Process Instruments
and Controls Handbook

OTHER McGRAW-HILL HANDBOOKS OF INTEREST

Process Instruments and Controls Handbook

THIRD EDITION

DOUGLAS M. CONSIDINE • Editor-in-Chief
Registered Professional Engineer (California)
in Control System Engineering

GLENN D. CONSIDINE • Managing Editor

• McGRAW-HILL BOOK COMPANY •

New York   St. Louis   San Francisco   Auckland   Bogotá   Hamburg
Johannesburg   London   Madrid   Mexico   Montreal   New Delhi
Panama   Paris   São Paolo   Singapore   Sydney   Tokyo   Toronto

Library of Congress Cataloging in Publication Data
Main entry under title:

Process instruments and controls handbook.

    Includes bibliographies and index.
      1. Process control—Handbooks, manuals, etc.
2. Automatic control—Handbooks, manuals, etc.
3. Engineering instruments—Handbooks, manuals, etc.
I. Considine, Douglas Maxwell. II. Considine, Glenn D.
TS156.8.P764 1985   629.8   84-10044
ISBN 0-07-012436-1

234567890   HAL/HAL   89876

ISBN 0-07-012436-1

The editors for this book were Harold B. Crawford and Rita Margolies,
the designer was Mark E. Safran, and the production supervisor was Ter-
esa F. Leaden. It was set in Baskerville by University Graphics, Inc.

Printed and bound by Halliday Lithograph.

Contents

For the detailed contents of any subsection, consult the title page of that subsection. The alphabetical index follows Section 19.

Index follows Section 19

ABOUT THE EDITORS

Douglas M. Considine, a registered professional engineer (California), received a chemical engineering degree from Case-Western Reserve University. He is a Fellow of both the Instrument Society of America and the American Association for the Advancement of Science, and a Senior Member of the American Institute of Chemical Engineers.

Among the technical management positions held by Mr. Considine were Process Design Engineer for The Lummus Company; Director, Process Engineering Applications, for Honeywell, Inc.; Manager, Technical Planning, for P. R. Mallory (Dart & Kraft, Inc.); and Director, Advanced Control Engineering, for the Hughes Aircraft Company. As a consultant Mr. Considine has served numerous firms, including Beckman Instruments, Varian Associates, Ralph M. Parsons, General Signal, Milton Roy, and Dun & Bradstreet. He has also chaired numerous symposia on chemical process control and instrumentation.

Mr. Considine has written over 60 technical articles and edited numerous handbooks and encyclopedias, including *Energy Technology Handbook* and *Chemical and Process Technology Encyclopedia* (both McGraw-Hill). He holds several patents for temperature and liquid level process controllers, is a coinventor of the laser fabric cutter, and was a pioneer in the development of the automated supermarket checkout stand.

Glenn D. Considine, a registered construction engineer (Georgia), is a member of both the American Society for Metals and the Institute of Food Technologists. He specializes in the management of information systems serving science, engineering, and technology.

Contributors[1]

R. K. ABELE. *Oak Ridge National Laboratory, Oak Ridge, Tennessee.* (NUCLEAR RADIATION DETECTORS)

PHILLIP J. AGARD. *Bowmar/ALI, Incorporated, Division of Bowmar Instrument Corporation, Acton, Massachusetts.* (INFORMATION DISPLAY AND STORAGE SYSTEMS)

JAMES D. BARTLETT, JR. *SLM Instruments, Inc./American Instrument Company, Urbana, Illinois.* (PHOTOMETRIC AND REACTION-PRODUCTS ANALYZERS)

MEL BERMAN. *Waugh Controls Corporation, Chatsworth, California.* (AUTOMATIC BLENDING SYSTEMS)

C. E. BERNSTEIN. *Process Control Division, Honeywell Inc., Fort Washington, Pennsylvania.* (DIGITAL CONTROL PROGRAMMERS)

ROBERT BICKING. *Micro Switch, a division of Honeywell Inc., Freeport, Illinois.* (PIEZORESISTIVE PRESSURE SENSING ELEMENT)

RICHARD W. BORUT. *Engineering Department, The M. W. Kellogg Company, Houston, Texas.* (CONTROL CENTERS)

L. L. BRACKESBUSCH. *Baird-Atomic, Inc., Bedford, Massachusetts.* (NUCLEAR RADIATION DETECTORS)

E. H. BRISTOL. *Research Department, The Foxboro Company, Foxboro, Massachusetts.* (CONTROL ALGORITHMS)

RICHARD A. BROWN. *Honeywell Inc., Test Instruments Division, Denver, Colorado.* (INFORMATION DISPLAY AND STORAGE SYSTEMS)

STEPHEN C. BUCHANAN. *Daytronic Corporation, Miamisburg, Ohio.* (LINEAR VARIABLE DIFFERENTIAL TRANSFORMER) (DIGITAL TECHNOLOGY)

WILLIAM H. BURTT. *The Foxboro Company, Foxboro, Massachusetts.* (RESONANT WIRE PRESSURE TRANSDUCERS)

C. A. CARLSON. *Milltronics Ltd., Peterborough, Ontario, Canada.* (SPEED AND VELOCITY MEASUREMENTS)

WILLIAM CARLSON. *DeZurik, Unit of General Signal Corporation, Sartell, Minnesota.* (CONSISTENCY MEASUREMENTS)

R. H. CHERRY. *Consultant, Huntingdon Valley, Pennsylvania.* (THERMAL-CONDUCTIVITY GAS ANALYZERS)

PAUL M. CHRISTENSEN. *Engineering Department, Fisher Controls Company, Marshalltown, Iowa.* (ELECTRONIC LEVEL CONTROLLERS)

WILSON A. CLAYTON. *Director of Engineering, Hy-Cal Engineering (A Unit of General Signal), El Monte, California.* (THERMORESISTIVE SYSTEMS)

---

[1]Persons who authored complete articles, subsections of articles, or who otherwise cooperated in an outstanding manner in furnishing information and helpful counsel to the editorial staff.

SHELDON CONNOLLOY. *Fisher Scientific Company, Pittsburgh, Pennsylvania.* (TITRATION SYSTEMS)

DOUGLAS M. CONSIDINE. *Editor-in-Chief.* (VARIABLES)

P. M. CORRAO. *Houston Atlas, Inc., Houston, Texas.* (REAGENT-TAPE ANALYZERS)

VINCENT B. CORTINA. *Environmental Equipment Division, EGEG International, Inc., Waltham, Massachusetts.* (HUMIDITY MEASUREMENT)

R. G. CRANE. *Division Manager, Research, Inc., Minneapolis, Minnesota.* (PROGRAMMERS)

JAMES A CROWSON. *Project Engineer, Robertshaw Controls Company, Industrial Instrument Division, Knoxville, Tennessee.* (MECHANICAL PRESSURE ELEMENTS)

LEE E. CUCKLER. *Consultant, Formerly Assistant to General Manager, Robertshaw Controls Company, Anaheim, California.* (MOISTURE MEASUREMENT SYSTEMS)

Z. C. DOBROWOLSKI. *Chief Development Engineer, Kinney Vacuum, A Unit of General Signal, Canton, Massachusetts.* (VACUUM MEASUREMENT)

HARRY A DOWNER. *Bindicator Company, Port Huron, Michigan.* (BULK LEVEL SYSTEMS)

EARLE DUNCAN. *Cascade Control Company, Roswell, Georgia.* (OPEN-CHANNEL FLOW MEASUREMENTS)

RODNEY M. DURHAM. *Manager, Instruments Division, Infrared Industries, Inc., Santa Barbara, California.* (INFRARED PROCESS ANALYZERS)

L. S. DYSART. *Retired, Formerly of Robertshaw Controls Company, Industrial Instrument Division, Knoxville, Tennessee.* (MECHANICAL PRESSURE ELEMENTS)

C. JOHN EASTON. *President, Sensotec, Inc., Columbus, Ohio.* (STRAIN GAGES)

P. A. ELFERS. *Retired, Bemidji, Minnesota. Formerly, Fisher Controls Company, Marshalltown, Iowa.* (FLOAT AND HYDROSTATIC LEVEL SYSTEMS)

ENGINEERING STAFF. *Bell Laboratories, Murray Hill, New Jersey.* (VOICE RECOGNITION AND SPEECH SYNTHESIS) (OPTICAL TRANSMISSION SYSTEMS)

ENGINEERING STAFF. *Extranuclear Laboratories, Inc., Pittsburgh, Pennsylvania.* (QUADRUPOLE MASS SPECTROMETRY)

ENGINEERING STAFF. *The Foxboro Company, Foxboro, Massachusetts.* (VARIABLES)

GEORGE F. ERK. *Sun Tech, Inc., Philadelphia, Pennsylvania.* (TRENDS IN ANALYTICAL INSTRUMENTATION)

ALLEN C. FAGERLUD. *Research Specialist, Fisher Controls International, Inc., Marshalltown, Iowa.* (CONTROL VALVE NOISE)

M. R. FARUKHI. *The Harshaw Chemical Company, Solon, Ohio.* (NUCLEAR RADIATION DETECTORS)

CHARLES E. FEES. *Product Manager, Fischer & Porter Company, Horsham, Pennsylvania.* (VARIABLE AREA FLOWMETERS)

D. L. FOWLER. *The Foxboro Company, Foxboro, Massachusetts.* (VORTEX FLOWMETERS)

ERIC L. GEIGER. *Eberline Division, Thermo Electron Corporation, Sante Fe, New Mexico.* (NUCLEAR RADIATION DETECTORS)

H. C. GERY. *Application Consultant, Leeds & Northrup (A Unit of General Signal), North Wales, Pennsylvania.* (ELECTRIC ACTUATORS)

DAVID B. GIACOMELLI. *Fluidyne Instrumentation, Oakland, California.* (POSITIVE-DISPLACEMENT FLOWMETERS)

DAVID T. GOLDMAN. *Assistant Director for Planning, National Measurement Laboratory, National Bureau of Standards, Washington, D.C.* (UNITS AND STANDARDS OF MEASUREMENT)

ALLEN M. GOLDSTEIN *Technical Associates, Canoga Park, California.* (NUCLEAR RADIATION DETECTORS)

B. P. GROPP. *Research and Development, Weston Instruments Division, Sangamo/Weston, Newark, New Jersey.* (ELECTRIC METERS)

ALEXANDRA B. GULDBERG. *Kaye Instruments Inc., Bedford, Massachusetts.* (INFORMATION DISPLAY AND STORAGE SYSTEMS)

G. A. HALL, JR. *Westinghouse Electric Corporation, Pittsburgh, Pennsylvania.* (FUNDAMENTALS OF AUTOMATIC PROCESS CONTROL)

C. HAMMERSTEIN. *Teledyne Analytical Instruments, City of Industry, California.* (THERMAL-CONDUCTIVITY GAS ANALYZERS)

R. L. HANES. *Rochester Instrument Systems, Rochester, New York.* (ALARMS AND ANNUNCIATORS)

NELS S. HANSSEN. *Aquatrol Corporation, St. Paul, Minnesota.* (INFORMATION DISPLAY AND STORAGE SYSTEMS)

JOHN N. HARMAN III. *Senior Chemist, Beckman Industrial Corporation, a subsidiary of Emerson Electric Company, Fullerton, California.* (OXIDATION-REDUCTION POTENTIAL MEASUREMENTS) (PH AND PH MEASURING SYSTEMS)

THOMAS J. HARRISON. *IBM Corporation, Boca Raton, Florida.* (DIGITAL TECHNOLOGY)

ROBERT L. HAYES. *Bindicator Company, Port Huron, Michigan.* (BULK LEVEL SYSTEMS)

W. A. HESKE. *Industrial Instrument Operations, Dresser Industries, Stratford, Connecticut.* (NON-CONTACTING OPTICAL SENSORS)

G. W. HICKENLOOPER. *The Ohmart Corp., Cincinnati, Ohio.* (NUCLEAR RADIATION METHODS)

STEPHEN P. HIGGINS, JR. *Retired, Senior Principal Software Engineer, Advanced Systems, Honeywell Inc., Process Management Systems Division, Phoenix, Arizona.* (PROCESS CONTROL TECHNIQUES)

E. H. HIGHAM. *Foxboro Great Britain Limited.* (MAGNETIC FLOWMETERS)

E. C. HIRTH. *Application Engineer, Moore Products Company, Spring House, Pennsylvania.* (PNEUMATIC PRESSURE TRANSMITTERS)   (PNEUMATIC LEVEL SYSTEMS)

EDWIN A. HOUSER. *Cyberdesign, Inc., Fullerton, California.* (SAMPLE CONDITIONING FOR PROCESS ANALYZERS)

DAVID W. HOWARD. *Brookfield Engineering Laboratories, Inc., Stoughton, Massachusetts.* (VISCOSITY MEASUREMENTS)

STEPHEN M. HOWARD. *StreeterAmet Measurement Systems Division, Mangood Corporation, Grayslake, Illinois.* (INDUSTRIAL SCALES AND BATCHING SYSTEMS)

T. S. IMSLAND. *Engineering Department, Fisher Controls Company, Marshalltown, Iowa.* (MEASUREMENT TERMS AND CHARACTERISTICS)

DANIEL A. JACKSON. *Omega Engineering, Inc., Stamford, Connecticut.* (THERMOCOUPLES)

BILL JOHNSON. *Kenco Plastics Company, Inc., Necedah, Wisconsin.* (OPEN-CHANNEL FLOW MEASUREMENTS)

C. M. JOHNSON. *Engineering Department, Fisher Controls Company, Marshalltown, Iowa.* (HYDROSTATIC LEVEL SYSTEMS)

RICHARD E. KEHRLI. *Tektronix, Inc., Beaverton, Oregon.* (INFORMATION DISPLAY AND STORAGE SYSTEMS)

DONALD B. KENDALL. *Retired, Formerly Toledo Scale Division, Reliance Electric Company, Worthington, Ohio.* (INDUSTRIAL SCALES)

RICHARD H. KENNEDY. *The Foxboro Company, Foxboro, Massachusetts.* (FUNDAMENTALS OF AUTOMATIC PROCESS CONTROL)

JOHN P. KING. *The Foxboro Company, Foxboro, Massachusetts.* (DISTRIBUTED CONTROL)

RALPH KLUCKEN. *Technical Association of the Pulp and Paper Industry, Atlanta, Georgia.* (CONSISTENCY MEASUREMENTS)

RIAD KOCACHE. *Taylor Servomex Limited, Crowborough, Sussex, England.* (OXYGEN ANALYZERS)

JAMES H. KOEGEL. *Director of Engineering, Robertshaw Controls Company, Industrial Instrumentation Division, Anaheim, California.* (CAPACITANCE LEVEL MEASUREMENTS) (ELECTRONIC CONTROLLERS)

JEROME KOHL. *Department of Nuclear Engineering, North Carolina State University, Raleigh, North Carolina.* (NUCLEAR RADIATION DETECTORS)

RICHARD J. KOTALIK. *Robertshaw Controls Company, Industrial Instrumentation Division, Anaheim, California.* (ELECTRONIC CONTROLLERS)

ANDREW L. KURYLCHEK. *Consultant, Wayne, New Jersey.* (BELT CONVEYOR SCALES AND WEIGH FEEDERS)

ALEXANDER KUSKO. *President, Alexander Kusko, Inc., Needham Hills, Massachusetts.* (ELECTRIC MOTOR DRIVE CONTROLS)

R. KUTKO. *Gaertner Scientific Corporation, Chicago, Illinois.* (DIMENSIONAL METROLOGY)

R. W. LALLY. *Engineering Department, PCB Piezotronics, Inc., Depew, New York.* (PIEZOELECTRIC PRESSURE TRANSDUCERS)

STEPHEN F. LANG. *Mine Safety Applicances Company, Pittsburgh, Pennsylvania.* (GAS ANALYZERS)

MICHAEL LATHAM. *Hardy Instruments, Inc., San Diego, California.* (INDUSTRIAL SCALES)

JOHN R. LAVIGNE. *The Foxboro Company, Foxboro, Massachusetts.* (CONSISTENCY MEASUREMENTS BIBLIOGRAPHY)

RICHARD E. LAWONN. *Giddings & Lewis Electronics Company, Fond Du Lac, Wisconsin.* (PROGRAMMABLE MACHINE CONTROLS)

G. ROBERT LEAVITT. *Manager, Product Reliability, Taylor Instrument Company, Division of Sybron Corporation, Rochester, New York.* (FILLED-SYSTEM THERMOMETERS)

MARA LEVINE. *Process Instrument Division, Panametrics, Inc., Waltham, Massachusetts.* (ALUMINUM OXIDE MOISTURE SENSORS)

A. J. LIVINGSTON. *The Ohmart Corporation, Cincinnati, Ohio.* (RADIATION-TYPE DENSITY METERS) (CONTINUOUS NUCLEAR WEIGH SCALES)

R. LYALL. *Brooks Instrument Division, Emerson Electric Company, Statesboro, Georgia.* (POSITIVE-DISPLACEMENT FLOWMETERS)

EDMOND E. LYNCH. *BIF, A Unit of General Signal Corporation, West Warwick, Rhode Island.* (OPEN-CHANNEL FLOW MEASUREMENTS)

S. A. MacNEILL. *Wallace & Tiernan Division, Pennwalt Corporation, Belleville, New Jersey.* (AMPEROMETRY)

E. C. MAGISON. *Process Control Division, Honeywell Inc., Fort Washington, Pennsylvania.* (SAFETY IN INSTRUMENTATION AND CONTROL SYSTEMS)

C. L. MAMZIC. *Moore Products Co., Spring House, Pennsylvania.* (FLUIDICS)

CRAIG McINTYRE. *Kay-Ray, Inc., Arlington Heights, Illinois.* (MASS FLOW MEASUREMENT) (CONTINUOUS NUCLEAR WEIGH SCALES) (MOISTURE MEASUREMENT SYSTEMS)

ARDMORE M. MILLER. *Love Controls Corporation, Wheeling, Illinois.* (ELECTRIC CONTROLLERS)

R. W. MILLER. *Flow Consultant, The Foxboro Company, Foxboro, Massachusetts.* (DIFFERENTIAL PRODUCERS)

JAMES E. MURPHY. *Instrument Division, Mine Safety Appliances Company, Decatur, Georgia.* (THERMAL-CONDUCTIVITY GAS ANALYSIS)

JOHN J. MURRAY. *The Superior Electric Company, Bristol, Connecticut.* (STEPPING MOTORS)

ALLEN E. MUSHIN. *Omega Engineering, Inc., Stamford, Connecticut.* (THERMOCOUPLES)

JOHN NAGY. *Section Manager, The Cedar Grove Division of Beckman Industrial Corporation, a subsidiary of Emerson Electric Company, Cedar Grove, New Jersey.* (ELECTROLYTIC CONDUCTIVITY AND RESISTIVITY MEASUREMENTS)

GREGORY NEEB. *Formerly, DeZurik, Unit of General Signal Corporation, Sartell, Minnesota.* (CONSISTENCY MEASUREMENTS)

JOE M. NELSON. *Section Head, Network Applications Systems Division, Honeywell Information Systems, Honeywell Inc., Billerica, Massachusetts.* (PROCESS CONTROL TECHNIQUES)

ROBERT A. NORCROSS. *Norcross Corporation, Newton, Massachusetts.* (VISCOSITY MEASUREMENTS)

EUGENE NORMAN. *Green Bay, Wisconsin.* (CONSISTENCY MEASUREMENTS)

JOHN H. NOYES. *Du Pont Instrument Products, Scientific and Process Division, Wilmington, Delaware.* (PHOTOMETRIC ANALYZERS)

ROBERT L. OSBORNE. *Manager, Diagnostics and Controls, Westinghouse Electric Corporation, Orlando, Florida.* (FUNDAMENTALS OF AUTOMATIC PROCESS CONTROL)

JAMES E. OVERALL. *Georgia-Pacific Corporation, Atlanta, Georgia.* (CONSISTENCY MEASUREMENTS)

RAY PEACOCK. *Vice President of Engineering, Land Instruments Inc., Tullytown, Pennsylvania.* (RADIATION THERMOMETRY)

CHRIS PERKINS. *Automatic Timing & Controls Co., King of Prussia, Pennsylvania.* (TIMERS)

DAVE POSTETTER. *John Fluke Mfg. Co., Inc., Everett, Washington.* (INFORMATION DISPLAY AND STORAGE SYSTEMS)

MOHAN PRASAD. *Manager, Technical Review, The Foxboro Company, Foxboro, Massachusetts.* (VARIOUS TOPICS)

ROY A. RATCLIFFE. *Vice President, Process Instruments, Gould Inc., Measurement Systems Division, Oxnard, California.* (STRAIN GAGE PRESSURE TRANSDUCERS)

GENE REBUCCI. *Schenck Weighing Systems, Totowa, New Jersey.* (BELT CONVEYOR SCALES AND WEIGH FEEDERS)

R. G. REIP. *Senior Development Engineer, GPE Controls, Division of Brunswick Corporation, Morton Grove, Illinois.* (HYDRAULIC CONTROLLERS)

WILLIAM L. RICKETSON. *Toledo Scale, Division of Reliance Electric, Atlanta, Georgia.* (INDUSTRIAL SCALES)

MARC L. RIVELAND. *Senior Research Engineer, Fisher Controls International, Inc., Marshalltown, Iowa.* (CONTROL VALVE CAVITATION)

ROBERT B. ROSS. *Gould Inc., Industrial Controls Division, Westminister, Maryland.* (RELAYS AND CONTACTORS)

BRUCE R. RUSCH. *President and General Manager, Gould Inc., Programmable Control Division, Andover, Massachusetts.* (PROGRAMMABLE CONTROLLERS)

E. R. SABINASH. *Cutler-Hammer Products, Eaton Corporation, Milwaukee, Wisconsin.* (LIMIT SWITCHES)

R. S. SALTZMAN. *Instrument Products Division, E. I. du Pont de Nemours & Co., Inc.* (ULTRAVIOLET-ABSORPTION ANALYSIS)

ANTHONY SCIACCA. *National Sonics Division—Xertex, Hauppauge, New York.* (ULTRASONIC METHODS)

G. B. SELLERS. *Engineering Department, Robertshaw Controls Company, Industrial Instrumentation Division, Anaheim, California.* (CAPACITANCE LEVEL MEASUREMENTS)

WINSTON G. SHEQUEN. *Manager, International Support, Applied Physics Laboratory, Division of Bausch & Lomb, Sunland, California.* (X-RAY ANALYTICAL METHODS)

EUGENE B. SHERWOOD. *Micro Motion, Inc., Boulder, Colorado.* (MASS FLOW MEASUREMENT)

W. H. SHIRK, JR. *Formerly Manager, Engineering Coordination and Services, Leeds & Northrup Company, A Unit of General Signal, North Wales, Pennsylvania. Presently, General Manager, Pocono Lake Preserve, Pennsylvania.* (BRIDGES AND POTENTIOMETERS)

WALTER A. SKOWRON. *Hewlett-Packard, Loveland Instrument Division, Loveland, Colorado.* (INFORMATION DISPLAY AND STORAGE SYSTEMS)

GARY SNYDER. *Micro Motion, Inc., Boulder, Colorado.* (MASS FLOW MEASUREMENT)

RICHARD F. SPEICHER. *Micro Switch—Honeywell Inc., Freeport, Illinois.* (PIEZORESISTIVE PRESSURE TRANSDUCERS) (PROXIMITY SWITCHES) (PANEL, KEYBOARD, AND CONTROL SWITCHES)

ELMER SPERRY. *Chief Chemist, The Cedar Grove Division of Beckman Industrial Corporation, a subsidiary of Emerson Electric Company, Cedar Grove, New Jersey.* (ELECTROLYTIC CONDUCTIVITY AND RESISTIVITY MEASUREMENTS)

H. A. STEINHERZ. *CTI-Cryogenics Division, Helix Technology Corp., Waltham, Massachusetts. Formerly with Varian Associates, Palo Alto, California and President, American Vacuum Society.* (HIGH-VACUUM MEASUREMENTS)

FRED TIPPING. *Taylor Servomex Limited, Crowborough, Sussex, England.* (OXYGEN ANALYZERS)

WALTER A. TROEGER. *Engineering Department, Weston Instruments Division, Sangamo/Weston, Newark, New Jersey.* (ELECTRIC POWER AND ENERGY MEASUREMENT)

RICHARD VILLALOBOS. *The Foxboro Company, Plymouth, Massachusetts.* (CHROMATOGRAPHY)

R. R. VOKOUN. *Seiscor Division, Seismograph Service Corporation, Tulsa, Oklahoma.* (RHEOLOGICAL MEASUREMENTS)

JAMES R. WALKER. *Chief Electrical Engineer, DeVlieg Machine Company, Royal Oak, Michigan.* (POSITION AND DISPLACEMENT MEASUREMENT)

T. J. WALSH. *Engineering Department, Leeds & Northrup Company, A Unit of General Signal, North Wales, Pennsylvania.* (SELF-BALANCING ELECTRICAL INSTRUMENTS)

J. D. WARNOCK. *Manager, Systems Engineering, Moore Products Company, Spring House, Pennsylvania.* (PNEUMATIC TRANSMISSION AND CONTROL)

ROBERT M. WHITTIER. *Manager of Development Engineering, Endevco Corporation, San Juan Capistrano, California.* (VIBRATION MEASUREMENTS)

PIETER R. WIEDERHOLD. *President, General Eastern Instruments Corporation, Watertown, Massachusetts.* (HUMIDITY MEASUREMENT)

JOSEPH WILDER. *Senior Scientist, Object Recognition Systems, Inc., Princeton, New Jersey.* (MACHINE VISION)

GEORGE WILLIS. *The Bristol Saybrook Company, Old Saybrook, Connecticut.* (TIMERS)

PAUL WING. *Consultant, Hingham, Massachusetts.* (CONTROL VALVES) (CONTROL VALVE SIZING) (CONTROL VALVE CHARACTERISTICS)

W. D. WOOD. *Formerly Director, Standards Laboratory, Taylor/Sybron, Rochester, New York.* (MEASUREMENT TERMS AND CHARACTERISTICS)

JOHN YAEGER. *Communications Department, Fischer & Porter Company, Warminster, Pennsylvania.* (ELECTRONIC PRESSURE TRANSMITTERS)

A. W. YOUNG. *Rank Precision Industries, Inc., Des Plaines, Illinois.* (CLINOMETER; GONIOMETER; THEODOLITE; SURFACE FINISH)

Preface

The industrial process instrumentation and control field has always been characterized by above-average technological progress. The decade just past was no exception. Each period of progress has certain "buzz" words which tend to highlight progress. Included among these contemporary words are "distributed control," "programmable controllers," and "microprocessors," to mention a few. Computer control, a popular term a few years ago, continues to find usage, but it has lost its identity—because computers and data processing techniques have invaded nearly all aspects of industrial instrumentation.

So much progress has been made since the 1974 publication of the Second Edition of this Handbook that well over 80 percent of the information is entirely new or, at a minimum, revised in a major way. This is attested by the increase in the printed pages required. The Third Edition contains 1565 illustrations and 177 tables—representing the input of 143 contributors.

Because the scope and complexity of process instrumentation/control has doubled and possibly trebled in a time span of less than a decade, commencing in the mid-1970s, the original planned content for this edition required alteration several times during the preparation of the manuscript. To hold the *Handbook* within a reasonable length, considering production cost and hence price of the bound book, it became necessary to withhold coverage of a number of topics of secondary, but real, importance to process control—in the interest of providing thorough coverage of the fundamental topics. Consequently, some topics included in prior editions are not found in this edition. Users who have copies of the Second Edition are urged to retain them for reference on such topics as heat-transfer effects in thermal measuring systems, wells and protecting tubes, bimetal thermometers, liquid-in-glass thermometers, temperature-sensitive material indicators, pressure switches, pressure regulators, some of the chemical analytical instrumentation methodologies, nondestructive testing, pollution instrumentation, volumetric feeders, balances and weighing, counters, dynamometers and torque measurements, compression and tensile testing machines, tension measurement and control systems, pattern recognition theory, appearance measurements, magnetic quantities. Also, special articles originally scheduled on microstructure fabrication, optical isolators, information theory, and computer-aided manufacturing/robotics (for discrete-piece handling industries) had to be deleted in the interest of using all available space for coverage of the hard-core aspects of process instrumentation and control.

Rather than weary the reader with a philosophical preface, we have elected to take a walk through the contents of the 83 separate articles.

As in past editions, every effort has been made to include a comprehensive alphabetical index and numerous cross references within the text in the interest of reader convenience.

DOUGLAS M. CONSIDINE
Editor-in-Chief

Introduction:
Descriptions of Articles

SECTION 1—MEASUREMENT FUNDAMENTALS

Variables The variable (or changing quantity) is the object of all instrumentation. An understanding of the variables is fundamental to measurement, display, and control engineering. There are dozens of variables encountered in the manufacturing and process industries. In this article, considerable order is brought to this complex topic by a scientific classification of the variables and, in terms of measurement means, a matrix is included that relates variables to measurement techniques.

Units and Standards of Measurement The quantification of measurements is based upon the establishment of physical units for each variable. Instrumental precision depends upon the formulation and maintenance of reliable standards. In this article, the basic units and standards of measurement are described, along with graphics which illustrate the derivation and traceability of standards that are maintained by the National Bureau of Standards in the United States and by similar organizations in other parts of the world.

Measurement Terms and Characteristics Technologies have unique lexicons. Instrumentation is no exception. In this article, several of the most important and frequently mentioned terms, such as range, resolution, accuracy, hysteresis, among many others, are precisely defined and fully explained with the aid of examples and diagrams.

Glossary of Units Frequently Used. There are over 100 fundamental measurement units that are encountered almost on a daily basis by instrumentation and control engineers. Those units which are mentioned most frequently are precisely defined in this article. Further, a very helpful listing of numerous organizations that participate in the responsibility for establishing and defining standards is included.

SECTION 2—TEMPERATURE SYSTEMS

Temperature Conversion Table Although in recent years there has been a marked trend to expressing temperature in degrees Celsius rather than degrees Fahrenheit, nevertheless the average instrument and process control engineer is called upon very frequently to convert from one scale to the other. The table included here is possibly the most convenient of its type and condensed into a relatively small space. Temperatures between 0° and 100° (either commencing with C or F) are given in precise 1° increments. Above 100°, temperatures are given up to 3000° in 10° increments, with a convenient table of interpolation factors for in-between readings. The table also covers, in 10° increments, temperatures from 0° to absolute zero.

Temperature Scales, Standards, and Calibration   The common (Celsius and Fahrenheit) scales, as well as some of the lesser-used temperature scales are carefully defined. Considerable attention is given to the establishment and characteristics of the International Practical Temperature Scale and how fixed points are used in describing that scale. An important table that gives approximate differences in kelvins between the values of the IPTS is included. Primary and secondary temperature standards are described. Of considerable importance, the levels of temperature measurement accuracy are delineated, ranging from level 1 (general purpose) to level 5 (international standards). A synopsis of comparison calibration and transfer standards is also included.

Thermocouples   This long-established and ubiquitous method of detecting temperature is reviewed from all important engineering angles. Following terse descriptions of the fundamental thermocouple effects (Seebeck, Peltier, Thomson), the principal compensation methods (cold junction, electronic-bridge, thermoelectric refrigeration, heated oven) are described in considerable detail. The fundamental physical laws that govern thermocouple characteristics are defined and illustrated. Important thermocouple characteristics are tabulated for each of the major (and some minor) thermocouple types (iron-constantan, platinum alloys, etc.). Insulation of thermocouples and extension wire is described, aided by tabular summaries. Thermocouple applications are evaluated from numerous practical situations—need for fast response, situations with high shock and high temperature, and importance of small size, probe flexibility, and cost among other factors. Practical instructions are given concerning wiring and installing thermocouples. The various methods for protecting thermocouples and tabular summaries of materials of construction used are included. An extensive description of fabricating thermocouple elements (instead of procuring finished assemblies) is included. This article features 27 illustrations and 9 tables.

Thermoresistive Systems (RTDs)   Once considered mainly for laboratory and research applications, the RTD, through the improvement of materials and electronics, now finds major application under rugged process conditions. The principal resistance materials used (platinum, copper, nickel, etc.) are described and supported by tabular data and graphs. Circuits used with RTDs are outlined and diagrammed in considerable detail.

Filled-System Thermometers   For many years, the filled-system thermometer was commonly used throughout the process field and, because the technology has gained from continued engineering refinements, the method continues to find extensive use for many applications. The advantages and disadvantages of the method are clearly outlined. Filled-system thermometers are classified, and the characteristics of each class (liquid-filled, vapor-pressure, and gas-filled) are described. A convenient table that compares the various thermal systems is included. Factors that affect selection—capillary and case compensation, speed of response, barometric and bulb-elevation errors, and overrange protection—are delineated, including specific examples and sample calculations.

Radiation Thermometry   Significant progress has been made in this field during the past decade, at which time the term "radiation pyrometry" was widely used. This article thoroughly appraises the field, ranging from theory to practical applications. Advantages of the method are outlined, followed by descriptions of the major classes of radiation thermometers [wideband and narrowband instruments, ratio (two-color) thermometers, optical pyrometers, and fiber-optic instruments]. The elements of radiation thermometer systems (optics, detectors, signal conditioners) are carefully delineated. Theoretical considerations (blackbody concept, effective wavelength, and N-value) are described. Much attention is given to factors affecting the performance (transparent objects, emissivity and reflection errors, interference in sighting path, moving objects, sighting over very long distances) of radiation thermometers, including descriptions of solutions for each of these problems. The article ends with a discussion of standards used for calibrating radiation thermometers. In all, the article includes 31 illustrations and 11 tables of detailed information.

SECTION 3—PRESSURE SYSTEMS

Mechanical Pressure Elements A wide range of mechanical pressure elements are described, including liquid-column elements, elastic elements (bellows, bourdon springs, diaphragms). Much information helpful to design and application is included in both tabular and graphic formats. The importance of materials of construction and sources of error is stressed. This article includes 29 diagrams and 9 tables.

Pressure Transducers and Transmitters This area of process instrumentation has essentially undergone a revolution during the past decade—with significant assists from electronics and solid-state technology, as well as better materials of construction and methods of fabrication. Much stress has been given to miniaturization and ease of connecting sensing elements into total systems. The new (and old) pressure transducers are classified and individually described in considerable engineering detail [strain gage transducers, capacitive, piezoelectric, frequency-change, potentiometric, reluctive, linear-variable differential transformers (LVDTs), pneumatic, and interferometric manometers are included]. Featured is a tabular comparison of characteristics of the several kinds of transducers. The extensive coverage of this article required 54 illustrations.

Vacuum Measurement This topic, always a specialty, is reviewed in terms of numerous advances in materials, circuitry, and measurement techniques which have occurred in recent years. Units of vacuum measurement and nomenclature are stressed. A graphical matrix which shows principal vacuum measurement methods versus torr (range) and another matrix which plots methods versus characteristics of both method and process being measured can be very helpful in vacuum gage selection. All important measurement methods (bourdon gage, capacitance manometer, liquid manometer, McLeod gage, thermal-conductivity gage, hot-filament ionization gage, cold-cathode ionization gage, spinning rotor friction gage) are described in terms of principles used, characteristics, and applications. Leak detection using vacuum gages and partial pressure analyzers are also covered. These descriptions are augmented by 23 diagrams and graphs.

Resonant Wire Pressure Transducers Although not new in concept, the principle of resonant frequency detection in pressure measurement has gained prominence only since the late 1970s. This principle is described, along with current reductions to operating hardware, including the main characteristics of these transducers. Sectional views of current sensors are included.

SECTION 4—FLOW SYSTEMS

Differential Producers Although inroads have been made by other methodologies of flow measurement, the differential-producing flowmeter (head flowmeter) continues to enjoy wide application throughout the process industries. The numerous refinements which have been made in this technology during the past decade or so are delineated, including progress in standardization and in the understanding of the physics of the various differential producers (orifices, venturis, flow nozzles). Several refinements in hardware are reported. A detailed discussion of the fundamental unit flow rate equations is included. Sizing versus flow rate determination is also discussed in detail. These descriptions are assisted by tabular data and equations and calculations. The article features 25 illustrations and 6 tables.

Variable Area Flowmeters The design and application of variable area flowmeters has been fully updated. Each of several types of meters (rotameter, orifice and tapered plug meter, piston meter) are described in terms of characteristics and applications. Correction factors for variable area meters are described with the support of several tabular summaries. Typical variable area meter calculations are included.

Magnetic Flowmeters   The use of magnetic flowmeters has been markedly expanded during the past decade. This expansion has been accompanied by numerous design and application refinements. A description of operating principle is followed by detailed hardware discussions, including sectional views of contemporary magnetic flowmeters, and functional and block diagrams.

Velocity, Vortex, Mass, Ultrasonic, and Other Flowmeters   Assisted by advancements in materials, solid-state technology, and circuitry, among other technological innovations, the "nonconventional" flowmeters have gained much prominence during the past decade. Described in this article are vortex shedding and its application to metering, turbine flowmeters, new concepts in mass flow measurements, radiation-type mass flowmetering, angular momentum meters, gyroscopic flowmeters, ultrasonic flowmeters (Doppler effect and transit-time meters) that feature clamp-on, noninvasive designs, thermal flowmeters, and radioisotope tracer flowmeters.

Positive-Displacement Flowmeters   This article reports the numerous design refinements which have occurred in recent years in this well-established and widely used principle of flow measurement. Descriptions include nutating-disk meters, oscillating-piston meters, fluted rotor meters, rotating meters, and bellows meters. These descriptions are supported by numerous illustrations and tabular data summaries as an aid to application selection decisions.

Open-Channel Flow Measurements   Weirs, flumes, and open flow nozzles are concisely described. Dimensional and capacity specifications are included. A one-piece fiberglass Parshall flume is described and illustrated.

SECTION 5—LEVEL SYSTEMS

Level Systems   The many ways of measuring and controlling liquid and bulk solid levels are described. In addition to systems descriptions (characteristics, application data), the article features concise appraisals of the importance of level control to process operations, level measurement system mathematics, replete with equations, capacity versus level height in various vessels, volume and weight measurement from level, and errors in measurement of quantities in storage tanks. Liquid-level system descriptions include point-contact methods, visual methods, electrode or probe systems, buoyancy methods, float-type mechanisms, magnetic float gages, displacer-type mechanisms, hydrostatic pressure methods, electric capacitance gages, sonic and ultrasonic methods, nuclear radiation methods, and thermal methods. Solid-level system descriptions include diaphragms, paddle wheels, tilt switches, vibration sensors, plumb bob system, capacitance sensors, and nuclear devices. Included is a tabular summary of density and angle of repose of over 40 common bulk commodities. In support of these descriptions are 85 illustrations and several tables.

SECTION 6—ANALYTICAL AND TESTING INSTRUMENTATION

In the rapidly expanding instrumentation and control technology, one of the most active and advancing segments has been analytical and testing instrumentation. To cope with reporting on the many advancements, this Third Edition includes many more pages on this topic as compared with the Second Edition (which, at that time—about a decade ago—was considered extensive).

Analysis Instruments   To assist the Handbook user in proceeding in an orderly manner through what might aptly be termed a "jungle of methodologies," this article classifies analysis instruments in terms of four interactive phenomena: (1) electromagnetic radiation, (2) chemical affinity or reactivity, (3) electric or magnetic fields, and (4) thermal or mechanical energy. Most analysis instruments fall into one of these categories in terms of their interaction with external energy. In addition to summarizing these categories in the text, a useful table on the generalized relationships between matter and energy that can be measured to ascertain chemical composition is included. The general applications for chemical composition measurements are summarized. A convenient table of factors for interconversion of concentration units of gases and vapors is included. At the end of the article, trends in analytical instrumentation are summarized, including the ever-growing trend to *on-line* applications.

Sample Conditioning for Process Analyzers   There is a common observation that an analysis instrument can be no better than its sampling system. The functions of sampling systems are described, with emphasis on isokinetic sampling. Particular attention is given to sampling with fast-loops. Direct sampling as well as preconditioning of samples (vaporizing, cooling, drying) are delineated. The article also features multiple-stream switching, calibration sample management, controlling analyzer pressure, instrument packaging and installation, and estimating transport time lag. These descriptions are supported by 14 diagrams.

X-Ray Analytical Methods   This well-established analytical technique is thoroughly described, highlighting modern hardware. The principles of x-ray crystallography, x-ray fluorescence analysis, and x-ray electron-microprobe analyzers are described. Particular attention is given to slurry, dry powder, and pulp density analyses. The article incorporates 22 illustrations and 8 tables.

Quadrupole Mass Spectrometry   Quite recently, these spectrometers have gained inroads into some of the more traditional process analysis applications that have been dominated by infrared-absorption and gas chromatography techniques. The quadrupole mass spectrometer is gaining acceptance because of its comparatively low cost, reliability, and ease of computer-controlled operation. The theory and principles of operation are described, including examples of techniques for determining which compound peaks and relationships provide the best results.

Ultraviolet-Absorption Analysis   Popular for many years, ultraviolet (UV) analyzers have experienced numerous engineering refinements in recent years, and their acceptance for process stream analysis, including problems encountered in pollution control, has increased. Among topics discussed in this article are the ultraviolet-absorption spectrum, the relation of component concentration to absorption, design criteria, components, sources of UV radiation, and the single-beam, split-beam, and dual-beam versions of these analyzers. Several examples of applications including the sample handling systems needed are described.

Infrared Process Analyzers   The most recent advances in the design and application of non-dispersive infrared analyzers (NDIR) using either microphone or solid-state detectors are delineated. Tabular summaries include typical specifications and representative gases and concentrations which are commonly measured by NDIR analyzers.

Photometric and Reaction-Product Analyzers   There has been considerable expansion in recent years in the use of photometric and reaction-product analytical methods. Such techniques include fluorometers, luminescence photometers, colorimetric analyzers, nephelometers, turbidimeters, titrators, amperometers, and reagent-tape analyzers. Each of these methods is discussed in considerable detail, supported by the inclusion of 32 illustrations.

Oxygen Analyzers   A quiet revolution in oxygen measurement instrumentation has occurred during the past decade. Oxygen measurements, important to combustion control and hence to energy conservation, have gained increased acceptance. In addition to the older methods for O_2 determinations, the more recent zirconium oxide sensors are described.

Oxidation-Reduction Potential Measurements   Impressive progress has been made in the development and application of selective ion electrodes. The basic principles of oxidation-reduction potential (ORP) methods are described.

pH and pH-Measuring Systems   Pioneered nearly a half-century ago, the measurement of pH provides one of the most common of the analysis instrumentation variables. Improvements continue to be made in the design and composition of glass and reference electrodes. In this article, pH is concisely defined and described against a backdrop of acid-base theory. Buffer solutions and asymmetry potential are explained. pH instrumentation, like so many other traditional measurement technologies has felt the impact of microprocessors in recent years.

Electrolytic Conductivity and Resistivity Measurements   Numerous applications for electrolytic measurements including the gauging of water quality and detection of contamination are described. Units, measuring circuits (including an electrodeless circuit), and recent advancements in conductivity cell design are described. A summary of sources of errors in conductivity measurement is included.

Thermal-Conductivity Gas Analysis   For certain gases and gas mixtures, thermal conductivity has been the analytical method of choice for many years. The principles of the method, along with a tabular summary of the thermal conductivities of various gases, are given. Other topics covered include typical applications and practical considerations in continuous measurements. Emphasis is placed on conditioning gas samples and sample purity.

Combustibles, Total Hydrocarbons, and Carbon Monoxide Analyzers   Operating principles and sources of measurement errors are described. Included is a convenient table that lists the heats of combustion of representative process materials.

Chromatography   During the past decade, there appeared to be no limitation on the applications of chromatography in process measurements. Chromatographs are close to occupying the number one position among analysis instruments. In addition to routine process control applications, gas and liquid chromatographs are widely used for waste disposal and personnel safety–area monitoring. Great stress has been placed on the automation of data display and storage. These factors, plus new refinements in valve and detector design, are described in this article.

Density and Specific-Gravity Measurement   Systems for measurement of density and specific gravity continue to be applied widely for a number of processes as criteria of composition. The several specific gravity scales are defined. The numerous instrumental approaches used (photoelectric, inductance-bridge, balanced-flow vessel, displacement meter, chain-balanced float, liquid-purged differential-pressure method, air bubbler system, viscous-drag instruments, force-balance horizontal U-tube meter, vibrating sensors, ultrasonic sensors, radiation-type meters, among others) are described and illustrated in this article. A convenient specific gravity conversion table is included.

SECTION 7—WEIGHING, FEEDING, AND BATCHING SYSTEMS

Industrial Scales   Few fields of instrumentation have changed as radically as industrial weighing—and during a comparatively short period of 10 to 15 years. The strain gage and other forms of load cells have largely displaced levers and counterweights in weighing systems. The capacities and functions of the principal industrial scale configurations are described. These include portable and bench scales, floor and deck scales, monorail scales, hopper and tank scales, motor truck scales, railway track scales, crane scales, counting scales, drum-filling scales, and bulk weighers. Examples of weigh-batching systems are given. The automation of weight data handling is described. Because of the large variety of equipment offered today, 45 diagrams and photos were required to illustrate current hardware.

Bonded Strain Gage Load Cells   This article concentrates on the principles, types, and applications of load cells with respect to weighing and other force measurements. Included are descriptions of columnar load cells, bending beams, and shear elements. In recent years, load cells which have capacities of 500 lb (227 kg) to 300,000 lb (136,080 kg) and which incorporate shear element designs have revolutionized scale designs. Cross-section, circuitry, and application diagrams are included.

Belt Conveyor Scales and Weigh Feeders   As with industrial scales, belt conveyor scales have been affected by strain gage, electronic, and data-processing technologies. Conveyor scales are evaluated in terms of their accuracy, testing, and calibration. Typical applications are illustrated. The application of microprocessors to gravimetric feeders is stressed. Attention is given to problems resulting from free-flowing flushy-type materials. The text is supported by 30 diagrams of equipment installations.

Continuous Nuclear Weigh Scales   Nuclear weigh scales are frequently the method of choice, particularly for certain severe process environments and materials characteristics where the measurement of the mass of bulk substances on the move, as on a conveyor, is required. In addition to operating principles, radioactive sources, and problems encountered in terms of equipment accuracy, loading, and the nature of the product weighed are described.

Automatic Blending Systems   Electronics and computer technology have affected this field in a major way during the past 10 to 15 years. The article includes discussions of batch versus continuous blending, typical blend configurations, the sizing of a blending system, the selection of the best-suited flowmeter for blending systems, factors in specifying equipment for blending applications, and a detailed description of the system-operator interface (displays, automatic rate control, formula or recipe storage).

SECTION 8—FORCE SYSTEMS

Strain Gages   In recent years, the strain gage has permeated a wide variety of sensors and detectors and has taken on a certain universality as a transducer. In this article, the principles of the strain gage are described. Numerous types of strain gages (unbonded, bonded metallic, flame spray, welded, bonded semiconductor, diffused semiconductor, and thin film), along with their materials composition and major characteristics are described and tabulated. Basic bridge circuits are delineated. Strain gage transducer applications are described with the aid of 14 diagrams.

SECTION 9—MOTION AND GEOMETRIC SYSTEMS

Position, Displacement, Motion, and Object Detection Systems   The rapidly expanding field of computer-aided manufacturing and robotics has resulted in large demands for improved and new ways to measure position, displacement, and other geometric variables. This article is a rather exhaustive description of such devices. The article commences with a discussion of machine conditions and requirements, followed by specific descriptions of position-measuring systems (resolvers and synchros, inductive plates, rotary encoders, inductive-bridge, pulse-type, and magnetic-pin or tooth transducers, optical binary scales, magnetic scales). Displacement-measurement systems described include linear variable differential transformers (LVDTs), linear potentiometers, thin-film potentiometers, and linear variable reluctance transducers (LVRTs). Other discussions include numerical and automatic-positioning control systems, the basic classes of N/C systems, and the impact of programmable controllers and other relatively new developments. Object detection and proximity switches are described in considerable detail, including photoelectric devices, scanning techniques, the use of fiber optics and light-emitting diodes (LEDs) as radiation sources for certain applications. Other equipment described includes electromechanical switches, inductive proximity sensors, magnetic proximity sensors (variable-reluctance sensors, dry-reed switches, Hall-effect switches, Wiegand-effect switches), capacitive proximity switches, and magnetostrictive limit switches. Because of the wide variety of hardware currently offered, a total of 59 illustrations was required to support the text of this article.

Dimensional Metrology   In the discrete-piece manufacturing industries, precision dimensional measurements are required in the laboratory and on-line. Included in this article are descriptions of interferometers, optical gratings, gage blocks, autocollimators, goniometers, clinometers, theodolites, optical flats, sine bars, spherometers, ellipsomers, optical comparators, and optical benches. Automatic and semiautomatic measurement machines are described. Pneumatic, electrical, and electronic gaging systems for on-line use are described. Attention is given to late-generation gauging systems, including laser gauging and image sensing systems involving charge coupled devices (CCDs). A separate section of the article is devoted to thickness gauging systems, including nuclear-radiation, ultrasonic, and x-ray systems. Instrumentation used for determining and evaluating surface finish is described.

Speed and Velocity Systems   This article includes descriptions of direct-current and alternating-current tachometers, as well as frequency-responsive tachometer systems. There are discussions of speed differential measurements, bearingless tachometer generators, analog versus digital transducers, pulse-type transducers, proximity switches adapted for speed measurement, capacitor-type switches, impulse tachometers, optical encoders, stroboscopic tachometers, variable-reluctance instruments, vibration pickups, and vibrating-reeds, eddy-current and velocity-head or hydraulic tachometers. Within the last few years, an interesting approach to regulating the speed of a motor without a tachometer has emerged. This is concisely described. Governors and air-velocity measurement are also covered.

Machine Vision   This is a rapidly expanding segment of industrial instrumentation, particularly in view of the marked trend to more automated production systems. Electronic vision systems combined with machine artificial intelligence are enhancing productivity and improving quality control in various electronic and mechanical assembly operations.

Vibration Measurements   These measuements are required during the development, construction, and operation of many kinds of machines. Described is the use of piezoelectric accelerometers, strain-gage sensors, servo accelerometers, and signal conditioning for accelerometers. The environmental effects on vibration sensors are delineated. Velocity transducers and relative motion sensing with the eddy current probe are also described. The text is augmented by 11 illustrations.

SECTION 10—HUMIDITY AND MOISTURE MEASUREMENTS

Humidity Measurement   Commencing with definitions and common humidity parameters, including a psychrometric chart (English and metric units), this article follows with detailed discussions of wet bulb/dry bulb measurements, the Dunmore element and the Pope cell (for relative humidity), dew point hygrometry, saturated salt dew point sensors, condensation-type hygrometers, electrolytic hygrometers, and aluminum oxide moisture sensors.

Moisture Measurement   This article commences with various definitions, including wet basis and dry basis. Principal moisture-sensing devices—including electrical-conductivity methods, electric-capacitance methods, aluminum oxide impedance sensors, and radio frequency absorption, microwave absorption, combined microwave and gamma absorption, and infrared absorption meters—are outlined. The relatively new vibrating quartz crystal moisture sensor is concisely described. The article also covers various laboratory moisture determinations, including the Karl Fischer technique. A total of 16 diagrams illustrate the various moisture-sensing methodologies.

SECTION 11—RHEOLOGICAL SYSTEMS

Viscosity Systems   Various substances are classified in terms of their rheological behavior—true plastics, pseudoplastics, thixotropic, and rheopectic materials. Fundamental viscosity units and relationships are defined. Numerous specific techniques used in viscometers are described, including timed fall of ball or rise of bubble, drag torque on stationary element in rotating cup, timed fall of piston in cylinder, pressure drop through a friction tube, and torque viscometers. The relatively recent trend toward increasing numbers of on-line process viscometers is emphasized. Representative viscosity control applications are described and illustrated. Rheological variables, such as viscosity, have grown in importance in several industries in recent years, notably in food processing.

Consistency Measurements   The professional literature is sparse in connection with consistency measurements which are so important in the pulp industry. The article commences with terse definitions and explanations of flowability, the relationship of consistency to percent solids, and to such factors as density and freeness. Specific applications are cited. While emphasis is on in-plant consistency measurement and control systems, attention is also given to laboratory-type devices.

SECTION 12—RADIATION SYSTEMS

Nuclear Radiation Instrumentation   The detection and measurement of nuclear radiation finds many uses in addition to applications in the operation of nuclear reactors. For example, beta-ray and x-ray absorption are used in analytical instruments, there are radiation-effects studies of living processes, applications in nuclear medicine and high-energy and low-energy physics, as well as dating techniques, and geological and mineral exploration. Radioactivity and the types of radiation are described in considerable detail, nuclear radiation units of measurement are defined, followed by descriptions of specific nuclear radiation detectors (ionization chambers, Geiger counters, gas ionization tubes, proportional counters, scintillation counters, phoswich detectors, solid-state detectors). Personal dosimetry for beta, gamma, and x-radiation, including film badges, thermoluminescent dosimetry, electrets, and electroscopes, is detailed. The article closes with a summary of the use of radioisotopes in process instrumentation for a variety of applications, including the measurement of moisture, density, level, and thickness. The article incorporates numerous diagrams and tables.

SECTION 13—ELECTRICAL ENERGY MEASUREMENTS

Electric Meters   These instruments have been undergoing continuous development for at least 150 years. Contemporary current, voltage, resistance, frequency, and electric charge measuring instruments are detailed. Measurement methodologies described include permanent-magnetic moving-coil instruments, moving-iron instruments, rectifier instruments, electrodynamic instruments, direct current–ratio indicating mechanisms, thermocouple instruments, and electronic instruments. The various approaches used in digital voltmeter design are described, including successive approximations, dual-slope techniques, ramp and counter, and voltage-to-frequency conversion. The article incorporates 28 illustrations.

Electric Power and Energy Measurement   As the author of this article points out, traditionally the term "power" in association with electricity has tended to lose its true meaning. Sometimes power is used where actually the correct term energy should be used. The author develops this observation and includes brief descriptions of power theory and average power. Wattmeters of various designs (dynamometer, thermal watt converter, and low power factor wattmeter) are described in some detail. Power measurement and the Blondel theorem are delineated. Other topics include power factor and its measurement, varmeters, and instrument transformers.

SECTION 14—CIRCUITS, COMPONENTS, AND SUBSYSTEMS

A number of circuits, components, and subsystems are used over and over in various type of industrial instruments. To avoid repetitive descriptions throughout the Handbook, we have grouped many of these elements in this special section.

Bridges and Potentiometers.   This section has been updated to reflect electronic technology of the past decade. The article features equations and diagrams for a large number of bridge types, including the simple bridge circuit, the unbalanced dc bridge, the dc Kelvin bridge, and variations of dc Wheatstone and Kelvin bridges and their use. Included are fault location test sets, the Varley loop test, the Murray and modified Murray loop test, the Fisher loop test, the three Varley method, determination of open faults, the universal ratio set (URS), the direct reading ratio set (DRRS), and the Julie ratiometric bridge. Attention is given to bridges for very high resistance measurements, the guarding and shielding of bridges, and the characteristics of resistance materials. Discussed in some detail are ac bridges (Maxwell bridge, impedance bridge, Wein bridge, Owen, Hay, Schering, Anderson, conductivity, Heydweiller, Heaviside, Campbell, and ratio transformer bridge). Direct current null detectors used in bridges (galvanometers, electronic null detectors) and ac null detectors are described. The article closes with a detailed discussion of potentiometers, including standardization sources (saturated and unsaturated cells and the zener regulated supply). There are 49 diagrams of bridge circuits.

Self-Balancing Electrical Instruments   In self-balancing bridge and potentiometer circuits, the servo amplifier functions with a motor to position a calibrated feedback device to accomplish null balance. The recording industrial potentiometer is a widely used form of self-balancing servo device. This article is mainly addressed to this type of instrument. Topics include operating principles, measuring circuits, conversion (dc to ac and ac to dc), rebalancing (feedback) devices, voltage references, and damping. The use of potentiometer instruments for temperature measurements is stressed. Normal-mode and common-mode interference is described in some detail. The article closes with an appraisal of technological advancements that have and will continue to affect the design of self-balancing electrical instruments.

Digital Technology   There are few, if any, aspects of industrial and process control instrumentation that have not been penetrated in some way by developments in digital technology over the past decade. Computers of numerous sizes and complexities, digital displays, printouts, digital transmission of data appear in systems, simple and complex, and run the gamut of the process variables. The term *computer control* no longer has the specific connotations of a decade or less ago—because of the great variety of computer and solid-state hardware available.

In this article is centralized the larger body of information that fundamentally makes up digital technology—computer fundamentals, computer and digital system characteristics, computer hardware, computer storage, digital codes, units, and language, computer programs and programming, operations and procedures, digital input/output subsystems and components, data converters, and digital system amplifiers. This is a long, but orderly article, accompanied by 83 diagrams of systems, circuits, hardware, and applications.

Safety in Instrumentation and Control Systems Much progress has been made in this field during the past 5 to 10 years. The coverage of this article is fully updated as of the mid-1980s. The article commences with a review of areas and materials, including definitions of the National Electrical Code Area Classification System. Techniques used to reduce explosion hazards are explored, including intrinsic safety, explosion-proof housings, hermetic sealing, and pressurization systems are described in considerable detail. The procedures for certification of systems is thoroughly discussed, as is the installation of equipment.

SECTION 15—INFORMATION AND DISPLAY SYSTEMS

Information Display and Storage Systems Rather dramatic changes have occurred during the past decade in information display and storage. This article highlights not only the newer techniques, but also reviews the time-honored methods still widely used. The article is divided into two major portions—indicating means and graphic recorders. Some of the topics covered under indicating means include moving-pointer indicators, moving-scale indicators, multipoint and multirange indicators, digital indicators, illuminative display elements, with description of luminescence, fluorescence, and phosphorescence, light-emitting diodes, and liquid-crystal display elements. Cathode-ray tubes are described in considerable detail, with special mention of storage CRTs, character generation tubes, and scan conversion tubes. Alarms and annunciators, which have become much more sophisticated in recent years, are dealt with in considerable detail, including an evaluation of alarm information. Diagrams are given for annunciator operational sequences. Other topics include graphic displays and computer graphics.

In the portion of the article on graphic recorders, the latest generation of hardware is described. Included is a convenient table that lists numerous factors in the classification and selection of information storage devices. The types of graphic coordinates most frequently used in industrial instruments are described. Considerable emphasis is given to strip-chart recorders, with details on marking-medium combinations, pen and ink, impact printing, thermal writing, electric writing, light-beam methods, and electrostatic writing. Traditional and contemporary standard strip-chart recorders are described and illustrated. Developments in hybrid recorders are delineated. Circular-chart recorders, which are still popular for many applications, are described, along with X-Y recorders, alarm information recorders, sequential events recorders, oscillographic recorders, and cathode-ray tube recorders, including the fiber-optic CRT recorder. The article closes with a description of loggers and printers. The text is supported by 78 diagrams and photos.

Panel, Keyboard, and Control Switches During the past decade, in addition to the continuing refinement in design of traditional switching formats, developments and trends have included greater emphasis on human engineering factors (for example, the tactile response of the operator in manipulating a switch), the application of solid-state switching circuitry, and the incorporation of such principles as the Hall effect, expansion of the role of switches in the total system (for example, encoding switches), and the integration of switches in large groupings in complex data-input and data-output systems (for example, microcomputer-based touch panel keyboards).

This article commences with a review of traditional switch configurations, giving detailed coverage of hardware (switching action and circuitry, contact rating, effective display area, service life, termination, etc.). Considerable detail follows on membrane switches which have become quite popular during the past 5 to 10 years. Mercury switches and electromechanical relays are also covered. The article incorporates 30 diagrams of traditional and very advanced switch designs.

Voice Recognition and Speech Synthesis The application of these techniques to industrial instrumentation still remains somewhat debatable. Mainly, it is the timing as to when these techniques can become well established. Under voice recognition, this article explains the differences between voiced and unvoiced sounds and how, to date, voice and speaker identification systems have been designed and applied. Under speech synthesis, the concept of the basic phonological element of speech, the *phoneme,* is described. The major approaches to speech synthesis explored and applied to date (adaptive differential pulse-code modulation, formant synthesis, and text synthesis) are described and diagrammed.

Control Centers—From Spec to Job Site This is a very practical article which summarizes control center design objectives, discusses the formulation of basic specifications, including sample specification forms, the concept of the general arrangement drawing and loop diagrams. Also covered are guidelines for handling inquiries and changes during the construction stage, evaluation of quotations, and the inspection of functional tests of panels at the factory and on site. The impact of distributed control on panel layouts is briefly summarized.

SECTION 16—DATA COMMUNICATION SYSTEMS

Telemetry Telemetry may be defined as *measuring at a distance* and applies to industrial, commercial, military, and space operations. This article commences with a broad classification of telemetry systems, followed by descriptions of specific systems [voltage telemetering, current telemetering, position telemetering, frequency telemetering, pulse telemetering, including pulse duration systems, pulse count systems, and pulse code (digital) systems]. Other topics include binary coded decimal (BCD) codes, Baudot code, baud versus bits, parallel and serial transmission, interface standards, code security and error detection, the Hamming code, telemetering communication channels (two-wire, privately owned, leased channels, wideband channels, a tabular summary of the principal channels of the National Communication Networks in the United States, microwave channels, and power line carrier channels). The use of time-division multiplexing and frequency-division multiplexing is delineated. The article closes with a discussion of modulation. The text is accompanied by 14 system diagrams and 9 tabular summaries.

Pneumatic Transmission and Control In this article, pneumatic transmission and control are lumped together because of the great similarities in hardware used for both objectives. Covered are the transmission signal range, transmission system dynamics, and the use of volume boosters. Descriptions of pneumatic devices include the baffle-nozzle amplifier, pilot relays, booster relays, nonbleed relays, pressure divider circuits, 1:1 repeaters, force-balance feedback amplifiers, and motion-balance feedback amplifiers. The modes of control as effected by pneumatic controllers are discussed.

Optical Transmission Systems Transmission systems and circuits based upon fiber-optics technology are attractive for instrumentation and process control applications for a number of reasons, including immunity to interference, high information-carrying capacity, effectiveness in alleviating ground loop problems, enhancement of safety in hazardous locations, and adaptability for handling digital data. These systems are progressively entering the process instrument field, but actually only a start has been made. This article reviews light wave transmission, digital light wave systems, the basic types of fibers used (modal and chromatic), optical-fiber performance and characteristics, light sources and detectors, cables and connections, the advantages of digital transmission, and imaging requirements.

SECTION 17—PRINCIPLES OF AUTOMATIC CONTROL

Fundamentals of Automatic Process Control As observed by readers of earlier editions of this Handbook, this is one of the most widely used articles of the Handbook. For the Third Edition, the article has been fully updated to reflect technological changes of the past decade. The article is intended not only for individual study, but also for use by groups of instrument engineers for college and in-plant training programs. The article commences with the nontheoretical analysis of a typical process. Discussed are process time lags, types of process disturbances, process reaction curves, manual versus automatic control, the basic functions of control, the anatomy of typical control systems, the impact of microprocessors in control instruments, the principal modes of control, selection of control action, stability of control loops, a brief summary of frequency-response analysis, criteria of control quality, nonlinearity of control-loop components, the adjustment of automatic controllers, multielement control (cascade, ratio, override controls), feedforward control, multiple variable-interactive control, self-tuning controllers, and general control system design procedures. The article features 56 diagrams.

Process Control Techniques   This article summarizes those basic control concepts and techniques which have been successfully applied to the design and analysis of analog and digital process control systems. The article is divided into five parts—system characteristics and representation, basic control concepts, continuous and sampled-data control systems, stability, and system design and control strategies. This article is the logical next step for the instrumentation engineer who understands the fundamentals of process control, but who desires to approach the complex subject in a more mathematical, theoretical manner. Nevertheless, the article provides very practical counsel for those persons charged with the responsibility of solving difficult control problems.

Control Algorithms   The implementation of a control function on a process requires either an analog control system, natural to the process but with its own problems, or a digital computation. The process of sampling an analog value, converting the resulting value to a (quantized) number, computing a control action, and outputting that value involves issues removed from the abstractions of the design control theory. This article deals with algorithm design. Some topics included are number systems and the effects of the basic arithmetic operations, fixed point format, fixed point scaling, error in fixed point arithmetic, fixed point multiplication and division, digital integration for control, floating point format, generalized multiple precision floating point, specification of fixed point algorithms, basic control algorithms, the lag calculation, lead/lag calculation, PID controller calculation, quantization and saturation effects, operational effects, and identification and matrix oriented operations. The article is replete with examples.

SECTION 18—AUTOMATIC CONTROLLERS

Electronic Controllers   Since their introduction in the late-1940s, electronic controllers have experienced several stages in their advancement, but none equal to the changes that have occurred during the past decade. This article is devoted to descriptions of the most recent electronic control configurations. The article commences with a review of controller requirements, including modes of control. Considerable attention is devoted to controller applications and common problems encountered in their application. A comparison is made between analog and digital controllers and a modern microprocessor-based controller is described, including functional block diagrams. The operation of digital controllers is described in some detail. Fifteen diagrams include process reaction curves, reaction recovery curves, ratio control loop, cascade control loop, override control loops, typical analog controller, typical digital controller, effective deviation for three control algorithms, and a block diagram of the microprocessor section of a digital controller.

Electric Controllers (Traditional)   The distinction between electric and electronic controllers is not sharp. The difference tends to be characterized as much by chronology as by technology. A number of traditional electric controllers, which have been used in industry for several decades and which remain in place in many installations, are described in this article.

Hydraulic Controllers   Self-contained closed-loop hydraulic controllers continue to be used for certain types of process control problems, but as the computer, with its electrical output, expands in process control applications, the electrohydraulic servo valve is gaining in usage. This combination adds the advantages of hydraulic control to the versatility of the computer. Also contributing to the expanding use of hydraulics is the steady improvement in fire-resistant fluids. This article commences with a review of the elements of hydraulic controllers, hydraulic relays, and hydraulic fluids. The comparison of these elements is accomplished through a series of tabular summaries. Other topics include power considerations, conservation, and a summary of the relative advantages and limitations of hydraulic controllers.

Fluidics   This is the technology of sensing, controlling, and information processing with devices that use a fluid medium and whose operation is based solely upon the interaction between fluid streams. Topics covered in this article include the characteristics of fluidic devices, principle of operation, various types of fluidic amplifiers, and applications, including fluidic control valves and fluidic flowmeters.

Programmable Controllers   These instruments have gained increasing popularity as control tools for industrial applications since their inception in 1969. Their power, flexibility, and ease of use have won acceptance by design engineers, operators, and maintenance personnel. This article describes the functions of a programmable controller, pointing out, for example, that unlike numerical control (NC) and computer numerical control (CNC), which are used to control position, the programmable control is used for sequence control. The principles and applications of PCs are described.

Time-Base Controllers   This article describes a large grouping of controllers which have in common a time-measurement system. Included are atomic clocks, crystal oscillators, pendulum clocks, electric clocks, synchronous motors, induction motors, regulated dc motors, air motors, capacitor discharge and charge circuits, flux-decay relays, delay lines, thermal devices, dashpots, pneumatic devices, electronic timers, and spring-actuated devices. In the portion of the article on timers, such devices as time switches, time-delay relays, interval timers, percentage timers, repeat-cycle timers, reset timers, and time-cycle controllers and step programmers are described. Both electromechanical and solid-state configurations are described. Another portion of the article describes time-scheduling controllers, including contoured cams and followers, strip-chart recorder-controllers with motor-driven setpoint, and digital program controllers. Because of the great variety of hardware available, the text is accompanied by 45 diagrams.

Distributed Instrument Systems   This article reviews the concept of distributed instrumentation/control from a practical viewpoint. Emphasized is the flexibility that can be designed into such systems. The architecture of microprocessor-based distributed instrument systems is reviewed, and the functional characteristics of the communications subsystem required to achieve practical objectives are summarized. Operations workcenters are described. The three major categories of controller subsystems are delineated—the dedicated loop controller, the programmable controller, and the multifunction, multiloop controller. The advantages and problems of combining auxiliary data collection are explained. Specific examples of problems/solutions facing the designer of a distributed instrument system are covered. The article includes a summary of a suggested systematic approach to avoid pitfalls in the design of a distributed instrument system.

SECTION 19—FINAL CONTROL ELEMENTS

Control Valves   This article is a thoroughly updated summary of the physical characteristics and functions of control valves. Included are discussions of body styles (sliding stem globe, single-port globe, top and bottom guided invertible globe, rotary stem, butterfly, and special valve designs for specific applications). The latter include small flows, cryogenic service, high pressure, slurry handling, mixing and proportioning, angle and Y pattern valves, and digital valves. Described are valve plug designs and body design details, body materials, valve trim, actuator design and applications, typical positioner performance, and auxiliary air relays. The article includes 52 diagrams and charts.

Control Valve Sizing   The selection of proper port and body size for a control valve is based on calculation of the correct valve flow coefficient C_v. This article is devoted to this proposition and includes discussions of pressure drop across the valve, the flowing quantity, specific gravity, and critical flow factors. Control valve flow formulas are discussed. Other topics include cavitation, the effect of pipe reducers, and the handling of high viscosity, laminar flow of liquids.

Control Valve Characteristics   The flow characteristic of a valve defines the flow behavior as the valve operates through rated stroke. Discussed in this article are inherent flow characteristics, linear characteristic, equal percentage characteristic, quick-opening characteristic, straight-sided V-port (parabolic) characteristic, three-way valve characteristic, rotary valve characteristics, installed flow characteristics, rangeability, and guidelines to selection of characteristics to suit the process.

Control Valve Noise   Fluid and transmission systems are a major source of industrial noise. A major component of process noise is the control valve. The article explains how control valve noise is the result of the turbulence introduced into the flow stream. The article includes a brief review of noise terminology, identification of sources of valve noise, the relation of gross flow parameters to noise potential of the valve, and techniques for predicting control valve noise and methods of noise control.

Control Valve Cavitation   Although cavitation as a hydrodynamic phenomenon has been known since the early-1900s, the actual mechanisms involved have been poorly understood until relatively recently. Cavitation is of major concern in control valves because not only is the effectiveness of a control valve decreased, but there are highly undesirable attendant side effects, including noise, vibration, and material damage. The three types of cavitation—vaporous, pseudocavitation, and gaseous—are defined and described in this article. The article includes the basic equations required in the analysis of cavitation. The primary causes of valve cavitation—mechanical and chemical attack—are detailed. The three principal methods for controlling cavitation—system design, material selection, and anticavitation products—are described.

Electric Actuators   This article presents a terse review of contemporary hardware, including reversible electric motor drives, saturable-core reactors, electropneumatic converters, and solid-state power switches.

Electric Motor Drive Controls   Much progress has been made during the past decade in the design and application of controls for electric motor drives, resulting in a marked increase in the use of these kinds of final controlling elements. Constant-speed and adjustable-speed drives are described. Power conversion technology and regulators are detailed. The article provides a succinct update on major developments in this field.

Stepping Motors   A stepping motor is an electromechanical device which rotates a discrete step angle when energized electrically. The principal available configurations (solenoid-operated ratchet, permanent magnet, and variable reluctance) are described. The most common applications for stepper motors include printers, tape readers, memory devices, valve actuators, positioning tables, counters, and indexers.

Process Instruments and Controls Handbook

Measurement Fundamentals *

Douglas M. Considine. *Editor-in-Chief. (Variables)*

Engineering Staff. *The Foxboro Company, Foxboro, Massachusetts. (Variables)*

David T. Goldman. *Associate Director for Planning, National Measurement Laboratory, National Bureau of Standards, Washington, D.C. (Units and Standards of Measurement)*

T. S. Imsland. *Engineering Department, Fisher Controls Company, Marshalltown, Iowa. (Measurement Terms and Characteristics)*

W. D. Wood. *Formerly Director, Standards Laboratory, Taylor/Sybron, Rochester, New York. (Measurement Terms and Characteristics)*

*Persons who authored complete articles or subsections of articles, or who otherwise cooperated in an outstanding manner in furnishing information and helpful counsel to the editorial staff.

Variables

**by
Douglas M. Considine***

A variable is that quantity or characteristic which is the object of measurement in an instrumentation/control system. Synonymous terms include measurement variable, instrumentation variable, and process variable. The last term is used commonly in the manufacturing (process) industries.

Variables generally are classified in two ways: (1) by their physical characteristics, such as those described in Table 1, and (2) by the measurement signals involved, such as those given in Table 2.

---

*Editor-in-Chief. Portions of this section were contributed by engineers of The Foxboro Company, Foxboro, Massachusetts.

Table 1. Classification of Variables by Physical Characteristics

Class of variable	Examples
Thermal variables—relate to the condition or character of a material dependent on the material's thermal energy.	Temperature, differential temperature Specific heat Thermal energy variables, such as enthalpy and entropy Calorific value
Radiation variables—relate to the emission, propagation, reflection, and absorption of energy through space or through a material in the form of waves; and, by extension, corpuscular emission, propagation, and absorption.	Nuclear radiation Electromagnetic radiation, such as radiant heat; infrared, visible, and ultraviolet light; x- and cosmic rays; gamma radiation Photometric variable Acoustic variable
Force variables	Total force Moment or torque Stress as force per unit area Pressure, differential pressure, vacuum or barometer as force per unit area Weight at local gravity Liquid level as a pressure
Rate variables—concerned with the rate at which a body, or a measured variable, is moving toward or away from a reference point or the repeat rate of some phenomenon. *Time* is always a component of a rate measurement.	Fluid volume flow rate Fluid mass flow rate Acceleration Frequency Linear velocity Angular velocity Vibration
Quantity variables—relate to the total quantity of material that exists within specific boundaries or passes some point in a given time period.	Mass Weight at local gravity Integrated flow rate Volume Thickness Moles of material Distance
Physical-property variables—concerned with the physical properties of materials with the exception of properties related to mass or chemical composition.	Density and specific gravity Humidity Moisture content Viscosity Consistency (e.g., of paper pulp) Structural characteristics, such as hardness, ductility, and lattice structure
Chemical-composition variables—relate to the chemical properties and analysis of substances. When measured at the output of a process, these variables can give a final measure of the performance of a plant. *Traditional laboratory instruments* are being replaced by on-line analyzers using techniques such as optical, infrared, and mass spectrometry, etc., as well as automated analysis following the old laboratory procedures, using microprocessors for calculations in real time.	A very abridged list of analysis variables would include the quantitative measurement of CO_2, CO, H_2S, NO_x, S, SO_x, C_2H_2, CH_4, $H^+(pH)$, water and air quality, and various solvents and chemicals.

Table 1. Classification of Variables by Physical Characteristics (*Continued*)

Class of variable	Examples
Electrical variables	Voltage
Capacitance and conductance can be used for product quality inference.	Electric current
	Resistance
Resistance most often is used as an indication of the temperature of a resistance element.	Conductance
	Inductance
Inductance and impedance are often used to measure the mechanical motion of a mechanical sensor.	Capacitance
	Impedance
Voltage and current are normally used for transmission of the output of a sensor to remotely located control equipment.	

Table 2. Classification of Industrial Variables and Measurement Methods

Variable	Mechanical motion	Liquid displacement	Light or electron beam	Force	Pressure	Voltage or amperage	Voltage/current	Pulse duration	Frequency	Pulse code	Temperature
Temperature	×	×	×	×	×	×	×		×		O
Nuclear radiation						×	×			×	×
Electromagnetic radiation	×					×	×				×
Total force	×	×	×	O	×	×	×		×		
Moment or torque	×	×	×	×		×	×				
Pressure or vacuum	×	×	×	×	O	×	×		×		
Flow	×	×		×	×	×	×	×	×		×
Speed or velocity	×			×	×	×		×	×		
Acceleration	×		×	×	×	×	×				
Mass[a]				×							
Weight	×			×	×		×		×		
Elapsed time	×					×	×			×	
Frequency	×					×	×		O	×	
Position	×	×	×	×	×	×	×	×	×	×	
Dimension[b]	×	×							×		
Contour[b]	×										
Level	×	×		×	×		×		×		
Density or specific gravity	×	×		×	×		×				
Humidity[c]	×						×				×
Moisture content	×			×			×				
Viscosity	×			×	×			×			
Chemical composition[d]					×	O	×				×
Electric voltage	×		×	×		O	×	×	×	×	×
Electric current	×		×	×		O	×	×	×	×	×
Electrical resistance				×		×	×	×	×		

Table 2. Classification of Industrial Variables and Measurement Methods (*Continued*)

Variable	Measuring methods										
	Mechanical motion	Liquid displacement	Light or electron beam	Force	Pressure	Voltage or amperage	Voltage/current	Pulse duration	Frequency	Pulse code	Temperature
Electric inductance						X	X	X	X		
Electric capacitance						X	X	X	X		
Electrical impedance						X	X	X	X		
Specific heat[e]											X
Thermal energy variables[e]											X
Calorific value[e]											X
Photometric variables[f]			X			X	X				
Acoustic variables[f]					X	X			X		

[a]Mass is almost invariably measured by the gravitational effect, that is, by weight.

[b]Dimension and contour, defined as the relative position between two points, are almost always measured by a position measurement, with a selected point of the dimension or contour maintained in a predetermined relationship to a reference point on the position measuring system.

[c]Humidity is measured by dewpoint temperature. This applies to direct measurement of the temperature of pure water whose vapor pressure is equal to the vapor pressure to be measured, and also to saturated lithium chloride vapor pressure in saturated lithium chloride elements.

[d]For most chemical composition measurements, no simple direct transducer exists. Measuring equipment in general is elaborate. In most cases, the measurement signal is an electric voltage or current, or an electric voltage-current relationship.

[e]No simple transducers exist for the measurement of specific heat, calorific value, entropy, enthalpy, and the like. Either measurements are derived from computations based on measurements of other variables, or specialized equipment is used to maintain a number of conditions constant so that one variable, usually temperature, varies in a predetermined relation to changes in the measured variable.

[f]Photometric variables and acoustical variables both include several distinct variables. The sensor for most photometric variables is some type of photocell; the sensor for most acoustic variables is some type of microphone. Both have electric outputs.

NOTE: X denotes that a standard transducer is commercially available for measuring the variable with the method indicated; O denotes that the variable is already in measurement method form.

Process Variable

Specifically, a process variable may be defined as a physical or chemical quantity, the variation of which will indicate a desirable or undesirable change in the operation of a manufacturing process. Process variables are measured and controlled for a number of reasons, including the following.

1. **Product Quality.** Most products are designed to meet requirements for certain specified physical or chemical properties. As the demand for higher quality material increases, the tolerances governing deviation from specifications become narrower. As tolerances grow narrower, the need for instrumental control increases.

2. **Material Savings.** By the consistent maintenance of adequate product quality, rejects are cut to a minimum, thus effecting savings of raw and in-process materials as well as the productive time of machinery and labor.

3. **Cost Accounting.** Instrumental measuring systems provide accurate raw data for efficiency and material balance calculations, as contrasted with rough manual estimates and arbitrary cost allocations.

4. *Production Inspecting and Testing.* Generally instruments can provide faster and more accurate day-in and day-out determinations of materials and product quality than are obtainable by manual means.

5. *Plant Safety and Comfort.* Instruments detect excessive heat and pressure, noxious vapors, inadequate ventilation, excessive noise, the presence of human limbs in dangerous machine areas, electric short circuits and leakage, flame failure, and numerous other unsafe or uncomfortable conditions.

Units and Standards of Measurement

by
David T. Goldman*

The International System of Units, officially abbreviated SI, is a modernized version of the metric system. The system was established by international agreement to provide a logical and interconnected framework for all measurements in science, industry, and commerce. The system is built upon a foundation of six base units of measurement, each of which is defined in the following paragraphs. All other SI units are derived from these base units. Multiples and submultiples are expressed in a decimal system. The use of metric weights and measures was legalized in the United States in 1866, and the customary units of weights and measures have been defined in terms of the meter and the kilogram since 1893. The only legal units for electricity and illumination in the United States are SI units. Common equivalents and conversions between customary units and SI units are given in Table 1. Multiples and submultiples, prefixes, and official symbols for SI units are given in Table 2.

Length

Length is expressed in terms of the *meter,* abbreviated m. The meter is defined as 1 650 763.73 wavelengths in vacuum of the orange-red line of the spectrum of krypton 86. An interferometer is used to measure length by means of light waves.

A new definition of the *meter* based on a very accurate measurement of the speed of light was adopted by the General Conference on Weights and Measures at its meeting in 1983: The meter has been defined as the length of path traveled by light in vacuum during a time interval of 1/299 792 458 of a second. Practical realization of this definition will be through time-of-flight measurements, by frequency comparison with laser radiations of known wavelengths, or through interferometric comparisons with radiations of stated wavelengths, such as that from krypton 86.

Other SI length-related units include:

Area, expressed in terms of the *square meter* (m^2). Land is often measured by the *hectare* (10,000 square meters, or approximately 2.5 acres).

Volume, expressed in terms of the *cubic meter* (m^3). The liter is a special name for the cubic decimeter (0.001 m^3).

*Associate Director for Planning, National Measurement Laboratory, U.S. National Bureau of Standards, Washington, D.C. (Several NBS scientists furnished information on specific measurement quantities.)

Table 1. Common Equivalents and Conversions

Approximate common equivalents		Conversions accurate to parts per million	
1 inch	= 25 millimeters	inches × 25.4*	= millimeters
1 foot	= 0.3 meter	feet × 0.3048*	= meters
1 yard	= 0.9 meter	yards × 0.9144*	= meters
1 mile	= 1.6 kilometers	miles × 1.60934	= kilometers
1 square inch	= 6.5 square centimeters	square inches × 6.4516*	= square centimeters
1 square foot	= 0.09 square meter	square feet × 0.0929030	= square meters
1 square yard	= 0.8 square meter	square yards × 0.836127	= square meters
1 acre	= 0.4 hectare	acres × 0.404686	= hectares
1 cubic inch	= 16 cubic centimeters	cubic inches × 16.3871	= cubic centimeters
1 cubic foot	= 0.03 cubic meter	cubic feet × 0.0283168	= cubic meters
1 cubic yard	= 0.8 cubic meter	cubic yards × 0.764555	= cubic meters
1 quart (liq)	= 1 liter†	quarts (liq) × 0.946353	= liters
1 gallon	= 0.004 cubic meter	gallons × 0.00378541	= cubic meters
1 ounce (avdp)	= 28 grams	ounces (avdp) × 28.3495	= grams
1 pound (avdp)	= 0.45 kilogram	pounds (avdp) × 0.453592	= kilograms
1 horsepower	= 0.75 kilowatt	horsepower × 0.745700	= kilowatts
1 millimeter	= 0.04 inch	millimeters × 0.0393701	= inches
1 meter	= 3.3 feet	meters × 3.28084	= feet
1 meter	= 1.1 yards	meters × 1.09361	= yards
1 kilometer	= 0.6 mile	kilometers × 0.621371	= miles
1 square centimeter	= 0.16 square inch	square centimeters × 0.155000	= square inches
1 square meter	= 11 square feet	square meters × 10.7639	= square feet
1 square meter	= 1.2 square yards	square meters × 1.19599	= square yards
1 hectare†	= 2.5 acres	hectares × 2.47104†	= acres
1 cubic centimeter	= 0.06 cubic inch	cubic centimeters × 0.0610237	= cubic inches
1 cubic meter	= 35 cubic feet	cubic meters × 35.3147	= cubic feet
1 cubic meter	= 1.3 cubic yards	cubic meters × 1.30795	= cubic yards
1 liter	= 1.057 quart (liq)	liters × 1.05669	= quarts (liq)
1 cubic meter	= 250 gallons	cubic meters × 264.172	= gallons
1 gram	= 0.035 ounces (avdp)	grams × 0.0352740	= ounces (avdp)
1 kilogram	= 2.2 pounds (avdp)	kilograms × 2.20462	= pounds (avdp)
1 kilowatt	= 1.3 horsepower	kilowatts × 1.34102	= horsepower

*Exact.
†Based on the U.S. survey foot (= 0.3048006 meter).

Table 2. Prefixes That May Be Applied to All SI Units

Symbol	Prefix	Pronunciation	Multiples and submultiples
E	exa	ĕx′á	10^{18}
P	peta	pĕt′á	10^{15}
T	tera	tĕr′á	10^{12}
G	giga	ji′gá	10^{9}
M	mega	mĕg′á	10^{6}
k	kilo	kĭl′ŏ	10^{3}
h	hecto	hĕk′tŏ	10^{2}
da	deka	dĕk′á	10
d	deci	dĕs′ĭ	10^{-1}
c	centi	sĕn′tĭ	10^{-2}
m	milli	mĭl′ĭ	10^{-3}
μ	micro	mĭ′krŏ	10^{-6}
n	nano	năn′ŏ	10^{-9}
p	pico	pĕ′kŏ	10^{-12}
f	femto	fĕm′tŏ	10^{-15}
a	atto	ăt′tŏ	10^{-18}

Time

Time is expressed in terms of the *second,* abbreviated s. The second is defined as the duration of 9 192 631 770 cycles of the radiation associated with a specified transition of the cesium atom. This duration is realized by tuning an oscillator to the resonance frequency of the cesium atoms as they pass through a system of magnets and a resonant cavity into a detector. Other SI time-related units include:

Frequency is the number of periods or cycles per second. The SI unit for frequency is the *hertz* (Hz) which is one cycle per second. Some nations broadcast standard frequencies and correct time. In the United States, radio stations WWV, WWVB, and WWVH, operated by the National Bureau of Standards, and U.S. Navy radio stations broadcast this information. Many shortwave receivers pick up WWV on frequencies of 2.5, 5, 10, and 20 MHz. The standard radio broadcast band extends from 535 to 1605 kHz.

Speed is determined by dividing distance by time. The SI unit for speed is the *meter per second* (m/s) which approximates 3 feet per second.

Acceleration is the rate of change in speed. The SI unit for acceleration is the *meter per second squared* (m/s^2).

Mass

Mass is expressed in terms of the *kilogram* (kg). The standard for the unit of mass is a cylinder of platinum-iridium alloy kept by the International Bureau of Weights and Measures at Paris. A duplicate in the custody of the National Bureau of Standards serves as the mass standard for the United States. This is the only base unit of measurement still defined by an artifact—as contrasted with a natural phenomenon which can be duplicated anywhere provided the required scientific equipment is at hand.

Force is closely allied to the concept of mass. The SI unit of force is the *newton* (N). A force of 1 newton, when applied for 1 second, will give to a 1-kilogram mass a speed of 1 meter per second (an acceleration of 1 meter per second squared). One newton equals approximately two-tenths of a pound of force.

The SI unit for *work and energy* of any kind is the *joule* (J).

$$J = N \times m$$

The SI unit for *power* of any kind is the *watt* (W).

$$W = \frac{J}{s}$$

Temperature

Temperature is expressed in terms of the *kelvin,* abbreviated K. The thermodynamic or Kelvin scale of temperature used in the SI has its origin or zero point at absolute zero and has a fixed point at the triple point of water defined as 273.16 kelvins. The Celsius scale is derived from the Kelvin scale. The triple point is defined as 0.01°C on the Celsius scale, which is approximately 32.02°F on the Fahrenheit scale. See article Temperature Scales, Standards, and Calibration in Sec. 2 of this Handbook. The triple-point cell, an evacuated glass cylinder filled with pure water, is used to define a known fixed temperature. When the cell is cooled until a mantle of ice forms where the temperature at the interface of solid, liquid, and vapor is 0.01°C,

$$\text{Temp. F} + 40 = 1.8 \, (\text{temp. C} + 40)$$

$$\text{Temp. F} = 1.8 \, (\text{temp. C}) + 32$$

$$\text{Temp. C} = (\text{temp. F} - 32)/1.8$$

NOTE: For a discussion on the derivation of the International Practical Temperature Scales, see J. F. Schooley, "Toward a New Scale of Temperature," *Dimensions/NBS,* September 1978, p. 21.

Electric Current

Electric current is expressed in terms of the *ampere* (A). The ampere is defined as the magnitude of the current which, when flowing through each of two long parallel wires separated by one meter in free space, results in a force between the two wires (due to their magnetic fields) of 2×10^{-7} newton for each meter of length.

The SI unit of *voltage* is the *volt* (V).

$$V = \frac{W}{A}$$

The SI unit of *electrical resistance* is the ohm (Ω).

$$\Omega = \frac{V}{A}$$

Amount of Substance

The amount of a substance is expressed in terms of the *mole* (mol). The mole is defined as the amount of substance that contains as many elementary entities as there are atoms in 0.012 kilogram of carbon 12.

Luminous Intensity

Luminous intensity is expressed in terms of the *candela* (cd). The candela is defined as the luminous intensity, in a given direction, of a source that emits monochromatic radiation of frequency 540×10^{12} hertz and that has a radiant intensity in that direction of 1/683 watt per steradian.

The SI unit of *light flux* is the *lumen* (lm). A source having an intensity of 1 candela in all directions radiates a light flux of 4π lumens. A 100-W incandescent bulb emits about 1700 lm.

Development of Units and Standards

If we were to commence anew today to develop a set of measuring units and standards, unquestionably we would proceed differently from the long, tortuous path that has been followed over several centuries. The process of establishing more precise and more easily achievable and reliable standards has no end in view. Progress must be reasonably slow, simply because so much is at stake when even a slight alteration is made in the definition of a standard. The changes that must take place in commerce and science to reflect alteration of a measurement standard are staggering and can have vast economic repercussions because of the basic resistance of the public to changes in traditions and habits. The continuing resistance to practical adoption of the metric system by the United States, the sole remaining industrial country that is not predominantly metric, is a manifestation of resistance to change.

The development of standards has, in general, proceeded from (1) the initial establishment of an arbitrary artifact, such as the yard of Henry II; and (2) the finding by scientists of natural phenomena that require no artifact for comparison. Several different artifacts have served as standards of length even within the boundaries of a single country. Gabriel Mouton, a vicar of Lyons, France, in the late seventeenth century, proposed that the standard of length be based on the length of a pendulum whose period was a second. Since this length depends on gravitational acceleration and therefore varies depending on altitude and location, this proposal was not accepted by the National Assembly of France in their deliberations establishing a new system of measurement. Instead, in 1791, the basic unit of length, the meter, was defined as one ten-millionth of the quadrant of the earth's meridian. By measuring the distance between Dunkurque, France, and Barcelona, Spain, which lie on the same meridian, the length of the meter could be determined. A standard of sintered platinum representing the meter was constructed and formed the basis for the start of the metric system. Following the signing of the Treaty of the Meter in 1875, platinum-iridium bars of modified X cross section were constructed and calibrated against the platinum standard. The bar closest to the platinum standard was selected to serve as the international standard of length. The other bars were distributed to other countries to ensure an internationally compatible system of length measurement. This meter bar continued until 1960 as the international standard of length. At that time the advance of scientific measurement, especially the accuracy with which the wavelength of light could be determined by spectroscopic means, led to a new definition of the meter that was not dependent on an artifact. In this definition, the meter is defined as 1 650 763.73 wavelengths in vacuum of the orange-red line of the spectrum of krypton 86. As measurement capability, particularly in the use of lasers, progressed, together with the need for even higher accuracy in the determination of lengths and distances, a new definition of the meter became both possible and desirable. Based on the very accurate measurement of the speed of light, 299 792 458 m/s, a new definition of the meter has been approved—as previously defined in this Handbook.

The continuing refinement of measurement standards is one of humankind's better examples of international cooperation. The Treaty of the Meter, which was signed by 17 nations including the United States, established an International Bureau of Weights and Measures to disseminate international standards of length and mass. Meetings of the General Conference on Weights and Measures have been held most recently at four-year intervals. Further, an International Committee for Weights and Measures was established to implement the recommendations of the general conference. The committee meets at present every year.

Examples of recent changes in measurement standards include new definitions of the meter and the candela, and extending the International Practical Temperature Scale to much lower temperatures. A list of actions taken in modifying the International System can be found in *NBS Special Publication 330,* "The International System of Units (SI)," 4th ed., 1981. Other decisions have been made, increasing international cooperation in radio frequency measurements, accurate time dissemination, and standard reference materials coordination.

Electrical Quantities

Electrical instruments are calibrated in terms of the prime standards of the quantity measured. These prime standards are established and maintained by various government standards organizations throughout the world—sometimes referred to as legal standards. The national caretaker of standards in the United States is the National Bureau of Standards. The method of tracing the accuracy of voltmeters, ammeters, and wattmeters to prime standards is outlined schematically in Fig. 1. Sophisticated standards and calibration services also are available from a limited number of private firms.*

Manufacturers of standard cells and precision electrical instruments and systems usually maintain their own standards laboratories which contain local standards that are calibrated or recalibrated on a periodic scheduled basis by an official organization, such as the National Bureau of Standards. Laboratory secondary standards and the working standards used in various stages of manufacturing and assembly are, in turn, calibrated against these local standards.

Various types of ratio devices are used to extend the standard-cell voltage (approximately 1 V) to cover many ranges of commercial instruments and in terms of current in amperes in addition to voltage through the use of standard resistors. Extension of voltage values is accomplished through the use of what is commonly referred to as a volt box. Effectively, this is a high-valued resistor,

---

*Government laboratories generally calibrate only primary standards in selected areas; private firms generally must be relied on to calibrate instruments, test equipment, or standards in other areas. See "A Directory of Standards Laboratories," prepared by the National Council of Standards Laboratories, available from Secretariat, National Conference of Standards Laboratories, c/o National Bureau of Standards, Boulder, Colorado 80303.

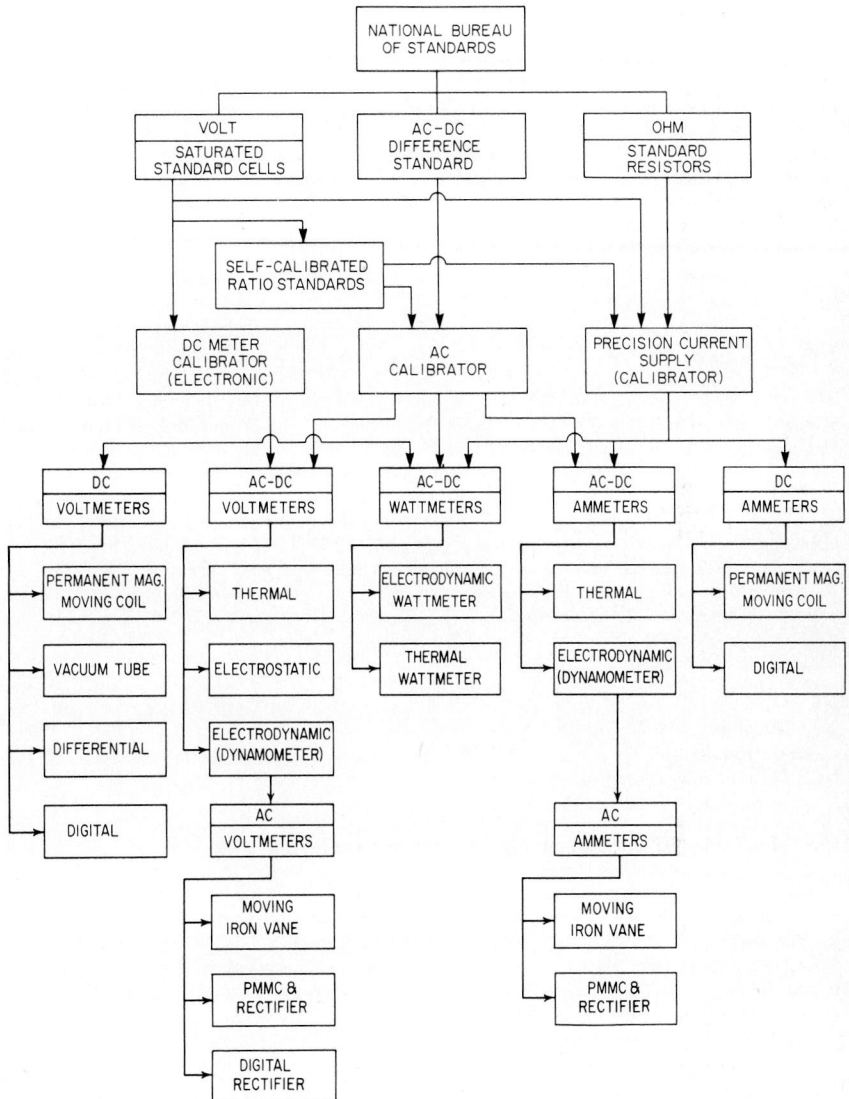

Fig. 1 Traceability of electrical instruments back to prime standards maintained by the National Bureau of Standards.

accurately tapped at exact proportions of its total value. If the total voltage is applied to the total resistance, then the value at the tapped points is low enough to compare accurately with the standard-cell voltage.

Extension of current values is accomplished through the use of very low-valued resistors; passing the unknown current through them results in voltage which, in turn, can be compared with the standard-cell voltage, and through the use of Ohm's law and the resistance value, the value of the current can be established. Such low-valued resistors are called shunts, and standard units of this kind are maintained with much care. These shunts can be calibrated—as by the National Bureau of Standards.

The direct voltages and currents established at this level are used to calibrate electronic calibrators which, in turn, are used to calibrate dc voltmeters, ammeters, and ac-dc voltmeters, ammeters, and electrodynamic wattmeters. Also shown in Fig. 1 are ac-dc difference or transfer standards. These are high-accuracy instruments, such as the thermal- and electrodynamic-type voltmeters, ammeters, and wattmeters, which are sent to the National Bureau of Standards for calibration. In this case, NBS does not perform an accuracy check but rather determines the difference between the ac calibration at various frequencies and that on direct current. These standards are then used for determining the performance of ac-dc instruments when used on alternating current.

Alternating-current voltmeters and ammeters are usually calibrated against electronic ac calibrators which have been calibrated against dc standards by using thermal transfer standards.

Voltage Standards

The cadmium cell, developed by Edward Weston in 1893 and universally known as the Weston standard cell, was used as a world standard of emf from the time of its acceptance by the International Committee on Electrical Units and Standards in 1908 until the advent of Josephson voltage standards in the 1970s. Standard cells are still the prevalent primary voltage standard in industry. The unit of voltage at the International Bureau of Weights and Measures (BIPM) and the legal voltages of most industrial nations are now maintained through the ac Josephson effect.* The Josephson effect is realized through weak connections (commonly through a dielectric tunnel barrier a few nanometers thick) between superconductors. Such connections are referred to as Josephson junctions. For a wide range of frequencies, the current-voltage characteristics of a Josephson junction irradiated with electromagnetic radiation of frequency f will exhibit zero-resistance parts (constant voltage steps) at voltages given by

$$V = \frac{nhf}{2e}$$

where n = an integer
h = Planck's constant
e = magnitude of the charge of an electron

The units of voltage are defined through the equation above by assigning a value to $2e/h$.† In general, the ac Josephson effect is used periodically to determine the emf of a group of standard cells that then embody the unit of voltage between determinations.

A standard cell is made in two types: (1) a normal cell containing a saturated cadmium sulfate solution, and (2) a type used as a working standard in which the solution is less than saturated above 4°C. The saturated cell is the basic standard for maintaining the value of the volt and is used in this manner in all national laboratories. Its rather high temperature coefficient must always be taken into account. See Fig. 2.

The emf of the unsaturated cell is within 1.0188 and 1.0198 abs V, the exact voltage of each cell being established by comparison with the normal or saturated cell. This cell is a useful working standard because of its negligible temperature coefficient.

Electrochemical cells of the type described have a relatively high internal resistance and thus should be used under zero load conditions. The cells should not be short-circuited, because several weeks may be required for full recovery of a stable reference voltage—if the cell has not been permanently damaged.

Highly precise reference signals are available from electronic standards which employ temperature-compensated zener diodes. A wide variety of precision black box instruments, arranged for flexibility and convenience of use, are marketed. For a survey of these, see Ref. 4 at end of this article.

---

*See "Josephson Effect" in *The McGraw-Hill Encyclopedia of Science & Technology*, 5th ed., vol. 7, McGraw-Hill, New York, 1982, pp. 438–441.

†For the United States legal volt, V_{NBS}, the assigned value is $2e/h$ = 483 593.42 GHz/V_{NBS}; for the international volt, V_{69-BI}, the assigned value is $2e/h$ = 483 594.0 GHz/V_{69-BI}.

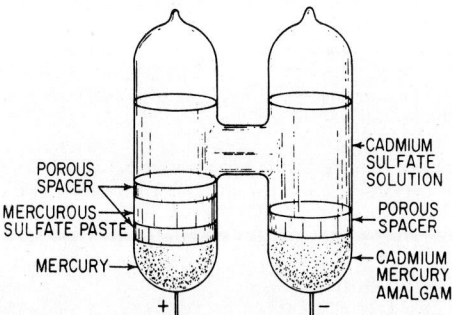

Fig. 2 Weston standard cadmium cell.

Standard Resistors

The hermetically sealed, wire-wound 1-Ω resistors designed by J. L. Thomas of the National Bureau of Standards are among the highest quality resistance standards ever produced.* The United States legal ohm, Ω_{NBS}, is the mean four-terminal resistance of a group of Thomas-type standard resistors immersed in oil at 25°C under a power dissipation of 0.01 watt. In the Thomas-type standard (Fig. 3), the container is made of coaxial cylinders only slightly different in diameter, with the space between the cylinders sealed. The resistance element (carefully annealed Manganin† wire) is mounted in this space in good thermal contact with the smaller cylinder, which serves as the inner wall of the container.

A subgroup of five Thomas 1-Ω resistors, together with resistors at different values and/or different design, are used as working standards at NBS.‡ These standards allow calibrations to be made over a wide range of resistance values (10^{-4} to $10^{15}\ \Omega$) and under conditions differing from those

*J. L. Thomas, *J. Res. NBS,* vol. 36, 1946, pp. 107–111.
†Trademark, Wilbur B. Driver Co., Newark, N.J.
‡N. B. Belecki, B. L. Dunfee, and O. Petersons, "The National Measurement System for Electricity," *NBSIR 75-935,* Septermber 1978.

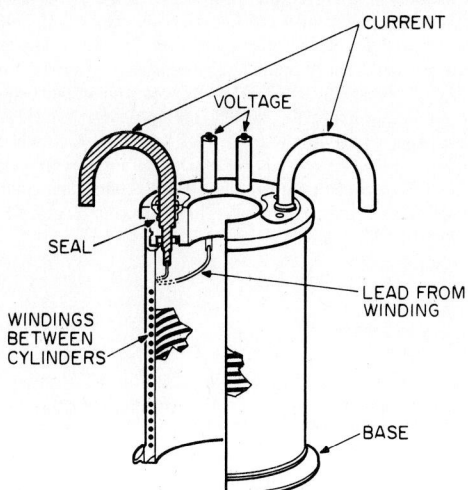

Fig. 3 Thomas-type resistance standard.

which the legal unit is maintained. Examples of the latter are standard resistors especially designed for stable characteristics at very high currents.

Temperature Standards

Temperature standards are described in the article Temperature Scales, Standards, and Calibration in Sec. 2 of this Handbook.

Dimensional and Geometric Quantities

Because of the extensive scope of the metalworking field, there are numerous industrial metrology laboratories which supplement the services of the National Bureau of Standards on a private basis. With instrumentation and environment comparable with the NBS to which all standards trace, the well-equipped industrial standards laboratory provides measurement and inspection services to private industry and others in doing such things as calibrating and certifying gage blocks and end measures, as well as plain and taper plug and ring gage calibration, thread gage calibration, hardness testing, and surface finish analysis. The degree to which the environment is controlled is one of the main differences between the average industrial gage laboratory and a metrology laboratory. For example, interferometric measurement of gage blocks should be performed at $68 \pm \frac{1}{20}°$F and at a relative humidity of 50% or less. Some of the tools used are described in the article Dimensional Metrology, Sec. 9, of this Handbook.

Calibration Services

With well over a score of major measurement quantities of concern to scientists and process engineers, the requirements for calibration services are extremely wide and well beyond the scope of this Handbook. The National Bureau of Standards publishes a catalog of its services, "Calibration and Related Measurement Services of the National Bureau of Standards," *NBS Special Publication 250 (SP250)*, which is updated every two years. An appendix to *NBS SP250* lists current prices and is updated every six months. Several hundred different calibration services are listed in *NBS SP250*, with typical fees ranging from a few hundred dollars to several thousand dollars. Many of the services listed are on a fixed-price basis, while others are listed "at cost." The latter are those services for which costs vary considerably, depending on the particular instrument submitted for calibration and the particular conditions of test desired. NBS distinguishes between three categories of service: (1) calibration services, (2) measurement assurance program services, and (3) special tests. The distinctions between these categories are explained in *NBS SP250*. Except for government priority work, tests are scheduled on the basis of order of receipt. A few services are performed only at certain times of the year, and in such cases communication with NBS regarding current schedules is essential. Questions regarding NBS calibration services should be referred to the NBS Office of Measurement Services, Washington, D.C. 20234.

Standard Reference Materials

Reliable measurement is one of the cornerstones of today's technically complex society. The need for accurate, consistent measurements is rapidly increasing—to improve efficiency in industry, monitor the environment, and improve health care.

Standard reference materials (SRMs), produced by the National Bureau of Standards since 1905, play a major role in increasing and ensuring the accuracy of these measurements. Standard reference materials are well-characterized, homogeneous materials with specific chemical or physical properties measured and certified by NBS. Standard reference materials are used widely throughout the

United States and the world to help develop test methods of proven accuracy, to calibrate instruments and measurement systems used to maintain quality control of the production of materials and goods, to help ensure equity in buyer-seller transactions, and to ensure the long-term reliability and integrity of the measurement process.

The National Bureau of Standards issues SRMs in 70 major categories, and over 800 different SRMs are available. During 1982, NBS distributed 40,000 SRM units to over 10,000 patrons throughout the world. Some of the major categories are

Metals	*Chemical Properties and Analytical*
Carbides	Calorimetric
Cast irons	Electron probe microanalytical
Cast steels	Ion activity
Gases in metals	Isotopics
High-purity metals	Metalloorganic compounds
Nonferrous alloys	Minerals
Steels	Ores
Steelmaking alloys	Primary chemicals
Physical Properties	Reference fuels
Color	Surface flammability
Freezing points	Trace elements
Magnetic	X-ray diffraction
Mechanical and metrology	*Products*
Melting points	Biologicals
Nuclear materials	Botanicals
Optical	Cements
Permittivity	Clinicals
Radioactivity	Computer tapes
Resistivity	Fertilizers
Sizing standards	Gas transmission
Smoke density	Glasses
Thermal conductivity	Photographic
Thermal expansion	Refractories
Thermal resistance	Rubber materials
Thermocouple materials	*Other*
Vapor pressure	Environmentals
Water vapor permeance	Industrial hygiene

REFERENCES

1. "Brief History of Measurement Systems with a Chart of the Modernized Metric System," *NBS Spec. Pub. 304A,* revised, August 1981.

2. Braudaway, D. W., and R. E. Kleimann: "A High-Resolution Prototype System for Automatic Measurement of Standard Cell Voltages," *IEEE Trans. Instrum. Meas.,* vol. IM-23, no. 4, December 1974.

3. Eicke, W. G., and J. M. Cameron: "Designs for Surveillance of the Volt Maintained by a Small Group of Saturated Standard Cells," *NBS Tech. Note 430,* October 1967.

4. Staff: "Voltage Standards Survey," *Instrum. Control Sys.,* vol. 44, no. 10, October 1971.

5. Benedict, R. P.: *Fundamentals of Temperature, Pressure and Flow Measurements,* 2d ed., Wiley, New York, 1977.

6. Schooley, J. F.: "Progress Toward a New Scale of Temperature," *Dimensionless/NBS,* November 1980, p. 10.

7. Heydemann, P. L.: A Primer on Pressure, *Dimensions/NBS,* July 1977, p. 3.

8. Croarkin, C., Beers, J., Tucker, C., and J. Cameron: "Measurement Assurance for Gage Blocks," *NBS Monog. 163,* 1979.

9. Youden, W. J.: "Uncertainties in Calibration," *IRE (IEEE) Trans.,* vol. 11, nos. 3 and 4, December 1962.

10. Eisenhart, C.: "Realistic Evaluation of the Precision and Accuracy of Instrument Calibration Systems," *J. Res. NBS,* vol. 67C, no. 2, April–June, 1962, p. 161.

11. Natrella, M. G.: "Experimental Statistics," *NBS Handb 91,* August 1, 1963.

12. Ku, H. H. (ed.): "Precision Measurement and Calibration: Statistical Concepts and Procedures," *NBS Spec. Publ. 300,* vol. I, February 1969.

13. "Calibration and Related Measurement Services of the National Bureau of Standards," *NBS Spec. Publ. 250* (issued periodically).

14. "Standard Reference Materials," *NBS Spec. Publ. 260* (issued periodically).

15. Mollet, P.: *Optics in Metrology,* Pergamon Press, New York, 1960.

16. Astin, A. V.: "Standards of Measurement," *Sci. Am.,* vol. 218, no. 6, June 1968.

Measurement Terms and Characteristics

Over a span of years, several technical societies and professional organizations that exert cognizance over the instrument and control field have been concerned with the development of a body of definitions and terms to describe in a consistent manner the many characteristics and specifications that apply to instruments and controllers. In the second edition of this Handbook (1974), some of the measurement terms drafted by the Scientific Apparatus Makers Association (SAMA) were presented. The full list of both measurement and control terms was released in the SAMA publication *PMC20-2-1970* ("Process Measurement and Control Technology"). More recently, the earlier work of SAMA committees was melded into the standards program of the Instrument Society of America (ISA), working in cooperation with the American National Standards Insititute and other standards organizations.

Although *ISA-S51.1* ("Standard Process Instrumentation Terminology") is easily available and procurement of it is recommended to all interested users of this Handbook, the review included in the second edition of this Handbook has been updated and somewhat expanded. The commentary* that provides some additional insight into the definitions is included. In the following paragraphs, the official definitions appear in bold type, whereas the commentary appears in standard type.

Signal Terms

In measurement devices that indicate, record, or transmit, it is important to know exactly what quantity is being measured. The following terms will help to clarify this area of measurement.

Measured Variable

The physical quantity, property, or condition which is to be measured.

Note 1: It is sometimes referred to as the measurand.

Note 2: Common measured variables are temperature, pressure, rate of flow, thickness, speed, etc. (Ref. 8).

---

*By Wallace D. Wood, formerly of Taylor/Sybron, Rochester, New York, and Theodore S. Imsland, Fisher Controls Company, Marshalltown, Iowa.

Measured Signal

The electrical, mechanical, pneumatic, or other variable applied to the input of a device. It is the analog of the measured variable produced by a transducer (when such is used).

Note 1: In a thermocouple thermometer, the measured signal is an emf which is the electrical analog of the temperature applied to the thermocouple.

Note 2: In a flowmeter, the measured signal may be a differential pressure which is the analog of the rate of flow through the orifice.

Note 3: In an electric tachometer system, the measured signal may be a voltage which is the electrical analog of the speed of rotation of the part coupled to the tachometer generator. See Table 1.

A *measured signal* is normally thought of as the signal produced by the primary element and applied to the input of the secondary element. However, it is also possible to consider that the output of the secondary element (usually a standard signal, such as 3 to 15 psig, (20 to 100 kPa), 10 to 50 mA, etc.) is another form of measured signal—sometimes called the *transmitted signal*.

The measured signal can be the result of output of more than one primary element. For example: the measurement of average temperature using more than one thermocouple, or the measurement of total flow by combining the outputs of more than one flow transmitter.

Input Signal

A signal applied to a device element, or system (Ref. 4).

The pressure applied to the input connection of a pressure transmitter is an input signal.

Table 1. Illustrations of the Use of the Terms Measured Variable and Measured Signal (English Units Used in Examples)

TYPICAL RANGES	TYPE OF RANGE	RANGE	LOWER RANGE-VALUE	UPPER RANGE-VALUE	SPAN
(1) THERMOCOUPLE 0 0-2000°F TYPE K T/C	MEASURED VARIABLE	0 TO 2000°F	0°F	2000°F	2000°F
-0.68 +44.91 mV	MEASURED SIGNAL	-0.68 TO +44.91 mV	-0.68 mV	+44.91 mV	45.59 mV
0 20 X100 = °F	SCALE AND/OR CHART	0 TO 2000°F	0°F	2000°F	2000°F
(2) FLOW METER 0 10,000 lb/h	MEASURED VARIABLE	0 TO 10,000 lb/h	0 lb/h	10,000 lb/h	10,000 lb/h
0 100 in H$_2$O	MEASURED SIGNAL	0 to 100 in H$_2$O	0 in H$_2$O	50 in H$_2$O	50 in H$_2$O
0 10 1000 lb/h	SCALE AND/OR CHART	0 TO 10,000 lb/h	0 lb/h	10,000 lb/h	10,000 lb/h
(3) TACHOMETER 0 500 rpm	MEASURED VARIABLE	0 TO 500 rpm	0 rpm	500 rpm	500 rpm
0 5V	MEASURED SIGNAL	0 TO 5V	0 V	5V	5V
0 80 ft/s	SCALE AND/OR CHART	0 TO 80 ft/s	0 ft/s	80 ft/s	80 ft/s

Output Signal

A signal delivered by a device, element, or system (Ref. 4).

Range-Related Terms

Terms such as *range* and *span* are commonly used to describe the region over which a quantity is being measured. The following definitions will clarify the relationship between these terms.

Range

The region between the limits within which a quantity is measured, received, or transmitted, expressed by stating the lower and upper range-values.

Note 1: For example:
    a.   0 to 150°F
         −20 to +200°F
         20 to 150°C

Note 2: Unless otherwise modified, input range is implied.

Note 3: The following compound terms are used with suitable modifications in the units: measured variable range, measured signal range, indicating scale range, chart scale range, etc. See Tables 1 and 2.

Note 4: For multirange devices, this definition applies to the particular range the device is set to measure (Ref. 8).

Span

The algebraic difference between the upper and lower range-values.

Note 1. For example:
    a. Range   0 to 150°F, span 150°F
    b. Range −20 to 200°F, span 220°F
    c. Range   20 to 150°F, span 130°C

Table 2. Illustrations of the Use of Range and Span Terminology

TYPICAL RANGES	NAME	RANGE	LOWER RANGE VALUE	UPPER RANGE VALUE	SPAN	SUPPLE-MENTARY DATA
0 + 100	−	0 TO 100	0	+ 100	100	−
20 + 100	SUPPRESSED ZERO RANGE	20 TO 100	20	+ 100	80	SUPPRESSION RATIO =.25
−25 0 + 100	ELEVATED ZERO RANGE	−25 TO + 100	− 25	+ 100	125	−
−100 0	ELEVATED ZERO RANGE	− 100 TO 0	−100	0	100	−
−100 −20	ELEVATED ZERO RANGE	−100 TO −20	−100	− 20	80	−

*Suppression ratio 0.25.

Note 2: The following compound terms are used with suitable modification in the units: measured variable span, measured signal span, etc.

Note 3: For multirange devices, this definition applies to the particular range that the device is set to measure (Ref. 3). See Table 2.

Overrange

Of a system or element, any excess value of the input signal above its upper range-value or below its lower range-value.

Note that it is the signal that is overrange, not the system or element.

On multirange instruments, such as the differential-pressure cell, it is often necessary to distinguish between the limits to which the device can be adjusted versus the limits to which the device is adjusted. The following terms describe these limits in detail.

Lower Range-Limit

The lowest quantity that a device can be adjusted to measure.

Note: The following compound terms are used with suitable modifications to the units: measured variable lower range-limit, measured signal lower range-limit, etc. See Tables 1 and 2 (Ref. 8).

Upper Range-Limit

The highest value of the measured variable that a device can be adjusted to measure.

Note: The following compound terms are used with suitable modifications to the units: measured variable upper range-limit, measured signal upper range-limit, etc. See Tables 1 and 2 (Ref. 8).

Lower Range-Value

The lowest value of the measured variable a device is adjusted to measure.

Note: The following compound terms are used with suitable modifications to the units: measured variable lower range-value, measured signal lower range-value, etc. See Tables 1 and 2 (Ref. 8).

Upper Range-Value

The highest value of the measured variable a device is adjusted to measure.

Note: The following compound terms are used with suitable modifications to the units: measured variable upper range-value, measured signal upper range-value, etc. See Tables 1 and 2 (Ref. 8).

Some confusion has existed in the use of the terms *elevated-zero range, suppressed-zero range, zero elevation, zero suppression,* and *suppression ratio.* The following definitions place these terms in a perspective that is logical and easy to understand.

Elevated-Zero Range

A range in which the zero value of the measured variable, measured signal, etc., is greater than the lower range-value. See Table 2.

Note 1: The zero may be between the lower and upper range-values, at the upper range-value, or above the upper range-value.

Note 2: The terms suppression, suppressed range, and suppressed span are frequently used to

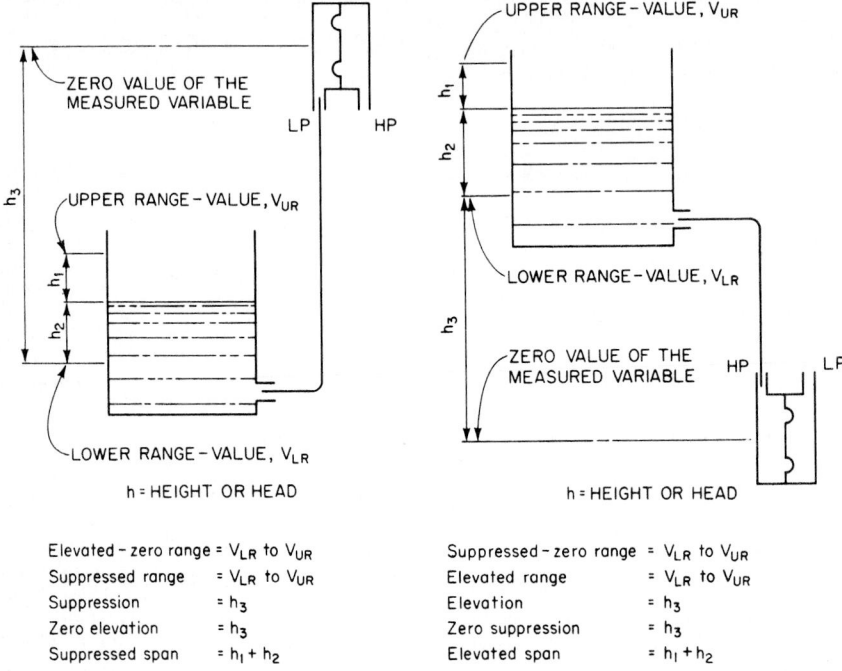

h = HEIGHT OR HEAD

Elevated – zero range	= V_{LR} to V_{UR}		Suppressed – zero range	= V_{LR} to V_{UR}
Suppressed range	= V_{LR} to V_{UR}		Elevated range	= V_{LR} to V_{UR}
Suppression	= h_3		Elevation	= h_3
Zero elevation	= h_3		Zero suppression	= h_3
Suppressed span	= $h_1 + h_2$		Elevated span	= $h_1 + h_2$

Fig. 1   Typical level application illustrates relationship between elevation-related terms.

Fig. 2   Typical level application illustrates relationship between suppression-related terms.

express the condition in which the zero of the measured variable is greater than the lower range-value. The term elevated-zero range is preferred (Ref. 8).

The relationship between the elevated-zero range-related terms is shown in Fig. 1.

Suppressed-Zero Range

A range in which the zero value of the measured variable is less than the lower range-value. (Zero does not appear on the scale.) See Table 2.

Note 1: For example: 20 to 100.

Note 2: The terms elevation, elevated range, and elevated span are frequently used to express the condition in which the zero of the measured variable is less than the lower range-value. The term suppressed-zero range is preferred (Refs. 2 and 8).

The relationship between suppressed-zero range-related terms is shown in Fig. 2.

Zero Elevation

For an elevated-zero range, the amount the measured variable zero is above the lower range-value. It may be expressed either in units of the measured variable or in percentage of span (Ref. 8).

Zero Suppression

For a suppressed-zero range, the amount the measured variable zero is below the lower range-value. It may be expressed either in units of the measured variable or in percentage of span. (Ref. 8).

Suppression Ratio

(Of a suppressed-zero range)—the ratio of the lower range-value to the span.

Note: For example:
      Range 20 to 100
      Suppression ratio = 20/80 = 0.25
      See Table 2 (Refs. 2 and 8).

Readability Terms

The fineness by which the measured variable can be observed depends on a number of factors. The length of the scale and the number of scale graduations greatly influence observer readability. Obviously a longer scale and a larger number of scale graduations will result in a more accurate observation. The performance of the measuring mechanism will also influence observation accuracy. A high degree of resolution and a small dead band and hysteretic error will improve the observed indication. The following terms relate the various factors involved in observation.

Indicator Travel

The length of the path described by the indicating means or the tip of the pointer in moving from one end of the scale to the other.

Note 1: The path may be an arc or a straight line.

Note 2: In the case of knife-edge pointers and others extending beyond the scale division marks, the pointer shall be considered as ending at the outer end of the shortest scale division marks (Refs. 2 and 8).

Pen Travel

The length of the path described by the pen in moving from one end of the chart scale to the other. The path may be an arc or a straight line (Ref. 8).

Resolution

The least interval between two adjacent discrete details which can be distinguished one from the other (Ref. 4).

The smallest change in the input signal that will result in a measurable change in the output signal is often called the *threshold* and is closely related to the resolution concept.

Sensitivity

The ratio of a change in output magnitude to the change in input which causes it after the steady state has been reached.

Note 1: It is expressed as a ratio with the units of measurement of the two quantities stated. (The ratio is constant over the range of a linear device. For a nonlinear device the applicable input level must be stated.)

Note 2: Sensitivity has frequently been used to denote the dead band. However, its usage in this sense is deprecated since it is not in accord with accepted standard definitions of the term (Ref. 4).

Accuracy-Related Terms—Static

These terms relate to the question, How good are measurements under steady-state conditions?

Calibrate

To ascertain outputs of a device corresponding to a series of values of the quantity the device is to measure, receive, or transmit. Data so obtained are used to (Ref. 3).:

1. Determine the locations at which scale graduations are to be placed.
2. Adjust the output, to bring it to the desired value, within a specified tolerance.
3. Ascertain the error by comparing the device output reading against a standard.

This term has come to have three meanings in usage. The first two describe a manufacturing operation. The third meaning relates to the checking of an instrument after it is completed. The checking may be as simple as a rough calibration check at 0, 50, and 100% span. Or, if a full complement of performance information is desired, there may be two or more cycles of up and down runs at 5 to 10 steps.

The third meaning is of primary interest here. The document on which results are recorded is referred to as a *calibration report.*

Usually a calibration report is limited to the recording of the average of output readings for a series of input values over the range of the instrument. The series of input values are applied both up-scale and down-scale. The output readings are averaged for each input value. The maximum difference found in readings, considering all input values, is the maximum hysteresis and is reported as *hysteresis,* discussed in detail later. Frequently a third column of information shows the difference between average output and true values in terms of instrumental errors or corrections.

In testing an instrument against a standard, the standard should have a history of known accuracy for the characteristic under test. The uncertainty should not exceed one-tenth to one-third that allowable in the instrument under test. There are also certain conditioning and environmental requirements. For a discussion of these, refer to page 32 in *ISA-S51.1* ("Standard Process Instrumentation Terminology").

Accuracy

Degree of conformity of an indicated value to a recognized accepted standard value, or ideal value (Refs. 4 and 8).

Conformity may be conceived as the maximum difference, over the range of the instrument, between indicated value and the true value being measured. It is often referred to as the maximum uncertainty or limit of uncertainty. The operating conditions for which the accuracy applies are stated in any reputable calibration report.

A handy means for representing the maximum uncertainty is an accuracy envelope, as illustrated in Fig. 3. The envelope encloses all variations that occur as the result of a full set of accuracy-related performance tests.

Accuracy Rating (sometimes called Reference Accuracy)

A number or quantity that defines a limit that errors will not exceed when a device is used under specified operating conditions. See Fig. 3.

Note 1: When operating conditions are not specified, reference operating conditions shall be assumed.

Note 2: As a performance specification, accuracy (or reference accuracy) shall be assumed to mean accuracy rating of the device when used at reference operating conditions.

Note 3: Accuracy rating includes the combined effects of conformity, hysteresis, dead band, and repeatability errors. The units being used are to be stated explicitly. It is preferred a ± sign precede the number or quantity. The absence of a sign indicates a + and a − sign.

Accuracy rating can be expressed in a number of forms. The following five examples are typical:

1. Accuracy rating expressed in terms of the measured variable. Typical expression: The accuracy rating is ±1°C, or ±2°F.

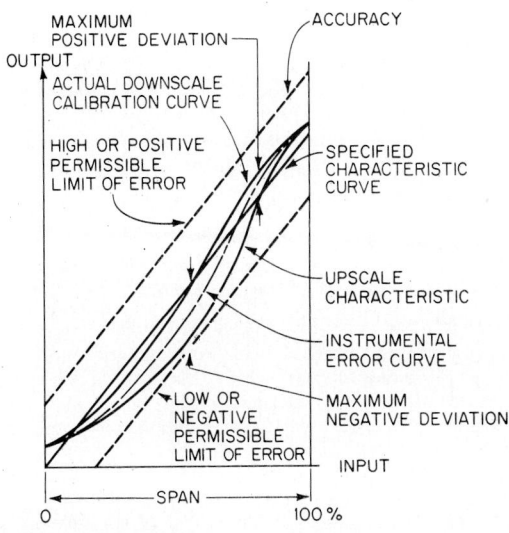

Fig. 3 Accuracy rating.

2. **Accuracy rating expressed in percentage of span.** Typical expression: The accuracy rating is ± 0.5% of span. (This percentage is calculated using scale units such as °F, psig, etc.)

3. **Accuracy rating expressed in percentage of the upper range-value.** Typical expression: The accuracy rating is ± 0.5% of the upper range-value. (This percentage is calculated using scale units such as kPa, °F, etc.)

4. **Accuracy rating expressed in percent of scale length.** Typical expression: The accuracy rating is ± 0.5% of scale length.

5. **Accuracy rating expressed in percentage of actual output reading.** Typical expression: The accuracy rating is ± 1% of actual output reading.

The term *errors* used in the first and second paragraphs of the foregoing definition might better be termed *uncertainties,* and the word *error* should be retained to mean the average difference between readings and true value.

Typical examples of accuracy ratings are: The pressure gage has an accuracy of 0.1% F.S. The accuracy is expressed in terms of span or full scale for a range starting at zero. *Example:* The 0- to 2000-psi pressure gage has a maximum uncertainty of 2 psi for any reading. Or the voltmeter has an accuracy of 0.2% R. The scale starts at zero, and the accuracy is in percent of reading. For a 0- to 100-V instrument, the maximum uncertainty at 50 V is 0.1 V, and at 100 V, 0.2 V.

Error

The algebraic difference between the indication and the ideal value of the measured signal. It is the quantity which algebraically subtracted from the indication gives the ideal value.

Note: A positive error denotes that the indication of the instrument is greater than the ideal value (Refs. 2 and 8).

$$\text{Error} = \text{indication} - \text{ideal value}$$

It is helpful to use the word *error* in this concept only and to describe it as an instrumental error,[*] meaning the difference between the average of a series of up and down readings, as indicated by the

---

[*]This concept is traditionally called bias or systematic error.

instrument output, and the corresponding ideal values of input. The central line (characteristic curve) in Fig. 3 represents an instrumental error curve.

Zero Error

The error of a device operating under the specified conditions of use when the input is at the lower range-value (Ref. 8).

The term *zero shift* is often used to represent a change or drift in zero error with time.

Span Error

The difference between the actual span and the ideal span.

Note: It is usually expressed as a percent of ideal span (Ref. 8).

The term *actual span* should be thought of as the span observed or measured when the input signal changes through the true or ideal span for which the instrument is designed. For illustrations of *span* refer to the foregoing section, Range-Related Terms.

With reference to Fig. 3, the measured span is a little less than the ideal span. In the following example, the span error is −2%.

Input, %	Output error, %
0	+4
100	+2

Correction

The algebraic difference between the ideal value and the indication of the measured signal. It is the quantity which algebraically added to the indication gives the ideal value.

Note: A positive correction denotes that the indication of the instrument is less than the ideal value (Refs. 4 and 8).

$$\text{Correction} = \text{ideal value} - \text{indication}$$

This term, rather than error, is generally the one included in a calibration report.

It is desired to know the ideal value of current when a meter reads 16.7 mA. The correction is +0.4 mA. Therefore, the ideal current is 17.1 mA.

A transmitter with a milliampere output signal is being adjusted to give an ideal output value of 20.00 mA for 100% input. The milliammeter correction is +0.34 mA. The instrument then is to be adjusted so that the output meter reads 19.66 mA.

Hysteresis

That property of an element evidenced by the dependence of the value of the output, for a given excursion of the input, on the history of prior excursions and the direction of the current traverse.

Note 1: It is usually determined by subtracting the value of the dead band from the maximum measured separation between up-scale going and down-scale going indications of the measured variable (during a full range traverse, unless otherwise specified) after transients have decayed. This measurement is sometimes called hysteresis error or hysteretic error. See Fig. 4.

Note 2: Some reversal of output may be expected for any small reversal of input; this distinguishes hysteresis from dead band.

An element that demonstrates hysteretic error only is a helical or cantilever spring. The error curve is generated by plotting deflection versus force.

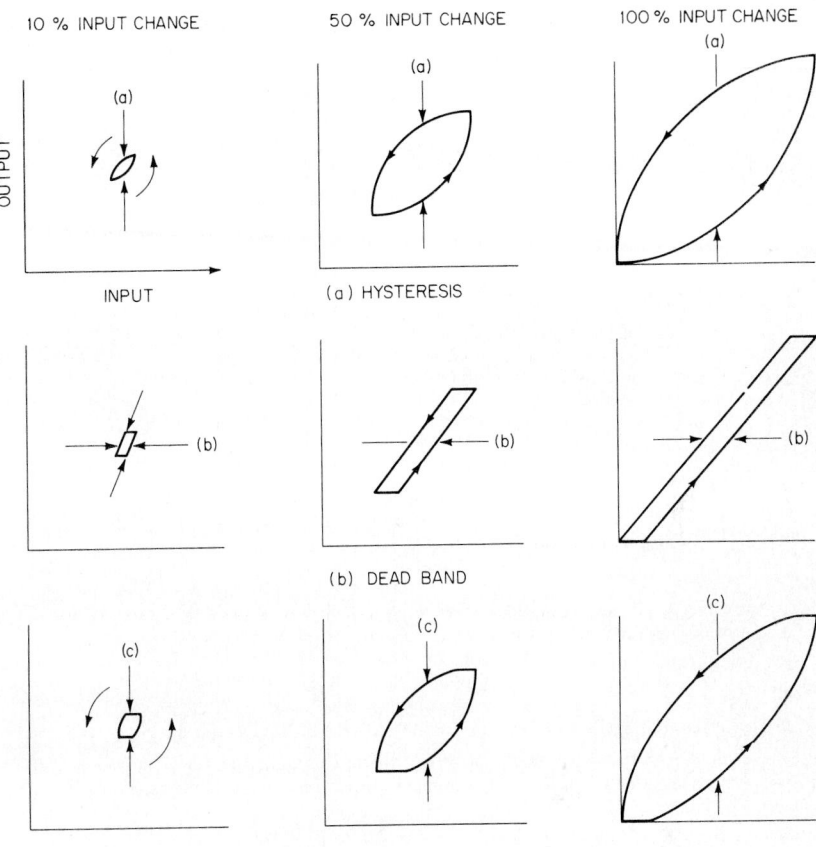

Fig. 4   Hysteresis and dead band.

Dead Band

The range through which an input can be varied without initiating observable response. See Fig. 4b.

Note: Dead band is usually expressed in percentage of span (Refs. 4 and 8).

Friction or "play" between elements of an instrument are direct causes of dead band, since the input may change with no force or motion available to the "driven" element. The term *sensitivity* is still frequently used to represent dead band.

Repeatability

The closeness of agreement among a number of consecutive measurements of the output for the same value of the input under the same operating conditions, approaching from the same direction for full range traverses (Refs. 2 and 8). See Fig. 5.

Note: It is usually measured as a nonrepeatability and expressed as a repeatability in percent of span. It does not include hysteresis.

The average of the up-scale curves would become the actual up-scale characteristic of Fig. 3.

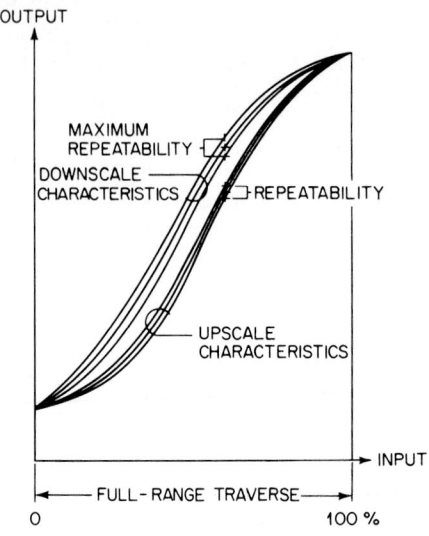

Fig. 5   Repeatability.

Similarly, the average of the down-scale curves would become the actual down-scale characteristic. The accuracy envelope, as previously described, must enclose all the curves of Fig. 5.

Deviation

Any departure from a desired or expected value or pattern (Refs. 4 and 8).

It may also be described as the difference between measured value and true value for a particular input value. The deviation is given a plus or minus value, depending on whether the indications are above or below the true value.

Linearity

The closeness to which a curve approximates a straight line.

Note 1: It is usually measured as a nonlinearity and expressed as a linearity, e.g., a maximum deviation between an average curve and a straight line. The average curve is determined after making two or more full range traverses in each direction. The value of linearity is referred to the output unless otherwise stated.

Note 2: As a performance specification, linearity should be expressed as independent linearity, terminal-based linearity, or zero-based linearity. When expressed simply as linearity, it is assumed to be independent linearity (Refs. 4 and 8).

The great majority of measuring instruments have a linear scale. Manufacturers put a great deal of effort into designing instruments so that an input change in the upper part of the range produces the same output change as a similar change in the lower part of the range.

Terminal-based linearity is commonly used in test work, since the "straight line" is most quickly located and thus the computing time is minimized.

Independent Linearity

The maximum deviation of the calibration curve (average of up-scale and down-scale readings) from a straight line so positioned as to minimize the maximum deviation (Ref. 8). See Fig. 6.

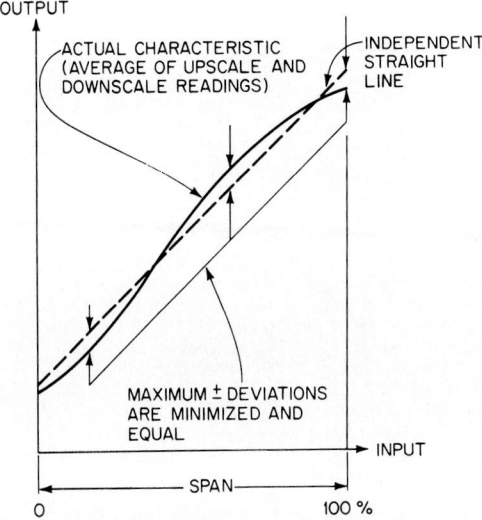

Fig. 6 Independent linearity.

Terminal-Based Linearity

The maximum deviation of the calibration curve (average of up-scale and down-scale readings) from a straight line coinciding with the calibration curve at upper and lower range-values (Ref. 8). See Fig. 7.

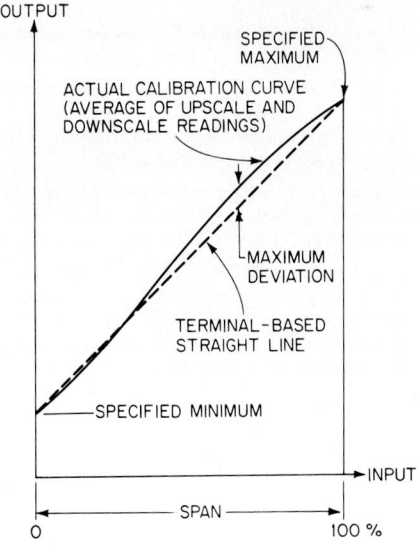

Fig. 7 Terminal-based linearity.

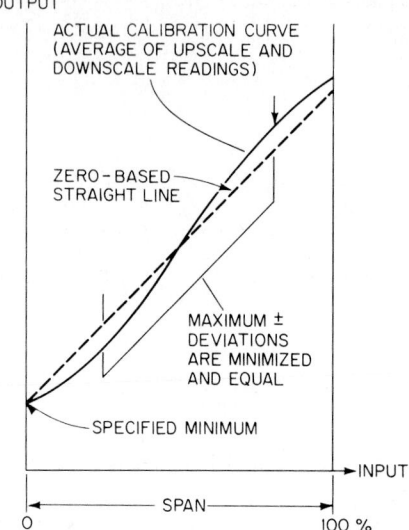

Fig. 8   Zero-based linearity.

Zero-Based Linearity

The maximum deviation of the calibration curve (average of up-scale and down-scale readings) from a straight line so positioned as to coincide with the calibration curve at the lower range-value and to minimize the maximum deviation (Ref. 8). See Fig. 8.

Conformity

Of a curve, the closeness to which it approximates a specified curve (e.g., logarithmic, parabolic, cubic, etc.).

Note 1: It is usually measured in terms of nonconformity and expressed as conformity; e.g., the maximum deviation between an average curve and a specified curve. The average curve is determined after making two or more full range traverses in each direction. The value of conformity is referred to the output unless otherwise stated.

Note 2: As a performance specification conformity should be expressed as independent conformity, terminal-based conformity, or zero-based conformity. When expressed simply as conformity, it is assumed to be independent conformity (Refs. 4 and 8).

Conformity has the same three types as linearity. The concept is of value in nonuniform scale instruments. An example is a millivoltmeter scaled to read temperature from a thermocouple whose output emf is nonuniformly related to temperature.

For definitions of each type of conformity and associated graphic illustrations, refer to pp. 8 and 9 of *ISA-S51.1* ("Standard Process Instrumentation Terminology").

Drift

An undesired change in the output-input relationship over a period of time.

Point Drift

The change in output over a specified period of time for a constant input under specified reference operating conditions.

Note: Point drift is frequently determined at more than one input, as for example, at 0, 50, and 100% of range. Thus, any drift of zero or span may be calculated.

Dynamic Terms

These terms relate to the question, How good are measurements under dynamic conditions?

There are two realms by which an instrument is evaluated: static and dynamic. Other categories of this description have dealt with static terms. Here, dynamic terms are covered. It is not enough to know that a device is "fast" or "stable." One must know "how fast" or "how stable." The following terms will provide a better understanding of the dynamic realm of instrument terms.

Damping

(1) (noun). The progressive reduction or suppression of oscillation in a device or system. (2) (adjective). Pertaining to or productive of damping.

Note 1: The response to an abrupt stimulus is said to be *critically damped* when the time response is as fast as possible without overshoot, *underdamped* when overshoot occurs, or *overdamped* when response is slower than critical.

Note 2: Viscous damping uses the viscosity of fluids (liquids or gases) to effect damping (Refs. 4, 8, and 11).

Note 3: Magnetic damping uses the current induced in electrical conductors by changes in magnetic flux to effect damping (Refs. 4, 8, and 11).

The condition described above as critically damped is actually an ideal or desired condition. The negative sense often associated with the term *critical* does not apply in this case.

Damping Factor

For the free oscillation of a second-order linear system, a measure of damping expressed (without sign) as the quotient of the greater by the lesser of a pair of consecutive swings of the output (in opposite directions) about an ultimate steady-state value (Refs. 4 and 8). See Fig. 9.

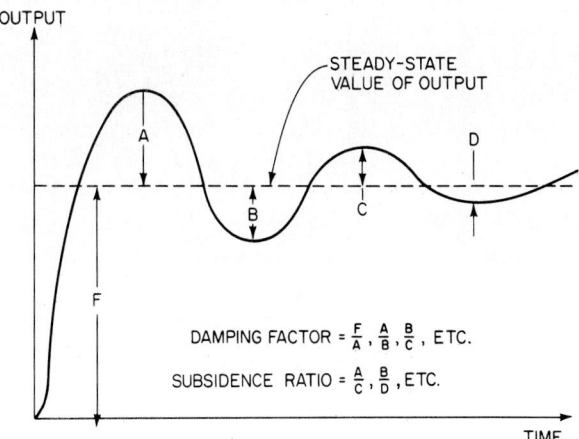

Fig. 9   Underdamped response of a system with second-order lag.

Noise

An unwanted component of a signal or variable which obscures the information content (Refs. 4 and 8).

Signal-to-Noise Ratio

Ratio of signal amplitude may be peak or rms. For nonsinusoidal signals, peak values should be used (Ref. 8).

Dynamic Response

The behavior of the output of a device as a function of the input, both with respect to time (Refs. 2 and 8).

Ramp Response

The total (transient plus steady-state) time response resulting from a sudden increase in the rate of change from zero to some finite value of the input stimulus (Ref. 4).

Step Response

The total (transient plus steady-state) time response resulting from a sudden change from one constant level of input to another (Ref. 4).

Step Response Time

Of a system or an element, the time required for an output to change from an initial value to a large specified percentage of the final steady-state value either before or in the absence of overshoot, as a result of a step change to the input. See Fig. 10.

Note: Usually stated for 90, 95, or 99% change (Refs. 4 and 8).

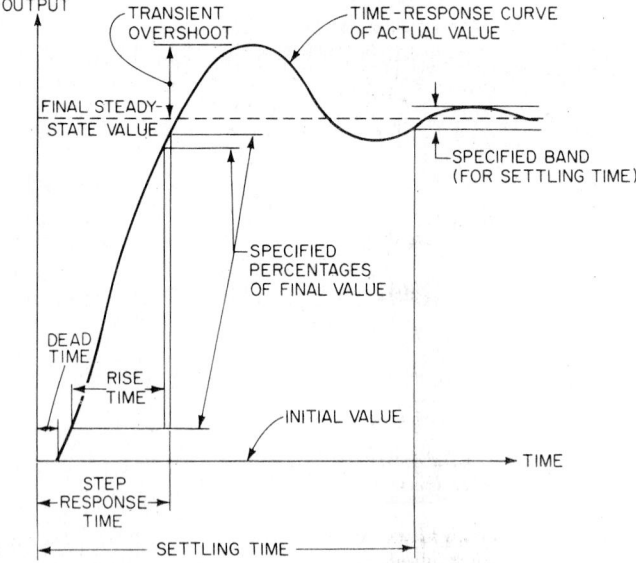

Fig. 10   Typical time response of a system to a step increase in input.

Response Time

An output expressed as a function of time, resulting from the application of a specified input under specified operating conditions (Refs. 4 and 8). See Fig. 10.

Rise Time

The time required for the output of a system (other than first order) to make the change from a small specified percentage (often 5 or 10) of the steady-state increment to a large specified percentage (often 90 to 95), either before or in the absence of overshoot. See Fig. 10.

Note: If the term is unqualified, response to a unit step stimulus is understood; otherwise the pattern and magnitude of the stimulus should be specified (Refs. 4 and 8).

Settling Time

The time required, following the initiation of a specified stimulus to a system, for the output to enter and remain within a specified narrow band centered on its steady-state value. See Fig. 10.

Note: The stimulus may be a step impulse, ramp, parabola, or sinusoid. For a step or impulse, the band is often specified as $\pm 2\%$. For nonlinear behavior both magnitude and pattern of the stimulus should be specified (Refs. 4 and 8).

Time Constant

The value T is an exponential response term $A \exp(-t/T)$ or in one of the transform factors $1 + sT$, $1 + j\omega T$, $1/(1 + sT)$, $1/(1 + j\omega T)$,

where s = complex variable
  t = time, s
  T = time constant
  j = $\sqrt{-1}$
  ω = angular velocity, rad/s

Note: For the output of a first-order (lag or lead) system forced by a step or an impulse, T is the time required to complete 63.2% of the total rise or decay; at any instant during the process, T is the quotient of the instantaneous rate of change divided into the change to be completed. In higher order systems, there is a time constant for each of the first-order components of the process. In a Bode diagram, breakpoints occur at $\omega = 1/T$ (Refs. 4 and 8).

Frequency, Undamped (Frequency, Natural)

1. Of a second-order linear system without damping, the frequency of free oscillation in radians or cycles per unit of time.

2. Of any system whose transfer function contains the quadratic factor $s^2 + 2z\omega_n s + \omega^2 n$, the value ω_n

   where s = complex variable
     z = constant
     ω_n = natural frequency, rad/s

3. Of a closed-loop control system or controlled system, a frequency at which continuous oscillation (hunting) can occur without periodic stimuli.

Note: In linear systems, the undamped frequency is the phase crossover frequency. With proportional control action only, the undamped frequency of a linear system may be obtained in most cases by raising the proportional gain until continuous oscillation occurs (Refs. 4 and 8).

Resonance

Of a system or element, a condition evidenced by large oscillatory amplitude, which results when a small amplitude of a periodic input has a frequency approaching one of the natural frequencies of the driven system (Refs. 4 and 8).

A classical example of resonance is the case of soldiers breaking step in marching across bridges to avoid setting up resonant and possibly destructive forces due to the rythmic beat of their feet.

Limit, Velocity

A limit which the rate of change of a specified variable cannot exceed.

Energy-Related Terms

These terms relate to the question, What power does the instrument require, and how much is consumed?

Supply Pressure

The pressure at the supply port of the device.

A typical transmitter air supply pressure specification follows:

20 psi recommended (limits: 18 to 25 psi)

Air Consumption

The maximum rate at which air is consumed by a device within its operating range during steady-state signal conditions.

Note: Air consumption is usually expressed in cubic feet per minute (ft^3/min) or cubic meters per hour (m^3/h) at a standard (or normal) specified temperature and pressure (Ref. 8).

Supply Voltage

The voltage at the electric supply terminals of the device.

Typical voltage specifications follow:

Example 1. 117 V $\pm$ 10%, 50 or 60 Hz

Example 2. DC: 24 V recommended (limits: 23 to 27 V)

Power Consumption, Electrical

The maximum power used by a device within its operating range during a steady-state condition.

Note 1: For a power factor other than unity, power consumption shall be stated as maximum volt-amperes used under the above-stated condition.

Note 2: For a device operating outside its operating range, the maximum power might exceed that experienced within the operating range (Ref. 8).

Operational Related Terms

The following terms relate to the question, What factors beyond the primary measurement may influence the instrument readings?

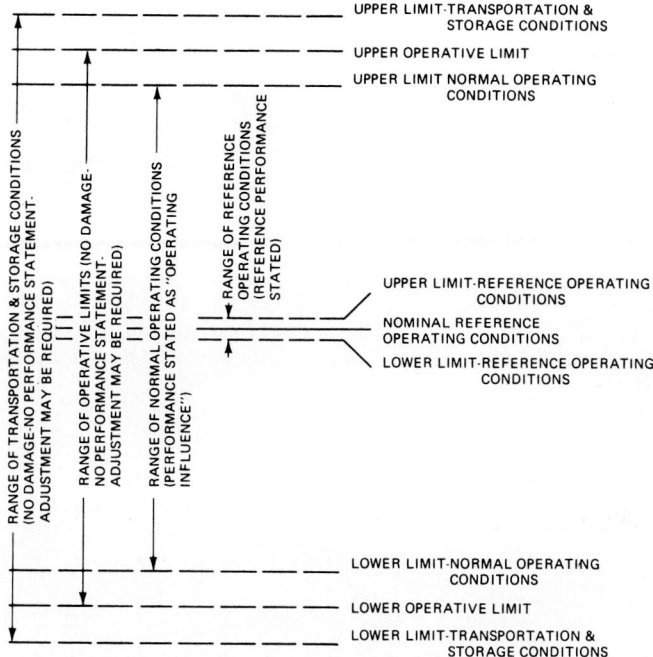

Fig. 11 Operating conditions.

Operating Conditions

Conditions to which a device is subjected, not including the variable measured by the device. Examples of operating conditions include ambient pressure, ambient temperature, electromagnetic fields, gravitational force, inclination, power supply variation (voltage, frequency, harmonics), radiation, shock, and vibration. Both static and dynamic variations in these conditions should be considered (Refs. 2 and 8).

Instruments located outdoors are subjected to the extremes of weather conditions, such as wind, sun, and rain, so that ambient temperature may be only one of several conditions.

The process may influence an instrument such as a pressure element immersed in a high-temperature fluid.

A designer continually attempts to design instruments so that conditions, other than the primary variable to be measured, have no or minimal influence on the reading. The art of design includes Ni-Span C* capsules in pressure instruments, temperature compensation bimetal strips on bourdon spring takeoff arms, and temperature-controlled compartments for precision resistors in high-accuracy resistance bridges.

Reference Operating Conditions

The range of operating conditions of a device within which operating influences are negligible. See Fig. 11.

Note 1: The range is usually narrow.

Note 2: They are conditions under which the reference is stated and the base from which the values of operating influences are determined (Ref. 8).

---

*Trademark, Huntington Alloy Products Division, International Nickel Co., Inc., Huntington, W.Va.

Operative Limits

The range of operating conditions to which a device may be subjected without permanent impairment of operating characteristics. See Fig. 11.

Note 1: In general, performance characteristics are not stated for the region between the limits of normal operating conditions and the operative limits.

Note 2: On returning to within the limits of normal operating conditions, a device may require adjustments to restore normal performance (Refs. 2 and 8).

Transportation and Storage Conditions

The conditions to which a device may be subjected between the time of construction and the time of installation. Also included are the conditions that may exist during shutdown.

Note: No permanent physical damage or impairment of operating characteristics shall take place under these conditions, but minor adjustments may be needed to restore performance to normal (Ref. 8).

Operating Influences

The change in a performance characteristic caused by a change in a specified operating condition from reference operating condition, all other conditions being held within the limits of reference operating conditions.

Note: The specified operating conditions are usually the limits of the normal operating conditions (Refs. 2 and 8).

Operating influence may be stated in either of two ways:

1. As the total change in a performance characteristic from reference operating condition to another specified operating condition.

   Example.   Voltage influence on accuracy may be expressed as: 2% of span based on a change in voltage from a reference value of 120 V to a value of 130 V.

2. As a coefficient expressing the change in a performance characteristic corresponding to unit change in the operating condition from reference condition to another specified operating condition.

   Example.   Voltage influence on accuracy may be expressed as

   $$\frac{2\% \text{ of span}}{130 \text{ V} - 120 \text{ V}} = 0.2\% \text{ of span per volt}$$

Note: If the relation between operating influence and change in operating condition is linear, one coefficient will suffice. If it is nonlinear, it may be desirable to state more than one coefficient such as 0.05% per volt from 120 to 125 V, and 0.15% from 125 to 130 V.

The error in a device due to an operating change is often unpredictable in direction. A typical example is the case of a transmitter influenced by ambient temperature. The structure is designed so that expansions of various connecting internal materials offset one another. However, the overall effect may be over- or undercompensation due to normal variations in materials, tightness of assembly, and other factors. The designer can only state the influence as an uncertainty within limits. For example:

Ambient temperature influence: ±0.5% for 50°F change in case temperature between 50 and 150°F.

Ambient Pressure

The pressure of the medium surrounding a device.

Ambient Temperature

The temperature of the medium surrounding a device.

Interference, Electromagnetic

Any spurious effect produced in the circuits or elements of a device by external electromagnetic fields.

Note: A special case of interference from radio transmitters is known as radio frequency interference (RFI).

Interference, Common-Mode

A form of interference which appears between measuring circuit terminals and ground (Refs. 3 and 8).

Interference, Normal-Mode

A form of interference which appears between measuring circuit terminals (Refs. 2 and 8).

ABRIDGED LIST OF REFERENCES APPLICABLE TO FOREGOING DEFINITIONS

1. ANSI: "Direct-Acting Electrical Recording Instruments (Switchboard and Portable Types)," *ANSI C39.2-1964.*
2. ANSI: "Specifications for Automatic Null-Balancing Electrical Measuring Instruments," *ANSI C39.4-1966.*
3. ANSI: "Dictionary of Electrical and Electronic Terms," *ANSI C42.100-1972.*
4. ANSI: "Terminology for Automatic Control," *ANSI C85.1-1963.*
5. ANSI: "Supplement to *C.85.1-1963.*"
6. ANSI: "Supplement to *C.85.1b-1966.*"
7. ANSI: "Temperature Measurement Thermocouples," *ANSI MC96.1-1975.* Foregoing available from American National Standards Institute, 1430 Broadway, New York, N.Y. 10018.
8. SAMA: "Process Measurement and Control Terminology," *PMC20.1-1973,* Scientific Apparatus Makers Association, 370 Lexington Avenue, New York, N.Y. 10017.
9. ISA: "Electrical Instruments in Hazardous Atmospheres," *ISA-RP12.1-1960.*
10. ISA: "Intrinsically Safe and Non-Incendive Electrical Instrument," *ISA-RP12.2-1965.*
11. ISA: "Electrical Transducer Nomenclature and Terminology," *ISA-S37.1-1969.*
12. ISA: "Control Center Construction," *ISA-S60.7-1975.* Foregoing available from Instrument Society of America, P.O. Box 12277, Research Triangle Park, N.C. 27709.
13. IEEE: *Publ. IEEE 279,* Institute of Electrical and Electronics Engineers, 345 E. 47th Street, New York, N.Y. 10017.
14. NFPA: *Publ. NFPA 493,* National Fire Protection Association, Boston, Mass.
15. NFPA: *Publ. NFPA 501,* National Fire Protection Association, Boston, Mass.
16. IEC: "Automatic Controlling and Regulating Systems," (International Electrotechnical Vocabulary Publication), *50(37)-1966.*
17. IEC: "Automatic Control and Regulation-Servomechanisms," (International Electrotechnical Vocabulary Publication), *351 (1037)-1972.*
18. IEC: "Service Conditions," Committee 65-Working Group 2, *IEC/TC65/WG2.* Foregoing available from International Electrotechnical Commission, 1 rue de Varembe, 1211 Geneva, Switzerland.

Glossary of Units Frequently Used

For convenience, the following definitions are presented in alphabetical order.

Ampere (A).   *The constant current that, if maintained in two straight parallel conductors, of infinite length and negligible cross section and separated from each other by a distance of 1 meter in a vacuum, will produce between these conductors a force equal to 2×10^{-7} newton per meter of length.* (The SI unit of electric current.)

Ampere per Meter (A/m).   *The magnetic field strength in the interior of an elongated uniformly wound solenoid excited with a linear current density in its winding of 1 ampere per meter of axial distance.* (The SI unit of magnetic field strength.)

Ampere-Hour (Ah).   *The quantity of electricity represented by a current of 1 ampere flowing for 1 hour.*

Angstrom (Å).   *A unit of length equal to 10^{-10} meter.* (The tiny circle over the Å is sometimes omitted, but its use distinguishes the symbol from that for ampere.)

Apostilb (asb).   *A unit of luminance. One lumen per square meter leaves a surface whose luminance is 1 apostilb in all directions within a hemisphere.* (The candela per square meter is the preferred unit of luminance.)

Atmosphere (atm).   *A unit of pressure. One standard atmosphere equals 101,325 newtons per square meter.*

Atomic Mass Unit, Unified (u).   *The atomic mass unit (unified) is 1/12 the mass of an atom of the ^{12}C nuclide.* [Use of the prior atomic mass unit (amu), defined by reference to oxygen, is no longer preferred.]

Bar (bar).   *A unit of pressure. One bar equals 100,000 newtons per square meter.*

Barn (b).   *A unit of nuclear cross section. One barn equals 10^{-28} square meter.*

Barrel (bbl).   *A unit of volume. One barrel equals 9702 cubic inches; or 0.15899 cubic meter.* (This is the standard barrel used for petroleum. A different standard barrel is used for fruits, vegetables, and dry commodities.)

Baud (Bd).   *A unit of signaling speed. One baud equals one element per second.* (The signaling speed in bauds is equal to the reciprocal of the signal element length in seconds.)

Bel (B).   *A dimensionless unit for expressing the ratio of two values of power, being the logarithm to the base 10 of the power ratio.* [The more commonly used unit, decibel (dB), is 10 times the logarithm to the base 10 of the power ratio. Thus 1 bel = 10 decibels.]

Bit (b).   *A unit of information generally represented by a pulse. A bit is a binary digit, i.e., a 1 or 0 in computer technology.* (In information theory, the bit is the smallest possible unit of information.)

Bit per Second (b/s).   *A unit of signaling speed. A transference rate of 1 bit per second.*

British Thermal Unit (Btu).   *A unit of heat. The heat required to warm 1 pound of water through an interval of 1 degree Fahrenheit.*

Calorie, International Table (cal_{It}).   *A unit of heat. One international table calorie equals 4.1868 joules.* (The 9th *Conférence Générales des Poids et Mesures* adopted the joule as the unit of heat.)

Calorie, Thermochemical (cal).   *A unit of heat. One calorie equals 4.1840 joules.* (See foregoing definition.)

Candela (cd).   *The luminous intensity of 1/600,000 of a square meter of a radiating cavity at the temperature of freezing platinum (2042 K).* (The SI unit of luminous intensity. The unit formerly was called the *candle.*)

Circular Mil (cmil).   *The area of a circle whose diameter is 0.0001 inch. One circular mil equals $\pi/4 \times 10^{-6}$ square inches.*

Coulomb (C).   *The quantity of electric charge which passes any cross section of a conductor in 1 second when the current is maintained constant at 1 ampere.* (The SI unit of electric charge.)

Curie (Ci).   *The unit of activity in the field of radiation dosimetry. One curie equals 3.7 × 10^{10} disintegrations per second.* (The activity of 1 gram of ^{226}Ra is slightly less than 1 curie.)

Cycle (c).   *An interval or space of time in which is completed one round of events or phenomena.*

Cycles per Second (Hz, c/s).   *The number of cycles per second.* [The name Hertz (Hz) is the accepted international term. The abbreviation Hz is preferred to c/s.]

Darcy (D).   *A unit of permeability of a porous medium. One darcy equals 1 cP (cm/s) (cm/atm) equals 0.986923 square micrometers.* (A permeability of 1 darcy will allow the flow of 1 cubic centimeter per second of fluid of 1 centipoise viscosity through an area of 1 square centimeter under a pressure gradient of 1 atmosphere per centimeter.)

Day (day).   A unit of time, the exact definition of which is dependent on which system of time measurement is referred to—apparent solar time, mean solar time, universal time, apparent sidereal time, ephemeris time, or atomic time. With the exception of atomic time, the time base is referenced to rotation of the earth. For general purposes, a day is considered the period taken for one revolution of the earth about its axis. The SI unit of time is the *second,* which see later in this list.

Decible (dB).   One-tenth of a *bel,* which see earlier in this text.

Degree Celsius (°C).   *One unit of temperature on the Celsius temperature scale, which is derived from the thermodynamic or Kelvin scale of temperature.* [Temperature (degrees Celsius) equals temperature (Kelvin units) minus 273.15. See Temperature Scales, Standards, and Calibration in Section 2 of this Handbook. In some modern scientific literature, the degree symbol before the C is omitted.]

Degree Fahrenheit (°F).   *One unit of temperature on the Fahrenheit temperature scale.* [Temperature (degrees Fahrenheit) equals 1.8 × (degrees Celsius) plus 32. See Temperature Scales, Standards, and Calibration in Section 2 of this Handbook. In some modern scientific literature, the degree symbol before the F is omitted.]

Degree Rankine (°R).   *One unit of temperature on the Rankine temperature scale.* [Temperature (degrees Rankine) equals (degrees Fahrenheit) plus 459.69. See Temperature Scales, Standards, and Calibration in Section 2 of this Handbook.]

Dyne (dyn).   *A unit of force. One dyne equals the force necessary to give 1 gram mass an acceleration of 1 centimeter/(second) (second).* (The dyne is the unit of force in the CGS system.)

Electronvolt (eV).   *A unit of energy. One electronvolt equals the energy acquired by an electron when it passes through a potential difference of 1 volt in a vacuum.* (One electronvolt equals 1.602 × 10^{-12} erg.)

Erg (erg).   *A unit of energy. One erg equals 10^{-7} joule.* (Also, 1 erg equals the work done when a force of 1 dyne is applied through a distance of 1 centimeter. One foot-pound equals 13,560,000 ergs. The erg is the unit of energy in the CGS system.)

Farad (F).   *The capacitance of a capacitor in which a charge of 1 coulomb produces a potential difference of 1 volt between the terminals.* (The SI unit of capacitance.)

Footcandle (fc).   *A unit of luminance. One footcandle equals 1 lumen per square foot.* [The name *lumen per square foot* is recommended for this unit. The SI unit, *lux* (lumen per square meter), is preferred.]

Footlambert (fL).   *A unit of luminance. One lumen per square foot leaves a surface whose luminance is 1 footlambert in all directions within a hemisphere.* (If luminance is measured in English units, the candela per square inch is preferred. However, use of the SI unit, the candela per square meter, is generally preferred.)

Footpound (fp).   *A unit of work equal to the work done when a mass of 1 pound is raised a distance of 1 foot.* (One footpound equals 14.59390 newton meters.)

Gal (Gal).   *A unit of acceleration. One Gal equals 1 centimeter per second per second.* (The milligal is frequently used because it is approximately 1 part in 1000 of the normal gravity of the earth.)

Gallon (gal).   Because the gallon, quart, and pint differ in the United States and the United Kingdom, the use of this term is generally discouraged for scientific and engineering purposes. An imperial gallon equals 1.20095 U.S. gallons. One U.S. gallon equals 3.785 × 10^{-3} cubic meter.

Gauss (G).   *A unit of magnetic flux density or magnetic induction. The ratio of the flux in any cross section to the area of that cross section, the cross section being taken normal to the direction of flow. One gauss equals 1 maxwell per square centimeter.* (The gauss is a unit of the CGS system; the SI unit, the *tesla,* is preferred.)

Gilbert (Gb).   *A unit of magnetomotive force. One gilbert equals 0.4π (ni), where (ni) is an ampere-turn.* [The gilbert is a unit of the CGS system. Use of the SI unit, the ampere (or ampere-turn), is preferred.]

Grain (gr).   *A unit of mass. One grain equals 0.06480 gram.* (One ounce avoirdupois equals 437.5 grains; 1 ounce troy equals 480 grains; 1 ounce apothecaries equals 480 grains. One pound avoirdupois equals 7000 grains.)

Gram (g).   *A unit of mass. One gram equals 1/1000 kilogram.* (The kilogram is the SI unit of mass.)

Henry (H).   *A unit of inductance. The inductance of a circuit in which a current of 1 ampere induces a flux linkage of 1 weber.* (The SI unit of inductance.)

Hertz (Hz).   *A unit of frequency. One hertz equals a frequency of one cycle per second.* (The CGS and SI unit of frequency.)

Horsepower (hp).   The horsepower is considered an anachronism in science and technology. Use of the SI unit of power, the *watt* is preferred. When used, 1 horsepower equals (1) 42.44 Btu/minute; (2) 33,000 footpounds/minute; and (3) 550 footpounds/second.

Hour (h).   *A unit of time. One hour equals 60 minutes, or 3600 seconds.*

Inch (in).   *A unit of length. One inch equals 2.540×10^{-2} meter.*

Inch of Mercury (inHg).   *A unit of pressure. One inch of mercury equals 3386.4 newtons per square meter.* [An inch of mercury also equals (1) 0.03342 atmosphere, (2) 1133 feet of water, (3) 345.3 kilograms/square meter, (4) 70.73 pounds/square foot, and (5) 0.4912 pounds/square inch.]

Inch of Water (inH$_2$O).   *A unit of pressure. One inch of water equals 249.09 newtons per square meter.* [An inch of water also equals (1) 2.458×10^{-3} atmosphere, (2) 0.07355 inch of mercury, (3) 2.540×10^{-3} kilogram/square centimeter, (4) 0.5781 ounce/square inch, (5) 5.204 pounds/square foot, and (6) 0.03613 pound/square inch. These figures hold for a temperature of 4°C.]

Joule (J).   *A unit of energy. The work done by 1 newton acting through a distance of 1 meter.* (The SI unit of energy. One joule equals 1 watt-second, 10^7 ergs, 10^7 dyne-centimeters.)

Joule per Kelvin (J/K).   A unit of heat capacity and entropy.

Kelvin (K).   *The basic unit of thermodynamic temperature. One kelvin is the fraction 1/273.16 of the thermodynamic temperature of the triple point of water.* (The term degree Kelvin was officially dropped in 1967. Thus, the symbol is K and not °K. The relationship of the Kelvin scale to the Celsius scale was given earlier in this list under Degree Celsius. Also see Temperature Scales, Standards, and Calibration in Section 2 of this Handbook.)

Kilogram (kg).   *A unit of mass and is based on a cylinder of platinum-iridium alloy kept by the International Bureau of Weights and Measures in Paris.* [A duplicate in the custody of the National Bureau of Standards in Washington is the mass standard for the United States. The kilogram is the only base unit still defined by an artifact. A kilogram equals (1) 1000 grams, (2) 2.205 pounds, (3) 9.842×10^{-4} long tons, and (4) 1.102×10^{-3} short tons. 1 ton *(tonne)* equals 1000 kilograms.]

Knot (kn).   *A unit of speed. One knot equals 1 nautical mile per hour.* (A knot also equals 6080.2 feet/hour and 1.151 statute miles/hour.)

Lambert (L).   *A unit of luminance. One lumen per square centimeter leaves a surface whose luminance is 1 lambert in all directions within a hemisphere.* (The candela per square meter is the preferred unit of luminance.)

Liter (L).   *A unit of volume. One liter equals 10^{-3} cubic meter.* [A liter also equals (1) 1000 cubic centimeters, (2) 0.03531 cubic foot, (3) 61.02 cubic inches, (4) 1.308×10^{-3} cubic yard, (5) 0.2642 U.S. liquid gallon, (6) 1.057 U.S. liquid quart, and (7) 0.22 imperial gallon.]

Lumen (lm).   *A unit of luminous flux. The flux through a unit solid angle (steradian) from a uniform point source of 1 candela.* (The SI unit of luminous flux.)

Lumen per Square Foot (lm/ft^2). *A unit of illuminance and also a unit of luminous excitation.* (Use of the SI unit, lumen per square meter, is preferred.)

Lumen per Square Meter (lm/m^2). *A unit of luminous excitation.* (The SI unit of luminous excitation.)

Lumen per Watt (lm/W). *A unit of luminous efficacy.* (The SI unit of luminous efficacy.)

Lumen Second (lm·s). *A unit of quantity of light.* (The SI unit of quantity of light.)

Lux (lx). *A unit of illuminance. One lux equals 1 lumen per square meter.* (The SI unit of illuminance.)

Maxwell (Mx). *A unit of magnetic flux. The flux through a square centimeter normal to a field at 1 centimeter from a unit magnetic pole.* (The CGS unit of magnetic flux.)

Meter (m). *A unit of length. Defined as the length of path traveled by light in vacuum during a time interval of 1/299 792 458 of a second.* (The SI unit of length.) A meter also equals (1) 100 centimeters, (2) 3.281 feet, (3) 39.37 inches, (4) 0.001 kilometer, (5) 5.396×10^{-4} nautical mile, (6) 6.214×10^{-4} statute mile, and (7) 1.094 yard.

Mho (mho). *A unit of conductance (and of admittance). The conductance of a conductor whose resistance is 1 ohm.* [The name *siemen* (S) is also used for this quantity.]

Micrometer (μm). *A unit of length. One micrometer equals one-millionth of a meter.* (The term *micron* formerly used for this unit is no longer preferred.)

Micron. See micrometer above.

Mil (mil). *A unit of length. One mil equals 1/1000 inch.*

Mile, Nautical (nmi). *A unit of length. One nautical mile equals 1.1516 statute miles.* [One nautical mile also equals (1) 6080.27 feet, (2) 1.853 kilometers, and (3) 2027 yards.]

Mile, Statute (mi). *A unit of length. One mile equals 5280 feet.* [One statute mile also equals (1) 1609 kilometers, (2) 1760 yards, (3) 6.336×10^4 inches, and (4) 0.8684 nautical mile.]

Minute, angle ('). One-sixtieth of a degree of an arc. A minute also equals (1) 1.8519×10^{-4} quadrant, (2) 2.90888×10^{-4} radian, (3) 1/60 or 0.016667 degree, and (4) 60 seconds.

Minute, time (min). *A unit of time. One minute equals 60 seconds.* (Time also may be designated by means of superscripts, as in $9^h46^m30^s$, in cases where otherwise there would be no confusion in the use of these symbols.)

Mole (mol). *A unit of amount of a substance. One mole is an amount of a substance, in specified mass units, equal to the molecular weight of that substance.* (The SI unit for amount of substance. Examples are the gram mole and the pound mole.)

Neper (Np). A dimensionless unit for expressing the ratio of two voltages, two currents, or two power values in a logarithmic manner. The number of nepers is the natural (napierian) logarithm of the square root of the ratio of the two values being compared. Thus, the neper uses the base of 2.71828, in contrast with the bel (or decibel) which uses the common logarithm base of 10. One neper equals 8.686 decibels.

Newton (N). *A unit of force. One newton is the force that will impart an acceleration of 1 meter per second per second to a mass of 1 kilogram.* (The SI unit of force. One newton equals 10^5 dynes.)

Nit (nt). *A unit of luminance that is synonymous with candela per square meter.*

Oersted (Oe). *A unit of magnetic field strength. The magnetic field produced at the center of a plane circular coil of one turn and of radius 1 centimeter, which carries a current of* ($\frac{1}{2}\pi$) *abamperes.* (An abampere equals 10 amperes. The oersted is the CGS unit of magnetic field strength. Use of the SI unit, the *ampere per meter*, is preferred.)

Ohm (Ω). *A unit of resistance (and of impedance). The resistance of a conductor such that a constant current of 1 ampere in it produces a voltage difference of 1 volt between its ends.* (The SI unit of resistance.)

Pascal (Pa). *A unit of pressure or stress. One pascal equals 1 newton per square meter.* (One pascal also equals 0.00014504 pound/square inch. Because the unit is so small, the larger kilopascal (kPa) and megapascal (MPa) are more commonly used.)

Phon (phon). *A unit of loudness level. The pressure level in decibels of a pure 1000-Hz tone.*

Phot (ph). *A unit of illuminance. One phot equals 1 lumen per square centimeter.* (The phot is the CGS unit of illuminance. The SI unit *lux* is preferred.)

Pint (pt). Because the gallon, quart, and pint differ in the United States and the United Kingdom, the use of this unit is generally discouraged for scientific purposes. One U.S. pint equals (1) 473.2 cubic centimeters, (2) 0.01671 cubic foot, (3) 28.87 cubic inches, (4) 4.732×10^{-4} cubic meter, (5) 6.189×10^{-4} cubic yard, (6) 0.125 U.S. gallon, (7) 0.4732 liter, and (8) 0.5 U.S. liquid quart.

Poise (P). A unit of dynamic viscosity. The unit is expressed in dyne second per square centimeter. The centipoise (cP) is more commonly used. The formal definition of viscosity arises from the concept put forward by Newton that, under conditions of parallel flow, the shearing stress is proportional to the velocity gradient. If the force acting on each of two planes of area A parallel to each other, moving parallel to each other with a relative velocity V and separated by a perpendicular distance X, be denoted by F, the shearing stress is F/A and the velocity gradient, which will be linear for a true liquid, is V/X. Thus, $F/A = \eta V/X$, where the constant η is the viscosity coefficient or dynamic viscosity of the liquid. The poise is the CGS unit of dynamic viscosity.

Pound, Mass. *A mass equal to 0.45359237 kilogram. Also a unit of measure of the inertial property equal to the mass of a body weighing 1 pound at the standard acceleration of gravity [980.665 centimeters/(second) (second)].*

Pound, Weight (lb). *A unit of force equal to the earth's attraction for a mass of 1 pound.* [In English units, this force, acting on a 1-pound mass, will produce an acceleration of 32.1747 feet/ (second)(second).]

Poundal (pdl). *A unit of force. One poundal equals the force required to give a standard 1-pound body an acceleration of 1 foot/(second) (second).*

Quart (qt). Because the gallon, quart, and pint differ in the United States and the United Kingdom, the use of this unit is generally discouraged for scientific purposes. One U.S. quart equals (1) 946.4 cubic centimeters, (2) 0.03342 cubic foot, (3) 57.75 cubic inches, (4) 9.464×10^{-4} cubic meters, (5) 1.238×10^{-3} cubic yard, (6) 0.25 U.S. gallon, and (7) 0.9463 liter.

Rad (rad). *A unit of plane angle. One radian equals the angle subtended at the center by a circular arc equal in length to the radius of the circle.* (The SI unit of plane angle.)

REM (rem). *A unit of dose equivalent in the field of radiation dosimetry. One rem equals the amount of ionizing radiation of any type which produces the same damage to humans as 1 roentgen of approximately 200 kilovolts x-radiation.* (The unit is an abbreviation of *roentgen equivalent man*.)

Revolution per Minute (r/min). Although use of rpm as an abbreviation is common, it is not intended for use as a symbol.)

Roentgen (R). A unit of exposure in the field of radiation dosimetry. That quantity of x- or gamma-radiation such that the associated corpuscular emission per 0.001293 gram of dry air (equals 1 cubic centimeter at 0°C and 769 millimeters of mercury pressure) produces in air ions carrying 1 electrostatic unit (esu) of electricity of either sign. The emu is a unit in the CGS system in which the statcoulomb is the charge that repels an exactly similar charge in a vacuum with a force of 1 dyne. One statcoulomb equals 3.3356×10^{-10} coulomb.

Second, angle ("). This is 1/3600 of a degree of arc. A second also equals (1) 4.84814×10^{-6} radian, (2) 2.7778×10^{-4} degree, and (3) 0.016667 minute.

Second, time (s). *A unit of time. The duration of 9 192 631 770 periods of the radiation corresponding to the transition between the two hyperfine levels of the ground state of the ^{133}Cs (cesium) atom.* (The SI unit of time.)

Slug (slug). *A unit of mass. One slug equals 14.5959 kilograms.*

Steradian (sr). *A unit of solid angle. One steradian equals the solid angle subtended at the center by ¼π of the surface area of a sphere of unit radius.*

Stilb (sb). *A unit of luminance. One stilb equals 1 candela per square centimeter.*

Stokes (St). *A unit of kinematic viscosity. The centistoke (cSt) is more commonly used.* (Kinematic viscosity is the dynamic viscosity divided by the density. See Poise given earlier in this list.)

Tesla (T). *A unit of magnetic flux density (magnetic induction). The magnetic flux density of a uniform field that produces a torque of 1 newton-meter on a plane current loop carrying 1 ampere and having a projected area of 1 square meter on the plane perpendicular to the field.* T = N/ A·m. (The SI unit of magnetic flux density.)

Therm (thm). *A unit of heat. One therm equals 100,000 British thermal units.*

Ton (ton). *A unit of weight.* If not otherwise specified, a *short ton* equal to 2000 pounds is assumed. A *long ton* equals 2240 pounds. A *metric ton* equals 1000 kilograms (2205 pounds); also called *tonne* (t).

Var (var). *A unit of reactive power. The reactive power at the port of entry of a single-phase two-wire circuit when the product of (1) the rms (root mean square) value in amperes of the sinusoidal current, (2) the rms value in volts of the voltage, and (3) the sine of the angular phase difference by which the voltage leads the current is equal to 1.* (The SI unit of reactive power.)

Volt (V). *A unit of voltage. The voltage between two points of a conducting wire carrying a constant current of 1 ampere when the power dissipated between these points is 1 watt.* (The SI unit of voltage.)

Voltampere (V-A). *A unit of apparent power. The apparent power at the port of entry of a single-phase two-wire circuit when the product of (1) the rms (root mean square) value in amperes of the current and (2) the rms value in volts of the voltage is equal to 1.* (The SI units of apparent power.)

Watt (W). *A unit of power. The watt equals 1 joule per second.* (The SI unit of power. One watt equals (1) 3.4192 Btu/hour, (2) 0.05688 Btu/minute, (3) 10^7 ergs/second, (4) 44.27 footpounds/ minute, (5) 0.7378 footpounds/second, (6) 1.341×10^{-3} horsepower, and (9) 0.001 kilowatt.)

Watt per Meter Kelvin (W/m·K). The SI unit of thermal conductivity.

Watt per Steradian (W/sr). The SI unit of radiant intensity.

Watt per Steradian Square Meter (W/Sr·m²). The SI unit of radiance.

Watthour (Wh). *A unit of power.* One watthour equals 3600 joules. [One watthour also equals (1) 3.413 Btu, (2) 3.60×10^{10} ergs, (3) 2656 footpounds, (4) 859.85 gram-calories, (5) 1.341×10^{-3} horsepower-hour, (6) 0.8598 kilogram-calorie, (7) 367.2 kilogram-meters, and (8) 0.001 kilowatt hour.]

Weber (Wb). *A unit of magnetic flux. The magnetic flux passing through an area of 1 square meter placed normal to a uniform magnetic field of magnetic flux density equal to 1 tesla.* (Wb = T·m². (The SI unit of magnetic flux. If the flux linked by a circuit changes at a uniform rate of 1 weber/second, a voltage of 1 volt is induced in the circuit. Wb = V·s.)

INFORMATION SOURCES

Numerous organizations throughout the world are active in sponsoring, formulating, and, in most cases, publishing standards that apply to the technology of measurement and control. Most of these documents undergo constant revision and updating. Because there are scores of such documents with changing contents, it is not practical in a reference handbook to attempt to list all of them. Only a selected few are included in the list of references at the end of this article. The user of this Handbook can obtain lists of available documents by contacting the following organizations, among others.

American Association for the Advancement of Science (AAAS)
   1515 Massachusetts Ave., N.W.
   Washington, D.C. 20005

American Chemical Society (ACS)
   1155 Sixteenth St., N.W.
   Washington, D.C. 20036

American Institute of Chemical Engineers (AIChE)
345 E. 47th Street
New York, N.Y. 10017

American National Standards Institute (ANSI)
1430 Broadway
New York, N.Y. 10018

American Nuclear Society (ANS)
555 N. Kensington
LaGrange Park, Ill. 60525

American Society for Testing and Materials (ASTM)
1916 Race Street
Philadelphia, Pa. 19103

American Society of Mechanical Engineers
345 E. 47th Street
New York, N.Y. 10017

British Standards Institution (BSI)
2 Park Street
London W1A 2BS, England

Deutsches Institue für Normung Burggrafenstr. (DIN)
4-7 D1000 Berlin 30
Federal Republic of Germany

Electronic Industries Assocation (EIA)
2001 Eye Street, N.W.
Washington, D.C. 20006

Institute of Electrical and Electronic Engineers (IEEE)
345 East 47th Street
New York, N.Y. 10017

Institute of Measurement and Control (IMC)
20 Peel Street
London W8 7PD, England

Instituto Argentino del Petroleo (IAP)
Maipu 645, 3er Piso
Buenos Aires, Argentina

Instrument Society of America
67 Alexander Drive
Research Triangle Park, N.C. 27709

International Bureau of Weights and Measures
Paris, France

International Electrotechnical Commission (IEC)
1, rue de Varembe
1211 Geneva, Switzerland

International Purdue Workshops on Industrial Computer Systems
Purdue University
West Lafayette, Ind. 47907

International Standards Organization (ISO)
1, rue de Varembe
1211 Geneva, Switzerland

National Bureau of Standards (U.S.) (NBS)
Washington, D.C. 20234

Scientific Apparatus Makers Association (SAMA)
1101 16th Street, N.W.
Washington, D.C. 20036

Society of Automotive Engineers (SAE)
    400 Commonwealth Drive
    Warrendale, Pa. 15096

Standards Council of Canada (SCC)
    Meadowvale Corporate Centre
    2000 Argentina Road, Suite 2-401
    Mississouga, Ontario L5N 1V8, Canada

Underwriters Laboratories (UL)
    333 Pfingsten Road
    Northbrook, Ill. 60611

SECTION 2

Temperature Systems

Wilson A. Clayton. *Director of Engineering, Hy-Cal Engineering, A Unit of General Signal, El Monte, California. (Thermoresistive Systems)*

Daniel A. Jackson. *Omega Engineering, Inc., Stamford, Connecticut. (Thermocouples)*

G. Robert Leavitt. *Manager, Quality Assurance, Taylor Instrument Company, A Division of Combustion Engineering, Inc., Rochester, New York. (Filled-System Thermometers)*

Allen B. Mushin. *Omega Engineering, Inc., Stamford, Connecticut. (Thermocouples)*

Ray Peacock. *Vice President of Engineering, Land Instruments, Inc., Tullytown, Pennsylvania. (Radiation Thermometry)*

Temperature Conversion Table

Temperature Conversion Table

General formula: $°F = (°C \times \tfrac{9}{5}) + 32$; $°C = (°F - 32) \times \tfrac{5}{9}$

C	value	F
-273.1	**-459.4**	
-268	**-450**	
-262	**-440**	
-257	**-430**	
-251	**-420**	
-246	**-410**	
-240	**-400**	
-234	**-390**	
-229	**-380**	
-223	**-370**	
-218	**-360**	
-212	**-350**	
-207	**-340**	
-201	**-330**	
-196	**-320**	
-190	**-310**	
-184	**-300**	
-179	**-290**	
-173	**-280**	
-169	**-273**	-459.4
-168	**-270**	-454
-162	**-260**	-436
-157	**-250**	-418
-151	**-240**	-400
-146	**-230**	-382
-140	**-220**	-364
-134	**-210**	-346
-129	**-200**	-328
-123	**-190**	-310
-118	**-180**	-292

C	value	F
-17.8	**0**	32
-17.2	**1**	33.8
-16.7	**2**	35.6
-16.1	**3**	37.4
-15.6	**4**	39.2
-15.0	**5**	41.0
-14.4	**6**	42.8
-13.9	**7**	44.6
-13.3	**8**	46.4
-12.8	**9**	48.2
-12.2	**10**	50.0
-11.7	**11**	51.8
-11.1	**12**	53.6
-10.6	**13**	55.4
-10.0	**14**	57.2
-9.44	**15**	59.0
-8.89	**16**	60.8
-8.33	**17**	62.6
-7.78	**18**	64.4
-7.22	**19**	66.2
-6.67	**20**	68.0
-6.11	**21**	69.8
-5.56	**22**	71.6
-5.00	**23**	73.4
-4.44	**24**	75.2
-3.89	**25**	77.0
-3.33	**26**	78.8
-2.78	**27**	80.6
-2.22	**28**	82.4
-1.67	**29**	84.2

C	value	F
10.0	**50**	122.0
10.6	**51**	123.8
11.1	**52**	125.6
11.7	**53**	127.4
12.2	**54**	129.2
12.8	**55**	131.0
13.3	**56**	132.8
13.9	**57**	134.6
14.4	**58**	136.4
15.0	**59**	138.2
15.6	**60**	140.0
16.1	**61**	141.8
16.7	**62**	143.6
17.2	**63**	145.4
17.8	**64**	147.2
18.3	**65**	149.0
18.9	**66**	150.8
19.4	**67**	152.6
20.0	**68**	154.4
20.6	**69**	156.2
21.1	**70**	158.0
21.7	**71**	159.8
22.2	**72**	161.6
22.8	**73**	163.4
23.3	**74**	165.2
23.9	**75**	167.0
24.4	**76**	168.8
25.0	**77**	170.6
25.6	**78**	172.4
26.1	**79**	174.2

C	value	F
38	**100**	212
43	**110**	230
49	**120**	248
54	**130**	266
60	**140**	284
66	**150**	302
71	**160**	320
77	**170**	338
82	**180**	356
88	**190**	374
93	**200**	392
99	**210**	410
100	**212**	413
104	**220**	428
110	**230**	446
116	**240**	464
121	**250**	482
127	**260**	500
132	**270**	518
138	**280**	536
143	**290**	554
149	**300**	572
154	**310**	590
160	**320**	608
166	**330**	626
171	**340**	644
177	**350**	662
182	**360**	680
188	**370**	698
193	**380**	716

C	value	F
260	**500**	932
266	**510**	950
271	**520**	968
277	**530**	986
282	**540**	1004
288	**550**	1022
293	**560**	1040
299	**570**	1058
304	**580**	1076
310	**590**	1094
316	**600**	1112
321	**610**	1130
327	**620**	1148
332	**630**	1166
338	**640**	1184
343	**650**	1202
349	**660**	1220
354	**670**	1238
360	**680**	1256
366	**690**	1274
371	**700**	1292
377	**710**	1310
382	**720**	1328
388	**730**	1346
393	**740**	1364
399	**750**	1382
404	**760**	1400
410	**770**	1418
416	**780**	1436
421	**790**	1454

C	value	F
538	**1000**	1832
543	**1010**	1850
549	**1020**	1868
554	**1030**	1886
560	**1040**	1904
566	**1050**	1922
571	**1060**	1940
577	**1070**	1958
582	**1080**	1976
588	**1090**	1994
593	**1100**	2012
599	**1110**	2030
604	**1120**	2048
610	**1130**	2066
616	**1140**	2084
621	**1150**	2102
627	**1160**	2120
632	**1170**	2138
638	**1180**	2156
643	**1190**	2174
649	**1200**	2192
654	**1210**	2210
660	**1220**	2228
666	**1230**	2246
671	**1240**	2264
677	**1250**	2282
682	**1260**	2300
688	**1270**	2318
693	**1280**	2336
699	**1290**	2354

C	value	F
816	**1500**	2732
821	**1510**	2750
827	**1520**	2768
832	**1530**	2786
838	**1540**	2804
843	**1550**	2822
849	**1560**	2840
854	**1570**	2858
860	**1580**	2876
866	**1590**	2894
871	**1600**	2912
877	**1610**	2930
882	**1620**	2948
888	**1630**	2966
893	**1640**	2984
899	**1650**	3002
904	**1660**	3020
910	**1670**	3038
916	**1680**	3056
921	**1690**	3074
927	**1700**	3092
932	**1710**	3110
938	**1720**	3128
943	**1730**	3146
949	**1740**	3164
954	**1750**	3182
960	**1760**	3200
966	**1770**	3218
971	**1780**	3236
977	**1790**	3254

C	value	F
1093	**2000**	3632
1099	**2010**	3650
1104	**2020**	3668
1110	**2030**	3686
1116	**2040**	3704
1121	**2050**	3722
1127	**2060**	3740
1132	**2070**	3758
1138	**2080**	3776
1143	**2090**	3794
1149	**2100**	3812
1154	**2110**	3830
1160	**2120**	3848
1166	**2130**	3866
1171	**2140**	3884
1177	**2150**	3902
1182	**2160**	3920
1188	**2170**	3938
1193	**2180**	3956
1199	**2190**	3974
1204	**2200**	3992
1210	**2210**	4010
1216	**2220**	4028
1221	**2230**	4046
1227	**2240**	4064
1232	**2250**	4082
1238	**2260**	4100
1243	**2270**	4118
1249	**2280**	4136
1254	**2290**	4154

C	value	F
1371	**2500**	4532
1377	**2510**	4550
1382	**2520**	4568
1388	**2530**	4586
1393	**2540**	4604
1399	**2550**	4622
1404	**2560**	4640
1410	**2570**	4658
1416	**2580**	4676
1421	**2590**	4694
1427	**2600**	4712
1432	**2610**	4730
1438	**2620**	4748
1443	**2630**	4766
1449	**2640**	4784
1454	**2650**	4802
1460	**2660**	4820
1466	**2670**	4838
1471	**2680**	4856
1477	**2690**	4874
1482	**2700**	4892
1488	**2710**	4910
1493	**2720**	4928
1499	**2730**	4946
1504	**2740**	4964
1510	**2750**	4982
1516	**2760**	5000
1521	**2770**	5018
1527	**2780**	5036
1532	**2790**	5054

Temperature Conversion Table

The boldface center column is the temperature to be converted. The left column gives the Celsius equivalent (°C); the right column gives the Fahrenheit equivalent (°F).

°C	Temperature	°F
−112	**−170**	−274
−107	**−160**	−256
−101	**−150**	−238
−95.6	**−140**	−220
−90.0	**−130**	−202
−84.4	**−120**	−184
−78.9	**−110**	−166
−73.3	**−100**	−148
−67.8	**−90**	−130
−62.2	**−80**	−112
−56.7	**−70**	−94
−51.1	**−60**	−76
−45.6	**−50**	−58
−40.0	**−40**	−40
−34.4	**−30**	−22
−28.9	**−20**	−4
−23.3	**−10**	14
−17.8	**0**	32
−1.11	**30**	86.0
−0.56	**31**	87.8
0	**32**	89.6
0.56	**33**	91.4
1.11	**34**	93.2
1.67	**35**	95.0
2.22	**36**	96.8
2.78	**37**	98.6
3.33	**38**	100.4
3.89	**39**	102.2
4.44	**40**	104.0
5.00	**41**	105.8
5.56	**42**	107.6
6.11	**43**	109.4
6.67	**44**	111.2
7.22	**45**	113.0
7.78	**46**	114.8
8.33	**47**	116.6
8.89	**48**	118.4
9.44	**49**	120.2
26.7	**80**	176.0
27.2	**81**	177.8
27.8	**82**	179.6
28.3	**83**	181.4
28.9	**84**	183.2
29.4	**85**	185.0
30.0	**86**	186.8
30.6	**87**	188.6
31.1	**88**	190.4
31.7	**89**	192.2
32.2	**90**	194.0
32.8	**91**	195.8
33.3	**92**	197.6
33.9	**93**	199.4
34.4	**94**	201.2
35.0	**95**	203.0
35.6	**96**	204.8
36.1	**97**	206.6
36.7	**98**	208.4
37.2	**99**	210.2
199	**390**	734
204	**400**	752
210	**410**	770
216	**420**	788
221	**430**	806
227	**440**	824
232	**450**	842
238	**460**	860
243	**470**	878
249	**480**	896
254	**490**	914
427	**800**	1472
432	**810**	1490
438	**820**	1508
443	**830**	1526
449	**840**	1544
454	**850**	1562
460	**860**	1580
466	**870**	1598
471	**880**	1616
477	**890**	1634
482	**900**	1652
488	**910**	1670
493	**920**	1688
499	**930**	1706
504	**940**	1724
510	**950**	1742
516	**960**	1760
521	**970**	1778
527	**980**	1796
532	**990**	1814
704	**1300**	2372
710	**1310**	2390
716	**1320**	2408
721	**1330**	2426
727	**1340**	2444
732	**1350**	2462
738	**1360**	2480
743	**1370**	2498
749	**1380**	2516
754	**1390**	2534
760	**1400**	2552
766	**1410**	2570
771	**1420**	2588
777	**1430**	2606
782	**1440**	2624
788	**1450**	2642
793	**1460**	2660
799	**1470**	2678
804	**1480**	2696
810	**1490**	2714
982	**1800**	3272
988	**1810**	3290
993	**1820**	3308
999	**1830**	3326
1004	**1840**	3344
1010	**1850**	3362
1016	**1860**	3380
1021	**1870**	3398
1027	**1880**	3416
1032	**1890**	3434
1038	**1900**	3452
1043	**1910**	3470
1049	**1920**	3488
1054	**1930**	3506
1060	**1940**	3524
1066	**1950**	3542
1071	**1960**	3560
1077	**1970**	3578
1082	**1980**	3596
1088	**1990**	3614
1260	**2300**	4172
1266	**2310**	4190
1271	**2320**	4208
1277	**2330**	4226
1282	**2340**	4244
1288	**2350**	4262
1293	**2360**	4280
1299	**2370**	4298
1304	**2380**	4316
1310	**2390**	4334
1316	**2400**	4352
1321	**2410**	4370
1327	**2420**	4388
1332	**2430**	4406
1338	**2440**	4424
1343	**2450**	4442
1349	**2460**	4460
1354	**2470**	4478
1360	**2480**	4496
1366	**2490**	4514
1538	**2800**	5072
1543	**2810**	5090
1549	**2820**	5108
1554	**2830**	5126
1560	**2840**	5144
1566	**2850**	5162
1571	**2860**	5180
1577	**2870**	5198
1582	**2880**	5216
1588	**2890**	5234
1593	**2900**	5252
1599	**2910**	5270
1604	**2920**	5288
1610	**2930**	5306
1616	**2940**	5324
1621	**2950**	5342
1627	**2960**	5360
1632	**2970**	5378
1638	**2980**	5396
1643	**2990**	5414

NOTE: The numbers in **boldface** type refer to the temperature (in either Celsius or Fahrenheit degrees) which it is desired to convert into the other scale. If converting from degrees Fahrenheit to degrees Celsius, the equivalent temperature is in the left column, while if converting from degrees Celsius to degrees Fahrenheit, the equivalent temperature is in the column on the right.

Interpolation factors

C		F	C		F
0.56	**1**	1.8	3.33	**6**	10.8
1.11	**2**	3.6	3.89	**7**	12.6
1.67	**3**	5.4	4.44	**8**	14.4
2.22	**4**	7.2	5.00	**9**	16.2
2.78	**5**	9.0	5.56	**10**	18.0

Temperature Scales, Standards, and Calibration*

Temperature, broadly defined, is the degree of hotness or coldness of a body or an environment. In a narrower sense, temperature is the degree of hotness or coldness referenced to a specific scale. Temperature is an intensive quantity independent of the size of the system. Knowledge of the temperature of two bodies makes possible the prediction of the direction of heat flow when they are brought into contact. Heat will flow from the body at the higher temperature. This behavioral mode is the basis of the second law of thermodynamics.

TEMPERATURE SCALES

There are two approaches that can be used in realizing the thermodynamic temperature scale. One is based on a heat engine operating according to the Carnot cycle. The other is based on the behavior of a perfect gas. The heat engine approach is impractical and is considered only in the development of theory. The perfect gas method is practical and has been used extensively for fundamental measurements ever since the concept was developed by Lord Kelvin (Sir William Thomson, 1824–1907) in 1848. Since there is no perfect gas to be found in nature, real gases are used and corrections applied for their departure from perfect gas behavior. The constant volume gas thermometer is used, but because of the complexities and very taxing experimental difficulties involved, such thermometers are only to be found in large university or governmental standardizing laboratories, such as the National Bureau of Standards.

From the outset the value of a degree on the thermodynamic scale was established by assigning a span of 100° to the temperature difference between freezing and boiling water. This made the value of a degree agree with the Celsius (then called centigrade) degree. The Celsius scale was already in common use in science. However, Kelvin recognized the significance of using only one fixed point to establish the scale. It was not until 1954 that his suggestion was officially adopted.

Although the thermodynamic concept is only about 100 years old, thermometer makers had been building instruments for many years before that. In the early days there were as many arbitrary scales as there were makers. The confusion so created has persisted down to the present time. Material progress in resolving this state of confusion has been made through adoption of the International Temperature Scale which is now in the fourth generation. These various scales and their development are discussed in more detail in the sections that follow.

Fahrenheit Scale

The first empirical scale of temperature developed is named for its inventor, Daniel Gabriel Fahrenheit (1686–1736), who described it in a paper published in 1724. Fahrenheit used the ice point, designated 32°, and the body temperature of a healthy person, designated 96°, as fixed points in

---

*Much of the material in this article has been retained from the second edition of this Handbook, as prepared by Dr. R. D. Thompson, then with Taylor/Sybron, Rochester, New York. Cooperation in updating and expanding the article was provided by J. A. Wise (U.S. National Bureau of Standards) and Robert D. Collier, formerly with Taylor/Sybron, Arden, North Carolina.

making scales for his thermometers. The fundamental interval, ice point to steam point, turned out to be 180°.

In the 1724 report of his invention Fahrenheit referred incorrectly to a third temperature as a fixed point which he designated 0°. He attained it with a mixture of ice, water, and sal ammoniac or sea salt. The temperature realized with such a system is actually a function of the relative proportions of the three ingredients and of the type of salt used.

The Fahrenheit scale is used extensively in English-speaking countries for meteorological, medical, and industrial purposes. Temperatures are denoted by the term degrees Fahrenheit and by the symbol °F.

Réaumur Scale

This scale evolved from the invention prior to 1730 of René-Antoine Ferchalt de Réaumur (1683–1757). The initial concept was to use diluted wine as the thermometric liquid and to designate the ice point as 1000 and the boiling point of the liquid as 1080. In time this was changed so that the fundamental interval, ice point to steam point, while still 80°, was based on an ice-point designation of 0° and a steam-point designation of 80°. At present the scale is little used except in the brewing and liquor industries. Temperatures are denoted by the term degrees Réaumur and by the symbol °R.

Celsius (Centigrade) Scale

In 1742 Anders Celsius of Uppsala University in Sweden reported on the use of thermometers in which the fundamental interval, ice point to steam point, was 100°. Celsius designated the ice point at 100° and the steam point as 0°. Subsequently Christin (1743) in Lyon, France, and Linnaeus (1745) in Uppsala independently interchanged the designations. For many years prior to 1948 it was known as the centigrade scale. In 1948 by international agreement it was renamed in honor of its inventor and to avoid an inconsistent connotation. The scale is used worldwide in scientific work and in medicine and for general purposes in non-English-speaking countries. Temperatures are denoted as degrees Celsius and by the symbol °C.

Thermodynamic Celsius Scale

This scale, called originally the Thermodynamic Centigrade Scale, was developed from thermodynamic considerations but was extremely difficult to reproduce in practice. The constant volume gas thermometer with its extensive and elaborate ancillary equipment was used.

By international agreement the temperatures of melting ice and condensing water vapor, both under the pressure of one standard atmosphere and numbered 0° and 100°, respectively, were chosen to define the scale. Temperatures are denoted by degrees Celsius thermodynamic and by the symbol °C (therm).

Thermodynamic Kelvin Scale

The currently accepted theoretical scale was named for Lord Kelvin who first enunciated the principle on which it is based. Thermodynamic temperature is denoted by the symbol T and the unit is the kelvin, symbol K. The kelvin is the fraction $1/273.16$ of the thermodynamic temperature of the triple point of water. The triple point is realized when ice, water, and water vapor are in equilibrium. It is the sole defining fixed point of the thermodynamic Kelvin scale and has the assigned value 273.16 K.

Rankine Scale

This scale is the equivalent of the thermodynamic Kelvin scale but is expressed in terms of Fahrenheit degrees. Thus, the temperature of the triple point of water on the Rankine scale, corresponding to 273.16 K, is very nearly 491.69°Rankine.

The International Practical Temperature Scale

Over the years international agreements have been made to adopt scales which reproduce the thermodynamic Celsius and Kelvin scales as closely as possible within the limits of the state of the art. The scales have been given various names. The first was the Normal Thermometric Scale (1887) based on the constant volume hydrogen gas thermometer with a fundamental interval of 100° between the fixed points of melting ice, 0°C, and of condensing water vapor, 100°C. Both points were realized at one standard atmosphere pressure. This scale, commonly referred to as the International Hydrogen Scale, was reproduced for international comparisons by means of four liquid-in-glass thermometers. The range was limited to −35 to +100°C.

Improvements in materials and techniques made possible extension of the range with a simultaneous increase in accuracy attainable. 1927 marked the adoption of the International Temperature Scale covering the range from the boiling point of oxygen to that of luminous incandescent bodies. Values were assigned to six fixed points, along with specifications for interpolation instruments, the platinum resistance thermometer, and the platinum 10% rhodium-platinum thermocouple. Radiation constants for optical pyrometry were also adopted. This was a practical or working scale and was intended to reproduce as closely as possible the Thermodynamic Centigrade Scale. If, in practice, it was desired to establish temperatures on the thermodynamic scale, measurements would be made on the International Temperature Scale and experimentally determined difference corrections could then be applied. These differences, however, were generally so small as to be of significance only in work of the very highest order of accuracy.

In 1948 a revision was made and the designation degree Celsius in place of degree centigrade was officially adopted. The scale changed slightly from the 1927 version, but at higher temperatures. Then in 1954 it was decided to define the scale by only one fixed point as had been recommended by Lord Kelvin a hundred years previously. The triple point of water was chosen, 273.16 K. A text revision was issued in 1960 covering these various changes, but it did not affect the values of the 1948 scale by as much as the experimental error of measurement. The name of the scale was changed to International Practical Temperature Scale of 1948.

In 1968, The International Practical Temperature Scale (IPTS 68) was adopted. Two changes of major significance were made at that time. The scale was extended below the oxygen point to effect standardization in the range 10 to 90 K. An updating over the range −183 to +1063°C based on more refined gas thermometer measurements made since 1927 was accomplished.

The fixed points for the 1968 scale are given in Table 1 which is reproduced from the approved English translation of the official text. Although the maximum change in the 0 to 100°C portion of the scale is only 0.01° Celsius, at higher temperatures the changes are much more significant. For example, at the gold point the value is changed by 1.4° Celsius. The radiation constant was assigned a new value to bring the optical pyrometer portion of the range into closer accord with the thermodynamic scale. Table 2 shows the approximate differences between IPTS 68 and IPTS 48.

STANDARDS

IPTS 68 as well as its predecessor international scales provided specifications for the interpolation instruments to be used at other than the fixed points in realizing the scale in practice. These instruments are the platinum resistance thermometer for use up to the antimony point, 630.74°C, and the platinum 10% rhodium-platinum thermocouple for use from 630.74 to 1064.43°C, the gold point. Above the gold point optical pyrometry is used, and the Planck radiation constant is defined in the text of the scale. The official text has been published in *Metrologia,* volume 5, no. 2, April 1969.

Table 1. Defining Fixed Points of the IPTS 68*

Equilibrium state	Assigned value of International Practical Temperature	
	T_{68}, K	t_{68}, °C
Equilibrium between the solid, liquid, and vapor phases of equilibrium hydrogen (triple point of equilibrium hydrogen)	13.81	−259.34
Equilibrium between the liquid and vapor phases of equilibrium hydrogen at a pressure of 33 330.6 N/m² (25/76 standard atmosphere)	17.042	−256.108
Equilibrium between the liquid and vapor phases of equilibrium hydrogen (boiling point of equilibrium hydrogen)	20.28	−252.87
Equilibrium between the liquid and vapor phases of neon (boiling point of neon)	27.102	−246.048
Equilibrium between the solid, liquid, and vapor phases of oxygen (triple point of oxygen)	54.361	−218.789
Equilibrium between the liquid and vapor phases of oxygen (boiling point of oxygen)	90.188	−182.962
Equilibrium between the solid, liquid, and vapor phases of water (triple point of water)†	273.16	0.01
Equilibrium between the liquid and vapor phases of water (boiling point of water)‡	373.15	100
Equilibrium between the solid and liquid phases of zinc (freezing point of zinc)	692.73	419.58
Equilibrium between the solid and liquid phases of silver (freezing point of silver)	1235.08	961.93
Equilibrium between the solid and liquid phases of gold (freezing point of gold)	1337.58	1064.43

*Except for the triple points and one equilibrium hydrogen point (17.042 K) the assigned values of temperature are for equilibrium states at pressure $p_0 = 1$ standard atmosphere (101 325 N/m²). In the realization of the fixed points small departures from the assigned temperatures will occur as a result of the differing immersion depths of thermometers or the failure to realize the required pressure exactly. If due allowance is made for these small temperature differences, they will not affect the accuracy of realization of the scale.

†The water used should have the isotopic composition of ocean water.

‡The equilibrium state between the solid and liquid phases of tin (freezing point of tin) has the assigned value of $t_{68} = 231.9681$°C and may be used as an alternative to the boiling point of water.

For routine measurements instruments of the same type may be used as working or secondary standards with their calibration corrections being determined by comparison with primary standard instruments. Other secondary standards such as liquid-in-glass thermometers similarly calibrated are commonly employed.

Platinum Resistance Thermometer

The text of IPTS 68 gives important details of construction of a standard platinum resistance thermometer. Such thermometers may be purchased from manufacturers who specialize in the production of instruments of this type. A thermometer to be used as a primary standard is calibrated by measuring its resistance at various fixed points. This can be done by the user or by a standardizing laboratory such as the National Bureau of Standards in the United States. Where traceability to establish accuracy of manufactured product is required, the national laboratory method is generally used.

Table 2. Approximate Differences ($t_{68} - t_{48}$) in Kelvins between IPTS 68 and IPTS 48 Values

t_{68}, °C	0	−10	−20	−30	−40	−50	−60	−70	−80	−90	−100
−100	0.022	0.013	0.003	−0.006	−0.013	−0.013	−0.005	0.007	0.012		
−0	0.000	0.006	0.012	0.018	0.024	0.029	0.032	0.034	0.033	0.029	0.022

t_{68}, °C	0	10	20	30	40	50	60	70	80	90	100
0	0.000	−0.004	−0.007	−0.009	−0.010	−0.010	−0.010	−0.008	−0.006	−0.003	0.000
100	0.000	0.004	0.007	0.012	0.016	0.020	0.025	0.029	0.034	0.038	0.043
200	0.043	0.047	0.051	0.054	0.058	0.061	0.064	0.067	0.069	0.071	0.073
300	0.073	0.074	0.075	0.076	0.077	0.077	0.077	0.077	0.077	0.076	0.076
400	0.076	0.075	0.075	0.075	0.075	0.074	0.074	0.075	0.076	0.077	0.079
500	0.079	0.082	0.085	0.089	0.094	0.100	0.108	0.116	0.126	0.137	0.150
600	0.150	0.165	0.182	0.200	0.23	0.25	0.28	0.31	0.34	0.36	0.39
700	0.39	0.42	0.45	0.47	0.50	0.53	0.56	0.58	0.61	0.64	0.67
800	0.67	0.70	0.72	0.75	0.78	0.81	0.84	0.87	0.89	0.92	0.95
900	0.95	0.98	1.01	1.04	1.07	1.10	1.12	1.15	1.18	1.21	1.24
1000	1.24	1.27	1.30	1.33	1.36	1.39	1.42	1.44			

t_{68}, °C	0	100	200	300	400	500	600	700	800	900	1000
1000		1.5	1.7	1.8	2.0	2.2	2.4	2.6	2.8	3.0	3.2
2000	3.2	3.5	3.7	4.0	4.2	4.5	4.8	5.0	5.3	5.6	5.9
3000	5.9	6.2	6.5	6.9	7.2	7.5	7.9	8.2	8.6	9.0	9.3

Platinum-10% Rhodium-Platinum Thermocouple

The text of IPTS 68 gives important details of construction of a standard thermocouple. As with the resistance thermometer, calibration can be accomplished using fixed points or by having the service performed at a national standardizing laboratory.

Optical Pyrometer (Radiation Thermometer)

National standardizing laboratories maintain their own standard optical pyrometer. Calibration of a user's instrument is best accomplished by comparison with such an instrument. When so calibrated, the pyrometer may be used for actual measurements or for the calibration of other working standards.

Secondary Standards and Calibration

Since a primary standard thermometer is an expensive instrument, the integrity of which is important, it is not commonly used for routine purposes. Instead, secondary standards are generally employed which in turn are calibrated against a primary standard. These secondary standards may be of exactly the same design as the primary standards, but if damaged or destroyed, replaceability is not as significant. They may also be instruments of different design such as noble metal thermocouples of different composition, base metal thermocouples or liquid-in-glass thermometers. In all cases they should be of a quality commensurate with their use. Calibrations should be conducted periodically against the primary standards to permit correction for aging effects and to evaluate the magnitude of alteration due to possible misuse.

*Levels of Temperature Measurement Accuracy**

The standards and calibrating procedures used are determined by need and generally fall into four categories.

Level 1—General Purpose

For many applications, temperature measurements need not be highly accurate. Frequently, errors of a few degrees are not troublesome. If a home thermostat, for example, has a slight error, the setting can be adjusted to compensate. Numerous industrial temperature measurements are not highly critical. Millions of thermometers are used in heating and air-conditioning systems primarily to indicate whether steam or chilled water is flowing properly, and whether the regulating equipment is functioning. Reasonable cost is paramount; the instruments used are relatively inexpensive. More accurate instruments can be used when trouble arises; attention should be given to more accurate instruments during installation and servicing.

Level 2—Industrial Grade

In many industrial process control uses and in connection with scientific experiments, there is often little margin for significant temperature errors. Returning to the air-conditioning application just mentioned, it is one thing to know whether the system is operating properly by observing thermometers in chilled water lines—but an entirely different matter to balance the chiller system for maximum economy. Here, an error of 0.5°C can mean thousands of dollars a year in energy loss. Similarly, in rubber curing, food processing, and medical sterilization, among many other applications, there is a critical time-temperature relationship, frequently with safety and health depending on knowing the temperature within an uncertainty of 1° or less. For these uses, a wide variety of equipment is available to indicate, record, and control temperature. In some devices, such as radiation

*Based on a paper by Collier (1978). See reference list.

thermometers, the sensing device need not contact the object to be measured. The principal temperature sensors are described in other articles in this Handbook section.

Generally, industrial systems do not have accuracy built in as a feature of basic design, but rather all systems must be checked for accuracy by reference to some standard. Thermocouples, for example, may vary in output slightly from one junction to another, and systems using them can have undetected errors as a result of poor connections, malfunctioning indicators, or junction deterioration with time. Most of these devices can be calibrated to operate with greatly reduced uncertainty if regularly checked against a reliable reference standard having a very high accuracy.

Level 3—Secondary Reference Standards

To calibrate, or to check the calibration of industrial-grade instruments, a more accurate thermometer is needed as a reference standard. It is important to know the range of uncertainty of equipment indication or control, both when new as a check on the manufacturer's specifications and correctness of installation, and after a period of use to check for deterioration or malfunction.

As a general rule, when an instrument is calibrated, the device used to calibrate it should be *very much more accurate*. Ideally, it should have an uncertainty an order of magnitude less, or in other words, be roughly 10 times as accurate. In thermometry, therefore, to calibrate an instrument with an uncertainty of a few tenths of a degree, the reference standard must have an uncertainty of only a few hundredths of a degree. Further, multiple readings, cross-referencing, and other techniques can be used to minimize error, particularly in connection with Level-4 standards.

Although more recent equipment, such as quartz crystal systems, is becoming available, most devices used as temperature reference standards are of two types—the precision platinum resistance thermometer, previously described, or a set of very carefully made and calibrated mercury-in-glass etched-stem thermometers, observed through a microscope. Both must be used with a precision temperature bath, and both must be checked frequently against a primary standard. Because it costs many times as much as the mercury-in-glass set, can be easily operated improperly, and needs expensive maintenance to ensure proper operation, the resistance bridge system is not as widely used as glass thermometers. Mercury-in-glass thermometers have long been used as reference standards because they offer a number of advantages: (1) extremely simple, trouble-free construction; (2) relatively low cost; (3) simple operation—a direct reading scale; (4) long-term stability; (5) easy visual check of malfunction or damage; and (6) very small uncertainty when properly calibrated.

Unfortunately, inexpensive and inaccurate glass thermometers appear at first glance to be little different from those used as reference standards. They are sometimes even described in sales literature as "precision," when, in fact, they may offer very poor accuracy. Furthermore, the manner in which they are used has a great deal to do with the final accuracy attained.

Level 4—Primary Standards Calibration System

It is possible to achieve still another order of magnitude in reducing uncertainty in temperature measurement—to a level of a few thousandths of $1°$. The cost of both the initial purchase of equipment and the continued effort needed to ensure the accuracy of the system restrict its use to a very few locations (about 25 in the United States, for example). In the larger industrial countries, primary calibration facilities are usually found in the nation's standards laboratory, such as the U.S. National Bureau of Standards, in a few factories where thermometers are made, and perhaps in a few commercial testing laboratories. However, most thermometer manufacturers do not maintain such facilities and rely on thermometers calibrated at the National Bureau of Standards and other similar laboratories in various parts of the world.

A primary standards system usually includes

1. A set of metrology-grade platinum-wire-in-glass resistance elements
2. A high-quality precision resistance bridge and electronic null detector
3. A calibrated standard resistor (by the National Bureau of Standards or an equivalent facility) to verify resistance bridge operation
4. A set of primary standard melting-freezing point cells, including a glass–distilled water triple-point cell

5. A set of precision stirred-liquid calibration baths whose capability covers the range of temperatures required

6. A regular program of three-way verification checks between the laboratory, the equipment manufacturer, and the standards laboratory to ensure continued accuracy of the resistance element and standard resistor unit calibration

7. A trained technician who maintains records, accomplishes calibrations under controlled conditions, and ensures that verification check programs are rigorously followed.

Level 5—International Standards Coordination

In much the same way that coordination between laboratories is done to ensure the accuracy achieved in Level 4, the National Bureau of Standards coordinates with its counterparts in other countries. In general, the procedures are built around reference to primary standards and a system of cross-verification between laboratories, but techniques are even more rigorous than those described for Level 4.

Much more detail on procedures is given by Collier (1978, 1982).

Comparison Calibration

Apparatus and techniques for comparison calibration are described in detail in the literature. *NBS Monograph 150* ("Liquid-in-Glass Thermometry"); *NBS Circular 590* ("Methods of Testing Thermocouples and Thermocouple Materials"); *ASTM Standard E77* ("Standard Method for Verification and Calibration of Liquid-in-Glass Thermometers"); and *ASTM Standard E220* ("Standard Method for Calibration of Thermocouples by Comparison Techniques"); are comprehensive, authoritative treatises on the subject.

Transfer Standards

It is desirable in maintaining a temperature scale to have transfer standards in the form of fixed-point apparatus available for occasionally checking the standard instruments. The simplest type of

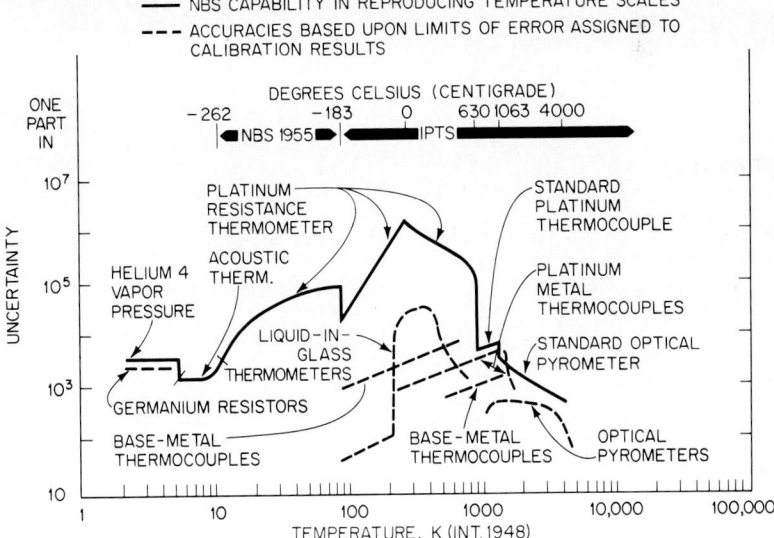

Fig. 1 Calibration of temperature measuring instruments.

Table 3. Estimated Uncertainties of the Assigned Values of the Defining Fixed Points in Terms of Thermodynamic Temperatures

Defining fixed point	Assigned value	Estimated uncertainty, K
Triple point of equilibrium hydrogen	13.81 K	0.01
17.042 K point	17.042 K	0.01
Boiling point of equilibrium hydrogen	20.28 K	0.01
Boiling point of neon	27.102 K	0.01
Triple point of oxygen	54.361 K	0.01
Boiling point of oxygen	90.188 K	0.01
Triple point of water	273.16 K	Exact by definition
Boiling point of water	100°C	0.005
Freezing point of tin	231.9681°C	0.015
Freezing point of zinc	419.58°C	0.03
Freezing point of silver	961.93°C	0.2
Freezing point of gold	1064.43°C	0.2

apparatus is an ice bath. If greater accuracy in that region is desired, a triple point of water cell may be used. Another temperature realized quite conveniently is the benzoic acid triple point, 122.37°C. This is a secondary fixed point on IPTS 68. Other materials commonly used for this purpose are zinc and organic compounds such as phenol, naphthalene, and phthalic anhydride. In these cases the freezing point is employed. A list of approved secondary standards is given in the text of IPTS 68.

Another form of transfer standard for use in the high-temperature range is the tungsten ribbon-filament lamp. These lamps can be made with a very stable lamp current-brightness-temperature relation and are particularly used for temperatures above 1050°C.

The reliability of a calibration is a function of a number of factors. Obvious elements such as the quality of the equipment and the care used in making the measurements enter in. There are some fundamental limitations, however, associated with the uncertainties in the values of the fixed points of IPTS 68 and the ability of laboratories such as the National Bureau of Standards to reproduce the scale. Figure 1, even though it relates to IPTS 48 rather than IPTS 68, gives a good overview of the reliability of calibration at the Bureau as a function of thermometer type and temperature. It remains for the user to relate this to each specific situation. If it is desirous to translate measurements on IPTS 68 to the thermodynamic scale, to the uncertainties shown in Fig. 1, the uncertainties of the IPTS 68 fixed points in terms of thermodynamic temperatures should be added. See Table 3.

REFERENCES

Buckingham, E.: "The Correction for Emergent Stem of the Mercurial Thermometer," *Bull. BS 8,239(S170)*, National Bureau of Standards, Washington, D.C., 1912.

Collier, R. D.: "Temperature Measurement Accuracy," ISA Conference, October 15–19, 1978, Instrument Society of America, Research Triangle Park, N.C.

Collier, R. D.: "Temperature Instrument Calibration Fundamentals," ISA Temperature Symposium, March 1982, Instrument Society of America, Research Triangle Park, N.C.

Furukawa, G. T.: "A Measurement Assurance Program—Thermometer Calibration," Unpublished paper, ASTM, June 25, 1980. Contact author at National Bureau of Standards, Washington, D.C.

Guillaume, C. E.: *Traité Practique de la Thermométrie*, Gauthier-Villars et Fils, Paris, 1889.

Liberatore, L. C., and H. J. Whitcomb: "Density Changes in Thermometer Glasses," J. Am. Ceram. Soc., vol. 35, 1952, pp. 67.

Martin, W. I., and S. C. Grossman: "Calibration Drift with Thermometers Repeatedly Cooled to −30°C," *ASTM Bull. 231*, American Society for Testing and Materials, Philadelphia, 1958, p. 62.

Osborne, N. S., and C. H. Meyers: "A Formula and Tables for the Pressure of Saturated Water Vapor in the Range 0 to 374°C," *J. Res. NBS* vol. 13 (RP691), 1934.

Riddle, J., G. Furukawa, and H. Plumb: *NBS Monogr. 125*, National Bureau of Standards, Washington, D.C., 1973.

Ruh, E. L., and G. E. Conklin: "Thermal Stability in ASTM Thermometers," *ASTM Bull. 233*, American Society for Testing and Materials, Philadelphia, 1958, p. 35.

Scott, R. B., and F. G. Brickwedde: "A Precision Cryostat with Automatic Temperature Regulation," *J. Res. NBS*, vol. 6 (RP284), 1931, p. 401.

Staff: *Comptes Rendus des Séances de la Treizième Conférence Générale des Poids et Mesures (1967-1968)*, *Resolutions*, International Bureau of Weights and Measures, Paris, 1967-1968.

Staff: "The International Practical Temperature Scale of 1968," *Metrologia*, vol. 5, April 1969, pp. 2, 35.

Staff: "Calibration and Test Services of the National Bureau of Standards," *NBS Spec. Publ. 250*, National Bureau of Standards, Washington, D.C., 1970.

Staff: "Calibration at Temperatures Other than Fixed Points," *ASTM E77-72*, American Society for Testing and Materials, Philadelphia.

Staff: "Evolution of the International Practical Temperature Scale of 1968," *ASTM STP 565*, American Society for Testing and Materials, Philadelphia, 1982.

Stimson, H. F.,: "The International Practical Temperature Scale of 1948," *J. Res. NBS*, vol. 65A, no. 3, May-June 1961.

Thomas, J. L.: "Reproducibility of the Ice Point" in *Temperature, Its Measurement in Science and Industry*, Reinhold, New York, 1941.

Thompson, R. D.: "Recent Developments in Liquid-in-Glass Thermometry" in *Temperature, Its Measurement in Science and Industry*, Reinhold, New York, 1962.

Wise, J.: "Liquid-in-Glass Thermometry," *NBS Monogr. 150*, 1976, sec. 2: "Thermometer Calibration Services at the NBS," 1-5; sec. 3: "Definitions," 5-6; sec. 4: "Temperature Scales and Standards," 6-8; sec. 5: "Calibration," 8-16.

Thermocouples*

by
Daniel A. Jackson and Allen E. Mushin†

The thermocouple principle was discovered in the early nineteenth century when experimenters found that bonding wires of two dissimilar metals together to form a closed circuit, as shown in Fig. 1, produced an electric current which would flow in the circuit when a difference in temperature was imposed between the end junctions. These experiments showed that a thermal emf was generated in that earliest version of a thermocouple. Today the thermocouple is the most widely used transducer for temperature measurement.

Simply stated, a thermocouple is a device that converts thermal energy directly into an electric voltage when a temperature gradient exists between the two end junctions of a pair of dissimilar metal wires. One end is fused together to form a measuring junction, the *hot junction*, and the other end, the *cold junction*, is connected to a measuring device. Temperature at the hot junction is determined by measuring the voltage appearing at the cold junction. The open circuit voltage developed

*This article has been copyrighted (1982) by Omega Engineering, Inc., Stamford, Connecticut.
†Omega Engineering, Inc., Stamford, Connecticut.

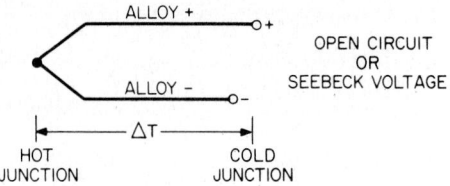

Fig. 1   Thermocouple circuits.

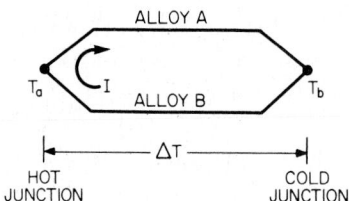

Fig. 2   Seebeck's circuit.

is a function of the Seebeck coefficient of the two metals and the difference in temperature between the hot and cold junctions. See Fig. 2.

Since the voltage produced by the thermocouple is a function of the difference in temperature between the hot and cold junctions, the temperature of one of the junctions (namely, the cold junction) must be accurately known.

Thermocouple Effects

The Seebeck Effect

Seebeck discovered, in 1821, the existence of thermoelectric currents while observing electromagnetic effects associated with bismuth-copper and bismuth-antimony circuits. His experiments showed that, when the junctions of two dissimilar metals forming a closed circuit are exposed to different temperatures, a new thermal electromotive force is generated which induces a continuous electric current.

The *Seebeck effect* concerns the net conversion of thermal energy into electric energy with the appearance of an electric current. The Seebeck voltage refers to the net thermal electromotive force (emf) set up in a thermocouple under zero-current conditions. The direction and magnitude of the Seebeck voltage E_S depend on the temperature of the junctions and on the materials making up the thermocouple. For a particular combination of materials, A and B, for a small temperature difference

$$dE_S = \alpha A,B \, dT$$

where $\alpha A,B$ is a coefficient of proportionality called the Seebeck coefficient.

The Peltier Effect

When an electric current flows across a junction of two dissimilar metals, heat is liberated or absorbed. When the electric current flows in the same direction as the Seebeck current, heat is absorbed at the hotter junction and liberated at the colder junction. The *Peltier effect* is defined as the change in heat content when a quantity of charge (1 coulomb) crosses the junction. See Fig. 3. This is the basis for thermoelectric refrigeration and heating.

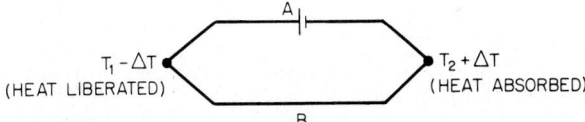

Fig. 3 Peltier effect.

The Thomson Effect

The *Thomson effect* is defined as the change in the heat content of a single conductor of unit cross section when a unit quantity of electricity flows through it along a temperature gradient of 1 kelvin. See Fig. 4.

Consider a single conductor which has been heated at one point to some temperature T_2. See Fig. 4a. A thermal gradient will exist on either side of the heated point. Two points, P_1 and P_2, of equal temperature, $T_1 < T_2$, will be found on either side of T_2. If current flows through the single conductor, the temperature at P_1 and P_2 will change. See Fig. 4b. The changes are the result of a motion of current with respect to the direction of the temperature gradient. Electrons moving against the increasing temperature gradient (from P_1) will absorb energy and increase their potential energy. Electrons traveling in the same direction as the gradient (toward the decreasing temperature at P_2) will give up energy and thus decrease their potential energy.

Heat will thus be absorbed at P_1, where the electron current direction is opposite that of the heat flow, while heat will be liberated at P_2, where the electron current direction is the same as that of the heat flow. The changes in the heat content of the conductor are known as the Thomson effects.

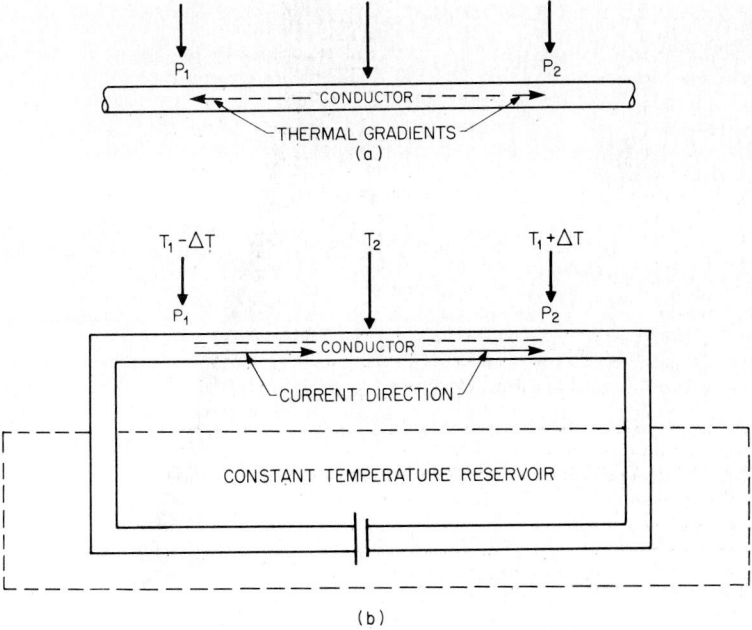

Fig. 4 Thomson effect: (a) single conductor heated at one point—a thermal gradient exists on either side of the heated point; (b) current flowing through a single conductor, causing temperature at P_1 and P_2 to change.

The Thomson effects are equal and opposite and thus cancel each other. The law of homogeneous conductors, which states that a thermoelectric current cannot be maintained solely by the application of heat to a single homogeneous conductor, regardless of any variation, has merit here. The configuration is ideal for connecting thermocouples to measuring devices using extension wires, because no electromotive force (emf) is added to the circuit.

Compensation Methods

Principal compensation methods include the (1) cold junction method, (2) electronic bridge method, (3) thermoelectric refrigeration method, (4) heat oven method, and (5) cryogenic method—as described in the following paragraphs.

Cold Junction Method

As a differential output transducer, the voltage output of a thermocouple is dependent on the temperature at both the hot and cold junctions. Historically, the freezing point of water, 32°F (0°C), was selected as a convenient cold junction reference.

To construct an ice bath reference junction, both legs of the thermocouple are fused to copper wire to form a transition junction. The leads are then waterproofed, and the transition junction is immersed in an ice bath. See Fig. 5. The open circuit voltage appears across the copper leads exiting from the ice bath.

The copper leads are used to connect to the emf readout device. This procedure avoids the generation of thermal emf at the terminals of the readout instrument. Voltages measured in this way may be directly converted into temperature by using NBS millivolt-temperature reference tables *(NBS Monograph 125)*. See Ref. 5.

Preparation of a Distilled Water Ice Bath

The following steps are required: (1) Drill a hole in the cap of a Dewar flask or Thermos bottle to accommodate the transition junction; (2) firmly pack the flask with small ice chips made from distilled water and then fill the flask with distilled water; the use of tap water may cause inaccuracies because of contaminants; (3) replace the melted ice with more ice while removing the excess water; (4) place the cap on the flask and insert the transition junction into the flask; allow 20 min for temperature stabilization.

Because ice baths are often difficult to maintain, and not always practical, several alternate methods of cold junction compensation are often employed.

Electronic Bridge Method

Because changes in the reference junction temperature influence the output signal of thermocouples, practical instruments must provide a means to cancel this potential source of error.

An alternate to the ice bath method is an electronic compensating bridge network. The temperature at a "floating" cold junction is monitored, and voltage corrections are applied to the incoming signals to make it appear as if the thermocouple has been referenced to some fixed temperature.

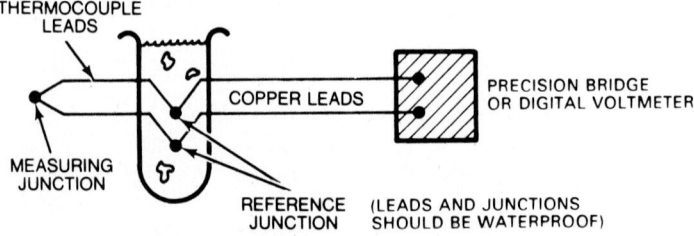

Fig. 5 Ice bath circuit.

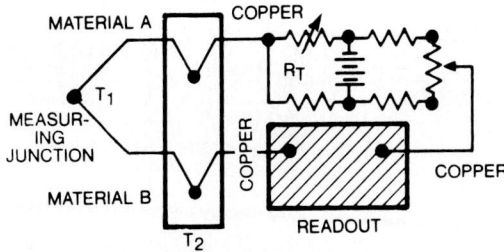

Fig. 6 Electronic bridge circuit.

This method usually employs a self-compensating electronic bridge network, as shown in Fig. 6. A temperature-sensitive element (R_T) is incorporated into a bridge network which is in series with one leg of the thermocouple pair. The temperature-sensitive element regulates the voltage applied by the battery or stable dc supply to the incoming thermocouple signal.

Most electronic cold junction compensators condition the signal so that the thermocouple appears to have been referenced to the ice point without resorting to the physical bath. They can operate over a wide range of ambient temperatures with high accuracy.

Thermoelectric Refrigeration Method

The ice point thermocouple reference chamber relies on the equilibrium of ice and distilled, deionized water at atmospheric pressure to maintain six reference wells at precisely 32°F (0°C). The wells extend into a sealed cylindrical chamber containing distilled, deionized water. The outer walls are cooled by the thermoelectric cooling elements. The increase in volume produced by the creation of ice within the cell is sensed by the expansion of a bellows, which operates a circuit controlling the cooling process. The alternate freezing and thawing of the ice accurately maintains a 32°F (0°C) environment around the reference wells. See Fig. 7.

The automatic operational features of this instrument eliminate the need for frequent attention required by common ice baths. Any combination of thermocouples may be used simply by inserting the reference probes in the reference wells. Calibration of other types of temperature sensors at 32°F (0°C) may be performed as well.

Heated Oven Reference Method

This method takes advantage of the thermoelectric effect using fixed temperature ovens instead of a battery to generate a correction voltage applied to the thermocouple. See Fig. 8.

Assume that we do away with the second oven shown in Fig. 8 and that a transition is made from thermocouple wire to copper wire in the first oven at 150°F (65°C). The output of the thermocouple will be based on the difference in temperature between the bead and the reference oven at 150°F (65°C). We then invoke the law of intermediate temperatures, and the millivolt equivalent of a thermocouple at 150°F (65°C) referenced to the ice point may be added to make the total output of the thermocouple appear as if it has been directly referenced to the ice point. A type-K junction at 150°F (65°C) yields 2.65 mV. Adding this voltage to the thermocouple signal coming out of the 150°F (65°C) oven effectively simulates a thermocouple referenced directly to the ice point.

This voltage may be produced by a battery or by creating another thermocouple junction to generate the voltage. In Fig. 8, thermocouple wire has been added to the leads from the first oven

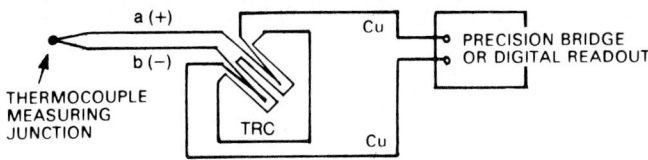

Fig. 7 Thermoelectric circuit.

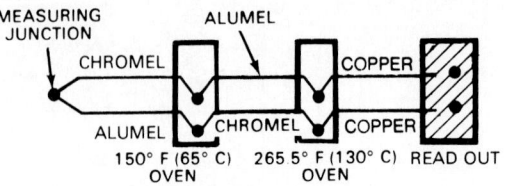

Fig. 8   Heated oven circuit.

and run into a second oven. If the temperature 265.5°F (130°C) of the second oven is selected so that the voltage produced between the two ovens is 2.65 mV, then the final output of the thermocouple appears as if it has been directly referenced to the ice point.

Cryogenic Method

Another less frequently used form of thermocouple reference junction is employed in the study of cryogenics for extremely low temperatures.

The thermocouple is made up of gold-atomic iron (Au–0.07% FE) versus Chromel-P, and the reference junction is boiling liquid helium which has a temperature of 4.26 K.

Depending on the operating conditions, reference junctions may vary from the 32°F (0°C) standard without sacrificing accuracy. Electronic-bridge-type compensators can be supplied with junction reference temperature up to 150°F (65°C). The user must generate a millivolt-versus-temperature table for the reference temperature from the 32°F (0°C) reference tables by

1. Looking up the millivolts for the reference temperature in the 32°F (0°C) reference tables
2. Subtracting that millivolt value from all voltages in the 32°F (0°C) reference tables

The result will be a new table referenced to the new reference temperature. NOTE: The reference temperature cannot be subtracted from the 32°F (0°C) reference indicated temperature. The correction must be made in the millivolt domain.

Laws Governing Thermocouples

Laws of Thermoelectric Circuits

Numerous investigations of thermoelectric circuits in which accurate measurements were made of the current, resistance, and emf have resulted in the establishment of three basic laws. These laws have been confirmed experimentally beyond a reasonable doubt and may be accepted in spite of lack of theoretical development.

Law of Homogeneous Metals

A thermoelectric current cannot be sustained in a circuit of a single homogeneous material, however, varying in cross section, by the application of heat alone.

A consequence of this law is that two different materials are required for any thermocouple circuit. Experiments have been reported suggesting that a nonsymmetrical temperature gradient in a homogeneous wire gives rise to a measurable thermoelectromotive force. Evidence indicates, however, that any emf observed in such a circuit arises from the effects of local inhomogeneities. Furthermore, any current detected in such a circuit when the wire is heated is taken as evidence that the wire is not homogeneous.

Law of Intermediate Metals

The algebraic sum of the thermoelectromotive forces in a circuit composed of any number of dissimilar materials is zero if all of the circuit is at a uniform temperature.

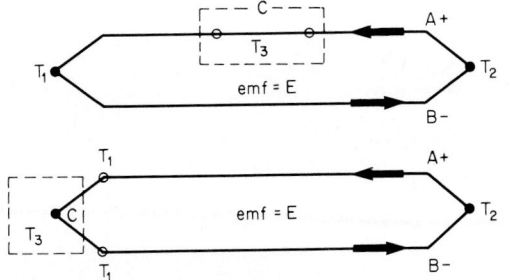

Fig. 9 Emf is unaffected by a third metal.

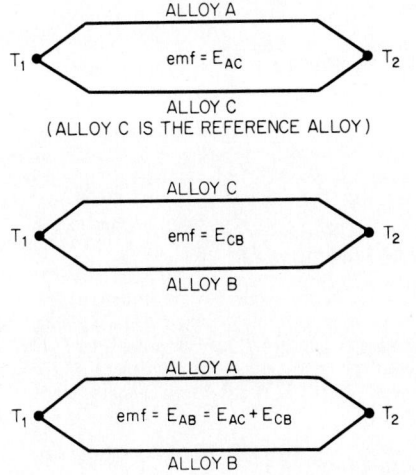

Fig. 10 Emf algebraic sums.

The result of this law is that a third homogeneous material can always be added to a circuit with no effect on the net emf of the circuit as long as its extremities are at the same temperature. Therefore, a device for measuring the thermoelectromotive force may be introduced into a circuit at any point without affecting the resultant emf, provided all the junctions added to the circuit by introducing the device are at the same temperature. It also follows that any junction whose temperature is uniform and which makes a good electrical contact does not affect the emf of the thermocouple circuit regardless of the method employed in forming the junction. See Fig. 9.

When applying this law, it follows that, if the thermal emfs of any two metals with respect to a reference metal are known, then the emf of the combination of the two metals is the algebraic sum of their emf's against the reference metal. See Fig. 10.

Law of Successive or Intermediate Temperatures

If two dissimilar homogeneous metals produce a thermal emf of E_1 when the junctions are at temperatures T_1 and T_2, and a thermal emf of E_2 when the junctions are at T_2 and T_3, the emf generated when the junctions are at T_1 and T_3 will be $E_1 + E_2$.

The application of this law permits a thermocouple calibrated for a given reference temperature to be used with any other reference temperature through the use of a suitable correction. See Fig. 11 for a schematic example.

Another example of this law is that extension wires having the same thermoelectric characteristics as those of the thermocouple wires can be introduced into the thermocouple circuit (from region T_2 to region T_3 in Fig. 11) without affecting the net emf of the thermocouple.

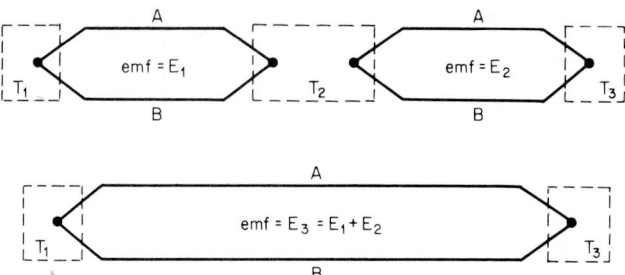

Fig. 11 Emf's are additive for temperature intervals.

Thermocouple Types

Iron-Constantan* (ANSI Symbol J)

The iron-Constantan J-curve thermocouple with a positive iron wire and a negative Constantan wire is recommended for reducing atmospheres. The operating range for this alloy combination is 1600°F (871°C) for the largest wire sizes. Smaller wires should operate at correspondingly lower temperatures. It is not recommended for use below 32°F (0°C) because of possible rusting and embrittlement of the iron.

Copper-Constantan (ANSI Symbol T)

The copper-Constantan T-curve thermocouple, with a positive copper wire and a negative Constantan wire, is recommended for use in mildly oxidizing and reducing atmospheres up to 750°F (399°C). It is suitable for applications where moisture is present. This alloy is recommended for low-temperature work, since the homogeneity of the component wires can be maintained better than that of other base metal wires. Therefore, errors due to inhomogeneity of wires in zones of temperature gradients are greatly reduced.

Chromel-Alumel† (ANSI Symbol K)

The Chromel-Alumel K-curve thermocouple, with a positive Chromel wire and a negative Alumel wire, is recommended for use in clean oxidizing atmospheres. The operating range for this alloy is 2300°F (1260°C) for the largest wire sizes. Smaller wires should operate at correspondingly lower temperatures. Type-K thermocouples are subject to green rot in oxygen-starved atmospheres at high temperatures. This condition causes the chromium to migrate in the material and destroys the calibration. This usually occurs in metal protection tubes containing insulation leaving no air.

Chromel-Constantan (ANSI Symbol E)

The Chromel-Constantan thermocouple may be used for temperatures up to 1600°F (871°C) in vacuum or in inert, mildly oxidizing, or reducing atmospheres. At subzero temperatures, the thermocouple is not subject to corrosion. This thermocouple has the highest emf output of any standard metallic thermocouple.

*The following trade names are used in this article: Inconel, Huntington Alloys, Inc.; Nextel, 3M Co.; Chromel, Hoskins Manufacturing Co.; Alumel, Hoskins Manufacturing Co.; Platinel, Engelhard Industries, Inc.; Constantan, Wilbur B. Driver Co.; Tophel, Wilbur B. Driver Co.; Nial, Wilbur B. Driver Co.; Thermokanthal KP, Kanthal Corporation; Thermokanthal KN, Kanthal Corporation; Teflon, E. I. DuPont de Nemours & Co., Inc.; Thermos, Thermos Division, King-Seeley Thermos Co.; Kapton, E. I. DuPont de Nemours & Co., Inc.; Refrasil, H. I. Thompson Co.

†Trademark of Hoskins Mfg. Co. Equivalent material is now available from several sources. ANSI designation KP is Tophel from the Wilbur B. Driver Company, and Thermokanthal KP from the Kanthal Corporation. ANSI designation KN is Nial from the Wilbur B. Driver Company, and Thermokanthal KN from the Kanthal Corporation.

Platinum-Rhodium Alloys (ANSI Symbols S, R, and B)

Three types of noble metal thermocouples are in common use: (1) type S (international transfer standard), a positive wire of 90% platinum (Pt) and 10% rhodium (Rh) used with a negative wire of pure Pt; (2) type R (Japanese standard), a positive wire of 87% Pt and 13% Rh used with a negative wire of pure Pt; and (3) type B, a positive wire of 70% Pt and 30% Rh used with a negative wire of 94% Pt and 6% Rh.

These thermocouples have a high resistance to oxidation and corrosion. However, hydrogen, carbon, and many metal vapors can contaminate a Pt-Rh thermocouple. The recommended operating range for Pt-Rh alloys is 2800°F (1538°C), although temperatures as high as 3270°F (1800°C) can be measured with the type B Pt–30% Rh versus Pt–6% Rh alloy combination.

Platinel Alloys

Platinel was designed to duplicate the type-K thermocouple curve because of greater stability at higher temperatures than Chromel-Alumel. Platinel is produced in two versions, Platinel I and Platinel II. The negative lead in both versions is a 65% gold–35% palladium alloy. The positive lead for Platinel I is 83% palladium, 14% platinum, and 3% gold, and the positive lead in Platinel II contains 55% palladium, 31% platinum, and 14% gold.

Platinel is generally used when the type-K thermocouple is desired but cannot be used because of a potential green rot atmosphere or other environmental conditions. Its temperature range is the same as that of the type-K thermocouple, to 2300°F (1260°C).

Tungsten-Rhenium Alloys (Industry Symbols G, C, and D)

Three types of refractory metal thermocouples are in common use for measuring temperatures up to 5000°F (2760°C). These thermocouples are made from tungsten-rhenium alloys and have inherently poor oxidation resistance. They should be used only in vacuum, hydrogen, or inert atmospheres.

Nicrosil-Nisil (Industry Symbol N)

Nicrosil-Nisil thermocouples were developed to complement type-K thermocouples by working at high temperatures under conditions unsuitable for a type-K thermocouple. Most notably, type-N thermocouples are not subject to green rot when used in oxygen-starved atmospheres at temperatures up to 2300°F (1260°C).

Thermocouple Characteristics

The principal characteristics of the thermocouples just described are summarized in Table 1.

Limits of Error

All wire, both thermocouple-grade and extension-grade, is manufactured to meet the industry-accepted standard ANSI MC 96.1-(1975), which specifies the maximum allowable thermoelectric deviation over a broad range of temperatures. (See Tables 2 to 4.)

Base metal thermocouple extension-grade wire is made from the same materials as thermocouple wire; however, its use is restricted to a lower range of temperatures. Within its range, extension-grade wire maintains the same limits of error as thermocouple-grade wire.

Noble and refractory metal thermocouple extension-grade wire is made from relatively inexpensive proprietary alloys created to simulate the thermoelectric behavior of the actual thermocouple elements over a limited range of temperatures (see Tables 2 to 4).

Tables 2 to 4 give the standard and special limits of error for thermocouple wire, extension-grade thermocouple wire, and thermocouple compensating extension wire. The limits of error for each type of thermocouple apply only over the temperature range specified. (The limits of error in the table do not include installation or system errors.) Furthermore, these limits of error should be applied only to standard published wire sizes. Where limits of error are given in percent in the tables, the percentage applies to the degrees Celsius temperature being measured. For example, if a type-J

Table 1. Thermocouple Characteristics

ANSI T/C	Symbol single	Generic and trade names	Color coding — Single	Color coding — Overall T/C wire	Color coding — Overall extension-grade wire	Magnetic Yes	Magnetic No	Maximum useful temp. range	Emf (MV) over useful temp. range, MV	Average sensitivity, μV/°C	Environment (bare wire)
T	TP	Copper	Blue	Brown	Blue		X	−328 to 662°F	−5.602 to 17.816	40.5	Mild oxidizing, reducing, vacuum, or inert; good where moisture is present
	TN	Constantan, Cupron, Advance	Red				X	−200 to 350°C			
J	JP	Iron	White	Brown	Black	X		32 to 1382°F	0 to 42.283	52.6	Reducing, vacuum, inert; limited use in oxidizing at high temperatures; not recommended for low temperature
	JN	Constantan, Cupron, Advance	Red				X	0 to 750°C			
E	EP	Chromel, Tophel, T^1 Thermokanthal KP	Purple	Brown	Purple		X	−328 to 1652°F	−8.824 to 68.783	67.9	Oxidizing or inert; limited use in vacuum or reducing
	EN	Constantan, Cupron, Advance	Red				X	−200 to 900°C			
K	KP	Chromel, Tophel, T^1 Thermokanthal KP	Yellow	Brown	Yellow		X	−328 to 2282°F	−5.973 to 50.633	38.8	Clean oxidizing and inert; limited use in vacuum or reducing
	KN	Alumel, Nial, T^2 Thermokanthal KN	Red			X		−200 to 1250°C			
S	SP	Platinum–10% rhodium	Black		Green		X	32 to 2642°F	0 to 14.973	10.6	Oxidizing or inert; atmospheres; do not insert in metal tubes; beware of contamination
	SN	Pure platinum	Red				X	0 to 1450°C			
R	RP	Platinum–13% rhodium	Black		Green		X	32 to 2642°F	0 to 16.741	12.0	
	RN	Pure platinum	Red				X	0 to 1450°C			
B	BP	Platinum–30% rhodium	Gray		Gray		X	32 to 3092°F	0 to 12.426	7.6	
	BN	Platinum–6% rhodium	Red				X	0 to 1700°C			
C*	CP*	Tunsten–5% rhenium	White/red trace		White/red trace		X	32 to 4208°F	0 to 37.066	16.6	
	CN*	Tungsten–26% rhenium	Red				X	0 to 2320°C			
G*	GP*	Tungsten	White/blue trace		White/blue trace		X	32 to 4208°F	0 to 38.564	16.0	Vacuum, inert, hydrogen atmospheres; beware of embrittlement
	GN*	Tungsten–26% rhenium	Red				X	0 to 2320°C			
D*	DP*	Tungsten–3% rhenium	White/yellow trace		White/yellow trace		X	32 to 4208°F	0 to 39.506	17.0	
	DN*	Tungsten–25% rhenium	Red				X	0 to 2320°C			

*Not ANSI symbol.

2.26

Table 2. Limits of Error for Thermocouples

Reference junction 0°C

Metal	Thermocouple type	Temperature range, °C	Limits of error	
			Standard (whichever is greater)	Special (whichever is greater)
Base metal	T	0 to 350	±1°C or ±0.75%	±0.5°C or 0.4%
	J	0 to 750	±2.2°C or ±0.75%	±1.1°C or 0.4%
	E	0 to 900	±1.7°C or ±0.5%	±1°C or ±0.4%
	K	0 to 1250	±2.2°C or ±0.75%	±1.1°C or ±0.4%
Noble metal	R or S	0 to 1450	±1.5°C or ±0.25%	±0.6°C or ±0.1%
	B	800 to 1700	±0.5%	
Base metal (low-temperature)	T*	−200 to 0°	±1°C or ±1.5%	†
	E*	−200 to 0°	±1.7°C or ±1%	†
	K*	−200 to 0°	±2.2°C or ±2%	†
Tungsten Rhenium	G, C, or D‡			
0.003 in wire diam		0 to 1760	±13.3°C or ±1%	†
0.005 in wire diam		0 to 1982	±13.3°C or ±1%	†
0.010-0.020 in wire diam		0 to 2315	±13.3°C or ±1%	†

*Thermocouples and thermocouple materials are normally supplied to meet the limits of error specified in the table for temperatures above 0°C. The same materials, however, may not fall within the subzero limits of error given in the "low temperature" section of this table. If materials are required to meet the subzero limits, this requirement should be specified, including selection of materials.

†Little information is available to justify establishing special limits of error for subzero temperatures. Limited experience suggests the following limits for types E and T thermocouples:

Type E −200 to 0°C ±1°C or ±0.5%; type T −200 to 0°C ±0.5°C or ±0.8%.

‡Not ANSI thermocouple types.

thermocouple is at a temperature of 200°C, then the limit of error is 200°C × ¾% = ±1.5°C. If the measurement temperature is given in degrees Fahrenheit, then the following formula applies: (°F − 32) × % = limits of error. For example, if the temperature is 400°F, then the limit of error is (400 − 32°F) × ±¾% = ±2.76°F.

Insulation

The purpose of insulation is to impose a dielectric material between adjacent thermocouple wires (primaries) to prevent shorts. Good insulators are Teflon, asbestos, fiberglass, etc. Wire supplied in duplex form includes an outer jacket which may be extruded, wrapped, or stranded over the insulated primaries. This outer layer is referred to as the overall insulation.

Table 3. Limits of Error—Thermocouple Extension Wire

Reference junction 0°C

Extension wire type	Temperature range, °C	Limits of error	
		Standard, °C	Special, °C
KX	0 to 200	±2.2	—
JX	0 to 200	±2.2	±1.1
EX	0 to 200	±1.7	—
TX	−60 to +100	±1.0	±0.5

Table 4. Limits of Error for Thermocouple Compensating Extension Wire

Reference junction 0°C

Thermocouple type	Compensating wire type	Temperature range, °C	Limits of error
R, S	SX*	0 to 200	±0.057 mV (±5°C†)
B	BX‡	0 to 100	+0.000 mV (+0°C†) −0.033 mV (−3.7°C†)
G	GX-W/W26% Re	0 to 260	±0.14/mV
C	CX-W5% Re/W26% Re	0 to 870	±0.11 mV
D	DX-W3% Re/W25% Re	0 to 260	±0.11 mV

*Copper (+) versus copper nickel alloy (−).

†Because of the nonlinearity of the types R, S, and B temperature-emf curves, the error introduced into a thermocouple system by the compensating wire will be variable when exposed in degrees. The degrees Celsius limits of error given in parentheses are based on the following measuring junction temperatures:

Wire type	Measuring junction temperature
SX	Greater than 870°C
BX	Greater than 1000°C

‡Copper-versus-copper compensating extension wire, usable to 100°C with maximum errors as indicated, but with no significant error over the 0 to 50°C range. Matched proprietary alloy compensating wire is available for use over the range 0 to 200°C with claimed limits of error of ±0.033 mV (±3.7°C).

Selection Criteria—Duplex Wire

When recommended environmental constraints for insulation are exceeded, shorts develop between conductors, resulting in erroneous thermocouple readings. A thermocouple will indicate the temperature at the short and not the temperature at the end measuring junction. In most cases, this situation will result in lower than expected readings. When several shorts occur, the temperature at the short closest to the instrumentation will be indicated.

When selecting thermocouple wire, the user should bear in mind the temperature limitations of the thermocouple material itself, as well as the limitations of the insulation.

Thermocouple Wire Insulation

Insulations are rated for a maximum temperature both for continuous usage and for a single exposure. It is important to observe the temperature limits when selecting an insulating material. At elevated temperatures even insulations which remain physically intact may become conductive. Under these conditions, the output of the thermocouple may be a function of the highest temperature to which the insulation is exposed, rather than the temperature of the measuring junction. The change in insulation resistance may be permanent if it is caused by the deterioration of organic insulants or binders which leave a carbon residue. In considering the temperature to which the insulation is exposed, it should not be assumed that this is the temperature of the measuring junction. A thermocouple may be attached to a massive specimen which is exposed to a high-temperature source to achieve a rapid heating rate. Parts of the thermocouple wires not in thermal contact with the specimen can be overheated severely, while the junction remains within safe temperature limits. With this in mind, high-quality insulation should be used when rapid heating rates are expected. Very little factual information is available on actual deterioration rates and magnitudes, but the condition is real, so a conservative approach is a requisite of good engineering practice. Insulation characteristics are given in Table 5.

The basic types of flexible insulations for elevated temperature usage are ceramic fiber, fiberglass, fibrous silica, and asbestos. Of the four materials, ceramic fiber has the highest temperature rating. Modern technology has led to the development of ceramic fibers which markedly increase the upper use temperature of flexible insulations. These insulations, if properly applied and handled, allow base metal thermocouples to be used to their maximum temperature limits within the limitations of exposure to the environment in which they are placed.

Table 5. Insulation Wire Characteristics

Material	Abrasion resistance	Flexibility	Water submersion	Temp., °F Max.	Temp., °F Min.	Temp., °C Max.	Temp., °C Min.	Resistance Solvent	Resistance Acid	Resistance Base	Resistance Flame	Resistance Humidity
Polyvinyl chloride	Good	Excel.	Good	+221	−40	+104	−40	Fair	Good	Good	Good	Good
Nylon	Excel.	Good	Poor	+300	−65	+149	−53	Good	Poor	Good	Poor	Good
Kapton	Excel.	Good	Good	+600	−450	+316	−268	Good	Good	Good	Good	Excel.
Teflon (PFA)	Excel.	Good	Excel.	+500	−450	+260	−268	Excel.	Excel.	Excel.	Excel.	Excel.
Teflon (FEP)	Excel.	Good	Excel.	+400	−450	+204	−268	Excel.	Excel.	Excel.	Excel.	Excel.
Silicone rubber	Fair	Excel.	Good	+392	−100	+200	−78	Fair	Poor	Good	Poor	Good
Asbestos	Good*	Good	Poor	+1000	−100	+538	−78	Excel.	Excel.	Excel.	Excel.	Poor
Glass	Poor	Good	Poor	+900	−100	+482	−78	Excel.	Excel.	Excel.	Excel.	Fair
Refrasil	Poor	Good	Poor	+1600	−100	+871	−78	Excel.	Good†	Good†	Excel.	Poor
Nextel	Fair	Good	Poor	+2600	0	+1427	−17.8	Excel.	Good†	Good†	Excel.	Fair

*Below 500°F (260°C).

†Attacked by hydrofluoric acid, phosphoric acid, and strong alkalies.

2.29

Fibrous silica also has good high-temperature electrical properties, but, because this insulation normally is not impregnated, its handling and abrasion characteristics also leave something to be desired. The next best high-temperature insulation is asbestos. Because this material has very poor mechanical properties, a carrier fiber or an impregnating material is added. In some instances, this carrier is cotton or another organic compound which leaves a carbon residue after exposure to a temperature at which it burns. This results in a breakdown of electrical insulation. Asbestos loses its mechanical strength after exposure to elevated temperatures and may break away from the wire with little or even no handling. A more commonly used insulation is fiberglass. This material can be impregnated to provide improved moisture and mechanical characteristics within the temperature limitations of the impregnating compound. The most frequently used type of fiberglass has an upper temperature limit of approximately 900°F (500°C) for continuous use. If one is willing to sacrifice the handling characteristics, nonimpregnated fiberglass insulations are available which withstand higher temperatures.

Types of Insulation

The following descriptions of the different types of insulation are available industrywide. The letter designations are simply those of a leading manufacturer.

AG and AHG conductors are covered with felted asbestos nominal 0.010 to 0.015 in (0.25 to 0.38 mm) thick (AG) or a nominal thin-wall thickness of 0.005 in (0.12 mm) (AHG). An overall glass braid nominal 0.006 in (0.15 mm) thick is then woven over the insulated primaries and impregnated with silicone varnish. This insulation is suitable for continuous operation up to 900°F (482°C) but may be used up to 1400°F (760°C) for a single reading. The silicone impregnation is permanently destroyed above 400°F (204°C). The flexure strength of asbestos is severely limited above 600°F (315°C). AA felted asbestos is applied over each conductor. Asbestos yarn is overbraided onto the insulated primaries. This construction will withstand continuous temperatures up to 1000°F (537°C) or one reading at 1400°F (760°C). Flexure strength is severely limited above 600°F (315°C).

AX felted asbestos is applied to each wire. The two wires are then twisted, and no overall insulation is applied. This material is suitable for continuous use to 1000°F (537°C) but can be used for a single reading up to 1400°F (760°C). Flexure strength is severely limited above 600°F (315°C).

XR is a braid of vitreous silica fiber applied to each conductor and then over the insulated primaries. Suitable for use to 1600°F (871°C), depending on the temperature limit of the wire, it will withstand temperatures of 1800°F (982°C) if not subjected to flexure or abrasion. It is not color-coded.

XC insulation refers to Nextel, a high-temperature alumina-boria-silica ceramic fiber, which is braided on each conductor and then over the insulated primaries. Nextel's continuous service rating is 2600°F (1426°C), with a short-term rating to 3000°F (1648°C). For base metal thermocouple applications, construction is limited by the temperature of the thermocouple wire rather than by the ceramic braiding. Nextel ceramic fiber is *not* recommended for use with platinum thermocouples or in applications exposed to molten tin and copper, hydrofluoric or phosphoric acid, or strong alkalies. It is not color-coded.

Teflon (TT) is extruded on each bare wire, and then a Teflon jacket is extruded over the insulated primaries. This material has superior abrasion and moisture resistance up to 400°F (204°C) for FEP Teflon and 500°F (260°C) for PFA Teflon. It also resists most acids and vapors and can be used for a single reading up to 600°F (315°C).

TG Teflon (PFA) is extruded over each conductor. Silicone-impregnated glass approximately 0.005 in (0.12 mm) thick is overbraided onto the insulated primaries. It has excellent moisture resistance to 500°F (260°C) and resists most acids and vapors. It has good abrasion resistance and may be used for a single reading to 650°F (343°C).

GG Braid

Each conductor is covered with a single braid of glass yarn approximately 0.005 in (0.12 mm) thick and then impregnated with a modified silicone varnish. The insulated primaries are then covered with a braid of glass yarn 0.005 in (0.12 mm) thick and impregnated with the silicone varnish. This braid is suitable for continuous use to 900°F (482°C). The impregnation improves resistance to abrasion and moisture but is burned off above 400°F (204°C).

Wrap: Glass yarn approximately 0.005 in (0.12 mm) thick is applied to each conductor as a double wrap instead of a single braid and then impregnated. The wrap provides tight covering for the individual conductors and resists fraying; it is commonly used on fine-gage wire.

PP color-coded polyvinyl is extruded over each bare wire, and a polyvinyl jacket is applied over the insulated primaries. This material is flexible within its temperature range, has good abrasion resistance, and is unaffected by petroleum, alkalies, acids, and most solvents except ketones and esters.

NN nylon is extruded over each bare wire. A nylon overall jacket is applied over the insulated primaries. Nylon is hygroscopic and over a long term moisture will degrade the insulation.

KK fused Kapton tape approximately 0.006 in (0.15 mm) thick is applied to each conductor. A 0.004-in (0.10-mm) jacket of fused Kapton tape is then applied over the insulated primaries. Primary and overall insulation is color-coded. Kapton has excellent moisture and abrasion resistance and high dielectric strength (about 7000 V/mil) and retains much of its physical integrity after exposure to gamma radiation. A coating of Teflon is added to the tape as an adhesive binding agent. The Teflon binder melts at approximately 500°F (260°C).

Thermocouple Applications

Thermocouples are best applied in measurement situations where the following criteria are critical.

Fast Response

Thermocouples can be made from wire pairs as small as 0.0005 in in diameter for millisecond response.

High Shock

Except for the refractory group (tungsten versus tungsten-rhenium), thermocouples are rugged and will withstand high shock.

High Temperature

Of the base metal thermocouples, type K will provide good continuous service up to as high as 2300°F (1250°C) under oxidizing conditions in bare wire form. In Inconel-sheathed form, they are rated to 2100°F (1150°C). Platinum alloy thermocouples can be used in clean oxidizing environments to 3000°F (1650°C) in wire form. They are good up to the same temperature with a platinum-rhodium alloy or high-purity alumina sheath. Tungsten-rhenium alloy thermocouples can be used only in vacuum or in an inert atmosphere and will go as high as 4500°F (2500°C) in wire form. Tantalum-sheathed tungsten-rhenium will withstand temperatures up to 4200°F (2300°C) under the same vacuum or inert atmosphere conditions.

Small Size

Thermocouples can be drawn in metal-sheathed form to as small as 0.010 in (~0.25 mm) O.D. In wire form, 0.0005-in (~0.013-mm) thermocouples can be made. Thermocouples can be made in foil form to a thickness of 0.0005 in (~0.013 mm).

Probe Flexibility

Thermocouple wire is available in solid or stranded wire form. Sheathed probes with compacted oxide insulation can be bent to a radius of curve three times the sheath outer diameter.

Heavy Duty Construction

Thermocouples can be made from heavy-gage wire or even round stock to withstand the most severe applications.

Cost

On a probe-for-probe basis, thermocouples are the least expensive temperature transducers. Exotic sheath materials or designs bring the price up accordingly. System costs can also escalate for long extension runs.

Low Temperature

Types T and E can be used to 50 K ($-223\,^\circ$C). Below this temperature, use gold–atomic iron (Au–0.07% Fe) paired with Chromel-P.

Limitations of Thermocouples

Some limitations are listed below.

Accuracy

Limits of error are wider than those of resistance-temperature detectors or thermistors.

Resolution

Instrumentation must resolve tens of microvolts per degree.

Drift

Thermocouples should be checked periodically when used at elevated temperatures.

Leads

Only thermocouple wire can be used. Type B may use copper. A cold junction reference is required, usually to 32°F (0°C).

WIRING, INSTALLATION, AND APPLICATION RULES FOR THERMOCOUPLES

Because thermocouples are differential transducers based on complex thermoelectric properties, they have their own rules and codes which must be followed to ensure sound, repeatable results.

Rule 1 Thermocouple wire must be used from the hot junction leads back to the point of reference; e.g., at the computer interface card, recorder, controller, or digital indicator.

Rule 2 Be consistent with the thermocouple type from start to finish. If the system calls for type-K thermocouples, then the probe, connectors, extension leads, and indicator must all be for type-K calibration.

Rule 3 Switches may be used which are manufactured with silver- or gold-plated contacts provided that (1) the switch is designed to be isothermal, and (2) thermocouple wire is used from the common output to the reference junction.

Rule 4 Avoid splitting the output of a thermocouple between two readout devices, e.g., a recorder and a controller. This arrangement may work, but usually there are impedance matching problems or possible generation of unwanted ground paths. A better choice would be to use dual ungrounded thermocouples.

Rule 5 Thermocouple wire should be routed away from large current carrying or voltage wires. If run in a conduit, it should be run with signal carriers only.

Rule 6 Thermocouple wire may be limited in use by temperature constraints on the insulation and by wire diameter.

Rule 7 Extension-grade thermocouple wire has certain temperature limitations regardless of the insulation characteristics. Base metal extension is good between 32 and 392°F (0 to 200°C). Types RX and SX are good between 0 and 200°C ambient. Type BX is good between 32 and 212°F (0 to 100°C).

Rule 8 Feedthroughs not made of thermocouple material are a potential source of error if a temperature gradient exists through the wall.

Rule 9 Immersion-type thermocouples must be installed to an adequate depth (10 to 20 times sheath outer diameter minimum immersion) to reduce conduction losses.

Rule 10 Protection tubes may sag unless supported in high-temperature environments [1500°F (815°C) and higher].

Rule 11 The negative leg of a thermocouple is always color-coded red per ANSI standards.

Rule 12 Do not use thermocouples in microwave- or radio frequency-heated ovens. Measurements may be taken only with the heat source off.

Rule 13 Grounded and ungrounded probe configurations refer to the relationship between the sheath and the thermocouple junction. An ungrounded probe is manufactured so that the junction is electrically isolated from the metal sheath, which may be a source of outside noise or unwanted voltages. In a grounded probe, the junction is welded directly to the metal sheath and hence is electrically grounded to it. Grounded junction probes have better response times than ungrounded probes. See Fig. 12.

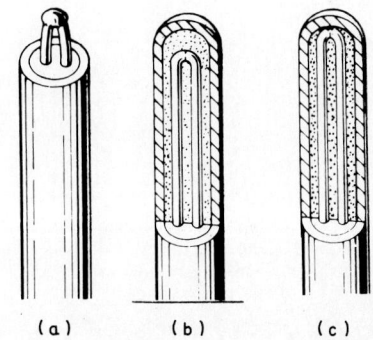

(a)   (b)   (c)

Fig. 12 Measuring junctions: (a) exposed junction; (b) ungrounded junction; (c) grounded junction.

Rule 14 Be aware of a hysteresis effect in type-K thermocouples used at high temperatures above 1000°F (537°C), and green rot in oxygen-starved atmospheres.

Rule 15 Avoid direct flame measurement with bare wire thermocouples.

Measuring Junctions

The principal types of measuring junctions include (1) exposed junctions, (2) ungrounded junctions, and (3) grounded junctions, as shown in Fig. 12.

Exposed Junction

This junction is recommended for the measurement of static or flowing noncorrosive gas temperatures where the response time must be minimal. The junction extends beyond the protective metallic sheath to give a fast response. The sheath insulation is sealed at the point of entry to prevent penetration of moisture or gas.

Ungrounded Junction

This type of junction is recommended for the measurement of static or flowing corrosive gas and liquid temperatures in critical electrical applications. The welded wire thermocouple is physically insulated from the thermocouple sheath by soft magnesium oxide (MgO) powder.

Grounded Junction

This junction is recommended for the measurement of static or flowing corrosive gas and liquid temperatures and for high-pressure applications. The junction of this thermocouple is welded to the protective sheath, giving a faster response than the ungrounded junction type.

Table 6. Additional Protection Tube Materials

Material	Recommended max. temp., °F	Application atmosphere				Applications*
		Oxidizing	Hydrogen	Vacuum	Inert	
304SS	1700 (927°C)	Very good	Good	Very good	Very good	Recommended for general chemical applications, food applications, oil refinery use, and steam lines
Inconel 600	2100 (1150°C)	Very good	Good	Very good	Very good	Recommended for gas furnaces, lead baths, and bath mixtures containing cyanide. Do not use in salt baths contaminated by sulfur.

*Although sheath material has excellent pressure ratings (up to 50,000 psi; 345 MPa), a thermowell is generally used in high-pressure applications.

Table 7. High-Temperature Sheath Materials

Sheath material	Maximum operating temperature	Workability	Working environment	Approx. melting point	Remarks
Molybdenum*	4000°F (2205°C)	Brittle	Inert, vacuum, reducing	4730°F (2610°C)	Relatively good hot strength; sensitive to oxidation above 930°F (500°C); resists many liquid metals and most molten glasses
Tantalum†	4500°F (2482°C)	Malleable	Inert, vacuum	5425°F (3000°C)	Resists most acids and weak alkalies; very sensitive to oxidation above 570°F (300°C)
Platinum-rhodium alloy	3050°F (1677°C)	Malleable	Oxidizing, inert vacuum	3400°F (1875°C)	No attack by SO_2 at 2000°F (1093°C); silica is detrimental; halogens attack at high temperatures
Inconel 600	2100°F (1149°C)	Malleable	Oxidizing, inert, vacuum	2570°F (1410°C)	Excellent resistance to oxidation at high temperature; do not use in presence of sulfur above 1000°F (538°C); hydrogen tends to embrittle

*Refractory metals are extremely sensitive to any trace of oxygen above approximately 500°F (260°C). They must be used in vacuum or in very pure inert gases such as helium and argon.
†Suitable for exposure to certain reducing atmospheres as well as inert gases and vacuum.

Table 8. High-Temperature Insulations and Wires

Material	Approx. upper useful temperature	Approx. melting point	Remarks
Insulation*			
Magnesia (MgO)	3000°F† (1650°C)†	5070°F (2800°C)	Hygroscopic, compacts well
Alumina (Al_2O_3)	2800°F† (1540°C)†	3660°F (2015°C)	Requires considerable volume reduction to compact satisfactorily
Beryllia (BeO)‡	4200°F† (2315°C)†	4620°F (2550°C)	Compacts well; high thermal conductivity; see warning
Sensing Wires			
Pt–6% Rh versus Pt–30% Rh ⎫ Pt–10% Rh versus Pt ⎬ Pt–13% Rh versus Pt ⎭	3000°F (1650°C)	3200°F (1770°C)	Some decalibration at continued high-temperature use because of rhodium volatization
W versus W–26% Re ⎫ W–5% Re versus W–26% Re ⎬ W–3% Re versus 25% Re ⎭	4200°F (2320°C)	5600°F (3100°C)	Brittle; avoid flexing

*At temperatures above 1800°F (982°C) all insulating materials experience a substantial decrease in resistivity with increasing temperature.

†Values given are for compacted insulation. For uncompacted hard-fired insulators, useful temperature range can be 100 to 200°F (38 to 95°C) higher.

‡*Warning:* Beryllium oxide and thorium oxide are toxic.

Protection of Thermocouples

Metal sheaths are frequently used to protect thermocouples. Although sheath materials have excellent pressure ratings (up to 50,000 psi, 345 MPa), a thermowell is generally used in high-pressure applications. See Tables 6 and 7.

Insulation and wires constructed to withstand high temperatures are frequently required. See Table 8.

Thermocouple Assembly

This assembly consists of a thermocouple element and one or more associated parts, such as a terminal block, connection head, or protection tube or thermowell.

Terminal Block

This is a block of insulating material used to support and join the termination of conductors. See Fig. 13.

Connection Head

A connection head is a housing enclosing a terminal block for an electrical temperature sensing device and is usually provided with threaded openings for attachment to a protection tube and for attachment of a conduit. See Fig. 13.

Connection Head Extension

A connection head extension is a threaded fitting or an assembly of fittings extending between the thermowell or angle fitting and the connection head.

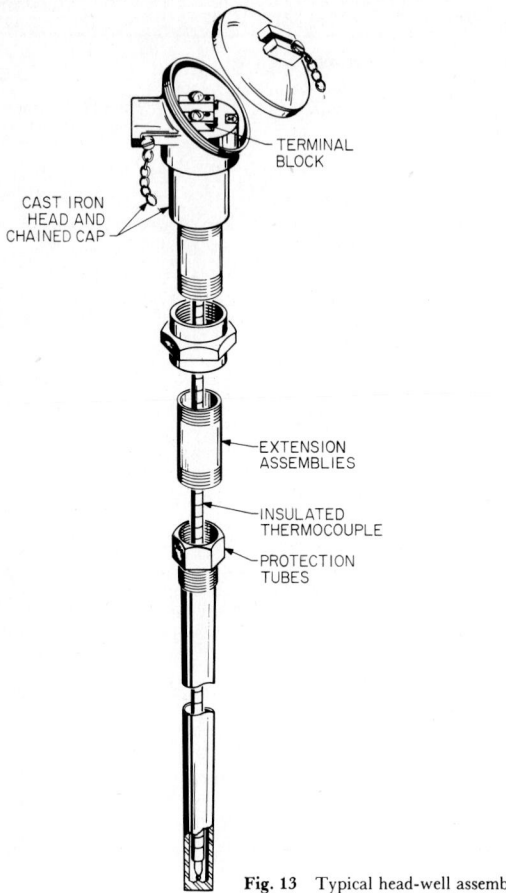

Fig. 13 Typical head-well assembly.

Protection Tube

A protection tube is a tube designed to enclose a temperature sensing device and protect it from the deleterious effects of the environment. It may provide for attachment to a connection head but is not primarily designed for pressuretight attachment to a vessel. A bushing or flange may be provided for the attachment of a protection tube to a vessel (Fig. 13).

Representative thermowell assemblies are shown in Fig. 14.

Ceramic Protection Tubes

Table 9 lists the properties of the two materials most commonly used for ceramic protection tubes, mullite and high-purity alumina. Although ceramic tubes are used primarily in high-temperature applications, they may also be used at lower temperatures where corrosive atmospheres can harm metal protection tubes. See Fig. 15. Ceramic is also used for insulation of bare thermocouple wire. See Fig. 16.

High-purity alumina tubes should always be used with platinum thermocouples above 2200°F (1200°C) to ensure long life and maximum accuracy. Mullite tubes contain impurities which can contaminate platinum above 2200°F (1200°C). See Table 9.

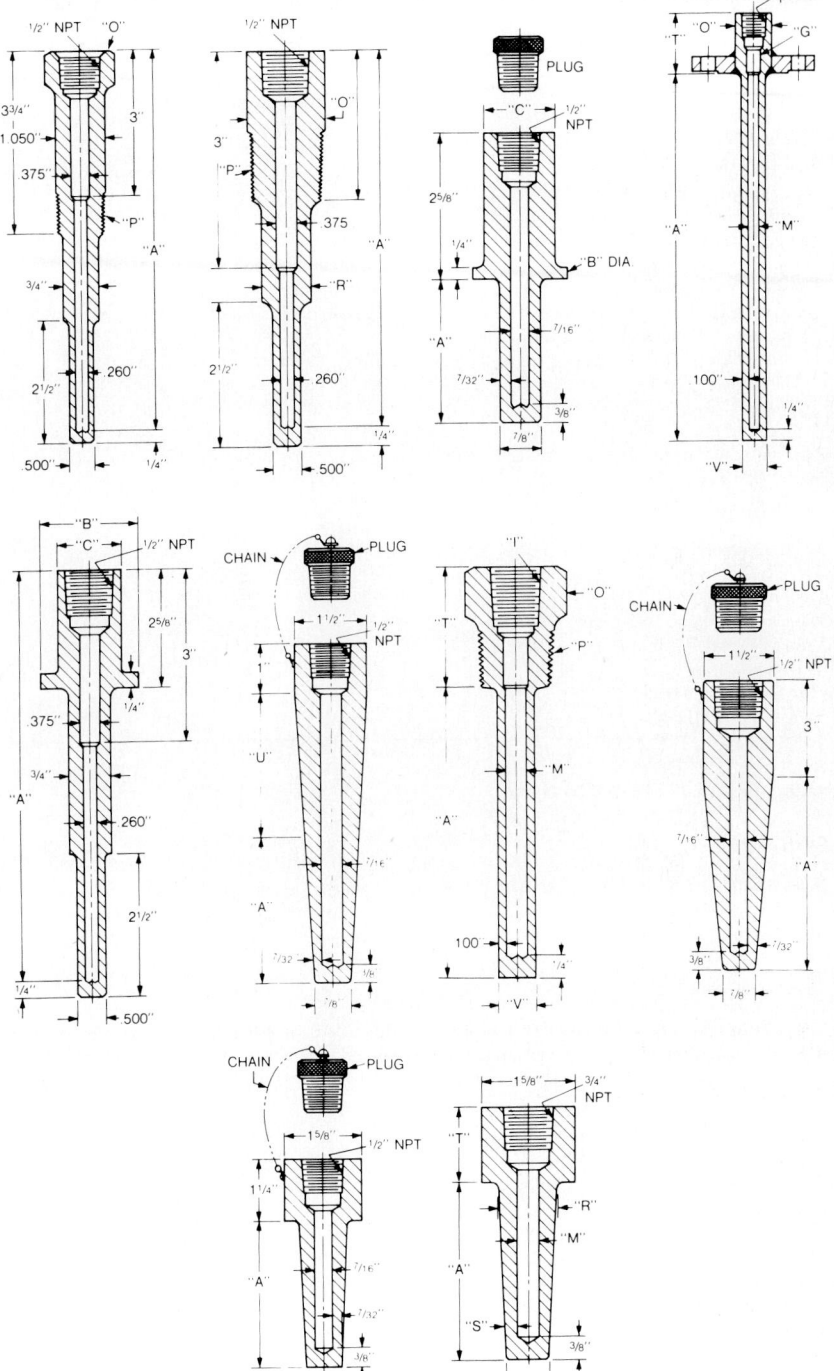

Fig. 14 Representative well asssemblies.

Table 9. Typical Physical Properties of Ceramic Materials

Property	High-purity alumina	Mullite
Composition	99.8% Al_2O_3	85% mullite, 15% glass
Water absorption	0.00	0.00
Specific gravity	3.85	2.8
Gas permeability	Gastight	Gastight
Compressive strength, psi (MPa)	>300,000 (2068)	>190,000
Tensile strength, psi (MPa)	30,000 (207)	18,000
Transverse strength, psi (MPa)	55,000 (380)	27.000
Coefficient of linear thermal expansion		
24–250°C	6.2×10^{-6}	3.3×10^{-6}
24–500°C	7.4×10^{-6}	4.0×10^{-6}
24–1000°C	8.5×10^{-6}	5.0×10^{-6}
24–1500°C	9.8×10^{-6}	—
Approximate thermal conductivity (btu) (in)/ (hr) (ft^2) (°F)		
24°C	230	40
800°C	60	25
Maximum service temperature	1950°C	1750°C
	3540°F	3180°F
Dielectric constant 1 MHz	9.29	5.80
Dielectric strength, (volts/mil)	230	250
Volume resistivity, Ω cm	10^{13}	10^{13}
Te value	800°C	800°C
Hardness (Mohs scale)	9	7.5
Fabrication method	Cast	Cast

Other Types of Thermocouples

Surface Probes

Special commercially available ANSI types K and E surface probes are shown in Fig. 17. Type E is preferred because of its high accuracy in most low-temperature applications. Type K is used where high temperatures must be measured. Types J and T are uncommon in surface probes.

Cement-On Styles

Special fast-responding construction techniques include thin-foil thermocouples with plastic laminates for cementing directly onto equipment to be monitored. Cement-on style I thermocouples are

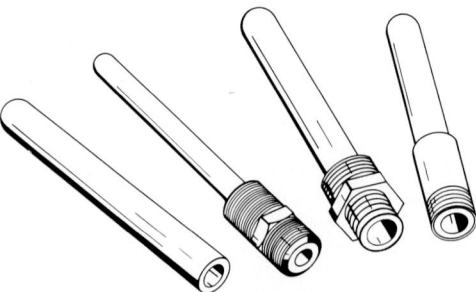

Fig. 15 Representative ceramic protection tubes.

BORE DIAMETER, in	1/8	3/32	5/64	1/16	3/64	1/32	0.020	1/64	0.005
MAXIMUM WIRE DIAMETER	0.102	0.064	0.064	0.040	0.032	0.025	0.016	0.010	0.003

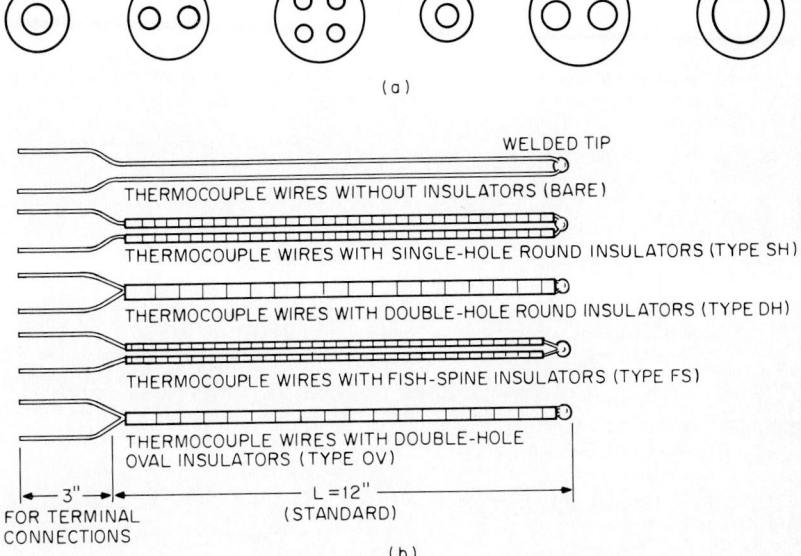

(a)

WELDED TIP

THERMOCOUPLE WIRES WITHOUT INSULATORS (BARE)

THERMOCOUPLE WIRES WITH SINGLE-HOLE ROUND INSULATORS (TYPE SH)

THERMOCOUPLE WIRES WITH DOUBLE-HOLE ROUND INSULATORS (TYPE DH)

THERMOCOUPLE WIRES WITH FISH-SPINE INSULATORS (TYPE FS)

THERMOCOUPLE WIRES WITH DOUBLE-HOLE OVAL INSULATORS (TYPE OV)

|←— 3" —→|← ————— L =12" ————— →|
FOR TERMINAL   (STANDARD)
CONNECTIONS

(b)

Fig. 16 (a) Various styles and sizes of ceramic insulators; (b) application of insulators to various styles of thermocouples.

easy to install. The foil sensor is embedded between two thin, glass-reinforced, high-temperature polymer laminates which both support and electrically insulate the foil section as well as provide a flat surface for cementing. The polymer-glass laminate, in general, determines the maximum temperature of the construction which is 500°F (260°C) continuous and up to 698°F (370°C) for short duration. Each style I, 30-gage thermocouple wire is bonded to the foil and strain relieved by the laminate. See Fig. 18.

Style II

Cement-on style II thermocouples are recommended where an extremely fast surface temperature measurement response time is a requirement. Intimate thermal coupling for this style is achieved by directly bonding the foil junction area to the surface to be measured. For ease of handling, the foil leads are fastened to a polyimide film frame which is a tough, flexible, dimensionally stable material rated for 500°F (260°C) continuous service. During application, the foil thermocouple can be peeled from the frame or released by the application of heat. As an alternative, portions of the frame may be cut away by scissors or knife.

The 6-in (152.4-mm) uninsulated foil leads are of 0.002-in (0.05-mm) material and thus are fragile and should be handled with care during installation (Fig. 18).

Style III

Cement-on style III thermocouples are constructed of 30-gage (0.010-in) (~0.25-mm)-diameter "standard limits of error" wire. The thermocouple is bead-welded and embedded in a paper-thin

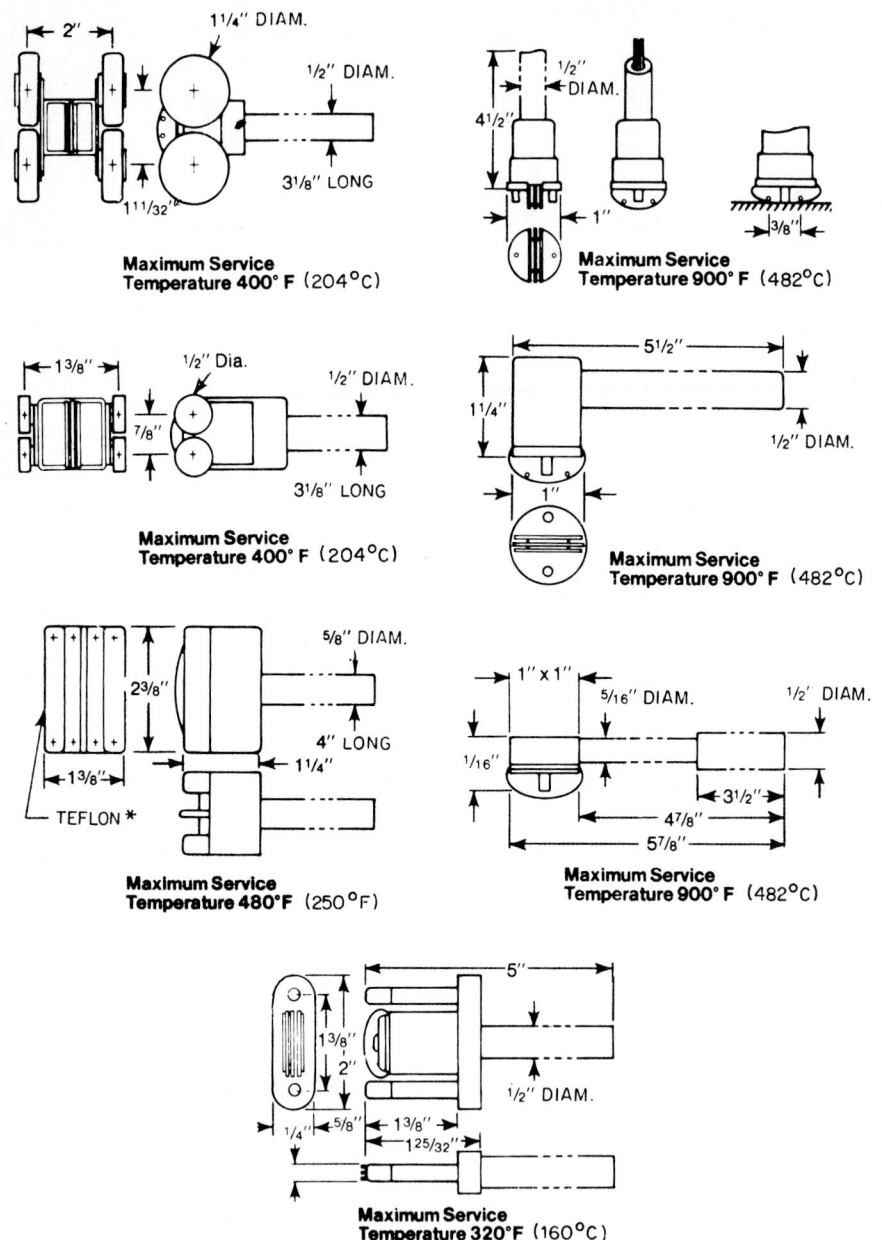

Fig. 17 Various styles of surface probes.

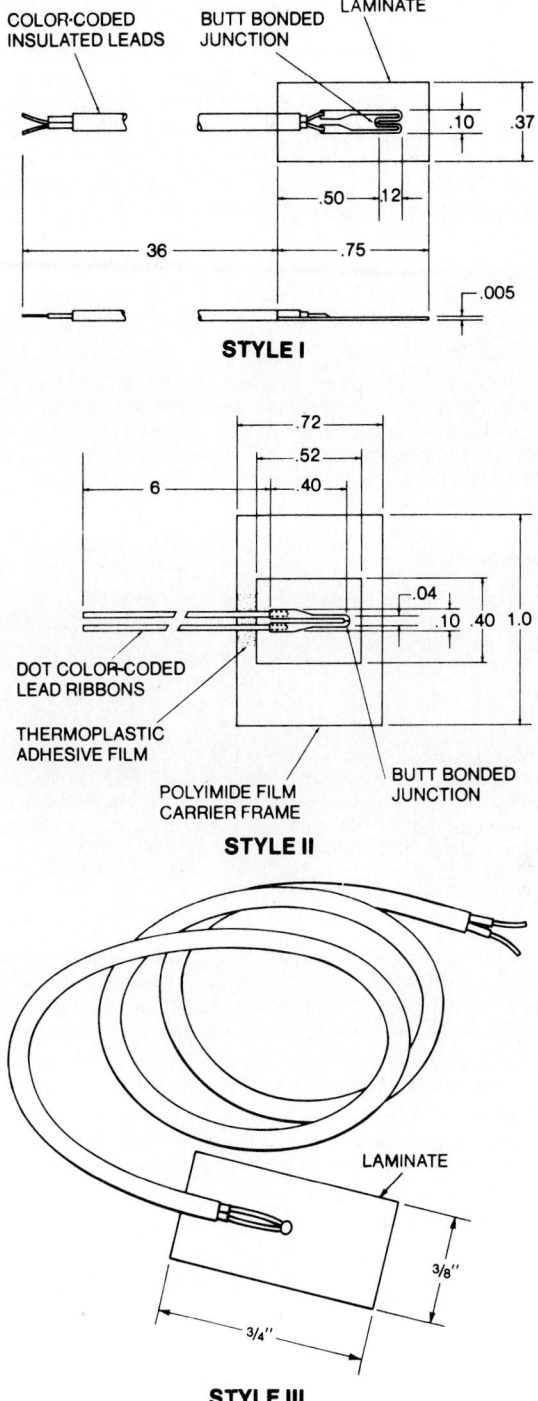

Fig. 18 Cement-on-style thermocouples. Dimensions are in inches.

laminate intended for surface applications by adhesive bonding. Color-coded, glass braid insulated leads are supplied. Style III thermocouples can be used in the range of −310 to +698°F (−190 to +370°C) depending on the time of exposure and the environment (Fig. 18).

Bare Thermocouple Element Fabrication

While completely fabricated thermocouples are available commercially, this section is intended to assist those who desire to fabricate their own thermocouples.

Thermocouple Wires

Carefully selected and tested pairs of thermocouple wires are available commercially in standard AWG diameters. When purchased as a pair simultaneously from a supplier, the pair will conform to the specified calibration limits and is referred to as a matched pair.

Interchange of a common wire between two types of thermocouples (e.g., Constantan from types J to T), or even between different matched pairs of the same type, may yield a thermocouple that will not conform to the specified calibration limits.

Joining Thermocouple Wires

The dissimilar wires of a thermocouple must be joined at the temperature measuring junction by a joint of good electrical and thermal conductivity without destroying the mechanical and metallurgical properties of the thermocouple wires at this joint.

1. For use below 1000°F (500°C) most base metal thermocouple wires may be silver-soldered using borax as a flux.

2. Above 1000°F (500°C), experience has shown that properly welded thermocouple junctions provide long life and excellent thermal and electrical properties. Welded thermocouple junctions are used in practically all industrial applications. Noble metal thermocouples should always be joined by welding. Common methods of welding thermocouples are gas, electric arc, resistance, tungsten inert gas (TIG), and plasma arc welding.

Preparation of Wires

1. Often the matched wires must be straightened prior to joining to facilitate stringing of insulators in the final thermocouple assembly but, where possible, excessive bending of thermocouple wires should be avoided because cold-working may alter the emf output of thermocouple wire. Hammering, stretching, and excessive twisting should be avoided for the same reason.

2. The thermocouple wires are cut to the length desired, allowing for one or two attempts at welding and for any forming that must be done at the junction.

3. Fixtures are usually used to shape the wires prior to welding, except for butt-welded thermocouples which are bent around a mandrel after welding. Care should be exercised to avoid nicking or damaging the wire during the forming operation, as damage may shorten thermocouple life.

Gas or Arc Welding Types E, J, K, and T Thermocouples

In preparation for welding, the wires may be twisted as shown in Fig. 19, or positioned in a "V" as shown in Fig. 22. The twisted construction adds strength and facilitates welding.

For larger twisted wire, 1 in of each wire should be prepared by removing any oxide or surface finish with abrasive paper. For smaller twisted wire, the prepared length need be only ½ in. The prepared ends are either twisted together to yield one and one-half turns as shown in Fig. 19 or positioned in a "V" as shown in Fig. 22 and then welded.

Fig. 19   Method of twisting wires for gas and electric arc welding.

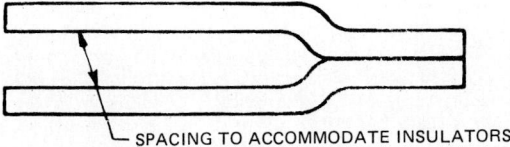

Fig. 20   Method of forming metal wires for resistance welding.

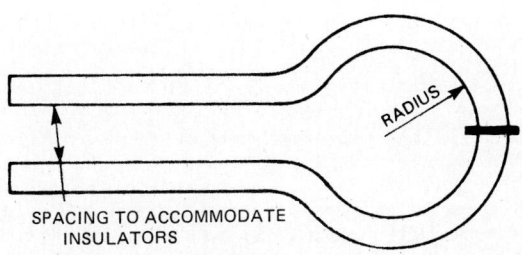

Fig. 21   Formed butt-welded thermocouple.

Resistance Welding Types J and K Thermocouples

This method is recommended only for larger gage wires. Approximately ½ in of each wire should be sanded with abrasive paper in preparation for welding.

The sanded ends should be formed to produce longitudinal contact as shown in Fig. 20.

Butt Resistance Welding Types E, J, and K Thermocouples

This method is recommended for larger gage wires and requires a commerically available wire butt welder of current capacity suitable for the gage wire being welded. Approximately ½ in of each wire should be sanded with abrasive paper in preparation for welding. The sanded ends are butted together in the spring-loaded butt welder jaws and spring pressure applied to the jaws. The weld is performed, and the flash is removed by grinding. The wires are then bent as shown in Fig. 21.

Resistance Welding Types B, R, and S Thermocouples

Extreme care should be taken to avoid cold-working and contamination by oils, perspiration, dirt, etc. Sanding is not required.

The ends should be formed to produce a longitudinal contact of about ⅛ in as shown in Fig. 20.

Arc Welding Types B, R, and S Thermocouples

Extreme care should be taken to avoid cold-working and contamination by oils, perspiration, dirt, etc. Sanding is not required.

The ends of the wires are positioned as shown in Fig. 22.

Fig. 22 Method of forming metal wires for electric arc welding.

Gas Welding

A neutral flame, as shown in Fig. 23, is essential for gas welding. The neutral flame is obtained by increasing the oxygen until the excess gas flame just vanishes. Overshooting the vanishing point gives an oxidizing flame. An oxidizing flame is injurious and should never be used. See Fig. 23.

The smallest tip that will readily heat the wires to fusion temperature should be used. Continued heating at welding temperatures yields a poor weld.

Heat the ends of the twisted or "V"-positioned wires to redness with the tip of the cone and plunge them into the flux. Reheat the wires to fusion temperature simultaneously and rotate the weld to form a ball at the tip. Quench the welded junction in water to remove excess flux.

Attainment of simultaneous fusion requires that the heating of the lower melting wire be delayed.

The junction should be smooth; a weld with a pitted surface must be rejected because it is burned and repair of the weld is not feasible. An unsatisfactory weld must be cut off, and the procedure repeated.

Electric Arc Welding

Welded junctions may also be produced by an arc between two soft carbon electrodes or between one carbon electrode and the thermocouple wires as the second electrode. Only larger gage thermocouples should be used as the second electrode. Finer wires require two carbon electrodes. Direct current or alternating current may be used, but direct current is preferred. With direct current and a single carbon electrode, the thermocouple is connected to the positive lead. The ends of the twisted or "V"-positioned wires are moistened, dipped into flux, and clamped upright in a vise with copper jaws. (Avoid surface damage in clamping.)

A brief welding cycle is best, because excessive current will result in burning. The bead should be small and solid. Bridges or gaps between wires are weak and unsatisfactory. If an unsatisfactory weld results, it must be cut off and the procedure repeated.

Types J and K in small sizes including AWG 14 and type T including AWG 20 have been arc-welded using submerged mercury as one electrode and the wires as the second electrode. The mercury is connected through a carbon electrode and covered with a silicone oil or water. On immersion in the mercury, the thermocouple is welded. Type-K thermocouples are welded using 30 V of alternating current and 15-A fuses.

Electric Resistance Welding

Heating and fusion of the wires are accomplished by resistance heating of the wires and by contact resistance at their junction. This method is recommended only for types B, R, and S, and the larger gage sizes of types J and K. The junction of Fig. 20 is placed between the electrodes of a resistance welder. A suitable pressure-current-time cycle must be established by trial and error on identical

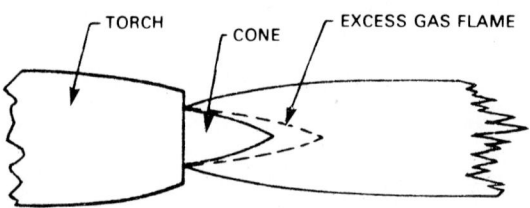

Fig. 23 Neutral flame for gas welding.

scrap wires. Visual and destructive examination are required to establish proper welding conditions. Excessive pressure will produce a good-looking weld but only peripheral fusion.

TIG and Plasma Arc Welding

The TIG welding process and plasma arc welding process are rapidly gaining in importance for welding thermocouple junctions. These welding processes use an inert gas envelope to protect the weld from oxidation. Welding using the TIG or plasma arc process is done following the same routine as welding with one carbon electrode, except that flux is not used. The plasma arc has distinct advantages: no tungsten inclusions in the weld, extremely high temperatures in the plasma arc, and a more controllable arc. These processes are especially recommended for welding junctions in sheathed thermocouple wire. The procedures outlined above may also be used on types B, R, and S.

Alternate Thermocouple Circuits

This article has concentrated on the use of thermocouples one at a time. However, as shown in Figs. 24 through 27, multiple thermocouples can be used in numerous configurations.

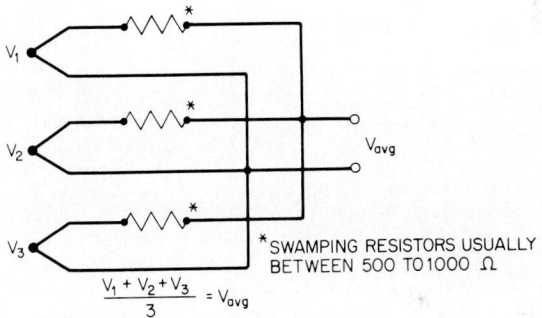

$$\frac{V_1 + V_2 + V_3}{3} = V_{avg}$$

Fig. 24 Thermocouples in parallel.

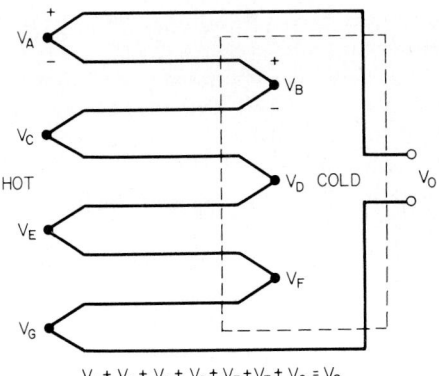

$$V_A + V_B + V_C + V_D + V_E + V_F + V_G = V_O$$

Fig. 25 Thermocouples in series (thermopile). Note: V_B, V_D, and V_F are negative thermoelectric voltages compared with V_A, V_C, V_E, and V_G. However, the alloys are also reversed, creating a net voltage.

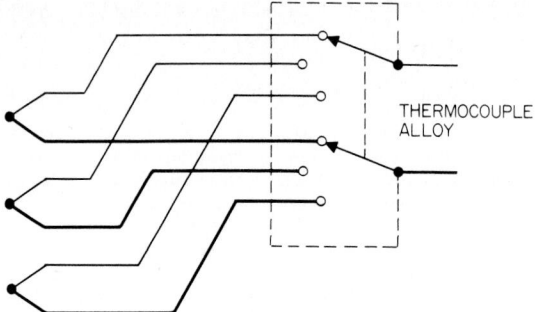

Fig. 26 Switch circuit. Note: Switch must be isothermal or made of the same thermocouple alloy material.

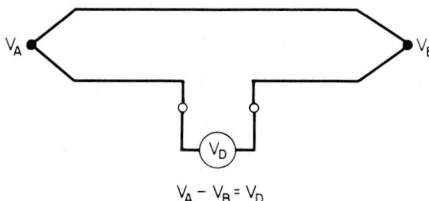

$$V_A - V_B = V_D$$

Fig. 27 Differential circuit. Note: Output voltage cannot be accurately cold junction-compensated because of the non-linearity of thermocouple emf versus temperature. Approximations can be made if the absolute temperature is known.

REFERENCES

1. Staff: "The Theory and Properties of Thermocouple Elements," *ASTM Spec. Tech. Publ. 492,* Omega Press, Stamford, Conn., 1979.
2. Staff: *Temperature Measurement Handbook,* Omega Press, Stamford, Conn., 1982.
3. Staff: "Manual on the Use of Thermocouples in Temperature Measurement," *ASTM Spec. Tech. Publ. 470B,* American Society for Testing and Materials, Philadelphia, 1975.
4. Staff: *The Nicrosil versus Nisil Thermocouple,* U.S. National Bureau of Standards, Washington, D.C., 1978.
5. Staff: "American National Standard for Temperature Measurement Thermocouples," *ANSI MC 96.1,* American National Standards Institute, New York, 1975.
6. Baker, H. D., Ryder, W. A., and N. H. Baker: *Temperature Measurement in Engineering,* vol. 1, Omega Press, Stamford, Connecticut, 1975.
7. Staff: "Thermocouple Reference Tables, Based on the IPTS-68," *NBS Monogr. 125,* Omega Press, Stamford, Conn., 1975.
8. Staff: "Manual on the Use of Thermocouples in Temperature Measurement," *STP 470B,* American Society for Testing and Materials, Philadelphia, 1981.

Thermoresistive Systems

by
W. A. Clayton*

The science of measuring temperature by utilizing the characteristic relationship of resistance to temperature has been advanced periodically since the early work of Faraday (circa 1835). Certain suitably chosen and prepared materials which vary in resistance, in a well-defined and calibrated manner, with temperature became readily available around 1925, prompting the use of resistance thermometers as primary elements for industrial applications where reproducibility and stability in measurement of process temperatures are of critical importance.

Platinum resistance thermometers are the international standard for temperature measurements between the triple point of hydrogen at 13.81 K and the freezing point of antimony at 630.75°C. Recent commercial availability of thin-film platinum resistance thermometers offers stable, broad-range performance in devices of small size and cost, combined with higher resistance values than previously obtainable. New applications exploit this capability to realize low-cost, high-performance temperature measuring systems incorporating rapid advances in electronic signal conditioning circuitry.

RESISTANCE THERMOMETRY

Resistance thermometry utilizes the characteristic relationship of electrical resistance with temperature to measure temperature. For pure metals, this relationship may be expressed by

$$R_t = R_0 (1 + at + bt^2 + ct^3 + \ldots) \tag{1}$$

where R_0 = resistance at reference temperature (usually at ice point, 0°C), Ω
  R_t = resistance at temperature t, Ω
  a = temperature coefficient of resistance, Ω/Ω (°C)
  b and c = coefficients calculated on the basis of two or more known resistance-temperature (calibration) points

For alloys and semiconductors, the relationship follows a unique equation dependent on the specific material involved. Whereas most elements constructed from metal conductors generally display positive temperature coefficients with an increase in temperature resulting in increased resistance, most semiconductors display a characteristic negative temperature coefficient of resistance.

Only a few pure metals have a characteristic relationship suitable for the fabrication of sensing elements used in resistance thermometers. The metal must have an extremely stable resistance-temperature relationship so that neither the absolute value of the resistance R_0 nor the coefficients a and b drift with repeated heating and cooling within the thermometer's specified temperature range of operation. The material's specific resistance in ohms per cubic centimeter must be within limits that will permit fabrication of practical-size elements. The material must exhibit relatively small resistance changes for nontemperature effects, such as strain and possible contamination which may not be totally eliminated from a controlled manufacturing environment. The metal's change in resistance with temperature must be relatively large in order to produce a resultant thermometer with inherent sensitivity. The metal must not undergo any change of phase or state within a reasonable

*Director of Engineering, Hy-Call Engineering, A Unit of General Signal, El Monte, California.

temperature range. Finally, the metal must be commercially available with essentially a consistent resistance-temperature relationship to provide reliable uniformity.

Industrial resistance thermometers, often referred to as resistance-temperature detectors (RTDs), are commonly available with elements of platinum, nickel, 70% nickel–30% iron (Balco), or copper. The entire resistance thermometer is an assembly of parts which includes the sensing element, internal leadwires, internal supporting and insulating materials, and protection tube or case. See Fig. 1 and Table 1.

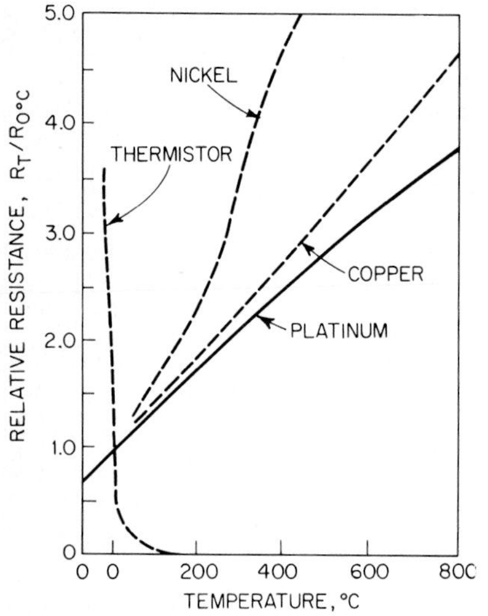

Fig. 1 Resistance-temperature characteristics of thermoresistive materials at elevated temperatures.

Platinum Resistance Thermometers (Wire-Wound)

Of all materials currently utilized in the fabrication of thermoresistive elements, platinum has the optimum characteristics for service over a wide temperature range. Although platinum is a noble metal and does not oxidize, it is subject to contamination at elevated temperatures by some gases, such as carbon monoxide and other reducing atmospheres, and by metallic oxides.

The metal is available commercially in pure form, providing a reproducible resistance-temperature characteristic. Platinum with a temperature coefficient of resistance equal to 0.00385 Ω /Ω (°C) (from 0 to 100°C) has been used as a standard for industrial thermometers throughout the United Kingdom and Western Europe since World War II and has gained prominence in recent years in the United States in the absence of a defined and commonly accepted standard coefficient. Platinum has a high melting point and does not volatilize appreciably at temperatures below 1200°C. It has a tensile strength of 18,000 psi (124 MPa) and a resistivity of 60.0 Ω/(cir mil) (ft) at 0°C (9.83 $\mu\Omega \cdot$ cm).

Platinum is the material most generally used in the construction of precision laboratory standard thermometers for calibration work. In fact, the laboratory grade platinum resistance thermometer (usually with a basic resistance equal to 25.5 Ω at 0°C) is the defining standard for the temperature range from the liquid oxygen point ($-$ 182.96°C) to the antimony point (630.74°C) as defined by the International Practical Temperature Scale of 1968 (IPTS 68).

The resistance-temperature relationship for platinum resistance elements is determined from the Callendar equation above 0°C;

$$t = \frac{100 \ (R_t - R_0)}{R_{100} - R_0} + \delta \left(\frac{t}{100} - 1 \right) \frac{t}{100} \tag{2}$$

where t = temperature, °C
    R_t = resistance at temperature t, Ω
    R_0 = resistance at 0°C, Ω
    R_{100} = resistance at 100°C, Ω
    δ = Callendar constant (approximately 1.50)

The fundamental coefficient (temperature coefficient of resistance) α is defined over the fundamental interval of 0 to 100 C°

$$\alpha = \frac{R_{100} - R_0}{100 R_0} \tag{3}$$

Table 1. Resistance versus Temperature of Various Metals

Metal	Resistivity, $g\Omega \cdot cm$	Relative resistance R_t/R_0 at $0°C$										
		−200	−100	0	100	200	300	400	500	600	700	800	900
Alumel*	28.1			1.000	1.239	1.428	1.537	1.637	1.726	1.814	1.899	1.982	2.066
Copper	1.56	0.117	0.557	1.000	1.431	0.862	2.299	2.747	3.210	3.695	4.208	4.752	5.334
Iron	8.57			1.000	1.650	2.464	3.485	4.716	6.162	7.839	9.790	12.009	12.790
Nickel	6.38			1.000	1.663	2.501	3.611	4.847	5.398	5.882	6.327	6.751	7.156
Platinum	9.83	0.177	0.599	1.000	1.392	1.773	2.142	2.499	3.178	3.178	3.500	3.810	4.109
Silver	1.50	0.176	0.596	1.000	1.408	1.827	2.256	2.698	3.616	3.616	4.094	5.586	5.091

*Trademark, Hoskins Manufacturing Co., Detroit, Mich.

Substituting Eq. 3 into Eq. 2, the Callendar equation becomes

$$t = \frac{R_t - R_0}{\alpha R_0} + \delta \left(\frac{t}{100} - 1\right) \frac{t}{100} \tag{4}$$

It should be noted that Eq. 4 must be modified to conform to IPTS 68:

$$t_{68} = t' + \Delta t \tag{5}$$

where t' = temperature by the Callendar equation (Eq. 4)
Δt = correction term given by

$$\Delta t = 0.045 \frac{t'}{100}\left(\frac{t'}{100} - 1\right)\left(\frac{t'}{419.58} - 1\right)\left(\frac{t'}{603.74} - 1\right) \tag{6}$$

Furthermore the Van Dusen modification of the Callendar equation previously applied below the ice point has been replaced by the new IPTS 68 reference function which extends down to the boiling point of hydrogen ($-259.34°C$). Although a thorough mathematical discussion of IPTS 68 is not within the scope of this presentation, the interested reader should consult Refs. 4 and 5.

IPTS 68 modifications to the Callendar equation result in temperature differences of less than $\pm 0.1°C$ at temperatures below $500°C$. However, these differences approach $+1.0°C$ at $850°C$. Industrial RTD constructions are commonly limited to 550 or $650°C$ by the methods and materials of assembly so that Eqs. 5 and 6 can be safely ignored in most applications.

Calculated results of the resistance-temperature relationship for a typical platinum industrial thermometer are given in Table 2.

The limitations imposed on the performance characteristics of industrial platinum resistance thermometers result more from the limitations of the element mounting and methods of construction than from the characteristics of the platinum itself. The thermometer's actual performance is a function of its critical design features. Since the sensitive winding of a resistance element usually comprises a small-diameter wire which must be protected from the medium being measured, it is generally mounted in a metal protective housing and in such a manner that the thermal resistance path from the outer surface of the housing to the wire is held to a minimum. It is imperative that the resistance winding be well insulated electrically from its housing, since shunt resistance will distract from the thermometer's performance.

Modern application requirements for high sensitivity, fast response, and extreme ruggedness at relatively low cost have led to the most versatile, most reliable, and widely used basic platinum resistance thermometer design—the fully encapsulated RTD. The platinum wire, usually 0.001 in (0.025 mm) O.D. or less, is wound into a coil and inserted into a multibore high-purity ceramic tube, or is wound directly on the outside of a ceramic tube. The most commonly used ceramic material is aluminum oxide (99.7% Al_2O_3). The winding is completely embedded and fused within or on the ceramic tube utilizing extremely fine granular powder.

The resultant fully encapsulated element, with only two noble metal leadwires exposed, provides maximum protection for the platinum resistance coil. Although special fusing techniques are used, such elements are not completely strain-free, but the effects of existing strains are fairly constant

Table 2. Resistance-Temperature Relationship of 100-Ω Platinum Resistance Thermometer.

Temperature below 0°C in Descending 10° Increments, °C (IPTS 68)

°C	0	-10	-20	-30	-40	-50	-60	-70	-80	-90	-100
						Resistivity, Ω					
0	100.000	96.083	92.154	88.213	84.259	80.291	76.309	72.313	68.302	64.275	60.230
-100	60.230	56.166	52.082	47.975	43.842	39.681	35.488	31.264	27.010	22.720	18.386
-200	18.386										

Temperatures above 0°C in ascending 10° increments, °C (IPTS 68)

°C	0	10	20	30	40	50	60	70	80	90	100
						Resistivity, Ω					
0	100.000	103.904	107.795	111.675	115.543	119.399	123.243	127.075	130.895	134.703	138.500
100	138.500	142.285	146.058	149.820	153.570	157.308	161.035	164.750	168.454	172.146	175.827
200	175.827	179.496	183.154	186.800	190.433	194.058	197.669	201.270	204.859	208.436	212.002
300	212.002	215.556	219.099	222.631	226.151	229.659	233.156	236.642	240.216	243.578	247.029
400	247.029	250.469	253.896	257.313	260.717	264.110	267.491	270.861	274.219	277.565	280.899
500	280.899	284.222	287.533	290.832	294.119	297.394	300.657	303.908	307.148	310.375	313.590
600	313.590	316.793	319.984	323.163	326.330	329.484	332.626	335.756	338.873	341.987	345.071
700	345.071	348.151	351.218	354.273	357.316	360.345	363.362	366.366	369.358	371.336	375.302
800	375.302	378.255	381.195	384.122	387.035	389.936					

Data Relevant to Table 2

Input data		Output data	
Temperature, °C	Resistance, Ω		Coefficients
-182.962 (b.p. oxygen)	25.747	$\alpha =$	0.385×10^{-2}
0.0 (f.p. water)	100.000	$\delta =$	1.508
100.00 (b.p. water)	138.500	$A_4 =$	-0.760×10^{-4}
419.580 (f.p. zinc)	253.753	$C_4 =$	-0.134×10^{-12}

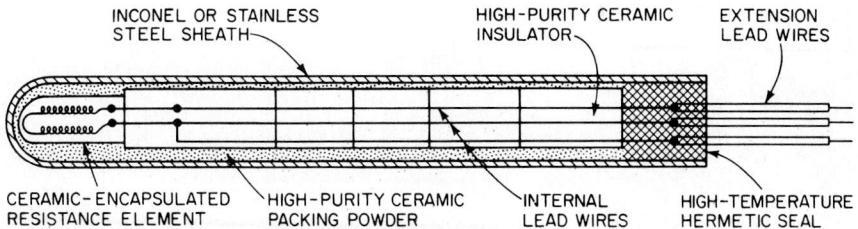

INCONEL OR STAINLESS STEEL SHEATH

HIGH-PURITY CERAMIC INSULATOR

EXTENSION LEAD WIRES

CERAMIC-ENCAPSULATED RESISTANCE ELEMENT

HIGH-PURITY CERAMIC PACKING POWDER

INTERNAL LEAD WIRES

HIGH-TEMPERATURE HERMETIC SEAL

Fig. 2   Platinum industrial resistance thermometer assembly.

with resulting errors well within the permissible limits for industrial applications. The intimate contact between the platinum winding and the ceramic encapsulation permits a rapid speed of response with the thermal conductivity of ceramic adequate for heat transmission through the protecting layer. Advanced manufacturing techniques can produce platinum elements with a base resistance of 100 Ω and physical dimensions as small as 0.080 in O.D. $\times$ 0.75 in long (2 $\times$ 19 mm).

The fully encapsulated design technique is maintained as the criterion for the entire thermometer assembly. Internal leadwires, usually a base metal alloy with high-temperature capabilities, such as *Inconel** or nickel-clad copper for applications to 550°C, are welded to the sensitive element's noble metal leadwires in close proximity to the element, with three and four lead circuits the most commonly utilized. Measuring circuits are described later in this article. A multibore ceramic tube serves to insulate the internal leadwires throughout the entire length of the higher-temperature assemblies. The thermometer's external extension leadwires, usually Teflon†-insulated copper, are then attached to the internal leadwires just above the ceramic insulators. The entire internal assembly is inserted into a stainless steel or Inconel sheath which has been treated to remove all forms of contamination. High-purity fine-grained ceramic powder is then packed into the assembly, using vibration techniques—after which the entire unit is hermetically sealed at the lead end with high-temperature epoxy materials. A platinum industrial resistance thermometer assembly with a maximum temperature range is shown in Fig. 2.

An assembly simpler than that shown in Fig. 2 gives enhanced ruggedness, including some capacity for bending without damage. The extension leadwires are connected directly to the resistance element, and the entire element and joint are again encapsulated in an aluminum oxide cover tube. This eliminates the points subject to strain as in the case of the assembly shown in Fig. 2. The armored element is then placed directly in the sheath without ceramic insulating spacers. High-purity ceramic powder packing and an end seal are used as previously described. Maximum temperature for these assemblies is a function of the internal leadwire insulation—260°C for TFE Teflon, and 550°C for fiberglass.

Industrial platinum resistance thermometers are available with a variety of basic resistances, ranging from 100 to 500 Ω, and in a wide variety of physical dimensions. Most commonly used sheath dimensions range from ⅛ through ¾₆ to ¼ O.D. (3.1 through 4.8 to 6.4 mm) and from a few inches to several feet in length. The resistance thermometer assembly commonly is provided with associated hardware as required by specific applications. Where physical size and/or time response are of critical importance, the thermometer assembly is provided with direct insertion hardware, including adjustable mounting fitting and an aluminum or miniature terminal connecting head.

For applications where high-temperature requirements are combined with high pressure, high flow, and high vibration, the resistance thermometer assembly includes a thermowell protecting tube. Aluminum or cast iron terminal connecting heads or the mounting fitting are designed with internal spring-loading mechanisms to provide improved thermal contact between the thermometer and the well, while at the same time eliminating relative vibration between thermometer sheath and thermowell assembly. Figure 3 shows the most widely used modern configuration with a spring-loaded fitting suitable for use with any thermowell and connecting head. Special design features in the fitting prevent rotation of the RTD to eliminate twisting and straining the leads in high-vibration environments.

*Trademark, Huntington Alloy Products Division, International Nickel Co., Inc., Huntington, W. Va.
†Trademark, E. I. du Pont de Nemours & Co., Wilmington, Del.

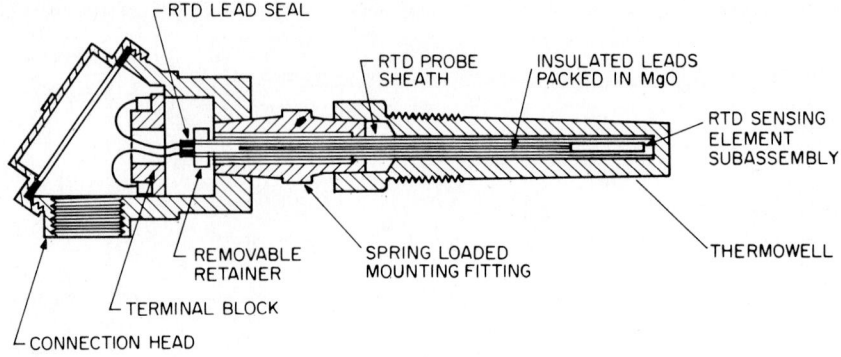

RTD LEAD SEAL

RTD PROBE
SHEATH

INSULATED LEADS
PACKED IN MgO

RTD SENSING
ELEMENT
SUBASSEMBLY

REMOVABLE
RETAINER

SPRING LOADED
MOUNTING FITTING

THERMOWELL

TERMINAL BLOCK

CONNECTION HEAD

Fig. 3 Industrial RTD-thermowell assembly.

Platinum Resistance Thermometers (Thin-Film)

New processing advancements have provided the capability to produce thin-film platinum RTD elements that are indistinguishable from wire-wound elements in reproducibility and stability. The final industrial configuration is the same as that shown in Fig. 3. The resistance table for 100-Ω elements is the same as that of Table 2 but is limited to 550°C. Thin-film platinum RTD elements are trimmed to the final resistance value on automated equipment. Wire-wound platinum resistance elements have a much higher labor content and require more platinum. Thin-film elements have a growing cost advantage that becomes especially significant at higher resistance values where the difference in platinum content is greatest. The final assembly methods are the same for both kinds of elements once the initial pigtail leads are attached.

Thin-film techniques have made high-resistance platinum elements a practical reality. One-thousand-ohm elements are routinely available. These high-resistance elements have, in many cases, made obsolete the need to consider base metal wire-wound sensors of nickel or nickel-iron. Thermistors also may often be replaced with a sensor of significantly greater stability and temperature range. As is the case with all resistance thermometers, higher resistances are used to swamp out lead errors in two-wire or long three-wire connections and to allow precise measurement with simplified signal conditioning circuits. With the use of thin-film elements, high-resistance 1000-Ω platinum sensors can be made without increasing the sizes over those commonly available at the 100-Ω level.

The technology involved in thin-film platinum sensors offers the greatest cost advantage if the film area is kept small. This has led to the introduction of 1000-Ω thin-film elements with a slightly lower fundamental coefficient of 0.00375 Ω/Ω(°C). The value 0.00385 Ω/Ω (°C) has become the defacto industrial standard coefficient (now being proposed as a U.S. and an international standard) for 100-Ω elements. A slightly lower coefficient allows a thin-film size reduction of several times; in fact, 1000-Ω thin-film elements are produced in exactly the same size as 100-Ω thin-film elements. Because the 1000-Ω film can be thinner in the patented application discussed here, it has a lower platinum content and cost of production than the 100-Ω element. With efficient assembly designs, 1000-Ω platinum thin films combine very high performance and low cost—formerly associated only with lower performing base metal wire-wound RTDs.

A slightly lower fundamental coefficient to obtain the advantages noted above does imply lower performance. In fact the 0.00385 Ω/Ω(°C) coefficient accepted for wire-wound platinum RTDs is a result of the combination of slightly doped platinum wire, controlled by the resistance wire manufacturer and strain associated with a fully supported element on alumina substrates. The effect of winding on alumina alone is shown by the reduction in the coefficient of reference-grade platinum from 0.003927 Ω/Ω(°C), a requirement for IPTS 68 standard thermometers which are relatively unsupported but strain-free, to 0.003902 Ω/Ω(°C), which is produced in industrial constructions using reference-grade wire. The combined effects that result in the 0.00385 Ω/Ω(°C) coefficient are reproducible by manufacturers everywhere to provide the so-called international-grade platinum curve in Table 2. The reduction in sensitivity is insignificant to gain the advantage of rugged construction. For accurate measurements, the user need only be aware of or specify the actual temperature coefficient. Similarly, the further slight reduction in coefficient to realize significant size and cost advantages in 1000-Ω thin-film sensors is readily accommodated. In the case of the thin film, only the purest materials equivalent to reference-grade platinum are used—so even the dependence on the platinum supplier for specific coefficient alterations is eliminated.

The higher order constants $b, c, \ldots$ in Eq. 1 or 8, or δ in Eq. 2 and 4 are the same for all types of platinum sensors within better than 0.1% of temperature, or about one-fifth of conventional interchangeability tolerances of 0.25°C, or 0.5% of temperature, whichever is greater. This allows higher coefficient sensors to be padded with a parallel resistor and a resistor in series with the combination to match the curve for lower coefficient sensors. Platinum sensors of different coefficients can be made to work through multiplexed inputs to common signal conditioning provided that the signal conditioning is calibrated for the lowest coefficient.

Nickel Resistance Thermometers

For industrial temperature measurements in the range of -70 to $150°C$, resistance thermometers with nickel sensing elements of 10 to 1000 Ω have had extensive and satisfactory use. The important characteristics of nickel include a tensile strength of 120,000 psi (828 MPa) and a resistivity of 38.36 $\Omega/(\text{cir mil})$ (ft) at $0°C$ (6.38 $\mu\Omega \cdot \text{cm}$). The upper temperature limit of the thermometer is imposed by the materials used in insulating the nickel wire and by the state of anneal which is affected at $250°C$. Above $300°C$, nickel undergoes transformations which make its resistance-temperature curve irregular.

The temperature coefficient of pure nickel is approximately 0.0066 Ω/Ω ($°C$), whereas that of platinum is less than 0.00393 Ω/Ω ($°C$) (refer to Fig. 1). Thus the use of nickel in place of platinum in resistance thermometry offers a promise of high sensitivity. However, in establishing standard tables for resistance versus temperature, engineers have had to take into account the variations in coefficient values for commercially available nickel. Since different lots of nickel differ in resistivity, it becomes necessary to adjust the nickel resistance thermometer element to match the theoretically determined function. To accomplish this adjustment a compensating coil constructed from a negligible temperature coefficient alloy, such as Manganin* or Constantan,* is connected in series (or shunt) within the same protecting sheath with the higher coefficient nickel element. See Fig. 4. This procedure results in allowable interchangeability of elements without necessitating the selection of nickel wire having a definite and highly reproducible temperature coefficient. However, the resultant coefficient and corresponding resistance-temperature function has been effectively lowered from that of the pure nickel and may turn out to be approximately the same as platinum. Hence, the anticipated improvement in sensitivity has been lost.

Another method of achieving interchangeable nickel sensors is to establish the specific anneal temperature required to obtain the design coefficient from each lot of nickel. The method of element construction and probe assembly can then be similar to that for platinum sensors except that no temperature exposure above $200°C$ can be allowed. Full sensitivity is retained.

Recent widespread use of the fully encapsulated platinum RTD with its superior performance characteristics, often at lower cost, has taken precedence over the use of nickel thermometers in almost every industrial application. Current production of nickel thermometers is primarily a component replacement function for already existing industrial measuring and control systems.

Copper Resistance Thermometers

Electrolytic copper of the highest purity is readily available commercially, having a highly consistent temperature coefficient of resistance value equal to approximately 0.00425 Ω/Ω ($°C$), which is slightly higher than that of platinum (refer to Fig. 1). Resistance elements of copper are fabricated to utilize the full temperature coefficient and are highly interchangeable with respect to established resistance-temperature relationships. The usable temperature range of copper resistance thermometers is confined to -200 to $+150°C$, with copper having a tendency toward oxidation at higher

*Trademark, Wilbur B. Driver Co., Newark, N.J.

temperatures. Copper has a tensile strength of 30,000 psi (207 MPa). The resistivity of copper is 9.38 Ω/(cir mil) (ft) at 0°C (1.56 $\mu\Omega \cdot$ cm), a value much lower than that of platinum or nickel.

Construction methods for copper resistance thermometers are very similar, and in many cases identical, to those utilized in the manufacture of nickel thermometers. Since the copper wire is highly reproducible from available stock, it is not necessary to provide the element with any type of zero-temperature-coefficient compensating resistors as is true of nickel elements. The relatively low resistivity property of copper suggests that fine wire must be used in element construction to avoid massiveness and consequent slowness in response to temperature change. Copper thermometer elements are therefore manufactured with base resistances of 10 and 25 Ω and generally do not exceed 100 Ω.

The stability of calibration is considered to be satisfactory, while being superior to that of nickel. The resultant resistance-temperature relationship is highly linear, within narrow limits, from approximately -50 to 150°C. This straight-line characteristic of copper thermometers is useful in allowing two sensors to be applied directly for temperature difference measurement applications.

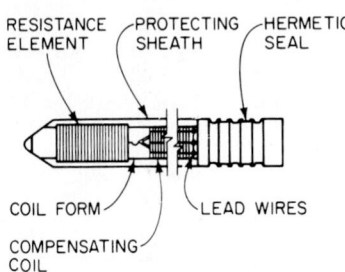

Fig. 4   Nickel resistance thermometer.

Special Resistance Thermometers

The development of resistance thermometers of the copper, nickel, and platinum types has advanced in recent years consonant with industrial process requirements. Standard thermometer designs are therefore able to meet the majority of these application requirements. However, many special purpose resistance thermometers have been designed and developed to meet particular and highly specialized application requirements.

Since the sensitive winding on most standard thermometers typically varies from approximately ¾ to 1½-in (19 to 38 mm), allowing for a factor of 2 for conductive heat losses propagated up the sheath of the assembly, the resultant temperature measurement is confined to the medium immediately surrounding the tip of the sensor sheath. Special copper, nickel, and platinum thermometers have been constructed to measure *average* temperature in large enclosures such as air, water, and oil baths; drying ovens; air-conditioning and heating ducts; pipelines; and other vessels. In applications of this type the resistance winding is distributed on a coil form over a large proportion of the protecting sheath's length. Elements of this type are known as distributive windings and have been fabricated in lengths varying from 3 to 240 in (76 mm to 6.1 m) or more. The resultant measurement is an effective average temperature of the medium surrounding the entire length of the distributive thermometer.

Many special resistance thermometers have utilized the basic wafer design, having the characteristics of small mass combined with good thermal contact and resulting in an extremely fast time response. Fine insulated wire of copper, nickel, or platinum usually is sandwiched between two protecting sheets of insulating material and sealed. Most wafer-type assemblies are fragile as compared with standard designs. However, for applications demanding fast time responses, such as air temperature measurements and surface temperature measurements, the wafer design is often an effective alternative to standard sheathed thermometers. Recent advances in film technology, both thin and thick, show promise for new wafer-type resistance thermometers with added ruggedness and high precision at relatively low production cost. See Fig. 5.

Performance Characteristics and Testing Procedures

Most manufacturers and users of resistance thermometers, as well as several societies, have developed a series of test methods for the evaluation of RTD performance characteristics. In recent years, much effort has been made toward standardization of these test methods so that resistance thermometers of various types and manufacturers may be compared in order to determine the proper selection and

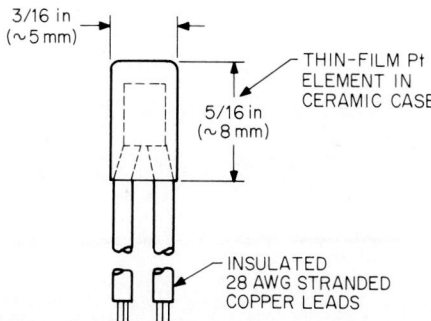

3/16 in
(~5 mm)

THIN-FILM Pt
ELEMENT IN
CERAMIC CASE

5/16 in
(~8 mm)

INSULATED
28 AWG STRANDED
COPPER LEADS

Fig. 5 RTD surfaces temperature sensor. *(Hy-Cal Engineering.)*

how to best meet application requirements. Most general test methods utilize equipment commonly available in temperature measurement and test evaluation laboratories to formalize resistance thermometer specifications. For specific applications involving conditions not commonly duplicated in the lab (for example, combined conditions of extremely high temperature, flow, and pressure), test methods are left to the interested individual or interested group.

In addition to the temperature range, temperature coefficient, and interchangeability characteristics as previously discussed, the following performance characteristics are of primary concern: (1) accuracy, (2) stability, (3) repeatability, (4) time response, (5) self-heating effect, (6) insulation resistance, and (7) vibration resistance. For the following test methods, sample specifications are given for a typical 100-Ω platinum industrial RTD. Reference 14 is recommended for description of detailed test procedures.

Accuracy

The accuracy of adjustment or calibration over the specified temperature range or operation is a prime consideration in selecting a resistance thermometer for a given application. The *accuracy* of calibration is defined as the ability of a thermometer to conform to its predetermined resistance-temperature relationship and commonly is expressed in terms of percent of actual temperature reading. The majority of industrial RTDs fall within the 0.1 to 0.5% accuracy range. Two test methods are used for determining sensor accuracy.

The *comparison technique* consists basically of measuring simultaneously the resistance of the test sensor at a desired temperature and the temperature itself, using a certified standard thermometer. To obtain consistent results, the sensor to be calibrated and the standard thermometer must reach thermal equilibrium before actual measurement. Thermal equilibrium is maintained with the use of a stirred, constant temperature bath. For elevated-temperature calibration an equalizing block (multibore metal block having large thermal mass) within a calibration furnace or fluidized sand bath is used to maintain thermal equilibrium.

The *fixed-point technique* consists of measuring the thermometer under test while the temperature is "defined" by the change in phase of some medium. The method provides a calibration of the test thermometer at one and only one distinct temperature. This temperature is an intrinsic property of the medium which is undergoing a phase change, usually freezing.

The following equilibrium states (fixed point) as defined by IPTS 68 commonly are used:

Fixed point	Temperature, °C
Triple point of water	0.01
Boiling point of water	100.00
Freezing point of tin	231.9681
Freezing of zinc	419.98
Freezing point of gold	1065.43

Both the comparison and fixed-point calibration techniques require an adequate number of sample test thermometer calibration results to determine a statistically valid general specification. A sample accuracy specification may read $\pm 0.5°$F or $\pm 0.5\%$ of temperature reading, whichever is greater.

Stability

In many applications, the stability of an industrial resistance thermometer is the foremost performance consideration. *Stability* is defined as the ability of a thermometer to maintain and reproduce its specified resistance-temperature characteristics for long periods of time within its specified temperature range of operation. The degree of stability exhibited by a thermometer is expressed in terms of "drift," which is defined as an undesirable change in resistance over a period of time which is unrelated to the actual operating temperature. The ice point is generally used as the defining reference temperature. The maximum drift usually is experienced at elevated temperatures. Hence, stability is specified as the drift in ice-point resistance over a given interval of operating time at the thermometer's upper temperature limit. The stability test method requires that the thermometer maintain stability under cyclic test conditions within its usable temperature range.

The test thermometer is immersed into the ice bath (a Dewar flask containing a mixture of finely divided pure ice and distilled water), and the initial ice-point resistance is recorded. The thermometer is removed from the bath and allowed to reach equilibrium with room temperature. The RTD then is inserted into a test furnace where the controlled temperature is equivalent to the thermometer's upper temperature limit. The test thermometer is allowed to "soak" for a given interval of time, typically 24 to 48 h. The thermometer is taken out of the furnace and allowed to reach equilibrium at room temperature. Another ice-point reading is taken. This resistance reading now is recorded and compared to the initial ice-point measurement. The difference between the two measurements is determined as the drift. This procedure is repeated at regular intervals for a prolonged period of time, typically one year. A drift-versus-time plot then can be maintained. The stability of the test thermometer is specified as the maximum recorded drift over the total period of test time. Sample stability specification: Drift in ice-point resistance after 10,000 h of operation at 1000°F is less than 0.05 Ω (approximately 0.25°F).

Repeatability

The repeatability performance characteristic is closely related to stability. *Repeatability* is defined as the conformity of consecutive temperature measurements for an individual test thermometer at selected temperatures within its specified temperature range of operation. The repeatability test method requires that a number of consecutive temperature readings (usually three) be recorded at selected temperature intervals of approximately 100°F. The maximum recorded conformity in reading at any given temperature within the usable range is determined to be the repeatability of a test thermometer. Sample repeatability specification: Conformance of consecutive temperature readings is better than 0.02 Ω (approximately 0.10°F).

Time Response

This is an especially important consideration in applications where the temperature of the medium being measured is subject to rapid changes. *Time response* is that time required for the thermometer to react to a step change in temperature and reach the resistance corresponding to 63.2% (one time constant) of the total temperature change. In order to assure a repeatable step change in temperature, test methods call for a reference point and an elevated fixed point subject to controlled thermal conditions. Usually the ice point and a constant temperature, rotating pan water bath, respectively, meet test requirements.

Test procedures require that the test thermometer be properly connected to a high speed recorder. The test sensor is allowed to reach equilibrium in the ice bath, after which it is rapidly plunged into the constant temperature water bath. The thermal response time τ is determined graphically from the recorded plot to be equal to 63.2% of the temperature step change encountered. The time response specification is the maximum recorded value over a statistical average of test thermometers. Sample time response specification: The time required for the sensor to reach 63.2% of a step change in temperature is less than 6.0 s. Conditions: 0 to 50°C in water moving at approximately 3 ft/s (1 m/s) transverse to the sensor sheath.

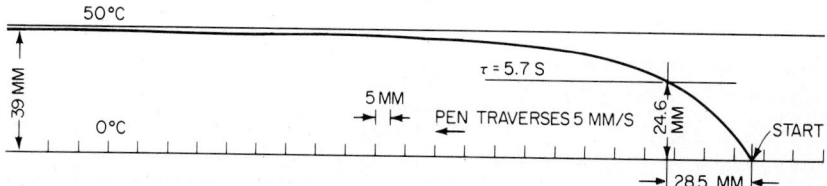

Fig. 6 A time-response test recording.

Figure 6 is a recording for the sample specification. Note that the temperature span of 0 to 50°C is equivalent to 39 mm on the Y axis. The time constant τ is plotted to be 63.2% of the total 39 mm, equal to 24.6 mm from the reference temperature at 0°C. τ intersects the X axis (time) at 28.5 mm from the starting point and, with the recorder chart paper moving at 5 mm/s, is determined to be 5.7 s.

Figure 7 shows representative values for the time constant versus the sensor diameter. Note that the time constant approximately doubles with an additional layer in the assembly when elements are inserted in the sheath to make probes. High-temperature probes, such as shown in Fig. 2, are even slower because of poorer thermal transfer through the packing.

Self-Heating

The operation of a resistance thermometer depends on the flow of current through the element, which consequently produces heat. The heat generated can be a source of error resulting from a "false" change in resistance which is not indicative of the temperature being measured. This effect is known as *self-heating* and is defined as the temperature change experienced by a resistance thermometer

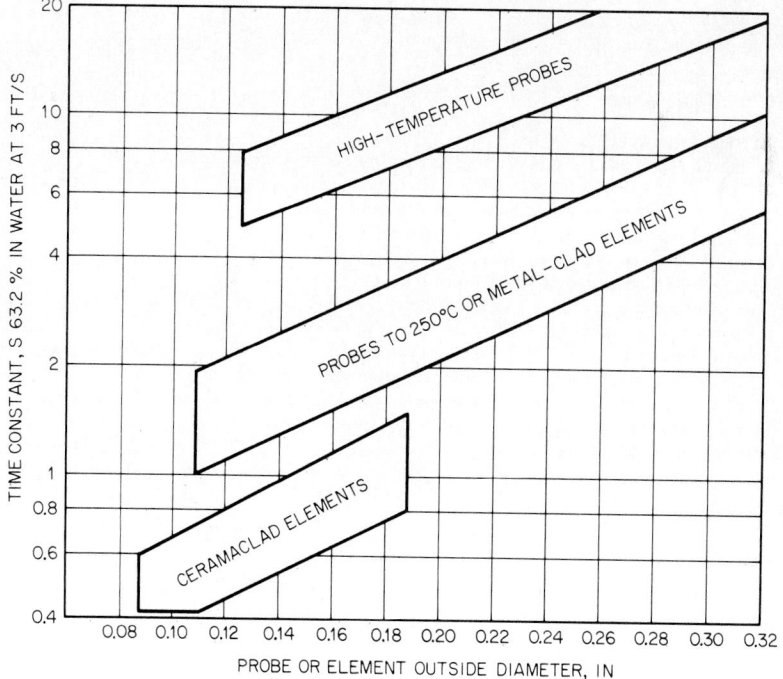

Fig. 7 Response time of typical resistance temperature sensors. *(Hy-Cal Engineering.)*

due to I^2R heating when normal operating current is applied to the thermometer immersed in a uniform heat-transfer medium.

The self-heating effect will vary, depending on the sensor construction characteristics. The well-designed resistance thermometer will properly dissipate self-developed heat into the measured medium, resulting in extremely small errors. It is evident that the self-heating error is dependent on the heat-transfer characteristics associated with the measured media. In still air, for example, where the heat transfer is poor, the error will be considerably greater than in a stirred water bath where the heat-transfer coefficient is relatively good. Therefore, any self-heating specification must define fully the working test medium and hence the heat-transfer coefficient.

Standard test methods to determine self-heating effects require the test thermometer to be immersed in a uniform heat-transfer medium. The measured operating current is then increased through the test sensor incrementally (usually 1 to 2 mA) within the specified range of excitation current. With each increment in excitation current the resistance of a thermometer is measured and recorded. A resistance-versus-current curve can be maintained with values below 1 mA extrapolated from the test plot. The self-heating error then is specified as the rise in indicated temperature due to the power dissipated through the sensor over the full range of excitation current. The unit associated with the self-heating error is milliwatts per degree Celsuis as calculated by

$$H = S \frac{P_2 - P_1}{R_2 - R_1} \tag{7}$$

where H = self-heating error, $mW/°C$
  S = average temperature coefficient of resistance, $\Omega/°C$
  P_1 = power dissipation at maximum current, mW
  R_1 = resistance at minimum current, Ω
  R_2 = resistance at maximum current, Ω

Sample self-heating specifications: Maximum self-heating error over excitation current range of 0 to 10 mA is less than $0.1°C$ based on a minimum dissipation factor of $100 \ mW/°C$. Conditions: water at $28°C$ moving approximately 1 ft/s transverse to the sensor sheath. Note: Self-heating is greater on thin-film elements.

Insulation Resistance

Insulation resistance between the sensitive element and all other subassemblies of the thermometer (sheath, thermowell, terminal, and connecting head) is ideally infinite under operating conditions. However, the shunt effect of leakage resistance often is a degrading factor on actual performance. Factors contributing to poor insulation resistance include (1) flaws in design or construction resulting in random physical contact between element and sheath, (2) leakage of a conducting medium through flaws in the sensor sheath or sheath welds, (3) mechanical deformation of the sheath under adverse conditions, and (4) moisture sealed in the sheath during construction. Of all the factors contributing to insulation breakdown, moisture effects are the most common, particularly in fully encapsulated designs where ceramic packing powder readily absorbs moisture. See Fig. 2.

Insulation resistance therefore may be a function of temperature, and test methods call for checks at a suitable lower temperature (usually ambient) and an elevated temperature above the steam point (usually the maximum specified operating temperature). Measurements of insulation resistance need not be of high precision, in that low insulation readings are grounds for rejection. Test voltages may vary from as high as 300 V dc to as low as 1.5 V dc, depending on the test equipment and application requirements. Sample specification: The insulation resistance between any leadwire and the thermometer sheath is greater than 100 MΩ at $72°F$ and 1 MΩ at $1000°F$.

Vibration Resistance

The ability of industrial resistance thermometer assemblies to withstand vibration and shocks within specified limits while retaining other performance characteristics is of unqualified importance. Test apparatus includes a vibration shaker capable of producing the sinusoidal vibration level over the specified frequency range while the test thermometer is mounted on the shaker head. Special mounting fixtures of aluminum or magnesium are usually necessary. Piezoelectric accelerometers are used to monitor and/or control the frequency response, amplitude, and cross-axial sensitivity of a test fixture. For the duration of the test the normal output of the thermometer is monitored, usually on an oscilloscope/oscillograph, such that momentary circuit interruptions or short circuits are detected. On completion of the vibration test, the thermometer is determined to show no visual defects and no breakdown in insulation resistance while maintaining ice-point resistance specifications.

Sample vibration resistance specifications: The sensor will withstand a vibration harmonic amplitude of 0.06-in (1.5-mm) total excursion over approximately 10 to 55 Hz up and back in approximately 1 min for 2 h total time in each of the three mutually perpendicular directions (*MIL Standard 202C, Method 201A*). In addition, the sensors will withstand a severe high-frequency vibration of 25 to 1000 Hz up and back at 20 g over a 15-min interval in each of the three mutually perpendicular directions.

Resistance Thermometer Measuring Methods

Three methods are commonly used for making electric connections from the resistance thermometer assembly to the measuring instrument, namely, the use of two-, three-, or four-lead circuits.

The two-lead circuit shown in Fig. 8 is the simplest, consisting of two relatively low-resistance leads a and b connecting the resistance thermometer with the measuring apparatus. For this example, the thermometer leadwires are copper and a wheatstone bridge is shown. Note that the leg R_x comprises the resistance of the element plus the resistance of the leads a and b. Therefore, leads a and b, unless of very low resistance, can add appreciably to the resistance of the element. Even though the resistances of the leads may be known and allowed for in the measurement, they are subject to ambient temperature changes and do not, of course, follow a precise resistance-temperature function. Therefore, the two-lead measuring method should be used only where leadwire resistance can be kept to a minimum and only where a moderate degree of accuracy is required. The extent of leadwire error is related to the resistance of the element and obviously will be proportionately greater with windings of low resistance than with windings having a higher resistance.

The three-lead circuit shown in Fig. 9 is the most widely used method in industrial resistance thermometry. In this circuit leads a and c are connected in close proximity to the resistance thermometer element at a common node. The third lead, b, is connected to the opposite resistance leg of the element. The resistance of lead a is now added to bridge arm R_3, while the resistance of lead b

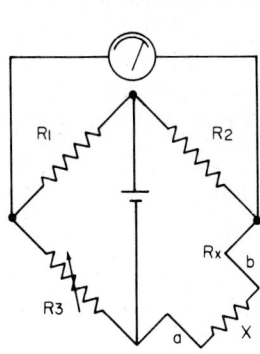

Circuit equations:

$$R_1 + R_3 = R_2 + a + b + X \qquad (8)$$
$$R_1 = R_2$$
$$\therefore \quad R_3 = a + b + X \qquad (9)$$

Fig. 8 Two-lead measuring circuit.

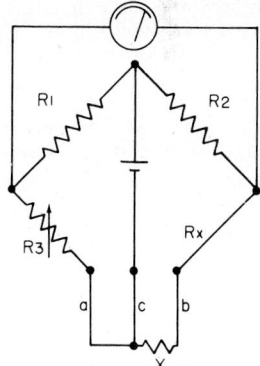

Circuit equations:

$$R_1 + R_3 + a + c = R_2 + b + X + c \qquad (10)$$
$$R_1 = R_2$$

If $a = b$ (lead resistance equal)

Then $R_3 = X \qquad (11)$

Fig. 9 Three-lead measuring circuit.

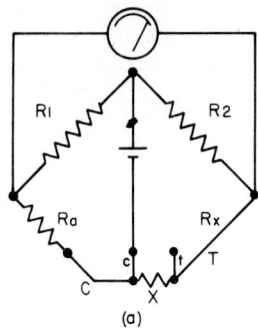

(a)   (b)

Circuit equations:

$R_1 + R_a + C + c = R_2 + T + X + c$ (12) $R_1 + R_b + T + t = R_2 + C + X + t$ (14)

$R_1 = R_2$ $R_1 = R_2$

$R_a + C = T + X$ (13) $R_b + T = C + X$ (15)

Summing eqs. (13) and (15) yields:

$R_a + C = T + X$

$R_b + T = C + X$

$\overline{R_a + R_b + C + T = T + C + 2X}$ (16)

$\therefore \quad X = (R_a + R_b)/2$ (17)

Fig. 10 Four-lead measuring circuits.

remains on bridge arm R_x, thereby dividing the lead resistance and retaining a balance in the bridge circuit. Lead resistance c is common to both the left and right loops of the bridge circuit. Although this method compensates for the effect of lead resistance, the ultimate accuracy of the circuit depends on leads a and b being of *equal* resistance. Special matching techniques are used on leads a and b, particularly when distances between the sensor and measuring equipment are relatively great.

The four-lead circuit, as shown in Fig. 10a and b, is used only where the highest degree of accuracy is required, as in the case of platinum resistance thermometers used as laboratory standards. A single measurement involves first taking a reading per Fig. 10a and then taking a reading per Fig. 10b. These are known as normal and reverse bridge measurements. The sum of the two measurements is divided by 2, yielding an average reading independent of leadwire resistances. Because of the inconvenience and time required in carrying out this procedure, this method is used only when the ultimate in accuracy is required. Referring to Fig. 10a, note that two separate leads are connected at a common node in close proximity and on both sides of the resistance element. These leads are designated C and c on one side and T and t on the opposite side. For the normal measurement, leads C and T are in opposite legs of the bridge with lead c common to both loops of the bridge as previously described in the three-lead method. In Fig. 10b, note that leads T and C are reversed in opposite legs of the bridge and that lead t is now common to both loops of the bridge. The circuit equations define the averaging technique used to cancel out leadwire effects. A triple-pole, double-throw mercury contact switch usually is used to transfer the resistance thermometer leads from normal to reverse circuits.

Recent developments in the electronics industry have resulted in highly stable constant

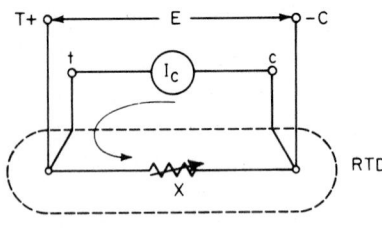

Circuit equations:

$E = IX$

$I \cong$ constant

$E = f(X) = f'$ (temperature) (18)

Fig. 11 Four-lead constant current measuring circuit.

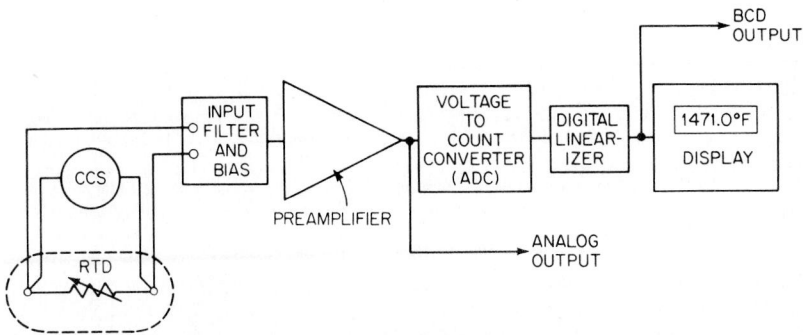

Fig. 12   Resistance-temperature detector system with both digital and analog outputs.

current source (CCS) power supplies available in miniature packages at relatively low cost. Techniques utilizing CCS power supplies offer an effective alternative to null balance bridge-type instruments, particularly in industrial process systems requiring scanning (both manual and automatic) of up to 100 or more individual RTD points usually located at varying distances from the central control panel.

The resistance thermometer circuit is the same basic four-lead circuit as previously discussed, with two leads joined in close proximity to each side of the resistance element. See Fig. 11.

The CCS is connected across leads t and c supplying a constant current I_c across the resistance element X. The value for I_c is usually kept to a minimum of 2 mA or less, to avoid excessive self-heating errors . The voltage drop across the resistance element is then measured between T and C. The resultant voltage drop across the thermometer element, in the constant current mode, varies with resistance directly as a function of temperature.

The advantage of CCS resistance thermometer circuits should now be apparent to the reader already versed in bridge measuring techniques. The CCS power supply will continue to maintain the fixed constant current (within its compliance voltage limitations) across the thermometer element, thus making costly matching techniques previously associated with leadwires unnecessary. In addition, leadwire contact resistances associated with manual or high-speed automatic switching are reduced to minimal concern.

An added feature of the CCS measuring circuit is its ability to interface directly with a wide variety of voltage measuring instruments. Sophisticated state-of-the-art digital linearizers can be

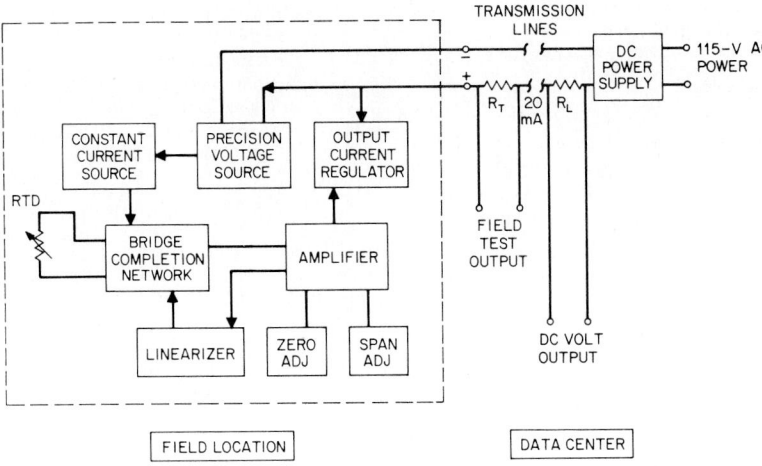

Fig. 13   Block diagram of 4 to 20 mA two-wire RTD transmitter. *(Hy-Cal Engineering.)*

applied to operate on the nonlinear platinum resistance thermometer function to display out directly in engineering units with linearization conformities to 0.1°F. Furthermore, digital instruments when applied in conjunction with CCS measuring circuits with resistance thermometers provide added features of analog output signals for recording and trending and linearized binary coded decimal (BCD) output for interface with digital printers as well as process control computers. Figure 12 illustrates the functional schematic of the digital temperature system with the resistance thermometer as a primary element.

When platinum RTD leads are limited in resistance relative to the sensing element by the use of short leads or high-resistance thin-film elements, a two-wire or three-wire bridge completion circuit can be used. Linearizing analog feedback from the bridge output amplifier gives results approaching those obtainable from the four-wire system of Fig. 12. The final output is converted to a linear 4 to 20 mA signal in the most common industrial configurations. Figure 13 is a typical block diagram.

THERMISTORS*

Thermistors are included in the class of solids known as semiconductors, having electric conductivities between those of conductors and those of insulators. The name "thermistors" is derived from *thermally* sensitive re*sistors,* since the resistance of a thermistor varies as a function of temperature. Although semiconductors have been known as a class for many years, only in recent years have sufficient theory and manufacturing techniques been developed to spur their use as primary temperature elements. Since about 1935, reproducibility and stability characteristics of certain of these materials have been brought into practical bounds. Usable electrical devices dependent on thermistor properties have become available, and industrial applications in both the United States and Europe utilize these primary elements in moderate quantities.

A thermistor is an electrical device made of a solid semiconductor with a high-temperature coefficient of resistivity which would exhibit a linear voltage-current characteristic if its temperature were held constant. When a thermistor is used as a temperature sensing element, the relationship between resistance and temperature is of primary concern. The approximate relationship applying to most thermistors is

$$R_t = R_0 \exp B \left(\frac{1}{T} - \frac{1}{T_0} \right) \tag{8}$$

where R_0 = resistance value at reference temperature T_0 K, Ω
  R_T = resistance at temperature T K, Ω
  B = constant over temperature range, dependent on manufacturing process and construction characteristics (specified by supplier)

$$B \cong \frac{E}{K} \tag{9}$$

where E = electronvolt energy level and K = Boltzmann's constant (8.625×10^{-5} eV/K).

A second form of the approximate resistance-temperature relationship is written in the form

$$R_T = R_\infty e^{B/T} \tag{10}$$

where R_∞ = thermistor resistance as temperature approaches infinity, Ω.

Both Eqs. 8 and 9 are only best approximations and, therefore, are of limited use in making highly accurate temperature measurements. However, they do serve to compare thermistor characteristics and thermistor types.

The temperature coefficient usually is expressed as a percent change in resistance per degree of temperature change and is approximately related to B by

*Based on Ref. 12.

$$a = \frac{dR}{dT}\left(\frac{1}{R}\right) = \frac{-B}{T_0^2} \tag{11}$$

where T_0 is in kelvin. It should be noted that the resistance of the thermometer is solely a function of its absolute temperature. Furthermore, it is apparent that the thermistor's resistance-temperature function has a characteristic high negative coefficient as well as a high degree of nonlinearity. The value of the coefficient a for common commercial thermistors is on the order of 2 to 6% per kelvin at room temperature. This value is approximately 10 times that of metals used in the manufacture of resistance thermometers.

Resultant considerations due to the high coefficient characteristic of thermistors include inherent high sensitivity and high level of output, eliminating the need for extremely sensitive readout devices and leadwire matching techniques, respectively. However, limitations on interchangeability (particularly over wide temperature ranges), calibration, and stability—also inherent to thermistors—are quite restrictive. The high degree of nonlinearity in the resistance-temperature function usually limits the range of the readout instrumentation. In many applications, special prelinearization circuits must be used before interfacing with related system instrumentation. The negative temperature coefficient also may require an inversion (to positive form) when interfacing with some analog and/or digital instrumentation.

Thermistor Construction Techniques

A number of metal oxides and their mixtures, including the oxides of cobalt, copper, iron, magnesium, manganese, nickel, tin, titanium, uranium, and zinc, are among the most common semiconducting materials used in the construction of thermistors. Usually compressed into the desired shape from the specially formulated powder, the oxides are then recrystallized by heat treatment, resulting in a dense ceramic body. The leadwires are then attached while electric contact is maintained, and the finished thermistor is then encapsulated.

There are many mechanical configurations for the thermistor as a primary temperature element. Beads are made (Fig. 14a) by forming small ellipsoids of material suspended on two fine leadwires approximately 0.10 in (2.5 mm) apart. The material is sintered at elevated temperatures, and the leadwires become tightly embedded within the bead, making electric contact with the thermistor material. For more rugged applications, the bead thermistor is sealed into the tip of a glass, ceramic, or suitable metal sheath.

Disk thermistor configurations (Fig. 14b) are manufactured by pressing the semiconductor material into a round die to produce a flat, circular probe. These pieces are sintered and then silvered on

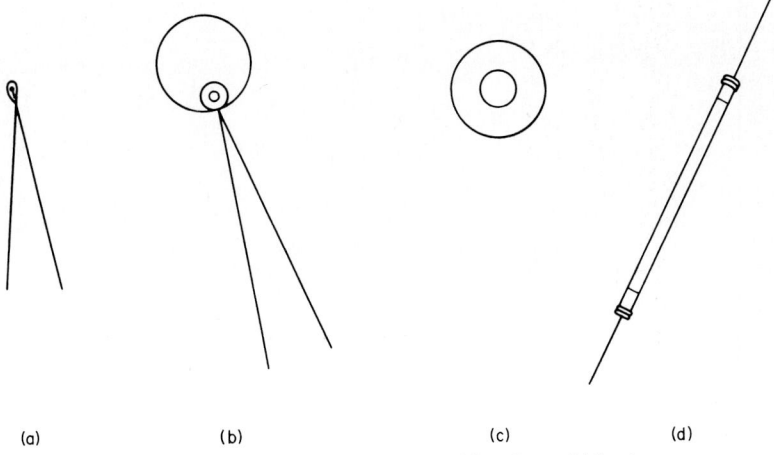

(a)   (b)   (c)   (d)

Fig. 14.   Some thermistor configurations: (a) bead, (b) disk, (c) washer, and (d) rod.

Table 3. Typical Thermistor Properties

Description	Temp. coefficient at 25°C, % per°C approx.	Unheated resistance, Ω			Max. continuous ambient temp, °C	Dissipation factor, mW/°C*	Thermal time constant, °C
		At 0°C approx.	At 25°C	At 50°C approx.			
Rods, disks, and washers with leads or mountings							
Bolted pileup, 4 washers connected in parallel	−3.8	7	2.5 ± 10%	1	150	30	300
0.4-in-diam disk, radial leads	−3.8	283	100 ± 10%	41	125	7	160
0.4-in-diam disk, radial leads	−4.4	3,270	1,000 ± 10%	360	125	7	160
0.11- × 1⅛-in rod, axial leads	−3.8	28,500	10,000 ± 10%	4,000	150	6	95
0.055- × ½-in rod, radial leads	−4.4	327,000	100,000 ± 10%	36,000	150	2.5	20
Unmounted beads							
Glass-coated bead, 0.015 in diam; 0.001-in leads, opposite	−3.4	5,000	2,000 ± 25%	900	150	0.1	1
Glass-coated bead, 0.045 in diam; 0.004-in leads, adjacent	−3.8	5,800	2,000 ± 20%	810	300	0.7	2
Glass-coated bead, 0.045 in diam; 0.004-in leads, adjacent	−4.7	355,000	100,000 ± 15%	34,000	300	0.7	2
Beads on glass probes							
Bead on 0.07- × 2-in glass probe	−3.8	5,800	2,000 ± 20%	810	300	1	25
Bead on 0.07- × 2-in glass probe	−4.7	355,000	100,000 ± 15%	34,000	300	1	25

*Mounting by the leads in still air is assumed. Rating increases for mountings having greater thermal conduction to surroundings.

†Rating is based on same mounting as in preceding footnote and decreases with increased thermal conduction to surroundings.

the two flat surfaces. Thermistor disks range from 0.1 to 1.0 in (2.5 to 25 mm) in diameter and 0.02 to 0.5 in (0.5 to 13 mm) in thickness. Disk thermistors are commonly applied where a moderate degree of power dissipation is required.

Washer-type thermistors (Fig. 14c) are manufactured like disk thermistors, except that a hole is formed in the center of the sensor to provide for bolt mounting. Normal washer configurations are approximately ¾ in (19 mm) in diameter and are applied where high power dissipation is a primary requirement.

Rod-type thermistors (Fig. 14d) are extruded through dies, resulting in long, cylindrical probes commonly varying from 0.05 to 0.175 in (1.3 to 4.4 mm) in diameter and from 0.25 to 2.0 (6.4 to 51 mm) in length, with leadwires usually attached at the ends of the rod. Rod configurations are generally of high terminal resistance and are generally applied where power dissipation is not a principal concern.

Table 3 summarizes thermistor configurations and their respective properties.

Thermistor Performance Characteristics

Evaluation of thermistor performance characteristics is in many cases similar to that of resistance thermometers. Test methods, therefore, are identical in some cases to those already outlined. The following description concentrates on thermistor characteristics not explicitly common to resistance thermometers.

Figure 15 demonstrates the logarithm of the specific-resistance-versus-temperature relationship for three typical thermistor materials as compared with platinum metal. The specific resistance of the thermistor represented by curve 1 *decreases* by a factor of 50 as the temperature is increased from 0 to 100°C. Over the same temperature range, the resistivity of platinum will *increase* by a factor of approximately 1.39.

Thermistors range in terminal resistance at room temperature from about 1 Ω to the order of 10^8 Ω, depending on composition, shape, and size (refer to Table 4). Within a given type they commonly vary 10 to 30% in resistance from the nominal at the reference temperature. Some types may be specially manufactured and specially selected, bringing resistance values within closer limits.

The range in power dissipation factor varies from approximately 10^{-5} W to several watts per degree Celsius of resultant rise in temperature. This value varies inversely with the degree of thermal isolation of the thermistor element.

The time constant varies from a few tenths of a second to several minutes among general purpose thermistors. Naturally the shorter times are associated with the smaller bead configurations, and the longer times with the larger rods and disks as well as larger beads. The time constant varies directly with the thermal capacity of the thermistor and inversely with the dissipation factor.

The stability with time of the resistance-temperature function depends on thermistor construction as well as the particular application conditions. Changes may occur either in the resistivity of the semiconductor or in the contact medium and its relation with the semiconductor. Resistivity may change through chemical changes in composition, caused by decomposition or diffusion processes which are generally accelerated at higher temperatures. Certain oxide compositions are inherently more stable than others. Thermistors enclosed in glass coverings with sintered-in noble metal leadwires usually exhibit relatively good stability with time. If subject to cycling at relatively high tem-

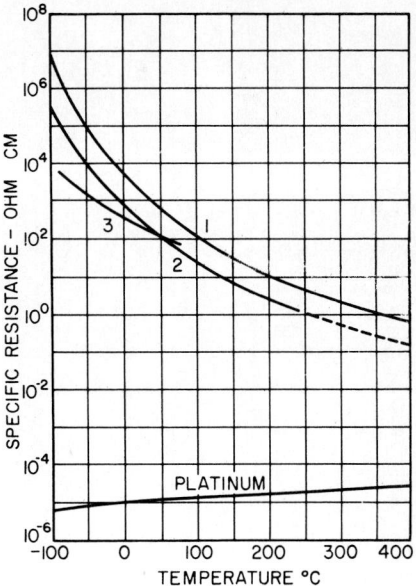

Fig. 15 Resistance-temperature characteristics of thermistors.

Table 4. Comparison of Various Temperature Sensors

Evaluation criteria	Platinum RTD 100-Ω wire-wound and thin-film	Platinum RTD ultra-7 1000-Ω thin-film	Nickel RTD 1000-Ω wire-wound	Balco RTD 2000-Ω wire-wound	Thermistor	Thermocouple	Semiconductor devices
Cost, OEM quantity	High	Low*	Medium	Medium	Low*	Low*	Low*
Temperature range	Wide −400°F to +1200°F*	Wide −320°F to +1000°F*	Medium −350°F to +600°F	Short −100°F to +400°F	Short to medium −100°F to +500°F	Very wide −450°F to +4200°F†	Short −57°F to +257°F
Interchangeability	Excellent†	Excellent†	Fair	Fair	Poor to fair	Good*	Fair
Long-term Stability	Good*	Good†	Fair	Fair	Poor	Poor to fair	Good to fair
Accuracy	High*	High*	Medium	Low	Medium	Medium	Medium
Repeatability	Excellent†	Excellent†	Good*	Fair	Fair to good	Poor to fair	Good*
Sensitivity (output)	Medium	High*	High*	Very high†	Very high†	Low	High*
Response	Medium	Medium to fast	Medium	Medium	Medium to fast*	Medium to fast*	Medium to fast*
Linearity	Good*	Good*	Fair	Fair	Poor	Fair	Good*
Self-heating	Very low to low*	Medium	Medium	Medium	High	NA‡	Very low to low*
Point (end) sensitive	Fair	Good*	Poor	Poor	Good*	Excellent†	Good*
Lead effect	Medium	Low*	Low*	Low*	Very low†	High	Low*
Physical size packaging	Medium to small	Small to large*	Large	Large	Small to medium*	Small to large*	Small to medium*

*Good rating.
†Best rating
‡Not applicable.

2.66

peratures, some types of thermistors may exhibit a behavior characteristic known as relaxation. Resulting resistance changes may be as much as several percent following the cyclic temperature change, with recovery time of a few hours. When thermistors are held at a fixed temperature with negligible current flow, resistance changes on the order of 1% per year are typical.

The upper operating temperature limit of thermistors is set by physical changes in the semiconductor material or by calibration considerations. This limit will depend on the construction of the thermistor and the degree of precision required of it for a given application. Typical temperature ranges are on the order of 100 to 400°C. The low-temperature operating limit is usually determined by the resistance, reaching such a high value as to make convenient measurements difficult. For these applications only thermistors of relatively low coefficient and very low room-temperature resistance are practical.

Thermistor Applications

The application of thermistors as primary temperature elements follows the usual principle of resistance thermometry. Conventional bridge or other resistance measuring circuits, as well as constant current circuits, as outlined in the previous section are employed. Special application considerations must be given to the negative and highly nonlinear resistance-temperature relationship, as previously noted. Common to resistance thermometers, consideration must be given to keeping the measuring circuit small enough to avoid significant heating in order that the element resistance shall be solely dependent on the temperature of the measured medium.

The use of thermistors as temperature sensing elements in industrial applications has increased in recent years. Although many applications for thermistors are restricted to the laboratory, increased use in industrial applications, resulting from advanced controlled manufacturing techniques and development of related process instrumentation, will continue.

Comparison of Various Temperature Sensors

Various RTDs are compared with each other as well as with thermistors, thermocouples, and semiconductor devices in Table 4.

REFERENCES

1. Robertson, D., and K. A. Walch: "Calibration Techniques for Precision Platinum Resistance Thermometers" *Leeds & Northrup Co. Tech. Publ. A1.1201,* North Wales, Pa., 1961.

2. Macan, W. A.: Electrical Sensors of Temperature, *Res. Develop.,* vol. 21, no. 6. June 1970. pp. 22–29.

3. Adams, E. F.: "Resistance Thermometry" in *Process Instruments and Controls Handbook,* Douglas M. Considine (ed.), McGraw-Hill, New York, 1957, pp. 47–59.

4. "The International Practical Temperature Scale of 1968: Amended Edition of 1975," *Metrologia,* vol. 12, July 17, 1976.

5. Riddle, J. L., Furukawa, G. T.. and H. H. Plumb: "Platinum Resistance Thermometers," *NBS Monogr.* 126, National Bureau of Standards. Washington, D.C., 1973.

6. Rubin, L. G.: "Temperature: Concepts, Scales and Measurement Techniques," *Leeds Northrup Tech. Publ. A1.0001RP,* North Wales, Pa.

7. Walker, W. T.: "Resistance Thermometry" in *Encyclopedia of Instrumentation and Control,* Douglas M. Considine (ed.). McGraw-Hill, New York, 1971.

8. Warner, F. B.: "Some Recent Developments in Applied Platinum Resistance Thermometry" in *Temperature: Its Measurement and Control in Science and Industry.* vol. 3, VanNostrand Reinhold, New York, 1962, pp. 297–304.

9. Green, C. B.: "Thermistors as Primary Temperature Elements" in *Process Instruments and Controls Handbook,* Douglas M. Considine (ed.), McGraw-Hill, New York, 1957, pp. 60–67.

10. Droma, C. R.: "Thermistors for Temperature Measurements" in *Temperature: Its Measurement and Control in Science and Industry,* vol. 3, VanNostrand Reinhold, New York, 1962, pp. 339–346.

11. Sachse, H. B.: "Measurement at Low Temperatures with Thermistors" in *Temperature: Its Measurement and Control in Science and Industry,* vol. 3, VanNostrand Reinhold, New York, 1962, pp. 347–353.

12. White, R. B.: "Thermistors" in *Process Instruments and Controls Handbook,* 2d ed., Douglas M. Considine (ed.). McGraw-Hill, New York, 1974.

13. Standard DIN 43760, "Fundamental Values of Measuring Resistors for Resistance Thermometers," German Standards Committee (DNA).

14. ASTM: *Standard Methods for Testing Industrial Thermometers,* American Society for Testing and Materials, Philadelphia, Pa., 1974.

15. ASTM: *Standard Specifications for Industrial Platinum Resistance Thermometers,* American Society for Testing and Materials, Philadelphia, Pa. 1984.

Filled-System Thermometers

by
G. Robert Leavitt*

Filled-system thermometers are used to provide a remote signal or indication of the temperature of a material for purposes of measurement and/or control. Filled-system thermometers appeared as a method of measurement following the invention of the bourdon spring which is still widely used as an element for converting a pressure or volume change to a usable motion in industrial measurement and control loops. Any reference to "bourdon" in this article is intended to also include other measuring elements, such as diaphragms and bellows.

Principle of Operation

The filled-system thermometer is designed to provide an indication or record of temperature at some distance from the point of measurement. See Fig. 1. The system usually consists of an element sensitive to temperature (bulb), an element sensitive to pressure or volume changes (bourdon, bellows, or diaphragm), a means of connecting these elements (capillary tubing), and a device for indicating or recording a signal related to the measured temperature. In exceptional cases, the capillary or even the bourdon may be designed to be the temperature-sensitive element, thus eliminating the use of a capillary and/or bulb.

The temperature sensing element (bulb) contains a fluid which changes its volume or pressure with temperature. The pressure-sensitive element (bourdon) responds to these changes by delivering a motion or force to a device which transduces the signal to a usable form. This is commonly a mechanical linkage which drives a pointer or pen, but may be a pneumatic or an electrical device which transmits the temperature signal over long distances. These temperature signals frequently are used for process control purposes.

*Manager, Quality Assurance, Taylor Instrument Company, Division of Combustion Engineering, Inc., Rochester, New York.

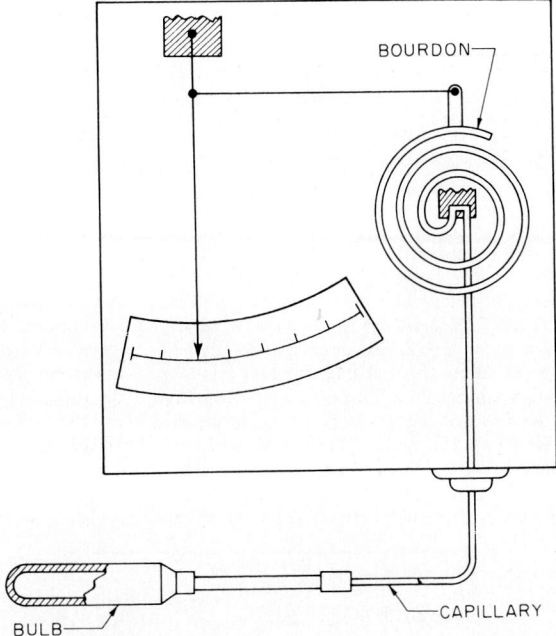

Fig. 1   Filled-system thermometer.

Relative Advantages and Limitations

Advantages

The wide use of filled-system thermometers in the process industries may be explained by the following reasons:

1. Fundamental simplicity of the system allows rugged construction, minimizing the possibility of damage or failure in shipment, installation, and use. The amount of upkeep is generally minor.

2. Simplicity of the system allows inexpensive design.

3. As used in the process industries, sensitivity, response time. and accuracy are generally the equal of any other temperature measuring instruments available.

4. The capillary allows considerable separation between the point of measurement and the point of indication. Although capillary lengths of 400 ft (120 m) have been used successfully, it is usually more economical to employ transducers for signal transmission of 100 ft (30 m) or more.

5. The measuring system is self-contained. It does not need auxiliary power unless it is combined with a pneumatic or an electric transmission system.

6. The system can be designed to deliver significant power if necessary to drive indicating or controller mechanisms, including valves.

Limitations

Thermal systems are limited as follows:

1. The bulb size may be too large to fit the available space.

2. The performance characteristics vary considerably with the type of filling fluid, and the user must be certain that he does not misapply a particular type of system.

3. The maximum temperature is more limited than that in some electrical measuring systems.

4. In case of system failure, the entire unit must be replaced or repaired.

5. Separation of sensing and indicating elements may be limited, depending on other characteristics, such as filling liquid and accuracy requirements.

Classification of Filled-System Thermometers

Filled-system thermometers may be separated into two fundamental types: those in which the bourdon responds to (1) volume changes and (2) pressure changes. Those that respond to volume changes are completely filled with a liquid. The liquid expansivity with temperature is greater than that of the bulb metal, the net volume change being communicated to the bourdon. An internal-system pressure change is always associated with the bourdon volume change, but this effect is not of primary importance. The systems that respond to pressure changes are either filled with a gas or partially filled with a volatile liquid. Changes in gas or vapor pressure with changes in bulb temperature are communicated to the bourdon. The bourdon will increase in volume with increase in pressure, but this effect is not of primary importance.

Filled-system thermometers have been classified* as follows:

Volumetric principle	
Liquid-filled (other than Hg)	Class I
Mercury-filled	Class V
Pressure principle	
Vapor-filled	Class II
Gas filled	Class III

This classification will be used throughout this article.

Liquid-Filled Thermal Systems

These systems operate on the principle that the temperature signal is generated by and proportional to a volume change in the bulb. They often are referred to as liquid-expansion type. They are divided into subclasses to designate the means of providing temperature compensation:

1. *SAMA Class IA (Full Compensation)*. The compensating means is a second thermal system minus the bulb, or an equivalent means of compensation. See Fig. 2.

2. *SAMA Class IB (Case Compensation)*. Here the only compensation is a means provided within the instrument case. See Fig. 3.

3. *SAMA Class IB (Case Compensation)*. The compensating means is a second thermal system minus the bulb, or an equivalent means of compensation.

4. *SAMA Class VB (Case Compensation)*. Here the only compensation is a means provided within the instrument case. See Fig. 3.

General Design Considerations

A manufacturer of liquid-filled systems generally will provide a single bourdon design to cover all possible temperature spans. The bulbs of all Class I or Class V thermal systems from one manufac-

---

*SAMA Stand. RC6-10-1963, Process Measurement and Control Section, Scientific Apparatus Makers Association, Washington, D. C.

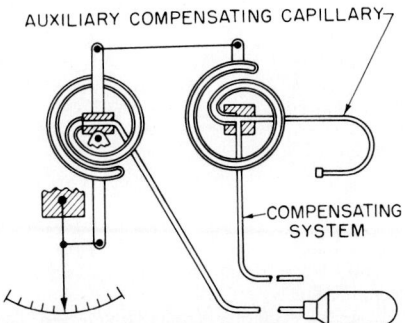

Fig. 2 Fully compensated liquid-, mercury-, or gas-filled thermal system (Class IA, IIIA, or VA).

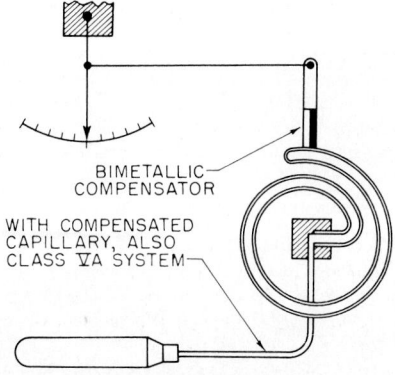

Fig. 3 Case-compensated liquid-, mercury-, or gas-filled thermal system (Class IB, IIIB, or VB).

turer will deliver the same volume change to the bourdon, regardless of range span. Bulb volume, therefore, will vary with range, in accordance with Eq. 1:

$$\Delta V = V_b \delta R = \text{constant} \tag{1}$$

where ΔV = bourdon volume change, cm^3
φ = differential expansivity between filling liquid and enclosing bulb, cm^3/(cm^3) ($^\circ$C)
V_b = bulb volume, cm^3
R = range span, $^\circ$C

Rearranging Eq. 1 yields

$$V = \frac{K}{\varphi R} = \frac{K'}{R} \tag{2}$$

where K and K' are constants. Equation 2 shows that the bulb volume is small for large-range spans, and vice versa. The ΔV utilized by different manufacturers varies considerably but is generally between 0.04 and 0.15 cm^3.

$$\varphi = 10.8 \times 10^{-4} \text{ cm}_3/(\text{cm}^3)\ (^\circ\text{C}) \qquad \text{for organic liquids}$$

$$\varphi M = 13.4 \times 10^{-5} \text{ cm}^3/(\text{cm}^3)\ (^\circ\text{C}) \qquad \text{for mercury}$$

Therefore:

$$V = \frac{\Delta V}{R} = \frac{0.04 \text{ to } 0.15}{10.8 \times 10^{-4}R} = \frac{65 \text{ to } 250}{R} \text{ cm}^3 \qquad \text{for organic liquids} \tag{3}$$

$$V = \frac{\Delta V}{\varphi R} = \frac{0.04 \text{ to } 0.10}{13.4 \times 10^{-5}R} = \frac{550 \text{ to } 1{,}400}{R} \text{ cm}^3 \qquad \text{for mercury} \tag{4}$$

Maximum and Minimum Temperatures and Range Spans

The minimum temperature is limited to the freezing point of the organic liquid employed as a filling medium. This is usually between -75 and -200°C (-103 and -328°F), depending on the liquid used.

The maximum temperature to which the liquid is stable is approximately 300°C (572°F). The minimum range span is determined by the maximum bulb size that is practical, generally 12 to 25°C (22 to 45°F), although spans as short as 5°C (9°F) are used. The maximum span sometimes is limited by nonlinearity of expansivity and compressibility of the fluid fill to approximately 150°C

(270°F), so that standard accuracy may be met with linear dials and charts. Some liquids, however, deliver a linear signal for spans as large as 250°C (450°F). Mercury-filled thermal systems may be used between −38 and +650°C (−39 and +1202°F). Organic liquids freeze at a much lower temperature and are commonly used between −75 and +300°C (−103 and +572°F). Because of the higher expansivity, organic-filled systems are more adaptable to short spans, while the lower compressibility of mercury makes it easier to use on long spans. In either case, the minimum span usually is limited by the largest practical bulb size. See Fig. 4.

Capillary and Case Compensation

The nonmetallic liquid systems are particularly vulnerable to errors resulting from capillary temperature changes because of the high expansivity of the fill. Only if the capillary volume is very small can these errors be ignored and a Class IB system used. See Fig. 3.

To neutralize the capillary temperature error an auxiliary thermal system is employed. See Fig. 2.

The mercury-filled system may be compensated fully by an alternative method. This employs a capillary containing low controlled expansion nickel steel wire which can, by proper selection of dimensions, cancel any net volume changes due to the expansion of the mercury. It should also be noted that, because of the low expansivity of mercury, Class VB systems can have a capillary up to 50 ft (15 m) with good performance.

A VA system using a compensated capillary, an IB, or a VB often is preferred to a Class IA because of the relative simplicity of construction, lesser vulnerability to abuse, and lower cost.

Since the bourdon volume changes with measured temperature changes, the case compensation can be perfect at only one bulb temperature. In practice, this compensation is adjusted to a specified tolerance with the bulb at midrange. This results in the thermometer being slightly overcompensated at range bottom and slightly undercompensated at range top. Attention must be paid to these details if precise measurements are desired.

Speed of Response

The speed of response of liquid-filled thermometers ranges from 1 to 10 s for 63% responses of a bare bulb in well-stirred water. See Fig. 5. The effect of capillary resistance on speed of response is

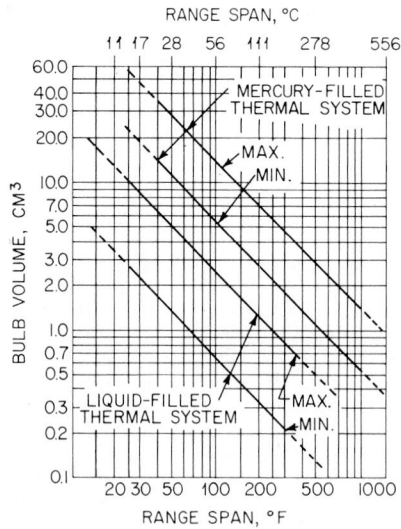

Fig. 4 Bulb volume versus range span, Class I and Class V systems (maximum and minimum values show extremes among manufacturers).

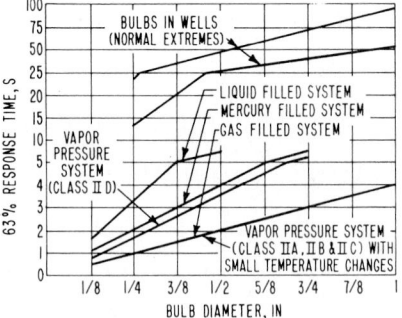

Fig. 5 Bulb response versus bulb outside diameter in well-stirred water. *Note:* ⅛ in = 3.2 mm; ¼ in = 6.4 mm; ⅜ in = 9.5 mm; ½ in = 12.7 mm; ⅝ in = 15.9 mm; ¾ in = 19.1 mm; ⅞ in = 22.2 mm.

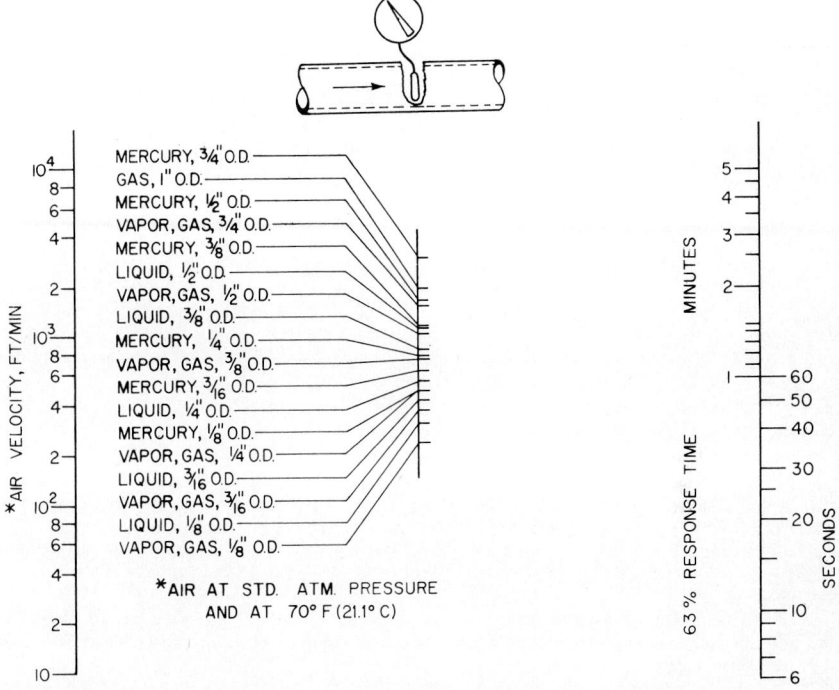

Fig. 6 Bulb response rate in air at various air velocities. *Note:* 1 ft/min = 0.3 m/min; ⅛ in = 3.2 mm; ³⁄₁₆ in = 4.8 mm; ¼ in = 6.4 mm; ⅜ in = 9.5 mm; ½ in = 12.7 mm; ¾ in = 19.1 mm; 1 in = 25.4 mm.

generally negligible. The low thermal conductivity of the organic fills is somewhat balanced by their smaller bulb size, but for a given diameter mercury systems are faster. To improve response in poor heat-transfer mediums, such as air, bulbs often are manufactured from lengths of capillary tubing which may be installed in straight lengths or preformed into relatively tight coils. See Fig. 6.

Barometric Errors

This effect is negligible in Class I and Class V systems.

Bulb Elevation Error

When a bulb of a liquid-filled system is elevated above or lowered below the case, a pressure difference equal to the liquid head develops between the bulb and bourdon. Most of the internal pressure change takes place at the stiff or bulb end of the system. A pressure change, for instance, causes the bulb to expand or contract and the liquid to compress or to expand, depending on whether the bulb is below or above the bourdon, thus causing a flow of fluid into or out of the bourdon and thereby causing a bulb-elevation error. If the bulb is elevated above the case, the instrument indication will rise. If the case is elevated above the bulb, the indication will be lowered by an equal amount. If the bulb is to be elevated above the case, it will be necessary for the manufacturer to increase the pressure within the system so that the bulb pressure will not drop to zero after bulb elevation, thereby reducing the indicated system temperature. This increase in pressure will reduce the overrange possibilities of the system. Since the bulb is usually very stiff and the liquid is highly incompressible, the error is generally small. The error in percent of range versus bulb elevation in feet may be approximated by Eq. 5:

$$E = \frac{P_h(K_b + K_f)V100}{\Delta V} = \frac{0.43hd(K_b + K_f)100}{\varphi R} \tag{5}$$

where E = error, % of span
h = bulb elevation, ft
P_h = head pressure, psi
d = density of fluid, g/cm^3
K_b = bulb stiffness, cm^3/(cm^3) (psi)
K_f = fluid stiffness, cm^3/(cm^3) (psi)
V = bulb volume, cm^3
ΔV = volume change across range, cm^3
φ = differential expansivity of fluid and bulb, cm^3/(cm^3) (°C)
R = span, °C

It is noted that this effect will be larger for mercury-filled systems. If the installed height changes, the instrument should be rezeroed. Functional limits usually are ± 30 ft (9 m) unless specially built to permit greater differences in height between bulb and case.

Overrange Protection

Maximum overrange temperature is defined as the maximum temperature to which the bulb of a filled system may be exposed indefinitely without damage to the system. It is usually expressed in percent of span above the range-top temperature.

Classes IB, VA, and VB systems generally are provided with overrange protection of 100% of range span, sometimes higher, depending on the design of the bourdon spring or an equivalent element.

On fully compensated systems using two bourdons and a double capillary, especially Class IA, the bourdon expansion capacity is taken up by the potential expansion of liquid in the capillary, and less overrange protection is available.

Vapor-Pressure Thermal System

This system is defined by SAMA as "a thermal system partially filled with a volatile liquid and operating on the principle of vapor pressure." Four types are employed:

1. **Class IIA.** Designed to operate with the measured temperature above the temperature of the rest of the thermal system. See Fig. 7.
2. **Class IIB.** Designed to operate with the measured temperature below the temperature of the rest of the thermal system. See Fig. 8.
3. **Class IIC.** Designed to operate with the measured temperature above and below the temperature of the rest of the thermal system. This type normally requires a larger sensitive portion than Class IIA or IIB. See Fig. 9.
4. **Class IID.** Designed to operate with the bulb temperature above, below, and at the temperature of the rest of the thermal system. See Fig. 10. In this type, the volatile liquid is confined to the sensitive portion and a second relatively nonvolatile liquid is used to transmit the vapor pressure to the expansible device.

General Design Considerations

The pressure within the system is the vapor pressure at the interface of the liquid and vapor. It is required that the system be filled so that this interface will always exist within the bulb. The fill will always be in a liquid state at the coolest part of the system. This combination of requirements and circumstances is responsible for the four basic designs just defined, each best for particular conditions of use.

The Class IIA system, the most common type of vapor-filled thermal system is designed for the condition that the bulb temperature is always above the temperature of the rest of the system. The bourdon, capillary, and part of the bulb, therefore, always will be liquid-filled. The bulb must be

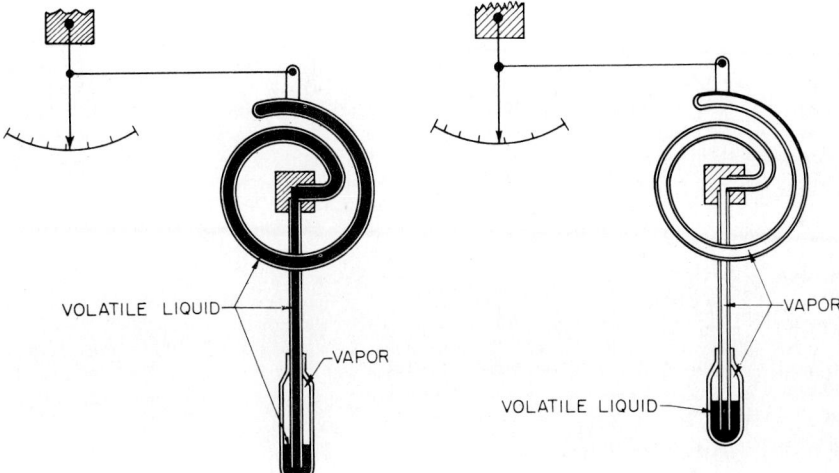

Fig. 7 Vapor-pressure thermal system (Class IIA).   **Fig. 8** Vapor-pressure thermal system (Class IIB).

large enough (1) to accommodate the volume change of the fill within the bourdon and capillary due to ambient temperature changes, (2) to accommodate the volume change with temperature of item 1 within the bulb, (3) to accommodate the volume change of the bourdon across the range, and (4) to provide some minimum liquid and vapor space within the bulb. A larger bulb, therefore, may be required for long systems ($\geqq$50 to 75 ft) ($\geqq$15 to 23 m).

The Class IIB system may have the smallest bulb, since the design does not need to accommodate the effect of items 1 and 3 just given. See Fig. 8.

The Class IIC system may operate above or below the ambient temperature of the case and capillary, as shown in Fig. 9. The bulb volume must accommodate, in addition to the requirements of the Class IIA system, the complete volume of the bourdon and capillary. This system will not give

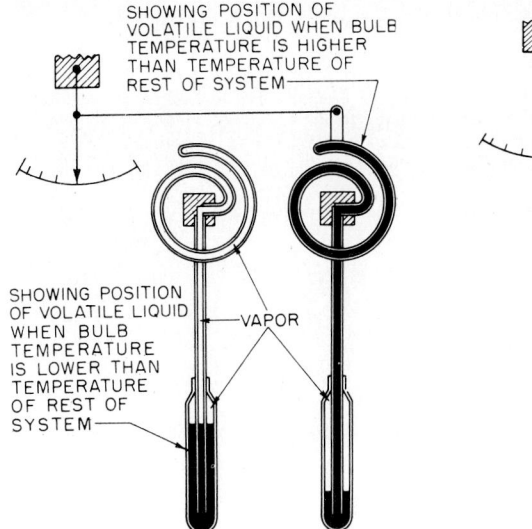

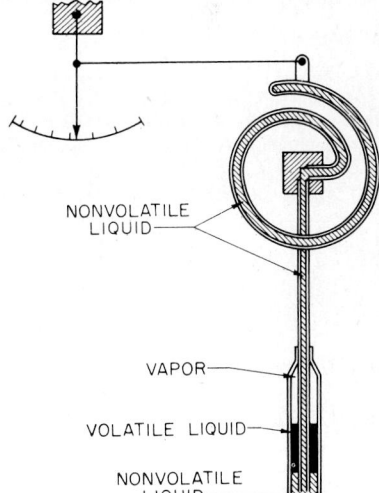

Fig. 9 Vapor-pressure thermal system (Class IIC).   **Fig. 10** Vapor-pressure thermal system (Class IID).

a reliable temperature indication during transfer of the liquid into or out of the capillary and bourdon, which occurs when the bulb temperature crosses the ambient temperature of the bourdon and capillary.

The Class IID system is specifically designed to measure temperature accurately at the ambient temperature of the bourdon or capillary as well as at other temperatures. The actuating volatile liquid is confined to the bulb by a transmitting liquid of low vapor pressure which fills the capillary and bourdon. The means of confinement or trap is generally designed to be large so that actuating liquid getting into the trap cannot enter the capillary under any conditions of use.

The bulb size required is generally somewhat larger than that of Class IIA, since it must accommodate this larger trap and additional volume of transmitting liquid. See Fig. 10.

The vapor-system indication is a nonlinear function of bulb temperature, the vapor pressure increasing at an increasing rate with rise in the absolute temperature. This is an advantage where the user desires more reading sensitivity toward the top of the range. At least one manufacturer incorporates a varying linkage ratio in order to linearize the pen or pointer motion.

It is desired that the minimum vapor-pressure change across the range be 100 psi to minimize barometric pressure effects to provide adequate bourdon energy to overcome frictional effects. Therefore, various actuating liquids are chosen for different range spans and different temperature levels.

Maximum and Minimum Temperatures

The maximum temperature is limited by the critical point of the liquid employed and by the tendency of most known organic liquids to change chemically at 316°C (600°F) or higher. The minimum temperature is generally limited to approximately -40°C (-40°F) because of loss in reading sensitivity at lower temperatures coupled with the requirement that the bourdon must be able to withstand the vapor pressure of the liquid fill at room (or possibly shipping) temperatures.

Capillary and Case Compensation

The capillary of Class II systems is insensitive to temperature changes. It is necessary, however, that the capillary temperature of Class IIA systems, when the capillary and bourdon are filled with actuating liquid, not be at a point that exceeds the bulb temperature. The internally contained fluid or vapor within the bourdon is also insensitive to temperature changes qualified as above. However, the modulus of elasticity will decrease with increasing temperature, causing a shift of only 1½ to 2% for a 56°C (100°F) case temperature change. In most uses, this effect is not important and no compensation is provided.

Speed of Response

The 63% response of Class II systems also varies from 1 to 10 s with the bulb in well-stirred water and a short capillary length. The characteristics affecting response include capillary length and conductivity, viscosity of fluid in the capillary, the size of the system relative to the available heat (especially with Class IIC where reversal occurs), and the heat transfer internally between the bulb and the filling liquid.

Class IID systems are generally slowest because of the bulb size and the viscosity of the fluid.

Barometric Errors

Class II systems are sensitive to barometric changes. They are generally designed to have a minimum of 100 psi (690 kPa) across the range so that errors will be kept to ¼% of the range.

Bulb Elevation Error

If the bourdon is above the bulb, the pressure within the bourdon equals the vapor pressure in the bulb minus the liquid pressure head along the capillary, or

$$E = \frac{100 P_h}{\Delta P} = 0.43 h \times \frac{100}{\Delta P} \qquad \text{(see Eq. 5)}$$

where E = elevation error, % of pressure span
  h = height of bourdon above bulb, ft
  P_h = liquid pressure head along capillary, psi

In the following example, assume:

1. Range is 80 to 200°F. See methyl ether curve of Fig. 11.
2. Case is installed above bulb by 50 ft (15 m).

Elevation error will be

$$E = 0.43 \times 50 \times 0.658 \times \frac{100}{329} = 4.3\% \text{ of pressure span}$$

This effect generally can be taken care of by the manufacturer if the user provides installation elevations for the bulb and case so that the instrument will read properly after installation. The Class IIC system will have liquid within the capillary only over part of the range span and is, therefore, not recommended when the bulb and case are at appreciably different elevations, since the instrument cannot read correctly for both situations, i.e., bulb temperature above or below capillary and case temperature ambients.

Overrange Protection

In vapor pressure thermal systems, overrange protection is generally more limited than in other systems because of the increasing rate of vapor-pressure rise with temperature rise. A specific overrange temperature for each range offered should be obtained from the manufacturer. This is generally appreciably less than 100% of the range span above the range top. Under some limited conditions, it is possible to fill the system so that all the liquid will be exhausted from the bulb at a bulb temperature above the instrument range, in which case the safe overrange temperature can be increased.

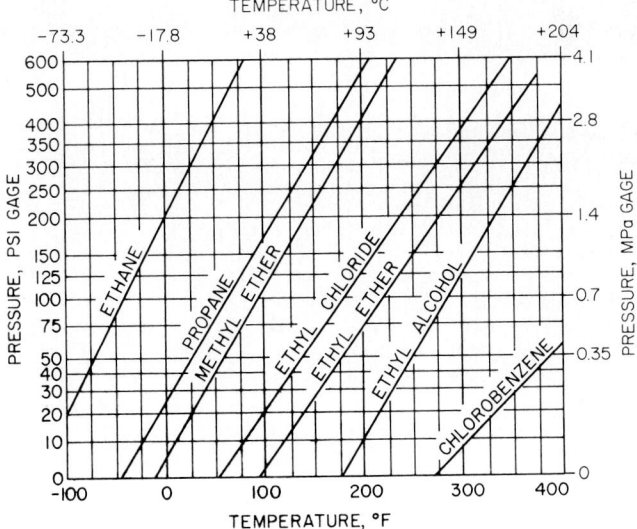

Fig. 11 Vapor pressure–temperature curves for various thermal system fills.

Gas-Filled Thermal System: Class III

This system is defined by SAMA as "a thermal system filled with a gas and operating on the principle of pressure change with temperature change." The system is usually compensated for ambient temperature effects in one of two ways:

1. **Class IIIA.** With a second thermal system minus the bulb, or an equivalent means of compensation. See Fig. 2.
2. **Class IIIB.** With compensating means within the case only. See Fig. 3.

General Design Considerations

The gas-filled system operates on the principle of Charles's law, which states that the absolute pressure of a confined gas is proportional to the absolute temperature. This law is approximated by keeping the capillary and bourdon volume small compared with the bulb volume, thereby preventing an appreciable percentage of the gas mass within the bulb from flowing out as the bulb temperature is increased across the range.

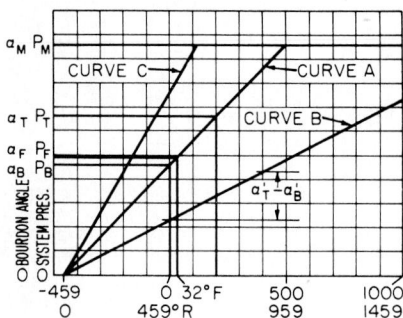

Figure 12 shows the system operation in which it is assumed that Charles's law is exactly followed. The manufacturer finds it expedient to keep the angle change across the range a_T-a_B (or a corresponding system pressure change across the range P_T-P_B) constant for all range spans. Curve A for the 0 to 200°F span is established by filling the system to a particular pressure P_F or angle a_F with the bulb at a particular temperature (say 32°F). The slope of curve B for a 0 to 400°F span will accordingly be one-half that of a 0 to 200°F span, i.e., $a_T'-a_B' = a_T-a_B$. The system may be expressed mathematically by Eq. 6:

Fig. 12   Function of a gas-filled thermal system.

$$
\begin{array}{cc}
\text{Condition 1} & \text{Condition 2} \\[4pt]
\dfrac{V_b P_1}{T_{b1}} + \dfrac{V_c P_1}{T_{c1}} + \dfrac{V_B P_1}{T_{B1}} = \dfrac{V_b P_2}{T_{b2}} + \dfrac{V_c P_2}{T_{c2}} + \dfrac{V_B P_2}{T_{B2}}
\end{array}
\tag{6}
$$

where V_b = volume of bulb, cm³
V_c = volume of capillary, cm³
V_B = volume of bourdon, cm³
T_{b1} = absolute temperature of bulb (condition 1)
T_{b2} = absolute temperature of bulb (condition 2)
T_{c1} = absolute temperature of capillary (condition 1)
T_{c2} = absolute temperature of capillary (condition 2)
T_{B1} = absolute temperature of bourdon (condition 1)
T_{B2} = absolute temperature of bourdon (condition 2)
P_1 = absolute system pressure (condition 1)
P_2 = absolute system pressure (condition 2)

System linearity and bulb and case temperature errors may be calculated from Eq. 1. This equation does not include the effect on modulus change of the bourdon. The theory previously described for Class II systems may be utilized to calculate this effect.

Maximum and Minimum Temperature and Range Spans

Gas thermal systems are able to cover the widest range of temperature of any of the filled systems. They are usually limited on the low side by the critical temperature of the gas used, and on the high side by the bulb materials (commonly 5 K and 925 K). The maximum span is limited only by the above conditions of use and the nonlinearity due to mass flow from the bulb. The minimum span is limited by the pressure at which the bourdon becomes overstressed.

The gas system lends itself to use with a transducer with biasing springs, making many more ranges, especially with short spans (25 K), available.

Capillary and Case Compensation

Equation 7, describing capillary temperature error in Class III systems, is derived from Eq. 6,

$$E = \frac{100 V_c \Delta T_c T_b^2}{(V_b T_c + V_c T_b) T_c R} \tag{7}$$

where T_c = mean capillary temperature, °R
T_b = absolute temperature of bulb, °R
ΔT_c = capillary temperature change, °F
R = span, °F

It can be observed from the equation that the capillary error increases rapidly as the bulb temperature increases and can be reduced by increasing the bulb volume and/or the span.

Speed of Response

The speed of response of Class III systems is generally fast because the ratio of bulb mass to external surface is more favorable than in other system types. This would be especially true in low conducting media. See Fig. 8. The 63% response time is generally between 1 and 4 s in well-stirred water.

Barometric Errors

Class III systems are sensitive to barometric changes. They are generally designed to have a 400-psi (2.8 MPa) change across the range, and so errors are usually negligible. Special compensation is usually required when they are used with transducers.

Bulb Elevation Error

Since the density of the gas used is very small, the bulb elevation errors are negligible.

Overrange Protection

The maximum temperature for all gas thermometers is that which produces the maximum allowable pressure in the bourdon or the maximum permissible bulb temperature, whichever is lower. See Fig. 12. It is noted that short spans reach this limiting pressure at a substantially lower temperature.

Averaging Thermal Systems

Class III systems lend themselves uniquely to applications requiring the average temperature of a large enclosure such as an oven or drying chamber. This is so primarily because bulb volume is not critical and capillary-type bulbs that can be easily installed will provide adequate volume and minimum sensitivity to bending or uncoiling.

Comparative Performance of Thermal Systems

Table 1 is intended as an approximate guide for comparison of the different classes of filled-system thermometers. The values apply principally to bourdon-type systems actuating an indicating pointer or pen. When they are used with transducers which provide biasing spring elements, many of the performance figures such as span, bulb size, response, and overrange may be changed for greater ease of use.

Table 1. Comparison of Thermal Systems*

Operating principle	Liquid expansion		Pressure change	
	Liquid-filled	Mercury-filled	Vapor pressure	Gas-filled
Type	Liquid-filled	Mercury-filled	Vapor pressure	Gas-filled
Class	Class IA and IB	Class VA and VB	Class IIA, B, C, D	Class IIIA and IIIB
Low temperature limit	−125°F (−87°C)	−38°F (−39°C)	−40°F (−40°C)	−450°F (−268°C)
High temperature limit	600°F (316°C)	1200°F (650°C)	600°F (316°C)	1200°F (649°C)
Longest span	350°F (194°C)	1200°F (650°C)	300°F (167°C)	1200°F (667°C)
Shortest span	25°F (14°C)	40°F (22°C)	80°F (44°C)	150°F (83°C)
Typical bulb size, diameter and length, in	⅜ × 5	1⁄16 × 4	11⁄16 × 4	1⁄16 × 4
Dial or chart divisions	Equal	Equal	Nonlinear, larger at range top	Equal
Maximum standard capillary length, ft	200, Class IA 15, Class IB	200, Class VA 50, Class VB	200	200
Capillary temperature compensation	Dual bourdon, Class IA None, Class IB	Compensated capillary, Class VA None, Class VB	None necessary	Generally none
Case temperature compensation	Dual bourdon, Class IA Bimetal strip, Class IB	Bimetal strip or core-type compensator	Usually none	Bimetal strip
Bulb elevation error	Negligible	Generally small	Often large	Negligible
Overrange capability in % of range	0–200%, Class IA 100%, Class IB	100% minimum	Generally small	0–300%, varies with span
Barometric errors	Negligible	Negligible	Usually small	Usually very small
Speed of response	....	See Fig. 5	....	....

*Above characteristics are to be used only as a guide. Dimensions and functional values will vary with manufacturer.

Radiation Thermometry

by
Ray Peacock*

Radiation thermometry is a practical application of the Planck law and Planck radiation formula†
and makes it possible to measure the temperature of an object without making physical contact with
the object. Many contemporary and special laboratory instruments are, in essence, sensors that
respond to a limited wavelength region of the thermal radiation spectrum, converting received radiant
energy into an electrical signal. Instrument calibration is based on the amount of radiation emitted
by a reference blackbody source, which is the ideal or perfect emitter of radiation that reliably follows
Planck's law. Interpretation of received radiation in terms of temperature is relatively simple, especially in laboratory measurements, but in industrial practice a number of factors can make the interpretation difficult.

Advantages of the Method

In radiation thermometry, the sensor does not have to be in thermal equilibrium with the object.
Thus, very high temperatures can be measured. Planck's radiation law is the basis of the International Practical Temperature Scale of 1968 (IPTS-68) at temperatures above the gold point
(1064.43°C). Realization of the temperature scale above the gold point with carefully designed
metrological instruments, while virtually without an upper limit, is possible to within a precision of
$\pm 0.01°C$ or better. But, radiation thermometry is not limited to high temperatures. Modern instruments are commercially available that can measure well below $-18°C$ (0°F). Typical industrial
precision lies in the range of ± 0.5 to 1% of absolute temperature.

The advantages and limitations of radiation thermometry are summarized in Table 1. Because
of the wide selection of instruments offered, determining the instrument best suited for a given application can be difficult. Two of the major criteria are (1) wavelength response and (2) target size.
Speed of response also may be a primary factor in selection. Factors which also contribute to the
difficulty of selection include lack of standards and precise terminology. Critical parameters of
waveband, target size, or field of view and calibration uncertainty are not always stated explicitly in
the commercial literature.

Perspective

Early radiation thermometers, called radiation and optical pyrometers, were radically different from
one another in both design and use. The simple radiation pyrometer was compact, designed for fixed
installation, and served only as a transducer. In contrast, the portable optical pyrometer was a com-

*Vice President of Engineering, Land Instruments, Inc., Tullytown, Pennsylvania.

†Planck's law of thermal radiation predicts very accurately the radiant power emitted per unit area per unit
wavelength by a blackbody, or complete radiator. Mathematically it can be written as

$$M^b(\lambda, T) = \frac{C_1}{\lambda^5} \frac{1}{e^{C_2/\lambda T} - 1} \; W \cdot m^{-3}$$

where $C_1 = 2\pi h C^2 = 3.7415 \times 10^{16} \; W \cdot m^2$ is called the first radiation constant
$C_2 = Ch/k = 1.43879 \times 10^{-2} \; m \cdot k$ is the second radiation constant

This radiation formula can be written in other forms, such as using wavenumbers instead of wavelengths. In this
article, the form above expresses the radiant exitance in terms of wavelength and absolute temperature. The units
used are SI, and the nomenclature is that recommended by the Optical Society of America.

Table 1. Relative Advantages and Limitations of Radiation Thermometers

Advantages

- Can measure:
    Very high temperatures
    Moving objects
    Large areas
    Inside vacuum or pressure vessels
    Inside semitransparent objects
- Does not contact (hence mar) object of measurement
- Instrument not physically exposed to temperature it measures (as are devices which require physical contact)
- Rapid response
- High differential sensitivity

Limitations and Disadvantages

- Relatively high cost
    Initial
    Installation
    Required maintenance
- Application engineering required to solve some problems
- No uniform calibration tables

plete measuring system and a much more sophisticated instrument. Developments in integrated circuits, transducer or detector devices, and optical technology have had a profound impact on both fixed and portable instrument design.

TYPES OF RADIATION THERMOMETERS

A convenient classification of commercial radiation thermometers is:

1. Wideband instruments
2. Narrowband instruments
3. Ratio (two-color) thermometers
4. Optical pyrometers
5. Fiber-optic instruments

Both portable and fixed-installation instruments are available in each class. See Table 2.

For those readers who may not be fully familiar with the theoretical background of radiation thermometry, this subject is covered later in this article.

Wideband Instruments

Wideband radiation thermometers are the simplest and are of the lowest cost. Available devices respond to radiation with wavelengths from 0.3 μm to between 2.5 and 20 μm, depending on the lens or window material used. These instruments also have been called *broadband* or *total radiation pyrometers* because of their relatively wide wavelength response and the fact that they measure a significant fraction of the total radiation emitted by the object of measurement. Historically, these

Table 2. Principal Types of Commercial Radiation Thermometers

Type	Temperature range, °C	Relative cost	Type of use
Wideband			
Fixed	0–4000	Low	General
Portable	0–2000	Low–medium	General
Narrowband			
Fixed	−50–2500	Medium–high	Special
Portable	0–2500	Low–medium	Special
Ratio (two-color)			
Fixed	1000–2500	High–very high	Special
Portable	700–2200	High	Special
Optical			
Fixed	800–2500	Very high	Special
Portable	800–2500	Medium	General
Fiber optic			
Fixed	100–2500	Medium–high	Special
Portable	250–800	High	Special

devices were the earliest fixed or automatic units. They still find use and continue to be produced. See Table 3.

Narrowband Instruments

A narrowband radiation thermometer usually has a carefully selected, relatively narrow wavelength response, often selected to meet a special application requirement. The detector, lens, window, and filter(s) are selected to provide the particular wavelength response desired. Optical pyrometers can be considered a subset of this class. The development of many different narrowband thermometers in recent years has been prompted by application needs matched by improved designs—to overcome the limitations of the earlier wideband models and to provide more accurate temperature measurement in numerous different applications. See Tables 4 to 6. Also see Figs. 1 to 3.

Ratio Thermometers

A ratio thermometer is essentially two radiation thermometers contained within a single housing. Several internal components, such as the lens and detector, may be shared. The unique characteristic of the ratio thermometer is that the output from the two thermometers, each having a separate wavelength response, is ratioed. See Table 7.

Table 3. Characteristics of Some General Purpose Wideband Radiation Thermometers

Manufacturer*	Model	Temperature range limits, °C	Waveband limits, μm	Approximate N-value range
Honeywell	RH	500–1800	0.4–2.6	9.5–4.0
Land	ORO	600–1900	0.4–2.6	8.8–4.0
Mikron	M65A	0–1000	7–20	5.3–1.5
Leeds & Northrup	18890	825–1800	0.4–2.6	6.5–4.0

*Manufacturers and locations: Honeywell Inc., Fort Washington, Pa.; Land Instruments, Inc. Tullytown, Pa.; Mikron Instrument Co., Inc., Ridgewood, N. J.; Leeds & Northrup Co., North Wales, Pa.

Table 4. Characteristics of Some General Purpose Narrowband Radiation Thermometers

Manufacturer*	Model	Temperature range limits, °C	Mean effective wavelength, μm	Approximate N-value range
Barnes	0710	600–3000	0.9	18–5
	2026	100–1500	2.3	17–4
	8014	−40–300	11.0	5–2
Ircon	2000	500–3000	0.9	21–5
	6000	80–1500	2.2	19–4
	4000	0–500	11.0	5–2
Land	QQ	1100–1700	0.6	17–12
	QO	600–2500	0.9	18–6
	CGA	−50–600	11.0	5–2
Mikron	M65H	800–1700	0.9	15–8
	M65B	0–1000	11.0	5–1
Raytek	S/L 400	800–1700	0.9	15–8
	S/L 300	250–1500	2.2	13–4
	LC-814	0–1000	11.0	5–1
Vanzetti	TM	250–1000	1.9	15–6
Williamson	4100	500–2000	1.0	19–11

*Manufacturers and locations: Barnes Engineering Co., Stamford, Conn.; Ircon, Inc., Skokie, Ill.; Land Instruments, Inc., Tullytown, Pa.; Micron Instrument Co., Ridgewood, N.J.; Raytek, Inc., Mountain View, Calif.; Vanzetti Systems, Inc., Stoughton, Mass.; Williamson Corp., Concord, Mass.

The concept behind the ratio thermometer is that the ratio signal is also a function of temperature, and so long as the ratio value is unchanged, the temperature measurement is accurate. Since the ratio is measured, the target size is relatively unimportant because the ratio of signal from a small target is the same as that from a large target.

Emissivity is one source of attenuation of radiation, and it is argued that, if the spectral emissivity in one waveband changes the same amount as in the second band, the ratio thermometer will be unaffected. This is a reasonable argument for some materials, but not for oxidizing metals, since emissivity as a rule is not a strong function of an object's temperature but is mostly affected by the material's composition, phase, and surface roughness. Oxidizing metals, however, change emissivity rapidly and quite differently at different temperatures.

Table 5. Applications and Advantages of Some Narrowband Radiation Thermometers

Waveband, μm	Applications and advantages
0.65	Accurate measurements of objects in the open at high temperatures
0.4–1.1	Ability to sight through nonluminous products of fossil fuel combustion; metals
0.9	in the open
3.43 ± 0.2	Thin-film plastics of the polyethylene type
3.6 ± 0.2	Subsurface glass temperatures
3.9 ± 0.15	Sighting through nonluminous combustion-produced gases
4.6 ± 0.1	Measuring "flame temperature," an absorption-emission band in hot carbon dioxide
5.0 ± 0.2	Silica, glass surface temperature
7.9 ± 0.1	Thin-film plastics of the polyester type
8–14	Atmospheric window for "low" temperature

Table 6. Characteristics of Some Special Purpose, Narrow Waveband Radiation Thermometers

Manufacturer*	Model	Temperature range limits, °C	Waveband, μm†	Range of N-values	Use
Barnes	3335	50–400	3.43 ± 0.1	13–6	Plastics
	4854	50–1700	5.1 ± 0.3	9–1	Glass
	7781	30–300	7.9 ± 0.2	6–3	Plastics
Ircon	3400	0–800	3.43 ± 0.07	15–4	Plastics
	7000	50–1300	5.0 ± 0.2	9–2	Glass
	8000	20–400	7.9 ± 0.15	6–3	Plastics
	G	250–1100	1.64 ± 0.08	16–8	Metals
	Q	800–2400	4.5 ± 0.05	9–2	Flames
Land	GGV	200–2500	5.0 ± 0.2	6–I	Glass, silica
	CGR	400–1600	3.9 ± 0.13	10–3	Furnaces
	GM	300–900	1.3 ± 0.5	16–8	Metals
	VH	200–900	1.7 ± 0.5	16–7	Metals
	VX	350–1100	1.2 ± 0.2	18–9	Metals
Micron	M65E	200–1100	6 ± 1.0	6–3	Glass
Williamson	4200	25–1000	3.05 ± 0.25	16–4	Low temp.
	4300	40–800	3.43 ± 0.05	13–4	Plastics
	4500	40–1370	5.1 ± 0.1	9–2	Glass
	4400	300–1650	NS	—	Petrochemical
	4800	50–375	7.95 ± 0.10	6–3	Plastics

*Manufacturers and locations: Barnes Engineering Co., Stamford, Conn.; Ircon, Inc., Skokie, Ill.; Land Instruments, Inc., Tullytown, Pa.; Micron Instrument Co., Ridgewood, N.J.; Williamson Corp., Concord, Mass.
†NS, Not stated.

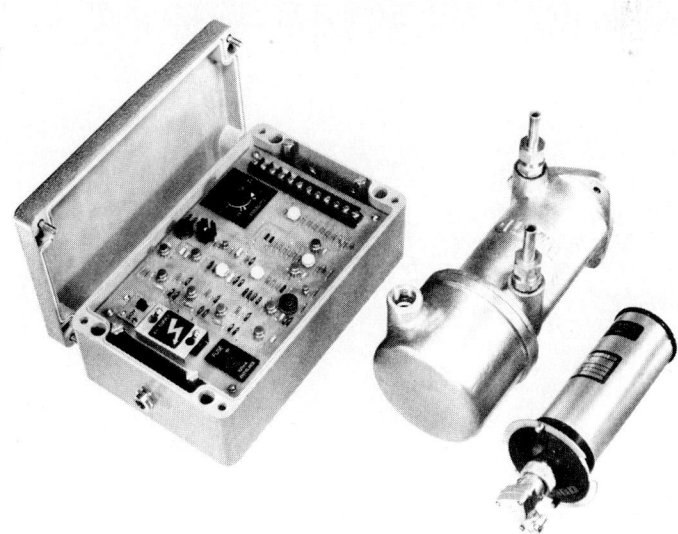

Fig. 1 Industrial thermometer system. *(Land Pyrometers Ltd. Unicard System 2.)*

Fig. 2   Industrial thermometer system. *(Williamson Corp. System 4000.)*

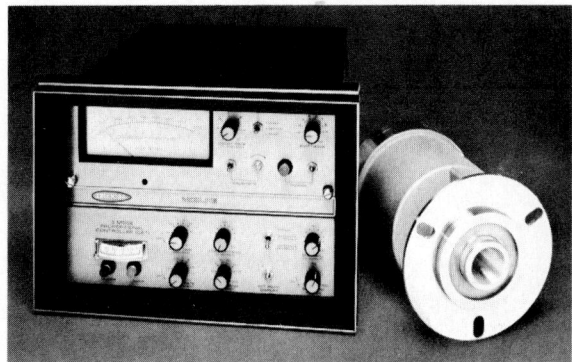

Fig. 3   Industrial thermometer system. *(Ircon Corp. Modline.)*

Table 7. Characteristics of Some Ratio (Two-Color) Radiation Thermometers

Manufacturer*	Model	Temperature range limits, °C	Waveband centers, μm	Equivalent wavelength, μm
Capintec	Red Eye	175–1250	1.65 and 2.2	6.6
	Ratio Scope	750–1750	0.81 and 0.45	5.5
	Thermoscope	1000–2000	0.55 and 0.70	2.6
Ircon	R	700–3500	0.95 and 1.05	10.0
Land	NQY	800–1900	0.75 and 0.88	5.1
	NQN	1200–2500	0.64 and 0.88	2.3
Williamson	8100	800–2200	0.71 and 0.81	5.8

*Manufacturers and locations: Capintec Instruments, Inc., Montvale, N.J.; Ircon, Inc., Skokie, Ill.; Land Instruments, Inc., Tullytown, Pa.; Williamson Corp., Concord, Mass.

Table 8. Characteristics of Some Optical Pyrometers

Manufacturer*	Model	Temperature range limits, °C	Waveband, μm	Notes
Leeds & Northrup	8630	775–2800	0.653	Two-piece unit
	8627	775–1750	0.65 ± 0.04	Rechargeable
Pyro	Optical	775–2800	0.65	One piece
	Microptical	700–5000	0.65	Many options

*Manufacturers and locations: Leeds & Northrup Co., North Wales, Pa.; Pyrometer Instrument Co., North-vale, N.J.

Optical Pyrometers

These instruments utilize a unique method of measurement; i.e., a photometric match is made between the brightness of the object and an internal lamp. Optical pyrometers are sensitive only in a very narrow wavelength range. The most popular instruments are manually operated; i.e., the operator performs the photometric match visually.

The manual optical pyrometer or visual optical pyrometer is the earliest and most respected portable radiation thermometer system available. It enjoys a reputation, unique among radiation

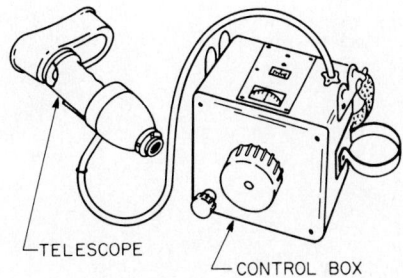

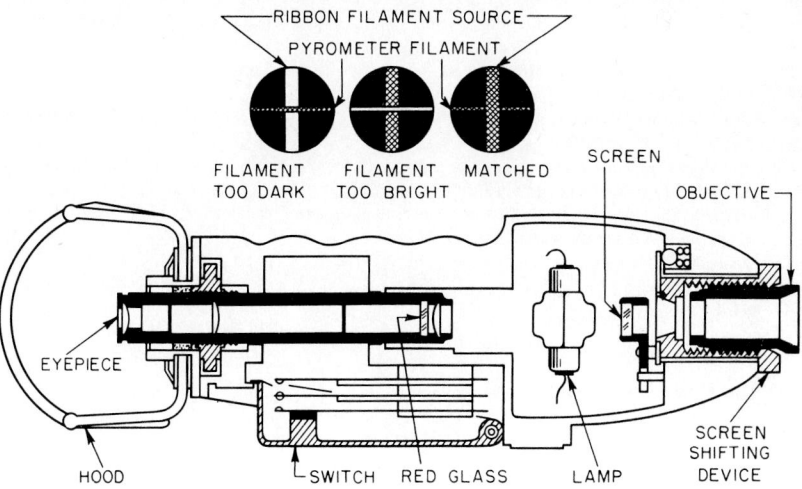

Fig. 4 Optical pyrometer telescope. *(Leeds & Northrup.)*

thermometers, for outstanding accuracy. The instrument occupies a special historical place in radiation thermometry because it was the first instrument widely accepted in both research and manufacturing and it demonstrated that excellent temperature measuring performance is possible with properly designed and maintained radiation thermometers. See Table 8.

Optical pyrometers differ from other radiation thermometers in both the type of reference source used and the method of achieving the brightness match between the object and reference. Figure 4 is a schematic view of a typical visual optical pyrometer and the indicators for over, under, and matching conditions as seen by the operator. The reference source used is an aged and calibrated tungsten strip lamp. In use, the operator views the object to be measured and the lamp filament simultaneously through an optical filter. The combination of filter characteristics and the response of the average human eye produces a net instrument wavelength response band that is very narrow and centered near 0.65 μm. By adjusting the current through the lamp or varying the intensity of the object radiation, the operator can produce a brightness match over at least a portion of the lamp filament, according to relative target size. Under matched conditions, the two "scenes" merge into one another, or the filament apparently vanishes.

Automatic optical pyrometers function in much the same manner, in the same waveband, but utilize a photomultiplier tube or other detector element in place of the human eye. An electronic circuit with feedback drives the system to a null point, thus determining the temperature. The optical field of view of an automatic unit is specified by the manufacturer, while that of the manual instrument is limited to the optical resolution of the eye, augmented by the telescope and optical lenses furnished with the instrument.

The temperature range of optical pyrometers is limited at the lower end by the need for an incandescent image of the filament to about 800°C. The upper temperatures are limited only by applicational needs. Temperature indications on manual units are analog scales or meters.

Fiber-Optic Thermometers

Fiber optics enable near-infrared and visible radiation to be transmitted around corners and away from hot, hazardous environments to locations more suitable for the electronics associated with modern radiation thermometers. Fiber optics also make possible measurements in regions where access is restricted to larger instruments and where large electric or radio frequency fields would seriously affect an ordinary sensor. Fiber-optic transmission devices have helped to solve a number of application problems, such as hot turbine blade temperature measurement in gas turbine engines and the temperature of hot metal inside induction heating coils or inside vacuum vessels.

Conceptually, a fiber-optic system differs from an ordinary thermometer system by the addition of a fiber-optic light guide, with or without a lens. The optics of the light guide define the field of view of the instrument, while the optical transmission properties of the fiber-optic elements form an integral part of the thermometer spectral response.

Present fiber optics transmit well to wavelengths as long as about 2.0 μm and thus the thermometers are limited to measuring temperatures upward from about 200°F (93°C). Fiber optics must be maintained in a clean condition just as an ordinary thermometer lens. It is important to observe manufacturers' recommendations in this regard.

Contemporary fiber-optic thermometers are shown in Figs. 5 and 6.

Portable Radiation Thermometers

A wide variety of portable single-waveband instruments are available, ranging from instruments of poor quality to some very sophisticated models that compare favorably with the optical pyrometer in terms of accuracy. The properties of some of these instruments are given in Table 9. See also Figs. 7 and 8.

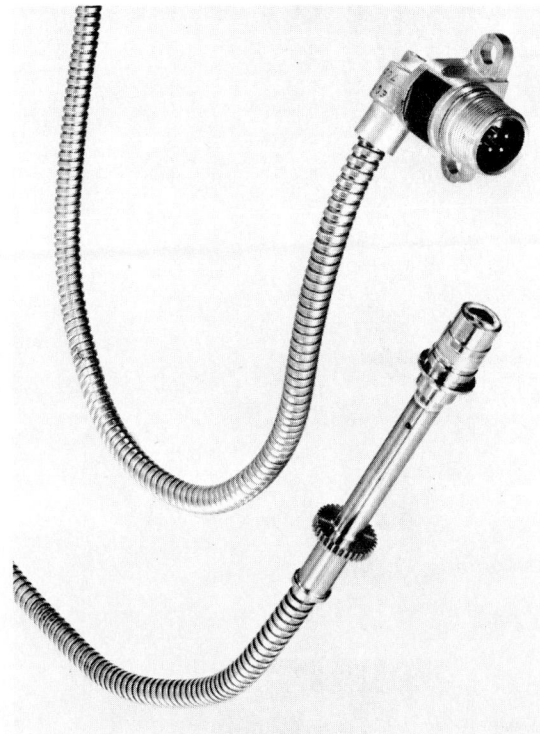

Fig. 5 Rugged fiber-optic thermometer for use in airborne gas turbine engines. *(Land Pyrometers Ltd.)*

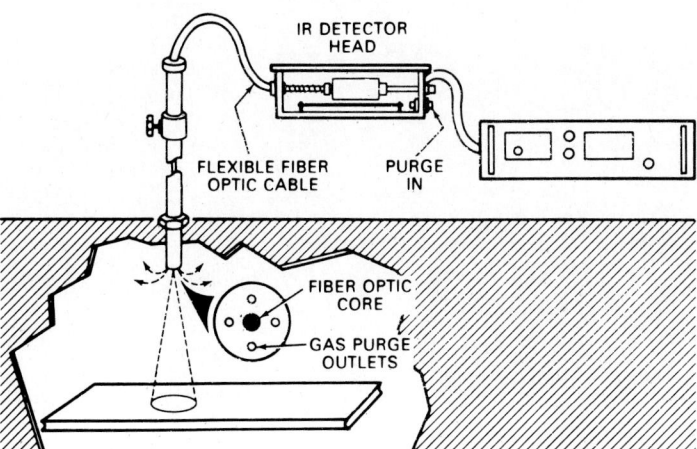

IR DETECTOR
HEAD

FLEXIBLE FIBER
OPTIC CABLE

PURGE
IN

FIBER OPTIC
CORE

GAS PURGE
OUTLETS

Fig. 6 Use of a fiber-optic system to allow measurement through a pressurized boundary. *(Vanzetti Infrared.)*

Table 9. Characteristics of Some Portable Radiation Thermometers

Manufacturer*	Model	Waveband, μm	Temperature span limits, °C	Range of N-values	Optics distance/target	
Land	Cyclops 33	8–14	−50–1000	4.8–1.0	30:1	General purpose-low temp.
	Cyclops 39	3.9 ± 0.13	600–1400	4.2–2.2	105:1	Furnace temperatures
	Cyclops 51	0.9 ± 0.2	600–3000	17.7–3.0	165:1	General purpose-high temp.
Mikron	5	1–20	40–1700	11.5–1.8	30:1	General purpose
	57	0.5–1.1	750–1750	15.6–7.9	75:1	General purpose-high temp.
	80B	8–14	0–315	4.8–2.2	30:1	General purpose-low temp.
Raytek	Ranger 2LT	8–14	−30–1100	5.2–0.8	30:1	General purpose
	Ranger 2HT	2.2 ± 0.1	400–3000	9.7–2.0	60:1	General purpose
Wahl	Heat Spy HSA-8	1.8–40	0–320	13.2–6.1	10:1	Small target-low temp
	Heat Spy HSA-7	1.8–4.0	600–1100	5.5–2.2	75:1	General purpose-high temp.
	Heat Spy DHS-14	7–20	0–500	5.3–2.7	20:1	General purpose-low temp.
Williamson	2214	2.8–3.3	50–1400	14.6–2.8	80:1	General purpose
	2310	3.43 ± 0.05	50–375	13.0–6.5	24:1	Plastics
	2410	Not specified	425–1650	—	144:1	Furnace temperatures

*Manufacturers and locations: Land Instruments, Inc., Tullytown, Pa.; Mikron Instrument Co, Inc., Ridgewood, N.J.; Raytek, Inc., Mountain View, Calif.; Wahl Instruments, Inc., Culver City, Calif.; Williamson Corp., Concord, Mass.

Fig. 7 Modern portable thermometer. Specifications are given in Table 9. *(Land Instruments.)*

Fig. 8 Modern portable thermometer. Specifications are given in Table 9. *(Williamson Corp.)*

ELEMENTS OF RADIATION THERMOMETER SYSTEMS

Optical System

The function of the thermometer optical system is to collect thermal radiation, in a known wavelength region, from a known external optical path and deliver it to the detector. There are several optical designs used in commercial instruments. See Figs. 9 and 10. Note that some devices are essentially dc instruments which are calibrated to absolute levels of output. Other designs make use of ac techniques of amplification, requiring some form of chopping means to produce an ac modu-

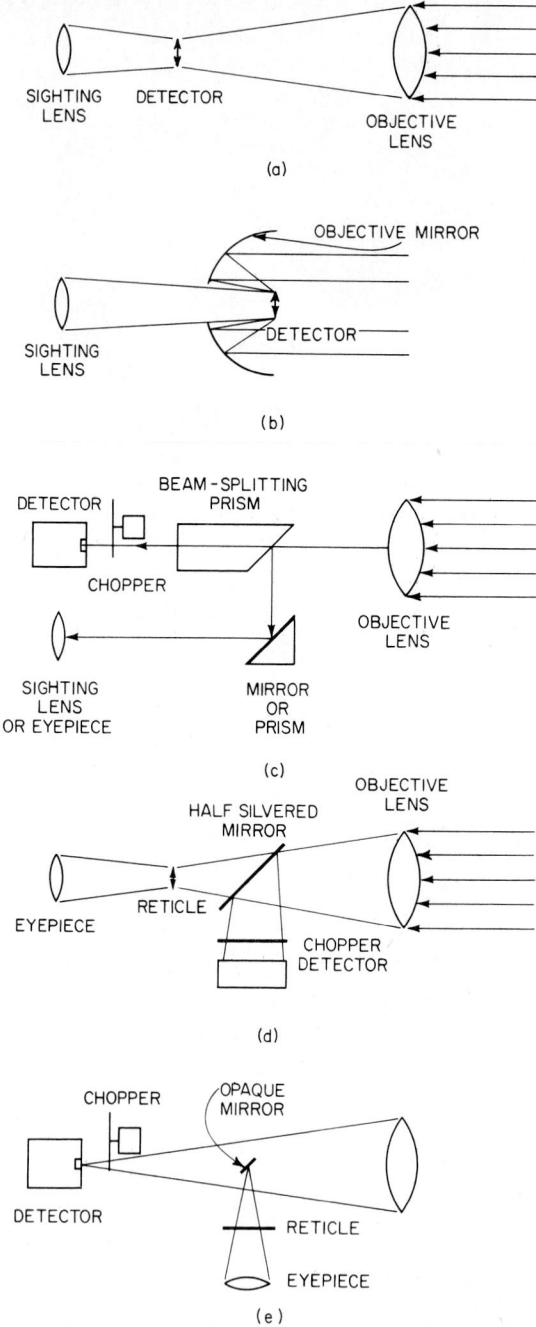

Fig. 9 Typical optical systems in industrial radiation thermometers: (*a*) fixed-focus lens optics without chopping; (*b*) fixed-focus mirror optics without chopping; (*c*) optics used where detector or modulator is too large to sight around; (*d*) variation of (*c*) with half-silvered mirror that transmits visible radiation and reflects infrared radiation; (*e*) opaque mirror in main radiation path which provides energy for sighting.

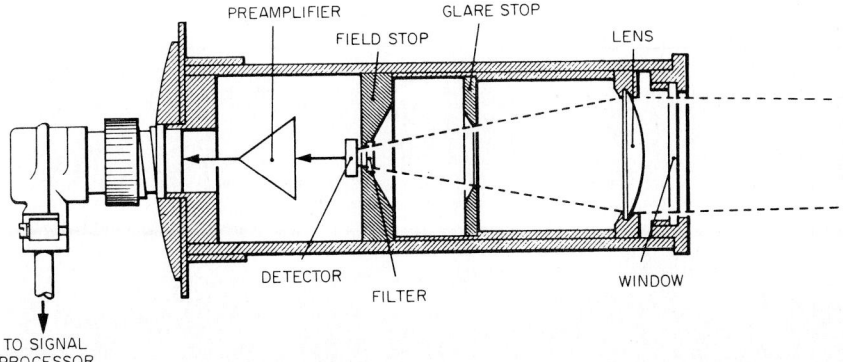

PREAMPLIFIER GLARE STOP LENS
FIELD STOP
DETECTOR WINDOW
FILTER
TO SIGNAL
PROCESSOR

Fig. 10 Cutaway view of typical industrial radiation thermometer. Optical path is shown by dashed lines.

lation on the incoming radiation. In these designs, the chopping device can also act as a local reference temperature source. Still other designs provide a built-in calibration or reference source; the detector acts as a comparator between the reference and the unknown. Thus variations in the detector with time or temperature do not affect accuracy.

Some instruments provide an ancillary optical system within the instrument to aid in aligning the instrument with the object of measurement. While sighting is valuable in some cases, it must be carefully centered with respect to the actual optic axis of measurement in order to be reliable. Often, even in systems with built-in sighting, the user must make final optimizing adjustments to make certain of the alignment.

In all thermometer designs with windows or lenses, the materials used for these components are an integral part of the wavelength determining portion of the instrument. If windows or lenses need replacement or change, the material and any coating specification should be carefully checked. Sometimes the best practice is to return the unit to the manufacturer for repair and a calibration check. Some of the infrared window materials are coated or are softer than ordinary glass optics, and thus particular care is required in cleaning to avoid changing the calibration.

Detectors

The transducer elements used to detect and convert thermal radiation into an electrical signal can be grouped into one of two classes: (1) *thermal detectors,* which produce an output because they absorb radiation and undergo slight temperature changes relative to radiation input variations; or (2) *photon detectors,* which produce an output because the received radiation releases electric charges in the body of the detector. The long wavelength limit of the latter detectors is confined to the photon energy which is just sufficient to release these charges.

The most common thermal detectors are thermopiles, metallic and thermistor bolometers, and pyroelectric devices. All three types are characterized by having a flat response versus wavelengths and having relatively lower sensitivity than photon detectors. Thermopiles and bolometers are limited in response speed by their thermal (heat) capacity. The pyroelectric detector is unique in that it does not have to attain a uniform temperature to respond. It is a rate-sensitive detector, and the radiation reaching it must be chopped in order to be measured. This type of detector requires considerable electronics to be useful. Its introduction as a radiation thermometer detector is relatively recent.

Photon detectors are usually semiconducting devices which can be grouped into two classes: (1) *photoconductive,* which change resistance as a function of temperature; and (2) *photovoltaic,* which produce an output as a function of absorbed radiation. Photoemissive devices, such as photomultiplier tubes, were used in earlier instruments but are no longer a significant factor.

The most common photoconductor cells are lead sulfide and lead selenide. These cells are sensitive in the 1 to 3 μm wavelength range, as shown in Fig. 11. They are used either in special narrowband thermometer detectors with filters, or in wideband, medium-temperature measurement requirements in the 200 to 800°F (93 to 427°C) region.

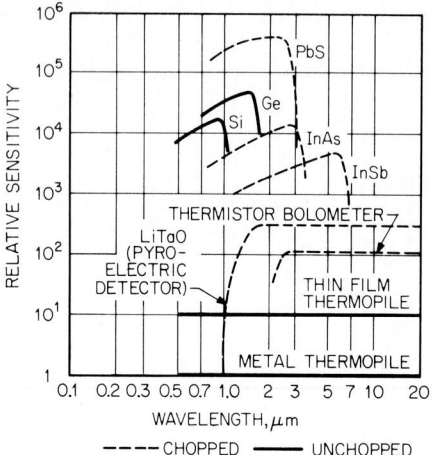

Fig. 11   Relative sensitivity versus wavelength for some common detectors.

The photovoltaic cells most often used are silicon, germanium, and indium antimonide. The wavelength response of these cells is also shown in Fig. 11. Silicon is the most widely used because its wavelength properties match the short wavelength thermal emission of objects at relatively high temperatures. The silicon cell finds use in thermometers covering temperature ranges from 750 to 7000°F (400 to 3870°C). Germanium cells are ideal for medium-temperature measurements as an adjunct to or a replacement for lead sulfide cells. Germanium is inherently more stable, reproducible, and faster in response than lead sulfide.

Indium antimonide is useful for midinfrared measurement—in special applications, such as hot glass surface temperature measurements. However, it has poor sensitivity at ambient temperatures much higher than normal room temperature.

Pyroelectric detectors are finding increasing use as replacements for indium antimonide and other less-sensitive thermal detectors in a number of new instrument designs, including those for the important 8 to 14 μm band.

SIGNAL CONDITIONING

One of the major advancements in radiation thermometer design since the late 1960s has been the widespread emergence of auxiliary signal conditioning. These new techniques have simplified the problem involving the interface between the measuring instrument and indicators, recorders, and controllers and have permitted closed loop control in situations previously considered impractical.

The principal signal conditioning features available are (1) linearization, (2) precise emissivity correction, (3) time modification of the raw signal, and (4) interfacing and control. The ready availability of low-cost, small, precision integrated circuits has largely been responsible for these advancements. Also, as of the early 1980s some low-cost microprocessors have appeared in radiation thermometer systems, and this trend will increase.

Linearization, Interfacing, and Display

Nearly all manufacturers of continuous radiation thermometers and most manufacturers of portable devices provide instruments with linear outputs. Nonlinear scales and charts are no longer required. Instrument outputs now follow acceptable industry standards, including 0 to 1 V dc, 0 to 20 mA, 4 to 20 mA, and 1 mV/degree plus offset. Multiple buffered outputs are available either as standard or optional features for interfacing with computers, digital panel meters, and recorders. Digital outputs are also available in most instruments. Most systems include integral temperature display directly in digital or analog form.

Precision Emissivity Correction

Compared with earlier radiation thermometers in which an uncalibrated knob was used to effect emissivity correction, contemporary instruments utilize precision amplifiers and this, coupled with a more knowledgable approach to applications, results in a precise emissivity correction setting—thus providing an accurate temperature measurement, usually of ±1% or better under many conditions. Typically, modern instruments include either precision analog or digital emissivity correction.

Time Modification

Four major time modification actions are used to condition the raw signal to provide a smoothed output for display and control. The main functions used are sample and hold, peak and valley picking, and averaging.

Sample and Hold

Triggered by external timing or logic signals, the sample-and-hold function enables the system to pick a specific window in time with respect to some datum to make a measurement. The system output thus disregards all other signals, and the output is a series of constant levels of temperature. Figure 12a shows the recovery of an important temperature measurement surrounded by extraneous signals as the result of using a sample-and-hold function.

Peak Picking

In many measurements, the desired temperature is the hottest temperature that appears to the thermometer. The object may be intermittent for some reason, such as obscuration by smoke. The peak picker is an amplifier with a fast rise time and an adjustable, slower decay time. Modification of the raw, noisy signal to produce a smoother signal by using the peak picking method is shown in Fig. 12b.

Valley Picking

This is the inverse of peak picking just described. This method is often used where the inside of a furnace environment at the lowest temperature seen is the desired criterion for recording and control.

Averaging

In attempting to control or monitor the average temperature of a series of objects, or of an object whose apparent temperature is varying rapidly, the simplest method is to slow down the response of

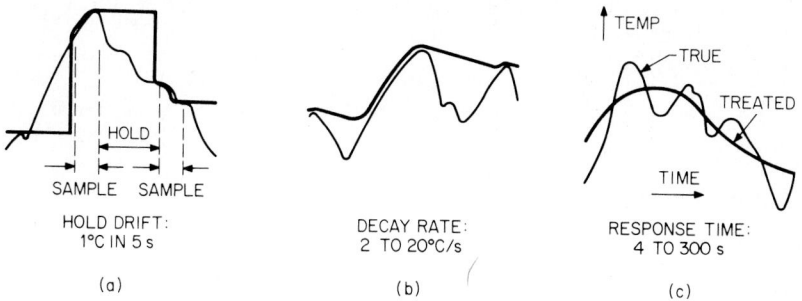

Fig. 12 Three signal conditioning time functions, showing typical specifications: (a) sample and hold; (b) peak picker; and (c) averager.

the thermometer to measure the average. An averaging function is a unity gain amplifier with a variable response. The results of applying an average to a rapidly fluctuating signal are shown in Fig. 12c.

Other Signal Conditioning Functions

Among other useful signal conditioning functions are the following.

1. **Hot Object Detection.** Indicates that the thermometer has received a signal well in excess of its minimum for a certain period of time (adjustable on-delay) and has then been absent for a minimum period of time (adjustable off-delay). Useful in activating and resetting a peak picker or sample-and-hold function for counting the number of objects passing, or for activating other logic functions.
2. **Delta T.** Calculates the difference between the maximum and minimum temperatures measured in a set of readings. Used in portable instruments for heat loss calculations.
3. **Maximum-Minimum.** Records the maximum or the minimum temperature in a series of readings.

Alarms and Controls

Some radiation thermometer manufacturers produce complete control systems. There is considerable variation in type and number of control features offered. Some may only offer an alarm feature, but a nonlatching alarm is in essence a simple on-off control function. More complex process control capabilities are offered in some packaged systems, including time proportional control and full three-term proportional control.

PRINCIPAL RADIATION THERMOMETER PROPERTIES AND CHARACTERISTICS

Irrespective of different design configurations and features, all radiation thermometers must meet certain basic conditions in order to provide repeatable and accurate readings: (1) The optical field of view must be filled; (2) the response time must be adequate; and (3) the instrument must remain stable during measurement. To fully appreciate the importance of these parameters and how they affect a measurement, particularly the optical field of view, a knowledge of two thermometer parameters is required: (1) effective wavelength and (2) N-value.

NOTE: Prior to a discussion of these properties, a short section on radiation thermometry theory and symbology is included for those readers who are interested in a refresher review of basic concepts.

Theoretical Considerations

Symbols, units, and a few fundamental definitions of terms commonly used in radiation thermometry are given in Table 10.

Blackbody Concept

The concept of the blackbody is central to radiation thermometer technology. The energy radiated by an object as a result of its temperature is quantitatively expressed in terms of a perfect radiating body, which is traditionally designated a *blackbody*. The blackbody concept has been described in several ways, e.g., a body which absorbs all the radiation it intercepts, and a body that radiates more thermal energy for all wavelength intervals than any other body of the same area, at the same tem-

Table 10. Symbols, Terms, Values, Units, and Definitions of Terms Used in Radiation Thermometry (SI Units)

Symbol	Term*	Value	Units
M	Radiation emitted per unit area, radiant exitance; a superscript b indicates radiation is from a blackbody		$W/m^2 \cdot \mu m$
τ	Stefan-Boltzman constant	5.6697×10^{-6}	$W/m^2 \cdot 4$
T	Temperature		K or °C
C_1	First radiation constant	3.7415×10^{-16}	$W \cdot m^2$
C_2	Second radiation constant	1.43879×10^{-2}	$m \cdot K$
λ	Wavelength		μm
λ_m	Wavelength of maximum radiant exitance		μm
b	Constant of Wein's displacement law	2.8978×10^{-2}	$m \cdot K$
ϵ_T	Total emissivity		Dimensionless
$\epsilon(\lambda)$	Emissivity at wavelength (spectral emissivity)		Dimensionless
n	Index of refraction		Dimensionless
r_t	Total reflectivity (coefficient of reflection)		Dimensionless
$r(\lambda)$	Reflectivity at wavelength		Dimensionless
t_T	Transmission coefficient (total)		Dimensionless
$t(\lambda)$	Transmission coefficient at		Dimensionless
e	Base of the natural logarithms	2.718	Dimensionless
$V(T)$	Calibration function (table)		V or A
N	N-value of a radiation thermometer		Dimensionless

Emissivity is a measure of the ratio of thermal radiation emitted by a non-blackbody to that of a blackbody at the same temperature. *Total emissivity* is the ratio of the total amount of radiation emitted. (By definition, a blackbody has maximum emissivity and its value is unity. Other bodies have an emissivity less than unity. The ratio is designated total emissivity.) *Spectral emissivity* is the ratio at a specified wavelength. *Emittance* is an alternate term for emissivity. *Gray body* refers to an object which has the same spectral emissivity at every wavelength, or one which has its spectral emissivity equal to its total emissivity.

perature. Physical realization of the blackbody includes a spherical cavity and the wedge-shaped cavity described by Mendenhall. See Fig. 13. The radiation leaving a small hole in the sphere or the inside of the wedge, for example, will fulfill the definition of blackbody radiation, provided the walls of the cavity are opaque and the cavity is uniform in temperature. Because the geometry of these models promotes multiple internal reflection of the radiated energy, the wall material may be an imperfect radiator and thus a practical substance. Laboratory blackbody sources must be stable and uniform when used as reproducible sources in the calibration of secondary standards and of radiation thermometers.

Many different blackbody source designs have been made, and several are available commercially. Typical shapes include spherical furnaces with small openings, as well as conical cavities and cylinders having a large length-to-diameter ratio. The emissivity of laboratory sources ranges from 0.98 to 0.9998. In some sources, a temperature uniformity of better than 1 K is obtainable. Thus,

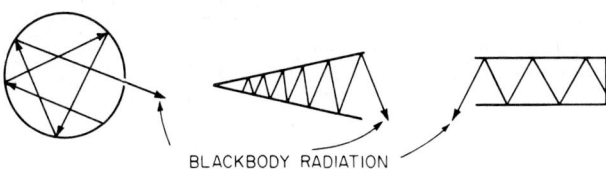

BLACKBODY RADIATION

Fig. 13 Classical blackbody models.

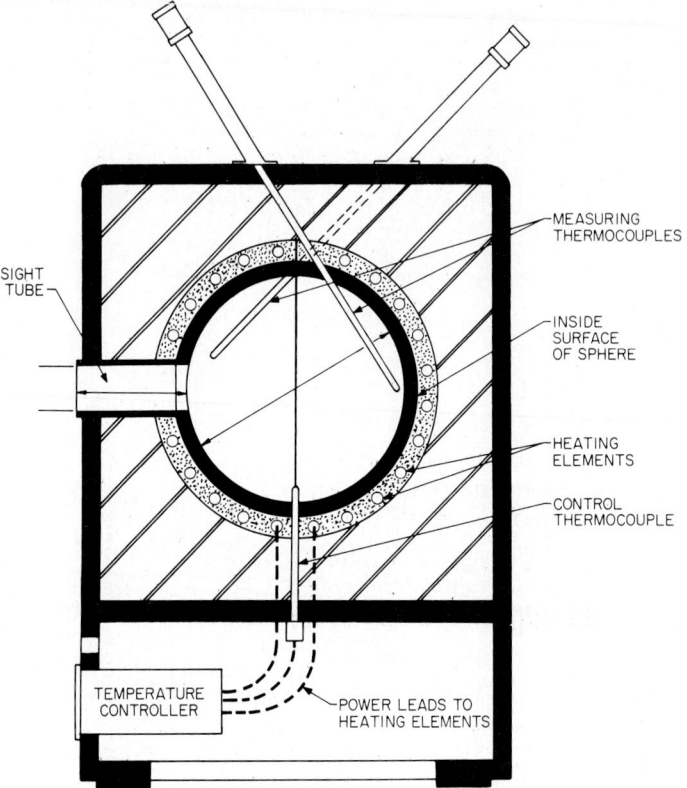

Fig. 14   Schematic view of commercial blackbody furnace. Entrance sight tube is shown to left of the hollow sphere. Note that two thermocouples are used; either can determine the sphere's temperature, but two allow internal temperature gradients to be measured as well.

reproducible calibration conditions of $\pm 1\%$ are readily attainable, and $\pm 0.1\%$ have been achieved for industrial instruments. One form of commercial blackbody furnace is shown in Fig. 14.

Relation between Blackbody Temperature and Radiation

As previously mentioned, Planck's radiation law predicts precise levels of radiation emitted per unit surface area of a blackbody at each wavelength. This may be written

$$M^b(\lambda, T) = c_1 \lambda^{-5} (e^{c2/\lambda T} - 1)^{-1} \tag{1}$$

The total radiation emitted per unit surface area is the integral of Eq. 1 over all wavelengths and is

$$M^b(T) = c_1 \int_0^{+\infty} \lambda^{-5} (e^{c2/\lambda T} - 1)^{-1} \, d\lambda \tag{2}$$

or simply

$$M^b(T) = \tau T^4 \tag{3}$$

This is known as the Stefan-Boltzman law. It is used extensively in the calculation of radiant heat transfer.

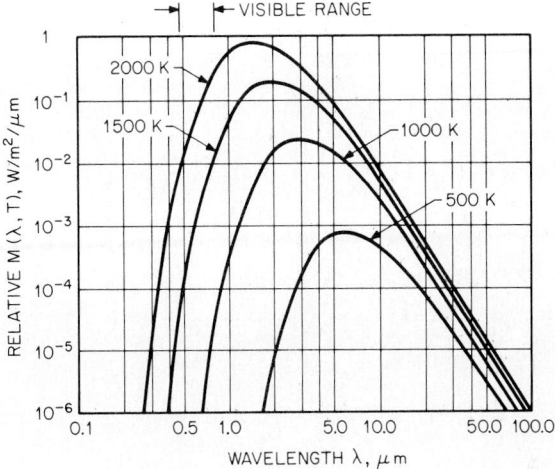

Fig. 15 Radiant emission of thermal radiation from a blackbody at various temperatures (in Kelvins).

In situations where the exponential term in Planck's law is much greater than one, i.e., $e^{c2/\lambda T} \gg 1$, Planck's equation can be approximated by

$$M^b(\lambda, T) = c_1 \lambda^{-5} e^{-c2/\lambda T} \tag{4}$$

which is known as Wein's law.

A graphical representation of the radiant exitance* as predicted by Planck's law, for several temperatures, is given by Fig. 15. Note that for each temperature there is a peak value of emission and that the peak value shifts to shorter wavelengths as the temperature is increased. This shift can be expressed as

$$\lambda_m T = b \tag{5}$$

which is known as Wein's displacement law.

Measuring Temperatures of Non-Blackbodies

Temperature measurements of real, non-blackbody objects can be complicated by three major factors:

1. Non-blackbodies emit less radiation than blackbodies, and often this difference is wavelength-dependent. Often but not always, an emissivity correction must be made. Without the needed emissivity correction of the measured signal, the apparent temperature will be lower than the actual temperature.

2. Extra radiation from other radiant sources may be reflected from the object's surface, thus adding to the measured radiation and thereby increasing the apparent temperature. In certain cases, the reflected amount may compensate for that needed to correct the object's emissivity.

3. The intensity of the emitted radiation may be modified in passing through media between the object and the instrument, thus resulting in a change in the apparent temperature. If radiation is lost, the apparent temperature will be low; if radiation is added, the temperature will read high.

In addition to these practical problems, there is often a need to measure an object that is transparent. Additional problems include added radiation produced by hot objects behind the object of measurement.

*A coined word not to be confused with excitance.

Effective Wavelength and *N*-Value

The calibration function, or output, of a radiation thermometer is a nonlinear voltage or current. Mathematically, it is an equation involving the spectral characteristics of the optical system and the detector response, integrated over all wavelengths. Once an instrument design is fixed, the relation between the thermometer output and a blackbody source temperature can be written

$$V(T) = K \int_0^{+\infty} M^b(\lambda, T)S(\lambda) \, d\lambda \tag{6}$$

where $S(\lambda)$ = net thermometer wavelength sensitivity
K = calibration constant

The calibration function is generated simply by aligning a unit with a blackbody source and measuring the output at different temperatures. Under conditions where Wein's law may be substituted for Planck's law, this equation becomes

$$V(T) = Kc_1 \int_{\lambda_1}^{\lambda_2} \lambda^{-5} e^{-c2/\lambda T} \, d\lambda \tag{7}$$

and a wavelength λ_e can be found such that

$$\lambda_e^{-5} e^{-c2/\lambda_e T} = \int_{\lambda_1}^{\lambda_2} \lambda^{-5} e^{-c2/\lambda T} \, d\lambda \tag{8}$$

The single wavelength λ_e, the effective wavelength, is representative of an instrument at temperature T. Thus the calibration function $V(T)$ can be written as

$$V(T) = K' e^{-c2/\lambda_e T} \tag{9}$$

from which it can be shown that the effective wavelength can be expressed in terms of the rate of change of the thermometer calibration function at T as

$$\lambda_e = \frac{c_2}{T_2} \left[\frac{\Delta V(T)}{V(T) \, \Delta T} \right]^{-1} \tag{10}$$

Over a range of temperatures, say from T_1 to T_2, the effective wavelength will change.

Also, at a given temperature, the calibration function can be approximated as a single-term power function of temperature:

$$V(T) = KT^N \tag{11}$$

Making use of the effective wavelength concept, it can also be shown that

$$N = \frac{c_2}{\lambda_e T} \tag{12}$$

where the power N is called the N-value of the thermometer. A ratio thermometer can be described in terms of effective (or equivalent) wavelength, even though it has two distinct and possibly widely separated wavelength response bands.

At high temperatures, $c_2/\lambda T \gg 1$, the calibration function of a ratio thermometer is essentially the ratio of the two calibration functions of the individual "channels," $V_1(T)$ and $V_2(T)$:

$$V_R(T) = \frac{V_1(T)}{V_2(T)} = A \frac{e^{-c2/\lambda_1 T}}{e^{-c2/\lambda_2 T}} \tag{13}$$

where λ_1 and λ_2 are the effective wavelengths of the two channels. When a net effective or equivalent wavelength λ_E is defined as

$$\frac{1}{\lambda_E} = \frac{1}{\lambda_1} - \frac{1}{\lambda_2} \tag{14}$$

then the *form* of the calibration function $V_R(T)$ is nearly identical to that of a single-wavelength thermometer with an effective wavelength of λ_e. The nearer the two effective wavelengths of the ratio thermometer channels are, the longer the equivalent wavelength λ_E.

A mean equivalent wavelength and N-value can also be used to characterize a given ratio thermometer.

It should be noted that, in using the foregoing equations, the temperatures must be expressed as absolute temperatures.

Optical Field of View

Inasmuch as Planck's law deals with the radiation emitted per unit surface area, radiation from a *known area* must be measured in order to establish a temperature measurement. This property, called the *field of view* of the instrument, is simply the target size that the thermometer "sees" at a given distance. Thus, a target-size-versus-distance table, chart, or formula is essential for the correct use of a radiation thermometer. At a minimum, the object of measurement must fill the required target size at a given distance. The optical field of view of a radiation thermometer is shown in Fig. 16. Similarly, in calibrating a thermometer, the radiation source must fill the field of view in order to generate or check the calibration output. If the field of view is not filled, the thermometer will read low. If a thermometer does not have a well-defined field of view, the output of the instrument will increase if the object of measurement is larger than the minimum size.

A useful criterion for a well-defined field of view is as follows: When the incremental increase in thermometer output versus the size of the target being measured is less than or equal to 0.1% per unit target area increase, i.e., the radiation collected from the defined viewing area is 99% or more of the radiation reaching the detector when viewing an extended object. But this is not essential so long as the pertinent collection percentage is known.

The target size is not precisely defined in most thermometer manufacturers' literature. It can easily be determined by viewing a uniform temperature source of large size through a small aperture. Varying the aperture sizes permits a graph similar to that of Fig. 17 to be generated. From this graph, the actual target size and the error due to viewing extended sources can be estimated quite accurately.

Such tests are most important when a measurement is to be made on objects about the same size as the stated target size—to determine what actual field of view is being filled. Also, they prove very useful when the best accuracy is required, or in comparing instruments with nominally the same specifications from two or more manufacturers.

Although the optical field of view of an instrument is never quite perfect, the effect of target size errors can be minimized by selecting an instrument which has the shortest wavelength response, or the largest N-value, consistent with the desired temperature span. For example, if a thermometer's field of view is only 95% filled, the output $V(T_A)$ will be approximately $0.95\ V(T_0)$, where $V(T_A)$

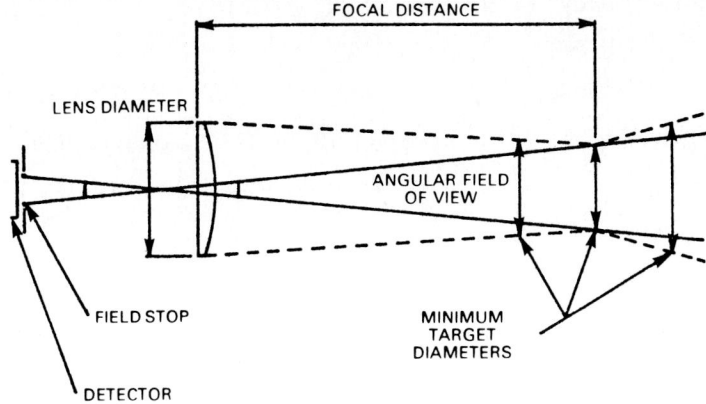

Fig. 16 Schematic representation of a radiation thermometer field of view.

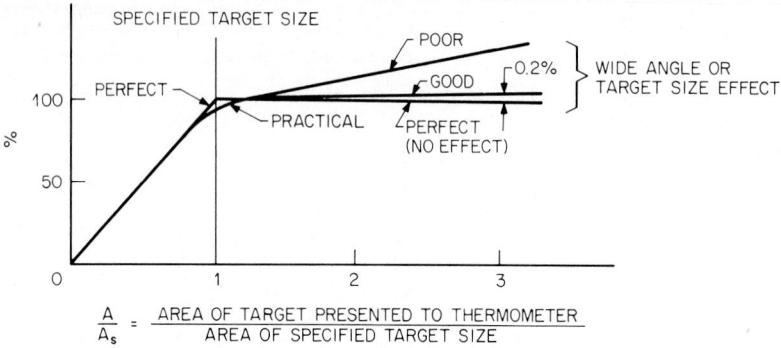

Fig. 17   Output of a thermometer as a function of the size of the hot object.

is the calibration function corresponding to an apparent temperature T_A and T_O is the true object temperature. The fractional error in temperature measurement can be shown to be

$$\frac{T_O - T_A}{T_O} = 1 - 0.95^{1/N} \tag{15}$$

In general, the error due to any factor that reduces the radiation input to a thermometer by a factor F can be written

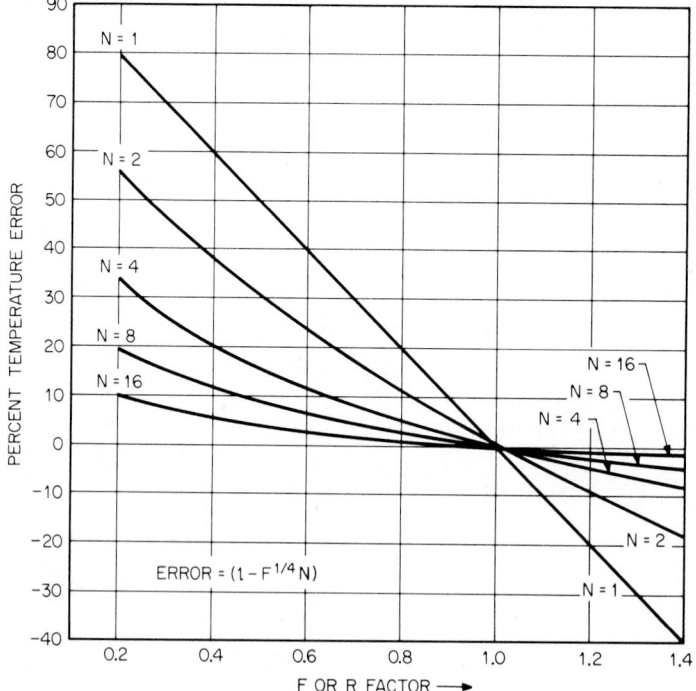

Fig. 18   Temperature error versus F (or R factor) for radiation thermometers with various N-values. See text for definitions.

$$\frac{T_O - T_A}{T_O} = 1 - F^{1/N} \tag{16}$$

Thus, the advantage of a large N-value, or short wavelength thermometer, is that the associated error in temperature measurement is minimized because of a factor F reduction in the actual target area versus the required target area. It should also be noted that a reduction in received radiation can result from a number of other factors (absorption or scattering radiation in the line of sight, dirt accumulation on thermometer lens, or change in emissivity of the object). The percent error in temperature measurement versus F for several values of N is shown graphically in Fig. 18.

It is also important to note that—because of the trend toward longer wavelength response thermometers, especially in general purpose instruments for furnace use or relatively low-temperature process and quality control uses—there is a greater need for optics with well-defined fields of view in order to obtain reasonably accurate measurements. Since the N-value of a given thermometer is inversely proportional to the effective wavelength, the measuring errors due to poorly defined optics are much greater at longer wavelengths than at shorter wavelengths. It is possible that an instrument specified as 1% accurate is barely 1% in the laboratory and is 2% or poorer in industrial use because of optical field of view limitations.

Optical Field of View—Ratio Thermometers

Since a ratio thermometer essentially measures the ratio of two thermometer signals, anything that tends to reduce the actual target size will not upset the ratio. In principle, ratio thermometers should be immune to changes in target size or not seriously affected if the object does not fill the field of view. There are, however, two limitations to this: (1) The optical field of view in the two wavebands must be the same; and (2) the net signals to be ratioed must be larger than any internal amplifier drifts, offsets, and noise.

The requirement that the two optical fields of view be the same can be expressed in the following equation for a ratio thermometer calibration function:

$$V_R(T) = kR \frac{T^{N_1}}{T^{N_2}} = kRT^{N_R} \tag{17}$$

where k = calibration constant
R = ratio of two fields of view (ideally $R = 1$)
N_1 = N-value of channel 1
N_2 = N-value of channel 2
$N_R = N_1 - N_2$ = net effective N-value for the ratio thermometer

Thus, the effect of the two fields of view not being the same is nearly identical to the effect of an unfilled field of view for a single-wavelength-response instrument, where F is replaced by R. The difference is that R can be greater than 1.0. The seriousness of this effect is magnified, much like the case of the long wavelength thermometer, as a result of the net effective N-value of a ratio thermometer being smaller than the N-values of either individual waveband.

While this effect is not a real problem when the object of measurement is larger than the required field of view, it is critical when the object is about the same size or smaller than that required. Furthermore, the error can be significant and of either sign, depending on the value of R. Thus, it is very important that a ratio thermometer, if it is to fulfill one of its major benefits, have fields of view that are the same in both wavebands.

Response Time

The response time required for a thermometer to attain a new value on an instantaneous change of object temperature is usually given in the manufacturer's specifications. Various manufacturers use different time constants, such as 90, 95, 98, or 99%. These are all related, or can be related to the exponential time constant of the system and can be expressed as

$$V(T,t) = V(T)(1 - e^{t/\tau})\qquad\qquad(18)$$

where t = time

τ = exponential time constant, the time required to reach 63.2% of the final value

In general, to achieve measurement within 1% of the final value, t/τ must be greater than or equal to 5. Figure 19 shows the curve and the relation between the more commonly used response times.

If the response time is inadequate to obtain the final value, an error will occur that is very similar to the error due to a partially filled field of view. The factor F, pertinent to a time-response-limited system, is approximately

$$F_t = 1 - e^{t/\tau}\qquad\qquad(19)$$

where τ = thermometer exponential time constant

t = time within which measurement is to be performed

A more detailed analysis of fast-response radiation thermometers is described in Ref. 7 in relation to measurements on rotating gas turbine components.

Ambient Temperature Stability

There are a number of differences in the environment in which a radiation thermometer may be used. Notable among these are the extremes of temperature that may occur in a plant environment. To obtain accurate readings under practical conditions, most instruments include ambient temperature compensation over a specified range of temperature. Two types of ambient temperature errors can occur: (1) transient, and (2) steady state. Each installation may experience these in differing amounts.

The transient error is at a maximum when the instrument is not at the same temperature as its local environment. Because of its thermal mass, there is always a lag in its response to local temperature changes. It is difficult to design an instrument to minimize these errors under such conditions other than to use a very large thermal mass so that internal temperature changes will occur slowly.

Most ambient temperature compensation methods are designed to minimize the error under steady-state conditions. This error can be measured with appropriate equipment, such as an environmental test chamber and a constant reference source of radiation.

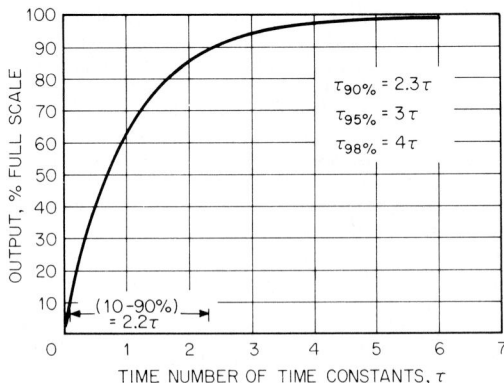

Fig. 19   Various definitions of response time in terms of the exponential time constant.

FACTORS AFFECTING PERFORMANCE OF RADIATION THERMOMETERS

Several factors external to the instrument can interfere with measurement performance: (1) The object is transparent; (2) the object has unknown or varying emissivity; and (3) attenuation or enhancement of radiation occurs in the sighting path.

Transparent Objects

The law of conservation of energy requires that, at every wavelength, the coefficient of transmission, reflection, and emission (absorption) of radiation add up to 1:

$$\epsilon(\lambda) + r(\lambda) + t(\lambda) = 1 \tag{20}$$

There are several ways to deal with transparent objects such as glass, plastic, semiconductor materials, and gases. The first and simplest is to select a waveband in which the object is opaque. For example, nearly all ordinary glasses are opaque at wavelengths longer than about 5.0 μm provided the glass thickness is 3.0 mm or more. Similarly, most thin polyethylene plastics are opaque in a narrow wavelength band centered at 3.43 μm. As a simple guide in determining if an object is opaque, one must examine the spectral absorption coefficient which can be roughly deduced from spectral transmission curves using the relationship

$$\alpha = -\frac{1}{x} \ln \frac{T}{[1 - r(\lambda)]^2} \tag{21}$$

where $r(\lambda)$ = surface reflection coefficient which, at normal incidence, is obtained from the index of refraction n
x = thickness
T = transmission

$$r(\lambda) = \left(\frac{n-1}{n+1}\right)^2 \tag{22}$$

More precise information can be obtained by measuring samples of different thickness in the laboratory and plotting T versus thickness on a log-log graph. The slope of the line will be α. An object can be considered opaque if T is less than about 1%.

Flames and hot gases are special cases because they have no sharply defined boundary—hence no reflection coefficient. In this case, Eq. 22 becomes $r(\lambda) = 0$.

Often there are situations where one cannot or prefers not to measure in a wavelength region where the object is opaque. In many such cases a similar analysis enables one to calculate the emissivity and correct for the transmission and reflection effects. Caution must be taken to ensure that the background beyond the object is either much cooler than the object temperature or is accounted for in the correction.

A variation of the Schmidt method for determining the "average" temperature of flames illustrates how this can be done. Transparent flames and hot gases are unique in that it is difficult to calculate the emissivity; both emissivity and temperature are unknown. Therefore, two measurements are needed, one with a cold background and a second with a background at a measured apparent temperature T_B. All measurements must be performed with the same instrument, or with two having the same calibration function.

The first measurement, with a cold background, yields a value of voltage (or current) V_1, where

$$V_1 = \epsilon(\lambda)V(T_F) \tag{23}$$

The second measurement yields a voltage V_2 which includes a fraction of the contribution from the hot background, which is known to have a temperature corresponding to a voltage V_3, or

$$V_2 = \epsilon(\lambda)V(T_F) + [1 - \epsilon(\lambda)]V_3 \tag{24}$$

$$\epsilon(\lambda) = 1 - \frac{V_2 - V_1}{V_3} \tag{25}$$

In measurements of some transparent objects, it is desired to look below the surface and determine the temperature without disturbing its distribution. Two well-known examples of measurements in the glass industry are described here, but the concepts apply to any similar applications.

If the object is optically thick and radiation can pass from regions well below the surface to an external radiation thermometer, it is known that for linear temperature gradient conditions the thermometer will indicate a temperature at a depth y_α where α is the absorption coefficient. Typical uses of this technique involve glass temperature in melting furnaces and forehearths where the temperature below the surface is more useful as a control parameter than the surface temperature.

Similarly, in several annealing and forming operations, the bulk temperature of glass objects is less affected than the surface by transient errors, such as spray and air currents, and has proven to be more reliable for process and quality control.

In the tempering of glass, signals from two thermometers, one of which monitors only surface temperature while the second looks into the glass, have been used to analyze the critical surface-to-center temperature gradient. Measurements have been used both in the development of process parameters and in tempering furnace start-up and monitoring.

Emissivity and Reflection Errors

In most applications, opaque objects are measured. Thus the equation relating emissivity and transmission becomes

$$\epsilon(\lambda) + r(\lambda) = 1 \tag{26}$$

Emissivity Data

A major practical problem that most users face is obtaining a reliable emissivity value for use in correctly setting the emissivity compensation. Similarly, a ratio thermometer user must correctly set the emissivity ratio compensation (nongrayness or chromatic compensation). Several approaches may be suggested.

1. Use representative table values from either manufacturers' literature or handbooks.

2. Measure the object's temperature, or at least that of a representative sample, by another accurate and reliable means and adjust the thermometer emissivity control to make the thermometer match.

3. Have the emissivity of a representative sample determined by a qualified laboratory. NOTE: It is important to specify the waveband, but not usually the temperature (only the range at which the emissivity value will be needed).

It is important to note that the emissivity of an object can be different from the usual values if the angle of incidence with which the thermometer views the object is too large. Figure 20 illustrates this situation. Note that the target area on the surface changes with the angle. As a general guideline, the emissivity of electrically conductive materials is uniform from angles of incidence of 0 to ~40°, whereas nonconductors are uniform to ~50°.

Classification of Measurements by Temperature of Surroundings

All measurements made by radiation thermometers can be classified into one of the three following categories:

1. Object and surroundings are at about the same temperature.

2. Object is in cooler surroundings.

3. Object is in hotter surroundings.

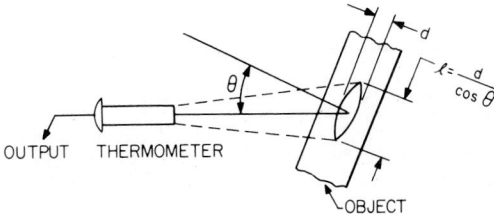

Fig. 20   Effect of the angle of viewing on target size and shape.

Measurements can be described by a simple equation which accounts for both the emissivity of the surface and the effect of thermal radiation from other sources reflected from the surface into the thermometer:

$$V(T_A) = \epsilon(\lambda)V(T_O) + [1 - \epsilon(\lambda)]V(T_B) \tag{27}$$

where $V(T_A)$ = indicated apparent blackbody temperature
  $V(T_O)$ = blackbody temperature of object
  $V(T_B)$ = blackbody temperature of net surroundings (assuming for simplicity that a reasonably uniform effective background exists)
  $\epsilon(\lambda)$ = spectral emissivity
  $r(\lambda)$ = average reflectivity of object over same wavelength span, equal to $1 - \epsilon(\lambda)$

Objects in Surroundings of the Same Temperature (Category 1)

In measurements where $T_O = T_B$

$$V(T_O) = V(T_B) \tag{28}$$

and

$$V(T_A) = V(T_O) \tag{29}$$

Where the actual emissivity is not very important, nor are minor variations in emissivity, the object looks like a blackbody.

But, the question arises, How close must the surroundings be to be considered equal? This depends on several parameters, the most important of which is the desired accuracy. A region of temperature T_B surrounding an object at temperature T_O having emissivity $\epsilon(\lambda)$ will produce a thermometer output $V(T_A)$, where T_A is the apparent temperature according to the relation

$$V(T_A) = \epsilon(\lambda)V(T_O) + (1 - \epsilon(\lambda))V(T_B) \tag{30}$$

Given a thermometer calibration function $V(T)$, one can calculate easily the allowed ranged of T_B such that $|T_B - T_A| \leq \Delta T$, where ΔT is the desired accuracy or systematic error allowed. This error must be factored into the overall error but will be systematic if T_B and $\epsilon(\lambda)$ are reasonably constant.

Target Tubes

Radiation thermometers frequently have been used in place of noble metal thermocouples, sighted into refractory tubes which are inserted into the heated region about 5 to 10 diameters, according to material properties. In this application, the thermometer measures the tube temperature and, since it is a reasonably good approximation of a blackbody with such an insertion length, no emissivity correction is required. See Fig. 21.

In these cases, the advantage of the thermometer is (1) a lower replacement cost if the tube breaks (i.e., tube replacement instead of tube and thermocouple replacement), or (2) negligible drift in calibration at high temperatures (compared with a thermocouple which can shift calibration with time and temperature), or (3) electrical isolation. Figure 22 illustrates a glass application where a target

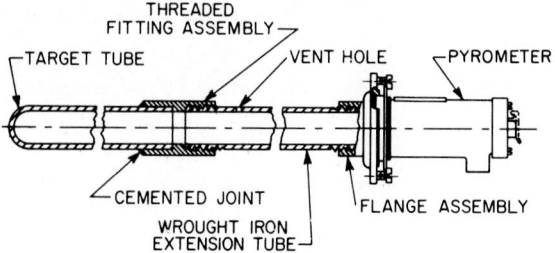

Fig. 21 Radiation thermometer with a closed-end target tube.

tube can be inserted into electrically heated, molten glass wherein all three of the aforementioned advantages are realized.

Product Temperature in Furnaces

Many heating processes have a normalizing or soaking temperature portion of the cycle where the object is kept at a constant temperature for a reasonable period of time. The heat is usually delivered by the radiant brickwork of the furnace which is kept hot by heaters or burners. Nearly any type of radiation thermometer can be used in this type of measurement with the emissivity correction set at 1.0. There are definite advantages in using the highest N-value instrument, because it will be least affected by dirt or soot buildup on the window or lens and requires less frequent cleaning. In some cases, ratio thermometers offer a measuring advantage—for the same reasons—but usually at a significantly higher initial cost.

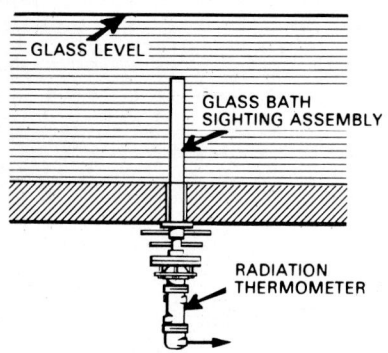

Fig. 22 Use of a target tube to measure the subsurface temperature of molten glass.

In selecting a thermometer for sighting inside furnaces fired by fossil fuels, it is important to select one with a wavelength response that avoids the emission wavebands of water vapor, carbon dioxide, and carbon monoxide, which are the major gaseous products of combustion. The effects of water vapor are discussed later. Also it is important to choose a sighting path that avoids evident flames. Although the surroundings in the furnace are nearly the same temperature as the object, the same is not true of gases and carbon particles in the vicinity of flames. Flames can be several hundred degrees hotter than the surroundings and produce little effect on the measurement—if (1) they are optically thin at all wavelengths, and (2) measurements are made in wavelength regions where the flames have the lowest emissivity, i.e., outside the strong emission band mentioned.

Measurements near Ambient Temperatures

Some widely used portable radiation thermometers are capable of measuring in the region of 0 to 100°F (−18 to 38°C) and have proven useful for many energy conservation and maintenance tasks. In the 65 to 95°F (18 to 35°C) measurement region, one often encounters objects which are at about the same temperature as the surrounding ambient conditions. Here, just as in high-temperature furnace measurements, the emissivity must be set at 1.0 in order to obtain the correct temperature.

Objects in Cooler Surroundings (Category 2)

In cases where $T_O > T_B$, $V(T_O)$ is usually much greater than $V(T_B)$ and the second term in Eq. 27 can be neglected (for materials where $\epsilon(\lambda)$ is greater than 0.5), or

$$V(T_A) = \epsilon(\lambda)V(T_O) \qquad (31)$$

and emissivity is important.

If no correction is made for $\epsilon(\lambda)$, the apparent temperature is called the *brightness temperature* because the object appears to the thermometer to be as "bright" as a blackbody at the apparent temperature. Modern instruments provide an emissivity correction which, in analog circuits, consists of an amplifier of gain G, where

$$G \propto \frac{1}{\epsilon(\lambda)} \qquad (32)$$

Multiplying the measured signal $V(T_A)$ by G yields

$$V(T) = GV(T_A) = G\epsilon(\lambda)V(T_O) \qquad (33)$$

When the gain is chosen so that $G = 1/\epsilon(\lambda)$, then the corrected signal is equal to the blackbody calibration function value. However, if there is a difference, then an error in measurement results. This error may be expressed as a fraction of the measured signal:

$$F_G = 1 - [G\epsilon(\lambda)]^{1/N} \qquad (34)$$

which affects the measurement similarly to previous error sources discussed. As described earlier, it is desirable to select a radiation thermometer with a very nonlinear response to minimize measuring error for this category of measurement, i.e., a thermometer having a large N-value or a short wavelength response.

Many measurements, especially of very hot objects, are made in a process where the object is in much cooler surroundings. The use of high-N-value or short wavelength thermometers is recommended for such situations. Figure 23 illustrates the relation between the emissivity error, temperature, and wavelength response of some typical radiation thermometers. The optimum measuring accuracy and span capabilities can easily be noted from this chart.

Numerous applications in this category encompass such diverse manufacturing operations as those involving the processing of iron and steel, nonferrous metals, glass and ceramics, semiconductors, chemicals, plastics, foods, rubber, textiles, cement, and minerals. Most radiation thermometer manufacturers provide detailed application specifics. The majority of applications are relatively simple—particularly when the objects are opaque and the emissivity is high (0.8 or more). Thermometer selection involves simply matching the requirements to the most suitable short wavelength instrument available and providing the correct emissivity correction.

Some of the more difficult measurements involve aluminum and other low-emissivity metals, tin- and zinc-coated steel, very small objects, and measurements that involve sighting through visible flames, such as in flame brazing and glassworking. Experienced counsel should be sought in specific applications where emissivities are low or there are unusual measuring conditions.

Objects in Hotter Surroundings (Category 3)

In cases where $T_O < T_B$ and $V(T_A)$, the sum of reflected and emitted radiation, is much higher than even the radiation emitted by a blackbody at the temperature of the object, problems can be quite difficult. In these situations, the apparent temperature is much higher than the actual temperature. Traditionally, this has been one of the most difficult measurement problems in radiation thermometry but has been solved in recent years through effective use of a second measurement to correct for the background radiation.

The Specular Solution

Where the object has a relatively smooth surface and is very close to being perpendicular to the sighting axis of the thermometer, it is possible to consider the background source in terms of *reverse optics,* i.e., as mostly a specular reflection. See Fig. 24. Here the thermometer sight path is shown being reflected from the surface of the object. If the sighting cone of the thermometer intercepts a cool surface rather than a hot surface, then the error will be due to the diffuse reflected component rather than the total reflected view of a hot region. Flat glass in tempering furnaces is an excellent example of this case. Glass is very flat and nearly a perfect specular reflector. Thus, as shown, a thermometer viewing itself by reflection will show no error from the surroundings.

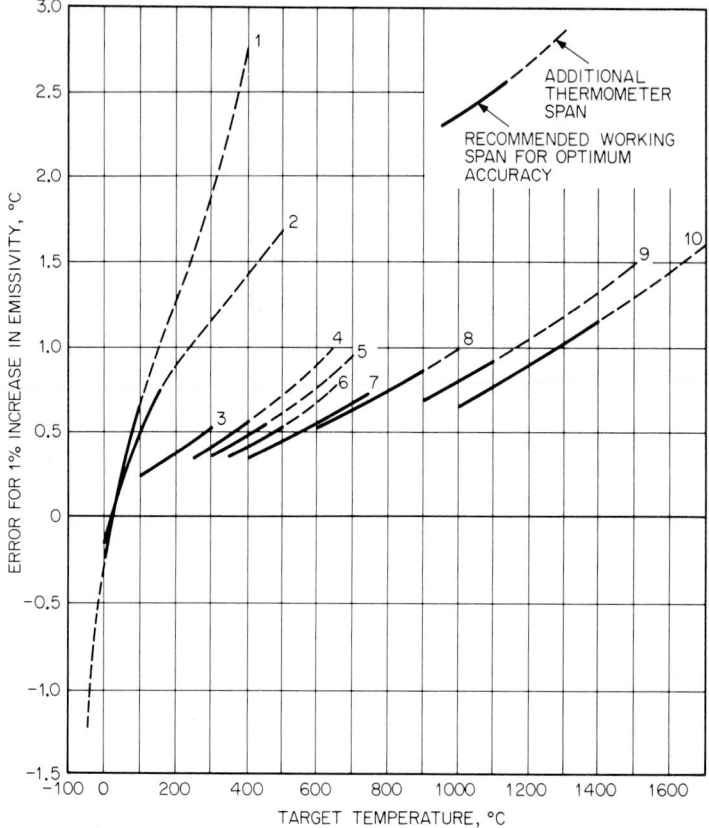

Fig. 23   Errors in measurements made in cooler surroundings due to uncertainty of change in object's emissivity over the range −40 to 1700°C. The effective wavelengths of the thermometers are shown below in micrometers:

Thermometer	μm	Thermometer	μm
1	10	6	1.3
2	7	7	1.0
3	2.3	8	0.9
4	1.8	9	0.75
5	1.5	10	0.6

Similarly, rolled steel sheet used for automotive or appliance purposes is often processed in high-temperature ovens for annealing or cleaning. These specialty products are often highly specular, up to 90% or more, and can be measured with very high-N-value thermometers using this arrangement.

Shielding

On rapidly moving objects, it is sometimes possible to solve the reflection problem for diffuse, rough, or varying-position objects by constructing a cooled shield similar to that shown in Fig. 25. It is essential that the shield be cooled, or else it will eventually reach some temperature between that of the object and the surroundings and introduce its own reflected component into the measurement.

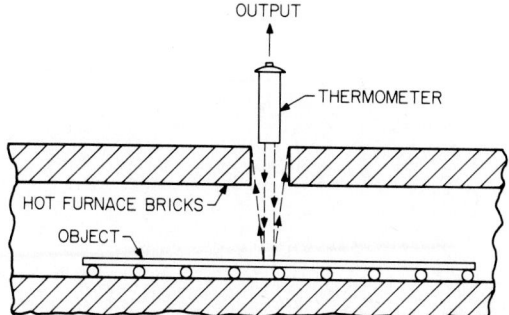

Fig. 24 Specular targets. The optical sight path is selected so that the thermometer "sees" a cool background rather than hot surfaces by reflection from the object.

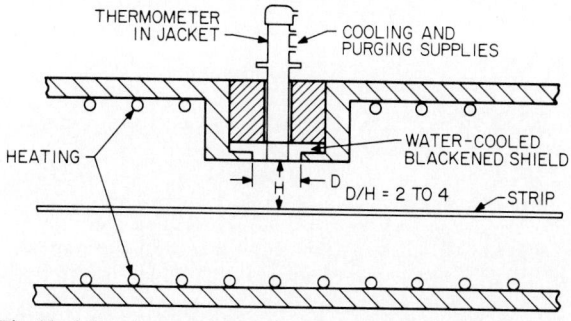

Fig. 25 Schematic representation of the shielded target method. $D/H =$ diameter of shield/spacing of strip.

Obviously, this technique would not be suitable in very high-temperature furnaces or on stationary or slowly moving objects because of the large cooling effect of the shield.

Background Compensation

Several years ago, Barber (Ref. 9) suggested a solution to the problem. This consisted of using a second thermometer to measure the background or surrounding temperature, as shown in Fig. 26, and then combining the two signals.

From the first thermometer,

$$V_1 = V(T_A) = \epsilon(\lambda)V(T_O) + (1 - \epsilon(\lambda))V(T_B) \tag{35}$$

and from the second thermometer,

$$V_2 = V(T_B) \tag{36}$$

Combining the two signals

$$V(T_O) = \frac{1}{\epsilon(\lambda)}[V_1 - (1 - \epsilon(\lambda))V_2] \tag{37}$$

the calibration function for the object temperature can be obtained. It is necessary that V_2 have the same calibration function as V_1, or if a furnace thermocouple is used, its signal should be scaled and made nonlinear to conform to the same scale shape as V_1. It has been shown recently that this also

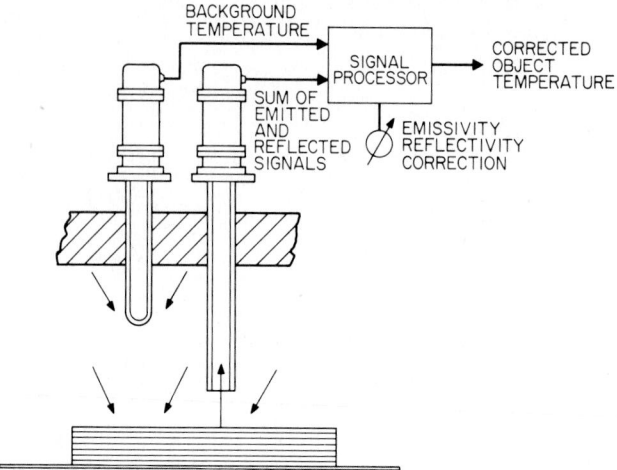

Fig. 26 Dual thermometer method for measuring the temperature of an object in surroundings that are hotter.

can be done by using the second thermometer to measure the term $[1 - \epsilon(\lambda)]V(T_B)$ directly by measuring the reflected component from a cooled piece of material having the same emissivity as the object in the furnace; i.e., V_2 becomes $[1 - \epsilon(\lambda)](T_B)$.

In both measurements, the errors are minimized by using a thermometer with a relatively low N-value, or long wavelength response. While not obvious from the foregoing equations, this has been shown from detailed calculations as summarized in Fig. 27. Here the errors in measurement are compared for two different thermometer systems, one with a short wavelength and the other with a significantly longer one. The examples given are for the typical product and background temperatures found in a steel slab or billet reheat furnace.

Dual Wavelength Method

Another, different, dual method has also been developed for gas turbine measurements. It utilizes a special two-wavelength thermometer, which in essence provides the following information:

$$V_1(T'_A) = \epsilon_1 V_1(T_O) + (1 - \epsilon_1)V_1(T_B) \tag{38}$$

$$V_2(T_A) = \epsilon_2 V_2(T_O) + (1 - \epsilon_2)V_2(T_B) \tag{39}$$

If ϵ_1 and ϵ_2 are known, then in principle it is possible to solve the equations using an iterative technique for the set of values of T_O and T_B which are common to both. The errors in this method are very sensitive to choice of wavebands, and slight errors in either channel at short wavelengths can produce large errors. Both the dual-wavelength and dual-thermometer methods are relatively new, and not much information on them has been published to date. They are indicative, however, of the significant applications that can be solved by using novel approaches.

Interference to Radiation in the Sighting Path

All radiation emitted must reach the thermometer for the measurement to be interpreted correctly, unless a ratio thermometer is used, which is excellent for extreme cases of attenuation in the sighting path. Principal causes of interference include:

1. Attenuation of radiation over part or all of the sighting path

2. Reflection and absorption losses on passing through windows

3. Aperturing or blockage of a portion of the sight path

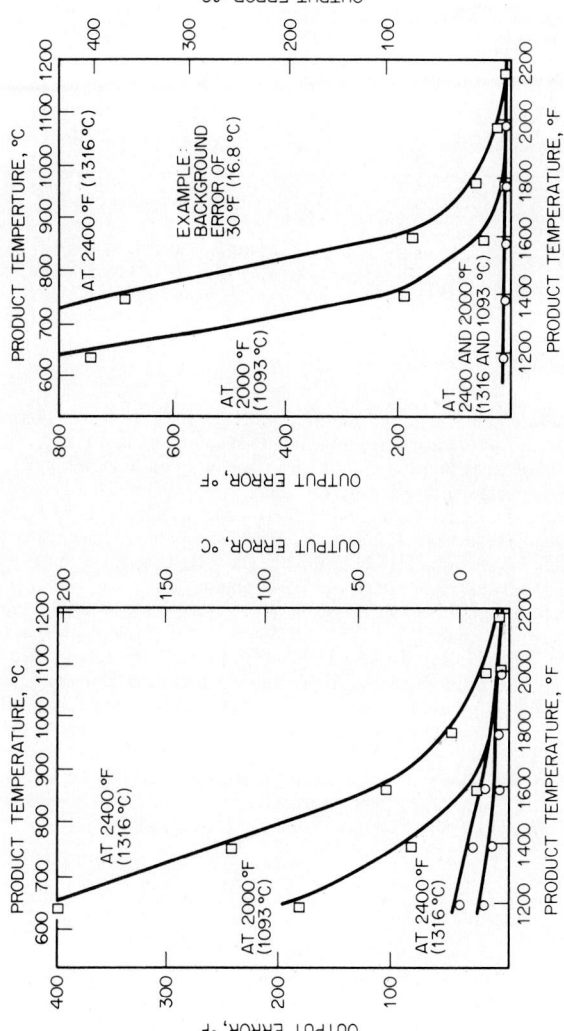

Fig. 27 Errors likely to occur in the measurement of objects in hotter surroundings due to (*a*) uncertainty or a change in emissivity of 25% and (*b*) inaccurate background temperature measurement for dual thermometer systems of two effective wavelengths. □, 0.9 nm; ○ 3.95 nm. Values shown are representative of temperatures in steel billet and slab reheat furnaces with backgrounds of 2000 and 2400°F (1093 and 1316°C).

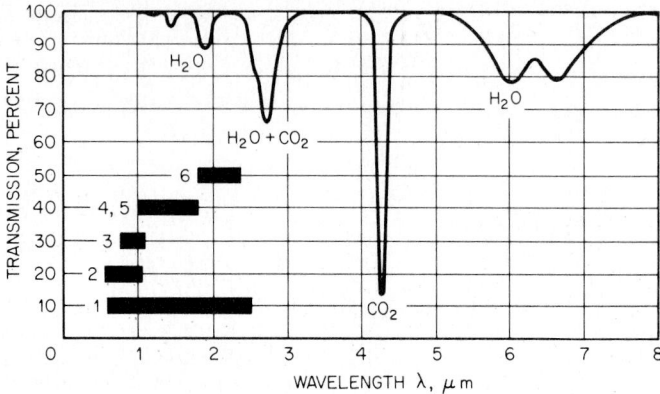

Fig. 28   Infrared radiation transmission through 1 m of air at 68°F (20°C), 55% relative humidity, and the response bands of some thermometers. 1, Wideband (WB); 2, silicon (Si); 3, silicon-filtered (Si); 4, optical pyrometers; 5, germanium (Ge); 6, lead sulfide (Pbs).

Attenuation

Water vapor is the major cause of errors in measurements with the earlier wideband thermometers because their wavelength-sensitive region overlaps some water vapor absorption bands. The principal absorption bands in the atmosphere are shown in Fig. 28. Higher output and more spectrally selective detectors enable modern radiation thermometers to avoid these problems and have essentially eliminated distance-dependent errors due to this source of error over nominal sighting distances.

Radiation may also be attenuated by (1) physical obscuration by dust, smoke, or condensing steam; (2) absorption by other gases in the sighting path; or (3) blocking on an intermittent basis by a colder opaque object which obscures all or part of the sight path.

Minor attenuation produces little error when large N-value thermometers or ratio thermometers are used and, if present, can be estimated from Fig. 18. However, the effects of small amounts of water on a surface or of water or steam in the sight path can be subtle because of their varying spectral absorption. In general, it is important to evaluate any unusual material in the sight path in terms of its absorption within the spectral response band of the instrument. Experimental errors for water and steam in the sight path of three common high-temperature thermometers and a ratio thermometer are given in Table 11.

Solutions to the problem of readily apparent obscuring sources are easily implemented through the use of sighting tubes or air jets and/or signal conditioning in the instrument electronics so that the thermometer output will ignore reductions in the signal level (peak picking).

Table 11.   Errors Due to Water Vapor and Steam in the Thermometer Sighting Path

Source at 1050°C (1900°F)

Radiation thermometer waveband	Water 6 mm, ¼ in		Steam 300 mm, 12 in	
	°C	°F	°C	°F
0.4 to 2.8	−420	−755	−150	−270
0.5 to 1.1	−7	−13	−56	−99
0.67	0	0	−29	−52
Ratio 0.67/0.90	+31	+56	+7	+13

The use of large-N-value thermometers in hotter surroundings (Category 3) can cause a situation where a slight error in either of the two measurements will produce a large error in the temperature calculated from the two readings. Thus a relatively small N-value or long wavelength response sensors provide the least sensitivity to sight-path absorption errors in this special case.

Reflection

Sighting through pressure sealing windows involves only a fixed correction which can easily be calculated. In the wavelength region of 0.5 to 2.5 μm, for example, glass windows introduce a reflection loss of about 4% per surface, and thus a net attenuation of 8% is incurred. This effectively means that only 92% of emitted radiation reaches the thermometer. The emissivity correction control on most instruments allows this correction to be set.

If the emissivity value without the window is $\epsilon(\lambda)$, then the value that should be set is $0.92\epsilon(\lambda)$. The reflection losses from other uncoated optical windows can be readily found from the index of refraction of the window material at the effective wavelength of the thermometer. If the index of refraction varies more than 10 to 20% from its nominal value over the full waveband of the thermometer, a more careful calculation is needed. However, this is seldom the case in practice. For coated optics, the transmission is usually specified by the window supplier and can be used directly.

Spurious reflections from sunlight or nearby incandescent or high-pressure lamps can significantly bias a measurement if a specular reflection enters the thermometer, even though it may only be periodic. The simplest solutions include shielding the instrument from the light, replacing the lamps with other lamps that do not affect the radiation thermometer, and the use of a periodic measurement, possibly through a sample-and-hold command to perform the measurement at that time window where no reflection exists.

Measurement of Moving Objects

Moving objects require a radiation thermometer with a response time short enough to compensate for the speed of motion. Modern thermometers can respond as rapidly as milliseconds, but response times much faster than a few milliseconds are difficult to obtain with industrial devices. Rotating blades in a gas turbine engine, for example, can require response times down to a few microseconds. A very careful circuit design and assembly are necessary to achieve such performance, but they are possible.

Sighting over Very Long Distances

In industrial plants, typical distances between instrument and object range from a few inches (centimeters) to 50 to 60 ft (15 to 18 m). Ordinarily, such installations do not require correction for atmospheric attenuation (except those instruments which are sensitive to water vapor absorption), because the effects of the atmosphere are usually small enough to be neglected. However, at longer distances—say 100 ft (30 m)—attenuation due to the atmosphere can become significant. Such applications require special analysis (Ref. 20 may be useful in this regard).

CALIBRATION OF RADIATION THERMOMETERS

Two types of secondary standards are used in radiation thermometer calibration: (1) a tungsten strip lamp and (2) secondary standard radiation thermometers with transfer sources. A transfer source is usually a gray body or a blackbody that is sufficiently stable and uniform to allow the secondary standard thermometer and the unknown to be compared under identical conditions.

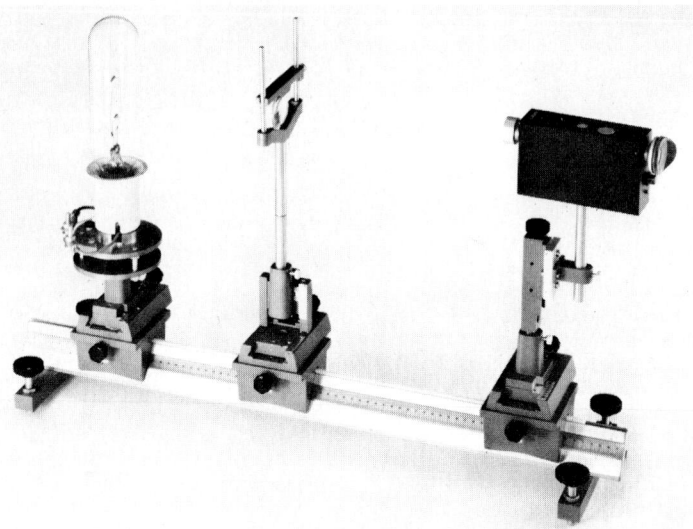

Fig. 29   Secondary standard lamp for optical pyrometer calibration. *(Land Pyrometers, Ltd.)*

Secondary Standard Lamps

Special flat-ribbon tungsten filament lamps, both vacuum and gas-filled, are available. These lamps can be calibrated against an absolute source and remain stable for use on a daily basis for nearly one year without serious changes. The concept is old and well proven for the calibration of optical pyrometers. The U.S. National Bureau of Standards offers a service which calibrates such lamps at 0.65 nm. It is also possible to use the lamps to calibrate other types of radiation thermometers with a different wavelength response at slightly degraded accuracy.

Typical lamp calibrations are made using a table of dc filament current versus brightness temperatures, and a moderate-cost system is capable of achieving accuracies or uncertainties of $\pm 0.5\%$ of absolute temperature or better. See Fig. 29, which shows both the familiar cylindrical lamp and a special high-stability lamp which can provide calibration capabilities over the range 800 to 2350°C.

Secondary Standard Thermometers

Although long practiced by many instrument manufacturers, the concept of using radiation thermometers as secondary standards has only recently been described as a desirable metrological practice. A well-characterized radiation thermometer can serve as a useful and practical secondary standard as well as a short-term transfer device. When used as a secondary standard, a blackbody source is not required. A lower quality furnace that can provide a reasonably uniform temperature can act as a gray body to provide reliable and accurate calibration. See Fig. 30, which shows a commercial reference, gray body source useful from 200 to 1450°C. This, when used with a good secondary standard thermometer, can provide calibration accuracy of $\pm 0.5\%$ of absolute temperature or better.

Process Setup and Checking

The initial setup of a radiation thermometer on a manufacturing process often requires checks to ensure that the correct emissivity value has been used, that reflected radiation is not interfering with the reading, and a number of other factors. Also, it is very convenient for the plant instrument engineer to be able to periodically verify that an installed instrument is still working properly without having to remove the device and return it to the shop or manufacturer for a calibration test.

Portable radiation thermometers and optical pyrometers play an important role in these two situations, particularly in measurements which require no emissivity correction, such as furnace mea-

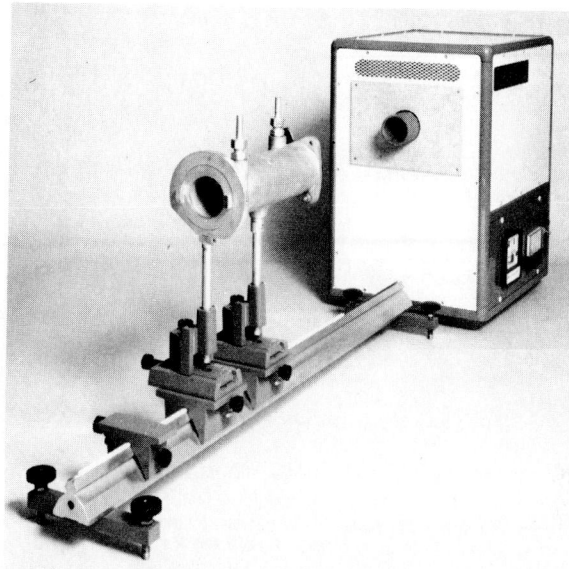

Fig. 30   Gray body calibration (reference) source.

Fig. 31   Portable emissivity-free thermometer with hemispherical gold-coated mirror. *(Land surface thermometer.)*

surements or those involving closed-end sighting (target) tubes. The Land surface thermometer is unique among radiation thermometers used for these purposes since, because of its hemispherical mirror, it requires no emissivity correction on objects with an emissivity greater than 0.6. The instrument is shown in Fig. 31. In use, it must be placed on the surface to be measured in order to be effective. A clip-on shield allows the instrument to be used to measure the spectral emissivity within its wavelength response band as well.

REFERENCES

1. Driscoll, W. G., and W. Vaughn: *Handbook of Optics,* McGraw-Hill, New York, 1978.
2. "Radiation Thermometers/Pyrometers—Annual Buyers Guide Survey," *Meas. Control,* vol. 16, no. 1, February 1982, pp. 233–243.

3. Nutter, G. D.: "Radiation Thermometry—A Review," *Mech. Eng.,* vol. 94, no. 6, 1972, p. 16.

4. Rucklidge, J. M. *A Beginner's Guide to Infrared Thermometry,* Land Instruments Inc., Tullytown, Pa., 1977.

5. Kostkowski, H. J., and R. D. Lee: "Theory and Methods of Optical Pyrometry," in *Temperature: Its Measurement and Control in Science and Industry,* vol. 3, pt. 1, Van Nostrand Reinhold, New York, 1962, p. 449.

6. Reynolds, P. M.: "Emissivity Errors of Infrared Pyrometers in Relation to Spectral Response," *Br. J. Appl. Phys.,* vol. 12, no. 8, 1961, pp. 401–405.

7. Beynon, T. G. R.: "Radiation Thermometry Applied to the Development and Control of Gas Turbine Engines," 6th International Temperature Symposium, March, 1982, Washington, D.C.

8. Barber, R.: "Non-Contact Temperature Measurements of Metal Surfaces in the Open," *Adv. Instrum.,* vol. 33, pt. 1, 1978, pp. 417–434.

9. Barber, R.: "Furnace Load Temperature Measurement during the Heating Process," *Ind. Process Heat.,* February 1967.

10. Peacock, G. R.: "Infrared Thermometers: A Comparison of Single Band and Ratio Types," *Meas. Control,* vol. 13, no. 4, September 1979.

11. Iuchi, T., et al.: "Temperature Measurement System of Steel Strips in a Continuous Annealing Furnace," *Tetsu-To-Hagame,* vol. 61, 1975.

12. Roney, J. E.: "Radiation Measurement of a Product Temperature in a Furnace," U.S. Letters Patent 4,144,758, 1978.

13. Atkinson, W., and R. R. Strange: "Pyrometer Temperature Measurement in the Presence of Reflected Radiation," *ASME Paper 76-HT-74,* American Institute of Mechanical Engineers, New York, 1976.

14. Barber, R.: "Review of Infrared Thermometry for the Glass Industry and the Interpretation of Their Readings," *J. Non-Cryst. Solids,* vols. 38 and 39, 1980, pp. 903–908.

15. Beattie, J. R.: "Application of Pyrometry to Glass Tempering," *IEEE Conf. Record, 6th IGA Meeting,* 1971, pp. 721–724.

16. Barber R., and M. E. Brown: "Calibration Methods and Standards for Infrared Thermometers," *ISA Trans.,* vol. 17, no. 3, 1978, pp. 63–69.

17. Banfield, A. R.: "Evaluating the Performance of Optical and Radiation Pyrometers," *SPIE Proc.* vol. 234, 1980, pp. 62–65.

18. Sakums, F., and S. Hattori: "Establishing a Practical Temperature Standard by Using Silicon Narrow Band Radiation Thermometer," 6th International Temperature Symposium, March 1982, Washington, D.C.

19. Warnke, G. F.: "Commercial Pyrometers" in *Temperature: Its Measurement and Control in Science and Industry,* vol. 4, pt. 1, Instrument Society of America, Research Triangle Park, N.C., 1972, pp. 503–518.

20. Wolfe, W. L., and G. J. Zissis (eds.): *The Infrared Handbook,* U.S. Government Printing Office, Washington, D.C., 1978.

Pressure Systems*

Robert Bicking. *Micro Switch, A Division of Honeywell Inc., Freeport, Illinois. (Piezoresistive Pressure-Sensing Element)*

William H. Burtt. *The Foxboro Company, Foxboro, Massachusetts. (Resonant Wire Pressure Transducers)*

James A. Crowson. *Project Engineer, Robertshaw Controls Company, Tennessee Division, Knoxville, Tennessee. (Mechanical Pressure Elements)*

Z. C. Dobrowolski. *Chief Development Engineer, Kinney Vacuum, A Unit of General Signal, Canton, Massachusetts. (Vacuum Measurement)*

L. S. Dysart. *Retired, Formerly of Robertshaw Controls Company, Tennessee Division, Knoxville, Tennessee. (Mechanical Presure Elements)*

W. A. Heske. *Industrial Instrument Operations, Dresser Industries, Stratford, Connecticut. (Noncontacting Optical Sensors)*

E. C. Hirth. *Application Engineer, Moore Products Company, Spring House, Pennsylvania. (Pneumatic Pressure Transmitters)*

R. W. Lally. *Engineering Department, PCB Piezotronics, Inc., Depew, New York. (Piezoelectric Pressure Transducers)*

Roy A. Ratcliffe. *Vice-President, Process Instruments, Gould, Inc., Measurement Systems Division, Oxnard, California. (Strain Gage Pressure Transducers)*

Richard F. Speicher. *Micro Switch-Honeywell, Freeport, Illinois. (Piezoresistive Pressure Transducers)*

---

*Persons who authored complete articles or subsections of articles, or who otherwise cooperated in an outstanding manner in furnishing information and helpful counsel to the editorial staff.

H. A. Steinherz. *CTI-Cryogenics Division, Helix Technology Corporation, Waltham, Massachusetts. Formerly with Varian Associates, Palo Alto, California, and President, American Vacuum Society. (Vacuum Measurement)*

John Yaeger. *Communications Department, Fischer & Porter Company, Warminster, Pennsylvania. (Electronic Pressure Transmitters)*

Mechanical Pressure Elements

by
James A. Crowson*

Mechanical pressure elements are often used as sensors in recorders, indicators, transmitters, and similar devices. Applications include measurement of vacuum, gage pressure, absolute pressure, and differential pressure. When equipped with suitable adapters, these elements may also be used to measure liquid level and flow.

The two basic types of mechanical elements are the liquid-column type, such as the manometer; and the elastic type, such as the bellows.

Liquid-Column Elements

Because of their inherent accuracy, liquid-column elements are widely used for direct measurement of pressure and vacuum, and also as standards for calibrating other pressure instruments. Several of the basic elements are described in this article.

U-Tube Manometer

As shown in Fig. 1, a glass U tube is partially filled with liquid and both ends are initially open to the atmosphere. When a gage pressure P_2 is to be measured, it is applied to the top of one of the columns and the top of the other column remains open. When the liquid in the tube is mercury, for example, the indicated pressure h is usually expressed in inches or millimeters of mercury. To convert to pounds per square inch (or kilograms per square centimeter),

$$P_2 = dh \qquad (1)$$

*Project Engineer, Tennessee Division Robertshaw Controls Company, Knoxville, Tennessee. Several tables were contributed by L. S. Dysart, formerly of Robertshaw Controls Company.

where P_2 = pressure, psig (kg/cm^2)
  d = density, lb/in^3 (kg/cm^3)
  h = height, in (cm)

For mercury, the density is 0.490 lb/in^3 at 60°F (15.6°C), and the conversion of inches of mercury to pounds per square inch becomes

$$P_2 = 0.490h \tag{2}$$

The density of water at 60°F (15.6°C) is 0.0361 lb/in^3, and if water is used in a manometer, the conversion of inches of water to pounds per square inch becomes

$$P_2 = 0.0361h \tag{3a}$$

The same principles apply when metric units are used. For example, the density of mercury at 15.6°C (60°F) may also be expressed as 0.0136 kg/cm^3, and the conversion of centimeters of mercury to kilograms per square centimeters

$$P_2 = 0.0136h \tag{3b}$$

For measuring differential pressure and for static balance,

$$P_2 - P_1 = dh \tag{4}$$

The U-tube manometer principle is also utilized in industry in an instrument usually called a *differential-pressure manometer*. In this device (Fig. 2), the tubes are expanded into chambers and a float rides on top of the liquid in one of the chambers. The float positions an outside pointer through a pressuretight bearing or torque tube.

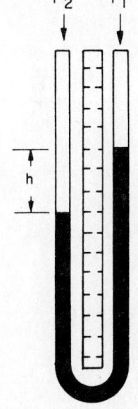

Well Manometer

In industrial well manometers (Fig. 3) one leg is replaced by a large-diameter well so that the pressure differential is indicated only by the height of the column in the single leg. The ratio of the diameters is important and should be as great as possible to reduce the errors resulting from the change in level in the large-diameter well.

Fig. 1 Manometer U tube.

The pressure difference can be read directly on a single scale. For static balance,

$$P_2 - P_1 = d \left(1 + \frac{A_1}{A_2} \right) h \tag{5}$$

where A_1 = area of smaller-diameter leg
  A_2 = area of well

If the ratio of A_1/A_2 is small compared with unity, then the error in neglecting this term becomes negligible, and the static balance relation becomes

$$P_2 - P_1 = dh \tag{6}$$

On some manometers, this error is eliminated by reducing the spacing between scale graduations by the required amount.

Inclined-Tube Manometer

In this device (Fig. 4), so as to read small pressure differentials more accurately, the smaller-diameter leg is inclined. This produces a longer scale so that

$$h = L \sin \alpha \tag{7}$$

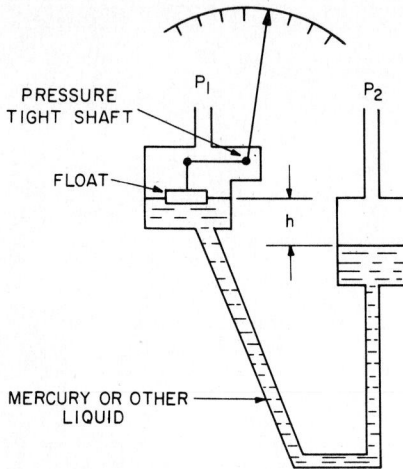

Fig. 2 Differential-pressure manometer.

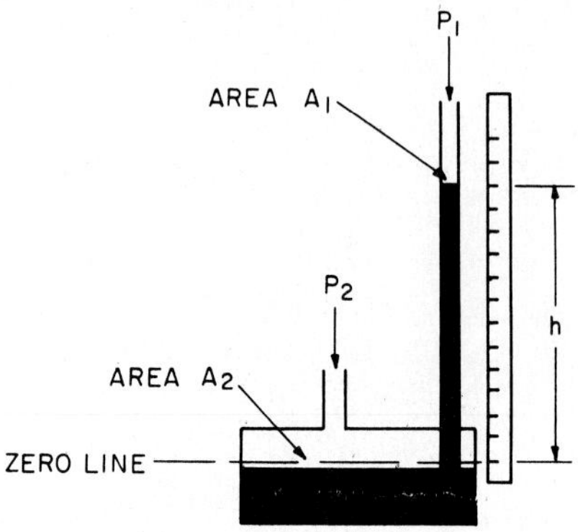

Fig. 3 Well manometer.

where L = length of scale corresponding to height h
α = angle of inclination of small-diameter leg

For static balance,

$$P_2 - P_1 = d\left(1 + \frac{A_1}{A_2}\right) L \sin \alpha \qquad (8)$$

or
$$P_2 - P_1 = dL \sin \alpha \qquad (9)$$

for a small ratio of A_1/A_2. By reducing the angle α, the scale length can be greatly increased. For good operation, however, the ratio of L to h should not exceed $10:1$.

Liquid-Sealed Bell

This device (Fig. 5) utilizes a container immersed in a sealing liquid. The pressure to be measured is applied to the inside of the bell, the motion of which is opposed by a restricting spring. For static balance, neglecting the buoyant effect of the liquid on the thin-walled bell,

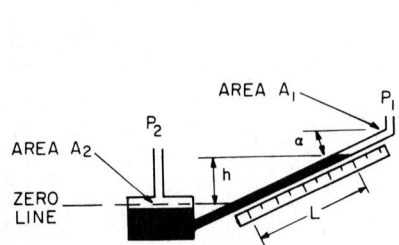

Fig. 4 Inclined-tube manometer.

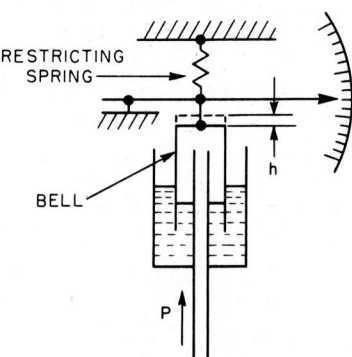

Fig. 5 Liquid-sealed bell.

$$P = \frac{F_c h}{A} \tag{10}$$

where P = pressure change inside bell
F_c = spring constant of restricting spring
h = motion of bell
A = area of interior of bell

Bell-Type Differential-Pressure Gage

This device (Fig. 6) consists of a bell sealed within another container, with pressures applied to both the outside and the inside of the bell. The motion of the bell is restricted again by an opposing spring. The motion must be transferred to the pointer by means of a pressure-tight shaft. For static balance,

$$P_2 - P_1 = \frac{F_c h}{A} \tag{11}$$

Double-Bell-Type Differential-Pressure Gage

This device (Fig. 7) consists of two identical bells immersed in a sealed liquid, the motion of the bells being restrained by a weight W suspended on a balance beam as shown. For static balance,

$$P_2 - P_1 = \frac{Wd \sin \alpha}{lA} \tag{12}$$

where d = weight radius about pivot of beam
l = lever arm of bell about pivot
A = area of bells
α = angle of weight from vertical

Liquid Barometer

A simple barometer may be constructed from a glass tube which is closed at one end and open at the other. The length of the tube must be greater than 30 in (76.2 cm). The tube is first completely filled with mercury, the open end temporarily plugged, and then the plugged end placed in a container partially filled with mercury.

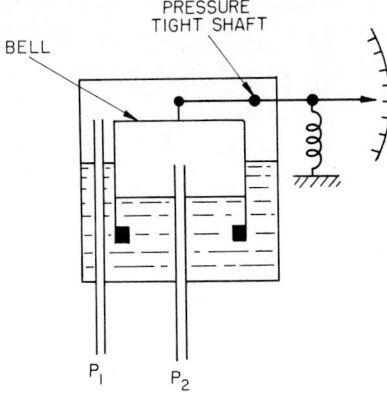

Fig. 6 Bell-type differential-pressure gage.

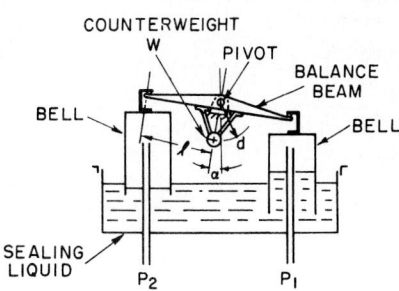

Fig. 7 Double-bell-type differential-pressure gage.

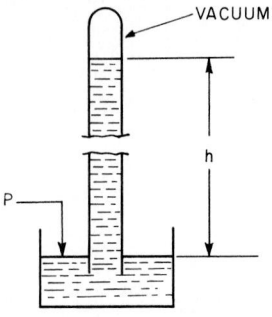

Fig. 8 Liquid barometer.

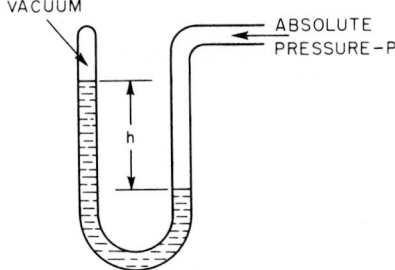

Fig. 9 Absolute-pressure gage.

When the plug is removed, the mercury in the tube will drop by a certain amount, creating a vacuum at the top of the tube. The height of the column, as measured in Fig. 8 and expressed in inches or millimeters of mercury, will then be proportional to atmospheric pressure.

Absolute-Pressure Gage

This gage, shown in Fig. 9, consists of a glass U tube partially filled with mercury, with the top of one leg evacuated and sealed off. The pressure to be measured is applied to the other leg, and h may be read in inches of mercury absolute. To convert to pounds per square inch absolute (psia),

$$P = 0.490h \tag{13}$$

where $P =$ absolute pressure, psia. If h is indicated in centimeters, this value may be converted to kilograms per square centimeter absolute by multiplying by 0.0136.

Manometers with Digital Indication

When using a basic manometer, the resolution and accuracy of the measurements are limited to some extent by the operator's ability to visually measure the height of the liquid column. Vernier scales and magnifying eyepieces are sometimes added to the instrument to facilitate this measurement.

Even greater accuracy may be achieved by using sonar techniques to determine the column height, and displaying the resulting measurement on a digital meter.

Manometer Liquids

Several liquids with a wide range of densities are available for use with manometers. The choice depends primarily on the magnitude of the pressure to be measured and the required resolution. Transposing Eq. 1, we have

$$\frac{h}{P} = \frac{1}{d} \tag{14}$$

From this it is seen that the column height of a water manometer, for example, would be 13.6 times greater than that of a mercury manometer, with equal pressure applied to both instruments.

Liquids used with manometers are classified according to their specific gravity. Commonly available specific gravities are 0.827, 1.000, 1.750, 2.950, and 13.560. The density of a liquid varies with temperature, and this error must be compensated for when making high-precision measurements.

Manometer scales may be graduated in any desired units. For example, mercury manometers often have dual scales, with one scale graduated in inches of mercury and the other scale graduated in pounds per square inch. Also, a manometer filled with a liquid having a specific gravity near that of water is usually equipped with a scale that is read directly in inches of water. Equivalent metric scales are also used.

Table 1. Ranges of Elastic Elements

Element	Application	Minimum range	Maximum range
Capsule	Pressure	0–0.2 in (0.5 cm) H_2O	0–1000 psig (70.3 kg/cm^2)
	Vacuum	0–0.2 in (0.5 cm) H_2O	0–30 in (76.2 cm) Hg vacuum
	Compound vacuum and pressure	Any span within pressure and vacuum ranges with a total span of 0.2 in (0.5) cm H_2O	—
Bellows	Pressure	0–5 in (12.7 cm) H_2O	0–2000 psig (141 kg/cm^2)
	Vacuum	0–5 in (12.7 cm) H_2O	0–30 in (76.2 cm) Hg vacuum
	Compound vacuum and pressure	Any span within pressure and vacuum ranges with a total span of 5 in (12.7 cm) H_2O	—
Bourdon	Pressure	0–5 psig (0.35 kg/cm^2)	0–100,000 psig (7030 kg/cm^2)
	Vacuum	0–30 in (76.2 cm) Hg vacuum	—
	Compound vacuum and pressure	Any span within pressure and vacuum ranges with a total span of 12 psi (0.84 kg/cm^2)	—

Elastic Elements

The four principal elements in this category are the bellows, bourdon tube, capsule, and diaphragm. The first three items are often used in similar applications. Table 1 shows the pressure ranges for which each type is usually used.

Bellows Elements

A bellows element is a thin-wall metal tube with deeply convoluted sidewalls which permit axial expansion and contraction.

Bellows vary in size from approximately ⅟₁₆-in (1.5 mm) to 6 ft (1.8 m) in diameter, but those used for pressure elements are limited to a maximum diameter of approximately 6 in (15 cm). Figure 10 shows common types of bellows and methods of attaching fittings. Note that the ends may be trimmed to receive either a disk- or a ring-type fitting. Also, bellows may be formed with only one end open or with both ends open.

Manufacture

Most bellows are made from seamless tubes, and the convolutions are either hydraulically formed or mechanically roll-formed. Combinations of these two processes are sometimes used in order to obtain required characteristics.

Bellows may also be manufactured by welding a series of formed plates together at their inner and outer diameters. Elements fabricated by this method usually have more free plates per unit length than those formed by the hydraulic process.

For several applications, bellows may also be formed by electrodepositing metal on an expendable mandrel which is later removed by melting.

Materials

Seamless bellows are made of brass, phosphor bronze, beryllium copper, Monel, stainless steel, Inconel, and other metals. Welded bellows are usually made of stainless steel, Monel, Inconel-X, or Ni-

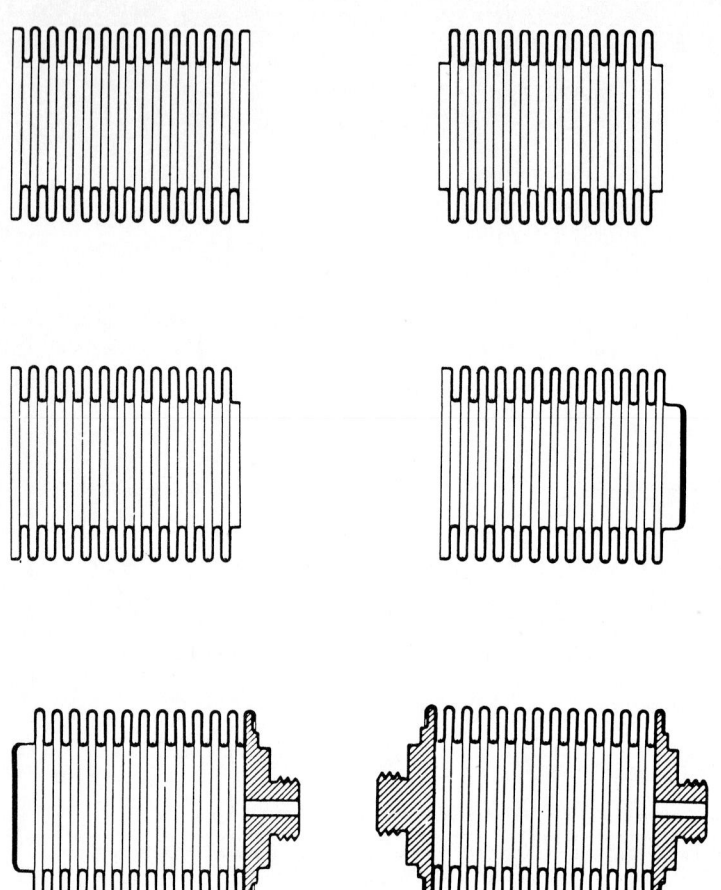

Fig. 10 Common types of bellows and methods of attaching fittings.

Span C. The choice of material depends on pressure and deflection specifications and also on the corrosion resistance required.

Bellows Spring Rate (Force Constant)

This is defined by

$$R_B = \frac{\Delta F}{\Delta d} \tag{15}$$

where R_B = bellows spring rate, lb/in (kg/cm)
ΔF = force increment applied to free end of bellows, lb (kg)
Δd = deflection due to force increment, in (cm)

Deflection with Force

This can be expressed by transposing Eq. 15 as

$$\Delta d = \frac{\Delta F}{R_B} \tag{16}$$

Deflection with Pressure

If pressure is applied to an unrestrained bellows assembly, the resulting deflection of the free end is

$$\Delta d = \frac{\Delta P A_e}{R_B} \tag{17}$$

where ΔP = pressure increment, psi (kg/cm^2)
    A_e = effective area of bellows, in^2 (cm^2)

If an opposing spring is used with the bellows, deflection is

$$\Delta d = \frac{\Delta P A_e}{R_B + R_s} \tag{18}$$

where R_s = spring rate of spring, lb/in (kg/cm).

Effective Area

An empirical formula often used to calculate the effective area of bellows is

$$A_e = \frac{\pi(A + B)^2}{16} \tag{19}$$

where A = outside diameter, in (cm)
    B = inside diameter, in (cm)

Characteristics

A complete listing of bellows characteristics for various designs can be obtained from the manufacturer. A sample group of bellows is represented in Table 2. Note that the maximum stroke and the maximum pressure are given, along with several other variables.

Maximum Pressure

Consider a bellows which is closed at one end, has pressure applied to the other end, and is clamped in a fixture to prevent deflection. The applied pressure will distort the convolutions and develop stress in the bellows. Under these conditions, the value of pressure that develops the maximum allowable stress for a given bellows is referred to as the *maximum pressure rating* for the bellows.

Maximum Stroke

Stress is also developed in the convolutions of a bellows by applying a force to the free end, causing the bellows to deflect in accordance with Eq. 16. The magnitude of stroke which develops the maximum allowable stress in a given bellows is referred to as the *maximum stroke rating* for the bellows.

Probable Life

If a bellows has its free end closed and pressure applied at the fixed end, the free end will deflect in accordance with Eq. 17, and stresses will be developed by both the pressure and the deflection. Given the maximum pressure rating and the maximum stroke rating for a specified bellows, the probable life of the bellows may be estimated for any combination of pressure and stroke, as shown in Fig. 11.

Bellows Applications

Bellows pressure elements are often used in applications that require relatively long strokes and high developed forces. They are well suited for input elements for recorders and indicators and for feedback elements in pneumatic control applications. Figures 12 and 13 show typical assemblies for these kinds of applications.

Table 2. Characteristics of Single-Ply Brass Bellows*

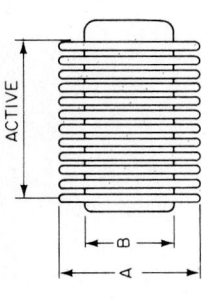

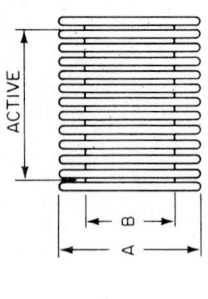

A, outside diam, in	B, inside diam, in	Approx. length per active convolution, in	Effective area, in²	Maximum stroke per active convolution, in	Spring rate per active convolution, lb/in	Stroke per active convolution, in per psi pressure	Maximum pressure, psi
15/32	5/16	0.045	0.12	0.0067	600	0.0002	315
9/16	23/64	0.045	0.16	0.0044	685	0.00023	235
3/4	1/2	0.057	0.31	0.0128	480	0.00 065	165
1 1/64	5/8	0.091	0.53	0.0285	125	0.0 0424	55
1 1/2	63/64	0.151	1.2	0.0293	825	0.00146	170
2	1 23/64	0.141	2.22	0.0405	346	0.0064	60
2 7/16	1 37/64	0.195	3.18	0.0517	528	0.00603	70
2 15/16	2 3/4	0.164	5.07	0.055	388	0.0131	50
4 3/4	3 1/8	0.19	9.92	0.1025	168	0.0592	20
5 5/8	4 15/16	0.231	23.0	0.0872	300	0.0769	20
8 5/8	7 57/64	0.22	55.5	0.0409	1218	0.0449	25
12	10 3/16	0.225	102.4	0.0676	530	0.198	10
Notes regarding use of table		Multiply by number of active convolutions to obtain normal free length of active portion of bellows.		Multiply by number of active convolutions to obtain maximum stroke of active portion of bellows.	Divide by number of active convolutions to obtain spring rate of active portion of bellows.	Multiply by number of active convolutions to obtain flexibility of active portion of bellows.	Do not use with maximum stroke. See Fig. 16, life expectancy chart.

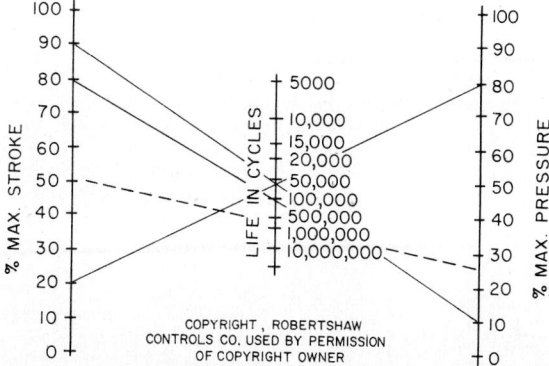

Fig. 11 Chart for determining service-life expectancy of bellows. This chart shows the relation between stroke, pressure, and life. It is based on the results of hundreds of tests conducted on various sizes and types of bellows under many different conditions.

Using a straight-edge, connect the *stroke*, expressed in percentage of the maximum stroke, and the *pressure,* expressed in the percentage of the maximum pressure. This line intersects the central scale at the *probable life,* expressed in cycles.

EXAMPLE:

A pressure element consisting of a brass bellows and a range spring is to make a stroke of 0.205 in with an applied pressure of 42.5 psi. The bellows has an outside diameter of 1½ in, an inside diameter of ⁶⁹⁄₆₄ in, and 14 active corrugations. What will be its probable life?

Solution: (Refer to Table 2)
Maximum stroke rating = 14 × 0.0293 = 0.410 in; percent of maximum stroke rating = (0.205/0.410) × 100 = 50%; percent of maximum pressure rating = (42.5/170) × 100 = 25%. The line connecting 50% maximum stroke and 25% maximum pressure intersects the central scale at 1 million cycles, which is the probable life under the stated conditions.

Note: The chart is equally applicable to metric or any other system of consistent units.

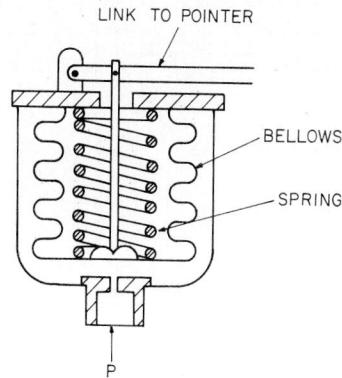

Fig. 12 Spring-loaded bellows gage.

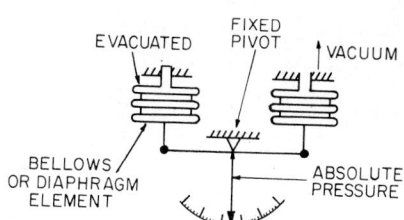

Fig. 13 Bellows absolute-pressure gage.

Differential-Pressure Elements

Figure 14 shows a compact element which utilizes a single small-diameter bellows. This assembly is used on widespan applications and where any overrange is relatively small.

The assembly illustrated schematically in Fig. 15 contains a pair of large-diameter oil-filled bellows which makes the element suitable for measuring low values of differential pressure at high values of static pressure. Differential-pressure spans down to 10 in (25.4 cm) H_2O are typical, and static pressures up to 6000 psi (421.8 kg/cm^2) can be tolerated. Note that the total volume enclosed by the two bellows is completely filled with oil, so that the pressure applied to the outside of the bellows will cause an equivalent oil pressure to be developed inside the bellows. This action maintains the bellows stress at a low level, making it possible to use thin-wall bellows with high sensitivity. The oil fill and overrange valves make it possible for this element to withstand high values of differential-pressure overrange.

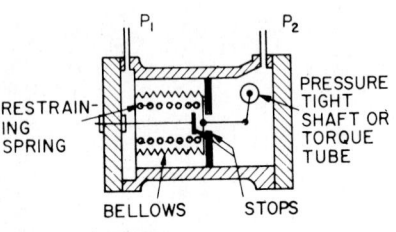

Fig. 14   Bellows differential gage.

Another function of the oil fill is to help provide pulsation damping. Movement of the bellows assembly will cause oil to be forced from one bellows to the other, and this displaced oil must flow through the damper valve. This valve may be adjusted to provide the desired amount of damping action.

Pneumatic Servomechanism

This device is sometimes used as a substitute for a single capsule or bellows to operate recorders and similar devices that require precision positioning of high-friction loads. For small increments in input pressure, a servomechanism equipped with a given motor bellows can develop up to 100 times more

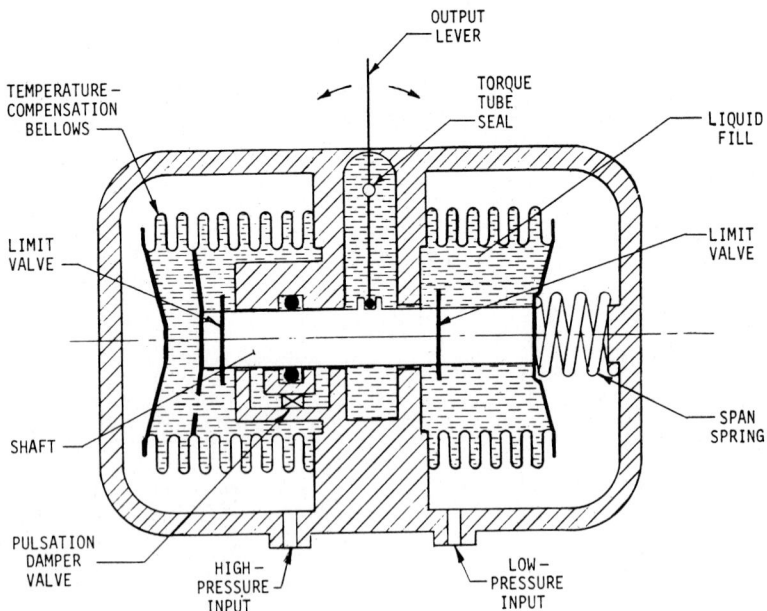

Fig. 15   Double-bellows differential-pressure gage.

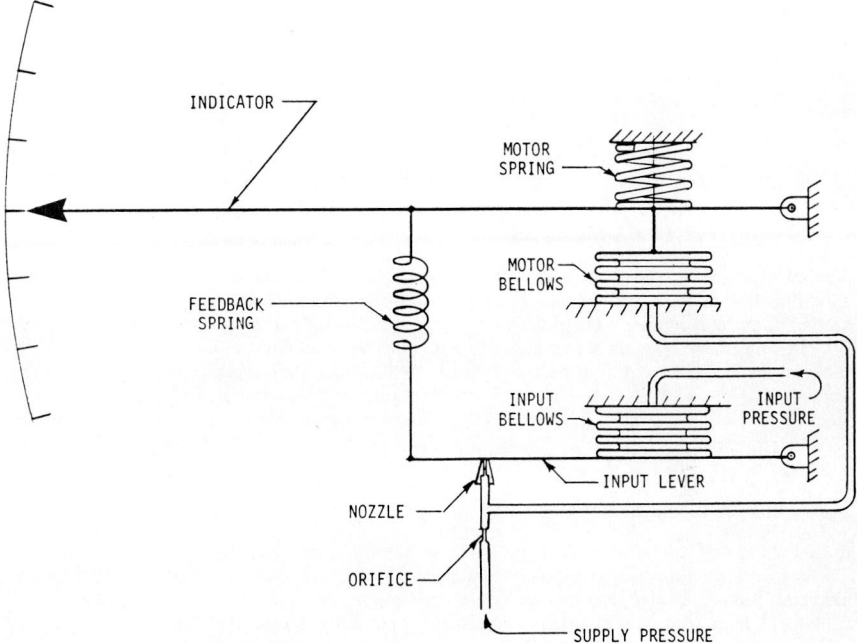

Fig. 16   Pneumatic servomechanism.

force than can be developed by the bellows alone. A typical servomechanism is shown schematically in Fig. 16.

Operation of the device is as follows. With the input pressure at zero, the force developed by the input bellows is low, allowing the feedback spring to pull the input lever away from the nozzle, resulting in a low nozzle back-pressure. This low pressure in the motor bellows allows the motor spring to force the indicator to a low position. When input pressure is applied, the input lever is forced toward the nozzle, causing the nozzle back-pressure to increase. This pressure increase in the motor bellows will force the indicator to move upward, stretching the feedback spring. The indicator will continue to move upward until the feedback spring pulls the input lever away from the nozzle to the point where a force-balance condition is obtained between the input pressure and the feedback spring.

Bourdon Tube Elements

In the original patent of E. Bourdon in 1852, the bourdon tube was described as a curved or twisted tube whose transverse section differed from a circular form. In principle, a tube closed at one end with an internal cross section that is not a perfect circle, if bent or distorted, has the property of changing its shape with internal pressure variation. An internal pressure increase causes the cross section to become more circular and the shape to straighten, resulting in motion of the closed end of the tube provided the open end is held rigid. This motion, commonly called "tip travel," is shown by arrow T in Fig. 17 (rising pressure). The amount of tip travel is a function of tube length, wall thickness, cross-section geometry, and modulus of the tube material.

The tube, when formed in a specified manner, becomes a bourdon spring element. Common forms are C, helical, and spiral.

The C type is formed by winding the tube circularly to form a segment of a circle (Fig. 17a).

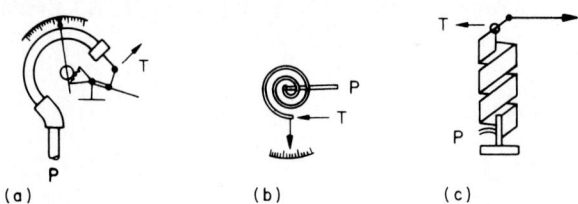

(a)   (b)   (c)

Fig. 17 Types of bourdon springs: (a) C type, (b) spiral, (c) helical.

The spiral type is formed by winding more than one turn in the shape of a spiral about a common axis (Fig. 17b). The helical type is formed by winding more than one turn in the shape of a helix (Fig. 17c). Forming of both cross section and shape frequently is done in the same winding operation.

The type of bourdon spring used is primarily a design requirement. The allotted space in an instrument case is one of the main considerations. What can be accomplished by one type can be duplicated by the other types, provided the proper length of tubing material can be used. In general, spiral and helical types, having greater tube lengths, are used where more tip travel is needed or where a reduction in stress level in the tube is required.

Bourdon Spring Materials

A bourdon spring can be made from any metal or alloy that exhibits satisfactory elastic qualities. The basic requirements for tube materials and the classification of alloys for bourdon springs are discussed in the literature (Ref. 3). The several paragraphs that follow are on (1) classification of alloys by temper, (2) characteristics of alloys, and (3) corrosion resistance.

Classification of Alloys by Temper

The large number of alloys used for bourdon purposes can be classified conveniently into groups on the basis of the treatment necessary to obtain the temper or hardness level required for a specific application. Such a grouping may be made as follows:

Group I. Strain-hardened alloys (Table 3):

1. Cartridge brass
2. Trumpet brass
3. Phosphor bronze
4. Silicon bronze
5. Austenitic stainless steel (AISI Types 304 and 316)
6. Monel

Group II. Precipitation-hardened alloys (Table 4):

1. Beryllium copper
2. K Monel
3. Ni-Span C
4. Inconel X

Group III. Heat-treated (quenched and drawn) alloys (Table 5):

1. Low-alloy steels (AISI 4130, 8630)
2. Martensitic stainless steel (AISI Type 403)

Table 3. Characteristics of Group I (Strain-Hardened) Alloys Used in Bourdon Spring Elements

Group I alloys	Nominal composition, %		Modulus of elasticity × 10^6	Temper usually supplied from mill	Rockwell B hardness	Nominal grain size, min.	Usual heat treatment	Common joining			Relative ease of forming	Application and remarks
								Weld	Braze	Soft solder		
Trumpet brass	Cu 81 Zn 18 Sn 1		15	3–4 B & S hard	80	0.035	None	—	X	X	Excellent	Widely used in low-pressure field under 500 psi (35.1 kg/cm^2)
Cartridge brass	Cu 70 Zn 30		15	3–4 B & S hard	80	0.035	None	—	X	X	Excellent	Low-pressure gages where drift and hysteresis are not critical
Phosphor bronze, grade A	Cu 95 Sn 5 (P 0.035)		15.5	3–4 B & S hard	80	0.035	None	—	X	X	Good	Low-pressure gages. Good resistance to corrosion and fatigue
Silicon bronze, grade B	Si 1.50 Mn 0.25 Cu Bal.		15	3–4 B & S hard	80	0.035	None	X	X	—	Excellent	Low-pressure gages. Difficult to solder because of surface oxide
AISI Type 304 stainless steel	Cr 18 Ni 9 C 0.08 max		28–29	5–10% reduction area	92	0.050	Stress relief 600–700°F	X	— (316–371°C)	—	Fair	Low- to medium-pressure gages. High corrosion resistance. Work-hardens rapidly
AISI Type 316 stainless steel	Cr 17 Ni 13 Mo 3 C 0.08 max		27.5–28.5	5–10% reduction area	92	0.060	Stress relief 600–700°F	X	— (316–371°C)	—	Fair	Same as above, except somewhat better corrosion resistance than Type 304 especially for sulfates and chlorides

Table 4. Characteristics of Group II (Precipitation-Hardened) Alloys Used in Bourdon Spring Elements

Group II alloys	Nominal composition, %		Modulus of elasticity $\times 10^6$	Temper usually supplied from mill	Rockwell B hardness	Nominal grain size, min.	Usual heat treatment	Common joining			Relative ease of forming	Application and remarks
								Weld	Braze	Soft solder		
Beryllium copper	Be Co Ni Cu	1.80 0.25 0.30 Bal.	18.5–19.0	5–10% reduction area	75	0.035	1–2 h at 600–650°F (316–343°C). Furnace- or air-cooled	—	X	X	Good	Medium- to high-pressure (over 500 psi) gages. Low hysteresis and relaxation limits. High endurance limits
Monel K	Ni Al Ti Cu	63–67 2.0–4.0 0.50 Bal.	26	5–10% reduction area	90	0.020	16 h at 1100°F (593°C). Slow cool 25°F (14°C)/h to 900°F (482°C) or 8 h at 1100°F (593°C), 4 h at 1000°F (538°C), 4 h at 900°F (482°C). Air-cooled	X	X	—	Good	Medium- to high-pressure gages. High fatigue strength. Low hysteresis and good corrosion resistance
Inconel X	Ni Cr Ti Cb Al	70 min. 15.0 2.5 1.0 0.40–1.00	31	5–10% reduction area	—	Not usually specified	16 h at 1350°F (732°C) recommended. However, satisfactory results obtained with 4–8 h at 1350°F (732°C)	X	X	—	Fair	For high-pressure gages, 1000 psi, or higher. Excellent mechanical and corrosion properties
Ni-Span C	Ni Cr Ti Mn C Si Al	42 5.5 2.5 0.4 0.06 0.5 0.4	24–27 Depending on cold work and heat treatment	5–10% reduction area	85	Not usually specified	8 h at 1100°F (593°C) or 4.5 h at 1250°F (677°C). Air-cooled	—	X	X	Good	Medium- to high-pressure gages. Constant modulus alloys, having good resistances

Table 5. Characteristics of Group III (Quenched and Tempered) Alloys Used in Bourdon Spring Elements

Group III alloys	Nominal composition, %	Modulus of elasticity × 10^6	Temper usually supplied from mill	Rockwell B hardness	Nominal grain size, min.	Usual heat treatment	Common joining			Relative ease of forming	Applications and remarks
							Weld	Braze	Soft solder		
AISI 4130	C 0.28–0.33 Cr 0.80–1.10 Mo 0.15–0.25	28.5	Annealed	85	Not specified	Oil quench from 1600°F (871°C). Temper 750–950°F (400–510°C)	X	X	—	Good	Medium- to high-pressure gages. Low hysteresis and relaxation
AISI 8630 steel	C 0.28–0.34 Cr 0.40–0.60 Ni 0.40–0.70 Mo 0.15–0.25	28.5	Annealed	85	Not specified	Oil quench from 1600°F (871°C). Temper 750–950°F (400–510°C)	—	X	—	Good	Medium- to high-pressure gages. Low hysteresis and relaxation
AISI 403 steel	C 0.15 max Cr 11.00–12.50 Ni 0.35–0.55 Mo 0.40–0.60	28.5	Annealed	86	Not specified	Oil quench 1750–1800°F (954–982°C) Temper 450–750°F (232–400°C)	X	X	—	Good	Medium- to high-pressure gages. Good combination of mechanical and corrosion properties

Although this grouping shows the common alloys used for bourdon tube application, it is by no means exclusive because many other alloys used for the same application fall into one or the other of the foregoing groups.

Characteristics of Alloys

Group I alloys are supplied in the strain-hardened (cold-worked) condition for bourdon-type purposes. Either a certain hardness range or a given cold reduction is specified by the gage manufacturer. In addition, close grain size control is maintained to provide uniform performance from one bourdon tube to the next.

Ordinarily this group does not require heat treatment after coiling, although stress relieving at low temperatures can be used to advantage in improving the precision of the gages. The copper base alloys in this group are relatively easy to form (coil); trumpet brass and cartridge brass are probably the easiest of all bourdon materials to form.

Group II alloys, preshaped in random or cut-length tubes, are generally supplied either in the slightly strain-hardened (approximately one-eighth hard) condition or to a specified hardness range. After forming, this group must be precipitation-hardened to obtain optimum operating characteristics. Generally, the heat treatment for precipitation hardening consists of holding the material within a specified temperature range for a predetermined time period and either furnace cooling or air cooling as indicated.

Group II alloys produce high-precision bourdon tubes possessing good creep resistance, low hysteresis, and low relaxation. These alloys are more difficult to form than the alloys of Group I and, therefore, should be obtained from the tube mill in the softest possible condition. Fatigue life of alloys in this group is usually excellent, although corrosion can cause early failure.

Group III alloys generally are supplied in soft annealed condition; however, for extremely light-wall tubing it is desirable to obtain tubing "as shaped," in order that specified dimensional tolerances can be maintained. In this condition the material has undergone a slight amount of cold work, which does not interfere with subsequent forming or coiling operations.

Alloys in Group III are not difficult to form, being intermediate between Groups I and II in workability. The martensitic stainless steels (Types 403 and 420) are somewhat more difficult to form than the low-alloy tubing (such as 4130) because of higher base hardness or strength.

All these alloys must be quenched and drawn (tempered) after forming to obtain the necessary mechanical properties. They can be used in high-pressure applications and show exceptionally good characteristics with respect to creep, hysteresis, and zero shift. Endurance limits of these alloys also are high—provided the surface of the original tubing is free of imperfections such as seams and gouges. Although this qualification applies to all seamless bourdon tubing, particular precautions as to surface quality must be taken with the low-alloy steels of Group III because of certain defects invariably occurring during piercing and hot-working of the seamless-tube rounds.

Corrosion Resistance

One of the important factors in determining the proper selection of bourdon tube materials is frequently the resistance to corrosion by a specific agent. Scores of effective corrosion resistance guides are available from manufacturers and found in the literature. Exemplary of the latter is sec. 23, Materials of Construction, in Perry's *Chemical Engineers' Handbook,* 5th ed., McGraw-Hill, 1973.

Bourdon Tube Design

An exact analysis of the behavior of a bourdon tube under pressure is extremely complicated. At the present time, no analytical developments have progressed to the point where a sufficiently reliable determination of bourdon tube deflections has been possible. Bourdon tube design in practice has been based on empirical equations derived from practical observations.

The deflection of a bourdon tube (Ref. 4), shown schematically in Fig. 18, is expressed by

$$\Delta a = K \frac{aP}{E} \int (A,B,t,R) \qquad (20)$$

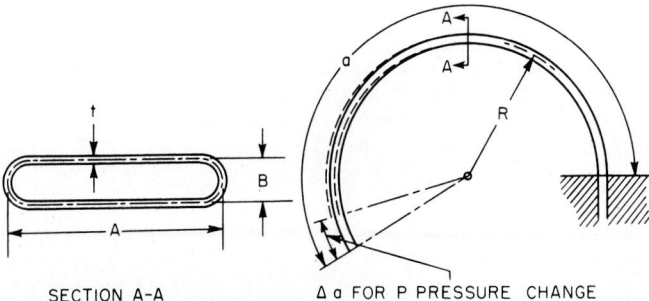

SECTION A-A   Δ a FOR P PRESSURE CHANGE

Fig. 18 Bourdon dimensions.

where Δa = angular deflection of element tip, deg
K = empirical constant determined by test on a number of bourdon tubes
a = total angle subtended by bourdon tube element, deg (see Fig. 18)
P = differential pressure between inside and outside of tube, psi
E = modulus of elasticity of tube material
A,B = cross-sectional length (see Fig. 18) and width of tube, in
t = thickness of tube, in
R = radius of curved tubing, in (in C- and helical-type elements, this is a constant; in a spiral type, a variable)

One form this relationship can take is

$$\Delta a = K \frac{aP}{E} \left(\frac{R}{t}\right)^x \left(\frac{A}{B}\right)^y \left(\frac{A}{t}\right)^z \qquad (21)$$

where x, y, and z, as well as K, are constants which must be determined by running tests on a number of bourdon tubes.

A typical equation found for a flat, cross-sectional type of tube as shown in Fig. 18 is

$$\Delta a = 0.05 \frac{aP}{E} \left(\frac{R}{t}\right)^{1/5} \left(\frac{A}{B}\right)^{1/3} \left(\frac{A}{t}\right)^3 \qquad (22)$$

The constants were determined from test results on nine bourdon tubes according to data given in Table 6.

Table 6. Design Constants for Circularly Bent Bourdon Tubes*

Bourdon tube no.	Dimensions, in				Observed $\Delta a E/ap$	Calculated $\Delta a E/ap$	Error, %
	A	B	t	R			
1	0.639	0.115	0.014	1.462	21,050	21,800	−3.5
2	0.649	0.135	0.031	2.025	1,900	1,810	+5
3	0.649	0.136	0.031	2.031	1,850	1,810	+2
4	0.649	0.126	0.031	1.444	1,660	1,732	−4
5	0.635	0.186	0.042	2.031	563	566	−0.5
6	0.635	0.193	0.042	1.998	552	556	−0.5
7	0.638	0.176	0.042	1.480	586	561	+4.5
8	0.655	0.200	0.056	1.995	237	243	−2.5
9	0.655	0.199	0.056	2.027	233	244	−4.5

*Ref. 1.

Table 7. Helical Element with 2½ Turns (Usable Deflection 25°)
$A = 0.604 + t, B = 0.027 + t, R = 0.469$

Material	Thickness t, in	Pressure, psi per angular degree deflection $P/\Delta a$	Torque gradient, g cm per angular degree deflection G	Kt	Empirical constant K	Torque-gradient constant C
Type 316 stainless steel	0.011	4.83	79	0.0069	0.63	16.4
	0.019	17.0	270	0.0115	0.61	15.9
	0.025	30.0	485	0.0160	0.64	16.1
Trumpet metal	0.012	3.30	35	0.0074	0.62	10.6
	0.0165	6.80	83	0.0104	0.63	12.2
	0.023	14.3	177	0.0146	0.63	12.4
	0.0265	20.0	235	0.0162	0.62	11.7

Performance of the elements of one manufacturer follows the formula

$$\Delta a = (0.63t)\frac{aP}{E}\left(\frac{R}{t}\right)^{1/5}\left(\frac{A}{B}\right)^{1/3}\left(\frac{A}{t}\right)^{3} \tag{23}$$

Table 7 lists data from bourdon elements designed for industrial instruments.

Life

In designing an element, the life of the element must be considered. The final design is based on practical observations of pressure-deflection relationships and pulsation or overpressuring tests. Elements can initially be overstressed to take a residual set and then can be worked to this overstressed pressure without any additional creep. Only experimentation will determine this allowable pressure. A useful criterion that has been found satisfactory is to apply pressure to the element and then remove the pressure. The pressure at which a zero set of 1% of the deflection is obtained is considered a satisfactory safe operating limit.

All bourdon tubes should be tested in some manner as previously indicated, as a reference on which the life of an element can be based.

Torque-Gradient Constant

This factor can be defined as the moment of the force applied at the tip of the spring (perpendicular to the direction of motion) times its distance from the instantaneous center of rotation required to give one angular degree of rotation. If the element is pivoted about a center shaft, its torque gradient is easily ascertained.

The torque gradient of a bourdon tube is very important, because it is a determining factor in the design of a complete instrument, as to what mechanisms can be operated with desired accuracy and sensitivity. The torque gradient will increase with pressure rating of the bourdon element for similarly designed cross-sectional tubes. Specifically,

$$G = C\frac{P}{\Delta a} \tag{24}$$

where G = torque gradient, lb-in per degree of deflection
  C = torque-gradient constant

Power Rating

Frequently, the power rating of the bourdon element is utilized:

$$W = FM \tag{25}$$

where F = force acting at tip, parallel to direction of motion, required to restrain element with application of pressure corresponding to pressure rating of element, lb
M = motion of element, unrestrained with same pressure applied, in
W = power rating or work factor of element

For a bellows or capsule element,

$$W = (RM)M = RM^2 \tag{26}$$

where R = spring rate of element, lb/in
M = full-scale movement, in

Hysteresis

As in the other types of pressure elements, bourdon tube material possesses some hysteresis in a pressure cycle. The amount of hysteresis can be kept to a minimum by proper heat treatment and by utilizing proper materials. In a bourdon tube, the amount of hysteresis is on the order of 0.1 to 0.5% for the maximum pressure cycle. For smaller cycles normally encountered in practice the hysteresis may be negligible.

Sensitivity

The sensitivity of a bourdon tube without any type of restraining member (no center pivot or attached mechanism) is excellent—response to less than 0.01% of the maximum pressure is possible.

With a center pivot attached, the response level can be as high as 0.1% of the maximum pressure. When the bourdon tube is driving other mechanisms, the sensitivity of the system is further reduced.

Nonlinearity

The nonlinearity of a bourdon tube may be 0.5% of full scale or higher. When the element is installed in a recorder or indicator, however, the error can be eliminated by adjusting the linkage as explained later in the paragraph Compensation by Mechanical Adjustments. Also, if the nonlinearity is consistent from one element to another, the indicator scale can be graduated to compensate for the error.

Static Errors

The modulus of elasticity of most elastic materials decreases with increasing temperature. Hence, the deflection rate of a bourdon tube made of one of these materials will increase with increasing temperature. The error is approximately

$$E = 0.02t \frac{P}{P_S} \tag{27}$$

where E = error, % of full scale
t = temperature change from reference, °F
P = applied pressure, psi
P_S = pressure span, psi

This error can be reduced by the use of bimetal which will change the amount of multiplication as the temperature changes. A compensating device is shown in Fig. 19.

The modulus of Ni-Span C is virtually constant with changing temperature. Elements made of this material do not require temperature compensation.

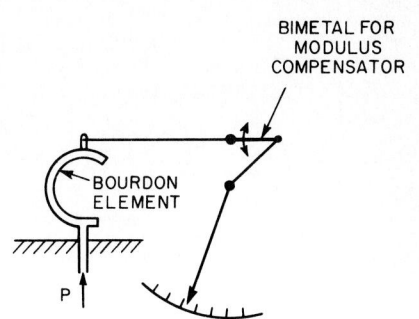

Fig. 19 Compensating device for ambient temperature changes.

Elevation Error

Elevation error in a pressure device may be present if the connecting line from the point of measured pressure to the measuring element is partially filled with a liquid. The error is

$$E = \frac{hd}{Ps} 100 \qquad (28)$$

where E = error of pressure device, %
 h = height of liquid between measuring element and point of measured pressure, in (cm)
 d = density of liquid in connecting line, lb/in^3 (kg/cm^3)
 Ps = pressure span of pressure device, psi (kg/cm^2)

Overrange and Underrange Protection

Overrange protection is required on most pressure measuring instruments; underrange protection is required on instruments calibrated for partial ranges (for example, a range of 20 to 50 psi). The element can be rigidly stopped, as shown in Fig. 27 which appears later in this article, or the indicator alone can be stopped, as shown in Fig. 20. In both examples, the element itself must be designed to withstand the overrange.

It should be noted that, when an element is used for a partial range, the torque gradient of the instrument is greatly affected. Whenever the multiplication of the motion of a given element is changed, the torque gradient changes inversely as the square of the multiplication. Assuming that the full-scale pen deflection is the same in each case,

$$\frac{G_e}{G_e^1} = \left(\frac{P - P_0}{P - P_1}\right)^2 \qquad (29)$$

where G_e = effective torque gradient at driven mechanism for pressure span
 G_e^1 = effective torque gradient at driven mechanism for partial pressure span, for same deflection of pointer
 $P - P_0$ = designed pressure span
 $P - P_1$ = partial pressure span

Example. A pressure element is designed to give a 50° deflection of the pen for a 100-psi change. It is desired to use this element for a partial range of 75 to 100 psi or a 25-psi span with the same 50° deflection of the pen. Then

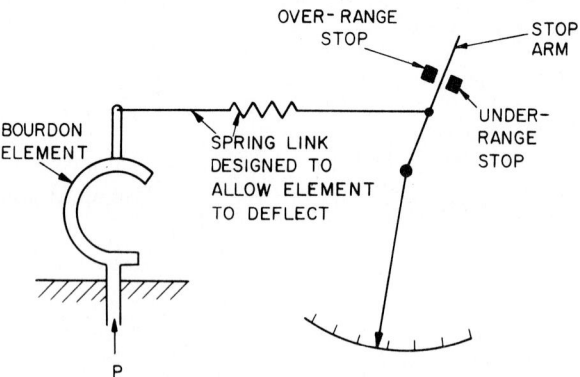

Fig. 20   Overrange and underrange protection.

$$\frac{G_e}{G_e^1} = \left(\frac{P - P_0}{P - P_1}\right)^2 = \left(\frac{100 - 0}{100 - 75}\right)^2 \tag{30}$$

or

$$G_e^1 = \frac{1}{16}Ge \tag{31}$$

Time Response

For a typical pressure element, the full-scale response to a gas pressure change is generally less than 100 ms if no flow restriction is encountered in the connecting tubing. Usually, the flow resistance of the connecting line is great enough to make the system time response slower, however, and this is particularly true for low-pressure measurements.

Diaphragms

Metallic Diaphragms

These elements are flexible disks, usually with concentric corrugations, that are used to convert pressure to deflection and also to serve as fluid barriers in transmitters, seal assemblies, and similar devices. They are also the basic component used in the fabrication of metallic capsules, which are discussed later in this article.

Materials

Metals commonly used for diaphragms are trumpet brass, phosphor bronze, beryllium copper, stainless steel, Ni-Span C, Monel, Hastelloy, titanium, and tantalum. Diaphragms made from different materials vary in stiffness according to their respective moduli of elasticity. Usually the metal is heat-treated before forming to produce a maximum elastic limit. After forming, the diaphragms are heat-treated to relieve stresses and then chemically cleaned before being assembled into capsules.

Design

A diaphragm is usually designed so that the deflection-versus-pressure characteristics are as linear as possible over a specified pressure range, and with a minimum of hysteresis and minimum shift in the zero point. When required by the application, however, a diaphragm may be designed with nonlinear characteristics, a type of element often used in altitude sensor capsules.

Both linearity and sensitivity are determined to a great extent by the depth and number of corrugations, and the angle of formation of the diaphragm face. The sensitivity can be increased by increasing the number of corrugations and by decreasing the depth of corrugations, with a sacrifice in linearity. The maximum sensitivity for extremely small motions is obtained by using a flat, uncorrugated diaphragm.

Deflection with Pressure

This deflection is dependent on a variety of factors: (1) diameter, (2) metal thickness, (3) shape of corrugations, (4) number of corrugations, (5) modulus of elasticity, and (6) applied pressure. An empirical formula often used to calculate this deflection is

$$\Delta d = \frac{\Delta P K D^4}{t^{1.5}} \tag{32}$$

where Δd = deflection, in
ΔP = pressure increment, psi
K = constant involving Young's modulus of elasticity and corrugation design
t = metal thickness, in

Table 8. Values of E and K for Common Materials

Material	Modulus of elasticity E		Capsule constant K
	psi	kg/cm^2	
Phosphor bronze	16×10^6	1.1×10^6	0.24×10^{-6}
Beryllium copper	20×10^6	1.4×10^6	0.19×10^{-6}
Ni-Span C	24–27.5×10^6	1.7–1.9×10^{65}	0.15×10^{-6}
Type 316 stainless steel	28×10^6	2.0×10^6	0.14×10^{-6}

It will be noted that the deflection varies as the fourth power of the diameter. Doubling the diameter increases the deflection 16 times for a given pressure change.

K is inversely proportional to the modulus of elasticity and varies with the design of the corrugation. Practical corrugated shapes produce two- to threefold variations in K. K values typical of one design are given in Table 8.

Application in Pressure Transmitters

Diaphragms are commonly used as pressure sensors in both pneumatic and electronic transmitters. In both applications, the full range deflection is usually limited to approximately 0.002 in (0.05 mm). In an electronic instrument, the deflection is converted to an electrical signal by a variety of methods, the most common employing a strain gage. The strain gage may be either a metallic or a semiconductor type.

Diaphragm Seals

In many pressure measuring applications, it is desirable to prevent the process fluid from contacting or seeping into the pressure measuring element or connecting line for the following reasons: (1) to prevent errors in measuring-element response due to effects of static head, (2) to protect the measuring element from corrosive fluids, congealing or viscous fluids, or clogging by entrained materials in the measured fluid, (3) to make possible cleaning of critical process systems (as in food and pharmaceutical production), and (4) to prevent loss of expensive or hazardous process fluids in case of failure of the measuring element or the need to remove the pressure instrument for service.

Such protection generally involves the use of diaphragm seals between the process fluid and the measuring system.

Construction

These devices, also known as *chemical seals,* use a flexible diaphragm made of a suitable corrosion-resistant material. The diaphragm is attached to or sealed within a suitable "body" which provides attachment to the process pressure on one side of the diaphragm and to the measuring element on the other side. See Figs. 21 through 24.

Diaphragms and bodies are available in a wide variety of materials. Selection is on the basis of a material suitable for the service at lowest cost. Corrugated diaphragms are used to achieve maximum flexibility and linearity with sufficient stroke to provide the displacement needed to produce full excursion of the sensing element.

The upper body member (usually of steel since it is not exposed to the process) generally is contoured to the shape of the diaphragm. See Fig. 21. This construction protects the diaphragm in event of loss of fill fluid or failure of the instrument element. It also permits removal of the instrument or gage for service without shutting down the process.

Metal diaphragms generally are welded to the top body member. This provides maximum protection against leaks. A clamped construction, where the diaphragm is held between two body members (Fig. 22) by bolts, is used for nonweldable diaphragms or where ready replacement of the diaphragm is desired.

Diaphragm seals are classified according to the method of attaching them to the process. The *threaded connection* (Figs. 21 and 22), with female pipe thread in the lower body, is the most com-

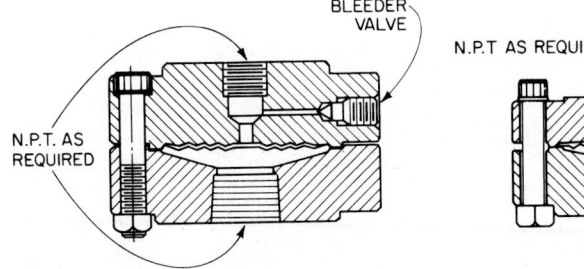

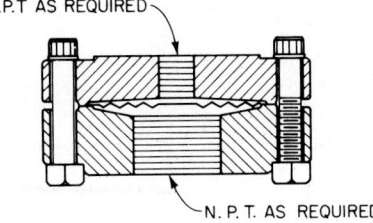

Fig. 21 Screwed-type diaphragm seal with bleeder valve.

Fig. 22 Screwed-type diaphragm seal with clamp construction.

mon design. The *flange connection* (Fig. 23) provides a means for bolting the seal to a flange provided by the user. Suitable gasketing is required. *Saddle types* (Fig. 24) allow the lower body section to be welded into a process pipe or vessel and provide a way of eliminating troublesome cavities.

Coatings of Teflon, Kel-F, and other synthetic materials sometimes are used on the exposed parts of the flange-connected seal to prevent erosion or buildup of process material on the diaphragm.

Fill Fluid

This fluid, which completely fills the measuring system and provides for pressure transmission between the seal diaphragm and the sensing element, must be chosen carefully to provide (1) minimum thermal expansion, (2) fluidity (low viscosity) and chemical stability under all temperatures to which exposed, and (3) minimum or no contamination of process fluid in the event of diaphragm rupture. Fills used include silicone oils (-50 to $100°F$, -46 to $38°C$), special mineral oils (100 to $300°F$, 38 to $150°C$), mercury (-30 to $700°F$, -34 to $370°C$), and sodium-potassium eutectic (20 to $1500°F$, -7 to $816°C$). One of the most popular fills is a 70 to 30% mixture of glycerin and water.

A filling procedure required to ensure a complete fluid fill of the measuring system without any air or gas bubbles is provided by the gage or instrument manufacturer and by many local instrument shops at modest cost. "Bleeder" valves at the protector (Fig. 21) and at the sensing element (where the element is connected by a capillary) simplify the filling. The use of vacuum in filling also is common.

The spring rate of a metal diaphragm often is sufficient to require calibration corrections in lower ranges (below 50 psi, 3.5 kg/cm^2). Also it should be noted that diaphragm seals may not have sufficient displacement to provide required travel of some long-stroke bellows elements.

Nonmetallic Diaphragms

This type of diaphragm is frequently used in low-range pressure and vacuum applications, since it is very flexible and has a low pressure constant. Typical applications are in pneumatic force-balance

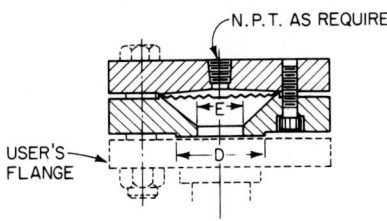

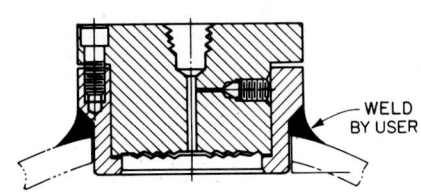

Fig. 23 Flanged-type diaphragm seal.

Fig. 24 Saddle- or welded-type diaphragm seal.

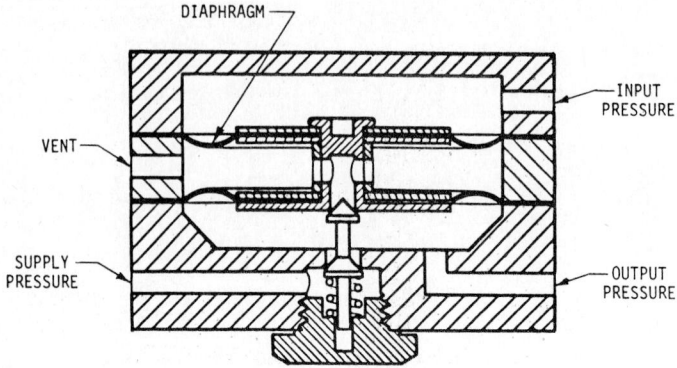

Fig. 25   Volume-booster relay with nonmetallic diaphragms.

instruments, such as relays and stack-type controllers. Figure 25 shows a relay which uses a nonmetallic diaphragm. Figure 26 shows a low-range differential-pressure indicator.

Materials

Nonmetallic diaphragms are made from a variety of materials, including most synthetic rubbers. The choice of material is influenced considerably by such factors as the temperature and composition of the medium contacting the diaphragm. Elastomers with fabric reinforcing are often used, and typical reinforcing materials are cotton, nylon, and dacron.

Force Due to Pressure

The force developed by pressure acting on a diaphragm is

$$F = PA_e \tag{33}$$

where F = force, lb (kg)
P = pressure, psi (kg/cm^2)
A_e = effective area, in^2 (cm^2)

Effective Area

The formula usually used to calculate the effective area of a nonmetallic diaphragm is

$$A_e = \frac{\pi}{4} D_o^2 \tag{34}$$

where D_o = diameter of the centerline of the convolution, in (cm).

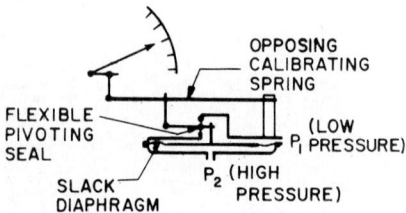

Fig. 26   Slack diaphragm differential unit.

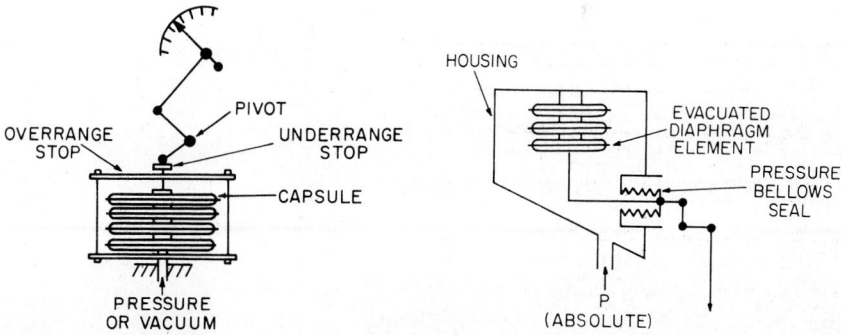

Fig. 27   Capsule element.   **Fig. 28**   Absolute-pressure gage.

Metallic Capsules

A capsule consists of two metallic diaphragms joined at their peripheries by soldering or welding.

Two or more capsules may be joined together as shown in Fig. 27, and the total deflection of the assembly is then equal to the sum of the deflections of the individual capsules.

These elements are often used in absolute-pressure gages, as shown in Fig. 28. Table 9 shows the characteristics and constants for a group of typical capsules designed for relatively low-pressure ranges. The high force constants obtainable with capsules also make them usable at higher ranges, with applications up to 1000 psi (70.3 kg/cm^2). Aircraft applications are also common, with 150 psi (10.5 kg/cm^2) ranges being typical.

Capsule Spring Rate (Force Constant)

This is defined by the equation

$$R = \frac{\Delta F}{\Delta d} \tag{35}$$

where R = spring rate, lb/in (kg/cm)
ΔF = force increment applied to free end of capsule, lb (kg)
Δd = deflection due to force increment, in (cm)

Deflection with Force

This can be expressed by transposing Eq. 35

as
$$\Delta d = \frac{\Delta F}{R} \tag{36}$$

Table 9. Constants for Typical Capsule Elements

Material	Thickness of metal, in	Diam. of capsules, in	No. of capsules	Range of capsule, inH$_2$O	Deflection for range, in	Pressure constant, psi/in	Force constant, lb/in	K
Phosphor bronze	0.0031	3$\frac{3}{16}$	2	0–15	0.255	2.12	8.0	0.25×10^{-6}
Phosphor bronze	0.0031	3	2	0–20	0.162	4.45	12.0	0.24×10^{-6}
Phosphor bronze	0.0042	3	4	0–35	0.304	4.17	11.8	0.21×10^{-6}
Phosphor bronze	0.0051	2$\frac{3}{16}$	4	0–50	0.203	8.90	19.0	0.24×10^{-6}
Phosphor bronze	0.0105	1$\frac{13}{16}$	5	0–360	0.185	70.4	92.1	0.22×10^{-6}
Ni-Span C	0.0030	3	2	0–20	0.104	7.5	22.8	0.15×10^{-6}

Deflection with Pressure

If pressure is applied to an unrestrained capsule, the resulting deflection of the free end is

$$\Delta d = \frac{\Delta P A_e}{R} \tag{37}$$

where ΔP = pressure increment, psi (kg/cm^2)
A_e = effective area of capsule, in^2 (cm^2)

Effective Area

The formula usually used for calculating the effective area of a capsule is

$$A_e = 0.1 D_E^2 \tag{38}$$

where D_E = diameter of the portion of the capsule exposed to pressure.

Pressure Constant

The pressure constant is the pressure required per unit deflection and may be calculated by

$$\frac{\Delta P}{\Delta d} = \frac{R}{A_e} \tag{39}$$

Pressure Differential of a Pressure Switch

For a capsule or bellows operating a snap-acting electrical switch, the formula for calculating the differential between the cut-out pressure and the cut-in pressure is

$$\Delta P = \frac{\Delta F_s + \Delta M_s R}{A_e} \tag{40}$$

where ΔP = pressure differential, psi (kg/cm^2)
ΔF = force differential of snap switch, lb (kg)
ΔM = movement differential of snap switch, in (cm)

Hysteresis

This is defined as the difference between the up-scale deflection and the down-scale deflection, measured at the same applied pressure and expressed as a percentage of the full-scale pressure. To test hysteresis, the total excursion of the applied pressure should be between zero and full scale. The maximum hysteresis of most capsules is usually 0.25 to 0.5% of full scale.

Nonlinearity

The term "nonlinearity" is often described as being the maximum deviation between a given curve and the straight line connecting the end points of the curve, expressed as a percentage of full scale. The nonlinearity of capsules is dependent on several design factors, as explained earlier under Diaphragms and may vary from 0.1 to 2% or higher. If the capsule is installed in an instrument, however, this nonlinearity may be compensated for as described below.

Compensation by Microprocessor

When a capsule is used as a sensor in an electronic pressure measuring system, predetermined values of nonlinearity may be compensated for by means of microprocessor technology. Correction factors are stored in memory and recalled as required.

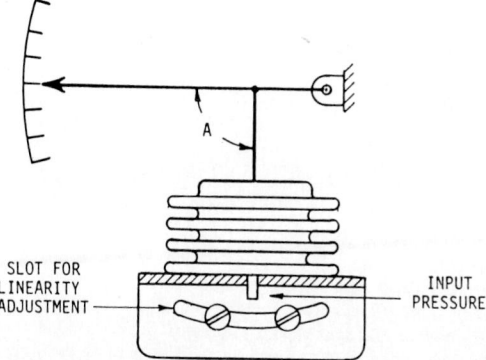

Fig. 29   Pressure indicator with linearity adjustment.

Compensation by Mechanical Adjustments

When performing the initial calibration of mechanical pressure indicators and recorders, adjustments should be made so that the angles between the various rotating arms and their associated links are 90° when the indicator or pen is at midscale. If subsequent tests then reveal that the calibration curve is not linear, this nonlinearity can usually be reduced by increasing or decreasing one of these angles.

Figure 29 illustrates a method of making this adjustment on a single-link indicator. For example, if a calibration curve for this indicator is plotted with pressure on the horizontal axis and reveals an upward bow in the center portion of the curve, the magnitude of this upward bow may be reduced by rotating the input element to the right. Angle A will then be greater than 90° when the indicator is at midscale. If the bow is in the downward direction, the correction can be made by rotating the input element to the left. A relatively large change in angle A is required for a small reduction in nonlinearity; hence a micrometer screw adjustment is not required here.

REFERENCES

1. "Manometer Tables: Recommended Practices," *ISA Res. Pap* 2.1, Sept. 6, 1952; revised 1962, Instrument Society of America, Pittsburgh, Pa.

2. Krell, G. H.: "Sensing Pressure through Diaphragms," *Chem. Eng.*, July 1968, pp. 88–93.

3. Giacobbe, J. B., and A. M. Bounds: "Selecting and Working Bourdon-Tube Materials," *Instrum. Mfg.*, July–August 1952.

4. Goitein, K.: A Dimensional Analysis Approach to Bourdon Tube Design, *Instrum. Pract.*, September 1952, pp. 748–755.

Pressure Transducers and Transmitters

As described in the preceding article, Mechanical Pressure Elements, several relatively simple pressure-sensitive elements are used as the interface between the source of pressure, as in a vessel or pipe, and the instrument. Pressure sensors are sometimes called *pressure summing* elements. Actual values of pressure are inferred by their effects on the pressure sensors, such as a change in force or position, so that pressure values may be transduced and expressed in terms of an electric, pneumatic, or hydraulic signal that can be transmitted and serve as the basis for indicating, recording, and controlling pressure. It should be emphasized that there is a marked distinction between the pressure sensor and the pressure transducer. The *sensor* provides the basis for measurement; the *transducer* converts energy from one form to another. In very simple cases, a spring may furnish the restoring force and thus, by way of a mechanical link-and-level system, indicate pressure. Or, an elastic bellows may be restored to a null postition by the application of pneumatic pressure. Transducers fall into two major categories: (1) *electronic,* in which the transducer may employ a strain gage, a change in capacitance, reluctance, or changes in piezoelectric or optical effects, among other phenomena used; and (2) *pneumatic,* in which a counter air pressure becomes a measure of the actual pressure on the process side of a diaphragm, bellows, bourdon, or other elastic element. Transducers may be used for local measurements or, very frequently, are incorporated into transmitters. The basic difference between a transducer and a transmitter is the incorporation in the transmitter of zero and span adjustments. Pressure transducers, notably strain gages, find application in other areas of force measurement as, for example, in the form of load cells commonly used in weighing systems and in compression and tensile testing machines. See also Section 8 of this Handbook.

As of the mid-1980s, digital and microprocessor concepts continue to penetrate the pressure measurement and transmitter field. Analog technology, however, continues to dominate. Proponents of analog methodology point to its relative simplicity and the large fund of experience that has been gained over the years in applying analog devices efficiently. For example, in a period where the traditional 4 to 20 mA loop continues in wide usage, digital standards largely remain unrefined. Analog devices require comparably simple power units, and transmission only two wires. It is further stressed by some users that analog sensor multiplexing is relatively free from problems. The long-range trend appears to be toward a pressure transducer packaged in a single feedback loop chip. Future transducers may incorporate on a single chip the pressure sensor, analog-to-digital input signal conversion and conditioning, a microprocessor, memory, and digital-to-analog output signal conversion. Several combinations of components ultimately will become available.

An abridged summary of pressure transducers, described in more detail later in this article, is given in Table 1.

STRAIN GAGE PRESSURE TRANSDUCERS

Strain gages are the most common pressure transducers, considering all process-related uses, and during the past decade they have made serious inroads on pneumatic pressure transducers and transmitters for use in many applications. In practice, strain gages usually are mounted directly on the pressure sensor or force summing elements, as described under Mechanical Pressure Elements earlier in this Handbook. Strain gages may be supported directly by sensing diaphragms or bonded to cantilever springs acting as restoring agents. The large majority of transducers continue to be manufactured by unbonded wire, foil, or semiconductor techniques. However, diffused semiconductor and deposited thin-film manufacturing techniques are revolutionary steps being researched by the strain gage industry. Strain gages are discussed in considerable detail in Sec. 8 of this Handbook.

A full survey of strain-gage-type pressure transducers and transmitters currently available is impractical for inclusion here. Several types are illustrated in Figs. 1 through 5. The units shown

Table 1. Representative Pressure-Transducing Mechanisms

Mechanism	Nominal accuracy*	Pressure-sensitive or summing element	Operating range†
Strain gage	0.1	Diaphragm with or without beam	0–20,000 psig
	0.1	Piezoresistive	0–5000 psig
Capacitative	0.1	Diaphragm	200 mmHg–10,000 psig
Piezoelectric	0.1	Distortion of crystal	0.1 mmHg–100,000 psig
Frequency change	0.2	Resonant wire	5–7 inH$_2$O
	0.025	Resonant crystal	0–20 psia
Potentiometric	0.5	Bourdon tube	10–10,000 psig
	0.5	Diaphragm capsule	5–10,000 psig
Reluctive	0.25	Diaphragm	0 inH$_2$O–10,000 psig
	0.25	Bourdon tube	5 inH$_2$O–100,000 psig
Linear-variable differential transformer (LVDT)	0.15	Diaphragm	5 inH$_2$O–100 psi
	0.3	Bourdon tube	5 inH$_2$O–10,000 psig
Pneumatic	0.5	Bellows—null balance	0–15 psi
	0.5	Bellows—force balance	0–15 psi
Interferometric	‡	Manometer	>0.029 psi

*Plus or minus percent full scale.
†Instruments available with a number of ranges within this broad range. psi, pounds per square inch (1 psi = 6.8947 kPa (kilopascals); psia, pounds per square inch absolute; psig, pounds per square inch gage.
‡Very small fraction of 1 psi.
NOTE: Characteristics of high-vacuum instruments are given in High-Vacuum Measurement.

Fig. 1 (*Left*) Differential-pressure transmitter *(PD3000)*. (*Right*) Same unit with local indication. *(Gould Inc., Measurement Systems Division, Oxnard, Calif.)*

Fig. 2 Flush-mount gage pressure transmitter *(PGF3000). (Gould Inc., Measurement Systems Division, Oxnard, Calif.)*

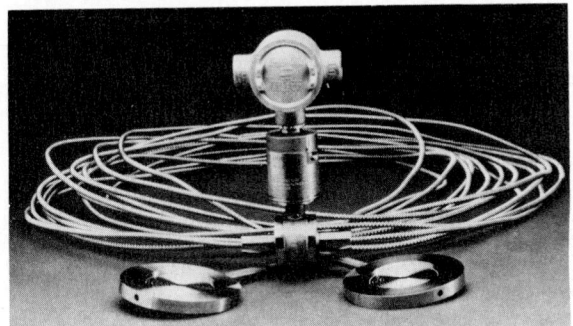

Fig. 3 Remote diaphragm transmitter *(PD3018) (Gould Inc., Measurement Systems Division, Oxnard, Calif.)*

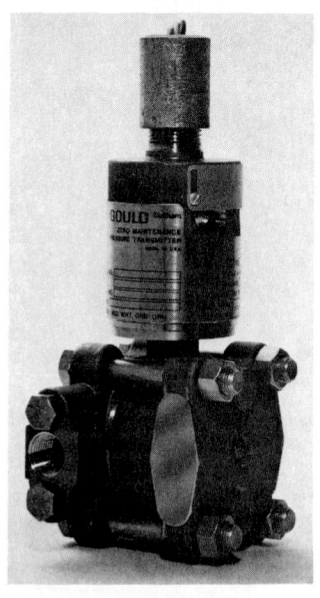

Fig. 4 Gage pressure transmitter *(PG3040). (Gould Inc., Measurement Systems Division, Oxnard, Calif.)*

3.34

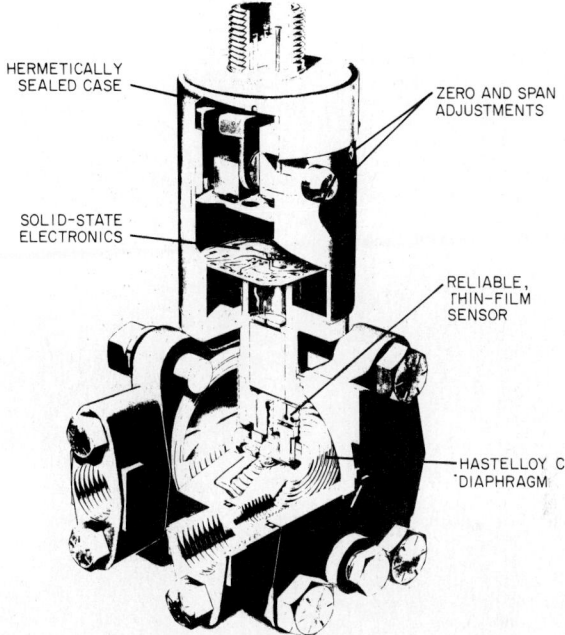

HERMETICALLY
SEALED CASE

ZERO AND SPAN
ADJUSTMENTS

SOLID-STATE
ELECTRONICS

RELIABLE,
THIN-FILM
SENSOR

HASTELLOY C
DIAPHRAGM

Fig. 5 Sectional view of the *P3000* series strain-gage pressure transducer. Magnetic zero and span adjustments permit calibration without violating the hermetic seal of the housing. Zero and span are totally noninteracting. *(Gould Inc., Measurement Systems Division, Oxnard, Calif.)*

have an accuracy of ±0.25% of calibrated span, including linearity, hysteresis, and repeatability. Generally, the stability of the units is estimated at ±0.25% of the upper range for 6 months. Outputs are 4 to 20 mA (milliamperes), maximum 25 mA (limited). The power supply is 12 to 55 V dc (reverse polarity protection). The enclosures are totally hermetically sealed. Zero elevation is −100% of full range or to full vacuum for certain models and applications. Zero suppression is +80% of full range. Generally, maximum possible process temperatures range between −65 and +250°F (−54 to +121°C). Wetted parts may be Type 316 stainless steel with diaphragms of Hastelloy C, 316 stainless steel, Monel, Inconel, or tantalum. Fill fluid is silicone oil or Fluorolube. Depending on models and applications, screwed or flanged connections are available.

Digital Storage of Calibration Data

In late 1983, Honeywell introduced a so-called "smart" pressure transmitter (ST 3000) which incorporates a unit that communicates digitally with a special hand-held terminal. The unit stores factory calibration data characterized to each transmitter and sensor and, in an EEPROM (electrically erasable programmable read-only memory) located in the transmitter case, makes the nonlinearity corrections permanent and fully insensitive to temperature and static pressure variations that create the need for such corrections. Analog techniques do not make it possible to correct such errors. It has been claimed that, even though the ultimate accuracy is still limited by the analog nature of a bridge in the sensor, accuracy is improved 100% over prior designs to 0.1% of the span or upper-range value, whichever is greater. Gain in precision and stability of nonlinearity correction achievable in the field permits turndown ratios that are as large as 400:1 to be specified. Typical conventional analog turndown ratios are about 5:1. As a result, the ST 3000 line requires only three models to cover d/p range from 0 to 3000 psi. The device is explained in considerable detail in the Kompass

(1983) reference listed at end of this article. As observed by Kompass, "The fact that ST 3000 has a digital communicator certainly will permit an extension to direct communication from the console in distributed digital control systems, although that capability has not yet been offered; and future models might well include transmission of digital signals for the measure variable (our own speculation)."

(a)   (b)

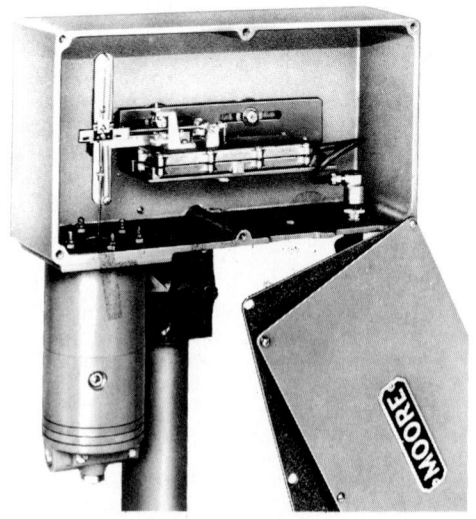

(c)

Fig. 6   Pneumatic pressure transmitters: (*a*) helix type; (*b*) capsular type; and (*c*) slack diaphragm type. *(Moore Products, Spring House, Pa.)*

PNEUMATIC PRESSURE TRANSMITTERS

Three types of pneumatic pressure transmitters are shown in Fig. 6. These instruments combine a pressure sensing element with a standard motion transmitter which detects the position of the pressure sensing element and provides a linear 3 to 15 psi (~20 to 100 kPa) output signal. Maximum accuracy is achieved by the fact that the pressure sensing element develops many times the 2-g force required for full-scale operation of the transmitter pilot. The unit shown in Fig. 6a combines a helix-type sensing element with a motion transmitter. The pressure sensing element is connected to the pilot rod of the motion transmitter. The position of the sensing element is thus converted into a linear pressure output. This type of instrument is often used for large-span pressure measurement applications where a standard pneumatic linear signal is desired. For maintaining a constant compressor discharge pressure under varying conditions, as shown in Fig. 7, this instrument may be used to transmit actual discharge pressure as a standard 3 to 15 psi (20 to 100 kPa) signal to a pressure controller.

The unit shown in Fig. 6b combines a stacked capsule pressure sensing element with a motion transmitter. The sensing element is also connected to the pilot rod of the motion transmitter to provide a linear pressure output, which is proportional to the position of the pressure sensing element. Among numerous other applications, this unit is frequently used in conjunction with an air-purge-type liquid-level system, as shown in Fig. 8, to provide a standard 3 to 15 psi level signal to the level controller. See also Sec. 5 of this Handbook.

The unit shown in Fig. 6c combines a slack diaphragm pressure sensing element with a motion transmitter. These transmitters are used primarily for general combustion or steam generating applications. The most common measurements made include forced draft, induced draft, first- and last-pass draft, stack draft, firebox pressure, and air or fuel gas flow. The general range for these instruments (positive pressure or differential pressure) is from 0 to 0.16 (minimum) to 3.5 in water [0 to 0.4 (minimum) to 8.9 cm water]. The instruments can also be used for negative-pressure applications, where a reverse-acting transmitter will provide an output which increases with a decrease in the measured pressure. A typical application is shown in Fig. 9.

In operation and as shown in the diagram of Fig. 10, supply air flows through a restriction into the motion transmitter's bellows assembly, past the wire pilot, to atmosphere. The pressure within the bellows (pilot pressure) depends on the position of the pilot with respect to its annular nozzle,

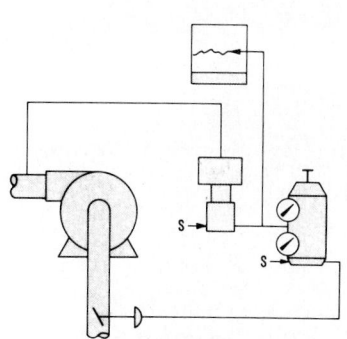

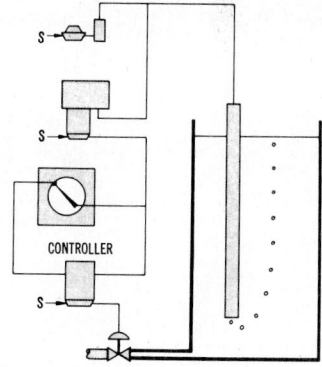

Fig. 7 Transmission of actual discharge pressure in the form of a standard pneumatic signal to a pressure controller for maintaining a constant compressor discharge under varying load conditions. S, instrument supply pressure. *(Moore Products, Spring House, Pa.)*

Fig. 8 Transmitter pressure sensing element used in liquid-level measurements for liquids in which sediments are likely to collect on the bottom of the vessel, or where the liquid would contaminate the transmitter pressure sensing element. In this diagram, a pressure transmitter is used in conjunction with an air purge system to provide a standard pneumatic signal to the level controller. S, instrument supply pressure. *(Moore Products, Spring House, Pa.)*

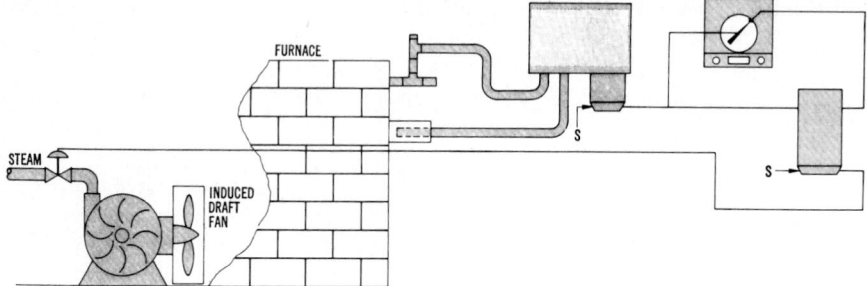

Fig. 9   Using the pressure outside the furnace as a reference, a pressure transmitter transmits a standard pneumatic signal, which is proportional to the furnace pressure, to a process controller which controls the speed of the induced draft fan. Accurate draft control in a furnace is necessary to maintain constant furnace pressure. S, instrument supply pressure. *(Moore Products, Spring House, Pa.)*

as it is displaced by the pressure sensing element. There is a tapered step on the diameter of the wire pilot. If the wire pilot moves down so that only the small diameter restricts the nozzle, the backpressure will decrease. Conversely, when the pilot moves up so that the large diameter restricts the nozzle, the back-pressure increases. For this reason, the system will always come to equilibrium with the annular nozzle aligned with the tapered portion of the pilot.

The pilot pressure is also present in the chamber above the booster-pilot diaphragm. As the pilot pressure increases, the booster-pilot diaphragm is forced down, closing the exhaust port and opening the supply port and thus increasing the transmitted output pressure. Rebalance occurs when the output increases enough to force the bellows assembly upward, bringing the annular nozzle back to its original position with respect to the pilot. The amount of change in output required to accomplish rebalance depends on the range spring in the motion transmitter. Helix elements are constructed of

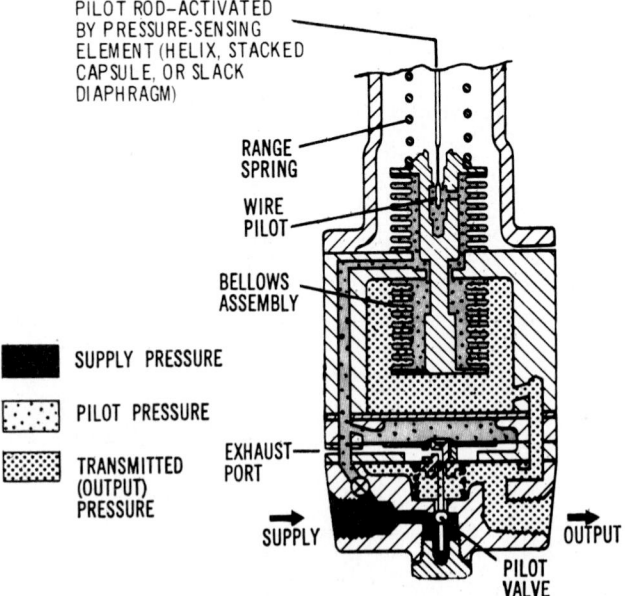

Fig. 10   Motion transmitter with a built-in booster-pilot valve. *(Moore Products, Spring House, Pa.)*

beryllium copper, Ni-Span C, or 316 stainless steel, depending on the model selected. In instruments using a capsular sensing element, the capsular element is constructed of phosphor bronze, Ni-Span C, or 316 stainless steel, depending on the model selected. The general ambient temperature limits of the units described range from $-40°F$ ($-40°C$) to $+180°F$ ($+82°C$).

For the measurement of high static pressures (up to 6000 psi, 42 MPa) and for differential-pressure measurement at low-, medium-, or high-pressure ranges, the instrument shown in Fig. 11 can be used. Available in several models, these differential-pressure cells transmit a pneumatic signal of 3 to 15 psi ($\approx$20 to 100 kPa). The units are available with threaded connections or for flange mounting. Instruments with threaded connections are used to measure low differentials. Their static working pressure limit is 500 psi ($\approx$3.4 MPa). Body and flange materials for the various models include cadmium-plated steel and 316 stainless steel. Optional elevation or suppression range kits are available. Instruments with flange mounting can be bolted directly onto a tank for providing liquid-level measurement of open or closed tanks. Depending on the corrosiveness of the application, Duranickel diaphragms and Monel trim may be used. Generally the high ambient temperature limitation is 250°F (121°C), but some units can be used up to 350°F (177°C).

Pressure measurement ranges fall into three principal categories, as indicated in Table 2.

Compact Differential-Pressure Transmitter

An instrument of the type shown in Fig. 12 can be used to monitor differential pressure, or absolute pressure. For example, a unit of this type can be used in conjunction with an orifice plate, a venturi, or another primary flow element to measure flow; and to transmit a pneumatic signal. These units also can be used to measure density or specific gravity, interface levels between liquids, and liquid levels in closed vessels. See also Sec. 5 of this Handbook. Each instrument consists of a compact meter body (Barton) and a motion transmitter (Moore).

In operation (flow application), as shown in Fig. 13, differential pressure is applied against opposing bellows in the meter body. These bellows, enclosed in separate pressuretight chambers, are connected by a common center shaft. The differential pressure causes the bellows assembly to move a proportional amount determined by the elasticity of the bellows, range spring, and torque tube. The two bellows are filled with a liquid (ethylene glycol is standard) and connected internally via a passage in the center plate. As the differential increases, liquid is displaced from one bellows to the other. If the differential exceeds the operating range of the meter, the bellows will continue to move until the O-ring seal on the center shaft seats against the center plate. This traps the liquid in the bellows, allowing the differential to continue to increase up to the full static pressure rating of the meter body without causing damage—because the liquid fill, being incompressible, prevents additional movement.

The motion transmitter operates in the same manner as previously described (see Fig. 10).

1:1 Pressure Transmitters

These are diaphragm-type transmitters used to measure process pressures or liquid levels in open vessels and to transmit a pneumatic signal in a 1:1 proportion to the measured variable. As shown in Fig. 14, process pressure is exerted against the process side of the metal diaphragm. The resultant force is opposed by the force of the pneumatic output pressure, exerted on the opposite side of the

Table 2. Differential-Pressure Cell Pressure Ranges

Range	Differential pressure, in H_2O, kPa	Static pressure, psi, MPa
Low-range capsule	0–5 to 0–25	500
	(0–1.2 to 0–6)	(3.5)
Medium-range capsule	0–20 to 0–250	1500
	(0–5 to 0–62)	(10.5)
High-range capsule	0–200 to 0–850	6000
	(0–50 to 0–210)	(42)

SOURCE: Moore Products Company.

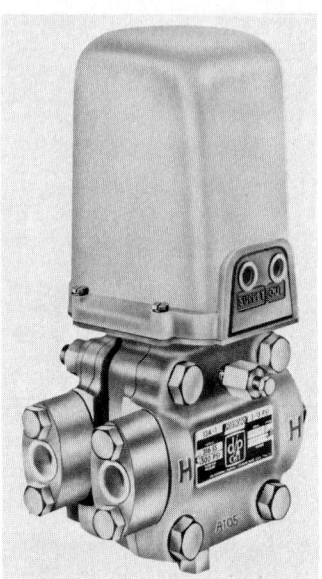

Fig. 11 Differential-pressure (D/P) cell used for numerous process pressure differentials and also in liquid-level and density measurements. *(Moore Products, Spring House, Pa.; unit manufactured by Foxboro.)*

Fig. 12 Compact differential-pressure transmitter used for various applications to monitor differential pressure or absolute pressure. *(Moore Products, Spring House, Pa.)*

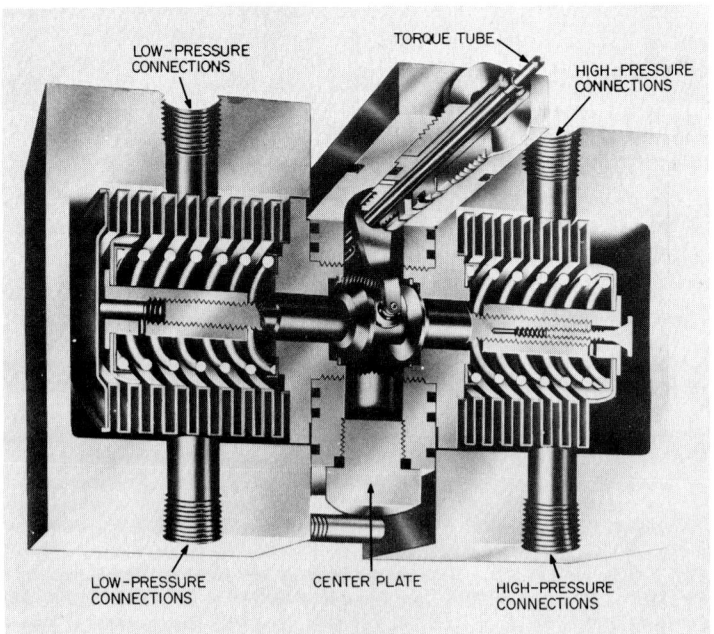

Fig. 13 Differential-pressure transmitter can be applied against opposing bellows in meter body for measuring flow as shown here. *(Moore Products, Spring House, Pa.)*

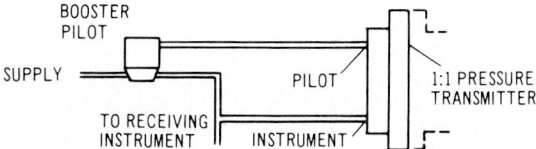

Fig. 14 Application of a 1:1 pressure transmitter. *(Moore Products, Spring House, Pa.)*

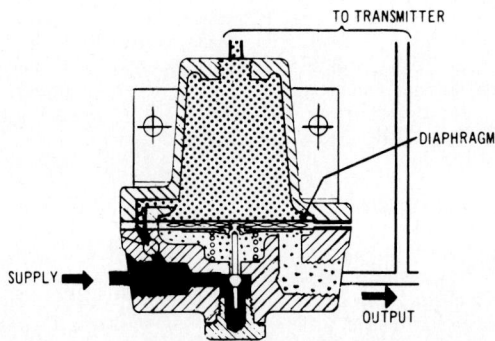

Fig. 15 Booster pilot used with 1:1 pressure transmitter. *(Moore Products, Spring House, Pa.)*

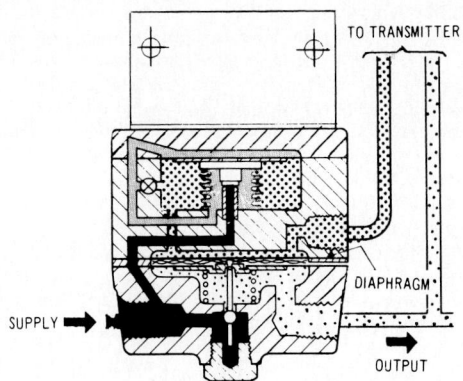

Fig. 16 Pilot valve featuring an automatic orifice to maintain a constant flow of pilot air through the transmitter nozzle. *(Moore Products Nullmatic.)*

diaphragm. Any unbalance in the opposing forces will change the clearance between the diaphragm and the detecting nozzle, producing a change in the pilot pressure. A separate booster-pilot valve is required with this system. The booster pilot (Fig. 15), which operates on the force-balance principle, changes its output in response to changes in the transmitter's pilot pressure. The output of the booster pilot is fed back to the transmitter to rebalance the process pressure in a 1:1 ratio.

The pressure change within the balance bellows depends on the action of the booster-pilot valve. For example, with the booster pilot illustrated, filtered and regulated supply air flows through a restriction to the top of the booster pilot's diaphragm, out of the chamber above this diaphragm, and up to and through the nozzle in the transmitter. An increase in differential pressure moves the differential bellows and nozzle seat toward the nozzle, increasing back-pressure on the nozzle. The increase in nozzle back-pressure forces the diaphragm in the pilot valve downward—closing the exhaust ports to the porous fabric center layer of the diaphragm and opening the supply-plunger port—to increase the transmitted output pressure and thus rebalance the system. A decrease in differential pressure will reverse the operation, exhausting the transmitted pressure through the exhaust port to rebalance the system.

When the transmitter is in equilibrium, the forces exerted within the booster-pilot valve will also be in balance. The nozzle pressure above the flexible diaphragm will equal the transmitted pressure plus the force exerted by the loading spring (below the diaphragm). The force of the loading spring, therefore, determines the pressure drop across the transmitter nozzle. Since the pressure drop remains essentially constant, undesirable nozzle characteristics are eliminated. For high-precision applications, a pilot valve of the type shown in Fig. 16 may be used. This pilot valve may be required for low-range applications. This valve features an automatic orifice which maintains a constant flow of pilot air through the transmitter nozzle, making the pilot insensitive to supply pressure variations.

CAPACITATIVE PRESSURE TRANSDUCERS

In a typical capacitance-type transducer, a measuring diaphragm moves relative to one or two fixed plates. Changes in capacitance are detected by an oscillator or bridge circuit. Generally, capacitive transducers have the advantages of low mass, high resolution, and good frequency response. Limitations include the need for sophisticated signal conditioning circuitry and some sensitivity to temperature and the effects of stray noise on the sensor leads. In some recent designs, the use of ceramic bellows has made it possible to extend the useful temperature range and to decrease the size of the devices. Design improvements have also included the minimization of hysteresis and nonlinearity, as effected, for example, through the use of a Franklin oscillator. Considerable research in recent years has gone into broadening the application of capacitance-type transducers, as exemplified by a design (Oxford University, United Kingdom) where researchers built a transducer with a resolution greater than 0.19 for pressures of 1 to 2 bars. Designed for operation at liquid helium temperatures, the device modulates the output frequency of a 10-MHz oscillator.

In a typical capacitative pressure transducer, as pressure is applied and changes, the distance between two parallel plates varies—hence altering the electric capacitance. This capacitative change can be amplified and used to operate into phase-, amplitude-, or frequency-modulated carrier systems. A frequency-modulated system using a tuned resonant circuit is shown in simple form in Fig. 17. In this electric circuit, the capacity C_3 is part of a tuned resonant circuit $L_2C_2C_3$. L_1C_1 forms part of a stable high-frequency oscillator circuit. The tuned circuit $L_2C_2C_3$ is loosely coupled to the circuit L_1C_1. The high-frequency potential induced in circuit $L_2C_2C_3$ is rectified, and the dc output current of the rectifier is indicated on a microammeter. The response of the tuned circuit $L_2C_2C_3$ to a constant frequency is shown in Fig. 18 as a function of the capacity ($C_2 + C_3$) of this circuit. Peak output occurs at point A when the circuit is tuned to resonate at the oscillator frequency. This circuit is tuned to its operating point B by increasing capacitor C_2 until the rectifier meter reads approximately 70% of maximum. Any small change in pressure transducer capacity C_3, due to pressure on the diaphragm, affects the response of the circuit according to Fig. 18.

In order to eliminate the effect of cable capacity between the transducer C_3 and the tuned circuit L_2C_2, a circuit of the type shown in Fig. 19 can be used. In this circuit, a coil L_3 is built as an integral part of the capacitative-transducer assembly. The coil L_3 is connected in parallel with the

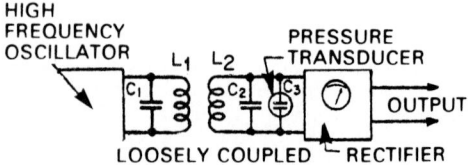

Fig. 17 Simplified diagram of a tuned resonant circuit that may be used in a capacitative pressure transducer.

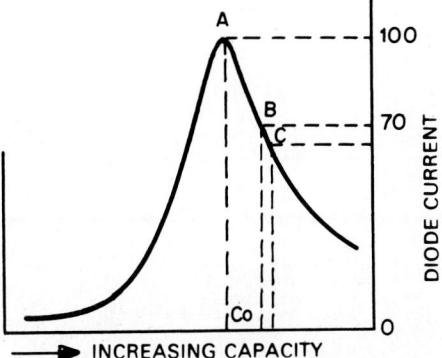

Fig. 18   Response of a resonant circuit to a constant frequency shown as a function of the capacity of the circuit.

transducer capacity (C_3) to form a tuned circuit with a resonant frequency (for example, 600 kHz). The tuned circuit L_3C_3 is close-coupled to the tuned circuit L_2C_2 by means of the link coils L_4 and L_5, which are connected by means of a low-impedance (70-Ω) untuned cable. The change in cable capacity, such as that produced by vibration, is negligible when reflected into the high-impedance tuned circuit. In this way, long cables can be used between the transducer and the electronic unit. The tuning characteristics of a link-coupled circuit are shown in Fig. 20. The operating range is the linear section noted midway between the maximum and minimum readings obtained with changing capacity.

A phase-modulated carrier system can be used in combination with transducers which incorporate a radio frequency matching transformer which is tuned with the fixed condenser plate and stray capacitances in the pickup to approximately the oscillator frequency and is properly matched to the transducer connecting coaxial cable. When pressure is applied to the diaphragm, the increase in capacity lowers the resonant frequency of the circuit. The resulting change in reactance is coupled back to the indicator by a suitable transmission line, producing a phase change in the discriminator.

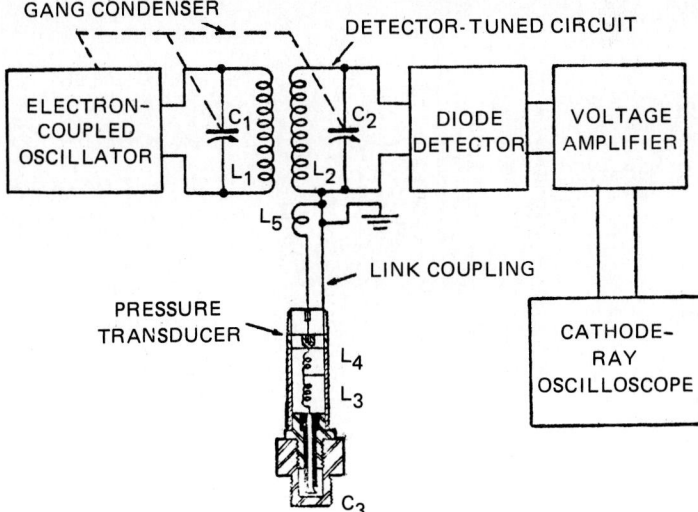

Fig. 19   One type of circuit used to eliminate the effect of cable capacity between transducer C_3 and tuned circuit L_2C_2.

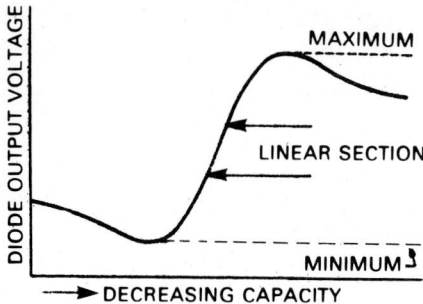

Fig. 20   Tuning characteristic of a link-coupled circuit.

This, in turn, produces an output voltage which is a function of the pressure on the diaphragm. The voltage can be indicated or recorded by the usual methods.

The measuring system of one modern (Fischer & Porter 50DPF100) capacitative differential-pressure transmitter is shown in Fig. 21. Both process diaphragms, high and low side, are mechanically attached to the connecting rod. In the middle of the connecting rod the movable electrode is attached and held in position by the spring diaphragm. The differential pressure is balanced by the restoring force of the spring diaphragm. Hence, the spring diaphragm represents the measuring element. When applying a differential pressure on the system, the movable electrode is shifted and the distances d_1 and d_2 to the fixed electrodes are changed simultaneously. As a result of the change in distance between the fixed and movable electrodes, the capacitances of the differential capacitor are also changed. This change is electronically amplified and transduced to a 4 to 20 mA dc output signal directly proportional to the differential pressure.

The internal construction of the measuring element is shown in Fig. 22. The measuring diaphragms are subdivided in the inner circular area and the outer circular ring area. At the circumference, the inner circular areas are welded to the connecting rod. Because of this connection of the inner diaphragms to the connecting rod, additional forces are absorbed—because of a volumetric

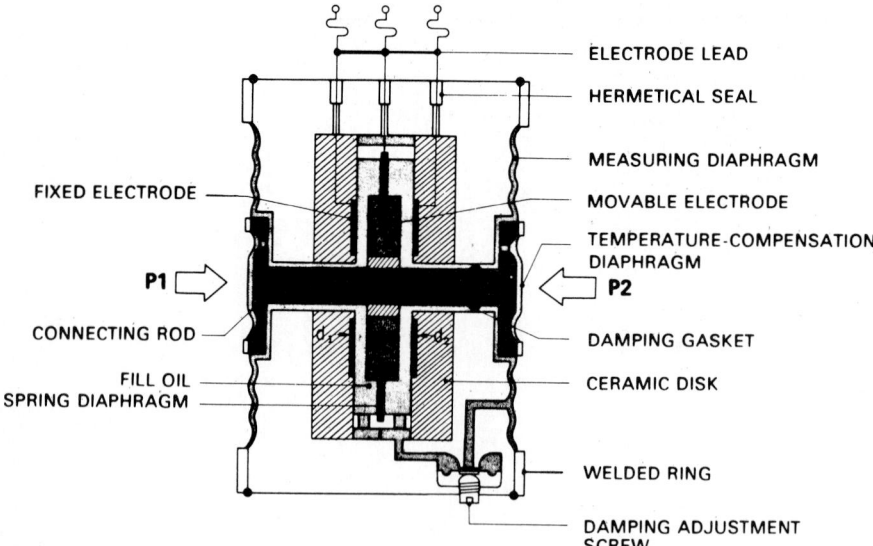

Fig. 21   Schematic diagram of the measuring element of one type (*50DPF100*) of modern capacitative pressure and differential-pressure transducer-transmitter. *(Fischer & Porter, Warminster, Pa.)*

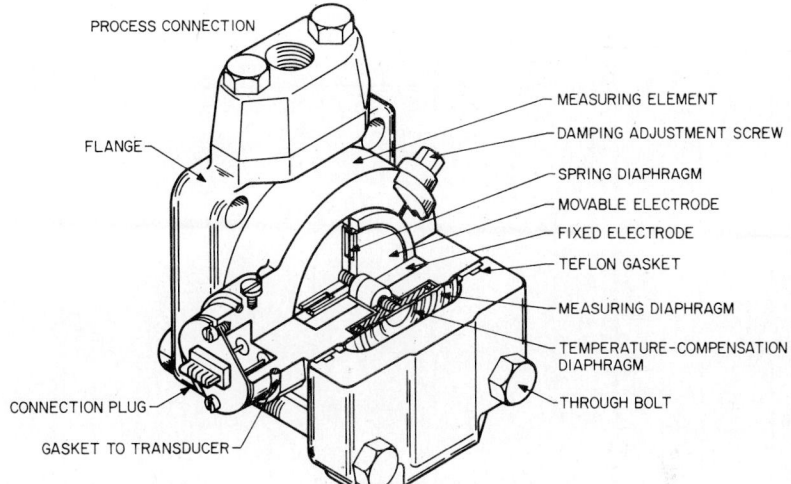

PROCESS CONNECTION

MEASURING ELEMENT
DAMPING ADJUSTMENT SCREW
FLANGE
SPRING DIAPHRAGM
MOVABLE ELECTRODE
FIXED ELECTRODE
TEFLON GASKET
MEASURING DIAPHRAGM
TEMPERATURE-COMPENSATION DIAPHRAGM
CONNECTION PLUG
THROUGH BOLT
GASKET TO TRANSDUCER

Fig. 22 Sectional view of measuring element of *50DPF100*. *(Fischer & Porter, Warminster, Pa.)*

change of the fill fluid. Thus, the inner diaphragms serve for temperature compensation, eliminating effectively zero drift of ambient and process fluid temperatures.

The adjustable restriction for damping is mounted in the fill fluid passage located between the control chamber and clearance space behind the low-pressure-side diaphragm. Since the direct passage, except a capillary opening, is sealed with the damper ring, the response of the movable electrode assembly to differential pressure is controlled by the adjustable restriction.

In the measuring element, the internal oil-filled space is a single-chamber system, an important part of the design required for maintaining the differential-capacitance constant. The voltage output is independent of the dielectric constant only when there is exactly the same oil between the electrodes on both sides. After prolonged service, the dielectric property of the oil may change to some degree. The single chamber permits the oil to homogenize. It is also claimed that the single-chamber design provides increased stability when the transmitter body is exposed to uneven temperatures.

Fundamental Relationships of Capacitative Transducer

A simplified diagram of the measuring circuit of the device shown in Figs. 21 and 22 is given in Fig. 23. Assume the gaps between the movable electrode and two fixed electrodes are both equal to d_0. When differential pressure $(P_1 - P_2)$ is applied, the connecting rod moves a distance of Δd. Then,

$$d_1 = d_0 + \Delta d \qquad (1)$$

$$d_2 = d_0 - \Delta d \qquad (2)$$

$$\Delta d = K_1(P_1 - P_2) \qquad (3)$$

where d_1 and d_2 represent the interelectrode gaps, respectively, on the high and low sides; K_1 is a proportional constant.

The capacitances of the gaps, C_1 and C_2, respectively, are

$$C_1 = \frac{K_2}{d_1} = \frac{K_2}{d_0 + \Delta d} \qquad (4)$$

$$C_2 = \frac{K_2}{d_2} = \frac{K_2}{d_0 - \Delta d} \qquad (5)$$

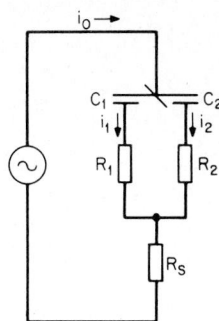

Fig. 23 Measuring circuit used in capacitative pressure transducer-transmitter *(50DPF100)*. *(Fischer & Porter, Warminster, Pa.)*

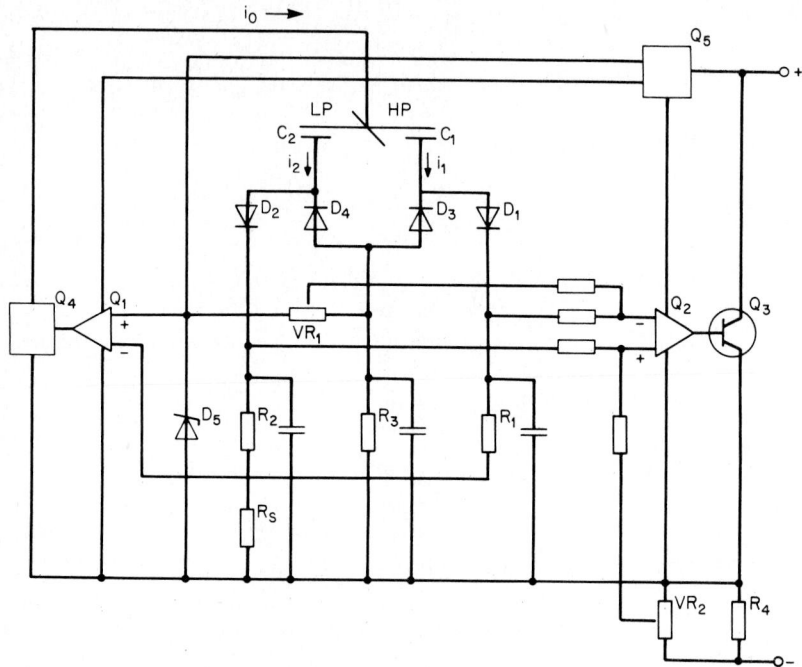

Fig. 24 Schematic diagram of an electronic circuit used in a capacitative pressure transducer-transmitter (*50DPF100*). (*Fischer & Porter, Warminster, Pa.*)

where K_2 is a proportional constant and depends on the electrode area and dielectric constant of the material filling the gaps. The capacitances are connected to the electronic measuring circuit shown in Fig. 24 to detect the capacitance change.

When, in this circuit,

$$\frac{1}{j\omega C_1}, \frac{1}{j\omega C_2} \gg R_1, R_2, R_s$$

where $j = \sqrt{-1}$

ω = frequency, rad/s

the branch currents i_1 and i_2 are expressed as follows:

$$i_0 = i_1 + i_2 \tag{6}$$

$$i_1 = (i_0 - i_1)\frac{C_1}{C_2} = \frac{i_0 C_1}{C_1 + C_2} \tag{7}$$

$$i_2 = (i_0 - i_2)\frac{C_2}{C_1} = \frac{i_0 C_2}{C_1 + C_2} \tag{8}$$

Therefore, the voltage drops on R_1, R_2, and R_s are, respectively,

$$e_1 = i_1 R_1 = \frac{i_0 R_1 C_1}{C_1 + C_2} \tag{9}$$

$$e_2 = i_2 R_2 = \frac{i_0 R_2 C_2}{C_1 + C_2} \tag{10}$$

$$e_s = i_0 R_s \tag{11}$$

And the difference between e_2 and e_1 is

$$e_2 - e_1 = \frac{i_0(R_2 C_2 - R_1 C_1)}{C_1 + C_2} \tag{12}$$

Furthermore, suppose $R_1 = R_2$, then from Eqs. 11 and 12,

$$e_2 - e_1 = \frac{C_2 - C_1}{C_1 + C_2} e_s \tag{13}$$

Using Eqs. 3 to 5,

$$e_2 - e_1 = \frac{e_s K_1}{d_0} (P_1 - P_2) \tag{14}$$

Hence,

$$e_2 - e_1 \sim P_1 - P_2 \tag{15}$$

is independent of the dielectric constant, provided that e_s is kept constant. Voltage e_s remains constant by controlling automatically current i_0.

In the electronic circuit of Fig. 24, the differential capacitors (C_1, C_2) are excited by the high-frequency oscillator O_4 through diodes (D_1, D_2, D_3, D_4) and develop dc voltages on R_1 and R_2, respectively, thus yielding input voltages to the amplifier. Both currents also flow through R_s. The comparator functions to compare the voltage on R_s with the zener voltage D_5 and controls the oscillator to keep the excitation current constant. The other half-cycle is used to develop a bias for zero adjustment. These circuit features significantly reduce interaction between zero and span adjustments. Even when the transmitter span is changed from minimum to maximum, only a single adjustment is usually required.

Various configurations of the sensing elements for different applications are shown in Fig. 25. Transmitters are available in the following approximate ranges:

Range	
psi	Pa
0.2–1.8	1.25–12.5 kPa
0.7–7	5–50 kPa
6–30	40–200 kPa
30–145	0.2–1.0 MPa
100–510	0.7–3.5 MPa
290–1160	2–8 MPa
1090–4350	7.5–30MPa

Units are furnished for both differential and absolute pressure in certain ranges. The output signal is 4 to 20 mA dc, or 10 to 50 mA dc (two wires). The accuracy of the span setting is $\leq 0.25\%$. A variety of materials of construction are used, depending on the process and ambient environment. Temperature limits for standard units are:

	Temperature limits	
	°C	°F
Process fluid	−40 to +120	−40 to +248
Transducer	−30 to +100	−22 to +212
Indicator	−10 to +60	+14 to +140

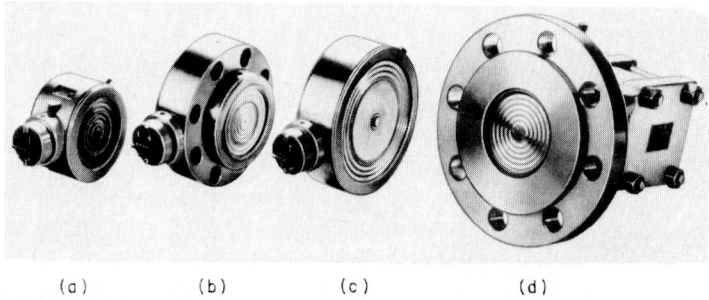

(a)   (b)   (c)   (d)

Fig. 25 Types of measuring elements used for different applications of an electronic capacitative-type pressure and differential-pressure transducer-transmitter (*50DPF100*). (*a* and *b*) Low-range, very-low range, and medium-range differential pressure; (*c*) high-range differential pressure, high working pressure, and absolute pressure; (*d*) liquid level. *(Fischer & Porter, Warminster, Pa.)*

A variety of other types of capacitative pressure transducer-transmitters is available in the marketplace.

PIEZOELECTRIC PRESSURE SENSORS*

When certain asymmetrical crystals are elastically deformed along specific axes, an electric potential produced within the crystal causes a flow of electric charge in external circuits. Called the *piezoelectric effect,* this principle is widely used in transducers for measuring dynamic pressure, force, and shock or vibratory motion. In a piezoelectric pressure transducer, as shown in Fig. 26, the crystal elements form an elastic structure which functions to transfer displacement caused by force into an electrical signal proportional to the pressure applied. Pressure acting on a flush diaphragm generates the force.

Piezoelectric pressure transducers historically have employed two different types of crystals: (1) natural single crystals, such as quartz and tourmaline; and (2) synthetic polycrystalline ceramic materials, such as barium titanate and lead zirconate. With the relatively recent development of artificially cultured quartz crystals, the foregoing distinction is no longer clear-cut.

Cultured quartz has the advantage of being readily available and reasonably priced. Near-perfect elasticity and stability, combined with an insensitivity to temperature, make quartz an ideal transduction element. Ultrahigh insulation resistance and low leakage allow static calibration, accounting for the popularity of quartz in pressure transducers.

Natural tourmaline, because of its rigid, anisotropic nature, offers submicrosecond response in pressure-bar-type blast transducers. Artificial ceramic piezoelectric crystals and *electret* (perma-

*Based on a description furnished by R. W. Lally, Engineering Department, PCB Piezotronics, Inc., Depew, New York.

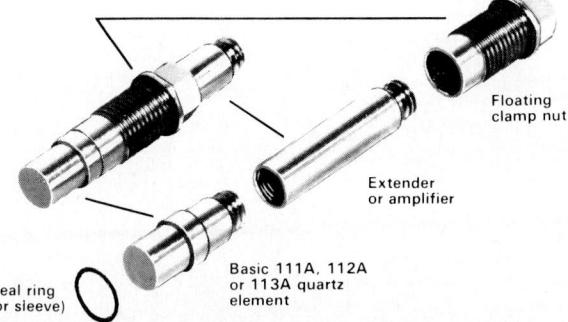

Fig. 26   Modular assembly of a piezoelectric dynamic pressure transducer. *(PCB Piezotronics, Inc., Depew, N.Y.)*

nently polarized dielectric material, the analog of a magnet) materials are readily formed into compliant transducer structures for generating and measuring sound pressure.

The charge signal from a piezoelectric pressure transducer is usually converted into a voltage-type signal by means of a capacitor, according to the law of electrostatics: $E = Q/C$, where E is the voltage signal, Q is the charge, and C is the capacitance. This circuit is shown in Fig. 27.

In response to a step function input, the charge signal stored in the capacitor will exponentially leak off through the always finite insulation resistance of the circuit components, precluding static measurements. The initial leakage rate is set by the circuit discharge time constant: $R \times C$, where R is the leakage resistance value, which can be as high as 10^8 MΩ in quartz crystals.

Because of the automatic rezeroing action of the discharge circuit, piezoelectric sensors measure relative pressures, sometimes noted as *psir*. They measure pressure relative to the initial level for transient events and relative to the average level for repetitive phenomena. Sometimes the slow action of these circuits is mistaken for zero drift by impatient operators.

To prevent the charge signal from quickly leaking off through the recorder or oscilloscope input resistance, a special isolation amplifier is required between the crystal and recorder. As illustrated in Fig. 28, if the charge converting capacitance is located at the input of this isolation amplifier, the amplifier is called a *voltage amplifier*. If the capacitor is in the feedback path, it is called a *charge amplifier*. Amplifiers are further classified as electrostatic (dc-coupled) or vibration (ac-coupled). The ac coupling circuitry behaves similarly to the sensor discharge circuit.

The high-frequency response of piezoelectric sensor systems depends on the resonant behavior of the sensor's mechanical structure, or on electronic low-pass filters in the sensor, amplifier, or recorder.

The advent of microelectronics and charge-operated field-effect transistors (JFET and MOS-FET) has and is continuing to profoundly change the design of piezoelectric sensors. The current trend is to package the isolation amplifier and signal conditioning circuitry inside the sensor. These integrated circuit piezoelectric (ICP) sensors with built-in microelectronics, which operate over a

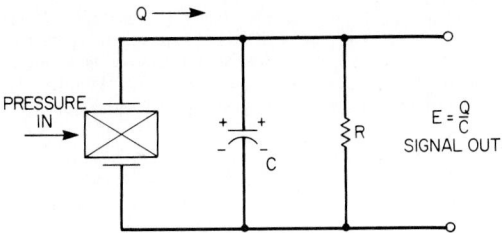

Fig. 27   Piezoelectric crystal circuit. *(PCB Piezotronics, Inc., Depew, N.Y.)*

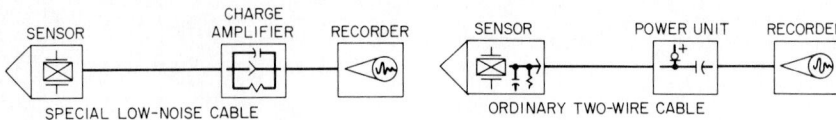

Fig. 28 Conventional charge mode system. *(PCB Piezotronics, Inc., Depew, N.Y.)*

Fig. 29 Modern voltage mode ICP system. *(PCB Piezotronics, Inc., Depew, N.Y.)*

simple two-wire cable, have sometimes been referred to as "smart" sensors. A typical ICP pressure sensor is shown in Fig. 29. A small unit (0.1 in, 2.5 mm in diameter) may have a sensitivity as high as 40 mV/psi (5.8 mV/kPa).

To eliminate spurious signals caused by environmental effects, such as temperature and motion, the mechanical structures of some piezoelectric pressure sensors are quite sophisticated. A typical acceleration-compensated pressure sensor containing an integrated accelerometer to cancel out motion signals is shown in Fig. 30. Durable conformal coatings of the sensor case and diaphragm provide electrical and thermal insulation. Hermetic seals are electron-beam-welded.

Piezoelectric pressure sensors offer several advantages for measuring dynamic pressures. They are generally small in size, lightweight, and very rugged. One transducer may cover a measuring range of greater than 10,000 to 1 and a frequency range from less than 1 Hz to hundreds of kilohertz, with little or no phase shift (time delay). As previously mentioned, piezoelectric sensors cannot measure static or absolute pressures for more than a few seconds, but this automatic elimination of static signal components allows unattended, drift-free operation.

Because of their unusual ruggedness, piezoelectric pressure sensors are widely used in difficult applications, such as ballistics, blasts, explosions, internal combustion, fuel injection, flow instabilities, high-intensity sound, and hydraulic or pneumatic pulsations—in connection with problems which may be encountered in connection with guns, shock tubes, closed bombs, rocket motors, internal-combustion engines, pumps, compressors, pipelines, mufflers, and oil exploration imploders.

Pressure-Sensitive Wire

Usually a material sensitive to linear strain is also sensitive to applied hydrostatic pressure. The two effects are called, respectively, strain sensitivity and pressure sensitivity, and both appear to be somewhat larger for small sizes of conductors than for larger sizes.

For pressure measurement, the sensitive element may be a helically wound coil mounted on a suitable form and protected from the liquid in which it is immersed. Since the effect of the pressure is exerted on the winding form and on the protective insulation as well as on the wire, care must be taken in the design of the assembly to ensure that deformations of these parts do not deform the wire coil. Precautions must be taken, too, in bringing the connecting leads out of the pressure chamber, in the selection of materials, and in the operation of the instrument to minimize errors due to temperature changes. The change in resistance of most materials due to a few degrees' change in temperature is as great as that resulting from fairly high pressure. Therefore, to reduce this error, Manganin wire has been employed in winding the coil. Manganin has an extremely low temperature coefficient of resistivity. Some researchers also have found that a properly treated gold-chromium wire is superior to Manganin and can be used for test pressure measurements to extremely high values, such as 200,000 psi (1380 MPa). Research at Harvard University several years ago showed that the coefficient of resistance with pressure is quite linear over an extremely wide range. The

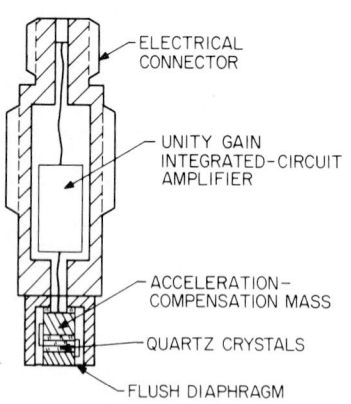

Fig. 30 Acceleration-compensated quartz pressure sensor with a built-in microelectronic unity gain isolation amplifier. *(PCB Piezotronics, Inc., Depew, N.Y.)*

pressure-sensitive wire concept is one of the few methods known for measuring pressures above the highest ranges capable of measurement with bourdon tubes.

The pressure-sensitive wire concept should not be confused with *vibrating-wire* technology, described later in this article.

RESISTIVE PRESSURE TRANSDUCERS

Resistive pressure transducers can be classified according to the method used by which a change in pressure results in a change in the resistance of an electrical element. Included are (1) a change in the dimensions or internal structure of the element, (2) movement of contacts (changing either the area or the length of the resistance element), (3) a change in the temperature of the element, among others. While variations in the design of transducers using these basic principles are limited only by the ingenuity and imagination of the user, a few representative designs are described briefly here. Some of these designs no longer are popular, but they are included to demonstrate the wide variety of approaches that can be taken.

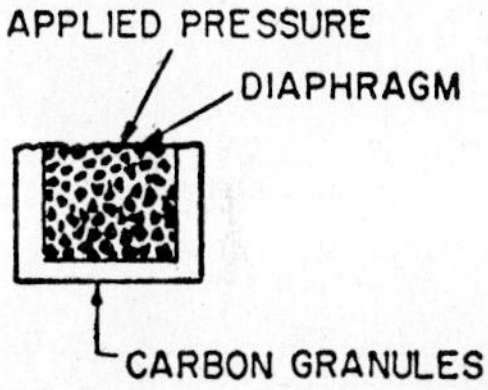

Carbon Pile

A variation in applied pressure on a confined mass of carbon particles will change its electrical resistance, primarily because of the variation in the area of the contracting surface between the adjacent granules. The carbon microphone (Fig. 31) is an example of the use of this principle for transducing sound pressures. While also feasible for measuring fluid pressure, the irregular carbon granules (which may be carbon, graphite, or of a molded carbonaceous material) have been replaced by smooth, thin disks (Fig. 32) for improved performance. Such a stack changes its resistance with applied pressure. The change for a single stack is not accurately proportional to the change in applied pressure, and to improve linearity the stacks are used in pairs, in a mechanically and electrically push-pull arrangement. The assembly of carbon stacks and the coupling of the stacks to a pressure-responsive diaphragm are shown in Fig. 33. Because the carbon pile element is sensitive to vibration, temperature changes, and other extraneous phenomena, it has largely been superseded by more stable types. An advantage of the carbon pile concept in some applications is that it is a low-resistance device capable of providing sufficient output current to actuate electrical indicating instruments without amplification.

Fig. 31   Carbon pile pressure transducer.

Moving-Contact Resistance Elements

Traditionally most variable resistance pressure transducers have incorporated moving contacts. Inasmuch as nearly every form of primary pressure (summing) element can be configured to position a movable contact along a resistance element, many designs have been manufactured or proposed. The variable resistance element may be (1) a single, continuous wire, (2) an assembly or coiled or folded wire, or (3) a tapped resistance coil. The resistance material may be nonmetallic, or it may be liquid

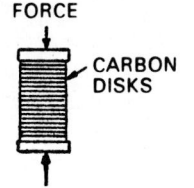

Fig. 32   Stacked carbon disk pressure transducer.

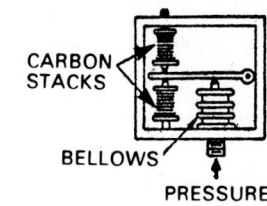

Fig. 33   Carbon stacks with bellows coupling.

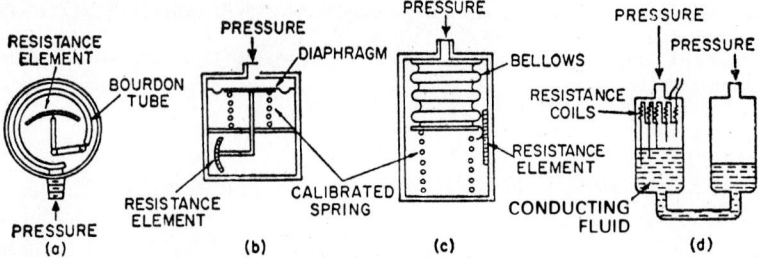

Fig. 34 Highly schematic diagrams of transducers using moving-contact resistance elements: (a) bourdon tube, (b) diaphragm, (c) bellows, (d) differential-pressure coil. In recent years, such devices have been highly miniaturized, in most cases through application of solid-state electronics.

(mercury or electrolyte). A number of pressure transducers are manufactured which incorporate a diaphragm, bellows, or bourdon tube. These devices are designed to respond to changes in either absolute, differential, or gage pressure. The resultant motion actuates one or more precision potentiometers, thus providing outputs of resistance ratio (by application of a voltage, ac or dc, across the ends of the potentiometer) or voltage ratio proportional to the input pressure. Some transducer arrangements are shown in Fig. 34.

Temperature-Resistance Elements

Devices of this type have been used to measure or compare pressures of gases and for determining the proportions of gaseous mixtures at known pressures. Thermal conductance of the gas is the measurement criterion. If a thin wire filament is suspended within a pressurized chamber and heat is supplied in the form of an electric current, the temperature of the filament will decrease as the gas pressure increases. Such an arrangement is shown in its simplest form in Fig. 35. The device comprises a pair of lamps each containing a fine wire filament, one lamp having been exhausted to a moderate vacuum and the other connected to the chamber in which the pressure is to be measured. Low-voltage electric lamps have been used for pressures less than 2in Hg(0.7 kPa). One lamp is fitted with a tubulature to permit access of the gas; while the other (unchanged) lamp is used as a reference. When there is no gas flow or only a very small flow, the loss of heat from such a filament takes place largely by conduction through the gas to the tube wall (and then through the wall to the open air), so that the temperature of the filament decreases as the gas pressure increases. If the filament material has a positive thermal coefficient, the resistance will decrease likewise. A bridge circuit can be used to measure the ratio of the resistance of the two filaments; if this ratio is the

PRESSURE

REFERENCE   WORKING
LAMP   LAMP

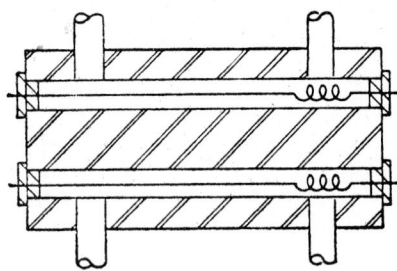

Fig. 35 Operating principle of a transducer for measuring low pressures. Each lamp contains a fine wire filament. One lamp is evacuated, while the other is exposed to the pressure being measured. The same principle can be used in a miniaturized version. See also the article Vacuum Measurement in this Handbook section.

Fig. 36 Transducer for measuring low pressures comprising two straight filaments stretched concentrically in small-diameter tubes.

criterion, changes in temperature of the entire assembly will have little or no effect on the measurement.

For very precise measurements, instruments have been designed to force the heat flow through the path, which gives the method its greatest sensitivity and accuracy. These devices use straight filaments stretched concentrically in small-diameter tubes. These tubes may simply be holes drilled in a massive metal block. The conducting path through the gas may be just a few millimeters. The construction is such as to facilitate the transfer and eventual loss of heat to the air. In operation (Fig. 36), the flow of gas is stopped during a reading, or the flow rate is maintained constant at all times.

Pressure Switches

Some pressure switches also operate on the principle of resistance change.

Piezoresistive pressure transducers are discussed later in this article.

MAGNETIC PRESSURE TRANSDUCERS

Magnetic pressure transducers comprise two groups: (1) those which operate through a change in inductance, and (2) those which utilize the change in reluctance of part of a magnetic circuit. Demarcation of the two types is not sharp.

Magnetic transducing elements must be supplied with energy, the variable inductance types being energized by electromotive forces and the variable reluctance types by magnetomotive forces. Electric circuits are usually low-impedance and are capable of carrying relatively heavy currents. In some designs, electronic amplification is not required. Transducing elements which fall within this class have been made in forms too numerous to cover in this discussion. However, operating principles are reviewed briefly here.

Inductive Elements

The simplest form of magnetic transducing element is the variable inductance unit as depicted in Fig. 37. This device consists of a coil of many turns of wire wound on a tube of insulating material with a movable core of magnetic material. As the magnetic core enters the solenoid coil, the inductance of the coil increases in a manner roughly proportional to the amount of metal within the coil. Figure 38 shows the inductance ratio element in which the movement being measured causes a simultaneous change in two similar inductors, increasing the inductance of one and, at the same time, decreasing the inductance of the other. It is therefore a push-pull device. It is quite similar to the device shown in Fig. 37, but instead of a single winding it has a center-tapped winding. Its iron core is always entirely within the coil, and the ratio of the inductance of the two halves of the coil is controlled by the position of the core. This construction provides double sensitivity, and the linearity

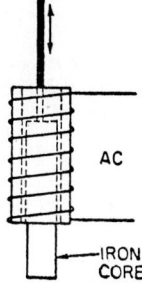

Fig. 37 Schematic representation of a variable inductance unit.

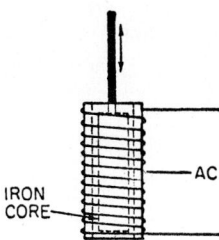

Fig. 38 Schematic representation of an inductance-ratio element.

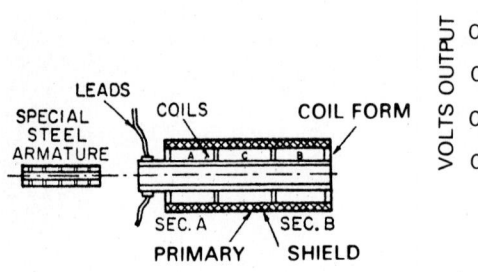

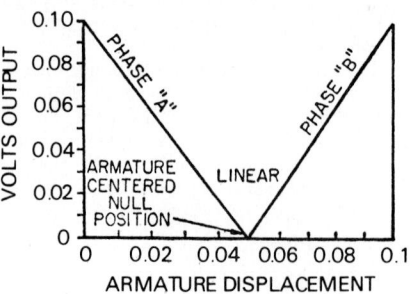

Fig. 39   Mutual-inductance element.

Fig. 40   Phase relationship in a mutual-inductance element.

of response is improved over that of a variable inductance element. Compensation for temperature, lead resistance, and other factors is far easier, and the iron core does not exert an appreciable force on the measuring element other than that due to its weight, as the forces due to the magnetic effect of the exciting current balance each other.

The mutual-inductance element (sometimes called a differential transformer) is a form of inductance ratio device in which the effective impedance of a coil is altered by changing the mutual inductance between it and another circuit. The cross section of such an element is shown in simplified form in Fig. 39. The device consists of three coils wound on a single spool with a free-moving armature of magnetic material mounted inside the spool. Alternating current is supplied to the center or primary coil *C,* and the magnetic flux generated by this coil is distributed by the armature so that a voltage is induced in the secondary coils *A* and *B*. If the armature is symmetrically located (centered), the induced voltages will be equal. In normal operation, coils *A* and *B* are connected in a series bucking relationship so that, if the armature is centered and both coils have equal voltages induced in them, the resulting output will be zero, and if the armature moves to the left a voltage of one phase (*A*) will predominate and if the armature moves to the right a voltage of the other phase (*B*) will predominate. Phase *A* will differ from phase *B* by 180°. This relationship is shown in Fig. 40. In a highly schematic way, the system is shown in Fig. 41.

A position repeating bridge with an amplifier used in a null-balance circuit is shown schematically in Fig. 42. Here, when the armatures of both mutual-inductance elements *A* and *A'* are in their proper position, their output will be equal and opposite in phase and the resultant input to the amplifier will be zero. No voltage will be applied to the amplifier phase of the motor, and it will be dormant.

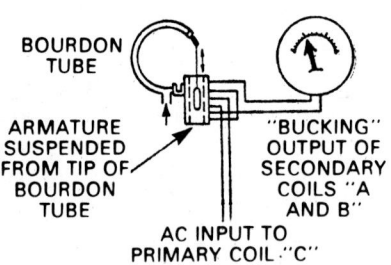

Fig. 41   Circuit connections for a simple pressure indicating system using a mutual-inductance element.

When the pressure measuring device raises the armature of *A,* the *A* phase will predominate and the input to the amplifier and subsequent input to the motor will be in the proper phase relationship with the line voltage to run the motor and raise the armature of *A'* a like amount and balance the system. The opposite takes place when the armature of *A* is lowered. A pointer or pen coupled to the motor shaft will thus indicate or record a motion exactly proportional to the motion of the pressure measuring device. The variable inductance or linear differential transformer has several advantages: (1) No friction is placed on the measuring device; (2) few or no reaction forces are exerted on the measuring device; (3) there is a linear electrical response when actuated by a linear mechanical motion; (4) operation on 60-cycle current requires no special power supply; and (5) the device is small, rugged, and dependable and can operate under water. The sensitivity of some systems is about 1 part in 12,000, and overall accuracy is on the order of 0.25% above 10% of full scale. Temperature sensitivity is minimal, as are voltage and frequency variations. See Fig. 43.

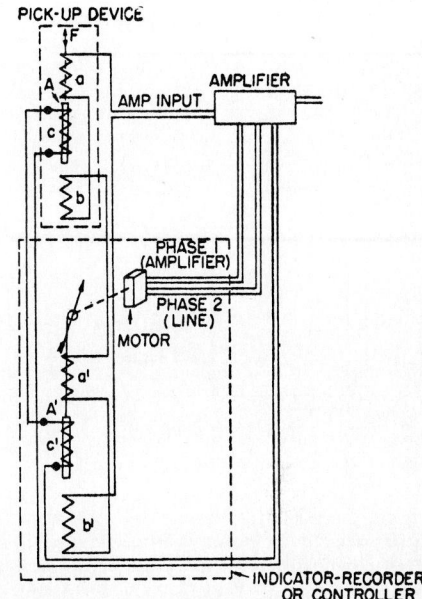

Fig. 42 Type of position repeating bridge that may be used with amplification in a null-balance circuit.

Reluctive Elements

Variable reluctance pressure transducers are distinguished primarily by the manner in which the exciting electric energy enters the system. In these devices, the energy is introduced as a magneto-motive force which may be derived from either a permanent magnet or an electromagnet assembly. Electromagnetic induction, roughly speaking, is the production of an electric current (or voltage) by

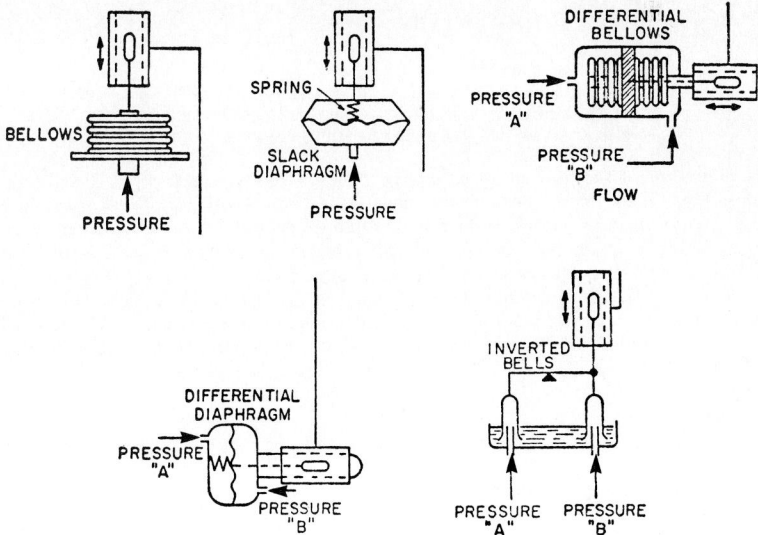

Fig. 43 Variations in measuring system that may employ a null-balance position repeating bridge.

the movement of a conductor through a magnetic field or by the changing of the strength of that field while the conductor is within it. In either case, it is the interaction of a magnetic field and an electric conductor in such a manner as to produce a current; a change in position or a change in flux is always involved.

A device of this type is shown schematically in Fig. 44. Here the coil is attached to the magnet, and the magnetic circuit is completed by a movable armature of magnetic material which in this case is positioned by a pressure measuring element. Changes in the position of the armature change the reluctance of the magnetic circuit, while the magnetomotive force remains constant; thus the magnetic flux varies, and a current is induced in the coil. This current is proportional to the rate at which the magnetic flux varies; therefore, it is proportional to the rate at which the armature is displaced. In some cases, the measuring element may be in the form of a torque tube consisting of a length of hollow tubing which has been flattened and twisted about its longitudinal axis. One end is free to rotate as pressure is applied. The rotation is directly analogous to the straightening of a bourdon tube. The tube untwists when the pressure on it is positive, and twists when it is negative or when the pressure outside is positive. Thus, the tube responds to pressure or vacuum applied either internally or externally and to the difference between the internal and external pressures. Devices of this kind may have a range from 0 to 5 to 0 to 10,000 psi (0 to 35 to 0 to 70,000 kPa) or higher. Variations in circuitry used with this type of device are shown in Fig. 45.

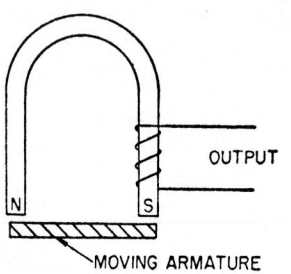

Fig. 44 Schematic representation of a variable reluctance pressure transducer.

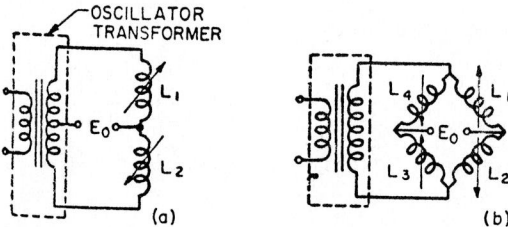

Fig. 45 Variations in circuitry used in connection with a torque tube-type measuring element: (*a*) inductance bridge with two active and two fixed arms; (*b*) bridge with four active arms.

A block diagram of a measuring system incorporating a variable reluctance pressure transducer is shown in Fig. 46. The inductance coils are connected as two arms of a bridge, as shown in Fig. 45*a*. The output of the bridge drives a discriminator type of demodulator unit. A carrier oscillator supplies the excitation voltage for the transducer and the bias current for the demodulator. Excitation may be 20 V at 3000 Hz, and the nominal open circuit output may be 0.1 V per volt full scale. Linearity may be 0.5%, and hysteresis 0.2 to 0.3% of the pressure range. The natural frequency can be up to 10,000 Hz, depending on the application. Zero drift and sensitivity change with temperature may be on the order of 0.02% of the pressure range per degree Fahrenheit (slightly higher per degree Celsius)—from -25 to $+180°F$ (-31 to $+82°C$).

See also other force measuring devices in Sec. 8 of this Handbook.

OPTICAL PRESSURE TRANSDUCERS

Over the years, a number of optical methods for detecting movement of the diaphragm, bellows, or other summing elements of a pressure sensor have been suggested. This problem, of course, is similar

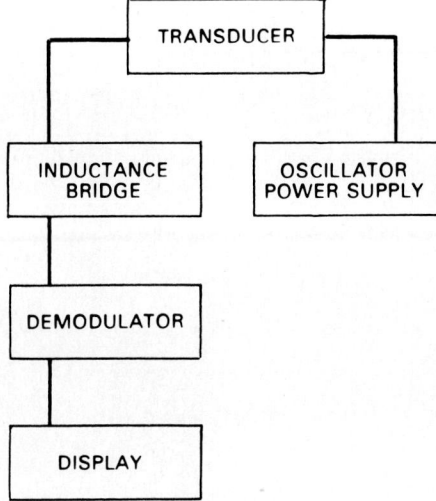

Fig. 46 Measuring system incorporating a variable reluctance pressure transducer.

to that of detecting the position of moving elements for other purposes. See also Sec. 9, Motion and Geometric Systems, in this Handbook.

One of the more recently developed noncontacting optical sensors for use with pressure sensors is shown in Figs. 47 and 48. The diaphragm or small helical bourdon tube element moves only 0.020 in (0.5 mm) during a full-scale excursion. Fixed to the element is an opaque vane (a kind of mini-

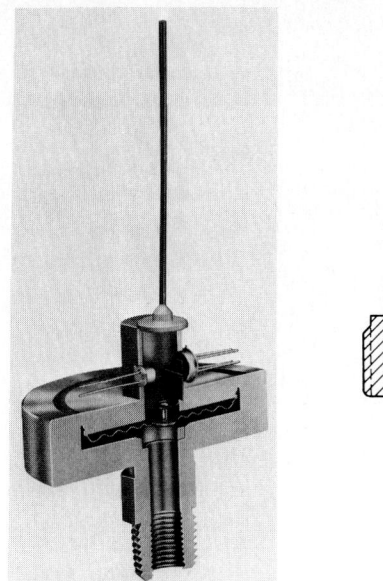

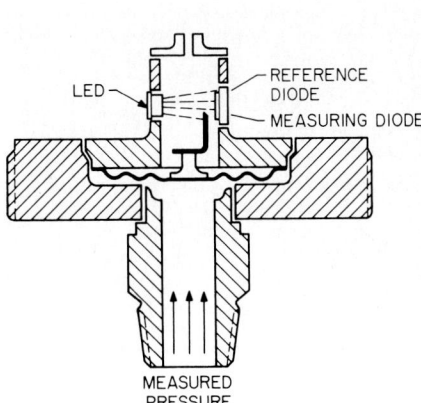

Fig. 47 Cutaway view of a noncontacting optical pressure sensor. *(Industrial Instrument Operations, Dresser Industries, Newtown, Conn.)*

Fig. 48 Cutaway view of the Heise noncontacting optical sensor. *(Industrial Instrument Operations, Dresser Industries, Newtown, Conn.)*

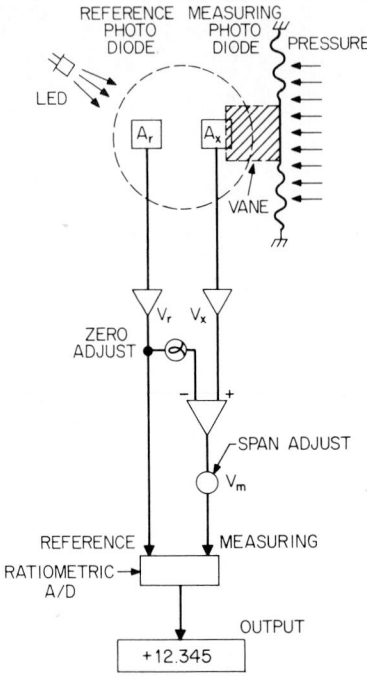

Fig. 49 Fundamental operation of the Heise optical sensor. Since both photodiodes are affected equally by temperature changes, because of their monolithic structure, temperature drift of the sensor can be eliminated. *(Industrial Instrument Operations, Dresser Industries, Newtown, Conn.)*

Fig. 50 Enlarged close-up of a noncontacting optical pressure sensor, showing the diaphragm at the lower center, opaque vane, and optical path. *(Industrial Instrument Operations, Dresser Industries, Newtown, Conn.)*

ature windowshade) which blocks near-infrared light impinging on one of two monolithic photodiodes. The other photodiode continuously measures the intensity of the light source. A light emitting diode (LED) serves as a reference against which the measured light is compared.

Two voltage signals, V_r and V_x, are generated by the reference and measuring photodiodes, respectively. Let k equal the photodiode light sensitivity factor, h the light intensity, and A_r and A_x the respective diode areas, and then

$$V_r = hkA_r \quad \text{and} \quad V_x = hkA_x$$

Each signal by itself will vary with light intensity h and with the sensitivity factor k. The ambient temperature can affect both h and k, but since a single light source is used and the k for each diode is the same because of the monolithic structure, both V_r and V_x are affected by temperature in the same proportion. See Fig. 49. An enlarged view, showing a cutaway of the diaphragm, optical path, and opaque vane is given in Fig. 50.

A portion of α of V_r is used to bias V_k to provide zero adjustment. The signal is then proportioned by a factor β representing a span adjustment. The resulting signals V_r and V_m are applied to a ratiometric, integrating-type analog-to-digital (A/D) converter.

The output of the converter is represented by

$$\frac{V_x}{V_r} \quad \text{or} \quad \beta \left(\frac{A_x}{A_r} - \alpha \right)$$

The ratiometric processing essentially eliminates any drift errors associated with the photodiodes and the LED, and the output is a function of only the potentiometer settings α and β and the ratio

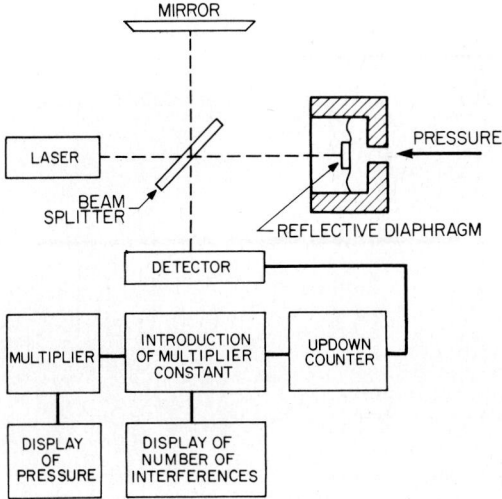

Fig. 51 Supersensitive and accurate pressure sensor that utilizes optical interferometry.

of the exposed photodiode areas A_x and A_r. Second-order drifts in the amplifiers are compensated for electronically, and an autozero technique is used in the converter.

The small motions of the pressure elements result in almost negligible hysteresis, generally on the order of 0.02% full scale or less, and repeatability of better than 0.02% full scale. By using a digitally controlled, variable frequency clock, the A/D converter accumulates pulses during the integration of V_r in accordance with a lookup table resident in a pair of PROMS* which are programmed at the time of calibration. The technique causes frequency changes between zero and full scale, resulting in a 40-segment linearization. Each sensor is thus characterized by the PROMS, which can be replaced if necessary to recalibrate the pressure sensor. The calibration reference temperature is 73°F (23°C).

The range of the system is from 0 to 50 inH2O (0 to 12.5 kPa) through 60,000 psi ($\approx$413 MPa); also, 5 psia through 60,000 psia (34 kPa through 413 MPa). Pressure elements are constructed of Inconel X718, and other wetted parts of AISI 316 stainless steel.

Optical Interferometry

For measuring surface variations on the order of millionths of an inch or less in the machine tool industry, optical interferometry is frequently used. This technique has been applied by Swiss engineers (Etel S.A., Renens, Switzerland) to a pressure sensor capable of measuring pressure differentials in the 200-Pa range with a resolution of 1 Pa (0.000145 psi). With reference to Fig. 51, a monochromatic light source emits a beam which passes through a beam splitter. Part of the beam is reflected to a fixed mirror which is exactly the same distance from the splitter as the diaphragm. The returning reflection from the splitter and the diaphragm create interference fringes at the splitter. These fringes are a function of the displacement of the diaphragm. The detector picks up the interference fringe, and a signal passes through an up-or-down counter. By introducing a multiplier constant, with the results passed through the multiplier stage, data are displayed by an LED readout. Linearity is rated at $\pm$0.3%, and hysteresis is limited to one digit on the readout. The system permits the use of very small pressure diaphragms, which are so fragile that the use of a conventional transducer would be impractical.

*Programmable read-only memory.

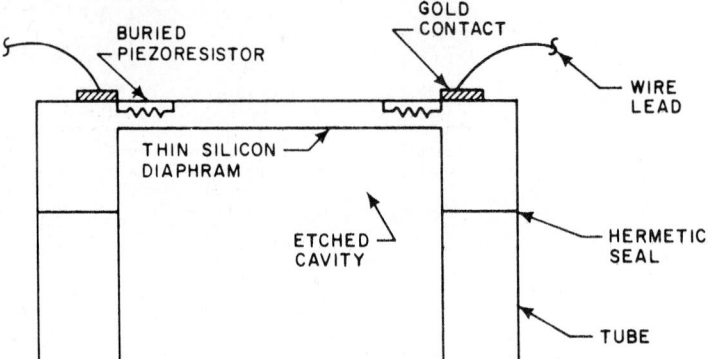

Fig. 52 Cross section of a piezoresistive sensing element with wire leads bonded to metal contacts. *(Micro Switch.)*

Vibrating-Wire Pressure Transducer

Although vibrating-wire technology is not new to instrumentation, it has not been widely applied until relatively recently. In essence, locating a wire under tension in the field of a permanent magnet and applying feedback causes the wire to vibrate continuously at its resonant frequency. In an ideal situation, the resonant frequency is a function of wire length, the square root of the tension, and the mass of the wire. Through electronic circuitry, a squaring operation can be performed on the resonant frequency, resulting in an output signal proportional to the differential pressure. See the article Resonant Wire Pressure Transducers later in this Handbook section.

PIEZORESISTIVE PRESSURE TRANSDUCERS

Stemming from research* in the 1950s on the piezoresistive properties of silicon diffused layers and the development of a piezoresistive device for a solid-state accelerometer, the first piezoresistive pressure transducers were developed as pressure inputs for a commercial airliner in the 1960s. Although piezoresistive transducers have been available for other applications over an input pressure range of 1 to 680 atm, the principal application developed in the early 1970s was in the automotive field. Since that time, uses for piezoresistive pressure transducers in process control and industrial applications have increased.

The sensing element consists of four nearly identical piezoresistors buried in the surface of a thin, circular silicon diaphragm. Gold pads attached to the silicon diaphragm surface provide connections to the piezoresistors and serve as pads for probe-type resistance measurements or for bonding of wire leads. The thin diaphragm is formed by chemically etching a circular cavity into the surface opposite the piezoresistors. The unetched portion of the silicon slice provides a rigid boundary constraint for the diaphragm and a surface for mounting to some other member. A cross-sectional view of the sensing element with wire leads bonded to the metal contacts is shown in Fig. 52.

Pressure causes the thin diaphragm to bend, inducing a stress or strain in the diaphragm and also in the buried resistor. The resistor values will change, depending on the amount of strain they undergo, which depends on the amount of pressure applied to the diaphragm. Hence, a change in pressure (mechanical input) is converted to a change in resistance (electrical output). The sensing element converts energy from one form to another. The resistor can be connected to either a half-bridge or a full Wheatstone bridge arrangement. For pressure applied to the diaphragm using a full bridge, the resistors can theoretically be approximated as shown in Fig. 53 (nonamplified units). The

*Honeywell Inc.

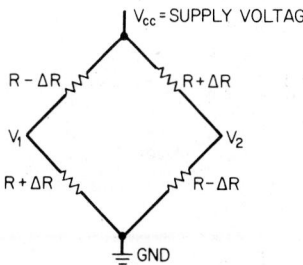

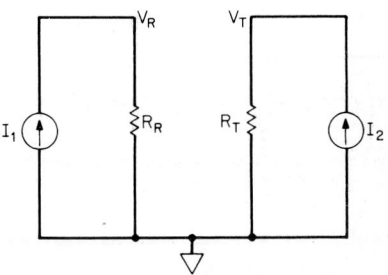

Fig. 53 Full-bridge arrangement of a piezoresistive transducer. $R + \Delta R$ and $R - \Delta R$ represent the actual resistor values at the applied pressure. R represents the resistor value for the undeflected diaphragm ($P = 0$) where all four resistors are nearly equal in value. ΔR represents the change in resistance due to an applied pressure. All four resistors will change by approximately the same value. Note that two resistors increase and two decrease, depending on their orientation with respect to the crystalline direction of the silicon material. The signal voltage generated by the full-bridge arrangement is proportional to the amount of supply voltage (V_{cc}) and the amount of pressure applied which generates the resistance change ΔR. *(Micro Switch.)*

Fig. 54 Half-bridge configuration used in signal-conditioned version of piezoresistive pressure transducer. Voltage across the piezoresistors: $V_R = L_1 R_R = I_1$ $(R_{RO} + kP)$ for radial resistor; and $V_T = I_2 R_T = I_2$ $(R_{TO} - kP)$ across the tangential resistor. I_1 and I_2 are adjusted at zero pressure to obtain: $V_R = V_T$ or $I_1 R_{RO}$ $= I_2 R_{TO}$. At other temperatures, when R_{RO} and R_{TO} vary, the equality will hold provided that the temperature coefficients of R_{RO} and R_{TO} are equal. I_1 and I_2 increase with temperature to compensate for the chip's negative temperature coefficient of span. The temperature coefficient of null, which may be of either polarity, is compensated for by summing a temperature-dependent voltage $V_N(T)$ with the piezoresistor voltage so that the output $V_O = V_R - V_T \pm V_N(T)$, with the polarity of $V_N(T)$ selected to provide compensation. *(Micro Switch, Series 110PC.)*

signal voltage generated by the full-bridge arrangement is proportional to the amount of supply voltage V_{cc} and the amount of pressure applied which generates the resistance change ΔR.

A half-bridge configuration used in a signal-conditioned version of the piezoresistive pressure transducer is shown in Fig. 54.

Among the pressure ranges of the transducers most frequently used are 0 to 1, 0 to 15, 3 to 15, 0 to 30, 0 to 100, and 0 to 250 psi (1 psi = ~6.9 kPa). Repeatability and hysteresis effects are typically less than 0.1% of full scale, and combined linearity and hysteresis do not exceed ±1% of full-scale output. The operating temperature range for standard units is from −40 to +125°C (−40 to +252°F).

REFERENCES

Bailey, S. J.: "Pressure Sensors 1982," *Control Eng.*, vol. 29, no. 5, 1982, pp. 89.

Bailey, S., J.: "Pressure Sensors and Transmitters Affected by Technological Change," *Control Eng.*, vol. 31, no. 1, 1984, pp. 97–102.

Bicking, R.: "P-to-V Transducer Available in Control Market after Automotive Baptism," *Control Eng.*, vol. 25, no. 11, 1978, pp. 51.

Easton, C. J.: "Electronics Pushes Strain Gage Technology," *Control Eng.*, vol. 28, no. 12, 1981, p. 180.

Jutila, J. M.: "Electronic Pressure Transducers: A Status Report," *Instrum. Technol.*, vol. 27, no. 1, 1980, p. 9.

Kompass, E. J.: "*Smart* Transmitter Stores Calibration Digitally," *Control Eng.*, vol. 30, no. 11, 1983, pp. 80–81.

McGowan, M. J.: "Multiple Pressure Transmitters Speak Digital on Two Wires," *Control Eng.*, vol. 26, no. 12, 1979, p. 32.

Moore, R.: "Pressure Transmitters: One Manufacturer's Approach to Reliability," *Control Eng.*, vol. 27, no. 5, 1980, p. 93.

Morris, H. M.: "D/P Cell Outputs Either Current or Frequency," *Control Eng.,* vol. 26, no. 2, 1979, p. 53.

Shapiro, B. H.: "Strain Gage Transducers Show Long Term Stability," *Control Eng.,* vol. 31, no. 2, 1984, pp. 160–163.

Staff: "Manometer Measures Low Pressures with 1 Pa Accuracy," *Control Eng.,* vol. 27, no. 7, 1980, p. 54.

Wolny, R. F.: "Applying Electronic Delta P Transmitters," *Instrum. Technol.,* vol. 25, no. 7, 1978, p. 47.

Vacuum Measurement

by
Zbigniew C. Dobrowolski*

This article is concerned with the measurement of gas pressures below atmospheric pressure. Subatmospheric pressure is usually expressed in reference to perfect vacuum or absolute zero pressure. Like absolute zero temperature (the concept is analagous), absolute zero pressure cannot be achieved, but it does provide a convenient reference datum. Standard atmospheric pressure is 14.695 psi absolute, 30 in of mercury absolute, or 760 mmHg of density 13.595 g/cm^3 where acceleration due to gravity is $g = 980.665$ cm/s^2. One millimeter of mercury, which equals one *torr,* is the most commonly used unit of absolute pressure. Derived units, the mullitorr or micrometer, representing 1/ 1000 of 1 mmHg or 1 torr, are also used for subtorr pressures.

In the MKS system of units, standard atmospheric pressure is 750 torr and is expressed as 100,000 Pa (N/m^2) or 100 kPa. This means that 1 Pa is equivalent to 7.5 millitorr (1 torr = 133.3 Pascal). Vacuum, usually expressed in inches of mercury, is the depression of pressure below the atmospheric level with absolute zero pressure corresponding to a vacuum of 30 in of mercury.

Since atmospheric pressure is not constant and since it also varies with elevation above sea level, the use of vacuum units can lead to a misunderstanding.

Fundamentals of Major Gage Types

Vacuum gages can be either direct or indirect reading. Those which measure pressure by calculating the force exerted by incident particles of gas are direct reading, while instruments which record pressure by measuring a gas property which changes in a predictable manner with gas density are indirect reading.

The range of operation for these two classes of vacuum instruments is given in Table 1.

Summary of Gage Properties

Since the pressure range of interest in present vacuum technology extends from 760 to 10^{-13} torr (over 16 orders of magnitude), there is no single gage capable of covering such a wide range. Although hundreds of types of gages have been used in past years, this discussion is limited to the

*Chief Development Engineer, Kinney Vacuum Co., Unit of General Signal Corp., Canton, Massachusetts.

Table 1. Range of Operation of Major Vacuum Gages

Principle	Gage type	Range, torr
Direct reading	Force measuring:	
	Bourdon, bellows, manometer (oil and mercury),	$760–10^{-6}$
	McLeod capacitance (diaphragm)	$760–10^{-6}$
Indirect reading	Thermal conductivity:	
	Thermocouple (thermopile)	$10–10^{-3}$
	Pirani (thermistor)	$10–10^{-4}$
	Molecular friction	$10^{-2}–10^{-7}$
	Ionization:	
	Hot filament	$10–10^{-10}$
	Cold cathode	$10^{-2}–10^{-15}$

relatively limited number of gages currently commercially available and in engineering use. See Fig. 1.

When using indirect reading gages, it is pertinent to remember that their calibration is based on dry air or nitrogen as the test gas. For other gases or vapors, a calibration factor must be given. This means that, in order to determine the correct pressure, the composition of the gas mixture measured must be known.

In practice, this is not known or at best only as a rough approximation. In most instances, it is sufficient to recognize whether light (hydrogen) or heavy (oil, organics) molecules predominate.

For example, pressure reading based on the hot-filament ionization gage is 10^{-5} torr, but the residual gas is known to be mainly diffusion pump oil. The gage reading is an order of magnitude too high because the calibration is based on nitrogen, and hence the ion pressure is less than 10^{-6} torr. Ion gage calibration factors are described later in this article. The pressure in the rough vacuum range (760 to 1 torr) can be measured with an accuracy of better than 10% by the use of direct reading gages. In fact, this can be extended to high vacuum (10^{-3} to 10^{-6} torr) by the use of a capacitance manometer without regard to gas composition or a McLeod gage for noncondensable gases. With indirect reading gages, particularly at pressures below 10^{-6} torr (very high-vacuum, 10^{-6} to 10^{-9} torr, and ultrahigh-vacuum, below 10^{-9} torr), the pressure is known to within an order of magnitude only if the gas composition is known at the same time.

At all pressures, gage location with respect to the vacuum pump needs to be considered, as throttling valves and line pressure drops can provide sharp differencs between the process chamber and pressure at the pump. Small air leaks into the gage line or gage connection can also produce large pressure reading errors on account of the small conductance of these lines. In such instances, pressure readings can be one or two orders of magnitude higher than the system pressure.

In high- and ultrahigh-vacuum regions, the relationship is even more complicated. Inadequate gage outgassing can result in too high a pressure reading, while a depressed pressure value can be recorded with a very well-outgassed gage. Gage pumping can further complicate the measuring process. The pressure equalization period between the process chamber and gage can be long.

The effect of the gaseous environment on the gage and that of the measuring mechanism on the gas composition needs to be noted. When using high-compression-ratio gages (McLeod), condensable vapors in a gas mixture can be liquefied, making for complicated pressure interpretations.

Contamination of vacuum gage sensors, interfering electric and magnetic fields, and environmental temperature and pressure excursion all can introduce undesirable errors. The selection of a suitable vacuum gage should take into account the effect of these parameters as well as specific operating requirements such as corrosion, vibration, and ease of operational interfacing.

Figure 2 summarizes the pertinent characteristics of the gages under discussion. Four different answer symbols are used in the chart, ranging from "yes" through "qualified yes" and "qualified no" to "no." For easy and uniform reference, these symbols are made to appear in shaded squares ranging from white (blank) to nearly black (heavy hash marks). This chart allows one to determine at a glance if the number of disadvantages or gage limitations is high or low. The assigned answers are unavoidably somewhat arbitrary and may be clarified by reading the descriptions of individual gage characteristics which follow this general discussion.

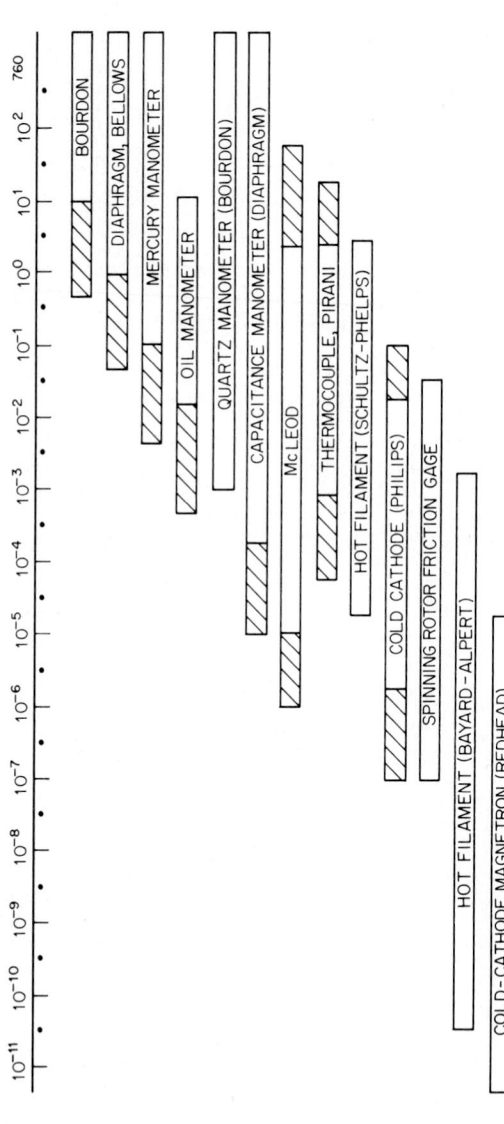
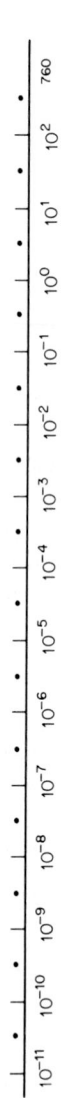

Fig. 1 Ranges of major types of vacuum gages.

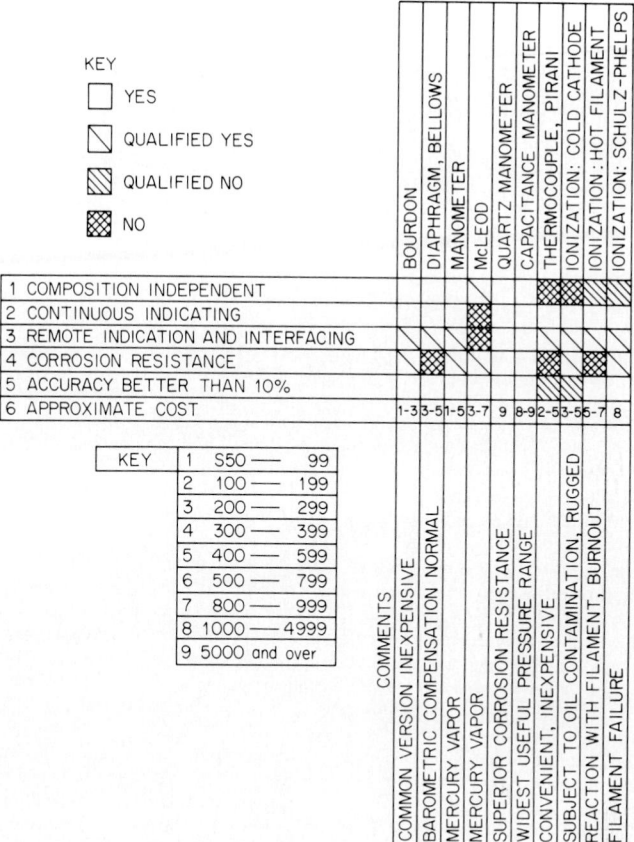

Fig. 2 Summary of vacuum gage properties.

Bourdon Gage

This type of gage, preeminently useful in the range of pressures above 1 atm, is also available commercially in units useful down to 10 torr. Figure 3 is a schematic drawing of a bourdon gage. As indi-ated in Fig. 1, a bourdon gage gives a composition-independent indication, does not chemically affect the gas being measured, and is not affected by the gas (apart from corrosion problems overcome by the use of stainless steels or other nonreactive metals or by the use of inert liners). To the extent that the elastic properties of the element are unaffected by temperature or are affected in a known and reproducible way, the gage can be used at a temperature considerably higher or lower than room temperature. The relationship between pressure and scale deflection is most frequently made to be quite closely linear, being controlled by adjustments of a simple linkage. Generally, the linkage is coupled to the pointer by a small pinion and a segment gear. In another design this gearing is replaced by a helical groove and a sliding inclined shoe, an arrangement for which greater useful life is claimed.

The principal disadvantage of the bourdon gage is its relatively poor accuracy below 20 torr. A multiturn spiral structure and thin-walled construction are required to push sensitivity below this limit. The bourdon-type structure is inherently strong and well adapted to high stresses; this feature is what makes it so satisfactory for high-pressure measurements but limits its usefulness for low pressures. With conventional metal-walled bourdon gages, remote indication or controlling action is most easily achieved pneumatically, although electromagnetic methods have also been used.

Common bourdon gages are not usually compensated barometrically, which means their readings are affected by altitude and by barometric changes. In an instrument designed for absolute pressure measurements a reference bourdon tube sealed at zero absolute pressure is installed in addition to the sensing tube. With this infrequently used arrangement, pressure readings in the subtorr range are possible.

An interesting version of a compensated gage using spiral bourdon tubes made of quartz, with an attached mirror and an optical amplifier for observing small deflections, is described in more detail later under Quartz Manometer.

Bellows and Diaphragm Gages

Like the bourdon gage, the bellows gage employs mechanical deflection of an impermeable elastic container in response to the internal and external pressure difference as an operating principle. A bellows made of copper-beryllium alloy is a more flexible structure with greater deflection per unit pressure difference and accordingly can be more easily made sensitive and accurate.

For measurement of absolute pressure, independent of ambient pressure variations, two techniques have been used, as shown in Fig. 4.

In the single-bellows version (Fig. 4a) the metal bellows capsule is evacuated to a pressure below 1 millitorr (considerably lower than the gage sensitivity of 100 millitorr) and is then permanently sealed.

The gage case is made vacuum tight and is fitted with a second flexible metal bellows in the wall, allowing mechanical connection to the external indicating or controlling linkages. The unknown pressure is introduced into the gage chamber, where it surrounds the sensing capsule and compresses it elastically. The bellows element is quite strong in compression and can thus be made quite flexible and sensitive, meanwhile retaining the necessary ability to withstand atmospheric pressure without damage. Gages embodying this construction are available with a full-scale reading of less than 10 torr and are useful down to pressures of a few tenths of 1 torr. Other gages with higher pressure ranges are readily available.

A similar pressure response can be obtained from the popular diaphragm vacuum gage. A barometer with an aneroid capsule is the most common type of diaphragm gage. The sensing element is a hermetically sealed, evacuated, thin-walled, flattened, circular capsule made of copper-beryllium alloy. It is housed in an enclosure similar to that shown for the single-bellows gage (Fig. 4a), except that the indicating mechanism is located in the space under vacuum (Fig. 4c). As the pressure is reduced, the diaphragm capsule expands and conveys its movement to a pointer through a mechanical amplification (levers) system.

Gages constructed on this principle have a nearly linear scale and are independent of altitude and barometric pressure by virtue of design. In the two-bellows gage for measurement of absolute pressure, atmospheric air surrounds two flexible bellows (Fig. 4b) which are mounted in a common frame and are mechanically opposed. One bellows is permanently sealed after being evacuated to a pressure two to three orders of magnitude lower than gage sensitivity. The other bellows communicates with the vacuum system. Perfect compensation requires that the effective areas of the two bellows be identical.

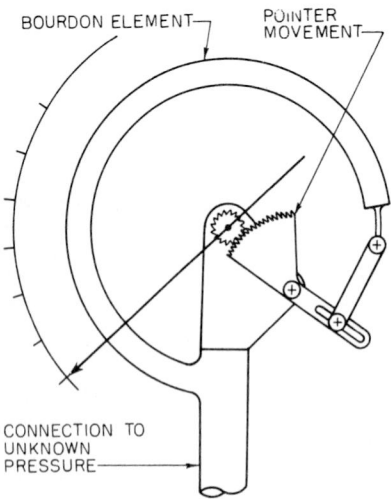

BOURDON ELEMENT

POINTER MOVEMENT

CONNECTION TO UNKNOWN PRESSURE

Fig. 3   Bourdon gage.

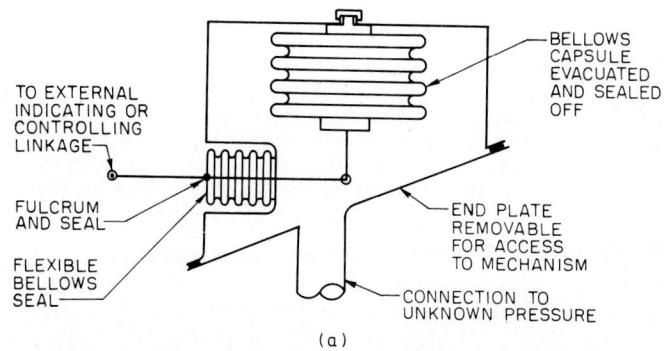

BELLOWS CAPSULE EVACUATED AND SEALED OFF

TO EXTERNAL INDICATING OR CONTROLLING LINKAGE

FULCRUM AND SEAL

FLEXIBLE BELLOWS SEAL

END PLATE REMOVABLE FOR ACCESS TO MECHANISM

CONNECTION TO UNKNOWN PRESSURE

(a)

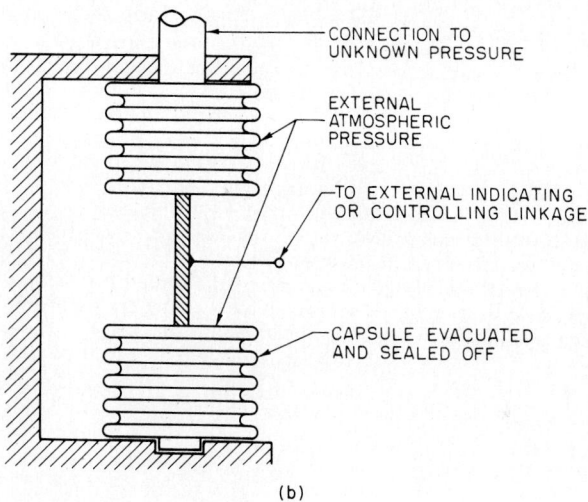

CONNECTION TO UNKNOWN PRESSURE

EXTERNAL ATMOSPHERIC PRESSURE

TO EXTERNAL INDICATING OR CONTROLLING LINKAGE

CAPSULE EVACUATED AND SEALED OFF

(b)

Fig. 4 Methods used for making a bellows gage read absolute pressure. (*a*) Sensitive single-bellows gage. (*b*) Two-bellows gage. (*c*) Gage in which indicating mechanism is located in space under vacuum.

3.67

Two-element gages of the bellows style are less sensitive than the single-element gages previously described. Remote indication is possible through the use of a bonded strain gage or transducer, or use of the sensor element converted to an electrical signal by a differential transformer. All mechanical vacuum gages are sensitive to vibrations, particularly those with high sensitivity, because of complicated mechanical linkages.

Unintended liquid flooding of the gage must be prevented by suitable location and self-draining orientation should liquid enter the instrument. Because material selection is limited, corrosion protection may be practical only by keeping gases and vapors in contact with the gage mechanism in an anhydrous state through temperature control, trapping, inert gas blanketing, vacuum storage, and similar operating techniques. Particularly troublesome is the behavior of elastomers in the presence of halogens. This is also true of vacuum-exposed gage scales made of plastic or paper. Teflon is preferred as the seal material with a gage scale made of metal or a known compatible alternate. If heating is applied, temperature compensation may be necessary.

Capacitance Manometer

The state of the art of the now widely used diaphragm gages has dramatically improved with the advent of electronic capacitance manometers. The wide pressure range, corrosion resistance, operational stability, ease of interface with electromechanical controls and with microprocessors, and reasonable cost all make it one of the most valuable process tools.

Diaphragm deflections as small as 10^{-9} in inch can be resolved, which in the most sensitive, thinnest membrane model, corresponds to an applied pressure of 10^{-6} torr. The useful measuring range of such an instrument extends from about 10^{-5} to 1 torr. In thicker membrane units, the measuring range extends from 10^{-2} torr to above 1 atm. Systems are available where analog and digital readouts are provided for continuously monitoring or controlling pressure. In the range of 10^{-5} to 100 torr, these are the most accurate continuous reading, total pressure measuring instruments available. Typical accuracies range from 0.05% at 1 torr and above to 3% at 10^{-4} torr.

Input pressure to the sensor is applied to a thin, radially tensioned metal membrane welded to massive supporting rings which deflect in approximate proportion to the applied pressure. See Fig. 5a. Stationary capacitor plates consisting of metal films deposited on ceramic insulating disks are

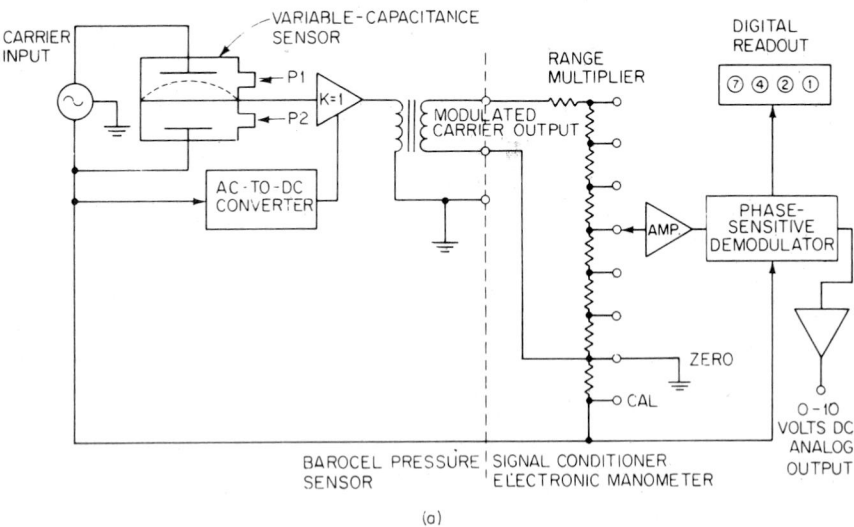

(a)

Fig. 5   Variable capacitance pressure transducing systems. (a) Input pressure to sensor is applied to a thin, radially tensioned metal membrane welded to massive supporting rings which deflect in approximate proportion to applied pressure.

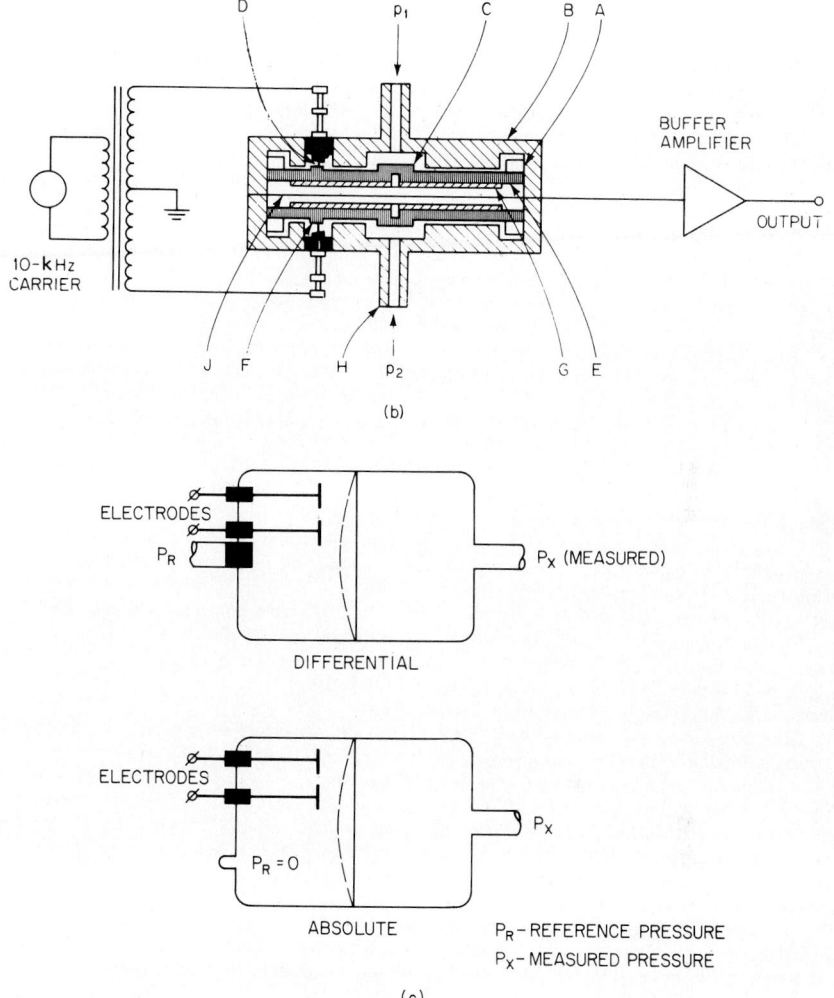

Fig. 5 Continued (b) Sensor output signal feeds into a metering console which accurately scales the pressure signal over three or four decades. (c) Single-sided capacitance sensors. A, Substrate support; B, sealed housing; C, screen; D, jumper wire; E, ceramic substrate; F, film termination; G, gold film; H, pressure port; J, diaphragm.

mounted adjacent to the membrane, one on either side, so that the diaphragm becomes the variable element of a capacitor. When the element is arranged in a bridge circuit excited by a frequency of 10 kHz, the capacitance change caused by the diaphragm motion unbalances the bridge and produces an output at 10 kHz of voltage amplitude precisely proportional to pressure.

The sensor output signal feeds into a metering console (Fig. 5b) which accurately scales the pressure signal over three or four decades, provides a meter display of pressure value (either analog or digital), and changes the 10 kHz into a dc voltage analog of pressure using phase-sensitive, synchronous, demodulation techniques.

The materials used in the sensors are metals and ceramics, fully compatible with the most critical vacuum applications. Some models can be baked to 450°C and actually measure pressure at temperatures up to 300 to 400°C.

The largest single problem with the sensor shown in Fig. 5b is the need to continuously pump on the reference side in vacuum applications. In addition, the dielectric constant on the process side of the diaphragm may not be identical with that on the reference side.

More recent advances in gettering and outgassing technology have permitted permanent evacuation of the reference cavity so that continuous pumping is no longer needed for absolute pressure measurements. See Fig. 5c. Sensitivity to corrosion and changes in the dielectric constant have been eliminated by placing the electrodes in the reference or sealed cavity.

Since the diaphragm is usually made of Monel, Inconel, or another high-nickel alloy, inherent corrosion resistance of the sensor is high. In compatible applications procedural techniques described in connection with bellows and diaphragm gages can be used. With the one-sided electrode arrangement, it is also possible to cover the same three, four, or even five pressure decades using a single sensor.

Like most instruments, capacitance manometers suffer from several fundamental sources of error; most of these are, however, negligible in engineering terms. The fundamental errors can be identified as hysteresis, nonlinearity, temperature effect, effect of media composition, and line pressure effect.

Hysteresis of the diaphragm is inherently very low on account of structural stretching and built-in tension (approximately 0.01% of reading).

Nonlinearity is the highest single source of error, representing 80% of all accumulated inaccuracies. It is composed of variations in sensor performance as well as variations in the electronic components used. The error is typically less than 0.07% of reading.

Temperature effects are second in importance, typically 0.06% of reading for a 1°C change in temperature. Automatic temperature compensation is complicated and, in cases where required accuracy demands, a dedicated sensor heater, set typically for 45°C $\pm$ 0.1% with a 10:1 control ratio (1°C sensor temperature change for 10°C change in ambient temperature) is commonly used.

Media composition effect is eliminated with the one-sided electrode location. For a two-sided electrode arrangement, a 1% change in the dielectric constant may mean a 0.5% error of reading.

Line pressure effect is on the order of 0.01% of full scale for 1 psi of line pressure change. This error is not significant in vacuum measurements.

The high-accuracy versions of the instrument have a usable resolution capability on the order of 1 part in 100,000 with noise levels an order of magnitude lower. Even the least expensive instruments have resolution and noise values only one order of magnitude worse than that of the high-precision units. For a resolution of 1:1,000,000, the high-accuracy instrument must resolve a capacitance change on the order of 10^{-5} pF corresponding to a deflection of 5×10^{-9} inch (0.5 Å). These values are for a 1-torr, full-scale sensor. All this is accomplished in the presence of typical interference associated with process environment which may include high-temperature gradients, radio frequency and magnetic fields, and perhaps some radiation.

The overall gaging transduces a measured absolute or differential pressure into a useful, stable dc voltage or current output which can be interfaced with data acquisition systems while being able to accept input command signals from microprocessors or mainframe process computers.

Quartz Manometer

While capacitance manometers push the diaphragm pressure response characteristics to present-day technological limits, quartz manometers do the same with the bourdon-tube principle.

The inherent accuracy of the measuring system is based on several factors: quartz is a perfect spring material, no-contact optical sensing of the quartz tube deflection, repeatable digital resolution, and control of the sensor environment.

Two quartz bourdon tubes (one at reference pressure or sealed at full vacuum) are formed into a helix which rotates as a pressure difference between the tubes is established. Since the quartz pressure sensor (QPS) is maintained at a constant temperature, it repeats the same angular deflection whenever the same pressure difference appears.

To preserve the inherent accuracy of quartz and prevent friction effects, optical sensing is employed based on an electrooptical differential which generates an electrical output proportional to the angle difference between the QPS rotation Θ_P and the instrument counter reading Θ_R. See Fig. 6.

Resolution of the instrument is 1 part in 100,000 with a 1-millitorr resolution for a 100-torr

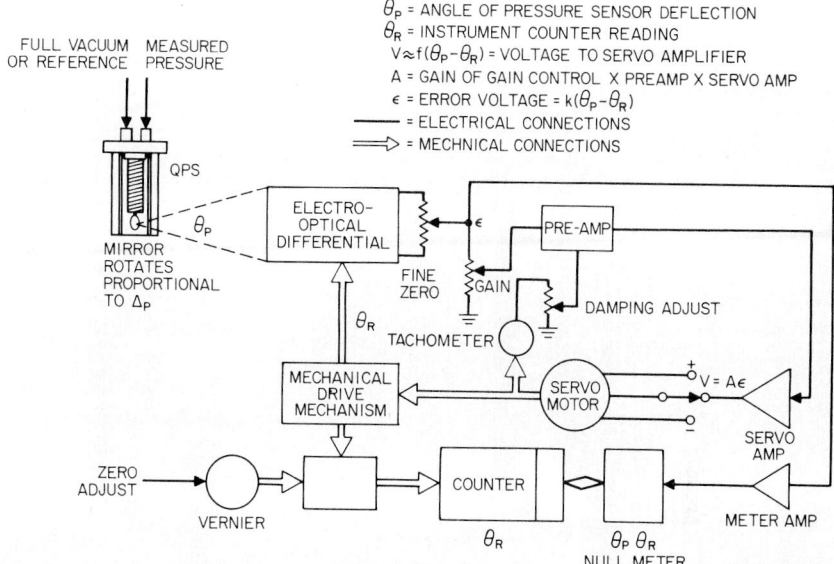

Fig. 6 Operating principle of a quartz manometer.

full-scale sensor. The quartz manometer can be interfaced with process control loops in a fashion similar to that used with a capacitance manometer. The corrosion resistance of quartz is superb (etching by hydrofluoric acid is the only danger), with helium permeability a potential complication when that gas pressure has to be measured. The instrument is typically more expensive and less convenient to use than a capacitance manometer, with which it shares many performance and operational characteristics.

Liquid Manometer

The simplest vacuum pressure gage is a transparent U tube filled with mercury or another low-vapor-pressure liquid. One branch of the tube is connected to the vacuum system, and the other is sealed under vacuum or is maintained at low reference pressure by an auxiliary vacuum source. With low reference pressure, the unknown pressure is given directly by the difference between the two columns of liquid.

A difference of 0.1 mm can be detected by eye, which sets the minimum detectable pressure change at 0.1 torr for mercury. The sensitivity can be increased about 15-fold by the use of low-vapor-pressure oil. This extrapolates to a pressure difference detection limit of 10 millitorr (0.1 mm liquid-column difference). Instrument errors are common to all liquid manometers, mainly involving temperature, density, capillary depression, gas solubility, and light refraction.

Temperature corrections are predictable, and using a micrometer readout it is possible to measure pressures at the 1 millitorr level for the purpose of calibrating other instruments. Manometers are insensitive to gas composition (except for reactions with the manometer liquid), direct reading, inexpensive, and readily field-fabricated. Within their range of application, manometers are considered to be primarily pressure standards.

McLeod Gage

This is a special case of a liquid manometer used as a pressure amplifier. If, for example, 100 cm³ (volume V in Fig. 7) of permanent gas is compressed into a section of capillary having a volume of

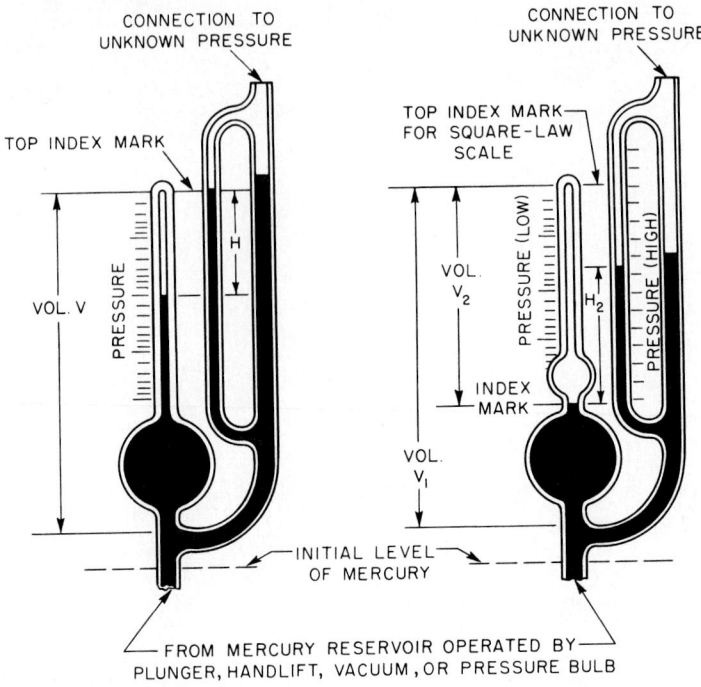

(a) SINGLE-RANGE, SQUARE-LAW SCALE

$$P = \frac{AH^2}{V} \text{ (VERY NEARLY)}$$

(b) DOUBLE-RANGE, LINEAR AND SQUARE-LAW SCALES

$$P = \frac{V_2}{V_1} H_2 \text{ (VERY NEARLY)}$$

Fig. 7   Two versions of the McLeod gage.

0.1 cm^3, the resulting pressure reading is amplified 1000 times. This principle allows pressure measurements into the 10^{-6}-torr region, considerably below the 10^{-2}-torr range of precision manometers.

If we assume that the volume V of gas trapped at the unknown pressure p (in centimeters of mercury for convenience) obeys Boyle's law, then $pV = (p + H)A$, where A is the cross section of the closed capillary in square centimeters, and

$$P = \frac{AH^2}{V - HA}$$

In practice, HA is quite negligible when compared with volume V, with $p = 10AH^2/V$ torr, and with other values expressed in centimeters.

A conveniently small McLeod gage may have a volume V of 200 cm³, with a capillary cross section A of 0.02 cm² and a length of 10 cm. Thus, for $H = 0.1$ cm, $p = 5 \times 10^{-6}$ torr, which would be the limit of unaided visual resolution and the reading could be wrong by 100%. At $H = 1$ cm, $p = 5 \times 10^{-4}$ torr, the possible error becomes 10%. For various reasons, the only significant improvement in accuracy can be achieved by an increase in volume V.

A carefully constructed nonportable gage with a 1300-cm³ volume gives reproducible readings of ± 0.5, ± 0.6, ± 2, and $\pm 6\%$ at pressures of 10^{-2}, 10^{-3}, 10^{-4}, and 10^{-5} torr, respectively. The errors for other volumes can be estimated to be no lower than those based on volume proportionality. Thus, in the previous example with $V = 200$ cm³ and $p = 10^{-4}$ torr, percent error = $(1300/200) \times 2 = 13\%$, which is in good agreement with the previous rough estimate of 10%.

Since the measured pressure in a McLeod gage is derived from basic (linear) dimensions, it is the industrial pressure standard with reference to which all other vacuum gages are calibrated. However, it should be emphasized that only the pressure of permanent gases is measured correctly. On account of the high compression ratio employed, vapor pressure of a substance of several tenths of a torr would not be detected with condensed liquid occupying negligible volume and not being visible to the eye.

The presence of condensable vapors in an unknown gas-vapor mixture is explained with the help of Fig. 8.

In a normal reading of a McLeod gage, h_1 is made equal to zero. Consider now what would happen if we raised the mercury to successive levels and measured the heights h_1 and h_2 in millimeters. The pressure of the compressed gas at the top of the closed capillary is $(h_2 - h_1)$ mmHg. If the system contains permanent gas only (that is, gases for which $PV = RT$), then

$$PV = pv \qquad (1)$$

where P, p = pressures before and after compression, respectively

V = volume of McLeod gage bulb

v = volume contained in length h_2 of the capillary, the volume to which the gas has been compressed

Since

$$v = \frac{\pi d^2}{4} h_2 \qquad (2)$$

then

$$PV = (h_2 - h_1)\frac{\pi d^2}{4} h_2$$

or

$$(h_2 - h_1)h_2 = \text{constant}$$

for all values of h_1 and h_2 as the mercury is raised in the process of compressing a particular sample of gas.

On the other hand, if a condensable material of vapor pressure P at room temperature is present in the system with practically no permanent gas, then, since $h_2 - h_1$ is the pressure on the compressed gas,

$$h_2 - h_1 = P \text{ a constant} \qquad (3)$$

as long as $h_2 - h_1$ is as large as P (sufficient compression to cause condensation).

Thus, for a wet system in which the remaining gas present is all vapor, the McLeod gage characteristic behavior follows $h_2 - h_1 = P$, a constant, whereas for a system containing permanent gases only, the McLeod gage characteristic behavior follows $(h_2 - h_1)h_2 = \text{constant}$. In a system involving a mixture of permanent gases and condensable vapors, a behavior somewhat intermediate is to be expected.

The criterion of whether a McLeod gage reading is being compromised by the pressure of a condensable vapor is, therefore, to raise the mercury to three or four different levels, measuring h_1

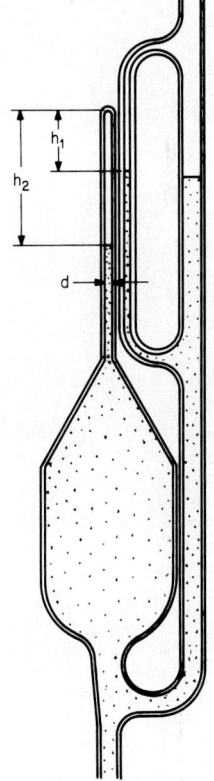

Fig. 8 Diagram explaining the presence of condensable vapors in an unknown gas-vapor mixture when using a compression (McLeod) gage.

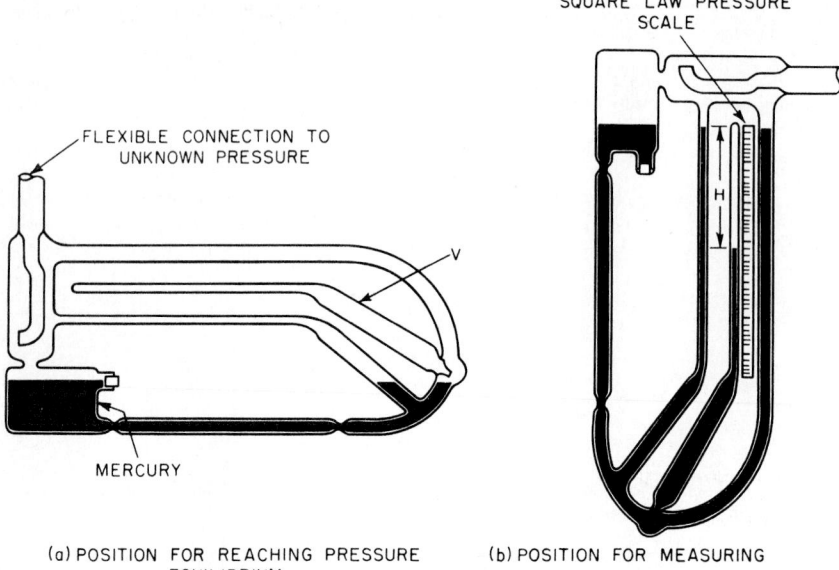

SQUARE LAW PRESSURE
SCALE

FLEXIBLE CONNECTION TO
UNKNOWN PRESSURE

V

H

MERCURY

(a) POSITION FOR REACHING PRESSURE   (b) POSITION FOR MEASURING
EQUILIBRIUM

Fig. 9   Tilting-type McLeod gage.

and h_2 for each level, and see whether the product $(h_2 - h_1)h_2$ is a constant for all levels. If this is the case, then the reading is not being compromised by the presence of condensable vapor. If, on the other hand, the departure of this product from a constant value is quite marked, then a condensable vapor must be influencing the readings. In the extreme case, in which the vapor pressure is very high as compared with the pressure of the permanent gas, it may even be found that $h_2 - h_1$ is constant and independent of the mercury level.

There are several methods of bringing the mercury level to the set mark in the reference capillary. In Fig. 7, the initial mercury level is a barometric mercury leg with an extra liquid head provided by a plunger or air pressure. The atmospheric leg can be shortened by keeping the reservoir under vacuum. Air admission controls the desired head. Magnetic coupling with handlift is also used. Figure 9 shows a tilting McLeod gage, while Fig. 10 shows a highly portable metal McLeod gage where precision-ground glass capillaries can be readily and inexpensively replaced in case of breakage.

Small oil McLeod gages are available, with a sensitivity to 10^{-4} torr quite common. To be usable these gages must be kept under vacuum; otherwise the oil outgassing becomes lengthy and laborious. Apart from scale magnification, the absence of mercury vapors can be desirable. Oily environments can be handled easily, as contaminating oil vapors can be made soluble in the gage fluid (dibutyl pthalate, dibutyl sebacate, light oil).

Thermal Conductivity Gages (Pirani and Thermocouple Gages)

Commercial thermal conductivity gages should not ordinarily be thought of as precision devices. Within their rather limited but industrially important pressure range they are outstandingly useful. The virtues of these gages include low cost, electrical indication readily adapted to remote readings, sturdiness, simplicity, and interchangeability of sensing elements. They are well adapted for uses where a single power supply and measuring circuit are used with several sensing elements located in different parts of the same vacuum system or in several different systems.

The working element of the gages consists of a metal wire or ribbon exposed to the unknown pressure and heated by an electric current. The temperature attained by the heater is such that the total rate of heat loss by radiation, gas convection, gas thermal conduction, and thermal conduction through the supporting leads equals the electric power input to the element. Convection is unimportant and can be disregarded, but the heat loss by thermal conduction through the gas is a function of pressure. At pressures of approximately 10 torr and higher, the thermal conductivity of a gas is high and roughly independent of further pressure increases. Below about 1 torr, on the other hand, the thermal conductivity decreases with decreasing pressure, eventually in linear fashion reaching zero at zero pressure. At pressures above a few torr, the cooling by thermal conduction limits the temperature attained by the heater to a relatively low value. As the pressure is reduced below a few hundred millitorr, the heater temperature rises, and at the lowest pressures, the heater temperature reaches an upper value established by heat radiation and by thermal conduction through the supporting leads.

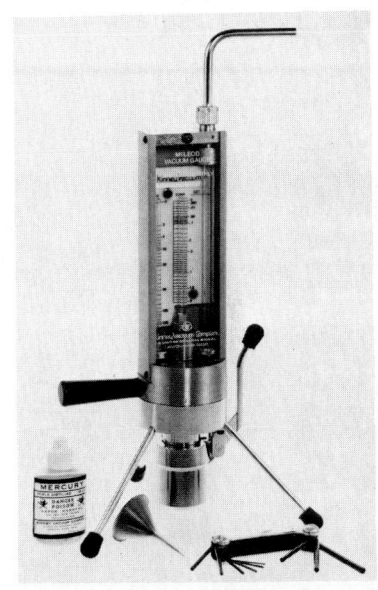

The principal sources of inaccuracy in thermal conductivity gages arise from the difficulty in maintaining constant surface conditions. The first of these is the accommodation coefficient, which is a measure of the closeness with which a gas molecule rebounding from a heated or cooled surface has acquired the temperature of that surface. More troublesome is the emissivity of the surface, which controls the relation between temperature and rate of radiant energy emission. The emissivity of a new, clean heater is low and results in high temperatures at low gas pressures, but after a period of service the metal surface becomes oxidized, carbonized, or otherwise soiled. When this occurs, the low-pressure range is particularly subject to error, frequently reading high by as much as 20 millitorr.

Fig. 10 Highly portable metal McLeod gage permitting precision-ground glass capillaries to be readily and inexpensively replaced in case of breakage. *(Kinney Vacuum Company, Unit of General Signal.)*

As previously indicated, the response of a thermal conductivity gage (Pirani) can conveniently be described in terms of three pressure ranges: low, medium, and high. The low-pressure range is characterized by a response quite closely proportional to the pressure. The response in the medium-pressure range is roughly logarithmic so that, for example, equal changes in meter current result from each successive doubling of the pressure. The high-pressure range is characterized by very small changes in reading, even when the pressure is increased manyfold. The boundaries between the low- and medium-pressure ranges and between the medium- and high-pressure ranges can be shifted by the details of construction, temperature of operation, and parameters of the indicating circuit.

The two types of thermal conductivity gage mentioned utilize different means for converting the temperature variations of the heater into meter readings. As shown in Fig. 11, Pirani gages are customarily used in a Wheatstone bridge circuit having two scales; the lower scale, covering from 0 to perhaps 20 or 40 millitorr, is approximately linear, while the higher scale, covering from a few millitorr to nearly 1 torr, is roughly logarithmic. The two scales of the meter are customarily calibrated directly in terms of air.

Two similar Pirani elements are used in adjacent arms of the bridge, one being exposed to the unknown pressure and the other being evacuated and sealed off. Although the second of these elements is usually considered as providing compensation for variations in ambient temperature and is usually mounted adjacent to the measuring element, this precaution is ordinarily unnecessary. Partial compensation of the voltage coefficient is the principal function of the compensating element. Without the compensator, even quite small fluctuations in the bridge voltage would give large, spurious pressure indications; the effect of ambient temperature is much less important than this voltage effect.

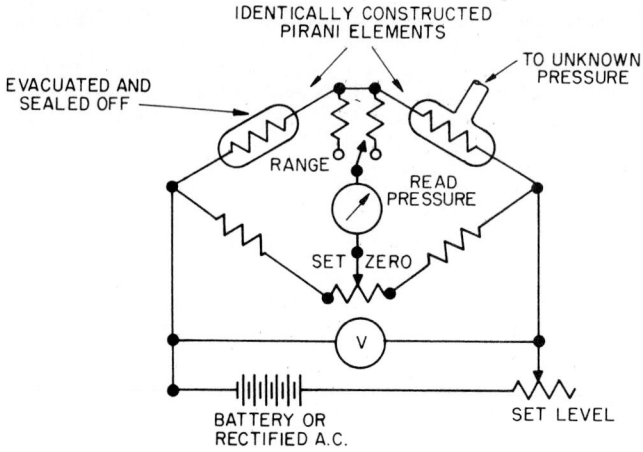

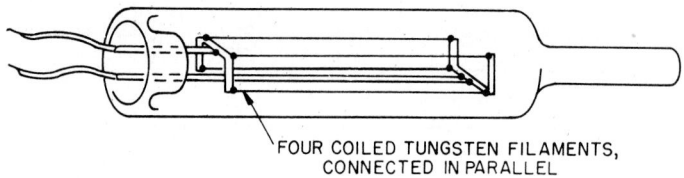

FOUR COILED TUNGSTEN FILAMENTS,
CONNECTED IN PARALLEL

Fig. 11 Pirani gage: (*top*) gage in fixed-voltage Wheatstone bridge; (*bottom*) sensing element.

The compensating element can thus often be mounted at some convenient distance from the sensing element, although if leads many yards long are employed it may be desirable to make the connections to the compensating element with leads of similar length and resistance.

A typical Pirani gage characteristic, showing meter current dependence on the pressure of dry air, is illustrated in Fig. 12. Since this is plotted on a linear scale, it is evident that at low pressures sensitivity is high but at high pressures the change in indication is relatively small even for large increases in pressure.

As indicated, the ways in which the output indications of the Pirani gage and of the thermocouple gage depend on pressure are governed by the same laws and are basically similar. It is useful, therefore, to compare the linear plot of Fig. 12 with the semilogarithmic plot of Fig. 16, which shows the response of a typical thermocouple gage and which makes it evident that in the medium-pressure range (10 to 200 millitorr) the response is roughly logarithmic.

Gases other than air have a different thermal conductivity and accommodation coefficient, and the reading of a Pirani gage with such a gas is generally somewhat different from the reading for air at the same pressure; i.e., the Pirani gage is composition-dependent. Figure 13 shows typical curves for several gases, giving the relationship between true pressure and reading observed on a scale calibrated in terms of the pressure of air.

The composition dependence of Pirani and thermocouple gages makes it possible to use them for locating large- and medium-sized leaks in vacuum systems, as will be discussed later.

A common type of thermocouple gage is shown in Fig. 14. The temperature attained by the midpoint of the heater is sensed by a single-thermocouple hot junction welded to the midpoint of the heater. This gives sufficient current to be read on a low-resistance dc microammeter whose single range is usually calibrated in terms of pressure, the gas being dry air. The inherent inaccuracy of these gages, as ordinarily constructed, does not make it worthwhile to correct for variations in ambient temperature.

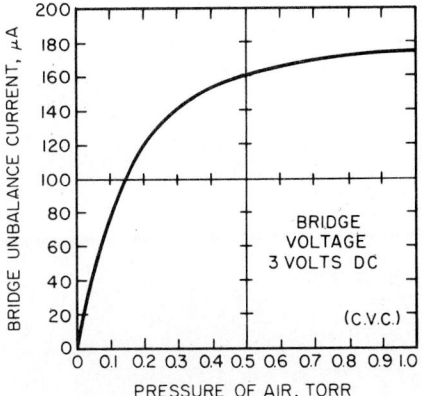

Fig. 12 Linear scale plot of typical Pirani gage response versus pressure of air.

Another variation of the thermocouple principle is shown in Fig. 15. In this gage the functions of the heater and the thermoelectric generator are combined in a single element constructed as a thermopile. The junctions are made alternately large and small, and a heating current is passed in series through the composite thermopile structure. The heating effect, and consequently the attained temperature, are greater in the portions having a small cross section; thus these become the hot

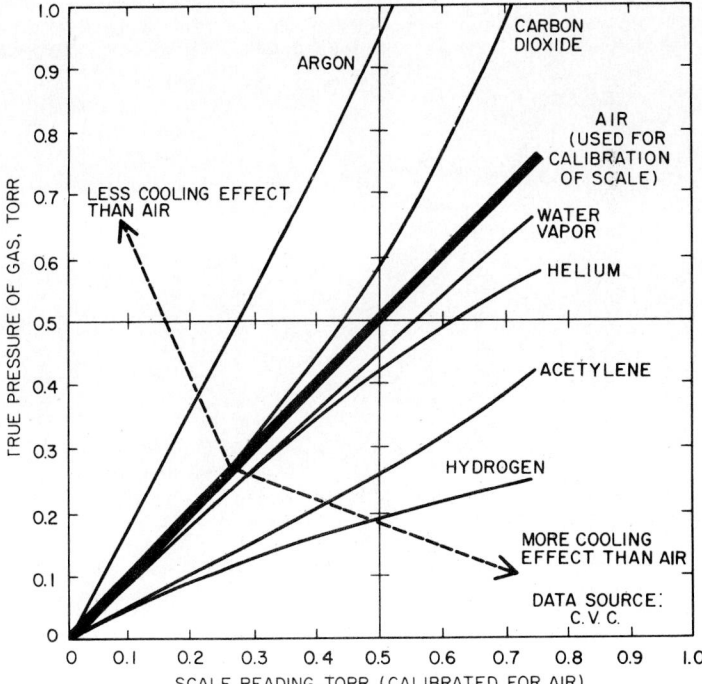

Fig. 13 Linear scale plot showing typical Pirani gage response for gases other than air, in terms of air calibration.

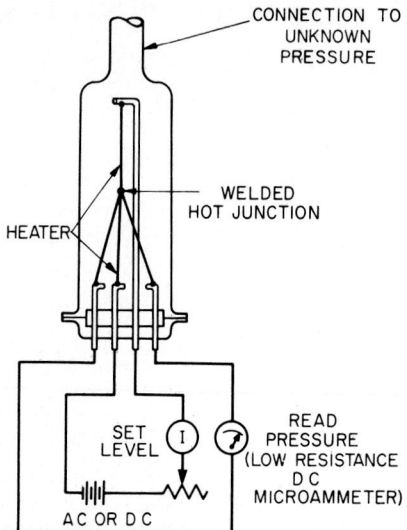

CONNECTION TO
UNKNOWN
PRESSURE

WELDED
HOT JUNCTION

HEATER

SET
LEVEL   I

READ
PRESSURE
(LOW RESISTANCE
D C
MICROAMMETER)

AC OR DC

Fig. 14   Thermocouple gage.

junctions of the thermopile. With the use of a thermopile, which produces higher voltages than a single thermocouple, it is possible to use a less sensitive indicating meter. The arrangement of Fig. 15 shows that an alternating current can be used for heating without interfering with the dc indicating circuit.

Figure 16 shows a semilogarithmic plot of the output meter current as a function of pressure for a typical thermocouple gage. The solid curve shows the calibration for dry air, and the meter scale

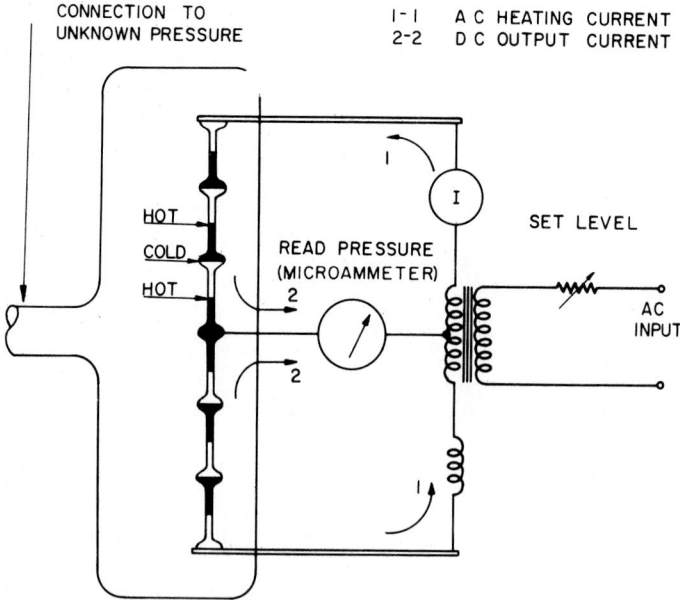

CONNECTION TO
UNKNOWN PRESSURE

1-1   A C HEATING CURRENT
2-2   D C OUTPUT CURRENT

HOT

COLD

HOT

READ PRESSURE
(MICROAMMETER)
2

2

I

SET LEVEL

AC
INPUT

Fig. 15   Modified thermal gage with a double-function thermopile.

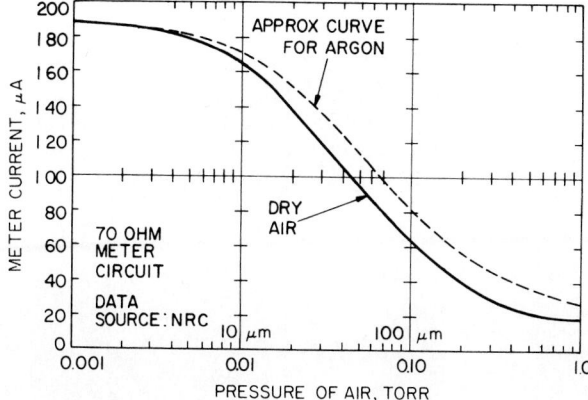

Fig. 16 Semilog plot of a typical thermocouple gage response versus pressure of air and of argon.

is marked in accordance with this calibration. The dashed curve gives approximately the response for argon. It should be noted that, while the low-pressure limiting value coincides with that for dry air, the high-pressure values approach a different asymptote, since the asymptotic high-pressure thermal conductivities of argon and air are different.

Figure 17 gives typical curves of the response of a thermocouple gage for gases other than air. These curves are comparable with those of Fig. 13, which shows the corresponding relationships for a Pirani gage, except that Fig. 17 is a log-log plot so as to cover a considerably larger pressure range and show more clearly the asymptotic behavior at high pressures. The fact that many of the curves cross the curve for air indicates that for these gases the low-pressure thermal conductivity is greater than that of air but the high-pressure thermal conductivity is less than the value for air. It is of some importance in metallurgical applications to note that the curve for carbon monoxide coincides with that for air over a considerable range.

Hot-Filament Ionization Gage

The hot-filament ionization gage is the most widely used pressure measuring device for the region from 10^{-2} to 10^{-11} torr. The operating principle of this gage is illustrated in Fig. 18.

A regulated electron current (typically about 10 mA) is emitted from a heated filament. The electrons are attracted to the helical grid by a dc potential of about $+150$ V. In their passage from filament to grid, the electrons collide with gas molecules in the gage envelope, causing a fraction of them to be ionized. The gas ions formed by electron collisions are attracted to the central ion collector wire by the negative voltage on the collector (typically -30 V). Ion currents collected are on the order of 100 mA/torr. This current is amplified and displayed using an electronic amplifier. The control circuit for a hot-filament ionization gage (Bayard-Alpert type) is shown in Fig. 19.

This ion current will differ for different gases at the same pressure; i.e., a hot-filament ionization gage is composition-dependent. Over a wide range of molecular density, however, the ion current from a gas of constant composition will be directly proportional to the molecular density of the gas in the gage.

Calibration of the indicating unit used with a hot-filament ionization gage is practically always understood to be with dry nitrogen or dry air. The response of the sensing element for other gases follows a reasonably well-verified empirical law. Of two different gases at the same pressure and temperature, the one having the higher molecular weight (or more strictly the higher total number of electrons in the molecule) will, in general, yield a higher ion current.

Dushman discusses this relationship and shows a linear plot of the experimental data fitted approximately with a straight line not passing through the origin. Figure 20 presents Dushman's

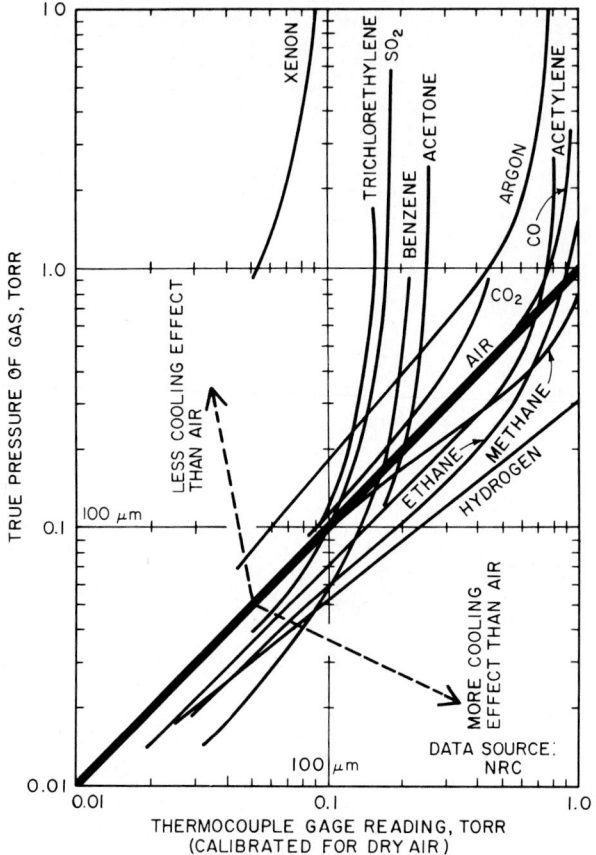

Fig. 17   Log-log plot showing a typical thermocouple gage response for gases other than air, in terms of calibration.

data on a log-log plot, showing that a satisfactory fit can also be obtained by use of an empirical power-exponent expression.

$$S_A = \left(\frac{N_c}{18}\right)^{0.8} = 0.099(N)^{0.8} \tag{4}$$

or

$$S_{\mathrm{air}} = 0.118(N)^{0.8} \tag{5}$$

In these equations N_c is the so-called net atomic number of the molecule (i.e., the sum of the atomic numbers of the component atoms, equal to the number of electrons in the neutral molecule), and S_A and S_{air} are the ion gage sensitivities relative to argon and to air, respectively. The relationship fails badly for hydrogen and neon and is poor for helium; for gases not shown in Fig. 20, it might be reasonable to guess that the calculated value will be accurate within 15 or 20%. The relation has no presently known theoretical basis, and since the ionizing ability of low-voltage electrons depends drastically on the gas and on the electron energy, it is surprising that such a relation exists at all. It is useful, nevertheless.

The hot-filament ionization gage is extremely useful but does have several undesirable features:

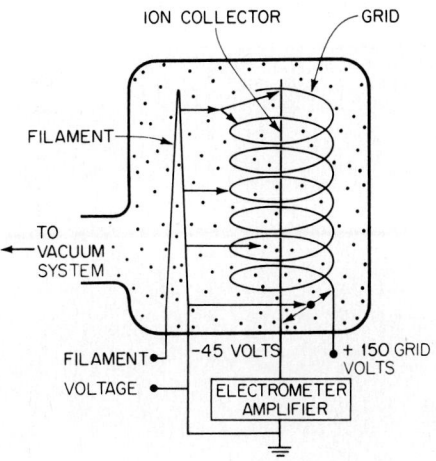

ION COLLECTOR   GRID

FILAMENT

TO
VACUUM
SYSTEM

FILAMENT   −45 VOLTS   + 150 GRID
                                              VOLTS
VOLTAGE   ELECTROMETER
                    AMPLIFIER

Fig. 18   Hot-filament ionization gage (Bayard-Alpert type).

1. Since an incandescent filament or cathode is required to obtain the necessary electron emission, and since air pressures higher than a few millitorr cause rapid deterioration of these thermionic emitters, it is not possible to use ionization gages at higher pressures (preferably below 0.5 millitorr). In addition, means must be provided for preventing burnout if accidental sudden increases in pressure occur. Oxidation-resistant electron emitters make this limitation less severe (e.g., thoria).

2. The hot filament is susceptible to attack by some gases and may cause undesirable chemical reactions in others. Typical filament temperatures range as high as 2200°C.

3. Since the filament operates at such a high temperature and since the other electrodes assume relatively high temperatures, the sorption of gas by these electrodes is important. Ideally, the

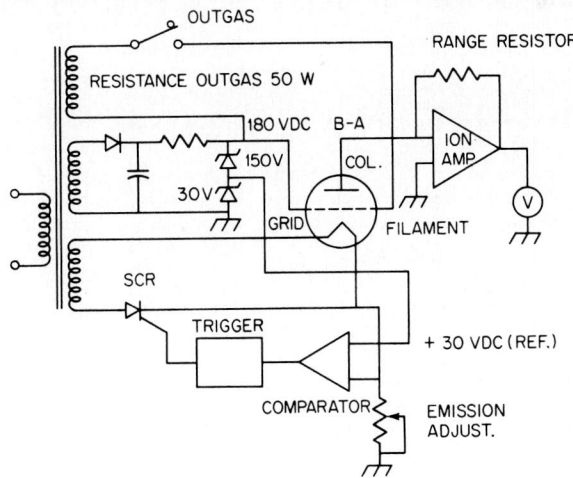

OUTGAS

RANGE RESISTOR

RESISTANCE OUTGAS 50 W

180 VDC   B-A

150V   ION
AMP
COL.
30V
GRID   FILAMENT
V

SCR

TRIGGER   + 30 VDC (REF.)

COMPARATOR   EMISSION
                        ADJUST.

Fig. 19   Control circuit of hot-filament ionization gage.

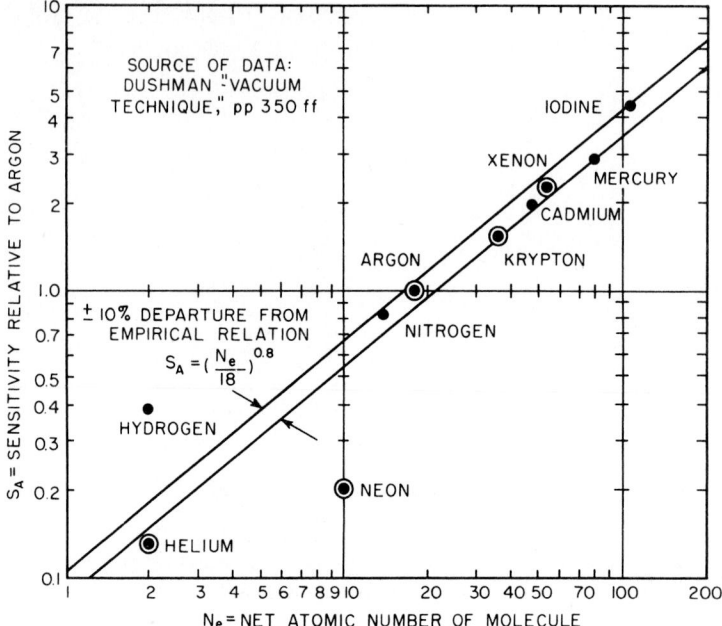

Fig. 20 Calibration of a typical hot-filament ionization gage for gases other than air.

electrodes would neither trap gas nor give it off, but this is not the situation. Various means of heating these electrodes are in use, including resistance heating with an electric current (particularly applicable to a spiral grid structure), electron bombardment, and baking out in an oven. After an ionization gage has been degassed by one or more of these means, the electrodes will take up more gas, so that readings made immediately after an outgassing procedure are likely to be erroneously low. For industrial purposes a reasonable compromise is to keep outgassing to a judicious minimum and to employ a gage having a connecting tubulation as large in diameter and as short as possible. Whatever sorption and desorption processes occur will then cause a minimum of inequality between the pressure in the gage and the pressure in the system. Gages mounted entirely within the vacuum system have also been used ("nude gages"). Although these gages avoid the conductance limitations, electrical behavior can be erratic.

Ion gage sensing elements can be made sufficiently alike so that it is not necessary to calibrate them individually for ordinary industrial applications; they are thus directly interchangeable.

Schulz and Phelps developed a modified hot-filament ionization gage for use in the 1 to 10^{-5} torr range. This gage is commercially available in one envelope together with a conventional Bayard-Alpert gage.

Cold-Cathode Ionization Gage

This ingenious gage, invented by Penning, possesses many of the advantages of the hot-filament ionization gage without being susceptible to burnout. Ordinarily, an electrical discharge between two electrodes in a gas cannot be sustained below a few millitorr pressure. To simplify a complicated set of relationships, this is because the "birthrate" of new electrons capable of sustaining ionization is

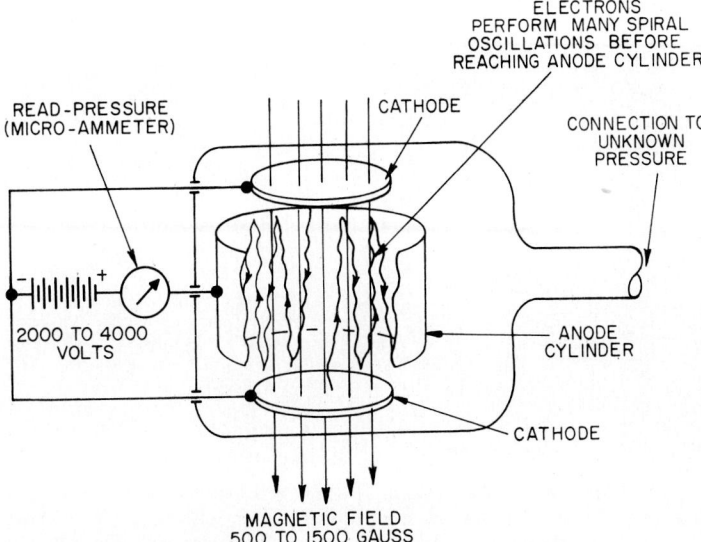

ELECTRONS
PERFORM MANY SPIRAL
OSCILLATIONS BEFORE
REACHING ANODE CYLINDER

READ-PRESSURE
(MICRO-AMMETER)

CATHODE

CONNECTION TO
UNKNOWN
PRESSURE

2000 TO 4000
VOLTS

ANODE
CYLINDER

CATHODE

MAGNETIC FIELD
500 TO 1500 GAUSS

Fig. 21   Philips cold-cathode ionization gage.

smaller than the "death rate" of electrons and ions. In the Philips gage this difficulty is overcome by the use of a collimating magnetic field which forces the electrons to traverse a tremendously increased path length before they can reach the collecting electrode. In traversing this very long path, they have a correspondingly increased opportunity to encounter and ionize molecules of gas in the interelectrode region, even though this gas may be extremely rarefied. It has been found possible by this use of a magnetic field and appropriately designed electrodes, as indicated in Fig. 21, to maintain an electric discharge at pressures below 10^{-9} torr.

Comparison with the hot-filament ionization gage reveals that, in the hot-filament gage, the source of the inherently linear relationship between gas pressure (more exactly molecular density) and gage reading is the fact that the ionizing current is established and regulated independently of the resulting ion current. In the Philips gage this situation does not hold. Maintenance of the gas discharge current involves a complicated set of interactions in which electrons, positive ions, and photoelectrically effective x-rays all play a significant part. It is thus not surprising that the output current of the Philips gage is not perfectly linear with respect to pressure. Slight discontinuities in the calibration are also sometimes found, since the magnetic fields customarily used are too low to stabilize the gas discharge completely. Despite these objections, a Philips gage is a highly useful device, particularly where accuracy better than 10 or 20% is not required.

The Philips gage is composition-sensitive but, unlike the situation with the hot-filament ionization gage, the sensitivity relative to some reference gas such as air or argon is not independent of pressure. Leak hunting with a Philips gage and a probe gas or liquid is a useful technique. Unlike the hot-filament ionization gage, the Philips gage does not involve the use of a high-temperature filament and consequently does not subject the gas to thermal stress. The voltages applied in the Philips gage are on the order of a few thousand volts, which is sufficient to cause some sputtering at the high-pressure end of the range, resulting in a certain amount of gettering or enforced take-up of the gas by the electrodes and other parts of the gage. Various design refinements have been used to facilitate periodic cleaning of the vacuum chamber and electrodes, since polymerized organic molecules are an ever-present contaminant.

The conventional cold-cathode (Philips) gage is used in the range from 10^{-2} to 10^{-7} torr. Redhead has developed a modified cold-cathode gage useful in the 10^{-6} to 10^{-12} torr range (Fig. 22). The operating voltage is about 5000 V in a 1-kG magnetic field.

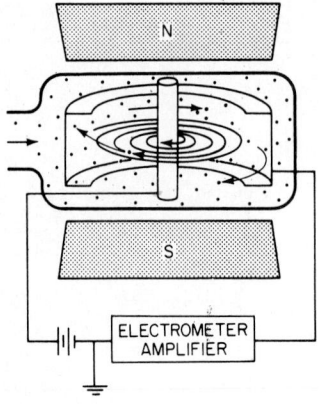

Fig. 22 Inverted magnetron, a cold-cathode gage, produces electrons by applying a high voltage to unheated electrodes. Electrons spiraling in toward the central electrode ionize gas molecules, which are collected on the curved cathode.

Spinning Rotor Friction Gage

While liquid manometers (U tube, McLeod) serve as pressure standards for subatmospheric measurements (760 to 10^{-5} torr) and capacitance (also quartz) manometers duplicate this range as useful transfer standards, calibration at lower pressures depends on volume expansion techniques and presumed linearity of the measuring system. The recently marketed friction gage allows extension of the calibration range directly down to 10^{-7} torr. The gage measures pressure in a vacuum system by sensing the deceleration of a rotating steel ball levitating in a magnetic field (Fig. 23).

It can be shown that

$$p = \frac{\pi r d}{10\theta t}\, \bar{v} \ln \frac{w(t)}{w_0}$$

where

r = rotor radius

d = rotor density

$\bar{v}$ = mean molecular velocity ($\sqrt{8RT/\pi M}$)

θ = rotor surface geometry factor (≈ 1.0)

w = initial angular speed of rotor (400 Hz)

$w(t)$ = rotor speed after elapsed time (t)

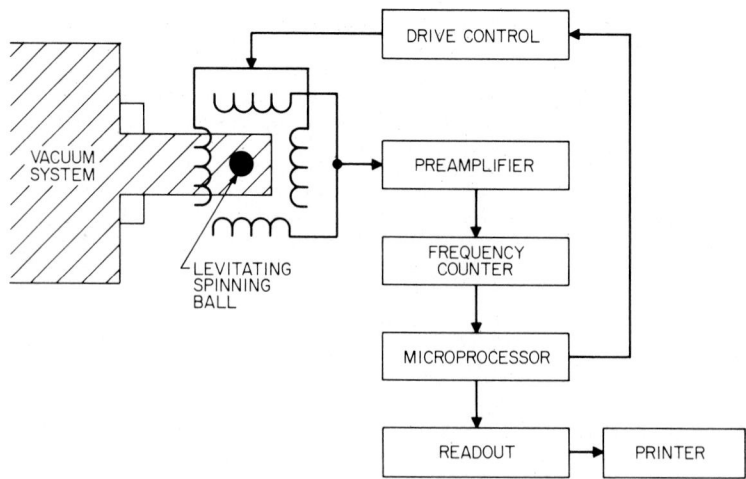

Fig. 23 Schematic diagram of spinning rotor friction gage.

or in terms of control unit interpretation

$$P \equiv \Delta t$$

where a typical reading takes only a few seconds. Calibration coefficients for gases based on

$$P = \text{coeff. ln } \frac{w(t)}{w_0}$$

for a steel ball with a radius of 2.25 mm and material density of 7.79 g/cm^3 are tabulated

Hydrogen	9756	(3.738)
Deuterium	6903	(2.638)
Helium	6324	(2.417)
Methane	3455	(1.320)
Neon	3083	(1.178)
Nitrogen	2617	(1.000)
Oxygen	2446	(0.935)
Argon	2181	(0.933)
Krypton	1513	(0.578)
Xenon	1208	(0.462)

Long-term reproducibility (several years) and accuracies of 1% from 10^{-2} to 10^{-5} torr and 3 to 5% for 10^{-7} torr have been documented.

Leak Detection Using Vacuum Gages

All vacuum gages can be used as leak detectors by the simple expedient of recording reduced or increased pressure when a leaking segment of a vacuum system is exposed to the pump and a gage located near its inlet. Similarly, gages located in strategic locations of a system can help in measuring the leak-up rate and hence the vacuum tightness of a particular section of the installation. Leak elimination can be based on the isolation method or on using plastic compound (Duxseal) for temporary repairs of suspected areas.

In addition, composition-dependent instruments such as ionization gages, Pirani gages, and thermocouple thermal conductivity gages are useful in the detection of leaks in vacuum systems and apparatus. The gage is installed at a suitable point on the vacuum system, which is continuously pumped, and the suspected points on the exterior of the system are probed with a gas or volatile liquid. When an important leak is probed, the stream of gas entering the system changes composition and a marked shift in gage reading can be observed (a sharp increase). When several leaks of different sizes are present, it is usually necessary to find and repair the larger ones first so that their presence will not mask the smaller leaks. This technique of leak detection is particularly useful on an operating system for maintenance purposes; at the present state of development, location of the smallest leaks requires the use of mass spectrometer leak detectors (their limit of sensitivity is about 10^{-10} st. cm^3/s).

Partial Pressure Analyzers*

Many applications of high-vacuum technology are more concerned with the partial pressure of particular gas species than with total pressure. Also, "total" pressure gages generally give accurate readings only in pure gases.

*Also known as residual gas analyzers (RGAs).

For these reasons partial pressure analyzers are finding increasing application. These are basically low-resolution, high-sensitivity mass spectrometers which ionize a gas sample in a manner similar to that of a hot-filament ionization gage. The resulting ions are then separated in an analyzer section, depending on the mass-to-charge ratio of each ion. The ion current corresponding to one ion type is then collected, amplified, and displayed. Partial pressure gages are very valuable diagnostic tools in both research and production work. The major types are listed in the accompanying table.

Type	Minimum partial pressure, torr	Resolution,† au	Magnetic field
Magnetic sector	10^{-11}	20–150	Yes
Cycloidal	10^{-11}	100–150	Yes
Quadrupole	10^{-12}	100–300	No
Time of flight	10^{-12}	200	No

†Maximum mass number at which a mass number difference of one can be observed.

REFERENCES

Benedict, R. P.: *Fundamentals of Temperature, Pressure and Flow Measurements,* Wiley, New York, 1969.

Bromberg: "Accurate Pressure Measurements in the Region of 0.1–1,000 Torr: The Intercalibration of Two McLeod Gages and a Capacitance Manometer," *J. Vacuum Sci. Technol,* September/October 1969.

Denison, D. R.: "Vacuum Physics and Technology," in *Methods of Experimental Physics,* vol. 14, Academic Press, New York, 1979.

Dobrowolski, Z. C.: "Fluid Flow-Low Pressure," in *Encyclopedia of Science and Technology,* McGraw-Hill, New York, 1981.

Drinkwine, M. J., and D. Lichtman: *Partial Pressure Analyzers and Analysis,* American Vacuum Society, New York, 1980.

Dushman, S.: *Scientific Foundations of Vacuum Technique,* 2d ed., Wiley, New York, 1962.

Leck, J. R.: *Pressure Measurement in Vacuum Systems,* The Institute of Physics, London, 1957.

McCracken, G. M.: "Small Magnetic Mass Spectrometers as Residual Gas Analyzers," *Vacuum,* vol. 19, no. 7, July 1969.

O'Hanlon, J. F.: *A User's Guide to Vacuum Technology,* Wiley, New York, 1980.

Pike, E. W., and N. E. Gibbs: A Study of Aneroid Capsules," *J. Appl. Phys.,* vol. 19, (1948), pp. 106–108.

Pirani, M., and T. Yarwood: *Principles of Vacuum Engineering,* Chapman and Hall, London, 1961.

Redhead et al: *The Physical Basis of Ultrahigh Vacuum,* Chapman and Hall, London, 1968, pp. 253–265.

Schulz and Phelps: *Rev. Sci. Instrum.,* vol. 28, 1957, p. 1051.

Sellenger, F. R.: "A Review of Vacuum Gauges and Methods for High Vacuum Gauge Calibration," *Vacuum,* vol. 18, no. 12, December 1968, pp. 645–650.

Staff: "Procedures for Hot Filament Gauge Calibration AVS Tentative Standard 6.4", *J. Vacuum Sci. Technol.,* July 1970.

Staff: *Quartz Manometer and Controller,* Mensor Corp., San Marcos, Texas, 1981.

Sullivan, J. J.: "Modern Capacitance Manometers," *Transducer Technol.,* July/August 1978.

Sullivan, J. J.: *A New Vacuum Gauge Transfer Standard,* MKS, Inc., Burlington, Mass., 1979.

Utterback and Griffith: "Reliable Submicron Pressure Readings with Capacitance Manometer," *Rev. Sci. Instrum.* July 1966.

Van Atta, C. M.: *Vacuum Science and Engineering,* McGraw-Hill, New York, 1965.

Van Der Pyl, L. M. M.: "Bibliography on Bourdon Tubes and Bourdon Tube Gauges," *Mech. Eng.,* vol. 75, 1953, p. 915.

Wilson, N. G., and L. C. Beavis: *Handbook of Vacuum Leak Detection,* American Vacuum Society, New York, 1976.

Resonant Wire Pressure Transducers

by
William H. Burtt*

While not a new technology, the principle of resonant frequency detection in pressure measurement was not generally available to the process industries until 1978. Late that year, a differential-pressure transducer using a resonant wire pressure element (sensor) was introduced. Similar transducers for gage and absolute pressure and liquid-level measurements followed. The advantages of a resonant wire transducer are high accuracy, long-term stability, reduced ambient temperature effect, and virtually nonhysteretic performance.

Principle of Operation

A wire under tension is caused to oscillate at its resonant (or natural) frequency, and changes in pressure are converted to changes in this frequency.

The application of an oscillating wire as a primary means of detecting force is based on fundamental principles initially outlined by Rayleigh's equations for a bar vibrating in a vacuum. Holst et al. (1979) modified Rayleigh's equations to fit an oscillating wire.

Their approximation of the resonant frequency f_n of a wire in a vacuum is

$$f_n = \frac{1}{2l} \sqrt{\frac{T}{\rho A} + (12 + \pi^2)\frac{EK^2}{\rho l^2}} + \frac{1}{l} \sqrt{\frac{EK^2}{\rho}} \tag{1}$$

where ρ = density of wire material
A = cross-sectional area of wire
T = tension in wire
E = modulus of elasticity of wire material
K = radius of gyration
l = length of wire

A practical resonant wire pressure sensor (Fig. 1) requires that the wire be in a nonvacuum environment. Fairly rigorous modifications and refinements of Eq. 1 were performed by Holst et al. to account for this. Simplifying, and assuming the length, density, and area remain constant in the range of tension applied to the wire, the result of these modifications can be approximated as

$$f_n \propto T^2 \tag{2}$$

A typical resonant wire sensor for differential pressure or liquid-level measurement is illustrated in Fig. 1. A wire under tension is located in the field of a permanent magnet. The wire is an integral part of an oscillator circuit that causes the wire to oscillate at its resonant frequency.

One end of the wire is connected to the closed end of the metal tube. The tube is fixed to the sensor body by the electrical insulator. The other end of the wire is connected to the low-pressure diaphragm and loaded in tension by the preload spring.

The spaces between the diaphragms and the backup plates, the fluid transfer port, and the metal tube are filled with fluid. An increasing pressure on the high-pressure diaphragm tends to move the diaphragm toward its backup plate. The fluid thus displaced moves through the fluid transfer port

*The Foxboro Company, Foxboro, Massachusetts.

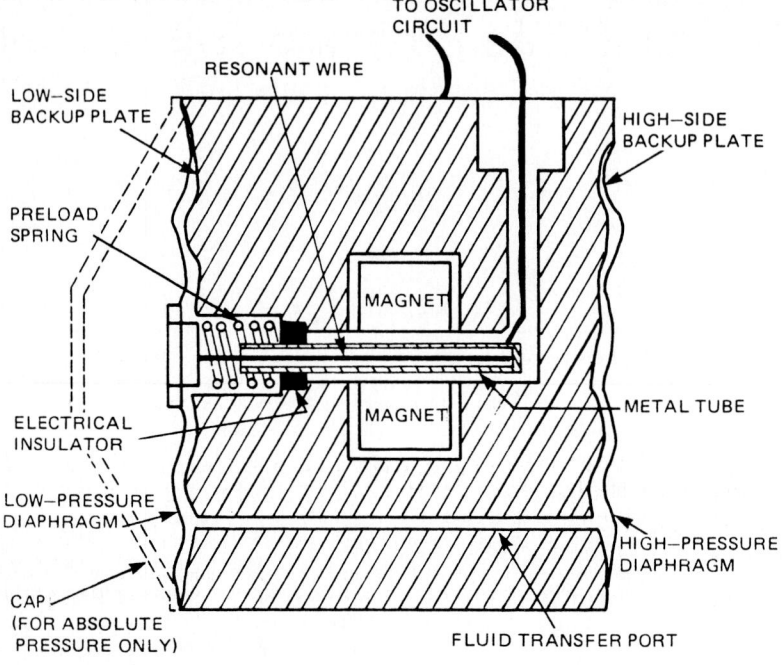

Fig. 1   Typical differential-pressure sensor. *(The Foxboro Company.)*

and tends to push the low-pressure diaphragm away from its backup plate. This increases the tension on the wire, raising its resonant frequency, and increases the output signal of the transducer.

Overrange protection for the wire is provided by an overrange spring (not shown) selected to limit the maximum tension on the wire to about two-thirds of its yield strength. The diaphragms are protected from overrange by the backup plates.

The "cap" over the low-pressure diaphragm is a modification for absolute pressure measurement. The absolute zero reference is established by evacuating the cap to 0.53 Pa (0.004 mmHg).

The same sensor is used in a liquid-level transducer. A resonant wire sensor for measuring gage pressure is illustrated in Fig. 2.

Developing the Output Signal

Figure 3 is a block diagram of the electronic circuitry. The oscillator circuit passes an ac excitation current through the wire. The forces acting on the wire cause it to oscillate at its resonant frequency. Because the wire is in a magnetic field, an emf is generated having a frequency equal to the resonant frequency of the wire. This output signal of the oscillator circuit is the input to a pulse shaper circuit. This circuit produces two complementary frequency signals which are applied in cascade to two frequency converter stages.

Each converter has two capacitors that are alternately charged and discharged by two opposite phase electronic switches operated at the resonant frequency of the wire. For example, during each half-cycle of the resonant frequency, the two capacitors in the first converter are alternately charged and discharged to the reference voltage. As one capacitor is charging, the other is being discharged by its corresponding switch. The output voltage of this converter is thus a function of the reference voltage and the frequency of the wire. The capacitors in the second converter are similarly being charged and discharged to the output voltage of the first converter. The output voltage of the second converter is thus proportional to the square of the frequency and, therefore, proportional to the

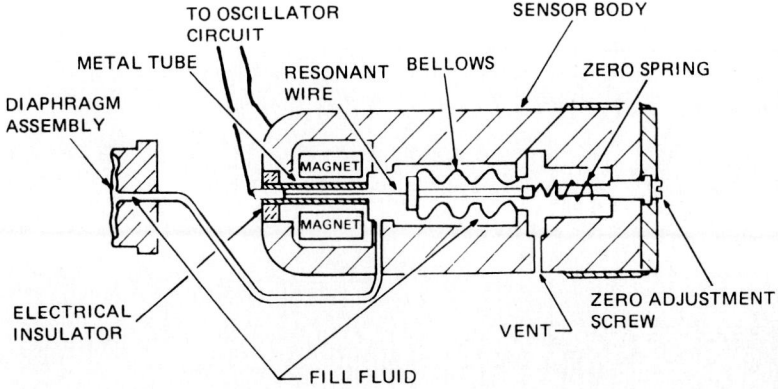

Fig. 2 Typical gage pressure sensor. *(The Foxboro Company.)*

tension on the wire. Because the actual relationship between pressure and frequency differs slightly from the ideal relationship, a compensating circuit is included in the second converter stage to ensure that the output voltage is linear with the pressure.

This voltage, directly proportional to the pressure, is converted to a 4 to 20 or 10 to 50 mA dc signal by the output amplifier.

Adjustments

The adjustments shown in Fig. 3 are made at the transducer. Zero, balance, and fine span are screw-driver adjustments, while coarse span, damping, and output action adjustments are made with jumpers. (The transducer using the gage pressure sensor illustrated in Fig. 2 does not have an electrical zero adjustment. The zero adjustment is mechanical, as shown in the figure.)

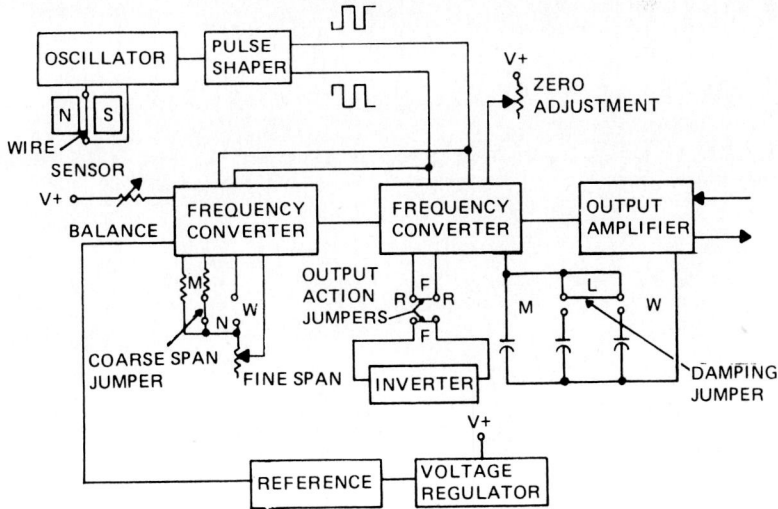

Fig. 3 Block diagram of the electronic circuitry of a resonant wire pressure transducer. *(The Foxboro Company.)*

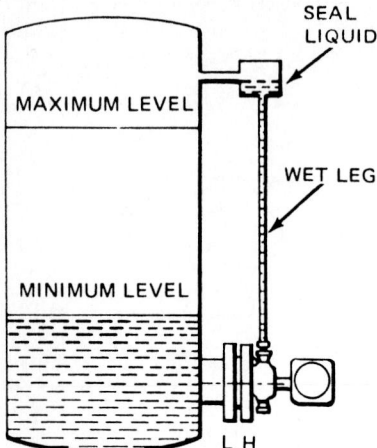

SEAL
LIQUID

MAXIMUM LEVEL

WET LEG

MINIMUM LEVEL

L H

Fig. 4 Typical elevated-zero range application of a resonant wire pressure transducer. *(The Foxboro Company.)*

Suppressed-Zero and Elevated-Zero Ranges

A suppressed-zero range is one that starts above the normal zero input signal to the transducer. The transducer is adjusted for this range with the zero adjustment screw.

An elevated-zero range is one that starts below zero and has a negative lower range-value. A typical elevated-zero range application is a wet leg, liquid-level measurement on a closed, pressurized tank (Fig. 4). Since the higher pressure must always be applied to the high-pressure diaphragm (to ensure tension on the wire), the low-pressure diaphragm must face the tank. This is necessary because the wet leg head is always greater than the tank head. So that increasing level will provide an increasing output signal, it is also necessary to place the output action jumpers in the reverse (R) position.

In effect, then, an elevated-zero range application is handled by transforming it to a suppressed-zero range application and reversing the output signal of the transmitter.

REFERENCES

Holst, P. A., Olsen, E. O., Simpson, D. C., Lindquist, R. P., and Karas, E. L., "Resonant Wire Technology Applied to Process Control Instrument Design," *Proc. 8th IMEKO Conf., Moscow, USSR,* May 1979.

Snodgrass, F. E.: "Precision Digital Tide Gauge," *Science,* October 9, 1964, pp. 198–208.

Flow Systems*

Earle Duncan. *Cascade Control Company, Roswell, Georgia. (Open-Channel Flow Measurements)*

Charles E. Fees. *Product Manager, Fischer & Porter Company, Horsham, Pennsylvania. (Variable Area Flowmeters)*

D. L. Fowler. *The Foxboro Company, Foxboro, Massachusetts. (Vortex Flowmeters)*

David B. Giacomelli. *Fluidyne Instrumentation, Oakland, California. (Positive-Displacement Flowmeters)*

E. H. Higham. *Foxboro Great Britain Limited, Redhill, Surrey, England. (Magnetic Flowmeters)*

Bill Johnson. *Kenco Plastics Company, Inc., Necedah, Wisconsin. (Open-Channel Flow Measurements)*

R. Lyall. *Brooks Instrument Division, Emerson Electric Company, Statesboro, Georgia. (Positive-Displacement Flowmeters)*

Edmond B. Lynch. *BIF, A Unit of General Signal Corporation, West Warwick, Rhode Island. (Open-Channel Flow Measurements)*

Craig McIntyre. *Kay-Ray, Inc., Arlington Heights, Illinois. (Mass Flow Measurement)*

R. W. Miller. *Flow Consultant, The Foxboro Company, Foxboro, Massachusetts. (Differential Producers)*

Eugene B. Sherwood. *Micro Motion, Inc., Boulder, Colorado. (Mass Flow Measurement)*

Gary Snyder. *Micro Motion, Inc., Boulder, Colorado. (Mass Flow Measurement)*

*Persons who authored complete articles or subsections of articles, or who otherwise cooperated in an outstanding manner in furnishing information and helpful counsel to the editorial staff.

Differential Producers

**by
R. W. Miller***

Differential producing flowmeters, sometimes referred to as *head* class flowmeters, are frequently selected because this type of flowmeter has had a long history of use in many applications. There are numerous primary elements in this class. The concentric orifice, venturi, flow nozzle, Lo-Loss† tube, target flowmeter, pitot tube, and Annubar‡ are all differential producers.

Other types of flowmeters are usually selected because of an obstructionless feature, wider rangeability, or no freezing or condensate buildup in lead lines, or because the fluid is abrasive, dirty, or multicomponent (slurry).

DIFFERENTIAL-PRODUCER FLOWMETER

Figure 1 illustrates a differential-producer flowmeter. It consists of two separate devices that act in combination to measure flow rate. The first is the primary device. The geometry of the differential-producer element, the length of the upstream and downstream piping, flow conditions, and pressure tap location with respect to the primary element are all individual elements of the primary device.

Differential pressure is measured by transmitting pressure through lead lines to a secondary

---

*This article is a condensation of the material presented in R. W. Miller, *Flow Measurement Engineering Handbook*, McGraw-Hill, New York, 1982. This material has been copyrighted by R. W. Miller and is reproduced here by permission of copyright holder. Mr. Miller is Flow Consultant for the Foxboro Company, Foxboro, Massachusetts.
†Trademark, Badger Meter, Inc.
‡Trademark, Dieterich Standard Corporation.

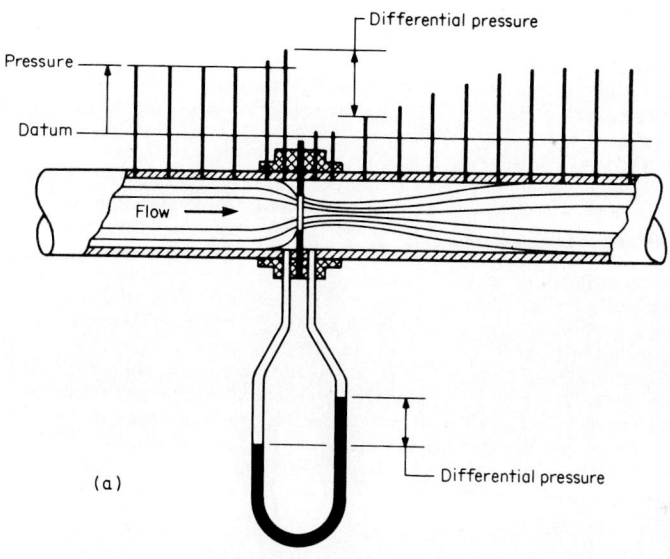

(a)

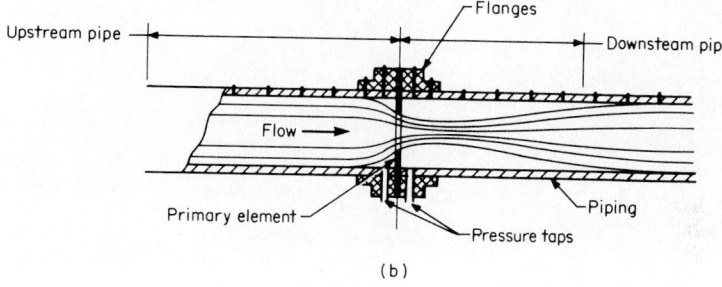

(b)

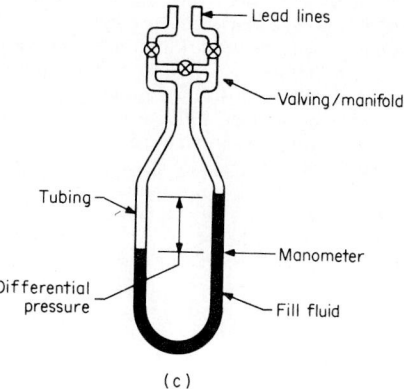

(c)

Fig. 1 Differential-producer flowmeter. (a) Flowmeter, (b) elements of the primary device, (c) elements of the secondary device.

differential-pressure measuring element. The secondary device in this case consists of three elements: lead lines, valving, and a manometer. Since flow can be calculated by visually observing the differential, no further readout equipment is required.

If flow rate were calculated by a central processor, by an on-site computer, or from a chart record, these would be additional elements of the secondary device. In combination, the two devices, primary and secondary, are referred to as the flowmeter.

As shown in Fig. 1, differential pressure depends on pressure tap location and on whether the contraction is abrupt or gradual—orifice or flow nozzle. The relationship between measured differential and flow rate is then a function of tap locations, primary element design, and associated upstream and downstream piping. These effects are included in the discharge coefficient which relates the actual flow rate to the theoretically calculated flow rate.

Many types of differential producers are in use, and for convenience they are separated into two broad categories. In the first, the bore of the primary element is "sized" to produce a given differential (normally 100 in, 25 kPa) at the upper range design flow rate. In the second type, differential pressure at the upper range flow rate cannot be varied because the primary element is of a fixed geometry. These differential producers are referred to as *fixed-geometry devices,* and the differential-pressure transmitter is calibrated to a value corresponding to the upper range flow rate.

The flow rate equations are identical for both types. However, in the first case sizing is required, while in the second case the upper range value for the differential-pressure transmitter must be determined. Orifices, nozzles, and venturis are usually considered type 1, while Annubars, elbows, and target flowmeters are considered type 2.

HISTORICAL BACKGROUND

Record of the first application of the differential-producer principle in the measurement or regulation fluid flow is lost in antiquity. It is known, however, that the Romans of Caesar's time used the orifice to measure water distributed to householders.

The developments that led to the design and theory used in present-day meters are of more recent origin. In 1732, Henry Pitot invented the pitot tube. In 1791, Venturi did his basic work on the venturi tube and developed the basis for much of the theory used in modern flowmeter computations. It was not, however, until 1887 that Clemens Herschel, using Venturi's basic work, developed the commercial venturi tube.

Impetus was added to the development and use of differential producing flowmeters by two twentieth-century industrial trends, namely (1) the handling of large volumes of flowing fluids and (2) the importance of flow rate in continuous industrial processes. Large volumes flowing at high velocities made most other then available types of flow measuring devices bulky and uneconomical. Probably the most important stimulant to the use of the differential producer, however, has been its adaptability in the control of flow rates in continuous processes.

The demands of the natural gas industry in 1903 led Thomas R. Weymouth to experiment with the use of a thin-plate, sharp-edged orifice for measuring large volumes of natural gas. Weymouth used flange pressure taps, 1 in upstream and 1 in downstream from the face of the orifice, which later were to become the predominant tap locations in the United States. Weymouth also developed empirical coefficient data correlated with the ratio of orifice bore to pipe diameter (beta ratio, β).

Vena contracta taps are those located one pipe diameter upstream and at the point of minimum pressure a short distance downstream from the orifice. In April 1916, Horace Judd presented a paper at a meeting of the ASME on the use of vena contracta taps. This paper also covered the first use of eccentric and segmental orifices for use with dirty air or air entrained in liquids. Taps one pipe diameter upstream and one-half pipe diameter downstream frequently are used as substitutes for vena contracta taps in industries where frequent orifice changes are necessary.

Although the thin, square-edged orifice was widely used with many different fluids, it was not until 1930 that a joint AGA/ASME/NBS test program obtained sufficient data to develop a coefficient prediction equation. The Ohio State University (1935) test report and the Buckingham fitting equations for the various tapping arrangements have been used by ASME and AGA since 1935. The ability to predict a coefficient from measured dimensions led to full commercialization of the orifice flowmeter.

In the late 1950s, the United States work was combined with European practice, and International Standards R541 (1967) for orifices and nozzles and R781 (1968) for venturis were issued.

In the mid 1960s, the ASME Fluid Meters Research Committee initiated a study to reevaluate the Ohio State data and to add additional coefficient data. The objective was to derive, by regression analysis, a simpler and more accurate coefficient prediction equation for flange-tapped orifices. The results (Dowdell and Chen, 1970) were not encouraging, with the Ohio State data having a ±1.85% deviation (two sigma) from the Buckingham equation. By limiting line size and beta ratios, the ASME Fluid Meters Research Committee (1971) changed its previous ±0.55% coefficient accuracy statement to ±1%.

In 1975, J. Stolz proposed a universal orifice equation to the ISO Orifice Flowmeter Subcommittee. His proposal was to combine, based on logical rules, the Ohio State data into a single dimensionless expression suitable for flange, corner, or D and $D/2$ taps. This equation was presented in a 1978 paper by Stolz. This orifice prediction equation appears in ISO Standard 5167 (1980) which combines R541 and R781 into a single differential-producer standard. Based on the Dowdell and Chen investigation and subsequent papers by Miller et al. (1979), the ASME Fluid Meters Research Committee (1981) adopted the ISO 5167 (1980) equation. This equation is increasingly being used by industry because of its simplicity, improved accuracy, and applicability over a wider Reynolds number range. The ANSI/ASME Measurement of Fluid Flow in Closed Conduits Committee is presently preparing an ANSI standard that includes this equation (MFC-draft 4, 1982).

For the measurement of natural gas, AGA 3 (ANSI/API 2530, 1978) is usually required for contractual purposes, and in it the Buckingham equations for flange and pipe taps are presented.

The first investigations of a flow nozzle date back to the nineteenth century (Froude, 1847). In 1930, Germany standardized on an ISA (1932) nozzle geometry. The International Federation of National Standardizing Organizations (ISA) was later replaced by the present International Organization for Standardization (ISO). In the United States, the long-radius flow nozzle was developed at Ohio State University (1948) primarily for the measurement of steam flows.

The need for improved accuracy when testing steam turbines led to development of the ASME throat tap nozzle. The downstream pressure tap in this nozzle is in the cylindrical throat section.

The commercial success of the orifice, venturi, and nozzle led to the development of continually improved secondary measuring elements. This, coupled with test work and user familiarity, further led to the development and use of other primary elements, such as the segmental orifice, eccentric orifice, Lo-Loss tube, elbow, and annular orifice. These primary elements, combined with a wide variety of available secondary elements, are used to indicate, record, compute, measure, and control the majority of flow measurements made today.

PRIMARY ELEMENTS

Orifice

Square Edge

In 2-in (50-mm) and larger line sizes, the concentric orifice (Fig. 2) is the most common restriction for clean liquids, gases, and low-velocity vapor (steam) flows. It is a sharp, square-edged hole bored in a flat, thin plate. The ratio of hole diameter (d) to pipe diameter (D) defines the beta ratio (β). For most applications this ratio should be between 0.2 and 0.75, depending on the desired upper range differential, a high beta ratio orifice producing less of a differential for the same flow rate than a small beta ratio.

Depending on upstream and downstream tap locations (Fig. 3), the flowmeter is referred to as a corner tap, a flange tap, a D and $D/2$ tap, a pipe tap ($2\frac{1}{2}D$ and $8D$), or a vena contracta tap orifice flowmeter. Corner and D and $D/2$ taps are widely used in Europe, while flange taps predominate in the United States. Pipe taps ($2\frac{1}{2}D$ and $8D$) are sometimes used as bypass pump restrictors for natural gas or where other tapping arrangements would require drilling too close to the plate. Vena contracta taps are replaced by D and $D/2$ taps when future changes in the orifice bore will require no tap relocation.

Accuracy ranges from ±0.8 to ±5% depending on fluid, upstream piping configuration, measurement equipment, and whether corrections for the Reynolds number, gas expansion factor, measured pipe diameter, or other effects are included in the computation. While standards differ for the minimum pipe Reynolds number, a value of 10,000 is considered acceptable by most. The maximum Reynolds number may be as high as 3.3×10^7 (Gorter and Rooij, 1977).

Quadrant and Conic Edge

When the pipe Reynolds number is below 10,000, the upstream orifice edge is either rounded (quadrant) (Fig. 4) or conical (Fig. 5), these contours having a more constant and predictable discharge coefficient at lower Reynolds numbers. At low Reynolds numbers the coefficient of a square-edged orifice may change by as much as 30%, but for these geometries the effect is only 1 to 2%, making it the more usable flowmeter for viscous fluids. The quadrant plate is widely used in the United States, while the conic plate predominates in European applications.

Fig. 2 Square-edged concentric orifice.

Small Line Sizes

In ½- to 1½-in (12- to 40-mm) line sizes, the effects of pipe roughness, plate eccentricity, and edge sharpness are magnified, resulting in unpredictable coefficients. All indications are that coefficients must be experimentally determined, with selection of a square or quadrant edge depending on the Reynolds number.

In these line sizes, corner pressure taps are preferred. Special honed meter tube assemblies (Fig. 6) are available that have predictable coefficients (Filbran et al., 1960), and these should be used if accuracy is required.

Integral Orifice

When the pipe size is ½ in (12 mm) or smaller and the fluid clean, it is common to select an orifice installed integrally (Fig. 7) with the differential-pressure transmitter. This provides a compact installation in which overall accuracy, usually ±2 to 5% uncalibrated, has been predetermined based on flow calibrating reproducible fixed-bore orifices.

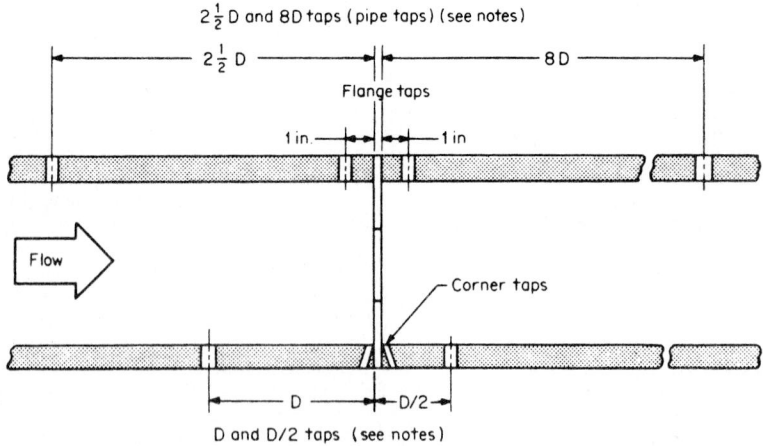

Fig. 3 Orifice pressure tap locations and their names.

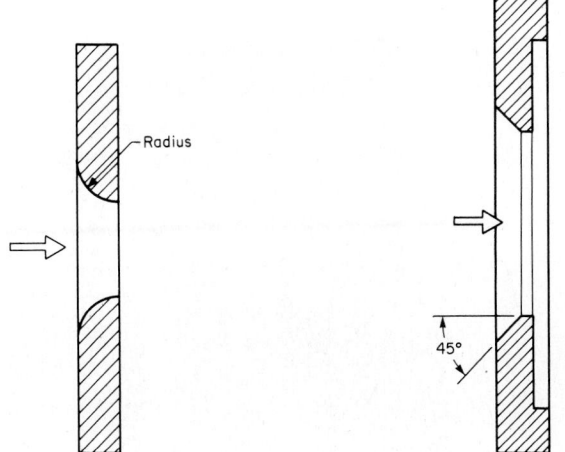

Fig. 4 Quadrant-edged orifice.   **Fig. 5** Conical orifice.

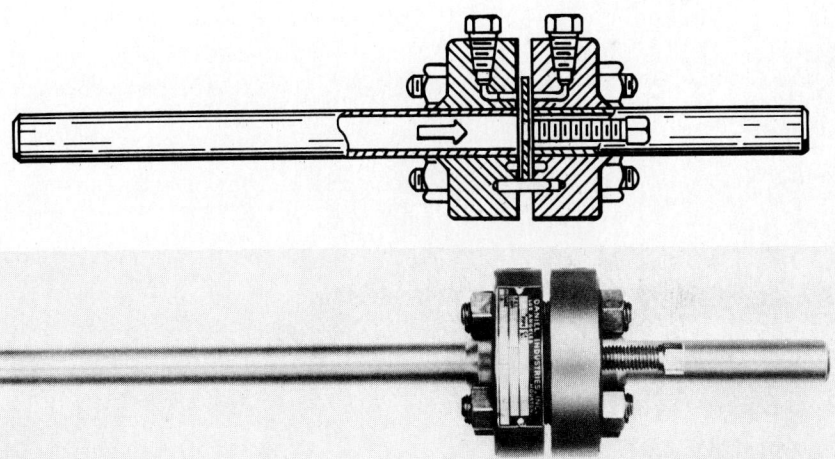

Fig. 6 Honed meter run. (*a*) Internal construction, (*b*) external view. (*Daniel Industries.*)

Eccentric and Segmental Orifice

By placing the orifice hole at the bottom (Fig. 8) of the pipe for gases and at the top for liquids, entrained water or gases flow through the plate rather than building up in front of it. If a segmental opening (Fig. 9) is machined in the plate, the passage of liquids, air, or particulate matter is also possible. Data for both orifice geometries are limited, but both are a low-cost alternative for troublesome applications. Accepted uncalibrated coefficient accuracy is ±2%.

Venturi Tube

The tapered inlet and diverging exit of the classical Herschel venturi (Fig. 10) substantially reduces permanent pressure loss. Since dirt will not build up as it passes through the contoured sections, like it does in front of an orifice, this differential producer can be used in dirty flow applications.

(a)

(b)

Fig. 7 Integral orifice. (*a*) In-line, (*b*) U-bend. *(Foxboro.)*

Fig. 8 Eccentric orifice.

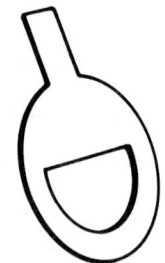

Fig. 9 Segmental orifice.

4.10

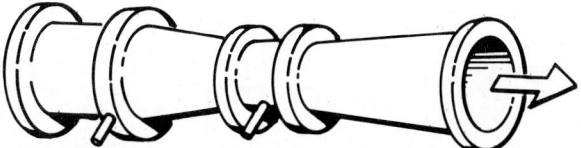

Fig. 10   Classical venturi.

Initially designed for large line size [6-in (150 mm) or larger] water and waste applications, the venturi has become more popular in smaller line sizes with the introduction of the proprietary universal venturi tube (UVT) (Fig. 11), a reduced weight and shorter overall length flowmeter having the advantages of the venturi principle (Halmi, 1974).

For pipe Reynolds numbers greater than 100,000, discharge coefficients for venturis are constant and predictable to within ±0.5 to 2%, depending on design. Venturis are not normally used for lower Reynolds numbers without accepting a decrease in accuracy. When pumping cost and/or shorter upstream installation lengths are important, the additional expense for a venturi design is usually warranted.

Flow Nozzle

The flow nozzle (Fig. 12) has an elliptical (ASME) or a radius (ISA) entrance and is generally selected for steam (vapor) flows operating at high pipe line velocities (100 ft/s, 30.5 m/s). Because of improved rigidity, it is dimensionally more stable at higher temperatures and velocities than an orifice.

The initial cost is substantially higher than that of an orifice but lower than that of venturi; however, permanent pressure loss is significantly higher than for a venturi. When sized to create the same differential at the same flow rate, pressure loss is approximately the same as for the orifice.

Flow nozzles or critical nozzle venturis (Fig. 13) are sometimes operated at critical (choked) flow for flow limiting, or as secondary flow standards (Jones, 1976). The most widely used nozzles are the ASME elliptical inlet long-radius wall tap nozzle and the ASME throat tap nozzle for steam turbine testing. In Europe, the ISA 1932 nozzle is widely used.

Fig. 11   Universal venturi tube. *(BIF.)*

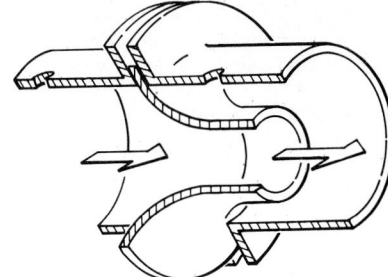

Fig. 12   Flow nozzle.

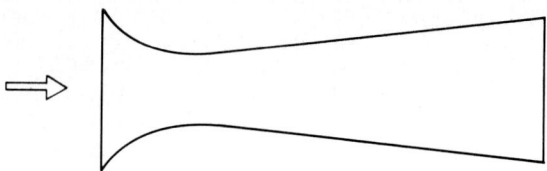

Fig. 13 Critical-flow venturi nozzle.

Lo-Loss Permanent Pressure Loss Devices

Several proprietary designs (Fig. 14) have been introduced that produce high differentials with lower permanent pressure loss, lower weight, and shorter overall laying lengths.

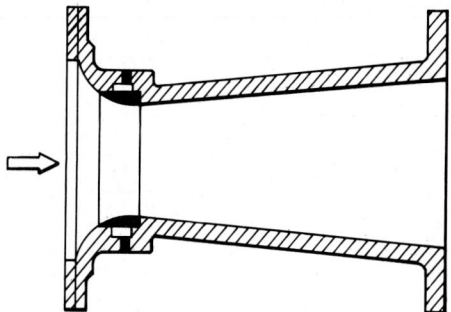

Fig. 14 Lo-Loss tube. *(Badger Meter.)*

Information on coefficients at a high Reynolds number is not generally available. Also, little theoretical work, required for good extrapolation, is available, and the user is cautioned to stay within calibrated ranges. Support information remains proprietary, and the user should contact the manufacturer for the latest information.

Elbow Flowmeter

As a fluid passes through a pipe elbow, the pressure increases at the outside radius of the elbow as a result of centrifugal force (Fig. 15). If pressure taps are located at the outside and inside of the elbow at either 22½ or 45°, a reproducible measurement can be made. Taps located at angles greater than 45° are not recommended, because flow separation may cause erratic readings.

Elbows are inexpensive flowmeters, since many installations have elbows. Even when the elbow is measured, differences between elbows limit accuracy to ±4% (Murdock et al., 1963), but precision (repeatability) is good (±0.2%). Flow is, however, unobstructed with no additional permanent pressure loss. Several manufacturers offer proprietary machined elbows for improved accuracy, but few data have been published. The major disadvantage of the elbow flowmeter is the very low differential produced, particularly for gas flows.

Pitot Tube and Annubar

The pitot tube (Fig. 16) is used for large pipe sizes when the fluid is a clean liquid or gas (vapor) and an inexpensive measurement is required. The difference between total (stagnation) pressure and

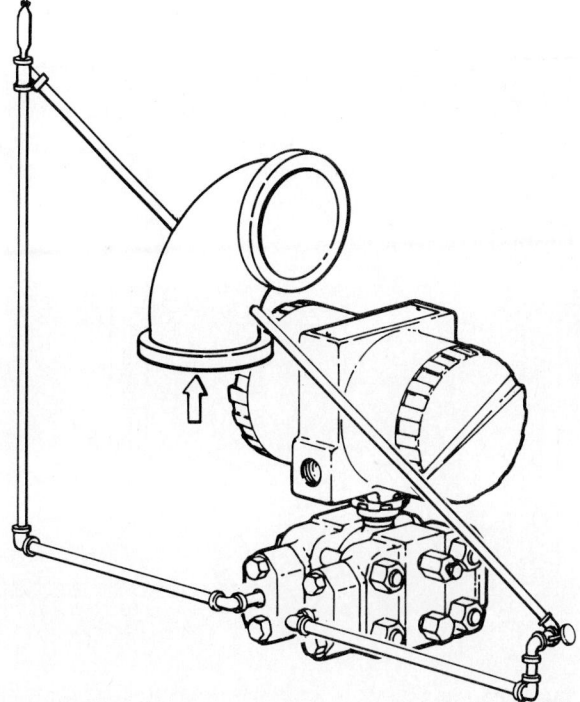

Fig. 15 Elbow flowmeter.

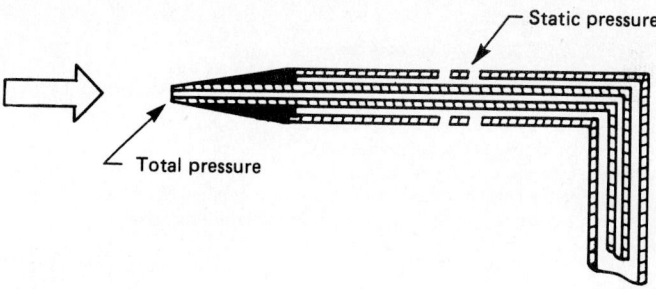

Fig. 16 Pitot tube.

static pressure follows the square root relationship, with velocity being sensed at the insertion depth only. By traversing, an average velocity point can be located and used to measure flow rate.

The Annubar (Fig. 17) is a multiple-ported pitot tube that spans the pipe. Pressure ports are located at mathematically defined positions based on published axisymmetric pipeline velocity profiles. These are claimed to average the differential, thereby eliminating the need to locate the average velocity point as is necessary for pitot tubes.

Ease of installation, low cost, very low permanent pressure loss, and insertability into existing piping make these devices convenient for ducts and large line size measurements. The Annubar has essentially replaced the pitot tube for clean liquids, gases, and vapors (steam). Because total pressure ports face the flow, a purging flow is suggested for dirty stream applications.

A variety of multiported designs are now available. The user should review individual manufac-

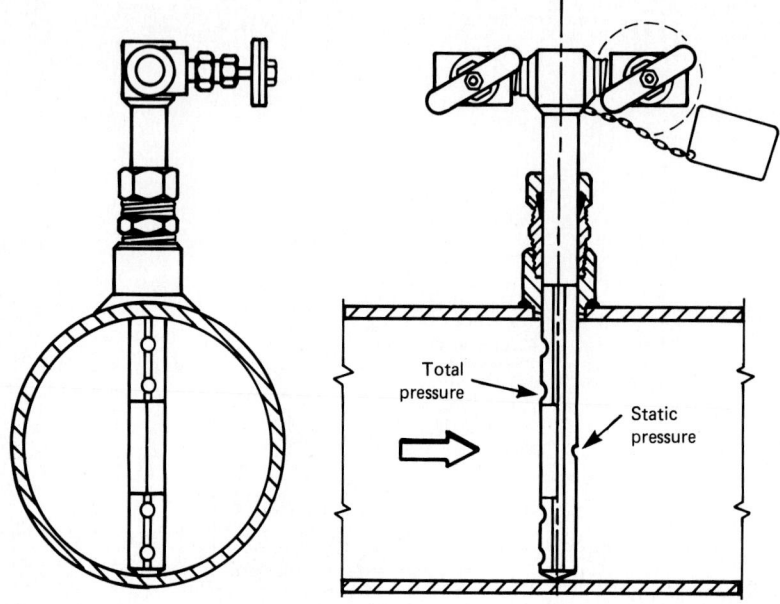

Fig. 17   Annubar. *(Dieterich Standard.)*

turer's support data before extending the devices outside the available data ranges. Neither ANSI nor ISO has standardized on any of the available designs.

Annular Orifice and Target Flowmeter

The annular orifice (Fig. 18) was developed to overcome the problem of dirt buildup in front of an orifice in liquid streams and of liquid buildup in a moist gas stream. Total (stagnation) pressure taps and rearward-facing taps produce a high differential for a given beta ratio (β), redefined as the ratio of disk diameter to pipe diameter. Few data have been presented for line size correlation. Only air data are available for the normally used beta ratios. A design that slips between flanges has reportedly been successfully used for air in 24-in (600 mm) and larger line sizes.

The target flowmeter (Fig. 19) has the features of the annular orifice without the disadvantages of freezing or plugging lead lines. The primary element consists of a sharp leading edge disk (target)

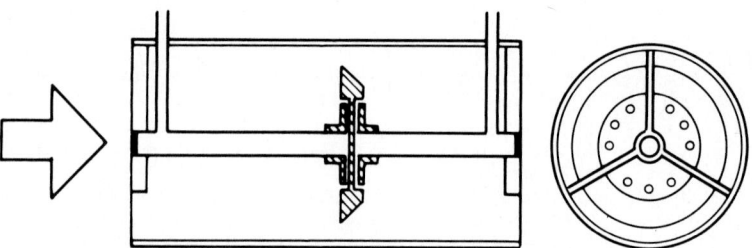

Fig. 18   Annular orifice.

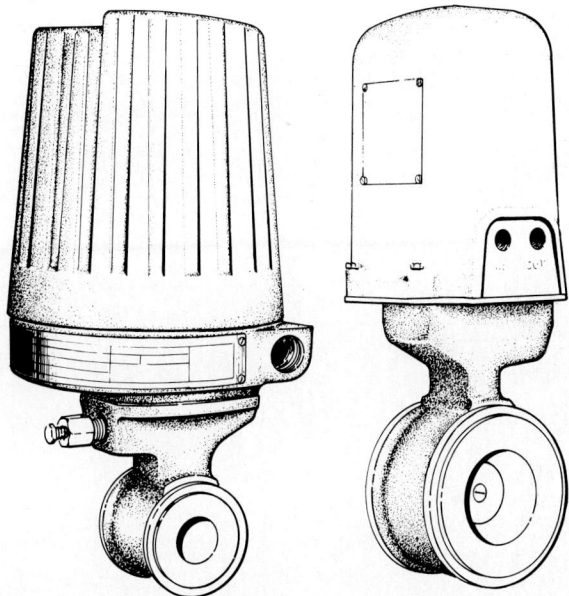

Fig. 19   Target flowmeter. *(Foxboro.)*

fastened to a bar. Differential pressure produced by the reduced annular area creates a disk *drag* force. This force is transmitted through a bar to a suitable force measuring secondary device, and the flow rate is calculated as the square root of this output.

Target flowmeters are particularly well-suited for dirty and low-Reynolds-number flows but are also used with clean fluids and natural gas (Reinecke et al., 1966). Uncalibrated accuracy ranges from ± 1 to $\pm 5\%$ depending on line size, beta ratio, and Reynolds number.

Energy Cost

In many instances, the energy cost resulting from permanent pressure loss is a factor in flowmeter selection. Pumping costs are sometimes significant in larger line sizes and may justify the selection of a more expensive flowmeter with either a lower permanent pressure loss coefficient or one that is obstructionless. Figure 20 shows the permanent pressure loss in percentage of differential pressure for some differential producers.

INSTALLATION

It is important that installation of the primary element approach the *standard* or reference conditions which prevailed when the coefficient information was obtained. The condition of the pipe, mating of pipe sections, pressure tap design, straight lengths of pipe preceding and following the primary element, and lead lines that transmit the differential pressure to the secondary measuring element all affect measurement accuracy. While some of these factors may have a minor effect, others can introduce 5 or 10% bias errors. In general, these errors are not predictable, and attempts to *adjust* a coefficient for *non*standard effects have not been successful.

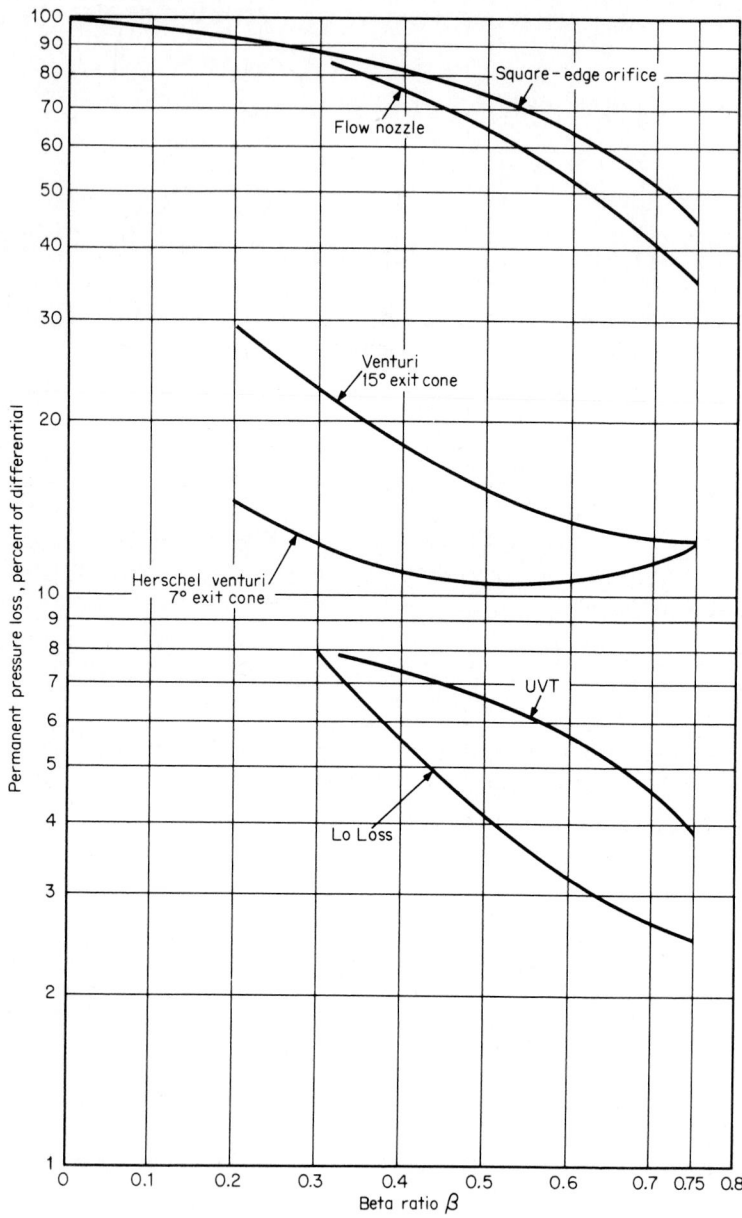

Fig. 20 Permanent pressure loss curves for differential producers.

PIPING

Reference Piping

ISO 5167 (1980) gives the following requirements for reference piping:

1. Visual condition of the outside of the pipe for both straightness and circularity
2. Visual condition of the internal surface of the pipe
3. Reference condition of relative roughness for the internal surface
4. Location of measurement planes and the number of measurements for determination of the average pipe diameter (D)
5. Circularity specifications for specified length of pipe preceding the primary element
6. Maximum allowable step height for mating section of upstream pipe

Pressure Taps

Both pressure taps should be of the same diameter, and where the hole breaks through they should be square with no roughness, burrs, or wire edges.

Protrusions or Pockets

If a pocket or obstruction is present in the upstream or downstream pipe, the velocity profile will be affected. Depending on the diameter of the disturbance, these obstructions should be located a sufficient distance from the primary element to ensure a fully recovered flow condition. It is good practice to locate thermal wells downstream to ensure minimum profile distortion.

Gaskets or weld beads that extend into the pipe increase fluid turbulence and alter the velocity profile. Welds should be ground smooth and gaskets trimmed so as not to visually protrude in any way.

Straight Pipe Lengths

The required length of pipe preceding and following the primary elements depends on the primary element and the type of upstream disturbance. Several U.S. and European test programs have obtained data for specific upstream disturbances on orifices, nozzles, venturi nozzles, and venturis. This information has been used to establish *recommended* minimum lengths. Minimum lengths differ in U.S. recommendations (ANSI 2530; ASME, *Fluid Meters*) and ISO 5167, with the ISO recommendation being more conservative.

Proprietary Devices

Only the straight lengths for venturis, nozzles, venturi nozzles, and orifices have been standardized. For proprietary devices, the user should consult the manufacturer for recommendations.

Lead Lines

Lead lines connect and transmit the process differential pressure to the secondary measuring element. Prior to the introduction of dry-type differential-pressure transmitters, lead line design was quite complex. Seal pots, condensation chambers, sediment chambers, and purge flows were required to ensure that volume changes within the wet-type meters would not affect the measurement or bring in sediment.

Dry-type meters have completely replaced mercury-filled meters in all but a few industries. The recognized hazards of mercury have led to a phasing-out program in the natural gas industry, and special safety precautions are required for laboratory use. Today many instrument manufacturers have discontinued the sale and repair of mercury-filled meters and offer only the dry type. Special high-accuracy mercury-filled manometers are, however, still available, but they are usually used only in the laboratory.

FUNDAMENTAL UNIT FLOW RATE EQUATIONS

If it is assumed that fluid is flowing through an inclined pipe, as shown in Fig. 21, the relationship between differential (pressure) and velocity can be derived from Bernoulli's theorem. For an ideal

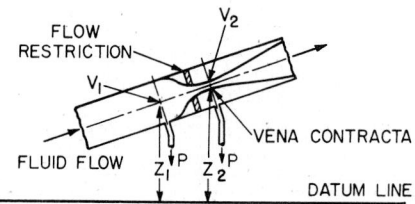

Fig. 21 Flowthrough restriction in a pipe.

incompressible fluid where there is no density difference between taps due to gas expansion, $\rho_{f1} = \rho_{f2} = \rho_f$, this takes the form

$$\frac{P_1}{\rho_f} + \frac{V_1^2}{2g_c} + \frac{g_l}{g_c}Z_1 = \frac{P_2}{\rho_f} + \frac{V_2^2}{2g_c} + \frac{g_l}{g_c}Z_2 \tag{1}$$

where Z = elevation above a datum of pipe centerline, ft, m
P = static pressure, absolute, lb_f/ft^2, Pa
ρ_f = fluid density, lb_m/ft^3, kg/m^3
V = average stream velocity, ft/s, m/s
g_c = conversion constant, 32.17405 $lb_m \cdot ft/lb_f \cdot s^2$, 1 $kg \cdot m/N \cdot s^2$
g_l = local acceleration due to gravity, ft/s^2, m/s^2
1, 2 = upstream and downstream tap locations

In Fig. 21, these factors are identified for upstream (subscript 1) and downstream (subscript 2) locations when an orifice plate is used with the downstream tap at the vena contracta, where the fluid is constricted to a minimum area (point of assumed minimum pressure). Various tap locations are employed with different types of primary elements. The difference in elevation $(Z_2 - Z_1)$ is zero for horizontal pipes or for inclined installation is compensated for by bringing the pressure connecting leads to the same elevation.

The following relationship between average pipe line velocity V_f and differential pressure is obtained:

$$\frac{\Delta P}{\rho_f} = \frac{P_1 - P_2}{\rho_f} = \frac{V_2^2 - V_1^2}{2g_c} \tag{2}$$

For steady flow the mass flow rate q_M (lbm/s, kg/s) between upstream and downstream locations is the same, and

$$q_{lbm/s} = \rho_{f1}A_1V_1 = \rho_{f2}A_2V_2 \tag{3}$$

where A_1 and A_2 are the areas (ft^2, m^2) at sections 1 and 2 and the subscript M is lbm/h or kg/h for U.S. and SI units, respectively.

The relationship between pipeline velocity V_1 and downstream velocity V_2 for a constant density fluid ($\rho_{f1} = \rho_{f2}$) is then

$$V_1 = \frac{A_2}{A_1} V_2 \tag{4}$$

Substituting Eq. 4 into Eq. 2 yields

$$V_2 = \sqrt{\frac{2g_c \, \Delta P}{[1 - (A_2/A_1)^2]\rho_f}} \tag{5}$$

Flow rate is calculated in volume or mass units rather than in the velocity units of Eq. 5. To make this conversion, it is assumed that the area at which the velocity is V_2 is the area at the primary element. Then the volumetric flow rate, at flowing conditions, is calculated as

$$q_v = A_2 V_2 = A_2 \sqrt{\frac{2g_c \, \Delta P}{\rho_f(1 - \beta^4)}} \tag{6}$$

where β = diameter ratio
d = primary element bore, ft, m
q_v = volumetric flow rate at flow conditions, ft^3/s, m^3/s
ΔP = differential pressure, lb$_f$/ft^2, Pa

In terms of mass flow rate, q_M

$$q_M = \rho_f q_v = \frac{\pi}{4} d^2 \sqrt{\frac{2g_c \rho_f \, \Delta P}{1 - \beta^4}} \tag{7}$$

where q_M = mass flow rate, lbm/s, kg/s.

Equations 6 and 7 are referred to as the theoretical flow rate equations in fundamental units (feet, pounds-mass, pounds-force, seconds or meters, kilograms, seconds). Only for contoured inlet devices, such as venturis and flow nozzles, will these equations calculate a liquid flow rate to within 2%. Energy loss across the primary element and the expansion of a gas (vapor) to the lower downstream pressure P_2 requires that two correction terms be introduced: the discharge coefficient and a gas expansion factor.

The empirically determined discharge coefficient C corrects the equation from theoretical to true flow rate based on experimental data obtained in a hydraulic laboratory. The discharge coefficient is defined as

$$C = \frac{\text{true flow rate}}{\text{theoretical flow rate (Eq. 6 or 7)}} \tag{8}$$

where the true flow rate is determined by collecting liquid in a mass or volume receiver over a measured time interval. The theoretical flow rate is calculated by Eq. 6 or 7 using the average differential ΔP and liquid density over the collection interval.

Experimental evidence shows that the discharge coefficient changes with the pipe line velocity profile. This change is correlated with the pipe Reynolds number by the generalized equation of the form

$$C = C_\infty + bR_D^n \tag{9}$$

where the constants C_∞ and b may be beta-dependent. Table 1 lists some of these equations.

The pipe Reynolds number is a dimensionless ratio of inertia to viscous flow forces and is calculated as

$$R_D = 22{,}737 \frac{q_{\text{lbm/s}}}{\mu_{\text{cP}} D} = 127{,}3239 \frac{q_{\text{kg/s}}}{\mu_{\text{cP}} D} \tag{10}$$

Table 1 Equations and Values for C_∞, b, and n of Eq. 9

Primary device	Discharge coefficient C_∞ at infinite Reynolds number	Reynolds number term	
		Coefficient b	Exponent n
Venturi			
Machined inlet	0.995	0	0
Rough cast inlet	0.984	0	0
Rough welded sheet-iron inlet	0.985	0	0
Universal venturi tube†	0.9797	0	0
Lo-Loss tube‡	$1.005 - 0.471\beta + 0.564\beta^2 - 0.514\beta^3$		
Nozzle			
ASME long radius	0.9975	$-6.53\beta^{0.5}$	0.5
ISA	$0.9900 - 0.2262\beta^{4.1}$	$1708 - 8936\beta + 19.779\beta^{4.7}$	1.15
Venturi nozzle (ISA inlet)	$0.9858 - 0.196\beta^{4.5}$	0	0
Orifice			
Corner taps	$0.5959 + 0.0312\beta^{2.1} - 0.184\beta^8$	$91.71\beta^{2.5}$	0.75
Flange taps (D in in) $D \geq 2.3$	$0.5959 + 0.0312\beta^{2.1} - 0.184\beta^8 + 0.09\,\dfrac{\beta^4}{D(1-\beta^4)} - 0.0337\,\dfrac{\beta^3}{D}$	$91.71\beta^{2.5}$	0.75
$2 \leq D \leq 2.3$	$0.5959 + 0.0312\beta^{2.1} - 0.184\beta^8 + 0.039\,\dfrac{\beta^4}{1-\beta^4} - 0.0337\,\dfrac{\beta^3}{D}$	$91.71\beta^{2.5}$	0.75
Flange taps (D in mm) $DD \geq 58.4$	$0.5959 + 0.0312\beta^{2.1} - 0.184\beta^8 + 2.286\,\dfrac{\beta^4}{D(1-\beta^4)} - 0.856\,\dfrac{\beta^3}{D}$	$91.71\beta^{2.5}$	0.75
$50.8 \leq D \leq 58.4$	$0.5959 + 0.0312\beta^{2.1} - 0.184\beta^8 + 0.039\,\dfrac{\beta^4}{1-\beta^4} - 0.856\,\dfrac{\beta^3}{D}$	$91.71\beta^{2.5}$	0.75
D and $D/2$ taps	$0.5959 + 0.0312\beta^{2.1} - 0.184\beta^8 + 0.039\,\dfrac{\beta^4}{1-\beta^4} - 0.0158\beta^3$	$91.71\beta^{2.5}$	0.75
$2\frac{1}{2}\,D$ and $8D$ taps	$0.5959 + 0.461\beta^{2.1} + 0.48\beta^8 + 0.039\,\dfrac{\beta^4}{1-\beta^4}$	$91.71\beta^{2.5}$	0.75

†From BIF CALC-440/441; the manufacturer should be consulted for exact coefficient information.
‡Derived from the Badger Meter, Inc., Lo-Loss tube coefficient curve; the manufacturer should be consulted for exact coefficient information.

where μ_{cP} = absolute viscosity, cP, g/cm·s
$\quad\quad D$ = pipe diameter, in, mm
$q_{lbm/s},q_{kg/s}$ = mass flow rate, lbm/s, kg/s

The assumption of a constant density between the two pressure taps is not valid for compressible fluids (air, nitrogen, etc.). Density decreases when a gas expands to the lower pressure measured at the downstream tap. A gas expansion factor is introduced into the equation to correct for this expansion. This factor is based on experimental data (orifice) or derived from the thermodynamic general steady flow energy equation to correct for this density difference. Assuming that the liquid-determined discharge coefficient applies to gas flows, the gas expansion factor is defined by

$$Y_1 = \frac{\text{true flow rate of gas}}{\text{flow rate calculated by liquid equation}} \tag{11}$$

where subscript 1 denotes that the gas expansion factor is based on an upstream tap location density determination; if a downstream tap is used, a subscript 2 is inserted. The relationship between these two gas expansion factors is

$$Y_2 = (1 - x_1)^{-0.5} Y_1 \tag{12}$$

where $x_1 = (P_1 - P_2)/P_1$.

Table 2 presents the gas expansion factor equation for orifices for contoured inlet devices. The *true* volumetric flow rate equation (ft³/s, m³/s) is then calculated by

$$q_v = \frac{\pi}{4} CY_1 d^2 \sqrt{\frac{2g_c \, \Delta P}{(1 - \beta^4)\rho_{f1}}} = \frac{\pi}{4} CY_2 d^2 \sqrt{\frac{2g_c \, \Delta P}{(1 - \beta^4)\rho_{f2}}} \tag{13}$$

and the *true* mass flow rate equation (lbm/s, kg/s) for all differential producers is then

$$q_M = \frac{\pi}{4} CY_1 d^2 \sqrt{\frac{2g_c \rho_{f1} \, \Delta P}{1 - \beta^4}} = \frac{\pi}{4} CY_2 d^2 \sqrt{\frac{2g_c \rho_{f2} \, \Delta P}{1 - \beta^4}} \tag{14}$$

where the subscript M is lbm/s in the U.S. system of units and kg/s in SI units.

ENGINEERING EQUATIONS

Equations 13 and 14 are written in a consistent set of fundamental units. In practice, however, flow rate units and measured differential-pressure units are seldom, if ever, the units of measurement. Also, density (lbm/ft³, kg/m³) in gas flows is more commonly calculated from temperature and pressure measurements and a knowledge of the properties of the flowing gas (vapor). Equations 13 and 14 must then be modified through suitable unit conversions to the units of measurement used in calculating the desired flow rate unit.

Flow rates are usually calculated in mass or volumetric units. The most common mass flow rate unit is pounds mass per hour of kilograms per hour for the flow of steam, ammonia, and ethylene. Gas flow rates of natural gas, nitrogen, or air usually have units of standard cubic feet per hour or standard cubic meters per hour. Liquid flow rates in the United States may have two volume units. The first is flowing gallons per minute (gpm) and the second is referred to as base gallons (or barrels) per minute (GPM). Both standard and base volumetric flow rates, whether for a gas or a liquid, refer the flowing volume to a reference condition of pressure and temperature. Usually these are 60°F (15.6°C) and 14.7 psia (101.3 kPa).

When calculating gas (vapor) density (lbm/ft³, kg/m³) from pressure and temperature measurements, the units of pressure and temperature must be defined in order to maintain consistency. The equations relating density to these measurements are

$$\rho_f(\text{lbm/ft}^3) = 2.698825 \frac{P_f G}{Z_f T_f} \quad\quad \rho_f\,(\text{kg/m}^3) = 3.483407 \frac{P_f G}{Z_f T_f} \tag{15}$$

Table 2 Summary of Gas (Vapor) Expansion Factor Equations

Equation	Pressure relationships	
	U.S. units	SI units
Contoured primary elements (nozzle, venturi nozzle, Lo-Loss† etc.)		
Upstream measurements $$Y_1 = \left\{ \frac{(1-\beta^4)(k/k-1)(p_{f2}/p_{f1})^{2/k}[1-(p_{f2}/p_{f1})^{(k-1/k)}]}{[1-\beta^4(p_{f2}/p_{f1})^{2/k}][1-p_{f2}/p_{f1}]} \right\}^{1/2}$$	$\dfrac{p_{f2}}{p_{f1}} = 1 - x_1$	$\dfrac{p_{f2}}{p_{f1}} = 1 - x_1$
Downstream measurements $$Y_2 = Y_1\sqrt{1+x_2}$$		
Orifice		
Corner, flange, D and $D/2$ taps Upstream measurements $$Y_1 = 1 - (0.41 + 0.35\beta^4)\frac{x_1}{k}$$		
Downstream measurements $$Y_2 = \sqrt{1+x_2} - (0.41 + 0.35\beta^4)\frac{x_2}{k\sqrt{1+x_2}}$$	$x_1 = \dfrac{h_w}{27.73 p_{f1}}$	$x_1 = \dfrac{\Delta p}{p_{f1}}$
$2\frac{1}{2}D$ and $8D$ Upstream measurements $$Y_1 = 1 - [0.333 + 1.145(\beta^2 + 0.7\beta^5 + 12\beta^{13})]\frac{x_1}{k}$$		
Downstream measurements $$Y_2 = \sqrt{1+x_2} - [0.333 + 1.145(\beta^2 + 0.7\beta^5 + 12\beta^{13})]\frac{x_2}{k\sqrt{1+x_2}}$$	$x_2 = \dfrac{h_w}{27.73 p_{f1}}$	$x_2 = \dfrac{\Delta p}{p_{f1}}$

†Registered trademark of Badger Meter, Inc. Manufacturers should be consulted for recommendations.

where p_f = absolute pressure, psia, absolute kPa
    G = ratio of the molecular weight of the gas to that of air
    Z_f = compressibility factor of the flowing gas (vapor)
    T_f = absolute flowing temperature, °R, K

The relationships between engineering flow rate units and the fundamental unit given in Eqs. 13 and 14 are

$$q_{\text{lbm/h}} = 3600(q_{\text{lbm/s}})_{\text{Eq.14}} = 3600\rho_f (q_{\text{ft3/s}})_{\text{Eq.13}} \tag{16}$$

for mass flow (U.S. units) and

$$q_{\text{kg/h}} = 3600(q_{\text{kg/s}})_{\text{Eq.14}} = 3600\rho_f (q_{\text{m3/s}})_{\text{Eq.13}} \tag{17}$$

for mass flow (SI units)

The relationship for volumetric flow at flowing conditions in U.S. units is

$$q_{\text{gpm}} = 448.83(q_{\text{ft3/s}})_{\text{Eq.13}} \tag{18}$$

For gas flows at a standard cubic foot, defined by ISO 5024, at 14.696 psia (101.33 kPa) and 59° (15°C), it is derived as

$$q_{\text{SCFS}} = \frac{\rho_f}{\rho_b} (q_{\text{ft3/s}})_{\text{Eq.13}} = 35.29 \frac{Z_b p_f}{Z_f T_f} (q_{\text{ft3/s}})_{\text{Eq.13}} \tag{19}$$

for U.S. units, and

$$q_{\text{SCMS}} = \frac{\rho_f}{\rho_b} (q_{\text{m3/s}})_{\text{Eq.13}} = 2.844 \frac{Z_b p_f}{Z_f T_f} (q_{\text{m3/s}})_{\text{Eq.13}} \tag{20}$$

where the subscript b refers to base (or standard) conditions.

The base gallon volumetric flow rate unit is similarly derived using the continuity of mass flow as

$$q_{\text{GPM}} = \frac{\rho_f}{\rho_b} (q_{\text{gpm}})_{\text{Eq.18}} = \frac{G_f}{G_b} (q_{\text{gpm}})_{\text{Eq.18}} \tag{21}$$

where G_f = flowing liquid specific gravity, ratio of flowing liquid density at its pressure and temperature to that of water at 60°F (15.6°C) and 14.7 psia (101.3 kPa)
    G_b = base specific gravity at its pressure and temperature, ratio of flowing liquid density to that of water when both are at 60°F (15.6°C) and 14.7 psia (101.3 kPa)

In U.S. practice the differential pressure unit is inches of water at 68°F (20°C), 14.7 psia (101.3 kPa), and standard gravity, 32.174 ft/s² (9.807 m/s²). The relationship between the fundamental unit and the measured unit is then

$$\Delta P \ (\text{lb}_f/\text{ft}^2) = 5.193hw \tag{22}$$

Substituting these conversion equations into the fundamental unit equations (Eqs. 13 and 14) develops the commonly used flow rate equations listed in Table 3. The U.S. and SI units for these equations are given in Table 4. In the above equations the bore of the primary element is at the flowing temperature. When the primary element and pipe diameter are measured at 68°F (20°C), assuming the pipe and primary element are of the same material, the bore used in these equations is calculated as

$$D^2 = F_a d_{\text{meas}}^2 = [1 + 2\alpha(T_F - 68)] \, d_{\text{meas}}^2 \quad \text{(for U.S. units)} \tag{23}$$

and

$$d^2 = F_a d_{\text{meas}}^2 = [1 + 2\alpha(T_{°C} - 20)] \, d_{\text{meas}}^2 \quad \text{(for SI units)} \tag{24}$$

where α = thermal expansion factor of the primary element, in/in·°F, m/m°C.

Table 3 Flow Rate Equations

	Liquid	Gas vapor
	Mass flow rate	
U.S. units		$q_{lbm/h} = 358.93 CY_1 d^2 \sqrt{\dfrac{p_{f1}\, hw}{1 - \beta^4}}$
SI units		$q_{kg/h} = 0.12645\ CY_1 d^2 \sqrt{\dfrac{p_{f1}\, \Delta p}{1 - \beta^4}}$
	Volumetric flow rate at flowing conditions	
U.S. units	$q_{gpm} = 5.6665 Cd^2 \sqrt{\dfrac{hw}{(1 - \beta^4)G_f}}$	
SI units	$q_{lpm} = 0.0668 Cd^2 \sqrt{\dfrac{\Delta p}{(1 - \beta^4)G_f}}$	
	Volumetric flow rate at base conditions	
U.S. units	$q_{GPM} = 5.6665 \dfrac{Cd^2}{G_b} \sqrt{\dfrac{G_f\, hw}{1 - \beta^4}}$	$q_{SCFH} = 7711\ CY_1 Z_b d^2 \sqrt{\dfrac{p_{f1}\, hw}{(1 - \beta^4)Z_{f1} T_{f1} G}}$
SI units	$q_{LPM} = 0.06668 \dfrac{Cd^2}{G_b} \sqrt{\dfrac{G_f\, \Delta p}{1 - \beta^4}}$	$q_{SCMH} = 0.19267 CY_1 Z_b d^2 \sqrt{\dfrac{p_{f1}\, \Delta p}{(1 - \beta^4)Z_{f1} T_{f1} G}}$

SIZING VERSUS FLOW RATE DETERMINATION

In using a differential producer, two distinctly different problems exist. First, the designer must establish the bore of the primary element required to produce a desirable differential at a design flow rate. And second, after the device has been fabricated, the differential pressure and other selected quantities must be measured and then substituted into the flow equation to calculate the flow rate.

To establish the bore, a desired differential pressure is first selected (usually 100 in, 25 kPa), and the transmitter calibrated to that range. Usually the pipe bore is not initially known, and a nominal standard diameter is selected. After the meter run is fabricated, the pipe and primary element diameters are known and, although the bore and differential-pressure transmitter's range is exactly as calculated, the beta ratio is different because of the pipe diameter measurement. This means that the dimensional diameter constant in the flow equation is not the same as that used in the sizing. This can result in bias errors of 3 to 4% for 2-in (50-mm) line sizes, and the measured pipe diameters should be used in the flow rate equation.

It is seen from the flow rate equations listed in Table 3 that in order to size the primary element bore the discharge coefficient must be known. Unless the coefficient is constant, as in the case of a venturi, the discharge coefficient is a function of the bore diameter ($\beta = d/D$) (see Table 1). For gas (vapor) flow the gas expansion factor is also bore-dependent (see Table 2). Therefore an iterative solution is required to properly size the bore such that the design conditions of flow rate, pipe size, and differential pressure satisfy the relationships given in Table 3.

A simplified sizing procedure is developed (Miller, 1983) by noting that the discharge coefficient is essentially constant over a wide Reynolds number range and that $d = \beta D$. By equating all bore-dependent terms to be known design conditions,. A sizing equation for a gas flow rate can be written as

$$S_M = \frac{CY_1 \beta^2}{\sqrt{1 - \beta^4}} = \frac{1}{7711} \sqrt{\frac{Z_{f1} T_{f1} G}{h_w p_{f1}}} \frac{q_{SCFH}}{Z_b} \tag{25}$$

where the sizing factor S_M is constant for the design conditions. The left hand side of the equation contains the selected design values of flow rate, pipe diameter, flowing pressure, temperature, etc.

Table 4 Measurements Units for Table 3 Equations

Symbol		U.S.	SI
d	Bore diameter of primary element	in	mm
hw	Differential pressure, inches of water at 68°F, $g_0 = 32.174$ and 14.7 psia	in	
C	Discharge coefficient	Dimensionless	
D	Pipe diameter	in	mm
G	Gas specific-gravity, ratio of the molecular weight of the gas to that of air	Dimensionless	
G_b	Liquid base specific gravity	Dimensionless	
G_f	Liquid flowing specific gravity	Dimensionless	
T_{f1}	Flowing absolute temperature measured at upstream tap gas	°R	K
Z_{f1}	Gas compressibility factor, at flowing conditions, determined at upstream tap	Dimensionless	
Z_b	Gas compressibility factor at a base pressure of 14.696 psia (101.32 kPa) and temperature at 59°F (15°C)	Dimensionless	
S_M	Sizing factor, $S_M = CY\beta^2/\sqrt{1 - \beta^4}$	Dimensionless	
Δp	Differential pressure		kPa
p_f	Absolute pressure measured at upstream tap	lbf/in^2	kPa
$q_{lbm/h}$	Mass flow rate	lbm/h	kg/h
q_{gpm}, q_{lpm}	Volumetric flow rate at flowing conditions of pressure and temperature	gal/min	L/min
q_{GPM}, q_{LPM}	Volumetric flow rate at 60°F (15.6°C) and 14.7 psia (101.3 kPa)	gal/min	L/min
q_{SCFH}, q_{SCMH}	Gas standard volumetric flow rate at 59°F (15°C) and 14.696 psia (101.325 kPa) (150, 5024 base)	ft^3/h	m^3/h

It is noted then that the S_M factor is constant and identically equal to the beta- (or bore)-dependent functions. This identity allows the following equation to be written:

$$\beta = \frac{d}{D} = \left[1 + \left(\frac{CY}{S_M}\right)^2 \right]^{-1/4}$$

(26)

If the discharge coefficient C is assumed constant and the gas expansion factor Y set equal to 1.0, then an initial estimate for beta (beta zero) can be calculated directly, and hence the bore. In most cases this estimate is within 0.1 to 1% of the solution found by iteration. The bore then becomes

$$d = \beta_0 D$$

(27)

Table 5 lists the S_M equations, and Table 6 the β_0 equations for several primary elements.

INFLUENCE QUANTITIES

Accuracy statements for flowmeters are based on the steady flow of a homogeneous, single-phase newtonian fluid with an approach velocity profile that does not alter the coefficient obtained in long,

Table 5 Sizing Factor Equations—S_M

	Liquid	Gas (vapor)

Mass flow rate

	Liquid	Gas (vapor)
U.S. units		$S_M = \dfrac{q_{\text{lbm/h}}}{358.93D^2\sqrt{\rho_{f1}hw}}$
SI units		$S_M = \dfrac{q_{\text{kg/h}}}{0.12645D^2\sqrt{\rho_{f1}\,\Delta p}}$

Volumetric flow rate at flowing conditions

	Liquid	Gas (vapor)
U.S. units	$S_M = \dfrac{q_{\text{gpm}}}{5.6665D^2}\sqrt{\dfrac{G_f}{hw}}$	
SI units	$S_M = \dfrac{q_{\text{lpm}}}{0.06668D^2}\sqrt{\dfrac{G_f}{\Delta p}}$	

Volumetric flow rate at base conditions

	Liquid	Gas (vapor)
U.S. units	$S_M = \dfrac{G_b q_{\text{GPM}}}{5.6665D^2\sqrt{G_f hw}}$	$S_M = \dfrac{1}{7711Z_bD^2}\sqrt{\dfrac{Z_{f1}T_{f1}G}{p_{f1}hw}}\,q_{\text{SCFH}}$
SI units	$S_M = \dfrac{G_b q_{\text{LPM}}}{0.06668D^2\sqrt{G_f\,\Delta p}}$	$S_M = \dfrac{1}{0.19267Z_bD^2}\sqrt{\dfrac{Z_{f1}T_{f1}G}{p_{f1}\,\Delta p}}\,q_{\text{SCMH}}$

straight runs of pipe. Departures from these reference conditions are called the flowmeter influence quantities. Velocity profile deviations, nonhomogeneous flow, pulsating flow, and cavitation are the four major influence quantities affecting all flowmeters. The errors associated with a particular influence quantity depend on the sensitivity of a particular flowmeter to that quantity and whether or not a calculation correction can be made. For newtonian fluids, velocity profiles can usually be brought into acceptable limits by the installation of sufficient straight pipe or, for shorter lengths, the installation of pulsating dampers or the use of a less sensitive flowmeter to achieve the desired degree of accuracy.

Cavitation

When decreased line pressure approaches the vapor pressure of a liquid in the line, cavitation begins. In essence, cavitation is boiling of the liquid caused by decreasing pressure rather than by increasing temperature. It is the formation and collapse (implosion) of vapor cavities. This imploding is responsible for the audible noise associated with cavitation, which can range from occasional popping to the sound of moving sand. Extensive cavitation destroys piping, restricts flow, ruins turbine blades, and produces unacceptable noise levels. Cavitation occurs in a system wherever pressure has been reduced sufficiently by either friction, flow separation, or restrictors such as valves, vortex elements, or differential producing flowmeters. Even in a well-designed piping system, cavitation can occur if control or relief valves are suddenly opened.

Dissolved gases and gas bubbles in liquids provide nucleative points and assist in the onset of cavitation. With gas concentrations in the range of 40 ppm, fluids will cavitate at higher static pressures. Generally, cavitation begins at higher static pressures and lower velocities in larger line sizes; once started, it will continue at higher static pressures than the initiating pressure.

The number used for correlating cavitation data is the dimensionless cavitation number, defined

Table 6 β_0 Approximate Sizing Equations

Type	Equation
Venturi	
Machined inlet	$\beta_0 = \left[1 + \left(\dfrac{0.995}{S_M} \right)^2 \right]^{-1/4}$
Rough-cast inlet	$\beta_0 = \left[1 + \left(\dfrac{0.984}{S_M} \right)^2 \right]^{-1/4}$
Rough-welded sheet iron	$\beta_0 = \left[1 + \left(\dfrac{0.985}{S_M} \right)^2 \right]^{-1/4}$
Universal venturi tube*	$\beta_0 = \left[1 + \left(\dfrac{0.9797}{S_M} \right)^2 \right]^{-1/4}$
Lo-Loss tube†	$\beta_0 = \left[1 + \left(\dfrac{0.92}{S_M} - 0.31 \right)^2 \right]^{-1/4}$
Nozzle	
ASME long radius	$\beta_0 = \left[1 + \left(\dfrac{0.9975}{S_M} \right)^2 \right]^{-1/4}$
ISA	$\beta_0 = \left[1 + \left(\dfrac{0.9944}{S_M} - 0.118 \right)^2 \right]^{-1/4}$
Venturi nozzle (ISA inlet)	$\beta_0 = \left[1 + \left(\dfrac{0.989}{S_M} - 0.09 \right)^2 \right]^{-1/4}$
Orifice	
Corner, flange, D and $D/2$ taps	
$R_p < 200,000$	$\beta_0 = \left[1 + \left(\dfrac{0.6}{S_M} + 0.06 \right)^2 \right]^{-1/4}$
$R_p > 200,000$	$\beta_0 = \left[1 + \left(\dfrac{0.6}{S_M} \right)^2 \right]^{-1/4}$
$2\frac{1}{2}D$ and $8D$ taps	$\beta_0 = \left[1 + \left(\dfrac{0.61}{S_M} + 0.55 \right)^2 \right]^{-1/4}$
Eccentric, all taps	$\beta_0 = \left[1 + \left(\dfrac{0.607}{S_M} + 0.088 \right)^2 \right]^{-1/4}$
Segmental, all taps	$\beta_0 = \left[1 + \left(\dfrac{0.634}{S_M} - 0.062 \right)^2 \right]^{-1/4}$
Quadrant($\beta \leq 0.6$)	$\beta_0 = \left[1 + \left(\dfrac{0.76}{S_M} + 0.26 \right)^2 \right]^{-1/4}$
Conic, corner ($\beta \leq 0.3$)	$\beta_0 = \left[1 + \left(\dfrac{0.734}{S_M} \right)^2 \right]^{-1/4}$

*From BIF CALC 440/441; the manufacturer should be consulted for exact coefficient information.
†Derived from Badger Meter, Inc., Lo-Loss flowtube coefficient curve.

as
$$\sigma_c = \frac{2g_c(P_{f2} - P_{v2})}{\rho_f V_{f2}^2} \tag{28}$$

where the pressures P_{f2} and P_{v2} are in pounds-force per square foot and the subscript 2 refers to the location of the maximum velocity within the flowmeter. In Eq. 28 the difference between the static pressure P_{f2} and vapor pressure P_{v2} can be viewed as the force required to collapse the vapor bubbles. The velocity-squared term is the dynamic pressure required to initiate bubble formation. The cavitation number σ_c is then the ratio between the collapsing and forming forces.

In general, the incipient cavitation number ranges from 1.0 to 2.5 for abrupt obstructions and orifice, vortex, and flow nozzles, where downstream pressure recovery is abrupt. For contoured inlet and exit devices, such as the venturi, venturi nozzle, and Lo-Loss tube, the incipient cavitation number ranges from 0.2 to 0.5.

Pulsation and Unsteady Flow

The differential producing flowmeter is basically a steady-rate measuring device. Because of the square root relationship between differential and flow rate, the average of a fluctuating differential does not give a true measure of flow; that is, the square root of the integral of the differential does not equal the integral of the square root of the differential except when the differential is constant. This error is quite large (10 or 20%) only when variations in the differential from the average reach large proportions.

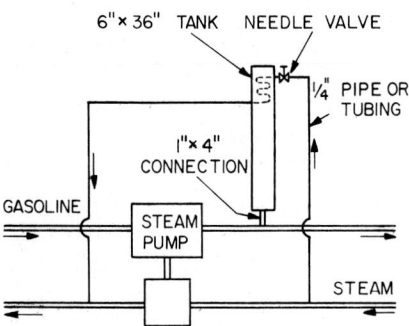

Fig. 22   Flow pulsation dampener for gasoline pumps.

The error is greater for a rectangular waveform than for a corresponding sine waveform. There is a theoretical basis for the belief that at high frequencies the rate of change of the differential also becomes a factor contributing to the error. The most frequent sources of pulsation errors are reciprocating pumps, compressors, and steam engines.

With liquid flows, the usual remedy is to install a cushion chamber partially filled with gas or vapor between the source of the pulsation and the head meter. A scheme for maintaining the cushion chamber properly charged when operating on the discharge of a pump handling a volatile fluid is shown in Fig. 22.

For gas or vapor flow, no way has yet been determined to predict the error from measurements of the usual parameters of static and/or differential pressure. Effects due to pulsation are found to be minimized by the use of high operating differentials and high beta ratios. Frequently, in order to accomplish this combination, it is necessary to use a meter run smaller than the adjoining pipe. Pressure loss or additional capacity applied between the source of pulsation and the meter, or a combination of both, is also found to be effective. When the two are used in combination, the throttling should be done between the capacity chamber and the meter. Orifice plates or valves can be used as throttling devices. If the system is equipped with a pressure regulator or flow controller, frequently satisfactory measurement can be provided by proper location of the primary device with respect to the regulating valve already in the system.

Flashing of Flowing Fluid

It is advisable to measure liquids at temperatures such that the pressure downstream from the orifice remains a reasonable amount above the boiling point. Tests of the flow of saturated water through an orifice plate indicate that appreciable time is involved in the change from the liquid to the vapor state and that a discharge coefficient calculated on the basis of the upstream conditions applies.

No tests have been run to prove whether this applies to mixtures containing liquids with different boiling points, but it has been found that most of the problems in measuring volatile liquids commercially arise from vapors which form upstream or downstream from the orifice or from vapors which accumulate in the piping. The latter trouble can be eliminated by proper use of seals or purges, and the former by use of one of the following: (1) segmental or eccentric orifices in horizontal lines, (2) a drain hole in the orifice plate flush with the top of the pipe when the quantity of vapor is small, or (3) a vertical installation with the flow in an upward direction.

COMPUTING, PLANIMETERING, AND INTEGRATING

Since the differential producing flowmeter is a rate measuring device, total flow is obtained by integrating the rate over a time period. The integration may be automatic, by an analog or digital computer, mechanical, or by a pneumatic integrator which is part of the secondary; or it may be done by computation from the chart record. Special configurations of planimeters are available for tracing chart records. The inherent accuracy of the planimeter is very high, but in use the integration accuracy depends on several factors: (1) care of the operator in following the chart record, (2) uniformity of the rate at which the chart is rotated during planimetering (nonuniformity may cause the roller of the countermechanism to slip), and (3) change in chart dimensions due to change in relative humidity. In view of these factors, an overall accuracy of $\pm 1\%$ is considered reasonable. When a planimeter is used, the discharge coefficient should be computed for the period of complete rotation of the chart, such as 24 h or 7 days.

The accuracy of mechanical integrators, depending on the specific type, is usually comparable to or better than that of hand planimetering procedures. The use of automatic integrators essentially is an economic consideration because addition of an integrator to a flowmeter increases the cost by a significant percentage. Generally, in an industrial installation where there are just a few flowmeters, it is less costly to invest in automatic integrators. Where many flowmeters may be involved, as in large oil and gas handling systems, a central, semiautomated setup for planimetering the charts or dedicated microprocessors are more economical. Certainly, the rapidity with which total flow information is required also is an important consideration.

METERS USED TO MEASURE DIFFERENTIAL PRESSURE

The secondary devices used to measure differential pressure are commonly divided into two types—wet and dry. In the dry type (Fig. 23a), the process fluid is isolated by a diaphragm; in the wet type, mercury is usually the separator. Mercury meters were the mainstay of the process and natural gas industries for many years. However, they have been completely replaced by the dry type in the process industry, and dry-type bellow meters (Fig. 24), which require no external power source, are rapidly replacing mercury meters in natural gas applications.

Dry Type

Dry-type devices are conveniently separated into motion and direct measuring types. In the motion type (Fig. 24), the pressure difference across a diaphragm causes a bellows to move against a restraining spring. The motion, which can be recorded directly, is proportional to the differential pressure. In the direct measuring type (Fig. 25), small deflections of a diaphragm are either measured or restrained by a feedback force. The deflections may be detected via induction, capacitance, a strain gage, or a taut resonating wire. In the force feedback device, the differential pressure is proportional to the feedback force. When the low-pressure side is evacuated, then the absolute pressure is measured. Dry-type transmitters have an accuracy in the range of ± 0.2 to $\pm 1\%$ of the upper range value (URV).

Wet Type

Shown in Fig. 23b are the two basic wet-type meters—the inverted bell and the float type. In the inverted bell, the force developed by the differential pressure acting on the bell is opposed by a spring force. The motion of the bell is a direct measure of the differential pressure; and, with a sizable bell,

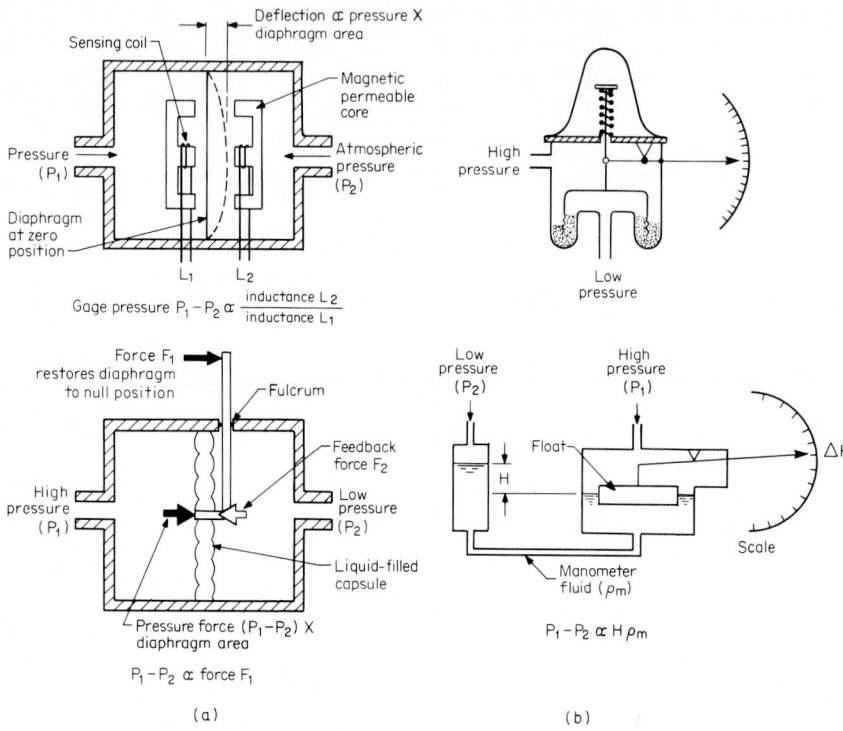

Gage pressure $P_1 - P_2 \propto \dfrac{\text{inductance } L_2}{\text{inductance } L_1}$

$P_1 - P_2 \propto$ force F_1

$P_1 - P_2 \propto H \rho_m$

(a)   (b)

Fig. 23 Dry and wet pressure or differential-pressure measuring devices. *(a)* Dry types, *(b)* wet types.

Fig. 24 Bellows dry-type differential-pressure measuring device *(Foxboro.)*

Fig. 25 Dry-type differential-pressure transmitter. *(Foxboro.)*

a considerable force can be developed, hence low differential pressures can be measured. Typical inverted-bell designs operate in the range 0 to 10 in H_2O (0 to 2.5 kPa).

In the many float types, a cylindrical steel float chamber forms one side of a manometer and contains a steel disk floating on mercury. A cylindrical steel reference chamber forms the second side of the manometer. The area of this chamber is selected to give the desired mercury level change for the operating differential pressure, which usually is from 20 to 200 in H_2O (5 to 50 kPa).

REFERENCES

AGA/ASME: "The Flow of Water Through Orifices," *Ohio State Univ. Stud. Eng. Ser., Bull. 89,* vol. IV, no. 3, 1935.

ANSI/API 2530, *Orifice Metering of Natural Gas,* ANSI, New York, 1978.

ANSI/ASME MFC: *Differential Producers Used for the Measurement of Fluid Flow in Pipes (Orifice, Nozzle, Nozzle Venturi, Venturi),* Draft 8, New York, December 1983.

ASME: "The ASME-ISO Orifice Equation," *Mech. Eng.,* vol. 103, no. 7, 1981.

ASME: *Fluid Meters,* 6th ed., ASME, New York, 1971.

Bailey, E. G.: "Steam Flow Measurement," *Power,* vol. 38, 1916, p. 25.

Beitler, S. R., and H. S. Bean: "Research on Flow Nozzles," *Ohio State Univ. Eng. Exp. Stn. Bull. 131,* 1948.

Dowdell, R. B., and Yu Lin Chen: "A Statistical Approach to the Prediction of Discharge Coefficients for Concentric Orifice Plates," *J. Basic Eng.,* vol. 92, no. 4, 1970, p. 752.

Filban, T. J., and W. A. Griffin: "Small Diameter Orifice Metering," *J. Basic Eng.,* vol. 82, no. 3, 1960, p. 735.

Froude, W.: "Discharge of Elastic Fluids under Pressure," *Proc. Inst. Civ. Eng.,* vol. 6, 1847.

Gorter, J. and D. G. Rooij, *An Investigation on Widening the Reynolds Number Range for Flow Measurements in Closed Conduits by Means of Orifice Plates,* Proceedings of a Conference on Fluid Flow Measurement in the Mid-1970s, vol. 1, Her Majesty's Stationery Office, Edinburgh, 1977, pp. 3–23.

Halmi, D.: "Metering Performance Investigation and Substantiation of the Universal Venturi Tube," pt. 1, "Hydraulic Shape and Discharge Coefficient," *J. Fluids Eng.*, ser. 1, vol. 96, no. 2, 1974, pp. 124–131.

Halmi, D.: "Metering Performance Investigation and Substantiation of the Universal Venturi Tube," pt. 2, "Installation Effect, Compressible Flow, and Head Loss," *J. Fluids Eng.*, ser. 1, vol. 96, no. 2, 1974, pp. 132–138.

Herschel, C.: "The Venturi Water Meter," *Trans. Am. Soc. Civ. Eng.*, vol. 17, 1887, p. 228.

Hickstein, E. O.: "Flow of Air Through Thin-Plate Orifices," *Trans ASME*, vol. 37, 1915, p. 765.

ISA Bulletin 12, International Federation of National Standardizing Organizations, Geneva, 1932.

ISO Standard R541, *Measurement of Fluid Flow by Means of Orifice Plates and Nozzles*, ISO, Geneva, 1967.

ISO Standard R781, *Measurement of Fluid Flow by Means of Venturi*, ISO, Geneva, 1968.

ISO Standard 5167, *Measurement of Fluid Flow by Means of Orifice Plates, Nozzles and Venturi Tubes Inserted in Circular Cross-Section Conduits Running Full*, ISO 5167, 1980(E), Geneva, 1980.

Jones, J. T.: *Field Experience with Sonic Nozzle*, AGA Operating Section Proceedings, American Gas Association, Arlington, Va., 1976, pp. 315–319.

Miller, R. W.: "The Stolz and ASME-AGA Orifice Equation Compared to Laboratory Data," *J. Fluids Eng.*, vol. 101, no. 4, 1979, pp. 483–490.

Miller, R. W.: *Flow Measurement Engineering Handbook*, McGraw-Hill, New York, 1983.

Miller, R. W., and G. A. Koslow: "The Uncertainty Values for the ASME-AGA and ISO 5167 Flange Tap Orifice Equation," ASME Annual Winter Meeting, Paper FM-5, 1979.

Murdock, J. W., C. J. Foltz, and C. Gregory: "Performance Characteristics of Elbow Flowmeters," ASME Annual Winter Meeting, Paper 63-WA-17, 1963.

Stolz, J.: An Approach Toward a General Correlation of Discharge Coefficients of Orifice Plate Flowmeters, ISO/TC30/SC2 (France 6) 654, 1975.

Stolz, J.: "A Universal Equation for the Calculation of Orifice Plates," in *FloMeko 1978, Gröningen, 1978*, North-Holland, Amsterdam, 1978.

Weymouth, Thomas, R.: "Measurement of Natural Gas," *Trans. ASME*, vol. 34, 1912, p. 1091.

Variable Area Flowmeters

by
Charles E. Fees*

The variable area flowmeter operates on the same basic principle as other differential head meters, such as those using orifices. In the orifice meter there is a fixed aperture and the flow rate is indicated as a function of differential pressure. In the variable area meter, there is a variable orifice and a relatively constant pressure drop—thus flow rate is indicated by the area of the annular opening through which the fluid must pass. This area is generally read out as the position of a float or obstruction in the orifice.

There are three general types of variable area meters:

1. **Rotameters.** In the usual rotameter, a float (usually metal and therefore denser than the fluid) contained in an upright tapered tube, large end up, is lifted to the position of equilibrium between

*Fischer & Porter Company, Horsham, Pennsylvania.

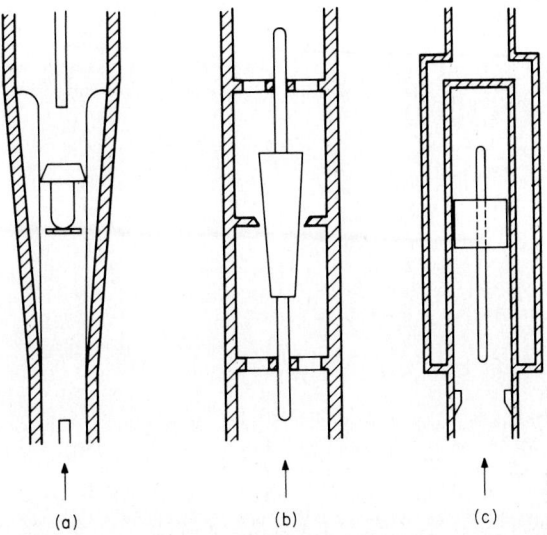

Fig. 1   Types of variable area flowmeters. *(a)* Tapered tube (rotameter),
(b) orifice and tapered plug, *(c)* cylinder and piston.

the downward force of the float and the upward force of the fluid flowing past the float through
the annular orifice. In its simplest form the tube is made of glass, which is graduated so that the
flow rate can be read directly by observing the position of the float. See Fig. 1*a*.

2. *Orifice and Tapered Plug Meters.* This type is equipped with a fixed orifice mounted in an
upright chamber. The float has a tapered body with the small end at the bottom and is allowed
to move vertically through the orifice. The fluid flow causes the float to seek the equilibrium
position. See Fig. 1*b*.

3. *Piston-Type Meters.* In these meters the piston is accurately fitted inside a sleeve and is lifted
by fluid pressure until sufficient port area in the sleeve is uncovered to permit passage of the flow.
The flow is indicated by the position of the piston. See Fig. 1*c*.

   All types of variable area flowmeters are available for remote indication, using electronic or pneumatic signals for transmission.

Performance Characteristics

Linearity

The flow rate (volume) through a variable area meter is essentially proportional to the area and, as
a result, most of these meters have essentially equal-scale increments. A typical indicating rotameter
scale is nonlinear by about 5%. Transmitting rotameters are commonly equipped with built-in linearity devices to ensure an output that will be linear with the flow rate within 1% of full scale.

Differential

An important characteristic of the variable area meter is that the pressure loss across the float is a
constant. The overall differential across the meter, however, will increase at higher flow rates because
of friction losses through the fittings.

Accuracy

This varies with scale length and degree of calibration. The most common accuracy is ±2% of full-scale reading. Individual meter calibration and the use of longer scales make possible accuracies as high as ±¼% of reading. Repeatability is excellent.

Capacity

Variable area flowmeters are the most commonly used means for measuring low-flow rates. Full-scale capacities range from 0.5 cm³/min of water and 30 std cm³/min of air in the smallest units to over 300 gpm (~1135 L/min) of water and 1000 std ft³/min (1700 m³/h) of air in 3-in (~7.6-cm) meters. Although larger size variable area meters are available, orifice plate installations are most common in pipe sizes of 4 in (~10.2 cm) and higher.

Pressure Drop

By placing very light floats in oversized meters, flow rates can be handled with a combination of very low pressure loss (often 1 in of water column or less) and a 10:1 flow range.

Generally, a variable area meter has only one moving part. Since that part and the fluid can be seen in a glass tube meter, this design enjoys a high degree of reliability and operator confidence. Because the variable area meter has no static pressure tap lines and because the orifice or annulus is in a vertical chamber, the meters tend to be self-cleaning.

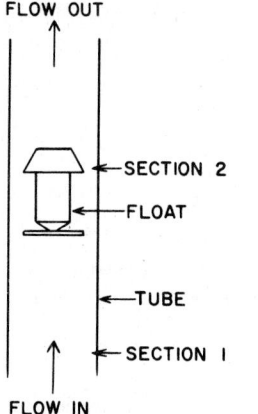

Fig. 2 Fundamental operation of a variable area flowmeter.

Basic Equations

Development of the following flow equations is based primarily on liquids. However, the resultant working equations can be used equally well on gas service. The analysis is approximate and is included here to show how the various factors enter into variable area meter calculations.

The float is assumed to have a sharp metering edge. The tapered tube is imagined to be replaced by a cylindrical tube having an inside diameter equal to the inside diameter of a metering tube measured at the metering edge of the float. The tube and float are shown in Fig. 2. Since the taper of an actual tube usually is small, the analysis of this substituted system with the cylindrical tube is approximately correct. Section 1 of this example is selected far enough upstream of the float so that the pressure at that section is not influenced by the float. Section 2 of the example is the minimum flow area between the float and the tube (vena contracta). Also, it is assumed that the streamlines crossing each of these sections are normal to the sections.

Definitions of the terms in the following analysis are written with reference to Fig. 2. According to Bernoulli's theorem:

$$V_2 - V_1 = \sqrt{2g(h_1 - h_2)} \tag{1}$$

where V_1 = velocity of fluid at section 1, ft/s
V_2 = velocity of fluid at section 2, ft/s
g = acceleration due to gravity, ft/s²
h_1 = hydraulic head at section 1, ft of fluid
h_2 = hydraulic head at section 2, ft of fluid

The hydraulic head drop is expressed in terms of pressure drop as

$$h_1 - h_2 = \frac{p_1 - p_2}{\gamma} \qquad (2)$$

where p_1 = pressure at section 1, lb/ft^2
p_2 = pressure at section 2, lb/ft^2
γ = specific weight of fluid, lb/ft^3

The continuity of flow equation may be written as

$$w = A_2 V_2 \gamma \qquad (3)$$

where w = flow, lb/s
A_2 = area at section 2, ft^2

To allow for factors not included in this analysis, however, a factor C, called the coefficient of discharge, is introduced. Then, Eq. 3 becomes

$$w = C A_2 V_2 \gamma \qquad (4)$$

Neglecting V_1 in Eq. 1 and combining Eqs. 1 and 4, the expression becomes

$$w = C A_2 \sqrt{2g\gamma(p_1 - p_2)} \qquad (5)$$

Dropping V_1 from Eq. 1 is justified on the basis that the area A_1 is much larger than A_2, causing V_1 to be small compared to V_2. To express the flow in volumetric units, since $q = w/\gamma$, Eq. 5 becomes

$$q = C A_2 \sqrt{2g \frac{p_1 - p_2}{\gamma}} \qquad (6)$$

where q = flow, ft^3/s.

For weight-loaded meters of either the rotameter or piston type, the simplest assumption that may be made regarding the pressure is that the differential pressure $(p_1 - p_2)$ equals the weight of the float or of the complete assembly, with proper allowance for buoyancy, divided by the float or piston area. This assumption ignores forces associated with momentum flux variations from section 1 to section 2 which require different analyses for rotameters and piston meters. Nevertheless it is dimensionally correct and leads to precise corrections for variations in specific weight of float and fluid under all circumstances. In the spring-loaded, piston-type meter, the same pressure component is present and, in addition, there is a component due to the spring force.

Rotameters

The term "rotameter" was derived from the fact that floats originally were produced with slots to give them rotation for the purpose of centering and stabilizing the float. The present trend, however, is toward guided nonrotating floats.

The essential elements of any rotameter are shown in Fig. 3. In addition to suitable inlet and outlet connections, they comprise (1) a metering tube and (2) a float.

Metering Tubes

At one time all glass metering tubes were made in Europe (principally Germany) from tapered scrap ends of drawn tubing. In modern practice they are formed on a mandrel and annealed to prevent internal stresses so that strong, uniform tubes result. This method also permits the forming of tubes with greater reproducibility and interchangeability and forming special shapes, such as nonconical tubes with curved elements designed to spread out the graduations at the lower end of the range (better readability at low-flow rates). It is possible to modify the conical form slightly in order to give the exact linear relationship between aperture and float position which is not quite achieved with a purely conical tube.

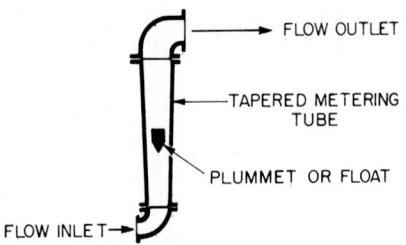

Fig. 3   Basic components of a rotameter.

The most important special shape is a modified conical section having internal beading or lands which serve to guide the float. See Fig. 4.

The manufacturer's recommendations regarding maximum safe working pressures always should be followed. However, as a preliminary guide in tube selection, Table 1 gives the parameters for standard wall thickness metering tubes.

Floats

Floats can be made from several materials to obtain corrosion resistance or capacity modification. Ratings are generally in terms of meter capacity, using a stainless steel float. If the only alteration in the meter is the float material (float dimensions remaining constant), a correction factor based on the preceding equations will be reliable. The densities of common float materials are given in Table 2.

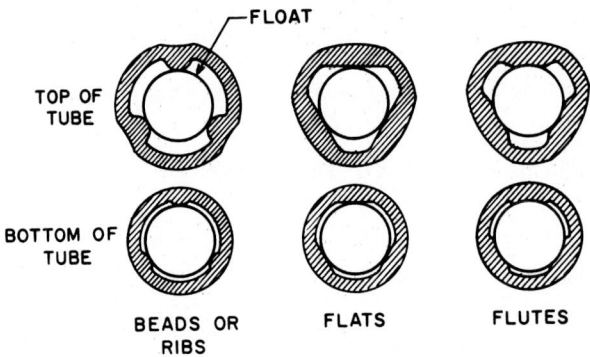

Fig. 4   Several types of glass rotameter tubes with ribs or beads for float guides.

Float Design and Effect of Viscosity

It has been found that float shape determines to a large extent how much a rotameter will be influenced by changes in the viscosity of the measured fluid. Floats having sharp edges have been found

Table 1 Approximate Safe Working Pressures for Heavy-Wall Borosilicate Glass Rotameter Tubes

Nominal tube size	Approximate maximum safe pressure
⅛ in (~ 3.2 mm)	550 psi (~3.79 MPa)
¼ in (~ 6.4 mm)	450 psi (~3.10 MPa)
⅜ in (~ 9.5 mm)	375 psi (~2.59 MPa)
½ in (~12.7 mm)	300 psi (~2.07 MPa)
¾ in (~19.1 mm)	230 psi (~1.59 MPa)
1 in (~ 2.5 cm)	180 psi (~1.24 MPa)
1½ in (~ 3.8 cm)	130 psi (~896 kPa)
2 in (~ 5.1 cm)	100 psi (~689 kPa)
3 in (~ 7.6 cm)	60 psi (~414 kPa)
4 in (~10.2 cm)	40 psi (~276 kPa)

Table 2 Specific Gravity of Float Materials (Approximate)

Material	Specific gravity	Material	Specific gravity
Aluminum	2.79	Lead (10% antimony)	10.67
Brass	8.46	Monel¶	8.80
Bronze (90-10 or 88-10-2)	8.80	Nickel	8.85
Dowmetal*	1.80	Porcelain	2.41
Durimet†	8.02	Stainless steel (303 or 304)	7.92
Everdur‡	8.54	Stainless steel (316)	8.02
Glass (borosilicate)	2.54	Steel (carbon)	7.80
Hastelloy B§	9.24	Titanium	4.5
Hastelloy C	8.94	Tantalum	16.6

*Trademark, Dow Chemical Co., Midland, Mich.
†Trademark, Duriron Co., Inc., Dayton, Ohio.
‡Trademark, Anaconda American Brass Co., Waterbury, Conn.
§Trademark, Union Carbide Corp., New York, N.Y.
¶Trademark, Huntington Alloy Products Division, International Nickel Co., Inc., Huntington, W. Va.

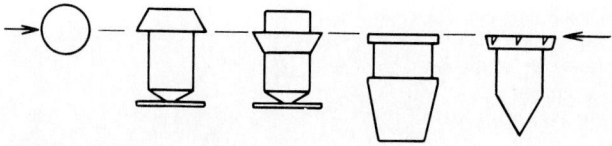

Fig. 5 Partial representation of typical rotameter float shapes.

to be relatively insensitive to viscosity changes over a considerable viscosity range. Some typical float shapes are shown in Fig. 5.

Rotameter Calibration Characteristics

Scales are essentially linear. Capacity scales on standard rotameters often are graduated in terms of percent of maximum flow, the graduations reading from 100 down to 8%. A factor tag is supplied with each instrument showing the number by which the percentage scale reading must be multiplied to determine the flow rate in particular engineering units. A change in fluid or metering conditions can be accommodated simply by changing the factor. In glass tube meters, the graduations, whether

Table 3 Approximate Upper Temperature Limits beyond Which Rotameter Float Diameters Should Be Reduced Because of Errors Due to Thermal Expansion Differences between Glass and Float

316 Stainless steel used in float head

Meter size	Temperature	
	°F	°C
¼ in (~6.4 mm)	200	~93
½ in (~12.7 mm)	180	~82
1 in (~2.5 cm)	160	~71
1½ in (~3.8 cm)	150	~66
2 in (~5.1 cm)	140	60
3 in (~7.6 cm)	130	~54

percentage or direct reading, may be fused into the glass tube or engraved on interchangeable metal scales mounted adjacent to the glass tube. In all-metal, or armored, meters the graduations are engraved on an externally mounted scale.

Thermal Expansion Limitations

When a borosilicate glass metering tube is designed for a given temperature, the differential thermal expansion between the metal float and the glass tube must be considered. The upper limit of temperature above which the head diameter of the float must be reduced from the standard value is given in Table 3. The values listed are based on borosilicate glass metering tubes and Type 316 stainless steel floats. The manufacturer will make recommendations as to the correct head diameter for higher temperatures.

Capacities of Rotameters

See Table 4.

Types of Rotameters

Factors affecting the selection of a rotameter include function required (indication, transmission, totalization), flow rate, pressure rating, temperature rating, corrosion resistance required, accuracy, connection size and type, availability of accessories, and cost.

Direct Viewing Rotameters

With the exception of handmade rotameters occasionally found in laboratories, some form of housing is necessary to contain the meter and tube and to provide for convenient connection. A common, general purpose design is shown in Fig. 6. The graduated scale is visible through sight windows mounted in the protective metal housing. The float is viewed directly—no extension rod is required. The maximum capacity of typical glass tube rotameters is given in Table 4. These meters can be used where (1) the pipe size and pressure do not exceed the values of Table 1, (2) the fluid is not so dark or opaque that direct viewing will be difficult, and (3) the fluid flows freely at ordinary temperatures.

Table 4 Maximum Capacities of Typical Glass Rotameters*

Nominal tube size, in	Water		Air		Approximate pressure loss at maximum flow†	
	L/min	gpm	m³/h	Std ft³/min	inH₂O	cmH₂O
⅟₁₆ and ⅛	0.001–0.32	0.0003–0.08	0.002–0.6	0.001–0.35	0.5–10	1.3–25.4
¼	0.1–2.0	0.03–0.5	0.24–3	0.14–1.8	1–50	2.5–127
½	0.95–19	0.25–5.0	1.7–34	1–20	1–32	2.5–81.3
¾	7.6–45	2–12	12–90	7–53	4–36	10.2–91.4
1	15–114	4–30	31–207	18–122	8–100	20.3–254
1½	49–204	13–54	82–384	48–226	9–100	22.9–254
2	76–340	20–90	133–653	78–384	12–120	30.5–305
3	230–450	60–120	348–850	205–500	20–40	50.8–102

*From a study of four manufacturers. Lower capacities are available from most manufacturers using special "low-pressure-drop" floats.
†Pressure losses for the same maximum flow can vary widely because of meter internal design, connection size, and orientation.

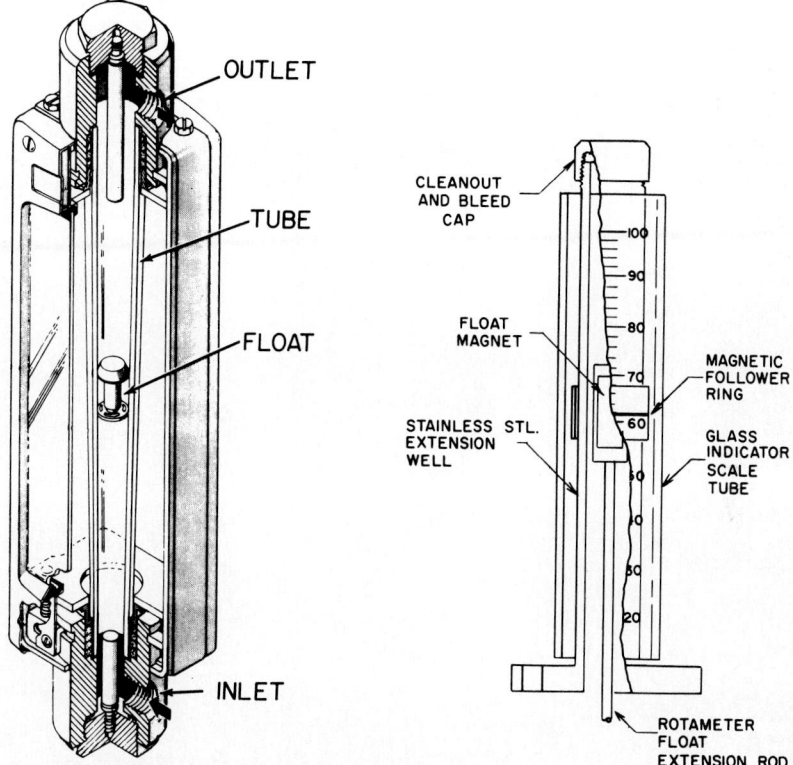

Fig. 6 Direct indicating glass tube rotameter with tube enclosures and metal end fittings.

Fig. 7 Magnetic indicating extension for rotameter.

Magnetic Couplings

To perform secondary functions (transmission, recording, controlling), magnetic couplings were developed to sense float position. Such couplings also make possible the use of all-metal metering tubes. A magnetic-follower arrangement for indicating the float position on all-metal meters is shown in Fig. 7. This type cannot be used for transmission or other functions.

The direct magnetic coupling of Fig. 8 can be used for secondary functions because of the higher torque available to sense the float position. The magnetic coupling of Fig. 9 generally is used with through-flow-type rotameters, with either glass or metal tubes. Because the follower is a passive ferromagnetic material of relatively small cross section, the coupling has a faster response, but lower torque, than that of Fig. 8. Because of the lower torque, a servooperated follower mechanism must trace the edge of the cam in order to provide sufficient power for such functions as transmission, recording, and totalization.

Rotameters for Transmission

Most transmitters are pneumatic or electronic. Both types can be adapted to the couplings of Figs. 8 and 9. Pneumatic transmitters have the usual full-scale output of 15 psig ($\sim$103.5 kPa). A 10% of flow rate output is 4.2 psig ($\sim$29 kPa). At no flow (float against the bottom float stop), the output is 3 psig ($\sim$21 kPa). Electronic transmitters most commonly are arranged for a linear output, ranging from 1 to 5, 4 to 20, or 10 to 50 mA dc—covering the flow range from 0 to 100%. The signal output from 0 to 10% is indeterminate. Actual measured output begins at 10% and from this point

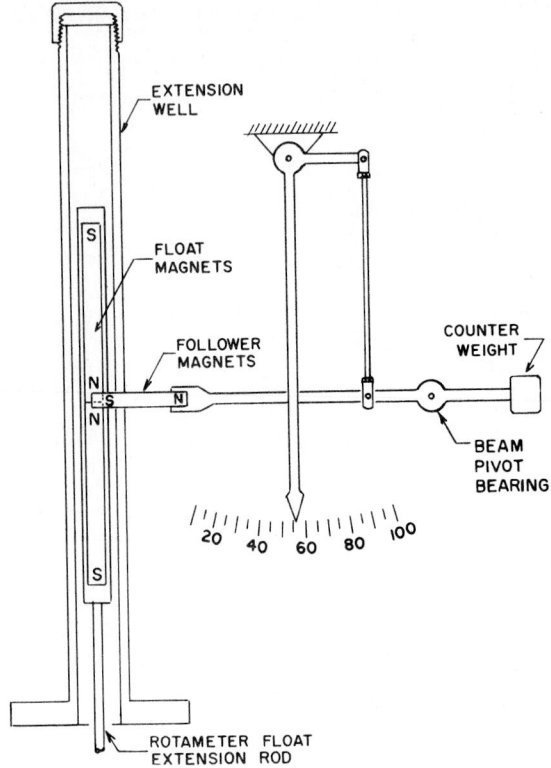

Fig. 8 Magnetically coupled rotameter extension for transmission, recording, and totalizing.

is linear to 100%. Although not as common, null balance inductance bridge and differential transformer types of transmitters are available.

Alarms

Rotameters are widely used for alarming purposes: high and low flow limits. The most widely used variety employs a magnet in a float extension which trips a magnetic dry reed switch mounted external to the pressurized extension well. Other types utilize magnetically coupled followers to operate switches, photoelectric sensors on glass tube meters where the float interrupts a light beam, and the impedance bridge coil which is used where flow rates are so low that only a small ball-type float can be used.

Other Secondary Functions

Magnetically coupled instruments also can be equipped with mechanical integrators, circular-chart recorders, and feedback controllers.

Special Designs

These include steam-jacketed meters, glass-lined or plastic-lined bodies, inverted meters (where fluids are heavier than metal floats), and series-mounted meters in which each tube covers a portion of an overall range, particularly useful for aircraft engine testing.

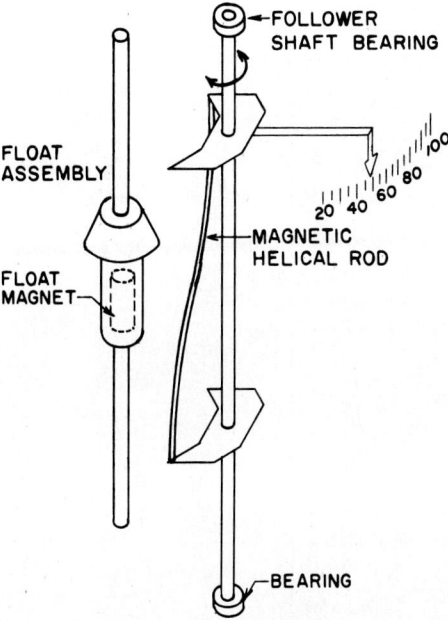

Fig. 9 Magnetic coupling for through-flow-type rotameters.

Installation and Maintenance Factors

(1) It is desirable to install rotameters to within 5° of vertical to avoid a reduction in the gravity component and to prevent float wear; (2) chipping or cracking of glass tubes that will reduce pressure rating must be avoided; (3) quick-opening valves should be avoided; (4) periodic cleaning is recommended where the float tends to be dirty or where substances precipitate on metal parts; (5) no metal should be removed from the metering edge of the float—a very small change in diameter or sharpness of a sharp-edged float can result in a significant error.

Orifice and Tapered Plug Meters

Typical construction of an orifice and plug-type meter is shown in Fig. 10. This type of design is used in the more compact, lower cost versions of all-metal meters. These meters represent a special form of a rotameter, and all standard rotameter principles and formulas apply. In the generally used configuration, the ratio of float weight to plug diameter is quite high, resulting in flow rates generally higher than those obtained with a standard glass tube rotameter of equal pipe size. The taper on the plug offers a considerable viscous drag surface, causing the plug to be more sensitive to viscosity than the standard sharp-edge rotameter float. Meters of this type are used on purge-water lines and other applications where high accuracy is not required and where compact size and cost are important.

Piston-Type Meters

The usual piston-type meter comprises (1) a sleeve or cylinder which is held rigidly in a cast body, and (2) a well-fitted piston or metering plug. Orifices, usually rectangular, are cut into the sleeve. These orifices are uncovered by the piston or plug until sufficient area is opened to permit passage

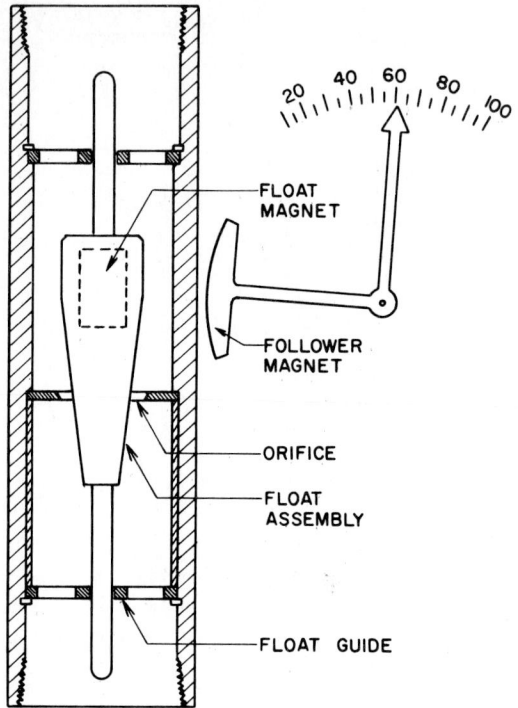

Fig. 10 Typical arrangement of an orifice and plug-type variable area flowmeter.

of the flow being measured. The metering edges consist of the port edges and the bottom edge of the plug. The position of the piston or metering plug, therefore, provides a direct indication of the orifice area and, consequently, of the rate of flow. Two piston meter designs are shown in Figs. 11 and 12. The position of the piston can be sensed visually through a glass tube mounted above the assembly—through a magnetic follower similar to those previously described, or by a transmitted signal using an impedance bridge or differential transformer coil.

The body may be of cast iron, steel, or other metal, depending on corrosion resistance and strength required. Body designs generally conform in dimensions to those of standard 250- or 600-lb (1.7- or 4.1-MPa) valve bodies with ASME or ANS flanges. These units are available only for liquid or gases at relatively high pressures and for pipe sizes up to 4 in (~10.2 cm). Every piston-type variable area meter has an extension to house the spring (for spring-loaded units) or to suspend the weight (if weight loading is substituted for the spring).

Sleeves usually are of heavy construction for rigidity and made of stainless steel for long life. They are accurately honed so that the plug may be very closely fitted without sticking. Metering capacity is directly influenced by the area of the ports cut into the sleeve, but the linearity of calibration may be destroyed if certain limits are exceeded. For this reason and because of the danger of distorting the sleeve, it is not generally recommended that capacity increases be attempted by enlarging the port area of a sleeve once it has been honed.

The piston or metering plug usually is made of stainless steel, but it may have Stellite* guide rings and a Stellite metering edge. The bottom face generally is a single flat plate except for a central stem fastening. The metering plug may be loaded by a tension spring or a weight extending into a cylindrical section attached to the body. The piston may be loaded by weights placed inside the body as shown in Fig. 12. Either method permits adjustment of capacity within limits. For marine or

*Trademark, Union Carbide Corporation.

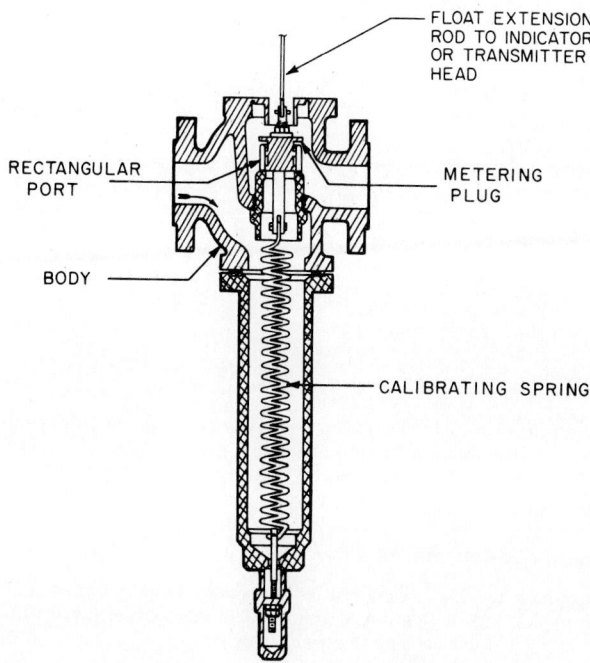

Fig. 11   Spring-loaded piston-type variable area flowmeter.

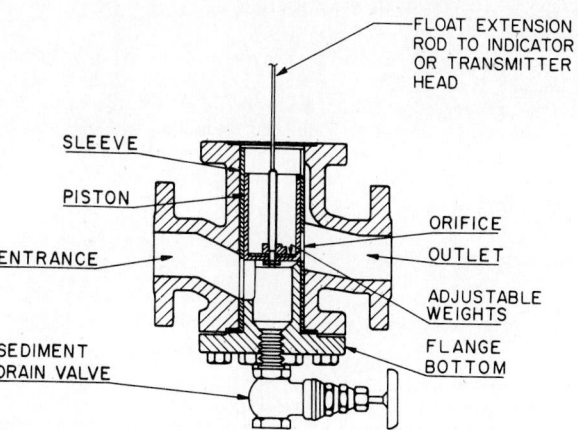

Fig. 12   Weight-loaded piston-type variable area flowmeter.

mobile use, the spring-loaded meter may be preferred, since it is not essential that such installations be absolutely vertical.

Viscosity Effects

All piston-type variable area meters appear to be influenced to some extent by changes in the viscosity of the metered fluid. Compared with the rotameter, in which the viscous drag is parallel to float motion, the viscous drag in the piston meter is approximately normal to the movement of the piston, thus exerting a smaller influence. In measuring very viscous fluids a steam-jacketed meter may be required.

Piping Requirements

As with rotameters, upstream piping configuration is not critical. It is generally recommended that throttling valves be located no closer than 10 pipe diameters upstream of the meter. Also, unusual arrangements such as a pipe discharging into a meter of larger nominal size directly through a reducer should be avoided.

Capacity

See Table 5.

Advantages and Limitations

Although piston meters are not used nearly so extensively as rotameters, they do offer certain advantages. They may be somewhat less expensive for large pipe sizes (3 and 4 in). However, they must be used with clean fluids free of foreign particles. They are most commonly used to meter such fluids as water, kerosene, gasoline, oil, tar, and other clean petroleum-base liquids.

Installation and Maintenance Factors

(1) Use of a strainer in the flow line upstream is recommended unless the metered fluid is absolutely free of foreign particles; (2) weight-loaded piston meters should be mounted within 5° of vertical; spring-loaded meters will tolerate a much greater angle.

Correction Factors for Variable Area Meters

A variable area meter generally is calibrated for one specific fluid. When it is used with a different fluid, correction factors always will be necessary for the density changes and possibly may be required for the viscosity change. The corrections required for viscosity are quite complex, and data would have to be obtained by way of an individual calibration with the new fluid, or perhaps obtained from the manufacturer. Assuming a meter that is insensitive to viscosity (usually the case), corrections are necessary for (1) specific gravity or specific weight of the fluid, and (2) the buoyancy of the piston assembly or float.

Specific-Gravity or Specific-Weight Correction

The correction for specific gravity or specific weight may be inferred from Eq. 6. Assuming that there are no changes other than specific gravity, the following equations apply to gas flow where buoyancy of the float is negligible.

Table 5 Maximum Capacities of Typical Piston-Type Meters

Nominal pipe size	Type of loading	Approximate capacity (water at 70°F, 21.1°C)			Approximate pressure drop
		gal/h	lb/h	k/h	
1 in (2.5 cm)	Weight	250–540	2,250–4,500	1,021–2,041	2 psi (13.8 kPa)
2 in (5 cm)	Weight	150–2,000	1,250–16,600	567–7,530	½–2 psi (3.5–13.8 kPa)
1 in (2.5 cm)	Spring	600–2,530	5,000–21,000	2,268–9,526	10 psi (68.9 kPa)
2 in (5 cm)	Spring	2,590–5,120	21,500–42,500	9,752–19,278	10 psi (68.9 kPa)
4 in (10 cm)	Spring	6,025–13,855	50,000–115,000	22,680–52,164	6 psi (41.4 kPa)

In terms of weight units,

$$\frac{w_2}{w_1} = \sqrt{\frac{\gamma_2}{\gamma_1}} = \sqrt{\frac{(\text{sp gr})_2}{(\text{sp gr})_1}} \qquad (7)$$

In terms of volume units,

$$\frac{q_2}{q_1} \sqrt{\frac{\gamma_1}{\gamma_2}} = \sqrt{\frac{(\text{sp gr})_1}{(\text{sp gr})_2}} \qquad (8)$$

The special case of gas flows, in terms of standard volume units of a perfect gas, is covered by Eq. 9. Typical standard volume units are scfm (standard cubic feet per minute) and slpm (standard liters per minute).

$$\frac{Q_2}{Q_1} = \sqrt{\frac{(\text{sp gr})_1 P_2 T_1}{(\text{sp gr})_2 P_1 T_2}} \qquad (9)$$

where w_1 = flow of original fluid, weight/unit of time
$\quad\quad w_2$ = flow of new fluid, weight/unit of time
$\quad\quad \gamma_1$ = specific weight of original fluid, weight/unit volume
$\quad\quad \gamma_2$ = specific weight of new fluid, weight/unit volume
$\quad (\text{sp gr})_1$ = specific gravity of original fluid (air = 1.00)
$\quad (\text{sp gr})_2$ = specific gravity of new fluid (air = 1.00)
$\quad\quad q_1$ = flow of original fluid, actual volume/unit of time
$\quad\quad q_2$ = flow of new fluid, actual volume/unit of time
$\quad\quad Q_1$ = flow of original gas, standard volume/unit of time
$\quad\quad Q_2$ = flow of new gas, standard volume/unit of time
$\quad\quad P_1$ = original operating absolute pressure, psia
$\quad\quad P_2$ = new operating absolute pressure, psia
$\quad\quad T_1$ = original operating absolute temperature, °R
$\quad\quad T_2$ = new operating absolute temperature, °R

It will be noted that any flow rate, specific weight, absolute pressure, and absolute temperature units may be used, so long as the units are consistent in numerator and denominator.

Buoyancy Correction

If the change is from metering a gas to metering a liquid, or from a liquid to a gas, it is best to work in terms of specific weight rather than specific gravity, since the latter will involve different reference fluids. For such changes, as well as a change from liquid to liquid, the effect of liquid specific weight on the apparent weight of the buoyed float must be considered.

To correct for buoyancy, the apparent weight of the float or piston assembly in each fluid should be determined by subtracting the weight of the displaced fluid from the weight of the float in a vacuum, the latter being approximated by ordinary weighing in air. The following equations apply:

$$Wa_1 = \frac{\gamma_f - \gamma_1}{\gamma_f} W \qquad (10)$$

$$Wa_2 = \frac{\gamma_f - \gamma_2}{\gamma_f} W \qquad (11)$$

where W = float weight, lb
$\quad Wa_1$ = apparent weight of float in original fluid
$\quad Wa_2$ = apparent weight of float in new fluid
$\quad \gamma_f$ = specific weight of float

The part of the correction resulting from buoyancy is

$$\frac{q_2}{q_1} = \frac{w_2}{w_1} = \sqrt{\frac{Wa_2}{Wa_1}} = \sqrt{\frac{\gamma f - \gamma_2}{\gamma f - \gamma_1}} \tag{12}$$

Combined Specific-Weight Corrections for Weight-Loaded Area Meters

Equations 7 and 12 combine to give

$$w_2 = w_1 \sqrt{\frac{(\gamma f - \gamma_2)\gamma_2}{(\gamma f - \gamma_1)\gamma_1}} \tag{13}$$

Equations 8 and 12 combine to give

$$q_2 = q_1 \sqrt{\frac{(\gamma f - \gamma_2)\gamma_1}{(\gamma f - \gamma_1)\gamma_2}} \tag{14}$$

Equations 13 and 14 are always applicable but are usually employed only in conversions from liquid to liquid or from liquid to gas, and vice versa. Conversions from gas to gas, where buoyancy effects are usually negligible, are often calculated using Eqs. 7 and 8. No knowledge of float weight is required in either case.

Combined Specific-Weight Corrections for Spring-Loaded Area Meters

In a spring-loaded meter, the spring load F must be added, and it is necessary to know the piston weight as well as the value of F (corresponding to each meter reading if a change in F with spring extension is a significant fraction of $F + Wa$) before the specific-weight correction can be calculated.
Equation 12 is then replaced by the following buoyancy equation:

$$\frac{q_2}{q_1} = \frac{w_2}{w_1} = \sqrt{\frac{Wa_2 + F}{Wa_1 + F}} \tag{15}$$

where Wa_2 and Wa_1 are calculated from Eq. 11. Equations 7 and 15 combine to give

$$w_2 = w_1 \sqrt{\frac{(Wa_2 + F)\gamma_2}{(Wa_1 + F)\gamma_1}} \tag{16}$$

Equations 8 and 15 combine to give

$$q_2 = q_1 \sqrt{\frac{(Wa_2 + F)\gamma_1}{(Wa_1 + F)\gamma_2}} \tag{17}$$

Capacity Changes

A weight-loaded variable area meter may be altered for higher or lower flow rates by changing the apparent weight of the float or piston. This may be achieved by duplicating the original float shape in materials of higher or lower specific weight, using an increased volume, or both. Regardless of the units of flow, if shape change has an insignificant effect on the flow coefficient, the adjustment will alter the capacity as

$$\frac{w_2}{w_1} = \frac{q_2}{q_1} = \sqrt{\frac{Wf_2(\gamma f_2 - \gamma)\gamma f_1}{Wf_1(\gamma f_1 - \gamma)\gamma f_2}} \tag{18}$$

where Wf_2 and Wf_1 are the original and new float weights in vacuum.

Spring-loaded area meters may be altered for a higher or lower capacity by changing the piston or float specific weight or volume, or more easily by adjusting the spring load F. Regardless of the units of flow, if shape change has an insignificant effect on the flow coefficient, the adjustment will alter the capacity as

$$\frac{w_2}{w_1} = \frac{q_2}{q_1} = \sqrt{\frac{F_2 + Wf_2(\gamma f_2 - \gamma)/\gamma f_2}{F_1 + Wf_1(\gamma f_1 - \gamma)/\gamma f_1}} \tag{19}$$

where again corrections must be calculated for the values of F_2 and F_1 at each scale reading if the spring force for either assembly varies significantly with the piston position or meter reading.

Metering head losses will vary significantly when capacities are so altered and the square root relationship limits the flow-range changes that can be made. Changes in the shape of the metering-edge portion of the float in either type is not recommended, since the change in flow coefficient generally is not predictable.

TYPICAL VARIABLE AREA METER CALCULATIONS

The following examples are designed to illustrate typical day-to-day measurement problems and to present methods for their solution. The tables, figures, and equations are numbered in accordance with those already given in this text.

Problem A

A variable area meter is to be selected for a liquid other than water. What capacity in terms of gallons per minute of water should be selected?

Known Data: Desired capacity of other liquid   10 gpm
              Specific gravity of other liquid   1.24
              Specific gravity of water   1.00
              Float specific gravity (316 stainless steel)   8.02

Step 1: Using Eq. 14:

$$q_2 = 10 \sqrt{\frac{(8.02 - 1.00)1.24}{(8.02 - 1.24)1.00}} \qquad q_2 = 10 \sqrt{\frac{(7.02)1.24}{6.78}} = 11.33$$

ANSWER: 11.33 gpm of water equivalent

Problem B

Consider the same meter capacity and fluid as in Prob. A except that the meter is to be supplied with a Hastelloy* C float. What is the equivalent water capacity for meter selection from a manufacturer's capacity table based on Type 316 stainless steel floats?

Additional Known Data: Hastelloy C float specific gravity   8.94
                       316 stainless steel float specific gravity   8.02

Step 1: Using Eq. 14:

$$q_2 = 10 \sqrt{\frac{(8.02 - 1)1.24}{(8.94 - (1.24)1}} \qquad q_2 = 10 \sqrt{\frac{(7.02)1.24}{(7.70)1}} = 10.63$$

ANSWER: 10.63 gpm water equivalent

*Trademark, Union Carbide Corporation.

Problem C

A variable area meter was originally supplied for a maximum flow rate of 2000 pph, liquid specific gravity 0.72. It is desired to use the meter on a new liquid specific gravity 0.91. Determine the correction factor and new maximum flow rate.

Known Data: Original fluid specific gravity   0.72
New fluid specific gravity   0.91
Float specific gravity (316 stainless steel)   8.02
Original maximum flow   2000 pph

Step 1:   Using Eq. 13:

$$\frac{w_2}{w_1} = \sqrt{\frac{(8.02 - 0.91)0.91}{(8.02 - 0.72)0.72}} \qquad \frac{w_2}{w_1} = \sqrt{\frac{(7.11)0.91}{(7.30)0.72}} = 1.109$$

ANSWERS: Correction factor   1.109
New maximum flow   2218 pph

Problem D

A spring-loaded piston-type area meter of known characteristics is measuring water at a given capacity. It is desired to determine the capacity of the meter for the measurement of motor gasoline.

Known Data: Type of meter: spring-loaded piston-type area meter, insensitive to viscosity changes.
Now measuring: water at 70°F, quantity = 10,540 lb/h
Weight of complete piston assembly, including armature, piston, spring, and tension adjusting stem: 1.22 lb
Assembly materials (excepting armature and stem): 304 stainless steel with specific gravity 7.92
Spring tension at zero flow: 8.94 lb

Note: For the purposes of this problem, it can be assumed that the armature and stem have the same specific gravity as other parts of the assembly.

Find: What is the capacity (in weight units) of this meter in terms of motor gasoline with specific gravity 0.738?

Step 1: Neglecting the buoyant effect of the atmosphere on the original weighing, the apparent weight Wa of the piston assembly in each fluid can be calculated, using Eq. 11, as follows:

$$Wa_1 = \frac{1.22(7.92 - 1.00)}{7.92} \qquad Wa_2 = \frac{1.22(7.92 - 0.738)}{7.92}$$

$$= 1.066 \text{ lb in water} \qquad\qquad = 1.106 \text{ lb in gasoline}$$

Step 2: The factor for spring loading can be determined from Eq. 15 as follows:

$$\frac{w_2}{w_1} = \sqrt{\frac{1.106 + 8.94}{1.066 + 8.94}} = 1.002$$

Step 3: The specific-gravity factor in Eq. 15 can be determined as follows:

$$\frac{w_2}{w_1} = \sqrt{\frac{0.738}{1.000}} = 0.8588$$

Step 4: The capacity in terms of motor gasoline can be determined by multiplying the capacity for water by the factors determined in steps 2 and 3, as follows:

Gasoline capacity = (10,540)1.002(0.8588)

ANSWER: 9070 lb/h at 70°F

Problem E

A meter is to be selected to handle a gas other than air. What capacity in terms of free air should be selected?

Known Data:	Desired capacity of carbon dioxide	100 scfm
	Flowing conditions of the gas	50 psig and 100°F
	Specific gravity of carbon dioxide relative to air at standard temperature and pressure	1.529
	Float specific gravity (assuming stainless steel)	8.02
	Standard conditions	14.7 psia and 70°F

Step 1: Using Eq. 9:

$$(\text{scfm})_2 = 100 \sqrt{\frac{1.529(14.7)560}{1.00(64.7)530}}$$

Note: $P_1 = 50 + 14.7 = 64.7$ psia
$T_1 = 460 + 100 = 560°R$
and $P_2 = $ (standard pressure) $- 14.7$ psia
$T_2 = $ (standard temperature) $- 70°F$ or $530°R$
$(\text{scfm})_2 = 100 \sqrt{0.3675} = 100(0.6056) = 60.56$

ANSWER: 60.56 scfm air equivalent

Problem F

For the same meter as in Prob. E what will be the maximum flow rate in terms of gallons per minute of water?

| *Additional Known Data:* | Specific gravity of water | 1.0 |
| | Specific gravity of air at standard temperature and pressure referred to water | 0.0012 |

Step 1: Using Eq. 14:

$$q_2 = 60.56 \sqrt{\frac{(8.02 - 1)0.0012}{(8.02 - 0.0012)1}}$$

$$q_2 = 60.56 \sqrt{\frac{(7.02)(0.0012)}{8.02}} = 60.56 \sqrt{0.00105}$$

$$q_2 = 60.56(0.0324) = 1.96 \text{ cfm of water}$$

Step 2: $1.96 \text{ cfm} \times 7.48 \text{ gal/ft}^3 = 14.65 \text{ gpm}$

ANSWER: 14.65 gpm water

Note: A convenient factor can be used when the float is Type 316 stainless steel ($\gamma f = 8.02$). Scfm air metered at 14.7 and 70°F $= 4.124 \times$ gpm water.

Problem G

A meter has a maximum flow of 50 std ft^3/min propane metered at 25 psig and 70°F. What will the maximum flow be if the operating pressure is changed to 50 psig?

| *Known Data:* | Original maximum flow | 50 scfm propane |
| | Original operating pressure | 25 psig or 39.7 psia |

Original operating temperature   70°F
New operating pressure   50 psig or 64.7 psia
Specific gravity of propane   1.56

Step 1:   Using Eq. 9:

$$(\text{scfm})_2 = 50 \sqrt{\frac{1.56(64.7)530}{1.56(39.7)530}} = 50(1.277) = 63.85$$

ANSWER: 638.85 scfm propane metered at 50 psig and 70°F

Problem H

For the same meter as in Prob. G what will be the maximum capacity for metering natural gas at 25 psig and 70°F?

Additional Known Data: Specific gravity of propane   1.56
                             Specific gravity of natural gas   0.6

Step 1:   Using Eq. 9:

$$(\text{scfm})_2 = 50 \sqrt{\frac{1.56(39.7)530}{0.60(39.7)530}}$$

$$= (50)1.61 = 80.5$$

ANSWER: 80.5 scfm natural gas metered at 25 psig and 70°F.

Magnetic Flowmeters

**by
E. H. Higham***

The *magnetic flowmeter* was developed for measuring the flow rate of fluids at installations where the more common methods of measurement are unsuitable. Its principal feature is that it presents an unobstructed passage to the flow of fluid and, provided that the fluid is an electrolyte, it is relatively insensitive to changes in fluid density, viscosity, and the flow velocity profile. It also has an essentially linear response and is particularly suitable for measuring the flow of hostile acids and alkalis as well as slurries with coarse or fine suspended material.

Principle of Operation

The operation of a flowmeter is based on Faraday's law of electromagnetic induction, which states that the voltage E induced in a conductor of length d is proportional to the velocity v with which it

*Foxboro Great Britain Limited. Redhill, Surrey, England.

traverses a magnetic field H. The vector of the generated voltage is perpendicular to both the velocity vector and the magnetic field vector.

As a first approximation, for example, a 1-mV signal is developed in a flowtube 0.1 m in diameter if the field strength is 0.1 T and the fluid velocity is 1 m/s.

The essential components of a practical magnetic flowmeter are shown in Fig. 1.

The tube is constructed of nonmagnetic stainless steel and lined with polytetrafluoroethylene (ptfe), polyurethane, or another similar insulating material to minimize short-circuiting or diversion of the relatively small flow signal to the metal tube.

The transverse magnetic field is generated by coils located on opposite sides of the tube, and the voltage induced by motion of the fluid through the magnetic field is detected by a pair of electrodes located diametrically opposite each other with their axes perpendicular to both the magnetic field and the axis of the tube.

The kinetics of the interface between the metallic electrodes and the flowing fluid are not only complex, but also unpredictable and variable. However, the interelectrode potential induced by motion of the fluid through the magnetic field changes at the same rate as the magnetic field itself. Hence, the only method of reducing the effect of these spurious signals is to measure the change in interelectrode potential

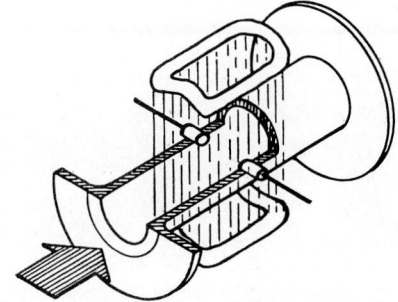

Fig. 1   Essential components of a practical magnetic flowmeter. *(Foxboro.)*

resulting from a known change in magnetic field strength. The simplest form uses the main supply to energize the coils of the electromagnet directly, in which case the flow signal is an alternating voltage of the same frequency in phase with the magnetic flux.

When sinusoidal excitation of the coils is used, then the induced signal is

$$e = Bdv \sin \omega t + K\omega B \cos \omega t$$

where B = mean flux
$\omega = 2\pi f$
v = mean velocity
ωt = angular frequency
d = diameter of the flowtube
K = constant

The first term is flow-dependent and in phase with the magnetic flux. The second term, which is in phase quadrature and independent of flow, is the transformer voltage. This latter voltage arises from the fact that the signal leads to the measuring circuit, combined with the effective current path through the fluid, form a complete loop that is cut by the alternating flux. Although careful attention to the positioning of the signal leads can reduce this signal appreciably, it is still necessary to use measurement circuits which reject this unwanted component. In particular, attention must be paid to the changes in eddy currents and iron losses in the magnetic circuit due to changes in ambient and/or process temperature.

Component Layout and Materials of Construction

Figure 2 shows one layout of the essential components. The nonmagnetic stainless steel tube provides the strength for the conduit through which the liquid flows. It is lined with insulating material such as ptfe and two flush electrodes, located diametrically opposite each other at the center of the flowtube, extending through the wall. The tube and connections are usually flanges to facilitate mounting in a pipeline.

Mounted outside the tube is a pair of coils powered directly from the mains supply to create the

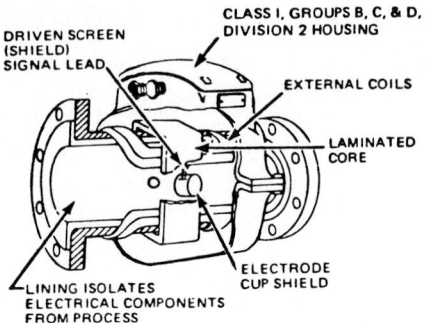

DRIVEN SCREEN
(SHIELD)
SIGNAL LEAD

CLASS I, GROUPS B, C, & D,
DIVISION 2 HOUSING

EXTERNAL COILS

LAMINATED
CORE

ELECTRODE
CUP SHIELD

LINING ISOLATES
ELECTRICAL COMPONENTS
FROM PROCESS

Fig. 2 Layout of essential components of a magnetic flowmeter. *(Foxboro.)*

electromagnetic field in the tube. A laminated steel core mounted around the coils completes the magnetic circuit.

A cast-aluminum housing provides environmental protection for the components and includes a two-compartment junction box in which the terminals for the connections to the coils and to the electrodes are located. The transmitter may be mounted integral with the flowtube or located in a remote position, in which case the connections to the electrodes are made via screened leads which include driven screens to reduce the loading effect of the cables on the flow signal.

The transmitter is also housed in a cast-aluminum enclosure and is designed to convert the small (millivolt) signal developed at the electrodes into a proportional standard transmission signal of 4 to 20 mA. Pulse and dc voltage outputs are also available.

Functional Diagram

A functional diagram is shown in Fig. 3. The signal from the electrodes is applied to the high-impedance input stage that is characterized by a high common mode rejection. This circuit also provides the drive for the screened signal leads so that the effect of cable capacitance is minimized.

After amplification, the flow signal is combined with a signal derived from the supply voltage to balance the residual signal at no flow. It then passes via a synchronous rectifier to the divider stage where it is divided by a signal derived from the reference voltage. The divider stage comprises a dual-ramp oscillator whose square wave output signal has a varying mark/space ratio applied via an optical isolator to a square wave generator with amplitude limiting.

At zero flow, the mark/space ratio from this stage is unity; as the flow increases, the mark/space ratio increases proportionally so that the averaging amplifier generates a voltage proportional to the flow rate. This voltage is applied to a voltage-to-current amplifier to generate a 4 to 20 mA output current for transmission or to a voltage-to-pulse rate converter to generate a pulse output to drive a counter or similar device.

An alternative approach for minimizing the spurious components in the flow signals is exemplified by the Fischer & Porter Mag X system. In this system, the electromagnet is energized by applying the rectified mains and then, after a few cycles during which the current stabilizes at a value determined by the inductance of the electromagnet and the mains voltage, the signal at the electrodes is sampled and stored. The excitation is then switched off, and when the electromagnet current falls to zero, the electrode signal is sampled again. Subtraction of this signal from that obtained and stored when the magnetic field is applied provides a measure of the fluid flow rate. Errors due to the variable kinetics of the electrode-liquid interface are largely eliminated, and a zero control is not required.

The flowtubes required for this system are essentially the same as those used for an ac-excited system, except that with this mode of excitation the transformer effect signal on the measuring circuit is virtually eliminated because the measurement signal is sampled only when the electromagnet current is constant except for the ripple at twice the mains frequency.

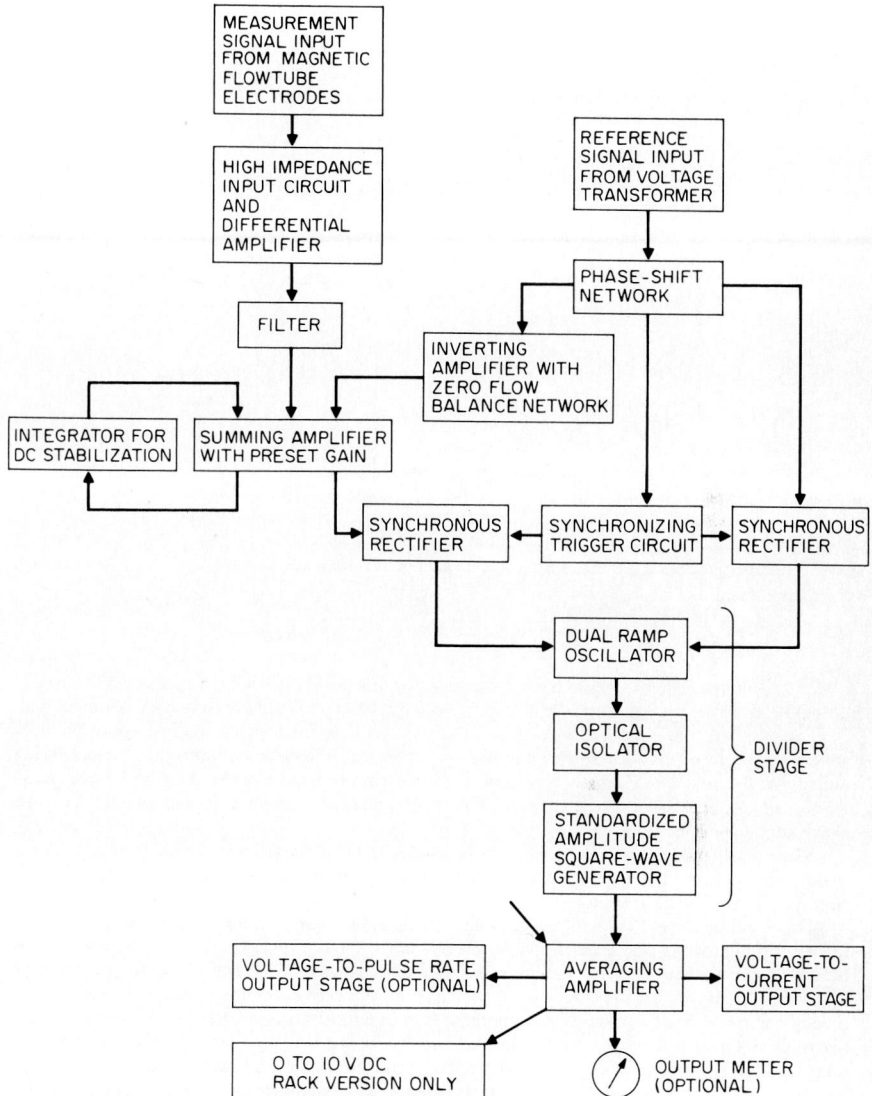

Fig. 3   Functional diagram of a magnetic flowmeter. *(Foxboro.)*

Transmitter Block Diagram

See Fig. 4. Power to energize the electromagnet is derived from a bridge rectifier connected across the secondary winding of a transformer, the mains supply to which is controlled by a triac. A small resistor included in series with the electromagnet provides a signal proportional to the current flowing and hence to the magnetic field. This signal is applied to a circuit that forms a component in the internal feedback loop, combining the output frequency with the reference signal to generate an input to the summing junction. The other input to the junction is the amplified signal from the electrodes.

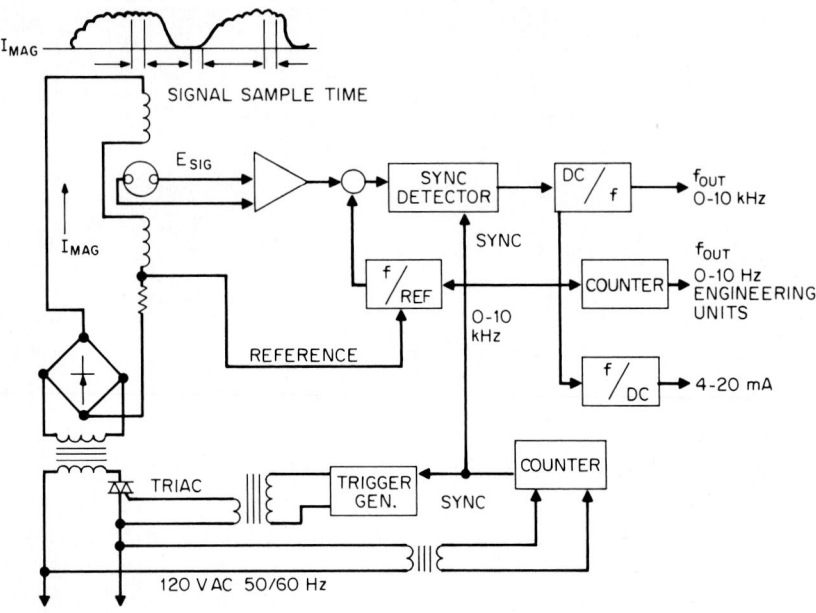

Fig. 4 Block diagram of a magnetic flowmeter transmitter. *(Fischer & Porter.)*

The combined signal is applied to the synchronous detector that also includes the "hold" circuits to enable the zero excitation signal to be subtracted from the flow signal to derive the real flow signal. The latter signal is applied to the dc-to-frequency circuit that operates the frequency that provides not only the basic output but also the input to a counter that allows an output signal in engineering units to be derived. The signal is also used to drive a circuit that generates the 4 to 20 mA transmission signal. Operation of the entire unit is synchronized by a counter driven directly from the mains supply.

A further refinement of the method of operation described in the previous section involves reversal of the polarity of the magnetic field during alternate cycles. It is thought that this reduces the spurious effects of electrode polarization.

The Altoflux DIGITIV T900 signal converter is an example of a transmitter based on square wave excitation of the magnetic circuit in the flowtube. A block diagram is shown in Fig. 5. The converter comprises two sections. One provides the square wave current of ± 0.125 A to the magnetic circuit, with galvanic isolation from the other circuits. The second section provides the signal processing and comprises the usual high-impedance input amplifier with an associated follower amplifier to drive the screens of the input cables and so reduce the loading effect. The output from this stage is applied to a summing junction where it is combined with the signal from the feedback loop. The output from this junction passes via a two-stage bucket chain memory to a lowpass filter and on to a precision synchronized rectifier. From the rectifier the signal is applied to an integrator, the output of which drives a voltage-to-frequency converter to generate a frequency of 4 kHz when the input signal from the flowtube corresponds to a fluid flow rate of 1 m/s (irrespective of the flowtube size).

This frequency signal is also applied to a frequency-to-voltage converter to generate an output signal that passes via a modulator to the summing point. Thus, in operation, fluid flowing through the flowtube generates a small signal which, after amplification, is applied to a summing junction where it is combined with a signal proportional to the frequency output signal of the transmitter. If the feedback signal is less than the flow signal, the integrator causes the voltage-to-frequency converter to increase its frequency until balance is established. If the feedback signal is greater than the flow signal, the converse applies. It should be noted that the memory, rectifier, and modulator are gated so that only the flow signals generated during periods of a steady magnetic field are processed.

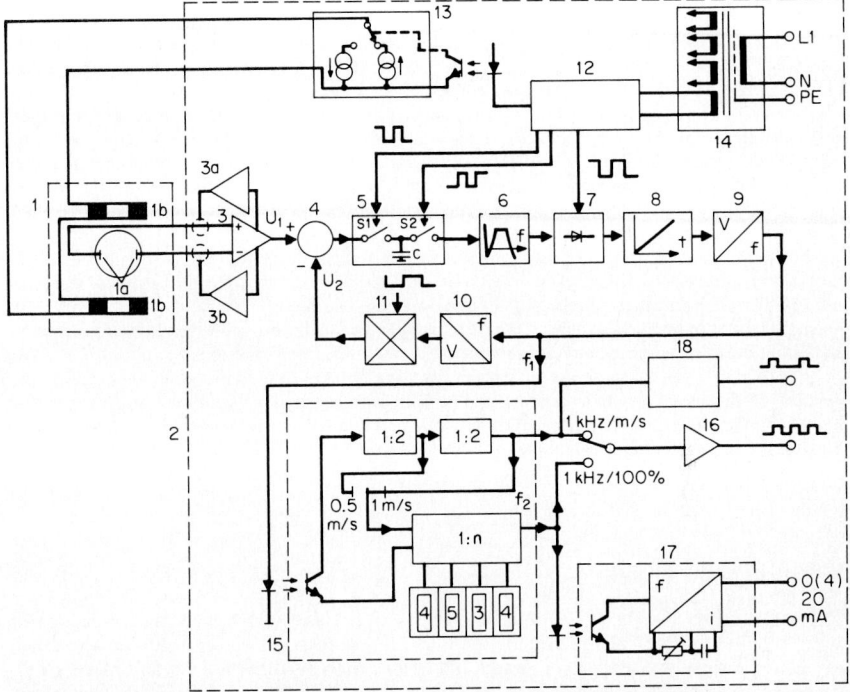

Fig. 5   Block diagram of a magnetic flowmeter transmitter. *(Krohne.)*

The output signal may be a frequency derived from the previously mentioned feedback frequency via an opto-isolator, two binary dividing stages, and an output amplifier, so that it is 1 kHz m^{-1}s^{-1}.

A second output is derived from the same input frequency, but in this case it passes via a present counter that allows scaling so that 1 kHz then equals 100% of the flow rate. This latter signal is also passed via a further opto-isolator to a frequency-to-current converter to generate an output current or 0 to 20 mA or 4 to 20 mA into a 1000-Ω load. This converter also includes means for adjusting the damping between 2 and 6 s.

An alternative method of excitation is to use square wave or trapezoidal signals. In both instances the duration of the switching transient can be controlled and the magnetic flux held constant during the measurement period so that the error due to transformer effect is essentially eliminated. However, the increased cost and complexity of the circuits which control the magnetic flux and process the measurement signals may be a disadvantage. A further advantage of these schemes compared with the ac excitation systems is the reduced reactive power requirement, and whereas ac systems require a different set of coils for each voltage and frequency, systems using pulsed or keyed dc excitation can be set to operate from any of the usual supply voltages and frequencies.

As mentioned previously, the flow rate signal is derived from the ratio of induced voltage to magnetic field strength. In practice, variations in the characteristics of the magnetic circuit from one unit to the next, as well as those resulting from changes in ambient or process conditions, make it necessary to determine the field strength.

In some instances this is done by measuring the voltage applied to the coils, but this voltage is sensitive to changes in coil resistance due to temperature variations. In some systems the coil current is measured, and this current is less sensitive to coil resistance changes, but both are affected by changes in the eddy currents in the lined metal tube as well as in the rest of the magnetic circuit.

Another method is to use a flux coil or a Hall effect probe, but all these techniques suffer from the disadvantage that eddy currents in the lined tube modify the magnetic field so that the signal measured in this way does not necessarily represent the flux through which the fluid is flowing.

Flowtube Geometry and Materials

From an application viewpoint, there is a persistent requirement to reduce the size of the flowtubes, particularly their length. Hitherto there has been a reluctance to reduce the overall length because of the adverse effect this would have on the overall system performance.

As mentioned previously, the flowtube usually comprises a nonmagnetic stainless steel tube lined with insulating material so that the flow signal is not attenuated by being diverted to the metal tube rather than to the electrode. If the adjacent pipework is similarly lined with insulating material or fabricated in nonconducting material, the flow signal will not be attenuated. If, however, the adjacent pipework and flanges are metallic, then the flow signal will be attenuated progressively as the length of the flowtube is reduced.

Furthermore, if the pipe and flange material is magnetic, then the basic magnetic field of the flowtube will be distorted, particularly as the length is reduced and the flanges become a more significant part of the magnetic circuit. In addition to modifying the flow constant of the flowtube, this may also make the instrument more sensitive to installation conditions. However, by using a suitable configuration of the magnetic circuit with respect to location of the electrodes and the earth or guard rings necessary to establish the mean potential of the flowing fluid, short form tubes have been devised which function well in process plants. But their calibration factor may differ from that obtained under reference conditions, and they may be more sensitive to distorted flow profiles than those of the corresponding conventional optimized design.

Velocity Profile Effect

The essential characteristic of a magnetic flowmeter is the linear relation between the induced voltage and the fluid flow rate. This is attributable to the fact that the small voltage developed across each element of fluid is precisely proportional to the velocity at which the element moves through the magnetic field. However, the electrical path from an element to the electrodes involves adjacent elements each of which has developed its own potential but, being a conductor, also acts as a partial shunt. Consequently, the contribution of each element to the output signal depends not only on the mean velocity of that element through the magnetic field, but also on the field strength at each particular point. In an ideal flowtube the magnetic field would be shaped so that the combined effect of the field strength and elemental velocity would be constant throughout each cross section. This would involve complex coil configurations which are difficult to implement in practice. But practical designs can lead to significant improvements compared with a uniform magnetic field, although the extent of the improvement is restricted when point electrodes are used.

Numerical analysis shows that performance can also be enhanced by using large area electrodes, but contamination of the electrode's surface displaces its effective electrical center, and this not only causes a change in the zero flow signal and calibration factor but also increases the measurement errors when distorted flow profiles are involved.

In the majority of instances, the calibration factor for a flowtube is determined in a flow rig, using water. However, careful control of all the manufacturing tolerances can provide a statistical basis for estimating the calibration factor, probably within 1%. A more accurate dry calibration can be calculated from measurements of the magnetic field strength at a sufficient number of points within the flowtube.

Because magnetic flowmeters involve no moving parts or mechanical linkages, they are free of hysteretic effects and should have a repeatability and reproducibility better than 0.1%.

As mentioned previously, magnetic flowmeters are essentially linear devices, but in practice they have a distinct velocity profile characteristic as a result of which their accuracy and linearity are influenced by the configuration of the associated pipework in the installation. If this configuration is such that the velocity distribution varies appreciably over the cross section at the flowtube, then it will almost certainly affect both accuracy and linearity. However, in any particular installation the flow pattern should remain reasonably constant and therefore the linearity, repeatability, and reproducibility should not be affected, but it would be fortuitous if the accuracy were not adversely affected.

Velocity, Vortex, Mass, Ultrasonic, and Other Flowmeters

Flowmeters may be classified in numerous ways. For practical purposes, the major categories of flowmeters in terms of frequency of use—differential, area, positive-displacement, magnetic, and open-channel meters—are described separately in this section of the Handbook. There still remain a number of important flowmetering methods, and these are described briefly here.

Velocity Meters

Velocity meters employ a measuring element, the action of which is proportional to stream velocity which, in turn, is a measure of rate of flow in a filled, closed pipe.

Turbine Meters

A turbine wheel (rotor) is set in the path of a fluid stream. The flowing fluid impinges on the turbine blades, imparting a force to the blade surface and setting the rotor in motion. When a steady rotational speed has been reached, the speed is proportional to fluid velocity. In one design (Fig. 1), a multiblade rotor is mounted at right angles to the axis of the flowing fluid. The rotor is supported by ball or sleeve bearings on a shaft which is retained in the flowmeter housing by a shaft-support section. This section may support the rotor shaft at one end only (commonly referred to as cantilevered or single-ended construction); or it may be in two sections, thereby supporting both ends of the shaft (double-ended construction). There are still other variations for affixing the rotor to the shaft. The shaft-support section also usually acts as a flow straightener (straightening vanes). An exception is the "Pelton wheel" design used in some low-flow turbine meters. This latter concept is analogous to the overshot or undershot waterwheel with the rotor axis at right angles to the flow axis. In each case, a count of the number of turbine wheel revolutions per unit time will be a measure of flow

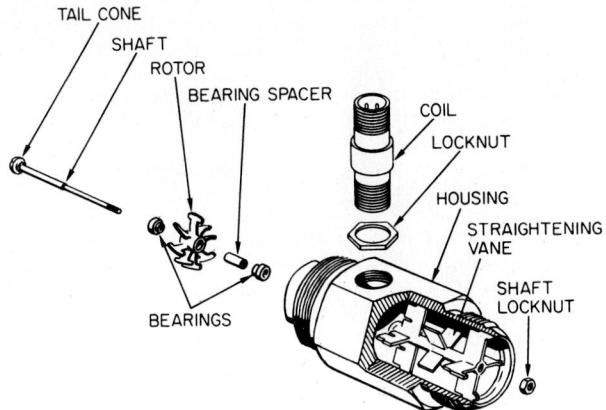

Fig. 1 Cutaway and exploded view of a turbine flowmeter transmitter. *(Foxboro.)*

rate, and a count of total revolutions will determine total volume flow. The overall compactness of design is shown in Fig. 2.

Rotor speed may be transmitted through the meter housing by a mechanical shaft with a magnetic coupling to an external shaft or through a suitable gland in the housing. These mechanical output types are adjusted to desired rotation-versus-flow ratios by use of selected gear trains. Another widely used method of signal generation is a magnetic pickup coil, consisting of a permanent magnet with coil windings, mounted in close proximity to the rotor but external to the fluid channel. The passage of each rotor blade past the base of the pickup coil changes the total flux through the coil, inducing 1 cycle of voltage. The frequency of the generated pulses is a measure of flow rate, and the total number of pulses a measure of total flow.

Usually, the coefficient of each transmitter is determined by actual flow calibration prior to shipment by the manufacturer. This coefficient, commonly referred to as the calibration factor or K factor, is expressed by

$$K = \frac{T_k f}{Q}$$

where f = frequency, Hz
Q = volumetric flow rate (such as gpm)
T_k = a time constant (such as 60 for gpm)
K = pulses per volume unit (such as pulses/gal)

Turbine flowmeters provide accurate flow measurement over wide flow ranges. The devices are particularly effective in aerospace and airborne applications for energy fuel and cryogenic flow measurements. Primary uses include flow totalizing for inventory control and custody transfer, precision automatic batching for loading and batch mixing, and automatic flow control using digital electronic techniques. Secondary instrumentation includes signal amplifiers, frequency-to-analog converters, totalizers, totalizer-batchers, and frequency (flow rate) indicators. The most commonly used sizes range from reduced-bore meters [usually ½ in (∼13 mm) nominal size] for fractional flow rates to 12-in (∼305 mm) nominal size meters for 8,000 to 9,000 (303 to 341 hectoliters) full-scale flow. End connections include NPT, male flared tube, flanged, and plain for slip-on couplings.

Vortex Flowmeter

In fluid flow technology, the principle of vortex shedding has been known for many years, but only during the last several years has it been applied in a practical way to flow measurement. Central to the operation of a vortex flowmeter is the generation of distinct vortices (whirlpools, in essence) in the flow stream by an obstruction. Detection of these vortices can serve as an attractively accurate measurement of flow rate. One configuration of the vortex flowmeter can be used to measure the flow rate of liquids, gases, and steam, as contrasted with the need to customize traditional flowmeters for different services. Vortex flowmeters are constructed of stainless steel parts and thus can handle corrosive fluids as well as dirty fluids because they incorporate no moving parts and have no cavities that can plug or foul. Process fluid temperatures up to 200°F (95°C) can be tolerated. Temperatures in excess of this value may have the effect of increasing the minimum measurable flow rate (low-flow cut-in). Vortex flowmeters are not affected by viscosity for flowing conditions where the Reynolds number (R_D) is greater than 10,000. But, the minimum linear flow rate for each size of flowmeter will be the flow rate which results in a Reynolds number of 10,000 at the flowing viscosity. The area between a Reynolds number of 3000 and 10,000 is generally considered the nonlinear region. Because of numerous advantages and few limitations, the vortex flowmeter in recent years has become an attractive alternative for traditional flowmeters for many applications.

This phenomenon of fluid flow can occur when a nonstreamlined obstruction is placed in a flowing stream. The shedder (obstruction) must be designed and shaped to produce vortices which create a strong differential pressure to enable a sensor to pick up a clear signal from this pressure. The geometry of one vortex flowmeter shedding element (Foxboro) is shown in Fig. 3. This particular T-shaped element resulted from hundreds of tests and extensive data analysis on possible elemental shapes. The shape of the tail is very important in the control of stable vortex shedding. The tail also

Fig. 2   Turbine flowmeter transmitter. *(Foxboro.)*

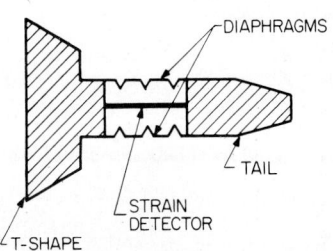

Fig. 3   Shedding element cross section. *(E83 Series, Foxboro.)*

provides a good location for the vortex detector because the signal-to-noise ratio at this location is high. The location also provides good protection for the detector from any objects that may be in the fluid stream.

As shown in Fig. 4, as the fluid flows through the flowmeter, it is divided into two paths by the shedding element, which is positioned across the flowmeter body. High-velocity fluid parcels flow past the lower velocity parcels in the vicinity of the element to form a shear layer. There is a large velocity gradient within this shear layer, thus making it inherently unstable. After some length of travel, the shear layer breaks down into well-defined vortices. Differential-pressure changes occur as the vortices are formed and shed. This pressure variation is used to actuate a sealed detector at a frequency proportional to vortex shedding.

With further reference to Fig. 3, the detector is made up of a double-faced, circular, diaphragm capsule filled with a liquid. A piezoelectric crystal is located in the center of the capsule in such a manner that the vortex-produced pressure changes are transmitted through the fill liquid to the crystal, whereupon the piezoelectric crystal produces a voltage output when a pressure change is applied. The alternate vortex generation reverses the differential, and a voltage of opposite polarity is generated. Thus, a train of vortices generates an alternating voltage output with a frequency identical to the frequency of vortex shedding. Depending on specifications, a flow output signal of either 4 to 20 mA dc or frequency is provided on the same pair of wires as used for the dc power. Again, depending on the particular model, a built-in scaling module and counter may permit totalizing of the flow while still operating in the two-wire mode without affecting the analog output signal or display. Through a dry-contact closure, a remotely mounted second counter may be driven. Depend-

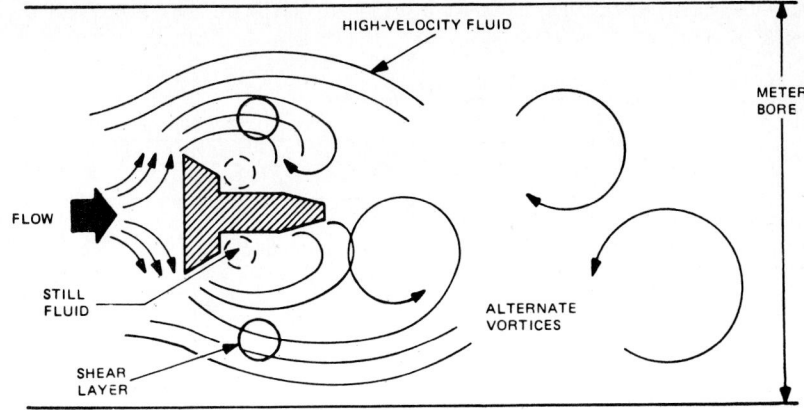

Fig. 4   Vortex shedding phenomenon. *(As depicted in the Foxboro E83 Series.)*

ing on the particular model, the electronics may be integrally mounted on the primary flowmeter or remotely mounted up to about 33 ft (10 ms) from the primary.

Analog Output

With reference to Fig. 5, the amplifier input stage accepts the voltage pulse signal from the vortex sensor and produces a conditioned rectangular pulse at a fixed amplitude. This pulse is coupled to

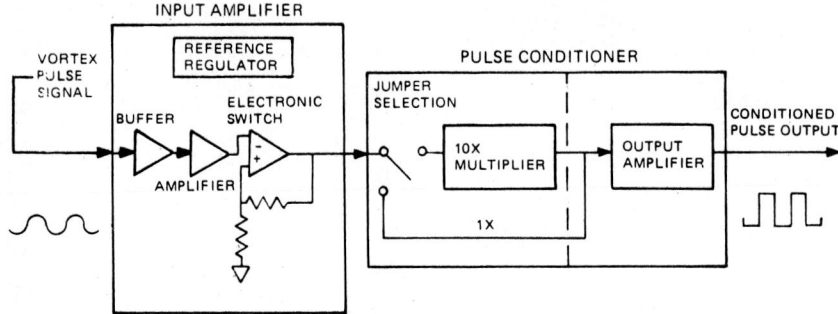

Fig. 5 Simplified block diagram—pulse output. *(Vortex flowmeter, E83 Series, Foxboro.)*

the pulse conditioner where, after buffering, it is connected to either a 10-times multiplier or directly to the output amplifier. The output pulse amplitude is related to the supply voltage.

The output is one pulse for each complete cycle of vortex shedding. There is no pulse output below the low-flow cut-in (minimum flow rate). Means are provided to adjust the minimum signal to which the amplifier will respond, which is particularly useful for gas flow. The pulse multiplier, which multiplies the basic pulse rate by a factor of 10, is typically used with equipment which demands a high-pulse-rate input, such as that normally received from turbine flowmeters. Pulse scaling to engineering units is accomplished in the receiving device.

With reference to Fig. 6, the amplifier input stage accepts the voltage pulse signal from the vortex detector and produces a conditioned rectangular pulse of fixed amplitude. These conditioned pulses are the input to a prescaler (frequency divider) which is set, by jumper selections, to normalize the frequency to a range that the pulse rate-to-current converter can accept. The output of the prescaler drives switches which are in the converter.

A capacitor in the converter is alternately charged and discharged at each zero crossing of the input signal. The charge level is determined by the attenuator reference voltage. The charging current of the capacitor is the output of this stage. It is directly proportional to this frequency of the rectangular pulses and, therefore, to the flow rate. The minimum frequency that can be set is 25

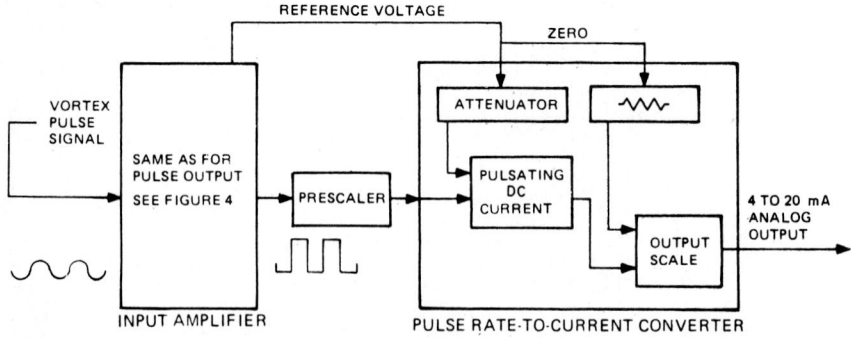

Fig. 6 Simplified block diagram—analog outtput. *(Vortex flowmeter, E83 Series, Foxboro.)*

pulses per second, which corresponds to the minimum upper range value of the flowmeter with an analog amplifier.

The output amplifier is of the two-wire type, with all power supplied by the signal lines. A zero circuit forces the amplifier to produce a 4-mA output signal for any flow velocity less than the low-flow cut-in. As flow increases, the signal current from the pulse rate–current stage is proportionally amplified to produce the 4-20 mA output signal.

K Factor

The *K factor* is used in describing and specifying the performance of the vortex flowmeter. This term defines the relationship between input and output and is expressed as the ratio of pulses per unit volume, e.g., pulses per gallon, per cubic meter, and per liter. When the K factor is measured at a series of flow rates for a given flowmeter and plotted against a parameter (pipe Reynolds number R_D), a "signature curve" for that flowmeter is generated. See Fig. 7.

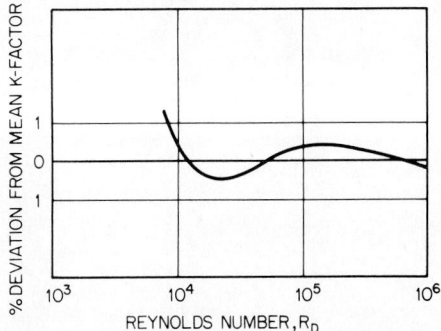

Fig. 7 Typical signature curve for a vortex flowmeter. *(E83 Series, Foxboro.)*

The Reynolds number is a dimensionless parameter which specifies the ratio of inertial effects to viscous effects in a flowing stream. For a given flowmeter and fluid, under a given set of conditions, the Reynolds number varies directly as the velocity. The use of the Reynolds number allows a prediction of the flow performance of various fluids based on the calibration of a specific fluid, such as water. Flowing fluids of different velocities and viscosities having an identical Reynolds number should perform identically.

As can be seen from Fig. 7, there is a certain minimum Reynolds number below which there is no pulse output. This minimum measurable flow is established by either one of two factors. One is the Reynolds number at which the vortex shedding phenomenon ceases; the other is the point at which the differential pressure across the detector becomes so low that it cannot be sensed by the detector.

Vortex flowmeters of the type just described are available in two body styles. The *wafer body* type is designed to be clamped by studs between pipeline flanges of the same size as the flowmeter. This style is furnished in 2-, 3-, and 4-in (50-, 80-, and 100-mm) sizes. The *flanged body* type is provided with flanges integral to the flowmeter and is furnished in 6- and 8-in (150- and 200-mm) sizes. Minimum and maxiuum upper range values (water flow) for the 2-in (50-mm) vortex flowmeter range between 62 and 209 gpm (3.9 and 13.2 L/s), and for the large 8-in (200-mm) vortex flowmeter, between 2600 and 2850 gpm (164 and 180 L/s). For the measurement of air flow, the minimum and maximum upper range values for the 2-in (50-mm) vortex flowmeter range between 6080 and 21,000 ft^3/h (172 and 594 m^3/h), and for the large 8-in (200-mm) vortex flowmeter, between 82,800 and 286,000 ft^3/h (2340 and 8100 m^3/h). For dry, saturated steam, the minimum and maximum upper range values for the 2-in (50-mm) vortex flowmeter range between 615 and 9710 lb/h (279 and 4410 kg/h), and for the large 8-in (200-mm) vortex flowmeter, between 8390 and 133,000 lb/h (3810 and 60,100 kg/h). Process fluid velocities up to 250 ft/s (76.5 m/s) can be accommodated by these meters. Accuracy claims for the meters are within 0.75% of actual flow for liquids, and to within 1.5% of flow rate for gas and steam. As compared with an orifice plate–differential-pressure transmitter, the vortex flowmeter has 5 times more rangeability.

Vortex flowmeters are widely used for flow measurement of water, benzene, formaldehyde, gasoline, ammonia, beer, and acetic acid, among numerous other process liquids, as well as for natural gas, air, nitrogen, propane, and methane. They are widely used for measuring steam in utility plants, cookers, digesters, dryers, and bleach plants, among others.

Vortex Velocity Meters

As illustrated by Fig. 8, this type of meter consists of a short section of pipe with an offset chamber in which a machined and balanced rotor is mounted. The rotor is fed by a portion of the main flow

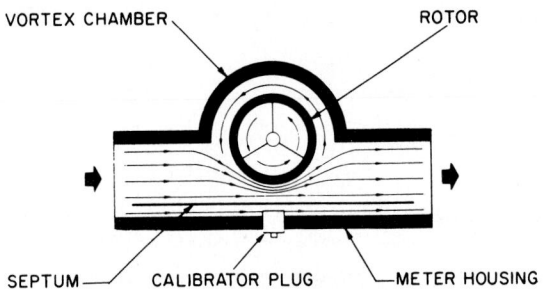

Fig. 8 Principle of vortex velocity meter.

so that the total number of rotor revolutions is directly proportional to the total gas flow through the meter. This principle compensates for gas density changes over wide ranges. The meter is divided into two sections by a septum or thin, flat plate. At the bottom of the meter, an adjustment plug screws into the lower section. Positioning the plug within the lower section forces more or less gas through the upper section with a proportional increase or decrease in rotor speed for the same total flow through the meter. Since meter registration is directly related to rotor speed, external adjustment of the plug position readily changes the meter registration.

Mass Flow Measurements

Most often, the user of a flowmeter essentially is interested in the weight rather than the volume or the linear flow rate of a fluid flow being metered. Flowmeters based on the measurement of volume are subject to ambient and process conditions, such as density, which changes with temperature and pressure. Viscosity changes also affect many volumetric flow measurement methods.

Traditionally, mass flow readings have been obtained through mathematically adjusting volumetric measurements (manually or via computer), although some conventional flowmeter designs are available where, within reasonable accuracy parameters, compensation is accomplished automatically within the meter. The advent of the microprocessor during the past few years has greatly reduced the difficulties of this task. Traditional schemes for compensating volumetric flowmeters to provide mass measurements are shown in Fig. 9. Such schemes of course add considerably to the cost of the basic meter. Further, as the measured liquid may change in composition (or different liquids are measured, as in a pipeline from time to time), additional mathematical correction factors must be entered into the automatic compensating system.

Numerous examples of automatic compensating methods could be cited. Because of space limitations, only one is given here. In one gas mass flowmeter, for example, a vortex gas velocity meter and a gas densitometer are combined. The densitometer is located upstream of the flow device. The densitometer produces a pressure difference that is linearly proportional to the density of the flowing gas at line conditions. Thus, the unit automatically corrects for variations in pressure, temperature, specific gravity, and supercompressibility.

The gas sample from the pipeline passes across a constant speed centrifugal blower and returns to the pipeline. The pressure rise across the blower varies directly with the gas density. A differential-pressure signal from the densitometer is combined in readout instrumentation with a flow rate signal from the gas meter to provide mass flow. Magnetic coupling is used for operating the readout devices. The cost of the gas meter–densitometer combination is severalfold that of a gas meter alone in the lower pipeline sizes (1 and 2-in, ~ 2.5 to 10 cm). As the pipeline and meter sizes go up, the cost of the densitometer-computer portion of the total system becomes less.

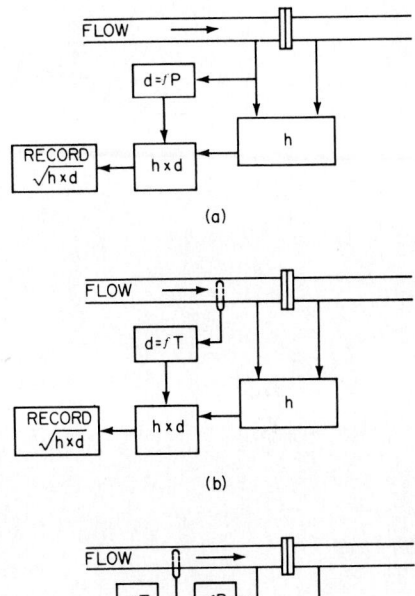

(a)

(b)

True Mass Flow Measurement

Over the years, the incentive for measuring mass flow directly has been high, and a variety of approaches have been taken. One of the most recent of these approaches is one that takes advantage of the Coriolis effect* and provides mass flow measurement for both liquids and gases. The principal elements of the flowmeter are shown in Fig. 10. A U-shaped sensor tube is vibrated at its natural frequency. The angular velocity of the vibrating tube, in combination with the mass velocity of the flowing fluid, causes the tube to twist. The amount of twist is measured with magnetic position detectors, producing a signal which is linearly proportional to the mass flow rate of every parcel and particle passing through the U-shaped sensor tube. See also Fig. 11. The output is essentially unaffected by variations in fluid properties, such as viscosity, density, pressure, temperature, pulsations, entrained gases, and suspended solids.

Nothing comes in contact with the flowing fluid except the inside wall of the tube, which is made of 316L stainless steel or other corrosion- or erosion-resistant materials as may be specified. Two magnetic position detectors, one on each side of the U-shaped sensor tube, generate signals which are routed to the associated electronics compartment for processing into a usable output. No solid-state components are in proximity with the U-shaped sensor, nor are there any moving parts in the sensing assembly. The fully assembled flowmeter is shown in Fig. 12.

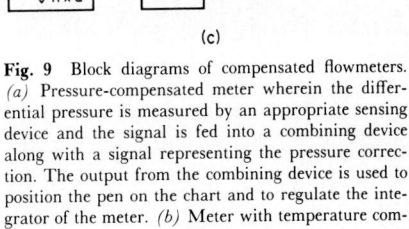

(c)

Fig. 9 Block diagrams of compensated flowmeters. *(a)* Pressure-compensated meter wherein the differential pressure is measured by an appropriate sensing device and the signal is fed into a combining device along with a signal representing the pressure correction. The output from the combining device is used to position the pen on the chart and to regulate the integrator of the meter. *(b)* Meter with temperature compensation. *(c)* Combined pressure-temperature compensation.

*Any object moving above the earth with constant space velocity is deflected relative to the surface of the rotating earth. This deflection was first discussed by the French scientist Coriolis about the middle of the last century and is now usually described in terms of the Coriolis acceleration or the Coriolis force. The deflection is found to be to the right in the Northern Hemisphere and to the left in the Southern Hemisphere. The Coriolis effects must be considered in a great variety of phenomena in which motion over the surface of the earth is involved. These include: (1) Rivers in the Northern Hemisphere should scour their right bank more severely than their left, and the effect should be more evident for rivers at high latitudes. Studies on the banks of the Mississippi and Yukon rivers indicate the predicted results. (2) The motions of air over the earth are governed to an appreciable extent by the Coriolis force. (3) A term, due to the Coriolis effect, must be included in the equation for exterior ballistics. (4) Any level bubble being carried on a ship or plane will be deflected from its normal position, and the deflection will be perpendicular to the direction of motion of the ship or plane. The correction for this effect may amount to several miles in the determination of a ship's position by methods of celestial navigation if a bubble octant is used in making observations.

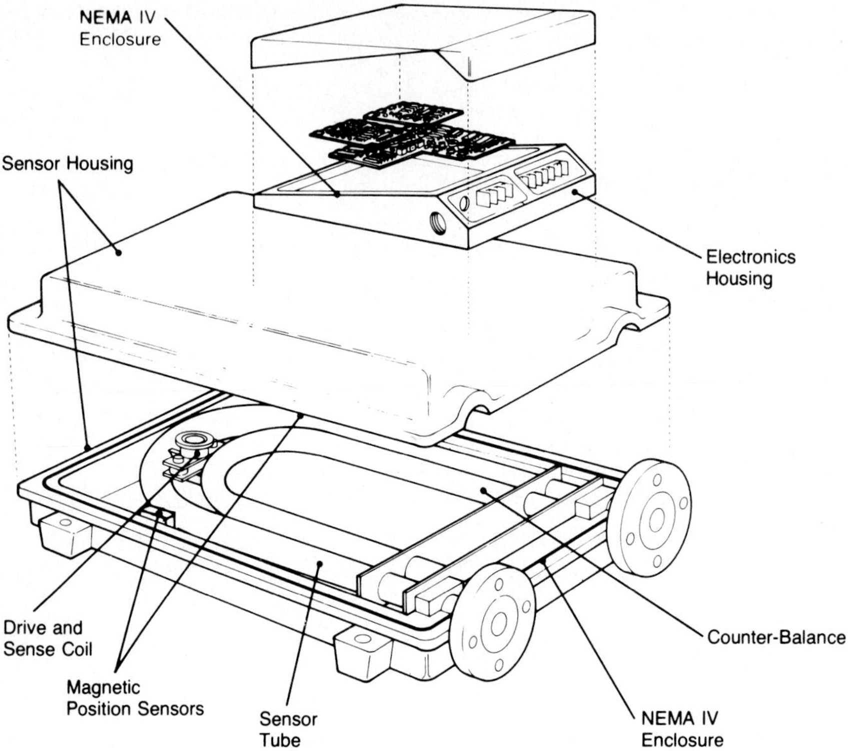

NEMA IV
Enclosure

Sensor Housing

Electronics
Housing

Drive and
Sense Coil

Counter-Balance

Magnetic
Position Sensors

Sensor
Tube

NEMA IV
Enclosure

Fig. 10 Principal elements of the Micro Motion Model C mass flowmeter. *(Micro Motion.)*

Meters have been designed to measure the flow rate of liquids, mixtures, foams, and slurries. In addition to measuring fluids that may be handled by conventional flowmeters, the mass flow unit also measures numerous difficult-to-handle substances, such as peanut butter, oil-sand mixtures, thick, sticky viscous fluids, and liquids containing gases. Since only the mass in motion is measured, material buildup in the sensor tube does not affect the calibration of the meter. Designed for easy cleaning, this mass flowmeter has been well received in the food processing field. Provided there is sufficient mass velocity, gases also can be measured by the unit.

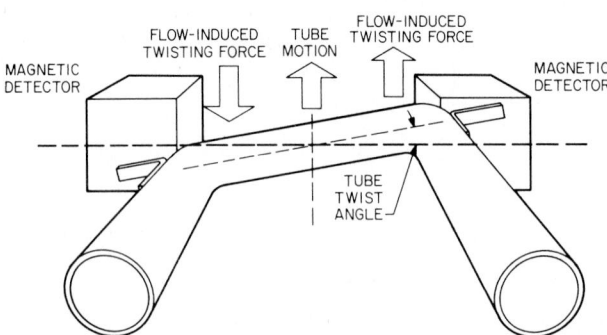

Fig. 11 Tube configuration showing the twist angle and manner in which magnetic detectors sense tube motion. *(Micro Motion.)*

Fig. 12 Fully assembled mass flowmeters. *(left)* Large configuration with flanged connections, *(right)* smaller meter with threaded connections. *(Micro Motion.)*

The mass flowmeter is currently available in seven models, ranging from 0 to 2.2 lb/min (0 to 1.0 kg/min) for a meter with ¼-in FNPT process connections 0 to 3200 lb/min (0 to 1450 kg/min) for a large model with 2-in flange connections. The maximum line pressure ranges from 2000 psi (138 bars) for the small meter to 750 psi (104 bars) for the largest model.

The rated accuracy for the meter is ±0.4% of rate, but long-term accuracy on the order of ±0.2% has been reported by some users. The fluid temperature range is −67 to +257°F (−55 to +125°C), and the ambient temperature range is −40 to +185°F (−40 to +85°C). The meter's output is linear from zero flow to a maximum rate limited only by the acceptable system pressure drop. The frequency output (digital), always proportional to the mass flow rate, is infinitely adjustable between 0 to 3 Hz and 0 to 15,000 Hz. The output can be scaled to any engineering unit. There is also an analog output available that provides 4 to 20 mA as well as other commonly used analog outputs. Two basic circuit boards cover the full range of either analog or frequency outputs. Calibration is performed at the factory and holds for any fluid. Verification or recalibration is easily accomplished in the field. Optional features include remote electronics, an expanded fluid temperature range of −400 to +400°F (−240 to +200°C), lined sensor tubes for corrosion-resistant operation, and UL-approved meters for hazardous area applications.

Radiation-Type Mass Flowmeter

In one system (Kay-Ray), gamma-ray-based density measurement is combined with magnetic flow measurement with a computer to yield a measurement of mass flow. See Fig. 13. This combination provides an obstructionless flow and noncontacting mass measurement in piping systems ranging from 2 in (∼5 cm) to 36 in (91.4 cm) in diameter. The flow sensor is operable up to 6500°F (3593°C). Features include pushbutton calibration, ability to tie in with supervisory computer communication, and a wide variety of materials of construction.

Angular Momentum Meters

In this approach, angular momentum is imparted to the stream of gas or liquid being measured. This is accomplished by using an impeller turning at constant speed. The angular momentum of the stream of fluid is removed by a turbine located just downstream of the impeller. The turbine is restrained (cannot rotate). The reaction torque produced by the turbine as it removes the angular momentum of the fluid becomes the meter output. The torque, under proper calibrated conditions, is directly proportional to the mass flow rate. In one design, the flow rate is totalized by applying

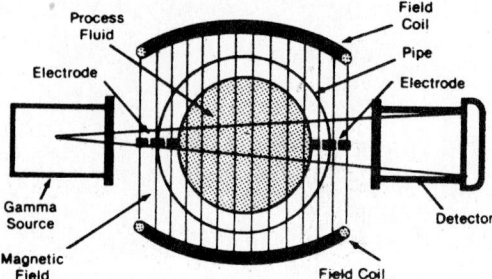

Fig. 13 Mass flowmeter that combines gamma-based density measurement with magnetic flow measurement and a computer. *(Kay-Ray.)*

the torque produced to the minor axis of a gyroscope. Meters of this type sometimes are referred to as axial flowmeters.

Gyroscopic Flowmeters

In this design, the flowmeter resembles a gyroscope, and its operation is most readily explained in gyroscopic terms. In Fig. 14, the *C* axis is equivalent to a gyroscope's spin axis, fluid motion in the pipe section perpendicular to the *C* axis replacing the spinning wheel. The entire pipe assembly is rotated about the *A* axis, which corresponds to the precession axis of a gyro. The mass flow rate produces a corresponding torque about the *B* axis. The pipe loop parallel to the *C* axis corrects for centrifugal force produced when deflection occurs about the torque axis. Flexure pivots connect the pipe elements to the sensing element. Sensing-element deflection, proportional to torque or mass flow rate, can be picked up by various forms of transducers. Meters of this type tend to be bulky.

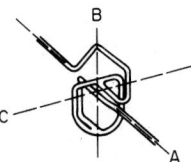

Fig. 14 Pipe configuration used as a measuring element in a gyroscopic mass flowmeter.

A mass flow measurement system designed especially for engine test installations is shown in the article on digital technology in Sec. 14 of this Handbook.

Ultrasonic Flowmeters

Ultrasonic or acoustic flowmeters are of two principal types: (1) *Doppler-effect meters* and (2) *transit-time meters*. In both types, the flow rate is deduced from the effect of the flowing process stream on sound waves introduced into the process stream. In *clamp-on* designs, these meters make it possible to measure the flow rate without intruding into the stream and thus are classified as *noninvasive*. But even in configurations where transducers are contained in shallow wells, the flowmeters are essentially nonintrusive.

The principles of ultrasonic flow measurement have been known for many years, but only within the last decade have these meters made measurable penetration of the flowmeter field. This lag in acceptance has been variously explained, but the general consensus is that too many designs were introduced too soon—prematurely and without testing them against adverse plant environments—to the point that for several years ultrasonic flowmeters had somewhat of a tarnished image. Ultrasonic flowmeters have numerous innate advantages over most of the traditional metering methods—linearity, wide rangeability without an induced pressure drop or disturbance to the stream, achievable accuracy comparable to that of orifice or venturi meters, bidirectionality, ready attachment to the outside of existing pipes without shutdown, comparable if not overall lower costs—attractive features which can make a product prone to overselling. Within the last few years, ultrasonic flow measurement technology has found a firmer foundation, and most experts now agree that these meters will occupy a prominent place in the future of flowmetering.

Doppler-Effect Meters

In 1842, Christian Doppler predicted that the frequencies of received waves were dependent on the motion of the source or observer relative to the propagating medium. His predictions were promptly checked for sound waves by placing the source or observer on one of the newly developed railroad trains. Over a century later, the concept was first considered for application in the measurement of flowing streams.

For the principle to work in a flowmeter, it is mandatory that the flowing stream contain sonically reflective materials, such as solid particles or entrained air bubbles. Without these reflectors, the Doppler system will not operate. In contrast, the transit-time ultrasonic flowmeter does not depend on the presence of reflectors.

The basic equations of a Doppler flowmeter are

$$\Delta f = 2f_T \sin \theta \, \frac{V_F}{V_S} \tag{1}$$

and, by Snell's law

$$\frac{\sin \theta_T}{V_T} = \frac{\sin \theta}{V_S} \tag{2}$$

Simultaneous solving of Eqs. 1 and 2 gives

$$V_F = \frac{\Delta f}{f_T} \, \frac{V_T}{\sin \theta_T} = K \, \Delta f \tag{3}$$

where V_T = sonic velocity of transmitter material
$\quad \theta_T$ = angle of transmitter sonic beam
$\quad K$ = calibration factor
$\quad V_F$ = flow velocity
$\quad \Delta F$ = Doppler frequency change
$\quad V_S$ = sonic velocity of fluid
$\quad f_T$ = transmission velocity
$\quad \theta$ = angle of f_T entry in liquid

Doppler-effect flowmeters use a transmitter that projects a continuous ultrasonic beam at about 0.5 MHz through the pipewall into the flowing stream. Particles in the stream reflect the ultrasonic radiation which is detected by the receiver. The frequency of the radiation reaching the receiver is shifted in proportion to the stream velocity. The frequency difference is a measure of the flow rate. The configuration shown in Fig. 15 utilizes separated dual transducers mounted on opposite sides of the pipe. Other possible configurations are shown in Fig. 16. In essence the Doppler-effect meter measures the beat frequency of two signals. The *beat frequency* is the difference frequency obtained when two different frequencies (transmitted and reflected) are combined.

When the measured fluid contains a large concentration of particles and/or air bubbles, it is said to be *sonically opaque*. The more opaque the liquid, the greater the number of reflections that originate near the pipe wall, a situation exemplified by heavy slurries. It can be noted from the flow profile of Fig. 15 that these reflectors are in the low-flow-rate region. In contrast, the preponderance of particle reflectors will occur in the center of the pipe (where the flow rate is highest) when the fluid is less sonically opaque. Where there are relatively few reflective particles in a stream, there is a tendency for the ultrasonic beam to penetrate beyond the centerline of the pipe and to detect slow moving particles on the opposite side of the pipe. Because the sonic opacity of the fluid may be difficult to predict in advance, factory calibration is difficult.

It will be noted from Fig. 15 that the fluid velocity is greatest near the center of the pipe and lowest near the pipe wall. An average velocity occurs somewhere between these two extremes. Thus,

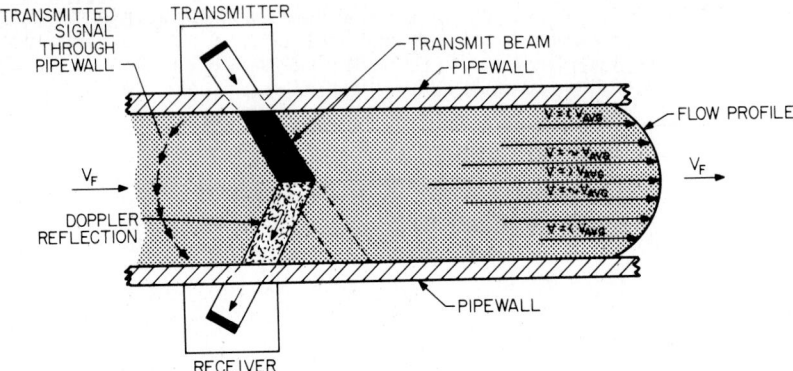

Fig. 15   Principle of the Doppler-effect ultrasonic flowmeter with separated opposite-side dual transducers.

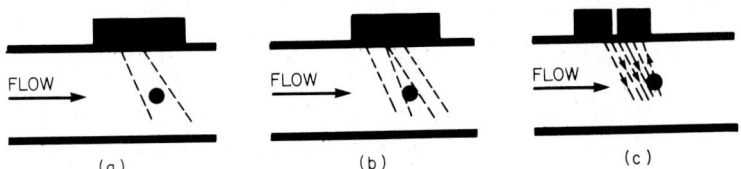

Fig. 16   Configurations of Doppler-effect ultrasonic flowmeters. *(a)* Single transducer, *(b)* tandem dual transducer, *(c)* separate dual transducers installed on same side of pipe. See also Fig. 17.

there are numerous variables, characteristic of a given fluid and of a specific piping situation, that affect the interactions between the ultrasonic energy and the flowing stream. Should a measured fluid have a relatively consistent flow profile and include an ideal concentration and distribution of particles, these qualities add to the fundamental precision of measurement and thus simplify calibration. Various designs are used to minimize interaction inconsistencies. For example, separation of transmitters and receivers makes it possible to restrict the zone in the pipe at which the principal concentration of interaction occurs. This zone usually occurs in the central profile region of the pipe less affected by variations in sonic opacity than near the pipe wall.

Transit-Time Flowmeters

With this type of meter, air bubbles and particles in the flowing stream are undesirable because their presence (as reflectors) interferes with the transmission and receipt of the ultrasonic radiation applied. However, the fluid must be a reasonable conductor of sonic energy. See Fig. 17. At a given

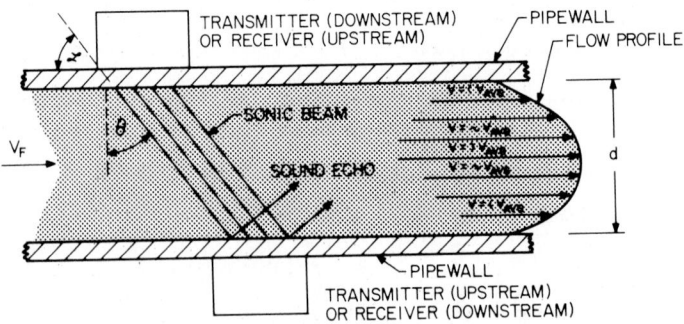

Fig. 17   Principle of the transit-time ultrasonic flowmeter. Clamp-on type is shown. Transducers alternately transmit and receive bursts of ultrasonic energy.

temperature and pressure, ultrasonic energy will travel at a specific velocity through a given liquid. Since the fluid is flowing at a certain velocity (to be measured), the sound will travel faster in the direction of flow and slower against the direction of flow. By measuring the differences in arrival time of pulses traveling in a downstream direction and pulses traveling in an upstream direction, this ΔT can serve as a measure of fluid velocity. Transit-time or ΔT flowmeters transmit alternately upstream and downstream and calculate this time difference. The operation is illustrated by the following equations.

$$V_F = \frac{(T_U - T_D)V_S}{\sin \theta} \frac{V_S \cos \theta}{d} = \frac{\Delta t V_S}{\sin \theta} \frac{1}{T_L} \tag{4}$$

By Snell's law

$$\frac{V_S}{\sin \theta} = \frac{V_C}{\sin \alpha} = K \tag{5}$$

$$V_F = \frac{K \Delta t}{T_L} \tag{6}$$

where T_U = upstream transit time
T_D = downstream transit time
T_L = zero flow transit time
V_S = liquid sonic velocity
d = pipe inside diameter
V_R = liquid flow velocity
α = angle between transducer and pipe wall
V_C = transducer sonic velocity

In the clamp-on transit-time flowmeter, where the transducers are strapped to the outside of the pipe, the sonic echo is away from the receiver and thus the device can retransmit sooner and operate faster. Clamp-on meters are installed on a normal section of piping, which is particularly attractive for retrofit applications. Other designs make use of what are called *wetted* transducers, which are mounted within the pipe. In both designs, of course, the tranducers are installed diagonally to the flow, i.e., *not* directly across from each other. Wetted transducers are usually installed in a shallow well, but because there are no projections beyond the pipeline wall, they are still considered nonintrusive. There is no significant disturbance to the general flow profile. However, slight, localized eddy currents may form in the vicinity of the wells. To avoid eddy currents, at least one manufacturer puts the transducers within the wells, forming what is termed an epoxy window. This results in completely filling the stream side of the well, making is essentially flush with the inner pipe wall.

In another design (dual-path ultrasonic flowmeter), two pairs of transducers are installed in the piping. The upstream and downstream propagation times between each pair of transducers are integrated in a microprocessor-based electronics package to determine flow rate.

Open Channels

Ultrasonic flowmeters are also applicable to open-channel flow measurement. The transducer is installed above the channel. Pulses emitted are reflected by the liquid surface back to the transducer. In a form of sonar, the time required for the pulse to return from the liquid surface is related to the height of the liquid surface. By knowing the cross-sectional properties of the open channel, height measurements can be converted to a measure of flow.

Numerous Variations

Presently, ultrasonic flowmeters are characterized by numerous configurations, adding to the difficulties of first determining whether an ultrasonic flowmeter is basically suited to an application and then selecting the best possible configuration. Careful checking with several suppliers is suggested by many experts. The decision favoring a Doppler effect or a transit-time ultrasonic flowmeter has been simplified by at least one manufacturer who offers an optional plug-in board which enables the unit to select between differential-time or Doppler techniques. A circuit (Foxboro Model 480) selects which technique should be used at any given instant. Thus, the instrument itself decides how to measure the flow in a given pipe—a decision based on actual conditions in the pipe. The Doppler sensing circuitry is calibrated by the differential-time circuitry. Thus, when the Doppler circuitry is being used, it will read the correct flow rate without a need for operator intervention.

Microphonic Sensor

In this device, which is useful only for detecting the presence or absence of flow, a microphone detects the shear noises (acoustic energy) emitted by the flowing material. This energy is converted to an electric signal and amplified over a wide band, usually 10 to 100 kHz or 10 to 35 kHz.

Thermal Flowmeters

In thermal flowmeters, heat in a measured quantity is added to the flowing stream upstream of the measurement. The cooling effect (heat dissipation) of the flowing stream becomes a measure of flow. Either the temperature rise of the stream at a point downstream or the amount of energy required to maintain a heated element at a constant temperature can be measured.

Radioisotope Tracer Flowmeters

The two principal techniques used for measuring flow rate utilizing radioisotopes are peak timing and dilution, as illustrated in Fig. 18.

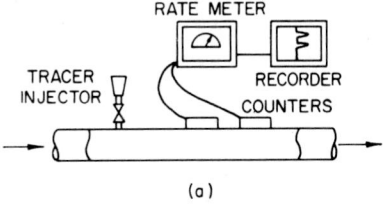

(a)

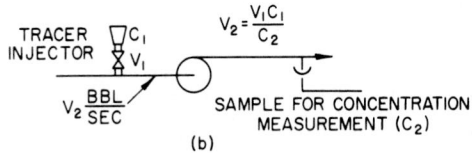

(b)

Fig. 18 Radioisotope tracer techniques for measuring flow. *(a)* Peak timing method, *(b)* dilution method.

Peak Timing

In this method, a gamma emitter such as ^{60}Co or ^{124}Sb is injected quickly at a point close to the section of the pipe in which the velocity is to be determined. The time of passage of the peak of the tracer wave is determined by using two detectors (counters or scintillation detectors), located a known distance apart and external to the pipe. The detectors usually are connected to a rate meter and a recorder. The time interval required for the fluid to travel between the two detectors is calculated from the distance on the recorder chart between the two radiation-level peaks; or it is determined by having the arrival of the activity at the first detector start a timer which is stopped by the arrival of the radioactive material at the second detector. The measured time interval is divided by the distance between the two observation points to calculate the linear flow rate. Although this is a seemingly complex approach to measuring flow, applications do arise where conventional approaches do not suffice.

Dilution Method

This technique is similar in principle to the thermal flowmeter and involves introducing a known concentration of radioisotope material at an upstream point and measuring the dilution at a point downstream. Thorough mixing between the two points is mandatory.

Laser Doppler Flowmeter

As shown in Fig. 19, fluid flow is determined by measuring the Doppler shift in laser radiation scattered from particles in the moving fluid stream. No sensor or transducer is necessary in the

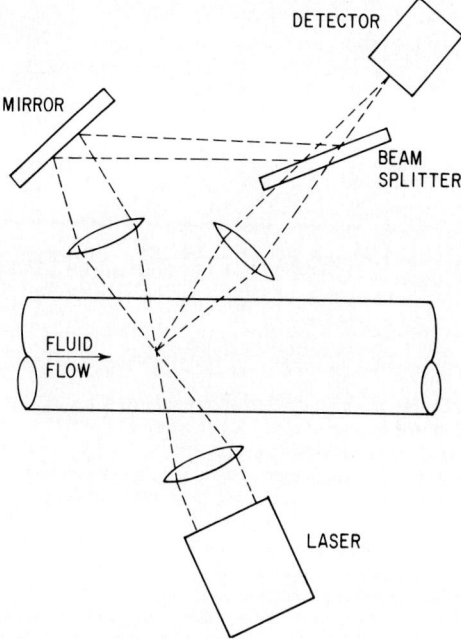

Fig. 19   Laser Doppler flowmeter.

moving fluid stream. The laser radiation focal point can be moved across the flow tube to measure velocity profiles. Fluid flows from 0.01 to 5000 in (0.25 mm to 127 m)/s have been measured. Contaminants, such as smoke, may have to be added to gases to provide scattering centers for the laser beam.

REFERENCES

Bailey, S. J. "Flowmeters," *Control Eng.*, vol. 25, no. 1, 1978, p. 49.

Bailey, S. J.: "Choosing a Flowmeter," *Control Eng.*, vol. 27, no. 4, 1980, p. 75.

Bailey, S. J.: "Fit Meter to Stream, Flow Loop for Top System Performance," *Control Eng.*, vol. 29, no. 6, 1982, p. 87.

Bailey, S. J.: "Computer Enhanced Metering Results in Better Flow Control," *Control Eng.*, vol. 30, no. 2, 1983, pp. 71–78.

Lynnworth, L. C.: "A Checklist of Ultrasonic Flowmeters," *Instrum. Technol.*, vol. 26, no. 11, 1979, p. 62.

Morris, H. M.: "What's Available in Ultrasonic Flowmeters," *Control Eng.*, vol. 26, no. 8, 1979, p. 41.

Morris, H. M.: "Ultrasonic Flowmeter Uses Wide Beam Technique to Measure Flow," *Control Eng.*, vol. 27, no. 7, 1980, p. 99.

Schmidt, T. R.: "What You Should Know About Clamp-on Ultrasonic Flowmeters," *Instrum. Technol.*, vol. 28, no. 5, 1981, p. 59.

Slomiana, M.: "Selecting Pressure and Velocity Head Primary Elements for Flow Measurement," *Instrum. Technol.*, vol. 26, no. 11, 1979, p. 40.

Staff: "Ultrasonics," *Instrum. Technol.*, vol. 24, 5, 1977, p. 11.

Staff: "Vortex Flowmeter for Steam Service," *Control Eng.*, vol. 26, no. 8, 1979, p. 9.

Thurston, C. W.: "Fundamental Changes in Flow Control Are Ahead," *Control Eng.*, vol. 27, no. 10, 1980, 69.

Waller, J. M.: "Guidelines for Applying Doppler Acoustic Flowmeters," *Instrum. Technol.*, vol. 27, no. 10, 1980, p. 55.

Postive-Displacement Flowmeters

Positive-displacement (PD) meters measure flow directly in quantity terms instead of indirectly or inferentially as rate meters do. In several rather ingenuous ways, the PD meter separates the flow to be measured into different discrete portions or volumes (not weight units) and, in essence, counts these discrete volumes to arrive at a summation of total flow. Fundamentally, PD meters do not have a time reference. Their adaptation to rate indication requires the addition of a time base. Similarly, by making appropriate corrections for changes in density resulting from pressure, temperature, and composition changes, PD meters can serve to indicate flow in terms of mass or weight.

Although inferential rate-type flowmeters appear most frequently in process control loops, departments of facilities concerned with materials accounting (receiving, shipping, inter- and intra-plant transfers, distribution, and marketing) largely depend on PD flowmeters. Also, in recent years, there has been growing acceptance of PD flowmeters for use in automatic liquid batching and blending systems. (See the article on fluid blending systems in Sec. 7 of this Handbook.) Utilities and their consumers are among the largest users of PD flowmeters, with millions of units in use for distributing and dispensing water, gas, gasoline, and other commodities. Where PD meters are used to dispense these commodities, they are subject to periodic testing and inspection by various governmental weights and measures agencies. Accuracy requirements vary with type of meter and service. Usually, water meters for domestic service and small industrial plants require an accuracy of only $\pm 2\%$, and gas meters $\pm 1\%$, whereas meters associated with retail gasoline and diesel fluid pumping have a tolerance (when new or after repair) of only 3.5 cm^3 in 5 gal (19.9 L). Routine tests permit a tolerance of 7 cm^3 in some states.

Nutating-Disk Meters

The nutating-disk meter (sometimes called a "wobble" meter), a cross section of which is shown in Fig. 1, is a very widely used meter and of particular importance in the measurement of commercial

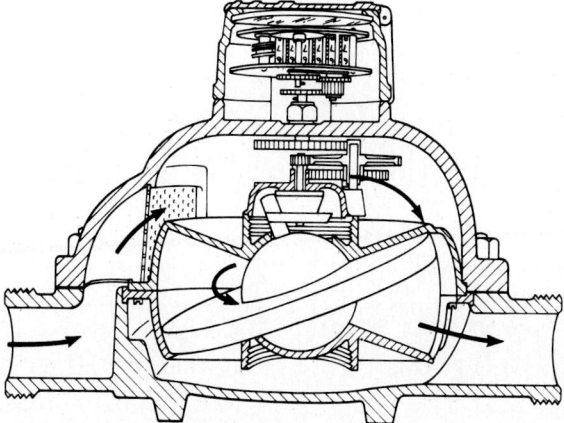

Fig. 1. Sectional view of a representative nutating-disk meter.

and domestic water. Although there are design differences from one proprietary make to the next, the fundamental operation of the nutating-disk meter is shown in Fig. 2. Each cycle (complete movement) of the measuring disk displaces a fixed volume of liquid. Note that there is only one moving part in the measuring chamber, namely, the disk. The liquid enters through the inlet port and fills the spaces above and below the disk, which fits closely and precisely in the measuring chamber. The advancing volume of liquid moves the disk in a nutating motion until the liquid discharges from the outlet port. The motion of the disk is controlled by a cam which keeps the lower face in contact with the bottom of the measuring chamber on one side, while the upper face of the disk is in contact with the top of the chamber on the opposite side. Thus, the measuring chamber is sealed off into separate compartments which are successively filled and emptied, each compartment holding a definite volume. The motion is smooth and continuous with no pulsations. The liquid being measured forms a seal between the disk and the chamber wall through capillary action, thus minimizing leakage or slippage and providing accuracy even at low flow rates.

In water metering installations, the meter commonly incorporates a built-in local reading dial and/or "speedometer-type" digital integrated readout. The motion of the nutating disk is transmitted

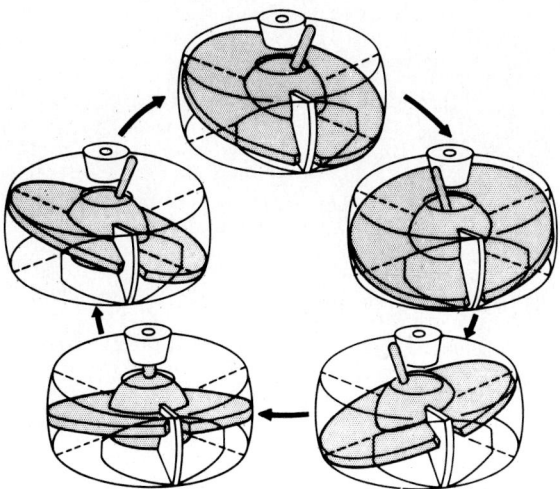

Fig. 2 Operating principle of a nutating-disk meter.

Table 1 Capacities of Nutating-Disk Meters*

Meter size, in (cm approx.)	Flow rate, gpm (α L/min, approx.)			
	Up to 180°F (82°C)		Above 180°F (82°C)	
	Maximum	Minimum	Maximum	Minimum
½ (1.3)	20 (76)	2 (7.6)	10 (38)	2 (7.6)
¾ (1.9)	30 (114)	3 (11.4)	15 (57)	3 (11.4)
1 (2.5)	50 (189)	5 (18.9)	25 (95)	5 (18.9)
1½ (3.8)	100 (379)	10 (38)	50 (189)	10 (38)
2 (5.1)	160 (606)	16 (61)	80 (303)	16 (61)

*Maximum working pressure, 150 psi (1034 kPa).

mechanically to the readout by way of a gear train. A variety of accessories can be added to the fundamental mechanism, including hand setbacks which permit reset of zero for use in batching operations, one register which can be reset along with a totalizer which cannot be reset in the field without dismounting (for example, where sealing is desired for accounting purposes), and impulse contactors which can be used for electrical transmission of readings. Meter sizes, flow capacities, and working pressures are summarized in Table 1.

Industrial versions of these meters provide accurate measurements for low flow rates and are easy to install and maintain. They can handle a wide range of chemicals, including caustics, with the proper materials of construction. Depending on the materials, the temperature range is from −300 to +250°F (−150 to +120°C).

Oscillating-Piston Meters

In principle, the oscillating-piston meter is similar to the nutating-disk meter with the important difference that mechanical motion takes place in one plane only (no wobble). In one design* the measuring chamber consists of five basic parts: (1) top head, (2) bottom head, (3) cylinder, (4) division plate, and (5) piston. The only moving part in the measuring chamber is the piston which oscillates smoothly in a circular motion between the two plane surfaces of the top and bottom heads. The division plate separates the inlet ports A and the outlet ports B. The piston is slotted to clear the division plate, which also guides the travel of the piston in its oscillating motion. A gear train transmits the piston motion to the register. The major components and operation of the meter are shown in Fig. 3.

The piston has a significantly smaller circumference than the chamber. This provides for maximum liquid area displacement for each oscillation. A small differential pressure across the meter produces motion of the piston within the measuring chamber. In order to obtain oscillating motion, two restrictions are placed on the movement of the piston. First, the piston is slotted vertically to match the size of a partition plate fixed to the chamber. This plate prevents the piston from spinning around its central axis and also acts as a seal between the inlet and outlet ports of the chamber. Second, the piston has a center vertical pin which is confined to a circular track that is part of the chamber. Differential pressure across the meter causes the piston to sweep the chamber wall in the direction of flow. This oscillating motion displaces liquid from the inlet to the outlet port in a continuous stream.

To further prevent unmeasured liquid from passing through the chamber, the piston has a horizontal partition or web. This web is perforated to promote balanced buoyancy of the piston within the chamber and a linear flow pattern of liquid through the meter. A drive bar and shaft are positioned through the top of the chamber so that, as the piston oscillates, the piston pin drives the bar and shaft in a circular or spinning motion. This rotating shaft is the driving link between the piston and the register or readout unit.

*Brooks Instrument Division, Emerson Electric Company, Hatfield, Pa.

TOP HEAD PISTON BOTTOM HEAD

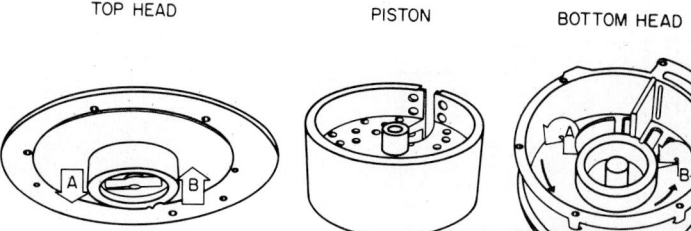

WITH CYLINDER AND
DIVISION PLATE

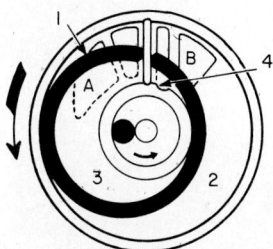

DIAGRAM I
SPACES I AND 3 ARE RECEIVING
LIQUID FROM THE INLET PORT, A,
AND SPACES 2 AND 4 ARE DIS-
CHARGING THROUGH THE OUT-
LET PORT B.

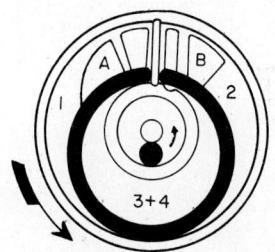

DIAGRAM 2
THE PISTON HAS ADVANCED AND
SPACE I, IN CONNECTION WITH
THE INLET PORT, HAS ENLARGED,
AND SPACE 2, IN CONNECTION
WITH THE OUTLET PORT, HAS
DECREASED, WHILE SPACES 3
AND 4, WHICH HAVE COMBINED,
ARE ABOUT TO MOVE INTO PO-
SITION TO DISCHARGE THROUGH
THE OUTLET PORT.

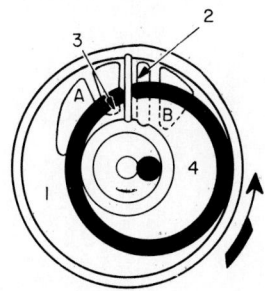

DIAGRAM 3
SPACE I IS STILL ADMITTING
LIQUID FROM THE INLET PORT
AND SPACE 3 IS JUST OPENING
UP AGAIN TO THE INLET PORT,
WHILE SPACES 2 AND 4 ARE DIS-
CHARGING THROUGH THE OUT-
LET PORT.

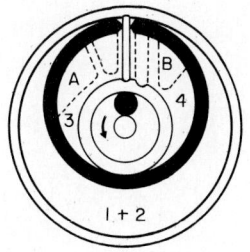

DIAGRAM 4
LIQUID IS BEING RECEIVED INTO
SPACE 3 AND DISCHARGED FROM
SPACE 4, WHILE SPACES I AND 2
HAVE COMBINED AND ARE ABOUT
TO BEGIN DISCHARGING AS PIS-
TON MOVES FORWARD AGAIN TO
OCCUPY POSITION AS SHOWN IN
DIAGRAM I.

Fig. 3 Major components and principle of operation of an oscillating-piston meter. *(Brooks Instrument, Division of Emerson Electric.)*

Table 2 Characteristics of Oscillating-Piston Meters*

Case material	Connections	Chamber material	Piston material	Max. safe working pressure	Max. safe working temperature
Cast iron	Screwed connections (½, ¾, 1 in)	Bronze	Bronze	150 psi (1034 kPa)	200°F (93°C)
			Aluminum	150 psi (1034 kPa)	200°F (93°C)
	Flanged connections 125 b F.F. (1½, 2 in)	Bronze	Bronze	125 psi (862 kPa)	200°F (93°C)
			Aluminum	125 psi (862 kPa)	200°F (93°C)
Cast bronze	Screwed connections (½, ¾, 1 in)	Bronze	Bronze	150 psi (1034 kPa)	200°F (93°C)
			Aluminum	150 psi (1034 kPa)	200°F (93°C)
	Flanged connections, 125 lb F.F. (1½, 2 in)	Bronze	Bronze	125 psi (862 kPa)	200°F (93°C)
			Aluminum	125 psi (862 kPa)	200°F (93°C)
Cast steel	Screwed connections ½, ¾, 1 in)	Bronze	Bronze	400 psi (2758 kPa)	300°F (149°C)
			Aluminum	400 psi (2758 kPa)	250°F (121°C)
		Ni-Resist	Ni-Resist	400 psi (2758 kPa)	300°F (149°C)
	Flanged connections (1½ or 2 in, 150 lb, ANSI R.F.)	Bronze	Bronze	275 psi (1896 kPa)	300°F (149°C)
			Aluminum	275 psi (1896 kPa)	250°F (121°C)
		Ni-Resist	Ni-Resist	275 psi (1896 kPa)	300°F (149°C)
	Flanged connections (1½ or 2 in, 300 lb ANSI R.F.)	Bronze	Bronze	400 psi (2758 kPa)	300°F (149°C)
			Aluminum	400 psi (2758 kPa)	250°F (121°C)
		Ni-Resist	Ni-Resist	400 psi (2758 kPa)	300°F (149°C)

As furnished by Brooks Instrument Division, Emerson Electric Company.
NOTE: When an all-aluminum piston is used, the maximum recommended temperature is 250°F (121°C). When a bronze piston is used on hot-water service over 180°F (82°C), the maximum rate of flow should be reduced by one-half the values shown. Minimum rates remain the same. For approximate metric equivalents of dimensions shown for pipe connections and meter sizes, see Table 1.

The principal specifications of this type of meter are summarized in Table 2. Because of the variety of materials available for casing, chamber, and piston, this meter can handle numerous chemicals and industrial liquids. In addition to the materials listed in Table 2, stainless steel meters are also available.

A variety of registers are available for use on the meters—with various calibrations in both English and metric units. See Fig. 4. For continuous or batching operations, all totalizers repeat after completion of a numerical cycle.

In addition to registers, a number of accessories are available with these meters, including ticket printers and interfaces for control operations. For example, an impulse contactor can actuate electrically controlled injection devices at regular intervals proportional to the volume of meter throughput. This contactor opens or closes an electric circuit at desired intervals proportional to the units measured by the meter, such as one impulse for every gallon or every tenth gallon. The contactor also can be used to actuate a variety of other electrical devices, such as electromagnetic counters, predetermining counters, pumps, solenoid valves, alarms, sampling devices, and totalizers, among others. A cam-actuated switch can provide electrical shutoff of auxiliary equipment after preset liquid volumes have been delivered. The meter is also available with a pneumatic pulser for interfacing with a pneumatic batch controller.

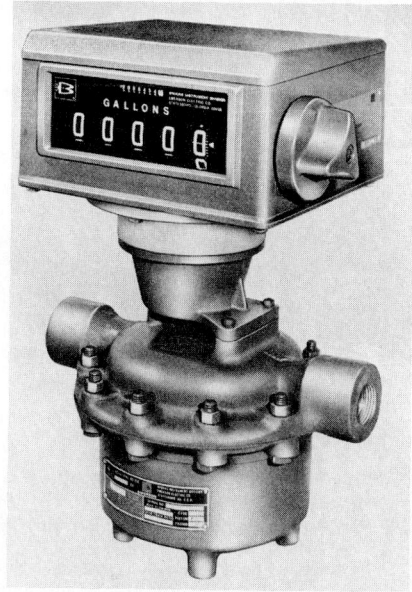

Fig. 4   A 1-in NPT bronze oscillating-piston flowmeter with a large dial register. *(ER-113, Brooks Instrument, Division of Emerson Electric.)*

There is a variety of designs of PD flowmeters based on the piston principle. One unit* consists of a self-porting, four-piston flowmeter. Four radial pistons drive a vertical crankshaft by connecting rods. A cylindrical permanent magnet is affixed to the crankshaft, which in turn drives a hollow cylindrical magnet attached to the drive shaft of the transmitter, thus providing isolation of the fluid and electronic portions of the meter, eliminating the need for shaft sealing devices. The capacity of the meters ranges from 1 cm^3/min to 80 L/min. The output is linear, and an accuracy of $\pm 0.5\%$ over a 500:1 to 1000:1 turndown ratio is claimed. The meters are machined from solid blocks of Type 303 stainless steel. Radial clearances between piston and cylinder bore are held to 0.002 to 0.003 in (5 to 8 μm). Such clearances limit the maximum size of particulate contamination of the measured fluid and thus require care in selecting the proper filtration system. It is claimed that these meters are particularly useful where the viscosity of the measured fluid varies over a wide range, or where a high turndown ratio would normally require the use of two or even three conventional meters.

Fluted Rotor Meters

Meters of this type are widely used in the flow measurement of crude and refined petroleum products and a variety of other commercial liquids. Meters of this type are frequently used on product loading racks and small pipelines. Ticket printers, available with most designs, provide authentically recorded documents for liquid transfer transactions.

In one design†, two spiral fluted rotors within the measuring chamber are dynamically balanced,

*Fluidyne Instrumentation, Oakland, Calif.

†Brooks Bi Rotor Meter.

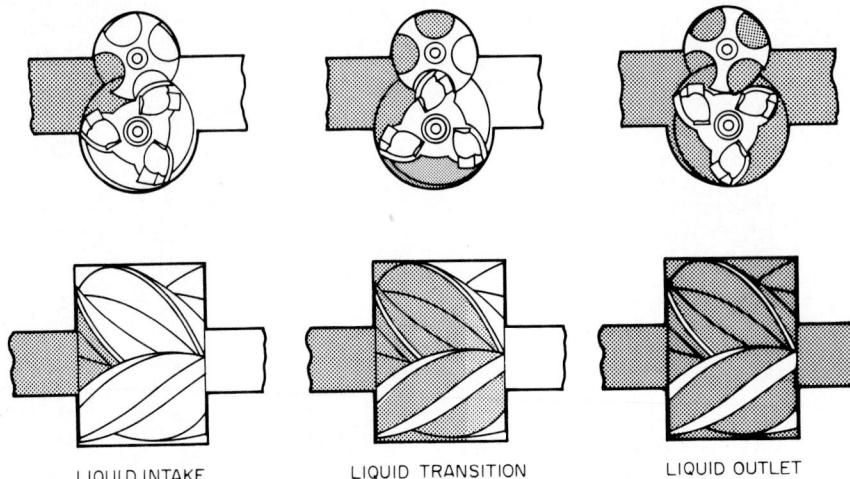

LIQUID INTAKE   LIQUID TRANSITION   LIQUID OUTLET

Fig. 5 Operating principle of a Brooks BiRotor positive-displacement flowmeter. *(Brooks Instrument, Division of Emerson Electric.)*

but hydraulically unbalanced. With reference to Fig. 5, as the product enters the intake of the measuring unit chamber, the two rotors divide the product into precise segments of volume momentarily and then return these segments to the outlet of the measuring unit chamber. During what may be referred to as a "liquid transition," the rotation of the two rotors is directly proportional to the liquid throughput. A gear train, located outside the measuring unit chamber, conveys mechanical rotation of the rotors to a mechanical or electronic register for totalization of liquid throughput. See also Fig. 6.

Fig. 6 Axial flow BiRotor positive-displacement flowmeter with a large dial register and printer. *(Brooks Instrument, Division of Emerson Electric.)*

Meter sizes range from 2 to 16 in ($\sim$5 to 41 cm). The rotors are nickel-cast iron or heat-treated aluminum; rotor shafts are of ground and polished nitralloy; rotor bearings are stainless steel; body and end covers are of nickel-cast iron. The operating pressure, depending on the particular model, ranges from 275 to 1440 psi (1896 to 9928 kPa), and the operating temperature, from -20 to 150°F (-29 to 66°C). Capacities range from 100 to 1000 U.S. gal/min, which is equivalent to 83 to 833 imperial gal/min, 143 to 1430 bbl/h, or 378 to 3785 L/min, to much higher capacities of 875 to 8750 gpm (729 to 7290 imperial gallons; 1250 to 12,500 bbl/h; 199 to 1986 m³/h).

These meters require lubrication. As shown in Fig. 7, the principal components of the lubricating

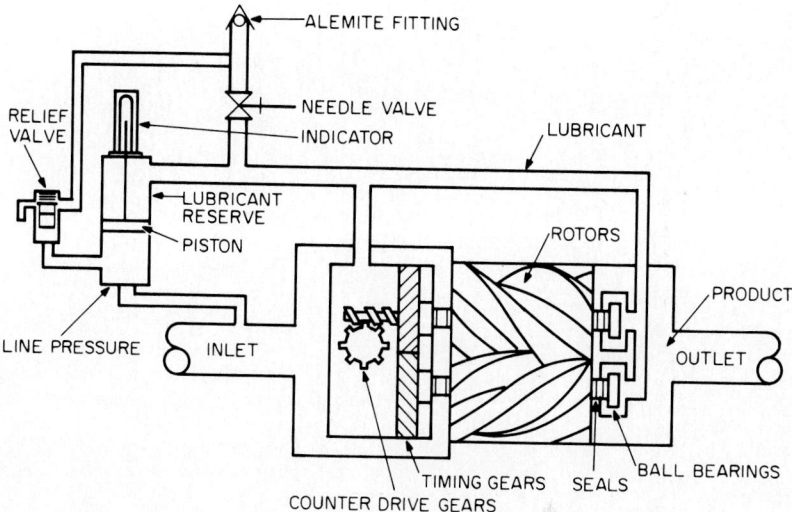

Fig. 7 Schematic diagram of lubricating system used with a BiRotor positive-displacement flowmeter. *(Brooks Instrument, Division of Emerson Electric.)*

system are a hydraulic cylinder and piston with a differential-pressure relief valve, a needle valve and lubricant fitting, and seals for isolating the lubricant from the flow. The hydraulic cylinder, in addition to serving as a lubricant reservoir and segregating the flowing stream from the lubricant, ensures that the lubricant pressure is at all times equal to or greater than the internal meter pressure. This positive pressure is maintained by the application of line pressure to the side of the piston opposite the lubricant supply. The external pressure relief valve protects the meter from excessive pressure when recharging the system with oil. This valve will open, releasing some lubricant to the atmosphere, should lubricant pressure exceed pipeline pressure by 30 psi (207 kPa). The mechanical isolating seals, located between the rotor and rotor bearing, effectively segregate the lubricant from the flowing stream. A hand pump may be attached to the lubricant fitting for recharging the system with a lightweight oil, even if it is under pressure. The differential relief valve in the hydraulic cylinder prevents overpressurization during this operation.

In another design,* a pair of radially pitched helical measuring elements is used (Fig. 8). The two nested helixes trap a relatively large pocket of liquid which progresses from the inlet to the outlet of the meter in a fashion similar to that used in progressive cavity pumps. The performance of the meters is essentially independent of viscosity. The large dimensions of the progressive cavity permit the passage of gels, undissolved solids, fines, and other agglomerates which normally would cause

*Fluidyne Instrumentation, Oakland, Calif.

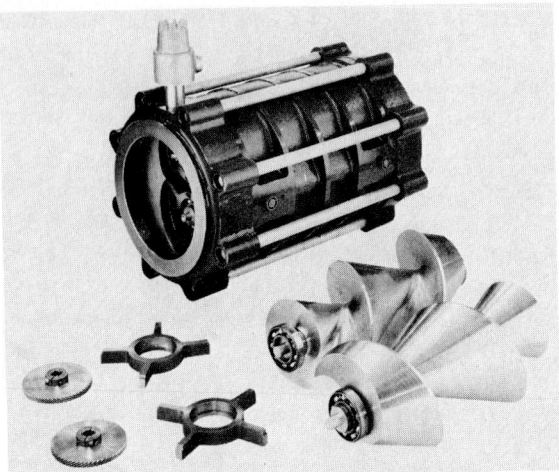

Fig. 8 Disassembled 10 in (~25.4 cm) helix-type flow transducer using radially pitched helical measuring elements. *(Helix 235/240, Fluidyne Instrumentation.)*

blockage in some designs. Machining tolerances of the cylinder bore and helical rotors are tight. An overall accuracy (uncertainty) of ±0.5% over a 150:1 flow range (turndown ratio) with a repeatability of 0.2 to 0.3% is claimed. The meter design can handle flow rate capacities from less than 0.5 to over 4000 gpm (1.9 L to ~151 hL) in sizes from 1½ to 10 in (~3.8 to 25.5 cm). By using heavy, raised-face flanges, operating pressures as high 3500 psi (24 MPa) can be achieved. Although higher temperatures can be obtained as an option, the normal operating temperature is up to 450°F (250°C).

Rotating Meters

Several types of rotating PD meters are offered, including lobed-impeller, sliding-vane, retracting-vane, and oval-shaped gear meters.

Lobed-Impeller Flowmeter

In the meter illustrated in Fig. 9, two rotors revolve with a fixed relative position inside a cylindrical housing. The measuring chamber is formed by the wall of the cylinder and the surface of one-half of one rotor. When the rotor is in a vertical position, a definite volume of fluid is contained in the measuring compartment. As the impeller turns, owing to the slight pressure differential between inlet and outlet ports, the measured volume is discharged through the bottom of the meter. This action takes place 4 times for a complete revolution, the impeller rotating in opposite directions and at a speed proportional to the volume of fluid flow. These meters are available in pipe sizes from 1½ to 24 in (~3.8 to 61 cm) for handling from 8 gpm (~30 L/min) to 25,000 bbl/h with an accuracy of about ±0.20% and a repeatability of ±0.015%. Advantages include performance at relatively high temperatures (400°F, 204°C) and pressures (1200 psi, 8276 kPa); low pressure loss; availability of numerous materials of construction, applicable to gases and a wide range of light to viscous fluids, including asphalts; and good accuracy, particularly at high rates. Limitations include susceptibility to damage from entrained vapors, a tendency to be bulky and heavy in larger sizes, relatively high cost, and slip at low flow rates; hence they are best used at high rates.

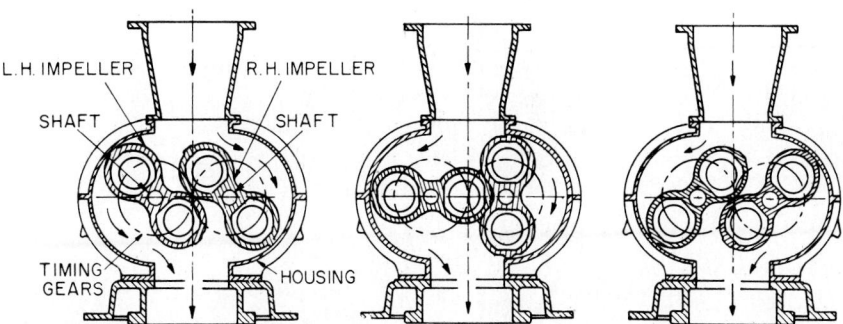

Fig. 9 Lobed-impeller flowmeter.

Sliding-Vane and Retracting-Vane Rotary Flowmeters

As shown in Fig. 10 vanes are moved radially as cam followers to form the measuring chamber or are articulated in another version, the retracting-vane type, shown in Fig. 11. The sliding-vane type

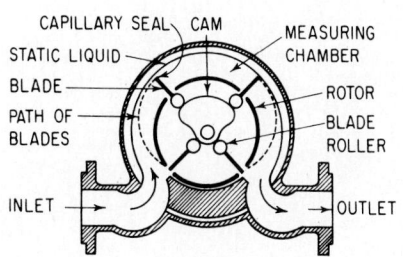

Fig. 10 Sliding-vane rotary meter.

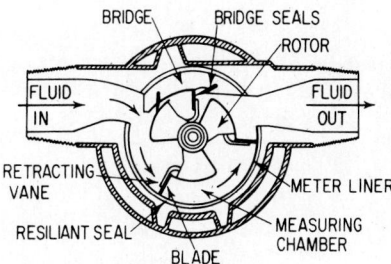

Fig. 11 Retracting-vane rotary meter.

contains a cylindrical rotor that revolves on ball bearings around a central shaft and stationary cam. As fluid flows against an extended blade, the resulting rotation of the rotor and the action of the cam cause the blades to act as cam followers, creating measuring chambers that measure fluid throughput. Capillary action of the metered fluid seals the blades to form measuring cavities. These meters are quite accurate and are available up to 16 in (41 cm). Liquids should be clean and below 400°F (204°C) and 900 psi (6207 kPa). In the retracting-vane rotary meter, fluid entering the meter is deflected downward against the extended blade, causing rotation of the measuring element. Sealing of the metering chamber is accomplished by incorporating features of the piston-ring seal. The vanes or blades, coated with a resilient material, are spring-loaded to provide the wiping action of a piston ring. Unlike the cam action of the sliding-vane design, the vanes of the retracting type are positioned not only by the configuration of the meter internals, but also by any foreign materials inside the meter. The meter is suitable for handling highly viscous (100,000 SSU) or slurry fluids. The meter is available in pipe sizes from 1 to 4 in (~2.5 to 10 cm) and can handle fluids up to 400°F (204°C) and 500 psi (3448 kPa).

Oval-Shaped Gear Flowmeters

In these meters, precision-matched, oval-shaped gears are used as metering elements. The principle of operation is illustrated in Figs. 12 and 13. A cutaway close-up, with one rotor removed from the meter housing to show the measuring cavity, appears in Fig. 14.

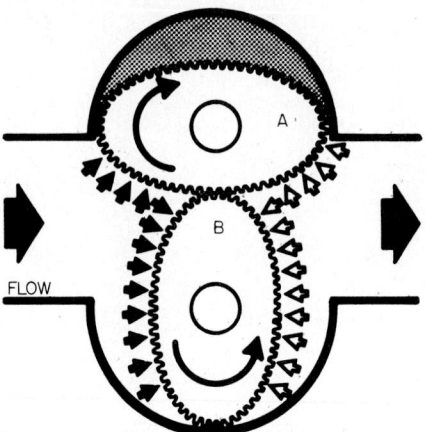

Fig. 12 Sectional schematic of oval-gear flowmeter, showing how a crescent-shaped gap captures the precise volume of liquid and carries it from inlet to outlet. *(Oval flowmeter, Brooks Instrument, Division of Emerson Electric.)*

Meter sizes (connections) range from ½ to 1, 1½, 2, 3, and 3 in (~1.3 to 3.8, 5.1, 7.6, and 10.2 cm), with capacity ranges for different fluids as follows:

Cold water	0.2–1.4 to 110–705 gpm (U.S.)
	0.8–5.3 to 416.4–2668.7 L/min
Hot water	0.3–1.0 to 132–484 gpm
	1.1–3.8 to 499.7–1832.1 L/min
Liquefied petroleum gas	0.4–1.6 to 176–837 gpm
(LPG)	1.5–6.1 to 666.2–3168.4 L/min
Gasoline (0.3–0.7 cP)	0.3–1.6 to 132–837 gpm
	1.1–6.1 to 499.7–3168.4 L/min
Kerosene (0.7–1.8 cP)	0.2–1.6 to 110–837 gpm
	0.8–6.1 to 416.4–3168.4 L/min
Light oil (2–4 cP)	0.1–1.9 to 705–1010 gpm
	0.4–7.2 to 2668.7–3826.3 L/min
Heavy oil (5–300 cP)	0.04–1.9 to 44–1010 gpm
	0.2–7.2 to 166.6–3826.3 L/min

In meter application engineering, three basic viscosity classifications are taken into consideration: (1) standard-viscosity class from 0.2 to 200 cP, (2) medium-viscosity class from 300 to 500 cP, and (3) high-viscosity class above 500 cP. Tests have shown that a meter calibrated on a 1-cP product and then applied to a 100-cP product does not shift more than 1.2% above the initial calibration. Normally, where the viscosity is 100 cP or greater, there is no significant shift in accuracy. Oval-gear PD meters are sized for maximum flow so that the pressure drop is less than 15 psi (103 kPa).

Normal operating pressure ranges from 255 to 710 psi (1758 to 4895 kPa), depending on whether steel or stainless steel flanges are used. The maximum operating temperature is 230°F (110°C), but special meters are available for higher temperatures. The lower limit is 0°F (−18°C). Housings, rotors, and shafts are constructed of Type 316 stainless steel or Type Alloy 20 (CN-7M stainless steel). Bushings are of hard carbon.

Meters are available with numerous accessories, including an electric impulse contactor, which is used to open and close an electric circuit at intervals proportional to the number of units measured.

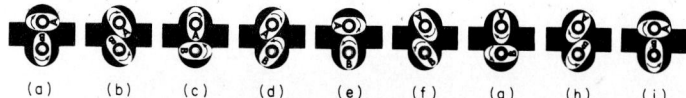

(a)　(b)　(c)　(d)　(e)　(f)　(g)　(h)　(i)

Fig. 13 Principle of operation of the Oval flowmeter. The meter measures liquid flow by using a slight pressure differential to rotate a pair of oval gears. The meshed gears seal the inlet from the outlet flow, developing the pressure differential. When in the position shown in *(a)*, gear *A* receives torque from the pressure difference, the torques on gear *B* cancel one another, and gear *A* drives gear *B* as shown in *(b)*. When gear *A* rotates to the position shown in *(c)*, it loses torque, but gear *B* obtains torque and drives gear *A*. This alternate driving action provides a smooth rotation of nearly constant torque without dead spots. As shown in *(d)* through *(i)*, the principle continues through a complete cycle, bringing gear *A* back to its original position as shown in *(a)*.

As the gears rotate, they trap precise quantities of liquid in the crescent-shaped gaps or measuring chambers. The total quantity of flow for one rotation of the pair of oval gears is 4 times that of the crescent-shaped gap. The rate of flow is proportional to the rotational speed of the gears. Because the amount of slippage between the oval gears and the measuring chamber wall is minimal, the meter is essentially unaffected by changes in viscosity and lubricity of liquids.

An output shaft is rotated in direct proportion to the oval gears by means of a magnetic coupling. The output shaft drives a gear train that provides meter registration in engineering units. *(Oval flowmeter, Brooks Instrument, Division of Emerson Electric.)*

Fig. 14 Gear-type positive-displacement flowmeter with one rotor removed from the housing to show the measuring cavity. *(Oval flowmeter, Brooks Instrument, Division of Emerson Electric.)*

It may actuate electromagnetic counters to indicate meter throughput; electromagnetic predetermining counters to start or stop pumps, to open or close solenoid valves after a predetermine volume has passed through the meter; electrically operated devices for injecting additives into the stream at regular intervals; or electronic counters, totalizers, recording printers, and sampling devices.

A solid-state pulse transmitter is available for providing signaling for remote indicating, totalizing, and data monitoring systems. A pneumatic pulser is available to provide a pulse signal for use with batch controllers. A mechanical predetermined batching system is shown in Fig. 15.

Bellows Gas Meter

This type of meter, widely used in both commercial and domestic gas service, comprises four measuring compartments which operate simultaneously. Some of the compartments are filling while oth-

Fig. 15 Oval positive-displacement flowmeter equipped for mechanical predetermined batching. The system comprises a two-stage preset meter register and a two-stage quantity control valve with mechanical linkage. In operation, the preset is returned to zero and the desired quantity is set. The quantity control valve is opened manually. As the desired volume is passing through the meter, the preset counts down toward zero. When the first-stage trip point is reached, the valve actuating ring (part of the preset) rotates to partially close the valve (low-flow condition). When the final stage (zero) is reached, the valve actuating ring continues its movement to allow the valve to fully close. An electrical version of the system is also available. *(Brooks Instrument, Division of Emerson Electric.)*

ers are emptying, but all always conform to set conditions, thereby ensuring uniform delivery of gas. The unit shown in Fig. 16 is constructed so that the number of times each measuring chamber is filled and emptied is registered, thus indicating the total volume in cubic feet on the index. The register or meter is operated from a crank that is rotated by movement of the diaphragms. The mechanism generally used is referred to as a standard Glover-type mechanism. Synthetic rubber diaphragms are used to ensure that displacement is directly proportional to stroke. Two valves of the D-slide type are actuated by a central crank through a suitable linkage. Motion of the meter mechanism cannot occur unless a pressure differential (hence gas flow) takes place. Normally, a pressure differential of 0.10 in (2.5 mm) of water will commence actuation of the meter. Maximum capacities range from 150 to 17,000 ft^3/h (4.2 to 481 m^3/h).

Fig. 16 Bellows-type gas meter. *(American Meter.)*

Meter Proving

Gas meters are tested by passing a known volume of gas at a known rate through them and comparing this volume indication with the standard. A prover is shown in Fig. 17. As the bell is lowered, it displaces a known volume in the container, which is forced through the meter under test. The rate at which the bell drops, hence the rate of flow through the meter, can be carefully controlled.

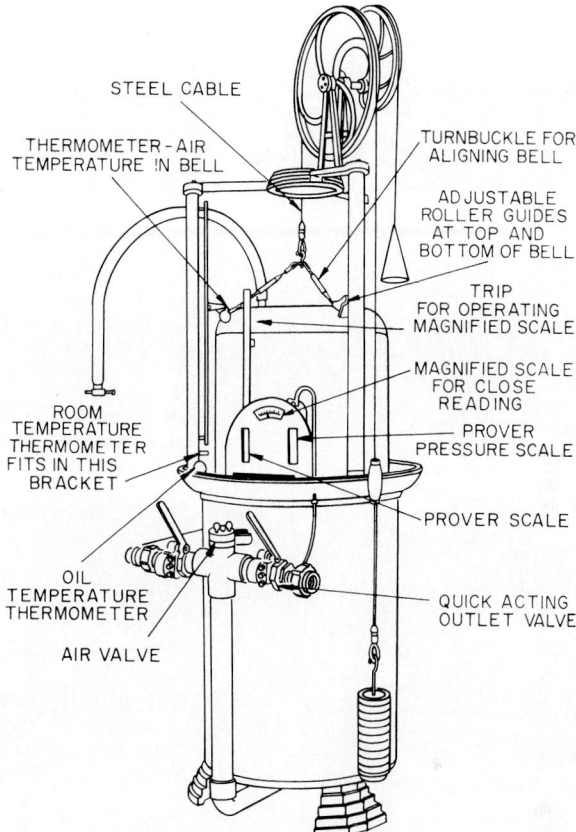

STEEL CABLE

THERMOMETER-AIR
TEMPERATURE IN BELL

TURNBUCKLE FOR
ALIGNING BELL

ADJUSTABLE
ROLLER GUIDES
AT TOP AND
BOTTOM OF BELL

TRIP
FOR OPERATING
MAGNIFIED SCALE

MAGNIFIED SCALE
FOR CLOSE
READING

ROOM
TEMPERATURE
THERMOMETER
FITS IN THIS
BRACKET

PROVER
PRESSURE SCALE

PROVER SCALE

OIL
TEMPERATURE
THERMOMETER

QUICK ACTING
OUTLET VALVE

AIR VALVE

Fig. 17 Bell-type prover for checking gas meters.

Open-Channel Flow Measurements

An open channel may be defined as any conduit in which a liquid, such as water, flows with a free surface. Immediately evident examples are rivers, canals, flumes, and other uncovered conduits. Certain closed conduits, such as sewers and tunnels when flowing partially full, not under pressure also may be classified as open channels. Two types of units are used in measuring open-channel flows: (1) units of discharge and (2) units of volume. *Discharge,* or *rate of flow,* is defined as the volume of liquid that passes a certain reference section in a unit of time. This may be expressed, for example, as cubic feet per second, gallons per minute, cubic meters per minute, or millions of gallons per day, among others. The unit of volume commonly used in irrigation work is the *acre-foot,* which is defined as the amount of water required to cover one acre to a depth of one foot (43,560 ft^3), or the *hectare-meter* (10,000 m^3).

To determine open-channel flows, a calibrated device must be inserted in the channel to relate the free surface level of liquid directly to the discharge. Primary devices used include weirs, flumes, and Kennison nozzles, among others. These devices fall into the general category of head-area meters and are used extensively in the measurement and allocation of irrigation water as well as primary devices for municipal sewage and industrial wastewater applications. With these devices, the discharge at any given instant can be directly related to the level. The relationship between discharge and level is termed *rating,* and this relationship is specific to a given primary device. Rating tables are available for all standard weirs and flumes, as, for example, the *Stevens Hydrographic Data Book* (see reference list).

Weirs

A weir consists of a partition or bulkhead of timber, concrete, sheet metal, or other fabrication having at its top edge an opening of fixed dimensions through which a stream can flow. The opening is called the weir notch; its bottom edge is the crest; the depth of water passing over the crest is the head (always measured at a definite point upstream from the weir); the sheet of water flowing through the notch and over the weir crest is called the nappe. When the weir has a sharp upstream edge so that the nappe springs clear of the crest, it is called a sharp crested weir. The nappe, immediately after leaving a sharp crested weir, suffers a contraction along the horizontal crest called crest contraction. If the sides of the notch have sharp upstream edges, the nappe also is contracted in width, and the weir is said to have end contractions. With sufficient side and bottom clearance dimensions of the notch, the nappe undergoes maximum crest and end contractions, and the weir is said to be fully contracted.

Of the various types of weirs, the three most commonly used are the V-notch, rectangular, and Cipolletti.

V-Notch Weir

This device is especially recommended for metering flows less than 1 ft^3/s (28, 320 cm^3/s) equivalent to 0.65 million gal/day (2500 m^3/day) and is suitable for measuring slowly changing flows up to 10 ft^3/s (0.28 m^3/s). Extensive experiments have been made to determine the calibration data for V-notch weirs with included angles of 60 and 90°; two acceptable formulas for calculations are given in Fig. 1.

Rectangular Weir

This weir is capable of high-capacity metering and is simple and inexpensive to construct. To ensure complete contraction of the nappe, the side and bottom clearance dimensions of the notch must equal or exceed those shown in Fig. 2.

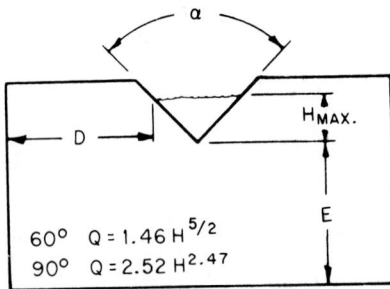

$$60° \quad Q = 1.46\,H^{5/2}$$
$$90° \quad Q = 2.52\,H^{2.47}$$

Fig. 1 V-notch weir. $D \geqq 2.5 \times$ maximum head (H_{max}); $E \geqq 2 \times$ maximum head (H_{max}); $Q =$ actual discharge, ft^3/s; $H =$ measured head, ft.

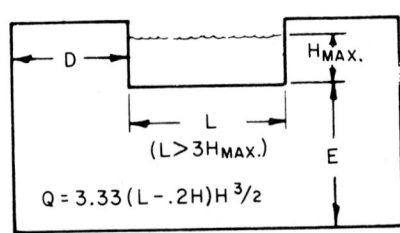

$$(L > 3H_{MAX}.)$$
$$Q = 3.33\,(L - .2H)H^{3/2}$$

Fig. 2 Rectangular weir. $D \geqq 2.5 \times$ maximum head (H_{max}); $E \geqq 2 \times$ maximum head (H_{max}); $Q =$ actual discharge, ft^3/s; $L =$ crest length, ft; $H =$ measured head, ft.

Cipolletti Weir

This weir, shown in Fig. 3, is similar to the rectangular weir except for sloping sides (1 horizontal to 4 vertical) of the notch. The design has the advantage of a simplified discharge formula which is more convenient to work with.

Other Weirs

Occasionally, hyperbolic (Sutro), broad crested, and round crested weirs are used. Check references.

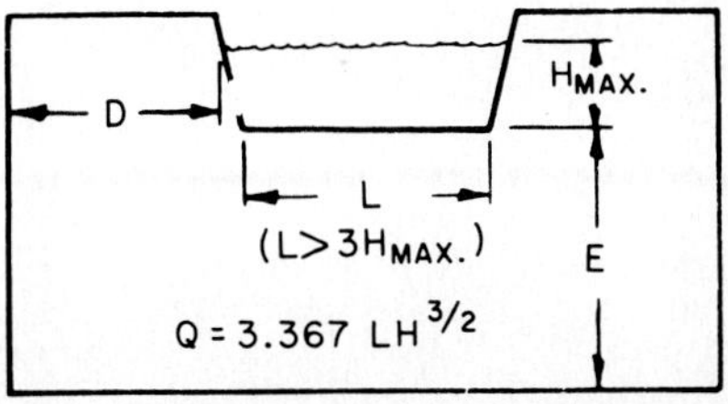

$$(L > 3H_{MAX.})$$

$$Q = 3.367 \, LH^{3/2}$$

Fig. 3 Cipolletti weir. $D \geqq 2.5 \times$ maximum head (H_{max}); $E \geqq 2 \times$ maximum head (H_{max}); $Q =$ actual discharge, ft³/ s; $L =$ crest length, ft; $H =$ measured head, ft.

Parshall Flumes

The Parshall venturi-type flume, shown in Fig. 4, consists of a converging upstream section, a downward sloping throat, and an upward sloping, diverging downstream section. It is usually constructed of reinforced concrete but may be of wood. Stainless steel and fiberglass-reinforced plastic liners have been used for metering corrosive solutions. Surfaces are true planes, finished smooth with close adherence to specified dimensions.

Sizes

Parshall flumes have been constructed in sizes with throat widths ranging from 1 in (2.5 cm) to 40 ft (12 m), for measuring flows up to 1500 million gal/day (5.7 million m³/day). The size of a flume is the width W of the throat. General economy of construction dictates that the smallest standard

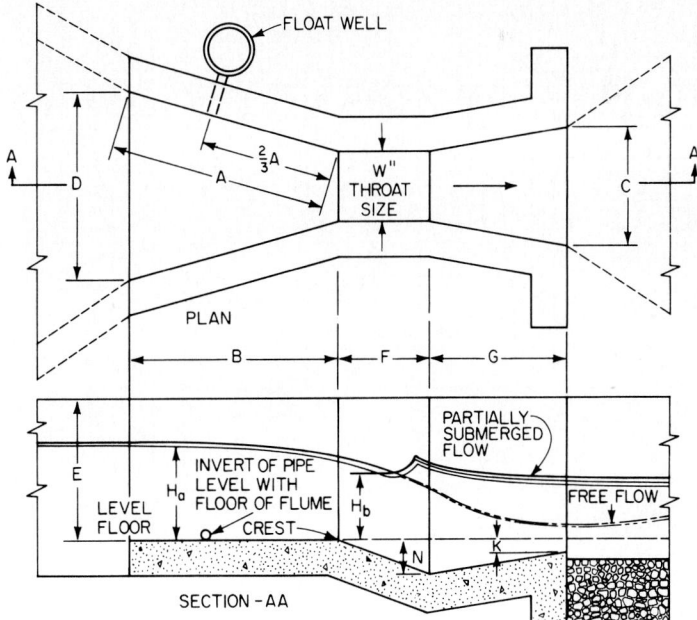

Fig. 4 Parshall flume dimensions. (See also Table 1.)

Table 1 Parshall Flume Dimensions*

W	*A*	⅔*A*	*B*	*C*	*D*	*E*	*F*	*G*	*K*	*N*	Flume flow extremes, million gal/day† Min.	Max.
0′3″	1′6⅜″	1′¼″	1′6″	0′7″	0′10³⁄₁₆″	2′0″	0′6″	1′0″	0′1″	0′2¼″	0.02	1.23
0′6″	2′7⁄₁₆″	1′4⁵⁄₁₆″	2′0″	1′3½″	1′3⅝″	2′0″	1′0″	2′0″	0′3″	0′4½″	0.03	2.52
0′9″	2′10⅝″	1′11⅛″	2′10″	1′3″	1′10⅝″	2′6″	1′0″	1′6″	0′3″	0′4½″	0.06	5.75
1′0″	4′6″	3′0″	4′4⅞″	2′0″	2′9¼″	3′0″	2′0″	3′0″	0′3″	0′9″	0.07	10.41
1′6″	4′9″	3′2″	4′7⅞″	2′6″	3′4⅜″	3′0″	2′0″	3′0″	0′3″	0′9″	0.10	15.90
2′0″	5′0″	3′4″	4′10⅞″	3′0″	3′11½″	3′0″	2′0″	3′0″	0′3″	0′9″	0.27	21.39
3′0″	5′6″	3′8″	5′4¾″	4′0″	5′1⅞″	3′0″	2′0″	3′0″	0′3″	0′9″	0.39	32.57
4′0″	6′0″	4′0″	5′10⅝″	5′0″	6′4¼″	3′0″	2′0″	3′0″	0′3″	0′9″	0.84	43.88
5′0″	6′6″	4′4″	6′4½″	6′0″	7′6⅝″	3′0″	2′0″	3′0″	0′3″	0′9″	1.03	55.32
6′0″	7′0″	4′8″	6′10⅜″	7′0″	8′9″	3′0″	2′0″	3′0″	0′3″	0′9″	1.68	66.89
7′0″	7′6″	5′0″	7′4¼″	8′0″	9′11⅜″	3′0″	2′0″	3′0″	0′3″	0′9″	1.94	78.46
8′0″	8′0″	5′4″	7′10⅛″	9′0″	11′1¾″	3′0″	2′0″	3′0″	0′3″	0′9″	2.26	90.16

*Dimensions are given in feet and inches. See also Fig. 4.

†Extreme minimum and maximum capacities of flumes; the actual measuring range depends on the type of instrument selected. 1 million gal/day = 3785 m^3/day.

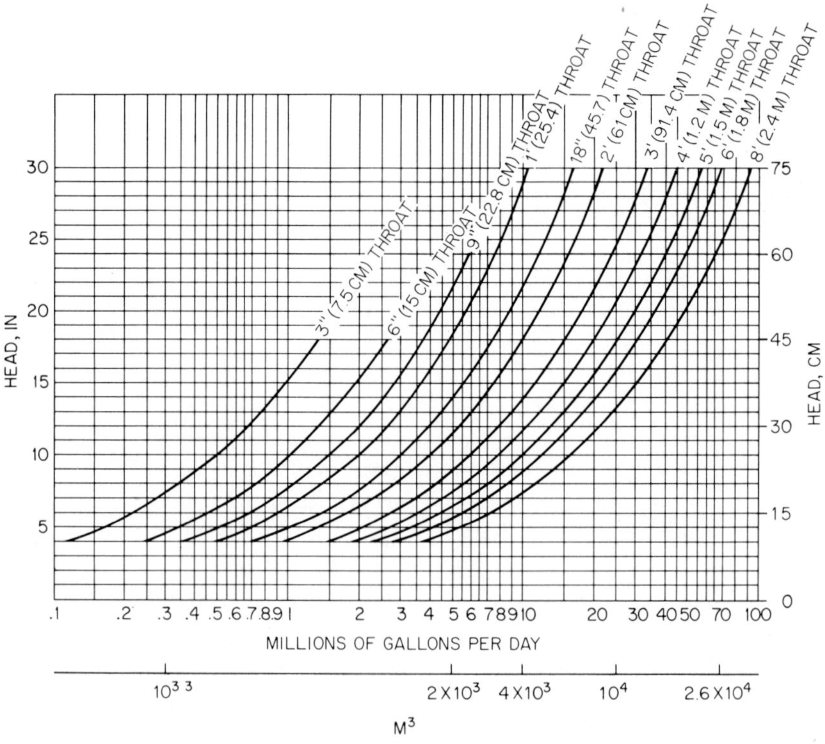

Fig. 5 Capacity curves for Parshall flumes.

throat size be selected, provided it is consistent with the depth in the channel at maximum flow and permissible head loss. As a general rule, the throat size should be one-third to one-half the width of the channel. See Table 1.

Free-Flow Conditions

Free-flow discharge, the condition under which the rate of discharge is dependent only on the width of the throat and the depth of water at the gage point H_a in the converging section, can occur at two different stages: (1) where the liquid moves at high velocity in a thin sheet conforming closely to the dip at the lower end of the throat, and (2) where the backwater raises the water surface to elevation H_b, causing a ripple or standing wave to form at or just downstream from the end of the throat. Under the latter condition the flume is said to operate under partial submergence, but the free-flow rate of discharge is not retarded as long as the ratio of H_b to H_a does not exceed the values given below.

Submerged Flow

When the ratio of H_b to H_a exceeds the values given below, the flume is said to operate under submerged flow and the rate of discharge is reduced. The operation of flumes under submerged flow conditions is not recommended since two gage points are required to determine the negative correction factor to apply to the free-flow calibration data. There are no instruments available for direct and accurate measurement of submerged flow. Ratings for the various flume sizes up to 8 ft (2.4 m) are given in Fig. 5.

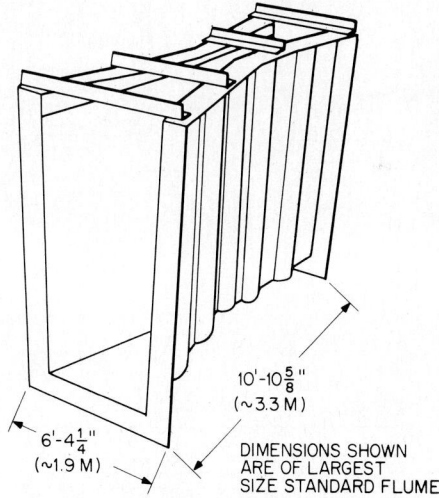

$10'\text{-}10\frac{5}{8}''$
(~ 3.3 M)

$6'\text{-}4\frac{1}{4}''$
(~ 1.9 M)

DIMENSIONS SHOWN
ARE OF LARGEST
SIZE STANDARD FLUME

Fig. 6 A rigid one-piece fiberglass Parshall flume constructed of fiberglass-reinforced polyester resins layered up over white Gel-cote. Advantages claimed for flumes of this type include (1) smoothness of construction which prevents debris buildup and hence head loss, (2) dimensional accuracy and stability—does not require external bracing, (3) long life in measurement of sewage and industrial wastes, (4) complies with U.S. Department of Interior requirements, (5) is lightweight—hence easy to ship, and (6) reduction in installation time. Flumes of the type illustrated are designed with 2½-in ($\sim$ 6.4-cm) flanges at the bottom and sides of the approach and discharge ends to connect with fluid channels. *(Kenco Plastics.)*

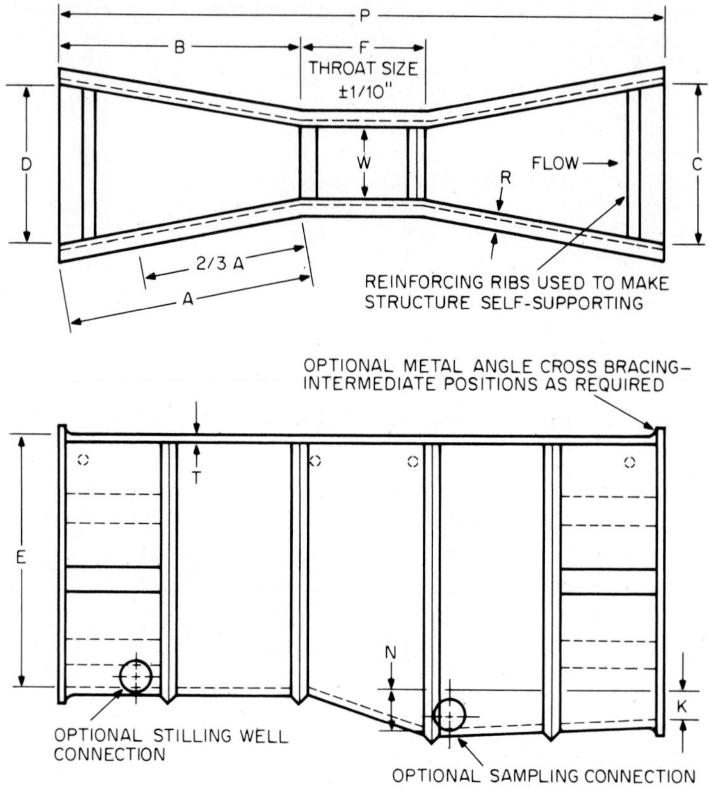

Fig. 7 Configuration and principal dimensional parameters of a rigid, one-piece Parshall flume. *(Kenco Plastics.)*

Table 2 Rigid One-Piece Parshall Flume Dimensions*

W	*A*	⅔*A*	*B*	*C*	*D*	*E*	*F*	*P*	*K*	*N*	*R*	*T*
1″	1′2 9/32″	9 17/32″	1″2″	3 21/32″	6 19/32″	1′8″	3″	2′1″	¾″	1⅛″	1¾″	3/18″
2″	1′4 5/16″	10⅞″	1′4″	5 5/16″	8 13/32″	9″	4½″	2′6½″	⅞″	1 11/16″	1¾″	3/16″
3″	1′6⅜″	1′¼″	1′6″	7″	10 3/16″	2′0″	6″	3′0″	1″	2¼″	2½″	3/16″
6″	2′7/16″	1′4 5/16″	2′0″	1′3½″	1′3⅝″	2′0″	1′0″	5′0″	3″	4½″	2½″	3/16″
9″	2′10⅝″	1′11⅛″	2′10″	1′3″	1′10⅝″	2′6″	1′0″	5′4″	3″	4½″	2½″	3/16″
1′0″	4′6″	3′0″	4′4⅞″	2′0″	2′9¼″	3′0″	2′0″	9′4⅞″	3″	9″	2½″	¼″
1′6″	4′9″	3′2″	4′7⅞″	2′6″	3′4⅜″	3′0″	2′0″	9′7⅞″	3″	9″	2½″	¼″
2′0″	5′0″	3′4″	4′10⅞″	3′0″	3′11½″	3′0″	2′0″	9′10⅞″	3″	9″	2½″	¼″
3′0″	5′6″	3′8″	5′4¾″	4′0″	5′1⅞″	3′0″	2′0″	10′4¾″	3″	9″	2½″	5/16″
4′0″	6′0″	4′0″	5′10⅝″	5′0″	6′4¼″	3′0″	2′0″	10′10⅝″	3″	9″	2½″	⅜″

*Dimensions are in feet and inches. See also Fig. 7. Data furnished by Kenco Plastics Company, Inc., Necedah, Wis.

Rigid One-Piece Parshall Flumes

For many applications, particularly for measuring liquids containing solids that settle under gravity head conditions, a one-piece fiberglass-reinforced polyester resin flume may perform better and may be easier to install and maintain. Such a flume is shown in Fig. 6. The wall thicknesses and integral support ribs are designed to prevent deformities. Extruded angles are bolted to the flange to aid in maintaining dimensional integrity. Dimensions are given in Fig. 7 and Table 2.

Flume throat	H_b/H_a
3–9 in (7.5–22.9 cm)	0.6
1–8 ft (0.3–2.4 m)	0.7
10–50 ft (3–15.2 m)	0.8

Open-Flow Nozzles

The Kennison nozzle,* shown in Fig. 8, is an unusually simple device for measuring flows through partially filled pipes. The nozzle successfully copes with low flows, wide flow ranges, and liquids containing suspended solids and debris. Because of its high accuracy, nonclogging design, and excellent head-versus-flow characteristics, the Kennison nozzle is well suited for the measurement of raw sewage, raw and digested sludge, final effluent, and trade wastes. Capacities are given in Table 3.

The unobstructed flow path (with a self-scouring action) prevents clogging by debris. Comprehensive calibration data from authoritative hydraulic laboratories form the basis of design calculations for particular flow conditions. A typical rating curve (Fig. 9) illustrates the relatively large changes in head (float movement) for equal flow rate increments.

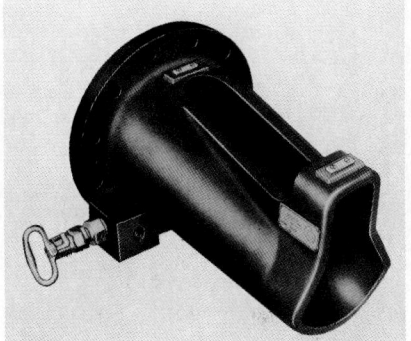

Fig. 8 Kennison nozzle.

Maintaining Approach Conditions

Like other open-flow metering devices, the Kennison nozzle is most accurate when the installation is provided with satisfactory, prescribed approach conditions. When laboratory-calibrated, the nozzle is mounted level at the end of a horizontal, uninterrupted, eight-diameter length of pipe. Highest accuracy can be expected when the installation most nearly duplicates this test condition. A typical installation for measuring flow from a chamber or basin is shown in Fig. 10.

Should the slope of the approach pipeline exceed the values given in Table 3, the velocity can be reduced to a permissible rate by interposing an adequate length of straight pipe of the same diameter as the nozzle, laid horizontally. The horizontal pipe ahead of the nozzle should have a length equivalent to five pipe diameters, plus a length equivalent to three pipe diameters for each 0.0010 of slope in excess of the values given in Table 4.

Where the main line is smaller than the nozzle and there is excessive velocity, the line must be reduced to the recommended value. Possible methods for reducing excessive velocity are shown in Fig. 11.

There are many schemes for converting the level signal from head area meters to flow readings. Most commonly employed is the float, using a separate float well. Others are float in-stream and bubble pipe. In all instances, the level signal must be characterized for linear and direct readout at the secondary line.

One essential consideration with any float-operated meter is to select a device that will give a

*Named after the designer, Karl R. Kennison, a hydraulic engineer.

Table 3 Capacities of the Kennison Nozzle

Inlet diameter	gpm	million gal/day	m³/day
6 in (15 cm) ½ section	90	0.13	492
6 in (15 cm)	100	0.29	1,098
8 in (20 cm)	320	0.45	1,703
10 in (25 cm)	560	0.8	3,028
12 in (30 cm)	850	1.2	4,542
16 in (40.6 cm)	1,900	2.7	10,220
20 in (50.8 cm)	3,100	4.5	17,032
24 in (61 cm)	5,200	7.5	28,388
30 in (76.2 cm)	8,500	12.5	47,313
36 in (91.4 cm)	14,000	20	75,700

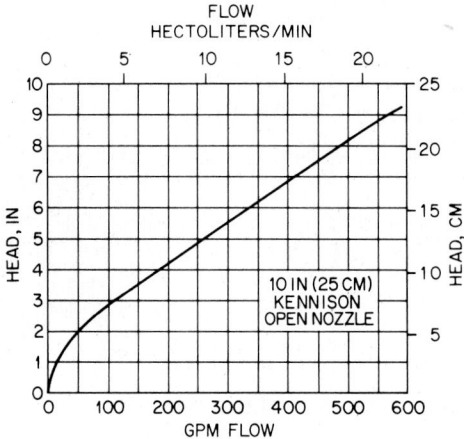

Fig. 9 Typical rating curve for a 10-in (25-cm) Kennison nozzle. Note linear relationship from ⅒ to maximum flow.

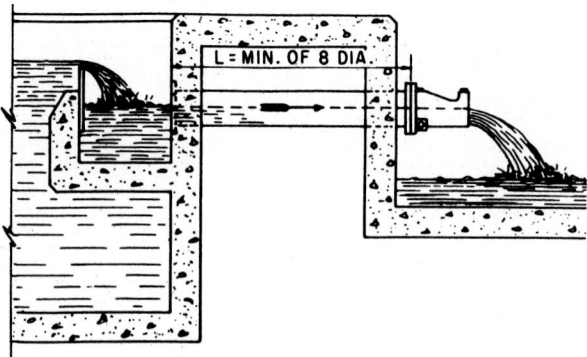

Fig. 10 Method for measuring flow from a chamber through a Kennison nozzle.

Table 4 Limiting Slope for Approach Piping for Kennison Nozzle

Nozzle size	Slope, ft/ft, m/m, etc.
6 in (15 cm)	0.0070
8 in (20 cm)	0.0050
10 in (25 cm)	0.0040
12 in (30 cm)	0.0033
16 in (40.6 cm)	0.0027
20 in (50.8 cm)	0.0023
24 in (61 cm)	0.0021
30 in (76.2 cm)	0.0020
36 in (19.4 cm)	0.0020

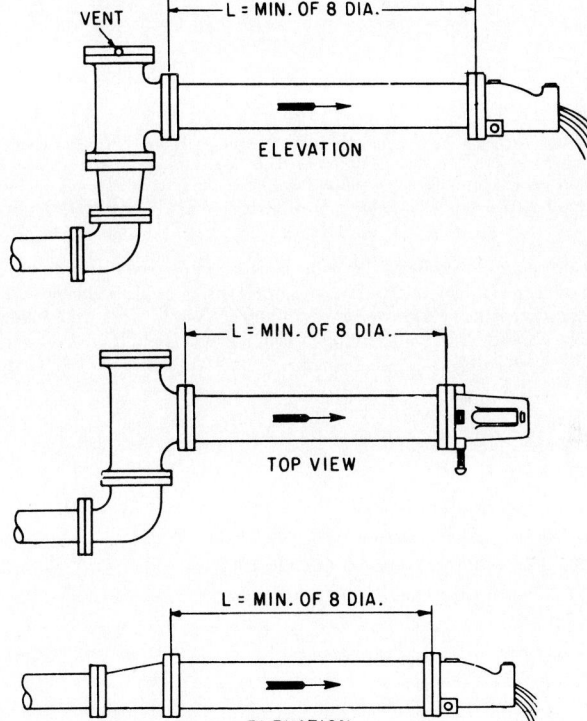

Fig. 11 Methods of reducing the velocity in a line smaller than the Kennison nozzle.

total level rise of at least 4 in (10 cm) at maximum flow. For low discharges this usually means selecting a 22½° V-notch weir or a 1- or 2-in (2.5- or 5-cm) Parshall flume. Flow cams can be designed to conform to the rating of any weir or flume. Flow gears can be designed to accommodate various ranges of flows. Level changes through the weir or flume cause the float to rise or fall. This float movement rotates, in one design,* the float pulley and through the flow gears changes the cam

*Leupold & Stevens, Inc., Beaverton, Oregon (U.S. Patent 3,219,772).

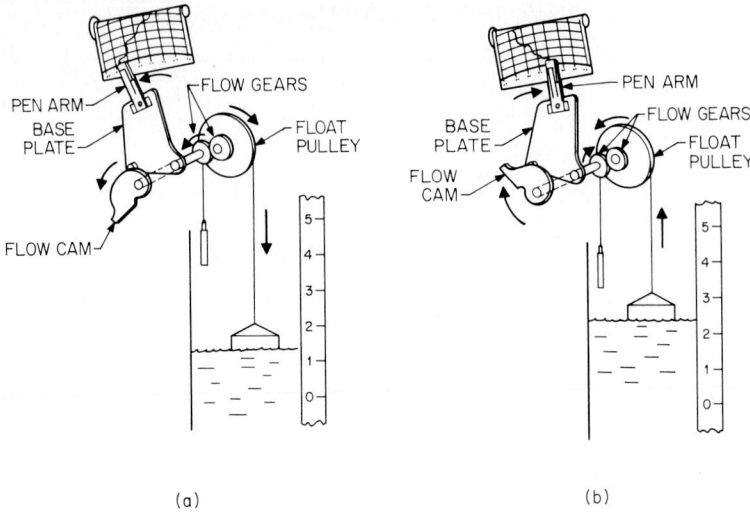

(a)   (b)

Fig. 12   Float-type open-channel flow measuring instrument. *(a)* Float movement rotates the float pulley and through the flow gears changes the cam position. As the cam rotates, it positions a pointer or pen to indicate instantaneous discharge at that moment. *(b)* Effect of a rising stage. A mechanical integrator determines the total volume. *(Model 61, U.S. Patent 3,219,772, Leupold & Stevens.)*

position. As the cam rotates, it positions the pointer or pen to correctly indicate the instantaneous discharge at that moment. See Fig. 12*a*. The effect of a rising stage is shown in Fig. 12*b*. The instruments permit convenient field conversion to different flow ranges through the use of interchangeable flow gears and cams. An integrating mechanism provides totalized volume.

REFERENCES

Baumeister, T. (ed.): *Standard Handbook for Mechanical Engineers,* 8th ed., McGraw-Hill, New York, 1978.

Chow, V. T.: *Open-Channel Hydraulics,* McGraw-Hill, New York, 1959.

Chow, V. T. (ed.): *Handbook of Applied Hydrology,* McGraw-Hill, New York, 1964.

Davis, C. V., and K. E. Sorensen (eds.): *Handbook of Applied Hydraulics,* 3d ed., McGraw-Hill, New York, 1970.

Morris, H. M., and J. M. Wiggert: *Applied Hydraulics in Engineering,* 2d ed., Wiley, New York, 1972.

Shoop, C. F., and G. L. Tuve: *Mechanical Engineering Practice,* 5th ed., McGraw-Hill, New York, 1956.

Staff: *Flow—Its Measurement and Control in Science and Industry,* vol. 1 (1974), vol. 2 (1982), Instrument Society of America, Research Triangle Park, North Carolina.

Staff: *Stevens Hydrographic Data Book,* Leupold & Stevens, Beaverton, Oregon (revised periodically).

Level Systems*

Paul M. Christensen. *Engineering Department, Fisher Controls Company, Marshalltown, Iowa. (Electronic Level Controllers)*

Harry A. Downer. *Bindicator Company, Port Huron, Michigan. (Bulk-Level Systems)*

P. A. Elfers. *Retired, Bemidji, Minnesota. Formerly, Fisher Controls Company, Marshalltown, Iowa. (Float and Hydrostatic Level Systems)*

Robert L. Hayes. *Bindicator Company, Port Huron, Michigan. (Bulk-Level Systems)*

G. W. Hickenlooper. *The Ohmart Corporation, Cincinnati, Ohio. (Nuclear Radiation Methods)*

E. C. Hirth. *Application Engineering, Moore Products Company, Spring House, Pennsylvania. (Pneumatic Level Systems)*

C. M. Johnson. *Engineering Department, Fisher Controls Company, Marshalltown, Iowa. (Hydrostatic Level Systems)*

James H. Koegel. *Director of Engineering, Robertshaw Controls Company, Industrial Instrumentation Division, Anaheim, California. (Capacitance Level Measurements)*

Anthony Sciacca. *National Sonics Division—Xertex, Hauppauge, New York. (Ultrasonic Methods)*

G. B. Sellers. *Engineering Department, Robertshaw Controls Company, Industrial Instrumentation Division, Anaheim, California. (Capacitance Level Measurements)*

*Persons who authored parts of this section or who otherwise cooperated in an outstanding manner in furnishing information and helpful counsel to the editorial staff.

Level Systems

Measuring and frequently controlling the level of materials contained in storage and processing vessels, such as tanks, wells, reservoirs, bins, and hoppers, is one of the most common procedures of industrial instrumentation. There is a great diversity of liquids and bulk and fluidized solids which require level measurement—ranging from pure, clear water to viscous, sticky, and corrosive fluids and slurries, and from free-flowing, dry crystals to moist, lumpy solids. The processing environments for level sensors extend from vacuum to high-pressure service, and from subzero to elevated temperatures. With the possible exception of chemical composition measurement, the available means for measuring level exceed, both in number and diversity, those of any other process variable. Largely because of the many technologies involved, the number of level system suppliers exceeds 300, with relatively few manufacturers offering what might be termed a complete line of level sensors.

Like other industrial process variables, level systems date back well over a century, representing a progression of measuring and controlling approaches—mechanical, pneumatic, electrical, and electronic. A number of level systems still in common use have been improved, but with little or no alteration in basic concept. Other level systems represent the application of relatively new technologies developed during the past few decades. Also, like other industrial instrumentation systems, level measurement and control have been markedly impacted during recent years by solid-state electronics and digital techniques.

CLASSIFICATION OF LEVEL SYSTEMS

System Objectives

The manner in which level data are to be used determines whether a single, fixed level is to be sensed, a series of multiple, fixed levels is to be sensed, or a continuous range of levels between a low and a high level is to be sensed. The principal need may be for information, as in the case of inventory control, on a continuous or periodic basis; the need may be mainly or exclusively concerned with controlling a level—to ensure an even supply of material to a process without problems of over- or underfeed; or the need may involve maintenance of a proper dynamic balance between numerous vessels in a continuous process. The need for information and/or control also may apply to only one fluid or to two or more fluids (essentially immiscible) where the locations of the interfaces between fluids bear importantly on the efficiency of a given process.

Characteristics of Measurands and Process Environment

Factors such as chemical composition (corrosiveness), viscosity (liquids), flowability (solids), and the process environment (temperature, pressure, absence or presence of agitation, among others) markedly influence the selection not only of the level measurement method but also the design and location of the sensing devices. Interfaces encountered include gas (commonly air) and liquid, air and foam, air and powder, air and granular solids, foam and liquid, liquid and liquid, solid floating on a liquid, solid beneath a liquid, and vapor and liquid.

Frequently, in selecting the most appropriate level measurement technique, this is the most convenient starting point. Suppliers of equipment tend to fall into measurement technique categories.

Direct and Inferential Methods

Direct methods involve a direct measurement of the distance (usually height) from the substance level to a datum line. Direct methods include (1) direct visual observation of distance on a suitably calibrated scale, as with a gage stick, hook gage, or gage glass; (2) determination of the position of a detecting member which rides on the material surface, such as a ball or other type of float (liquids) or a plumb bob (solids); (3) contact of electrode probes with the material; (4) interruption of a light beam to a photoelectric cell; and (5) reflection of radio and radar frequency waves or of sonic and ultrasonic energy from a material surface.

Inferential methods utilize effects other than the changing position of a material surface for determining the level in a vessel. Inferential methods include (1) measurement of the fluid or hydrostatic head developed by the material (liquids); (2) measurement of the buoyant force created when a detecting element is partially or completely immersed (liquids); (3) attenuation of radiation; (4) utilization of electrical properties of materials, such as capacitance, conductance, and resistivity; and (5) utilization of thermal and other physical differences between the material and the gas or vapor space above the material.

In contrast to direct methods of measurement, most inferential methods have in common certain inherent errors because of changes in density due to varying composition and temperature, among other factors.

IMPORTANCE OF LEVEL CONTROL IN PROCESS OPERATION

In many processes involving liquids contained in vessels, such as distillation columns, reboilers, evaporators, crystallizers, and mixing tanks, the particular level of liquid in each vessel can be of great importance in process operation. A level which is too high, for example, may upset reaction equilibria, cause damage to equipment, or result in spillage of valuable material. A level that is too low may have equally bad consequences. Combined with such basic considerations, there is the advantage in continuous processing of reducing stroage capacity throughout the process. This reduces the initial cost of equipment, but less storage also accentuates the need for accurate and sensitive level control. Effective measurement and control of level can usually be justified in terms of economy and/or safety. To the operator, knowledge of this variable provides data on the (1) quantity of raw material available for processing, (2) available storage capacity for products being manufactured, and (3) satisfactory or unsatisfactory operation of the process. In the following paragraphs, a few examples are given.

Efficient Operation of Equipment

Control at One Height

In many process applications, the level must be maintained accurately at a predetermined height, irrespective of load conditions on the process. Several examples will serve to illustrate this point.

In a steam or vapor generator, such as a boiler, it is desired to maintain the level at a predetermined value in order that two sets of conditions will be present at all times regardless of the output from the generator. The first conditions will be present at all times regardless of the output from the generator. The first condition requires that the quantity of liquid inventory in the vessel be maintained in order to provide feed for the evaporation process. The second condition requires that a vapor volume space be maintained in order to have available storage capacity for the vapor, plus a volume space which will prevent carryover of entrained liquids in the vapor.

In continuous processes, a correct level head in certain equipment is of considerable importance. In evaporators, for example, the heating medium may be inside a tube bundle which must at all times be covered to an optimum depth, thereby requiring precise level control. Too low a level will uncover the heating surface, lowering the efficiency of the process. Too high a level will require a greater heat input as the head pressure increases, which may result in damage to throughput material or unsatisfactory evaporation.

Proportional (Wideband) and Averaging Level Control

In continuous processes, accumulators or storage vessels are introduced between various stages of the process in order to provide storage (inventory). Process upsets or disturbances are absorbed in such accumulators, and only a minimum of them are passed on to the next phase of the process.

In such process applications, control of the level at a constant height is not always desirable. It is more important that the outflow of the storage vessel does not change suddenly and cause an upset in the subsequent process stage. Any sudden increase in input to the storage vessel should be absorbed in the vessel. To accomplish this, "averaging" liquid-level control is used, wherein a wideband proportional-plus-reset mode of control is incorporated in the level control instrument.

With this type of control, if the uncontrolled input suddenly increases, the wide proportional band permits the level to rise temporarily, with little change in the controller output regulating outflow from the vessel. If the input remains at its higher value for a period of time, the automatic reset functions to return the level to the set point, and gradually changes the outflow rate. Conversely, if the uncontrolled input suddenly decreases, the level is permitted to drop; the outflow is not similarly affected but is changed at a gradual rate if the decreased input continues.

The size of the vessel required becomes a function of the (1) maximum process upset to be expected in the system and to be absorbed by the accumulator, (2) duration of the upset, and (3) elapsed time allowed before the full value of the continued upset is to be passed along to the next stage of the process.

The term "holding time" is often used to describe the function just explained. The size of the vessel also may be limited by physical height considerations, by the change in level height allowable, and by the economics of vessel cost. All these factors become variables for design and instrument engineers to evaluate in determining the size of vessels and accumulators.

The actual performance of the liquid-level controller in smoothing out minor upsets and in absorbing them in the accumulator can be further adjusted by proportional-band settings and reset rate adjustments.

Level Control Permits Smaller Vessels

In simple single-capacity systems, the utilization of level measurement and control can be advantageously applied to keep the capacity of processing vessels within practical limits. If level measurement and control are used, the size of a mixing or reaction vessel may be small. It is not necessary that large vessels be used to handle all available liquid to be processed, since the liquid-level control device will feed only the fluid required to keep the liquid concentration or its height at a predetermined value. Large or bulky vessels thus can be eliminated, with accompanying economy. This also means that a small amount of process material is under reaction or in process, thereby reducing attendant hazards, potential losses, or spoilage.

Protection of Centrifugal Pumps

Where it is desired to maintain a head pressure against the suction of a centrifugal pump, the level of the liquid in the storage tank must be maintained at an optimum value. If the level drops too low, flashing and cavitation will occur in the pump suction, with resultant erratic pump discharge and extreme wear on pump impellers. If the level rises too high, there may be a loss of accumulator volume in the vessel, thereby affecting the process from an operating viewpoint.

Product Quality Control

Warp sizing in the textile industry is a good example of how close control of the liquid level directly affects product quality. The warp yarn is run through a bath of sizing solution which adds a protective coating to the yarn. The amount of size absorbed by the yarn is a function of the time during which the yarn is in contact with the size solution. As the yarn passes through the size solution in a prescribed path (usually around a large cylinder rotating at a fixed number of revolutions per minute), the time of contact is a function of the level height of the size solution. A variation in solution level, therefore, will change this contact time and thus destroy warp uniformity, later causing breakage of threads on the loom.

Cost Accounting

The flowmeter, weighing scale, and liquid-level gage are the process cost accountant's principal tools for obtaining facts concerning quantities of liquid raw materials and finished products in storage, and of liquids in process. Thousands of liquid-level meters of all types, from the simplest gage stick, often still used for taking inventories at tank farms, to the most sophisticated remote-level indicating gages are in daily use, principally for cost accounting needs.

LEVEL MEASUREMENT SYSTEM MATHEMATICS

Important to the engineer who designs and specifies liquid-level instrumentation systems are such factors as (1) the relationship of flow to level in vessels, and (2) capacity versus level height in variously shaped vessels.

Constant Head for Steady Process Flow

Often, steady process flows, such as the introduction of a raw material to a process, are maintained by holding a constant head pressure on the feed line. This can be achieved by control of the liquid level in the feed tank whose feed line exits from the bottom of the tank, as shown in Fig. 1.

As regards the theory of this arrangement, in the free flow of liquid through an opening or orifice, or across a weir plate, the quantity of liquid discharged is a function of the level height above the orifice or weir. Briefly, the flow is governed by the equation

$$Q = CA\sqrt{2gH}$$

where Q = quantity, ft^3/s
$\quad C$ = orifice constant
$\quad A$ = area of flow, ft^2
$\quad g$ = acceleration of gravity, 32.2 ft/s^2
$\quad H$ = height of liquid, ft

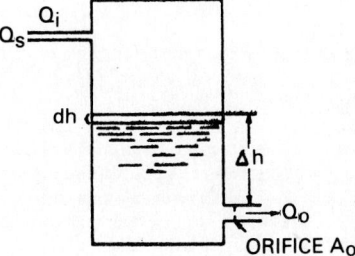

Fig. 1 Basic flow versus level factors.

Although English units are usually used, the relationship, of course, also applies where appropriate metric units are used.

Thus, with the area of flow A constant and the other factors constant, steady process flow Q will be obtained if the head of liquid H above the orifice or weir is accurately measured and controlled.

Relationship of Flow to Level in Process Vessels

In designing process equipment and determining the requirements of liquid-level controlling equipment, an engineer often must calculate in advance how a change in liquid inflow will affect the level in the vessel. In any given system of vessels, a stable quantity of inflow to the vessel equals the outflow. With a stable inflow, the height of the level H above the drawoff increases until the head pressure developed causes flow through orifice A by an amount Q_o which is equal to the inflow Q_i.

From the above equation and reference to Fig. 1, this relation may be expressed by

$$Q_i = Q_o = 0.897 C A_o \sqrt{2gH}$$

where Q_i = flow rate into vessel, gal/min
$\quad Q_o$ = flow rate out of vessel, gal/min
$\quad C$ = orifice constant or coefficient
$\quad A_o$ = orifice area, in^2
$\quad g$ = acceleration of gravity, 32.2 ft/s^2
$\quad H$ = height of liquid level above top of outlet orifice, in

Although English units are usually used, the relationship, of course, also applies where appropriate metric units are used.

Variation in the inflow or in the orifice area, with the other variables remaining constant, causes a level change. A control valve substituted for the fixed orifice provides an easy means of varying the orifice area. If the orifice area is increased, the level will fall until, at the new area, the level stabilizes at a point where its head effect on the orifice causes outflow to equal inflow.

If the orifice area is decreased, the level will rise until the product of a smaller orifice value and a larger head effect causes outflow to equal inflow. The rate at which the level rises or falls can be expressed by the following relationships:

$$Q_o = 0.897 CA_o \sqrt{2gH} \qquad \text{(standard flow formula)}$$

$$\frac{dH}{dt} = (Q_i - Q_o)\frac{231}{A} = \text{height change (units/min)}$$

$$= (Q_i - Q_o)\frac{231}{0.785d^2}$$

$$= \frac{294}{d^2}(Q_i - Q_o)$$

$$= \frac{294}{d^2}(Q_i - CA_o\sqrt{2gH})$$

$$= \frac{294}{d^2}(Q_i - 7.20CA_o\sqrt{H})$$

where dH = change in liquid-level height, in
dH/dt = rate of change in liquid-level height, in/min
A = transverse tank area, in^2
d = diameter of tank (cylindrical), in

Although English units are usually used, the relationship, of course, also applies where appropriate metric units are used. Where English units are used, it should be noted that 1 gal = 231 in^3 and 1 psi = 27.7 in of water.

To find height H_2 at a time t_2, when Q_i is initially equal to Q_o and is instantaneously changing at time t_1,

$$\frac{dH}{dt} = \frac{294}{d^2}(Q_i - 7.20CA_o\sqrt{H})$$

$$dt = \frac{dH}{(204/d^2)Q_i - (294/d^2)(7.20CA_o\sqrt{H})}$$

For easy manipulation, let

$$B = \frac{294}{d^2}7.20CA_o \qquad \text{and} \qquad D = \frac{294}{d^2}Q_i$$

Then

$$dt = \frac{dH}{D - B\sqrt{H}} \qquad \text{and} \qquad dt\int_{t_1}^{t_2} = \int_{H_1}^{H_2}\frac{dH}{D - B\sqrt{H}}$$

From a table of integrals after transformation into standard form,

$$2x\left[-\frac{\sqrt{H}}{B} - \frac{D}{B^2}\log_e(D - B\sqrt{H})\right]_{H_1}^{H_2} = t_2 - t_1$$

or
$$\frac{2}{B^2}[\sqrt{HB} + D \log_e (D - B\sqrt{H})]\big|_{H_2}^{H_1} = t_2 - t_1$$

It should be noted that, if there is a change in inflow Q_i, the resultant liquid-level height H_2 may be calculated when outflow Q_o again equals inflow by use of the standard flow formula previously given. By use of the last equations given, the amount of time required for the liquid level to reach any height between H_1 and H_2 can be calculated. Also, by substituting a figure for t_2 in the equation, the change in height for any given change in time can be calculated. Also, if an uncontrolled flow is introduced into a vessel and a level controller is placed on the outflow, the rate of flow and level changes can be calculated. Further details of this procedure are given by Smith in *Automatic Control Engineering,* McGraw-Hill, New York, 1944.

Capacity versus Level Height in Various Vessels

Many processing and storage vessels where liquid-level measurement is a factor are cylindrically shaped and mounted vertically on end. Thus, the content for any level height can be simply calculated by

Content = cross section area × level height

For example, in English units, $ft^3 = \pi r^2 h$, where r = inside radius of tank (in ft) and h = level height (in ft).

This relationship becomes more complex where cylindrical tanks are mounted horizontally. This is best illustrated by calculating a typical problem, the dimensions of which are in English units and illustrated in Fig. 2.

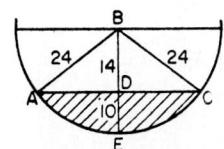

Fig. 2   Dimensions used in calculation of the capacities of horizontal cylindrical tanks.

Example.   Determine the volume in a cylindrical tank (flat ends) mounted horizontally, with the following factors known:

Diameter of tank = 48 in
Depth of liquid = 10 in
Length of tank = 120 in
Area ACE (shaded portion) = area $ABCE$ − area ABC
Area $ABCE$ = (2$LABD$/360) × area of circle

and $\angle ABD$ is found from its cosine which is $^{14}\!/_{24}$.

$$\therefore \angle ABD = 54.25°$$
$$\text{Area } ABC = 14 \times 24 \times \sin ABD$$

or
$$14 \times 24 \times \sin 54.25 = 14 \times 24 \times 0.8116 = 272.7$$
$$\text{Area } ABCE = \frac{108.50}{360} \times \pi \times (24)^2 = 545.4$$
$$\therefore \text{Area } ACE = 545.4 - 272.7 = 272.7 \text{ in}^2$$
$$\text{Volume (U.S. gallons) per foot of length} = \frac{272.7 \times 12}{231} = 14.17 \text{ gal/ft}$$
$$\therefore \text{Total volume} = \frac{14.17 \text{ gal} \times 120}{12} = 141.7 \text{ gal}$$

An alternative method of calculating the capacity of a horizontal cylindrical tank employs the formula

$$A = R_2 \left[\text{arc vers } \frac{H}{R} - \frac{\sqrt{2RH - H^2}}{R^2} (R - H) \right]$$

$$\text{Volume} = A \times \frac{L}{231}\,\text{gal}$$

where A = area in^2
$\qquad H$ = height of liquid in tank, in
$\qquad R$ = radius, in
$\qquad L$ = length, in
arc vers H/R = angle in radians whose versine is H/R

Values of A are given by Hewes and Seward in *Design of Diagrams for Engineering Formulas,* McGraw-Hill, New York, 1923. See also Musick reference listed at the end of this article. A system utilizing a computer-optimized linearizer is shown in Fig. 3.

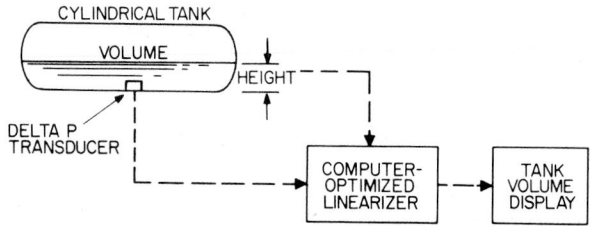

Fig. 3   Means for translating liquid height to volume (from a plot or an equation) to provide component values for a computer-optimized linearizer. In one configuration *(AP4155, Action Instruments)* a plug-in 12-segment linearizer accepts input from a height-of-level transducer and provides an output proportional to the tank volume.

Spherical Tanks

Useful relationships for determining the volume of a sector of a sphere are given in Fig. 4.

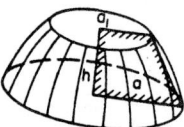

Fig. 4   Volume of a partially filled spherical tank equals one-half the total volume of the sphere (if the tank is 50% full) plus the volume of the spherical sector filled with liquid above the midline of the tank. If the tank is less than half full, the volume is that of the spherical sector filled with liquid above the bottom of the tank. Volume of a sphere = ⅘ × π × radius^3. With reference to the diagram, volume of a spherical sector = ⅙ × π × h × $(3a^2 + 3a_1^2 + h^2)$.

Volume and Weight Measurement from Level

Liquid-level measurement resolves itself into position measurement, namely, the position (height) of a liquid surface above a datum line. Measurement, however, need not always be expressed in terms of inches, feet, or meters above the datum line but, with a knowledge of the dimensional and contour characteristics of the containing vessel, can be conveniently interpreted (hence calibrated) in terms of the volume of liquid contained—and further, with information concerning the specific gravity of the liquid, can be expressed in terms of the weight of the liquid in the vessel.

Volume Determinations

If the purpose of level measurement is to determine the volume of liquid contained in the vessel, then direct measurement of level height is preferable because

$$V = A \times H$$

where V = volume of vessel
A = area of vessel
H = height of level

Thus, the volume measured is independent of liquid density.

If measurement of the pressure due to hydrostatic head must be used because of specific requirements, volume is determined by the relation

$$V = \frac{A \times P}{D}$$

where V = volume in vessel at given level
P = pressure due to hydrostatic head
D = density of liquid in vessel

In this case, the volume measurement depends on the density of the liquid.

Weight Determinations

If the purpose of level measurement is to determine the weight of the liquid contained in the vessel, then measurement of the pressure due to hydrostatic head has advantages because

$$W = H \times D \times A = A \times P$$

where W = weight of liquid in vessel
H = height of level
D = density of liquid
A = area of vessel
P = pressure due to hydrostatic head

Thus, the measurement is independent of liquid density.

If direct measurement of level height is used, then

$$W = A \times D \times H$$

and the weight measurement depends on knowing the density of the liquid.

Errors in Measurement of Quantities in Storage Tanks

Where highly accurate measurements of the quantities of liquids stored in tanks are required, it is the usual practice to correct the fluid density for temperature changes and neglect the other temperature and pressure effects. Generally, this is an adequate approach, but it is reassuring to know the magnitude of the uncorrected errors. The following equations allow measurements to be easily and completely corrected to any desired reference conditions.

It is assumed that the true volume of a container, empty and at its reference temperature, is accurately known. Also, it is assumed that the measuring element is without error. Under these ideal conditions, the remaining errors are:

1. The tank may not be perfectly round (if cylindrical) or geometrically true (if some other shape).
2. The tank wall is stretched, owing to hydrostatic pressure.
3. The fluid is not at its reference temperature.

Corrections for each of these unavoidable conditions are presented as relatively simple, dimensionless expressions in which any consistent data can be substituted. The results are expressed as relative errors e, defined as

$$e = \frac{\text{observed value} - \text{true value}}{\text{true value}}$$

The error e is therefore a decimal and can be converted to a percentage by multiplying by 100. It is positive or negative, depending on whether the observed value is high or low.

Also included is a method by which the temperature error can be completely and exactly compensated for. The correction allows for expansion of the fluid and the tank at a temperature T_1 below the liquid level, and for the tank and float wire (configuration selected as an example) at another temperature T_2 above the liquid level. It allows for an average liquid temperature consisting of one or an infinite number of striations, each at a different temperature. The only requirement is that each horizontal level be uniform.

Necessity for Corrections

The results of these calculations are given in summary form. For a tank that is not perfectly round but appears round to a person with normal vision, it is likely that the error from this cause is essentially insignificant and generally can be disregarded. Similarly, for a tank withstanding hydrostatic pressure that appears to be structurally sound, it is likely that any errors due to stretching are insignificant and generally can be disregarded.

In the case of temperature errors, however, it is seldom prudent to neglect corrections. The greatest errors are involved when the fluid density enters the calculations and is not corrected to the *actual* liquid temperature. This occurs when a pressure measuring system, as shown later (Fig. 7a), is used to report volumetric contents, or when a volume measuring system (Fig. 7b and c) reads in weight units.

Error for Tanks "Out of Round" (Fig. 5)

Assume that the tank is measured by strapping and that the area is calculated from perimeter P:

$$A' = \frac{P^2}{4\pi} = \text{calculated area}$$

$$A = \pi ab = \text{true area}$$

$$e = 0.298 \left(\frac{a}{b} - 1\right)^{1.952}$$

where a and b = measured dimensions (Fig. 5).

Error Due to Pressure Expansion (Fig. 6)

$$e = \frac{\pi S}{E} \times \frac{H_1^2 - H_0^2}{H_1(H_1 - H_0)}$$

and

$$S = \frac{6PD}{t}$$

where H_0 = pressure above liquid
H_1 = pressure at bottom of liquid
S = average wall stress
P = average pressure
E = modulus of elasticity
D = tank diameter
t = wall thickness

For open tanks, $H_0 = 0$ and $e = \pi S/E$.

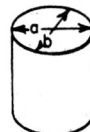

Fig. 5 Error due to out-of-round tank. See text.

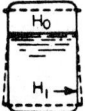

Fig. 6 Error due to pressure expansion. See text.

Error Due to Temperature (Fig. 7)

Expressions for errors in measurements in volumetric or weight units are given in Table 1 for three methods of level measurement. These relatively simplistic sensing methods are still numerous, particularly where cost is a major factor. An elementary mechanical temperature compensating system is shown in Fig. 8.

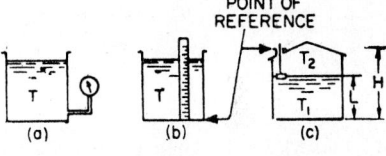

Fig. 7 Errors due to temperature (with various measuring systems). (a) Pressure-type system, (b) rod, (c) float and tape. See Table 1.

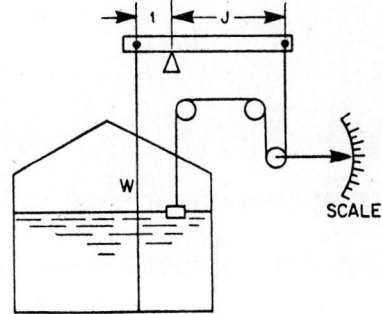

Fig. 8 Simplistic mechanical temperature compensating system. The wire W is subjected to the same depth of liquid and to the same temperature as the liquid itself. A change in the length of the wire is therefore proportional to the rise and fall of the liquid level due to temperature change. To compensate exactly for temperature, add to the motion of the float wire the motion of the compensating wire multiplied by a factor J, where $J = (3g - f)/g$, where f and g are expansibility factors for the liquid and tank material, respectively.

Sample Calculations

Examples are given for pressure and temperature errors. Microprocessors and other electronic techniques are built into some modern level measuring systems and involve making auxiliary measurements of temperature and density and integrating these values into the overall system of measurement prior to display of data or control actions.

Table 1 Error Equations for Some Level Measuring Systems*

Type of measuring system	Expressions for error	
	Reading, volumetric units	Reading, weight units
Pressure (Fig. 7a)	$e = -T(f + 2g)$	$e = -2Tg$
Rod (Fig. 7b)	$e = -T(2g + c)$	$e = T(f - 2g - c)$
Float and tape (Fig. 7c)	$e = 3gT_1 - \dfrac{L}{H} T_2(g - c)$	$e = T_1(f - 3g) - \dfrac{L}{H} T_2(g - c)$

*T = fluid temperature; f = volume expansion of fluid; g = linear expansion of wall; c = linear expansion of rod or float wire.

Example 1.   A tank 20 ft in diameter contains 30 ft of liquid of specific gravity 1.7. If the wall is ⅜-in steel plate, what is the error due to pressure?

$$P = 30 \times 1.7 \times 0.433 = 22 \text{ psi}$$

$$S_{max} = \frac{6PD}{t} = \frac{6 \times 22 \times 20}{⅜} = 7040 \text{ psi}$$

$$e = \frac{\pi S}{E} = \frac{\pi \times 7040}{30,000,000} = 0.00074 = 0.074\%$$

Example 2.   A manometer reads the pressure at the bottom of a steel tank. What error is introduced if the water temperature is 80°F instead of 25°F?
   In weight units of calibration

$$e = -2gT = -2(80 - 25) \times 0.0000132 = -0.00145 = -0.145\%$$

In volumetric units of calibration

$$e = -T(f + 2g) = -55(0.000207 + 0.000026) = -0.0128 = -1.28\%$$

Example 3.   A tank has a maximum diameter of 10 ft 6 in and a minimum diameter of 10 ft 4 in. What error is introduced by assuming it to be a circle of the same perimeter?

$$\frac{a}{b} = \frac{125 \text{ in}}{124 \text{ in}} = 1.0161$$

$$e = 0.298 \times 0.0161^{1.952} = 0.000097 = 0.01\%$$

POINT-CONTACT METHODS

It is most likely that liquid-level control was a forerunner of liquid-level measurement. Centuries ago, humans controlled water level height by building dams or bulwarks. A further step led to leaving an opening in the wall and filling it with logs or timbers which could be manually changed in number to control the height of water in a reservoir. Later, conveniences such as valves, sluice gates, and coffer dams were added to the system. The need for point-level measurements probably arose when liquids of value, such as oil, were stored in vessels, precipitating the need for inventory control.

In early reservoirs, a *notched stick* served as a crude level measuring device, the gager noting the interface line between a wetted and a dry stick. With the millions of *dipsticks* used in transportation vehicles and stationary engines, this unsophisticated tool appears to be the most common form of level measurement despite its very early roots.

An early refinement was the *hook gage*, shown in Fig. 9a. The hook gage point is moved up from beneath until it just pierces the liquid surface. The scale is usually engraved on a square brass rod. Another early device was the *plumb bob* or plummet gage, shown in Fig. 9b. This gage is lowered from above until a depresssion shows contact with the liquid surface. A variation of these point-contact devices is the bronze or steel tape, used with a plumb bob gage, as shown in Fig. 9c. For some applications, these remain inexpensive, reliable, and unsophisticated approaches where convenience and accuracy are not a major consideration.

Over the years, particularly since the advent of solid-state electronics a few decades ago, these basic approaches have been modernized and greatly improved in terms of convenience and absence of human error, as well as permitting centralized measurement of numerous vessels at remote locations. One system, for example, consists of a plummet, perforated tape, measuring sprocket, counter, take-up reel, drive motor, and detecting and timing circuits. As the level changes, the drive motor moves the plummet to correspond to the level. The tape drive turns a measuring sprocket shaft and counter and synchro indicating the level. The counter, reading in inches (or centimeters), tenths, and

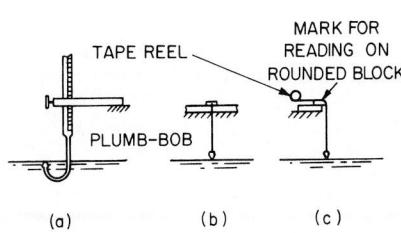

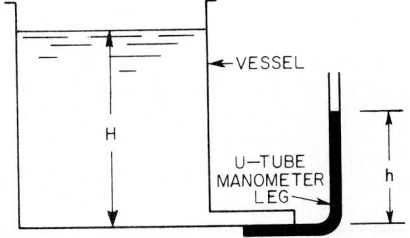

Fig. 9 Point-contact level measuring devices. (*a*) Hook gage, (*b*) plumb bob gage, (*c*) tape-and-plumb-bob gage.

Fig. 10 Schematic diagram of U-tube manometer for open vessels.

hundredths, provides a local readout for checking and calibration purposes. The synchro provides a transmission signal (FIC Industries). Tape-and-plummet systems are also used to measure the depth of sludge in clarifying or settling tanks. In one such device, high-intensity infrared beams illuminate a phototransistor detector until such time that sludge blocks the probe gap (Markland Specialty Engineering). Other kinds of transducers can become a part of the plummet and cause a change in the signal when the plummet engages a liquid or solid surface.

VISUAL METHODS

The dipstick requires visual reading of a mark on a scale, and visual methods are also represented by the *gage glass*. A desire for a clear indicating scale and a more convenient location for the scale inspired the introduction of a simple open-end manometer or tube, as shown in Fig. 10. The liquid height in the tube equalizes with the level in the vessel and, by means of a transparent tube material, such as glass, allows the level in the vessel to be read at any outside point. Gage glasses for pressure and vacuum vessels are shown in Fig. 11. Capillary attraction in small-diameter tubing causes water, for example, to rise an extra height of about $0.046/d$, where d is the internal diameter of the tube in inches; this extra height due to capillarity would be over ¼ in (≈ 6.4 mm) for a ⅛-in (≈ 3.2 mm) diameter tube. Larger diameter tubes are therefore normally employed in gage glasses.

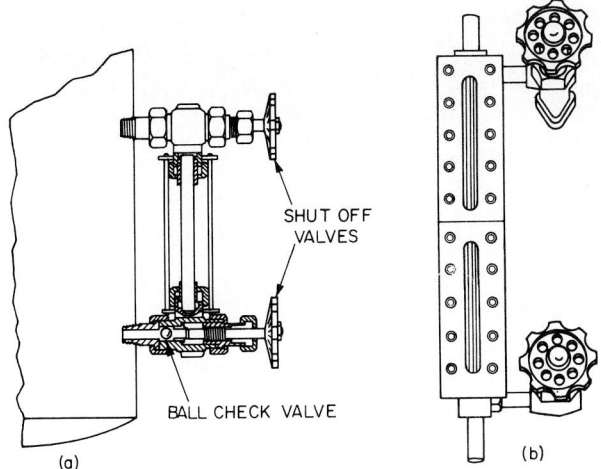

Fig. 11 Gage glasses. (*a*) Detail of a typical type of column, (*b*) reflex type.

The gage glass (Fig. 11a) can be considered a manometer in which the level in the glass seeks the same position as the level within the vessel. These simple instruments remain popular where local direct reading of level is satisfactory and where the application falls within their limitations.

Gage glasses are used with transparent tubes made of glass or plastic with sufficient strength to withstand the pressure in the vessel. To this end, tubular gage glasses are generally limited to 150 lb (1.04 Mpa) at 400°F (204°C) service, although they are available for service at 450 lb (3.1 Mpa) at 450°F (232°C).

In the type illustrated, the liquid chamber is commonly of steel. The reflex gage is of the prismatic or reflex type, in which the inside face of the glass is provided with 90° prisms running lengthwise, but still retaining a suitable gasket bearing surface. Rays of light normal to the face of the glass strike the prisms at an angle of 45°. If no liquid is in contact with the prisms, total reflection takes place, since the critical angle for a ray passing from glass to air is 42°; thus the gage appears silvery white. The critical angle of a ray passing from glass into water is 62°; thus the light ray passes on through the water, making it possible to see the inside of the chamber. The inside is usually painted black; so the viewer sees the liquid portion in the gage as black, though the liquid may be colorless. The reflex type does not permit observation of the color of a liquid or an interface between two different liquids.

Variations of this basic design include models for:

1. Low-temperature liquids which require a transparent frost preventing unit around the vision slot.
2. Acids and other corrosive liquids which require a resistant lining or coating of all chamber surfaces in contact with the liquid.
3. Provision for circulating a heat-transfer medium around the liquid chamber of the gage for (a) heating viscous liquids to enhance their flowability, and (b) cooling high-temperature liquids to avoid any tendency to boil.

Where it is desired to note the color, characteristics, or interface of a liquid, a transparent-type gage, similar in construction to the reflex type, can be used. In this design heavy sight glasses are placed on opposite sides of the liquid chamber; neither glass has the prism on one side.

Photoelectric Readout

Sometimes a light source and a light-sensitive receiver are used to enhance gage glass reading and for remote indication. The basic circuitry permits analog or digital displays. See Fig. 12.

Frost-Plug Detection

This method is usable for liquid detection where the liquid is below 32°F (0°C), such as propane and butane. With sufficient moisture in the air, frost will form on all plugs below the surface level of the liquid. See Fig. 13.

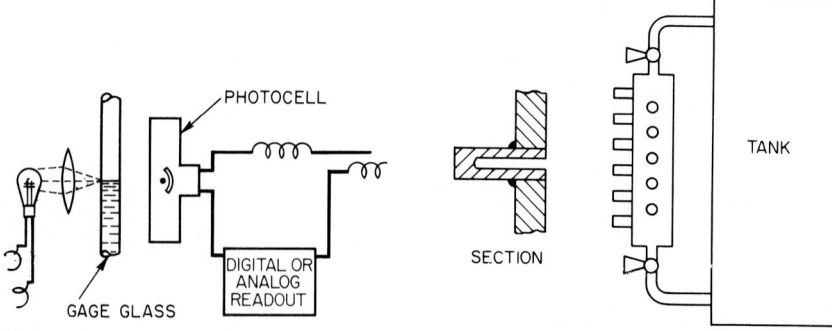

Fig. 12   Gage glass with photoelectric cell detector.   **Fig. 13**   Frost plug gage.

Dip Tube

A level gage commonly used in liquefied petroleum gas (LPG) and anhydrous ammonia tanks under pressure consists of a dip tube installed through a pressuretight fitting in the vessel. The outer end of the tube has an orifice to restrict flow and a shutoff valve which, when opened, allows observation of the state of the discharged fluid, liquid, or vapor, thereby indicating whether the end of the dip tube is submerged in liquid or vapor.

A variation of this consists of a dip tube through a packing box in the side of the vessel. The inner end of the tube is bent at an angle so that a half rotation of the tube in the packing box will traverse the inner end of the tube through the vertical range of liquid level. The angular position of the dip tube as it starts to discharge liquid indicates the liquid level.

ELECTRODE OR PROBE SYSTEMS

If the liquid level is to be controlled at one specific point or between predetermined limits, and no indication or record is required, simple and inexpensive electrode or probe systems are often suitable. Such systems depend on the electric conductivity of the vessel liquid, employing electrodes or probes installed vertically, as shown in Fig. 14, at the points of maximum and minimum level, but insulated

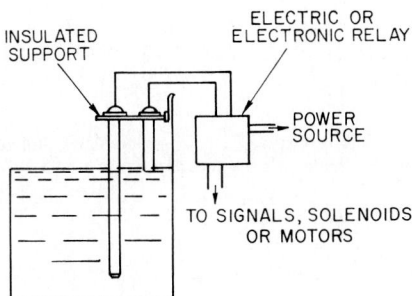

Fig. 14   Electrode or probe-type level system.

from the vessel. Low-voltage current flows between the vessel wall and the electrodes when they are covered by the liquid. This current flow is used to operate sensitive electric relays or, if the liquid conductivity is low, electronic relays.

Both electric and electronic probe systems can be used for interface-level control where one liquid is a conductor and the other is not. Both are also applicable for open tank or pressure vessel service. Because the vessel is a part of the electric sensing circuit, it must be made of a conducting material, or a plate of conducting material must be installed in it to provide a continuous circuit with the electrodes and liquid.

Electric Type

For most aqueous solutions of electrolytes or other liquids that have resistivities lower than 20,000 $\Omega \cdot cm$ and will not be ignited by arcing, a relatively simple electrode system employing a sensitive electric relay can be used. When the liquid surface is relatively calm, a single electrode can be used to provide high- or low-level control and alarm. If the surface is turbulent, however, two electrodes are employed for level control at one point, the electrodes being installed with a small vertical distance between them and interlocked in the electric circuit so that, in effect, a "dead" neutral zone is provided in the on-off control action.

Electronic Type

With liquids of low electric conductivity, electronically operated relays are employed to increase the sensitivity of the electrode system to a very small current. Such devices are available for operation at resistivities up to 20 million $\Omega \cdot$ cm, with a knob adjustment for selection of the most suitable relay sensitivity. Even distilled water or alcohol has sufficient sensitivity to operate such relays.

System for Molten-Glass Level (Hartford Empire System)

Operation of this system as diagramed in Fig. 15 is as follows: An electric motor-driven mechanical relay system lowers a probe electrode until contact is made with the molten glass. This contact completes a circuit through the glass which is conducting at elevated temperatures. An electromechanical relay stops the probe at the point of contact and holds it in position for approximately 2 s. During this 2-s interval, the electric relay also makes contacts which energize the recording balancing circuit in an electronic recorder. This instrument records the position of the glass level as determined by the position of the slidewire in the mechanism during this interval. After the 2-s recording interval, the electromechanical relay raises the probe until the contact is broken. The cycle then repeats itself. The total time of an average cycle of operation is approximately 12 s.

The gaging unit has a maximum span of approximately 3 in (7.6 cm). Its sensitivity is on the order of 0.005 in/1-in span (0.13 mm/25.4-mm span). Maintenance of the system is low, calibration is comparatively easy, and accuracy over long periods of time is obtainable. The system may be used to control glass-level height by electric or pneumatic means feeding proper ingredients into the melt pot by control of the batch charger.

BUOYANCY METHODS

Some liquid-level detectors employ floating, partially submerged, and fully submerged elements. These systems use buoyancy (or force of buoyancy) as the fundamental sensing principle.

Theory of Buoyancy

Archimedes' principle states that the resultant pressure of a fluid on a body immersed in it acts vertically upward through the center of gravity of the displaced fluid and is equal to the weight of the fluid displaced. With reference to Fig. 16, it will be noted that, by measuring the difference in

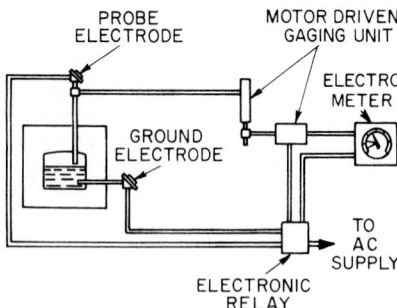

Fig. 15 Electrode system for molten-glass level measurement.

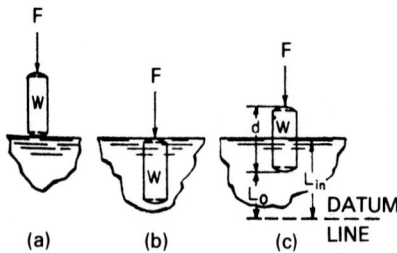

Fig. 16 Relation between level position and displacer element. (*a*) Level position at the bottom or below the displacer, (*b*) level position at the top or above the displacer, (*c*) level position between the top and bottom of the displacer.

weight of a partially submerged element at various degrees of submergence, the level of the liquid in which the displacer element is submerged can be determined. The following equations are useful for determining the forces exerted by the displacer element.

Condition 1

Level position at bottom or below displacer (Fig. 16*a*):

$$F = W$$

where F = force or weight to be supported
W = weight of displacer

Condition 2

Level position at top of displacer (Fig. 16*b*):

$$F = W - \frac{V \times G}{D}$$

where V = volume of displacer
G = specific gravity of liquid at a reference temperature (usually 60°F, 15.6°C)
D = density of fluid

In the case where V is given in in³, F is in lb, W is in lb, and G is in lb/in³, which for water is 62.4 lb/ft³, or 1/27.7 lb/in³. The relationship, of course, also applies where appropriate metric units are used.

Condition 3

Level position in intermediate position (Fig. 16*c*):

$$F = W - \frac{V \times G}{D} \times \frac{L_{in} - L_0}{d}$$

where L_{in} = level position in intermediate position
L_0 = level position at bottom or below displacer
d = length of displacer

In the English units given for condition 3, L_{in}, L_0, and d are given in inches.

With a cylindrical displacer, F varies as the level position around the displacer varies. The value of F is measured by suitable means, such as a torsion spring, pneumatic force balance, or electronic transducer. Thus, this value becomes a function of the level position above the datum line and is governed by the relationship shown in Fig. 17.

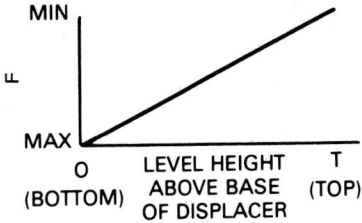

Fig. 17 Force F versus level height.

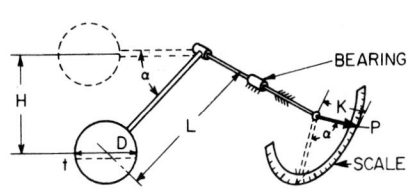

Fig. 18 Ball-float mechanism operating principle.

Float-Type Mechanisms

Float mechanisms use the principle of a buoyant member which floats on the surface of the liquid and changes position as the liquid level varies.

Ball-Float Mechanism

A physical member, floating on a liquid and referred to a datum point, provides a direct means of liquid-level measurement. Mechanically, a hollow metal ball is one of the most practical designs, as its nonabsorbent qualities eliminate buoyancy changes and a sphere provides maximum buoyancy for its weight.

A ball, attached to a rod and thence to a rotary shaft operating in a bearing, trunnion, or packing gland, with a pointer and scale is shown in Fig. 18.

Requirements for this device are simple. For maximum sensitivity, the ball float should be weighted so that it will sink to its largest (center) section. As the level rises or falls around the sphere, this produces the largest amount of available power in both directions needed to overcome friction of shafts and bearings plus inertia of component parts.

Range of Level Measurement

The effective travel of a float ball and shaft measuring device is limited by practical considerations. Angle α, shown in Fig. 18, should not exceed 60° of the shaft rotation ($\pm 30°$ from the horizontal) in order to obtain a relatively satisfactory measurement and response.* The maximum practical range of level measurement can, therefore, be approximately expressed as

$$\frac{H}{2} = L \sin \frac{\alpha}{2}$$

$$H = 2L \sin \frac{\alpha}{2}$$

where H = range of level measurement
$\quad\quad L$ = length of rod connecting ball float to shaft
$\quad\quad \alpha$ = angle of shaft rotation, degrees

From the foregoing, it can be seen that height of level measurement may be mechanically limited by the length of the rod L.

Power Availability

The power developed by a ball-float unit is a function of the allowable displacement of liquid around the circumference of the float ball. For example, if the ball is held in a fixed position, by friction or other means, a rise in level at the circumference of the float will cause a displacement of liquid equal to the product of the segment of the float and the level rise. A buoyant force equal to the weight of displaced liquid is produced. On the other hand, if the liquid level falls around the circumference of the float, the weight of liquid equivalent to the volume of the displaced segment will be the value of the negative buoyancy, or the amount of force acting in a downward direction.

As shown in Fig 18, this force can be expressed as

$$F = 0.0361 \, AtG$$

where F = force of float due to buoyancy effect
$\quad\quad A$ = area of float segment covered by level rise or exposed by level fall
$\quad\quad t$ = amount of level rise or fall
$\quad\quad G$ = specific gravity of liquid

---

*When 60° is substituted for α, H becomes equal to L.

English units are usually used: F is in lb, A is in in^2, t is in in. The relationship, of course, also applies where appropriate metric units are used. In the foregoing equation, 0.0361 is lb/in^3 of water.

The area A for a ball float is $D^2/4$, where D is the diameter of the float in inches. Therefore, the foregoing equation becomes

$$F = 0.0284D^2tG$$

The available torque required to overcome friction, resistance, and other forces in the system at any point p on the measurement arm as illustrated can be expressed as

$$T = \frac{FL}{k}$$

where T = torque at a point on measuring arm a distance k from rotating shaft, in·lb
F = force of float, lb
L = length of rod connecting float to rotating shaft, in
k = length of measuring arm to point p, in

Specifically, for a ball float, the above equation becomes

$$T = \frac{0.0284D^2tGL}{k}$$

Float Valves

From measurement it is a simple transition to a closely coupled controller such as the float valve shown in Fig. 19. In this device, the float directly and mechanically positions a valve mechanism to open and close it and allow more or less flow of liquid into the vessel. These devices have been used for many years for open tanks, reservoirs, sumps, and domestic plumbing fixtures, among other applications of a relatively simplistic nature. A different configuration of a mechanically operated control valve is shown in Fig. 20. A cage-type float-operated gage linked to a control valve is shown in Fig. 21. Devices with these configurations have been widely used for the measurement and control of levels in hot wells, storage tanks, and stills, among other similar applications. Where a float measuring device is located within the vessel, repair or maintenance while the unit is in operation cannot be made. Level measurement and control may be upset and erratic if ebullient fluid surfaces exist in the vessel. Gage-type devices (Fig. 21) generally are limited to 6-, 8-, and 10-in ($\approx$15-, 20-, and 25-cm) float sizes to avoid excessive size housings, and level measurement distance is normally limited to 1.5 times the float diameter. Power limitations imposed on the measurement and control of liquid level by direct mechanically operated ball-float devices created the need for a pilot- or relay-operated device using an external source of energy for measurement and control action. See Fig. 22.

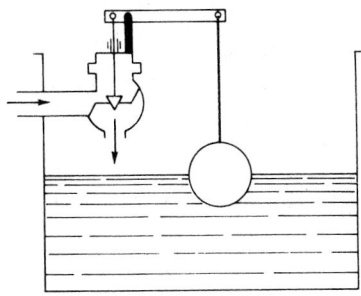

Fig. 19 Schematic diagram of a mechanically operated float valve.

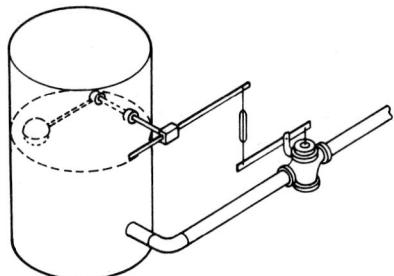

Fig. 20 Ball-float mechanism linked to a control valve.

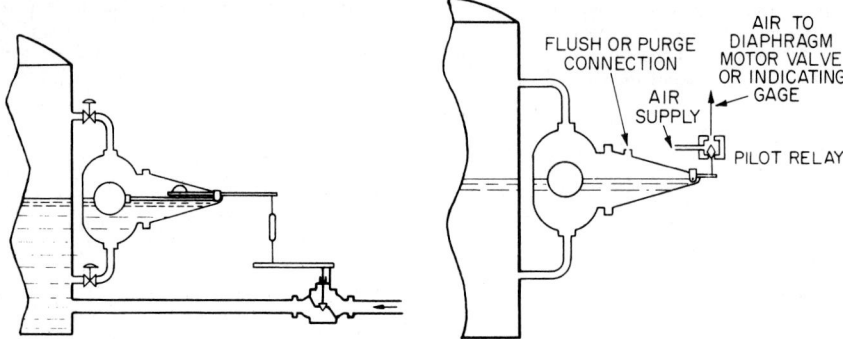

Fig. 21 Cage-type float-operated gage linked to a control valve.

Fig. 22 Float-cage-type gage with a pilot relay.

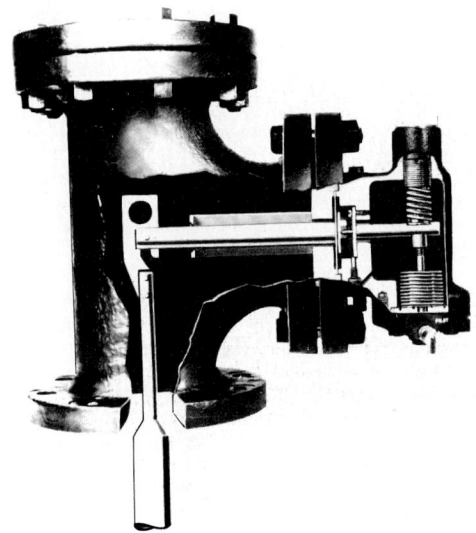

Fig. 23 Liquid-level transmitter. *(Moore Nullmatic.)*

Float-Type Transmitters

Liquid-level transmitters (Fig. 23) provide an accurate and dependable solution for many problems. These devices also can be used for the measurement of density or interface level. The transmitter shown features a packless design which permits the use of a variety of corrosion-resistant materials, permits mounting on top of a vessel above the liquid (Fig. 24), and is well suited to small-diameter tanks, as well as for hortonspheres and other vessels. The transmitters operate on a force-balance, weighbeam principle to measure the change in buoyant force resulting from a change in liquid level. The buoyant force is transmitted to the sensing bellows through the float arm, which is supported by a packless, flexible disk.

A booster-pilot valve, furnished with each transmitter, maintains all members in equilibrium at essentially their original positions. When the transmitter is in equilibrium, the forces exerted at the booster-pilot valve are also in balance. See Fig. 25. In this null-balance condition, the nozzle back-pressure will equal the transmitted pressure plus the pressure caused by the differential spring. The force of the differential spring, therefore, determines the pressure drop across the nozzle in the transmitter. Since this pressure drop remains essentially constant throughout the measuring range, the operation of the instrument is essentially linear.

A rise in liquid level changes the buoyant force on the float and tends to move the bellows and the nozzle seat toward the nozzle, thus increasing nozzle back-pressure. This increase in nozzle back-pressure forces the diaphragm in the pilot valve downward—closing the exhaust port and opening the supply port—to increase the transmitted output pressure until the change in output rebalances the change in buoyant force. A fall in liquid level reverses the operation, exhausting transmitted pressure through the exhaust port to rebalance the system. With each of these transmitters, the transmitted output pressure balances the buoyant force on the float at all times. The fulcrum distance, the float cross section, and the bellows area are all fixed elements. Once established, the calibration and measuring span remain essentially constant.

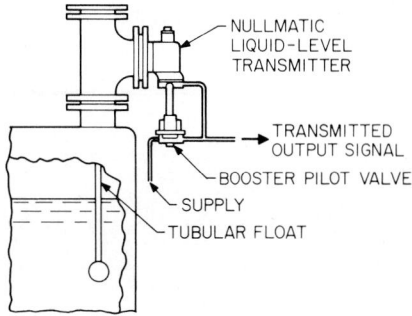

Fig. 24 Top-mounted installation of liquid-level transmitter. *(Moore Nullmatic.)*

Standard spans range from 18 in (46 cm) to 60 ft (18.33 m) of the liquid-level change. The standard supply pressure is 125 psi (646.3 cm Hg), and the output pressure range is 3 to 15 psi (15.5 to 77.6 cm Hg). The response level is 0.08% of full scale, and the reproducibility is 0.15% of full scale. Ambient temperature limits for the transmitter are −40 to 450°F (−40 to 232°C), and for the booster-pilot valve, −40 to 180°F (−40 to 82°C). The unit is furnished in a variety of construction materials, depending on the corrosive characteristics of the measured liquid.

Float-Type Controllers

The liquid-level controller shown in Fig. 26 consists of a float, mounted on a short float arm connected to the free end of a patented packless, flexible shaft. The fixed end of this flexible shaft is mounted on a standard flange. The shaft is tubular, with a flattened center section which permits only vertical float motion. This flexible part is similar to a bourdon tube in cross section. A stiff tongue within the shaft transmits the float motion to the air pilot for operation of a diaphragm-type motor control valve, or to a single-pole double-throw contact switch for operation of some other final controlling element. The device operates on the change in force which results from the buoyancy of the float—not on the motion of a float riding on a changing liquid level. The required float motion is approximately ¹⁄₃₂ in (0.8 mm). Various configurations to accommodate specific tank requirements are illustrated in Fig. 27.

In a somewhat different configuration, a solid float (polypropylene or Teflon)* is mounted on a short float arm. The device operates on a buoyancy change of less than 1 oz (28.3 g) to provide full actuation of a microswitch. See Fig. 28. The unit operates from full vacuum to 600 psig (≈ 4.1 MPa). Internal and external installations of the device are shown in Fig. 29.

Float-Operated Hydraulic Level Gage

The device shown in Fig. 30 incorporates a float member and arm which operate a dual-bellows member mechanically connected together, each a member of a separate hydraulic system terminating in a receiving bellows member. The two hydraulic circuits are completely filled with a liquid whose temperature-viscosity relationship is relatively constant, thus providing two separate hydraulic systems mechanically connected for transmission of float motion. The fundamental principle has been in use for years in the measurement of all kinds of liquids, corrosive and noncorrosive. As the float moves downward, the float arm operates the stroke link. This causes bellows A to expand and take in liquid from the dial or indicating end bellows D, which compresses. Simultaneously, bellows B compresses and displaces liquid into bellows C. Since bellows C and D are linked, their combined

*DuPont.

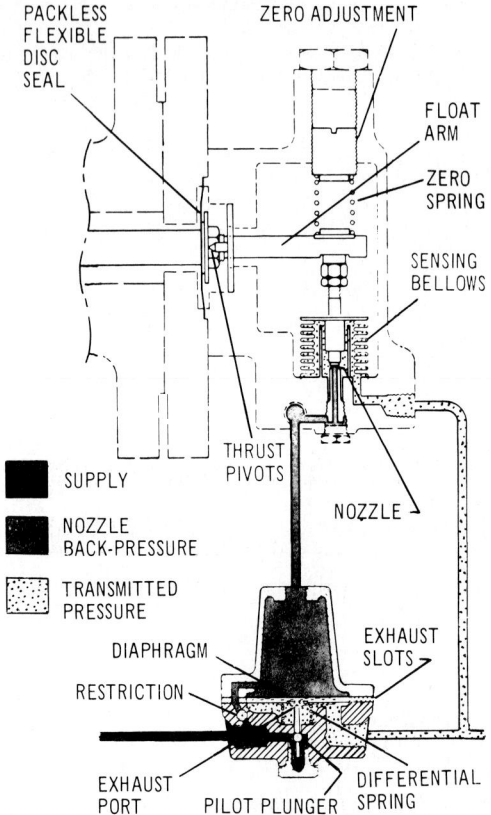

SUPPLY

NOZZLE BACK-PRESSURE

TRANSMITTED PRESSURE

Fig. 25 Booster-pilot valve used with liquid-level transmitter. *(Moore Nullmatic.)*

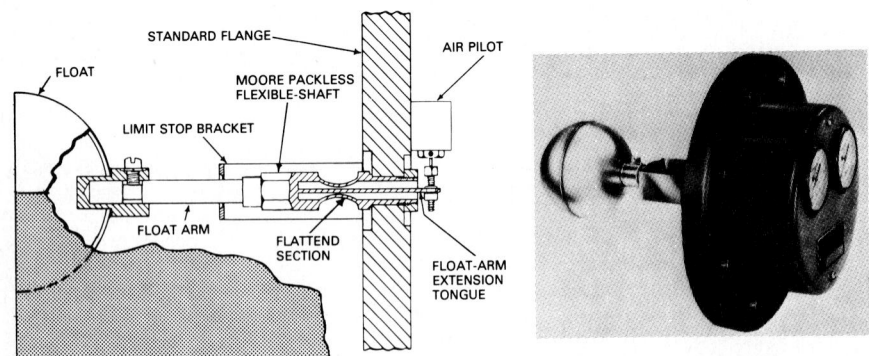

Fig. 26 Force-balance liquid-level controller. *(Left)* Mounting arrangement in side of tank, *(right)* complete pneumatic assembly. This controller is also available for electrical operation. *(Moore Products.)*

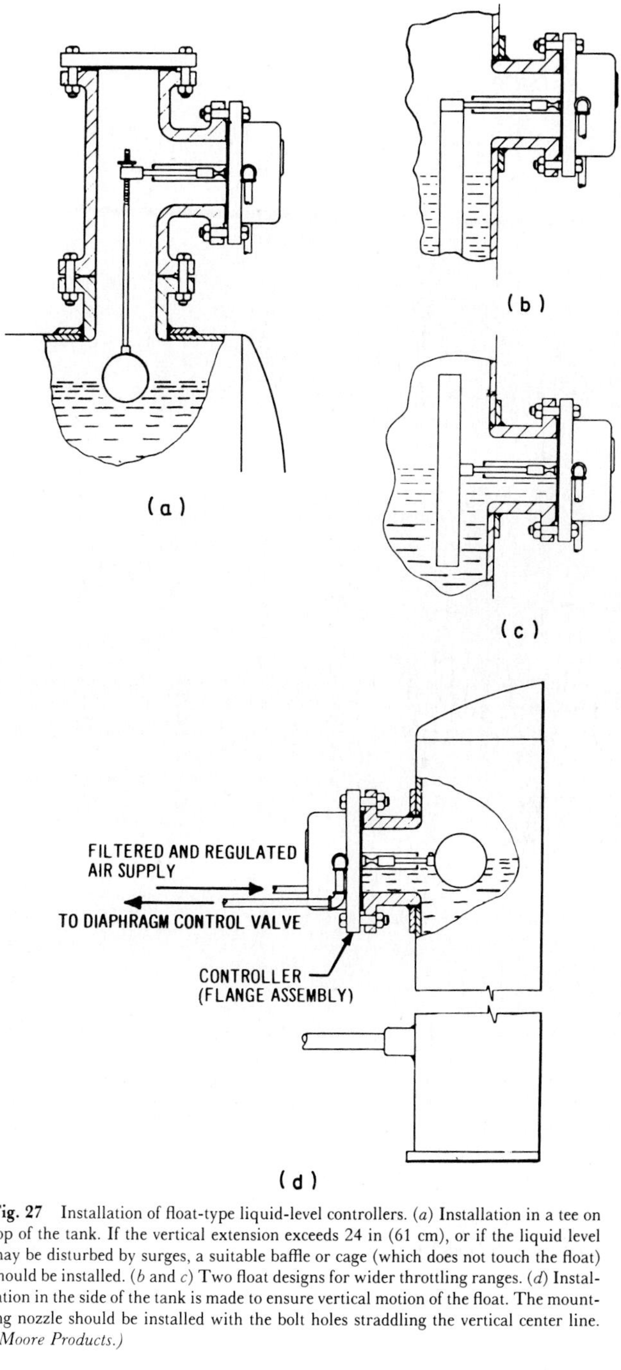
(b)
(a)
(c)
FILTERED AND REGULATED AIR SUPPLY
TO DIAPHRAGM CONTROL VALVE
CONTROLLER (FLANGE ASSEMBLY)
(d)

Fig. 27 Installation of float-type liquid-level controllers. (*a*) Installation in a tee on top of the tank. If the vertical extension exceeds 24 in (61 cm), or if the liquid level may be disturbed by surges, a suitable baffle or cage (which does not touch the float) should be installed. (*b* and *c*) Two float designs for wider throttling ranges. (*d*) Installation in the side of the tank is made to ensure vertical motion of the float. The mounting nozzle should be installed with the bolt holes straddling the vertical center line. *(Moore Products.)*

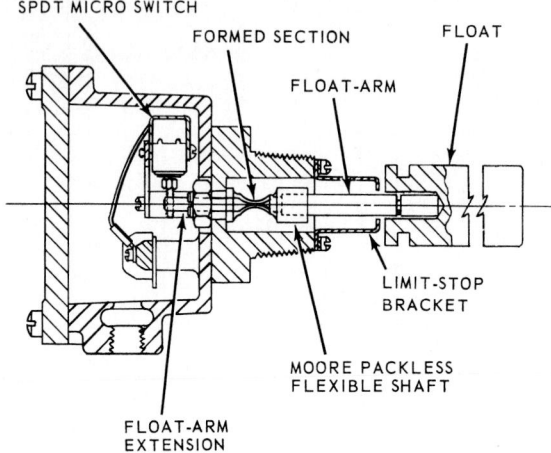

Fig. 28 Sectional diagram of an electric liquid-level controller equipped with a solid float. *(Moore Products.)*

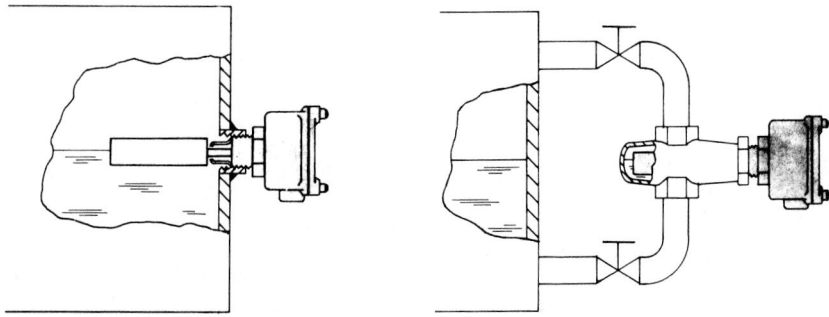

Fig. 29 Electric liquid-level controller. *(Left)* Direct-mounted in the side of a vessel. The float passes through a 1½-in (≈3.8-cm) NPT coupling. *(Right)* For external mounting, the controller can be furnished in a float chamber ready for installation. Standard connections on the chamber are 1 in (≈2.5 cm) NPT. *(Moore Products.)*

action causes a pointer or dial to move and indicate the liquid level in the vessel. When the liquid level increases, the float moves up and the action is reversed.

A link arrangement at the indicating or recording end provides compensation for differences in temperature to which the system and the communicating tubing are exposed—and avoids any impairment in accuracy due to temperature variations. This is accomplished by offsetting bellows C and D and mechanically connecting them with a fulcrum link member having its axis at K.

These devices are generally limited to a distance between vessel and indicator of 250 ft (76 m) and to a vessel pressure of 200 psi (≈1.38 MPa).

Magnetic-Type Float Gages

The problems of stuffing boxes in ball-float gages led to designs which employ magnetic forces to follow or sense the float position. Two such designs are described below: (1) the magnet-bond method and (2) the magnetically operated float switch.

From the operating principle, several advantages of a magnetically operated device are apparent.

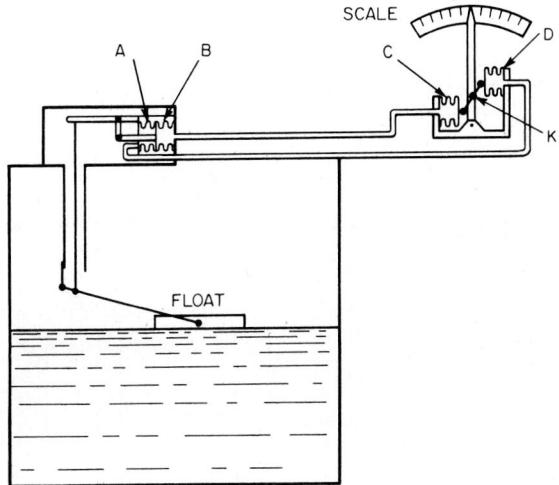

Fig. 30   Float-operated hydraulic-type gage.

Because no seal is required around the shaft actuated by the float, any possible application problems from this source are eliminated. Further, the unit involves a minimum of moving parts, so that friction and wear of parts is negligible; sensitivity of control action is also high.

Magnetic-Bond Method

A float mechanism employing the magnetic-bond method of measuring the liquid level utilizes a magnetic member which floats on the surface of the liquid. A strong magnetic field is transmitted from the magnetic float to a suitable pickup device which actuates the indicator or controlling device.

A typical construction is shown in Fig. 31. Although a doughnut-type float is illustrated, this member may be a ball, disk, or other shape. Other means of conveying magnetic bond action to the pickup member may also be employed; and mechanical, pneumatic, hydraulic, or electric coupling to the receiver may be used.

No stuffing boxes are required, thus providing a leakproof construction suitable for toxic, explosive, or flammable liquids.

The float member is free-floating—with no provision for manually checking its position—therefore, it must be ensured that it will be free-moving and not stick, freeze, or bind.

A variation of the method is shown in Fig. 32, where an indicating scale and steering device are fastened to the outside of the liquid or indicating chamber.

The float-controlled actuating magnet has greater magnetic force than the edge-magnetized bicolor wafers in the scale. As the actuating magnet passes the wafers, they are rotated 180° and present the other color for indication.

In another version (Columbia Controls), continuous level indication is provided by a series of reed switches within the central column of the unit. A magnet contained in the float ring closes the respective reed switch. The reed switch, in turn, is connected to a particular leg of a voltage divider. The varying voltages can be used to represent the level.

Magnetically Operated Float Switch

Another type of ball-float mechanism employing magnetic force to operate a mercury switch is illustrated schematically in Fig. 33. In this device the ball float positions a magnetic piston, attached to the float rod which moves up and down within a nonmagnetic enclosing tube, as shown. Outside the enclosing tube is a high-strength permanent magnet attached to a pivoted arm with a mercury switch mounted on it, as illustrated.

When the level is up, the magnetic piston is in the magnetic field and the magnet is drawn against

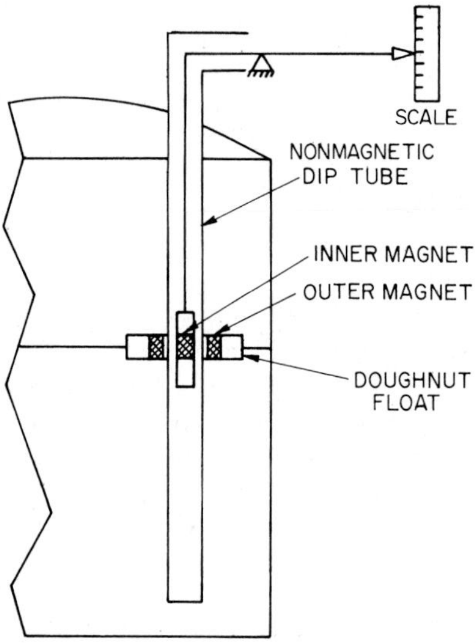

SCALE

NONMAGNETIC
DIP TUBE

INNER MAGNET

OUTER MAGNET

DOUGHNUT
FLOAT

INDICATOR

MAGNETIC
STEERING
DEVICE

MAGNET

FLOAT

COLORED
INDICATOR
WAFERS

Fig. 31   Principle of a magnetic-bond-type ball-float gage.

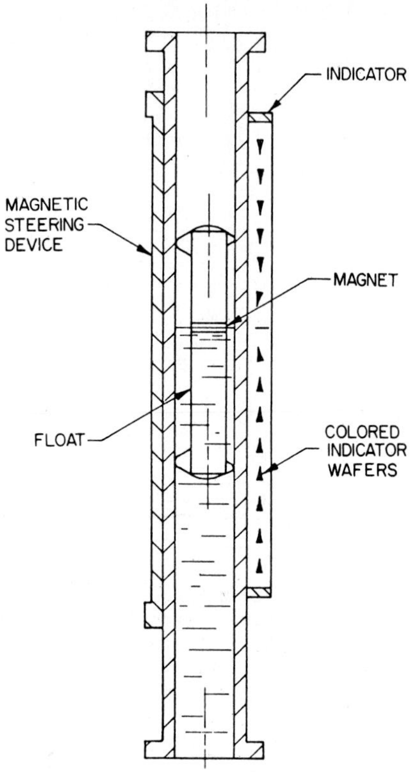

Fig. 32   Magnetic indicator gage.

the enclosing tube, thereby tilting the mercury switch to one position (to open or close a circuit as desired). When the level drops to a predetermined point, the piston is moved down out of the magnet field and the magnet is pulled out by the tension spring, thereby tilting the mercury switch to its other position. Figure 33*a* and *b* illustrates these two conditions with an SPST mercury switch open for high level and closed for low level, where, for example, an electrically operated valve might be

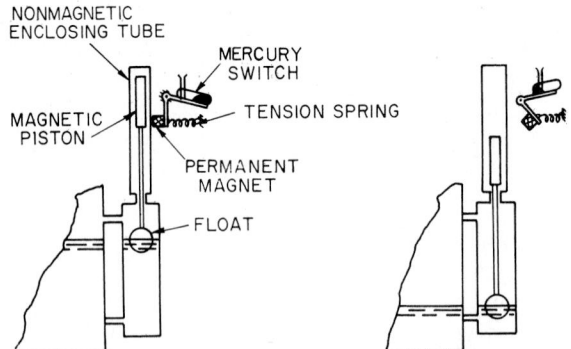

NONMAGNETIC
ENCLOSING TUBE

MERCURY
SWITCH

MAGNETIC
PISTON

TENSION SPRING

PERMANENT
MAGNET

FLOAT

Fig. 33   Principle of a magnetically operated float switch. (*a*) High or normal level (mercury switch open), (*b*) low level (mercury switch closed).

energized to open and admit more liquid. A variety of switching actions, however, can be obtained by selection of the type of switch, and multiple switches can be installed on the float assembly.

Chain or Tape Float Gages

In gages of these types the float is connected by means of a flexible chain or tape to a rotating member (which is connected to the indicating mechanism). A counterweight, as shown in Fig. 34, is employed to keep the chain or tape taut as the float rises or falls. In a chain-type gage, the chain engages a sprocket which turns the rotating member; in the tape type, the tape wraps around a drum.

These gages can be installed either within a tank or vessel or in a long pipe located adjacent to and connected to the vessel by suitable connections into the liquid and vapor phases of the vessel.

Displacer-Type Mechanisms

Since its introduction, the displacer measuring element has played an important role in measuring and controlling the liquid level in process control. The device is simple, reliable, accurate, and adaptable to measurement over a very wide range of level variation; it can be used at very high pressures and temperatures and can be installed in many different ways.

The displacer element is usually of cylindrical shape, of a length to correspond to the level variation expected, and weighted to sink in the liquid being measured. The maximum force change developed is equivalent to the weight of the liquid displaced when the displacer is completely submerged minus the weight of the vapor displaced when completely out of the liquid.

Various methods have been used to transmit the buoyant force change on the displacer to a motion or force outside the pressure wall. The most successful has been the torque tube which over-

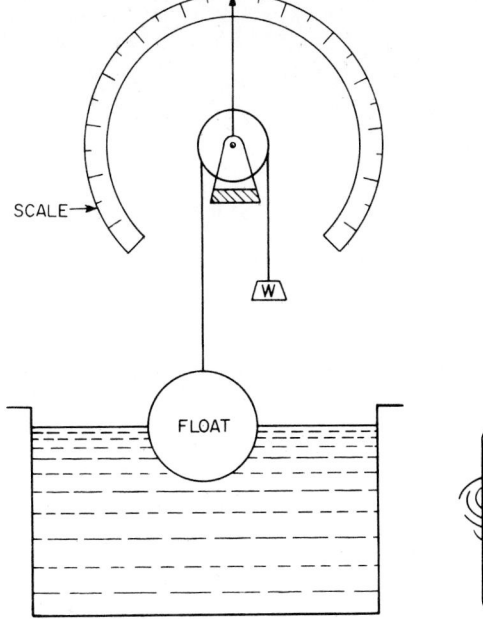

Fig. 34 Chain- or tape-type float gage.

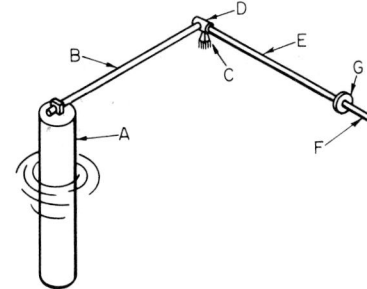

Fig. 35 Schematic diagram of a torque tube displacer unit.

comes the leakage and friction problems of the stuffing box and yet provides adequate motion to drive either pneumatic or electronic transmitters and controllers. Torque tubes can be designed for very high pressures and temperatures and are virtually insensitive to pressure changes within the vessel.

A schematic diagram of this design is shown in Fig. 35. The displacer A is supported at one end of a support rod B, while the other end of the rod is carried by a pivot bearing assembly C. A torque tube E is located at 90° to the support rod and is attached to it at the pivot point. This tube has a pressuretight inner end fitting D and a flange member G bolted or suitably held in place to form a pressure seal with a housing which encloses the above parts. A shaft F is inserted into the hollow torque tube and is attached to the fitting D. Before the outer torque tube flange G is bolted or sealed, it is manually rotated by an amount necessary to twist the tube, rotating the support rod about its pivot point and completely supporting the weight of the displacer.

Operation of the displacer type is based on Archimedes' principle which states that a body immersed in a liquid will be buoyed up by a force equal to weight of liquid it displaces. A rise in liquid about the displacer causes it to become effectively lighter by the amount of the weight of the liquid being displaced. This change in weight then permits the twisting spring action of the torque tube to raise the displacer and rotate the support rod B about its pivot point along with the torque end fitting D. Since the operating shaft F is attached to fitting D, it too is rotated by a similar amount, thus giving a measurable shaft rotation which is directly proportional to the liquid-level change.

Most of the present designs use approximately 2 to 7° of rotary movement produced by the torque tube shaft to actuate the flapper of an air relay pilot mechanism which usually is attached directly to the cage housing enclosing the torque tube. There are several methods of converting the shaft movement into a proportional air-pressure signal to be used for level indication or control. Electrical systems are also available.

Span or Proportional-Band Adjustment

The proportional band of a liquid-level indicator or controller is defined as the percentage of full level range necessary to change the output pressure by an amount to actuate the indicator over its range or to fully travel its companion control valve. Thus, with a 14-in (35.6-cm)-long displacer and a 7-in (17.8-cm) liquid-level change being required to obtain a full output signal, the level indicator or controller would have a 50% proportional band.

When proportional plus reset (integral) control action is desired, it can be easily added to the pilot relay system.

Level-Position Adjustment

Most of the liquid-level indicators and controllers available today provide an adjustment for the starting-point or level set positon.

Specific Gravity Correction

Since the displacer movement is a function of the weight of liquid being displaced, it is obvious that the torque tube shaft and flapper rotation will vary in total range, depending on the specific gravity of the liquid involved. Hence, it is necessary to correct for this difference in order to provide a uniform output pressure signal from the pilot relay. The adjustment may be a mechanical change in the linkage of flapper-arm travel, or a means to incorporate it in the proportional-band adjustment and make two corrections in specific gravity and proportional band together so as to produce a flapper-nozzle travel relation which will produce a full signal output with unit level height change but with a different rotation of torque tube and shaft. A calibrated scale and suitable adjustment are provided for this purpose.

Force-Balance Unit

In this design shown in Fig. 36, a change in buoyant force is transmitted through (1) a float arm and flexible disk or (2) flexure tube or torque tube pressure seal or (3) magnetic-bond method to a force balancing pilot system.

A rise in liquid level causes the displacer to become effectively lighter by the amount of the additional liquid displacement, which in turn permits the balance spring to move a flapper arm closer

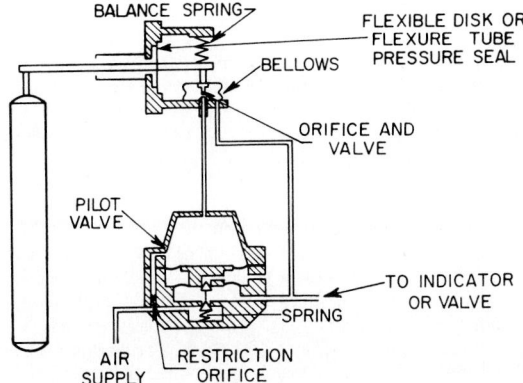

Fig. 36 Schematic diagram of a force-balance displacer unit with a flexible disk pressure seal.

to a nozzle. This action increases the nozzle pressure to an air relay, causing an increase in output pressure of magnitude sufficient to balance the force of the loading spring with a feedback bellows and return the displacer to its original position.

Mounting of Displacer-Type Units

Many units of this kind are of the cage type and are attached to the side of a vessel with equalizing connections containing block valves, as shown in Fig. 37. With this type of installation it is possible to close the block valves and remove part or all of the measurement unit for servicing. These units are also available for mounting on the side or top of a vessel; when so used, the cage portion is eliminated and the displacer is hung from its supporting rod directly inside the vessel. When this internal-type unit is used, and there is a possibility of surging or turbulence, a stilling chamber should be provided around the displacer to minimize the effect of erratic displacer action.

Construction of Displacer-Type Units

Cage-type units are available with 1½- and 2-in (≈3.8- and 5.1-cm) screwed or flanged equalizing connections with a choice of locations at the top, sides, or bottom of the cage. Many commercial units are of iron or steel with pressure standards up to 2500 psi (17.2 MPa). Internal parts in contact

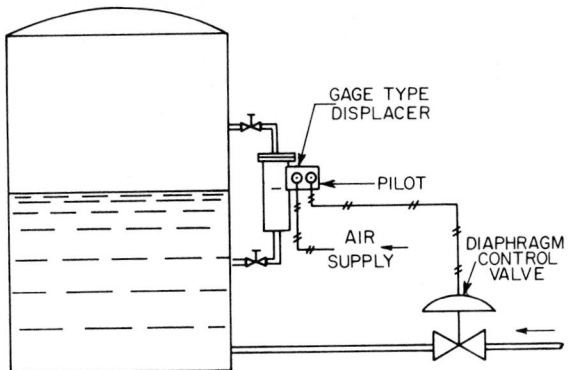

Fig. 37 Typical installation of a cage-type displacer unit.

with the liquid are made of corrosion-resistant materials, such as stainless steel, Monel*, nickel, bronze, Hastelloy†, and other alloys.

To accommodate the various level range requirements, standard cages are made to fit displacer lengths up to 15 ft ($\approx$4.6 m) and, for special needs, even longer. Common range lengths are 14, 32, 48, 60, 72, 84, 96, 108, 120, and 180 in ($\approx$35.6, 81.3, 121.9 cm; 1.5, 1.8, 2.1, 2.4, 2.7, 3.0, 4.6 m). For longer displacer lengths, it is usually more practical to use a control unit flanged to the side or top of a vessel and hang the displacer directly in the liquid being measured.

Interface Measurements with Displacer-Type Units

Interface liquid-level measurement is the measurement of the location of the contact point between two immiscible liquids. All displacer level measurements are actually interface measurements, since the variable is the level point between two fluids having different specific gravities. It follows that these two immiscible fluids may be liquid and gas, liquid other than water and water, or two liquids other than water. For interface applications having relatively close specific gravities, it may be necessary to use a larger than standard displacer in order to obtain a buoyant force change of sufficient magnitude to operate the pilot relay mechanism properly. A special displacer must have the proper weight to sink in the heavier of the two liquids; also, when hanging in the lighter fluid, it cannot have a net weight of a greater value than the standard load rating for a given torque or flexure tube.

HYDROSTATIC PRESSURE METHODS

In application, hydrostatic methods of liquid-level measurement for open and closed vessels are of several configurations: (1) direct connection of hydrostatic head to measuring device, (2) diaphragm-box system, (3) air-trap system, (4) air-bubble tube or purge system, (5) force-balance system, and (6) opposed-diaphragm types, among others.

Theory of Hydrostatics

Hydrostatic head may be defined as the weight of liquid existing above a reference or datum line and can be expressed in various units, such as pounds per square inch, grams per square centimeter, and feet or meters of liquid measured. The head is a real force, due to liquid weight, and, as shown in Fig. 38, is exerted equally in all directions. It is independent of the volume of liquid involved or the shape of the containing vessel. Measurement above the datum line may be expressed by the relationship

$$H = \frac{P}{D} = \frac{P}{D_W G} = \frac{P \times M}{G}$$

where H = height of liquid above datum line
P = pressure due to liquid head
D = density of liquid at operating temperature
D_W = density of water at a reference temperature
G = specific gravity of liquid at operating temperature
M = multiplying factor, depending on units of measurement used. For example, if H is in in, P is in psi, D is in lb/in^3, G is in 62.4 lb/ft^3 (0.036 lb/in^3), and the reference temperature is 60°F (15.6°C), M = 27.70.

From this relationship, it is seen that a measurement of pressure P at the datum or reference point in a vessel provides a measure of the height of the liquid above that point, provided the density

*International Nickel.
†Union Carbide.

or specific gravity of the liquid is known. Also, this relationship shows that changes in the specific gravity of the liquid will affect liquid-level measurements by this method, unless corrections are made for such changes. During the past relatively few years, compensation for environmental changes which affect measurement accuracy has been automated in some systems through the use of microprocessors and sensors that continuously or intermittently detect changes in such factors as liquid temperature and/or density.

When a pressure greater than atmospheric is imposed on the surface of the liquid in a closed vessel, this pressure adds to the pressure due to the hydrostatic head and must be compensated for by a pressure measuring device which records liquid level in terms of pressure. See also Pressure Transducers and Transmitters in Sec. 3 of this Handbook.

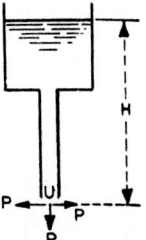

Fig. 38   Basic elements of hydrostatic head.

Pressure-Gage Systems (Open Vessels)

A conventional indicating recording pressure gage requires only calibration of its scale or chart in terms of liquid level in order to be employed for liquid-level measurement. Calibration is based on the relationship previously given.

Normally, these systems are a function of the total head and thus are not suitable for interface or duolevel measurements. The minimum range of measurement with conventional pressure gages employing a sensitive spring-and-bellows assembly is about 0 to 5 in (0 to 12.7 cm) of water. The maximum range has no practical limit.

Direct Connection of Gage to Open Vessel

The simplest system of installing a pressure gage is illustrated in Fig. 39. Within limits, however, the gage can be mounted below the minimum vessel level and the instrument zero-adjusted for the head due to the added liquid column.

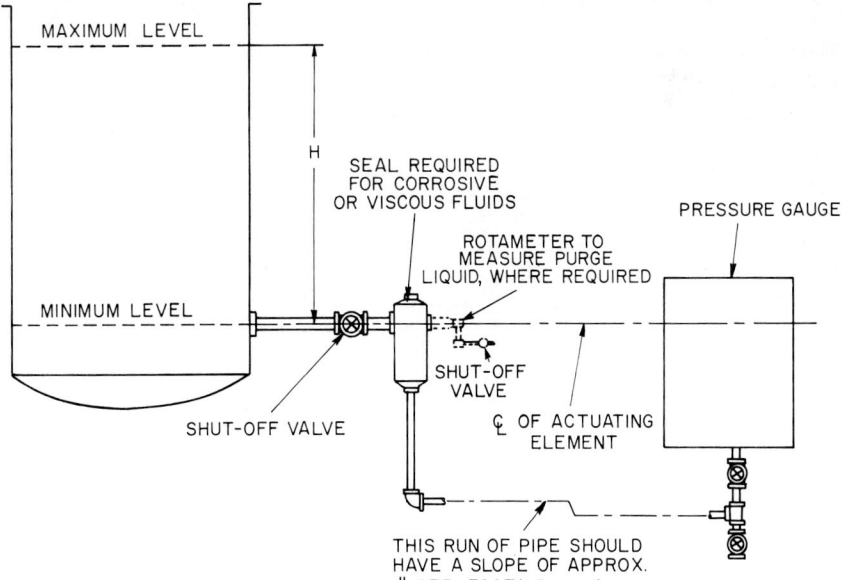

Fig. 39   Pressure gage system for open vessels.

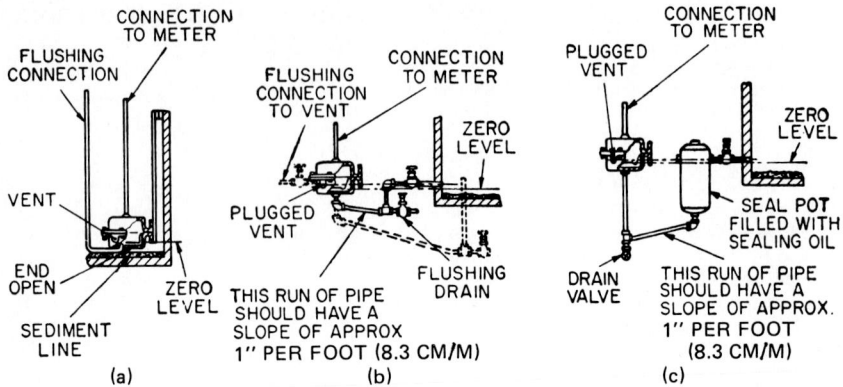

Fig. 40 Diaphragm-box system used for liquid-level measurement in open vessels. (*a*) Open-type diaphragm box submerged in the liquid, used for measurement of noncorrosive fluids under conditions that permit installation of the box in the vessel at approximately the minimum level. The box must be at least 2 to 3 in (5 to 7.5 cm) above any sediment in the vessel bottom to avoid clogging of openings on the box. (*b*) Closed-type diaphragm box located outside the tank, used where the liquid is noncorrosive but conditions do not permit installation within the vessel. (*c*) Closed-type diaphragm box for use with corrosive liquids.

Diaphragm-Box System

Where it is not feasible to locate the gage at or below the minimum vessel level as previously discussed, a diaphragm arrangement can be used. General configurations are shown in Fig. 40. Basically, this system operates by trapping air in the gage connecting line and in the upper portion of a box which has a flexible diaphragm. Hydrostatic pressure acts on the underside of the diaphragm, causing it to move upward and compress the entrapped air in direct relation to the head of liquid in the vessel. At least one manufacturer offers a choice of flush diaphragm, extended diaphragm, or remote seal diaphragm. See Fig. 41.

Air-Bubble Tube or Purge System

If air is connected into a tube system similar to that described for the air-trap system, and its pressure is regulated at a value slightly greater than the maximum head of liquid in the vessel, air pressure

Fig. 41 Remote seal continuous range level sensor for vented vessels. (*Sybron-Taylor.*)

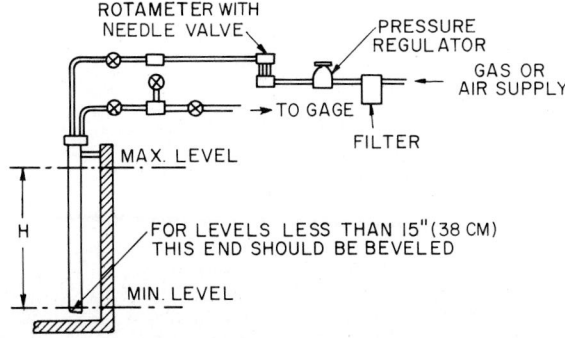

Fig. 42   Gas or air purge system for measurement of the liquid level in open vessels.

in the system will be equal to the pressure due to the hydrostatic head of the vessel liquid at any level because any excess air pressure will bubble out of the bottom of the tube. Figure 42 illustrates the principle of such a system as embodied in a common industrial arrangement employing a 2-in (5-cm) standpipe and cap to act as an air bell. Two connections are made to this pipe, one from the regulated air supply with a tube extending down inside the pipe to within about 3 in (7.5 cm) of the pipe bottom. The other connection is for the pressure gage.

Because air or some other gas is continuously bubbling from the bottom of the pipe, keeping liquid out, this system is also known as a purge type and is especially suited for measuring the level of corrosive liquids, liquids that become viscous on cooling, and liquids containing entrained solids. For corrosive liquids only the standpipe need be made of corrosion-resistant material. A further advantage of this system is that it permits location of the gage not only at any height with respect to the vessel but also at distances up to 1000 ft (305 m) from the vessel.

As shown in Fig. 42, the range of the gage for this system is based on the head for maximum level above the bottom of the immersed pipe. The pipe bottom is kept at least 3 in (7.5 cm) above the vessel bottom to avoid clogging from any sediment.

In typical applications, a rotameter with capillary-type restrictor, a pressure regulator, and an air filter are used. Air pressure is set at a value slightly greater than the hydrostatic pressure for maximum level, and its rate of flow adjusted by the needle valve so that about 0.2 to 2.0 std ft^3/h (57 × 10^{-4} to 57 × 10^{-3} m^3/h) are emitted from the bottom of the standpipe. For periodic level check methods, a hand pump may be used for the air supply.

Hydrostatic Differential-Pressure Meters

The basic operating principle of a differential-pressure-type liquid-level measurement system is illustrated in Fig. 43. The system as applied to an open vessel illustrates how a simple U-tube manometer can serve to measure the pressure due to hydrostatic head by connection to a minimum-level tap. As shown, the liquid head H in the vessel causes the manometer liquid to rise until its head h counterbalances the vessel head. The balance of pressure is expressed as

$$h \times D_W \times G_m = H \times D_W \times G_v$$

where h = manometer scale reading, units of height
D_W = density of water (at standard reference temperature)
G_m = specific gravity of manometer liquid (such as mercury)
H = height of liquid above tap, units of height
G_v = specific gravity of vessel liquid

Thus,

$$h = H \frac{G_v}{G_m}$$

For many years, the traditional application of the differential-pressure method was a basic mercury manometer (wet-type) system. See Fig. 44. For open vessels, the manometer was installed as

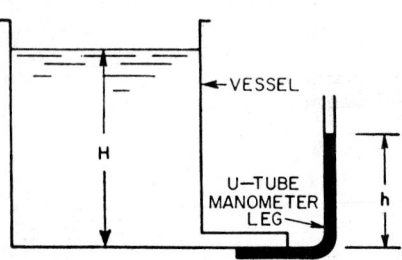

shown, with the float chamber side connected to the minimum-level tap and the range tube side connected to a reservoir pot located as shown to provide a liquid head equivalent to the minimum-level head in the vessel and open to the atmosphere. Operation of the system depended on the balance of heads, just as in a simple manometer. Here, however, vertical movement of the float is a fixed value related to full-scale movement of the instrument pen or pointer, and a range tube had to be selected to accommodate the change in head due to liquid level. Frequently, a special tube had to be designed to permit full-scale pen or pointer travel, or an output signal, with the given change in liquid level. The

Fig. 43 Operating principle of a U-tube manometer for measuring the liquid level in an open vessel.

wet-type system was largely replaced by the dry-type differential-pressure meter several years ago. Consequently the calculation of head measurements for mercury manometers is not included here.

Dry-Type Differential-Pressure Meter

In applications for indication, recording, and control of level, changes in differential pressure are sensed directly by a diaphragm or bellows. The resulting force or motion produced by the diaphragm or bellows may be converted into an electric or pneumatic transmission signal. For open vessels a dry-type differential-pressure meter may be installed as shown in Fig. 45, with the high-pressure side of the measuring element connected to the minimum-level tap. For a closed vessel application, separate sensors placed near the bottom and in the vapor spaces at the top of the tank may be used. Where provisions are made for the handling of corrosive and clogging fluids and where the specific gravity of the fluid is known, differential-pressure devices are relatively easy to calibrate. Differential-pressure cells are available with casings and diaphragms in numerous combinations of corrosion-

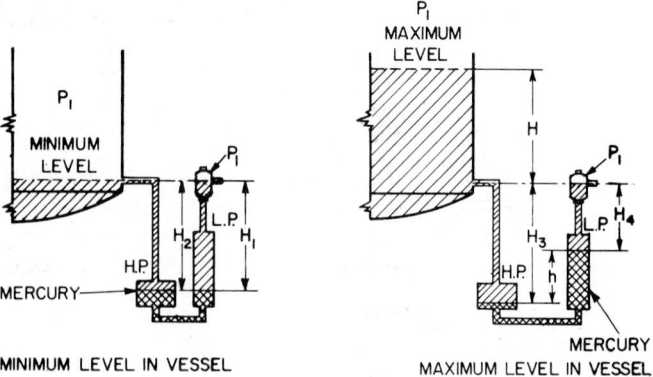

MINIMUM LEVEL IN VESSEL   MAXIMUM LEVEL IN VESSEL

Fig. 44 Principle of a wet-type mercury manometer for measuring the liquid level. Prior to the advent of the dry-type system, this configuration was widely used.

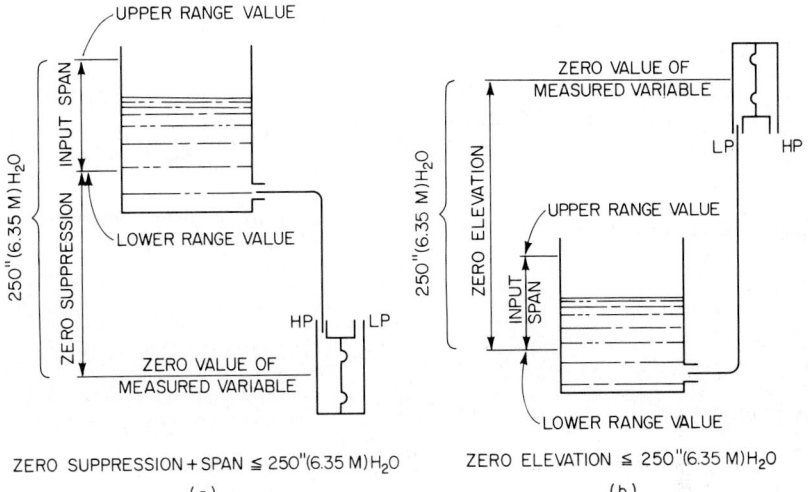

Fig. 45 For some applications it is necessary to either suppress the measured variable zero or elevate it. (*a*) Suppressed zero range, (*b*) elevated zero range.

resistant materials and coatings, including stainless steel, Hastelloy, Monel, Kel-F,* and Teflon, among others. In some situations, flanged differential-pressure cells may be a useful alternative. Although isolating valves are required, the flanged cells can be connected directly to the tanks, thus eliminating the requirement for piping on the high-pressure side of the sensor.

Differential-pressure calculations for open tanks are given in Fig. 46. For a closed tank with a

*3M Co.

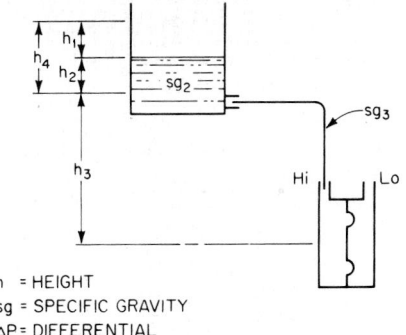

h = HEIGHT
sg = SPECIFIC GRAVITY
ΔP = DIFFERENTIAL

Any level or head:
$$\Delta P = h_1 sg_1 + h_2 sg_2 + h_3 sg_3$$
Lowest level or minimum head, where $h_2 = 0$, $sg_1 = 0$:
$$\Delta P_l = h_3 sg_3$$
Highest level or maximum head, where $h_1 = 0$, $h_2 = h_4$:
$$\Delta P_h = h_4 sg_2 + h_3 sg_3$$
Span $= \Delta P_h - \Delta P_l$
$\quad\quad\; = h_4 sg_2 + h_3 sg_3 - h_3 sg_3$
$\quad\quad\; = h_4 sg_2$

Fig. 46 Differential-pressure calculations (open tank).

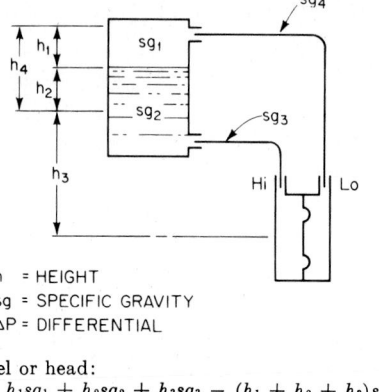

h = HEIGHT
sg = SPECIFIC GRAVITY
ΔP = DIFFERENTIAL

Any level or head:
$$\Delta P = h_1 sg_1 + h_2 sg_2 + h_3 sg_3 - (h_1 + h_2 + h_3) sg_4$$
Lowest level or minimum head, where $h_2 = 0$, $h_1 = h_4$:
$$\Delta P_l = h_4 sg_1 + h_3 sg_3 - (h_4 + h_3) sg_4$$
Highest level or maximum head, where $h_1 = 0$, $h_2 = h_4$:
$$\Delta P_h = h_4 sg_2 + h_3 sg_3 - (h_4 + h_3) sg_4$$
$$\begin{aligned}\text{Span} &= \Delta P_h - \Delta P_l \\ &= h_4 sg_2 + h_3 sg_3 - h_4 sg_4 - h_3 sg_4 - h_4 sg_1 - h_3 sg_3 \\ &\quad + h_4 sg_4 + h_3 sg_4 \\ &= h_4 (sg_2 - sg_1)\end{aligned}$$

Fig. 47 Differential-pressure calculations (closed tank with a noncondensable atmosphere over the liquid).

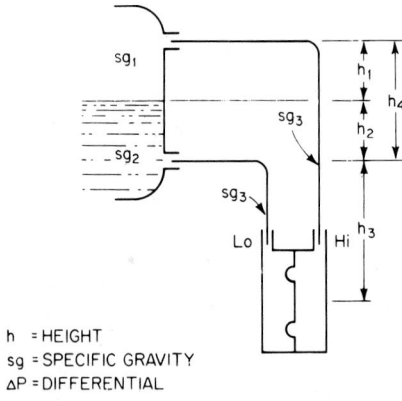

h = HEIGHT
sg = SPECIFIC GRAVITY
ΔP = DIFFERENTIAL

Any level or head:
$$\Delta P = (h_1 + h_2 + h_3) sg_3 - h_1 sg_1 - h_2 sg_2 - h_3 sg_3$$
Lowest level or maximum head, where $h_2 = 0$, $h_1 = h_4$:
$$\Delta P_l = h_4 sg_3 + h_3 sg_3 - h_4 sg_1 - h_3 sg_3 = h_4 (sg_3 - sg_1)$$
Highest level or minimum head, where $h_1 = 0$, $h_2 = h_4$:
$$\Delta P_h = h_4 sg_3 - h_3 sg_3 - h_4 sg_2 - h_3 sg_3 = h_4 (sg_3 - sg_2)$$
$$\begin{aligned}\text{Span} &= \Delta P_l - \Delta P_h \\ &= h_4 sg_3 - h_4 sg_1 - h_4 sg_3 + h_4 sg_2 \\ &= h_4 (sg_2 - sg_1)\end{aligned}$$

Fig. 48 Differential-pressure calculations (condensable atmosphere over the liquid).

noncondensable atmosphere over the liquid, see Fig. 47. For a closed tank with a condensable atmosphere over the liquid, see Fig. 48.

Piping Arrangements

See Fig. 49 for a piping arrangement for a closed tank with a noncondensable atmosphere above the liquid. In an open tank installation, only the high-pressure side piping is required.

For a vessel with condensable vapor above the liquid, such as a still or boiler, piping per Fig. 50 may be considered.

Sealing Liquid in Connecting Piping

Level measurements of liquids which are free of excessive solids but must be sealed from the meter body are often handled by the use of seal pots located at the minimum and maximum connections with the meter body connecting lines filled with a sealing liquid. This system is almost universally used in oil refinery service where the measurement of liquid levels in pressure vessels is required.

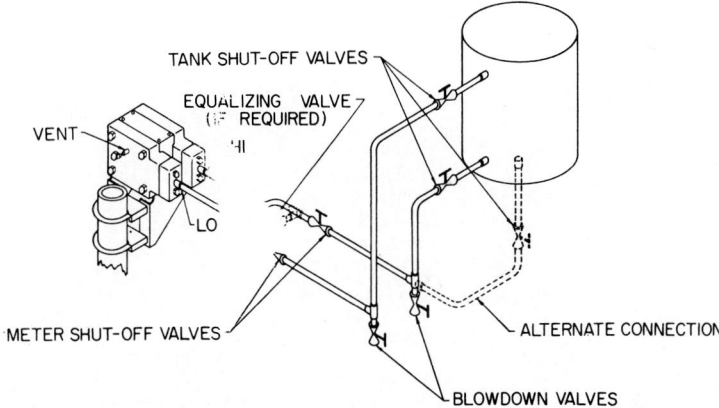

Fig. 49   Piping arrangement sometimes used for the differential-pressure method of measuring the liquid level in a closed tank with a noncondensable atmosphere.

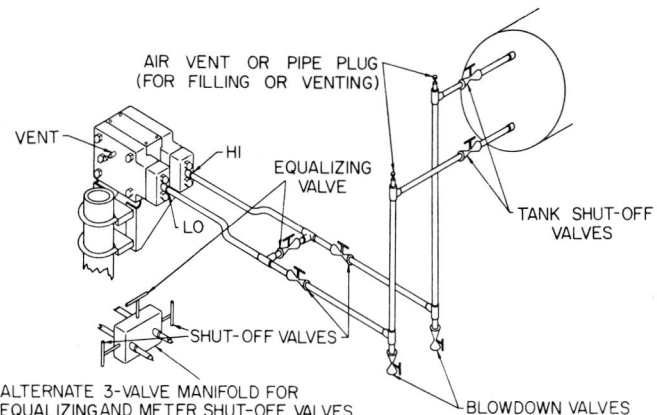

Fig. 50   Piping arrangement sometimes used for the differential-pressure method of measuring the liquid level in a closed tank with a condensable atmosphere.

In this service, the sealing liquid is generally heavier than the vessel liquid, quite often being a 50-50 mixture of glycerin and water, which has a specific gravity of 1.125 at 60°F (15.6°C). The system is also applied frequently in measuring the level of light volatile liquids under pressure, such as butane.

Sealing Liquids and Chamber

A schematic diagram is shown in Fig. 51.

The formulas for the required differential range of the instrument (including suppression) are exactly the same as those given above for the case where the outer-leg liquid has a different specific gravity. See Figs. 46 to 48. In this case sg_2 represents the specific gravity of the sealing liquid at its operating temperature (usually about 80°F).

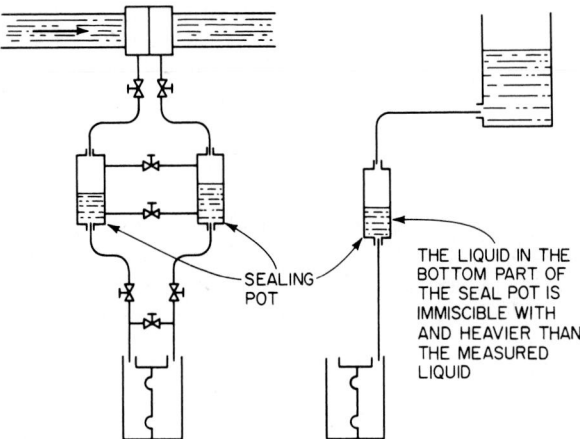

SEALING POT

THE LIQUID IN THE BOTTOM PART OF THE SEAL POT IS IMMISCIBLE WITH AND HEAVIER THAN THE MEASURED LIQUID

Fig. 51   Schematic view of a sealing chamber. In some cases it is desirable to keep the measured fluid out of the connecting piping and the measuring element body. Reasons include: (1) the liquid must be kept from freezing, (2) the measured liquid is too viscous at ambient temperatures, and (3) the measured liquid is corrosive. A sealing chamber is placed in the connecting piping, and the connecting lines are filled with a liquid that is immiscible with the measured liquid.

Liquid Purge System

The purge system employs a suitable liquid constantly flowing from a source of supply through the level tap connections, as shown in Fig. 52, and into the vessel. The manometer, connecting piping, and outer leg are filled with the purge liquid. Needle regulating valves with rotameters in the two lines to the legs are adjusted so that the flow of liquid into the vessel is at as low a rate as possible, yet sufficient to prevent the vessel liquid from entering the connecting lines. The flow through each line must be the same.

Applications of this sytem are for corrosive liquids, for liquids having entrained solid particles, and for liquids which when cooled in the manometer and piping would solidify or precipitate solids.

Gas or Air Purge System

This system, illustrated in Fig. 53, is fundamentally the same as the liquid purge system, the scavenging medium being air or gas instead of liquid. Likewise, the system can, in general, be used in the same applications as outlined for the liquid purge system. The pressure in the vessel must, of course, be less than the available gas or air pressure. Differential-pressure regulators are usually installed across the rotameter, with a needle valve to ensure constant flow of the purge medium.

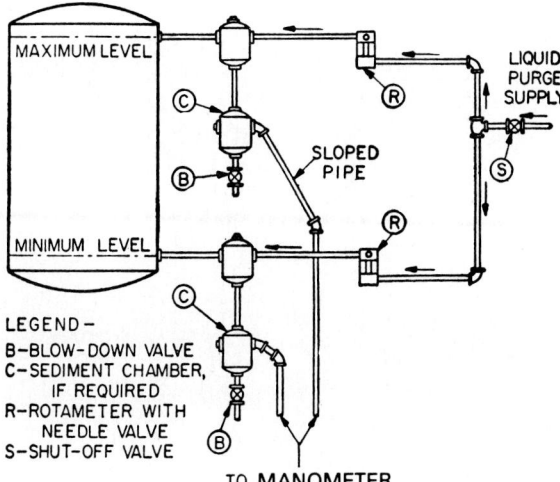

Fig. 52   Liquid purge system for manometers (closed-vessel service).

In operation, the flow of gas or air is regulated so that there is no appreciable velocity head. The manometer, connecting piping, and outer leg are filled with the gas, and since the gas density can be neglected, the only head effect to consider is that due to the liquid in the vessel.

Drip legs, also called sediment traps or settling chambers, are shown in Fig. 54 in a typical manner. Condensation or moisture in piping would create measurement errors and thus should be drained at frequent intervals. Also, in liquid seal systems, sediment or foreign material collects in the chambers and can be drained therefrom.

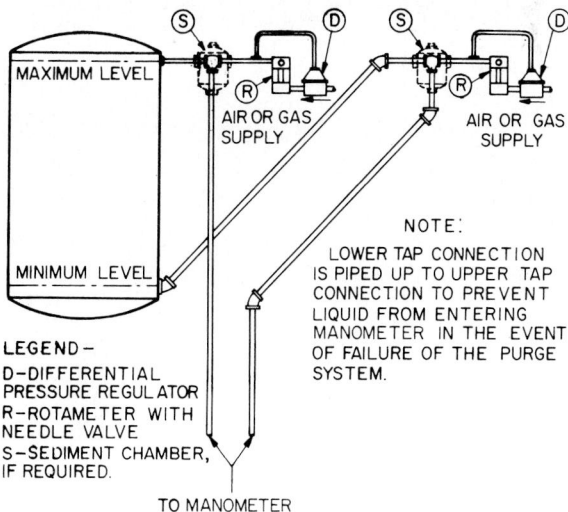

Fig. 53   Air or gas purge system for manometers (closed-vessel service).

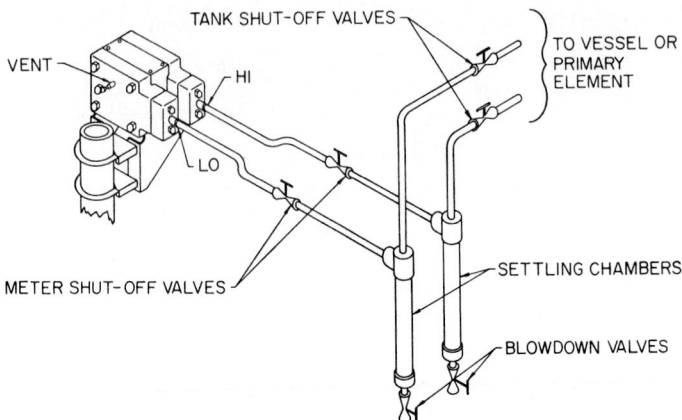

Fig. 54   Typical piping arrangement for settling chambers.

Force-Balance Diaphragm System

A force-balance system is one in which hydrostatic head acting against a diaphragm is balanced by air pressure on the opposite side of the diaphragm. In this system the balancing air pressure can be connected to an indicating or recording pressure gage or manometer with one leg open to the atmosphere for evaluation of the level height. For measurements in closed vessels, a second compensating unit can be installed in the vessel space above the maximum-level line and connected to the low-pressure side of any type of differential-pressure measuring instrument.

This sytem is particularly desirable for service where air or gas bubbling through the vessel liquid is objectionable for such reasons as contamination or crystallization. Further, it eliminates the need for any seals or purges where the vessel liquid contains suspended solids. The diaphragm can be of an alloy or another material resistant to corrosive liquids.

For the open-vessel application, the installation can be either as shown in Fig. 55, or from the top as illustrated in Fig. 56, with the location at the point of minimum-level measurement.

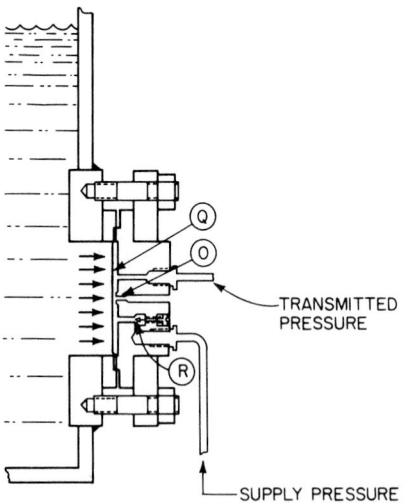

Fig. 55   Pneumatic liquid-level transmitter.

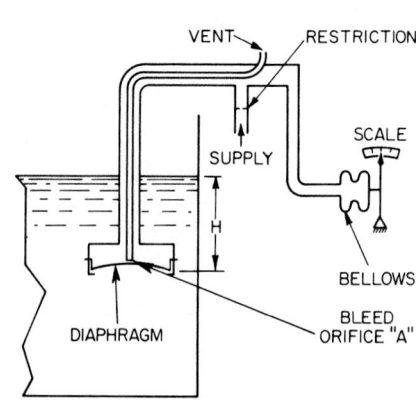

Fig. 56   Force-balance diaphragm system for open vessels.

Air supply at a constant pressure is fed to the diaphragm via the restrictor or dampening orifice *R* (Fig. 55). As the level in the vessel rises, the head pressure working against the diaphragm *Q* acts to close the vent orifice *O* nozzle. Pressure builds up on the transmitting side of the diaphragm until it exerts a force equal to the head pressure in the vessel. At that time the vent nozzle *O* opens slightly until the air pressure stabilizes at a value equal to the liquid head pressure in the vessel.

The head pressure picked up by the measuring instruments is expressed by the basic equation previously given in the introduction to the discussion on hydrostatic pressure methods.

This type of transmitter reproduces the liquid-level head pressure or process pressure without multiplication (1:1 ratio) and, therefore, has no range or calibration inherent in its construction.

In a variation of the simple one-to-one force-balance transmitter, a welded wafer-type diaphragm sensing element is mounted on a tank, connected by a capillary and filled system to a sensing diaphragm in the transmitter, and provided with a force balance, either pneumatic or mechanical, to offset and balance the liquid head pressure exerted on the tank sensing diaphragm. This force is measured and connected to level height measurement by appropriate mechanical and pneumatic means—used as indication of level height and/or control action for liquid-level control.

A design modification to permit more convenient access to a differential-pressure sensor without interrupting the process flow is shown in Fig. 57.

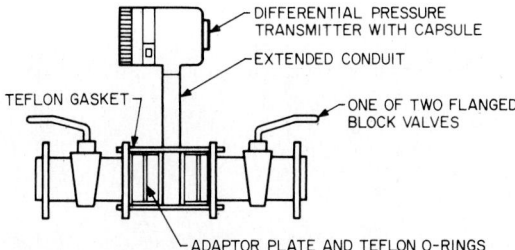

Fig. 57 Modification of differential-pressure liquid-level sensor to permit convenient removal for maintenance without interruption of the process under control. *(Beckman Instruments.)*

The ASME Boiler and Pressure Code states, "When the direct reading of gage glass water level is not readily visible to the operator in his working area, two dependable indirect indications shall be provided, either by transmission of the gage glass image or by remote level indicators." An electric manometer primary sensor (Yarway) designed to meet one of the foregoing alternatives is shown in Fig. 58. The pressure from a constant head pipe (compensated as required to ensure accuracy) and the actual pressure from the vessel are applied on opposite sides of a diaphragm. The displacement

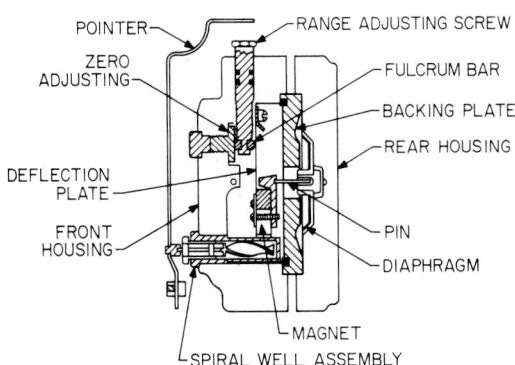

Fig. 58 Cross section of an electric manometer primary sensor. *(Yarway.)*

of the diaphragm due to the differential pressure results in the motion of a pin which deflects a plate anchored at one end and fitted with a permanent horseshoe magnet at the other. The poles of the magnet straddle a spiral armature inside the pressure isolating well which is threaded into the housing. A counterbalanced pointer is fastened to the other end of this armature. When motion on the deflection plate changes the position of the horseshoe magnet, its movement along the well causes rotation of the spiral armatrue which, in turn, actuates the pointer to respond to the changes in liquid level.

Open Channels

A differential linear integrated circuit pressure transducer can be submerged in a flow stream for measuring the liquid level. A system is available which measures and stores level data and provides for periodic, rapid collection of the data by means compatible with direct transfer to a computer (ISCO).

Static Head Type

A typical static head level device is illustrated by an altitude valve (Fig. 59). It is a self-contained, self-regulating controller which may be quite remote from the vessel or tank. The static measuring line usually supplies the operating fluid in addition to sensing the liquid level.

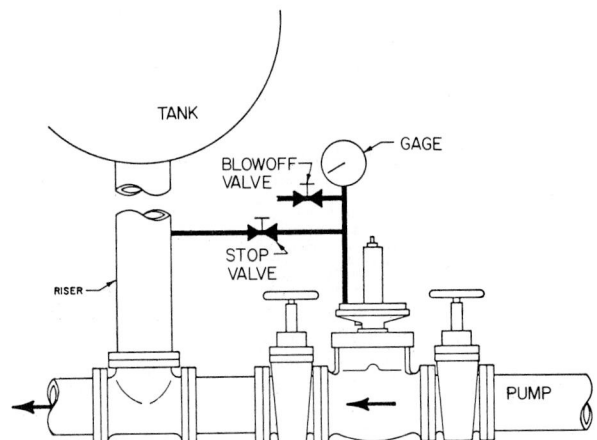

Fig. 59 Altitude valve arrangement for measuring static head.

LEVEL MEASUREMENT BY ELECTRICAL CAPACITANCE*

Liquid level, interface level, and the level of granular solids can be measured using the electrical capacitance effect. The capacitance of a suitable electrical capacitance sensing element varies with the level of the material, and electrical measurement of this capacitance gives a direct reading of the level. Most measurements use low-level radio frequency energy on the probe. A variety of electronic measurement instruments are available for single-point on-off control or for continuous measurement with current outputs of 4 to 20 mA dc.

---

*This section was prepared by J. H. Koegel and G. B. Sellers, Robertshaw Controls Company, Industrial Instrumentation Division, Anaheim, California.

Advantages and Limitations

The primary sensing element is very simple and rugged and has no moving parts. Capability for temperature, pressure, and corrosion resistance is easily obtained. The sensing elements are easily cleaned, and sanitary standards are readily met. Intrinsically safe elements with explosion-proof instruments are readily available. The cost of most capacitance systems is competitive with that of the simple mechanical or pneumatic units; however, more costly special purpose capacitance systems are available for special difficult-to-measure applications.

The capacitance system has certain limitations: (1) If the dielectric constant of the measured medium changes with the temperature, a measurement error will result unless a dielectric compensated detector is used; (2) viscous conducting liquids which coat the sensing element can cause erroneous or false readings unless a detector which compensates for coatings is used; (3) air bubbles in the liquid or foam on top of the liquid can give erroneous readings; and (4) sensing the interface level between two conducting liquids is difficult, depending on the magnitude of the conductivity.

Sensing Elements

A capacitor consists of two conducting plates separated by a nonconducting material or dielectric. The capacitance of a parallel plate capacitor is determined by the equation

$$C = \frac{KA}{d}$$

where C = total capacitance
K = dielectric constant of material between plates (see Table 2)
A = area of plates
d = distance between plates

Table 2 Dielectric Constants of Common Process Materials

Material	Temperature, °C	Dielectric constant
Hydrogen peroxide	0	84.2
Water	25	78.5
Aqueous solutions	—	50–80
Formic acid	15.6	58.5
Glycerin	25	42.5
Glycol	25	37
Benzoyl chloride	0	29
Ethanol	25	24.3
Ammonia	25	16.9
Sulfur dioxide	0	15
Magnesium titanate	25	13.9
Glass	—	3.7–10
Sodium chloride	25	5.9
Alkyd resin	25	5.1
Sulfur	118	3.52
Butyric acid	20	2.97
Polystyrene	25	2.6
Styrene	25	2.43
Benzene	25	2.27
Carbon tetrachloride	25	2.22
Cyclohexane	25	2.015
Chlorine	−50	2.1
Carbon dioxide	20	1.6
Propane	0	1.61
Argon	−191	1.53
Nitrogen	−203	1.454
Air	—	1.0

The most common sensing element used in capacitance-level systems is the capacitance probe. See Figs. 60 and 61. The probe shown in Fig. 60 is composed of three elements: a probe gland, an electrode (inner conductor or plate), and insulation on the electrode to electrically insulate it from the gland.

The capacitance of an insulated probe used with a conductive liquid is approximated by the equation

$$C = \frac{K(0.613l)}{\log_{10}(D/d)}$$

where C = total capacitance
 K = dielectric constant of insulation (Teflon 2.1, polypropylene 1.5)
 l = length of active portion of probe
 d = diameter of probe rod
 D = diameter of insulation

Refer to Fig. 62. If the probe shown in Fig. 60 is screwed into a metal tank, the tank itself will be electrically connected to the probe and thus will become the outer conductor. As the level rises on the probe, the capacitance will increase as a result of the increase in dielectric constant between the

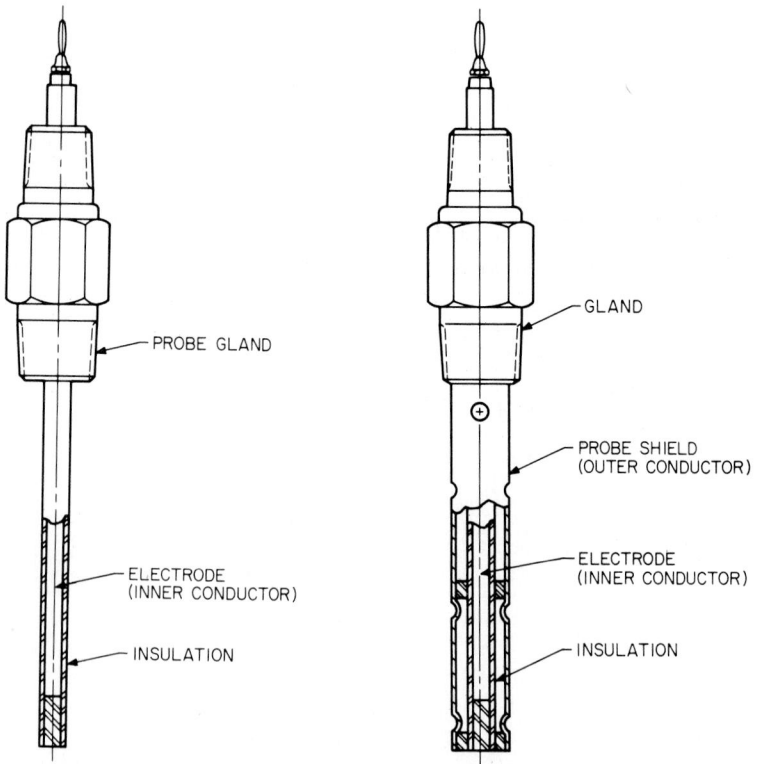

Fig. 60 Capacitance probe without a shield. *(Robertshaw Controls.)*

Fig. 61 Capacitance probe with a shield. *(Robertshaw Controls.)*

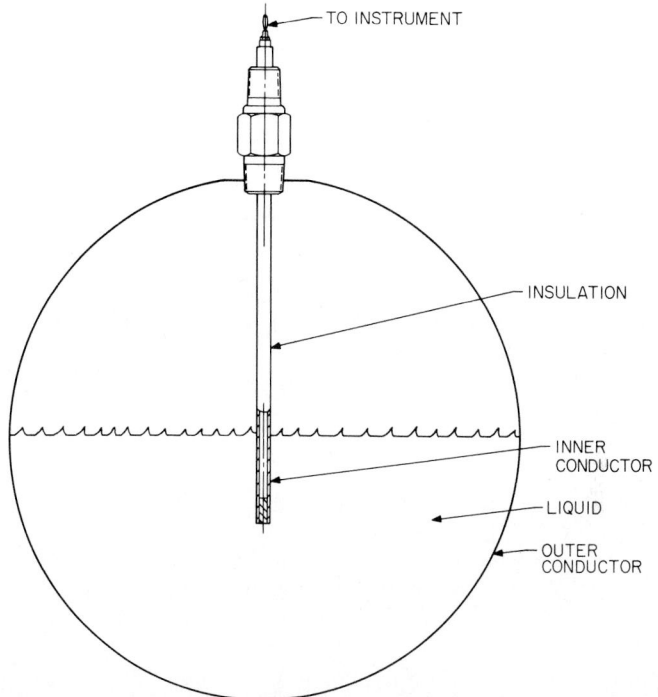

Fig. 62 Capacitance probe installation. *(Robertshaw Controls.)*

plates. If the fluid is a conducting liquid, this has essentially the same effect as moving the tank wall (outer conductor) against the outside surface of the insulation.

A somewhat different probe construction can be employed when the liquid being controlled is a dielectric or nonconductive liquid. A shield can be added, as shown in Fig. 61, which then becomes the outer conductor. This increases the gain of the probe (capacitance change per inch change in level) and is also used to obtain a linear capacitance change with level on a nonlinear geometric configuration, as shown in Fig. 62.

Bare or uninsulated probes can be made for dry, nonconductive granular solids. For extremely low-dielectric-constant, nonconducting fluids, a bare probe with a shield may be used to increase further the gain of the probe.

Many other probe configurations are possible. An insulated cable suspended in a storage bin and insulated from the bin is used for grain storage, as shown in Fig. 63.

The probe gland, element, and insulation can be constructed of materials to resist almost any corrosive effects. Tetrafluoroethylene (TFE) insulation and stainless steel metal parts are most commonly used and fulfill most corrosion and temperature requirements.

Cages are available for capacitance probes similar to those employed in displacement and float-type level systems. See Fig. 64.

Application Considerations

An insulated probe must be used for all level measurements when the material being measured is conductive or there is a possibility of moisture being present in the material being measured. Any moisture present in a nonconducting liquid raises the effective dielectric constant. In low-dielectric-constant liquids, an addition of 1% water raises the dielectric constant of the mixture or emulsion about 3%.

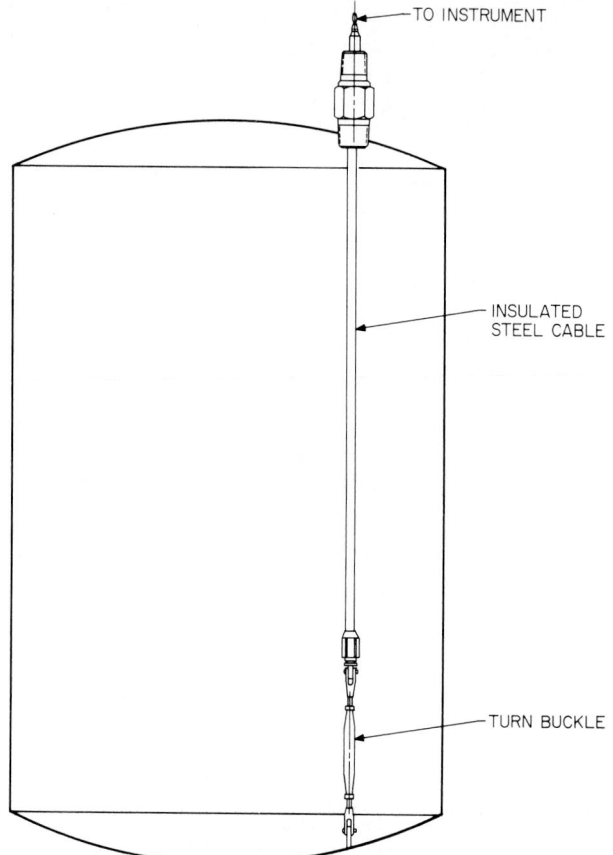

Fig. 63 Electrode formed from insulated steel cable for measurements in tall vessels or bins [up to 125 ft (38 m)]. *(Robertshaw Controls.)*

Dielectric-Compensated Systems

By utilizing a dual-element probe or two separate probes and a microprocessor-based capacitance detector, variation in dielectric can be monitored. The microprocessor can then make an adjustment to the level indication to correct for dielectric shift, thus eliminating error due to dielectric changes. This system also minimizes errors due to conductivity changes.

Compensation for Coating and Buildup on the Probe

When a bare probe is used with a nonconductive viscous material, coating is usually not of consequence, since the effective spacing of the capacitance plates is much greater than the coating thickness. In this application, the resistance of the coating is very high, the average dielectric of the medium between the capacitance elements approaches 1, and the detector circuit does not recognize the coating.

When an insulated probe is used with a viscous conductive material, this coating can be a cause of measurement error. There are a number of techniques used in detector circuits which are very effective in minimizing errors due to this type of coating. In this case, the conductivity of the coated portion of the probe is used to cancel the effect of the capacitance of the coated portion, for negligible

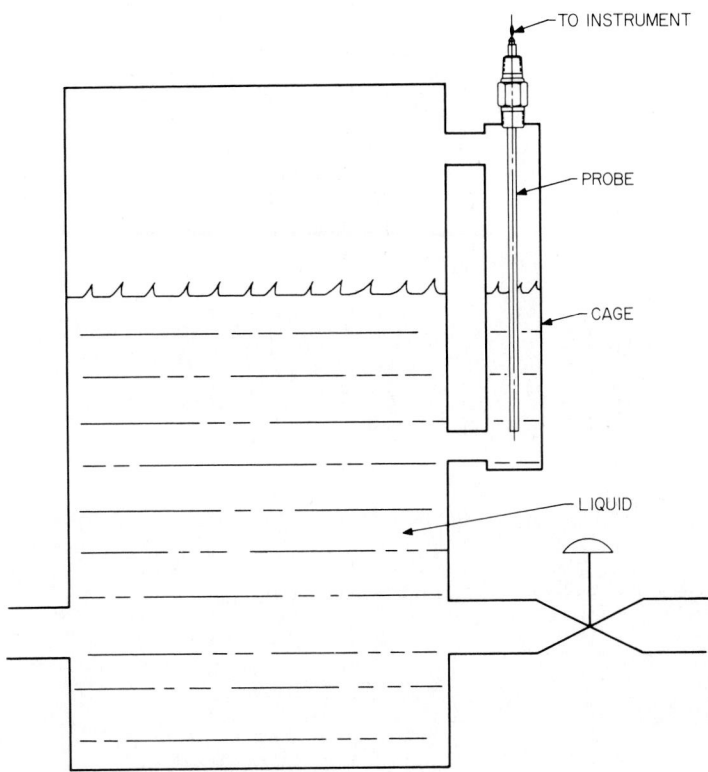

Fig. 64 Cage-mounted capacitance level probe. *(Robertshaw Controls.)*

error due to coating. Usually, a phase sensor is employed where the detector output is attenuated when a phase shift in the probe signal is sensed as compared to the reference oscillator. The phase shift is a manifestation of the resistance and capacitance of the coating portion of the probe signal. The inherent phase of the measured media with respect to the reference is taken into consideration when the system is calibrated. A change in this inherent phase is normally the result of coating and is used to compensate for the output.

Detector Configuration

Modern capacitance detector circuits are available for direct mounting on the probe or for remote mounting. Probe-mounted detectors are more cost-effective; however, remote-mounted units are often required because of temperature conditions at the vessel, space consideration, or accessibility for calibration and service.

Figure 65 illustrates a remote-mounted on-off detector. Note the optional indicator lights on the remote detector, which indicate if the set point of the level has been reached or exceeded. Also, a probe-mounted on-off detector with a relay output is shown in Fig. 65.

Figure 66 illustrates a remote-mounted continuous monitor with an indicator to reflect the level in the vessel, and Fig. 67 illustrates a probe-mounted continuous level monitor with a 4- to 20-mA output.

Figure 68 shows a probe-mounted continuous level detector with its cover removed.

Figure 69 shows a dielectric-compensated microprocessor-based level detector with its electronic assemblies on display. This instrument has options which include data output for transmission to a master computer and relay outputs for alarm.

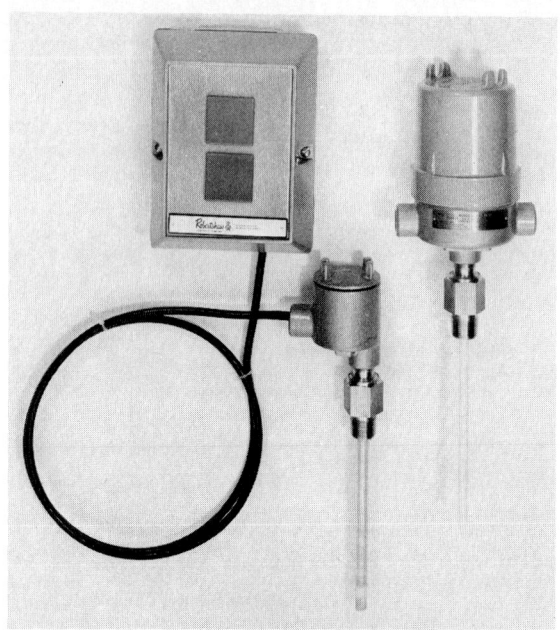

Fig. 65 On-off capacitance level instrument with remote-mounted detector and probe-mounted detector. *(Robertshaw Controls.)*

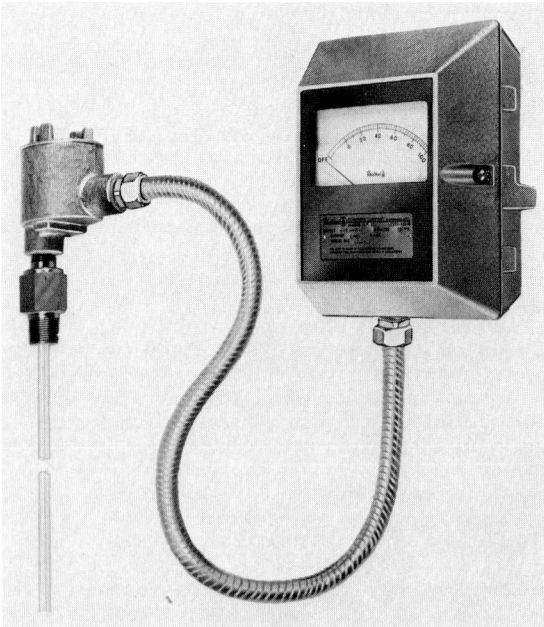

Fig. 66 Remote-mounted, continuous capacitance level monitoring system. *(Robertshaw Controls.)*

On-off instruments are usually offered with high or low-level failsafe features, switchable on the circuit card. Probe-mounted detectors are usually explosion-proof and weathertight, while remote-mounted instruments are optionally available with either explosion-proof or weathertight enclosures. The probe circuit should be designed to be intrinsically safe so that it can be used in hazardous applications without fear of explosion.

One very cost-effective on-off configuration available is a probe-mounted detector designed with two independent set point adjustments corresponding to two separate output relays. With this arrangement, one instrument can take the place of two by utilizing one probe when two independent trip points are required.

Normally, on-off detectors are supplied with adjustable deadband and time delay features. Narrow deadband settings are used to desensitize the trip setting and prevent excessive output switching due to splashing or rippling of the process. Wide deadbands can be implemented for pump-on/pump-off cycling of a vessel. Normally, the deadband is adjustable from 0.5 to ≈ 200 pF.

The time delay feature can be adjusted from several milliseconds to many seconds. This feature is also designed to minimize nuisance tripping due to splashing on the probe, etc.

Modern electronic and microprocessor technology makes level measurement by electrical capacitance attractive for many applications which were not practical in the past.

Fig. 67 Probe-mounted, continuous capacitance level monitoring system. *(Robertshaw Controls.)*

SONIC AND ULTRASONIC METHODS

Sonic and ultrasonic technology has been applied to the problem of level measurement of liquids and bulk solids for many years. Generally, these methods have been used for point (rather than continuous) level measurement and control. There are two main categories: (1) measurement of the transit time between a sound wave transmitter and a receiver which detects the returning sound wave pulse (echo), and (2) the absorption or attenuation of acoustic energy, such as occurs when a material (at a given height in a container) interferes with the transmission of energy from the transmitter to the receiver. In another system, a vibrating member, placed at some height in a container, displays a frequency change when it is covered by another medium. Sonic methods also have been found useful in making interface level measurements—this method is based on the alteration in the speed of sound in different media. Like probes (capacitive or conductive), sonic methods are particularly useful when the more traditional methods do not work well or at all—as exemplified by foaming liquids.

A continuous sonic-type level measuring unit is shown in Fig. 70. The equipment includes a transmitter that periodically sends a sound pulse to the surface from the transducer, a receiver (which is included in the transducer) that amplifies the returning pulse, and a time interval counter that measures the time elapsing between the transmission of a pulse and receipt of the corresponding pulse echo. Echo pulses are ordinarily reflected back from the surface of the liquid; however, the line of demarcation or interface between immiscible liquids also reflects sufficient energy to allow the system to gauge these obscure interface levels. One transducer and its single coaxial cable constitute the only installation equipment usually required to gauge an individual tank. The receiving indicator may be switched to gauge as many transducer-equipped tanks as needed. Designers of such systems must take into very serious consideration the effect of environmental changes on the measurements. Notably, these changes are in temperature, pressure, and chemical composition—all factors which affect the velocity of acoustic propagations and on which the measurement is fundamentally based.

Fig. 68 Probe-mounted, continuous capacitance detector with cover removed. *(Robertshaw Controls.)*

Fig. 69 Microprocessor-based dielectric-compensated capacitance level system. *(Robertshaw Controls.)*

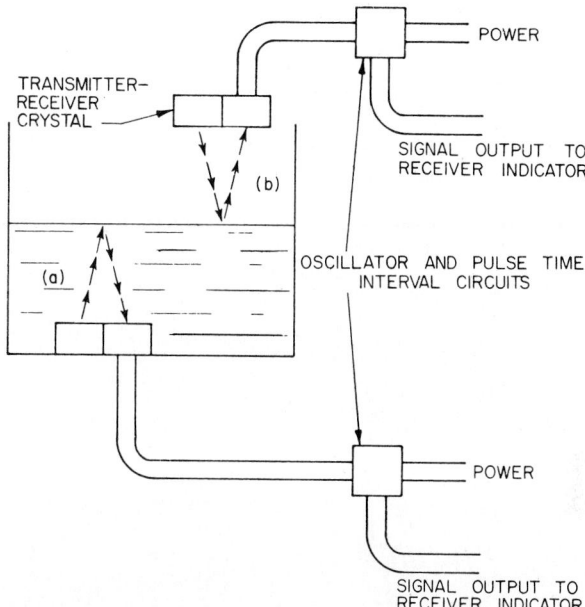

Fig. 70 Continuous sonic-type level measuring units. (*a*) Liquid phase, (*b*) vapor phase.

In some processing operations, for example, where chemical-composition changes may be expected, such changes can severely affect calibration unless additional electronic means are incorporated in the system to anticipate and correct for such changes. The velocity of sound at $0°C$ in air is 1087.42 ft (331.45 m)/s. In ammonia, the velocity is 1361 ft (415 m)/s; in carbon dioxide, 1106 ft (337 m)/s; in chlorine, 674 ft (205 m)/s; in helium, 3182 ft (970 m)/s.

A representative single-sensor liquid-level indicator is shown in Figs. 71 and 72. The sensor typically is a small, hermetically sealed probe whose front face oscillates at a relatively high frequency (35,000 to 40,000 Hz). When rising liquid covers at least half the face of the sensor (if mounted horizontally), or the entire face (if mounted vertically), the oscillating action of the sensor will be damped out. This damping is recognized by the amplifier and causes a relay in the amplifier to drop out and actuate either an on or an off signal (high level) and/or suitable control action through relays or other actuators. The sensors usually are about 1 in (2.5 cm) in diameter, are made of stainless steel or another corrosion-resistant alloy, and can easily be placed in a fitting or tank. These sensors are not limited by the physical properties of various liquids, such as pressure, conductivity, and density, and can be used with flammable liquids.

In the sonic system shown in Fig. 73, one transmitting sensor creates the sonic beam, and sound waves are picked up by a receiving sensor. This can be accomplished by a direct path or by reflective waves from the transmitter, which strike a solid surface and are reflected back to the receiving sensor. Sensors can be spaced as close as ¼ in (6 mm) or as far apart as 10 ft (3 m) in a direct beam path. They can pipe their beams through tubing where sensor beams cannot be direct, but generally the tubing length is limited to several feet. The sensor's sound beam is unaffected by mist, smoke, dust, or fumes, since it is interrupted only by a solid or liquid entering the beam path. This type of system is generally used for dry bulk solids.

NUCLEAR RADIATION METHODS

Nuclear radiation from a selected source can be related to the liquid or solids level in a vessel and is employed in several practical industrial systems. As a detector for converting nuclear gamma ray

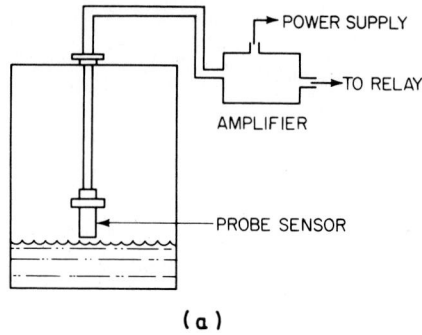

(a)

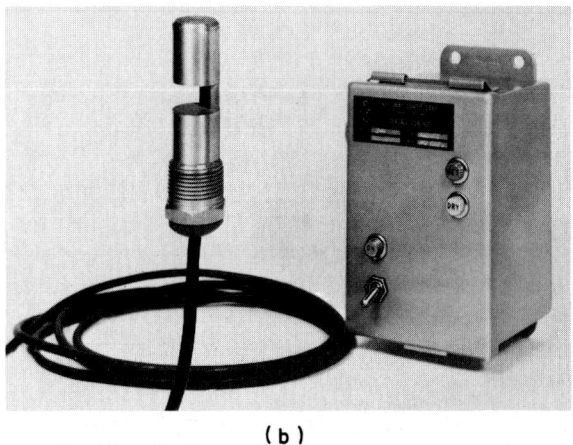

(b)

Fig. 71. (a) Single-sensor sonic-type liquid-level indicator using top-mounted sonic probe. *(Delavan Industrial Controls.)* (b) Slot configuration sensor. *(National Sonics Division, Xertex.)*

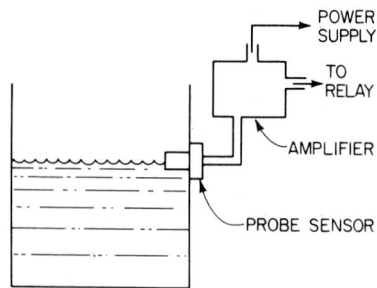

Fig. 72 Single-sensor sonic-type liquid-level indicator using side-mounted sonic probe. *(Delavan Industrial Controls.)*

5.54

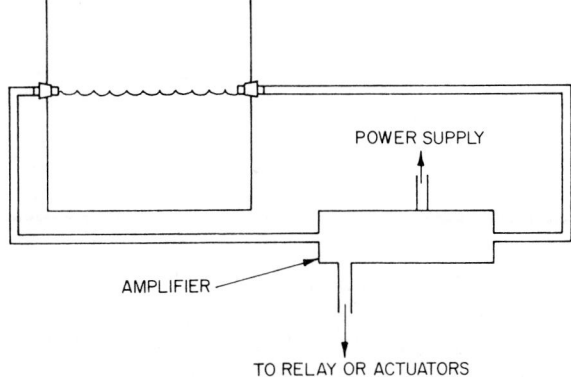

Fig. 73 Sonic probe level detector utilizing two sonic probes.

radiation into electrical quantities related to level, one system employs a Geiger counter (Fig. 74), while the other utilizes a specially designed gas ionization cell.

Geiger-Mueller Detector

Two variations of this system depend, respectively, on two basic relationships regarding the intensity of radiation received by a detector:

1. Intensity varies in proportion to the thickness of any material interposed between the source and the detector.
2. Intensity varies in inverse proportion to the square of the distance between the source and detector.

Thus, for liquid- or solid-level measurement, the source can be located so that the liquid or solid itself forms a path of varying thickness through which the gamma rays must pass to reach the detector. Alternatively, for liquids the source can be placed in a float which rises and falls with the liquid level, thereby varying the intensity of radiation picked up by the detector. A substantial strength of gamma ray penetration power is utilized—the source being generally located outside and on the vessel and the rays passing through the steel or other material of the vessel.

Figure 75 illustrates a typical installation in which the emission source, usually a minute quantity of capsulated radioactivity isotope, such as cobalt 60, cesium 137, radium 226, has a fixed location in or on the side of the vessel. The detector, with single or multiple detector elements (Geiger-Mueller tubes), is fixed to the outside of the vessel directly across from the source location so that the liquid, as the level varies, will intercept the emission path and change the quantity and radiation intensity

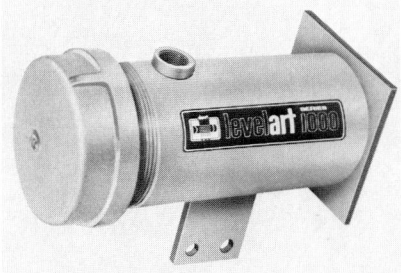

Fig. 74 Point-level detector for rapid response, non-contact, and high- or low-level alarm systems, as well as for material flow detection. The design permits easy field installation and adjustment. *(Ohmart.)*

received by the detector. See Fig. 76. The gamma ray absorption of the vessel wall is constant, and the absorption in the gas or vapor space above the liquid is negligible; therefore, proper readout of the radiation value received will define the level in the vessel. The readout range depends on the active length of detector elements exposed to radiation.

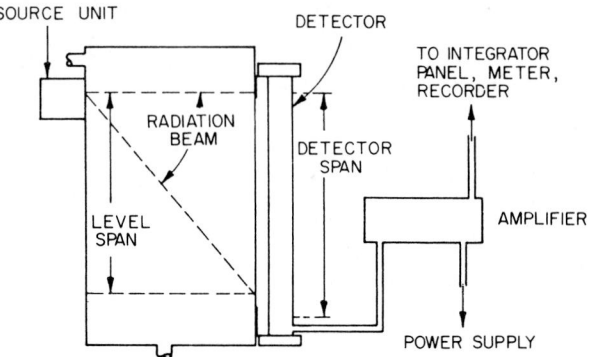

Fig. 75 Typical installation of a radiation-type level detector.

The detector (Geiger-Mueller tubes) converts the gamma radiation received into an output, a series of small current pulses which when fed to the circuit of the amplifier are converted to voltage pulses and can be transmitted to a recorder and/or controller. Ionization chambers are also used as radiation detectors.

The form of a system employing a float, which contains the radiation source, is shown in Fig. 77. This form has the same limitation as other float-type detectors in that its calibration can be altered by material clinging to the float, thereby changing its position with respect to liquid surface. Both forms have the advantage that they require no stuffing box for connections through the vessel wall. This is of particular value in high-pressure service or where corrosive or dangerous liquids are involved.

Accuracy. From ± ⅟₁₆ in up to 1% of range or span. Accuracy directly affected by variations in density of material.

Fig. 76 Assembly of multiple detectors installed on the side of a tank. *(Ohmart.)*

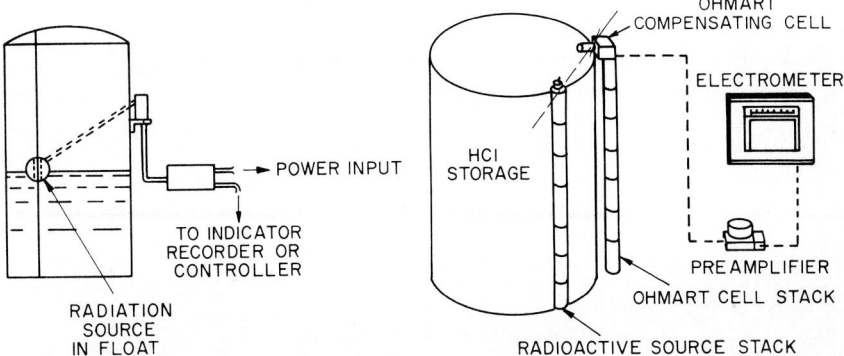

Fig. 77 Gamma-ray emission-type gage. Radiation source is in a float. *(Ohmart.)*

Fig. 78 Typical application of an Ohmart cell for level measurement. *(Ohmart.)*

Span or Range. From ½ in to 20 ft (12 mm to 6 m) with single units. More with multiple installations.

Temperature. Source unit limited only be materials of construction. Detector units: −40 to 140°F (−40 to 60°C).

Time Constants. Up to 10 s for continuous level measurement.

For certain applications where other common methods are impractical or unreliable, these systems provide high accuracy of measurement. The use of radioactivity isotopes cesium 137 and cobalt 60 requires a license in most countries. No license is needed for radium 226, although it is more expensive than cesium 137. Radiation source devices are furnished with built-in safety features for transport, installation, and use, with shuttering and shielding so that the radiation field intensity is below the regulatory tolerance level and thus noninjurious to personnel.

Gas Ionization Cell

The Ohmart* gages employed for liquid- or solids-level measurement are based on a patented Ohmart cell which converts gamma radiation energy directly into electric energy. See Fig. 78. This cell contains two electrodes which have different work functions. They are separated by a filling gas which, when forcibly ionized by exposure to nuclear radiation, attracts positive ions to the electronegative electrode and electrons to the electropositive electrode; thus an electric current is generated which can be amplified.

Another measuring method involves a halogen-filled Geiger-Mueller tube and either a thyratron or a cold-cathode trigger tube, with no vacuum tubes in the measuring circuit. The output of the Geiger-Mueller tube is a series of pulses with the repetition rate proportional to the radiation field intensity. These pulses are integrated and appear as positive dc voltages at the tube.

The radioactive source is placed in one or multiple positions on a tank or vessel (a column or cell stack is common) at a location where the material whose level or interface position is being measured interposes itself between the source and the cell. The gamma source emitting radioactivity is generally cesium 137 or cobalt 60.

The amount of radiation reaching the measuring cell is an inverse function of distance, liquid between source and cell, or density of liquid. All electrical output is, therefore, proportional to the level-position or material density.

Applications are common such as:

One Level Position. One radiation source, one measurement cell.

High-Low Level. Two radiation sources, two measurement cells.

*Ohmart Corp., Cincinnati, Ohio.

Continuous Level. Stack or strip source and detection. See Fig. 78. Source must be shielded in accordance with regulations and licenses.

Zero Suppression. Can be either electrical or by a compensating cell.

Amplification. A number of available types. Generally furnished with an output of 0 to 20 mV or 0 to 50 mV.

Time Constant. From 2 to 10 s.

Accuracy. Liquids, interface, or solids ±⅛ in (3 mm) in range up to 50 ft (15 m) or more. Continuous level. Motor drive up to ±1% of range or scale.

THERMAL METHODS

In the instrumentation literature of the mid-1980s, thermal methods for measuring liquid level are seldom mentioned. Thermal systems were devised for handling difficult measurement problems prior to the advent of electronic and current sophisticated approaches to liquid-level instrumentation. Because of the numerous very difficult liquid-level measurement problems which continue to arise, the editors believe that, at least for the third edition of this Handbook, it remains in order to describe these earlier, essentially very simple systems.

Expansion-Tube Unit

In a process where the liquid phase is at a different temperature than the vapor phase, the liquid level can be measured and controlled through utilizing the expansion of a metal tube in which both liquid and vapor phases are present. This method has been in use for more than 60 years to control and measure levels in vapor generators such as steam boilers.

A metal tube is connected to the vessel (Fig. 79): the upper end to the vapor space, and the lower end to the liquid phase. The level is at the same reference point in both tube and vessel. The temperature of the liquid is below that of the vapor space because of radiation—whereas that of the vapor space is always in accordance with the vapor pressure of the evaporated liquid.

As the level in the vessel lowers, less liquid is in the tube, and greater tube surface is in contact with the high-temperature vapor—thus, the tube expands. As the level rises, more lower-temperature liquid enters the tube and the tube contracts.

In order to obtain the largest physical change in level in the tube, it is inclined at an angle—practically determined to be between 15 and 30° from horizontal. The tube length must be of such value as to extend over the complete range of level to be measured and controlled.

The expansion tube must be firmly anchored at the lower (liquid) end. The expansion or contraction movement takes place at the upper end and can be measured directly or multiplied through mechanical leverage. A leverage ratio of 60:1 is sometimes favored. This movement can be used to

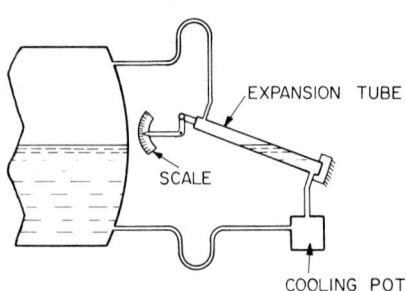

Fig. 79 Expansion-tube-type level indicator.

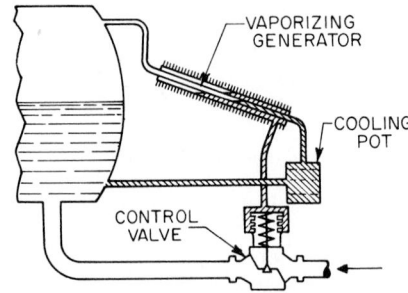

Fig. 80 Thermal-hydraulic-type level controller.

operate a pointer on a scale for level measurement. Also, it can be directly used to operate a control valve mechanically or to operate a pilot relay valve for pneumatic or hydraulic action of a diaphragm motor valve.

The assumption is that the ambient temperature is reasonably constant; otherwise erratic measurement and control result.

The materials of construction can be any metal that will withstand pressure and temperature and/or corrosion action if present; however, metals with the highest expansion coefficients are selected in order to obtain large tube motion. Copper alloys are useful in this respect.

Thermal Hydraulic System

In a primary phase system—where liquid and vapor of the same composition exist under conditions of common pressure and temperature—the liquid level may be measured and controlled by a thermal hydraulic system. This system has been utilized for more than 70 years for controlling and measuring level in such units as vapor generators and boilers.

The theory employed is that the vapor-phase temperature is always that of the pressure-temperature value. If pressure in a vessel is held constant, its vapor temperature also remains constant, either in the vessel or in a separate system outside the vessel. The liquid phase does not so react. Away from its heat source, its temperature drops. In a system outside the vessel, therefore, a difference between the temperature in the vapor phase and in the liquid phase can be obtained.

Figure 80 shows a thermal hydraulic system which operates as follows: A generator, consisting generally of two concentric tubes, is installed as shown with the inner tube connected to the vessel in which the level is to be controlled. One end of the tube is above the liquid level and the other end below, resulting in the tube being subject to the same pressure and liquid level as in the vessel. A closed system is formed by the annular spacer between the inner and outer tubes and is connected to an indicating or controlling device.

Before vapor is admitted in the inner tube, the closed system is filled with a liquid similar to the vessel liquid, with characteristics such that it will vaporize when the system is put in operation. This vaporization of liquid in the closed system is used to expand a bellows or similar member to obtain measurement or control. In the system illustrated, it is used to operate a control valve directly by means of a bellows unit.

The relative amount of liquid and vapor in the inner tube determines the quantity of vapor generated in the annular space. If there is a rise in level, cool liquid will come into the inner tube from the reservoir leg. The amount of heat available for vaporization in the generator is a direct function of the length of internal tube exposed to vapor conditions. With rising level the heat is reduced, thereby causing condensation of vapor and a reduction in the vapor pressure in the annular space. This change in pressure causes a change in the position of the measuring and controlling bellows. A falling level in the primary vessel reverses this procedure.

Although normally applied with the indicating or control device close to the measurement element, this portion of the system can be located up to distances limited in practice to usually not over 600 ft (183 m). The measuring element (generator) must be placed close to or on the primary vessel and should not be subject to sudden and wide ambient temperature variations.

Thermometer-Bulb Immersion Method

Where the temperature of the fluid in open or closed vessels is above or below ambient and a simple single-point cutoff or alarm is desired, the conventional pressure-type thermometer system can be employed. The thermometer bulb is installed, usually through the vessel wall, at the desired point for level control or alarm (high or low level). As liquid rises above or falls below the bulb, the temperature change is detected and used to initiate an electric or pneumatic action for alarm or control purposes.

For example, the thermometer instrument is commonly available with electric contacts which can be set to close above a temperature set point selected in accordance with the temperature of the vessel liquid. The system can be installed and adjusted so that with a vessel liquid at 150°F (66°C) a set point of, say, 100°F (38°C) (sufficiently above ambient) will be exceeded when the vessel liquid covers the bulb which is installed, say, midway up the height of the vessel. Closing of the electric contact can be used to energize a visible or audible alarm signal or a solenoid valve installed in the

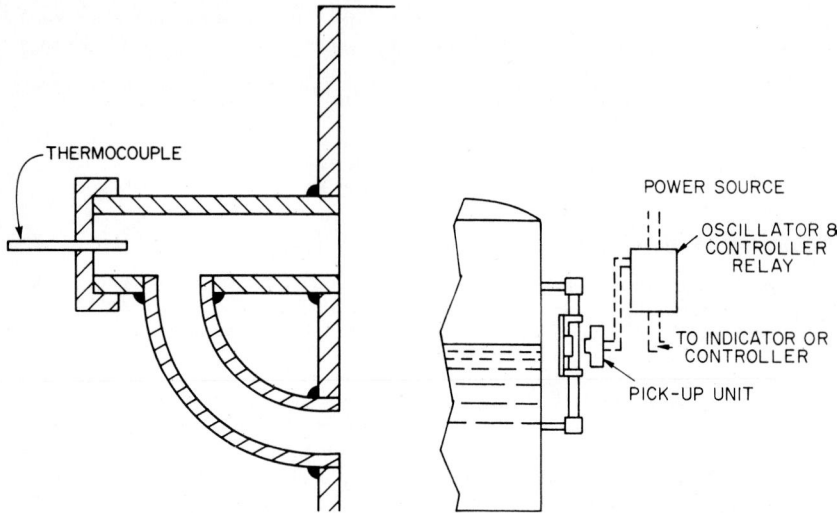

Fig. 81 Ram-horn-type level detector for severe service.

Fig. 82 Oscillator-type pickup on gage glass.

supply line to the vessel. Usually nonindicating types of thermometers will suffice for applications of this method, although recorders can be used to show periods when high- or low-level limits are exceeded.

Ram's Horns

As shown in Fig. 81, these devices are used in dirty, waxy, sludgy, corrosive service which is too severe for most other instruments. The large, curved leg permits liquid to flow up to the temperature sensing device, here illustrated as a thermocouple covering it and indicated by a temperature change that the level is at that point.

OTHER LIQUID-LEVEL MEASUREMENT METHODS

Oscillator Type

A typical oscillator liquid measurement unit is shown in Fig. 82. These units are generally used to indicate or measure a specific point of level.

An electronic oscillator generates a high-frequency alternating current, which is conducted to a pickup unit clamped to a clean gage glass as illustrated.

Any change in characteristic of material, such as level change, in the field of pickup detunes the oscillator. The resultant change in current when amplified can operate relays, contacts for remote indication and control, solenoids, and motorized valves. The accuracy of operation can be as close as 0.001 to 0.125 in (0.025 to 3.18 mm) of level change.

Rotating-Paddle Method

When a paddle such as that shown in Fig. 83 is placed in a tank, its power requirement when rotating in air is low. Power consumption increases as the liquid level rises and starts to cover the paddle.

Weighing Method

The weight of a vessel can be used as a measure of the liquid level where the vessel shape provides a uniform relationship with changes in level and where all other pertinent factors, such as density and temperature, are fully known. Numerous kinds of scales (load cells or levers) can be used. Such installations are expensive and usually are more often applied to bulk solids than to liquids. Where severe corrosion, stickiness, and other sensor fouling conditions are present, this method can provide improved accuracy with less maintenance. The system can be used in combination with separate density and temperature probes which can be integrated into the system's circuitry.

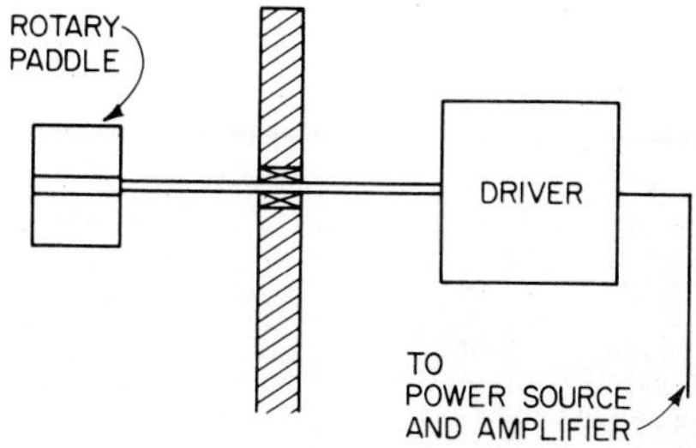

LEVEL MEASUREMENT OF BULK SOLIDS*

Fig. 83 Rotating-paddle-type level detector.

Emphasis on the measurement of bulk solids in bins, tanks, and other kinds of containers has increased markedly during the last few decades. In addition to the conventional handling of bulk materials in the various processing industries, other applications have increased. These include use by food commodity traders who hold products for higher prices and by plastic molding firms that find it economical to purchase raw ingredients in large volumes, among others. Bulk materials come in a variety of shapes, sizes, and densities. Products range from prilled substances that may weigh as little as 4 to 5 lb/ft^3 (64 to 80 kg/m^3) to large, lumpy materials that may have a density well over 100 lb/ft^3 (1600 kg/m^3).

Like the liquid-level control applications previously described, solids-level instrumentation takes the following forms: (1) *Point control,* the most frequently used form, where a sensing element indicates the presence or absence of material at a specific location (level). Point control systems usually provide operator notification by means of a switch closure corresponding to high- or low-level storage conditions. Controls of this type prevent bin run-outs and overflows which interrupt the normal materials handling procedure. (2) *Continuous sensing and control,* where a sensing element which produces a continuous indication (often in the form of voltage or current output) is used. The output is proportional to level and is displayed on a meter (analog or digital). (3) Inventory control can be considered another subcategory. This type of equipment provides level information over a continuous measurement range, either on demand or in a predetermined time sequence. Inventory control devices satisfy operator requirements for determining material consumption, material on hand, and reorder levels.

Bulk Solids–Level Sensors

There are several parallels between the level sensors for bulk solids and those used for liquids. In fact, considerable adaptation of devices from one technology to the other has occurred during the last few decades. Progess has been accelerated by solid-state circuitry and digital methods. Several detection principles are applied in both solids- and liquid-level measurements, including the employment of electronic field effects, sonic and ultrasonic principles, photoelectric transmission, radio- and microwave absorption, nuclear methods, and diaphragms and tilt switches. Diaphragm sensors may utilize capacitance, reluctance, and strain gage transduction, among other principles.

*The assistance of R. L. Hayes, Bindicator Co., Port Huron, Michigan, in preparing parts of this section is gratefully acknowledged.

Importance of Proper Mounting Location

A characteristic of bulk solids, referred to as the *angle of repose,* is a very important consideration. See Table 3. If this factor is not properly evaluated in location determination, level control devices will not respond properly. The angle of repose may be defined as the maximum angle with the horizontal at which an object on an inclined plane will retain its position without tending to slide. The tangent of the angle of repose equals the coefficient of static friction. The term is used in a

Table 3 Density and Angle of Repose of Some Bulk Commodities

Commodity	Density		Angle of repose, degrees	Tangent angle of repose
	lb/ft^3	kg/m^3		
Alfalfa	48	769	45	1.00
Alfalfa, pellets	41–43	657–689	45	1.00
Alum, fine	45–50	721–801	30–45	1.00
Alumina, fine	55	881	55	1.40
Asphalt, crushed	45	721	30–45	1.00
Asphalt, lump	70–95	1121–1522	30–45	1.00
Barley	37–48	593–769	28	0.50
Bauxite, ground	68	1089	35	0.70
Bauxite, mine raw	75–85	1202–1362	35	0.70
Borax	60–75	961–1202	30–45	1.00
Calcium carbide	70–80	1121–1282	30–45	1.00
Calcium oxide (lime)	27	433	45	1.00
Cement, portland	75–95	1202–1522	30–45	1.00
Cement, clinker	75–80	1202–1282	30–45	1.00
Coal, anthracite	50–55	801–882	30–45	1.00
Coal, anthracite ($-\frac{1}{8}$ in, -3 mm)	60	961	45–50	1.20
Coal, bituminous	50	801	25–35	0.70
Flour	30–45	481–721	30–45	1.00
Fuller's earth	35–40	561–641	35	0.70
Lime, hydrated	15–35	240–561	25–30	0.60
Oats	25–35	401–561	28–32	0.65
Phosphate, rock	75–85	1201–1362	30–45	1.00
Phosphate, pulverized	60	961	40–52	1.30
Phosphate, sand	90–100	1442–1602	30–45	1.00
Polyethylene, pellets	30–35	481–561	23–27	0.50
Polyethylene, powder	45	721	52	1.30
Polystyrene, beads	45	721	30–45	1.00
Polystyrene, powder	42	673	30–45	1.00
Polyvinyl chloride, pellets	36	577	30–45	1.00
Polyvinyl chloride, powder	45	721	30–45	1.00
Rye	44–48	705–769	26	0.50
Salt, coarse	45–50	721–801	35–45	1.00
Salt, fine	70–80	1121–1282	30–45	1.00
Sand, foundry	90–100	1442–1602	30–35	0.70
Sand, silica	90–100	1442–1602	25–30	0.60
Soybeans	46–48	737–769	20–30	0.60
Soybean, flour	27–30	433–481	35–40	0.85
Soybean, meal	36–50	577–801	30–45	1.00
Wheat	48	769	30	0.60
Zinc oxide, heavy	30–35	481–561	45–50	1.20
Zinc oxide, light	10–15	160–240	45–50	1.20

closely related way to describe the maximum angle with the horizontal at which loose materials, such as grain, sand, coal, and stone, will retain a position without tending to slide. The moisture content and distribution of fine and coarse particles have a marked effect on the value of this angle.

Other factors which enter into consideration in designing instrumentation systems for bulk solids include *arching* and *flushing,* both related to the angle of repose. Many materials, such as lampblack, activated carbon, zinc oxide, titanium oxide, fine soda ash, and hydrated lime, tend to arch when placed in containing vessels such as hoppers. Arching is best overcome by keeping the individual solid particles continually in motion. Electric vibrators are commonly used for this purpose. Very few materials are crushed or pulverized so well that all particles are of equal size. In the absence of hopper vibration, the smaller particles and fines tend to segregate and roughly collect in a center cone, while the larger particles roll to the side. Consequently, when a hopper is discharged, the fines drop out first, while the larger particles discharge later, the end result of which is formation of an arch. Flushing is a condition caused by the sudden breaking of an arch or otherwise clogged state. Vibrators also help to eliminate flushing.

Other conditions which affect selection and location of a level sensor include moisture content of the material and physical characteristics such as stickiness, temperature, and pressure.

Diaphragm Controls

This type of device makes use of a pressure-sensitive diaphragm which when depressed by the presence of material produces a correspondig switch actuation. A technique that has been used for about a half-century, it is considered quite reliable. Diaphragm controls are available in a wide variety of materials, making it possible to use them over a wide range of product densities. The controls are usually externally mounted and offer no restriction to normal product flow. Other than the switching circuitry, the devices require no external power for their operation. They can be used interchangeably for both high- and low-level detection. Advantages include simple installation and maintenance and comparatively low cost. The devices are not applicable to positive-pressure environments such as those encountered in pneumatic conveying systems.

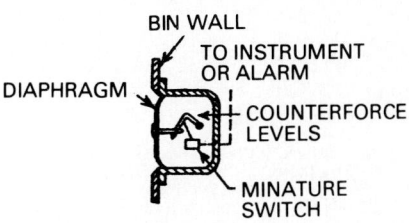

Fig. 84 Schematic diagram of a diaphragm-type solids-level detector.

The added pressure results in a false indication. Special attention should be given to mounting diaphragm controls, as they can be position-sensitive, especially when used on hopper underslopes or when mounted horizontally. The general operating principle of the diaphragm is shown in Fig. 84.

Paddle Wheel Controls

Very commonly used, these controls make use of a motor which slowly rotates a paddle in the absence of material, and rotates itself to actuate a switch when the paddle is prevented from turning by the presence of material. These controls are modestly priced, are versatile, and are available with a variety of paddle options corresponding to various product densities. They may be used for both high- and low-level point control. Paddle wheel controls are not position-sensitive, and their operation is not adversely affected by positive-pressure environments, but checking with the specific manufacturer on this point is suggested. The electrical components are contained in either weatherproof or explosion-proof enclosures and are easily serviced in the field, requiring no special skills. For installation, see Fig. 85.

Tilt Switch

This is probably the simplest device used for point level detection of solid materials. These controls make use of a free-hanging sensor which produces a switch contact when the bulk material tilts the unit off its vertical plane. Other than switching circuitry, tilt switches require no external power for their operation. They are usually quite rugged and reliable. Because of mounting requirements, tilt switches are not quite as versatile as most level devices and are usually limited to high-level control.

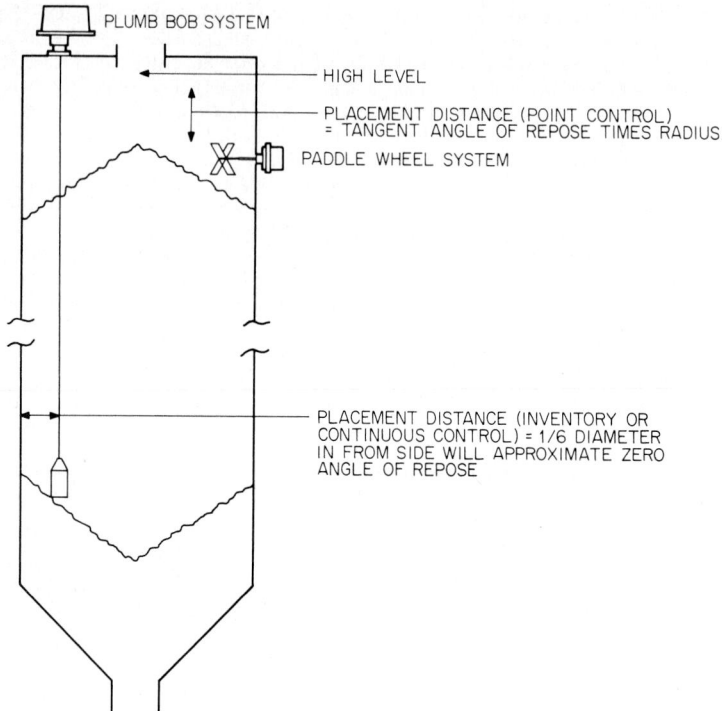

PLUMB BOB SYSTEM

HIGH LEVEL

PLACEMENT DISTANCE (POINT CONTROL)
= TANGENT ANGLE OF REPOSE TIMES RADIUS.

PADDLE WHEEL SYSTEM

PLACEMENT DISTANCE (INVENTORY OR
CONTINUOUS CONTROL) = 1/6 DIAMETER
IN FROM SIDE WILL APPROXIMATE ZERO
ANGLE OF REPOSE

Fig. 85   Mounting of plumb bob and rotating-paddle types of solids-level detectors. *(Bindicator.)*

Vibration Sensors

This type of control features a tuning fork as a sensing probe. A piezoelectric crystal vibrates the probe when there is no material surrounding it. When material contacts the probe, the circuitry detects its presence and operates a contact closure. Vibratory controls can be used for both high- and low-level detection. Solid-state circuitry adds to their cost, but they are considered very reliable. A skilled technician is required at the time of a failure. Vibration controls work well with free-flowing materials and can tolerate both temperature and pressure extremes. Because vibration may cause cavitation and subsequent false signals in connection with moist materials, prepurchase experimentation may be indicated.

Plumb Bob System

Sometimes called "yo-yo" controls, these devices are useful for generating inventory control information. In operation, a weight attached to a cable is lowered into a storage vessel. When the plumb bob strikes the material, the slackened cable triggers retraction of the cable and weight. The amount a cable is paid out or withdrawn is measured and indicated by an appropriate readout. The devices are easy to install, simple to operate, and accurate to ±0.1 ft (3 cm). They have the advantage of being able to operate in vessels well over 100 ft (30.5 m) deep. They are not affected by dust, moisture, or most ambient conditions, and the system operates on demand (manual or sequentially timed). They are not to be regarded as continuous measurement devices. For installation, see Fig. 85.

Capacitance-Type Sensors

These systems are among the fastest growing methods of solids-level control. A variable capacitor is formed by a probe mounted in the vessel and in the vessel wall. Variations in this measured capacitance can be correlated with material level. Capacitance controls make use of solid-state electronics and have no moving parts within the vessel. They are not position-sensitive and work well in the presence of relatively high temperatures and pressures. They are available for both point control and continuous level measurement. The systems, however, are sensitive to moisture, and thus continuous measurement of solids is usually avoided. The principal limitation is materials which tend to adhere to and build up on the probe. The principles of capacitance measurement are described in more detail earlier in this article in connection with liquid-level applications.

Nuclear Devices

In these systems, a source transmits a beam of radiation across a bin to a detector on the other side. The nuclear radiation is capable of penetrating the bin wall, and no holes have to be cut to mount the transmitter and receiver. The nuclear beam is partially absorbed by the material in the bin, and the sensitivity of the receiver can be set to detect when material occupies the gap between the transmitter and receiver. Nuclear devices provide a comprehensive approach to level control, incorporating many advantages. Costs are high, and skilled, licensed technicians are required for installation and servicing. The principles of nuclear level detecting systems are described in more detail earlier in this article in connection with liquid-level applications.

REFERENCES

Bailey, S. J.: "Level Sensors: A Key Partner in Productivity," *Control. Eng.,* vol. 27, no. 10, October 1980, p. 75.

Bailey, S. J.: "Analysis of Material-Sensor Interaction Essential to Proper Level Meter Selection," *Control Eng.,* vol. 28, no. 13, December 1981, p. 67.

Buckley, P. S.: *Techniques of Process Control,* Wiley, New York, 1964.

Buckley, P. S.: "Material Balance Control in Recycle Systems," *Instrum. Technol.,* vol. 21, no. 5, May 1974, p. 29.

Cho, C. H.: *Measurement and Control of Liquid Level,* Instrument Society of America, Research Triangle Park, North Carolina, 1982.

Khandheria, J., and J. P. Shunta: "Adaptive Sampling Increases Sampling Rate as Process Deviations Increase," *Control Eng.,* vol. 24, no. 2, February 1977, p. 33.

Luyben, W. L., and P. S. Buckley: "A Proportional-Gage Level Controller," *Instrum. Technol.,* vol. 24, no. 12, December 1977, p. 65.

Mariam, P. L.: "Measuring Level in Hostile or Corrosive Environments," *Instrum. Technol.,* vol. 26, no. 4, April 1979, p. 45.

Morris, H. M.: "Level Instrumentation from Soup to Nuts: Figuratively and Literally," *Control Eng.,* vol. 25, no. 3, March 1978, p. 56.

Morris, H. M.: "Sensing Interface Levels Poses Many Challenges," *Control Eng.,* vol. 25, no. 8, August 1978, p. 40.

Morris, H. M.: "D/P Cell Outputs Either Current or Frequency," *Control Eng.,* vol. 26, no. 2, February 1979, p. 53.

Morris, H. M.: "Solid State Electronics Dominates Level Measurement," *Control Eng.,* vol. 26, no. 11, November 1979, p. 47.

Musick, J. E.: "Linearization of the Level-to-Volume Relationship," *Instrum. Technol.,* vol. 25, no. 9, September 1978, p. 141.

Slimiana, M.: "Using Differential Pressure Sensors for Level, Density, Interface, and Viscosity Measurements," *Instrum. Technol.,* vol. 26, no. 9, September 1979, p. 63.

Soisson, H. E.: *"Instrumentation in Industry,"* Wiley, New York, 1975.

Spenser, J.: "Using Monolithic Analog Level Detectors," *Control Eng.,* vol. 27, no. 5, 1980, p. 97.

Staff: "Monitor Temperature and Measure Liquid Level Using a Thermistor," *Control Eng.,* vol. 24, no. 7, July 1977, p. 22.

Staff: "Copper Mine Uses Ultrasonic Level Sensors," *Control Eng.,* vol. 27, no. 8, August 1980, p. 38.

Analytical and Testing Instrumentation *

James D. Bartlett, Jr. *SLM Instruments, Inc./American Instrument Company, Urbana, Illinois. (Photometric and Reaction Products Analyzers)*

R. H. Cherry. *Consultant, Huntingdon Valley, Pennsylvania. (Thermal Conductivity Gas Analyzers)*

Sheldon Connolloy. *Fisher Scientific Company, Pittsburgh, Pennsylvania. (Titration Systems)*

P. M. Corrao. *Houston Atlas, Inc., Houston, Texas. (Reagent Tape Analyzers)*

Rodney M. Durham. *Manager, Instruments Division, Infrared Industries, Inc., Santa Barbara, California. (Infrared Process Analyzers)*

Engineering Staff. *Extranuclear Laboratories, Inc., Pittsburgh, Pennsylvania. (Quadrupole Mass Spectrometry)*

George F. Erk. *Sun Tech, Inc., Philadephia, Pennsylvania. (Trends in Analytical Instrumentation)*

C. Hammerstein. *Teledyne Analytical Instruments, City of Industry, California. (Thermal Conductivity Gas Analyzers)*

John N. Harman, III. *Senior Chemist, Beckman Industrial Corporation, a subsidiary of Emerson Electric Company, Fullerton, California. (Oxidation-Reduction Potential Measurements) (pH and pH Measuring Systems)*

Edwin A. Houser. *Cyberdesign, Inc., Fullerton, California. (Sample Conditioning for Process Analyzers)*

*Persons who authored complete articles or subsections of articles, or who otherwise cooperated in an outstanding manner in furnishing information and helpful counsel to the editorial staff.

Riad Kocache. *Taylor Servomix Limited, Crowborough, Sussex, England.* *(Oxygen Analyzers)*

Stephen F. Lang. *Mine Safety Appliances Company, Pittsburgh, Pennsylvania. (Gas Analyzers)*

A. J. Livingston. *The Ohmart Corporation, Cincinnati, Ohio. (Radiation-Type Density Meters)*

S. A. MacNeill. *Wallace & Tiernan Division, Pennwalt Corporation, Belleville, New Jersey. (Amperometry)*

James E. Murphy. *Instrument Division, Mine Safety Appliances Company, Decatur, Georgia. (Thermal Conductivity Gas Analysis)*

John Nagy. *Section Manager, The Cedar Grove Division of Beckman Industrial Corporation, a subsidiary of Emerson Electric Company, Cedar Grove, New Jersey. (Electrolytic Conductivity and Resistivity Measurements)*

John H. Noyes. *Du Pont Instrument Products, Scientific and Process Division, Wilmington, Delaware. (Photometric Analyzers)*

R. S. Saltzman. *Instrument Products Division, E. I. du Pont de Nemours & Company, Inc. (Ultraviolet Absorption Analysis)*

Winston G. Shequen. *Manager, International Support, Applied Physics Laboratory (Division of Bausch & Lomb), Sunland, California. (X-Ray Analytical Methods)*

Elmer Sperry. *Chief Chemist, The Cedar Grove Division of Beckman Industrial Corporation, a subsidiary of Emerson Electric Company, Cedar Grove, New Jersey. (Electrolytic Conductivity and Resistivity Measurements)*

Fred Tipping. *Taylor Servomex Limited, Crowborough, Sussex, England. (Oxygen Analyzers)*

Richard Villalobos. *The Foxboro Company, Plymouth, Massachusetts, (Chromatography)*

Analysis Instruments

Numerous attempts have been made to categorize the impressive variety of analysis instruments commercially available. The system proposed by C. M. Albright, Jr. (Du Pont) in 1956 and first described in an earlier edition of this Handbook remains one of the most useful. This classification is based on the *interactions between energy and matter* used as the basic operating principles of analysis instruments.

Matter is made up of complex, systematic arrangements of particles that are characterized by their mass, their electric charge, or both. From a practical standpoint, these particles consist of *neutrons,* having mass, but with no charge; *protons,* having essentially the same mass as neutrons, but with a unit positive charge; and *electrons,* having negligible mass, but with a unit negative charge. Neutrons and protons comprise the nuclei of atoms, and each nucleus ordinarily is provided with sufficient orbital electrons, in a progressive shell-like arrangement of different energy levels, to neutralize the net positive charge on the nucleus. The total number of neutrons plus protons determines the atomic weight. The number of protons, which in turn fixes the number of electrons, determines the chemical and physical properties, except the mass, of the resulting atom.

The chemical combination of atoms into molecules involves only the electrons and their energy states. Chemical reactions involving both structure and composition usually occur through loss, gain, or sharing of electrons among atoms. Every configuration of atoms in a molecule, crystal, solid, liquid, or gas can be represented by a definite system of electron energy states. Moreover, the particular physical state of the molecules, as represented by their mutual arrangement, also is reflected in these electron energy states. These energy states, which are characteristic of the composition of any particular substance under consideration, can be most readily inferred by observing the consequences of interaction between the substance and an external source of energy. This external energy source may be in any of the following basic groups:

1. Electromagnetic radiation
2. Chemical affinity or reactivity

3. Electric or magnetic fields

4. Thermal or mechanical energy

These groups differ fundamentally in their mode of interaction with matter. Moreover, the types of information which these interactions afford may vary considerably in specificity or uniqueness, a situation which sometimes can be controlled by combining techniques. Many properties can be measured or inferred by more than one type of interaction, as can be readily observed by inspection of Table 1.

The philosophy of Table 1, with especial regard to the definition of the four basic energy groups, is based on the considerations given in the following paragraphs:

Electromagnetic Radiation

Interaction of electromagnetic radiation with matter affords information of a most basic kind owing to the fact that photons of electromagnetic radiation are emitted or absorbed whenever changes occur in the quantitized energy states occupied by the electrons associated with atoms and molecules. For example, x-rays, which consist of photons or electromagnetic wave packets having relatively high energy, penetrate deeply into the electron orbits in an atom and provide, upon absorption, the large amount of energy required to excite one of the innermost electrons. The pattern of x-ray excitation or absorption is related to the identity of the atoms whose orbital electrons are excited, making the x-ray technique useful for determining the presence of atoms and elements in dense samples. However, because of their great penetrating power, x-rays are not adapted to the excitation or observation of low-energy states corresponding to outer-shell or valence electrons or of interatomic bonds involving vibration or rotation. In such applications, the use of x-rays would be analogous to attempting to stop a windmill with a rifle bullet. On the other hand, electromagnetic energy at longer wavelengths, in the infrared region, is made up of photons having relatively low energy, corresponding to the energy transformations involved in the vibration of atoms in a molecule due to the stretching or twisting of the interatomic bonds. Using the windmill model to represent a molecule, infrared radiation would correspond more to a breeze, which can indeed have a profound effect on windmill operation.

Chemical Affinity or Reactivity

In a very real sense, chemical reactions permit recognition of certain substances because, by mixing, the two reactants can be brought into rather intimate contact so that the valence potential or activity coefficient, or what might be called the "potential driving force toward reaction," can come into play. This situation can be characterized by a high degree of specificity and permits composition determinations in liquids, slurries, and the like, where the information afforded by electromagnetic radiation absorption would be somewhat less significant.

Electric or Magnetic Fields

This is a powerful method for the determination of chemical composition when it is possible to rely on some inherent or conferred electrical or magnetic distinction between the sought-after components. It is employed by incorporating the sample in a suitable electric or magnetic circuit so that the distinguishing features can be sorted out and measured. The mass spectrometer, which sorts out the constituent ions in a sample—according to their mass and conferred charge—in a combination of electric and magnetic fields, can produce a complete, although empirical, chemical analysis of gas or vapor samples. Techniques have also been devised for the analysis of both liquids and solids. A more commonly encountered system for the determination of ions in solution is electric-conductivity measurement. In this case, however, no distinction is afforded between different ions.

Thermal or Mechanical Energy

This technique involves interactions of a gross nature compared with the other three techniques just described. For example, the distinguishing ability of some gas molecules to become highly excited during vibration, twisting, and rotation enables them to conduct larger amounts of heat away from heated bodies with which they collide than other gas molecules. The gross cooling effect on the heated body can be used to determine the quantity of the particular molecule present in a mixture. The simple and widely used thermal-conductivity analyzers that depend on this principle are indeed limited in their specificity or ability to recognize just one molecule, but where a gas of high thermal conductivity, such as H_2, occurs in a gas of lower thermal conductivity, such as N_2, the method is an excellent choice. Another example of gross energy transfer is measurement of the viscosity of a substance by doing work on it with such devices as rotating disks or paddles or dropping a weight through it. Here again the measurement affords an insight regarding the actual intermolecular forces that must be overcome whether the measurement is used for determining concentration, degree of polymerization, or composition.

A table of conversion units is frequently helpful in evaluating chemical analysis problems. See Table 2.

Application of Chemical-Composition Measurements

There are very few phases of industrial operations where chemical-composition variables are not important. Following are some of the applications of major significance:

Raw Materials:

1. Composition analysis to check purchase specifications.
2. Detection of contamination by trace impurities.
3. Analysis check on materials priced on an active-ingredient basis.
4. Continuous analysis of materials delivered by pipeline; water analysis.

Process Control:

1. Speed up and improve control by automatization of, or replacement of, control laboratory tests on "grab" samples.
2. Improve control by replacing or augmenting inferential measurements, such as temperature or pressure, with more significant composition data.
3. Permit use of continuous processes that could not be controlled except by continuous analysis instrumentation.

Process Troubleshooting:

1. Temporary use of analysis instruments for process studies aimed at overcoming occasional upsets.

Yield Improvement:

1. Continuous analysis of process streams to measure effects of variables influencing yields.
2. Analysis of overflow or purge streams, recirculated material, sumps, and the like, to determine product losses and detect buildup of undesirable by-products that affect yield.

Inventory Measurements:

1. Analytical monitoring of material flowing between process steps and plant areas to establish consumption and in-process inventory on the basis of active or essential ingredients.

Product Quality:

1. Determination of product composition.
2. Assess structurally dependent attributes, such as color, melting or boiling point, and refractive index.
3. Assist in adjustment of product to meet specifications.

Table 1 Generalized Relationships between Matter and Energy That Can Be Measured to Ascertain Chemical Composition

Phenomenon to be measured	Group I interaction with electromagnetic radiation	Group II interaction with other chemicals	Group III reaction to electric and magnetic fields	Group IV interaction with thermal or mechanical energy
Definition	Measurement of the quantity and quality of electromagnetic radiation emitted, reflected, transmitted, or diffracted by the sample.	Measurement of the results of reaction with other chemicals in terms of amount of sample or reactant consumed, product formed, or thermal energy liberated or determination of equilibrium attained.	Measurement of the current, voltage, or flux changes produced in energized electric and magnetic circuits containing the sample.	Measurement of the results of applying thermal or mechanical energy to a system in terms of energy transmission, work done, or changes in physical state.
Relations of measurements to chemical variables	Electromagnetic radiation varies in energy with radiation frequency, that of the highest frequency or shortest wavelength having the highest energy and penetration into matter. Radiation of the shortest wavelengths (gamma rays) interacts with atomic nuclei, x-rays with the inner shell electrons, visible and ultraviolet with valence electrons and strong interatomic bonds, and infrared radiation and microwaves with the weaker interatomic bonds and with molecular vibrations and	The selectivity inherent in the chemical affinity of one element or compound for another, together with their known stoichiometric and thermodynamic behavior, permits positive identification and analysis under many circumstances. In a somewhat opposite sense, the apparent dissociation of substances at equilibrium in chemical solution gives rise to electrically measurable valence potentials, called oxidation-reduction potentials, whose magnitude is indicative of the concentration and composition of the substance. While individually all the above effects are unique for each element or compound, many	The production of net electric charge on atoms or molecules by bombardment with ionizing particles or radiation or by electrolysis or dissociation in solution or the induction of dipoles by strong fields establishes measurable relationships between these ionized or polarized substances and electric and magnetic energy. Ionized gases and vapors can be accelerated by applying electric fields, focused or deflected in magnetic fields, and collected and measured as an electric current in mass spectroscopy. Ions in solution can be transported, and deposited if desired, under the influence of various applied potentials for coulometric or polarographic analysis and for electrical conductivity measurements. Inherent and induced magnetic properies	The thermodynamic relationship involving the physical state and thermal energy content of any substance permits analysis and identification of mixtures of solids, liquids, and gases to be based on the determination of freezing or boiling points and on the quantititive measurement of physically separated fractions. Useful information can often be derived from thermal conductivity and viscosity measurements, involving the transmission of thermal and mechanical energy, respectively.

rotation. Most of these interactions are structurally related and completely unique. They may be used to detect and measure the elemental or molecular composition of gas, liquid, and solid substances within the limitations of the available equipment.

Emitted Radiation

1. Thermally excited:
   a. Optical emission spectrochemical analysis
   b. Flame photometry
2. Electromagnetically excited:
   a. Fluorescence (see Sec. 6, X-Ray Analytical Methods)
   b. Raman spectrophotometry
   c. Induced radioactivity
   d. X-ray fluorescence (see Sec. 6, X-Ray Analytical Methods)

are readily masked by the presence of more reactive substances so they can be applied only to systems of known composition limits.

Consumption of Sample or Reactant

1. Orsat analyzers
2. Automatic titrators

Measurement of Reaction Products

1. Impregnated-paper-tape devices (see Sec. 6, Photometric Reaction Product Analyzers)
2. Continuous chemical reaction types (see Sec. 6, Photometric and Reaction Product Analyzers)

Thermal Energy Liberation

1. Combustion types (see Sec. 6, Combustibles, Total Hydrocarbons, and Carbon Monoxide Analyzers)

give rise to specialized techniques, such as oxygen analysis based on its paramagnetic properties and nuclear magnetic resonance, which is exceedingly precise and selective for determination of the compounds of many elements.

Mass Spectroscopy (see Sec. 6, Quadrupole Mass Spectrometry)

Electrochemical (see Sec. 6, Reaction Product Analyzers)

Electrical Properties

1. Electrical conductivity (see Sec. 6, Electrical Conductivity and Resistivity Measurements)
2. Dielectric constant and loss factor
3. Oscillometry
4. Gaseous conduction

Magnetic Properties

1. Paramagnetism (see Sec. 6, Oxygen Analyzers)
2. Nuclear magnetic resonance
3. Electron paramagnetic resonance

Effects of Thermal Energy

1. Thermal conductivity (see Sec. 6, Thermal Conductivity Gas Analyzers)
2. Melting and boiling point determinations
3. Ice point (crystallization) (see Sec. 10, Humidity Measurement)
4. Dew point (see Sec. 10, Humidity Measurement)
5. Vapor pressure
6. Fractionation
7. Chromatography (see Sec. 6, Chromatography)
8. Thermal expansion

Table 1 Generalized Relationships between Matter and Energy That Can Be Measured to Ascertain Chemical Composition (*Continued*)

Phenomenon to be measured	Group I interaction with electromagnetic radiation	Group II interaction with other chemicals	Group III reaction to electric and magnetic fields	Group IV interaction with thermal or mechanical energy
	Transmission and Reflection Measurements 1. X-ray analysis (see Sec. 6, X-Ray Analytical Methods) 2. Ultraviolet spectrophotometry (see Sec. 6, Ultraviolet Absorption Analysis) 3. Conventional photometry (transmission colorimetry) 4. Colorimetry 5. Light scattering 6. Optical rotation (polarimetry) 7. Refractive index 8. Infrared spectrophotometry (see Sec. 6, Infrared Process Analyzers) 9. Microwave spectroscopy 10. Gamma ray spectroscopy 11. Nuclear quadrupole moment	Equilibrium Solution Potentials (Oxidation-Reduction) 1. Redox potentiometry (see Sec. 6, Oxidation-Reduction Potential Measurements) 2. pH (hydrogen ion concentration) (see Sec. 6, pH and pH Measuring Systems) 3. Metal ion equilibria		Effects of Mechanical Energy or Forces 1. Viscosity (see Sec. 11, Viscosity Systems) 2. Sound velocity 3. Density (see Sec. 6, Density and Specific Gravity Measurements)

Table 2 Factors for Interconversion of Concentration Units of Gases and Vapors

At 27°C, 3 in H_2O gage pressure

Desired units	Present units						
	Percentage by volume	Parts per million by volume	Moles per liter	Milligrams per cubic centimeter	Milligrams per liter	Milligrams per cubic meter	Milligrams per cubic foot
Percentage by volume	—	10^{-4}	2450	$24,500/M$	$2.45/M$	$2.45 \times 10^{-3}/M$	$0.0863/M$
Parts per million by volume	10^4	—	24.5×10^6	$24.5 \times 10^6/M$	$24,500/M$	$24.5/M$	$863/M$
Moles per cubic centimeter	4.1×10^{-4}	4.1×10^{-8}	M	$1/M$	$10^{-3}/M$	$10^{-6}/M$	$35.3 \times 10^{-6}/M$
Milligrams per cubic centimeter	$4.1\,M \times 10^{-7}$	$4.1\,M \times 10^{-8}$	—	—	10^{-3}	10^{-6}	35.3×10^{-6}
Milligrams per liter	$0.41\,M$	$4.1\,M \times 10^{-5}$	$M \times 10^3$	10^3	—	10^{-3}	0.0353
Milligrams per cubic meter	$410\,M$	$0.041\,M$	$M \times 10^6$	10^6	10^3	—	35.3
Milligrams per cubic foot	$11.6\,M$	$1.16\,M \times 10^{-3}$	$28,300\,M$	$28,300$	28.3	0.0283	—

To use the table:
1. Locate column along top of table which gives present unit.
2. Locate row along left of table which gives desired unit.
3. Read down and across to locate multiplying factor.
4. Multiply present quantity by factor.
Example: Given 700 ppm to convert to moles per liter: $700 \times 4.1 \times 10^{-8} = 2.87 \times 10^{-5}$ mol/L

NOTE: In table, M is molecular weight of the gas or vapor.

Safety:

1. Detection of leaks in equipment.
2. Survey operating areas for escape of toxic materials from leaks or spills, especially materials not readily detected by human senses.
3. Detection of flammable or explosive mixtures in atmosphere or process lines.

Waste Disposal:

1. Monitoring plant stacks for accidental discharge of toxic or nuisance gases, vapors, or smoke.
2. Analysis of waste streams for toxic or other objectionable materials.
3. Control of waste treatment or product recovery facilities.

Research and Development:

1. Continuous analysis to speed up research and optimize results.
2. Provide structural and compositional information not otherwise obtainable.
3. Produce results in a more directly usable form.

In considering the application of instrumental methods for the determination of chemical composition, it is important to bear in mind that measurement is the first step toward control and that the closer the information can be brought to the process, the better will be the control. This is true whether the information is merely presented continuously to the operator or actually used to control the process itself automatically. Complete automatic control becomes more desirable as process throughput is increased and holdup is decreased in modern highspeed processing equipment. This trend also places a premium on high speeds of response in analytical instruments and their associated control equipment.

Practical Considerations in the Measurement of Chemical Composition

Any practical appraisal of the merits of chemical-composition variables for process control purposes must recognize certain inherent physical limitations in their measurement. Generally speaking, these limitations are the following:

1. **Sample Must Be Representative.** Although this requirement may appear obvious, it is a factor that is very frequently overlooked. In the first place, the sample must be gathered or drawn off in such a fashion that it will be of the same composition as the body of the processed material. Moreover, there must be assurance that any change in conditions, such as temperature or pressure, between the sampling and measuring points cannot influence sample composition. In addition, in nearly all cases the probable composition of the sample must be known ahead of time through some independent method before an analysis technique can be selected.

2. **Physical State of Sample.** The technique must provide for interaction between the applied energy and the entire sample, as well as for observation of the total result. This can seldom be accomplished. It is for this reason that a large majority of techniques are applicable to gases, where the molecules are widely spaced and free to react in a characteristic manner, and that fewer techniques are applicable to liquids and still fewer can be applied to solids.

3. **Uniqueness of Specificity of Method.** The selection of the method must be tailored to the sample composition and to the information requirements. Some methods or techniques involving atomic and molecular structure are rather universal in that they permit exact identification and measurement of every elemental or molecular constituent present in the sample. These methods are usually the most complex and costly. They are sometimes considerably less sensitive than simple methods whose only drawback is an inability to distinguish between related substances having similar gross interactions with energy. Where the related substances are known not to be present in the sample, these simpler, less specific methods should always be considered.

Trends in Analytical Instrumentation*

Much progress has been made in analytical instrumentation in the past, and further progress is expected in the future. It is therefore important for anyone investigating the state of the art to check the latest literature and vendor information available.

Several factors favor continuing developments and refinements in this field.

1. Conservation of energy has increased efforts to use *on-line* analytical instruments. Fuel costs have risen over the last decade to a point where on-line instrumentation has become even more cost-effective. One area immediately affected by this is the application of an oxygen and combustibles analyzer in determining optimum fuel-to-air ratios in the combustion process and the resulting heat generation. Likewise, calorimetric analysis of fuel quality is becoming even more important, as are density and specific gravity analyses.

2. Requirements for monitoring pollution generated additional needs for high-quality analytical devices with the capability to provide good records of various pollutants and particulate emissions. Stack gas monitors are widely used for pollution control and monitoring and have design features permitting long-term unattended operation within stacks or through sampling systems.

3. Pressing demands for more accurate, thorough, and rapid means of testing materials and products—from receipt of raw materials and inspection throughout manufacturing to the completion of production, warehousing, and distribution.

Just as the availability of better analysis instruments since the 1950s has moved the formerly isolated chemical control laboratory on-stream in terms of the use of continuous instrumental analyzers, the advent of improved nondestructive inspection techniques and fully automated testing procedures will complete a revolution already underway in the quality control of production, particularly in discrete piece manufacturing industries.

In the past 5 to 10 years, the replacement of mechanical timers and logic sequencers with microprocessors for automated sequencing and data reduction has not only improved reliability, accuracy, and ease of operation and maintenance, but has also allowed further development of enhanced production strategies through interfacing with hierarchical computer systems, such as data acquisition and management information systems. Automated calibration, self-diagnostics of electronics and sequencing operations have further improved reliability and have elevated the importance of on-stream analyzers as primary measuring devices for process control. Microprocessors allow easy programming of all the functions required for each individual application and can, of course, be reprogrammed for a much wider field of applications, as compared with a single analyzer with a conventional analog control function, timer, cams, sequencers, and the like.

Analyzer Systems

Many users turn away from buying just the analyzer proper, and instead they specify and purchase fully integrated, preassembled analyzer systems using modular sample system panels, standardized calibration, and sample disposal methods—all incorporated in preassembled shelters provided with heat, light, ventilation, and air-conditioning, among other features. The advantages of modular and preassembled systems include:

1. Systems designed and built by a specialist supplier often may be superior to those built at the site.

2. Factory construction is independent of weather and labor conditions existing at the site.

3. Systems can be fully tested at the factory under simulated operating conditions, and major design, equipment, and construction faults can be corrected before shipment.

Even with preassembled analyzers and specialist suppliers, the task of engineering and managing a stream analyzer installation is not simple. Figure 1 outlines a typical sequence of events that occur during an analayzer installation.

*Prepared by George F. Erk, Sun Tech, Inc., Philadelphia, Pa.

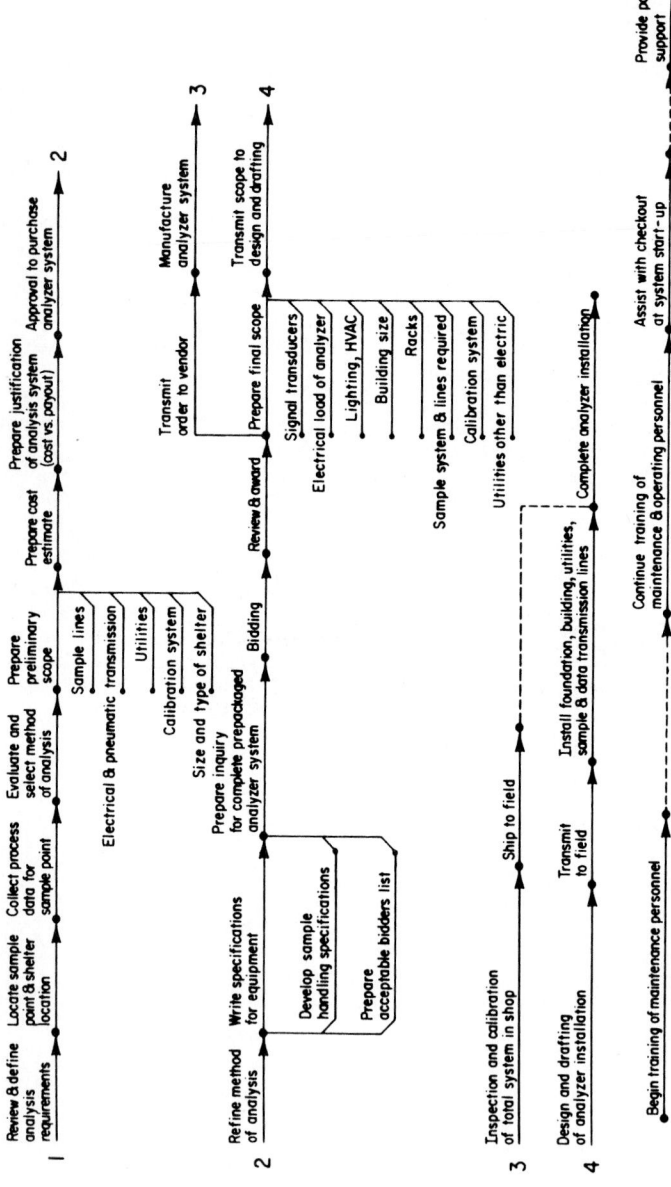

Fig. 1 Planning diagram shows major activities involved in specifying, bidding, manufacturing, and installing a prepackaged analyzer system. Depending on the application, certain items can be accomplished at different times than as shown. The sequence shown does not imply a degree of importance for any one factor.

Investment Parameters

Computer input data

Heater size: 30 MMBTH
Heater efficiency: 88%
Fuel heat value: 6x10^6 Btu/FUEB
Yearly operating time: 8,400 h
Increased heater efficiency: 1%
Analyzer installed cost: $11000.
Yearly maintenance cost: $2000.
Investment tax credit: 10%
Analyzer life: 10 years

Computer output data

Total investment = $11000.
Taxes = 50.80% paid 4 times/year
Project Life = 10.00 years
Lead time = 0. years
Undepreciated book value at end of project = $0.
Interest rate of return = 22.75%
Total return on investment = 126.29%
Average return on investment = 12.63%/year

Discount rate	Present worth	Payout (years)
0.00%	$13892.	3.64
10.00%	$ 5561.	4.68
15.00%	$ 2918.	5.66

End year	Investment	Book value	Deduct. expense	Depreciation	Taxes	Investment tax credit	Cash flow
0	11000	11000	0	0	0	0	0
1	11000	8800	3700	2200	762	1100	4038
2	11000	7040	3700	1760	986	0	2714
3	11000	5632	3700	1408	1164	0	2536
4	11000	4506	3700	1126	1307	0	2393
5	11000	3604	3700	901	1422	0	2278
6	11000	2884	3700	721	1513	0	2187
7	11000	2163	3700	721	1513	0	2187
8	11000	1442	3700	721	1513	0	2187
9	11000	721	3700	721	1513	0	2187
10	11000	0	3700	721	1513	0	2187
Total		0	37000	11000	13208	1100	24892

Fig. 2 Example of a relatively straightforward computer program that can be written to calculate investment parameters for any analyzer system. The data can then be used to help economically justify the analyzer.

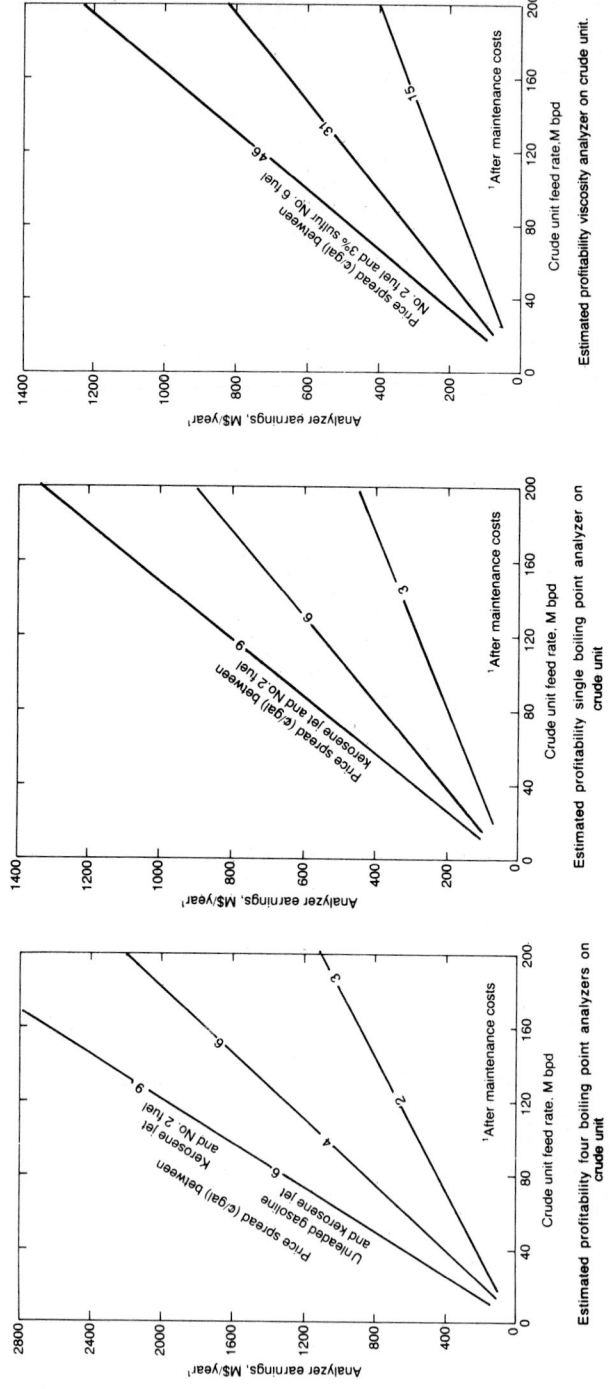

Fig. 3 Estimated profitability of applying various forms of analytical instrumentation on a refinery computer model.

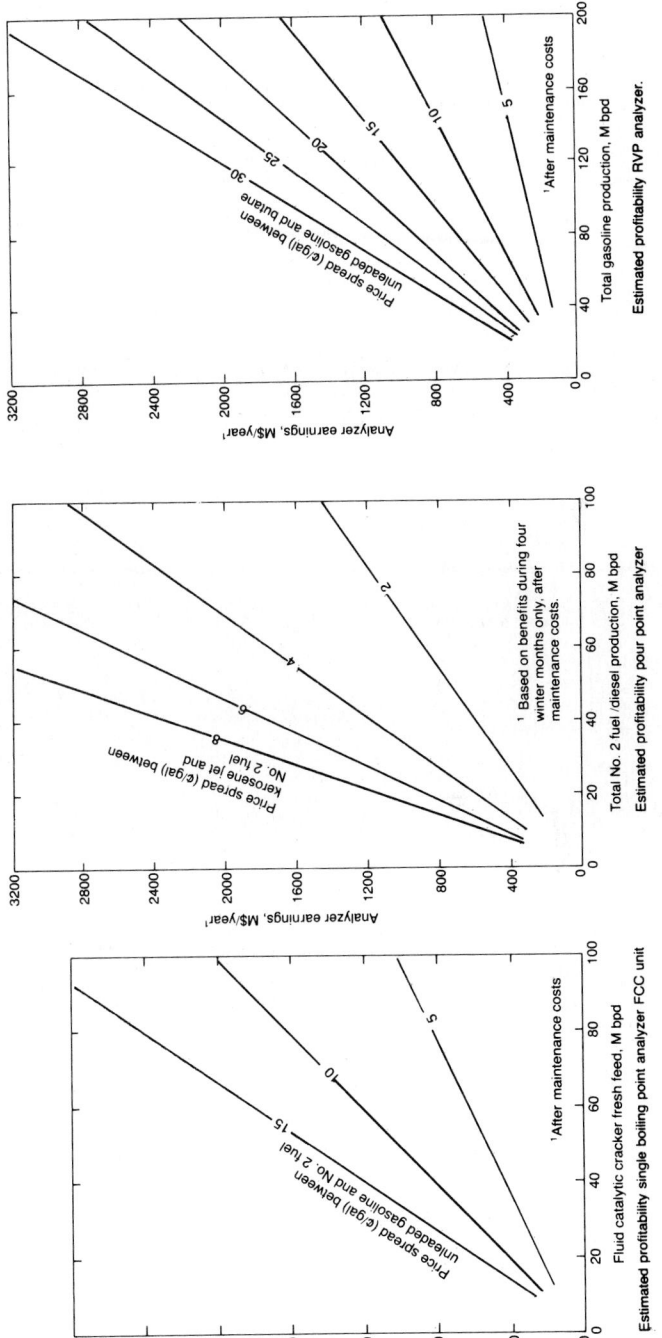

Fig. 3 (*Continued*)

6.19

Economic Justification

Process analyzers usually require justification on an economic basis. For example, the installation of an oxygen and combustibles analysis program can be established to calculate investment parameters (Fig. 2) based on amount of fuel saved, increase in yield, efficiency, or decreased "off-spec" material. Studies performed on a refiner computer model developed for the U.S. Department of Energy by Turner Mason and Associates indicate large payouts on boiling point, pour point, Reid vapor pressure and/or viscosity analyzers. See the series of graphs in Fig. 3.

REFERENCES

Editor's Note: See reference lists at the end of each of the over 20 separate articles on analysis instruments in Sec. 6 of this Handbook.

Sample Conditioning for Process Analyzers

by
Edwin A. Houser*

Process stream analyzers may be classified as (1) in situ or (2) extractive, according to whether there is a need to extract (remove) a sample of the process fluid and transport it to a remotely located analyzer. "In situ" refers to an analysis device within the process itself, or at least to a device that senses process fluid characteristics without actually withdrawing a sample. Although some sample conditioning actions may occur within an in situ analyzer, the term "sampling system" nearly always refers to extractive analyzers. One usually considers the elimination of a sampling system the main advantage of an in situ analyzer. However, there are relatively few analyzer technologies which lend themselves to in situ application, and thus the principal emphasis has been on extractive systems.

Functions of Sampling System

The main functions of a sampling system are (1) representative sampling, (2) transporting, and (3) conditioning, all of which must be accomplished within the context of the parameter to be measured and the characteristics of the analyzer to which the system is connected. For example, ideally, a fast analyzer should have a fast sampling system. Generally, it does not matter if the composition of the sample is modified in the system, so long as the desired parameter is measured with sufficient accuracy. Sometimes sample composition is deliberately changed to achieve the desired measurement; for example, removing a large variable quantity of water vapor places the measurement on a dry basis.

*Cyberdesign, Inc., Fullerton, California.

Most process stream analyzers are designed to provide automatic, unattended operation for long periods without operator intervention, and sampling systems should be similarly designed. However, many difficult sampling systems require considerably more maintenance than analyzers.

In designing a sampling system, three components must be considered: (1) the process, (2) the environment, and (3) the analyzer. To obtain a sample from the process in such a form that it is representative of the material in the process is the primary objective of a sampling system. In an extractive system, the sample must be transported a significant distance through an environment which is usually different from that of the process. The effects of temperature, pressure, time, and materials of construction must be evaluated to ensure sample integrity. Obviously, the sample must be fully compatible with the analyzer—in terms of conditions such as temperature, pressure, flow rate, viscosity, density, phase, and degree of contamination. Failure to achieve this compatibility leads to an inaccurate analysis and, in a worst case, to damage to the analyzer. It is evident, then, that the analyzer manufacturer's specifications are an important factor in sample system design. The choice of analyzer location may or may not be decided by the sampling system designer. With a knowledge of the analyzer and its location, the sampling system design must cope with such factors as environmental conditions, distance, elevation changes, and transport time requirements, among others.

The first task in sampling system design is to gather needed application data. Several organizations have developed fill-in forms which provide an organized format for considering problem factors. The ISA analyzer system application data form, for example, has been prepared by the Instrument Society of America and can be very helpful. Often, the final system represents a compromise. In any event, experience has shown that, where there is a choice, the simpler the system the better.

Sample Takeoff Probes

Probe is a term that generally applies to any type of apparatus which protrudes into the interior of the process equipment being sampled. *Equipment* may refer to a vessel, a pipe, or even a room. A probe is preferred over a surface connection for a number of reasons. The probe makes it possible:

1. To collect samples from the interior regions of the equipment
2. To avoid picking up contamination which might creep along the inside surface
3. To reduce the internal flushing volume (and therefore transport lag time) while providing a structurally strong connection
4. To achieve a degree of separation (rejection) of particulates or droplets which otherwise would contaminate the sample
5. To achieve proportional sampling of mixed phases, i.e., avoid selection of one phase over another (isokinetic sampling)
6. To collect and mix equal proportions across a duct or flue (average)
7. To stop a reaction (quenching) by sudden cooling and/or pressure reduction
8. To create a pressure differential to obtain flow through the system (by velocity head)
9. To dilute the sample before it can cool as it flows out of the equipment

Many probes combine two or more of the aforementioned functions in one unit. Figure 1 shows a simple probe that can be fabricated in the field. It reduces internal volume, avoids pipe wall contamination, and provides a strong connection. Figure 2 shows a commercially available plain pipe probe for clean, low-temperature flue gases. Note the sloping installation, which provides self-draining for the condensate. Figure 3 shows a water-cooled, water-spray-washed type of probe for hot, dirty flue gases. The sample is drawn into the probe opening through a curtain of water spray to knock back fly ash. This type of probe is not suitable for the measurement of water-soluble gases.

Factors to be considered when designing a probe include:

Mechanical Strength. Mechanical strength must be considered, with due allowance for temperature, length, weight, material, fluid velocity, and vibration. This requires considerable knowledge of process conditions.

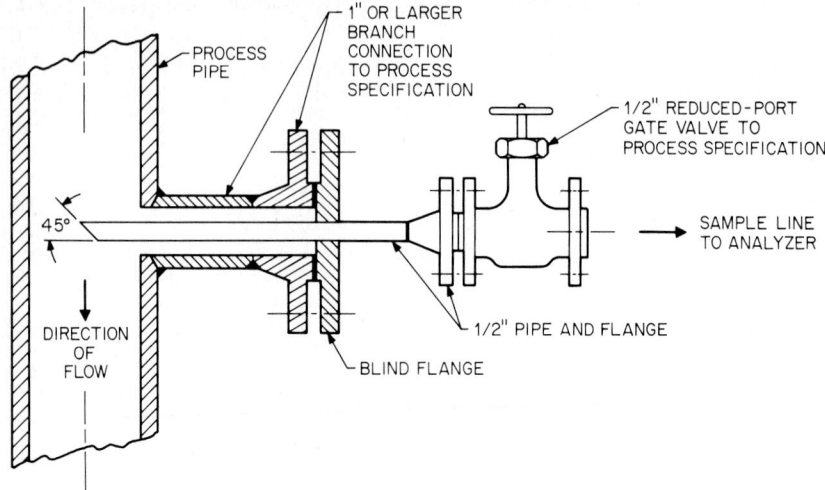

Fig. 1 Sample probe used to avoid pipe-wall contamination *(Cyberdesign.)*

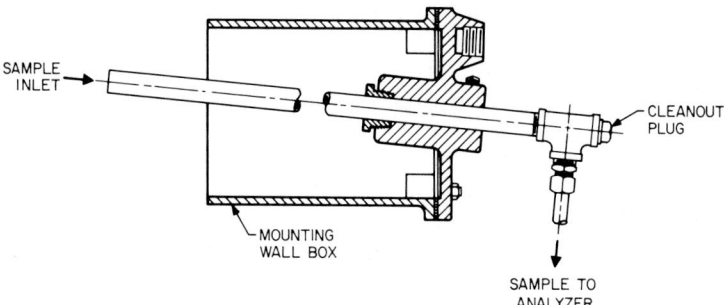

Fig. 2 Sample probe for clean flue gases. *(Hays-Republic Division, Milton-Roy.)*

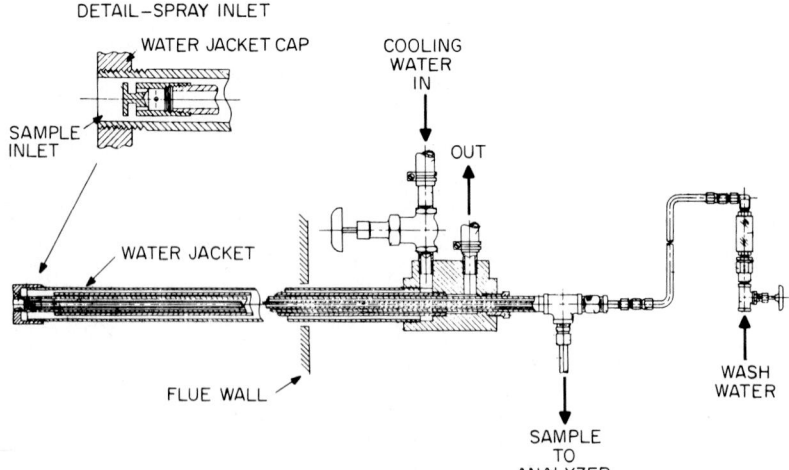

Fig. 3 Sample probe for dirty flue gases. *(Hays-Republic Division, Milton-Roy.)*

Corrosion Resistance. Corrosion resistance is a difficult problem because specific boundary layer conditions may vary drastically over very short distances. For example, when sampling hot gases containing sulfuric acid vapors, the first liquid condensing can be quite high in sulfuric acid, whereas later more water condenses to dilute the sample but not before excessive local corrosion has occurred.

Effect on Sample Composition. Some probes make use of water washing sprays to clean the sample. The relative solubilities of the sample constituents should be evaluated to ensure that the measured component is not adversely affected.

Ease of Maintenance. Probes which contain filters or small passageways may require cleaning or filter renewal at unpredictable intervals. Here one must consider the following: Can the probe be serviced without shutting down the process? Can personnel access the probe location during operation?

Further, the probe as an integral part of the process line or vessel must follow the same design standards in terms of class and service.

Sampling with Fast Loops

Fast loop sampling refers to the practice of creating a relatively large flow across a pump, control valve, or other process equipment for the purpose of reducing transport lag time without wasting an unacceptable quantity of sample. Because of their much higher mass/volume ratio, liquid samples are used with fast loops more commonly than gas samples. A return point in the process must be chosen to have a sufficiently lower pressure to ensure a sufficient differential to obtain reliable and adequate flow. The need for pressure gages and a flowmeter to monitor fast-loop conditions should not be overlooked.

A common practice with fast loops is to terminate them (at the analyzer location) with a type of slipstream or bypass filter. Two representative types are shown in Figs. 4 and 5, using a tubular, porous, stainless steel element (Fig. 4) and a thin, microporous (membrane filter) element in which the fast-loop flow washes the upstream side of the element (Fig. 5). In both cases, only the smaller flow to the analyzer passes through the element. Figure 6 shows a typical fast-loop arrangement, with bare-minimum pressure and flow instrumentation.

The fast-loop flow rate can be predicted using accepted turbulent and laminar flow formulas for the pressure drop, distance, and line size proposed. Care must be taken to allow for variations in process pressure with load changes. Pressure drops estimated across passive process equipment should be reduced by at least 75% if the sampling system is to maintain performance when the process is operating at reduced capacity. Usually, differentials across pumps and control valves increase when process flows are reduced, so sizing under normal conditions should suffice. Unless it is obvious, the sample flow as a fraction of the process flow should always be checked.

In the case of an insufficient process differential (or where the sample is to be returned to the same line) a pump located close to the takeoff point to ensure adequate net positive suction head (NPSH) can provide a fast-loop differential. It is important to size the pump correctly, allowing for anticipated variations in flow and fluid parameters.

In a few cases, the supply pressure will be much larger than necessary to obtain flow, and it may be advantageous to reduce the pressure to protect downsteam conditioning equipment or, in the case of gases, to reduce the dew point or increase the effective velocity at the same mass flow rate. In such cases, the pressure regulator should be located close to the takeoff point.

Direct Sampling

Direct sampling refers to the practice of venting all of the sample to the atmosphere or in a flare, for gases; or into a drain (water or oil), for liquids. Of course, the drain can include a collection or sample recovery system which eventually returns the sample to the process.

Direct sampling is used mostly for gases, where the mass of material taken from the process is usually small even for fairly high sample line velocities. Frequently, the choice for high-pressure gas samples is to locate a pressure regulator close to the takeoff or to use a fast loop so as to avoid wasting a large flow of material. Assume for the moment a fixed-quantity flow rate to be taken from the process. Because the velocity in the sample line is inversely proportional to the absolute pressure in the line, reducing the pressure at the analyzer location, instead of at the takeoff, can increase tremendously the transport lag from process to analyzer. Clearly, it is important to reduce the volume between the sample tap and the pressure regulator.

Figure 7 shows a typical direct system for high-pressure gases. Heat should be applied between the takeoff and the regulator if the dew point there can be above ambient temperature. A similar check should be made for the dew point after the regulator to determine whether it is necessary to heat all the way to the analyzer location. Usually, a reduced pressure of about 10 to 15 psig (70 to 105 kPa) is enough to develop a reasonable velocity, but very long lines may need more.

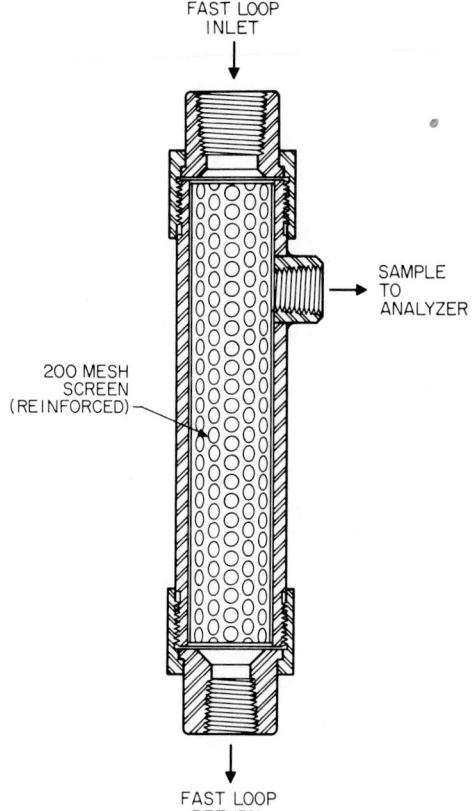

Fig. 4 In-line strainer for liquid fast loops. *(Precision Scientific Development, GCA.)*

Vaporizing Liquid Samples

There are many occasions where the sample exists as a liquid at the takeoff but may be analyzed in the vapor phase. Although special factors may dictate otherwise, if the vapor pressure of the sample is low enough so that it can be analyzed as a liquid, it probably should be, even though it is vaporizable. Doing so may eliminate the need for heat-tracing sample lines and conditioning systems and

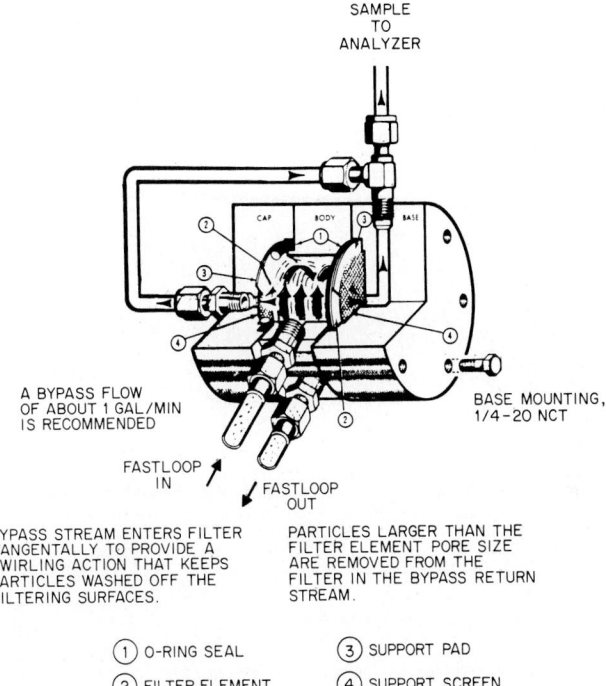

Fig. 5 Membrane-type fast-loop liquid filter. *(Collins Products.)*

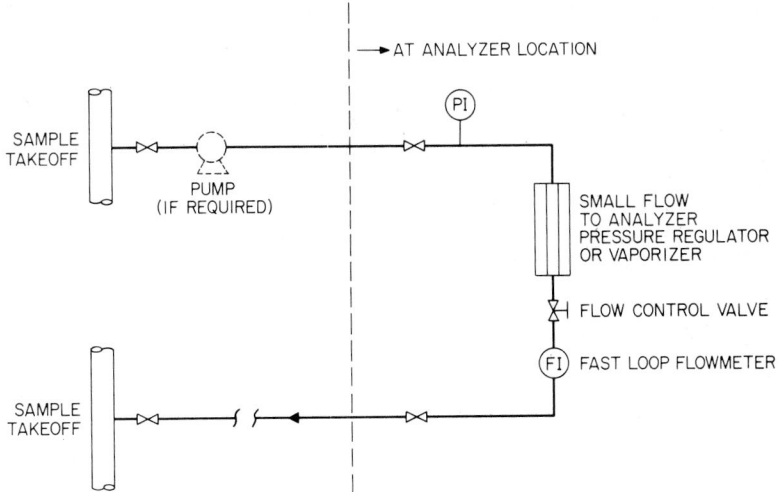

Fig. 6 Typical fast loop for a liquid. *(Cyberdesign.)*

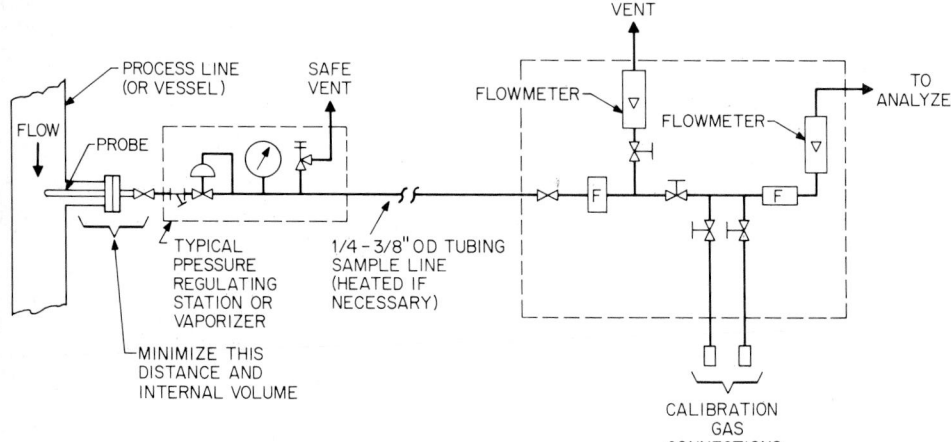

Fig. 7 Typical direct system for high-pressure gases or vaporized liquids. *(Cyberdesign.)*

avoid problems resulting from the presence of even small amounts of nonvolatile materials. Also, vaporizers are notorious for introducing long time lags. The most common reason for vaporizing a liquid sample is that it is difficult if not impossible to keep the sample a liquid at the analyzer temperature.

Vaporization of liquid samples is usually done by devices designed for the purpose called vaporizing pressure regulators. These are basically spring-loaded pressure reducing regulators having an integral heating chamber close to the valve orifice to provide the heat of vaporization. A typical proprietary steam-heated vaporizing regulator is shown in Fig. 8. This type of device automatically reduces and controls pressure as well as vaporizes. Other methods involve the use of capillary tubing or larger tubing packed with some kind of material to stabilize the liquid phase. In any case, the objective is to convert the liquid to a vapor in as small a space as possible to avoid fractionation, or physical separation of the sample into its components of various volatility. Such fractionation may be smoothed out or remixed, especially if the analyzer is continuous, but this is difficult if not impossible when the analyzer is a chromatograph, which is usually the case. The best course is to convert the sample as quickly as possible.

As with pressure regulators, the choice *may* be between direct sampling (vaporizer at takeoff) or using a fast loop. The latter is preferred *if* there is enough pressure to keep the sample a liquid all the way to the vaporizer, and if there is a fast-loop return available.

If direct sampling is chosen, the internal volume between the process and the vaporizer must be kept to an absolute minimum, because of the very low liquid volume/vapor volume ratio. Special small-diameter probes and reduced port block valves are essential to minimize volume while still providing a mechanically strong attachment to the process line.

The direct and fast-loop schematics shown in Figs. 6 and 7 apply to vaporizer systems also. In either case, an important objective is to minimize the internal volume upstream of the vaporizer. The probe arrangement (Fig. 7) minimizes the internal volume between the line and the vaporizer; this is essential if the purge-time delay is to be kept reasonable. It is important to evaluate carefully the pressure drop and sample temperature in the supply line relative to the sample bubble point so as to ensure a completely liquid sample at the inlet to the vaporizer.

After the vaporizer, it may be necessary to heat the system to keep the sample vaporized, depending on the dew point of the mixture at the reduced pressure.

Consideration should be given to keeping the flow rate through the vaporizer within its design capability. A flow rate too low will result in partial vaporization upstream of the reducing valve. A rate too high will cause incomplete vaporization at the valve, and fractionation as the heavier fractions vaporize later.

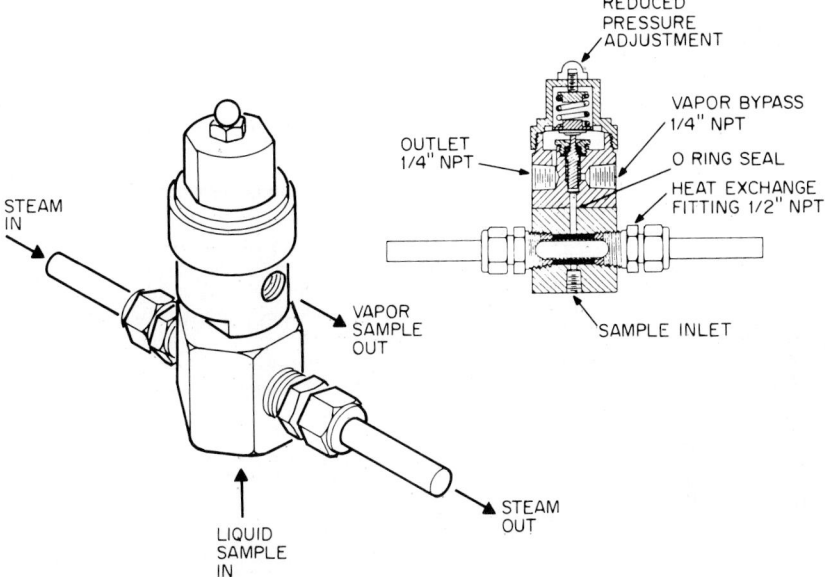

Fig. 8   Liquid sample vaporizer. *(Go., Inc)*

Drying Samples

Probably the most common unwanted sample constituent is water, which can occur in the same phase as the sample or as free liquid in vapor and liquid samples.

When the sample is a liquid, free water can create serious problems in the analyzer, depending on its type. Liquid-inject gas chromatographs are particularly susceptible because the size of an injected sample can vary greatly if entrained water droplets are present. Also affected are cut point, cloud point, and many types of optical analyzers.

The removal of free water from liquid hydrocarbon streams is nearly always accomplished by passing the sample through a coalescing device. This device is usually a type of filter in which small droplets of water coalesce into larger ones as the sample flows through the porous medium, followed by a region of low velocity where the water is separated by gravity. Figure 9 shows a cross section of a commercially available coalescing filter for hydrocarbons.

Actually, passage through practically any porous medium will cause water to coalesce, although a "depth-type" medium is perhaps more effective. Media which are hydrophilic (water-wetted) are preferred. In any case, it is essential to provide a continuously flowing purge stream from the bottom of the unit to drain off the water.

Because water is less soluble at lower temperatures, it is important to locate the coalescer at the coldest point in the system, hopefully also

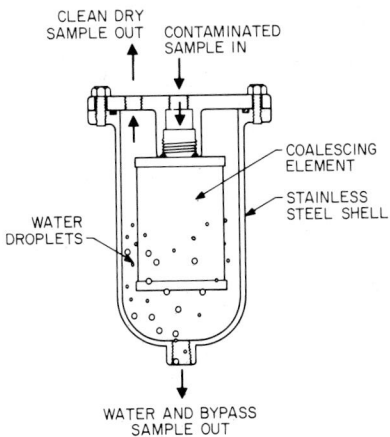

Fig. 9   Coalescing filter for liquid hydrocarbons. *(Precision Scientific Development, GCA.)*

close to the analyzer. The manufacturer's flow ratings should be checked to make sure that water will not break through under normal system operation.

Water can be removed from gas samples by any one or a combination of the following: (1) cooling, condensing, and separating, (2) adsorption or absorption on a suitable desiccant, and (3) diffusion drying.

Cooling, condensing, and separating are attractive for relatively high concentrations of water, say above 5 to 10%, depending on the pressure. The concentration of water remaining afterward depends on the temperature and pressure, which of course fix the partial pressure and mole fraction of water in the vapor phase. Calculation of the water remaining and the water separated is straightforward when detailed steam tables are used.

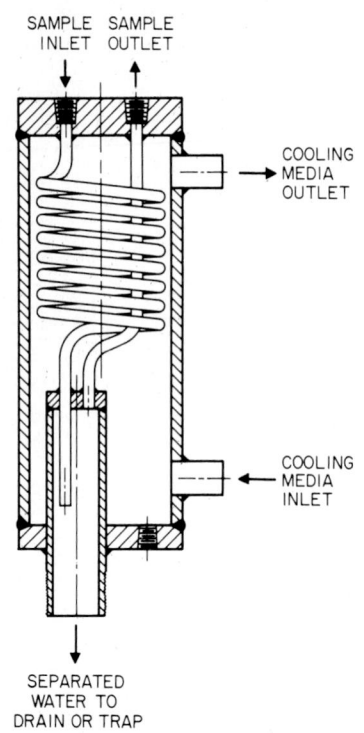

SAMPLE SAMPLE
INLET OUTLET

COOLING
MEDIA
OUTLET

COOLING
MEDIA
INLET

SEPARATED
WATER TO
DRAIN OR TRAP

Fig. 10 Liquid-cooled separator. *(Adapted from Houser.)*

It is important to evaluate the effect of removing water for two reasons. First, the component of interest may have significant solubility in the condensed water, resulting in an analysis value lower than the actual value. Second, if the concentration is on a wet basis, it will naturally increase if significant amounts of water are removed.

The first effect can be difficult to estimate, especially if the system does not follow Henry's law of solubility. The pH of the condensed phase can greatly affect the solubility of acidic or alkaline gases. Sometimes, two or more gas phase components react together when a liquid water phase is present. Considerable research into the chemistry of the system may be necessary.

Cooling, condensing, and separating can be readily accomplished with a variety of hardware designed for the purpose. Small concentric tube or coil-in-shell heat exchangers are best; the latter should disassemble for cleaning if tube wall temperatures can cause scaling. Small, low-volume, stainless steel float traps which operate at fairly high pressure can be obtained. At low pressures, manometer traps operate without moving parts—which is always helpful. High-pressure situations should be fully evaluated with respect to purge delay, because actual volumetric gas flow is inversely proportional to pressure for the same mass flow.

Figure 10 shows a cross section of a water-cooled separator for removing water by cooling and condensing. Note that the separator section is contained within the cooling jacket so as to provide separation at the lowest temperature. A ball-float trap, manometer leg, or continuous blowdown may be connected at the bottom.

Drying by adsorption and/or absorption on desiccant beds has some application in sampling systems, although there are usually serious disadvantages. If the component of interest is strongly polar, it probably will also be adsorbed to some extent. Sometimes, a major nonmeasured component is adsorbed significantly and then breaks through, causing a shift in the background composition. And, finally, the problems of limited life and replacement requirements must be anticipated.

The last method, diffusion drying, is becoming more popular because of two factors. First, when properly designed for the application, it requires little attention and has a very long life. If the sample contains no free solids or condensables (other than water), the service life can be nearly unlimited.

The other advantage is that diffusion drying is highly selective. It has been demonstrated that many highly polar components such as SO_2, H_2S, and NO_2 are not lost with the water. It is very important to keep all liquids out of a diffusion dryer, especially water. If there is any possibility that liquid water can reach the dryer, the inlet half should be heated to a temperature above the maximum sample dew point. Careful attention is also needed to provide the proper dry gas flow and purge-to-sample pressure ratio.

Figure 11 shows a commercially available diffusion dryer consisting of a stainless steel jacket surrounding a bundle of multiple small-diameter polymer tubes. Normally, the sample flows inside the tubes and dry purge gas in the jacket. Since the purge gas picks up water from the sample, care should be taken to evaluate the exit purge gas dew point. It may be necessary to heat or dilute the purge gas or to provide a trap for the condensate.

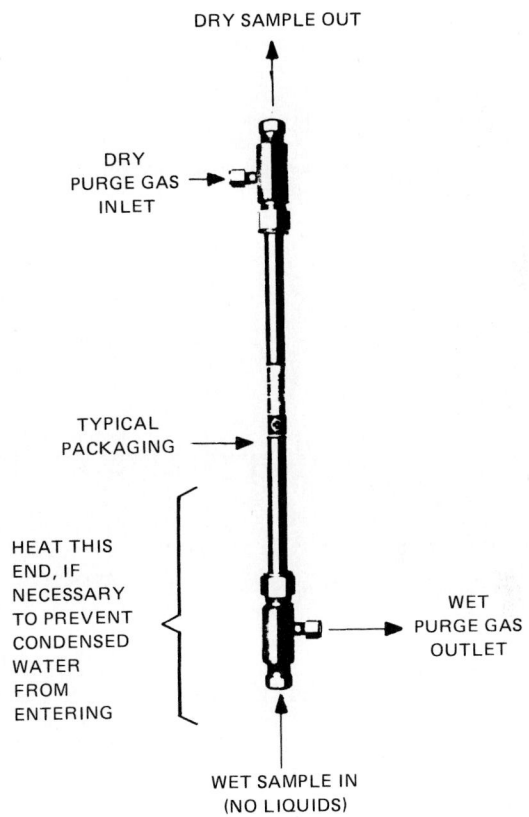

DRY SAMPLE OUT

DRY PURGE GAS INLET

TYPICAL PACKAGING

HEAT THIS END, IF NECESSARY TO PREVENT CONDENSED WATER FROM ENTERING

WET PURGE GAS OUTLET

WET SAMPLE IN (NO LIQUIDS)

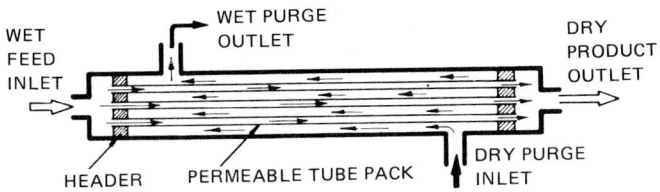

WET FEED INLET

WET PURGE OUTLET

DRY PRODUCT OUTLET

HEADER   PERMEABLE TUBE PACK   DRY PURGE INLET

Fig. 11 Permeation tube dryer. *(Permapure Products.)*

Multiple-Stream Switching

When the analyzer must be shared by more than one sample, careful consideration should be given to the following:

1. Provide positive control of switching so that unwanted samples are prevented from contaminating the selected sample if a selector valve should leak.

2. Make sure that the selected sample is "fresh," that is, up to date with respect to the timing cycle used.

3. Evaluate fully the adequacy of the sampling rate, e.g., whether each sample is being seen frequently enough with respect to application requirements.

Positive control of switching usually means some type of valve arrangement in which valve leakage is always from the selected sample to an unwanted sample rather than the reverse. This may be accomplished by pressure, flow balancing, or a combination.

Figures 12 and 13 show two methods for preventing contamination by unwanted samples using the pressure-balance method. The use of two valves in series connected so that the space between is

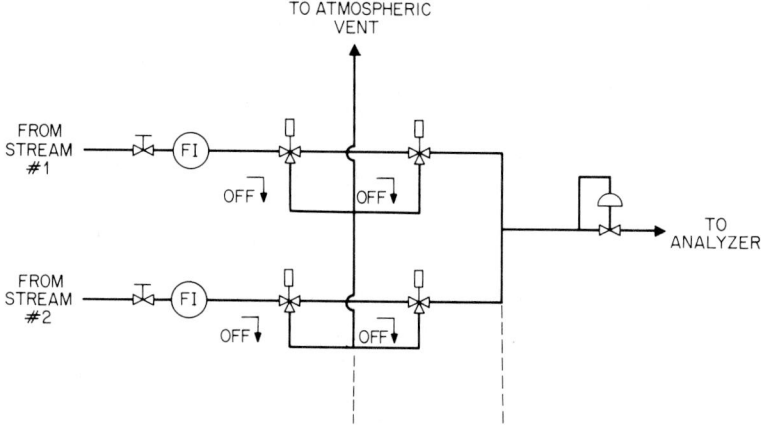

Fig. 12 Multistream block-and-bleed system. *(Cyberdesign.)*

vented to a lower pressure is usually called "block and bleed." It is clear that, if either valve leaks, the flow will be toward the lower pressure. The two valves may be remotely operated shutoff valves or one three-way valve in series with a check valve, since normal flow is always in one direction.

In Fig. 13, the valves in the second set are three-way connected so as to eliminate the dead-ended volumes that occur in a conventional tee-type manifold. Figure 13 is better, but switching signals to the valves is more complicated.

To determine timing, it is necessary to evaluate three time frames accurately:

1. Transport delay time from the sample tap to the switching valve in the stream selection system

2. Delay time from the switching valve to the analyzer sensor

3. Response time of the sensor itself, to at least 99% of the new sample composition

Clearly, the sum of time frames 2 and 3 must in all cases be less than the time per stream allotted by the sequencing system. If the unselected samples are kept flowing (usually desirable), time frame 1 could conceivably have any value; it does not necessarily need to be less than the time per stream. However, the actual age of each sample may be very important from an application standpoint, so the effect of not including time frame 1 in the total should be fully evaluated.

The selection of desirable switching valves is important for reliable system operation. Packless solenoid valves are most commonly used because they are relatively inexpensive and compact. When used in a block-and-bleed arrangement, they do not need to shut off tightly. Other considerations, such as heat buildup, corrosion resistance, and internal dead volume, can be more important.

Solenoid valves dissipate considerable heat, much of which is absorbed by the sample and can accelerate decomposition or polymerization reactions.

Direct-operated and packless solenoid valves must necessarily contain magnetic materials which are not as corrosion-resistant as 18-8 stainless steel. Further, ac packless solenoid valves must have a shading ring (exposed to the sample) to short-circuit bucking currents induced in the valve armature. This ring must be a good electrical conductor and is usually copper or silver, neither being particularly corrosion-resistant. Sometimes aluminum shading rings can be specified.

Where solenoid valves are unacceptable, miniature air-operated valves find good use in multiple-stream systems. These valves are available in a variety of two-way and three-way designs with all-stainless-steel wetted parts and high-temperature operators. Air service solenoid valves can be obtained from proven pneumatics equipment manufacturers without compromises, since these valves are exposed only to air at room temperature.

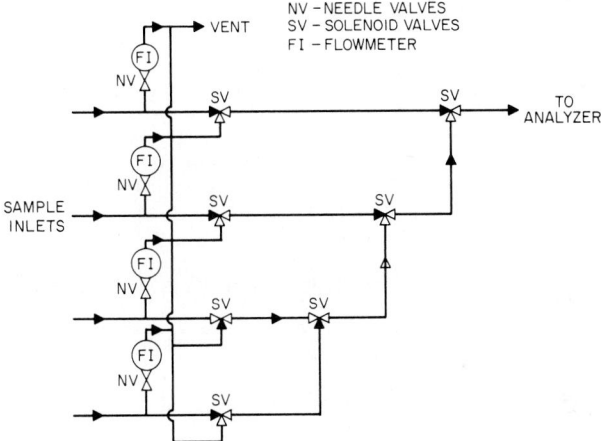

Fig. 13 Block-and-bleed manifold—no dead-ended volume. *(Adapted from Houser.)*

Calibration Sample Management

When designing a sampling system, one must not overlook facilities for the introduction of calibration standards for analyzers requiring standardization on one or more samples of known concentration. Some analyzers, such as processs chromatographs, have inherent zeroing and require only one calibration point. Others, such as infrared analyzers, require at least two calibration points, close to zero and up-scale. More than two points may be required if the analyzer response is sufficiently nonlinear.

A *calibration standard* is usually a flowing sample of material which either contains a known quantity of the substance for which the analyzer is calibrated or creates a known analyzer response. The usual approach is to shut off the normal sample and substitute a calibration standard for a length of time sufficient to establish output equilibrium and adjustment.

Calibration standards can be either stored or dynamically prepared at the time of use. Storage stability is a clear requirement of the former, although a stored sample can be "updated" by a spot laboratory determination if there is no other way to obtain a constant value. The stability of a stored sample can be improved by blanketing it (if it is a liquid) with nitrogen to exclude air. Internal reactions are harder to cope with and usually result in an attempt to create a synthetic mixture containing the measured component (or a proxy) in a nonreactive, noninterfering background. For example, an infrared analyzer for the measurement of carbon dioxide in a heat-treating furnace atmosphere might be calibrated using a mixture of carbon dioxide in nitrogen, on the assumption that other components in the sample do not absorb or interfere.

Gaseous calibration samples are usually stored in pressure cylinders, having been mixed by partial pressure techniques and verified by some type of referee determination. The selection of storage container and gas handling equipment is clearly important if calibration standard integrity is to be maintained. Some common causes of composition changes include:

1. Reaction of calibration sample with wetted materials in the container and piping

2. Adsorption or desorption of the component of interest by materials in the system

3. Diffusion of the component of interest into the calibration standard from the atmosphere through permeable materials in the system

4. Physical separation (segregation) of components, such as by molecular weight

All but the last item may be corrected by selection of materials, assuming they are available. Separation can frequently be prevented by heating the bottom of the container to develop convection circulation.

Where significant interfering components exist in the sample, it may be desirable to include an average concentration of them in the calibration sample so as to provide a comparable signal.

After creating a storage-stable calibration standard with the same concentration as the actual sample, it remains to present the sample to the analyzer under conditions of flow, temperature, and pressure sufficiently like those existing when the sample is analyzed. The calibration point of many analyzers will be shifted away from that desired if the analyzer is quite sensitive to temperature, pressure, and/or flow rate.

Generally, calibration samples for gas analyzers should be vented to the same point and at the same flow rate as when analyzing the sample. The temperature of a gas sample is usually not a problem, because gases equilibrate rapidly to the temperature of the analyzer. Liquid calibration samples should be maintained at a temperature near that of the process sample or heat-exchanged against fast-loop or bypassed flow. The effect of temperature may be estimated from manufacturer's specifications or determined by testing.

Storage of volatile hydrocarbon calibration samples in the vapor phase may not be feasible because storage pressure (capacity) is too low. Liquid mixtures may be stored and then vaporized at the time of use, as shown in Fig. 14. The item labeled "vaporizer" is nothing more than a ¼-in tube packed with firebrick and heated by steam or electricity.

Where stored calibration samples are not feasible, two other approaches may be considered, laboratory analysis of an immediate spot sample or dynamic blending. The former requires only a convenient tap for sample collection.

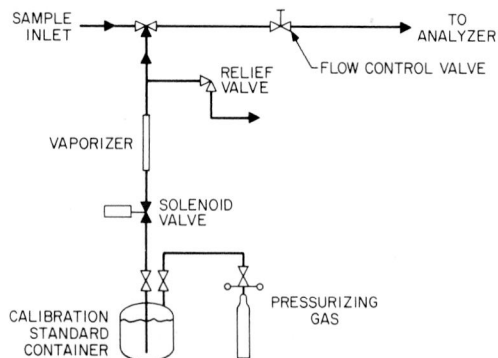

Fig. 14 Method for vaporizing liquid calibration standards. *(Adapted from Houser.)*

Dynamic blending is used frequently for the preparation of calibration samples having very low concentrations of highly reactive or strongly polar components, as in air pollution monitoring. Nitrogen dioxide standards are a good example. A storage-stable mixture of nitric oxide and nitrogen is dynamically mixed with air containing an excess of ozone, instantaneously and quantitatively oxidizing the NO to NO_2. Other examples include diffusion cells wherein a constant flow of carrier gas (usually nitrogen) flows at a constant temperature and pressure over a permeable membrane containing the component of interest, such as SO_2. The important consideration is usually what happens to the mixture after it is prepared and before it reaches the detector. Usually, these mixtures are highly polar and reactive and adsorb or break down rapidly, so it is necessary to prepare them as close to the detector as possible and use materials as inert as possible.

Controlling Analyzer Pressure

Gas analyzers are frequently (but not always) affected by absolute pressure in the sensing device or cell. This is generally true where the analyzer response is proportional to the number of molecules contained within the detecting device. Common examples include all photometer-type analyzers, such as infrared, ultraviolet, and visible spectrum absorption analyzers, as well as gas chromatographs, in which the mass of sample injected into the carrier gas is directly proportional to the absolute pressure in the sample loop. Practically any gas analyzer based on the measurement of a gross physical property falls into this category.

The most common method of controlling sample pressure is to vent the analyzer to the atmosphere, on the assumption that either atmospheric pressure variations are insignificant or calibration is carried out with sufficient frequency to follow these changes.

However, in some cases, the barometer reading may vary as much as ± 2 to 3%, causing this error to be added to the other errors in the system.

Another reason for controlling gas analyzer pressure may be to obtain greater sensitivity (higher pressure) or to allow the handling of high dew point samples without possible condensation in the cell (lower pressure). Of course, one affects the other, so it is important to evaluate the overall result.

Usually, controlling other than the ambient pressure is accomplished by locating a back-pressure regulator immediately downstream of the sensor and regulating the flow upstream, as in Fig. 15.

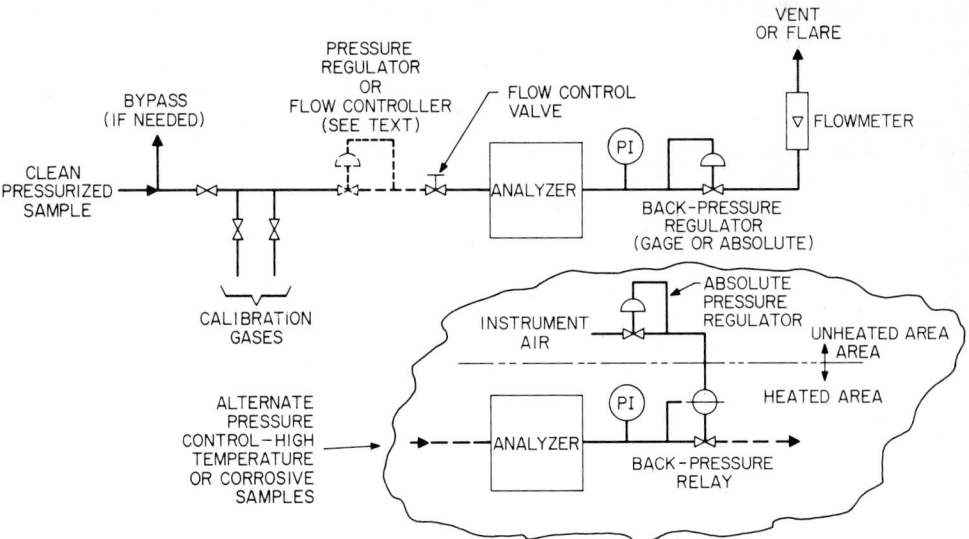

Fig. 15 Methods for controlling analyzer pressure for gas samples. *(Cyberdesign.)*

When the calibration gas pressure is significantly different from that of the sample, a reducing regulator or flow controller is needed upstream of the analyzer.

The choice of regulator is usually dictated by the accuracy required and materials-of-construction availability. If the pressure is higher than atmospheric and barometric pressure variations are acceptable, a regulator referenced to the atmosphere is suitable. Of course, barometric changes are proportioned to the higher pressure and are therefore less significant.

Absolute pressure regulators eliminate the effects of the barometer but are frequently not available in suitable materials. One method of solving this problem is to use an absolute pressure regulator to control air pressure to a pneumatic relay (air-loaded regulator) which is made of suitable materials. This arrangement is also shown in Fig. 15 (inset). Another advantage of this method is that it allows the use of a reducing absolute pressure regulator with a back-pressure relay. Yet another plus can be obtained by locating the regulator in a protected environment and the relay in a severe environment (hot, corrosive, etc.). Care must be taken to ensure that the required pressure differentials are maintained across supply and pilot connections, if used, and that necessary minimum flows are used.

Although the emphasis has been on back-pressure regulators following the analyzer, one may also use a reducing regulator upstream and control the flow downstream. It is important that the regulator be of the nonbleed type unless venting the sample to the immediate atmosphere is permissible.

Packaging and Installation

A sampling system generally consists of hardware such as valves, tubing, fittings, gages, flowmeters, filters, pumps, among other items.

On the assumption that the locations of the various subsystems have been determined and that due consideration has been given to ambient temperature extremes, weather exposure, hazardous area classifications, and the like, primary consideration should be given to protection of the sample, i.e., maintaining its integrity as it passes from the process to the analyzer. If a minimum temperature above ambient is required to prevent condensation, several methods can be considered, including (1) direct heat tracing, (2) indirect heat sourcing, and (3) air bath packaging.

Direct heat tracing involves wrapping of, or wiring to, the components and sample lines, using small-diameter tubing (steam or hot fluid) or insulated resistance wire (electric) and then insulating overall. Where the number of components is small and where easy access to them for servicing is not important, the direct method is adequate and inexpensive. Steam tracing or self-limiting electric tracing is preferred to ordinary resistance wire. Careful workmanship to ensure a good thermal contact between tracer and components is very important. The stiffness and circular shape of steam trace tubing makes this difficult, but there is vast improvement when heat transfer mastic is puttied into the voids.

Unfortunately, all the results of this careful effort may be destroyed when components must be accessed for repair or adjustment. Frequently, not nearly as much care is taken the second time around. If only warming is needed, major components can be mounted on a flat plate which is then heated. The operating temperature of the components is determined by conductance and to some extent by radiation between the components and the plate. Painting everything flat black helps, and using an insulated cover is even more effective. Where later access is required, the cover should be easily removable. A window in the cover will allow inspection of controls and visual detection of the leakage of liquids before they soak into the insulation.

Where the minimum temperature required is close to the temperature of the heating medium, an air bath can be used to prevent hot and cold spots. Although air circulation venturis are available, a piece of ⅛ O.D. tubing flattened at the end and supplied with 25 to 50 psig (138 to 172 kPa) air will ensure circulation at a lower cost. The benefits of good circulation overcome the small cooling effect of introducing cold air. Access for making operating adjustments during system operation should be provided.

Temperature control can be difficult. On-off control of electric heating elements or steam can be successful if the right sensor is coupled at the right place. Heated plate designs should sense the plate rather than the air to avoid overheating the plate during cycling. Air bath designs in which moving air couples heat from the heater to the components should sense the air (with a fast-response sensor).

Temperature control with steam is best obtained by controlling steam pressure, and providing good insulation and good circulation. Self-operated temperature regulators are often too large for sampling system enclosures and tend to wire-draw the seat if run continuously at low loads.

Where maintaining a high sample temperature is critical, consider the difficulty of bringing the sample through the wall of an enclosure without creating a cold spot. Cast-aluminum explosion-proof enclosures are notorious in regard to this problem. The enclosure wall provides a strong heat sink that will drain heat away even if the heat tracer is brought right to the fitting. Sometimes concentric-tube steam jackets are required.

Estimating Transport Time Lag

System response time is usually a key objective in any sampling system. Whether the purpose is operator assistance, closed-loop control, or safety monitoring, an effort should always be made to estimate the system response time. Although some analyzers may be the limiting factor, the sampling system is usually the most significant factor.

The designer should begin by breaking the system into sections which have readily calculated or estimated transport lags, preferably starting at the takeoff. Add up the calculated (or estimated) internal volume of the fitting or probe, including the block valve. Next, determine the actual flow of sample in this section. For gases, use simple absolute pressure and temperature ratios unless you are a purist and compressibility factors are available. Estimation of transport times in sampling systems is seldom done accurately, so assuming the sample is an ideal gas is usually acceptable.

When estimating the lag upstream of a vaporizing regulator, be especially careful to consider all the volume, including an estimate of the internal volume in the vaporizer upstream of the reducing valve. Usually, the flow rate is known as the vapor flow after the vaporizer, requiring that the equivalent liquid flow be calculated. First calculate the average molecular weight of the sample, and then find the weight rate of flow from the molar volume of an ideal gas (22.4 L/g · mol). Then, the volume rate of flow of the liquid can be calculated from its estimated density.

Having obtained the purged volume and volumetric rate of flow, can we calculate the time lag by simple division? Well, maybe. If the phenomenon known as "plug flow" is close enough to the real situation, yes. Two things should be considered: flow volumes of large cross section and the possibility of adsorption on internal surfaces. While advanced mathematics can be used to further evaluate these effects, they are usually calculated by multiplying the plug-flow time lag by an estimated factor. Large cross-section volumes are increased by a factor of 2 or 3. Factors for samples adsorbed on surfaces are more difficult to obtain and usually amount to educated guesses.

The estimated time lags for the individual sections of the system are then added to obtain a total value. This quantity should closely approximate the observed dead time if carefully calculated; however, the time constant or integrating lag for control system purposes should be obtained experimentally.

ADDITIONAL READING

Erk, G. F.: "Engineering Analyzer Systems," *Instrum. Technol.,* vol. 24, no. 8, August 1977, p. 39.

Franczak, D. F., and Paprocki, T. J.: "Locating and Installing Equipment for Continuous Emission Monitoring," *Instrum. Technol.,* vol. 27, no. 11, November 1980, p. 38.

Houser, E. A.: "Principles of Sample Handling and Sampling Systems Design for Process Analyzers," *ISA Monogr.,* Instrument Society of America, Research Triangle Park, North Carolina, 1972.

Lester, D. J.: "EPA Requirements for Stack Monitors," *Instrum. Technol.* vol. 25, no. 2, February 1978, p. 53.

Nishioka, K., and R. Brooks: "Automated In-Situ Water Quality Monitoring," *Instrum. Technol.* vol. 26, no. 10, October 1979, p. 47.

Strauss, R.: "Sample Filters for Analyzers," *Instrum. Technol.* vol. 24, no. 8, August 1978, p. 53.

Trawick, E. G., and G. L. Baker: "Organic Spill Detector," *Instrum. Technol.* vol. 24, no. 4, April 1977, p. 71.

Yamashita, S.: "On-Line Sensor Cleaning Keeps the Process Flowing," *Control Eng.,* vol. 25, no. 3, March 1978, p. 66.

X-Ray Analytical Methods

by
W. G. Shequen*

X-rays occupy the portion of the electromagnetic spectrum between $\approx$ 0.01 and 100 Å in wavelength. Their range of approximate quantum energy is 2×10^6 to 2×10^{-10} erg, or from 10^6 to 100 eV. Important x-ray analytical methods are based on (1) fluorescence, (2) emission, (3) absorption, and (4) diffraction and are used to determine qualitatively and quantitatively the elemental content of complex mixtures and to determine exactly the atomic arrangement and spacings of crystalline materials.

Sources of X-Rays

X-rays are emitted by atoms which are bombarded with energetic electrons. This results from two separate effects: (1) deceleration of highspeed electrons as they pass through matter, and (2) ionization of individual atoms which abruptly stop the electrons. The first effect results in a continuous-type spectrum; the second results in characteristic line spectra.

Continuous Spectrum

The bulk of x-radiation arising from electron bombardment is in the continuous spectrum. If an individual electron is abruptly decelerated, but not necessarily stopped, in passing through or near the electric field of a target atom, the electron will lose some energy ΔE, which appears as an x-ray photon of frequency $\Delta = \nu E/h$, where h is Planck's constant. An electron may experience several such decelerations before it is finally stopped, emitting x-ray photons of widely different energy and wavelength. A few electrons will be stopped in a single process, losing their energy and emitting an x-ray photon having the exact energy of the incident electron.

X-Ray Spectral Lines

These lines result when incident electrons knock orbital electrons out of an atom. If an ejected electron is from one of the inner orbits of the atom (Fig. 1), an electron from an outer shell will fall to the inner orbit to fill the vacancy. The increase in potential energy of this electron in approaching the nucleus results in the emission of an x-ray photon having an energy exactly equal to that lost by the electron. The wavelength λ of such photons is related to ΔE by $\lambda = ch/\Delta E$, where c is the velocity of electromagnetic energy, and h is Planck's constant. Because the energy of orbital electrons is quantized, x-ray photons can have only certain definite wavelengths which are characteristic of the atom. This situation is somewhat analogous to the more familiar ultraviolet (uv) and visible emission spectra of materials, the difference being that the optical spectra are the result of electron transitions between energy levels of just the outermost electrons of the atom.

    X-rays resulting from an electron transition filling an electron vacancy in the innermost shell of an atom are known as K x-rays, or K lines. X-rays formed by electron transitions to the second or L shell are known as L lines, and so on. Transitions to the K shell in very light elements or to the M or N shell of heavier elements represent a smaller energy change. Consequently, the photons

*Manager, International Support, Applied Research Laboratories (Division of Bausch & Lomb), Sunland, California.

emitted may have an energy and a wavelength corresponding to the uv or visible region. To fill a vacancy in a K shell, an electron can come from either an L or an M shell; the transitional energy will be slightly different, depending on the source of the electron, giving rise to K lines of slightly different energy or wavelength. These are known as K_α and K_β lines when they come from the L and M shells, respectively. The K_β line has a shorter wavelength than the K line, since it corresponds to a greater change in the potential energy of the electron. Under high resolution, these lines can be shown to consist of several other lines (denoted $K_{\alpha 1}$, $K_{\alpha 2}$, $K_{\beta 1}$, $K_{\beta 2}$, $K_{\beta 3}$, and so on) corresponding to the exact initial and final electron positions or energy levels within the shells.

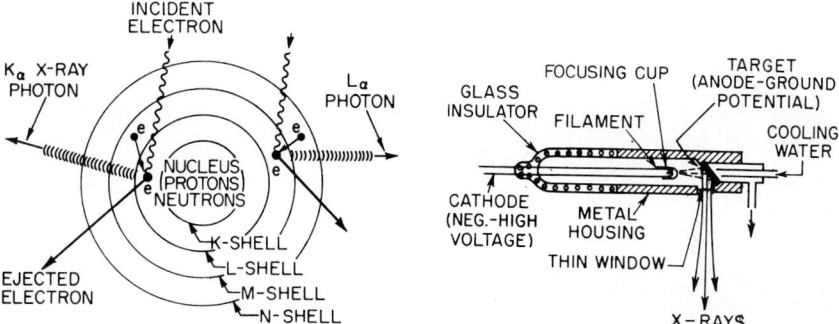

Fig. 1 Origin of x-ray spectra due to electron bombardment.

Fig. 2 High-voltage, high-vacuum (Coolidge-type) x-ray tube.

Generation of X-Rays

A high-vacuum, Coolidge-type tube, wherein electrons are emitted from a heated tungsten filament and accelerated by high voltage to an anode (target), is the most common source of x-rays. The x-rays are emitted by the process previously described when the electrons strike the target. The target usually is placed at a slight angle to the impinging electron beams so that the focal spot can be viewed from either a side or end window in the tube. This window through which the x-rays are emitted is commonly made of a thin sheet of beryllium foil, mica, aluminum, or special low-absorption glass (see Fig. 2). Only a portion of the x-rays emitted by the target actually fall on the exit window.

By proper choice of target material and voltage, it is possible to generate either an essentially continuous spectrum (Fig. 3) or a continuum with superimposed line spectra (Fig. 4). If tube voltage is sufficiently high, the K lines of the target material will be superimposed on the continuum. At lower tube voltages, the K line will not be excited, but the L and M lines, occurring at longer wavelengths, will still be present. To obtain an essentially continuous spectrum over a fairly wide spectral range, it is thus common practice to employ a target material of high atomic number, such as tungsten, and to operate the tube at a voltage lower than that required to excite the K lines (Fig. 3). It has been shown that the total intensity of the continuum is directly proportional to the atomic number Z of the target material and to the square of the applied voltage.

Figure 3 shows that there is a sharp lower limit to the wavelengths generated in this manner. This short wavelength cutoff varies with the applied tube voltage; at higher voltages, the cutoff is a shorter wavelength. This results from the fact that, with increased tube voltage, higher energy electrons strike the target material; a few of these electrons are completely stopped in a single encounter with an atom, giving up all their kinetic energy by emitting a single x-ray photon. The energy of photons generated in this fashion increases with the energy of the incident electrons. The wavelength λ_0 of the shortest wavelength x-rays emitted is related to the tube voltage by

$$\lambda_0 = \frac{12,398}{V} \quad \text{Å} \tag{1}$$

where V = tube voltage, V

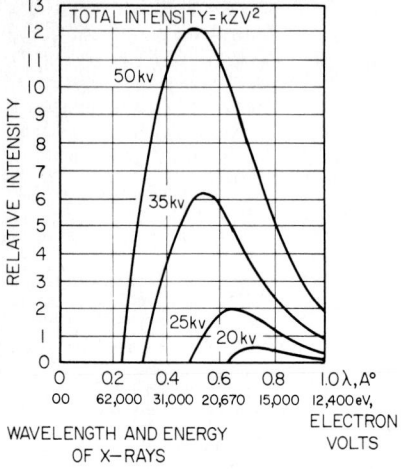

Fig. 3 Continuous x-ray spectrum of tungsten ($Z = 74$) at various tube voltages.

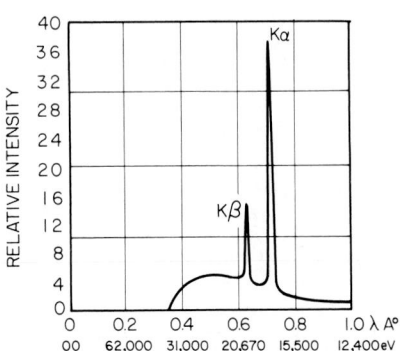

Fig. 4 Intensity of x-rays from molybdenum target tube at 35 kV.

The wavelength of peak intensity of the x-ray continuum spectra has been shown to be related to λ_0, the cutoff wavelength, by

$$\lambda_m = \tfrac{3}{2}\lambda_0 \qquad (2)$$

where λ_m is the wavelength of maximum intensity

When the x-ray tube voltage is increased sufficiently to excite the innermost or K-shell electrons of the target, the K-line spectra are superimposed on the continuum (see Fig. 4). The wavelength of the K lines emitted by a number of typical target materials is given in Table 1, together with the necessary minimum potential required to excite these lines. A convenient relationship between the wavelength λ of the K or L lines and the atomic number of the element is given by Mosley's law,

Table 1 Characteristic Wavelength of Typical X-Ray Tube Targets

Based on kX units $1.00202 = 1$ Å, where kX unit $= \tfrac{1}{3}$, 029.45 spacing of calcite cleavage planes

Target	Angstroms				Minimum potential required to excite K lines, kV
	$K_{\alpha1}$	$K_{\alpha2}$	$K_{\alpha avg}$	$K\beta1$	
Chromium	2.28962	2.29352	2.2909	2.08478	5.98
Copper	1.54050	1.54434	1.5418	1.39217	8.86
Tungsten	0.209	0.213	0.211	0.184	69.48
	$L_{\alpha1}$	$L_{\alpha2}$	$L_{\alpha avg}$	$K\beta_1$	Minimum potential required to excite L lines, kV
Rhodium	4.597	4.605	—	—	3.42

which in its approximate form is

$$\lambda_K = \frac{1218}{(Z - 1)^2} \tag{3}$$

and

$$\lambda_L = \frac{6560}{(Z - 7.4)^2} \tag{4}$$

where λ_K and λ_L = approximate wavelengths of K and L emission lines, respectively, Å

A similar equation can be derived to fit the M, N, and other series for all elements which have such lines.

Commercial X-Ray Tubes

A wide variety of tubes are available. All high-power (high-current) commercial tubes employ a water-cooled anode, although some special high-current tubes have rotating anodes so that no one spot on the anode becomes unduly heated. Tubes are built with ratings up to 10 kW. Tubes are available for the generation of either very weak, long wavelength x-rays, or extremely hard x-rays, such as the case of the 2-million-V tubes. Special pulse-type x-ray tubes generate extremely intense beams of very short duration. For personnel safety, all tubes must be adequately shielded. Typical operating voltages are 20 to 75 kV with currents up to 60 mA. A tube normally rated for 1000 h may have a life of 10,000 h if operated at one-half the nominal power. Windows of x-ray tubes are most commonly of Lindemann glass or beryllium. Beryllium is more transparent at shorter wavelengths, sections (0.01 in thick) having a transmission of 85% in contrast with 33% for Lindemann glass at 2 Å. At wavelengths less than 0.7 Å, glass is nearly as good as the beryllium relative efficiency varies with the target material and atomic number of the element being analyzed. See Fig. 5.

Detectors include (1) Geiger-Mueller tubes, (2) ionization chambers, (3) scintillation counters, (4) proportional counters, (5) electron multiplier tubes, and (6) nondispersive detectors using cooled lithium-drifted Si detectors. See also the article on nuclear radiation instrumentation in Sec. 12 of this Handbook.

X-Ray Crystallography

X-rays penetrating below the surface of crystalline materials are scattered by the individual parallel layers of atoms; each atomic layer acts as a new, although weak, source of x-rays. To be reinforced in a given direction at an angle θ (Fig. 6), the spacing d between crystal planes must be rigorously related to the wavelength of the radiation. At a given angle, x-rays of one definite wavelength will be constructively reinforced. These variables are related by Bragg's law of x-ray diffraction:

$$n\lambda = 2d \sin \theta \tag{5}$$

where λ = wavelength
  θ = angle of incidence and diffraction of x-rays from a crystal whose lattice spacing is defined by d
  n = order or harmonic, of the diffraction. For almost all fluorescence analysis, $n = 1$; i.e., the characteristic prime x-ray line is used.

The usable x-ray spectrum normally extends from approximately 0.1 to 10 Å. The x-ray transmission curves of Fig. 7 illustrate the need for providing a helium, hydrogen, or vacuum path for the x-ray channel for elements with atomic numbers lower than $Z \cong 24$, i.e., at wavelengths longer than approximately 2.3 Å.

Early x-ray instruments generally consisted of a single goniometer mechanism that could be manually or motor driven through a broad 2θ range. Wide wavelength coverage was obtained by repeat-

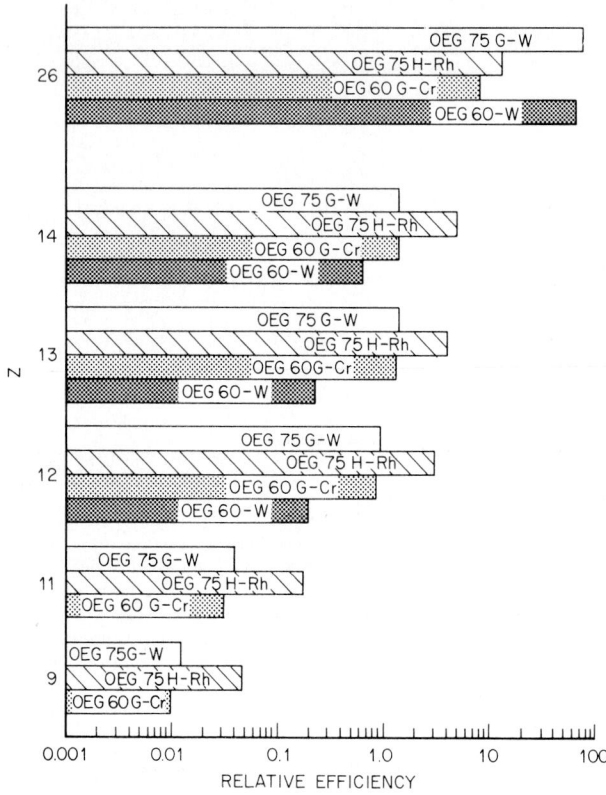

Fig. 5   Comparison of x-ray tubes.

edly driving the goniometer through its range, using each of a limited variety of crystals, collimators, and detectors provided in the basic instrument design. Less sophisticated designs required manual replacement of one or more components to obtain full wavelength coverage. Either approach was satisfactory in that the researchers who used these instruments were not particularly interested in high analytical speed and, at the same time, were interested in keeping the instrument cost as low as possible.

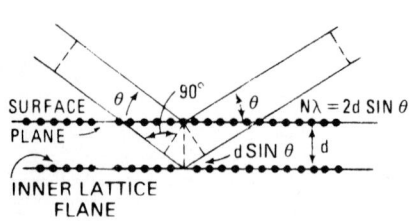

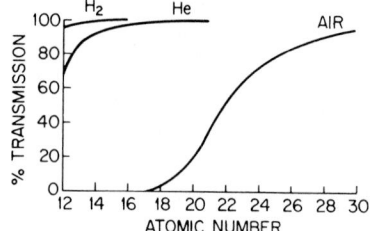

Fig. 6   Reflection of x-rays from an internal crystal plane.

Fig. 7   Transmission of K_α radiation.

The goniometer has become a supplement to the production control or process control instrument, which is normally mounted with a number of carefully selected fixed x-ray channels (monochromators) to provide optimum analytical data for the elements of major interest. Figure 8 illustrates schematically the fully focusing Johansson curved and ground crystal spectrometer. Fully focusing crystals, as opposed to flat crystals, ensure highest possible x-ray diffraction intensities.

Selection of an optimum diffraction crystal involves four basic considerations: (1) crystal d spacing, (2) diffraction efficiency at the wavelength of interest, (3) resolution at the specific wavelength, and (4) discrimination provided against higher order interfering radiation (see Table 2). Normally, there are several crystals which satisfy the d-spacing requirements for any given element. Figure 9 illustrates the intensity resolution characteristics of various crystals for the diffraction of Al, Si, Ca, and Mg. Of the crystals illustrated, ethylenediamine d-tartrate (EDT) provides not only the highest diffraction intensities but also the best resolution (half-width) for the analysis of Al and Si. LiF and ammonium dihydrogen phosphate (ADP) provide similar characteristics, respectively, for Ca and Mg diffraction. Higher-order discrimination may be obtained by either crystal selection or energy discriminating pulse-height analysis. In some circumstances, the selection of an appropriate crystal may save the user the expense of purchasing pulse-height analysis equipment. Selective use of high-resolution crystals eliminates the problem of higher order interference.

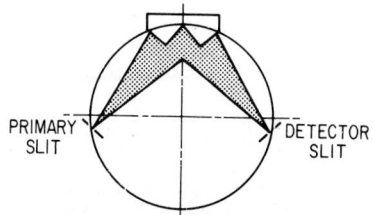

Fig. 8 Optical diagram of Johansson curved crystal spectrometer.

X-Ray Fluorescence Analysis

One of several types of spectrochemical techniques used in the laboratory, x-ray fluorescence analysis is nondestructive. The characteristic x-ray spectrum of each element bears a simple direct relationship to the element's atomic number. The relation of the wavelength λ to the atomic number Z is

$$\frac{1}{\lambda} \propto Z^2 \quad \text{(Mosley's law)}$$

Table 2 Discrimination Factors of Crystals for Orders Greater than $n = 1$

Crystal	Plane	λA	Discrimination factor			
			$n = 2$	$n = 3$	$n = 4$	$n = 5$
ADP	(101)	9.889 (Mg K_a)	—	—	1.0	1.0
Gypsum	(020)	—	—	—	4.6	1.7
Mica	(002)	—	—	—	<1.0	<0.1
ADP	(101)	8.339 (Al K_α)	—	1.0	1.0	—
EDT	(020)	—	—	4.0	6.0	—
ADP	(101)	7.126 (Si K_α)	—	1.0	1.0	—
EDT	(020)	—	—	3.2	8.8	—
SiO$_2$	(01.0)	—	—	0.9	1.6	—
LiF	(200)	3.360 (Ca K_α)	1.0	1.0	1.0	1.0
NaCl	(200)	—	0.9	0.4	0.2	0.8
Ge	(111)	—	>60.0	1.1	0.1	13.0

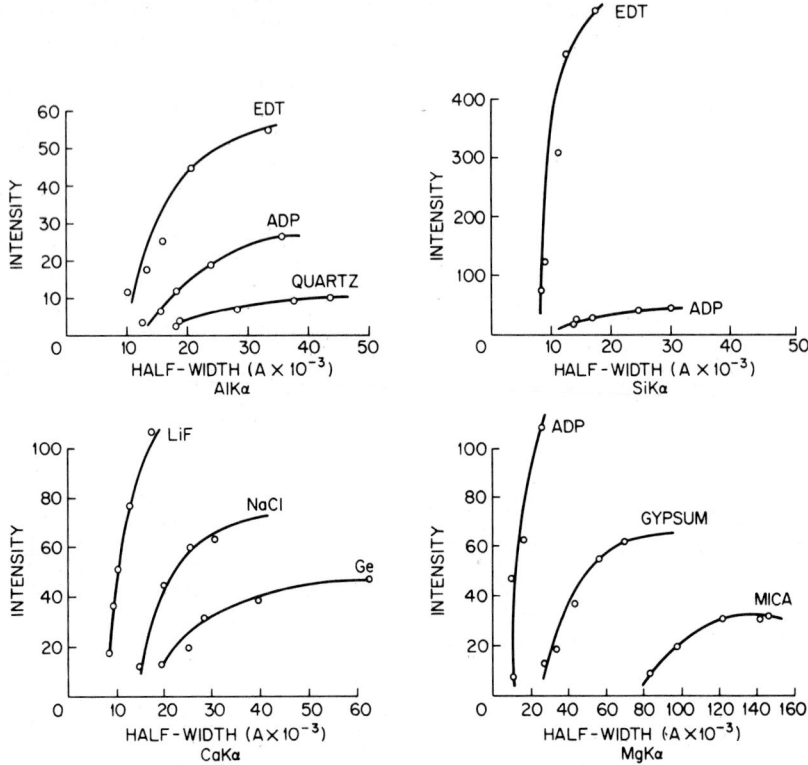

Fig. 9 Intensity resolution function of crystals and slit width. ADP, ammonium dihydrogen phosphate; EDT, ethylenediamine *d*-tartrate.

Since x-ray spectral lines come from the inner electrons of atoms, they are not related to the chemical properties of the elements or to the compounds in which they reside. Because the characteristics of x-ray spectra are associated with energies released through transitions of electrons within the inner shells of an atom, the spectra are simple. Most practical x-ray fluorescence analysis involves the detection of radiation released through electron transitions from outer shells to the K shell (K spectra), from outer shells to the L shell (L spectra) and, in a very few cases, from outer shells to the M shell (M spectra).

The simplest type of energy source available for commercial instrumentation is that obtained from an x-ray tube. For samples containing predominantly low-atomic-number elements, as in cement raw mix, the most efficient excitation is accomplished by using an x-ray tube target material of relatively low atomic number, such as chromium. Elements having higher atomic numbers are most effectively excited by high-atomic-number targets, such as tungsten and platinum. An optimum target material is rhodium for the analysis of a broad range of elements. The x-ray tube irradiates the sample which, in turn, emits characteristic fluorescent radiation of its atoms.

Once x-ray fluorescence is produced from the sample by means of an x-ray tube, appropriate components of the instrument separate this radiation into its characteristic wavelengths, detect the energy emitted from each excited atom, and produce a signal representative of the number of atoms (concentration) of the element in the sample. Typical excitation conditions (x-ray tube) are 50 kV, 35 mA. Bragg's law of x-ray diffraction is satisfied by the condition $N\lambda = 2d \sin \theta$, where λ is the wavelength and θ is the angle of incidence and diffraction of x-rays from a crystal whose lattice spacing is defined by d and N, the harmonic order of the diffraction. For almost all fluorescence analyses, $N = 1$. The usable x-ray spectrum normally extends from 0.1 to ~20 Å. A helium or vacuum path is required for x-ray analysis of elements with atomic numbers lower than $Z \cong 24$ (wavelengths longer than 2.3 Å).

Instrumentation can provide simultaneous elemental analysis using fixed, preselected x-ray detection channels and scanners or a goniometer to supply one or more channels that can be tuned to a wide wavelength coverage. Up to 30 monochromator positions are possible. Typical crystal materials covering the practical wavelength range of 1 to 20 Å are lithium fluoride, silicon oxide, sodium chloride, EDT, and ADP. Optimum analytical data for the elements of interest are obtained by using fully focusing Johansson curved and ground crystals.

An optical diagram of a Johansson curved-crystal spectrometer was previously given in Fig. 8. Each spectrometer of an x-ray quantometer may be equipped with optimum crystal detector combinations for specific determinations in a wide variety of matrixes, including steel, aluminum, copper-base materials, ores, cement, and slags—in both liquid and solid states.

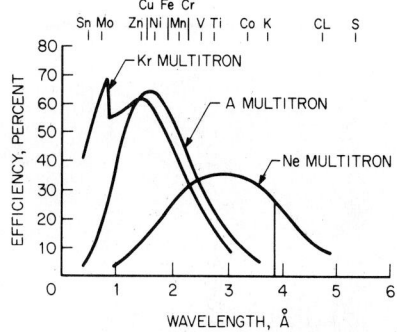

Fig. 10 Relative efficiency of x-radiation detectors.

The diffracted x-radiation is detected by Geiger, proportional, or semiproportional detectors. See Fig. 10. The detector of each monochromator generates pulses which are a measure of the intensity of radiation of each wavelength. The pulses are filtered through a discriminator in order to avoid undesired interferences. The shrinkage of pulses due to an increase in pulse frequency is automatically compensated. Collected pulses are transferred to a computer for processing output. See Fig. 11.

The transition from laboratory to automated instrument to achieve highspeed continuous analysis

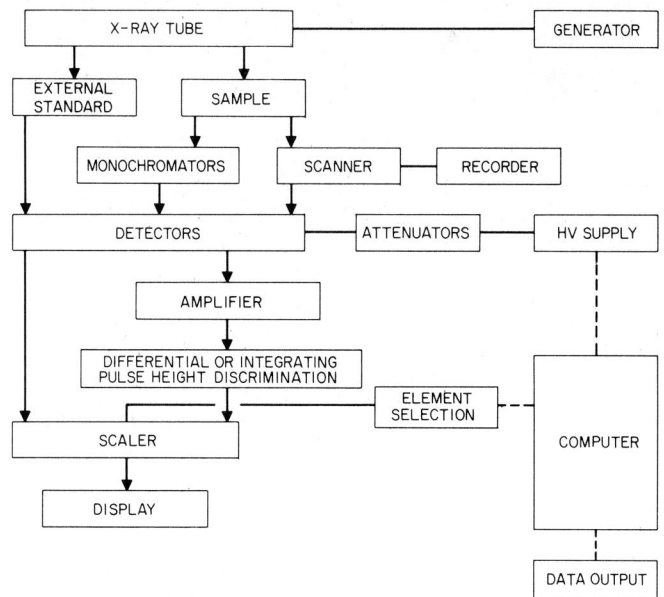

Fig. 11 System configuration of x-ray fluorescence spectrometer. *(Bausch & Lomb.)*

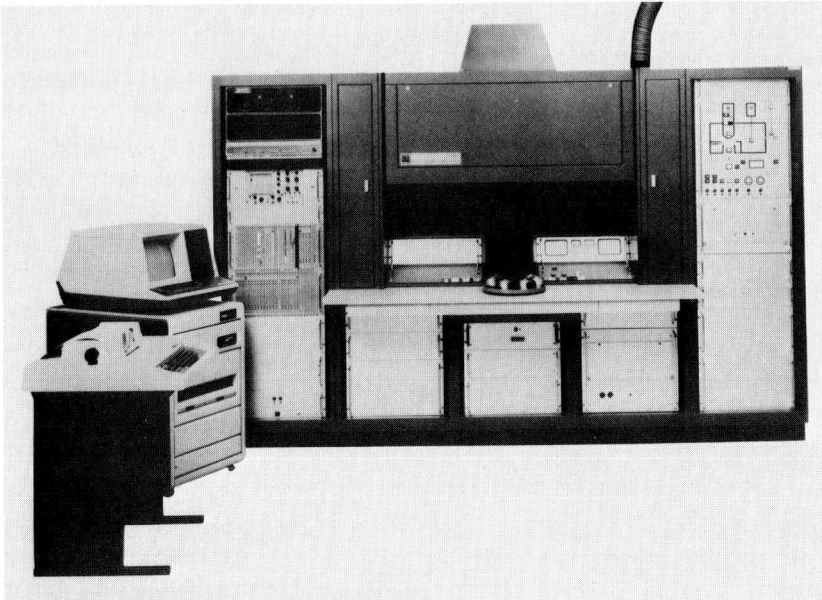

Fig. 12   X-ray fluorescence quantometer provides a high order of sensitivity in the analysis of a variety of materials and products, ranging from precision bearings to the steel and cement used in buildings. Among the elements which can be determined are sodium, magnesium, silicon, aluminum, sulfur, calcium, iron, and fluorine. X-ray fluorescence instruments are widely used in the cement, glass, and refractory industries, among others. *(Bausch & Lomb.)*

of dry or wet materials primarily involves the sample handling and presentation hardware. See Fig. 12.

Limits of detectability for the desired elemental analyses vary, depending on the matrix, elements, methods of sample preparation, and quality of instrumentation applied. Generally, these are on the order of 1 to 100 ppm. The limit of detectability, however, is only one criterion used in evaluating methods of analysis. The time of analysis is important, particularly in production and process control laboratories. In multielement spectrometers, it is possible to perform as many as 30 simultaneous elemental determinations in from 20 to 120 s, depending on the material being analyzed.

X-Ray Analyzer (Electron Microprobe)

The advantages of this analytical instrument include the following: (1) Analysis can be confined to very small amounts of materials (microsamples); (2) the particular material to be analyzed need not be physically separated from its surrounding materials as is often required with some other analytical methods; and (3) through the development of associated instrumentation, diagnostic techniques, and information displays, the method can be quite fast. Limits of detection in solid solutions are from approximately 0.005 to 0.5%, depending on the elements and sample matrixes involved. See Fig. 13. Concentrations as low as 10^{-16} g may be measured. In addition, magnified images from 30× to 150,000× may be obtained on a cathode ray tube (CRT) display. The CRT display may also be recorded photographically in the x-ray, backscattered electron, or secondary-electron mode. A back-scattered electron image of aluminum, heat-treated and etched, is shown in Fig. 14.

Although this technique is mainly used for metallurgical studies, nonmetallics also may be analyzed when samples are properly prepared. Biological applications include tooth and bone samples, cytochemical problems and staining techniques, physiochemical problems, and studies in pathology.

Relative weight-fraction-detection limits for most elements in biological specimens are in the general range of 0.01 to 0.1%. Electronics industry applications include studies on diffusion phenomena, electrical contact surfaces, interfaces on transistors, and microcircuitry analysis.

As shown in Fig. 15, electrons from an electron gun are directed toward the sample through an electron optical system. Once the electron beam strikes the sample, a number of signal sources are

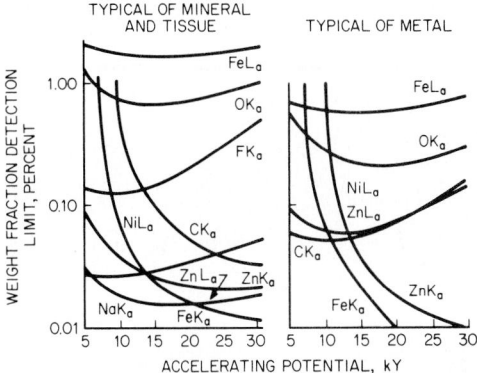

Fig. 13   Typical weight fraction detecting limits of an electron microprobe x-ray analyzer.

activated, including (1) high-energy backscattered electrons, (2) low-energy secondary electrons, (3) cathodoluminescence, and (4) x-rays. Some heat is also generated within the sample. Volume d_3 in Fig. 15 is the *volume from which x-rays are emitted.*

The x-rays produced may be detected nondispersively by a proportional counter whose output may be separated as a function of energy by a pulse-height analyzer into various wavelength components. Better detection sensitivity, however, can be obtained through the use of a fully focusing diffracting-crystal spectrometer in conjunction with a proportional detector and pulse-height analyzer. As shown in Fig. 16, the necessary condition for fully focusing optics is to have the x-ray source, the crystal, and the detector slit all placed in a common circle. The geometry requires that the diffracting-crystal planes be bent to the diameter of the Rowland circle.

Fig. 14   Backscattered electron image of aluminum heat-treated and etched. *(Bausch & Lomb.)*

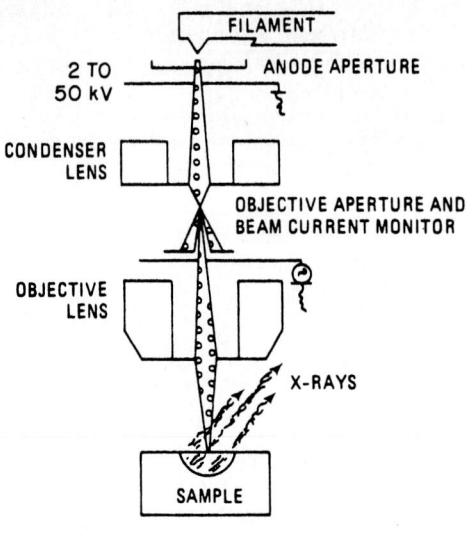

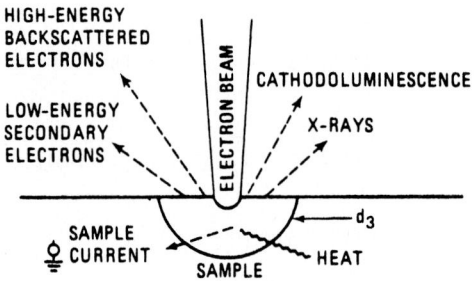

Fig. 15 Electron gun and probe forming lens system of an integrated electron probe. *(Bausch & Lomb.)*

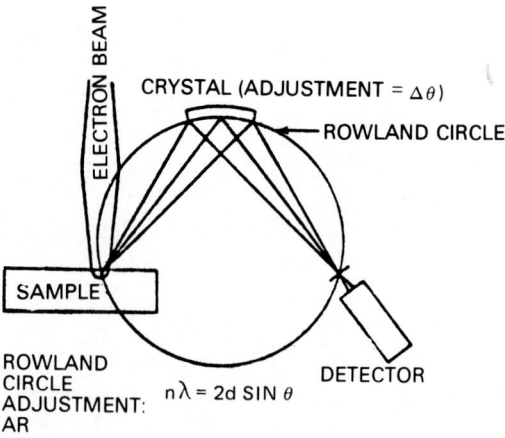

Fig. 16 Geometry of a fully focusing diffracting crystal spectrometer. *(Bausch & Lomb.)*

6.46

Spherical aberration at the detector slit is minimized by further grinding the crystal surface to fit the radius of the Rowland circle. With the resultant Johansson optics, the crystal radius is fixed, and the 2θ range of the spectrometer is scanned by moving the crystal radially away from the source and, at the same time, rotating it into the detector to achieve a true focus throughout the spectrometer range. X-rays, particularly those of wavelength greater than 2 Å, and electrons are highly absorbed in an air atmosphere. Thus, the spectrometer must be enclosed in a vacuum on the order of 10^{-5} torr. The present wavelength range of interest extends from ~1 to 100 Å. Diffracting crystals to cover this range must provide broad wavelength coverage, high diffraction efficiency for high peak count intensity, good resolution, and good resulting peak-to-background ratios. Crystals that meet these requirements include lithium fluoride, ADP, EDT, quartz, and sodium chloride.

Detectors

Of the three commonly used x-ray detectors—(1) Geiger counter, (2) scintillation counter, and (3) proportional counter—the latter is used most frequently for electron probe microanalysis. In the wavelengths from 1 to 10 Å, sealed proportional counters may be used. For longer wavelength analysis—in the range 10 to 93 Å —the thinnest possible detector window is required to limit spectral attenuation. See Table 3. Nitrocellulose windows have proved successful. Nondispersive detection systems using cooled Li-drifted Si are also applicable.

Readout

An optical microscope is required in the system to provide the analyst with a means of reference to identify various sample areas for analysis. Sample stages may hold single or multiple samples and are provided with means for moving the sample in $x, y,$ and z planes without breaking the system vacuum. After the point of interest is located on the specimen, the data may be read out in a number of ways. As many as 12 wavelength-dispersive spectrometers may render simultaneous multielement analyses. These data are computer-processed to yield a direct percent concentration printout for the elements of interest.

On-Stream Analysis

The polychromator design of x-ray analyzers allows the rapid acquisition of analytical data with a high degree of precision. Because of the short analytical time required, many streams or circuits may be analyzed on an alternating basis.

From 1 to 20 slurry streams may be analyzed on a single instrument, and as many as 9 elements may be determined simultaneously as each stream is cycled for analysis.

Slurry Analysis

In slurry analysis, one prime requisite is that pressure and flow remain constant throughout the analytical cell. As a result, thin, wear-resistant cell windows must have the added requirements of high resistance to x-radiation, ease of replacement, and the ability to transmit x-rays with a minimum absorption of x-ray energy. In a typical system, slurry samples flow into a head tank at a controlled rate (up to 30 L/min for cement). A preselected amount of slurry is passed through the analytical cell of the instrument, and the remainder is permitted to overflow into the return line or launder. Cell window life is evaluated for each application, and the windows are changed at intervals to provide adequate safety against failure. Both Mylar and Kapton, ¼ to 1 mil in thickness, have been used successfully.

The sequence of the slurry streams may be controlled by manual, automatic, or computer means.

Table 3 Crystal and Detector Wavelength Coverage

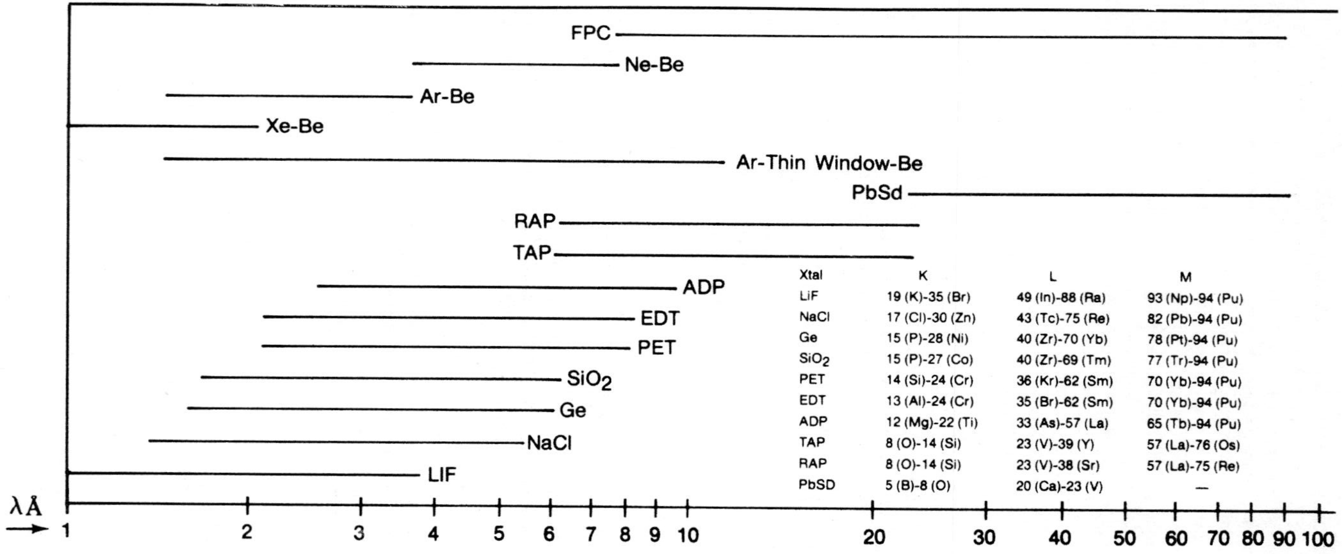

FPC
Ne-Be
Ar-Be
Xe-Be
Ar-Thin Window-Be
PbSd
RAP
TAP
ADP
EDT
PET
SiO2
Ge
NaCl
LIF
λÅ
1
2
3
4
5
6
7
8
9
10
20
30
40
50
60
70
80
90
100

Xtal	K	L	M
LiF	19 (K)-35 (Br)	49 (In)-88 (Ra)	93 (Np)-94 (Pu)
NaCl	17 (Cl)-30 (Zn)	43 (Tc)-75 (Re)	82 (Pb)-94 (Pu)
Ge	15 (P)-28 (Ni)	40 (Zr)-70 (Yb)	78 (Pt)-94 (Pu)
SiO_2	15 (P)-27 (Co)	40 (Zr)-69 (Tm)	77 (Tr)-94 (Pu)
PET	14 (Si)-24 (Cr)	36 (Kr)-62 (Sm)	70 (Yb)-94 (Pu)
EDT	13 (Al)-24 (Cr)	35 (Br)-62 (Sm)	70 (Yb)-94 (Pu)
ADP	12 (Mg)-22 (Ti)	33 (As)-57 (La)	65 (Tb)-94 (Pu)
TAP	8 (O)-14 (Si)	23 (V)-39 (Y)	57 (La)-76 (Os)
RAP	8 (O)-14 (Si)	23 (V)-38 (Sr)	57 (La)-75 (Re)
PbSD	5 (B)-8 (O)	20 (Ca)-23 (V)	—

Table 4 Typical Slurry Analysis*

Component	Concentration, %	Matrix	ν, %
Al_2O_2	5.0	Cement slurry	1.52
So_3	2.0	Cement slurry	2.05
Ca	40.0	Cement slurry	0.08
SiO_2	1.40	Fe–Cu–Zn slurry	1.10
Ni	2.85	Cu–Ni–solution	0.35
Cu	0.0004	Cu solution	4.20
Cu	0.09	Zn–Fe–SiO_2 slurry	2.80
Zn	0.40	Zn–Fe–SiO_2 slurry	0.25
Ag	0.50 oz/ton	Cu–Pb–Zn slurry	5.30

*ν = Relative standard deviation.

Double safety interlocks prevent x-ray generation when an analytical cell is removed. A dual leak detection system will shut off the slurry flow should either slow or burst-type leaks occur in the flow cell system. Sampling and sample preparation techniques for on-stream slurry analysis tend to be less critical than those of laboratory or batch sample analysis methods because of the large averaging factor provided by the continuous flow system. The results shown in Table 4 indicate the variety of matrices which have been tested and the analytical precision obtained in the determination of several elements.

Dry Powder Analysis

Sampling and sample preparation techniques are somewhat more critical for the on-stream analysis of dry powders than for slurry stream analysis. The best results, both in making briquettes and in analysis, are obtained when the sample is ground to a uniform particle size, preferably less than 325 mesh. This can be accomplished in a continuous manner where the ground material is fed into a hopper, or by sampling means where the flow is not continuous. Measured amounts of the ground, dry material are fed to the automatic briquetting system. With an analysis time of 90 s, up to 1000 samples or 9000 elements can be determined within a 24-h period.

The results shown in Tables 5 and 6 indicate the typical analytical precision obtained in the analysis of raw mix cement samples in dry powder form. Table 5 lists the data obtained with a 3000-W rhodium target x-ray tube. Table 6 lists the data obtained with a 1400-W chromium target x-ray tube. For applications requiring the determination of elements having low atomic numbers, such as sodium and fluorine, the sample must be in a dry powder form to overcome cell window radiation absorption losses. Typical data for the determination of sodium are shown in Table 7.

Table 5 Typical Raw Mix Analysis (Dry Powder): OEG-75H Rhodium Target X-Ray Tube*

Component	Concentration range, %	Concentration, %	ν, %
MgO	2–5	3.07	0.59
Al_2O_3	3–8	4.66	0.56
SiO_2	9–21	13.22	0.28
CaO	31–45	40–35	0.08
Fe_2O_3	1–3	1.83	0.10

*ν = Relative standard deviation. Integration time = 40 s.

Table 6 Typical Raw Mix Analysis (Dry Powder): OEG-60G Chromium Target X-Ray Tube*

Component	Concentration range, %	Concentration, %	ν, %
MgO	2–5	3.07	2.37
Al_2O_3	3–8	4.66	0.94
SiO_2	9–21	13.22	0.43
CaO	31–45	40–35	0.03
Fe_2O_3	1–3	1.83	0.16

*ν = Relative standard deviation. Integration time = 40 s.

Table 7 Sodium Analytical Results (Dry Powder)*

Matrix	Na_2O, %	ν, %
Coal	0.11	5.3
Coal	0.39	3.8
Polymer	0.25	4.0
Bauxite	13.9	0.54
Bauxite	0.028	5.64

*ν = Relative standard deviation. Integration time = 100s.

Pulp Density, Matrix, and Interelement Corrections

Pulp density measurements may be obtained simultaneously with the elemental analytical data by means of the scattered radiation technique. Scattered radiation, measured at predetermined wavelengths, is related to the percentage of water in the slurry stream. This measurement then is used as a correction factor in determining the analysis.

Matrix changes in either dry powder or slurry stream analyses, which may alter the element analytical data, can be corrected in many cases by use of the scattered radiation technique. In a manner similar to that applied for pulp density correction, the scattered radiation, measured at a predetermined wavelength, is used as a correction factor.

Interelement effects, which may be caused by the absorption or enhancement of one element by another, may be corrected by computer programming. Pulp density, matrix effects, and interelement effects are related and require highspeed digital computers to make all three corrections, thus providing accurate analytical values.

Another type of instrumentation for dry material analysis receives material from the process flow line and directs it through a pipe to the powder presenter wheel. This wheel has a rubber groove in its center. In the groove are flanges which catch the powder flowing from the pipe presenter. The powder is compressed by a shoe to present a uniform sample area to the x-ray beam. As the powder sample rotates under the beryllium window, it is exposed to the x-ray beam and irradiated. Exposure to the beam causes the sample to fluoresce and emit secondary radiation, which becomes the basis for determining the concentration of the elements in the sample.

Trend Analysis versus Composition Analysis

Trend analysis, an indication of changes within the process, can be misleading and incorrect. With little effort, trend analysis can be converted to composition analysis for a more reliable evaluation of process control. For processes that exhibit minor composition fluctuations, a slope factor may be sufficient to convert trend data to analytical data. For processes that exhibit relatively wide variations in pulp density, dry specific gravity, matrix changes, and/or composition, trend analysis can indicate an inverted trend for any of the elements. Only by use of composition analysis corrected for the various effects can accurate process control be achieved.

Analytical hardware for the construction of closed-loop process control systems utilizing x-ray fluorescence as the elemental detecting and measuring medium is available. When interfaced with highspeed digital computers, this type of instrumentation makes possible (where the dictates of a process are of an elemental control nature) a true closed loop.

X-Ray Absorption Analytical Methods

Like the absorption of visible or uv radiation, the absorption of x-radiation at any given wavelength follows Lambert's law, which for the visible and uv region is usually given by

$$I = I_0 e^{\mu_1 x} \tag{6}$$

where I = intensity of radiation passing through material
  I_0 = intensity of incident radiation
  e = natural logarithm base
  μ_1 = fraction of energy absorbed per centimeter of absorber traversed
  x = thickness of absorbing material

Because of one striking difference between the absorption of x-rays and optical radiation, this equation is commonly modified for the x-ray region. Because the absorption of x-rays when passing through a fixed mass of a given material is completely independent of the physical state of the material, the absorption coefficient of Lambert's law usually employed for the x-ray region is the mass absorption coefficient μ. This coefficient is constant for any given element or mixture of elements irrespective of whether they are present in the gas, liquid, or solid state. For example, the absorption in passing through 18 g of water is the same as if the rays pass through 18 g of ice, 18 g of steam, or a mixture of 16 g of oxygen and 2 g of gaseous hydrogen. Hence, as applied to x-rays, Lambert's law usually is written

$$I = I_0 e^{-\mu x p} \tag{7}$$

where μ = mass absorption coefficient μ_1/p, p being the density of the absorbing material

Lambert's law holds only if monochromatic x-rays are used. As is the case in the optical region, the value of the absorption coefficient varies widely with the wavelength of the radiation, and μ must be specified for that wavelength. However, there are abrupt changes in absorption coefficients at certain wavelengths characteristic of the elements in the sample. These abrupt changes occur at wavelengths somewhat shorter than, but related to, the K, L, and M (and so on) emission lines, i.e., near the characteristic x-ray wavelengths emitted when elements are subjected to electron bombardment. These discontinuities are called *absorption edges* and are due to the photoelectric ejection of electrons from the K, L, and M shells of atoms subjected to x-rays. If the incident x-rays have sufficient energy (sufficiently short wavelength), they will photoelectrically eject electrons from the K shell. The energy of the photon is transferred to the electron ejected, hence that photon appears to be absorbed. Absorption edges are given Roman numeral subscripts, such as L_I and L_{II}, to distinguish them from the emission lines. A typical absorption curve as a function of wavelength is shown in Fig. 17.

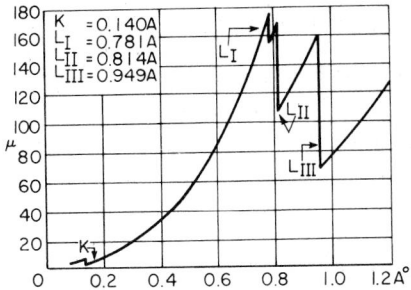

Fig. 17 Mass absorption coefficient for lead as a function of wavelength.

The absorption coefficient for x-rays on the shorter side of the K absorption curve as a function of wavelength is shown in Fig. 17.

The absorption coefficient for x-rays on the shorter side of the K absorption edge is much larger than for the long-wavelength side in the immediate vicinity of the edge. This is explained by the fact that x-rays having a wavelength just longer than that of the edge are not sufficiently energetic to eject a K electron, while the probability that they will eject an L electron or be lost by scattering is not significantly different from that for x-rays on the short-wavelength side of the edge. From this, it can be concluded that, if an x-ray has more than sufficient energy to eject a K electron, the probability is that it will either eject a K electron or pass through the material without any photoelectric action at all. It is very improbable that it will eject an L or M electron. Similarly, if the x-ray photon has more than enough energy to eject an L electron but not enough to eject a K electron, the probability is that is will either eject an L elecron or pass through the material without any action.

It can be shown that the mass absorption coefficient μ for a given element varies nearly linearly with the cube of the wavelength, except in the region of the absorption edges, according to

$$\mu = C\lambda^3 + \sigma \tag{8}$$

where C = constant for a given element between any two absorption edges
σ = "mass scattering coefficient" of same element

Photoelectric ejection of electrons is responsible for the $C\lambda^3$ term which is generally much larger than σ. Equation 8 is approximate because the mass scattering coefficient σ is not strictly constant over a wide wavelength range. Both $C\lambda^3$ and σ increase with the increasing atomic number Z of the absorber-scatterer. Hence, heavier elements like lead absorb x-rays much more strongly than lighter elements. $C\lambda^3$ has been found to vary approximately as the third or fourth power of the atomic number. σ varies nearly linearly with $Z\lambda$ between atomic numbers 4 (beryllium) and 20 (calcium) at wavelengths longer than 0.2 Å. For heavier elements, σ is proportional to $Z^m\lambda^n$, where m varies between 1 and 2 and n varies between 1 and 3.

It is important to note in connection with the K, L, and M critical absorption edges that the wavelength at which an absorption edge is observed is somewhat shorter than the wavelength of x-rays emitted when electrons from outer shells move to fill the vacancy. This is to be expected, since the photons responsible for the electron ejection must have sufficient energy to eject completely the electron from the atom, while in filling the vacancy an electron travels only from one of the atom's outer shells, thus undergoing a smaller energy change than for electron ejection. For example, the K absorption edge for magnesium is at 9.53 Å, while the $K_{\beta 1}$ emission line for magnesium is at 9.559 Å.

Total Absorption Spectrometers

Techniques employing both the mass absorption coefficients of the elements and their critical absorption edges have been used for chemical analysis. The simplest instrument is one that continuously measures the total amount of absorption from an x-ray beam as it traverses the sample. An instrument of this type, applicable to liquids, gases, or solids, is a photometer that compares the absorption of the sample with that of a reference sample. X-rays from the source pass through a chopper which permits the scintillation-type detector to receive alternately the radiation transmitted by the sample material and by the reference material.

Critical-Edge Technique

The most selective and a potentially useful analytical technique based on x-ray absorption is the critical-edge technique. This entails making absorption measurements just above and just below the critical-edge wavelength specific to the element to be determined (see Table 8). It has been shown that the difference in absorption between a wavelength just shorter than this critical-edge wavelength and another wavelength just longer is proportional to the concentration of that element in the sample. If the wavelengths are selected sufficiently close to the critical edge, the method becomes specific for one particular element. To obtain maximum sensitivity, it is important to select the proper edge, which may be either the K edge or one of the L or M edges. Maximum sensitivity is obtained when the difference between the mass absorption coefficients across the edge is as large as possible. Figure 18 shows the difference in mass absorption coefficient as a function of atomic number for the K, L_{III},

Table 8 Critical-Edge Wavelengths for Common Elements in Angstroms

Element	K edge	L_I edge	L_{II} edge	L_{III} edge
S	5.009	—	—	75.7
I	4.384	—	—	60.9
Cr	2.066	—	—	—
Fe	1.739	—	—	—
Cu	1.374	—	1212 .9	13.15
Mo	0.618	4.290	4.712	4.904
Ag	0.485	3.247	3.507	3.691
W	0.178	1.020	1.071	1.212
Pb	0.140	0.781	0.814	0.949

and M_v edges. In general, when the L_I edge can be used, it is stronger than the K edge. Similarly, the L_{II} edge is preferred to the L_I edge whenever that line falls at a practical working wavelength. The M lines are still stronger than the L lines.

Usually the choice of available edges is really not great, since the stronger edges occur at wavelengths that fall in the region where it is difficult to generate x-rays or where all practical window materials are too opaque. To obtain the highest selectivity with the absorption edge method, it is apparent that the wavelengths used should be as near as possible to the critical edge; for this reason, the most accurate work is conducted with a crystal monochromator, which can isolate extremely narrow wavelength bands on each side of the edge.

The critical-edge method employing an x-ray spectrometer (Fig. 19) has been investigated by a number of workers. This method has been used for the analysis of elements ranging from sulfur to

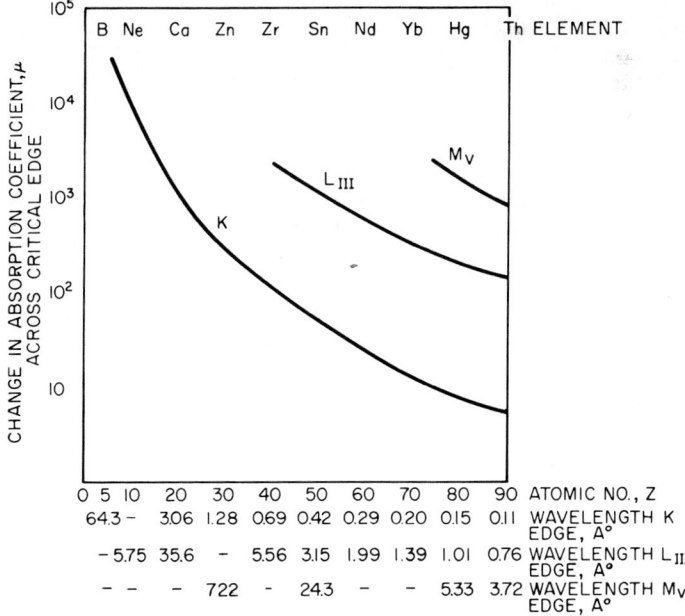

Fig. 18 Height of absorption edge $\Delta\mu$ as a function of atomic number.

lead. It is not particularly applicable for the determination of elements lighter than sulfur because the K absorption edge for all lighter elements occurs at wavelengths greater than 5 Å, where air and common window materials absorb quite strongly. For elements where the method is applicable, it is possible to detect as little as 50 ppm of a given element in complex mixtures. If the proper edge is selected for this type of measurement, the sensitivity of the method will be about the same for all elements heavier than titanium.

In many instances, it is possible to obtain sufficient selectivity using the critical-edge method by employing filters or x-rays tubes having targets that provide narrow emission lines in the wavelength region of interest. By proper selection of filters and tubes, the selectivity and sensitivity of the method can approach that obtained with spectrometers. It is also possible to use x-rays generated by fluorescence for critical-edge measurements.

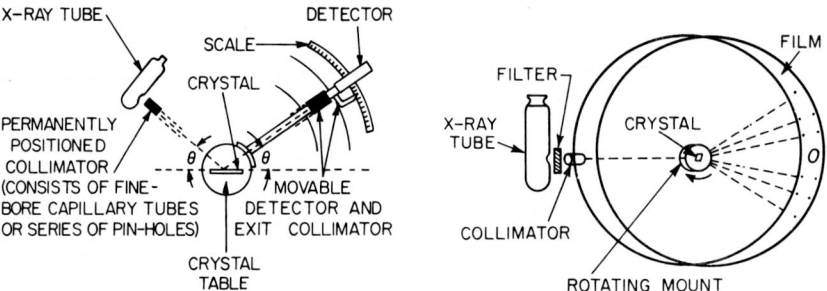

Fig. 19   Crystal monochromator for x-rays. Crystal table is geared to detector arm so that the angular motion of crystal is one-half that of the detector.

Fig. 20   Circular camera as used for rotation spectra.

X-Ray Diffraction Analytical Methods

Basically, it is known that, because of diffraction effects, x-rays appear to be reflected by lattice planes. By using crystals of known lattice spacing it is possible to determine the wavelength of reflected energy (using Bragg's law). As a corollary, if the wavelength is known, the spacing of an unknown single crystal, mounted as shown in Fig. 19, can be determined. This is the theoretical basis of much x-ray diffraction work. However, to use this method for the study of crystal structure, two requirements must be met: (1) A single crystal must be available; and (2) the crystal must be properly oriented in the holder. In view of these limitations, other techniques are often employed.

When single crystals are available, the problem of proper orientation can be minimized by rocking the crystal, commonly through $\pm 15°$, about a principal axis (found empirically). During the rocking, whenever a major crystal plane is properly oriented, monochromatic x-rays incident on the crystal will be reflected. Thus, during rotation, a series of spot reflections occur. These can be conveniently recorded on film, either held in a flat holder or preferably bent in the form of a cylinder along the circumference of a circle centered on the crystal (see Fig. 20). If the crystal is turned 90° in the holder, again rotated, then turned 90° in a third direction, and again rotated, all unit cell dimensions of most crystalline forms can be obtained. Crystals only a fraction of a millimeter in size can be studied in this way.

For single crystals, it is also possible to use the Laue technique, which consists of recording the reflections obtained when the crystal is irradiated with a collimated beam of polychromatic radiation. Since radiation of all wavelengths between 0.2 to 2 Å is generally present in the incident x-rays, all important sets of planes in the crystal will give mirrorlike reflections of some monochromatic wavelength present in the incident x-rays. If the crystal is oriented with a principal axis parallel to the primary beam, the dot pattern formed will be symmetrical; if not, the pattern will either be distorted or be asymmetrical.

Since samples are frequently available only in powder form, more convenient techniques for measuring lattice spacing have been developed. The Hull-Debye-Scherrer method is shown in Fig.

21 where the sample is irradiated with collimated monochromatic x-rays. Because powder samples consist of many small, randomly oriented crystals, the patterns obtained as a result of any one lattice spacing are not single spots but rather a series of spots overlapping to form a ring on a photographic film. Each spot, and hence the ring, is located at a characteristic angle of 2θ. This simplifies the translation of data from the photographic record. In modern cameras, the film circles the specimen in a full 360° (Fig. 22) so that the back-reflections are also recorded. Geiger tubes are commonly used in place of film to record the angle and intensity of reflections.

To obtain diffraction patterns, powder samples are commonly pressed in a wedge shape and mounted so that the edge of the wedge bisects the x-ray beam. The powder can alternately be formed

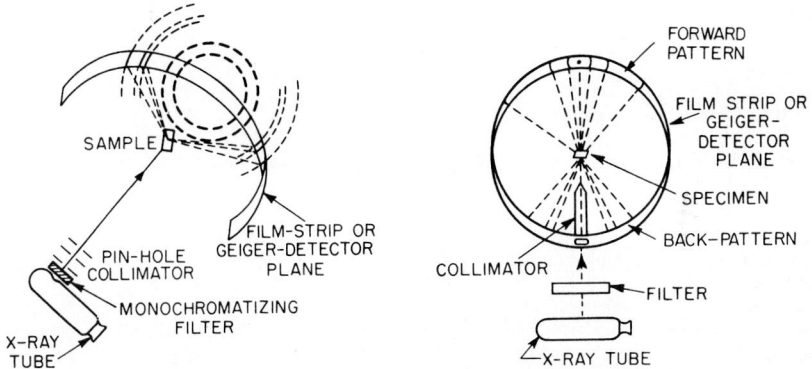

Fig. 21 Arrangement for producing transmission diffraction patterns.

Fig. 22 Cylindrical arrangement for obtaining simultaneous transmission and reflection spectra.

into a rod, held on a thin fiber with adhesive, or held in a thin-walled plastic or glass capillary. For some studies, the sample is best pressed into a small hole in a plastic disk.

Liquids and gases composed of large or highly polar molecules also exhibit definite x-ray diffraction patterns, indicating that the molecules are not randomly distributed but rather are arrayed in some preferred orientation. Furthermore, even completely unoriented materials will have distinctive scattering characteristics, causing poorly defined halos. These are a result of the interaction of the electronic fields of neighboring atoms, giving rise to a preferred direction of scattering.

Because of the complexity of diffraction studies and because the interpretation of patterns is highly involved, a thorough coverage of this subject is beyond the scope of this Handbook.

X-ray diffraction can be used to obtain a detailed understanding of the structure and spacing of single crystals or of powdered crystalline materials. Now that the crystal arrangements and spacings for many materials are known, the x-ray diffraction technique is a good tool for qualitative identification of an unknown. The x-ray diffraction characteristics of common materials have been classified according to the interplanar spacings that produce the three most intense diffraction lines. Numerous groups have cooperated in collecting and indexing these data. ASTM has collected data and published patterns for more than 5000 materials. Details of crystal patterns have been published. With this information, a crystallographer generally finds it possible to identify most of the materials in complex mixtures. However, in some cases, overlapping of the lines makes identification uncertain.

Diffraction techniques can be used for quantitative analysis of relatively simple mixtures. The method is generally applicable only to high concentrations and is of little value when the mixture consists of more than three constituents. The technique is based on measurement of the absolute intensity of the diffraction line in various patterns and is most reliable only when the intensities are compared with those of synthetic mixtures made up of the same materials. The method is plagued with numerous complications, and it is suggested that it be avoided if other more direct quantitative methods are applicable.

X-ray diffraction makes it possible to determine rapidly whether a given material is amorphous or crystalline. This involves simply looking at a diffraction pattern to note whether the lines are sharply defined, as for crystalline materials, or broad and indistinct, indicating amorphous substances. It is usually possible to determine the degree of crystallinity of polymeric materials.

An inspection of a transmission diffraction pattern also can provide an estimate of the grain size of many materials. For grain sizes above 5×10^{-3} mm, the grain size can be estimated by counting the number of spurious spots near the center of the pattern. Grain sizes between 2×10^{-4} and 5×10^{-3} mm give rise to sharp rings. For smaller grain sizes, the rings lose their sharpness and broaden out, finally disappearing if the crystal grains are sufficiently small. Because of this dependence on grain size, it is important that powdered samples used in x-ray diffraction work be carefully prepared and finely ground.

Diffraction methods may be used to measure stress and strain; to determine the amount of preferred orientation in metals; to calculate the size, shape, and arrangement of molecules; and to determine the molecular weight of crystalline materials. One particular advantage of x-ray diffraction methods is the small quantity of sample required; frequently, quantities as small as a fraction of a milligram are sufficient.

X-Ray Emission Analytical Methods

It is possible to analyze for the elements in an unknown material qualitatively and semiquantitatively by bombarding the material with electrons in an x-ray-type tube. The sample is placed on the anode, where, on bombardment, it emits characteristic x-rays at definite wavelengths. The wavelength of the emitted x-rays can be compared with those emitted by known materials, or the atomic number of the material can be calculated using Mosley's law. The wavelength of the emitted x-rays is determined with a conventional x-ray spectrometer as previously described. The method can be used for rough quantitative measurements by comparing the emission of the sample with that of an internal standard, similar to methods employed in visible and uv emission spectroscopy. Suitable standards are other elements of adjacent atomic number. It can be assumed that the intensity of the emitted x-rays is approximately the same for elements of adjacent atomic number at a fixed tube potential.

REFERENCES

Abelmann, R. A., W. G. Shequen, and A. H. Smallbone: "Dynamic Analysis with X-Ray Fluorescence," *Instrum. Control Syst.*, vol. 40, no. 8, August 1967.

Andermann, G., and J. W. Kemp: "Scattered X-Rays as Internal Standards in X-Ray Emission Spectroscopy," *Anal. Chem.*, August 1958.

Buerger, Martin J.: *Crystal-Structure Analysis*, Wiley, New York, 1960.

Clark, G. L.: *Applied X-Rays*, 4th ed., McGraw-Hill, New York, 1955.

Cooper, H. R.: "Using On-Stream X-Ray Fluorescence for Slurry Composition Analysis," *Instrum. Technol.* vol. 28, no. 7, July 1981, p. 45.

Davidson, E., and A. H. Smallbone: "Automated Sample Analysis: Wet or Dry," 10th Annual Rocky Mountain Spectroscopy Society Meeting, August 1968, Denver, Colo.

Dryer, H. T.: "The Effects of X-Ray Tube Parameters on Fluorescence Analysis," Conference on Analytical Chemistry and Applied Spectroscopy, March 1963, Pittsburgh, Pa.

Dryer, H. H., and J. A. Anzelmo: *Evaluation of ARL 72000S X-Ray Quantometer for the Analysis of NBAS/ BAS Iron Base Alloy Standards* Bausch & Lomb/ARL, Dearborn, Mich., 1978.

Herglotz, H. K. and L. S. Birks, (eds.) *X-Ray Spectrometry, Practical Spectroscopy Series*, vol. 2, Dekker, New York, 1978.

Jutila, J.: "Multicomponent On-Stream Analyzers for Process Monitoring and Control," *Instrum. Technol.* vol. 26, no. 7, July 1979, p. 38.

Roy, P., and W. Wittwer: *Evaluation of the 72000S X-Ray Quantometer for the Analysis of Alloyed Aluminum Products*, Bausch & Lomb, Ecublens, Switzerland, 1979.

Smallbone, A. H.: "New X-Ray Fluorescence Analytical Techniques and Material Handling Methods," Conference on Analytical Chemistry and Applied Spectroscopy, March 1965, Pittsburgh, Pa.

Smallbone, A. H.: "Briquetting, X-Ray Techniques Refine On-Stream Analysis," *Rock Prod.*, December 1965.

Smallbone, A. H., and R. Lathe: "A Simplified Approach to Ore Analysis Using an On-Stream X-Ray Analyzer and a Digital Computer," Conference on Analytical Chemistry and Applied Spectroscopy, March 1969, Pittsburgh, Pa.

Staff: "High-Speed X-Ray Analysis," *Instrum. Technol.* vol. 25, no. 8, August 1978, p. 26.

Quadrupole Mass Spectrometry*

Mass spectrometry has long been recognized as a valuable tool for both the qualitative and quantitative analysis of volatile inorganic and organic compounds. Until only recently, however, the use of mass spectrometers in process analysis has not been widespread because of its perceived complexity and the high cost of the instrumentation. Technological advancements over the past several years have reduced costs and simplified mass spectrometry to the point where it is very suitable for many process analytical applications in place of infrared absorption or gas chromatography (GC) techniques. The quadrupole mass spectrometer is well accepted in such applications because of its low cost, reliability, and ease of computer-controlled operation.

Simply, a quadrupole mass spectrometer can be divided into three parts, (1) the ionizer, (2) the mass filter, and (3) the detector, all of which are contained in a vacuum chamber maintained at a low pressure. When a gaseous sample is introduced into the system's ionizer, it is bombarded with a stream of electrons, producing positively charged parent ions (ions with the same molecular weight as the neutral molecule) and fragment ions. The ionizer has a series of lenses which serve to collimate the cloud of sample molecules toward the mass filter.

A quadrupole mass filter is a set of four rods disposed parallel and symmetrically with one another; opposite rods are electrically connected. A radio frequency and dc voltage of equal potential, but opposite charge, is applied to each set of rods (Fig. 1). By varying the absolute potential applied to the rods, it is possible to stop all ions except those of a given mass-to-charge (m/e) ratio.

Finally, the ions flowing down the quadrupole strike a Faraday plate detector. In some cases, the signal is amplified further by an electron multiplier. Thus, there is obtained a spectrum of signal intensity versus m/e value. Each molecule has a unique fragmentation pattern, so that a spectrum can be used as a fingerprint for compound identification. In addition, it is possible to quantitate the amount of a particular compound by comparing the sample signal intensity with the intensity produced by a known amount of the compound.

Many applications of the quadrupole mass spectrometer use a gas or liquid chromatograph to introduce the sample into the ionizer. When using the spectrometer in this manner, it is most common to scan a wide mass range (50–1000 atomic mass units, amu) at rates on the order of 1000 amu/s for compound identification. For process analyses, it is most common to introduce the sample directly into the ionizer and scan a narrower mass range. For both applications, computer systems are needed to collect the enormous amounts of data produced.

A computer is also used to control the potential applied to the quadrupole in order to scan the mass range of interest. Alternatively, the computer can command the appropriate potential required to focus a particular mass, a technique referred to as *selected ion monitoring* (SIM). Process analysis is essentially continuous SIM to achieve on-line quantitation of the components in the process stream.

*Engineering Staff, Extranuclear Laboratories, Inc., Pittsburgh, Pennsylvania.

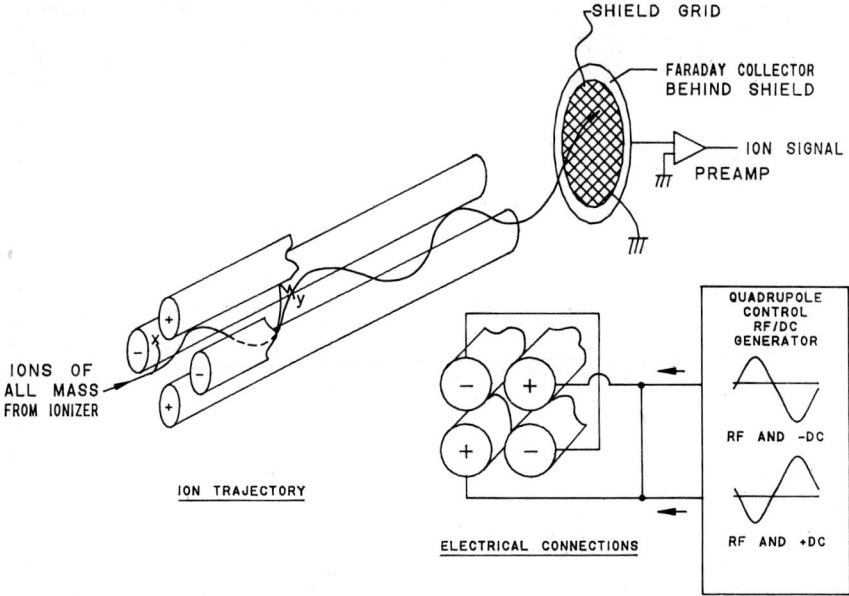

Fig. 1   Schematic diagram of quadrupole mass spectrometer. *(Extranuclear Laboratories.)*

Theory of Operation

Once the stream components to be analyzed have been identified, the mass spectrum of each component must be compared to the spectra of all other components to find unique mass peaks. Not all constituents of the mixture will have unique peaks; some may have intense peaks that might be useful in measuring the compound except that the mass peak also exists in another compound. For a stream

Table 1   Mass Peaks in Each Component's Spectrum and Their Relative Intensities, Using the Largest Peak from Each Compound as 100% in Each Case

Sample described in test

Mass	H_2O	CO	N_2	CO_2
1	6.00	—	—	—
12	—	4.00	—	9.00
14	—	—	5.00	—
15	—	—	0.03	—
16	4.00	1.00	—	10.00
17	28.00	—	—	—
18	100.00	—	—	—
28	—	100.00	100.00	11.00
29	—	1.00	0.73	0.09
30	—	—	0.02	—
44	—	—	—	100.00
45	—	—	—	1.00
46	—	—	—	0.04

containing N_2, CO, H_2O, and CO_2, Table 1 shows the mass peaks in the spectrum of each component and their relative intensities, using the largest peak from each compound as 100% in each case. Inspection of the table shows that m/e 18 and m/e 44 are unique peaks for H_2O and CO_2, respectively. However, the parent molecular ions of both N_2 and CO appear at m/e 28. Thus, some other method of measuring these components must be applied.

One might choose to measure CO at m/e 12 even though CO_2 has a fragment at this mass. Since the CO_2 intensity can be determined by measuring the peak at m/e 44, the CO_2 contribution to mass 12 is known and can be subtracted. The resultant at m/e 12 is then a measurement of the CO present in the mixture.

Two possibilities exist for the N_2 measurement; measure m/e 28 and subtract the CO contribution, since the CO intensity is known from the step above, or measure the intensity at m/e 14 and subtract the contribution from CO.

In practice, the spectral information used for calibration is obtained by actually introducing the compound of interest into the mass spectrometer and measuring the spectrum. This is usually done by using a known binary mixture of the particular compound in nitrogen or argon. The ratio of each mass peak compared to that of some standard ion, for example, m/e 28 of nitrogen, is measured. The fragmentation of the peaks of interest is then programmed into a fragmentation matrix. For each compound to be measured, only one mass peak is programmed.

For the analysis of N_2, CO, H_2O, and CO_2, the fragmentation matrix becomes:

	12	18	28	44
CO	4.00*	0.00	100.00	0.00
H_2O	0.00	100.00*	0.00	0.00
N_2	0.00	0.00	100.00*	0.00
CO_2	9.00	0.00	11.00	100.00*

The peaks to be measured for each component are indicated by an asterisk. The subtraction of interfering mass peaks is accomplished by inverting the matrix to yield:

	12	18	28	44
CO	25	0	0	−2.25
H_2O	0	1	0	0
N_2	−25	0	1	2.14
CO_2	0	0	0	1

In the actual analysis, the amplitudes of peaks at masses 12, 18, 28, and 44 are measured and multiplied by the inverted matrix. For example:

	12	18	28	44	
CO	25	0	0	−2.25	I(12)
H_2O	0	1	0	0	I(18)
N_2	−25	0	1	2.14	I(28)
CO_2	0	0	0	1	I(44)

Therefore, CO intensity is determined by:

$$25[Int(12)] + 0[Int(18)] + 0[Int(28)] - 2.25[Int(44)]$$

Values for H_2O, N_2, and CO_2 are similarly calculated. Intensities may be converted to percentage concentration by dividing by the component's sensitivity and normalizing to 100%.

Fig. 2 Quadrupole mass spectrometer. *(EL-150, Extranuclear Laboratories.)*

The formulas for these calculations are therefore

$$INT(I) = IBKG(I) - BKG(I)$$
$$CONC(I) = \text{Sum over } J \; [\, INVER(I,J) \cdot INT(J)\,] \; / \; SENS(I)$$
$$\%CONC(I) = CONC(I) \cdot 100 \; / \; \text{Sum over } J \; [\, CONC(J)\,]$$

Process Analyzer System

One process analyzer* (Fig. 2) has a rapid response time, a wide dynamic range, accuracy, and ease of computer operation. Because the mass spectrometer is a nonspecific detector, it is useful in detecting almost any gaseous compound that produces ions in the mass spectrometer's range (typically 200 amu). The quadrupole mass spectrometer is particularly adaptable to process analysis because it is easily and rapidly directed to monitor any mass ion peak within its range. The instrument shown is controlled by a computer system which commands masses, gathers and reduces data, and presents useful information almost instantly to the operator. The instrument was developed specifically for the process analyzer to provide the flexibility required for a variety of applications, while keeping to a minimum the amount of operator training needed to effectively run the analyzer. The instrument enables the operator to design analysis, calibration, and sampling sequence methods to optimize conditions for a particular application. An analysis method is designed by specifying the molecular weight and the mass to be monitored for each of the components to be analyzed. The data system

---

*Extranuclear Laboratories EL-150.

then searches a library stored on a floppy disk for the spectra and sensitivity of the compounds of interest. The fragmentation matrix used for the analysis is thereby constructed. Default values are set for the various mass spectrometry operating parameters such as dwell and settling times, input resistors, background, and alarm limits. The operator can change parameters for any or all compounds as desired. Once designed, the method can be stored on a floppy disk, allowing the user to design a different procedure for each stream to be analyzed.

Once a suitable calibration procedure has been designed for the stream of interest, the system can be programmed for automatic calibration. In a calibration sequence, the operator can specify:

Delay	Sets the time required to clear the lines before calibration is initiated
Concentration	Sets the concentration of each component to be calibrated
Scan	Sets the number of scans to be averaged for calibration calculations
Background	Measures the background and makes the appropriate adjustments in the background values used in the analysis
Ratio	Adjusts the fragmentation matrix to reflect any changes that may have occurred.
Sensitivity	Adjusts sensitivity values used in concentration calculations

Complex streams may require more than one calibration gas to properly calibrate all stream components. A separate sequence for each gas can be programmed, and the procedures chained together for automatic operation. Different calibration procedures can be programmed for the different streams analyzed, and the procedures stored on a disk.

Finally, a sampling sequence method can be designed for automatic switching of up to 16 valves.

Fig. 3   In the manual mode, data are presented on the CRT in the form shown here. *(Extra-nuclear Laboratories.)*

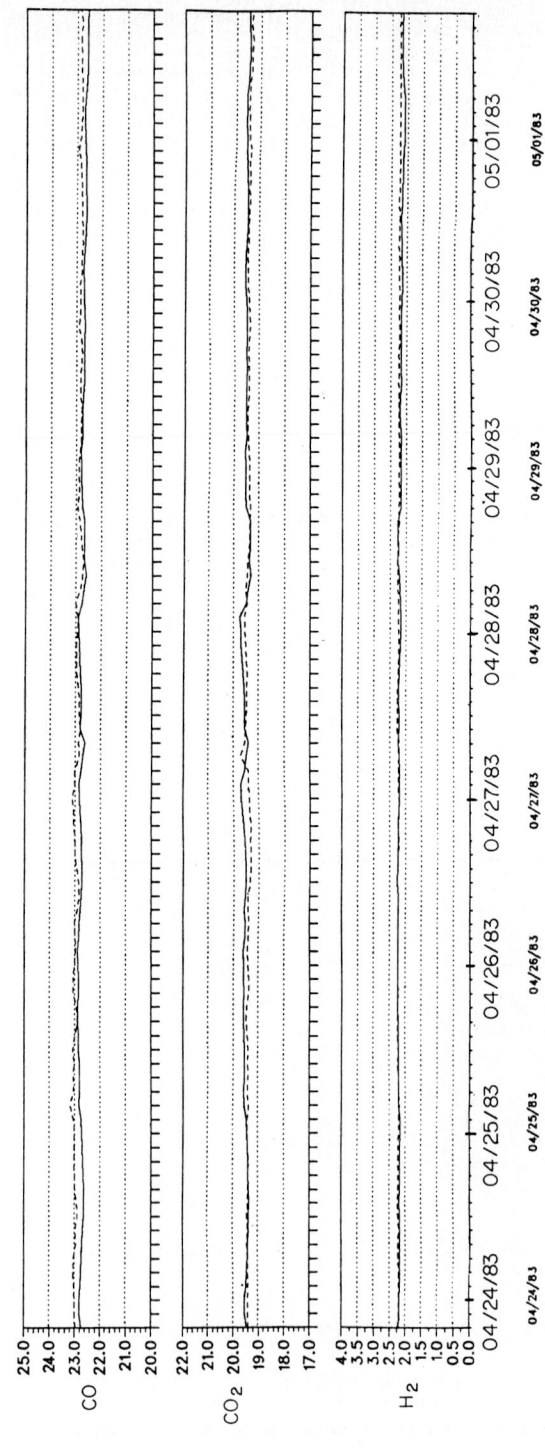

Fig. 4 Demonstration of the stability of the system. *(EL-150, Extranuclear Laboratories.)*

In a sampling sequence method, among the parameters which can be specified are valve number, analysis or calibration file name, time for stream clearing, time for stream analysis, and calculation method. When the method is run, the data system reads the sampling sequence method, automatically opens and closes the specified valves, loads in mass spectrometry analysis parameters and scans the mass spectrometer for the specified length of time before going on to the next stream. If the next entry in the method is calibration, the procedure previously defined is performed. Each time a calibration is performed, the new analysis parameters are stored on the floppy disk to ensure that operating procedures are preserved.

The process analyzer has three modes of operation. In the sweep mode it can be commanded through the cathode ray tube (CRT) keyboard to sweep the quadrupole over a specified range with the data displayed on an oscilloscope. This mode is useful for tuning or to determine the masses present in a particular stream.

In the manual mode, data are presented on the CRT in the form shown in Fig. 3. During real-time operation, the operator can display and change all mass spectrometry parameters in order to fine-tune a particular analysis. Data can be presented as intensity (raw signal voltages), as percent concentration, or as a ratio to a specified component. Optionally, data can also be printed as hard copy or stored on the disk.

The third mode of operation provides automatic analysis as described previously.

Figure 4 shows the stability of the system. The figure shows percentage concentration for CO, CO_2, and H_2 as measured by the instrument (dashed lines) and an infrared (ir) analyzer (solid lines) for an 8-day period. The mass spectrometer was not tuned during this period, and an automatic calibration was performed at 2-h intervals. All signals showed a long-term accuracy of $\pm 0.5\%$ of their concentrations, i.e., 0.1% and better full-scale accuracy.

Ultraviolet Absorption Analysis*

Process stream analyzers based on the measurement of ultraviolet (uv) radiation absorption are used throughout the process industries for monitoring (and controlling) the concentrations of components in both gas and liquid streams. In contrast to a laboratory uv spectrophotometer, a plant analyzer is designed to be used in a plant environment for the continuous analysis of a *specific* component. The spectrophotometer, which reads absorbance (or transmittance) versus wavelength, is a very useful laboratory tool both for identifying the components of a mixture and for determining the concentration of each of the components.

The uv absorption pattern of a compound is not as distinctive a fingerprint as its ir counterpart, and fewer compounds absorb in the uv region than in the ir region. However, several important classes of compounds absorb strongly in the uv region, whereas water and the usual components of air do not absorb in this region. As a result, uv absorption analyzers may be more selective and sensitive than ir and other types of analyzers used in many plant stream analysis applications.

*Based partially on information furnished by R. S. Saltzman, Instrument Products Division, E. I. Du Pont de Nemours & Company, Inc., Wilmington, Delaware.

Ultraviolet Absorption Spectrum

The region of the electromagnetic spectrum covered by the uv absorption analyzers described here is from 200 to 400 nm and excludes the "vacuum" uv, below 200 nm. Absorption in this region, corresponding to photon energies between 6.12 and 3.06 eV, is due to excitation of the more loosely bound valence electrons, including the "unsaturation" (π) electrons of multiple bonds and the unpaired electrons of free radicals.

The spectra of molecules generally are composed of broad bands when compared with the line spectra of atoms because each electronic state has associated with it numerous vibrational and rotational energy levels. The vibrational-rotational fine structure of the bands is not resolved except for gas-phase spectra of light, simple molecules at low total pressure. For example, nitrogen dioxide and sulfur dioxide at low pressure show a fine structure when studied at high resolution. Also, the absorption spectra of simpler aromatics, such as benzene and cymene, show a distinct fine structure, even in liquid phase. In general, however, pressure broadening and solvent interaction make the spectrum appear as an envelope of the otherwise expected fine-structure spectrum.

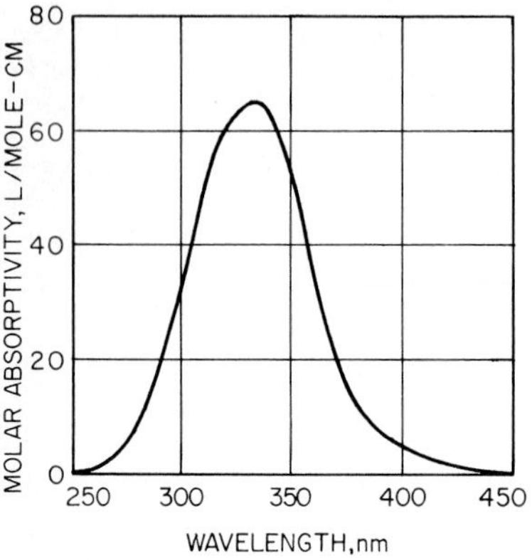

Fig. 1   Spectral absorptivity of chlorine.

Relation of Concentration to Absorption

Uv absorption data usually are tabulated and plotted in terms of absorptivity as a function of increasing wavelength (Fig. 1). The unit of wavelength in the uv region is the nanometer (nm) [1 nm (10^{-9} cm) equals 10 Å], which is the same as the millimicron (mμ), a unit which was previously used extensively in analytical chemistry literature. The most common unit of absorptivity is liters per mole-centimeter. A material with absorptivity of unity at a specific wavelength has an absorbance of 1.0 in a path length of 1 cm when the sample concentration is 1.0 mol/L.

The absorbance of a substance is directly proportional to the concentration of the material which causes the absorption in accordance with the Lambert-Beer law, more commonly referred to simply as Beer's law:

$$A = abc = \log \frac{I_0}{I} = \log \frac{1}{T} \tag{1}$$

where A = absorbance

a = molar absorptivity, L/mol·cm

b = pathlength, cm

c = concentration, mol/L

I_0 = intensity of radiation striking detector with nonabsorbing sample in light path

I = intensity of radiation striking detector with concentration c of absorbing sample in light path b

T = transmittance = I/I_0

For the vapor phase,

$$A = \frac{abc'}{2450} \tag{2}$$

at 25°C and 760 torr pressure where c' = volume percent or mole percent, or

$$A = \frac{abc'}{2450} \frac{P + 14.7}{14.7} \frac{298}{t + 273} \tag{3}$$

at any temperature or pressure where P = pressure, psig and t = temperature, °C.

For the liquid or solid phase,

$$c = \frac{c'' \times d}{MW} \times 10 = \text{mol/L} \tag{4}$$

where c'' = weight percent in liquid
   d = density of liquid
   MW = molecular weight of material to be measured

$$A = \frac{10abc''d}{MW} \tag{5}$$

Ultraviolet Absorption Analysis Techniques

The basic uv absorption analyzer consists of a radiation source, optical filters, a sample cell, a detector, and an output meter (Fig. 2). A transmittance measurement is made by calculating the ratio of the reading of the output with the sample in the cell to the reading with the cell empty (of uv absorbing materials). The concentration can be calculated from the known absorptivity of the substance by means of Beer's law (Eq. 1) or may be obtained by comparison with known samples.

The source must provide the desired uv wavelength and may be either a line source, such as a mercury arc, or a continuous source, such as a hydrogen or deuterium arc. Tungsten and tungsten-iodine lamps also may be used at longer uv wavelengths. Optical filters or a spectral dispersing system are used for screening out radiations of unwanted wavelength emitted by the source. The sample cell is equipped with windows that are transparent at the chosen wavelengths, and the path length between the windows must be fixed. Vacuum phototubes, photomultipliers, and solid-state cells are used as uv detectors, and their output is read with a sensitive meter. Plant stream uv analyzers incorporate elaborate variations of these basic elements.

Design Criteria

Experience has shown that simple modifications of laboratory instruments are not adequate to meet the special requirements for continuous monitoring and control of process streams. A few basic design requirements must be met in nearly every case. For example, plant instruments must be rugged and

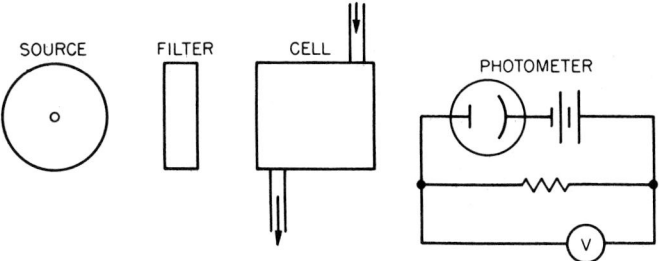

Fig. 2   Elementary uv analyzer.

serviceable; they usually cannot be, and are not, treated as carefully as laboratory types. Instruments in the field will be exposed to plant fumes and vibrations and must meet the electrical classification of the area. Maintenance periods, except for routine checks and adjustments, should be months apart, and the design should be as simple as possible with components readily accessible for maintenance.

The output reading of the instrument, in most cases, should be linear in concentration units. Straight-line calibrations are less confusing and preferred by operating personnel.

Absorbance readings should be accurate to $\pm 1\%$ of the full-scale reading. For gas analyzers, absolute pressure control may limit the accuracy with respect to concentration to $\pm 2\%$. High accuracy can be maintained by frequent and automatic standardization which may include zero setting and range or sensitivity adjustment. If the analyzer is used for automatic control, high reproducibility with a minimum of dead zone may be essential.

The instrument must rapidly respond to changes in concentration. A time constant of less than 1 min usually is required. Uv analyzer lags generally are small, but delays through long sampling lines must be avoided. Greater lags sometimes can be tolerated, but where the analysis is used for automatic control or for safety shutoff, a faster response (sometimes less than 10 s) often is required.

The analyzer must be easy to calibrate. Calibration based on known values of absorptivity is easiest and accurate if the analytical radiation is monochromatic and unchanging in wavelength.

Over 80% of the problems associated with analyzer installations are found in the sampling systems. With an analyzer designed to minimize sample handling problems, difficult analyzer applications can often be handled reliably with simplified sampling systems.

Specific Design Requirements

The specific analysis will dictate many of the features of the analyzer. Some points usually requiring consideration are:

1. *Absorbance Range.* In general, the absorbance range for a specific application should be as broad as possible for high stability but is limited to ensure good linearity and a reasonable cell length. A range of 0 to 1 absorbance (90% loss of the radiation at the analytical wavelength) is often a good compromise.

2. *Analytical Wavelength.* This must be carefully selected from among those obtainable with standard photometric components. Ideally, it should be a wavelength where the desired component absorbs strongly and all others are transparent, and the cell path length is reasonable.

3. *Cell Length.* This requirement, subject to mechanical limitations, can be determined from Beer's law after the absorbance range and analytical wavelength have been specified. A very thin cell may be required in recording the concentration of a strongly absorbing major component of a liquid mixture, and a very long cell may be required in recording the concentration of a trace constituent of a gas.

Ultraviolet Analyzer Components

A uv analyzer usually is designed for a specific application but can be modified for a new analysis. Information provided here on several of the basic components should be of help in evaluating the feasibility of a new application and in understanding the operation of uv analyzers. The source, filters, and detectors all must be considered in determining the spectral distribution of analytical radiation in the analyzer. The source emits radiation of a limited range or number of wavelengths, some of which are blocked by a selected filter combination, and the detector responds selectively to one or more of the wavelengths transmitted.

Good performance of a uv absorption analyzer can be maintained easily only when the analytical radiation is of high purity (not contaminated with other wavelengths). The potential absorbance range of the analyzer varies with the purity of the analytical radiation. If the radiation is 99% pure, i.e., with 1% of radiation not absorbed, then a sample whose true absorbance is 2.0 will be indicated as having an absorbance of only 1.7, or for a true absorbance of 3.0, the indicated absorbance will be 2.0 when the analyzer is calibrated in the usual way by Beer's law. These figures show the loss in sensitivity and departure from linearity that occur owing to stray light or unabsorbed impurity radiation. Instability results from variation in the impurity content, particularly at high-absorbance readings.

Inaccurate analyses may be expected if the absorptivity of the material analyzed varies markedly within a narrow range of wavelengths that might be used as the analytical radiation. This occurs because the effective analysis wavelength shifts to the wavelength of lowest absorptivity as the concentration increases. For highest accuracy and linearity, monochromatic radiation is preferred. Even with the best available uv interference filters, continuous spectrum sources cannot provide the narrow bands of wavelengths comparable to the spectral emission line isolated with filters. As a result, the linearity and accuracy of analyzers using continuous spectrum sources are limited except where the absorptivity is constant over the range of wavelengths.

Sources of Radiation

The uv sources most useful for uv absorption analysis include tungsten and tungsten-iodine lamps and mercury, zinc, cadmium, hydrogen, deuterium and other arc or gas discharge lamps. The source best suited depends on the wavelength, stability, life, and other requirements.

Tungsten filament incandescent lamps are the most readily obtainable, inexpensive lamps used for uv photometry. They provide a continuous spectrum of energy that is relatively feeble in the uv region but adequate in an efficient instrument above 350 nm. *Tungsten-iodine cycle lamps* have quartz envelopes and may provide adequate emission above 300 nm in an analyzer with highly sensitive detectors.

Mercury vapor lamps are the most useful for uv analyzers. They usually have long life expectancies. Useful discharge lines are highly intense, usually three to five orders of magnitude greater than the background emission, and are at fixed, unchangeable wavelengths. Isolation of a line may often be provided with simple blocking filters, while still maintaining the high spectral purity possible from the narrow spike of emission.

Low-pressure mercury lamps have a highly intense and narrow emission line at the mercury resonance wavelength at 254 nm. Temperature regulation of the source is required for long-term stability.

Medium-pressure mercury lamps generate high-intensity emission of several mercury lines. The usual glass envelope limits its usefulness to lines above 300 nm. The life expectancy of this lamp with a glass envelope is exceptionally long. (Some have operated continually for over 5 years.) These lamps must be carefully selected to obtain stable arcs.

For use of the mercury lines below 302 nm, such as 265, 275, 280, and 289 nm, a lamp with a "uv transmitting" envelope is required. These lamps are generally considerably more expensive.

The zinc discharge lamp is valuable for the 214-nm emission line, and the cadmium discharge lamp for its 228-nm emission line. The power to the lamp and the lamp temperature must be closely regulated for stable operation. These lamps, even when carefully selected, are not usually as stable as mercury vapor lamps.

Several other types are available but have limited applicability. Hydrogen and deuterium lamps are widely used in the laboratory but are too delicate, expensive, and short-lived for widespread use in process stream uv analyzers.

Optical Filters

The analytical radiation in an uv analyzer must be as nearly monochromatic as possible in order to facilitate the calibration of the instrument and to ensure high linearity, long-time stability, and sustained accuracy. Monochromatic radiation is obtained by proper selection of source, filters, and phototubes, each of which is selective in regard to the wavelengths that it respectively emits, transmits, or responds to.

Monochromators, employing prism or grating dispersing systems, provide monochromatic radiation beams, but those that have appeared to date are complex and inefficient for use in a plant stream analyzer.

Uv interference filter technology has made great strides in recent years. Uv interference filters are still comparatively expensive, but stable filters with half-bandwidths of less than 2.0 nm are now available and are being used in uv analyzers.

For isolation of spectral lines from a gas discharge source or for a relatively broad emission band, combinations of bandpass filters often provide the least expensive and most stable means of wavelength isolation. These filters may be in the form of special optical glass plates, films, or sealed quartz bottles of gases or liquid solutions. For sustained use in uv absorption analyzers, filters are selected to have good stability in the intense uv radiation and heat of the light source.

Single-Beam Analyzer

The simplest analyzer is the *single-beam* type, with a functional diagram essentially the same as that of the general absorption analyzer in Fig. 2. The output of this type of instrument is affected by fluctuations and drift of the light source, dirt or bubbles in the sample cell, and any drift in the detector or detector circuit. As a result, single-beam instruments must operate on relatively low sensitivity (high-absorbance) levels to provide reasonably stable analyses. Stabilized light sources and improved detector circuits, in recent years, have improved stability to some extent, and these simplified units have found application where only a "go-no-go" or broad-range measurement is required.

Split-Beam Analyzer

This type of design is based on a differential absorption measurement at two wavelengths and has certain advantages, including compensation for dirt and bubbles in the sample cell.

Radiation from the source, generally a gas discharge lamp, is partially absorbed in passing through the sample. The radiation leaving the sample is divided into two beams by a semitransparent mirror. Each beam passes through an optical filter to a phototube. The filter removes radiation at all wavelengths except the one to be measured.

Radiation at the analytical wavelength striking one phototube is absorbed strongly by the component whose concentration is being measured. Radiation at the reference wavelength, directed to the second phototube, is absorbed weakly or not at all by the component. Each phototube develops a current directly proportional to the intensity of radiation striking the phototube.

The phototube current is converted to a dc voltage directly proportional to the negative logarithm of the phototube current by a special logarithmic amplifier. The output voltages from the measuring and reference amplifiers are subtracted in the control station to produce a final output voltage. Hence, the output is proportional to the difference between the logarithms of the phototube currents and proportional to the sample concentration (based on Beer's law).

This split beam photometric analyzer has separate light source, sample, and photometer housings. The three housings of the analyzer are easily separated. Because of this, the analyzer is readily adapted for monitoring process gas and liquid streams in pipeline cells installed directly in the process line and for monitoring the thickness of film or film coatings.

Dual-Beam Analyzers

A *dual-beam* analyzer, using a measuring and reference beam at the same wavelength, is fundamentally the most stable design from strictly an instrumental standpoint, provided it has the equivalent stability of optical and electronic components of the single-beam and split beam instruments.

It does not compensate for dirt and bubbles in the sample cell, although the addition of a short pathlength reference cell tends to compensate for dirt accumulation on cell windows.

APPLICATIONS OF ULTRAVIOLET ABSORPTION ANALYZERS

Their high sensitivity, accuracy, reliability, and precision make uv absorption analyzers suited to many process stream analysis problems. Applications of these analyzers run into the thousands throughout the process industries.

Elemental Halogens

Elemental halogens all broadly absorb uv and/or visible radiation, and analytical wavelengths within these absorption bands usually can be chosen for selective analyses. Historically, uv analyzers have been most widely used in monitoring elemental chlorine (Cl_2). These analyzers often provide the primary measurement for closed-loop control of chlorination processes.

Sample streams with elemental halogens may be highly corrosive, and materials of construction must be carefully chosen.

Chlorine is photochemically activated by radiation of $\sim$400 nm. It reacts very rapidly with hydrogen and slowly with CO and hydrocarbons if this radiation is not blocked from the sample. This phenomenon is utilized, however, in monitoring the hydrogen (H_2) concentration in Cl_2 with a dual-beam analyzer by measuring the difference in the chlorine absorption before and after the Cl_2 photochemically reacts with any hydrogen in the stream on exposure to the 405-nm radiation emitted by the source.

Uv absorption analyzers for the monitoring of F_2 concentrations find application in uranium recovery processes. Figure 3, showing the absorption spectra of F_2 and uranium hexafluoride (UF_6), illustrates how a split beam, two-wavelength measurement can be used to compensate for interfering compounds. Since UF_6 absorbs nearly equally at 334 and 405 nm, the difference in absorbance measurement from a split beam analyzer using these two wavelengths, from UF_6 concentration changes, is practically nil. The F_2, however, has an appreciable difference in absorbance at the two wavelengths, and the output is calibrated for F_2 concentration.

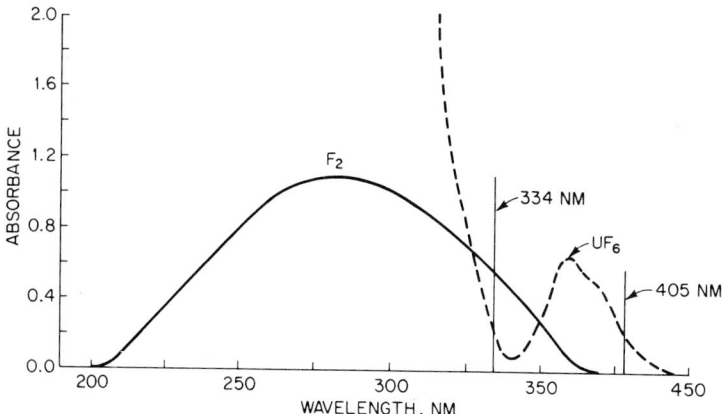

Fig. 3   Absorption spectra (vapor phase) of F_2 and UF_6, illustrating how a split beam, two-wavelength measurement can be used to compensate for interfering compounds.

Aromatic Compounds

The benzene ring, with its conjugated double bonds, is a strong absorber of uv radiation. In general, the greater the substitution for the hydrogen in the ring, the greater the absorbance of the aromatic compound. Aromatic compounds with radicals containing double bonds, such as nitro ($-NO_2$) or sufonyl ($-SO_2-$) groups, are particularly strong uv absorbers—as are naphthalene, anthracenes, and related compounds.

The absorption band of a para isomer of an aromatic compound usually extends to higher wavelengths than those of the ortho and meta isomers. As a result, the para isomer concentration can often be selectively monitored by measuring the absorption at a wavelength on the high side tail of the absorption band.

Aromatics, including phenols, are monitored in wastewater streams for pollution control.

Color Formers

Water-white products, both natural and synthetic, may "yellow" with time despite elaborate processing to remove color. Impurities causing color degradation often are strong uv absorbers, and their presence can be sensitively monitored with on-stream uv analyzers. Examples of products monitored for color formers include p-xylene, terephthalic acid, bisphenol A, and corn syrup.

Compounds with UV Absorption Shifting with pH

The uv absorption bands of a comparatively few, but important, chemical species shift significantly with pH. This phenomenon can be utilized to provide highly selective and sensitive analyses for these compounds with uv absorption analyzers. Examples include monitoring concentrations of phenols in the parts per billion (ppb) range in wastewater, of peroxide in bleaching operations, and of sulfite ions in process and waste streams.

Oxidizing Agents

Oxidizing agents, in general, have strong uv absorption bands. Examples of such materials being monitored with uv absorption analyzers include sodium hypochlorite, chlorine dioxide, hydrogen peroxide, ozone, and potassium permanganate. "Bleaching" steps are often controlled by monitoring the oxidizing agent concentration.

Inorganic Compounds

Salts of iron, nickel, manganese, copper, and other transition metals may be strong absorbers in both the uv and visible regions. Anions with strong uv absorption include nitrates, sulfites, and chromates. For example, the concentration of nitric acid is often monitored with uv absorption analyzers.

Sulfur Compounds

Sulfur-containing compounds often strongly absorb uv radiation. In sulfur recovery plants, two uv absorption analyzers are used to monitor the concentrations of H_2S and SO_2 for control of the H_2S/SO_2 ratio for maximum recovery efficiency.

Pollutants

Uv absorption analyzers are particularly well-suited for source monitoring of many industrial pollutants. Sulfur dioxide, nitrogen oxides (NO_x), chlorine, phosgene, mercury, and ozone are examples of compounds monitored in the gas phase for pollution control. Aromatics, Cr^{6+}, ions, "oil," phenols, and sulfites, as well as mercury compounds, are monitored in waste streams for pollution control.

Lower Explosive Limit (LEL) of Solvents

The high speed of response and the reliability of uv absorption analyzers make them well-suited for the monitoring and control of the concentration of many flammable solvents slightly below the lower explosive limits. As a result, highly efficient operation of a process can be maintained under safe conditions. Examples of solvents monitored include toluene, benzene, xylene, acetone, and dimethylformamide.

Table 1 Representative Gases, Vapors, and Liquids That Do Not Absorb Visible or Ultraviolet Radiation

Do not absorb above 210 nm		
Inorganics	Saturated Straight-Chain Hydrocarbons	Lower Alcohols
Argon		
Carbon dioxide	Butane	Ethanol
Carbon monoxide	Ethane	Methanol
Helium	Methane	n-Butanol
Hydrochloric acid	Propane	n-Propanol
Hydrogen		
Krypton		
	Unsaturated Straight-Chain Hydrocarbons	
Neon		
Nitrogen		
Oxygen	Acetylene	
Water	Ethylene	
Xenon	Propylene	

Do not absorb at 254 nm or at higher wavelengths		
Chlorides	Acids	Lower Alcohols
Ethyl chloride	Acetic	Ethylene glycol
Ethylene dichloride	Butyric	Isobutanol
Methyl chloride	Propionic	Isopropanol
Methylene dichloride		
Vinyl chloride		
Esters	All Ethers	
Butyl acetate		
Cellosolve*		
Dimethyl sulfate		
Ethyl acetate		
2-Ethyl butyl acetate		
Vinyl acetate		

*Trademark, Union Carbide Corporation, New York, N.Y.

Other Types of Compounds

Other materials not falling in the applications categories above but commonly monitored with uv absorption analyzers include 1,3-butadiene, furfural, caffeine, ketones, aldehydes, and proteins.

Compounds that absorb uv radiation and hence are potential candidates for uv absorption analyses are of course numerous. Also, the number of readily available uv wavelengths for uv absorption analyzers has greatly expanded in the past 20 years, and any table of uv absorption data inclusive enough to be of real value is beyond the scope of this article.

Materials that do not absorb uv radiation are listed in Table 1. These compounds are transparent when they are highly purified, and any absorption they exhibit usually can be traced to impurities in the sample.

Many different materials absorb the same uv wavelengths; hence, complex mixtures of materials are sometimes difficult or impossible to analyze unambiguously. Often a wavelength might be selected near the fringe of an absorption band that will provide a more accurate and selective analysis than is obtainable at the wavelength of highest absorptivity. Working away from the peak of an absorption band often is considered not acceptable by analytical chemists who have often observed what appear to be deviations from Beer's law when measuring off the peak absorption with laboratory spectrophotometers. Usually, however, deviations from Beer's law usually result from the limitations of the spectrophotometer with respect to wavelength purity and reproducibility. With high-purity measuring wavelengths, deviations from Beer's law are rare. Deviations that do appear usually result from "pressure broadening," changed chemical structure, and solvent effects. The first two of these sources of deviation usually can be eliminated by proper sample conditioning (temperature and pressure).

The uv absorbance of a liquid is stronger than an equal path length of its vapor roughly by an amount proportional to the density of the ratio of the two phases. The peak absorptivities of the liquid in the wavelengths of peak values are shifted various amounts by the interaction effects of various solvents. Polar solvents, such as water and acids, produce the greatest shifts. Because of such shifts, the sensitivity of uv absorption analyzers to liquids cannot be predicted with accuracy on the basis of absorptivity values of the pure material. There are, however, several uv-transparent solvents, such as cyclohexane, hexane, and isooctane, which cause little shift in the absorptivity of the solute. Hence, absorptivity values of materials in these solvents can be used to predict concentration ranges of gas analyzers reliably to about ±20% in many cases.

Infrared Process Analyzers

by
Rodney M. Durham*

The infrared portion of the electromagnetic spectrum extends from just beyond the visible to the microwave region. This is nominally from 0.75 μm (micrometer) to 1000 μm. The manner in which electromagnetic radiation interacts with matter is a function of the wavelength of the radiation. Infrared wavelengths interact at the molecular level. It is this molecular interaction which makes the infrared radiation a valuable probe in studying the properties of molecules. The infrared wavelength region from 3 to 10 μm is especially valuable in detecting the presence of molecular species. Infrared

*Manager, Instruments Division, Infrared Industries, Inc., Santa Barbara, California.

radiation at these wavelengths has sufficient energy to excite molecular vibration. The frequency of the vibration is a function of the weights of the atomic elements bound in the molecule and the strength of the molecular bond. This characteristic vibration frequency results in an infrared absorption band which is indicative of the presence of a particular molecular bond. For example, the carbon-hydrogen bond is excited by infrared radiation at the wavelength of 3.4 μm, while the carbon-oxygen bond in the carbon monoxide molecule is excited by 4.7-μm radiation.

Infrared Spectroscopy

Scientists have long used infrared absorption as a means of probing the structure of molecules. Studying the manner in which specific wavelengths of infrared energy excite vibration and rotation in molecules reveals information about the molecule that can be used to determine what and how many molecules are present. The *infrared spectrophotometer* is the principal instrument used by scientists for these measurements. Most laboratory spectrophotometers are of a dispersive design; i.e., a prism or grating is used to separate the spectral components in the source radiation. Modern infrared spectrophotometers have a wide wavelength range from 2 to 50 μm. They find use in research, quality control, and analytical service laboratories.

Infrared Process Analyzer

The infrared process analyzer has evolved from the laboratory spectrophotometer to satisfy the specific needs of industrial process control. While dispersive instruments continue to be used in some applications, the workhorse infrared analyzers in process control are predominantly nondispersive infrared (NDIR) analyzers. The NDIR analyzer can be used for either gas or liquid analysis. For simplicity, the following discussion addresses the NDIR gas analyzer, but it should be recognized that the same measurement principle applies to liquids. The use of infrared as a gas analysis technique is certainly aided by the fact that molecules which consist of two like elements, such as nitrogen (N_2) and oxygen (O_2), do not absorb in the infrared spectrum. Since nitrogen and oxygen are the primary constituents of air, it is frequently possible to use air as a zero gas.

Many different analyzer configurations have been developed to address the diverse needs of the industrial process control industry. The basic constituents of a NDIR analyzer are (1) a source of infrared radiation, (2) a means of restricting the wavelength range of the source radiation, (3) a means of detecting the infrared radiation, (4) a sample chamber to hold the gas or liquid to be measured, (5) a means of modulating the source radiation, and (6) electronics to process the signal generated by the source energy falling on the detector.

NDIR Analyzers Using Microphone Detectors

The majority of infrared process analyzers installed over the last 20 years utilize microphone detectors. These detectors are generally the Veingerov single-sided microphone system and the Luft balanced condenser microphone system, shown schematically in Figs. 1 and 2. (See Hill-Powell reference.) The microphone detector uses an absorbing gas as its detecting medium. When radiation reaches the detector that the sensitizing gas will absorb, the gas heats up and expands. This causes a diaphragm to distend. The diaphragm movement varies the condenser microphone capacity which is part of an electric circuit which generates an electrical output signal. Both analyzers use dual sources which are chopped to alternately allow energy to pass through a sample cell and a reference cell. When the sample cell contains a nonabsorbing zero gas, such as nitrogen, the modulated beams reaching the detector through the two paths are of equal amplitude. In the case of the Veingerov single-sided detector, the chopper is configured so that at any given time the sum of the cross-sectional areas of the two beams as seen by the detector equals the total cross-sectional area of a single beam so that, when no absorbing sample is present, a constant signal is produced and the output is zero. When a sample is present, the sample and reference path signals become unbalanced and a signal at the chopper frequency is developed. The amplitude of this signal is a function of the concentration of the gas present in the sample cell.

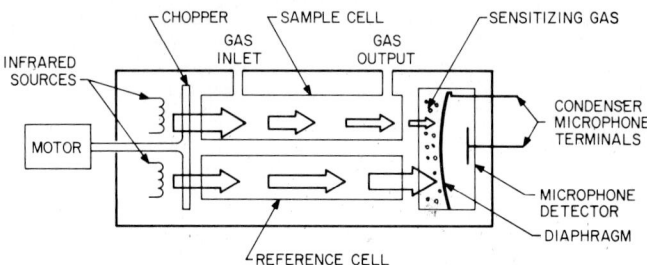

Fig. 1 NDIR analyzer with a Veingerov-type detector.

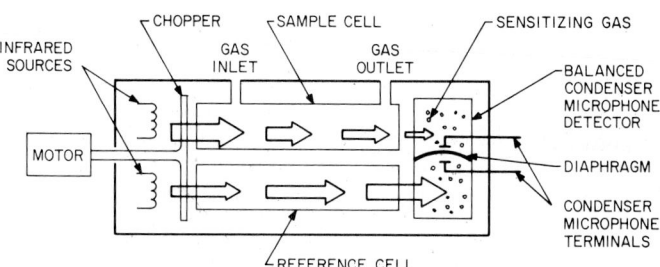

Fig. 2 NDIR analyzer with a Luft-type detector.

The Luft detector operates similarly but has two chambers separated by a diaphragm. The signal generated by the presence of an absorbing gas in the sample cell is at twice the chopping frequency. This arrangement has an advantage over the single-sided microphone detection system, since it is less susceptible to vibration caused by imbalance of the chopper motor. Having separate chambers does, however, allow for the possibility of a change occurring in one half of the detector and not in the other, thus resulting in zero drift. Recently, infrared process analyzers have been introduced which use a Luft-type detection system but have replaced the diaphragm with flow sensors. The flow of gas from one chamber to another is sensed by the flow sensor rather than by using a capacitance detection technique. This is claimed to eliminate one of the major causes of detector failure—failure of the thin diaphragm. For a given path length, process analyzers which use a microphone detector are more effective than those which use solid-state detectors and optical filters in measuring low concentrations of gases with a lot of structure in their absorption band. This structure results from the molecular rotation spectrum being superimposed on the vibration spectrum and is easily resolved in simpler molecules, such as carbon monoxide, methane, and ammonia (Fig. 4).

Disadvantages of the use of microphone detection in NDIR analysis include sensitivity to vibration and the high cost of detector replacement. A Luft-type detector typically costs over $600. Figures 1 and 2 show that two sources are required in these systems. To avoid zero drift, these sources must have a stable, balanced output. Also note that no lenses or mirrors are used to direct the energy from the sources to the detector. These systems rely on reflection of the infrared beams off the interior walls of the sample and reference cells. To achieve high reflectivity, these walls are highly polished and are frequently lined with gold foil. This makes the replacement of a sample cell expensive. Contamination of the cell walls by the sample also results in zero drift. Since a microphone detector uses a gas as its detecting agent, NDIR analyzers with this detection mechanism are generally available for only a limited number of gas species. Unstable or corrosive gases cannot be used to charge the cell, although it is possible to use an alternative gas with similar infrared absorption properties.

NDIR Analyzers Using Solid-State Detectors

A new generation of NDIR analyzers has evolved to satisfy the frequently harsh industrial environment. These analyzers utilize solid-state sensors for the detection of infrared radiation. The sensors most frequently used are lead selenide (PbSe), thermopiles, and pyroelectric detectors. The gas analyzers generally are configured as single-path, dual-beam with a reference path, or dual-channel with a reference filter.

The single-beam instrument finds application where low cost is important but stability requirements are not stringent. Generally, this is true when the measurement period is short and frequent rezeroing is practical. Changes in source intensity due to power variations or changes in detector sensitivity due to temperature fluctuations are reflected directly in the output as zero drift. To avoid this, a reference path is commonly used. The dual-beam configuration is shown in Fig. 3. The source

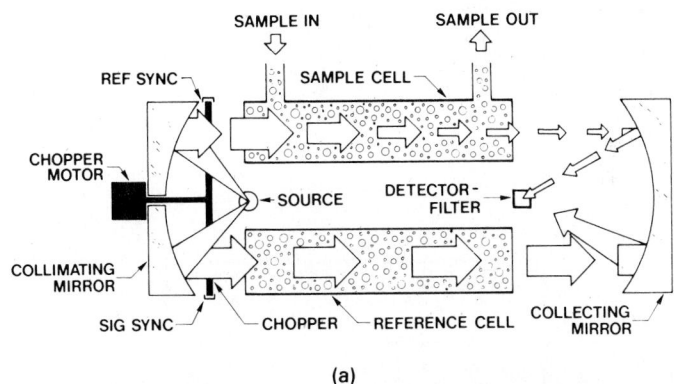

(a)

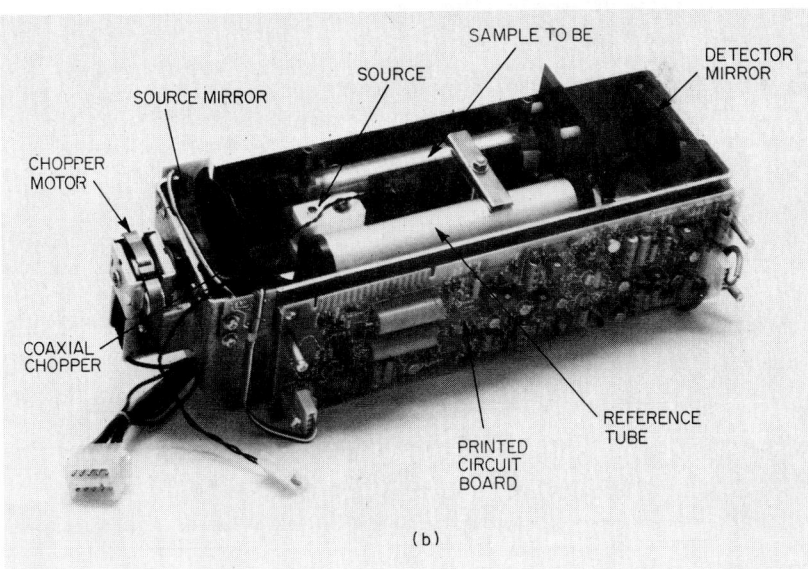

(b)

Fig. 3   Dual-beam NDIR analyzer with a solid-state detector. (*a*) Optical paths, (*b*) optical bench assembly. *(Infrared Industries.)*

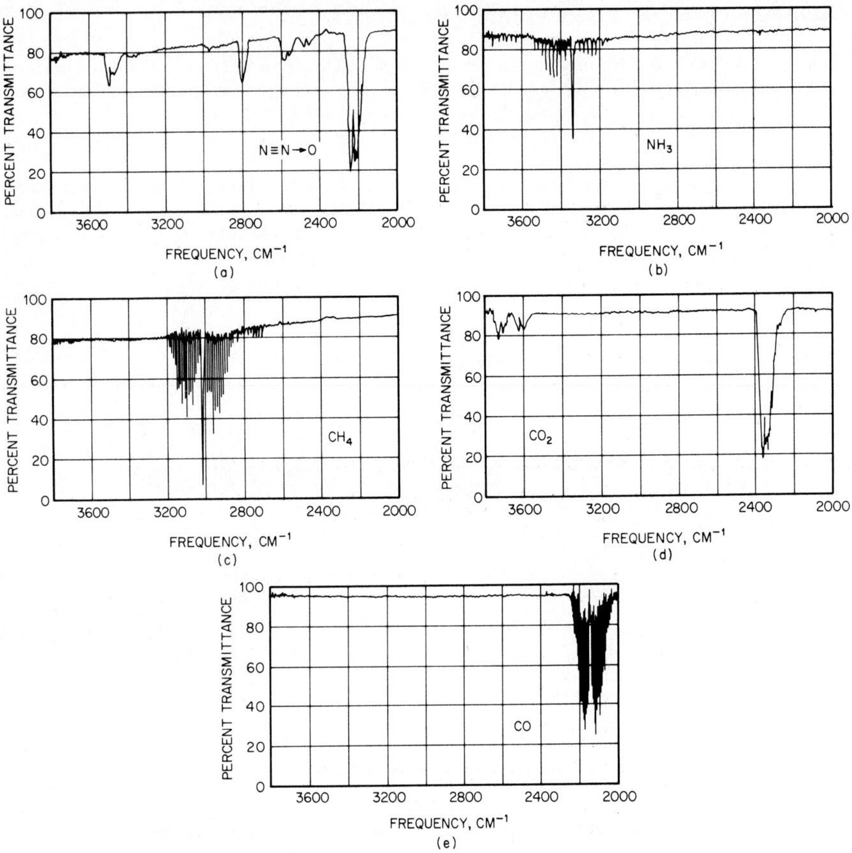

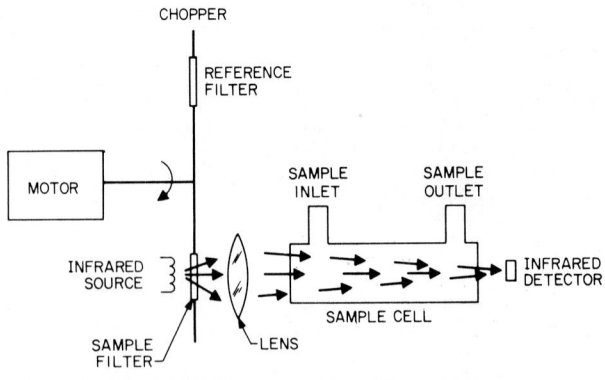

Fig. 4 Infrared spectra. (*a*) Nitrous oxide, (*b*) ammonia, (*c*) methane, (*d*) carbon dioxide, (*e*) carbon monoxide. (*Sadtler Research Laboratories.*)

Fig. 5 Dual-channel NDIR analyzer with a solid-state detector.

energy is modulated by a chopper blade, which allows the source to pass alternately through the reference and sample paths. The reference path is always free of absorbing gas, so the detector is exposed to the source through a path unaffected by the presence of the sample. This signal level is monitored by an automatic gain control (AGC) circuit which holds the reference signal level constant. If the source intensity or detector sensitivity should change, the AGC would correct for it in both the reference and sample channels. Sync pickups monitor the chopper position and inform the electronics when the sample or reference path is irradiated. A narrow bandpass optical filter is located in the front of the detector to limit the infrared energy and sensitize the analyzer to a particular gas absorption band. The signals generated by the two optical paths are synchronously demodulated. The amplitude of the reference signal is used by the AGC circuit. When an absorbing gas is introduced, the signal reaching the detector through the sample path is attenuated and the magnitude of the detected signal corresponds directly to the concentration of sample gas present in the sample cell. The use of a reference path results in stable operation with effectively reduced drift due to power line or temperature fluctuations. Synchronous detection is an effective means of providing very narrow bandpass filtering of the chopped signal, resulting in good noise rejection. The use of collimating optics eliminates the reliance on internal reflection in the sample and reference cells to transmit the source energy to the detector. The sample cell construction is thus simplified, and the cell is easier to clean, less susceptible to damage, and less costly to replace.

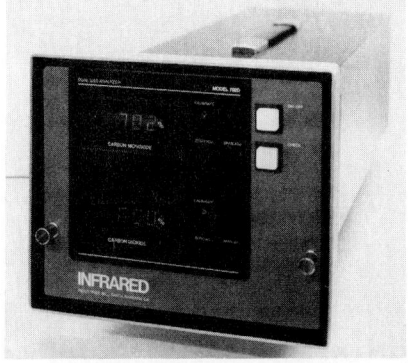

A reference optical filter can be used as an alternative to the reference path. This requires that a spectral window exist where there is no interference from the sample. The 3.8 to 3.9-μm spectral region is frequently used in NDIR gas analyzers for this purpose. The spectral curves of Fig. 4 show that this spectral window would work well as a reference for measuring carbon monoxide, carbon dioxide, methane, or ammonia. It is unsuited for nitrous oxide.

Fig. 6 Two-gas NDIR process analyzer. *(Infrared Industries.)*

The dual-channel (reference filter) configuration is outlined in Fig. 5. The reference filter and the sample filter, which define the spectral region of interest, are mounted on a spinning chopper wheel. As the chopper spins, it alternately positions the filters in the optical path. The signal is demodulated in a manner similar to that employed in the dual-beam approach. In order for the reference filter to be effective, it is important that the performance of the source and detector be similar in the sample and reference spectral regions. Unfortunately, this is not normally the case. The spectral properties of the source and detector are both strong functions of temperature. They must be controlled precisely, or zero and span drift will result. Similarly, the optical interference filters used to isolate the spectral region of interest will shift with temperature. These too, should be temperature-controlled. An advantage of this approach is that drift due to accumulation of dirt or film on the windows is avoided, since this is generally nonspectral in nature and corrected by the AGC electronics. The analyzer is effectively zeroed at each revolution of the chopper wheel.

Another attribute of both dual-beam and dual-channel (reference filter) process analyzers is that they can be configured for measuring more than one gas. Figure 6 shows an NDIR gas analyzer for two gases.

Properties and Features

The infrared process analyzer is frequently used on-line as a continuous monitor of process stream parameters. A sample conditioning system is generally required to remove particulates and condensable vapors from the sample and transport the sample to and away from the analyzer. The process analyzer should have corrosion-resistant materials in contact with the stream. Typical materials of construction are stainless steel and Teflon. Special materials are available for exceptional requirements.

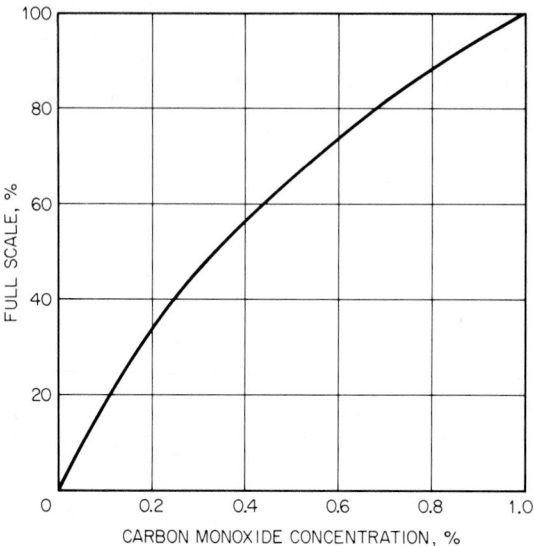

Fig. 7 Typical nonlinear gas calibration curve.

To accommodate the different absorption properties of the various gases and liquids in the infrared, the sample cell length can be selected. The magnitude of absorption is a function of the concentration and the path length. A high concentration generally requires a shorter path length than a low concentration. Certain gases are strong absorbers (CO_2, N_2O), but others (CO, SO_2) are relatively weak absorbers in the infrared.

The change in signal as a function of concentration is usually nonlinear. A typical calibration curve is shown in Fig. 7. The sample cell path length is selected for a given gas in order to minimize the nonlinearity while providing an adequate degree of absorption. Some process analyzer manufacturers provide instruments with nonlinear outputs and a calibration chart. Others provide internal linearizers as standard equipment or as options. An important consideration when specifying a process analyzer is what gases other than the gas to be monitored are found in the process stream. Two gases that have overlapping absorption bands cannot be discriminated. However, it is possible that a second absorption band exists that is unique for the gas of interest or that the overlap is only partial and some discrimination can be achieved. Process analyzer manufacturers can use negative gas filters or special optical filters to address these cross-talk problems only if complete applications data are provided by the process engineer.

Internal span calibration is available from many manufacturers. This feature allows the span of an instrument to be checked without using calibration gas. When activated, an optical attenuator is introduced into the sample path. This is done when zero gas is present in the sample cell. A fixed-amplitude signal results from the attenuation and is used to check the gain of the span electronics. An alternate approach does not involve the optical system but just introduces a fixed electronic signal into the span circuit to check the gain.

Typical specifications for an NDIR process analyzer are given in Table 1.

Applications

The availability of rugged NDIR process analyzers at moderate cost has resulted in their wide acceptance as process control tools. Table 2 lists typical gases and concentration ranges generally available from NDIR analyzer suppliers. Literally hundreds of different gases can be monitored with infrared process analyzers. The petrochemical industry utilizes infrared process analyzers to monitor CO_2 in

Table 1 Typical Specifications for an NDIR Process Analyzer

Repeatability ±1% of full-scale *Zero drift* ±1% of full-scale per 24 h *Span drift* ±1% of full-scale per 24 h *Linearity* ±0.5% of full-scale fitted to theoretical curve *Noise level* Less than 1% of full-scale	*Warm-up* 15 min *Internal span check* Electrically activated *Temperature range* 0–50°C (32–122°F) *Ambient humidity* Up to 95% relative humidity *Power requirements* 115 V ac ± 10%, 50/60 Hz

Table 2 Typical Gases and Concentrations Measured by NDIR Analyzers

	Range (full-scale concentration)	
Gas	Minimum, ppm	Maximum, %
Ammonia	1000	100
Butane	300	100
Carbon monoxide	500	100
Carbon dioxide	200	100
Ethylene	1000	100
Ethane	500	100
Ethylene oxide	1500	100
Hexane	400	5
Methane	400	100
Nitrous oxide	150	100
Propane	400	100
Sulfur dioxide	400	100
Water vapor	1500	10

the manufacture of ethlyene oxide and ammonia. Acetylene is monitored during the manufacturing of acetylene and vinyl chloride. The metals industry uses infrared analyzers to monitor CO_2 in steel converting, in soaking pits, and in the heat-treat process. Carbon monoxide is monitored during heat-treating, aluminium powder processing, and tin plant annealing. The food industry monitors CO_2 in greenhouses, in storage facilities, and during fermentation. The analyzers find use in monitoring for explosive and toxic hazards. Infrared analyzers are used to monitor stack gases for pollution control and are becoming increasingly important in flue gas monitoring for combustion efficiency.

REFERENCES

Hill, D. W., and T. Powell: *Non-Dispersive Infrared Gas Analysis in Science, Medicine, and Industry,* Plenum Press, New York, 1968.

Perkins, W. D.: "Infrared Spectroscopy," in *Process Instruments and Controls Handbook* (D. M. Considine, ed.), McGraw-Hill, New York, 1974.

Sadtler: *The Sadtler Standard Spectra: Gases and Vapors,* Sadtler Research Laboratories, Inc., Philadelphia, 1972.

Photometric and Reaction Product Analyzers

Determination of chemical composition by measurement of a reaction product is the basis for most classical methods of qualitative and quantitative chemical analysis. Several instrumental methods have been used for this purpose.

Basically, the technique involves two steps: (1) promotion of a desired chemical reaction, and (2) measurement of a reaction product as a means for determining the presence and quantity of a particular constituent in the sample.

In general, determination of a constituent in a sample or process stream by measurement of a reaction product may be represented by

$$X + R \rightarrow M$$

where X is the constituent to be determined,
R is the reactant, and
M is the reaction product to be measured.

The reactant R may be present in the sample, in which case it is necessary only to expose the sample to reaction conditions suitable to produce M. Otherwise it will be required to add the reactant to the sample, or vice versa. Frequently, the reaction of X and R is spontaneous under normal instrument operating conditions; sometimes, however, appropriate conditions must be established either to promote the desired reaction or to ensure a suitable reaction rate.

In many instances it is possible to measure the reaction product as it forms in the reaction zone. Occasionally the products and sample residue must be removed from the zone before a measurement can be made. Finally, the reaction product may be measured directly, or its presence may be inferred from a secondary reaction. For example, CO (X) in air or O_2 (R) may be determined by combustion to CO_2 (M), which in turn may be measured *directly* by thermal conductivity, or *inferentially* by absorbing it in a solution such as barium hydroxide and measuring the change in electrolytic conductance.

Instrumental Equipment Used

Many elements of the instrumentation used are common to other techniques. These include (1) fluorometers, (2) luminescence photometers, (3) colorimetric analyzers, (4) nephelometers, (5) turbidimeters, (6) titrators, (7) amperometers, and (8) reagent-tape analyzers.

Fluorometers

In fluorescence analysis, the amount of light emitted characteristically under suitable excitation is used as a measure of the concentration of the responsible material under observation. Thus, the method is closely related to colorimetric or spectrophotometric analysis, in which the amount of light absorbed characteristically is used to measure the concentration of the dissolved species.

The main advantage of fluorescence methods is their high sensitivity, about 1 part in 10^8, in many determinations of both inorganic and organic substances. This is two or three orders of magnitude greater than the sensitivity of absorption methods, where sensitivity is limited by the necessity for detecting a very small fractional decrease in the light transmitted by the solutions.

In fluorescence, the situation is inherently more favorable. Inasmuch as zero concentration corresponds to darkness (neglecting reagent blanks) and sensitivity depends on detecting the first faint emission of light as the concentration is increased, advantage can be taken of highly sensitive detectors, such as photomultipliers, and high-intensity ultraviolet sources of excitation. Combining these with sophisticated electronic and optical techniques has led to a remarkable achievement—under favorable conditions it is possible to detect rhodamine 5DGN down to the extremely low concentration of 1 part in 10^{12} using an instrument designed for tracing ocean currents with a fluorescent dye marker.

The use of fluorescent methods requires that the substance to be determined is fluorescent under suitable irradiation, or can be made so by a chemical reaction. Among organic substances, fluorescence is shown mainly by aromatic compounds, including such hydrocarbons as benzene, naphthalene, anthracene, and their derivatives, rather than the aliphatic series. Among the metal ions, only a few show intrinsic fluorescence, such as uranium and thallium, but many others can be determined fluorometrically by adding a specific reagent which reacts with the metal to form a fluorescent complex. Thus, aluminum is complexed with the dye Pontachrome BBR, beryllium with morin, and zirconium with flavenol.

Various sources are used for exciting fluorescence, including proprietary lamps, mercury vapor lamps, and xenon arc lamps. See Fig. 1. A tungsten lamp may be used for substances having a strong excitation band above 450 nm. Both the desired excitation band and emission band may be isolated

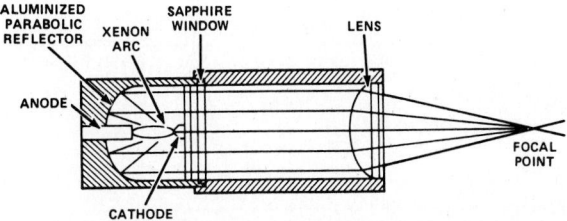

Fig. 1 Optical diagram of a xenon lamp. The excitation source is a 250-W xenon arc lamp with an integral parabolic reflector. The arc output is collimated by the reflector and then focused by a lens. The area between the lamp and lens is an ozone trap and filter which reduces ozone producing wavelengths from the arc. No venting hoses or deozonation is needed. The lamp is of ceramic and metal construction, which permits finned heat sinks to be bolted directly to it. *(SLM Instruments, American Instrument.)*

by means of interference filters. However, since the desired excitation band is usually in the ultraviolet region, tungsten lamps are not in general use, but they may have specific applications. These lamps give a band spectrum and do not have the sharp line limitation of mercury vapor lamps. The location of mercury lamp spectral lines is shown in Fig. 2.

If the emission is in the visible spectrum, it may be estimated by visual comparison with standards. In any range of the spectrum, the intensity of the emission may be measured with a phototube, a barrier layer cell, or a photographic plate and densitometer. By far, the most common procedure is to use a phototube or an electron multiplier phototube attached to a microphotometer or a recorder.

A fluorometer constructed with two monochromators is called a *spectrofluorometer*. With a spectrofluorometer, two types of information can be obtained easily—the wavelength of best excitation and the wavelength of strongest emission. Two curves are generally plotted on the recorder for each fluorescing material—an excitation curve and an emission curve. The excitation spectrum is a plot of the wavelength of the exciting source against the intensity of the emission. The excitation wavelength producing the greatest intensity of emission seems to be the best exciting wavelength. However, this statement is true only for the particular light source and grating used. See Fig. 3.

A diagram of a specific spectrofluorometer is shown in Fig. 4. The optical system uses lenses $L1$ and $L2$ and mirrors $M1$ and $M2$ to focus an image of the light source on entrance slit 1 of the excitation monochromator. The excitation monochromator selects light of the monochromatic wavelength from the light source and focuses it on the sample cell. Light focused on entrance slit 1

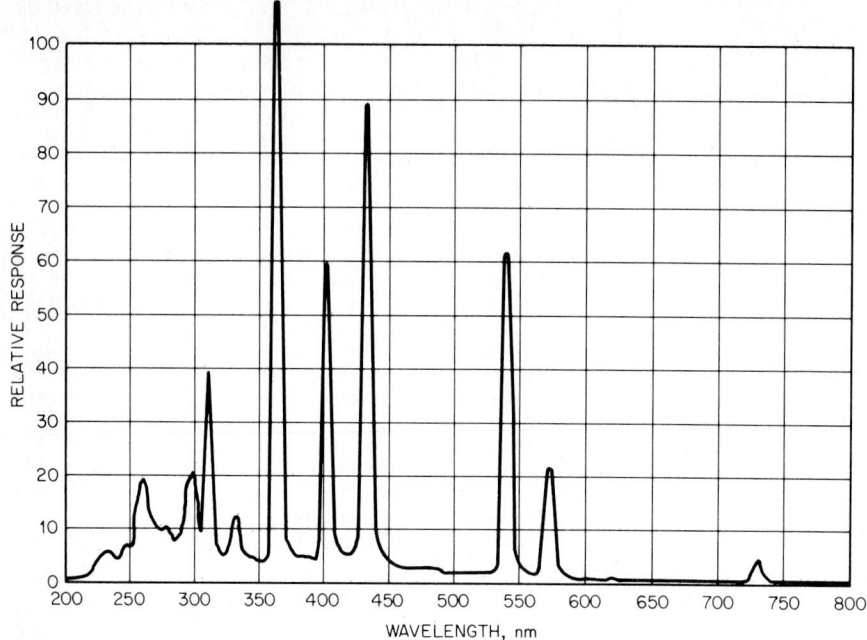

Fig. 2 Locations of a mercury lamp spectral lines.

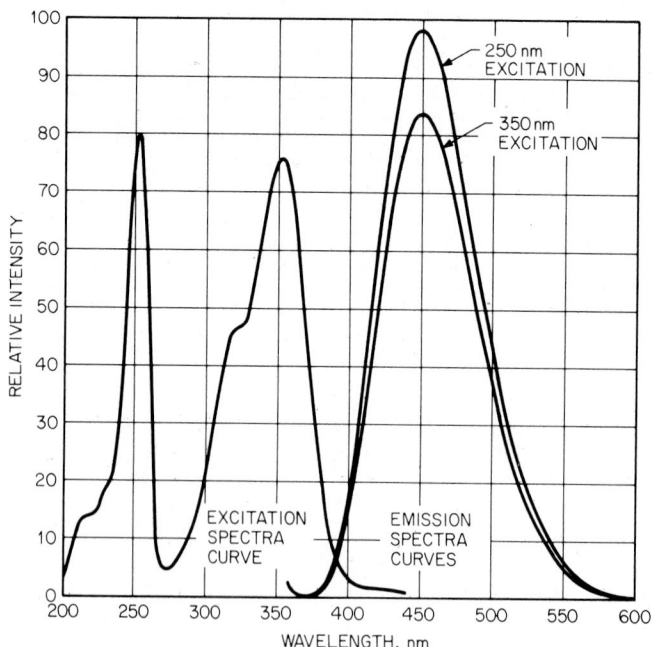

Fig. 3 Plot of excitation and emission spectra curves made by a spectrophotofluorome-
ter. *(Aminco-Bowman.)*

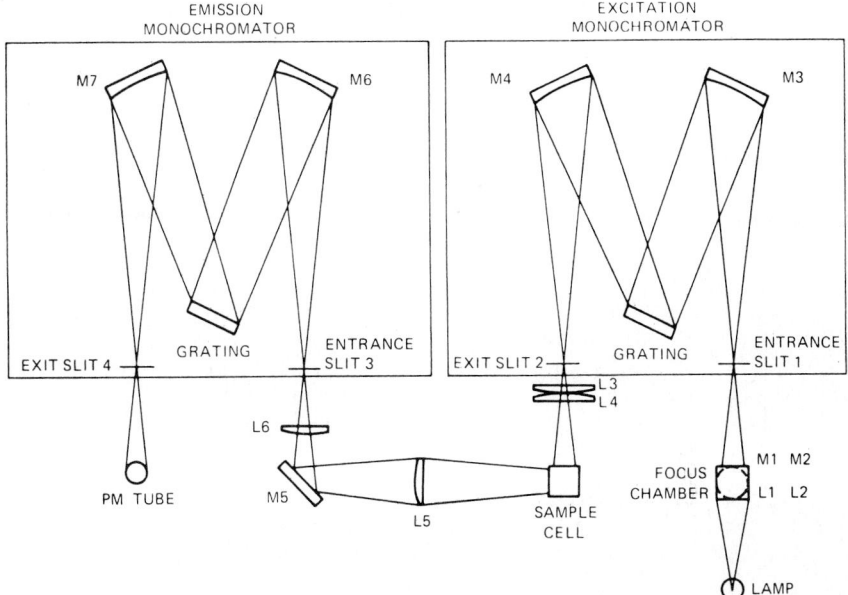

Fig. 4 Optical system diagram of a spectrofluorometer. L, Lens; M, mirror. *(Model SPF-125, SLM Instruments, American Instrument.)*

impinges on mirror *M*3. This mirror changes this diverging light into parallel rays which are reflected to the plane grating. This grating is blazed to produce maximum intensity at 300 nm (first order). Monochromatic light from the grating is directed to mirror M4 which focuses a monochromatic image of the lamp on exit slit 2. Lenses $L3$ and $L4$ then focus a reduced image in the center of the sample cell. The spectral bandwidth of the excitation monochromator is determined by the entrance and exit slit width. These slits are mechanically coupled to maintain equal widths. The spectral bandpass of the excitation monochromator can be incrementally varied from 1.5 to 44 nm (1 mm slit width equals 11 nm bandpass).

The excitation energy is absorbed by the sample solution and emitted as energy of a longer wavelength, i.e., reemitted as fluorescence. The fluorescent light emitted is focused by lenses $L5$ and $L6$ on entrance slit 3 of the emission monochromator. The operation of the emission monochromator is the same as that of the excitation monochromator except that the emission monochromator's plane grating is blazed at 500 nm. The monochromatic fluorescent light is applied via exit slit 4 to the photomultiplier tube. Light energy at any wavelength other than that set on the emission monochromator is blocked by that monochromator.

The photomultiplier tube converts the incident light into a corresponding electrical signal which is applied to the photometer's measuring circuit. The photometer essentially provides noise-free, stable amplification of the photomultiplier's output signal and a stable high voltage to operate the photomultiplier tube.

The high-voltage output of the photometer is applied to a series of dynodes in the photomultiplier tube. When light impinges on the photomultiplier tube's cathode, electrons are emitted which are accelerated to the first dynode by the potential difference between the dynode and the cathode. Each electron striking the dynode liberates from two to five secondary electrons. These secondary electrons are accelerated to the second dynode, and the process is repeated until the anode is reached. The resulting cathode current amplification is on the order of several hundred thousand.

The configuration of another type of fluorometer is shown in Fig. 5.

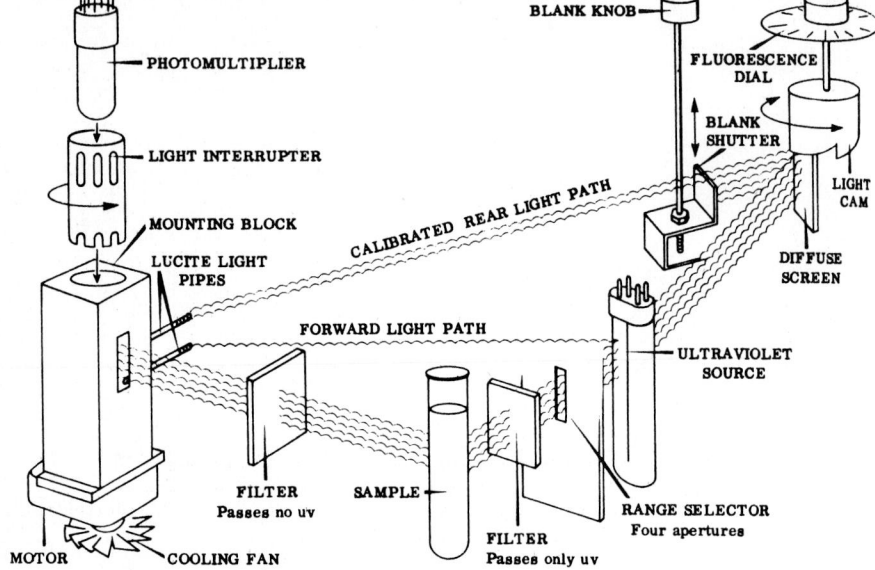

Fig. 5 Schematic of a Turner fluorometer. Primary filters pass uv light to the sample for excitation. Secondary filters transmit visible light emitted by a sample for measurement. The intensity of the emitted light is proportional to the sample concentration even when exciting and measured light are not at optimal wavelength. Filter selection, therefore, is not as critical as in colorimetric procedures. Latitude in the choice of wavelengths permits considerable selectivity, so that the extraneous fluorescence of a reagent blank can be minimized without sacrificing sensitivity.

Fluorescence Linearity Curve

A linearity curve indicates the accuracy with which analytical standard solutions have been prepared and whether the fluorescence can be treated simply. The fluorescence emitted by the sample is defined by

$$F = k_1 P_0 (1 - e - abc)\Phi$$

where F = observed fluorescence intensity
k_1 = constant determined by instrumental parameters
P_0 = intensity of incident energy
a = molar absorptivity (absorbance divided by product of molar concentration c of substance and cell path length)
b = path length
c = concentration (molar) of fluorescent solute (fluorophor)
Φ = quantum efficiency (ratio of quanta emitted to quanta absorbed)

However, when the absorption of incident energy by the sample is less than 0.05 absorbance, the equation simplifies to

$$F = k_1 P_0 \Phi abc$$

Under this assumption, the fluorescence of dilute or low absorbing sample solutions is directly proportional to the concentration of the sample solution. Changes in the parameters in this equation are directly related to the reproducibility of the results obtained with the instrument and also to the chemical nature of the solutions.

Ratio Spectrofluorometer

This type of instrument is used to obtain fluorescence spectra and percent transmittance measurements of samples for chemical assays. Generally, fluorescence spectra cannot be used for positive identification of compounds because the width of the fluorescence bands renders them comparatively nonspecific. However, the measurement of fluorescence spectra can provide identification when materials are similar in both bioassay and spectrophotometry but differ markedly in fluorescence properties. Quantitative assays can be performed using either the linearity curve method or the proportionality method.

With reference to Fig. 6, light from a xenon lamp is dispersed by a grating in the excitation monochromator into monochromatic radiation incident on the sample. The fluorescent light produced by the sample is dispersed by a similar grating in the emission monochromator. Both the excitation and emission radiation are monitored by separate photomultiplier tubes. These tubes convert the radiation into reference and fluorescence electrical signals. When the instrument is in the *ratio mode*, these signals are amplified and applied to an electronic divider which performs the following operation to stabilize the signal:

$$\frac{E_{\text{fluorescence}}}{E_{\text{reference}}} = E_{\text{ratio-compensated}}$$

If the excitation energy to the sample changes, its fluorescent output will change. But since the excitation energy is monitored by the reference photomultiplier tube, both the numerator and denominator of the electronic divider will change by equal percentages, leaving the output constant. The

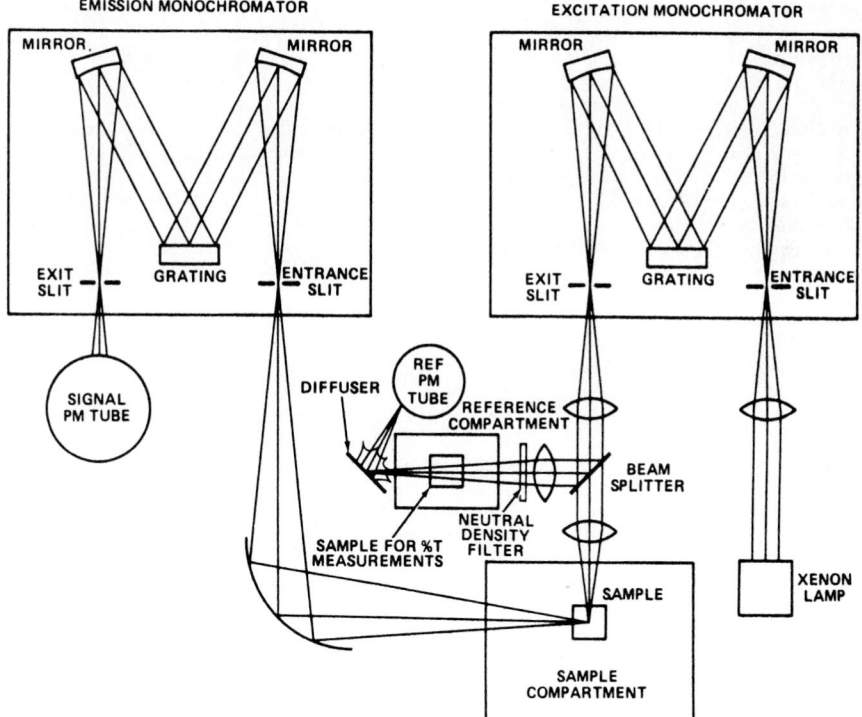

Fig. 6 Simplified optical diagram of a ratio spectrofluorometer. *(Model SPF-500, SLM Instruments, American Instrument.)*

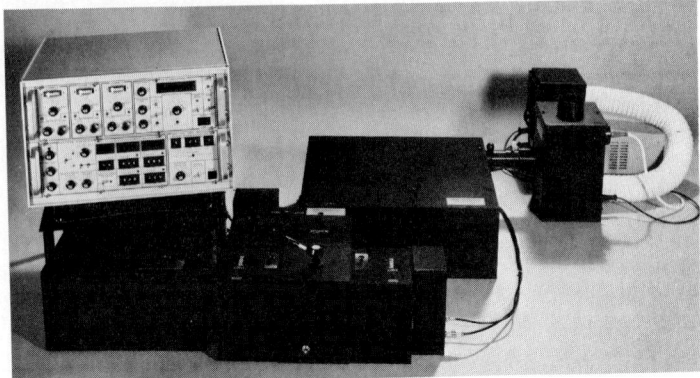

Fig. 7 Photon counting spectrofluorometer with microprocessor-based electronics. Two data acquisition channels (a third is optional) are included in the total package, each with its own photomultiplier tube, power supply, logarithmic photon rate meter, and switch-selectable analog or photon counting modes. *(Model SLM 8000, SLM Instruments, American Instrument.)*

ratio-compensated signal is displayed as relative fluorescence on a digital readout meter. A wavelength-versus-relative-fluorescence graph spectrum can be obtained by applying the signal to an XY recorder. An excitation spectra is obtained by scanning the excitation wavelength while maintaining a constant emission wavelength. An emission spectrum is run by scanning the emission wavelengths while maintaining a constant excitation wavelength.

Modern spectrofluorometers are available with microprocessors and calculator accessories to facilitate the manipulation and interpretation of measurement data. See Fig. 7. Among the many features are expansion of scale to increase detectability; curve smoothing; a differential spectra function which calculates the difference between two spectra and automatically adjusts the display to plot both positive and negative differences, thus eliminating background signal and scatter interference from the resulting spectra; first-, second-, third-, and higher order derivatives of spectra can be calculated, permitting more precise measurements when complex compounds produce spectra with overlapping bands.

Luminescence Photometers

A relatively new but rapidly growing field of analysis involves the measurement of bioluminescence and chemiluminescence as a qualitative and quantitative indication of substances which exhibit these phenomena. In bioluminescence, for example, photometers can measure adenosine triphosphate (ATP) in such applications as bacteria counting, assay for cyclic adenosine monophosphate (AMP), assay for flavin mononucleotide, *Chlorella* determinations, evaluation of the viability of stored blood and cell counts, biomass determination in sewage, and evaluation of the effects of microbial cells in the environment. In chemiluminescence, for example, hydrogen peroxide determinations, which are based on copper-catalyzed chemiluminescent reactions between luminol and hydrogen peroxide, can be made. The same technique has proven useful for the determination of bacterial levels.

Luciferase in the presence of luciferin, oxygen, magnesium, and ATP catalyzes the formation of adenyl-luciferin. Oxidation of adenyl-luciferin in the presence of atmospheric oxygen produces adenyl-oxyluciferin and *light*.

$$\text{ATP} + \text{luciferin} \underset{\text{luciferase}}{\overset{\text{Mg}^{2+}}{\rightleftharpoons}} \text{adenyl-luciferin} + \text{pyrophosphate}$$

$$\text{Adenyl-luciferin} \underset{\text{O}_2}{\rightarrow} \text{adenyl-oxyluciferin} + \text{water} + \text{light}$$

Consequently, when a sample containing ATP is added to a suitably buffered solution of luciferase and luciferin, the flash of light that occurs is directly proportional to the quantity of ATP present. Since the total population of microorganisms of a given size class, e.g., bacteria, in a sample is proportional to the total quantity of ATP present, the intensity of the light flash when measured by a luminescence meter can be calibrated in terms of the total number of living organisms present (biomass). From a process control standpoint, this instrumentation can be used to optimize sewage and sludge feed rates in an activated sludge treatment plant, as well as to measure the effects of toxic agents introduced into the process.

One type of luminescence photometer is shown in Fig. 8. It consists of a reaction chamber mounted on top of a photomultiplier microphotometer. The reaction chamber contains a photomultiplier tube surrounded by a rotating head which, when moved by a lever to the loading position, permits access to a cell holder cavity for insertion of a sample cuvet. When the head is rotated to the operating position, the sample cuvet is positioned in front of the photomultiplier tube and directly under the injection port. The injection port is fitted with a knurled nut which holds a rubber septum and serves as a guide to direct the injection needle into the center of the sample cuvet. The cell holder cavity has a curved mirror lining to reflect emitted light from the luminescent reaction to the photomultiplier tube. The tube converts the light to corresponding electrical signals.

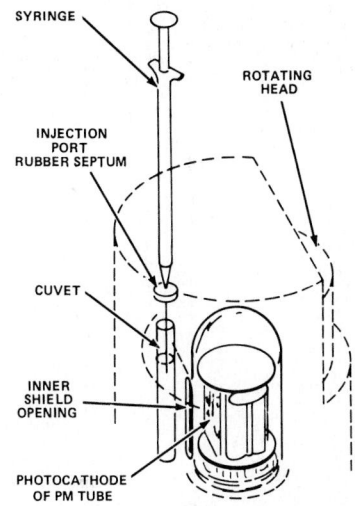

The microphotometer amplifies the electrical signals from the photomultiplier tube and indicates the relative intensity of the signal on a meter. The microphotometer also supplies the high voltage needed for the photomultiplier tube. Outlets are usually furnished for a 0 to 50 mV recorder or oscilloscope.

The assay of ATP with extract from the luminous organ of the firefly was first used in the early 1970s. This extract is supplied in a desiccated form in vials and when reconstituted with distilled water provides a suspension with a pH value of 7.4. Before using the reconstituted firefly extract for the assay, a diluent containing

Fig. 8 Reaction chamber of a luminescence photometer showing the rotating head in an operating position. *(Chem-Glow Photometer, SLM Instruments, American Instrument.)*

equal volumes of sodium arsenate and magnesium chloride is required. The buffer used is Trisborate. In use, the operator prepares standard solutions and a standard curve plotting relative intensity versus concentration of ATP.

Colorimetric Analyzers

Color measurement in analysis is used in basically two ways: (1) as a direct measure of the concentration of one substance, or a combination of substances, whose color is indicative of product purity in terms of desired product characteristics—as exemplified by the continuous on-stream color monitoring of various organic and inorganic gases and liquids, lube oils, kerosene, and other petroleum products, rosins, waxes, corn syrup—or of impurities of certain substances in the environment (water and air) whose color (absorption of light and possibly ultraviolet radiation) is a reliable indication of concentration and composition; and (2) as a measure of the concentration of a reaction product, i.e., the end point color of a reaction. In this latter use, the color indicates when sufficient reagent has been added to a measured volume of a sample to complete a chemical reaction, and thus the amount of reagent required indicates the concentration of substance in the sample, a procedure sometimes referred to as *colorimetric titration*. In execution colorimetric analysis ranges from manual and visual comparison of sample and reference colors to continuous on-stream measurements.

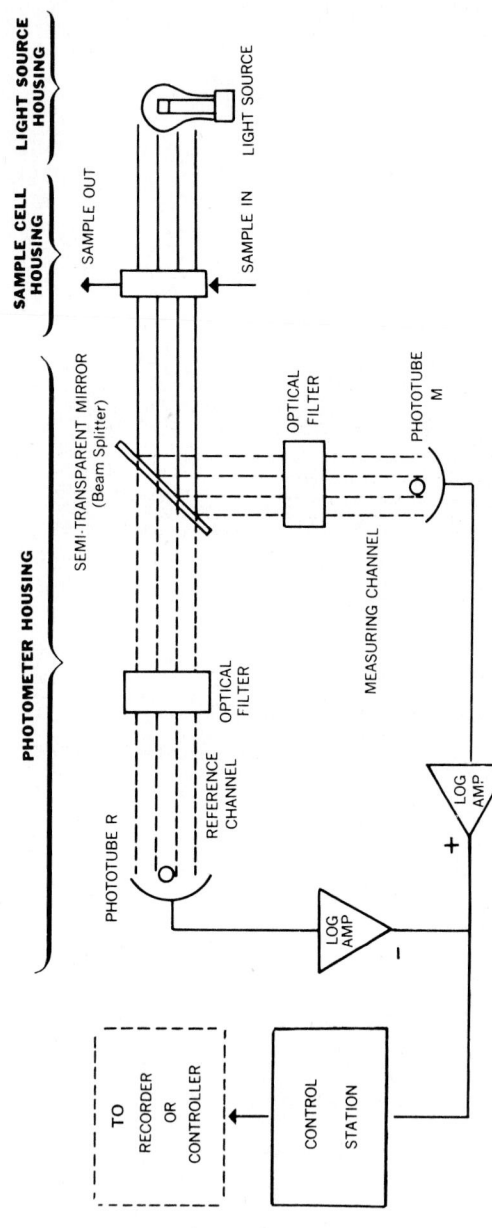

Fig. 9 Functional diagram of a photometric analyzer. (*Du Pont 400.*)

Colorimetry Fundamentals

The fundamental principle of colorimetry states that the amount of light absorbed by a given substance in solution is proportional to the intensity of incident light and to the concentration of the absorbing species. This is expressed mathematically in the Lambert-Beer law:

$$\log I_0/I = abc$$

where I_0 = intensity of incident light
I = intensity of transmitted light
a = absorptivity of substance
b = light path length
c = concentration of colored substance
I/I_0 = transmittance
$\log I_0/I$ = absorbence

Visual Methods

Visual colorimetry is a simple method and is fairly precise. Essentially it requires matching of the color of a standard solution with that of an unknown sample so that, when they become identical, they contain the same amount of colored substance in columns of equal cross section. At this point

$$C_x b_x = C_s b_s$$

$$C_x = \frac{C_s b_s}{b_x}$$

where C_x = concentration of unknown solution
b_x = length of light path of unknown solution
C_s = concentration of standard solution
b_s = length of light path of standard solution

A standard series of solutions may be prepared, each with a known concentration, having the same volume as the unknown and each contained in identical flat-bottomed tubes of equal diameter (Nessler tubes). The solutions should be compared in daylight and examined against a white background.

A more refined method uses a Duboscq colorimeter. This instrument features a dual-matched optical system. Uniformly intense light is incident on both colorimeter tubes, and the difference in absorption of the standard and unknown solutions is compensated for by adjusting the thickness of a solution through which light passes. When the two colors match, $C_x = C_s b_s/b_x$.

Visual colorimetric methods are also used in connection with the determination of pH. See the article on pH and pH measuring systems in this Handbook section.

Continuous Monitoring of Product Color

Complete color definition requires multidimensional analysis, which in a liquid process stream is complex and costly. There are, however, simpler, one-dimensional color standards or scales which have been developed and accepted for grading liquid products in most process industries.

Frequently, a split beam analyzer is used to measure the difference in absorption of radiation by the sample at two wavelengths in the ultraviolet-visible region. In certain applications (e.g., chlorine purity), the analyzer can be supplied for dual-beam operation to measure the absorption of light at a single wavelength.

In one type of split beam configuration (Du Pont patent), radiation from a selected light source passes through the sample cell and into the photometer where it is split by a semitransparent mirror into two beams. See Fig. 9. One beam is directed to the measuring wavelength phototube through an optical filter which excludes all wavelengths except the measuring wavelength (the wavelength absorbed strongly by the sample constituent under analysis). The other beam falls on the reference wavelength phototube after passing through an optical filter which transmits only the reference wavelength. Light at the reference wavelength is absorbed weakly (compared to light at the measuring wavelength) by the constituent being analyzed.

The phototubes convert both light signals to electric current signals proportional to the light intensity of each beam. The phototube signals provide inputs to the amplifier system, which compares the logarithm of each beam intensity signal to provide a single photometer output signal (in accordance with Beer's law) that is linearly proportional to the concentration or thickness of the sample. Changes in light intensity, or foreign matter in the sample cell, have a minimum effect on accuracy of the analysis. Any such variation affects both beams in a similar manner and is automatically corrected. Variations in the concentration of the measured constituent primarily affect the measuring beam light intensity.

For gas and liquid analyses, the sample passes through a cell in the optical beam. The sample chamber is isolated from the two housings containing the optical and electronic components. This permits analyses to be made at controlled temperatures.

In addition to using the photometric analyzer for color measurement, the thickness of light absorbing coatings on solid transparent or translucent film materials also can be measured by locating the photometer and lamp housings on opposite sides of the moving film, completely omitting the cell and cell housing.

Also, in connection with certain materials, such as sugars and some antibiotics, the photometric analyzer can be used for optical rotation measurements. These materials have the ability to rotate the plane of a beam of polarized light passing through the sample to a degree proportional to the concentration of the component. When the instrument is so used, it is generally referred to as a *polarimeter*.

Among the practical applications for this type of photometric analyzer are those listed in Table 1.

ASTM Color Scale

One of the most commonly used one-dimensional color scales is that described in ASTM* method D-1500. This method is used primarily for grading the color of lubricating oils in the petroleum refining industry. It is of value because it is often used as the primary quality measurement in the purchase and sale of partially refined oils. Crude oil, before it is refined, is nearly opaque black-brown in color. As the oil is refined, there is a good correlation between the extent of refining and the "lightness" of the oil. The color changes from a very dark red-brownish black, to brownish-orange, orange, yellow, and eventually water white.

The ASTM color scale takes into account broad color changes resulting from changes in absorbance over a wide band of wavelengths, as indicated in Fig. 10. In determining ASTM color, the measuring channel of the instrument shown in Fig. 9 responds to a similarly broad band of wavelengths and, in effect, approximates the human eye or a standard observer. The dotted curve in Fig. 10 shows the approximate phototube-filter combination response of the measuring channel in an analyzer which provides a linear output when monitoring a product ranging from 0–6 ASTM color. The reference channel response is at a wavelength of 750 nm where the absorbance change is small compared to that of the measuring channel. In effect, the analyzer measures the amount of light absorbed by the sample, which is related to the area between the filter cutoff and the ASTM curve of interest. It compares this absorbance to the light absorbed at 750 nm, which remains relatively constant. For example, in measuring an ASTM 2 number, the analyzer compares the light absorbed at the measuring channel (indicated by the hatched area in Fig. 10) to the light absorbed at the reference channel, 750 nm.

Saybolt Color Scale

This is another widely accepted one-dimensional standard. It is used primarily within the petroleum and pharmaceutical industries to grade the yellowness of pale products. Color grading with the Saybolt color standard is described under ASTM procedure D156, "Test for Color of Petroleum Products." It is based on matching the color of a column of liquid sample with the Saybolt standards, which are color-controlled glass disks. The depth of the liquid column is adjusted to correspond in

*American Society for Testing and Materials, Philadelphia, Pa.

Table 1 Representative Applications for Photometric Analyzers

Gas and liquid compositions	Range*
Acetaldehyde in water	0–100% w/w
Acetone	0–2% v/v
Benzene in air	0–4% v/v
Benzoic acid in water	0–7500 ppm w/w
Bisphenol A in caustic	0–2000 ppm w/w
Bromine	0–150 ppm v/v
Carbon bisulfide	0–100% v/v
Chlorine in HCl	0–0.2% w/w
Chlorine dioxide	0–15% v/v
Chromium^{6+} (etching baths)	0–10 g/L
Ferrous ion (pickling baths)	0–20 g/100 ml
Fluorine in uranium hexafluoride	0–1% v/v
Hydrogen peroxide	0–1% w/w
Hydrogen sulfide	0–1000 ppm v/v
Mercury in air	0–400 ppb v/v
Naphthalene in water	0–10 ppm w/w
Nitric acid	40–60% w/w
Nitrogen dioxide	0–1500 ppm in air v/v
Phenols in wastewater	0–100 ppb to 0–10 ppm w/w
Sulfur dioxide	0–500 ppm v/v
Toluene	0–1% v/v
Xylenes and other aromatics	0–1% v/v

Color Analyses	Optical Rotation (polarimetry)
Purified organics	Amino acids
Wastewater	Antibiotics
Corn syrup (yellowness)	Dextrose equivalent
Films (blueness)	Penicillin
Kerosene	Plant nutrients
Lube oils	Starches
Rosin	Sugars
Wax	Vitamins
Tea	

Opacity, Turbidity, Haze	Thickness (films and coatings)
Aluminum particles in mineral spirits	Nitrocellulose film coatings
Film haze	Polyethylene terephthalate (polyester film)
Solids in certain liquors	
Turbidity in aqueous systems	Resin coating on cellophane

*v/v = Volume on volume; w/w = weight on weight.
Source: Du Pont Analytical Instruments Division.

"darkness" to one of the three Saybolt color standards. The absorbance-versus-wavelength curves for a product graded by Saybolt standards are shown in Fig. 11. The instrument shown in Fig. 9 monitors Saybolt color grades by measuring the difference in absorbance at 436 nm and 546 nm (blue and green). A mercury vapor lamp and isolation filters are used to provide these high-purity lines. A typical calibration curve is shown in Fig. 12.

Rosin Color Scale

Rosin has been graded by color for many years. In several respects, the rosin color standard scale is similar to the ASTM scale. Analogous to the change in color of petroleum as it is refined, rosin also

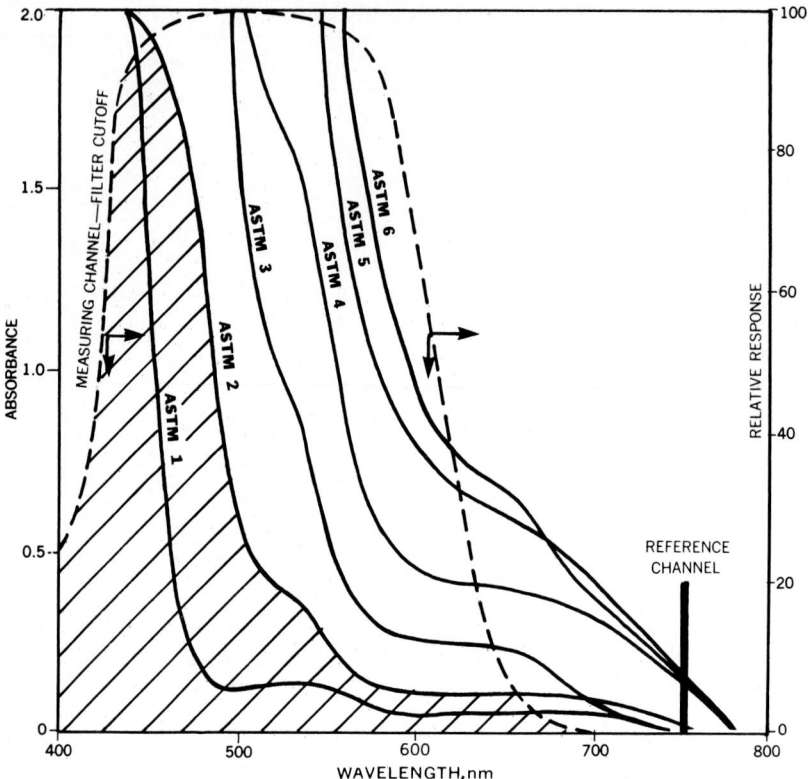

Fig. 10 Absorbance-wavelength curves for a product graded by ASTM color standards. *(Du Pont Analytical Instruments Division.)*

becomes progressively lighter as it goes through various steps of refining. The rosin scale is made up of 12 standards, ranging from a dark reddish-orange, the D curve of Fig. 13, to an X standard which is a pale yellow. Rosin refiners usually can grade rosin to within one-quarter of a color grade. Again, the photometric analyzer can be set up for rosin color grade monitoring.

APHA Color Scale

The American Public Health Association color standards are known concentrations of platinum-cobalt solutions. These standards, like the Saybolt standards, are used mainly to grade the degree of yellowness in pale products. The color of the product under test is compared with the APHA standards in matched Nessler tubes. These also can be monitored instrumentally, as are the other colored substances previously mentioned.

Wastewater Sampling

The photometric analyzer described can be combined with a special sampling system for automatic monitoring of wastewater streams for low levels of phenols, using the characteristic wavelength shift with pH. Figure 14a shows the shift toward higher wavelengths and the marked increase in ultraviolet absorption of basic phenol solutions compared to acidic solutions. The system (Fig. 14b) uses time-sequenced acid and base additions to the sample, generally operated on a 10-min cycle. Initially,

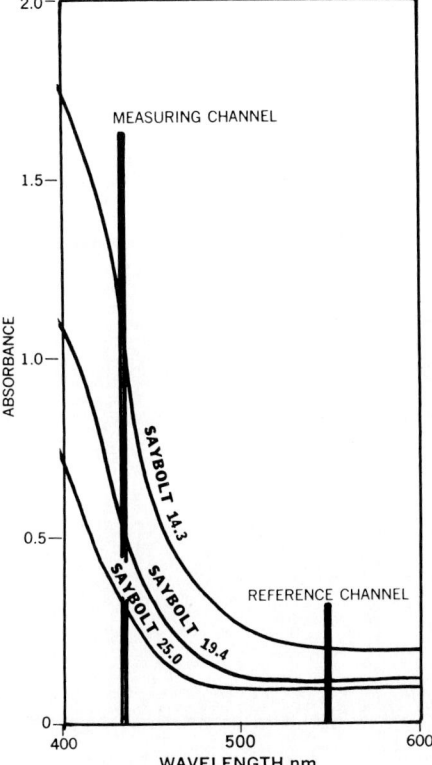

Fig. 11 Absorbance-wavelength curves for a product graded by Saybolt color standards in a 25.4-cm (10-in) cell. *(Du Pont Analytical Instruments Division.)*

water is flushed through the system for 20 s. Then the sample, normally made basic by continuous caustic addition, is acidified with acid drawn with the sample through the analyzer cell. This acidified sample acts as the "zero" standard. After approximately 90 s, the analyzer is automatically adjusted to provide a zeroed output for recording. After zero standardization, the acid flow is stopped, and phenol in the basic solution selectively absorbs light at the measuring wavelength (289 nm). At the end of 10 min, the cycle repeats itself. The photometer output signal is directly proportional to phenol concentration, even in the presence of high concentrations of materials that would normally cause background interference.

Nitric Oxide Analyzer

This instrument uses a differential photometer and is specific for extremely small amounts of NO (down to 0.05 ppm full-scale) in gas streams and atmospheres, particularly in coke oven and water gas. A schematic diagram of the device is shown in Fig. 15.

The NO is oxidized in the presence of a catalyst, butadiene, after which it is brought into contact with Griess solution, with which it reacts to form an azo dye. The process is continuous, requiring good flow control for both sample and reagent. A color comparison of the reagent before and after the reaction with the oxidized sample is made with a differential photometer. The photometer has a

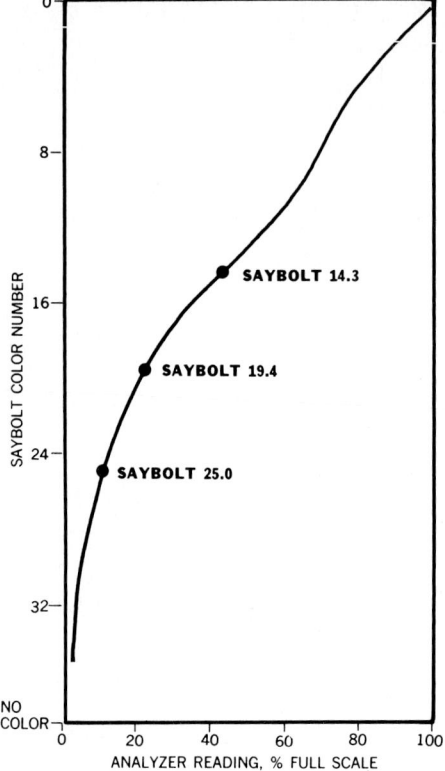

Fig. 12 Calibration curve for no color to zero Saybolt color in a Du Pont 400 photometric analyzer.

bridge-type measuring circuit which is automatically balanced by a null recorder. The instrument is calibrated empirically by the manufacturer, and complete data and charts are supplied.

Nephelometry

Sir John Tyndall noted that particles which are invisible when directly in the path of a strong light become discernible when viewed from the side. Noted about a century ago and known as the Tyndall effect, the phenomenon derives from reflection of part of the incident light by the particles. The reflected light is directly proportional to the number of particles in suspension. An instrument for measuring the intensity of reflected light so produced is called a *nephelometer* and is used for the quantitative determination of small amounts of diverse materials which have the ability to reflect light when in liquid suspensions. Examples include the measurement of traces of silver, wherein chloride ion is added to a solution of material containing silver. Insoluble silver chloride particles are suspended in the solution. Small amounts of calcium in titanium alloys may be determined by measuring suspensions of the stearate formed in a suitable medium. Nephelometry also finds application in the measurement of bacterial growth rates, in the analysis of cholesterol, glycogen, and enzymes, in controlling the clarity of beverages, water, and wastewater, in solution control in tanning operations, and in many measurement situations where an unknown composition may be transformed into or related to a form of suspension.

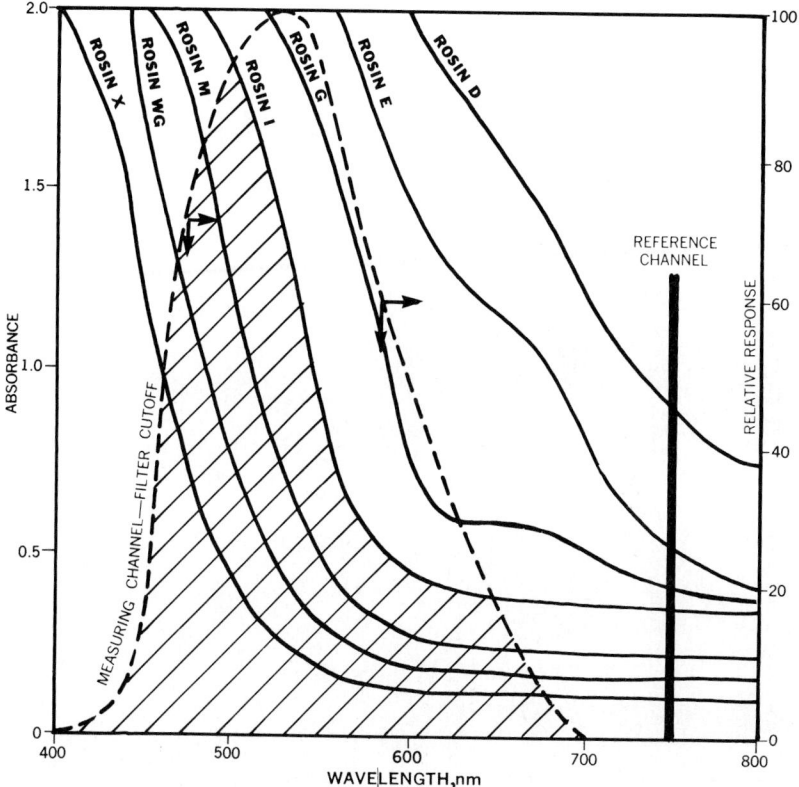

Fig. 13 Absorbance-wavelength curves for a product graded by a rosin color sample. *(Du Pont Analytical Instruments Division.)*

This method of instrumental analysis lacks a universally accepted standard. However, one instrument manufacturer in the field proposed a number of years ago that measurement of the Tyndall effect in suspensions be based on so-called nephelos units, defined as the particle concentration that will reflect up to 2% of the total light passing through the sample. In this system, the region between 2% and 0% (no reflection) would be divided into 100 uniform units.

Nephelometric methods are similar to fluorometric methods in that both involve measurement of scattered light. However, the scattering is inelastic in nephelometry and elastic in fluorometry. Thus, the scattered light measured in fluorometry is of a longer wavelength than the incident light, and both incident and scattered light are of the same wavelength in a nephelometric determination. In fact, the two functions sometimes are combined in one instrument, which may be called a nefluorophotometer. When the instrument operates as a nephelometer, it utilizes two Tyndall windows located opposite each other in a cylindrical sample cell with their common axis perpendicular to the path of the entering light. The concentration of suspended particles is determined by summing the photocurrents of the two cells. When used as a fluorometer, the instrument measures light emitted by a sample which is excited by incident radiation in the appropriate spectral band. Further, the same instrument can be set up for use as a photometer to measure light transmitted by the sample. Three light sources may be used—an incandescent (tungsten) source for colorimetric or nephelometric applications; a mercury arc source for fluorometry; and a sodium arc source with principal emissions at 320 and 590 nm when a sharp peak at either of these wavelengths is required. See Fig. 16.

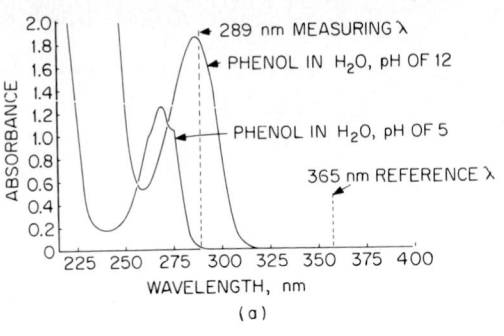

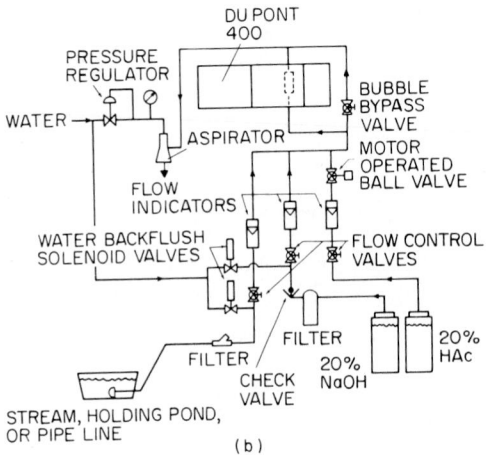

Fig. 14 Wastewater analysis system. (*a*) Phenol absorption in acid and basic water solutions, (*b*) flow diagram of a phenol monitoring system. (*Du Pont Analytical Instruments Division.*)

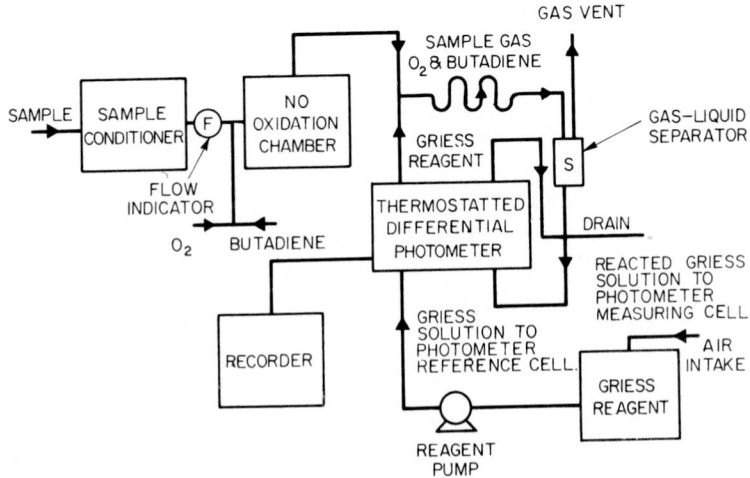

Fig. 15 Nitric oxide analyzer using the colorimetric principle. (*Penn Airborne Products.*)

Turbidimetry

In essence, turbidimetry is a special case of nephelometry. The cloudiness of a liquid is caused by the presence of finely divided suspended material and is called *turbidity*. Turbidimetric methods are similar to colorimetric procedures in that both involve measurement of the intensity of light transmitted through a medium. They differ in that the light intensity is attenuated by scattering in turbidimetry and by absorption in colorimetry. Both determinations utilize similar instrumentation.

Turbidity may be due to a single chemical substance, or it may result from a combination of several components. For example, silica may be determined in approximate concentrations of 0.1 to 150 ppm, expressed as SiO_2. Sometimes, composite turbidities are expressed as equivalent to silica. Higher concentrations may be determined by dilution. Numerous applications are possible in which turbidity is developed from the test sample under controlled conditions. A widely used determination is that of sulfate after the addition of barium chloride to an unknown sample to form a suspension of barium sulfate. The procedure is particularly applicable within the concentration range of 0.2 to 100 ppm sulfate. Routine procedures have been developed for the determination of sulfur in coal, oil, and other organic materials in which the sample is first fused in a sodium peroxide bomb prior to precipitation of the sulfur as barium sulfate.

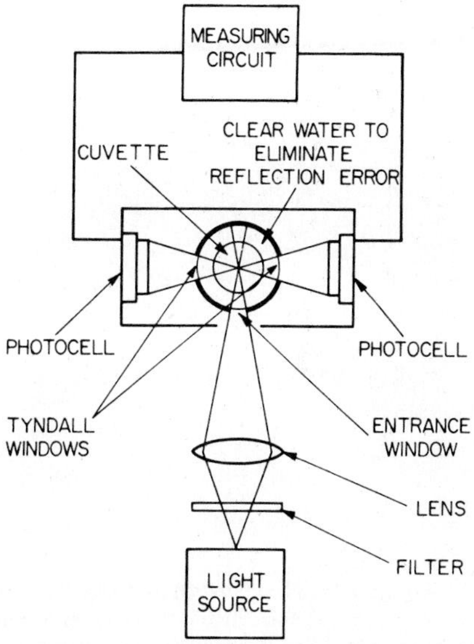

The *Parr turbidimeter* is an extinction-type instrument which consists of a cylinder to contain the turbid suspension, a lamp filament of fixed intensity at the base, and an adjustable plunger through which visual observation is made. Measurement is made of the depth of turbid medium necessary to extinguish the image of the lamp filament. Standard suspensions are used to prepare a calibration curve, which is a plot of depth versus concentration.

The *Hellige turbidimeter* is also a variable-depth-type instrument using visual detection. A combination of vertical and horizontal illumination of the sample and a split ocular permit the eye to compare the intensities of two images appearing simultaneously in the ocular. This is a special form of double-beam operation, and the intensity of the light source need not be extremely constant from one sample to another. Adjustment is made of a slit in the path of the direct or vertical illumination of the sample, and the calibration curve consists of a plot of this slit opening versus concentration.

Fig. 16 Nefluorophotometer, which can function as a nephelometer, fluorometer, or photometer. When operated as a nephelometer, it utilizes a split beam to obtain twin optical paths, which permits a direct comparison of solutions in the sample and reference cells, thus reducing the time required to complete an analysis. When used as a fluorometer, the instrument measures light emitted by a sample which is excited by incident radiation in the proper spectral band. The apparatus can also be set up as a photometer to measure light transmitted by the sample.

The *Sargent-Welch turbidimeter,* shown in Fig. 17, contains two basic elements, an optical system and an electronic system. See Fig. 18. The light source is a low-voltage tungsten filament lamp powered by an adjustable, regulated dc power supply. A biconvex lens and mechanical aperture serve to collimate the light beam before it enters the bottom of the sample tube. The sample compartment is designed to accept a 25-mm diameter sample tube and contains two apertures, the main beam entrance aperture and the orthogonal exit aperture for scattered light. The light shield stops extraneous light from entering the sample tubes through the top or the exposed sides. Both the sample compartment and the light shield have black, nonreflective finishes to minimize stray light effects. The scattered light is detected by a high-vacuum photodiode with a bialkali photocathode. The useful response of the detector covers the spectral region of 300 to 630 nm. The detector has high luminous sensitivity and very low dark current. The photocurrent generated in the detector is amplified by a solid-state, direct-coupled, negative feedback amplifier. The instrument is furnished with turbidity standards for four ranges: (1) less than 1 nephelometric turbidity unit (NTU), (2) less than 10 NTU, (3) less than 100 NTU, and (4) less than 1000 NTU.

Titrators

Titration is the procedure by which an unknown quantity of a particular substance is determined by adding to it a standard reagent with which it reacts in a definite and known proportion. With a knowledge of the reaction taking place and the amount of standard reagent needed to reach an end point (completion of the reaction), the concentration of the other reactant (unknown in terms of quantity) is readily determined. Titrations are almost always performed with solutions, but they can be performed with other phases when the needed equipment is available.

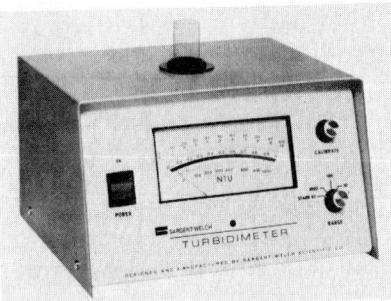

The point in a titration where the amount of titrant added is just sufficient to combine in a stoichiometric or empirically reproducible proportion with the substance to be determined is referred to as the *equivalence point*. The experimentally observed signal for termination of a titration is referred to as determination of the *end point*. The end point should coincide or be very close to the equivalence point. The difference between the equivalence and end points is sometimes referred to as the *indicator blank*.

Fig. 17 Sargent-Welch turbidimeter.

Titrations can be classified in a number of ways—for example, by the kind of reaction, such as acid-base, oxidation-reduction (redox), precipitation, or complexation, or by the manner in which the end point is determined. In the latter regard, there are colorimetric titrations, where the end point is noted by a marked change in the color of the solution and for which a large number of organic dyes are available. For example, pH-sensitive dyes are very useful in acid-base titrations. Instrumental methods of detecting the end point are based on the measurement of a number of different physical properties of the solution. The most generally used electrochemical methods of end point detection utilize various types of electrodes in the titration vessel across which changes in either voltage, current, or resistance are measured. These are known as *potentiometric* (voltage), *amperometric* (current), and *conductometric* (resistance) titrations. Spectrophotometric or photometric titrations utilize the changes in absorbed radiation to indicate the end point. Less commonly used instrumental titration methods utilize changes in such physical properties as solution temperature, dielectric constant, refractive index, viscosity, sound velocity, radioactivity, and scattered light, among others. *Coulometric titrations* are named for the mode of preparation of the titrant. Titrant is prepared either in the reaction flask in the presence of the substance to be determined or externally by electrolysis of a suitable solution, and from the measured number of coulombs required to reach the end point, the quantity of the sought-for substance is calculated.*

Automatic Titrations

These analyses can be performed whereby either the titration curve is automatically recorded or the delivery of titrant is automatically terminated at the end point. Such systems release the operator for sample preparations and can greatly decrease the expense per sample determination. The majority of modern automatic titrators are equipped to measure pH (0 to 14) or millivolts (0 to ~2000 mV, ±). Thus, they fall in the potentiometric titration class.

*Karl Fischer titrations for determining moisture content are described in the article on moisture measurement in Sec. 10 of this Handbook.

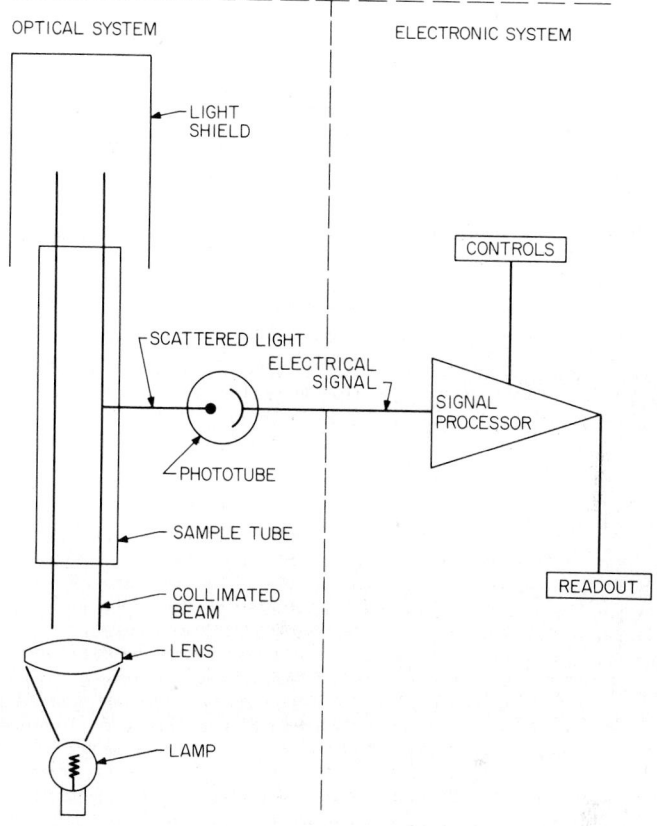

OPTICAL SYSTEM

ELECTRONIC SYSTEM

LIGHT SHIELD

CONTROLS

SCATTERED LIGHT

ELECTRICAL SIGNAL

SIGNAL PROCESSOR

PHOTOTUBE

SAMPLE TUBE

READOUT

COLLIMATED BEAM

LENS

LAMP

Fig. 18   System diagram of a Sargent-Welch turbidimeter.

Potentiometric Titration Fundamentals

Potentiometric titration is based on the principle of the Nernst equation, which may be written

$$E = E^0 - \frac{0.059}{n} \log \frac{A_{ox}}{A_{red}}$$

where E = measured electromotive force (emf)
  E^0 = standard value of emf (electrode potential) when substances of electrochemical reaction are in their standard states
  n = valence change (change in number of electrons per mole of reactants)
  A = activities of oxidized and reduced forms of reactants

Activities are proportional to concentrations, so that the concentration of one of the reactants may be determined if that of the other is known. Thus, the concentration of Cu^{2+} ions in a solution can be found by using an electrode of metallic copper (unstrained metals are assumed to be at unit activity); or the concentration of Cl^- ions in a solution can be found by using an electrode (insoluble) of silver chloride, AgCl deposited in an electrode of metallic silver; or the concentrations in a solution containing two ions in different states of oxidation, such as Fe^{2+} and Fe^{3+}, can be found by using an inert electrode, such as one of platinum. The three foregoing reactions and corresponding forms of the Nernst equation (using the approximation of substituting concentrations for activities) are

$$Cu^0 \rightleftharpoons Cu^{2+} + 2e^-$$

$$E = E^0 - \frac{0.0591}{2} \log c_{Cu^{2+}}$$

$$Ag^+ + Cl^- \rightleftharpoons AgCl\downarrow$$

$$E = E^0 - (0.0591)\log K_{sp} + (0.0591)\log c_{Cl^-}$$

where K_{sp} = solubility product of AgCl

$$Fe^{2+} \rightleftharpoons Fe^{3+} + e^-$$

$$E = E^0 - (0.0591) \log \frac{c_{Fe^{3+}}}{c_{Fe^{2+}}}$$

In constructing an electric cell for potentiometric titrations, it is necessary to use a second electrode to complete the circuit in addition to the measuring electrode (sometimes called the *indicator electrode*). Ideally, the second electrode should be a hydrogen electrode, which is the standard reference electrode for which the potential, in equilibrium with its ions, is defined as *zero*. Since a hydrogen electrode is awkward to use, other electrodes of known potential are used.

Instrumentally Assisted Manual Titration

A simple apparatus for manually feeding the titrant, but with instrumental end point indication, is shown in Fig. 19. The procedure is to determine the potential of the indicator electrode after each addition of titrant. This is done by closing the switch just long enough to read the potential on a sensitive electrometer. Of course, very small additions of titrant are made as the expected end point is approached. Then the readings of potential are plotted against the volume of titrant added, as shown in Fig. 20. Reactions suitable for determinations by this method show sharply defined end points, as illustrated. To correspond to the true stoichiometric end point, certain conditions should be met, including reversibility of the reaction and allowing time for the electrodes to reach equilibrium before closing the switch to read the potential difference. Types of titrations for which potentiometric methods are particularly useful include titrations of halide mixtures, various metal solutions, and alkaloids, and various titrations with oxidizing agents such as permanganate, dichromate, iodate, and ceric sulfate solutions.

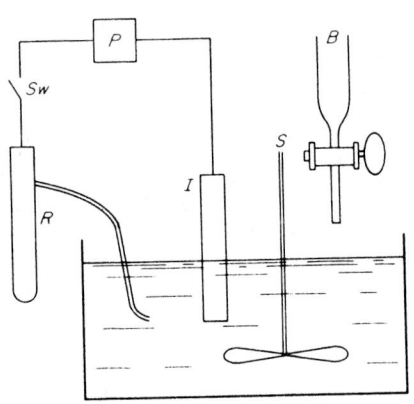

Fig. 19 Apparatus for potentiometric titration. R, reference electrode; I, indicator electrode; P, potentiometer; B, buret; S, stirrer; Sw, switch.

Automatic Configurations

Certain parts of an automatic titrating system appear relatively easy to automate, e.g., flow control of the titrant. One problem involves the need to greatly slow the speed of titrant addition as the end point is approached. In one design, a demand module provides a recorder output signal that is proportional to dE/dV (the derivative of electrode output with respect to titrant volume dispensed) without restricting the need to dispense titrant at a constant rate. Instead of the titration curve shown in Fig. 20, a curve of the type shown in Fig. 21 is obtained. This type of curve provides peaks for easy detection of equivalence points but is not distorted by time-dependent experimental variables.

Titrations are also complicated by reactions that have two end points, as illustrated in Fig. 22. Also, depending on the particular application, a titrator may be desired with an end point seeking

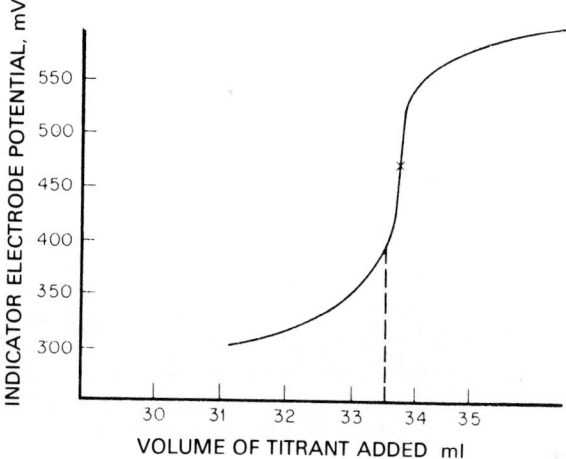

Fig. 20   Potentiometric titration curve.

mode, as contrasted with preset end points. Thus, during the past few years, there has been considerable opportunity to apply modern microprocessor and other electronic techniques to automated titration systems. See Figs. 23 and 24.

Amperometry

A device in this category, as illustrated in Fig. 25, is used for the continuous analysis of residual chlorine in aqueous solution in low concentrations (milligrams per liter). Infrared can be used to record either free or total available chlorine. Since chlorine does not exist as Cl_2 in water, free avail-

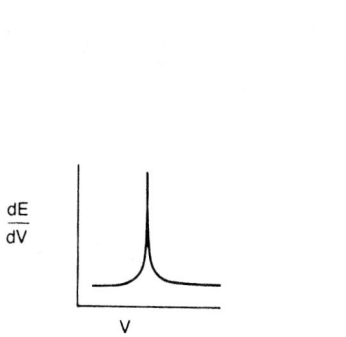

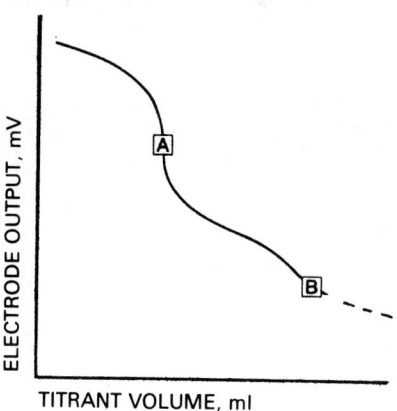

Fig. 21.   Titration curve obtained when the volume-based derivative signal dE/dV, which is a function of titration chemistry rather than titrant delivery rate, is calculated. *(Model 383 AEP demand module, Fisher Scientific.)*

Fig. 22   Titration with two end points. When so programmed, the instrument will output values of equivalence points at both locations on the curve and stop automatically after reaching the second end point. *(Fisher Scientific.)*

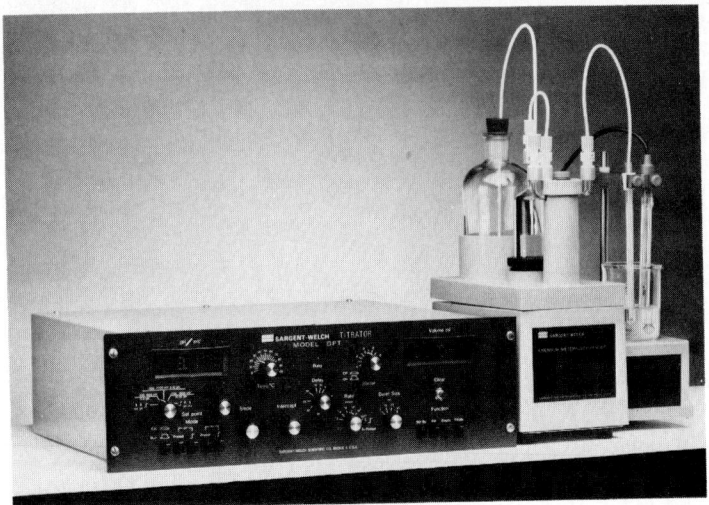

Fig. 23 Automatic titrator that will perform pH and potentiometric titrations to a preset pH or millivolt value. There is an output available for the recorder to provide a complete titration curve. On the left is a control module with two digital readouts, one for pH or millivolts and the other for titrant volume. On establishment of the parameters of the titration, no further operator assistance is required. The unit in the center is the dispenser. A stepper motor provides a choice of six different rates. The initial delivery can be varied from 40% of buret volume per minute down to a 4% rate. There is an automatic decrease in titrant delivery as the selected set point is approached. The rate of delivery is determined by the voltage difference between the end point and the cell voltage and the rate of change in the cell voltage. The solution being titrated and the electrode assembly are shown on the right. *(Model DPT, Sargent-Welch Scientific.)*

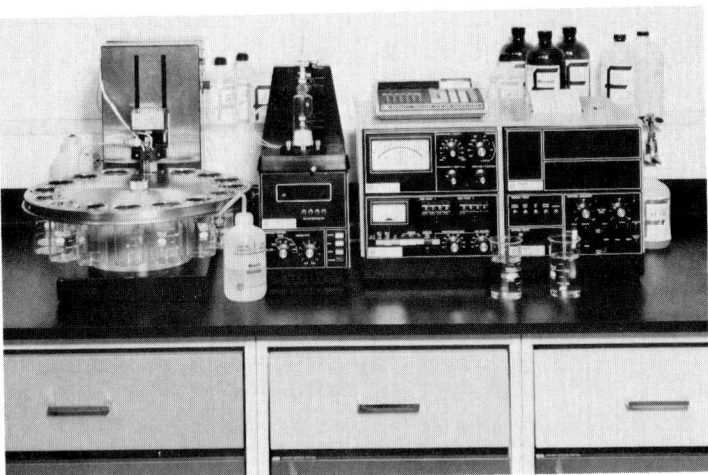

Fig. 24 Automated multisampler instrument for automatic pH or millivolt titrations. A 16-place turntable, designed to hold 200-ml beakers, is shown on the left. Such systems are used where there are large workloads involving fixed end point requirements. A similar apparatus is available with a linear rack for holding samples instead of a turntable. The display includes an LED readout for digital indication of dispensed titrant (0.01-ml units), a BCD output for connection to a computer, and data processing equipment. A microprocessor control refills a buret syringe as needed to deliver the required titrant volume. A memory retains a cumulative total of delivered volume for continuous readout and for printout at the end of the procedure. *(Titralyzer II, Fisher Scientific.)*

able chlorine is defined as that present in the form of HOCl. Total available chlorine includes all forms available for sterilization purposes, including HOCl, NH_2Cl, $NHCl_3$, and NCl_3.

Current produced in a depolarizing cell by the reduction of a free halogen is proportional to the halogen residual (milligrams-per-liter range). When free chlorine, as HOCl, is passed through the cell, the current produced can be conditioned to represent the concentration. Since pH affects the dissociation constant of HOCl, the sample is conditioned by carbon dioxide or an acetate buffer to the pH range 4.4 to 5.5. A thermistor in the cell circuit compensates for temperature change effects.

When total available residual chlorine is to be measured, an iodide solution is added to the sample. The iodide is oxidized to HOI in stochiometric proportion to the available chlorine. Current from the cell produced by the HOI is proportional to the equivalent amount of chlorine present in the sample. An excess of iodide of 4 to 10 times the molar concentration of chlorine ensures a complete reaction.

The cell utilizes a copper anode and a platinum cathode. A unique grit bombardment system keeps the electrode surfaces clean. The analyzer is calibrated by adjusting the output value to that obtained by amperometric titration of the sample. A diagram of the system is shown in Fig. 26.

The range of the instrument is in steps from 0 to 1 to 0 to 20 mg/L, the sensitivity is 0.01 mg/L, and the speed of response is 10 s after the sample enters the analyzer.

Impregnated-Tape Devices

Instruments are available which utilize reagent-treated tapes to analyze for gases or suspended particles in industrial processes or atmospheres. They can be used to monitor exhaust or smoke

Fig. 25 Indicating residual chlorine analyzer. *(Wallace & Tiernan Division, Pennwalt Corporation.)*

in combustion control and dusts, aerosols, or corrosive and toxic gases in air pollution control; and to analyze and control specific components in gas mixtures, over wide ranges of concentrations from fractions of a part per million to several percent.

The principle of operation of a tape unit for the detection of hydrogen sulfide (H_2S) is shown in Fig. 27. The gas sample flows into the reaction window of the sample chamber where it passes over only on one surface of a moving sensing tape, causing the surface to darken in proportion to the H_2S concentration. Measuring and reference photocells sense the tape's color change. Electronics provide a meter deflection in units of sample concentration and can be calibrated in parts per billion (ppb), parts per million (ppm), or percentage ranges, depending on application and needs. The instrument with door open is shown in Fig. 28. Reproducibility and accuracy are $\pm 2\%$ of full scale, and the instrument is calibrated by using standard samples. Typical applications include environmental monitoring, personnel protection, and process monitoring as found in amine and caustic treaters, pipeline product analysis, stacks, fuel, gas to boilers, and reformer cycle gas, among others.

The apparatus shown in Fig. 29 converts sulfur-containing compounds into H_2S to obtain a specific measurement of the total sulfur in gas or liquid hydrocarbons. This conversion unit is used in connection with a detector of the type previously described. Typically the converter and detector are used for continuous on-stream analysis for total sulfur in samples ranging from C_1 through C_{15}, including alcohols, ketones, amines, water, gasoline, kerosene, finished and intermediate products, and paraffinic and olefinic hydrocarbon gas mixtures.

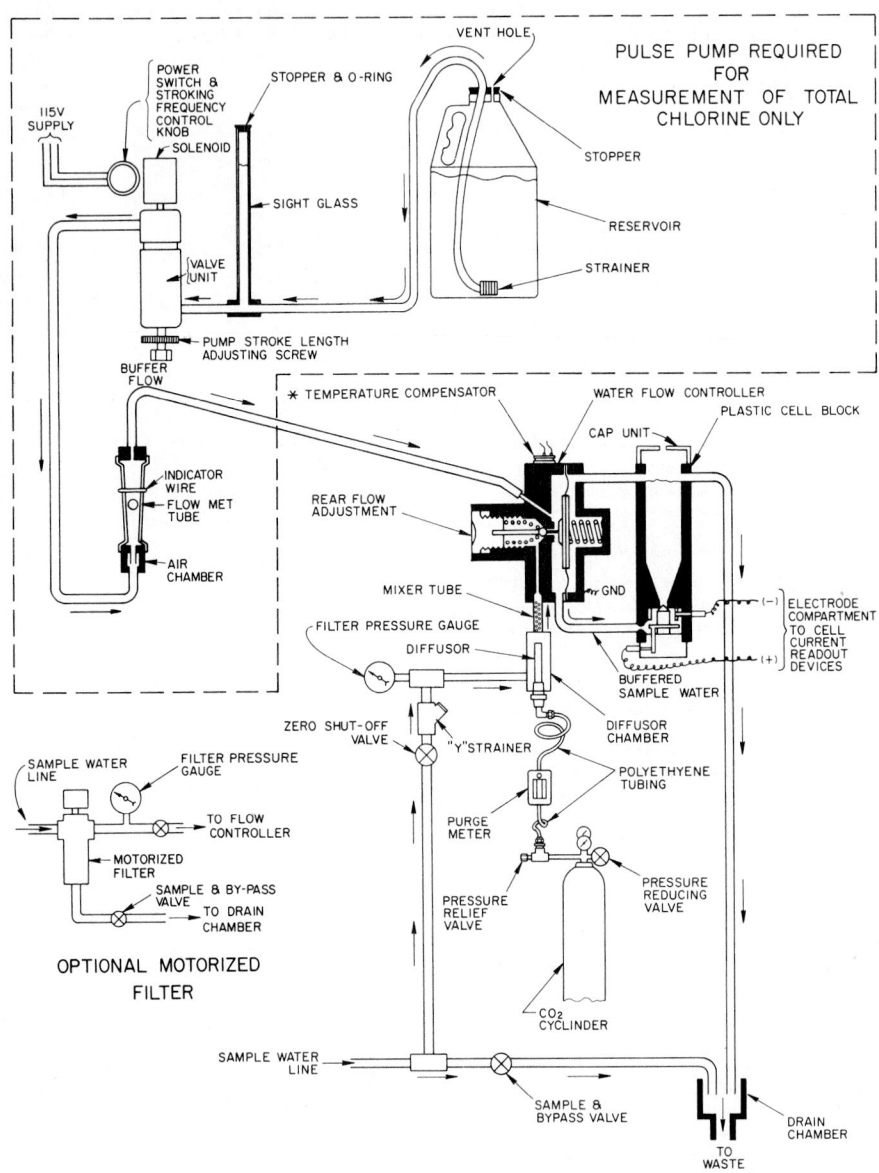

Fig. 26 Residual chlorine analyzer using the amperometric principle. Readout devices include indicators and recorders with an optional controller and transmitter. *(Wallace & Tiernan Division, Pennwalt Corporation.)*

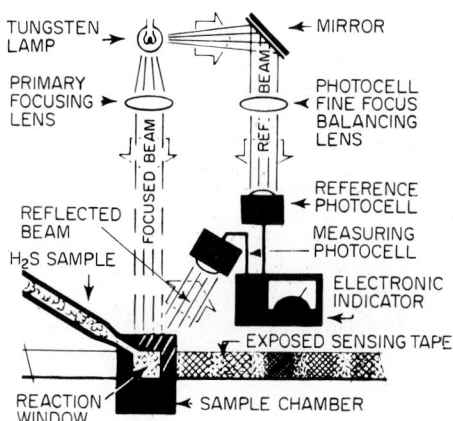

Fig. 27 Moving reagent tape analyzer for hydrogen sulfide measurement. *(Houston Atlas.)*

Fig. 28 Reagent-tape-type hydrogen sulfide gas analyzer. *(Houston Atlas.)*

In Fig. 27, the labeled components are:

- TUNGSTEN LAMP
- PRIMARY FOCUSING LENS
- REFLECTED BEAM
- H₂S SAMPLE
- REACTION WINDOW
- MIRROR
- PHOTOCELL FINE FOCUS BALANCING LENS
- REFERENCE PHOTOCELL
- MEASURING PHOTOCELL
- ELECTRONIC INDICATOR
- EXPOSED SENSING TAPE
- SAMPLE CHAMBER
- FOCUSED BEAM
- REF. BEAM
- BEAM

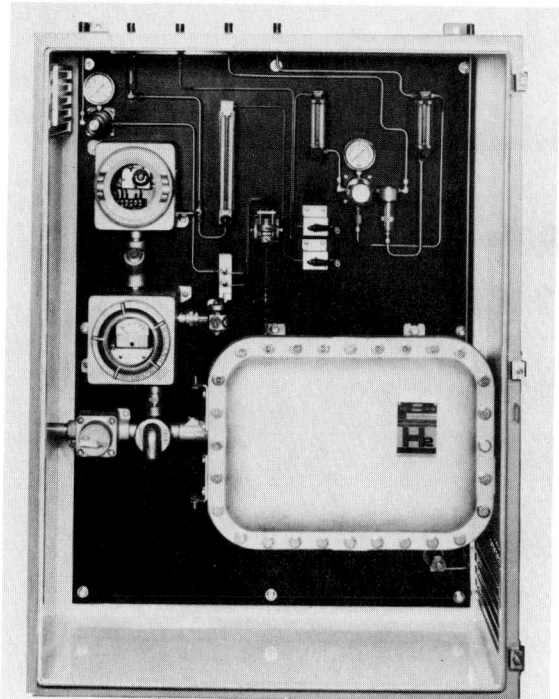

Fig. 29 Total sulfur hydrogenerator for gases or liquids. *(Houston Atlas.)*

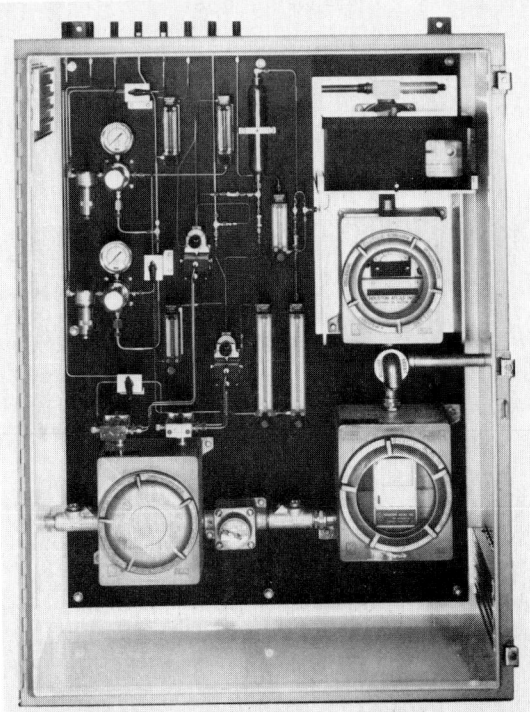

Fig. 30 Hydrogen sulfide analyzer system specially designed for the measurement of H_2S in fuel gas. *(Houston Atlas.)*

Fig. 31 Hydrogen sulfide monitor mounted in the field for wide-area gas monitoring. *(Houston Atlas.)*

An apparatus designed specifically for ensuring compliance with environmental standards is shown in Fig. 30. Hydrogen sulfide is a major sulfur contaminant in fuel gas undergoing combustion. When combined and diluted with excess air and combustion products, it establishes the maximum concentration of sulfur dioxide (SO_2) that can leave the process and flow from the stack. Measurement of H_2S enables immediate corrective action to protect the process downstream of the sampling point and preventing excess contamination of the atmosphere. The instrument shown in Fig. 31 is designed for wide-area H_2S monitoring. The unit continuously samples ambient air simultaneously at many points within or along boundary areas where H_2S is suspect or being processed. A reference gas generator is shown in Fig. 32. The device volumetrically meters precise microliter quantities of a known high-purity primary standard gas into a suitable carrier gas stream with a known flow rate. Thorough mixing and uniform blending of the standard gas establishes the quantitative concentration (parts-per-million range) in the gas stream. A schematic operating diagram is shown in Fig. 33. The rangeability is from 1 to 12,000 ppm and is reproducible to ±2% of the generated value.

Fig. 32 Reference gas generator with a range of 1 to 12,000 ppm. *(Houston Atlas.)*

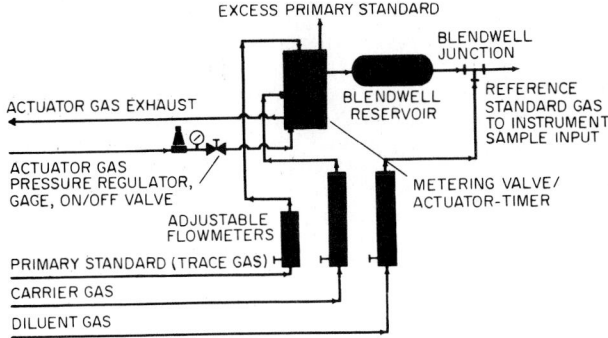

Fig. 33 Principal elements of a reference standard generator. *(Houston Atlas.)*

REFERENCES

ASTM: *Standard Test Method for Sulfur in Petroleum Products by Hydrogenolysis and Rateometric Colorimetry,* American Society for Testing and Materials, Philadelphia (current).

Bailey, S. J.: "On-Line Analysis 1979: A Microprocessor Market," *Control. Eng.,* 26, no. 3, March 1979, p. 61.

Bailey, S. J.: "Stream Analysis '80; System Concepts Gain," *Control. Eng.,* vol. 27, no. 6, June 1980.

Cines, M. R., Haskell, D. M., and C. G. Houser: "Molecular Sieves for Removing H_2S from Natural Gas," *Chem. Eng. Prog.,* August 1976, pp. 89–93.

Drushel, H. V.: "Trace Sulfur Determination in Petroleum Fractions," *Anal. Chem.,* vol. 50, no. 1, January 1978, pp. 76–81.

Jutila, J.: "Multicomponent On-Stream Analyzers for Process Monitoring and Control," *Instrum. Technol.,* 26, no. 7, July 1979, p. 38.

Krigman, A.: "Turbidity Instrumentation," *Instrum. Tech.,* vol. 31, no. 2, February 1984, pp. 31–36.

Oxygen Analyzers

Although numerous concepts have been suggested for the measurement of oxygen concentrations in air and other gaseous mixtures, the phenomenon of magnetic susceptibility of oxygen was the predominant principle used in industrial instruments for many years. During the past 10 or 15 years, technological advancements in other fields have affected oxygen measurement techniques to the extent that a number of relatively new sensors, such as zirconium oxide probes, galvanic cells, and fuel cells, among others, have received much acceptance. Further, oxygen analysis has taken on added importance in recent years, spurred by two principal factors—energy conservation and environmental protection—both of which are intimately associated with increased combustion efficiency and cleanliness of combustion effluents. In this regard, oxygen analyzers are frequently associated with analyzers for carbon monoxide and carbon dioxide. Like most other areas of analytical instrumentation, oxygen analysis in the recent past has benefited from solid-state electronics and computing and data processing technology, where oxygen measurements are infrequently made in isolation but most often are integrated into larger, broad-based control systems.

Zirconium Oxide Sensors

At relatively high temperatures, zirconium oxide becomes a conductor of oxygen ions (O^{2-}). These cells produce a millivolt signal in response to an oxygen partial pressure differential—for example, that in a stack or that of the outside ambient air. The response is logarithmic, hence these cells have the greatest sensitivity at low concentrations of oxygen. The cells are sensitive to temperature change, and thus many sensors are equipped with thermocouples or other temperature sensors to provide the required temperature compensation. Or, in some designs, heating elements are integral to the sensors, thus maintaining a constant temperature. The desired temperature ranges from 1500 to 1550°F (816 to 843°C). When properly equipped with heaters and temperature compensation, zirconium oxide sensors can be used over a wide range of actual stack temperatures—0 to 1500°F (−18 to 816°C) as reported by at least one supplier.

There are numerous sizes and configurations. Frequently the cells are enclosed in porous tubes which provide some filtration of particulates from the gaseous stream. Some systems are equipped for purging to keep the sensors clean. In some designs, porous platinum coatings on the inside and outside of the tube serve as the electrodes. Probe lengths range from about 8 in (20 cm) up to about 9 ft (2.7 m) for uses ranging from packaged boilers and small furnaces to very large ducts found in large boiler installations. Transmitters are often incorporated with the sensors. Averaging modules are available which provide a single signal based on the outputs of several probes.

Cell response varies from ~3 s down to 1 s, and common range is 0.1 to 21% oxygen. The oxygen content of normal ambient air at sea level is 23.15% (weight) or 20.95% (volume). However, some sensors respond to 0.1 to 100% oxygen, as may be required in chemical processing.

Fuel Cell Sensors

A fuel cell may be defined as an electrochemical device which directly combines hydrogen and oxygen from air to produce electricity and water. The concept of the fuel cell has been known for well over a century but was not utilized in a practical sense until the advent of the various space exploration programs. Cells developed by Bacon in the late 1930s provided the basis for the *Apollo* power plant and led to a sophisticated technology of cells using an alkaline electrolyte and pure hydrogen and oxygen as reactants. The shortcomings of the early alkaline cells were overcome by new types of cells, of which the phosphoric acid cell is among the most advanced. The principle of operation of a hydrogen-air cell with an acid electrolyte is shown in Fig. 1.

Thus, it required only a little imagination to apply the fuel cell principle to oxygen concentration measurement. The key to development of a practical fuel-cell-type sensor for oxygen measurement is not an understanding of the basic principles, but—somewhat akin to semiconductor design and production—the required infratechnology that is gained only from experience, and this explains the proprietary nature of most fuel cells and the relative scarcity of detailed information in the literature.

With further reference to Fig. 1, the electrode has a central function in cell operation. In its catalyzed layer, it provides a large number of sites where gases and electrolyte can react. Its porous

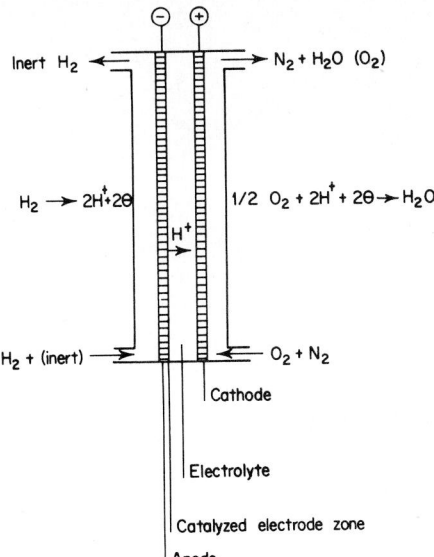

Fig. 1   Operation of a hydrogen-air cell with an acid electrolyte, which illustrates the basic concept of a fuel cell, numerous adaptations (various electrolytes, electrodes, etc.) of which are now employed in a large variety of fuel cell oxygen sensors.

configuration permits fast reactant transport and removal of inerts and product moisture. The electrode also provides a path for current to flow to the terminals and serves to contain the electrolyte. The latter not only provides ionic conduction, but also ensures separation of the reactants.

Cell Voltage

The cell voltage and the free energy of the underlying reaction are defined by

$$U = \frac{\Delta F}{nF}$$

where U = theoretical cell voltage
n = number of electrons transferred in the reaction
F = Faraday constant

Since $\Delta F = \Delta H - T\Delta S$, it follows that, depending on the value of ΔS, the electric energy to be derived from the cell can be larger or smaller than the energy ΔH obtained by direct combustion of the fuel. ΔH is the reaction enthalpy for the current-generating reaction, ΔS is the entropy change, and T is the absolute temperature. For a hydrogen-oxygen couple, the corresponding theoretical cell voltage at 25°C is 1.23 V if a liquid coater is formed, or 1.18 V if the product water is vaporized.

The thermodynamically possible conversion efficiency, however, is only partly realized in a practical fuel cell. Two basic losses are encountered: (1) the ohmic loss, and (2) the electrode polarization, i.e., the deviation of the actual from the thermodynamic electrode potential. The polarization is the result of the irreversibility of the electrode process, i.e., the activation polarization and the voltage loss which develops from the concentration gradients of the reactants. This leads to the current-voltage characteristics shown in Fig. 2.

Electrolytes available include aqueous potassium hydroxide (KOH), where the OH^- ion provides current transport. The general operating temperature range is 20 to 90°C (68 to 194°F). Electrode catalysts include nickel, silver, and platinum metals. The reactants are hydrogen (fuel) and an oxidant (oxygen). Concentrated phosphoric acid (H_3PO_4) is another available electrolyte, where H^+ ions serve as the current transport. The general operating temperature range is higher than that

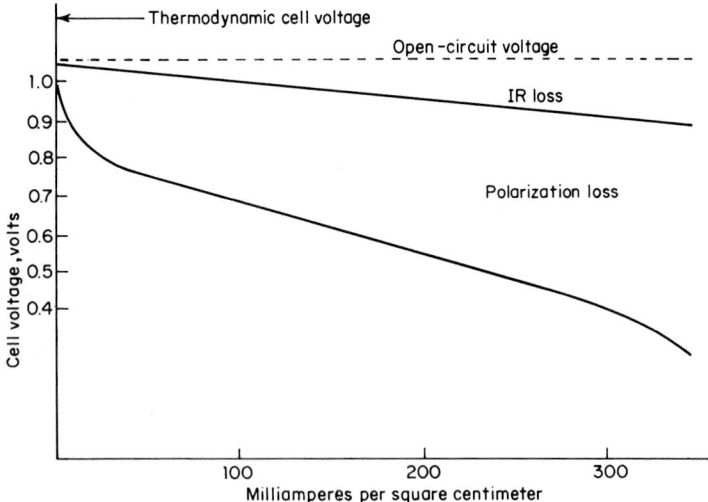

Fig. 2   Illustrative of the thermodynamic characteristics of fuel cells is this current-voltage relationship of a hydrogen-air cell with a phosphoric acid electrolyte at an operating temperature of 125°C.

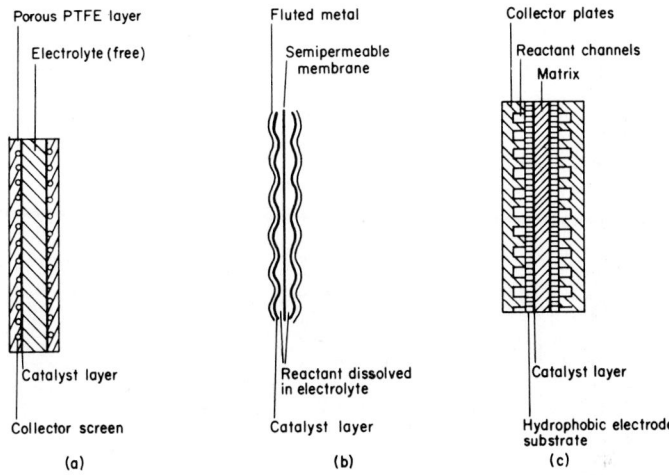

Fig. 3 Types of cell structures. (*a*) Free-electrolyte cell with screen reinforced electrodes with a polytetrafluoroethylene (PTFE) liquid barrier; (*b*) cell with nonporous electrodes and semipermeable membranes; (*c*) electrodes and matrix sandwiched between groove collector plates.

of aqueous potassium hydroxide. Electrode catalysts are platinum metals. Stabilized zirconium oxide (ZrO_2), where the O^{2-} ion is the current transport, can also serve as the electrolyte of a fuel cell. Operating temperatures are much higher, 700 to 1000°C (1292 to 1832°F), with base-metal oxides used as the electrode catalyst.

An adaptation of the fuel cell for power generation can be used for oxygen measurement. Phosphoric acid electrolyte cells are among the most advanced. A matrix-type cell construction can be used, and in this configuration a limited amount of electrolyte is trapped in a microporous structure by capillary forces. See Fig. 3. As a result, thin, highly porous, comparatively low-cost electrodes can be used, inasmuch as the electrodes are not required to contain the electrolyte. For electrolyte absorption, a thin layer of Teflon-bonded silicon carbide powder may be directly applied to the electrode surface. The electrodes may be made of 0.3-mm-thick Teflon-impregnated carbon fiber sheets which are activated on one surface with a layer of platinum dispersed on carbon black. Single cells may be sandwiched between carbon plates with a suitable pattern for current collection and grooves for reactant distribution. The direction of flow channels may be perpendicular to each other. In order to minimize temperature gradients, heat sinks are required and may be temperature-controlled or compensated.

Unlike a paramagnetic-type oxygen analyzer, for example, which does not require replenishment or replacement of materials and parts periodically, fuel cells are expected to require periodic replacement (not repair). In one design,* continuous process life varies from 6 to 18 months, depending on the class of cell. This cell is designed to detect oxygen concentrations from 1 ppm to 100% O_2. Oxygen surrounding the fuel cell diffuses through a Teflon membrane and is reduced at the cathode. A corresponding oxidation occurs at the anode internally, resulting in the generation of a current directly proportional to the partial pressure of oxygen.

In another design,† a fuel cell of the stabilized type responds to the O_2 partial pressure difference between the atmosphere inside and outside the cell. A high-level signal is collected by a platinum coating applied inside and outside a ceramic cell. The signal is logarithmic, increasing readability for low O_2 concentrations. To operate properly, the cell is heated to approximately 760°C in a heated

*Teledyne Micro-Fuel cell.
†MSA Model 803 oxygen analyzer.

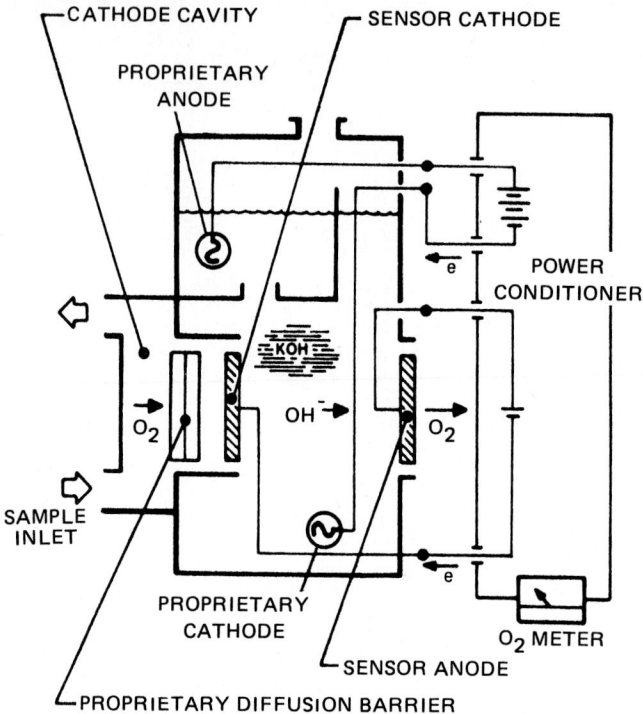

Fig. 4 Coulometric-type oxygen analyzer. *(Infrared Industries.)*

box which maintains all gases above their dew points. A thermocouple detects cell temperature and provides a signal which may be switched to the panel meter to read out the cell temperature. The system measures the entire content of flue gas, for example, including moisture. Thus, it is claimed that errors associated with "dry" conventional techniques are avoided. Such instrumentation is available for permanent mounting or portable meters. The scale range is 0.1 to 21% O_2, with a response time of about 3s.

Numerous other fuel cell–based oxygen meters are available.

Other Electrochemical Sensors

The use of a variety of galvanic cells and electrochemical means in the sensing and measurement of oxygen dates back many years. With the availability of improved production techniques borrowed from other areas of the electronics industry, particularly in terms of miniaturization, electrochemical O_2 sensors have reappeared and are now important in oxygen measurement technology.

In one type* of instrument, oxygen is sensed by a galvanic cell containing two dissimilar electrodes in a basic electrolyte. The cell is encapsulated in inert plastic, and the cell face is a fluorocarbon polymer that protects the sensing electrode. As oxygen diffuses through the cell face, it initiates an oxidation-reduction (redox) reaction which, in turn, generates a small current that is proportional to the partial pressure of the oxygen. The current is amplified through a solid-state electronic circuit and converted to a digital signal which indicates the percentage of oxygen. The range is 21% O_2 and 2000 ppm CO, and the response time is about 15 s for O_2.

*MSA Model 803-P analyzer.

The oxygen analyzer shown in Fig. 4 operates through a simple coulometric process whereby O_2 in the sample stream is reduced in an electrochemical cell. The gas stream enters the cathode cavity, and oxygen is metered to the cathode through the diffusion barrier. Oxygen is electrochemically reduced at the cathode:

$$O_2 + 2H_2O + 4e^- \rightarrow 4OH \qquad \text{(cathode reaction)}$$

The electrolyte solution contains potassium hydroxide (KOH) which assists in the migration of hydroxyl ions (OH^-) to the anode where they are oxidized to reform elemental oxygen:

$$4OH \rightarrow O_2 + 2H_2O + 4e^- \qquad \text{(anode reaction)}$$

An emf of approximately 1.3 V dc, applied to the sensor electrodes, is the driving force for the reduction and oxidation reactions. The resulting cell current, which is directly proportional to the oxygen concentration in the gas stream, is measured accurately by a solid-state electronic circuit and conditioned for display. Operation of the cell is compatible with gas compositions containing CO_2, H_2S, Cl_2, NO_x, SO_2, and inert and passive gases such as N_2, H_2, CO, and hydrocarbons, among others. Above $150°F$ ($66°C$) cell cooling is required and below $32°F$ ($0°C$) cell heating is required. Accuracy is claimed to be $\pm 2\%$ full-scale at $72°F$ ($22°C$). Standard ranges: for Trace analyzers, 0 to 100 ppm, 0 to 1000 ppm, 0 to 10,000 ppm; for percent analyzers, 0 to 1%, 0 to 10%, 0 to 25% O_2. It is claimed that, unlike some sensors, the electrochemical cell is not significantly affected by "oxygen shock" when exposed to high oxygen levels. The standard display is digital.

Polarographic Principles

As early as 1930, it was suggested that the current generated by an oxygen-depolarized primary battery could be used as a measure of oxygen concentration. Several workers experimented and developed cells with totally immersed electrodes. Dry forms of such cells have overcome a number of the disadvantages of wet types. In these cells, the sample gas, electrolyte, and cathode form a three-phase boundary where the oxygen molecules are reduced. The percentage of oxygen molecules reduced in the cell depends on the cell design. At low flow rates, it is possible to approach a state where all oxygen molecules are reduced and the cell thus operates coulometrically. This improves stability and reduces the temperature coefficient. Some cells have a thin membrane fixed near the electrode surface which allows oxygen to diffuse to the electrode. In another form (polarographic), a constant polarizing voltage is applied to the sensor.

Magnetic Susceptibility

When atoms and molecules are placed in a strong nonuniform magnetic field, either they are attracted to the strongest part of the field (in which case they are termed *paramagnetic*—ferromagnetism is a special case), or they tend to go to the weakest part of the field (termed *diamagnetic*).

The ratio of the intensity of magnetism induced in a unit volume of a substance to the magnetic intensity of the field acting on it is called the *susceptibility per unit volume*. The symbol is $K =$ volume susceptibility of gas mixture, cgs. This ratio is positive for paramagnetic substances and negative for diamagnetic materials. The susceptibility per unit mass X is equal to K/ρ, where ρ is the density of the substance. The atomic susceptibility X_A is equal to the product of X and the atomic weight, or the product of K and the atomic volume. The molar susceptibility X_M is equal to the product of X and the molecular weight, or the product of K and the molecular volume.

Diamagnetism is a universal property of matter; all substances, even though classified as paramagnetic, have a diamagnetic component. The mass susceptibility of diamagnetic substances usually is independent of temperature. The volume susceptibility of a diamagnetic substance is dependent on its density and hence at constant pressure is inversely proportional to the absolute temperature. In general, the mass susceptibility of a paramagnetic substance is inversely proportional to the absolute temperature. When volume susceptibilities are considered, paramagnetic volume susceptibility is inversely proportional to the square of the absolute temperature.

Table 1 Relative Volume Magnetic Susceptibilities of Gases at 20°C Expressed as Equivalent Percentage of Oxygen Normalized to the Scale

Nitrogen = 0; oxygen = 100

Gas	%
Ammonia	−0.26
Argon	−0.22
Acetylene	−0.24
Butadiene	−0.65
n-Butane	−1.3
Isobutane	−1.3
Carbon dioxide	−0.27
Carbon monoxide	+0.01
Chlorine	−0.77
Ethane	−0.46
Ethylene	−0.26
Hydrogen	+0.24
Hydrogen chloride	−0.30
Methane	−0.20
Nitrous oxide	−0.20
Nitric oxide	+43.0
Oxygen	+100.0
Propane	−0.86

The volume susceptibilities of some common gases are given in Table 1. The values for oxygen and nitric oxide are remarkably high. *This property of oxygen is utilized in a variety of ways for the measurement of oxygen concentration in gas mixtures.*

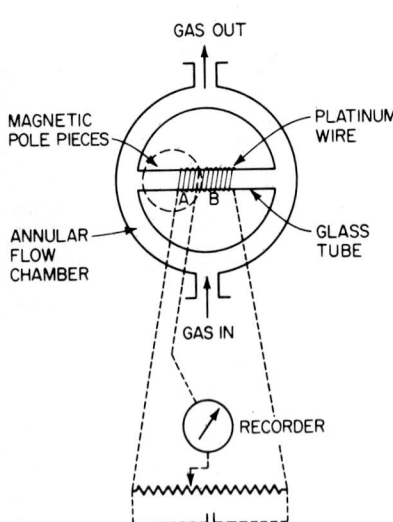

Fig. 5 Schematic of a Lehrer-type analyzer.

Thermomagnetic-Type Analyzers

In 1930 Senftleben observed that the thermal conductivity of paramagnetic gases appeared to decrease under the influence of a magnetic field. These results led Rein and later Klauer et al. to carry out further investigations which resulted in the discovery of a much larger effect in which the thermal conductivity of a paramagnetic gas increases when the detector is placed in certain positions under the nonuniform magnetic field.

These investigators proposed that if a paramagnetic gas comes in the vicinity of a hot wire inside a nonuniform magnetic field, then the gas will be heated, reducing its magnetic susceptibility and causing it to be displaced by a cooler gas experiencing a strong magnetic force. Thus, a form of *magnetic wind* is created. This tends to cool the hot wire.

Lehrer, in 1942, developed an industrial oxygen analyzer based on these effects. The instrument (Fig. 5) consisted of a brass annular flow chamber with inlet and outlet gas connections diametrically opposed. The two semicircular sections

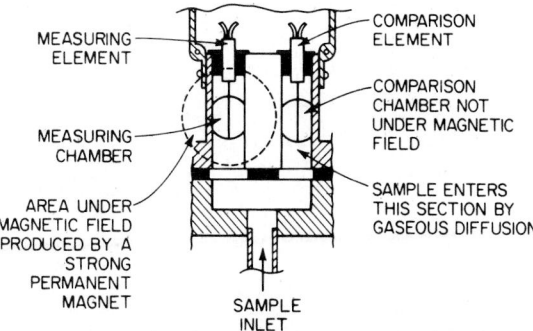

MEASURING ELEMENT

MEASURING CHAMBER

AREA UNDER MAGNETIC FIELD PRODUCED BY A STRONG PERMANENT MAGNET

COMPARISON ELEMENT

COMPARISON CHAMBER NOT UNDER MAGNETIC FIELD

SAMPLE ENTERS THIS SECTION BY GASEOUS DIFFUSION

SAMPLE INLET

Fig. 6 A measuring element and comparison element form two legs of a Wheatstone bridge and are heated by bridge current. Measuring and comparison chambers are exposed to the same sample gas.

of the flow chamber were connected by a glass tube positioned symmetrically with respect to the inlet and outlet connections. A platinum wire was wound on the glass tube and divided into two sections by a central tapping point, forming the arms of a Wheatstone bridge. The wire was wound on the outside of the tube to ensure that its temperature would be determined largely by constant external heat losses and be only slightly affected by the heat conductivity of the gas in the measuring chamber. A current through the coil raised its temperature.

A strong nonuniform magnetic field was produced by an electromagnet, with the tube between the pole pieces and arranged so that only one section of the coil was within the field. The heater between the pole pieces heated any oxygen molecules attracted to that area by the magnetic forces. As these molecules lost their paramagnetism, they were displaced by cold oxygen molecules and moved toward the second coil. Thus, there was a flow of oxygen molecules, drawing other molecules with them and causing cooling of one coil and heating of the other. The change in the resistance of the coils was monitored and calibrated as percentage oxygen. The lowest range of the instrument was 1% full scale. Higher ranges and supressed zero were possible. A zero stability of about 1% full scale was observed over a period of several months. A functional disadvantage of the instrument was the high degree to which the thermal conductivity of the gas influenced its sensitivity. For ranges greater than 2% oxygen, considerable deviations from linearity were observed. See Figs. 6 through 10.

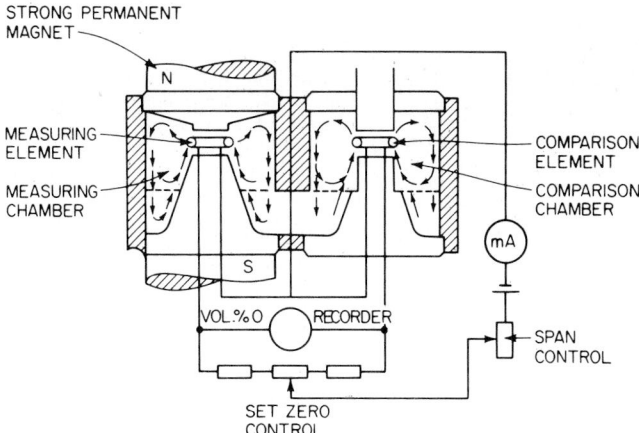

STRONG PERMANENT MAGNET

N

MEASURING ELEMENT

MEASURING CHAMBER

COMPARISON ELEMENT

COMPARISON CHAMBER

mA

S

VOL.% O   RECORDER

SPAN CONTROL

SET ZERO CONTROL

Fig. 7 Main part of a Hartmann & Braun analyzer. A measuring element and comparison element form half of a Wheatstone bridge. Measuring and comparison chambers are exposed to the same sample gas.

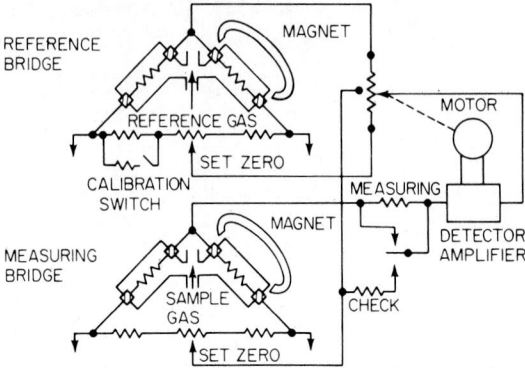

Fig. 8 Schematic arrangement of a double-bridge analyzer. Recorder output is a ratio of the percentage of oxygen in the sample to that in the reference gases. This equipment makes measurements independent of barometric pressure variations.

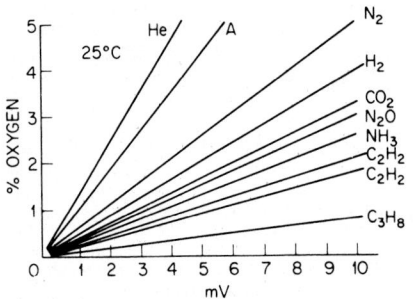

Fig. 9 Calibration curve for an oxygen analyzer.

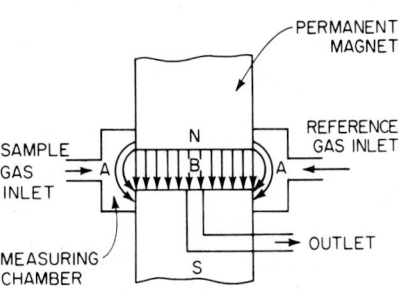

Fig. 10 Schematic of a Quincke-type analyzer.

1. Carrier Gas Composition. The output of the bridge is dependent on the ratio Cp/η, where Cp = specific heat, joules/$(cm^3)(°C)$ at a given temperature and pressure, and η = viscosity of gas mixtures at mean temperature of gas in heated tube, cgs units. This ratio is given for several gases in Table 2. The main effect of this dependence is to alter the slope of the calibration curve (Fig. 9). For example, if an analyzer is calibrated for N_2 (90%) and O_2 (10%) and then used to analyze a flue gas containing CO_2 (10%), N_2 (80%), and O_2 (10%), the reading must be multiplied by 0.948, as indicated in Table 2. This correction has to be altered each time the carrier gas changes in composition.

The chemical composition of the carrier gas also affects the zero position. Diamagnetic gases produce a reverse flow through the heated tube, since the pressure differential is dependent on the nature and concentration of the various gases. In addition, as the tube is heated, any secondary

Table 2 Cp_0/η_0 Ratio for Various Gases*

Gas	N_2	O_2	H_2	A	He	CO_2	N_2O	C_2H_4	C_2H_2	C_3H_8	NH_3
Cp_0/η_0	1.00	0.876	1.92	0.576	0.636	1.55	1.56	2.52	2.58	5.57	2.35

Cp_0 = specific heat at 0°C and 760 torr, J/$cm^3 \cdot$ (°C).
η_0 = viscosity of gas at 0°C, cgs units.

chemical action initiated by the heat, whether dissociative or reactive, will influence the reading and result in error. Hydrocarbon vapors, for example, may react on the heater filament, producing large errors.

2. *Temperature Effects.* The thermal conductivity of the sample gas affects the temperature distribution inside the tube and hence the sensitivity of the instrument.

3. *Diameter of Glass Tube and Tube Wall Thickness.* Any variations in these parameters result in distortion of thermal symmetry, leading to temperature differences and bridge unbalances. The effect also is a function of the thermal conductivity of the carrier gas. These errors result in a shift in the zero position.

4. *Ambient Temperature.* The output of the bridge decreases as the ambient temperature increases, the change depending on the nature of the carrier gas. A typical coefficient is about 1.5% per degree Celsius. Usually the analyzer is temperature-controlled and includes means for temperature compensation.

5. *Other Operational Factors.* Water vapor must be excluded from the sample. The response of the instrument to other paramagnetic gases will be similar to its response to oxygen, that is, proportional to their volume susceptibility. The bridge output is proportional to the square of the magnetic field strength, an effect which is small because the instrument is temperature-controlled. As the differential pressure across the heated tube is very small, a slight tilt of the tube produces a thermosyphon effect. The magnitude of this error depends on the gas and is larger for denser gases. Each instrument has slightly different flow characteristics. Hence a flow regulator usually is included. The output of the bridge is proportional to the second power of the pressure. Thus, pressure changes alter the sensitivity of the instrument, hence the instrument usually is pressure-controlled or compensated. Line voltage and frequency regulation are included because variations affect the temperature of the hot wire.

Quincke-Type Analyzers

When a paramagnetic gas is placed in the vicinity of a magnetic field, as shown in Fig. 10, the gas experiences a force driving it from the nonuniform part of the field A to the uniform part B. The Quincke method relies basically up measurement of the pressure differential betwen the two regions.

In 1951 Luft reported the development of an analyzer based on the Quincke method. See Fig. 11. The gas sample is admitted through a capillary K into two identical measuring tubes, M_1 and M_2. These tubes are located within the poles of a strong permanent magnet, which has one pole piece made of two halves. One half is formed of soft iron, producing a strong field on the tube under it. The other half is of a nonmagnetic material, making the field on the second tube a weak one.

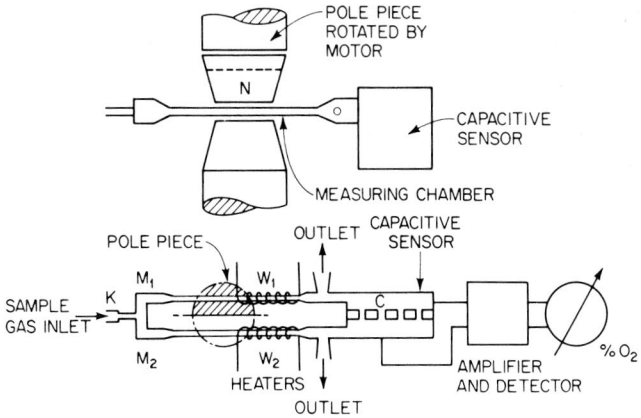

Fig. 11 Main parts of a Phlogor analyzer.

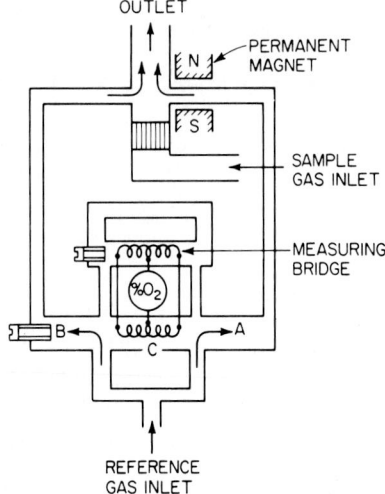

Fig. 12   Main part of a Maihak Oxygor analyzer.

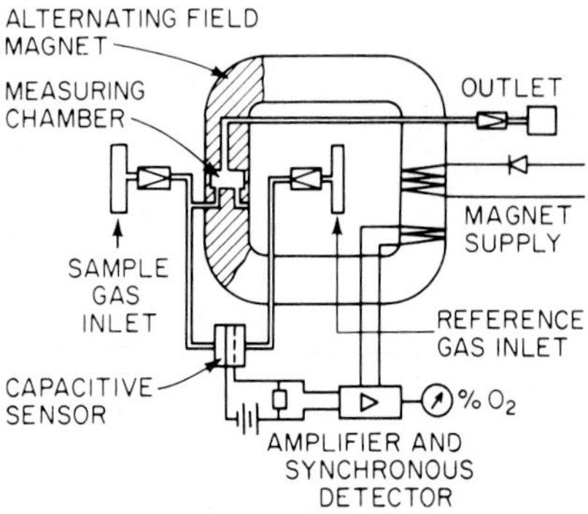

Fig. 13   Main part of a Hartmann & Braun Oxytest S analyzer.

The sample gas which enters the tube under the strong nonuniform field, at room temperature, experiences a force action on its paramagnetic component. This force is reduced at the other end by heating the gas with heaters W_1 and W_2. The sample in the second tube experiences a very small force. The pressure differential betwen the two tubes is measured by a capacitive pressure sensor C. A motor rotates the pole piece at constant speed and thus modulates the pressure sensor. The modulator signal has an amplitude proportional to the magnetic susceptibility of the gas mixture.

A later development was reported by Luft and Mohrmann in 1967. See Fig. 12. A reference gas stream having a weak magnetic susceptibility is split into two identical branches, A and B. These join together and with a pipe through which the sample gas is introduced. There is a strong magnet at the end of one of the branches. If the sample gas contains components of high magnetic susceptibility, these will experience a force at the end of the branch A, producing a pressure differential between A and B. This results in a reference gas flow in C. This flow is measured by a Wheatstone bridge. The bridge imbalance can be calibrated to read percentage oxygen.

Hummel in 1968 reported the development of a fast-response analyzer, shown in Fig. 13. The sample gas and a reference gas are admitted into a chamber which is within a magnetic field. The pressure differential between the streams due to the magnetic forces acting on the high-susceptibility component in the sample gas is measured using a capacitive pressure sensor. The magnetic field is switched on and off, modulating the signal whose amplitude is proportional to the magnetic susceptibility of the sample. A synchronous detector is used for demodulation. Another form is shown in Fig. 14. In the latter instrument, the reference gas flow due to the differential pressure is measured with a miniature flow sensor.

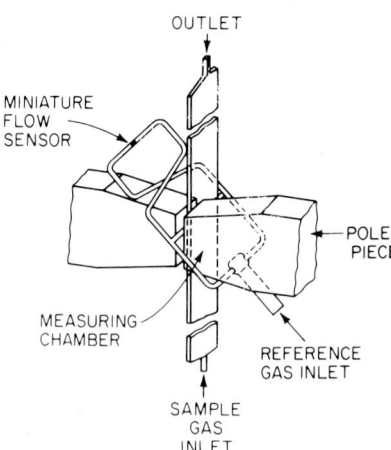

Fig. 14   Main part of a Siemens Oxymat II analyzer. The differential pressure across the measuring chamber due to the susceptibility of the sample gas causes the reference gas to flow over the flow sensor.

The foregoing instruments have been developed to overcome the dependence of the magnetic wind designs on the thermal conductivity and viscosity of the sample gas and, in this regard, have been successful. The instruments require a reference gas at constant flow.

Magnetodynamic Oxygen Analyzers: Faraday Type

Early work on the magnetic susceptibility of matter was done by Faraday. He developed a simple, sensitive method of whereby the sample to be tested is suspended from the arm of a very sensitive balance in a nonuniform magnetic field at a position where the gradient of the field is maximum. The sample experiences a force which is proportional to the field strength at that position, to the gradient of the field, and to the difference in magnetic susceptibility between the sample and the surrounding medium.

Pierre Curie used this method extensively, and later many investigators improved the technique. Havens experimented with a variety of test bodies and finally selected the dumbbell as the most suitable for his experiments in determining the magnetic susceptibility of nitrogen dioxide.

In 1946 Pauling et al. reported the design of an instrument for determining the partial pressure of oxygen based on the Faraday method. Modern magnetodynamic instruments are refinements of the Pauling instrument. Two spheres filled with nitrogen form a dumbbell which is suspended within a sealed cell having a sample gas inlet and outlet under a symmetrical nonuniform magnetic field produced by suitable pole pieces. See Fig. 15. The dumbbell has a resultant diamagnetic component and, when the cell is filled with a gas of low magnetic susceptibility, such as nitrogen, the dumbbell is deflected away from the intense part of the field. It takes an equilibrium position whereby the magnetic forces are balanced by the torsional torque produced by the suspension.

In some cells the suspension material is quartz fiber. In the later Munday cell the suspension utilizes platinum-iridium. When a gas is admitted to the cell, the dumbbell assumes a new position, one further out if the sample has a net paramagnetic component. In a simple, nonlinear instrument, the position can be measured by means of a mirror fastened to the dumbbell which reflects a beam of light onto a graduated scale.

A more accurate technique involves always restoring the dumbbell to its initial zero position. This is achieved by electrostatic forces in one instrument. The dumbbell is sputter-coated with a conducting metal, and two electrodes are placed adjacent to it and held at constant potential—one above and one below ground potential. A voltage is applied to the dumbbell from a suitable source to provide the required electric field. The voltage required to restore the dumbbell to the zero position is directly proportional to the volume magnetic susceptibility of the sample.

In the Munday cell (Fig. 16), a plantinum coil wound around the dumbell is supplied with a

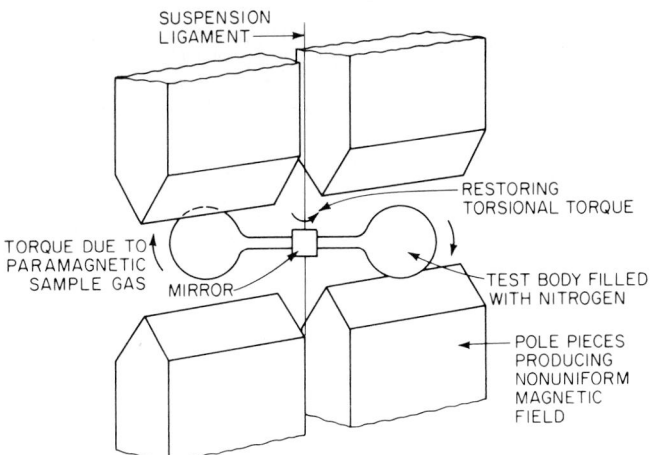

Fig. 15   Basic measuring element in a magnetodynamic analyzer.

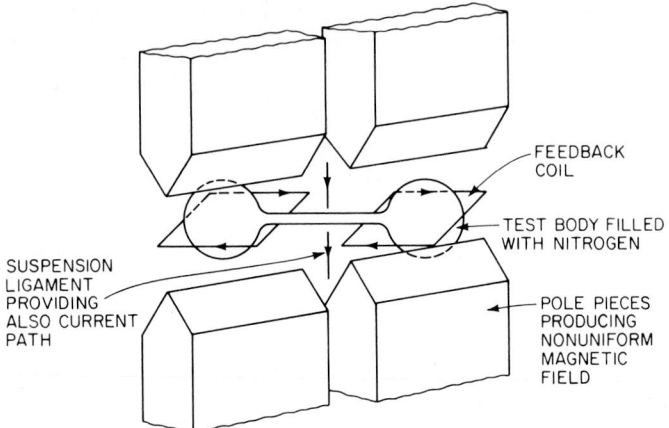

Fig. 16   Basic measuring element in a magnetodynamic analyzer with current feedback.

small electric current which interacts with the magnetic field and provides a restoring torque. The current required to restore the dumbbell to the zero position is directly proportional to the volume magnetic susceptibility of the sample. A self-nulling automatic instrument (Fig. 17) is made by the inclusion of two photocells to detect the deflection of the light beam. Feedback circuitry maintains the dumbbell at the null position.

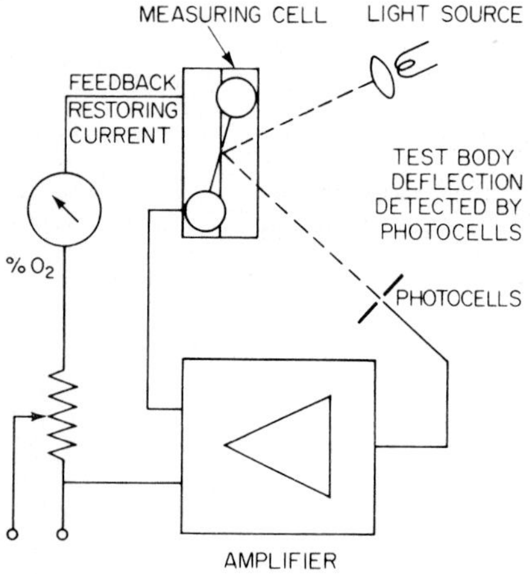

Instruments of the foregoing type offer high sensitivity in measuring the volume magnetic susceptibility of the sample. Provided that the sample does not contain large amounts of gases, such as nitric oxide, nitrogen dioxide, and chlorine dioxide (which are strongly paramagnetic), the errors that may be introduced by other common gases (with low susceptibilities) can be estimated accurately from a prior knowledge of sample composition. Water vapor introduces some errors. A dry sample usually is used. With a sample containing common gases and oxygen, the instrument can be calibrated to read directly in percentage oxygen. Or, the calibration may be in terms of oxygen partial pressure, in which case the instrument reading is directly proportional to the atmospheric pressure variation.

One instrument,* which is independent of atmospheric pressure variations, ratios the output of two channels. Each channel acts as a separate oxygen analyzer. One is a reference channel; the other is a sample channel. The instrument also can be used in a difference mode.

Fig. 17   Schematic of a magnetodynamic analyzer. The amplifier generates current to keep the dumbbell in a null position.

The instrument reading is inversely proportional to the square of the temperature. Temperature compensation and control are provided in the industrial range. The use of feedback yields exceptional linearity over the whole range (0 to 100% O_2). Feedback also increases the natural frequency of the system. The time response is dependent on the length of the sample lines and the response of the

*Servomex.

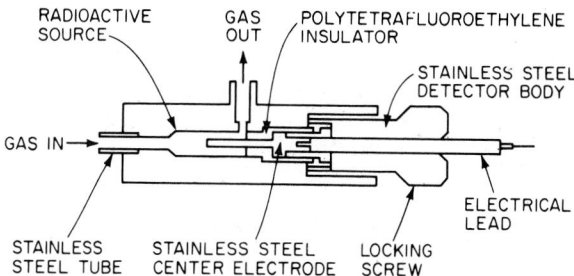

Fig. 18 Main part of an electron capture analyzer.

cell. In one cell,* the sample is diffused, giving a 90% response time of 40 s. However, special faster cells are available. The Munday cell has a special flow design, giving it a 90% response time of 7s. The errors resulting from changing the flow rate from 0 to 150 ml/min are less than 0.1% O_2. Tilt has a small effect on these instruments—about 0.005% O_2 per degree of tilt.

Electron Capture Analyzers

Oxygen absorbs low-energy beta radiation, whereas gases such as nitrogen and argon do not. The apparent reduction in the radiation of a given source is dependent on the number of oxygen molecules present. This principle can be used as a means of measuring oxygen to parts-per-million levels. Figure 18 shows a detector of this type.

The device uses a low-energy beta source (tritium). There are both inlet and outlet ports for passage of the carrier gas. Centrally placed in the cylinder is an electrode. The body of the detector is made of a stainless steel cylinder. When a gas such as nitrogen flows through the detector cell, a cloud of electrons is formed between the electrode and the wall of the cell as a result of beta emission from the radioactive source.

An applied voltage between the electrode and the body gives a standing current. When an electron-capturing material, such as oxygen, mixes with the carrier gas flowing through the detector, some of the flowing electrons are captured and the standing current falls. The amount of fall is proportional to the concentration of the electron capturing material. A commercial instrument† with a detector of this type and ranges from 0 to 5 to 0 to 250 ppm oxygen in nitrogen is available.

Catalytic Combustion

This method is suitable for the determination of oxygen content in noncombustible gases and saturated hydrocarbons. The method is based on measurement of the change in temperature due to total catalytic combustion of the oxygen in the sample (or a known proportion of it) with hydrogen.

The sample gas enters the analyzer at a constant rate, after which it is mixed with a constant stream of hydrogen, heated, and admitted to the measuring cell. The measuring cell is maintained at a constant temperature suitable for catalytic combustion. The measuring cell consists of two sections, each containing a platinum sensor. The measuring sensor is exposed to the sample; the other is a compensating sensor. Only a small amount of the sample enters the compensating sensor. Combustion of the mixture takes place at both sections, raising the temperature of the measuring sensor. The sensors are connected in a bridge arrangement which drives a recorder or meter calibrated in percentage oxygen.

*Beckman.
†Analytical Instruments, Ltd.

Inferential Thermal Conductivity

Hydrogen has a very high thermal conductivity, and it also reacts catalytically with oxygen. These properties of hydrogen are the basis of the inferential thermal conductivity method. The sample gas, at room temperature, is mixed with a constant stream of hydrogen. The mixture is admitted to one chamber of a dual thermal conductivity detector, after which the oxygen in the sample is reacted with the hydrogen in a suitable oven to form water. The remaining mixture, after removal of its water content, is admitted to the second chamber of the detector. The imbalance in the thermal conductivity bridge is directly proportional to the oxygen concentration in the gas stream.

REFERENCES

Adlhart, O. J.: "Fuel Cells," in *Van Nostrand's Scientific Encyclopedia,* 6th ed., Van Nostrand Reinhold, New York, 1982.

Arthur, R. M.: "On-Line Measurements Improve Activated Sludge Process Control," *Instrum. Techol.,* vol. 27, no. 9, September 1980, p. 103.

Bailey, S. J.: "Stream Analysis '80: System Concepts Gain," *Control. Eng.,* vol. 27, No. 6, June 1980, p. 71.

Bockris, J. O., and T. Srinivasan: *Fuel Cells: Their Electrochemistry,* McGraw-Hill, New York, 1969.

Breiter, M. W.: *Electrochemical Process in Fuel Cells,* Springer-Verlag, Berlin, 1969.

Meagher, R. F., and J. R. Grinker: "Sensors for Wastewater Plant Control," *Instrum. Technol.,* vol. 28, No. 5, May 1981, p. 51.

O'Meara, J. E., Jr.: "Oxygen Trim for Combustion Control," *Instrum. Technol.,* vol. 26, No. 3, March 1979, p. 47.

Staff: Papers presented at the Intersociety Energy Conversion Engineering Conference, Atlanta, Georgia, 1981.

Staff: Papers presented at National Fuel Cell Seminar, Norfolk, Virginia, 1981.

Williams, K. R.: *Introduction to Fuel Cells,* Elsevier, Amsterdam, 1966.

Oxidation-Reduction Potential Measurements

by
J. N. Harman III*

All chemical reactions involving an exchange of electrons are considered oxidation-reduction reactions. These reactions produce measurable and predictable potentials. The relative strengths of the oxidants and reductants involved can be measured by determining the oxidation-reduction potential (ORP) prevailing. This can be done by inserting an unattackable electrode, such as platinum, rhodium, or gold, into the solution together with a reference electrode and measuring the resulting electromotive force (emf) by means of a high-impedance potentiometer or amplifier. The emf or voltage measured is the difference between the individual voltages developed at each electrode. If a hydrogen gas reference electrode could be used, the potential at the measuring electrode would be the true ORP of the solution since the hydrogen gas reference electrode potential, by convention, is zero. Generally, a silver-silver chloride or calomel reference electrode is employed instead of hydrogen, and the potential developed by either of these electrodes requires a correction in the ORP meter reading to relate the measuring electrode potential to the hydrogen zero potential.

To understand and properly relate the ORP values and equations it is necessary to appreciate the rather confusing sign conventions prevailing. There are two existing conventions, and they differ only by being opposite in polarity. The *American Convention* expresses the potential with the polarity as it truly exists in the solution *surrounding* the noble measuring electrode. The *European Convention* concerns itself with the opposite potential which exists *on* the measuring electrode. Instrument makers and many users are concerned only with the sign of the electrode since the meter is only aware of this polarity. Thus, most ORP meters and recorders read out in signs followed by the European Convention.

Specific oxidation-reduction values for a wide variety of reactions involving an exchange of electrons can be found in handbooks and in the literature. The values, in volts, typically are referenced to the hydrogen-hydrogen ion couple (having *zero* potential) at unit activity and at 25°C. These values (for the reaction being measured) can be inserted into the general form of the Nernst equation:

$$E = E_0 - \frac{0.0591}{n} \log \left(\frac{\text{oxidant}}{\text{reductant}} \right)$$

where E = measured emf expressed in volts opposed to normal hydrogen electrode (zero potential is by definition)

E_0 = standard emf as found in the literature for oxidation-reduction reaction involved under the conditions specified

n = number of electrons involved in oxidation-reduction reaction

From the foregoing relationship, the expected potential can be predicted, depending on the ratio between oxidant and reductant. Conversely, by reading the measured potential, the prevailing ratio of oxidant to reductant can be determined.

*Senior Chemist, Beckman Industrial Corporation, a subsidiary of Emerson Electric Company, Fullerton, California. Some of the earlier work on the fundamentals of this topic was preserved from the second edition article authored by T. J. Kehoe and is gratefully acknowledged.

Table 1 ORP of Quinhydrone Solutions

| | Solution pH and temperature, °C | | | | | |
| | pH = 4 | | | pH = 7 | | |
Electrode	20°C	25°C	30°C	20°C	25°C	30°C
Calomel (mercury-mercurous chloride)	223	218	213	47	41	34
Silver–silver chloride	268	263	258	92	86	79
Hydrogen	470	462	454	295	285	275

Limitations of Methods

Many types of chemical reactions and processes can be controlled by ORP measurement, but with the following qualifications:

1. The measuring electrode senses oxidants and reductants in a solution. For example, hexavalent chromate may be the component requiring measurement. However, other oxidants and reductants, such as iron salts and sulfides, may be present, and these can influence the net potential.

2. In many systems, variations in the pH of the solution cause variations in the net potential of the solution, since many oxidation-reduction couples involve the hydrogen ion.

3. Temperature influences the potential slightly. The necessary correction usually is less than 1 mV/ °C but cannot be predicted theoretically for most systems since mixed oxidation-reduction reactions may be present.

Calibration

If the electrodes and amplification stage are performing properly, the measured potential is a true measurement not requiring calibration. On the other hand, the measuring electrode can become coated (and thereby insensitive) or can be "poisoned" by prior exposure to very strong oxidants or reductants and still retain a "memory" of this exposure. Thus, some users find it opportune to calibrate the system periodically. This can be done by placing the electrodes first in a pH 4 buffer solution containing a few grams of quinhydrone and then placing them in a pH 7 buffer solution also containing quinhydrone. The solutions should be freshly prepared just before use. Table 1 lists the millivolt values that can be expected from such solutions at several temperatures and using various reference electrodes. The values are all *plus* and follow the European convention.

Applications

Water treatment probably employs ORP measurement to a greater extent than any other group of processes. ORP is relied on almost universally both to monitor and to control the cyanide oxidation and chromate reduction processes so frequently required in industrial waste treatment of metal treating wastes before discharge into sewers or streams. Regulatory agencies often monitor freshwater streams with ORP analyzers to determine the relative "health" of receiving waters and to detect unwanted dumping of strong oxidants or reductants upstream. Some sewage plants employ ORP to monitor sewage influent, digester sludge, and to a limited extent, to control chlorination for in-plant odor control.

ORP is used to control the flow of chlorine or other oxidizing agents in various bleaching processes in the pulp and paper industry. A wide variety of applications exist throughout the chemical industry.

Measurement of Specific ions

Another type of ORP measurement involves the use of a metal-metal salt as the measuring electrode instead of a noble metal electrode. As an example, a silver-silver chloride measuring electrode can measure chloride concentrations ranging from a few to over 10,000 ppm. This essentially is an oxidation-reduction potential measurement. Thus, other oxidants and reductants in the solution being measured can, if in sufficient concentration, introduce a secondary potential, although most applications are not concerned with this aspect. Applications include the monitoring of streams for chlorides as an index of water quality, for salt intrusion detection, and to detect dumping of wastes of high chloride content. Other metal-metal salt combinations, such as silver-silver sulfide for measuring sulfide levels in aqueous solutions, have been used as process analyzers.

Selective ion electrodes have found wide application in laboratory determinations of many species of industrial significance. Continuous process analyzers incorporating the electrodes have been successfully applied in industrially important analyses, but significant sample pretreatment or preconditioning to afford useful operation is required.

Selective ion electrodes measure the activity of anionic species in solution by developing a potential which is related to the activity of the ion of interest and may be determined experimentally by measuring the voltage difference between the selective ion electrode and an appropriate reference electrode. The Nernst relationship expresses the potential of the selective ion electrode as a function of the ionic activity of interest:

$$E = E_0 - \frac{0.0591}{n} (\log a_{ion})$$

where E = measured emp expressed in volts compared to a normal hydrogen electrode (zero potential by definition)

E_0 = standard emf for oxidation-reduction reaction

n = number of electrons involved in reaction

Selective ion electrodes are usually calibrated against known standards to account for behavioral anomalies and to ensure that the system is functional for a given determination. Commercially available selective ion electrodes are classified by species in Table 2, in which the detection limit is given in parentheses after the species.

The reader is referred to the publications of Orion Research Corporation for information on selective ion electrode application and theory. See references listed. These electrodes have found widespread use in laboratory determinations but have been less frequently used for continuous process analysis because of the problems of short life of liquid ion-exchanger-based selective ion electrodes, the problems of cross-interferences which are more difficult to deal with in the continuous analysis setting than the laboratory setting, and the lack of long-term demonstrated electrode stability.

Table 2 Commercially Available Selective Ion Electrodes

Detection limit is given after the species

Glass	Solid state	Liquid membrane	Gas sensing
H^+ (10^{-14} M)	Br^- (5×10^{-6} M)	Cu^{2+} (10^{-5} M)	NH_3 (10^{-6} M)
Na^+ (10^{-7} M)	Cd^{2+} (10^{-7} M)	Cl^- (10^{-5} M)	CO_2 (10^{-7} M)
K^+ (5×10^{-5} M)	Cl^- (10^{-5} M)	NO_3^- (10^{-5} M)	NO_2 (2×10^{-5} M)
NH_4^+ (10^{-5} M)	Cu^{2+} (10^{-7} M)	ClO_4^- (10^{-5} M)	SO_2 (10^{-7} M)
Ag^+ (10^{-2} M)	Cr^- (10^{-6} M)	K^+ (10^{-5} M)	
	F^- (10^{-7} M)	$Mg^{2+} + Ca^{2+}$ (10^{-10} M)	
	I^- (1.5×10^{-10} M)		
	Pb^{2+} (10^{-7} M)		
	Ag^+ (10^{-14} M)		
	CNS^- (10^{-5} M)		

REFERENCES

Kehoe, T. J., and R. Jones: "ORP Measurement in Waste Treatment," *J. Water Sewage Works,* vol. 107, no. 8, August 1960.

Latimer, W. M.: *The Oxidation States of the Elements and Their Potentials in Aqueous Solutions,* Prentice-Hall, Englewood Cliffs, N.J., 1952.

Lingane, J. J.: *Electroanalytical Chemistry,* Wiley, New York, 1953.

Orion: *Orion Research Analytical Methods Guide,* Orion Research, Inc., Cambridge, Mass. (current).

Orion: *Applications Bulletin,* Orion Research, Inc., Cambridge, Mass. (current).

pH and pH Measuring Systems

by
J. N. Harman III*

One of the most important applications of solution potential measurements involves determination of the effective concentration of acids and bases in solution. Because acids and bases are of vital significance in such diverse fields as biology, medicine, food technology, and water and sewage treatment, as well as the chemical industry, measurements of the concentration and strength of acids and bases are the most common generally made on chemical systems. The pH measurement technique employs special selective electrodes that develop an emf proportional to the hydrogen ion concentration in the solution in which they are immersed. By definition,

$$pH = \log \frac{1}{\text{hydrogen ion concentration, mol/L}} \qquad (1)$$

All water solutions of acids and bases owe their chemical activity to their relative hydrogen and hydroxyl ion concentrations. In water, the equilibrium product of the hydrogen and hydroxyl ion concentrations is a constant, 10^{-14} at 22°C, and the pH scale is uniquely related to water at this temperature. Thus, when the concentrations of H^+ and OH^- in pure water at 22°C are equal, the H^+ ion concentration must be 10^{-7}, and pH $= \log (1/10^{-7}) = 7.0$.

The pH scale covers the range of both acid and alkaline solutions. Pure water is neither acidic nor basic. Acid solutions increase in strength as the pH value falls below 7. Alkaline solutions increase in strength as the pH rises above 7. The scale is not linear with concentration. A 1-unit change in pH represents a 10-fold change in the effective strength of an acid or base, e.g., a solution of pH 3 is 10 times as strong an acid solution as a solution of pH 4.

It is important to recognize that pH measures only the concentration of hydrogen ions actually dissociated in a solution, and not the total acidity or alkalinity. It is this factor that is responsible for the observed pH change in pure water with temperature. As the water temperature increases, the

*Senior Chemist, Beckman Industrial Corporation, a subsidiary of Emerson Electric Company, Fullerton, California. Some of the earlier work on the fundamentals of this topic was preserved from the second edition article authored by T. J. Kehoe and is gratefully acknowledged.

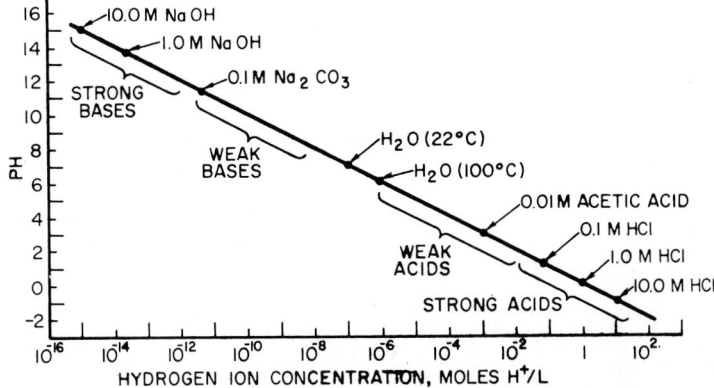

Fig. 1   Relationship among pH, hydrogen ion concentration, and various aqueous solutions of acids and bases.

amount of dissociation increases and the quantity of hydrogen and hydroxyl ions increases equally. Since pH is related to the concentration of hydrogen ions alone, the pH actually decreases, although the water is still neutral. Unless the relationship between dissociation constant and temperature is known, therefore, it is not possible to predict the pH of a solution at a desired temperature from a known pH reading at some other temperature.

pH meters generally cover a range of 0 to 14 pH units. However, it is possible to make measurements beyond these limits in very concentrated solutions. Figure 1 shows these relationships and indicates how the pH scale encompasses measurements on solutions of any strength.

Acid-Base Theory

Ionization and Neutralization in Solution

Acids dissolve in water to produce conductive solutions because charged ions are formed. The equation for the ionization of hydrochloric acid may be written

$$
\begin{array}{cccc}
\text{HCl} & = & \text{H}^+ & + & \text{Cl}^- \\
\text{hydrochloric} & & \text{hydrogen} & & \text{chloride} \\
\text{acid} & & \text{ion} & & \text{ion}
\end{array}
\tag{2}
$$

In the same manner, sodium hydroxide forms ions when dissolved in water:

$$
\begin{array}{ccccc}
\text{NaOH} & = & \text{Na}^+ & + & \text{OH}^- \\
\text{sodium hydroxide} & & \text{sodium} & & \text{hydroxyl} \\
& & \text{ion} & & \text{ion}
\end{array}
\tag{3}
$$

When these two solutions are mixed, a neutralization reaction occurs; i.e., the acidic and basic properties of the solution are lost because of the reaction of the hydrogen and hydroxyl ions to form water.

$$
\text{Na}^+ + \text{OH}^- + \text{H}^+ + \text{Cl}^- = \text{HOH} + \text{Na}^+ + \text{Cl}^-
\tag{4}
$$

The resulting solution is still conductive because of the presence of sodium and chloride ions but is neither acid nor alkaline if the quantities of the original acid and base were equivalent.

Normality and Concentration

The characteristic properties of acids are due to the presence of hydrogen ions. The characteristic properties of bases are due to the presence of hydroxyl ions. A solution of acid that contains 1 g/L of ionized hydrogen is of the same strength insofar as the neutralization reaction is concerned no matter what acid is used. Such an acid solution is said to be 1 N in acidity. Similarly a solution containing 17 g of hydroxyl ion, or 1 g/L ionic weight, is 1 N in alkalinity. This should not be confused with the more common usage of normality for solution concentrations on a gross weight basis.

Hydrochloric acid is referred to as a strong acid and sodium hydroxide as a strong base because both are almost completely dissociated into ions in aqueous solutions. Thus, the concentration of active hydrogen ions in the solution of a strong acid is approximately equal to the normality, and the same is true for the hydroxyl ion concentration of a strong base. This is not true for acids and bases which do not ionize completely and are consequently termed weak acids and bases.

Generalized Theory

It is convenient to write HA for the general form of an acid and BOH for the general form of a base. Acidic and basic properties are related to the behavior of water, since to a limited extent water is amphoteric and acts as either an acid or a base. Both hydrogen and hydroxyl ions are produced by the dissociation of water:

$$HOH = H^+ + OH^- \qquad (5)$$

The rate of ionization of water molecules is exactly equal to the rate of recombination of the ions, and the equilibrium expression can be written

$$K = [H^+][OH^-] = 10^{-7} \times 10^{-7} = 10^{-14} \text{ at } 22°C \qquad (6)$$

where the brackets indicate the concentration of the ions in moles per liter. K is called the ion product constant.

Actually hydrogen ions, or protons, do not exist free in solution but are associated with solvent molecules. The ionization of water would thus more properly be written

$$HA + HOH = H_3O^+ + A^- \qquad (7)$$

H_3O^+ is called the *hydronium ion* and, in aqueous systems, is the ion responsible for acidic properties; equations are more simply written using H^+, and this practice will be followed in this discussion. Strong acids such as hydrochloric, nitric, sulfuric, and perchloric all appear to be of the same strength in water, because each completely dissociates to form hydronium ion which acts as the acid.

pH Theory

In the foregoing discussion, it has been assumed that strong acids ionize completely, producing hydrogen ion concentrations equal to the concentration of the acid present. When the concentration of hydrogen ions exceeds approximately 0.01 N, the properties begin to deviate from the ideal behavior of ions predicted by thermodynamic theory. The active hydrogen ion concentration appears to be lower than the value corresponding to complete ionization.

This apparent concentration is called the *activity* of the ion and is obtained by multiplying the ion concentration by the activity coefficient. The value of the activity coefficient approaches 1 in very dilute solutions. The difference in the ion concentration and the ion activity is largely accounted for by the interaction of the electric fields associated with ions when they are present at high concentrations. Hydrogen ion activity is the most useful measure of effective acid strength. For many years the general method of expressing acid strength was to give the hydrogen ion concentration in moles per liter. This was awkward because of the wide concentration range to be covered and the fact that acid strength and concentration are easily confused.

The pH Scale

Sorenson, in 1909, proposed that the expression pH be adopted for hydrogen ion concentration and defined pH as an operator standing for "power of hydrogen" in the following manner:

$$pH = -\log [H^+] \tag{8}$$

The pH of a solution may be determined by measuring the voltage of a concentration cell comprising two platinum electrodes, one immersed in a solution of known pH and the other immersed in the sample, with the two solutions separated by a salt bridge and with hydrogen gas at known pressure contacting the two electrodes. The voltage of such a cell may be expressed as

$$E_H = \frac{RT}{F} \ln \frac{(a'_{H+})}{(a'_{H_2})^{1/2}} - \frac{RT}{F} \ln \frac{(a''_{H+})}{(a''_{H_2})^{1/2}} \tag{9}$$

where (a'_{H+}) and (a''_{H+}) are the activities of the hydrogen ions in the known solution and in the sample, respectively, and (a'_{H_2}) and (a''_{H_2}) are the activities of the hydrogen gas. If (a'_{H+}) and (a'_{H_2}) are unity, the reference electrode is a standard hydrogen electrode and the first term in the equation becomes zero. If the hydrogen pressure over the sample is 1 atm, then (a''_{H_2}) becomes unity and the voltage is dependent solely on the hydrogen ion concentration. Then

$$E_H = -\frac{RT}{F} \ln (a'_{H+}) = -0.0591 \qquad E_H \text{ (at 25°C)} = 0.0591 \text{ pH} \tag{10}$$

Thus,

$$pH = \frac{E_H}{0.0591} \text{ at 25°C} \tag{11}$$

and is a useful expression of acidity.

As stated at the beginning of this discussion, the pH scale includes alkaline solutions. This is possible because of the relation

$$K_w = [H^+][OH] = 10^{-7} \times 10^{-7} = 10^{-14} \text{ at 22°C} \tag{12}$$

Thus in a solution in which hydroxyl ion activity is unity,

$$a_{H+} = \frac{10^{-14}}{1} = 10^{-14} \qquad pH = 14 \tag{13}$$

The general practice is to speak of the hydrogen ion concentration as corresponding to a given pH when actually the effective concentration or activity is meant. This practice will be followed in the remainder of this discussion.

Table 1 lists pH values and some other characteristics for a variety of acids and bases. For a more complete treatment of acid-base theory and of activity, the reader is referred to standard textbooks on physical chemistry.

Buffer Solutions

These are solutions that resist a change in pH despite the additions of acid or base. Buffer action is exhibited by any solution that contains substantial concentrations of both a weak acid and the salt of that acid. When dissolved, the salt yields a substantial concentration of anions (A^-) and the weak acid yields mostly undissociated molecules (HA). Mixtures of weak bases and their salts also act as buffers in alkaline solutions. The hydrogen ion concentration, $[H^+]$, of the solution is determined by the mass action law:

$$\frac{[H^+][A^-]}{[HA]} = K_A \quad \text{(ionization constant)} \tag{14}$$

Table 1 pH Values and Characteristics of Various Materials

Compound or material	Total normality	Effective H^+ (as normality)	pH (at 25°C)
Acids and common acidic solutions			
Hydrochloric acid	1.0	0.8	0.1
	0.1	0.083	1.08
	0.001	0.001	3.00
Sulfuric acid	1.0	0.48	0.32
	0.1	0.68	1.17
Acetic acid	1.0	0.0043	2.37
Lemon juice	—	0.01–0.0063	2.0–2.2
Acid fruits	—	10^{-3}–3×10^{-5}	3.0–4.5
Vegetables, including melons	—	10^{-5}–10^{-7}	5.0–7.0
Jellies, fruit	—	10^{-3}–3×10^{-4}	3–3.5
Fresh milk	—	3×10^{-7}–2.2×10^{-7}	6.50–6.65

Temperature, °C	pH
Water at various temperatures	
0	7.472
22	7.00
25	6.998
50	6.631
100	6.13

Compound or material	Total normality	Effective OH^- (as normality)	pH (at 25°C)
Bases and basic solutions			
Sodium hydroxide	1	0.57	13.73
	0.1	0.071	12.84
Ammonia (10% NH_3)	5.9	0.006	11.8
	0.1	0.0018	11.27
Lime water [$Ca(OH)_2$ sat.]	—	0.04	12.4
Trisodium phosphate $Na_3PO_4 \cdot 10H_2O$, 2%	—	0.009	11.95
Blood plasma, human	—	—	7.3–7.5

Acids	Bases
Common characteristics	
Sour taste	Bitter taste
Neutralize bases to yield water and salts	Neutralize acids to yield water and salts
Catalyze some reactions	Ionize in water to yield hydroxyl ions
Ionize in water to yield hydrogen ions	

The constancy of K_A over a wide range of concentrations, and thus the ratio of the free acid times the anion concentration divided by the undissociated acid concentration, forces the buffering action when the system is perturbed.

If a strong acid is added to this solution, the resulting hydrogen ion concentration will be less than if the acid were added to the same amount of pure water, because the hydrogen ion concentration cannot increase and the A^- and HA concentrations remain unchanged. This would be in violation of the mass action law. Conversely, if something should occur, such as dilution or neutralization, to alter the mass action relationship to decrease the hydrogen ion concentration, the HA molecules will dissociate until the law is again satisfied. In the measurement of pH, buffer solutions are used to calibrate the electrodes and the system to ensure measurement system accuracy. The composition of some standard buffer solutions is given in Table 2.

METHODS OF pH MEASUREMENT

Two methods are in general use for the direct determination of pH in chemical systems: chemical indicators and potentiometric instruments, i.e. pH meters.

Chemical Indicators

Early in the history of chemistry it was found that the color of certain compounds in solution was dependent on the hydrogen ion concentration. These compounds are in general weak acids and bases and their salts. When the salt of a weak acid is different in color from the nonionized acid, the resultant color of the solution will depend on the ratio of the concentrations of the two forms. The pH range in which the color change takes place with an acid indicator, for example, is dependent on the ionization constant of the acid. When the ratio of acid form to salt is 1, then $K = [H^+]$; this is the midpoint of the pH range of color change.

Indicators have been found for most ranges of pH and for many years offered the only available means of determining pH. pH determinations are made with indicators by adding a small quantity of indicator solution to the sample and comparing the color with a color standard. When good color standards are available in steps of 0.2 pH unit and observations are made in a "comparator" by a skilled observer, 0.1 pH unit is considered the limit of accuracy for the method. For rough measurements the colorimetric method is convenient and inexpensive. Turbid and colored solutions cannot be observed with accuracy, and indicators are not stable in many strongly oxidizing or reducing solutions.

Electrometric pH Measurement

The basic measuring system comprises (1) a pH-responsive electrode, (2) a reference electrode, and (3) a potential measuring instrument.

Hydrogen Electrode

Considered to be the standard for the electrometric measurement of pH, the hydrogen electrode was used experimentally to follow changes in acidity as long ago as 1897. (W. Böttger, *Z. Phys. Chem.*, vol. 24, 1897, p. 253.) The hydrogen electrode is employed by bubbling hydrogen gas past a wire or foil which is able to catalyze the reaction $H^+ + e \rightleftharpoons \frac{1}{2}H_2(g)$ and thus establish an equilibrium between molecular hydrogen and the hydrogen ions. The metal wire or foil usually is platinum which has been pretreated to provide a "platinum black" catalyzing surface.

Table 2 Buffer Solutions*

Standard buffer systems: pH values 2.2–10.0 at 20°C

pH	0.2 M KH phthalate, ml	0.2 M HCl, ml	Dilution, ml
	Potassium phthalate plus hydrochloric acid		
2.2	50	46.60	To 200
2.4	50	39.60	200
2.6	50	33.00	200
2.8	50	26.50	200
3.0	50	20.40	200
3.2	50	14.80	200
3.4	50	9.95	200
3.6	50	6.00	200
3.8	50	2.65	200

pH	0.2 M KH phthalate, ml	0.2 M NaOH, ml	Dilution, ml
	Potassium phthalate plus sodium hydroxide		
4.0	50	0.40	To 200
4.2	50	3.65	200
4.4	50	7.35	200
4.6	50	12.00	200
4.8	50	17.50	200
5.0	50	23.65	200
5.2	50	29.75	200
5.4	50	35.25	200
5.6	50	39.70	200
5.8	50	43.10	200
6.0	50	45.40	200
6.2	50	47.00	200

Quinhydrone Electrode

This electrode consists of a gold or platinum electrode surrounded by a saturated solution of quinhydrone. Quinhydrone consists of equimolecular amounts of benzoquinone and hydroquinone. These two substances, together with the hydrogen ion to be measured, form a reversible oxidation-reduction system. Thus, the quinhydrone electrode will assume a potential based on the activities of these three substances. The electrode is limited to pH measurements in the acidic region, hence is impractical for continuous process measurements.

Antimony Electrode

This electrode simply consists of antimony metal coated with antimonous oxide. The potential of the electrode varies with pH and thus serves as a means of measuring pH. The potential is an oxidation-reduction type developed by the metal-metal salt. The electrode is simple to construct and rugged. However, the potential is influenced by (1) the level of oxygen dissolved in solution, (2) any oxidants or reductants present, (3) a variable temperature coefficient over its 0 to 11 pH span of useful measurement, and (4) the need for frequent standardization due possibly to change in electrode surface characteristics with time.

Glass Electrode

It was not until development of the glass electrode that pH measurement became a simple, reliable tool for all types of users. See Fig. 2. The possibility that a thin glass membrane of special compo-

Table 2 Buffer Solutions* (*Continued*)

Standard buffer systems: pH values 2.2–10.0 at 20°C

	Monobasic potassium phosphate plus sodium hydroxide		
pH	0.2 M KH$_2$PO$_4$, ml	0.2 M NaOH, ml	Dilution, ml
5.8	50	3.66	To 200
6.0	50	5.64	200
6.2	50	8.55	200
6.4	50	12.60	200
6.6	50	17.74	200
6.8	50	23.60	200
7.0	50	29.54	200
7.2	50	34.90	200
7.4	50	39.34	200
7.6	50	42.74	200
7.8	50	45.17	200
8.0	50	46.85	200

	Boric acid plus potassium chloride plus sodium hyroxide		
pH	0.2 M H$_3$BO$_3$ + 0.2 M KCl, ml†	0.2 M NaOH, ml	Dilution, ml
7.8	50	2.65	To 200
8.0	50	4.00	200
8.2	50	5.90	200
8.4	50	8.55	200
8.6	50	12.00	200
8.8	50	16.40	200
9.0	50	21.40	200
9.2	50	26.70	200
9.4	50	32.00	200
9.6	50	36.85	200
9.8	50	40.80	200
10.0	50	43.90	200

*From Daniels, Mathews, and Williams, *Experimental Physical Chemistry*, 3d ed., McGraw-Hill, New York, 1941, p. 441.

†This solution is prepared by dissolving 12.41 g of purified H$_3$BO$_3$ and 14.91 g of purified KCl in redistilled water and diluting to 1000 ml.

sition could develop a potential was described as early as 1909 by the German chemist Fritz Haber. However, little progress with this approach was made until the mid-1920s. Developments since have accelerated to the point that both laboratory and continuous process pH measurements today are almost as common as temperature and pressure measurements. This acceptance primarily is due to development of the glass electrode together with improved means of extremely stable electronic amplification of the potential developed by the electrode.

The mechanism whereby certain types of glass develop potentials depending on the pH level is an extremely complex one. It has been experimentally observed that a glass electrode produces a predictable potential directly related to the hydrogen ion concentration of the solution in which it is immersed. The glass electrode responds in a predictable fashion throughout the normally accepted 0 to 14 pH range, developing 59.2 mV per pH unit at 25°C consistent with the classical Nernst equation. The recognition that actual pH electrodes seldom exhibit the theoretically predicted millivolt-pH sensitivity is a relatively recent event observed in many modern analyzers which allow slope adjustment to calibrate the analyzer for electrodes with less than the theoretical slope. Unlike the earlier types of pH electrodes, the glass electrode is not influenced by oxidants or reductants in the solution. With suitable temperature compensation, pH measurements can be made from 0 to 100°C, and to even higher temperatures under controlled conditions. The glass electrode pH measurement, by its obvious character, involves high resistances in the measuring loop. This demands that a high-impedance amplifier be employed. Glass electrodes have resistances varying from a few to a thousand megohms. In recent years, this has ceased to be a significant problem with use of modern electronic circuitry.

The *asymmetry potential* (AP) of a glass electrode can be observed by coupling the glass electrode to a stable reference electrode of the same internal half-cell system as used in the glass electrode in a buffer and noting the voltage developed by the electrode pair. Typically, most pH glass electrode–reference electrode pairs are disposed to put out 0 mV in a pH 7 buffer at 25°C—deviations in values are termed the AP. By definition, asymmetry potential implies that the two surfaces of the membrane may not respond in an identical fashion for various reasons. The possibility of one side of the membrane responding differently, even to the same type of solution, increases as the age and use of the electrode increases. Most pH meters have an asymmetry potential adjustment control, which is simply a variable potentiometer, to correct for any drift which does occur.

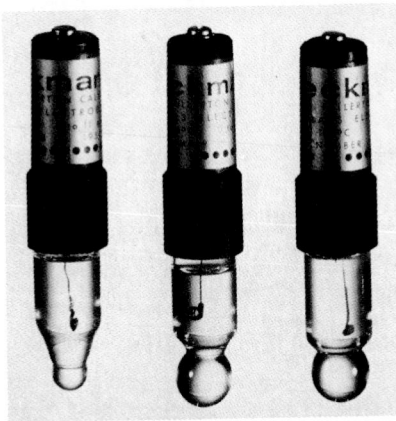

The asymmetry potential drift, when it does occur, is not a sharp or sudden change but almost can be regarded as a constant of the system which requires periodic adjustment by standardization with a buffer solution. It is the author's experience that weekly standardization will keep a continuous pH measurement within ±0.05 pH unit accuracy, assuming that amplifier drifting and electrode fouling do not occur.

Fig. 2 Configurations of glass electrode. *(Beckman Industrial Corporation.)*

The most likely cause for asymmetry potential drift in continuous pH measurements is prolonged chemical or abrasive attack of the stream on the glass membrane. When either of these occurs, the outer surface becomes altered to the extent that the response of the membrane to the presence of hydrogen ions gradually alters.

Dehydration of the outer surface by prolonged exposure to alcoholic solutions also can cause asymmetry potential drift. When glass electrodes have been stored dry for long periods of time, the pH readings continue to drift for periods up to 24 to 48 h until a full equilibrium condition is attained. The adsorption of surface-active agents and a greasy film can cause an apparent asymmetry potential drift as the exchange capacity of the surface is upset. Exposure of the glass membrane to strong acids and caustic solutions can permanently alter the external surface of the glass. Hydrofluoric acid can quickly etch the membrane to the point of complete destruction. The term "asymmetry potential change" attempts to summarize the several effects of this usually slight decay.

Reference Electrode

In pH measurement, a second electrode is required in the solution simply to complete the circuit. A good, stable reference electrode (1) must produce a predictable potential compatible with the glass measuring electrode, (2) must be linear with respect to temperature change, and (3) must be simple to use.

Although various reference electrodes have been employed, only mercury-mercurous chloride (calomel) and, more recently, silver-silver chloride reference electrodes are in common use today. The calomel electrode consists of an inner glass tube packed with mercury mixed with mercurous chloride. A hole in the bottom of the tube communicates with a saturated potassium chloride solution, and this in turn is contained in a relatively large glass chamber. At the bottom of this chamber one of various types of junctions is employed to permit the potassium chloride to diffuse or leak into the solution being measured. To complete the circuit, a wire is inserted into the packed column and goes to the amplifier along with the shielded cable from the glass electrode. With this arrangement, the potential developed at the calomel-KCl interface is constant at a given temperature since the solution surrounding it remains the same.

The silver-silver chloride electrode can be nearly identical to the calomel electrode. Instead of the packed column, only a silver wire need be employed if coated with a heavy coating of silver chloride. The silver-silver chloride electrode finds broader acceptance since it remains stable and reproducible at much higher temperatures than the calomel electrode. The latter electrode is limited to about 80°C maximum, particularly for prolonged or continuous use.

Aside from the type of internal employed, the other important parameter of the reference electrode is its type of junction. Different junction styles have differing advantages and limitations. Various schemes are employed, such as an asbestos fiber embedded in the glass; the use of a glass bead with a temperature coefficient different from that of the glass body surrounding it to permit a controlled crack around the bead for liquid communication; a palladium wire embedded in a near molten glass body with, again, an annular space surrounding the wire resulting after the glass cools; and a ground glass sleeve mated to a ground glass portion of the body. Reference electrodes are now available fabricated from plastic and incorporate a polymeric diffusion junction; these junctions have found successful application in the process industries.

Temperature Compensation

Examination of the Nernst equation reveals that the magnitude of the Nernst slope factor is a function of the temperature at the measurement electrodes. Most process pH measurement systems incorporate a temperature compensating element in the process sample to normalize the temperature-dependent slope to standard conditions by appropriate modifications of the gain of the pH amplifier; devices used for temperature compensation are generally suitably packaged resistors, thermistors, or platinum resistance thermometers. It is important to point out that the only function of the temperature computation is to correct for the change in the Nernst slope with temperature—it does not correct for changes in the reference electrode or pH electrode absolute potential with temperature or for a change in the pH of a test solution due to temperature changes.

Developments in Measurement Hardware

Significant effort has been expended in the development of novel electrode packaging configurations. Combination electrodes, incorporating the pH and reference electrode together with the temperature compensation device in one package, suitable for threading into a pipe fitting for flowing measurement applications or submersion measurement applications, are available from many manufacturers. A rather useful development is the "live" insertion probe, available from many manufacturers, which allows a combination electrode to be inserted into and withdrawn from a valve fitting in a tank wall or process line, allowing periodic calibration or maintenance of the probe without shutting down the process.

Ultrasonic probe cleaners have been successful in many applications where electrode life has been limited as a result of coating problems. This technique seems to work best for coating with hard crystalline precipitates, although some success has been reported in preventing the coating of electrodes with films of grease and oil by use of the ultrasonic cleaning technique. In use, the electrodes are continuously or periodically exposed to ultrasonic excitation from a transducer mounted in the solution close to the pH and reference electrodes.

Use of Microprocessors

The advent of the relatively low-cost microprocessor chip has had a significant impact on the features now available in modern process pH analyzers. It is possible to use the computational capability to calculate parameters of interest to the end user, such as the actual pH electrode slope (millivolts/pH unit) obtained when it has been calibrated in buffer solutions; to program in corrective factors to normalize the temperature dependency of the pH of solutions due to changes in solution composition; and to have the ability to simulate pH changes by keying in inputs to the microprocessor to verify any pH-dependent function, such as alarms, operation of a final control element and recorder, or computer inputs. The accuracy of temperature compensation is improved by computing the amount of temperature compensation required with a mathematical equation solved by the microprocessor.

Allied to the development of microprocessor-based pH-ORP instrumentation (see previous article on ORP measurements) is the development of pH controllers based on the same methodology. These controllers allow greater flexibility and accuracy than the earlier analog generation. Both on-off control and pulse duration control schemes are commercially available, and one such controller allows the user to program into the instrument the titration curve characteristics of a particular process to specifically tune the instrument to the particular end application.

Applications

There are more continuous pH analyzers in use than all other continuous analytical-type instruments combined. Continuous pH analyzers can be found in practically every industry that uses water within its processes. Installations of pH analyzers throughout the free world probably are in excess of 10,000. A sizable portion of these form the primary measuring element within closed-loop automatic control systems. Applications range from industrial water and waste treatment to pH control for flotation processes in the mining industry.

Numerous pH applications can be found in the pulp and paper metals and metal treating fields, and in petroleum refining, synthetic rubber manufacturing, power generation plants, pharmaceuticals, chemical fertilizer production, and a broad spectrum of the chemical industry. See Fig. 3.

When properly applied, pH measurements offer the following benefits: (1) Many chemical reactions can be controlled better, resulting in greater processing efficiency and better quality of the ultimate product, and sometimes contributing to a safer process. (2) Records are provided to assure management that optimum processing operations prevail. In the case of industrial waste discharge into public streams, lakes, or collection systems, the recorded measurement provides evidence of compliance from a regulatory standpoint. (3) In certain cases, continuous pH measurement has permitted certain processes to be converted from a batch to a continuous basis, e.g., certain chemical fertilizer processes and some food manufacturing processes.

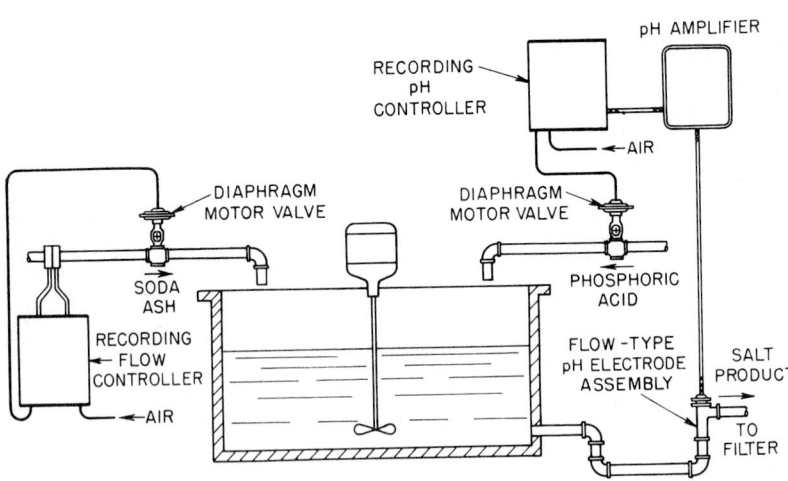

Fig. 3 Continuous, automatic neutralization in the manufacture of disodium phosphate is made possible through the use of automatic flow and pH controllers. The control afforded by this system protects the purity of the salt produced and prevents the waste of reagents.

REFERENCES

Bates, R. G.: *Determination of pH Theory and Practice,* Wiley, New York, 1964.

Dole, M.: *The Glass Electrode,* Wiley, New York, 1941.

Eiserman, George: *Glass Electrodes for Hydrogen and Other Cations: Principles and Practices,* Dekker, New York, 1967.

Ives, D. J. G., and G. J. Janz: *Reference Electrodes,* Academic Press, New York, 1961.

Mattock, G.: *pH Measurement and Titration,* MacMillan, New York, 1961.

Skinskey, F. G.: *pH and pION Control in Process and Waste Streams,* Wiley, New York, 1973.

Electrolytic Conductivity and Resistivity Measurements

by
Elmer Sperry* and John Nagy†

Industrial interest in the measurement of electrolytic conductivity (of which electrolytic resistivity is the reciprocal) arises chiefly from its usefulness as a measure of ion concentrations in water solutions, together with the fact that the equipment is simple and inexpensive. Water itself is a very poor conductor. The very purest water, such as may be obtained by passage through a mixed-bed ion exchanger, has a conductivity approaching very closely the theoretical minimum of approximately 0.05 μS/cm (18 M$\Omega \cdot$cm) at 25°C, which is due to the dissociation products of water itself. The conductivity of a water solution is, in practice, almost exclusively due to a dissolved electrolyte rather than to the solvent (water) ions, and thus a criterion for electrolyte concentration can be established.

Solutions of strong electrolytes follow a rather uniform pattern of change in conductivity with concentration, which is almost linear at low concentrations, rising more gradually to a maximum (usually about 20 to 30% by weight) and then falling as the concentration rises further. A series of such conductivity concentration curves is shown in Fig. 1.

Range of Conductivity Measurements

The range of practical conductivity measurement extends from a fraction of a microsiemen per centimeter to more than 1 S/cm. Although electrolytic conductivity measurements are widely used in the very low-concentration region, where they are almost uniquely sensitive, many applications have

*Chief Chemist

†Section Manager, The Cedar Grove Division of Beckman Industrial Corporation, a subsidiary of Emerson Electric Company, Cedar Grove, New Jersey.

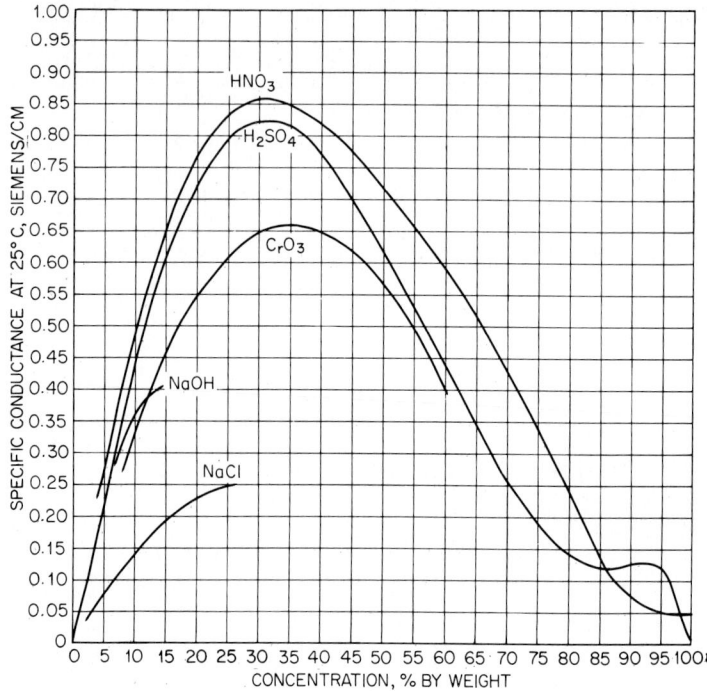

Fig. 1   Conductivity-concentration curves for certain electrolytes.

been established in medium- and high-concentration regions. Since water is so poorly ionized, addition of the slightest trace of electrolytic material causes a large increase in conductivity. For example, distilled or demineralized water with a conductivity of 2 μS/cm at 25°C will double its conductivity on addition of 1 ppm of a typical salt, and 1 ppm of a strong acid will increase its conductivity by as much as 500%.

At very high concentrations, because of the steep negative slope of the conductivity-concentration curve, conductivity measurements in such systems as concentrated sulfuric, hydrofluoric, and nitric acids have proved to be exceedingly sensitive to small concentrations of water. Differences in acid concentrations of 0.05% in the concentration span 95 to 100% acid are easily measured.

Most practical applications of electrolytic conductivity measurement fall into one of the following categories.

Concentration in Simple Water Solutions

Common examples are sodium chloride, sodium hydroxide, and sulfuric acid. In such cases, the conductivity-concentration curve must be known in advance or else the system must be experimentally calibrated.

Gauging the Quality of Pure Water

Examples are distilled or demineralized water and condensed boiler steam. The exact nature of the dissolved electrolyte is usually of less importance than an estimate of the magnitude of its concentration, and so an average value of about 2 S/cm at 25°C for each part per million of salt is assumed. Ultrapure water is discussed a bit later.

Measuring the Extent of Reactions

Examples include neutralizations, precipitation, and the washing of soluble electrolytes from insoluble materials. Such procedures require calibration or sometimes simply a comparison between the conductivities of the incoming and outgoing streams.

Detecting Contaminations

Conductivity measurements have been widely used to detect leaks in heat exchangers and resultant contamination of heating or cooling media in such equipment as acid coolers, condenser coils, and steam coils. Any sudden increase in conductivity of the heat-exchange medium is taken to indicate a leakage of electrolyte. A break in steam condenser tubes results in an increase in condensate conductivity.

Natural Waters

In fresh water and saltwater, conductivity measurements can be related to total dissolved solids content and salinity. Both laboratory and field instruments are in wide use. In fresh water, conductivity provides a rapid quality check and is useful in the field for locating sources of pollution which generally increase the dissolved solids content and, therefore, conductivity. In saltwater, conductivity provides a rapid and accurate method of determining salinity. Inductive-type laboratory salinometers may be used to determine salinity to an accuracy of $\pm 0.01\%$. Saltwater intrusions in streams and wells can be detected by conductivity measurements.

Interfaces, Suspensions, and Emulsions

Simple two-part systems in which one part is conductive and the other nonconductive, such as water and oil, provide applications for conductivity measurements. The interface between oil and water may be located easily by probing with a conductivity cell. Generally, an electrodeless or inductive-type cell is preferred, since oil may coat the electrodes and reduce sensitivity. Suspended droplets or particles of a nonconducting material will reduce the ability of a given volume of electrolyte to conduct electricity and provide an application for conductivity measurements in suspensions and slurries. Oil-in-water and water-in-oil emulsions may be distinguished by the fact that the former are conductive and the latter are essentially nonconductive.

Enhanced Conductivity

A sample may be pretreated to enhance the conductivity of certain ions which are of interest. Thus, very conductive hydroxyl ions may be removed from boiler water by the addition of a weakly ionized organic acid to reveal a conductivity which is more nearly proportional to that of the remaining dissolved salts. In a similar manner, conductive ammonium hydroxide may be removed from steam condensate by passage through a hydrogen form cation exchanger to reveal the conductivity of the remaining dissolved salts which are converted to their corresponding mineral acids. Similarly, samples may be boiled or sparged to remove conductive dissolved gases.

Fundamentals of Electrolytic Conductivity Measurement

The flow of electricity through matter is accomplished by the movement of electric charges, which in metallic conductors are electrons and in electrolytic conductors are positive and negative ions. Conducting solutions in general are electrolytic conductors. In electrolytic conductors, current is usually introduced and leaves the system through metallic electrodes on the surface of which chemical reactions occur. It is possible when using alternating current to cause current to flow by inductive coupling as well as by direct contact between electrode and electrolyte. Inductive coupling is discussed later under measuring circuits.

Positive ions or cations move toward the cathode, where reduction takes place, and negative ions or anions move toward the anode, where oxidation takes place. The conductivity of a solution depends on the concentration and mobility of all the ions present. The ion mobility in turn depends on ion size and charge, as well as the dielectric constant of the solvent and the solution temperature and viscosity.

The determination of electrolytic conductivity presently consists of measuring the ac electrical conductance of a column of solution. Although ac measurement methods greatly reduce errors associated with electrolysis, when electrodes are used, they introduce other errors associated with series and shunting capacitance which must be compensated for in the design of the measuring instrument. A precision of 0.01% can be obtained in laboratory measurements following the bridge techniques first discussed by Grinnell Jones and his coworkers. Industrial conductivity meters are capable of providing accuracies of 1% of the actual conductivity under ideal conditions.

Electrolytic conductivity is most often measured by placing electrodes in contact with the electrolytic solution which is contained in such a way that the measured electrical conductance between the electrodes can be related to the conductivity of the solution. The conductivity cell most commonly comprises an enclosure made of electrically insulating material, such as glass or plastic, which serves to hold or isolate a portion of the electrolytic solution and to accommodate the two electrodes. The cell constant of such a device is then used to relate the measured electrical conductance between the electrodes to the actual electrolytic conductivity.

Two electrodes, 1 cm^2, located on opposite interior faces of a hollow cube, 1 cm on an edge, would have a cell constant of 1/cm, and a measured conductance of 100 μS at 25°C would indicate a conductivity of 100 μS/cm (10 mS/m) at 25°C.

Definitions and Units

Electrolytic conductivity is often defined as the electrical conductance of a unit cube of solution as measured between opposite faces. It is expressed in the same units as electrical conductivity, namely, reciprocal ohms per unit length. Most commonly we find: mho/centimeter (mho/cm), siemens/centimeter (S/cm), and siemens/meter (S/m)

$$1 \text{ mho/cm} = 1 \text{ S/cm} = 100 \text{ S/m}$$

Few solutions exhibit conductivities as great as 1 S/cm. The most commonly used decimal submultiples are: micromho/centimeter (μmho/cm), microsiemens/centimeter (μS/cm), and millisiemens/meter (mS/m)

$$1 \text{ } \mu\text{mho/cm} = 1 \text{ } \mu\text{S/cm} = 0.1 \text{ mS/m}$$

Electrolytic resistivity (the reciprocal of conductivity) is similarly defined as the electrical resistance of a unit cube of solution. It is expressed in the same units as electrical resistivity, namely, ohms times a unit of length. Most commonly we find:

$$\text{ohm} \cdot \text{cm } (\Omega \cdot \text{cm}) \text{ and ohm} \cdot \text{meter } (\Omega \cdot \text{m}) \qquad 100 \text{ } \Omega \cdot \text{cm} = 1 \text{ } \Omega \cdot \text{m}$$

Again, decimal multiples commonly encountered are:

$$\text{megohm-centimeter } (\text{m}\Omega \cdot \text{cm}) \text{ and megohm} \cdot \text{meter } (\text{M}\Omega \cdot \text{m}) \qquad 100 \text{ M}\Omega \cdot \text{cm} = 1 \text{ M}\Omega \cdot \text{m}$$

Resistivity units are used almost exclusively to describe ultrapure water in the 10 to 18 M$\Omega \cdot$cm (0.1 μS/cm to 0.05 μS/cm) range generated by mixed-bed ion exchange and used as boiler feed water and in certain critical washing applications.

The cell constant of a conductivity cell is defined as a factor which relates the measured conductance between the cell terminals to the conductivity of the electrolyte being measured. It is generally expressed in reciprocal units of length (although occasionally in units of length by certain European manufacturers). Most commonly we find:

$$1/\text{centimeter } (\text{cm}^{-1}) \text{ and } 1/\text{meter } (\text{m}^{-1}) \qquad 1 \text{ cm}^{-1} = 100 \text{ m}^{-1}$$

The conductance measured between the cell terminals is multiplied by the cell constant given in reciprocal units of length to calculate the conductivity. The measured resistance between the cell terminals is divided by the cell constant to calculate the resistivity. Although the cell constant in reciprocal units of length can be calculated from the dimensions of the conductivity cell by dividing the length of the electrical path through the solution by the cross-sectional area of the path, in practice these measurements are difficult to make and are only used to approximate the cell constant which is determined by the use of standard solutions of known conductivity or by comparison with other conductivity cells which have been so standardized. See the section below on cell constant determination.

Measuring Circuits

Although several circuits are used for measuring electrical conductivity, as described below, the ac Wheatstone bridge is widely applied and is potentially the most stable and accurate.

AC Wheatstone Bridge

A typical system, shown in Fig. 2, comprises the bridge, including the voltage source, the null indicator, and the conductivity cell. In Fig. 2, D represents an ac voltage-sensitive device called a detector. The ac source may be a low-voltage tap on a line-frequency-operated transformer or battery or a line-powered electronic oscillator for higher frequencies. The magnitude of the bridge voltage necessarily is related to the sensitivity of the detector and also to the general characteristics of the electrolytes to be tested.

The usual industrial measuring and control equipment is supplied with bridge voltages of 1 to 10 V. The frequency of this ac source in commercial units is commonly 60 Hz, and more rarely 1000 Hz. Where measurements are to be made on high-resistance electrolytes, such as in distilled water or steam condensate, lower bridge source frequencies are preferable. For measurements in high-conductivity solutions, higher bridge frequencies are of advantage.

R_S is the so-called standard arm of the bridge and is generally variable, as a device either to change the range of the instrument by selecting one of a number of resistors differing in resistance by powers of 10 or to correct for the temperature coefficient of resistance of the electrolyte. R_3 and R_4 are end resistors whose function is to establish the limits of the bridge calibration. R_5 is the

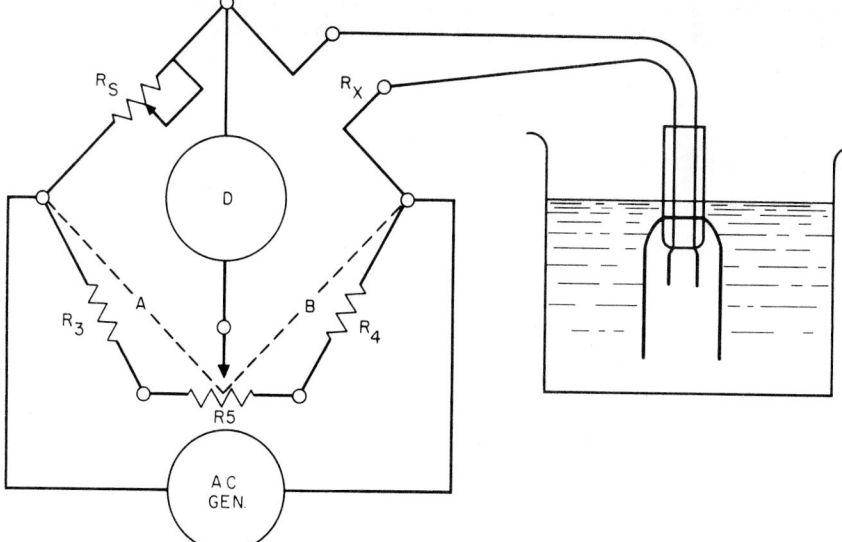

Fig. 2 Alternating current Wheatstone bridge used in electrolytic conductivity measurements.

calibrated slidewire potentiometer. With R_3 and R_4 short-circuited, the range of the bridge would be zero to infinity in resistance or conductance. Increasing the values of R_3 and R_4 compared with the value of R_5 will reduce the range covered. It should be noted that the slidewire contact resistance is in series with the detector, and thus variable values of this resistance cause no error in bridge readings. R_X is effectively the resistance of the electrolyte measured between the two electrodes of the conductivity cell immersed in the liquid under test. The condition for balance of the Wheatstone bridge is that $A/B = R_S/R_X$, and this condition is indicated by no current flow through the detector D.

While most laboratory conductivity bridges are manually balanced, the Wheatstone bridge circuit also finds use in a variety of conductivity monitors, controllers, and recorders, where it is mechanically rebalanced by a servomechanism operated by the detector. Generally, in these devices advantage is taken of the phase shift which occurs in the detected signal as the bridge is driven through balance by the servomotor.

Conductivity Meter

A second system of conductivity measurement utilizes a simple ohmmeter circuit, shown in Fig. 3. A meter, a transformer secondary winding, and a conductivity cell are connected in series so that the current is a function of the cell conductance. The meter may be calibrated in resistivity or conductivity units.

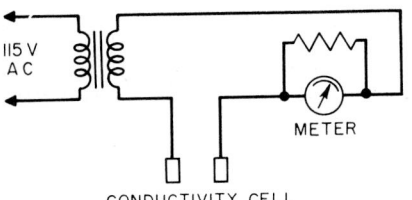

While early circuits of this type suffered from inaccuracies due to line voltage variations, the addition of a regulated power supply to drive the transformer has brought this relatively simple and inexpensive circuit into wide use. Complete isolation may be achieved by interposing a second transformer between the cell and meter. Generally, a stage of amplification is added to increase sensitivity and to reduce nonlinearity caused by meter resistance. This, combined with

Fig. 3   Conductivity measurement system utilizing a simple ohmmeter circuit.

gated detection, reduces the polarization errors associated with series capacitance at the electrodes. Driven shields are employed to reduce the errors associated with the shunt capacitance of long cell leads. The addition of automatic temperature compensation, alarm contacts, and electrical outputs makes the conductivity meter the most widely used instrument for industrial measurement and control applications.

Electrodeless Circuit

As previously mentioned, an electric current may be caused to flow in an electrolyte by means of induction without the use of contacting electrodes. In such electrodeless systems, the electrolyte is contained in an electrically insulating tube which passes through the cores of two transformers (Fig. 4) in such a way that the electrolyte forms a closed loop linking the flux in both cores. In the first transformer, this loop of electrolyte serves as a single-turn secondary winding in which an alternating voltage is induced. In the second transformer, the loop forms a single-turn primary winding, providing a means for measuring the resulting current which is directly proportional to the specific conductance of the electrolyte comprising the loop. Alternatively, both transformers may be located about an insulated tube immersed in the electrolytic solution.

Variations of these systems employing glass tubes were in use before 1907. However, more recently, the introduction of chemically resistant, high-temperature electrical insulators, such as fluorocarbons, has simplified the design of the insulated tube comprising the electrodeless conductivity cells. Since no contacting electrodes are used, all the problems associated with electrodes, such as polarization and electrode surface maintenance, simply disappear. Wide application of these systems is found in highly conductive electrolytes, such as strong mineral acids and bases—often in conjunction with abrasive slurries or materials containing entangling fibers.

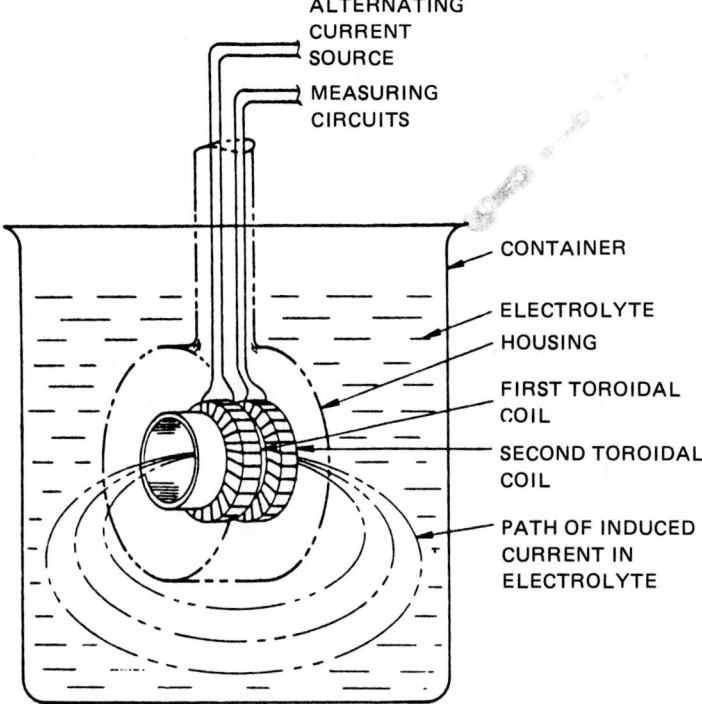

Fig. 4 Inductive electrolytic conductivity measuring circuit.

Four-Electrode Circuit

The four-electrode method avoids errors caused by polarization and fouling by using a set of measuring electrodes located between a set of current producing electrodes. See Fig. 5. The measuring electrodes are used to determine the voltage drop in the electrolyte caused by the current. The conductivity of the electrolyte between the measuring electrodes is proportional to the current divided by the potential. Laboratory measurements may be made with either alternating or direct current. However, process instrumentation almost always utilizes alternating current. In practice, the alternating current through the entire cell is varied to maintain a constant potential between the measuring electrodes. When this is done, the conductivity of the electrolyte between the measuring electrodes is proportional to the cell current. Changes external to the measuring electrodes, such as may be caused by current electrode polarization or fouling, will not cause a change in the cell current, which will be maintained at such a value that the potential between the measuring electrodes is constant. Systems have been designed that can accommodate a 10-fold increase in impedance at the current electrodes due to polarization and fouling. Fouling and polarization errors do not occur at the measuring electrodes, since the measurement there is essentially potentiometric with no current flowing through the measuring circuit. Size and orientation of the measuring electrodes are also chosen to avoid such errors.

Two-Wire Transmitter

These devices, widely used throughout industry for measurement and control, are also provided for conductivity. They can accommodate both electrode and electrodeless circuits and provide a current output of 4 to 20 mA dc with direct current excitation of 15 V or more. The power to operate the transmitter is derived solely from the dc excitation voltage and the resulting current flow. These transmitters are usually located close to the conductivity cells and interface with a remote computer. Although generally blind, local and remote readouts can be supplied. Such devices, since they are low power, are most often designed to be intrinsically safe and can be used in hazardous areas. Since they are located near the conductivity cells, lead length problems are minimized. The current output can be transmitted over long distances without loss or interference.

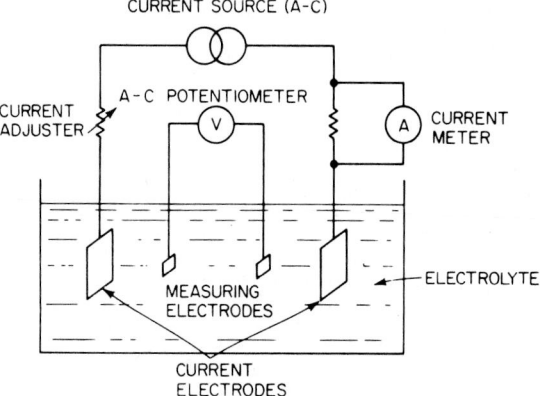

Fig. 5   Four-electrode conductivity circuit.

Conductivity Cells

Conductivity cells are simple in basic structure, consisting typically of two metal plates or electrodes spaced within an insulating chamber. Examples are shown in Figs. 6 and 7. This arrangement permits isolation and measurement of a portion of the solution and serves to make the measured independent of sample volume and proximity to conductive or nonconductive surfaces.

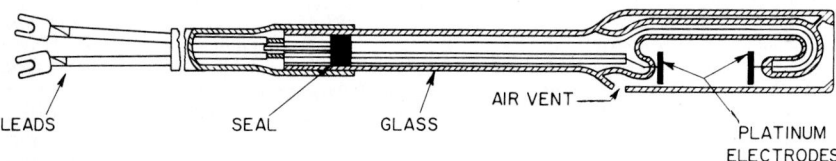

Fig. 6   Dip-type conductivity cell.

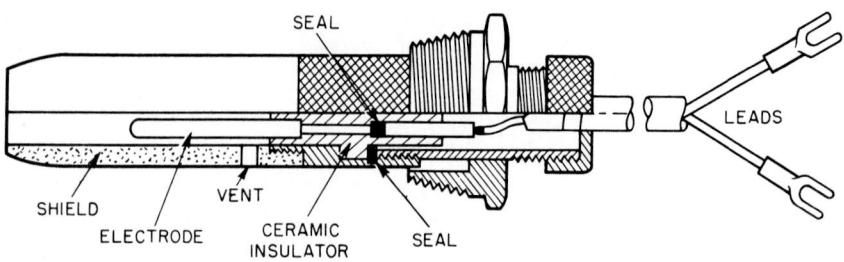

Fig. 7   Screw-in conductivity cell for high-pressure service.

In laboratory cells, platinum electrodes mounted in glass structures are commonly employed for their excellent chemical resistance. For work of the highest accuracy, fill-type cells, such as those designed by Jones and Bollinger to eliminate capacitive shunting, must be used.

For all practical work where errors up to ½% can be tolerated, design considerations are much less critical, and dipping or immersion cells are almost universally employed. These cells possess the advantage of greater ease in handling and permit measurements without transferring the solution. A cell of this type with a cell constant close to 1.0 cm^{-1} is shown in Fig. 6. For plant use, where rugged construction and heavier electrodes are desirable, structural parts of a suitable resistant metal, plastic, or ceramic and electrodes of stainless steel or nickel are utilized.

Conductivity cells suitable for plant and control work can be conveniently grouped into four types.

1. **Dip Cells.** As the name indicates, these cells are designed for dipping or immersing in open vessels. Materials of construction differ widely and include glass, rubber, epoxy, polystyrene, and Teflon for the insulating shield and body, and stainless steel, nickel, platinum, gold, and platinum-plated metals for the electrodes. Figure 6 shows a typical dip cell. The shield is usually perforated both to increase circulation and to provide a means of venting air to ensure its filling with the liquid under test. During both calibration and use a clearance of at least ¼ in (6½ mm) should be allowed between all parts of the conductivity cell and the containing vessel. In addition, care must be exercised to ensure immersion of the uppermost air vent at least ¼ in (6½ mm) below the surface of the liquid. Temperature and the chemical and mechanical characteristics of the materials of construction establish the service limits.

2. **Screw-In Cells.** For permanent installation in pipelines and tanks, cells provided with threaded fittings are used. Temperature and pressure limitations vary with the materials and the construction. Heavy-walled, glass-bodied conductivity cells, sealed by means of compressed rubber or neoprene gaskets in 1-in N.P.T. stainless steel fittings, are suitable for continuous service up to about 50 psi (345 kPa) at 220°F (104°C). For higher pressure service the construction shown in Fig. 7 permits operation at 500 psi (3.5 MPa) and 300°F (149°C).

3. **Insertion Cells with Removal Devices.** These cells are designed to permit removal of the element without closing down or depressurizing the line in which they are installed. For inspection or repair the conductivity cell element is drawn through a packing gland past the normally opened gate valve and reaches a stop, whereupon the gate valve is closed by hand and the element is removed. Typical construction, shown in Fig. 8, for service up to 200 psi (1.4 MPa) at 200°F (93°C) employs brass and bronze fittings, a plastic or ceramic insulator, and gold-sheathed or nickel electrodes. For higher pressures or more corrosive service, similar cells of stainless steel with fluorocarbon insulation are suitable up to 300 psi (2.1 MPa) at 250°F (121°C).

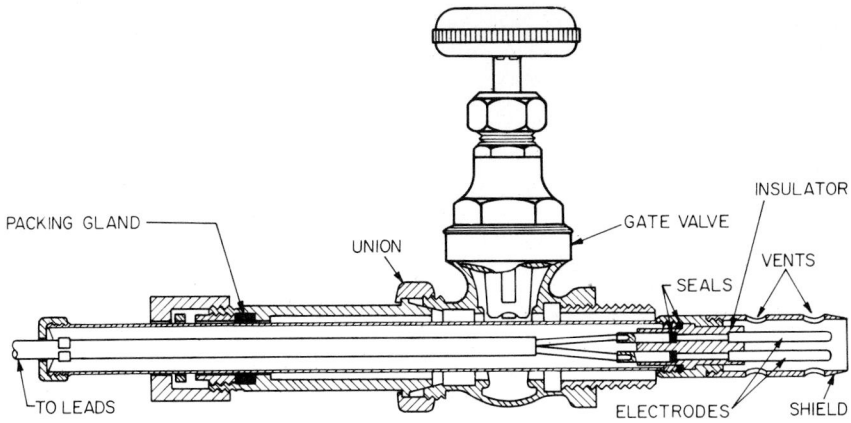

Fig. 8 Insertion conductivity cell with removal device.

4. *Flow Cells.* Built in sections of plastic or glass tubing with bores from several millimeters to 1 in (25.4 mm) and more, these cells have internal electrodes, usually metallic or carbon rings, mounted flush with the wall to offer little resistance to flow. It is frequently found necessary to utilize three electrodes, connecting the outer two in common and to ground. This eliminates almost completely the pickup of spurious ac voltages resulting from current leakage from electric equipment which may be part of the system and also eliminates the effect of any external electrical shunt path that might exist. In the smaller sizes, these cells are connected to the piping system with rubber or plastic tubing, and in the larger sizes by standard pipe flanges. A small glass flow cell is shown in Fig. 9.

Proper Location of Cell

Most industrial conductivity cells are provided with a means for permanent mounting in a pipeline or through a tank wall. The following requirements govern proper cell location:

1. Circulation must be good.
2. Representative sampling is necessary.
3. Linear velocity at the point should not be great enough to cause cavitation or damage to the cell.
4. Where linear velocity is low, it is advisable to use a flow-type conductivity cell or else to mount the conductivity cell so that the flow impinges on the open end.
5. Provision should be made nearby for measuring the solution temperature. A location with small temperature variations is preferable.
6. The cell may be mounted at any angle, provided that it cannot become air-bound or gradually fill with particles of sediment.

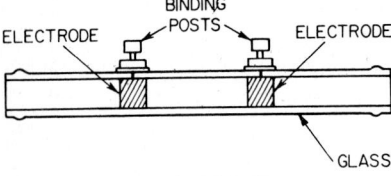

Fig. 9   Flow-type conductivity cell.

Design of Conductivity Cells

The design of conductivity cells falls into three categories: (1) electrical, (2) mechanical, and (3) chemical. All three factors must be taken into careful consideration when designing or using a cell to obtain optimum measurement results.

Electrical Factors

As previously discussed, water solutions of electrolytes exhibit electrical conductivities from less than 0.1 μS/cm for deionized water to over 1 S/cm for concentrated acids. This enormous range of more than seven orders of magnitude presents problems in the design of accurate measuring instruments, which fortunately can be solved by using conductivity cells of different cell constants to reduce the range in conductance that these instruments must measure. By using four conductivity cells covering four orders of magnitude (0.01, 0.10, 1.0, and 10.0 cm^{-1}), it is only necessary for the conductivity measuring instrument to measure slightly more than three orders of magnitude in conductance—say 5 to 10,000 μS. Cell constants of 0.0001 cm^{-1} and 100 cm^{-1} are occasionally employed to reduce the measuring range even more.

Conductivity cells with large-area electrodes close together have low cell constants and are used to measure low conductivities (high resistivities). Electrodes spaced far apart with the path between them restricted in cross-sectional area provide high-cell-constant cells which are used to measure high conductivities (low resistivities).

Electrode Surface

In addition to having a cell constant of magnitude commensurate with the range of specific conductance to be covered, a conductivity cell with relatively large electrode surfaces in relation to the electrolyte cross section performs better than one with small surfaces. A most useful means of increasing the effective electrode surface, called platinizing, was discovered by Kohlrausch. Platinum black may be deposited on platinum, gold, or nickel surfaces as follows:

1. The conductivity cell is cleaned in aqua regia or chromic acid cleaning mixture and rinsed thoroughly in running tap water.

2. The conductivity cell is immersed in platinizing solution containing 3 g platinic chloride and 0.02 g lead acetate per 100 ml distilled water.

3. Electrode leads are connected to two 1½-V dry cells in series or to any source of 3 to 6 V direct current. Polarity is reversed every 15 s. Plating is continued for 3 min or until a dense black coating covers the entire electrode. The platinizing solution should not be discarded until it has been exhausted.

4. The conductivity cell is rinsed for ½ h in running tap water. For cells with a low cell constant to be used for measurement of high-purity waters, this rinse should be followed by soaking for 15 min in several changes of distilled water, preferably hot.

By roughing or etching the base-metal surface, the deposit of black platinum can be made reasonably resistant to mechanical removal. For usual measurement and control purposes, the deposit is not affected by materials which poison platinum-hydrogen electrodes used in pH measurements, nor does it deteriorate appreciably on drying.

Well-platinized electrode surfaces are most necessary in the measurement of highly conductive solutions. With a well-designed cell having an adequately high cell constant and heavily platinized electrodes, measurements even at 60-Hz bridge current are free of the difficulties and errors associated with polarization. More commonly, electrodeless conductivity instruments are used for highly conductive solutions, thus eliminating polarization and the need for platinization. For measurements in poorly conductive media, such as condensed steam or demineralized water where measured resistances are high, polarization does not occur to an appreciable extent. Therefore, the platinum black coating can be made much thinner or completely eliminated. This is frequently desirable in the measurement of nonflowing samples, since platinum black is quite absorptive and many times serves to introduce electrolyte into otherwise highly resistive solutions. Black platinum is easily removed by electrolysis in aqua regia for several minutes, with occasional current reversal.

Mechanical Factors in Design of Cells

Significant factors include pressure, temperature, velocity of flow, piping size, and the presence of solids. Only when all these are taken into consideration can the user expect a long, satisfactory performance with a minimum of maintenance.

Sampling Arrangements

Where the temperature and pressure in the system are higher than the maximum rating of available conductivity cells, it is common practice to draw a continuous small sample of the liquid through a cooling coil of adequate rating to cool and condense it, through a throttling valve to reduce the pressure, and only then to bring the sample into contact with a conductivity cell. Alternatively, by running the sample through a proper length of suitable capillary tubing its temperature and pressure can be reduced without the use of a throttling valve.

Effects of Temperature Changes

Regardless of the temperature at which the conductivity cell operates, efforts should be made, in the interest of accuracy, to hold that temperature as constant as practically possible. As shown below, the conductivity of most solutions increases about 2½% for each degree Celsius rise in temperature. Uncontrolled or uncompensated, this factor overwhelmingly limits the overall accuracy of the measurement. A convenient method of controlling sample temperatures is to use a large excess of cooling water to bring the sample temperature to cooling-water temperature, which is usually constant except for gradual seasonal variations.

Although solution conductivity is influenced by pressure, the magnitude of the effect is negligible at ordinary pressures.

Velocity Effects

Up to the point at which cavitation occurs in the conductivity cell, high velocities of flow do not affect conductivity measurements, although it is usually necessary, in order to improve adhesion, to deposit the coating of platinum black on a roughened or matte surface, which may be obtained by sandblasting or chemical etching. When flow velocity is extremely low, it is particularly important that the conductivity cell be selected and installed in a manner to ensure circulation of liquid between the electrodes. For example, the screw-in type of cell should be introduced at a turn in the piping and mounted in a tee with the open end facing the flow.

Where the piping is smaller than the available conductivity cells, it is necessary to bush up the piping size in its immediate area. A tee is the most common means of mounting the cell. It should be born in mind that the conductivity cell occupies space in the fitting and offers resistance to flow of fluid. Piping at least one size larger than the fitting of the conductivity cell is recommended. In the larger sizes, from about 4 in (~ 10 cm) up, threaded fittings are conveniently welded on, to provide a means of mounting the conductivity cell. Where the removal type of insertion cell shown in Fig. 8 is used, sufficient clearance must be left to pull the element out. The electrode chamber must be positioned well within the flow to ensure adequate sampling.

Effects of Undissolved Solids

The presence of undissolved solids in the system should influence the choice and location of the conductivity cell. Where velocities are high, entrained solids increase the scouring effect. To avoid stripping the platinum black coating from the electrodes, the cell should be placed to avoid direct impingement of flow on the electrodes, but in no case should the mounting be such that solids may settle and eventually plug the cell chamber.

Where velocities are low and suspended solids are present, the cell is best mounted facing the flow, with the open end pointing downward to avoid accumulation of solids. Where there is a strong tendency for solids to deposit out, as in ore and paper pulping, starch suspensions, and gypsum slurries, the usual type of cell construction with platinized platinum electrodes inside an insulating sheath presents great operating difficulties. For such service, electrodeless and four-electrode systems find wide application. In severe service, where metal electrodes are destroyed by erosion, electrodeless systems predominate.

Chemical Considerations in Cell Design

The requirement of resistance to chemical action is common, so that a conductivity cell of properly resistant materials must be chosen. The problem is sometimes complicated by the fact that ac potentials are imposed on the electrode material. An example showing the need for considering this factor is measurements in strong hydrochloric acid. Although platinum is resistant to hydrochloric acid, it will dissolve with the application of 60-Hz current unless the conductivity cell is designed with a cell constant sufficiently high to reduce the current density at the platinum surfaces.

With the expanding industrial use of hydrofluoric acid, the measurement of conductivity as an index of its concentration has assumed importance. The change in specific conductance with concentration is quite steep and almost linear up to about 70% HF by weight, after which the conductivity curve descends at a sharp angle. For all concentrations cells constructed of fluorocarbon and platinum have proved suitable, even at elevated temperatures. For lower concentrations and temperatures, polyethylene and polystyrene have been used.

Conductivity measurements are almost universally used to gauge the quality of pure waters such as demineralized or distilled water and condensed steam. Borosilicate glass, dense ceramic, and epoxy plastics have proved to be satisfactory insulators and to stand up indefinitely under continuous flow. Types 304 and 316 stainless steel, nickel, titanium, gold, and platinum structural parts and electrodes are suitable. Care must be exercised to avoid the use of fluxes in the fabrication of electrodes to prevent subsequent contamination by leaching of electrolytic material. It is particularly important in the measurement of exceedingly pure water effluent from mixed-bed ion exchangers. Similarly, after chemical cleaning or replatinization of conductivity cells in such service, particularly thorough washing is necessary. Pipe joint compounds and dopes must be employed sparingly or replaced by fluorocarbon thread tape.

The design of electrodeless cells is somewhat simpler. Two types of electrodeless cells are in wide use: *flow* type and *dip* type. Flow-through cells are fabricated most frequently from plastic pipe, such as polyvinyl chloride or glass-reinforced polyester or epoxy. For applications requiring greater chemical resistance, the more common fluorocarbons, including vinylidene fluoride, find wide application. Greater economy, strength, and dimensional stability can be achieved by using linings made of these chemically resistant materials. Flow-through cells exhibiting excellent chemical resistance at high temperatures and pressures may be fabricated from lined steel pipe, provided that the pipe flanges are insulated to prevent flux linkages in the electrodeless circuit. In flow-through electrodeless cells, the transformers are located around a portion of the cell tube or pipe.

In dip-type cells, the transformers are contained in an immersible housing in such a way that they surround a short tube or pipe which runs through the housing and is connected at both ends to the outside so that it fills with electrolyte when the cell is immersed. All dip-type electrodeless cells are designed to be immersed completely in the solution to be measured. Materials of construction are similar to those used in flow-through cells, favoring polyvinyl fluoride and epoxies. However, coated steel and anodized aluminum housings have performed well under high pressures in deep sea measurements.

Inspection and Maintenance of Conductivity Cells

The surfaces of the electrodes of conductivity cells, which are normally supplied coated with platinum black, operate satisfactorily only when this coating is present and in good condition. The only common exception to this is in the case of cells of low constant in service in very poorly conductive solutions. Under these conditions the operation will not be affected adversely by removal of the platinum black coating.

It is recommended that conductivity cells in service be inspected at regular intervals. Conditions of service determine the frequency of inspection. Any unusual behavior of the conductivity measuring and recording instrument unattributable to known variations in the system under measurement should be taken as a possible indication that the cell requires inspection and cleaning or replatinization. A gradual appearance of and increase in sluggishness in recording action or bridge balance are almost always identified with deteriorated electrode surfaces. In the absence of all indications, an inspection interval of 1 month is suggested.

Periodic Inspection Checklist

Periodic inspection once a month should include a check of the following points:

1. Are there cracks or chips in the cells, or do the cells show any appearance of wear or deterioration?
2. Is the platinum black deposit still present over most of the electrode surface?
3. Is there any adherent coating on the electrodes or any discoloration?
4. If there are vent holes, are they clear and free from obstruction?
5. Is there any indication that excessive liquid velocities have caused changes in the position of the electrodes?
6. Is the electrode shield intact?
7. Do the electrodes show signs of corrosion or deterioration?
8. Is leakage resistance of dry cells above 50 MΩ?

Check of Cell Constant

At less frequent intervals it is advisable to check the constant of the conductivity cell. This can be done most readily with existing equipment and without the necessity of making up special test solu-

Table 1 Standard Reference Solutions

Approx. normality	Composition KCl, g/L	Temp., °C	Specific conductance, μmhos/cm	Suitable for cell constants, cm^{-1}	Approx. cell resistance, Ω
0.001	0.0744	25	147	0.01–0.1	70–700
0.01	0.7440	18	1.221	0.1–10	70–7,000
		25	1.409		
0.1	7.437	18	11.170	10	8,000
		25	12.860		

tions by merely comparing the instrument reading under identical conditions with the cell in question and with another cell of the same constant. If instrument readings (specific resistance, measured resistance, or specific conductance) are not the same, the cell constant can be calculated as follows:

$$K = K_{std\ cell} \times \frac{R_{cell}}{R_{std\ cell}} \qquad (3)$$

or

$$K = K_{std\ cell} \times \frac{cond_{std\ cell}}{cond_{cell}} \qquad (4)$$

To check the cell constant by independent means it is necessary to measure the resistance of the conductivity cell immersed in a carefully prepared solution of known specific conductance, usually 0.001, 0.01, or 0.1 N potassium chloride in water. Refer to ASTM Standard Test Method D 1125. The choice of standard solution to be used for checking a conductivity cell depends largely on the magnitude of the cell constant, since it is desirable to hold the measured resistance as close as possible to the optimum range of 500 to 10,000 Ω. The values in Table 1 are suggested.

The values in the table do not include the conductivity of the water used in making up the solutions. Naturally, since this must be added to the values given above, water with as high resistance as possible should be used in making up the standard solution, particularly 0.001 N KCl.

In measuring the cell constant, the conductivity cell should be well cleaned and rinsed in several changes of distilled water, and then rinsed at least twice with portions of the standard solution. Several measurements using different portions of the solution should be made of the resistance of the conductivity cell, and the temperature of the solution should be measured to the nearest 0.1°C. The cell constant K is calculated as follows:

$$K = R_x \times \frac{S.C.\ KCl + S.C.\ H_2O}{1,000,000} \qquad (5)$$

where R_x = measured resistance, Ω
S.C. KCl = specific conductance of KCl as in Table 1 at temperature of measurement
S.C. H_2O = specific conductance of distilled water used in making up solution at same temperature.

Temperature Compensation

A family of conductivity curves for sodium chloride at several temperatures is shown in Fig. 10. This electrolyte is typical, in that the conductivity increases uniformly with temperature at a rate of about 2½%/°C. The magnitude of the temperature dependence is large enough for all common electrolytes to be of significance in electrolytic conductivity measurements. Consequently, the matter of a suitable means of temperature compensation must be considered in the requirements for each installation. The most common means of temperature correction is inclusion in the measuring circuit of a manual adjustment calibrated in temperature units. The calibration is based on a suitable average temperature coefficient of conductivity.

The manual setting of a knob to correspond to the temperature reading at the conductivity cell serves to balance out electrically the influence of temperature on the conductivity of the solution. The instrument will thus read in terms of microsiemens per centimeter or ohm-centimeters at some reference temperature (usually 25°C) regardless of the cell temperature, as long as the manual temperature compensator is properly adjusted. This means of temperature compensation is adequate where actual temperature variations in the system are small or the rate of change in temperature is low, for example, in the measurement of raw water conductivity.

When the temperature of the liquid is erratic, it is necessary to reset the temperature compensator at frequent intervals, which in addition to being burdensome yields a record of questionable validity. Where sampling and cooling equipment are employed, as in the measurement of steam purity or boiler water conductance, automatic valves are occasionally used to throttle the cooling-water input to maintain a constant sample output temperature.

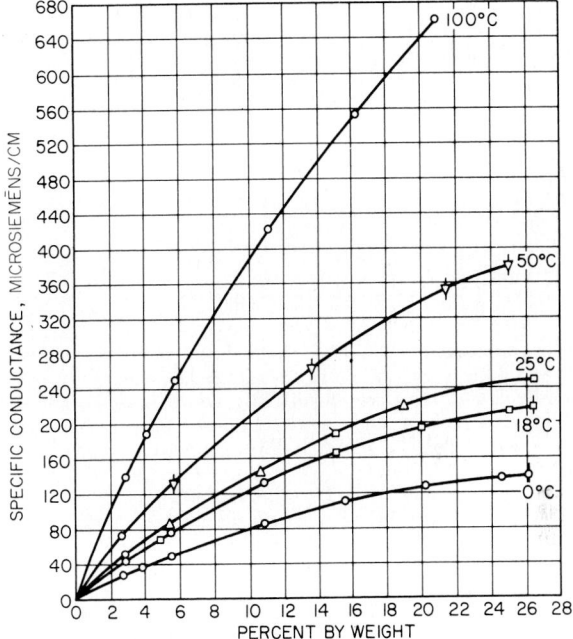

Fig. 10   Specific conductance of sodium chloride.

Another expedient is to feed thermostatically controlled cooling water to a small secondary heat exchanger and so maintain a still more uniform sample output temperature. In many cases, however, as in the measurement of hot-well condensate, it is not possible or practical to draw a continuous sample and bring it to constant temperature by external means. For this reason, automatic temperature compensation methods have received much attention.

At present, thermistors are used universally for automatic temperature compensation in electrolytic conductivity measuring instruments. Usually the high negative temperature coefficient of the thermistor is attenuated by being incorporated into a network containing fixed resistors. However, some electrolytes, such as ultrapure water, may require thermistors with extremely high temperature coefficients.

Although many solutions, such as those of neutral salts, have closely similar temperature coefficients, the automatic temperature compensator coefficient for best performance must be matched to the characteristics of the solution being measured. Strong acids and bases have coefficients differing markedly from those of salts. Temperature coefficients of electrolytes also change with concentration and temperature. Different compensator coefficients may be required for the same electrolyte at dif-

Table 2 Effect of Temperature on Conductivity of Various
Materials

	Ratio to conductivity at 25°C				
Solution	0°C	25°C	50°C	75°C	100°C
18-MΩ·cm water	0.22	1.00	3.11	7.46	14.2
NaCl	0.54	1.00	1.53	2.15	2.73
5% NaOH	0.57	1.00	1.43	1.87	2.32
5% H$_2$SO$_4$	—	1.00	1.24	1.42	1.52
Sodium orthoscilicate	—	1.00	1.46	1.90	2.31
Dilute NH$_3$	0.50	1.00	1.47	1.83	2.05
Dilute HNO$_3$	0.65	1.00	1.31	1.58	1.80
4% acetic acid	—	1.00	1.30	1.52	—
Black liquor	—	1.00	1.44	1.83	2.19
White liquor	—	1.00	1.50	2.03	—
Brown stock	—	1.00	1.55	2.07	2.55
Washer filtrate	—	1.00	1.55	2.07	2.55
Drilling mud	—	1.00	1.58	2.10	—
0.01% H$_3$PO$_4$	—	1.00	1.30	1.52	1.67
Sugar syrup	0.34	1.00	2.41	4.40	6.93
10% HCl	0.64	1.00	1.33	1.63	1.87
98% H$_2$SO$_4$	—	1.00	1.71	2.56	3.50
25% oleum	—	1.00	1.92	2.82	3.75

ferent concentrations or temperatures. Table 2 shows the ratio of the conductivity at the tabulated temperature to the conductivity at 25°C for some common solutions.

The measurement of ultrapure water (2 to 18 MΩ·cm) resistivity presents special temperature compensation problems because the temperature coefficient changes both with resistivity and temperature. Figure 11 shows a family of curves describing this changing temperature coefficient. The coefficient of 18-MΩ·cm water is given in Table 2, and it can be seen from the figure that the coefficient of 2-MΩ·cm water more nearly resembles that of sodium chloride, as shown in Table 2 and Fig. 10. Automatic temperature compensation circuits used for ultrapure water must be designed to apply a correction which is dependent on both resistivity and temperature instead of simply dependent on temperature, as are most other compensators. Several methods of accomplishing this have been devised and are currently offered by manufacturers.

For automatic temperature compensators to operate satisfactorily, the thermistor must be in good thermal contact with the electrolyte. Often the thermistor is built directly into the conductivity cell where it assumes the same temperature as the cell and its contents. Separate compensator probes must be located near the cell, where temperatures are the same as those of the cell. Separate probes for thermistors simplify service, since they may be made entirely of strong, chemically resistant materials such as stainless steel and are left in service almost indefinitely—whereas the cells must be removed periodically for servicing and occasional replacement.

Sources of Error in Conductivity Measurements

Possible sources of error are as follows:

1. *Circulation is insufficient.*
   SYMPTOM—Sluggishness of response to concentration changes.
   REMEDY—Relocation or more agitation.
2. *Conductivity cell is fouled.*
   SYMPTOM—Great sluggishness in response to large concentration changes.
   REMEDY—Clean cell and relocate if necessary.

3. *Conductivity cell requires replatinization.*

SYMPTOM—Broad null point in manually balanced bridge, stepwise change in recorder action.

REMEDY—Clean and replatinize.

4. *Electrical leakage in conductivity cell.*

SYMPTOM—Erratic results, discordant checks under apparently identical conditions, change in reading with shift in position or angle of immersion of conductivity cell.

TEST—Finite resistance reading when cell is removed from solution and rinsed. Resistance rises when cell is thoroughly dried.

REMEDY—Electrical leakage which lessens on rising and drying may be indicative of leakage of solution past electrode seals or through cracks in electrode insulators. Tightening of seals or replacement of defective insulators is indicated.

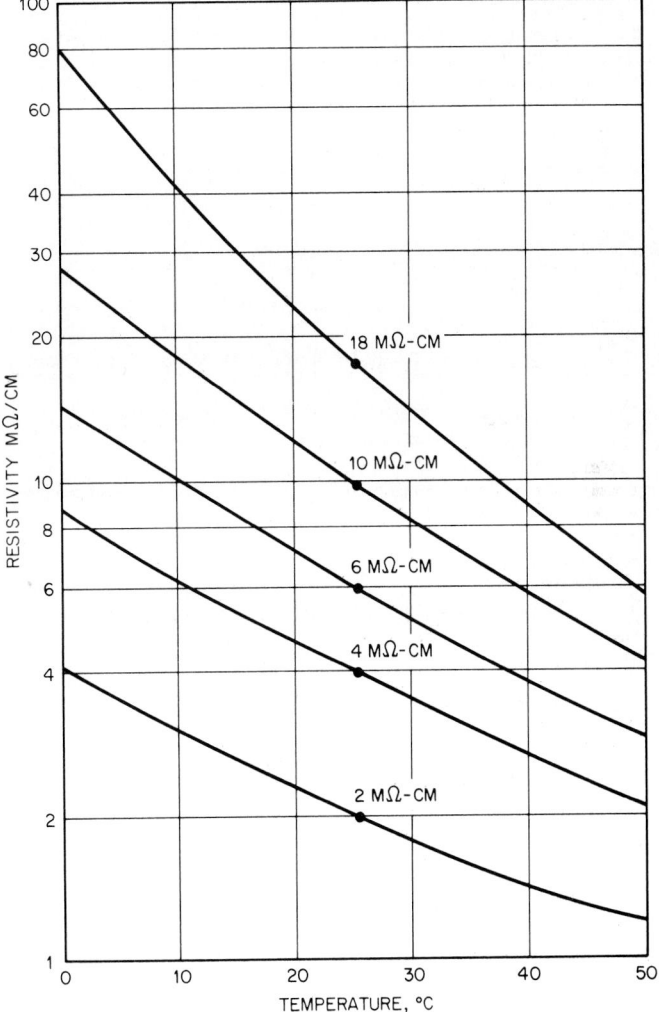

Fig. 11 Resistivity-temperature relationship of ultrapure water contaminated with sodium chloride.

5. *Leaching of electrolytes from cell or container.*
   SYMPTOM—Drift of reading (typically with pure water at low flow rates) toward higher conductance.
   TEST—On agitating conductivity cell under liquid level, resistance rises, only to fall again on standing.
   REMEDY—Remove conductivity cell and wash with several changes of hot distilled water. Reduce density of platinum black coating. Clean the container.

6. *Temperature errors.*
   SYMPTOM—Instrument readings vary somewhat with time even though solution is known to remain constant in composition.
   TEST—Checking temperature of the solution and resetting manual compensator more frequently improve record.
   REMEDY—Provide better temperature control or use automatic temperature compensator.

7. *Reference temperature.*
   SYMPTOM—Cannot obtain check reading from two different instrument systems even though each bridge and cell check against manufacturer's test data.
   REMEDY—Examine instrument specifications and ascertain the so-called reference temperature of each. For example, an instrument referred to 25°C will read about 1175 μmhos/cm under the same conditions that a comparable instrument referred to 18°C will read 1000 μmhos.

REFERENCES

ASTM: *Standard Test Method D-1125,* American Society for Testing and Materials, Philadelphia (periodically updated).

Fuoss, R. M.: "Conductance-Concentration Function for the Paired Ion Model," *J. Phys. Chem.,* vol. no. 82, 22, 1978, p. 2477.

ICT: *International Critical Tables,* vol. 6, McGraw-Hill, New York, 1929, pp. 229–309.

Janz, G. J.: "Conductance Cell Calibrations: Current Practices," *J. Electrochem. Soc.,* vol. 124, no. 2, 1977, p. 57C.

Jones, Grinnell, et al.: Series of papers on precise electrolytic conductivity measurements and equipment. Last paper in series appeared in *J. Am. Chem. Soc.* vol. 57, 1935, p. 280.

Rosenthal, R., and R. Kidder: "Solution Conductivity Handbook," *Beckman Instrum. Bull.* 4090, Beckman Instruments, Cedar Grove, N.J., 1969.

Thermal Conductivity Gas Analysis

For certain gases and gas mixtures, thermal conductivity has been the analytical method of choice for many years. The method is based on the fact that various gases differ considerably in their ability to conduct heat. Although this approach was first suggested as early as 1880 by Leon Somzec, it was not until 1908 that Koepsal developed a practical instrument for the continuous indication of hydrogen content in producer gas. Koepsal made use of the hot-wire method for comparing the thermal conductivities of two gases. Since that time, numerous adaptations have been made of the original hot-wire system.

The thermal conductivity of a gas mixture depends on the nature and concentration of the con-

Table 1 Thermal Conductivities of Gases

Gas	K_{gas}*		K_{gas}/K_{air}	
	0°C	100°C	0°C	100°C
Air	2.23	2.854	1.000	1.000
Acetone	0.906	1.558	0.406	0.546
Ammonia	2.00	3.10	0.897	1.086
Argon	1.58	2.07	0.709	0.725
Carbon dioxide	1.37	2.069	0.614	0.725
Carbon monoxide	2.15	—	0.964	—
Chlorine	0.718	—	0.322	—
Ethane	1.80	3.204	0.807	1.123
Ethylene	1.64	2.624	0.735	0.919
Ethyl alcohol	1.11	1.96	0.498	0.687
Helium	13.9	16.68	6.233	5.844
Hydrogen	15.9	20.03	7.130	7.018
Hydrogen sulfide	1.20	—	0.538	—
Methane	2.94	—	1.318	—
Methyl alcohol	1.32	2.033	0.592	0.712
Neon	4.44	5.44	1.99	1.91
Nitrogen	2.28	2.896	1.022	1.015
Nitric oxide	2.08	—	0.933	—
Nitrous oxide	1.44	2.09	0.646	0.732
Oxygen	2.33	3.006	1.045	1.053
Sulfur dioxide	0.768	—	0.344	—
Water vapor	—	2.17	—	0.760

*K is given in the units kiloergs per second per square centimeter per centimeter per degree Celsius. To convert to other units multiply the value by 41.833 to obtain calories per second per square centimeter per centimeter per degree Celsius.

stituents. Dependence of thermal conductivity on gas composition is treated extensively in the literature (Bennett and Vines; Daynes; Palmer and Weaver; Vines) but in general it is not profitable to make elaborate calculations of thermal conductivity. Furthermore, the available data on thermal conductivity of gases generally cannot be relied on to an accuracy of better than ±5%. In gas analysis the only practical reason for computing the thermal conductivity of a gas mixture is to obtain a rough estimate of probable sensitivity over a limited range of composition. For this purpose, the following simple linear relation usually is adequate:

$$K_1, K_2, \ldots, K_n = (K_1 p_1 + K_2 p_2 + \cdots + K_n p_n)/100$$

where $K_1, K_2, \ldots, K_n$ = thermal conductivity values of each constituent of the mixture
$p_1, p_2 \cdots, p_n$ = respective concentrations of constituents, vol%

Thermal conductivity data for a few gases of interest in industrial gas analysis are given in Table 1. Among the most common gases and mixtures measured are the following: argon in nitrogen or oxygen; carbon dioxide in air, flue gases, or nitrogen; Freon in air; helium in air or nitrogen; hydrogen in air, blast furnace top gases, carbon monoxide, carbon dioxide, nitrogen, argon, or hydrocarbons, such as reformer gases; methane in air; oxygen in argon, nitrogen, or hydrogen; and propane in air.

Principle of Operation

A thermal conductivity gas analysis cell usually consists of an electrically conductive, elongated sensing element mounted coaxially within a cylindrical chamber containing the gas. The element in the cell is maintained at an elevated temperature with respect to the cell walls by passing an electric current through it. An equilibrium temperature is attained by the element when its electric power input is equalized by all thermal losses from the wire. When material having a suitable temperature coefficient of resistance is used, the sensing element can serve a dual function of heat source and sensor of equilibrium temperature. The magnitude of the temperature difference between the element and the cell walls (i.e., the temperature rise of the element) at equilibrium is a function of the electric power input and the combined rate of heat loss from the wire by gaseous conduction, convection, radiation, and conduction through the solid supports of the element.

By proper design and cell geometry and by limiting the temperature rise of the heated element, it is possible to maximize heat loss due to gaseous conduction. Under these circumstances, the temperature rise of the element at constant electric power input is inversely related to the thermal conductivity of the gas confined within the cell.

Thermal Conductivity Measurement Systems

A Wheatstone bridge is the usual form of network employed to measure the resistance change of the sensing element. The current used to energize the bridge also serves to heat the wire. It is impractical to employ a single hot-wire cell in a bridge measurement because of the extreme sensitivity of such a measurement to changes in bridge supply voltage and ambient temperature. Instead, it is common practice to use two cells in adjacent arms of a Wheatstone bridge, one of the cells containing a reference gas and the other containing the gas to be analyzed. The bridge then responds to the difference in temperature rise (i.e., resistance) of the two sensing elements and thus is dependent on the difference in the thermal conductivities of the gases in the two cells.

A wide variety of thermal conductivity gas analyzers are available commercially. These devices embody a variety of cell types, geometries, and Wheatstone bridge configurations. For example, the measuring cell may receive sample gas by diffusion, convection, or a combination of these. In some instances, as in gas chromatography, the cell cavity may be a part of the gas channel, with all the sample gas passing through the cavity. This minimizes response time but requires the capability to establish a relatively low, but reproducible and constant, gas flow rate to maintain calibration integrity and minimize signal noise.

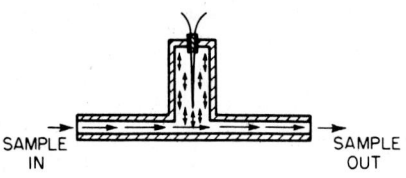

Fig. 1 Thermal conductivity cell of the diffusion exchange type.

Generic cell types are depicted in Figs 1 through 3. Figures 4 through 7 show basic Wheatstone bridge arrangements for single-pass gas analyses, i.e., analyses in which the sample gas affects the resistance of only one arm of the bridge and is compared with a reference cell containing gas of constant composition in the adjacent bridge arm. Figure 8 shows the basic double-pass configuration in which the sample gas influences the cell in one arm of the bridge and then is subjected to modification and finally passed through the cell in the adjacent arm. In this case the unmodified sample gas usually is considered the reference to which the modified sample gas is compared. Figure 9 shows modified Wheatstone bridge circuitry to compensate for small variations in bridge supply voltage. Resistors A and B in all cases and resistor M in Fig. 9 have substantially a zero temperature coefficient of resistivity. Views of modern thermal conductivity analyzers are shown in Figs. 10 through 12.

Scope of the Thermal Conductivity Gas Analysis Method

Thermal conductivity gas analysis is a nonspecific, nonabsolute method which depends on empirical calibration. Nevertheless, the simplicity, reliability, relative speed, and easy adaptation to continuous recording and control have made this method one of the most widely used means of industrial gas analysis. As a laboratory tool, the method has been used extensively in conjunction with analytical

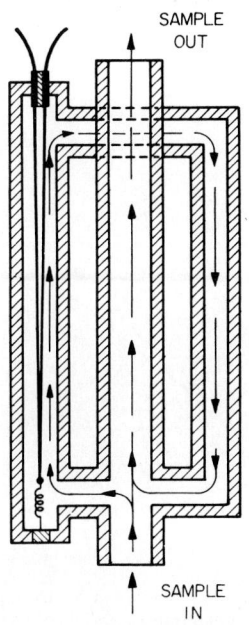

Fig. 2 Thermal conductivity cell of the convection exchange type.

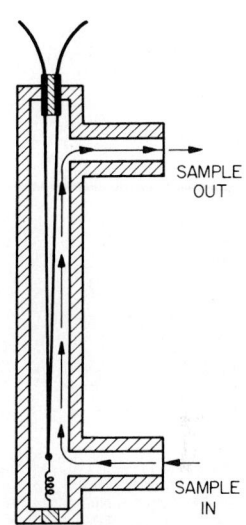

Fig. 3 Thermal conductivity cell of the direct flow type.

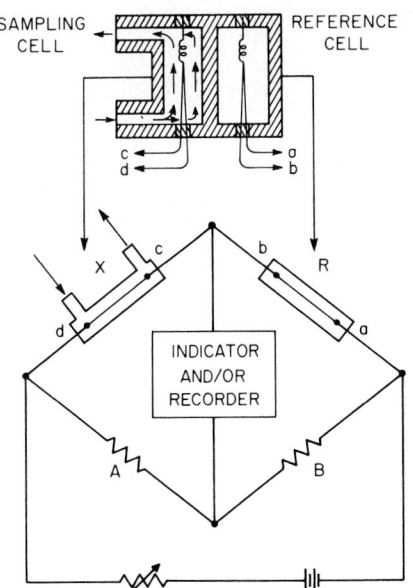

Fig. 4 Thermal conductivity bridge for single-pass analysis.

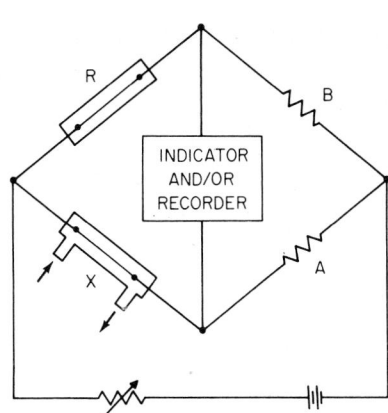

Fig. 5 Alternate form of thermal conductivity bridge for single-pass analysis.

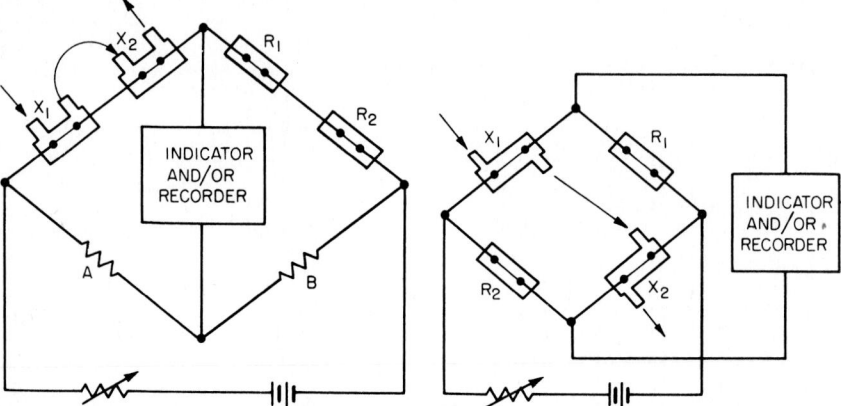

Fig. 6   Four-cell thermal conductivity bridge for single-pass analysis.

Fig. 7   Alternate form of four-cell thermal conductivity bridge for single-pass analysis.

techniques, such as gas chromatography where the sample gas is essentially a binary mixture for which the thermal conductivity method is ideally suited. For quantitative analysis of a binary gas mixture, provided the full-scale change in thermal conductivity is not less than about 2%, it is possible to obtain a useful sensitivity of better than 1% of full-scale and comparable zero stability. The approximate practical limits of this method, as applied to several binary mixtures, are given in Table 2. In each case, the range shown is equal to a 2% change in thermal conductivity. The tabulated values are accurate only to approximately 10% because of inherent limitations in available thermal conductivity data and the assumptions made with respect to operating conditions.

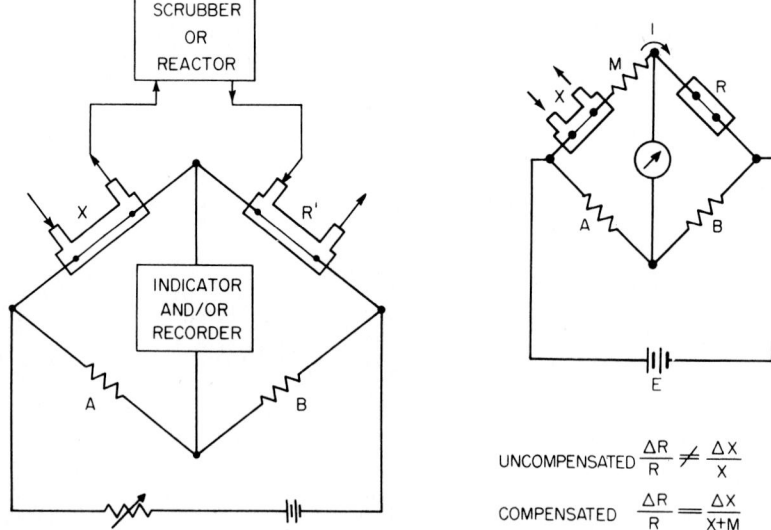

$$\text{UNCOMPENSATED } \frac{\Delta R}{R} \neq \frac{\Delta X}{X}$$

$$\text{COMPENSATED } \frac{\Delta R}{R} = \frac{\Delta X}{X+M}$$

Fig. 8   Thermal conductivity bridge for double-pass analysis.

Fig. 9   Compensation for variation in supply voltage.

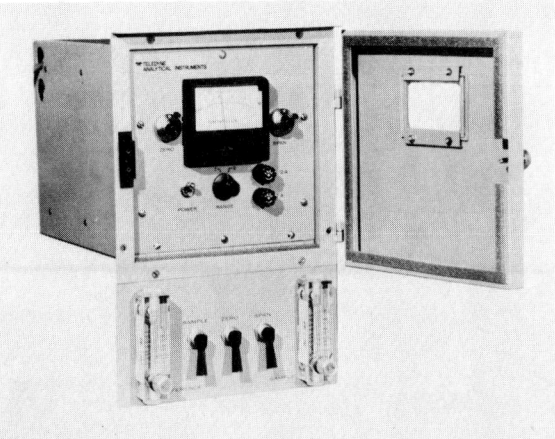

Fig. 10 Modern thermal conductivity analyzer. Two pairs of temperature-sensitive resistance elements are mounted in separate chambers in a cell block. These elements, arranged in a Wheatstone bridge configuration, comprise the measuring and reference components of the analytical system. A reference gas, such as air, with a known thermal conductivity, flows into one side of the cell block. The gas to be analyzed is fed into the other side. The reference gas absorbs heat from the reference elements in a direct ratio to its thermal conductivity. The amount of heat absorbed by the reference gas remains constant, since it has a constant thermal conductivity. Since the temperature of the reference elements remains constant, the resistance stays constant. The sample gas fed into the measuring side of the cell block absorbs heat from the measuring elements in a direct ratio to its thermal conductivity. As the thermal conductivity of the sample changes, as a result of changes in its concentration, the resistance of the measuring elements changes also, unbalancing the Wheatstone bridge. The resultant output signal is directly proportional to the impurity in the sample gas stream. *(Teledyne Analytical Instruments.)*

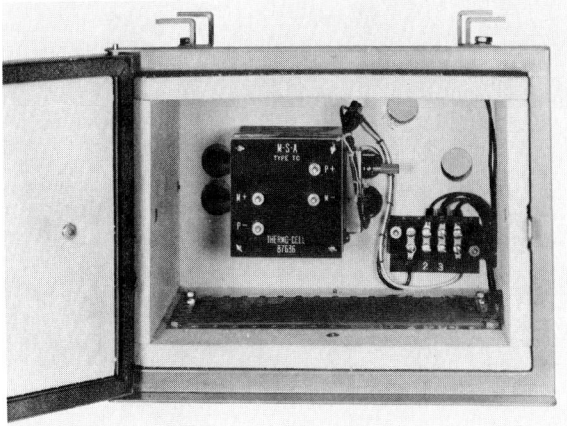

Fig. 11 Detector unit of a Thermatron analyzer. *(MSA Instruments.)*

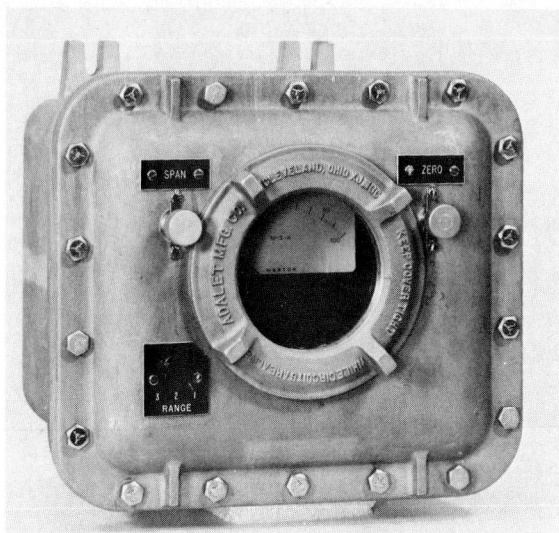

Fig. 12 Explosion-proof housing for a Thermatron analyzer where required. *(MSA Instruments.)*

Table 2 Approximate Practical Range of Thermal Conductivity Method Applied to the Analysis of Certain Binary Gas Mixtures

Gas mixture	Approximate practical full-scale range
Air*–oxygen	0–40% air in O_2
	0–38% O_2 in air
Air–carbon dioxide	0–5.3% air in CO_2
	0–7.3% CO_2 in air
Air–sulfur dioxide	0–1% air in SO_2
	0–3% SO_2 in air
Air–helium	0–2.4% air in He
	0–0.4% He in air
Air–helium	0–2.4% air in He
	0–0.4% He in air
Nitrogen–oxygen	0–55% N_2 in O_2
	0–52% O_2 in N_2
Nitrogen–carbon dioxide	0–5% N_2 in CO_2
	0–7% CO_2 in N_2
Nitrogen–hydrogen	0–2.3% N_2 in H_2
	0–0.3% H_2 in N_2
Nitrogen–argon	0–5% N_2 in A
	0–7% A in N_2
Carbon dioxide–oxygen	0–6.4% CO_2 in O_2
	0–4.4% O_2 in CO_2
Hydrogen–helium	0–10% H_2 in He
	0–12% He in H_2

*Where air appears in the table, the air may be regarded as a pure gas because of its substantially fixed composition.

In some cases, the variation in the thermal conductivity of binary mixtures does not follow the simple linear law. Ammonia in air and water vapor in air are typical examples. Information pertinent to application of thermal conductivity gas analysis for the determination of water vapor in air is given in the three Cherry references listed. Information pertinent to other binary mixtures exhibiting conductivity maxima and nonlinearities may be found in the Angerhofer and Dewey; Bennett and Vines; Daynes, and Vines references.

Complex Gas Mixtures

When a gas mixture contains three or more constituents, all of which are subject to independent variation, it is necessary to evaluate the individual effect of each constituent to determine the feasibility of applying the thermal conductivity method. There are many important industrial gas analysis problems involving such complex gas mixtures. In some of these it is known that variations in the concentration of individual constituents are interdependent. In such cases the problem may be equivalent to that of a simple binary mixture, although the matter of proper calibration may be more complex.

Applications of Thermal Conductivity Gas Analyzers

Table 3 lists several commercial and a few special applications. Part A of the table shows applications for which a single-pass instrument is normally used and indicates which comparison gases are suitable. Part B includes applications involving pretreatment of the gas sample followed by a single-pass measurement against a comparison gas. Part C lists applications usually accomplished by a double-pass measurement and indicates the nature of treatment of the gas sample between passes. Information on some applications not covered by Table 3 is given in the references Angerhofer and Dewey; Berl; Daynes; Minter et al.; Palmer and Weaver; and Phillips and Bulger.

Practical Considerations in Continuous Measurements

A thermal conductivity installation to be used for continuous automatic gas analysis must meet certain requirements if it is expected to operate reliably and with minimum maintenance.

Conditioning Gas Sample

Water vapor is the most common variable component found in industrial gases to be analyzed. Unless the purpose is to analyze for water vapor concentration, it is necessary either to remove the water vapor from the gas stream or in some other way to make its thermal conductivity effect negligible. In a few applications, drying the gas sample is desirable to avoid or minimize corrosion, but generally it is most convenient to fix the concentration of water vapor and eliminate it as a variable. This can be done by saturating the sample gas at a constant temperature. Several practical gas saturators requiring very little maintenance are now standard or optional equipment with commercially available analyzers. The use of saturated gas presupposes that the analyzer cells and associated parts exposed to the gas will withstand the more corrosive atmosphere caused by saturation. When the sample gas must be dried, there is a decided maintenance advantage in selecting an analyzer that will operate satisfactorily with a minimum flow rate of sample.

Gas Sampling and Cleaning

Use of any gas analyzer is futile unless the analyzer receives an adequate and representative sample of the gas to be analyzed. Proper location of the sampling probe or the use of multiple sampling probes must be considered in relation to the process being monitored and the desired results. This is especially important when automatic control of gas composition is involved. It is always desirable to keep sampling lines as short as possible. Particular care should be exercised to avoid low points where

Table 3 Typical Applications of Thermal Conductivity Gas Analysis

A. Binary mixtures (or equivalent) by direct comparison in single-pass instrument

Gas mixture	Suitable comparison gas
H_2 in O_2	O_2, air, or H_2
H_2 in Cl_2	H_2 or Cl_2
H_2 in N_2	H_2, N_2, or air
H_2 in air	H_2 or air
H_2 in CH_4	H_2, CH_4, or H_2 + CH_4
H_2 in CO_2	H_2, CO_2, or H_2 + CO_2
H_2 in water gas (H_2 + CO)	H_2 or H_2 + N_2
He in air, N_2, or O_2	He, air, H_2 or O_2
Ne in air	Air
Cl_2 in air	Air
HCl in air	Air
Acetone in air	Air
NH_3 in air	Air
O_2 in enriched air	Air
SO_2 in air or N_2	Air or N_2
Water vapor in air, N_2, or O_2	Air, N_2, or O_2
CO_2 in air, N_2, or flue gas	Air
A in N_2, air, or O_2	N_2, air, or O_2

B. Pretreatment of sample gas followed by direct comparison in single-pass measurement

Gas mixture	Pretreatment (combustion)	Comparison gas
Acetone in air	$CH_3COCH_3 + 4O_2 \rightarrow 3CO_2 + 3H_2O$	Air or N_2
Benzol in air	$2C_6H_6 + 15O_2 \rightarrow 12CO_2 + 6H_2O$	Air or N_2

C. Analysis by double-pass measurement

Gas mixture	Treatment between passes
NH_3 in N_2 and H_2	Absorb NH_3
O_2 in N_2, or O_2 in flue gas	Convert O_2 to H_2O
Add H_2 before first pass	
CO in flue gas or air	Convert CO to CO_2
Water vapor in variable gas mixture	Absorb H_2O
O_2 in H_2 and CO	Convert O_2 to H_2O (under controlled conditions of combustion)

*Selection of the comparison gas depends on the range of composition to be covered and the ability of the reference cell to retain highly diffusive gases, such as hydrogen, or to withstand the corrosive action of such gases as chlorine.

condensed vapors can collect and trap the line. The sampling line should have a minimum number of elbows, have a smooth inner surface, and be of a material that can withstand the thermal and corrosive action to which it will be subjected. Plastic tubing, when it meets the requirements, is the most convenient form of sampling line. Heavy-wall glass tubing and corrosion-resistant metal pipes frequently are used.

Gases from industrial processes are rarely free of suspended solids, smoke, and other particulates. The nature of the process and the type of particulate contaminants will determine the necessary measures for obtaining a satisfactory gas sample for analysis. Alundum and similar porous ceramic materials are commonly used to remove suspended solids; but these are of little value in dealing with smoke, fly ash, mists, and volatile materials like tars which subsequently condense in the sampling line. Water spray washers, impingers, and steam injectors are used with some success, particularly when high sampling rates are necessary.

At the high sampling rates desired in automatic control applications to reduce response lag, it is very difficult to clean the entire gas sample. Instrument maintenance, however, can be minimized by the use of secondary filters. This is frequently done by using one of the primary means mentioned above for at least partially cleaning the main sample stream. A portion of the main sample is then bypassed to the analyzer cells through a secondary filter usually located in the analyzer housing. By locating the secondary filter in the analyzer housing it is possible to avoid unfavorable temperature gradients that might cause condensation of volatile contaminants. Secondary filtering materials that have proved useful include filter paper, washed long-fibered asbestos, porous sintered metal plates, and a variety of paperlike materials containing fibers of one or more types such as cellulose, glass, quartz, and asbestos. Asbestos, of course, must be considered a hazardous material.

Sample Purity

Assuming that a representative sample of gas is obtained at the primary sampling point, it is also necessary to avoid significant changes in composition of the sample as it passes through the sampling, cleaning, and conditioning systems. This may impose necessary limitations on the type of cleaning operation feasible, and it may be necessary to accept additional maintenance in order to deliver a chemically significant sample for analysis. Perhaps the most frequent cause of delivery of a contaminated or diluted sample is air infiltration. Since most samples are obtained by applying suction at some point in the sampling system, there is always a danger of faulty lines admitting air to the gas stream. Even systems operating at static pressures higher than atmospheric are not immune to this difficulty. It is possible, at high velocities of gas flow in such a system, to have air infiltrate by aspirator action at restricted points or turns in the sampling line. Regular inspection and maintenance of sampling lines and periodic leak tests are the best assurance of satisfactory operation.

REFERENCES

Angerhofer, A. W., and B. M. Dewey: *Instruments,* vol. 26, pp. 580–583, 1953.

Bennett, L. A., and R. G. Vines: "Thermal Conductivities of Organic Vapor Mixtures," *J. Chem. Phys.,* vol. 23, 1955, p. 1587.

Berl, W. G.: *Physical Methods in Chemical Analysis,* Academic Press, New York, 1951, pp. 387–437.

Cherry, R. H.: "Determination of Water Vapor by Thermal Conductivity Methods," *Anal. Chem.,* vol. 20, 1948, p. 958.

Cherry, R. H.: U.S. Patent 2,501,377, March 21, 1950.

Cherry, R. H.: *Proceedings of the 1963 International Symposium on Humidity and Moisture,* p. 539, Reinhold, New York, 1965.

Daynes, H. A.: *Gas Analysis by Measurement of Thermal Conductivity,* Cambridge University Press, London, 1933.

Hebler, William O.: U.S. Patent 2,116,239, May 3, 1938.

Koepsal, A.: *Ber. Phys. Ges.,* vol. 10, 1908, p. 814; vol. 11, 1909, p. 237.

Linde, H. W., and L. B. Rogers: *J. Chem. Educ.,* vol. 28, 1951, pp. 576–577.

Minter, Clark C., and Lyle M. J. Burdy: "Thermal Conductivity Bridge for Gas Analysis," *Anal. Chem.,* vol. 23, 1951, p. 143.

Neumann, R. K., et al.: *Am. Rocket Soc. Paper* 64–52.

Palmer, P. E., and E. R. Weaver: "Thermal Conductivity Method for the Analysis of Gases," *NBS Technol. Paper* 249, 1924.

Phillips, G. L., and J. W. Bulger: *U.S. Dept. Agr. Bull.* E-851.

Vines, R. G.: "The Thermal Conductivity of Organic Vapors: The Influence of Molecular Interaction," *Aust. J. Chem.,* vol. 6, 1953, p. 1.

Combustibles, Total Hydrocarbons, and Carbon Monoxide Analyzers

The measurement of combustible gases is important in controlling their concentrations within safe limits below the lower explosive limit (LEL), optimizing concentrations for efficient reactivity, and detecting process system upsets. The principal method used for combustible gas or vapor analysis employs self-heated hot wire detectors of a material such as platinum. These detectors are mounted in a Wheatstone bridge circuit. The hot catalytic detector causes the combustibles in the sample to burn in the presence of oxygen. If the sample does not contain an excess of oxygen, then oxygen or air must be added in controlled amounts to provide an excess such that the reaction will be limited by the amount of combustibles present.

The combustion of the mixture releases heat (see Table 1), which is detected as a temperature rise in the catalytic detector spiral. Since the spiral's resistance is a function of its temperature, an electric unbalance occurs in the Wheatstone bridge proportional to the concentration of the combustibles. Measurement of this electric signal provides a measure of the combustibles concentration.

Total Combustibles

Instruments for measuring the concentration of total combustibles are widely used for continuously analyzing usually inert atmospheres for the presence of a wide range of combustible gases or vapors in the interest of preventing explosions. The instruments are commonly calibrated in terms of the LEL, specifically 0 to 1%, 0 to 2%, 0 to 5%, 0 to 10%, 0 to 20%, 0 to 25%, and 0 to 100% LEL. The instruments usually have high-low contacts which actuate panel lights and can direct output relays or recorder or controller signals. These instruments find frequent application in controlling inert gas

Table 1 Heats of Combustion of Representative Process Materials*

		Heat of combustion Hc° at 25°C and constant pressure to form					
		H_2O (liq) and CO_2 (gas)			H_2O (gas) and CO_2 (gas)		
Material	Formula	kcal/mol	cal/g	Btu/lb	kcal/mol	cal/g	Btu/lb
Hydrogen	H_2	68.3174	33,887.6	60,957.7	57.7979	28,669.6	51,571.4
Carbon monoxide	CO	67.6361	2,414.7	4,343.6	—	—	—
Methane	CH_4	212.798	13,265.1	23,861	191.759	11,953.6	21,502
Ethane	C_2H_6	372.820	12,399.2	22,304	341.261	11,349.6	20,416
Propane	C_2H_8	530.605	12,033.5	21,646	488.527	11,079.2	19,929
n-Butane	C_4H_{10}	687.982	11,837.3	21,293	635.384	10,932.3	19,665
Isobutane	C_4H_{10}	686.342	11,809.1	21,242	633.744	10,904.1	19,614
n-Pentane	C_5H_{12}	845.16	11,714.6	21,072	782.04	10,839.7	19,499
Isopentane	C_5H_{12}	843.24	11,688.0	21,025	780.12	10,813.1	19,451
Neopentane	C_5H_{12}	840.49	11,649.8	20,956	777.37	10,775.0	19,382
n-Hexane	C_6H_{14}	1002.57	11,634.5	20,928	928.93	10,780.0	19,391
n-Heptane	C_7H_{16}	1160.01	11,577.2	20,825	1075.85	10,737.2	19,314
n-Octane	C_8H_{18}	1317.45	11,533.9	20,747	1222.77	10,705.0	19,256
n-Nonane	C_9H_{20}	1474.90	11,500.2	20,687	1369.70	10,680.0	19,211
n-Decane	$C_{10}H_{22}$	1632.34	11,473.0	20,638	1516.63	10,659.7	19,175
Benzene	C_6H_6	789.08	10,102.4	18,172	757.52	9,698.4	17,446
Toluene	C_7H_8	943.58	10,241.4	18,422	901.50	9,784.7	17,601
Ethylene	C_2H_4	337.234	12,021.7	21,625	316.195	11,271.7	20,276
Acetylene	C_2H_2	310.615	11,930.2	21,460	300.096	11,526.2	20,734

*Additional values are obtainable from the National Bureau of Standards and American Petroleum Institute.

generators, solvent evaporating ovens, heat-treating atmospheres, and analyzing flue gas. Instruments designed for mounting in a permanent location usually are equipped for drawing a sample from a distance up to 100 ft (30 m). The sample gas is mixed with an equal amount of ambient air to ensure a sufficient supply of oxygen for combustion. The general arrangement is shown in Fig. 1. The response time is 5 s for 90% of full scale, and the repeatability is ±2% of full scale. The capabilities of one instrument for detecting samples of high-combustibles content and those in the area of slight excess air are shown in Fig. 2.

Customized Combustibles Analyzers

In addition to the availability of instruments for permanent mounting, semiportability, and portability (including battery-operated, hand-held meters), there are numerous models specifically designed for specific applications. Practically all of these instruments operate on the catalytic filament principle previously described. Examples include instruments designed for use in testing atmospheres that may be oxygen-enriched (more than 21% O_2). The rate of flame propagation of such mixtures is much higher than that of other combustibles in air. Therefore, such meters are equipped with heavy duty flashback arrestors capable of confining explosions of hydrogen or acetylene and oxygen within the combustion chamber. Instruments of this type are calibrated on hydrogen and acetylene. Special designs are available for the detection of vapors of leaded gasoline. When a hot-wire indicator is used with these vapors, the oxidation of tetraethyllead can produce a solid lead combustion product which condenses on the filament and reduces its catalytic activity, especially with respect to combustibles such as natural gas which have high ignition temperatures. In this case, the filament temperature is maintained sufficiently high to prevent the condensation of lead contaminants. In an alternative approach, an inhibitor filter may be used in the filter chamber. This filter promotes a chemical reaction with tetraethyllead vapors, which yields a more volatile combustion product and thus prevents filament contamination.

Silanes, silicones, silicates, and other compounds containing silicon in the tested atmosphere may seriously impair the response of combustibles meters because these materials can rapidly poison the filament. Where such materials are suspected, the instrument manufacturer will suggest that the instrument be checked after every fifth test.

Specially designed combustibles meters are available for use in detecting, measuring, and pinpointing leaks. Portable instruments are equipped with probe tubes or rods ranging from 3 to 4 ft

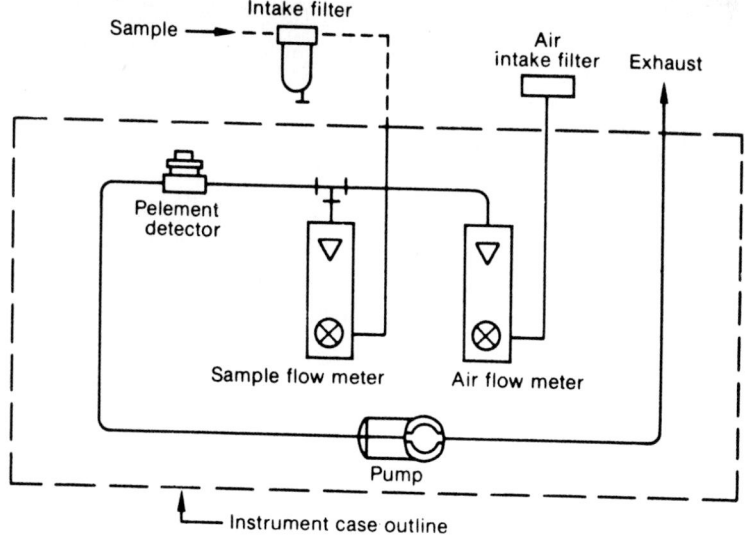

Fig. 1   General configuration of total combustibles analyzer. *(Model 513, MSA Instruments.)*

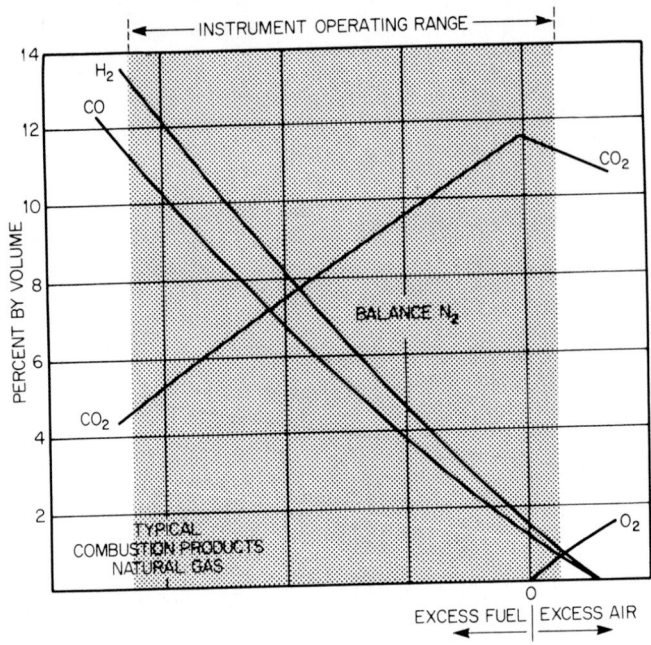

Fig. 2   Within the standard available operating ranges of 0 to 1 to 0 to 25% (shaded area), samples of high combustibles content as well as those in the area of slight excess air can be detected and measured. *(Model 513 total combustibles analyzer, MSA Instruments.)*

(0.9 to 1.2 m) in length, which are particularly effective for sampling from manholes or barholes and for testing tanks or other vessels which may contain liquids. The probe rod prevents the drawing of liquids into the flow system. A meter designed especially for gas utilities in routine testing for methane-in-air concentrations in manholes, sewers, curb boxes, and other street openings may be calibrated on methane in air. A similar meter for general industrial use may be calibrated on pentane in air. Pentane calibration is used because it is representative of petroleum vapors. Applications include testing tank and vessel interiors, locating leaks in pipelines and process streams, and checking confined areas in sewage disposal plants, refineries, paint factories, chemical plants, and iron and steel mills, among others. To prevent errors caused by stray magnetic fields, the instrument may use a core magnet-type meter movement. Where it is necessary to distinguish between condensable hydrocarbon vapors and natural gas, a charcoal cartridge which absorbs hydrocarbon vapors may be used. The difference in the readings with and without the charcoal cartridge indicates that vapor, gas, or a combination of the two is present in the sample.

In some instruments, two different types of filaments are used—a catalytic combustion filament for low ranges (percentage LEL) and a thermal conductivity filament for high ranges. In connection with the latter, combustibles in the sample cool this filament, decreasing its resistance and unbalancing the bridge. The imbalance, proportional to the gas concentration, is measured by the meter and read as percentage by volume.

Total Hydrocarbons Analyzers

Analyzers for determining total hydrocarbons in various concentrations are widely used for pollution monitoring and control programs, particularly in connection with internal combustion engines (auto-

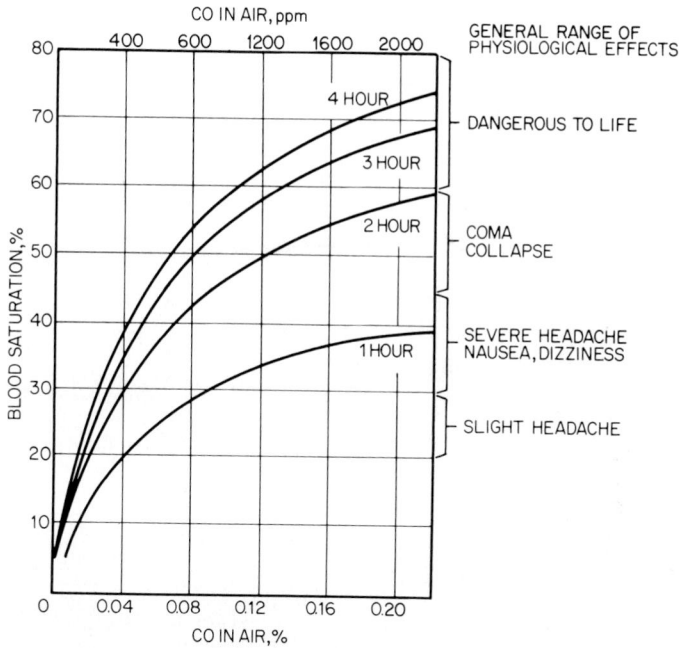

Fig. 3 Effects of carbon monoxide on human beings. This chart can be considered only a general guide, because the percentage of CO blood saturation varies with exertion, excitement, fear, depth of respiration, anemia, and general physical condition of the individual.

mobile exhausts) and fossil-fueled stationary combustion installations. They are also used for leak detection in aerosol packaging, to detect breakthroughs from carbon adsorption beds, and for leak detection in refrigerant systems, among other applications.

Some total hydrocarbon analyzers are based on the ionization of carbon atoms in a hydrogen flame. Normally, a flame of pure hydrogen contains an almost negligible number of ions. The addition of organic compounds (even traces) results in a large number of ions in the flame. In one analyzer,* the sample is mixed with a hydrogen fuel and passed through a small jet. Air is supplied to an annular space around the jet to support combustion. Any hydrocarbon carried into the flame results in the formation of carbon ions. An electric potential across the flame jet and an "ion collector" electrode suspended above the flame produce an ion current proportional to the hydrocarbon count. This current is measured by an electrometer circuit whose output then provides an analog analysis signal for a direct reading meter, graduated from 0 to 100, or for an optional potentiometric recorder. The analyzer features an electrical range attenuator, with change factors of ×1, ×3, ×10, ×30, ×100, ×300, ×1000, and ×3000. The stabilized zero setting is unaffected by range change factors. Calibration is by a span potentiometer—for greater accuracy than is provided by adjustment of the sample flow. The analyzer features sintered metal filters for the sample, air, and fuel capillary tubing. The instrument includes a flame-out alarm and an automatic fuel shutoff, for maximum safety, and an optional integral sampling pump. The full-scale range of the instrument is 0 to 4 to 0 to 12,000 ppm by volume, expressed as methane, the sensitivity is 1% of full-scale reading in 1 s or less, the response is 90% of final reading in 1 s or less, and the warmup time is 4 h from a cold start to stability; and 1 h after a temporary shutdown.

Another version of this instrument uses a dual-flame detector and continuously monitors ambient air for methane and nonmethane hydrocarbons. The instrument can measure (1) total hydrocarbons,

---

*MSA total hydrocarbon analyzer.

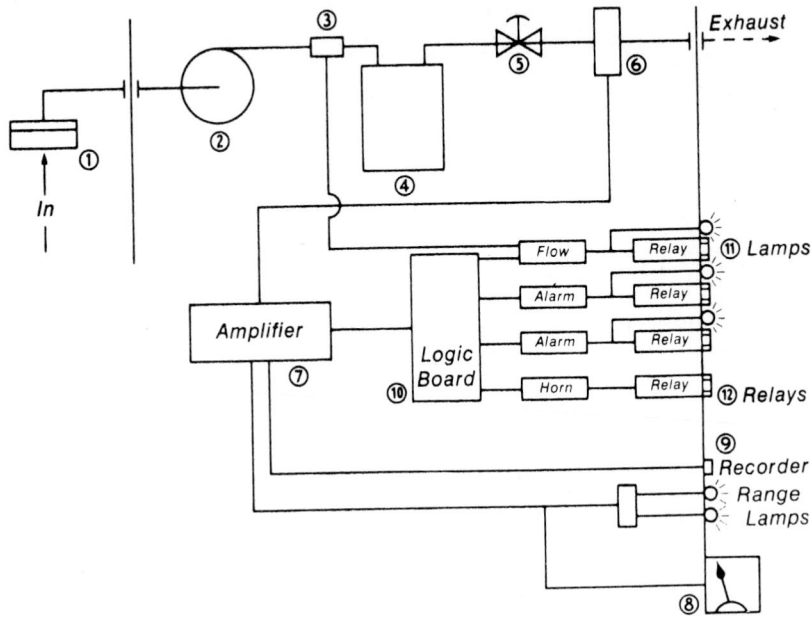

Fig. 4 Flow diagram of carbon monoxide alarm. *(Model 704, MSA Instruments.)*

(2) total hydrocarbons less methane, and (3) methane only. The full-scale range is 0 to 5 and 0 to 20 ppm. The instrument is used for pollution monitoring and frequently for checking the effectiveness of hydrocarbon control equipment.

Carbon Monoxide Measurement

Numerous uses for carbon monoxide (CO) detection include chemical, metallurgical, and many other types of manufacturing plants, gas and utility properties, sewers, flight test centers, mines, bus terminals, and garages. The instruments are also used for testing compressed air intended for delivery to respiratory equipment and for controlling ventilating equipment (e.g., in a vehicular tunnel).

In addition to indication, CO meters almost always incorporate both audible and visible alarm circuits. The threshold limit value for CO adopted by the American Conference of Governmental Industrial Hygienists is 50 ppm. See Fig. 3. As with the combustibles analyzers previously described, a large variety of configurations are available—from small, portable units to permanently mounted indicators and recorders, some of which are equipped to handle inputs from several detectors at various locations.

In one instrument (Fig. 4), the CO detector cell operates on the principle of catalytic oxidation. In the cell, the sample passes through an inactive chemical bed and then through an active catalytic bed of hopcalite.* Each bed contains a thermistor, making up part of a Wheatstone bridge circuit. Any CO present is immediately oxidized in the hopcalite half of the cell. This raises the temperature of the hopcalite thermistor and its electrical resistance, which causes a resistance imbalance in the bridge and thus a signal proportional to the CO concentration in the sample.

In another system, the sensor contains an electronic interface and an electrochemical polaro-

---

*A mixture of oxides of copper, cobalt, manganese, and silver as a catalyst for conversion of carbon monoxide to carbon dioxide.

graphic cell containing a sulfuric acid electrolyte. Air samples diffuse through a gas-porous membrane and a sintered metal disk and enter a sample area within the cell. The cell electrooxidizes CO to CO_2 in proportion to the partial pressure of CO within the sample. The oxidation generates an electrochemical signal proportional to the concentration of CO in the ambient air. The resulting electrical signal is monitored, temperature-compensated, and amplified to drive a meter. Under normal operation, the service life of the sensor assembly is usually a minimum of 6 months. Certain easily oxidized compounds in the sample atmosphere act as interferents in CO readings. These mainly include methane, ammonia, and sulfur dioxide. Hydrogen, hydrogen sulfide, nitrogen dioxide, propane, nitric oxide, ethylene, ethyl alcohol, and acetylene affect readings to a much lesser extent.

Chromatography

by
Richard Villalobos*

Chromatography is an instrumental procedure based on physical absorption principles for separating various components from a mixture of chemical substances. In its broad interpretation, chromatography is a combination of separation, identification, and quantitative measurements.

The procedure is broadly applied to mixtures of organic and inorganic materials and is particularly useful for mixtures of compounds whose chemical characteristics (composition and molecular structure) and physical properties (boiling point, density, etc.) are so nearly identical as to make other separation and analytical techniques difficult or impractical. Some forms of chromatography also offer the distinct advantage of requiring only very small samples—in terms of grams. Consequently, chromatography is used widely in the laboratory for organic chemical, biological, and medical studies—and in industry, notably in chemical, petroleum, and petrochemical plants, for quality and process control.

The recovery of pure compounds from mixtures in the laboratory often is as important as or more important than making quantitative determinations; hence, chromatography is widely used to separate and collect pure compounds for study by other methods.

For industrial applications a chromatograph is a complex transducer which not only puts out a signal that identifies the types and amounts of given substances but also must first separate the target substances from a stream that may contain numerous other substances, some of which may be closely related physically or chemically. The transducer output can be used for off-line quality control, or on-line so that the chromatograph becomes part of the total control loop.

In 1903, a Russian botanist, Mikhail Tsvet, used the chromatographic principle to separate plant pigments. He filled a vertical glass tube with an absorbent; and as a sample of the pigments was washed through the tube with a solvent, a series of colored absorption bands was produced, in essence a graphic presentation of colors—thus the name chromatography—literally "color writing," the term first used by Tsvet to describe this procedure. The term no longer is meaningful in this way, however, because chromatography now is used with colorless materials and most often with the operator no longer visually witnessing the separation process.

*Prepared at Beckman Industrial Corporation, Process Instruments and Controls Division, a subsidiary of Emerson Electric Co., Fullerton, California.

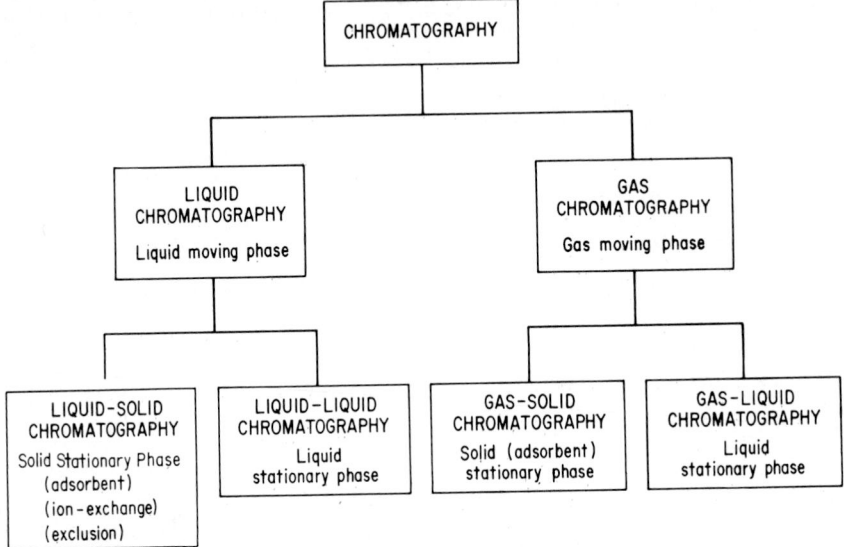

Fig. 1   Chromatography techniques as classified according to moving and stationary phases.

Basic Principle

Chromatography depends on selective retardation and separation of substances by a stationary bed of porous sorptive media as they are transported through the bed by a moving fluid. The degree of retardation, and hence the rate of migration of each substance, is determined by its relative affinity for the sorbent.

The sorbent bed is referred to as the *stationary phase,* and the moving fluid as the *moving phase.* The moving phase may be a liquid or a gas. The stationary phase may be liquid dispersed on and supported by a porous, inert solid in which the sample components are soluble and are partitioned in equilibrium with the moving phase—hence the *solvent** or *partition liquid.* Alternatively, the stationary phase may be a solid which sorbs sample components on its surface and inner structure by various reversible physical or chemical mechanisms, namely, absorption, ion exchange, or exclusion by molecular size.

Nomenclature

Chromatography may be classified by a number of schemes:

1. *The form of technique:* (a) frontal analysis, (b) displacement, (c) development, and (d) elution
2. *The nature of the moving and stationary phases* as shown diagramatically in Fig. 1, e.g., gas-liquid and liquid-liquid chromatography
3. *The nature of the interaction—whether physical or chemical—between the components and the stationary phase,* e.g., partition, absorption, and exclusion (permeation) chromatography
4. *Some distinctive feature of the apparatus or method,* most commonly used for referring to a specific technique, e.g., high-performance liquid chromatography (HPLC) and column, paper, and thin-layer (TLC) chromatography

---

*The term "solvent" is used both in gas chromatography in reference to the stationary liquid phase and in liquid chromatography in reference to the moving liquid phase.

Frontal analysis and displacement chromatography are not discussed here because they are of limited value and use.

Most procedures and techniques are often and variously referred to using two or more of the above nomenclature or classification systems, as shown diagrammatically in Fig. 2. Hence, the widely used gas chromatography can, for different purposes, be referred to as gas, solid (or liquid), column, or elution chromatography.

Development Chromatography

This technique is used only with liquid moving phases. The sample is introduced onto a dry column or sorbent bed and washed through the bed with a solvent that is less strongly sorbed than the sample components. The solvent washing process is continued until the solvent reaches a point just short of the exit or opposite end of the bed. The migration rate of each sample component is dependent on its *partition coefficient* or distribution between sorbent and solvent. The most useful forms of development chromatography are paper chromatography and TLC. The use of columns for development chromatography has declined because of the difficulties in visualizing and documenting the developed column and in recovering pure compounds from the sorbent bed.

Elution Chromatography

In this method, the moving phase is passed through the sorbent bed until all sample components have been washed or eluted from the bed. In contrast to development chromatography, elution chromatography uses columns exclusively. In gas elution systems and in most types of liquid elution chromatography, the column can be reused many times and without interruption of the flow of the moving phase. Components move through the column at rates dependent on the partition coefficient and are separated into bands which elute at characteristic times.

A principal value of elution systems is the precision with which elution times can be reproduced from one analysis to another—an important element in identification and quantitation.

A detector at the end of the column can generate an analog signal proportional to the concentra-

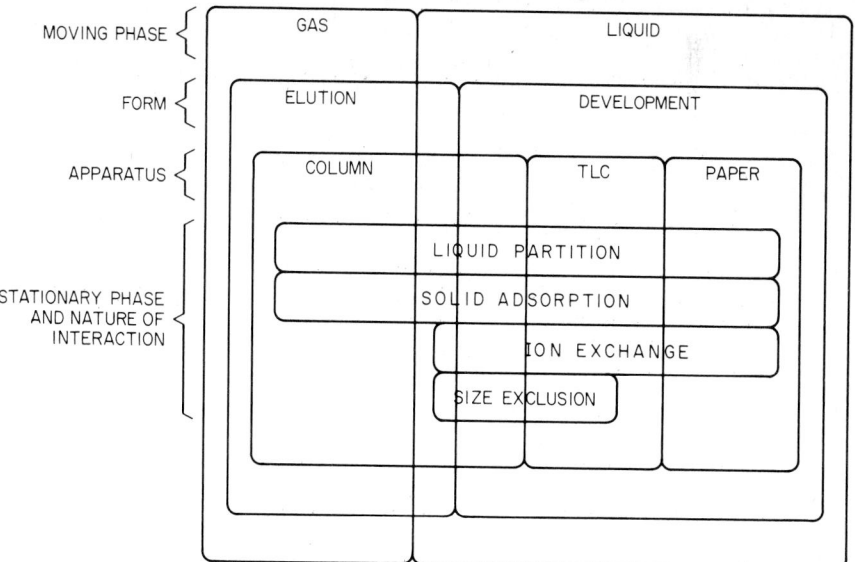

Fig. 2   Relationship of various chromatographic techniques.

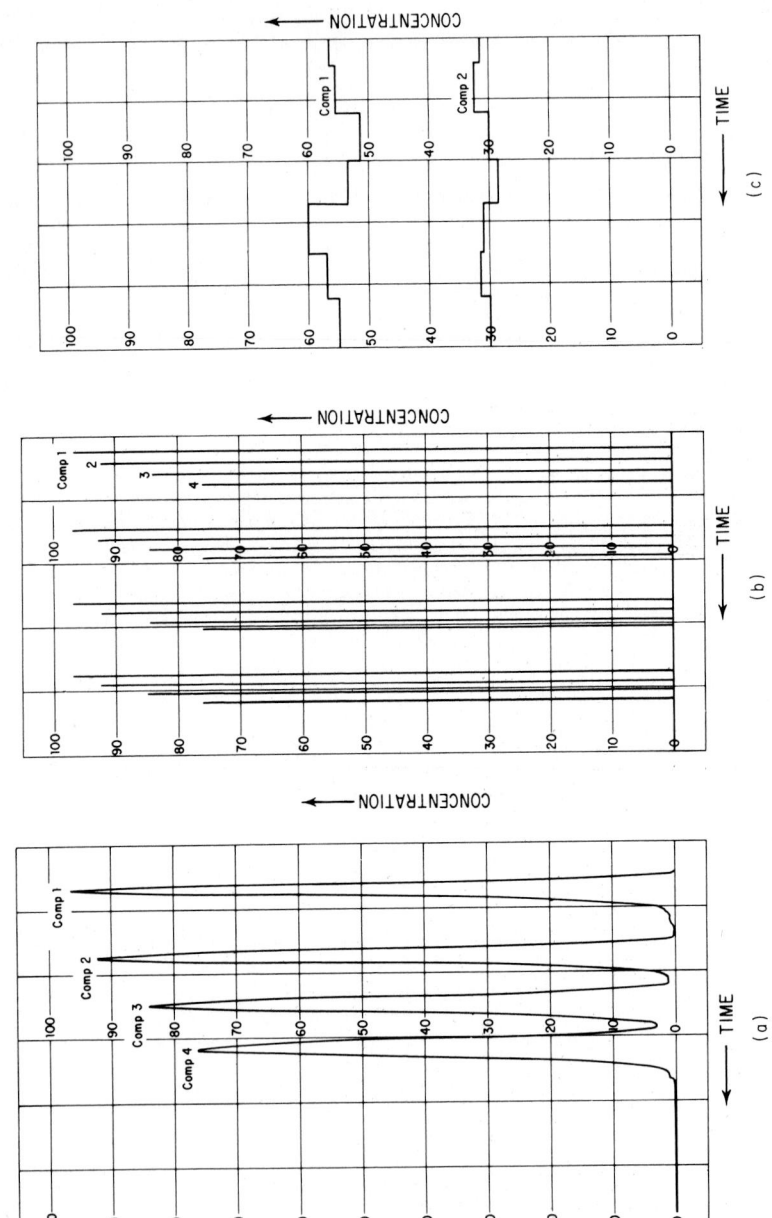

Fig. 3 Typical chromatograph readouts. (*a*) Chromatogram, (*b*) bar graph, (*c*) trend.

tion of the sample component in the moving phase. A time record of the detector signal is called a *chromatogram* and, as shown in Fig. 3*a*, contains the characteristic gaussian peaks.

Gas and liquid chromatography are the most popular forms of column elution chromatography. The theory, technique, and apparatus for both methods have been highly developed. Completely automated versions have been developed and widely applied to continuous on-line analysis of chemical and petroleum-refining process streams.

Gas Chromatography

Almost any organic or inorganic compound that can be vaporized can be separated and analyzed with a gas chromatograph. As shown in Fig. 4, the minimum requirements for a system include (1) a column which contains the substrate or stationary phase, (2) a supply of inert carrier gas (moving phase) which is continually passed through the columns, (3) a means for maintaining pressure and

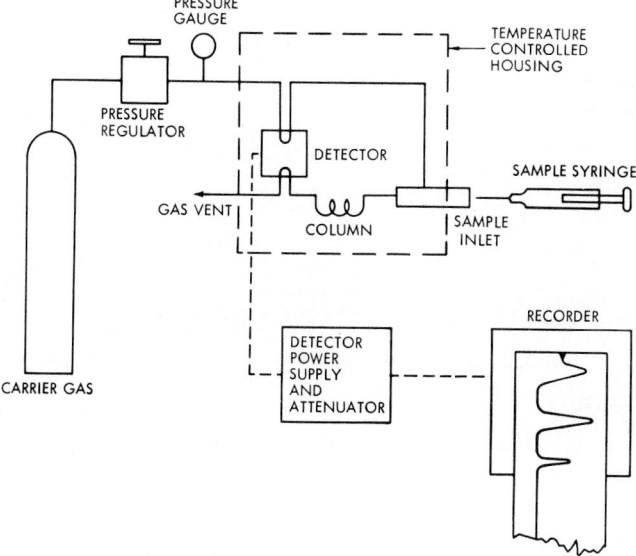

Fig. 4 Basic elements of a gas chromatograph.

flow constant, (4) a means of admitting or injecting the sample into the carrier gas stream, (5) a detector which senses the sample components as they elute, and (6) a recorder. The carrier gas may be any gas that does not react with the sample or adversely affect the detector. Helium, hydrogen, nitrogen, and argon are most commonly used

Gas Chromatography in the Laboratory

A typical application is illustrated and described in Fig. 5. Gas chromatography is so flexible that it has many end uses and apparatus configurations in the laboratory, some of which include:

1. *Programming of column temperature* to permit analysis of wide boiling range fractions. A low starting temperature (sometimes subambient) allows the separation of low-boiling components, whereas a later, higher column temperature reduces the elution time of high-boiling components and, overall, reduces the analysis time. This technique is used widely for complex samples such

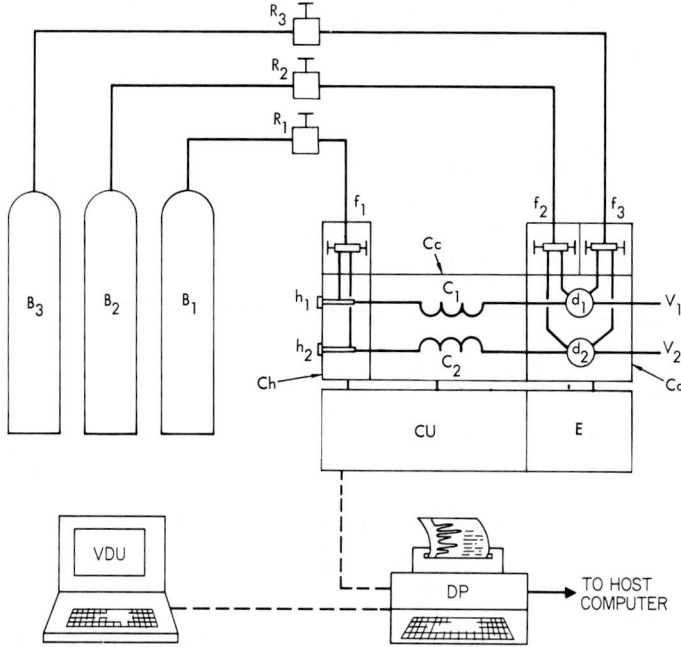

Fig. 5 Basic elements of a gas chromatograph for research and laboratory purposes. The dual column/dual detector gas chromatograph shown is suitable for the analysis of complex organic mixtures. Columns C_1 and C_2, with different characteristics, are enclosed in a temperature-controlled column compartment C_c, which may be isothermal and/or temperature-programmed during analysis for rapid elution of high-boiling compounds. Individual flash vaporizations of liquid samples are enclosed in temperature-controlled inlet compartment Ch. Carrier gas, supplied from a high-pressure cylinder B_1, is maintained constant by pressure regulator R_1 and flow controller f_1. Sample components are detected by hydrogen-flame ionization detectors d_1 and d_2 in constant temperature detector compartment Cd. Hydrogen fuel B_2 and combustion air B_3 are supplied to detectors through pressure regulators R_2 and R_3 and flow controllers f_2 and f_3. Combustion products are vented through V_1 and V_2. Control CU gives individual temperature control of sample inlets and detectors and programs the temperature of the column compartment at the rate and over the range required. Amplifier E amplifies the detector signal and transmits it to a microprocessor-based data processor and programmer DP with a keyboard and built-in printer. The data processor digitizes the signal, processes data, calculates the composition in accordance with an analytical program stored in the microprocessor RAM, and prints out a chromatogram and analytical data on a printer-plotter. Analytical method parameters are entered by the operator through the keyboard and stored in the methods file in RAM. All temperature settings and a programmed temperature profile are set in CU by DP. An auxiliary video display unit VDU displays a chromatogram and allows postrun data manipulation for methods development simultaneously with data reduction by DP. Data and programs can be transmitted via the RS232 port and serial link to the host computer.

as essential oils and flavor and aroma concentrates, and in the petroleum industry for simulated distillation analysis of crude oil and petroleum fractions.

2. *Multiple columns and column switching* to effect separations where a single column would be impractical, and also to reduce analysis time.

3. *Fraction collecting or preparative chromatography* where separated compounds are collected for analysis by other methods.

4. *Pyrolysis chromatography* where the thermal degradation of high-molecular-weight materials which yield characteristic fragment patterns can be studied.

5. *Reaction chromatography* where compounds may be converted to other related compounds in a reaction zone in the column system to facilitate detection or further separation.

6. *Subtraction chromatography* where a reaction zone in the system preferentially reacts with and removes compounds of a certain class.

7. *Physiochemical measurements,* such as the determination of activity coefficients, partition coefficients, K values, second virial coefficients, heat and entropy of solutions, absorption, vaporization, and other physical and thermodynamic properties.

8. *Hyperpressure (supercritical) chromatography* in which a fluid above the critical point is used as the moving phase and which enhances the volatility of sample components by as much as 10^4. This permits the analysis of compounds of extremely high molecular weight (MW = 1000) and of high boiling point ($1000°C$) at column temperatures below $300°C$.

Numerous types of detectors have been used, many of which are highly specific and of limited application. The most widely used are thermal conductivity (TCD), hydrogen flame ionization (FID), argon ionization, electron capture (ECD), flame photometric (FPD), and conductivity detectors.

Microprocessor-Based Automation

Microprocessor-based systems are now widely used to carry out functions previously performed manually by the operator. These systems may take many forms: (1) integral to and dedicated to a single chromatograph, and (2) as a separate unit which controls one or several chromatographs. The latter are also available for upgrading older manually operated chromatographs to provide automated data reduction and output.

Systems are equipped with read only memory (ROM) for the operating system and random access memory (RAM) for storage of user-specific methods, operating instructions, and analytical data acquisition and manipulation.

Depending on user requirements, one or more of the following functions can be provided:

1. Individual temperature control and indication of column, ovens, detectors, and sample inlets

2. Dynamic control of programmed temperature profiles, starting and ending temperatures, rate of temperature increase with one or more time-controlled segments of isothermal temperature, and cooling-down cycle

3. Automatic integration of peak areas or peak height measurements, with a choice of several calibration methods for calibration of results

4. Raw data storage with postrun calculation and manipulation of data for optimizing variables and methods development

5. Integral printer-plotter for hard copy of chromatograms and analytical results

6. Keyboard for input of analytical parameters and operator-selected variables

7. Video terminal for display of chromatograms and analytical results

8. Cassette or floppy disk extended memory for storage of files, methods, and user-specific applications programs

9. High-level language for user programming of special calculations such as physical properties and comparison with alarm levels

10. Serial communications via RS-232 ports with a central computer or other peripheral devices
11. Timed output control signals to peripheral and ancillary devices such as sample valves, column switching valves, and other external functions which are part of the automated procedure
12. On-board diagnostics to aid in troubleshooting and isolating and identifying failures

Chromatograph Data Reduction—Qualitative Analysis

For a given substrate under given conditions, each compound has a characteristic retention time which can be used for tentative identification. However, two or more compounds may have the same elution time on a particular column. In such cases, the compound may be rerun on a different column with other characteristics to reduce ambiguity. Extensive compilations of individual compound retention times on different substrates are available for reference. Positive identification can be made only by collecting the compound or transferring it as it elutes directly into another apparatus for analysis by other means, such as infrared or ultraviolet spectroscopy, mass spectrometry, or nuclear magnetic resonance. Commercially available apparatus is available which combines in a single unit both a gas chromatograph and an infrared, ultraviolet, or mass spectrometer for routine separation and identification. The ancillary system may also be microprocessor-based, with an extensive memory for storing libraries of known infrared spectra or fragmentation patterns (in the case of mass spectrometers); such systems allow microprocessor-controlled comparison and identification of detected compounds.

Quantitative Analysis

Quantitative analysis is based on the proportionality of detector response to the amount of component in the elution band. The most widely used measure of detector response is the area under the chromatogram peak. However, the peak height (amplitude of the detector signal at peak maxima) also may be used. Early methods for measuring the peak area included direct measurement on the chart with a planimeter, calculating from the height and width of the peak, and using a ball-and-disk integrator attached directly to the recorder. Electronic integrators have also been used to sense the detector signal directly, or after it is amplified, and to provide a digital printout of the peak area and time of peak maxima.

These methods of quantitation have been largely replaced by microprocessor-based data systems. The detector signal is directly coupled to a highspeed analog-to-digital converter which samples it at predetermined rates as high as 40 Hz. The digitized values are stored in memory for subsequent manipulation. Slope sensing algorithms find the beginning and end of peaks, peak maxima, and valley points between incompletely resolved peaks and can differentiate between peaks and the baseline (absence of peaks) to correct the digitized signal for baseline drift. Tangent skimming algorithms can locate and digitize small "rider" peaks on tailing edges of larger peaks.

The microprocessor sums the digitized values for a given peak to obtain the peak area. Results are then calculated by various methods:

Area Percentage, Normalized. In this method, the area of each peak is determined as a percentage of the total of the areas of all peaks. Results must add up to 100%; therefore, all components are detected.

Concentration by Relative Response Factors, Normalized. Similar to area percentage, but each area is corrected by a relative response factor characteristic for a given compound.

Internal Standard. Wherein results are calculated relative to a standard added to the sample in a known amount. Results are independent of sample size. The internal standard must be a compound not normally in the sample and well separated from other components in the sample.

Standard Addition. In which the sample is analyzed with and without the addition of a known amount of a compound which is also in the sample (spiking). The concentration is calculated from the observed increase in area.

External Standard. Wherein the area of each component is compared to the area of a separately run standard with a known concentration of each component.

Table 1 Principal Uses of Process Gas Chromatographs

Process Control: Use information to adjust the process through open- or closed-loop control
Process Study: Obtain information about the process to improve yield or throughput; correlate process variables with reaction products and yields
Process Development: Obtain information about process characteristics, as in pilot plants; correlate process variables with reaction products and yields
Material Balance: Use information to calculate the material balance for process units
Product Quality Specification Monitor: Monitor impurities in outgoing or incoming product for conformance to specifications
Waste Disposal Monitoring: Monitor liquid or gas effluent wastes for loss of valuable product or for the presence of toxic compounds
Personnel Safety Area Monitoring: Monitor ambient air for the presence of toxic compounds

---

Special Calculations. Analytical results can be used to calculate other properties or characteristics of the sample, such as average molecular weight, specific gravity, heating value, octane number, Reid vapor pressure, and boiling point distribution (simulated distillation).

Process Gas Chromatography

This is a system for continuous, repetitive, and fully automatic on-line analysis of process streams which is similar to the laboratory chromatograph in all essential elements of basic technique but different in design and appearance. Factors affecting design include the need (1) to comply with the National Electrical Code for operation in hazardous atmospheres, (2) to automate the procedure, and (3) for ready adaptability to closed-loop process control and communication with process control systems and computers. The principal uses of process gas chromatographs are listed in Table 1. The demand for maximum reliability and minimum maintenance has emphasized simplicity of hardware and methodology. Emphasis is placed on analyzing for a few rather than a number of components and on minimizing analysis time. These design targets have resulted in the extensive use of multicolumn techniques for rapid separation of selected components, with large portions of the sample being discarded. As shown in Fig. 6, the major components of a microprocessor-based process gas chromatograph system are (1) the analyzer, (2) the data processor, (3) the sample conditioner system, (4) one or more recorders, and (5) analog and serial outputs to peripheral devices and/or process control sytems and computers. A typical analyzer and process are shown in Fig. 7.

Analyzer

This equipment is usually located close to the sample point, enclosed in a shelter for weather protection. An analyzer is typically designed to comply with NEC Class I, Groups B, C, and D, Division 1, requirements for operation in hazardous areas by combination of explosion-proof enclosures, air purging, and intrinsically safe electric circuits. Sections of the analyzer include a controlled temperature compartment (heated air bath) for the columns, sample and column switching valves, and a detector. A pneumatics section for pressure or flow controllers for the carrier gas and other auxiliary gases (such as hydrogen and combustion air for a FID), as well as service air for the heater, electronics purge, and valve actuation. The electronics compartment contains a microprocessor with a central processing unit (CPU) and RAM and ROM for program control, data acquisition and reduction, output, and all communications functions. The RAM is battery-backed to prevent loss of applications programs due to power failure.

In its most usual form, the microprocessor performs these functions:

1. Controls all sequenced analyzer functions, such as sample injection and valve switching, by means of the applications program stored in RAM.

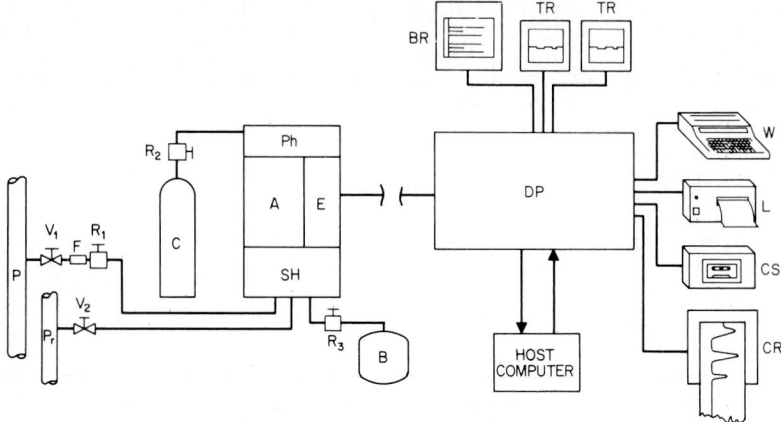

Fig. 6 Basic elements of a process gas chromatograph. The vapor sample is continuously withdrawn at a high rate from process line P, circulated through sample conditioner SH, and returned to lower pressure point P_r through shutoff valves V_1 and V_2. Particulate matter is removed by filter F, and the pressure reduced to a constant low level by regulator R_1. The sample conditioner contains flow control and other conditioning components and a valve for switching to synthetic calibration blend B through pressure regulator R_3. A sample slipstream is circulated to the sample valve in analyzer A, which also contains columns, detectors, and a temperature control system. Carrier gas C is controlled by regulator R_2 and pneumatics control section P. A microprocessor in electronics module E stores an analytical program in RAM and controls analyzer functions and data acquisition and reduction. Analytical results are transmitted over a serial link to data processor DP which converts results to an analog signal for presentation to bar graph recorder BR and as many as 30 to 40 trend recorders TR. Real-time constructed chromatogram is presented for maintenance on the recorder, CR. Serial outputs (RS232) flow to writer or panel-mounted line printer L for data logging and to cassette recorder CS for storing applications programs. Results and alarm messages flow to the host computer via serial link. An applications program is entered via data processor and downloaded to an analyzer RAM for execution in the analyzer. The processor controls several analyzers.

2. Samples and digitizes the detector signal at up to 40 Hz; performs peak area integration or peak height measurement with baseline correction and deconvolution of incompletely resolved peaks. Figure 8 illustrates several methods of peak deconvolution and baseline error correction.

3. Identifies components by comparing elution times with values stored in memory.

4. Calculates the composition with a choice of several calibration and calculation methods stored in ROM.

5. Controls the sequencing of sample conditioner in multistream systems.

6. Performs automatic calibrations by analyzing calibration standards at user-selected intervals and automatically updating calibration factors.

7. Performs auxiliary calculations such as determining average molecular weight, specific gravity, heating (thermal units) value, or other properties based on the calculated sample composition.

8. Monitors electromechanical sensors in the analyzer and sample conditioner system to detect abnormal conditions: oven temperature out of limits, carrier gas flow failure, sample flow failure, etc.

9. Performs software diagnostics on the detector signal and analytical results to detect abnormal conditions: change in elution time, total peak area out of limits, excessive baseline noise or drift.

10. Communicates with data processor over a serial link to transmit analytical data and calculations and receives new or modified applications programs. A digitized form of the detector signal may also be transmitted for remote reconstruction of a real-time chromatogram.

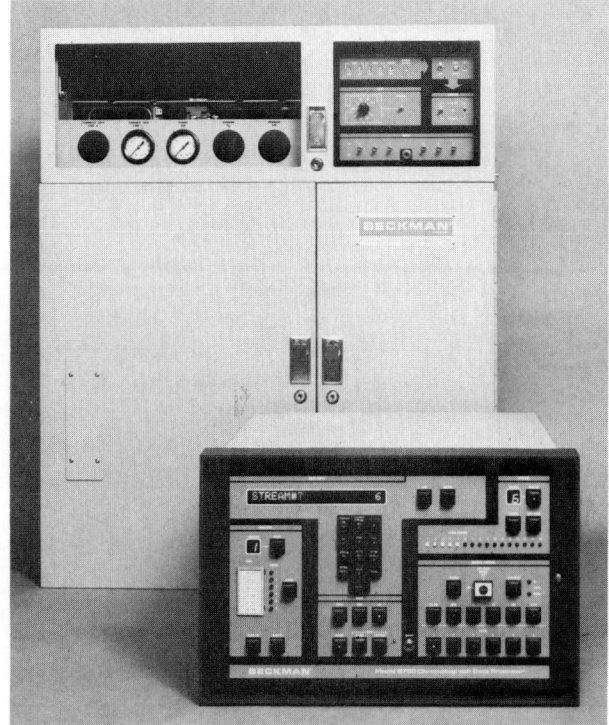

Fig. 7 Data processor controlling up to six analyzers at a remoted location. The processor keyboard permits a user-designed applications program, which is stored in the analyzer microprocessor memory. *(Model 6750 microprocessor process gas chromatograph, Beckman.)*

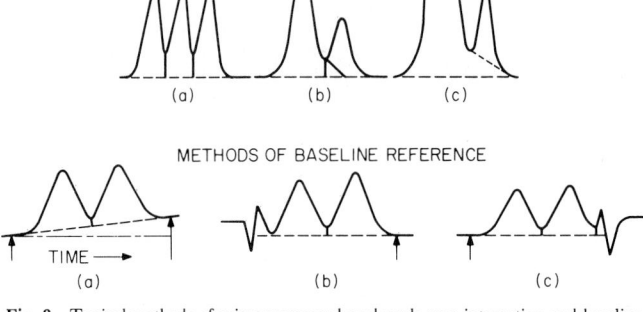

Fig. 8 Typical methods of microprocessor-based peak area integration and baseline correction for a process chromatograph. Peak Area Integration: (*a*) Forced integration, (*b*) perpendicular drop from a valley minimum for unresolved peaks of approximate equal area, (*c*) perpendicular drop with area correction for peaks of unequal area. Note tangential skimming for rider peaks on the tail of a large peak. A slope sensing algorithm finds the peak beginning, end, and maxima, and the valley minima. Baseline correction methods to correct for baseline drift and shifts are (*a*) trapezoidal reference to baseline before and after peak, (*b*) look-forward for baseline shift before peak, and (*c*) look-back for shift after peak.

6.179

Additionally, the analyzer can accept analog signals from other field-mounted analyzers or sensors such as flowmeters and pressure transducers. The signal can be scaled, digitized, and incorporated into special calculations to determine mass flow, therms per day, reactor yields, etc.

In alternative forms of these systems, the applications program is stored in the processor; the analyzer microprocessor digitizes only the detector signal and transmits the digitized values to the data processor. Applications program event commands are received in real time from the data processor and converted at the analyzer to electrical and pneumatic signals for sample valve actuation, column switching, sample conditioner control, etc.

Data Processor

This unit is commonly located near or in the control room in a nonhazardous environment, and as much as 2000 to 3000 ft (610 to 914 m) from the analyzers. The data processor also has its own microprocessor with a CPU and a complement of ROM and RAM in which the operating system and user-specific applications programs are stored. Communication with the analyzers is by a serial link.

In its most usual form, the processor is a special purpose microprocessor and can control up to six or eight analyzers. In other forms the processor may be a microprocessor-based minicomputer and control as many as 32 analyzers.

The processor has several main functions:

1. Input of applications programs by means of a special purpose keyboard, with an alphanumeric display and interactive dialogue. Prompting of the operator and screening of the input data ensures the input of all necessary parameters and prevents conflicting data inputs. The program is down-loaded into the analyzer memory and can be recalled for editing and modifications. In some versions, input may be accomplished through an ancillary video terminal with a keyboard and menu-driven operator communications.

2. Receives data from the analyzers and distributes it to the various output devices in analog or digital form as required.

3. Monitors the status of the analyzers and displays alarms and transmits them to peripherals.

4. Provides manual control of all analyzer functions during setup and maintenance and acts as a diagnostic center for troubleshooting and corrective maintenance. A real-time chromatogram is available.

Except for the initial setup and maintenance of the analyzer, all operations, including programming, manual operation, calibration, etc., take place at the processor location.

In some versions the applications program for all analyzers may be stored in the processor memory instead of the analyzer. Event commands are sent to the analyzer in real time over the serial link and converted to analog commands by the analyzer. Digitized data are received from each analyzer, with all data acquisition and reduction accomplished in the processor.

In yet other versions the processor is part of the analyzer and may be dedicated to, and integral to, a single analyzer at the analyzer location with all analyzer and processor functions described above performed locally.

Data Outputs

The simplest form of record is a bar graph, as shown in Fig. 3b. The record consists of a series of bars, one for each component, of height proportional to the component concentration. The output for each component is scaled to give a full-scale reading equivalent to a convenient concentration value. Each component may have a different full-scale range. A large number of components in one or many streams may be recorded on the same instrument. Different streams may be identified by the height of a flat-top bar preceding the series of bars for that stream.

Trend Outputs

Trend outputs consist of a continuous electrical signal (0 to 10 V or 4 to 20 MA) from the processor. As many as 30 to 40 such outputs may be available from a single processor. Each output represents the concentration of a particular component in one of the sample streams on a given analyzer scaled to some convenient range. Component identity and scale factors for each output channel are user-assigned from the processor keyboard.

The output value is held constant at the last value until a new value is determined in a subsequent analysis, causing a stepwise change in the record as the signal is updated at the end of each analytical cycle. See Fig. 3*c*. A separate recorder pen is required for each trend output recorded.

Trend outputs may also be used to input analytical data to a process control system. A separate two-wire 4 to 20 mA output line is typically required for each component input.

Chromatogram

A real-time chromatogram for any selected analyzer may be obtained at the data processor for setup and maintenance. The chromatogram is received in digital form at a rate of 10 or more data points per second and reconstructed into analog form and scaled at the processor.

Serial Outputs

Serial output ports (RS232) are usually provided for serial transmission of data—usually in ASCII code—to peripherals. See Fig. 6. These include:

1. **Printer.** Provides output for logging of analytical results data, and alarms. It also prints out, on command, a record of applications programs, analog and channel assignments, calibration data, and other user-selected parameters, with the time and data for maintenance record keeping. Some systems have two printers at different locations—one for data logging (in the control room) and the other for testing and maintenance.

2. **Panel-Mounted Printer.** Outputs same data as printer but on narrow paper tape.

3. **Host Computer.** Transmits all data, alarms, and analyzer status reports, to host computer for data logging and historical archiving for subsequent input to process control systems. Special software may be required to be written for either the data processor or the host computer for compatibility of communications protocol and message structure.

4. **Magnetic Tape Cassette Recorder.** Provides for archiving analyzer and processor applications programs on cassette tapes, usually with a conventional portable recorder. The tape may be played back to reload the program in the event of loss due to system failure.

Data Reduction

Qualitative analysis is less important in process gas chromatography because at the outset stream composition is usually defined within narrow limits both as to compounds present and range of concentration. The need for maximum accuracy dictates a maximum separation of measured components and an accounting of all components to avoid errors due to interferences. Therefore, qualitative aspects are limited to matching observed elution times to a table of expected elution time windows for identification and also for detecting and alarming elution time changes due to system malfunction.

Quantitative Analysis

Methods are in general similar to laboratory methods (described above), with an emphasis on accuracy. Time slices of a detector signal are digitized and stored in RAM. Slope sensing algorithms find the beginning and end of peaks, valleys, and rider peaks by tangent skimming. Baseline corrections are made by look-forward, look-back, or a combination of both to compensate for a drifting baseline due to column switching. See Fig. 8.

Composition is calculated by different methods which are user-selected at the processor keyboard.

Concentration Percentage, Normalized. Uses relative response factors and internal normalization. Does not require calibration standard.

External Standard. Compares area or peak height to a value in memory for a calibration standard of known concentration. Only components of interest need to be measured. Normalization can be carried out if all components in the sample are measured.

Reference to Key Component. Calibrates the system on a single key component in calibration standard and relative response factors. The key component may be a pure compound and need not be present in the sample. This method permits calibration for samples which are unstable, reactive, or hazardous by use of a safe key component.

Automatic Calibration. Automatic introduction and analysis of a calibration standard at user-selected intervals for updating calibration factors. User input limits prevent updating if an incorrect calibration occurs as a result of a depleted or contaminated standard.

Valves

Electrically or pneumatically operated valves are used for the injection of liquid or gas samples and column switching. Rotary, spool-and-O-ring, diaphragm, and sliding-plate valves with from 4 to 12 ports are used. Vaporizing liquid sampling valves mounted through the wall of the analyzer transfer liquid sample from the cold exterior zone to the heated vaporizing zone within the oven. The sample valve meters a fixed volume of liquid or gas into the column with a repeatability of $\pm 0.25\%$. The sample size may vary from approximately 0.1 μl to 50 ml (with an external sample loop).

Detectors

The need for ruggedness and reliability has limited the types of detectors suitable for on-line process chromatographs. TCDs (for general use) and hydrogen FIDs (for trace organic analysis) are the most widely used detectors. The flame photometric detector is used for detecting trace quantities of sulfur-containing compounds. See Table 2 for applications for the detection limits of typical process chromatograph detectors.

Liquid Chromatography

This method is particularly useful for the separation and analysis of high-molecular-weight compounds beyond the range of gas chromatography. The most widely used forms are paper chromatography and TLC, forms of development chromatography, and the liquid column chromatography form of elution chromatography, the latter frequently being referred to as HPLC.

Liquid Chromatography in the Laboratory

Paper and Thin-Layer Chromatography

Both methods are similar in apparatus and technique. The sorbent bed is in the form of a thin sheet of paper or a thin layer of a finely divided sorbent material deposited on a supporting metal, glass, or plastic plate. The sample is spotted near one end of the bend, which is then brought in contact with a source of solvent. As the solvent moves through the bed by capillary action, the sample components are washed through the bed at different rates and are separated into spots of pure compound, which can be recovered after the solvent is allowed to evaporate. Spots can be detected visually if the substances are colored, or with ultraviolet light if they are fluorescent. Components can be reacted to give colored or fluorescent derivatives by spraying with reagents. Conventional quantitative deter-

Table 2 Characteristics of Process Gas Chromatograph Detectors

Detector	Compounds detector	Minimum detectability*	Advantages	Disadvantages
Thermal conductivity	All	5 ppm	Rugged, reliable universal	Limited sensitivity
Gas density balance	All	100 ppm	Similar to thermal conductivity, wide choice of carrier gases; sample does not contact filaments	Low sensitivity; greater internal volume
Hydrogen flame ionization	All organic except formic acid and formaldehyde	20 ppb	Rugged, reliable, ultrahigh sensitivity; wide choice of carrier gases	Cannot detect inorganic compounds and inert gases; complex electronics
Flame photometric	Sulfur containing	25 ppb	Rugged, reliable; selectively sensitive to sulfur	Nonlinear response, limited dynamic range

*Referred to concentration in process sample with nominal 1-cm^3 sample size; component peak approximately 0.25 min wide at base; He or H_2 carrier. Under special conditions, sensitivity can be increased up to 50 times by use of large samples. See "Process Gas Chromatography Detection Systems," *Chem. Eng. Prog.*, vol. 64, no. 4, p. 55, 1968.

minations can be made after recovery of the substance of interest with a solvent, or roughly estimated by comparison with spots produced with known quantities and concentrations. Recording densitometers are used for quantitative measurements in situ. The paper is drawn automatically past a slit in front of a photocell, and the transmitted or reflected light is recorded. Radioactive compounds are detected by contact exposure with x-ray film, which after development can be measured with a densitometer in which a radiation detector is used in place of a photocell.

These techniques have gained wide acceptance because of ease and convenience of use and the modest cost and availability of materials. Moreover TLC affords a rapid and inexpensive means for preliminary screening and selecting solvent–stationary phase systems for liquid chromatographs.

Liquid (Column) Chromatography

This method is generally classified according to the stationary phase or to the nature of its interaction with sample components:

1. Liquid-liquid partition chromatography, wherein the sample components are partitioned between a moving liquid phase and a stationary liquid phase deposited on an inert solid. The two solvent phases must be immiscible. The stationary phase may be a large molecule chemically bonded to the surface of a solid (bonded liquid phase) to prevent loss by solubility in the moving phase. This method can also be subdivided into *normal-phase* systems, in which the moving phase is less polar than the stationary phase, and *reverse-phase* systems, in which it is more polar.

2. Liquid-solid or absorption chromatography, in which the sample components are absorbed on the surface of an adsorbent such as silica gel.

3. Ion exchange, in which ionic sample components interact with functional groups on a permeable ionic resin.

4. Exclusion or gel permeation, in which compounds are separated by molecular size by a range of pore sizes in a polymeric gel. This method is useful for measuring the molecular weight distribution of polymers.

Isocratic elution uses a solvent of constant composition throughout the analysis.

Gradient elution is a modification of the technique in which the solvent is a mixture of two solvents which differ in solvent strength. The composition or ratio of the two is changed during the analysis in accordance with a predetermined program. The change may be continuous, linear or nonlinear, stepwise, or a combination.

Apparatus for Liquid Chromatography

The need for more efficient columns and faster separations has led to the development of stationary-phase packings with particles as small as 2 μm, operating at pressures as high as 10,000 psig (69 MPa) and with solvent flow rates as low as 1 ml/min or less. This has in turn led to the development of detectors with internal volumes of only a few microliters and special fittings and connectors with minimum dead volume to prevent band spreading and loss of resolution. These improvements in apparatus and technique have resulted in an ability to achieve complex separations with speeds comparable to those of gas chromatography.

The solvent is moved through the system by constant flow or constant pressure pumps which are driven mechanically (screw-driven syringe or reciprocating) or by gas pressure with pneumatic amplifiers. For gradient elution two pumps may be synchronized and programmed to provide a controlled, reproducible composition change.

Samples may be introduced by syringe directly into the column through a septum or by means of valves with a fixed volume which has been prefilled with the sample. Valves may be of a rotary, sliding plug, or diaphragm design and of stainless steel and fluoroplastic construction for inertness. Auto samplers are used for unattended injection of samples loaded into vials and sequentially rotated into the injection mechanism.

The differential refractive index (RI) is probably the most widely used detector; other detectors are based on photometry. Fixed-wavelength and variable-wavelength ultraviolet detectors are the most commonly used. The dielectric constant (DC) detector and electrical conductivity (EC) detector are also used and are especially suitable for process liquid chromatography. Transport detectors transfer the eluate to a moving wire or belt through a solvent evaporator and into a pyrolyzer which fragments and transfers the sample components to a hydrogen FID.

Microprocessor-Based Automation

As in gas chromatography, microprocessor-based systems are now widely used to automate the technique and the data acquisition and reduction. Because of the similarity to gas chromatography, the microprocessor technology employed is virtually identical. Some multichromatograph data systems interface simultaneously with both liquid and gas chromatographs. The microprocessor controls all the variables critical to analytical reproducibility and precision, including gradient elution profiles, solvent pumping rates, etc.

Data Reduction—Qualitative Analysis

For a given solvent-substrate system, each compound has a characteristic retention time which can be used for tentative identification. Compilations of retention times are less readily available and more limited than for gas chromatography. Greater reliance on other analytical methods for the identification of compounds is therefore necessary.

Quantitative Analysis

All the methods of calculating composition in gas chromatography are applicable to liquid chromatography. Relative response factors are more sensitive to operating conditions and individual apparatus characteristics than in gas chromatography and should be determined with pure compounds for a given method.

Process Liquid Chromatography

Liquid chromatography, like gas chromatography, lends itself to automation for in-line process analysis. See Table 3. The constraints on design are the same:

1. Operation in hazardous atmospheres
2. Need for automation
3. Compatibility with process control systems and computers

Except for the elements of the analytical hardware, a process liquid chromatograph is virtually identical to a process gas chromatograph in (1) appearance, (2) data acquisition and reduction techniques, and (3) information and data output. Except for the analyzer, other system components such as the data processor and communications and output peripherals are the same and may be shared

Table 3 Representative Process Liquid Chromatography Applications

Food Industry	Macromolecule Industries
Additives	Polymers and copolymers
Preservatives	Rubbers and resins
Flavor components	Silicones
Starches	
Heat exchanger fluids	Dyes and Dye Intermediates
Triglycerides	
Cooking oils	Anthraquinones
	Benzidine
Beverage Industries	Naphthols
	Water analysis
Carbohydrates	
Sugars	Petrochemicals and Monomers
Wines	
	Acrylic acid
Agricultural Products	Acrylic esters
	Methacrylic acid
Pesticides	Terephthalic acid
Herbicides	Acrylonitrile
Insecticides	
Wood preservatives	Petroleum Products
Water analysis	
	Water analysis
Inorganic Ions	Fuels and hydrocarbon
	analysis
Pharmaceuticals	
Water Pollution	
Phenols	
Petrochemicals	
Platicizers	
Nitroaromatics	
Organic acids	
Polyaromatics	
Chlorinated organics	

simultaneously with process gas chromatographs. Data output is also as shown in Figs. 3a to c. Emphasis of technique is placed on analyzing for a few rather than for all components and minimizing the analysis time. This has resulted in extensive use of column switching and back-flushing. Isocratic methods for optimizing the separation of the target components are preferred to gradient elution.

Analyzer

Typical process liquid chromatography analyzer systems are shown in Figs. 9 and 10. The choice of individual components is influenced by considerations of reliability, safety, and ease of maintenance. For example, columns are sized for optimum separation in 10–15 min cycles at solvent flow rates of 1 to 2 ml/min. Bonded-phase packings for liquid-liquid partition methods are preferred for long-term column stability. Reverse-phase techniques with water-containing carriers avoid the cost of expensive organic solvents.

Solvent pumps using pneumatic amplifiers eliminate safety problems and are preferred to motor-driven mechanical versions. Carrier pressures of less than 1000 psig (6.9 MPa) are employed to lessen the strain on valves and other components. Sample injection and column switching are done with sliding plate, rotary, and diaphragm plunger valves with air operators. Back-flushing is frequently used to rapidly remove heavy compounds from columns. Exceptionally tenacious compounds may be displaced and removed by back-washing to a waste vent with a different solvent.

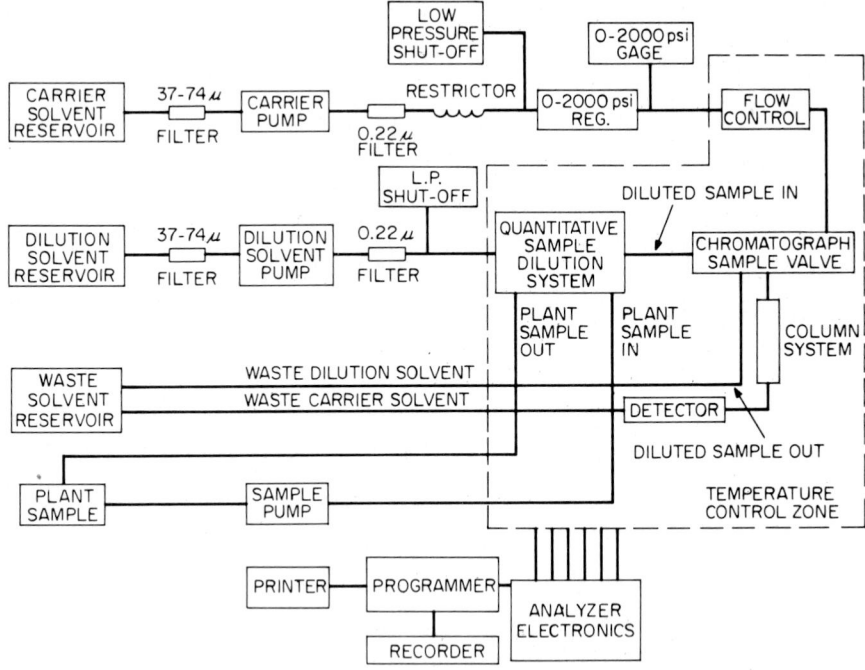

Fig. 9 Typical process liquid chromatograph system suitable for size exclusion analysis of polymers using isocratic elution. Solvent carrier is filtered and pumped through a pulse dampening restrictor and through pressure and flow controllers to the chromatograph sample valve. A low pressure shutoff protects the pump when the solvent is depleted. The process sample is circulated to a dilution system for blending with dilution solvent and circulation to the sample valve. The column system may include several columns with additional switching valves. Detector output is monitored and digitized by analyzer electronics and transmitted to a microprocessor in the programmer for data acquisition and reduction. Results are output on a printer or transmitted over a serial link to the host computer.

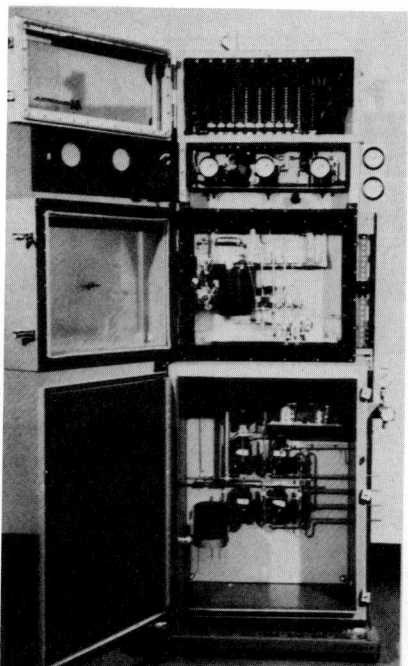

Fig. 10 Microprocessor-based process liquid chromatograph analyzer. Class I, Groups C and D, Division 1, for operation in hazardous areas. Electronics in the top section control analyzer functions and digitize detector signal for transmission to the data processor (not shown). Pressure and flow controls in the second section control solvent flow to an analyzer thermostatically controlled oven (middle unit) which houses columns, sample and column switching valves, and the detector. A sample conditioner (bottom section) conditions and dilutes the sample. *(Optichrom 2100, Applied Automation.)*

The detectors most often used in process liquid chromatography include fixed-wavelength ultraviolet, refractive index, dielectric constant, and electrical conductivity detectors.

REFERENCES

Bailey, S. J.: "On-Line Analyzers Prevent Plant Operating Pitfalls," *Control Eng.*, vol. 30, no. 8, 1983, pp. 57–61.

Heftmann, E.: *Chromatography*, 3d ed., Van Nostrand Reinhold, New York, 1975.

Kobayashi, R.: Chappelear, P. S., and H. A. Deans: "Physico-Chemical Measurements in Gas Chromatography," *Ind. Eng. Chem.*, vol. 59, no. 10, 1967, p. 63.

Krigman, A.: "On-Stream Analyzers: Progress with New Technologies," *In Tech*, vol. 30, no. 10, 1983, pp. 9–10.

Krigman, A.: "Process Chromatography: Difficult Becomes Routine," *InTech*, vol. 30, no, 10, 1983, pp. 34–40.

Mowery, R. A.: "Process Liquid Chromatography," in *Automated Stream Analysis for Process Control*, vol. 1 (D. P. Manka, ed.), Academic Press, New York, 1982.

Mukerji, A.: "Specifying Process Chromatography Systems," *Hydrocarbon Processing*, May 1981, pp. 187–192.

Purnell, J. H.: *Gas Chromatography*, Wiley, New York, 1962.

Villalobos, R.: "Process Gas Chromatography," *Ana. Chem.*, vol. 47, 1975, p. 983A.

Wolfe, T. C., Kinlen, P. J., and J. F. Combs: "A Computer-Based Evaluation System for Process Analyzers," *Advances in Instrumentation*, vol. 19, 1981, pp. 103–109.

Density and Specific Gravity Measurement

During the past decade, except for the increasing sophistication resulting from advances in solid-state and computer and digital technology, fluid density and specific-gravity measurement fundamentals and methodologies have remained pretty much the same. There are a few exceptions—the horizontal U-tube densitometer, vibrating U-tube and vibrating cylinder densitometers, and a comparatively few installations of equipment utilizing ultrasonic technology. Although control of the more common variables, such as temperature and pressure, is the basic criterion for process control, there are frequent cases where the measurement of density or specific gravity is the best method of determining and controlling the concentration of a solution or mixture. This can be done in open tanks or in vessels under pressure. In some instances density and specific gravity provide a good means for direct measurement of product quality.

Definitions

Density. For a fluid, density is defined as the mass per unit volume and usually is expressed in units of grams per cubic centimeter, pounds per cubic foot, or pounds per gallon.

Specific Gravity. In a fluid, the specific gravity is the ratio of the density of the fluid to the density of water. For critical scientific work, the reference is to double-distilled water at 4°C. In engineering practice, the reference is commonly to water at 60°F.

Since the specific gravity of water is unity, the specific gravities of oils and spirits are below this figure; and those of all aqueous solutions are above it.

Specific Gravity Scales

Early specific-gravity measuring devices used a float immersed in the solution so that more or less of the scale was submerged according to the specific gravity of the solution. The scale was graduated in specific-gravity units or in units selected arbitrarily for their convenience to a particular industry. In industry, several different scales are used.

API. A scale selected in 1921 by the American Petroleum Institute, the U.S. Bureau of Mines, and the National Bureau of Standards is now standard for petroleum products in the United States. It is represented by the formula

$$\text{Degrees hydrometer scale (at } 60°F) = \frac{141.5}{\text{sp gr}} - 131.5 \qquad (1)$$

Balling. Principally used to estimate the percentage of wort in the brewing industry, but also used to indicate percentage by weight of either dissolved solids or sugar liquors. Graduated in percent weight at 60°F or 17.5°C.

Barkometer. Used in the tanning and tanning-extract industry. Water equals zero and each scale degree equals a change of 0.001 in specific gravity, so that the following formula can be used.

$$\text{Sp gr} = 1.000 \pm 0.001n \qquad (2)$$

where n = degrees Barkometer.

Baumé. Widely used to measure acids and light and heavy liquids such as syrups. Proposed by Antoine Baumé, a French chemist, in 1768, this scale has found widespread acceptance because of the simplicity of the numbers representing liquid gravity. Two scales are used: (1) for liquids heavier than water and (2) for liquids lighter than water.

To calibrate the instrument for heavy liquids, Baumé prepared a solution of 15 parts by weight of salt in water. Zero on the scale was the point to which the float submerged in distilled water; a 15 mark was made at the point to which the float submerged in the salt solution. The distance between was then divided into 15 equal degrees, and degrees of the same length continued above 15.

For liquids lighter than water, a 10% salt solution was used in the same way. The point to which the float submerged in this solution was 0; in distilled water, it was 10. Thus, a scale was created with increasing numbers as density decreases.

It was found that oil of vitriol generally stood at 66 on the Baumé scale. Therefore, many early manufacturers fixed the 66 mark by immersing the instrument in oil of vitriol. As a result of these various calibration methods, there were many different scales bearing the name Baumé. In 1904, after a careful survey, the National Bureau of Standards adopted the following as standard Baumé scales:

For light liquids

$$°Bé = \frac{140}{\text{sp gr}} - 130 \tag{3}$$

For heavy liquids

$$°Bé = 145 - \frac{145}{\text{sp gr}} \tag{4}$$

(60°F is the standard temperature for these formulas.)

Brix. Used almost exclusively in the sugar industry. The degrees represent percent sugar (pure sucrose) by weight in solution at 17.5°C.

Quevenne. Used for milk testing, this scale represents a convenient abbreviation of specific gravity: 20° Quevenne means a specific gravity of 1.020; 40° Quevenne, a specific gravity of 1.040; and so on. One lactometer unit is approximately the equivalent of 0.29° Quevenne.

Richter, Sikes, and Tralles. Three alcoholometer scales which read direct in percent ethyl alcohol (C_2H_5OH) by weight in water.

Twaddle. This scale represents an attempt to simplify the measurement of industrial liquors heavier than water. The range of specific gravity from 1.000 to 2.000 is divided into 200 equal parts, so that one degree Twaddle equals 0.005 sp gr.

Variables Which Affect Specific Gravity

The temperature, the pressure, and the nature of a solution all affect its specific gravity. Industrial specific-gravity measurement is usually made to determine the nature of a liquid, such as the percent suspension, solution, and concentration. Therefore, the effect of the other two variables must be considered. Pressure has only a slight effect and can usually be ignored, but often the effects of temperature must be compensated. For practical purposes, however, specific gravity is commonly measured at the existing temperature of the liquid, providing a basis for control, particularly where the temperature is reasonably uniform.

Hand Hydrometer

This instrument (Fig. 1) consists of a weighted float with a small-diameter stem proportioned so that more or less of the scale is submerged according to the specific gravity of the fluid in which it is floated. Since the hydrometer weight is constant, this is equal to measuring the volume of liquid equivalent to a given weight.

Hydrometers may be calibrated to any of the scales previously described and are accurate to three or four decimal places. They are widely employed where automatic operation is not required and can be used in a standpipe with overflow at reading level to permit visual readings of a continuously flowing liquid.

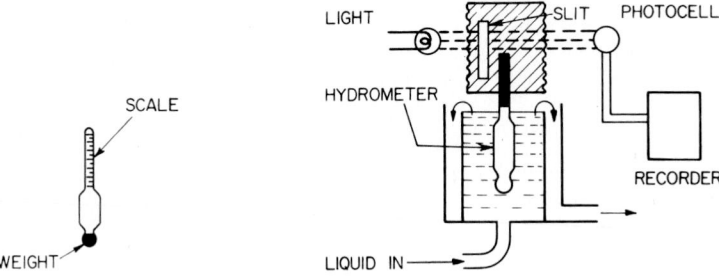

Fig. 1   Hand hydrometer.   **Fig. 2**   Photoelectric hydrometer.

Photoelectric Hydrometer

A glass hydrometer, similar to the hand-type described previously, is placed in a continuous-flow vessel (see Fig. 2). The instrument stem is opaque, and as the stem rises and falls, it affects the amount of light passing through a slit to a photocell. Thus, photocell output is made proportional to specific gravity and is recorded by a potentiometric-type instrument. This system is suitable for most specific-gravity recording applications not harmful to glass and is accurate to two or three decimal places.

Inductance Bridge Hydrometer

In this instrument (Fig. 3) the level of the measured liquid is held constant at an overflow tube. A glass hydrometer either rises or falls in the liquid as the specific gravity varies. The lower end of the hydrometer supports an armature in an inductance coil; any movement of this armature is duplicated by a similar coil in the recording instrument. With this system, the temperature of the liquid is usually recorded along with the value of specific gravity, so that corrections can be made.

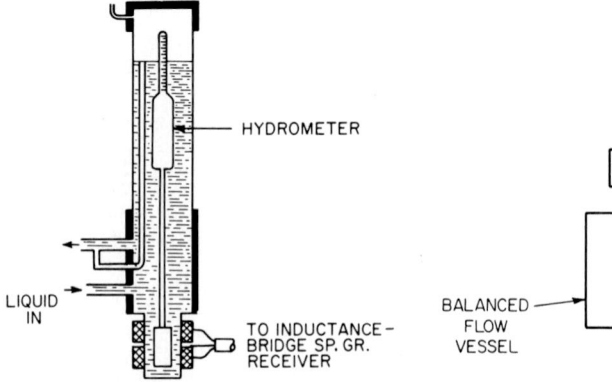

Fig. 3   Inductance bridge hydrometer.   **Fig. 4**   Balanced-flow vessel for measuring specific gravity and density.

Balanced-Flow Vessel

A fixed-volume vessel (Fig. 4) is employed, through which the measured liquid flows continuously. This vessel is weighed automatically by a scale, spring balance, or pneumatic force-balance transmitter. Since the weight of a definite volume of the liquid is known, the instrument can be calibrated to read directly in specific gravity or density units. Either open or closed vessels with flexible connections can be used. A high-accuracy measurement results, which is especially useful in automatic density control.

Displacement Meter

In a displacement meter (Fig. 5) liquid flows continuously through the displacer chamber. An upward force acts on the balance beam because of the volume of liquid displaced by the float. A pneumatic system, similar to the one shown, balances this upward force and transmits a signal proportional to the density of the liquid. Liquids with specific gravities 0.5 and higher can be measured with this equipment as long as suitable materials are used to prevent damage from corrosion. If the temperature of the flowing liquid changes, a thermostatic heater may be used to hold it constant.

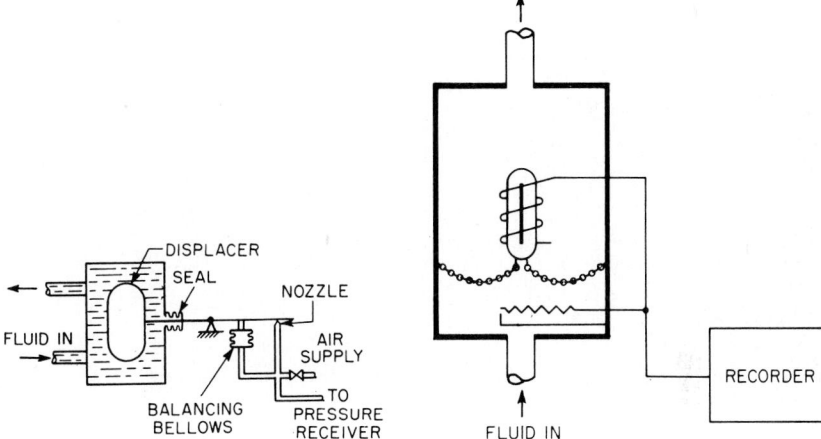

Fig. 5 Displacement meter for measuring specific gravity and density.

Fig. 6 Chain-balanced, float-type, density-sensitive element.

Chain-Balanced Float

This instrument (Fig. 6) uses a submerged plummet which is self-centering, operates without friction, and is not affected by surface contamination. The volume of the body is fixed and remains entirely under the liquid surface. As the plummet moves up and down, the effective chain weight acting on it varies, and for each density within the range of the assembly, the plummet assumes a definite equilibrium point.

In order to transmit these changes in density, the plummet contains a metallic transformer core, whose position is measured by a pickup coil. This voltage differential, as a function of plummet displacement, is a measure of specific-gravity change. To compensate for a density change due to temperature, a resistance thermometer bridge notes the temperature change and impresses a voltage across the recorder, which is equal to and opposite the voltage transmitted to it by the pickup coil as a result of the temperature-induced density change.

Liquid-Purged Differential-Pressure Method

Two taps are installed in the side of a tank or at different elevations in a vertical pipe, as shown in Fig. 7. These taps are led to a pressure-differential measuring device and are purged with a reference

erence fluid, usually water. In this way an automatically suppressed range is obtained, as well as freedom from plugged taps. In effect, the measured differential pressure is created by two equal columns, one of water and the other of sample fluid. The purge rate is small, so that dilution is negligible.

This system is frequently used in the pulp and paper industry to measure and control the density of green liquor, heavy black liquor, clay or starch slurries, lime slurries, and lime-mud slurries.

Air Bubbler—Reference Column Method

The reference column method (Fig. 8) displaces a known head of sample liquid and of water from their respective bubbler pipes and compares the two with a differential-pressure measuring device. The instrument can be calibrated directly in units of specific gravity. By varying the depth of immersion of the pipes, a great number of ranges can be obtained.

When it is necessary to correct the specific gravities to a standard temperature, the temperatures of both columns must be known. However, if the columns are arranged so that the sample liquid flows around the reference column, a single temperature measurement will suffice.

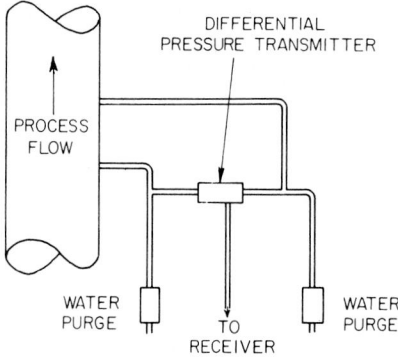
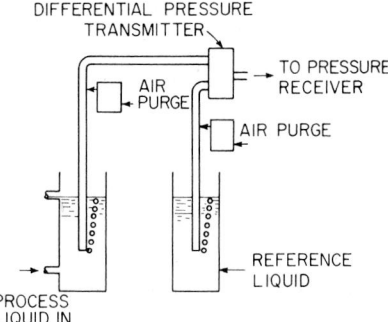

Fig. 7   Liquid-purge differential-pressure device for measuring specific gravity and density.

Fig. 8   Differential bubbler–reference column method for determination of specific gravity and density.

Air Bubbler—Single-Vessel Method

One of the simplest and most widely used methods of density measurement (Fig. 9) is to install two bubbler tubes in the sample fluid so that the end of one tube is lower than that of the other.

The pressure required to bubble air into the fluid is equal to the pressure of the fluid at the ends of the bubble tubes. Since the outlet of one is lower than that of the other, the difference in pressure will be the same as the weight of a constant height column of the liquid. Therefore, the differential-pressure measurement is equivalent to the weight of a constant volume of the liquid and can be represented directly as density.

This method is accurate to within about 0.3 to 1% specific gravity when used with liquids that do not tend to crystallize in the measuring pipes. Light black liquor, white liquor, and bleach are frequently measured by this method.

Air Bubbler—Range Suppression Type

This method (Fig. 10) is essentially the same as the single-vessel method just described. However, since it is often advantageous to read from a scale with a minimum marking greater than zero, a constant pressure drop range suppression chamber is connected in series with the low-pressure side, as shown.

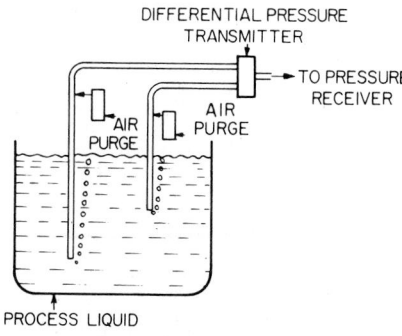

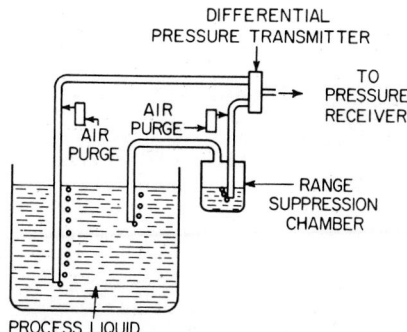

Fig. 9 Differential bubbler–single-vessel method for determination of specific gravity and density.

Fig. 10 Differential bubbler–range suppression method for determination of specific gravity and density.

Boiling-Point Rise

The temperature of a boiling solution is compared with that of water boiling at the same pressure (see Fig. 11). For a particular solution, the boiling-point elevation can be calibrated in terms of specific gravity at standard temperature. This method, which usually uses resistance thermometers to record the difference between the two temperatures, is commonly employed in evaporators to determine the end point of evaporation. High accuracy is realized in the measurement of one dissolved component or mixtures of fixed composition.

Gas Specific Gravity Balance

A tall column of gas is weighed by the floating bottom of the vessel (see Fig. 12). This weight is translated into the motion of an indicating pointer, which moves over a scale graduated in units of density or specific gravity. This system is suitable for any gas and can be used for recording.

Buoyancy Gas Balance

In this instrument (Fig. 13) a displacer is mounted on a balance beam in the vessel. The displacer is balanced for air, and the manometer reading is noted at the exact balance pressure. The air is then displaced by gas, and the pressure is adjusted until balance is restored. The ratio of pressure with

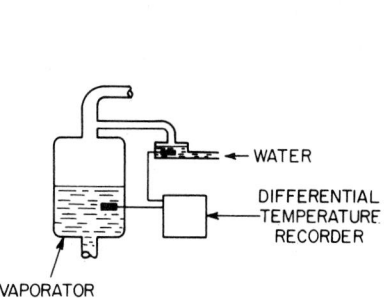

Fig. 11 Boiling-point rise apparatus for determination of specific gravity and density.

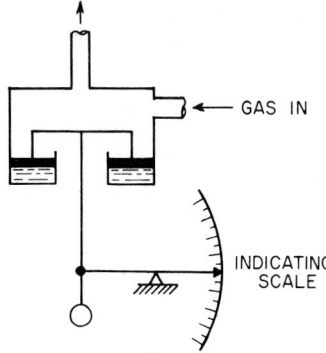

Fig. 12 Gas specific-gravity balance.

air to pressure with gas is the density of the gas relative to air. Primarily used for laboratory measurements, this device cannot be adapted for continuous measurement.

Viscous Drag Type

Impellers are driven in standard and test gas chambers and produce opposite rotation in gas columns (see Fig. 14). Nonrotating impellers on the opposite side of the chambers are coupled together by a linkage and measure the relative drag shown by the tendency of the impellers to rotate. The balance point is a function of relative density, and the instrument can be calibrated to read in density units.

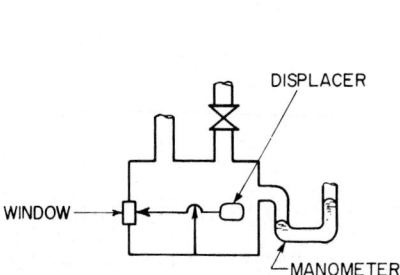

Fig. 13   Buoyancy gas balance.

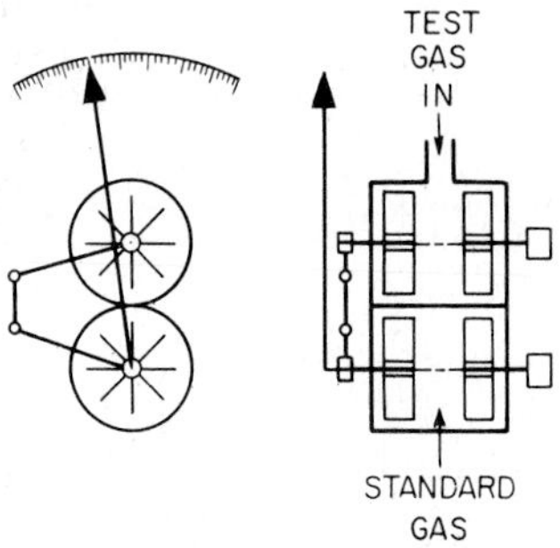

Fig. 14   Viscous-drag-type gas density apparatus.

Force-Balance Horizontal U-Tube Meter

In this scheme, a fixed volume of fluid flowing through a horizontal U tube is essentially weighed pneumatically. This comparatively large U tube tends to deform (sag or press down) below its null (empty) position when it is filled with the fluid to be measured. Increases in fluid density increase the tendency to be deformed (i.e., pulled downward by gravity). The air pressure required to return the U tube to its null position thus becomes a measure of fluid density. The device is, in essence, a pneumatic density transmitter. Increases in fluid weight tend to close the air nozzle, and this results in an increase in the output signal. The principal limitation of the method is its sensitivity to mounting attitude and vibration. Determining the mass of a fixed volume of fluid is a very direct approach to specific-gravity measurement. See Fig. 15. All wetted parts of the device are constructed of stainless steel. Claimed accuracy is ±1% of span for spans larger than 0.2 sp gr. For streams with wide temperature ranges, automatic temperature compensation is required to correct for thermal expansion of the liquid being measured. Span is factory-adjustable from 0.05 to 1.0 sp gr full scale. The range, however, is adjustable for 0.6 to 1.8 sp gr.

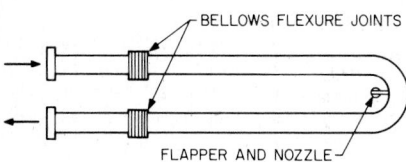

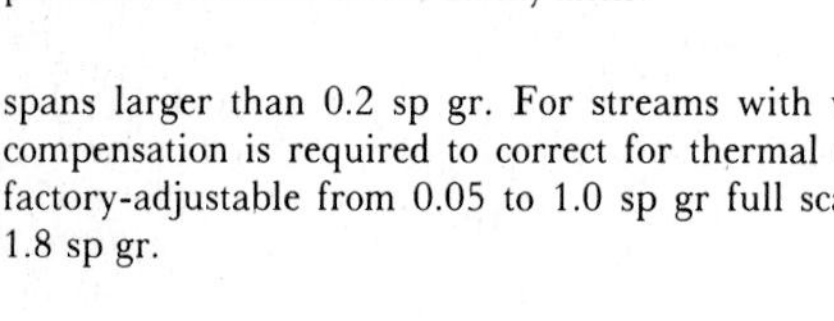

Fig. 15   Very schematic top view of force-balance pneumatic horizontal U-tube density meter.

Vibrating U-Tube Density Sensor

This device depends on a change in the natural oscillating frequency with a change in the process fluid density. When the fixed-volume tube is filled with a fluid that increases in density, it follows that the mass of the tube increases and this in turn decreases the amplitude of vibration. An accuracy of ±0.5% has been claimed for this type of device. Factors which affect performance of the device include extraneous mechanical vibrations, changes in flow rate, and blockage that may occur with slurries or liquids with a high fiber content.

Vibrating Cylinder Sensor

A thin-walled cylinder is located concentrically and inside the sensor housing. The process fluid flows around the thin cylinder, and the entire mass is excited into circumferential resonance electrically. The oscillating frequency changes as the mass of the process fluid changes. The device is useful for both liquids and gases. As presently designed, the instrument is limited to an operating viscosity range up to 20 S. An advantage is that the device can handle gas or gas-liquid combinations; however, it is susceptible to blockage by certain kinds of fluids.

Vibrating Vane Sensor

With this device, a vibrating vane, oscillating at its natural frequency when inserted directly into the process, measures fluid density in terms of the energy required to maintain the natural frequency of the sensor—more energy being required as the fluid density increases, and vice versa. Use with a wide range of densities and use on pipes of nearly any size are advantages of the device, as well as its apparent insensitivity to flow noise and piping vibrations. Although operation is not affected by nominal pressure changes, it is sensitive to temperature changes.

Vibrating Single-Tube Sensor

Quite similar to end-to-end tuning forks, this device has two cantilevered masses. As shown schematically in Fig. 16, the active portion of the center consists of a flow-through tube for the process fluid, with the cantilevered masses mounted on the same structure, but beneath the flowtube. In

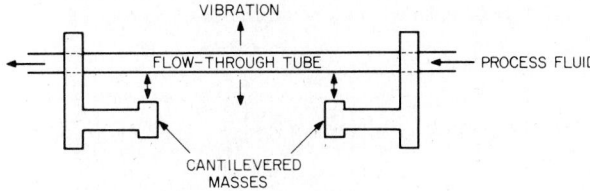

Fig. 16 Vibrating single-tube fluid density sensor.

operation, the entire structure is driven electronically into oscillation at a frequency determined by the combined mass of the device and the process fluid contained within it at any given instant. Any changes in the density (mass) of the flowing fluid alter the frequency—thus an indirect measure of density or specific gravity is obtained. The device requires both pressure and temperature compensation. Currently the unit is not available in many pipe sizes. The device is claimed to be insensitive to external vibrations and changes in flow rate.

Vibrating Twin-Tube Sensor

In a vibrating twin-tube sensor there are two parallel sensor tubes. Process fluid is passed through both tubes, which converge into a single flow passage at each end. See Fig. 17. Similar to a tuning

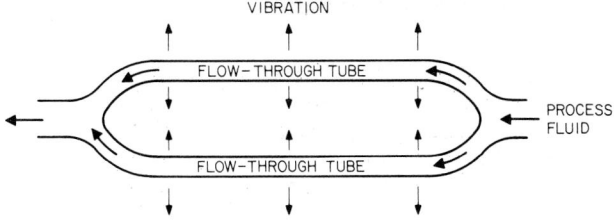

Fig. 17 Vibrating twin-tube fluid density meter.

fork, the tubes are driven at their mean natural frequency. This frequency varies inversely with changes in fluid density. Two drive coils maintain the tubes in resonance.

Ultrasonic Density Sensors

As in several other areas of process variable measurement, ultrasonic technology has been used in connection with difficult density measurements. One application is in the density measurement of lime slurries used to adjust the pH in acid water neutralization processes. The slurry is very difficult to handle, having a strong tendency to settle out and to coat all equipment with which it comes in contact. An ultrasonic density control sensor can be fully immersed in an agitated slurry, thus avoiding coating and clogging. Use of the device for controlling the specific gravity of slurries within $\pm$ 0.01% of the desired value has been demonstrated. In this application, the specific gravity ranges between 1.05 and 1.10. In actuality, the ultrasonic device measures the percentage of suspended solids in the slurry—providing a close approximation of the specific gravity. The suspended solids attenuate the beam, and the resultant electronic signal is proportional to the "specific gravity."

Radiation-Type Density Meter

Nuclear density gages measure and control slurries and liquids moving in process through pipelines. Accuracy is as good as $\pm$0.03%. The gages can be mounted on pipes from 1 through 30 in (76 cm) in diameter.

As shown in Fig. 18, a radioisotope source in a lead-shielded, shutter-controlled holder is clamped to one side of the pipe with a detector cell clamped to the other side of the pipe and in line with the radioisotope source. Transmitted radiation is in proportion to the specific gravity of the material in the pipeline. The detector cell converts transmitted radiation into minute electric current.

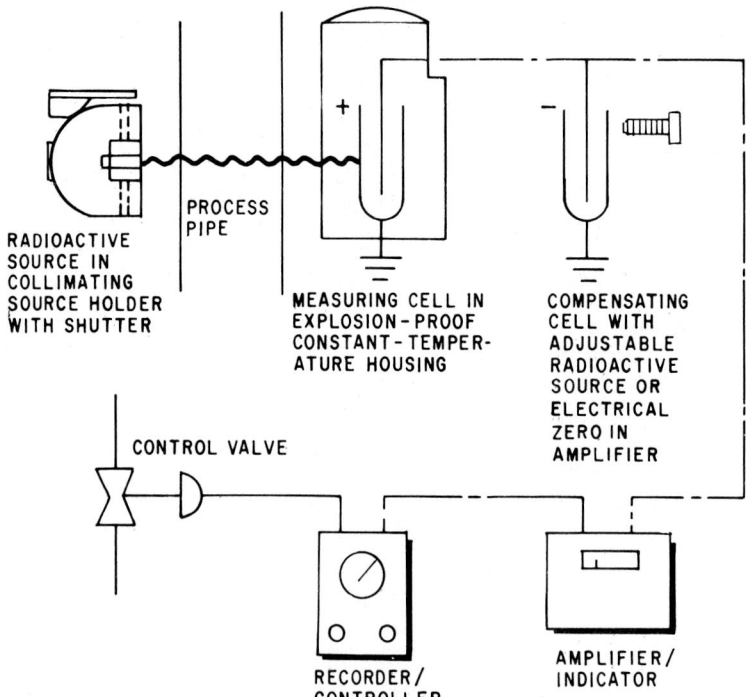

Fig. 18   Density control system of the radiation type. *(Ohmart.)*

Current is amplified for the readout and to operate control functions. Standard transducers can be used. Span and repeatability are determined by pipe size. The final control servo or the indicating panel may be several hundred feet away from the measuring point.

An ac or a dc amplifier may be used. The ac type is used for accurate density determination with very narrow spans. The dc type is used with larger spans and where less precision is required. Both types have industrially accepted drift rates and spans. Zero suppression is supplied by either a compensating cell or an electrical zero. The compensating cell virtually eliminates the zero drift rate caused by radioisotope source decay. The radioisotope, with the same half-life as the measuring source, is mounted integrally with the cell. Opposing current from the compensating cell decreases at the same rate as the measuring current. The electrical zero suppression uses zener diode-regulated voltage and is best suited where long time stability and span requirements are less strict.

Predetermined calibration is accomplished in most applications by determining zero and span settings from calculations of known absorption values of a given process material. Periodic rechecking is done to standardize the unit for source decay by emptying the pipe or by filling the process pipe with water (known factor of 1.0). No access to the gage head is required to recheck zero and span. Equivalent absorbers are used for on-site calibration of known parameters. Absorbers accurately represent the absorption of radiation by process material. This method is more accurate than predetermined calibration, since absorbers are made for specific measurement. Absorbers are used for restandardizing the gage when the pipe cannot be emptied for long periods.

Specific-Gravity Conversion Tables

Useful information for converting the various units of specific-gravity and density measurement will be found in Tables 1 and 2.

REFERENCES

Magaris, P.: "On-Line Density Measurement Is Fast and Accurate," *Control Eng.,* vol. 28, no. 6, June 1981, p. 98.

Slomiana, M.: "Using Differential Pressure Sensors for Level, Density, Interface, and Viscosity Measurements," *Instrum. Technol.,* vol. 26, no. 9, September 1979, p. 63.

Staff: "Ultrasonic Sensors Solve Slurry Control Problems," *Control Eng.,* vol. 25, no. 3, March 1978, p. 35.

Staff: "Aircraft Fuel Indication System," *Control Eng.,* vol. 27, no. 7, July 1980, p. 61.

Weast, R. C.: *Handbook of Chemistry and Physics,* CRC Press, Boca Raton, Fla. (published annually).

Table 1 Specific Gravity, Degrees Baumé, Degrees API, Degrees Twaddell, Pounds per Gallon, Pounds per Cubic Foot—Conversion Table*

$$°Bé = 145 - \frac{145}{\text{sp gr}} \text{ (heavier than } H_2O); \quad °Bé = \frac{140}{\text{sp gr}} - 130 \text{ (lighter than } H_2O);$$

$$°Tw = \frac{\text{sp gr } 60°/60°F - 1}{0.005}$$

Sp gr 60°/60°	°Bé	°API	lb/gal at 60°F wt in air	lb/ft³ at 60°F wt in air	Sp gr 60°/60°	°Bé	°API	lb/gal at 60°F wt in air	lb/ft³ at 60°F wt in air
0.600	103.33	104.33	4.9929	37.350	0.850	34.71	34.97	7.0775	52.943
0.605	101.40	102.38	5.0346	37.662	0.855	33.74	34.00	7.1192	53.255
0.610	99.51	100.47	5.0763	37.973	0.860	32.79	33.03	7.1609	53.567
0.615	97.64	98.58	5.1180	38.285	0.865	31.85	32.08	7.2026	53.879
0.620	95.81	96.73	5.1597	38.597	0.870	30.92	31.14	7.2443	54.191
0.625	94.00	94.90	5.2014	39.910	0.875	30.00	30.21	7.2860	54.503
0.630	92.22	93.10	5.2431	39.222	0.880	29.09	29.30	7.3277	54.815
0.635	90.47	91.33	5.2848	39.534	0.885	28.19	28.39	7.3694	55.127
0.640	88.75	89.59	5.3265	39.845	0.890	27.30	27.49	7.4111	55.438
0.645	87.05	87.88	5.3682	40.157	0.895	26.42	26.60	7.4528	55.750
0.650	85.38	86.19	5.4098	40.468	0.900	25.56	25.72	7.4944	56.062
0.655	82.74	84.53	5.4515	40.780	0.905	24.70	24.85	7.5361	56.374
0.660	82.12	82.89	5.4932	41.092	0.910	23.85	23.99	7.5777	56.685
0.665	80.53	81.28	5.5349	41.404	0.915	23.01	23.14	7.6194	56.997
0.670	78.96	79.69	5.5766	41.716	0.920	22.17	22.30	7.6612	57.310
0.675	77.41	78.13	5.6183	42.028	0.925	21.35	21.47	7.7029	57.622
0.680	75.88	76.59	5.6600	42.340	0.930	20.54	20.65	7.7446	57.934
0.685	74.38	75.07	5.7017	42.652	0.935	19.73	19.84	7.7863	58.246
0.690	72.90	73.57	5.7434	42.963	0.940	18.94	19.03	7.8280	58.557
0.695	71.44	72.10	5.7851	43.275	0.945	18.15	18.24	7.8697	58.869
0.700	70.00	70.64	5.8268	43.587	0.950	17.37	17.45	7.9114	59.181
0.705	68.58	69.21	5.8685	43.899	0.955	16.60	16.67	7.9531	59.493
0.710	67.18	67.80	5.9101	44.211	0.960	15.83	15.90	7.9947	59.805
0.715	65.80	66.40	5.9518	44.523	0.965	15.08	15.13	8.0364	60.117
0.720	64.44	65.03	5.9935	44.834	0.970	14.33	14.38	8.0780	60.428
0.725	63.10	63.67	6.0352	45.146	0.975	13.59	13.63	8.1197	60.740
0.730	61.78	62.34	6.0769	45.458	0.980	12.86	12.89	8.1615	61.052
0.735	60.48	61.02	6.1186	45.770	0.985	12.13	12.15	8.2032	61.364
0.740	59.19	59.72	6.1603	46.082	0.990	11.41	11.43	8.2449	61.676
0.745	57.92	58.43	6.2020	46.394	0.995	10.70	10.71	8.2866	61.988
0.750	56.67	57.17	6.2437	46.706	1.000	10.00	10.00	8.3283	62.300
0.755	55.43	55.92	6.2854	47.018	1.005	0.72	1	8.3700	62.612
0.760	54.21	54.68	6.3721	47.330	1.010	1.44	2	8.4117	62.924
0.765	53.01	53.47	6.3688	47.642	1.015	2.14	3	8.4534	63.236
0.770	51.82	52.27	6.4104	47.953	1.020	2.84	4	8.4950	63.547
0.775	50.65	51.08	6.4521	47.265	1.025	3.54	5	8.5367	63.859
0.780	49.49	49.91	6.4938	48.577	1.030	4.22	6	8.5784	64.171
0.785	48.34	48.75	6.5355	48.889	1.035	4.90	7	8.6201	64.483
0.790	47.22	47.61	6.5772	49.201	1.040	5.58	8	8.6618	64.795
0.795	46.10	46.49	6.6189	49.513	1.045	6.24	9	8.7035	65.107
0.800	45.00	45.38	6.6606	49.825	1.050	6.91	10	8.7452	65.419
0.805	43.91	44.28	6.7023	50.137	1.055	7.56	11	8.7869	65.731
0.810	42.84	43.19	6.7440	50.448	1.060	8.21	12	8.8286	66.042
0.815	41.78	42.12	6.7857	50.760	1.065	8.85	13	8.8703	66.354
0.820	40.73	41.06	6.8274	51.072	1.070	9.49	14	8.9120	66.666
0.825	39.70	40.02	6.8691	51.384	1.075	10.12	15	8.9537	66.978
0.830	38.67	38.98	6.9108	51.696	1.080	10.74	16	8.9954	67.290
0.835	37.66	37.96	6.9525	52.008	1.085	11.36	17	9.0371	67.602
0.840	36.67	36.95	6.9941	52.320	1.090	11.97	18	9.0787	67.914
0.845	35.68	35.96	7.0358	52.632	1.095	12.58	19	9.1204	68.226

*1 lb/gal (U.S.) = 119.826 kg/m³, 1 lb/ft³ = 16.0185 kg/m³.

Table 1 Specific Gravity, Degrees Baumé, Degrees API, Degrees Twaddell, Pounds per Gallon, Pounds per Cubic Foot—Conversion Table *(Continued)*

$$°Bé = 145 - \frac{145}{\text{sp gr}} \text{ (heavier than H}_2\text{O); } °Bé = \frac{140}{\text{sp gr}} - 130 \text{ (lighter than H}_2\text{O);}$$

$$°Tw = \frac{\text{sp gr } 60°/60°F - 1}{0.005}$$

Sp gr 60°/60°	°Bé	°API	lb/gal at 60°F wt in air	lb/ft³ at 60°F wt in air	Sp gr 60°/60°	°Bé	°API	lb/gal at 60°F wt in air	lb/ft³ at 60°F wt in air
1.100	13.18	20	9.1621	68.537	1.355	37.99	71	11.2884	84.443
1.105	13.78	21	9.2038	68.849	1.360	38.38	72	11.3301	84.755
1.110	14.37	22	9.2455	69.161	1.365	38.77	73	11.3718	85.067
1.115	14.96	23	9.2872	69.473	1.370	39.16	74	11.4135	85.379
1.120	15.54	24	9.3289	69.785	1.375	39.55	75	11.4552	85.691
1.125	16.11	25	9.3706	70.097	1.380	39.93	76	11.4969	86.003
1.130	16.68	26	9.4123	70.409	1.385	40.31	77	11.5386	86.315
1.135	17.25	27	9.4540	70.721	1.390	40.68	78	11.5803	86.626
1.140	17.81	28	9.4957	71.032	1.395	41.06	79	11.6220	86.938
1.145	18.36	29	9.5374	71.344	1.400	41.43	80	11.6637	87.250
1.150	18.91	30	9.5790	71.656	1.405	41.80	81	11.7054	87.562
1.155	19.46	31	9.6207	71.968	1.410	42.16	82	11.7471	87.874
1.160	20.00	32	9.6624	72.280	1.415	42.53	83	11.7888	88.186
1.165	20.54	33	9.7041	72.592	1.420	42.89	84	11.8304	88.498
1.170	21.07	34	9.7458	72.904	1.425	43.25	85	11.8721	88.810
1.175	21.60	35	9.7875	73.216	1.430	43.60	86	11.9138	89.121
1.180	22.12	36	9.8292	73.528	1.435	43.95	87	11.9555	89.433
1.185	22.64	37	9.8709	73.840	1.440	44.31	88	11.9972	89.745
1.190	23.15	38	9.9126	74.151	1.445	44.65	89	12.0389	90.057
1.195	23.66	39	9.9543	74.463	1.450	45.00	90	12.0806	90.369
1.200	24.17	40	9.9960	74.775	1.455	45.34	91	12.1233	90.681
1.205	24.67	41	10.0377	75.087	1.460	45.68	92	12.1640	90.993
1.210	25.17	42	10.0793	75.399	1.465	46.02	93	12.2057	91.305
1.215	25.66	43	10.1210	75.711	1.470	46.36	94	12.2473	91.616
1.220	26.15	44	10.1627	76.022	1.475	46.69	95	12.2890	91.928
1.225	26.63	45	10.2044	76.334	1.480	47.03	96	12.3307	92.240
1.230	27.11	46	10.2461	76.646	1.485	47.36	97	12.3724	92.552
1.235	27.59	47	10.2878	76.958	1.490	47.68	98	12.4141	92.864
1.240	28.06	48	10.3295	77.270	1.495	48.01	99	12.4558	93.176
1.245	28.53	49	10.3712	77.582	1.500	48.33	100	12.4975	93.488
1.250	29.00	50	10.4129	77.894	1.505	48.65	101	12.5392	93.800
1.255	29.46	51	10.4546	78.206	1.510	48.97	102	12.5809	94.112
1.260	29.92	52	10.4963	78.518	1.515	49.29	103	12.6226	94.424
1.265	30.38	53	10.5380	78.830	1.520	49.61	104	12.6643	94.735
1.270	30.83	54	10.5797	79.141	1.525	49.92	105	12.7060	95.047
1.275	31.27	55	10.6214	79.453	1.530	50.23	106	12.7477	95.359
1.280	31.72	56	10.6630	79.765	1.535	50.54	107	12.7894	95.671
1.285	32.16	57	10.7047	80.077	1.540	50.84	108	12.8310	95.983
1.290	32.60	58	10.7464	80.389	1.545	51.15	109	12.8727	96.295
1.295	33.03	59	10.7881	80.701	1.550	51.45	110	12.9144	96.606
1.300	33.46	60	10.8298	81.013	1.555	51.75	111	12.9561	96.918
1.305	33.89	61	10.8715	81.325	1.560	52.05	112	12.9978	97.230
1.310	34.31	62	10.9132	81.636	1.565	52.35	113	13.0395	97.542
1.315	34.73	63	10.9549	81.948	1.570	52.64	114	13.0812	97.854
1.320	35.15	64	10.9966	82.260	1.575	52.94	115	13.1229	98.166
1.325	35.57	65	11.0383	82.572	1.580	53.23	116	13.1646	98.478
1.330	35.98	66	11.0800	82.884	1.585	53.52	117	13.2063	98.790
1.335	36.39	67	11.1217	83.196	1.590	53.81	118	13.2480	99.102
1.340	36.79	68	11.1634	83.508	1.595	54.09	119	13.2897	99.414
1.345	37.19	69	11.2051	83.820	1.600	54.38	120	13.3313	99.725
1.350	37.59	70	11.2467	84.131	1.605	54.66	121	13.3730	100.037

Table 1 Specific Gravity, Degrees Baumé, Degrees API, Degrees Twaddell, Pounds per Gallon, Pounds per Cubic Foot—Conversion Table *(Continued)*

$$°Bé = 145 - \frac{145}{sp\ gr}\ (\text{heavier than } H_2O);\quad °Bé = \frac{140}{sp\ gr} - 130\ (\text{lighter than } H_2O);$$

$$°Tw = \frac{sp\ gr\ 60°/60°F - 1}{0.005}$$

Sp gr 60°/60°	°Bé	°API	lb/gal at 60°F wt in air	lb/ft³ at 60°F wt in air	Sp gr 60°/60°	°Bé	°API	lb/gal at 60°F wt in air	lb/ft³ at 60°F wt in air
1.610	54.94	122	13.4147	100.349	1.810	64.89	162	15.0824	112.824
1.615	55.22	123	13.4564	100.661	1.815	65.11	163	15.1241	113.136
1.620	55.49	124	13.4981	100.973	1.820	65.33	164	15.1658	113.448
1.625	55.77	125	13.5398	101.285	1.825	65.55	165	15.2075	113.760
1.630	56.04	126	13.5815	101.597	1.830	65.77	166	15.2492	114.072
1.635	56.32	127	13.6232	101.909	1.835	65.98	167	15.2909	114.384
1.640	56.59	128	13.6649	102.220	1.840	66.20	168	15.3326	114.696
1.645	56.85	129	13.7066	102.532	1.845	66.41	169	15.3743	115.007
1.650	57.12	130	14.7483	102.844	1.850	66.62	170	15.4160	115.318
1.655	57.39	131	13.7900	103.156	1.855	66.83	171	15.4577	115.630
1.660	57.65	132	13.8317	103.468	1.860	67.04	172	15.4993	115.943
1.665	57.91	133	13.8734	103.780	1.865	67.25	173	15.5410	116.255
1.670	58.17	134	13.9150	104.092	1.870	67.46	174	15.5827	116.567
1.675	58.43	135	13.9567	104.404	1.875	67.67	175	15.6244	116.879
1.680	58.69	136	13.9984	104.715	1.880	67.87	176	15.6661	117.191
1.685	58.95	137	14.0401	105.027	1.885	68.08	177	15.7078	117.503
1.690	59.20	138	14.0818	105.339	1.890	68.28	178	15.7495	117.814
1.695	59.45	139	14.1235	105.651	1.895	68.48	179	15.7912	118.126
1.700	59.71	140	14.1652	105.963	1.900	68.68	180	15.8329	118.438
1.705	59.96	141	14.2069	106.275	1.905	68.88	181	15.8746	118.740
1.710	60.20	142	14.2486	106.587	1.910	69.08	182	15.9163	119.062
1.715	60.45	143	14.2903	106.899	1.915	69.28	183	15.9580	119.374
1.720	60.70	144	14.3320	107.210	1.920	69.48	184	15.9996	119.686
1.725	60.94	145	14.3737	107.522	1.925	69.68	185	16.0413	119.998
1.730	61.18	146	14.4153	107.834	1.930	69.87	186	16.0830	120.309
1.735	61.34	147	14.4570	108.146	1.935	70.06	187	16.1247	120.621
1.740	61.67	148	14.4987	108.458	1.940	70.26	188	16.1664	120.933
1.745	61.91	149	14.5404	108.770	1.945	70.45	189	16.2081	121.245
1.750	62.14	150	14.5821	109.082	1.950	70.64	190	16.2498	121.557
1.755	62.38	151	14.6238	109.394	1.955	70.83	191	16.2915	121.869
1.760	62.61	152	14.6655	109.705	1.960	71.02	192	16.3332	122.181
1.765	62.85	153	14.7072	110.017	1.965	71.21	193	16.3749	122.493
1.770	63.08	154	14.7489	110.329	1.970	71.40	194	16.4166	122.804
1.775	63.31	155	14.7906	110.641	1.975	71.58	195	16.4583	123.116
1.780	63.54	156	14.8323	110.953	1.980	71.77	196	16.5000	123.428
1.785	63.77	157	14.8740	111.265	1.985	71.95	197	16.5417	123.740
1.790	63.99	158	14.9157	111.577	1.990	72.14	198	16.5833	124.052
1.795	64.22	159	14.9574	111.889	1.995	72.32	199	16.6250	124.364
1.800	64.44	160	14.9990	112.200	2.000	72.50	200	16.6667	124.676
1.805	64.67	161	15.0407	112.512					

*1 lb/gal (U.S.) = 119.826 kg/m³, 1 lb/ft³ = 16.0185 kg/m³.

Table 2 API Gravity-Specific-Gravity Conversion Table*

Formula: Specific gravity $60°F/60°F$ (vac./vac.) $= \dfrac{141.5}{\text{API} + 131.5}$

°API	Sp gr	lb/gal	°API	Sp gr	lb/gal	°API	Sp gr	lb/gal	°API	Sp gr	lb/gal
0	1.076	8.96	40	0.825	6.87	80	0.669	5.568	120	0.563	4.68
1	1.068	8.90	41	0.82	6.83	81	0.666	5.54	121	0.56	4.66
2	1.06	8.83	42	0.816	6.79	82	0.663	5.52	122	0.558	4.64
3	1.052	8.76	43	0.811	6.75	83	0.659	5.49	123	0.556	4.63
4	1.043	8.70	44	0.806	6.71	84	0.656	5.46	124	0.554	4.61
5	1.037	8.62	45	0.802	6.67	85	0.654	5.44	125	0.552	4.59
6	1.028	8.55	46	0.797	6.64	86	0.65	5.42	126	0.55	4.575
7	1.02	8.50	47	0.793	6.60	87	0.647	5.39	127	0.547	4.56
8	1.014	8.45	48	0.788	6.56	88	0.645	5.36	128	0.545	4.54
9	1.007	8.37	49	0.784	6.53	89	0.642	5.34	129	0.543	4.52
10	1.00	8.33	50	0.779	6.49	90	6.639	5.32	130	0.541	4.51
11	0.993	8.27	51	0.775	6.46	91	0.636	5.29	131	0.539	4.49
12	0.986	8.21	52	0.771	6.42	92	0.633	5.27	132	0.537	4.47
13	0.979	8.16	53	0.767	6.39	93	0.63	5.25	133	0.535	4.45
14	0.973	8.10	54	0.763	6.35	94	0.628	5.22	134	0.533	4.43
15	0.966	8.04	55	0.759	6.32	95	0.625	5.19	135	0.531	4.42
16	0.959	7.99	56	0.755	6.28	96	0.622	5.18	136	0.529	4.40
17	0.953	7.94	57	0.75	6.25	97	0.619	5.15	137	0.527	4.38
18	0.946	7.88	58	0.747	6.22	98	0.617	5.13	138	0.525	4.37
19	0.94	7.83	59	0.743	6.18	99	0.614	5.11	139	0.523	4.35
20	0.934	7.78	60	0.739	6.151	100	0.611	5.086	140	0.521	4.34
21	0.928	7.73	61	0.735	6.12	101	0.608	5.06	141	0.519	4.32
22	0.922	7.68	62	0.731	6.09	102	0.606	5.045	142	0.517	4.30
23	0.916	7.63	63	0.728	6.06	103	0.604	5.03	143	0.515	4.28
24	0.910	7.58	64	0.724	6.03	104	0.602	5.015	144	0.514	4.27
25	0.904	7.53	65	0.72	5.99	105	0.598	4.98	145	0.512	4.26
26	0.898	7.48	66	0.716	5.96	106	0.596	4.96	146	0.51	4.245
27	0.893	7.43	67	0.713	5.93	107	0.594	4.94	147	0.508	4.23
28	0.887	7.39	68	0.709	5.90	108	0.591	4.92	148	0.506	4.21
29	0.882	7.34	69	0.706	5.87	109	0.589	4.90	149	0.504	4.19
30	0.876	7.30	70	0.702	5.845	110	0.586	4.88	150	0.502	4.18
31	0.871	7.25	71	0.699	5.82	111	0.584	4.855	151	0.501	4.17
32	0.865	7.21	72	0.695	5.79	112	0.581	4.845	152	0.499	4.16
33	0.860	7.16	73	0.692	5.76	113	0.579	4.82	153	0.497	4.14
34	0.855	7.12	74	0.689	5.73	114	0.576	4.80	154	0.496	4.13
35	0.85	7.08	75	0.685	5.70	115	0.574	4.78	155	0.494	4.11
36	0.845	7.03	76	0.682	5.67	116	0.572	4.76	156	0.493	4.10
37	0.840	6.99	77	0.679	5.65	117	0.570	4.745	157	0.49	4.09
38	0.835	6.95	78	0.675	5.62	118	0.567	4.72	158	0.489	4.07
39	0.83	6.91	79	0.672	5.59	119	0.565	4.70	159	0.487	4.06

*1 lb/gal (U.S.) $= 119.826$ kg/m^3.

Weighing, Feeding, and Batching Systems*

Mel Berman. *Waugh Controls Corporation, Chatsworth, California. (Automatic Blending Systems)*

Stephen M. Howard. *StreeterAmet Measurement Systems Division, Mangood Corporation, Grayslake, Illinois. (Industrial Scales and Batching Systems)*

Donald B. Kendall. *Retired, formerly Toledo Scale Division, Reliance Electric Company, Worthington, Ohio. (Industrial Scales)*

Andrew L. Kurylchek. *Consultant, Wayne, New Jersey. (Belt Conveyor Scales and Weigh Feeders)*

Michael Latham. *Hardy Instruments, Inc., San Diego, California. (Industrial Scales)*

A. J. Livingston. *The Ohmart Corporation, Cincinnati, Ohio. (Radiation-Type Density Meters)*

Craig McIntyre. *Kay-Ray, Inc., Arlington Heights, Illinois. (Continuous Nuclear Weigh Scales)*

Gene Rebucci. *Schenck Weighing Systems, Totowa, New Jersey. (Belt Conveyor Scales and Weigh Feeders)*

William L. Ricketson. *Toledo Scale, Division of Reliance Electric, Atlanta, Georgia. (Industrial Scales)*

Staff. *StreeterAmet, Division, Mangood Corporation, Grayslake, Illinois. (Industrial Scales)*

James R. Story. *Marketing Director, Transducers, Inc., Cerritos, California. (Bonded Strain Gage Load Cells)*

*Persons who authored complete articles or subsections of articles, or who otherwise cooperated in an outstanding manner in furnishing information and helpful counsel to the editorial staff.

Industrial Scales

The measurement of weight predates many of the other important measurements encountered in industrial instrumentation. In fact, the weighing of commodities for purposes of trade goes back to antiquity. One of the first reasonably refined scales was the steelyard, originally known as the Roman balance. As shown by Fig. 1, a *steelyard* is an unequal-arm, portable weighbeam having a single bar whereon the applied load (hung directly from the load pivot) is counterpoised by positioning a poise on the long arm. A steelyard is sometimes furnished with a plurality of load pivots, fulcrum pivots, or poises. Later, the Danish steelyard was developed. This is a type of steelyard in which the applied load is balanced by the weight of the long arm and a fixed counterpoise; the fulcrum is movable, and its position when equilibrium is obtained indicates the weight value of the applied load. Later, the term "weighmaster's steelyard" was used to describe a refinement of the steelyard, a scale used officially for the determination or arbitration of weights. Unquestionably, the steelyard was popular because of the simplicity of its lever system—hence construction, portability, and inherent accuracy. The steelyard, however, lacked convenience—the load to be weighed had to be lifted into a pan a meter or so above the ground or attached to a hook above ground level. The load limits of the steelyard were determined by the strength of the individuals who operated it.

An Evolving Scale Technology

For centuries, the underlying principle of the steelyard, i.e., determining weight through the use of levers and counterweights, persisted. The most modern of industrial and commercial scales as recently as a few decades ago relied on levers, pivots, bearings, and counterweights. The improvements in this technology over the years represent impressive achievements in mechanical and materials engineering. For example, mechanical scales measured the variable (weight) over an exceedingly wide range—from grains or grams to several thousand tons (tonnes) with an accuracy (heavy-duty equipment) of 1 part in 10,000. This measurement achievement rivaled and bettered by far (an order of magnitude in some instances) the measurement of the majority of other process variables. Although pneumatic and hydraulic load cells appeared at about the same time, the great impact on weighing technology was made by the strain gage load cell, which has dramatically changed technology during the past decade or so.

The major shortcoming of mechanical scales was not their accuracy or ruggedness (they performed over a lifetime of 30 or 40 years under severe conditions with only moderate maintenance), but rather their bulky and heavy nature (larger scales weighing many hundreds of pounds) and their consequent need for costly, sturdy foundation structures, pits for the lever systems if weighing were to be conducted at floor, road, or track level, and the flexibility required for integration into complex

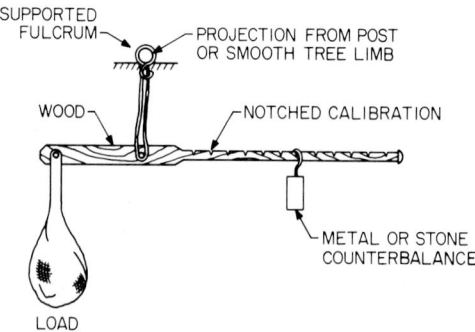

Fig. 1 Steelyard used for weighing since ancient times.

industrial systems. Despite this latter difficulty, however, as early as the 1940s scale engineers pioneered many of the principles used today in modern batching systems. Through a combination of electrical and magnetic equipment and ladder logic (prior to the programmable controller of today), and an ingenious blending of mechanical and electrical solutions, rugged and accurate automatic batching and bulk weighing, as well as continuous weighing for mass production lines, were developed. Some of these methods are still in use, but the trend since the 1960s has been toward adapting existing mechanical systems to electronic scales, or even more commonly, the abandonment of mechanical scales and full replacement with electronic systems. It is also interesting to observe that some of the first sophisticated data displays (tape and ticket printers and optical digital readouts) were first used in connection with mechanical scales. It is also notable that it has only been since the 1950s that weight as a process variable has received deserved recognition, along with temperature, pressure, flow, etc., as a really valid ingredient of industrial instrumentation. For many decades, scale professionals had their own organizations, considerably apart from other organized instrumentation groups, but the strain gage load cell changed all this.

Components of Weighing Devices

1. A load receiving (containing, holding, etc.) element

2. An indicating, recording, and/or controlling means

3. A sensing or measuring and transmitting apparatus to couple components 1 and 2

In either a mechanical or electronic scale, the load receiver—depending on the characteristics of the load—may be a platter, scoop, platform, hopper, tank, hook, section of conveyor (roller, belt, or overhead), section of railway track, section of roadway (for trucks), bag holder (for filling operations), or one of many other devices that at least momentarily "captures" the load to be measured.

Depending on the form of scale used, the load may be indicated by (1) the positioning of a poise on a beam (manually or electrically), (2) outputs from strain gage or other types of load cells, (3) the position of an indicator on a circular or straight graduated chart (either the indicator or the chart may be in motion), (4) optical projection and magnification of a section of moving chart, or (5) conversion of values to digits. Today, the display aspects of weighing can take full advantage of the digital and solid-state techniques developed during the last few decades. Even prior to modern electronics, certain industrial and commercial scales were among the first instruments to incorporate analog computing devices to convert weight into other terms (as, for example, the total cost of so many ounces of meat at a given price, a feature of retail scales that dates back many years). Today, with microprocessors, there hardly appears to be a limit to the manner in which basic weight data can be manipulated and displayed.

Mechanical Scale Engineering

Because there still are thousands of mechanical scales in daily use throughout the world, a significant portion of which have been converted to electronics (hybrids), it seems in order to review briefly the fundamental engineering lore of the once traditional scale.

Levers

In a mechanical scale, the load is reduced by levers so that smaller counterbalancing forces, such as poises on a beam, a bent lever and pendulum, or a spring, can be used. Figure 2 shows the three classes of levers and how the multiple of a lever of given length varies with its class. For a lever in equilibrium,

$$\text{Load} \times \text{load arm} = \text{power} \times \text{power arm}$$

The *load arm* is defined as the distance from the fulcrum to the point of application of the load, measured perpendicularly to the direction of the load force. The *power arm* is defined as the distance from the fulcrum to the point of application of the power (counterbalancing force in a scale), measured perpendicularly to the direction of the power force. In Fig. 2, a lever of 10 units in length is shown in each case. Figure 3 shows a lever system wherein a class 2 lever is connected to a class 1 lever. Here, the multiple of the system is the product of the multiples of the individual levers.

$$\text{Multiple} = \frac{\text{power arm}}{\text{load arm}} \qquad \text{mechanical advantage} = \frac{\text{load}}{\text{power}}$$

Since the main purpose of levers in a scale is to reduce the load to a smaller force which can be easily measured, class 3 levers, which have multiples of less than 1, are seldom used. Class 1 levers are frequently used to change the direction of application of a force and also can be used to extend the distance between the load and the indicating system. These uses are shown in Fig. 4. Quite often in such cases the levers may be equal-arm, i.e., with a multiple of 1.

The number of levers required (a hanging scale has no levers) in a mechanical scale varies from 1, for an even-arm scale, to as many as 13 (in the underground system) in a four-section railway track scale. In the latter instance, in addition to the 13 levers, there may be three or four more levers between the underground levers and the indicating system.

As used in mechanical scales, levers are fundamentally of two designs: (1) straight levers, as shown in Figs. 2 and 5, and (2) torsion or pipe levers, as shown in Fig. 6. A straight lever normally has one fulcrum pivot, one load pivot, and one power pivot. Some straight levers have a secondary load pivot coupled to the power pivot of another lever in the system. Construction of this type is shown in the long levers of an A-lever assembly (Fig. 5) used for built-in platform scales, hopper scales, and tank scales. The common terminology used for the additional levers is also indicated in Fig. 5.

The torsion lever has two fulcrum and two load pivots, with the power pivot on an arm which may be located between the fulcrum pivots or an extension of the pipe or shaft beyond one of the fulcrums. In some cases, there may be a secondary load pivot in the arm, which is connected to the power pivot of another lever system. Torsion levers are used for weighing materials in hoppers or tanks (Fig. 6) because they can be arranged to surround the hopper and support it at four points for stability. A torsion lever without an extension arm also is sometimes used as an extension lever (Fig. 6).

In a mechanical scale where the load is supported on more than one load pivot (Figs. 5 and 6), the multiple of the lever system for each load pivot to the indicating mechanism must be precise within the accuracy limits of the scale. If this were not true, a load placed near one load pivot would show a different weight than when it was placed near another load pivot. An error of this type is termed a *shift error*.

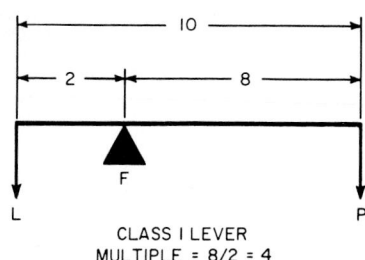

CLASS I LEVER
MULTIPLE = 8/2 = 4

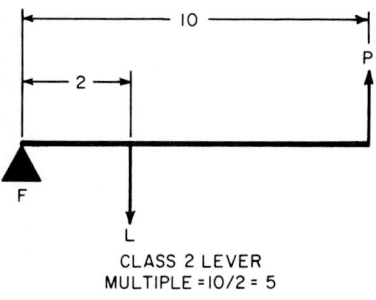

CLASS 2 LEVER
MULTIPLE = 10/2 = 5

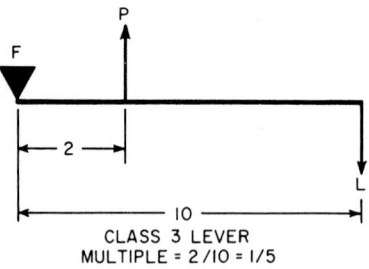

CLASS 3 LEVER
MULTIPLE = 2/10 = 1/5

F FULCRUM POINT
L LOAD POINT
P POWER POINT

Fig. 2   Classes of levers.

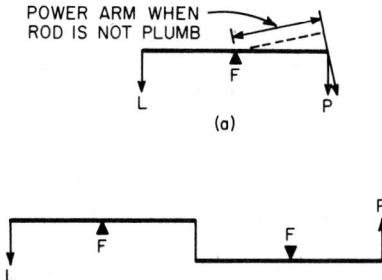

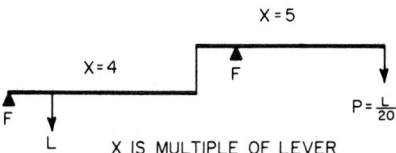

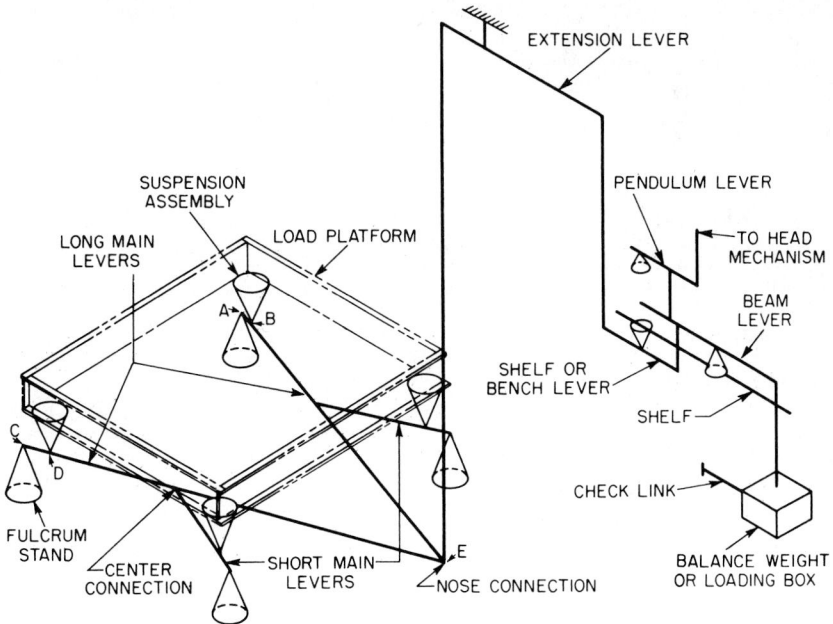

Fig. 3 Combination of levers made up of one class-1 and one class-2 lever.

Fig. 4 Uses of class-1 levers. (*a*) To change the direction of force (note the importance of plumbness in the application of forces), and (*b*) to change the direction as well as to extend the distance between points of application of forces.

A lever or lever system must transmit a definite fraction of the load to the indicating system with a minimum of friction. Knife-edge bearings were traditional for many years for the fulcrum, load, and power connections. In current mechanical designs, the most common form of bearing is a V groove, to receive a knife-edge for holding the lever in place. In some scales, ball bearings are used at some connection points. Some scales incorporate vertical plates in compression instead of pivots in bearings. Such instruments are called *plate-fulcrum scales*. Plates and bands also can be used in tension, as shown in Fig. 7. A horizontal member generally is added to stabilize the lever.

Fig. 5 Mechanical-type straight- or A-lever scale widely used prior to the introduction of electronic scales.

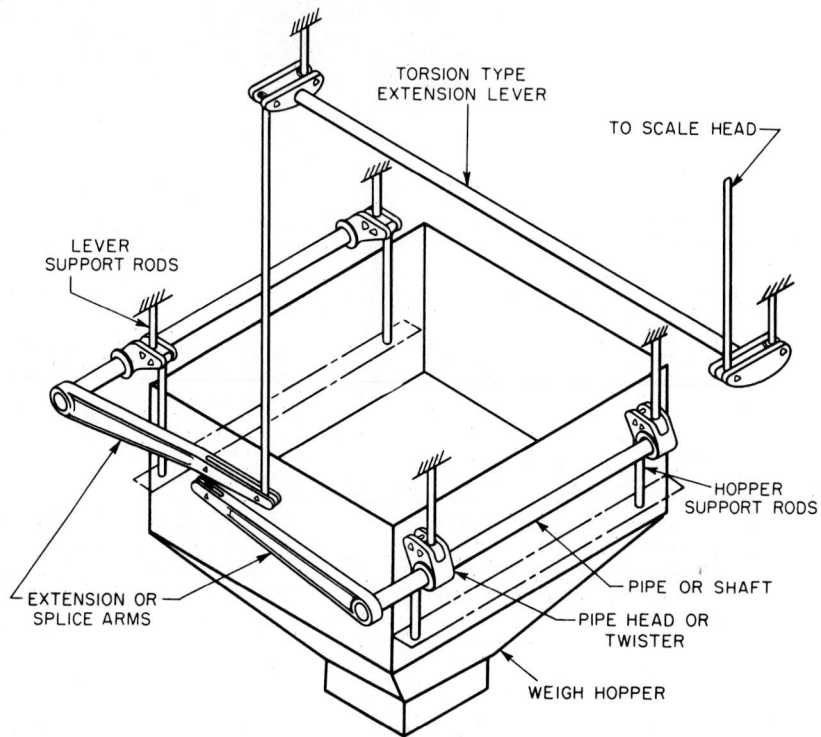

Fig. 6   Mechanical-type pipe or torsion lever hopper scale widely used prior to the introduction of electronic scales.

Counterbalancing Means

In mechanical-type beam or weighbeam scales, the load is measured by counterbalancing the force exerted by the load, this force generally being reduced through a lever system by moving one or more poises on a bar or weighbeam until the weight of the poises times their distance from the fulcrum equals the force due to the load times its distance from the fulcrum. Figure 8 shows a weighbeam unit with a large poise which can be moved manually on a notched bar to determine large increments in weight, and a smaller poise which is moved on a graduated bar and acts as a vernier for intermediate weights. An auxiliary poise which is moved on a graduated bar can be set to the tare weight of the container or vehicle containing the load.

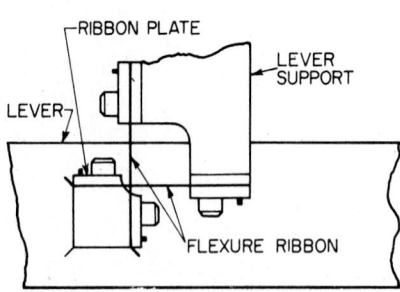

Fig. 7   Flexure ribbon construction.

Figure 9 shows a simpler unit in which loose counterpoise weights are hung from a pivot on the weighbeam to counterbalance large increments in weight. The load is determined by adding the amounts counterbalanced by each of the poises and/or counterweights. In some later models of this scale, the poises are motor-driven. In this construction, the bar generally is a screw driven by a motor and the poise is a "nut" which is prevented from turning. The revolutions of the screw are counted to determine the position of the poise at the time counterbalance is achieved. The revolution counter may be calibrated in terms of weight.

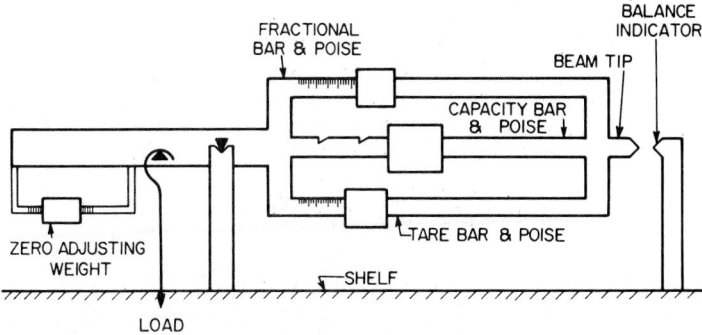

Fig. 8 Weighbeam with sliding poises used in earlier mechanical scales.

Pendulum Scales

A very successful means of *automatically* bringing a scale into balance with the load is well over a half-century old in principle. In a pendulum-type automatic indicating scale, the force due to the load is counterbalanced by the rotation of a bent lever. As shown in Fig. 10, rotation continues until the weight of the pendulum times its distance from the fulcrum equals the force due to the load times its distance from the fulcrum. A common design is a rotating indicator which moves over a stationary chart.

Figure 11 shows a double-pendulum automatic indicator. Here the bent levers take the form of cams or sectors designed so that rotation of the indicator is proportional to the load. Such an indicator can be the head mechanism for a scale of the type shown in Fig. 5. When one or two weighbeams are mounted on the beam lever, the tare weight of the container can be offset so that the automatic indicator reads the net weight. One or more unit weights can be added to the balance weight to provide zero suppression in multiples of dial capacity. Development of the automatic pendulum (or bent lever) scale opened up many possibilities for greater use of the scale in an automatic way. For example, photoelectric or magnetic detectors can be placed at various locations around the periphery of the circular dial and cut in or out when the indicating pointer reaches a given location.

Spring Scales

In spring scales, the deflection of a spring, generally helical in form, is used to measure the load. A straight-face spring scale employs an indicator directly attached to the end of the spring, pointing to a row of graduations to indicate the load. In a circular dial-type spring scale, shown in Fig. 12, a rack and pinion are used to magnify the extension of the spring for finer reading.

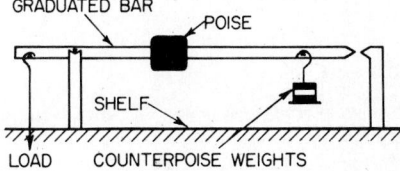

Fig. 9 Weighbeam with sliding poises and loose weights commonly used in some earlier scales.

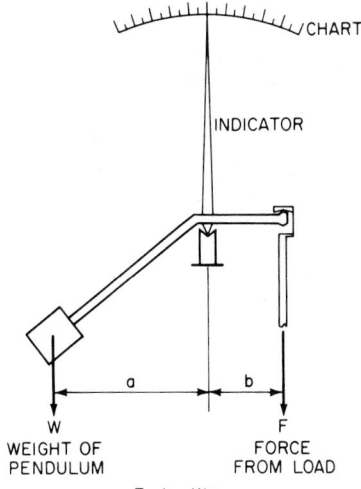

$$F \times b = W \times a$$

Fig. 10 Pendulum-type indicating scale popular prior to electronic scales.

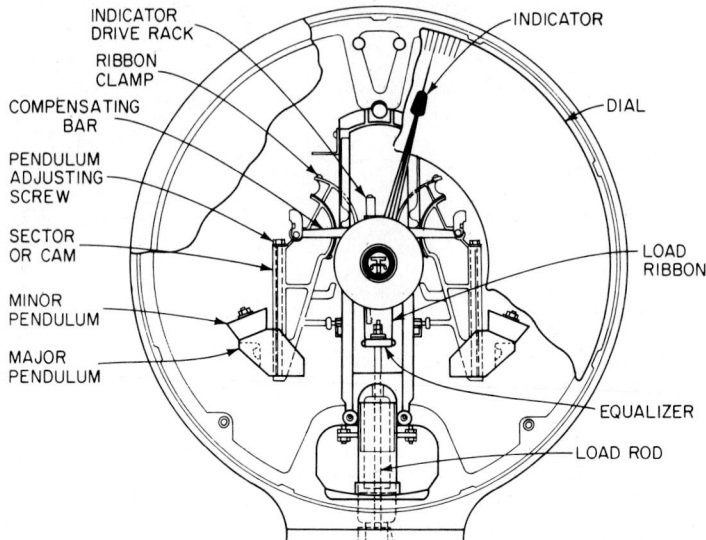

Fig. 11 Double-pendulum dial indicator, a principal display device prior to the digital displays of electronic scales.

Dashpots

Generally, a dashpot is used in both pendulum- and spring-type automatic indicating scales. The dashpot consists of a plunger moving in an oil-filled cylindrical housing. An adjustable orifice controls the rate of oil past the plunger so that the oscillation of the scale system will be damped and thus speed up the weighing operation. The dashpot also protects the indicating mechanism from the shock of suddenly applied loads. Magnetic or eddy-current damping also has been used.

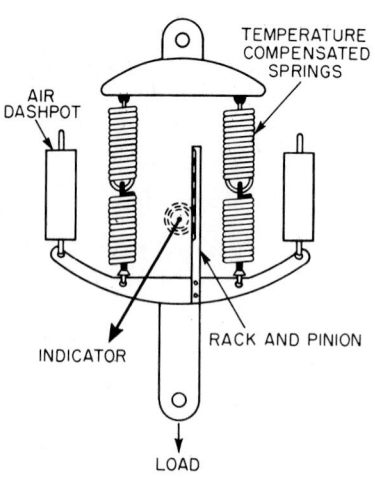

Fig. 12 Dial-type spring scale.

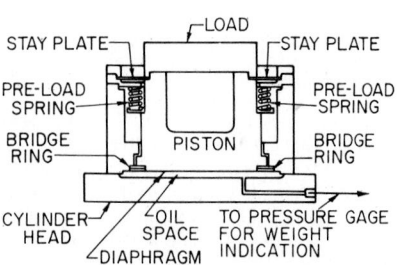

Fig. 13 Hydraulic scale using a load cell.

Hydraulic and Pneumatic Systems

Hydraulic pressure can be used to measure weight. A properly designed capsule containing oil can be connected to a bourdon tube or other pressure gage, as shown in Fig. 13. The hydraulic system serves the same purpose as the level system of a traditional scale, reducing the force due to the load by the use of a large capsule so that the weight is reduced to a measurable hydraulic pressure. In some scales of very large capacity, levers may be used to initially reduce the large force prior to final balancing by the hydraulic counterforce. Pneumatic systems in which air is the force-transfer medium operate on similar principles. See Fig. 14.

Electrical and Electronic Force Sensors

Particularly for scales and balances of low capacity, including laboratory balances, electromagnetic force may be used as a counterbalancing force. Measurement of this force can easily be translated into terms of weight (load).

Although other electrical means have been infrequently applied to weighing (for example, linear differential transformers), by far the majority of force (weight) transducers used in scales today are strain gage load cells. Because these load cells are used for other force measuring applications, they are discussed separately in the next article on load cells. Likewise, the theory and application of strain gages are described in the article on strain gages in Sec. 8.

Load cells used in modern industrial scales are precision strain gage devices which measure minute deflections of the load cell column or beam caused by applied weight or force. Factors which must be considered in selecting from a variety of load cell designs include (1) accuracy required, (2) scale capacity, (3) mode of loading (tension or compression), (4) number of cells required, (5) loading conditions, (6) environment, (7) output characteristics preferred, and (8) available space. Installation assemblies fall into two general categories—*compression* and *tension*. Check rods are usually used to provide lateral restraint while still permitting unrestricted vertical movement.

In their construction, strain gage load cells take advantage of all the characteristics of bonded resistance strain gages. Hermetically sealed within the load cell are one or more sets of matched strain gages bonded to a high-strength steel element machined to close tolerances. The strain gages are connected to form a balanced Wheatstone bridge. See Fig. 15. In some designs, a fixed excitation voltage is applied so that, when no load is present, the output signal of the cell will nominally be ±1% of its rated output. As force is applied, minute deflections of the strain member occur. This results in a change in the cross section of the gage which in turn creates an unbalanced bridge, causing a millivolt output to be generated. This output is proportional to the force applied.

The total weight of the structure (platform, tank, hopper, etc.) plus the anticipated "live" load and the number of support points determine cell capacity. The total sum capacity of the load cells

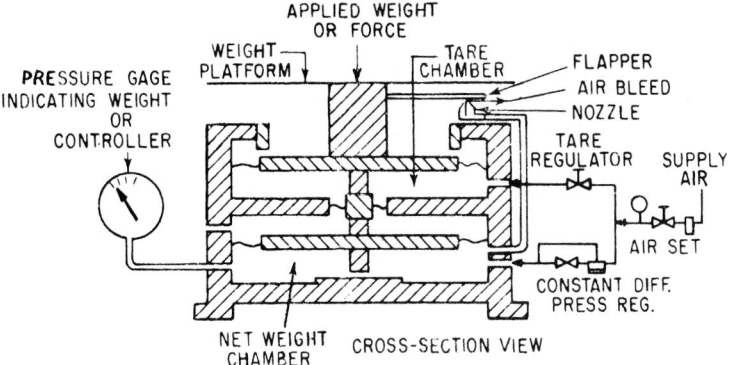

Fig. 14   Pneumatic load cell of the type used in some scales.

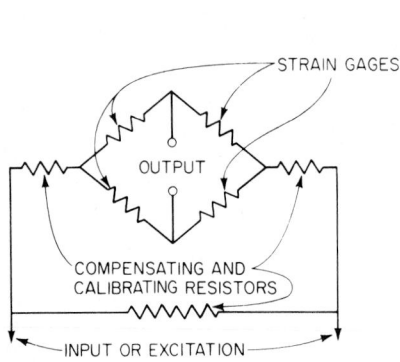

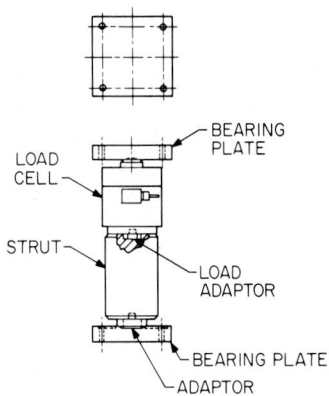

Fig. 15   Strain gage load cell circuit.

Fig. 16   A compression installation assembly showing the use of a strut between the load cell and the lower plate. This strut, by increasing the overall height of the assembly, reduces errors caused by lateral movement by reducing the incurred angle. Designed primarily for use on horizontal vessels. *(Hardy Instruments.)*

selected should be as close as possible to the gross load anticipated so as to maximize the available signal.

Selection of tension or compression support is largely determined by structural factors. For example, tension support for very large tanks or hoppers may be difficult and more costly than compression support. Tension load cells require greater vertical clearance than comparable compression units.

Most scale suppliers offer load cells in various capacity ranges—for example, 10 to 1000 lb (~5 to 450 kg), up to 10,000 to 500,000 lb (4536 to 226,800 kg) and even higher. Maximum deflection is usually but a few thousandths of an inch (a small fraction of a millimeter). One of a number of possible assembly and mounting configurations is shown in Fig. 16. To prevent side loads from reaching and possibly damaging the load cell, the loading arrangement (Hardy) shown in Fig. 17

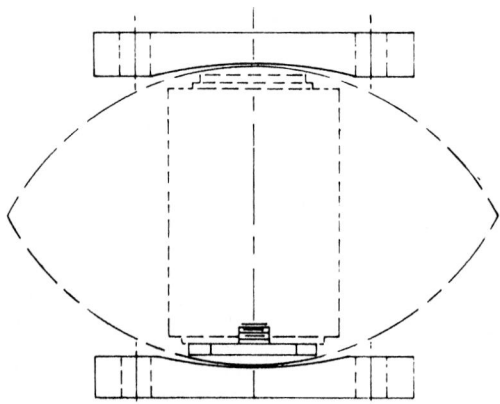

Fig. 17   Double convex loading arrangement to prevent side loads from reaching and possibly damaging the load cell. Lateral movement of the weigh bridge due to expansion or contraction causes the load cell to "rock." *(Hardy Instruments.)*

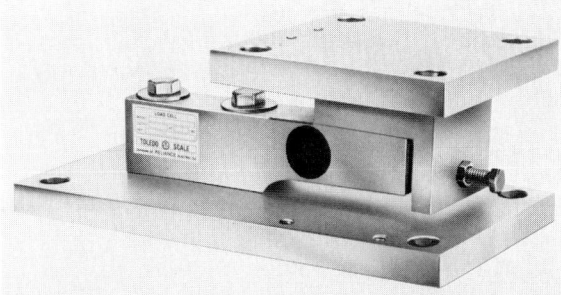

Fig. 18 Self-checking shear beam load cell designed to eliminate the need for check rods and restraints. The design is widely used in connection with tank and conveyor scales and as replacement hardware for existing mechanical scales. *(Toledo Scale, Division of Reliance Electric.)*

may be used. A self-checking shear beam load cell (Toledo) is shown in Fig. 18. Most scale manufacturers offer kits for converting existing mechanical scales to electronic weighing operations. See Fig. 19.

Applications and Types of Scales

There are six fundamental and, in many respects, quite different roles scales play in commerce and industry:

1. *Basic Weighing* of a vast variety of objects in connection with receiving, shipping, and transporting operations, where weight per se is the variable of interest. Also in this category are scales used to enforce load limit regulations in connection with highway and railway carriers.
2. *Counting,* particularly of small parts, where weight can be a useful indication of the number of pieces.
3. *Calculating,* where basic weight data can be automatically converted to price or value, density, specific gravity, etc.

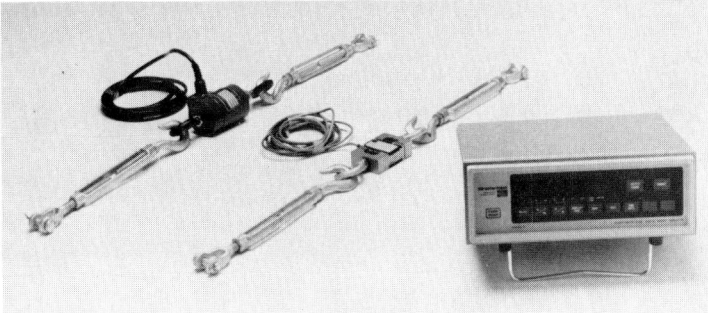

Fig. 19 Elements of a kit for converting mechanical scales to electronic weighing, showing how an electronic load cell can be incorporated in a steelyard rod, on the left, and a programmable weight indicator, on the right. *(StreeterAmet Division, Mangood Corporation.)*

4. Inspecting, to determine if containers are properly filled.

5. Preweighing and Filling.

6. Batching and Formulating, weighing the ingredients for formulation in a process.

   Scales may also be classified in terms of their physical structure, as dictated by the type of load to be weighed. Various formats include (1) portable and bench scales, (2) floor and deck scales, (3) overhead track or monorail scales, (4) hopper and tank scales, (5) motor truck scales, (6) railway track scales, (7) crane scales, and a number of special purpose scales, such as check weighing, counting, and drum and bag filling scales.

Portable and Bench Scales

These general purpose scales find many hundreds of uses in commerce and industry. Portable scales are equipped with casters or wheels for convenience in moving from one location to another. Bench scales are designed for semipermanent location. Some scales are constructed of stainless steel for use in demanding environments as may be found, for example, in meat packing and other food processing plants. Mechanical and lever-type scales are still available, but most users prefer electronic load cell designs. In the latter designs, automatic pounds-to-kilograms switching, pushbutton tare, and printer output are usually furnished as standard equipment. A stainless steel bench scale with a digital indicator is shown in Fig. 20. Some concept of the range of size and capacity of these scales is indicated in Table 1.

Floor and Deck Scales

A floor scale is designed so that the platform is essentially flush with the floor. The platform of a floor scale, with frame [∼10 in (25 cm) in depth] and appropriate foundation, is shown in Fig. 21. The platform is made of heavily ribbed cast iron with a removable steel center plate. The main levers, both long and short, are equipped with self-centering, self-gaging pivots and bearings, and with double parallel-link suspension. The removable center plate provides access to the lever system and load cell. Scales of this type are available in the following capacities:

   600 lb (0.2-lb graduations),   300 kg (0.1-kg graduations)
   1500 lb (0.5-lb graduations),   600 kg (0.2-kg graduations)

Table 1 Bench and Portable Scales—Capacity and Platform Size*

Type of scale	Total capacity		Approximate platform size	
	Pounds	Kilograms†	Inches	Centimeters
Bench	25	10	14 × 14 × 4	36 × 36 × 10
	50	25	14 × 14 × 4	36 × 36 × 10
	100	50	14 × 14 × 4	36 × 36 × 10
	150	60	20 × 20 × 6	51 × 51 × 15
	200	100	14 × 20 × 6	36 × 51 × 15
	300	150	20 × 20 × 6	51 × 51 × 15
Portable	800	400	19 × 28 × 8	48 × 71 × 20
	1000	500	19 × 28 × 8	48 × 71 × 20
	2400	1000	24 × 30 × 8	61 × 76 × 20
	2500	1000	24 × 30 × 8	61 × 76 × 20
	2500	1250	30 × 30 × 8	76 × 76 × 20

*Accuracy may be expressed as 1 part in 5000 (usual), or in sensitivity, such as 0.01 kg.
†Kilogram values are not an exact conversion of pound values at left because scales are calibrated to provide rounded ranges for either pounds or kilograms.

3000 lb (1-lb graduations), 1500 kg (0.5-kg graduations)

6000 lb (2-lb graduations), 2950 kg (1-kg graduations)

A computerized weighing system utilizing a floor scale is shown in Fig. 22.

A deck scale is similar to a floor scale, except that the frame of the scale rests on the existing floor rather than requiring a pit and special foundation. Thus the platform of the scale ranges from 8 to 10 in (20 to 25 cm) above floor level. This requires lifting of the load up to the level of the platform, a considerable inconvenience in many applications. To avoid this problem, low-profile scales of the type shown in Fig. 23 were developed. This type of scale is movable, and no pit is required. The platform of the scale shown is only 2 in (5 cm) above floor level. Scales of this construction are available with a capacity of up to 5000 lb (2000 kg). Another version of this design, called a self-contained scale, has a platform that is only 3 in (7.5 cm) above floor level and can weigh up to 20,000 lb (10,000 kg). Both designs are equipped with side rails for accommodating oversize loads or for attachment of conveyors, or for other platforms on top of the scale. Platform sizes range from 6 × 4 ft (1.8 × 1.2 m) to 7 × 5 ft (2.1 × 1.5 m).

For highly specialized requirements, customized scales can be constructed. For example, the shock-compensated scale shown in Fig. 24 was designed to withstand the impact from a 40-ton coil of steel dropped at 22.5 ft (6.9 m)/min by an overhead crane. The scale has 2-in (5-cm) spring coils under each corner to absorb the shock. The pressure of a 21-ton (19-metric-ton) load compresses the coils only 1 in (2.5 cm). The load cells are further protected by a built-in mechanical stop. The capacities of scales of this general design range from 5 to 275 tons (4.5 to 248 metric tons).

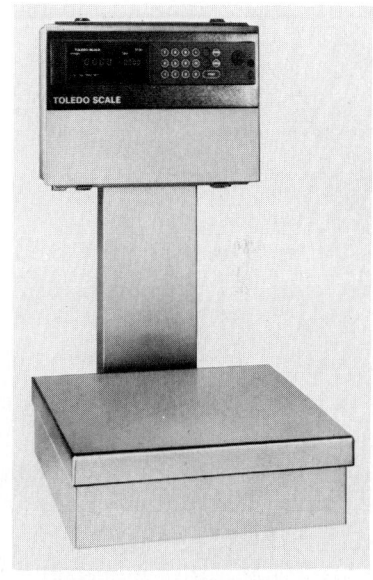

Fig. 20 Stainless steel bench scale with digital readout, with a pushbutton digital tare, a continuous display of tare, automatic pounds-to-kilograms switching, and data output to a printer. *(Toledo Scale, Division of Reliance Electric.)*

Monorail Scales

These scales generally fall into two classes—light- and heavy-duty. Typically, a light-duty scale is used in the meat packing industry for motion or static weighing of carcasses. The scale consists of a rugged, self-contained weighing unit usually supported by existing monorail systems. Modern electronic scales contain no moving parts, and weight accuracy is not affected by swinging loads. Heavy-duty monorail scales are typically used within steel mills, foundries, distribution warehouses, and other locations for weighing material in transit. The scale consists of preassembled, heavy-duty load cell units designed to be incorporated with the user's existing monorail system.

Light-duty monorail scales have capacities of up to 1000 lb (450 kg), and heavy-duty monorail scales, up to 10,000 lb (4500 kg).

Hopper and Tank Scales

These scales are widely used in the chemical and process industries, often in connection with highly automated batching systems, or in the continuous receiving and shipping of bulk materials. A pipe- or torsion-lever-type hopper scale was shown in Fig. 6. These systems have largely been converted to electronic systems, using strain gage load cells. Several arrangements for mounting tanks and hoppers are shown in Fig. 25. Hopper scales range up to over 60,000 lb (30,000 kg) in capacity,

Fig. 21 Floor scale with a removable center plate which provides access to a lever system and load cell. Flush mounting permits easy roll-on and roll-off of vehicles and wheeled containers. *(Toledo Scale, Division of Reliance Electric.)*

Fig. 22 Computerized weighing system utilizing a floor scale installed at a John Deere plant. By operating a pushbutton control panel, loads can be weighed in a few seconds without requiring the forklift operator to leave the vehicle. *(StreeterAmet Division, Mangood Corporation.)*

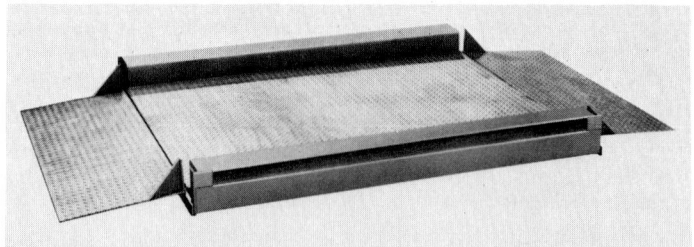

Fig. 23 Low profile scale, unlike a standard floor scale, requires no pit. Platform is only 2 in (5 cm) above the floor level and thus by way of ramps can be accessed by wheeled vehicles. The live side rails make it possible to weigh items somewhat larger than the platform. *(Toledo Scale, Division of Reliance Electric.)*

Fig. 24 Specially designed shock-compensated, self-contained platform scale with large springs under each corner to absorb shock. *(StreeterAmet Division, Mangood Corporation.)*

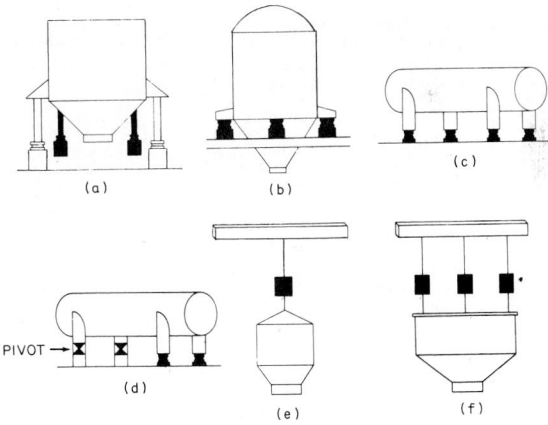

Fig. 25 Arrangement of load cells and mounting assemblies (shown in solid black) for various hopper and tank configurations. (*a*) Four-load-cell compression arrangement (vertical tank built above the floor) commonly used in batching and vehicle loading, (*b*) three-cell compression arrangement (vertical tank built through the floor or supported from beams) commonly used for storage of liquid or powders and for vehicle loading, (*c*) four-cell compression arrangement (horizontal tank with pivot arrangement) for storage of liquids, powders, or solids, (*d*) two-cell compression arrangement (horizontal tank with pivot arrangement) for storage of liquids, (*e*) single-load-cell suspension arrangement (vertical hopper) for process weighing or batching of small loads, (*f*) three-cell suspension arrangement (vertical hopper) for process weighing or batching of large loads. *(StreeterAmet Division, Mangood Corporation.)*

and tank scales up to 125,000 lb (60,000 kg). Kilogram values are not an exact conversion of the pound values because scales are calibrated to provide rounded ranges for either pounds or kilograms.

Bulk Weighing

For weighing large quantities, such as the contents of a railway car, very large scales may be required. The car can be placed on a railway track scale long enough and heavy enough to support the entire car. The gross weight is then determined and then the car is emptied and the tare weight determined. The net weight, of course, is the difference between the two readings. In another method, the contents may be discharged into a single very large hopper and weighed in one draft. However, with modern large hopper cars, such a scale would necessarily be large and of high capacity. A simpler and the most common method is to use a succession of drafts in a system of the type shown in Fig. 26. This is known as a bulk weighing system (or simply bulk weigher). With reference to Fig. 26, an automatic system accomplishes the following sequential steps:

A. *Filling.* The upper garner gate opens, filling the weigh hopper to the preset draft weight.

B. *Gross Weighing and Recording.* The upper garner gate closes, and the gross draft weight is recorded as a plus (+) weight along with an "A" indicating automatic operation. If the time-and-date feature is in use, the time will be printed.

C. *Discharging.* The weigh hopper gate opens, discharging the product.

D. *Tare Weighing and Recording.* When the discharge is complete, the weigh hopper gate closes. The tare weight of the material in the hopper (residual or heel) is recorded as a minus (−) weight with an "A" indicating automatic operation. If the sequential numbering feature is in use, a sequence number will be printed.

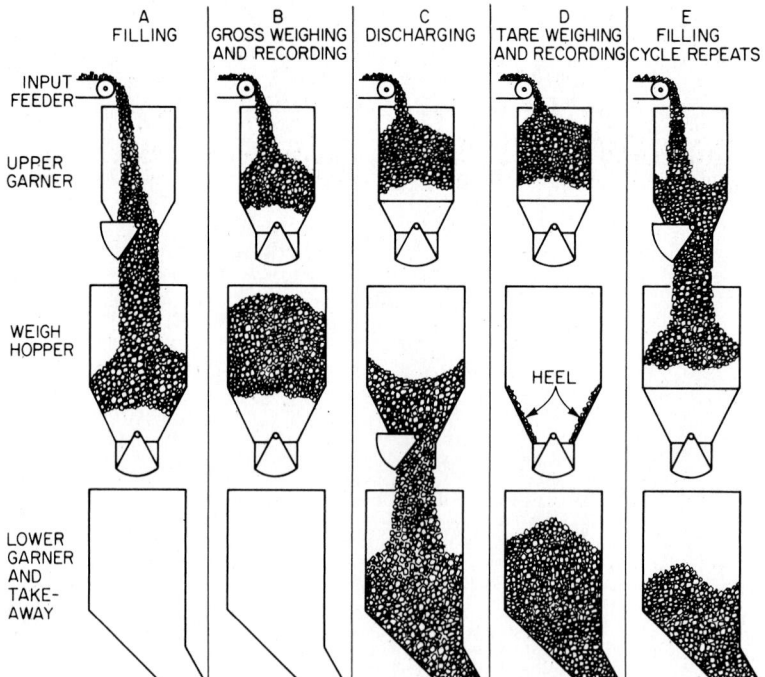

Fig. 26 Bulk weigher. Sequence of steps, *A* through *E*.

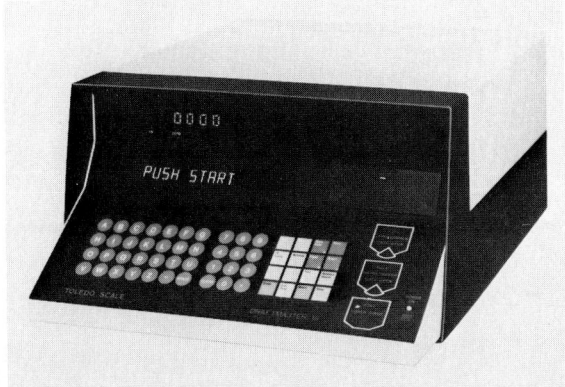

Fig. 27   Keyboard–information display unit used in connection with a bulk weigher. *(Toledo Scale, Division of Reliance Electric.)*

E. Filling Cycle Repeats. The upper garner gate opens, and the operation is repeated until a predetermined quantity is delivered, all the material is received, or an automatic sequence is manually completed. Subtotals are recorded, and the operator is instructed to either push END or CONTINUE.

If the weigh hopper has a capacity of 10,000 lb (5000 kg) it will weigh the contents of a 100-ton hopper car in 20 drafts instead of in one 200,000-lb draft. The keyboard-information display unit of this type of system is shown in Fig. 27. A piece of the printed tape output is shown in Fig. 28. The printer is alphanumeric, dot matrix, with 34 characters per line. Such systems can handle most free-flowing bulk materials and liquids.

A bulk weighing system equipped with a continuous graphic display and installed on a grain elevator is shown in Fig. 29.

Motor Truck Scales

Many of the earlier mechanical-type truck or highway scales have been replaced or converted in recent years to electronic systems which utilize strain gage load cells. As has been the case over many years, the newer designs require heavy-duty steel and concrete construction. Weighbridge foundations are of several configurations. The modern motor truck scale is available in two basic forms: (1) static weighing, and (2) in-motion weighing. The general advantages of electronic scales are faster weighing and self-diagnostic features, which reduce maintenance, downtime, and operator error. Electronic weighing can provide simultaneous weights for individual and tandem axles, as well as gross weights. Additional electronics supply information for billing, inventory, and other accounting functions. Truck scales must be constructed and installed to withstand the most severe weather, including dust, rain, ice, and snow. The most common truck scales have deck sizes ranging from 10 × 10 ft (3 × 3 m), with a capacity of 30 tons (27 metric tons), up to 100 × 12 ft (30 × 36.6 m), with a capacity of 100 tons (90 metric tons). At least one firm supplies a standard 300-ton- (270-metric-ton)-capacity truck scale with a platform of 80 × 12 ft (24 × 3.7 m). In-motion motor truck scales are of lower capacity, ranging between 20 and 50 tons (18 and 45 metric tons).

Some manufacturers incorporate structural hardware to enhance scale operation, for example, to prevent damaging nonaxial forces from reaching the load cells. A combination of flexures and load cell mounting struts affords unrestricted vertical movement while maintaining axial alignment. Shear forces caused by thermal expansion or contraction of the weighbridge and by vehicle movement across the scale are also essentially eliminated. Design details of a motor truck scale of this type are shown in Fig. 30.

Motor truck scales are also made in a low-profile configuration, as shown in Fig. 31. These scales are available with 60- and 100-ton (54- and 90-metric-ton) capacities with platforms of 60 × 10 ft

Fig. 28   Section of printed tape produced by a bulk weighing system. *(Toledo Scale, Division of Reliance Electric.)*

(18 × 3 m) or 70 × 10 ft (21 × 3 m). The weighbridge is suppled in two weldments which are bolted together at the site. Usually excavation is required only for the load bearing piers. The approaches at either end are 10 ft (3 m) long.

For in-motion weighing, scales are designed to measure the individual axles of a vehicle moving across the scale and then electronically combine the axle, tandem axle, and gross weights at an accuracy in excess of 90% for vehicle speeds up to 30 miles (48.3 km)/h.

Railway Track Scales

As with motor truck scales, standard pit and low-profile static scales and in-motion scales are available for weighing railway cars, locomotives and, in fact, entire trains through a sequenced procedure. Railway track scale capacities range between 100 and 200 tons (90 and 180 metric tons) with platforms 8 to 10 ft (2.4 to 3 m) wide and 60 to 125 ft (18 to 76 m) long. For static or in-motion weighing,

Fig. 29 Continuous graphic display, printer, and keyboard control unit used with bulk weighing system for a grain elevator installed at Cargill, Inc. *(StreeterAmet Division, Mangood Corporation.)*

a number of features (often proprietary) offer various advantages; for example, self-balancing and self-checking eliminate the need for manual attendance to monitor round-the-clock train weighing. A detection system may allow elimination of locomotive and caboose weights from the total train weight. Or, another feature may provide automatic collection of gross and net weight information for individual cars and total trains. Scales also may be designed for single draft or multidraft and coupled or uncoupled weighing with a capability for two-, four-, six-, or eight-axle detection. Built-in features may eliminate sequence errors and duplicate car weights.

For example, one system* obtains complete data on each railcar, including weight of lading and total car weights, and transmits this information to a central computer for recording and billing. The

*LoadWEIGH, StreeterAmet, Grayslake, Illinois.

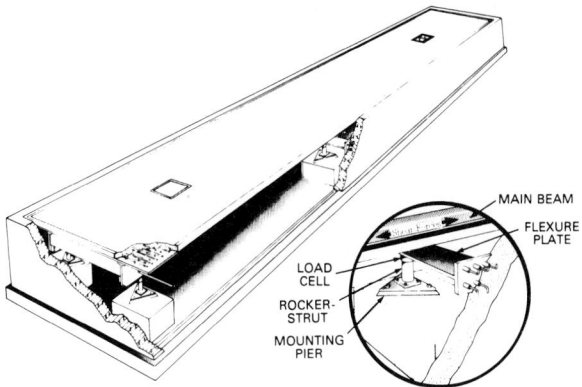

Fig. 30 Structural details of a motor truck scale designed to prevent damaging nonaxial forces from reaching the load cells. *(Accutruck, Hardy Instruments.)*

Fig. 31   Low-profile motor truck scale. *(Toledo Scale, Division of Reliance Electric.)*

loading and measuring operation takes place while the train moves up to 1.5 miles (2.4 km)/h. This type of system is particularly applicable in coal loading operations in connection with unit trains. Other materials that can be handled in this manner include ores, minerals, grains, and aggregates. A control center for one of these systems is shown in Fig. 32. The sequence followed in this system is shown in Fig. 33. A typical loading report is shown in Fig. 34. The system was designed to be used in conjunction with flood-loading-type loadout systems. The system can be modified to operate with most current loadout arrangements. A combination of weighbridges beneath a fload loading chute acts in conjunction with vehicle detection equipment to control the opening and closing of the chute precisely, to obtain final, predetermined weights. The system can be applied to single- or double-loading chute operations and typically incorporates the five following basic elements: (1) a multiplatform scale system, (2) a process control computer, (3) car position detectors, (4) cathode ray tube (CRT) terminals, and (5) a printer supplying hard copy of weight and product information.

Fig. 32   Control center for in-motion rail car loading and weighing system. *(LoadWeigh, StreeterAmet Division, Mangood Corporation.)*

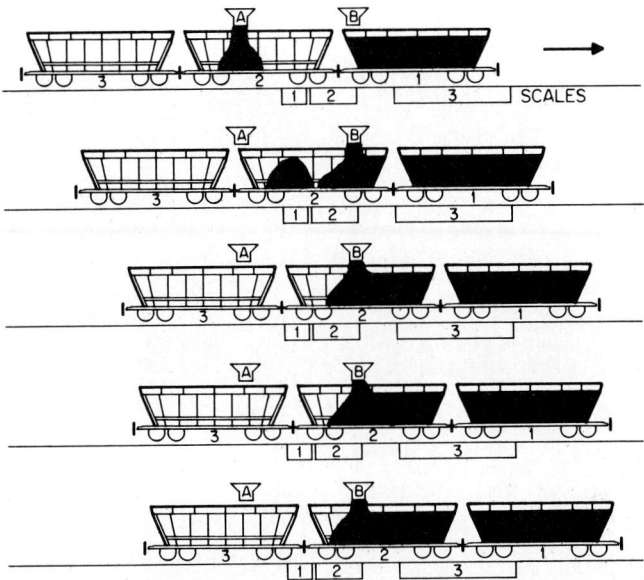

Fig. 33 Sequence of operations of in-motion rail car loading and weighing system. Each car is weighed twice. First, each car is weighed single draft during the loading. The unit controls the loading chute cutoff in order to fully load, but not overload each car. Second, each car is weighed again after loading is complete. This weighment is taken when the loading disturbances are no longer present. This weighing is accomplished in a two-draft mode. *(LoadWeigh, StreeterAmet Division, Mangood Corporation.)*

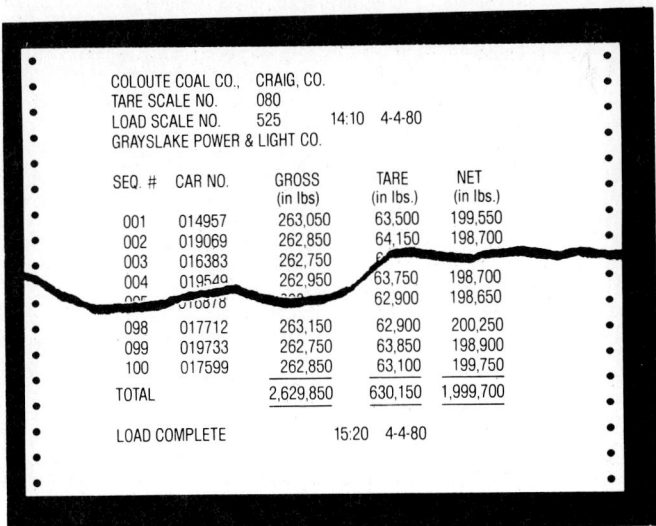

COLOUTE COAL CO., CRAIG, CO.
TARE SCALE NO.   080
LOAD SCALE NO.   525   14:10   4-4-80
GRAYSLAKE POWER & LIGHT CO.

SEQ. #	CAR NO.	GROSS (in lbs)	TARE (in lbs.)	NET (in lbs.)
001	014957	263,050	63,500	199,550
002	019069	262,850	64,150	198,700
003	016383	262,750		
004	019549	262,950	63,750	198,700
005	016878		62,900	198,650
098	017712	263,150	62,900	200,250
099	019733	262,750	63,850	198,900
100	017599	262,850	63,100	199,750
TOTAL		2,629,850	630,150	1,999,700

LOAD COMPLETE   15:20   4-4-80

Fig. 34 A typical loading report produced by the LoadWeigh system. Reports may be in pounds, kilograms, or tons. *(LoadWeigh, StreeterAmet Division, Mangood Corporation.)*

Fig. 35   Full load cell railway track scale. *(Toledo Scale, Division of Reliance Electric.)*

The full load cell railway track scale shown in Fig. 35 is designed for use where flow of rail and truck traffic requires a combination railway track and truck scale. Standard designs are available in 60-ft (18-m) and 72-ft (22-m) platform lengths.

Crane Scales

These scales eliminate double handling by weighing and moving materials at the same time. Generally, the configuration is a load cell contained in a strong, ruggedized assembly with a hook at the top connected to the crane pulley and a hook at the bottom connected to the load. Crane scales range in capacity from 1000 lb (450 kg) up to 60,000 lb (27,000 kg). Where strain gage load cells are used, reading can be remote. Local reading can be effected by using a hydraulic load cell, as shown in Fig. 36.

Counting Scales

Counting of identical parts, for shipping, interdepartmental transfer, or inventory, is an operation which can be expedited by the use of scales. In the mechanical scale, this has been traditionally accomplished through the use of an unequal-arm or fan-type scale equipped with 9:1 and 99:1 ratio counting scales. When this scale is used to count an unknown number of parts, the parts are put in a large scoop on a platter. See Fig. 37. Sufficient parts are transferred from the large scoop to the 99:1 scoop until the indicator is just to the left of zero (if it stops at zero, the count is completed). If the indicator is to the left of zero, one part from the 99:1 scoop is placed in the 9:1 scoop until the indicator stops at zero or just to the right of zero. If it is to the right of zero, parts are taken from the large scoop and held until the indicator is at zero. The total count is then 1 times the units in the hand plus 10 times the units in the 9:1 scoop plus 100 times the units in the 99:1 scoop. For instance, with 3 parts in the hand, 7 parts in the 9:1 scoop, and 5 parts in the 99:1 scoop, the count is 573. To issue 573 parts, 5 parts are put in the 99:1 scoop and 7 in the 9:1 scoop, and parts are then added to the large scoop until the indicator reads zero. Then 3 parts plus those in the ratio scoops are issued with those in the large scoop to make 573. For repeated issuing of a quantity of identical parts, counting scales are available with 10:1 and 100:1 ratio scoops. Scales are also available with 999:1 or 1000:1 scoops (triple ratio).

Another type of counting scale uses the variable ratio principle. A counting scoop is placed on a

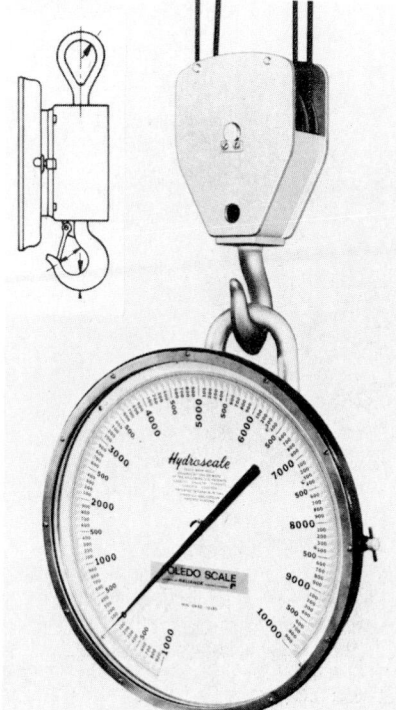

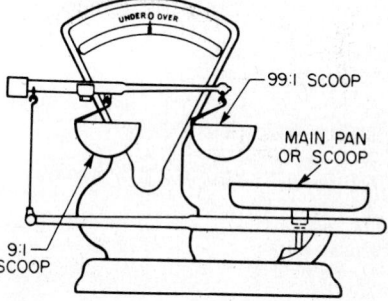

Fig. 36   Crane scale with a dial indicator. Side view is shown in inset. *(Hydroscale crane scale, Toledo Scale, Division of Reliance Electric.)*

Fig. 37   Double-ratio counting scale.

lever connected to the scale so that it can be moved along the lever and its position indicated on a graduated bar. Normally, the bar has several rows of graduations, such as for 1, 2, 5, or 10 pieces in the scoop, and the total count is read directly from the corresponding graduated bar.

Counting also can be performed with one or two scales coupled to a computer. The weight per piece is determined either on a separate more sensitive scale or on the main scale, possibly by the use of 5 or 10 more pieces in the sample. The weight of the parts to be counted is then entered into the computer which determines the number of pieces by dividing total weight by weight per piece.

In the scale illustrated in Fig. 38, a digital readout is used. One or two additional scale bases can be used with the scale contained in the control console, resulting in a total of three possible weighing or counting scales. Any one or two of the three scale bases can be used for a specific counting operation. All count calculations are based on an internal resolution of 200,000 increments. One hundred memory locations are available for storage of tare and average piece weights, as well as for the accumulation of counts. Performance depends on uniformity of weight per piece, number of pieces in the sample, individual piece weight, and the percentage of rated load placed on the scale. The most significant variable is uniform weight of the parts to be counted.

Drum Filling Scale

The scale shown in Fig. 39 consists of a scale, a lance assembly, and a microprocessor controller. Several material handling equipment accessories are also available. A brief drum summary report is printed automatically for each drum as it is filled, including net weight, drum count, product identification, and time and date. Totalized reports for a shift or a day are also produced.

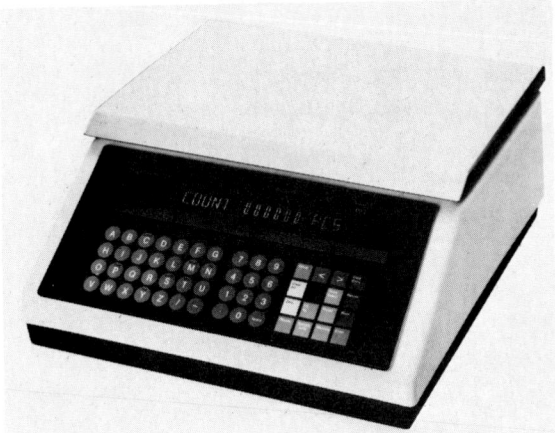

Fig. 38 Digital parts counting scale. *(Toledo Scale, Division of Reliance Electric.)*

Weigh-Batching Systems

Weighing the ingredients for formulation in a process can be done (1) by continuous blending, using continuous weigh feeders, or (2) by batching where definite amounts of each ingredient are weighed and accumulated. Generally, more precise results can be obtained with batch weighing, particularly when production is large enough to justify a scale for each ingredient. For the manufacture of a

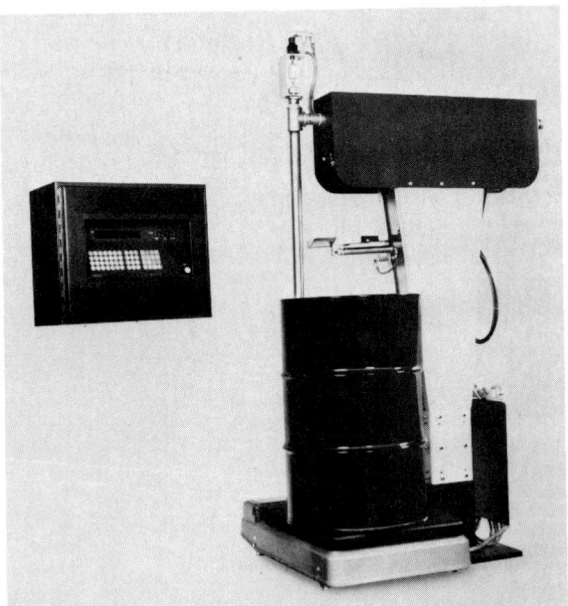

Fig. 39 Microprocessor-based automatic drum filling scale. *(Toledo Scale, Division of Reliance Electric.)*

product like plate or window glass, where large tonnages of material are used per day and where radical changes in the formula are seldom made, it is possible to use a scale of the proper design and capacity for each material and weigh the ingredients quite accurately. Feeding equipment that is particularly suitable for the material being weighed should be selected. Closely associated with the weighing operations per se are the hoppers, hopper feeders, and hopper discharging apparatus which, with some materials, can cause severe difficulties in terms of bridging and flooding.

The use of continuous weigh-feeding systems is described in detail later in this Handbook section in the article on belt conveyor scales and weigh feeders. In some cases where raw materials are quite consistent in density, with a uniform moisture content and uniform handling characteristics, volumetric rather than weigh feeders may be selected. See the article on volumetric feeders later in this Handbook section.

A modern weight- and volume-based batch controlling system* control panel is shown in Fig. 40. Following is a typical automatic sequence of operation:

1. The operator establishes the formula, selects the *automatic sequence mode,* and presses the *automatic control start.*

2. The first bin (ingredient) selected starts to feed into the weigh hopper, and the bin number is displayed on the panel.

3. The *full flow* lamp illuminates, indicating that the ingredient is feeding at full flow. *Dribble* and *full flow* may illuminate if this option has been programmed.

*Quantobatch, StreeterAmet, Grayslake, Illinois.

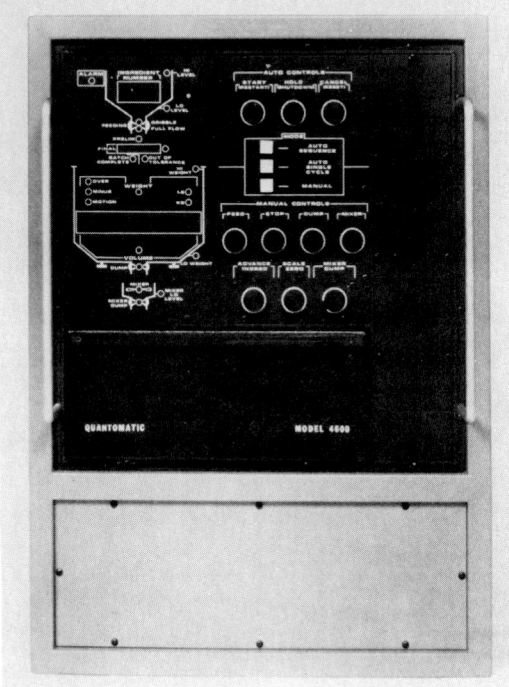

Fig. 40 Control panel for modern weight- and volume-based batch controlling system. *(Quantobatch, StreeterAmet Division, Mangood Corporation.)*

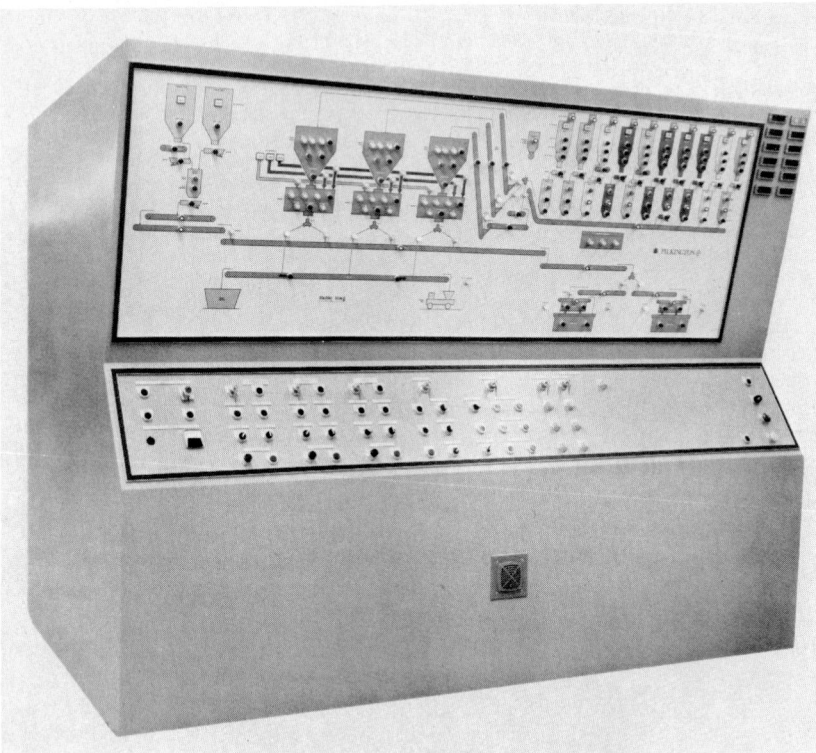

Fig. 41   Control center for complex batching system involving numerous ingredient hoppers, collecting hoppers, and conveyors. *(Toledo Scale, Division of Reliance Electric.)*

4. The *prelim* lamp illuminates, the *full flow* lamp goes off, and the *dribble* lamp illuminates, indicating the ingredient is feeding at a slow flow.

5. The ingredient weight or volume is continuously updated on the large digital display.

6. The *final* lamp illuminates when the ingredient weight minus the *preact* is reached.

7. The automatic tolerance check is made and, if within tolerance, the *tolerance* lamp will flash, indicating acceptance.

8. The remaining bins (ingredients) are then fed into the weigh hopper sequentially, according to the parameters selected in the setup mode.

9. The formula is discharged from the weigh hopper to final disposition.

10. Print outputs are provided as an integral part of the operation for use with an optional printer. Outputs include total batch, weight/volume, total weight/volume of each ingredient dispensed, and hopper residual weight after dumping.

The system can weigh, measure, and control 15 ingredients in a single operation and is capable of automatically repeating a batching cycle up to 99 times.

Control centers for weigh-batching systems may be of several configurations. A complex batching system involving numerous ingredient hoppers, collecting hoppers, and conveyors is shown in Fig. 41. Additional keyboard and display elements are shown in Fig. 42. Another batch controller is shown in Fig. 43.

A process where weighing is important in five separate areas is shown in Fig. 44.

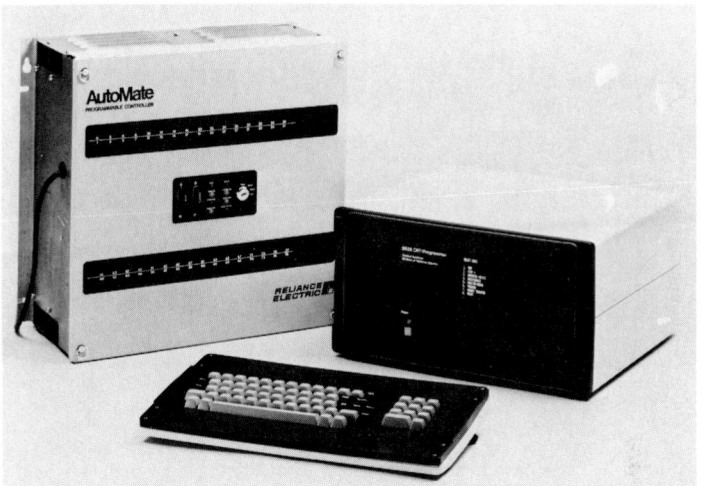

Fig. 42 Electronic equipment used in connection with batching systems of the type shown in Fig. 41. On the left is a programmable controller; on the right, a CRT display and programmer; and at the bottom, an entry keyboard. *(Toledo Scale, Division of Reliance Electric.)*

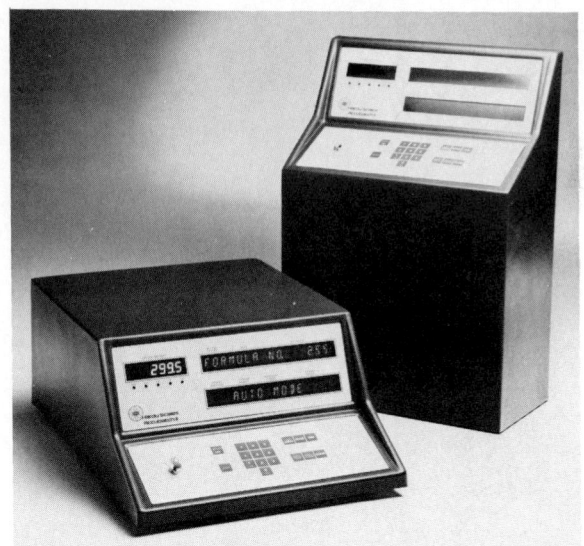

Fig. 43 Batch controller designed to store and process up to 255 formulas with up to 32 ingredients each. Measurement inputs can be accepted from both strain gage load cells and flowmeters. Process programming includes time-based and maintained interrupts, tolerance verifications, multiple discharges within a formula, and discharging a batch to an ordered weight. Two-speed feed control, data logging, serial communication ports, batch load calculation, autojog, and self-adjusting preact compensation also can be included. *(Accubatch II, Hardy Instruments.)*

7.29

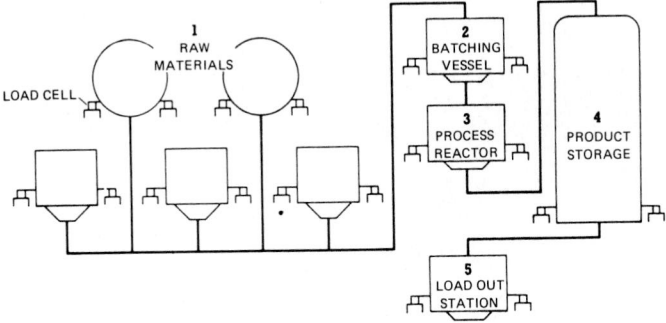

Fig. 44   Process involving nine weighing points in five areas: (1) material storage, (2) batching, (3) process reactor, (4) product storage, and (5) load-out station. *(Hardy Instruments.)*

A programmable batch controller, combining the flexibility of a programmable controller with the application expertise found in batching controllers, is shown in Fig. 45. This system controls all phases of multi-ingredient batching operations with one or two weigh hoppers. The unit is capable of virtually unlimited formula memory and features an active graphic display of critical operation processes.

In the more demanding processes where speed of formula changeovers is critical to production, state-of-the-art batching systems provide instant recall for the formulas stored in memory, plus an array of reports for management. See Figs. 46 and 47.

Fig. 45   Programmable batch controller for multi-ingredient batching operations with one or two weigh hoppers. *(ProBatcher, StreeterAmet Division, Mangood Corporation.)*

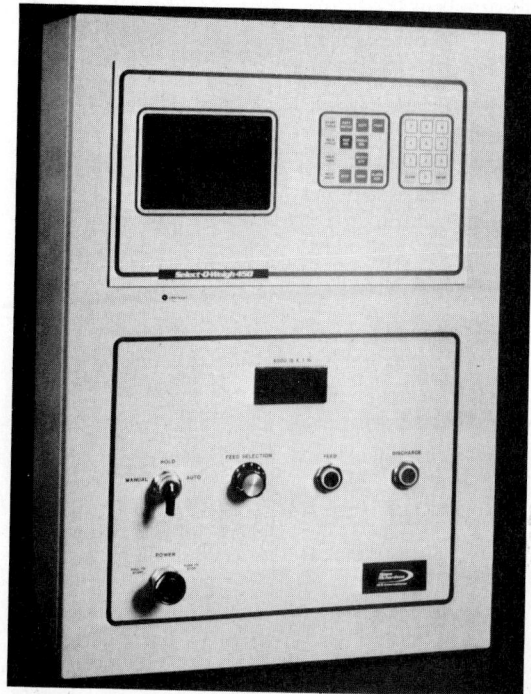

Fig. 46 Typical single scale with up to 96 formulas of 24 materials stored in memory. *(Select-O-Weigh 450, Howe Richardson Corp., Mangood Corporation.)*

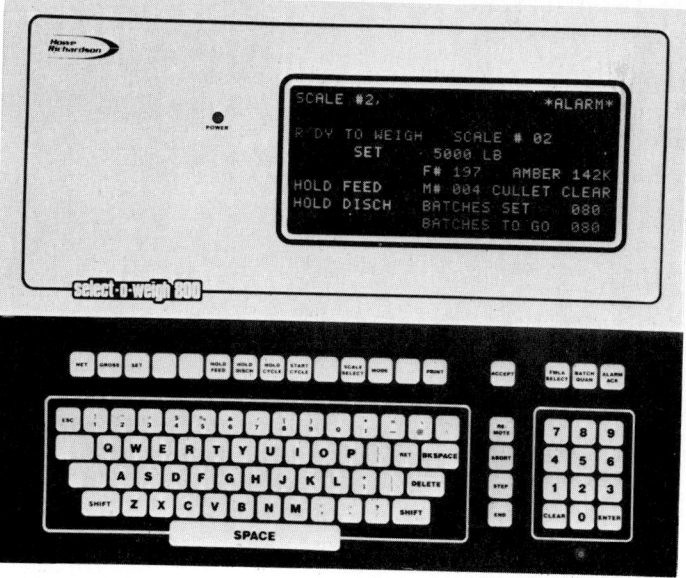

Fig. 47 Typical keyboard and display of a more powerful system than that shown in Fig. 46. The system accommodates 200 formulas in memory for larger processes with as many as 16 scales and 16 mixers working simultaneously. *(Select-O-Weigh 800, Howe Richardson Corp., Mangood Corporation.)*

REFERENCES

Bailey, S. J.: "Microcomputer Improves Resolution in Mass-Selective Scale," *Control Eng.*, vol. 25, no. 11, 1978, p. 45.

Fraade, D. J.: "Load Cells for Batch Weighing," *Instrum. Technol.*, vol. 24, no. 12, 1977, p. 53.

Kompass, E. J.: "Microprocessor Enhances Automatic Weigh-Batching Flexibility," *Control Eng.*, vol. 24, no. 4, 1977, p. 42.

Kurylcheck, A. L.: "Bin Discharging Problems and Solutions," *Chem. Eng. Prog.*, vol. 74, no. 7, 1978, p. 84.

McIntyre, C. E.: "Selecting and Applying Nuclear Belt Scales," *Control Eng.*, vol. 25, no. 7, 1978, p. 28.

Morris, H. M.: "Measuring Force in Difficult or Varying Environments," *Control Eng.*, vol. 25, no. 9, 1978, p. 78.

NBS: "Specifications, Tolerances, and Other Technical Requirements for Weighing and Measuring Devices," *Handb. 44*, National Bureau of Standards, Washington, D.C., 1979.

Regits, R. T., and R. E. Bergeron: "PC-Based Weigh Batching System Features Time Sharing Matrix Data Transfer," *Control Eng.*, vol. 25, no. 2, 1977, p. 77.

Staff: "Gyroscopic Scale Weighs by Measuring Precession Rate," *Control Eng.*, vol. 26, no. 1, 1979, p. 23.

Van den Berge, H., et al: "Weighing On-the-Fly Keeps the Process Moving," *Control Eng.*, vol. 24, no. 9, 1977, p. 52.

Wilson, D.: "Weigh-Feeder Control Needs Favor the Digital Route," *Control Eng.*, vol. 23, no. 1, 1976, p. 48.

Zupko, R. E.: *British Weights and Measures*, University of Wisconsin Press, Madison, Wisc., 1977.

Bonded Strain Gage Load Cells

**by
James R. Story***

A strain gage load cell is an electromechanical device which converts weight or force into an electrical signal. The heart of a strain gage load cell is the load (weight or force) receiving element (Fig. 1a) to which the strain gage Wheatstone bridge network, shown in Fig. 1b, is bonded. Excitation voltage, ac or dc, is applied to two legs of this bridge. The readout device, which accepts a millivolt input, is attached to the other two legs. Under a no-load condition, the Wheatstone bridge is balanced. As the load is applied, the bridge becomes unbalanced and a millivolt output is measured by the readout device. Typically, this output is 2 or 3 mV per volt of excitation. At full scale (full capacity), excitation voltage for the bridge is usually 10 to 20 V. Therefore, the maximum signal output of the load cell is typically 20 to 40 mV at full scale. Higher voltage can damage the strain gage or cause the performance of the load cell to deteriorate.

*Marketing Director, Transducers, Inc., Cerritos, California.

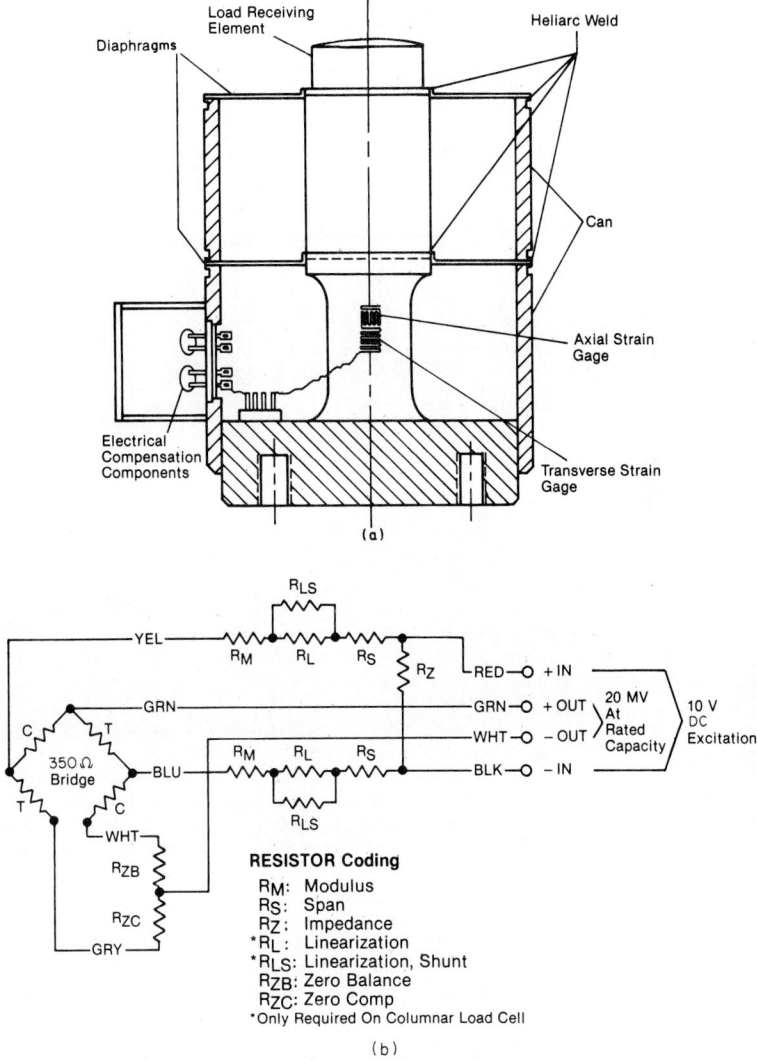

Fig. 1 (a) Compression load cell, (b) Wheatstone bridge wiring diagram (tension positive). *(Transducers, Inc.)*

Types of Load Cells

Strain gage load cells can be classified by the element design they incorporate. They usually consist of one of three basic configurations: (1) column, (2) cantilevered beam, or (3) shear design, all incorporating many modifications to meet the specific characteristics required for the application.

The Column

A typical design for a columnar-type load cell is shown in Fig. 1. This design is usually used in load cells having a capacity of 5000 lb (2268 kg) or greater. Two strain gages are bonded axially along

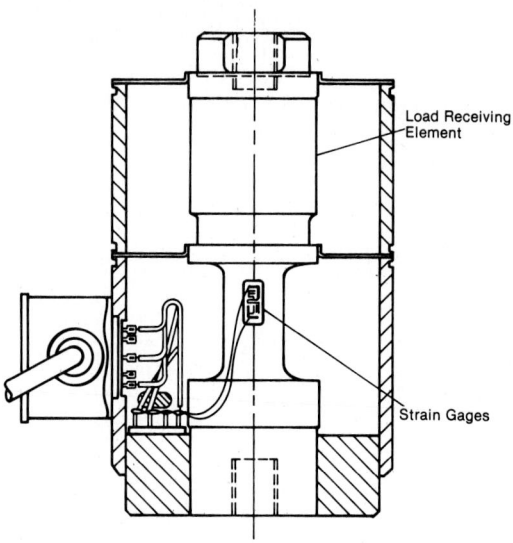

Fig. 2   Bidirectional load cell. *(Transducers, Inc.)*

the column and are referred to as the *active gages.* Two additional gages are mounted 90° to the axially positioned strain gages to measure transverse strain. These are referred to as the *Poisson gages.* In Fig. 1, a load cell is depicted which can be used only in compression, since it has only a fitting at the bottom base for attachment to the support structure and a load receiving button at the top.

In construction of the load cell, the load receiving element is shrunk-fit into the base in a compression-type load cell. The bottom section of the load cell housing, commonly referred to as the *can,* is welded onto the base structure. A diaphragm is welded to the top of this lower can. The diaphragm is welded both to the edge of the can and the center column. The upper can is then welded to the connection between the diaphragm and the lower can, thereby completing the basic outer shell or can of the load cell. Finally, a top diaphragm is welded to the upper lip of the top can and the center column, thereby sealing the internal portion of the load cell and making it impervious to gas or moisture. All wiring for the load cell, i.e., the wires from the strain gages to the external cable, is through a glass-to-metal seal located in the wall of the lower can.

In Fig. 2, the tension-type cell has an integral load receiving element (one piece) with either female or male fittings at both ends for the attachment of loading hardware. This type of cell is commonly referred to as a *bidirectional* or *universal* load cell.

Bending Beams

For force and weight measurements below 500 lb (227 kg), some type of variation is usually made in the bending beam design. The most common is the dual-guided cantilever beam, incorporating four strain gages mounted in the corners of a stabilized rectangle (Fig. 3). As a load is applied to the free end of this dual-guided cantilever beam, the strain gages bonded to the element undergo a resistive change proportional to the load applied. In this design, two gages go into tension and two gages go into compression. The location of these gages and their loading phase are shown in Fig. 3.

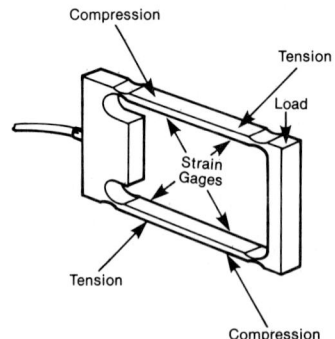

Fig. 3   Dual-guided cantilever beam load cell. *(Transducers, Inc.)*

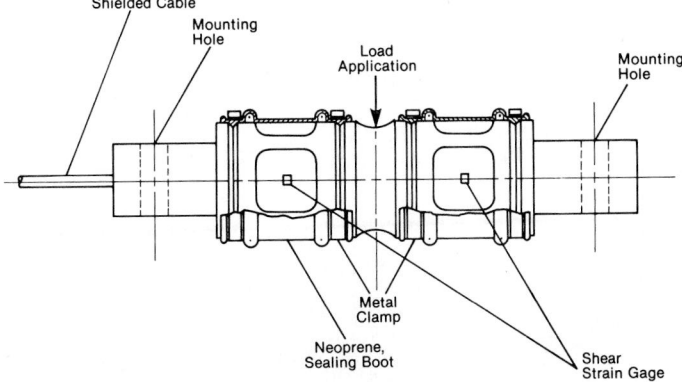

Fig. 4 Center-loaded (end-mounted) shear beam. *(Transducers, Inc.)*

Shear Element

A recent design of a strain gage load cell incorporates a shear-type load element. In this type of design, either four or eight gages are bonded to a specifically designed element and wired to form a Wheatstone bridge. These strain gages differ from the type used in columnar and dual-guided cantilever designs. They are designed to activate when they are placed in shear (Fig. 4). The advantages of this type of element are (1) higher capacities in smaller load cell sizes and (2) less sensitivity to side load error (loads not placed in the proper orientation).

New Load Cell Configurations

In recent years, load cells from 500 lb (227 kg) to 300,000 lb (136,080 kg) incorporating shear element designs have revolutionized scale designs. Formerly, a high-capacity truck and track scale utilized elaborate load cell mounting arrangements and weighbridge designs to obtain optimum performance. The shear beam load cell has markedly changed the design of medium- and high-capacity scales. To a great extent, this has been possible because the shear beam load cell tends to be insensitive to side load error and lends itself to very simple mounting arrangements, yet provides optimum performance.

Shear beam load cells may be classified as of two types: (1) a single-ended beam, and (2) a double-ended beam. In Fig. 5, a single-ended medium-capacity shear beam [500 lb (227 kg) to 10,000 lb (4535 kg)] is shown. This type of load cell lends itself to medium-capacity low-profile industrial scales. The load cell can be loaded by means of a flexible high-tensile-strength cable attached to a scale weighbridge or by a load button.

Another variation of the shear beam is the center-loaded end-mounted design commonly used for

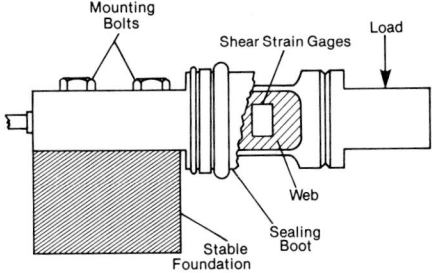

Fig. 5 Single-ended shear beam *(Transducers, Inc.)*

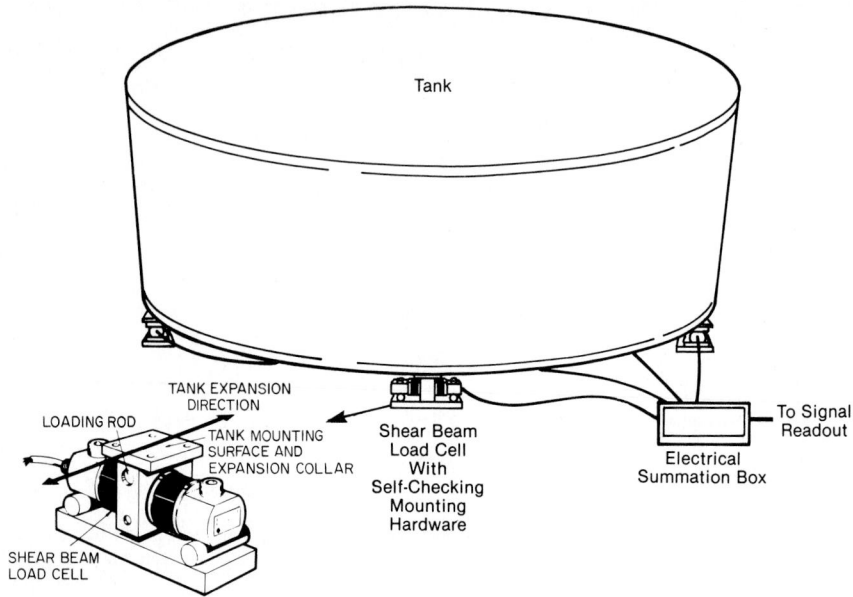

Fig. 6 Tank weighing installation. *(Transducers, Inc.)*

tank weighing, hopper weighing, truck scales, and track scales. Figure 6 shows the load cell used in a tank weighing installation with the mount bolted directly to the superstructure of the item to be weighed. As with the single-ended shear beam, this unit is insensitive to side load error and is well suited for industrial applications. It does not require an exceptional knowledge of scale technology to install. As indicated by Fig. 6, a valuable feature of the center-loaded end-mounted beam is its self-checking capability. The mounting hardware encircles the basic element and restricts the move-

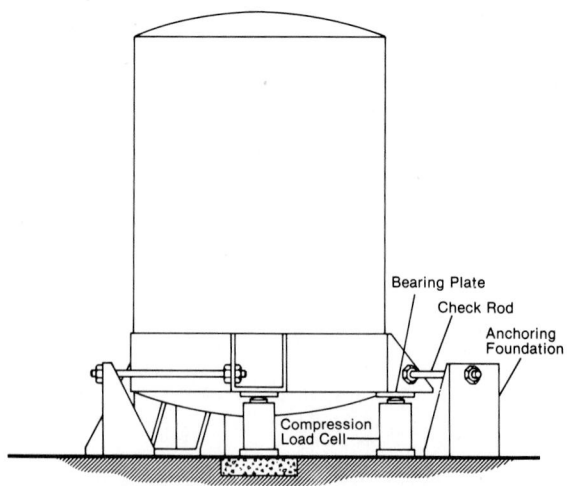

Fig. 7 Compression cell tank weighing installation. *(Transducers, Inc.)*

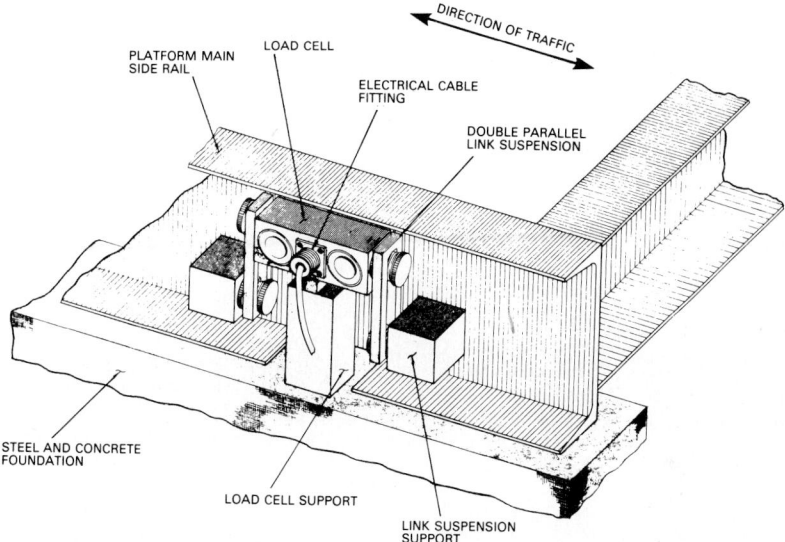

Fig. 8   End-loaded (center-mounted) shear beam. *(Transducers, Inc.)*

ment of the vessel to be weighed, thereby eliminating external check rods to secure the tank on the load cells. Prior to this development, a typical compression cell installation required special bearing plates and check rod assemblies connected to concrete foundations to secure the tank on the load cells. See Fig. 7.

Another variation is a center-mounted end-loaded version (Fig. 8). In this design, the load can either be hung in tension by means of links attached to the ends or in compression by means of a loading rod at the top of each end.

Both the center-loaded and end-loaded versions lend themselves to high-capacity truck and track scales. Since this load cell is insensitive to side load conditions, it can be mounted externally to the weighbridge of the truck or track scale and loaded through some type of link arrangement. This mounting design allows for pitless truck and track scales and weighbridges utilizing less steel, since the flexing of the weighbridge does not detract from the accuracy of the load cells.

Load Cell Classification by Environmental Protection

Load cells can also be classified by their environmental protection. These classes include (1) hermetically sealed, (2) environmentally protected, and (3) controlled environment. True hermetic sealing requires that the load cell construction incorporate only metal-to-metal or glass-to-metal seals to protect the internal construction of the load cell. (See the columnar load cell shown in Fig. 1.) By simply incorporating welded or soldered connections, the load cell can be made totally gastight and waterproof.

Environmentally protected load cells employ some type of potting material, usually a flexible rubber-type material and/or a waterproof covering, such as a neoprene boot over the strain gage area of the load cell. See Fig. 4.

Controlled environment load cells require protection from the environment in that this design does not incorporate any covering of the electrical portion of the cell. Usually this type of cell has only a dust cover over the gaged area.

Belt Conveyor Scales and Weigh Feeders

by
Andrew L. Kurylchek* and Gene Rebucci†

A belt conveyor scale may be looked on as a primary element for use in the measurement of the flow rate of bulk materials. Flow measurement is expressed in weight (such as pounds or kilograms) per unit of time (minutes or hours) rather than in units of volume (cubic feet or cubic meters). In this light, the belt conveyor scale is a true mass flowmeter. There are many parallels in the use of belt scales for solids applications and flowmeters for liquid systems.

Basic Belt Conveyor Scale

Belt scales are applied and installed on practically any size, length, and shape of conveyor—from 12-in ($\approx$30.5-cm)-wide belts handling as little as 1 lb (0.45 kg)/min at a minimum belt speed of 1 ft (30.5 cm)/min to 120-in (3-m)-wide belts handling as much as 20,000 tons (18,145 metric tons)/h at speeds up to 1000 ft (305 m)/min. The basic design and application of a conventional belt conveyor scale are illustrated in Fig. 1. A weigh platform (or weighbridge), on which are mounted standard conveyor idlers, is used to sense the weight of the material passing over it. A belt speed pickup system measures the speed of the conveyor belt. Weight and speed signals are transmitted to a multiplier whose product or true rate output signal is integrated and displayed on a continuous totalizer. The rate, of course, is equal to weight times speed.

The basic components used to sense the weight and speed may be mechanical, electrical, elec-

*Consultant, Wayne, New Jersey.
†Schenck Weighing Systems, Totowa, New Jersey.

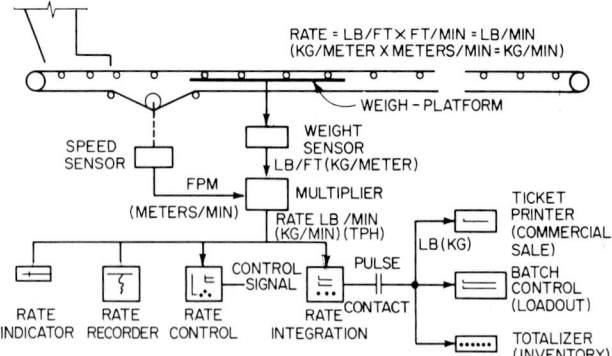

Fig. 1 Basic belt conveyor scale illustrating a variety of display and control instrumentation.

tronic, pneumatic, or hydraulic. This information can be used for local or remote presentation in several ways:

Rate Indication. To show rate of material flow at any given instant

Rate Recording. For production records, downtime, over- or underloading of the system, material stoppages due to breakdown or plugged hoppers, and monitoring overall system efficiency

Rate Control. Whose output signal to a final operator, such as a prefeed device, control the feed to any particular process demand or set point

Rate Integration. To display on a continuous totalizer the weight conveyed and to produce digital pulses, equivalent to a specific increment of weight, to:

> *Ticket printers* which display weight, date, and time for commercial sale of the conveyed product

> *Batch controllers* for predetermined loading of trucks, railcars, ships, barges, and containers—and to certain processes that demand batch feeding

> *Totalizers* for remote weight readout and inventory control

Radiation Gage Belt Conveyor Meter

See the article Continuous Nuclear Weigh Scales later in this Handbook section.

Belt Conveyor Scale Accuracy

Wherever possible, the scale should be installed where it will not be subjected to conveyor influences that cause a lifting effect of the belt off the weigh platform section. Belt tensions and belt lift have more influence on accuracy than any other factor. Therefore, the scale should be installed where the belt tensions produce the least effect. A tail-end scale location is shown in Fig. 2. For both an empty and a fully loaded belt, the least belt tension is in this area, immaterial of its length or shape, as evidenced by the graphic distribution of the tensions involved.

Theoretically, belt scale accuracy is directly proportional to (1) the vertical misalignment of the weighing idlers with the adjacent fixed idlers, (2) the belt stiffness factor (relative to troughing angle, belt thickness, and the modulus of elasticity), and (3) belt tension. The accuracy is inversely proportional to the length of the weigh platform. Thus, designing the scale with a minimum weigh platform deflection and increasing the length of the platform tends to overcome the inherent adverse effects of belt tension and belt stiffness. Multiidler weigh platforms, therefore, are utilized for applications that demand extremely high accuracy, whereas single-idler designs are suitable for less stringent accuracy requirements.

It is also important that the material (load) travel at the same velocity as the belt when crossing the weigh platform and not be in a turbulent state as it is when leaving the feed hopper or chute. The distance from feed point and weigh platform is designated the Q distance and generally should be

3 to 4 ft for speeds up to 300 ft/min   (1 to 1.3 m for speeds up to 100 m/min)

6 to 8 ft for speeds from 300 to 450 ft/min   (2 to 2.5 m for speeds from 100 to 150 m/min)

9 to 12 ft for speeds in excess of 450 ft/min (2.7 to 3.7 m for speeds in excess of 150 m/min)

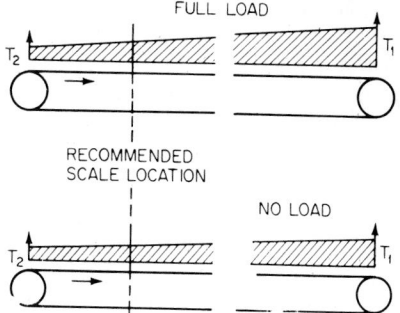

Fig. 2   Relative tension in the upper portion of a belt for (1) full load and (2) no-load conditions, indicating that the least belt tension in both conditions occurs at the tail-end location (at left of figure).

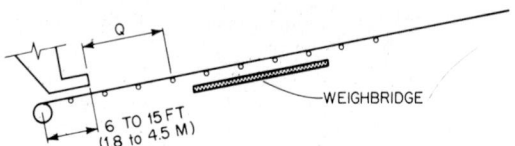

Fig. 3   Distance Q between point of feed on belt and weighbridge location is critical to maximum weighing accuracy.

See Fig. 3. The conveyor inclination should not exceed the angle where material will slide back down the belt and across the weigh platform and, consequently, be weighed a second time.

Accurate weighing results can be achieved on a curved belt provided that the scale is a minimum of ~40 ft (~12 m) from the point of tangency (section A of Fig. 4). If the available room is less than required, alternate section B can be used for the installation, but the accuracy guarantee will be less because of the aforementioned effects of belt tension. At or near a curve the belt does not fully rest in the idler trough, and thus the conveyed material is not being totally weighed. A portion of the weight is supported by the belt because of tension effects. For ideal weight sensing, the belt must ride completely in the trough for both an empty and a fully loaded condition. Also see Fig. 5 and Ref. 1.

An accuracy guarantee curve for load-cell-type belt scales is shown in Fig. 6. Load cell belt scale guarantees are based over a 3:1 to 4:1 range and normally read, "Guaranteed to weigh within plus or minus ½% of the totalized weight when installed in accordance with installation and operational instructions." Figure 6 shows accuracy curves for two- and four-idler weighbridges. Belt scale manufacturers do not adhere to a common or standard accuracy statement. Thus, the purchaser of a belt conveyor scale should make certain to specify exactly what is acceptable in terms of accuracy guarantee or, at a minimum, fully understand the manufacturer's interpretation of the accuracy terms. See Ref. 2.

Testing and Calibrating Belt Scales

Three methods are used to test and calibrate a belt scale:

1. **Material Test.** This is the most accurate and reliable method because it duplicates actual operating conditions. The material from no fewer than two circuits of belt travel must be weighed on a reliable platform, truck, or track scale, and the results compared with the readings obtained from the belt scale. Adjustments are made, and a second test is run to verify the adjustments. This is a time-consuming test method and often not practical in a busy plant.

2. **Test Chain Method.** This dynamic test is second best to the material test method. Chains are designed to weigh approximately the same as the material normally conveyed. See Fig. 7. The product of the known (pounds per foot, kilograms per meter) of chain and the number of feet or

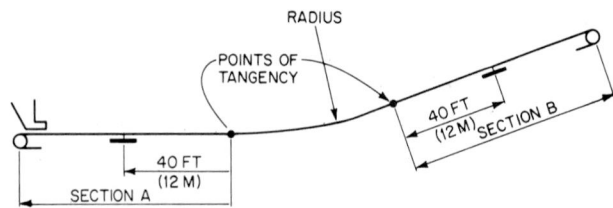

Fig. 4   A minimum distance of 40 ft (12 m) is recommended for the installation of a belt conveyor scale between the weighbridge and the point of tangency of a concave curve as shown.

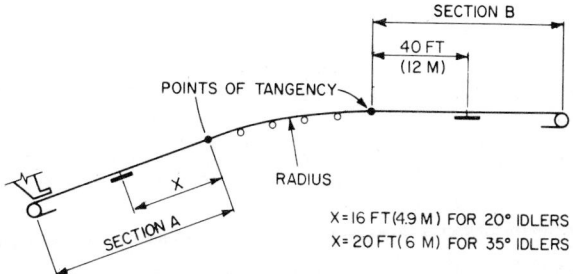

Fig. 5 Recommended distance between a weighbridge and the point of tangency of a convex curve as shown.

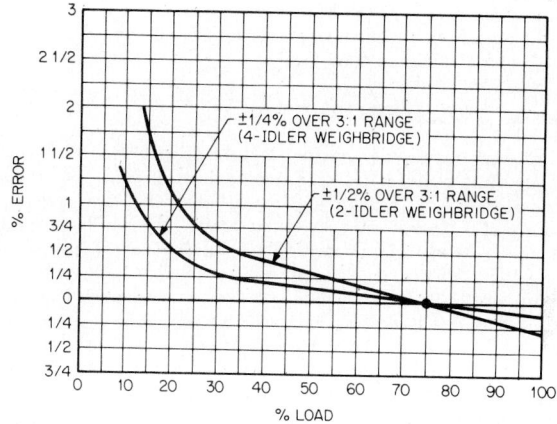

Fig. 6 Accuracy guarantee curve for load-cell-type belt scales when calibrated at 75% of maximum load.

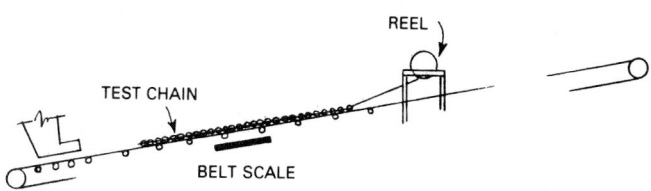

Fig. 7 Use of a test chain for calibrating a belt conveyor scale.

meters of belt travel per test provides the weight figure that the scale totalizer should read. Percent error is calculated, and necessary calibration adjustments are made. A verification check is always desirable.

3. *Calibrating Weight Test.* Although the easiest and quickest test to perform, this method does not take into account the effects of belt tension and weighbridge deflection as the other two test methods do. In this method, a known static weight simply is applied to the scale mechanism, causing a deflection or voltage output change (depending on the type of scale) equivalent to the normal weight per length of material conveyed. The equivalent weight per length of the static weight must be determined by accurately measuring the weight per length and the lever ratio and computing the equivalent weight per length. Unless this calculation is performed accurately, large errors can be *calibrated into* the system. The weight per length equivalent to the calibrating weight times the distance of belt travel per test provide the weight figure against which the scale totalizer should check. It is recommended that actual material tests occasionally be performed to verify the accuracy of the calibrating *weight-pounds-per-foot* (or other weight-per-length) equivalent.

Tare Adjustment

Because a conveyor belt varies in weight per running foot or meter, no less than one circuit of the belt must be run before a tare adjustment is made. On completion of one or more circuits, the integrator totalizer must display the same readout as before start-up, indicating that the varying weight of the conveyor belt will not introduce an error into the flow rate output. For this reason, an integrator must have the capability of readout on the minus side for lightweight sections of the belt, as well as on the plus side for heavier sections.

Certification of Belt Conveyor Scales

In 1967 the U.S. National Bureau of Standards adopted a belt conveyor scale code (NBS Handbook H-44) for use in certifying conveyor scales in commercial applications for the legal sale of bulk materials. The loading control and configuration of this system (Fig. 8) is typical. A ticket printer is used to record the weight of the material delivered to the car. The system may be automated further to load and print out the car's weight automatically. Some of the advantages of using belt conveyor scales for this application include: (1) Weighing fees are eliminated, representing a cost savings of several dollars per car. (2) The system inherently provides net weight, whereas track scales normally are used for gross weighing of loaded cars only wherein stenciled car tare weights (sometimes in error) are used to compute net weight. Furthermore, car repairs and general wear and tear may alter the tare weight by as much as 10%. (3) Because of faster loading, demurrage charges are eliminated. (4) There no longer is a need to unload overloaded cars, and fewer cars per day are required because underloading is prevented.

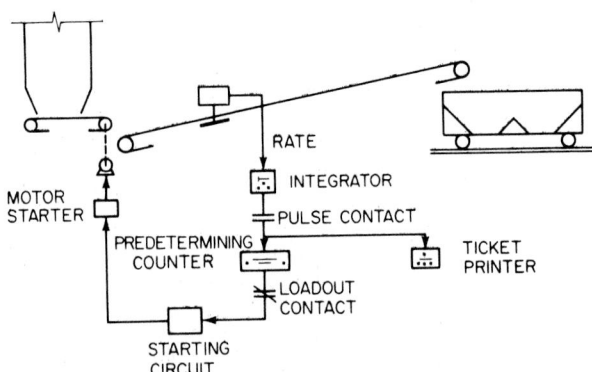

Fig. 8 Loading control and configuration (typical) of a belt conveyor scale system for commercial materials handling.

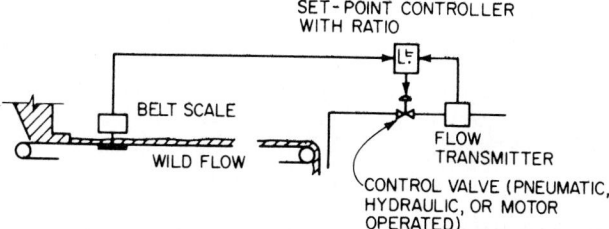

Fig. 9   System in which the flow rate of a liquid additive is proportioned to "wild" or uncontrolled flow rate of a bulk solid. A signal from the belt scale feeds the ratio input circuit of the set-point controller that establishes the ratio of liquid additive to solid flow. More than one additive material can be handled by a system of this type.

Typical Applications of Belt Conveyor Scales

Only a few representative applications can be described here, including (1) systems for proportioning liquids with bulk solids (Fig. 9), (2) "dry weight" measuring systems (Fig. 10), (3) process feed control (Fig. 11), (4) inventory control and railcar and truck loading as previously discussed for Fig. 8.

GRAVIMETRIC FEEDERS

The basic function of a belt scale is to continuously weigh material as it leaves a hopper, chute, or bin for conveyance to a given location. The scale accurately weighs the material *without* concern for the *amount* delivered per unit of time, whereas a gravimetric feeder is used to accurately deliver a *desired amount* of material *within* a given time period. In short, gravimetric feeders are used to *control* the flow rate to the process.

With gravimetric belt feeders, capacities range from 1 lb/min to 4000 tons/h ($\sim$0.5 kg/min to 3630 metric tons), and ranges up to 100:1 are readily available. These devices are furnished as self-

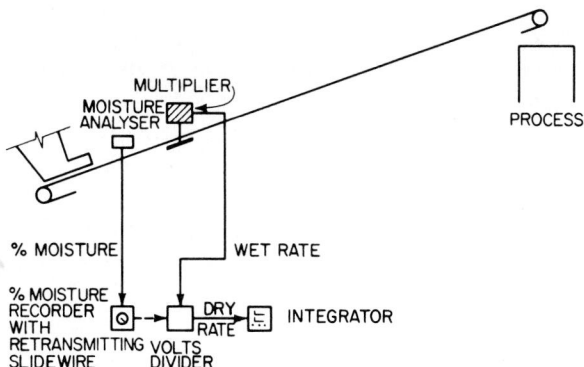

Fig. 10   Use of a belt conveyor scale in connection with a moisture analyzer to deliver a controlled dry weight of material to a process. The belt scale and moisture analyzer have compatible output signals which are fed to a voltage divider that calculates the rate of material passing over the scale on a dry basis. This information is fed to the integrator that provides the total weight readout of dry material to the process.

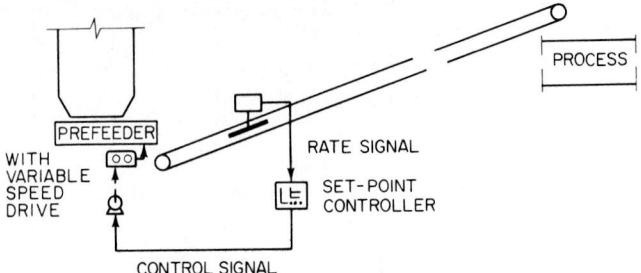

Fig. 11 Use of a belt conveyor scale to automatically control the material flow rate to a process and thus ensure a maximum capacity output. Where the feed rate is set and controlled manually, it must be set below the norm to take into account the occurrence of surges and density variations.

contained units, complete with drive, scale assembly, and controls, as shown in Fig. 12. Variable lengths and belt widths are available to suit plant layouts and required capacities and ranges.

Control System Variations

Of limited application, the system shown in Fig. 13 varies the material weight to maintain or control the rate of flow by employing a *constant speed feeder belt drive*. Normally, an electrical weight signal (weight per length) is transmitted to a weight controller preset for a particular rate of material to the process. By utilizing the constant speed belt drive, speed can be assumed constant, thereby inferring that weight is proportional to rate. In one configuration, the controller output correction signal is transmitted to a vibratory feeder; in another scheme, the signal goes to a motorized gate which

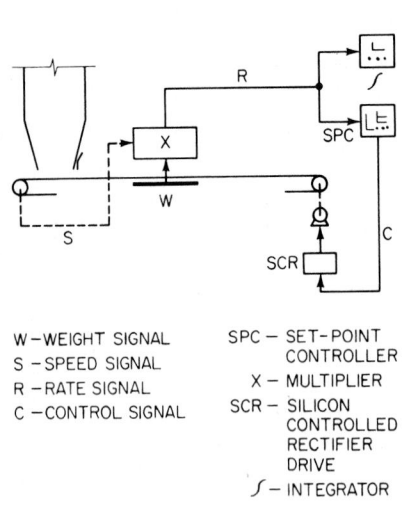

W — WEIGHT SIGNAL
S — SPEED SIGNAL
R — RATE SIGNAL
C — CONTROL SIGNAL

SPC — SET-POINT CONTROLLER
X — MULTIPLIER
SCR — SILICON CONTROLLED RECTIFIER DRIVE
∫ — INTEGRATOR

Fig. 12 Basic weigh feeder system.

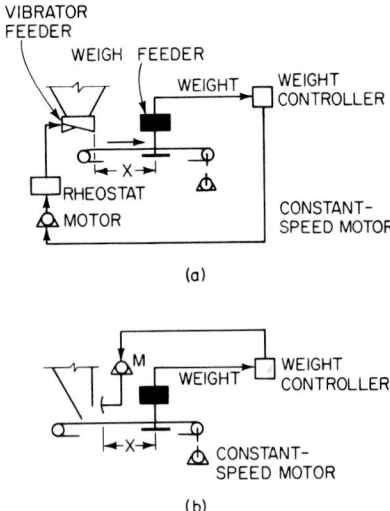

Fig. 13 Weigh feeders incorporating a constant speed feeder belt drive. (*a*) Feed rate controlled by a vibratory feeder, (*b*) feed rate controlled by a motorized gate.

permits more or less material to cross the weigh platform until the preset weight (or rate) as set on the controller is satisfied.

In both cases, material depth of load must be changed to satisfy a rate change—a factor that limits the range of the feeder to 3:1 or 4:1, because the accuracy of a belt scale or constant speed gravimetric feeder begins to deteriorate when operated below one-third of maximum capacity. Furthermore, a normally accepted design criterion for accurate weighing is that the minimum material bed depth on the weigh belt be three times the maximum lump size. For a maximum lump size of 1 in (2.5 cm), the minimum bed depth would be 3 in (7.6 cm). To increase the range to the upper limit (assumed 3:1), the maximum bed depth would be 9 in (22.9 cm). Obviously, to go above a 3:1 range, the maximum design limits of the gravimetric feeder would be exceeded. The height of the skirts used to contain the material also would exceed an acceptable design limit. Also, if a troughed belt design is used, overloading and spillage would occur with excessive bed depths.

Control response with this concept also is affected by the transport lag between the point of measurement and the point of control (indicated as X in the diagram). This distance, normally a few feet ($\approx$1 meter), creates a transport lag that tends to cause control system oscillation. However, because of its inherent simplicity, the system can be used satisfactorily for applications that do not require precision control or a maximum range of over 4:1.

Rate Control System

In the system shown in Fig. 14, there are no inferred constants. Weight is measured by a sensitive weigh platform and transmitted to either a mechanical or an electronic multiplier. Speed also is measured, either mechanically or electrically, and transmitted to the multiplier. The rate output from the multiplier is transmitted to a set-point controller where the rate is compared with the desired rate. When deviations occur, the controller corrective signal is transmitted to a silicon controlled rectifier dc drive which alters the speed of the feeder belt—before the material has left the end of the weigh platform—thereby providing almost instantaneous and continuous maintenance of the set rate.

A fixed gate adjacent to the infeed hopper is manually set at a predetermined height to maintain a fixed material bed depth on the belt—hence a relatively constant weight per length loading which permits high accuracy over very wide ranges—up to 100:1. This is possible because changes in weight are due primarily to changes in bulk density, which usually do not exceed ±10%.

Feeder Response to Density Changes

The control responses of the three systems described to density changes are illustrated by Fig. 15. The differences in response are ascribed to the factors previously mentioned.

Accuracy of Gravimetric Belt Feeders

Normally, accuracy guarantees are specified as ±½% or ±1%, dependent on specific designs, of set rate over the entire operating range. This approach is valid provided that a rate control feeder design is used, or when the range does not exceed 4:1 for "weight control" feeder designs.

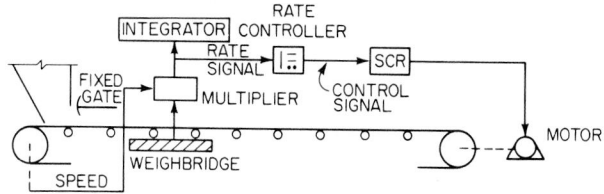

Fig. 14 Weigh feeder with a true rate control system wherein there are no inferred constants. Weight × speed = flow rate.

A more scientific approach to accuracy statements involving a sigma rating is preferred to the more generalized statements. For example, in this system:

1 sigma means that 68% of all tests performed must fall within ±½%.

2 sigma means that 95% of all tests performed must fall within ±1%.

3 sigma means that 99% of all tests performed must fall within ±1.5%.

Using the sigma rating, a typical accuracy statement will read: "±½% of the set rate at 1 sigma based on a minimum sample of 1 min or the time equivalent for two circuits of the weigh belt, whichever is greater."

Microprocessor Instrumentation

Microprocessor technology has been available for continuous weighing for several years and has now established itself as the present generation for proprietary instruments. The microprocessor provides accuracy, automation, and flexibility features not feasible with previous analog and digital systems.

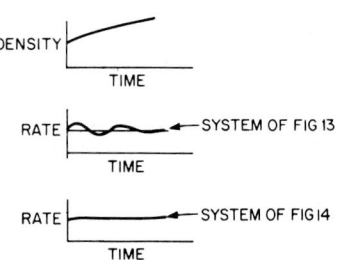

Fig. 15   Comparison of feeder response to density changes.

The basic difference is that the microprocessor does things sequentially using its memory and an instruction set (software), whereas analog and digital systems do things continuously using circuits (hardware). Incorporating functions and features into software is limited mainly by the ingenuity of the programmer and involves a large investment of time and test equipment. Once all the features are worked out they are all available all the time in every instrument. Previous analog and digital equipment required more hardware to incorporate added features. Therefore, these instruments required cost options, black boxes, external annunciators, etc., because of the high cost of putting all features in each piece of equipment.

Early microprocessor instruments tended to be hybrids utilizing a microprocessor chip to perform some logic but retaining digital circuits for basic operation. Second-generation systems can utilize the microprocessor kits presently available consisting of complete standard boards and a high-level language system for programming ease. However, full utilization of microprocessor capability requires the use of basic machine level programming to achieve 100% software implementation with the resultant hardware economy. A microprocessor instrument provides external possibilities for operator interfaces because it has an alphanumeric display and a keyboard where the operator and instrument can "talk to each other." Figure 16 is a typical front panel utilizing two displays and two keyboards. In general, the upper display and the left keyboard provide normal operation, while the lower display and the right keyboard are for communication. A single instrument as shown provides the following capabilities as a base unit:

1. One instrument serves as a system for several basic functions—weigh feeder, belt scale, or solids flowmeter—using the same hardware and software package.

2. Automatic calibration is possible. Given the calibration instructions, the instrument automatically puts the machine through a prescribed calibration test and corrects for any error within a normal range. The amount of the correction is retained for future reference.

3. There are three independent operating modes for feeding applications. One mode is a fixed volumetric mode for an emergency or for troubleshooting. The other two modes are normally defined at start-up or may be varied at any time as desired, including any combination of set-point source, start-stop source, display units, batching, damping, resetting source, and output signal, and as a ratio or cascade control system. Input and output signals can be either analog, binary-coded decimal (BCD), or serial.

4. Microprocessor systems inherently provide extensive possibilities for logic interlocks and self-

Fig. 16 A microprocessor-based instrument for continuous weighing systems. *(Schenck Weighing Systems.)*

testing. Standard schemes usually include permissives from upstream and downstream equipment interlocking as desired to start or output only. Self-testing includes the load cell and associated analog electronics, the speed sensor, and any serial data links.

5. All faults may be programmed independently to either shut down or just annunciate.

6. The displays of rate, weight, and speed are individually adjustable (damped) to suit the operator.

7. The outputs of rate, weight, and speed are available as analog, BCD, or serial with individually adjustable damping. The set point is also available as output.

8. Set-point and rate displays, as well as two independent totalizers, can be set to any units and registration.

9. Maximum and minimum trip points are available for rate, weight, and speed. Each point is individually adjustable and annunciated and can function as a warning or alarm.

10. Rate deviation limits from set point and the normally associated time delay are fully adjustable.

11. The control constants of proportional band and reset are implemented in software, providing exact mathematical modules for direct keyboard entry.

12. Sophisticated prefeed control is provided, including speed proportioning with weight override correction. Volumetric or preset start-up modes provide total automatic operation on long-lag systems.

13. Batch feed control includes internal or external set point, remote reset and start, fast feed and dribble feed adjustments, dribble point, and cutoff point settings.

This list of possible functions and features will be extended as new software is developed. As previously mentioned, software development is a very expensive undertaking for the manufacturer. The software determines exactly what features are available, and "minor" alterations for a given job are simply out of the question. Thus the flexibility of the microprocessor must be in the software; it is not inherent.

Calibrating and Testing Gravimetric Feeders

These methods are essentially the same as those used for belt conveyor scales. Material testing is considered the most accurate.

Applications of Gravimetric Feeders

Proportioning Systems

Some situations require adding a small quantity of additive material to an uncontrolled flow of bulk material. A belt scale may be used to measure the "wild flow" stream whose rate signal feeds the

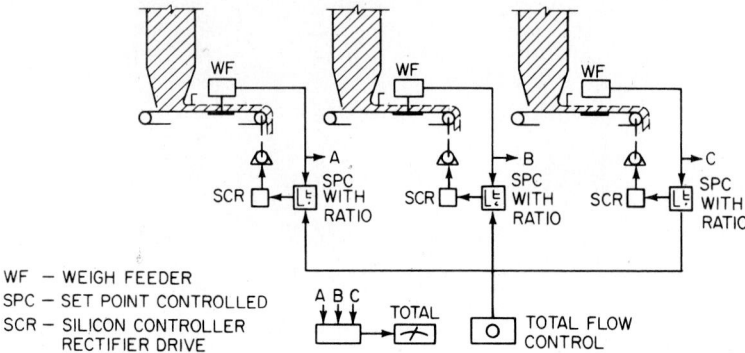

Fig. 17 Multiunit proportioning system involving the simultaneous blending of three bulk materials.

ratio input circuit of a set-point controller which incorporates a ratio adjustment that establishes the ratio of additive to wild flow. An increase or decreaes in the wild flow rate produces a corresponding change in the additive flow rate of the feeder, thereby maintaining the correct proportion. More than one additive may be involved in such a system. The wild flow may be of a solid or a liquid; the additive materials may be solids or liquids.

Multiunit Proportioning Systems

The system shown in Fig. 17 incorporates a master control that permits increasing or decreasing the total flow rate of the system. This may be accomplished in several ways: (1) a manually adjusted power supply that provides a master reference signal to feed the set-point controllers for each feeder, (2) a demand signal generated by a primary control loop designed to maintain a process variable that is affected by the total feed from the system, or (3) a total flow control loop that maintains a uniform total flow by summing the individual feed rates. The last mentioned provides feedback to a total flow set-point controller which in turn provides the demand signal to the individual feeder controllers.

Set-point controllers often are graduated in percent of total feed, permitting the operator to preset and control the system on a *percent formulation*. A summing amplifier normally is used with a percent system to sum the outputs of the individual feeders and to provide a total rate signal that is displayed on a total rate indicator.

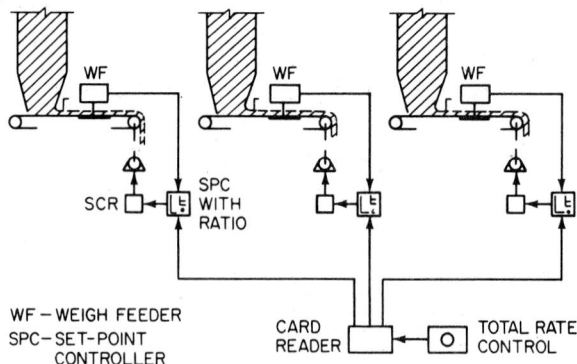

Fig. 18 Multiunit proportioning system with a card reader.

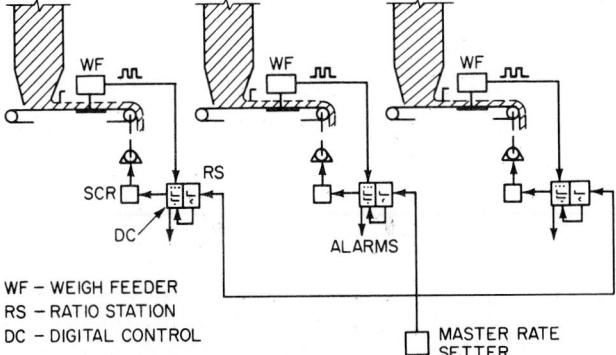

Fig. 19   Multiunit digitally controlled bulk blending system with a memory.

Multiunit Proportioning System with Card Reader

In the system shown in Fig. 18 card readers are used. This situation involves a large number of formulations for custom blending applications. Prepunched cards establish the percent of each ingredient in the blend. The outputs from the card reader are cascaded into each of the set-point controllers as the set-point signal for the individual feeder. An advantage of this system is elimination of operator error in adjusting the controller set points.

Multiunit Digital Control with Memory

Digital blending systems provide the most accurate means for automatic, continuous control flow of bulk materials and are dependent entirely on the accuracy of the digital pulses transmitted from the individual weigh feeders—with a control accuracy of plus or minus one pulse. In the system shown in Fig. 19, the output signal from the weigh feeder is a series of pulses whose value is a specific increment of weight (such as 2 lb, 20 lb, and so on). The pulse frequency (pulses per minute) is directly proportional to the flow rate. Thus, 200 pulses per minute, with a pulse value of 10 lb, equals 2000 lb/min.

These weighted pulses are compared with a set-point frequency established by a ratio station which receives digital pulses from a master rate setter, the latter establishing the total flow rate of the system. The ratio adjustment normally is calibrated in percent and is adjusted by the operator for the desired percentage of total flow. The controller compares, pulse for pulse, the set point with the actual rate frequencies, accumulating and storing the error in a memory bank. Any count difference between the measured and demand quantities is converted into a signal for controller action. The controller sends an output analog signal to a silicon controlled rectifier drive to speed up or slow down the feeder belt. Continuous accumulation of an error count in the controller memory triggers preset memory points for alarms or shutdown of the system. The net effect is a reduction in the total error to zero over a limited time period, as shown in Fig. 20. The comparison is between analog and digital control.

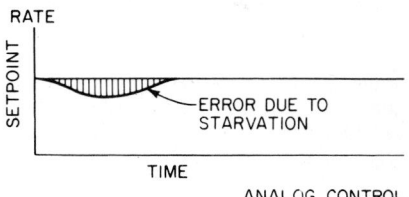

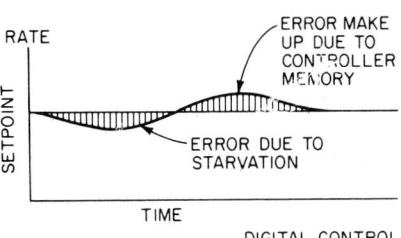

Fig. 20   Comparison of analog and digital blending systems showing how starvation errors are made up in a digital system by the use of a memory bank.

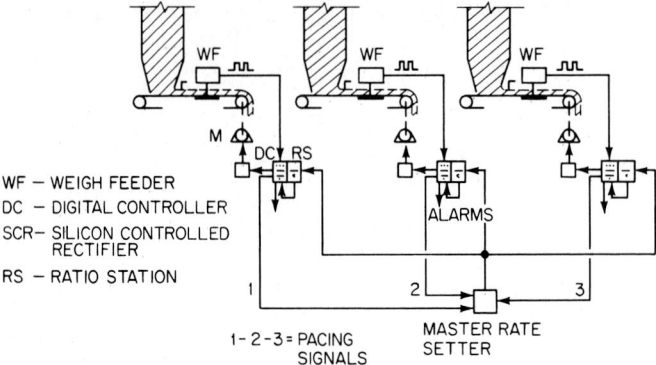

Fig. 21 Multiunit digitally controlled bulk blending system with an automatic pacing feature to accommodate a lagging feeder.

Multiunit Digital Control with Automatic Pacing

Automatic pacing provides a solution in systems where one or more materials may lag the remainder of the system at start-up because the material characteristics may produce a resistance to flow from the storage bins. This system, shown in Fig. 21, is identical to that just described, except that the controllers are equipped with a special pacing circuit that detects an underfeed condition and provides a control signal feedback to the master rate setter to reduce the master reference signal, thus slowing down the rest of the feeds and permitting the lagging feeder to keep pace and thus prevent "off spec" blends.

Multiunit Control with Computer-Operated Controllers

Some systems incorporate a combination of analog and digital control. Analog controllers are used to control the process—with a computer backup for computing and optimizing set-point settings. Operation of the individual control loops is identical with that of the previously described analog systems, except that the controllers are designed with motorized set-point potentiometers which are adjusted by the computer and equipped with slidewires to provide computer feedback to indicate the setting made. The total feed rate may be adjusted independently of the computer through a master rate setter or separate demand signal from a primary control loop. The weigh feeder transmits an analog signal to the controller and a digital signal to the computer. The set-point controller has a computer bypass switch that permits set-point settings independent of the computer. Some of these systems incorporate continuous analyzers that feed back composition information to the computer.

INTEGRATED CONCEPT OF FEEDER AND BIN DESIGN

An effective feed control system obviously requires a constant supply of material. Too often, bin designs are based on storage capacity requirements without regard for material flow characteristics. The only way to ensure an accurate controlled flow is to design an integrated system that considers both the bin and the gravimetric feeder design. The familiar flow problems of doming or arching of material over the bin discharge opening and of piping or rat-holing which occurs when the material forms about a stable void in the shape of the bin outlet are the result of failure to consider material characteristics in the initial design of the bin.

Careful study and evaluation of the following material properties is a prerequisite to the design of any bulk material blending system:

Density. The material may have a wide range of bulk densities which may be caused by variations in particle size, chemical composition, moisture content, and a packed or aerated condition.

Size. A screen analysis should be made to determine a percentage breakdown of the particle size.

Moisture Content. High moisture content and a large percentage of fines indicate that the material may tend to arch or rat-hole in the storage bin.

Temperature. Many materials are very free flowing at high temperatures, but when allowed to cool at rest, they may bridge and cease flowing owing to material arching.

Special Properties. Some materials require special consideration with regard to construction of system components—because they are corrosive or abrasive, or stick, or are of a flammable or explosive nature.

Shear Tests

Generally, if a material is coarse (+60 mesh), primary consideration should be given to maximum lump size because it is doubtful that a coarse material will bridge or rat-hole in the bin. If the material contains a high percentage of fines, a shear test should be made to determine the critical bin dimensions. To simulate actual bin storage conditions, the material sample is preconditioned and allowed to remain under pressure for periods of time equal to the at-rest conditions expected in actual operation. This procedure simulates the conditions that cause the material to gain strength and to approach the no-flow conditions of doming or piping. The net result is determination of the material flow function which, when plotted against the flow factor of a specific hopper design, will establish the parameters required to ensure flowability of the solid. Hopper factors have been computed and are published in Ref. 7.

Slotted-Type Weigh Feeder versus Table Feeder

The effective choice of equipment is illustrated by a "before-and-after" case history. Because the requirements of the process under consideration called for a controlled flow to feed a wet ore concentrate, a belt scale was installed on the conveyor following a table feeder, as shown in Fig. 22. To reduce the transport lag between the scale and the table feeder, a relatively small-width conveyor was selected and run at a fairly high speed. Because of the high speed of the conveyor and the slow revolution of the table, there resulted a series of material hills and valleys on the belt. This produced a noisy rate signal which, for the purpose of control, required damping to smooth out the output rate signal of the belt scale. This damping of the scale signal virtually eliminated the initial intent of reducing the transport lag and improving the system response.

The situation was improved by the use of a slotted-type gravimetric feeder whose hopper configuration is shown in Fig. 23. Tests determined that the minimum outlet dimension for a rectangular opening is 18 in (~46 cm) and that the hopper slope angles must be 8 and 20° from the vertical to achieve unobstructed flow. Note that the length of the outlet is 3 times greater than the width and is tapered in the direction of feed with a relationship greater than 2:1. The taper is required to

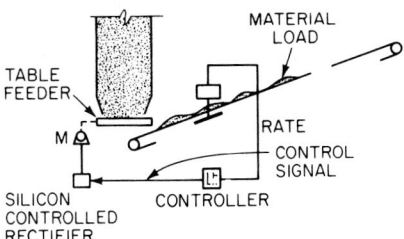

Fig. 22   Table feeder and scale control of wet ore concentrate. A system of this configuration has several inherent drawbacks.

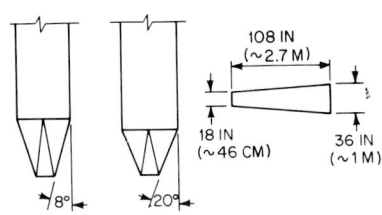

Fig. 23   Slot feeder hopper dimensions.

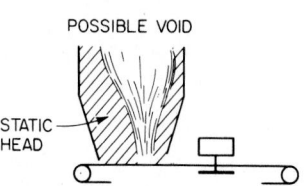

Fig. 24 Flow pattern without a tapered infeed.

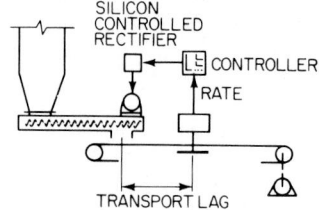

Fig. 25 Typical prefeed control system that may have several inherent limitations for handling very dry, flushy materials.

ensure uniform withdrawal across the entire length of the bin outlet. If the outlet were not tapered to provide an increasing volumetric capacity in the direction of flow, the material would flow from the front or rear section of the infeed, causing a dome or pipe to form, as illustrated in Fig. 24. This would result in a nonuniform flow and also in a static head of material that would increase the horsepower required to drive the feeder belt. The horsepower required for the slot-type gravimetric feeder in many cases may be only one-fifth of that required for a comparable table feeder. Advantages of the redesign included greater accuracy, wider range, improved system response, reduced cost, and lower maintenance and operating costs. Because downtime caused by plugged bins was eliminated, production was increased.

Free-Flowing Flushy-Type Materials

Aerated materials often flush through most standard bin designs. These materials usually are bone dry and very fine (−200 mesh) and are in many cases conveyed pneumatically by the bulk handling system. Unlike the sluggish wet ore concentrate of the prior example, these materials are very free flowing and often behave like water, having no angle of repose. Typically very free flowing materials are ground phosphate rock, ground petroleum coke, raw cement, and bentonite.

Unsuccessful attempts to control these materials were made by the use of rotary-vane and screw feeders. The problems encountered were not caused by these feed devices but rather were directly related to bin design.

Mass-Flow Bin versus Screw Prefeeder (See Fig. 25)

Consider a typical control system that utilizes a prefeed screw to control the flow of material to a gravimetric feeder. The screw feeder will be sized on the basis of capacity without regard for the minimum bin opening. The transport lag between the discharge of the prefeeder and the weigh platform will hinder the response of the control system. Because the bin design did not take into account the material flow factor, the material, once deaerated after a short duration, will tend to form a pipe or dome over the discharge of the bin. This causes a starved condition, and the system reacts by speeding up the screw feeder to obtain more material. This sets the stage for grave problems.

The material may not remain in the no-flow condition indefinitely, and the dome or pipe will fail, causing the material to become aerated—flushing through the screw and onto the weigh belt. The situation is aggravated because the screw speed now is many times that required to deliver the proper rate of flow. When the flushing condition is sensed by the weigh platform, the controls respond by severely reducing the screw speed to the point where the screw comes to a complete stop. Hopefully, the flushing will subside and the entire content of the bin will not be emptied onto the weigh feeder and the surrounding area. The result of the slowdown of the screw produces a deficiency of material which causes the controller to again increase the speed of the screw. This cycle may be repeated indefinitely until the operator resorts to manual control or the controller eventually gains control until the next no-flow condition occurs.

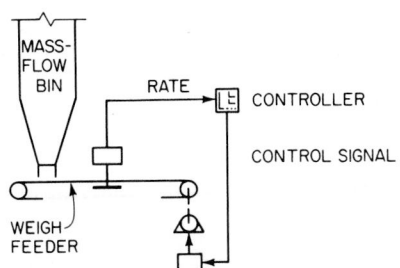

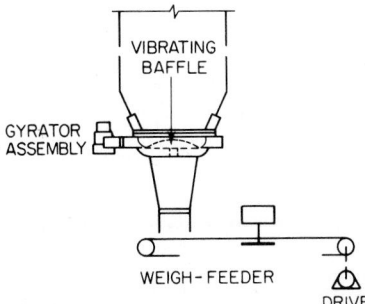

Fig. 26 Integrated concept of feeder design using a mass-flow bin.

Fig. 27 Use of bin activator with a weigh feeder.

At this point, several attempts to modify the system are made in an effort to eliminate or retard the flushing condition. Modifications of the screw usually are the first remedy considered. Changing the pitch of the flighting or additional flights sometimes reduces the flushing action, but this also reduces the capacity and does not solve the no-flow problem. Vibrators applied to the bin may help, unless it is found that they worsen the condition by packing the material. High-amplitude low-frequency vibrators usually work best. Air pads also help if they do not introduce excessive amounts of air. They may require readjustment from time to time as changes in the material characteristics occur.

The same material can be fed directly from a mass-flow bin to the weigh feeder, as shown in Fig. 26. A prior flow factor test would have indicated the required minimum opening and the slope of the hopper required to achieve mass flow. No prefeeder is required, and the transport lag in the system is eliminated. Improved system response eliminates the oscillations in the rate of flow. The mass-flow bin deaerates the material and also produces a more uniform density at the outlet of the bin, which reduces the burden placed on the control system.

Although the mass-flow bin appears to be a simple solution, certain aspects of the design must be considered. The slopes required usually are steep, 10 to 30° from the vertical. If the required storage capacity is large, the bin will have to be very high. However, the capacity of the mass-flow bin does not represent dead storage. The minimum retention time in the hopper should be at least 15 min to allow the material to deaerate. A cutoff gate should be provided between the discharge of the bin and the gravimetric feeder inlet to permit filling the empty bin.

If storage capacity is a problem, a flow promoting device, such as a bin activator, shown in Fig. 27, may be considered. A primary advantage of this is overcoming the problem of piping or doming by providing a large outlet in the bin. Introduction of vibration into the material also tends to deaerate it. The bin activator shown in Fig. 27 does not control the flow, but aids it. The bin activator shown in Fig. 28 serves as a flow aid device and a control feeder. The unit, operating on the principal of controlled vibration, increases or decreases its flow rate on command from the weigh feeder.

Loss-in-Weight Feeder

The gravimetric feeder design shown in Fig. 29 utilizes the loss-in-weight principle of operation. The weigh hopper and the metering screw are supported on load cells. As the metering screw delivers material from the system, the resultant load cell signal decreases proportionally with time; the differential dG/dt produces the true feed rate F as graphically illustrated. Thus the flow rate of material to the process may be controlled by varying the speed of the metering screw to maintain the true feed rate equal to the demand.

When the material in the weigh hopper reaches a predetermined lower level G_{min}, a fill cycle is initiated. During the fill cycle the system runs volumetrically. The filling system requires special engineering to provide fast filling cycles to keep the filling time (volumetric feed) short relative to the normal gravimetric feeding cycle. A minimum ratio is about 10:1 to maintain accuracies of ±1% or better.

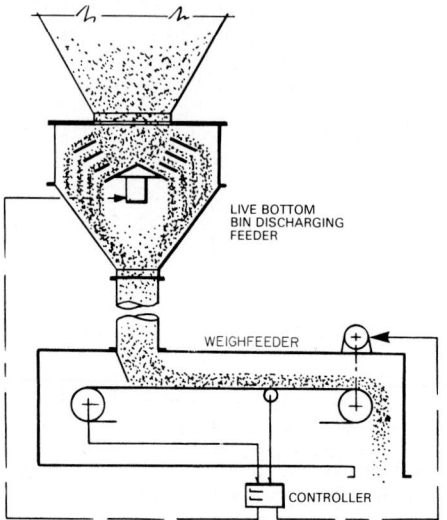

Fig. 28 Bin discharging feeder controls the flow rate on demand from the weigh feeder. *(Solids Flow Control.)*

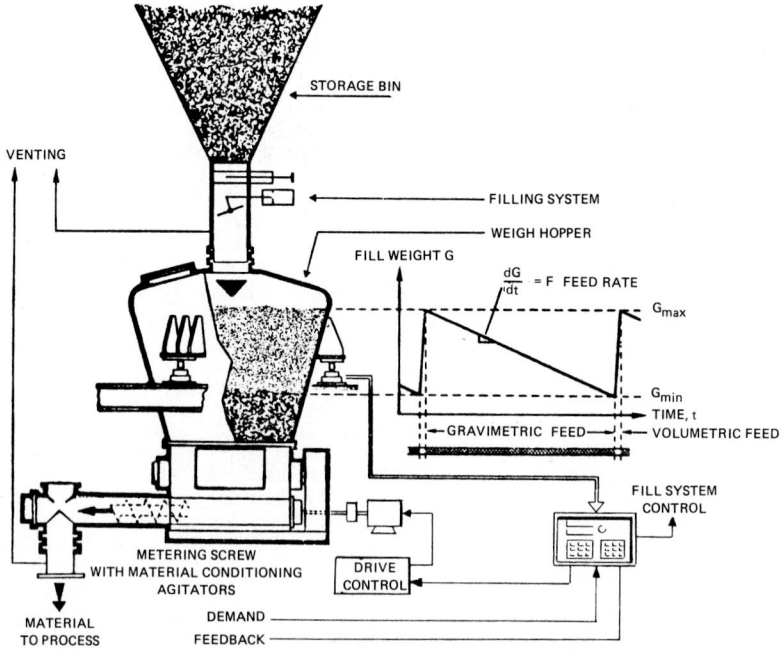

Fig. 29 Gravimetric feeder using the loss-in-weight principle. *(Schenck Weighing Systems.)*

Loss-in-weight systems provide excellent static accuracy and isolation from the atmosphere. Therefore a major application for these systems is in feeding pulverized coal in indirect firing systems. The basic feed system is easily engineered to meet the Boiler Code (85°F); however, the required flexible connections at the inlet and discharge are critical to weighing accuracy and are not commercially available. Venting is critical to fast filling and also with regard to settling times for the resumption of gravimetric feed after filling.

Microprocessor instrumentation is inherently adaptable to loss-in-weight feeding systems because of the basic cyclical operation, and the availability of this instrumentation has increased the flexibility and application of these systems.

Solids Flowmeter

Solids flowmeters are not true gravimetric devices but rather utilize impact or centrifugal force to generate a signal proportional to material flow. The basic requirement for a solids flowmeter arises from operating problems encountered with a belt scale or weigh feeder on dry, flushy (fluidlike) materials involving high flow rates, heat, environment, or explosion-proof requirements. A solids flowmeter is a sealed system which works best with flushy materials acting like fluids and is not inherently limited in upper capacity.

Typical materials for flowmeter applications are cement, raw meal, pulverized coal, bentonite, filter dust, and fly ash. The solids flowmeter is essentially an incline system with the measurement taken as the material falls through an opening. There is a height requirement to establish a stable flow velocity. The flowmeter requires a delivery system or prefeeder such as an air slide or screw, and a feed control system can be implemented by closing the loop with a valve or variable speed drive.

The unit is capable of providing long-term repeatability in the 2 to 3% range. Since the flowmeter is not a true gravimetric device, calibration to actual material to better than 5% requires a material test sample. For this reason, a static weighing system is frequently used in conjunction with a flowmeter to eliminate the difficulty of material sampling and provide in-line calibration on a periodic basis.

Figure 30 illustrates the flowmeter system with a closed loop for rate control and in-line calibration.

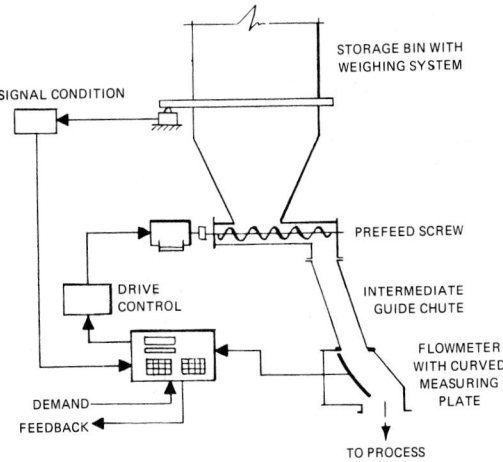

Fig. 30   Flow rate control system utilizing a solids flowmeter with an in-line calibration system *(Schenck Weighing Systems.)*

REFERENCES

1. Gilmore, D. R., and A. L. Kurylchek: "Maintenance of Belt Conveyor Scales and Weigh-Feeders," Minerals Processing Conference, Chicago, September 1966.

2. Small, R. L., and H. Colijn: "Belt Scale Weighing Can Be Accurate," ISA 19th Annual Conference, Paper 11.3-1-64, New York, October 1964, Instrument Society of America, Pittsburgh, Pa.

3. Gilmore, D. R.: "Weigh-Feeder Control Systems and Design," Rock Prod., 2d Annual Cement Conference, Chicago, November 1966.

4. Gilmore, D. R., and A. L. Kurylchek: "Efficient Weigh-Feeding," *Automation,* vol. 17, no. 2, pp. 77–79, Febraury 1970.

5. Gilmore, D. R., and A. L. Kurylchek: "Maintenance of Weigh-Feeders for the Cement Industry," IEEE Cement Industry Technical Conference, St. Louis, May 1968, Institute of Electrical and Electronic Engineers, New York.

6. Gilmore, D. R.: The Integrated Concept of Feeder Design, ISA 22d Annual Conference, Paper 41-2-MCID-67, Chicago, September 1967, Instrument Society of America, Pittsburgh, Pa.

7. Jenike, A. W.: "Storage and Flow of Solids," *Utah Eng. Exp. Sta. Bull.* 123, University of Utah, Salt Lake City, November 1964.

8. Colijn, H., and P. W. Chase: "Belt-Scale Instrumentation for High-Tonnage Conveyors," *Instrum. Technol.,* vol. 14, pt. 1, no. 5, pp. 64–68, May 1967; pt. 2, no. 6, pp. 75–80, June 1967.

9. Marhauer, H. H.: Weighbelt Dynamics, *Chem. Eng. Mag. Feature Rep.,* August 17, 1964, and September 14, 1964.

10. Considine, D. M.: Process Weighing, *Chem. Eng. Mag. Feature Rep.,* August 17, 1964, and September 14, 1964.

11. Kurylchek, A. L.: "Feeding the Feeder," Conference on Coal Feeding Systems, June 21–23, 1977, Jet Propulsion Laboratory, California Institute of Technology, Pasadena, Calif.

12. Kurylchek, A. L.: "Problems and Solutions in Bin Discharging and Feeding Difficult to Handle Powdery Materials," 70th Annual Meeting of the American Institute of Chemical Engineers, November 13–17, 1977, New York.

13. Wilson, D.: "Weigh-Feeder Control Needs Favor the Digital Route," *Control Eng.,* vol. 23, no. 1, 1976, p. 48.

14. Van den Berge, H., Klasaens, H. A., and A. Kopmels: "Weighing On-the-Fly Keeps the Process Moving," *Control Eng.,* vol. 23, no. 9, 1977, p. 52.

Continuous Nuclear Weigh Scales

Particularly for certain severe process environments and materials characteristics where it is required to measure the mass of bulk materials on the move, as on a conveyor, nuclear weigh scales are frequently the method of choice. These mass measuring systems require no contact with the moving materials. Representative of the materials and processes with which a nuclear scale may be used are hot, abrasive, dusty, corrosive, and difficult-to-handle materials encountered in the mining and benefician field (coal, iron ore, copper concentrates); in the pulp and paper industry (wood chips, sawdust, and fines); in the chemical and fertilizer industry; in the building materials field (sand, rock, lime, cement); and in some food processing plants (grain, vegetables, fruits). As with other modern weighing devices, nuclear scales are compatible with computers and sophisticated electronics and can tie into complex batching, formulating, ratioing, blending, and materials inventory control systems as may be required. Although pioneered for use in connection with belt conveyors, nuclear scales are used today with other conveyor configurations (screw, apron, air slide, drag chain, vibrating).

Principle of Operation

The mass of the material being measured is determined by the quantity of gamma ray radiation absorbed by the moving material. The applicable equation is

$$I = I_R \exp \frac{-\mu L}{W}$$

where I = radiation transmitted to detector
I_R = radiation transmitted to detector with empty belt
μ = radiation absorption coefficient
L = loading on belt in mass per unit belt length
W = belt width

The simplified diagram in Fig. 1 illustrates the relationship between belt loading and the radiation level at the detector. The actual configurations of the nuclear scale vary somewhat from one supplier to another, but the steps or elements of the measurement system are similar and include:

1. *A radioactive source emits gamma energy.* The radiation source used most commonly is cesium 137 (^{137}Cs), but a combination of cobalt 60 and americium 241 (^{60}Co and ^{241}Am) may be used in some cases. A license to use these radioactive substances in the United States must be obtained from the U.S. Nuclear Regulatory Commission. In the case of ^{137}Cs, a scale manufacturer can obtain a general license and, if the device is used on any standard conveyor, the user does not have to obtain a separate license. Also, in the case of ^{137}Cs, in the United States, the U.S. Food and Drug Administration permits such scales to be used in connection with food products.

2. *Part of the emitted radiation energy is absorbed by the mass of material on the conveyor.*

3. *The remaining energy is transmitted to the detector.* Thus, less absorption by the material results in greater energy received at the detector.

4. *The radiation received at the detector is inversely proportional to the mass on the conveyor.*

5. *The detector emits an electrical signal to an electronics package where it is amplified, scaled, and calibrated in terms of weighing units.*

Two principal configurations are used: (1) the A-frame (Fig. 2) and (2) the C-frame (Fig. 3). In both configurations, the gamma radiator is mounted above the conveyor and the detector is mounted below the conveyor.

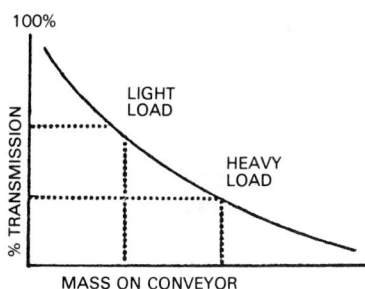

Fig. 1   Loading signal versus conveyor loading.

Fig. 2   A-frame configuration with a gamma radiator above the belt and a detector below the belt. *(Kay-Ray.)*

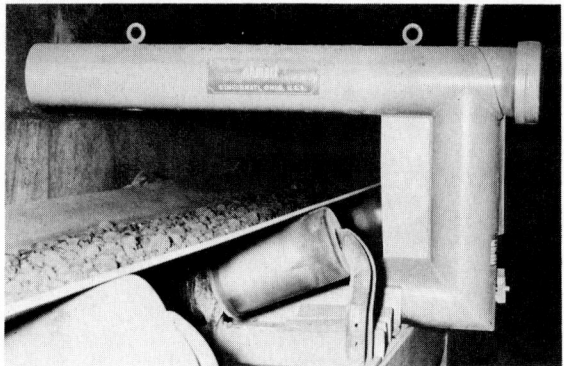

Fig. 3 C-frame configuration with a gamma radiator above the belt and a detector below the belt. *(Ohmart.)*

To relate the gamma radiation absorption measurement to mass per unit time, a conveyor speed transducer (usually a tachometer) measures the belt speed. These measurements then can be fed into a totalizer to indicate or record the total mass that has passed through the scale. The system is shown schematically in Fig. 4. A more detailed description of one type of system is given in Fig. 5.

Accuracy

Typically, under representative industrial conditions, accuracy ranges from ± 0.50 to $\pm 0.75\%$ of full scale. Major factors that affect accuracy include conveyor loading (principally sufficient loading on the belt to avoid high signal-to-noise ratios) and uniformity of the material being conveyed.

Loading*

A minimum 2.6 lb/ft^2 (12.7 kg/m^2) loading establishes that the material on the belt will provide a great enough radiation change for the system to accurately measure. This minimum loading will produce a radiation change of approximately 5%; a lower belt loading will degrade repeatability.
Conveyor loading is determined by the following equations:

$$L = \frac{33.3 \times T}{R \times W} \qquad L = \frac{0.274 \times T}{R \times W}$$

	lb/ft^2	kg/m^2
L = belt loading	tons/h	metric tons
T = maximum mass flow on belt	ft/min	m/s
R = maximum speed of belt	ft	m
W = width of belt		

Maintaining a belt load between 70 and 100% of full load ensures that errors due to system response time and unrepeatable profile changes are kept to a minimum. Operational belt loadings that constantly change more than 30% require that the specified repeatability be derated.
The system integrates the mass passing through the measurement point with respect to time. When belt runs become short, errors due to system time reponse become more significant. If belt runs are typically less than 10 min long, the scale manufacturer should be consulted.

*Extracted from a paper by Craig McIntyre, Kay-Ray, Inc., Arlington Heights, Illinois.

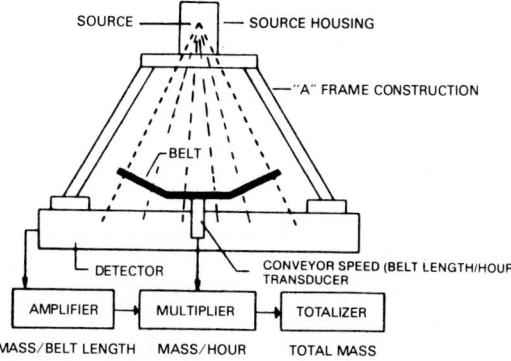

Fig. 4   Principal elements of a continuous nuclear weighing system.
Mass/belt length × belt length/hour = mass/hour. *(Kay-Ray.)*

Nature of Product

It is important to determine whether only one product will be weighed by the scale. Dissimilar products can have radiation absorption characteristics sufficiently different to affect scale calibration. There are cases, for example, where a belt may convey limestone and coke (or two other different materials) at different times. The scale can be calibrated to measure more than one product accurately, but this requires factory consultation.

The nuclear belt scale is sensitive to large variations in product profile. As the average particle size of a product varies, so does its profile for a given belt loading. Since particle size changes are usually related to bulk density changes, the following equation may be used to estimate the effects on basic belt scale accuracy due to particle size variations.

$$A_p = 1 - \left(\frac{D_{min}}{D_{max}}\right)$$

where A_p = the repeatability factor to be added to the basic ±0.5% repeatability
D_{min} = minimum bulk density
D_{max} = maximum bulk density

Other factors that can greatly influence product profile (such as large feed variations and large product lumps greater than 15% of belt width) require examination by scale factory engineers.

A change in product moisture content may also result in an erroneous scale reading. This error is caused by the change in radiation absorption characteristics as water displaces material and is easily calculated. (Up to a ±5% change in moisture content has been included in the basic ±0.5% total accuracy figure.)

$$A_m = \frac{(M_{max} - M_{min}) - 5}{15}$$

where A_m = repeatability factor to be added to basic ±% repeatability
M_{max} = maximum percent moisture by weight
M_{min} = minimum percent moisture by weight

Standard and Auxiliary Equipment

The basic nuclear weigh scale can provide several signal outputs, including:

1. **Loading Signal (*mass/length*).** All weigh scales have an analog output that is proportional to the belt loading. For applications where there is no speed transducer input (belt speed is assumed constant), the loading signal is also the mass-flow (mass/time) signal.

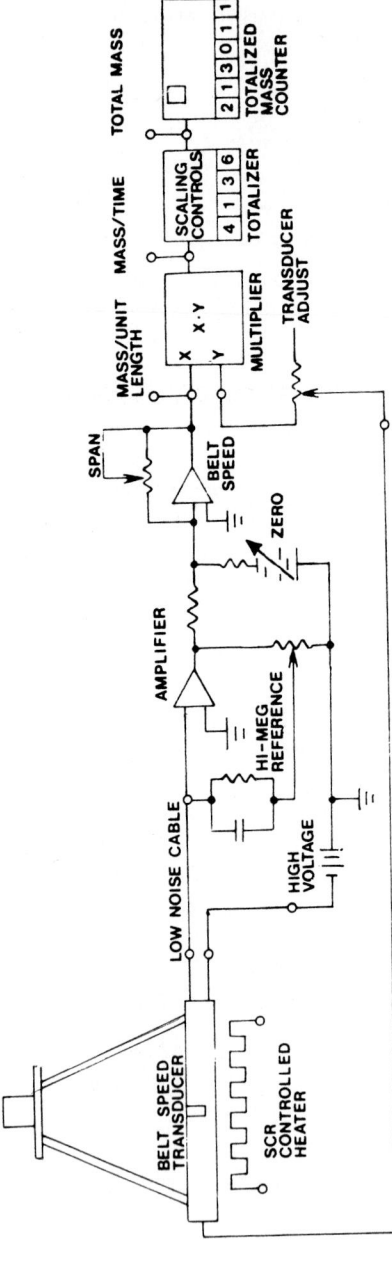

Fig. 5 Detailed diagram of a nuclear weigh scale system. The ion chamber detector under the belt reacts to the changing intensity by producing a small proportional current. This current, typically 10 A, is fed via a special low-noise coaxial cable to a Hi-Meg resistor-amplifier. Because the Hi-Meg resistor is many times greater than the reference feedback pot, essentially all of the resulting voltage appears across the Hi-Meg. The weigh scale is referenced with the belt running empty. The reference control is set to produce 10 V out of the first stage. The second control, zero, is adjusted to add an equal but opposite voltage, giving a net input (and output) of 0 V at the second stage. The transducer adjust control is set to input 10 V into the multiplier; this represents the top speed of the belt. The multiplier outputs 0 V into the totalizer. The up-down totalizer mass counter with nothing to count remains at zero.

When material flows on the belt, more radiation is absorbed, causing the ion chamber current to decrease. This results in less voltage across the Hi-Meg, so that a net positive voltage appears at the input to the second amplifier. The gain of this stage is set by the span pot so that 10 V is input to the multiplier. The multiplier outputs 10 V into the totalizer where the scaling controls are set to cause the 10 V to be translated into the desired units for the up-down totalized mass flow counter. When the belt is stopped and/or empty, there is no output from the multiplier, thus no totalized flow. For optimum stability, the ion chamber is heated to a constant 60°C by an SCR-controlled heater. *(Kay-Ray.)*

Fig. 6 Display used on a digital nuclear weigh scale. *(Kay-Ray.)*

2. Mass-Flow Signal *(mass/time)*. Weigh scales provided with a belt speed transducer (generally where the belt speed variation is greater than ±0.25%) have an additional analog mass flow signal corrected for belt speed variations.

3. Totalizer Output Mass *(weight)*. A totalizing output giving the amount of mass that has passed through the weigh scale.

Additionally, some nuclear weigh scales can be equipped with pneumatic outputs, high or low alarm outputs, resettable counters, and predetermining electronic counters.

A scale with a digital readout is shown in Fig. 6. The electronics enclosure for wall mounting of one type of scale is shown in Fig. 7.

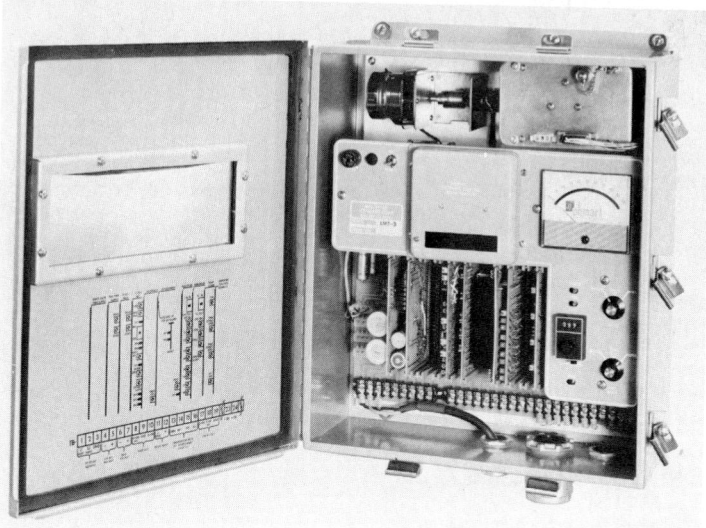

Fig. 7 Electronics enclosure for a wall mounting furnished with one type of nuclear weigh scale. *(Ohmart.)*

Nuclear radiation detectors are described in some detail in the article Nuclear Radiation Instrumentation in Sec. 12 of this Handbook.

Automatic Blending Systems

by
Mel Berman*

The need to blend various ingredients in pots or vats dates back to antiquity. Over the years, various means of combining liquid or powder components in preprogrammed sequences and amounts were devised, including bucket trains, sprocket gears, water wheels, and variable speed gearing techniques. Later came pneumatic controllers which could control the flow rate of a particular component. Proportioning techniques were developed utilizing mechanical and pneumatic devices. Electronics introduced analog amplifiers-comparators coupled with accurate metering devices and preset electromechanical counters for multiple-component flow control and ratioing. Further developments brought forth digital techniques with integrated circuits performing precise measurement, comparisons, and control of multiple-stream blending processes. Present microcomputer technologies and sophisticated programmable controllers find wide application in a variety of industries for bringing together multiple components in a precise and controlled manner. Applications are found in the petroleum, petrochemical, food and beverage, building materials, pharmaceutical, automotive, and chemical fields, among others.

Batch versus Continuous Blending

In a batch-type process, a recipe is followed by adding specific amounts of ingredients in a predefined sequence with mixing, stirring, and brewing, or processing times between the addition of each component. This practice, for example, was followed for many years in the manufacture of lubricating oils through programmed mixing, stirring, and heating of various hydrocarbon components and additives in large batch vats and allied equipment. This procedure has largely been replaced in many industries by *in-line blending* which essentially refers to a process whereby component streams (liquids, gases, powders, or aggregates) are measured and controlled in a precise relationship or ratio to each other. All components flow together simultaneously to a central collection point where they combine to form the finished product, such as a lubricating oil. The obvious advantage of in-line blending over the batch-type process is that large vessels for the mixing of components are eliminated. The blend header, augmented by static or active in-line mixers, is all that is required to form the final product. The finished product can go directly to a canning line, to a finished product storage tank, or into a pipeline for distribution.

Typical Blend Configuration

A modern in-line blending scheme is shown in Fig. 1. The blend controller nominally utilizes microprocessor technology with a cathode ray tube (CRT) display. Each fluid component is pumped from

*Waugh Controls Corporation, Chatsworth, California.

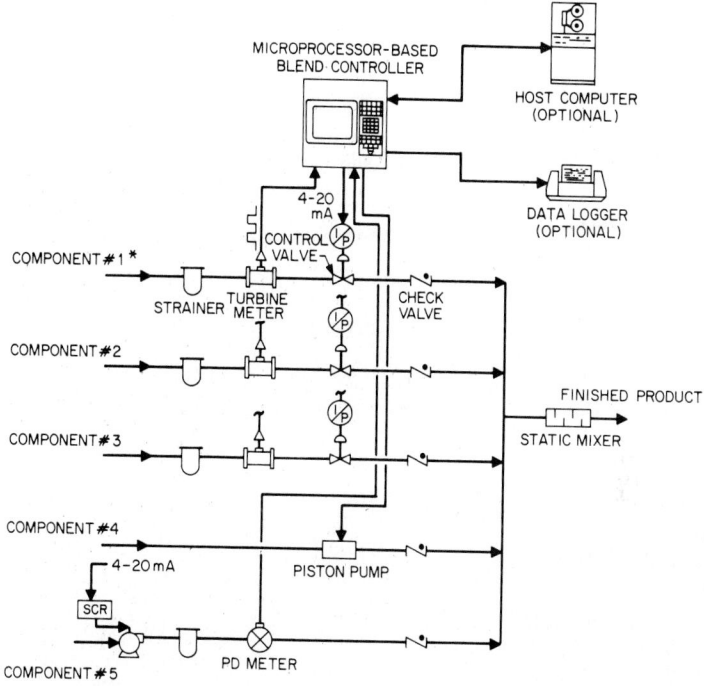

* Up to 16 closed or open-loop configurations may be used.

Fig. 1 Typical blender configuration. *(Waugh Controls.)*

a storage tank, through a strainer, and then through a flowmeter, with meters and valves selected for prevailing process conditions (viscosity, temperature, pressure, flow rates, etc.). The signal from the flowmeter is fed to the blend controller which compares the actual flow rate to the desired flow rate. If the actual flow rate is incorrect, the blend controller will adjust the control valve via the 4 to 20 mA signal to the valve. In this way, each component is controlled in a closed-loop fashion relative to its flow rate and ratio to the other component streams. For minor components, such as dyes or additives, it is sometimes most practical to control the flow rate by means of proportioning pumps which inject a precise amount of the fluid when a pulse signal from the blend controller is received. This type of open-loop control is cost-efficient, but some means for ensuring flow (not a dry line) should be considered inasmuch as no direct fluid measurement device is used.

Other variations of measurement and control involve the use of variable speed pump motor controllers (silicon controlled rectifiers) for flow control, adding a flowmeter in series with an injection pump, and the use of weigh belt feeders with variable feed-speed control and tachometer-load cell outputs (for powders and aggregates). See the prior article in the Handbook section on weigh feeders.

Liquid or Powder Blending

Solid materials, or combinations of solid and fluid materials, such as feeds to cement kilns, asphalt aggregates, and concrete aggregates, are readily controlled through the use of techniques similar to those employed for fluids. Flow input, in the form of pulses representing the weight of the material, is obtained from weigh belt feeders, while the 4 to 20 mA control output is used to operate a gate, regulating the amount of material fed from a hopper onto the weigh belt. Many weigh belt feeders require multiplication of belt speed by the mass on the belt in order to obtain the mass flow rate. This computation is performed by external computing modules mounted in a rack. Many food and

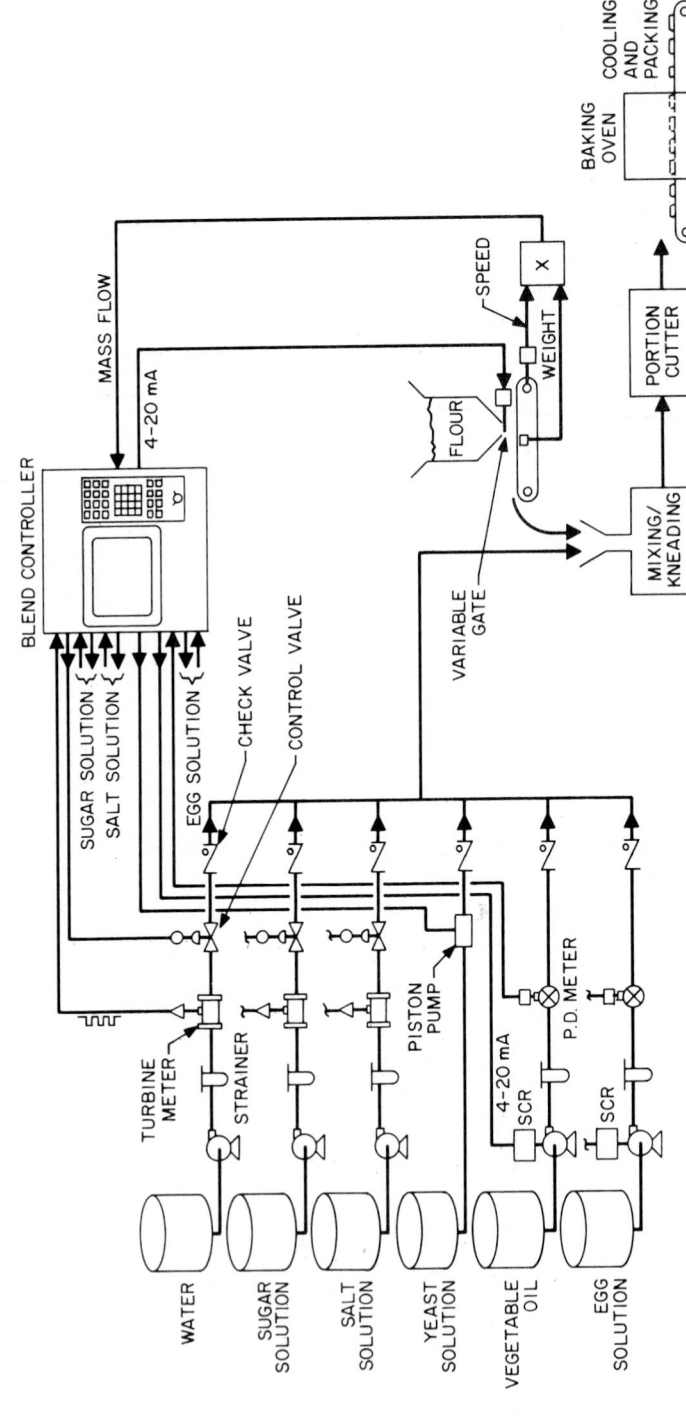

Fig. 2 Blending system for preparing bread and pastry dough. (*Waugh Controls.*)

7.64

chemical products require a combination of both liquid and powder ingredients, blended together to form the finished product or an intermediate feed to additional processing. A combination-type blending system used to make bread or pastry dough is shown in Fig. 2.

Sizing a Blending System

The first step in designing a blending system is to construct a list of all the components required to form the various products. Next, after each component, list the ratio ranges for these fluids as they relate to the final products. The combined blend delivery rate should be such that the system will fulfill all daily production needs in an 8- to 9-h shift, thus providing time for maintenance when required. Once the overall maximum delivery rate has been determined, each component stream can be sized to provide its percentage range of the total blend. The rangeability of the component streams then must be considered. This should not exceed 10:1 if turbine meters are used, or 20:1 in the case of positive-displacement (PD) meters. Possible future production rate increases should enter into sizing the component streams. However, caution should be exercised to avoid excessively oversizing control valves for the current flow rates.

Flowmeter Selection

Blend controllers are compatible with a broad variety of flowmeters and mass-flow measurement transducers used for the measurement of gases, liquids, and solid products. Most commonly used are flowmeters that produce a digital pulse frequency in direct proportion to the flow rate. These include turbine, PD, vortex shedding, ultrasonic, and magnetic flowmeters. Additionally, analog flowmeter elements, such as orifice flowmeters (using differential-pressure transmitters) and magnetic flowmeters (employing analog signal conditioners) are sometimes used.

When specifying flowmeters for blending applications, process conditions as well as mechanical requirements must be considered. See Table 1.

Positive-Displacement versus Turbine Flowmeters

Both PD and turbine flowmeters are commonly used for blending applications. PD meters are best used in applications involving very viscous fluids up to 2200 cSt or more. Turbine meters are useful

Table 1 Factors to Be Considered When Specifying Flowmeters for Blending Applications

Process conditions
Fluid or gas (common and chemical name)
Flow rate range
Pressure range
Temperature range
Specific gravity
Viscosity
Corrosive properties
Other fluid characteristics (slurries, contaminants, etc.)
Hazardous classifications
Electromechanical considerations
Flowmeter repeatability and accuracy
Line size
Flange rating
Other mechanical accessories (meter registers)
Pulser type and power requirements
Meter calibration factor (K factor)
Pulse output levels and frequency range

with viscosities at or below 20 cSt, depending on meter size. Thus, in the blending of lubricating oils, for example, PD meters are the primary measurement devices, while in the blending of gasoline and other light products, turbine meters are extensively used.

Because PD meters are constructed with very close mechanical tolerances, they are easily damaged by suspended abrasives and other foreign matter. A suitable strainer with a mesh size based on the meter manufacturer's recommendations is a mandatory requirement. Similarly, the bearings of the turbine meter are susceptible to excessive wear or damage by suspended particles and must be protected by strainers.

Positive-displacement meters do not differentiate between airflow and liquid flow. Thus, it is important to consider the use of air eliminators prior to the fluid entering the flowmeter in installations where a substantial amount of air may enter the system. Pump cavitation or air start-up pockets in the piping can cause registration in PD meters unless eliminators are used.

Turbine meters should be installed with straight pipe runs. Straightening vane sections upstream and downstream from the meter should be considered. Typical recommendations are 20 pipe diameters of upstream pipe length and 5 pipe diameters of downstream pipe length. Preamplifers mounted on the turbine meters ensure good noise immunity and a pulse transmission capability.

PD meters typically have contactors or solid-state pulse generators as output devices. Contactors should be specified as "mercury-wetted" to prevent excessive contact bounce which can be miscounted by the blend controller. Practical speed limitations of contactors are in the 30 to 60 Hz range. Solid-state pulsers provide higher pulse counts than contactors, but they are more expensive and require a dc power supply (typically $+12$ V dc). Pulsers are capable of outputting pulses in the 1- to 4000-Hz frequency range, thus providing excellent resolution of the metered volume.

Another type of PD meter pulser uses a pickup coil similar to that of a turbine meter. The coil is excited by a revolving gear in the meter stack. Care should be exercised when using this type of pickup, since it is sometimes susceptible to vibration. The meter registration gear train can be affected by local vibration (from external sources) and inadvertently cause the pickup coil to provide false output pulses. A preamplifier similar to that used with a turbine meter must be used with this type of PD meter pickup.

In the case of either a PD or a turbine meter, the meter should be used within the flow rate, temperature, and pressure ranges recommended by the manufacturer if meter damage is to be avoided. If a meter is to be used beyond its specified limitations, the manufacturer's advice should be sought.

Control Valve Selection

In most blending applications, one control valve is installed in each component stream. The position of the valve is based on a 4 to 20 mA control signal produced by the blend controller. Control valve actuators provide the mechanical forces required to throttle the valve accordingly. Selection of the appropriate control valve depends on a number of criteria, including fluid characteristics, corrosive character of the fluid, flow rate range, allowable pressure drop, shutoff pressure, and the availability of effective actuators.

Commonly used flow control valves include globe body plug valves, butterfly valves, eccentric disk valves, and ball valves. Each valve offers unique advantages for particular applications. The details of various valve designs are given in Sec. 19 of this Handbook.

Standard specifications for blending system control valves require that the actuator be integrally assembled on the valve body, together with a current-to-pressure converter, input air filter-regulator, pressure gages, and a shaft positioner if the valve size exceeds 2 in.

Valves are usually sized for a 15-psig pressure drop when fully open and at maximum flow rate, with either linear or equal-percentage characteristics. The objective in choosing the characteristic, which is the relationship between valve C_v and stem position, is to ensure a proportional change in the flow rate with the valve control current so as to maintain a constant loop gain. Linear characteristics are selected if the pressure at the input remains constant with changes in the valve opening.

Soft seats are selected when bubble-tight shutoff is required. This condition is typically encountered when the blender operates as a batch-size controller with automatic shutdown. The valve actuator must be sized to ensure complete shutoff at maximum line pressure and to allow stable control with maximum differential pressure.

Table 2 Factors to Be Considered When Specifying
Control Valves for Automatic Blending Systems

Process conditions
    Fluid or gas (common and chemical name)
    Flow rate range
    Pressure range
    Shutoff pressure requirements
    Temperature range
    Specific gravity
    Viscosity
    Corrosive properties
    Fluid contaminants
    Hazardous classifications
    Allowable differential pressure
    Desired control profile (linear, equal-percentage, etc.)
Power source
    Instrument air pressure
    Mains power supply
Mechanical considerations
    Line size
    Flange rating
    Valve type (plug, butterfly, etc.)
    Actuator style and size
    Mechanical accessories (gages, filters, positioners, etc.)

The valve time constant and slewing speed should be minimized to achieve loop stability with minimum response and correction times. Boosters are recommended when stroking times exceed 6 s. These boosters should include integrally adjustable feedback pilots, which can be used to stabilize their gain.

When specifying control valves for blending applications, the process conditions, source of power, and mechanical requirements must be considered. See Table 2.

Skid Assemblies

It is sometimes cost-effective to preassemble the flow control and measurement elements of a blending system on skid assemblies which are constructed and flow-tested prior to delivery at the blending site. These assemblies are then piped into a common blend header on the outlet side and to the component selection manifold on the inlet side. A typical skid assembly layout is shown in Fig. 3.

TYPICAL BLEND CONTROLLER

A single 19-in rack-mounted housing contains a complete blending controller for up to 16 components. Field installation is thus simplified, requiring only wiring to field equipment and to line power in order to put the system in operation.

A few circuit modules comprise the basic controller—a microprocessor module containing the processor and related circuits, a support module containing additional memory, one or two loop input-output modules containing inputs and outputs for eight control loops, and a video interface module. An analog input module is optionally used for temperature or other 4 to 20 mA inputs.

Simplicity of maintenance is ensured by easy removal and replacement of major subassemblies comprising the power supply, the video display, the plug-in circuit modules, the data cassette, and plug-in field interfacing relays. All are replaceable without removal of the blend controller from its panel mounting. A block diagram of a typical blend controller is shown in Fig. 4.

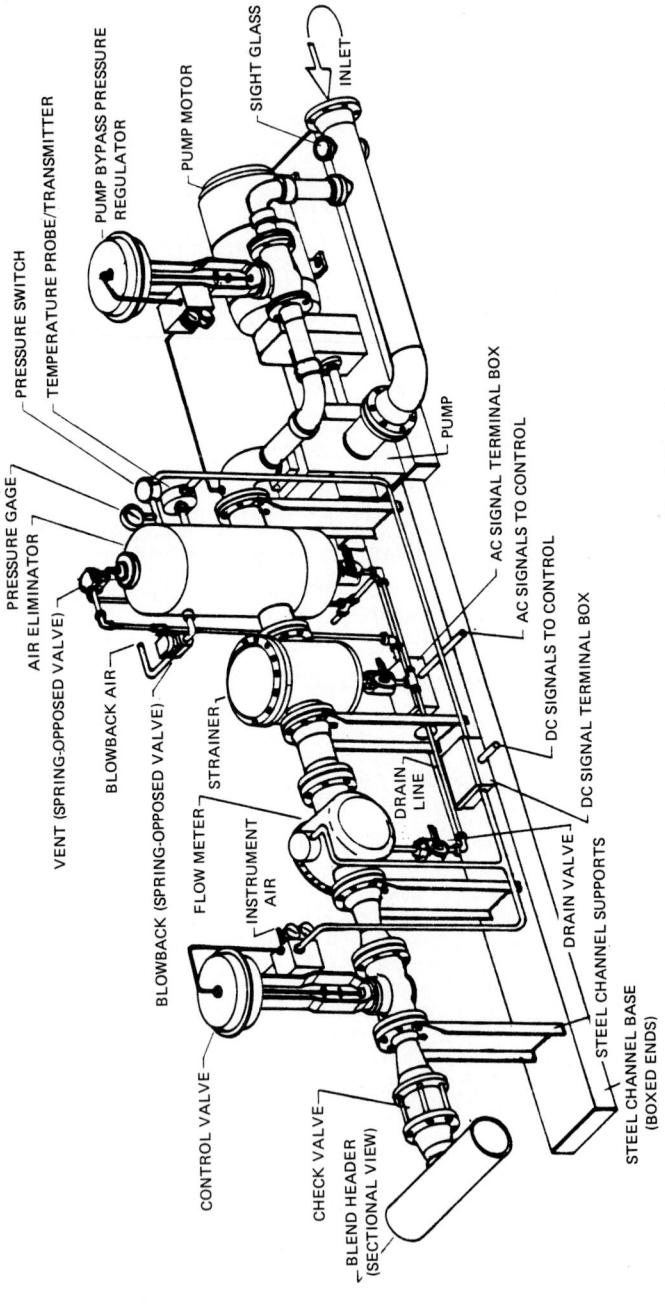

Fig. 3 Typical automatic blending system skid assembly. Only one component stream is shown. *(Waugh Controls.)*

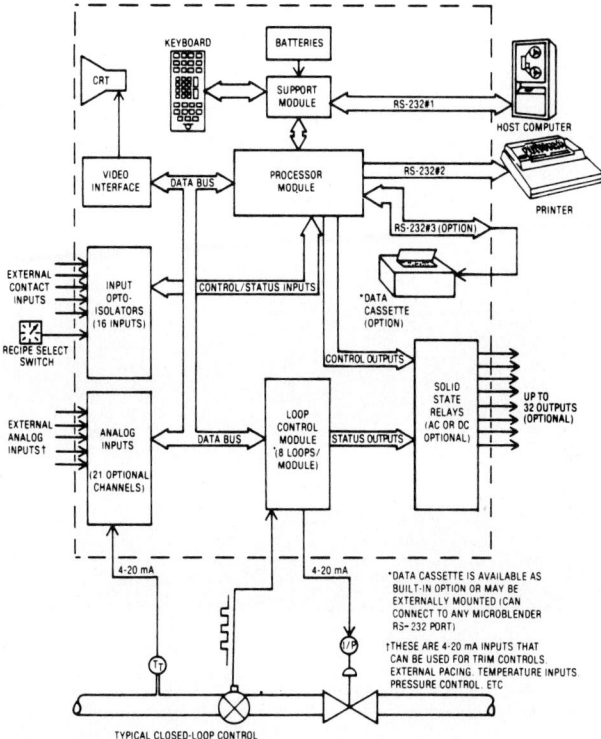

Fig. 4 Blend controller block diagram.

Blending Operation

With the assumption that compatible flow control and measurement elements have been selected and properly installed in the field, the system is ready to go on line. A typical CRT blend controller display is shown in Fig. 5. This display gives the operator all the pertinent information required to oversee the blending operation. Such data as product names, loop numbers, ratio set points, com-

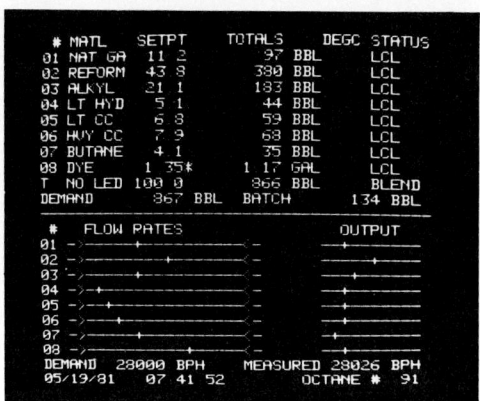

Fig. 5 Typical CRT (video) blend controller *General* display.

Fig. 6 Two tabletop blend controllers. *(Waugh Controls.)*

ponent totals, component temperatures, control status, batch total, actual combined flow rates, and totals (as well as demanded flow rates and totals) are available. In addition, a graphic display of the components' instantaneous flow rates and the associated control current outputs (4 to 20 mA) are included.

Keyboard Control

The operator configures the blend controller via the control keyboard. Such items as meter K factors, flow rate and totalization units, and product names are entered through the keyboard. Blending start-stop control and call-up of specific displays ("pages") are also accomplished by keyboard-operator interface. See Fig. 6. The details of the blender controller and keyboard panel are given in Fig. 7.

Data Displays

The operator can completely monitor the status of the current blending operation or examine the blender data base that defines the blender's operation characteristics. This is done by selecting one of the many available display formats and associated display pages. There are three types of displays: (1) general, (2) master, and (3) loop.

General Display

The general display gives a complete picture of the current blending operation status and is the display the operator selects most of the time during a blending operation, as previously shown in Fig. 5.

Master Display

This display consists of display pages of data which define master status and data base items. See typical master display pages shown in Fig. 8.

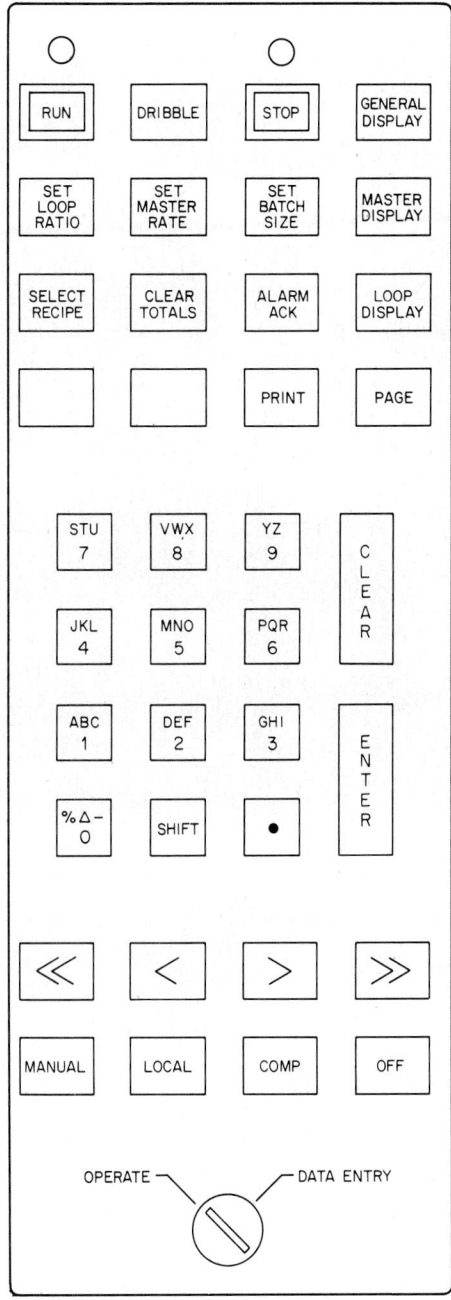

Fig. 7 Detail of blender control and keyboard panel.

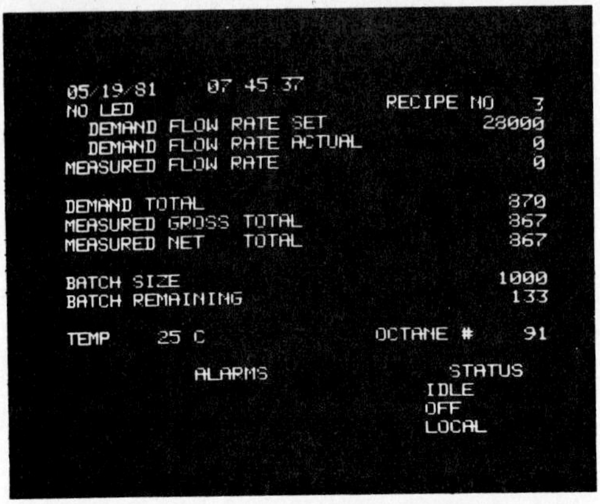

```
05/19/81    07 45 37
NO LED                       RECIPE NO   3
   DEMAND FLOW RATE SET              28000
   DEMAND FLOW RATE ACTUAL               0
MEASURED FLOW RATE                       0

DEMAND TOTAL                           870
MEASURED GROSS TOTAL                    867
MEASURED NET    TOTAL                   867

BATCH SIZE                            1000
BATCH REMAINING                        133

TEMP   25 C               OCTANE #      91

          ALARMS              STATUS
                              IDLE
                              OFF
                              LOCAL
```

(Page 1)

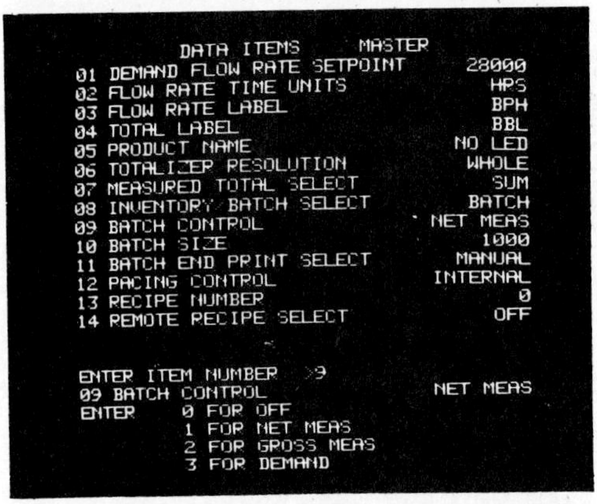

```
          DATA ITEMS      MASTER
   01 DEMAND FLOW RATE SETPOINT    28000
   02 FLOW RATE TIME UNITS           HRS
   03 FLOW RATE LABEL                BPH
   04 TOTAL LABEL                    BBL
   05 PRODUCT NAME                NO LED
   06 TOTALIZER RESOLUTION         WHOLE
   07 MEASURED TOTAL SELECT          SUM
   08 INVENTORY BATCH SELECT       BATCH
   09 BATCH CONTROL            * NET MEAS
   10 BATCH SIZE                    1000
   11 BATCH END PRINT SELECT      MANUAL
   12 PACING CONTROL            INTERNAL
   13 RECIPE NUMBER                    0
   14 REMOTE RECIPE SELECT           OFF


   ENTER ITEM NUMBER   >9
   09 BATCH CONTROL               NET MEAS
   ENTER    0 FOR OFF
            1 FOR NET MEAS
            2 FOR GROSS MEAS
            3 FOR DEMAND
```

(Page 2)

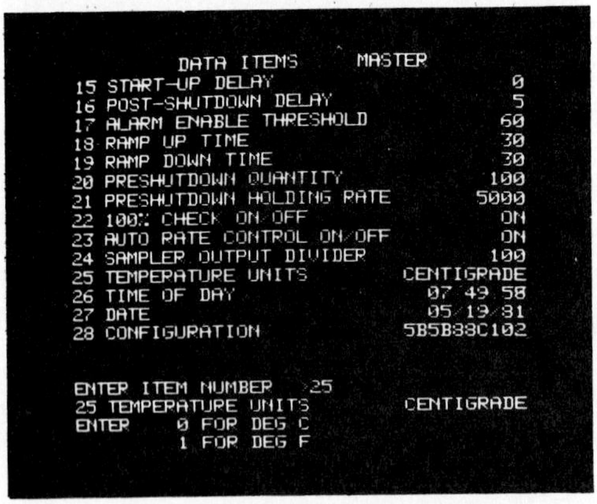

```
          DATA ITEMS      MASTER
   15 START-UP DELAY                    0
   16 POST-SHUTDOWN DELAY               5
   17 ALARM ENABLE THRESHOLD           60
   18 RAMP UP TIME                     30
   19 RAMP DOWN TIME                   30
   20 PRESHUTDOWN QUANTITY            100
   21 PRESHUTDOWN HOLDING RATE       5000
   22 100% CHECK ON/OFF                ON
   23 AUTO RATE CONTROL ON/OFF         ON
   24 SAMPLER OUTPUT DIVIDER          100
   25 TEMPERATURE UNITS         CENTIGRADE
   26 TIME OF DAY                07 49 58
   27 DATE                       05 19 81
   28 CONFIGURATION            5B5B88C102


   ENTER ITEM NUMBER   >25
   25 TEMPERATURE UNITS         CENTIGRADE
   ENTER    0 FOR DEG C
            1 FOR DEG F
```

(Page 3)

Fig. 8   Blender control *Master* display, pages 1 through 3.

7.72

Flow Control Loop Displays

For each (up to 16 available flow control loops), there are three pages of data: page 1 displays the present status of all dynamic data associated with the selected loop. Pages 2 and 3 display all flow loop data base parameters for the selected loop. See Fig. 9.

Blend Controller Functions

Modern microprocessor-based blend controllers can perform numerous tasks associated with the blending process. Some of these tasks are described in the following paragraphs.

Automatic Rate Control

If, in the course of a blending operation, one or more streams are unable to maintain a sufficient flow rate to achieve the desired blend ratios, the blender will automatically *ramp down* to a rate at which the selected ratios can be maintained. On restoration of flow in the lagging streams, the blender will automatically return to its original set blend rate.

When an alarm signal is received from any control loop, indicating that its valve output exceeds 95% of its maximum, the blender ramps the master flow rate down until the alarm vanishes, whereupon it ramps the master rate up until either the original set rate is achieved or the alarm reappears. Thus, the blender always operates either at its set master demand flow rate or at the maximum rate allowed by a lagging stream.

A rate limiting condition is indicated on the video display and through a discrete output. The auto rate control function may be disabled to allow operation in a memory mode.

Memory Mode

In some blending applications, it is necessary to maintain blender operation even though the setpoint ratio of one or more components is not maintained. In this mode, the automatic rate control feature is disabled, and the quantity of material either short or in excess of the set amount is stored in RAM. When flow in the affected loop is restored, the stored value decreases until it reaches zero and the blender is controlling in a normal manner. Sufficient memory is provided for each loop to retain a positive or negative error accumulation of up to a million counts.

Batch Quantity Control

The blender is capable of automatic shutdown after attaining a selected total blend quantity. The quantity at which shutdown is to be made is entered by the operator, and the blend is started in a normal manner. Near completion of the batch, the blender typically ramps down to a dribble flow and then shuts down completely at the selected total quantity.

Batch control is selected to function on either gross or net total and on either the sum of individual streams or on the blend total as determined by a tally meter in the combined stream. The batch size as entered by the operator is displayed as well as the amount remaining to be blended in each batch.

Closed-Loop Pulse Input Control

For each component stream, closed-loop control is maintained by comparing the scaled total pulse input from the flowmeter with a demand total derived from the master demand total and the percent set point for that loop. Any error, i.e., a difference between the measured and the demand total, results in an output change to the control device in a direction that will reduce the error to zero.

Sequential Pump Control

Solid-state relays for interfacing with motor starters, solenoid valves, or other field equipment may be provided in specially designed plug-in panels mounted on the rear of the blender housing. Terminals on each relay provide direct field wiring connections. Both ac and dc relays are available.

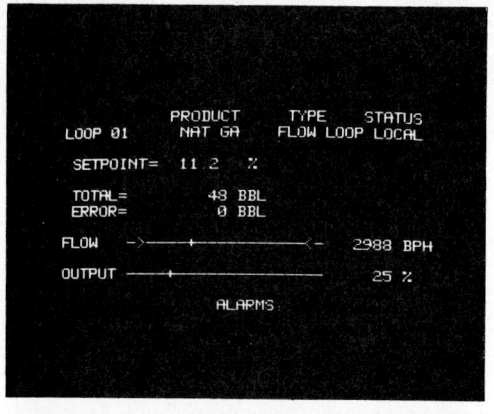

```
                    PRODUCT     TYPE    STATUS
        LOOP 01     NAT GA    FLOW LOOP LOCAL

        SETPOINT=  11.2    %

         TOTAL=          48 BBL
         ERROR=           0 BBL

        FLOW   ->----+-------<-   2988 BPH

        OUTPUT ------+---------    25 %

                    ALARMS:
```

(Page 1)

```
            DATA ITEMS      LOOP 01
    01 LOOP USED/UNUSED              USED
    02 OPERATIONAL MODE        FLOW LOOP
    03 PUMP START DELAY                 0
    04 COMPONENT NAME             NAT GA
    05 TOTAL LABEL                   BBL
    06 FLOW RATE LABEL               BPH
    07 LOOP RATIO                   11.2
    08 LOOP RATIO LABEL                %
    09 RATIO CONVERSION FACTOR         1
    10 TOTALIZER RESOLUTION       WHOLE
    11 METER K-FACTOR              200
    12 TOTAL UNITS FACTOR            1
    13 FLOW RATE TIME UNITS         HRS
    14 FULL SCALE FLOW RATE        9000

    ENTER ITEM NUMBER >2
    02 OPERATIONAL MODE        FLOW LOOP
    ENTER    0 FOR FLOWMETER
             1 FOR CLOSED LOOP PROP PUMP
             2 FOR OPEN LOOP PROP PUMP
```

(Page 2)

```
            DATA ITEMS      LOOP 01
    15 LOW FLOW ALARM SETPOINT         900
    16 HIGH FLOW ALARM SETPOINT       8100
    17 ERROR SETPOINT                    1
    18 ERROR SHUTDOWN SETPOINT          10
    19 TEMP. COMPENSATION ON/OFF       OFF
    20 MINIMUM TEMPERATURE              0
    21 MAXIMUM TEMPERATURE            100
    22 REFERENCE TEMPERATURE           50
    23 COEFF OF EXPANSION(%/DEG)        0
    24 LOOP GAIN                      0.2
    25 LOOP RESET TIME (SECS)          10
    26 DISCRETE OUTPUT ALARM      SHUTDOWN

    ENTER ITEM NUMBER >26
    26 DISCRETE OUTPUT ALARM      SHUTDOWN
    ENTER    0 FOR LOW FLOW
             1 FOR HIGH FLOW
             2 FOR ERROR
             3 FOR ERROR SHUTDOWN
```

(Page 3)

Fig. 9 Blender control *Flow Loop* display, pages 1 through 3.

7.74

Sequential starting of pumps to avoid sudden large increases in electrical load is readily accomplished by sequencing pump-start relay closures at the beginning of each blend. A different time delay may be entered for each relay closure, thus permitting any desired time sequence of pump starting.

Additive or Dye Pump Control

Additive pumps of the diaphragm or piston type, which inject a fixed amount of fluid into the blended stream with each stroke, may be actuated by the blender to proportion the additive to the total blend. Solid-state relay closures for actuating each pump stroke are paced by the master demand flow rate and bear a keyboard-selectable ratio to that rate. Solid-state relays, mounted on the blender, provide the interface with the pump.

Sampler Drives

A proportional sampler, which extracts samples from the blended stream in proportion to the total quantity blended, may be actuated in a manner identical to the additive pump pacing feature previously described. This signal is ratioed from the actual measured flow rate for the blend.

Ramp Up or Down

At a keyboard-selected interval following receipt of a blend-start signal, the blend rate of flow increases linearly until the selected rate is achieved. Ramp rates, 0 to 255 s, are selected by the keyboard. On receipt of any shutdown signal other than *Emergency Stop,* the blender reduces the blend rate to zero. In a similar manner, an *Emergency Stop* command causes all valves to go immediately to a fail-safe (normally closed) condition.

High- or Low-Blend Error Alarm

If a blend error occurs, e.g., a difference between the total amount blended in any loop and the total demand for that loop, an error alarm is initiated, causing a flashing display in the *Status* column and an audible alarm. At a second, larger, accumulated error, the blender shuts down immediately. High- and low-error and shutdown alarm points for each loop are entered as data for that loop.

High- or Low-Flow Alarm

Alarms indicating high or low flow in each stream are provided to warn of operation at rates outside the linear range of flowmeters, or for other purposes. High and low trip points are provided for each loop, with the setting displayed as an arrow on the graphic flow rate presentation of the *General* and *Loop* displays. The trip points are also shown numerically in each *Loop* display.

When an alarm condition occurs, a flashing *LF* or *HF* appears in the *Status* column opposite the affected loop. The condition is also indicated by flashing of the pointer on the graphic flow rate display, a flashing message in the lower part of the screen, and an audible alarm. The audible alarm is silenced by the acknowledge button on the keyboard, *Alarm Ack.*

The low-flow alarm is inhibited during ramp-up, ramp-down, and dribble flow, and when external pacing signals are below a preset point.

Internal or External Blend Pacing

All control loops are paced by a master demand flow rate and master demand total, which may be either internally generated or obtained from an external source.

The source for external pacing may be a flowmeter in an uncontrolled or wild stream, producing a pulse train input to the system. In this mode, the master demand signal is derived by the system directly from the master flow input.

External pacing may also be in the form of a 4 to 20 mA analog input from a flowmeter or from other sources. A typical analog input other than flow might be from a level controller on a surge tank receiving the blended output. The controller regulates the blend flow rate in order to maintain a constant level in the surge tank.

Temperature Compensation

When blending lubricating oils, fuel oils, or other heavy products, it is often necessary to correct the apparent gross volumes to a net basis. This can be done electronically within the blend controller.

Each flow control loop possesses the capability to correct the volume, as measured by the flow-meter, to a standard volume at a selected reference temperature. Fluid temperature is measured by a 100-Ω platinum probe with a 4 to 20 mA transmitter. A correction is made in accordance with a linear, keyboard-entered expansion coefficient for each stream.

When temperature compensation is used, blend ratios are controlled on the basis of net quantities. For batch control, the operator may select a batch size based on either the gross or the net total. This latter feature is most valuable for blending into vessels having a known volume, such as trucks, where the batch setting is determined by the capacity of the receiving vessel in gross units and the blend ratios are controlled in net units.

Analyzer Trim Control

In gasoline and other blending applications, it is frequently desirable to increase or decrease the percentage of certain components during the blending process to meet quality requirements of the blended product. To this end, the blender is capable of accepting inputs from up to three on-stream analyzers and adjusting the ratio of any one or more streams to control the analyzed quality of the finished blend. Typical applications in gasoline blending involve control of vapor pressure and octane number.

The 4 to 20 mA input from the analyzer may be either directly proportional to the analyzed quality or proportional to the deviation of the quality from a desired value. The analyzer signal acts as the process variable input to an internal proportional integral derivative (PID) controller function, whose output is then cascaded into selected loop controllers, serving to adjust percentage set points up or down. Changes in the flow of the selected streams affect the quality of the blend, thus closing the cascaded control loop.

Combined Quality Computation

When inputs from on-stream analyzers are provided, the system calculates (and shows on the *Master* display), the combined *Quality Number* (or *Quality Deviation*) of the total amount blended to that point, using the expression

$$N_t t = \sum \frac{(N \times \Delta Q)}{Q} \quad \text{or} \quad E_t = \sum \frac{(E \times \Delta Q)}{Q}$$

where $N_t t$ = combined quality number of total blend
E_t = combined quality deviation of total blend
N = quality number input at time of calculation
E = quality deviation input at time of calculation
ΔQ = incremental volume blended since last calculation
Q = total volume blended at time of calculation

This feature is typically used for octane and RVP* and analyzers. However, any on-stream analyzer with a 4 to 20 mA output can be used.

Recipe Storage

This feature provides storage on tape or in random-access memory (RAM) of repetitively used blend formulations that are readily called up and entered. Entry of a complete blend formulation is thereby greatly simplified, involving the entry of a single blend formulation number rather than a series of percentage set points. Lube oil and asphalt blending, where standard formulations are frequently used, are typical applications for this feature.

---

*RVP = Reid Vapor Pressure

Up to 15 formulations of 16 components each can be stored in the system's RAM. The formulations are originally entered via the *Select Recipe* function key and the alphanumeric keyboard. Any formulation can be called up subsequently by depressing the *Select Recipe* key and entering the recipe number. Alternatively, the formulations may be selected by a remote rotary switch.

An unlimited number of formulations may be stored in an optional built-in data cassette deck. Formulations are entered and called up, utilizing the *Select Recipe* and alphanumeric keys in the same manner as for RAM-stored formulations. Each data cassette tape can store up to 255 recipes.

Data Base Protection

All entered data are stored in RAM and backed up by an internal battery to preserve data during power outages of up to a week. A pair of terminals at the rear provides connections for a larger external backup battery to extend the backup time to any desired period. In addition, the optional data cassette can store the operating data for quick reentry in the event that the data base is erased for any reason.

Backup Totals

The system provides solid-state pulse outputs for each stream to actuate externally mounted backup totalizers or ticket printers if so desired.

Flowmeter Factor Scaling

Most pulse producing flowmeters used in blending systems require scaling of the input pulses to *engineering units* so that the blender may produce total and flow rate readings in meaningful values. The meter K factor, expressed in pulses per unit volume, together with the units in which the volume is to be measured, are entered through the keyboard into the data base for the particular loop in which the meter is installed. The system then automatically converts input pulses to actual flow totals and flow rates, expressed in the engineering units that have been entered.

Flow Rates and Totals

All individual loop flow rates and totals are shown graphically on a CRT display. Demand and measured total blend flow rates and totals are also displayed. The actual demand may differ from the set demand rate during the ramp-up or ramp-down phase of operation.

Remote Computer Control

A remote host computer can control the operation of one or more blenders through an RS-232 port provided on the blender for this purpose. It can, on command, effect the setting (and reading) of ratios or set points, flow rates, totals, and batch quantities—and can start or stop a blend.

When in the computer mode, all data base items may be entered and read back by the host computer. In addition, alarms and end-of-batch messages are initiated by the blender and sent to the host computer via the RS-232 computer port.

Data Printout and Logging

Essential operating data are printed out via an RS-232 data link connected directly to an optional printer. Printing is initiated automatically at the end of a batch, or manually by pressing the *Print* key on the *Blender* front panel. The printout assumes the same format as that shown on the video display. A printout may also be initiated by remote contact input or through the RS-232 data link.

Trends

With the introduction of microcomputers for use in blending systems, many ancillary functions in the blend controller mainframe became available at a relatively small additional cost. Thus, many microprocessor-based controllers include high- and low-flow alarms, independent analog controllers (e.g., temperature or pressure controls), computer interface, printer outputs, tape or disk storage (for multiple recipes), programmable solid-state relay outputs, and computation blocks that can be tailored for a specific process control task. Future features may include large-screen color displays, automatic loop tuning capabilities, and integral programmable controller functions for complex manifold lineup, control, and process sequencing.

REFERENCES

Baker, P. D.: "Positive Displacement Liquid Meters/Turbine Meters for Liquid Measurement," *Tech. Pap. 101A, 103A,* Smith Meter Division, Geosource Inc., Erie, Pa., 1977.

Bauman, D. E.: "Automate Small Refinery Blending Operations," *Hydrocarbon Process.* November 1981, pp. 247–251.

Hutchison, J. W.: *ISA Handbook of Control Valves,* 2d ed., Instrument Society of America, Research Triangle Park, N.C., 1976.

Krigman, A.: "Guide to Selecting Weighing, Batching, and Blending Systems," *Instrum. Technol.* October 1982, pp. 43–55.

Swientek, R. J.: "Microprocessor Controller Blends up to 16 Ingredients," *Food Process.* October 1982, pp. 70–72.

Waugh: *Microblender,* Model 2200 Brochure, Waugh Controls Corporation, Chatsworth, Calif., 1982.

Force Systems

C. John Easton. *President, Sensotec, Inc., Columbus, Ohio. (Strain Gages)*

Strain Gages

by
C. John Easton*

A strain gage is a device for measuring mechanical surface strain. The uses of strain gages fall into two broad areas: (1) applications, as in the case of stress analysis, where the gage measures strain as the primary objective of measurement, and (2) uses where the measurement of strain is employed in tranducers as a measure of another parameter, such as pressure, load, acceleration, or another force associated variable. Strain can be measured by different methods, including electrical, mechanical, and optical techniques. The electrical-type strain gage is based on the measurement of an impedance change that is proportional to strain. Over the years, the electrical-resistance strain gage has become the most widely used device, and this is what is usually meant when the term "strain gage" is used. The measurement of resistance change is the method most commonly used commercially to determine strain, because it is easy to apply to a wide variety of surfaces and because it is reliable, accurate, and easily interfaced with electronics. In order to measure strain, most commercially available resistance strain gages are attached directly to the surface being measured (such as a structural member or an element of a transducer).

Background

The resistance strain gage is based on the principle that the electrical resistance of a conductor depends on its shape. The basic relationships between resistance change and shape are shown in Fig.

*President, Sensotec, Inc., Columbus, Ohio.

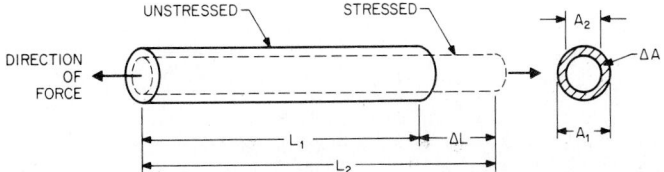

Fig. 1 Basic relation between resistance change and strain in a resistance-type strain gage. When under stress, the wire changes in length from L_1 to L_2 and in area from A_1 to A_2.

$$R = p\frac{L}{A} = \text{resistance}$$

where L = conductor length
A = cross section area
p = resistivity constant

$$S = \frac{\Delta R/R}{\Delta L/L} = \text{strain sensitivity (gage factor)}$$

$$\frac{\Delta R}{R} = \text{resistance change}$$

$$\frac{\Delta L}{L} = \text{strain}$$

1. The electrical resistance of a uniform cross section conductor is given by the equation

$$R = P\frac{L}{A} \tag{1}$$

where R = resistance, Ω
L = conductor length
A = cross-sectional area
P = resistivity constant of conductor material

The resistivity constant changes for different alloys. If a conductor is stretched elastically, the length will be increased and the cross-sectional area will be reduced as a result of Poisson's effect. Inspection of Eq. 1 shows that these two effects are additive in causing the resistance of the wire to increase when the resistivity constant of the material does not change. The term *strain sensitivity* is used to describe the resistance change ΔR of a conductor in relation to the length change ΔL. By definition,

$$\text{Strain sensitivity (gage factor)} = \frac{\Delta R/R}{\Delta L/L} \tag{2}$$

The strain sensitivity is commonly called the *gage factor* when referring to a specific strain gage material. Poisson's ratio is ~0.3 for most wire. The strain sensitivity or gage factor is ~1.6 when considering only the dimensional change aspect. This means that a 0.1% increase in length within the elastic range should produce a resistance increase of 0.16%. When actual tests are performed, a metal or alloy exhibits different values of strain sensitivity for different temperatures. The resistivity constant does not remain constant for all conditions.

The ideal strain gage would change resistance in accordance with deformations of the surface to which it is bonded and for no other reason. Unfortunately, gage resistance is affected by other factors (including temperature). Any resistive change in the gage not caused by strain is referred to as *apparent strain*. Apparent strain may be caused by a change in the gage factor due to temperature (thermal coefficient of the gage factor), by a change in resistance due to temperature (thermal coefficient of the resistance), by the stability of the metal, and even by the properties of the adhesive that bonds the gage to the surface being measured.

Table 1 shows some common types of resistance strain gages and their characteristics.

Table 1 Major Types of Resistance Strain Gages

Common name	Basic material of strain gage	Method of attachment of strain gage to surface	General application
Unbonded	Wire	Connected at ends	Transducer
Bonded metallic	Wire or foil	Epoxy	Stress analysis and transducer
Flame spray	Wire	Spray-coated	Stress analysis
Welded	Foil	Spot-welded	Stress analysis
Bonded semiconductor	Silicon or germanium	Epoxy	Stress analysis and transducer
Diffused semiconductor	Silicon	Semiconductor diffusion	Transducer
Thin film	Metal alloy	Sputtering or deposition	Transducer

Components of a Bonded Strain Gage System

Key parameters that need to be optimized for an ideal strain gage include high gage factor, minimal temperature effects, small size, and high ohmic resistance. Increasing the gage factor makes the gage more sensitive to strain but often increases the undesirable effects of temperature. A small size is preferred, so that the gage can be placed close to the high-strain area. A high resistance permits a larger input voltage excitation and thus a larger millivolt output with a lower power consumption.

Bonded foil strain gages are made using special metal alloy conductors with high resistivities, high gage factors, and low temperature coefficients. Wire strain gages are not widely used because in order to obtain 350 Ω by using no. 28 copper wire (0.000126 in^2, 0.08 mm^2 in cross section), 5360 ft (1633.7 m) of wire would be needed [350 Ω/(65 Ω/1000 ft)]. The metal alloy of a bonded foil strain gage is formed into a back-and-forth grid to decrease the overall length of the strain gage system (Fig. 2). The length of the grid versus the width is designed to concentrate the strain sensing

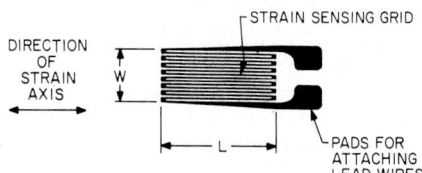

Fig. 2 A typical single bonded-foil strain gage. Dimensions can be as small as 0.031 × 0.062 in (0.79 × 1.57 mm). The thickness of a single gage is 0.0022 in (0.056 mm).

grid over the high-strain area. Foil strain gages with gage resistance values of 120, 350, and 1000 Ω are common, with special gages for use with 4 to 20 mA electronic transmitters having resistances as high as 5000 Ω.

The sensing grid is tiny and fragile in comparison to the structure to which it is usually attached; therefore, pads for connecting leadwires must be manufactured as part of the strain gage. The strain gage is bonded to the specimen surface by a thin layer of epoxy adhesive (Fig. 3), and care must be taken to ensure a thin, uniform, strong bond. A uniform bonding force applied by a contoured gaging block is used to exert a constant, even pressure against the strain gage. In summary, when installed and ready for use, the strain gage system consists of the specimen surface, an effective bond between the gage and the specimen, the strain gage, appropriate leads and connectors, and, if needed, a protective waterproof coating.

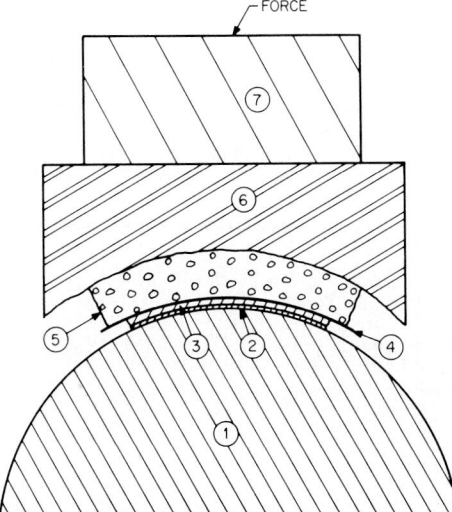

Fig. 3 Installation of a foil strain gage on a nonplanar surface. (1) Cylindrical specimen surface, (2) thin adhesive layer (typically 0.001 in, 0.025 mm), (3) strain gage, (4) Teflon sheet to prevent the rubber pressure pad from sticking, (5) rubber pressure pad, (6) metal gaging block that conforms to the specimen surface, (7) weight or clamp to apply pressure while the adhesive is curing.

Metallic Strain Gage Materials

All electrical conductors exhibit a strain gage effect, but only a few meet the necessary requirements to be useful as strain gages. The major properties of concern are (1) gage factor, (2) resistance, (3) temperature coefficient of gage factor, (4) thermal coefficient of resistivity, and (5) stability. High gage factor materials tend to be more sensitive to temperature and less stable than the lower gage factor materials.

Some of the most common strain gage materials are listed below.

1. ***Constantan.*** Constantan or Advance (copper-nickel alloy) is primarily employed for static strain measurement because of its low and controllable temperature coefficient. For static measurements, under ideal compensation conditions, or for dynamic measurements, the alloy may be used from -100 to $+460°$F (-73.3 to $+283°$C). Conservative limits are 50 to 400°F (10 to 204°C).

2. ***Karma.*** Karma (nickel-chrome alloy with precipitation-forming additives) provides a wider temperature compensation range than Constantan. Special treatment of this alloy gives minimum drift to 600°F (316°C) and excellent self-temperature compensation characteristics to $\sim$800°F (427°C).

3. ***Nichrome V.*** Nichrome V (nickel-chrome alloy) is commonly used for high-temperature static and dynamic strain measurements. Under ideal conditions, this alloy may be used for static measurements to 1200°F (649°C) and for dynamic measurements to 1800°F (982°C).

4. ***Isoelastic.*** Isoelastic (nickel-iron alloy plus other ingredients) is used for dynamic tests where its larger temperature coefficient is of no consequence. The higher gage factor is a distinct advantage where dynamic strains of small magnitude are measured.

5. ***479PT.*** 479PT (platinum-tungsten alloy) shows an unusually high stability at elevated temperatures. It also has a relatively high gage factor for an alloy. A gage of this material is recommended for dynamic tests to 1500°F (816°C) and static tests to 1200°F (649°C).

Table 2 Properties of Strain Gage Materials

Material	Composition, %	Gage factor	Thermal coefficient of resistivity, °C^{-1} × 10^{-5}
Constantan (Advance)	Ni 45, Cu 55	2.1	±2
Isoelastic	Ni 36, Cr 8, Mn-Si-Mo 4, Fe 52	3.52 to 3.6	+17
Karma	Ni 74, Cr 20, Fe 3, Cu 3	2.1	+2
Manganin	Cu 84, Mn 12, Ni 4	0.3 to 0.47	±2
Alloy 479	Pt 92, W 8	3.6 to 4.4	+24
Nickel	Pure	−12 to −20	670
Nichrome V	Ni 80, Cr 20	2.1 to 2.63	10
Silicon	p-Type	100 to 170	70 to 700
Silicon	n-Type	−100 to −140	70 to 700
Germanium	p-Type	102	—
Germanium	n-Type	−150	—

Semiconductor Strain Gage Material

Semiconductor material has an advantage over metals because its gage factor is approximately 50 to 70 times higher. However, this desirable increase in gage factor is offset by its greater thermal coefficient of resistivity (the common term is "temperature effect"). Semiconductor strain gages are used to manufacture miniature transducers (0.125 in, 3.2 mm in diameter). See Fig. 7. They are also used in stress analysis when a high electrical output is required from low-stress applications. The most commonly used semiconductors are silicon and germanium. A comparison of common metal and semiconductor gage materials is given in Table 2.

Bonding Agents

The importance of the adhesive which bonds the strain gage to the metal structure under test cannot be overemphasized. An ideal adhesive should be suited to its environment, transmit all strain from the surface to the gage, have high mechanical strength, high electrical isolation, and low thermal insulation, and be very thin. Also, it should not be affected by temperature changes. The adhesive must provide a strong bond while electrically isolating the gage from the surface to which it is attached. Electrical isolation is needed because most of the structures to which gages are bonded would electrically short out the elements if no electrical isolation existed. In a typical strain gage installation, the electrical isolation between the gage and the specimen surface should be at least 1000 MΩ at room temperature and 50 V dc. Electrical isolation (leakage) becomes a problem with bonding agents at high temperatures and in high-moisture environments. At high temperatures, even ceramic materials begin to exhibit a loss of electrical isolation. This is one of the most severe limitations on strain gage performance at temperatures above 1200°F (649°C). Some properties of common adhesives are given in Table 3.

Because of the wide variation in properties obtainable with different resin and hardener combinations, epoxy resins have become a very important class of strain gage adhesives. Phenolics are still employed occasionally, and some polyamides are used for temperatures over 600°F (316°C). Fast-cure adhesives, such as cyanoacrylates, are very popular for routine stress analysis work. Between 600 and 700°F (316 and 371°C), organic plastics soften rapidly. Therefore, strain gages operating above these temperatures are usually installed with inorganic (ceramic) bonding agents.

An alternate method for attaching free filament wire gages is the flame spray technique. Installation time per gage is approximately 15 to 20 min, and as soon as the gage is attached it is ready

Table 3 General Properties of Selected Strain Gage Adhesives

Characteristic	Eastman 910	Duco	M-Bond 43-B	EPY-500	M-Bond 610	PLD-700	M-Bond 6A-100
Base	Acrylic	Nitrocellulose	Epoxy	Epoxy	Epoxy-phenolic	Polyimide	Ceramic
Temperature range °F	−100 to +150	−100 to +150	To +250	To +400	To +450	To +750	To +1300
°C	−73 to +66	−73 to +66	To +120	To +205	To +230	To +399	To +705
Strain limit at room temperature, %	>10	>10	4	>5	3	>2	0.5
Shelf life, months	6	24	9	12	12	4	12
Cure time	1–5 min	12–48 h	2 h	1–30 h	2 h	2½ h	1 h
Cure pressure, psi	Contact	1–5	40–50	15–30	30–40	35–45	None
General application remarks	Fast cure, deteriorates with time	Good general purpose	Resistant to moisture, thin glue line	Long-term stability, transducers	Wide temp, range, transducers, thin glue line	Fast cure for high temperature, organic adhesive	For free filament gages, difficult to use

for use. In this method of attachment, a ceramic spray is created by atomizing a solid rod. The ceramic spray solidifies on the fine wire filaments of the strain gage and produces a homogeneous encapsulation without damaging the specimen or sensor. The flame spray technique produces a bond with an operating range of $-452°F$ $(-269°C)$ to $+1500°F$ $(816°C)$.

Another method for attaching strain gages is to first epoxy the gage to a thin piece of metal. These "weldable gages" are then attached to the specimen with low energy spot welders. The type of sensing element material, design, and attachment technique used on the thin metal shim (usually 0.005 in, 0.12 mm thick) is selected for the temperature range of operation. The shim material is selected for welding compatibility and a coefficient of expansion to match that of the specimen material. Weldable gages can be used in adverse environments and when the surface of the specimen cannot be properly polished for epoxy bonding procedures (as typically seen in field testing applications). They are also used in elevated temperature testing and when a minimum installation time is required.

Basic Bridge Circuit

In order to make use of the basic operating principle of the bonded resistance strain gage (i.e., change in resistance proportional to strain), the strain gage must be connected to an electric circuit capable of measuring small changes in resistance. Since the strain induced resistance changes are small (typically 0.2% for full-scale output in one active gage), the gages are wired into a Wheatstone bridge. A Wheatstone bridge is a circuit designed to accurately measure small changes. It can be used to determine both dynamic and static strain gage readings. The Wheatstone bridge also has certain compensation properties.

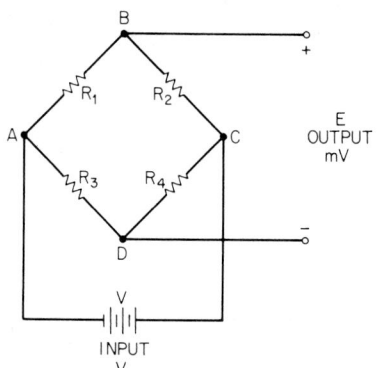

Fig. 4   Four-arm Wheatstone bridge circuit. Strain gages are inserted as R_1, R_2, R_3, and R_4.

The Wheatstone bridge detects small changes in a variable by comparing its value to that of a similar variable and then measuring the difference in magnitude, instead of measuring the magnitude directly. For instance, if four equal resistance gages are wired into the bridge (Fig. 4) and a voltage is applied between points A and C (input), then there will be no potential difference between points B and D (output). However, any small change in any one of these resistances will cause the bridge to become unbalanced, and a voltage will exist at the output in proportion to the imbalance.

In the simplest Wheatstone bridge configuration, a strain sensing grid is wired in as resistance R_1. For this circuit, the output voltage (E_o) can be easily derived. In reference to the circuit shown in Fig. 4, the voltage drop across R_1 is denoted V_{ab} and given as

$$V_{ab} = \frac{R_1}{R_1 + R_2} V \tag{3}$$

Similarly, the voltage drop across R_4 is denoted V_{ad} and given by

$$V_{ad} = \frac{R_4}{R_3 + R_4} V \tag{4}$$

The output voltage from the bridge, E, is equivalent to V_{bd}, which is given by

$$E = V_{bd} = V_{ab} - V_{ad} \tag{5}$$

Substituting Eqs. 3 and 4 into Eq. 5 and simplifying gives

$$E = \left(\frac{R_1}{R_1 + R_2} - \frac{R_4}{R_3 + R_4}\right) V \tag{6}$$

$$= \frac{R_1 R_3 - R_2 R_4}{(R_1 + R_2)(R_3 + R_4)} V$$

The voltage E will go to zero, and thus the bridge will be considered to be balanced when

$$R_1 R_3 - R_2 R_4 = 0$$

or

$$R_1 R_3 = R_2 R_4 \tag{7}$$

Therefore, the general equation for bridge balance and zero potential difference between points B and D is

$$\frac{R_1}{R_4} = \frac{R_2}{R_3} \tag{8}$$

Any small change in the resistance of the sensing grid will throw the bridge out of balance and can be detected by a voltmeter.

When the bridge is set up so that the only source of imbalance is a resistance change in the gage (resulting from strain), then the output voltage becomes a measure of the strain. From Eq. 6, with a small change in R_1,

$$E = \frac{(R_1 + \Delta R_1)R_3 - R_2 R_4}{[(R_1 + \Delta R_1) + R_2](R_3 + R_4)} V$$

Most Wheatstone bridge circuits are produced with all four arms serving as active strain gages.

Types of Resistance Strain Gages

Many different types of resistance strain gages have been developed since the first bonded strain gage was introduced in 1936. Currently, the most commercially used gage is the bonded foil gage, but this could change in the future. This gage uses a thin foil of metal alloy to sense strain. The bonded semiconductor strain gage uses silicon or germanium to sense strain.

Two other types of gages eliminate the problems of adhesive bonding by molecularly bonding the gage to the surface being measured. The diffused semiconductor strain gage diffuses an impurity such as boron into a semiconductor transducer diaphragm to form a strain gage, and the thin-film strain gage molecularly deposits metal alloy directly on a metallic structure such as the beam or diaphragm of a transducer element.

Bonded Foil Strain Gages

Foil strain gages first received serious commercial attention in the United States in the mid-1950s. Before that, all strain gages were made from wire. Foil strain gages are produced by a printed circuit process or by being stamped from selected alloys that have been rolled into a thin foil. Depending on the manufacturer and the gage type, foil thicknesses range from 0.0001 to 0.0002 in (0.00254 to 0.00508 mm). The foil is often specially heat-treated before use in order to optimize mechanical properties and the thermal coefficient of resistivity. For a given cross-sectional area, the foil conductor displays a large surface area. The large ratio of surface area to cross section provides superior mechanical stability under prolonged strain and high-temperature conditions. The large surface area also provides a good heat transfer surface between the grid and the specimen; therefore, high-input-voltage levels are possible without developing severe temperature gradients across the insulating matrix.

The photoetching process used to make foil strain gages permits the manufacture of sensing grids in virtually any two-dimensional pattern. Normally, a geometric pattern is developed that will provide maximum electrical and mechanical efficiency from the sensing element. The most common shape for the measurement of uniaxial strains is that of a long, narrow sensing element of the type shown in Fig. 1. This configuration maximizes the sensitivity in the direction of interest and minimizes the effects of strain in different orientations.

For measuring strain in special applications, different foil patterns, as shown in Fig. 5, can be utilized.

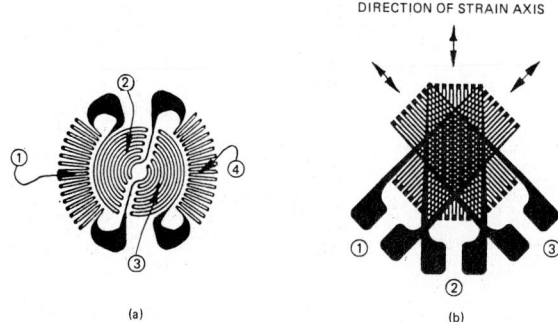

Fig. 5   Gage pattern variations. (*a*) Pattern used to measure strain in a diaphragm. Elements 1 and 4 are subjected to compressive radial strains, while elements 2 and 3 are subjected to tensile tangential strains. (*b*) Rosette gage that measures strain in three directions at one time.

Bonded Semiconductor Strain Gages

Semiconductor strain gages are used in applications similar to those employing conventional metallic gages in transducers and stress analysis. The principal difference between foil and semiconductor gages is the greater response of semiconductor gages to both strain and temperature. The large resistance-versus-strain characteristic of a properly oriented semiconductor crystal is due primarily to the piezoresistive effect. Gage behavior may be accurately described by the equation

$$\frac{\Delta R}{R_0} = \left(\frac{T_0}{T}\right)(E)(GF) + \left(\frac{T_0}{T}\right)^2 (E)^2(C_2) \tag{9}$$

where R_0 = unstressed gage resistance at T (changes as T changes)
   ΔR = change in gage resistance from R_0
   T = temperature, K
   T_0 = 298 K (24.9°C)
   E = strain
GF and C_2 = constants of particular gage in question

The resistance change due to strain is a parabola for high-resistivity p-type silicon. Pure material of this resistivity is not used to produce gages because of this severe nonlinearity. As can be seen in Eq. 9, the linearity can be improved by reducing the nonlinearity constant C_2. Figure 6 shows the behavior of a typical p-type semiconductor strain gage for a material that has been doped so that C_2 is low and the slope is more linear. Equation 9 also shows that large tensile strains on the gage filament and higher temperatures increase gage linearity. As temperature T rises, the value of both terms on the right-hand side of Eq. 9 decrease, as does gage sensitivity. The nonlinearity coefficient, however, decreases faster than the gage factor coefficient, thus improving the linearity.

n-Type semiconductor strain gages are similar in behavior to p-type gages, except that the gage factor is negative.

The high output obtainable from semiconductor piezoresistive elements makes them particularly attractive for transducers that are ⅛ in (3.2 mm) and smaller. Miniature transducers are formed by attaching individual silicon strain gages to a force collecting diaphragm or beam (Fig. 7). When the diaphragm or beam deflects, the surface strains are sensed by the semiconducting elements. Output levels typically ~100 mV full-scale are availa-

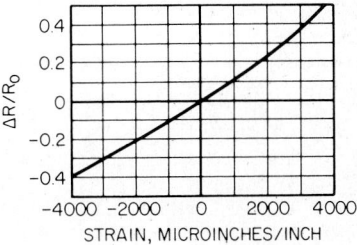

ble from full bridge transducers with ~10 V input.

A typical semiconductor gage, unlike a bonded foil gage, is not provided with a backing or carrier. Therefore, bonding the gage to a surface requires extreme care in order to obtain a thin epoxy bond. The same epoxies used for foil gages are used for semiconductor gages, but clamping techniques cannot be used because of the brittleness of the silicon semiconductor strain gages.

Fig. 6   Gage sensitivity versus strain level for a p-type semiconductor gage.

Diffused Semiconductor Strain Gages

A major advance in transducer technology was achieved with the introduction of diffused semiconductor strain gages. The gages are diffused directly into the surface of a diaphragm, utilizing photolithographic masking techniques and solid-state diffusion of an impurity element such as boron. Since the bonding does not use an adhesive, no creep or hysteresis occurs.

The diffusion process does not lend itself to the production of individual strain gages and also requires that the strained member (diaphragm or beam) be made from silicon. Therefore, diffused semiconductors are used for manufacturing transducers (primarily pressure) instead of for stress analysis. Typically, a slice of silicon 2 to 3 in (5 to 7.5 cm) in diameter is selected as the main substrate. From this substrate, hundreds of transducer diaphragms 0.1 to 0.5 in (2.5 to 21.7 mm) diameter with full four-arm Wheatstone bridges can be produced. A silicon pressure transducer diaphragm with a diffused semiconductor strain gage is shown in Fig. 8. The entire Wheatstone bridge circuitry is diffused into the diaphragm (strain gages and connection solder areas for leadwires).

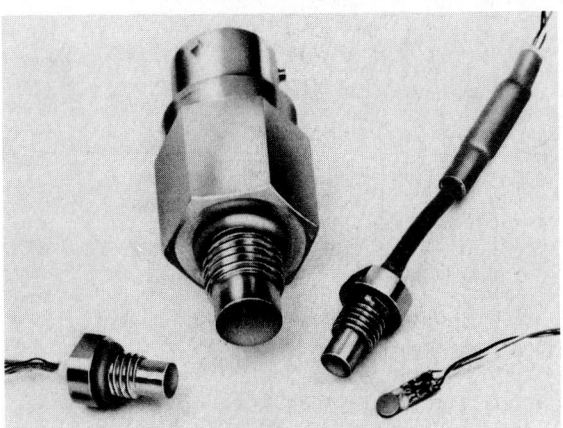

Fig. 7   Flush diaphragm subminiature pressure transducers. Shown (left to right) are models with (1) a straight thread and a differential pressure tube, (2) a hermetically sealed electrical connector, (3) a 0.125-in (3.2-mm) diameter, and (4) a wafer-thin thickness of 0.035 in (0.9 mm). *(Sensotec.)*

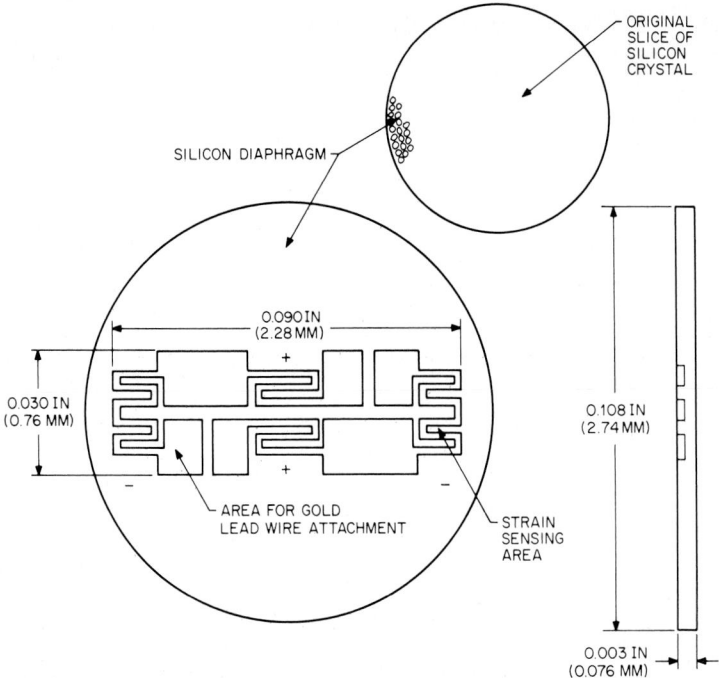

Fig. 8   Typical silicon pressure transducer diaphragm with a diffused Wheatstone bridge circuit. Hundreds of diaphragms are produced from a single slice of silicon crystal. The strain gage elements are situated to measure the compressive radial strains and the tensile tangential strains.

A diffused silicon sensor is unsuitable for high-temperature measurements because the gage-to-gage resistance decreases sharply as a function of temperature. Two mechanisms combine to produce this effect. First, isolation from the substrate is accomplished by a *p-n* junction, and its effectiveness is sensitive to increased heat. Second, the diaphragm (essentially an insulator) becomes increasingly conductive as the temperature is raised.

Refinements in the semiconductor diffusion process have allowed manufacturers to produce the entire strain gage transducer diaphragm, strain gage, temperature compensation elements (i.e., thermistors), and amplifier circuits with semiconductor technology. The introduction of very high volume, extremely low cost transducers is now practical.

Thin-Film Strain Gages

Another widely used, relatively new type of strain gage technology is the thin-film process. The thin-film technique independently controls the major strain gage properties (i.e., strain gage and electrical isolation). This technique uses the inherent advantages of metal strain gages (low-temperature effects and high gage factors) and the process advantages available with the diffused semiconductor technique (no adhesive bonding). The thin-film strain gage is potentially capable of producing an ideal strain gage system.

A thin-film strain gage is produced by depositing a thin layer of metal alloy on a metal specimen by means of vacuum deposition or sputtering. This technique produces a strain gage that is molecularly bonded to the specimen, so the disadvantages of the epoxy adhesive bond are eliminated. Like the diffused semiconductor process, the thin-film technique is used almost exclusively for transducer applications.

To produce thin-film strain-gage transducers, first an electrical insulation (such as a ceramic) is deposited on the stressed metal member (diaphragm or beam). Next, the strain gage alloy is deposited on top of the isolation layer. Both layers may be deposited either by vacuum deposition or by sputtering.

In vacuum deposition, the material to be deposited is heated in a vacuum and vapor is emitted. The vapor deposits on the transducer diaphragm in a pattern determined by substrate masks.

The sputtering technique also employs a vacuum chamber. With this method, the gage or insulating material is held at a negative potential and the target (transducer diaphragm or beam) is held at a positive potential. Molecules of the gage or insulating material are ejected from the negative electrode by the impact of positive gas ions (argon) bombarding the surface. The ejected molecules are accelerated toward the transducer diaphragm or beam and strike the target area with kinetic energy several orders of magnitude greater than that possible with any other deposition method. This produces superior adherence to the specimen.

In order to obtain maximum bridge sensitivity (millivolt output), to minimize heating effects, and to obtain stability, the four strain gages, the wiring between the gages, and the balance and temperature compensation components are all integrally formed during the deposition process. This ensures the same composition and thickness throughout.

The thin-film strain gage transducer has many advantages over other types of strain gage transducers. The principal advantage is long-term stability. The thin-film strain gage circuit is molecularly bonded to the specimen, and no organic adhesives are used which could cause drift with temperature or stress creep. The thin-film technique also allows control of the strain gage resistance value. A resistance as high as 5000 Ω can be produced in order to allow increased input and output voltages with low power consumption.

Bridge Correction Circuits

When static strains or the static component of a varying strain are to be measured, the most convenient circuit is the Wheatstone bridge (Fig. 4). The bridge is balanced (i.e., $E = 0$) when

$$\frac{R_1}{R_4} = \frac{R_2}{R_3}$$

Now, consider a bridge in which all four arms are separate strain gages. Assume that the bridge is initially balanced, so that $R_1R_3 = R_2R_4$ and $E = 0$. A strain in the gages will cause a change in each value of resistance, R_1, R_2, R_3, and R_4, by an incremental amount, ΔR_1, ΔR_2, ΔR_3, and ΔR_4. The voltage output ΔE of the bridge can be obtained from Eq. 6, which becomes

$$\Delta E = \frac{(R_1 + \Delta R_1)(R_3 + \Delta R_3) - (R_2 + \Delta R_2)(R_4 + \Delta R_4)}{[(R_1 + \Delta R_1) + (R_2 + \Delta R_2)][(R_3 + \Delta R_3) + (R_4 + \Delta R_4)]} \text{ V} \tag{10}$$

After considerable simplification, Eq. 10 becomes

$$E = \frac{R_2/R_1}{(1 + R_2/R_1)^2}\left(\frac{\Delta R_1}{R_1} - \frac{\Delta R_2}{R_2} + \frac{\Delta R_3}{R_3} - \frac{\Delta R_4}{R_4}\right) \text{ V} \tag{11}$$

Equation 11 shows that, if all four gages experience the same strain, the resistance changes will cancel out and the voltage change ΔE will equal zero. On the other hand, if gages R_1 and R_3 are in tension (ΔR_1 and ΔR_3 positive) and gages R_2 and R_4 are in compression (ΔR_2 and ΔR_4 negative), then the output will be proportional to the sum of all the strains measured separately. All four-arm Wheatstone bridge transducers are wired to give two gages in tension and two gages in compression. An example of a four-gage setup for the diaphragm of a pressure transducer is shown in Fig. 9. This design takes full advantage of the tensile tangential strains developed at the center of the diaphragm and the compressive radial strains present at the edge.

Another advantage of using a four-gage bridge, besides the increased output, is the effect on the temperature sensitivity. If the gages are located close together, as on a pressure transducer diaphragm, they will be subjected to the same temperature. Therefore, the resistance change due to

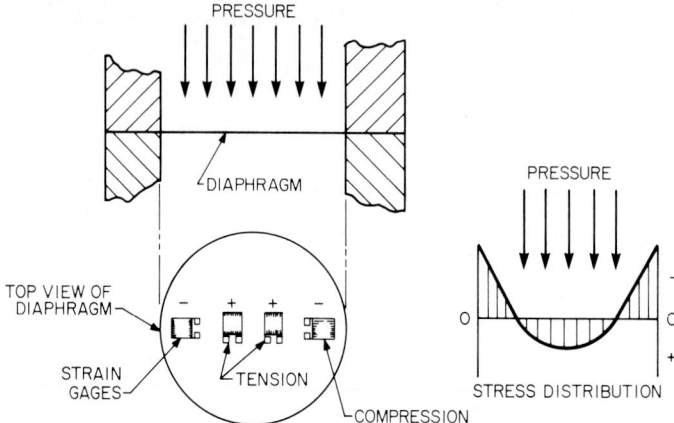

Fig. 9 Typical strain gage positions on a pressure diaphragm. The orientations take advantage of the stress distribution. The gages are wired into a Wheatstone bridge with two gages in tension and two in compression.

temperature will be the same for each arm of the Wheatstone bridge. If the gage resistance changes due to temperature are identical, the temperature effects will all cancel out and the output voltage of the circuit will not increase or decrease due to temperature.

The output voltage of the Wheatstone bridge is expressed in millivolts output per volt input (mV/V). For example, a transducer rated at 3.0 mV/V at 500 psi (~73 kPa) will have an output signal

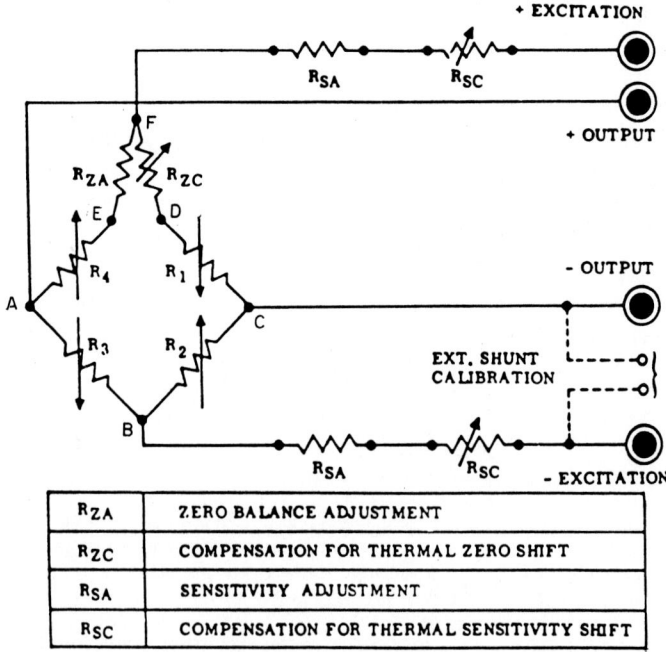

R_{ZA}	ZERO BALANCE ADJUSTMENT
R_{ZC}	COMPENSATION FOR THERMAL ZERO SHIFT
R_{SA}	SENSITIVITY ADJUSTMENT
R_{SC}	COMPENSATION FOR THERMAL SENSITIVITY SHIFT

Fig. 10 Strain gage transducer circuit with four active strain gage elements (R_1, R_2, R_3, and R_4). Balance, sensitivity, and thermal compensation resistors are also shown.

of 30.00 mV for a 10-V input at 500 psi ($\sim$73 kPa) or 36.00 mV for a 12-V input. Any variation in the power supply will directly change the output of the bridge. Generally, power supply regulation should be 0.05% or better.

In production the four strain gages in a Wheatstone bridge never come out to be exactly equal for all conditions of strain and temperature (even in the diffused semiconductor process). Therefore, various techniques have been developed to correct the differences in the individual strain gages and to make the strain gage bridge easier to use with electronic instrumentation. Four main values normally need adjusting (Fig. 10): (1) electrical bridge imbalance, (2) balance shift with temperature, (3) span or sensitivity shift of bridge output with temperature, and (4) standardization of the bridge output to a given millivolts-per-volt value. Other transducer characteristics such as accuracy, linearity, hysteresis, acceleration effect, and drift are part of the transducer element design (beam or diaphragm) and cannot be corrected after the strain gage bridge has been produced.

Figure 10 shows the circuit diagram of a Wheatstone bridge circuit with adjusting resistors. One corner of the bridge (points D and E) remains "open," so that the bridge can be adjusted electrically. This means that five leads come from the four gages. The zero-balance adjustment compensates for the electrical imbalance in the bridge caused by unequal resistances of the strain gages. Depending on which leg is unbalanced, R_{za} is placed between point E and F or between points D and F. The zero balance changes with temperature, and R_{zc} is inserted inside the bridge to correct for this change. A small piece of nickel wire is selected to provide a resistance change opposite the resistance change of the bridge. R_{sc} is also a temperature thermistor or sensor which changes resistance with temperature to adjust the excitation to the bridge. The values for R_{zc} and R_{sc} have to be selected by running each bridge over its desired temperature range (usually -65 to $+300°$F, -54 to $+149°$C).

R_{sa} is a non-temperature-sensitive resistor, and it is used to adjust the output to a precise millivolts-per-volt value once all the balance and temperature-sensitive resistors have been inserted within the bridge. This means that five balance and temperature-sensitive resistors have been inserted within the bridge. The user of transducers is not affected because all this circuitry is contained within the transducer and does not interfere with connections to amplifiers, power supplies, or computers.

A Wheatstone bridge can also be used in applications that require only one or two active strain gages. To compensate for temperature in two-gage applications, the gages must be located in adjacent arms of the bridge, as shown in Fig. 11. In placing gages, one must only recognize

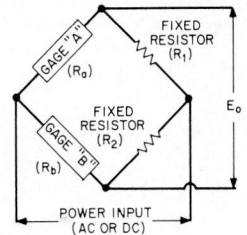

Fig. 11 Wheatstone bridge circuit utilizing two strain gages.

that the bridge is unbalanced in proportion to the difference in the strains of the gages located in adjacent arms and in proportion to the sum of the strains of gages located in opposite arms.

Transducer Applications

A strain gage is used in many transducer applications in addition to stress analysis. Since the electrical resistance strain gage is small, precise, inexpensive, fast responding, and easy to apply to different surfaces, it is used in a wide variety of transducers.

The heart of all strain gage transducers is the sensing element that detects the load, pressure, torque, acceleration, or other value being measured. The basic transducer element designs, such as the diaphragm, the diaphragm-beam, the column, and the shear web, use the advantages of strain gages to produce transducers with wide ranges of performance (accuracy to 0.02%) and physical characteristics (measuring ranges from 1 lb to 10 million lb, 0.5 to 5 million kg).

For different capacities (for instance, pressure ranges) of the same model strain gage transducer, the elements are designed so that for each separate capacity the sensing element will have approximately the same strain at the transducer's rated maximum output. For example, each pressure transducer model utilizes a specific diaphragm diameter and thickness plus element beam width and length to set the same microstrain; thus the output voltage for a 5-psi (0.7-kPa) transducer at capacity output is the same as that for a 1000-psi (145-kPa) transducer at capacity output, normally

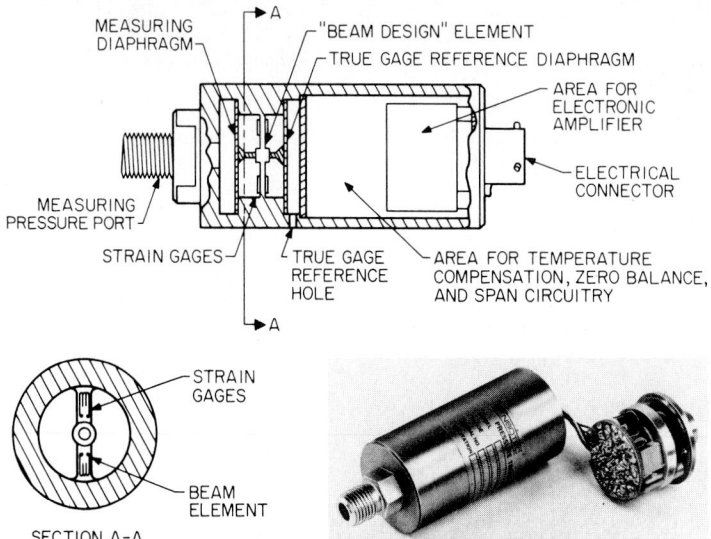

Fig. 12 Pressure transducer (diaphragm-beam design) that compares the measured pressure to atmospheric (reference) pressure. Welded stainless diaphragms permit its use in corrosive environments because the strain gages are in an environmentally protected chamber. Notice the internally installed electronic amplifier. *(True Sense, Sensotec.)*

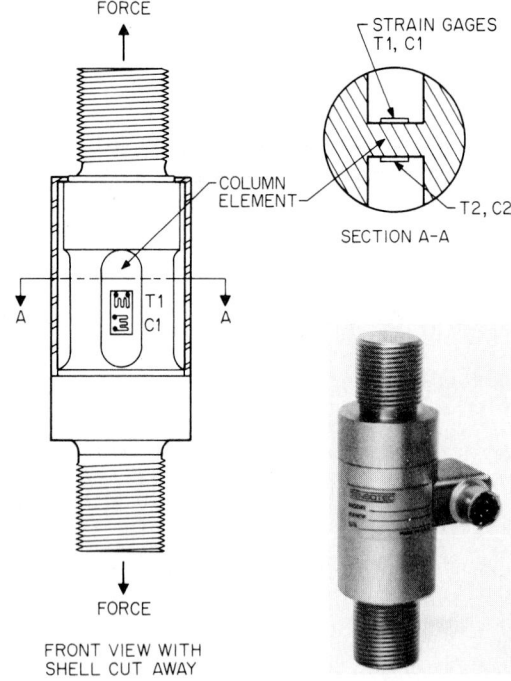

Fig. 13 Tension (column design) load cell. The special I-shaped cross section of the sensing element is designed to minimize the effects of any bending forces. *(Sensotec.)*

3 mV/V (1500 microstrain—from Fig. 1 with a gage factor of 2). A limitless number of transducer element designs are possible because of the flexibility of application to various element materials and shapes and the reliability and accuracy of the strain gage.

Several of the more common combinations of transducer elements and bonded strain gages are illustrated. Figures 7 to 9 show the basic flat diaphragm, and Fig. 5 shows a bonded foil strain gage design often used on diaphragms. The diameter and thickness of the diaphragms are varied for different pressure ranges.

Figure 12 shows the detail of a precision pressure transducer that utilizes a diaphragm to apply force on a beam element. Any pressure on the diaphragm is transmitted as a force on the beam. Strain gages are located on the beam to detect the strain resulting from the force.

A column transducer element design is shown in Fig. 13. This is a simple load tension measuring transducer. The element design is varied for different capacities by changing the dimensions of the cross-sectional area (section *A-A*).

A more complex type of transducer is the shear web element load cell. One use of the shear element design is in making high-capacity (50,000-lb, 22,650-kg), thin (1.5-in, 3.81-cm) load cells (Fig. 14). A shear web connects an outer, stationary hub to an inner, loaded hub. Strain gages detect the shear strain produced in the web. The large diameter of the 50,000-lb (22,650-kg) shear element

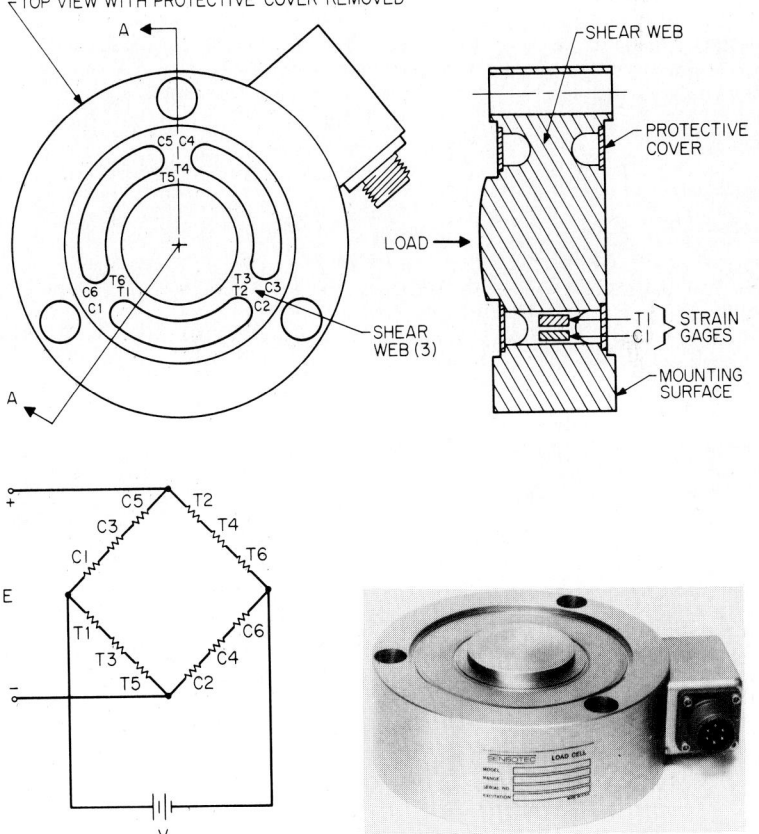

Fig. 14 Pancake (shear web) load cell. Each arm of the Wheatstone bridge circuit contains one strain gage from each of the three shear webs. The microstrains from the three webs are added together in one bridge circuit to determine the load. *(Sensotec.)*

requires that the strain be measured at more than one location (note the three webs). This is common in all types of large transducer elements in order to obtain an average of the total strains on the element and to eliminate errors caused by minor off-center loading. The strain gages are wired so the user sees only one 350-Ω bridge (Fig. 14). Each separate gage has a resistance of 120 Ω.

Electronics for Transducers and Transmitters

The full-scale output of a typical bonded-foil, four-active-element, strain gage bridge with all compensating and adjusting resistors connected is ∼20 to 30 mV at 10-V excitation. An amplifier must be used to obtain the 0 to 5 V or the 4 to 20 mA outputs used in control instrumentation. As a result of the advances in integrated circuitry, many transducers now have amplifiers that are internally installed within the transducer body (Fig. 12).

High-gain, low-noise, instrumentation quality, differential operational amplifiers such as the OP-07 make amplification of the strain gage bridge output for standard 0 to 5 V transducers and for 4 to 20 mA transmitters reliable and easy. These integrated circuit amplifiers have a high common mode rejection ratio and are thus well suited for use with Wheatstone bridge circuits. They are also inherently well compensated in order to deliver a constant output irrespective of temperature changes. The operational amplifiers used in instruments have controllable gains and zero-balance adjustments. Since the offset of the instrumentation channel's output is equal to the sum of the offsets in the bridge and in the amplifier, the combined offset can be adjusted at the amplifier so that the channel delivers 0 V at zero stimulus (pressure, load, torque, etc.) for 0 to 5 V output transducers or 4 mA at zero stimulus for 4 to 20 mA output transmitters.

REFERENCES

Dally, J. W., and W. F. Riley: *Experimental Stress Analysis,* McGraw-Hill, New York, 1978.

Easton, C. John: "Internal Transducer Electronic Amplifiers," 10th Transducer Workshop, Colorado Springs, Colo., June 12–14, 1979, Sensotec, Inc., Columbus, Ohio, 1979.

Mallon, J., and D. Germanton: *Advances in High Temperature Ultraminiature Solid State Pressure Transducers,* Kulite Semiconductor Products, Inc., Ridgefield, N.J., 1982.

Perino, P. R.: "Thin-Film Strain Gage Transducers," *Instrum. Control Syst.,* December 1965.

Perry, C. C., and H. R. Lissner: *The Strain Gage Primer,* 2d ed., McGraw-Hill, New York, 1962.

Shapiro, B. H.; "Strain Gage Transducers Show Long Term Stability." *Control Eng.,* vol 31, no. 2, 1984, pp. 160–163.

Weymouth, L. J., Starr, J. E., and J. Dorsey: "Bonded Resistance Strain Gages," *Exp. Mech.,* March 1979.

Motion and Geometric Systems*

Stephen C. Cuchanan. *Daytronic Corporation, Miamisburg, Ohio. (Linear Variable Differential Transformer)*

C. A. Carlson. *Milltronics Ltd., Peterborough, Ontario Canada. (Speed and Velocity Measurements)*

R. Kutko. *Gaertner Scientific Corporation, Chicago, Illinois. (Dimensional Metrology)*

Richard E. Lawonn. *Giddings & Lewis Electronics Company, Fond Du Lac, Wisconsin. (Programmable Machine Controls)*

E. R. Sabinash. *Cutler-Hammer Products, Eaton Corporation, Milwaukee, Wisconsin. (Limit Switches)*

Richard F. Speicher. *Micro Switch Division, Honeywell Inc., Freeport, Illinois. (Proximity Switches)*

James R. Walker. *Chief Electrical Engineer, DeVlieg Machine Company, Royal Oak, Michigan. (Position and Displacement Measurement)*

Robert M. Whittier. *Manager of Development Engineering, Endevco Corporation, San Juan Capistrano, California. (Vibration Measurements)*

Joseph Wilder. *Director of Research, Object Recognition Systems, Inc., Princeton, New Jersey. (Machine Vision)*

A. W. Young. *Rank Precision Industries, Inc., Des Plaines, Illinois. (Clinometer; Goniometer; Theodolite; Surface Finish)*

---

*Persons who authored complete articles, subsections of articles, or who otherwise cooperated in an outstanding manner in furnishing information and helpful counsel to the editorial staff.

Position, Displacement, Motion, and Object Detection Systems

Among the variables mentioned in the title of this article, position is the key factor. *Displacement* is the difference between two positions—frequently the difference between an actual position and a reference position. *Motion* is the result of going from one position to another. Although exact coordinates may not have to be determined, *objects are detected* (or not detected) by virtue of the position they occupy. *Dimension,* like displacement, also represents the difference between two positions—frequently the difference between two fixed points on a calibrated scale of some kind. But, because dimension measurement is a large subclass of position measurement technology, it is described separately—in the next article in this Handbook section.

It is interesting to note that there are several analogous situations involving the aforementioned variables as they apply to discrete pieces and variables by other names that apply to fluids and bulk materials. Thus, using a photoelectric device, for example, to count objects moving along a conveyor is analogous to determining the flow of a liquid or gas through a pipe. Level sensing of liquids in a tank or of bulk solids in a bin is fundamentally a measurement of position. And, particularly in connection with mechanical and pneumatic instruments, but also applying to electronic recorder servo systems, for example, position matching can be of importance.

Control systems involving position measurement tend to fall into one of two categories: (1) systems, as exemplified by machine tool and transfer line control, where achieving a position as described by exact coordinates is desired, i.e., where control actions are taken to attain a given position; and (2) systems, as exemplified by object detection systems, where control actions are taken because an object occupies (even for an instant) a definite (or approximate) position (location). In most metrological applications, the position is established and precedes the measurement of position in checking and inspection operations.

MACHINE CONDITIONS AND REQUIREMENTS

The degree of positioning accuracy and repeatability required by various production machines varies over a wide range: ± 0.0001 in (0.0025 mm) for boring machines, ± 0.001 in (0.025 mm) for drilling and contour machining, ± 0.005 in (0.127 mm) for armature insulator assembly, ± 0.010 in (0.254 mm) for tube bending and frame welding, ± 0.015 in (0.381 mm) for automobile seat cushion spring welding, and $\pm 0.1\%$ of full-scale reading for asphalt batching machines. In the manufacture of solid-state circuits and components in the electronics industry, involving laser and electron beam manipulation, accuracy is in terms of a fraction of a micrometer. The complexity of the control system varies with the type of machine, particularly with the number of axes that must be controlled—ranging from two up to six or more axes. The speed of response needed is related to the total cycling time of the machine and ranges from seconds and fractions of seconds up to 2 min or greater.

Backlash

To position a machine member with acceptable accuracy, it is necessary to establish the extent of the backlash or dead band region for the positioning mechanism used. The measuring transducer and its attendant dead band characteristics, when used with the machine member, determine the amount of dead band or backlash to be included in the total control system loop. For all types of repeat-back devices whose mechanical input is provided by rotating a shaft, an important consideration is the means of coupling the device to the positioned machine element. Frequently, a gear train is required to reduce member travel to one revolution or less of the transducer shaft. It is necessary to determine

9.5

whether any backlash in this train is comparable, when expressed as an arc of the transducer shaft, with the positional accuracy requirement of the machine member. Satisfactory results on the basis of this comparison rest on the assumption that the electrical and mechanical error factors for the transducer are small ($\frac{1}{10}$ or less) compared with the machine member positioning tolerances. Sometimes a transducer that does not require a mechanical input shaft should be considered. Where this is impractical, a separate rack and gear train, both exhibiting minimal backlash, may be used to position the mechanical input shaft of the associated transducer.

Stable Machine Base

The controlled member may have undesired movement with respect to the machine base—in a direction transverse to the controlled axis of travel. If the slide and table ways wear nonuniformly, variation in the transverse position of a point on the table of a machine may cause a variation in the air gap of a magnetic slot transducer system, for example. The same problem will result in misalignment of optical transducer systems if the table motion becomes crablike after wear of the slides has progressed.

Vibration

Nondata components of both a cyclic and a random nature may be superimposed on the true data because of machine-induced vibrations of the transducer. Thus, every effort should be made to reduce these effects and to take the residue effects fully into account when designing the total positioning system. See the article on vibration measurements later in this Handbook section.

POSITION MEASURING SYSTEMS

Position measurements fall into two broad classifications: (1) measurements concerned with rotary motions where the position of rotary machine driving members, such as shafts and screws, is translated into terms of linear motion or differences between two or more positions, each with its own coordinates; and (2) measurements concerned directly with linear motions.

Resolvers and Synchros

For many years, resolvers and synchros have been used in machine position control systems. These electric motorlike devices generally are rugged, reliable performers with sufficient resolution for many machine applications. Because resolvers and synchros produce an analog output, converters are required to provide a digital format for interfacing with digital numerical control systems—which is very important today as the strong trend to digital systems continues. At one time, these converters were comparatively large and expensive. Technological advances in electronics have solved these problems in recent years, so that these devices, when equipped with synchro-to-digital (S/D) converters, are a practical approach to overall digital positioning systems and are preferred over encoders for many applications.

A resolver or synchro is a rotary transformer, the primary of which is a winding on the shaft (rotor) much like a motor; the secondaries (stators) are wound in the case, again much like a motor. The difference betwen a resolver and a synchro is in the number of stator windings: A resolver has two stator windngs 90° apart, whereas a synchro has three windings 120° apart. In both devices, as the shaft turns, the relative positions of the rotor and stator windings change, and the root mean square (rms) voltage output of the stator winding varies as the sine of the angle between them. Only the ratio of the outputs is used. It should be pointed out that, although a synchro is a three-wire motorlike device, it is not a three-phase device. The ac outputs are either in phase or 180° out of phase. Phase shift does not change with angle except to reverse the phase at certain angles.

The resolver or synchro is coupled to the shaft to be measured, such as a pinion gear, lead screw, or robot arm, and then wired directly to an S/D converter. The S/D converter then outputs a digital word for further processing. Some S/Ds are available with three-state outputs for direct electrical interfacing to a microprocessor.

The operating principle of a resolver and a synchro is illustrated in Fig. 1. By having two stator windings 90° apart (resolver) or three windings 90° apart (synchro) and only considering the ratio of the outputs, the variations due to voltage or frequency changes of the reference become unimportant. The signals are a ratio of high-level ac voltages (generally 11.8 or 90 V line to line) that are relatively insensitive to noisy environments and can be transmitted thousands of feet (meters) with negligible loss of accuracy.

The resolver shaft is connected to the positioned machine member inside a closed control loop where it is moved in a direction to reduce its error voltage essentially to zero. This minimal output generally denotes the exact machine position given by the input data. Zero voltage occurs at a readout position corresponding to chosen data. But zero voltage also may occur 180 shaft degrees from this position. Thus, there is the possibility of a false data region. To eliminate this possibility, the designer chooses a gear ratio for use between the machine member and its transducer that will permit only 180 mechanical degrees of transducer shaft rotation during full member travel, or in some other fashion a nonambiguous data input system utilizing two 90° related sine and cosine windings is designed.

Synchro-to-Digital (S/D) Converters

These are of two basic types: (1) *Tracking converters,* wherein the output tracks the input in real time. They are solid-state servo loops with feedback. Tracking converters have good noise rejection characteristics. Pickup on the synchro leads can be minimized by using twisted shielded cable. (2) *Sampling converters* that sample and hold the signals, generally on the carrier peaks, and then perform the actual conversion. The conversion technique may be a successive approximation routine, or a harmonic oscillator may be used. The successive approximation method is usually faster and lends itself well to multiplexed systems. Sampling S/D converters are more susceptible to noise than tracking units. Where long runs in noisy environments are required, careful shielding must be used. The multiplexed S/D system does have a cost advantage. See Fig. 2.

Inductive Plates

The rotary form of the inductive plate position transducer is essentially a two-phase synchro or resolver whose windings have been projected onto a linear medium. As shown in Fig. 3, an inductive plate includes an etched stator winding that has been projected on a dimensionally stable nonconducting surface. The rotor associated with this transducer is constructed in a like manner. Variations in inductor displacement are averaged over a large number of inductors by summing the voltages from a like number of coils located on the rotor plate. Thus, the reproduced rotor and stator inductors need not be printed with a positional accuracy equivalent to that of the final transducer. Either a trigonometrically related transformer amplitude analog system or a clock-pulse-derived phase analog system can be used to supply sine and cosine voltages to the two windings contained in the stator plate. This is the same arrangement used with rotating resolvers and synchros. Essentially, this transducer is a two-phase synchro or resolver whose windings have been projected onto a linear medium.

The advantage of the device is elimination of gear backlash between the positioned machine member and the movable element of the transducer. A related type of device uses bifilar windings on a rod and sensing sleeve.

Widely used is the linear version of the inductive plate, as shown in Fig. 4. In this configuration, a slider moves across the scale, but with an air gap between the two. Since there is no physical contact, there is no apparent wear on the feedback device. The slider is attached to a movable push rod which can traverse up to 500 in (1270 cm)/min and attain a total of 2.5 million travel cycles without replacing the seal. Each scale is laser-checked to ensure an accuracy within ±0.001 in (±0.0025 mm). The scales are manufactured in 10-in and 250-mm lengths and can be placed adjacent to one another for long travels up to 200 ft (60 m). Transducers of this type are used on jig borers, horizontal boring machines, contouring machines, boring and turning machines, drilling machines, milling machines, positioning tables, vertical turret lathes, grinders, and horizontal turret

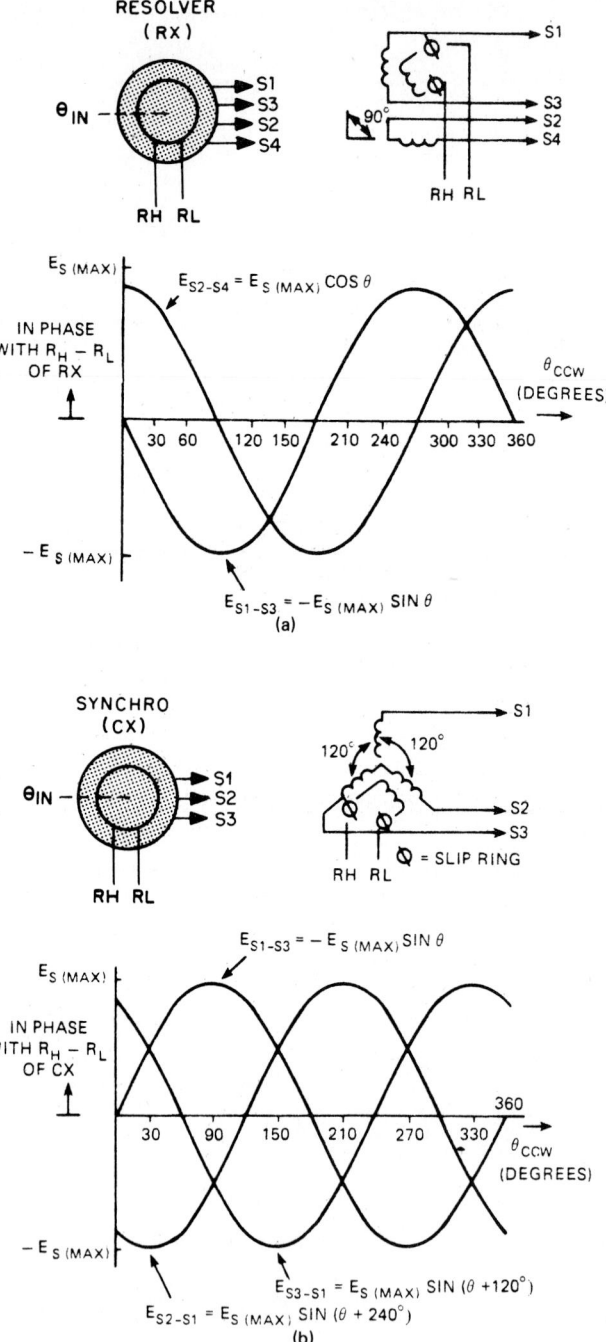

Fig. 1 (a) Mechanical angle-to-resolver generator, with output signals shown at right, (b) mechanical angle-to-synchro generator. (*After Muth, ILC Data Device.*)

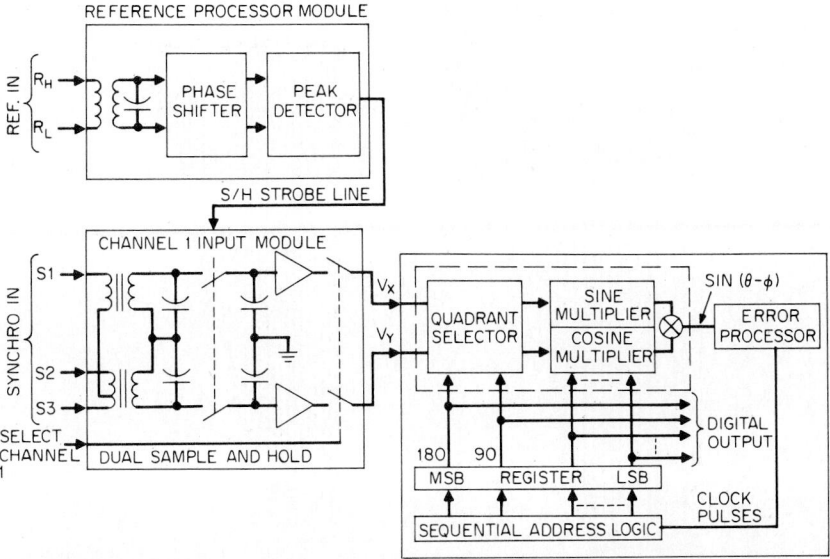

Fig. 2 Sampling S/D converter samples and holds signals on the carrier peaks and performs a conversion before the next peak. *(After Muth, ILC Data Device.)*

lathes, among others. The induction plate transducer is also available* in the form of a tape for ready mounting on machines. The tape yields a bidirectional linear accuracy of ±0.002 in (0.05 mm) when installed on slides, arms, or machine members.

The tape can be mounted on a nonmachined surface and aligned with self-contained leveling screws. The self-contained unit incorporates the tape, movable slider assembly, and built-in scale amplifier. The tape comprises a copper scale pattern etched on spring steel tape.

The use of a programmable controller with an inductive plate transducer is shown in Fig. 5.

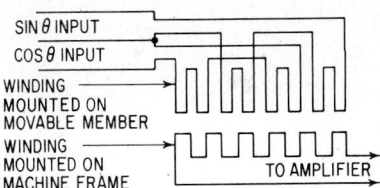

Fig. 3 Schematic circuit for a position transducer utilizing inductive plates.

Rotary Encoders

Encoders are of two basic types: (1) *Absolute encoders,* which provide a unique output signal for each single or multiple revolution of shaft gearing. An absolute encoder outputs a complete binary code (digital output) for each position. These devices are generally used in applications where position information rather than change in position is important. Absolute encoders have an individual digital address for each incremental move, and thus the position within a single revolution can be determined without a starting reference. By gearing two or more absolute encoders together, so that the second advances one increment for each complete revolution of the first (reminiscent of a mechanical counter), the range of absolute position can be extended. (2) *Incremental encoders* produce a symmetrical pulse for each incremental change in position. Pulses from the incremental encoder are counted for each incremental movement from a calibrated starting point in an up-down counter to track position.

Absolute encoders usually incorporate optical principles. A typical optical encoder consists of a light source, usually a light emitting diode (LED), and a detector separated by a transparent encod-

---

*Numerislide Spar, Giddings & Lewis Electronics Company, Fond Du Lac, Wisconsin.

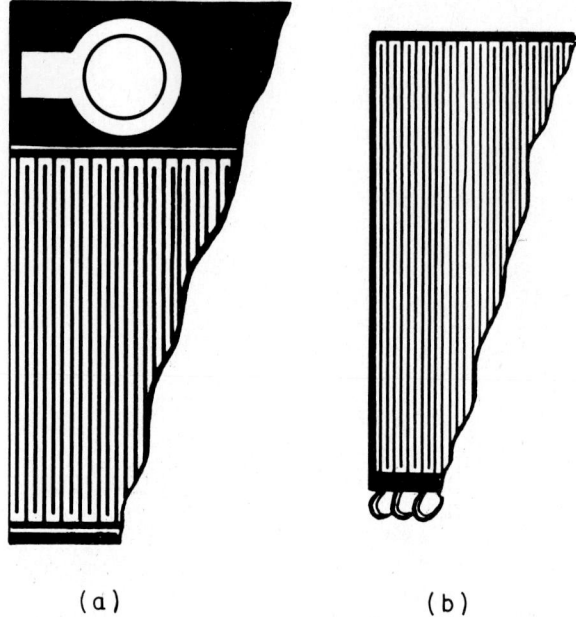

(a)   (b)

Fig. 4 Linear version of inductive plate transducer. (*a*) Scale. Standard sections of the scale are 10 in (250 mm) long, 2.3 in. (58.4 mm) wide, and 0.375 in (9.5 mm) thick. They are made up of a copper pattern bonded to heat-treated steel. (*b*) Slider is 4 in (101.6 mm) long, 2.875 in (73 mm) wide, and 0.375 in (9.5 mm) thick. (*Linear Inductosyn, licensed under patents of Farrand Industries to Giddings & Lewis Electronics.*)

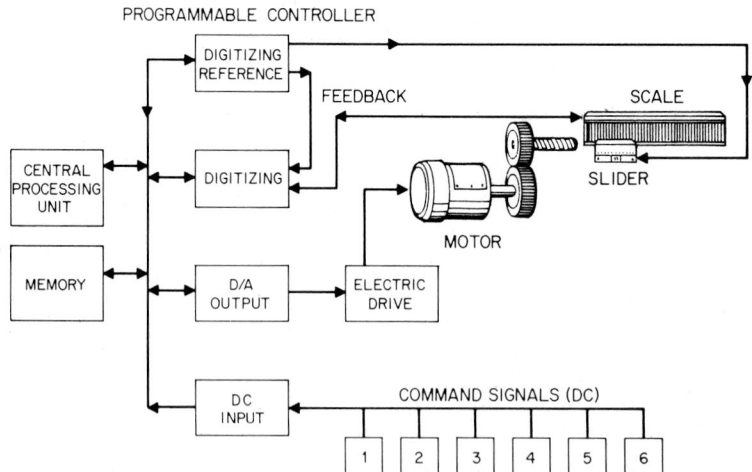

Fig. 5 Programmable controller for machines and processes, utilizing inductive plates. (*Giddings & Lewis Electronics.*)

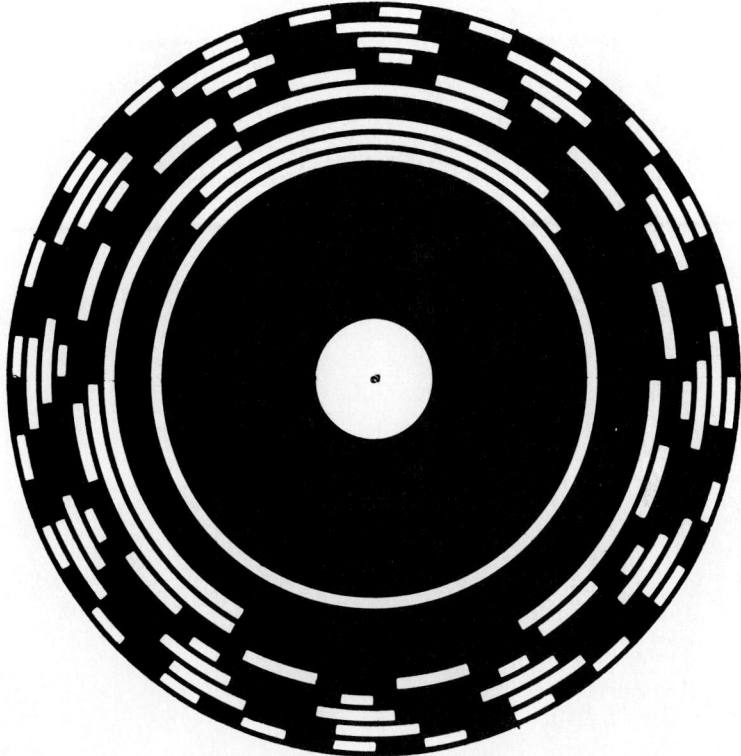

Fig. 6   Reasonable facsimile of a transparent encoding wheel etched with opaque marks.

ing wheel etched with opaque marks. Built-in electronics convert the on-off response of the detector to a digital output as the disk turns and the opaque marks pass between source and detector. See Fig. 6. The light sensor may be a photocell or a phototransistor.

During the last few years many issues of professional trade journals have described innovations in the design of encoders, particularly ways to make them fully compatible with computer control systems. See the list of references at the end of this article.

Inductive Bridge Transducers

The device shown in Figs. 7 and 8 is used in high-performance production machines with restricted axis motion. Operation is based on the use of a fixed inductive member B slightly longer than the axis to be measured, and a movable member A approximately half the length of B. Selectable taps are placed on B in a successive decade with externally located inductors to provide a bridge configuration that may be externally unbalanced by placing A (coil) and N (point) across a pair of tap points; then the coil is moved until equal voltage prevails between the two ends of the coil as evidenced by the occurrence of a small voltage at O and O'. A disadvantage of this system is the relatively large number of wires that must be taken from the device through the machine to the control system. An advantage is the high output voltage per unit of displacement (5 mV/0.001 in, 2 mV/0.01 mm). The supply frequency usually is between 400 and 1500 Hz.

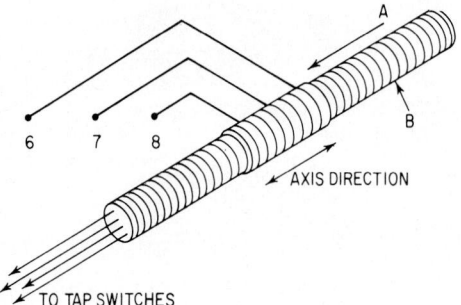

Fig. 7 Inductive bridge transducer for position measurement.

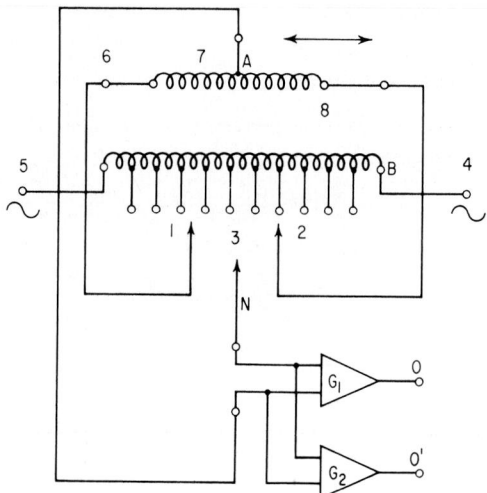

Fig. 8 Schematic circuit diagram of an inductive bridge transducer.

Pulse-Type Transducers

Transducers in this category function in one of the two following ways:

1. Transducers that require counting of least-dimension units for machine member positional measurement. An optical grating with a photocell, a toothed magnetic drum with a reluctance pickoff, and a magnetic scale are examples of pulse counting systems.

2. Transducers that use logic networks for readout of machine member position. A digital contactor using a geared sequence of printed circuit drums contacted by multiple brushes and a sequence of magnetic drum sections with reluctance pickoffs are examples of systems using logic networks to read out positions. A single-speed linear resolver with decade counters and an accumulator register can be used both to drive a visual readout and to position sensitive contacts.

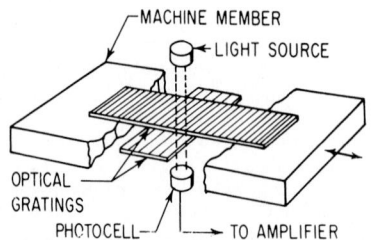

Fig. 9 Position transducer utilizing optical gratings.

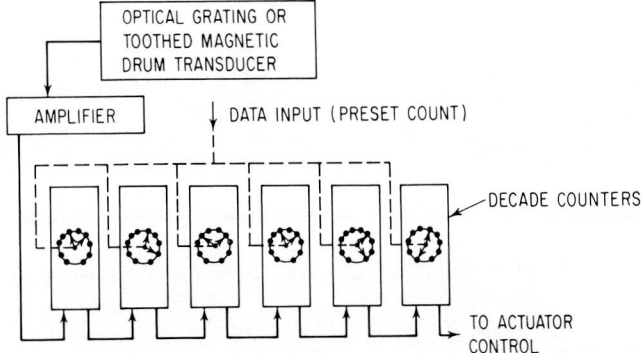

Fig. 10   Tap-switch circuit for optical gratings or toothed magnetic drums.

Optical Grating

A transducer of this type is shown in Fig. 9. The device requires an amplifier to raise the power level of the pulses from the associated photocell or phototransistor so that position readout displays or relays used in conjunction with other logic elements can be operated. Also see Fig. 10. Transducers of this type are used in some precision measuring machines.

Digital Contactors

These transducers do not require an auxiliary relay or amplifier circuit. Digital contacts have the disadvantage that they must employ at least one printed circuit drum and brush combination where the drum rotates at 1000 times the input shaft speed of the transducer—if the unit is to have positional accuracy comparable with that of other transducers. In some machines this has required the highspeed drum to be rotated up to 16,000 rpm. Life expectancy at this rate is short.

Designers of digital drum-brush transducers have minimized this difficulty by using a brush lifting means to remove the brush from the high-speed drum during coarse positioning of the transducer shaft. However, generally unacceptable reliability of operation has attended the use of this type of transducer in control systems requiring four or more significant digits. See Figs. 11 to 13.

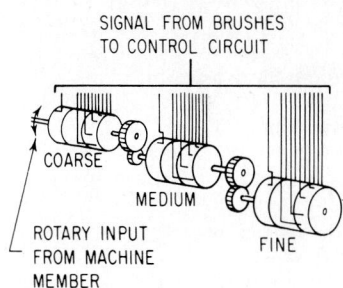

Fig. 11   Digital-contacter-type position transducer.

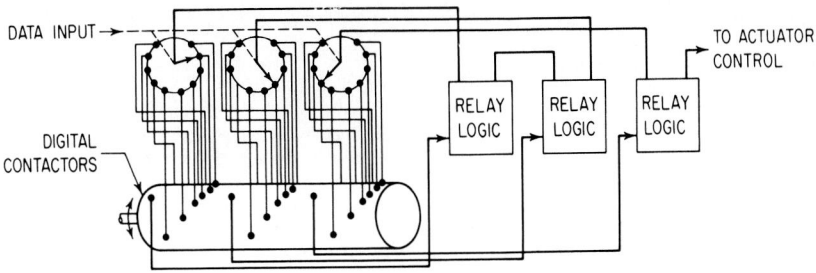

Fig. 12   Switch and relay logic circuit for digital contactors.

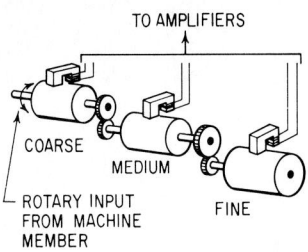

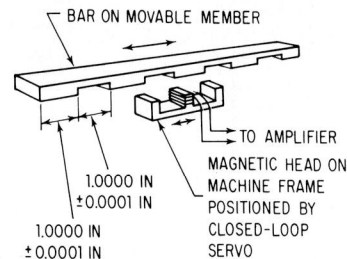

Fig. 13 Position transducer utilizing digital magnetic drums.

Fig. 14 Principle of the magnetic tooth as used in a position transducer.

Magnetic-Pin or Tooth Transducers

Trandsucers of this type are shown in Figs. 14 and 15. A comparison of positional resolution for these transducers with that for other position transducers is given in Table 1. A magnetic-pin or tooth transducer has a useful range of ±0.150 in (±3.8 mm) from the readout point. The transducer must be used with a synchro, so that the measuring range can be extended outside the restricted range. A switching means must be provided to switch successively from one transducer to another. A transducer using a toothed magnetic drum is shown in Fig. 16.

Potentiometers

Multiple-turn potentiometers are used in machine control systems that do not require an accuracy greater than the positional equivalent of ±0.001 in (±0.025 mm). Circuitry for such a transducer system is shown in Figs. 17 and 18. A disadvantage of the potentiometer is unreliability related to frictional wear in the slidewire.

Optical Binary Scale

The transducer shown in Fig. 19 is a direct-position readout that uses a scale as indicated. A binary-coded decimal (BCD) pattern is deposited on a glass plate that is slightly greater than 1 m in length. This scale is used with a plurality of LEDs and photodiodes arranged in fixed relation to register the bit condition of at least a 14-bit metric pattern. Accordingly, it is possible to measure a length of 3 m with a three-scale array. In combination with this device, a logic network of NAND and NOR gates can be used with appropriate input devices to provide position coincidence logic for linear position readout and/or position-contact actuation.

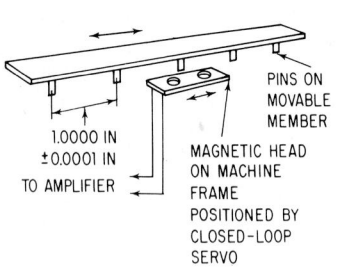

Fig. 15 Use of magnetic pins in a position transducer.

Magnetic Scale

A system for magnetizing the surface of an essentially square flux-current loop ferromagnetic-alloy tape provides alternate magnetized and unmagnetized linear bits that are 0.02 mm in length. The location of a magnetized bit can be determined by "reading" the surface with a dual-pole, second-harmonic modulator that gives an output when its poles are subject to a steady field of 0.0005 Oe or greater. A series of decade counters and storage registers can be used to store the bits read per unit of reading-head traverse. A disadvantage of the system is the necessity for a high-

Table 1 Representative Position Transducer Tolerances

Tranducer type	Tolerance
Synchro (three-phase)	4′ of shaft arc
Resolver	4′ of shaft arc
Optical grating	0.00001 in (0.000254 mm)
Magnetic toothed drum	15′ of shaft arc
Digital drum	35′ of shaft arc
Magnetic pin or tooth	0.00005 in (0.00127 mm)
Linear or printed circuit resolver (single-speed)	0.1000 in (2.54 mm) cycle = ±0.0050 in (±0.0127 mm)
	0.2000 in (5.1 mm) fine cycle = ±0.0002 in (0.005 mm)
Linear or printed circuit resolver (three-speed)	10 in (254 mm) cycle = ±0.020 in (0.51 mm)
	400 in (1016 cm) cycle = ±0.800 in (±20.3 mm)
Rotary printed circuit resolver	180° cycle; 2° per cycle = 1.5″
Magnetic scale	0.02 mm/cycle = ±0.002 mm
Optical binary scale	1 m/section = ±0.01 mm
Inductive bridge transducer	30 in (762 mm) max. = ±0.001 in (0.025 mm)

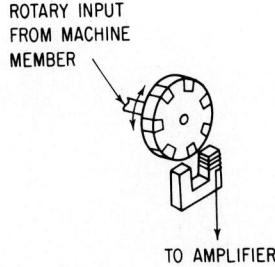

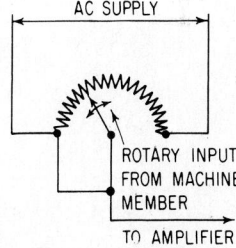

Fig. 16 Principle of position transducer employing a toothed magnetic drum.

Fig. 17 Basic circuit of a multiturn potentiometer as used in a position transducer.

permeability magnetic shield that must be located on three sides of the tape. Also, the tape must be maintained in a prestressed condition in order to exhibit acceptable magnetic characteristics. See Fig. 20.

DISPLACEMENT MEASUREMENTS

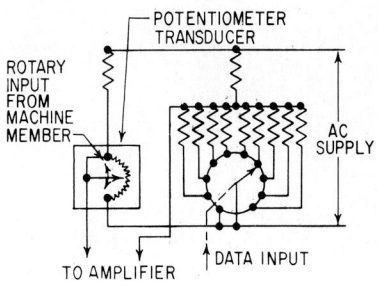

Fig. 18 Decade switch circuit for a multiturn potentiometer.

Instrumental techniques are available for the measurement of displacements as small as a wavelength of light to several feet (meters). Generally, displacement is thought of in terms of a motion of a few millimeters or less. Frequently, a measurement of displacement is made to relate to some other measurand, and hence displacement transducers are fundamental components of many instrumentation systems. Examples include the measurement of motion of the free end of a bourdon, the movement of a scale beam, and deflection in an accelerometer. Displacement obviously is associated closely with motion and position. Displacement implies motion from one point to another; and it also implies position, that is, a change from one position to the next. Displacement generally implies the establishment of a new position as related to a stable or normal position.

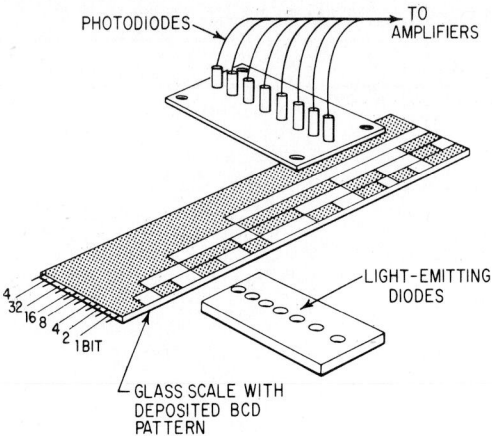

Fig. 19 Optical binary scale used for position measurement.

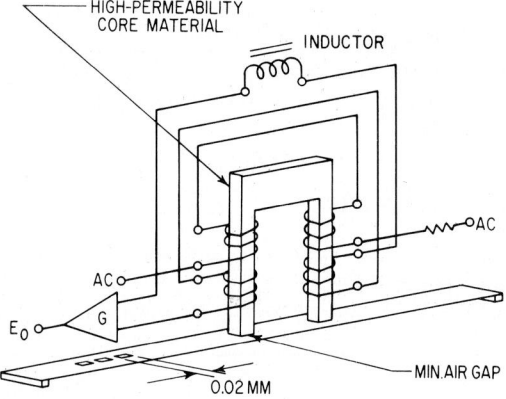

Fig. 20 Magnetic scale used for position measurement.

Displacement can be measured by a number of transducers—capacitance, electronic, inductance, magnetoelectric, photoelectric, piezoelectric, radioactive, and resistive.

Linear Variable Differential Transformer (LVDT)

These transducers depend on inductance effects. As shown in the sectional view of Fig. 21, a magnetic nickel-iron core, supported by a nonmagnetic push rod, moves axially within the cylinder in exact accordance with mechanical displacement of the probe tip. As shown by Fig. 22, with ac excitation of the primary coil, induced voltages appear in the secondary coils. Because of the symmetry of the magnetic coupling to the primary, these secondary induced voltages are equal when the core is in the central ("null" or "electric zero") position. When the secondary coils are connected in series opposition, the secondary voltages cancel and (ideally) there is no net output voltage. If, however, the core is displaced from the null position, one secondary voltage will increase while the other decreases. Since the two voltages no longer cancel, a net output voltage results. If the transducer is properly designed, this output will be exactly proportional to the magnitude of the displacement,

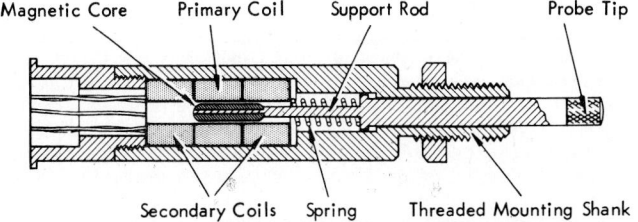

Fig. 21 Sectional view of a LVDT linear displacement transducer. *(Daytronic.)*

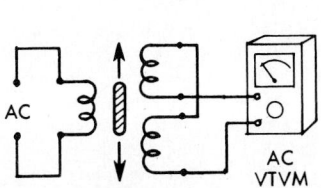

Fig. 22 Schematic of the operating principle of a LVDT linear displacement transducer. *(Daytronic.)*

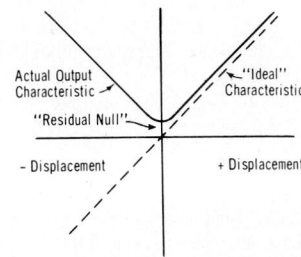

Fig. 23 Graph showing the "null" or "electric zero" position of a LVDT linear displacement transducer. *(Daytronic.)*

with a phase polarity corresponding to the direction of displacement, as shown in Fig. 23. A miniaturized displacement transducer designed for measurement applications in ranges up to ± 1 mm (± 0.04 in), where minimum transducer size is a prime requirement, is shown in Fig. 24. Small size also reduces thermal expansion errors.

Probes of this general design for longer stroke applications have linear ranges up to $\pm$ 25.4 mm (± 1 in). They have a useful temperature range of -40 to $+100°C$ (-40 to $+212°F$). Depending on the particular design, the excitation frequency ranges from 0.4 to 10 kHz. Again, depending on specific design, linearity ranges from 0.1 to 0.5% of range.

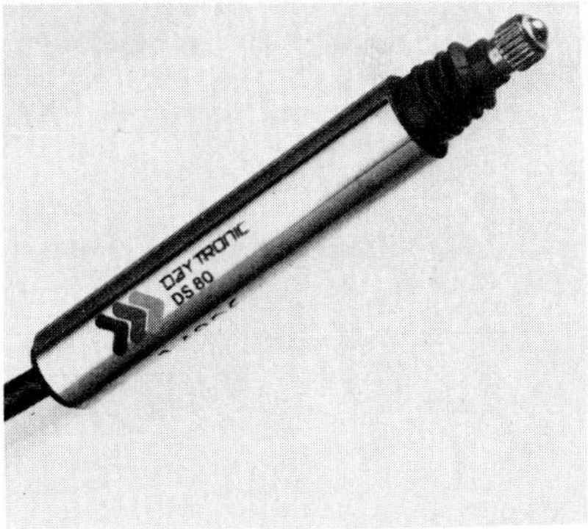

Unlike strain gages, LVDTs cannot be furnished with meaningful calibration data. System sensitivity is a function of excitation frequency, cable loading, and amplifier phase characteristics, among other factors. It is a general practice to calibrate each LVDT-cable-instrument system after installation using known input standards. LVDTs have a wide variety of applications, ranging from industrial gaging (thickness, taper, etc.) through structural testing. LVDT probes are furnished with several tip configurations.

Although the LVDT is used to electronically measure linear displacement, such as actuator or

Fig. 24 Subminiature LVDT linear displacement transducer. Probe (unextended) is 14.0 mm (0.55 in) long; extended, 51 mm (~ 2 in) long. The diameter of the tip is 3 mm (0.12 in). Transducer parts are made of stainless steel. *(Daytronic.)*

mechanism positioning, the device also is employed in conjunction with pressure sensing devices. In the latter case, the pressure sensor should deflect linearly proportional to pressure, thus providing a linear motion that can be sensed by a LVDT. The LVDT also can be used in an acceleration measuring device where displacement of a seismic mass produces a linear displacement of a magnetic core within the windings.

Linear Potentiometers

Linear potentiometers for the measurement of position and displacement take numerous forms dependent on intended application. The simplest, least costly form is a single length of wire along which a slider or other form of moving device contacts the wire. The position of the slider determines the effective length of the conductor. Hence a change in electrical resistance or voltage drop is related to the position or displacement of the slider. This simple device is useful for laboratory demonstration but seldom is used for military and industrial applications. The latter make use of wire-wound, thin-film, or printed circuits because these techniques make possible a considerably greater length of resistor within the same linear length. In summary, the voltage drop per unit length is larger, thus producing stronger, more useful signals.

One example of a linear displacement transducer of the wire-wound potentiometric type used for the high environmental and reliability requirements of aircraft and missiles is shown in Fig. 25. The basic design is cylindrical because the device generally is employed in connection with hydraulic actuators. The device comprises the following major elements: (1) two resistance elements j which are mounted (molded) along with slide bars d in the element block s, which in turn is contained within the outer case n. Wiper assemblies h are attached to wiper carrier e and are aligned to coincide with the resistance elements and contact bars immediately opposite each other. See sectional view. The wiper carrier is secured to the actuating shaft a in such a manner as to eliminate backlash and yet allow the shaft to rotate freely 360° when installed. This is necessary in many cases because the end fitting f provides a thread for attachment to a moving member and free rotation of the shaft is necesary to avoid deflection of the wiper system or dislocation of the wipers from the element and contact bars. The forward end of the transducer provides a bush b into which are inserted a pair of O-ring seals, providing adequate sealing against intrusion of salt spray, sand, dust, and moisture. These transducers are ½ and ¾ in (12.7 and 19.1 mm) in diameter, dependent on application—the larger size is generally used where motion is in excess of 6 in (152 mm).

Applications of these transducers include aircraft control surface indication, such as rudder, aileron, and elevator—either combined into a screw jack or separated from it—landing gear retractor systems, missile stage separation, and industrial uses, including machine tools and valve indicators.

Thin-Film Potentiometers

Three basic types include:

1. **Cermet.** A ceramic-metal mix which provides extremely long life and is capable of taking high temperatures (250°C). The material provides continuous or stepless output and is used primarily for trimming potentiometers where linearity is not a problem.

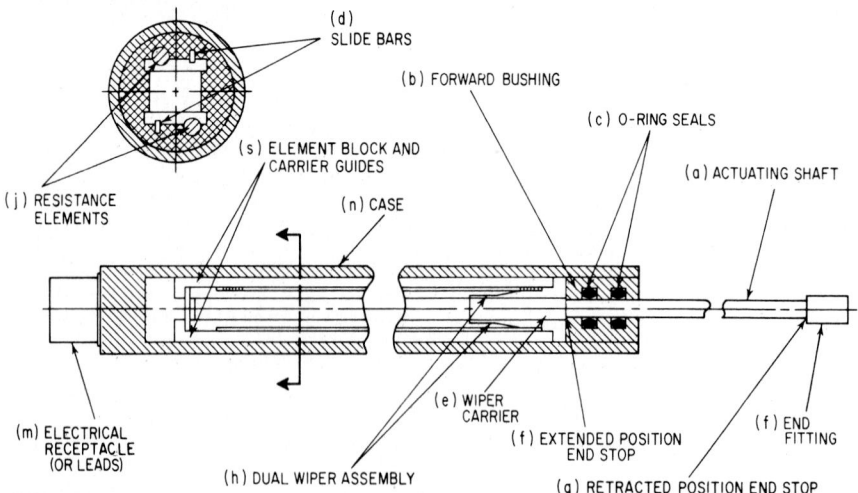

Fig. 25 Cylindrical wire-wound potentiometric-type linear displacement transducer. *(Gulton.)*

2. *Conductive Plastic.* Generally a carbon particle and plastic mix which provides continuous, stepless output and comparatively long life (in excess of 3 million cycles).

3. *Metal-Deposited Thin Film.* A homogeneous metal generally vacuum-deposited onto a glass substrate. The deposit may be extremely thin—on the order of 5000 Å, giving 0.001- to 0.002-in (0.025- to 0.05-mm) resistances of 20,000 to 1 million Ω, or thicker films giving resistances of 100 to 1000 Ω. These types generally are not associated with long life, but they do provide a continuous output.

Linear Variable Reluctance Transducer (LVRT)

This device is useful for laboratory measurements of growth, stress, thickness, vibration, shock, and contour and is employed to provide electrical feedback for valves and actuators in servo control systems for missiles, aircraft, and industrial machinery. As shown in Fig. 26, the variable reluctance unit forms half of a four-arm bridge. For many servo applications, a center-tapped transformer is used to form the reference half of the bridge. The output signal then is produced between the transformer center tap and the transducer centerpin. The signal phase changes 180° at midrange position. A variable resistive ratio divider or a variable inductive ratio divider may be used to complete the bridge. The null position is adjustable to any point within the transducer range.

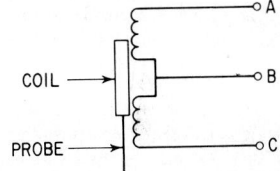

Fig. 26   Circuit of a LVRT. *(Gulton.)*

The winding is continuous over the length of a spool-type bobbin. There are no intermediate partitions that divide the bobbin segments as is the case in LVDTs. The advantages of the LVRT include excellent linearity over long strokes—up to 24 in (61 cm) and good temperature stability. A disadvantage is that the overall length must be double the total stroke. The devices can withstand rapid temperature changes from −160 to +450°F (−107 to +232°C). Encapsulation of the coils in an inert epoxy resin hermetically seals the unit and protects it against vibration and shock. The working portion of the probe is made of magnetic stainless steel; nonmagnetic stainless steel is used for the extension. Standard probe diametric clearance is 0.001 to 0.020 in (0.025 to 0.05 mm). Carrier frequencies from 1000 to 20,000 Hz may be used. Optimum results are obtained with a carrier frequency of 3000 to 5000 Hz. Operation with other frequencies results in slight changes in the maximum excitation voltage, phase shift, and linearity. The size of the devices (less probe extension, but including the connector) ranges from ¼ × 2.2 in to ½ × 10.2 in (6.4 × 56 mm to 12.7 × 259 mm) in standard configurations. The linearity is ±1% or better.

Linear Transformer (LT)

This is a specialized synchro consisting of a salient-pole rotor and a single-phase stator, distributively wound. The winding on the stator is designed to produce an output voltage that varies linearly with rotor position. This linear function is valid only within a restricted band about the zero position—generally ±50° or ±85°—which is known as the *excursion range.* Past the excursion range the output voltage no longer is linear and tends to become sinusoidal.

The LT is used in many cases as a replacement for a potentiometer. It is an infinite resolution device and also has the advantage of being constructed in the same manner as other servo system components. Thus, service behavior of the LT is the same as that of the other elements in the system.

NUMERICAL AND AUTOMATIC POSITIONING CONTROL SYSTEMS

During the decade of the 1970s and continuing at an accelerated pace into the 1980s, the traditional concepts of numerical control (NC) have been added to and significantly modified—mainly as a result of the appearance of microprocessors, programmable controllers, and improvements in transducer and servo systems. Prior to describing briefly some of the technological factors that continue to affect the control of machine tools and other automated (often quite complex) industrial production systems, it is in order to review the technology of traditional NC. For several decades, NC advanced rather slowly.

Traditional NC

What might be termed plain NC generally connotes a positioning control system for machine tools and similar equipment wherein numerical values corresponding to desired positions of tools and symbolic information corresponding to auxiliary functions are recorded in some convenient input form (punched paper tape, punched cards, or magnetic tape) where the information can be stored indefinitely. Readers convert this information into signals which operate servomechanisms on each axis of the machine whose motions are to be controlled.

Because perforated plastic or paper tape has been used so widely, NC often is referred to as *tape control.* Although comprising a limited fraction of the total installations, dial-input NC systems also are in use. NC originally was applied to metalworking machinery, such as mills, drills, boring machines, and punch presses, and these applications remain predominant. In recent years, NC has expanded into other areas of metalworking, including applications on cutoff machines, broaching machines, grinders of several types, tube benders, gear cutters, electrochemical machining, flame cutting, and welding. Outside of metalworking, NC is used for inspection, automatic testing systems, drafting systems, electronics assembly machines, step-and-repeat photography, laser cutting and bonding processes, woodworking machinery, and garment cutting equipment. The direct coupling of digital computers with production machine control systems is broadening both the applications for and the basic definition of NC.

Objectives of NC

As originally put forth, the basic objective for the development of NC for the metalworking field was to reduce the economic lot size for machined parts production. NC achieved this objective by reducing machining time, reducing fixture costs, increasing cutter life, reducing tooling storage, and lowering the skills required. *(In the quest to integrate NC with computer control systems and with the other advancements in technology, there appears to be a tendency to concentrate in the literature on the details of the new technology and to overlook some of the fundamental reasons that have brought NC to where it is today—or perhaps it is now believed that the fundamentals have been learned and do not require repetition.)* Other advantages of NC include product uniformity, reduced setup time, and a general reduction in other direct costs. Historically, NC has permitted closer shop scheduling, more accurate cost prediction and control, faster cash turnover, higher facilities utilization, and a more rapid return on equipment investment—as compared with manual skills or less sophisticated machine control techniques. Computer-aided manufacturing and robotics have stemmed from earlier knowledge gained in the application of traditional NC techniques.

NC brought a marked improvement over earlier instrumental attempts to direct machines automatically. Briefly, these included:

1. *Tracer Controls.* This type of control employs a form of mechanical analog input, usually a machined template or a three-dimensional model, in which the analog is scanned and sensed by an electric or hydraulic probe, and the probe's motions are then repeated by a servocontrolled tool.

2. *Automatic Machines.* These include automatic screw machines and large turning machines as found in high-production plants. They use manually preset hard stops to determine the limits of slide motions during cuts. Although relatively unsophisticated, the term "automatic" for machines of this type was preempted many years ago for automatic screw machines and has persisted in usage.

3. *Graphic Input Systems.* In this type of system an optical line follower can trace or follow curves in graphic form and hence guide machine spindles or, more commonly, flame cutters.

As contrasted with the foregoing types of control systems, the symbolic inputs of NC offer the following advantages:

1. The inputs usually are expressed in BCD format and hence can be processed easily by digital logic circuits.
2. A physical model is not required. Sometimes even an accurate drawing of the desired part may not be needed. Thus, skills for preparation prior to machining are conserved. Information storage on tapes or cards for future production runs also requires less space than that required for the storage of models and templates.
3. Symbolic control lends itself well to the expression of nondimensional commands, such as spindle speed changes, tool changes, drill cycles, and such functions as turning the coolant on and off.
4. With NC the accuracy of the input medium is easy to verify. It should be stressed that both mechanical analog input and graphic input control systems have advantages for certain applications.

Basic Classes of NC Systems

Point-to-Point Control

This is the simplest form of NC. It is used to guide a machine to particular programmed points, as in drilling, with no particular regard for path or velocity control. The system is also called *positioning control*. Several versions of point-to-point control systems have the ability to control straight cuts, usually parallel to machine axes or at 45° to the axes. Such systems have velocity control as well as end-point and limited-path control.

Continuous-Path Control

This system contains computing elements (interpolators) which permit computation of successive points on desired line segments or curves, starting with minimal input data such as end points, radii, and center coordinates. Interpolation may be linear, circular, or parabolic. Continous-path controls also are referred to as *contouring controls*.

Closed-Loop Control

These systems use position and often velocity feedback to control the dynamic behavior and final position of machine slides (or equivalent machine members in other automated systems). As previously described, a variety of position transducers (analog and digital) are available.

Open-Loop Control

These systems have no feedback to the control console. They rely on the drive motor to provide both the actuation and measurement functions. A form of feedback, integral to the motor, may be used to ensure command execution. Such controls may employ stepping motor drives, sometimes called *digital* or *pulse motors*. Stepping motors are described in Sec. 19 of this Handbook.

Absolute NC Systems

The term *absolute control* generally identifies controls that use absolute rather than incremental feedback. *Absolute feedback* connotes the assignment of a unique value to each possible position of the machine slide or equivalent actuating member. *Incremental feedback,* in contrast, simply gives the control a pulse each time the slide or member travels one increment of resolution. Absolute systems do not require rezeroing after shutdown.

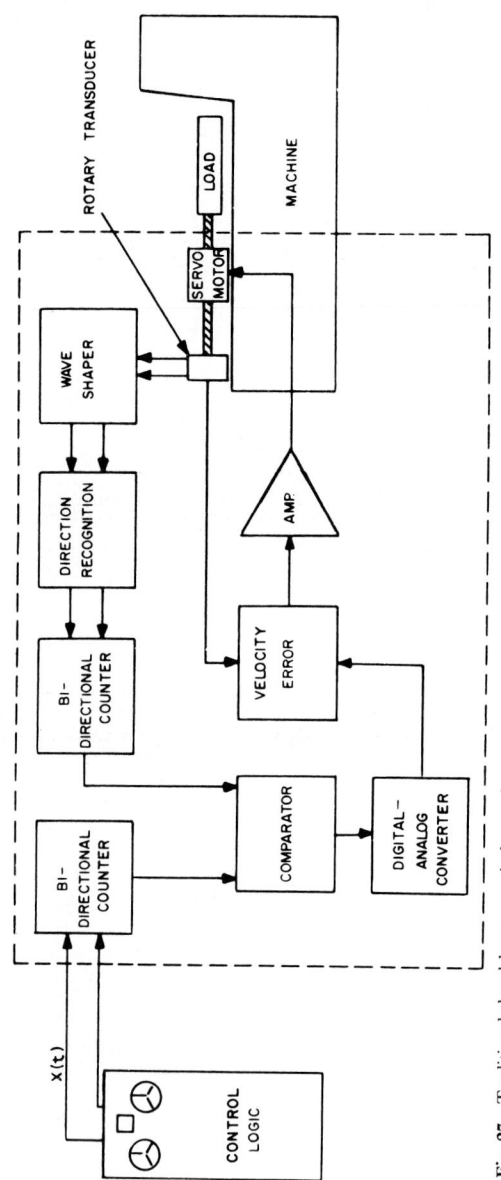

Fig. 27 Traditional closed-loop numerical control.

9.22

Computer-Based Numerical Control (CNC)

This area is a broad one, ranging from the direct coupling of a small general purpose digital computer all the way to direct integration of all production functions of a manufacturing operation, including machine axis control, workpiece changing, machine scheduling, and cost control. As described later, programmable controllers have been effectively used in machine control systems during the last few years.

NC Operating Principles

A common form of closed-loop NC (sometimes called semi-closed-loop control because the position of the slides is only inferred from a rotary device) is shown in Fig. 27. In this type of system, the control logic unit includes the operator controls and tape reader, along with circuits for reading, decoding, and timing commands. The output of the control logic unit can be a train of pulses that correspond in length and rate to the desired distance and velocity of motion.

Some closed-loop controls employ velocity feedback as well as position feedback for two reasons: (1) to permit precise control of cutting feed rates, and (2) to permit use of low-gain servos which thus reduce end point overshoot, hunting, and other undesirable effects of high-gain servos. Either rotary or linear position transducers can be used, depending on performance requirements and user preference.

The servomotor may be electric or hydraulic. In some machines, the actuator is not a motor on a lead screw, but a hydraulic positioner. The latter may cause difficulty in dynamically controlling a variable volume oil column. As mentioned later, there is a trend toward direct electrical actuation of highly automated machines, such as represented by transfer lines.

An open-loop numerical control system is shown in Fig. 28. The control logic sector is about the same as that used in closed-loop systems. In the open-loop system, however, the drive motor (usually a stepping motor) and its amplifier assume all the functions of the elements enclosed within the dashed box of Fig. 27. The open-loop system offers equal accuracy and performance, but there is no inherent position verification. Both systems depend equally on the accuracy of the lead screw and other mechanical elements for final workpiece accuracy.

Unlike the closed-loop system, the open-loop system is relatively indifferent to machine dynamics and resonances over a wide range. The control is never told what the machine is doing. Any correction is made within the motor. For this reason open-loop control has been popular for retrofitting where machine dynamics may be unknown. Conversely, closed-loop control is used extensively on new NC machine designs.

Programming

In traditional NC, programming is the process of converting the desired tool motions and auxiliary commands to an appropriate input, such as a tape or cards. Programming may require the use of a general purpose computer with special purpose software for complex parts. In *absolute programming,* all dimensions are referenced to a single point. In *incremental programming,* each dimension is simply referenced to the preceding point in the program. In recent years programmable controllers have impacted on these procedures.

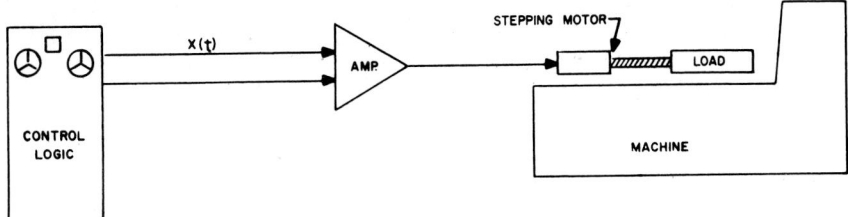

Fig. 28 Open-loop numerical control system.

Certain features have been offered to assist the programming function, including:

Tool Length Offset. This permits on-floor correction for worn or reground tools without changing tapes or presetting tools.

Manual Data Input. This feature allows manual (nontape) entry of dimensions and auxiliary functions into the system.

Position Readout. This is a display, usually digital, of slide positions.

Cutter Radius Offset. This feature permits on-floor compensation for different diameter cutters. In some controls the same tape can be used to make both male and female parts.

Mirror Image. This permits changing the sign of commands on one axis so that one tape can make a right and a left part.

Tape Search. This permits the operator to find the way to a particular block of information on a tape. It often is accompanied by a numerical display of the current tape position.

Shared Axes. This allows one data channel to move two or more different machine axes, one at a time. It is used to minimize the control cost for a multiaxis machine. The term "timed-shared axes" also is used to describe this feature.

Impact of Programmable Controllers and Other Changes

The National Electrical Manufacturer's Association (NEMA) defines a *programmable controller* (PC) as a digital electronic device that uses a programmable memory to store instructions to implement specific functions, such as logic, sequencing, timing, counting, and arithmetic to control machines and processes. Unlike NC and CNC units, which are used to control *position,* the programmable controller is used for *sequence* control. In operations that require periodic tool changes, the PC eliminates the need for a special control system. To change assembly line operations controlled by a PC, the new instructions are written into the PC's internal memory by pushing a few buttons on a portable programming panel. The task requires no special computer or computer programming skills.

In developing the relationship between PCs and NC, it is well to consider how the following principal functions of a PC can contribute to the overall task of controlling industrial machinery. These capabilities include relay logic, latches, timing, counting, ASCII interface, proportional integral derivative (PID) loops, shift register, data highway communications, arithmetics, comparison, computer interface, matrix manipulation, BCD, binary conversion, analog data manipulation, and printer interface, among others. PCs are described in some detail in an article in Sec. 18 of this Handbook.

The PC* illustrated schematically in Fig. 29 has a programmed memory which stores instructions. These instructions provide the intelligence to control a machine or process by performing logic, sequencing, timing, counting, mathematics, and positioning. The PC receives or provides data through digital or analog input and output devices. The PC serves as a replacement for the traditional relay controller. Programming is accomplished by the use of relay ladder diagrams. The more complex functions of relays can be provided with greater simplicity. Mathematical functions, once considered the tasks of computers, can be provided in a simplified format. The PC shown in Fig. 29 is available in two different arrangements. The first utilizes an enclosed rack with front terminal connections at the individual card. The second is an open rack with circuit card connectors wired to separately mounted terminals.

The processor section includes a microprocessor-based central processing unit, the memory, a timer card, and programmer console board. Available memory types are random-access memory (RAM) and/or programmable read-only memory (PROM), or complementary metal oxide semiconductor (CMOS) with a battery backup. The basic input-output (I/O) section may include ac and dc input and output boards, digital-to-analog (D/A) boards, and resolvers and Inductosyn (previously described) digitizing and driver boards. A power supply services the processor and I/O section.

---

*PC-400, Giddings & Lewis Electronics Company.

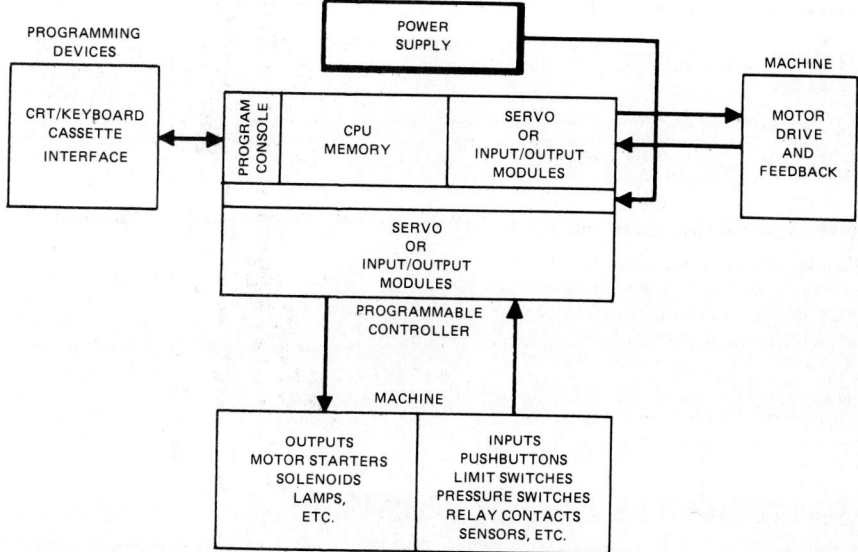

Fig. 29 Block diagram of programmable controller designed for machine and process control. *(PC-400, Giddings & Lewis Electronics.)*

Programs are loaded into the PC through a portable program panel. This may be, for example, a special terminal, which also provides hard copy of the ladder diagram, or a cathode ray tube or a (CRT)-keyboard programming unit. A third device is a tape cassette which can record existing programs and then simplify the reprogramming of identical controllers.

The input modules receive either ac or dc signals from machine devices such as pushbuttons, limit switches, and other indicating sensors. The output modules provide signals to motor starters, solenoid valves, or other similar devices. The ac and dc I/O modules have status indicators and are optically isolated. A PC ladder diagram is shown in Fig. 30.

```
!
!
! 1        2         3         4         5         6         7          104
!--] [--¡--] [--¡--] [--¡--] [--¡--] [--¡--] [--¡--] [--¡--] [--¡-( )--
! 8      !                   !         !         !         !          !
!--]/[--+         !         !         !         !         !          !
! 9       10       !         !         !         !         !          !
!--]/[-----]/[--+         !         !         !         !          !
!11       12        13       !         !         !         !          !
!--]/[-----]/[-----]/[--+         !         !         !          !
!14       15        16       17       !         !         !          !
!--] [-----] [-----] [-----] [--+         !         !          !
!18       19        20        1         2       !         !          !
!--]/[-----]/[-----]/[-----]/[-----]/[--+         !          !
! 3        4         5         6         7       !         !          !
!--] [-----] [-----] [-----] [-----] [-----] [--+         !          !
! 9       10        11        12        13        14        15        !
!--]/[-----]/[-----]/[-----]/[-----]/[-----]/[-----]/[--+
```

Fig. 30 Representative ladder diagram used with programmable controller shown in Fig. 29. In this example, the first rung dictates the position of the branching, and the other rungs branch up to the predetermined branch lines. The maximum number of relay contacts per line is 7, and the maximum number of rungs is 8. With this configuration, the maximum number of relay contacts for each ladder is 56, and the minimum is 1. There is a total of 2000 internal control relays available in the controller system. Relay contacts can be duplicated many times, and there is no restriction in normal use. *(Giddings & Lewis Electronics.)*

Other Technological Changes

As of the mid-1980s, there are two other trends which are impacting on machine control: (1) the use of all-electric drives in place of hydraulic systems, and (2) the use of more sophisticated position sensing transducers (i.e., as compared with conventional limit switches) where there is continuous, accurate feedback of position data. In one of the first all-electric systems, an ac servo-driven workpiece clamping system is used in place of conventional hydraulic clamps, thus eliminating the piping and other accessory equipment requirements of hydraulic systems. The motor-clamping device can be programmed to provide a precise and repeatable torque to produce clamping forces over a wide range. In the system, the lead screw has sufficient friction to prevent back-driving—acting as it were as a wedge between plunger and machine. Thus, power is required only momentarily to clamp or unclamp the workpiece.

In the system, transfer slide positioning is effected with ac servos. Position information is provided by a shaft-mounted 625-line optical encoder. However, the "home position" is detected by a traditional proximity switch. Each device includes an erasable progammable read-only memory (EPROM) chip which plugs into the microprocessor module. Each EPROM can store up to 80 program steps. It is originally programmed by the transfer line manufacturer but is reprogrammable by the user, as may be required by later production line changes.

OBJECT DETECTION AND PROXIMITY SWITCHES

As a general observation, in object detection—as contrasted with position measurement and control—the position (location of an object) *initiates* a control action. In position control, a control action is taken to establish a position. Some of the same measurement techniques apply to both systems. In object detection, the exact coordinates of position are seldom of interest, the primary purpose being that of detecting the presence or absence (no presence) of an object. The principal categorization of object detection systems places them in two categories: (1) contacting types, where the sensor makes actual physical contact with the object; and (2) noncontacting types, where the object only need be in the vicinity of the sensor. Depending on the type of sensing technology used, the vicinity can range from a millimeter or so up to several hundred feet (meters), although these are extreme cases.

Objects are detected for several reasons: to control the flow of discrete pieces as they move along a conveyor or other type of transport system; to detect human limbs (hands and fingers) where machines are manually loaded or unloaded, thus providing immediate machine reactions to avoid accidents; to count discrete pieces; to detect whether containers are filled to proper levels; to determine if all operations have been satisfactorily completed (e.g., inspection of assemblies to determine if all parts are present); to provide registration control; to read bar codes; to count gear teeth or other rotating elements as a means of determining revolutions per minute; among many others.

Noncontacting types of sensors are frequently called *contactless* or *proximity* sensors or switches. Depending on their configuration object detection devices can generate analog or digital output signals.

Object detection devices have been key components of many automated systems, particularly in the discrete piece manufacturing industries, for many years. As the trend toward computer-aided manufacturing and robotics accelerates, these devices will assume even larger roles.

Photoelectric Devices

Although the engineering of photoelectric devices has developed over several decades, refinement of these devices continues because of the impact of other technologies—fiber optics and new approaches to control circuitry are but two examples. It is estimated that about one-fifth of the position or presence (no presence) sensors specified are photoelectric devices.

Background

The *photoelectric effect,* discovered by Hertz in 1887, was not fully explained until 1921 when Einstein formulated the photoelectric law. The photoelectric effect is manifested by changes in electrical characteristics of certain substances when they are subjected to radiation, generally in the form of light or the near visible. The effect, first understood in terms of metals, causes the irradiated material to lose bound electrons, which are given off with a maximum velocity proportional to the frequency of the radiation, i.e., to the entire energy of the photon. The Einstein law, first verified by Millikan, states

$$E_k = h\nu - \omega$$

where E_k = maximum kinetic energy of emitted electron
   h = Planck constant
   ν = frequency of radiation (frequency associated with absorbed photon)
   ω = energy necessary to remove electron from system, i.e., the photoelectric work function for the surface of the emitting substance

An inverse photoelectric effect results from the transfer of energy from electrons to radiation.

The principal aspects of the photoelectric effect of interest industrially are (1) *photoconductivity,* evidenced by an increase in electrical conductivity of a material on the absorption of light (or other electromagnetic radiation), and (2) the *photovoltage* or *photovoltaic effect,* wherein the energy of photons is converted to electrical voltage by the substance receiving the radiation. Photoconductivity is the basis of the operation of photocells and phototransistors utilized in industrial photoelectric switches. Photovoltaic cells have found numerous applications in the electronic aerospace, and solar energy fields and, for example, have been used in satellites for instrument power.

Scope of Usage

Photoelectric controls are found in many kinds of applications because they respond to the *presence* or *absence* of either opaque or translucent materials at distances from a fraction of an inch (a few millimeters) up to 100 or even 700 ft (30 to 210 m). Photoelectric controls need no physical contact with the object to be triggered—which is very important in some cases, such as those involving delicate objects and freshly painted surfaces. Some of the more common applications include thread break detection, edge guidance, web break detection, registration control, parts ejection monitoring, batch counting, sequential counting, security surveillance, elevator control, conveyor control, bin level control, feed and/or fill control, mail and package handling, and labeling, among others. Several of these applications are described later.

Photoelectric Control System Configurations

A *self-contained* control includes a light source, a photoreceiver, and the control base function, which amplifies and imposes logic on the signal to transform it into a usable electrical output. A *modular* control uses a light source–photoreceiver combination or reflective scanner separate from the control base. Self-contained retroreflective controls require less wiring and are less susceptible to alignment problems, while modular controls are more flexible in allowing remote positioning of the control base from the input components and hence are more easily customized.

Photoelectric controls are further classified as nonmodulated or modulated. *Nonmodulated* devices respond to the intensity of visible light. Thus, for reliability, such devices should not be used where the photosensor is subject to bright ambient light, such as sunlight. Modulating controls, employing LEDs, respond only to a narrow frequency band in the infrared. Consequently, they do not recognize bright, visible ambient light.

Controls typically respond to a change in light intensity above or below a certain value of threshold response. However, certain plug-in amplifier-logic circuits cause controls to respond to the rate of light change (transition response) rather than to the intensity. Thus, the control responds only if the change in intensity or brightness occurs very quickly, not gradually.

Operating Mode

Both modulated and nonmodulated controls energize an output in response to:

1. A light signal at the photosensor when the beam is not blocked (light-operated, LO), or

2. A dark signal at the photosensor when the beam is blocked (dark-operated, DO)

Although some controls have built-in circuitry that determines a fixed operating mode, most controls accept a plug-in logic card or module with a mode selector switch that permits either light or dark operation.

In addition to a light source, light sensor, amplifier (in the case of modulated LED devices), and power supply, a complete system includes an electrical output device (in direct interface with logic level circuitry—the output transistor of a dc-powered modulated LED device or of an amplifier-logic card).

Scanning Techniques

There are several ways to set up the light source and photoreceiver to detect objects. The best technique is the one that yields the highest signal ratio for the particular object to be detected, subject to scanning distance and mounting restrictions. Scanning techniques fall into two broad categories: (1) thru (through) scan and (2) reflective scan.

Thru Scan

In *thru (direct) scanning,* the light source and photoreceiver are positioned opposite each other, so light from the source shines directly at the sensor. The object to be detected passes between the two. If the object is opaque, direct scanning will usually yield the highest signal ratio and should be the first choice. See Fig. 31.

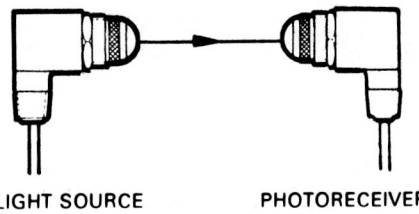

LIGHT SOURCE   **PHOTORECEIVER**

Fig. 31 In direct (or thru) scan configuration, the light source is aimed directly at the photoreceiver. Sometimes the configuration is referred to as the transmitted beam system. *(Micro Switch.)*

As long as an object blocks enough light as it interrupts the light beam, it may be skewed or tipped in any manner. As a rule, the object size should be at least 50% of the diameter of the receiver lens. To block enough light when detecting small objects, special converging lenses for the light source and photoreceiver can be used to focus the light at a small, bright spot (where the object should be made to pass), thereby eliminating the need for the object to be half the lens' diameter. An alternative is to place an aperture over the photoreceiver lens in order to reduce its diameter. Detecting small objects typically requires direct scan.

Because direct scanning does not rely on the reflectiveness of the object to be detected (or a permanent reflector) for light to reach the photosensor, no light is lost at a reflecting surface. Therefore, the direct scan technique permits scanning at farther distances than reflective scanning. Direct scanning, however, is not without limitations. Alignment is critical and difficult to maintain where vibration is a factor. Also, with a separate light source and photoreceiver, there is additional wiring, which may be inconvenient if the application is difficult to reach.

Reflective Scan

In *reflective scanning* the light source and photoreceiver are placed on the same side of the object to be detected. Limited space or mounting restrictions may prevent aiming the light source directly at the photoreceiver, so the light beam is reflected either from a permanent reflective target or surface, or from the object to be detected, back to the photoreceiver. There are three types of reflective scanning: (1) *retroreflective scanning,* (2) *specular scanning,* (3) *diffuse scanning.*

Retroreflective Scanning

With retroreflective scanning the light source and photosensor occupy a common housing. The light beam is directed at a retroreflective target (acrylic disk, tape, or chalk)—one which returns the light along the same path over which it was sent. See Fig. 32. Perhaps the most commonly used retro

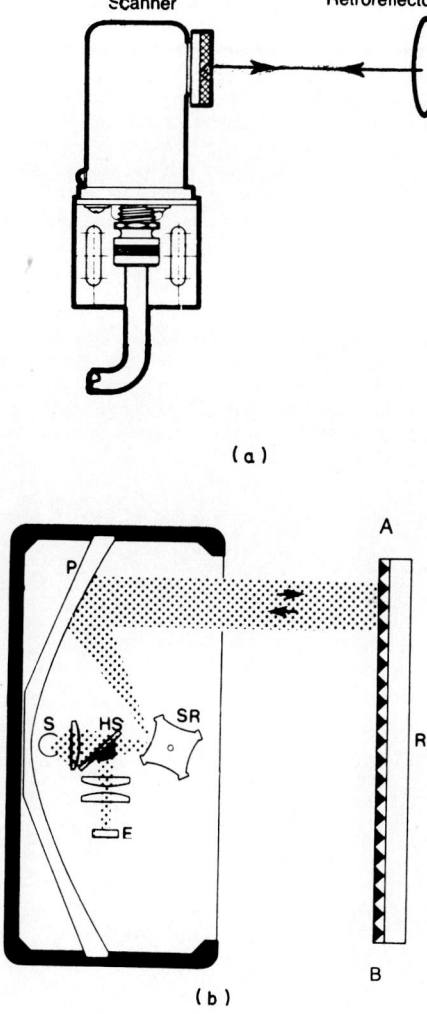

Fig. 32 (*a*) Reflected beam (retroreflective scan) system in which the light source and photoreceiver are contained in a single enclosure. This simplifies wiring and avoids critical alignment of the source and sensor. *(Micro Switch.)* (*b*) By adding a rotating-mirror wheel (*SR*), a parabolic reflector (*P*), and a semitransparent mirror (*HS*), a parallel-scanning beam can be obtained. This beam moves at high speed from *A* to *B*, thus forming a "light curtain," any interruption of which is detected and signaled by a relay. S, light source; E, photoreceiver. *(Sick Optik Elektronik.)*

target is the familiar bicycle-type reflector. A larger reflector returns more light to the photosensor and thus allows scanning at a further distance. With retro targets, alignment is not critical. The light source–photosensor can be as much as 15° to either side of the perpendicular to the target. Also, inasmuch as alignment need not be exact, retroreflective scanning is well suited to situations where vibration would otherwise be a problem.

Retroreflection from a stationary target normally provides a high signal ratio so long as the object passing between the scanner and the target is not highly reflective and passes very near the scanner. Retroreflective scanning is preferred for the detection of translucent objects and ensures a higher signal ratio than is obtainable with direct scanning. With direct scanning, the "dark" signal may not register very dark at the photosensor, because some light will pass through the object. With retroreflective scanning, however, any light that passes through the translucent object on the way to the reflector is diminished again as it returns from the reflector. The system is also useful where retroreflective tape or chalk coding can be placed on cartons for sorting. Retroreflective scanning normally can be used at distances up to 30 ft (9 m) in clear air conditions. As the distance to the target increases, the retro target should be made larger so that it will intercept and return as much light as possible. Single-unit wiring and maintenance are secondary advantages of retroreflective scanning. See Fig. 33.

Specular Scanning

The specular scan technique uses a very shiny surface, such as rolled or polished metal, shiny plastic, or a mirror to reflect light to the photosensor. See Fig. 34. With a shiny surface, the angle at which light strikes the reflecting surface equals the angle at which it is reflected from the surface. Positioning of the light source and photoreceiver must be precise (mounting brackets which fix the light source–photoreceiver relationship should be used), and the distance of the reflecting surface from the light source and photoreceiver must be consistently controlled. The size of the angle between the light

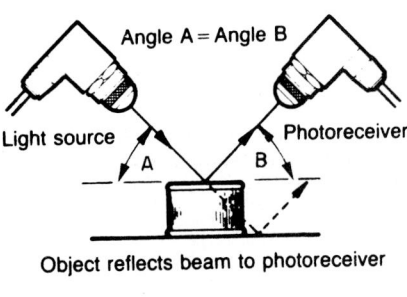

Object reflects beam to photoreceiver

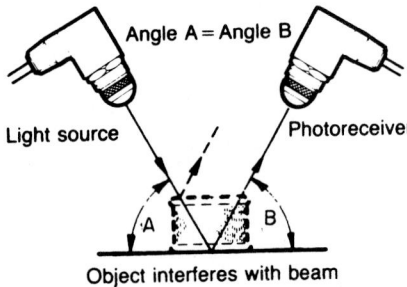

Object interferes with beam

Fig. 33 Retroreflective phototransistor scanner. Unit combines light source and receiver in a single case. One-piece lens-prism rejects ambient light that is not close to its optical axis. *(Automatic Timing & Controls.)*

Fig. 34 Specular scan technique employs a very shiny surface, such as rolled or polished metal, shiny plastic, or a mirror to reflect light to the photosensor. *(Micro Switch.)*

source and photoreceiver determines the depth of the scanning field. With a narrower angle, there is more depth of field. With a wider angle, there is less depth of field. For a fill-level detection application, for example, this means that a wider angle betwen the light source and photoreceiver allows detection of the fill level more precisely.

Specular scanning can provide a good signal ratio when required to distinguish between shiny and nonshiny (matte) surfaces, or when using depth of field to reflect selectively off shiny surfaces of a certain height. When monitoring a nonflat, shiny surface with high and/or low points that fall outside the depth of field, these points appear as dark signals to the photosensor.

Diffuse Scanning

Nonshiny (matte) surfaces, such as kraft paper, rubber, and cork, absorb most incident light and reflect only a small amount. Light is reflected or scattered nearly equally in all directions. In diffuse scanning, the light source is positioned perpendicularly to a dull surface. Emitted light is reflected back from the target to operate the photoreceiver. See Fig. 35. Because the light is scattered, only a small percentage returns. Therefore, the scanning distance is limited (except with some high-intensity modulated LED controls), even with very bright light sources. It is often difficult to obtain a sufficient signal ratio with diffuse scanning when the surface to be detected is almost the same distance from the sensor as another surface (for instance, a nearly flat or low-profile cork liner moving along a conveyor belt). Contrasting colors can help in such situations.

Diffuse scanning is used in registration control and to detect material (corrugated metal, for example) with a slight vertical flutter—which might prevent a consistent signal with specular scanning. Alignment is not critical in picking up diffuse reflection.

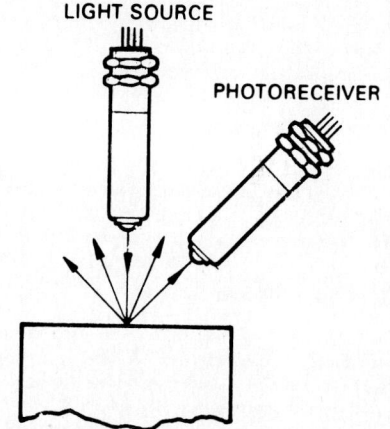

Fig. 35 Diffuse scan is used in registration control and to detect material (corrugated metal, for example) with a slight vertical flutter—which might prevent a consistent signal with specular scan. Alignment is not critical in picking up diffuse reflection. *(Micro Switch.)*

Color Differentiation (Registration Control)

In distinguishing color, as in registration mark detection, contrast is the key. High contrast (dark color on light, or vice versa) provides the best signal ratio and control reliability. Therefore, if possible, bright, well-defined, contrasting colors should be considered in the interest of the registration control system.

Diffuse scanning is normally used to detect color change. Table 2 gives some of the common color

Table 2 Factors That Determine Selection of the Photosensor and Scan Technique in Registration Control Applications

Background	Mark	Photosensor	Scan technique
Clear film	Black, blue, red	Any	Direct scan
White (kraft paper, metal foil)	Black, blue	Phototransistor or CdSe photocell	Diffuse scan
	Red	CdS photocell with blue-green filter	Diffuse scan
Black, blue, or other dark colors	Red	Any	Diffuse scan
Red	Black, blue	Any	Diffuse scan

combinations that must be distinguished in registration control, plus the most suitable type of photosensor and scan technique.

When the background is clear (transparent), the best method is to detect any color mark with direct scanning. When the background is a second color, contrasts such as black against white usually ensure a sufficient signal ratio (difference between dark and light signals) to be handled routinely with diffuse scanning. Red, or a color that contains considerable red pigment (yellow, orange, brown), on a white or light background is a special case. In such instances, a photoreceiver with a cadmium sulfide cell for detecting red marks is preferred because it makes red appear dark on a light background.

A retroreflective scanner with a short-focal-length lens (but without a retrotarget) can be used to detect registration marks. It is placed near the mark and is actually used in the diffuse scan technique. If a retroreflective scanner is employed to detect marks on a shiny surface, the scanner should be cocked somewhat off the perpendicular to make certain that only diffuse reflection will be picked up. Otherwise, the shiny surface of the mark could mirror-reflect so brightly that it would overcome the dark signal a CdS cell normally receives from red. This would mean a light signal from both the background and the mark. In detecting colors, a rule of thumb is to use diffuse (weakened), rather than specular (mirror), reflection.

Sensitivity Adjustment

Most photoelectric controls have a sensitivity adjustment that determines the light level at which the control will respond. Conditions which may require an adjustment of sensitivity include (1) detecting translucent objects, (2) a high speed of response, (3) a high cyclic rate, (4) line voltage variation, and (5) a high electrical noise atmosphere.

Light Sources and Sensors

Early photoelectric control systems used incandescent light sources and traditional photocells—a combination still widely used. A *photocell* changes its electrical resistance with the amount of light that falls on it. A number of photocells have been used over the years for different applications (photoelectric controls, copying machines, television pickup tubes, etc.). Widely used for photoelectric controls are cadmium sulfide and cadmium selenide cells. During the relatively recent past, phototransistors and photodiodes have become available as sensors—and LEDs have been used as light (infrared) sources. The advantages of the newer components in certain types of applications are described shortly.

Photocells

The *sensitivity* of a photocell can be defined in two ways. *Static sensitivity* expresses the resistance of the cell at a given light intensity; the lower the resistance, the more sensitive the photocell. So long as the resistance falls within the range of the control unit with which the photocell is used, static sensitivity is usually not an important consideration. *Dynamic sensitivity* is an expression of the ratio of photocell resistance at one light level to its resistance at a different level; the greater the ratio, the higher the sensitivity. This is a much more useful expression of photocell sensitivity.

The *speed of response* of a photocell is the time it requires to produce a change in resistance in response to a given change in light intensity. Although all photocells are fast, they require a finite amount of time to respond to changing light. And, their speed of response depends on the amount of light falling on the cell; the greater the light intensity, the faster the response.

Light history effect refers to a characteristic of a photocell that has been kept dark or light for extended periods. Such a cell overresponds to a change in light before returning to its normal response (somewhat analogous to the response of the human eye to sudden changes in light). Although in some applications it is well to know about this effect, it normally is not a significant consideration in industrial uses.

The *effect of temperature* on a photocell is to increase its resistance and thus decrease its current with increases in temperature for a given level of illumination. The temperature effect is smaller when the level of illumination is high than when it is low.

Photocells also respond differently to different *colors* of light. Photocells generally used in industrial applications have a far greater response to colors in the red and infrared range than in the blue-violet range. Except in certain applications, as previously described for registration control, the color response is not a significant factor.

Phototransistors

A phototransistor produces a collector current that is a function of both base current and light. Since the base lead of a phototransistor is usually left unconnected, only variations in light intensity produce variations in current output. There are a number of differences between the phototransistor and the photocell: (1) The current output of a phototransistor is largely independent of the voltage across it, whereas that of a photocell is not. As a result, controls designed to work with photocells will not necessarily work well with phototransistors, and vice versa. (2) The response of phototransistors is affected by changes in temperature, but in a way opposite to that of photocells; the higher the temperature, the higher the current output. (3) Phototransistors have a polarity which must be observed, while photocells do not. (4) Phototransistors respond to light much faster than photocells but typically have a lower sensitivity.

Photodiode response is narrower than that of the phototransistor, making the diode more effective in blocking stray light from incandescent, sun, or other light sources.

LED Sources

The introduction of LEDs as radiation sources brought a number of advantages for certain applications. The useful life of an LED is estimated at 100,000 h, which is about 10 times that of an incandescent lamp. However, incandescent lamps are frequently used because they have a spectrum

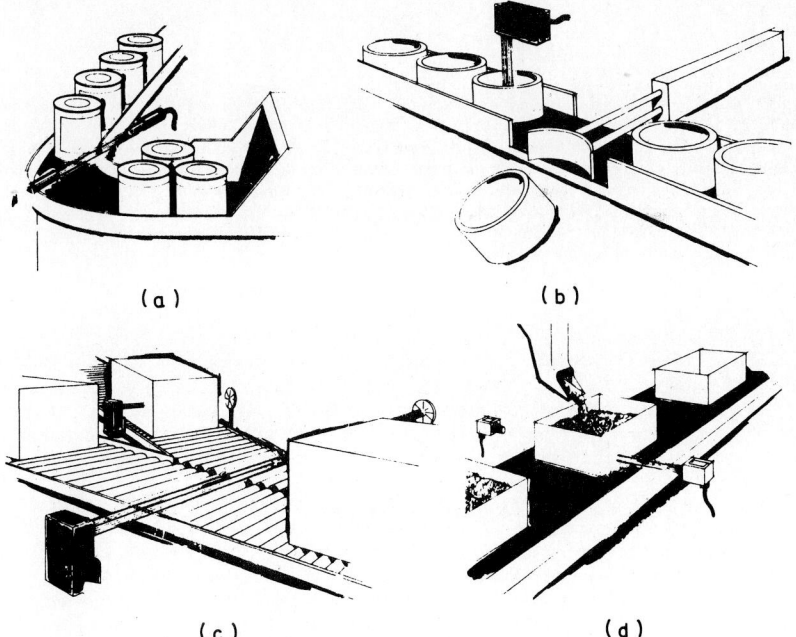

(a)

(b)

(c)

(d)

Fig. 36 Conveyor-associated applications of photoelectric control systems. (*a*) Counting products is a common application. As each can breaks the light beam, one count is added to total-count register. (*b*) Dark caps of containers are checked for white liners by a photoelectric scanner. The scanner activates a mechanism that rejects caps without liners. (*c*) To prevent collisions where two conveyors merge, each conveyor is monitored by a control that powers the other conveyor when its own is cleared. (*d*) Using logic for a one-shot pulse output, a photoelectric control slows a conveyor and fills the carton which has interrupted the light beam. (*Micro Switch.*)

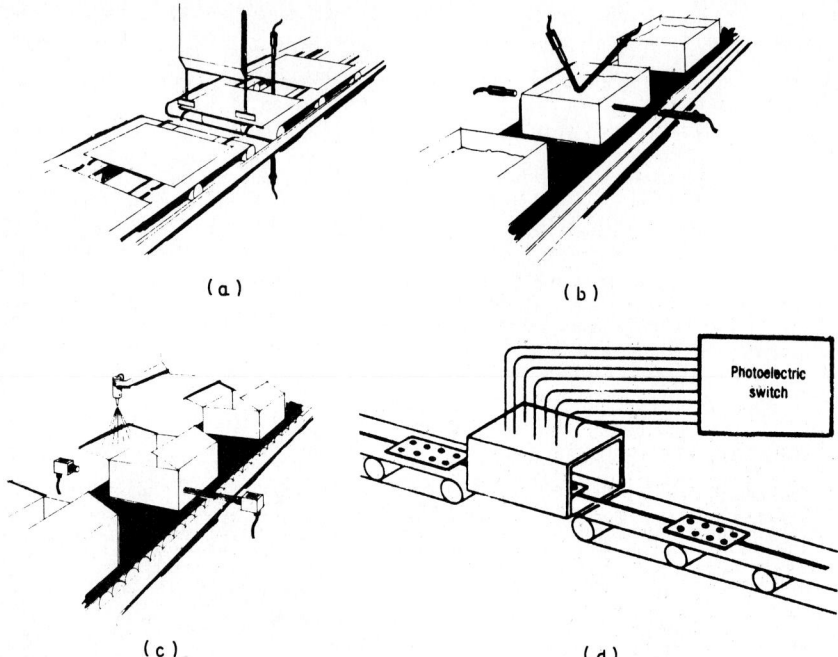

Fig. 37 Conveyor-associated photoelectric control applications. (*a*) A light source and photoreceiver placed near a guillotine are used to detect products and operate the blade for cutting the link between products. (*b*) Two light source–photoreceiver pairs work together to check the fill level. The box detecting pair turns on, or enables, the fill inspection pair, thereby preventing the inspection pair from mistaking the space between boxes as an "improper fill." (*c*) Turning a glue nozzle on and off is an example of process control easily effected with a photoelectric control. *(Micro Switch.)* (*d*) A matrix of fiber-optic scanners checks each package for complete fill (in this case, specifically located discrete pieces). Any missing part will pass light, which activates the photoelectric switch to emit an output signal—an example of AND logic. *(After Krueger.)*

from the ultraviolet to the visible to the infrared, allowing a wide range of colored targets to be detected. LEDs have the advantage that they can be modulated directly, whereas incandescent lamps require a mechanical chopper. Silicon phototransistors and photodiodes are excellent matches for infrared LEDs because their greatest sensitivity peaks almost match precisely at the transmitter's (LED) wavelength.

Use of Fiber Optics with Photoelectric Switches

Fiber-optic bundles can be added to existing photoelectric switches to provide object sensors, and these can be combined to implement logic functions. A system has been developed using fiber optics for applications that require several sensing inputs and one or more outputs to interface with microcomputers or minicomputers. Program selection permits use of the LO or DO mode and allows operation of any channel for a predetermined time, thus avoiding sequential channel operation in fixed time frames. These systems find application where a PC is not warranted because of cost or complexity. Input can be from a relay or switch contacts, transducers, memory devices, CMOs, or TTL. The output section provides a channel signature for each emitter and detector pair, resulting in the capacity of actuating one or more output devices. The actual architecture can be configured to suit the application.

Applications of Photoelectric Controls and Other Optical Pickups

Photoelectric sensors and controls are among the most versatile tools found in industrial instrumentation. They have been used in numerous ingenious ways—far too many to report here. Probably the most common applications involve either discrete pieces moving on conveyors, or moving webs, and sheet goods. A number of representative conveyor-associated applications are illustrated and described briefly in Figs. 36 and 37. As will be evident, some of the supporting logic and instrumentation goes well beyond the simplest source-photoreceiver application. Applications associated with continuously moving webs and sheets are shown in Fig. 38, and applications associated with code reading in Fig. 39. Textile and apparel industry applications of photoelectric controls are shown in Fig. 40. Various other applications of photooptical devices are shown in Fig. 41.

Electromechanical Switches

Commonly referred to as mechanical limit switches (or simply limit switches), these devices have been available for decades and, in fact, for many years mechanical limit switches were key elements in automating machines and conveyor systems and, in large measure, continue to be. Even prior to the appearance of solid-state circuitry, great advancements were made toward manufacturing and improving the electromechanical characteristics of these switches. Stress on design for miniaturization commenced in the 1940s. Because there are so many repetitive operations in automated systems, limit switches must be capable of reliable performance over millions of actuations, often in relatively dirty and severe industrial environments.

Limit switches have been refined over the years to the point where most problems of system geometry in matching switch to machine have been solved. The main categories of limit switch actuation are (1) rotary operating heads, (2) plunger operating heads, and (3) wobble lever operating heads. These configurations permit switch actuation by moving parts and by moving machine members of various sizes and shapes and approaching the switch from different angles and directions. Side and top rotary switches are shown schematically in Fig. 42. Various kinds of levers used with rotary switches are shown in Fig. 43. The adjustability of rotary switches is illustrated in Fig. 44. Plunger switches are shown in Fig. 45, and wobble actuators are illustrated in Fig. 46.

Depending on the particular configuration, limit switches vary in dimension: height, 2 to 6 in (51 to 152 mm); width, 0.6 to 3.2 in (15 to 81 mm); length, 1.6 to 3.4 in (41 to 86 mm). The upper operating temperature range, depending on the particular design, is 160°F (71°C) to 250°F (121°C), and the lower operating temperature range is −25°F (−32°C) to 35°F (2°C). Housing materials usually are zinc or aluminum.

Electrical characteristics of some heavy-duty plug-in, watertight and oil-tight limit switches are shown in Table 3. Additional characteristics include:

Standard. Overtravel 60° minimum, pretravel 15° maximum, differential travel 3° (single-pole) and 7° (double-pole) maximum.

Low-Differential Travel Design. Overtravel 68° minimum, pretravel 7° maximum, differential travel 3° (single-pole) and 4° (double-pole) maximum.

Low-Operating Torque Design. Overtravel 60° minimum, pretravel 15° maximum, operating torque 1.7 in · lb (0.19 N · m) maximum.

Low-Torque, Low-Differential Travel Design. Overtravel 68° minimum, operating torque 1.7 in · lb (0.19 N · m) maximum, differential travel 3° (single-pole) and 4° (double-pole).

Sequence Action Design. Delayed action between operation of two poles, in each direction. Overtravel 48° minimum.

Center Neutral Design. One set of contacts operates on clockwise rotation, and another set on counterclockwise rotation. Overtravel 53° minimum.

Maintained Contact Design. Operation maintained on counterclockwise rotation, reset on clockwise rotation, and vice versa. Overtravel 20° minimum.

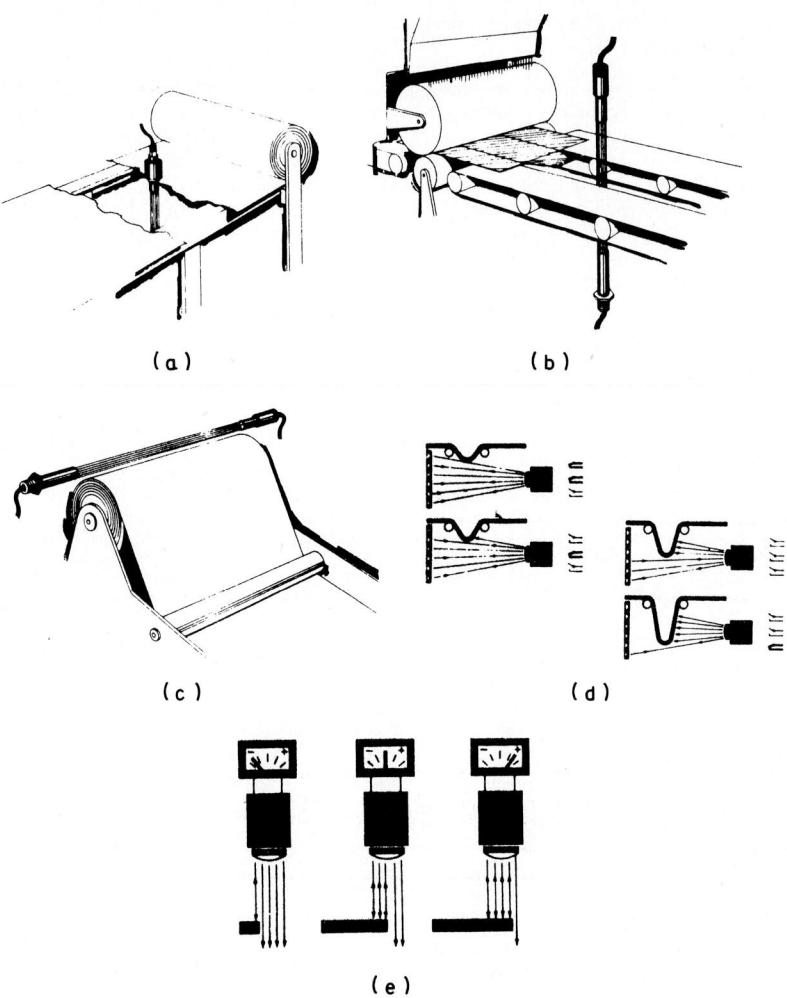

(a)

(b)

(c)

(d)

(e)

Fig. 38 Applications of photoelectric controls associated with moving sheets and webs. (a) A photoelectric control operating on reflected light is a simple way to detect a web break. An alternative is to put a light source above the web and a photoreceiver below. (b) Gluing, buffing, or flattening can be done efficiently by controlling the pressure rollers or buffer with a photoelectric light source and photoreceiver that detects the product to be processed. (c) The size of a paper or fabric roll can be controlled by positioning a light source and a photoreceiver so that roll diameter blocks the beam. *(Micro Switch.)* (d) A loop control monitors the loop of opaque materials and provides control outputs if the loop is above or below the desired level. The center of the desired value is aligned with the optical axis. Upper and lower desired value limits are automatically set so that correct reference values are always provided even where material feed is not continuous, as occurs on power presses and punch presses. (e) The edge of a moving web (paper, textiles, plastics, metals) consists of an optic head and a control amplifier. Control amplifiers are available with adjustable limit values for three-point control, or with a proportional power output for regulation and control purposes, and provide an indication of edge position. *(Sick Optik Elektronik.)*

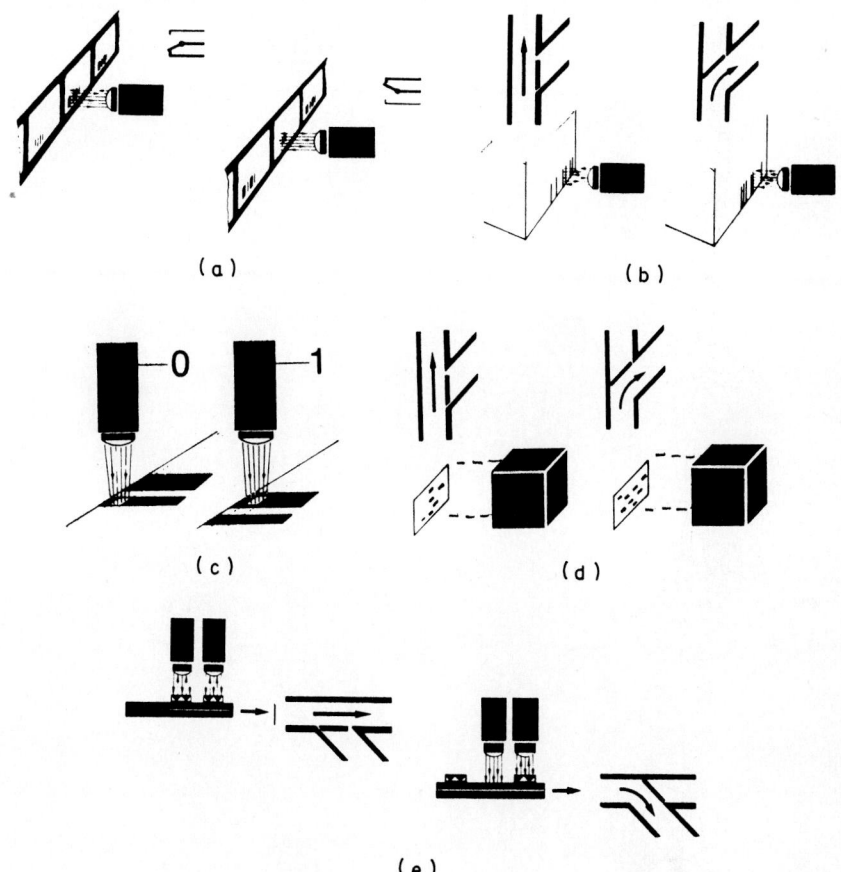

Fig. 39 Bar code systems utilizing photoelectric devices. (*a*) A label inspection system inspects roll labels. The system reads a very simple and reliable code printed along with the regular information during product manufacture. Before the labels are applied, they are rewound on a label inspection machine for high speed checking. The maximum winding speed of the machine is 16.4 ft/s (5 m/s). (*b*) A photoelectric scanner reads bar codes on paperboard in a material handling and distribution system. The scanner detects code bars with varying contrast to the surface of the packaging material. The code may be printed along with other printing, without requiring a special procedure. Data can be entered into computer systems for accounting or further processing operations. (*c*) An identification scanner on a conveyor line recognizes code marks from retroreflective material through a punched card mask. The identification is normally permanently affixed to the container. Further devices for data processing, monitoring, and routing can be connected into the system. (*d*) A code reader monitors packaging components on-line to ensure that all components in a process are correct. The code reader consists of an optic head and a microprocessor-controlled evaluation unit which performs counting and control functions. The system is capable of operating with material speeds up to 33 ft/s (10 m/s). Along with the regular information or design printed on the packaging material during manufacture, a very simple and reliable bar code is printed for subsequent reading during the packaging operation. Bar codes may be printed in color on a colored background or even on highly reflective materials. Readability is determined by the contrast between the bar color and the background color. During operation the codes are read and compared with the programmed desired code. Incorrect codes can be signaled immediately or delayed on-line by a built-in shift register until a programmed station is reached when an ejection signal is given. Through the use of additional sensors, the unit can verify that the incorrect item has been ejected and provide a stop signal if the ejection has not been fully executed. Built-in counters maintain and display a count of correct, incorrect, and sum of all codes read. A stop signal is also triggered in the event that five consecutive incorrect codes are counted. The system can be connected with external data terminals. Programming and operating data are stored at least 4 weeks with the power off. (*e*) A code bar system that provides destination information for automatic routing control. Code bars are secured to containers or pallets. The destination information is set by moving the reflector to the proper position and then reading while in motion on the conveyor by photoelectric scanners. *(Sick Optik Elektronik.)*

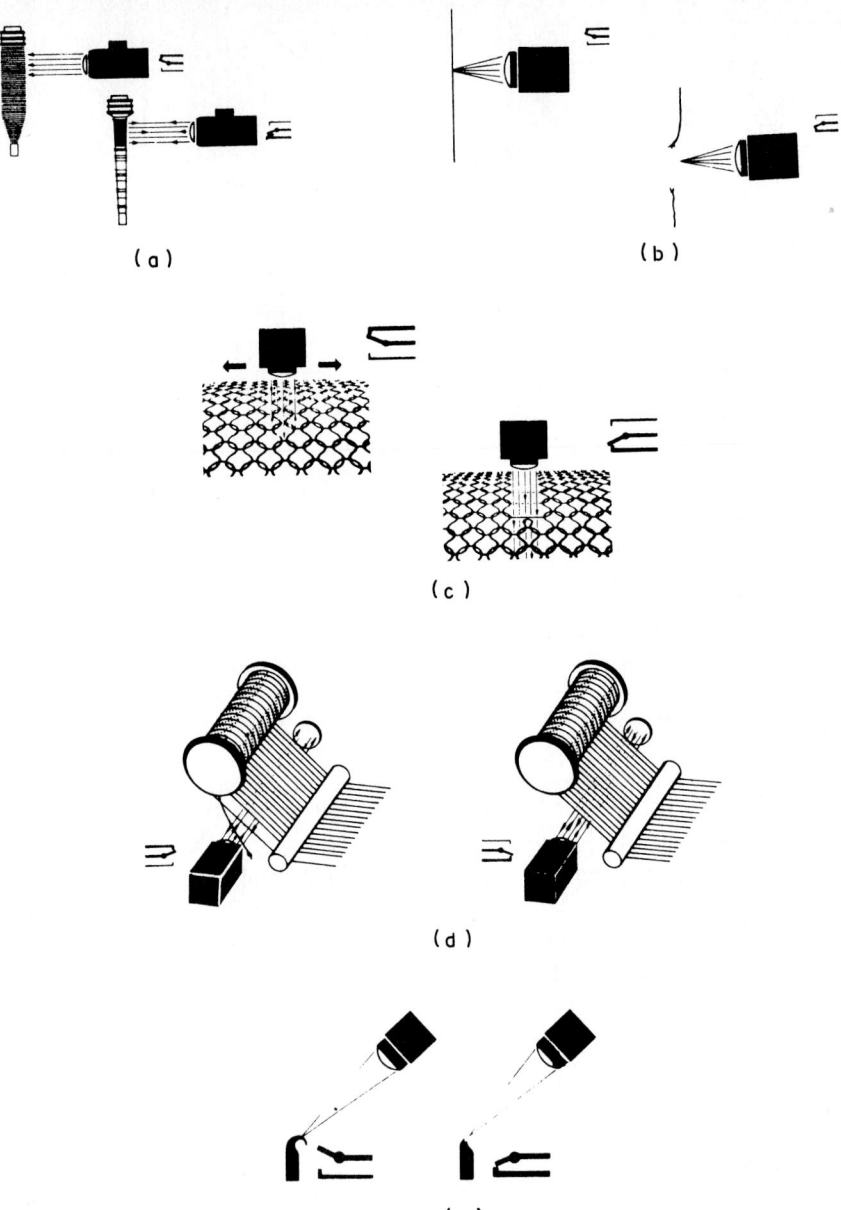

Fig. 40 Textile and apparel industry applications of photoelectric controls. (*a*) A bobbin monitor determines when the supply of thread or yarn on weaving loom bobbins has been exhausted. It is used to trigger bobbin change on high speed automatic equipment. Special optics are used for detection of high-gloss or metallic thread materials. (*b*) A thread scanner detects breaks or counts thin threads or filaments during production. The unit monitors weft threads on unprotected weaving looms with mechanical or pneumatic weft feed. (*c*) A system for monitoring material for knitting defects. An optic head continuously looks for defects while moving back and forth across the web. On detection of a defect, the machine is stopped and signal is given. (*d*) A thread break detector scans all types of thread groupings, e.g., warp or Raschel knitting machines. Breakage of any thread over 0.045 mm (0.0018 in) diameter is detected over a 20-ft (6-m) range. The device is also used for monitoring the windings of transformers and electrical coils. (*e*) A needle scanner for detecting broken and defective needles. *(Sick Optik Elektronik.)*

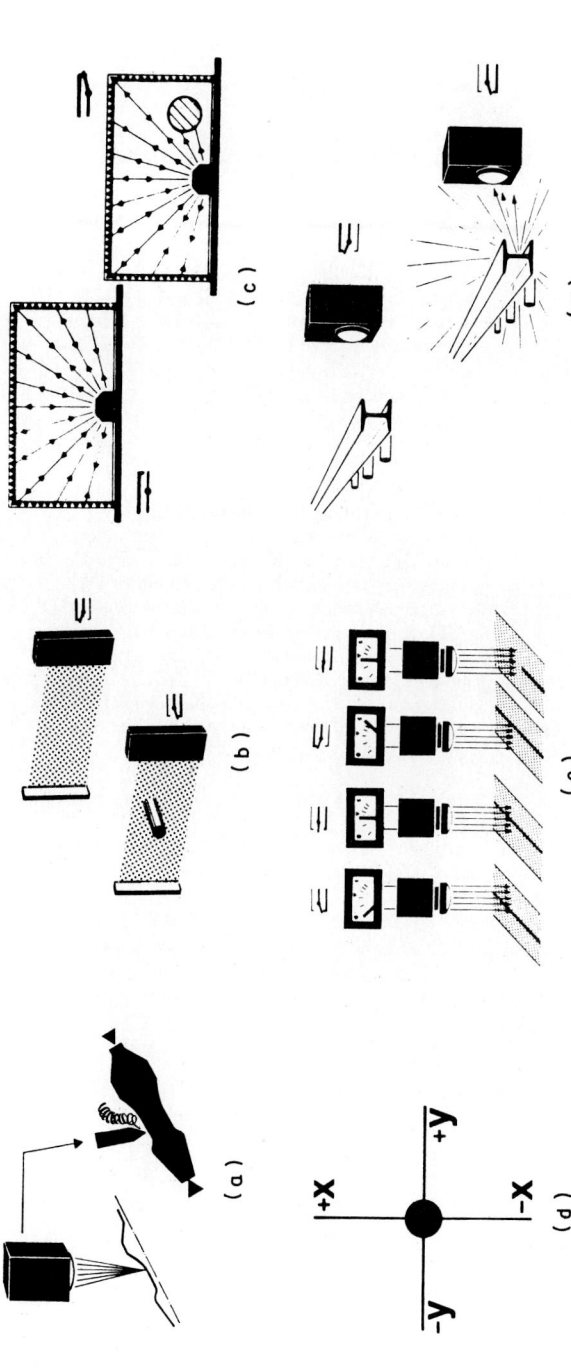

Fig. 41 Various applications of photooptical devices. (*a*) Optical curve tracers guide processing machines or tools directly by following a master drawing. They are particularly useful in the automatic production of irregularly shaped parts. The complete system includes automatic contour change and automatic depth setting of tools in response to code marks next to the line being followed. Some of the machines using curve tracers include milling, those used for electrodischarge (EDM), glass and foam cutting, laser cutting and woodworking. Units are capable of translating a master drawing to an NC format. The system has a scale reduction capability, used on pantograph machines for the production of small precision parts. (*b*) A light curtain system operates on the autocollimation principle using invisible modulated light. It can monitor an area approximately 7.9 in × 6 ft 6 in (200 × 2000 mm). It is used where parts cannot be detected by traditional photoelectric controls because of inaccurate guidance, such as in ejection monitoring of automatic punch or injection molding equipment; for the detection of parts of varying shape; and for control of movable safety shields. (*c*) A reflex scanning device used to detect objects anywhere within a given plan. Typical applications include monitoring doorways, and checking shelves, pallets, and moving transport containers for overhanging or projecting goods. A narrow reflective surface is mounted around the perimeter of the area to be monitored. The unit emits a light beam which scans the area and is reflected back to a photoreceiver. The angle of effective scan is adjustable from 5 to 220°. An object larger than 0.6 in (15 mm) anywhere within the operating range will interrupt the light beam and cause an output signal. (*d*) An *xy* axis final positioner controls drive mechanisms for obtaining a precision final position of conveyance devices, cranes, loading ramps, and robots. The unit focuses on a reflector attached to the target object. The scanning field of the unit is divided into four sectors corresponding to the quadrants of a rectangular coordinate system. One signal output and an indicator diode represent each of the four directions. A fifth output signals the presence of a reflector. The scan range is 12 in (300 mm), the scan area is 3.5 × 3.5 in (90 × 90 mm), and the depth of field is ±4 in (100 mm). (*e*) A line control unit follows light or dark lines and emits an analog signal in response to a deviation error; the position is then adjusted by a servomotor. The system can be used to follow a printed line and guide webs of paper, metal, or plastic with irregular edges being wound onto rolls; to control machine tools; or as a program transmitter for controlling processing operations. (*f*) A heat radiation receiver detects metal parts with a temperature above 450°C (842°F), such as ingots, slabs, sheets, and wires in rolling mills or continuous heating furnaces. Units respond to infrared radiation and thus are not affected by conveyors of a relatively lower temperature. Various attachments, including water cooling and air purging, are available. (*Sick Optik Elektronik.*)

Gravity-Return Side Rotary Switches

Unlike standard switches with spring return mechanisms, a switch is available with *gravity return*. The weight of the actuating lever must provide the force to restore it to the free position. The very small 5 in · oz (0.035 N · m) operating torque is useful in some conveyor applications because it permits operation with small or lightweight objects.

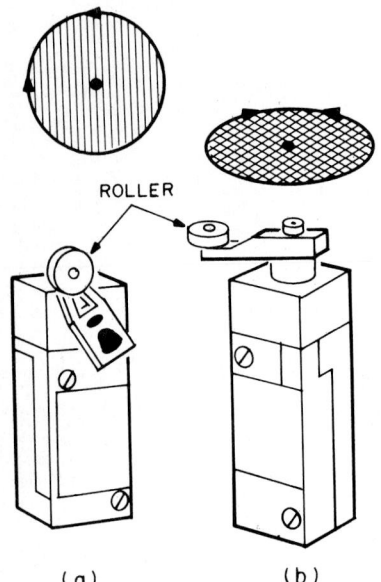

ROLLER

(a)   (b)

Fig. 42 Rotary operating heads on limit switches may be (a) side rotary, or (b) top rotary.

Solid-State Sensors

In addition to the traditional types of electromechanical switches just described, units employing Hall-effect position and vane sensors are available. See Fig. 47. The external design of these switches parallels that of the switches just described. The computer-compatible digital output of these switches makes it possible to interface directly with most electronic circuits, discrete transistor circuits, microprocessors, and integrated logic circuits (register-transistor logic, diode-transistor logic, transistor-transistor logic, and Hall-effect–transistor logic), and silicon controlled rectifiers (SCRs). Switches incorporating the Hall-effect principle* cannot be used in areas where high magnetic fields are present. Also, wiring for these devices should not be run in the same conduit with high (ac or dc) power lines. Use of the Hall-effect device eliminates any problems of contact bounce sometimes encountered with mechanical designs. The supply voltage is 5 V dc or 6 to 16 V dc.

Inductive Proximity Sensors

In an inductive proximity sensor, an electromagnetic field (radio frequency, rf) is generated by an oscillator circuit. When a metal object enters the effective field generated by the sensor, a countercurrent (eddy current) is set up in the metal object. This causes a voltage drop in the oscillator. This drop is sensed by the detector, which triggers the output. The output can be used for many industrial control purposes. See schematic diagram of a typical sensor given in Fig. 48. The sensors are available in three basic configurations (Fig. 49). Types of mounting are shown in Fig. 50. Nominal detection ranges from about 2 mm to about 20 mm, depending on the specific design. Some units are adjustable from 10 to 50 mm. Nominal detection range specifications assume that there is no interference from objects other than the single object being sensed and that the object is steel. A standard target is considered to be mild steel (45 × 45 × 1 mm, 1.77 × 1.77 × 0.04 in). When an object is made of other metals, appropriate conversion factors must be used, considering steel = 1.00.

Stainless steel, Type 430	1.03
Stainless steel, Type 302	0.85
Nickel	0.85
Brass	0.50 to 0.54
Aluminum	0.47 to 0.50
Copper	0.40 to 0.46

*When a steady current flows in a steady magnetic field, electromotive forces are developed which are at right angles both to the magnetic force and to the current and are proportional to the product of the intensity of the current, the magnetic force, and the sine of the angle between the directions of these quantities. This is known as the Hall effect.

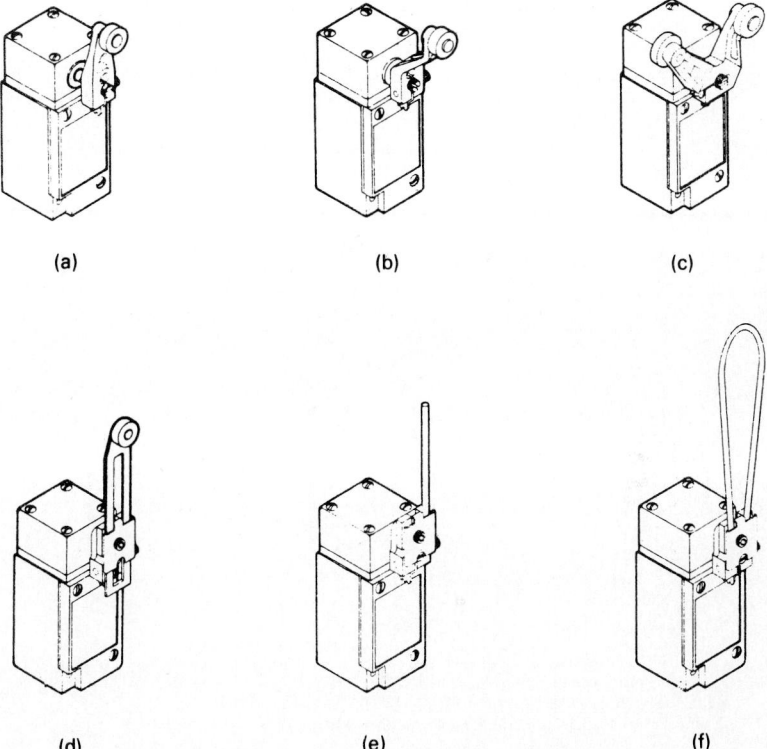

(a)   **(b)**   **(c)**

(d)   **(e)**   **(f)**

Fig. 43   Levers for use with side or top rotary actuated switches are available in a variety of sizes and materials. Rollers may be on either side of the lever to best match the external actuating mechanism. They permit a wide range of cam tracking possibilities. (*a*) A standard roller lever with a fixed 1½-in (38.1-mm) radius. (*b*) An offset roller lever with a fixed 1½-in (38.1-mm) radius. (*c*) A yoke roller lever used with side rotary maintained switches where a reciprocating actuator operates the switch in one direction and reverses it when moving in the other direction. (*d*) An adjustable-length roller lever with an adjustable radius from 1½ in (38.1 mm) to 3½ in (88.9 mm). (*e*) A rod lever which may be formed by the user. The hub permits lever length to be adjusted. A flexible spring or rod version is available. (*f*) An adjustable loop lever for accommodating certain types of external actuating mechanisms. *(Micro Switch.)*

The point at which a target will be detected is influenced by the type of metal, its size, and its surface area. Targets may approach axially (head on) or laterally (from left to right) and are detected at the point where they first touch the envelope of the sensing curve. Curves for standard and extended ranges are given in Figs. 51 and 52. The effect of temperature on the operating points is shown in Fig. 53. Output configurations are shown in Fig. 54.

Inductive proximity sensors have a number of characteristics in common with electromechanical and photoelectric sensors. Some advantages over mechanical switches for particular applications include greater immunity to shock and vibration, notably in connection with fast-moving parts and high-speed operations, and the handling of small and lightweight parts as well as large and rough parts. These sensors are widely used on machine tools, conveyor lines, transfer lines, material handling equipment, packaging machinery, and positioning controls, among other uses. The sensors can interface directly with programmable controllers and other solid-state logic systems or conventional relay systems.

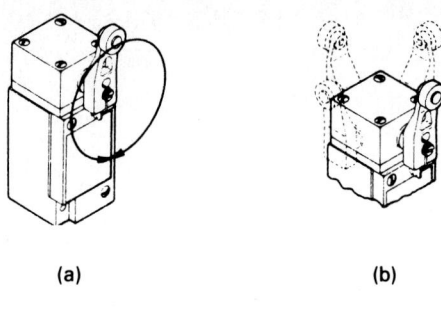

(a)   **(b)**

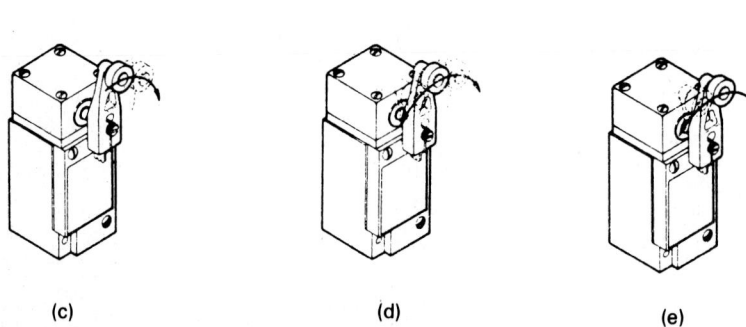

(c)   **(d)**   **(e)**

Fig. 44   The actuation of limit switches with rotary levers is adjustable for operation clockwise, counterclockwise, or in both directions. (*a*) The lever locks in any position, 360° around the shaft. (*b*) The head may be positioned and locked in any of four 90° positions. (*c*) Clockwise rotation. (*d*) Clockwise and counterclockwise rotation. (*e*) Counterclockwise rotation. *(Micro Switch.)*

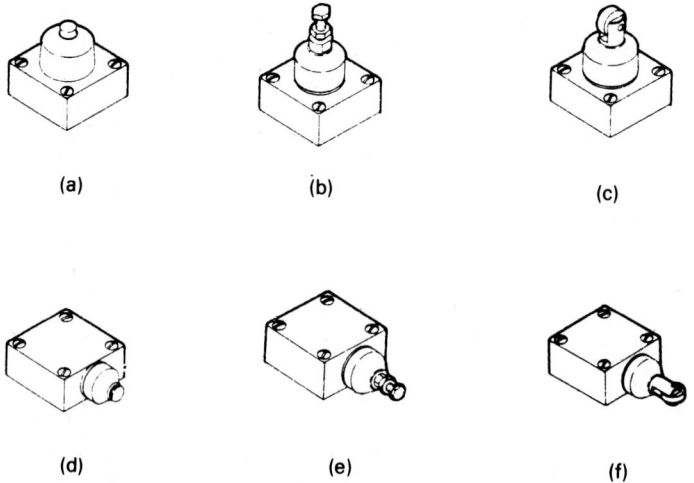

(a)   **(b)**   **(c)**

(d)   **(e)**   **(f)**

Fig. 45   Plunger-operated limit switches are available in several configurations to accommodate various external actuating mechanisms. (*a*) Top plunger, (*b*) adjustable top plunger, (*c*) top roller plunger, (*d*) side plunger, (*e*) adjustable side plunger, and (*f*) side roller plunger. *(Micro Switch.)*

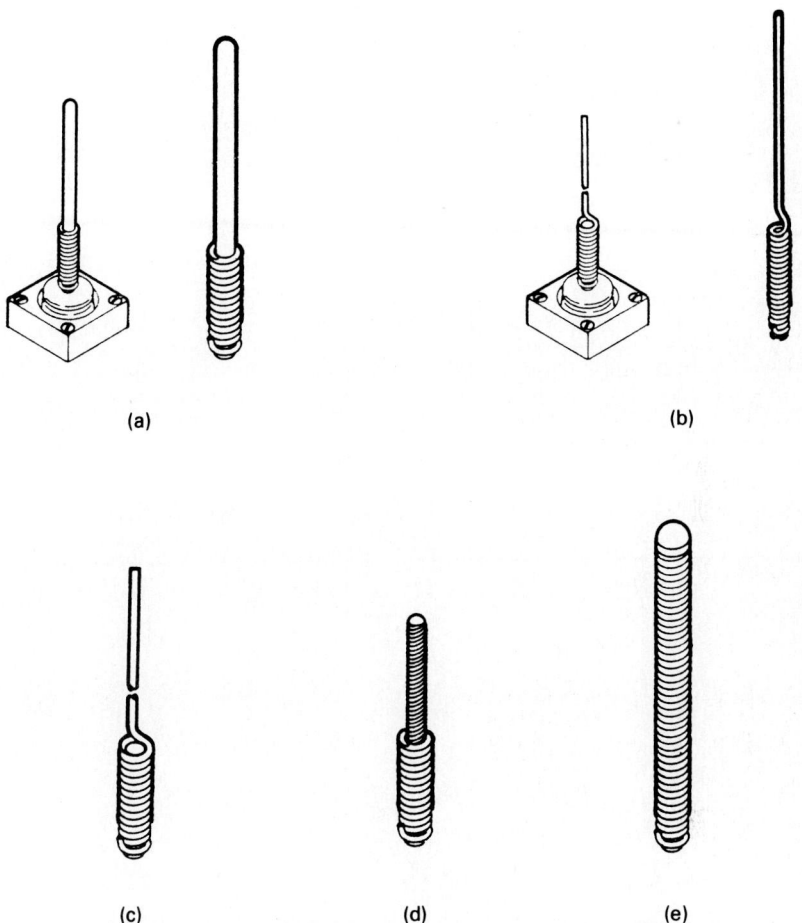

(a)   (b)

(c)   (d)   (e)

Fig. 46 Wobble actuators. (*a*) A wobble stick mounted on a switch head with detail of stick at right, (*b*) a cat whisker mounted on a switch head with detail of whisker at right, (*c*) spring wire, (*d*) cable, and (*e*) coil spring. *(Micro Switch.)*

Magnetic Proximity Sensors

There are four principal types of magnetic proximity sensors: (1) variable-reluctance-type sensors, the operation of which depends on the interruption of a fixed magnetic field (circuit) by a ferrous actuator; (2) magnetically actuated dry reed or mercury switches; (3) Hall-effect sensors; and (4) Weigand-effect sensors.

Variable-Reluctance-Type Sensors

The principle of operation of variable reluctance position (presence) transducers is shown very schematically in Fig. 55. These transducers convert motion (rotating, sliding, or oscillating) into electrical control signals. As shown in Fig. 55*a*, with no actuating object in the vicinity of the sensor (pole piece plus coil plus magnet), the path of magnetic flux is undisturbed. However, as an object

Table 3 Maximum Contact Ratings of Some Heavy-Duty Plug-in Limit Switches

AC volts	Current, A		Cont.	Voltamperes		DC volts	DC current, A
	Make	Break		Make	Break		

One normally open, one normally closed, contact on same polarity

NEMA A600 rating							
120	60	6	10	7200	720	120	0.2
240	30	3	10	7200	720	240	0.1
480	15	1.5	10	7200	720	240	0.1
600	12	1.2	10	7200	720	240	0.1

Two normally open, two normally closed, each pole on same polarity

NEMA B600 rating							
120	30	3	10	3600	360	120	0.1
240	15	1.5	10	3600	360	240	0.05
480	7.5	0.75	10	3600	360	240	0.05
600	6	0.60	10	3600	360	240	0.05

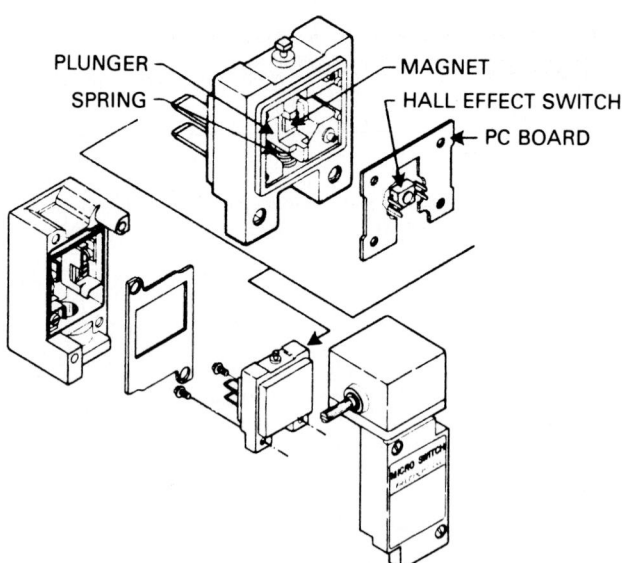

PLUNGER — MAGNET

SPRING — HALL EFFECT SWITCH

PC BOARD

MICRO SWITCH

Fig. 47 Solid state limit switch that incorporates a Hall-effect switch. *(Micro Switch.)*

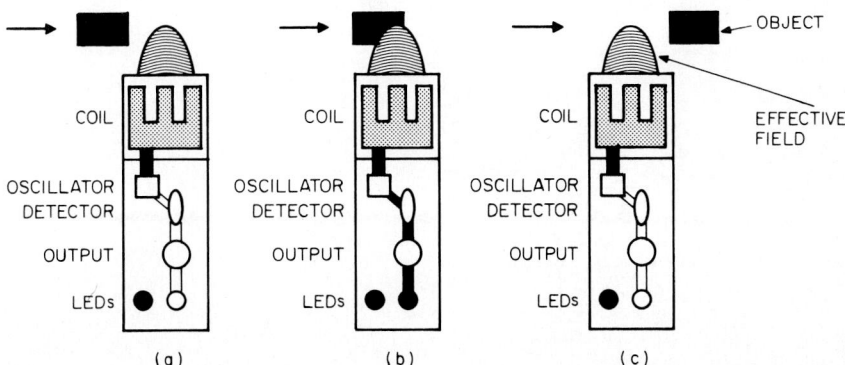

Fig. 48 Schematic diagram of inductive proximity sensor. (*a*) Object approaching sensor, (*b*) object in field of sensor, (*c*) object beyond sensor. *(Cutler-Hammer.)*

approaches and passes near the pole piece, the flux path is distorted. This alters the voltage output of the coil surrounding the pole piece. This system is most frequently used in connection with rotating equipment for speed measurement (tachometry) where discontinuities, such as gear teeth, shaft keyways, drilled holes in steel plates, etc., alter the magnetic flux in proportion to revolutions per minute. Specific applications include engine revolutions-per-minute measurement—aircraft, automotive vehicles; motor revolutions-per-minute measurement—drills grinders, lathes, among many other machines; process speed measurement in food, textile, paper, printing, and other industries; flow rate on turbine meters; revolutions-per-minute measurement of tape and disk drives. The sensors have a wide thermal operational range—from -300 to $+800°F$ (-184 to $+427°C$) at speeds up to 200,000 rpm. Sensors are available in passive and active configurations. Passive sensors require no external electric power. The output signal is an alternating current, the waveform of which is a function of the actuator, usually sinusoidal, for example, with typical gear actuators. The amplitude and frequency of the output signal are both proportional to the surface speed of the actuator as it passes the sensor's pole piece. The active configuration requires a dc power supply. The output signal is a pulse train whose amplitude is constant over the operating performance range for a fixed supply voltage level. Active magnetic sensors provide usable output signals at very low actuator speeds and at relatively large air gaps between the sensor pole piece and actuator. They produce a logic-level output signal directly compatible with digital instrumentation. An excellent detailed discussion of the design and application of variable reluctance sensors is given in the EC HDBK listed in the references at end of this article.

Magnetically Actuated Dry Reed Switches

Consisting of a thin reed (wire) contained in a hermetically sealed tubelike container, this type of switch is both inexpensive and rugged. Whenever an activating magnet approaches the critical range of the switch, a contact closure is made. Life expectancy generally is in excess of 20 million operations at contact ratings of about 15 VA. These switches generally can operate loads directly, thus avoiding the cost and complexity of comparable solid-state systems. Since the actuating magnet can be installed on a rotating or reciprocating object, the switch can be used in a wide variety of applications for counting, positioning, and synchronizing. Contact closure speeds can be up to 100 per second. See Fig. 56. Mercury switches with flexible electrodes that can be attracted by the proximity of a magnet also have been used.

Hall-Effect Switches

The adaptation of a Hall-effect switch as part of a traditional electromechanical switch was previously described. Self-contained units are also available where the Hall effect serves the principal

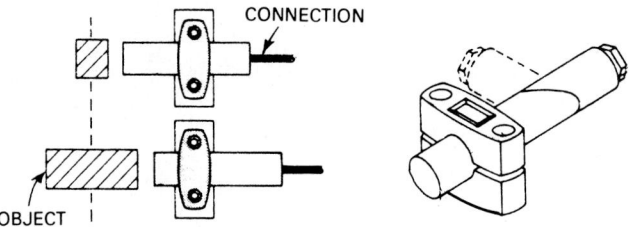

(a) CYLINDRICAL CONFIGURATION

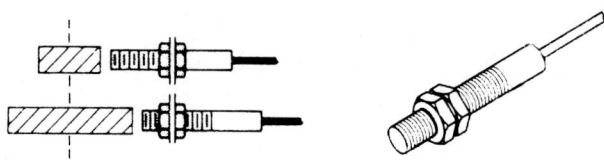

(b) THREADED CONFIGURATION

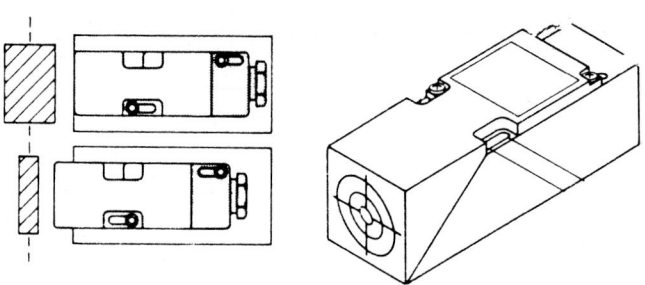

(c) RECTANGULAR CONFIGURATION

Fig. 49 Basic styles of inductive proximity sensors. (*a*) Cylindrical—a two-piece mounting clamp with socket head screws allows the installed sensor to be moved to the desired position. (*b*) Threaded—the installed sensor can be rotated to the desired position and held in place with two flat nuts. (*c*) Rectangular—the sensor has slots in mounting base, which allow adjustment to desired position after installation. (*Effector, Automatic Timing & Controls.*)

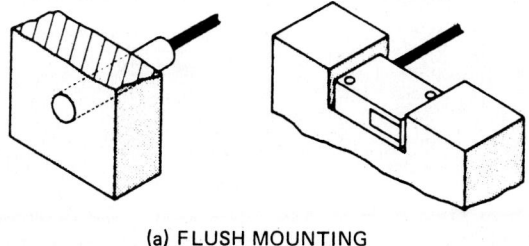

(a) FLUSH MOUNTING

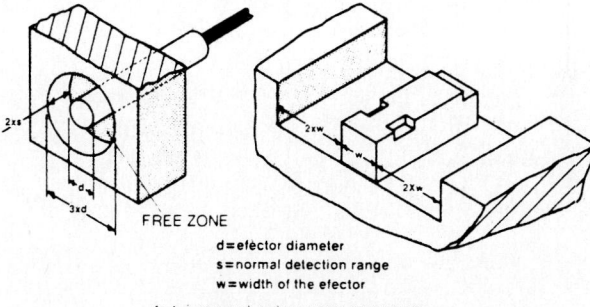

d = efector diameter
s = normal detection range
w = width of the efector

(b) NONFLUSH MOUNTING

Fig. 50 Inductive proximity sensor mounting. (*a*) Some sensors are flush-mountable, i.e., they can be mounted so that the active zone is completely surrounded by metal except for the face. (*b*) Some sensors are designated for nonflush mounting and must be mounted so that the area surrounding the active zone is free and open as shown. *(Effector, Automatic Timing & Controls.)*

switching function. Although magnetically actuated, these devices are capable of sensing through nonferrous metals. They can be interfaced directly with industrial logic PCs. Speed of operation is about 25 kHz. Operating temperature range is -40 to $+185°F$ (-40 to $+85°C$). A typical unit may be actuated by the south pole of any permanent magnet or electromagnet with sufficient flux density. The axis of the magnet should be parallel to the axis of the device. All ferromagnetic material must be at least 0.5 in (12.7 mm) away from the sensing face of the sensor and any part of the magnet.

Wiegand-Effect Switches

A Wiegand wire is a small-diameter wire that has been selectively work-hardened so that the surface and the core of the wire differ in magnetic permeability. When subjected to a magnetic field, the wire emits a well-defined pulse that requires little signal conditioning. This pulse induces a voltage in the surrounding sensing coil. The wire is insensitive to polarity and emits a pulse whether the magnetic field is flowing from north to south, or vice versa. A Wiegand proximity sensor senses the presence or absence of ferromagnetic material.

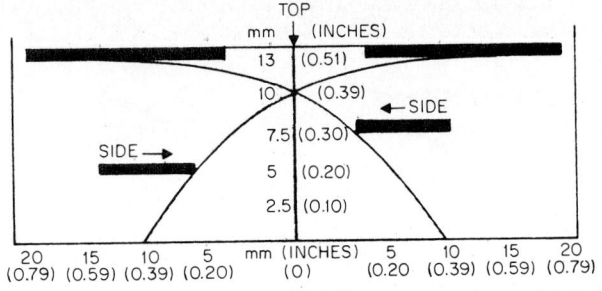

(a) STANDARD RANGE. (CAN BE MOUNTED FLUSH WITH METAL SURFACE.)

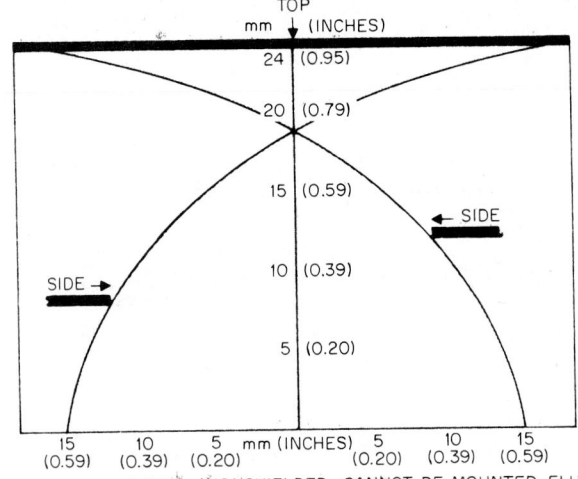

(b) EXTENDED RANGE. (NONSHIELDED–CANNOT BE MOUNTED FLUSH WITH METAL SURFACE.)

Fig. 51 Sensing distance—typical response curve for a target of mild steel (45 × 45 × 1 mm, 1.77 × 1.77 × 0.04 in). Black rectangular areas represent objects approaching the sensor from the top or sides. *(Cutler-Hammer.)*

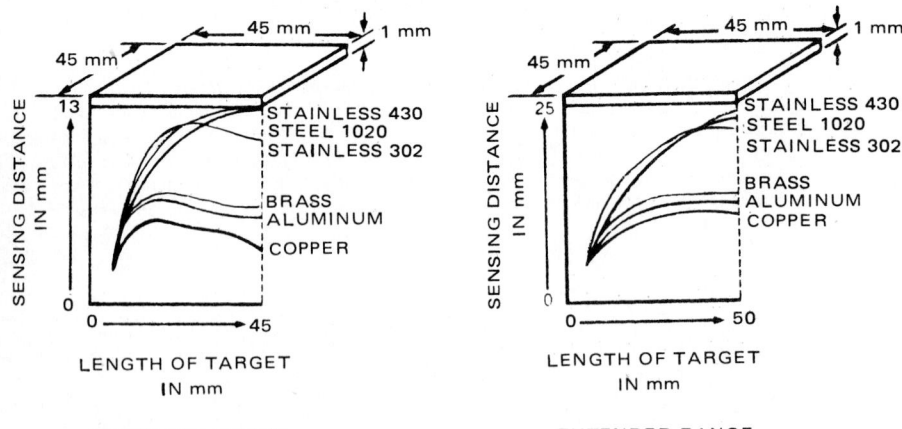

Fig. 52 Effect of target size and target material on the sensing distance for some inductive proximity sensors. *(Cutler-Hammer.)*

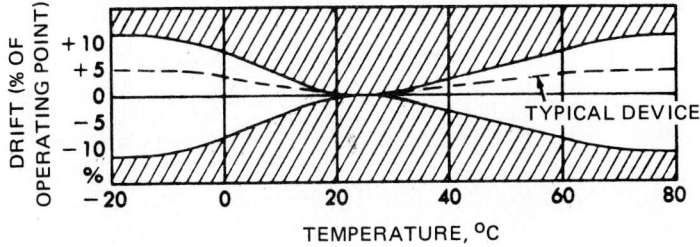

Fig. 53 Effect of temperature on the operating points of some inductivity proximity sensors. *(Cutler-Hammer.)*

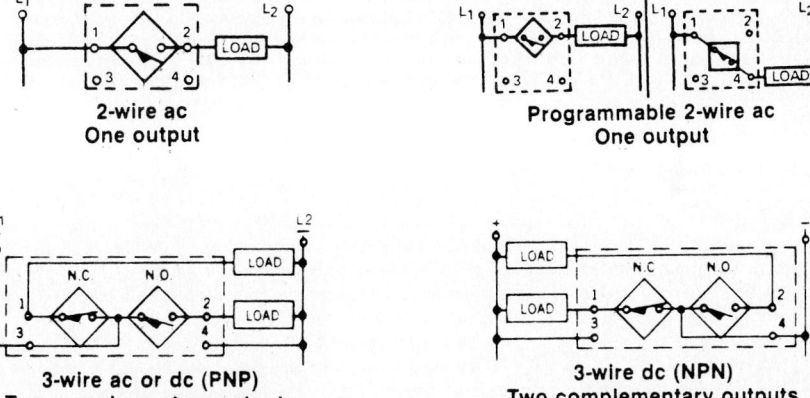

Fig. 54 Output configurations of some inductivity proximity sensors. *(Cutler-Hammer.)*

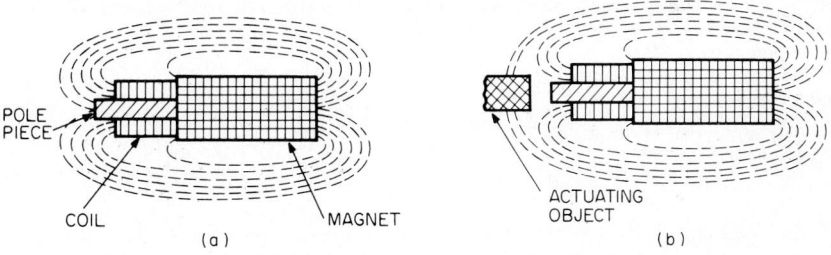

Fig. 55 Schematic diagram of the action of a variable reluctance object sensor. (*a*) A sensor with no actuating object in the magnetic field; (*b*) actuating objects in field alters the voltage generated at coil terminals. The voltage is proportional to the rate of change of magnetic flux.

Fig. 56 Magnetically actuated dry reed switch. *(Automatic Timing & Controls.)*

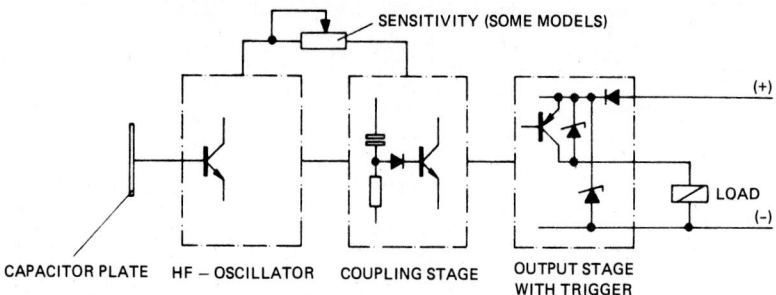

SENSITIVITY (SOME MODELS)

CAPACITOR PLATE   HF – OSCILLATOR   COUPLING STAGE   OUTPUT STAGE WITH TRIGGER

Fig. 57   Principle of operation of a capacitive proximity switch. The device shown incorporates an independent transitorized oscillator which is remote from the control amplifier. Amplifiers available can be operated from ac and dc power supplies with solid-state, reed, and electromechanical relay outputs. Sensing ranges available include 0 to 1, 0 to 2, 0 to 4, and 0 to 6. Switching rate is 500 Hz, temperature range is −15 to 140°F (−26 to 60°C), and connecting cable length is 6 ft (1.8 m). *(ProxSwitch, R. B. Dension.)*

Capacitive Proximity Switches

The basic element of a capacitive proximity sensor is a high-frequency oscillator containing a capacitor, one of the plates of which is built into the end of the sensor. See Fig. 57. When oscillating, a field is created around this free capacitor plate. When an object is placed in the field, the amplitude of the oscillator output changes. These oscillations are rectified and smoothed by an integrating coupling stage. The resulting dc signal is fed to a trigger circuit which switches an output transistor.

The sensing zone or envelope of the sensor is influenced by the physical properties of the object being sensed in the following ways: (1) Nonconducting materials, such as glass and plastics, are sensed by a change in dielectric characteristics. Since this change is small, the sensing ranges are necessarily limited. (2) Conducting materials are sensed by a change in dielectric characteristics as well as by an additional disturbance of the noise field caused by terminal conductivity. (3) Materials containing both conducting and nonconducting properties, especially if grounded, are sensed by a combination of the foregoing characteristics, as well as by absorption. These conditions produce the greatest switching distance for a given sensor. Because capacitive sensors are so markedly affected by material characteristics, an estimate of the switching range requires knowledge of the medium to be measured.

See also Fig. 58. Capacitive systems for measuring solid and liquid levels are described in Sec. 5 of this Handbook.

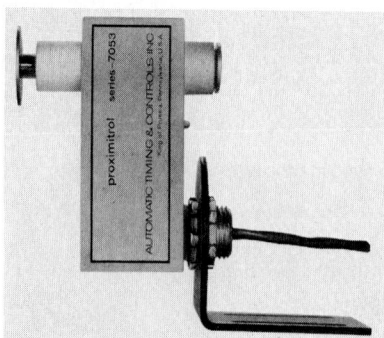

Fig. 58   Capacity-sensitive limit switch. A sensor disk is shown at the upper left. As the object approaches, the capacitance of the sensor increases. When the sensor capacitance surpasses the refrence level, the difference is amplified by an amplifier controller and the control relay contacts transfer. The reference level can be optimized for specific applications by means of a screwdriver adjustment. The standard disk sensor is 1³⁄₁₆ in (30 mm) in diameter. Disk size and shape can be altered to provide improved sensitivity, as determined by the shape of the object being detected, its dielectric constant, and the operating distance. The operating range is from 0.2 to 5 in (5 to 127 mm), speed of response is up to 200 operations per second, capacitance range is 0.1 to 2 pF, and a change in probe capacitance of 0.01 pF can be detected. The temperature range is 32 to 131°F (0 to 55°C). Minimum object size can be as small as 0.1 in³ (1.6 cm³). *(Proximitrol, Automatic Timing & Controls.)*

Magnetostrictive Limit Switch

Discovered in 1858, but rarely applied, magnetostriction represents a change in dimension of a material as the magnitude or direction of magnetization in a crystal is changed. The principle was introduced in the limit switch field in the mid-1970s. As shown in Fig. 59, the active element is a helical spring subjected to axial extension or compression caused by mechanical displacement. The spring may be made of a variety of ferromagnetic materials, inasmuch as most of these materials have sufficient magnetostrictive properties for this application. To obtain an optimum combination of linear range, sensitivity, temperature coefficient, and hysteresis, the spring is heat-treated. Figure 59 explains how any compression or extension of the spring can be reduced to a measurable output that indicates displacement. A helical spring is well suited to many applications, but the magnetostrictive element can be selected by design to fit specific application requirements for range of displacement and operating force. In some units, the spring material may be a nickel alloy wire about 1 mm in diameter with approximately 2000 turns of 40-gage wire solenoid wound before the helix is formed. In some designs—to avoid electrical connections to a movable point within the device, a double helix is used. The magnetostrictive transducer is considered to have good potential where high-resolution information is required over a substantial range. The principle of magnetostriction transducers is well explained by Garshelis (1977) in the reference listed.

Other Object Detection Devices

Just about all measurement techniques have been considered for or adapted to the problem of object detection. Several of these methods have been borrowed from other fields—for example, load cells

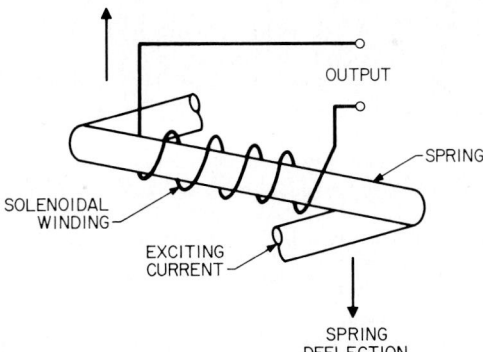

Fig. 59 Operating principle of a magnetostrictive limit switch transducer. Alternating current is passed through a wire element to establish a time varying circular magnetic field. A second winding on the wire element forms a long solenoid in which an emf is inducted proportionally to the time rate of change in any axial component of the magnetization caused by the applied field. In the absence of torsion, magnetization is circular and there is no output signal. If the spring is stretched or compressed, resulting torsional strain causes magnetization to move toward alignment with the direction of greatest strain. The direction of the principal strain is 45° to the axis of the wire. Thus, the time varying magnetization has a component coaxial with the solenoidal winding and induces in it a time varying emf. This is the transducer output signal.

can be used to detect (and count) objects passing over a conveyor, although the principal application of load cells is in connection with weighing operations. Similarly, alpha- and gamma-radiation detectors used in weighing and density measurement applications also can serve to detect discrete objects or masses as they move along a conveyor. The use of laser beam detectors tends to parallel the optical paths which serve as the basis of photoelectric controls previously described. However, the main uses of lasers are concerned with metrological measurements as discussed in the next article in this Handbook section. This also applies to pneumatic methods.

Ultrasonic ranging methods have been used. The presence of objects can be detected by return echoes from reflecting materials in a form of sonar. Similarly rf reflection (radar) can be applied. Fluidics has been applied to object detection problems. All of these currently uncommon methods are described in further detail by Morris (1980) in the reference listed.

REFERENCES

Andreiev, N.: "Machines Identify, Measure, and Inspect Yet Fall Short of Vision," *Control Eng.,* vol. 27, no. 11, 1980, p. 73.

Andreiev, N.: "On-Line Optical Gaging," *Control Eng.,* vol. 27, no. 11, 1980, p. 78.

Bailey, S. J.: "Precise Positioning: A Matter of Sensors and Loop Dynamics," *Control Eng.,* vol. 27, no. 12, 1980, p. 50.

Bailey, S. J.: "Digital Encoders Enhance Loop Design Options," *Control Eng.,* vol. 28, no., 4, 1981, p. 61.

Bailey, S. J.: "Incremental Motion Control," *Control Eng.,* vol. 28, no. 5, 1981, p. 93.

Bailey, S. J.: "Linear Motion: Control It by Electric or Fluid Power," *Control Eng.,* vol. 28, no. 6, 1981, p. 69.

Bailey, S. J.: "Optical Sensors Critical to Future Productivity," *Control Eng.,* vol. 29, no. 1, 1982, p. 72.

Brenza, R. M.: "Digital Tracking Controls Motor Rotor Positions," *Control Eng.,* vol. 29, no. 1, 1982, p. 144.

Bullock, T. B.: "A Programmable Controller for Servo Positioning," *Tech. Pap. MS79-935,* Society of Manufacturing Engineers, Dearborn, Mich. 1979.

Cullum, W., and H. Kratzer: "Measuring Shaft Position by Applying Synchro or Resolver Transducers," *Control Eng.,* vol. 29, no. 1, 1982, p. 148.

EC HDBK: *Magnetic Sensors—Technical Applications Handbook,* Electro Corporation, Sarasota, Florida, 1981.

Garshelis, I. J.: "Magnetostriction Put to Work in Industrial Limit Switching," *Control. Eng.,* vol. 24, no. 10, 1977, p. 60.

Guichard, R. C.: "Plastic Lens Used in Photoelectric Control," *Control Eng.,* vol. 29, no. 2, 1982, p. 134.

James, K. A., Quick, W. H., and V. H. Strahan: "Fiber Optics: The Way to True Digital Sensors?" *Control Eng.,* vol. 26, no. 2, 1979, p. 30.

Keehbauch, T. J.: "Programmable Position Control Uses Standard Induction Motor as Servo," *Control Eng.,* vol. 31, no. 1, 1984, pp. 108–110.

Krueger, A. H.: "Applying Fiber Optics to Photoelectric Switches," *Control Eng.,* vol. 27, no. 8, 1980, p. 61.

Laduzinsky, A.: "Nine Basic Ways to Sense Position," *Control Eng.,* vol. 24, no. 3, 1977, p. 41.

Levy, M. S.: "Optically-Coupled Triac Drivers Reduce Parts Count," *Control Eng.,* vol. 27, no. 9, 1980, p. 134.

Morris, H. M.: "Learn the Jargon Before Specifying Photoelectric Switches," *Control Eng.,* vol. 25, no. 11, 1978, p. 47.

Morris, H. M.: "Object Detection Techniques Range from Limit Switches to Lasers," *Control Eng.,* vol. 27, no. 11, 1980, p. 65.

Morris, H. M.: "Robotic Servo Systems Need Accurate Positional Feedback Inputs," *Control Eng.,* vol. 31, no. 1, 1984, pp. 90–93.

Morris, H. M.: "Computerized Numerical Control Evolves in Response to Changing Needs," *Control Eng.,* vol. 31, no. 2, 1984, pp. 76–79.

Murphy, T. F.: *Practical Considerations in Closed Loop Positioning Control with A Programmable Controller,* Giddings and Lewis Electronics Company, Fond Du Lac, Wis., 1982.

Muth, S.: *Getting Reliable Digital Data from Shaft Position Sensors,* ILC Data Device Corporation, Bohemia, N.Y., July 1981.

Staff: "Wafer Stepping System Uses Computer and Laser to Achieve One Micron Circuit Lines," *Control Eng.,* vol. 27, no. 1, 1980, p. 49.

Dimensional Metrology

Dimensional metrology is concerned with the measurement of length, distance, thickness, and angle and their associated geometric and trigonometric functions and configurations. The topic is closely related to the material in the preceding article of this Handbook, Position, Displacement, Motion, and Object Detection Systems, particularly in connection with automated systems, because position and displacement sensors and transducers must relate to a fundamental metrological base.

From a practical standpoint, industrial dimensional metrology falls into five main categories: (1) the basic instruments, tools, and standards normally found in the well-equipped metrology laboratory responsible for establishing and maintaining manufacturing quality standards; (2) instrumental methods for assisting inspectors in manually gauging parts, workpieces, jigs, fixtures, and final assemblies; (3) automated systems for gauging parts and pieces in high-volume production; (4) continuous thickness gages and control systems; and (5) dimensionally oriented control systems for machine tools and other production equipment.

Basic Instruments, Tools, and Standards

The standard of length is defined in Sec. 1 of this Handbook in the article Units and Standards of Measurement.

Interferometer

An interferometer is a precision instrument which uses the interference of light waves as the basis of measurement. The optics of an interferometer are designed so that the variance of known wavelengths and path lengths within the instrument permits accurate measurement of distances. The French physicist Jacques Babinet in 1827 suggested the possibility of using the wavelength of light as a standard for length. However, it was not until 1960 that the meter was officially defined by the International Bureau of Weights and Measures in terms of interferometry (the length of the path traveled by light, 299 792 458 m/s), thus replacing the prior 200-year-old definition based on a platinum-iridium bar. Thus, the prime standard of length no longer is confined to meter bar no. 27 at the National Bureau of Standards in Washington, D.C., similar bars at the International Bureau of Weights and Measures at Sèvres, France, and those held by several other nations. Now, with adequate facilities and skills, the standard of length may be duplicated anywhere in the world with equal precision.

The interferometer derives its name from the fact that it makes it possible to see light wave interference patterns as a series of bright and dark lines. If two light beams from a given source are directed optically over separate paths that may differ by as little as $\frac{1}{10}$ of a wavelength and then are recombined, detectable destructive interference will occur. Thus, the interferometer may serve two roles: (1) to measure distances in terms of known wavelengths, or (2) to make precise measurements

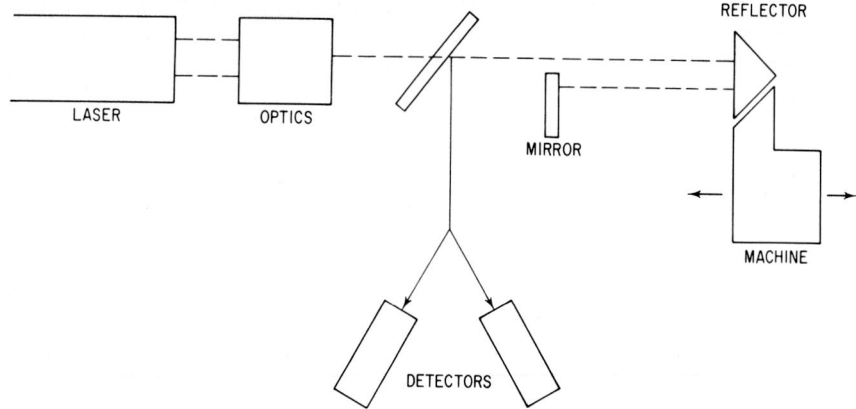

Fig. 1   Laser interferometer. In the early 1980s, scientists at **IBM** Corporation developed a compact device for measuring the wavelengths of tunable dye laser light. These lasers, invented in 1966, are sources of variable color coherent light and are used in the field of laser spectroscopy. Earlier devices used to measure laser wavelengths were limited to accuracies of about 1 part in 10 million (also true of the new device), but there is a major difference. Earlier devices required that the laser beam be reflected off a moving mirror that traveled along a track at least 1 m long. The new device requires that the mirror travel only a few centimeters.

of wavelengths. Currently, laser interferometers are being used industrially in the construction and assembly of precision machine tools where the laser can provide accuracies of better than twenty millionths of an inch (1.27×10^{-6} mm) in lengths up to 200 in (5080 mm). A highly schematic representation of a laser interferometer is given in Fig. 1.

Optical Gratings

Diffraction gratings are widely used in optical instruments associated with precise dimensional measurements. Close, equidistant, and parallel grooves are ruled on a polished surface, commonly a glass base coated with aluminum. Gratings can be (1) reflection or (2) transmission types. The number of grooves ranges from several hundred to many thousands per inch, depending on the dispersion required. The cross section of a diffraction grating is shown in Fig. 2.

The relative movement of two gratings produces optical patterns or fringes. Such patterns are sometimes referred to as moire patterns.

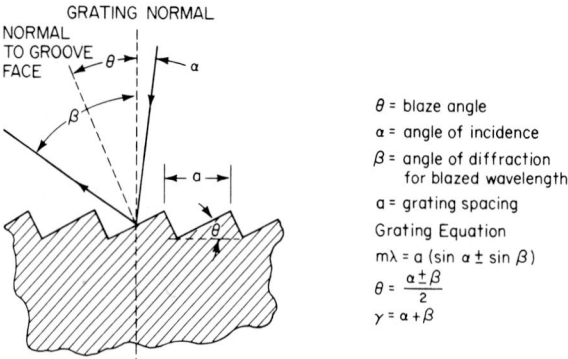

θ = blaze angle
α = angle of incidence
β = angle of diffraction for blazed wavelength
a = grating spacing

Grating Equation

$m\lambda = a\,(\sin \alpha \pm \sin \beta)$

$\theta = \dfrac{\alpha \pm \beta}{2}$

$\gamma = \alpha + \beta$

Fig. 2   Cross section of a diffraction grating showing the "angles" of a single groove, which are microscopic in size on an actual grating. *(Bausch & Lomb.)*

Gage Blocks

Usually furnished in sets of 81 blocks, ranging from 0.05 to 4.0 in (1.27 to 101.6 mm) gage blocks serve as secondary standards traceable to the prime length standard. The blocks are rectangular pieces of hardened steel or carbide that are ground and polished flat with a square or oblong cross section. The length of a block is the perpendicular distance between the two opposite, polished faces. By carefully sliding the gaging surfaces of two blocks one over the other, they can be "wrung" together and thus can be built up to provide a useful range of standard lengths. The dimensions of individual blocks are established by interferometric methods with apparatus set up along the lines shown in Fig. 3. Class AA blocks have a tolerance of ± 0.000002 in ($\pm 5.1 \times 10^{-5}$ mm), and are intended for reference purposes in temperature-controlled gage laboratories. These "masters" should be sent periodically to the National Bureau of Standards for checking. Class A blocks are used in inspection departments and also may serve as masters. Blocks of class B and C quality are used throughout the factory for accurate measurements and tool setting.

Gage blocks are termed *end standards*. Additionally, there is need for "line standards" in the form of precision scales. Before the development of laser interferometric techniques, line standards were compared with master scales visually by the use of a microscope. The National Bureau of Standards now uses an automatic fringe counting interferometer which employs a laser light source which measures line standards directly in lengths up to a meter with a precision of a few parts in 100 million.

Autocollimator

An autocollimator is an optical instrument for directly measuring small angles of tilt of a reflective surface. The instrument effectively combines the functions of a collimator and a viewing telescope

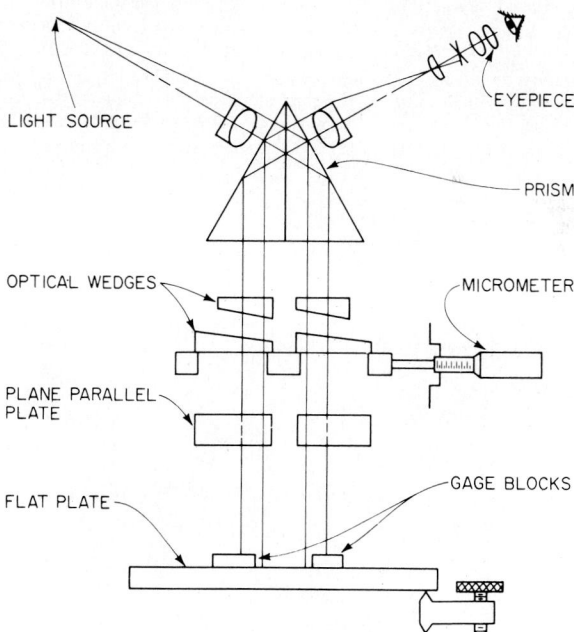

Fig. 3 Interferometric comparator for comparing length of one gage block with another. The difference in gage block lengths is measured by the horizontal distance through which optical wedges must be moved to bring the two interference patterns successively into coincidence with a cross hair in the eyepiece. Wedges and micrometer are rotated 90° for illustration. *(Tool Engineer.)*

into a single system. Diffused light is directed at a partially silvered mirror which diverts part of the beam forward past a target cross hair to the reflective surface to be observed. The reflected beam is passed back into the autocollimator and focused by the objective so that the shadow of the cross hair appears in a plane which coincides with the image of the cross hair. A measuring device, such as a filar micrometer, is used to determine the displacement of the shadow of the cross hair with respect to the cross hair per se. This displacement then is related to the angle of tilt of the reflector. Accurate readings can be taken direct to 0.5 second of arc, within a 10-minute range, up to a distance of 30 ft (9.1 m).

Goniometer*

This is an angular measuring device essentially similar to a circular table or a dividing head except that autocollimating telescopes are used to establish the datum values for the readings taken from the divided circle—either directly or through a micrometer eyepiece. The instrument is employed for determining circular divisions and utilizes the basic principle of the divided circle. See Fig. 4. The goniometer permits rapid inspection of plane surfaces, such as prisms and parallel plates. After one face is viewed, the telescope can be turned with the divided circle so that a second face of the prism can be viewed—the angle through which the telescope is turned is the measurable angle between the two faces. Some precision goniometers are capable of establishing values of 0.1 second of arc. The device is used to establish parallel faces and Brewster angles of laser rods and for production inspection of prisms and optical components.

Clinometer*

This device is essentially a divided-circle instrument which simplifies the transfer of angles between planes. Bubbles or electronic levels frequently are used to establish the null setting principle of a precision clinometer, while less precise instruments utilize the comparison of an angle measured to a datum surface. Angular indications using a clinometer are established by placing the instrument successively on mutually inclined surfaces. The difference between the two observed readings is the angle between the surfaces. Inasmuch as a clinometer is basically a divided-circle instrument, it along with autocollimating telescopes can form the basis of a goniometer. A clinometer is illustrated in Fig. 5.

*Information furnished by Rank Precision Industries, Inc., Des Plaines, Ill.

Fig. 4   Precision goniometer. *(Hilger & Watts.)*

Fig. 5 Clinometer. *(Hilger & Watts.)*

Fig. 6 One-fifth-second microptic theodolite. *(Engis Equipment.)*

Theodolite

Essentially, a theodolite is a surveying instrument that has been adopted for use in metrology. The instrument consists of a telescope in which rotation about the vertical axis is measurable. The telescope also can rotate about a horizontal axis in a true vertical plane. This rotation is capable of measurement. Theodolites are calibrated for (1) collimation value, that is, the centering of the cross hair to the telescope, (2) trunnion error—the truth of rotation of the telescope in a true vertical plane, and (3) the centering of the axis or rotation. Accuracy of the circle, position of the plate bubble, and accuracy of the micrometers also are measurable and correctable features. Theodolites are used for alignment of large jigs and fixtures. Through triangulation, the instruments also can be used for length measurements. A theodolite is shown in Fig. 6.

Optical Flats

An optical flat is a transparent disk, usually of quartz, two sides of which are parallel. One side is polished for clear vision of the surface of the gage, part, or tool on which the flat is placed. The other side of the disk is ground optically flat. Under proper conditions, the phenomenon of interference bands is created. These bands appear as a pattern of dark stripes on the illuminated work surface. Observation of these bands can be used with great accuracy for two purposes: (1) to determine the flatness of a surface i.e., to assess the amount of concavity or convexity and (2) to measure the linear difference between a reference gage and an inspection gage, or between a gage block and a highly accurate part. Any linear difference can be detected between approximately 0.002 in (0.051 mm) as a maximum and one to two millionths of an inch ($\sim$2.5 to 1.3 $\times$ 10^{-5} mm) as a minimum. The principle of the optical flat is illustrated in Fig. 7.

Sine Bar

A sine bar consists essentially of a bar serving as a straight-edge and two cylindrical buttons, which may be on the side or on the undersurface. If the side-button type is used, one button rests on a gage block, and the thickness of that gage block is added to the height of the gage block stack used to set

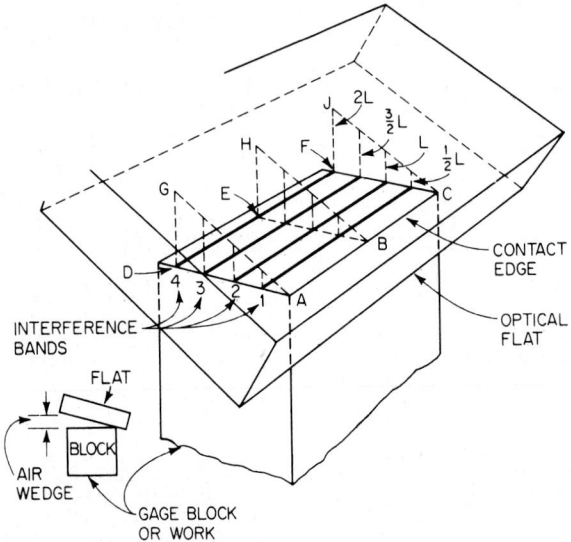

Fig. 7   When an optical flat is manipulated so that a wedge of air is created, dark bands (interfernce bands) result at locations where the separation of float and work equals a multiple of $L/2$, or one-half wavelength of the monochromatic light used.

the second button. If a base-button sine bar is used, the first button rests directly on the surface plate or master flat, and the second button rests on a stack of blocks equivalent to the sine of the wanted angle. Thus, the sine bar serves to reduce problems of laying out or checking angles to terms of the right triangle. The bar itself may be 5, 10, or 20 in (~127, 254, or 508 mm) between the centers of the buttons. The distance between the centers of the buttons is the hypotenuse of the angle; the height of the elevated button above the button in contact with the reference surface is the sine of the angle. A sine bar is illustrated in Fig. 8.

Spherometer

This is an instrument for determining the radius of curvature of lenses and other spherical surfaces. A design of wide acceptance consists of a depth measuring device, such as a micrometer screw, mounted at the center of a tripodlike support. The concave or convex surface to be measured is centered directly under the spherometer, and the micrometer screw is rotated until its tip just touches the spherical surface. The displacement of the tip of the micrometer screw above or below the plane of the three support points is a measure of the radius of curvature of the surface. A typical sphero-meter reads to 0.01 mm. A high-precision spherometer of the type shown in Fig. 9 reads to 1 μm.

Ellipsometer

This instrument is used to determine the thickness of very thin films of monomolecular dimensions. An ellipsometer is basically a polarization interferometer utilizing a photometer as a readout device. A production-type ellipsometer can measure surface film thickness from a few angstroms to 1 μm and permits rapid determinations of thickness and refractive index. In one type, a laser source oper-ates in a fully lighted room, providing a small, intense spot that permits the study of very small samples. A solid-state detector is used. Automatic ellipsometers are available. In operation, a speci-men is placed on the stage, and the start button depressed. A minicomputer displays and prints the film thickness in less than 4 s for most samples. High speed and simplicity make such an instrument ideal for monitoring rapidly changing films and for high-volume measurements in the semiconductor industry.

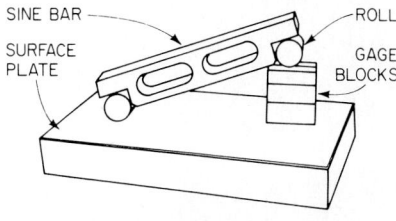

Fig. 8 The sine bar allows work to be located in an accurate relationship to a plane or surface plate.

Fig. **9** High-precision ring-type spherometer. *(Gaertner.)*

Optical Comparator*

Also known as a contour projector, the optical comparator is an instrument for accurate visual inspection and measurement of a part in which a magnified image of the part or portion thereof is projected onto a viewing screen. See Fig. 10.

The instrument operates by directing a light beam at the part to be inspected so as to cast a shadow which is relayed through a series of optical lenses and mirrors and finally projected onto the viewing screen. A contour projector can measure triaxial linear dimensions, radial dimensions, angles, and irregular as well as regular profiles. Two basic measuring techniques are used: (1) measurement by *comparison* and (2) measurement by *movement*.

Measurement by comparison is achieved by comparing the projected image with precise chart gages laid out on the viewing screen. The scale of the chart gages must correspond precisely with the magnifying power of the lens in use, and the lens must be distortion-free. Field diameter is also important in measuring by comparison. The field diameter (the area that a contour projector can measure at one time) is a function of the viewing screen diameter and the lens magnification. A 14-in (356-mm) screen, for example, when used in conjunction with a 20 × magnification lens, will yield a field diameter of 0.7 in (17.8 mm).

Measurement by movement is achieved through table motions which are measured by various devices, including micrometers, indicators, and digital readouts.

Contour projectors have many uses. For example, the instruments can be used (1) in *engineering and development* for photoelastic stress analysis, (2) in *toolrooms* to check cutter shapes and sizes, and (3) in *assembly operations* where dimensional relationships between mating parts are critical and require precise checking. A contour projector can be conceived as a variety of gages combined in a single unit.

Optical Bench

Although not compactly assembled in a black box, an optical bench (with accessories) may be considered a modular, highly flexible, disassembled instrument capable of convenient, customized assembly by a skilled user. An optical bench is a graduated support on which carriages for holding lenses,

---

*Information furnished by John A. Bernardini, Ex-Cell-O Corporation, Detroit, Michigan.

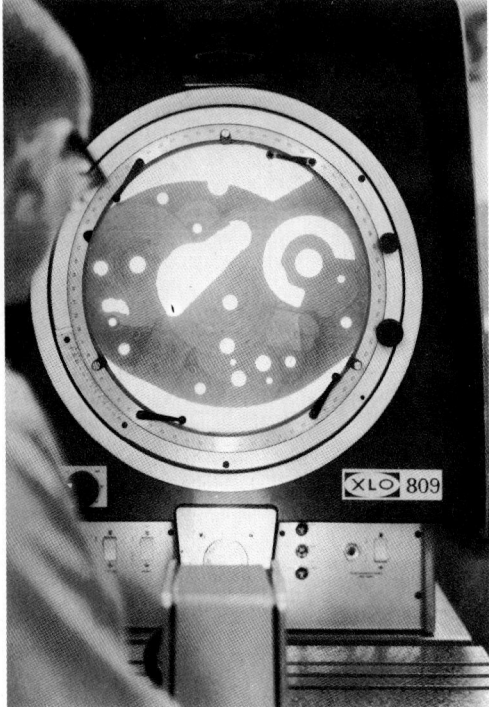

Fig. 10 Use of a contour projector to inspect a part by simultaneous direct and surface projection. The operator can search for surface defects and measure the diameters of holes at the same time. The smallest hole visible on the part illustrated is 0.019 in (0.48 mm).

mirrors, and other components can be mounted and positioned. The optical bench can be of a linear configuration, or rectangularly shaped as shown in Fig. 11.

The Linear Bench

The linear bench, which was the traditional shape for many years, consists of a long, graduated horizontal track on which the carriages (holding the optical components) move. The lathe bed type is in this category and is the most precise optical bench available. The bench is typically 1 to 2 m long and is strong, rigid, and extremely straight. Since the typical precision lathe bed bench does not deviate in straightness by more than 0.002 cm, it is excellent for research purposes. The single-rod bench consists of one graduated rod mounted on two end supports. The double-rod bench consists of two parallel rods, one of which is graduated, mounted on two and sometimes three supports. Rod-mounted benches, which are less expensive and support less weight, are commonly used for educational purposes.

The Rectangular Bench

The rectangular bench consists of several parallel coplanar tracks rigidly attached to a baseplate. The plate is mounted on a frame, usually in such a way that the system is isolated from room vibrations. Rectangular optical benches are useful for laser setups and holographic studies, or wherever the experiment makes extensive use of folded optical paths.

Fig. 11   Rectangular optical bench with related optical components. *(Gaertner.)*

Measuring Machines

These are highly instrumented systems used to measure and record linear dimensions and xyz coordinates of holes and surfaces in parts and tools. Various styles of machines operate under manual, motorized, or computer control. The mode of measurement may be direct, by deviation, or by a combination of these two methods. Direct measuring machines with digital readout have been the most widely used. These machines provide direct or absolute measurement of probe position or movement in three axes to 0.001- or 0.000050-in (0.02540- or 0.00127-mm) resolution.

Computer-aided measuring machines generate tapes for use on automatically controlled machine tools. Coordinate data from a prototype model measured on the machine or fed into the machine from a drawing are combined with tool control code information to produce input to the machine tool control system. Characteristics of some commercially available measuring machines include:

Measuring range	
x axis	30 in (762 mm) to 72 in (1829 mm)
y axis	20 in (508 mm) to 30 in (762 mm)
z axis	8 in (203 mm) to 24 in (610 mm)
Work height capacity	28 in (711 mm) to 36 in (914 mm)
Readout resolution	0.0001 in (0.0254 mm) (average)
Accuracy*	± 0.0005 in (0.0127 mm) to 0.001 in (0.025 mm)

Operation of Digital Readout Measuring Machine

The workpiece is mounted on the table in the conventional manner and the probe is moved in the x, y, z axes until it is positioned in a reference hole or against a reference surface. The readout then is cleared to zero by pushing a reset button. The probe then is manually moved to the first check location, and the amount of movement in each axis is displayed. The output signals can be interfaced directly with electronic computers, thus adding the use of storage banks, logic, and capability to perform mathematical and algebraic computations. The clear-to-zero feature allows measurement to begin at any point. Readout is automatic to any point within the measurement range. All readouts are progressive from zero and are preceded by a plus or minus sign to indicate position.

A typical measuring system consists of three basic components for each axis of measurement: (1) a steel grating, (2) a reading head, and (3) the readout. The grating and reading head are mounted so that the short glass grating in the reading head is superimposed over the steel grating at a slight angle. As shown in Fig. 12, a beam of light is passed through the glass grating, reflecting back from the steel grating onto four photocells. The relative movement of the two gratings produces optical patterns or fringes. As these fringes move, they interface with the beam of light being reflected from the steel grating.

---

*Over measuring range at a plane 8 in (203 mm) below y-axis carriage.

The fringe pattern travels across the grating at right angles to the physical motion. Photocells in the reading head detect the interference fringes (light and dark bands) and convert the changes in light intensity into electric signals, thus "counting" the fringes and indicating the precise amount of movement by digital display.

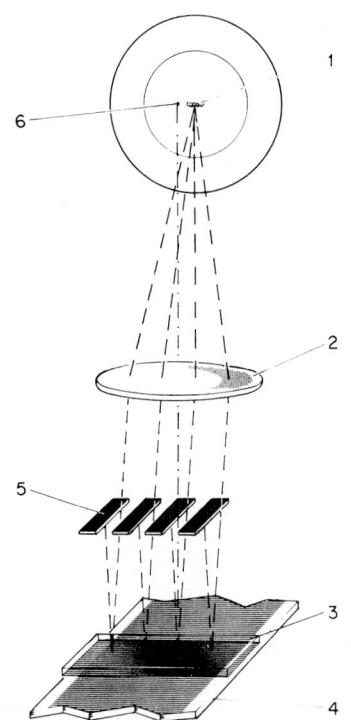

Fig. 12   Optical system of a measuring machine. (1) Line filament, (2) collimating lens, (3) index grating, (4) scale grating, (5) photocell strips, (6) principal focus of lens.

Accessory instrumentation available with digital readout measuring machines includes a wide variety of interchangeable probes, pneumatic drill attachment, optical viewing screen, layout punch, and microscope.

A computer-aided measuring machine collects, memorizes, and produces dimensional data in digital form from workpieces of simple to complex configuration. Some typical computer programs provided are (1) automatic alignment computation, (2) difference computation, (3) out-of-tolerance computation, (4) coordinate conversion (cartesian to polar), and (5) polar deviation (true position). With automatic alignment computation, part setup time is reduced, as no fine-positioning of the part to the machine is required. The computer calculates and compensates for the difference in the alignment of the workpiece and the axis of the measuring machine. A computerized gaging system can determine and print out how much an actual measurement is out of tolerance from specified size, translate linear coordinates into radial terms, and measure contours "on-the-fly" (the measuring probe scans the part's surface without interruption). Tapes for contouring work may also be generated from a prototype part. As the part surface is scanned by the probe, the computer monitors and stores points. From this information a tape is produced with the minimum number of points necessary to cut the contour within programmed tolerances.

Automated Gaging Systems

Generally, the engineering problems involved in automating a dimension inspection and control operation are considerably more difficult than those encountered in designing automatic control systems for such process variables as temperature, pressure, liquid flows, levels, and so on. For one thing, accuracy requirements generally are much higher. Further, with the exception of very high production runs that may extend over many months, the control setup must be much more flexible and adaptable to multiple use than, for example, a temperature or fluid-flow controller that may serve the same function year after year over the lifetime of a plant. But, despite difficult engineering problems and severe demands for economic payout, much progress has been made in the automatic gaging field. The foregoing conditions also account for the extensive use of semiautomatic gages and instrumentation systems which require but reduce manual skills and hence contribute markedly to greater inspection accuracy and lower overall inspection costs.

Dimensional variables, while capable of reduction to terms of length, distance, and angle, present themselves in numerous specific formats—diameter, depth, taper, center distance, thickness, contour, squareness, roundness, parallelism, camber, concentricity, surface finish; and certain types of geometric configurations, such as gears and threads, have unique characteristics, such as runout, pitch, profile, chamfer, and truncation, which for all practical purposes discourage attempts to create a universal approach to automatic gaging and inspecting problems.

Pneumatic Gaging Systems

For many years pneumatic detecting means predominated the automatic gaging field, and these methods continue to be important. A pneumatic gaging system consists essentially of components that provide a constant pressure air supply, an indicating means, and a metering orifice. The principle of operation is based on the effects of varying the flow of air from the metering orifice. For example, as flow from the orifice is obstructed, pressure in the system builds up to the regulated value, and flow through the system drops. Over a significant range of values in such a gaging system, there exists a linear relationship between flow or pressure and the size of the escape orifice. In many gaging applications, the linear relationship is equated to the clearance that separates a sized metering orifice and an obstruction. The indicating component shows the change in flow or pressure as a linear measurement.

The orifice obstruction can be either the workpiece being gaged or an integral part of the gage spindle. An example of the first is found in the gaging of a bore by inserting a gaging spindle containing two diametrically opposed orifices through which air is flowing. Here there is no contact between the gaging spindle and the part being measured; the closeness of fit serves as the obstruction. In the second kind of gage the gage spindle includes a component that mechanically contacts the work being measured. Movement of this component relative to a sized orifice affects air flow and establishes the basis for measurement.

Basic applications of pneumatic gages are shown in Fig. 13. Industrial installations range from the simple application of a single gaging spindle, for checking a single dimension, to a multiple sensing application in which, for example, 40 or more dimensions may be checked simultaneously. Pneumatic gages are well suited to postprocess gaging, in which parts are checked against specifications after machining. Pneumatic units also are applied to *in-process* gaging and, in some cases, are tied into complex computer control systems. Air-to-electric transducers frequently are used in automated systems.

Combinations of various gaging devices are used to provide signals and controls for a variety of applications, including (1) adjusting movement of cutting or forming tools, (2) signaling for replacement of worn tools, (3) stopping machines, (4) warning when part-size trend approaches minimum or maximum limits, (5) segregating parts at various stages of production, (6) weighing and checking weight relationships, (7) connecting or straightening parts automatically, (8) matching parts for selective assembly, (9) inspecting finished parts, (10) classifying and matching parts, and (11) selectively packaging parts.

Electric and Electronic Gaging Systems

The use of air jets in pneumatic systems has many advantages, because to some degree the concept does approach universality of application—not being severely affected by temperature, vibration, and other ambient conditions, or by the materials characteristics of the pieces being inspected. Going from pneumatic sensing to electric or electronic control circuitry poses no large problems. Practically every conceivable type of electric transducer—induction, reluctance, resistance, capacitance, and so on—has been applied to parts gaging applications. With the vast variety of transducers available, it is evident that the designer of an automatic gaging system is presented with almost limitless component possibilities and circuitry configurations. These matters are well beyond the intended scope of this volume. Frequently the techniques developed for successful application for manually operated table gages can be applied to automatic systems.

In one widely used system, the gage head contains two coils with a sintered iron core centered between them. This core is attached to the gage spindle and moves axially, as shown in Fig. 14. Its position relative to the coils affects the impedance of the coils. The coils and a symmetrical transformer in the oscillator form a bridge. When the core rests equally between the two coils, the bridge is balanced and the output signal from the gage head is zero. When the core is displaced, the impedance of the coils is changed. The impedance of one is reduced, while that of the other is increased, which generates a signal in proportion to the amount of displacement of the core. The signal then is amplified, rectified, and indicated on a dc meter calibrated in units of length. For portability, battery power can be used. Differential amplifiers are available to determine the difference between, or the sum of, two measurements, using two gage heads simultaneously. Typical applications include checking roundness, concentricity, parallelism, thickness, cam contours, tapers, flatness, and squareness—without requiring precision fixturing. This difference technique is well suited for the selection of

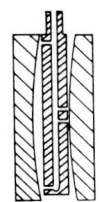

Two diametrically opposed open jets check true diameter of holes having tolerances of 0.005 in. or less

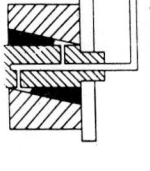
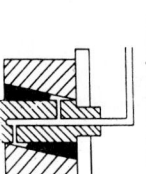

Contact-type gaging head is used in measuring diameters for interrupted bores and bores having a keyway

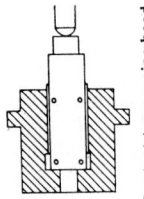

Spindle with opposed open orifices spaced longitudinally checks squareness of a bore axis with a face

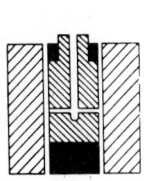

Camber or straightness of hole is checked by rotating through 180° a spindle having four jets

Outside diameter can be checked by two opposed standard jets; tolerance should be 0.002-in. or less

Parallelism of holes can be indicated by combining two spindles of type used in squareness checking

Squareness of surfaces for close-tolerance parts having good surface finish can be checked by a fixtured spindle

Concentricity can be indicated by one or more contact-type spindles mounted in a suitable gage fixture

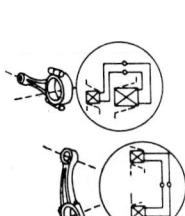

Out-of-round is indicated when any spindle is rotated through 90°

Center distance between holes can be checked by a fixture with two spindles, each having two opposed jets

Multiple-contact-type gage heads mounted in a suitable fixture can check contour

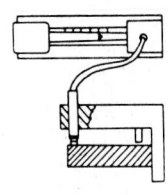
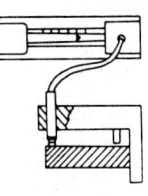
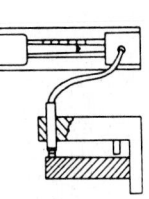
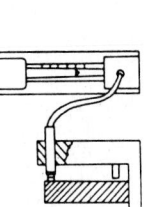

Taper is indicated as any spindle is passed through a bore

Thickness gage for thin parts and items having close tolerance incorporates two opposed standard jets

Height, width, or depth can be measured by a suitably fixtured contact-type gaging head

Fig. 13 Basic applications of pneumatic gaging.

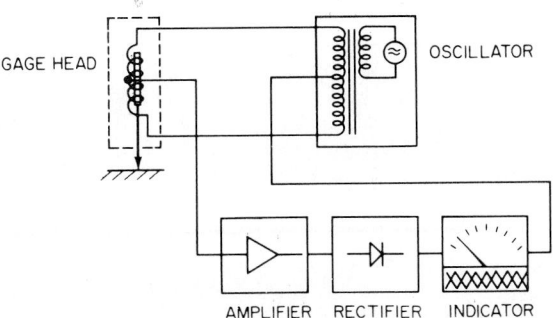

Fig. 14 Impedance-type gaging system. *(Brown & Sharpe.)*

parts with a specific clearance where the actual size itself is not critical; for example, in mating the inside diameter of a cylinder with the outside diameter of a piston. Examples of various measurements along these lines are given in Figs. 15 through 17.

Late-Generation Gaging Systems

With the advancements in solid-state components and circuits since the early 1970s and with much increased attention to automating many production and assembly operations—as in the automotive and aircraft fields—the character of automatic gaging systems is changing markedly.

Relatively recent innovations include the use of laser detectors, cameralike units for object detection, automatic leveling devices, laser holography, photodiode detectors, and imaging systems, including charge-coupled devices (CCDs). Automatic gaging is no longer limited to inspecting small parts, but full assemblies as well, and thus a number of nongaging instruments and devices, such as levelers and positioners, are required to ensure accurate gaging. The integrated system, for example, will most likely include object recognition devices. See the article Machine Vision later in this Handbook section. The emphasis in recent years has been on *noncontact* gaging.

Laser gaging, as pointed out by Jutila (1980), through the use of a scanning laser system has been available for on-line use since the early 1970s. In such systems, a fixed laser beam is directed at an oscillating or rotating mirror, whereupon the projected arc is collimated into a flat ribbon of light which is passed across the target to a vidicon tube or similar detector. An object in the light path causes a silhouette on the detector field and, as the beam scans across the object, detector circuitry determines the interval during which the light is blocked. Phase detection is combined with spatial modulation to measure the scan time across the shadow. Thus, time is related to the section length. Claims have been made that laser scanning can achieve resolutions better than 0.1 mil (0.0001 in) and can operate with the transmitter and detector separated by more than 6 ft (1.8 m). The response of laser scanners is frequently under 1 ms. Systems have been built that can scan several hundred objects per second. Common applications have included the monitoring of various metal, rubber, and plastic extrusions. Laser scanners also have been installed to replace contact instruments for detecting flaws in such parts as transmission gears. Laser triangulation has been used to determine distances from the light source to the surfaces of objects. Specific applications include parts positioning, machine tool control, product height, and level monitoring. The system is also applicable to web thickness measurement.

Photodiode detectors, also described by Jutila (1980), are widely used for noncontact gaging. Arrays comprise sets of solid-state detectors which are sensitive to the shadow cast by the object being measured. The diodes are scanned electronically, and the shadow dimension is determined by the number of dark elements. In addition to use for quality control of wire, cable, tubing, bar, and other extrusion cross sections, the system has been used for sorting and parts inspection.

Image sensing has commonly been accomplished by some form of television camera tube. Recently, CCDs have been used in memory arrays as alternatives to video imaging sensors. Invented in 1969 (Bell Laboratories), the CCD in its configuration as of the mid-1980s is a three-layered semiconductor device—one layer of metallic electrodes and another of silicon crystal separated by an insulating layer of silicon dioxide. CCDs can be fabricated with standard metal oxide semiconductor (MOS) processing techniques. Almost from the outset of its development, a very promising appli-

By placing two gage heads *A* and *B* parallel and on the same side of a workpiece and master with gage settings properly actuated, the indicator will show the difference between *A* and *B*. If both master and workpiece are affected by the same source of error *D* (such as temperature), the difference will remain the same. Such a setup can be used for measuring roundness, parallelism and flatness.

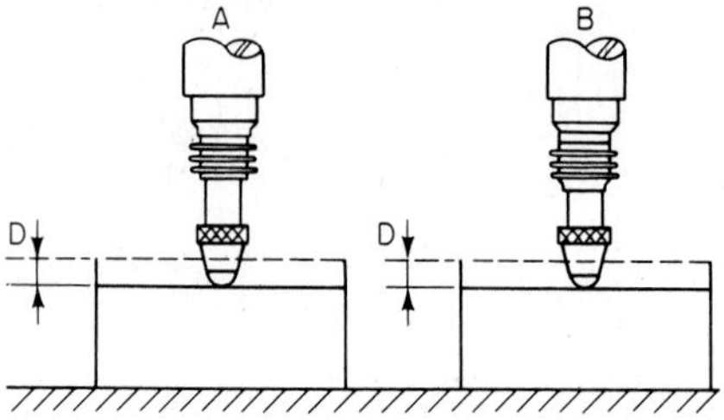

With this setup, only the difference in concentricity is shown on the indicator. If both parts are out of round in same amount, the reading is not affected.

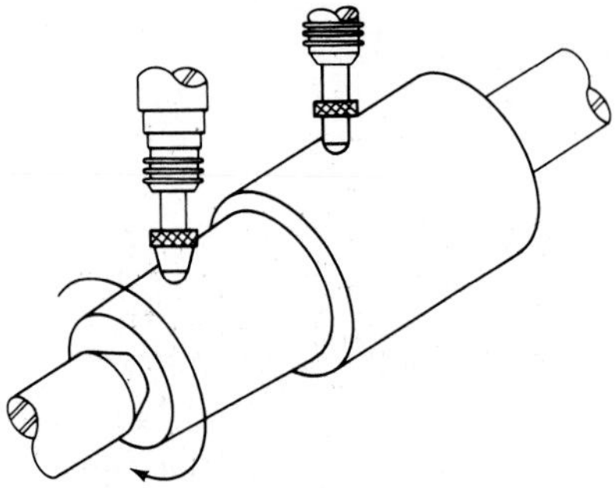

By placing two gage heads parallel on a tapered part, it is possible to check the degree of taper as compared with a master part, without regard to its diameter.

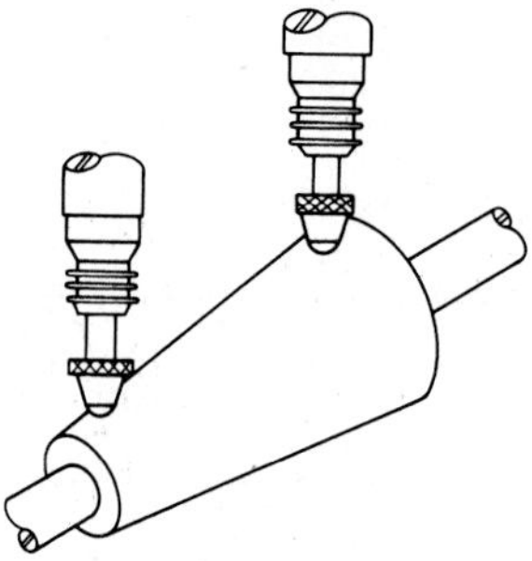

By placing two gage heads *A* and *B* perpendicular and opposite to the workpiece with gage settings properly actuated, the indicator will show the sum of *A* and *B*. If the workpiece is displaced either upward or downward, the indicator reading will not change. In this example, gage head *A* will move a distance +*D* while gage head *B* moves a distance −*D* so that the reading of the thickness of the part does not change because of such displacement of the part. Such a setup can be used to measure thickness or diameters without the need for precision fixtures.

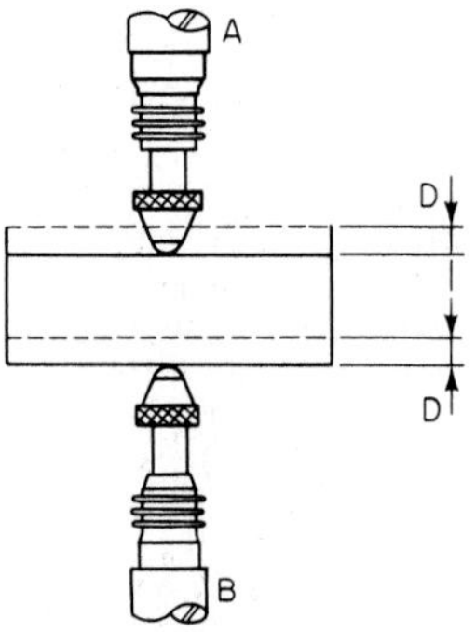

Fig. 15   Examples of the use of impedance-type dimension gages. The first three cases are concerned with "difference" measurements, whereas the last two cases deal with "sum" measurements. *(Brown & Sharpe.)*

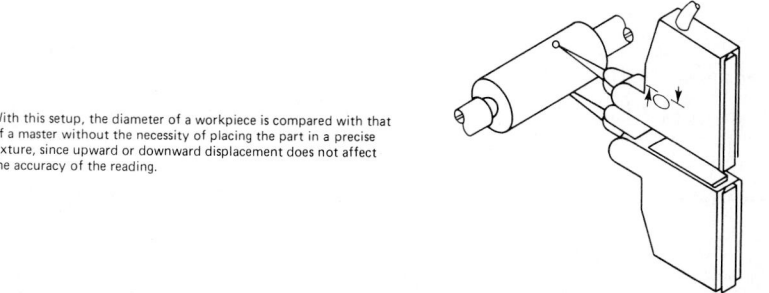

With this setup, the diameter of a workpiece is compared with that of a master without the necessity of placing the part in a precise fixture, since upward or downward displacement does not affect the accuracy of the reading.

Fig. 15   (*Continued*)

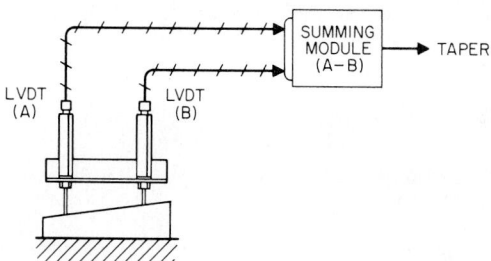

Fig. 16   Use of two LVDTs for the measurement of taper. (*Daytronic.*)

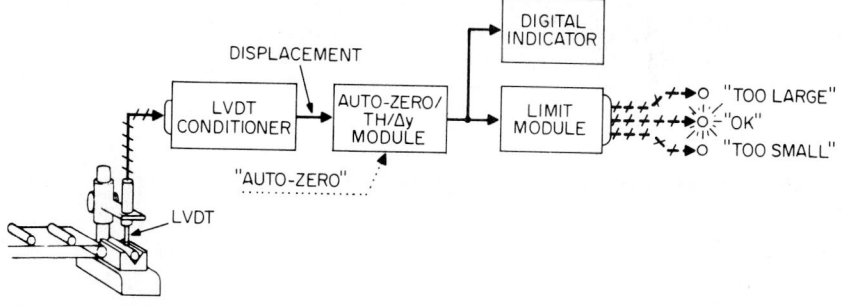

Fig. 17   Industrial comparator gaging requires precise zeroing on a standard "master" part in order to measure small deviations from "size" on subsequent production units. In the system shown, a LVDT is used in connection with a LVDT conditioning module, an auto-zero, track-hold, delta-wye module, a limit module, and a digital indicator. (*Daytronic.*)

cation has been the use of the CCD in a solid-state television camera. When light from an image or scene is focused on the CCD, a pattern of electric charges is created. The charges vary in proportion to the amount of light and thus serve as an accurate electrical representation of picture elements.

Thickness Gaging Systems

In addition to the types of thickness gages already described, there is wide application for continuous thickness measurement of sheets and webbed materials, both metallic and nonmetallic, such as paper and plastics. Many types of transducers are used or have been tried for such applications. Three of these are described here, namely, nuclear radiation, ultrasonic, and x-ray-type thickness gages.

Nuclear Radiation Thickness Gages

Noncontact gaging techniques are used. Several types are available: (1) single-measuring assemblies for sheet thickness control, (2) double-measuring assemblies for control of coatings, (3) a single-gage system (Prefcoat) for controlling coatings on paper and plastics, and (4) a single-gage system (Florescoat) for coating control on sheet steel.

The basic single-head gage, shown in Fig. 18, holds a beta radioisotope source beneath the sheet

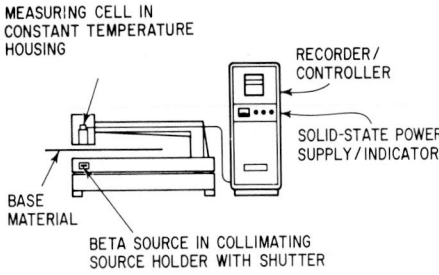

Fig. 18   Basic single-head beta gaging system for sheet thickness control. *(Ohmart.)*

to be measured and a detector cell above. The gage can measure a single point or scan the sheet automatically. Transmitted radiation is converted into an electric current and amplified for readout or automatic control. Measurement is basis weight or weight per unit area, such as ounces per square foot. The precision is $\pm 1\%$ of range.

The double-gage system shown in Fig. 19 is applied to measure and control coatings on paper,

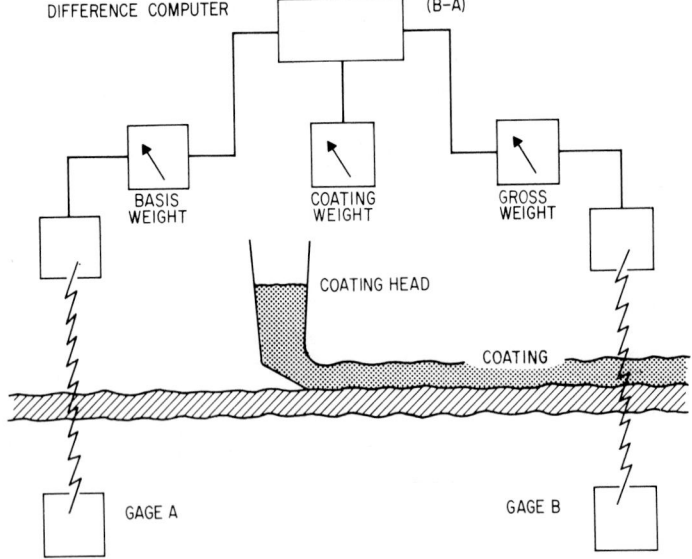

Fig. 19   Double-gate beta-radiation system for control of coatings on paper, textiles, and plastics. *(Ohmart.)*

textiles, and plastics. Measurement is made of the substrate before coating, and a second measurement is made after coating. The difference between the two weights determines the net coated weight. The system normally is used when the coating is 20% or more of the substrate.

Gages are matched on dwell cycles so that the same areas are read before and after coating. Signals are transmitted to a difference computer for readout. These can be recorded or used as the basis of automatic coating control. Uncontrolled short-term variations are not properly subtracted from the second signal unless a memory-delay unit is installed.

One system designed to overcome the disadvantages of a two-gage system measures and controls coatings on paper and plastics with a single gage. A photoelectric effect involving photoabsorption of energy is employed. A low-energy gamma-emitting source is utilized. The system can be used when the atomic number of the coating material is at least twice the atomic number of the base material. The preferential absorption principle detects only radiation absorbed by the ingredient with the highest atomic number—which must be the coating. Low energy (less than 0.5 meV) from the radioisotope source restricts the system to limited total mass—base material and coating combined. The total weight per unit area and mass coefficient of the materials combination must be calculated for acceptable absorption units.

Another system employs a phenomenon known as nuclear fluorescence to measure and control coatings on sheet steel or aluminum. This system measures only coating weight regardless of base material hardness or thickness. Control and readout are in absolute (not relative) terms. The system may be applied to single- or double-sided coating.

Nuclear fluorescence is produced when gamma radiation excites electrons in the metal of the coated strip. As shown in Fig. 20, excitation is continuous, but each occurrence is only temporary. Electrons returning to the original unexcited state produce a characteristic low-energy radiation called fluorescence. The coating material also produces fluorescence in some applications. The total radiation spectrum produced is returned to the detection system. Each fluorescence has its own energy level which is differentiated by two filters and two ion detection chambers at the measuring point. Detectors are of opposite polarity. One filter permits all backscattered radiation to reach an ion chamber; the other filter permits only radiation from the metal strip to reach the other chamber. Hence, the base-metal-strip signals cancel one another leaving only the coating signal as an output from the detectors. Precision is ± 0.02 oz/ft^2 (305 g/m^2). The area measured at any one point is 3 $\times$ 5 in (76 $\times$ 127 mm) but can be reduced to $\frac{3}{4} \times$ 5 in (19 $\times$ 127 mm) in some instances.

The foregoing nuclear systems bear a strong resemblance to nuclear systems used for weighing (see the article on Continuous Nuclear Weigh Scales in Sec. 7 of this Handbook) and for bulk-solids level detection (see the article on Level Systems in Sec. 7 of this Handbook).

Ultrasonic Thickness Gages

These gages find extensive application in the measurement of wall thickness and are useful in measuring most structural materials which transmit sound. Almost all metals, plastics, and ceramics, and

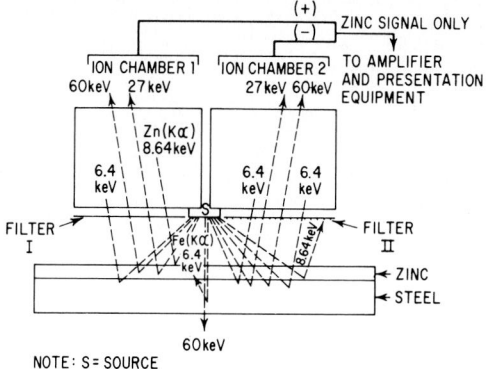

Fig. 20 Nuclear fluorescence gage for controlling coatings, such as zinc, on sheet metal. *(Ohmart.)*

various composite materials can be measured easily. Measurements are instantaneous and can be made from one side of the part. Suitable transducer configurations also permit readings at elevated temperatures for the determination of corrosion losses in chemical equipment while on-stream.

These gages utilize ultrasonic vibrations, most commonly in the portion of the frequency spectrum between 1 and 15 MHz. Sound waves in these frequencies have certain characteristics which make them suitable for making measurements and finding defects. The sound beam can readily pass through most structural materials. It reflects from acoustic boundaries, either internal flaws or geometric boundaries such as the back surface of the part. Sound travels at a characteristic velocity in various materials; this velocity is dependent on the elastic properties and the density of the material, and this permits timing of the wave propagation through the material.

The basic element of an ultrasonic gaging system consists of a *generator* to produce high-frequency electronic vibrations which in turn activate the crystal *transducer*. The transducer converts the electric signals to mechanical vibrations, or ultrasound, which is introduced into the part. Echoes from various boundaries in the part are picked up by the same transducer and are reversibly converted from mechanical vibrations to an electric signal. This is further amplified and processed, and the information is displayed in one of several types of data presentations.

There are two general categories of ultrasonic gages, *resonance* and *pulse-echo*. Each has distinct electronic characteristics and readout methods. Both are dependent on transmission of the acoustic waves, reflections, and velocity in the material.

Resonance ultrasonic gages produce a frequency-modulated continuous wave signal. This provides a corresponding swept frequency of sound waves which are introduced into the part. When the thickness of the part equals one-half wavelength, or multiples of one-half wavelength, standing wave conditions or mechanical resonances occur. The frequency of the fundamental resonance, or the difference frequency between two harmonic resonances, is determined in the instrumentation. The thickness is determined by the formula

$$\text{Th} = \frac{\text{Vel}}{2F}$$

where Th = thickness of part under transducer
    Vel = speed of sound in the material
    F = frequency, Hz

The most common readout for ultrasonic resonance gages is a large CRT. Frequency indications on the tube are compared to an overlaid scale calibrated in inches, and direct thickness readings can be obtained. The readings are instantaneous, and thickness variations can be monitored as the transducer is scanned over the part.

The resonance frequency gages generally produce accuracies of 1% of nominal thickness.

The *pulse-echo* technique operates somewhat like a refined sonar system. Very short electric pulses, usually at discrete frequencies, are generated in the electronic instrument. These produce short acoustic pulses from the transducer, which in turn pass through the material and reflect from the boundaries, just as in the resonance test. The transit *time* of the pulses through the material is measured by the instrumentation.

Thickness is determined by the formula

$$\text{Th} = \frac{\text{Vel} \times T}{2}$$

where Th = actual thickness of part under transducer or within sound beams
    Vel = speed of sound in the material
    T = transit time of sound pulse through one round trip in the material, s

The simplest and most common readout for this type of gage is a direct reading panel meter. This is incorporated into small, portable, battery-operated instrument packages. Accuracies between ½ and 1% of full scale are usually obtainable.

Numerical readout instruments are also available which incorporate specialized digital voltmeter readings with the ultrasonic electronic circuitry. On a standard three-digit readout accuracies of ⅒% of full scale are available. The digital readouts are very easy to read and also lend themselves to incorporation in more sophisticated systems.

One limitation of ultrasonic gaging is the requirement for continuous coupling of the sound beam between the transducer and the part.

Sound cannot pass across an air-solid or air-liquid boundary. Liquid coupling, either a continuous thin film or some other type, is required. In some instances complete immersion of the transducer and the material is feasible. In others, commonly used on production lines, a bubbler or partially contained water column provides the continuous coupling path.

The total accuracy of the test system depends on the reading accuracies noted above, as well as on the test conditions. Surface roughness, curvatures, and material property variations may slightly reduce the total accuracy.

Both testing methods can be applied to automatic gaging systems. Electronic accessories such as high- or low-limit switches and strip chart recorders to provide thickness profiles can be included.

X-Ray Thickness Gages

Noncontact x-ray thickness gages measure the thickness or density of hot or cold materials while in motion or when stationary. Steel, aluminum, brass, copper, glass, paper, and rubber—in a continuous strip or sheets—as well as plastic films, foils, and material coatings are typical of materials that can be gaged by this method. Different styles of x-ray gages span a thickness range from 0.00025 in (0.0064 mm) in plastic to 2 in (50.8 mm) in steel. The capacities of various gages are shown in Table 1.

All materials absorb x-rays to varying degrees, dependent on thickness and density. Thickness is determined by measuring the amount of x-ray energy a material absorbs as the material passes between an x-ray emitter and a receiver. See Fig. 21. The gage is set to the desired thickness standard, and deviation in material thickness from nominal is indicated in percent or thousandths or millionths of an inch on a meter or digital display. A sample piece of the material to be measured or a reference standard is used to calibrate the gage. The gaging signals can be used to operate a variety of accessory instrumentation, including automatic controllers, recorders, totalizers, classifiers, markers, and sorters.

An x-ray gaging system comprises three basic units: (1) a scanning unit that contains the x-ray generator and a detecting unit, (2) an operator control station, and (3) a power unit. The scanning unit generally is a C- or O-frame mounted in a stationary position or on a traversing track—as on a rolling mill, process, or inspection line.

X-Ray Gage Characteristics

Systems are accurate to ±0.25% of the thickness being measured—or better. Repeatability of readings is 0.01%. The speed or response time in which an x-ray gage reacts to a change in material thickness is important in keeping fast-moving strip and sheet stock within tolerance. For example, in a mill running 6000 ft/min (1829 m/min), a gage with a 50-ms response time reacts to a change in material thickness within 5 ft of material passing—in contrast with a 250-ms response time that permits 25 ft (7.6 m) of material to pass before indicating a thickness change.

Continuous process x-ray gaging is an integral part of a computer-operated rolling mill. The gaging system includes an interface that receives commands for thickness settings from the computer and, in return, verifies that such commands have been accomplished. Such systems may also include

Table 1 Typical Thickness Capacities of X-Ray Gages

Gage power, kV	Material measured, in*					
	Steel	Hot steel	Brass and copper	Aluminum	Plastic	Glass
25	—	—	—	0.0004–0.030	0.00025–0.100	—
40	0.0001–0.040	—	0.0001–0.030	0.001–1.0	0.100–0.500	0.050–1.0
65	0.005–0.1500	—	0.005–0.100	0.060 and up	0.500 and up	0.060 and up
100	0.005–1.000	0.010–1.000	0.005–0.500	—	—	—
200	0.005–2.000	0.010–2.000	0.010–1.000	—	—	—

*1 in = 25.4 mm.

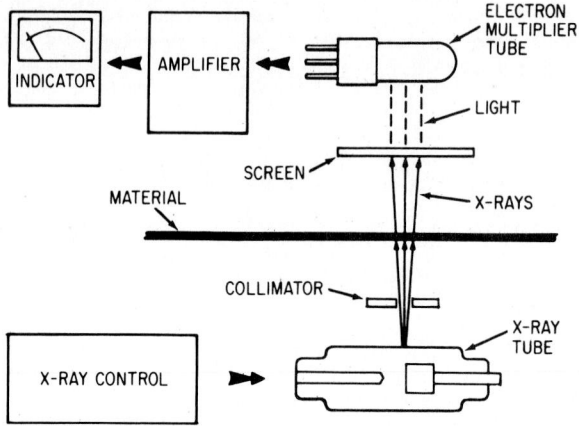

Fig. 21 Principal elements of an x-ray thickness gage. *(Bendix.)*

a digital display of the gage setting, compensation for varying temperatures and alloys of steel, fully automatic calibration to maintain accuracy, and electric motor-operated positioning systems for locating the scanning unit over the steel. A system may also record the difference in thickness between the edge and the center or "crown" of the steel. Anticipatory gaging systems anticipate an out-of-tolerance trend or change in material thickness and use the signals to actuate screw-down, speed, and tension controls to keep the material on gage. Automatic control permits rolling to close tolerances, maximum on-gage footage per ton, and less scrap or out-of-gage material at both ends of a coil or run.

Linear Variable Differential Transformers

A thickness measurement system that utilizes two LVDTs is shown in Fig. 22. The system can also be used to measure slope or taper.

Surface Finish*

Any surface produced by machining departs from the perfect form because of a variety of causes, such as inaccuracies in the machine tool, deformation of the work under the cutting force, and irregularities caused by vibration. Irregularities also may be caused by rupture of the material during separation of the chip. These factors in turn produce geometrical inaccuracies associated with errors of form, secondary texture of waviness, and roughness. In considering the typical surface, two features are of importance: (1) height of the irregularities and (2) amount of separation between the units. Roughness of surface is affected by both the size and the shape of the undulation. Wavelength spacing is just as important as height. A perfectly smooth surface can be changed into a very rough surface simply by changing the wavelength of the undulation. It is not necessary to make any changes in height to accomplish this increase in roughness. As the wavelength of the undulation becomes smaller, the quality of the surface deteriorates. This deterio-

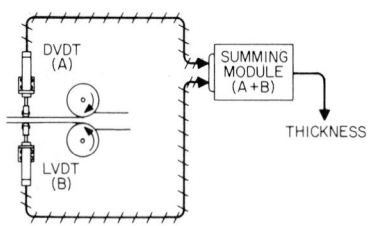

Fig. 22 Use of two LVDTs and a summing module for measuring thickness. *(Daytronic.)*

---

*Information furnished by Rank Precision Industries, Inc., Des Plaines, Ill.

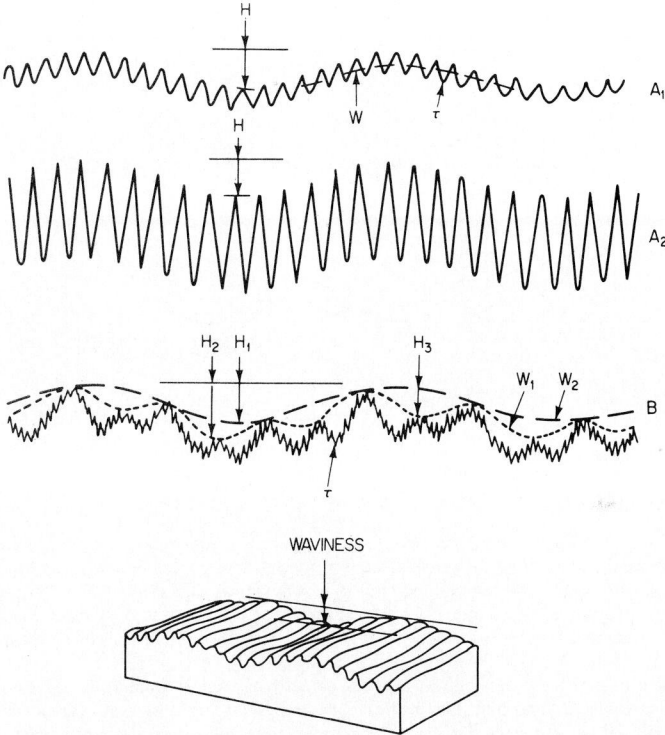

Fig. 23 Examples of waviness. When measuring across the roughness lay, several types of waviness can be found. The profile A_1 has a wavy undulation W on which is superimposed roughness r of smaller amplitude. The profile A_2 is of the same kind, but the roughness is larger in amplitude than the waviness. In these cases, the mean line and crest line waviness are substantially the same. The crest line is irregular and rarely identical with its mean line. Roughly equivalent representations of waviness often can be obtained by filtering. On ground surfaces, the crest spacing of closely spaced waviness surfaces often is in the region from 0.02 to 0.1 in (0.5 to 0.254 mm), and the height generally is less than half the overall height of the grinding texture. Surfaces also may be encountered where the spacing and height are the same, yet the surfaces are quite different.

ration becomes increasingly apparent to the eye as the wavelength becomes shorter, even though the height remains the same.

The patterns formed on the surface by waviness often are varied. Some patterns, like pronounced chatter marks and the coarse feed marks of a badly trued grinding wheel, can be identified at a glance; others may require an instrument to reveal their presence. In the case of surfaces of revolution, the marks extending along the lay of the roughness often become the circumferential departures of roundness. Waviness, as seen in a profile graph, often can be appraised both as an undulation of the mean line and as an undulation of a line drawn through the more prominent crests—thus the terms "crest line waviness" and "mean line waviness."

In measuring across the roughness lay, several types of waviness can be found, as indicated in Fig. 23. The most common specification used to control surface texture is average roughness expressed in microinches. This value represents the actual vertical distance from a datum line of every point of the profile occurring in the length of the sampled surface. The position of the datum line or center line is not a constant. The elevation varies with each specific length of profile being measured. By definition, the position of the center line is such that the total areas of the profile lying above and below the line are equal. Surface texture includes roughness, waviness, lay, and flaws. See Fig. 24.

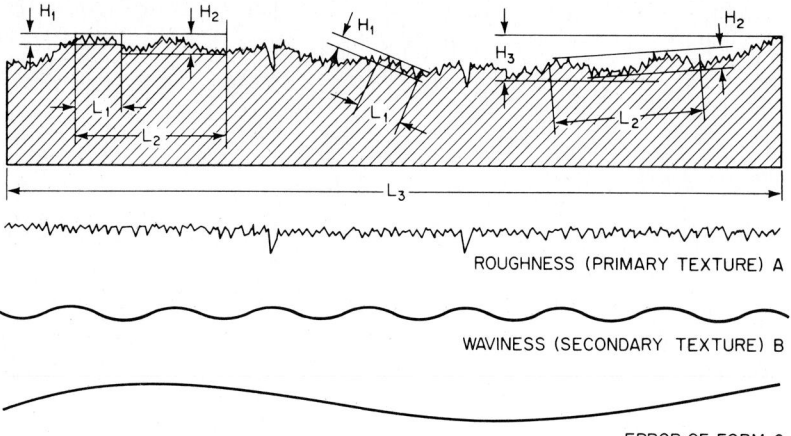

Fig. 24   A surface texture representing the combined effects of several causes.

Since the measurement of surface roughness involves determination of the average linear deviation of the actual surface from the nominal surface, there is a direct relationship between the dimension tolerance for a part and the permissible surface roughness. It is evident that a requirement for the accurate measurement of a dimension is that the variations introduced by surface roughness not exceed the tolerance placed on a dimension. If this were not the case, measurement of the dimension would be subject to an uncertainty greater than the required tolerance.

Surface geometry has a fairly direct bearing on metrology and fits and limits. Measuring instruments ordinarily have anvils and gaging tips which, because of their size, make contact only with the highest points of surface irregularities. The intervening valleys, however, may have appreciable depth and, if this depth amounts to a large proportion of the tolerance, it may affect the size of the part. Subsequent removal of the high spots may differ from that indicated by the measuring instrument by an appreciable amount.

In order to accomplish measurements functionally and to an average numerical value, stylus-type instruments generally have been used. Noncontact methods, usually optical interference means, are effective for the interpretation of fringe values—hence interpretation of the surface. New methods employing scanning electron microscopy are now being used as a means of establishing a three-dimensional evaluation. The essential elements of a stylus-type instrument are (1) a sharply pointed stylus for tracing the profile at a cross section, (2) a means for generating a datum, and (3) a way of amplifying and indicating the stylus movement. Of the stylus-type instruments, two are in general use: (1) a carrier-modulated device in which the magnitude of a carrier current is controlled at every instant of time in accordance with the position of the stylus relative to the datum—regardless of how long the stylus remains in a given position, and (2) a device in which a current or potential is generated in accordance with the motion of the stylus as the stylus is displaced from one level to another.

Carrier-modulated instruments are useful for obtaining graphs because, in acting like simple levers, these devices faithfully reproduce every movement of the stylus relative to the datum, regardless of the spacing. These instruments are calibrated by using gage blocks or interferometrically. The instruments behave in a manner similar to a mechanical lever with magnification ratios of up to 1 million times. The generating instruments reproduce only if the stylus is rising and falling at a rate above the low-frequency limit. Therefore, widely spaced irregularities over which a stylus may be rising and falling only slowly will not be reproduced. This instrument is not desirable for a profile recording but is suitable for numerical evaluation. It is apparent that no matter how valuable and necessary the profile graph may be, some form of numerical assessment is required even if only for purposes of establishing a print value. However, a numerical value cannot be readily established until sufficient data concerning measurement of the component have been established.

Roughness width cutoffs utilized with average values are regarded as the greatest spacing of repetitive surface irregularities to be included in the measurement of average roughness height. A roughness width cutoff is rated in inches and must always be greater than the roughness width. When no value is specified, the value of 0.030 in (0.8 mm) is assumed. The fundamental object of taking the length of the surface into consideration is based on the fact that different makes of instruments and different operators should obtain the same answer for any given surface. The profile graph on the other hand accurately defines and establishes all irregularities, giving values of height as well as width.

REFERENCES

Andriev, N.: "On-Line Optical Gaging Assures Perfect Fit of Jeep Doors," *Control Eng.*, vol. 27, no. 11, 1980, p. 78–80.

Beers, J. S.: "A Gage Block Measurement Process Using Single Wavelength Interferometry," *NBS Monogr. 152*, National Bureau of Standards, Washington, D.C., December 1975.

Beers, J. S., and J. E. Taylor: "Contact Deformation in Gage Block Comparisons," *NBS Tech. Note 962*, National Bureau of Standards, Washington, D.C., May 1978.

Beers, J. S., and C. D. Tucker: "Gage Block Flatness and Parallelis Measurement," *NBSIR 73-239*, National Bureau of Standards, Washington, D.C., August 1973.

Beers, J. S., and C. D. Tucker: "Intercomparison Procedures for Gage Blocks Using Electromechanical Comparators," *NBSIR 76-979*, National Bureau of Standards, Washington, D.C., January 1976.

Croarkin, C., Beers, J., and C. Tucker: "Measurement Assurance for Gage Blocks," *NBS Monogr. 163*, National Bureau of Standards, Washington, D.C., February 1979.

Higdon, R.: "Dealing with Noise in Net Coat Weight Control," *Control Eng.*, vol. 26, no. 6, 1979, p. 76.

R. Hocken et al.: "Three Dimensional Metrology," *Ann. CIRP*, vol. 26-1, 1977.

Jutila, J. M.: "New Dimensions in Optical Gaging," *Instrum. Technol.*, vol. 27, no. 6, 1980, p. 9.

Munsinger, R. A.: "Fiber Optic Probe Colorimeters Eliminate Manual Color Measurement," *Control Eng.*, vol. 28, no. 4, 1981, 136.

Reeve, C. P.: "The Calibration of Indexing Tables by Subdivision," *NBSIR 75-750*, National Bureau of Standards, Washington, D.C., July 1975.

Reeve, C. P.: "The Calibration of an Optical Flat by Interferometric Comparison to a Master Optical Flat," *NBSIR 75-975*, National Bureau of Standards, Washington, D.C., 1975.

Reeve, C. P., and R. C. Veale: "A Survey of the Stability of Optical Flats," *NBSIR 73-232*, National Bureau of Standards, Washington, D.C., June 1973.

Schoonover, R. M., Ku, H. H., Whetstone, J., and J. F. Houser: "Liquid Level Instrumentation in Volume Calibration," *NBSIR 75-900*, National Bureau of Standards, Washington, D.C., October 1975.

Staff: "Surface Roughness Averages for Common Production Methods," *Met. Prog.* vol. 118, no. 2, 1980, p. 51.

Staff: "New Device Measures Laser Wavelengths More Accurately," *Control Eng.* vol. 27, no. 1, 1980, p. 44.

Teague, E. C.: "Surface Finish Measurements: An Overview," *Tech. Pap. IQ750137*, Society of Manufacturing Engineers, 1975.

Young, R., Ward, J., and F. Scire: "The Topgrafiner: An Instrument for Measuring Surface Microtopography," *Rev. Sci. Instrum*, vol. 43, no. 7, 99901011 July 1972.

Young, R. D.: "Eight Techniques for the Optical Measurement of Surface Roughness," *NBSIR 73-219*, National Bureau of Standards, Washington, D.C., May 1973.

Speed and Velocity Systems

Speed is a scalar quantity equal to the magnitude of velocity. Velocity is a vector quantity denoting both the direction and the speed of a linear motion, or denoting the direction of rotation and the angular speed in the case of rotation. Industrially, linear speeds generally are inferred from rotational measurements simply because of the manner in which most machines are designed—with rotating shafts, wheels, and gears to which speed transducers can be conveniently attached.

DC and AC Tachometers

These devices, which utilize the technology of electric motors and generators, have been used in industry for decades and continue in wide usage. They generate an analog signal that is proportional to rotational speed. They have proven reliability and acceptable accuracy for most applications, and they enjoy the advantage of being technically mature. There are two fundamental types of dc tachometers: (1) brush-type and (2) brushless. Brush-type dc tachometers are of two constructions: (1) iron core and (2) moving coil. Fundamentally, the ac tachometer is a three-phase electric generator with a three-phase rectifier on the output. Each of these designs has relative advantages and limitations, the most important of which are developed in the text that follows.

DC Tachometers

Relative perpendicular motion between a magnetic field and a conductor results in voltage generation in the conductor. The magnitude of the voltage is a direct function of the strength of the magnetic field and the *speed* with which the conductor moves perpendicularly to it. Current will flow if the ends are connected to a load, such as an instrument. The polarity of the voltage and, therefore, the direction of current flow depend on the polarity of the field and the direction of conductor motion. See Fig. 1. This is the principle used in the dc or commutator-type tachometer. The same effect can be obtained by rotating the magnet and holding the conductor still—the principle of the ac or rotating-magnet tachometer.

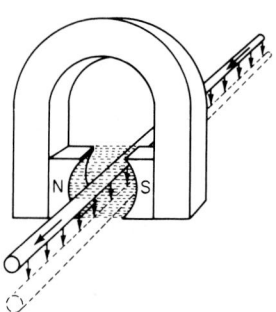

Fig. 1 Principle of the commutator-type dc tachometer generator.

A dc tachometer system consists of a dc generator and a dc indicator or recorder, as shown schematically in Fig. 2, and may be used when the top operating generator speed is at least 100 and not more than 5000 rpm. Special indicators may be used for top operating generator speeds as low as 100 rpm and special recorders for speeds as low as 10 rpm. Generally, the characteristics of the tachometer can be matched to the equipment whose operating speed is being determined by using suitable gearing for effecting speed reduction or multiplication. Several types of dc tachometer generators are available. The composite characteristics of a representative device are given in Table 1.

In the iron core brush-type dc tachometer, stamped iron laminations with an overlaid wire winding and commutator make up the rotor assembly. Limitations of this simple design include the large inductance of the armature due to the rotor laminations. The inductive spike from this type of construction causes erosion of the brushes. The output signal also contains considerably more ripple than may be found in other kinds of tachometers. There is also the problem of surmounting the extra inertia caused by the mass of the laminations.

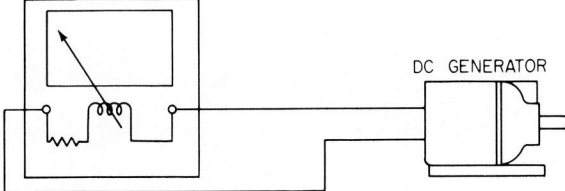

Fig. 2   Highly schematic circuit of a dc tachometer system.

Table 1 Composite Characteristics of a Representative DC Tachometer Generator

Voltage output at 1000 rpm	6 V ± 1%
Accuracy	±1%
Emf linearity	±0.15%
Permissible current drain	50 mA
Maximum rms value of ac ripple	2%
Allowable end play	0.005 in (0.13 mm)
Maximum operating temperature	250°F (121°C)
Internal resistance at 25°C	20 Ω ± 2%
Composition of brushes	Palladium-silver alloy
Armature	12 bars, 12 slots
High-potential test	500 V for 1 min
Bearings	Ball
Temperature compensation	$\frac{1}{10}$% per 10°C change
Normal continuous speed	2000 rpm
Minimum top speed	100 rpm
Maximum top speed	5000 rpm
Weight	~2 lb (0.9 kg) (dust-resistant models)
	~25 lb (11.3 kg) (weatherproof and explosion-proof models)
Shaft diameter	$\frac{3}{16}$–$\frac{3}{4}$ in. (5–13 mm)
Adjustable magnetic shunt range	±4%
Length/width/height (approximate)	4½–5 in/3 in/3 in (114–127 mm/76 mm/76 mm) (dust-resistant models)
	12½ in/6 in/5 in (318 mm/152 mm/127 mm) (weatherproof and explosion-proof models)

In the moving coil brush-type dc tachometer, the winding is in the form of a shell or cup. In this construction, there is a magnet on one side and an iron slug on the other. Thus, the magnetic field passes through the cup-shaped winding. This overcomes much of the inertia because only the winding is rotating. The electrical inductance is also markedly reduced.

Brush-type dc tachometers are usually limited to relatively clean environments. Brush life is shortened in many cases because of particulate and erosive contaminants. Some airborne contaminants may also build up in the form of films on the commutator. Sealed enclosures can be used, but these create a thermal problem because of entrapped heat. This heat is not generated within the tachometer per se because of the very low currents involved, but it can be conducted through the shaft. Magnets are sensitive to temperature (estimated to be 0.01 to 0.05% per degree Centigrade) and therefore, if the stability of the output is critical, temperature compensation may be required.

Speed Ratio Systems with DC Tachometers

By using two dc tachometer generators (Fig. 3) connected to a ratio meter mechanism, measurements that are dependent on differential processing speed, such as "percent stretch" and "ratio of draw," can be measured and controlled through additional elements in the system. The system shown for a textile application has wide use in the paper and steel industries, as well as where it is important to

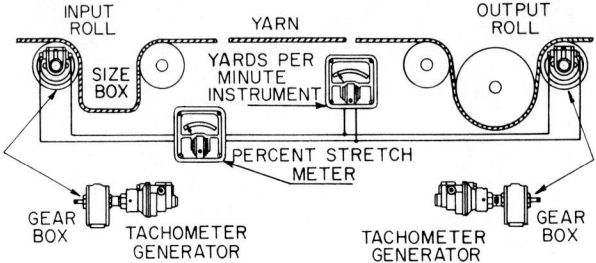

Fig. 3 Speed ratio tachometer system used in the textile industry.

know the ratio between two quantities expressible in terms of rotation. In these installations, the minimum generator speed must be 400 rpm because of voltage requirements of the indicator. Full-scale range limits are from 10 to 100% shrink. Percent stretch = ouptut − input × 100% input. If the input generator = 100 units/min and the output generator reads 125 units/min, the percent stretch = 125 − 100/100 = 25%. Through suitable switching arrangements, the outputs of several pairs of generators may be selectively fed through the ratio and production rate instruments to provide readings from various sections of multistage machines.

AC Tachometers

There are (1) voltage-responsive tachometer systems, and (2) frequency-responsive systems.

Voltage-Responsive Tachometer Systems

Consisting of an ac generator and a rectifier-type indicator, as shown in Fig. 4, these systems may be used in any installation where the generator speed for full scale is not less than 500 and not greater than 5000 rpm. With adequate attention given to bearings, conventional ac generators may be used at speeds up to 10,000 rpm. The ac tachometer generator embodies a stator surrounding a rotating Alnico* or equivalent permanent magnet. The output of the generator for voltage response systems is temperature-compensated and is proportional to speed.

Frequency-Responsive Tachometer Systems

This type of system consists of a dc indicator or recorder, a frequency-responsive network which may be contained in the recorder or a separate transformer box, and an ac tachometer generator of either the conventional or bearingless form. See Fig. 5.

Several types of ac tachometer generators are available. The composite characteristics of a representative device are given in Table 2.

*General Electric Co.

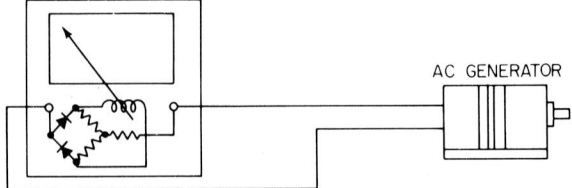

Fig. 4 Highly schematic circuit of an ac voltage-responsive tachometer system.

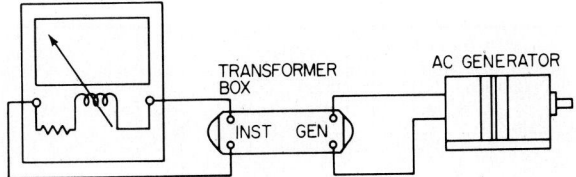

Fig. 5 Highly schematic circuit of an ac frequency-responsive tachometer system.

Table 2 Composite Characteristics of a Representative AC Tachometer Generator

Voltage output at 1000 rpm	10 V ± 1% open circuit
Accuracy	±1%
Permissible current drain	150 mA
Frequency at 900 rpm	60-Hz sine wave
Allowable end play	0.005 in (0.13 mm)
Maximum operating temperature	250°F (121°C)
Internal resistance at 25°C	100 Ω ± 1% (voltage-responsive units)
	32 Ω ± 20% (frequency-responsive units)
	8 poles
Stator	
Rotor	Alnico V, 8 poles
Bearings	Ball
Temperature coefficient	³⁄₁₀% per 10°C change (voltage-responsive units)
	No temperature compensation (frequency-responsive units)
Emf linearity	Depends on load and speed
Minimum top speed	500 rpm
Maximum top speed	5000 rpm
Weight	About 3 lb (1.4 kg) (dust-resistant models)
	About 25 lb (11.3 kg) (spray-resistant and explosion-proof models)
Shaft diameter	³⁄₁₆–¾ in (5–13 mm)
Length/width/height (approximate)	4½–5 in/3 in/3–3½ in (114–127 mm/76 mm/76–89 mm) (dust-resistant models)
	12½ in/6 in/5 in (318 mm/152 mm/127 mm) (spray-resistant and explosion-proof models)

Electrically Suppressed Zero Speed Measurement

A system of the type shown in Fig. 6 consists of one ac tachometer generator, a reference signal such as 110 V 60 Hz, and a transformer box that contains two frequency-responsive networks—plus an indicator or recorder. The upper 10% or more of the full-scale speed may be expanded over the complete scale of the instrument. Variations of this basic circuit may be obtained, such as multiple ranges including both zero left and suppressed zero, and the use of a reference signal for critical accuracy adjustment.

Speed Differential Measurements

A speed differential tachometer measures the difference between two speeds. The instrument may be calibrated in terms of revolutions per minute differential, feet per minute draw, inches per minute shrink, and so on. One or more additional indicators or recorders may be connected to one of the generators and calibrated in terms of production rate.

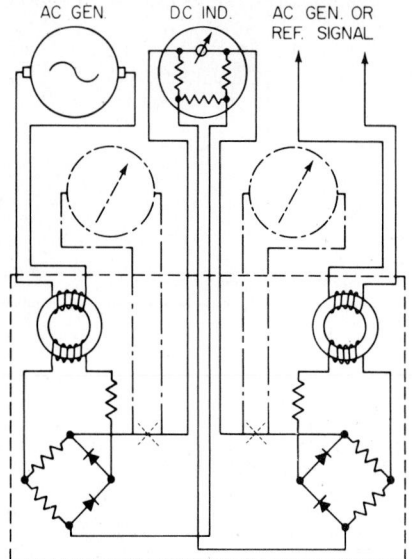

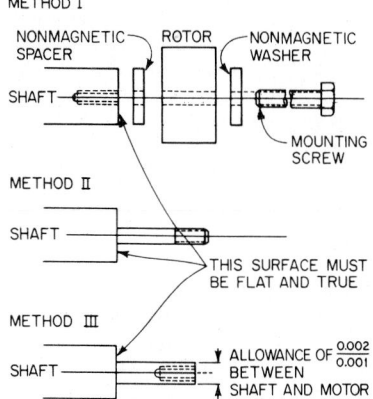

Fig. 6 Highly schematic circuit of a tachometer system with electrically suppressed zero speed measurement. *(Weston.)*

Fig. 7 Representative methods for mounting the rotor of a tachometer generator unit.

The first and most basic method for measuring speed differential is to arrange two dc generators so that their output voltages are in opposition. By connecting a dc instrument between the two generators, a measurement of their voltage output difference may be read directly in terms of revolutions per minute differential or whatever unit of measurement is required. Instruments may be connected in parallel to the generators to measure the speeds.

Another effective method is to use two ac generators and frequency-responsive networks arranged so that their outputs feed into a differential indicator or recorder.

Typical of the operations to which these systems can be applied is that of indicating the rpm difference between two engines driving a common shaft through a fluid drive. A tachometer generator is coupled to the driving shaft of each engine. Two wires from each generator are connected to either a differential indicator or a transformer box, depending on whether a dc or an ac frequency-responsive system is used. When a frequency-responsive system is used, four wires are connected from the box to the differential indicator so that, during operation of the engines, the instrument will then show the difference between their speeds. An additional instrument, calibrated in revolutions per minute, may be connected in the circuit to give the actual speed of either or both engines.

Bearingless Tachometer Generators

These devices are ac generators of the most basic form consisting of only a permanent magnet rotor and a stator. The devices have no bearings or brushes. They are designed to be impervious to oil, grease, and relatively high temperatures and, consequently, may be installed in inaccessible areas, such as gearboxes, which permits saving of space. They have very low torque burdens of less than 1 oz·in and are capable of speeds up to 100,000 rpm.

In general, when a bearingless generator is used, the frequency-responsive approach is employed. Since the system is solely dependent on the frequency output of the generator, voltage variations caused by reductions in the magnetic strength of the rotor due to handling, poor alignment of stator and rotor, or axial travel of the rotor with respect to the stator will not affect the overall accuracy. The rotor of the generator unit should be mounted to the true center of the shaft with extreme care, particularly in high-speed installations. A few representative methods are illustrated in Fig. 7. The relatively low inertia of the rotor makes it possible to secure it to the shaft with a right-hand thread regardless of the direction of rotation. It is recommended that a steel screw having an SAE thread be used to permit maximum tightening. The rotor should not be pressed onto the shaft because magnetic rotor material is brittle and may shatter.

Analog versus Digital Speed Transducers

The tachometers described thus far in this article generate an analog signal. For compatibility with digital systems, pulse-type transducers have grown in popularity in recent years. However, it should be pointed out that some pulse-type transducers were developed years before the major trend toward digital systems, which only dates back a few decades. For example, pulse generators have been used in connection with certain types of flowmeters for many years. There are parallels between speed transducers and position and motion transducers, often with the same basic principles applying. For this reason, reference to the article Position, Displacement, Motion, and Object Detection Systems in Sec. 9 of this Handbook is suggested.

Advantages of digital systems include (1) compatability with other digital equipment, (2) use of a high-accuracy time base, and (3) resolution to the number of digits in the display. However, where very low-speed indications to several decimal places are required, the analog approach may be the best choice because of the longer gate (update) time of the digital transducer.

Advantages suggested for analog systems include: (1) A moving pointer tells instantly whether speed is increasing or decreasing; (2) the cost is usually less than that of digital transducers; and (3) additional remote indicators may easily be added to the system. Analog systems also have the advantage of maturity of design and application experience developed over many years of usage.

Pulse-Type Transducers

The two principal categories of pulse generators are (1) magnetic, and (2) optical.

Magnetic Pickup

This device is essentially a coil wound around a permanent magnet probe. When ferrous objects, such as gear teeth or turbine blades, are passed through the probe's magnetic field, the flux density is modulated, inducing ac voltages in the coil. One complete cycle of voltage is generated for each interruption or object passed. If the objects, such as gear teeth, are evenly spaced on a rotating shaft, the total number of cycles will be a measure of the total rotation, and the frequency of the ac voltage will be directly proportional to the rotational speed of the shaft. A sectional view of a magnetic pickup is shown in Fig. 8. If a gear with 60 teeth is selected to measure the revolutions per minute of a

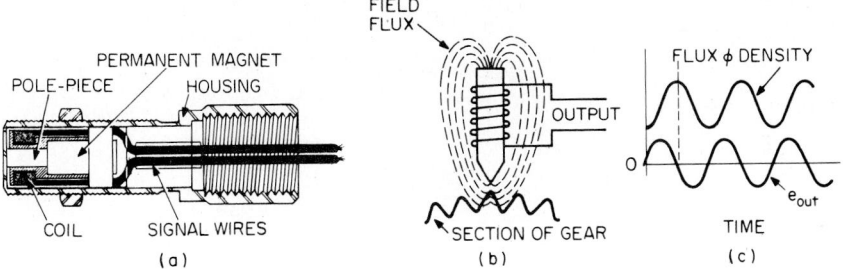

Fig. 8 Magnetic pickup. (*a*) Sectional view of the pickup; (*b*) placement of the pickup with a small air gap between the pickup and gear teeth; (*c*) an output wave form which is a function not only of rotational speed but also of gear tooth dimensions and spacing, pole piece diameter, and the air gap between the pickup and the gear tooth surface. The pole piece diameter should be less than or equal to both the gear width and the dimension of the tooth's top (flat) surface. The space between adjacent teeth should be approximately 3 times the diameter. Ideally, the air gap should be as small as possible, typically 0.005 in (0.13 mm). *(Daytronic.)*

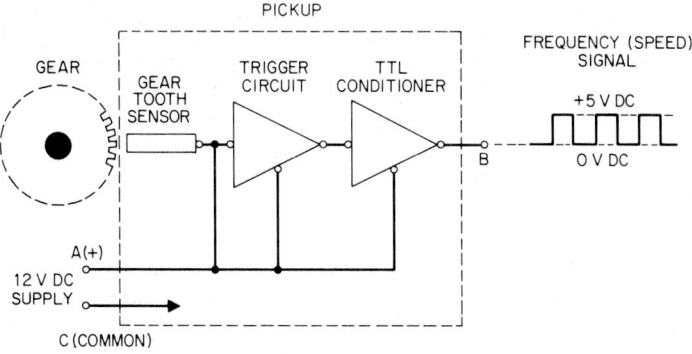

Fig. 9 Circuit of "zero velocity" magnetic pickup. *(Daytronic.)*

rotating shaft, the output frequency (in hertz) will be numerically equivalent to the revolutions per minute—a situation that may allow frequency meters to be used without calibration. For high rotational speeds, the use of a gear with a smaller number of teeth may be specified. The magnetic pickup circuit shown in Fig. 9 contains its own signal conditioning circuitry for generating a clean square wave output pulse ($+5$ V) for each ferrous discontinuity passing the head of the pickup. The output is either on or off, depending on the presence or absence of ferrous material. The unit senses motion down to "zero velocity" and always produces a pulse train of constant amplitude, irrespective of the rotational speed of the gear. The use of a magnetic pickup in connection with a turbine flowmeter is shown in Fig. 10.

Proximity Switches

Various kinds of proximity switches (described in the first article in this Handbook section) can be adapted to speed measurement. Sometimes these adaptations are referred to as speed switches. The unit shown in cross section in Fig. 11 is a magnetic noncontacting device used for speed and motion measurement. These units are particularly popular for *slowdown* indication for conveyors and other process machinery. See Fig. 12. On start-up, the incoming pulses from the probe are ignored for a fixed 5 s to allow equipment to accelerate to normal operating speed. In some systems, there is an adjustable time delay on the start-up feature, calibrated at 0 to 60 s. This is a rugged system especially configured for heavy-duty machinery in demanding atmospheres. The elements detected by the probe are rather large ferrous blocks measuring $2 \times 2 \times 1$ in ($5 \times 5 \times 2.5$ cm) mounted directly

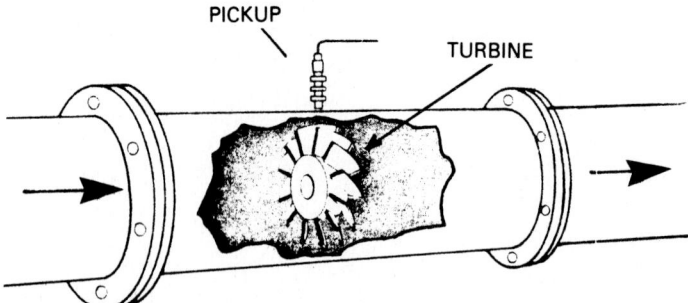

Fig. 10 Use of a magnetic pickup on a turbine flowmeter. In properly designed flowmeters, the output frequency produced by the magnetic pickup is a linear function of the volumetric flow rate. Each output cycle represents the passage of a known volume of fluid. A flowmeter can be calibrated in cycles per gallon, liters, etc. The rating is known as the K factor of the meter. *(Daytronic.)*

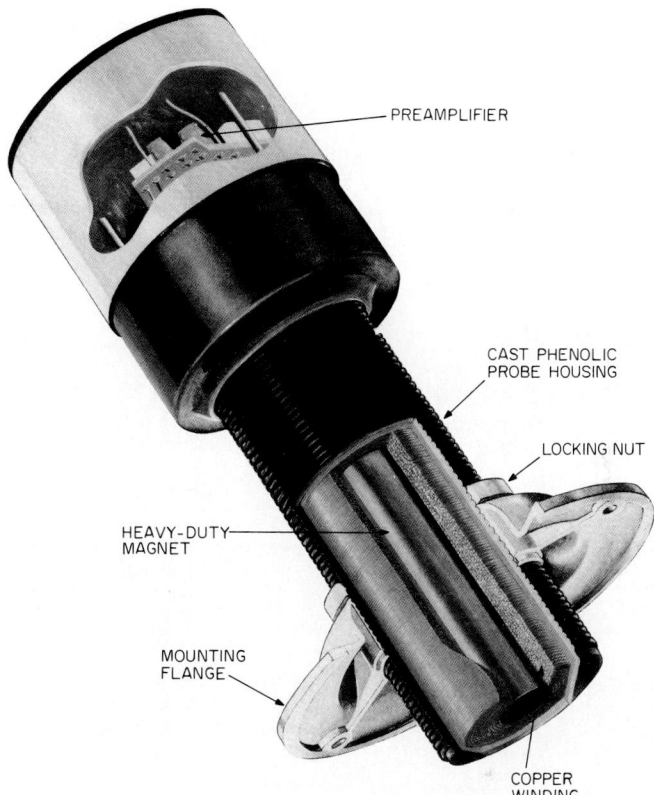

PREAMPLIFIER

CAST PHENOLIC
PROBE HOUSING

LOCKING NUT

HEAVY-DUTY
MAGNET

MOUNTING
FLANGE

COPPER
WINDING

Fig. 11   Sectional view of a magnetic speed switch. The probe shown is over 8 in
(20 cm) long and 2 in (5 cm) in diameter. The probe can detect ferrous moving objects
up to 5 in (12.5 cm) from the face of the probe. The devices are fully sealed for
protection against dust and moisture, and are operational up to 500° F (260°C). The
preamplifier boosts the signal for cable runs of up to 5000 ft (1250 m) with no. 18
AWG, and 25,000 ft (6250 m) with no. 12 AWG. Smaller probes are available for
mounting on smaller process machinery. *(Milltronics.)*

on the rotating conveyor or machine. The speed switch is equipped with a liquid crystal display unit,
which can be calibrated to suit need (percentage of desired speed, etc.). See Fig. 13.

Capacitor-Type Switches

The principle of these switches is described in the first article in this Handbook section. In the speed
sensing mode of this type of switch, a trigger cam is mounted on a rotating or reciprocating element
of a machine so that it appears within the range of the proximity switch at eve y revolution or stroke.
The distance between the cam and the switch must be no more than one-hal the nominal detection
range of the switch. The cam may be metallic or nonmetallic, depending on whether an inductive or
a capacitative switch is used. The switch generates a pulse each time the cam appears within range.
A controller measures the instantaneous rate between successive pulses and compares it with the set
point. Through the use of appropriate relays, the arrangement can be used for either overspeed or
underspeed correction.

Because these switches measure the time interval between two successive pulses, there is an inherent time lag in response when the pulse rate decreases below the set point; and the size of the lag

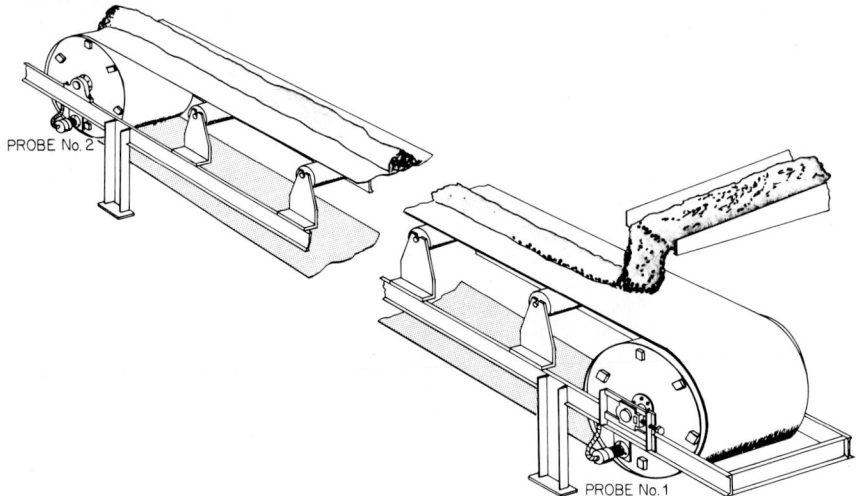

Fig. 12   Conveyor loss-of-motion detection by using two probes, one on the head pulley and another on the tail pulley. By computing the plus or minus speed relation of the tail to the head pulley and comparing this value with percentage slip set points, slip protection is ensured. A minimum percent speed feature of the system checks for mechanical failure between the motor and the tail pulley. Each of these features has its own time delay after start-up and one common delay to ignore nuisance alarms and/or shutdown. The four alarms are fed to a first-out annunciator and latch in the relay output circuit. Reset of the relay can be manual or automatic. *(Milltronics.)*

depends on the value of the set point. For example, with one switch design, at a set point of 5000 pulses per minute, the switch must wait 0.012 s (i.e., 60 divided by 5000) before it can determine that the pulse rate has dropped below the set point, and therefore it must wait that long before initiating relay action. At a set point of 5 pulses/min, the lag increases to 12 s (i.e., 60 divided by 5). Should this lag be too large to satisfy the needs of a particular application, it can be shortened by using several equally spaced trigger cams on the rotating or reciprocating machine element. This results in multiplying the number of pulses per cycle by the total number of cams used—and thus dividing the time lag by an equal number.

There is no time lag when the pulse rate increases above the set point. If a pulse arrives before the end of the interval determined by the set point, the switch detects the increase in speed as soon as the pulse arrives, and the control relay responds immediately.

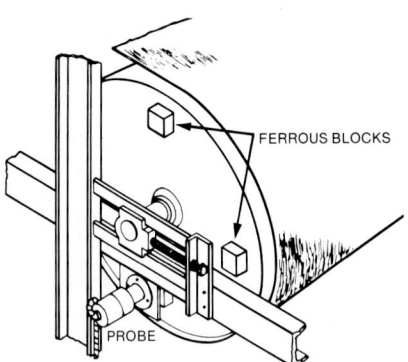

Fig. 13   Probe and ferrous block mounting detail. A noncontacting tachometer can be calibrated in the field—from 0 to 8 ppm through 0 to 720 ppm equal to 4 to 20 mA output. The parts-per-million input is equal to the number of ferrous objects per revolution times the revolutions per minute of the machine. *(Milltronics.)*

Impulse Tachometer

In the instrument shown in Fig. 14, the charging current of a capacitor is used. The pickup head usually contains a reversing switch, operated from a spindle, which reverses twice with each revolution. Thus, battery potential is applied to the capacitor in each direction, and with each impulse a current is passed through the milliammeter. The indicator responds to the average value of these impulses. Therefore, the indications are proportional to the rates of the pulses, which in turn are proportional to the rates of the spindle revolutions.

No current is drawn from the battery when the spindle is not revolving. The pulse current is approximately 1 mA. The spindle speed and battery voltage influence the indicator deflection. Thus, it is important to check and correct the battery voltage at frequent intervals. This is accomplished by means of an adjustable resistor placed in the circuit.

The oscillating switch may be connected directly for speeds of 200 to 10,000 rpm and, with suitable gears, speeds below or above these values can be measured. The readings of a properly standardized instrument are not affected by temperature, humidity, vibration, or magnetic fields. The indicator and head may be separated up to a distance of 1000 ft (300 m), and where suitably shielded connections are used, the distance may be increased. The indicator scale is uniform.

A high-accuracy instrument is also available wherein the capacitor and reversing switch are connected to one leg of a bridge circuit. The pulses from the periodically charging capacitor upset the balance of the bridge and thus cause an indication on the milliammeter. Multiple ranges are obtained by using different capacitor values.

Optical Encoders

Many modern position control systems use an incremental optical encoder for position determination, as explained in the first article in this Handbook section. By taking advantage of the calculating power of a microprocessor, the impulses from an optical encoder can be converted to a velocity measurement.

Optical encoders are available for handling very wide dynamic ranges, such as 10,000 to 20,000 to 1. Accuracies are claimed to be better than 0.01% per revolution.

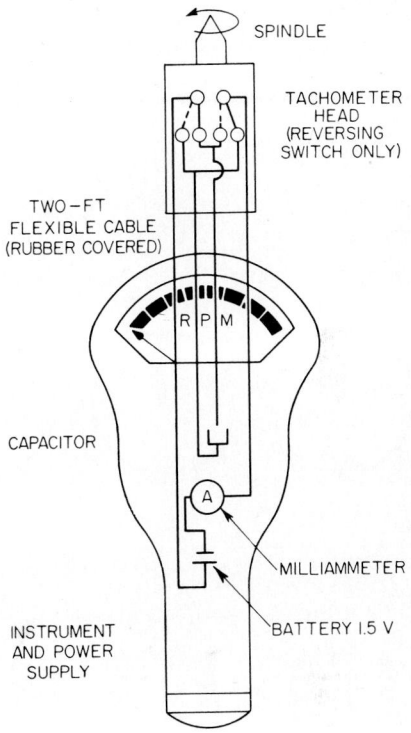

Fig. 14   Capacitor-type impulse tachometer.

Stroboscopic Tachometers

A stroboscope permits intermittent observation of a cyclically moving object in such a way as to produce an optical illusion of stopped or slowed motion. This phenomenon is readily apparent, for example, when rewinding a tape at many revolutions per minute when the tape deck is located under a 60-cycle incandescent lamp. Patterns on the reel tend to slow and then appear to stop before reversing their direction. Stroboscopic effects have been known for decades,* of course, one of the first scientific applications being found in very highspeed photography. Intermittency of observation can be provided by mechanical interruption of the line of sight (as with a motion picture camera) or by intermittent illumination of the object being viewed. The industrial stroboscope basically is a lamp plus the electronic circuits required to turn the lamp on and off very rapidly—at rates as high as 150,000 flashes per minute and higher.

The schematic diagram of an electronic stroboscope is shown in Fig. 15. The device includes a strobotron tube with its associated discharge capacitors, a triggering tube to fire the strobotron, an oscillator to determine the flashing rate, and a power supply. With the use of harmonic techniques, speeds up to 1 million rpm can be measured. Accuracy is nominally ±1% of the dial reading after calibration.

To serve as a tachometer, a stroboscope must have its own flashing-rate control circuits and

---

*Invented independently by Stampfer of Vienna and Plateau of Ghent in 1832. Stampfer chose the name "stroboscope," which is derived from the Greek words meaning "whirling watcher."

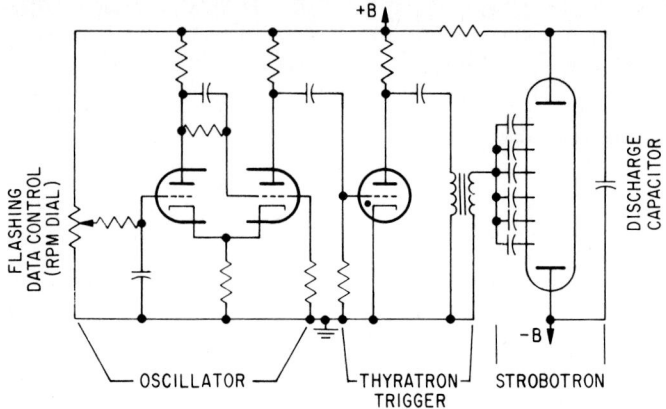

Fig. 15 Schematic diagram of an electronic stroboscope circuit.

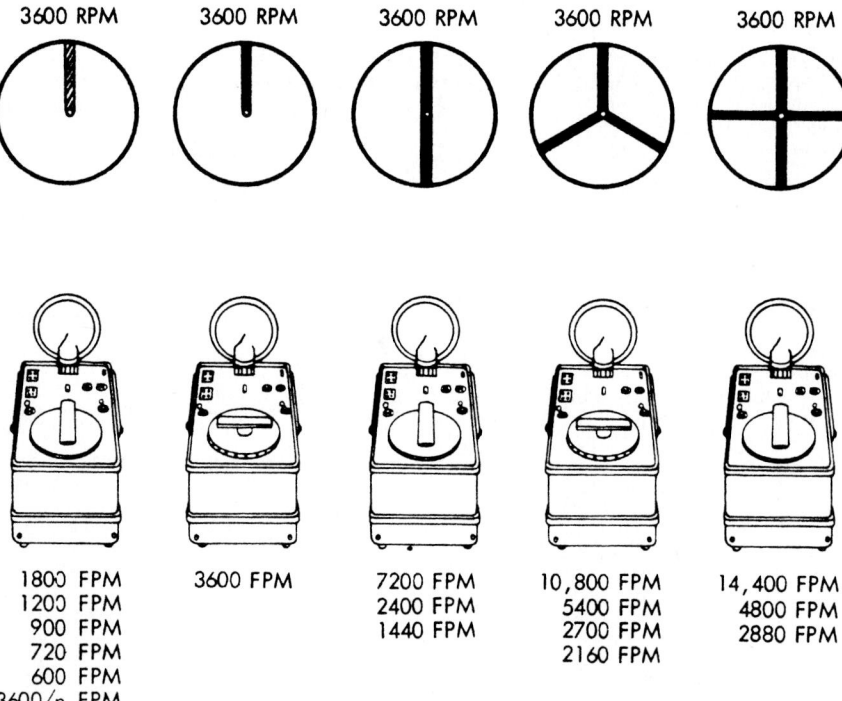

Fig. 16 Diagrams showing the images obtained at harmonic and subharmonic flashing rates of a stroboscope. Even with an asymmetrical object, the correct fundamental image is repeated when the stroboscope is flashing at one-half, one-third, etc., the speed of the object. Flashes are then occurring every other revolution, etc., and even though such submultiple images appear progressively dimmer, they can be confusing. The proper setting for a fundamental speed measurement is the highest setting at which a single stationary image can be achieved. This does not hold, however, if the fundamental is beyond the flashing-rate limit of the stroboscope. There are several ways to distinguish fundamental from submultiple images. The flashing rate can be decreased until another single image appears. If this occurs at half the first reading, the first reading was the actual speed of the device. If it occurs at some other value, then the first reading was a submultiple. Or the user can double the flashing rate and check for a double image. Or the user can flip the range switch to the next higher range. Because of the 6-to-1 relationship between ranges, a 6-to-1 pattern should appear. The 6-to-1 relationship between ranges also makes it convenient to convert speed readings from revolutions per minute into cycles per second. One simply flips to the next lower range and divides the new reading by 10. *(GenRad.)*

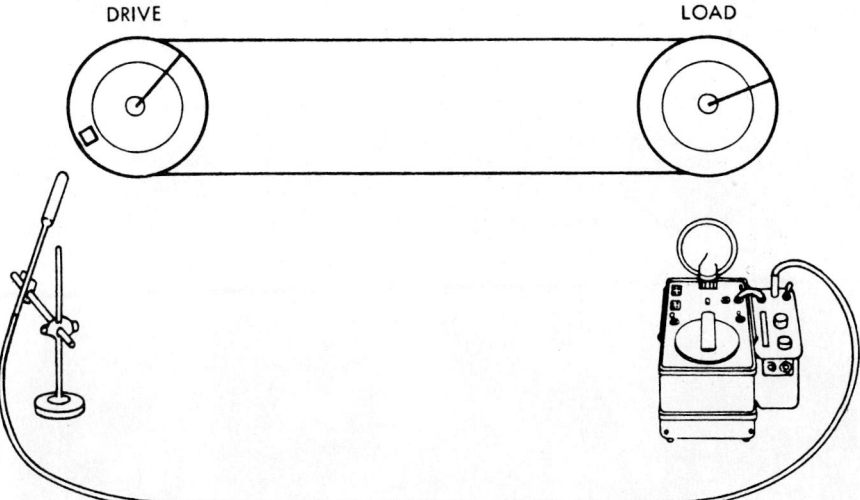

Fig. 17 A method of measuring belt slippage is to observe a load under stroboscopic flashes photoelectrically synchronized with the drive shaft. A simpler way is to adjust the stroboscope flashing rate for stopped images of the drive shaft and then carry a strobe to load to check slippage. *(GenRad.)*

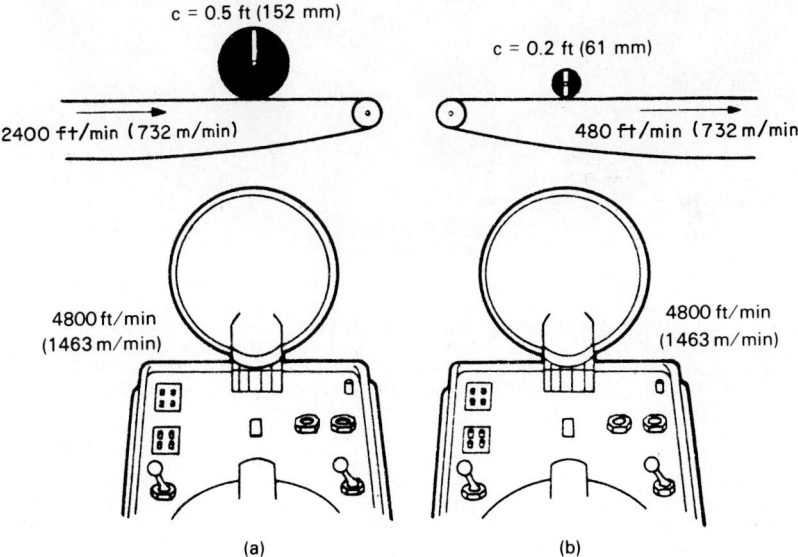

(a)   (b)

Fig. 18 At times it may be desirable to measure the linear speed of a device rather than the number of revolutions per minute. The surface speeds of drums, wheels, and rollers and the linear speeds of belts and pulleys can be measured stroboscopically with the aid of a surface speed wheel. (a) The larger of the two surface speed wheels gives a single stationary image when the flashing rate is twice the surface speed in feet per minute. Thus, in the example above, the belt speed is 2400 ft/min. (b) The smaller of the two surface speed wheels gives a double stationary image when the flashing rate is 10 times the surface speed in feet per minute. Thus, in this example, the belt speed is 480 ft/min. The surface speed wheel is a disk that can be held against a moving surface so that a point on the wheel's circumference will move at the surface speed. The diameter of the disk is chosen so that its rotational speed, indicated on the stroboscope dial, is simply related to the surface speed being measured. *(GenRad.)* Metric conversions: 0.2 ft = 61 mm, 0.5 ft = 152 mm, 480 ft/min = 146 m/min, 2400 ft/min = 732 m/min, 4800 ft/min = 1463 m/min.

Fig. 19   Stroboscopic tachometer test disk. These disks can be cut out and mounted on light cardboard or metal. The center must be carefully located and fitted onto the drive shaft, or possibly on a dummy tachometer plugged into one of the positions on a multiple-unit tachometer test stand. With the stroboscope set to flash at line frequency, as the speed of the test stand motor increases, the rings of the disk will appear successively to stand still. The outside ring will appear motionless at each even hundred revolutions per minute on the tachometer scale. The actual speed, in hundreds of revolutions per minute will be readable intermittently by stroboscopic combination of the outside ring figures. *(GenRad.)*

calibrated dial. In adjusting the stroboscope flashing rate for a single stationary image, one must avoid being confused by harmonics or subharmonics. Spurious images are especially confusing when the object being viewed is symmetrical. See Figs. 16 through 19.

Other Types of Tachometers

Variable Reluctance Instrument

A pickup of this type produces pulses that are proportional to speed and are amplified, are rectified, and control the direct current to a milliammeter. This type of instrument is rated at 10,000 to 50,000 rpm, with an accuracy of $\pm\frac{1}{2}\%$ of full-scale reading. The pickup is rated to withstand ambient temperatures from -60 to $+500°F$ (-51 to $+260°C$).

Vibration Pickups

These devices are also used to measure velocity. The pickup usually consists of a very lightweight coil attached to the end of a pivoted shaft and free to move in the field of a permanent magnet.

Vibrating-Reed Tachometer

Vibrating reeds provide a natural means of measuring the frequency of vibrating or revolving equipment. The reeds are of various lengths, in accordance with their natural period of vibration, and are so mounted on a base with a reference scale that observation of the reed which is vibrating forms a means of measuring the frequency of vibration. See Fig. 20.

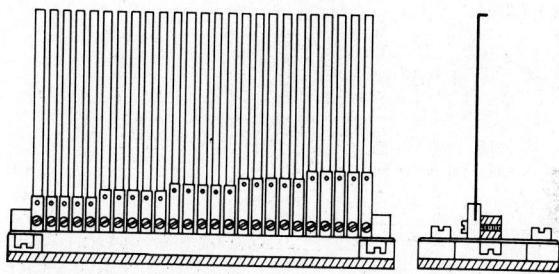

Fig. 20   Vibrating-reed tachometer.

Photoelectric Devices

In one instrument of this type, designed to measure speeds up to 3 million rpm, the movable part subject to measurement is arranged to provide reflecting and absorbing areas. The interrupted reflected light produces, by means of a photocell, electric impulses which are applied to a frequency meter which generates a square wave from the pulse voltage and applies it to a discriminating circuit. A fixed current pulse at each half-cycle is produced. These pulses are rectified and applied to a dc milliammeter which indicates the average value. Thus, the meter readings are proportional to the number of pulses per second, or the frequency.

Eddy-Current Tachometer

The eddy-current or drag-type tachometer has been widely used for certain types of speed measurements. A preponderance of these units has been employed in automobile speedometers, in which case a flexible shaft arrangement is used, but they also find industrial usage.

In its basic form, as shown in Fig. 21, the drag-type instrument employs a permanent magnet which is revolved by the source being measured. Close to the revolving magnet is an aluminum disk, pivoted so as to turn against a spring. A pointer attached to the pivoted disk is associated with a calibrated scale. As the permanent magnet is revolved, eddy currents are set up in the disk. The magnetic fields caused by these eddy currents produce a torque which acts in a direction to resist this action and turns the disk against the spring. The disk turns in the direction of the rotating magnetic field and turns (or is dragged) until the torque developed equals that of the spring. This torque is proportional to the speed of the rotating magnet. The instrument has a uniform scale.

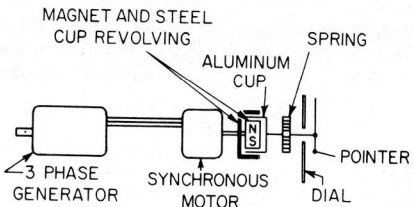

Fig. 21   Drag-type eddy-current tachometer.

The rotating field usually is produced by a permanent magnet but may be of any form which is steady. The disk may also take the form of a cup and may be of copper.

Remote indication is obtainable with one form of this tachometer. A three-phase generator is driven from the shaft whose speed is to be measured. The generator output is connected to a three-phase synchronous motor, attached to the indicator, which rotates the magnetic field. Several indicators, each with its own synchronous motor, may be connected to the three-phase generator and indicate in proportion to the speed of the generator. Since the synchronous motors keep in step with the generator frequency over a wide range, the indications are independent of voltage developed by the three-phase generator.

Velocity Head or Hydraulic Tachometers

In devices of this type, advantage is taken of the fact that pumps or blowers produce a velocity which can be converted into a static pressure. See Fig. 22. The hydraulic tachometer incorporates a rotary

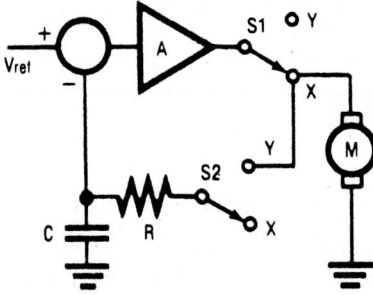

Fig. 22 Motor speed regulation without a tachometer. In the sample-and-hold diagram shown here, the motor is switched from a free running state where it charges capacitor C to a second state where it is driven by an error voltage produced by the last stored sample. When $S1$ is in position Y, the motor is disconnected from the amplifier and is free running. At the same time, $S2$ connects the free running motor to capacitor C, thereby charging C to the voltage nK_e (the back emf). *(After Geiger.)*

pump as the transmitter and a piston as the receiver. The pump, usually contained in the indicator case, is driven by a flexible shaft and a gear train which automatically handles reversed speed, but the instrument normally is not equipped to show the direction of rotation. Pump displacement, which is positive and free of pulsations, raises or lowers a counterweight piston. The piston operates a pointer through a rack and gear. The piston also drives a tape-marking stylus for recording. The recorder tape is driven by the flexible shaft from the speed pickup.

The indicator may be read to 0.4% of full-scale value and is claimed to be accurate to within ± 1% of full-scale reading. One application is in railroad locomotives where the instrument accuracy is affected very little by ambient temperature changes.

Tachometer-less Regulation of Servo Speed

Within the last few years, an interesting approach to regulating the speed of a motor without a tachometer has emerged. Basically, the arrangement consists of allowing a motor to coast for a very short interval, during which interval the back electromotive force (emf) is measured. In one technique, *sample and hold,* the motor armature is time-shared. About 90% of the time, it operates as a motor. During a 10% coasting period, the motor functions as a generator or dc tachometer. Thus, it can provide an output voltage that is directly proportional to its speed. The applicable sample-and-hold block diagram is shown in Fig. 22. The motor inductance must be sufficiently small, as in the case of a printed-circuit motor, to qualify for this approach. Equations and more details are given by Geiger (1979) in the reference listed.

Governors

A governor is an automatic controller used to maintain the rotative speed of a machine at a desired value. The governor measures the speed, compares the measured value with the desired value, and acts to correct any error between the two values—usually by adjusting the flow of energy to the machine. Governors may be divided into two main types: (1) devices in which the speed sensing element operates the energy metering device directly, and (2) devices that use one or more stages of power amplification between the speed sensing element and the energy control device. There is a natural distinction between these two types, arising from the fact that the first type usually gives stable control on an engine or other prime mover, whereas the second type requires the presence of some stabilizing factor to prevent continual oscillation of the speed (hunting).

Air Velocity Measurement

Kata Thermometer

This means of measuring air velocity employs an alcohol thermometer with a large bulb and two marks on the stem, corresponding to 100 and 95°F (38 and 35°C). In use, the bulb is heated above 100°F (38°C) and the time required for the column to fall from 100 to 95°F (38 to 35°C) is measured. Factors for converting from time units to velocity are available. The device is nondirectional and can be used for low velocities but is quite cumbersome and subject to radiation errors.

Heated Thermal Element

By using two thermometer elements, one of which is electrically heated to restore its reading to the uncooled value, a measure of air velocity can be obtained. Such measurements give average values, are nondirectional, and are subject to radiation errors. Thermocouples can be used instead of thermometers.

Pitot Tube Air Speed Indicator

The pitot tube air speed indicator consists of two elements: (1) a dynamic tube, which points upstream and determines the dynamic pressure, and (2) a static tube, which points normal to the air stream and determines the static pressure at the same point, as shown in Fig. 23. The tubes are connected to the two sides of a manometer or inclined gage so as to obtain a reading of velocity pressure, which is the algebraic difference between the total pressure and the static pressure.

The relationship between air velocity and velocity pressure is

$$V = \sqrt{2GH} \tag{1}$$

where V = velocity
G = acceleration due to gravity
H = velocity head or pressure

The pressure differential created is quite small in relation to air velocity. At 110 ft/min (33.5 m/min) the velocity pressure is only 0.0625 in (1.6 mm) water. Consequently, the instrument is not generally used for measuring velocities less than 1000 ft/min (305 m/min). Turbulence in the air

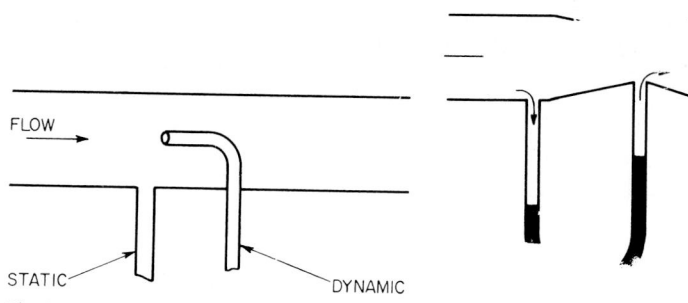

Fig. 23 Pitot tube air speed indicator.

Fig. 24 Venturi air speed indicator.

stream affects device accuracy, and the tubes are subject to clogging where dusty, unclean air is involved.

Venturi Air Speed Indicator

Limitations of the pitot tube led to the design of a venturi air speed indicator in which a greater differential pressure is created. The device is shown in Fig. 24 and requires individual calibration for best accuracy. Parts are not readily interchangeable.

Revolving-Vane Anemometer

This widely used device comprises a paddle wheel which is revolved by the moving air stream. See Fig. 25. The wheel is attached to a counter, and by selection of the proper gear ratios and vane pitch the counter can be calibrated to read directly in feet of air. The air velocity can be determined by measuring the time interval. The device is supplied with a curve for correction of the nonlinear relation between air velocity and rotational speed of the vanes. Air density also should be considered when high accuracy is needed. The measurement tends toward the average air speed. The range of the device usually is 300 to 3000 ft/min (91.5 to 915 m/min).

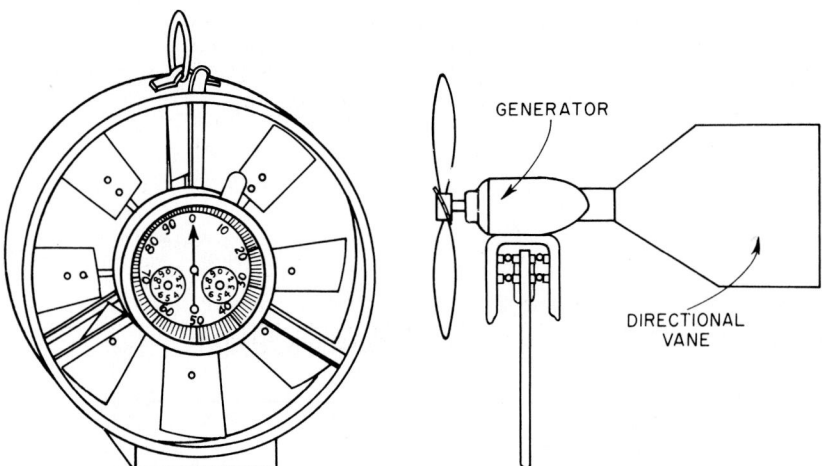

Fig. 25 Revolving-vane anemometer.   **Fig. 26** Propeller-type electric anemometer.

Propeller-Type Electric Anemometer

This is a version of the basic rotating-vane device. Figure 26 shows the propeller type in which the blades are fastened to the shaft of an electric generator which develops emf or frequency proportional to speed. The emf or frequency signal is fed to an indicator. The generator and propeller are pivoted so that the directional vane can keep the device headed directly into the direction of airflow. The device reads in average air velocity. The direction of airflow can also be indicated mechanically or electrically.

Revolving-Cup Electric Anemometer

One type is shown in Fig. 27. The generator is mounted on a vertical axis, and, like the propeller type, its emf or frequency output is proportional to the speed of the revolving cups. The speed readings are average, but the device is not directional.

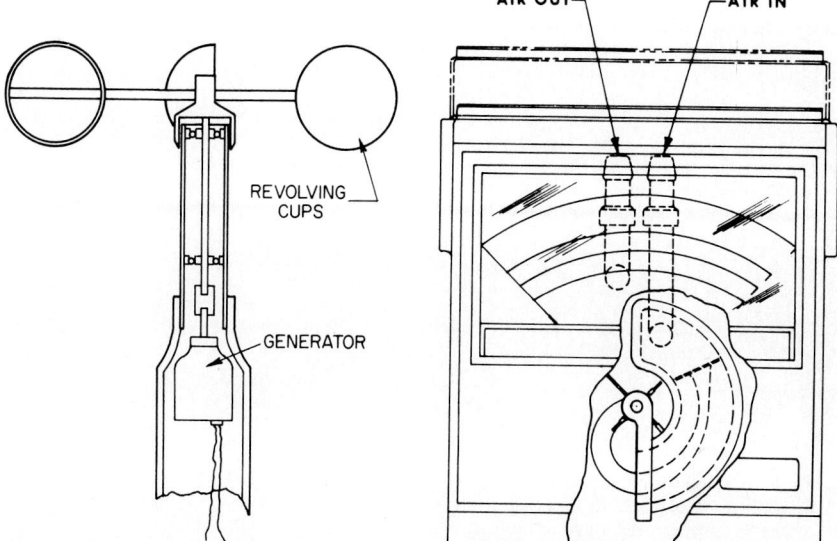

Fig. 27 Revolving-cup electric anemometer.

Fig. 28 Alnor Velometer. *(Alnor Instrument.)*

Alnor Velometer*

This instrument, shown in Fig. 28, measures static and total pressures as well as air velocity. It utilizes a taut-band suspension movement having a piston arm attached to an aluminum staff which locates an aluminum vane in a "tunnel" in the instrument case. As air flows through the tunnel, it impinges on the vane, thus deflecting against the taut-band suspension and moving the pointer across scale. The instrument is portable, and since the moving element is statically and dynamically balanced, it can be used in any normal position. Full-scale deflection is possible with a minimum airflow of 700 cm^3/min.

An assortment of probes and range selectors allows velocity pressures to be measured for various conditions, including (1) air velocity within ducts and tunnels, (2) supply and return velocities from an air duct to a room, and (3) air velocity in an open space or room.

A quantitative relationship applying to this instrument is

$$Q = KAV \qquad (2)$$

where Q = quantity of air, ft^3/m
  A = cross-sectional area, ft^2
  V = average velocity, ft/min
  K = coefficient of discharge

For an opening without baffler or grills, K may be considered unity. If a grill is involved, the following relationship applies:

$$Q = A'V \qquad (3)$$

where $$A' = \frac{\text{gross grill area} + \text{free grill area}}{2} \qquad (4)$$

*Trademark, Alnor Instrument Company, Niles, Illinois.

The instrument is standardized for 70°F (21.1°C) and 29.92 in (760 mm) Hg pressure. Under these conditions, normal air density is 0.0749 (1.2 kg/m³). For other than standard conditions, the indicated reading may be corrected for density by using the following relationship:

$$V_t = V_r \frac{\rho_{std}}{\rho_{actual}} \tag{5}$$

where V_t = true velocity, ft/min
V_r = indicated velocity, ft/min
ρ_{std} = density, standard = 0.0749 lb/ft³

$$\rho_{actual} = 1.324 \frac{\text{barometric pressure, in Hg}}{460 + \text{temp., °F}} \tag{6}$$

Changes in relative humidity have little effect on instrument accuracy. The reading may be corrected for temperature by Eq. 7.

$$V_t = V_r \sqrt{\frac{460 + t}{460 + 70}} = V_r \sqrt{\frac{460 + t}{530}} \tag{7}$$

where V_t = true velocity, ft/min
V_r = indicated velocity (reading), ft/min
t = temperature at time of measurement, °F

Equation 5 corrects for density, and when used, Eq. 7 need not be applied.

REFERENCES

Geiger, D.: "Regulating Servo Speed without a Tachometer," *Control Eng.,* vol. 26, no. 10, 1979, p. 73.

Morris, H. M.: "Rotary Tachometers Dominate the Velocity Sensing Market," *Control Eng.,* vol. 25, no. 12, 1978, 37.

Morris, H. M.: "The Many Roads to Measuring Speed Are Increasingly Digital," *Control Eng.,* vol. 27, no. 3, 1980, pp. 57–61.

Morris, H. M.: "Controlling a Web on the Fly," *Control Eng.,* vol. 29, no. 7, 1982, pp. 69–72.

Uram, R.: "Hardware and Software Team Up to Control Steam Turbine Speed," *Control Eng.,* vol. 23, no. 2, 1976, pp. 50–52.

Van Veen, F.: *Handbook of Stroboscopy,* GenRad, West Concord, Mass. (revised periodically).

Machine Vision

by
Joseph Wilder*

Machine vision for industry may be defined as the process of extracting information from visual sensors, enabling machines to make intelligent decisions. The need for automatic decision making is growing rapidly in industry as the speed and accuracy required for such decisions increase beyond the limits of human capabilities. Since many decisions concerning the identity, quality, dimensional integrity, and precise location of objects are made on the basis of visual information, machine vision has become an important part of many manufacturing processes. The advantages of machine vision include (1) increased productivity, (2) improved product quality, (3) substitution of 100% inspection for sampling, (4) reduction in errors, (5) inspection in hostile environments, and (6) automatic collection of data for statistical analysis and record keeping.

Recent reviews (Refs. 1 and 2) have described applications of image processing to specific tasks, such as automatic position detection of transistors and integrated circuits during wire bonding. However, some of the systems and techniques described to date have yet to be implemented in a factory environment. This article reviews several general categories of industrial applications of machine vision. Also included are descriptions of major components of machine vision systems and some new technology that will be incorporated into future systems. Some of the factors involved in incorporating machine vision systems into rugged machines for the factory floor are also mentioned toward the end of the article.

Principal Functions of Machine Vision

These include the following.

Identification and Sorting. These operations involve the determination of which of a known number of objects or patterns is present. An example of identification and sorting is reading serial numbers or bar codes on stamped metal, plastic parts, or cartons. Another example is sorting items by their graphic content. The content may be the graphics (designs, printing, etc.) on cartons being fed to automatic palletizers or, as illustrated in Figs. 1 and 2, on cigarette packs that have been returned to the manufacturer to be sorted by brand.

Inspection. Checking the total content of a scene as well as the orientation of patterns or objects within it. Examples of inspection include examining sheet steel for cracks and checking labels on bottles or cans for misalignment and tears.

Verification. Validation of a particular piece of information in a scene. An example might be checking a lot code or date for correctness or legibility, or establishing that a kit of parts or an assembly is complete. A system that verifies that a key is in the correct location on a computer keyboard is shown in Fig. 3.

Measurement. Checking that parts are the right size; i.e., that they are within their specified dimensional tolerance. In this regard, there are two different types of visual measurement tasks. One involves exhaustive, accurate, and precise testing in a quality control laboratory on a small sample of parts. Another involves a simple dimensional check at high speed as part of the production process. An example of the latter is measuring the depth of threads of the twist-off aluminum closures on bottles of carbonated beverages. This measurement cannot be performed by human inspectors inasmuch as bottling lines can run at rates in excess of 1000 bottles per minute.

*Director of Research, Object Recognition Systems, Inc., Princeton, New Jersey.

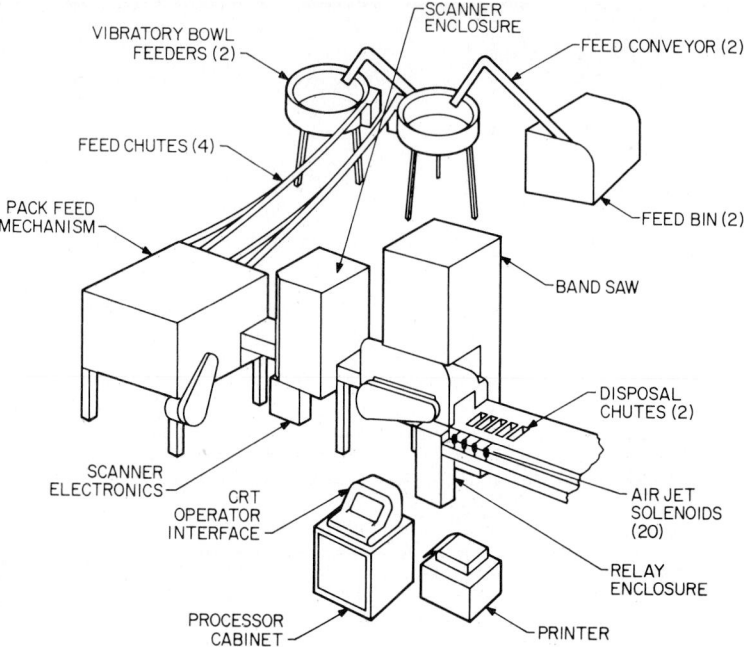

Fig. 1   Schematic diagram of a cigarette pack sorter. *(Object Recognition Systems.)*

Manipulation. The use of visual information to guide robots and other automated production equipment. Applications in this category may be classified as:

1. *Tracking.* Following seams to aid gluing or welding robots and guiding spray painting robots.
2. *Materials Handling.* Part acquisition, pose estimation, reorientation and transfer of objects from one work station to another. One type of part acquisition, *bin picking,* represents a difficult

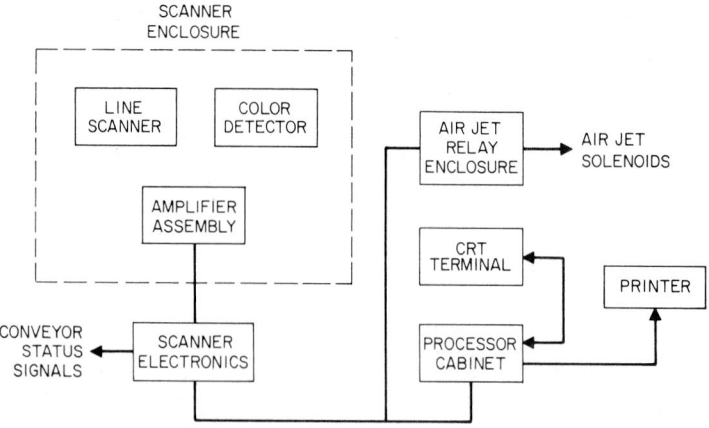

Fig. 2   Block diagram of a cigarette pack sorting system. *(Object Recognition Systems.)*

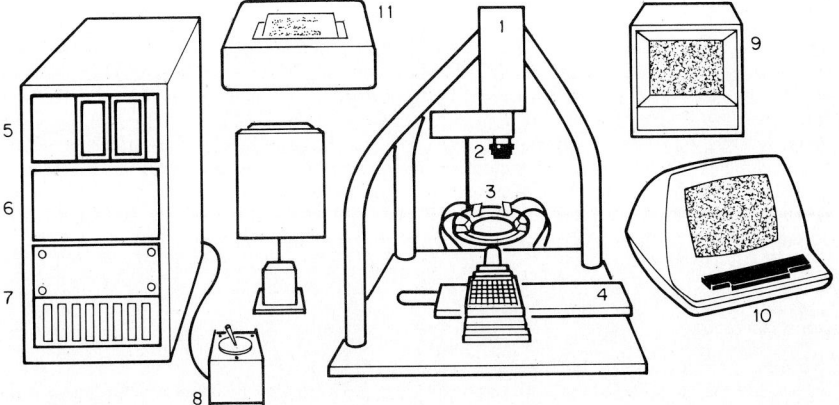

Fig. 3   Machine vision system for inspecting keyboards. (1) Television camera, (2) split image optics, (3) fiber-optic illuminators, (4) xy table, (5) microprocessor with dual floppy disk drives, (6) controller (xy table), (7) pre-processing and data acquisition electronics, (8) joystick, (9) monitor, (10) CRT terminal, and (11) printer. *(Object Recognition Systems.)*

and important problem in many factories where parts are jumbled in bins or only partially oriented.

3. Assembly.   Putting parts together. Vision systems can aid robots in a number of assembly operations, such as inserting screws and pins. A vision system can find the centers of holes and help to orient parts to be inserted in the holes.

The construction of a complete work cell in which objects are acquired, oriented, assembled, and inspected may require that several of the vision problems just discussed be solved.

Components of a Machine Vision System

Contemporary vision systems perform to some degree all of the following functions.

Visual Sensing

The sensors selected for a vision system must be carefully matched to the requirements of the vision problem addressed. Factors which must be considered include (1) field of view, (2) resolution, (3) signal-to-noise ratio, (4) spectral response, (5) dynamic range, (6) geometric distortion, (7) lag, (8) time and temperature stability, and (9) cost. Although there are special purpose photosensors in use, this article is confined to *scanning sensors.*

Scanning sensors may be classified as (1) line scanners, or (2) area-type scanners. Line scanners include solid-state linear arrays, flying spot scanners, and prism, mirror, or holographically deflected laser scanners. These scanners are fast, high-resolution devices that are relatively free of geometric distortion in one dimension. However, in order to capture a complete two-dimensional scene, the second dimension must be obtained either by motion of the object past the scanner or by mirror or prism deflectors. Mechanical motion slows down data acquisition considerably and, in some cases, produces geometric distortion.

The most common area-type scanners are closed-circuit television cameras with vidicon image sensors. These sensors are inexpensive and provide high resolution (300 to 500 television lines). However, they suffer from geometric distortion, time and temperature instability, lag (requiring several television frames for complete erasure), and sensitivity to nearby magnetic fields. Solid-state scanners do not suffer appreciably from the foregoing problems, but until recently have provided limited resolution at a high cost. The resolution of solid-state scanners has increased as costs have decreased, and now solid state scanners are displacing vidicons in most industrial applications.

Preprocessing

The preliminary stage of image processing in which the vast amount of information in a visual scene is reduced to a more manageable form is called *preprocessing*. This procedure can be carried out in real time (at video rates) by special purpose high-speed hardware. A more flexible but usually slower approach is to use a frame grabber or frame buffer, i.e., a memory block in which the digitized image is stored. Once the image is captured in memory, it can be processed in a variety of ways by a computer. The availability of increasingly faster microprocessors and lower cost memories is expected to result in flexible image processing systems that can compete in speed with dedicated hardware systems.

Some of the preprocessing steps incorporated in a machine vision system may include the following.

Noise Removal

It is frequently necessary to remove unwanted clutter from an image. It may also be necessary to reduce electrical noise generated in the scanner circuits. A variety of one- and two-dimensional filtering techniques have been developed for noise removal. Spatial noise is generally filtered by applying local operators that replace a picture element (pixel) with a function of its surrounding pixels, e.g., 3×3, 5×5 neighborhoods. Nonlinear operators, such as the median filter, are particularly useful, in that they remove speckle noise without affecting the location or sharpness of edges. However, it is expensive to achieve these operators with real-time hardware because accurate delay lines are required, e.g., four one-line delays for a 5×5 operator. It is possible, however, to construct large kernel operators from a pipelined sequence of less expensive 3×3 kernel operators. This approach has been implemented in several systems.

Segmentation

The purpose of segmentation is to separate the objects of interest from the rest of the scene or background. The most frequently used techniques are thresholding and edge detection. Thresholding involves separating the scene into regions of similar gray scale intensity. The intensity of each pixel is compared with one or more thresholds and assigned to the appropriate region, e.g., black or white. Fixed thresholds are functional only when there is high contrast that is constant over time and over the entire scene. A high signal-to-noise ratio is also required. Otherwise, some kind of adaptive or local-area thresholding technique must be used.

Edge detection techniques generally employ local operators that respond to the first or second spatial derivative of the gray scale intensity in the neighborhood of each pixel. An edge is detected when these derivatives exceed a given magnitude. Because edge detectors respond to signal changes, they are less sensitive to absolute values of intensity or contrast than intensity threshold detectors. Consequently, they are more robust. Edge detectors, however, are basically high-pass spatial frequency filters and, as such, are somewhat sensitive to high-frequency disturbances (speckles, scratches).

Location

Once the objects of interest have been segmented from the rest of the image, it is usually necessary to determine their location in the field of view. The location algorithm used in any application depends on the nature of the available reference marks, if any, and the accuracy of the initial mechanical registration. One technique for locating the objects of interest is to correlate the horizontal and vertical projections of the silhouette of an object with the corresponding projections of its reference pattern. Another approach that is useful when the object is constrained in translation and rotation is to compute the centroid and the major and minor axes of the part. In some instances, locating the bounding box may be sufficient.

Where there is more than one object in the field of view, some kind of clustering algorithm must be used to associate the object pixels with a particular object. Additional intelligence also must be incorporated in the clustering algorithm when objects can touch one another.

Feature Extraction

Once the initial preprocessing steps have been completed, the next step is to generate a set of descriptors or features on which a decision concerning the objects in the image can be based. The features that describe an object may be as simple as the pixels of the object or the sum of those pixels. However, in many cases, a higher-level description of the object is required in order to differentiate between nearly identical objects, find small defects in objects, or precisely determine the object's size and location.

One family of a higher-level descriptions of objects involves the object's geometrical properties, which may include area, perimeter, bounding box, major and minor axes, number of corners and holes, and moments and moment invariants. Sometimes functions of these properties, such as area/perimeter^2, are selected. The advantage of such features is that they are invariant to translation and rotation. Some are also invariant to changes in scale. The disadvantages of geometric features include a requirement for high-contrast silhouettes, the absence of extraneous or noisy pixels and, for some, a great deal of computation. Further, the features selected for a particular application are critically dependent on the shape of the objects to be inspected.

Another family of object descriptors makes use of data transformations. Such transformations may be geometric, e.g., rectangular to polar or log spiral, or may involve spatial frequency or sequence-related transformations. Of the latter, transforms such as the Fourier, cosine, and Walsh-Hadamard are linear, orthogonal, and conserving of both energy and entropy. They compact the bulk of the information in a highly correlated input image pixel array into a small number of relatively uncorrelated transformed array components. These transforms thereby increase the discrimination between classes for a given number of components. The effect of individual pattern noise components is also greatly minimized, because every point in the transformed array contains a contribution from every point in the pattern array. Finally, there are fast algorithms for these transforms that are simple to implement on a small computer.

Feature selection in the spatial frequency domain can be treated as a filtering problem. It may be desirable to omit the very lowest spatial frequency components which represent absolute light-level variations and gradual shading across the field of view. It may also be desirable to omit the very highest spatial frequency components, since they might represent unwanted details like scratches and other texture variations on the object or in the background. Spatial frequency feature selection is less critically dependent on the shape of the objects to be inspected than are geometric features.

Spatial frequency features are not invariant to translation, rotation, and scale to the same degree as geometric features, and this is a major disadvantage. They can, however, be quite useful for inspecting and classifying well-oriented and positioned objects.

An additional family of object descriptors makes use of relationships between parts of the objects. Such relationships may be syntactically constructed or based on principles of geometric modeling. There have been a few instances of the use of relational structures in industrial vision systems to date, but the present interest in three-dimensional vision, as it relates to robotics, probably will accelerate the development of this type of object description.

Decision Making

Most decision rules employed in industrial machine vision systems must be fast and easily implemented by small computers. Simple Euclidean and "city block" distance metrics on the features are commonly used. It is frequently possible to incorporate information that will improve performance by using a decision tree approach. One simple example is the sorting of cartons by size and by color prior to discriminating on the basis of patterns on the sides of the cartons.

Most general purpose vision systems operate on a "train-by-showing" basis in which prototypes are formed by having the system scan a set of training samples. The system can thus be trained and retrained for a wide variety of applications.

It is useful for a flexible machine vision system to be able to trade time for reliability, i.e., when time permits, feedback that modifies parameters of the earlier stages of image processing can improve the reliability of the decisions. Multiple scans and decisions involving the same scene may also be used to improve reliability.

There are two major considerations in designing reliable decision making systems: (1) It is important to include all available information that can contribute to the decisions; and (2) it is important to characterize, accurately, the variations in the populations of objects or patterns to be inspected. Such characterization may require either large training sets or careful study of the manufacturing processes that produce the objects of interest.

Trends

New Machine Architectures for Vision

Serial processing of image data by conventional microprocessors is far too slow for many industrial applications of machine vision. As a result, a great deal of research and development is presently underway to devise special purpose image processing machines (Refs. 3 and 4). Various combinations of pipeline and parallel processing architectures are being studied. One pipeline architecture concatenates* many programmable small-area pixel operators. Various stages of image processing on a particular pixel can be accomplished by "handing off" the pixel from one operator to another. A machine that uses such an architecture can perform sophisticated image processing algorithms at video rates. However, because each local area operator is quite expensive to build with discrete components, the practical realization of a system of this type for industrial use awaits the implementation of these operators in very-large-scale integrated circuits (VLSI). Similarly, VLSI will be necessary for practical implementation of parallel processing architectures. Such systems will employ multiple processing elements under common synchronous control to analyze all parts of an image simultaneously. Hybrid parallel and pipeline architectures will also become viable. However, complex problems of algorithm partitioning, resource allocation, and bus design must be solved before such systems can become practical.

Optical Computing

This is another area of active research which may ultimately affect the design of industrial image processing systems. Optical computing systems make use of light valves and deflectors, coherent fiber-optic arrays, optical feedback devices, and special purpose solid-state components. They can carry out "instantaneous" parallel processing operations on images. Examples of such operations include complex data processing functions like the log spiral, Fourier, and Walsh-Hadamard transforms. Although a small number of special purpose inspection systems using Fourier optics exist, optical computing components have not yet been incorporated into general purpose machine visions systems for factory use. Optical components impose stringent requirements for mechanical stability and, in some cases, control of temperature and humidity. Also, the costs of these components have been prohibitively high.

Advanced Processing Techniques

There are several innovative image processing techniques under investigation. Researchers are using color information to improve the ability of machine vision systems to inspect multicolored objects, such as electronic components. Range data are being combined with two-dimensional information to help characterize three-dimensional objects. Researchers are studying radiometric stereo (one scanner, multiple light sources) and stereo (two scanners) image processing systems as a source of information for modeling three-dimensional objects.

Vision systems are also being coupled with computer-aided design–computer-aided manufacturing (CAD-CAM) data bases to combine the conception, design, production, and inspection of manufactured goods in one coherent endeavor. The systems approach, combined with data-driven computations made possible by the new machine vision architectures, will ultimately bring true artificial intelligence into the manufacturing process.

---

*To join or link, as in a chain.

Incorporating Machine Vision in Factory Machines

When image processing technology is transferred from the laboratory to the factory floor, the first step is careful definition of the problem to be solved. In many cases, the particular visual inspection problem addressed is only one of a number of visual and manipulative tasks being performed by an operator. The operator must be relieved of a significant proportion of labor, or the quality of the inspection process must be significantly improved in order to justify the installation of a vision system.

A second important step is evaluation of the environmental conditions at the inspection site. The design of equipment that is sufficiently rugged to operate reliably must consider temperature, humidity, vibration, ambient lighting, power line variations and impulse noise, the presence of nearby electric and magnetic fields, and rf interference.

A third step is assessment of the requirements for and a careful design of the optics and illumination components of the vision system. These must be stable and must provide sufficient contrast, uniformity, and scanner signal-to-noise ratio to create consistently high-quality images for analysis. Image quality is usually the most important contributor to the performance of a machine vision system.

A fourth step is determination of the ease with which the system can be integrated into the existing factory. The system must be simple to train and run. No vision system can succeed without the support of the factory personnel who operate it.

REFERENCES

1. Kruger, R. P., and W. B. Thompson: "A Technical and Economic Assessment of Computer Vision for Industrial Inspection and Robotic Assembly," *IEEE Proc.,* vol. 69, no. 12, December 1981, pp. 1524–1538.

2. Zeller, H., and G. Doemens: "Industrial Application of Pattern Recognition," *Proceedings of the 6th International Conference on Pattern Recognition,* October 1982, pp. 202–213.

3. Danielson, P., and S. Levialdi: "Computer Architectures for Pictorial Information Systems," *Computer,* vol. 14, November 11, 1981.

4. IEEE: "Computer Architectures for Image Processing," *Computer,* vol. 16, no. 1, January 1983.

Vibration Measurements

by
Robert M. Whittier*

Vibration measurements are required during the development and construction of many kinds of machines and systems. Measurements are often made to determine levels of vibration which could be destructive or cause excessive noise. Once a system is operational, vibration monitoring is used as an important indicator of mechanical health. Excessive or increasing vibration is often an early indicator of mechanical degradation.

Vibration sensing is required over a wide range of amplitudes and frequencies. For example,

*Manager of Development Engineering, Endevco Corporation, San Juan Capistrano, California.

vibration from rotating machinery, such as pumps, motors, compressors, and turbines, occurs from about 1 Hz to over 20,000 Hz, the principal interest being from 10 to 2000 Hz. Vibration amplitudes vary widely depending on the equipment design. For example, a smooth-running motor may vibrate at 0.01 g (1 g = 386 in/sec², 980 cm/sec²), but a high-speed gearbox can easily vibrate at more than 100 g at a frequency of more than 10,000 Hz.

Ideally, a vibration sensor is attached to a body in motion and provides an output signal proportional to the vibrational input from that body. Sometimes it is not practical to attach a sensor directly to the moving body. In these cases, measurement is made by attaching the sensor to another body and making a measurement relative to the motion of that body. In any event, the performance of the measurement technique should not be degraded by the location of the sensor.

Inertial Motion Sensing

Inertial motion measurement is achieved by attaching a sensor to the moving structure. By doing so, the vibration is fundamentally changed because of the addition of the mass of the sensor; however, the changes are usually insignificant. The specific requirement is that the dynamic mass of the sensor be much less than the dynamic mass of the structure at the point of attachment.

Inertial motion sensors consist of a mass, a spring, a viscous damper, and a means of electrical pick-off. Figure 1 shows a simplified mechanical schematic of such a system. The response of this to vibration is well known. When the mass is small and the spring is stiff, the system can be used at frequencies below its resonant frequency where its response is constant for acceleration inputs. Thus, it is an accelerometer. When the mass is large and the spring is flexible, the device can be used to sense relative displacement at a high frequency. Damping is sometimes added to these, which limits the relative response at its resonance. However, damping does create phase shift. When damping is used, it is advisable to maintain its value at approximately seven-tenths of critical damping to achieve proportional or linear phase shift. When damping is not used in an accelerometer, the frequency range is limited to typically one-fifth of the resonant frequency.

Accelerometers

The most common type of vibration sensor is the accelerometer. It can be made small, lightweight, and rugged—all necessary attributes. Both self-generating accelerometers and those requiring electrical excitation are available. The most common is the self-generating piezoelectric device. Typical performance characteristics for accelerometers are listed in Table 1.

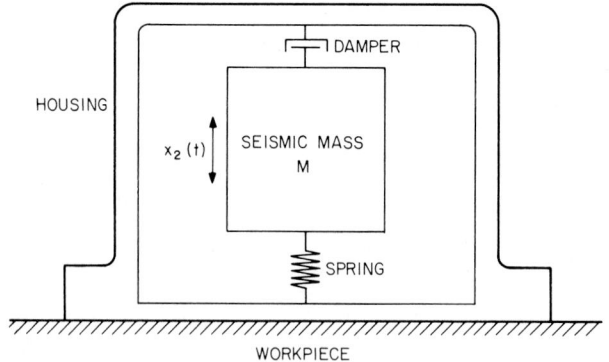

Fig. 1 Mechanical schematic of an inertial sensor.

Table 1 Typical Accelerometer Performance Characteristics

Characteristic	Piezoelectric accelerometers		Piezoresistive accelerometer	Servo accelerometer
Sensitivity, pC/g, mV/g	10	100	20	250
Frequency range, Hz	4–15,000	1–5000	0–750	0–500
Resonance frequency, Hz	80,000	20,000	2500	1000
Amplitude range, g	500	1000	25	15
Shock rating, g	2000	10,000	2000	250
Temperature range, °C	−50 to +125	−50 to +260	0 to +95	−40 to +85
Total mass, g	1	29	28	80

Piezoelectric Accelerometers

These devices utilize a mass in direct contact with a piezoelectric component, or crystal. When a varying motion is applied to the accelerometer, the crystal experiences a varying force excitation ($F = ma$), causing a proportional electrical charge q to be developed across it.

$$q = d_{ij}F = d_{ij}ma$$

where d_{ij} is the material's piezoelectric strain constant.

As the equation shows, the electrical output from the piezoelectric material is dependent on its properties. Two commonly used materials are lead-zirconate titanate ceramic (PZT) and quartz. As self-generating materials, they both produce a large electric charge for their size, although the piezo-electric strain constant of PZT is about 150 times that of quartz. As a result, accelerometers using PZT are more sensitive or are much smaller. The mechanical spring constants for the piezoelectric components are high, and the inertial masses attached to them are small. Therefore, these accelerometers are useful to extremely high frequencies. Damping is rarely added to these devices. Figure 2 shows a typical frequency response for such a device. Piezoelectric accelerometers have comparatively low mechanical impedance. Therefore, their effect on the motion of most structures is small. They are also rugged and have outputs that are stable with time and environment.

Two principal design configurations are used for piezoelectric accelerometers. One stresses the piezoelectric material in *compression* while the other stresses it in *shear*. Simple diagrams of these two types are shown in Fig. 3. When the accelerometer is accelerated upward, the mass is moved downward toward the bottom. Conversely, downward acceleration moves the mass element upward. With vibration motion, the resultant dynamic stress deforms the piezoelectric element. In the compression accelerometer, vibration varies the stress in the crystal which is held in compression by the preload element. In the shear accelerometer vibration simply deforms the crystal in shear. The mechanical construction of actual designs can be more complex, but the model is the same.

Piezoresistive Accelerometers and Strain Gage Sensors

Piezoresistive accelerometers are strain gage sensors which use semiconductor strain gages in order to provide much greater gage factors than are possible with metallic gages. Higher gage factors are achieved because the material resistivity changes with stress, not just its dimensions. The increased sensitivity is critical to vibration measurement in that it permits miniaturization of the accelerometer.

A typical piezoresistive accelerometer uses either two or four active gages in a Wheatstone bridge. It is more important to use multiple gages than when using metallic gages because the temperature coefficients of the semiconductor elements are greater than those of metallic gages. To control the electrical bridge balance and sensitivity variations with temperature, other resistors are used within the bridge and in series with the input.

The mechanical construction of an inertial system using piezoresistive elements is illustrated in Fig. 4, and the construction of a complete accelerometer is shown in Fig. 5. This design includes overload stops to protect the gages from high-amplitude inputs and includes oil to improve damping. Such an instrument is useful for acquiring vibration information at low frequencies (for example, below 1 Hz), and the device can be used to sense static acceleration.

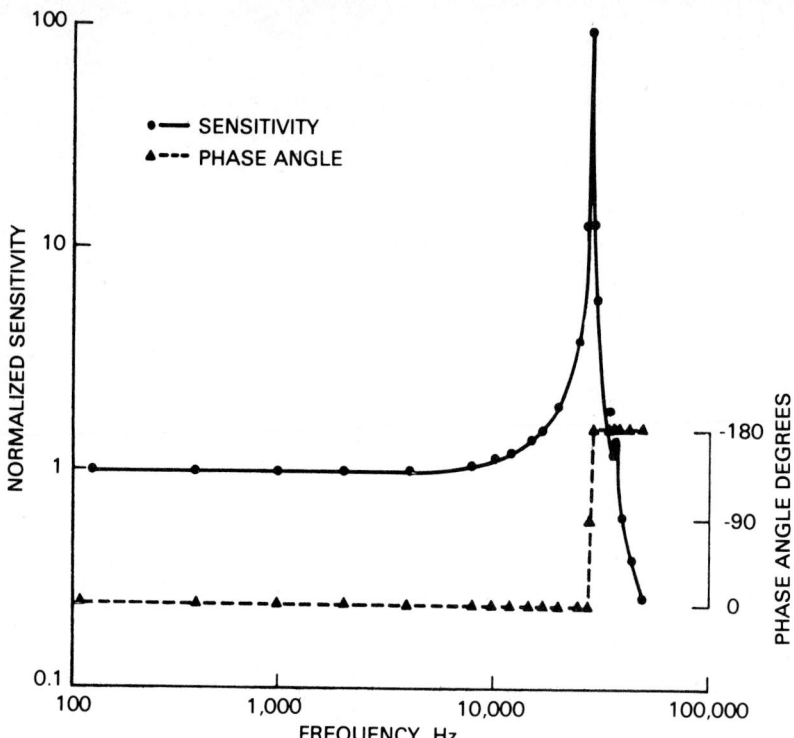

Fig. 2 Frequency response from a typical piezoelectric accelerometer.

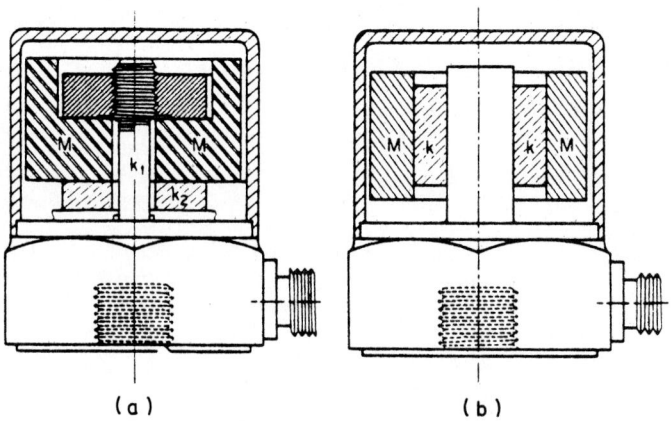

Fig. 3 Conceptual diagram for piezoelectric accelerometers. (*a*) Compression, (*b*) shear.

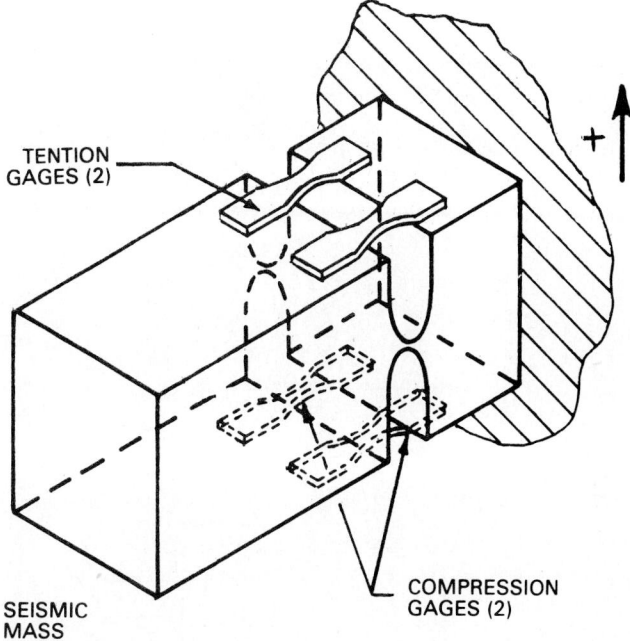

Fig. 4 Inertial system using piezoresistive elements.

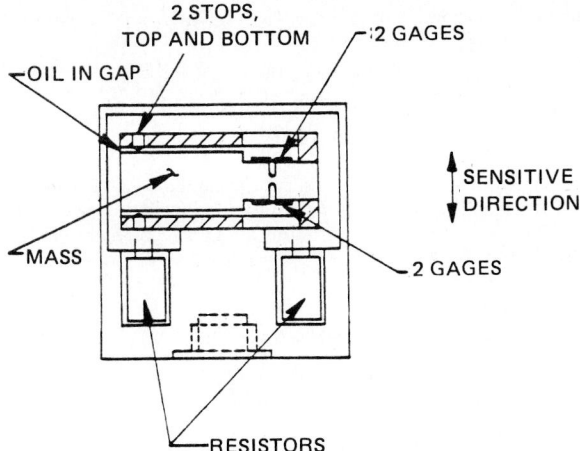

Fig. 5 Construction of typical piezoresistive elements.

Servo Accelerometers

The construction and operating principle of a servo accelerometer are illustrated in Fig. 6. When subjected to acceleration, the proof mass deflects relative to the base of the accelerometer, and the pick-off changes its capacitance as a result of changes in the damping gap. As this occurs, the servo supplies current to the coil which is located in the gaps of the permanent magnets. The resulting

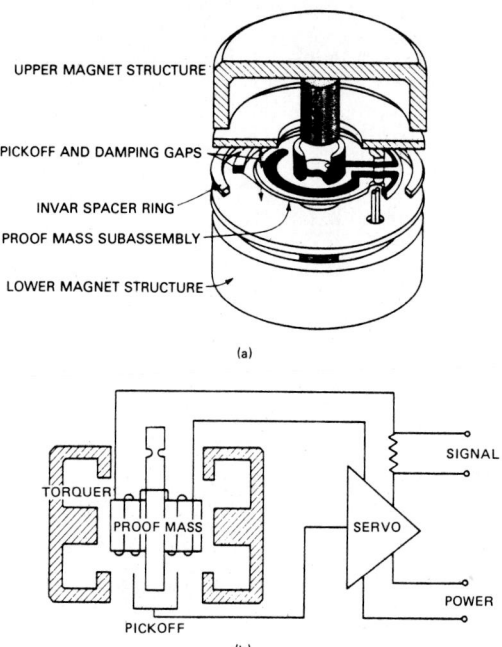

Fig. 6 Construction of a servo accelerometer.

force restores the coil to its equilibrium position. The output signal is a measure of the coil current and is proportional to the applied acceleration.

Signal Conditioning for Accelerometers

Signal conditioners interface accelerometers to readout and processing instruments by (1) providing power to the accelerometer if it is not self-generating, (2) providing proper electrical load to the accelerometer, (3) amplifying the signal, and (4) providing an appropriate filtering and drive signal. Piezoelectric and piezoresistive transducers both require conditioners with certain characteristics, as is now discussed.

Conditioning Piezoelectric Accelerometers

The piezoelectric accelerometer is self-generating and supplies a very small amount of energy to the signal conditioner. It presents a very high-source impedance, mainly capacitive, to the conditioner. Two forms of schematic representation of a piezoelectric accelerometer are shown in Fig. 7. It may be regarded as a voltage source in series with a capacitance, or as a charge source in parallel with a capacitance. The signal conditioner determines how the transducer is treated in a given system. Both voltage and charge sensing are used. The charge amplifier is by far the most common approach. The charge amplifier is advantageous because the system gain and low-frequency response are well defined and are independent of the cable length and accelerometer capacitance.

The charge amplifier consists of a charge converter and a voltage amplifier, as shown in Fig. 8. The system does not amplify charge per se. It converts input charge to a voltage and then amplifies the voltage.

A charge converter is essentially an operational amplifier with integrating feedback. The equivalent circuit is shown in Fig. 9. With basic operational-type feedback, the amplifier input is main-

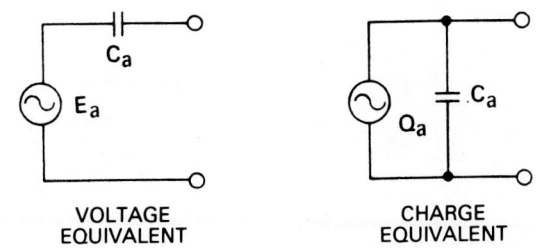

VOLTAGE EQUIVALENT   **CHARGE EQUIVALENT**

Fig. 7 Electrical schematic representation of piezoelectric accelerometers.

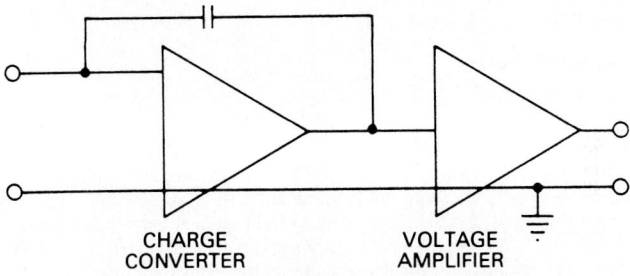

CHARGE CONVERTER   **VOLTAGE AMPLIFIER**

Fig. 8 Block diagram of a charge amplifier.

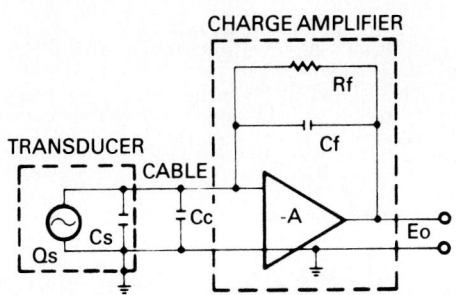

Fig. 9 Equivalent circuit of a charge converter.

tained at essentially zero volts and therefore looks like a short circuit to the input. The amplifier output is a function of the input current. Having integrating operational feedback, output is the integral of the input current—hence the name "charge amplifier," since

$$q = \int i\,dt$$

In operation, the charge converter output voltage which occurs as a result of a charge input signal is returned through the feedback capacitor to maintain the voltage at the input at, or close to, zero. Thus, the charge input is stored in the feedback capacitor, producing a voltage across it which is equal to the value of the charge input divided by the capacitance of the feedback capacitor. The transfer characteristic (charge gain) of this amplifier is therefore solely dependent on the value of the feedback capacitor (or network if a more complicated feedback is used), provided that the amplifier's open-loop gain and input impedance are sufficiently high. The midband charge gain (millivolts per picocoloumb) of this circuit is

$$\frac{E_o}{Q_i} = \frac{1}{C_f}$$

The complete circuit transfer function is given by

$$\frac{E_o}{Q_i} = \frac{-A}{1 + A} \cdot \frac{sR_f}{sR_f \left(\dfrac{[(Cs + C_c)]}{(1 + A)} + C_f \right) + 1}$$

where E_o = charge converter output
Q_i = accelerometer signal (charge)
s = accelerometer sensitivity (charge)
C_a = accelerometer capacitance
C_c = cable capacitance
C_f = feedback capacitance
R_f = feedback resistance
A = amplifier open-loop gain

To simplify, we can assume

$$-A/(1 + A) = -1,$$

since the amplifier open-loop gain A is very high and, therefore, $A = 1 + A$.

Because C_f is very large compared to $(C_s + C_c)/(1 + A)$ in most applications, system gain is independent of cable length. Only the unlikely combination of an extremely large source or cable capacitance and a very small feedback capacitance would influence system gain, and then only to a minor degree. Therefore, charge measuring systems are seldom calibrated end to end.

With the above simplifications, the denominator of the expression becomes $sR_fC_f + 1$. This is the characteristic form of a simple first-order rolloff at

$$f_{-3 \text{ dB}} = \frac{1}{2\pi \; R_fC_f}$$

with a terminal slope approaching 6 dB per octave. For all practical purposes, low-frequency response of a charge measuring system is a function of well-defined electronic components and does not vary with cable length. This is a very important feature when measuring low-frequency vibrations.

Conditioning for Low-Impedance Piezoelectric Accelerometers

Piezoelectric accelerometers are available with simple electronic circuits internal to their cases to provide signal amplification and low-impedance output. Some designs operate from low-current dc voltage supplies and are designed to be *intrinsically safe* when coupled by appropriate barrier circuits. Other designs have common power and signal lines and use coaxial cables.

The principal advantages of piezoelectric accelerometers with integral electronics are their relative immunity to cable-induced noise and spurious response, the ability to use lower cost cable, and a lower signal conditioning cost. In the simplest case, the power supply might consist of a battery, a resistor, and a capacitor.

These advantages do not come without compromise. Because the impedance matching circuitry is built into the transducer, gain cannot be adjusted to utilize the wide dynamic range of the basic transducer. Ambient temperature is limited to that which the circuit will withstand, and this is considerably lower than that of the piezoelectric sensor itself. In order to retain the advantages of small size, the integral electronics must be kept relatively simple. This precludes the use of multiple filtering and dynamic overload protection and thus limits their application. But when conditions are relatively benign, these accelerometers can economically provide excellent noise immunity and signal fidelity.

Conditioning Piezoresistive Transducers

Piezoresistive transducers are relatively easy to condition. They generally have high-level output, low-output impedance, and very low intrinsic noise. These transducers require an external power supply. This supply is usually direct current, but it may be alternating current providing the carrier frequency is at least 5 to 10 times the maximum frequency of interest.

Most transducers are designed for constant voltage excitation and are used with relatively short cables. With long cables, wire resistance is not negligible. Moreover, resistance changes with temperature, and the voltage drop along the line varies as the transducer resistance or load changes. For these applications, transducers should be calibrated for constant current excitation so their output will be less dependent on external effects.

Many piezoresistive transducers are full-bridge devices. Some have four active arms to maximize sensitivity. See Fig. 10. Others have two active arms and two fixed precision resistor arms to permit

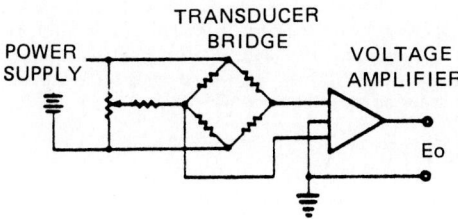

Fig. 10 Typical system and bridge circuit for a piezoresistive accelerometer.

shunt calibration by precision calibration resistors in the signal conditioner. Miniature transducers are usually half-bridge devices, with bridge completion accomplished in the signal conditioner.

Adjustment of the unbalanced output of an accelerometer can easily be performed in the signal conditioner. For full-bridge transducers, the balancing potentiometer R_1 is connected across the excitation terminals and a current limiting resistor is connected between the wiper arm of the potentiometer and the bridge. This is shown in Fig. 11a. For half-bridge transducers, a small balance potentiometer (typically 100 Ω) is connected between the bridge completion arms as shown in Fig. 11b.

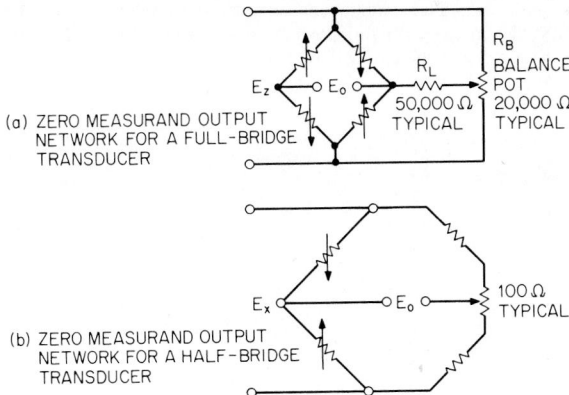

Fig. 11 Bridge balancing for a piezoresistive accelerometer.

Environmental Effects

Temperature

Accelerometers can be used over wide temperature ranges. Piezoelectric devices are available for use from cryogenic temperatures ($-270°$C) to over $650°$C. The sensitivity changes with ambient temperature, but the changes are systematic and can be calibrated. If the ambient temperature changes suddenly so that strains develop within the accelerometer and within the time response of the measurement system, further errors can occur. These are evaluated by testing the response of accelerometers to step function changes in temperature per industry standard test procedures. Errors usually appear as a wandering signal or a low-frequency oscillation.

Cable Movement

Cabling from the accelerometer to the signal conditioner can generate spurious signals when it is subjected to movement or dynamic forces. This is usually significant only for systems using high-impedance piezoelectric accelerometers. The major noise generating mechanism is triboelectric noise, which is caused by charge trapping due to relative motion, or localized separation between the cable dielectric and the outer shield around the dielectric. To reduce this effect, cabling is available which is "noise-treated." These cables have a conductive coating applied to the surface of the dielectric which prevents charge trapping. Another way to eliminate this effect is to use a sensor which includes an electronic circuit for reducing the impedance to $\sim 100\ \Omega$.

Dynamic Strain Inputs

In vibration environments, some structures may dynamically flex, stretch, or bend at the mounting location of the accelerometer. Being in intimate contact with this strained area, the base of the accelerometer can also be strained. A portion of this base strain is transmitted directly to the crystal sensing element and generates error signals. In addition to strains in the structure, it is also possible to induce errors from forces or pressures onto the case of the accelerometer. Outputs from these forces vary greatly, depending on the internal design of the accelerometer. The errors from these sources are usually checked against industry standard test procedures, and the results are included in specifications.

Electrostatic and Electromagnetic Fields

Electrostatic noise can be generated by stray capacitance coupling into the measurement system. It is important that the cabling between a high-impedance piezoelectric sensor and the signal conditioner be fully shielded. Ground loops can be avoided by grounding the system at a single point, usually at the output of the signal conditioner. Magnetically coupled noise can best be avoided by not placing signal cables in close proximity to power or high-current conductors and by avoiding electromagnetic sources when possible. Accelerometers should also be checked for their sensitivity to electromagnetic fields.

Velocity Transducers

Electrodynamic velocity pickups can be used directly into readout instruments. The self-generating voltage in the transducer is proportional to the velocity of the motion being measured and is usually of sufficient amplitude so that no voltage amplification is required. The disadvantages of velocity pickups are their large size and mass and their inability to be used for measurements at frequencies below ~ 10Hz. Also, the output at high frequencies, above ~ 1000 Hz, is quite small in most applications. Care must be taken in using these devices in strong magnetic field environments.

A typical velocity sensor consists of a seismically mounted, and usually critically damped, magnetic core suspended in a housing rigidly attached to the vibrating surface. A coil of wire attached to the housing surrounds the core. Relative motion between the magnetic core and the housing causes magnetic lines of flux to cut the coil, inducing a voltage proportional to the velocity. These sensors operate above their first natural frequency.

Relative Motion Sensing—Eddy-Current Probe

In some cases, it is not practical to place a sensor in contact with the moving part. Relative motion measurement approaches are then used. The most commonly used device is the eddy-current probe.

Noncontact eddy-current displacement measuring systems have achieved general acceptance for industrial machinery protection and condition monitoring. An eddy-current displacement probe, generally about -0.30 in (7.5 mm) in diameter, contains a small coil of fine wire at its tip which is excited by a remote rf oscillator to generate a magnetic field. As the tip of the probe is brought close to a conductive surface, such as a rotating shaft, eddy currents induced in the conductor by the probe's magnetic field oppose the field and reduce the amplitude of the carrier by an amount proportional to the change in promixity. A demodulator, usually encapsulated in the same enclosure as the oscillator, converts the change in carrier amplitude to a low-impedance, calibrated voltage output.

An eddy-current displacement sensor and its companion oscillator-demodulator therefore constitute a gap-to-voltage measuring system. The average gap, or the distance between the probe tip and the conductive surface, is represented by a dc bias or offset on which is superimposed an ac analog of the surface's dynamic motion. A typical linear amplitude range is 1 to 2 mm with a frequency response capability from static to more than 2000 Hz. The sensitivity changes for different target materials and with changes in cable length.

REFERENCES

Anon.: "ISA Recommended Practice for Specifications and Tests for Piezoelectric Acceleration Transducers," *ISA 37.2,* Instrument Society of America, Research Triangle Park, N.C. 1964.

Anon.: "ANSI Standard for the Selection of Calibration and Tests for Electrical Transducers Used for Measuring Shock and Vibration," *ANSI Rep. S2.11,* American National Standards Institute, New York, 1969.

Bouche, R. R.: "Accelerometers for Use in Nuclear Reactor Components," ASME Meeting, December 1970, New York, American Institute of Mechanical Engineers New York.

Eller, E. E.: "Introduction to Shock and Vibration Measurements," in *"Shock and Vibration Handbook,"* 2d Ed. (C. M. Harris and C. E. Crede, eds.), McGraw-Hill, New York, 1976, chap. 12.

Jacobs, E. D.: "New Developments in Servo Accelerometers," *Proc. Inst. Environ. Sci.* 1968, pp. 517–523.

Mitchell, R, S.: "The Use of Vibration and Other Operating Characteristics in a Comprehensive Engineered Monitoring System," 22nd International Instrumentation Symposium on Fundamentals of Aerospace Instrumentation," May 1976, San Diego, Calif.

Thomas, R. L.: "Signal Conditioning for Commonly Used Transducers," 14th ISA Aerospace Symposium, June 1968, Instrument Society of America, Research Triangle Park, N.C.

Whittier, R. M., and L. C. Ensor: "Accelerometers Developed for Vibration Measurement at High Temperatures," 25th ISA Conference, October 1970, Philadelphia, Pa., Instrument Society of America, Research Triangle Park, N.C.

Humidity and Moisture Systems

Vincent B. Cortina. *Environmental Equipment Division, EGEG International, Inc., Waltham, Massachusetts. (Humidity Measurement)*

Lee E. Cuckler. *Consultant, Formerly Assistant to General Manager, Robertshaw Controls Company, Anaheim, California. (Moisture Measurement Systems)*

Mara Levine. *Process Instrument Division, Panametrics, Inc., Waltham, Massachusetts. (Aluminum Oxide Moisture Sensors)*

Craig McIntyre. *Kay-Ray, Inc., Arlington, Heights, Illinois. (Moisture Measurement Systems)*

Pieter R. Wiederhold. *President, General Eastern Instruments Corporation, Watertown, Massachusetts. (Humidity Measurement)*

Humidity Measurement

by
Pieter R. Wiederhold*

Humidity can be expressed in a variety of different forms: wet bulb temperature, percent relative humidity (% RH), vapor pressure, mixing ratio, dew point or frost point, grains per pound, grams per kilogram, parts per million, among others. These parameters can be measured by a number of different instruments, each capable of accurate measurement under certain conditions and within specific limitations.†

*President, General Eastern Instruments Corporation, Watertown, Mass.
 †Humidity measurement is a special case of moisture measurement in gases. See also the following article, Moisture Measurement Systems.

Definition of Humidity

Unless one is routinely working with humidity measurements, there is a tendency to overlook the fact that humidity is water gas, behaving in accordance with the ideal gas laws. One of the easiest ways to put humidity in its proper perspective is through application of Dalton's law of partial pressures to the most commonly encountered gas—*air*.

Dalton's law states that the total pressure P_m exerted by a mixture of gases or vapors is the sum of the pressure of each gas if it were to occupy the same volume by itself. The pressure of each individual gas is called its *partial pressure*. The total pressure of an air-water gas mixture, containing oxygen, nitrogen, and water, is equal to the sum of the partial pressures of each gas:

$$P_m = P_{N2} + P_{O2} + P_{H2O} + \cdots$$

Therefore, the partial pressure of water vapor in air is directly related to the measurement of humidity. This vapor pressure varies from 1.22×10^{-3} millibar (mbar) of mercury (0.122 Pa), at the $-75°C$ frost point of "bone-dry" arctic or industrial dry air, to 1.013×10^3 mbar of mercury (0.1013×10^6 Pa) at the $100°C$ dew point of saturated hot air in a product dryer. This is a change of almost a million to one over the span of interest in industrial humidity measurement.

The ideal humidity instrument would be a linear, wide-range pressure gage, specific to water vapor and employing a primary or fundamental measuring method. Such an instrument, although physically possible, would be cumbersome. Most humidity measurements are made by a secondary instrument which is responsive to humidity-related phenomena.

Common Humidity Parameters

The humidity parameters most often encountered in scientific and industrial applications are shown in Table 1. In addition to these common parameters, numerous other formats exist for use in narrow applications or specific technologies. However, most of these are variations of the basic parameters given in Table 1.

Psychrometric Chart

The psychrometric chart (Fig. 1) provides a quick means for converting from one humidity format to another, because dew point, relative humidity, ambient temperature, and wet bulb temperature can be conveniently related to each other on a single sheet of paper. The psychrometric chart has long been the basic tool of air-conditioning engineers, and in Fig. 1 are shown the temperatures most often encountered in comfort control or product conditioning applications. Psychrometric charts are available for higher temperatures and humidities and are quite useful in dryer and condensation system design. Charts are also available for lower temperatures but tend to be less useful because wet bulb measurements are difficult to make with any accuracy at temperatures below $-7°C$.

Wet Bulb–Dry Bulb Measurements

Psychrometry has long been a popular method for monitoring humidity, primarily because of its simplicity and inherent low cost. A typical industrial psychrometer consists of a pair of matched electrical thermometers, one of which is maintained in a wetted condition. Water evaporation cools the wetted thermometer, resulting in a measurable difference between it and the ambient or dry bulb measurement. When the wet bulb reaches its maximum temperature depression, the humidity is determined by comparing the wet bulb–dry bulb temperatures on a psychrometric chart. In a properly designed psychrometer, both sensors are aspirated at an airstream rate between 4 and 10 m/s for proper cooling of the wet bulb, and both are thermally shielded to minimize errors from radiation.

A properly designed and utilized psychrometer, such as the Assman laboratory type, is capable

Table 1 Humidity Measurement Methods

Parameter	Description	Units	Typical applications
Wet bulb temperature	Minimum temperature reached by a wetted thermometer in an airstream	°F or °C	High-temperature dryers, air conditioning, meterology, test chambers
Percent relative humidity	The ratio of the actual vapor pressure to the saturation vapor pressure, with respect to water, at the prevailing dry bulb temperature	0–100%	Monitoring conditioning rooms, test chambers, pharmaceutical and food packaging
Dew point or frost point	Dew point is the temperature to which the air must be cooled to achieve saturation. If the temperature is below 32°F, it is called the frost point	°F or °C	Heat treating, annealing atmospheres, dryer control, instrument air monitoring, meteorological/environmental measurements
Volume or mass ratio	Parts per million (ppm) by volume is the ratio of the partial pressure of the water vapor to the partial pressure of the dry carrier gas. PPM by weight is identical to ppm by volume, but the ratio changes according to the molecular weight of the carrier gas	ppm_v, ppm_w	Used primarily to ensure dryness of industrial process gases such as air, nitrogen, oxygen, methane, hydrogen, etc.

of providing accurate data. However, very few industrial psychrometers meet these criteria and are limited to applications where low cost and moderate accuracy are the underlying requirements. The psychrometer has certain inherent advantages: (1) It is capable of highest accuracy near 100% RH. From an accuracy standpoint, it is superior to most other humidity sensors near saturation. Since the dry bulb and wet bulb sensors can be connected differentially, the wet bulb depression (which approaches zero as the relative humidity approaches 100%) can be measured with a minimum of error. (2) Although large errors can occur if the wet bulb becomes contaminated or improperly fitted, the simplicity of the device allows easy repair at minimum cost. (3) The psychrometer can be used at an ambient temperature above 100°C, and the wet bulb measurement is usable up to 100°C.

Major limitations of the psychrometer include: (1) As relative humidity drops below 20% RH, the problem of cooling the wet bulb to its full depression becomes difficult. The result is impaired accuracy below 20% RH, and few psychrometers provide reliable data below 10% RH. (2) Wet bulb measurements at temperatures below 0°C are difficult to obtain with any high degree of confidence. Automatic water feeds are not feasible because of freezing. (3) Because a wet bulb psychrometer is a source of moisture, it can be used only in environments where added water vapor from the psychrometer exhaust is not a significant component of the total volume. (4) Generally speaking, psychrometers cannot be used in small, closed volumes.

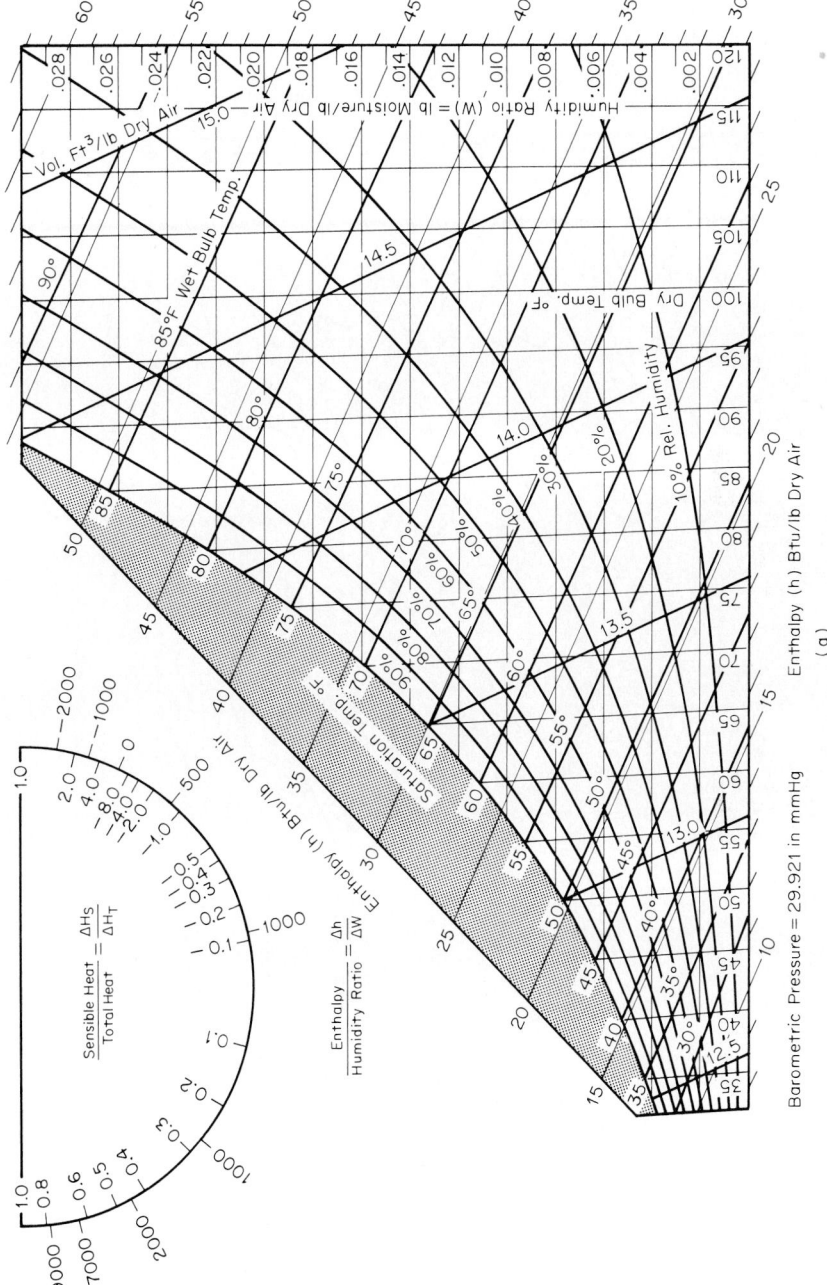

Fig. 1 (a) Psychrometric chart in conventional units. (b) Psychrometric chart in SI units. *Note:* These charts only apply at atmospheric pressure. *(American Society of Heating, Refrigerating, and Air Conditioning Engineers.)*

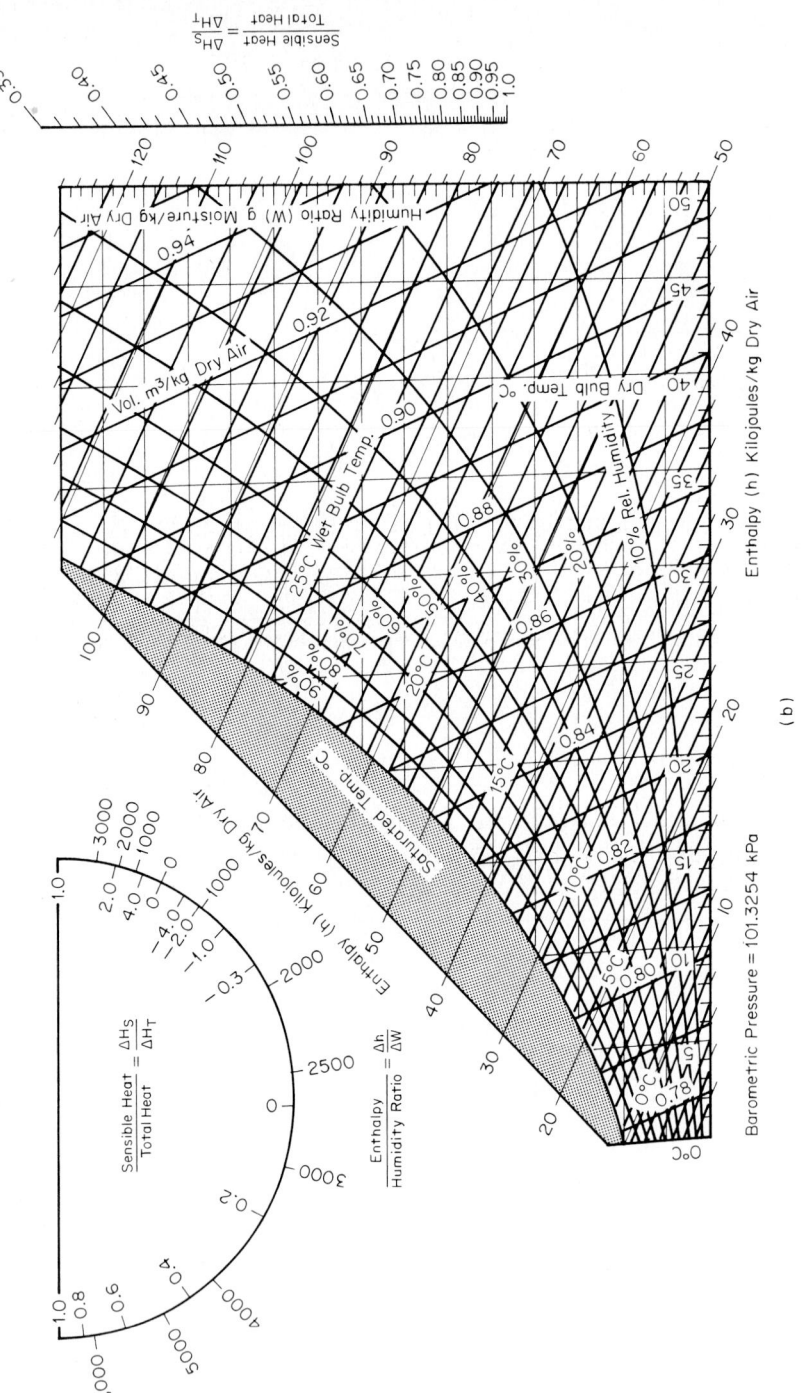

Fig. 1 (*Continued*)

10.6

Percent Relative Humidity

Percent relative humidity is the best known and perhaps the most widely used method for expressing the water vapor content of air. Percent relative humidity is defined as the ratio of the prevailing water vapor pressure e_a to the water vapor pressure if the air were saturated e_s multiplied by 100:

$$\% \ RH = \frac{e_a}{e_s} \times 100$$

The term "percent relative humidity" appears to have derived from the invention of the hair hygrometer in the seventeenth century. The hair hygrometer operates on the principle that many organic filaments, such as hair, goldbeater's skin, and even nylon, change length as a nearly linear function of the *ratio* of *prevailing water vapor pressure* to the *saturation vapor pressure*.

Basically, percent relative humidity is an indicator of the water vapor saturation deficit of the gas mixture, rather than an indicator of sorption, desorption, comfort, or evaporation. A measurement of RH without a corresponding measurement of dry bulb temperature is not of particular value, since the water vapor content cannot be determined from % RH alone.

Sensors for Measuring % RH

Over the years, devices other than the simple hair hygrometer have evolved, which permit a direct measurement of % RH. These devices are, for the most part, electrochemical sensors which offer a degree of ruggedness, compactness, and remote electronic readout ability not afforded by hair devices.

Two widely used electronic % RH sensors are the Dunmore element and the Pope cell. The Dunmore sensor employs a bifilar-wound inert wire grid on an insulative substrate coated with a lithium chloride (LiCl) solution of a controlled concentration. The hygroscopic nature of this salt causes it to take up water vapor from the surrounding atmosphere. The ac resistance of the sensor is an indication of the prevailing % RH. Dunmore cells are excellent RH sensors but, because of the characteristics of lithium chloride, are usually designed to cover a narrow range of interest. For example, a single sensor may cover from 40 to 60% RH, and the sensor output is usable only in that range. See Fig. 2a.

Wide-range Dunmore sensors can be made with a cluster of narrow-range sensors in a common housing, mated with an electrical matching network. This arrangement, however, usually results in a rather bulky sensor.

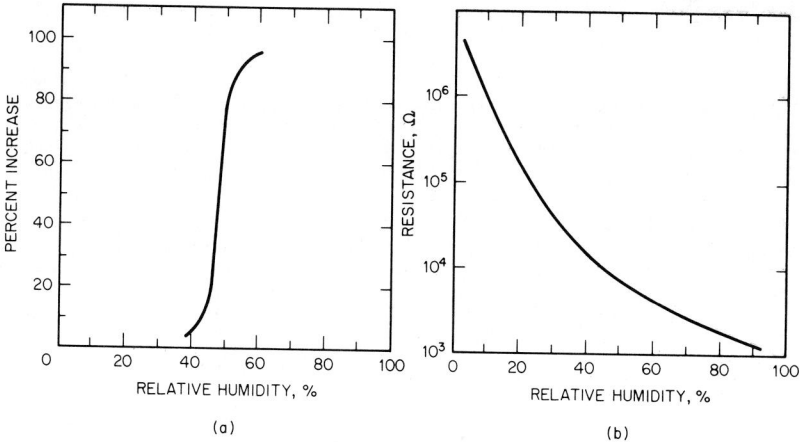

(a)   (b)

Fig. 2 Resistance characteristics of typical Dunmore and Pope sensors. (*a*) Dunmore sensors are limited to a narrow range of humidity. This sensor operates between 40 and 60% RH. (*b*) Pope sensors operate over a wide humidity range, but output impedance of the sensor varies from 1000 Ω (100% RH) to several megohms (10% RH), which complicates readout circuitry.

The Pope cell employs a similar bifilar conductive grid on an insulative substrate. In this sensor, the substrate is made from polystyrene which has been treated in a prescribed fashion with sulfuric acid. This results in sulfonation of the longer-chain polystyrene molecules. Because the sulfate radical (SO_4) is highly mobile in the presence of hydrogen ions (available from the water molecule in vapor form), the SO_4^{2-} ions can detach and take on H^+ ions, thereby altering the surface resistivity of the sensor as a function of humidity.

In both Dunmore and Pope sensors, the element is arranged in an ac-excited Wheatstone bridge so that only alternating current flows through the grid. Direct current excitation of either the Dunmore or the Pope element polarizes the sensor, eventually causing loss of calibration.

The Pope sensor has one significant advantage over the Dunmore sensor in that the Pope unit is a wide-range sensor, typically covering 15 to 99% RH in a single element. See Fig. 2b. Considerable attention must be given to readout circuitry for the Pope sensor because the resistance varies in a nonlinear fashion from 1000 Ω to several megohms.

Dew Point Hygrometry

Dew point measurements are widely used in scientific and industrial applications when precise measurement of water vapor pressure is needed. The dew point, the temperature at which water condensate begins to form on a surface, can be accurately measured from $-75°C$ to $+100°C$ across the entire range of humidity with a condensation (chilled mirror) hygrometer.

Three types of instruments have received wide acceptance in dew point measurement: (1) the saturated salt dew point sensor, (2) the condensation-type hygrometer, and (3) the aluminum oxide sensor. Many other instruments are used in specialized applications, including pressure ratio devices, dew cups, and fog chambers. The latter are manually operated.

Saturated Salt Dew Point Sensors

The saturated salt (lithium chloride) dew point sensor is a widely used sensor because of its inherent simplicity, ruggedness, and low cost. Both the U.S. and Canadian government weather services use this type of sensor for most official ground-based humidity measurements. However, some of these instruments are being converted to the more accurate condensation hygrometers.

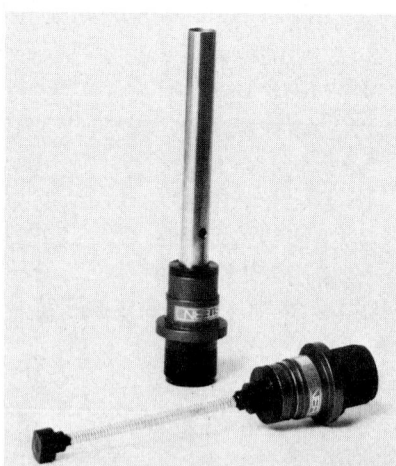

The principle of the saturated salt dew point sensor is that the vapor pressure of water is reduced in the presence of a salt. When water vapor in the air condenses on a soluble salt, it forms a saturated layer on the surface of the salt. This saturated layer has a lower vapor pressure than water vapor in the surrounding air. If the salt is heated, its vapor pressure increases to a point where it matches the water vapor pressure of the surrounding air and the evaporation-condensation process reaches equilibrium. The temperature at which equilibrium is reached is directly related to the dew point.

A saturated salt sensor is constructed with an absorbent fabric bobbin covered with a bifilar winding of inert electrodes coated with a dilute solution of lithium chloride. See Fig. 3. Lithium chloride is often used as the saturating salt because of its hygroscopic nature which permits application in relative humidities between 11 and 100%.

Fig. 3   Typical lithium chloride sensor. *(General Eastern Instruments.)*

An alternating current is passed through the winding and salt solution, causing resistive heating. As the bobbin heats up, water evaporates into the surrounding air from the diluted LiCl solution.

The rate of evaporation is determined by the vapor pressure of water in the surrounding air. When the bobbin begins to dry out, as a result of evaporation of water, the resistance of the salt solution increases. With less current passing through the winding, because of increased resistance, the bobbin cools and water begins to condense, forming a saturated solution on the bobbin surface. Eventually, equilibrium is reached and the bobbin neither takes on nor loses any water.

Properly used, a saturated salt sensor is accurate to ±1°C between dew point temperatures of −12 and +38°C. Outside these limits, small errors may occur as a result of multiple hydration characteristics of lithium chloride. These may produce ambiguous results at 41, −12, and −34°C dew points. Maximum errors at these ambiguity points are 1.4, 1.6, and 3.4°C, respectively, but actual errors encountered in typical applications are usually less.

Applications

The saturated salt sensor has certain advantages over other electrical humidity sensors, such as % RH instruments. Because the salt sensor operates as a current carrier saturated with Li and Cl ions, the addition of contaminating ions has little effect on its behavior compared to the situation for a typical RH sensor which operates "starved" of ions and is easily contaminated. A properly designed saturated salt sensor is not easily contaminated since, from an ionic standpoint, it can be considered precontaminated.

But, if a saturated salt sensor does become contaminated, it can be washed with an ordinary sudsy ammonia solution, rinsed, and recharged with lithium chloride. It is seldom necessary to discard a saturated salt sensor if proper maintenance procedures are observed.

Limitations of saturated salt sensors include (1) a relatively slow response time and (2) a lower limit to the measurement range imposed by the nature of lithium chloride. The sensor cannot be used to measure dew points when the vapor pressure of water is below the saturation vapor pressure of lithium chloride, which occurs at 11% RH. With certain gases, ambient temperatures can be reduced, increasing the RH to above 11%, but the extra effort needed to cool the gas usually warrants selection of a different type of sensor. Fortunately, a large number of scientific and industrial measurements fall above this limitation and are readily handled by the sensor.

Condensation-Type Hygrometers

The condensation-type dew point hygrometer is one of the most accurate and reliable instruments and offers the widest range of sensors for humidity measurements. These features are obtained, however, through increased complexity and cost. In the condensation-type hygrometer, a surface is cooled (thermoelectrically, mechanically, or chemically) until dew or frost begins to condense out. The condensate surface is maintained electronically in vapor pressure equilibrium with the surrounding gas, while surface condensation is detected by optical, electrical, or nuclear techniques. See Figs. 4 and 5. The surface temperature is then the dew point temperature, by definition.

The largest source of error in a condensation hygrometer stems from the difficulty in measuring the condensate surface temperature accurately. Typical industrial versions of the instrument are accurate to ±0.2°C over very wide temperature spans, and laboratory models offer accuracies up to ±0.1°C.

A wide span and minimal errors are the two main features. A properly designed condensation hygrometer can measure from dew points of 100°C down to frost points of −75°C.

The response time of a condensation dew point hygrometer is usually specified in terms of its cooling or heating rate, typically 1.5°C/s, making it considerably faster than a saturated salt dew point sensor and nearly as fast as most electrical % RH sensors. Perhaps the most significant feature of the condensation hygrometer is its fundamental measuring technique, which essentially renders the instrument self-calibrating. For calibration, it is only necessary to manually override the surface cooling control loop, causing the surface to heat, and witness that the instrument recools to the same dew point when the loop is closed. On the assumption that the surface temperature measuring system is calibrated, this is a reasonable and valid check on the instrument's performance.

Because of its fundamental nature and superior accuracy and repeatability, this kind of instrument is widely used as a secondary standard (National Bureau of Standards) for calibrating other lower level humidity instruments.

The inert construction of the condensation hygrometer makes it virtually indestructible. Although

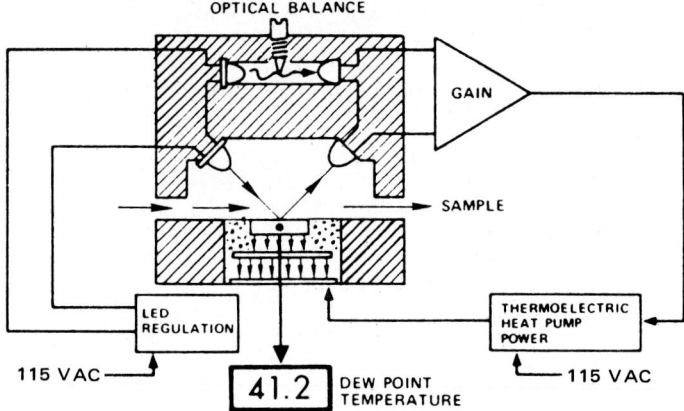

Fig. 4 Dew point is detected in a condensation hygrometer by cooling a surface until water begins to condense. Condensation is detected optically or electronically. The signal is fed into a control circuit which maintains the surface temperature at the precise dew point.

the instrument can become contaminated, it is easy to wash and return to service without impairment of performance or calibration.

The condensation (chilled mirror) hygrometer measures the dew or frost point temperature. Unfortunately, many applications require measurement of % RH, water vapor in parts per million, or some other humidity parameter. In such cases, the user must decide whether to employ the fundamental, high-accuracy condensation hygrometer and convert the dew or frost point measurement to the desired parameter, or to use lower level instrumentation to measure these parameters directly. In recent years, microprocessors have been developed which can be incorporated in the design of a

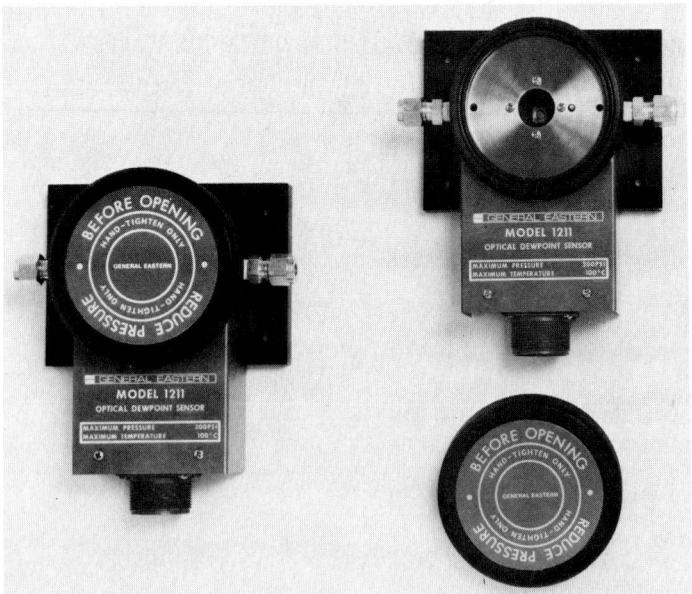

Fig. 5 Typical optical condensation (chilled mirror) detector. *(General Eastern Instruments.)*

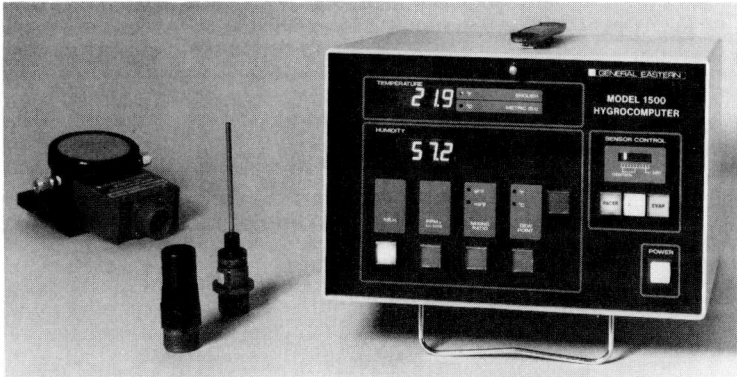

Fig. 6   Chilled mirror hygrometer with a built-in microprocessor to provide humidity data in various parameters, such as ppm$_v$, ppm$_w$, dew point, and % RH. *(General Eastern Instruments.)*

condensation hygrometer, thus resulting in instrumentation which can offer accurate measurements of humidity in terms of almost any humidity parameter. Two instruments of this type are shown in Figs. 6 and 7.

Electrolytic Hygrometer

A typical electrolytic hygrometer utilizes a cell coated with a thin film of phosphorus pentoxide (P_2O_5), which absorbs water from the sample gas. See Fig. 8. The cell has a bifilar winding of inert electrodes on a fluorinated hydrocarbon capillary. Direct current applied to the electrodes dissociates the water, which is absorbed by the P_2O_5, into hydrogen and oxygen. Two electrons are required to electrolyze each water molecule, and thus the current in the cell rep-

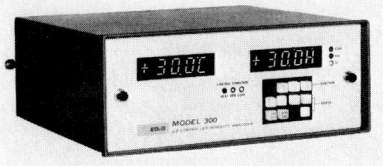

Fig. 7   Microprocessor-controlled humidity analyzer using a chilled mirror sensor. *(EG&G Environmental Equipment.)*

resents the number of molecules dissociated. A further calculation, based on flow rate, temperature, and current, yields the parts-per-million concentration of water vapor.

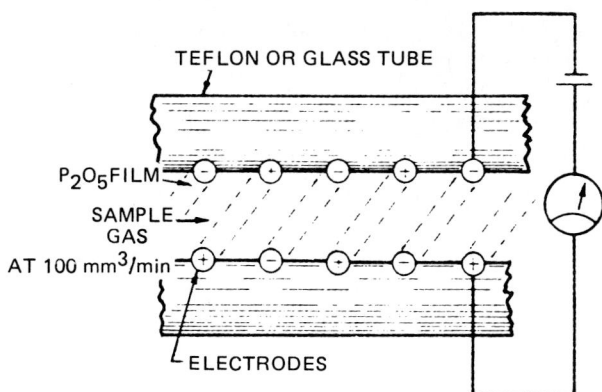

Fig. 8   An electrolytic hygrometer dissociates water, absorbed by P_2O_5, into hydrogen and oxygen by electrolysis. Since two electrons are required to electrolyze a molecule of water, the amount of current used by the hygrometer relates to parts per million of water vapor.

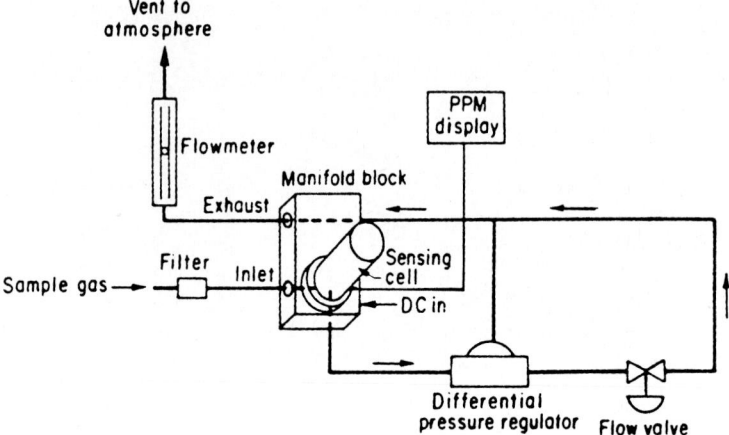

Fig. 9 Calculation of the water vapor content in an electrolytic hygrometer is dependent on precise control of the flow rate. This arrangement controls the sample pressure across the cell, ensuring correct flow regardless of input pressure fluctuations.

In order to obtain accurate data, the flow rate of the sample gas through the cell must be known and constant. Since the parts-per-million calculation is partially based on flow, an error in the flow rate causes a direct error in measurement.

A typical sampling system for ensuring constant flow is shown in Fig. 9. Constant pressure is maintained within the cell. Sample gas enters the inlet, passes through a stainless steel filter, and enters a stainless steel manifold block. It is very important that all components prior to the sensor be made of an inert material, such as stainless steel, to minimize contamination. After the sample gas passes through the sensor, its pressure is controlled by a differential-pressure regulator which compares the pressure of the gas leaving the sensor with the pressure of the gas venting to atmosphere through a preset valve and flowmeter. In this way, constant flow is maintained, even though there may be nominal pressure fluctuations at the inlet port.

A typical electrolytic hygrometer can cover a span from 0 to 2000 ppm with an accuracy of $\pm 5\%$ of the reading, more than adequate for most industrial applications. The sensor is suitable for most inert elemental gases and organic and inorganic gas compounds that do not react with P_2O_5.

Electrolytic hygrometers cannot be exposed to high water vapor levels for long periods of time, because this results in a high usage rate for the P_2O_5 and high cell currents.

Aluminum Oxide Moisture Sensor

This type of sensor is a capacitor, formed by depositing a layer of porous aluminum oxide on a conductive substrate and then coating the oxide with a thin film of gold. The conductive base and the gold layer become the capacitor's electrodes. Water vapor penetrates the gold layer and is absorbed by the porous oxidation layer. The number of water molecules absorbed determines the electrical impedance of the capacity, which is in turn a measure of water vapor pressure.

Advantages of the aluminum oxide sensor are (1) small size and suitability for *in situ* use; (2) economical use in multiple-sensor arrangements; (3) suitability for very low dew point levels without the need for sensor cooling (as required in condensation-type sensors); typically, dew points down to $-100°C$ can be measured without serious difficulty; and (4) a wide measurement span.

The aluminum oxide sensor has the following limitations. (1) It is a secondary measurement device and must be periodically calibrated to accommodate aging effects, hysteresis, and contamination, and (2) sensors require separate calibration curves, which are typically nonlinear.

Aluminum oxide humidity instruments are available in a variety of types, ranging from a low-cost single-point system, including portable battery-operated models, to multipoint microprocessor-based systems with an ability to compute and display humidity information in different parameters, such as dew point, parts per million, and % RH.

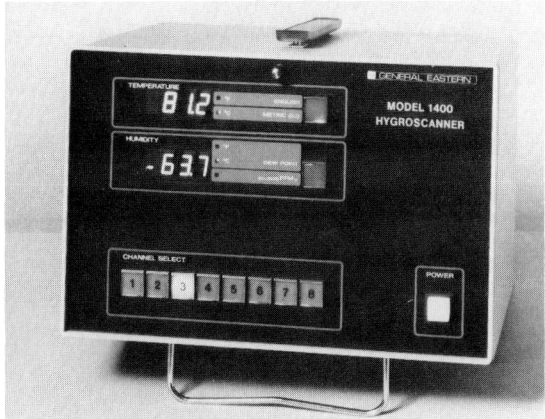

Fig. 10 Microprocessor-controlled multichannel humidity analyzer using aluminum oxide sensors. *(Hygroscanner, General Eastern Instruments.)*

The aluminum oxide sensor is also used for moisture measurements in liquids (hydrocarbons) and, because of its low power usage, it is suitable for use in explosion-proof installations. These sensors are frequently used in petrochemical applications where low dew points must be monitored on line and where the reduced accuracies and other limitations are acceptable. The advantages of the sensor must be weighed against the fact that its accuracy is lower than any of the fundamental measurement sensor types. As a secondary measurement device, it can provide reliable data only if kept in calibration and if damage due to incompatible contaminants is avoided.

REFERENCES

Arnold, J. H.: "The Theory of the Psychrometer," *Physics,* vol. 4, 1933, p. 255.

ASHVE: *Heating, Ventilating, Air Conditioning Guide,* American Society of Heating and Ventilating Engineers, New York, 1956, chap. 3.

ASRE: *Data Book,* 6th ed. American Society of Refrigeration Engineers, New York, 1949, chap. 4.

Carrier, W. H.: "Rational Psychrometric Formulae," *ASME Trans.,* vol. 33, 1911, p. 1005.

Carrier, W. H., and C. O., Mackey: "A Review of Existing Psychrometric Data in Relation to Practical Engineering Problems," *ASME Trans.,* vol. 59, 1937, p. 32.

Dunmore, F. D.: "An Improved Electrical Hygrometer," RP 1265, *J. Res. Natl. Bur. Stand.,* vol. 23, 1939, p. 701.

Harding, J. C., Jr.: "Dew-Point Hygrometer with Contaminant Error Immunity," *Meas. Control,* February 1979.

Hickes, W. F.: "Humidity Measurement by a New System," *Refrig. Eng.,* vol. 54, 1947, p. 351.

Irving, C. L., and C. T. Higgins: "HVAC System Design Narrows RH Control to ±1%," *Control Eng.,* vol. 27, no. 11, 1980, p. 150.

Ivory, J.: "On the Hygrometer by Evaporation," *Phil. Mag.,* vol. 60, no. 81, 1922.

Jennings, B. H., and A. Torloni: "Psychrometric Charts for Use at Altitudes Above Sea Level," *Refrig. Eng.,* vol. 62, 1954, p. 71.

Marvin, C. F.: "Psychrometric Tables," *Publ. 235,* U.S. Weather Bureau, 1900.

Ruskin, R. E., ed.: "Principles and Methods of Measuring Humidity in Gases," in *Proceedings of the 1st International Symposium on Humidity and Moisture Measurement and Control in Science and Industry, Washington, D.C.,* Van Nostrand Reinhold, New York, 1963.

Sherwood, T. K.: "The Curious History of the Wet Bulb Hygrometer," *Chem. Can.* June 1950, pp. 19–22.

Smithsonian Institute: *Smithsonian Meteorological Tables,* Washington, D.C., 1951.

Wexler, A., and W. G. Brombacher: "Methods of Measuring Humidity and Testing Hygrometers," *NBS Circ. 512,* National Bureau of Standards, Washington, D.C., 1951.

Whipple, F. W. J.: "The Theory of the Hair Hygrometer," *Trans. Phys. Soc.,* 1921–1922.

Wiederhold, P. R.: "Humidity Measurement. II. Hygrometry," *Instrum. Technol.* vol. 22, no. 8, 1975, p. 45.

Wildhack, W. A., ed.: "Fundamentals and Standards (Humidity)," in *Proceedings of the 1st International Symposium on Humidity and Moisture Measurement and Control in Science and Industry, Washington, D.C.,* Van Nostrand Reinhold, New York, 1963.

Zimmerman, O. T., and I. Lavine: *Psychrometric Tables and Charts,* Industrial Research Service, Dover, N.H., 1945.

Moisture Measurement Systems

With the exception of the close coupling of relative humidity measurement with moisture measurement, as in the case of some systems for moving-web moisture control, for example, the fundamental methods for measuring moisture in air and other gases are described in the preceding article Humidity Measurement in this Handbook section. This article is concerned principally with the measurement of moisture in solids and liquids.

Moisture content has a number of synonymous terms, several of which are specific to certain materials and applications. The water content of solid, granular, or liquid materials usually is referred to as moisture content on either a *wet basis* or a *dry basis*. For industrial measurements, the wet basis is most common. Wet basis refers to the quantity of water per unit weight or volume of wet material. The textile industry uses the dry basis in measuring the moisture content of textile fibers. Often referred to as the *regain* moisture content, the dry basis refers to the quantity of water in a material expressed as a percentage of the weight of bone dry material. The relationship between the wet and dry moisture content bases is shown in Fig. 1.

Instruments for moisture measurement may be considered *periodic* or *continuous*. Periodic instruments generally are automated versions of conventional laboratory moisture analysis procedures—with a measurement typically requiring 2 min or longer and sometimes up to 20 min. Thus, periodic instruments are quite impractical for automatic control systems. Moisture measuring instruments also may be classified by the operating principle involved. Instruments using electrical conductivity (dc or ac), absorption of electromagnetic energy (radio frequency regions), electrical capacitance (dielectric constant change), and infrared energy radiation are most adaptable to continuous measurement because of the comparatively fast speed of response. Instruments that use automatic oven drying, chemical titrations (Karl Fischer technique), equilibrium hygrometric methods, and distillation methods usually are of the intermittent periodic type.

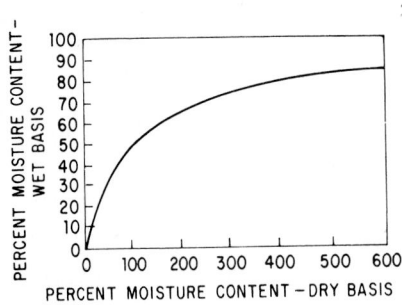

Fig. 1   Relation between dry and wet moisture content bases.

Electrical Conductivity Methods

The relationship between dc resistance and the moisture content of such materials as wood, textiles, paper, and grain is the physical basis of measurement. Specific resistance when plotted against moisture content results in an approximate straight line up to the moisture saturation point. Beyond saturation, where all the cells and intermediate spaces in the material are saturated with free water, electrical conductivity methods are not reliable. This point varies from about 12 to 25% moisture content, depending on the type of material. The sample under test is applied to suitable electrodes in the form of needle points for penetration into the material. Granular or fibrous materials may make use of electrodes in the form of a cup or clamp arrangement to confine the material to a fixed volume. The electrodes and sample under test comprise one arm of a Wheatstone bridge as shown in Fig. 2.

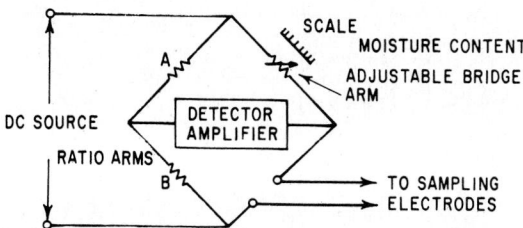

Fig. 2   Basic bridge circuit for the conductivity method of moisture content measurement.

The high sensitivity required of the detector dictates the use of electronic amplifiers. The range of resistance values corresponding to the normal moisture content varies from less than 1 to 10,000 MΩ or higher, dependent on material, electrode design, and moisture content. An increasing moisture content results in decreasing resistance values.

The effects of moisture on the electrical characteristics of various materials are shown in Figs. 3 through 6.

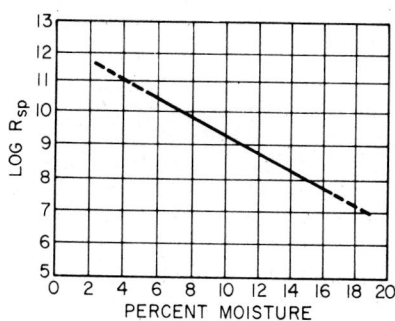

Fig. 3   Moisture content versus logarithm of specific resistance for spruce, basswood, fir, cedar, and hemlock. The plot results in practically a straight line. *(U.S. Forest Products Laboratory.)*

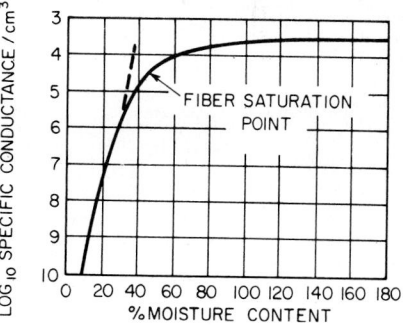

Fig. 4   Effect of moisture content on specific conductance of redwood. It will be observed that the change in the slope of the curve occurs at approximately 30% moisture content. This point is the fiber saturation point and correlates with other testing methods for fiber saturation, which have always been difficult to make. These data help to support the theory that a relatively linear relationship exists between moisture content and the logarithm of the specific resistance. *(U.S. Forest Products Laboratory.)*

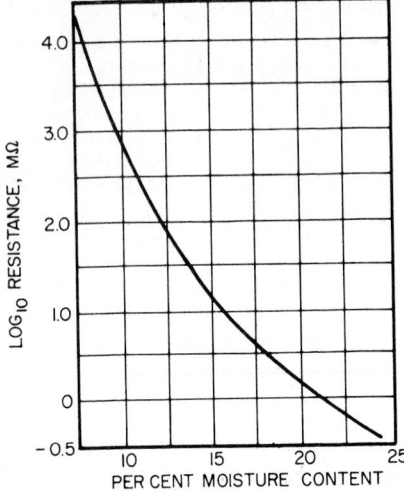

Fig. 5   Relation between moisture content and logarithm of resistance for Douglas fir. These data indicate some divergence from the straight-line relationship. *(General Electric.)*

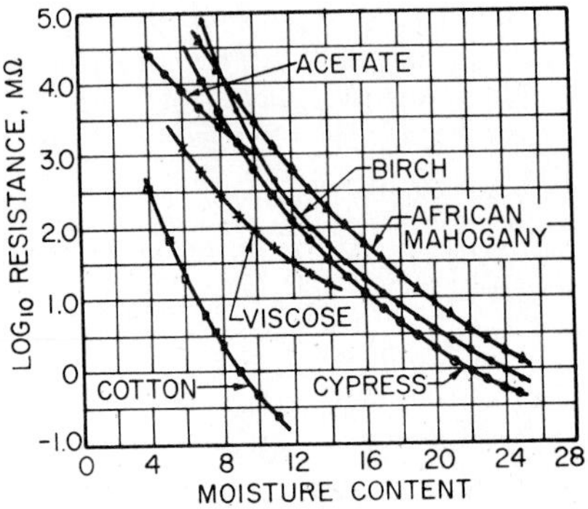

Fig. 6   Moisture content versus $\log_{10}(R_{SP})$ of woods and textile fibers. *(General Electric.)*

Electrical Capacitance Methods

The change in dielectric constant under dry and moist conditions of a material is the physical principle used. The dielectric constant of most vegetable organic materials is 2 to 5 when dry. Water has a dielectric constant of 80. Therefore, the addition of small amounts of moisture to such materials causes a considerable increase in the dielectric constant. The material being measured forms part of a capacitance bridge which has radio frequency (rf) power applied from an electronic oscillator as shown in Fig. 7. Electronic detectors measure bridge imbalance or frequency change, dependent on

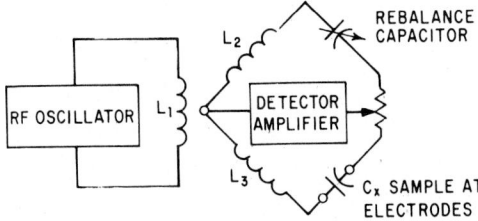

Fig. 7   Basic bridge circuit for capacitance method of moisture content measurement.

the method employed. Electrode design varies with the type of material under test. Parallel-plate electrodes are used for sheet materials, whereas cylindrical electrodes generally are used for liquids or powders. Moisture content measurable by electrical capacitance methods ranges from 2 to 3% to 15 to 20%, dependent on the product involved. See Figs. 8 and 9.

In an effort to increase the speed of measurement over that of mechanical scanning systems, a system* which uses an electronically scanned multielectrode array for determining the instantaneous

*F. P. Research Laboratory, Inc., Bellevue, Wash.

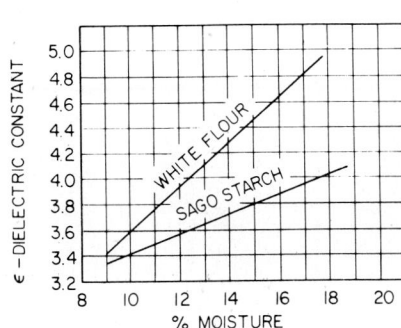

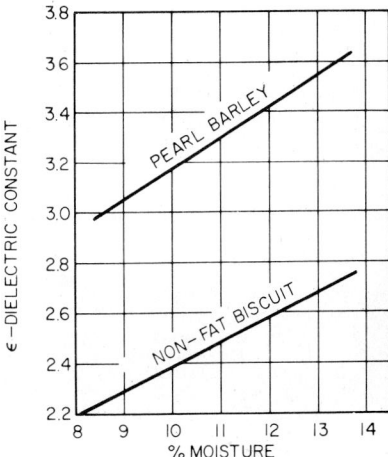

Fig. 8 Moisture content versus dielectric constant for starch and flour. *(J. Soc. Chem. Ind., London.)*

Fig. 9 Moisture content versus dielectric constant for biscuit and barley. *(J. Soc. Chem. Ind., London.)*

moisture profile in a paper web was developed. The system is typically used at the dry end of a paper machine or coater on most grades of paper from newsprint to boxboard and pulp sheets. The sensor array normally consists of 8 to 16 individual electrode elements which are in light contact with the paper web. Each sensing element consists of a center electrode, and the surrounding grounded electrodes are separated by an insulator. A high-frequency sensing field is applied to the center electrode to measure the amount of field coupling through the paper web to the grounded electrode. This value is basically related to the dielectric coefficient of the paper. It is claimed that the sensor can operate on materials up to 0.5 in (13 mm) in thickness without a serious drop in sensitivity. It is reported that the sensor measures the average moisture content of multilayer sheets, such as those produced on multicylinder machines.

As reported, the system scans up to 1000 points per second. It is possible to pick out a particular wet streak and use it in the control strategy, or to delete particular wet streaks as desired.

Because of the commerical importance of moisture content in agricultural products, much effort in recent years has been directed toward grain and other moisture meters, including battery-operated portable instruments that assist not only grain elevator operators and processors but growers as well, and in some instances moisture detectors have been attached to harvesting equipment.

Similarly, much research and development activity has been directed toward soil moisture measurement. In one system,* the sensor is basically a liquid crystal oscillator (30 MHz) arranged so that its frequency is sensitive to the capacitance of external electrodes which compose the walls of the case. A fringing field penetrates into the soil in the region of the insulating gap between the electrodes. See Fig. 10. The measurement region sensitive to soil moisture is indicated. Oscillator output is coupled to the coaxial line which also carries the dc power supply and control pulses. See Fig. 11.

Upon digital command (advance pulse), the electrode is electrically disconnected (SW-1) from the oscillator so that the oscillator base frequency can be measured. Long-term drifts in oscillator frequency are compensated for by recording the frequency difference as the soil capacitance is switched in. Both switches are controlled by a modulus four (2-bit) counter which is cleared by the reset pulse and clocked by the advance pulse. The sensor switch states produced by sensing one or more advance pulses (following reset) are indicated in the table shown in Fig. 11. The system is described in considerable detail by Wobschall (1978) in the paper listed at the end of this article. A typical calibration curve is given in Fig. 12.

*ICE/Electronics, Williamsville, New York.

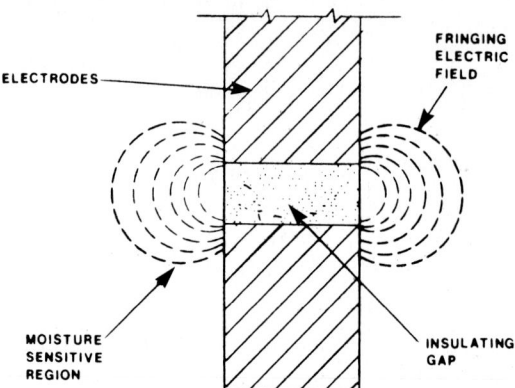

Fig. 10 Electrodes used in a soil moisture measurement system. Only moisture in the vicinity of the sensor insulating gap (½ in, 13 mm) is detected by the sensor. The sensor response corresponds to the average moisture within one or two gap diameters. For permanent installation, the hole is refilled with the cable left in place. The sensor will operate under water. A waterproof sheath (shrink tube) can be put over the connector and cable end to protect against long-term corrosion. Water seepage into the connector does not short out or affect operation of the sensor directly, and therefore waterproofing is not required for temporary installations. *(IS2-C1, ICE/Electronics.)*

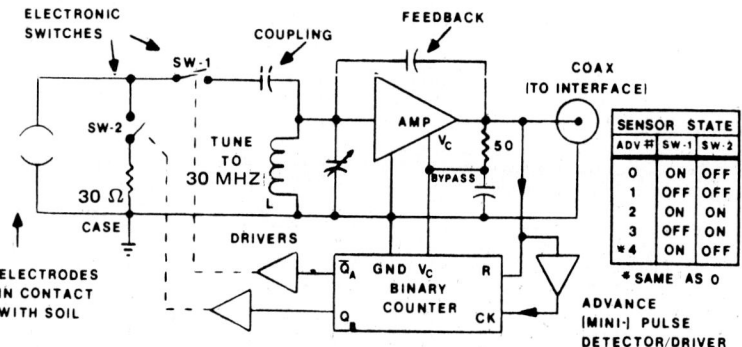

Fig. 11 Sensor electronics functional diagram. *(IS3-C1, ICE/Electronics.)*

Aluminum Oxide Impedance Sensor*

The use of an aluminum oxide sensor for the measurement of water vapor in air is described briefly in the preceding article in this Handbook section. These sensors are also effective detectors of moisture in liquids. See Figs. 13 and 14. The aluminum oxide sensor consists of an aluminum strip anodized to provide a porous oxide layer. A very thin coating of gold is evaporated over this structure. The aluminum base and the gold layer form the two electrodes of what is essentially an aluminum oxide capacitor. Water vapor is rapidly transported through the gold layer and adsorbed by the pore walls of the sensor. The number of water molecules so adsorbed is proportional to the fugacity† of the water in the liquid. At equilibrium conditions, the fugacity of the water in the liquid is equivalent

---

*Information furnished by Panametrics, Inc.

†In thermodynamics, fugacity is a measure of the tendency of a substance to escape by some chemical process from the phase in which it exists.

to the water vapor pressure above the liquid. In most cases it is directly proportional to the concentration of dissolved water in the organic liquid.

Since the pore openings are small in relation to the size of most organic molecules, admission into the pore cavity is limited to small molecules, such as water. The number of water molecules adsorbed determines the pore wall conductivity which influences the value of the sensor's electrical impedance. The surface of an aluminum oxide sensor can then be viewed as a semipermeable structure allowing the measurement of water vapor pressure in liquid organics in the same manner as is accomplished in a gaseous medium. In general, the aluminum oxide transducer provides identical values of electrical output whether directly immersed in the fluid or placed in the gas space immediately above; i.e., the measured vapor pressure is identical in both phases.

The measured vapor pressure can be used to determine the weight content of dissolved water within an organic liquid by application of Henry's law, which states that the mass of gas dissolved by a given volume of solvent, at constant temperature, is directly proportional to the pressure of the gas with which it is in equilibrium. The concentration of water in an organic liquid, expressed in parts per million by weight or in milligrams per liter, equals the partial pressure of water vapor times a constant:

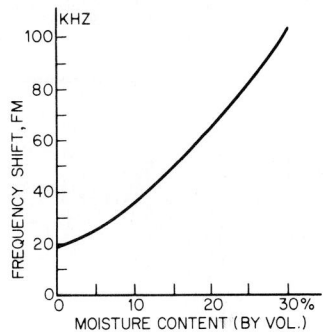

Fig. 12   Typical calibration curve for a soil moisture sensor. *(Series IS3, ICE/Electronics.)*

$$C_W = KP_W$$

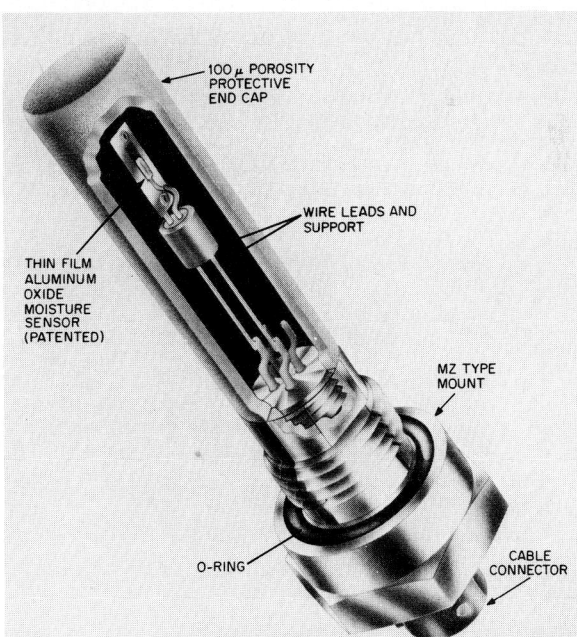

Fig. 13   Aluminum oxide sensor with the principal elements indicated. The impedance of the probe ranges from 2 MΩ to 50 kΩ at 77 Hz, depending on the vapor pressure of water. *(Panametrics.)*

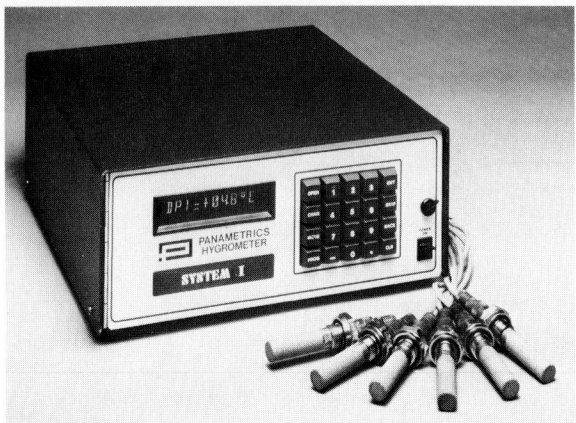

Fig. 14 Hygrometer employing an aluminum oxide sensor and capable of measuring moisture in gases and liquids. *(System I, Panametrics.)*

Thus a measurement of the water vapor pressure of the liquid and knowledge of Henry's law constant K permit direct calculation of the absolute moisture content. For liquids to which Henry's law is not applicable, an empirical calibration of the measured meter reading versus the known moisture content can be made.

It is generally possible to use an aluminum oxide sensor and Henry's law to determine directly either the absolute moisture content or the percent saturation of a nonpolar liquid. For nonpolar liquids, the above equation can be written

$$C_S = KP_S$$

where the subscript S refers to the saturation concentration and the saturation vapor pressure of water, i.e., the values of concentration and water vapor pressure such that, when additional water is added, free water forms.

After eliminating K from the prior equations, it can be seen that the ratio of the actual vapor pressure of water to its saturation vapor pressure at the same temperature can be expressed as the percent saturation of the liquid. Thus,

$$\% \text{ Saturation} = \frac{C}{C_S} \times 100 = \frac{P_W}{P_S} \times 100$$

A measurement of the dew point (vapor pressure of water) and knowledge of Henry's law constant K allows the moisture content of the liquid to be calculated directly.

Saturation values for water in a number of organic liquids are available in the literature. If the saturation value for a particular organic liquid cannot be found, it can be determined experimentally.

The only measurements required to determine the absolute moisture content of the liquid are the dew point as measured with an aluminum oxide sensor and the temperature of the liquid. The moisture can then be calculated from

$$C_W = KP_W = \frac{C_S}{P_S} P_W$$

As K is generally a function of temperature, the values of C_S and P_S should be ascertained at the temperature of measurement. For simple straight-chain hydrocarbons, the temperature coefficient of K is in most cases small enough to be ignored for measurements in the vicinity of room temperature.

A Karl Fischer titration analysis can be used as the transferable standard for comparison with an aluminum oxide sensor.

The principal application of aluminum oxide sensors is in the measurement of moisture concentration in liquid process streams. The sensor operates satisfactorily in a stream containing either a pure hydrocarbon or a mixture of hydrocarbons made up of only carbon and hydrogen atoms. For such liquids, an analysis based on Henry's law is generally applicable. The sensor also operates satisfactorily in various classes of organic liquids, including halogenated hydrocarbons, esters, amides, and alcohols, among others. The user should be aware of the fact that some of these compounds may hydrolyze or decompose under certain operating conditions, with possible harm to the probe. For example, perchloroethylene in the presence of water hydrolyzes readily to form hydrochloric acid which is harmful to the probe.

Common liquids in which the aluminum oxide probe performs well include:

Butadiene, butene propylene, ethylene dodecane, and styrene in polymerization reactions where moisture inhibits polymerization if present

Liquid natural gas, ethane, and propane as found in fuel transport systems where moisture can cause freeze-ups and reduce the heat content of the fuels

Cyclohexane, tetrahydrofuran, and diethyl ether, all of which are solvents for pharmaceuticals, polymer intermediates, and dyes—where the presence of moisture inhibits reactions

Periodic calibration of an aluminum oxide system is suggested at 6-month intervals unless the measured liquid contains acidic, basic, or other contaminants, in which case more frequent calibration may be required. For liquids to which a Henry's law analysis is applicable, calibration can be carried out in the gaseous phase, using dilution techniques, but if the liquids are conductive, calibration must be confined to the liquid phase.

These sensors are not recommended for use where acid or base concentrations exceed 10 ppm_w (weight), unless the water concentration is well below 10 ppm_w, and in the exceptional case of hydrofluoric acid (which is particularly corrosive to these sensors) above the 1 ppm_w level. Salts, mercury, and suspended conductive solid particles (carbon, iron, etc.) cause deterioration or a low-resistance path on the sensor surface, leading to erroneously high readings. Sensors should not be exposed to ambient temperatures above 70°C. Although the sensor is not flow-rate-dependent, mechanical damage can occur where flow rates exceed those of a linear velocity of 50 cm/s. Excessive flow rates are avoided by installing the sensor in a suitable bypass.

Pressurization of the liquid has no significant effect on the sensor. Probes that can physically withstand pressures up to 5000 psig (34.5 MPa) are available.

The range of the sensor for Henry's-law-type liquids is from saturation (100 to 1000 ppm_w) to less than 1 ppm_w. The theoretical lower limit for these liquids is much greater than the level to which they can be dried in practical process streams. In conductive liquids, the range is from approximately 1% by weight to 10 ppm_w. Actual range depends largely on the conductivity of the liquid.

The accuracy of the aluminum oxide sensor depends on the accuracy to which the saturation value of the liquid is known and/or the calibration accuracy. Accuracies of ± 20 to $\pm 25\%$ of reading can be achieved at moisture concentrations of 1 ppm_w or less.

RF Absorption

The attenuation of electromagnetic energy when passed through a material is the physical principle used. Radio frequency devices are operated at frequencies below 10 MHz, and for best results the products being measured must be made up of polar materials. The rf energy is passed through the polar material, and the water molecules absorb some of the energy in the form of molecular motion. A block diagram of a typical rf instrument is shown in Fig. 15. By means of suitable electrodes for the sampling process, the instrument can be applied to solid or sheet materials. Granular materials can be placed in cell-type electrodes. The test results will be affected by any polar material in the sample. Thus, the results are nonspecific to water. However, water of hydration in a product will not be detected. Instruments of this type must be calibrated to the particular product under test. Although these instruments usually are available as periodic devices, they can be adapted for continuous measurements on some products. The moisture content range is dependent on the material and ranges from 0.1 to 60%. For certain materials, the range is quite narrow.

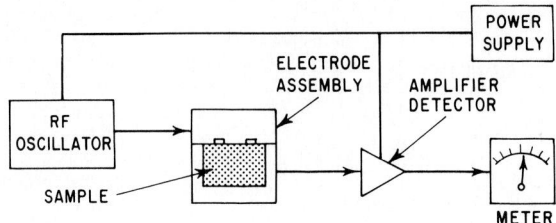

Fig. 15   Radio frequency absorption-type moisture indicator.

Microwave Absorption

The frequency of the electromagnetic energy used in these devices is in the region of 1000 MHz and higher. The 2.45-GHz (S band), the 8.9 to 10.68 GHz (X band), and the 20.3 to 22.3 GHz (K band) regions all have been used. Water molecules greatly attenuate the transmitted signal with respect to other molecules in the material in the S- and X-band frequencies. In the K-band frequencies the water molecule produces molecular resonance. There are no other molecules that respond to this particular resonant frequency. Thus, the frequency is most specific to the moisture (free water) content in paper products. The wavelengths of the K-band frequencies are approximately 1.35 to 1.5 cm. At these very short wavelengths, the energy may be guided or transmitted by waveguides. A block diagram of a microwave moisture instrument is shown in Fig. 16. Energy

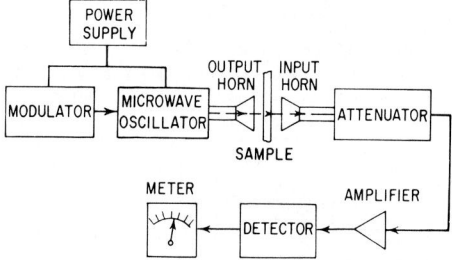

Fig. 16   Microwave absorption-type moisture indicator.

absorption can be detected by attenuation (loss) or by phase shift methods, or both measurements can be made simultaneously. The useful range of these devices is from less than 1 to 70% moisture and higher. At low moisture contents, S- and X-band devices require a minimum mass of material for practical operation. The accuracy of these instruments is within ±0.5% of the indicated value— up to approximately 15% moisture content.

Combined Microwave and Gamma Absorption

In one system,* a noncontacting, on-line moisture detector combines a microwave and gamma absorption signal. Microwave energy, strongly absorbed by water, is indicated by a power drop in the microwave signal. Absorption of the gamma energy accurately measures the mass of the material in the process. The percent moisture is then calculated by dividing the water units (the microwave signal) by the mass units (the gamma signal). See Fig. 17. The combined signal automatically compensates for changes in product density. The product temperature, having a second-order effect on the final calculation, is factored into the microwave signal to maintain accuracy. Outputs (standard and optional) available with this system include: (1) digital percent moisture and density display; (2) analog percent moisture and density signals (0 to 5 V, 5 mA maximum; 4 to 20 mA, 250 Ω maximum (floating); and isolated 4 to 20 mA, 500 Ω max; (3) three-mode (proportional integral derivative)

*Kay-Ray Inc., Model 8100, Arlington Heights, Ill.

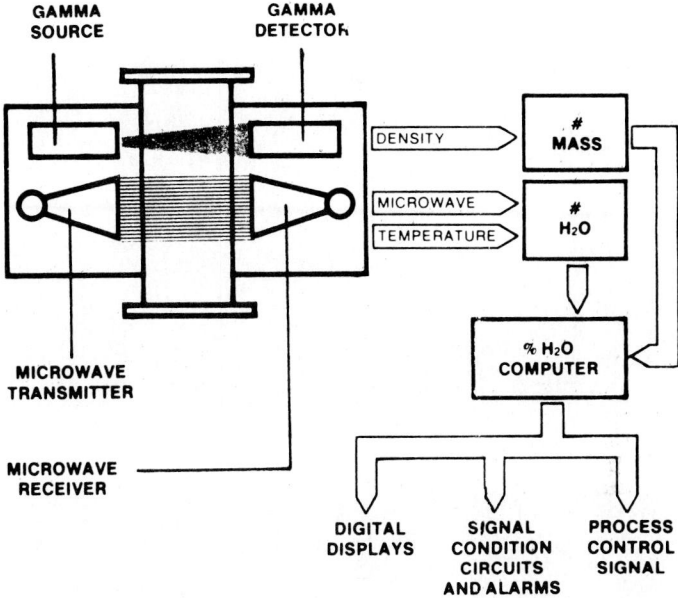

Fig. 17   Combined microwave and gamma absorption system which simultaneously measures moisture and density. Percent moisture is calculated by dividing the water units (microwave signal) by the mass units (gamma signals). *(8100 Systems, Kay-Ray.)*

controller (0 to 5 V or 4 to 20 mA floating); (4) high-low moisture alarm contacts; (5) high-low density alarm contacts; and (6) system alarm contact. Accuracy claimed is up to $\pm 0.2\%$ moisture (which varies with application). Product conditions are (1) 2 to 30% moisture, (2) 10 to 50 lb/ft^3 (160 to 801 kg/cm^3), and (3) 50 to 180°F (10 to 82°C) operating temperature range. The use of the system in three representative applications is shown in Fig. 18.

Infrared Absorption

The water molecule becomes resonant at certain infrared frequencies, and thus the amount of energy absorbed by the water absorption band is a measure of moisture content. One device for calculating the moisture content of sheet paper uses infrared energy at two different wavelengths and measures the differential absorption. Infrared (IR) radiation is passed through a rotating disk chopper that contains two filters. One filter passes a wavelength of 1.94 μm, which covers the water absorption band; the second filter passes a band of 1.80 μm, which is not significantly affected by moisture content. A block diagram of the instrument is shown in Fig. 19. The paper sheet is exposed to dual-frequency narrowband radiation, and the reflected radiation is collected in an integrating sphere where it activates a lead sulfide detector. The signals are amplified, and the ratio of the two narrowband signals is read out directly as moisture content. The system is quite immune to the effect of other paper variables, such as basis weight, coatings, density, and composition. The useful range is from 1 to 12%.

An infrared instrument useful for gas and liquid measurements operates on the principle of selective energy absorption of the sample when compared with a reference material having little or no absorption. The instrument contains two infrared sources—one source is directed through a cell that contains the sample material, and the second is directed through the reference cell. In the case of gas analysis the reference cell is filled with an inert gas or air. A detector unit consisting of two cavities separated by a metallic diaphragm, which acts as a variable capacitor with respect to a fixed plate within the cavity, is used. The major components of the instrument are shown in Fig. 20. Operation involves chopping their energy at 10 Hz to the sample and reference cells. The gas or fluid in the

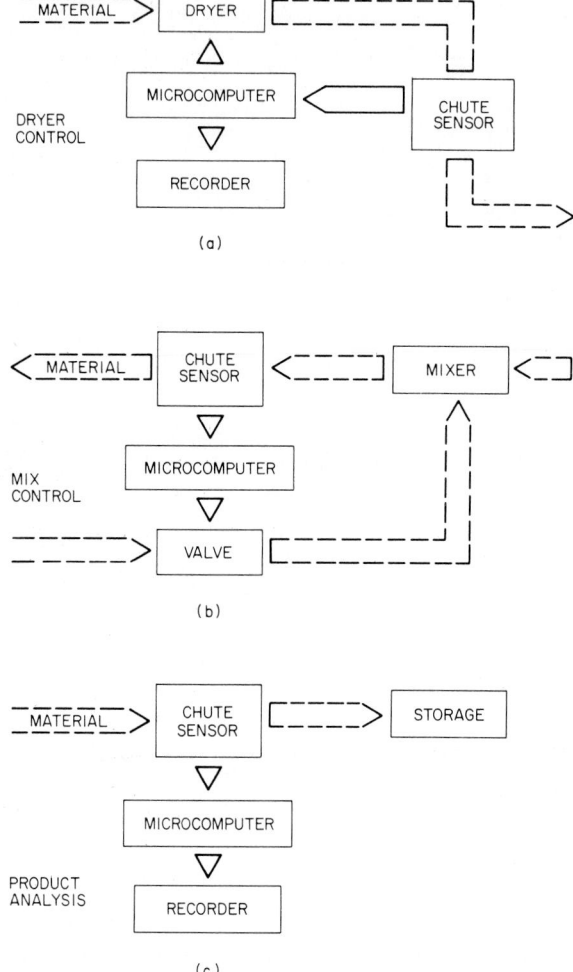

Fig. 18 Representative applications of a microwave–gamma absorption moisture measurement system. (*a*) In dryer control, the output is monitored for moisture content and the control signal is fed back to the dryer controller—useful for such products as dried pet foods, cereals, grains, nuts, and fertilizer. (*b*) In the mixer control, the addition of water to the process is controlled by measuring the input material and instantaneously feeding the percent moisture signal to a water addition valve. This is useful in chemical processes, wheat tempering, and ground roast coffee manufacture, among others. (*c*) Continuous indication of moisture content as well as density eliminates need for certain laboratory procedures—useful in soybean and grain storage, coking coal, and processing of green beans, among others. *(System 8100, Kay-Ray.)*

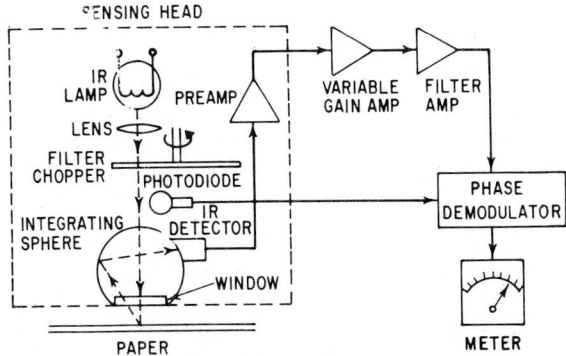

Fig. 19 Infrared backscatter moisture indicator.

sample cell absorbs light at a wavelength characteristic of the material, the amount being proportional to the concentration. The reference cell does not absorb any appreciable IR energy. The IR energy passes through the cells and into the detector, which causes the temperature of the gas in the detector to rise. However, if some of the energy has been absorbed by the sample cell, the sample side of the detector will receive less energy than the reference half of the detector. The temperature on either side of the diaphragm causes a definite pressure—with the diaphragm moving in a direction that tends to equalize the two pressures. This movement in turn causes the capacitance between the diaphragm and its sensing plate to change. This change is used to modulate an oscillator. The chopper alternately blocks and then passes the energy from each IR source. Each time the energy is blocked, the pressures equalize in the detector compartments. Then, alternately, the energy is permitted to enter the detector cavities. The modulating frequency is 10 Hz, and the magnitude of capacitance change is related directly to the amount of energy absorbed in the sample cell. The signal from the

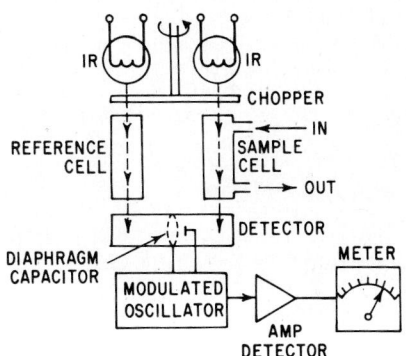

Fig. 20 Infrared gas or liquid analyzer.

oscillator is amplified and demodulated for meter indication or applied to a recorder. The range of moisture in gases is from 1000 ppm to 100% water vapor, and for fluids, from 6 to 100% water. The response speed is 0.5 s for 90% of full-scale output.

Vibrating Quartz Crystal

In one system,* the vibrational frequency changes of a hygroscopically sensitized quartz crystal are measured. The sample gas is divided into two streams. One stream is passed directly to the measuring crystal; the other is passed through a dryer and then to the same measuring crystal, alternating with the first stream every 30 s. The wet and dry vibrational frequencies are electronically monitored and compared to the frequency of an uncoated sealed reference crystal. These data are further processed, and the correct moisture level is computed. The instrument is available with a programmable microprocessor (programmable read-only memory) to furnish an up-to-date display and binary-coded decimal (BCD) output compatible with computer processing. The instrument is designed for on-line monitoring of noncorrosive process gases and measures moisture levels as low as 1 ppm in such gases as hydrogen, ethylene, refrigerants, and natural gas. The standard output is in BCD form, with

*DuPont Analytical Instruments Division, Wilmington, Del.

digital readout, directly in parts per million (volume) or pounds per cubic foot (kilograms per cubic meter). Optionally, the output can be converted to a continuous analog signal of 0 to 10 mV, or 4 to 20 mA direct reading, updated every 30 s for display on a strip chart or used as an alarm or control signal. The control module can be located up to 2000 ft (610 m) from the sensor.

Laboratory Moisture Determinations

Karl Fischer Technique

This method offers high sensitivity (5 ppm) over a wide range of moisture content (10 ppm to 100%). Accuracy to ±1% or better can be achieved. The method involves titration of the sample using Karl Fischer reagent (a solution of iodine, sulfur dioxide, and pyridine in methanol or methyl Cellosolve). As with other titrations (see the article Photometric and Reaction Product Analyzers in Sec. 6 of this Handbook), Karl Fischer titrations have become highly automated in recent years. In one system,* an amperometric system controls titrant delivery to a preset end point automatically. The system detects the end point by measurement of current through a solution, using a dual platinum electrode and a dc voltage source. A panel control adjusts the electrode polarization current from 0 to 50 mA. A reference button allows the use of a microammeter to reset the desired polarization current precisely. The circuitry used is all solid state. The buret can be controlled automatically or manually. A continuous light emitting diode (LED) readout digitally indicates the amount of dispensed titrant to the nearest 0.01 ml. The rate of change of the digital display gives an indication of buret delivery rate. With the use of a proportional end point approach, at a selected proximity to the end point, the titrant delivery rate tapers off to a preset minimum, thus minimizing overshoot. The apparatus accepts liquid, solid, or gas samples, and consecutive samples may be introduced manually into the reaction vessel. Several titrations, depending on volume, can be performed before evacuation of the reaction vessel is required. Reaction vessel capacity is 400 ml. An end point timer has a range of 0 to 100 s.

Distillation Methods

The water is separated from the liquid sample by virtue of the differences in boiling points of the other liquids and water. The water fraction is collected carefully and measured volumetrically.

Oven Drying

The oldest and still a very common analytical method is that of determining moisture content by heating the sample to ensure complete drying. Moisture is calculated on the basis of loss in weight between the original and the dried sample. Semiautomatic drying and weighing ovens are available in which the moisture content is indicated directly by a weighing scale built into the oven. Other volatile materials interfere with the method and must be accounted for to determine how much weight loss is due specifically to water evaporation.

Equilibrium Methods

The equilibrium moisture content of the air at the surface of a material is representative of the moisture within the material. This is particularly true of hygroscopic granular and fibrous materials. The instrumentation involves the use of humidity measuring devices.

REFERENCES

Bailey, S. J.: "Instrumentation for On-Line Moisture Measurement," *Control Eng.,* vol. 26, no. 10, October 1979, p. 34.
Bailey, S. J.: "Moisture Sensors," *Control Eng.,* vol. 27, no. 9, 1980, p. 112.

*K-F Titrimeter, Fisher Scientific Company, Pittsburgh, Pa.

Cucchiara, O.: "The Measurement of Dissolved Water in Organic Liquids Using an Aluminum Oxide Hygrometer," *Anal. Instrum.,* vol. 15, no. 1, 1977.

Cucchiara, O.: "Aluminum Oxide Hygrometry in Gases," *Anal. Instrum., Ser. 17,* "Measurement Technology in the '80s," 1979, pp. 182–186.

Gardner, R. C.: "Moisture Basis Weight Infrared Gage for Paper," *Instrum. Technol.,* vol. 15, no. 1, 1958, p. 51–54.

Glasstone, S.: *Thermodynamics for Chemists,* Van Nostrand Reinhold, New York, 1947.

Gumpert, R., and I. E. Pakulis: "Three Moisture Measurement Techniques," *Instrum. Technol.,* vol. 28, no. 8, 1981, p. 43.

Kraft, P. L.: "Nine Ways to Measure Moisture," *Control Eng.,* vol. 16, no. 7, 1969, pp. 86–89.

Polta, R. C., Anderson, C. T., and D. A. Stule: "Monitoring Moisture in Dewatered Sludge," *Instrum. Technol.,* vol. 27, no. 6, 1980, p. 45.

Reim, T. E.: "Nuclear Gage Controls Process Moisture," *Instrum. Technol.,* vol. 14, no. 7, 1967, p. 47–51.

Shinkskey, F. G.: "How to Control Product Dryness Without Measuring It," *Instrum. Technol.,* vol. 15, no. 9, 1968, p. 47–51.

Sinn, P. N. (Ed.): "Principles and Methods of Measuring Moisture in Liquids and Solids," in *Proceedings of the 1st International Symposium on Humidity and Moisture Measurement Control in Science and Industry, Washington, D.C.,* Van Nostrand Reinhold, New York, 1963.

Staff: "Electronically Scanned Moisture Meter Replaces Mechanically Scanned Version," *Control Eng.,* vol. 26, no. 6, 1979, p. 52.

Staff: "Moisture Analyzer Helps Extend Period Between Dryer Column Recharging," *Control Eng.,* vol. 27, no. 8, 1980, p. 31.

Wobschall, D.: "A Frequency Shift Dielectric Soil Moisture Sensor," *IEEE Trans. Geosci. Electron.,* GE-16, 1978, pp. 112–118.

Zagorzycki, P. E.: "Automatic Control of Conveyor Dryers," *Chem. Eng. Prog.,* vol. 75, no. 4, April 1979, p. 50.

Rheological Systems*

William Carlson. *DeZurik, Unit of General Signal Corporation, Sartell, Minnesota. (Consistency Measurements)*

David W. Howard. *Brookfield Engineering Laboratories, Inc., Stoughton, Massachusetts. (Viscosity Measurements)*

Ralph Klucken. *Technical Association of the Pulp and Paper Industry, Atlanta, Georgia. (Consistency Measurements)*

A. L. Landesman. *Woodstock, New York. (Consistency Measurements)*

John R. Lavigne. *The Foxboro Company, Foxboro, Massachusetts. (Consistency Measurements Bibliography)*

Gregory Neeb. *Formerly, DeZurik, Unit of General Signal Corporation, Sartell, Minnesota. (Consistency Measurements)*

Robert A. Norcross. *Norcross Corporation, Newton, Massachusetts. (Viscosity Measurements)*

Eugene Norman. *Green Bay, Wisconsin. (Consistency Measurements)*

James E. Overall. *Georgia-Pacific Corporation, Atlanta, Georgia. (Consistency Measurements)*

R. R. Vokoun. *Seiscor Division, Seismograph Service Corporation, Tulsa, Oklahoma. (Rheological Measurements)*

---

*Persons who authored complete articles or subsections of articles, or who otherwise cooperated in an outstanding manner in furnishing information and helpful counsel to the editorial staff.

Viscosity Systems

Rheology is the study of the internal mechanics of the flow of liquids and suspensions. Substances that undergo continuous deformation when subjected to a shear stress are said to exhibit fluid characteristics. There are two rheological variables which frequently must be measured in the laboratory and controlled in manufacturing processes—viscosity and consistency. The measurement and control of consistency is described separately in the next article.

Classification of Substances

Fundamentally, liquids or suspensions in liquids when subjected to a shear stress behave in two ways, and this is a basis for classification: (1) When a *newtonian substance* undergoes deformation, the ratio of shear rate (flow) to shear stress (force) is *constant*. (2) With a *non-newtonian substance*, this ratio is *not constant*. These two types of behavior are illustrated in Figs. 1 and 2.

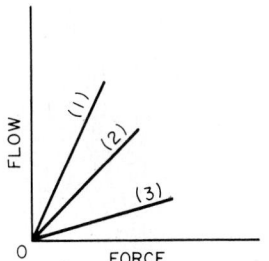

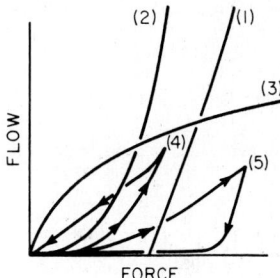

Fig. 1 Behavior of newtonian substances.

Fig. 2 Behavior of non-newtonian substances. (1) True plastic (Bingham body), (2) pseudoplastic, (3) dilatant, (4) thixotropic, (5) rheopectic.

Non-newtonian substances are further classified according to the general pattern of functional dependence of rate of shear on shear stress (and in some cases time, and occasionally frequency of vibrational agitation). Five principal subclassifications are illustrated in Fig. 2. Examples of specific classes of fluids are given in Table 1. These definitions include:

True Plastic (Bingham Body). This substance when subjected to a constantly increasing shear stress flows only after a definite yield point has been exceeded.

Pseudoplastic. This material appears to have a yield stress beyond which flow commences and increases sharply with an increase in stress. In practice, such substances are found to exhibit flow at all shear stresses, although the ratio of flow to force increases negligibly until the force exceeds the apparent yield stress.

Dilatant. The initial flow of this substance under a low shear stress is at a high rate; further increases in shear stress result in a lower flow rate. This substance is sometimes termed inverted plasic or inverted pseudoplastic.

Thixotropic. Flow rate of this substance increases with increasing duration of agitation, as well as with increased shear stress. When agitation is stopped, internal shear stress exhibits hysteresis. On reagitation, generally less force is required to create a given flow than is required for the first agitation.

Table 1 Examples of Common Fluids Exhibiting Diverse Rheologic Characterics

	Non-newtonian				
Newtonian	Pseudoplastic	Plastic	Thixotropic*	Rheopectic	Dilatant*
Water Most mineral oils	Catsup	Chewing gum Tar	Silica gel Most paints	Bentonite sols Gypsum in water	Quicksand Peanut butter
Gasoline	Printers' ink, paper pulp	High concentrations of asbestine in oil	Glue, molasses, lard, fruit juice concentrates		Many candy compounds
Kerosene Most salt solutions in water Light suspensions of dye stuffs			Asphalts		

*Some liquids may change from thixotropic to dilatant, or vice versa, as the temperature or concentration changes.

Rheopectic. If certain thixotropic suspensions are rhythmically shaken or tapped, they will "set" or build up very rapidly, a phenomenon termed *rheopexy*. Apparent viscosity of a rheopectic substance increases with time (duration of agitation) at any constant shear rate.

Viscosity Relationships

The basic relationship, as given by Newton, is

$$\frac{F}{A} = \mu \frac{V}{L} \tag{1}$$

where F/A = shear stress (force per unit area)
V/L = shear rate (velocity per unit thickness of layer)
μ = proportionality constant

Thus, viscosity came to be defined as

$$\mu = \frac{\text{shear stress}}{\text{shear rate}} \tag{2}$$

Hagen-Poiseuille Law

Hagen (Germany) and Poiseuille (France) described viscosity as the ratio of shear stress to shear rate at the wall of a capillary tube:

$$\mu = \frac{PR/2L \text{ (shear stress)}}{4Q/\pi R^3 \text{ (shear rate)}} = \frac{\pi PR^4}{8QL} \tag{3}$$

where P = pressure differential across liquid in tube
R = inside radius of tube
L = length of tube
Q = volume rate of flow of liquid
μ = viscosity

This equation is limited to conditions of laminar or viscous (streamline) flow.

Viscosity Units

From Eq. 3 came the unit of viscosity—originally in cgs units—a measure of the internal resistance of a material to flow: the ratio of the applied shear stress to the rate of shear. The SI unit of viscosity is pascal second ($Pa \cdot s$). The cgs unit is poise (P); 1 P = 1 $dyn \cdot s/cm^2$ and is equivalent to 0.1 $Pa \cdot s$. Values of viscosity are commonly expressed in centipoise (cP); 1P $\times$ 100 is equivalent to 1 millipascal second ($mPa \cdot s$). The viscosity of water at 68°F (20°C) is 1 cP.

Kinematic Viscosity Units

Kinematic viscosity is a measure of the internal resistance to flow of a liquid under gravity: the ratio of the viscosity to the density of a liquid. The SI unit of kinematic viscosity is square meter per second (m^2/s). The cgs unit is stoke (St); 1 St = 1 cm^2/s; 1 St $\times$ 100 equals 1 centistoke (cSt).

$$\text{Kinematic viscosity, St} = \frac{\text{viscosity, P}}{\text{density, g/cm}^3} \qquad (4)$$

Viscosity Index

This term has been used with reference to petroleum products. It is an empirical number that indicates the effect of a change in temperature on the viscosity of an oil. A low index number signifies that an oil changes viscosity by a large amount with a given temperature change.

Viscosity Scales

A few time-based viscosity scales are in common use: (1) Saybolt scales (in the United States), (2) Redwood scales (in Great Britain), and (3) Engler scales (in Europe). All three scales are based on the Hagen-Poiseuille law and indicate the time of efflux, under specified conditions, for a fixed volume of fluid through a specific capillary or aperture. Kinematic viscosity (and from it, absolute viscosity) can be determined from such scales by the empirical formula

$$\nu = At - \frac{B}{t} \qquad (5)$$

where A and B = constants applicable to the viscometer
t = time of efflux, s
ν = kinematic viscosity, cSt

Commonly accepted values of A and B are given in Table 2. Various viscosity conversion factors are given in Table 3.

Table 2 Commonly Accepted Values of A and B (Eq. 5)

	Value of constants		
Viscometer type	A	B	Efflux time t, s
Saybolt Universal	0.226	195	32–100
	0.220	135	Above 100
Saybolt Furol	2.24	184	25–40
Redwood no. 1 (Standard)	0.260	179	34–100
	0.247	50	Above 100
Redwood no. 2 (Admiralty)	2.46	100	32–90
	2.45	—	Above 90

Table 3 Viscosity Conversion Factors

		Multiply the number of	
To convert to	Poise by	$\dfrac{\text{Pound-force second}}{\text{Square foot}}$ by	$\dfrac{\text{Pound-mass}}{\text{Foot second}}$ by
Poise	1	478.8	14.88
Pound-force seconds per square foot	0.002089	1	0.03109
Pounds-mass per foot second	0.06721	32.174	1

1 poise = 0.01 centipoise
1 poise = 0.1 pascal second
1 pound-force second/square foot = 1 slug/foot second
1 pound-mass/foot second = 1 pound second/square foot
ν (stokes or square centimeter/second) $\times$ 0.001076 = ν (square feet/second)

Viscosity Measurement

Over the years, numerous techniques have been developed for measuring viscosity. Some methods are best applied in a laboratory setting, while a few can be adapted for on-line control.

Time to Discharge

In this method, the time required to discharge a given volume of fluid through an orifice or nozzle is used as a measure of viscosity. This principle is used in the Saybolt, Redwood, Engler, Scott, Ubbelohde, and Zahn viscometers, and in the Parlin and the Ford cups. A similar type utilizes a capillary tube in place of the orifice or nozzle, including the modified Ostwald, Bingham, and Zeit-fuchs viscometers.

   These are laboratory or batch-type instruments which have been used for many years in the petroleum and allied industries. The Zahn, Parlin, and Ford types have been used in the paint and varnish industry. With changes in orifice or capillary dimensions, these instruments are suitable for low- to moderately high-viscosity ranges and for viscosity approximations. Viscosity is reported in seconds and can be converted to absolute viscosity units only if the fluid is newtonian.

Timed Fall of a Ball or Rise of a Bubble

In this method, the time required for a ball to fall or a bubble to rise through liquid confined in a tube is proportional to the absolute viscosity, because in either case the liquid moves in viscous flow through a restriction. This is primarily a laboratory- or batch-type instrument and finds use in the petroleum industry for high-viscosity oils. The ball method can be timed with considerable accuracy by field coils located at the start and finish points.

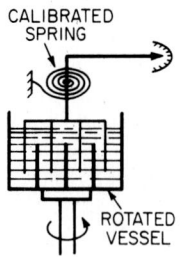

CALIBRATED SPRING

ROTATED VESSEL

Fig. 3 Viscometer employing the principle of drag torque on a stationary element in a rotating cup. *(Brabender.)*

Drag Torque of a Stationary Element in a Rotating Cup

In this method, a cup rotates at constant speed. The stationary element, which may be a cylinder or a paddle, is restrained by a calibrated spring. See Fig. 3. The spring deflection indicates torque which can be converted to absolute viscosity units. This is principally a laboratory- or batch-type instrument, but may be equipped to record. The bowl may be designed to control the sample temperature. The instrument can be used with both newtonian and non-newtonian fluids and, by a change in rotational speed, a shear stress versus shear rate diagram can be obtained. The instrument is suitable over a wide range of viscosities.

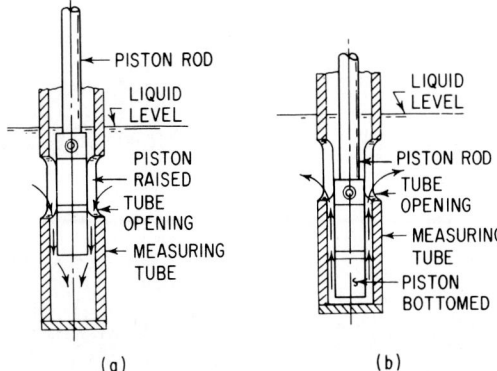

Fig. 4 Viscometer employing timed fall of a piston in a cylinder. The position of the piston during the filling phase is shown in (a). The piston is raised, drawing liquid into the space formed under it. The measuring phase is shown in (b). The piston falls, expelling the liquid through the same path that it entered. The time of fall is directly proportional to the viscosity. (Norcross.)

Timed Fall of a Piston in a Cylinder

In this system, a submerged plunger is raised automatically and then dropped in a cylinder, forming an orifice. The piston and cylinder of one type of instrument are shown schematically in Fig. 4. The time of fall may be recorded in terms of absolute viscosity. The instrument is available with a recorder and can be used in automatic control systems as found in the paint, oil, soaps, and plastics industries. The device is suitable for both open and pressurized service and can be designed to handle viscosity ranges from 0.1 to 1 million P with high repeatability at operating temperatures up to 800°F (427°C).

Similar types, utilizing a holed disk or a holed cone and designed for laboratory use, report results in seconds, representing the time required for the piston to fall a given distance through the sample.

Pressure Drop through a Friction Tube

In this system, the sample is pumped through a friction tube in viscous flow. The pressure drop across the tube is measured by a differential-pressure transmitter. The transmitter output is a direct solution to the Hagen-Poiseuille equation.

Torque Viscometers

In this system, a sensor (spindle) is rotated at some discrete speed through the same couple arrangement which allows the sensor to be displaced to an equilibrium position at which point the calibrated torque spring gives a reference to the viscosity of the fluid being measured. See Fig. 5. Different types of spindles and speeds may be used.

Laboratory Torque Viscometer

The laboratory unit shown in Fig. 6 operates on the torque principle. The light emitting diode (LED) digital display and a 0 to 10 mV output provide continuous readings which can be converted into centipoise units. A permanent record of viscosity can be obtained by connecting the unit to a stripchart recorder (Fig. 7). Recording permits the analysis of rheological processes (rheological profiles) that may occur rapidly, or slowly over a long period of time. The instrument is accurate within 1% of the range in use, and sensitivity and reproducibility are 0.2%. The selectability of several rotational speeds on each instrument allows non-newtonian flow properties, such as plasticity, pseu-

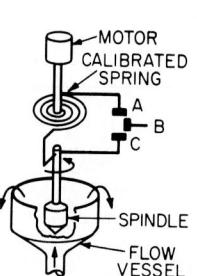

Fig. 5 Viscometer in which the torque required to rotate a torque element in a liquid is a measure of viscosity. *(Brookfield.)*

Fig. 6 Laboratory torque viscometer with digital readout. *(Model LVTD, Brookfield.)*

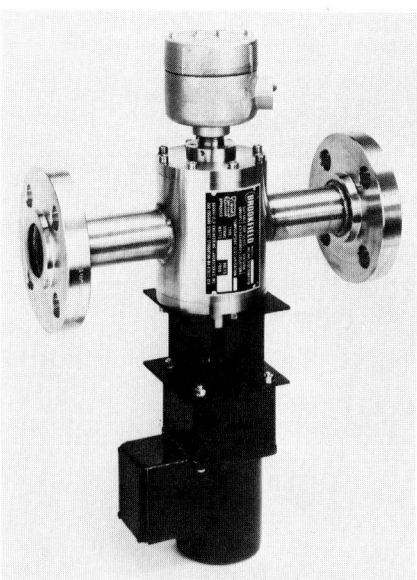

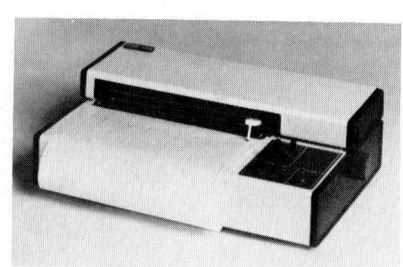

Fig. 7 Strip chart recorder used in connection with the laboratory torque viscometer shown in Fig. 6. *(Brookfield.)*

Fig. 8 Variable speed rotational viscometer—fully flooded in-line system. *(Model TT-100VS, Brookfield.)*

Table 4 Broad Classification of Fluids by Viscosity Range

Low-viscosity fluids (15–2 million cP)	
Adhesives (water and solvent base)	Paints
Chemicals	Paper coatings
Cosmetics	Plastisols
Hot waxes	Starches
Inks (Flexo and Roto)	Surface coatings
Latex	Toothpaste
Oils	Varnish
Paints and coatings	High-viscosity fluids (800–64 million cP)
Pharmaceuticals	Asphalt
Photo resist	Caulking
Rubber solutions	Chocolate
Solvents	Epoxies
Soups	Gels
Textile fibers (synthetic)	Inks (offset and thick film)
Water-based systems	Molasses
Medium-viscosity fluids (100–8 million cP)	Pastes
Adhesives (hot melt)	Peanut butter
Creams	Putty
Food products (dressings, cheeses, batters)	Roofing compounds
Gums	Tars
Inks (screen and offset)	
Organisols	

Source: Brookfield Engineering Laboratories, Inc., Stoughton, Massachusetts.

doplasticity, and dilatancy to be rapidly identified. Thixotropic flow curves are easily generated by a simple procedure.

Although specification of a particular model of this type of viscometer generally should be prepared in concert with the manufacturer, there are three broad viscosity ranges—low, medium, and high. Some of the fluids which nominally fall into these ranges are listed in Table 4.

On-Line Process Viscometers

In the instrument shown in Fig. 8, instead of rotating and sensing on the same shaft through the same couple arrangement, the operation has been divided into each end cap; i.e., a cylinder is rotated directly from the motor through one end cap, and another cylinder is positioned inside the rotating cylinder on a torque tube which goes through the opposite end cap. With this design, one need be concerned only with pressures and temperatures on the seals that are rotating and that are not influencing the friction associated with the torque measurement. The nonrotating member is sensed in this design. The instrument shown can measure a wide variety of fluids under various conditions of pressure and flow. A shear rate from 1 to 1200 s^{-1} can be manually selected in increments of 0.01 to 1.0 s^{-1}, depending on the range. Shear rate, shear stress (in dynes per square centimeter), and viscosity (in poise or centipoise) are continuously displayed on digital LED readouts. These values can also be recorded through back-panel recording outputs. The shear stress range is variable to allow measurement of a wide variety of viscosities at desired shear rates.

The system consists of three basic components: (1) a viscometer, which is mounted in the process pipeline and does the actual measuring, this being the only component that contacts the process fluid; (2) a control unit, which contains most of the system's electronic circuitry and houses the digital readouts and associated switches and controls (an external voltage stabilizer for this unit is supplied); and (3) the frequency generator, which controls the speed of the viscometer drive motor and provides the shear rate signal to the control unit. The system has the following characterisitics: viscosity, 1 to 150,000 cP; shear rate, 1 to 1200 s^{-1}; temperature, -40 to 300°F (-40 to 150°C); pressure, 0 to 200 psi (1380 kPa); flow rate, 20 gal/min (76 L/min maximum); accuracy, $\pm 0.5\%$ of span; repeatability, $\pm 0.5\%$ of span; signal output, 0 to 1.0 V; overall dimensions of viscometer, 5 × 5 × 20 in (127 × 127 × 508 mm). Units for higher temperatures and pressures are available.

Viscometer Controller

As shown in Fig. 9, this unit combines a viscometer and indicator-controller into a single, compact instrument. The instrument, including the valve for solvent addition, is pneumatically operated and is thus well adapted for hazardous areas. However, an electric synchronous motor (no arcing or sparking contacts) is optional for the viscometer drive. The unit provides on-off pneumatic control with 1% throttling action. A common application for the system is the controlled addition of solvent in the manufacture of such products as inks, coatings, and lacquers. In the illustration are shown the viscosity indicating dial and the set-point knob directly below. The case is stainless steel.

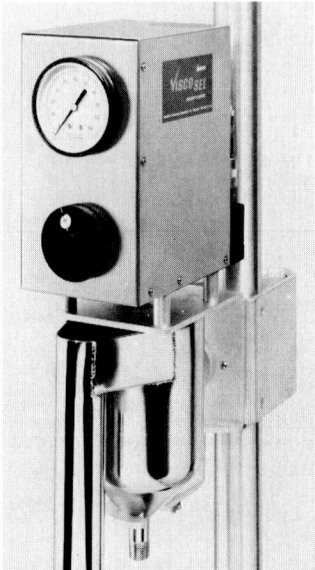

Fig. 9 Viscometer controller widely used for the controlled addition of solvent and related operations. The instrument is pneumatically operated and is thus well adapted to hazardous areas. *(Viscosel, Model VTA120, Brookfield.)*

Cone-Plate Viscometer

The instrument shown in Fig. 10 is designed for routine determination of the absolute viscosity of fluids in small sample volumes. The torque measuring system consists of a calibrated beryllium-copper spring connecting the drive mechanism to a rotating cone, which senses the resistance to rotation caused by the presence of sample fluid between the cone and a stationary flat plate. The resistance to the rotation of the cone produces a torque that is proportional to the shear stress in the fluid. See Fig. 11. Indication is by dial or digital display. Readings are converted to absolute centipoise units (millipascal seconds) from precalculated range charts. Alternatively, viscosity can be calculated from the known geometric constants of the cone, the rate of rotation, and the stress-related torque. Purge fittings can be installed to maintain a controlled atmosphere in the sample cup. A Luer fitting can be installed to allow sample material to be introduced without removing the sample cup.

The instrument is equipped with an eight-speed transmission with a 200:1 ratio of maximum to minimum speeds, thus providing a wide variety of shear rates and viscosity ranges. These ranges can be further extended by the use of interchangeable cone spindles. The small sample volume is advantageous to research involving samples of biological fluids where the supply may be limited or where costly materials are involved, such as thick-film coatings that contain precious metals.

Ultrasonic Probe

Over the years, a number of instruments utilizing ultrosonic energy have been designed for the measurement of viscosity.

Representative Viscosity Control Applications*

Continuous viscosity measurement and control systems are widely used (1) in the processes involving solvents where viscosity is a measure of solvent concentration; (2) in certain chemical reactions, such as polymerization, where viscosity is an indication of the *end point* of the reaction; (3) in processes

*This portion of this article was prepared by Robert A. Norcross, Norcross Corporation, Newton, Mass.

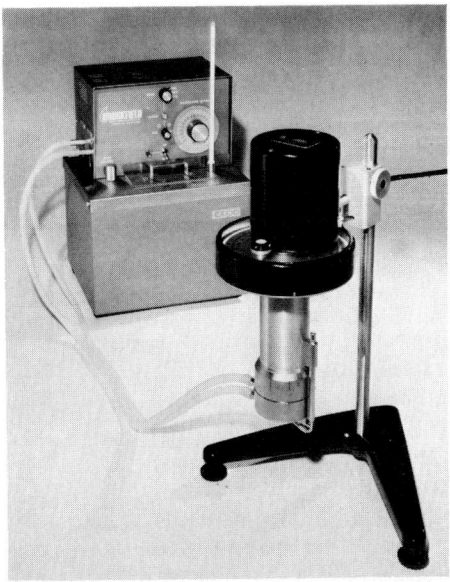

Fig. 10 Laboratory-type viscometer that operates on the cone-plate principle. A constant temperature water bath is shown in the background. *(Wells-Brookfield.)*

involving starch—as found in the paper and textile industries—where viscosity is a measure of the enzyme actions occurring; and (4) in the blending of petroleum and chemical materials to desired end specifications.

Solvent Control

Typical processes that involve solvents and the need to automatically correct for solvent evaporation losses include (1) flow coating, dipping, and spraying in finishing operations; (2) coating of wire with enamel, plastic, rubber, and various synthetic materials; (3) impregnating motors and transformers with varnish; (4) printing processes, particularly rotogravure and flexographic; and (5) sizing paper and textiles. Typically a solvent valve and a throttling valve are installed in the solvent feed line. When evaporation causes the viscosity to exceed the control point setting, the solvent valve is opened. The throttling valve is preset for load conditions to prevent overcorrecting. Changes in viscosity occur slowly, because the volume is large compared to the exposed evaporating surface. Thus, on-off control usually suffices, because corrections are made before the viscosity increases appreciably above the set-point value. See Fig. 12.

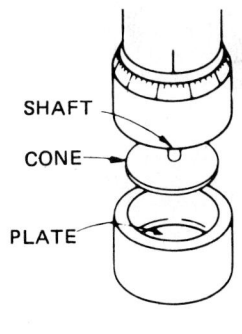

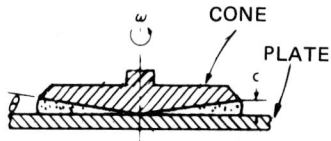

Fig. 11 Operating principle of the cone-plate viscometer shown in Fig. 10. *(Wells-Brookfield.)*

End Point Determination

In certain chemical reactions, such as are encountered in (1) the bodying of oils; (2) polymerization of resins, such as alkyd, epoxy, phenolic, polyester, polyurethane, polystyrene, silicones, and turpene

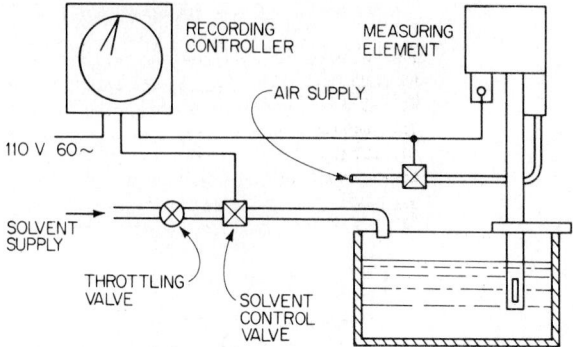

Fig. 12 Viscosity instrumentation system for automatically adding solvent to correct for evaporation losses. *(Norcross.)*

vinyl; (3) manufacture of viscose, acetate, and nylon; and (4) polymerization of polyvinyl acetate emulsions, the reaction end point can be very critical. Often, a high-viscosity alarm indicates the approach of the end point. The process can be terminated automatically, but this refinement seldom is used because operating personnel always follow the batch closely at this time. The operating conditions for the viscosity measuring element when applied directly to reactors often are severe. Reproducible measurements are required under operating pressures from full vacuum to 300 psig, and at temperatures of 700°F or higher. Usually, explosion-proof design is required, as well as provisions for purging, flushing, and disassembling for cleaning. Most processes of this type require accurate temperature control so that the viscosity measurements will be meaningful.

Starch Processes

Three processes, starch conversion, sizing, and coating, require viscosity measurement and control. For example, as more paper mills change over to computer control, the continuous measurement of viscosity becomes an important input for closed-loop operation. A batch-type enzyme starch conversion system which accurately finishes each batch to the desired viscosity is shown in Fig. 13.

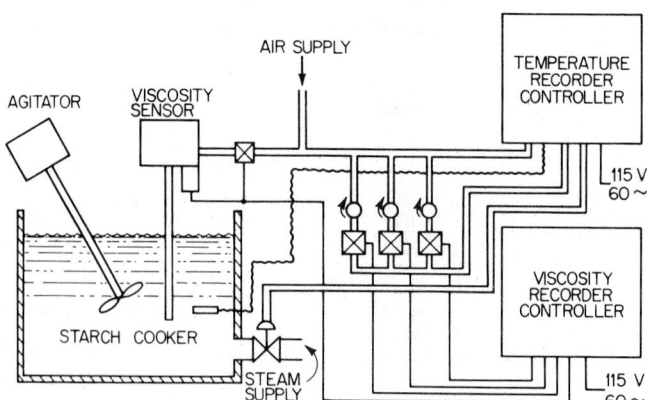

Fig. 13 Automatic batch-type starch conversion system utilizing automatic viscosity measurement and control. *(Norcross.)*

Blending

There are many applications in the petroleum and chemical industries in which liquids are blended continuously to make a product to specifications. The viscosity of the blended product can be measured with a viscometer installed either in a side stream or directly in the main line. The necessary sample flow can be maintained in the side stream either by a restriction in the line or by adding a small pump in the sample stream. In some cases the side stream can be connected across an existing pump in the main line to obtain the sample flow. Viscosity measurements also can be made by bleeding off a small sample continuously for measurement at atmospheric pressure. For most applications of this type the effect of temperature variations must be taken into consideration.

Temperature Corrections

These can be made (1) by controlling the sample temperature accurately to a predetermined value, or (2) by a means of temperature compensation. The viscosity of most fluids is temperature-dependent. Thus, both temperature and viscosity must be measured before the viscosity, with reference to a base or reference temperature, can be determined. An instrument that reads the viscosity of a fluid and automatically adjusts this reading to compensate for temperature variations is highly desirable in most continuous flow processes. The output from such a compensated viscometer can be used to provide a feedback signal which, in turn, will control temperature, rate of flow, or other variations in input in order to maintain the desired process uniformity.

REFERENCES

Considine, D. M., and G. D. Considine (eds.): *Foods and Food Production Encyclopedia,* Van Nostrand Reinhold, New York, 1982, pp. 1998–2004.

Deman, J. H., et al: *Rheology and Texture in Food Quality,* AVI, Westport, Conn., 1976.

Eirich, F. R. (ed.): *Rheology, Theory and Application,* 5 vols., Academic Press, New York, 1956–1969.

Kapsalis, J. G., and D. D. Hamann: "Trends in Food Texture Measurement," Institute of Food Technologists Symposium, St. Louis, Mo., 1979.

Matuski, F. J., and P. C. Scarna, Jr.: "Instrument Makes On-Line Viscosity Control of Slurries Possible," *Control Eng.,* vol. no. 28, 13, 1981, p. 116.

Norcross, A. S.: "Automatic Viscosity Control," *Instrum. Control Syst.* vol. 32, no. 6, June, 1959.

Reinher, M.: *Deformation, Strain, and Flow,* 2d ed., H. K. Lewis, London, 1960.

Reinher, M.: *Lectures in Theoretical Rheology,* 3d ed., Wiley, New York, 1960.

Rha, C. K.: "Viscoelastic Properties of Foods as Related to Micro- and Molecular Structures," *Food Technol.,* vol. no. 33, 10, 1979, pp. 71–76.

Sherman, P.: *Industrial Rheology,* Academic Press, London, 1970.

Slomiana, M.: "Using Differential Pressure Sensors for Level, Density, Interface, and Viscosity Measurements," *Instrum. Technol.,* vol. 26, no. 9, 1979, 63.

Consistency Measurements*

Substances that undergo continuous deformation when subjected to a shear stress are said to exhibit fluid behavior. The resistance offered by the substance to this deformation is termed *consistency*. The consistency is constant for gases and newtonian fluids if static temperature and pressure are fixed. When such fluids are referred to, the consistency is called *viscosity*.

The consistency of non-newtonian fluids is not constant but is a function of the applied shear stress and in some cases may vary with time. The term *apparent viscosity* is frequently applied with reference to the consistency of non-newtonian liquids.

Although consistency is a real property of non-newtonian fluids, its measurement is usually relative to arbitrary standards. Thus, the measurement can be stated in no definitive units such as those of absolute viscosity, temperature, and pressure.

Measurement and control of fluid consistency are of industrial concern because this variable has considerable effect on the efficiency of unit operations and on the quality of end products from many processes. Consistency control is occasionally the means of reducing the quantity of valuable material lost in the process.

Flowability

Through the years, the term "consistency" has acquired several connotations, largely as a result of the usage made of the measurement. When considering the operation of process machinery, one is interested in the characteristics of the fluid as a flowing medium, that is, its flowability. Here the term retains its true definition of resistance to deformation. For example, consistency (and temperature) control is necessary for proper operation of a sterilization method used in the manufacture of canned foods. This method employs the short-time high-temperature principle of sterilization in continuous agitating cookers, with induced convection instead of conduction of heat. This process does not allow increases in the consistency of the food slurry. Thus, control yields improvement in the color, flavor, and texture of the product. This method has proved successful for processing cream-style corn and similar food products.

Consistency of the slurry is also an important variable in the refining of wood pulps for paper-making. The purposes of refining are to brush out lumps or knots of fibers, to cut fibers to a shorter length, or to produce more hydration. The result of each form of mechanical treatment of cellulose pulp is influenced, in both extent and kind, by the consistency (flowability) of the pulp. If any form of mechanical treatment is to be so governed as to have a uniform effect, then consistency must be uniformly controlled.

Consistency of the material also influences most mechanical treatment operations, such as kneading of dough, mixing of oil and water and clay and water, and formulation of paint. In all mixing operations, the apparent viscosity of the mixture is a most important physical factor. A knowledge of the rheological properties of mixtures is essential to proper design of mixing equipment. A mixer, operating at constant speed and for a fixed length of time, yields a mixture whose physical properties depend largely on the consistency of the substances being treated.

Relationship of Consistency to Percent Solids

The apparent viscosity of a suspension of certain fibrous materials is related to the percentage of fibers in the suspension. This relationship provides the operating principle for many consistency regulating devices; that is, these devices employ measurement of apparent viscosity as the base for

*Information in this section was originally prepared by Arthur L. Landesman for the first edition of this book. For the second edition, the very active cooperation of James Overall, Georgia-Pacific Corporation, Portland, Oregon, and Gregory Neeb, DeZurik Corporation, Sartell, Minnesota, was obtained and is hereby gratefully acknowledged. For the third edition, several new descriptions and insights were provided by Eugene Norman, Green Bay, Wisconsin.

control of fiber concentration. This principle has, through constant usage, caused the term "consistency" to acquire the connotation of solids content and has caused its true scientific definition to be disregarded in some industries.

"Consistency" is the term used in the paper industry, as well as in some food processing and mining operations, to designate the concentration of weight of air-dry pulp (i.e., of total solids) in any combination of pulp and vehicle, the vehicle usually being water. Expressed as an equation,

$$\frac{\text{Air-dry weight of solids (per unit wt)}}{\text{Weight of solids + water (per unit wt)}} \times 100 = \text{consistency, \%}$$

Papermakers occasionally use the terms "bone-dry," "oven-dry," and "moisture-free" with reference to pulp fibers or paper from which all water has been removed by evaporation. Since such thorough drying is seldom practiced in the paper mill, the term "bone-dry" is primarily used in laboratory work and in process equipment design. Air-dry pulp is considered to contain 10% moisture, so that

$$\text{Bone-dry consistency, \%} = 0.9 \times \text{air-dry consistency, \%}$$

These equations find practical application as shown by the following problem.

Example. For what capacity (flow rate) must a stuff pump be designed in order to handle pulp slurries at 3.0% air-dry consistency in a system designed to produce 100 tons/day of bone-dry paper? Assume that 1 gal of 3.0% consistency pulp weighs 8.34 lb.

$$\frac{100 \text{ tons/day} \times 2000 \text{ lb/ton}}{24\text{h/day} \times 60 \text{ min/h}} = 139 \text{ lb/min bone-dry fibers}$$

and thus $$\text{Flow rate} = \frac{139 \text{ lb/min}}{0.03 \times 0.9 \times 8.34 \text{ lb/gal}} = 617 \text{ gal/min}$$

Other useful flow conversion formulas include:

$$\text{Gallons/minute} = \left(\frac{16.65}{\%\text{C}}\right) \text{tons/day}$$

$$\text{Liters/minute} = \left(\frac{100}{\%\text{C}}\right) \text{kilograms/minute}$$

When percentage solids is the connotation, uniform consistency is also important to the efficient operation of process machinery and to the control of end product quality. It should be noted that this meaning of consistency is more akin to density than to the concept of viscosity. Devices that determine apparent viscosity, rather than density, are frequently used to infer the solids content.

The end point in the evaporation of tomato products is determined by this type of consistency measurement. Tomato paste, for example, is now concentrated to 36% solids. The concentration is not indicated properly by density measurement but may be correlated reliably with apparent viscosity. Variations in the solids content of mineral-ore slurries may be determined by observing corresponding changes in the consistency. Papermakers are vitally concerned with basis weight (weight per unit area) of the sheet because this weight is the unit on which sales are frequently based. Consistency control is essential to basis-weight control because it is the means of delivering to the papermaking machine a constant volume of stock containing a uniform amount of fibers.

Efficient operation of other paper process machinery also depends on supplying the device with a definite and constant quantity of pulp. The operation of fine screens is a complex problem in which consistency is a large factor. An examination of Fig. 1 reveals that optimum screen operation is effected by control of the inlet load which, at constant volume of feed, varies with consistency.

The chart shows the mutual changes in the various important factors in screening. With increasing consistency the horsepower consumed per ton of material processed decreases, but the percentage of tailings or rejected material increases. Therefore, it is necessary to control the consistency of the feed at some point which results in the optimum balance between accepted and discarded material as well as low power consumption.

Relationship of Consistency to Moisture*

Moisture is the virtual reciprocal of consistency. It may be considered a measure of the liquid water content of the solid medium. This measurement is preferred for materials of higher consistencies ranging from 12 to 99%. However, it may be applied to consistency measurements outside this range. For example, a paper web in the fourdrinier section of a paper machine may vary from 0.3% at the breast roll to 20% at the press roll. The dryer consistency—entrance to exit—may vary from 10 to 99%.

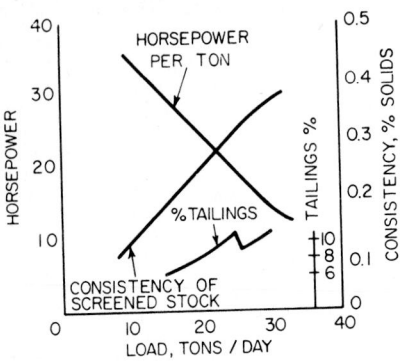

Fig. 1   Curves showing the effect of consistency on load, power, and tailings for fine screens in the paper industry.

Relationship of Consistency to Density

Density is the mass per unit volume of a substance, whereas consistency is the mass per unit mass of a substance. As an example, 3 g of solids in 100 g of mixed slurry is a 3% consistency.

Relationship of Consistency to Freeness

Freeness is the amount of water that leaves a sample of specific volume and consistency. The Canadian Standard Freeness Scale (CSF) indicates the amount of bypass water out of a calibrated funnel from a 1-l sample at 0.3% consistency. The freeness is read directly in cubic centimeters. The higher the number, the more readily the slurry will release the water.

Reasons for Consistency Variation

One of the general reasons for variation in fluid consistency at any point in the process is fluctuation of the composition of raw materials. In some mills paper stock is formed by slushing wet lap pulp, wastepaper, or broke with water in a beater or other mechanical pulper. In order to prepare a stock of uniform consistency it is necessary to weigh accurately the pulp charged to the beater and adjust the amount of water added to compensate for the varying moisture content of the pulp. Furthermore, since preparation of the batch is done manually, composition of the slush stock is also subject to human error. In industrial practice the slushed pulp is usually made up to approximately the desired consistency and then controlled accurately by means of a consistency regulator.

Nature is responsible for the varying composition of the raw materials for food processing operations. Consistency is affected by such factors as variations in starch content and size of food pulp particles.

Normal variations in the operation of process machinery also create a need for consistency control. When slime or other solid particles partially plug screens and other filters, both their overall capacity and their ability to pass fibrous materials are reduced and the concentration of fibers in the filtrate varies. In the preparation of mechanical wood pulp, consistency variations are caused by fluctuations in the supply of water to the grindstone. The apparent viscosity of mineral ore slurries varies with changes in particle size caused by gradual dulling of blades in attrition mills and other grinding machinery.

A few general observations concerning consistency instrumentation are given in the following paragraphs.

Instruments used to measure and control the absolute viscosity of newtonian liquids take advantage of the fact that temperature is the only factor with a significant effect on viscosity. Velocity of flow, history of treatment, temperature, and other variables exert influences on the apparent viscosity of non-newtonian fluids. However, the consistency of a suspension of fibrous materials varies primarily with the percentage of fibers in the slurry. As long as the nature of the suspended material is not greatly changed, the use of apparent viscosity as the control point gives proper consistency control.

*By Eugene Norman.

With respect to the operation of any regulator, the question of exact variation in the viscosity of the suspension is mathematically unimportant, as may be noted from a consideration of the viscosity-consistency curve of Fig. 2.

It can readily be seen that the maintenance of fixed viscosity involves a corresponding consistency, regardless of whether the relationship is linear or not.

It is evident that the usefulness of any apparatus based on the relation between consistency and shear rate (apparent viscosity) of paper stock is limited to the range in which the viscosity-consistency curve may show considerable slope.

Of course, the most accurate means of measurement should be a direct method, such as drying and weighing the amount of fibers in a pulp sample. The accuracy depends largely on the accuracy of sampling. For example, in some product vessels consistency varies from top to bottom, and a sample of the cross section must be taken. The weight-measurement method is essentially confined to laboratory experimental and control work.

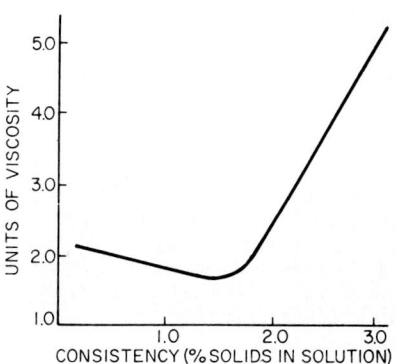

Fig. 2   Relation between shear value (apparent viscosity) and consistency.

Specific Variables in Consistency Measurement

Among process variables that affect the measurement of consistency are temperature, velocity, freeness, hydrogen ion concentration (pH), furnish, and inorganic materials content.

Temperature

Temperature changes affect the shear stress of fiber suspensions in that increasing temperatures provide decreasing forces on a sensor at a fixed consistency. Thus the sensor produces a change in signal indicating lowering consistency with the increasing temperature.

Velocity

The velocity effect is a pertinent but virtually unpredictable variable with respect to consistency measurement. A varying velocity at the sensor alters the forces acting on the sensor at a fixed consistency. If the fluid velocity is too low, laminar flow conditions can be created in which the consistency varies across the pipe diameter. If the fluid velocity is too high and the sensor is mounted in a long run of straight pipe, boundary layering may occur. This effect is the reverse of the laminar effect. In the latter case, the consistencies at the boundaries may be lower than the consistencies at the center of the pipe. Velocity effects vary for each particular type of sensor and for each installation. It is common practice, however, to avoid the velocity effect by mounting the sensor in a turbulent section of the pipe (e.g., directly after a pump) with a reasonably constant flow rate.

Freeness

Since freeness is one of the characteristics of the fibrous medium, a variation in freeness imposes a variation in consistency transmitter output. Increasing the freeness at a fixed consistency provides an increasing consistency signal. To compensate for the freeness effect, the freeness should be measured and included in the calibration of the sensor. Where the sensor is subjected to composite media of varying freeness, an on-line freeness tester can be used to compensate the consistency signal.

pH

As the alkalinity of the slurry increases at a fixed consistency, the sensor output increases. This is due to the effect of pH on the shear stress of the fibrous medium. Also, chemical agent additions, used for controlling the pH of the medium, may react with the fibrous mass, producing stress variations.

Furnish

The material in a fibrous medium pertaining to the type of fiber is the *furnish*. This furnish may vary from a short-fiber stock to a long-fiber stock. At a fixed consistency, an increasing fiber length results in an increasing sensor signal, as a result of the ability of long fibers to dewater rapidly, resulting in higher freeness.

Inorganic Materials Content

Since consistency transmitters and sensors respond essentially to fibers in a slurry, an increasing inorganic materials content results in a decreasing consistency signal at a given consistency level. The filler material does not affect or contribute to the shear stress as does the fiber. Thus there is an apparent reduction in the consistency. For example, a 3% slurry with 25% filler has a fiber consistency of 2.25%. If the use of these materials is fairly constant, the sensor-transmitter should be compensated during calibration.

Consistency Instrumentation

Consistency is measured and controlled with one or more of the following objectives in mind: (1) determination of rheological properties, that is, the characteristics of the flowing medium, (2) determination of the quantity of solids in suspension (closely related to density), and (3) determination of the in-process quality of non-newtonian liquids and suspensions.

Classification of Applications

The applications for consistency instruments may be broadly classified corresponding to the above basic reasons for their use as follows:

1. *To indicate the proper operating procedures for handling non-newtonian fluids in pipelines and processing equipment.* Most of these materials are easily handled as long as sufficient force is applied to keep them in motion. When at rest, previously flowing plastic or pseudoplastic substances can set and form "plugs" in pipelines and pumps. Awareness of this characteristic leads to provision for drain-off of such fluids from equipment before shutdown in order to facilitate subsequent start-up.
2. *To aid in control of the end point of unit operations and in control of final product quality.* For example, the end points of such unit operations as evaporation, cooking, and mixing are frequently controlled by consistency determination. Additionally, control of paper basis weight and canned-food texture are examples of applications relative to final product quality. The range of percent solids that can be determined by the various instruments depends very much on the physical characteristics of the solid particles and the limitations of pumps and other accessory equipment.
3. *To determine, under conditions of uniform solids content, fluctuation in apparent viscosity of non-newtonian fluids as a clue to intermediate product quality variations.* Here the term "quality" is used in a broad sense to connote the effects of composition, amount of hydration, particle size and shape, and length of time during which the material has been subjected to beating, mixing, or other mechanical treatment.

Laboratory-Type Devices

In addition to the drying-weighing direct method, a limited number of laboratory instruments are available for consistency determinations.

Consistency Cups

Several types of cups (see Fig. 3) are available for measuring the consistency of viscous liquids, solutions, and suspensions, such as paint, lacquer enamels, gelatin solutions, and drilling muds. Consistency is expressed at the time of efflux of a fixed volume of fluid through an orifice in the bottom of the cup. One such device consists of a one-piece plastic molding with an overflow well at the top, vertical sides, and a tapering bottom with a ⁵⁄₃₂-in hole in the tip. Capacity for overflow is about 100 ml. Consistency cups also are used for determining the consistency of nitrocellulose clear lacquers.

Rotational Viscometers

The rotational-type viscosimeter may be used to measure shear values (apparent viscosity) of

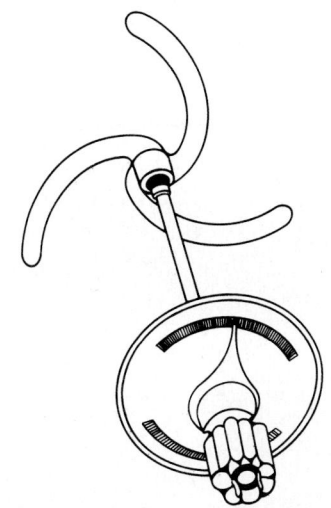

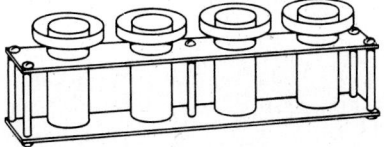

Fig. 3   Four-cup lacquer consistency tester.

Fig. 4   Device employing a small torque wrench to measure shear. *(DeZurik, Unit of General Signal.)*

non-newtonian substances. The instrument operates on the principle that the resistance to deformation offered by a fluid is directly proportional to the torque on a spindle rotating in a container with the fluid; or the container may rotate about the spindle. The torque is measured by various means, most of which may be used to position indicating pointers on arbitrarily calibrated scales. A very simple device, but one that provides accurate comparative results for developing a shear value–consistency curve (helpful in determining controllable ranges of consistency), is shown in Fig. 4.

Some applications for instruments of this type include (1) automatic regulation of aniline inks on rotogravure presses, (2) automatic control of starch sizing by means of regulating the orifice size of a homogenizer, (3) measurement of portland cement settling rates, (4) measurement of viscosity in resin condensation processes, and (5) measurement of catsup, paper coating, and crystal slurries.

Continuous Consistency Measurements

Because of stratification and nonuniformity of the fluid stream, it usually is preferred to have the consistency sensor exposed to the entire flow and not just to a sample. However, because of the tremendous volumes involved in many operations, this approach is not always practical and sampling techniques must be used. In order to provide effective control over consistency, a sample truly representative of the entire fluid flow must be taken into the measuring chamber of the sensing apparatus. There should be no separation or settling of solids from a suspension. Therefore, if samples are taken from a large tank of material, the entire volume of the tank should be well agitated. If the sample is taken from a pipeline, there should be provision for thorough mixing of diluent and the main stream prior to sampling for the consistency sensor. Sampling at a point immediately after a centrifugal pump is good practice. Detailed measurements and angles for one acceptable means of installing a consistency sampling device in a pipeline in a papermill are given in Fig. 5a. Another effective scheme is shown in Fig. 5b.

In controlling consistency, it usually is much easier and more practical to dilute, that is, to reduce the consistency, rather than vice versa. Therefore, the feed to the sensor should be at a consistency

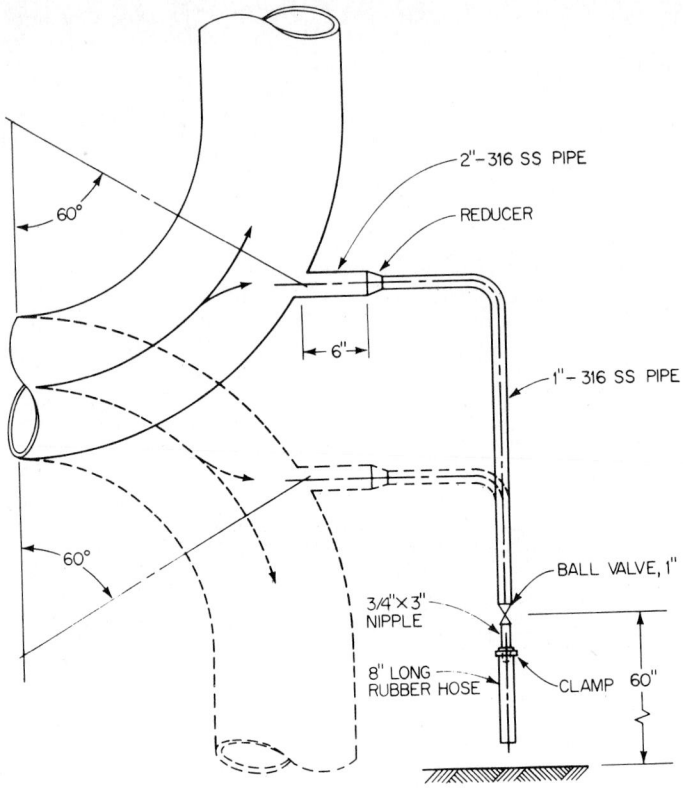

Fig. 5 (a) Suggested geometry for installing a consistency sampling line in a large pipe in a paper mill. *(Georgia-Pacific.)*

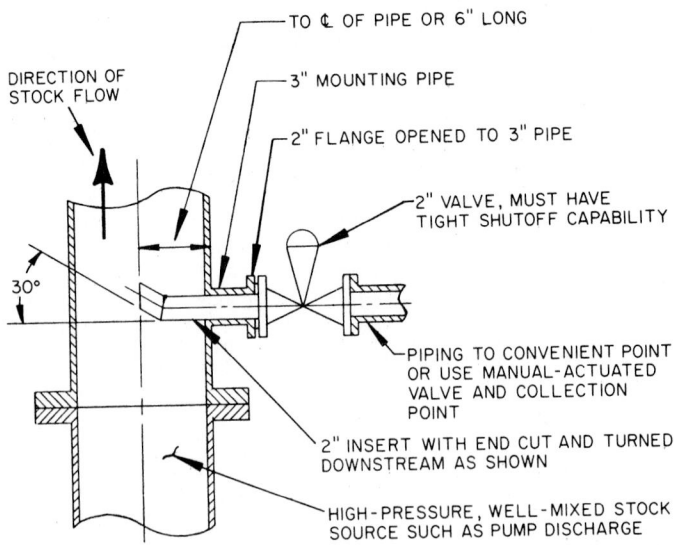

Fig. 5 (b) Suggested consistency sampling system piping for vertical lines. *(DeZurik, Unit of General Signal.)*

higher than the desired value. For best operation of the sensor, it is desirable to standardize operating techniques so that minimum variations occur in the density of the feed to the sensor. This is an important consideration for sizing the sensor and the automatic control valve for controlled addition of dilutent to the heavy material.

Blade-Type Consistency Transmitter

The transmitter* shown schematically in Fig. 6 operates on a shear rate principle by utilizing a blade sensor suspended in the pipeline. Consistency changes are measured by sensing the change in force required by the sensing blade *A* to shear through the flowing stock. A torque arm *B*, connected to the sensor, extends into a pneumatic torque transducer which converts torque arm movement into a pneumatic output signal proportional to consistency change. The sensor is so designed that it is not affected by changes in flow rate over a wide range of flows. The output signal is transmitted to a recording controller which records the consistency and positions an air-operated dilution valve.

The unit has a consistency range of 1.75 to 6.00% with a sensitivity capable of sensing consistency within ±0.02% for many applications. The unit has a 40:1 span adjustment ratio and a claimed repeatability of ±0.5% of chart reading. The flow velocity range is 0.75 to 5 ft (23 cm to 1.5 m)/s. The unit is available in pipe size dimensions from 4 to 30 (~10 to 76 cm). The 4-in size handles a minimum of 30 gal (~114 L)/min and a maximum of 195 gal (~738 L)/min, whereas the 30-in size handles a minimum of 1665 gal (~63 hL)/min and a maximum of 11,000 gal (~416 hL)/min. The maximum line pressure rating is 125 psi (862 kPa). The sensor, mounting neck, and torque arm are constructed of Type 316 stainless steel, and the body of the unit is cast aluminum. Typical horizontal and vertical installations are shown in Fig. 7.

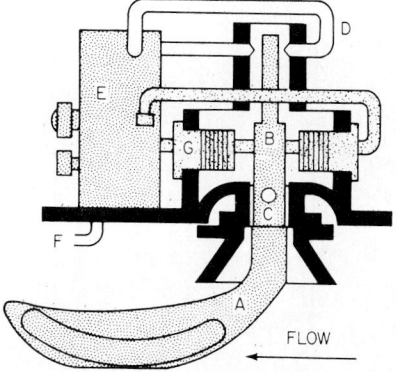

Fig. 6 Typical installation of a blade-type consistency transmitter. (*A*) Sensing blade; (*B*) torque arm; (*C*) flexure; (*D*) pneumatic bridge circuit; (*E*) booster pilot; (*F*) pneumatic signal; (*G*) force-balance feedback system. *(DeZurik, Unit of General Signal.)*

*DeZurik, Unit of General Signal. Middletown, N.Y.

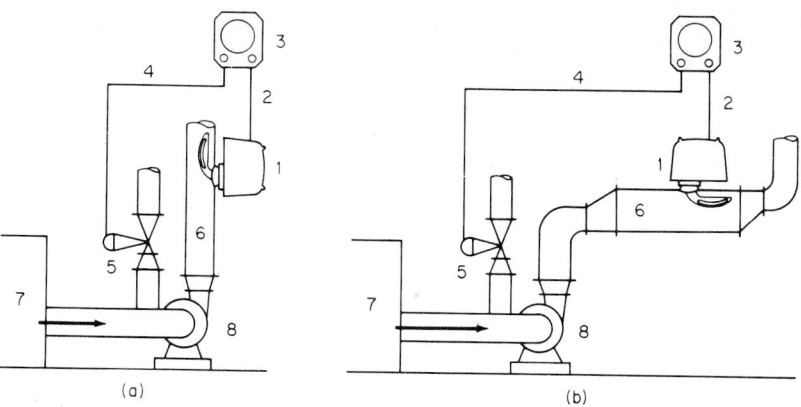

(a)   (b)

Fig. 7 Typical installation of a blade-type consistency transmitter. (*a*) Vertical, (*b*) horizontal. (1) Transmitter, (2) transmitter output, (3) controller-recorder, (4) controller output, (5) dilution valve, (6) stilling pipe, (7) stock chest, (8) pump. *(DeZurik, Unit of General Signal.)*

Rotational-Type Consistency Transmitter

Operation of the instrument* shown in Fig. 8 is similar to that of a rotational viscometer. The unit consists of a motor-driven sensor which rotates within the process line. The torque required to rotate the sensor at constant speed within the slurry is directly related to the slurry consistency. A change in consistency results in a change in torque, which produces a pneumatic output signal. The air signal normally is fed to a recorder-controller which then positions an air-operated dilution valve to maintain the desired consistency. These units also are available in an electronic configuration to provide output signals of milliamperes or millivolts.

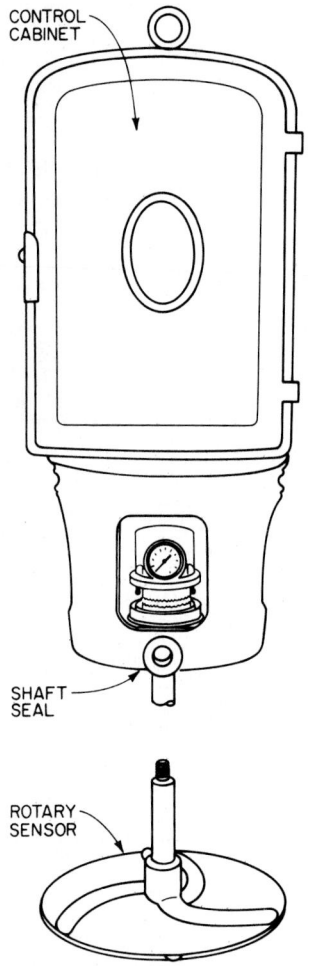

These assemblies (consistency range 0.75 to 6.00%) are available in several sizes, from a minimum flow for small units of 15 gal and a maximum flow of 180 gal (~57 and 681 L) per minute to the largest size with a minimum flow of 500 gal and a maximum flow of 10,000 gal (~19 and 379 hL) per minute. It is claimed that the sensor is capable of sensing consistency within ±0.01% for some applications. Sensing accuracy is maintained on stock as thin as 0.75% or as thick as 6.0%, and the maximum working pressures is 125 psi (862 kPa). All parts coming in contact with stock are constructed of stainless steel.

In order to reduce dependence on a virtually constant flow, the rotating sensing element† is positioned outside the direct flow path. The requirement that a representative sample of the material be measured is still critical for obtaining accurate results with this method. For this reason, the transmitter should be mounted immediately downstream of an in-line mixer (Fig. 9) or in the turbulent region of the discharge of a pump (Fig. 10).

Fig. 8 Rotational-type consistency transmitter. *(DeZurik, Unit of General Signal.)*

Probe-Type Consistency Transmitter

The sensor‡ is a cylindrical body in the flowing stream that senses force in the direction of flow by the imbalance of a strain gage bridge in the hollow neck of the probe (Fig. 11). This sensing method is dependent on the characteristics of consistency versus viscosity. If the fluid being measured is water, the probe will sense the flow rate. If the fluid flow rate is constant and the viscosity of the fluid increases, the probe will measure the increase. At low flow rates, however, the forces exerted on the probe by the water or viscous media are minor in comparison to the forces exerted by the consistency of the stock (Fig. 12). These forces are caused by the shearing required to separate the stock and make it flow around the probe. Thus, the forces exerted by the liquid carrier of the stock may be measured and calibrated out of the consistency measurement. The relationship of various stock furnishes measured by this probe is shown in Fig. 13.

*DeZurik, Unit of General Signal, Middletown, N.Y.
†Eur-Control. Decatur, GA.
‡Thompson Equipment Company.

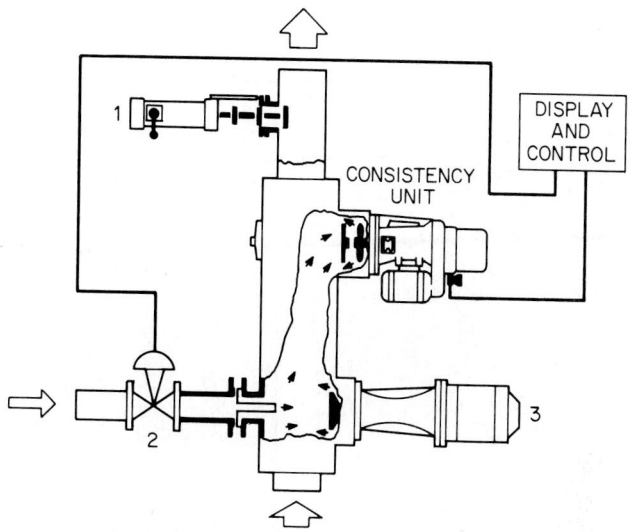

Fig. 9 Consistency control system with consistency transmitter, in-line mixer, dilution water valve, and sampling valve. (1) Sampling valve, (2) ball sector valve, (3) mixer. *(Eur-Control.)*

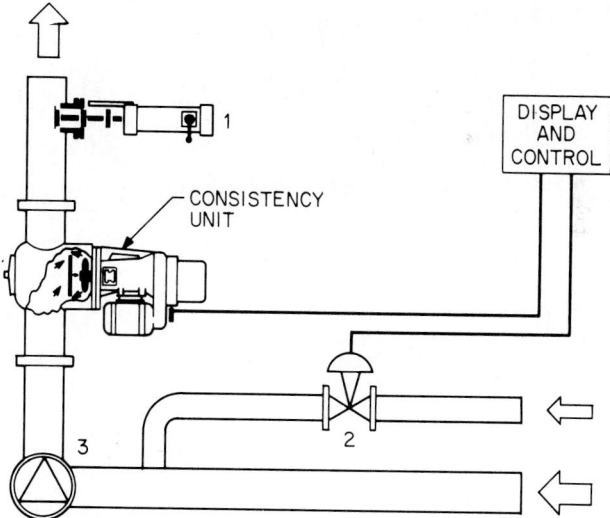

Fig. 10 Consistency control system with a pump as a mixer for dilution water. (1) Sampling valve, (2) ball sector valve, (3) pulp pump. *(Eur-Control.)*

Polarized Light Consistency Transmitter*

This is a sampling-type device for sensing low-consistency stocks. A representative portion of the main stream is passed through the sample cell through which a polarized light beam is transmitted. The beam is analyzed to determine the amount of depolarization created by the suspended particles.

*The Electron Machine Company. Umatilla, Fla.

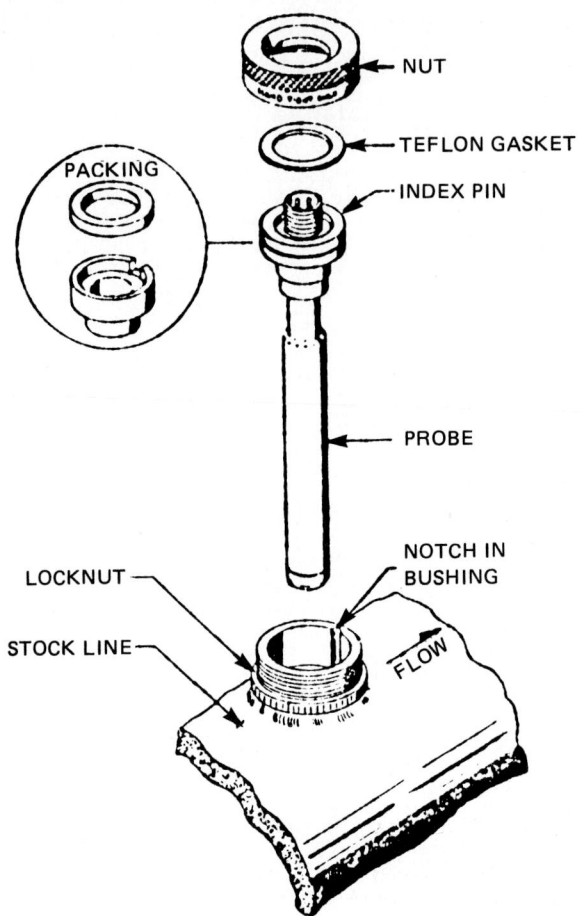

Fig. 11 Probe-type consistency transmitter. *(Thompson Equipment.)*

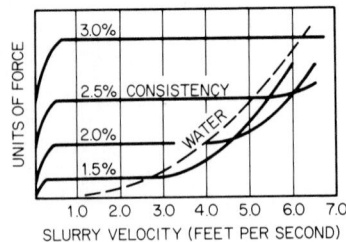

Fig. 12 Graph showing that the forces exerted on a probe by water or viscous media are minor in comparison to the forces exerted by the consistency of the stock.

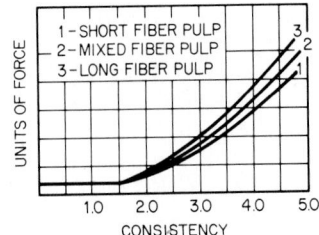

Fig. 13 Relationship between various stock furnishes measured by a probe.

The resultant signal is conditioned to indicate the percent concentration of the particles or consistency of the fluid.

Web Consistency Measurements[*]

The previously mentioned methodologies and sensing techniques have referred to stock forming and conditioning processes. When the material is formed and deposited on a moving web, the consistency of the material must still be measured and controlled to characterize the final product. The method of measurement is radically different from that of the stock forming process. While contact measurement is prevalent in the former process, noncontact measurement is essential once the material has been deposited. Where moisture, density, viscosity, and freeness are factors to be considered in the former measurement, only moisture is the consideration of the latter. Web consistency is the inverse of web moisture. This simplifies the measurement, as moisture sensing is relatively easy as compared to consistency sensing. However, even with these methods, inaccuracies described as loss factors can interfere with the measurements. In direct current, audio frequencies, and up to radio frequency, electrical conductivity is responsible for this loss factor. In microwave regions, the absorption of radiation is responsible as a result of molecular rotation. In the infrared region, the loss factor is related to the vibrational energy and the binding energies of the molecules. In the visual frequency regions, the loss factor is related to the absorption of energy as the electrons change their energy levels by shifting from one orbit to another.

See also Sec. 10 of this Handbook.

REFERENCES

Balls, B. W.: "Towards Better Understanding of Consistency Measurements," *Meas. Control,* vol. 1, no. 9, September 1968.

Britt, K. W. (ed.): *Handbook of Pulp and Paper Technology,* 2d ed., Van Nostrand Reinhold, New York, 1970.

Casey, J. P. (ed.): *Pulp and Paper—Chemistry and Chemical Technology,* 3d ed., vol. 1, Wiley, New York, 1980.

Clark, J. d'A: *Pulp Technology and Treatment for Paper,* Freeman, San Francisco, 1978.

Dykes, J. T.: "Consistency Installations and System Design Techniques," *Tappi,* vol. 46, no. 11, November 1963, p. 680.

Fjeld, M.: "Application of Modern Control Concepts on a Kraft Paper Machine," *Automatica,* vol. 14, 1978, p. 107.

Lavigne, J. R.: *An Introduction to Paper Industry Instrumentation,* rev. ed., Freeman, San Francisco, 1977.

Lavigne, J. R.: *Instrumentation Applications for the Pulp and Paper Industry,* Freeman, San Francisco, 1979.

McDonald, R. G. (ed.): *Pulp and Paper Manufacture,* 2d ed., vol. 2 (prepared under the direction of the Joint Textbook Committee of the Paper Industry), McGraw-Hill, New York, 1967.

McGill, R. J.: *Measurement and Control in Papermaking,* Adam Hilger, Bristol, England, 1980.

Staff: *A Consistency Manual,* Process Control Committee, Technical Section, Canadian Pulp and Paper Association, Montreal, Canada, June 1967.

Torborg, R. H.: "Fine Tuning of a Consistency Control System for Maximum Performance," *Pulp Paper,* March 1980, pp. 134–138.

[*]By Eugene Norman.

Radiation Systems *

R. K. Abele. *Oak Ridge National Laboratory, Oak Ridge, Tennessee. (Nuclear Radiation Detectors)*

L. L. Brackesbusch. *Baird-Atomic, Inc., Bedford, Massachusetts. (Nuclear Radiation Detectors)*

M. R. Farukhi. *The Harshaw Chemical Company, Solon, Ohio. (Nuclear Radiation Detectors)*

Eric L. Geiger. *Eberline Division, Thermo Electron Corporation, Sante Fe, New Mexico. (Nuclear Radiation Detectors)*

Allen M. Goldstein. *Technical Associates, Canoga Park, California. (Nuclear Radiation Detectors)*

Jerome Kohl. *Department of Nuclear Engineering, North Carolina State University, Raleigh, North Carolina. (Nuclear Radiation Detectors)*

---

*Persons who authored complete articles or subsections of articles, or who otherwise cooperated in an outstanding manner in furnishing information and helpful counsel to the editorial staff.

Nuclear Radiation Instrumentation

The detection and measurement of nuclear radiation finds many uses. An abridged list includes:

1. **Analytical Instrumentation.** Includes beta-ray absorption and backscattering techniques, gamma-ray scintillation spectrometry, radioactivation analysis, and radioimmunoassay. The fundamentals of some nuclear radiation detectors are also applied in x-ray (electromagnetic) spectrometers.

2. **Radiation Effects.** Finds use in studies of living processes, the environment, and materials, the latter being of particular importance for nuclear power equipment, armaments, and weapons.

3. **Nuclear Medicine.** Embraces research, diagnostics, and therapy.

4. **High- and Low-Energy Physics.** Includes subatomic particle research.

5. **Process Instrumentation.** Measures radiation from radioactive isotopes to determine liquid level, bulk density, materials thickness, among several other process variables.

6. **Monitoring, Surveillance, and Warning Systems.** Finds use in research laboratories and industrial plants where there is a potential hazard of accidental exposure to unsafe levels of radiation.

7. **Dating Techniques.** Relates emissions from radionuclides to age determination (trees, rocks, artifacts from archeological digs, fossils, and other objects and substances of interest to the paleosciences).

8. **Astrophysics and Other Pure Sciences.** Has uses, for example, in gamma-ray astronomy.

9. **Geological and Mineral Exploration.** Includes oil well logging.

Radioactivity and Types of Radiation

Radioactivity may be defined as the spontaneous disintegration of the nucleus of an atom with the emission of radiation. By distinction, electromagnetic radiation does not derive from a disturbance of the atomic nucleus. Although gamma radiation and x-radiation can be measured by many of the techniques which apply to nuclear radiation, gamma radiation and x-radiation are technically electromagnetic. Another useful distinction is that between *ionizing* and *nonionizing* radiation. Ionizing radiation, such as alpha and beta radiation, charged mesons, protons, and heavier charged nuclei, has sufficient energy to produce ionization directly in its passage through air and is capable of nuclear interactions in which sufficient energy is released to produce ionization. Gamma rays and neutrons, on the other hand, require conversion to ionizing radiation before they can be detected by ionization-dependent techniques.

Radioactivity was discovered by Becquerel in 1896, who observed that pitchblende (a uranium containing mineral) could produce an exposure effect on a photographic plate that was wrapped in black paper and kept in the dark. Soon after this, it was found that uranium minerals and uranium chemicals showed more radioactivity than could be accounted for by their uranium content. About the same time, the radioactivity of thorium minerals and thorium chemicals was also discovered. The excess radioactivity of mineral as compared to chemical uranium led Pierre and Marie Curie to experiment with the mineral. For these experiments, the first type of radiation detecting instrument, a gold-leaf electroscope, was used. In the relatively few years that followed, other naturally radioactive elements were reported, including radium, polonium, and radon. During these early years, it had not been established that several types of radiation were emitted.

Beginning in 1899 and continuing through the next two decades, E. Rutherford and associates conducted a rather thorough study of the radiations emitted by radioactive substances. During this study, the radiations were found to be of three types, alpha, beta, and gamma. In kind, they resemble anode rays, cathode rays, and x-rays, respectively. See Table 1.

Table 1 Principal Types of Radiation Emitted by Radioisotopes*

Type	Symbol	Description	Rest mass (O = 16)	Charge (electron = 1)	Range in air	Ion pairs (per cm in air)
Alpha particles	α	Nuclei of helium atoms	4	+2	2–9 cm (for 3–10 MeV)	30,000–70,000 (varies with distance from source)
Beta particles	β	Electrons ejected from a nucleus	1/1840	−1	160–2000 cm (for 0.5–5 MeV)	150–40 (for 0.1–5.0 MeV)
Gamma rays	γ	Electromagnetic radiations produced only in nuclear processes	None	None	15,000 cm ½ value thickness (for 1.5 MeV)	¹⁄₁₀₀ of number of pairs produced by same energy β

*X-rays resulting from the filling of nuclear shells and bremsstrahlung generated by the deceleration of beta particles are two additional sources of electromagnetic radiation produced by radioisotopes used in instrumentation. None of the above radiations produce detectable amounts of radioisotopes.

Alpha Rays

The name "alpha particle" was used in the earlier years of radioactivity investigations before it was fully understood what alpha particles are. It is known now, of course, that alpha particles are the same as helium nuclei. When a radioactive nucleus emits an alpha particle, its atomic number decreases by $Z = 2$ and its mass number by $A = 4$. It is, therefore, a spontaneous nuclear reaction of the form

$$A_Z \rightarrow {}^{A-4}Z - 2 + {}^4\text{He}$$

The process that occurs when alpha particles are emitted by radioactive nuclei is referred to as *alpha decay*.

Alpha rays have a definite velocity and a definite range for each radioactive nuclide. The velocity is from 5 to 7% that of light, and the range is defined as the distance traversed in a homogeneous medium before absorption. The penetrating power of alpha rays is the lowest of the three principal kinds of radiation, beta rays being on the order of 100 times, and gamma rays $\sim 10,000$ times more penetrating. Alpha rays may be defined as twice-ionized nuclei of helim (He^{2+}).

Ramsay and Royds (1909) experimentally determined that accumulated alpha particles, quite independently of the matter from which they have been expelled, consist of helium. They sealed radon in a glass tube with a wall so thin that the alpha particles could pass through the wall into a surrounding vessel. After 6 days, the optical spectrum of helium was observed. Therefore, alpha particles, on losing their positive charge, become ordinary helium.

All the energy released by the transition is carried away by the product nuclei. Therefore, a spectrum of alpha-particle numbers as a function of energy shows a series of distinct peaks, each corresponding to a single alpha-particle transition. To conserve both energy and momentum, the energy must be shared by the two product nuclei, with the daughter nucleus ($^{A-4}Z - 2$) recoiling away from the direction of emission of the alpha particle. If E_x and M_x are, respectively, the kinetic energy and mass of the alpha particle, and E_R and M_R the kinetic energy and mass of the recoiling product nucleus, the transition energy is $Q = E_a + E_R$; and the kinetic energy of the emitted alpha particle is $E_a = [M_R(M_a + M_R)]Q$.

The first alpha particle detector was the *spinthariscope*, developed by Crookes, in which the tip of a wire coated with a tiny amount of radium salts was placed near a screen coated with zinc blende. Viewed in the dark with a magnifying eyepiece, each alpha particle striking the zinc blende target was observed to produce a visible scintillation. The detection and counting of single alpha particles was accomplished by Rutherford and Geiger (1908) by the deflection of an electrometer needle on the arrival of each alpha particle in a gas at low pressure in an electric field somewhat below the sparking point.

Almost all radioactive nuclides that emit alpha particles are at the upper end of the periodic table, with atomic numbers greater than 82 (lead), but a few alpha-particle emitting nuclides have lower atomic numbers. The reason why alpha-particle emitters are principally limited to nuclides with larger mass numbers is that generally alpha-particle emission is energetically possible only in this region. Most radioactive nuclides with smaller mass numbers emit beta-particle radiation.

Beta Rays

Beta radiation is composed of negatively charged electrons. Beta rays are strongly deflected to electrical and magnetic fields and have varying velocities almost up to that of light. When beta particles are emitted by radioactive nuclei, the process is referred to as *beta decay*. The names "beta particle" and "beta radiation" were applied in the early years of radioactivity investigations before it was fully understood what beta particles are. It is now known that beta particles are electrons. When a radioactive nuclide undergoes beta decay, its atomic number Z changes by $+1$ or -1, but its mass number A is unchanged. When the atomic number is increased by 1, negative beta-particle *(negatron)* emission occurs; and when the atomic number is decreased by 1, there is positive beta-particle *(positron)* emission, or orbital electron capture.

Because atomic nuclei contain only protons and neutrons, beta particles must be created at the instant of emission, just as photons are created at the time of emission of electromagnetic radiation. Because of this creation process, the amount of energy equal to the rest energy $m_e c^2$ of an electron must be consumed when beta decay occurs. Any remaining energy can be given to the beta particle as kinetic energy. The nuclear transitions producing beta decay are between discrete energy states differing by a definite amount of energy W_0, so we expect the total energy of a beta decay transition to be W_0. However, emitted beta particles are experimentally found to have a continuous range of total (rest plus kinetic) energy W of such magnitude that $m_e c^2 < W < W_0$, rather than all having a single energy W_0. This distribution as a function of energy (or momentum) forms what is known as a beta-ray spectrum. The shape of the spectrum depends on the sign of the charge on the beta particle (positive or negative), the energy W_0, and the particular nucleus involved. Unless energy and momentum are *not* conserved in the process, the energy not carried away by the beta particle must be given to some other particle. Furthermore, since the beta particle has a spin quantum number ½, angular momentum cannot be conserved unless another ½ unit of angular momentum can be disposed of. Both of these possible discrepancies in the conservation laws have been taken care of in the Fermi theory of beta decay through postulation of a massless particle, a neutrino or an antineutrino, which has a spin quantum number ½ and also carries away the remaining energy and momentum. Neutrinos were difficult to find experimentally but, even before they were experimentally detected, so much evidence had been obtained to show their existence that the Fermi theory of beta decay was generally accepted.

Gamma Rays

Gamma radiation is made up of photons or quanta of electromagnetic radiation that are emitted when an atomic nucleus undergoes a transition from one of its excited energy levels to a lower level. Like the terms "alpha" and "beta," "gamma" ray was used in the earlier years of radioactivity investigations when the exact nature of radiation was still a mystery. Gamma radiation is much more penetrating than alpha or beta particles. The presence of gamma rays from 30 mg of radium can be observed in an electroscope after passing through 30 cm of iron, as first reported by Rutherford. Gamma-ray energies range from 10^4 to 10^7 eV. They are often emitted as part of a nuclear reaction when an atomic nucleus is left in an excited stage or during an isomeric transition. Gamma rays also can be emitted following alpha-particle decay, beta-particle decay, or orbital electron capture, if the daughter nuclide is left in an excited state.

In the strictest sense, the term "gamma ray" is applicable only to photons produced as a result of transitions in atomic nuclei. However, the term is also sometimes used to denote bremsstrahlung radiation produced when the high-energy electrons in the beam of an electron accelerator, such as an electrostatic generator, synchrotron, or linear accelerator, strike the target of the accelerator.

Gamma rays carry away the full energy of the transition with which they are associated. As a result, if detecting systems are used that are capable of absorbing the full energy of the gamma ray, a spectrum of gamma-ray numbers as a function of energy shows a series of distinct peaks. On the other hand, the discrete energy characteristics of gamma rays are more difficult to observe if the detecting system separates the effects of different types of gamma-ray interactions with matter, such as the Compton, photoelectric, and pair production interactions. Under certain circumstances, a transition that would normally be expected to emit a gamma ray may sometimes release its energy through an internal conversion process.

In addition to other areas of interest, an entirely new branch of astronomy called gamma-ray astronomy has emerged. This is part of a trend in recent years to extend the capabilities of traditional telescopes to the measurement of gamma radiation and x-radiation, as well as infrared and ultraviolet radiation.

Neutrons

The neutron is an elementary nuclear particle that has no charge and a mass of 1.088665 amu. Thus, it is heavier than the proton by 0.00139 amu. The discovery of the neutron by Chadwick in 1932 represented a great step forward in the investigation of atomic nuclei. Chadwick found that radiation emitted when alpha particles from polonium reacted with beryllium could project protons from a thin sheet of paraffin wax. Although the radiation itself produced no observable ionization when passing through a gas, the protons released from the paraffin were detected in an ionization chamber. Inability to produce ionization was interpreted as a lack of electric charge. Chadwick correctly deduced that the so-called beryllium radiation must consist of neutral particles with a mass very nearly equal to that of the proton. The kinetic energy of neutrons has an important bearing on their behavior when interacting with nuclei and may range from near zero to as much as 50 MeV. It is therefore logical to classify neutrons in terms of energy according to their properties in each range of energy.

Thermal Neutrons. Thermal neutrons are in thermal equilibrium with the substance in which they exist; most commonly, neutrons have a kinetic energy of $\sim$0.025 eV, which is about two-thirds of the mean kinetic energy of a molecule at 15°C.

Epithermal Neutrons. Epithermal neutrons have energies just above those of thermal neutrons and range in energy between a few hundredths electronvolts and $\sim$100 eV.

Slow Neutrons. A less definite classification is that of slow neutrons, which have energies up to $\sim$100 eV, but "slow neutron" is sometimes used as a synonym for "thermal neutron."

Intermediate Neutrons. Intermediate neutrons have energies in a range that extends roughly from 100 to 100,000 eV. This range is above that of epithermal neutrons and below that of fast neutrons.

Fast Neutrons. Fast neutrons have energies exceeding 10^5 eV, although sometimes a lower limit is given.

Prompt Neutrons. Prompt neutrons are neutrons released coincident with a fission process.

Delayed Neutrons. Delayed neutrons are neutrons released subsequently in a fission process or, more generally, neutrons emitted by excited nuclei formed in any radioactive process (beta disintegration, in all cases thus far known). The neutron emission itself is prompt, so that the observed half-life is that of the preceding beta emitter.

Neutron Detection

Because it is a neutral particle, the neutron is detected by means of a secondary charged particle which it releases in passing through matter, or by means of the radioactivity which the neutron induces in stable elements. Protons may be projected by collisions with neutrons in hydrogenous material, and the ionization from the protons can be measured in an ionization chamber. Also, fission may be utilized for the detection of neutrons by placing fissionable material inside an ionization chamber and observing the ionization generated by the fission fragments.

Specialized Radiation Detectors

Although nuclear radiation detectors are concerned with the measurement of alpha, beta, and gamma radiation (and neutrons by inference from these measurements), a number of specialized instruments, embraced by this general classification, are marketed for highly specific needs.

For example, there are deuteron detectors, tritium detectors, radon daughter analyzers, and radioactive iodine (in air) detectors, among others. Deuteron is the nucleus of deuterium (heavy hydrogen), a particle containing one proton and one neutron, and is encountered in some nuclear reactors. Similarly, there is tritium, a radioactive isotope of hydrogen with mass number 3. The uptake of radioactive iodine (^{131}I) by humans as a result of exposure to a serious nuclear accident has brought about the development of special instrumentation for detecting ^{131}I in the air (particulates). This is accomplished by the use of a single-channel scintillation detector that rejects other radiations.

Nuclear Radiation Units

Curie

The *curie* (Ci) is the SI unit for rate of radioactive decay and is defined as the quantity of any radioactive material having 3.7×10^{10} disintegrations per second (dps). The *microcurie* (μCi), 3.7 $\times 10^4$ dps, and the *millicurie* (mCi), 3.7×10^7 dps are commonly used.

Roentgen

The *roentgen* (R) is a measure of the intensity of ionizing radiation in air. It is defined as the quantity of gamma radiation or x-radiation which produces 2.083×10^9 ion pairs (one electrostatic unit of charge, esu) per cubic centimeter of free air at a temperature of 0°C and a pressure of 1 atm. Inasmuch as the same number of roentgens from various types of radiation produce different amounts of body damage, a term reflecting relative biological effectiveness (RBE) was created.

Roentgen Equivalent Man

One *roentgen equivalent man* (REM) is the amount of ionizing radiation of any type which produces the same damage in humans as 1 R of ~200 kV x-radiation (1 REM = 1 rad in tissue/RBE). When the physical dose is measured in roentgen equivalent physical (REP) units, the approximate definition 1 REM $\approx$ 1 REP/RBE is used.

Roentgen Equivalent Physical

One REP is the amount of ionizing radiation of any type which results in the absorption of energy at the point in question in soft tissue to the extent of 93 ergs/g. It is approximately equal to 1 R of ~200-kV x-radiation in soft tissue.

Rad

A *rad* is an ionizing radiation unit which corresponds to the absorption of energy in any medium of 100 ergs/g. One rad in tissue = 100/93 REP.

Inverse Square Law

Radiation emitted by radioactive substances is uniformly distributed in all directions in space; thus the number of particles or quanta passing through a unit volume at any point distant from the source varies inversely as the square of the distance from the source, or

$$I = \frac{I_0 r_0^2}{r_i^2}$$

where I_0 is the radiation intensity at distance r_0 from a source and I is the radiation intensity at distance r_i from same source. The foregoing equation neglects absorption effects and assumes a point source. It is most useful for gamma radiation, where the source-to-detector distance is usually much larger than any source or detector dimension.

NUCLEAR RADIATION DETECTORS

Ionization Chambers

In an ion chamber, a potential is applied across a volume of gas to collect the ions produced by ionizing radiation. In the integrating type of ion chamber where construction is often of the cylindrical type, the outer cylinder is usually (but not necessarily) at positive high voltage and the inner signal or collector electrode is insulated from ground by a high-impedance insulator. This insulator must have high surface resistivity as well as volume resistivity, and in most cases this high surface resistance must hold for high-humidity conditions. The resistance must be on the order of 10^{15} Ω or higher for measuring small currents. Some of the most effective materials for this insulator include Teflon, fluorothene, polystyrene, quartz, and ceresin wax.

Surrounding this insulator is a grounded guard electrode, or guard ring, which serves as an electrostatic shield for the signal insulator, protects it from high-voltage leakage problems, and may be used to define the collecting volume of the chamber (by defining the electrostatic field). The guard ring must, in turn, be insulated from the high-voltage electrode of the system.

The saturated current from an integrating-type ion chamber is directly proportional to the amount of radiation present. Where air is the gas, the response is proportional to the density of the air, which is affected by atmospheric pressure and temperature. The following equation is useful for calculating the approximate current output of an ionization chamber:

$$I = \frac{kVPR \times 10^{-13}}{1.08}$$

where I = output current, A
   k = proportionality constant for type of gas (air = 1)
   V = chamber volume, cm^3
   P = gas pressure, atm
   R = radiation intensity, R/h

Ion chambers may be operated with several atmospheres of gas pressure. Gases used for ion chamber filling include air, argon, nitrogen, krypton, xenon, and boron trifluoride. Sealed ion chambers can be filled with a gas other than air under pressure, but the usefulness of the instruments for measuring low-energy x-rays and beta radiation is diminished because the wall thickness must be sufficient to withstand the internal pressure. The advantages of gas-filled ion chambers include less sensitivity to humidity, smaller size, low weight, ruggedness, no requirement for warm-up, fast response, and no correction needed for temperature or pressure.

The major advantage of air ion chambers is that they directly measure the exposure rate (roentgens per hour or milliroentgens per hour) over a wide range of photon energies. The major disadvantage is associated with attempts to measure a voltage drop across a high-megohm resistor, especially at low exposure rates, when the ionization current is extremely small. Moisture condensate inside the chamber or in the electronics can cause erroneous readings and sometimes render the instrument ineffective. Air ion chambers that are not sealed may also give erroneous readings in the presence of a high noble gas concentration, because the noble gas inside the chamber produces a memory effect. In such cases, the chamber may be temporarily sealed to keep the noble gas out of the chamber proper.

Ionization chambers are commonly used to survey a work area for external exposure rates. When beta radiation absorbed dose rates are to be measured, the ion chamber is designed with a thin window covered by a removable beta shield. An ion chamber instrument used for survey work is shown in Fig. 1. The photon energy and beta response for the instrument are shown in Fig. 2.

A design for an ionization chamber for a fixed location in a laboratory or reactor installation is shown in Fig. 3. The unit is used for the measurement of beta and gamma radiation, or gamma radiation only. Approximately one-half of the wall area is cut away in the form of four windows fitted with 0.005-in (0.1-mm)-thick cellulose acetate. This design is used for soft radiation. The chamber consists of a bakelite tube 5 in (12.7 cm) in diameter and 18.75 in (47.6 cm) long, with bakelite end plates. One end plate holds the high-voltage input and signal output connectors. At the other end of the chamber, a space is provided to hold a desiccant, such as silica gel, for operation under conditions of high humidity. The collector is an aluminum tube 0.5 in ($\sim$1.3 cm) in diameter and 17% in (44.1 cm) long, which is supported by Teflon insulation at the back end and by polystyrene

Fig. 1 Portable air ion chamber instrument used to detect beta, gamma, and x-radiation, with four linear ranges of operation, from 5 to 5000 mR/h full-scale. The detector is 3 in (7.6 cm) in diameter and 208 cm^3 in volume, with 200 mg/cm^2 phenolic walls inside a 0.05-in (1.3-mm)-wall aluminum case. The sliding beta shield is 400 mg/cm^2 phenolic on the bottom of the case with a positive friction lock. The window is 7 mg/cm^2 Mylar. *(Eberline, Division of Thermo Electron.)*

insulation at the connector end. Leakage resistance is 10^9 Ω or higher to either ground or high voltage. The inside of the cylinder and inside of the end plates are coated with Aquadag to form a conducting surface.

Geiger Counters

Also called Geiger-Müller (GM) tubes or counters, these devices are widely used for detecting ionizing radiation. A typical configuration is shown in Fig. 4. When operating in the Geiger region, the tube produces an output voltage pulse of approximately constant magnitude for each ionizing event taking place within the cylindrical electrode. The development of this output pulse depends on the production of an avalanche of ionization along the central wire electrode, possible only if the central wire is of sufficiently small diameter (typically less than 0.010 in, 0.25 mm) and a very high field gradient exists in the immediate vicinity of the wire. This high field gradient causes electrons attracted toward the central electrode to attain kinetic energies that are high enough to allow ionization of additional atoms of the gas inside the tube. Additional electrons, probably produced by photons emitted when some of the ion electron pairs recombine, are attracted toward the central electrode and produce complete ionization of the entire region immediately surrounding the central wire in about 10 μs. Because of their low mobility, the positive ions produced near the central wire build up as a sheath to destroy the high voltage gradient and render the tube inactive. This action results in a pulse of approximately constant magnitude for each ionizing event.

After the Geiger counter discharges and produces a pulse, it remains inoperative for a period of time called the *dead time.* This is the time required for the positive-ion sheath to move out from the wire to a position where the electric field can recover so that another avalanche can form. The *resolving time* of the counter is greater than the dead time and is determined by the point at which the pulse size becomes large enough to again trigger the electronic equipment. See Fig. 5. The *recovery*

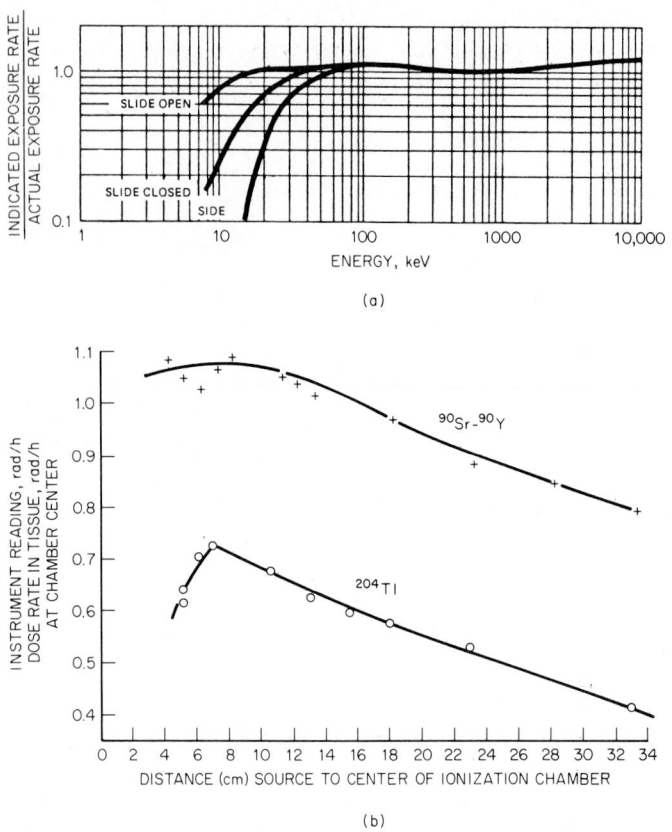

Fig. 2 Typical photon energy response (*a*) and typical beta response (*b*) for the portable air ion chamber instrument shown in Fig. 1. *(Eberline, Division of Thermo Electron.)*

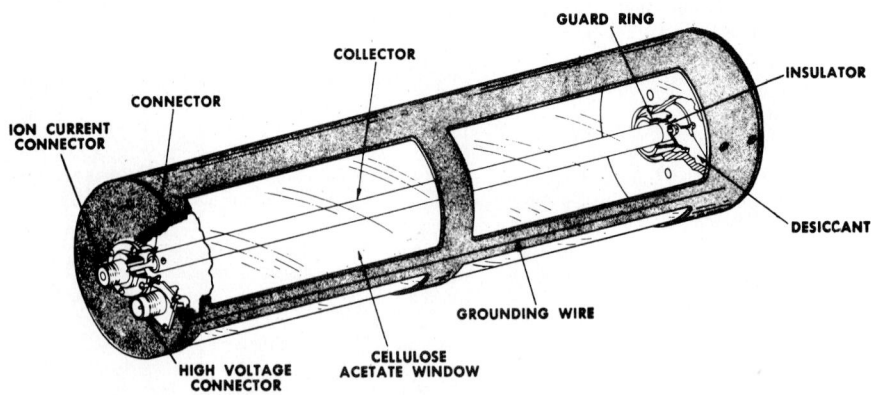

Fig. 3 Ionization chamber designed for fixed "hot" locations in a laboratory or reactor installation. The unit measures beta and gamma radiation (with a window) or gamma radiation only (solid wall construction). *(Technical Associates.)*

12.10

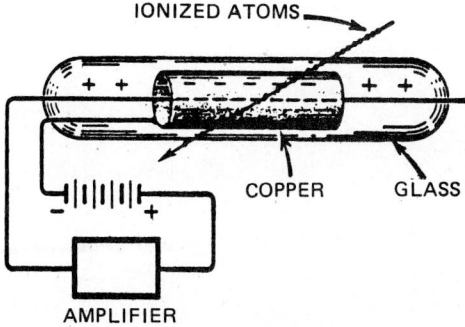

Fig. 4 Schematic illustration of a Geiger counter.

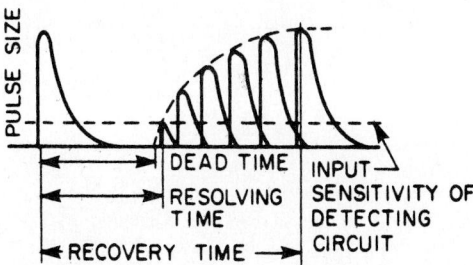

Fig. 5 Dead time and subsequent gradual recovery of pulse sizes in a Geiger counter.

time, greater still than the resolving time, is determined by the point at which the pulse again gains its original amplitude. All these factors determine the speed at which a counter can operate without losing a large number of counts. The dead time and recovery time are on the order of 100 to 200 μs for a typical Geiger counter.

Several different quenching vapors, such as amyl acetate, ether, and alcohol, can be used. Halogen gas, often mixed with an inert gas such as neon, is frequently used. The halogen molecule does not dissociate as does the polyatomic molecule. In actual practice, the useful life of an organic-quenched counter is on the order of 10^8 counts, whereas that of a halogen-quenched counter may be 10^{10} counts or more. Halogen counters, unlike organic counters, are not damaged when subjected to voltages above the plateau region.

Geiger counters are made in a variety of shapes other than the most common shape of the cylindrical shell and axial wire. Cleanliness cannot be overstressed in counter construction. The counter should be washed with a detergent or alcohol, followed with several rinses of distilled water, and then dried by heating. The central wire should have no sharp projections and be free of dust and lint. A typical counter with a 1-in (2.54-cm)-diameter shell and a 0.001-in (0.025-mm) wire, filled with ethyl alcohol to a pressure of 1 cm Hg and argon to a pressure of 9 cm Hg, will operate at approximately 1000 V. The pulse size varies with the counter dimensions and the voltage above the Geiger threshold. See Fig. 6.

All photons or beta particles that initiate ionization within the tube produce a pulse just as large as photons of high energy, and since each pulse has equal weight (each is one count), the Geiger counter tends to overrespond at photon energies between 40 and 200 keV when it has been calibrated for an exposure rate at higher photon energies, e.g., 600 keV. To compensate for this overresponse, Geiger-counter-type probes used for exposure rate measurements should be provided with energy compensating shields. Some commercial counters do not have this feature and thus should be used only to measure exposure rates for which they are specifically designed. For specific applications where the higher sensitivity of end window GM probes is required, an acceptable energy-dependent response characteristic can be achieved.

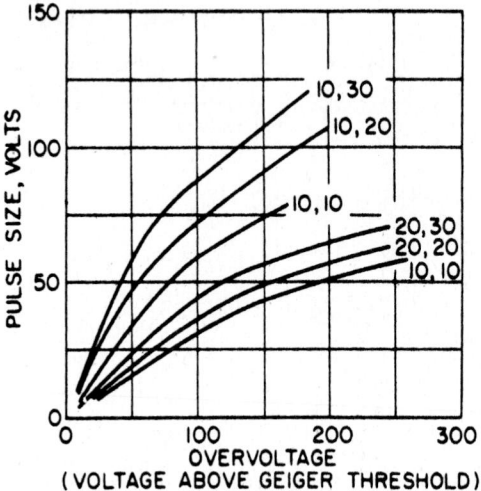

Fig. 6 Pulse size as a function of counter dimension. Tube dimensions are indicated as cathode radius (millimeters) and wire radius (thousandths of a millimeter).

The major advantages of GM instruments are their reliability, ruggedness, and relatively low cost. The pulses from GM tubes are large and easily counted. Changes in humidity, temperature, and atmospheric pressure do not affect the response, as is the case with ion chambers. Low-level exposure rates can be measured more reliably with an energy-compensated GM instrument than with an ion chamber. However, if the primary purpose is to measure exposure rates, the probe response should be relatively independent of photon energy between 40 keV and 3 MeV and the instrument should continue to read full scale in a radiation field that is 10 times the full-scale response. GM probes are not normally used for beta absorbed dose rate measurements because the

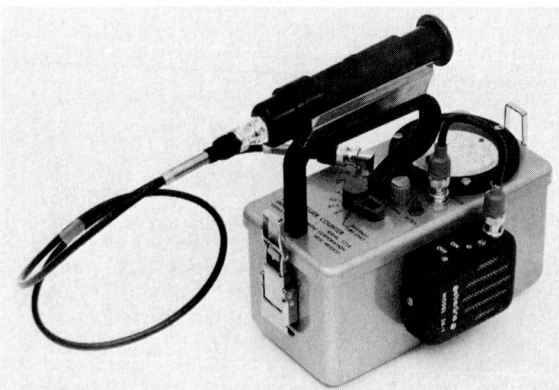

Fig. 7 A GM survey meter with five linear ranges: 0.2, 2, 20, 200, and 2000 mR/h. Response time varies from 2 to 10 s for 0 to 90% full-scale. The instrument is operational from −40 to 60°C. The meter is furnished with an internal, small, halogen-quenched GM tube with energy compensation and a sensitivity of approximately 100 cpm per milliroentgen per hour ^{137}Cs, and an energy-compensated hand probe. [*Model HP-270 with hand probe (HP-270) and speaker (SK-1), Eberline, Division of Thermo Electric.*]

response in terms of rads per hour is strongly dependent on the energy of the beta particles. In some cases, end window GM probes are used for this purpose, but the energy-dependent limitations must be understood and appropriate calibration provided.

A GM survey meter is shown in Fig. 7. The wide range (0 to 2000 mR/h) is obtained by the use of two detectors. An external hand probe is used for 0 to 200 mR/h, and an internally mounted GM detector is used for 0 to 2000 mR/h.

Some survey GM instruments are equipped with high-impedance headphones for detecting the clicks produced by the instrument.

A wall-mounted or portable laboratory monitor or ratemeter is shown in Fig. 8.

Fig. 8 A laboratory monitor-ratemeter for portable or wall-mounted use. The insturment is ac-coupled for GM detectors. There are 10 ranges to 3 million cpm and three response ranges (0.1, 1, and 10 s). A recorder output is provided for graphic data. The instrument is designed to accommodate a variety of detectors (including low-energy x-ray-sensitive detectors). Audible and visual alarms are provided. The alarm level can be preset in values from 10 to 100% full-scale for all ranges. A built-in loudspeaker produces a succession of audible clicks at a speed that varies with the count rate. The audio intensity may be set to loud, low, or off. The instrument is designed for battery or line voltage operation. *(Baird.)*

Gas Ionization Tubes

One of the simplest devices for radiation detection is a gas-filled diode (Fig. 9). Consider a gas-filled cylinder with a fine axial wire and an ionizing radiation particle traversing the tube volume. As the particle traverses the tube, the gas is ionized. This ionization consists of positive ions and electrons. As the fine central wire (anode) is at positive potential with respect to the cylinder (cathode), the

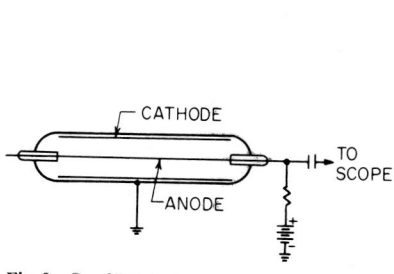

Fig. 9 Gas-filled diode.

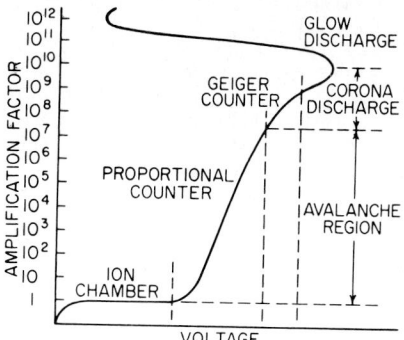

Fig. 10 Relation between voltage and amplification in a gas-filled diode.

electrons move to the wire in a region of increasing field strength and the positive ions move outward to the cathode. The electrons are collected in a short period of time during which the positive ions move only a short distance out toward the cathode.

At low voltages (ion chamber region) no additional ions are created by collision and the number arriving at the central wire is essentially the number produced during the initial ionizing event (neglecting recombination), as illustrated by Fig. 10.

The proportional region is reached when the counter voltage is raised to the point where secondary electrons are first formed by collisions of the primary electron with gas molecules. As the voltage on the counter is further increased, this multiplication of electrons increases until the phenomenon known as the *Townsend avalanche* occurs. This is a cumulative process in which the initial electrons and their progeny also produce more electrons. The pulse on a proportional counter is given by

$$V = \frac{Ane}{C}$$

where V = pulse size, V
  A = gas amplification
  n = number of electrons in initial ionizing event
  e = charge on electron, C
  C = distributed capacity of central wire system, F

The gas amplification A is defined as the total number of all electrons produced (including the original electron) as a result of each initial electron traveling from its origin to the central wire (A = 1 for an ion chamber). The proportional region is characterized by amplifications greater than 1, up to values of 10^5 or 10^6. In this region each discharge is a single avalanche, originates from a primary ion pair, and is limited to a small distance along the wire. The output voltage pulse is proportional to the number of ions formed during the initial ionizing event.

As the voltage is further increased, each avalanche produces new avalanches until the discharge spreads along the whole length of the central wire. The amplification increases until it is no longer dependent on the amount of primary ionization and all discharge pulses become equal in amplitude. This is the beginning (threshold) of the Geiger counting region.

Proportional Counters

In an ionization chamber, as previously described, the voltage is set high enough to collect all the ions produced in the chamber but low enough to avoid ion multiplication, i.e., production of additional ion pairs caused when the initial ions collide with molecules of gas. If the voltage is increased until additional ion pairs are formed, ion multiplication may be achieved and the output pulse will remain proportional to the energy of the initial ionization. Under these conditions, the ion chamber becomes a proportional counter, where the output pulse from the detector is proportional to the energy absorbed in the detector.

One type of proportional counter is the unsealed air proportional alpha counter used to measure alpha contamination. Until recently, air proportional counters did not have good alpha plateau characteristics or high counting efficiency. They were used only in areas of low humidity, because moisture in the probe and electronics caused erratic operation. Most of these problems have been resolved with improved electronics and new probe design. Good plateau characteristics, high counting efficiency, and good beta-gamma rejection have now been achieved. The problem with humidity is similar to that previously mentioned for ionization chambers. A major advantage of air proportional alpha counters is simplicity of operation coupled with low cost.

To avoid some of the problems encountered with air proportional counters, gases such as propane, methane, or methane-argon mixtures may be used. With use of the proper gas, the ion multiplication factor inside the probe is much greater and provides good plateau characteristics for alpha or beta contamination measurements. Excellent rejection of beta or gamma pulses is possible when measuring alpha radiation. Thus, some authorities consider the proportional alpha counter superior to a scintillation alpha counter for measuring alpha radiation in the presence of beta or gamma radiation. In a flowing gas configuration, the cylinder of counting gas must be periodically replaced.

Sensitive, fast-neutron area monitors which are not sensitive to gamma rays are important to nuclear power reactor, high-energy power accelerator, and medical cyclotron environs (Lucas, 1978). While many neutron detectors have been devised, few are directly sensitive to fast neutrons and, at the same time, insensitive to gamma rays. The helium 3 proportional counter has undergone continuous development for a number of years. Such counters can be fabricated which operate at pressures up to 10 atm with volumes of 3 to 4 L and useful lives of more than a decade. The counters are normally used to detect slow neutrons by way of the energy deposited in the counter as a result of stopping the reaction products of neutron capture:

$$n = {}^3He \rightarrow {}^3H + p + 0.76 \text{ MeV}$$

The helium 3 proportional counter can be used to infer the dose equivalent for neutrons having energies in the range of 0.2 to 20 MeV with sufficient accuracy for radiation safety purposes. Advantages claimed are durability, sensitivity, stability against temperature and high-voltage changes, and built-in calibration and stabilization on pulse height. The primary limitation is the inability to accurately infer the dose equivalent for intermediate-energy neutrons.

Scintillation Counters

A scintillation counter consists of a scintillation phosphor optically coupled to a photomultiplier tube, both enclosed in a lighttight shield. The particle to be detected loses energy in exciting and ionizing the molecules of the crystal. Energy in the form of light is radiated from these molecules, and some of this light falls on the photocathode of the photomultiplier tube.

Photoelectrons are ejected from the photocathode and accelerated to the first dynode where each photoelectron ejects several other electrons by secondary emission. This process of multiplication is repeated at each succeeding dynode by placing it at a higher potential than the preceding dynode. On arrival at the anode or collector, a multiplication on the order of 10^6 may thus have been produced. This charge (initiated by one particle) on arrival at the anode is then used to produce a voltage pulse on the anode circuit capacitance.

Scintillation Phosphors

The prerequisite property of a scintillation phosphor is that it be transparent to its own excited radiation. When an atom of any material is raised to an excited state by an ionizing particle, it emits light in the process of returning to the ground state. With most materials the atoms are equally capable of reabsorbing the light thus emitted. The peculiar property of phosphors useful for scintillation counting is that they are excited to a particular energy level by ionizing radiation such that the light energies emitted on return to the ground state do not correspond to the resonant absorption energies of that phosphor for light quanta. It is thus transparent to its own excited radiation. This property occurs naturally with certain phosphorescent crystals, particularly those of organic nature. Most inorganic crystals require "doping" with another element (notably thallium or europium) to produce this condition.

There are many properties to be considered in the selection of a phosphor, including conversion efficiency of energy-in to light-out, response to different particles and quanta, linearity of a phosphor to a wide range of energies of a particle or quanta, and decay time of light emission.

Several inorganic crystals possess characteristics desirable for general gamma-ray counting applications. Table 2 presents selected characteristics for the major inorganic crystals. NaI(Tl) is the most widely applied scintillation crystal and is regarded as the standard against which other crystals are compared. It offers high luminescence efficiency and good gamma absorption and is well suited for use with many types of photomultiplier tubes. Other crystals often selected for gamma counting include CsI(Na) and CsI(Tl). These high-Z crystals offer higher gamma absorption than NaI(Tl) and superior resistance to thermal and mechanical shock. Cesium fluoride (CsF) is a unique scintillation crystal combining both a high count rate capability ($\sim$10 MHz) and high gamma absorption and is slightly better than NaI(Tl). It possesses a decay constant of $\sim$5 ns, which is comparable to that of many plastic scintillators. CsF is hygroscopic and has a relatively low scintillation light output.

Table 2 Scintillation Crystal Materials Specifications

Material	Wavelength of maximum emission, nm	Decay constant, μs^a	Scintillation cutoff wavelength, nm	Index of refraction[b]	Density, g/cm^3	Hygroscopic	Gamma scintillation conversion efficiency %[c]
NaI(Tl)	410	0.23	320	1.85	3.67	Yes	100
CsI(Na)	420	0.63	300	1.84	4.51	Yes	85
CsI(Tl)	565	1.0	330	1.80	4.51	No	45
CsF	390	0.005	220	1.48	4.11	Yes	5
^{6}LiI(Eu)	470–485[d]	1.4	450	1.96	4.08	Yes	35
CaF_2(Eu)	435	0.9	405	1.44	3.19	No	50
$Bi_4Ge_3O_{12}$	480	0.30	350	2.15	7.13	No	8
BaF_2	325	0.63	134	1.49	4.88	No	10
TlCl(Be,I)	465	0.2	390	2.4	7.00	No	2.5
KI(Tl)	426	0.24/2.5[e]	325	1.71	3.13	Yes	24
$CaWO_4$	430	0.5–20[f]	300	1.92	6.12	No	50
$CdWO_4$	530	0.5–20[f]	450	2.2	7.90	No	65

[a]Room temperature, best single exponential decay constant, $I_0 e - \lambda t$.
[b]At wavelength of maximum emission.
[c]Referred to NaI(Tl) with S-11 photocathode response.
[d]Primarily used for neutron detection.
[e]KI(Tl) has two scintillation decay components for gamma excitation.
[f]Several decay components have been reported for tungstates.
Source: Harshaw Chemical Company, Solon, Ohio.

For applications requiring neutron detection, LiI(Eu) and ^{6}LiI(Eu) are widely used. These materials offer the convenience of scintillation techniques for thermal neutron detection. Because the neutron peak appears at a gamma equivalent energy of approximately 5.28 MeV, lithium iodide offers effective discrimination against gamma rays. This characteristic has made lithium iodide useful in oil and gas well logging.

CaF_2(Eu) is often used for alpha and beta detection because this low-Z crystal is nonhygroscopic and relatively inert, eliminating the need for an entrance window between the sample and detector in many applications. When machined into thin sections, calcium fluoride becomes increasingly transparent to high-energy gamma rays, yet it is efficient for the detection of alpha and beta particles. This characteristic has made CaF_2(Eu) useful in low background detectors, such as the *Phoswich*.

Bismuth germanate (BGO) is another high-Z crystal frequently used in gamma- and x-ray detection. Because it is nonhygroscopic and relatively inert, BGO offers many packaging options compared to crystals such as NaI(Tl) and CsI(Na) which must be hermetically sealed. The low-afterglow characteristics of BGO follow rapid changes in a high-flux x- or gamma-ray beam with a photomultiplier operated in the dc mode. The light output drops to ~0.1% within 3 ms after cessation of x-ray excitation.

Scintillators for specialized applications include BaF_2, TlCl(Be), KI(Tl), $CaWO_4$, and $CdWO_4$.

Phoswich Detectors*

High-efficiency detection of low-level radiation in the presence of an ambient background requires special design techniques. The Phoswich detector is composed of a thin front NaI(Tl) section optically coupled to a larger section of a different scintillation material [usually CsI(Tl) or CsI(Na)]. The entire detector package is optically coupled to a photomultiplier tube whose output is composed of signals originating within either one or both of the scintillators. See Fig. 11.

The two scintillation crystals viewed by the photomultiplier tube operate as an efficient low-background detector for the radiation of interest. The larger section acts as an anticoincidence shield, while the thin section transmits the unwanted high-energy radiation but has sufficient thickness to

*Developed by Harshaw Chemical Company, Solon, Ohio.

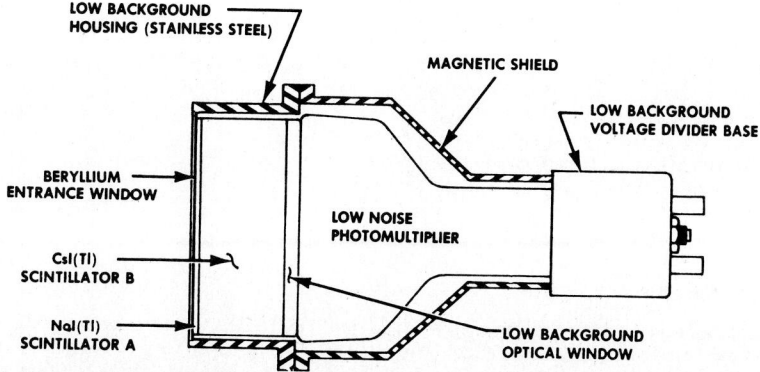

Fig. 11 Detector designed for high-efficiency detection of low-level radiation in the presence of an ambient background. The detector shown is of the Matched Window type for x-ray and low-energy gamma detection. *(Phoswich, Harshaw Chemical.)*

be essentially 100% efficient for the radiation of interest. See Fig. 12*a*. With this configuration, any high-energy radiation either passes through without interaction, as illustrated in Fig. 12*b*, or Compton scattering (Fig. 12*c*) occurs in the thin section. If Compton scattering occurs, the larger CsI(Tl) or CsI(Na) scintillator will absorb the scattered photon and supply an anticoincidence signal to allow electronic rejection of the event. Radiation entering from the rear or sides of the detector (Fig. 12*d*) is viewed by the anitcoincidence scintillator. If it undergoes Compton scattering into the thin detector, the event will still be recognized as spurious.

To ensure optimum performance of the system, the associated electronics must be able to separate the different signals arising from the two scintillation crystals. Separation is determined by the phos-

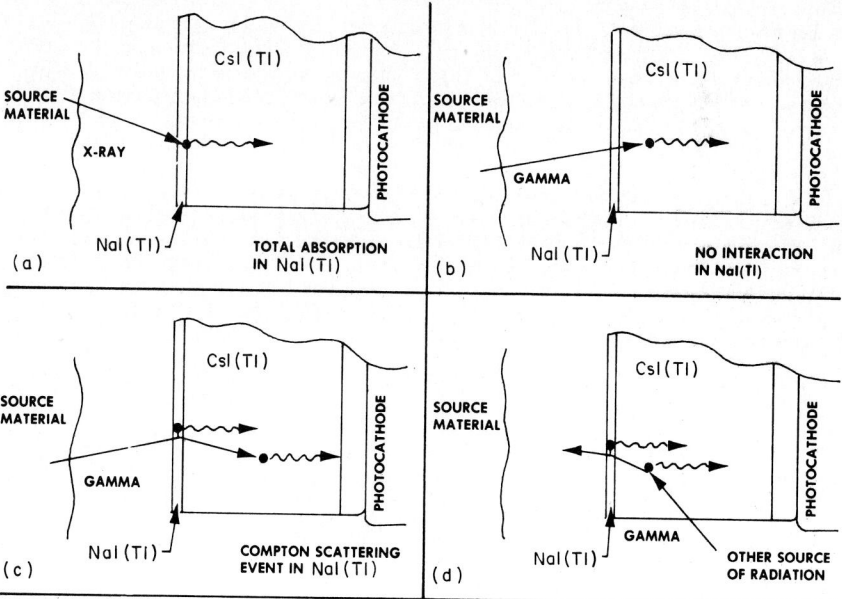

Fig. 12 Typical interactions in Phoswich detectors. (*a*) Total absorption in NaI(Tl), (*b*) no interaction in NaI(Tl), (*c*) Compton scattering event in NaI(Tl), (*d*) other source of radiation. *(Harshaw Chemical.)*

phor decay time of each material, and various techniques are used to separate these times. With "pure" signals (those originating in one or the other of the two crystals but *not* in both simultaneously) separation is relatively simple. However, a significant proportion of the events that should be rejected is composed of mixed signals originating from both crystals. Accordingly, the electronics must be capable of rejecting an event even if just a small fraction of the energy is scattered into the anticoincidence crystal. At the same time, the electronics must accept events of the desired energy that originate from total absorption in the primary NaI(Tl) crystal.

The Phoswich principle may be expanded to facilitate the detection of beta particles in the presence of a gamma-ray background by utilizing a $CaF_2(Eu)$ scintillator coupled to a larger NaI(Tl) anticoincidence crystal.

Massive Crystals

For detecting gamma rays and high-energy electrons in the gigaelectronvolt range, large-diameter crystals are required. In use, these detectors are coupled across the flats in a linear arrangement to permit side viewing by the photomultiplier tubes. The design also permits minimum crystal-to-crystal dimensions in the system.

Solid-State Detectors

The solid-state detector is an outgrowth of other advances in semiconductor materials and may be regarded as a parallel-plate ion chamber in which the plates separate proportionally to the applied voltage. The detector is shown in Fig. 13. The sensitive volume of the detector is known as the *depletion layer*. This is essentially the extent of the region at the junction of the semiconductor materials where a sharp potential gradient exists and consequently where ionization created by an external influence, such as nuclear radiation, is quickly removed with negligible recombination. The thickness of the depletion layer x is given by $x = \frac{1}{6} \sqrt{\rho V}$ approximately, where x is in micrometers, ρ is the conductivity of the semiconductor material in ohm centimeters, and V is the reverse bias voltage applied to the junction.

For measurement of the energy of ionizing particles, particularly alphas and protons, it is necessary that the particle dissipate negligible energy before entering the depletion layer. Two techniques are used. A very thin layer of p-doped semiconductor may be superimposed on an n-type semiconductor, thus forming a junction very near one outer surface; or a surface-barrier-type diode may be used. In the latter, the discontinuity of space charge within a semiconductor present at a surface is used to provide a high-potential-gradient region. The surface is coated with a thin layer of gold to provide a conductive contact.

The advantages of the solid-state detector are associated with its small size and the high specific ionization for these materials. This specific ionization, the number of ion pairs per electron volt or its reciprocal, the number of electronvolts of energy lost by the ionizing particle to create one ion pair, is 10 times more favorable for solid-state detectors ($\sim$3 eV per ion pair) than for gas chamber detectors. Hence the ionization intensity is an order of magnitude higher in the semiconductor for a given ionizing particle, and as a result the energy resolution of solid-state detectors is greater than that of gas ion chambers.

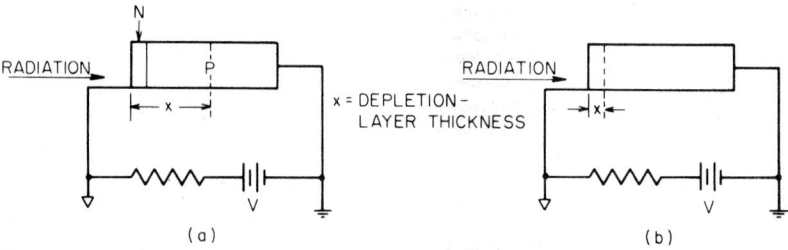

Fig. 13   (*a*) Junction diode solid-state detector, (*b*) surface barrier solid-state detector.

For integrated current measurement, the solid-state detector does not appear nearly so favorable when compared with the gas ionization chamber. The polarizing voltage determines the thickness of the active volume, so this detector cannot show a plateau characteristic similar to that for the gas ion chamber. Hence, a stable voltage is required for stable operation. More important than this is the fact that the semiconductor is a very low-resistance material compared with the gas in a conventional ion chamber. Consequently, a background current on the order of 10^{-6} to 10^{-7} A is passed by the solid-state detector compared with less than 10^{-16} A by a conventional ion chamber. It is very difficult to stabilize this ohmic current to permit satisfactory measurement of currents of 10^{-9} to 10^{-12} A, such as are associated with common radiation fields.

Personnel Dosimetry for Beta-, Gamma-, and X-Ray Exposure

Devices in this category are used (1) to control or limit personnel exposure and (2) for dose documentation (official records). Chirpers, alarming dosimeters, and self-reading pocket dosimeters fall into the first category. They measure only the photon (gamma- or x-ray) exposure or provide an audible indication of radiation intensity. Thermoluminescent dosimeters (TLDs) and film badges fall into the latter category and are used to estimate shallow (skin) and deep (whole body) dose equivalents from beta radiation, gamma radiation, and x-radiation. *Electrets* offer possible advantages in both categories for photon (x- or gamma-ray) radiation.

Available pocket gamma monitoring devices range from relatively simple to highly sophisticated devices. The least sophisticated are chirpers which respond to variations in radiation level by variations in the audible pulse rate. These instruments are not equipped with meters or other visual indicators and are effective in quiet locations as alerting or warning devices. Chirpers are relatively small and low cost and are the most widely used of the pocket gamma monitoring devices.

Film Badges

Personnel dosimetry for many years was based almost entirely on the use of small badges containing special x-ray film and various energy compensating shields. Although TLDs have largely replaced film badges at many larger facilities, especially where higher radiation exposures are expected, some low-risk users still rely on film badges because the associated dosimetry service is generally less expensive than that for TLDs. Some of the disadvantages of film include fogging as the result of mechanical pressure, elevated temperature, or exposure to light before development. Fading of the latent image results in a reduction in density, which is dependent on the time interval between exposure and development and the degree of exposure to moisture and heat. This may be minimized by using special packing to exclude moisture from the film and by storing the film under refrigeration prior to use. Film dosimeters also exhibit directional dependence, particularly for densities recorded behind metal filters. Microscopes are used for scanning and viewing the developed plates.

Thermoluminescent Dosimetry

Like film badges, TLDs must be read and thus require an in-house capability for this or use of an outside service. Lithium fluoride (LiF) is the most commonly used TLD material because its response to radiation is similar to that of human soft tissue (tissue equivalent) and thus it can be used to estimate more directly shallow and deep dose equivalents. The methods of annealing and readout of LiF can have a significant effect on such factors as fading and reproducibility. Therefore, the entire dosimetry system must be evaluated, along with the dosimeter. Other materials used or suggested for TLDs include lithium borate, calcium fluoride, and calcium sulfate. Generally these materials exhibit a strong overresponse to most x-rays and must be provided with energy compensating shields for estimating the deep dose equivalent. TDL readers range in complexity from manually operated to automated.

Electrets

An electret is a permanently polarized piece of dielectric material, produced by heating the material and placing it in a strong electric field during cooling. Barium titanate ceramics and certain polymer films can be polarized in this way. The electric field of an electret corresponds to the magnetic field of a permanent magnet. If a controlled volume of air surrounds an electret and this air is exposed to ionizing radiation, the charges created in the air will be attracted to the electret and injected into it. The apparent charge of the electret will decrease in proportion to the radiation exposure. In France, direct and indirect reading electret dosimeters have been developed and tested and, at some time, may impact on the self-reading pocket dosimeters, alarming dosimeters, and chirpers currently used.

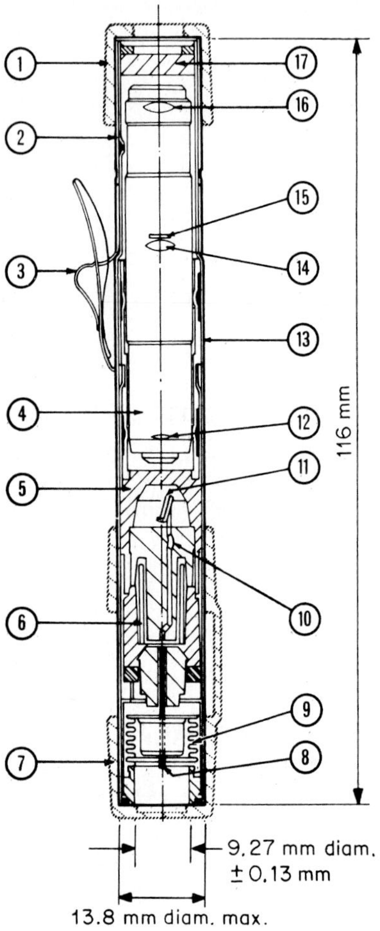

Electroscope

Like film badges, the electroscope has been used for many years for personnel monitoring. The Stephen dosimeter, for example, uses this principle. With reference to Fig. 14, an ionization chamber containing a moving quartz fiber electroscope is used. When the instrument is charged, positive charges distribute themselves over the fiber and the wire electrode on which it is mounted, causing the fiber to bend away from the electrode to a degree dependent on the amount of charge the system has received. The adjustable output in the charging unit allows the operator to set the fiber to a point corresponding to zero on the reticle. When the surrounding air in the ionization chamber is ionized, negative ions are attracted to the positively charged electrode, reducing its charge and producing a degree of movement directly related to the quantity of ionizing radiation. The rate of fiber movement is calibrated directly in roentgen units and is proportional to the level and duration of x- or gamma-ray exposure. Although primarily for personnel monitoring, the electroscope finds useful application in radiation measurement where the integrated dose at a point location is desired over a long exposure (hours or days).

116 mm

9.27 mm diam.
± 0.13 mm

13.8 mm diam. max.

Fig. 14 Simplified sectional diagram of R. A. Stephen quartz fiber dosimeter. (1) Protective cap, (2) marker sleeve, (3) pocket clip, (4) microscope body, (5) ionization chamber, (6) capacitor, (7) captive protective cap, (8) charging pin, (9) charging bellows, (10) electrode, (11) quartz fiber, (12) objective lens, (13) case tube, (14) field lens, (15) reticle, (16) lens eye, (17) window. Usual ranges are 0 to 200 mR, 0 to 500 mR, 0 to 1 R, 0 to 2 R, 0 to 5 R, and 0 to 10 R. Higher ranges (instrument contained in brass case) include 0 to 50, 0 to 100, 0 to 200, 0 to 500, and 0 to 1000 R. *(Baird.)*

Radiation Systems

Modern electronics has brought a high degree of sophistication to nuclear radiation monitoring and surveillance systems. Indicative of the manner in which the numerous types of radiation detectors previously described can be integrated into a total system is the very generalized block diagram of Fig. 15.

RADIOISOTOPES IN PROCESS INSTRUMENTATION

Industrial research and process control have found scores of uses for radioisotopes. Frequently a radiation source and detector form the basis for measuring a wide diversity of process variables—

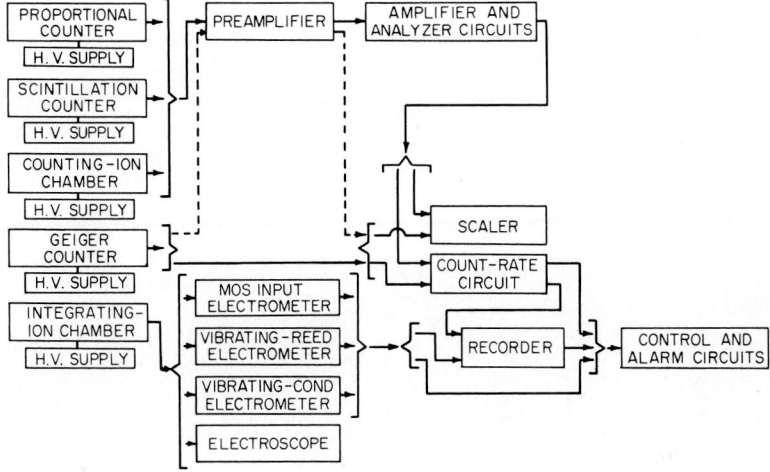

Fig. 15 Various types of nuclear radiation detectors and how they can be tied into allied equipment to achieve a complete recording, alarm, or controlling system.

weight and mass, particularly on moving conveyors, levels of slurries and bulk solids, thickness of materials, both stationary and moving, moisture content, and chemical composition. Radioisotopes are used in certain types of flowmeters, for measuring wear, for determining pathways in photosynthesis, and for establishing CO_2 fixation rates—not to mention their very extensive use in medical research, diagnosis, and therapy.

Isotope

The word "isotope" is derived from the Greek words *iso* (same) and *topos* (place). Atomic species of the same atomic number, i.e., belonging to the same element, but having different mass numbers, are called *isotopes*. Here, we are interested in radioisotopes. Isotopes distinguishable from other species of atoms with the same atomic number by radioactive transformation are known as *radioactive isotopes* or *radioisotopes*. Radioisotopes are useful as sources of radiation and as *tracers*. They can be used as tracers because they exhibit substantially the same chemical behavior as the stable isotope species, while emitting radiation, which permits determination of their identity and location. This tracer feature has been widely used in medical instrumentation. The principal radiations emanating from radioisotopes are alpha, beta, and gamma rays, described earlier in this article.

Half-Life

Radioisotopes are unstable; they decay in accordance with the relation

$$N = N_0 e^{-\lambda t}$$

where N = number of unchanged atoms at time t
  N_0 = number of atoms present when $t = 0$
  λ = constant characteristic of radioactive species, the decay constant

Lambda is related to the half-life $t\frac{1}{2}$ of the radioactive isotope by the relation

$$t\frac{1}{2} = \frac{0.693}{\lambda}$$

The constant λ relates the rate of disintegration A and the number of radioactive atoms N by the relation $A = N\lambda$, where A usually has units of disintegration per second. To date scientists have been unable to change the value of λ by varying such factors as temperature and pressure. The half-lives of the known radioisotopes cover the range from a fraction of a second to many years. For example, carbon 10 is a radioisotope with a half-life of 20 s, while carbon 14 has a half-life of 5100 years.

Radiation Absorption and Measurement of Density, Level, and Thickness

Measurements of density, level, and thickness all depend on a determination of the number of radiations per unit time penetrating the sample (a pipe filled with slurry, for example) and producing a measurable signal in the detector. An increase in the amount of matter between the source and the detector *usually* results in a decrease in the signal.

The exponential nature of the attenuation of beta radiation,* x-radiation,† or gamma radiation is indicated by the relationship

$$\frac{I}{I_0} = Be^{-\mu\rho t} \tag{4}$$

where I_0 = initial radiation intensity
I = radiation intensity through absorbing material
B = buildup factor dependent on energy and collimation of source, and on ρ and t. B accounts for radiation which has been "scattered" or changed in direction by intereactions that do not stop the radiation
μ = absorption coefficient, dependent on composition of absorbing material, energy of radiation, and source detector geometry
ρ = specific gravity of absorbing material
t = thickness of absorbing material

Optimum performance and accuracy are obtained for an absorption-type gage when the signal change is maximized for a given process or product change. The signal change is at a maximum for $\mu\rho t = 1$. For thickness gaging where ρ and t are fixed, proper selection of the source permits optimizing $\mu\rho t$. For density gaging the pipe diameter or path length t can be selected to optimize $\mu\rho t$.

Since all material between the source and detector contribute to scattering (a change in direction of the radiation) and absorption, errors in absorption-type gage readings can occur owing to changes in air density, pipewall thickness, and solids deposit. Gages must be regularly zeroed and calibrated. Simple solid standards usually are provided for calibration.

Scattering can produce a 180° change in direction of the incident radiation; measurement of this *backscattered* radiation, which is sensitive to the density and atomic number of the scattering material, provides a means of determining the density of compacted soils with a surface reflection gage. Backscattering also permits measuring the weight per unit area of a sheet on a roll or backup plate when the sheet (e.g., rubber) has a different atomic number than the roll (e.g., iron).

Neutron Moderation and Moisture Measurement

If a source of fast neutrons (Sb-Be, Pu-Be, ^{252}Cf) is placed on or in a medium containing hydrogen, some of the neutrons will be slowed down (moderated or thermalized) by collisions with the hydrogen atoms; the number of slow neutrons per unit area per unit time can be determined with a detector selectively sensitive only to slow neutrons (e.g., a ^{3}He- or BF_3-filled proportional or G-M detector). The relative response of the detector provides a measure of the hydrogen content of the medium surrounding or adjacent to the source. When water is the only or principal hydrogen-containing variable constituent of the medium, then the technique can be used to measure moisture content.

Gage readings are affected by changes in dry bulk density and to a lesser extent by composition changes of the medium. Calibration at regular intervals is essential.

Neutron moisture gages are used for highway, water content of snow, foundation, and agricultural purposes and are finding a place as process instruments in industry.

---

*Beta radiation has a finite range, whereas in theory x- and alpha rays are exponentially attenuated.
†X-radiation of each energy exhibits sharp changes in absorption coefficients for certain absorbers.

Statistics, Background, Time Constant, Response Time, and Accuracy

The inherent statistical nature of isotopic radiation gaging systems results from the fact that radioisotopes do not decay and emit radiations continuously but instead decay in a random manner. Because the disintegration rate is random in time, the *actual* number of radiations emitted in a given unit of time is related to the *average* value by a probability function.

The relationship between the actual or observed number of events (the measured signal) and the average becomes closer—in terms of percent deviation—as larger numbers of radiations are measured. A larger number of measurements can be attained by increasing the size of the source (number of millicuries), improving the geometry (moving the source closer to the detector), improving the efficiency of the detector (e.g., lining it with a high-atomic-number material such as platinum), or increasing the time taken for the measurement (related to system time constant*). An optimum gage design must consider all the above factors. Normally a change to improve the statistical response (or accuracy) involves additional cost or a slower response time.

*Time constant is the time needed by a gage to attain 63% of its new equilibrium value when subjected to a step input. For density gages 15 s is a common time constant with values from 2 to 60 s available.

Table 3 Properties of Common Radioisotope Radiation Sources

Radioisotope	Radiation of interest	Half-life	Energy of radiation	Principal use
Alpha Sources				
^{210}Po	α	138 days	5.3 MeV	Static elimination, thin-film thickness gaging
^{226}Ra	α	1620 years	5.15 MeV	Vacuum gage
Beta Sources				
^{14}C	β	5700 years	0.155 MeV	Thin plastic gaging
^{85}Kr	β	10.8 years	0.695 MeV	Light paper and plastic gaging
^{90}Sr-^{90}Y	β	28 years	0.61 and 2.18 MeV	Heavy paper, thin metal, rubber gaging
X-Ray and BS* Sources				
^{3}H	BS*	12.46 years	17.6 keV and down	S and Pb analysis
^{90}Sr-^{90}Y	BS*	25 years	2.18 MeV and down	Steel, Cu, Al sheet gaging
^{57}Co	X-ray	0.74 year	122, 136, and 14 keV	X-ray fluorescence analysis
^{210}Pb	X-ray	22 years	47 and 11–13 keV	X-ray fluorescence analysis
^{241}Am	X-ray	458 years	60 and 14–21 keV	X-ray fluorescence analysis
^{238}Pu	X-ray	86 years	12–20 keV	X-ray fluorescence analysis
^{55}Fe	X-ray	2.7 years	6 keV	X-ray fluorescence analysis and sulfur analysis
^{109}Cd	X-ray	1.3 years	88.2 and 22.2 keV	X-ray fluorescence analysis
Gamma Sources				
^{137}Cs	γ	33 years	0.66 MeV	Level density, thick plate gaging
^{60}Co	γ	5.2 years	1.2 and 1.3 MeV	
^{226}Ra and daughter	γ	1620 years	Up to 0.8 MeV	
Neutron Sources				
^{226}Ra-Be	Neutrons†	1620 years	Up to 13 MeV	Moisture and density gaging
^{210}Po-Be	Neutrons	138 days	Up to 11 MeV, avg. 4 MeV	Moisture and density gaging
^{124}Sb-Be	Neutrons†	60 days		Moisture and density gaging
^{239}Pu-Be	Neutrons	24,360 years		Moisture and elemental gaging
^{252}Cf (spontaneous fission source)	Neutrons†	2.65 years	2.3 MeV avg., 6.0 MeV max.	Activation analysis— moisture measurement
^{241}Am-Be	Neutrons	475 years		Moisture measurement

*Bremmsstrahlung.
†Also emits gamma radiation.

In most systems the detector responds to background radiation due to cosmic rays and naturally radioactive materials (mostly potassium, radium, thorium, and their decay products), as well as to radiation from the source. Statistical variations in this background radiation produce unwanted variations in the response of the system; background effects are minimized by shielding the detector with iron or lead and by using sources that produce radiation intensities several orders of magnitude above the background. Solenoid-operated "check" sources are a most desirable accessory for verifying operation of the detector readout system.

In making specific-gravity measurements for liquids or slurries, temperature changes can produce major signal generation changes; these can be balanced out electrically by using a temperature measuring device, or a correction can be calculated. Air bubbles, grease deposits, and pipewall corrosion can all cause major errors.

Radioisotope Radiation Sources

Important properties of the more commonly used radioisotope radiation sources are tabulated in Table 3.

There are literally hundreds of applications where radioisotopes play an important role in instrumental measurements. Several commercially available instruments employ these techniques. The highlights of proven radioisotope instrument applications are given in Table 4.

Table 4   Principal Applications for Radioisotope in Instruments

Variable	Range	Operating principle	Typical applications
Level	Off-on or wide range 10 ft (0.3 m)	Absorption of beta or gamma rays by liquid or solid; as level rises cuts radiation reaching the radiation detector	Slurries, viscous materials, solids, crushed rock, corrosive materials, aircraft and rocket fuel tanks, can filling level
Mass flow in pipes	Unlimited	Product of density measured by density meter $\times$ flow rate	Slurries, viscous or corrosive materials
Mass flow on belts	0–120 tons/h	Mass by gamma attenuation $\times$ belt velocity from tachometer	Pneumatic conveying of solids, coal, wood chips, potatoes, gravel on belt, vibrator, apron, drag chain conveyor
Density-specific gravity of fluids in ducts	0.0001 sp gr, units up	Relative absorption of radiation depends on mass of material between source and detector	Coal slurries, sewage sludges, food products, granular materials bulk density, black and green liquor density control, lime kiln feed consistency control, pipeline interface location
Density-specific gravity of solids	70–170 lb/ft^3 (1121–2723 kg/m^3)	Relative scattering or absorption of gamma rays	Compacted soil density for roads, footings, dams, asphaltic concrete
Thickness of a sheet or coating	0.0005-in (0.1mm) plastic film to 1-in steel (2.5 cm)	Relative absorption or back-scattering of radiation varies with thickness of sample beneath source	Plastic, paper, metal foil, rubber impregnation and coating weights of organic and inorganic materials and metals
Moisture content of soils and other solids	0–40 lb/ft^3 (0–640 kg/m^3)	Neutron slowing down affected by hydrogen atoms in water—measures relative number of slow neutrons when source of fast neutrons placed in or on sample	Soil moisture for construction of highways, airports; for agricultural purposes; water in concrete or coke
Composition Sulfur Metals, alloys, and minerals	0–5% 0 and up	X-ray absorption X-ray fluorescence or excitation—detect characteristic x-rays of elements of interest	S in fuel oil and other liquids, Identify alloys, ash content of coal, locate ore bodies, analyze powdered mineral samples

REFERENCES

Editor's Note: See the following articles and reference lists in this Handbook: Level Systems, in Sec. 5; X-Ray Analytical Methods, in Sec. 6; and Continuous Nuclear Weigh Scales, in Sec. 7.

Farukhi, M. R.: "Recent Developments in Scintillation Detectors for X-Ray CT and Positron CT Applications," *IEEE Trans. Nucl. Sci.*, Workshop on Positron Emission Tomography, October 21–23, 1981, San Francisco, Calif.

Geiger, E.L.: "Health Physics Instrumentation," Health Physics Society Meeting, Seattle, Wash., July 25, 1980. (Reprints available from Eberline, a Division of Thermo Electron Corporation, Santa Fe, N.M.)

ICRP: "Radiation Protection in Uranium and Other Mines," International Commission on Radiological Protection, *Rep. 24, Ann. ICRP,* vol. 1, no. 1, 1977.

Kathren, R. L., Selby, J. M., and E. J. Vallario: "A Guide to Reducing Radiation Exposure to as Low as Reasonably Achievable (ALARA)," DOE/EV/1830-T5, U.S. Department of Energy, Washington, D.C., April 1980.

Kroger, F. A.: *Some Aspects of the Luminescence of Solids,* Elsevier, Amsterdam, 1948.

Lucas, A. C.: "The Use of a Helium-3 Filled Proportional Counter as a Sensitive Dose or Dose Equivalent Meter," Nuclear Science Symposium, October 18–20, 1978, Washington, D.C.

Lucas, A. C.: "PIN Diode Neutron Dosimeters," *Tech. Pap.,* Harshaw Chemical Company, Solon, Ohio, 1978.

Lucas, A. C., and B. M. Shoffner: "A Stabilized Scintillation Crystal System for Monitoring Gamma Ray Exposure in the Environment," *IEEE Trans. Nucl. Sci.,* NS-23, 1976, p. 699.

Maisch, G., and G. Schmidt: "Methods for Measuring Engine Wear by Means of Radionuclides," *Kernitechnik,* vol. 33, 1979, p. 150.

NCRP: "Instrumentation and Monitoring Methods for Radiation Protection," *Rep. 57,* National Council on Radiation Protection and Measurement, March 1978.

NCRP: "A Handbook of Radioactivity Measurements Procedures," *Rep. 58,* National Council on Radiation Protection and Measurement, November 1978.

Sailer, S.: "State of the Industrial Isotope Technology," *Automobilindustrie,* vol. 3, 1976, p. 45.

Sailer, S.: "Wear Measurement with Radionuclides," *Met. Prog.,* April 1981, p. 67.

Upton, A. C.: "The Biological Effects of Low-Level Ionizing Radiation," *Sci. Amer.,* vol. 246, no. 2, 1982, pp. 41–49.

Williams, F.: "Theoretical Basis for Solid-State Luminescence," in *Luminescence of Inorganic Solids,* P. Goldberg (ed.), Academic Press, New York, 1966.

Electrical Energy Measurements

B. P. Gropp. *Research and Development, Weston Instruments Division, Sangamo/Weston, Newark, New Jersey. (Electric Meters)*

Walter A. Troeger. *Engineering Department, Weston Instruments Division, Sangamo/Weston, Newark, New Jersey. (Electric Power and Energy Measurement)*

Electric Meters

Ohm's law, $E = IR$, describes the relationship of the three basic electrical quantities—voltage, current, and resistance. Instruments used to measure these quantities are quite similar in many respects and generally fall into the overall category of *electric meters,* especially in terms of the industrial-commercial configurations. The fundamental electrical quantities are defined and described in the article Units and Standards of Measurement in Sec. 1 of this Handbook.

Because of its widespread use as a reference voltage, it is in order here to describe in more detail the zener diode. This is a type of silicon diode which acts like a rectifier until the applied voltage reaches the avalanche breakdown voltage *(zener voltage)*. At this point, the diode becomes conducting. The voltage drop across the diode remains essentially constant and independent of the current. The diode is used in voltage limiting circuits and power supplies. Zener diodes are also used as precise reference signals. A potentiometric instrument can be no more accurate than the calibrated (standard) source of electromotive force (emf) to which the unknown potential is compared. Standard cells and a variety of batteries calibrated against standard cells have been of prime use, but during the past decade or so batteries, such as the mercury cell, have been used as a standard when accuracy tolerances permit. Since the 1960s, the zener diode used in an appropriate circuit has become most useful by offering low cost, very long life, small size, and accuracies equivalent to or better than those of standard cells, particularly over long periods. A typical circuit using zener diodes is shown in Fig. 1. The resistors R_1, R_2, and R_3 and the dynamic resistance of CR_1 constitute a balanced Wheatstone

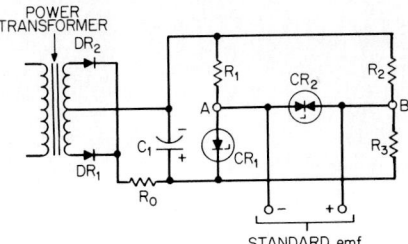

Fig. 1 Bridge-type circuit in which a zener diode is used as a constant voltage source.

bridge which tends to eliminate variations in V_{supply} from the nodes of A and B. CR_1 and CR_2 are in a conventional cascade configuration, because R_2 is inherently large compared to R_1 and R_2. CR_2 is also a temperature-compensated zener diode, and the resulting standard emf can be held to within 0.02% over a normal working range of supply voltage and ambient temperature.

Brief Review of Electric Meters

Electrical instruments have been undergoing continuous development for at least 150 years. During most of that time, the principal effort has been directed toward the perfection of pointer deflecting instruments. In these devices, the deflection angle of the pointer is a function of, and thus analogous to, the value of the electrical quantity measured. Therefore, the name "analog meters" was coined to distinguish them from the meters in which the value of the quantity is displayed in numerals (digital meters). As in most other areas of industrial instrumentation, there has been a trend in recent years toward the solid-state digital indicating electric meters, but analog indicators remain popular and, as of the mid-1980s, considering established as well as new installations, probably predominate, particularly for panel-mounted switchboard applications. Measurement principles are discussed in more detail later in this article.

Current Meters

Fundamental methods used for current measurement include the coulometer, the potentiometer (see Handbook Sec. 14), and industrial-commercial ammeters. Principal measurement methodologies used include permanent-magnet moving-coil instruments, moving-iron instruments, electrodynamic instruments, and electronic instruments.

Ammeters for panels and switchboards are obtainable with ranges from 0 to 1 to 0 to 500 A, milliammeters from 0 to 1 to 0 to 500 mA, and microammeters from 0 to 5 to 0 to 500 μA. The ranges for ac and dc instruments are similar. The meters are obtainable in both circular and rectangular front formats. Scale lengths from 1½ to 7½ in (3.8 to 19 cm) are available in stock models. Panel meters in which the scale is optically projected (moving scale) or in which the pointer is optically projected (moving pointer) are obtainable. The accuracy of stock panel ammeters ranges from ±2 to ±3% full-scale. Ammeters also are available in portable form for testing and laboratory use where the accuracy generally ranges from ±1 to ±2% full-scale, although special, higher-cost instruments are available for greater accuracy. Ammeter ranges also are obtainable in multimeter form where voltage and resistance as well as current may be measured and displayed. Most multimeters are portable. Like any other instrument signal, the input can be used for practically any kind of display (see Handbook Sec. 15). Shunts and instrument transformers frequently are used in connection with ammeters to stretch their range.

A well-designed dynamometer-type instrument is satisfactory for audio frequency measurements, while thermocouple ammeters can be used for all frequencies from power frequencies up to and including radio frequencies. Through careful design, the errors of a thermocouple ammeter due to frequency can be kept smaller than 1% for up to as high as 50 MHz.

For applications requiring that an instrument be carried to the place of measurement there are portable instruments. These generally are of higher accuracy than panel and switchboard instruments, so that the accuracy of the latter can be checked on location. Although very few electrodynamic instruments are found in panel instrument form for reasons of cost, they are quite frequently encountered in portable form where their greater accuracy and flexibility of use justify the higher cost. Laboratory instruments usually are of the permanent-magnet, moving-coil type or the electrodynamic type, both in the interest of highest achievable accuracy. Iron-vane-type instruments are not in the same accuracy class because of inherent limitations of their operating principle.

Voltage Meters

Fundamental methods used for voltage measurement include the potentiometer and electrical bridge (Handbook Sec. 14), the sphere-gap voltmeter, and industrial-commercial voltmeters. Principal measurement methodologies used include permanent-magnet, moving-coil instruments, moving-iron instruments, electrodynamic instruments, electrostatic instruments, and electronic instruments.

For emf's of 1.5 V downward into the millivoltage and microvoltage ranges, potentiometers may be used, particularly in the laboratory where high precision is required. However, very rugged potentiometer-type transducers have been developed for the measurement of millivoltage outputs from various sensors where the final display usually is not in terms of voltage but rather in terms of some other variable. (For example, see the article Pressure Transducers and Transmitters in Sec. 3 of this Handbook.)

Electrostatic voltmeters measure ac voltage from 150 to 3500 V.

For very high voltages, the measured distance between two electrodes and the beginning of a corona discharge provides a measure of voltage. This is the standard method for measurement of the crest value of alternating currents. Numerous investigators have calibrated sphere gaps with sphere diameters ranging from a few centimeters up to 2 m.

Commercial and industrial panel and switchboard voltmeters of the permanent-magnet, moving-coil type are obtainable for dc voltage measurements from 0 to 1.5 V to 0 to 1 kV, and for ac voltage measurement from 0 to 1 V to 0 to 500 V. Scale lengths from 1½ to 7½ in (3.8 to 19 cm) are obtainable in standard models with both circular and rectangular front formats. The accuracy of stock panel voltmeters ranges from ±2 to ±3% of full scale. Permanent-magnet, moving-coil voltmeters are also available in portable form for testing and laboratory use where the accuracy generally ranges from

±1 to ±2% full scale, although special higher cost instruments are available with greater accuracy. Voltmeters also are obtainable in multimeter form where current and resistance as well as voltage may be measured with the same meter.

Moving-iron voltmeters are available in formats, ranges, and characteristics similar to those just described. They are less complex in design but require more power. They are used mainly for ac voltage measurements of modest accuracy.

Thermocouple voltmeters employ a permanent-magnet moving-coil millivoltmeter to show the emf setup in the thermocouple attached to the heater wire (through which the unknown current is passing). With a suitably constructed resistance multiplier, the instrument may be used at frequencies up to 15 kHz or higher and for voltages from 10 V to several hundred volts.

Electronic voltmeters are used for measuring ac emf's from a few microvolts up to several hundred volts. With capacitance dividers, the range can be increased to even higher voltages. Such meters draw practically no current from the source. Input impedance is on the order of megohms. With careful design, such meters can give correct indications at frequencies up to 100 MHz.

Rectifier voltmeters are solid state and capable of working at high temperatures and operating frequencies. Full-wave ac rectification may be obtained by connecting a bridge network formed of separate half-wave rectifier units to a permanent-magnet moving-coil instrument (Fig. 2). Appropriate series resistances are used to provide the desired ranges. Deflections are proportional to the average value of the current flowing through the indicator, and thus a practically uniform scale is achieved. Errors result if the wave shape departs appreciably from a sine curve.

Voltmeters are further described later in this article.

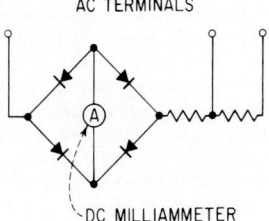

Fig. 2   Rectifier-type voltmeter.

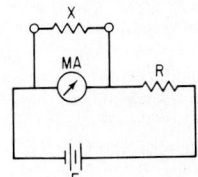

Fig. 3   Parallel ohmmeter.

Resistance Meters

An instrument or meter arranged to permit the measurement of resistance is termed an *ohmmeter*. An ohmmeter usually is actuated by a self-contained or external source of voltage, such as a battery. In contrast to resistance bridges which permit the comparison of a known with an unknown resistance, direct reading ohmmeters are milliammeters calibrated in ohms. The two simplest types are those in which the actuating current is a direct or an inverse function of the resistance connected to the instrument terminals. In both types, the actuating voltage is assumed either to remain constant under all conditions or to be adjustable so that the pointer can be set to a specific mark on the scale with the unknown resistance either open-circuited or short-circuited, depending on whether the instrument is of the parallel or series type.

The basic circuit of a *parallel ohmmeter* is shown in Fig. 3. The unknown resistance is connected in parallel with a milliammeter. The current through the milliammeter rises with increasing resistance X, so that 0 on the meter scale corresponds to 0 resistance, and full-scale corresponds to infinite resistance. The supply voltage E will remain reasonably constant if a resistance R is inserted to limit the total current that can be drawn from the source.

The circuit of a *series ohmmeter* is shown in Fig. 4. The unknown resistance X is connected in series with a milliammeter. As the resistance of X rises, the current through the meter diminishes and becomes 0 when X approaches infinity. In order to make full scale on the meter correspond to $X = 0$

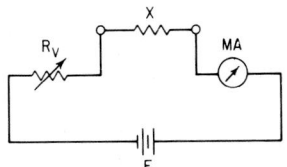

Fig. 4   Series ohmmeter.

without causing the low meter resistance to draw excessive current from the source E when X is low, a resistor R_v is inserted. Usually R_v is made adjustable so that the meter can be set at top mark (0 on the resistance scale) with the terminals for the unknown resistor short-circuited out, thus compensating for any variation in the supply voltage E.

The desire to make the readings of a direct reading ohmmeter independent of fairly wide variations in supply voltage without the need for manual adjustment has been fulfilled by *ratio meters* whose pointer deflection corresponds to the ratio of the voltage across an unknown resistance to the current flowing through it. Instruments of this type usually balance the torques produced by two moving coils attached to the same shaft so that the position of rest represents one specific ratio.

Most ohmmeters are offered in combination with a multiplicity of current and voltage ranges—in the form of multimeters or analyzers that serve a wide variety of purposes.

Frequency Measurements

Electric frequency is the number of periodic swings or cyles of voltage or current in a unit of time. Measurement can be accomplished by (1) counting the number of cycles in a given time, (2) measuring the time in which a given number of cycles occurs, (3) balancing an impedance bridge in which the impedances of the arms are known functions of frequency, (4) tuning electric circuits to a condition of resonance, (5) reasonating tuned reeds magnetically, and (6) using a deflection-type frequency meter.

Counting requires electronic means, including sources of pulse trains, pulse counters, and electronic switches (gates). Very wide ranges of frequency up to many millions of hertz are measurable in this manner. Time measurements require similar instrumentation. Methods 3 and 4 require manual adjustment of capacitors or other circuit elements and the observation of a balance or resonance detector in a bridge circuit. Tuned reeds are limited to a fairly narrow range of electromagnetically produced mechanical oscillations. Instruments based on this method usually extend little beyond the common power frequencies. The usual deflection-type frequency meter employs a rotatable iron vane whose position is determined by the superimposition of two stationary magnetic fields whose resultant changes its orientation as a function of frequency. These fields are produced by two coils, one of which is connected into a resistive circuit; and the other into an inductive circuit. As the frequency increases, the current in the latter decreases. Such instruments are widely used in electric power installations and are calibrated for the range of frequencies encountered in the field.

Oscilloscope Frequency Comparison

This method involves the generation of Lissajous figures on the face of a cathode ray tube. A source of known frequencies is connected to one set of deflection plates, and the unknown frequency is connected to the other set. If one frequency is a harmonic or integral multiple of the other, a stable and recognizable pattern will be displayed.

Electronic Frequency Meters

The input signal is amplified and converted to a square wave which charges a capacitor through a diode. A second diode discharges the capacitor through the meter circuit. The current through the meter is proportional to the rate of the charging pulses. Hence it is proportional to the frequency of the input signal. In another configuration, the input waveform drives a trigger circuit which generates a series of negative trigger pulses. These actuate a constant current source and a linear timing circuit. The output of the latter turns off the current source, thus determining the width of the current pulse delivered to the meter circuit. One such stable pulse is passed through the meter circuit for each input cycle. The meter averages the pulses it receives and presents an indication that is proportional to the average frequency.

Heterodyne Frequency Meter

This instrument essentially is a frequency-calibrated stable oscillator, a mixer, and an indicator. An amplifier for driving the beat indicator also is included. The output signal may be used to drive external indicators, such as headphones, oscilloscopes, or recorders. With the unknown frequency applied to the mixer, the oscillator usually is tuned for a zero beat, which is indicated by a minimum or zero output from the mixer. The oscillator frequency then is the same as, or a subharmonic of, the frequency of the applied signal. Heterodyne techniques also are used when the oscillator is tuned for a difference frequency in a specified range of frequencies instead of for a zero beat. The difference frequency then is determined by a separate frequency measuring circuit to obtain the desired information. By using oscillator harmonics to zero beat against the unknown, the basic range of the instrument may be extended by 50 to 100 times the highest fundamental frequency available from the oscillator. Many heterodyne frequency meters also include crystal calibrators which are used to check accuracy at various points on the oscillator frequency control dial. A heterodyne frequency meter with a crystal calibrator can provide measurement accuracy of up to 1 part in 10^6 or better.

Electronic Counter

This instrument consists of a time-base generator, a signal gate, and decade counting units. Frequency is measured by counting the number of input cycles over a precisely controlled period of time. The time-base generator develops control signals which are applied to the signal gate. When the first or "start" signal is received, the signal gate opens to pass input pulses from the unknown frequency source to the decade counting units. When the second or "stop" signal is received, the signal gate closes to prevent further input pulses from reaching the decade counting units. Totalization of input pulses by the decade counting units during the interval when the gate was open is a measure of the input signal frequency.

The frequency measurement accuracy of an electronic counter is ± 1 count plus or minus the accuracy of the time-base generator. In one standard arrangement frequencies up to 500 MHz or more are measured with a 10-MHz counter and a frequency converter. With a transfer oscillator or other harmonic generator-mixer arrangements, the range is further extended to at least 40 GHz. At lower frequencies the accuracy of measurement is the basic accuracy of the electronic counter. Inherent characteristics of transfer oscillators limit the measurement accuracy at higher frequencies to about 1 part in 10^7.

Slotted-Line Measurement of Frequency

An electronic counter and transfer oscillator can be used for very accurate measurements of frequency in the ultrahigh frequency and microwave bands. For some measurements, where an accuracy on the order of $\pm 0.5\%$ is sufficient, the frequency may be determined with a slotted line or slotted section by observing the standing-wave pattern.

Lumped-Constant Wavemeters

These instruments are used up to frequencies of 1200 MHz. A crystal detector is coupled to a resonant LC circuit. For frequencies up to 100 MHz, a fixed inductor and a variable capacitor are used to form the resonant circuit. Plug-in coils of various inductance values are used to cover the frequency range. The capacitor dial is calibrated to read the frequency directly. When the wavemeter coil is in the presence of the field of the unknown signal, the circuit is tuned to resonate at the same frequency as the unknown signal. The output of the crystal detector is indicated by the meter circuit. In wavemeters operating above 100 MHz a butterfly resonant circuit is used. This element varies both the inductance and capacitance simultaneously. The accuracy of a lumped-constant wavemeter depends on the Q of the resonant circuit. In general, accuracies between ¼ and 1% may be obtained.

Cavity-Type Wavemeters

These instruments are applicable for determining frequency in a waveguide system. Typically, a cylindrical cavity and piston are used. As the position of the wavemeter piston is varied, the distributed inductance and capacitance of the cavity, hence the resonant frequency, are changed. In a reaction-type wavemeter, there is a drop in the transmitted power when the piston is adjusted for resonance, because power is absorbed in the tuned cavity. For this reason, a wavemeter in a waveguide system should be detuned so that maximum power is transmitted to the load. Accuracy of a cavity-type wavemeter depends on the selectivity, or Q, of the resonant cavity. Commercial wavemeters are available with accuracies of 0.1 to 0.01%.

Measurement of Electric Charge

Instruments for measuring electric charge are not usually routinely grouped with electric meters, notably those of the switchboard type. Electric charge is measured by *electrometers* of various configurations. The familiar attraction and repulsion of the suspended-gold-leaf electroscope exemplifies the fundamental principle of the instrument. Numerous forms of electrometers have been devised. In the Wulf electrometer, two silvered or platinized quartz fibers are placed side by side so that they tend to bulge apart on being charged. Similar fibers are used in the string electrometer in which the fibers are lightly stretched and deflect laterally when in an electric field. The behavior of the fibers in either design can be observed with a low-power micrometer microscope. Devices of this type are most widely used in nuclear radiation and cosmic ray applications. See the article Nuclear Radiation Instrumentation in Sec. 12 of this Handbook.

MEASUREMENT METHODOLOGIES USED IN BASIC ELECTRIC METERS

The basic electrical analog instrument can be traced to Oersted's discovery (1819) of the relation between current and magnetism. Faraday (1821) learned that a current carrying conductor would rotate in a magnetic field (Fig. 5). Ampere (1821) demonstrated that a current in one conductor attracted or repelled another current carrying conductor. Sturgeon (1836) wound the current carrying wires into a coil and suspended it in a magnetic field. Kelvin (1867) placed a soft-iron core in

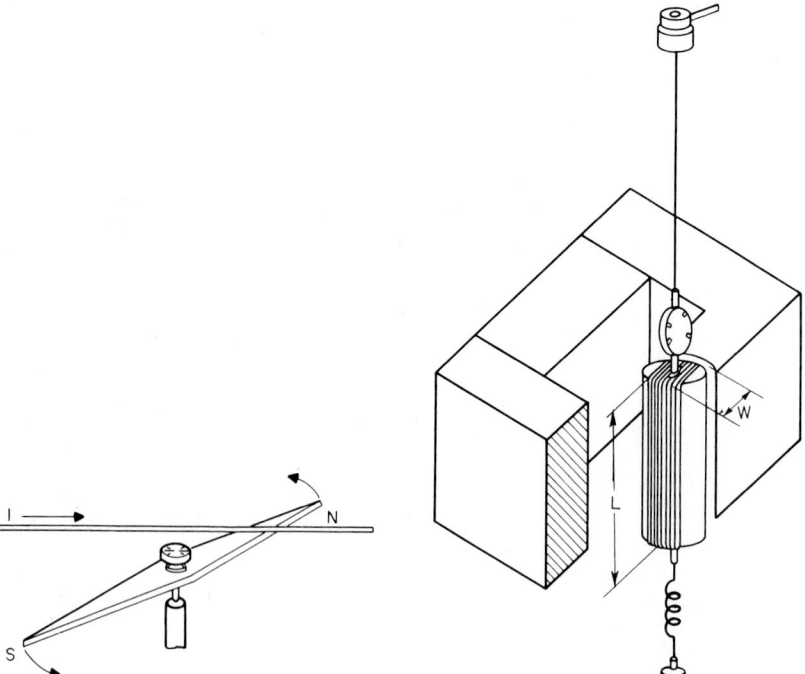

Fig. 5   Action of a compass needle when an electric current is passed through an adjacent wire.

Fig. 6   D'Arsonval galvanometer, the forerunner of industrial-type permanent-magnet moving-coil instruments.

the center of the coil, shortening the air gap, increasing the sensitivity, and improving the scale characteristics of the device. D'Arsonval (1881) perfected and patented an instrument of this type (Fig. 6). Weston discovered the factors required to produce a permanent-magnet system, added soft-iron pole pieces, devised current carrying control springs, and produced the first commercial double-pivoted permanent-magnet moving-coil instruments. From Oersted's and Kelvin's work also stem the principles on which polarized iron-vane and electrodynamic types of instruments evolved through the further work of Ayrtron and Perry (1881), Bruger (1886), and Weston (1889), among others. Digital instruments, particularly those based on solid-state technology, thus are relatively recent in the history of electric meters.

Permanent-Magnet Moving-Coil Instruments

The operating principle of these instruments is identical to that of D'Arsonval galvanometers. The principle of operation of the permanent-magnet moving-coil mechanism is shown in Fig. 7. The

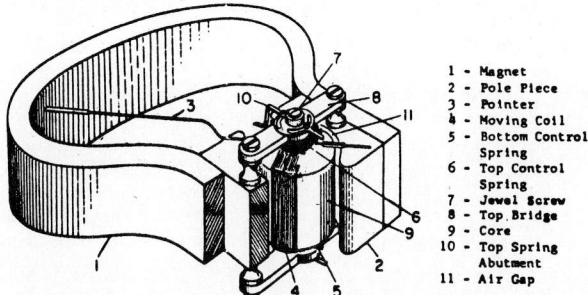

1 - Magnet
2 - Pole Piece
3 - Pointer
4 - Moving Coil
5 - Bottom Control Spring
6 - Top Control Spring
7 - Jewel Screw
8 - Top Bridge
9 - Core
10 - Top Spring Abutment
11 - Air Gap

Fig. 7   Earlier design of a permanent-magnet moving-coil mechanism in which an external magnet was used. *(Weston.)*

reaction of a permanent-magnet flux with the current establishes the torque. The induced electromagnetic torque is balanced by the mechanical torque provided by control springs attached to the movable coil. This permits the angular position of the movable coil to be indicated by a pointer against a fixed scalar reference. With the advent of Alnico and other improved magnetic materials, it became feasible to design a magnetic system in which the magnet serves as a core. Such magnets operate at their highest energy product with minimum lengths, thereby making the core magnet mechanism a practical reality. These mechanisms have the advantage of being relatively resistant to external magnetic fields, thus allowing the elimination of magnetic shunting effects in a panel and

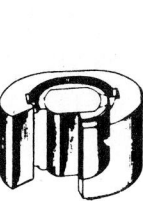

Fig. 8 Core magnet used in a permanent-magnet moving-coil instrument. *(Weston.)*

Fig. 9 Movement of a permanent-magnet moving-coil instrument which incorporates a core magnet. *(Weston.)*

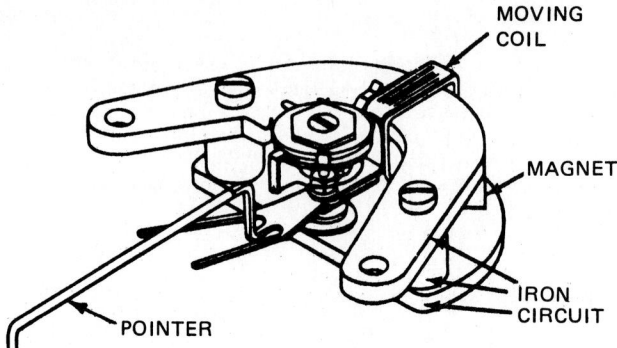

Fig. 10   Representative edgewise meter mechanism used in a permanent-magnet moving-coil instrument. *(Weston.)*

the need for magnetic shielding in the form of iron cases. See Fig. 8 and 9. A representative edgewise meter mechanism is shown in Fig. 10.

A typical moving coil for a permanent-magnet moving-coil system is shown in Fig. 11. Most voltmeter coils have metal frames for damping—a short-circuited turn in a magnetic field; most ammeter coils are frameless—the coil turns are shorted by the shunt. The pointers, springs, and pivots are assembled to the coil by means of pivot bases. The moving system is statically balanced for all positions by balance weights.

Torque Equation

Assume that the device (Fig. 12) has the following parameters:

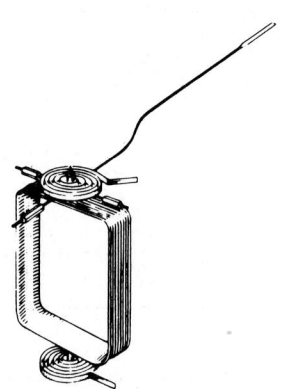

N = number of turns in coil
W = width of coil, cm
L = length of coil, cm
B = field strength in which coil operates, G*

Fig. 11   Moving coil for a permanent-magnet moving-coil instrument. *(Weston.)*

The force on a conductor 1 cm long carrying 1 abampere (10 A) in and normal to a magnetic field of 1 G or 1 line/cm² is 1 dyne at right angles to both the conductor and the field. The total force on one side of the coil of Fig. 6, when it is carrying a current of I' abamperes, is

$$F = I'BNL \tag{1}$$

Since the force is applied at a radius of $W/2$ cm, the turning moment, or torque, in dyne centimeters, is

$$T = \frac{I'BNLW}{2} \tag{2}$$

---

*These relationships also apply, of course, when SI terms, such as tesla for gauss, amperes per meter for oersted, and weber for maxwell are appropriately substituted.

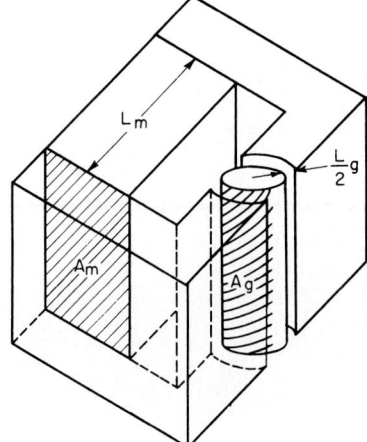

Fig. 12 Magnetic system for a galvanometer, with important dimensions indicated.

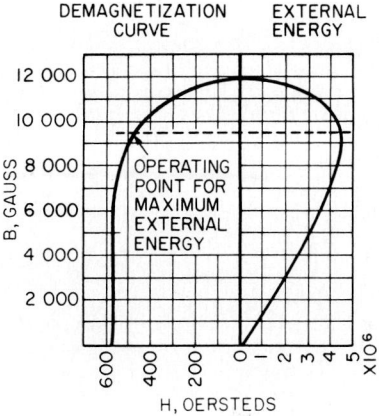

DEMAGNETIZATION CURVE   EXTERNAL ENERGY

B, GAUSS

OPERATING POINT FOR MAXIMUM EXTERNAL ENERGY

H, OERSTEDS

Fig. 13 Demagnetization and external energy curve for Alnico V.

But, since both sides of the coil are working, the total torque is twice that of Eq. 2, or

$$T = I'BNLW \tag{3}$$

The product of the width and length of the coil W is equal to the area. Therefore, A can be substituted for LW. Also, it is possible to write I (amperes)$/10 = I'$ (abamperes). The result is

$$T = \frac{IBNA}{10} \tag{4}$$

Equation 4 gives the torque in dyne-centimeters in terms of the dimensions of a moving coil, the ampere-turns in it, and the magnetic field through which it moves. This is the fundamental equation for all moving-coil electromagnetic instruments. Equation 4 applies to highly sensitive suspended galvanometers, as well as to industrial millivoltmeters, voltmeters, and ammeters. Also, Eq. 4 may be considered to apply to electrodynamometer wattmeters, in which case the magnetic field is that produced by the fixed field coils rather than by a permanent magnet.

The shape of the moving coil has been studied by many designers, and the optimum configuration, as may be expected, depends on its final intended use. If sensitivity to low currents is the only criterion, the coil should be square. However, when arranged for a high sensitivity with low restoring force and high flux density, the square coil responds very slowly because it has a relatively high moment of inertia and a high degree of damping.

The most sensitive instruments generally have a coil so proportioned that the length is three to four times the width. Indicating instruments, where ample energy is available, use coils more nearly square.

Magnetic System Characteristics

Since the current sensitivity is proportional to the magnetic field, the designer must consider the permanent-magnet system that produces the field. With reference to Fig. 12, the dimensions of the magnetic system are:

L_m = length of magnet
L_g = length of air gap
A_m = cross-sectional area of magnet
A_g = cross-sectional area of air gap

In designing a moving-coil instrument, whether a galvanometer or an ammeter, a certain flux density B_g is needed. This is the flux density in a working air gap which has an area A_g and a total air-gap length L_g (the sum of two gaps in series). This is to be obtained by means of a suitable magnet material, the magnet having a length L_m and a cross section A_m.

A permanent-magnet material can be described by (1) its demagnetization curve, which is the portion of the BH curve (gauss versus oersteds) in the second quadrant, and (2) an associated external energy curve comprising the product of the BH values. A typical curve for Alnico V is shown in Fig. 13. The BH product curve varies with the BH ratio and is important to magnetic system design, since it is desired that the magnet operate at, or close to, its point of maximum energy.

If it is desired to use this particular magnet material in a moving-coil instrument, it is necessary to solve for the dimensions of the magnet by means of the following equations:

$$\text{Magnet length } L_m = \frac{B_g L_g}{H} \tag{5}$$

where L_m = minimum length of magnet, cm
$\quad B_g$ = flux density in air gap, gauss
$\quad L_g$ = length of total air gap, cm
$\quad H$ = magnetizing force at operating point

$$\text{Magnet area } A_m = \frac{B_g A_g}{B} \tag{6}$$

where A_m = cross section of magnet, cm^2
$\quad B_g$ = flux density in air gap, G
$\quad A_g$ = cross-section area of air gap, cm^2
$\quad B$ = flux density in magnet, G

Thus, it is possible to obtain values for the magnet cross section and length. If the shape of the air-gap system is such that it is necessary to place the permanent magnet rather far away, more flux will be lost in leakage than would be the case with some other conformation where the magnet itself could be closely associated with the air-gap system. Thus, flux leakage may vary from as little as 10% to as much as 50%. Additional magnet cross section is required to furnish this leakage flux. The magnet length from Eq. 5 is a minimum for stable operation. A slight excess in length is desirable.

New magnetic systems usually defy complete analysis because of the unknown factor of leakage. An experienced designer is usually able to estimate the leakage within 20%. Some experimentation almost always is required to arrive at the optimum dimensions of a system involving permanent magnets.

Equation of Motion

The moving system of a permanent-magnet moving-coil instrument has mass, inertia, a restoring force, and an electrical winding. The balancing of all these factors, together with the magnet system, indicating means, and a housing, is the designer's problem. Since galvanometers are used for many purposes, many different designs result.

In a spring-controlled moving system where the controlling torque is proportioned to the angular deflection and where the damping associated with the system is proportional to the velocity of motion, the oscillation resulting is a harmonic motion. This motion was first represented in the form of a differential equation by Gauss and Weber:

$$K\frac{d^2\theta}{dt^2} + G\frac{d\theta}{dt} + S\theta = 0 \tag{7}$$

where θ = angular deflection from initial position, rad
$\quad t$ = time from initiation of motion, s
$\quad K$ = moment of inertia of system around axis, g·cm^2
$\quad G$ = damping coefficient, g·cm/rad/s
$\quad S$ = spring or control constant, g·cm/rad

When such a system is critically damped, however, that is, when its character of motion lies at the borderline between just oscillatory and just overdamped, the following relation holds:

$$G_0 = 2\sqrt{KS} \tag{8}$$

It is desirable to term the ratio of the actual damping coefficient G to the critical damping coefficient G_0 a new factor n. Thus Eq. 8 can be reduced to a simple dimensionless form based on two parameters. Expressed mathematically, the value of the damping coefficient for any damping is

$$G = nG_0 = 2n\sqrt{KS} \tag{9}$$

Also, for damping less than critical,

$$n = \frac{\delta}{\sqrt{\delta^2 + \pi^2}} \quad \text{and} \quad \delta = \log_e k \tag{10}$$

where δ is the logarithmic decrement in motion and k is the damping factor, defined as the ratio of a deflection or swing to the successive swing in the opposite direction. k may also be considered the reciprocal of the percent overthrow when an impulse is applied to the instrument. This value n is less than 1 for less than critical damping, is equal to 1 when damping is critical, and is greater than 1 for the overdamped condition. The term n is variable since, other factors being maintained the same, it varies directly as the square of the total flux through the coil, the coil area, and the number of turns of wire in the coil; it varies inversely as the coil circuit resistance and as the square root of the moment of inertia, the spring constant. More specifically, and where R is the circuit resistance,

$$n = \frac{B^2 A^2 N^2}{2R\sqrt{KS}} \tag{11}$$

Damping

The damping associated with the moving system usually is obtained in moving-coil permanent-magnet systems by induced currents in the moving-coil winding which, for example, flow through (1) the external closed circuit of a thermocouple or (2) a shunt resistance connected internally in the instrument which may or may not involve a more elaborate network for temperature compensation.

The moving coil also may be wound on an aluminum frame which contributes materially to mechanical strength and secures further damping arising from eddy currents in the frame itself. However, where great sensitivity is needed, which requires a relatively high flux density in the air gap, the damping obtained with the ordinary external circuit is usually such as to overdamp the instrument without the necessity for a frame. As a consequence, most millivoltmeters used as such, or as temperature indicating instruments, are made without frame damping. Indeed, with high flux density and many turns in the moving coil, an overdamped condition frequently exists, causing an unduly long time delay for the instrument pointer to reach its final position. In many instances, the circuit resistance is arbitrarily increased or the flux reduced in order to obtain less damping—even at the sacrifice of some degree of absolute torque. A good engineering compromise is required to balance these several factors as needed for any specific application.

Jewel Bearings

Jeweled bearings for fine mechanisms, particularly clocks and watches, were invented by Nicholas Facio, a Swiss watchmaker, about 1705. Such bearings in timepieces are used because of their low friction, but their form must be such as to keep the tiny teeth of the gears constantly in mesh. A ring jewel is, therefore, necessary to maintain alignment with the table jewel for end thrust. The jewels are mounted in watch plates as shown in Fig. 14a.

For instrument bearings, the ring jewel produces entirely too much friction, and the V jewel, as shown in Fig. 14b, is almost universally used. The pivot may have a radius at its tip from 0.0005 in to as high as 0.003 in (0.01 to 0.08 mm), depending on the weight of the mechanism and the vibration it will encounter. The radius of the pit in the jewel is somewhat greater so that contact is in the form of a circle a fraction of a thousandth across. The design shown in Fig. 14b has the least friction of any practical instrument bearing.

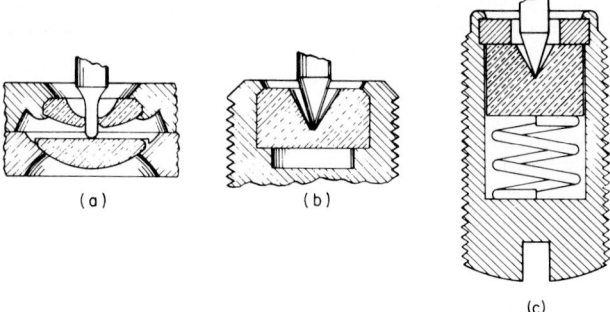

Fig. 14   (*a*) Ring and end-stone jewel bearing, (*b*) V jewel bearing, (*c*) spring-back jewel bearing.

Although the moving elements of instruments are designed to be of the lowest possible weight, the minute contact area between pivot and jewel results in large stresses for which the bearing must be designed. For example, a moving element weighing 300 mg resting on the area of a circle 0.0002 in (0.005 mm) diameter produces a stress of about 10 tons/in^2 (1407 kg/cm^2).

If the load is further increased, the contact area will not rise in equal measure, so that the stress will grow. Stresses set up by relatively moderate accelerations (jarring or dropping an instrument) may cause pivot damage (friction) except in an instrument specially protected.

The spring-back jewel bearing is now widely used and is made as shown in Fig. 14*c*. It is located in its normal position by the spring and may move axially when the shock to the mechanism becomes severe.

Temperature Compensation

A permanent-magnet moving-coil mechanism is not insensitive to temperature by itself, but may be made so by the appropriate use of proper series and shunt resistors of copper and Manganin. Magnets and springs decrease in strength and copper increases in resistance with an increase in temperature. The changes in the magnet and the copper tend to make the pointer read low on fixed voltage impressed, while the spring change tends to cause high readings. The effects are not identical, however, with the result that an uncompensated mechanism tends to read low by approximately 0.2% per degree Centigrade. For purposes of specification, an instrument is considered compensated when the change in accuracy due to a 10°C change in temperature is not more than one-fourth of the total allowable error.

Taut-Band Suspension

The suspension mechanism shown in Fig. 15 has been known for years and was utilized for many years strictly for laboratory equipment where high sensitivities were required and available torques were low so that it was desirable to eliminate even the low friction of pivots and jewels. The taut-band instrument, widely used in industrial-commercial meters for many years, enables one to obtain the advantage of a virtual absence of friction while also eliminating the sag by placing the ribbons under sufficient tension. This tension is provided by a tension spring so that the instrument can be used in any position. Generally, taut-band suspension instruments can be made with higher sensitivities than those using pivots and jewels. Common failures of pivot-and-jewel instruments are cracked jewels and/or blunted pivots, often brought on by subjecting the meter to shock. Taut-band meters can withstand shock up to 100 *g*. However, pivot-and-jewel instruments are generally superior for installations where there is a lot of vibration in the absence of shock.

Moving-Iron Instruments

An iron-vane instrument provides a means for measuring ac or dc current using an attraction-repulsion method. A field coil is wound, through which the current to be measured is passed. In the center

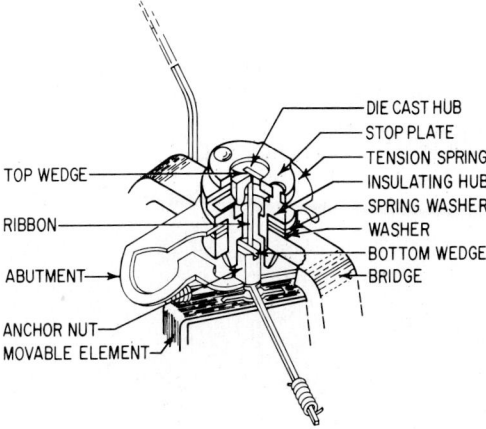

Fig. 15 Taut-band suspension.

of this coil are placed two iron vanes. One vane is stationary, and the other is made movable by connecting it to an axis, pointer, and torque spring. The magnetic field set up by the current passing through the coil induces like polarities in the iron vanes, causing them to separate. The amount of separation is indicated by a pointer. The principle is illustrated in Fig. 16, and an early magnetic vane mechanism is shown in Fig. 17.

The energy stored in the magnetic field can be written

$$E = \tfrac{1}{2}I^2L \tag{12}$$

where L is the inductance of the system, and its derivative, the torque T, is

$$T = \frac{\partial E}{\partial \theta} = \frac{1}{2} I^2 \frac{\partial L}{\partial \theta} \tag{13}$$

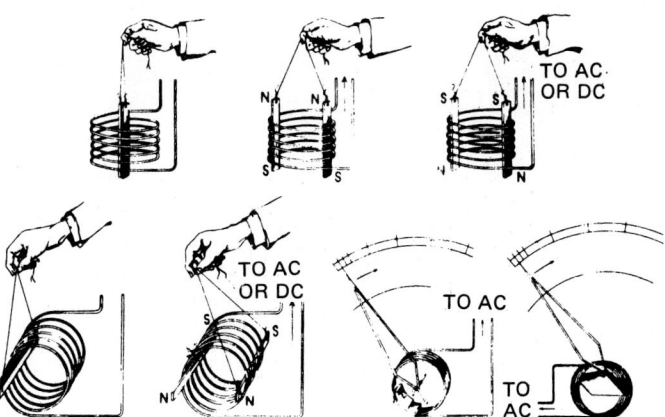

Fig. 16 When two similar adjacent iron bars are similarly magnetized, a repelling force is developed between them which tends to move them apart. In the moving-iron vane mechanism, this principle is used by fixing one bar in space and pivoting the second so that it tends to rotate when the magnetizing current flows. A spring attached to the moving vane opposes the motion of the vane and permits the scale to be calibrated in terms of the current flowing. *(Weston.)*

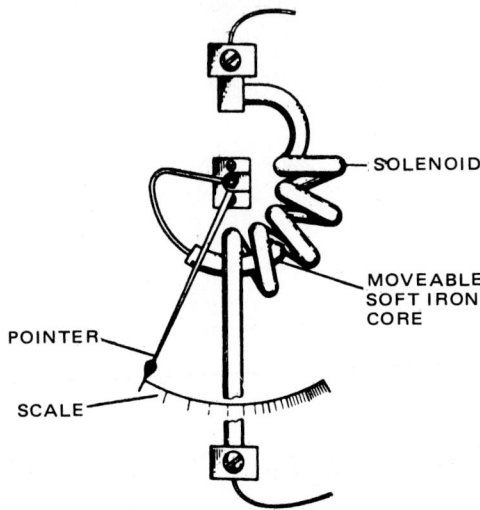

Fig. 17 Early magnetic vane mecnanism of the suction type. The opposing force or restoring torque is provided by gravity instead of the traditional spring used for many years. This method was widely used in older instruments. All gravity-controlled instruments had the major disadvantage of being subject to serious position errors. *(Weston.)*

The system's response is proportional to I^2. The instrument indicates root mean square (rms) current or voltage. By suitably shaping and arranging fixed and movable irons in the field of the coil, the instrument scale can be made very nearly linear over as much as 80% of its range. Alternatively, the upper range of the scale can be greatly compressed, and a small portion of the total range can be expanded to cover much of the scale. The latter arrangement is particularly appropriate in a voltmeter used to monitor a voltage which is nearly constant most of the time, such as a line voltage. Over the years, various vane configurations have been developed, two of which are shown in Fig. 18.

Rectifier Instruments

An instrument of this type makes use of four semiconductor rectifier elements arranged in a square with the input across one diagonal and a permanent-magnet moving-coil instrument connected across the other diagonal, the rectifier elements being arranged with their direction of conduction such that current in the permanent-magnet moving-coil instrument is in the same direction for either polarity at the input terminals. This arrangement can be used either for a voltmeter or for a low-range-current meter (milliammeter). Use of the arrangement for high-range-current meters (ammeters) is considered impractical. The rectifier-type instrument indicates the *average* values of current or voltage over a half-cycle of the alternating current. Scales are marked in terms of rms value of current or voltage for a sine wave input. Consequently, waveform errors are present for a nonsinusoidal input.

Rectifier-type meters provide a nominally linear (dc) scale rather than the conventional iron-vane distribution which has a slight crowding at the low end of the scale. Alternating current measurements of microamperes and milliamperes at minimum power consumption are permitted. Alternating current voltmeters of higher sensitivity (ohms per volt) than iron-vane types can be made. The frequency response of rectifier-type meters is essentially flat from 20 to 1000 Hz (1% effect from reference 60 Hz). In some designs, rectifier-type meters employ a full-wave copper oxide rectifier, available in a rating of 50 μA to 20 mA.

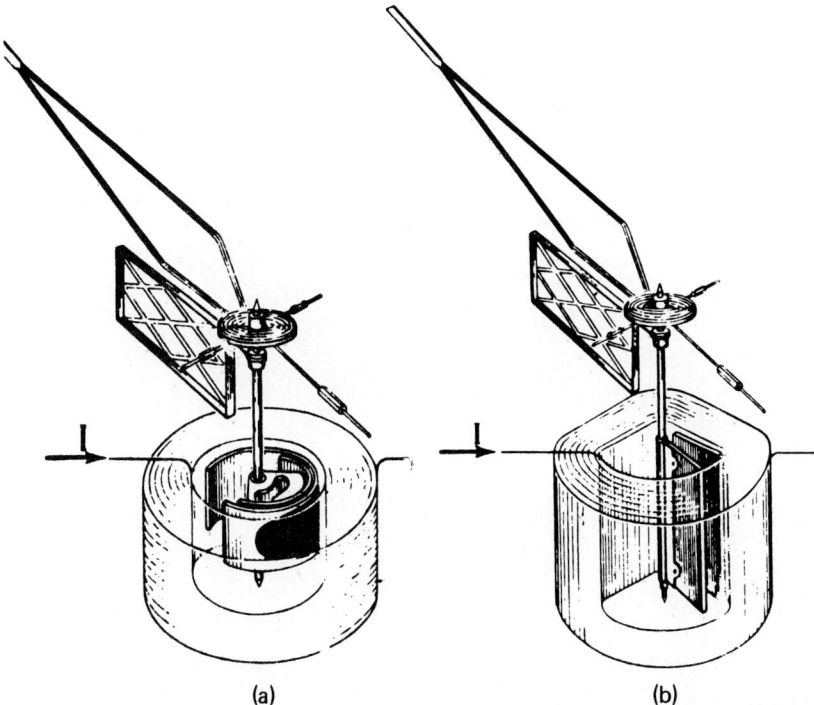

(a)   (b)

Fig. 18   Numerous vane configurations have been designed. (*a*) A concentric vane mechanism where vanes slip laterally under repulsion. This design has square law characteristics. Short magnetic vanes result in small dc current reversal and residual magnetism errors. However, with this mechanism, it is possible to shape the vanes to secure special characteristics, thereby opening the scale when needed. (*b*) A radial vane mechanism opens up like a book under repulsion. This device provides the most linear scale, but requires better design and better magnetic vanes for good grades of instruments. An aluminum damping vane, attached to the shaft just below the pointer, rotates in a close fitting chamber to bring pointer to rest quickly. *(Weston.)*

Electrodynamic Instruments

Electodynamic ammeters and voltmeters have both a fixed and a moving-coil system, each of which carries the current to be measured or a fraction of it. The interaction of their fields produces a torque proportional to the square of the current if the coils are connected in series (in a voltmeter), or proportional to the product of the fixed and moving-coil currents in a parallel arrangement (in an ammeter). Thus, the scale indication is of rms (effective) voltage or current, as in the moving-iron instrument. Here, however, eddy current errors are much less, and electrodynamic instruments are generally useful over an extended range of power and low audio frequencies. In addition, the dc response can be substantially error-free, and electrodynamic instruments are used extensively as ac-dc transfer standards to determine the ac performance of other instruments which cannot be calibrated reliably on direct current. This transfer function is important because the basic standards of resistance and emf, and the potentiometer techniques for accurately measuring current and voltage, are available only on direct current. The transfer function is generally limited to frequencies below 1000 Hz. Electrodynamic instruments are also used as wattmeters to measure power at low frequencies, generally below 1000 Hz. An electrodynamic mechanism is shown in Fig. 19. With a longer pointer and other minor changes, the mechanism shown can be used as a laboratory standard. With further modification of crossed moving coils, fundamentally the same mechanism can be used for measurements of power factor, phase angle, and capacity. In this form, the mechanism measures a ratio by balancing two torques.

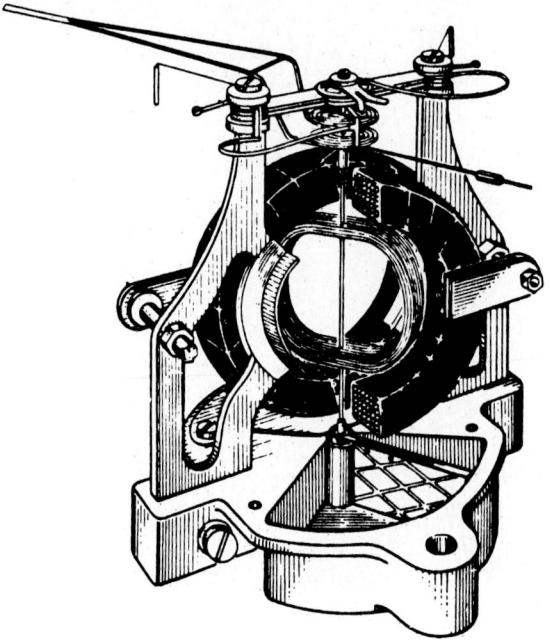

Fig. 19   Electrodynamic mechanism. *(Weston.)*

DC Ratio Indicating Mechanism

Shown in Fig. 20 is a mechanism in which two independent coils rigidly mounted on opposite sides of a common axis move in an air gap formed by a core mounted eccentrically between the cylindrical pole faces of a permanent magnet. No control springs are attached to the moving system, the current being fed to the coils by fine filaments. The ampere-turns (or flux) developed by each coil react with

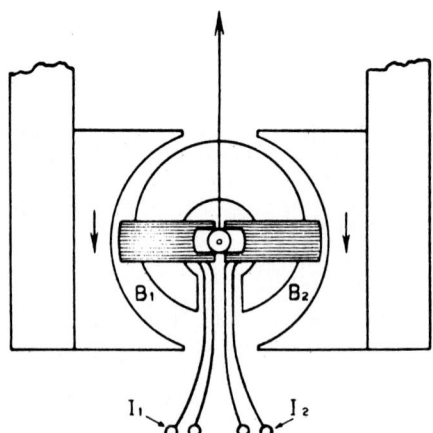

Fig. 20   Direct current ratio indicating mechanism. *(Weston.)*

the flux in its part of the air gap, causing the pointer to move until the product of ampere-turns and gap flux is equal on each side—so that a balance is reached.

Except for a multiplying factor to take care of the turns ratio and coil areas, the relationship can be written

$$I_1 B_1 = I_2 B_2 \qquad (14)$$

or

$$\frac{I_1}{I_2} = \frac{B_2}{B_1} \qquad (15)$$

where I_1 and I_2 are the currents in the two coils and B_1 and B_2 are the flux densities in the air gap at the points where the coils rest in the balanced state.

Equation 15 expresses a ratio between two currents in terms of a ratio between two flux densities. In view of the nonuniform and reversed flux distributions on the two sides of the air gap, a given ratio of flux densities occurs only once within the total angle of moving system rotation and is, therefore, associated with only one point on the scale. Each point on the scale thus represents only one current ratio, and the scale can be so calibrated. Such an instrument is a ratio meter capable of measuring any condition which can be expressed as a ratio of two currents.

Thermocouple Instruments

These instruments use the heating effect of an electric current. A fine wire or thin-walled tube is heated by the current to be measured. A thermocouple is attached to the heater element, and the temperature rise produces an emf whose value is indicated on a low-range permanent-magnet moving-coil millivoltmeter. The rate of heat production (and temperature rise) is proportional to the square of the current. Thus the indication is of *effective* (rms) value, and there is no waveform error. Such instruments may be used for current measurement (milliamperes and amperes) from dc to the radio frequency range without serious frequency errors. As a voltmeter, the instrument is generally restricted to audio frequencies, but with multipliers having a low distributed capacitance, the useful range can be extended upward to the megahertz level.

Electronic Meters

These devices employ electrical and electronic means to convert electrical quantities into digital outputs and readings. Almost always the input quantity is a voltage, other quantities being easily converted into voltages before being processed by the instrument. A digital output, in addition to use for display, can be transmitted over distances to actuate data processing equipment and inputted directly into computers.

The electrical means of conversion used in these instruments essentially are the same passive circuit elements also found in analog instruments, resistors, capacitors, inductors, transformers, and so on. Along with these passive elements are active electronic elements, such as transistors and diodes. The latter elements are not commonly found in conventional pointer deflecting instruments. The combination of active and passive elements is used to construct circuits which, in themselves, perform certain basic functions: (1) the *amplifier*—to provide an amplified version of the input; (2) the *pulse generator*—a circuit which produces electrical pulses at its output terminals, either continuously or on receipt of a command; (3) the *gate*—a circuit which has at least two inputs and one output, one input controlling whether a signal from the second input is fed through to the output; (4) the *counter*—a circuit whose output is indicative of the number of events that occurred on its input (frequently coupled with this circuit is a readout circuit which decodes the output of the counter circuit and displays it as a number); and (5) the *clock*—a circuit which provides electrical impulses at its output, at a fixed rate.

Modern electronic meters have taken full advantage of the latest developments in solid-state technology.

Digital Voltmeters

Several approaches are used in digital voltmeter design. Typical means include (1) the principle of successive approximations, (2) dual-slope integration, (3) a ramp and counter, and (4) voltage-to-frequency conversion.

Successive Approximations

In this technique, the unknown voltage is successively compared with standardized voltages generated in the instrument so that each voltage increment is one-half the amount of its predecessor. Whenever a standardized voltage exceeds the unknown voltage, the measurement is disregarded. Whenever the standardized voltage is less than the unknown, the former is "remembered." All the remembered voltages are added up until the difference between this sum and the unknown voltage can be neglected, and the sum is then displayed.

Dual-Slope Techniques

This requires a fixed-time integration interval during which the voltage to be measured is first integrated. This may be accomplished by charging a capacitor. The capacitor is then discharged by applying an opposed fixed reference voltage greater than the unknown. The time required to discharge the capacitor until its voltage reaches zero is used as a measure of the unknown voltage. During this discharge time, the output of a pulse generator is gated to a counter. At the end of the required time, the counter reads the voltage to be measured and the cycle repeats itself.

Ramp and Counter

In this type of instrument, a linear rise in voltage is generated and compared with the unknown voltage. A counter which receives clock pulses is started at the beginning of the process and stopped when the two voltages are equal. The time this takes is proportional to the unknown voltage. Therefore, the number of pulses counted during this interval is a measure of the unknown voltage.

Voltage-to-Frequency Conversion

In this method, the input voltage determines the rate or frequency at which a series of pulses is generated. This pulse train is gated to a counter for a prescribed fixed-time interval, and the number of pulses counted during this interval is a measure of the unknown voltage.

An important element in the conversion of voltage to a frequency is the operational integrator. This is an amplifier with a capacitor connected between the output and input. In front of the input is a resistor (Fig. 21). The amplifier is assumed to have a high gain. As a result of the action of this

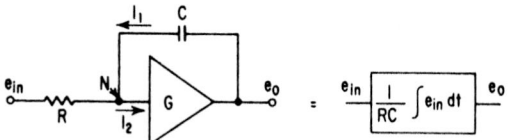

Fig. 21   Operational amplifier integrator.

configuration as an integrator, the application of a *step* input results in a linear *ramp* output in which the slope of the ramp is proportional to the input voltage step.

To generate a quantitative measure of the input voltage E_i, a relay or gate is energized by the output voltage E_o when the slope of the output voltage crosses a preset threshold level. This relay, when closed, permits a square voltage pulse of fixed amplitude and duration to be applied to the input to the integrator. The pulse is produced precisely by the pulse generator when triggered by the clock circuit. The pulse, being of polarity opposite that of the input and greater in amplitude than the input, results in a reversal of the integrator output (Fig. 22). In addition to triggering the pulse generator, the clock frequency controls the width of the reset pulses. The first clock pulse passing through turns the pulse generator on, and the next clock pulse gates it off.

The sequence of events is as follows. At time zero, the unknown input voltage is applied to the integrator, and the output begins to rise at a uniform rate. As soon as the output voltage reaches a fixed level, the gate is opened, thus permitting a clock pulse to trigger the pulse generator. The pulse generator then supplies a well-defined pulse of voltage to the input of the integrator. The summing of this pulse and the unknown input voltage results in the integrator "seeing," at its input, a potential with reversed sign. In order to maintain a balanced current condition at the summing mode, the integrator begins to decrease its output at the required rate. When the pulse ends, the integrator output resumes its uniform rate of rise, and the cycle of events just described is repeated. This loop is a voltage-to-frequency converter, since the number of pulses required by the integrator in a given time interval, in order to maintain a balanced condition at its input, is directly proportional to the magnitude of the input voltage.

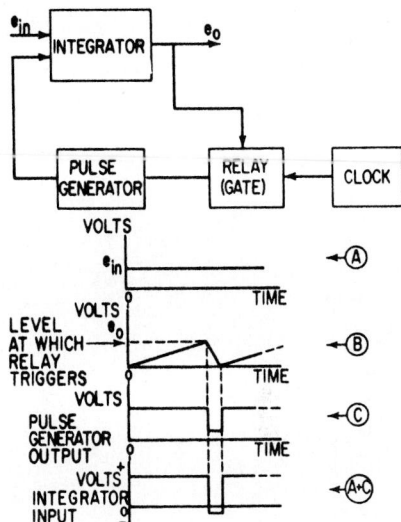

Fig. 22   Cyclic integrator.

Some refinements of this basic scheme are necessary because a high-gain operational amplifier with its associated large input summing resistors tends to degrade (lengthen) the rise time of the pulses supplied by the feedback loop. But, if as shown in Fig. 23, a dc amplifier is added ahead of the integrator and the summation of the input voltage and the voltage of the feedback pulses is made at this point, the requirement for perfect linearity of output voltage as a function of time is removed from the integrator. This summation is made so that the average difference between the two voltages appears at the output of the first amplifier.

Because the integrator now sees only an error signal as the output of the first amplifier and operates on this error (rather than the full input), the original stringent requirements placed on it are greatly relaxed. Without the requirement for accurate transformation, the integrator needs only to gate a clock pulse whenever the integral of the error signal exceeds a preset level. The pulse generator output then resets the integrator as previously described.

Accuracy is dependent primarily on the amplitude and width stability of the pulses which are smoothed and fed back through a potential divider for comparison with the input. Any difference between the average pulse level and the input results in an error signal which adjusts the rate at which the pulses are generated, thus maintaining a null balance.

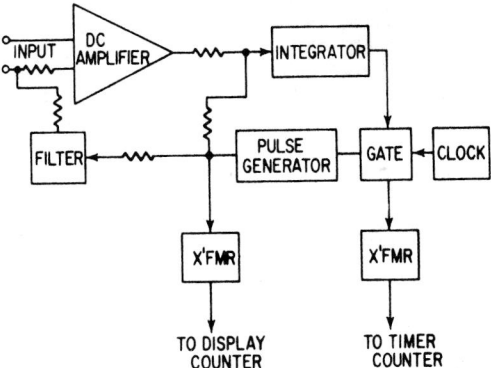

Fig. 23   Voltage-to-frequency converter.

Counting and Display

With a train of pulses available, it remains only to totalize these pulses for a fixed interval of time and to display this total as a direct measure of voltage. Because the timing or totalizing interval is controlled by the same clock which gates the pulse generator in the voltage-to-frequency converter

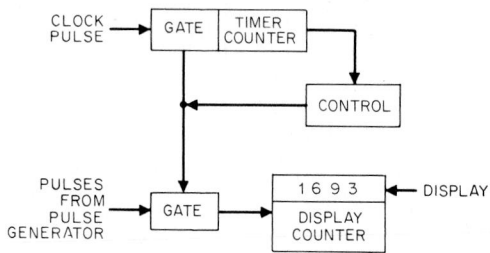

Fig. 24   Counting and display of pulse train.

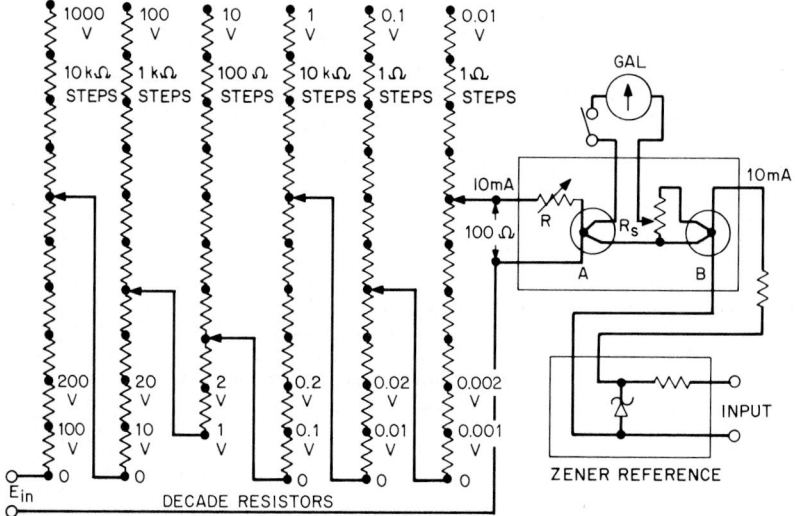

Fig. 25   Use of a thermal voltage standard for measuring ac voltages. The measuring element is a vacuum thermoelement A, which consists of a heater of a high-resistance alloy and a thermocouple thermally attached to the heater by means of an insulating bead. The output of thermocouple A is compared with that of a similar thermocouple B, the heater of which is connected to a zener reference. Both the measuring and reference thermoelements have a similar temperature dependence, which cancels the effects of ambient temperature changes. Also, both couples are mounted in a temperature lag box to reduce further the effects of ambient temperatures.

In use, resistor R_A is adjusted so that its resistance plus that of the heater of thermoelement A equals 100 Ω. The outputs of both thermocouples are next made equal by adjusting R_B for a null reading on the galvanometer with 10 mA from the zener reference flowing through both heaters in series. The decade resistors are selected to provide 10 mA to the heater of thermoelement A when connected as shown, and the switches indicate the input voltage E_{in} as they are adjusted for a null indication of the galvanometer.

Decade resistors and thermoelements are designed for a wide frequency response, such as 50 kHz, and both ac and dc readings can be made. The difference between the ac and dc readings (ac/dc difference) is checked on all instruments against known standards calibrated at a national standards laboratory. The zener reference contains zener diodes especially aged and tested for long-term stability. The line input is rectified, filtered, and regulated. Temperature-sensitive components are mounted in a temperature-controlled oven.

system and which controls the width of the pulses, no degradation of the measurement occurs if the clock is somewhat imprecise.

The sampling interval normally is ⅟₆₀ s. This time corresponds to 1 Hz of 60-Hz voltage and permits the influence of any 60-Hz voltage component at the input to integrate to zero.

The pulse train is measured and displayed as shown in Fig. 24. With both input gates open initially, the clock pulses and pulse generator output are permitted to pass into the timer counter and display counter, respectively. When the proper number of clock pulses have accumulated in the timer counter, the timer counter passes a signal to the control section, which closes the input gates. This action prevents further pulses from entering the counters. The number of pulses accumulated in the display counter during this totalizing interval establishes the output status of the display counter, and this status is gated to the display. The control section then resets the counters and reopens the gates to permit a fresh measurement to be made and the display to be updated.

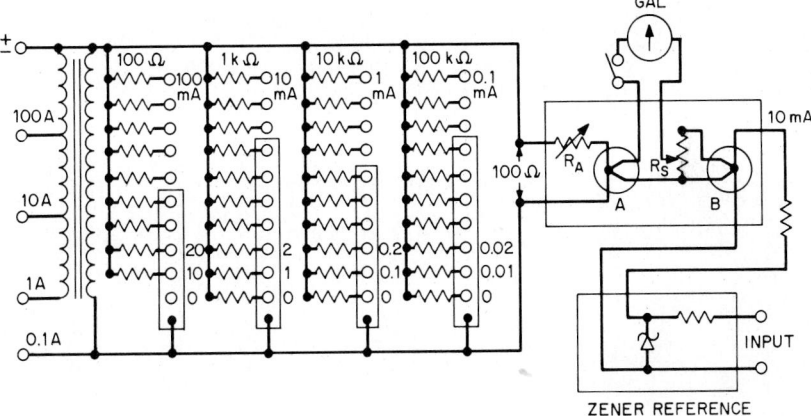

Fig. 26 Current measurements can be made in a manner similar to that shown in Fig. 25 by replacing the series decade resistors with decade shunts for the basic range of 10 to 100 mA with current transformers for the higher ranges.

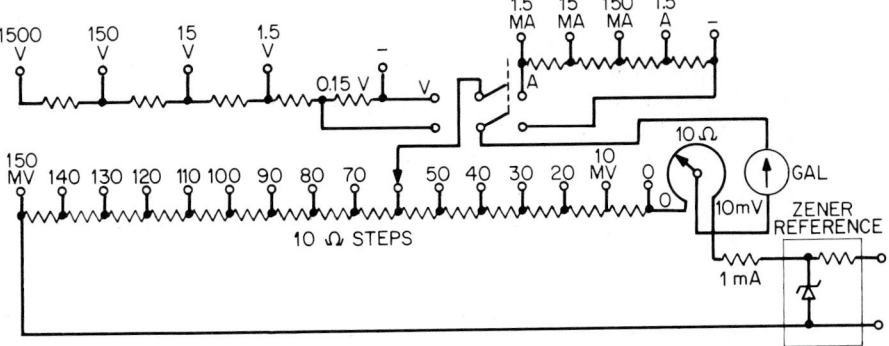

Fig. 27 The circuits of Figs. 25 and 26, plus a voltage divider, shunts, and potentiometer can be combined into one instrument to form a complete ac and dc precision standard. In its simplest form, the dc portion is shown here. The input voltage to the voltage divider or the input current to shunts provides a voltage drop of 0 to 150 mV, which is balanced by a voltage from a potentiometer consisting of a decade and slidewire. The voltage is obtained by passing current from the zener reference through the slidewire and decade resistors. In its final form, the entire instrument is protected by overload circuits, fuses, and relays to prevent the burnout of resistors and thermoelements.

Precision Voltage and Current Measurements

Measurements to an accuracy of better than 0.05% are beyond the capability of standard indicating instruments. The highlights of precision measuring methdologies are shown and briefly described in Figs. 25 through 28.

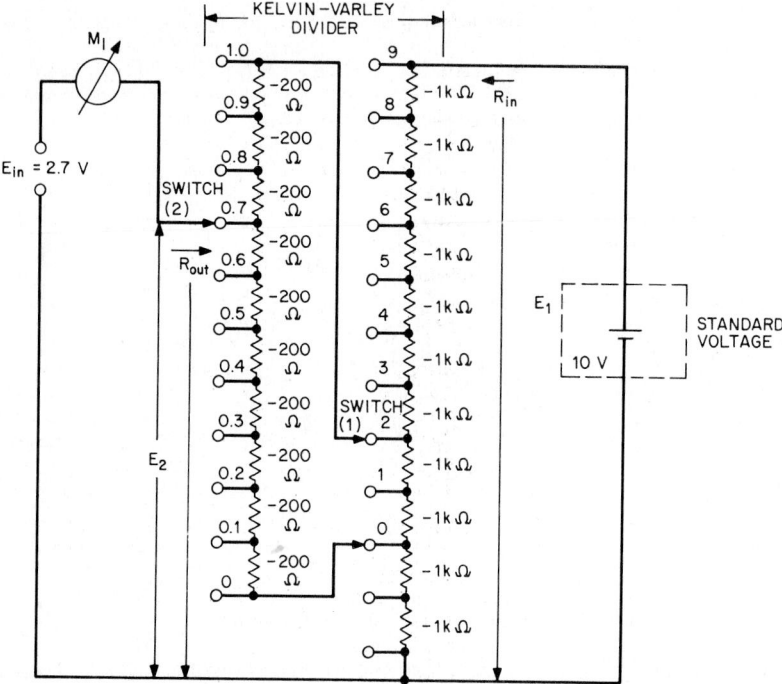

Fig. 28   Basic circuitry of a typical differential voltmeter. If the known voltage E_1 is adjusted through the use of a Kelvin-Varley divider to the exact potential of the unknown voltage E_{in}, the null detector will indicate zero and the value of the unknown voltage may be read from the position of the dials associated with the divider. In the example shown here, the unknown voltage is 2.7 V and the standard voltage is 10 V. The null detector will indicate zero when $E_2 = 2.7$ V or $E_2 = 2.7 = R_{out}/R_{in} \times E_{in}$. Since $R_{in} = 10$ kΩ (note that this is constant for any setting of the switches), this occurs when $R_{out} = 2.7$ kΩ. Examination of this diagram will show that this will occur when switch 1 is placed in its second position and when switch 2 is moved to its seventh position. Only two decades are shown. Others can be added whereby the total resistance of each added decade must equal the resistance of two steps of the preceding one. For high-voltage measurements with differential voltmeters, the value of the standard voltage must be greater than the unknown. This may be overcome by inserting a voltage divider between the unknown and the null detector.

Power and Energy Measurement

The use of wattmeters, varmeters, and power factor meters are described in the following article Electric Power and Energy Measurement.

REFERENCES

Brooks, P. B.: "Calibration Procedures for D-C Resistance Apparatus," *NBS Monogr. 39,* National Bureau of Standards, Washington, D.C., 1962.

Dziuba, R. F., and B. L. Dunfee: "A Resistive Voltage-Ratio Standard and Measuring Circuit," *IEEE Trans. Instrum. Meas.,* IM-19(4), November, 1970, pp. 266–267.

Eicke, W. G.: "Comments on Zener Diodes as Voltage Standards," Proceedings of the 10th Session Comité Consultatif d'Electricté du Comite International des Poids et Measures, Paris, 1963.

Eicke, W. G., and L. M. Auxier: "Regional Maintenance of the Volt Using NBS Volt Transfer Techniques," *IEEE Trans. Instrum. Meas.,* IM-23(4), December 1974, pp. 290–294.

Field, B. F., Finnegan, T. F., and J. Toots: "Volt Maintenance at NBS Via 2e/h: A New Definition of the NBS Volt," *Metrologia,* vol. 9 1973, pp. 155–166.

Gard, E. F., and T. E. Wells: "10-Kilohm Standard Resistors," *Natl. Bur. Stand. Tech. News Bull.,* vol. 52, no. 11, 1968, pp. 250–251.

Hermach, F. L.: "AC-DC Transfer Instruments for Current and Voltage Measurements," *IRE Trans. Instrum. Meas.* I-8, 1958, p. 235.

Hermach, F. L., and R. F. Dziuba (ed.): "Precision Measurement and Calibrations, Electricity-Low Frequency," *Spec. Publ. 300,* vol. 3, Government Printing Office, Washington, D.C., 1976.

Hermach, F. L.: "AC-DC Comparators for Audio-Frequency Current and Voltage Measurements of High Accuracy," *IEEE Trans. Instrum. Meas.,* IM-25(4), December 1976, pp. 489–494.

Houghton, S. R.: "Transfer of the Kilowatthour," *IEEE Trans. Power Appar. Syst.,* PAS-94(4), July–August, 1975, pp. 1232–1240.

Kahler, R. L.: "An Electronic Ratio Error Set for Current Transformer Calibrations," *IEEE Trans. Instrum. Meas.,* IM-28(2), June 1979, pp. 162–164.

Kieffer, L. J. (ed.): "Calibration and Related Measurement Services of the National Bureau of Standards," *NBS Spec. Publ. 250,* National Bureau of Standards, Washington, D.C., 1980.

Oldham, N. M.: "A Measurement Assurance Program for Electric Energy," *Tech. Note 930,* National Bureau of Standards, Washington, D.C. (1976).

Petersons, O. and W. E. Anderson: "A Wide Range High-Voltage Capacitance Bridge with One PPM Accuracy," *IEEE Trans. Instrum. Meas.,* IM-24(4), December 1975, pp. 336–344.

Schoenwetter, H.: "An RMS Digital Voltmeter/Calibrator for Very-Low Frequencies," *IEEE Trans. Instrum. Meas.,* IM-27(3), Septemeber 1978, pp. 259–267.

Schoenwetter, H.: "NBS Provides Voltage Calibration Service in 0.1–10 Hz Range Using AC Voltmeter/Calibrator," *IEEE Trans. Instrum Meas.,* IM-28(4), December 1979, pp. 327–331.

Souders, T. M.: "Wide-Band Two-Stage Current Transformers of High Accuracy," *IEEE Trans. Instrum. Meas.,* IM-21(4), November 1972, pp. 340–345.

Turgel, R. S.: "Sampling Techniques for Electric Power Measurement," *Tech. Note 870,* National Bureau of Standards, Washington, D.C., 1975.

Williams, E. S.: "Thermal Current Converters for Accurate AC Current Measurements," *IEEE Trans. Instrum. Meas.,* IM-25(4), December 1976, pp. 519–523.

Electric Power and Energy Measurement

by
Walter A. Troeger[*]

Traditionally, the term "power" in association with electricity has tended to lose its true meaning. Sometimes "power" is used where actually the correct term "energy" should be used. By definition, *power* is the rate at which energy is transformed, or made available, and is measured in *watts*. *Energy* may be defined as the time integral of power, or as the total energy supplied, and is measured in *watthours*.

From an economic point of view, the most important of all electrical measurements is the measurement of energy. The watthour meter in various forms can be found in nearly every home, factory, highway billboard, and other location where electrical energy is being purchased. Metering, installation, and wiring have been governed by national, industrial, and local codes for so many years that, at least in the United States, particular types of installations are nearly identical wherever located.

For homes or small stores where energy demand is low and fractional horsepower motors are used, the common supply is single-phase three-wire. Two voltage levels are generally available, 120 and 240 V, depending on which pair of wires is selected. Electric ranges and heavy-duty home air conditioner motors can take advantage of the high voltage to reduce line currents which reduce losses and permit smaller-size wiring.

The measurement of energy almost always involves fixed-installation metering. This provides safety through grounding of the meter enclosure and ease of reading through proper location and mounting. Tamperproof housings, which are also weatherproof where required, are common practice to ensure the integrity of readings.

In contrast, the measurement of power (watts) follows no such set of rigid rules. Frequently, considerable planning must go into a watt measurement to properly use existing metering or to purchase new equipment so that the test will be valid and results will be within the expected accuracy limits.

Whereas the measurement of energy is almost entirely restricted to 60 Hz, power measurements range from direct current to alternating current, including distorted waves, chopped waves, and missing pulses. A variety of circuits for connecting wattmeters to single-phase and polyphase systems has been developed over many years. Basic connection diagrams appear in most electrical engineering textbooks. However, the user of a wattmeter is hard pressed to find diagrams covering practical or unusual situations. Wattmeter manufacturers usually offer an instruction book or a bound set of connection diagrams with their instruments. Basic wattmeter connection diagrams appear later in this article.

Power Theory

Since energy is simply the total power over a time period, an understanding of the power equations will provide some information on both power and energy terminology. A direct current under steady-state conditions produces power, computed as the product of the voltage across the circuit and the current in amperes in the circuit. This also applies to ac circuits so long as instantaneous values of volts and amperes are used. The product of volts and amperes at any instant gives the instantaneous power in watts. However, such a measurement is unusual and difficult to make, and the resulting

---

[*]Engineering Department, Weston Instruments Division, Sangamo/Weston, Newark, New Jersey.

information is of limited use. Instantaneous power is of interest, of course, in the study of transient phenomena.

Average Power

Average power in an ac circuit is of far more interest, since it is equivalent to dc power and is a measure of mechanical work being done or heat liberated. Wattage or average power has an exact mathematical relation to horsepower or British thermal units or calories.

The most basic equation for power, relating voltage, current, power, and the phase angle between the voltage and current, is derived as follows. If both voltage and current are sinusoidal, the average power over a cycle is

$$P = \frac{1}{T} \int_0^T ei \, dt$$

$$= \frac{1}{2\pi} \int_0^{2\pi} E_m \sin \theta \, I_m \sin (\theta - \phi) \, d\theta$$

where E_m and I_m are maximum values and ϕ is the phase angle by which the current lags behind the voltage.

From $\sin (\theta - \phi) = \sin \theta \cos \phi - \cos \theta \sin \phi$,

$$P = \frac{E_m I_m}{2\pi} \left(\int_0^{2\pi} \sin^2 \theta \cos \phi \, d\theta - \int_0^{2\pi} \sin \theta \cos \theta \sin \phi \, d\theta \right)$$

$$= \frac{E_m I_m}{2\pi} \left(\left[\frac{\theta}{2} - \frac{\sin 2\theta}{4} \right]_0^{2\pi} \cos \phi - \left[\frac{1}{2} \sin^2 \theta \right]_0^{2\pi} \sin \phi \right)$$

$$= \frac{E_m I_m}{2} \cos \phi$$

The root mean square (rms) values of sinusoidal voltage and current are

$$E = \frac{E_m}{\sqrt{2}} \quad \text{and} \quad I = \frac{I_m}{\sqrt{2}}$$

Substitution in the previous equation yields

$$P = EI \cos \phi$$

An immediate concern is what happens to the indications of a wattmeter if the voltage or current or both are not sinusoidal. Since it is possible to synthesize an odd wave shape with higher harmonics of the fundamental frequency, a wattmeter will give correct indications if it is frequency-compensated over the span of harmonics. It is to be noted that, if a particular harmonic is present in either the current or the voltage but not the other, it does not contribute to the average or active power. Frequency compensation of the wattmeter is still necessary for an accurate measurement.

Wattmeter Construction

Dynamometer

All wattmeters of this type contain a fixed coil (usually divided into two coils), which carries the current, and a moving coil, which has series resistance connected for voltage, turning within the fixed coil. The torque on the moving system is proportional to the product of the currents in the fixed and moving coils:

$$\text{Torque} = K_1 i_m i_f \frac{dM}{d\Theta}$$

where M is the mutual inductance between the two sets of coils and remains constant over the usable scale range.

The period of the instrument is very long compared with the period of the alternating voltage. Since the instrument movement cannot follow the rapid variations in torque, it takes up a balance position where the driving torque equals the spring restoring torque. Deflection represents the average torque:

$$\text{Deflection} = \frac{K_2}{T} \int_0^T ei \, dt$$

which is identical to the average power equation

$$P = \frac{1}{T} \int_0^T et \, dt$$

multiplied by a constant.

It follows then that a dynamometer wattmeter is a "true rms wattmeter" and takes into account the magnitudes of voltage and current as well as the phase angle between them. Furthermore, a meter indication reverses when the flow of power reverses.

Thermal Watt Converter

Like the dynamometer, the thermal watt converter dates back to the early days of electricity. Heat produced by the voltage and current directly heats thermocouples arranged in a network to provide a dc output directly proportional to the wattage input. Figure 1 shows the essential parts of a thermal

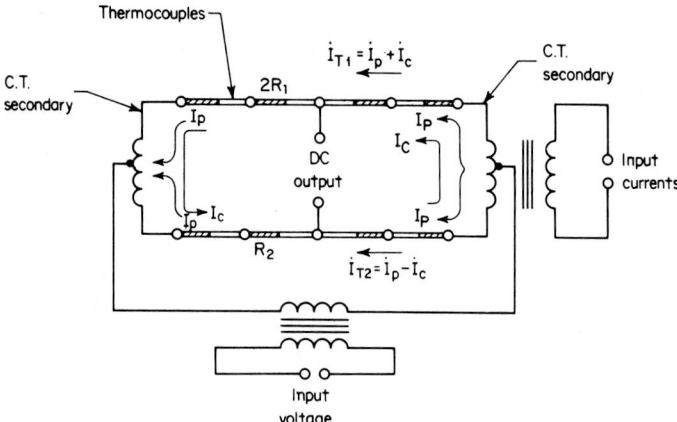

Fig. 1   Basic circuit of a single-element thermal watt converter. I_P, Current due to potential transformer (PT); I_C, current due to current transformer (CT); I_{T1}, vectorial sum of currents in R_1; I_{R2}, vectorial difference of currents in R_2; $R_1 = R_2$.

watt converter, namely, a potential transformer which is connected to monitor voltage, a current transformer which has a double-wound center-tapped secondary, and two sets of thermocouples.

The quantity of heat in R_1 is proportional to $R_1 (I_{T1})^2$ and in R_2 is equal to $R_2 (I_{T2})^2$. A vector diagram can be drawn for the sum and difference of I_p and I_c. From this diagram, equations can be developed which yield the difference in wattage in resistance R_1 and R_2 as $4RE1 \cos \phi$. The variable part of this term is $EI \cos \phi$ which is the expression for ac power.

Some early thermal wattmeters used bimetallic elements or liquid-filled thermometers to measure the difference in heat between resistors R_1 and R_2. Thermocouples are used in more recent designs. In one design,* the resistors and the thermocouples are one and the same. They act as both heating resistors and temperature sensing elements, and they show a very rapid response to power changes. Also, because the impedances of the several parts of the circuit are inherently balanced, there is little tendency for an interchange of currents or potentials between the ac and the dc portions of the network.

Used mostly by the power industry, especially in totalizing electrical system loads, the watt converter has also found widespread use in measuring the power taken by very large motors where a remote readout is needed. Classed as a true rms wattmeter, this device responds to magnitude of voltage and current as well as to the phase angle between them. Converter output will also reverse if the flow of power in the system reverses.

Specific Design Considerations

For clarity, it is in order to comment on the design factors as they affect a few specific commercial instruments. Available single-element and two-element watt converters have a response to 99% in 0.7 s. Because of this rapid response, some protection of the thermocouples is necessary during overload conditions. Some manufacturers use transformer cores which saturate at a moderate overload and minimize thermocouple burnout or other damage. If repeated overloads occur, the current circuit is usually operated below the 5-A normal level. Output is 50 mV open circuit per element when connected to a 500-W load. An available two-element watt converter has an output of 100 mV for 1000 W of circuit load. Provision is made for adjusting resistors in the output network so that the output can be reduced to achieve a particular ratio between input wattage and output millivolts.

Since thermal watt converters have no moving parts, they have been made very rugged by putting the entire circuit in a steel case. Polyphase wattmeters are easily made by assembling two or more single-element wattmeters and providing for a common output signal. An available three-element wattmeter has been designed for use on military motor-generator sets having three-phase four-wire distribution systems. Such systems can be expected to operate under unbalanced conditions, and to correctly read system power a three-element wattmeter must be used. If the internal potential and current transformers are of good quality, a thermal watt converter can be used on frequencies extending to 20,000 Hz. A high-frequency-type meter has a working frequency span of 180 to 20,000 Hz. In general, all thermal watt converters use internal transformers, which excludes their use on direct currents.

Low-Power-Factor Wattmeter

RMS wattmeters operate over the full power factor range from zero to unity. Many low-power-factor examples can be found in the laboratory, such as motor or transformer testing, core loss tests, and power supply circuits. At zero power factor, a wattmeter indicates zero watts even with the rated voltage and current flowing. It quickly becomes apparent that the major difficulty in making a low-power-factor measurement is the low indication obtained on the meter. For example, if a wattmeter indicates full-scale with a unit power factor load, it will indicate half-scale with a 50% power factor load for the same level of voltage and current. At 20% power factor, the pointer will move only one-fifth the distance up-scale.

Special wattmeters are available for use on low-power-factor circuits. They are commonly called 20% power factor meters, since the full-scale wattage is equal to the maximum voltage times the maximum current times 0.2. Both the accuracy and readability are improved through the use of this type of instrument. Since a 20% power factor wattmeter is designed to develop 5 times the torque of a unit power-factor-type instrument, care must be taken not to apply voltage or current above the maximum values shown on the instrument rating. Large overloads will soon burn out the resistors, fixed coils, or moving coils. Small overloads continuously applied will cause deterioration of the overheated insulation. A wattmeter designed for a low-power-factor load can be used on higher power factor circuits provided either the voltage or the current is sufficiently reduced to keep the

*Weston Instruments.

pointer on the scale. Likewise, normal unity power factor meters may be used at low circuit power factors provided maximums are not exceeded.

Power Measurement

Power in an ac or dc circuit may be determined indirectly by making appropriate measurements of voltage, current and, where necessary, the power factor. The power factor is usually expressed as a decimal value ranging between 0 and 1 and is derived from the cosine of the angle between the voltage and the current. The power factor is further designated lead or lag, depending on whether the current vector is ahead or behind the voltage vector based on counterclockwise rotation of the vectors. When making power calculations, it is not necessary to know if the power factor is lead or lag.

Therefore, calculating power is a valid procedure, but with some reservations. Results are accurate only if all quantities of voltage, current, and power factor are correctly measured. The most elusive quantity is the power factor. It is rare that a single-phase power factor meter can be used on other than 60 Hz. Even a three-phase power factor meter has the requirement that the load be balanced, although some designs cover several thousand hertz. Modern electronic phase angle voltmeters give excellent results on good sine waves over a wide frequency span. However, when distorted waves are encountered, results are questionable, since many instruments of this type operate on the zero crossing principle. Further, the power factor of a distorted wave has little meaning since, by definition, it is based on a sine wave. The conclusion is soon reached that the only way to measure power accurately is with a wattmeter.

In recent years, the phrase "true rms wattmeter" has been reserved for the description of the ultimate in wattmeters. This is because some of the types of wattmeters available today are accurate at only one frequency, must operate over a narrow voltage span, cannot operate on direct current, or must be worked at a high power factor. Although more descriptive than technically correct, the designation will probably remain in the literature.

The original, basic, true rms wattmeter was the dynamometer. Even until recently, this type of instrument was used as the standard wattmeter at the U.S. National Bureau of Standards. A dynamometer wattmeter can be calibrated very accurately on direct current and then used on alternating current. It is often the standard used to check other wattmeter devices, because it can be made to a high accuracy, is a passive device, and retains its accuracy for many years.

Blondel Theorem

Probably the most important theorem in electric power measurement is that proposed by Blondel. In essence, it states that, to correctly measure total system power, it is permissible to use one less wattmeter than current carrying conductors. Also, the common point for the potential circuits is the conductor without a wattmeter current connection. The circuit being so measured may be operated at any power factor or under any condition of current or voltage unbalance. Strict adherence to Blondel's theorem would require the use of three wattmeters or a 3-element-type meter to correctly measure a three-phase four-wire system. Since large commercial systems strive to maintain a good voltage balance, a less expensive wattmeter of the 2½-element design may be used and still achieve good accuracy.

Many questions often arise as to the proper wattmeter connections for various types of loads. For example, in a three-phase three-wire circuit, the load can be delta, wye, or some other configuration. The wattmeter is concerned only with the three wires. This leads to a simple pictorial concept for the connection of a wattmeter. Visualize a laboratory bench with an unknown power supply on the left, the connecting wiring across the bench, and an unknown load to the right. Without knowledge of the source or load, a true measurement of the total system power can be made by following Blondel's theorem.

Assume four wires are present and that all may be carrying current. Provide three wattmeters, making the wire without a meter the potential common. It is to be noted that one terminal of the meter current circuit and potential circuit carries an instantaneous polarity marking (usually plus or minus). That means that, if (+) of a dc supply is applied to each of these terminals, the meter will deflect up-scale. Likewise, if (−) is so applied, the meter will still go up-scale. If one terminal is made to have the opposite polarity, the meter will move down-scale. In a multimeter correction, the

($\pm$) current terminals should all be toward the source. Even so, because of load reactance, one wattmeter may produce a reversed indication. The total power then will be the algebraic sum of all the meter readings. The measurement has been made without any knowledge of the source or the load. Voltage and current levels must be within the range of the meter to avoid overheating damage.

If more knowledge of the load could be obtained, the immediate benefit would be reduced metering costs. If we still have a three-phase, four-wire system, but know that the neutral wire carries no current, then only two wattmeters are needed to give a true reading.

If we further know that both voltage and currents are balanced around the phases, then metering costs can be further reduced by using a single wattmeter and a wye box. When combined with the meter, the two arms of resistance in the wye box form a wye having a neutral point. The wattmeter will then measure a phase power which is a known fraction of the total power. The scales of switchboard meters are usually direct reading, but the indication from a portable wattmeter must be multiplied by the wye box multiplying factor.

If only voltages in this three-phase four-wire example are known to be balanced, then a so-called 2½-element wattmeter is satisfactory.

Adherence to Blondel's theorem always provides a true power measurement regardless of circuit conditions. A knowledge of the circuit often points the way to less expensive metering which can provide adequate results.

For the measurement of power in a two-wire circuit, Figs. 2 and 3 show the possible connections.

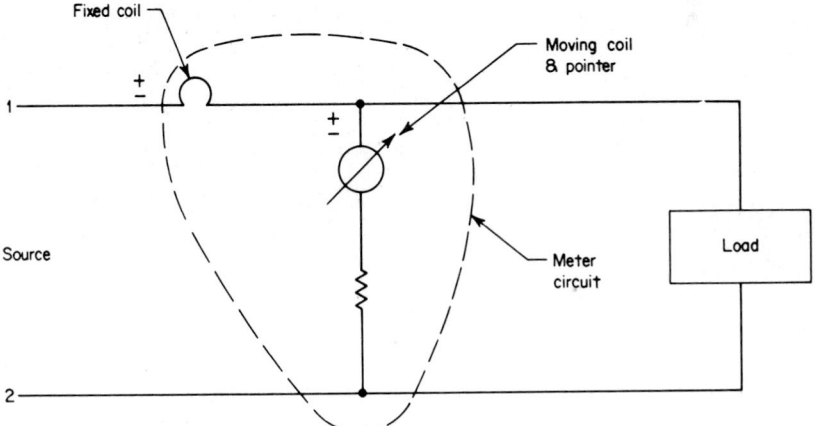

Fig. 2 Single-element wattmeter with the potential circuit connected on the load side. This connection is most often used.

The connection shown in Fig. 2, indicating the wattmeter potential circuit connected on the load side of the current coils, is most often used. Readings taken on small wattage loads may easily be corrected for meter loss by opening the load and reading the wattmeter. Although not truly correct under varying loads, this "tare" reading much improves the accuracy of the measurement.

When the wattmeter current coils are wound for low currents, such as 0.1 A, there is a sufficient voltage drop across them under load to make the connection shown in Fig. 3 and thus yield a more accurate result.

Figure 4 clearly demonstrates Blondel's theorem of two wattmeters in a three-wire circuit where the common potential circuit connection is made in the line without a current coil. There are some two-element wattmeters available which connect the two moving coils into line 2. Such an arrangement is adequate for moderate voltages and accuracy, but the mechanical force set up between the fixed and moving coil due to the electrostatic effect precludes the use of this connection where high accuracy is required.

The 2½-element wattmeter of Fig. 5 monitors all the current flowing and so is capable of a correct wattage measurement where current imbalance exists. Line-to-line voltages should be nearly balanced, and it is assumed that they are 120° from one another in vector rotation.

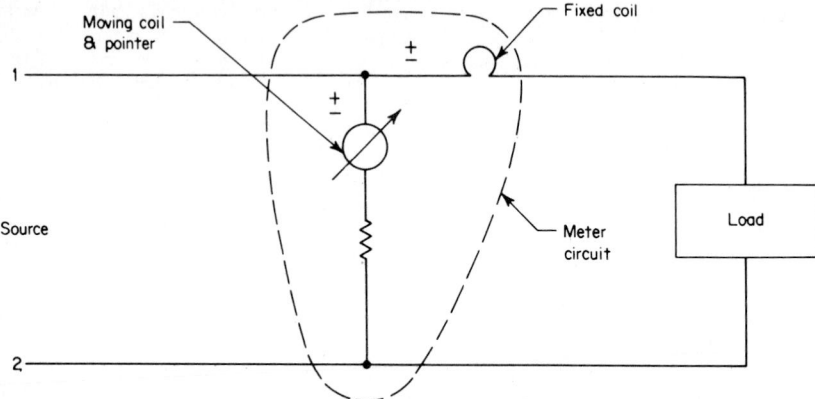

Fig. 3 Single-element wattmeter with the potential circuit connected on the source side. Generally used when fixed coil has a low current rating and a high voltage drop.

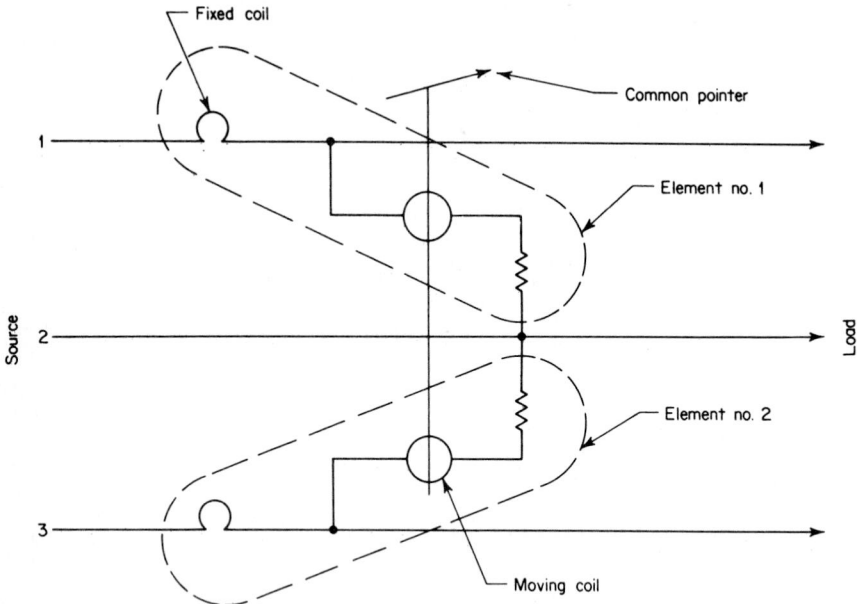

Fig. 4 Two-element wattmeter connected to a three-phase three-wire system.

Power Factor

Whenever the voltage and current are not in phase, a third term called the *power factor* must be introduced. The formula is

$$\text{Power factor} = \cos\left(\tan^{-1}\frac{Q}{P}\right)$$

where Q is reactive power and P is active power.

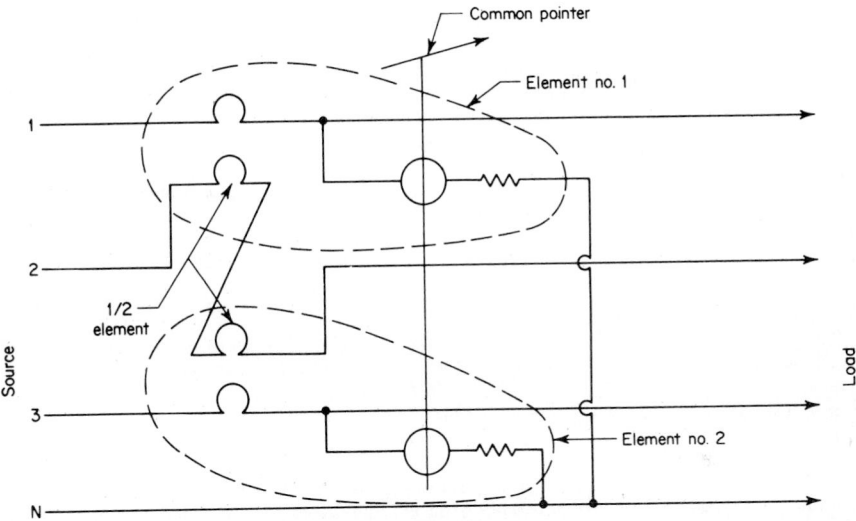

Fig. 5   A 2½-element wattmeter connected to a three-phase four-wire system.

This basic formula may be used on simple, two-wire circuits, as well as on a three-phase system, so long as the polyphase system is balanced in both voltage and current.

When a polyphase system is unbalanced in any manner, the system power factor will no longer have any specific physical meaning. A numerical ratio can be obtained and is defined as the interval power vector. This is not, in general, equal to the average value of the power during the interval.

The power factor is not usually one of the quantities measured when testing low-power circuits under laboratory conditions. Commercial users of bulk power often monitor the power factor so as to be able to make adjustments in condenser banks of synchronous motors to keep the power factor as high as possible. This in turn usually reduces the cost of the power purchased.

Power Factor Meter

Since the power factor meter can show at a glance the operating condition of a power system, it is most often found on switchboards of both consumers and suppliers of power.

Ratio-type movements are commonly found in both single- and polyphase power factor meters. The single-phase power factor meter measures the ratio of vars to watts, which corresponds to the tangent of the power factor angle when the voltage and current are sinusoidal. A scale can then be drawn for the power factor, which is the cosine of the angle.

The polyphase power factor meter also uses the same basic ratio mechanism as the single-phase meter and is connected to the three-phase three-wire circuit. This instrument indicates the vector power factor by measuring the angle between line current 2 and line voltages 3-2 and 1-2. The indication is the polyphase power factor only for balanced voltages and currents when both are sinusoidal.

Single-phase instruments can be scaled in a variety of combinations, such as 0-1 power factor lag or lead, or 0.3-1-0.3 power factor. Because of the principle of the instrument, not every range combination is possible in the polyphase power factor meter.

Varmeters

When compared to other electrical quantities, the *var* is a comparatively new term, having been recognized by international agreement in 1930. The letters were taken from volt-ampere-reactive and represent *power incapable of producing work*. Voltages used in this form of metering are always 90° in vector rotation from those used in wattage measurement. In any ac system having sinusoidal voltages and currents operating at other than unity power factor, the real power is less than the volt-ampere product and is related by the familiar right triangle of Fig. 6.

Since more iron and copper are required to deliver a given amount of power at a low than at a high power factor, design allowances must be made for any reactive volt-amperes in addition to the designed load power. Line losses are higher and voltage regulation poorer when the power factor is low. Any customer contributing to a low power factor must be expected to pay for this added loss in addition to the actual energy consumed. Var metering is generally used to obtain an estimate of the average power factor of a fluctuating load over a period of time. A recording-type meter is used for this measurement:

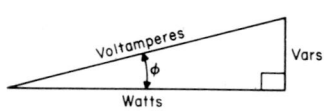

Fig. 6   Real power is less than volts × amperes. ϕ, Power factor angle.

$$Var = EI \sin \phi = EI \cos (90° - \phi)$$

A wattmeter can be converted to a varmeter if the voltage is shifted into quadrature with the line voltage at the load. In single-phase varmeters, this voltage shift is done by means of added reactance. In polyphase systems, the voltage shift is most easily done by reconnection if all necessary points are available. On three-phase three-wire systems, a var connection may be made to wye-connected resistors forming an artificial neutral. A special polyphase, phase shifting autotransformer, also called a potential converter or phasing transformer, is available for var service. When a varmeter is purchased from the manufacturer, the proper scaling, instrument adjustment (watts), and connection diagrams are provided. If it is desired to convert an existing wattmeter to a varmeter, several problems must be avoided (or at least understood).

Varmeter indication can be either up- or down-scale, depending on whether clockwise or counterclockwise rotation is selected for the new var voltages. Either is correct, but it should be determined that up-scale indication will occur for a leading or lagging power factor. Another more serious problem is voltage level. If the new var voltages differ from those used by the basic wattmeter, then, for the same power supplied to the load, the full-scale value and therefore all scale points will be in error. In order to correct this, the meter series resistance has to be changed (a task for a meter shop with wattmeter calibration equipment).

Varmeters are usually arranged to deflect to the right (up-scale) on leading power factor circuits with phase rotation 1-2-3. Zero center varmeters also have the words IN at the left and OUT at the right. The OUT refers to the flow of power from a source, considered to be located at the left of a diagram, to a load somewhere to the right of the diagram.

Instrument Transformers

For reasons of personnel safety and good instrument design, it becomes impractical to connect meters directly to circuits having voltages above 1000 V and currents above 50 A. Furthermore, instruments tend to become inaccurate when directly connected to high voltages because of electrostatic forces that act on the moving system. Shunts are often considered for large ac currents but have the disadvantages of large power loss at high current ratings and no voltage isolation. Unless a shunt is especially designed for ac use, its inductance, nonuniform split of current in the blocks, skin effect, proximity effect, and nonuniformity of the material all will contribute to ac shunt error.

In contrast, potential and current transformers offer a practical and safe means of reducing voltage and current by an exact ratio for instrument use. All instrument transformers have the following basic design objectives: (1) careful attention to an exact primary-to-secondary ratio, and (2) as small a phase angle as possible. Instrument transformers by design have a small load (burden) capability, because meters do not need large amounts of power and a low burden enhances the design for best possible accuracy.

In a power station, the instrument transformers used for station metering, relay operation, and control services are often several feet (a few meters) tall, resulting from a design to withstand very high voltages. For laboratory and shop testing at low distribution voltages, both potential transformers and current transformers are very small in comparison. They weigh perhaps 15 lb (~7 kg) and can be hand-carried. Quality potential transformers for laboratory use can support a 25-V/A burden, have a ratio accuracy from 0.1 to 0.5%, and have a phase angle of 10 min. A core-type design in which the winding surrounds the iron provides the necessary insulation for a high voltage range: 2300/1150 or 115 V. The iron is usually grain-oriented silicon steel from 0.012 to 0.025 in (0.3 to 0.6 mm) thick, depending on the quality of the transformer and frequency range. Current transformers of the toroidal type, using tape-wound high-nickel-iron cores, have phase angles of less than 2 min and a ratio error of ±0.02%. The burden capability is up to 25 VA. Primary-to-secondary insulation is difficult to provide in a toroidal transformer. A rating of 2500 V is common for stock transformers, with 5000 working volts pushing the practical limit for this construction in custom designs.

Editor's Note: See references listed at the end of the article on electric meters in this Handbook section.

Circuits, Components, and Subsystems *

Stephen C. Buchanan. *Daytronic Corporation, Miamisburg, Ohio. (Digital Technology)*

Thomas J. Harrison. *IBM Corporation, Boca Raton, Florida. (Digital Technology)*

E. C. Magison. *Process Control Division, Honeywell Inc., Fort Washington, Pennsylvania. (Safety in Instrumentation and Control Systems)*

W. H. Shirk, Jr. *Formerly Manager, Engineering Coordination and Services, Leeds & Northrup Company. A Unit of General Signal, North Wales, Pennsylvania. Presently, General Manager, Pocono Lake Preserve, Pennsylvania. (Bridges and Potentiometers)*

T. J. Walsh. *Engineering Department, Leeds & Northrup Company, A Unit of General Signal, North Wales, Pennsylvania. (Self-Balancing Electrical Instruments)*

*Persons who authored complete articles or subsections of articles, or who otherwise cooperated in an outstanding manner in furnishing information and helpful counsel to the editorial staff.

Bridges and Potentiometers

by
W. H. Shirk, Jr.[*]

In this article, an attempt is made to provide a broad overview of the many types of electrical bridge and potentiometer circuits that are so important to measuring instruments.

    A simple, passive bridge circuit is characterized by a network of four impedances having a voltage signal source applied to two opposite junctions and a null detecting device connected to the other two junctions. When the network contains impedances that are pure resistance, the signal source is generally a dc voltage, such as a battery, and the detector is a dc amplifier or galvanometer. If the network contains impedances that have reactive components, the signal source is an ac voltage, such as a signal generator, and the associated detector is designed to be sensitive to both in-phase and quadrature voltage or current.

    [*]Formerly Manager, Engineering Coordination and Services, Leeds & Northrup Company, A Unit of General Signal, North Wales, Pa. Presently, General Manager, Pocono Lake Preserve, Pennsylvania.

The advantages of the electrical bridge include:

1. An unknown impedance can be measured in terms of three other known impedances or one known impedance and an impedance ratio.
2. At null balance (no current flowing through the detector), the impedance values of all four bridge arms are independent of the characteristics of the signal source or null detector. In fact, the signal source and detector may be interchanged without affecting the balance equation.
3. One impedance bridge arm can be significantly greater or smaller in value than the other impedance arms, allowing measurement of an unknown impedance beyond the range of the known impedances.
4. If three of the bridge arm impedances are known in value and the fourth is to be determined, the maximum uncertainty in the accuracy of determining that value is equal to the sum of the uncertainties in the three known impedances.

Precautions to be taken when using an electrical bridge include:

1. The effects of lead and contact resistance on the four bridge arm impedances must be considered.
2. The secondary effect of self-heating bridge arm impedances is a determining factor when choosing the magnitude of the signal source voltage and the physical size of the components.
3. The characteristics of the null detector, including sensitivity, input impedance, and response time, in conjunction with the voltage signal source and magnitude of the bridge arm impedances, determine the overall sensitivity of measurement.
4. The effect of insulation resistance (and reactance for ac bridges) on all four bridge arms must be considered.

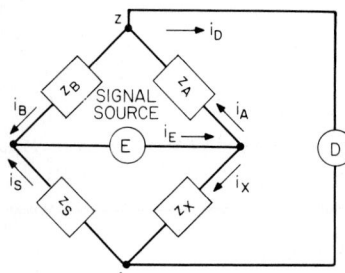

Fig. 1   Fundamental electrical bridge circuit.

Simple Bridge Circuit

The fundamental bridge circuit is shown in Fig. 1. As a result of the signal source E, current flow through the bridge arms Z_A, Z_B, Z_X, and Z_S is as shown in the diagram. But no current flows through the detector at null balance; $i_D = 0$. Therefore, $i_A = i_B$ and $i_X = i_S$.

Voltage at Node 1:

$$e_1 = \frac{i_X Z_X}{i_X Z_X + i_S Z_S} \times E$$

Voltage at Node 2:

$$e_2 = \frac{i_A Z_A}{i_A Z_A + i_B Z_B} \times E$$

Since at balance the two voltage drops across Z_S and Z_B are equal and $i_A = i_B$, and $i_X = i_S$, then

$$\frac{i_x Z_X}{i_x Z_X + i_x Z_X} \times E = \frac{i_A Z_A}{i_A Z_A + i_A Z_B} \times E$$

$$Z_X Z_A \times Z_X Z_B = Z_X Z_A \times Z_S Z_A \tag{1}$$

$$Z_X Z_B = Z_S Z_A$$

$$Z_X = Z_S \frac{Z_A}{Z_B}$$

In an ac bridge circuit, the impedance values are complex, containing both resistive and reactive components. It can be shown that, under this condition involving complex impedances, two balance equations apply and must be simultaneously achieved:

$$R_X = R_S \frac{R_A}{R_B} \tag{2}$$

$$X_X = X_S \frac{X_A}{X_B} \tag{3}$$

where, in Fig. 1, $Z_X = R_X + jX_X$ and $Z_S = R_S + jX_S$. Of course, in applications where the resistive component is inconsequential, only the relationships of Eq. 2 apply.

In general for a dc bridge, ratio arms R_A and R_B are adjustable in even exponential steps (1, 10, 100, 1000 Ω, etc.) and R_S is an adjustable rheostat consisting of four, five, or six decades with minimum steps of 1 Ω or less and a maximum range of 10,000 Ω or more. When the bridge is balanced, the unknown value of R_X is determined by the reading of the rheostat arm multiplied by the ratio R_A/R_B.

Unbalanced DC Bridge

Although the objective in an electrical bridge circuit is usually to achieve balance, it is the unbalanced bridge that is of most interest, and knowledge of the parameters affecting bridge sensitivity is essential in selecting the optimum values for the bridge arms, null detector characteristics, and signal source.

If the circuit in Fig. 1 is pure resistance (no reactive component to the impedances and commonly called a dc Wheatstone bridge), it can be shown, using Thevenin's theorem, that the current flowing through the detector in an unbalanced condition is

$$i_D = \frac{E(R_A R_S - R_B R_X)}{R_A R_S (R + X) + R_X (A + B) + D(A + B)(R + X)} \tag{4}$$

where E = voltage of signal source
$A, R, B,$ and X = resistance values of four bridge arms
D = input resistance of detector

For a small imbalance where the proportional imbalance is 10% or less, an excellent approximation of the unbalanced bridge voltage available to the detector is

$$V_{\text{DET}} = \frac{EKR_S R_X}{(R_S + R_X)^2} \tag{5}$$

where V_{DET} = voltage available to detector from a theoretical voltage source having an internal resistance $R_A + R_X$ in parallel with $R_B + R_S$
E = voltage of signal source
K = bridge imbalance in proportional parts
R_S and R_X = nominal values of bridge arms

If the detector and signal sources are interchanged, this relationship becomes

$$V_{\text{DET}} = \frac{EKR_AR_X}{(R_A + R_X)^2} \tag{6}$$

To illustrate this, assume a bridge circuit where $R_A = 10\ \Omega$, $R_B = 10,000\ \Omega$, $R_S = 1000\ \Omega$, and R_S is nominally $1\ \Omega$. If the maximum wattage dissipation in any one resistor were to be limited to 0.1 W, the maximum applied voltage would be

$$E = \sqrt{RW}$$

$$= \sqrt{(1000)(0.1)}$$

$$= 10\ \text{V}$$

Since 99.9% of the voltage will appear across R_A, it is assumed that the value of E is 10 V and not 10.01 V, the theoretical limit.

With a bridge imbalance of 1% (either $R_X = 1 \pm 0.01\ \Omega$ or $R_S = 1000 \pm 10\ \Omega$), the voltage available to the detector, from Eq. 5, is

$$V_{\text{DET}} = \frac{EKR_SR_X}{(R_S + R_X)^2}$$

$$= \frac{(10)(0.01)(1000)(1)}{(1000 + 1)^2}$$

$$= 0.0001\ \text{V (approximately)}$$

If the detector has an input resistance that is large with respect to $R_A + R_X$, the full 0.0001 V will be available to it. If the detector has an input resistance of 10,000 Ω and a sensitivity of 10 μV per division, it will deflect 10 divisions for a 1% unbalanced condition.

If the detector and power supply were interchanged, the applied voltage would be limited to

$$E = \sqrt{RW}$$

$$= \sqrt{(10)(0.1)}$$

$$= 1\ \text{V}$$

Using Eq. 6, the unbalanced bridge would result in a voltage available to the detector of

$$V_{\text{DET}} = \frac{EKR_AR_X}{(R + R_X)^2}$$

$$= \frac{(1)(0.01)(10)(1)}{(10 + 1)^2}$$

$$= 0.0008\ \text{V (approximately)}$$

With a detector with the same characteristics noted previously, the 1% imbalance would result in a deflection of approximately 80 divisions (actually 72 divisions taking into account the bridge resistance as seen by the detector), a more sensitive arrangement with less bridge voltage, but more power being dissipated by the unknown (0.01 versus 0.0001 W).

DC Kelvin Bridge

One of the precautions required when using a Wheatstone bridge is to consider the effect of lead and contact resistance, particularly when measuring low-value resistances. An error of 0.1% can be introduced by 6-in (15-cm) leads of B&S no. 10 copper wire connected to a 1-Ω resistor. One way of reducing this error is to connect the Wheatstone bridge as shown in Fig. 2.

In this arrangement, R_X is a four-terminal resistor adjusted so that the exact value of resistance

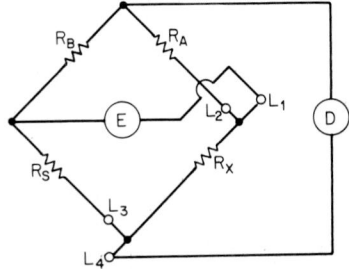

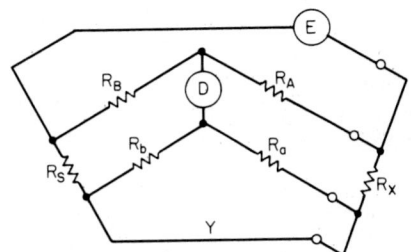

Fig. 2 Connection of a Wheatstone bridge for low resistance R_X.

Fig. 3 Kelvin bridge.

is defined between the L_1L_2 node and the L_3L_4 node. The lead resistance L_1 is in series with the signal source; L_2 is in series with R_A and, if R_A is larger than R_X, the effect of the lead resistance on the accuracy of measurement will be less than if L_2 were in series with R_X. L_3 is in series with R_S, which also results in a reduced lead resistance error; and L_4 is in series with the detector. The accuracy of measurement is unaffected by L_1 and L_4.

An alternate circuit which further reduces the effects of lead and contact resistance is a Kelvin bridge, shown in Fig. 3.

The balance equation for a Kelvin bridge is

$$R_X = \frac{R_S \times R_A}{R_B} + \frac{Y \times R_b}{Y + R_a + R_b} \left(\frac{R_A}{R_B} - \frac{R_a}{R_b} \right) \tag{7}$$

As can be seen, the only way to reduce this complex expression to the simple equation for a balanced Wheatstone bridge is to maintain the ratio R_a/R_b as nearly equal to R_A/R_B as possible and to keep the Y resistance (called the *Yoke resistance*) as low as possible. Since the ratios can never be exactly equal and the Yoke resistance is always finite, these terms contribute to the overall inaccuracy of measurement. However, the advantage of this circuit is that R_X can be very much smaller in value than the other bridge arm resistors, and the effect of lead and contact resistance associated with R_X minimized. Kelvin bridge measurements down to 1 $\mu\Omega$ are not uncommon.

Calculations of bridge sensitivity for a Kelvin circuit are based on the principle described in Eq. 5, except that the effect of the equivalent parallel resistances R_a and R_b, in series with the detector, must be considered. With electronic detectors, this effect is often negligible. A further consideration is the true voltage available from the voltage source, since a low value for R_S and R_X can result in a significant load on the voltage source and reduce the terminal voltage well below the expected value depending on the voltage source's internal resistance.

An example of a Kelvin bridge sensitivity calculation is as follows. Assume $R_X = 0.0001$ Ω, $R_S = 0.01$ Ω, $R_A = 100$ Ω, and $R_B = 10,000$ Ω. Also, R_a and R_b are 100 and 10,000 Ω, respectively.

The voltage source, based on the previous assumption of 0.1 V maximum dissipation per bridge arm will be

$$E = \sqrt{(R_S + R_X)W}$$

$$= \sqrt{0.0101 \times 0.1}$$

$$= 0.03 \text{ V}$$

For a 1% bridge imbalance, the voltage available to the detector, using Eq. 5, is

$$V_{\text{DET}} = \frac{EKR_SR_X}{(R_S + R_X)^2}$$

$$= \frac{3 \times 10^{-2} \times 10^{-2} \times 10^{-2} \times 10^{-4}}{(0.01 + 0.0001)^2}$$

$$= \frac{3 \times 10^{-10}}{1.02 \times 10^{-4}}$$

$$\approx 3 \times 10^{-6} \text{ or } 3 \ \mu V$$

The detector described above, with a 10,000-Ω input resistance and a sensitivity of 10 μV per division, would indicate only a one-third division deflection for a 1% imbalance.

For a Kelvin bridge, the required low value for R_S makes it impractical to have this arm as the adjustable rheostat; therefore, the common Kelvin design establishes R_S as a resistor whose value is a selectable power of 10, and the adjustable rheostat is R_A. Of course R_a must be closely matched in value to R_A.

In the above example, if R_A and R_a are matched in ratio to 1%, and Y is just below 0.01 Ω, the error contribution of the second term of the Kelvin bridge balance equation will be

$$\text{Error} = \frac{Y \times R_b}{Y + R_a + R_b} \left(\frac{R_A}{R_B} - \frac{R_a}{R_b} \right)$$

$$= \frac{0.01 \times 10^4}{0.01 + 10^2 + 10^4} \left(\frac{100}{10,000} - \frac{99}{10,000} \right)$$

$$\cong \frac{100}{10^4} (0.01 - 0.0099)$$

$$\cong 10^{-2}(10^{-4})$$

$$\cong 1 \ \mu\Omega \text{ or } 1\% \text{ of } R_X$$

Variations of DC Wheatstone and Kelvin Bridges

Many instruments are available with variations of the basic Wheatstone or Kelvin bridge circuit for specific applications.

Fault Location Test Sets

A series of Wheatstone bridge circuits, test methods, and balance equations has been developed for use in determining the distance to a fault (grounded or crossed wiring) in a communications cable by measuring the resistance of cable pairs. A few of the more common circuits, constraints, and balance equations for these applications are shown in Figs. 4 through 11.

Varley Loop Test

This method requires two good conductors joined at one end and is applicable for high-resistance loops. With reference to Fig. 4,

$$r = R_g + X_b + X_a$$

$$X_a = \frac{rB - AR}{A + B}$$

where R_g = resistance of a good wire
$\quad r$ = total resistance of loops
$\quad X_a$ = resistance to faulty wire from test set

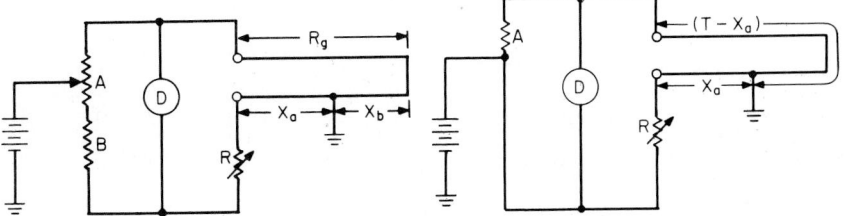

Fig. 4 Varley loop test.

Fig. 5 Murray loop test—for short sections and faults near the test location.

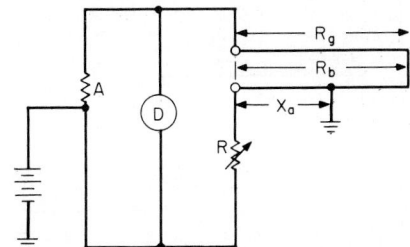

Fig. 6 Modified Murray loop—for use when only one conductor (not of the same size or resistance as the faulty wire) is available.

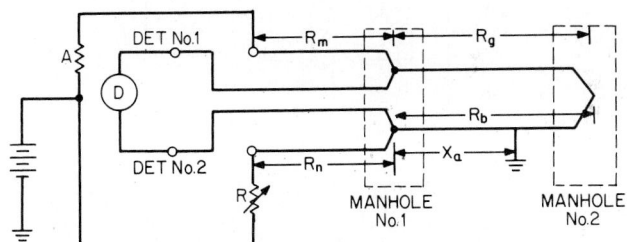

Fig. 7 Moody loop—for use when testing faults at locations distant from control cable locating desks.

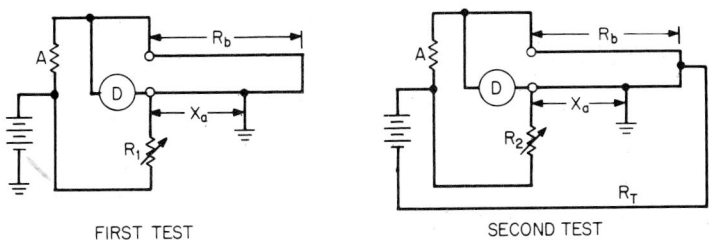

FIRST TEST SECOND TEST

Fig. 8 Fisher loop.

14.13

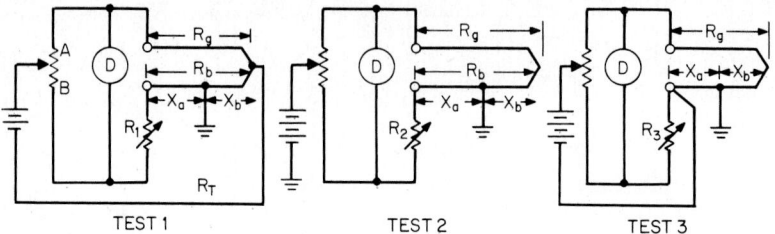

TEST 1   TEST 2   TEST 3

Fig. 9   Three-Varley method—useful when all conductors in a cable are defective.

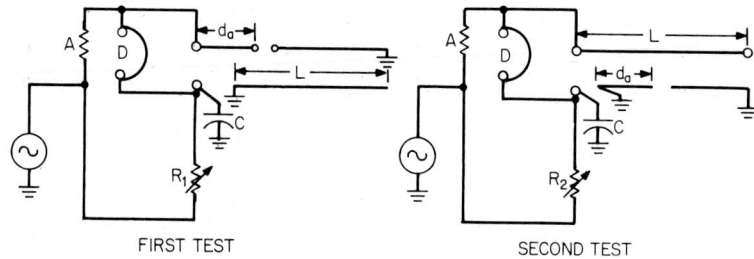

FIRST TEST   SECOND TEST

Fig. 10   Location of a break in a pair.

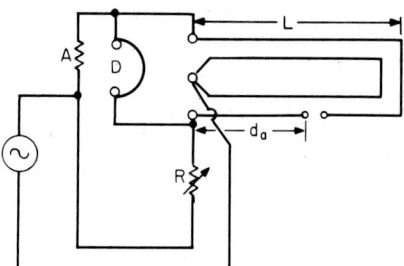

Fig. 11   Location of a break in a quadded cable.

Murray Loop Test

This method is used for short sections and faults near test locations. With reference to Fig. 5,

$$X_a = \frac{Rr}{A + R}$$

Modified Murray Loop

This test is for use when only one conductor (not of the same size or resistance as the faulty wire) is available. In this arrangement, as shown in Fig. 6, R_1 and A_1 are the test set readings when the bridge is connected as shown, and R_2 and A_2 are the test set readings when the left end of the pair of wires is shorted together and the bridge is reconnected to the open pairs of wires from the right.

$$X_a = \frac{R_1(A_2 + R_2)R_b}{R_2(A_1 + R_1) + R_1(A_2 + R_2)}$$

For a modified Murray loop, if $A_1 = A_2$ and $A + R$ is a constant, then $X_a = \dfrac{R_1}{R_2} R_b$.

Another modification of the Murray loop test is the Moody loop test shown in Fig. 7. This method is used when testing from control cable location desks and where the faults are at distant locations.

$$X_a = \frac{R_1 + R_n}{A + R_1 + R_n + R_m} r$$

where $r = r_g + R_b$.

Fisher Loop Test

This method is useful where all conductors in a cable are defective (grounded or crossed). Two good external conductors, R_b and R_T, are required, but need not be of either the same length or resistance as the faulty conductors. However, the resistance of R_b must be known. See Fig. 8.

$$X_a = \frac{(A + R_2)R_1}{(A + R_1)R_2} R_b$$

Three-Varley Method

Like the Fisher loop test, the three-Varley method, shown in Fig. 9, is useful when all conductors in a cable are defective. Two good conductors are required, R_g and R_T, of any resistance and length provided they can be connected sequentially to the far end and test set end of the faulty conductor.

$$X_a = \frac{A}{A + B} (R_3 - R_2)$$

$$X_b = \frac{A}{A + B} (R_2 - R_1)$$

$$R_b = \frac{A}{A + B} (R_3 - R_1)$$

If A/B can be made ⅛ or ¼, a function available in some test sets, the resolution of the measurement and resulting accuracy in locating the fault can be improved. Under these conditions

For $A/B = $ ⅛:

$$X_a = \frac{R_3 - R_2}{10} \qquad X_b = \frac{R_2 - R_1}{10} \qquad R_b = \frac{R_3 - R_1}{10}$$

For $A/B = $ ¼:

$$X_a = \frac{R_3 - R_2}{5} \qquad X_b = \frac{R_2 - R_1}{5} \qquad R_b = \frac{R_3 - R_1}{5}$$

Open Faults

If the fault is an open fault instead of a ground or a cross fault, the distance to the fault can be determined by measuring the capacitance between a good conductor and each section of the open conductor. This measurement, of course, requires an ac signal source and an ac null detector, usually a pair of sensitive headphones. Some typical arrangements are shown in Figs. 10 and 11.

When locating an open fault in a pair of cables, the capacitance of the cable is compared to a capacitor C (usually 1 or 2 μF). Two measurements are made, and the resulting ratio multiplied by the length of the good cable gives the distance to the open fault. See Fig. 10.

$$da = \left(\frac{R_1}{R_2}\right) L$$

where da = distance to fault and L = length of cable.

When four conductors are available in a cable, three of which are good, a simple measurement can be made by tying two of the conductors together at both ends. See Fig. 11. Under these conditions,

$$da = \frac{2A}{A + R} L$$

Universal Ratio Set (URS)

This arrangement is equivalent to a highly accurate continuously adjustable ratio arm of a Wheatstone bridge with a resolution of 0.001 Ω and a total range of 2111.110 Ω. The circuit is shown in Fig. 12. The ratio arms are so constructed that as resistance is removed from one ratio arm, it is added to the other, maintaining a constant total resistance.

The URS is particularly useful in measuring unknown resistance values in terms of high-accuracy fixed-decade standards within a 10:1 ratio of the unknown resistors. The effects of lead and contact resistance are eliminated by taking up to four readings and substituting these in the equation

$$\frac{X}{S} = \frac{R_4 - R_3}{R_2 - R_1}$$

Direct Reading Ratio Set (DRRS)

This arrangement has the same basic application as a URS but has only one adjustable arm, as shown in Fig. 13. This arm is adjustable from 99.445 to 100.555 Ω in 0.001-Ω steps. A DRRS has a very narrow range and is limited to deviations from a nominal 1:1 ratio of approximately $\pm 0.5\%$ with a resolution of 0.001%. When the standard and battery are connected to point D, the nominal ratio becomes 10:1 (or 1:10). Since only one measurement can be made, R_S and R_X must be of such magnitudes that lead and contact resistances are negligible.

Julie Ratiometric Bridge

Another variation of the dc Wheatstone bridge using a linear resistive (Kelvin-Varley) voltage divider (KVVD) is shown in Fig. 14.

In this arrangement a seven-decade resistive divider, normally having a 100,000-Ω total resistance, is connected in series with a zero adjustment resistance (R_0) and in parallel with a range adjustment (R_R). In operation, two calibration steps are required. In the first, the unknown resistance is a low-resistance shorting link and the voltage divider dials are set on zero. If A and B are nominally equal, R_0 will be adjusted to 100,000 Ω. The second step requires connecting to the X terminals a known resistance of nominal full-scale value, say 100,000 Ω. With the voltage divider set on 1.00000 (100,000 Ω) R_R is adjusted to 200,000 Ω. It can be shown that any value of resistance between 0 and 100,000 Ω can be measured with this circuit by reading the ratio setting of the voltage

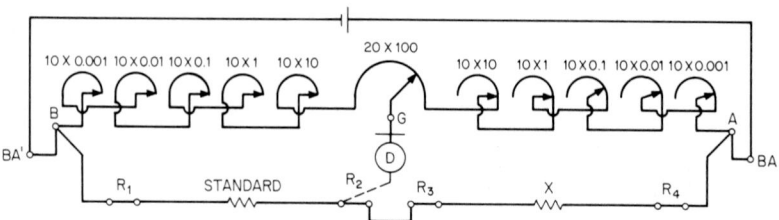

Fig. 12   Universal ratio set.

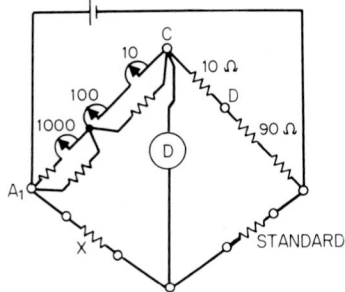

Fig. 13   Direct reading ratio set.

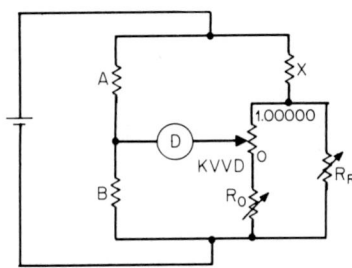

Fig. 14   Julie ratiometric bridge.

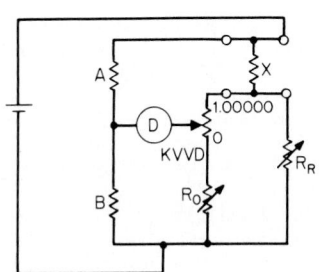

Fig. 15 Julie arrangement for low-resistance measurements.

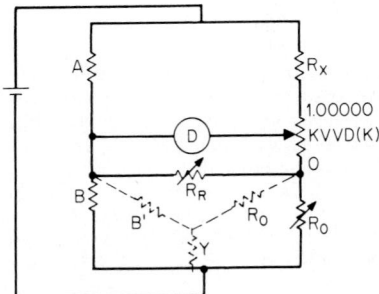

Fig. 16 Julie arrangement for high-resistance measurements.

divider and applying the proper multiplier. Changing the ratio A/B will necessitate selecting a different value of R_R, but the range of the KVVD will still be linear.

Low-resistance four-terminal measurements can be made by reconnecting the Julie bridge as shown in Fig. 15. With $A = 100\ \Omega$ and $B = 1000\ \Omega$, a full-scale range of $100\ \Omega$ can be achieved. Under these conditions, $R_0 = 10^6\ \Omega$ and $R_R = 1000.91\ \Omega$.

For a full-scale range above the range of the KVVD, say 100 MΩ, the bridge is reconnected as shown in Fig. 16. With a value of $10^6\ \Omega$ for A, $10^6\ \Omega$ for B, and $10^5\ \Omega$ for R_0, R_R is calculated to be 1102.2 Ω. Using a wye-delta transformation, B' becomes 1001 Ω and $R_0' = 100.1\ \Omega$. Y ($\sim$91 Ω) is in series with the battery and is of no significance in this example. Under some conditions, however, Y could require an increase in supply voltage to offset the voltage drop across Y and thereby improve sensitivity.

The general equations for Figs. 14 and 15 are

$$R_{X(\text{max})} = \frac{R_R(K + R_0)}{R_R + K + R_0} \times \frac{A}{B}$$

$$R_0 = \frac{KB}{A}$$

$$R_R = \frac{KBR_X(1 + B/A)}{KA + KB - BR_X}$$

The general equations for Fig. 16 are

$$R_{X(\max)} = \frac{A(K + R_0')}{B'}$$

$$R_0 = K \qquad A = B$$

$$R_R = \frac{K(B + K)}{R_{X(\max)} - 2K}$$

$$B' = \frac{R_R \times B}{R_R + B + R_0}$$

$$R_0' = \frac{R_0 \times R_R}{R_R + B + R_0}$$

Very High Resistance Measurements

Accurate resistance measurements above the 100-MΩ range involve problems of bridge sensitivity and the accuracy and stability of the high-valued bridge arm resistors. For maximum sensitivity, bridge arm resistors should be close to the unknown resistance in value, but high-accuracy high-stability gigohm (10^9-Ω) resistors are very difficult to obtain.

Several bridge circuits have been used successfully to overcome these problems. One involves a bridge circuit incorporating simulated high-valued resistances using a wye configuration for the high-valued bridge arm. With resistors no larger than 1 MΩ in value, simulated bridge arms of 10^{12} Ω are achievable. The basic bridge circuit is shown in Fig. 17.

For a range in X of 10^{11} to 10^{12} Ω, $R_{ao} = 10^6$ Ω, $R_{bo} = 10^6$ Ω, and $R_{co} = 10.2$ KΩ, the simulated value of A is 100 MΩ. The value of ratio arm B is selected so that when shunted with the delta equivalent of the wye circuit it will provide a nominal value of 1000 Ω. The rheostat arm is designed to have a maximum value of 10^7 Ω, with the highest decade consisting of ten 1-MΩ steps. Bridge sensitivity for this arrangement can be calculated from the equation

$$e_{\det} = V_{\text{DET}} \frac{R_D}{B_R + R_D}$$

where $e_{\det}$ = voltage available to detector
V_{DET} = unbalanced bridge voltage
R_D = detector input resistance
B_R = bridge resistance as seen by detector

A wye-connected circuit is also useful in simulating unknown resistances for checking bridge performance in high-megohm ranges. One such arrangement is shown in Fig. 18 where a simulated 100-MΩ resistor is used to calibrate a bridge.

The general equation for this wye-delta configuration is

$$X = d_o + a_o + \frac{d_o \times a_o}{R_o}$$

The shunt effect of the wye-delta configuration on the ratio arm of the bridge is

$$R_{\text{s}} = d_o + R_o + \frac{d_o \times R_o}{a_o}$$

The equivalent value of the rheostat arm when shunted becomes

$$Re = \frac{R_S \times R}{R_S + R}$$

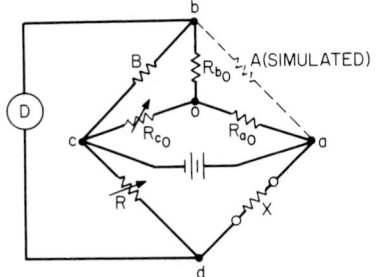

Fig. 17 Wye-delta bridge for high-resistance measurements.

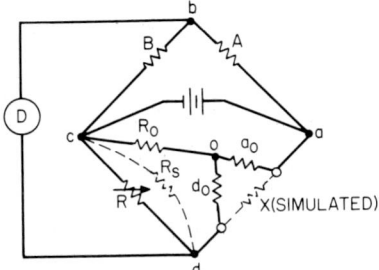

Fig. 18 Simulated wye-delta calibration resistor.

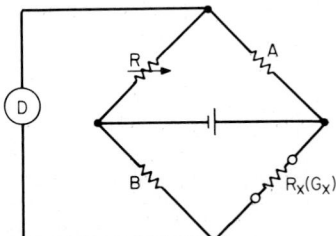

Fig. 19 Conductance bridge for high-resistance measurements.

Another arrangement used in measuring high-megohm resistance is a conductance bridge arrangement as shown in Fig. 19.

In this arrangement R, instead of being a resistance decade, becomes a conductance decade, and the bridge balance equation becomes

$$\frac{1}{R_X} = G_X = \frac{R}{A \times B}$$

If it is desired to measure 100 MΩ (equivalent to 0.01 μmhos) to a resolution of 1% (1 μmho), typical bridge arm resistance values will be 1 MΩ for the A and B arms and 10,000 Ω (100 μmhos) for the R arm adjustable to the nearest 100 Ω (10,000 μmhos). Note that the R arm will actually be constructed with conductance steps of 10,000, 5000, 3333.3, 2500, 2000, 1666.7, 1428.6, 1250, 1111.1, and 1000 Ω.

The advantage of this circuit is that the adjustable bridge arm is kept within a reasonable range of resistance values, while the fixed ratio arms A and B can become very large.

Guarding and Shielding

The measurement of high-valued resistances or even lower-valued resistances to a high accuracy requires the use of protective circuitry to prevent leakage currents from developing across imperfect insulators and thus degrading the accuracy of measurement. Such protective circuitry is referred to as *guarding*. Actually guarding does not eliminate leakage currents but rather diverts them to portions of the bridge circuit where the effect on the accuracy of measurement is significantly reduced. An example of an unguarded bridge circuit is shown in Fig. 20.

In this example leakage resistance between a (terminal 1) and d (terminal 2 or ground, if d is connected to ground) is directly across the unknown X resistance. If X is 1 MΩ, a leakage of 1000

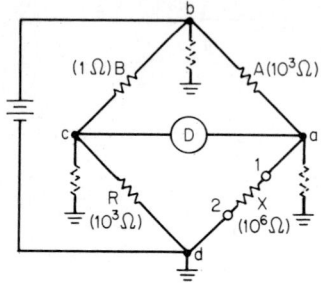

Fig. 20 Leakage paths in an unguarded bridge.

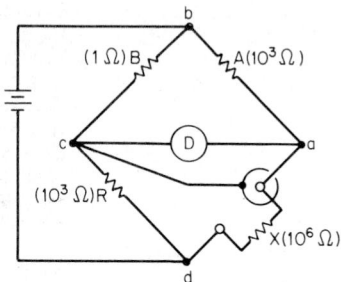

Fig. 21 Guarded Wheatstone bridge.

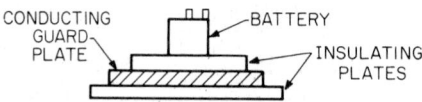

Fig. 22 Guarded battery arrangement.

$M\Omega$ will introduce an error of 0.1%. Most quality insulating materials can degrade to 1000 $M\Omega$ or less over a period of time as dust and dirt are deposited and in the presence of high relative humidity. This same leakage resistance can also occur across other bridge arms or between bridge arms and ground, but the effect is considerably less (only 0.0001% across the A and R arms).

A way of minimizing the effect of this leakage resistance by reducing its significance to an acceptable level is to surround terminal 1 with a conducting guard ring, as shown in Fig. 21, and connecting this guard ring to the c corner of the bridge.

In this arrangement, leakage resistance between the a corner (or terminal 1) to the guard ring is directly across the detector and is insignificant until its magnitude begins to approach the value of the input impedance of the detector, at which point the effect is a reduction in sensitivity. Leakage from the guard ring to terminal 2 is directly across R and has a negligible effect on the accuracy of the measurement. Guarding must be applied to all connections associated with the a bridge node including, in this example, the input circuit of the detector. If a guarded detector is not available, interchanging the detector and battery will permit placement of the battery on an insulating plate beneath which is mounted a guard plate and then on another insulating plate, as shown in Fig. 22.

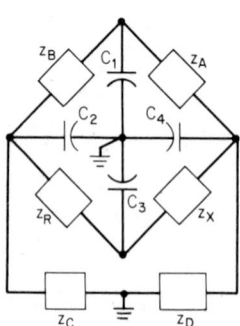

Fig. 23 Shielded impedance bridge with a Wagner ground.

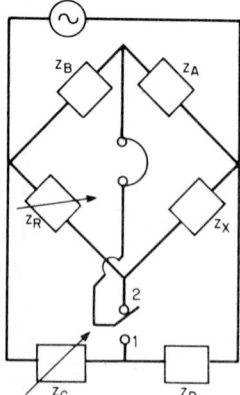

Fig. 24 Method of balancing a Wagner ground.

Guarding is different from shielding. Where guarding relates to protection from a dc leakage resistance, generally across the surface of an insulator, *shielding* relates to protection from an ac leakage conductance, generally the result of capacitance coupling between a source of signal noise and the measuring circuit. The purpose of shielding is to make these capacitances fixed in magnitude, location, and effect. The normal method of shielding is to totally surround the measuring circuit with a ground plane so that the capacitance coupling between the ground plane and measuring circuit remains stable. When shielding all the bridge arms in this manner, it is normally desirable to balance out the component to shield capacitance by the use of double shielding so that all capacitances are across portions of the bridge circuit where this effect is compensatable or insignificant. Another method sometimes used is to construct a circuit with only partial shielding and then use a Wagner ground circuit as shown in Fig. 23.

It can be seen that, by using the balance method shown in Fig. 24 of first connecting the detector to point 1 and balancing Z_C and then to point 2 and balancing Z_R (and repeating steps 1 and 2 until there is no change), the capacitance C_2 (Fig. 23) is in parallel with Z_C and C_4 is in parallel with Z_D, thus effectively removing them from the bridge circuit. The Wagner ground balance technique of Fig. 24 likewise eliminates the effect of C_1 and C_3 because, at null balance, both sides of the detector are at the equivalent of ground potential. This holds true as long as $Z_A/Z_B = Z_X/Z_R = Z_D/Z_C$.

Characteristics of Resistance Materials

When measuring resistance or selecting bridge designs for dc resistance measurements, consideration must be given to the characteristics of the unknown resistor and bridge component resistors. Ideally, a resistor should be stable with time and temperature and not generate spurious voltages when dissipating internally generated heat or when the instrument is subjected to a temperature change. Such spurious voltages are commonly called *thermal electromotive forces* (emf's). In addition, an ideal resistor should be sufficiently nonreactive that its time constant will not result in detector transients (kicks of the meter pointer) when voltage is applied to or removed from the bridge. These characteristics are all controlled by selecting the appropriate alloy material and manufacturing technique. The alloys most commonly used are copper-manganese, copper-nickel, and nickel-chromium alloys. Copper-manganese alloys, one having the trade name Manganin, have a parabolic temperature coefficient that can be controlled to peak at close to room temperature (25 to 30°C), so that the resulting temperature coefficient of resistance is within ±10 ppm/°C over the range 23 to 33°C. The alloy can be altered so that the peak can be higher for applications involving significant wattage dissipation. This high-peak material is referred to as shunt Manganin. Manganin has a very low thermal emf against copper (2 to 3 μV/°C), is inherently stable, providing the chemical constituents are properly controlled, and is easily soldered to copper leadwires, resulting in a stable termination. As with all alloy wires, ultimate stability can be achieved only if the stresses developed during winding are relieved by accelerated aging at high temperature (generally 24 to 48 hours at 140°C).

Constantan and Advance are trade names for two copper-nickel alloy wires. These materials have significantly flatter coefficient curves than Manganin but are characterized by very high thermal emf's against copper (approximately 40 μV/°C). Such thermals would be unacceptable in a dc bridge but are inconsequential in an ac circuit. As a matter of fact, one way of eliminating the thermal effects in a dc bridge is to perform a null balance measurement, reverse the polarity of the voltage supply, and take a second measurement. The average value of these two measurements eliminates the error of the thermal emf in the bridge circuit including the unknown resistor, providing the thermal emf remains constant during the two measurements. Karma and Evanohm are the trade names of two nickel-chromium alloys. These materials have high resistivity and tensile strength and are suitable for high resistance value applications. Nickel-chromium alloys are very susceptible to strain, and great care must be taken during the winding process to minimize strain and relieve induced strain by high temperature accelerated aging. Stable terminations of this alloy wire to copper leads are also difficult to achieve, and sophisticated welding or brazing techniques are required. This alloy has a very low temperature coefficient, however, and its parabolic curve is extremely flat, with coefficients of a few parts per million achievable in the 10 to 50°C range.

Minimizing the reactive component of a resistor can be achieved by selecting appropriate winding configurations. Low-valued resistors tend to be inductive, and noninductive windings are produced by using a bifilar configuration. This technique involves selecting a specified length of wire, determining the midpoint, folding the wire in half, and winding the two halves together. For high-valued resistors, this technique results in a capacitive winding, and therefore either a segmented spool with a unifilar winding is used, reversing the turns in each segment (commonly called a *pi winding*), or a reversed layer-wound resistor is used wherein each layer effectively cancels the inductance of the layer beneath it. The least reactive configuration using resistance alloy wire is a woven resistor in which the web pattern is chosen to reduce the inductive and capacitive effects to the lowest possible level.

Metal film resistors are far less reactive than wire-wound configurations, can be designed so that substrate and alloy temperature coefficients have a compensating effect, and can be made reasonably stable. Generally they are not as stable as wire-wound configurations and do not have the range of values available in wire-wound units. Nonmetallic deposited resistors are available with very high-resistance values but are even less stable.

All resistors must be protected against the adverse effects of humidity and other atmospheric contaminants. The most stable resistors are hermetically sealed in metal cans, in some instances in a high-grade, inert oil. Other resistors are protected by dipping them in plastic or waxlike materials or molding them in a plastic capsule. All plastics eventually allow moisture to penetrate. The only true hermetic seal is a soldered glass-to-metal seal. The degree of protection required depends, therefore, on the ultimate stability desired and the environment in which the resistor is used.

AC Bridges

An ac bridge differs from a dc instrument in that an ac signal source and ac null detector are required and, generally, two balance conditions must be met, the in-phase or resistive condition and the quadrature or reactive condition. Some ac bridges require only one balance, providing only the resistive or reactive component is desired, and utilize a phase-sensitive detector.

Examples of various ac bridge circuits, their applications, and bridge balance equations are given in the following several paragraphs. See Figs. 25 through 38.

Maxwell Bridge

As shown in Fig. 25, the Maxwell bridge is useful in measuring unknown inductors in terms of an adjustable resistance and inductance. Obtaining an adjustable inductor may be difficult.

$$R_X = \frac{R_A}{R_B} R_S \qquad L_X = \frac{R_A}{R_B} L_S$$

An alternative version of this circuit, shown in Fig. 26, measures the series inductance in terms of a parallel capacitance.

$$R_X = \frac{R_A R_S}{R_B} \qquad L_X = R_A R_S C_B$$

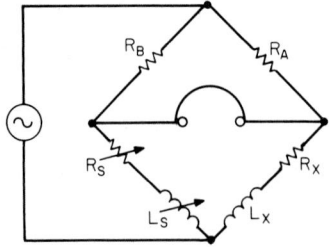

Fig. 25   Maxwell bridge.

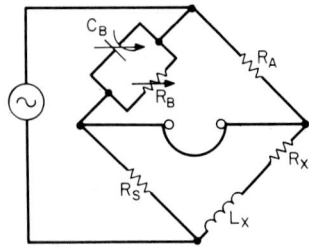

Fig. 26   Alternate form of a Maxwell bridge.

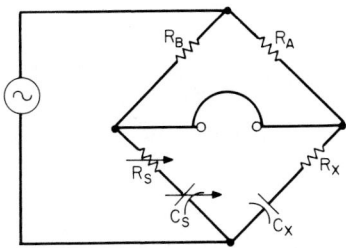

Fig. 27   Impedance bridge.

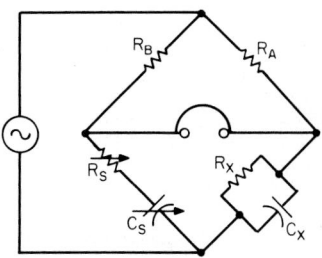

Fig. 28   Wien bridge.

Impedance Bridge

The bridge shown in Fig. 27 is useful in measuring capacitance.

$$R_X = \frac{R_A}{R_B} R_S \qquad C_X = \frac{R_B}{R_A} C_S$$

Wien Bridge

This bridge is shown in Fig. 28.

$$R_X = \frac{R_A}{R_B} \times \frac{1 + W^2 C_S^2 R_S^2}{W^2 C_S^2 R_S^2}$$

$$C_X = \frac{R_A}{R_B} \times \frac{C_S}{1 + W^2 C_S^2 R_S^2}$$

where $W = 2\pi f$.

Although a Wien bridge can be used to measure an unknown resistance and parallel capacitance in terms of a standard adjustable series resistance and capacitance, its more general application is in the measurement of frequency. If $R_B = 2R_A$, $C_S = C_X$, and $R_S = R_X$, the balance equation becomes

$$f(\text{frequency}) = \frac{1}{2\pi R_S C_S}$$

where f is the frequency of the ac source required to produce a null on the detector. This circuit is often used in frequency recorders.

Owen Bridge

As shown in Fig. 29, an Owen bridge is useful when measuring wide ranges of unknown inductances and using reasonable ranges of standard capacitances. This bridge also has been used for core loss and permeability measurements.

$$R_X = R_A \frac{C_B}{C_S} \qquad L_X = R_A R_S C_B$$

Hay Bridge

As shown in Fig. 30, this bridge is particularly useful in measuring Q [or dissipation factor D ($D = 1/Q$)] and is often used in general purpose impedance bridge circuits.

$$L_X = \frac{R_A R_S C_B}{1 + W^2 R_B^2 C_B^2}$$

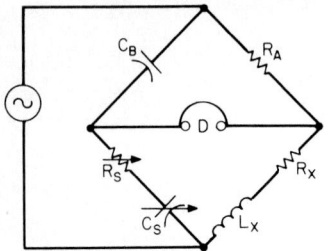

Fig. 29   Owen bridge.

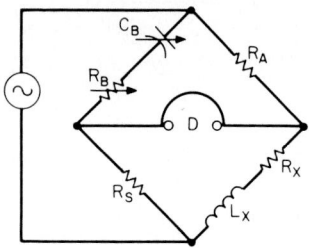

Fig. 30   Hay bridge.

$$R_X = \frac{R_B R_A R_S W^2 C_B^2}{1 + W^2 R_B^2 C_B^2}$$

$$Q_X = \frac{1}{W C_B R_B}$$

Schering Bridge

Shown in Fig. 31, a Schering bridge is particularly useful for insulation testing of high-quality capacitors, insulators, condenser bushings, insulating oils, and other insulating materials.

$$R_X = R_A \times \frac{C_B}{C_S}$$

$$C_X = \frac{R_B}{R_A} \times C_S$$

$$D_X = W C_B R_B$$

Anderson Bridge

Shown in Fig. 32, an Anderson bridge is useful in measuring extremely low-Q coils and has its greatest sensitivity when $R_B = R_S = \frac{1}{2}R_X$, $R_A = R_X$, and $L_X/C_0 = 2R_X^2$.

$$L_X = C_0 R_A \left[R_0 \left(\frac{R_S}{R_B} + 1 \right) + R_S \right]$$

$$R_X = \frac{R_A \times R_S}{R_B}$$

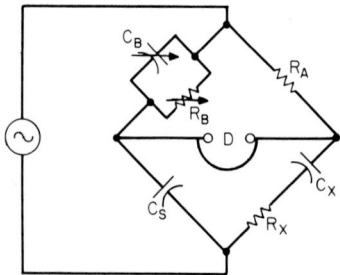

Fig. 31   Schering bridge.

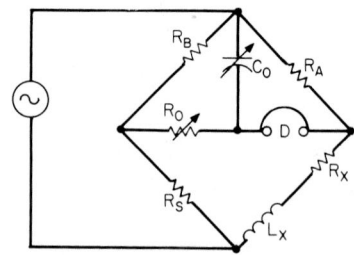

Fig. 32   Anderson bridge.

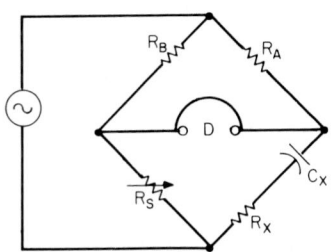

Fig. 33 Conductivity bridge.

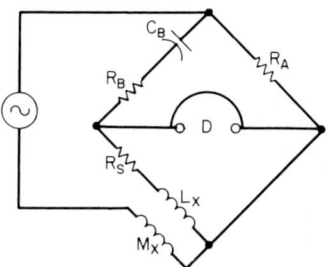

Fig. 34 Heydweiller bridge.

Conductivity Bridge

When measuring conductivity cells, as in liquid analysis applications, a circuit similar to Fig. 33 is used. A conductivity cell has the equivalent circuit of a series capacitance and series resistance. If the cell could be measured at direct current, the capacitance component of the cell would not be significant, but alternating current must be used to prevent polarization of the electrodes, resulting in false readings. The higher the frequency, the lower the polarization error but the higher the impedance-to-resistance ratio error. If a detector is used that is sensitive only to the in-phase component, R_X can be measured in terms of R_S, R_A, and R_B with only a small resultant error. When $R_A = R_B$, the resultant error of Z_X compared to R_X can be as high as 30%, but with a phase-sensitive detector this error can be reduced to less than 5%.

The balance equation of the conductivity bridge is

$$R_X = \frac{R_A}{R_B} R_S$$

Heydweiller Bridge

This bridge, shown in Fig. 34, is applicable in determining the mutual inductance M between two windings. The balance equation is

$$M_X = R_A R_S C_B \qquad L_X = C_B R_S (R_A + R_B)$$

Heaviside Bridge

This bridge, shown in Fig. 35, also is used for mutual inductance measurements. However, the inductance of the primary winding must be known.

$$M = \frac{R_A L_S - R_B L_X}{R_A + R_B} \qquad R_X = \frac{R_A}{R_B} R_S$$

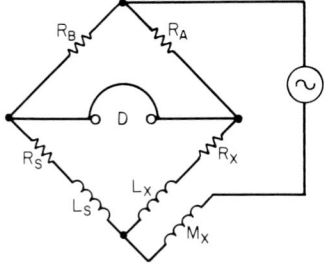

Fig. 35 Heaviside bridge.

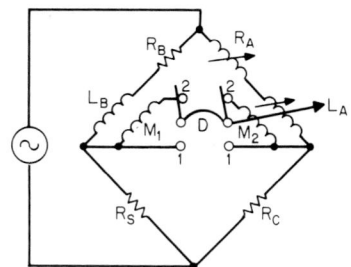

Fig. 36 Campbell bridge.

Campbell Bridge

This bridge, shown in Fig. 36, is initially balanced by adjusting R_A and L_A to obtain the relation shown below with the detector on 1. With the detector on 2, the mutual inductance of only L_A is rebalanced and the result is measurement of a mutual inductance M_2 in terms of a mutual inductance standard M_1.

With the detector on 1:

$$\frac{R_C}{R_S} = \frac{R_A}{R_B} = \frac{L_A}{L_B}.$$

With the detector on 2:

$$M_1 = \frac{R_S M_2}{R_C}.$$

Ratio Transformer Bridge

Another ac bridge arrangement makes use of the ratio transformer for the bridge ratio arms. See Fig. 37. In this circuit, T_1 is a voltage transformer and T_2 is a current transformer. The equation at balance is

$$Z_X = \frac{N_X}{N_s} \times \frac{m_X}{m_S} \times Z_S$$

This equation shows that two sets of ratio multipliers are available to achieve balance, providing greater flexibility in the selection of ratios should the reactive and resistive components be in different ratio relationships. N_S and m_S can be designed with selectable taps, with G_S and C_S connected to different taps. Adding further taps permits multiple standards of capacitance and conductance to be added for increased resolution.

Transformer Voltage Divider

This inductive voltage divider is similar to a resistive KVVD and can also be connected in an ac bridge configuration. One arrangement is used in the high-accuracy measurements of platinum resistance thermometers, as shown in Fig. 38. Under proper design conditions, a rather complex true balance equation can be reduced to

$$R_t = \frac{R_S \eta_0}{1 - \eta_0}$$

where η_0 = ratio reading on the primary side of the inductive voltage divider.

DC Null Detectors

While the basic "brain" of a dc resistance measurement is certainly the bridge, the heart of the measurement is the null detector. A complete discussion of null detectors is beyond the scope of this article, but no detector should be selected for use in a measurement system without some knowledge of the inherent characteristics of such a device. Galvanometers, electronic null detectors, and ac null detectors are described briefly here.

Galvanometers

These detectors were popular as relatively low-cost dc null detectors before the advent of dc operational amplifiers. However, the galvanometer still has an application where its natural damping and slower speed of response make it relatively immune to electrical interference problems and where a detector power source (battery or dc supply) is not available.

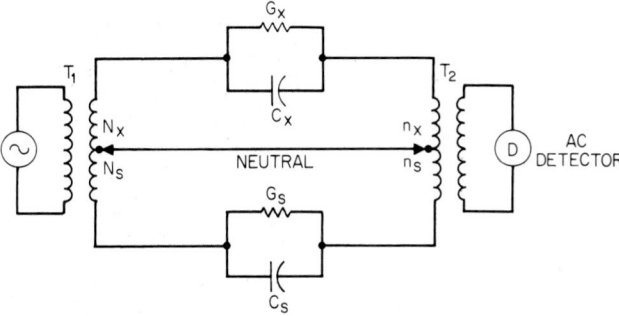

Fig. 37   Ratio transformer bridge.

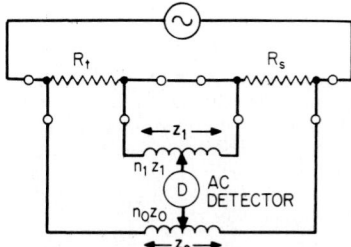

Fig. 38   Transformer voltage divider for platinum resistance temperature detector (RTD) measurements.

Sensitivity

The sensitivity of a galvanometer is generally stated in terms of microamps per division or microvolts per division. Since a galvanometer is a low-impedance device, it is thought of as a current-sensitive instrument. Voltage sensitivity is therefore current sensitivity times the sum of the internal resistance plus the external critical damping resistance, described shortly.

Internal Resistance

This value is normally a low number (less than 1000 Ω), and the voltage divider effect of the galvanometer's internal resistance compared to the bridge resistance must be taken into account when calculating bridge sensitivity.

Critical Damping Resistance, External (CDRX)

This is the optimum external circuit resistance for normal operation. If the external resistance significantly exceeds the CDRX, the galvanometer will oscillate (it will be underdamped). If it is below the CDRX, it will take an exceptionally long time to achieve balance (it will be overdamped). When critically damped, the galvanometer achieves its steady-state condition in the shortest time. Many operators prefer a slightly underdamped condition so that they can better observe the return to null.

Period

The undamped swinging of a galvanometer pointer in terms of the time required to complete one cycle is known as the period. A typical period is 3 to 5 seconds.

Electronic Null Detectors

The low-cost dc operational amplifier, low-noise field-effect transistor (FET) converter, and stabilized feedback circuit have made the electronic null detector the most popular detector for dc bridges and potentiometers. However, this design has also introduced other characteristics that must be considered when selecting a null detector for a given application. Some of these characteristics include the following.

Sensitivity

The ultimate sensitivity of any detector is limited by the thermal agitation or Johnson noise of the circuit. This limit is defined by the equation

$$En = \sqrt{4KTBR}$$

where En = ultimate root-mean-square (rms) noise voltage, V
$\quad K$ = Boltzman's constant (1.38×10^{-23} joules per kelvin)
$\quad T$ = absolute temperature, kelvin
$\quad B$ = bandwidth of noise [$1/(2\pi \times$ time constant)]
$\quad R$ = resistance, Ω

As an example, the limiting rms noise voltage for a 10^6-Ω resistor at 30°C with a bandwidth of 10 Hz is 0.6 μV. In addition, other sources of noise include thermal noise (the Peltier effect), transistor noise, flicker noise, popcorn noise, pink noise, white noise, and spot noise—all of which can affect the usable sensitivity of an electronic detector. Consequently, some detector performance specifications reference a threshold level (the minimum detectable signal change including noise interference). In general, the most sensitive commercial detectors are at least a factor of 5 or 10 less sensitive than the theoretical Johnson noise limit.

Stability

Electronic detectors must remain sufficiently stable to complete a measurement. Instability of a zero setting can result from time, temperature (an effect on voltage and current offset), changes in the source resistance (generally called current offset or the change in voltage offset with a change in source resistance, $I_0 = \Delta E_0/\Delta R_S$), and changes caused by inadequate power line regulation.

Input Impedance

Depending on the circuit used, an electronic detector can have an input impedance based on the true input filter and amplifier resistance or it can have a high-input impedance resulting from the use of a negative feedback circuit. The primary concern with feedback circuits is that the input impedance is high only at balance; it may be very low during transient conditions and therefore could cause a problem if high impedance is required for circuit protection purposes, as is the case in certain potentiometer applications when measuring standard cells.

Speed of Response

Several terms are used to describe the speed of response. *Response time* is the time required to achieve 100% deflection or the true steady-state condition. *Rise time* is the time required to achieve 90% of the response time. If the rise time of a detector is 1 s, it will take 2 s before the detector is within 1% of its final value. The *time constant,* of course, is the time required to achieve approximately 63% of the steady-state condition.

Isolation

Two types of isolation are specified for electronic detectors—dc isolation and ac isolation.
   Direct current isolation involves the use of high insulating materials and possibly a dc guard circuit surrounding the input of the detector. More common is ac isolation, or ac guarding required to achieve a high common-mode rejection ration (CMRR).
   The principle of ac guarding is illustrated in Fig. 39 which shows a detector connected to a test

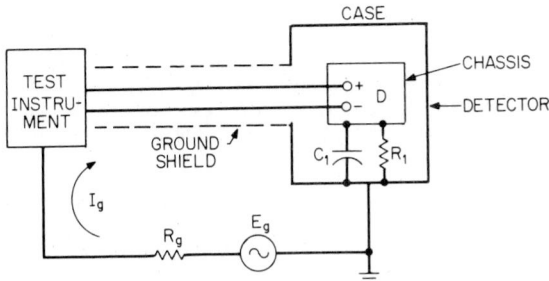

Fig. 39 Unguarded detector.

instrument. The common circuit of the detector is usually isolated from the case by insulation resistance R_1 and a filter capacitor C_1. The ground connection to the test instrument, R_g, involves a finite resistance, and E_g is an induced ac signal resulting from an external pickup. Even when the detector input leads are shielded, a circulating current I_g flows into the input of the detector and can result in an unstable voltage offset.

The same arrangement with an ac guarded input is shown in Fig. 40. In this circuit, the same I_g circulating current exists but flows through the guard shield to the guard chassis, effectively bypassing the input. Note that, to properly use an ac guard, both the test instrument and the detector must have provision for a guard connection.

The CMRR capability of a detector is expressed in decibels by

$$\text{CMRR} = 20 \log \frac{E_g}{E_0}$$

where E_g = common-mode-generated interference voltage and E_0 = resulting offset voltage.

If a generated interference voltage is 100 V and the resulting dc offset voltage is 100 μV, the CMRR will be 120 dB.

Another form of interference voltage is normal mode, or noise voltage that appears across the input terminals of the detector. The offset resulting from this type of interference is reduced by shielded input wiring, twisted input wiring, and input filtering. The problem with input filtering is that it tends to slow down the response time of the detector. The effectiveness of such filtering, the normal-mode rejection ratio (NMRR), is again expressed in decibels by the formula

$$\text{NMRR} = 20 \log \frac{E_g}{E_0}$$

where E_g = normal-mode interference voltage and E_0 = resulting detector offset voltage.

As an example, a 1-V ac normal-mode signal might result in a 100-μV dc offset, in which case the NMRR would be 80 dB.

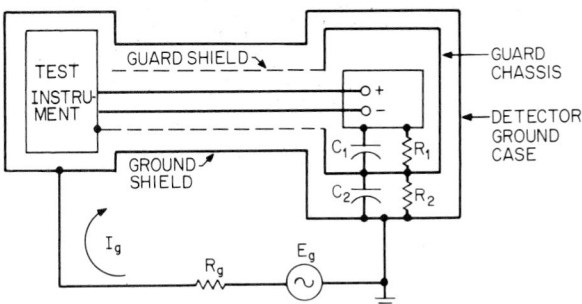

Fig. 40 Guarded detector.

AC Null Detectors

Alternating current null detectors come in a large variety of forms. For most bridge applications the detector is a tuned ac amplifier with a rectified dc output feeding a dc milliameter or oscilloscope indicator. Sensitivities of 1 μV with an input impedance of 1 MΩ are available. The tuned filter circuit is for the purpose of rejecting signals at frequencies other than the desired frequency. However, low-cost ac bridges may have relatively inexpensive ac detectors including ac galvanometers, ac amplifiers, tuning-eye tubes, or telephone headsets (under proper conditions of minimal background noise the current required to produce an audible signal at 1 kHz is less than 1 μA). The oscilloscope is also usable and indicates an ac signal as low as 1 mV.

POTENTIOMETERS

A *potentiometer* is an instrument which measures an unknown voltage or voltage difference in terms of a known voltage. The primary advantage of a potentiometer is that it is a null balance measuring instrument, inherently capable of high accuracy. *Null balance* means that at balance the potentiometer presents no load to the unknown voltage. The primary purpose of a potentiometer is in the measurement of dc voltages, generally in the region of 10 V and lower. As used in this discussion, the term "potentiometer" should not be confused with a three-terminal adjustable resistor which is not a measuring instrument. Also, this discussion does not include digital or differential voltmeters which are discussed in another section.

Potentiometer circuits fall into two basic categories, constant current and constant resistance. See Fig. 41.

In the constant current arrangement the R_P resistor is adjustable in increments equivalent to the resolution of the instrument. R_P can be a single slidewire with a resolution of only 1 part in 1000, or a multidecade device, using the Kelvin-Varley arrangement, with a resolution of 1 part in 10^7. In either case a linear scale or dial calibration is mechanically linked to the movable contact. In operation a known voltage is connected to the E_X terminals and the R_P resistor is adjusted to an equivalent dial setting. The standardizing resistor R_S is adjusted until the voltage drop E_S, caused by the circulating current I_S, is equal to E_X as indicated by a null on the detector. At this point the potentiometer is standardized. The known voltage may be removed, and any unknown voltage within the range of the instrument ($R_P \times I_S$) connected in its place. Readjusting the contact of R_P to a new null on the detector results in an indication of the value for the unknown voltage, providing I_S has not changed.

In the constant resistance arrangement the R_P resistor is fixed for a given range, and the resistor R_S adjusted until a null is achieved. The ammeter reads the value of I_S which, when multiplied by the value of R_P, indicates the value of E_X. A constant resistance potentiometer is particularly useful in measuring extremely low voltages, since there is no movable contact in series with the unknown voltage. Such contacts can generate significant thermal (Peltier) voltages. A 0 to 10 μA ammeter in conjunction with a 1-Ω resistor for R_P provides a potentiometric range of 0 to 10 μV.

Multirange constant current potentiometers can be designed using two approaches. Figure 42a shows a double-range instrument, and Fig. 42b a triple-range instrument. Both circuits meet the required conditions that (1) regardless of the setting of the range switch, the resistance as seen by the battery remains constant (necessary to preserve the constant current condition), and (2) the current flowing thru R_P (the parallel value of a resistor shunting a slidewire) is adjustable in exponential steps (factors of 1:1, 1:10, and 1:100).

A constant current potentiometer must be standardized against a known voltage to establish the proper current through the measuring circuit. One way of accomplishing this is to switch the reference voltage (usually a standard cell) across the input circuit as shown in Fig. 43. The problem with this arrangement is that the slidewire must be reset to the value of the standard cell each time the potentiometer is restandardized. To overcome this disadvantage, a circuit similar to that shown in Fig. 44 can be used.

In the circuit in Fig. 44 R_{SC} is a resistor that provides the major portion of the voltage drop for the standard cell, and S_{SC} is an adjustable voltage divider that can be set for the precise value of the standard cell. The detector is switched first to the "standardize" position, and the battery rheostat adjusted for null, and then to the "measure" position.

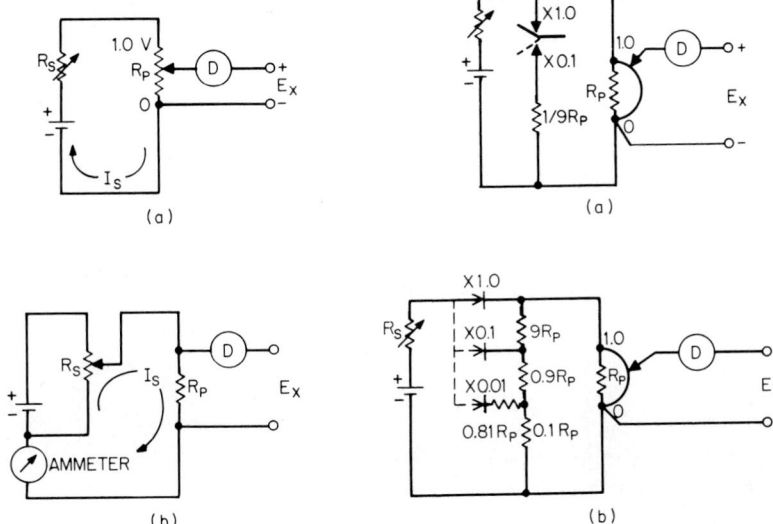

Fig. 41 Potentiometer circuits. (*a*) Constant current potentiometer, (*b*) constant resistance potentiometer.

Fig. 42 Potentiometer range-change circuits. (*a*) Double range, (*b*) triple range.

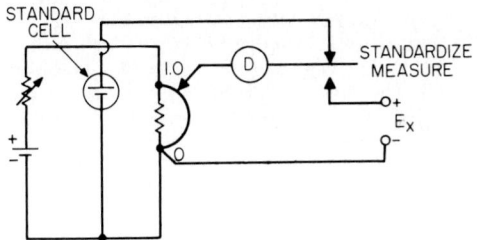

Fig. 43 Standardizing circuit.

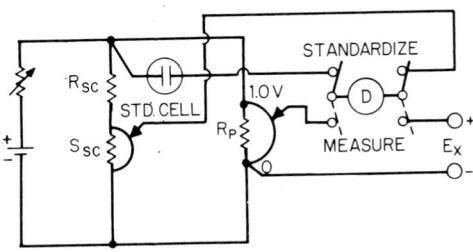

Fig. 44 Alternate standardizing circuit.

14.31

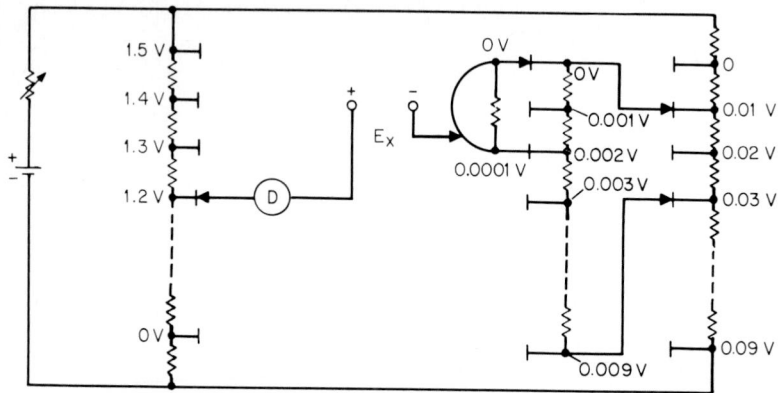

Fig. 45   Kelvin-Varley decades in a double-branch potentiometer.

If greater resolution is desired than is available in a single slidewire potentiometer, a double-branch circuit may be used employing the Kelvin-Varley arrangement as shown in Fig. 45. This arrangement can be extended to as many decades as desired and, of course, the standardizing circuits and range change circuits described previously may also be employed.

An alternate application of a Kelvin-Varley Voltage Divider (KVVD) in the measurement of dc voltages is shown in Fig. 46 and is known as a Julie universal potentiometer. It differs from the potentiometric circuits previously described in that it does not utilize a power source, such as a battery, which provides constant current at a constant resistance, but rather employs a true constant current source operating into an inverted KVVD, resulting in a varying resistance as seen by the current source. A mathematical analysis of this circuit shows that E_X is directly proportional to the setting of the KVVD. Typical circuit constants are 400 μA for the constant current source, 100,000 Ω for the KVVD, 25,000 Ω for the input resistance on the 10-V range, and 2500 Ω for the input resistance on the 1-V range. This circuit has the advantage of high accuracy and resolution, attenuated thermal effects generated by the switch contacts of the KVVD, and the capability to incorporate certain self-calibration features.

Various auxiliary devices may be added to a potentiometer to extend its usefulness.

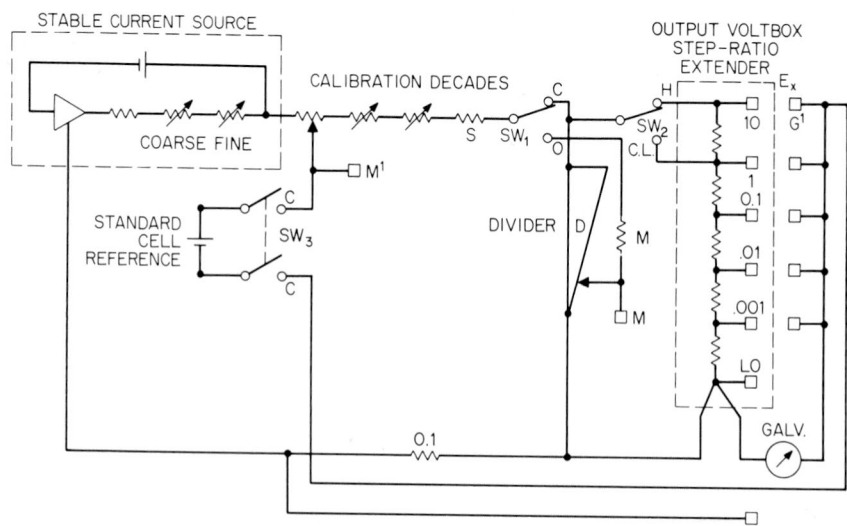

Fig. 46   Julie universal potentiometer.

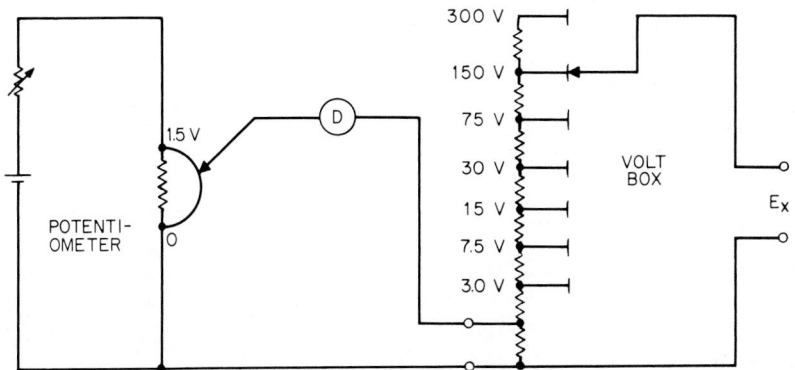

Fig. 47 Potentiometer with a volt box.

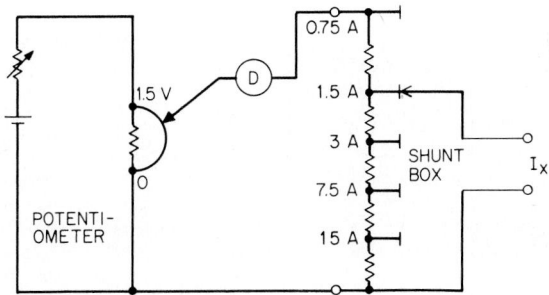

Fig. 48 Potentiometer with a shunt box.

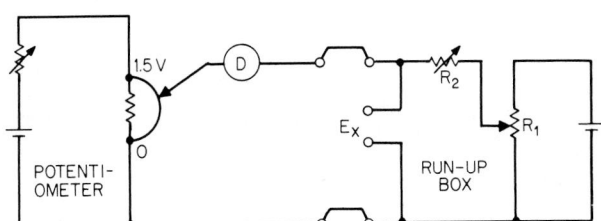

Fig. 49 Potentiometer with a run-up box.

Extension of the voltage range is accomplished by means of a volt box, a high-impedance voltage divider that provides convenient range multipliers. The basic volt box circuit is shown in Fig. 47. Volt boxes are designed to have as high an ohm-per-volt rating as possible to minimize loading on the unknown voltage and reduce self-heating. In the circuit shown, a 750-Ω/V circuit would result in a 225,000-Ω voltage divider on the 300-V tap and a 2250-Ω divider on the 3-V tap.

A potentiometer can be used to measure current by utilizing a shunt box which consists of a series of high-wattage resistors whose values provide voltage drops that are convenient multiples of the current being measured. Such a circuit is shown in Fig. 48. The resistance of the 15-A range in the circuit shown to provide 1.5 V output would be 0.1 Ω. However, the wattage dissipated by this resistor would be 22½ W, a large dissipation for an accurate resistor. Therefore a more practical shunt box design would be a 150-mV output, in which case the 15-A section would need to be only 0.01 Ω and the wattage dissipation 2¼ W.

Sometimes a potentiometer is used to calibrate a voltage measuring device which requires a finite input current at balance, such as a deflection meter, recorder, or millivoltmeter. Since a potentiometer is not accurate if it is delivering current to the unknown voltage, an auxiliary current source is required such as the run-up box shown in Fig. 49.

A run-up box contains its own voltage source, an adjustable potential divider R_1 that provides the proper voltage to the instrument under calibration, and an adjustable series resistor R_2 that provides for the designated circuit resistance specified for the instrument under calibration. In operation, the instrument under calibration is connected to E_X, the run-up box is adjusted for the desired deflection, and the resulting voltage is measured by the potentiometer.

Since a potentiometer is a comparative device, it requires a known reference voltage to measure an unknown voltage. It is appropriate, therefore, to include a brief discussion of the characteristics of various voltage reference devices. The three most common reference devices are compared in Table 1. The values given are representative for comparison purposes and may vary among manufacturers.

The saturated cell is presently the transportable reference standard for voltage and exhibits the

Table 1 Comparison of Standardization Sources

Characteristic	Saturated standard cell	Unsaturated standard cell	Zener-regulated supply*
Nominal voltage	1.018220 V $\pm$ 20 μV at 28°C	1.0193 V $\pm$ 200 μV at 25°C	Selectable, 7 V common
Accuracy	$\pm 2~\mu$V	$\pm 50~\mu$V	1–200 ppm, depending on stability
Stability	1–2 μV/year	20–40 μV/year	3–50 ppm/year, 30 ppm typical
Temperature coefficient	$-50~\mu$V/°C	-1 to $-5~\mu$V/°C	1–2 ppm/°C compensated, 0.1 ppm/°C in constant temperature chamber
Internal resistance	300–1 kΩ	300–500 Ω	Generally less than 0.1 Ω
Special considerations Noise	None	None	<0.5 ppm to >10 ppm (p-p)
Loading	$\ll 1~\mu$A to maintain stability	10 μA for short periods	No effect—may supply current
Shock/vibration	Sensitive	Somewhat sensitive	Insensitive
Storage temperature	4–40°C, temperature hysteresis 10–15 ppm/$\pm$10°C		Not sensitive
Regulation	Not applicable		2–3 ppm change for 10% line voltage shift
Constant temperature chamber	Required	Not required	May be required for highest stability units
Relative cost	Low but requires constant temperature bath; system cost is high	Low	Medium to high depending on desired stability

*Characteristics cover a wide range because of variety of supplies available.

highest accuracy and best long-term stability. It has a large temperature coefficient, however, and so must be maintained in a constant temperature bath (with air or oil). The unsaturated standard cell has a much lower temperature coefficient providing its two legs are maintained at about the same temperature, but it has poorer long-term stability. Both types of standard cells are mercurous sulfate-cadmium sulfate solutions in sealed glass cells and are therefore relatively fragile. The zener-regulated supply, however, requires an auxiliary power source (battery or isolated, regulated and rectified ac line operation). Zener supplies are available in a whole range of performance characteristics. In general a zener supply with characteristics equal to those of an unsaturated standard cell is considerably more costly. Selected zeners with characteristics approaching those of the saturated cell may cost as much as the saturated cell because both require a constant temperature enclosure.

REFERENCES

Calvert, R.: "The Transformer Ratio Arm Bridge," *Instrum. Control Syst.,* vol. 34, no. 1, January 1961.

Harris, F. R.: *Electrical Measurements,* Wiley, New York, 1952.

Hill, J. J., and A. P. Miller: "An AC Double Bridge with Inductivity Coupled Ratio Arms for Precision Platinum-Resistance Thermometry," *Proc. IEEE,* no. 110, vol. 2, February 1963.

Julie, L.: "Ratiometric DC Calibration Technique," *Instrum. Control Syst.,* vol. 39, no. 1, January 1966.

Julie, L.: "A Universal Potentiometer for the Range of 1 Nanovolt to 10 Volts," *IEEE Trans. Instrum. Meas.,* IM-16(3), September 1967.

Koep, K. J.: "Voltage Reference Standards," *Measurements and Control,* April, 1979.

L&N: "Fault Location in Cables," *Tech. Publ. 177737,* Leads & Northrup, North Wales, Pa., 1973.

Luppold, D. S.: *Precision DC Measurements and Standards,* Addison-Wesley, Menlo Park, Calif., 1969.

Parnell, J.: "The Measurement of Very High Resistance," ISA 18th Annual Conference, *Preprint 7. 3. 63,* September 1963.

Riley, J. C.: "AC Measurements Using Ratio Techniques," *ISA J.,* March 1965.

Sauer, H. A., and W. H. Shirk, Jr.: "A DC Wheatstone Bridge for Multi-Teraohm Measurements with High Accuracy Capability," *IEEE, Pap. No. CP 63-45,* 1963.

Stout, M. B.: *Basic Electrical Measurements,* Prentice-Hall, Englewood Cliffs, N.J., 1950.

Self-Balancing Electrical Instruments

by
T. J. Walsh*

In manually balanced potentiometers and bridges, the detection of a condition of imbalance is made visually from a zero-centered galvanometer or electronic null detector. A balanced or null condition is established by manually adjusting the value of one or more circuit parameters.

In self-balancing bridge and potentiometer circuits, the servo amplifier (or in some cases, a galvanometer) functions with a motor to position a calibrated feedback device to accomplish null balance. The recording industrial potentiometer is a widely used form of self-balancing servo device. This article mainly addresses this type of instrument.

*Engineering Department, Leeds & Northrup Company, A Unit of General Signal, North Wales, Pa.

Galvanometer self-balancing systems have generally been replaced by electronic self-balancing instruments because of the improved response and lower manufacturing cost of the latter.

Electronic Self-Balancing Systems

As pointed out later in this article, the technological advancements that have contributed to the design and construction of self-balancing electrical instruments, for both industrial and laboratory use, have been numerous, and progress continues.

Operating Principles

A self-balancing electrical instrument is a position-type linear servomechanism applied to a variety of potential and bridge-type circuits calibrated for many kinds of measurements. The basic arrangement of this type of servomechanism is shown in Fig. 1. A position-type linear servomechanism compares the unknown input signal to be measured with a reference voltage produced by a rebalancing device (feedback element) which is accurately calibrated with respect to a mechanical position on an output display scale. The detector or servo amplifier senses the dc imbalance or difference between the unknown and reference voltages and actuates a motor which positions an indicator on the output display scale for a zero-error or null voltage condition between the unknown and reference voltages. With the appropriate scale factor, the position of the indicator on the output display scale indicates the magnitude of the input signal being measured.

A position-type servomechanism (Fig. 1) must be provided with a forward gain block having adequate threshold sensitivity, zero stability, sufficient gain and power output, and judiciously selected noise and response characteristics; it must also meet the required consistency of operation and life under a variety of operating influences.

Low-level input signals create a severe zero-stability problem for self-balancing instruments, which limits the accuracy of measurement. Generally, the zero-stability problem for spans below 100 mV is circumvented by the use of a chopper-type amplifier. These amplifiers are used because low-level dc amplifiers drift. Chopper amplifiers convert low-level dc input signals into alternating current, amplify them, and then convert them back to direct current.

The range through which the instrument signal can vary without initiating a response on the output display scale is called the *dead band* and is twice the threshold value. The dead band is a prime factor affecting the accuracy of self-balancing instruments and must be kept within specified limits. In practice, the effects of servomotor load variations are minimized by sizing the motor and

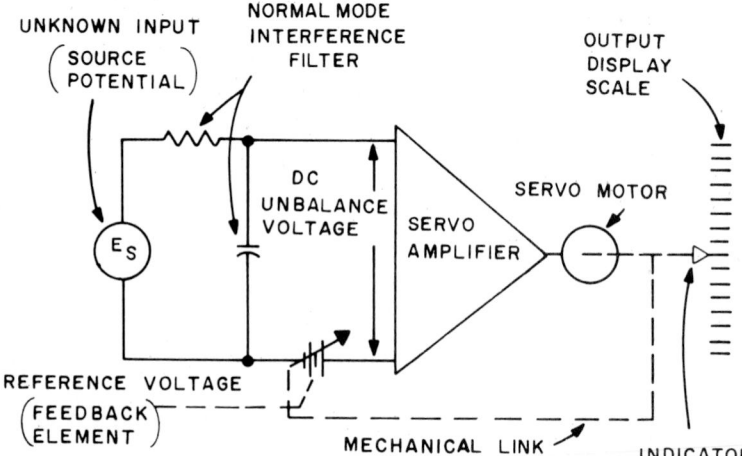

Fig. 1 Basic arrangement of a linear-position-type servomechanism used in self-balancing electrical instruments.

amplifier output so that the stall torque of the motor is an order of magnitude greater than the maximum torque load.

Measuring Circuits

The measuring circuit of a self-balancing instrument is defined as the circuitry that compares the unknown input signal to be measured to the reference or calibrated voltage, such as the voltage at the contact of a slidewire, and provides an output signal directly related to the dc imbalance or difference between the unknown and reference voltages.

A filter is often used in the measuring circuit to minimize the amplification of stray emf's (normal-mode interference). Interference such as line frequency stray on the input circuits is rejected by the filter.

Figure 1 also illustrates the classic measuring circuit for a self-balancing instrument. The majority of commercial self-balancing instruments employ measuring circuits of this form. This type of circuit has distinct characteristics in that the source resistance affects the response and dead band of the instrument, and the response and rejection of the input filter are related so that an increase in filter rejection can be achieved only at the expense of a slower response. Consequently, the source resistance must be kept within a specified range of values, and an amplifier gain adjustment and/or a damping adjustment may be required for the specified range of source resistance in order to bring the dead band of the instrument to within its specifications.

Figure 2 shows a typical measuring circuit for a self-balancing instrument that employs a conductive plastic slidewire operating at a voltage level of 6 V, a chopper-type preamplifier, and a servo system whose gain and damping are independent of source resistance.

Noise is filtered from the input signal as it passes to the preamplifier. The preamplifier amplifies the signal up to 6 V, calibrated by the span adjustment, and applies it to the servo amplifier. The servo amplifier compares it to the voltage from the slidewire, buffered through an amplifier.

If the signals are not the same, the servo amplifier drives the motor which moves the slidewire until they are the same. The damping network adds a small signal to make the servo amplifier "think" the slidewire is at the proper position before it is, to prevent overshoot and effect good damping.

Conversion (DC to AC and AC to DC)

With the development of solid-state technology, numerous methods of conversion, both dc-to-ac and ac-to-dc, have been used in commercial self-balancing instruments. The more common methods

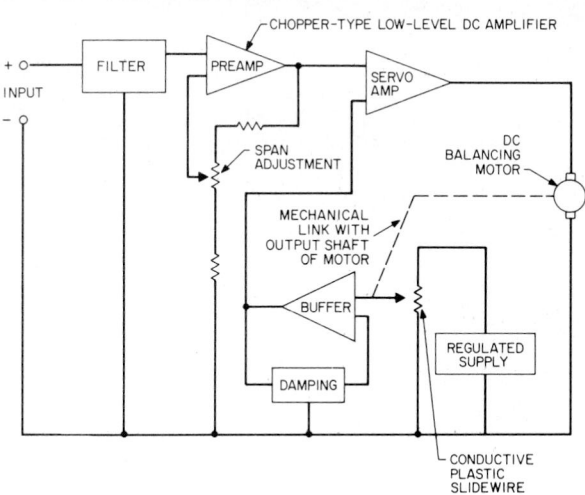

Fig. 2 Simplified diagram of a measuring circuit with a chopper-type preamplifier, span adjustment, high slidewire voltage, and high input impedance.

include (1) solid-state devices, such as field effect transistors (FETs) and bipolar transistors, which respond as "on" or "off" electric switches in series or shunt with the dc imbalance circuit, (2) photoconductive devices which cyclically vary the resistance in the dc imbalance circuit with the intensity of light from neon bulbs, (3) a vibrating reed which makes and breaks the dc imbalance circuit, (4) parametric devices, generally reactive in nature, which vary the capacitance or inductance in the dc imbalance circuit with the application of current, voltage, magnetic field, etc.

For many years dc-to-ac conversion was accomplished by either a vibrating contact-making converter or a converter employing saturable reactors or the Hall effect. Of these methods, the contact-making vibrator was by far the most commonly used device. However, more recently, solid-state devices have been utilized in commercial self-balancing instruments for conversion. FETs have been the more commonly used solid-state elements for conversion, and they have been developed to the point where they are being employed in many versions of self-balancing instruments. The conversion of direct current to alternating current and alternating current to direct current is accomplished with chopper-type amplifiers.

Figure 3 shows a chopper-type amplifier. A low-level dc signal passes through a filter to remove noise and goes to an FET switch $S1$. A feedback signal from the dc amplifier through resistors $R1$ and $R2$ is switched into the ac amplifier by FET switch $S2$. Since switches $S1$ and $S2$ are driven out of phase by the chopper drive oscillator, the input signal and the feedback signal are alternately switched into the ac amplifier. If the two signals are not equal (i.e., the output signal divided by $R1$ and $R2$ does not equal the input signal), an ac signal is produced at the input to the ac amplifier. This ac signal is amplified and applied to demodulator switch $S3$. Switch $S3$, driven at the same frequency and phase as $S1$, shorts half of the ac signal across it to the common. The other half which passes through is dc pulses. The dc pulses are amplified and filtered by the dc amplifier. The signal at the output of this amplifier is the output signal. In the presence of an input signal, the output signal increases until the feedback signal equals the input signal, at which time the ac signal at the input to the ac amplifier equals zero. At zero input to the ac amplifier, the dc amplifier stops changing, and the output becomes stable.

Rebalancing (Feedback) Devices

A measuring circuit compares the unknown input signal with a position-calibrated voltage to provide an error signal to the servo amplifier. The calibrated voltage is derived from the rebalancing (feedback) element which in classical form is a linear slidewire. This element (Fig. 4) is called a slidewire because a sliding contact is used to pick off the voltage.

A slidewire must have a sliding contact that is wear-resistant and essentially free of thermal emf's because the contact surface is heated by high speed rubbing. Resistance of the slidewire to chemical

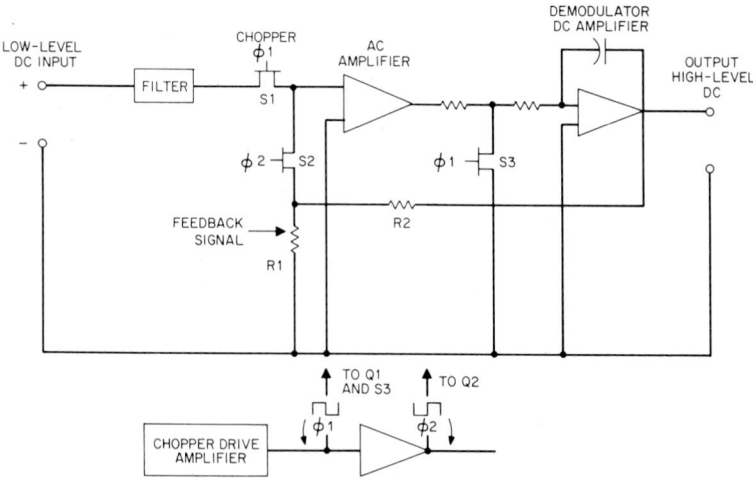

Fig. 3 Chopper-type amplifier.

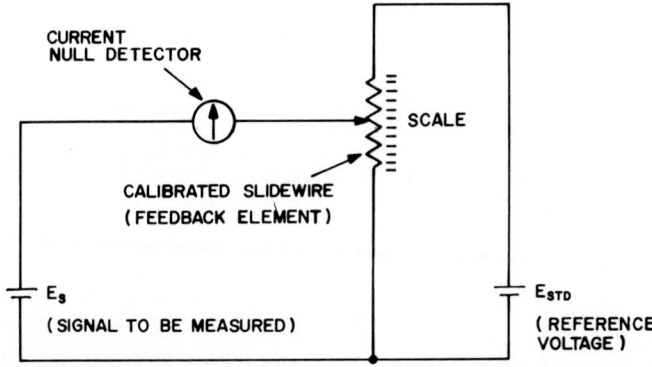

Fig. 4   Basic manual potentiometer.

attack in an industrial atmosphere and selection of the material to minimize mechanical polymerization are desirable; sealing of the assembly, lubrication, oil filling, and other factors all play a part based on the requirement.

Resolution of a good slidewire that is not convoluted will approach infinity, but if the wire is convoluted, the best that can be expected is the convolution step itself.

Conductive plastic slidewires have an expected life considerably greater than that of conventional wire-wound slidewires, and they can be constructed with linearity errors less than 0.1% of full scale and a temperature coefficient of less than 50 ppm/°C. High speed laser trimming of these slidewires guarantees this excellent linearity with no large step changes. The use of high-level voltage on the conductive plastic slidewire eliminates the old problem of having to contend with the thermoelectric voltage at the slidewire contact (particularly when the slidewire is in motion), which is large with respect to the slidewire voltage when it is operated at magnitudes near that of the low-level input signals to be measured.

Voltage References

A potentiometer cannot be more accurate than the calibrated (standard) source of emf against which the unknown potential is compared. Standard cells and a variety of batteries calibrated against standard cells have been of prime use. Since the 1960s, a zener diode used in an appropriate circuit has become most useful by offering low cost, very long life, small size, and accuracies better than those of standard cells when required. A typical circuit employing a zener diode is shown in Fig. 5. The power transformer, diodes $CR1$ and $CR2$, and capacitor $C1$ form a full-wave center-tapped rectifier power supply. The voltage across $C1$ causes a small current, about 6 mA, to flow through $R1$ into $VR1$, the zener diode. The voltage across $VR1$, usually about 6.2 V, is stable to about 0.002% per degree Celsius.

Damping

Because of inertia in the moving parts of the balancing system there is a tendency to overshoot the position of balance, which under some conditions may result in hunting (oscillation). This tendency is overcome in some instruments by manually adjusting a damping control. In other instruments it

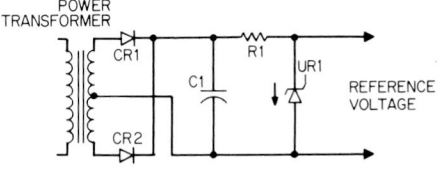

Fig. 5   Zener diode reference circuit.

is overcome by electronic circuits which stabilize the dynamic performance of the instrument so that no damping adjustments are required.

Generally, both methods introduce a small dc (damping) voltage in an appropriate circuit, which effectively is in opposition to the imbalance voltage and proportional to the speed of the motor or rebalancing element. When the indicator is approaching the balance point at full speed, the imbalance voltage, because of its position on the output scale, becomes less than the opposing voltage because of the speed of the motor; the applied voltage to the motor is then reversed to exert a breaking action, slowing the motor and decreasing the damping voltage. The circuit parameters of the instrument are selected (when properly adjusted) to prevent overshoot at balance but to allow the balance to be complete in the minimum time. In an instrument with manual damping controls the user is normally required to adjust the amplifier gain and damping controls in a prescribed manner in order to achieve the desired response for a particular circuit and/or measurement. In an instrument with circuitry which stabilizes the dynamic characteristic no adjustment is required by the user in order to achieve the desired response within the specified range of source resistances and spans of the instrument.

In some earlier electronic self-balancing instruments the damping voltage was supplied by a dc tachometer on the shaft of the balancing motor, and in some later instruments the damping voltage resulted from charging currents into or out of a capacitor. With the advent of inexpensive integrated circuitry and rapid advancements in the technology of the fields of servomechanisms and control theory, a considerable number of techniques for achieving improved damping characteristics for self-balancing instruments have come to the foreground. However, the nature of their complexity does not permit a discussion of the principles of even the more commonly used methods.

Temperature Measurements

Potentiometer instruments are extensively used for thermocouple, emf, and temperature measurements. Figure 6 shows the basic arrangement for thermocouples. The two junctions of metals 1 and 2 constitute a thermocouple. The potential appearing on terminals A and B is a function of the difference in temperature between the junction; this is the Peltier effect. By fixing one junction at a known temperature T_{REF}, the thermocouple potential becomes a function of the second junction temperature T_{MEAS} which is usable as a temperature measuring element. The reference junction T_{REF} may be held at a fixed and known temperature, but it may also be allowed to vary with the ambient temperature if the thermocouple potential is manually or automatically compensated.

The circuit shown in Fig. 6 illustrates how the basic arrangement for thermocouples is used in self-balancing electronic instruments for temperature measurements with reference junction compensation. The practical circuit of Fig. 10 shows metal 3 (copper) introduced in series with the thermocouple comprising metals 2 and 3, which gives rise to two more junctions at the junction terminal points A and B. When these terminals are maintained at the same temperature, this couple produces no net emf to affect the desired measurement. Thus, intermediate metals may be introduced anywhere in a thermocouple circuit, providing that no temperature difference exists between the junctions of any such couple. See the article Thermocouples in Sec. 2 of this Handbook.

The temperature T_{REF} varies with the ambient temperature and would produce a zero error in the thermocouple potential. To compensate for this error, temperature-sensitive resistor Rc, at the same temperature as T_{REF}, is used to shift the electrical zero of the system.

Interferences

Interference is any electric energy in an instrument's circuits that limits the precision with which the desired signal information can be discerned. Various terms used for interference are *noise, pickup,* and *hash.* Good instrument design demands that internal interference generation be minimized and that adequate protection be provided for expected levels of externally generated interference. In general, commercial self-balancing instruments are designed to give a specified level of protection for interference that can enter the instrument's signal circuits.

There are two modes of interference: normal mode and common mode.

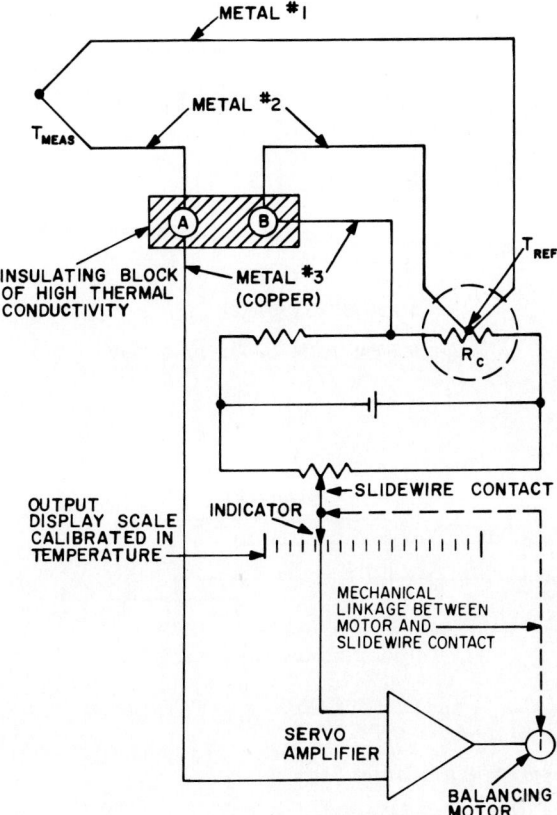

Fig. 6   Thermocouple-actuated potentiometer with T_{REF} compensation.

Normal-Mode Interference

This interference directly impresses a potential difference between the input terminals of the instrument in the same manner as a legitimate signal. As shown in Fig. 7*a*, the applied normal-mode voltage appears as a voltage in series with the input terminals of the recorder. Normal-mode interference acts directly on the instrument because its full magnitude is developed at the input terminals of the measuring circuit. Normal-mode voltages are frequently a great problem because they are usually large in relation to the input signal. For this reason input filters have a large attenuation ratio, and some recorders (see Fig. 10) have a speed control which helps minimize interference effects.

Common-Mode Interference

This type of interference causes both instrument input terminals to change potential simultaneously and by the same amount relative to ground. It is injected into the signal circuit by sources having one terminal grounded and the other coupled to the measuring circuit capacitively or conductively, as shown in Fig. 7*b*.

Common-mode interference acts indirectly on the instrument; a part of the common-mode energy incident on the measuring circuit must be converted to normal-mode interference. Normal-mode interference results from common-mode currents caused to flow through the source resistance in the measuring circuit and leakage impedance Z.

Interference can affect recorder performance in two ways:

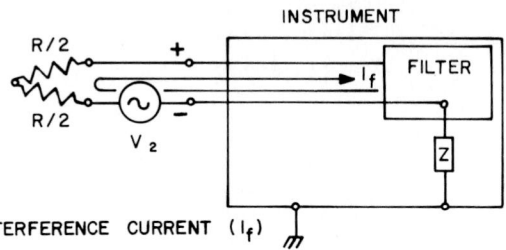

R= SOURCE RESISTANCE

V_2= NORMAL–MODE APPLIED VOLTAGE

(a)

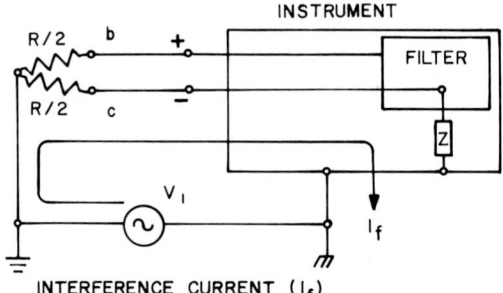

R= SOURCE RESISTANCE

Z=HIGH IMPEDANCE BETWEEN CASE (GROUND) AND MEASURING CIRCUIT
CAUSED BY LEAKAGE RESISTANCE AND STRAY CAPACITANCE

V_1= COMMON–MODE APPLIED VOLTAGE

(b)

Fig. 7   Types of interference. (*a*) Common mode, (*b*) normal mode.

1. To produce balancing motor torque and cause a shift in instrument calibration and/or produce oscillatory motion at the indicator (pen)

2. To produce no motor torque, but overload the amplifier, causing sluggish operation and excessive dead band

Both types of interference may exist simultaneously, resulting in a combination of these effects. In addition, if the interference source is variable or intermittent, the adverse effects will also be of an erratic nature.

The figure of merit for recorder performance under the influence of interference signals is called the rejection ratio. The rejection ratio is specified in decibels, defined as 20 times the logarithm of the ratio of the applied interference signal [root-mean-square (rms) volts for alternating current] in volts to the pen response expressed in equivalent volts of input. The rejection ratio expressed in decibels is only a figure of merit for recorder performance under the influence of interference signals when

1. The interference is specified for the worst phase condition
2. The source resistance or range of source resistances is specified
3. The frequency or range of frequencies is specified
4. The span or range of spans is specified
5. The arrangement of the connections to the input terminals (such as the point of connection for any guarded circuits) is specified

Generally, instruments with the better performance under the influence of external interference employ special techniques to suppress the interference.

Normal-mode interference is attenuated by low-pass or band rejection filters arranged in the instrument input circuits. Filters affect the instrument response, requiring a compromise between interference rejection and response specifications.

Because of the high manufacturing costs of other methods, shielding or guarded circuitry is commonly used to suppress interference; the extent of shielding or guarded circuitry throughout the instrument determines the level of external interference it can tolerate before it exhibits a detectable error in the measurement.

Figure 8 illustrates guarding against a common-mode generator appearing with the source potential. Current is drawn from the common-mode generator, but practically no current flows in a path to produce a normal-mode voltage drop, and thus the instrument system is essentially free of common-mode effects.

Adherence to proven circuit design techniques is used to suppress internal interference, for example, (1) twisting critical leads, (2) properly locating and spacing critical components such as low-level amplifiers, and (3) properly connecting the ground and return leads within the instrument. Some of the aforementioned design techniques also help to suppress external interference.

Technological Advancements

The advancement of solid-state technology and the development of new and improved materials have altered and will continue to change the design of commercial electronic instruments. In particular, such advancements are reflected in the overall design concepts of self-balancing instruments. Compared with earlier instruments, technological advancements have (1) reduced the size of electronic

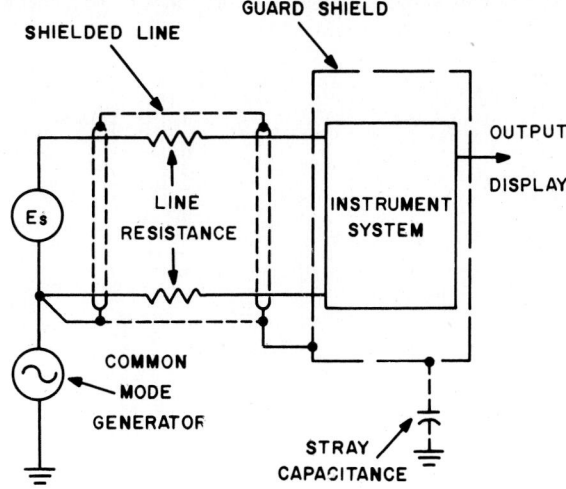

Fig. 8 Guard shield for an external common-mode generator.

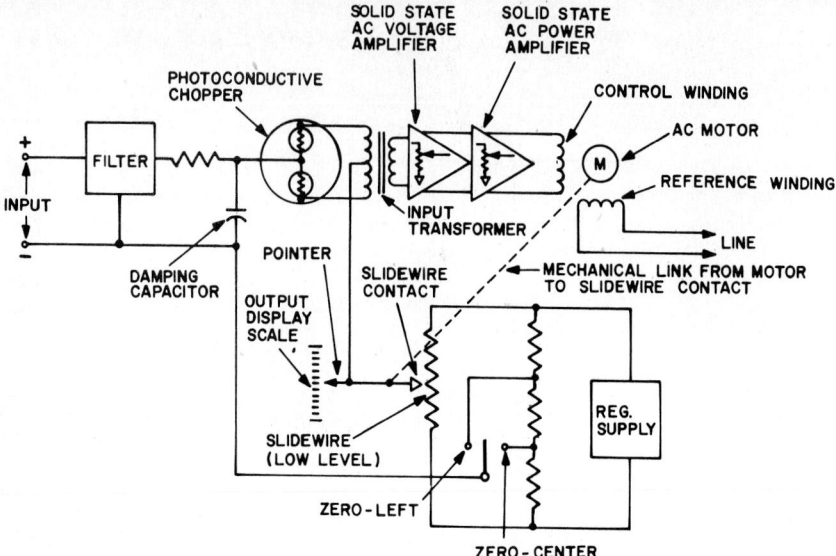

Fig. 9 Simplified circuit diagram of an electronic recorder using an ac balancing motor. *(Speedomax W/L recorder, Leeds & Northrup.)*

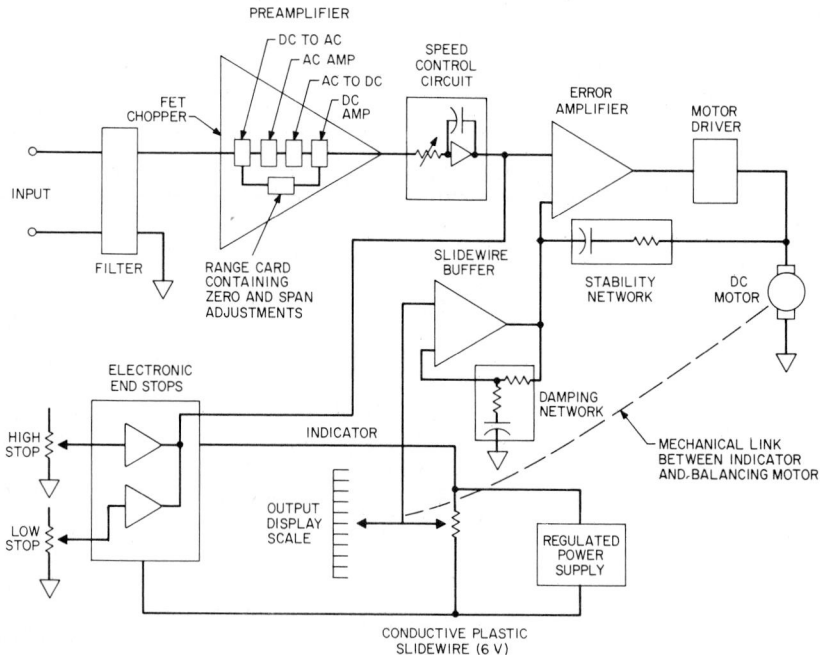

Fig. 10 Simplified circuit diagram of an electronic recorder using a dc balancing motor. *(Model 100, 165, and 250 recorders, Leeds & Northrup.)*

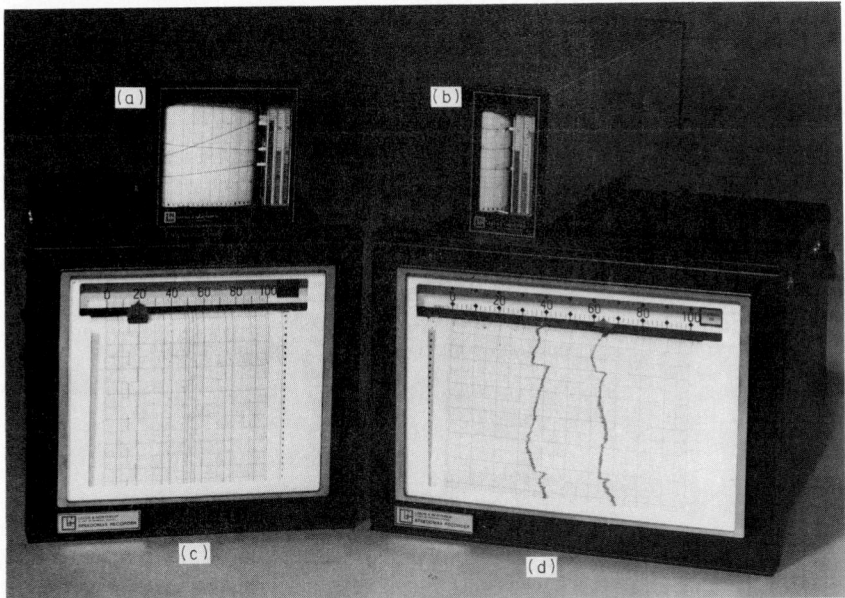

Fig. 11 Four modern strip chart recorders with the specifications tabulated below. *(Leeds & Northrup, Unit of General Signal.)*

	(a)	(b)
Dimensions	6 × 6 × 17 in (~15.2 × 15.2 × 43.1 cm)	3 × 6 × 23 in (~7.6 × 15.2 × 58.4 cm)
Chart	100-mm strip	
Chart speed	1–720 cm/h	
Type of record	Disposable fiber tip marker	
Input level	0.1–100 V spans 0.2–1000 mA spans 1.0–1000 mV spans Direct thermocouple, pH, conductivity	0.1–100 V spans 0.2–1000 mA spans
Step response	1–10 s, adjustable	
Input channels	1, 2, or 3 channels	
Other features	Alarm contacts, transmitter power supply, and multiple chart speeds	
	(c)	(d)
Dimensions	13 × 12 × 15 in (~33 × 30.5 × 38.1 cm)	17 × 12 × 15 in (~43.1 × 30.5 × 38.1 cm)
Chart	165-mm strip	250-mm strip
Chart speed	¼–900 in/h (~6.4 mm–22.9 m/h)	
Type of record	Disposable fiber tip marker, thermal pen, or multipoint printing	
Input level	1.0 mV–100 V dc spans 0.1–1000 mA dc spans Direct thermocouple, pH, and conductivity	

Fig. 11 (Legend continues on p. 14.46.)

Fig. 11 *(Continued)*

	(c)	(d)
Step response	1, 2, 5, or 10 s, selectable	
Input channels	1 or 2 with pen, up to 15 multipoint	1 or 2 with pen, up to 30 multipoint
Other features	2, 3, or 4 chart speeds, proportional control, alarm contacts, and alarm processor	

packages, (2) reduced the power requirement for the electronics, which generally results in the more efficient use of input power so that two or three instruments can be housed in one case without an excessive temperature rise, (3) improved the techniques for conversion from direct current to alternating current, and vice versa, (4) brought about the use of multiloop feedback circuits to improve dynamic performance of self-balancing instruments, (5) made it economically feasible and practical to use dc balancing motors in commercial self-balancing instruments, and (6) replaced the standard cell with a voltage-regulated power supply.

The smaller size of electronic packages has been a direct result of small solid-state devices and integrated circuitry (IC), and the reduced power requirement of electronic packages is mainly attributed to the absence of filament power in solid-state amplifiers and associated circuitry. The inherent characteristics of solid-state devices make them especially suitable for circuitry that converts direct current to alternating current, and vice versa, whereas the low cost of integrated amplifiers has made multiloop feedback circuits feasible. Individual loops built around high-gain IC amplifiers can perform assigned operational tasks such as stabilizing the dynamics of the instrument so that a damping adjustment for optimum response is not required or automatically regulating the gain with span changes to maintain optimal performance.

Improved brush materials have increased the expected life of dc balancing motors. Low-cost integrated amplifiers can be employed very effectively to protect dc balancing motors against damage for the blocked rotor condition at full applied voltage. Since dc balancing motors are more efficient than ac balancing motors, and the former require little or no power at balance because no power is needed for a line or reference winding, they are being used in increasing numbers in commercial self-balancing instruments.

Stable zener diodes have eliminated the need for standard cells in self-balancing instruments, and it follows that standardization on a frequent or an automatic schedule is no longer needed. Figure 9 shows the conventional type of electronic recorder that uses solid-state circuitry and an ac balancing motor. Figure 10 shows an electronic recorder that uses a dc balancing motor and IC in multiloop feedback circuits to provide automatic gain and damping with balancing time adjustment and electronic end stops to prevent damage if overdriven.

Four modern strip chart recorders are shown in Fig. 11.

Digital Technology

This article is divided into nine subsections: (1) digital computer fundamentals; (2) digital computer and digital system characteristics; (3) digital computer hardware; (4) digital computer storage; (5) digital codes, units, and languages; (6) computer programs and programming: operations and procedures; (7) input-output subsystems and components; (8) data converters; and (9) digital system amplifiers.

Most of these subsections are arranged alphabetically for convenience of reference. Most descriptions take the form of an expanded glossary of terms. The classification of topics and their appearance in various subsections is somewhat arbitrary because of multiple meanings of terms—and the several viewpoints which can be brought to any classification scheme.

Background

The progressive adaptation of digital technology to process instrumentation and automatic control is well documented in the literature. Most experts agree that, while the transition from analog to digital techniques occurred at a phenomenal rate during the 1970s and is continuing into the 1980s, analog methodologies will persist into an unpredictable future and, in particular, digital technology is only at the threshold as regards digitalization of many measurements (sensors) and of final controlling elements (actuators). Had digital computers been originally developed for process control rather than for general business data processing, the application of digital technology to industrial instrumentation may have been hastened. But the realistic economics of the situation intervened. The costs of digital equipment development may have been inordinately high without the massive business computing market to support the exceedingly high research and development costs required to perfect modern computers and digital systems.

The application of digital technology to process control and industrial instrumentation developed in a "secondhand" fashion, so to speak, thus requiring a melding of the disciplines of business data processing with the disciplines of process control. This commingling of interests required much time and patience. The presently achieved roles of computer manufacturers and old-line instrument manufacturers were not easy to establish and, in fact, still continue to shift. Digital technology has catalyzed the entry of hundreds of new firms, establishing a trend toward specialization in the various facets of the technology.

Analog Computers by Comparison

As a base of historical and technical reference, it is in order here to describe briefly the analog computer. This computer solves problems by physical analogy. The analog computer translates physical variables, such as temperature, flow, speed, and altitude, among many others, into related electrical quantities (predominantly) and uses electrical equivalent circuits as an analog for the physical phenomenon involved. But, it should be stressed that pneumatic, mechanical, and hydraulic analogs also have been widely used and continue to be used for numerous applications today. The advantages of analog methodology for certain classes of applications are described elsewhere in this Handbook, notably in Sec. 18.

Some of the acronyms now in popular use to identify relatively recent or enhanced computer applications in the manufacturing and processing field include the following: AI, artificial intelligence; AMS, advanced machine systems; CAD, computer-aided design; CAE, computer-aided engineering; CAM, computer-aided manufacturing; CIM, computer-integrated manufacturing; CNC, computer numerical control; FMS, flexible manufacturing system; MRP, materials requirements planning.

DIGITAL COMPUTER FUNDAMENTALS*

In contrast to an analog computer which operates on continuous data, a digital computer performs a series of discrete operations on numeric or alphabetic data in the solution of a problem. Compared to a calculator in which frequent human operator intervention is required, a computer accomplishes the data processing with little or no intervention by the human user. The following description provides a brief overview of the modern electronic digital computer.

A digital computer may be classified in a variety of ways. One way is to distinguish between *general purpose* and *special purpose* computers. A *general purpose* computer is designed to satisfy a wide variety of data processing needs with approximately equal speed and efficiency. A *special purpose* computer, in contrast, is optimized for the solution of problems associated with a restricted class of applications. With very few exceptions, a special purpose computer can provide all the functions of a general purpose computer, although it may not be as fast or as efficient for some problems.

*By Thomas J. Harrison, IBM Corporation, Boca Raton, Fla.

For example, a special purpose computer may be designed to efficiently handle arrays of data, such as those found in a two-dimensional Fourier transform problem. Although this computer could be programmed to perform accounting applications, its facilities would be inefficient. Similarly, a general purpose computer can be programmed to calculate the two-dimensional Fourier transform, but its solution undoubtedly would take more time.

Within the classification of general purpose computers, it is common for a manufacturer to provide several different models capable of executing the same program. The difference between models generally is related to the tradeoff between speed of processing versus the number of circuits and, therefore, cost. Between computers provided by different manufacturers, however, there usually is little compatibility at the basic instruction level. Programs written in assembly language for one manufacturer's machine generally cannot be run on a computer designed by another manufacturer. However, a high degree of program compatibility is possible if programs are written in a high-level language such as COBOL or FORTRAN.

A second classification of digital computers is roughly related to the physical size or cost of the computer. In commonly used terminology, computers are categorized as microcomputers, minicomputers, and mainframe or large-scale computers. The prefix "maxi" is sometimes used for the latter classification. Price often is used to distinguish the classes. However, computer prices continue to decline, and the functions available for a given price continue to increase. Thus, this means of defining the classes is time-dependent.

It is common for a microcomputer design to depend on a single-chip microprocessor as the primary processing element. Thus, a computer utilizing such a microprocessor, when combined with storage and input-output equipment, often is referred to as a *microcomputer*. In addition, a microcomputer generally is physically small and often may be packaged in a desktop structure.

To a large degree, but also diminishing with the passage of time, the *minicomputer* is recognized by its package and its variety of possible configurations. A typical minicomputer is mounted in an industry standard 19-in ($\approx$48-cm) rack. It is generally offered with a wide variety of separately packaged options and input-output equipment which also are rack-mounted. As a result, the minicomputer has been widely utilized to provide tailored systems for particular applications, such as those found in industrial process control. Comparatively recently, however, some models of these computers have been marketed in conventional packaging and with limited configuration options. At the same time, some microcomputers are being packaged in a rack. In addition, many current minicomputer designs utilize one or more microprocessors as the central processing unit. Thus, the distinction between the two classes appears to be lessening.

The *mainframe* or *maxicomputer* often is associated with large, centralized computing centers. At the lower range of cost and performance, the maxicomputer provides a capability similar to that of larger minicomputers. At the high end, the maxicomputer may provide computing power hundreds of times greater than that associated with minicomputers, with a correspondingly higher price. Mainframes typically are packaged in customized enclosures with little or no flexibility as to the location of a particular feature. Although many options may be available, the ability to efficiently mix features in an arbitrary manner usually is more limited than in the case of minicomputers.

As brought out by the foregoing observations, class distinctions among computers are imprecise. Across the classes there is a performance range, measurable in terms of instructions executed per second, ranging from about 50,000 instructions per second (ips) to over 5 million ips. Similarly, the price range is from a few thousand dollars to a few million dollars. All the machines—from micro to maxi—follow the same basic conceptual design. As a result, the popular use of the aforementioned classification is, in fact, of little practical value to the user and tends to be confusing.

Meaning of the Term "Computer"

It should be noted that the term "computer" and the phrases "electronic data processing system" (EDP) and "automatic data processing system" (ADP) are commonly used as synonyms. Initially, "to compute" meant to determine a quantity from other quantities by performing a calculation, an arithmetic operation, or a series of operations. Most data processing operations, however, actually do not require computing in this sense, but rather they involve collating, selecting, correlating, and otherwise manipulating information to or from an electronic storage unit. In modern language, these kinds of operations are ascribed to a computer whether or not any actual mathematical calculations are being performed on the data.

Elements of a Digital Computer System

Basically, a computing system comprises input and output equipment for physical handling of machine information and instructions, and a central unit which performs the actual electronic processing.

Central Unit

The central unit of a computer typically is made up of one or more *central processing units* (CPUs) and their associated storage. In larger systems having more than one CPU, the term *central electronic complex* (CEC) sometimes is used to describe the set of CPUs. A CPU, in turn, is made up of a control section, an arithmetic and logical unit (ALU), local storage *(registers)*, and, in some larger machines, an input-output channel controller. The control section controls the step-by-step operation of the system through fetching instructions from the storage, interpreting them, and providing the necessary signals to effect control of the ALU and other portions of the system. If a separate channel controller is used, this unit provides the detailed control of information flow to and from the input-output units. In smaller systems, where a separate channel controller is not used, this function is provided by the control section of the CPU. Storage holds both the instructions to be executed, called the *program,* and the data to be processed. This information is stored so that it may readily be obtained when needed. Data and instructions also are stored on external storage devices, such as magnetic disks and diskettes. The information stored on these devices is read into the main storage when needed. More detail is given later in this article under the definitions for *central processing unit* and for *storage.*

Actual processing of data is performed in the ALU according to the stored program instructions. Fundamentally, the ALU provides for the comparison of two data items, arithmetic operations (such as addition and subtraction) and logical operations (such as AND and OR). However, most computers provide several hundred different instructions which are combinations of these fundamental operations. For example, a single instruction may be provided which will fetch two operands from storage locations, add them together, and store them in a third storage location. A very important class of instructions alters the sequence of instructions based on the result of a calculation or comparison. For example, when comparing two numbers A and $B,$ if A is less than or equal to $B,$ the computer may *branch* or transfer control to a different set of instructions than if A is greater than $B.$ This provides alternatives in obtaining the solution of a problem and is the most significant difference between a computer and a calculator.

Input devices read data to be processed and programming instructions into the main storage unit or, sometimes, into the CPU itself. Common input devices include card readers, keyboards, magnetically encoded document readers, magnetic drums, disks, diskettes, and tape units. Output devices transcribe processed data into machine media, such as tapes, disks, or cards, and to devices providing human-readable information, such as cathode ray tube (CRT) displays and printers.

A computer also may accept input from, or provide output to, another computer. When two or more computers are interconnected in this fashion, are in close physical proximity (usually), and cooperatively participate in the solution of a single problem, the aggregate of computers often is referred to as a *multiprocessor.* The interconnection between the computers is by a channel-to-channel link or similar means to allow high speed interchange of information. Also, typically, a single cor.*rol program directs the operation of the interconnected computers.

A related concept, sometimes called *distributed processing* or *computer networking,* also involves interconnected computers. Although precise definitions remain to be created, computer networks generally involve two or more computers which are geographically separated and interconnected by serial communication facilities, such as telephone lines or dedicated coaxial cable. Information is interchanged between the computers, generally at a rate much slower than in the case of a multiprocessor. The computers in a network generally are under the control of individual, autonomous control programs, so that cooperation is effected in a manner similar to that encountered in a peer group of individuals.

Distributed processing may involve computers interconnected either by slow speed communication facilities or by high speed channel-to-channel connections. Distributed processing usually implies, however, that the interconnected processors work cooperatively on the *same* problem (as in the case of the multiprocessor), but not necessarily under control of a single control program; that is, the involved computers are at least semiautonomous. See also the article A Practical Look at Distributed Instrument Systems in Sec. 18 of this Handbook.

Computer Hardware

Computer *hardware* is the physical equipment which comprises a computer system. This is contrasted with *software,* which is the aggregate of the programs, procedures, and, sometimes, documentation necessary to effectively use the hardware. Taken together, the hardware and the software represent the *computer system.* It is not uncommon, however, to find the term "system" used in reference to only the hardware. The basic structure and cyclical operation of a computer has remained essentially unchanged since the invention of the early electromechanical and electronic computers in the 1940s. The basic sequence of operations is (1) fetch an instruction from storage, (2) decode (interpret) the instruction to determine the operation to be performed and the location of the operand(s), (3) fetch the operands from storage, if necessary, (4) perform the operation on the operand(s), and (5) fetch the next instruction from storage. There often is considerable data manipulation during each of these basic steps. For example, fetching operands from storage may require arithmetic operations to calculate the physical storage location from a relative address provided in the instruction.

The basic speed of a computer is determined by the characteristics of the circuits used to build the hardware. However, particularly in high-performance machines, a number of techniques are used to increase the effective processing speed beyond the limits imposed by circuit speed. One method is to overlap the basic steps outlined in the prior paragraph. As one example, unless the location of the next instruction depends on the outcome of the current operation, it is possible to overlap the operation (step 4) and the fetching of the next instruction (step 5). Another technique, called *pipelining,* utilizes a series of processing subsystems to operate on several instructions concurrently. For example, during the cycle in which the operation specified in instruction n is being executed, a separate processing unit is calculating the effective or physical address needed for instruction $n + 1$, while yet a third processing unit is fetching instruction $n + 2$. Occasionally, the instructions being preprocessed ($n + 1$ and $n + 2$) need to be "thrown away" since instruction n causes a branch to an instruction other than $n + 1$. On the average, however, a decrease in total processing time is effected by the concurrent operation.

It is also common to provide for concurrent transfer of information to and from input-output equipment. This is particularly effective, since many input-output devices are electromechanical in nature and operate at speeds several orders of magnitude slower than the CPU speed. One technique used in small computers is to utilize *cycle stealing* or *direct memory access* (DMA) between the input-output device and the main storage. In this approach, the CPU provides the channel equipment with the starting storage address of the data to be transferred (or, in the case of an input device, the starting location of an input storage area) and the number of data words to be transferred. The channel equipment then independently transfers the information, using *(stealing)* storage cycles not used by the CPU, while the CPU executes subsequent program instructions. In larger machines, the channel may have a significant functional capability which allows it to do more than merely sustain the information transfer. For example, it may perform error checking on the data or automatically *chain* to the next input-output request on completion of the current transfer operation.

Data Representation

Binary Notation

Virtually all modern computers utilize the binary number system and binary-coded alphanumeric representations for internal data manipulation. The binary number system utilizes 2 as the number base or radix, as compared to 10 used in the familiar decimal number system. Only two digits, 0 and 1, are needed in the binary system, as compared to the 10 symbols used in the decimal system. As in the decimal system, the quantity represented by a digit depends on its position in the numeral. Thus, the units position has a value of 1; the next position, a value of 2; the next 4; and so forth. As a specific example, the binary number 11010 represents the quantity $(1 \times 2^4) + (1 \times 2^3) + (0 \times 2^2) + (1 \times 2^1) (0 \times 2^0)$ or 26 (decimal). The binary equivalents of the first 20 decimal integers are:

Decimal	Binary	Decimal	Binary
1	00001	11	01011
2	00010	12	01100
3	00011	13	01101
4	00100	14	01110
5	00101	15	01111
6	00110	16	10000
7	00111	17	10001
8	01000	18	10010
9	01001	19	10011
10	01010	20	10100

Any quantity can be represented in the binary number system, although it requires considerably more digits than decimal notation. Binary representation is used in computers primarily because only two symbols are required and it is easy and economical to build electronic circuits having two states, "on" and "off," corresponding to 1 and 0.

The symbols 0 and 1 are called *bits*, a contraction of the words *bi*nary and dig*it*. In some usage, it is common to refer to the 1 as a bit and to the 0 as no bit, even though this is technically incorrect.

Numerical Data Types

Two numeric data types are provided for arithmetic operations on most computers. These are known as *integer* (or *fixed-point*) and *floating-point*. Integer data type provides for the representation and manipulation of whole numbers, . . ., $-3, -2, -1, 0, 1, 2, 3, \ldots$. Integer arithmetic can be used for dealing with fractions (e.g., $0.575 + 1.2$) by appropriate scaling of the numbers (i.e., multiplying by factors of 10 in the decimal system to align the decimal point). When utilizing integer arithmetic in this way, the computer programmer is responsible for the explicit scaling of the numbers and for keeping track of the radix point. The advantage of integer arithmetic is that it is faster than floating-point arithmetic and requires fewer circuits to implement. Its disadvantage is that the range of numbers that can be represented in a typical computer word is quite limited. For example, an 8-bit byte can be used to represent only 0 to 256 (or -128 to $+127$), and a 32-bit word can represent only 0 to 4 294 967 296.

The use of the floating-point data type provides a greater range of values by representing the quantity as a quantity *and* a scaling factor, in a manner similar to the use of scientific notation such as 0.424×10^{12}. In the computer, the fractional part and the exponent representing the scaling factor are stored separately. For example, the numeral might be stored in a computer word in the form:

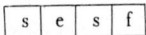

This is interpreted as $\pm 0 \cdot f \times 2^{\pm e}$, where s represents the sign, e is the binary exponent, and f is the binary fraction. In performing floating-point arithmetic, the computer manipulates the fraction and the exponent separately. Some small computers do not provide hardware for floating-point arithmetic, in which case the programmer is restricted to fixed-point arithmetic or must rely on programmed subroutines to perform the necessary manipulation.

The precision of a floating-point number is determined by the number of bits in the fraction, whereas the span or range is determined primarily by the number of bits in the exponent. Thus, a 1-byte fraction and a 1-byte exponent provide representation of quantities from -0.128×2^{128} to $+0.128 \times 2^{127}$.

The advantages of using floating-point numbers are the range of values that can be accommodated and the placement of the radix point by the hardware or software. The disadvantage is that floating-point operations are slower because the exponent and fraction are manipulated separately.

Character Coding

In addition to numbers, computers must deal with information composed of alphabetic characters and special symbols (e.g., @, #, $, *), and, since computer circuits are designed to manipulate only 0 and 1, these characters must be coded in a binary notation.

A number of different alphanumeric codes are used. One of the most common is ASCII (pronounced "as-key"), an acronym for American Standard Code for Information Interchange, which is defined by American National Standard X3.4-1977 and its international counterpart ISO 646-1973. This code provides for the representation of 128 alphabetic (upper- and lowercase), numeric, special symbol, and control characters in a 7-bit byte. Typically, a 1-bit parity check is added (see below), and the character is stored as an 8-bit byte in the computer. The following are sample ASCII codes:

Character	Binary notation
Numeric	
1	0110001
2	0110010
3	0110011
Alphabetic	
A	1000001
a	1100001
K	1001011
k	1101011
Special	
#	0100011
(	0101000
=	0111101
?	0111111
Control	
End of text (ETX)	0000011
Bell (BEL)	0000111
Acknowledge (ACK)	0000110
Cancel (CAN)	0011000

It is important to note that these codes are indistinguishable from the codes illustrated earlier for numeric quantities. Their interpretation as characters is determined by the intent of the programmer who writes the program for their manipulation. It is possible, and allowable in some computers, for example, to numerically add A to 6. Since this usually does not make sense, however, programs generally are designed to detect such operations and, as a minimum, to notify the programmer of the use of mixed data types.

Error Detection and Correction

Many codes are designed to assist in ensuring that an error has not occurred in transferring data from one location to another. Such an error might take place if an electrical noise pulse upset a circuit such that it was turned "off," thus converting a 1 into a 0. One of the simplest error detecting codes, often used internally in computers, is the *parity check*. In this scheme, the value of one bit in the code is selected such that the number of 1's, in the coded character is either even (even parity) or odd (odd parity). For example, in even parity, the 7-bit representation 0110101 of a character would be altered by adding a leading 0 to form 00110101, whereas 1010100 would be altered to 11010100 so as to provide an even number of 1's, in this case four. If a single-bit change error occurs, the number of 1's will no longer be even and, when tested by the checking circuit, the error will be detected. Note, however, that any even number of erroneous bit reversals does not provide a parity check.

More sophisticated error detection codes sometimes are used in portions of computers and in digital communications. These are capable of detecting multiple-bit errors and may be capable of determining the exact error so that it can be corrected automatically. Such codes require additional bits in the representation and, therefore, have a cost in terms of additional circuits and storage capacity.

The Computer Word and Byte

Although single bits of information sometimes are manipulated in a computer, it is common to handle a group of bits in parallel. Through common usage, an 8-bit group usually is the minimum number of bits manipulated in parallel. Although not technically correct, an 8-bit group usually is called a byte. Technically, a *byte* is any portion of a word (see below), and there is a growing tendency to utilize the more specific words *octet* (8-bit byte), *sextet* (6-bit byte), etc., to avoid confusion. An octet is a convenient size for representing a character set, since it allows 256 distinct representations, a number sufficient for the numerals 0 to 9, upper- and lowercase alphabetic characters, a generous number of special symbols, control codes, and parity.

A computer *word* typically is one or more bytes (usually 8-bit bytes). The internal circuits in the computer usually are designed to transfer or operate on a word, although instructions allowing the individual manipulation of bytes commonly are provided. In addition, a word is the usual amount of information obtained on a single storage access.

There is a correlation between the number of bits in a computer word and the number of circuits required in a particular design. Additionally, there is a correlation between the number of circuits and the classes of computer represented by the so-called microcomputer, minicomputer, and maxi-computer or mainframe computer. Microcomputers typically utilize a 1- or 2-byte word (8 or 16 bits), minicomputers usually use a 2- or 4-byte word (16 or 32 bits), and mainframes usually use 4- to 8-byte (32 to 64 bits) words. However, this is a time-dependent association since, for example, only a few years ago microprocessors used only 4-bit bytes and 32-bit (4-byte) microprocessors were designed in 1981.

Most instructions are contained in one word. Also, arithmetic and logical operations usually are performed on one-word operands. As previously indicated, however, most computers provide instructions for the manipulation of single bytes. This is particularly useful when manipulating characters coded in bytes. Since the length of the word determines the precision of simple arithmetic operations, the number of bits in the word may not be sufficient for some applications. As a result, many computers provide double-precision arithmetic operations which use two-word operands. These usually require additional computation time, since the implementation merely provides for handling the two words individually. In addition, some instructions are contained in more than one word. This often is caused by the need to include an operand or a physical address in the instruction itself, and the number of bits available often is not sufficient to express all the needed location addresses.

Instruction Representation

A key digital computer concept is that the instructions which control the computer are themselves stored in the computer storage as binary codes. The instruction code is stored in one or more words and is accessed by the control unit when needed. Depending on the particular operation, a variety of information may be contained in the instruction. For example, in an addition operation, the computer must be told where the addend and augend are located and what to do with the resulting sum. For this reason, the format of the stored instruction varies with the operation to be performed. In general, however, an instruction consists of an operation to be performed, one or more addresses indicating the location of the operands, and, possibly, other modifiers or data. Thus, an instruction might be stored in the following format:

OP	R1	R2	R3

where OP represents the operation code, and R1, R2, and R3 are three addresses indicating the sources and destinations of operands and results. With this format, the operation, "Add the contents of register 1 to the contents of register 3 and store the result in register 2," might be coded as

0110	0001	0011	0010

Here the operation code for ADD is coded as 0110, the address of register 1 is 0001, the address of register 2 is 0011, and the address of register 3 is 0010.

A typical computer is capable of performing several hundred different instructions of this type. Sequences of such instructions direct the detailed operation of the hardware on a step-by-step basis. Determining the exact sequence for the solution of a particular problem is part of the task of the programmer, as discussed below.

Although a particular computer may have several hundred instructions, they can be grouped into several major categories.

Arithmetic Instructions

These include arithmetic operations, such as addition, subtraction, multiplication, and division. There may be several different ADD instructions, for example, to provide for integer and floating-point data types or to allow the addition of quantities stored in particular locations in the computer; separate ADD instructions might provide for the addition of quantities located in registers or the addition of quantities stored in main storage.

Logical Instructions

These provide for the logical operations, such as AND, OR, and EXCLUSIVE OR. Logical operators may act on single bit positions or on whole words or bytes.

Data Movement Instructions

These provide for the movement of data between various part of the machine. They might include, for example, "Move the contents of storage location x into register y," "Replace the contents of register r with the contents of register s," or "Move the contents of storage locations x thru y to input-output port z."

Control Instructions

These provide for control of the sequence of operations. For example, a typical instruction might be, "Transfer control to the instruction located in storage location x," or "Execute instruction y if the contents of register 1 are negative, otherwise execute the next sequential instruction."

Basic Hardware Operation

The diagram of Fig. 1 is used to illustrate the basic operation of the computer. Although greatly simplified, the diagram shows the data paths between the functional units found in a typical CPU and its associated main storage. The CPU central control unit consists of circuits which "decode" (interpret) the operation code and subsequently provide the timing and actuation signals that control the other functional units in the CPU.

The main storage (MS) is used to store both the program (instructions) and the data on which it operates. Storage is organized as a sequence of locations, typically represented by a byte, which are assigned sequential addresses starting with 1. The amount of storage may vary from as little as a few thousand bytes in a small machine to millions of bytes in a large machine. The storage address register (SAR) holds the address of the storage location to be activated, either for the purpose of reading the contents of the location or for storing at the location. The storage data register (SDR) temporarily holds data being read into or out of storage. The arithmetic and logic unit (ALU) performs the specified arithmetic and logical operations on the data presented at its two inputs. The ALU output is routed to the register stack, the input-output control unit, or the MS by signals from the CPU. The register stack (RS) is a special purpose storage unit consisting of typically 16 to 64 locations that can be used for the temporary storage of data and addresses. It is used in lieu of the MS because it can be accessed much more quickly. The input-output control unit (IOCU) represents the channel or other circuits which provide the detailed control of input-output units such as video terminals, communications equipment, disk and diskette storage, and data acquisition equipment. The instruction address register (IAR) contains the location of the instruction currently being executed. Normally it is automatically incremented by 1 to access the next sequential instruction. The instruction register (IR) is a temporary storage location in which the current instruction is held during execution.

The basic operational sequence is as follows:

1. Fetch the next instruction from MS.
2. Fetch the operands (if any) from MS or RS.

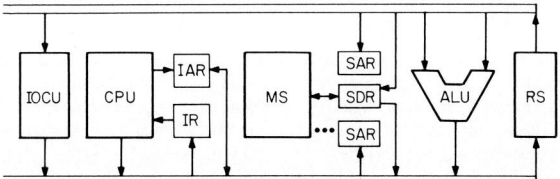

Fig. 1 Hypothetical computer data flow. CPU, Central processing unit; MS, main storage; SAR, storage address register; SDR, storage data register; ALU, arithmetic logic unit; IOCU, input-output control unit; RS, register stack; IAR, instruction address register; IR, instruction register.

3. Execute the operation.

4. Fetch the next instruction and repeat.

As a detailed example, consider the following sequence of instructions stored in sequential MS locations (abbreviations are used to represent the operation codes, but the actual representation in storage is a sequence of binary codes as described earlier).

Storage location	Instruction	Meaning
1	M,34,8	Move the contents of MS 34 into RS 8
2	M,87,6	Move the contents of MS 87 into RS 6
3	A,8,6,3	Add the contents of RS 8 to the contents of RS 6 and place the result in RS 3
4	BCN,3,93,65	If the contents of RS 3 are negative, go to the instruction in MS 93; otherwise, go to the instruction in MS 65

The execution of this program fragment proceeds as follows. The address in the IAR is set to 1, indicating that the instruction in MS location 1 is to be executed. The CPU causes the contents of location 1 to be transferred into the IR where it is decoded by the CPU. The CPU transfers the data address (34) to the SAR, and the contents of MS location 34 are transferred to the SDR and then routed to RS location 8. The operation is complete, and the CPU increments the IAR by 1, causing it to advance to 2.

Since the IAR is set to 2, the instruction in storage location 2 is transferred to the IR and decoded. The sequence is the same as above, except that the data in MS 87 is transferred to RS location 6. On completion of the transfer, the IAR is incremented to 3.

The instruction from storage location 3 is placed in the IR and decoded. This causes the contents of RS 6 to be routed to one input of the ALU and the contents of RS 8 to be routed to the other input. The two numbers are added together by the ALU, and signals from the CPU cause the sum to be stored in RS 3. The IAR is then incremented to 4.

The instruction in MS 4 is transferred to the IR and decoded. This causes the contents of RS 3 to be examined by the CPU to determine whether it is a positive or negative number. If it is found to be negative, the CPU replaces the contents of the IAR with 65, the address of the next instruction to be executed.

Control of the program has now been transferred to the instructions located at MS 65, and the basic fetch-execute-increment cycle continues. This example is simplified greatly, but it is conceptually accurate. In an actual computer, many operations may take place during a single instruction execution. For example, the address of the data may not be explicitly contained in the instruction, so that the CPU may have to fetch data from storage and perform some calculations to determine the actual address of the data. Similarly, a single instruction to send data to a printer may involve a number of data transfers between MS, registers, and the IOCU, all under control of the CPU.

In this brief entry, it only is possible to present basic concepts and definitions associated with digital computer hardware. Although similar in concept, there are many variations in computer hardware designs.

Computer Programming

Computer hardware is controlled by a sequence of instructions which are stored in the MS. This sequence of instructions is called a *computer program,* and the act of devising the particular sequence needed to solve a given problem is called *computer programming.* Collectively, the set of programs and associated documentation available for a particular computer is referred to as *software.*

Programming consists of all the activities associated with production of the actual program. This includes determining the solution to a problem, describing or organizing the solution in such a way that the computer can be used efficiently, coding the solution in a form acceptable to the computer through the use of a programming language, testing the resulting program, and maintaining the program and its documentation over its useful life.

The first phase, problem definition, involves a thorough understanding and description of what the computer is expected to do. This includes understanding what data are available, what errors are likely to be encountered (e.g., a person's name contains a numeral), what data manipulation is necessary, and what results are expected. It is also necessary to understand the characteristics of the data, such as whether certain parameters are positive or negative, the range of numerical inputs, the desired numerical precision, the number of data items to be handled or stored, and the desired format for the results.

Once the problem is defined, a procedure must be devised which provides the desired solution. This typically is done by breaking the problem into a set of subproblems, each of which can be addressed separately.

Frequently, this decomposition of the problem is facilitated by graphical or other techniques which are useful in organizing the solution. One popular technique is called flowcharting in which each subproblem is represented by a block and the relations between blocks illustrated by connecting lines. Blocks having different functions (e.g., input, output, decisions, calculations, etc.) often have distinctive shapes. American National Standard X3.5-1970 describes a set of such symbols and their use. Following the overall decomposition of the problem into subproblems, each of the major steps (blocks in the flowchart) is further decomposed into a more detailed flowchart.

Once the solution is planned, it must be translated into a sequence of computer instructions which represent the program. Although the actual program representation in storage is a series of bits, attempting to write a program in this representation is difficult and error-prone, because of the need to manipulate the inconvenient binary codes. As a result, programming languages have been devised to ease the task and reduce the chance for error.

There are many programming languages available for most computers. In general, however, they can be categorized as assembly languages, macroassembly languages, higher level or procedural languages, and problem-oriented languages. In *assembly language,* the binary code for each operation code is represented by a mnemonic code. For example, the binary code 01101 meaning ADD REGISTER might be represented by ADR. Similarly, address locations are represented by mnemonics chosen by the programmer; for example, NP may represent net pay. The program is coded using these symbols and might appear as

$$
\begin{array}{l}
\text{ADR,1,NP} \\
\text{M,34,86} \\
\text{CMP,3,RES}
\end{array}
$$

The resultant program, called the *source program,* is used as data for an assembly program, which substitutes binary codes for the mnemonic codes and assigns a physical storage addresss in place of the mnemonic address. The assembly program output is a sequence of binary words (bytes) which can be placed in storage for execution. This binary-coded program is called the *object program.*

It is soon discovered when using an assembly language that there are particular sequences of operations which are used frequently. For example, reading a character from a keyboard may require the same 100 assembly language instructions for each character read. Rather than writing this sequence each time it is needed, the whole sequence can be assigned a name such as KEYI, standing for key input. Whenever the programmer needs to read a keyboard input, the mnemonic KEYI is inserted in the assembly language program. The sequence of instructions KEYI is called a *macro.* When the program is processed using a macroassembler to produce the object program, the computer inserts the entire instruction sequence represented by KEYI whenever it appears. This technique can reduce significantly the effort needed to code a program.

Procedural languages go a step further and provide an even greater function for each statement in the source program. For example, a typical procedural language allows the programmer to write

$$A = B + C/D \times E$$

in order to express a mathematical calculation. Similarly, control statements can be expressed in a form such as

IF $\times$ .GT. y THEN z = x-y ELSE z = y-x

These statements are not executable directly by the computer and must be translated into a sequence of binary-coded instructions. This is done by a program called a *compiler* or an *interpreter* which analyzes each statement and substitutes the necessary sequence of instructions to accomplish the desired result. Each statement in the procedural language source program creates many machine instructions. As a result, the productivity of the programmer is greatly enhanced. A number of high-level procedural languages are in common use. These include, for example, FORTRAN, COBOL, Pascal, LISP, and BASIC. A majority of programming today is done using such languages to express the problem solution.

Problem-oriented langages are similar to procedural languages except that they are designed for the solution of a specific class of problems. They often use a vocabulary which is unique to a particular field. For example, a typical statement in a problem-oriented language for process control applications might be

AT 1500 HRS CLOSE VALVE V1 WAIT 10 SEC TEST

A source program written in such a language must also be processed by a compiler or interpreter program before it can be executed by the computer.

Program Debugging

Once the solution to a problem has been described in a programming language and processed by an assembler, compiler, or interpreter program, it must be tested for correctness before being used. Errors in programs are called "bugs," and the process of finding and correcting them is referred to as *debugging*. A variety of techniques are utilized in this process. Most assemblers, compilers, and interpreters include facilities which identify errors associated with the syntax and semantics of the programming language being used and other detectable errors such as an expression involving mixed data types (e.g., attempting to add a character to a number). Errors in the problem solution, of course, cannot be detected by these language processing programs. These are usually detected by visual inspection, sometimes by a second programmer who was not involved in the original writing of the program, and by using test data to determine if the results are as expected. Many errors are associated with the interaction of a particular program with other programs being used concurrently. These are detected when the program is integrated into the total set of programs being used on the computer.

Debugging a program is a difficult task, since all possible combinations of data must be considered. It is usually impossible to exhaustively test a program, so it is likely that some bugs will remain undetected until the program has been in use for some time. It is this phenomenon which gives rise to the need for maintenance over the life of the program.

Systems Programs

Programs can be separated into two major categories: *applications programs* and *systems programs*. *Applications programs* are programs directly involved in the solution of a user's problem and are typically related to the business or scientific purpose of the program. Thus, for example, accounting programs, data base inquiry programs, and airline reservation programs are applications programs.

Considerable programming, however, is required just to control the internal operation of the computer, independent of the particular application. The suite of programs used for this purpose often are called *systems programs*. In general, these provide facilities and services which are useful to applications programmers and serve to free them to concentrate on the details of the application. A major example of this is the so-called operating system, monitor, or executive program which handles many of the internal details associated with running an applications program.

In modern computers, it is common for several applications programs to be in main storage simultaneously and be executed concurrently. Two terms associated with this type of operation are time sharing and multiprogramming. In *time sharing*, each program is executed for a fixed period of time or until it needs a computer facility which currently is not available. At that time, the program is halted temporarily and another program is executed. After every active program has an opportunity to execute, the computer returns to the original program. Because of the speed of the computer, the individual user often is not aware that his or her program has been interrupted several times during its execution.

Multiprogramming is a generalization of time sharing where the computer facilities are shared among several programs based on the availability of resources, such as input-output devices, or some other criteria. For example, a priority scheme may be used where more important programs are allowed precedence over other programs. Less important programs are not allowed to run until higher priority tasks are completed.

The scheduling of programs in a time sharing or multiprogramming environment is one of the important functions of the operating system. In addition, the operating system often provides detailed control of input-output devices, such as disk files and printers, and security features that, for example, ensure that a program does not access data for which it is not authorized. The computer typically spends more time executing operating system code than it does applications programs. As a result, the design of the operating system is crucial to the performance perceived by the user.

In addition to the operating system, a suite of programs referred to as *utility programs* usually are provided. These include the assemblers, compilers, and interpreters needed to process the high-level language source programs. Other utilities are housekeeping programs for rearranging storage, programs for testing the correct operation of the computer and its peripheral devices, and accounting programs to aid in billing users for their portion of the computer's time.

Brief History of the Digital Computer

The earliest digital computer was the abacus, invented in the pre-Christian era and still is used in some parts of the world. Semiautomatic mechanical calculators did not appear until the seventeenth century, however, when Blaise Pascal, the son of a French tax collector, invented a calculator to assist in adding and subtracting the columns of figures associated with the assessment and collection of taxes. Although never totally successful, the toothed-wheel mechanism invented by Pascal was the basis for the modern mechanical adding machines. Other mechanical calculators followed, including those invented by Leibnitz and Napier, and during the latter half of the nineteenth century resulted in business equipment bearing the familiar names Burroughs and Monroe, to mention only two.

One of the most remarkable developments was that of Charles Babbage who, in the 1830s in England, proposed and partially built a "difference engine" for calculating mathematical and astronomical tables. This mechanical device was to be powered by a steam engine and, utilizing the method of successive differences, was to calculate numbers and punch them into copper printing plates. Babbage never succeeded in building his machine, partially because of the lack of adequate mechanical technology, but a scaled-down version was built by Scheutz in Sweden. Babbage then went on to design in the 1840s his "analytical engine" which embodied many of the fundamental principles found in the modern digital computer. It consisted of a "mill" which performed the actual calculations, internal storage, and, like the difference engine, a mechanism for punching results into copper printing plates. In addition, input instructions and data were fed into the machine using punched cards, an idea borrowed from the Jacquard loom of 1799. The machine was never intended to be built, although Babbage's son made several attempts to obtain support to build it, and the work was forgotten until after the invention of the modern digital computer in the 1940s.

The next significant development was the use of punched card equipment for the 1890 U.S. census. Herman Hollerith invented (reinvented?) the punched card as a means of recording data on each citizen and produced equipment for punching and reading the information. The invention allowed the census to be completed in a matter of a few years, rather than the more than 10 years originally predicted for hand tabulation methods. Hollerith founded a company which further developed the idea and, during the early 1900s, the use of punched card accounting equipment became common.

Concurrently, there were significant developments in electronics with the invention of the vacuum tube through the efforts of Edison, DeForest, and Fleming.

These efforts merged and culminated in the development of the first major all-electronic digital computer by John Mauchly and J. Prespert Eckert at the Moore School of Electrical Engineering at the University of Pennsylvania from 1942 to 1946. This machine, called ENIAC (electronic numerical integrator and calculator) was built at the request of the U.S. Department of War to assist in calculating ballistic tables. It consisted of about 18,000 vacuum tubes, measured 80 $\times$ 8 $\times$ 3 feet ($\approx$24 $\times$ 2.5 $\times$ 1 m), and consumed 120 kW of power, of which 80 kW was used to heat the tube filaments. The machine utilized punched card equipment for input and output and was programmed by means of cards and wired panels. Although powerful for its day, its computing power was significantly less than that found in a small desktop computer of the 1980s.

ENIAC lacked one essential concept found in today's computers—the stored program. Program storage was suggested by John Von Neumann of Princeton University and implemented in BINAC in 1948. The use of computer storage for storage of the program provided two advantages which contributed greatly to the flexibility of the machine. First, the program instructions were available at electronic speeds, rather than at the electromechanical speed dictated by the punched card equipment of ENIAC. Second, having the program in storage allowed it to alter instructions, thereby altering a calculation based on data or programmed controls. Although frowned on today, the ability to alter internally stored instructions was a major factor in realizing the true potential of the electronic computer.

Since these early machines, technology has allowed rapid advances in computers. The invention of the transistor in 1949 by researchers at Bell Telephone Laboratories provided a fast, inexpensive, and low-powered device which, when incorporated into computers in the late 1950s, eliminated the bulky, power consuming vacuum tubes of earlier designs. Progress in transistor technology has been phenomenal and now provides for the fabrication of entire computer processors, many times as powerful as ENIAC, on a tiny chip of silicon about 10 mm square.

This brief history omits thousands of advances and a detailed consideration of the many research scientists and engineers who have contributed to the modern computer. No single contribution or invention created the computer—rather it was a combination of many technical developments and the recognition of need on the part of creative businesses that provided the impetus for its past and present development.

DIGITAL COMPUTER AND DIGITAL SYSTEM CHARACTERISTICS*

Acceleration Time. The brief time span required for a magnetic tape transport or any type of mechanical device to attain operating speed; also referred to as *start time* acceleration. Several milliseconds may be required to start a tape in motion and attain a speed at which data can be written or read.

Access Time. The time required for a computer to locate data or an instruction in storage and transfer it to an arithmetic unit where the required computations are performed. Also, the time required to transfer information which has been operated on from the arithmetic unit to storage. In an *immediate-access* storage device, such as a magnetic core or semiconductor storage, access time is the interval that occurs from the start of the storage cycle to the availability of the addressed data at the output register of the storage. This corresponds to the read cycle of the storage unit. In a *disk storage unit,* access time is the interval required for the read/write head to be moved to the addressed track *(seek time),* plus the required rotation time of the disk *(latency).* The average latency is one-half the disk rotation time.

Asynchronous. A term used to designate the property of a device or action whose timing is not a direct function of the clock cycles in the system. In an asynchronous situation, the time of occurrence

*Terms are presented in alphabetical order for convenience of reference.

or duration of an event or operation is unpredictable because of factors such as variable signal propagation delay or a stimulus which is not under the control of the computer.

In terms of a computer channel, an asynchronous channel does not depend on computer clock pulses to control the transmission of information to and from the input or output device. Transmission of the information is under the control of interlocked control signals. Thus, when a device has data to send to the channel, the device activates a service request signal. Responding to this signal, the channel activates a SERVICE OUT signal. The latter, in turn, activates a SERVICE IN signal in the device and also deactivates the request signal. Information is then transferred to the channel in coincidence with SERVICE IN, and the channel acknowledges receipt of the data by deactivating SERVICE OUT.

Asynchronous operation also occurs in the operation of analog-to-digital (A/D) subsystems. The systems may issue a command to the subsystem to read an analog point and then proceed to the next sequential operation. The analog subsystem carries out the A/D conversion. When the conversion is complete, the subsystem interrupts the system to signal the completion.

Asynchronous also has a broader meaning—specifically unexpected or unpredictable occurrences with respect to a program's instructions.

Conversion Time.   The interval of time between the initiation and completion of a single analog-to-digital (A/D) or digital-to-analog (D/A) conversion operation. Also, the reciprocal of the conversion rate. In practice, the term usually refers to A/D converters or digital voltmeters. The conversion time required by an A/D converter is comprised of (1) the time needed to reset and condition the logic, (2) a delay to allow for settling time of the input buffer amplifier, (3) a polarity determination time, (4) the actual A/D conversion operation, and (5) any time required to transfer the digital result to an output register. Not all factors are always present. A unipolar A/D converter, for example, involves no polarity determination time.

Conversion time also is used in connection with data acquisition or analog input subsystems for process control computers. The more precise term in this case is *measurement time*. In this case, the measurement rate may not be the reciprocal of the measurement time, inasmuch as some of the operations performed may overlap with other operations. With some types of A/D converters, it is possible to select the next multiplexer point during the time the previous value is being converted from analog to digital form.

Conversion or measurement time also may include the time required for such operations as multiplexer address decoding, multiplexer switch selection, and settling time, and the time required to deactivate the multiplexer switches and permit the subsystem to return to an initial state.

Off-Line.   Pertains to the operation of a functional unit without the continual control of a computer. For example, a terminal operating in a "local" mode as a keyboard-printer is said to be off-line. Similarly, in computer process control, a process unit under manual control is off-line. Contrast with *on-line.*

On-Line.   Pertaining to a system or functional unit under the continual control of the central processing unit. The term is also used to describe a user's access to a computer via a terminal. Thus, if the terminal is connected to the computer and is able to communicate with it, it is said to be on-line. Contrast with *off-line.*

Real-Time Computing.   The computer was first used mainly to solve scientific problems and automate record keeping. In these applications, the problem description and solution format are presented to the computer in the form of a program, after which data are typically furnished on which the program operates. The results of program runs are reports which are distributed for use by people. Other than the time required for execution, there are no real-time constraints on tasks of this nature.

However, some applications require the computer to respond to external events and to perform computations and control functions within specified, often very brief, time limits. Systems of this type are referred to as *response-* or *real-time-oriented* systems. The time intervals involved may range from several seconds to several microseconds.

Airline reservation systems exemplify systems which must respond within a few seconds with certain information. The problem in this case is one of large numbers of data files (reservation schedules), constant changing of the data, and their use and modification by many sources (agents at various locations).

In process control applications, the primary task of the system may be to control information output as a result of sensor information read in. Petroleum refinery and chemical plant control, engine and transmission production, performance testing of products, and gas distribution control all typify uses for real-time computing. The tasks range from simple control algorithm calculations to complex process optimization and resource scheduling.

Resolution. In systems where either the input or output of the subsystem is expressed in digital form, the resolution is determined by the number of digits used to express the numerical value. In a digital-to-analog converter, the output analog signal takes on a finite number of discrete values which correspond to the discrete numerical input. The output of an analog-to-digital converter is discrete, although the analog input signal is continuous.

In digital equipment, resolution is typically expressed in terms of the number of digits in the input or output digital representation. In the binary system, a typical specification is that "resolution is x bits." As an example, if V_{fs} is the full-scale input or output voltage range, this specification states that the resolution is $V_{fs}/2^x$. If $x = 10$ and $V_{fs} = 5$ V, the resolution is $5/2^{10}$, or 0.00488 V. It is also common to express resolution in terms of parts. A four-digit decimal converter may be said to have a resolution of 1 part in 10,000, and a 10-bit binary converter may be said to have a resolution of 1 part in 1024. The term *least significant bit* (LSB) also is used. It may be stated, for example, that the binary resolution is $\pm\frac{1}{2}$ LSB. Also used is the term *least significant digit* (LSD). This term is used in relation to decimal or other nonbinary digital equipment.

Synchronous. A synchronous operation takes place in a fixed time relation to another operation or event, such as a clock pulse. When a set of contacts is sampled at a fixed time interval, the operation is termed synchronous. This situation is to be contrasted with that where the contacts may be sampled randomly under the control of an external signal. Generally, the read operation of a main storage unit is synchronous. The turning on of the X and Y selection drivers and the sampling of the storage output on the sense line are controlled by a fixed frequency. Contrast with *asynchronous* defined earlier in this section.

DIGITAL COMPUTER HARDWARE*

Accumulator. Generally, a register and associated equipment in the arithmetic unit of a computer in which arithmetical and logical operations are performed. The term also applies to a unit in a digital computer where numbers are totaled, i.e., accumulated. Often an accumulator stores one operand and on receipt of any second operand forms and stores the result of performing the indicated operation on both the first and second operands. Commonly, an accumulator works in conjunction with an accumulator extension, the latter being used in multiplication, division, and some shifting operations. In a division operation, for example, the quotient may appear in the accumulator—with the remainder appearing in the extension. Where a machine may use several accumulators, odd- and even-numbered accumulators may be paired for accomplishing multiplication and division operations.

More recent designs generally provide for multiple general purpose registers which perform the function normally associated with an accumulator in addition (typically) to other functions.

Adder. A digital circuit which provides the sum of two or more input numbers as an output. A 1-bit binary adder is shown in Fig. 2. In this diagram, A and B are the input bits and C and $\overline{C}$ are the carry and no-carry bits from the previous position. There are both serial and parallel adders. In a *serial adder*, only one adder position is required and the bits to be added are sequentially gated to the input. The carry or no-carry from the prior position is remembered and provided as an input along with the bits from the next position. In a *parallel adder*, all the bits are added simultaneously, with the carry or no-carry from the lower order position propagated to the higher position. In a parallel adder, there may be a delay due to the carry propagation time. See also *half-adder* later in this section.

An adder may perform the subtraction as well as the addition of two numbers. Generally, this is effected by complementing one of the numbers and then adding the two factors. The following is an example of a two's complement binary subtraction operation:

---

*Terms are presented in alphabetical order for convenience of reference.

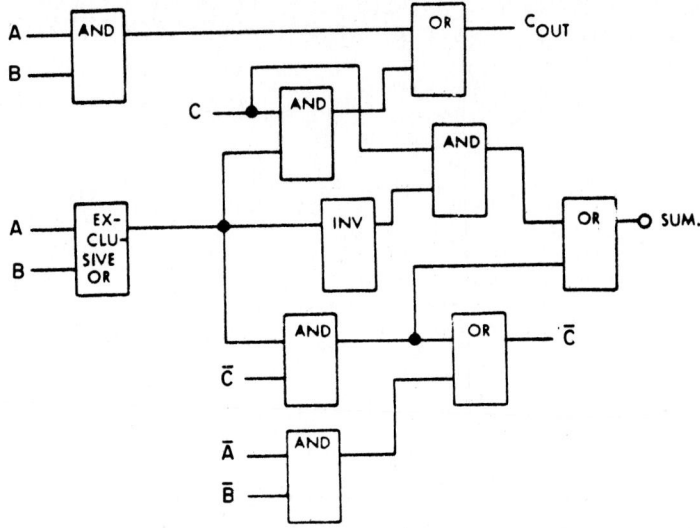

Fig. 2   Binary adder.

(a)　　　　0110　+6 (true)
　　　(+) 1010　−6 (complement)
　　　　10000 0 (true)

(b)　　　　0101　+5 (true)
　　　(+) 1010　−6 (complement)
　　　　1111　− 1 (complement)

(c)　　　　1111　(complement) = −0001 (true)

The two's complement of a binary number is obtained by replacing all 1's with 0's, replacing all 0's with 1's, and adding 1 to the units position. In (a) above, 6 is subtracted from 6, and the result is all 0's; the carry implies that the answer is in true form. In (b), 6 is subtracted from 5 and the result is all 1's with no carry. The no-carry indicates the result is in complement form and that the result must be recomplemented as shown in (c).

Thomas J. Harrison, IBM Corporation, Boca Raton, Fla.

Address Register.   A collection of single-bit logical storage elements for the temporary storage of an address in a computer. Typically, a computer contains a number of address registers, each with a specific purpose. An adjective modifier, such as "storage address register," is often used to indicate the specific use of the register. A *storage address register* contains the address of the data currently being accessed in storage. This register and its associated decoding logic provide the signals to the storage selection circuits while storage is being cycled.

An *instruction address register* contains the address of the instruction to be read from storage. When the instruction is called for, the contents of this register typically are transferred to the storage address register.

A *data address register* contains the address of the data to be obtained from storage, or the address at which the data are to be stored. When data are being read from or written to storage, the contents of this register are transferred to the storage address register. Two data address registers may be required in machines which perform the function of moving data from one storage location to another storage location with a single instruction. In this case, one register contains the address of the present location of the data, and the other contains the address of the location to which the data are to be transferred.

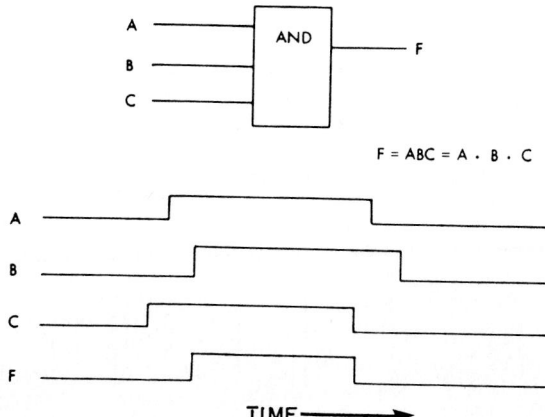

$$F = ABC = A \cdot B \cdot C$$

Fig. 3 AND circuit.

AND Circuit. A computer logical decision element which provides an output if and only if all the input functions are satisfied. A three-variable AND element is shown in Fig. 3. The function F is a binary 1 if, and only if, A and B and C are all 1's. When any of the input functions is 0, the output function is 0. This may be represented in Boolean algebra by $F = A \cdot B \cdot C$ or $F = ABC$. Diode and transistor circuit schematics for two-variable AND functions are shown in Fig. 4. In integrated circuits, the function of the two transistors or diodes may be fabricated as a single active device. In the diode AND circuit, output F is positive only when both inputs A and B are positive. If one or both inputs are negative, one or both diodes will be forward-biased and the output will be negative. The transistor AND circuit operates in a similar manner; i.e., if an input is negative, the associated transistor will be conducting and the output will be negative.

Generally referred to as "fan in," the maximum number of input functions of which a given circuit configuration is capable is determined by the leakage current of the active element. Termed "fan out," the number of circuits which can be driven by the output is a function of current that can be supplied by the AND circuit.

Thomas J. Harrison, IBM Corporation, Boca Raton, Fla.

Block Diagram. A graphical presentation of hardware and paths along which data and control information flow between various parts of a computer and/or data processing system. Through the use of numerous and varied symbols, much information can be condensed into a small space. Templates are available to facilitate the rapid construction of a block diagram. Some of the more com-

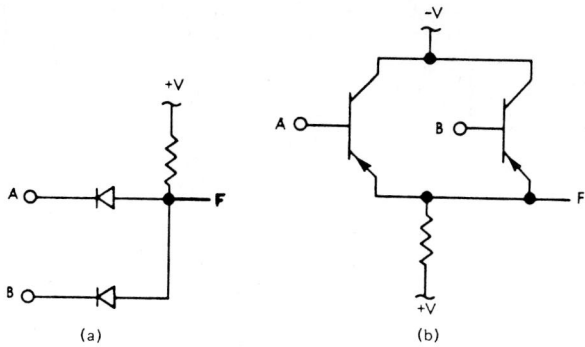

Fig. 4 (*a*) Diode-type AND circuit, (*b*) transistor-type AND circuit.

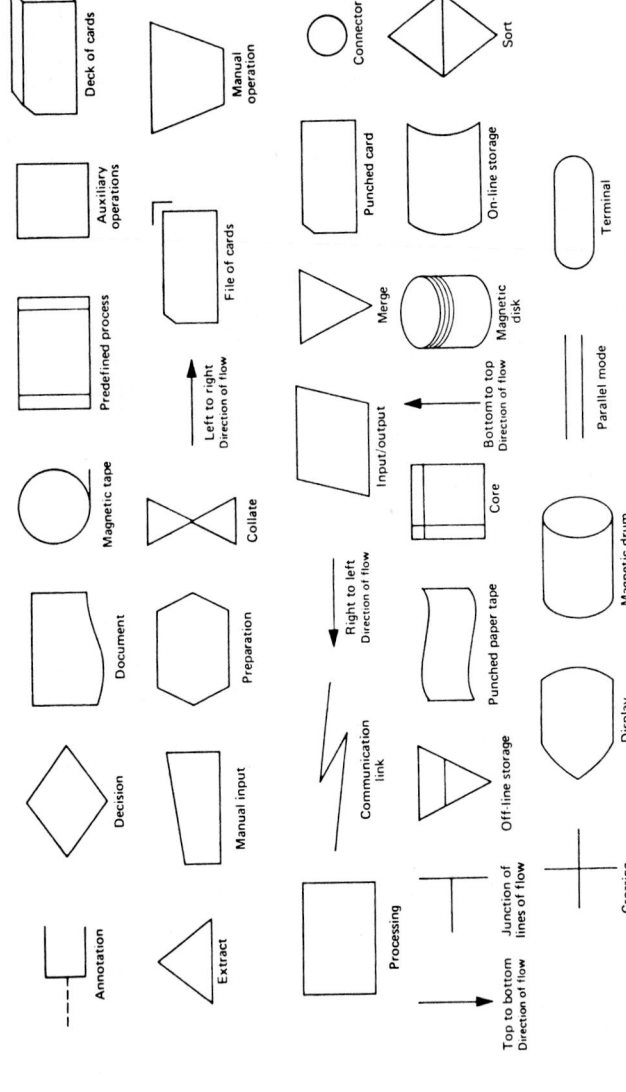

Fig. 5 Representative symbols used in preparing block diagrams.

monly used symbols are given in Fig. 5. The use of a block diagram to represent a central processing unit of a computer is shown in Fig. 6. This overall approach to the system does not require a detailed use of symbols. In a block diagram, each block generally represents a functional subdivision of the system under consideration. The function may range from a single AND block to a very complex representation of a total computer.

Buffer. An internal portion of a digital computer or data processing system which serves as intermediary storage between two storage or data handling systems with different access times or formats. The word *buffer* is also used to describe a routine for compensating for a difference in data rate, or time of occurrence of events, when transferring data from one device or task to another. For example, a buffer usually is used to connect an input or output device with the main or internal high speed storage. The term *buffer amplifier* applies to an amplifier which provides impedance transformation between two analog circuits or devices. A buffer amplifier may be used at the input of an analog-to-digital (A/D) converter to provide a low source impedance for the A/D converter and a high load impedance for the signal source.

Central Processing Unit (CPU). Also called the *mainframe,* the CPU is the part of a computing system exclusive of input or output devices and, sometimes, main storage. As indicated in Fig. 7, the CPU includes the arithmetic and logical unit (ALU), channels, storage and associated registers, and controls. Information is transmitted to and from the input or output devices via the channel. Within the CPU, data are transmitted between storage and the channel and between each of these and the ALU. In some computing systems, if the channel and storage are separate assemblages of equipment, the CPU includes only the ALU and the instruction interpretation and execution controls.

A simplified example of CPU data flow was previously given in Fig. 6. Generally, the operations performed in the CPU can be divided into (1) fetching instructions, (2) indexing (if appropriate to the instruction format), and (3) execution.

In the example shown, to initiate an operation, the contents of the instruction address register are transmitted to the storage address register (SAR), and the addressed instruction is fetched from

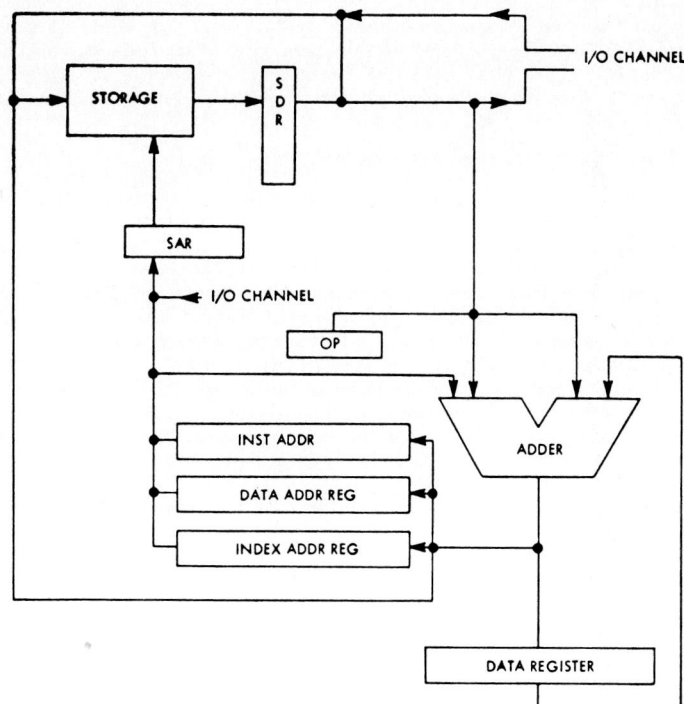

Fig. 6 Block diagram of a central processing unit of a computer.

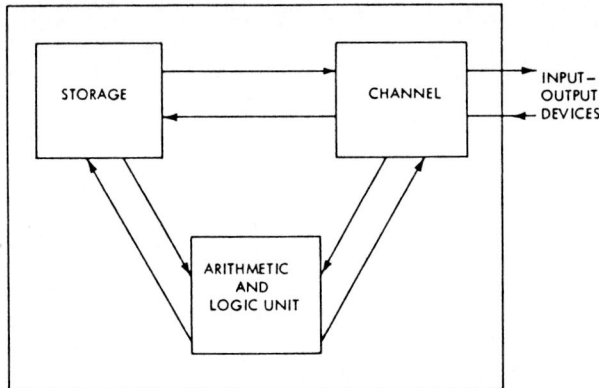

Fig. 7 Computer central processing unit.

storage and transferred to the storage data register. The operation code portion of the instruction is set into the operation code register (OP), and the data address part is set into the data address register. If indexing is defined in the instruction format, the indexing address is set into the index address register.

During the indexing phase of the operation, the index address register is transmitted to the SAR and the contents of the storage location are set into the storage data register. The contents of the storage data register and of the data address register are summed in the adder, and the result is placed in the data address register.

In connection with an arithmetic operation, the data address register is transmitted to the SAR, and the data in the referenced location are gated through the adder along with the contents of the data register. The result replaces the contents of the data register. If the arithmetic operation causes a carry-out of the high-order position of the word being operated on, the overflow indication is saved in a condition code trigger. This condition may be tested on a subsequent operation, such as a transfer on condition. On completion of the specified operation, the next instruction is fetched from storage and the previously described phases are repeated.

Thomas J. Harrison, IBM Corporation, Boca Raton, Fla.

Channel. The portion of the central processing unit (CPU) of a computer which connects input and output devices to the CPU. It may also execute instructions relating to the input or output devices. A channel also provides the interfaces and associated controls for the transfer of data between storage and the input-output devices attached to the CPU. The channel generally maintains the storage address for the device in operation and includes the buffer registers needed for synchronizing with storage. A serial or selector channel may be used for slower devices.

A *multiplex channel* has the capability of concurrently servicing several devices. The storage address for each operating device is controlled by the channel and is maintained in main storage or in the logic of the channel. If the address is maintained in the logic, this increases the maximum data rate of the channel and reduces the number of storage references needed to service the device. The storage-access channel is an adaptation of the multiplex channel. With this type of channel, the device transmits both the data and the storage address to the central processor when it requires servicing.

Where a *series channel* is used, one or more devices may be physically attached to the channel interface, but only one of them is logically connected at any given time. Thus, for the selected device, the full channel data rate capacity is available. For a given device data rate requirement, a series channel is less costly than either dedicated channels per device or a multiplex channel. In the latter instance, the saving is realized from the fact that the maximum required data rate is determined by the data rate of a single attached device. The maximum rate of a multiplex channel is the sum of the data rate requirements of several devices, plus the data rate required for device addressing.

Channel also may be used to describe other portions of some computers. For example, *analog input channel* refers to the path between the input terminals of an analog input subsystem and an analog-to-digital (A/D) converter. Similarly, it may describe the logical or physical path between the source and destination of a message in a communications system. In some applications relating to data acquisition in physics experiments, it is synonymous with the quantization interval in an A/D converter.

The tracks along the length of magnetic tape used for storing digital data are also referred to as channels.

Thomas J. Harrison, IBM Corporation, Boca Raton, Fla.

Collator. A data processing device used to combine sets or decks of cards or other information bearing units into a desired sequence. Typically, a card collator has two input feeds so that two ordered sets may enter into the process, and four output stackers so that four ordered sets can be generated by the process. Three comparison stations are used to route the cards to one stacker or the other on the basis of a comparison of criteria as specified by the collator controls. Collating is required where data from two or more physically separated files must be combined. Combining a file with names and addresses with another file containing one or more items of personal information is an example. The term *merge* usually signifies the combining of two similarly ordered sets of data into a single ordered set. The order, for example, may be alphabetical or numerical. A data set containing B, H, L, Q, and T may be combined with another set containing C, J, N, and S to produce the ordered set B, C, H, J, L, N, Q, S, and T. The combined set may be referred to as a *unified file*. The four basic operations involved in collating are merging, sequence checking, selection, and matching.

The term is also used for a program or routine which provides similar functions when applied to one or more files stored in a computer storage unit.

Counter. A physical or logical device capable of maintaining numeric values which can be incremented or decremented by the value of another number. A counter, located in storage, may be incremented or decremented under control of the program. An example of such use is recording the number of times a program loop has been executed. The counter location is set to the value of the number of times the sequence of instructions is to be performed. On completion of the sequence, the counter is decremented by 1 and then tested for zero. If the answer is nonzero, the sequence of instructions may be repeated.

A counter also may be in the form of a circuit that records the number of times an event occurs. A counter which counts according to the binary number system is illustrated by the truth table

	A_3	A_2	A_1
P1	0	0	1
P2	0	1	0
P3	0	1	1
P4	1	0	0
P5	1	0	1
P6	1	1	0
P7	1	1	1
P8	0	0	0

Fig. 8 Truth table for a binary counter. The situation illustrates a binary counter of three stages capable of counting up to eight pulses. Each trigger changes state when a pulse is gated to its input. In the instance of trigger A_2, it receives an input pulse only when trigger A_1 and the input are both 1. Subsequently, trigger A_3 changes its state only when both triggers A_1 and A_2 and the input pulse are all 1's. This table shows the value of each trigger after each input pulse. At the eighth pulse, the counter resets to zero.

shown in Fig. 8. Counters also may be used to accumulate the number of times an external event takes place. The counter may be a storage location or a counter circuit which is incremented as the result of an external stimulus.

Specific counter definitions include the following.

A *binary counter* is (1) a counter which counts according to the binary number system, or (2) a counter capable of assuming one of two stable states.

A *control counter* records the storage location of the instruction word which is to be operated on following the instruction word in current use. The control counter may select storage locations in sequence, thus obtaining the next instruction word from the subsequent storage location, unless a transfer or special instruction is encountered.

A *location counter* or *instruction counter* is (1) the control section register which contains the address of the instruction currently being executed, or (2) a register in which the address of the current instruction is recorded. It is synonymous with *program address counter*.

Thomas J. Harrison, IBM Corporation, Boca Raton, Fla.

Data Acquisition Computer. When a digital computer operates in real time to collect physical data, it often is referred to as a *data acquisition computer*. The instruction repertoire of a data acquisition computer typically includes special instructions to facilitate this application. Characteristic of

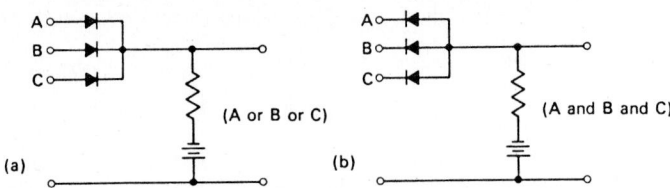

Fig. 9 Diode logic circuits. (*a*) OR circuit; (*b*) AND circuit.

the data acquisition computer is its ability to attach a wide variety of input or output devices. These include telemetering devices and flying-spot digitizers which are unique to the particular application. The data acquisition computer also can handle random demands from devices which are asynchronous with respect to other program activity. Although the average data rates may be reasonable, data acquisition input-output devices may have very high peak data rates. Exemplary of this is a flying-spot digitizer, which may transmit data to a computer at an average rate of 100,000 bytes/s, but with a peak rate of 1 million bytes/s. Generally, external devices transmit data randomly to a data acquisition computer. Hence, so that the necessary action may be taken by the control program, the data acquisition computer is designed with an ability to respond very quickly to an external interrupt signal.

Diode Logic (DL). A form of logic circuit using diodes and resistors. Circuits for performing the function OR and AND are shown in Fig. 9. Considering a positive signal to represent 1, one notes that a positive voltage connected to any one of the inputs A, B, or C in Fig. 9*a* will result in a positive signal at the output. Thus, the circuit performs the logical OR operation. In Fig. 9*b*, the AND operation is accomplished. To obtain a positive (1) output, A, B, and C must all be positive (1).

Diode-Transistor Logic (DTL). A form of logic circuit using diodes, transistors, and resistors. Circuits for performing the logical operations NOT OR (NOR) and NOT AND (NAND) are shown in Fig. 10. The arrangement is an extension of diode logic wherein the logic operation is effected by diodes and the transistor furnishes an inversion of sign and amplification. A positive signal represents the 1 logic state for the circuits in the figure. In Fig. 10*a* the transistor is brought into conduction with the collector voltage close to ground potential (0 output, corresponding to NOT 1) when A or B or C is positive (1 condition). In Fig. 10*b* the transistor is brought into conduction only when all inputs A, B, and C are in the 1 condition (positive signal). This state results in a transistor output corresponding to 0 (NOT 1). In both circuits, the output of the transistor is positive corresponding to 1 whenever it is cut off.

Emulator. A feature that enables a computer to execute instructions written for a different type of computer. This feature is particularly common where different computers are of similar but not identical configurations. An emulator implies the use of special logic circuits or a microcode to assist in execution of the instructions of the computer being emulated. The same functions, without the

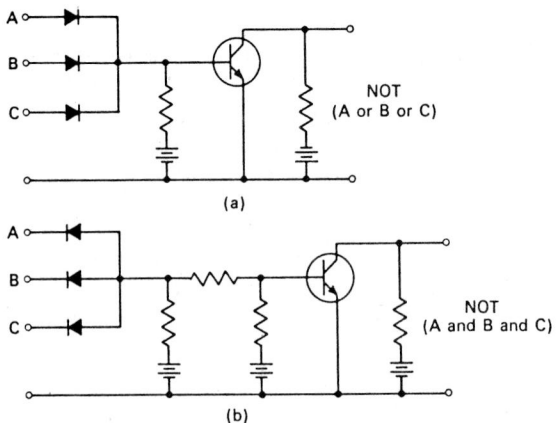

Fig. 10 Diode transistor logic circuits.

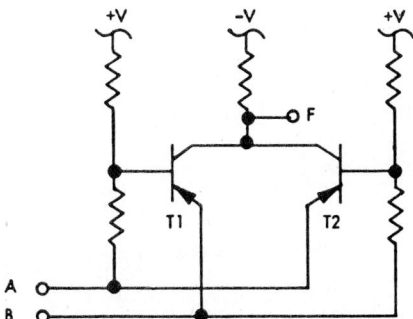

Fig. 11 EXCLUSIVE OR circuit.

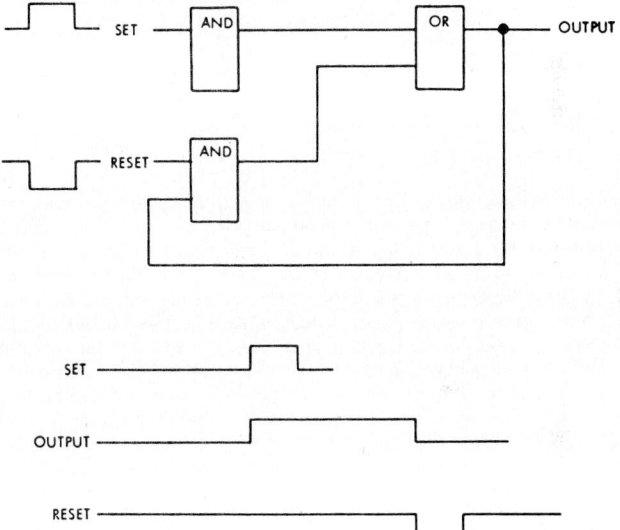

Fig. 12 Direct-coupled flip-flop or latch.

requirement for special circuits or a microcode, can be implemented as a special "simulator" program at the expense of storage and performance.

EXCLUSIVE OR Circuit A logical element which has the properties that, if either of the inputs is a binary 1, then the output is a binary 1. If both the inputs are a binary 1 or 0, the output is a binary 0. In terms of Boolean algebra, this function is represented as $F = AB' + BA'$, where the prime denotes the NOT function. With reference to the transistor EXCLUSIVE OR circuit shown in Fig. 11, the output is positive when either transistor is in saturation. When input A is positive and B is negative, transistor T_2 is in saturation. When B is positive and A is negative, transistor T_1 is in saturation. When A and B are either both positive or both negative, then both transistors are cut off and the output F is negative. Although shown as discrete devices in the figure, fabrication using large-scale integrated circuit technology may utilize other circuit and design configurations.

Fixed-Program Computer. A computer in which the sequence of instructions is permanently stored or wired in and performed automatically and not subject to change by either the computer or the programmer except by rewiring or changing the storage input.

Flip-Flop. A bistable device, the output of which assumes one of two stable states depending on the state of the most recently applied input signal; also known as a *toggle*. The flip-flop in computer

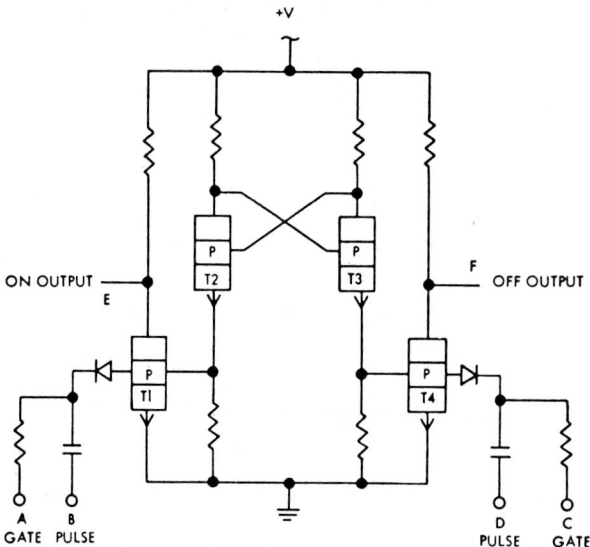

Fig. 13   Gated direct-coupled flip-flop.

systems is used for storing information. A direct-coupled flip-flop or *latch*, constructed of AND and OR logical elements, is shown in Fig. 12. The output is fed back to the input. A set pulse initially turns on the output, and the output then provides its own input even though the set pulse has been removed. The circuit remains "latched" in this condition until a reset pulse breaks the feedback loop.

A gated transistor flip-flop is shown in Fig. 13. When the flip-flop is in the OFF condition, transistor T_4 is OFF and the OFF output F is at + volts. This represents a binary 1 state. Transistor T_1 is ON and ON output E is at 0 V. This represents a binary 0 state. Transistor T_3 is in a state of low conduction, and its emitter voltage is not sufficient to forward-bias T_4. Transistor T_2, however, is in a state of heavy conduction and its emitter voltage forward-biases T_1 and thus holds it ON. To change the output state of the flip-flop, a conditioning gate of 0 V is applied to input C while the signal on input D is still positive. Thus, when the set pulse at input D goes to 0 V, a negative shift appears at the emitter of T_3. T_3 instantly goes into heavy conduction. The reduced collector voltage causes T_2 to go into a state of low conduction. The voltage at the emitter of T_2 is reduced and biases T_1 OFF. When the ac transient has receded, the T_3 emitter voltage holds T_4 ON. The flip-flop can be turned OFF by applying gate and pulse inputs to points A and B, respectively.

A flip-flop of this type may be connected for binary operation by connecting A to F and C to E and by applying the input pulse to both B and D. Since the gate must be applied before the pulse arrives, the output alternately changes state whenever a pulse is applied. The circuit may be used to build a shift register by connecting the OFF output of the previous state to gate C, connecting the ON output of the previous stage to gate A, and applying a pulse to inputs B and D.

Thomas J. Harrison, IBM Corporation, Boca Raton, Fla.

Gate.   A circuit having a binary output which is fully determined by the binary state of its input signals, such as in the AND and OR gate circuits. Also, a signal which permits an AND circuit to pass a signal. Usually the gate signal is of longer duration than the signal to make certain that coincidence occurs. In conditioning the set pulse of a flip-flop, for example, the gate must precede the set signal in order that the negative shift will be recognized by the transistor.

Gate Circuit.   A circuit which amplifies or passes a signal only in the presence of an appropriate synchronizing or "gating pulse" which "opens the gate." Also used to refer to the various logic functions and circuits used to realize computer designs, such as AND, OR, NOT, NOR, and NAND.

Half-Adder.   A circuit having two output points, S and C, representing a sum without carry and carry, and two input points, A and B, representing the addend and augend such that the output is related to the input according to the table:

Input		Output	
A	B	S	C
0	0	0	0
0	1	1	0
1	0	1	0
1	1	0	1

Two half-adders and an INCLUSIVE OR circuit, properly connected, can provide a full adder having two inputs (augend and addend) and a carry input which produces a sum output (without carry) and a carry output.

Hardware. The physical equipment or devices forming a computer and peripheral equipment. This is contrasted with the program, procedures, rules, and associated documentation which, collectively, are termed *software*.

Index Register. The contents of the index register of a computer are generally used to modify the data address of the instruction as the instruction is being read from storage. The modified address is called the *effective data address*. A particular index register is addressed by a specified field in the format of the instruction. The data address of the instructions would contain the address of the required data with reference to the start of the table. All instructions which reference table data are indexed by the specified index register which contains the address of the start of the table. Thus, when the program is to perform these operations on another table of data, the value in the index register is changed to the start address of the new data table. This effectively modifies all the indexed instructions in the sequence.

Index registers may be fixed locations in main storage, or they may be implemented in logic using flip-flops or triggers. In most modern computer designs, the function of index registers is provided through the capabilities of general purpose registers. In the case of storage resident index registers, the index register address contains only the number of bits required to specify the register number uniquely, and a fixed prefix is supplied by the system logic to provide the actual storage address. This technique minimizes the length of the computer instruction, inasmuch as most computers have only a few index registers. Thus, only 2 or 3 address bits are required in the instruction, even though a complete storage address may require 10 or more bits.

Index registers also are used as counters by the program. The same register may be used as both a counter and an address modification value to step through tables. The index register is initialized to contain the number of factors in the table to be operated on, and the instructions to be performed contain the table start address as a data address which is to be indexed by the same index register. Each time the sequence of instructions is performed, the index register is decremented by 1 and tested for zero value. Where the result of the test is nonzero, the sequence of instructions is repeated. However, since the index register value has been reduced by 1, the effective data address references the next lower table value. When the test result is zero, all the factors in the table have been operated on, and the program steps to the next sequential operation.

Thomas J. Harrison, IBM Corporation, Boca Raton, Fla.

Instruction Counter. A counter register in a computer which contains the address of the instruction to be accessed in storage. Also known as a *program counter* in some computer designs. Each time an instruction is executed, the register is incremented such that, at the completion of the operation, the instruction counter is able to address the next instruction. When a program is interrupted, the instruction counter address must be saved so that the program may resume at the point of interruption when the interrupt program is finished. If a BRANCH or CONDITION instruction is executed and the branch is taken, the contents of the instruction counter is replaced by the BRANCH TO address.

Logic. In hardware, a term referring to the circuits which perform the arithmetic and control operations in a computer. In designing digital computers, the principles of Boolean algebra are employed. The logical elements of AND, OR, INVERTER, EXCLUSIVE OR, NOR, NAND, NOT, and so forth, are combined to perform a specified function. Each of the logical elements is implemented as an electronic circuit which in turn is connected to other circuits to achieve the desired result. The word *logic* is also used in computer programming to refer to the procedure or algorithm necessary to achieve a result.

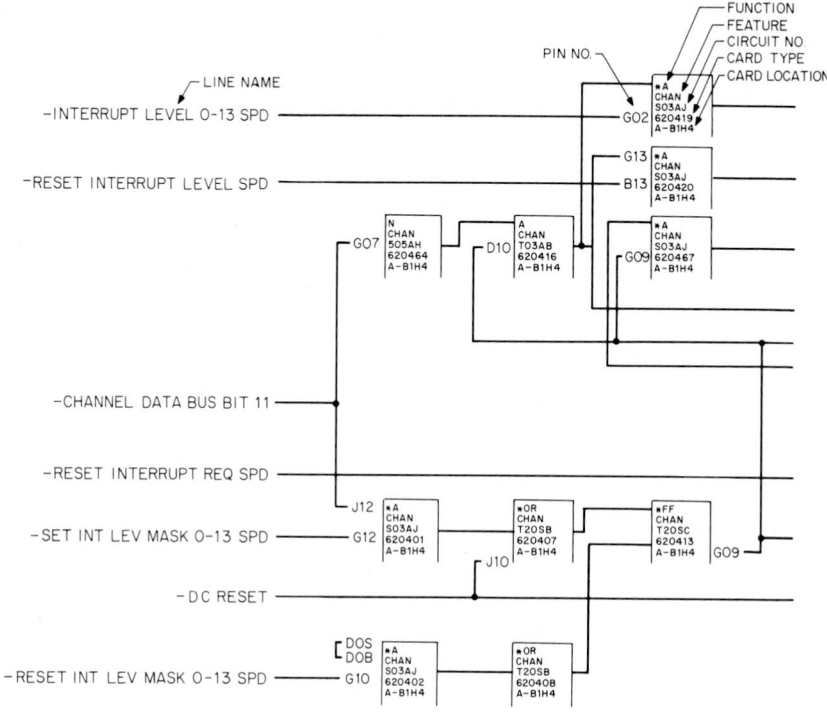

Fig. 14   Computer logic diagram.

Logic Diagram.   A drawing which indicates the interconnection of the individual logic elements in a computer. A logic diagram incorporates all the information needed for wiring a computer. The logic diagram shown in Fig. 14 indicates the logic blocks and electrical interconnections. The logic block in the figure contains the name of the function performed by it, such as AND, OR, and FLIP-FLOP, and the physical location of the circuit within the computer. Input and output signal connections for the function are given by the logic block. The logic diagram also shows the connection of a given logic block to other logic blocks. The computer manufacturer often provides wiring lists for the machine and a printed logic diagram. The logic diagram is used as an aid in troubleshooting the system. When utilizing large-scale integrated circuits, each block may contain a complex logic function, such as an adder or register, rather than single elemental logic functions. Although not needed for physical wiring or troubleshooting, these complex logic functions can, in turn, be expressed in terms of elemental logic similar to that shown in Fig. 14.

Thomas J. Harrison, IBM Corporation, Boca Raton, Fla.

MBM.   Abbreviation for *magnetic bubble memory.*

Microcomputer.   A computer which utilizes a microprocessor as its central processing unit (CPU). This CPU must perform two functions: (1) sequence through the instructions and (2) execute each instruction. A microcomputer requires two other fundamental elements: (1) memory for the program—a sequence of instructions to be performed, and (2) input-output circuits to tie it to external devices. A microcomputer is a digital logic device; i.e., all input-output signals are at digital logic levels, either 0 or 5 V. A significant amount of interface circuitry is required between the microcomputer and external devices.*

*E. R. Garen, *Control Eng.*, vol. 26, no. 6, 1978.

Microprocessor (MPU). Defined as a programmable controlled large-scale integrated (LSI) component, the latter usually referring to a LSI circuit, generally considered as being a single crystal of silicon on which is implemented a circuit involving a minimum of 1000 transistors. A microprocessor is a program-controlled component, implying that it is not the configuration of the circuit which determines its function but rather the sequence of instructions making up the program. The microprocessor sequences through these instructions, examining each in turn and executing each instruction. Executing each instruction means that the microprocessor performs some manipulation of data. In general, microprocessors have three categories of instructions which can be performed: (1) instructions to move data from one place in the system to another (e.g., from an input line to a storage register); (2) instructions which tell the microprocessor to perform an arithmetic or logical operation on data (e.g., to add, subtract, or perform AND or OR operations on two pieces of data); (3) instructions which tell the microprocessor where to go next in the program. By constructing sequences of instructions from these categories, the program is developed.* Microprocessors are mentioned in many places in this volume. Consult the alphabetical index.

Minicomputer. There is no widely accepted definition of this term, although it is frequently used in the information processing industry, as are other computer classifications, such as microcomputer, supermini, supercomputer, and mainframe computer. Nevertheless, computers classified as minicomputers often have one or more of the following characteristics: (1) They are called minicomputers by their manufacturers; (2) they utilize a 16-bit instruction and data word, often divided into two separately addressable 8-bit bytes; the supermini utilizes a 32-bit word, usually consisting of four separately addressable bytes; (3) they are often packaged using a rack-and-panel construction, although some may consist of only a single printed circuit board; (4) they are highly modular, having a wide variety of optional features, peripheral equipment, and adapters for attaching specialized peripheral devices, such as data acquisition equipment; (5) they are often utilized by original equipment manufacturers (OEMs) to provide computing capability in specialized equipment, such as electronic assembly testers; (6) they often provide as standard or optional features devices such as time-of-day clocks, interval timers, and hardware-implemented priority interrupts which facilitate their use in real-time applications; and (7) they often cost (1983) between $10,000 and $100,000, depending on the amount of storage, optional features, and peripheral equipment. It is important to recognize the lack of precision implicit in this definition and the fact that virtually all computers, from the microcomputer to the supercomputer, are conceptually similar in architecture and design.

Thomas J. Harrison, IBM Corporation, Boca Raton, Fla.

Modem. An acronym for *mo*dulator-*dem*odulator. A device for converting data from one code (or form) to another, mainly for transmission. A modem typically comprises the interface between a data sending (or receiving) device and common carrier lines.

Module. An incremental block of storage or other building block for expanding the capacity of a computer. In computer programs, a program unit that is discrete and identifiable and which usually can be assembled, compiled, and loaded as a unit. Also, *module* may refer to the individual package which contains an integrated circuit and which typically is mounted on a printed circuit board.

MOS. Acronym for *metal-oxide semiconductor*.

MOSFET. Acronym for *metal-oxide semiconductor field-effect transistor*.

NAND Circuit. A computer logical decision element which has the characteristic that the output F is 0 if, and only if, all the inputs are 1's. Conversely, if any one of the input signals A or B or C or the three-input NAND element shown in Fig. 15 is not a 1, the output F is a binary 1. Although the NAND function can be achieved by inverting the output of an AND circuit, the specific NAND circuit requires fewer circuit elements. A two-input transistor NAND circuit is shown in Fig. 16. The output F is negative only when both transistors are cut off. This occurs when both inputs are positive. The number of inputs, or fan-in, is a function of the components and circuit design. NAND is a contraction of NOT AND.

NOR Circuit. A computer logical decision element which provides a binary 1 output if all the input signals are a binary 0. This is the overall NOT of the logical OR operation. A Boolean algebra expression of the NOR circuit is $F = (AB)'$, where the prime denotes the NOT function. A two-input tran-

*E. R. Garen, *Control Eng.*, vol. 26, no. 6, 1978.

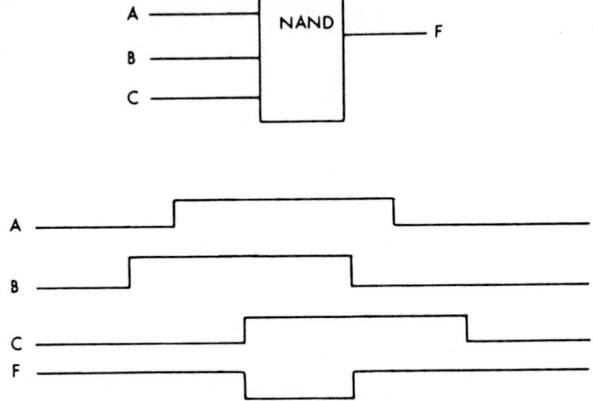

Fig. 15   Schematic of a NAND circuit.

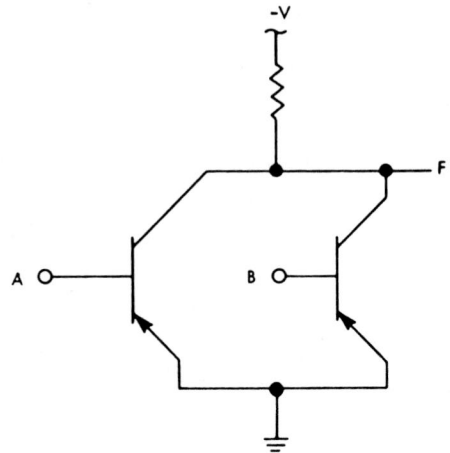

Fig. 16   Transistor-type NAND circuit.

sistor NOR circuit is shown in Fig. 17. Output F is a positive only when both transitors are cut off. This occurs when both inputs A and B are negative.

NOT Circuit.   Also known as an inverter circuit, this is a circuit which provides a logical NOT of the input signal. If the input signal is a binary 1, the output is a binary 0. If the input signal is in the 0 state, the output is in the 1 state. In reference to Fig. 18, if A is positive, the output F is at 0 V inasmuch as the transistor is biased into conduction. If A is at 0 V, the output is at $+$ volts because the transistor is cut off. Expressed in Boolean algebra, $F = A'$, where the prime denotes the NOT function.

OR Circuit.   A computer logical decision element which has the characteristic of providing a binary 1 output if any of the input signals are in a binary 1 state. This is expressed in terms of Boolean algebra by $F = A + B$. A diode and a transistor representation of this circuit is shown in Fig. 19. In a diode-type OR circuit, if either (or both) input signal A or B is positive, the respective diode is forward-biased and the output F is positive. The number of allowable input signals to the diode OR gate is a function of the back-resistance of the diodes. The input transistors of a transistor-type OR circuit are forced into higher conductivity when the respective input signal becomes positive. Thus, the output signal becomes positive when either or both inputs are positive.

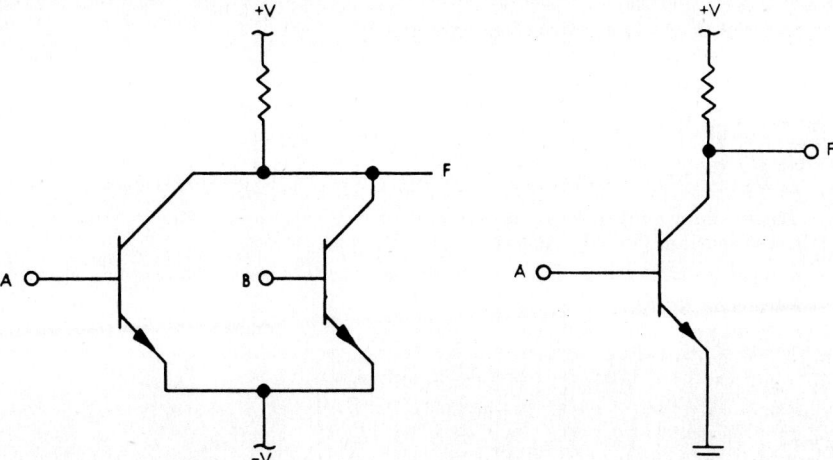

Fig. 17 Transistor-type NOR circuit.

Fig. 18 Inverter or NOT circuit.

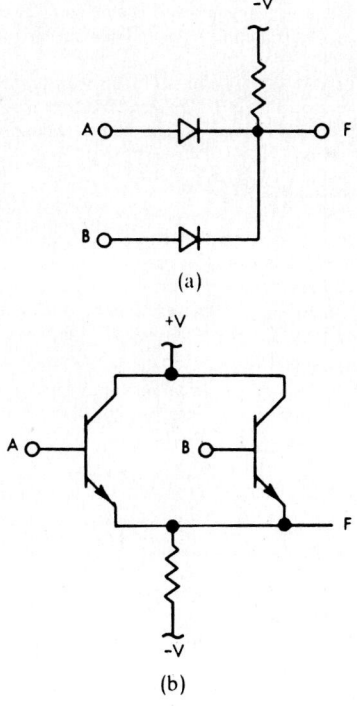

(a)

(b)

Fig. 19 OR circuits. (a) Diode or gate type, (b) transistor type.

Register. A hardware device used to store a certain amount of bits or characters and usually provided for a particular purpose. A register usually is constructed of semiconductor devices, such as transistors, and usually contains approximately one word or 1 byte of information.

Address Register. A register for the temporary storage of an address in a computer.

General Purpose Register. A register which may be utilized for several purposes, such as accumulation, address indexing, and shifting.

Index Register. A register which contains a quantity which may be used to modify addresses.

Instruction Register. A register which holds the identification of the instruction word to be executed next to the time sequence following the current operation. This register is often a counter which is incremented to the address of the next sequential storage location unless a transfer or other special instruction is specified by the program.

Program Register. A register in which the current instruction of the program is stored. Contrast with *control register.*

Shift Register. A register in which the characters may be shifted one or more positions to the right or left. In a right shift, the rightmost character(s) are lost. In a left shift, the leftmost character(s) are lost.

Storage Register. A register in the storage of a computer, in contrast to a register in one of the other units of the computer.

Resistor-Transistor Logic (RTL). A form of logic circuit using transistors and resistors. A circuit for performing the logical operation NOT OR (NOR) is shown in Fig. 20. Although the circuit values are chosen to bias the transistor beyond the cutoff when *A*, *B*, and *C* are all at ground potential (0 logic state), a change to the positive voltage associated with 1 operation of either of the three inputs will result in the transistor being brought into full conduction. Its collector will then be close to ground potential corresponding to a 0 (NOT 1) output. Thus, the circuit effects the operation NOT (*A* OR *B* OR *C*).

A variation of this arrangement, known as *resistor-capacitor-transistor logic* (RCTL), is shown in Fig. 21. Operation is the same as that of RTL except that higher speeds of operation are possible because the switching time of the transistor from one state to the other is reduced markedly by the use of the capacitors.

Thomas J. Harrison, IBM Corporation, Boca Raton, Fla.

Scaling Circuit. A circuit that produces an output pulse whenever a prescribed number of input pulses has been received. A binary scaler produces an output pulse whenever two input pulses have been received. By putting binary scalers in sequence, scales of 2, 4, 8, 16, and so forth, are obtained. A decade scalar produces an output pulse whenever 10 input pulses have been received. By putting decade scalars in sequence, scales of 10, 100, 1000, and so forth, are obtained.

Turing Machine. A mathematical abstraction of a device that operates to read from, write on, and move an indefinite tape, thereby providing a model for computerlike procedures. The behavior of a Turing machine is specified by listing an alphabet, i.e., a collection of symbols read and written, a

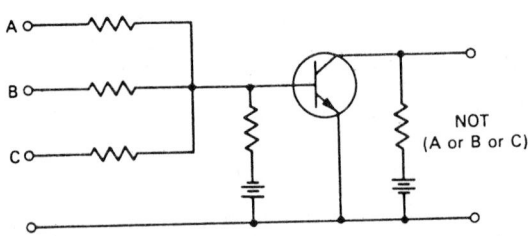

Fig. 20   Resistor-transistor logic (RTL) circuit effecting NOT OR (NOR) function.

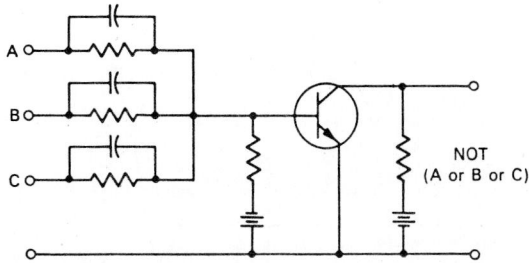

Fig. 21   Resistor-capacitor-transistor logic (RCTL) circuit effecting NOT OR (NOR) function.

set of internal states, and a mapping of an alphabet and internal states which determines what the symbol written and tape motion will be, and also what internal state will follow when the machine is in a given internal state and reads a given symbol. The Turing machine, proposed in 1936 by A. M. Turing, is the theoretical prototype of real general purpose computers and has had a profound impact in many areas of computer science.

Wired-Program Computer.   A computer in which the instructions that specify operations to be performed are determined by the placement and interconnection of wires. The wires usually are held by a removable control panel, allowing flexibility of operation. When the wiring is permanent, the machine is called a *fixed-program computer*. In this case, the program is often stored (wired) in a read-only storage (ROS) unit. Although no longer common in terms of general purpose computers, the wired-program computer is used in some special applications; for example, many programmed logic controllers (PLCs) utilized for elementary sequence control in industrial processes are wired-program computers.

DIGITAL COMPUTER STORAGE*

Core Storage.   Also termed *core memory* or simply *core,* core storage memory is a storage medium in which 1 bit of binary data is represented by the direction of magnetization in each unit of an array of ferrite toroidal rings. Although at one time the most widely used device for main storage, core storage has been largely displaced in practice by semiconductor storage, which is less costly, is smaller, utilizes less power, and is faster. Core storage continues to be used where retention of data in main storage is required when power is removed. Magnetic cores also find application in power and signal transformers and as transformerlike logical elements.

The ferrite material used in core storage has a rectangular hysteresis loop as shown in Fig. 22, and the magnetic flux is in one of two stable states, either A or A', when no current is flowing through the X and Y select lines, as shown in Fig. 23.

When a particular core is to be read out, current is passed through the cores via the X and Y select lines in such a direction as to set the particular core to zero. Thus, if $X0$ and $Y0$ are selected, the currents produce a combined magnetizing force, $-$ henrys (H), which sets the core at coordinates 00 to the flux state C'. If the core contained a 1, i.e., if the flux was at point A on the hysteresis loop, the change in flux generates a voltage in the sense winding which is interpreted as a 1 by discriminating logic in the sense amplifier circuit. If the core is in the 0 state or A', the change in flux is smaller, corresponding to transversal of the hysteresis loop from A' to C', and the voltage generated in the sense winding is ignored by the logic in the sense amplifier circuit. In this example, the cores at the other coordinates in the memory are, at most, only half-selected; i.e., the magnetizing force is -0.5 H which only sets the core to point B if the core is the 1 state. When the current is removed, the magnetic flux in the core returns to point A.

In order to write 1 in a core, current is passed through the X and Y select wires to produce a

*Terms are presented in alphabetical order for convenience of reference.

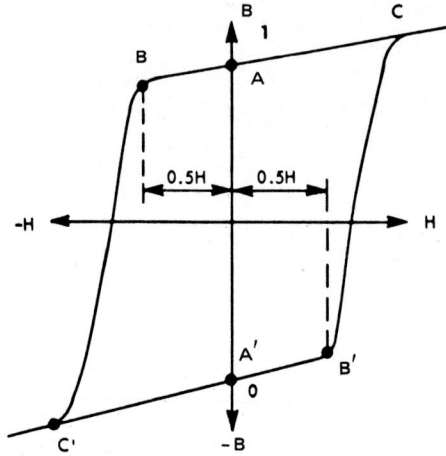

Fig. 22   Hysteresis loop of magnetic core.

combined magnetizing force, $+H$, on the core located at the intersection of the two select lines. This causes the core to be set to point C, and it subsequently stabilizes at point A when the current is removed. If a zero is to be set in the core, a current is driven through the inhibit winding which produces a magnetizing force of -0.5 H. This, in conjunction with the magnetizing force due to the current in the select lines, limits the net magnetizing force to 0.5 H and the flux moves only on the hysteresis loop to point B'. When the currents are removed, the magnetic flux of the core reverts to the 0 state or point A'.

Thomas J. Harrison, IBM Corporation, Boca Raton, Fla.

Data File.   A space for a set of related data organized in a specific manner; also called *data set*. As an example, an inventory file may consist of 12,000 records of 300 characters each, with each record representing the current inventory situation for a given part number and containing the part number, part name, automatic reorder point, economic reorder quantity, back-order quantity, current quantity in stores, current quantity in work in progress, unit cost, total value, date of last receipt, and so forth.

As files may be defined in a wide variety of ways on a large variety of media and devices, it has

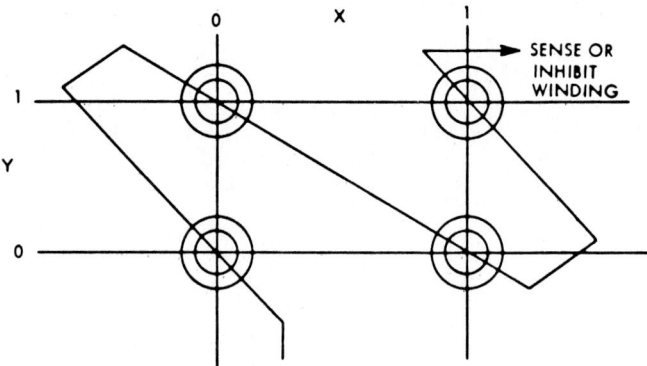

Fig. 23   Two-dimensional core array.

become important to structure them in some ordered way and to try to make them program-independent.

Two of the major problems in file management are to ensure that the file accessed actually is the correct one and to check if a particular program can be allowed access to the file. As the data processing field has expanded, the number of files available at a given installation has burgeoned—so that currently even rather small users may have hundreds of active files, some of which are confidential and should be accessible through special programs. Current methods of attaining a degree of program independence include moving a portion of the file definition from the program to the job control language where the definition may be changed without recompilation of the program. In an improved system the file definition material is included in the initial records of the file (called the *file label*) and is automatically recovered by the data management system when access to the file is attempted.

Density. Data density is the number of bits that can be stored per unit of recording medium, or the number of integrated circuits fabricated on a single chip or packaged in a single module. With magnetic tape, the density is expressed in terms of the number of bits per inch (or centimeter) of track. This also applies to disks, drums, and other mobile magnetic media. In semiconductor storage, *density* usually refers to the number of bits of storage on a single chip or in a single module. Logic circuit density is normally expressed in terms of circuits per chip.

In traditional longitudinal recording, two magnetic poles are pressed close to the recording surface and arranged so that the field is parallel to the motion of the surface. Density of information is limited to about 10,000 bits/in (4000 bits/cm). Several suppliers as of the mid-1980s are researching the concept of orienting the magnetic domains perpendicularly to the recording surface, with expectation that densities greater than 445,000 bits/in (175,000 bits/cm) may be achieved. With this concept, there may be as many as 2000 tracks/in (800 tracks/cm) of width of the recording medium.

Diskette. One form of disk storage utilizes a thin, flexible, plastic disk with one or two magnetic recording surfaces. This diskette, either 8 in ($\approx$ 20 cm) or 5¼ in ($\approx$13.3 cm) in diameter, is permanently housed in a lubricated protective paper envelope in which it rotates at speeds up to 360 rpm.

Originally conceived as a replacement for the keypunch, the first single-sided, single-density units had a capacity of 256,000 bytes, or the equivalent of 3200 80-column card images. Currently, a typical 8-in diskette drive has a formatted storage capacity of 1.3 million bytes (1.6 million bytes unformatted) in a double-density, double-sided configuration, an average access time on the order of 200 ms, and a data transfer rate of 62,500 bytes/s.

Diskettes are suited to data entry, word processing, and "personal" computer applications, for which data rates are low, response time requirements are modest, and required storage capacities are fairly low. Synonymous with *flexible disk* is *floppy disk*.

Disk Storage. In this form of storage, a magnetic coating on a rotating disk is used. Data are recorded on the surface of the rotating disk by magnetizing the surface in accordance with the pattern of binary data and are read by detecting the magnetic flux. A typical disk organization is shown in Fig. 24. Data are recorded on cylindrical tracks on the top and bottom surfaces of the disk. Track density is on the order of 500 tracks per radial inch (2.54 cm), and data density exceeds 8000 bits/in (3150 bits/cm) on each track. The disk spins at over 3000 rpm.

A magnetic head for each of the two data surfaces performs the reading and writing functions. The write portion of the read/write head consists of a coil of wire which magnetizes the magnetic surface of the disk when a current flows through the coil. The read portion of the head consists of a similar coil of wire into which a voltage is induced as it intercepts the magnetic flux on the disk.

The magnetic heads are attached to an access arm capable of positioning the head over any one of the cylinders on the disk. In addition, fixed leads may be positioned over tracks on one surface. In order to ensure reliable operation, the read/write head must be positioned very close to the surface of the disk without actually touching it. One method of accomplishing this is to shape the head surface such that the head "flies" over the surface at a height of about 10 microinches (0.24 μm). The disk may also be fitted with nonmovable (fixed) heads positioned over the recorded tracks. This provides minimum access time to recorded information on the track. The resulting characteristics are similar to those of drum storage.

Disk storage is used on a digital computer to store both data and programs. If the disk storage device has replaceable disk packs, they can be used as a medium of interchange of data with another computer. Although the access time on the disk is slower than either semiconductor or magnetic drum

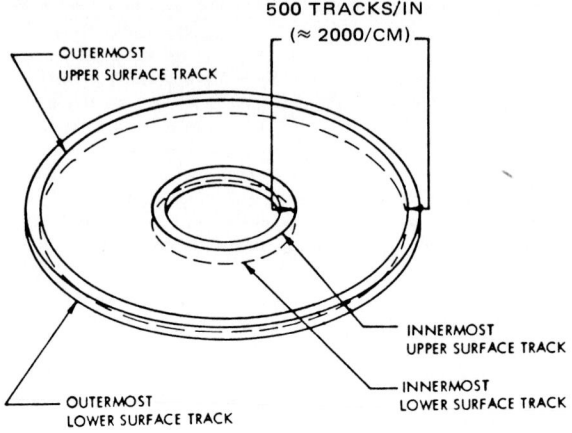

500 TRACKS/IN
(≈ 2000/CM)

OUTERMOST
UPPER SURFACE TRACK

INNERMOST
UPPER SURFACE TRACK

INNERMOST
LOWER SURFACE TRACK

OUTERMOST
LOWER SURFACE TRACK

Fig. 24   Disk cylinder concept.

storage (unless fixed heads are used), the cost per bit of information is considerably less. The access arm or carriage is moved forward or backward over the surface of the disk under control of the program in the computer. The data are written on the track in a serial-by-bit format; i.e., the data are read from the disk a bit at a time and formed into a character or word for transmission to the computer by associated logic circuits.

Thomas J. Harrison, IBM Corporation, Boca Raton, Fla.

Drum Storage.   A magnetic drum is a metal cylinder which revolves about its axis and on whose surface a magnetic material is deposited, which permits the recording of information. The information is recorded on the drum by passing current through a coil of wire in the write head, which magnetizes a track on the drum surface in accordance with the pattern of binary data. In order to read the information stored on the drum, a similar coil of wire in the read head is mounted near the drum surface where it intercepts the magnetic flux and transmits the induced voltage to a sense amplifier.

With reference to Fig. 25, the drum is divided into tracks around the drum circumference and a read/write head is associated with each track. Data may be stored on the drum in a serial-by-bit serial-by-digit format or in a parallel-by-bit serial-by-digit format, depending on the design selected by the manufacturer. A timing track is recorded on the drum to provide the sampling pulses for reading or writing the data and for address determination. The timing track signals increment counters which are used to determine the digit and word locations passing under the read/write head at any particular time. Other means for providing timing (e.g., optical) can also be used.

When data are to be written on the drum in a serial-by-bit serial-by-digit format, the drum storage address selects the appropriate write head. When the addressed location passes under the head, the information in the buffer register is gated to the write amplifier which drives current through the write head according to the bit pattern to be written. On a read operation, the selected read head gates the signal to the read amplifier as the addressed data pass under the head.

The average time required to read or write a selected address on the drum is one-half the revolution time of the drum. This may be shortened by placing multiple read/write heads on each track around the periphery of the drum and by providing logic in the drum storage unit to determine which head to select. The number of heads which can be mounted on the drum is a function of the allowable track-to-track density of the magnetic recording material and the physical dimensions of the read/write head. By arranging the heads along a helical path, increased track density can be achieved. Another factor which limits the number of read/write heads on the drum storage unit is the cost of the heads and their associated mounting hardware.

Drums make possible large-capacity data storage with the fastest access of any mechanical (mov-

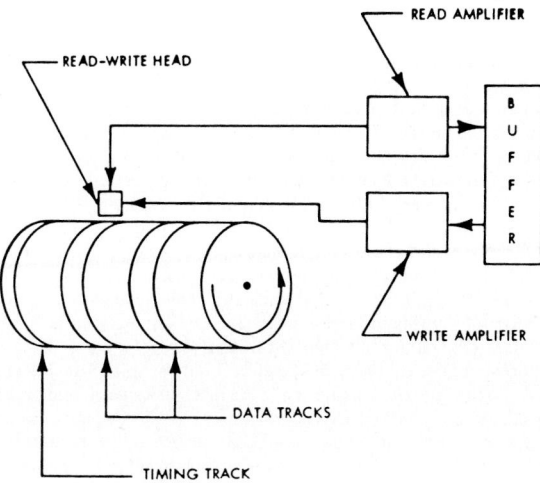

Fig. 25 Magnetic drum.

ing) storage configuration—with the exception of fixed-head disk files, which provide comparable access times. The cost per digit is relatively high because of the number of read/write heads required.

Thomas J. Harrison, IBM Corporation, Boca Raton, Fla.

EPROM. Acroynym for *erasable programmable read-only memory.*

GaAs. Symbol for gallium arsenide; used in microcircuits and chips.

Floppy Disk. See *diskette* earlier in this subsection.

Laser Bulk Storage. When a fine substrate is subjected to a very fine laser beam, a small hole is burned in the material. The presence or absence of a hole at a given position encodes the information. Theoretically, data density is ultimately limited to a hole that is limited by one wavelength on a side. Researchers are attempting to combine laser technology with that of the property of certain materials to resonate in response to external radiation. Rather than return to a lower energy state in the absence of radiation, some molecules reorient and create a persistent "hole" in the absorption spectrum of the material, with each hole representing one bit of information. See also *spectral memory* in this article.

Magnetic Tape Storage. Information is recorded on tape by magnetizing narrow (lengthwise) strips (termed *tracks*) in a pattern corresponding to a sequence of binary states (1's and 0's). Binary data, often corresponding to a single byte, are stored in column form across the width of the tape. A read/write head is usually associated with each row of magnetized material, so that one column can be read or written at a time as the tape traverses the head. The density of recorded data is most commonly 800, 1600, or 6250 bits per linear inch (approximately 315, 630, or 2461 bits per linear centimeter). A nine-track 2400-ft ($\approx$732-m) tape can carry from 20 to 156 million characters, dependent on the density. Magnetic tape storage is most often used to provide off-line archival storage of data. A drawback to tape storage is the fact that it must be read serially. Retrieval of randomly distributed information can be time-consuming, compared with disk storage.

Tape material usually is polyester plastic, one side of which is coated with a suspension of ferrite or magnetic oxide particles. Tapes are commonly ½ in (12.7 mm) in width, 1½ mils (0.0015 in, 0.4 mm) in thickness, and 2400 ft ($\approx$732 m) in length. Tape widths of ¼, ¾, and 1 in (0.64, 1.91, and 2.54 cm) and lengths of 200, 300, 600, 1200, and 3600 ft ($\approx$61, 91, 183, 366, and 1097 m) also are available for use at computer installations.

Recorded material can be removed from a tape by passing it through a strong, constant magnetic field *(dc erase)*, or through a high-frequency alternating magnetic field *(ac erase)*. Information is stored on the tape by magnetizing the magnetic film in one direction or the other in a pattern determined by the binary data. The information is recorded on the tape in a parallel-by-bit serial-by-digit format; i.e., the bits which comprise a digit are recorded across the width of the tape, and the digits are recorded sequentially along the length of the tape.

A magnetic tape is stored on reels and is transferred from one reel to another past two-gap read/ write heads in the course of reading or writing data. Vacuum columns or other tension mechanisms are provided so that the tape can be moved rapidly a few inches at a time without waiting for movement of the reels. The two-gap head allows automatic error checking of the tape data while it is being written. The first gap is used for writing, and the second gap for reading.

The reading or writing of data on a tape is controlled by clock circuits within the logic of the tape unit. In writing, the frequency at which the clock is stepped is a function of the tape speed and the recording density. In reading, the oscillator which drives the read clock is also gated by the first bit read. To compensate for the time it takes for the tape motion to get up to speed, read delay and write delay counters are designed into the logic of the tape unit.

In a tape read operation, the character is stored in a buffer register as it is read from the tape and is then transferred to the digital computer while the next character is being read. In a tape write operation, the next character is fetched from the computer while the previous character is being written. Since data are not written on the tape until the tape gets up to speed, the information to be written is blocked into records to minimize the number of gaps caused by the write delay.

Thomas J. Harrison, IBM Corporation, Boca Raton, Fla.

Magnetic Bubble Storage.   A storage technology whereby information is stored as a small magnetized region (bubble) having a magnetic polarity opposite the polarity of the surrounding region. The device consists of a garnet substrate on which is deposited a magnetic film only a few micrometers thick. The movement of bubbles is controlled by a magnetic field created by passing current through electrical conductors deposited over the magnetic film. Developed for memory circuits used in telephony, as of the early 1980s some computer firms are offering equipment in which this technology is incorporated. See Fig. 26.

Memory.   Although the word *memory* is widely used, international standards list the word as a deprecated term and prefer the word *storage*. Memory or storage is a means of storing information. A complete memory system incorporates means of placing information into storage and of retrieving it.

Optical Disk Storage.   As of the mid-1980s, optical digital disks for computer mass storage are under development by several suppliers. Considered one of the most promising developments in computer mass storage technology, the optical video disk was spawned by the home entertainment industry. The optical disk has high potential for on-line storage, random access graphics to complement on-line information systems, hybrid network architectures, automation systems, and archival storage.

PROM.   Acronym for *programmable read-only memory*.

RAM.   Acronym for *random-access memory*.

ROM.   Acronym for *read-only memory*.

Semiconductor Storage.   The basic storage element is a transistor circuit, typically containing from one to eight transistors. The circuit is the functional equivalent of a flip-flop. Semiconductor technology has made it possible to fabricate tens of thousands of these circuits on a single silicon chip. A semiconductor storage device has the capability of achieving faster access times than are obtainable with magnetic core units. Inasmuch as information is not erased when a bit position is read out, the information need not be written back into the location after a read operation. Thus, the storage cycle time is essentially the same as the read or write access time. In some designs, storage actually is accomplished by depositing an electronic charge in a small capacitance in the circuit. Since normal RC decay would eventually dissipate this charge, it is necessary to periodically "refresh" the storage by reading and then rewriting each storage location. This is usually done automatically by associated circuits and is not apparent to the user.

Semiconductor storage also makes possible the design of storage devices that use the same voltage levels as the logic circuits. Thus, the cost of smaller units can be minimized. It is important to note that, inasmuch as the basic storage element is a transistor circuit, the information in storage is lost

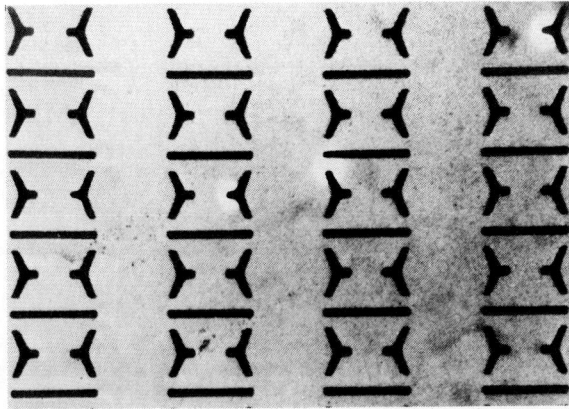

Fig. 26 Magnetic bubbles are tiny cylindrical magnetized areas less than ⅟₁₆ the diameter of a human hair. The areas can be moved about electronically so that they can be used for computing and storing information. In this illustration, magnetic bubbles, highly magnified and shown as light-colored circles, are seen moving through a circuit pattern formed on a thin epitaxial film of uniaxial garnet. The Y and bar circuit elements are part of an experimental shift register.One bubble, somewhat elongated, can be seen in transition from one pole to the next. The bubbles are 0.0003-in (0.008 mm) in diameter. *(Bell Laboratories.)*

when the power is turned off. Some designs provide battery power to retain the storage contents for short periods of time. Magnetic core storage, of course, retains its information when the power is properly sequenced off. With semiconductor storage, the information must be reloaded, usually from a diskette or disk storage unit, after the power is turned on. In a read operation, the selected bit is gated to the sense amplifier, and amplitude-discrimination logic in the sense amplifier interprets the output as a 0 or 1. In a write operation, the selected storage element is set to a 0 or a 1, depending on the data on the input lines.

Spectral Memory. Traditionally, the presence or absence of a mark on a sheet of paper can encode one bit of information. By allowing the marks to vary in color, the use of colors greatly enhances data storage density. Several industrial and university laboratories are extending this concept to the molecular level in which laser radiation causes distinct molecular colors. It is estimated that over 1000 colors could be obtained in this manner and that possibly a hundred million coded entities could be packed in one square centimeter of surface area.

Storage. Any medium capable of storing information. As generally defined, however, a storage unit is a device on or in which data can be stored, read, and erased. The major classifications of storage devices associated with computer systems are (1) immediate access, (2) random access, and (3) sequential access. As a general rule, the cost per bit of information is greater for immediate-access storage devices, but the access time is considerably faster than for the other two types.

Immediate-Access Storage. Devices in which information can be read in a microsecond or less. Usually an array of storage elements can be directly addressed, and thus all information in the array requires the same amount of time to be read.

Random-Access Storage. Devices in which the time required to obtain information is independent of the location of the information most recently obtained. This strict definition must be qualified by the observation that what is meant is *relatively* random. Thus, magnetic drums are relatively nonrandom access when compared with monolithic storage but are relatively random access when compared with magnetic tapes for file storage. Disk storage and drum storage units usually are referred to as random-access storage devices. The time required to read or write information on these units generally is in the 10- to 200-ms range but is dependent on where the information is recorded with respect to the read/write head at the time the data are addressed.

Sequential-Access Storage. Devices in which the items of information stored become available only in a one-after-the-other sequence, whether or not all the information or only some of it is desired. Storage on magnetic tape is an example.

Some other computer storage configurations are defined as follows:

Auxiliary Storage. A storage device in addition to the main storage of a computer, e.g., magnetic tape, disk, diskette, or magnetic drum. Auxiliary storage usually holds much larger amounts of information than the main storage, and the information is less rapidly accessed.

Buffer Storage. (1) A synchronizing element between two different forms of storage, usually between internal and external storage. (2) An input device in which information is assembled from external or secondary storage and stored ready for transfer to internal storage. (3) An output device into which information is copied from internal storage and held for transfer to secondary or external storage. Computation continues while transfers between buffer storage and secondary or internal storage, or vice versa, take place. (4) Any device which stores information temporarily during data transfers.

Circulating Storage. A device or unit which stores information in a train or pattern of pulses, where the pattern of pulses issuing at the final end are sensed, amplified, reshaped, and reinserted into the device at the beginning end.

External Storage. (1) The storage of data on a device which is not an integral part of a computer but in a form prescribed for use by the computer. (2) A facility or device, not an integral part of a computer, on which data usable by a computer is stored, such as off-line magnetic tape units and punch card devices.

Internal Storage. (1) The storage of data on a device which is an integral part of a computer. (2) The storage facilities forming an integral physical part of a computer and directly controlled by the computer. In such facilities all data are automatically accessible to the computer, e.g., magnetic core and magnetic tape on-line.

Magnetic Bubble Storage. See separate description of this storage mode in this section.

Main Storage. Usually the fastest storage device of a computer and the one from which instructions are executed.

Program Storage. A portion of the internal storage reserved for the storage of programs, routines, and subroutines. In many systems protection devices are used to prevent inadvertent alteration of the contents of the program storage.

Serial Storage. A storage technique in which time is one of the factors used to locate any given bit, character, word, or group of words appearing one after the other in a time sequence, and in which access time includes a variable latency or waiting time of from zero to many word times. Storage is said to be serial by word when the individual bits comprising a word appear serially in time; storage is serial by character when the characters representing coded decimal or other nonbinary numbers appear serially in time; e.g., magnetic drums are usually serial by word but may be serial by bit, parallel by bit, or serial by character and parallel by bit.

Working Storage. A portion of the internal storage reserved for the data on which operations are being performed. Synonymous with *working space* and *temporary storage* and contrasted with *program storage.*

Thomas J. Harrison, IBM Corporation, Boca Raton, Fla.

Storage Cycle Time. The interval of time needed for a complete storage operation; also known as *memory cycle time.* It is the minimum allowable time interval between two successive accesses to storage. Cycle time in the case of magnetic core storage is the interval needed to read the data from the core and to regenerate or write it back into the core. In the case of nondestructive readout storage, such as monolithic storage, it is the time needed either to read from the selected location or to write into the selected location and thus is equal to the access time.

Storage Protect. Several methods are used to effect storage protection in digital computers. The objective is to protect certain areas of code from alteration by other areas of code. Storage may be protected by areas, each area being given a different program-settable code or key. A master key is used to permit the control program to refer to all areas. In another system, the storage is protected on an individual word basis, providing finer resolution but increasing the time required to protect a given area size. In both cases, instructions are provided to set the protect status by programming. Some systems provide for a disable of the feature from the console. This permits a preset protection pattern to be established, but with the program capability to disable selected protection status. Protection is required so that stored information will not be altered accidentally by store-type instructions. In single-word schemes, any storage modification is prevented. In the multikey area systems storage approach, modification is limited to operation in areas with the same key.

DIGITAL CODES, UNITS, AND LANGUAGES*

ADA. A relatively recent computer language developed by the U.S. Department of Defense for use in "embedded computer systems." The language, designed by CII-Honeywell Bull of France with the assistance of numerous computer language experts, was selected from a set of four languages contracted for by the Department of Defense. It is based on a set of functional requirements developed and refined over the period 1974–1979, known as STRAWMAN (1975), WOODENMAN (1975), TINMAN (1976), IRONMAN (1978), and STEELMAN (1979), which defined approximately 100 requirements to be satisfied by the language. Although not a superset of Pascal, ADA is a descendant of Pascal and extends its strongly typed data abstractions. In addition, ADA provides extensive features for tasking and exception handling, making it appropriate for use in real-time applications.

As of the early 1980s, operational compilers for the language are being written. Concurrently, national and international standardization efforts have been initiated. It is expected that ADA will find acceptance in a broad range of applications during the 1980s.

The language is named in honor of Ada Augusta, Countess of Lovelace, the daughter of Lord Byron and a colleague of Charles Babbage. Ada devised sequences of "orders" for Babbage's "analytical engine" in the mid-nineteenth century and thus became the world's first computer programmer. A summary by J. G. P. Barnes of the history and the language itself can be found in "An Overview of Ada," *Software-Practice and Experience,* vol. 10 1980, pp. 851–887.

ALGOL. An acronym for *alg*ebraically oriented *l*anguage. This is a compiler language designed for ease of computation of algebraic expressions. It may be noted that ALGOL is also an acronym for *algo*rithmic *l*anguage. A version of ALGOL used in the programming of scientific problems has more frequently served as a model for the development of new artificial computer language systems and for the formal description of algorithms. Some features of ALGOL are indicated in the following example. Note that *comment* statements have no effect on the execution of the program.

```
begin
comment      This is a sample ALGOL program which illustrates
             some features of the language. Underlined words
             are 'keywords' in the language;
integer      A,B,C,EROR;
real         D;
             EROR  : = - 1;
             A     : =    1.0;
             B     : =    2.0;
             C     : =    3.0;
comment      The above statements declared the variable type
             for A,B,C,EROR and D and assigned values to A,B,C
             and EROR. Next D is calculated and some logic done
             testing its value;
        D    : = A/B × C;
if      D    = 4 then go to over;
if      D    = 3 then go to right
```

*Terms are presented in alphabetical order for convenience of reference.

<pre>
                      else go to Exit 1;
         Over: EROR: = 1.0;
         ‾‾‾‾   go to Exit 1;
         Right: EROR: = 0;
         ‾‾‾‾‾
         Exit1:   end
         ‾‾‾‾‾
</pre>

APT.   Acronym for *a*utomatically *p*rogrammed *t*ool. The programming of part production is required in connection with numerically controlled machine tools. Part programming is comprised of preparing written numerically coded instructions which, when followed by the machine control system, produce the desired part. Programming must be precise and is frequently intricate. APT accomplishes the majority of the detailed mathematical operations required for part programming and in this respect parallels for the part programmer the function of COBOL for business applications programming. Where APT is used, the programmer specifically describes the part in terms of geometric components, i.e., dimensions of straight lines, circles, ellipses, and other geometric conditions. The components are described in terms of English language statements. The programmer also instructs APT to position the cutting tool with reference to key intersections. The output from APT, sometimes punched into a paper tape, is used directly by the numerically controlled machine. APT is described by American National Standard X3.37-1980.

BASIC.   An acronym for *b*eginner's *a*ll-purpose *s*ymbolic *i*nstruction *c*ode. This language was developed at Dartmouth College by Kemeny and Kurz for time-shared use by students and other nonprofessional programmers. The language has been adopted by most computer firms and has been offered via time-sharing services. BASIC is extensively used for small desktop computer systems and by computer hobbyists. The language is characterized by a very simple statement form. It normally requires that the first word in the statement be one of the small number of keywords and that a restricted naming convention for variables be used. An example of a program prepared in BASIC is

```
 05  LET E = −1
 10  READ A, B, C
 15  DATA 1, 2, 3
 20  LET D = A/B*C
 30  IF D = 4 GO TO 60
 40  IF D = 3/2 GO TO 80
 45  PRINT A, B, C, D, E
 50  STOP
 60  LET E = +1
 70  GO TO 45
 80  LET E = 0
 90  GO TO 45
100  END
```

The nucleus of the language is described by American National Standard X3.60-1978 (minimal BASIC), and a standard for significant extensions of this nucleus has been developed.

Baud.   A traditional unit of telegraph signaling speed derived from the duration of the shortest signaling pulse. A telegraphic speed of one baud is one pulse per second. The term *unit pulse* sometimes has the same meaning. A related term, *dot cycle,* refers to an on-off or mark-space cycle in which both mark and space intervals have the same length as the unit pulse.

Binary Notation.   See subsection Digital Computer Fundamentals earlier in this article.

Bit.   A contraction of *b*inary dig*it*. A single character in a binary numeral, i.e., a 1 or 0. A single pulse in a group of pulses also may be referred to as a bit. The bit is a unit of information capacity of a storage device. The capacity in bits is the logarithm to the base 2 of the number of possible states of the device.

Parity Bit.   A check bit that indicates whether the total number of binary 1 digits in a character or word (excluding the parity bit) is odd or even. If a 1 parity bit indicates an odd number of 1

digits, then a 0 bit indicates an even number of 1 digits. If the total number of 1 bits, including the parity bit, is always even, the system is called an even parity system. In an odd parity system, the total number of 1 bits, including the parity bit, is always odd.

Zone Bit. (1) One of the two leftmost bits in a system in which 6 bits are used for each character; related to overpunch. (2) Any bit in a group of bit positions used to indicate a specific class of items, e.g., numbers, letters, special signs, and commands.

Boolean Algebra. Originated by George Boole (1815–1864), Boolean algebra is a mathematical method of manipulating logical relations in symbolic form. Boolean variables are restricted to two possible values or states. Possible pairs of values for the Boolean algebra variable are YES and NO, ON and OFF, TRUE and FALSE, and so forth. It is common practice to use the symbols 1 and 0 as the Boolean variables. Since a digital computer typically uses signals having only two possible values or states, Boolean algebra enables the computer designer to combine mathematically these variables and manipulate them in order to obtain the minimum design which realizes a desired logical function.

A table of definitions and symbols for some of the logical operations defined in Boolean algebra is shown in Fig. 27.

Byte. A group of binary digits, usually shorter than a word and usually operated on as a unit. Through common usage, the word most often describes an 8-bit unit. This is a convenient information size to represent an alphanumeric character. Particularly in connection with communications, the trend has been toward the use of *octet* for an 8-bit byte, *sextet* for a 6-bit byte, and so on. Computer designs commonly provide for instructions that operate on bytes, as well as word-oriented instructions. The capacity of storage units is often specified in terms of bytes. See also subsection Digital Computer Fundamentals earlier in this article.

Character. One symbol of a set of elementary symbols, such as those corresponding to the keys of a typewriter, that is used for organization, representation, or control of data. Symbols may include the decimal digits 0 through 9, the letters A through Z, and any other symbol a computer can read, store, or write. Thus, such symbols as @, #, $, and / are commonly used to expand character availability.

A *blank character* signifies an empty space on an output medium; or a lack of data on an input medium, such as an unpunched column on a punched card.

A *check character* signifies a checking operation. Such a character contains only the data needed to verify that a group of preceding characters is correct.

A *control character* controls an action rather than conveying information. A control character may initiate, modify, or stop a control operation. Actions may include the line spacing of a printer, the output hopper selection in a card punch, etc.

An *escape character* indicates that the succeeding character(s) is in a code that differs from the prior code in use.

A *special character* is not alphabetic, numeric, or blank. Thus, @, #, etc., are special characters.

COBOL. An acronym for *common business oriented language*. COBOL was developed in the late 1950s to assist in the processing of large quantities of character-oriented data, such as name and address files. The system is widely used in government and business, particularly in connection with payroll and inventory programs. The chief characteristics of COBOL are a good clear method of file and record description and powerful data manipulation facilities. An example of a program written in COBOL is given here. COBOL is described in American National Standard X3.23-1974, with a revision scheduled for the early 1980s.

LOGICAL OPERATION	SYMBOL	DEFINITION
AND	•	$A \cdot A = A, A \cdot 0 = 0, A \cdot 1 = 1, A \cdot \bar{A} = 0$
OR	$+$	$A + A = A, A + 0 = A, A + 1 = 1, A + \bar{A} = 1$
NOT	$-$	$A\bar{A} = 0, A + \bar{A} = 1$
EXCLUSIVE OR	$\oplus$	$A \oplus B = \bar{A}B + A\bar{B} = \overline{A \odot B}$
COINCIDENCE	$\odot$	$A \odot B = \bar{A}\bar{B} + AB = \overline{A \oplus B}$
NAND (OR SHEFFER STROKE)	/	$A/B = \bar{A} + \bar{B} = \overline{AB}$
NOR (OR PEIRCE)	$\downarrow$	$A \downarrow B = \bar{A}\bar{B} = \overline{A + B}$

Fig. 27 Boolean algebra symbols.

```
IDENTIFICATION DIVISION
  PROGRAM-ID. SMPLPRGM.
    REMARKS. THIS IS A SAMPLE PRINT PROGRAM.
    REMARKS. IT ILLUSTRATES ONLY SOME FEATURES OF COBOL.
    REMARKS. THERE ARE 4 DIVISIONS OF A COBOL PROGRAM
             IDENTIFICATION, ENVIRONMENT, DATA AND
             PROCEDURE.
    REMARKS. IN THE ENVIRONMENT DIVISION DETAILS ARE
             GIVEN OF THE CPU'S AND I/O DEVICES NEEDED.
    REMARKS. IN THE DATA DIVISION DETAILS ARE GIVEN OF
             THE FILES AND OF THE RECORD LAYOUTS FOR
             BOTH INPUT AND OUTPUT FILES.
    REMARKS. THE KEYWORD PICTURE IS USED TO DESCRIBE THE
             FORMAT OF A FIELD WITHIN A RECORD.
    REMARKS. FOR EXAMPLE IN THE # FILE "READ-IN", THE
             RECORD "SOURCE"HAS 3 FIELDS—"AUTHOR",
             "INITIAL", "TITLE", OF 20, 3 AND 57 CHARACTERS
             RESPECTIVELY.
    REMARKS. THE PROCEDURE DIVISION IS THE ACTUAL PROGRAM.

ENVIRONMENT DIVISION.
  CONFIGURATION SECTION
    SOURCE-COMPUTER.   IBM-360 F50.
    OBJECT-COMPUTER.   IBM-360 F50.
  INPUT-OUTPUT SECTION.
    FILE-CONTROL
    SELECT   READ-IN    ASSIGN TO UNIT-RECORD 2540 UNIT.
    SELECT   PRINTOUT   ASSIGN TO UNIT-RECORD 1403 UNIT.

DATA DIVISION.
  FILE SECTION.

FD READ-IN
    RECORDING MODE IS F, DATA RECORD IS SOURCE.
    01   SOURCE
        02   AUTHOR   PICTURE X(20).
        02   INITIAL  PICTURE X(3).
        02   TITLE    PICTURE X(57).
FD PRINTOUT
    RECORDING MODE IS F, DATA RECORD IS OBJECT.
    01   OBJECT
        02   FILLER   PICTURE A(20).
        02   AUTHOR   PICTURE X(20).
        02   FILLER   PICTURE A(3).
        02   INITIAL  PICTURE X(3).
        02   FILLER   PICTURE A(9).
        02   TITLE    PICTURE X(57).
        02   FILLER   PICTURE A(20).

PROCEDURE DIVISION.
  OPEN INPUT READ-IN. OUTPUT PRINTOUT.

  NOTE   THIS PROGRAM READS CARDS AND LISTS THEM IN A
  NEAT FORMAT.
  NOTE   AFTER THE LAST CARD IT SHUTS DOWN THE READER AND
  PRINTER THEN STOPS.

  LOOP. READ READ-IN RECORD AT END GO TO COMPLETE.
```

MOVE CORRESPONDING SOURCE TO OBJECT.
WRITE PRINTOUT RECORD, GO TO LOOP.

COMPLETE. CLOSE READ-IN, PRINTOUT, STOP RUN.

Code.   A system of symbols for representing data or instructions in a computer or data processing machine. A machine language program sometimes is referred to as a *code*.

Alphanumeric Code. A set of symbols consisting of the alphabet characters A through Z and the digits 0 through 9. Sometimes the definition is extended to include special characters. A programming system commonly restricts user-defined symbols to only those using alphanumeric characters and for the system to take special action on the occurrence of a nonalphanumeric character, such as $, %, or &. For example, a job currently in progress may be stopped should a given special character be encountered.

Binary Code. (1) A coding system in which the encoding of any data is done through the use of bits; i.e., 0 or 1. (2) A code for the 10 decimal digits 0, 1, . . . , 9 in which each is represented by its binary, radix 2, equivalent, i.e., straight binary.

Biquinary Code. A two-part code in which each decimal digit is represented by the sum of the two parts, one of which has the value of decimal 0 or 5, and the other the values 0 through 4. The abacus and soroban both use biquinary codes. An example follows:

Decimal	Biquinary	Interpretation
0	0 000	0 + 0
1	0 001	0 + 1
2	0 010	0 + 2
3	0 011	0 + 3
4	0 100	0 + 4
5	1 000	5 + 0
6	1 001	5 + 1
7	1 010	5 + 2
8	1 011	5 + 3
9	1 100	5 + 4

Column-Binary Code. A code used with punch cards in which successive bits are represented by the presence or absence of punches in contiguous positions in successive columns as opposed to rows. Column binary code is widely used in connection with 36-bit word computers where each group of three columns is used to represent a single word.

Computer Code or Machine Language Code. A system of combinations of binary digits used by a given computer.

Excess-Three Code. A binary-coded decimal code in which each digit is represented by the binary equivalent of that number plus 3; for example:

Decimal digit	Excess-3 code	Binary value
0	0011	3
1	0100	4
2	0101	5
3	0110	6
4	0111	7
5	1000	8
6	1001	9
7	1010	10
8	1011	11
9	1100	12

Gray Code. A binary code in which sequential numbers are represented by expressions which are the same except in one place and in that place differ by one unit:

Decimal	Binary	Gray
0	000	000
1	001	001
2	010	011
3	011	010
4	100	110
5	101	111

Thus in going from one decimal digit to the next sequential digit, only one binary digit changes its value. Synonymous with cyclic code.

Instruction Code. The list of symbols, names, and definitions of the instructions which are intelligible to a given computer or computing system.

Mnemonic Operation Code. An operation code in which the names of operations are abbreviated and expressed mnemonically to facilitate remembering the operations they represent. A mnemonic code normally needs to be converted to an actual operation code by an assembler before execution by the computer. Examples of mnemonic codes are ADD for addition, CLR for clear storage, and SQR for square root.

Numeric Code. A system of numerical abbreviations used in the preparation of information for input into a machine; i.e., all information is reduced to numerical quantities. Contrasted with *alphabetic code*.

Symbolic Code or Pseudocode. A code which expresses programs in source language; i.e., by referring to storage locations and machine operations by symbolic names and addresses which are independent of their hardware-determined names and addresses.

Two-out-of-Five Code. A system of encoding the decimal digits 0, 1, . . ., 9 where each digit is represented by binary digits of which two are 0's and three are 1's, or vice versa.

Thomas J. Harrison, IBM Corporation, Boca Raton, Fla.

Data Base. An accumulation of information, often well organized and assembled for easy retrieval, as by a data processing system. Data bases often are specialized in terms of information categories, such as business, process conditions, medical, and science, among others. The term *data bank* is no longer preferred.

Digit. A symbol used to convey a specific quantity either by itself or with other numbers of its set; e.g., 2, 3, 4, and 5 are digits. The base or radix must be specified, and each digit's value assigned.

Binary Digit. A number on the binary scale of notation. This digit may be zero (0) or (1). It may be equivalent to an On or an Off condition or a Yes or No condition. Often abbreviated *bit*.

Check Digit. One or more redundant digits carried along with a machine word and used in relation to the other digits in the word as a self-checking or error detecting code to detect malfunctions of equipment in data transfer operations.

Equivalent Binary Digits. The number of binary positions needed to enumerate the elements of a specific set. In the case of a set with five elements, it will be found that 3 equivalent binary digits are needed to enumerate the five members of the set 1, 10, 11, 100, and 101. Where a word consists of three decimal digits and a plus or minus sign, 1999 different combinations are possible. This set would require 11 equivalent binary digits in order to enumerate all its elements.

Octal Digit. The symbol 0, 1, 2, 3, 4, 5, 6, or 7 used as a digit in the system of notation which uses 8 as the base or radix.

Sign Digit. A digit incorporating 1 to 4 binary bits, which is associated with a data item for the purpose of denoting an algebraic sign. In most binary, word-organized computers, a 1-bit sign is used: $0 = +$ (plus); and $1 = -$ (minus). Although not strictly a digit by the foregoing definition, it occupies the first digit position and it is common to consider it a digit.

Thomas J. Harrison, IBM Corporation, Boca Raton, Fla.

Fixed-Point Arithmetic. A method of storing numeric data in a computer such that the data are all stored in integer form (or all in fractional form) and the user postulates a radix point between a certain pair of digits. Consider a computer whose basic arithmetic is in decimal and in which each computer word consists of seven decimal digits in integer form. If it is desired to add 2.796512 to 4.873214, the data are stored in the computer as 2796512 and 4873214, the sum of which is 7669726. It is recalled that a decimal point between digits 1 and 2 has been postulated. The result, therefore, represents 7.669726. Input and output conversion routines often are provided for convenience. These routines can add or delete the radix point in the external representation and align the data as required internally.

Fixed-point operations are fast and thus preferred over floating-point operations. It is important, of course, that the magnitude of the numbers be much better known than for floating-point numbers, since the absolute magnitude is limited by word size and the availability of double-length operations. For many applications, fixed-point calculations are practical and increase speed.

FORTRAN. An acronym standing for *for*mula *tran*slation. It is a programming language designed for problems which can be expressed in algebraic notation, allowing for exponentiation, subscripting, and other mathematical functions.

FORTRAN was introduced by IBM in 1957 after development by a working group headed by John W. Backus. It was the first computer language to be used widely for solving numerical problems and was the first to become an American National Standard. Numerous enhancements have been added to the language. The current version is described by American National Standard ANSI X3.9-1978. Real-time extensions are described in ANSI/ISA Sol. 1-1976 and ANSI/ISA S61.2-1978.

FORTRAN Source Program. When a problem is coded in FORTRAN, the programmer produces a FORTRAN source program consisting of a set of statements. A given statement may express an algebraic equation to be solved or a logic decision to be made. There are five types of statements in FORTRAN: (1) *arithmetic statements* which express a series of arithmetic operations to be performed in algebraic notation, (2) *input-output statements* which allow the programmer to control the transfer of data between the central processing unit (CPU) and the outside, (3) *control statements* which allow the programmer to govern the sequence in which the statements are to be executed (control may be exercised on a logical decision basis), (4) *specification statements* which allow the programmer to define the nature of the data being manipulated, and (5) *subprogram statements* which allow the user to define the subroutines which may be called at any point within the program.

FORTRAN Compiler. A FORTRAN compiler is a program for a specific computer which accepts as input a FORTRAN source program and translates it into a machine language object program. The latter then may be loaded into the computer. Some compilers yield a very rapid translation, but the object code produced is inefficient in terms of core requirements and execution speed. Other compilers require more time for translation, but the object code is much more efficient in both core and execution time required. In a process control situation where the user is coding control algorithms in FORTRAN, normally these programs are compiled only once but executed many times. Thus, one of the primary considerations when evaluating a digital computer and its associated FORTRAN compiler for a control application is the efficiency of the machine language object code generated by the compiler.

Thomas J. Harrison, IBM Corporation, Boca Raton, Fla.

Hartley. In information theory, a unit of logarithmic measurement of information equal to the decision content of a set of 10 mutually exclusive events, expressed by the logarithm with the base 10. For example, the decision content of a character set of eight characters equals $\log_{10} 8$, or 0.903 hartley. Synonymous with information content decimal unit.

Hollerith. Pertaining to a widely used system of encoding alphanumeric information onto cards (described by American National Standard ANSI X3.26-1970). The term *Hollerith cards* is synonymous with *punch cards*. Such cards were first used in 1890 for the U.S. census and were named after Herman Hollerith, their originator.

JOVIAL.   An acronym for *J*ules' (Schwartz) *version of the international algorithmic language*. JOVIAL is a procedure-oriented language originally derived from ALGOL. The language differs mainly from ALGOL in terms of the types of data that can be described and manipulated. Originally created for command and control applications, as used by the military and government agencies, JOVIAL has been applied to real-time programs, software, and selected commercial uses. Although the fundamental structure of a program is simple, the rules of the language are extensive. A number of extensions and variants have been developed. One of these, known as J-73, is a language approved for use by the U.S. Department of Defense and is utilized primarily by the U.S. Air Force.

Key.   (1) A group of characters which identifies or is part of a record or item; thus any entry in a record or item can be used as a key for collating or sorting purposes. (2) A marked lever manually operated for copying a character, e.g., a typewriter, paper tape perforator, card punch, manual keyboard, digitizer, or manual word generator. (3) A lever or switch on a computer console for the purpose of manually altering computer action.

Language.   A communications means for transmitting information between human operators and computers. The human programmer describes how the problem is to be solved using the computer language. A computer language consists of a well-defined set of characters and words, coupled with a series of rules *(syntax)* for combining them into computer instructions or statements. There is a wide variety of computer languages, particularly in terms of flexibility and ease of use. There are three levels in the hierarchy of computer languages: (1) machine languages, (2) procedure-oriented languages, and (3) problem-oriented languages.

> *Machine Language.*   (1) A language designed for interpretation and use by a machine without translation. (2) A system for expressing information which is intelligible to a specific machine, e.g., a computer or class of computers. Such a language may include instructions which define and direct machine operations, and information to be recorded by or acted on by these machine operations. (3) The set of instructions expressed in the number system basic to a computer, together with symbolic operation codes with absolute addresses, relative addresses, or symbolic addresses. In this case, it is known as an *assembly language*.
>
> *Procedure-Oriented Language.*   A machine-independent language which describes how the process of solving the problem is to be carried out. For example, FORTRAN, ALGOL, PL/I, and COBOL.
>
> *Problem-Oriented Language.*   A language designed for convenience of program specification in a general problem area. The components of such a language may bear little resemblance to machine instructions and often incorporate terminology and functions unique to an application. This type of language is also known as an *applications language*.

Other computer languages include:

> *Algorithmic Language.*   An arithmetic language by which numerical procedures may be precisely presented to a computer in a standard form. The language is intended not only as a means of directly presenting any numerical procedure to any appropriate computer for which a compiler exists but also as a means of communicating numerical procedures among individuals.
>
> *Artificial Language.*   A language specifically designed for ease of communication in a particular area of endeavor, but one that is not yet "natural" to that area. This is contrasted with a natural language which has evolved through long usage.
>
> *Comman Machine Language.*   A machine-sensible information representation which is common to a related group of data processing machines.
>
> *Common Business-Oriented Language.*   A specific language by which business data processing procedures may be precisely described in a standard form. The language is intended not only as a means for directly presenting a business program to any appropriate computer for which a compiler exists but also as a means of communicating such procedures among individuals.
>
> *Object Language.*   A language which is the output of an automatic coding routine. Usually, object language and machine language are the same. However, a series of steps in an automatic

coding system may involve the object language of one step serving as a source language for the next step, and so forth.

Thomas J. Harrison, IBM Corporation, Boca Raton, Fla.

LISP. An acronym for *list processing.* LISP is an interpretive language especially formulated for the manipulation of symbolic strings of recursive data. The language was developed at the Massachusetts Institute of Technology, commencing in 1959 under the direction of John McCarthy. Although developed to assist in the handling of symbolic lists, the language also finds uses in the manipulation of mathematical and arithmetic logic. A description and introduction to the system can be found in *The Programmer's Introduction to LISP,* W. D. Maurer, American Elsevier, New York, 1972.

Pascal. Invented by N. Wirth in 1971, Pascal is a comparatively recent high-level programming language named in honor of Blaise Pascal, the seventeenth-century French philosopher who invented the first workable mechanical adding machine at age 19. The language was designed for the systematic teaching of programming as a discipline based on fundamental concepts of structure and integrity. Although taught widely, Pascal is increasingly being used outside the classroom in non-business-oriented applications. The language is characterized by strong typing and a syntax which encourages readable programs and good programming practices. It is a descendant of ALGOL, although not a strict superset of it. Pascal compilers are available on many computers, ranging from microcomputers to mainframes. Efforts to develop national and international standards for this language commenced in 1981. The following is a program fragment as found in the excellent publication, *Introduction to Pascal,* C. A. G. Webster, Heyden & Sons, London, 1976.

```
procedure in 4 digits (var n: integer);
const zero = int ('0');
var s : char;
    i : integer;
begin
    read(s);
    if '0' ≥ s ∧ s ≤ '9' then
        begin
        n : = s − zero;
        for i : = 1 to 3 do
            begin
            read (s); n : = n * 10 + s − zero
            end
        end
end
```

PL/I. Jointly developed by IBM and SHARE (a users' group), PL/I is a multipurpose programming language used for scientific and commercial applications. The language is characterized by its modularity, because it possesses different subsets for various applications at different levels of complexity. This also applies for input-output. The program permits mixed expressions, adding to the simplicity and flexibility of the language. PL/I is a high-level language that is largely machine-independent. It is described in American National Standard ANSI X3.53-1976. Various standardized subsets and extensions are under development as of the early 1980s under the auspices of American National Standards Committee X3.

Quantization. When the range of a variable is divided into a finite number of distinct, nonoverlapping subranges, each of which is assigned a value within the subrange, the variable is said to be quantized. The process of quantizing is known as *quantization.* The distinct subranges sometimes are called *quanta.* The magnitude of the subranges need not necessarily be equal. Where the widths are nominally equal, they may be termed *quantization units.* Where they are not equal by intent, the term *quantization steps* is used. The degree used in temperature measurement is a quantization unit. A slide rule which performs quantization in terms of a nonlinear device where the subranges (considering the range of numbers 1 and 10) are unequal involves quantization steps.

It should be stressed that *quantize* is not synonymous with *digitize* or *digitalize*. As related to digital data acquisition and instrumentation systems, electronic devices which perform quantization are analog-to-digital (A/D) converters (encoders). Interpreted broadly, quantization also applies to the assignment of discrete values to a quantity obtained by reading the position of a pointer on a calibrated scale. Most quantizers of A/D converters are linear. Thus, the discrete subranges are properly called quantization units. Inasmuch as most of this equipment is based on binary representation, the common term for the quantization unit is the bit. In a 10-bit converter, the least significant bit is the quantization unit, which is equal to $V_{fs}/2^{10}$, where V_{fs} is the full-scale input to the A/D converter.

Scale Factor.   In digital computing, an arbitrary factor which may be associated with numbers in a computer to adjust the position of the radix so that the significant digits occupy specified columns. In analog computing, a proportionality factor which relates the magnitude of a variable to its representation within a computer.

Shannon.   In information theory, a unit of logarithmic measure of information equal to the decision content of a set of two mutually exclusive events expressed by the logarithm to base 2; e.g., the decision content of a character set of eight characters equals 3 shannons. Synonymous with *information content binary unit*.

Shannon Formula.   A theorem in information theory which states that a method of coding exists whereby C binary digits per second may be transmitted with arbitrarily small frequency of error where C is given by

$$C = B \log_2 \left(1 + \frac{S}{N} \right)$$

and no higher rate can be transmitted. B is the bandwidth and S/N is the signal-to-noise ratio.

SIMSCRIPT.   A programming language prepared by the Rand Corporation in the early 1960s for the U.S. Air Force. The language was designed to assist in programming simulation problems. SIMSCRIPT allows simulated systems to be described in terms of entities, attributes (i.e., properties associated with an entity), and sets (groups of entities). The language uses statements similar in appearance to those used in FORTRAN.

SNOBOL.   An acronym for *string*-oriented sym*bol*ic *l*anguage. This programming language developed by Bell Laboratories has been used in program compilation and in the generation of symbolic equations. A description of the current form of the language can be found in *The Programmer's Introduction to SNOBOL*, W. D. Maurer, American Elsevier, New York, 1976.

Word.   A character or bit string that traditionally has been an entity in computer technology. A word typically consists of one or more bytes. In small computers, there are usually 2 bytes per word. In large machines, there may be up to 8 bytes or more. Instructions are provided for manipulating words of data and, typically, most instructions occupy one word of storage. In addition, the internal data paths in a computer (parallel) are designed to transfer one word of data at a time. See also subsection Digital Computer Fundamentals earlier in this article.

COMPUTER PROGRAMS AND PROGRAMMING—
OPERATIONS AND PROCEDURES*

Absolute Address.   See *address* in this subsection.

Address.   An identification, represented by a name, label, or number, for a digital computer register, device, or location in storage. Addresses are also parts of an instruction word along with commands, tags, and other symbols. The part of an instruction which specifies an operand for the instruction may be an address.

Absolute address or *specific address* indicates the exact physical storage location where the referenced operand is to be found or stored in the actual machine code address numbering system.

*Terms are presented in alphabetical order for convenience of reference.

Direct address or *first-level address* indicates the location where the referenced operand is to be found or stored with no reference to an index register.

Indirect address or *second-level address* in a computer instruction indicates a location where the address of the referenced operand is to be found. In some computers, the machine address indicated can in itself be indirect. Such multiple levels of addressing are terminated either by prior control or by a termination symbol.

Machine address is an absolute, direct, unindexed address expressed as such or resulting after indexing and other processing have been completed.

Symbolic address is a label, alphabetic or alphameric, used to specify a storage location in the context of a particular program. Sometimes programs may be written using symbolic addresses in some convenient code, which then are translated into absolute addresses by an assembly program.

Base address permits derivation of an absolute address from a relative address.

Effective address is derived from applying specific indexing or indirect addressing rules to a specified address.

Four-plus-one address incorporates four operand addresses and a control address.

Immediate address incorporates the value of the operand in the address portion instead of the address of the operand.

N-level address is a multilevel address in which N levels of addressing are specified.

One-level address directly indicates the location of an instruction.

One-plus-one address contains two address portions. One address may indicate the operand required in the operation, and the other may indicate the following instruction to be executed.

Relative address is the numerical difference between a desired address and a known reference address.

Three-plus-one address incorporates an operation code, three operand address parts, and a control address.

Zero-level address permits immediate use of the operand.

Thomas J. Harrison, IBM Corporation, Boca Raton, Fla.

Assembler.   A computer program which operates on symbolic input data to produce machine instructions by carrying out such functions as (1) translating symbolic operation codes into computer instructions, (2) assigning locations in storage for successive instructions, and (3) assigning absolute addresses for symbolic addresses. An assembler generally translates input symbolic codes into machine instructions item for item and produces as output the same number of instructions or constants that were defined in the input symbolic codes.

Assembly language may be defined as computer language characterized by a one-to-one relationship between the statements written by the programmer and the actual machine instructions performed. The programmer thus has direct control over the efficiency and speed of the program. Usually, the language allows the use of mnemonic names instead of numerical values for the operation codes of the instructions and similarly allows the user to assign symbolic names to the locations of the instructions and data. For the first feature, the assembler contains a table of the permissible mnemonic names and their numerical equivalents. For the second feature, the assembler builds such a table on a first pass through the program statements. Then, the table is used to replace the symbolic names by their numerical values on a second pass through the program. Usually, dummy operation codes (or pseudocodes) are needed by the assembler to pass control information to it. As an example, an origin statement is usually required as the first statement in the program. This gives the numerical value of the desired location of the first instruction or piece of data so that the assembler can, by counting the instructions and data, assign numerical values for their symbolic names.

The format of the program statements is usually rigidly specified, and only one statement per input record to the assembler is permitted. A representative statement is: symbolic name, operation code (or pseudocode), modifiers and/or register addresses, symbolic name of data. The mnemonic names used for the operation codes usually are defined uniquely for a particular computer type with little standardization among computer manufacturers even for the most common operations. The programmer must learn a new language for each new machine.

An example of a program prepared in an assembler language follows. The explanatory comments following the REM (remarks) mnemonic and those to the right of the other program statements are ignored by the assembler program and thus do not affect execution of the program.

```
            ABS
            ORG    100
            REM    THIS IS AN EXAMPLE OF A PRO-
            REM    GRAM IN AN ASSEMBLER LAN-
            REM    GUAGE. "ABS", "ORG", "CALL",
            REM    "REM", "DC", "DEC" AND "END"
            REM    ARE PSEUDOCODES. THE OTHER
            REM    MNEMONICS IN THE SAME COL-
            REM    UMN ARE OPERATION CODES.
            REM    THE NAMES SUCH AS A. B.
            REM    START, AND OVER ARE SYM-
            REM    BOLIC ADDRESSES ASSIGNED BY
            REM    THE PROGRAMMER

START       SLT    32          THESE INSTRUCTIONS
            LD     A           COMPUTE A/B*C AND
            D      B           PUT THE VALUE IN D
            M      C
            SLT    1
            STO    D
            CMP    D4B1        THESE INSTRUCTIONS
            BSC    Z+          COMPARE D WITH 4 AND
            MDX    OVER        3.2 AND BRANCH OUT IF
            CMP    D32B1       IT DOES NOT LIE BE-
            BSC    Z-          TWEEN THEM
            MDX    UNDER

EXIT        CALL   EXIT        THE ASSEMBLER SETS
            REM                UP A BRANCH TO THE
            REM                SYSTEM EXIT SUBROU-
            REM                TINE (NAMED "EXIT")

OVER        LD     D1          THESE INSTRUCTIONS
            STO    EROR        SET THE VALUE OF
            MDX    EXIT        EROR TO 1 BEFORE
            REM                EXITING

UNDER       SRA    16          THESE INSTRUCTIONS
            STO    EROR        SET THE VALUE OF
            MDX    EXIT        EROR TO ZERO BEFORE
            REM                EXITING

            ORG    200         THIS SET OF PSEUDO-
A           DC     1           CODES IS USED BY THE
B           DC     2           ASSEMBLER TO SET UP
C           DC     3           THE DATA AND DATA
D           DC     0           WORKSPACE FOR THE
EROR        DEC    -1.0        ABOVE PROGRAM
D1          DEC    1.0
D4B1        DEC    4.0B1
D32B1       DEC    3.2B1
            REM                LAST STATEMENT IN
            REM                PROGRAM
            END
```

Thomas J. Harrison, IBM Corporation, Boca Raton, Fla.

Blocking of Records. See *record* in this subsection.

Bookkeeping Operation. An operation that does not contribute directly to the result, e.g., arithmetical, logical, and transfer operations used in modifying the address section of other instructions, in counting cycles, and in rearranging data. This is similar to a housekeeping operation but generally implies an enumeration and recording of actions taken.

Bootstrap. A technique for loading the first few instructions of a routine into storage, followed by using these instructions to bring in the remainder of the routine. The technique usually involves either entering a few instructions manually or, more commonly, using a special key on the console which invokes a routine stored in read-only storage (ROS). Also used as a verb, meaning use of a bootstrap.

Branch. A set of instructions that may be executed between a couple of successive decision instructions. Branching allows parts of a program to be worked on to the exclusion of other parts and provides a computer with considerable flexibility. The branch point is a junction in a computer routine where one or both of two choices are selected under control of the routine. Also refers to one instruction that controls branching.

Breakpoint Instruction. See *instruction* in this subsection.

Check. A process of partial or complete testing of the correctness of computer or other data processing machine operations. Checks also may be run to verify the existence of certain prescribed conditions within a computer or the correctness of the results produced by a program. A check of any of these conditions usually may be made automatically by the equipment; or checks may be programmed.

Automatic Check. A procedure for detecting errors that is a built-in or integral part of the normal operation of a device. Until an error is detected, automatic checking normally does not require operating system or programmer attention. For example, if the product of a multiplication is too large for the space allocated, an error condition (overflow) will be signaled.

Built-in Check. An error detecting mechanism that requires no program or operator attention until an error is detected. The mechanism is built into the computer hardware.

Checkpoint. A point in time in a machine run at which processing is momentarily halted to perform a check or to make a magnetic tape or disk record (or equivalent) of the condition of all the variables of the machine run such as the status of input and output devices and a copy of working storage. Checkpoints are used in conjunction with a restart routine to minimize reprocessing time occasioned by functional failures.

Duplication Check. Two independent performances of the same task are completed and the results compared. This is illustrated by the following operations:

12	31	84	127
9	14	43	66
21	45	127	193

Echo Check. For checking the accuracy of transmission of data, the transmitted data are returned to the sender for comparison with the original data. Essentially a reading test, the echo check guards against malfunctions of output operations.

Modulo n Check. Same as *residue check*. See below.

Odd-Even Check. Same as *parity check*.

Parity Check. A summation check in which the binary digits in a character or word are added, modulo 2, and the sum checked against a single, previously computed parity digit, i.e., a check which tests whether the number of 1's in a word is odd or even.

Programmed Check. A system of determining the correct program and machine functioning either by running a sample problem with similar programming and a known answer or by using mathematical or logic checks, such as comparing $A \times B$ with $B \times A$. Also, a check system built into the program for computers that do not have automatic checking.

Reasonableness Check. Same as *validity check*. See below.

Residue Check. An error detection system in which a number is divided by a quantity n and the remainder compared with the original computer remainder. This is also termed a *modulo n check*. See above.

Sequence Check. A data processing operation designed to check the sequence of the items in a file assumed to be already in sequence.

Summation Check. A check in which groups of digits are summed, usually without regard for overflow, and that sum checked against a previously computed sum to verify that no digits have been changed since the last summation.

Validity Check. A check based on known limits or on given information or computer results; e.g., a calendar month will not be numbered greater than 12, or a week does not have more than 168 h.

Thomas J. Harrison, IBM Corporation, Boca Raton, Fla.

Column. A character or digit position in a positional information format, particularly one in which characters appear in rows and the rows are placed one above another, e.g., the rightmost column in a five-decimal-place table or in a list of data.

Command. The specification of an operation to be performed. In terms of a control signal, a command usually takes the form of YES (go ahead), or NO (do not proceed). A command should not be confused with an instruction. In most computers, an instruction is given to the central processing unit (CPU), as contrasted with a command which is an instruction to be followed by a data channel. An input command, for example, my be READ, and an output command, WRITE.

Compiler. A program designed to translate a higher level language into machine language. In addition to its translating function, which is similar to the process used in an assembler, a compiler program is able to replace certain items of input with a series of instructions, usually called subroutines. Thus, where an assembler translates item for item and produces as output the same number of instructions or constants that were put into it, a compiler typically produces multiple output instructions for each input instruction or statement. The program which results from compiling is a translated and expanded version of the original.

Compiler language is characterized by a one-to-many relationship between the statements written by a programmer and the actual machine instructions executed. The programmer typically has little control over the number of machine instructions executed to perform a particular function and is dependent on the particular compiler implementation. Typically, the language is very nearly machine-independent and may be biased in its statements and features to a particular group of users with similar problems. Thus, these languages sometimes are referred to as problem-oriented languages (POLs). Slightly different implementations of a given language are sometimes called *dialects*. Some compiler languages in wide use are ALGOL, BASIC, COBOL, FORTRAN, and PL/I.

Computer Operating System. Generally defined as a group of interrelated programs to be used on a computer system in order to increase the utility of the hardware and software. There is a wide range in the size, complexity, and application of operating systems. The need for operating systems arose from the desire to obtain the maximum amount of service from a computer. A first step was simple monitor systems providing a smooth job-to-job transition. Modern operating systems contain coordinate programs to control input-output scheduling, task scheduling, error detection and recovery, data management, debugging, multiprogramming, and on-line diagnostics.

Operating system programs fall into two main categories: (1) control programs and (2) processing programs:

Control Programs
    Data management, including all input-output
    Job management
    Task management
Processing Programs
    Language translators

    Service programs
      Linking programs
      Sort or merge
      Utilities
      System library routines
    User-written programs

Control programs provide the structure and environment in which work may be accomplished more efficiently. They are the components of the supervisory portion of the system. Processing programs are programs that have some specific objective that is unrelated to controlling the system work. These programs use the services of supervisory programs rather than operating as part of them.

Data management is involved with the movement of data to and from all input-output devices and all storage facilities. This area of the system embraces the functions referred to as the input-output control system (IOCS), which frequently is segmented into two parts, the physical IOCS and the logical IOCS.

Physical IOCS is concerned with device and channel operations, error procedures, queue processing, and, generally, all operations concerned with transmitting physical data segments from storage to external devices. *Logical IOCS* is concerned with data organization, buffer handling, data referencing mechanisms, logical device reference, and device independence.

Job management involves the movement of control cards or commands through the system input device, their initial interpretation, and the scheduling of jobs so indicated. Other concerns are job queues, priority scheduling of jobs, and job accounting functions.

Task management is concerned with the order in which work is performed in the system. This includes management of the control facilities, i.e., central processing unit, storage, input-output channels, and devices, in accordance with some task priority scheme.

Language translators allow users to concentrate on solving logical problems with a minimum of thought given to detailed hardware requirements. In this category are higher level languages, report generators, and special translation programs.

Service programs are needed to facilitate system operation or to provide auxiliary functions. Sort programs, program linking functions, and file duplication, as well as system libraries, are within this group.

User-written programs are programs prepared specifically to assist the user in the accomplishment of his or her objectives.

Thomas J. Harrison, IBM Corporation, Boca Raton, Fla.

Concurrent Operation. The performance of several actions during the same interval of time, although not necessarily simultaneously. Compare with *parallel operation,* described later in this section. Multiprogramming is a technique that provides concurrent execution of several tasks in a computer. See also *multiprogramming* in this section.

Control Algorithm. A general computational procedure which may include instructions, limits, and equations representing functional relationships in the controlling elements. In control applications, an algorithm usually defines the functional relationship existing between the manipulated variable and the actuating or error signal. See article on *Control Algorithms* in Section 17.

CTS. Abbreviation for *clear to send.*

Cycle. (1) A set of operations repeated regularly and in the same sequence. (2) In data processing, the word *cycle* often refers to the storage cycle, i.e., the shortest time interval between one fetch (or store) and the next fetch (or store) within the same storage unit. Normally, the storage cycle establishes the pace for the entire computer. In some designs, access to more than one instruction per cycle is possible. Multiple independent storage units and overlapping cycle times also are possible. Storage cycle often is referred to as machine cycle or *major cycle.* The *clock cycle* of a synchronous computer establishes the commencement of each elementary subtask of the execution routine. This is a much faster cycle, internal to the computer hardware, and is sometimes referred to as the *minor cycle.*

Cycle Steal. A type of channel operation that has the ability to access main storage for data while the arithmetic unit is performing an operation which does not require storage access. When both the channel and the arithmetic unit request storage access simultaneously, the arithmetic operation typically is delayed until the channel storage request is serviced. Also known as a *direct storage access* (DSA) or *direct memory access* (DMA) channel.

Diagnostic Routine.   See *routine* in this subsection.

Diagnostics.   Programs provided for the maintenance engineer or operator to assist in discovering the source of a particular computer system malfunction. Diagnostics generally consist of programs which force extreme conditions (the worst patterns) on the suspected unit with the expectation of exaggerating the symptoms sufficiently for the engineer to readily discriminate among possible faults and to identify the particular fault. In addition, diagnostics may provide assistance in localizing the cause of a malfunction to a particular card or component in the system.

There are two basic types of diagnostic programs: (1) off-line and (2) on-line. Off-line diagnostic programs are those which require that there be no other program active in the computer system, sometimes requiring that there be no executive program in the system. Off-line diagnostics are typically used for central processing unit (CPU) malfunctions, very obscure and persistent peripheral device errors, or critically time-dependent testing. For example, it may be suspected or known that a harmonic frequency is contributing to the malfunction. Thus, it may be desirable to drive the unit continuously at various precise frequencies close to the suspected frequency to confirm the diagnosis and then to confirm the cure. Interference from other activities may well make such a test meaningless. Thus, all other activity on the system must cease.

On-line diagnostics are used mainly in a multiprogramming environment and are vital to the success of real-time systems. The basic concept is that of logically isolating the malfunctioning unit from all problem programs and allowing the diagnostic program to perform any and all functions on the unit. Many of the more common malfunctions can be isolated by such diagnostics, but there are limitations imposed by interference from other programs also using the CPU.

Thomas J. Harrison, IBM Corporation, Boca Raton, Fla.

Dummy.   An artificial address, instruction, or record of information inserted into a computer system solely to fulfill prescribed conditions, such as to achieve a fixed word length or block length, but without itself affecting machine operations except to permit the machine to perform the desired operations. The technique sometimes is used as an aid in debugging a system.

Error Procedures.   Computer errors fall into three principal categories: (1) central processor errors, (2) peripheral device errors, and (3) program errors. Program errors by far are the most common, but sometimes they may appear to be processor or peripheral errors. If an error cannot be duplicated through the use of other programs, it is almost certainly a program error.

The procedure for isolating peripheral device errors normally consists of retrying the operation several times and logging the failures. When the failure persists, the operator is informed and the computer may wait for operator intervention, or abort the current job and logically disconnect the device. Or, the error may be documented for the operator with a return to the current program with a special indicator to permit the applications programmer to determine the error procedure to be followed. This is a good approach when two different programs, each having different error procedure requirements, are running in the same machine.

Procedures for isolating central processor errors are usually more difficult. Often a "warm start" is attempted. All current jobs are aborted, all of the storage is initialized where possible, and a restart from a previously defined state is attempted. If the first warm start fails, another cycle usually is attempted. It is unlikely that a machine which cannot be warm-started can be safely used without operator intervention.

Executive Routine.   See *routine* in this subsection.

FIFO.   Abbreviation for *first in first out*.

Firmware.   See *software* in this subsection.

Flag.   A bit or character of information attached to a character or word to indicate the boundary of a field. Also an indicator used frequently to tell some later part of a program that some condition occurred earlier; or an indicator used to identify the members of several sets which are intermixed.

Heuristic Routine.   See *routine* in this subsection.

Housekeeping Operation.   A general term for computer operations which must be performed for a machine run but do not contribute directly to the solution of a problem; rather, they contribute to operation of the computer. Examples of housekeeping operations include establishing controlling marks, setting up auxiliary storage units, reading in the first record for processing, initializing, setup verification operations, and file identification.

Indirect Address. See *address* in this subsection.

Instruction. (1) A set of characters which defines an operation together with one or more addresses, or no address, and which, as a unit, causes the computer to perform the operation on the indicated quantities. The term *instruction* is preferable to the terms *command* and *order*. *Command* is reserved for a specified portion of the instruction word, i.e., the part which specifies the operation to be performed. *Order* is reserved for the ordering of the characters, implying sequence, or the order of the interpolation, or the order of the differential equation. (2) The operation or command to be executed by a computer, together with associated addresses, tags, and indices.

Alphanumeric Instruction. The name given to instructions that can be used equally well with alphabetic or numeric kinds of fields of data.

Branch Instruction (or Transfer Instruction). An instruction to a computer that enables the programmer to instruct the computer to choose between alternative subprograms depending on the conditions determined by the computer during the execution of the program.

Breakpoint Instruction. (1) An instruction that causes a computer to stop or to transfer control in some standard fashion to a supervisory routine which can monitor the progress of the interrupted program. (2) An instruction which, if some specified switch is set, will cause the computer to stop or take other special action.

Macro Instruction. (1) A pseudo instruction which causes a sequence of instructions to be inserted into the object routine for performing a specific operation. (2) A more powerful instruction which combines several operations into one instruction.

Micro Instruction. A basic or elementary machine instruction.

Multiple-Address Instruction. An instruction consisting of an operation code and two or more addresses. Usually specified as a two-, three-, or four-address instruction.

One-Address Instruction. An instruction consisting of an operation and exactly one address. The instruction code of a single-address computer may include both zero-address and multiaddress instructions as a special cases.

Pseudo Instruction (or Quasi Instruction). (1) A symbolic representation in a compiler or interpreter. (2) A group of characters having the same general form as a computer instruction but never executed by the computer as an actual instruction.

Two-, Three-, or Four-Address Instruction. An instruction consisting of an operation and two, three, or four addresses, respectively. The addresses may specify the location of operands, results, or other instructions.

Thomas J. Harrison, IBM Corporation, Boca Raton, Fla.

Interpreter. (1) An executive routine which translates a stored program expressed in some pseudocode into machine code and performs the indicated operations, by means of subroutines, as they are translated. Interpreters are used widely for translating some high-level languages such as BASIC and APL. An interpreter is essentially a closed subroutine which operates successively on an indefinitely long sequence of program parameters, pseudo instructions, and operands. (2) In punched card operations, a device that prints on a card the characters corresponding to hole patterns punched in the cards.

Interpretive Routine. See *routine* in this subsection.

Interrupt. A signal which causes the central processing unit (CPU) to change state as the result of a specified condition. An interrupt represents a temporary suspension of normal program execution and arises from an external condition, from an input or output device, or from the program currently being processed in the CPU. On recognition of an interrupt, the current program is suspended and replaced by another program. On completion of the new program, control of the CPU is returned to the interrupted program at the exact point where discontinuance occurred. As there is more than one possible interrupt condition in most systems, a priority may be established to determine the sequence for servicing programs. The priority may be established by hardware logic or programming. When the priority is established by programming, the interrupt condition causes the system to transfer control to an interrupt-service subroutine. This routine determines the specific cause of the interrupt. Based on the assigned priority of the interrupt condition, the routine schedules execution of the interrupt-service routine in the correct sequence. In the case of a hardware priority interrupt structure, the hardware is designed to prevent any interrupt from being recognized should it be of lower priority than the current program.

Job.   A set of data that completely defines a unit of work for a computer. A job usually includes all necessary computer programs, linkages, files, and instructions to the operating system. A job may comprise one or more tasks.

Location.   A storage position in the main internal storage which can store one computer word and which is usually identified by an address. More generally, it is any place in which data may be stored. A location address usually is expressed as a number, although it is referenced in many computer languages by using a unique name or acronym. An analogy can be made between a computed location or address and a post office box number—with the stipulation that the box can hold only one letter at a time.

A *protected location* is one to which access is denied under specified conditions. *Full protection* signifies that a memory location cannot be written into or read from without proper authorization through the use of a key or security code. *Write protection* denotes that a memory location may be read from, but not written into (thus altered), without proper authorization. *Open* or *no protection* signifies full and free accessibility to any program or user.

Logical Operation.   (1) A logical or Boolean operation on N-state variables which yields a single N-state variable, e.g., a comparison of the three-state variables A and B, each represented by $-$, 0, or $+$, which yields $-$ when A is less than B, 0 when A equals B, and $+$ when A is greater than B. Specifically, operations such as AND, OR, and NOT on two-state variables which occur in the algebra of logic, i.e., Boolean algebra. (2) Logical shifting, masking, and other nonarithmetic operations of a computer.

Loop.   A sequence of instructions that may be executed repeatedly while a certain condition prevails. The productive instructions in a loop generally manipulate the operands, while bookkeeping instructions may modify the productive instructions and keep count of the number of repetitions. A loop may contain any number of conditions for termination, such as the number of repetitions or the requirement that an operand be nonnegative. The equivalent of a loop can be achieved by the technique of straight-line coding, whereby the repetition of productive and bookkeeping operations is accomplished by explicitly writing the instructions for each repetition.

Linkage.   An address, instruction, or group of instructions that passes control and parameters between separate portions of a computer program, such as the main program and subroutines of tasks.

Machine Address.   See *address* in this subsection.

Macroassembler.   An assembler which permits the user to define pseudo computer instructions which may generate multiple computer instructions when assembled. Source statements which may generate multiple computer instructions are termed *macrostatements* or *macroinstructions*. With a process control digital computer, a macroassembler can be a most significant tool. By defining a set of macrostatements, for example, a process control engineer can define a process control programming language specifically oriented to the process.

Macrolibrary. A set of defined macrostatements and the associated computer instructions. A macrolibrary results, for example, when a process control engineer defines a language for a system. The effectiveness of a macroassembler depends largely on the ease of creating, manipulating, modifying, and linking various macrolibraries.

Mask.   A pattern of digits used to control the retention or elimination of portions of another pattern of digits. Also, the use of such a pattern. For example, an 8-bit mask having a single i bit in the ith position, when added with another 8-bit pattern, can be used to determine if the ith bit is a 1 or a 0; i.e., the ith bit in the pattern will be retained and all other bits will be 0's.

As another example, there are situations where it is desirable to delay the recognition of a process interrupt by a digital computer. A mask instruction permits the recognition of specific interrupts to be inhibited until it is convenient to service them.

Microprogram.   Microprogramming is a technique of using a special set of instructions for an automatic computer. Elementary logical and control commands are used to simulate a higher level instruction set for a digital computer. The basic machine-oriented instruction set in many computers is comprised of commands, such as ADD, DIVIDE, SUBTRACT, and MULTIPLY, which are executed directly by the hardware. The hardware actually implements each function as a combination of elementary logical functions, such as AND, OR, and EXCLUSIVE OR. The manner of exact implementation usually is not of concern to the programmer. Compared with the elementary logical functions, the ADD, SUBTRACT, MULTIPLY, and DIVIDE commands are a high-level language set in the same sense that a macrostatement is a higher level instruction set when compared with the machine-oriented language instruction set.

In a microprogrammed computer, the executable (micro) instructions which may be used by the (micro) programmer are comprised of a logical function, such as AND and OR, and some elementary control functions, such as shift and branch. The (micro) programmer then defines microprograms which implement an instruction set analogous to the machine-oriented language instruction set in terms of these microinstructions. Using this derived instruction set, the systems or applications programmer writes programs for the solution of a problem. When the program is executed, each derived instruction is executed by transferring control to a microprogram. The microprogram is then executed in terms of the microinstructions of the computer. After execution of the microprogram, control is returned to the program written in the derived instruction set. The microprogram typically is stored in read-only storage (ROS) and thus is permanent and cannot be changed without physical replacement.

An advantage of microprogramming is increased flexibility. This can be realized by adapting the derived, machine-oriented instruction set to a particular application. The technique enables the programmer to implement a function, such as square root, directly without subroutines or macrostatements. Thus significant programming and execution efficiency is realized if the square root function is commonly required. Also, the instruction set of a character-oriented computer can be implemented on a word-oriented machine where adequate microprogramming is provided.

Thomas J. Harrison, IBM Corporation, Boca Raton, Fla.

Multiprogramming. The essentials of a multiprogramming system in connection with digital computer operations are: (1) Several programs are resident in main storage simultaneously; and (2) the central processing unit (CPU) is time-shared among the programs. This makes for better utilization of a computer system. Where only one program may reside in the main storage at any given time, inefficient use of the CPU time results when a program requests data from an input-output device. The operation is delayed until the requested information is received. In some applications, such delays can constitute a large portion of the program execution time. In the multiprogramming approach, other programs resident in storage may use the CPU while a preceding program is awaiting new information. Multiprogramming practically eliminates CPU lost time due to input-output delays. Multiprogramming is particularly useful in process control or interactive applications which involve large amounts of data input and output.

Input-output delay-time control is a basic method for controlling the interplay between multiple programs. Where multiprogramming is controlled in this manner, the various programs resident in storage are normally structured in a hierarchy. When a given program in the hierarchy initiates an input-output operation, the program is suspended until such time as the input-output operation is completed. A lower priority program is permitted to execute during the delay time.

In a *time-slice* multiprogramming system, each program resident in storage is given a certain fixed interval of CPU time. Mutliprogramming systems for applications where much more computation is done than input-output operation usually use the time-slice approach.

Multiprogramming systems allow multiple functions to be controlled simultaneously by a single process control digital computer. A multiprogramming system allows a portion of the storage to be dedicated to each type of function required in the control of the process and thus eliminates interference among the various types of functions. In addition, it provides the means whereby asynchronous external interrupts can be effectively serviced on a timely basis.

Operand. An entity to which an operation is applied. The operand may be a portion of the computer instruction or it may be identified by the address part of the instruction.

Object Program. See *program* in this subsection.

Odd-Even Check. See *check* in this subsection.

One-Plus-One Address. See *address* in this subsection.

Operating System.   An integrated collection of service routines for supervising the sequencing of programs by a computer, which may provide debugging, input-output control, accounting, compilation, storage assignment, data management, and related services. Essentially synonymous with *monitor system* and *executive system*.

Overlay.   A technique for bringing routines into high speed storage from some other form of storage during processing, so that several routines will occupy the same storage location at different times. Overlay is used when the total storage requirement for instructions exceeds the available main storage. Also refers to the routine itself.

Parallel Operation.   The simultaneous performance of several actions, usually of a similar nature, through provision of individual similar or identical devices for each such action, particularly flow or processing of information. Parallel operation is performed to save time over serial operation. The decrease in the cost of multiple high-function integrated circuits has made parallel operation much more common in recent years. Because computer logic speeds already have approached a very high limit, the use of parallel execution holds the best promise for increasing overall computing speeds. Multiprocessors represent a case where parallel operation is achieved by providing two or more complete processors capable of simultaneous operation. Compare with *concurrent operation* described earlier in this section.

Parity Check.   See *check* in this subsection.

Patch.   A section of coding inserted into a computer routine to correct an error or to alter the routine. Often, it is not inserted into the actual sequence of the routine being corrected but rather is placed elsewhere, with an exit to the patch and a return to the routine provided. Also, the act of altering a routine by using a patch.

POL.   Abbreviation for *problem-oriented language.*

Postprocessor.   A program which operates in conjunction with another program as a successor. In the traditional numerical control field, a postprocessor is a program which takes the output of a general numerical control language compiler, such as APT (*automatically programmed tool*) and produces the specific machine-tool controller command information. The APT compiler generates tool path information from tool and part geometry descriptions and constraint information furnished by the programmer. The postprocessor adapts this output to a specific machine control format. In light of the foregoing description *postprocessor* is not to be confused with the postprocessing of data as occurs in the aerospace industry, for example, where information recorded during a run or flight is reduced and analyzed *(postprocessed)* at a later time.

Program.   (1) The complete plan for the computer solution of a problem, more specifically the complete sequence of instructions and routines necessary to solve a problem. (2) To plan the procedures for solving a problem. This may involve, among other things, analysis of the problem, preparation of a flow diagram, preparing details, texting and developing subroutines, allocation of storage locations, specifications of input and output formats, and the incorporation of a computer run into a complete data processing system.

> *Internally Stored Program.*   A sequence of instructions stored inside the computer in the same storage facilties as the computer data, as opposed to external storage on punched paper tape and pinboards.
>
> *Object Program.*   The program which is the output of an automatic coding system, such as an assembler or compiler. Often the object program is a machine language program ready for execution, but it may well be in an intermediate language.
>
> *Source Program.*   A computer program written in a language designed for ease of expression of a class of problems or procedures by humans, e.g., symbolic or algebraic. A generator, assembler, translator, or compiler routine is used to perform the mechanics of translating the source program into an object program in machine language.

See also earlier subsection Digital Computer Fundamentals.

Program Generator.   A program that permits a computer to write other programs automatically. Generators are of two types: (1) The *character-controlled generator,* which is like a compiler in that it takes entries from a library of functions but is unlike a simple compiler in that it examines control characters associated with each entry and alters instructions found in the library according to the directions contained in the control characteristics. (2) The *pure generator,* which is a program that writes another program. When associated with an assembler, a pure generator is usually a section of program which is called into storage by the assembler from a library and then writes one or more entries in another program. Most assemblers are also compilers and generators. In this case, the entire system is usually referred to as an *assembly program.*

Programming Flowchart.   A graphical representation of a program, in which symbols are used to represent data, flow, operations, equipment, and so forth. A digital computer program may be charted for two primary reasons: (1) ease of initial program design, and (2) program documentation. By coding from a flowchart, instead of coding without any preliminary design, a programmer usually conserves time and effort in developing a program. In addition, a flowchart is an effective means of transmitting an understanding of the program to another person.

A programming flowchart is comprised of function blocks with connectors between the blocks. A

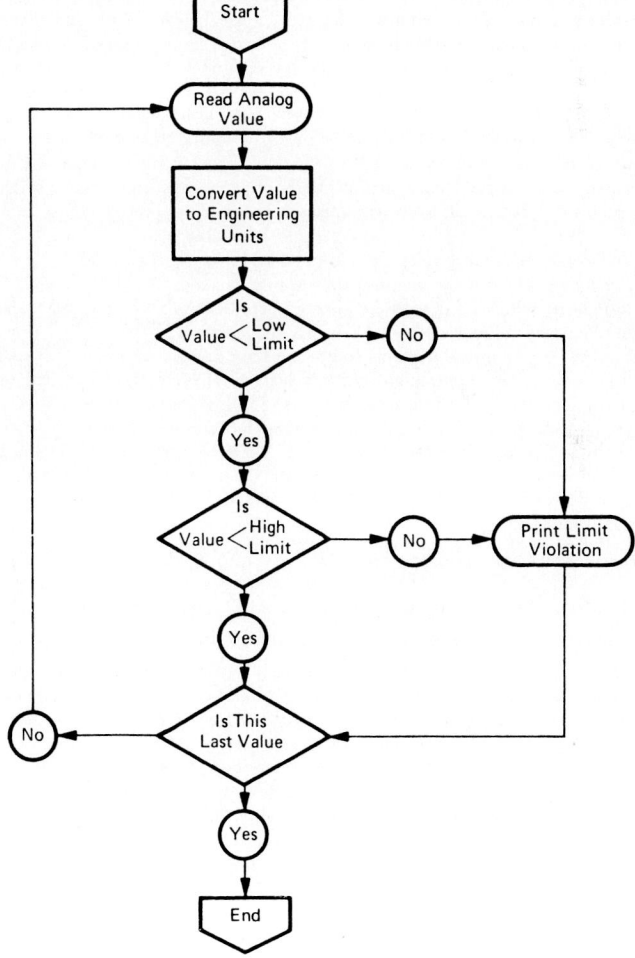

Fig. 28   Representative computer programming flowchart.

specific function box may represent an input-output operation, a numerical computation, or a logic decision. The program chart shown in Fig. 28 is of a program that reads values from the process, converts the values to engineering units, limit-checks the converted values and, if there is a violation, prints an alarm message on the process operator's typewriter. A set of symbols used in flowcharting is described in American National Standard ANSI X3.5-1970.

Various levels of detail are presented in programming flowcharts. A functional block in a low-level flowchart may represent only a few computer instructions, whereas a functional block in a high-level flowchart may represent many computer instructions. A high-level flowchart is used mainly for initial program design and as a way of informing a nonprogrammer of what the program does. A low-level chart usually appears in the last stage of flowcharting before a program is actually coded. It is used for documentation that may be needed for later modifications and corrections.

Thomas J. Harrison, IBM Corporation, Boca Raton, Fla.

Program Relocation. A process of modifying the address in a unit of code, i.e., a program subroutine, or procedure to operate from a different location in storage than that for which it was originally prepared. Program relocation is an integral part of nearly all programming systems. The technique allows a library of subroutines to be maintained in object form and made a part of a program by relocation and appropriate linking. When a program is prepared for execution by relocating the main routine and any included routines to occupy a certain part of the storage on a one-time basis, the process usually is termed *static relocation*. The resulting relocated program may be resident in a library and may be loaded into the same location in storage each time it is executed. Dynamic storage allocation schemes also may be set up so that each time a program is loaded it is relocated to an available space in storage. This process is termed *dynamic relocation*. Time-sharing systems may temporarily stop the execution of programs, store them in auxiliary storage, and later reload them into a different location for contined execution. This process is also termed dynamic relocation.

Software relocation commonly refers to a method whereby a program loader processes all the code of a program as it is loaded and modifies any required portions. Auxiliary information is carried in the code to indicate which parts must be altered. Inasmuch as all code must be examined, this method can be time consuming.

In *hardware relocation*, special machine components are used, as a base or relocation register, to alter addresses automatically at execution time to achieve the desired results. In dynamic relocation situations this is a fast method. Coding techniques also are used for relocation. The resulting code is caused to be self-relocating and executes in any storage location into which it is loaded. An index register may be used to make all references to storage. Double indexing is required to perform both the relocation and the normal indexing operations. Or, the code may actually modify all addresses as it executes to provide the correct reference. This method is slow compared with other methods and requires additional storage.

Thomas J. Harrison, IBM Corporation, Boca Raton, Fla.

Protected Location. See *location* earlier in this subsection.

Pseudo Instruction. See *instruction* earlier in this subsection.

Queue. When events occur at a faster rate than they can be handled in a computer system, a waiting line or *queue* must be formed. The elements of a queue typically are pointers (addresses) which refer to the items waiting for service. The items may be tasks to be executed or messages to be sent over communications facilities. The term is also used as a verb, meaning to place an item in a queue. Several methods of organizing queues are used. The *sequential queue* is common. As new elements arrive, they are placed at the end; as elements are processed, they are taken from the front. This is the first in–first out (FIFO) organization. In the *pushdown queue*, the last one in is the first one out (LIFO). The *multipriority queue* processes from the front, but in terms of priority of the elements waiting. Essentially, this is a modified sequential queue containing subsequences and is sometimes referred to as priority in first out (PIFO).

Record. In connection with computing systems, a group of related facts or fields of information treated as a unit.

Unit Record. (1) A card containing one complete record; a punched card. (2) Pertaining to equipment used in processing cards, such as readers, sorters, keypunches, and so forth.

Grouping of Records. The combining of two or more records into one block of information on a tape or disk to conserve tape or disk space. This is also called *blocking of records.*

Reentrant. Usually used with subroutines or procedures, this is a method of program coding which permits the code to be used concurrently by different calling programs. One copy of the code is resident in storage and can be used by several programs simultaneously. This conserves the storage that would be required for multiple copies of the code. The technique is commonly used in multiprogramming situations.

Relocation. See *program relocation* earlier in this subsection.

Residue Check. See *check* earlier in this subsection.

Restart. Normally, the restart function is activated by computer system failure or some other unusual occurrence in the area of the task being executed. Essentially, *restart* is the process of commencing again a computing job, task, or other program from a known point in the process. The restart function may be a separate task or program specified by the system designer in the event that the desired function cannot continue. A restart function may be specified for the whole system to provide for total system recovery in case of the detection of systemwide failures. This function may be referred to as the *reload* function and may be main-storage-resident or auxiliary-storage-resident.

Roll In–Roll Out. A function whereby a portion of the main storage is temporarily stored on auxiliary storage media to make storage available for a different function, usually one of higher priority. The later reloading of the stored portion for continuation of the execution is also part of the function. The function allows low-priority tasks or jobs to be run when the main or higher priority task does not require the storage but must have it on demand. A system which uses roll in–roll out must be able to tolerate the time required to accomplish the moves to and from the auxiliary storage.

Routine. A set of coded instructions arranged in proper sequence to direct a computer to perform a desired operation or sequence of operations. A subdivision of a program consisting of two or more instructions that are functionally related—hence a program.

Diagnostic Routine. A routine used to locate a malfunction in a computer or to aid in locating mistakes in a computer program. Thus, in general, any routine specifically designed to aid in debugging or troubleshooting.

Executive Routine. A routine which controls loading and relocation of routines and in some cases makes use of instructions not available to the general programmer. Effectively, an executive routine is part of the machine itself. Synonymous with *monitor routine, supervisory routine,* and *supervisory program.*

Heuristic Routine. A routine by which the computer attacks a problem not by a direct algorithmic procedure but by a trial-and-error approach frequently associated with the act of learning.

Interpretive Routine. A routine which decodes and immediately executes instructions written as pseudocodes. This is contrasted with a compiler which decodes pseudocodes into a machine language routine to be executed at a later time. The essential characteristic of an interpretive routine is that a particular pseudocode operation must be decoded each time it is executed. Synonymous with interpretive code.

Service Routine. A broad class of routines provided at a particular installation for the purpose of assisting in the maintenance and operation of a computer as well as the preparation of programs as opposed to routines for the actual solution of production problems. This class includes monitoring or supervisory routines, assemblers, compilers, diagnostics for computer malfunctions, simulations of peripheral equipment, general diagnostics, and input data. The distinguishing quality of service routines is that they are generally tailored to meet the servicing needs at a particular installation, independent of any specific production-type routine requiring such services.

Tracing Routine. A diagnostic routine used to provide a time history of one or more machine registers and controls during execution of the object routine. A complete tracing routine reveals the status of all registers and locations affected by each instruction each time the instruction is executed. Since such a trace is prohibitive in machine time, traces which provide information only following the execution of certain types of instructions are more frequently used. Furthermore, a tracing routine may be under the control of the processor or may be called in by means of a trapping feature.

RTS. Abbreviation for *request to send.*

Scheduler. The component of a programming system which determines the allocation of system resources in accordance with a predefined scheme. Usually, the scheduling function is broken down into several separate subfunctions, such as task or job scheduling and input-output scheduling. The scheduler acknowledges the requests for activity, ensures that any requisite conditions are met, and may initiate the requested action or put it in a queue to be acted on at a later time.

Sequence Check. See *check* earlier in this subsection.

Serial Operation. The flow of information through a computer in time sequence using only one digit, word, line, or channel at a time. Serial addition in character-oriented computers permits the formation of sums with low-cost hardware. Addition occurs from right to left. Parallel addition is used in faster word- or byte-organized computers.

Magnetic disk and drum storage units may access and record data on a serial-by-bit basis. Conversion to (or from) the parallel form utilized in the central processor (CPU) is performed in the associated control unit. Except in the case of short distances, most communications between computers or between computers and many types of terminals take place on a serial-by-bit basis.

Service Routine. See *routine* earlier in this subsection.

Simulator. A device, data processing system, or computer program which represents a system or phenomenon and which mirrors or maps the effects of various changes in the original, allowing the original to be studied, analyzed, and comprehended by means of the behavior of the model. The term is also used to designate a routine which is executed by one computer but imitates the operations of another computer.

Software. The totality of programs, procedures, rules, and (possibly) documentation used in conjunction with computers, such as compilers, assemblers, narrators, routines, and subroutines. References are made to the software and hardware parts of a system where the hardware comprises the physical (mechanical and electronic) components of the system. In some machines, the instructions are microprogrammed in a special control storage section of the machine, using a more basic code actually wired into the machine. This is contrasted with the situation where the instructions are wired into the control unit. The microprogram technique permits the economic construction of various size machines which appear to have identical instruction sets. However, microprograms generally are not considered software and are sometimes called *firmware.*

Subroutine. (1) The set of instructions necessary to direct a computer to carry out a well-defined mathematical or logical operation. (2) A subunit of a routine. A subroutine is often written in relative or symbolic coding even when the routine to which it belongs is not. (3) A portion of a routine that causes a computer to carry out a well-defined mathematical or logical operation. (4) A routine arranged so that control may be transferred to it from a master routine and so that, at the conclusion of the subroutine, control reverts to the master routine. Such a subroutine is usually a closed subroutine. (5) A single routine may simultaneously be both a subroutine with respect to another routine and a master routine with respect to a third. Usually control is transferred to a single subroutine from more than one place in the master routine, and the reason for using the subroutine is to avoid having to repeat the same sequence of instructions in different places in the master routine.

Closed Subroutine. A subroutine not stored in the main path of the routine. Such a subroutine is entered by a jump operation, and provision is made to return control to the main routine at the end of the operation. The instructions related to the entry and reentry function constitute a linkage.

Open Subroutine. A subroutine of which a replica must be inserted at each place in the computer program at which the subroutine is used. Although requiring more storage, this approach avoids linkage and housekeeping overhead.

Thomas J. Harrison, IBM Corporation, Boca Raton, Fla.

Summation Check. See *check* earlier in this subsection.

Tabulation. Preparation of output data in the format of a table, sometimes indicating differences and totals or simply totals. In order to reduce the demand on the programmer, utility programs, such as report generators, are used.

Task. The smallest unique unit that can contend for the control and storage resources of a computer. In a multiprogramming system, tasks are the elements in the system to which control is switched.

Time Sharing. The use of a device, particularly a computer or data processing machine, for two or more purposes during the same overall time interval, accomplished by interspersing component and subsystem actions in time. In the case of a digital computer, time sharing generally connotes the process of using main storage for the concurrent execution of more than one job by temporarily storing all jobs in auxiliary storage except the one in control. This technique allows a computer to be used by several independent program users. The method most often is associated with a computer-controlled terminal system used in an interactive mode.

The prime objective in using an interactive time-sharing system is to permit more efficient use of the computer system in proceeding from initial specification of a program to a final checked-out operating revision and also the more efficient use of people in the solution of problems by providing answers and services in a short time on demand. The traditional batch processing type of operation necessitates the coding and preparation of a program as a separate first step. Submission of the job to a computer control center for inclusion in a batch run can result in several hours of delay. The problem of writing and checking out a large, complex program in this fashion does not efficiently use a programmer's time because of loss of continuity in solving the problem. Time-sharing systems are designed to allow a user to operate the computer in an interactive manner, by way of a variety of console devices, and thus obtain immediate response to requests for answers while also receiving immediate indications of errors in programming.

Modern systems have the capability of handling many terminal users concurrently and still use available time to process batch jobs in the normal fashion. The magnitude of the capability for this type of operation depends on the computer speed, storage size, auxiliary storage size and speed, operating system used to manage and control these facilities, and type and amount of terminal service to be offered. *Time slicing,* described next, is a common approach to allocating time for each user in the terminal system.

Time sharing is also used in nonterminal systems for the purpose of sharing a portion of the system memory between jobs of different priority. This usually is called multiprogramming. See *multiprogramming* described earlier in this subsection. In some control and data acquisition systems, a portion of the main storage (a partition) is allocated to run non-process-related jobs (background jobs) in the conventional batch processing manner. However, when the system is notified that a process-related program requires the partition, the background job is stored in auxiliary storage until the higher priority program has been executed. The background then is restored and resumed if no other higher priority job is waiting for service.

Thomas J. Harrison, IBM Corporation, Boca Raton, Fla.

Time Slicing. A technique that allows several users to utilize a computer facility as though each had complete control of the machine. Several users can be serviced one at a time, unaware of each other, because of the relative speed between computer operation and human response. Essentially, time slicing is used by a software control system to control the allocation of facilities of a computer to tasks requesting service. The allocation basis is a fixed time interval. Each job is executed for the time period used in the time slice. The job is then temporarily stored on an auxiliary storage device (time sharing) or suspended (multiprogramming) while another job is being run. Each job, therefore, is run in an incremental fashion until complete.

Write Protection   See *location* earlier in this subsection.

Zero-level address   See *address* earlier in this subsection.

INPUT–OUTPUT SUBSYSTEMS AND COMPONENTS*†

Aliasing Error.   The Shannon-Nyquist sampling theorem states that, in order to reconstruct a signal containing frequency components in the spectrum 0 to *fm* hertz from sampled data, samples must be taken at a rate of at least $2f_m$ samples per second. Failure to obtain data at this rate converts high-frequency components into low-frequency components. Thus, the reconstructed signal contains low-frequency energy not present in the original signal. Figure 29 illustrates a 5-Hz signal sampled at a four-samples-per-second rate. The reconstructed signal has a frequency of 1 Hz. An *aliasing error* is an error which can be introduced into sampled data in a digital data acquisition system as a result of violation of the basic sampling theorem.

The term *aliasing* results from the interpretation that the high-frequency components take the "alias" of a lower frequency component. An equivalent term, *folding error,* arises from the interpretation that the frequency spectrum of the signal is folded such that high-frequency components appear in a lower frequency spectrum.

Sampling a continuous signal is equivalent to modulating the signal with a series of uniformly spaced impulses as indicated in Fig. 30. If a signal containing frequency components up to a frequency f_m is modulated by a carrier of frequency f_s, the frequency spectrum of the resulting sampled signal will consist of the original spectrum, plus harmonic spectra centered on the sampling frequency and its harmonics as shown in Fig. 30. This is the same effect as modulation of a radiofrequency carrier with an audio signal in amplitude-modulated radio, except that there are an infinite number of carrier frequencies because the frequency spectrum of a train of equally spaced impulses (the modulating signal) consists of an infinite number of frequency components at f_s and its harmonics.

Figure 30 shows that, if the sampling frequency f_s is less than $2f_m$, the lower sideband of the first-harmonic spectrum will overlap the original signal spectrum. Inasmuch as mathematical signal reconstruction methods are equivalent to low-pass filtering in the signal domain, it is evident that the signal energy from the first-harmonic spectrum overlapping the original spectrum will be passed by the low-pass filter and, therefore, will affect the nature of the reconstructed signal.

In the foregoing description, the signal was assumed to have no frequency components above the frequency f_m. In practice, signals are not definitively band-limited and high-frequency components are present in any real signal. In the majority of process control applications, the signals are nominally band-limited by the low-pass characteristics of the process of the signal transducers. In connection with such signals, the sampling theorem frequently is stated so as to require sampling at twice the highest *significant* signal frequency. In practice, selection of the sampling frequency depends on the interpretation of the term "significant." A rule of thumb is that the sampling rate should be 5 to 10 times the highest frequency of interest. Although not technically perfect, the rule is valuable inasmuch as signals are usually band-limited to reject frequencies above those of interest. The 5 or 10 multiplier provides a safety factor to account for the finite rolloff rate of most low-pass systems. Where accurate reconstruction of the sampled signal is needed, an estimate of the aliasing error should be made.

An estimate of the aliasing error can be made by using a graphical technique based on a frequency spectrum diagram of the type shown in Fig. 30. The frequency spectrum of the original signal and the spectra of the modulation harmonics are drawn as indicated. At each frequency in the bandpass of the reconstruction filter or its mathematical equivalent, the amplitude of the original frequency components and all contributions from aliased harmonics are summed. This sum is an estimate of the magnitude of this frequency component in the reconstructed signal. The estimate is conservative, inasmuch as the true contribution is represented by the vector sum of the components, whereas the estimate is based on the algebraic sum.

Thomas J. Harrison, IBM Corporation, Boca Raton, Fla.

*Terms are presented in alphabetical order for convenience of reference.
†Data converters and digital system amplifiers are described in subsequent subsections.

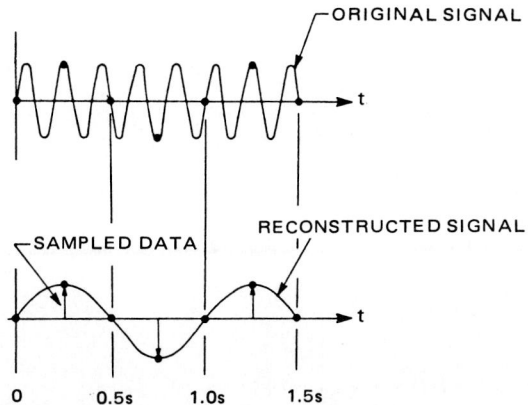

Fig. 29 Example of aliasing error in a sampled-data system.

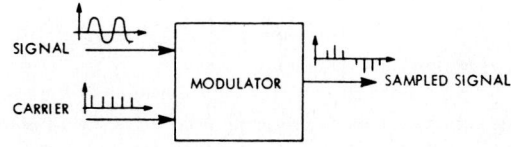

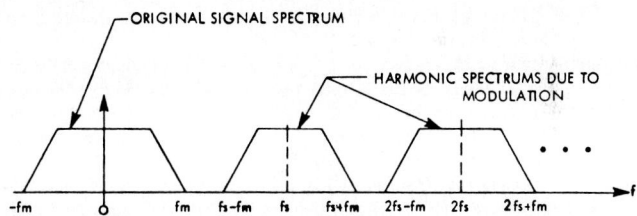

Fig. 30 Frequency domain sampled spectra.

Analog Input. An assembly of equipment (a subsystem) for performing the selection of the analog signal to be sampled, modification or conditioning of the analog signal, analog-to-digital (A/D) conversion to provide a digital representation, and the necessary digital controls for subsystem control and communication with a digital computer. Generally included in this assembly are signal conditioners, such as attenuators and filters, analog signal multiplexers, amplifiers, A/D converters, and the control logic.

For data acquisiton or process control computers, there are two fundamental analog input subsystem types. However, within each of these basic configurations, there are numerous variations possible as dictated by the use of different kinds of multiplexers, amplifiers, and A/D converters. Cost and performance are the usual predominating factors in selection. The two major subsystem types are (1) high-level and (2) low-level systems.

High-Level Analog Input Subsystems. High-level signals typically are in the range 0 to ± 10 V. They are connected to the subsystem input terminals of a configuration of the type shown

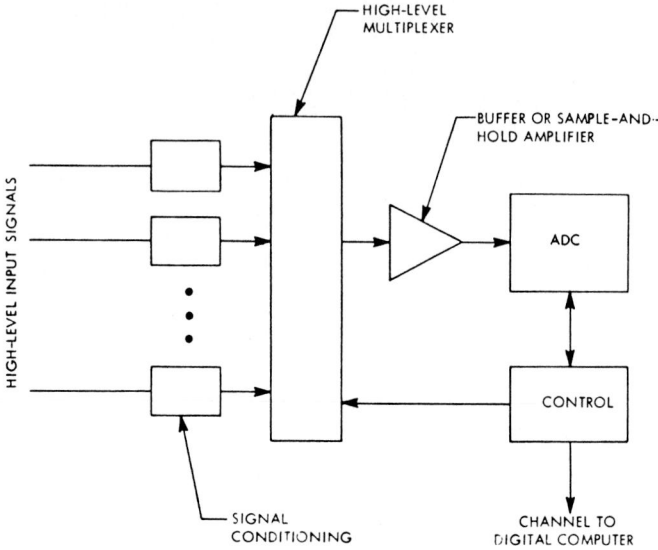

Fig. 31   High-level analog input subsystem.

in Fig. 31. Including filtering, limiting to protect the analog-input subsystem from overvoltage, and attenuation, the conditioning of signals is performed on a *per-channel* basis. The multiplexer selects the signal to be converted and is under the control of the subsystem control logic. Isolation between the A/D converter and the input signal source is provided by a buffer amplifier. A sample-and-hold amplifier may be used to reduce aperture-time errors. The buffer amplifier sometimes is included as an integral part of the A/D converter. The timing and sequencing of the subsystem operation is transmitted by the control logic, which also transmits the digital value to the digital computer.

Usually configurations of theses subsystems are single-ended, although some differential systems are obtainable. The latter are more costly but offer easier installation because noise and ground loops are less serious. Reduction of the latter problem is important, particularly in the case of high speed, high-resolution applications and where input signals must come from signal sources a considerable distance from the subsystem.

Solid-state switches, capable of sampling signals at rates in excess of 250,000 samples per second, normally are used in the analog multiplexer. Both bipolar transistors and field-effect transistors (FETs) are used. Bipolar transistor switches restrict levels to a high level becasue of voltage offset errors. The most common A/D converter used is of the successive approximation type, since its speed matches the solid-state multiplexer characteristics. For lower speed applications, ramp and integrating ramp A/D converters are used.

Timing and channel communications functions are provided by the control logic. The digital computer normally initiates a control word, specifying one or more input addresses, to commence operation of the subsystem. The control selects the appropriate multiplexer switches. After a predetermined delay time, the control unit sends a START CONVERT signal to the A/D converter. The latter operates asynchronously with respect to the computer and, at the end of the conversion, signals the control logic. Prior to transmitting the result back to the digital computer, the control unit may reformat the output word from the A/D converter; that is, it may add parity bits or convert from a parallel to a serial form. It may also collect some or all of the results in a buffer storage where they are available to the computer on demand.

The subsystem sampling speed may be as high as 100,000 samples per second at resolutions up to 14 bits. These characteristics are determined by the type of multiplexer, amplifier, and A/D converter used. Higher speeds are obtainable, but with reduced resolution or using special techniques. Under controlled operating conditions, a total measurement error of less than 0.05% of full scale is obtainable.

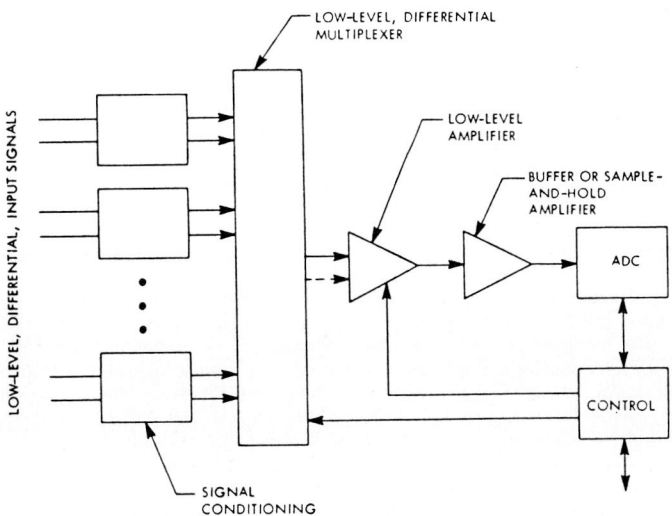

Fig. 32   Low-level analog input subsystem.

Subsystems of this type are found in numerous high speed, data acquisition applications inlcuding hybrid computation and rocket test stand monitoring, as well as for general research. These uses typically have wide channel bandwidth requirements, high-level signals, and/or a relatively small number of signals. Because the configuration is not adapted to the handling of low-level signals, it is not used widely for process control. For applications involving only a few data points, a high speed, low-level system can be achieved by adding a high-gain amplifier for each input signal. This is a costly approach and not practical where there are large numbers of low-level signals.

Low-Level Analog-Input Subsystems. Systems of this type are similar to the high-level system just described, with the exception of the use of a time-shared low-level amplifier. See Fig. 32. In operation, the input signals are conditioned and multiplexed by a low-level multiplexer. Before conversion to a digital representation by the A/D converter, the multiplexer output is amplified by a time-shared amplifier. Measurement accuracy for the total system for a 50-mV signal range typically is 0.1% with a resolution of 10 to 14 bits.

The low-level signals generally are less than 100 mV and often as low as 10 mV full scale. Thus, this type of subsystem usually provides a differential-signal input. The multiplexer and signal conditioning circuits are differential. The amplifier may or may not be differential, depending on the type of multiplexer used. A single-ended amplifier may be used where the multiplexer provides common-mode isolation, such as a transformer-coupled multiplexer. The output of the amplifer and the A/D converter input are single-ended in most cases.

Differential-multiplexer configurations for low-level uses most commonly are double-pole single-throw differential, flying capacitor, and transformer-coupled designs. Switching devices normally are dry reed relays, FETs, or mercury-wetted contact relays. FET switches can be applied for speeds in excess of 100,000 samples per second. Electromechanical devices are limited to about 300 samples per second or less. In any case, however, subsystem speed normally is limited to less than 10,000 samples per second because of amplifier performance. With the proper selection of multiplexer configuration and amplifier capabilities, common-mode voltages up to several hundred volts will not affect the acceptable performance of the system. However, where solid-state switches are used in multiplexers, the tolerance is limited to somewhat less than 30 V of common-mode voltage.

In the design of Fig. 32, a differential or single-ended amplifier may be used, depending on the multiplexer characteristics. In some cases, the amplifier will have a fixed gain. In other instances, the control logic may select the gain, either automatically or under control of the computer program. When a differential amplifier is used, the common-mode tolerance of the amplifier is critical in determining the common-mode rejection of the subsystem. Low-level applications also require that amplifier noise be kept to a minimum.

The design configuration just described is used extensively in process control and large data acquisition systems. Such uses typically have low channel bandwidth requirements and a large number of low-level signals. The time-shared amplifier provides the system with a cost advantage as compared with a system of the amplifier-per-channel type. The low-bandwidth application characteristic makes it possible to use the limited system sampling speed in numerous applications.

Thomas J. Harrison, IBM Corporation, Boca Raton, Fla.

Analog Input Module.   See *signal conditioners and logic modules* in this subsection.

Analog Multiplexer.   An array of analog switches used for the selection of one of several analog signals for transmission to subsequent devices in a data acquisition subsystem. Multiplexers most typically are used for signal selection prior to amplification and conversion to digital form in time-shared subsystems. An analog multiplexer used in the analog input subsystem of a data acquisition system is shown schematically in Fig. 33. Multiplexers are not confined to data acquisition systems but are used in audio switching networks at central telephone exchanges. The particular hardware configuration varies considerably with the nature of the signals and their manipulation.

The control units supervises switch selection. Where the multiplexer is differential, two or three switches are required for each input signal. Where the multiplexer is single-ended, only one switch per input is required, using a common ground to complete the circuit for all input signals. Although the control unit varies with particular applications, its primary function is to furnish the signals required to select or address the various switches in the multiplexer at the proper time. The logic in the control unit or a command from a computer determines the particular address to be selected. In some cases, the switches may be selected sequentially under the control of a ring counter circuit in the control unit. Or, the selection of each address may require a data word from the computer or microprogrammed control unit.

The control unit also furnishes timing for the multiplexer and associated equipment. Usually, the analog-to-digital (A/D) converter cannot commence a conversion instantly after selection of the multiplexer switch. A time delay is required to permit actuation time for the switch for settling of the amplifier and for the dissipation of other transients. Counters or delay circuits in the control unit usually furnish the required timing signals.

Several types of solid-state or electromechanical analog switches may be used in the multiplexer. Types of electromechanical analog switches used are mercury-wetted contact relays, dry reed relays, and crossbar switch assemblies. Of the solid-state swtiches, FETs are used most often. For special applications, diodes, bipolar transistors, and silicon controlled rectifiers may be used.

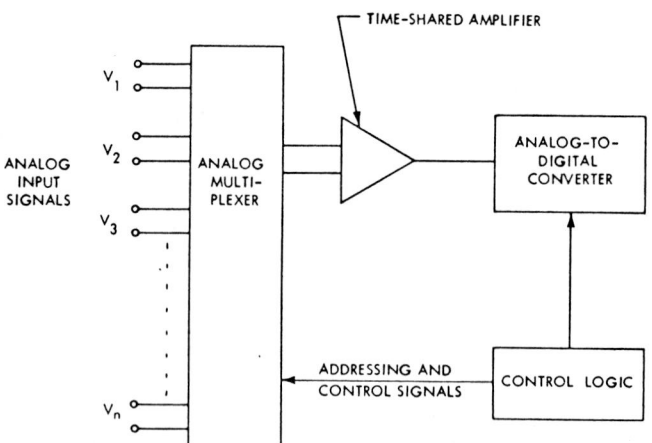

Fig. 33   Location of analog multiplexer in an analog input subsystem.

Among the major performance characteristics of an analog multiplexer are accuracy, sampling speed, and noise. The common-mode rejection ratio also is important in the instance of a differential multiplexer. The features of the multiplexer switching device mainly determine sampling speed. Electromechanical devices, such as mercury-wetted contact relays and dry reed relays, usually are limited to about 250 samples per second. Bipolar transistors and FETs can provide sampling speeds in excess of 100,000 samples per second. Although a switch closes rapidly, a time interval must be provided to permit transients to dissipate prior to conversion of the multiplexer output signal to digital form. In the case of the mercury-wetted relay, the actual switch closing time is in the range of 1 ms, but there may be a noise transient which persists at levels in excess of 10 μV for perhaps 5 ms or even longer. Where a noise level of this magnitude cannot be tolerated, a delay is required between switch selection and conversion initiation. Noise considerations also affect solid-state switches, but noise duration times are generally much shorter.

Careful consideration must be given to possible errors a multiplexer can contribute to the operation of a digital data acquisition subsystem. Leakage currents associated with the "off" switches will, in the case of solid-state switches, flow through the source impedance of the channel being sampled by the "on" switch and also through the multiplexer load impedance. Offset in the voltage being sampled thus can be caused by these currents. Even though leakage current from a single switch may be small (considerably less than a microampere), an appreciable error can result from the cumulative effect of all the switches connected to the common multiplexer output bus.

Inasmuch as the FET does not show an offset voltage in the "on" condition, it may be used for multiplexing low-level signals. It should be noted, however, that some FETs show a significant "on" resistance and, because of source loading effects, can introduce errors. Further, the "on" resistance increases the effect of leakage currents as previously described. This may increase the output signal rise time.

Where differential multiplexers are used, the common-mode rejection ratio is important. This factor is determined by leakage impedances between the swtich signal path and the drive circuit. Further, the maximum common-mode voltage which may be applied without causing damage to the multiplexer is determined by the breakdown voltage of the switch. Electromechanical devices have high drive-to-signal-path and contact-breakdown voltages. Hence, these devices can be used for multiplexers that will withstand several hundreds of volts of common-mode voltage. Generally, solid-

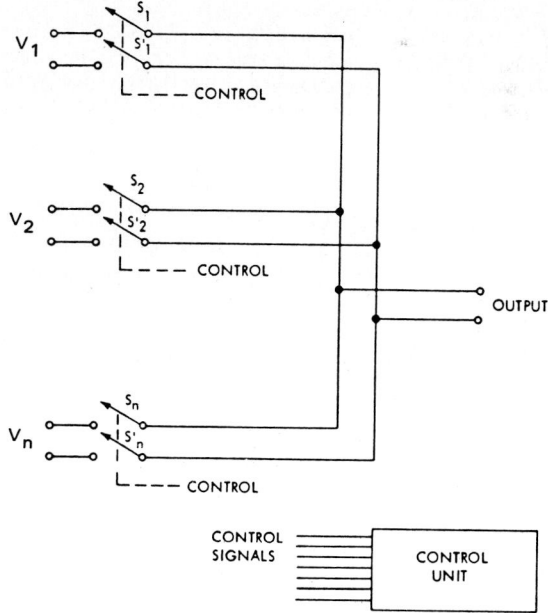

Fig. 34   Generalized configuration of an analog multiplexer.

state multiplexers are limited to a common-mode voltage less than 20 to 30 V. Improvement can be obtained through the use of isolation techniques, such as transformers.

The chances of multiplexer error increase as the number of input channels in the multiplexer increases. Additional channels decrease the common-mode rejection ratio and also increase the offset resulting from leakage currents. *Submultiplexing* or *block switching* are techniques frequently used to minimize these conditions. Block switching entails the use of two levels of multiplexing. The first level is the same as that indicated in Fig. 34. This level furnishes selection of an input signal. The second level of multiplexing is provided at the output of the first-level multiplexer. As an example, a second-level switch may be provided for each group of 16 first-level multiplexing switches. Of course, in the selection of an input, both the first-level switches and their associated second-level switch must be actuated. The advantage of second-level multiplexing is the isolation of each block from the leakage currents and other possible disturbances which may arise from other inputs. In a system of 16 blocks of 16 channels each, for example, the leakage errors are determined mainly by the 15 "off" switches in the same block as the addressed point and the 15 "off" block or second-level switches. Hence, these errors equal those encountered in a 30-channel single-level multiplexer, although a total of 256 input channels are serviced through the two-level multiplexer.

Thomas J. Harrison, IBM Corporation, Boca Raton, Fla.

Analog Output.   An assembly of equipment (a subsystem) and operations in a process control computer or data acquisition system with the capability to provide a continuous voltage or current output which can be controlled by a digital computer. Closed-loop control systems, for example, which utilize a digital process control computer, require current signals which may have ranges of 4 to 20, 1 to 5, or 10 to 50 mA. These signals, in turn, are used to control process actuators, such as valve positioners. For the generation of visual displays, recorders, and cathode ray tubes require similar analog signals.

A digital-to-analog (D/A) converter is the principal component of an analog output subsystem. Simply defined, a D/A converter is an electronically controlled attenuator network and a constant reference source. Digital input signals received by the D/A converter activate analog switches which determine the attenuation factor of a passive network. Either electromechanical or solid-state switches may be used. The input to the attenuator network is a constant voltage or current source. The output of the network is proportional to the attenuator switch settings and thus to the digital input signals.

Connection of the D/A converter to the digital computer is through the control logic. The latter provides addressing for each D/A converter and also the required control and timing signals. Shown in Fig. 35 is an analog output system of the kind used in process control computer or data acquisition systems. The digital computer, via digital control words, specifies which D/A converter is to be adjusted. Another digital data word carries the desired output value. The address information is decoded in the control logic. The digital data also are routed to the input of the addressed D/A

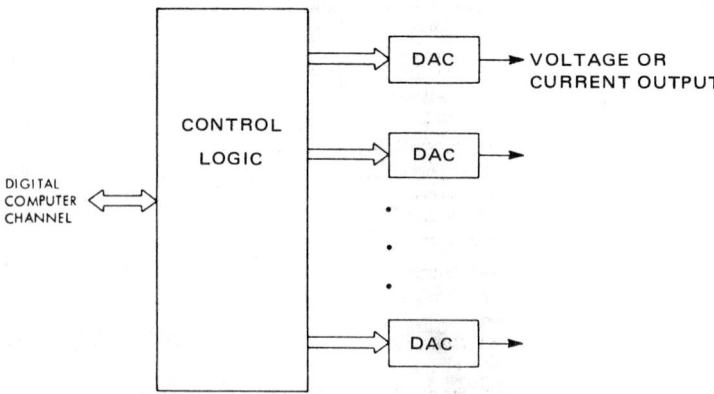

Fig. 35   Basic analog output subsystem.

converter. This setting of the input switches of the D/A converter provides the desired analog output value. The kind of subsystem shown is frequently used where the analog output value is changed often, at a rate in excess of 1000 samples per second. This subsystem configuration can adjust many D/A converters at rates exceeding 100,000 samples per second. The resolution of the D/A converters used for applications of this type usually ranges from 8 to 13 bits. The total error of the output voltage, under controlled operating conditions, may be less than 0.01%.

Thomas J. Harrison, IBM Corporation, Boca Raton, Fla.

Analog Switch. This term applies to a large family of switches designed for switching analog signals and normally implies high accuracy and high resolution. In data acquisition and instrumentation systems, the most important switch characteristics are speed, voltage, and current errors, "on" and "off" resistance, and noise.

Electromechanical analog switches, including mercury-wetted contact relays and dry reed relays, usually have a lower "on" resistance and a higher "off" resistance than most solid-state switches. Also, because of excellent isolation between drive and signal circuits, electromechanical switches have no inherent voltage offset and leakage currents. Disadvantages include their slow performance, as compared with solid-state switches, and the production of noise by the switching action, a condition which tends to persist for an appreciable period.

Solid-state switches on the other hand are high speed devices and with "on" and "off" resistances, compared with electromechanical switches. Where a *pn* junction is part of the signal path, as in a bipolar transistor, the switch shows an inherent voltage offset. The leakage currents between drive and signal paths also are larger in bipolar transistors as the result of less isolation between drive and signal. Field-effect transistors (FETs) display leakage comparable with that of electromechanical switches. Noise results from the switching action mainly because of coupling of the drive signal into the signal path by way of interelectrode capacitance. Even though the noise magnitude may be appreciable, decay to a negligible value occurs much more rapidly than in electromechanical switches.

Use of the major classes of analog switches may be summariezed as follows:

Relays and Other Electrochemical Devices. Used for low-level signals (less than 1 V full scale) and multiplexing speeds generally not exceeding 250 samples per second.

Field-Effect Transistors. Used for low-level applications at higher speeds. Limited in common-mode voltage handling capability because of lower breakdown voltages.

Bipolar Transistors. Used for high-level signals, but with appropriate circuitry for compensation of offset voltages; also may be used for low-level signals. They have been used less frequently in recent years as a result of improvements in FET technology.

Thomas J. Harrison, IBM Corporation, Boca Raton, Fla.

Analog-Zero Module. See *signal conditioners and logic modules* in this subsection.

Badge or Card Reader. Magnetic stripes or optical character recognition (OCR) characters on the back of credit cards and identification badges, in conjunction with special purpose terminals, are used to enter limited information into computers for identification or credit authorization purposes. In the case of magnetic stripes, a narrow band of magnetic material is applied to the back of the card and data are recorded using the same techniques used with magnetic tapes or disks. The optical techniques are identical to those used in commercial applications of OCR. In most cases, the recorded information is not altered and only the terminal can read the information. However, several rapid transit systems are issuing tickets having magnetic stripes on which the value of the ticket is recorded. Upon exiting from the train at the destination, the customer inserts the ticket in a terminal which debits the proper amount. If a value balance remains, the ticket is returned to the customer for subsequent use. This technology suggests many uses in industrial instrumentation applications.

Bar Code Scanner. The use of a bar code for identifying food and other products is rapidly gaining acceptance. The code, known as the Universal Product Code (UPC), provides for the encoding of 12 numeric characters, including two check characters. Each character code consists of two dark bars and two light bars (i.e., the absence of dark bars). Each individual character is encoded in a region consisting of seven rectanlinear areas called modules. The dark bars may occupy up to three modules, and the light bars may be up to four modules wide. In addition, two one-module dark bars at each end of the symbol and in the middle provide delimiters. The overall symbol has the appearance of

30 dark bars of varying width, separated by light bars. The encoded characters are also printed along the bottom edge of the symbol to provide human readability. For decoding, the symbol is scanned optically in an X-shaped pattern. The resulting binary signal is processed to identify the particular character.

The UPC symbol was developed for the Uniform Grocery Product Code Council, Inc., which is composed of representatives of grocery manufacturers and supermarket chains. See Fig. 36. It is an identification code only and does not incorporate pricing information. Pricing information, since it varies with location and time, is associated with each unique identification number by a computer. Although originally developed for groceries, the UPC is being expanded for use on other products.

Fig. 36   Example of a bar code. Shown here in full size (as used on grocery items).

Bipolar Transistors.   See *analog switch* earlier in this subsection.

Block Switching.   See *analog multiplexer* earlier in this subsection.

Bright Switch.   Invented by R. L. Bright, this device is a solid-state switch and consists of two bipolar transistors connected in the inverted connection shown in Fig. 37. Inasmuch as offset voltage is reduced by use of the inverted connection and the back-to-back transistor connections, a Bright switch more closely approximates an ideal voltage switch than a single transistor. When a bipolar transistor is driven into saturation by a signal applied between the base and emitter, an open-circuit (where the collector current is zero) voltage appears between the emitter and collector. Called the *offset voltage,* this voltage represents an error when the switch is used in analog switching applications. This is the case because the voltage appears in series with the signal source being switched. If, on the other hand, the transistor is driven into saturation by a drive signal applied between the base and collector, the offset voltage magnitude is substantially reduced. The difference between the two offset voltages is mainly a function of the difference between the normal β (common-emitter current gain with the drive connected between the base and emitter) and the inverted β (common-emitter current gain with the collector functioning as the emitter). The Bright switch configuration utilizes this phenomenon and the cancellation effect of connecting the transistors in a back-to-back configuration to achieve a low-offset switch. The offset of one transistor is partially canceled by the offset voltage of the other transistor. As contrasted with an offset voltage of several millivolts in a typical silicon transistor, net offsets of much less than 50 mV can be achieved with a Bright switch. The switch finds limited usage today.

Card Reader.   See *badge or card reader* earlier in this subsection.

Charge-Transfer Device.   Abbreviated CTD, a *charge-transfer device* is an array of closely spaced capacitors fabricated using metal-oxide semiconductor (MOS) technology. Operation of the device involves the movement of a charge packet which is stored on one capacitor to an adjacent capacitor

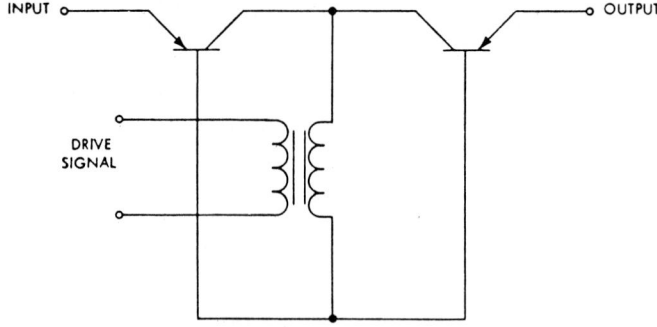

Fig. 37   Circuit configuration of a Bright switch.

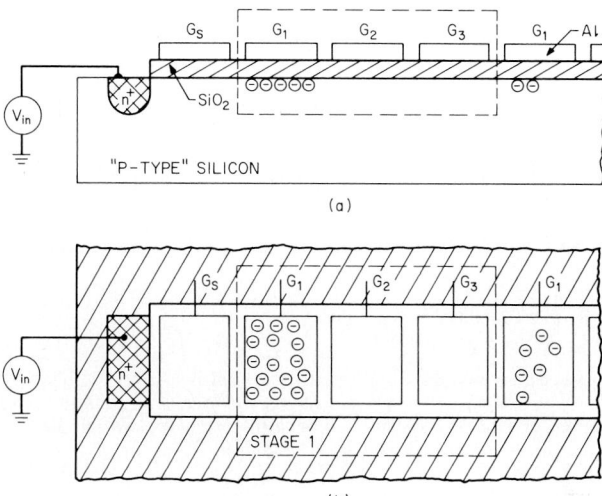

Fig. 38 (*a*) Cross section of a portion of a charge-transfer device (CTD) with an input composed of an n^+ diode and a sampling rate G_s which isolates the signal charge packet. The first stage of the device is enclosed by dashed lines. (*b*) Top view corresponding to (*a*).

The device is composed of a silicon dioxide insulating layer 1000 Å thick sandwiched between a metal layer (typically aluminum) and a semiconducting (silicon) substrate, which for CTDs is generally doped to be *p* type. A CTD is formed by selective etching of the metal, lcoalized oxidations, and diffusion of elements, such as boron and phosphorus, into the silicon. The features of the device can be extremely small (on the order of micrometers), and photographic techniques are used to define the circuit patterns. Application of a positive voltage to one of the metal gates depletes the *p*-type silicon underneath that gate of holes, which are its major charge carriers. However, this voltage attracts minority carrier electrons, because it produces an energy minimum (potential well) for electrons at the interface between the silicon and the insulator. It is these potential wells that are used for storing and transferring the signal charge.

in the array. The simplest application of a CTD is an electronically variable analog time delay. A CTD provides an inexpensive and convenient means of sampling and storing a data transient in analog form. Since the CTD sample period can be varied, after a number of samples have been taken at sufficiently short sample intervals (i.e., a sufficiently high clock rate $1/T_c$) the clock rate is reduced and the data are read out at a slower rate for analysis. Devices that can store 1000 samples taken at time intervals as short as 50 ns (a clock rate of 20 MHz) are commercially available, and clock rates as high as 130 MHz have been achieved with special CTD designs. An example of an application of a CTD delay line that makes use of the ability to vary the transfer rate of charge packets is transient data recording. Many physical phenomena occur so quickly that it is very difficult to analyze the data from a single event as it happens. It is therefore desirable to store the data and then perform the analysis at a more moderate rate. Unfortunately, most conventional storage techniques require that the data be converted first into a digital representation with a high speed, analog-to-digital converter, which can become very expensive for short sampling intervals. There are many other applications of analog time delay, including echo generation for electronic music, ghost cancellation in television, and filtering. Also, by the addition of input (or output) structures at each stage of delay, time division multiplexers (or demultiplexers) for radar and communication systems can be implemented. See Fig. 38.

CMRR.   Abbreviation for *common-mode rejection ratio.* See next term.

Common-Mode Rejection Ratio (CMRR).   A parameter used to express the ability of a differential subsystem or instrument to reject the effect of a common-mode voltage applied at its input

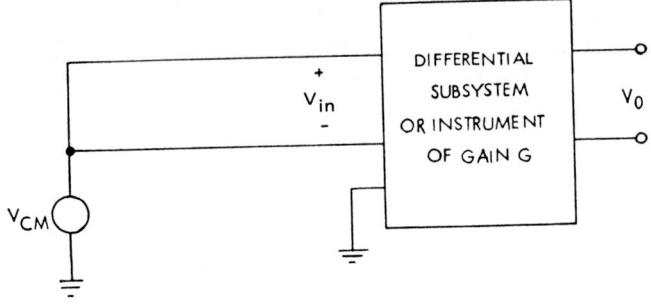

Fig. 39 Model of the common-mode rejection ratio.

terminals. An idealized model of the CMRR is shown in Fig. 39. Voltage V_{CM} is the common-mode voltage, V_{in} is the equivalent differential-input voltage due to V_{CM}, V_O is the output voltage (in the case of a digital measurement subsystem or instrument, the output value), and G is the gain or conversion factor of the differential subsystem or instrument.

The CMRR $= V_{CM}/V_{in}$. Inasmuch as it is difficult to measure V_{in} directly, an equivalent definition, CMRR $= V_{CM}G/V_O$, may be used as the basis of measurement.

If the common-mode voltage is ac, measurement of the CMRR is based on the ac component in the output. Where a digital output is involved, this is the increase in the repeatability or spread in output values. Consistent units must be used; i.e., if peak-to-peak values are used for V_{CM}, then the peak-to-peak increase in the spread of output readings must be used.

Frequently, the CMRR of an instrumentation subsystem ranges from 100 to 140 dB or, when expressed as a ratio, from 10^5:1 to 10^7:1, with 120 dB being the most common. A definition of the CMRR in decibel units follows the formula: CMRR (dB) $= 20 \log(\text{CMRR})$. A specification of 10^6:1 signifies that every volt of common-mode voltage results in the equivalent of 1 μV of normal-mode voltage at the input of the subsystem. Since the CMRR of most subsystems depends on the resistive source imbalance and, often, on the value of the common-mode voltage, a CMRR specification should include these parameters. A complete specification of the CMRR is CMRR $= 10^6$:1 at up to 1000 Ω source imbalance and 200 V dc.

Common-Mode Voltage. A voltage common to both inputs of a differential subsystem or instrument when measured with respect to a system reference point (usually ground). V_s is shown in Fig. 40. The point G_1 is the ground reference point for the measurement subsystem. It will be noted that the signal voltage is referenced to G_2. The latter may or may not be at the same potential as G_1. Voltages at input terminals A and B are $V_A = V_s + V_{CM}$ and $V_B = V_{CM}$. The voltage common to both A and B is V_{CM}, the common-mode voltage. Common-mode voltage sometimes is defined as the

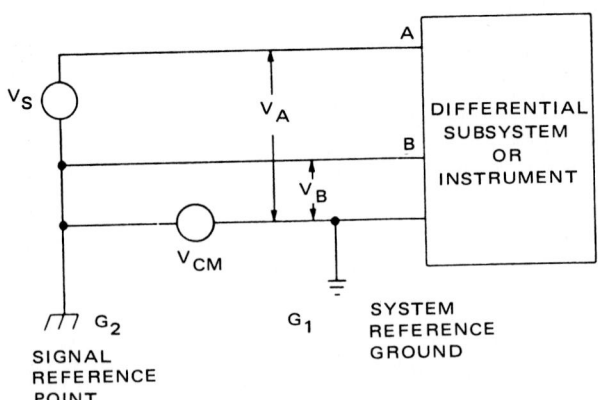

Fig. 40 Common-mode voltage.

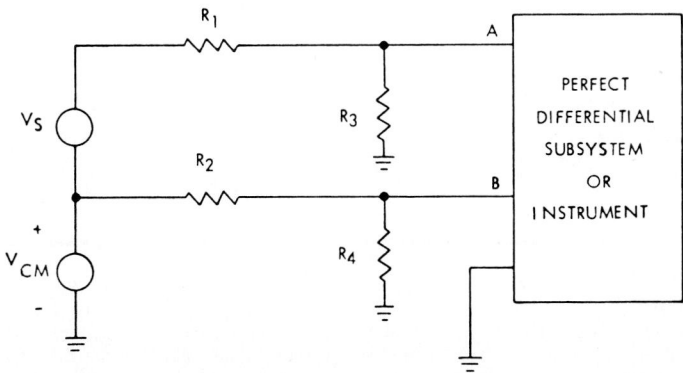

Fig. 41 Model of a conversion from common to normal mode.

average value of V_A and V_B, i.e., $(V_A + V_B)/2$. Inasmuch as common-mode voltage generally is of concern only when $V_{CM} \gg V_s$, there is little practical difference between the two definitions.

Common-mode sources cause problems in differential-measurement systems inasmuch as they can cause measurement error due to common-to-normal-mode (or differential-mode) conversion. In a perfect differential-measurement system, only the value of the signal voltage V_s would be measured—with no contribution arising from the presence of a common-mode voltage V_{CM}. However, in practice, some common-mode voltage usually appears as a signal voltage.

Conversion can occur as shown in Fig. 41. Resistances R_1 and R_2 are, respectively, representative of the input line resistance and the source impedance of V_s. Resistances R_3 and R_4 are leakage resistances to ground in both the input lines and the differential measurement subsystems. Resistances R_1 and R_3 comprise a voltage divider, as do resistances R_2 and R_4. If the input impedance of the instrument is neglected, the differential- or normal-mode voltage is

$$V_A - V_B = V_S \frac{R_3}{R_1 + R_3} + V_{CM}\left(\frac{R_3}{R_1 + R_3} - \frac{R_4}{R_2 + R_4}\right)$$

If one assumes that $R_3 \approx R_4 \gg R_1, R_2$, the expression becomes

$$V_A - V_B \approx V_S + V_{CM}\left(\frac{R_2 - R_1}{R_3}\right)$$

The desired measurement result is the value V_S. Consequently, the second term in the foregoing expression represents the error arising from common-to-normal-mode conversion, commonly referred to as the *common-mode error*. The common-mode rejection ratio, which is a measure of the ability of a system to reject the effect of common-mode voltage, is the inverse of the coefficient of V_{CM}, namely, $|R_3/(R_2 - R_1)|$.

The situation shown in Fig. 41 is only one possibility whereby common-mode voltage can be converted into a normal-mode signal. Where the common-mode source is ac, resistances R_3 and R_4 usually can be neglected. However, the leakage capacitance from input terminals A and B to ground must be considered. The diagram of Fig. 41 represents an idealized version of the actual phenomenon. Lumped parameters are assumed, but in an actual situation the resistances and capacitances are distributed along the length of the input lines.

Common-mode voltage can result from the way a system is used, or from entirely unintentional causes. Bonding of a thermocouple to a current carrying conductor and use of a subsystem to measure the voltage drop across an ungrounded resistor in a circuit are representative of error arising from the manner in which the system is used. Where the signal source is grounded, and where ground potential differs from that at the measurement point, the source of common-mode voltage error can be considered unintentional. Needless to say, however, the system does not differentiate between the two causes.

Thomas J. Harrison, IBM Corporation, Boca Raton, Fla.

Comparator. See *signal conditioners and logic modules* later in this subsection.

Computer Interface Module. See *signal conditioners and logic modules* later in this subsection.

CRT Devices. See *keyboard–cathode ray tube devices* later in this subsection.

CTD. Abbreviation for *charge-transfer device.*

Curve Plotter. A digital computer output device which draws curves of one or more variables as a function of one or more variables. The function is essentially that of an *xy* recorder. Data from the digital computer are translated into plotter actuating signals and converted to incremental plotter movements to produce a drawing, chart, or other graphic portrayal. Two configurations are common. In a *flat-bed plotter,* the paper is fixed to a flat surface and one or more pens are mounted on a carriage capable of moving in two dimensions, normally designated *x* and *y*. In a *drum plotter,* the paper is affixed to a cylindrical drum whose rotation provides one dimension (usually) of motion. A pen mounted above the drum surface and capable of moving axially with respect to the drum provides the *x* dimension. Control is also provided to raise or lower the pen from or to the paper surface. The information from the computer is decoded into fixed incremental movements of the drum and/or pen carriage. Each plotter command transmits information which is decoded into the number of pen-carriage increments to be moved (*y* axis) or the number of drum increments to be moved (*x* axis). Specified bit combinations also control the raising or lowering of the pen to the paper. See also Sec. 15 of this Handbook.

Data Link. (1) *RS-232* is a standard established by the Electronic Industries Association to define a widely used electrical interface for communication between digital data handling devices. The specification refers strictly to *physical* characteristics of the transmitting and receiving devices, along with the interconnecting cable. These characteristics include impedance values, voltage levels that define logic 1 and logic 0 states, electrical connectors, pin assignments, and the specific function of each cable conductor. The full RS-232 interface, rarely used, provides for 25 cable conductors. Most practical operations use various smaller subsets, depending on the functions being implemented. This bit-serial transmission requires only two wires, one for each direction of travel. Remaining lines are used for ground returns and for various "handshake" and "interrupt" functions. A minimum configuration (for one-way transmission to a simple printer, for example) may consist of only two wires. Full details are available from Electronic Industries Association, 2001 Eye St., N. W., Washington, D.C. 20006.

(2) *ASCII,* standing for *American Standard Code for Information Interchange,* is an American Standards Association specification for the language code used by a majority of manufacturers of data equipment. RS-232 and ASCII are separate standards; neither is necessarily associated with the other, since the interface can operate with other codes.

Seven binary bits determine 128 standard ASCII characters: numerals, upper- and lowercase letters, punctuation marks, and various special control characters. An optional eighth bit allows parity checking to detect transmission errors. In serial transmission, this 8-bit string is preceded by one "start" bit and followed by one or more (usually two) "stop" bits. This makes a total of 10 or more (usually 11) bits for the transmission of one character.

Transmission speed must fall within the common capability of both devices and, of course, each device must "know" the agreed-on rate. There are various standard speeds, called *Baud rates,* ranging from 50 to 19,200 bauds (or higher). One baud equals one bit transmitted per second. Dividing the bauds by the number of bits per character yields the maximum number of characters that can be transmitted per second.

When parity is used, 1 is transmitted (or not transmitted) in the eighth-bit position in order to make the total number of 1's in the character string either *even* or *odd* (as agreed). By counting the bits transmitted for each character, the receiving device can quickly recognize the occurrence of transmission errors.

Delta *Y* Module. See *signal conditioners and logic modules* later in this subsection.

Differential-Mode Voltage. The voltage appearing between two terminals or other points in a circuit, neither of which is necessarily at the system reference potential (usually designated as ground), is termed the differential- (or normal)-mode voltage. In Fig. 42, $V_A - V_B$ is the differential input voltage. A perfect differential system would indicate only the value of the differential signal voltage V_s. However, in practical installations, imbalances and inaccuracies in the system often result in conversion of some of the common-mode voltage into a differential-mode signal. See *common-mode voltage* described earlier in this subsection.

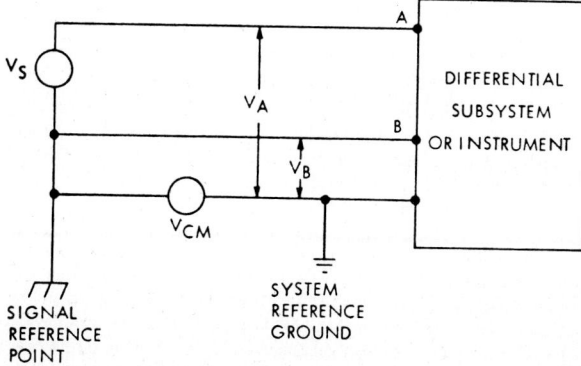

Fig. 42   Differential-mode voltage.

Digital Multiplexer.   A device that permits sharing a common information path between multiple groups of input or output digital signals. For example, a digital multiplexer can be used to transfer information between the central processing unit (CPU) of a computer and any one of several digital input or output devices. A form of digital multiplexer for use in reading information of groups of contact points is shown in Fig. 43. To read the status of contacts in group A, a positive signal is transmitted on the group A select line. Should the contact be closed, the positive signal will appear on the associated bit line. Bit lines are sampled during the period when the select line is active. The status of the group of contacts is stored in computer memory. On completion of the operation, the signal on the select line is removed. When it is desired to read group B, a positive signal is placed on the group B select line and the status of the contacts in group B is placed on the associated bit lines. Diodes in series with the contacts are needed to prevent back-circuits; the number of groups that can share the same input bit lines is a funciton of the back-resistance of the diode.

Digital Output.   The digital output information of a computer may be a contact-operate signal or a voltage-level signal. Under the control of the computer program, digital information is transferred from the central processing unit to a digital output register. The output of the register may be connected directly to the digital output signal lines to provide the voltage-level signal. To provide the contact-operate signals, the ouptut of the register is connected to a driver circuit which is connected

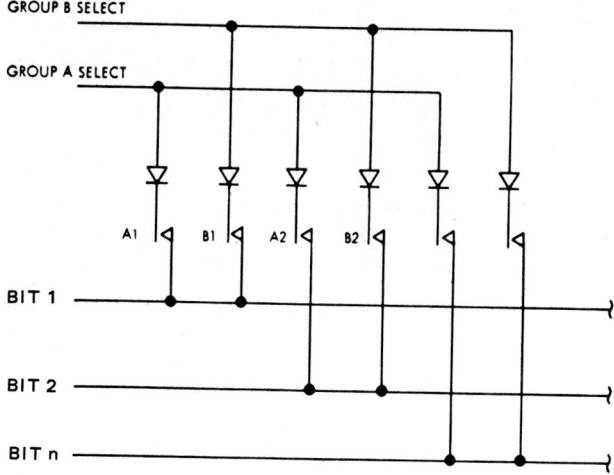

Fig. 43   Digital multiplexer.

to the digital output signal lines. When a register position is turned on, the drive circuit provides the current which actuates an external device, such as a relay. The information remains in the register until another computer command to the same group of digital output points is executed, or the register may be reset by a signal from the external device or after a fixed time interval has elapsed.

Dry Reed Relay.   An electromechanical, magnetically actuated, hermetically sealed switch used mainly for analog multiplexing and for controlling low-power loads (3 to 5 VA). The relay capsule is comprised of overlapped cantilevered contacts which are housed in a hermetically sealed glass capsule. An inert or reducing atmosphere under moderate pressure is contained in the capsule. See Fig. 44. The length of the capsule ranges from less than an inch (2.5 cm) to several inches (8 to 10 cm), depending on the power level and application for which the switch is designed. Diameters range from 0.25 to 0.1 in (0.6 to 0.25 cm). Nickel-iron alloy contacts, with relatively high permeability and low retentivity, usually are used. Frequently, the contact portion of the reed is electroplated with another metal for the reduction of contact resistance and to increase reliability.

The switch usually is actuated by a magnetic field produced by a coil wrapped around the capsule or by a magnet located near the capsule. Opening of the switch is accomplished by the restoring spring force of the reeds proper. Through the use of permanent magnets, the reeds can be biased to achieve normally closed, latching, and other contact actions.

The so-called pick time of the relay is dependent on the construction of the capsule and characteristics of the drive signal. Normally, pick times are less than 1 ms (including bounce). Drop times may be as small as a few tens of microseconds. A noise voltage is produced in the reed following closure as the result of magnetostrictive effects. The noise may persist at appreciable levels (greater than 20 μV) for several milliseconds. The reed relay generally is limited to sampling speeds of less than 250 samples per second because of the noise and pick time characteristics. Typically, relays are specified for a life well in excess of 10^6 operations at a full-rated resistive load.

Electrophotographic Printing.   See *printers* later in this subsection.

Encoder.   A device or subsystem which accepts an input and produces an output in coded form. This definition includes digital logic configurations which convert a digital input word in one code to an output word in a different code. Examples include decimal-to-binary and decimal-to-octal decoding circuits. *Encoder* also refers to a device or subsystem which converts an analog quantity into a digital representation by use of a quantization technique. Electronic and electromechanical analog-to-digital converters, such as shaft-to-digital converters, can be regarded as encoders. There is a trend toward restriction of the term to electromechanical converters, such as shaft-to-digital encoders. See also Sec. 9 of this handbook.

Field-Effect Transistor (FET) Switch.   A FET switch generally is used to control analog signals and is comprised of one or more FETs. In analog switching and multiplexing applications, a metal-oxide semiconductor FET (MOS FET) is used. The absence of an offset voltage makes the FET switch suitable for switching low-level signals. Further, the high input impedance provides excellent isolation between signal path and drive voltage.

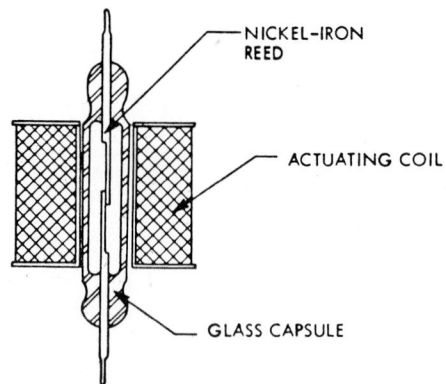

Fig. 44   Dry reed relay.

A FET switch can be used for switching at rates in excess of 100,000 samples per second even though they are not as fast as bipolar switches. A relatively high "on" resistance is the main disadvantage of the FET switch. "On" resistance ranges for 50 to 200 Ω. However, some devices are obtainable with less than 5-Ω "on" resistance.

There is no inherent offset voltage because there is no pn junction in the signal path as there is, for example, in a bipolar transistor. The FET is a majority carrier device in which the conductance of a conducting channel between two electrodes, the source and the drain, is controlled by a signal applied to a third electrode, the gate. A field established between the channel and the gate, insulated from the channel by a thin layer of material (usually silicon dioxide), controls the width of the conducting channel, hence resistance, in the MOS FET. Thus, there is no pn junction in the signal path from source to drain. Consequently, there is no offset voltage as in the case of a bipolar transistor.

The FET is a voltage-controlled device—hence has a very high input impedance. The input impedance in a MOS FET is the capacitor formed by the gate and channel, separated by the insulating layer. Thus, the input impedance is very high, usually in excess of 10^9 Ω. Leakage currents are very low and are determined in a MOS FET by the dielectric properties of the insulating layer. The high input impedance and low leakage thus provide excellent isolation between the drive signal and the signal being switched.

Thomas J. Harrison, IBM Corporation, Boca Raton, Fla.

Folding Error.   See *aliasing error* earlier in this subsection.

High-Level Signal Processing.   See *analog input* earlier in this subsection.

Horsepower Module.   See *signal conditioners and logic modules* later in this subsection.

Ink-Jet Printing.   See *printers* later in this subsection.

Integrator Module.   See *signal conditioners and logic modules* later in this subsection.

Josephson Tunnel-Junction.   As early as 1962, B. Josephson recognized the implications of the complex order parameter for the dynamics of the superconductor, in particular when one considers a system consisting of two bulk superconductors connected by a "weak link." The basic requirement for the weak link is that the amplitude of the order parameter at the link should be substantially smaller than in the bulk regions. In early experimentation, such a situation was realized in a variety of ways—two evaporated films separated between two bulk superconductors; a single hourglass-shaped evaporated film, with the constriction of dimensions small compared to the coherence length; or even a bare niobium wire with a pendant frozen blob of soft solder, where the weak links, indeterminate in number, are formed by solder bridges through pinholes in the surface oxide. Collectively, all such weak link junctions are referred to as *Josephson junctions.*

During recent years, considerable research has gone into investigating the use of Josephson superconducting devices of the tunnel-junction type (operated at near absolute zero) in connection with building superfast computers. In such devices, the electrical signals have only a millimeter or two to travel. Switching time is about 10^{-11} s. The power dissipation of Josephson devices permits high circuit density. It is estimated that the power dissipation of transistors is about 100 times that of Josephson devices. Different materials are being studied to improve the original lead-alloy thin-film materials. These include lead-indium-gold alloys and niobium-tin alloys.

Keyboard–Cathode Ray Tube (CRT) Devices.   Keyboards, often used in conjunction with CRT display devices, provide means to directly enter data into a computer. The depression of each key is translated into a unique code which is interpreted by the computer or the control electronics. For many purposes, the keyboard is similar to that used on typewriters with, typically, the addition of some extra keys which provide control signals. Special purpose keyboards also are widely used. They may consist of a numeric keypad and a certain number of dedicated functional keys. This is exemplified by the numeric and functional keys found on bank or cash dispensing machines.

The CRT is a means of outputting numeric or graphic data from the computer. In principle, it operates in a manner identical to its use in television receivers, except that the controlling signal is generated by the computer or controller electronics. Raster scan units are the least expensive and are used in most applications for the display of alphanumeric data. The display is written by the beam as it traverses the face of the CRT from left to right and top to bottom. In order to provide a flicker-free display, the image must be rewritten or refreshed at a rate in excess of 50 times per second. This is identical to the manner in which a television picture is generated. Because of the need to refresh and the signficant burden this would place on the central processing unit (CPU), CRT display units include controllers which provide the detailed control needed to generate characters and refresh the display. Data to be displayed are stored in a buffer accessible to the CPU. In graphics applications, a CRT having long persistence or storage capabilities is utilized to eliminate or minimize the need for refreshing the image between data changes. CRT displays are described in further detail in Secs. 15 and 18 of this Handbook.

Keypunch. A device, usually operated by a keyboard, for recording information on cards or tape by punching holes in the card or tape to represent letters, digits, and special characters. There is a wide variety of devices, ranging from manually operated portable devices, which operate one hole at a time, to semiautomatic designs which operate under the direction of a program contained on a drum card. Keypunches are available which print the punched information while punching the holes. Although once the primary data entry device for terminals, keypunches are being displaced in many applications by on-line interactive terminals and off-line key-to-diskette (disk or tape) units.

Light-Coupled Switch. A switch in which the switching signal is transmitted to the device in the form of light energy. The swtiching element may be a phototransistor, a photodiode, or a photo field-effect transistor (FET). The receipt of light energy by such a device changes the transmission characteristics of the device, permitting conduction between two terminals. When no light is present, the resistance of the device is high. When excited by photon energy, the resistance drops to a much lower value. Various light sources are used in connection with light-coupled switches. Although common incandescent sources may be used, gas discharge sources, such as neon lamps and solid-state devices (light emitting diodes, LEDs) and lasers, are more commonly used in instrumentation and communications systems. As compared with incandescent sources, the other aforementioned sources produce less heat and possess higher speed and reliability.

A major advantage of the light-coupled switch is the isolation that can be obtained between the signal and the drive source. Signals can be controlled without introducing errors due to the drive source. Further, isolation helps to maintain a high common-mode rejection ratio in analog signal handling equipment. See also Sec. 12 of this Handbook.

Low-Level Signal Processing. See *analog input* earlier in this subsection.

Magnetic Ink Character Recognition (MICR). See *printers* later in this subsection.

Mass-Flow Logic Module. See *signal conditioners and logic modules* later in this subsection.

Maximum-Minimum Logic Module. See *signal conditioners and logic modules* later in this subsection.

Mercury-Wetted Relay. Somewhat similar in construction and operation to a dry reed relay, a mercury-wetted relay is an electromechanical, magnetically actuated relay that operates within a hermetically sealed capsule. Mercury wetting has the advantage of providing low contact resistance and more reliable contact action. See Fig. 45. Nickel-iron alloy contacts with relatively high permeability and low retentivity usually are used. An inert or reducing atmosphere under moderate pressure is contained in the capsule. The armature base is immersed in mercury, and a coating of mercury is maintained on the armature and contacts because of the surface tension of the mercury and capillary grooves in the armature.

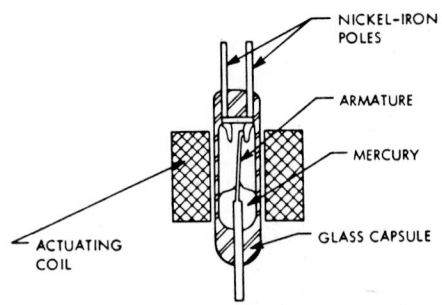

Fig. 45   Mercury-wetted relay capsule.

The switch is usually actuated by a magnetic field produced by a coil within which the relay capsule is enclosed. Small biasing magnets, as are also used in a dry reed relay, may be located near the capsule and thus change the switching characteristics, including contact actions such as latching and normally closed.The so-called pick time of the relay is dependent on the construction of the capsule and characteristics of the drive signal. Normally, pick times range from 0.75 to 5 ms. Drop times may be as high as 2 or 3 ms. Noise is produced by magnetostriction in the armature. For capsules of comparable size, however, the noise of a mercury-wetted relay is appreciably less than that of a dry reed relay. Because of the decay time of the noise, the speed of the relay for high-precision multiplexing operations is somewhat less than 300 samples per second.

A principal use of the mercury-wetted contact relay is in low-level multiplexing. It is also widely used in power switching. Particularly where high common-mode voltages are encountered, its low contact resistance, high leakage, high breakdown voltages, and long life make it attractive. At full-rated resistive loads, the life of this type of relay is in excess of 10^9 operations. For some applications, the nearly vertical mounting requirement is a disadvantage. The mercury-wetted relay may be considered a rugged device, suitable for withstanding the vibration and shock encountered in most industrial environments.

Multiplexer. See *analog multiplexer* and *digital multiplexer* earlier in this subsection.

Optical Character Recognition (OCR). See *printers* later in this subsection.

Output. The information transferred from the internal storage of a computer to secondary or external storage or to any device outside the computer. Also, the act of transferring such information. Collectively, all the devices required to accomplish the foregoing tasks. Analog signals for the control of a process or device can be obtained from a digital computer by means of an analog output subsystem. The analog output subsystem receives digital data from the central processing unit (CPU) and translates them into the required analog form by means of a digital-to-analog converter. Printers and plotters are output devices which provide means for obtaining hard copy of alphanumeric or graphic output from a digital computer.

Perforated Tape. Holes are punched into a continuous strip of material (typically paper or plastic) as the information storage mechanism. Data are recorded by means of a pattern of partly or completely punched holes across the width of the tape. Various tape widths are in use in different applications. The width most common in computer-related applications is 1 in (2.5 cm), with provision for eight data holes plus a feed hole. [American National Standards ANSI X3.6-1975 (R1973) and ANSI X3.18-1974.] In perforated tape readers and punches, the tape is mechanically fed by means of a sprocket drive wheel or friction rollers. The feed holes, usually punched at the same time data holes are punched, are used in the reader for timing purposes. Holes in the tape are typically sensed by photoelectric means, although mechanical brushes were common and are still used in some slower speed equipment. Although extensively used with early computers, paper tape input-output has generally been displaced by other media except in special applications as, for example, to provide data interchange with tape-controlled numerical control machines. It is interesting to note that an early use of paper tape to control a digital device was the Pianola player piano, patented in 1897 by E. S. Votey.

Plotter. See *curve plotter.*

Printers. These devices provide a means of outputting a permanent record in human-readable form. Printing techniques can be categorized as either impact or nonimpact. In *impact printing,* the image is formed by striking an inked ribbon placed near the surface of the paper. The technique, which is very old in concept, has the advantage that multiple copies can be produced simultaneously through the use of carbon or sensitive paper. Impact printing can be further categorized as either formed-character or matrix printing. In *formed-character printing,* the element which strikes the ribbon to produce the image consists of a raised mirror image of the desired character. This is similar to the method used in conventional typewriters. At slower printing speeds, mechanisms similar to those used in typewriters are often used. The character set is contained on an element in the form of a cylinder or ball or at the ends of a spoked wheel. A hammer or other striking motion causes the element to strike the ribbon. In higher speed printers, the raised images may be carried on a single band of material (typically metal) or individual elements may be linked together as in a chain. The band or chain contains one or more copies of the entire character set which is printable and moves continuously in a loop. When the desired character passes the print position(s), it is struck by a hammer, which transfers the ink to the paper.

In *matrix printing,* the character is formed as a series of closely spaced dots produced by striking a wire (typically 0.01 in, 0.25 mm in diameter) against the ribbon. In a typical printer of this type, the printing element consists of seven to nine such wires arranged in a vertical column and mounted on a carrier which traverses the paper horizontally. By selectively striking the wires, any character representable by a matrix of typically 5×9 dots can be produced. Although the printing does not have the aesthetic quality of formed-character printing, the matrix printer is mechanically simpler and usually less expensive; in addition, altering the printed character set is a matter of reprogramming the control unit, as opposed to mechanical replacement of the printing element.

The matrix method can also be used in nonimpact techniques. In a *thermal printer,* for example, the head consists of a vertical column of seven to nine heating elements, similar in principle to the arrangement of the wires previously described. The wires are selectively heated by electrical pulses, and the generated heat causes a mark to appear on specially treated paper. A similar technique, *electrostatic printing,* utilizes an electric arc between the printing element and a conductive paper to remove an opaque coating, thus exposing a sublayer of a contrasting color.

Ink jet printing is another form of nonimpact printing. In this technique, ink is forced through a tiny nozzle to form a droplet. The droplet is electrostatically charged and is attracted to an oppositely charged platen located behind the paper. Using electrically controlled deflection plates similar to those in a cathode ray tube (CRT), the trajectory of the droplet can be controlled to hit the desired spot on the paper. Unused drops are deflected away from the paper into a reservoir for recycling the ink. Because of the small size of the droplet and the precise trajectory control, ink jet printing quality can approach that of formed-character impact printing on untreated paper.

Electrophotographic printing techniques are based on principles similar to those used in many plain-paper copiers. A photoconductive material is selectively charged (or discharged) utilizing a light source, such as a *laser.* A powdered toner is attracted to the charged regions and, when placed in contact with a sheet of paper, transfers to the paper surface. The toner is then subjected to heat, which fuses it to the paper. As in the case of ink jet printing the quality rivals to some degree that of formed-character impact printing.

Nonimpact methods tend to provide higher printing speeds, primarily because they are electronic. This is contrasted with mechanical impact methods which are limited by the inertia of the hammer and print element, coupled with the resistance-inductance delays inherent in charging and firing an electromechanical device. In addition, the ink jet and electrophotographic methods, and to a lesser degree matrix printing techniques, provide the advantage of the ability to produce a spot anywhere on the paper. This makes them suitable for printing graphics and halftone images. It is possible, for example, to print simultaneously both the information on an invoice and the invoice itself (lines, logo, etc.) by nonimpact methods. The inability to print carbon copies may not be significant if the printing speed is high, since duplicate copies can be produced quickly.

Magnetic ink character recognition (MICR) and *optical character recognition* (OCR) input devices are utilized extensively in some applications. MICR, for example, is universally used in the banking industry for recording and reading data on checks and other bank documents. Stylized character sets, which ease the task of character recognition without sacrificing human readability, have been developed for use with these techniques [ANSI X3.3-1970 (R1976), X3.49-1974, and X3.2-1970 (R1976)]. For computer input, the character image is scanned optically or magnetically, resulting in an electrical signal. Controller electronics (typically) or a computer program then decodes the signal and identifies the particular character. The appropriate internal computer representation is then stored in memory.

Product-Ratio Module.   See *signal conditioners and logic modules* later in this subsection.

Punched Cards.   Information is stored as holes punched in a card. A commonly used card provides 960 hole positions arranged as 12 rows and 80 columns. A variety of codes are used to represent data, including pure binary in which punches in a row or column represent a 1 bit. The Hollerith punched card code (ANSI X3.26-1980) provides for coding numeric, alphabetic, and special symbols using one, two, or three punches per column. In representing alphanumeric data, it is common to print all or a portion of the data along the top edge of the card. This is known as *interpreting* and may be provided by a device called an interpreter or by the keypunch used to prepare the cards.

When cards are used with a computer, a card reader reads information from the cards and transmits the data to the computer. The reader typically uses photoelectric means to sense the holes as the cards pass from an input stacker to an output stacker. In addition to their use with a computer, punched cards are often processed by means of unit record equipment, such as collators and mechanical sorters. The punched card continues to be one of the most commonly used computer input media, despite the emergence of terminals and other media such as magnetic stripe cards. One reason is that the card provides a convenient, easily handled document which can be used as an invoice or a receipt in many transactions. It is interesting to note that the Jacquard loom, invented in 1799, utilized punched cards for controlling the weaving of a pattern. In 1832, Babbage, who invented an early computer-calculator known as an "analytical engine," proposed the use of punched cards as input to his machine, an idea he borrowed from Jacquard. Possible instrumentation applications are suggested by the relatively recent use of punched paper and plastic cards as "keys" for unlocking rooms and storage spaces as found, for example, in hotels and leased lockers.

Read Out, Readout.   As a verb, *read out* means to output information from a computer, generally to a display device. As a noun, *readout* refers to any of the several forms of display that may be used—moving indicators, dials, lights, printed or punched tape or cards, cathode ray tube (CRT) displays, and so forth. Although the term may be used in connection with information that automatically goes into some form of storage without a display interface, readout generally is considered output information that is visible or read out.

Ripple Module.   See *signal conditioners and logic modules* in this subsection.

Signal Conditoners and Logic Modules.   Signal conditioners modify an input signal prior to introducing it into an electronic system such as a digital data acquisition or instrumentation operation. There is no precise definition of signal conditioning, with considerable variation of meaning from one type and application to the next. Frequently, because of their intimate association in instrumentation systems, signal conditioners are often described along with various logic modules and, in fact, from a hardware standpoint the two "boxes" may be integrated in some instances. Modification of an analog signal prior to amplification may include attenuation (scaling), filtering, conversion (current to voltage, voltage to current, and so forth), impedance-level transformation, bridge or signal compensation and, in numerous instances, specialized operations. Commercially available signal conditioners may include both amplification and conversion from analog to digital form.

Included among numerous signal conditioning modules are strain gage conditioners, thermocouple and resistance temperature detector (RTD) conditioners (see Fig. 46), linear variable differential transformer (LVDT) conditioners, and frequency conditioners. The application of strain gage conditioners and associated logic modules are shown in Figs. 47 and 48. The use of a frequency-to-voltage conditioner is shown in Fig. 49. Brief descriptions of the associated logic modules are given next.

> *Horsepower Module.*   In one particular configuration (Daytronic), the module receives two inputs: (1) an unconditioned pulse signal (revolutions per minute) from an external sensor, and (2) a conditioned analog signal (torque) from a strain gage load cell. The output consists of two standard 5-V data signals: (1) a revolutions-per-minute signal proportional to the frequency of the pulse input, and (2) a horsepower signal proportional to the product of the revolutions-per-minute signal and the input torque signal. A principal application of the horsepower module is in engine dynamometer instrumentation. Here, the pulse input originates from a revolutions-per-minute sensing magnetic pickup, while the analog torque signal is produced by the conditioner module for a strain gage shaft-torque sensor in connection with a reaction sensing load cell. See Fig. 50. The horsepower module also is effective in determining the hydraulic power output of a pump, which is proportional to the product of the flow and the differential pressure across the device. In such applications, a turbine flowmeter generates the pulse input, while the analog input is supplied by the conditioner module for a differential-pressure transducer.

> *Analog Input Modules.*   This is an umbrella term sometimes used to describe modules for signals derived from other instrument systems, potentiometer-type sensors, dc-to-dc LVDTs, and Hall-effect transducers, among others. As shown in Fig. 51, the signal source configuration can be two-, three-, or four-wire. A regulated 10-V excitation source is included for use with potentiometers, dc-to-dc LVDTs, and similar devices. Some modules also feature zero adjustment for suppression of tare input values. Selectable low-pass filtering in some units removes normal-mode dynamic components that might prevent stable control action or digital readout.

> *Summing Module.*   A module of this type is designed to obtain the *sum* or *difference* of multiple analog signals. Accepting up to three 5-V data signal inputs (x, y, and z), one module design

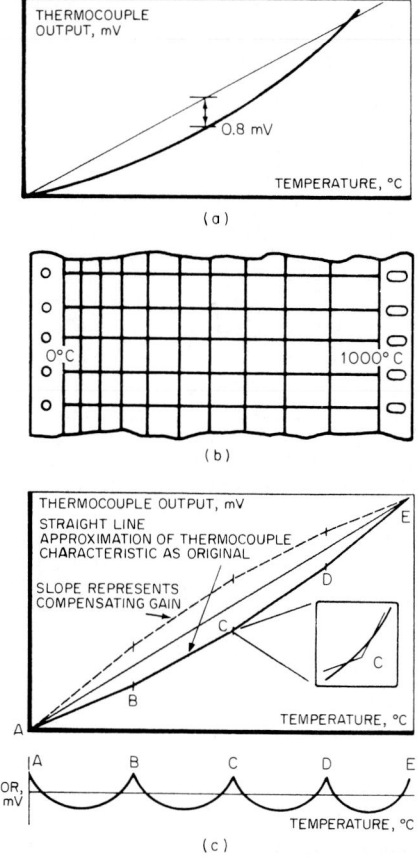

Fig. 46 Linearization of thermocouple signals. With a type-R (platinum–platinum–13% rhodium) thermocouple, the linearity error can be as much as 8% of span as shown (exaggerated) in (*a*). This indicates a maximum deviation from a straight line of 0.8 mV (equal to 8%) error over the 0 to 1000°C span. Thus, unless compensated in some way, the temperature instrument will have significant errors.

Traditionally, this problem was solved by specially calibrating indicating scales and recording charts. Although not shown to scale, the nonlinear recorder grid for the thermocouple output would appear as in (*b*). Linearization of the thermocouple signal can be accomplished electronically through the use of a variable-gain amplifier. The basis of the method (Chessell) shown in (*c*) is the approximation of a curve by a series of straight lines differing in slope. For practical use, excellent approximations are achieved in most cases by dividing the temperature range into four segments. Each of these segments is assigned a gain, depending on the curve's divergence from the straight line. With reference to (*c*), voltage discriminators in the signal conditioner switch the amplifier gain when the thermocouple output passes through nodes *B*, *C*, and *D*. The broken line represents the gains required to compensate for divergence of the thermocouple curve from line *AE*. In this example, divergence and, therefore, gain decrease progressively from *A* through *E*. Only segment *DE*, if required, can have a greater gain than the preceding segment. Intersections (inset) are chosen to minimize the small remaining error.

produces a 5-V data signal output conforming to the general expression $e_{out} = K(x \pm y \pm z)$. Internal switches allow the polarities of the y and z terms to be independently selected. Thus, each of these terms may be either *added to* or *subtracted from* the summed ouput. Unused input variables can be set to zero, while the value of the multiplier K can be set at 1, ½, or "variable." Selecting the "variable" setting permits the operator to adjust the coefficient value over a range of 0.25 to 100. Momentary front-panel pushbuttons are provided for the selection of individual

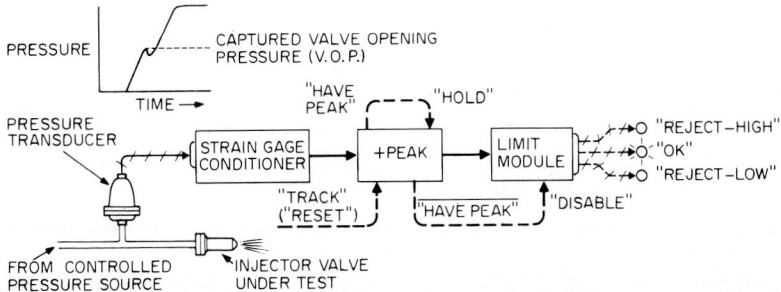

Fig. 47 System for testing the valve opening pressure of diesel engine fuel injectors or pressure relief valves. A strain gage conditioner picks up a signal from a pressure transducer. To start the test, the inlet pressure is caused to ramp upward from zero to a high final value. When it reaches the point of valve opening, there is a momentary dip (see graph) caused by the sudden increase in flow and partial restriction of the supply line. At this point, the peak-track-hold module captures the peak inflection point and immediately issues a "have peak" output to its own "hold" input, thereby freezing the captured value and making it unresponsive to subsequent, higher pressure values. The "have peak" output is used to disable the limit module output until a valid rest result has been obtained. The application of a momentary "track" command resets the system for the next part to be tested, the entire sequence taking only about 2 s. *(Daytronic.)*

input signals, useful for setup and checking purposes. The use of a summing module ($\pm$) is shown in Fig. 52.

Product-Ratio Module. This type of module may receive two 5-V data signal inputs (x and y) and produce a 5-V data signal output that can be selected by means of internal settings to provide any of the following functions: (1) $e_{out} = K(x - y)$; (2) $e_{out} = K(y/|x|)$; (3) $e_{out} = K(x^2)$; (4) $e_{out} = K\sqrt{|x|}$. Input signals can be either positive or negative, regardless of the mode of operation. In the *product* mode, the module is a full four-quadrant multiplier. However, in

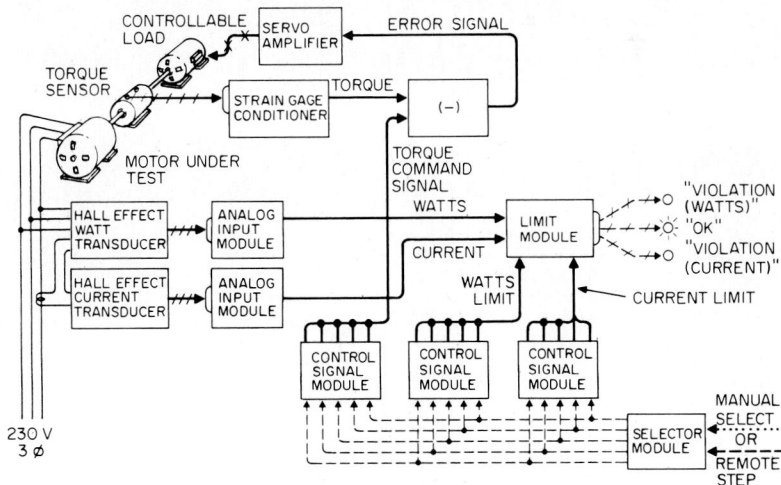

Fig. 48 Automatic "go–no go" motor test system. Analog input modules accept signals from a Hall-effect watt and Hall-effect current transducer; a strain gage conditioner operates on a signal from a torque sensor, and a summing module ($-$) generates an error signal by comparing the torque measured with the torque command signal. A limit module monitors both current and wattage consumption for conformance to individually preset limits at each of five sequentially selected torque loading test points. *(Daytronic.)*

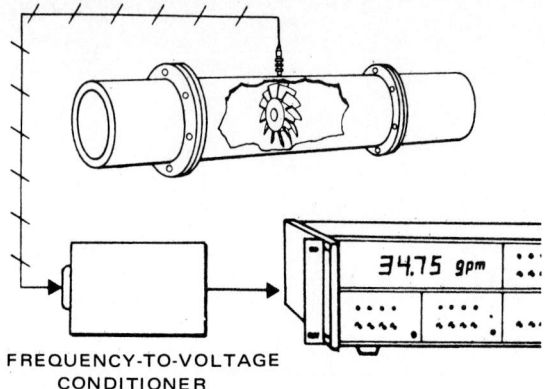

FREQUENCY-TO-VOLTAGE CONDITIONER

Fig. 49 Use of a frequency-to-voltage conditioner in a system for calibrating a turbine flowmeter, using a known K factor (cycles per gallon) of the flowmeter. *(Daytronic.)*

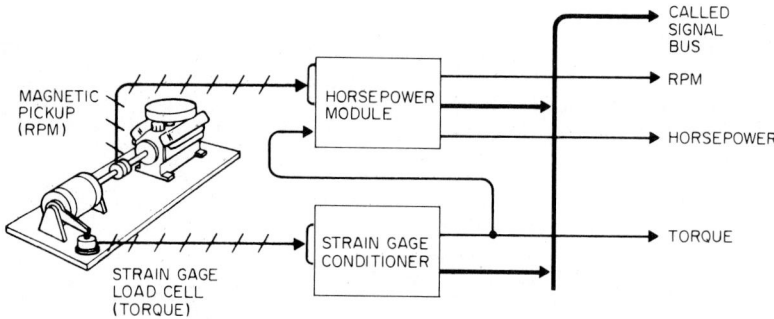

Fig. 50 System for measuring engine output, using a strain gage conditioner in connection with a load cell and a horsepower module. The pulse input to the horsepower module originates from a revolutions-per-minute sensing magnetic pickup, while the analog torque signal is produced by the conditioner module for a strain gage shaft torque sensor. *(Daytronic.)*

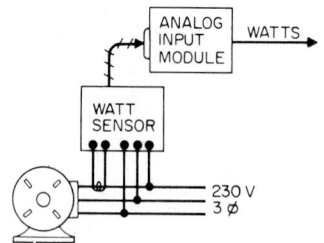

Fig. 51 Use of an analog input module with available Hall-effect transducers. Alternating current voltage, current, or power measurements (single- or three-phase) can be made. *(Daytronic.)*

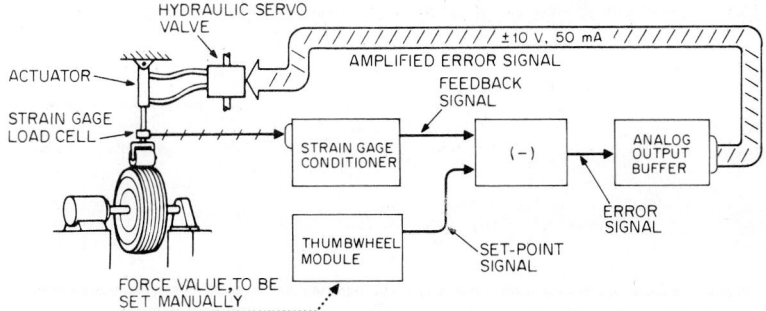

Fig. 52 Representative use of a summer module to obtain an error signal in a closed-loop servo system. This signal is the amplified difference between the set-point signal and the resulting follow-up signal. The summer module is indicated here by $(-)$. In this example, the loading of a tire under test is precisely controlled, regardless of tire wear, changes in hydraulic pressure, or other disturbing factors. The output gain of the summer (the K factor in the output expression is given in the text) can be adjusted to give "tight" yet stable servo action. *(Daytronic.)*

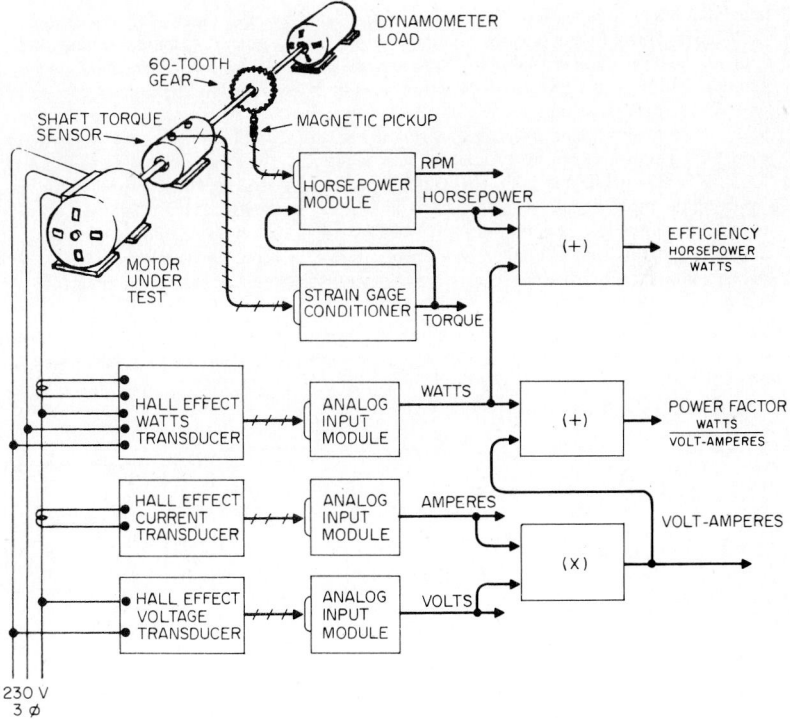

Fig. 53 Use of three product-ratio modules in a test-stand instrumentation system for computing the continuous value of the power factor and the efficiency of three-phase electric motors under various conditions of loading. It will be noted that the installation also incorporates three Hall-effect transducers (watts, current, voltage) with their associated analog input modules, a horsepower module, and a strain gage conditioner. *(Daytronic.)*

14.133

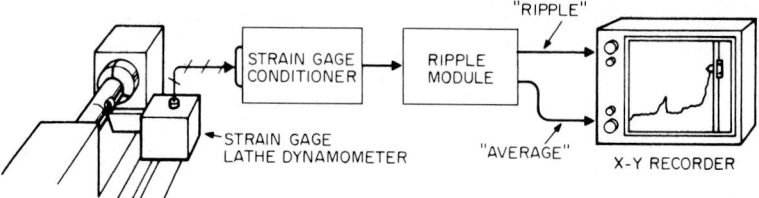

Fig. 54 In this system, newly manufactured automobile transmissions are spun up by an electric motor and checked for roughness of torque input. Uneven torque loading, resulting from improper fit, broken gear teeth, burrs, or foreign matter, is detected by the system and used to actuate a reject indicator. The system uses a strain gage conditioner and a ripple module. *(Daytronic.)*

the *ratio* and *square root* modes, it reads the *absolute* value of the x-input signal. Consequently, the output signal polarity (for the ratio mode) is that of the numerator signal, and it is possible to extract square roots from input signals of either polarity. For proper operation during ratio measurement, the denominator signal must be greater than 10% of full scale. If the signal should fall below an acceptable value, a front-panel "insufficient denominator" lamp will light automatically. An electric motor test stand using three product-ratio modules $(+)$, $(+)$, and $(\times)$ is shown in Fig. 53.

Ripple Module. In one type of ripple module, a 5-V data signal input produces a 5-V data signal ouptut proportional to the rectified average value of the dynamic component of the input signal that falls within a selectable frequency band. This module is useful for process monitoring and quality control testing in cases where the value of a dynamic component can imply deviation from normal or acceptable conditions. Typical applications include measurements of *conformance* ("thump") in tires, *tool chatter* in precision machining operations, *roughness* in torque output for transmissions or prime movers, *instability* ("hunting") in servo control systems, *hydraulic noise* in pump output, and *imbalance* in rotating machinery. The low end of the measured frequency band is determined by front-panel buttons offering selectable corner frequencies of 3, 10, 30, or 100 Hz—while the high-end cutoff depends on the active filter setting of the module supplying the input signal. Transfer gain of the ripple output signal is normally unity; however, internal switches allow gain multipliers of $\times 2$, $\times 5$, or $\times 10$ to be selected if needed. In one type, an auxiliary output signal equal to the average value of the input signal is supplied. This permits correlation of the ripple value with the average value as in the xy recording application shown in Fig. 54.

Auto-Zero Module. In this type of module, a 5-V data signal (x) is the input, and the output is a 5-V data signal equal to $x - x_t$, where x_t represents the *held tare value* of the input signal—

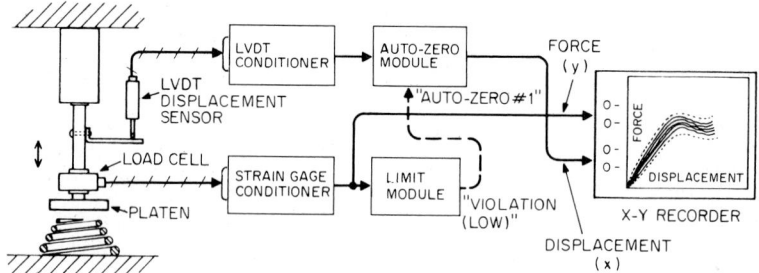

Fig. 55 In monitoring the production of critical nonlinear springs, quality control requires that the force displacement characteristics of successive lot samples be plotted on a single sheet of an xy recorder. Ideally, all plots should fall within the narrow band that defines the desired nonlinear characteristic. This method allows quick spotting of any aberrant units. However, such determinations often prove difficult (sometimes subjective) because any variation in spring height will cause the displacement plot to start at some point other than the origin of the xy graph. The use of an auto-zero module solves this problem by automatically rezeroing the displacement channel for each measurement. *(Daytronic.)*

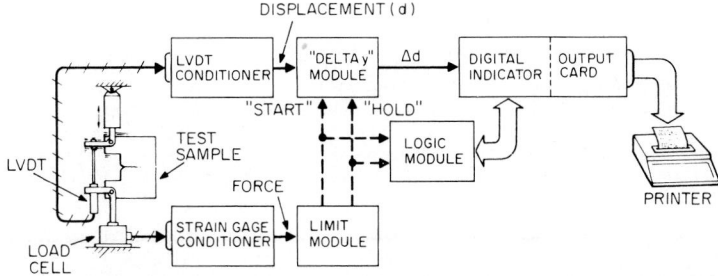

Fig. 56 System for measuring continuous crack growth in a material sample subject to cyclic loading in a fatigue testing machine. Force is the independent variable and is monitored by a limit module for the passage of two preset values, detection of which sends the "start" and "hold" commands to the delta y module. This module captures the change in displacement (delta y) corresponding to the known change in force (delta x). The output of the delta y module represents the "compliance" of the sample, a quantity directly related to the propagation of the crack at any given time. Other system elements cause the measurement to be printed automatically at periodic intervals or after a preset number of accumulated cycles. *(Daytronic.)*

i.e., the value at the instant when an input AUTO-ZERO logic command was applied. This module is useful when there is a need for quick, automatic establishment of an arbitrary zero reference. It also finds use in instrumentation systems involving automatic taring of container weights and adjustment of the zero baseline for graphic recording. Thus, the module is used in automatic batching systems. Another application is shown in Fig. 55.

Delta y Module. This unit accepts a 5-V data signal input (y) and produces a 5-V data signal output (delta y) equal to the *net change* in the input signal transpiring between the application of a logic START command and the application of a logic HOLD command. When both logic inputs are in the 0 logic condition, the analog output signal is zero, irrespective of the input signal value. On application of a logic START command, the output signal, *starting from zero,* will continuously reflect any subsequent change in the value of the input signal. On application of a HOLD command, the output signal instantly freezes at whatever value of delta y (i.e., whatever *net change in input signal*) has been achieved. When both logic commands are released, the module reverts to its original zero-output condition.

Based on a capacitor memory principle, module operation gives essentially instantaneous response to logic commands. Thus the module is capable of accurate capture of transient values in fast-moving dynamic situations. Configurations of this module can yield rate of change versus time ($\Delta y/\Delta t$) or rate of change versus an independent variable ($\Delta y/\Delta x$). An example of the latter is shown in Fig. 56.

Maximum-Minimum Module. This module accepts a standard 5-V data signal input. The output is a 5-V data signal equal to the *difference* between the maximum and the minimum excursions of the input signal that have been perceived since the removal of a logic RESET com-

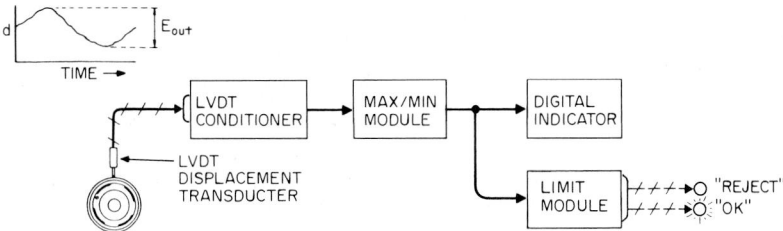

Fig. 57 Measurement of total indicated runout of an automobile wheel, using a maximum-minimum module in connection with a linear variable displacement transducer (LVDT), a limit module, and an LVDT conditioner. *(Daytronic.)*

mand. The actual polarity of either excursion may be either positive or negative—the maximum excursion being the most positive (or least negative) value experienced by the input signal and the minimum excursion being the least positive (or most negative) value.

A logic HOLD input will freeze the output, while a logic RESET input zeros the output until the command is removed, at which time the module is free to obtain a new maximum-minus-minimum reading. Both inputs can be supplied from external logic sources, but the RESET function is also available via front-panel pushbutton.

Maximum-minimum monitoring is useful in determining the precise range of an excursive phenomenon like the *runout, wobble,* or *looseness* of a rotating part. Figure 57 shows a system for the measurement display of an automobile wheel's total indicated runout. On detection of *excessive runout, a limit module produces a* NO-GO signal.

Integrator Module. This type of module accepts a 5-V data signal input and produces a 5-V data signal output proportional to the time integral of the applied signal. Instantaneously controllable by logic signals, the integration process can be commanded to START, STOP (with output held at the accumulated value), RESUME (from the existing value), or RESET TO ZERO. Integration periods from a fraction of a second to several minutes can be established by setting the input range from 0.003 to 3000 V·s full scale. Independent scaling of the output integral in any desired engineering units is made possible by a simple calibration procedure.

The integral of a variable with respect to time is represented by the area under the curve of its time plot. See Fig. 58. Such a quantity has many uses in engineering measurements. *Flow,* for example, integrates to *volume* (or *mass*) transferred; *velocity* integrates to *displacement;* and *power* integrates to *work* (or *energy*). The time integral of the product of two signals is also useful in many applications. Pressure times flow, torque times revolutions per minute, and force times velocity all integrate to yield energy consumed or delivered over the integration period.

In addition to time integrals, it is possible to obtain the integral of one variable (y) with respect to another (x)—provided that dx/dt (the time derivative of the x signal) is available. In the shock absorber testing system shown in Fig. 58 it is possible to compute the integral of force (y) with respect to displacement (x) for one complete actuation cycle of the testing machine. Represented by the shaded area in the graph, this integral is equivalent to the total energy absorbed by the shock absorber during the test cycle.

Mass-Flow Logic Module. In connection with engine test cell instrumentation, it is frequently desirable to determine fuel consumption in terms of fuel mass rather than fuel volume, the latter requiring the use of a specific gravity correction factor. In this special module, the application of

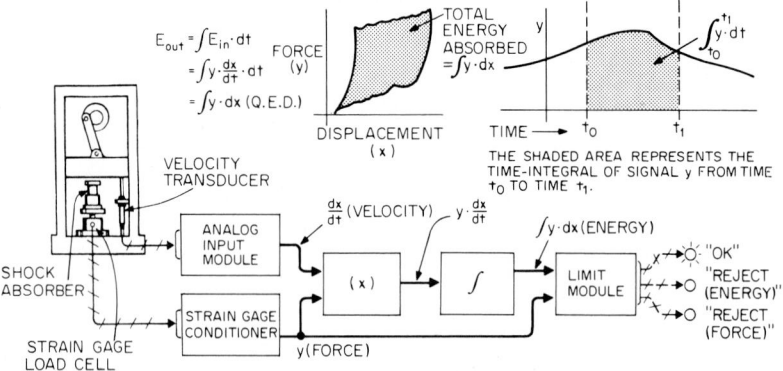

Fig. 58   System for testing automobile bumper shock absorbers for proper operation, measuring the total energy absorbed (area enclosed on the graph) during an impact cycle delivered by a flywheel and crank arm. Substandard viscous damping, the result of a low fluid level on a leaky port, will be reflected by a low amount of energy absorbed, while a plugged port will cause excessive force buildup during the stroke. In addition to an analog input module for accepting a signal from a velocity transducer, a strain gage conditioner, a product-ratio module (X), an integrator module, and a limit module are used in the system. *(Daytronic.)*

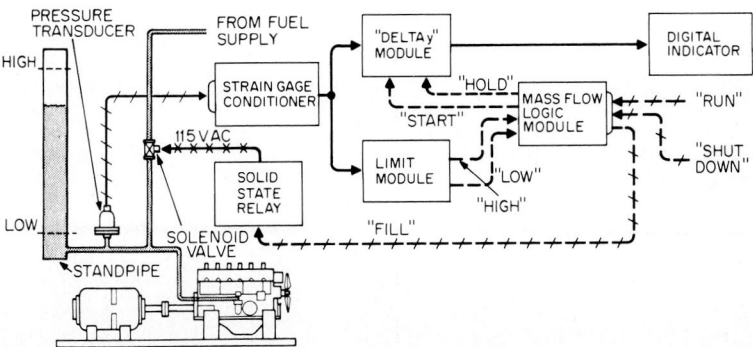

Fig. 59 System for determining mass flow of an engine test installation. In "weighing" the fuel consumed, the pressure head at the base of a fixed standpipe of constant cross section is measured. The output of the pressure transducer goes to a strain gage conditioner. The change in pressure (delta p) over a fixed time period (delta t) can be scaled in units of mass flow into or out of the standpipe. In addition to the strain gage conditioner, other important elements of the system include a solid-state relay, a delta y module (rate of change versus time), a mass-flow logic module, a limit module, and a digital indicator. *(Daytronic.)*

which is shown in Fig. 59, the measurement of standpipe pressure (delta p) over a fixed time period (delta t) can be scaled in units of mass flow.

Tract-and-Hold Module. This type of module may supply independently controllable TRACK and HOLD functions for up to five analog data channels. Through proper connections, the unit can be made to freeze five different analog signals at the same instant, to freeze five different values of one analog signal at five different instants, or to perform any required combination of these functions. A principal role of the track-and-hold module is the elimination of *time skew* in multichannel data acquisition. At the initiation of a scan cycle, the track-and-hold module receives a logic HOLD command from the system scanner controller and freezes all outputs simul-

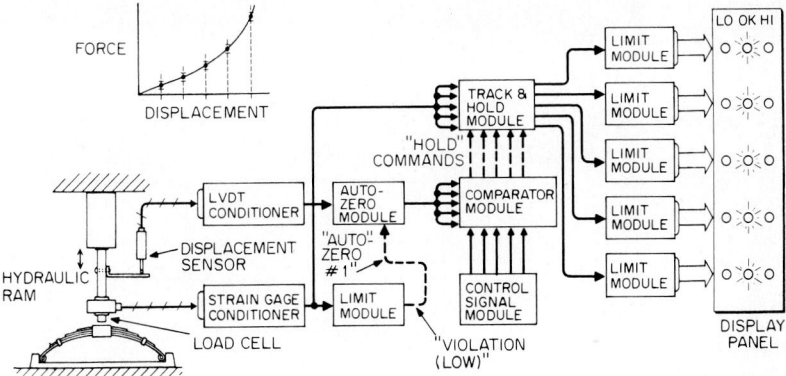

Fig. 60 Installation for testing an automobile spring assembly for proper force displacement characteristics. On the production line, only 2 s is required to make the test. A linear variable differential transformer (LVDT) conditioner accepts a signal from an LVDT; a strain gage conditioner accepts a signal from a load cell. When the descending ram makes initial contact with the part, the displacement channel is instantly auto-zeroed by the auto-zero module. As the displacement signal subsequently reaches each of five test points, which are preset on the control signal module, the comparator module issues a logic HOLD command to the track-and-hold module. In this way, the system captures the five individual values corresponding to the five limit modules for conformance with appropriate high and low limits at each point. The results are then displayed. *(Daytronic.)*

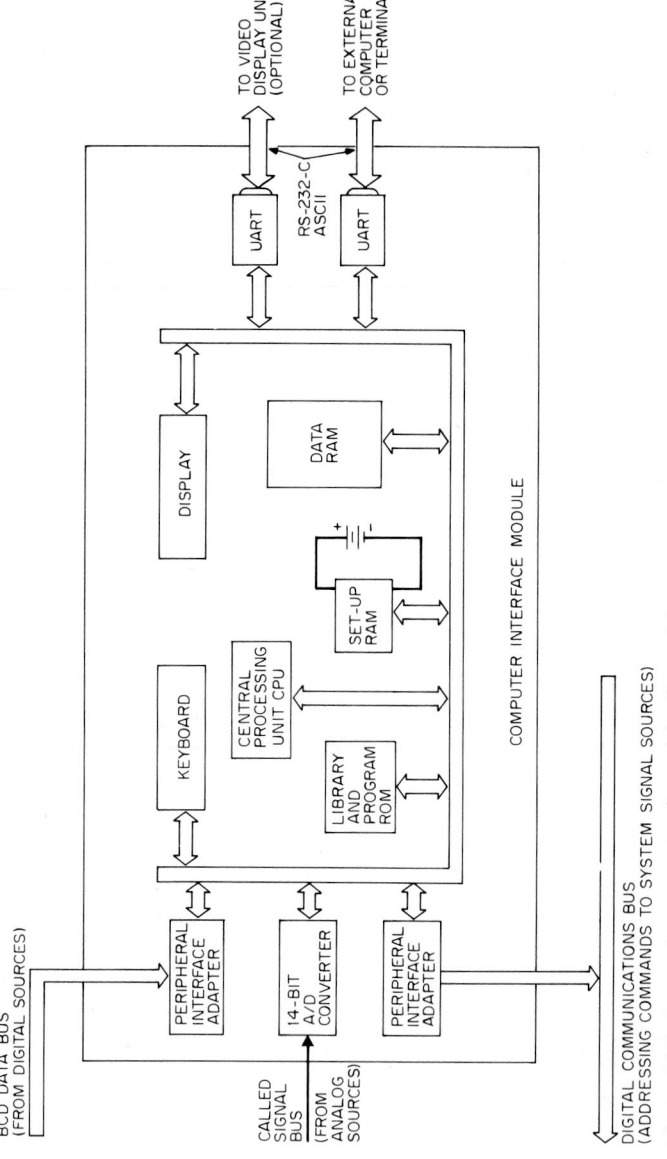

Fig. 61 Block diagram of a computer interface module. *(Daytronic.)*

BCD DATA BUS
(FROM DIGITAL SOURCES)

CALLED
SIGNAL
BUS
(FROM
ANALOG
SOURCES)

DIGITAL COMMUNICATIONS BUS
(ADDRESSING COMMANDS TO SYSTEM SIGNAL SOURCES)

PERIPHERAL
INTERFACE
ADAPTER

14-BIT
A/D
CONVERTER

PERIPHERAL
INTERFACE
ADAPTER

KEYBOARD

DISPLAY

CENTRAL
PROCESSING
UNIT CPU

LIBRARY
AND
PROGRAM
ROM

SET-UP
RAM

DATA
RAM

COMPUTER INTERFACE MODULE

UART

UART

RS-232-C
ASCII

TO VIDEO
DISPLAY UNIT
(OPTIONAL)

TO EXTERNAL
COMPUTER
OR TERMINAL

taneously. This action ensures that all data readings are taken at exactly the same instant—not skewed in time by finite delays inherent in the printing or recording process. A system for testing automobile spring assemblies is shown in Fig. 60.

Computer Interface Module. This module facilitates connection of the numerous parochial modules of the kinds just described to external computers. The detailed design of computer interface modules varies considerably from one manufacturer to the next. One representative design (Daytronic) is described here. A general block diagram showing how the computer interface module fits into the total scheme is given in Fig. 61. The elements of the module itself are shown in Fig. 62. The module is based on a microprocessor and includes all necessary programming for (1) setup and calibration of up to 398 data channels, (2) scanning and digitizing at a rate of over 1500 channels per second, (3) digital zeroing and scaling (the $y = mx + b$ function), (4) refreshing a "live" video display, and (5) rapid and efficient servicing of an external computer's requests for input data.

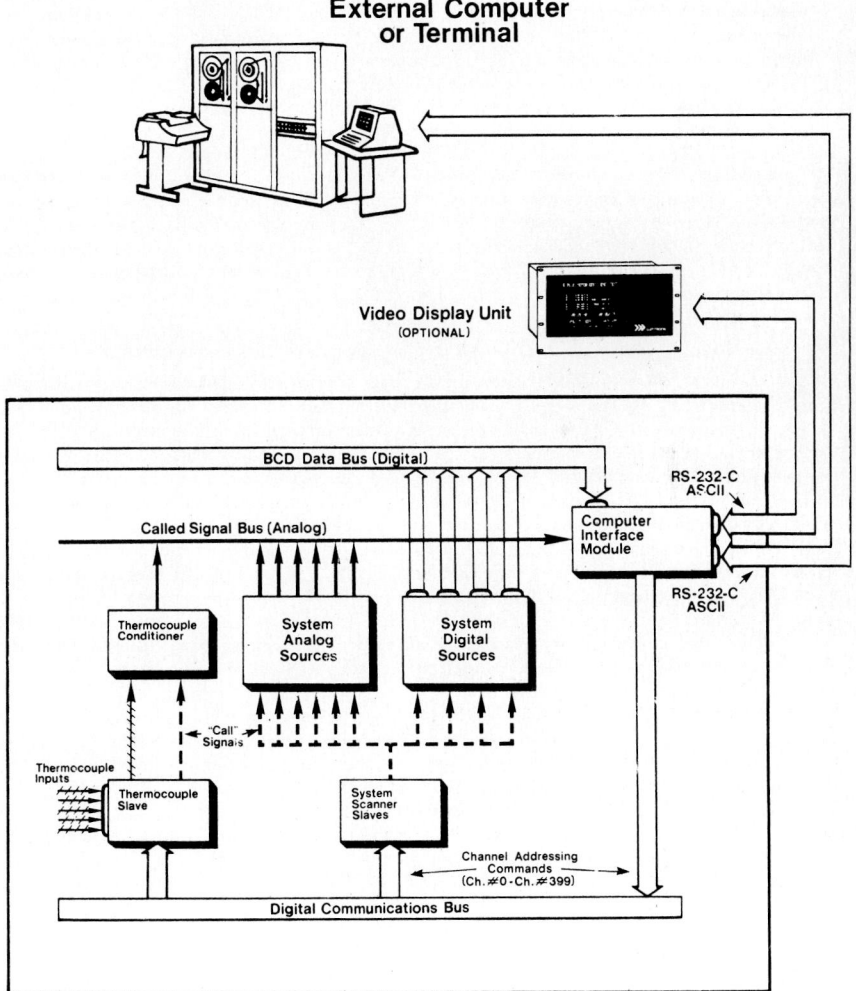

Fig. 62 Elements of computer interface module. *(Daytronic.)*

During operation, the computer interface module scans continuously at high speed, writing properly scaled measurement data into an internal buffer memory called the DATA RAM. Via an RS-232 port, an external computer can read the DATA RAM at any time. Simple prompting commands yield "random access," "dump," and other useful forms of response. These writing and reading operations are completely transparent to each other, with no mutual interference or discernible delays.

From the computer's point of view, the computer interface module is programmed to emulate a standard RS-232/ASCII data terminal—one that will respond instantly to simple requests for data. It can be addressed, therefore, by *any RS-232/ASCII-compatible computer operating in any conventional user programming language.* The module is designed to simplify the user's task of adapting and programming a selected computer for a given project. The computer (and its programmer) can concentrate on the process itself and not be concerned with managing the operation of the particular instrument configuration that performs the measurements.

With reference to Fig. 61, the module sends channel addressing commands to the system via the digital communications bus, receiving called analog signals via the called signal bus and called digital signals via the binary-coded decimal (BCD) data bus. Analog and digital signals may be intermixed as required. Setup and calibration operations are performed on a front-panel keyboard, assisted by prompting messages that appear on an alphanumeric light emitting diode (LED) display. All such entries are stored in a battery protected SET-UP RAM and designed to survive a power-off period of several weeks. The keyboard can also be used during normal operations to select live display of the readings of any system channel.

Stepping Motor.   An electromechanical device which rotates a discrete step angle when energized electrically. Typical step angles vary from as small as 0.72° to as large as 90°. Several means for electrically energizing stepping motors include dc pulses, square waves, and fixed logic sequence or multiple-phase square waves. Basic design types are (1) solenoid-operated ratchet, (2) permanent magnet, and (3) variable reluctance. Stepping motors provide a means for digitizing final controlling elements in process control and other industrial applications. More detail will be found in Sec. 19 of this Handbook.

Summing Module.   See *signal conditioners and logic modules* earlier in this subsection.

Surface Acoustic Wave (SAW) Device.   In a surface acoustic wave device information is represented by ultrasonic acoustic waves propagating freely along the surface of a planar solid. These waves, whose intensity is greatest at the surface, involve exceedingly minute motions of the particles of the solid on which they propagate; in typical SAW devices, particle displacements are at most a few angstroms. The wave velocity depends on the materials composing the path of propagation and on the structure employed. The surface wave velocity is independent of frequency for a uniform solid, but propagation on a layered solid is dispersive (frequency-dependent). A typical velocity for SAWs is 3×10^3 m/s, which is 100,000 times lower than the velocity of electromagnetic waves in a vacuum. Hence, at a given frequency a SAW wavelength is about 100,000 times smaller than an electromagnetic wavelength. For example, a 300-MHz signal in a SAW device would have a wavelength of only about 10 μm. The lowest frequency at which SAW devices are useful is limited to ~5 MHz by the dimensions of available crystals and the increase in wavelength as the frequency decreases, while wave attenuation and fabrication problems set an upper limit of a few gigahertz for room-temperature devices.

It is easy to convert electrical signals to SAWs (and vice versa) by the use of transducers based on the piezoelectric effect. The transducers consist of conducting electrodes evaporated onto the surface and shaped by photolithographic techniques like those used in making integrated circuits. The operation of SAW electrode transducers is illustrated in Fig. 63.

If the transducer on the right in Fig. 63a were identical to that on the left, this device would simply delay signals by 3.3 μs for each centimeter separating the two transducers and would operate well only in a band of frequencies near the design center frequency (300 MHz). Such frequency-selective transmission has been employed in millions of miniature SAW intermediate-frequency bandpass filters for use in television sets and other electronic equipment. See Fig. 63b. In the output transducer actually shown in Fig. 63a, the placement of fingers varies along the transducer, from larger spacings on the left side to smaller ones on the right, causing the left end of the transducer to respond most strongly to lower frequency waves and the right end to higher frequency waves. If a brief input signal composed of many frequency components is applied to the input transducer, the

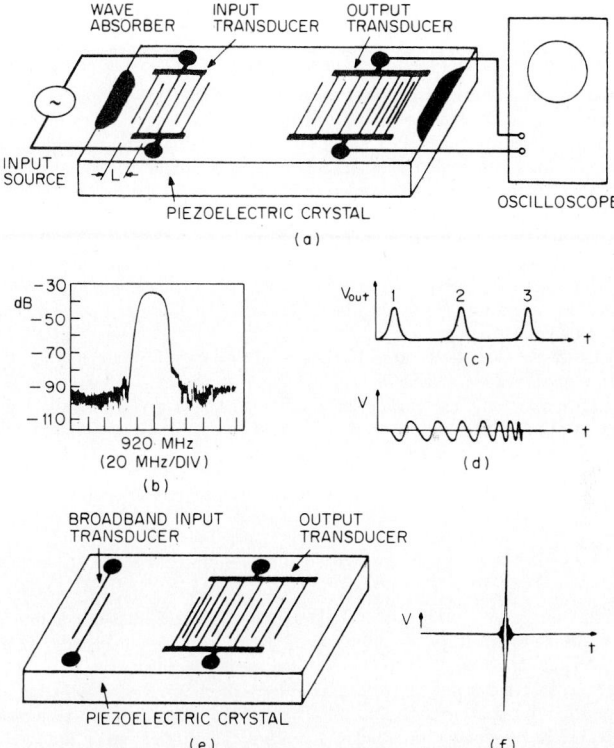

Fig. 63 Operation of surface acoustic wave (SAW) electrode transducers. (*a*) Schematic of a typical SAW device. (*b*) Bandpass filter characteristic (insertion loss versus frequency) resulting from frequency-dependent excitation and reception by transducers in the actual SAW filter. (*c*) Frequency analyzer application showing the output voltage *V* plotted against the time delay *t* in a device like that in (*a*) when low, medium, and high frequencies are input (outputs 1, 2, and 3, respectively). (*d*) Chirp output that would be obtained from the device in (*a*) if a voltage pulse of very short duration were applied to a broadband input transducer consisting of a single pair of electrodes. (*e*) Sketch of a SAW device having a broadband input transducer and an output transducer identical with the device in (*a*) but reversed right to left. (*f*) Sketch of the output when the expanded chirp pulse (*d*) is applied to the input of the device in (*e*). The voltage and time scales are the same as in (*d*).

low-frequency components will produce a voltage at the output transducer sooner than the high-frequency components because they arrive earlier at the portion of the output transducer responsive to them. Thus, a frequency analyzer can be made with this simple structure, in which frequency differences in an impulse excitation are converted to differences in arrival time.

Two other uses of SAW devices are pulse expansion and pulse compression.

SAWs provided the first practical and convenient way to make high-performance, compact, transversal filters which can operate at data rates of hundreds of megahertz. Most SAW devices used to date have been bandpass filters based on the transversal filter concept.

Terminal. A location where information can enter or leave a data processing system. Terminal equipment normally consists of one or many input or output devices (often both types) and the necessary electronics for interfacing with the remainder of the system. A terminal can be quite sophisticated and include data buffering and storage units, timing devices, and so forth. Often, a cathode ray tube (CRT) display, a keyboard for entering data, a small control console, and a buffer memory constitute a terminal. Many specialized terminals have been developed for particular applications. The Unviersal Product Code (UPC) scanner utilized in supermarkets is but one example.

Track-and-Hold Module.   See *signal conditioners and logic modules* earlier in this subsection.

Threshold Detector.   A circuit that provides an indication (digital) that a signal input is in excess of a predetermined magnitude. If a predetermined threshold value is 6 V, an input of 6.5 V will result in an output representing a binary 1 and an input of 5.5 V will cause a digital output representing a binary 0. Thus, a threshold detector is a form of comparator. The terms sometimes are used interchangeably. However, the term *comparator* may refer only to a circuit that detects whether an input signal has changed polarity.

From a practical standpoint, a band of uncertainty surrounds the predetermined threshold value. Thus, an input signal with a value that falls within the dead band will produce an indeterminant output, i.e., one that cannot be interpreted as either a 1 or a 0. Increasing the gain of the threshold circuit will reduce the magnitude of the dead band. However, stability and overload considerations place limitations on gain.

Threshold circuits are extensively used in data acquisition and instrumentation systems. The terms *voltage sense* and *level sense* sometimes are used to describe a threshold circuit in digital systems. The devices are commonly used in instrumentation alarm systems for monitoring circuits, for example, the output of a thermocouple to initiate a visual or audible alarm should the temperature rise too high or fall too low in a process.

DATA CONVERTERS*

Analog-to-Digital Converter.   Abbreviated A/D converter or ADC, this device provides a digital representation of an analog quantity. Examples of the latter include voltage, current, and position. There are two principal types of A/D converters used in data acquisition systems: (1) electromechanical converters and (2) electronic converters. The electromechanical types sometimes are referred to as shaft- or position-to-digital encoders. Generally, they are made up of a mask attached to the moving mechanical element, along with a means of reading the information on the mask. Magnetic, optical, and electrical means are used for reading. Where electrical sensing is used, the mask may consist of a conducting pattern on an insulated substrate. The code represented by the mask pattern is read by means of fixed conducting brushes which are in contact with the pattern. Optical sensors use a light source and photodetectors, whereas magnetic sensors employ inductive pickup coils.

The sensing means has a finite width. Consequently, there may be an ambiguity in the digital output. This results, for example, when a conducting brush is in contact with two adjacent portions of the mask pattern at the same time. Ambiguity can be avoided through the use of special codes and ingenious arrangements of the sensing detectors. The Gray code, where only one bit in the digital output changes at any given time as the position of the mask is varied, may be used. A V-scan technique also can be used. In the latter, two detectors are used for each track on the mask. With appropriate decoding of the outputs from the pairs of detectors, an unambiguous digital representation will result.

Electronic A/D Converters.   Two classes are used: (1) The input quantity, usually a voltage or current, is converted into another form, such as frequency or a pulse duration; this intermediate quantity then is measured to yield a digital representation of the input signal. (2) The input is compared directly with a known reference signal which can be varied under the control of the A/D converter logic. Several subclasses of A/D converters are obtainable within these two broad classes.

The principal techniques most frequently used in process control and data acquisition computers are (1) ramp, (2) integrating ramp, (3) voltage-to-frequency, (4) successive approximation, and (5) parallel-serial methods. With reference to the two broad classifications, the ramp, integrating ramp, and voltage-to-frequency methods involve conversion of the input signal to an intermediate quantity before measurement. The successive approximation and parallel-serial

---

*Terms are presented in alphabetical order for convenience of reference.

methods are direct comparison methods. These techniques are described in further detail in other specific terms listed in this section: *integrating-ramp A/D converter, parallel A/D converter, parallel-serial A/D converter, ramp A/D converter, successive approximation A/D converter,* and *voltage-to-frequency A/D converter.*

Numerous cost-performance tradeoffs are involved in selection of the most appropriate A/D converter for a given process control or data acquisiiton system. The characteristics are summarized as follows:

Ramp and Voltage-to-Frequency Converters. Relatively slow (require serial counting); used at speeds less than several thousand samples per second; usually a resolution of less than 12 bits

Successive Approximation Converters. Useful up to ∼100,000 samples per second at a resolution of 16 bits and useful up to more than 250,000 samples per second at a resolution of 8 bits or less

Parallel-Serial Converters. High speed uses requiring conversion rates in excess of 100,000 samples per second; resolution of 8 to 14 bits

Thomas J. Harrison, IBM Corporation, Boca Raton, Fla.

Average Value. See *integrating-ramp A/D converter* later in this subsecton.

Decoding Networks. See next term described.

Digital-to-Analog Converter. Abbreviated D/A converter or DAC, this is a device for generating an analog voltage or current proportional to the value of a digital input word. Frequently, in systems under the control of a digital computer, analog signals must be generated to actuate analog devices. The latter include positioning mechanisms such as valve positioners, graphic displays such as *xy* recorders, oscilloscope displays, and strip chart recorders. D/A converters also are used apart from computers to provide voltages proportional to the settings of input switches.

The majority of D/A converters consist of a D/A converter decoder network, an analog switch for each digital input bit, a buffer amplifier, and the necessary control logic in a configuration along the lines shown in Fig. 64. The decoder network is an electronically controllable attenuation network whose attenuation factor is proportional to the position of the input switches. The analog switches, under control of the logic, either connect the reference source to the decoder network input terminal or ground the terminal. This results in an output voltage proportional to the setting of the switches.

The control logic provides digital storage of the input data word and control of the switch posi-

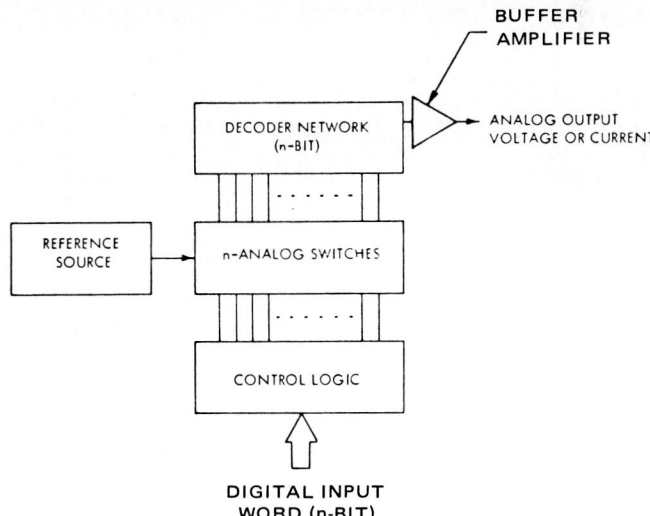

Fig. 64   Representative digital-to-analog converter.

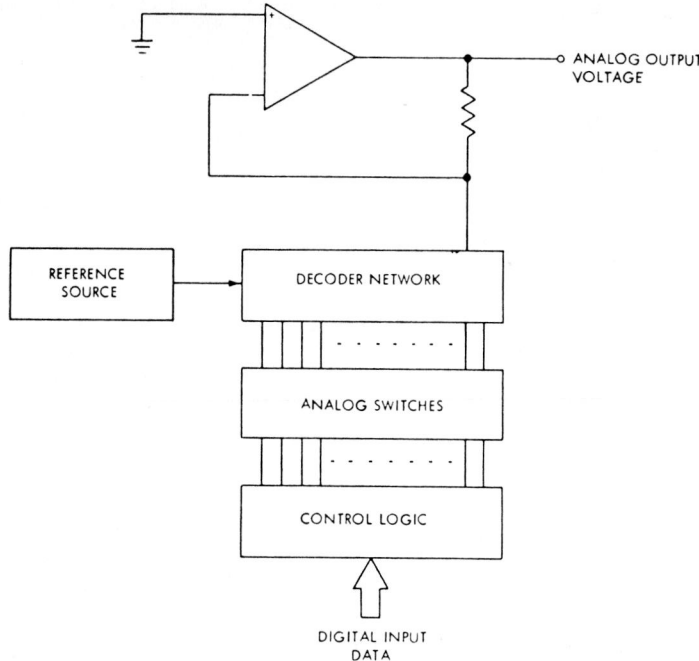

Fig. 65 Operational-amplifier-type digital-to-analog converter.

tion. Additionally, it may provide decoding of addresses and digital multiplexing of digital data to one of several D/A converters included in the subsystem. Impedance buffering and current driving capability are provided by the buffer amplifier. Generally, this is required because the output impedance of the decoder network is relatively high. Driving loads directly from the decoder network can result in loading errors which may be intolerable in a high-accuracy system.

The D/A configuration shown in Fig. 65 is comprised of control logic, a decoder network, a reference source, and an operational amplifier.

Decoding Networks. Although an ac decoder network is possible, the networks used in D/A converters generally are designed for direct current using precision resistors. The two major

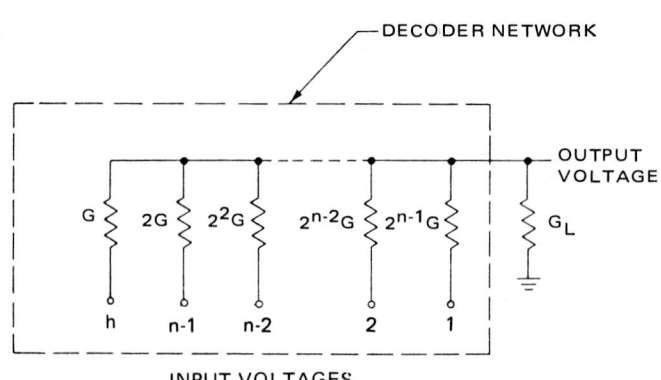

Fig. 66 Weighted-resistor decoder network.

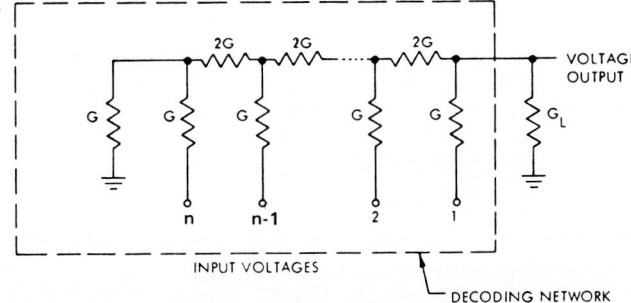

Fig. 67 Ladder decoder network of the R-$2R$ type.

decoder networks are shown in Figs. 66 and 67. Although the ladder networks shown are n-bit decoders, other codes can be realized by appropriate modifications. As an example, in a weighted-resistor binary-coded-decimal decoder, the configuration is identical but the resistors are R, $2R$, $4R$, . . .; in a ladder decoder, the resistors are R and $2R$, but an attenuation network providing an attenuation factor of 0.8 is inserted between each 3-bit section of the ladder decoder to provide the binary-coded-decimal ouptut.

The weighted-resistor network of Fig. 66 is used with a voltage reference source. Each of the terminals 1, 2, . . ., n either is connected to the reference voltage V_r or it is grounded. If the kth node is connected to V_r and the other $(n - 1)$ nodes are grounded, the ouptut voltage V_0 is

$$V_0 = \frac{V_r}{2^n - 1 + G_L/G} \sum_{k=1}^{n} A_k 2^{n-k} \qquad A_k = 1, 0$$

where A_k is an index which is 1 when the $(k - n)$ terminal is connected to the voltage source and 0 when it is grounded. When the conductance $G_L = G$, the expression reduces to

$$V_0 = V_r \sum_{k=1}^{n} A_k 2^{-k} \qquad A_k = 1, 0$$

which shows that the output voltage is binary and is determined by the conditions at the input terminals. The input conductance at the kth terminal when G_L is assumed equal to G is

$$G_{\text{in}} = 2^{n-k}(1 - 2^{-k})G$$

and the output conductance of the network is $2^n G$, a constant. Although this network is designed for use with a voltage reference source, a similar network can be designed for use with current reference sources.

The resistor network of Fig. 67 is commonly known as an $R2R$ ladder network. Either a reference voltage is connected to the kth terminal or the terminal is grounded. The expression for the output voltage is

$$V_0 = \frac{V_r}{1 + G_L/2G} \sum_{k=1}^{n} A_k 2^{-k} \qquad A_k = 1, 0$$

where $A_k = 1$ when the kth terminal is connected to the voltage V_r and 0 if the terminal is grounded. If the load resistance G_L is made equal to 0, the expression becomes

$$V_0 = V_r \sum_{k=1}^{n} A_k 2^{-k} \qquad A_k = 1, 0$$

which shows the binary contribution of each input. The output conductance is constant and is equal to $2G$ if $G_L = 0$. The input resistance is also constant and is equal to $2\,G/3$. A similar network can be used for current sources.

The weighted-resistor network uses a minimum number of resistances. The ratio of the largest to the smallest resistor is 2^{n-1} (or 512 for $n = 10$) as compared with only 2 for the ladder network. The type of resistor used in the network depends primarily on performance limitations and cost. Discrete wire-wound resistors are often used for high-precision networks which have a very low temperature coefficient. Film-type resistors can be used where the temperature coefficient and precision requirements are less.

Thomas J. Harrison, IBM Corporation, Boca Raton, Fla.

Dual-Ramp Integrating Converter.   See next term described.

Integrating-Ramp A/D Converter.   This type of instrument is designed to convert an unknown input signal to an equivalent pulse duration signal. The latter is measured by counting pulses generated by a precision clock pulse generator. Several design configurations are obtainable, one of the most common being a *dual-ramp* or *dual-slope* integrating A/D converter. The input signal is integrated over a precise time period; then the resultant integral signal is integrated over a variable length of time wherein a reference signal of opposite polarity is used. Inasmuch as the integral of the unknown input signal is proportional to the input signal, it follows that the duration of the second integration also is proportional to the input signal. The second integration period is measured by counting constant frequency pulses and thus yields a digital representation of the input signal.

A device of this type is shown schematically in Fig. 68. When the polarity of the integrator output changes, the comparator at the output of the integrator changes state. The signal to be integrated is determined by the switches at the input of the integrator. The operation sequence of the switches and the required gating of clock signals into the counter is provided by the control logic. The counter also is the output register for the A/D converter.

As shown in Fig. 68, the START CONVERT signal causes the counter to clear and closes switch S_1. As shown in Fig. 69, inasmuch as the input signal V_s may be considered essentially constant, the integrator output increases as a linear ramp signal, commencing from an initial integrator offset voltage $-V_i$. Then, for a fixed time interval $(0, t_1)$, an interval normally determined as the period required to fill the counter one time, the integration is continued. As indicated by an overflow pulse from the counter, when the interval is completed, switch S_1 is opened, while swtich S_2 is closed to apply a reference signal $(-V_i)$ to the integrator input. The integrator output decreases as a result of this action (Fig. 69). It is during the second integration that the interval clock pulses are counted

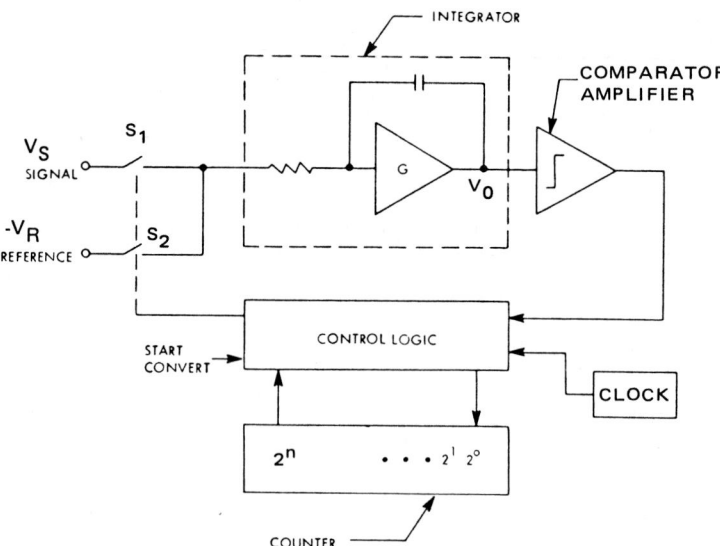

Fig. 68   Dual integrating ramp analog-to-digital converter.

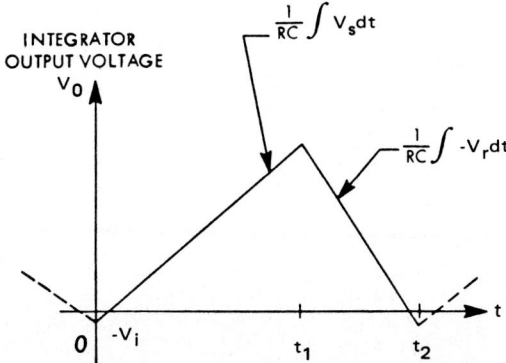

INTEGRATOR
OUTPUT VOLTAGE
V_0

$-\frac{1}{RC}\int V_s dt$

$-\frac{1}{RC}\int -V_r dt$

Fig. 69   Integrator output voltage of dual integrating ramp analog-to-digital converter.

by the counter. Once the integrator output reaches its initial level $(-V_i)$, the comparator changes state and thus pulses are kept from entering the counter. Inasmuch as the integrator output at time t_1 was proportional to the average value of the input signal [during time interval $(0, t_1)$], the length of the second integration also is proportional to V_r. Thus, the count shown by the counter at time t_2 is a digital representation of the input signal.

Typically, a dual-ramp integrating A/D converter is limited to conversion speeds of from 1000 to 2000 samples per second at a resolution of 10 or 12 bits as a result of limitations on the counter counting speed. The conversion rate can be increased to ~30,000 samples per second at a 14-bit resolution without markedly increasing the logic requirements by using a two-step integration during the second integration period.

The averaging characteristics and cancellation of errors that usually limit the performance of a ramp-type A/D converter are the principal advantages of an integrating-ramp A/D converter. The integration characteristic provides the *average value* of the input signal during the period of the first integration. Consequently, disturbances, such as spurious noise pulses, are minimized. The integration compares with a low-pass filter with a 6-dB roll-off and points of infinite attenuation at harmonics of $f = 1/t_1$.

Performance analysis of these converters also indicates that some other types of errors cancel out as well. Long-term drifts in the time constant, as may result from temperature changes or aging, do not affect conversion accuracy. Also, long-term alterations in clock frequency have no effect.

Thomas J. Harrison, IBM Corporation, Boca Raton, Fla.

Ladder Decoder Network.   See *digital-to-analog converter* earlier in this subsection.

Parallel A/D Converter.   Of the electronic analog-to-digital converters, this device probably has the simplest design. A comparator is provided for each quantization level in an n-bit converter. A digital representation of the input signal is obtained by appropriately decoding the output of the multiple comparators. See Fig. 70 where a 2-bit converter is shown. Comparators 1, 2, and 3 are biased with reference voltages of $0.75V_r$, $0.5V_r$, and $0.25V_r$, respectively. V_r is the full-scale reference voltage and is equal to the full-scale input range of the A/D converter. All comparators biased at a level less than the level of the input signal provide a binary 1 output when the input signal is applied. In the example shown, an input signal of $0.6V_r$ will result in a 1 output from comparators 2 and 3. The output of comparator 1 will be a 0. Through appropriate decoding, the binary digital representation of 10 is yielded. The binary state at the input and output of each logic block is included in the figure.

Even though the example provides only 2-bit resolution, the addition of more comparators and decoding logic can increase the resolution. However, each additional bit of resolution requires a doubling of the necessary hardware. Consequently, a parallel A/D converter normally is used only for converters of less than 5- or 6-bit resolution. Not only must the hardware be increased, but

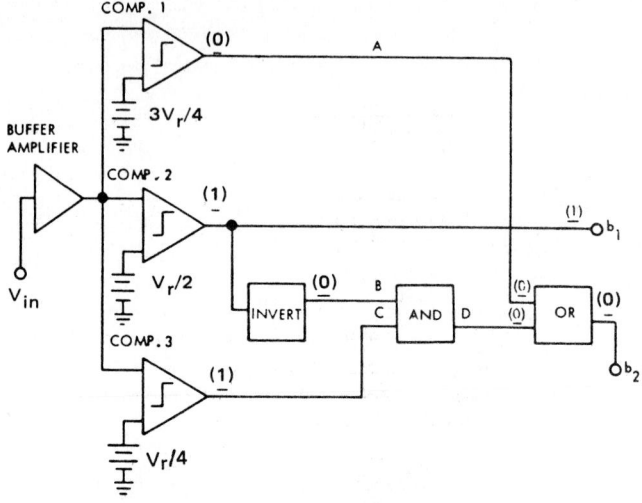

INPUT SIGNAL	A	B	C	D		b_1	b_2
$3V_r/4 < V_{in}$	1	0	1	0		1	1
$V_r/2 < V_{ir} < 3V_r/4$	0	0	1	0		1	0
$V_r/4 < V_{in} < V_r/2$	0	1	1	1		0	1
$V_i < V_r/4$	0	1	0	0		0	0

Fig. 70   Parallel analog-to-digital converter.

increasing the resolution makes marked demands on the threshold stability of the comparators in the interest of avoiding output ambiguity and excessive errors.

Simplicity for low-resolution requirements and high speed are the principal advantages of the parallel A/D converter. The settling times of the comparators and logic delays are the main factors that determine the speed. Even with simple hardware, speeds of over 50 million samples per second can be achieved.

Thomas J. Harrison, IBM Corporation, Boca Raton, Fla.

Parallel-Serial A/D Converter.   This type of analog-to-digital converter is somewhat similar to the successive approximation A/D converter because the conversion is accomplished in the form of a series of distinct steps. The speeds of parallel-serial A/D converters are not as high as those of parallel A/D converters, but practical attainment of resolution is considerably better, that is, 12- to 14-bit resolution at speeds of over 100,000 samples per second.

A parallel-serial A/D converter is shown schematically in Fig. 71. Here the A/D converter is comprised of four 3-bit digital-to-analog (D/A) converters and four amplifiers. Each has a gain of 8, which corresponds to the 3-bit resolution of each section of the A/D converter; i.e., $2^3 = 8$. A 3-bit conversion is accomplished by the seven comparators shown and in a fashion similar to that used in parallel A/D converters. The voltage that corresponds to the 3-bit converison is subtracted from the input signal by D/A converter 1 (DAC 1). This difference signal is then amplified by a factor of 8. The next 3 bits are then determined in a similar manner in the next stage of the A/D converter. Inasmuch as 3 bits are determined simultaneously (nearly), five steps yield a 15-bit quantization of the input signal. To accomplish this in a successive approximation A/D converter requires a 15-step procedure. See *successive approximation A/D converter* later in this section.

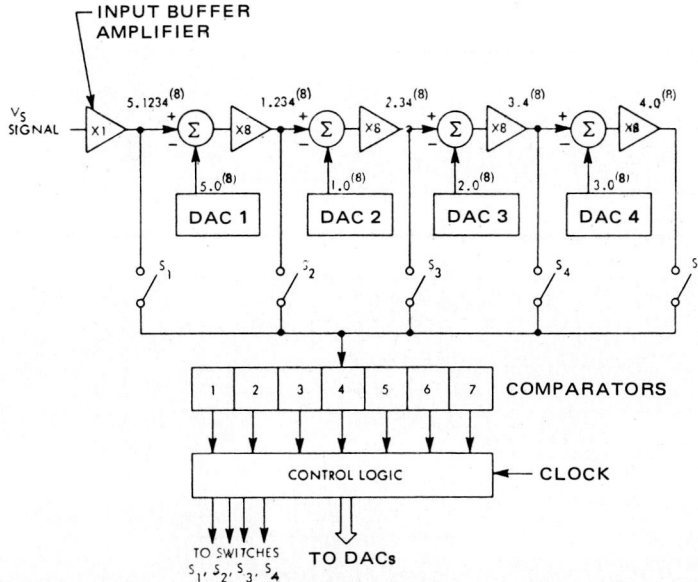

Fig. 71   Parallel-serial analog-to-digital converter (15-bit).

With reference to the diagram, assume an input voltage of 5.163086 V. Also assume a full-scale range of the A/D converter of 8 V. The operation of this A/D converter configuration is best explained in terms of octal numbers. Thus, the input voltage is 5.1234^8. The superscript 8 shows that the number is expressed in octal code. The conversion process commences with the closing of switch S_1 and applying the input signal to the parallel comparators. The output of the comparators shows that the input is in excess of 5^8 but is less than 6^8. Hence DAC 1 is set to 5^8 and a voltage of 0.1234^8 is applied to the first-gain-of-8 amplifier. By opening S_1 and closing S_2, the output of the amplifier, 1.234^8 V is applied to the parallel comparators. This parallel conversion yields the determination that the amplified 2 is set to 1.0^8 V to furnish a difference signal of 0.234^8 V. Other steps follow which involve the closing of switches S_3, S_4, and S_5. These actions result in the octal result 5.1234^8 or the 15-bit binary result 101001010011100.

In the device shown, it is required that each stage of the converter settle completely before the subsequent 3-bit conversion is commenced. Further circuitry is added for error correction so that, if a particular stage has not completely settled, with a resultant incorrect 3-bit converison, the conversion in the following stage will provide a signal to correct the prior 3-bit representation.

Although the parallel-serial A/D converter has relatively high speed conversion, the design requires greater hardware complexity than that of a successive approximation A/D converter with equal resolution. The parallel-serial A/D converter is not as fast as the parallel A/D converter, but it requires considerably less hardware for resolutions of more than 6 bits. Further, precise comparator adjustments, required in a high-resolution parallel A/D converter, are avoided with a parallel-serial A/D converter.

Thomas J. Harrison, IBM Corporation, Boca Raton, Fla.

Ramp A/D Converter.   This type of analog-to-digital converter quantizes an analog input signal through conversion of the signal to a time duration pulse. The latter is measured by a counter and a constant frequency pulse generator. A ramp A/D converter is shown schematically in Fig. 72. With an n-bit converter, a START CONVERT signal will reset the n-bit counter and initiate the operation of a function generator. The latter produces a linear ramp output signal, $V_R = kt$, where k is a constant and t is time. A comparator then continuously compares the reference ramp signal with the value of the unknown input signal V_S. The resulting ouptut of the comparator represents a binary

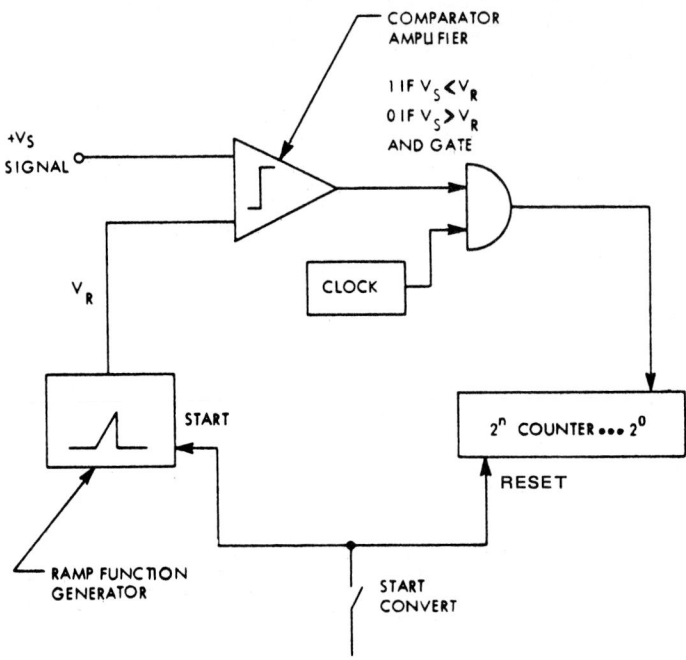

Fig. 72   Ramp-type analog-to-digital converter.

0 where V_R is greater than V_S, and a binary 1 where V_R is less than V_S. While the comparator output is 1, pulses from a clock pulse generator are counted by the n-bit counter. At such a time that the input signal V_S and the ramp signal V_R are equal, the comparator output changes to a binary 0. Also, by action of the AND gate, clock pulses are prevented from entering the counter. The time during which the comparator output remains in the 1 state is proportional to the magnitude of the input signal. Also, the count in the counter at the instant the comparator changes state is proportional to the time interval that the comparator output is 1. Thus, the count in the counter is a digital representation of the input signal.

This type of A/D converter is one of the simplest forms of electronic A/D converters and may be used for conversions of less than 10 to 12 bits at speeds that do not exceed several thousand conversions per second. Higher speeds can be obtained at lower resolution. Logic speed is the main factor in determining the speed of the device. It will be noted that the speed of the converter equals the clock frequency divided by the number of quantizing intervals. For example, 4096 pulses must be counted for a full-scale input signal in a 12-bit converter. At a clock frequency of 10 MHz, this necessitates 1024 pulses per 10 MHz, or ~0.4 ms, consequently providing a converter speed of 2500 samples per second. Added to simplicity is the advantage of excellent differential linearity, accounting for the very wide use of this type of A/D converter, particularly in such applications as the generation of histograms as encountered in the field of nuclear experimentation.

Thomas J. Harrison, IBM Corporation, Boca Raton, Fla.

Successive Approximation A/D Converter.   This type of analog-to-digital converter makes a direct comparison between an unknown input signal and a reference signal. The converter, as shown schematically in Fig. 73, includes a digital-to-analog (D/A) converter. This device provides an output voltage V_R, a precise fraction of the reference source voltage. The setting of analog switches, which are controlled by the digital outputs of the output register, determines the attenuation factor of the converter. Digital output information of the A/D converter also is stored in the output register. Two input AND logic gates are used as conditional reset gates. The shift register also functions as a step counter.

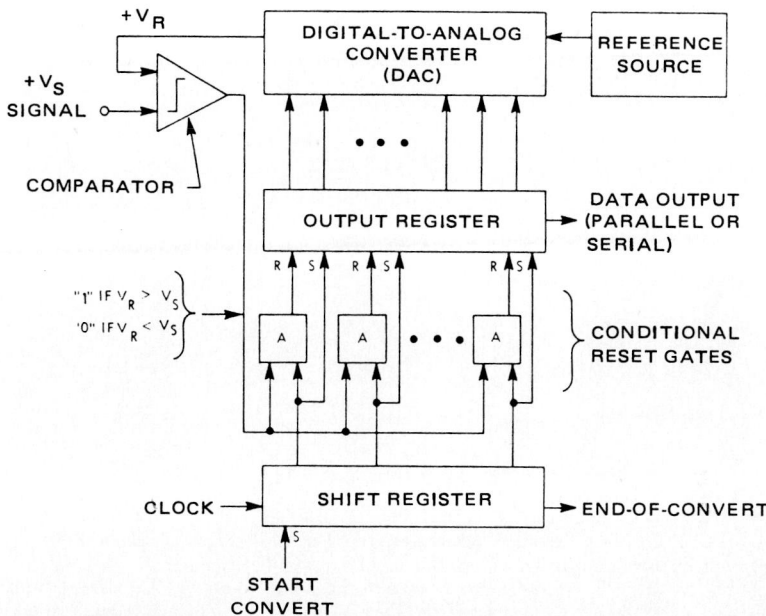

Fig. 73   Successive approximation analog-to-digital converter.

The A/D converter shown in the figure is a binary n-bit converter. A START CONVERT pulse causes all the registers to clear and sets a 1 into the first position of the shift register. This action also turns on the first switch in the A/D converter and sets the first stage of the output register to 1. The output of the D/A converter is equal to one-half the full-scale D/A converter output voltage inasmuch as the D/A converter is binary. The D/A output voltage is compared directly with signal voltage V_s by the comparator. The output of the latter is binary 1 if V_R is greater than V_s, and binary 0 if V_R is less than V_s.

The first bit in the output register may be reset to 0 or may remain as 1, depending on the comparator output. On appearance of the next clock pulse, the 1 in the shift register moves to the second bit, at which instant the comparison process is repeated—with the exception that the output of the D/A converter is either 0.25 or 0.75 of the full-scale D/A-converted ouput. This is dependent on whether the first bit was reset to 0 or remained as a 1, respectivley.

Codes other than binary can be used in successive approximation A/D converters. What is required is the production of an attentuation proportional to the weighting value of the code digit. A binary-coded decimal converter can be constructed by using a binary-coded decimal D/A converter. The D/A converter provides attenuation ratios of 0.8, 0.4, 0.2, 0.1, 0.08, and so on. Of course, appropriate coding of the output register is necessary so that the information can be stored in coded form.

Successive approximation A/D converters can be used at conversion speeds of up to approximately 100,000 samples per second at resolutions of up to 16 bits (not including sign). At lower resolutions, speeds of over 250,000 samples per second are practical. Factors to be considered in the design and application of these A/D converters include (1) stability and regulation of the reference source, (2) overload and recovery characteristics of the comparator, (3) analog switch characteristics, and (4) speed and response of the ladder network.

Thomas J. Harrison, IBM Corporation, Boca Raton, Fla.

Voltage-to-Frequency A/D Converter.   This type of analog-to-digital converter converts the magnitude of an analog input signal (usually a current or voltage) to a frequency which can be measured digitally. As an example, the analog input signal could control the bias of a varacter diode.

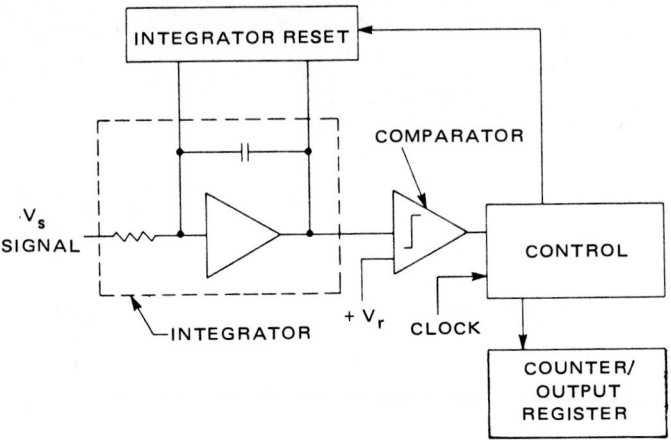

Fig. 74   Voltage-to-frequency analog-to-digital converter.

If the diode is used as the frequency determining component of a resonant circuit in an oscillator, then the frequency of oscillation will depend on the value of the input signal.

In Fig. 74, the voltage-to-frequency A/D converter uses an integrator. At commencement of the conversion, the counter is returned to zero and the integrator is reset. An input signal V_s is integrated until an output value V_r is reached by the integrator. A change in state of the comparator output detects this point. On this change of state of the comparator, the integrator is reset and the process repeated.

The rate of integration depends on the value of the input signal. Thus, the frequency of the sawtooth waveform generated by the integrator is a function of the input signal. If the number of integration cycles that take place in a fixed time period are counted, a digital representation of the input signal is yielded. Clock pulses in a separate counter determine the counting interval.

An advantage of the voltage-to-frequency A/D converter is that it provides a digital representation of the average value of the input signal during the conversion interval. Thus, input noise transients have a minimal effect on performance. As the result of serial counting, this A/D converter is comparatively slow. Speed usually is less than 1000 samples per second at a resolution of 10 bits or less.

DIGITAL SYSTEM AMPLIFIERS*

Amplifier.   A device for increasing the strength of a signal without appreciably altering other signal characteristics, such as waveform.

Bridge Amplifier.   An amplifier extensively used for instrumentation purposes. The commercial configuration generally is a direct-coupled amplifier, offering reasonably wide bandwidths up to 50 kHz at gains ranging from nearly unity to 1000. The use of four subamplifiers in a bridge amplifier configuration is shown in Fig. 75. The output voltage, assuming that the open-loop gains G_1, G_2, G_3, and G_4 of the separate amplifiers are quite large, is given by

$$V_0 = \frac{R_1 + R_2}{R_1} V_1 - \frac{R_2}{R_2'} \times \frac{R_1' R_2'}{R_1} \times V_2$$

*Terms are presented in alphabetical order for convenience of reference.

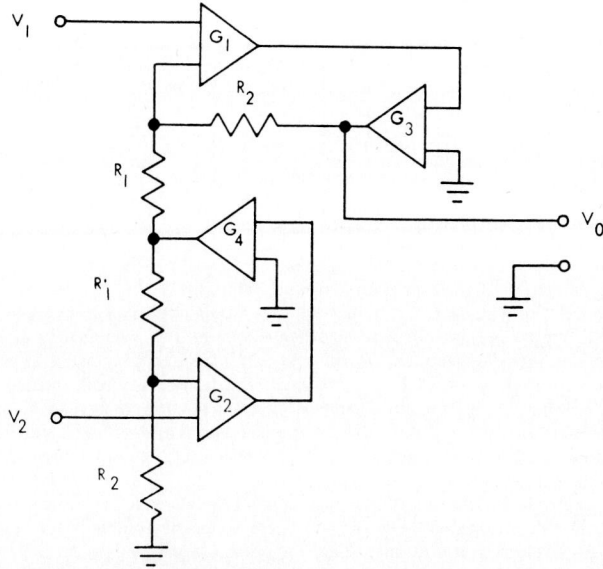

Fig. 75 Dynamic bridge amplifier.

Voltage V_1 is the sum of the differential voltage $V_{\text{signal}} = V_1 - V_2$ and the common-mode voltage. Voltage V_2 is the applied common-mode voltage V_{cm}. Substituting these factors in the foregoing expression, the output voltage is given by

$$V_0 = \frac{R_1 + R_2}{R_1} V_{\text{signal}} + \left(1 - \frac{R_2 R_1'}{R_2' R_1}\right) V_{cm}$$

The closed-loop gain of the amplifier thus is $(R_1 + R_2)/R_1$. The common-mode rejection ratio is $|G/[1 - (R_2 R_1')/(R_1 R_2')]|$. If $R_1'/R_1 = R_2'/R_2$, the condition for a balanced resistive bridge, theoretically infinite common-mode rejection, can be obtained. This analysis does not bring out the practical limitations of matching resistors and of other errors. Thus, the common-mode performance is finite. However, values in excess of 120 dB can be achieved. The common-mode rejection ratio (CMRR) of this type of amplifier is directly proportional to the gain. For most differential amplifiers, the CMRR is largely independent of the gain.

Thomas J. Harrison, IBM Corporation, Boca Raton, Fla.

Carrier Amplifier. A dc amplifier wherein the signal first is modulated, and then demodulated during amplification. Electronic switches or electromechanical devices are used in most cases to effect the modulation. Thus, the "chopping" action accomplishes the equivalent of square wave modulation of the signal.

The carrier technique is employed for two main purposes: (1) to reduce to a minimum the effects of zero-offset drift, which is a critical performance parameter in any dc amplifier, and (2) to provide isolation between the input and the output of the amplifier.

With reference to Fig. 76, a conceptual design is shown. The input signal is first modulated to produce an ac signal, after which the signal is amplified by an ac amplifier. Then, the output of the latter is demodulated to provide a dc output signal. Zero-offset drift in the amplification section of the amplifier does not affect the value of the output signal because only the ac component is amplified. However, offsets in the modulator can cause the equivalent of an offset in the output signal should they increase or decrease the magnitude of *both* the positive and negative peaks of the modulated signal. In most cases, if the input signal is greater than 1 V, such offsets do not create a serious problem. In the case of low-level amplifiers, however, they can cause significant errors. Because the

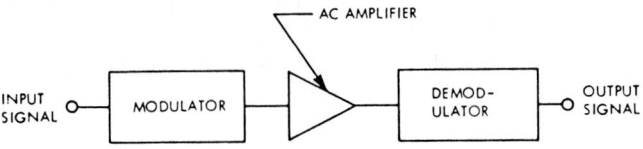

Fig. 76   Carrier amplifier.

output demodulator usually operates at a high level, demodulator offset is not considered an important limitation on overall amplifier performance. The use of carrier amplifiers designed mainly for the reduction of zero-offset drift is diminishing mainly as a result of the improvement of techniques and components for producing low-drift direct-coupled amplifiers.

In the instance of using a carrier amplifier to provide isolation between the input and output of an amplifier, the amplifier commonly is termed a *floating amplifier*. See *floating amplifier* later in this section. An amplifier of this design, incorporating an overall feedback path, is shown in Fig. 77. The basic carrier concept is used—the input signal is modulated and demodulated by a chopper circuit. In this example, since the main purpose is isolation of rather than reduction in drift, an ac amplifier is not used. By means of the four-terminal isolation characteristic of the transformer, the input signal can be referenced to a ground point independent of the output signal reference point. Floating-carrier amplifier designs of this type are used in digital data acquisition subsystems and instrumentation subsystems to accomplish amplification of signals under conditions where high common-mode voltages may be present. The common-mode voltage is essentially limited by the breakdown voltage of the coupling transformer. Thus, amplifiers of this design can function with up to several hundred volts of common mode, as contrasted with the usual 10- to 20-V limitation inherent in most direct-coupled amplifiers as a result of the breakdown limitations of most semiconductor devices.

Thomas J. Harrison, IBM Corporation, Boca Raton, Fla.

Chopper Amplifier.   Two connotations apply to the term *chopper amplifier*. In one type, the input signal is chopped (or modulated), amplified by an ac amplifier, demodulated, and filtered to provide a dc output signal. This type of amplifier was described earlier in this subsection. See *carrier amplifier*. In the second type, the error signal is chopped for the purpose of providing stabilization of gain and offset. Possibly a more apt term would be *chopper-stabilized amplifier*.

With reference to Fig. 78, the unit is a chopper-stabilized amplifier with an overall gain A. The amplifier output is attenuated by a factor of $1/A$ and then compared with the input signal. An error may exist as the result of zero-offset drift, a change in the gain of the dc amplifier, or simply noise. The error is chopped (modulated), amplified by an ac amplifier, demodulated, and then summed with the input to the dc amplifier—in a manner that compensates for the disturbance causing the error.

Excellent zero-offset gain characteristics are features of the chopper-stabilized amplifier, along with the favorable wide-bandwidth characteristics of the direct-coupled amplifier. The modulator-

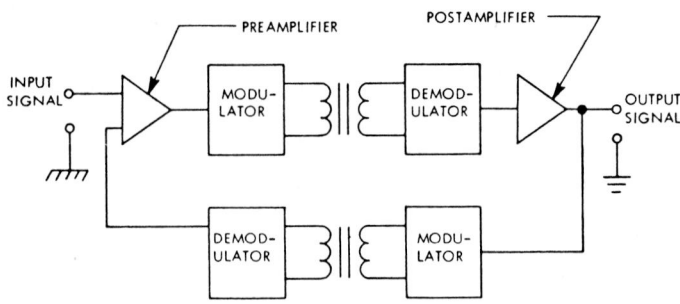

Fig. 77   Floating-carrier amplifier.

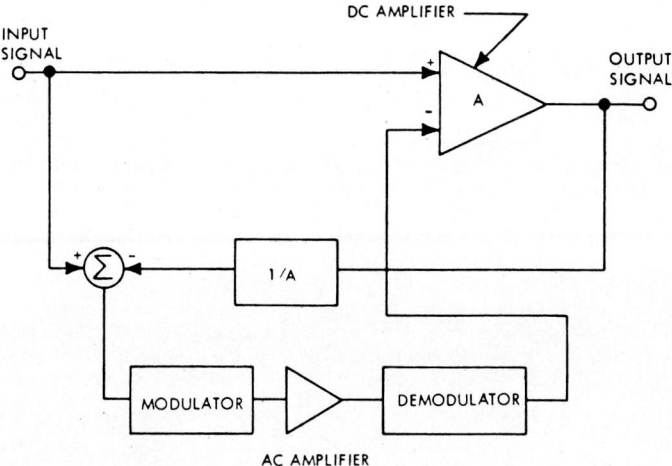

Fig. 78 Chopper-stabilized amplifier.

demodulator circuitry is relatively complex, and consequently the design is not extensively used in digital data acquisition and instrumentation amplifiers. Comparable performance can be obtained from conventional direct-coupled amplifiers. One negative feature of the chopper-stabilized amplifier is the frequently encountered long saturation recovery time, a deterrent for use in time-shared systems.

Thomas J. Harrison, IBM Corporation, Boca Raton, Fla.

Comparator Amplifier. A nonlinear amplifier for sensing either the polarity or the magnitude of an input signal. The amplifier output is one of two states, usually for representing the binary states 1 and 0. When the input signal is in excess of a predetermined level (frequently zero), the amplifier output is in one state. When the input signal is less than the predetermined level, the ouptut remains in the other state. In the case where the predetermined level is other than zero, a comparator amplifier may be termed a *threshold detector*. See also *threshold detector* described earlier in this section.

An ideal comparator and a practical comparator are compared in Fig. 79. The ideal characteristic is shown by the heavy line. In this example, the amplifier output is $+6$ V when the input signal is greater than zero, and zero when the input is less than zero. However, in practice, a linear region of operation exists between these two output levels, corresponding to a small uncertainty in the performance of the comparator. Consequently, the input level must be in excess of a threshold level V_t to make certian that the output will reach the $+6$-V level. Within the limitations of amplifier stability factors, the threshold voltage can be decreased by increasing the gain of the amplifier. Gains in excess of 10,000 are typical of comparator circuits.

The comparator is a high-gain single-ended or differential amplifier. Usually, the comparator is configured for fast recovery from saturation, thus permitting it to follow a rapidly varying input

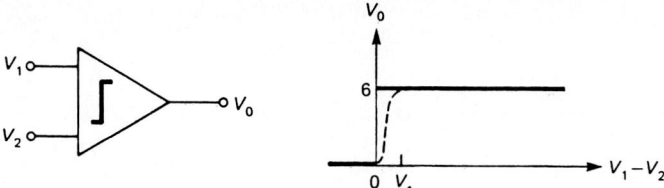

Fig. 79 Transfer characteristics of comparator amplifier.

signal. A low zero offset is required for precision applications inasmuch as zero offset is equivalent to a shift in the threshold level.

Thomas J. Harrison, IBM Corporation, Boca Raton, Fla.

Direct-Coupled (DC) Amplifier.   As used in digital data acquisition and instrumentation systems, the term *DC amplifier* has two connotations: (1) a *direct-coupled amplifier* (i.e., a low-resistance dc connection is used between each stage and the succeeding stage in the amplifier), and (2) an amplifier with a frequency response that extends to zero frequency (direct current). It should be pointed out that, although all DC (direct-coupled) amplifiers have a frequency response that extends to direct current, not all amplifiers with this type of frequency response are direct-coupled. A carrier amplifier, for example, has a response extending to direct current, but it is not direct-coupled.

One type of direct-coupled differential amplifier is shown in Fig. 80. Only the first stage of the amplifier is shown in detail, because the most important characteristics of any amplifier usually are determined by its first stage. The two or three remaining stages are shown as a single symbol. These stages provide high gain, typically greater than 1000, and also the output drive capability of the amplifier. Usually, the conversion from differential input to single-ended output is also effected in these stages.

As shown, the input stage is comprised of a pair of transistors, the bases of which are biased by current sources. A current source also determines the emitter current of the transistor pair. Feedback resistor R_{f1} and emitter resistor R_{e1} determine the gain of the stage. There is a direct coupling of the differential-output signal generated at the collectors of the first-stage transistors and the bases of the transistors in the second stage.

Usually, the gain of the first stage is on the order of 100. Thus, the first stage essentially determines the drift and noise characteristics of the overall amplifier. Because of the manner in which the first stage is configured, the drift characteristics are determined by the difference in the drifts of the

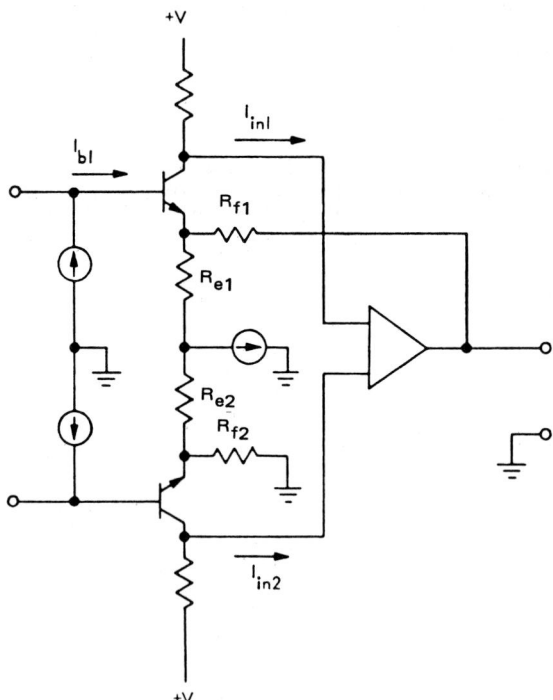

Fig. 80   Direct-coupled differential amplifier.

two input transistors. The base emitter voltage drop is the major contributor to the zero-offset temperature coefficient. The overall amplifier drift can be minimized (a few microvolts per degree Celsius) by using matched transistors in the differential configuration. Where additional temperature compensation is used in the first stage, a drift of less than 1 μV/°C can be achieved. Noise characteristics are mainly related to the intrinsic noise of the input transistors. Some control over this can be effected by selection of the biasing point. The usual procedure is to select low-noise transistors.

The design of the first stage also affects the common-mode characteristics. Where possible, the configuration should be made symmetrical with respect to ground—thus the inclusion of the dummy feedback resistor R_{f2}. Also, the input base bias sources must be balanced. The balance between the RC constants of the base-to-collector capacitance and the collector resistance determine high-frequency common-mode characteristics. Where a design may be critical, an adjustment of one of these parameters may be required. The common-mode performance of the amplifier also may be improved by common-mode feedback between the first-stage emitter and a common-mode point in a subsequent stage. Feedback is frequently applied to the ground side of the emitter bias source.

The type of amplifier just described schematically is used extensively in digital data acquisition and instrumentation systems. Direct-coupled design permits a wide bandwidth, typically 0 Hz to frequencies in excess of 50 kHz and at relatively high gains. As compared with some other designs of differential amplifiers, the DC amplifier is relatively inexpensive. For some applications, a disadvantage is the relatively low common-mode voltage capability. The common-mode voltage essentially is limited to 10 to 20 V because of the breakdown characteristics of the input transistors.

Thomas J. Harrison, IBM Corporation, Boca Raton, Fla.

Floating Amplifier.   Also known as an *isolated amplifier,* the design does not require that the input and output signals be referred to the same signal reference point (ground). Generally, differential-input amplifiers meet this definition. However, the term "floating" normally excludes an amplifier where the input reference point and the output reference point are common, i.e., either through a signal conductor or a power supply. Specifically, a floating amplifier is an amplifier which includes a four-terminal coupling device, such as a light-coupled signal transmission element or a transformer.

The most common floating amplifier used in digital data acquisition and instrumentation systems is the carrier amplifier. See also *carrier amplifier* described earlier in this section. The input signal is coupled to the output amplifier, demodulated, and filtered to produce an output signal. As shown in Fig. 81, there may be feedback from the output to the input by means of a similar modulator-demodulator combination.

The four-terminal isolation property of the transformer provides complete isolation of the amplifier input and output; i.e., the signal reference point (ground) for the input may be completely independent of the output signal reference point. Thus, even though the preamplifier and postamplifier may be single-ended, the amplifier performs as a differential amplifier. The breakdown voltage limitation of the transformer determines the maximum difference in the ground potential between the input and output circuits. Normally, this is quite high (hundreds of volts). Thus, this amplifier design

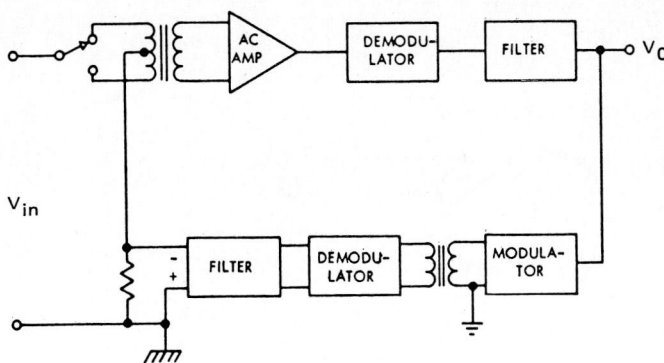

Fig. 81   Floating-carrier amplifier with feedback.

is particularly well adapted to uses that require the amplification of signals in the presence of high common-mode voltages. Several commercial designs are available.

The achievement of a high common-mode rejection ratio is made difficult in the design of these amplifiers because of the unbalanced nature of the preamplifier and imbalances in the coupling transformer. Typically used for these designs are well-shielded transformers with low, interwinding capacitances. Also, if optimum performance is to be achieved, the guard shielding technique usually is needed.

As compared with other differential-amplifier techniques, the isolated carrier amplifier is relatively expensive. A primary advantage is the input-output isolation. A major disadvantage is a reasonably narrow bandwidth (normally less than 10 kHz), which is imposed by carrier frequency limitations and by transformer characteristics. If the guard shield is not correctly connected to the source of common-mode voltage, the ac common-mode rejection ratio frequently is markedly degraded. Where optimum performance is to be achieved, the design is restricted to three-wire systems.

Operational Amplifier.   An amplifier with high dc stability and high immunity to oscillation, usually achieved by using a large amount of negative feedback. The operational amplifier is widely used to perform analog-computer functions. For example, an operational-amplifier integrator is used in the conversion of voltage to a frequency. In this configuration, a capacitor is connected between output and input. Application of a step input results in a linear ramp output in which the slope of the ramp is proportional to the input-voltage step.

Sample-and-Hold Amplifier.   Also known as a *track-and-hold amplifier,* this device has an output that is proportional to the input until a "hold" signal is received. On receipt of the signal, the amplifier output is maintained essentially constant even though there may be changes in the input signal. The input and output waveforms of a sample-and-hold amplifier are shown in Fig. 82.

As shown in Fig. 83, the conceptual design of a sample-and-hold amplifier comprises two independent amplifiers connected by a switch. With sampling switch S_1 closed, holding capacitor C is charged when an input signal is applied to the first amplifier. On receipt of the HOLD command, switch S_1 is opened, thus leaving capacitor C charged at the instantaneous value of the input signal. Capacitor C is not discharged because the second amplifier has a high input impedance. The output of the second amplifier remains essentially steady for a period of time. The HOLD signal may be generated by an external circuit (coupled to a process or experiment) or by a computer or digital control unit under control of a stored program.

Sample-and-hold amplifiers meet certain specialized needs in digital data acquisition systems. Usually one or both of the following requirements exist: (1) The value of a single signal must be determined at a precise instant in time; (2) the values of two signals must be compared at a precise instant in time. These requirements cannot be met by a time-shared amplifier and analog-to-digital converter system because these systems require a finite settling and conversion time. Where a sample-

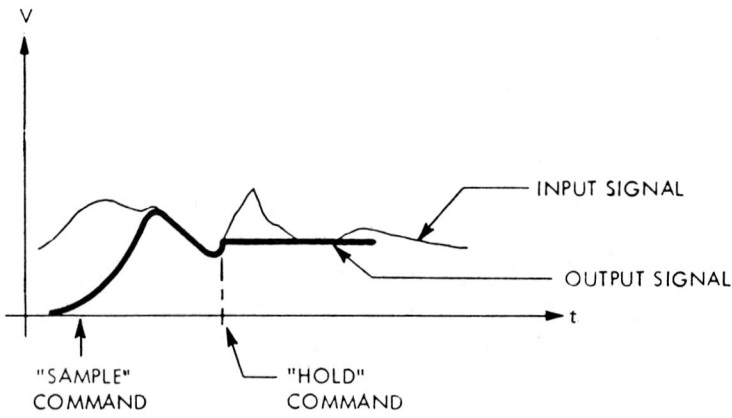

Fig. 82   Action of sample-and-hold amplifier.

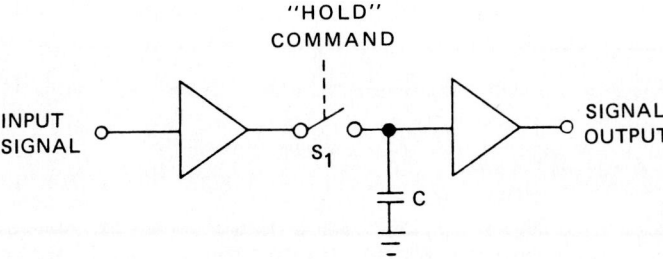

Fig. 83 Schematic circuit of sample-and-hold amplifier.

and-hold amplifier is used, it is possible to retain the instantaneous value of one or more input signals over the time interval required to convert them to digital values.

Thomas J. Harrison, IBM Corporation, Boca Raton, Fla.

Single-Ended Amplifier. A device designed to amplify signals between a single input terminal and the ground or signal reference point. Generally, the output signal is generated between a single output terminal and the same signal reference point.

A disadvantage of the design stems from installation difficulties caused by the common connection between the input and output circuits. Unfortunately, any potential that exists between the amplifier signal reference point (ground) and the reference point for the input signal appears in series with the input signal. Thus, this additional potential is amplified with the signal and cannot be distinguished from the true signal voltage. Since by definition a potential between two signal reference points is a common-mode voltage, the common-mode rejection of a single-ended amplifier is zero.

Where amplifiers of this kind are used in digital data acquisition and instrumentation systems, careful attention must be given to the grounding circuits. Obviously, a major requirement is for the input signal reference point to be at precisely the same potential as the signal reference point for the amplifier input. Thus, a low-impedance connection between these two reference points at all frequencies of concern is needed. This is a difficult condition to achieve when the signal source and the amplifier are separated by a significant distance. Thus, amplifiers of this type seldom are used when such separation exists. For small subsystems where the ground structure is well controlled, single-ended amplifiers may be used.

REFERENCES

Andreiv, N.: "Will the Real Data Acquisition System Please Stand Up?" *Control Eng.,* vol. 25, no. 11, 1977, p. 68.

Andreiv, N.: "Magnetic Bubble Memories Displace Drums in CNC Applications," *Control Eng.,* vol 28, no. 1, 1981, p. 57.

Aron, J. D.: *The Program Development Process, Part I: The Individual Programmer,* Addison-Wesley, Reading, Mass., 1974.

Avir, S., and G. J. Lauer: "The Characteristics and Decodability of the Universal Product Code Symbol," *IBM Syst. J.,* vol. 14, no. 1, 1975, p. 16.

Bailey, S. J.: "Put Memories in Control: Enhance Process Response," *Control Eng.,* vol. 24, no. 7, 1977, p. 7.

Bailey, S. J.: "LSI Subsystems Leading to Distributed Data Acquisition Systems," *Control Eng.,* vol. 25, no. 9, 1978, p. 69.

Bailey, S. J.: "Memory Developments Will Impact Control Systems," *Control Eng.,* vol. 26, no. 1, 1979, p. 37.

Bailey, S. J.: "Let's Get Control Experience into System Memory," *Control Eng.,* vol 27, no. 7, 1980, p. 71.

Bell, C. G., Mudge, J. C., and J. E. McNamara: *Computer Engineering,* Digital Press, Bedford, Mass., 1978.

Besant, C. B.: *Computer-Aided Design and Manufacture,* Wiley, New York, 1980.

Bohl, M.: *Information Processing,* 3d ed., Systems Research Associates, Chicago, Ill., 1980.

Chase, D.: "System Architecture Affects Data Acquisition Accuracy and Flexibility," *Control Eng.,* vol. 27, no. 7, 1980, p. 83.

Cox, G.: "Designing with Magnetic Bubble Memories," *Control Eng.,* vol. 26, no. 7, 1979, p. 54.

Garelick, E. L.: "Characterizing Signal Conditioners with Plug-In Memory," *Control Eng.,* vol. 24, no. 9, 1977, p. 62.

Ginn, P. L.: *An Introduction to Process Control and Digital Minicomputers,* Gulf Publishing, Houston, Texas, 1982.

Halevi, G.: *The Role of Computers in Manufacturing Process,* Wiley, New York, 1980.

Harrison, T. J.: *Minicomputers in Industrial Control,* Instrument Society of America, Research Triangle Park, N.C., 1978.

John, C. D.: *Process Control Instrumentation Technology,* Wiley, New York, 1982.

Klein, A.: "I/P, P/I Converters Link Computers with Process," *Control Eng.,* vol. 26, no. 1, 1979, p. 26.

Kompass, E. J.: "Environmentally Rugged I/O System Provides New Capabilities for Distributed Control," *Control Eng.,* vol. 25, no. 4, 1978, p. 51.

Kompass, E. J.: "Semiconductors for Control: The Control Engineer's Viewpoint," *Control Eng.,* vol. 28, no. 8, 1981, p. 62.

Laduzinsky, A. J.: "Multiplexers Provide Many Different Paths for Data," *Control Eng.,* vol. 28, no. 12, 1981, p. 72.

Loren, H., and H. M. Deitel: *Operating Systems,* Addison-Wesley, Reading, Mass., 1981.

Morris, H. M.: "Distributed Systems Makes Wide Use of Bubble Memories," *Control Eng.,* vol. 29, no. 1, 1982, p. 68.

Pratt, T. W.: *Programming Languages,* Prentice-Hall, Englewood Cliffs, N.J., 1975.

Simons, G. L.: *Computers in Engineering and Manufacture,* Wiley, New York, 1982.

Stire, T. G.: *Process Control Computer Systems,* Ann Arbor Science Publ., Woburn, Massachusetts, 1982.

Microprocessors

Alzos, W.: "Power Plant Optimization Using a Microcomputer," *Control Eng.,* vol. 26, no. 5, 1979, p. 126.

Barck, R. T.: "How to Select the Optimum Microprocessor for Consumer and Industrial Controls," *Control Eng.,* vol, 26, no. 7, 1979, p. 106.

Barnes, G. F.: "Single-Loop Microprocessor Controls," *Instrum. Technol.* vol. 24, no. 12, 1977, p. 47.

Beaston, J.: "Automatically Restarting a Microprocessor After Power Failure," *Control Eng.,* vol. 25, no. 12, 1978, p. 84.

Bibbero, R. J.: *Microprocessors in Industrial Control,* Instrument Society of America, Research Triangle Park, North Carolina, 1983.

Bottari, W.: "How to Design Single Chip Microcomputers into Control Systems," *Control Eng.,* vol. 26, no. 5, 1979, p. 69.

Bullock, T. B.: "Special CPU Design Makes Programming Logic More Efficient," *Control Eng.,* vol. 25, no. 5, 1978, p. 52.

Cassell, D. A.: *Microcomputers and Modern Control Engineering ,* Reston Publishing, Reston, Virginia, 1983.

Crater, K.: "Systems Approach Reduces Microprocessor to Pneumatics Interfacing Costs," *Control Eng.,* vol. 28, no. 8, 1981, p. 86.

Deshon, W. E.: "Microprocessors Improve Serial Communications," *Control Eng.,* vol. 25, no. 3, 1978, p. 127.

Garen, E. R.: "Microcomputers: The Ubiquitous Control Element," *Control Eng.,* vol. 25, no. 6, 1978, p. 53.

Gray, D. M: "Microprocessor Characterizes pH Ahead of Controller for Easy Tuning," *Control Eng.,* vol. 27, no. 1, 1980, p. 79.

Harrison, T. J.: "Micros, Minis and Multiprocessing," *Instrum. Technol.* vol. 25, no. 2, 1978, p. 43.

Hawkins, R. D.: "The Impact of Software on Future Microprocessor Distributed Architecture," *Instrum. Technol.,* vol. 27, no. 3, 1980, p. 49.

Hollister, A. L.: "A Primer on Microprocessor Software," *Instrum,. Technol.,* vol. 24, no. 10, 1977, p. 47.

Horbal, M. T., and D. Derrico: "Microcomputer System Controls Steel Angle Fabrication," *Control Eng.,* vol. 25, no. 6, 1978, p. 69.

Jutila, J. M.: "The Micros Are Restless," *Instrum. Technol.,* vol. 26, no. 10, 1979, p. 9.

Karl, L. J., and D. J. Jones: "Microprocessors: DDC for Large Electric Utility Boilers," *Instrum. Technol.,* vol. 27, no. 7, 1980, p. 47.

Kompass, E. J.: "Microprocessor-Based Controller Handles Analog and Digital I/O," *Control Eng.,* vol. 23, no. 7, 1976, p. 22.

Kompass, E. J.: "Microprocessor Enhances Automatic Weigh-Batching Flexibility," *Control Eng.,* vol. 24, no. 4, 1977, p. 42.

McGowan, M.: "A Microcomputer and A/D Converter on a Single Chip," *Control Eng.,* vol. 25, no. 6, 1978, p. 66.

McGowan, M. J.: "Designing and Assembling Microcomputer Systems Grows Easier," *Control Eng.,* vol. 26, no. 2, 1979, p. 34.

Raphael, H. A.: "Keyboard/Microprocessor Interface: Software or Hardware?" *Control Eng.,* vol. 24, no. 2, 1977, p. 41.

Sagues, P.: *Microprocessors for Measurement and Control,* Osborne/McGraw-Hill, Berkeley, California, 1982.

Williams, T. J.: "Two Decades of Change—A Review of the 20-Year History of Computer Control," *Control Eng.,* vol. 24, no. 9, 1977, p. 71.

Safety in Instrumentation and Control Systems

by
E. C. Magison[*]

Increased use of electrical control systems, analyzers, and computer systems in connection with process instrumentation has focused increasing attention on reducing the probability of fire or explosion due to electric instrument failure. In the United States, until the 1950s, explosion-proof housings were the common method of providing protection. Subsequently, more consideration has been given to other means which can provide the same or higher levels of safety, but with less weight and easier accessiblity for maintenance and calibration and at equivalent or lower costs.

Because instrument manufacturers serve an international market, increased activity within national jurisdiction is being matched by recognition that standardization must be accomplished at the international level as well.

Area and Material Classification

North America

In the United States, Articles 500 to 503 of the National Electrical Code (NEC) provide basic definitions of hazardous areas and the requirements for electrical installations. Article 500 defines the

[*]Manager, Regulatory Affairs, Honeywell Inc., Fort Washington, Pa.

Table 1 National Electrical Code Area Classification System

Nature of hazard		
Class I Gases and vapors	Class II Dusts	Class III Flyings
Group A—Acetylene Group B—Hydrogen or gases of similar hazardous nature, such as manufactured gas, butadiene, ethylene oxide, propylene oxide Group C—Ethyl ether, ethylene, cyclopropane, unsymmetrical dimethylhydrazine, acetaldehyde, isoprene Group D—Gasoline, hexane, naphtha, benzene, butane, propane, alcohol, acetone, benzol, lacquer solvent, natural gas, acrylonitrile, ethylene dichloride, propylene, styrene, vinyl acetate, vinyl chloride, *p-xylene*	Group E—Metal dusts Group F—Carbon black, coal, coke dusts Group G—Grain dust, flour, plastics, sugar	No group assigned: Typical materials are cotton, kapok, nylon, flax, wood chips—normally not in air suspension
Division 1		
For heavier-than-air vapors, below grade sumps, pits, et al., in Division 2 locations. Areas around packing glands; areas where flammable liquids are handled or transferred; areas adjacent to kettles, vats, mixers, et al. Where equipment failure releases gas or vapor and damages electrical equipment simultaneously	Cloud of flammable concentration exists frequently, periodically, or intermittently—as near processing equipment. Any location where conducting (resistivity less than 10^5 Ω-cm) dust may accumulate	Areas where cotton, spanish moss, hemp, et al., are manufactured or processed.
Division 2		
Areas adjacent to a Division 1 area. Pits, sumps containing piping, et al., in nonhazardous location. Areas where flammable liquids are stored or processed in completely closed piping or containers. Division 1 areas rendered nonhazardous by forced ventilation.	Failure of processing equipment may release cloud. Deposited dust layer on equipment, floor, or other horizontal surface.	Areas where materials are stored or handled.

NOTES:
In a Division 1 location, there is a high probability that a flammable concentration of vapor, gas, or dust is present during normal plant operation, or because of frequent maintenance. In a Division 2 location, there is only low probability that the atmosphere is hazardous—for example, because of equipment failure.

Until the 1971 revision, material classification in the NEC differed from the practice in almost all other countries except Canada. Material groupings were based on consideration of three parameters: autoignition temperature (AIT) (or SIT, spontaneous ignition temperature), maximum experimental safe gap (MESG), and the maximum pressure rise in an explosion test chamber.

classification of hazardous locations broadly in terms of kind and degree of hazard. The kind of hazard is specified by class and group. The degree of hazard is designed by division. Typical industrial locations, for example, might be classified as Class I, Group D, Division 1; or Class II, Group G, Division 2; and so on. The principal features of the NEC classification are summarized in Table 1. Many additional materials are listed in NFPA 497M.

Similar definitions are given in the Canadian Electrical Code.

International Electrotechnical Commission (IEC) Classifications

Most industrial nations are now adopting the area and material classification definitions of the IEC. Locations where a flammable concentration may be present are called Zone 0, Zone 1, and Zone 2. A Zone 0 location is a location where the atmosphere may be in the explosive range such a high percentage of the time (about 10%) that extraordinary measures must be taken to protect against ignition by electrical apparatus. Zone 1 locations have a probability of hazard between Zone 2 and Zone 0. A Zone 2 location is similar to North American Division 2. Taken together, Zone 1 and Zone 0 equate to North American Division 1.

In Zone 2, requirements are analogous to requirements in North America. The advantage of distinguishing the extraordinary hazards of Zone 0 from the lesser hazards of Zone 1 is that apparatus and installation requirements can be relaxed in Zone 1. For example, intrinsically safe systems in North America are judged on the basis of two faults because of the encompassing definition of Division 1. For use in Zone 1, consideration of only one fault is required, although two faults are assessed in Zone 0.

Material classification in most nations now uses IEC terminology.

A Group I hazard is due to methane (firedamp) in the underground works of a mine. The presence of combustible dust and other environmental aspects of mining works are assumed when preparing apparatus requirements for Group I.

Group II gases and vapors are flammable materials found in industrial, aboveground premises. They are divided into Groups IIA, IIB, and IIC, which are similar although not identical to North American Groups D, C, and B, respectively.

Classifying a Hazardous Location

The NEC definitions provide guidelines but do not give a quantitative method for classifying a specific hazardous location. Factors to consider include quantity of hazardous material which may be released, topography of the site, construction of the plant or building, and past history of fire and explosion (of a particular location or plant, as well as of an entire industry). Although authorities recognize the need, there are no concise rules for deciding whether a location is Division 1 or Division 2. The best guides to area classification known to the author are American Petroleum Institute (API) RP500A, B, and C for petroleum installations, and National Fire Protection Association (NFPA) 497 for installations in chemical plants. These documents are applicable to any industry.

Special Cases of Area Classification

It is common practice to pressurize instrument systems to reduce the area classification inside the case. The inside of an instrument case provided with a simple pressurization system, located in a Division 1 area, can be considered a Division 2 location because only by accidental failure of the pressurization system can the internal atmosphere become hazardous. If the pressurization system is designed to deenergize all equipment within the enclosure when pressurization fails, the interior can

---

In Europe, materials long have been grouped by AIT and separately by MESG. Pressure rise is not a material classification criterion. It is now recognized in the United States that there is no correlation between MESG and AIT. Hydrogen, for example, has a very small MESG and a very high AIT. Many Group C and D materials have lower ignition temperatures but wider, experimental safe gaps. Because the NEC classification was based on two uncorrelated parameters, Untied States experts could not use the results of experimental work on new material in other countries, or use other classification tables. The 1971 NEC revisions separate AIT from considerations of MESG. Explosion-proof housings now can be designed for MESG typical of a group of materials. External surface temperatures are limited by the AIT of the hazardous gas or vapor of concern.

be considered a nonhazardous location. Two failures are required—of the pressurization system and of the interlock system—before a hazard will exist within the case.

An important limitation of this philosophy is that, if any single failure can make the enclosure hazardous, the interior of the enclosure must not be classified less hazardous than Division 2 regardless of the pressurization system design. Such is the case with bourdon tube- or diaphragm-actuated instruments where process fluid is separated from the instrument interior only by a single seal, the bourdon or diaphragm. Unless the pressure is high enough or the enclosure air volume great enough to prevent a combustible concentration inside the enclosure should the measuring element fail, the interior should never be classified less hazardous than Division 2. Such systems often are referred to as singly sealed systems.

In a doubly sealed system, two seals are provided between the process fluid and the area being purged, and a vent to the atmosphere is provided between the seals. Failure of both seals is required to make the enclosure interior hazardous. Even so, pressurization can prevent the hazardous material from entering the compartment because the hazardous material is at atmospheric pressure.

Techniques Used to Reduce Explosion Hazards

The predisposing factors to fire or explosion are (1) the presence of a flammable liquid, vapor, gas, dust, or fiber in an ignitable concentration, (2) the presence of a source of ignition, and (3) contact of the source with ignitable material. The most obvious way to eliminate the possibility of ignition is to remove the source to a location where there is no combustible material. This is the first method recognized in the NEC, Article 500. Another method is to apply the principal of *intrinsic safety*. Equipment and wiring which are intrinsically safe are incapable under normal or abnormal conditions of igniting a specifically hazardous atmosphere mixture. For practical purposes, there is no source of ignition.

Figure 1 summarizes the techniques used to reduce explosion hazards. Methods based on allowing ignition to occur force combustion under well-controlled conditions so that no significant damage results. Explosion-proof housings contain an explosion so that flames do not spread into the surrounding atmosphere. Until relatively recently, this has been the most common technique. However, a continuous source of ignition, such as the continuous pilot in gas appliances, to localize combustion is commonplace.

There are several methods for reducing hazards by preventing the accumulation of combustible material in an explosive concentration or for isolating an ignition source from flammable material. Pressurization of instruments is common. In hydrogen annealing furnaces and hydrogen-cooled electric generators, the concentration is held above the upper explosive limit. Blanketing of tanks with nitrogen or CO_2 and rock dusting of coal mine galleries and shafts are examples of using inert materials to suppress a combustible mixture.

Several techniques are used to isolate the ignition source. Oil immersion prevents contact between the atmosphere and ignition source. In Europe sand-filled equipment sometimes is used. Sealing and encapsulation both provide a barrier to impede contact.

Increased safety is a technique used for transformers, motors, cables, etc., which are constructed with special attention to ruggedness, insulation reliability, and protection against overtemperature so that an ignition-capable failure is of very low probability. Increased safety, developed first in Germany and now accepted widely in Europe, may be used in Zone 1.

Nonsparking apparatus and nonincendive apparatus are suitable for use in Division 2 and Zone 2 because they have no normal source of ignition.

Restricted breathing is the technique of using a tight, but not sealed, enclosure in Zone 2 to allow only slow access of flammable vapors and gases to the source of ignition. This technique, developed in Switzerland, is slowly achieving acceptance in Europe.

Explosion-Proof Housings

Explosion-proof housings (termed *flameproof enclosure* in international English) remain the most practical protection method for motor starters and other heavy equipment which produces sufficient

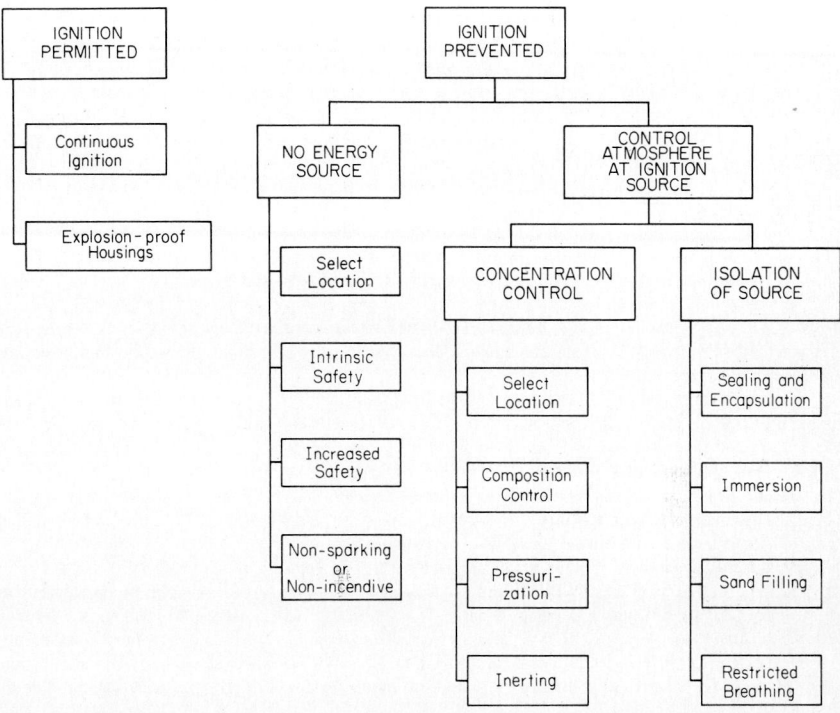

Fig. 1   Techniques used to reduce explosion hazards.

normal operation to ignite a flammable atmosphere. Explosion-proof enclosures are not vaportight; it is expected that a flammable atmosphere will enter the enclosure.

A pressure rise of 100 to 150 psi (690 to 1034 kPa) is typical for the mixture producing the highest explosion pressure. In small enclosures, loss of energy to the enclosure walls decreases the pressure rise. Because the enclosure must contain the explosion and also must cool escaping gases, cast or heavy metallic construction with wide, close-fitting flanges or threaded joints is typical. Non-metallic materials of construction are gaining recognition in some countries.

For specific design criteria in the United States, reference should be made to standards such as Underwriters Laboratories UL698—"Industrial Control Equipment for Use in Hazardous Locations"—or the Factory Mutual Standard. In general, in addition to explosion tests, the enclosure must withstand a hydrostatic pressure of 4 times the maximum pressure observed during the explosion test and must not have an external case temperature high enough to ignite the surrounding atmosphere. In Canada, the applicable standard is CSA C22.2-30.

The NEC requires that conduits entering enclosures containing arcing contacts or high-temperature devices be sealed. All conduits leaving Division 1 and 2 areas and all conduits larger than 2 in which enter junction boxes with splices or terminals must also be sealed. There may be no union, coupling box, or fitting between the sealing fitting and the point at which the conduit enters an area of lower classification. Conduits must be installed to prevent the accumulation of water in low spots.

In Europe, the applicable standards are CENELEC EN50014, "General Requirements," and CENELEC EN50018, "Flameproof Enclosure 'd'." These are available in English as British Standard BS5501, Parts 1 and 5.

European requirements emphasize special fasteners to prevent unauthorized opening of flame-proof enclosures. Requirements for flange gaps are less restrictive than North American standards. Routine testing of enclosures at lower pressures than those of the North American type test is common, although type testing is achieving recognition.

Encapsulation, Sealing, and Immersion

These techniques are not likely to be applied to a complete instrument. They serve to reduce the hazard classification of the instrument by protecting sparking components or subassemblies. Oil immersion and sand filling are applied to power handling apparatus, but neither technique has important applications in instrument systems, although oil immersion may be a convenient technique for some hydraulic control elements.

Hermetic Sealing

Paragraph 501-3(*b*) (1) of the NEC states that general purpose enclosures may be used in Division 2 locations if make-and break contacts are hermetically sealed against the entrance of gases or vapors. NEC provides no definition of a satisfactory hermetic seal, however. Seals obtained by fusion, welding, or soldering and, in some instances, plastic encapsulation are widely accepted. In reality, the leak rate of soldered or welded seals is lower than that required for protection against explosion in a Division 2 location.

The long-time average concentration inside a sealed enclosure approaches the average concentration outside. The function of a seal is to prevent transient excursions above the lower explosive limit (LEL) outside the device from raising the concentration inside the device to the LEL. Three mechanisms can force material through a seal: (1) changes in ambient temperature, (2) changes in barometric pressure—both effects tending to make the seal breathe, and (3) wind and strong air currents. The last-named mechanism usually can be ignored because a sealed device must be installed in a general purpose enclosure to protect it from such conditions.

Encapsulation involves enclosing a component or subassembly in a plastic material, a tar, a grease, with or without the additional support of a can. If the encapsulated assembly is robust and has mechanical strength and chemical resistance adequate for the environment in which it is used, it can be considered the equivalent of a hermetic seal. An external hazardous atmosphere must diffuse through a long path between the encapsulating material and the device leads to reach the interior. Standards for sealed devices are under consideration by the IEC and the Instrument Society of America (ISA).

Pressurization Systems

Lowering the hazard classification of a location by providing positive-pressure ventilation from a source of clean air has long been recognized in the NEC, and instrument users have pressurized control rooms and instrument housings for many years. The first detailed standard for instrument purging (pressurizing) installations was ISA SP12.4 (reissued in 1970 as S12.4). These requirements in essentially the same form make up the first section of NFPA 496 which also covers purging of large enclosures, ventilation of control rooms, and Class II hazards.

ISA S12.4 defines three types of pressurized installations:

Type Z. Pressurization to reduce the classification within an enclosure from Division 2 to nonhazardous

Type Y. Pressurization to reduce the classification in an enclosure from Division 1 to Division 2

Type X. Pressurization to reduce the classification within an enclosure from Division 1 to nonhazardous

Type Z Pressurization

This permits installation of ignition-capable equipment inside the enclosure. For an explosion to occur, the pressurized system must fail, and also, because the surrounding area is Division 2, there must be a process equipment failure which releases flammable material. Thus, there must be two independent failures, and no additional safeguards in the pressurization system are necessary.

Figure 2 shows a typical installation for Type Z pressurization from ISA S12.4. Only a pressurization indicator is required. The probability that a process fault will make the location hazardous

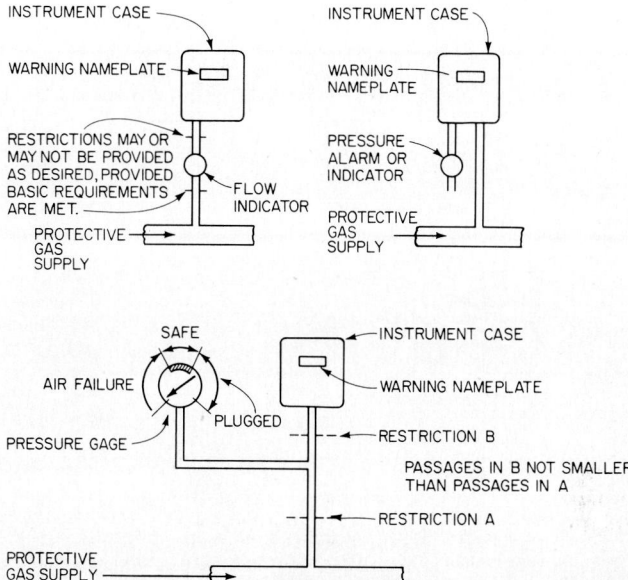

Fig. 2   Acceptable installations for Types Y and Z pressurization. (Reference: ISA S12.4.)

before any failure of the pressurization system is recognized and corrected is assumed to be extremely low. An electric indicator or alarm must meet the requirements of its location before pressurization. If a pressure indicator is used, no valve may be installed between the pressure indicator and the case. Any restriction between the case and the pressure device must be no smaller than the smallest restriction on the supply side of the pressure device. The case must be opened only if the area is known to be nonhazardous or if power has been removed. In normal operation, the pressurization system must maintain a minimum pressure of 0.1 in (2.5 mm) of water gage. The flow required to maintain this pressure is immaterial. The temperature of the pressurized enclosure must not exceed 80% of the ignition temperature of the gas or vapor involved when it is operated at 125% of rated voltage. A red warning nameplate must be placed on the instrument to be visible before the case is opened.

Type Y Pressurization

In this case, all requirements of Type Z pressurization must be met. Because the instrument is located in a Division 1 location, additional safeguards are required. Units must be fused so that, should a wire arc to an external case wall, the external surface temperature will not exceed the Type Z case temperature limit. Equipment inside the enclosure must be suitable for Division 2.

Type X Pressurization

In this system, the pressurization is the only safeguard. The atmosphere surrounding the enclosure is presumed to be frequently flammable. The equipment within the enclosure is ignition-capable. Pressurization system failure must automatically deenergize internal equipment. All requirements for Types Z and Y pressurization must be met. The interlock switch may be pressure- or flow-actuated. The switch must be suitable for Division 1 locations even if it is mounted within the instrument case, because it may be energized before purging has removed all flammable material. If it can be opened without using a tool, the door must be provided with an automatic disconnect switch suitable for Division 1. A timing device must prevent power from being applied until four case volumes of purge gas can pass through the instrument case with an internal pressure of 0.1 in (2.5 mm) water. The timing device also must meet Division 1 requirements even if inside the case. See Fig. 3.

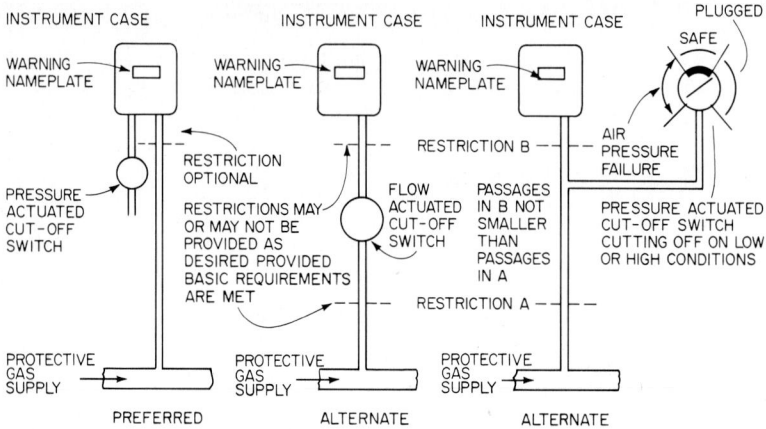

Fig. 3   Acceptable installations for Type X pressurization. (*Reference: ISA S12.4.*)

IEC and CENELEC standards for pressurization systems are similar to those of ISA and NFPA, although the requirements are not phrased in terms of reduction in area classification. The minimum pressure is 0.2 in (5.1 mm) of water gage.

The IEC is also readying a publication on continuous dilution. The hardware is similar to that required for pressurization, but the rationale for selecting the level of protection needed is based on the presence of a source of flammable material within the enclosure, as in an analyzer. The objectives are to prevent entry of an external flammable atmosphere (pressurization) and also to dilute any internal release to a low percentage of the lower flammable limit (continuous dilution).

INTRINSIC SAFETY

Experiment and theory show that a critical amount of energy must be injected into a combustible mixture to cause an explosion. If the energy provided is not greater than the critical ignition energy, some material will burn but flames will not propagate. An explosion occurs only when enough energy is injected into the mixture to ignite a critical minimum volume of material. The diameter of a sphere enclosing this minimum volume is called the *quenching distance* or *quenching diameter*. It is related to the MESG but is about twice as large. If the incipient flame sphere does not reach this diameter, it will not propagate.

The energy required for ignition depends on the concentration of the combustible mixture. There is a concentration at which the ignition energy is minimum. The concentration curve is asymptotic to limits of concentration commonly called the lower explosive limit (LEL) and the upper explosive limit (UEL). Figure 4 illustrates the influence of concentration on the critical energy required to cause ignition. A hydrogen-air mixture, one of the most easily ignited atmospheric mixtures, supports combustion over a wide range of concentrations. A propane-air mixture, which is typical of many common hazardous materials, is flammable only over a narrow range of concentrations. The amount of energy required to ignite the most easily ignited concentration of a mixture under ideal conditions is the minimum ignition energy (MIE).

Definition

The NEC defines intrinsically safe equipment and wiring as "incapable of releasing sufficient electrical or thermal energy under normal or abnormal conditons to cause ignition of a specific hazardous atmosphere mixture."

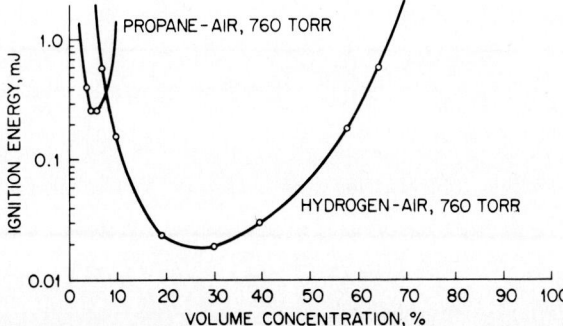

Fig. 4 Effect of concentration on ignition energy.

Early Developments

The British first applied intrinsic safety in dc mine signaling circuits. The first studies on intrinsic safety began about 1913. In 1936, the first certificate for intrinsically safe equipment for other than mining was issued. By the mid-1950s, certification in Great Britain for industrial applications was common. At the U.S. Bureau of Mines, work on intrinsically safe apparatus, although the term was not used, started about the same time as the British investigations. Rules for telephone and signaling devices were published in 1938.

During the 1950s, increased use of electric equipment in hazardous locations stirred worldwide interest in intrinsic safety, and by the late 1960s almost every industrial country either had published a standard for intrinsically safe systems or had drafted one. The major industrial countries also were active in the IEC Committee SC31G which prepared an international standard for intrinsically safe systems.

The first standards for intrinsically safe equipment intended for use by the instrument industry were published in ISA RP12.2, issued in 1965. The NFPA used ISA RP12.2 as a basis for the 1967 edition of NFPA 493.

During the years following the publication of ISA RP12.2 and NFPA 493-1967, the certification of intrinsically safe systems by independent approval agencies, such as Factory Mutual and Underwriters Laboratory in the United States, CSA in Canada, BASEEFA in the United Kingdom, and PTB in Germany, became a legal or marketing necessity in most countries. All standards for intrinsic safety have therefore become much more detailed and definitive. Adherence to the standard is the objective, not a judgment of safety. The work of the IEC has served as the basis for later editions of NFPA 493, as well as for Canadian Standard CSA C22.2-157 and CENELEC Standard EN50020. Any product marketed internationally must meet all these standards.

All the standards agree in principle, but differ not only in detail but also in the way they are interpreted by the approval agencies.

CSA and U.S. standards are based on safety after two faults, because in these cases Division 1 includes Zone 1 and Zone 0. European standards provide ia and ib levels of intrinsic safety for Zone 0 and Zone 1 application, based on consideration of two faults and one fault, respectively.

Standards for intrinsic safety can be less intimidating to the user if it is appreciated that most construction details are efforts to describe what can be considered a fault or what construction can be considered so reliable that the fault will never occur. When viewed in this light, creepage and clearance tables, transformer tests, and tests of protective components make much more sense. They are guidelines for making design decisions—not mandated values for design. They apply only if safety is affected.

Design of Intrinsically Safe Systems

The objective of any intrinsically safe design, whether produced by an equipment manufacturer or by a user attempting to assemble a safe system from commercially available devices, is the same—to ensure that the portion of the system in the Division 1 location is unable to release sufficient energy to cause ignition, either by thermal or electrical means, even after failures in system components. It is not necessary that the associated apparatus, i.e., the apparatus located in Division 2 or a nonhazardous location connected to the intrinsically safe circuit, be itself intrinsically safe. It is only necessary that failures, in accordance with the accepted standard for intrinsic safety, do not raise the level of energy in the hazardous area above the safe level.

Basic Techniques Used by Manufacturers

Techniques used by manufacturers in the design of intrinsically safe apparatus are relatively few in number, and all manufacturers use the same fundamental techniques.

Mechanical and Electrical Isolation

The most important and most useful technique is *mechanical isolation* to prevent intrinsically safe circuits and nonintrinsically safe circuits from coming in contact. Often mechanical isolation is achieved solely by appropriate spacing between the intrinsically safe and nonintrinsically safe circuits. In other cases, especially at field connections or in marshaling panels, partitions or wireways ensure that nonintrinsically safe wiring and intrinsically safe wiring are separate from one another. Encapsulation is sometimes used to prevent contact between intrinsically safe and nonintrinsically safe circuits.

Related to mechanical isolation is what can be called *electrical isolation*. Except in battery-operated systems, intrinsically safe systems have some connection to the power line, usually through a power transformer. The designer must consider the possibility of a transformer fault which connects the line voltage primary winding to the low-voltage secondary winding. In most systems, if one must consider the presence of line voltage on secondary circuits, the value and power rating of limiting elements would make design of an intrinsically safe system both functionally and economically impractical. Therefore, in modern standards for intrinscially safe construction, several varieties of transformer construction are recognized to be so reliable that one can assume that a primary-to-secondary short circuit will never occur. In one such reliable construction, a grounded shield between the primary and secondary ensures that any fault is from the primary to the grounded shield, so that the secondary winding potential is not raised to an unsafe voltage. In addition to special attention to transformer construction and testing, it is also necessary that the wiring layout prevent any accidental connection between wiring on the primary side of the transformer and wires connected to the transformer secondary.

Current and Voltage Limiting

Except in some portable apparatus, almost all intrinsically safe circuits require both current and voltage limiting to ensure that the amount of energy released under fault conditions does not exceed safe values. Voltage limiting is often achieved by use of redundant zener diodes to limit the voltage, but zener-triggered silicon controlled rectifier (SCR) crowbar circuits are also used to limit the voltage. Redundancy is provided so that, in the case of failure of a single diode or limiting circuit, the second device continues to provide voltage limiting. Current limiting in dc circuits and in most ac circuits is provided by metal film or wire-wound resistors of high reliability. See Fig. 5. Properly mounted resistors which meet the requirements for protective resistors in the applicable standard need not be redundant. They are of a level of quality that they will not fail in a way that allows current to increase to an unsafe level.

One very common use of current and voltage limiting is in the Redding barrier. See Fig. 6. The unique feature of Redding barriers is the incorporation of a fuse in series with the zener diodes, so that when a fault causes current to flow through the zener diode, the fuse will open before the power in the zener diode reaches a level at which the diode may open the circuit. In the design shown in Fig. 6, the 20-Ω resistor does not perform a safety function. It allows testing of the barrier to determine that the diodes are still intact. The current limiting function is performed by the 120-Ω resistor.

Devices to be connected to terminals 3 and 4 must be approved as intrinsically safe, but any device incapable of applying voltage to terminal 1 higher than the barrier rating may be connected. If the barrier is designed to limit against full power line potential, then the equipment in the nonhazardous area may be selected, connected, and intermingled without regard to safety in the field circuits.

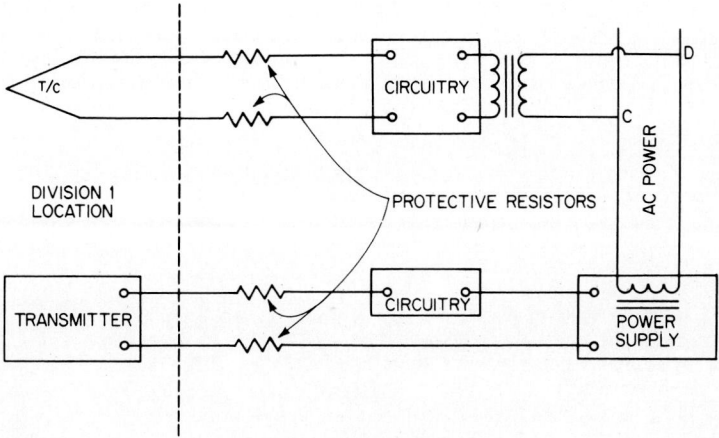

Fig. 5   Use of resistors to limit current to hazardous location.

In use, terminals 2 and 4 are both bolted to a busbar which is grounded through a very low (usually less than 1 Ω) ground resistance. The power supply also must be grounded. In operation, diodes $D1$ and $D2$ conduct only leakage current, which is small compared with the normal circuit current flowing between terminals 1 and 3. When high voltage is applied to terminal 1, the diodes conduct and limit the voltage at terminal 3 to a safe value. $R2$ limits the current into the hazardous area. Under fault conditions, the barrier looks like a low-voltage resistive source from the intrinsically safe side, and like a very low-impedance load at terminals 1 and 2. The values in Fig. 6 are for a 30-V 120-Ω barrier. Under fault conditions, the intrinsically safe circuit will appear to be driven from a nominal 30-V source with a source resistance of 120 Ω. Safety is provided by the diodes and resistors. The resistors can be presumed not to fail. The diodes are redundant. Should one fail, limiting would still take place. The fuse serves no purpose regarding safety and could be replaced by a resistor. Its function is to make the diode barrier economical. Should a fault occur, the zener diodes would connect heavily and, except for the fuse, would have to be impractically large and costly. The fuse is selected to blow at a current much lower than that which would damage the diode, permitting lower power, less costly diodes to be used.

Shunt Elements

These devices are used to absorb the energy which would otherwise be released by an inductor to the arc. The function of a shunt diode is shown in Fig. 7. Although capacitors and resistors can be used to shunt inductors, in dc circuits diodes are placed so that in normal operation they are back-biased and draw no current. The shunt elements must be redundant so that if one fails, the other

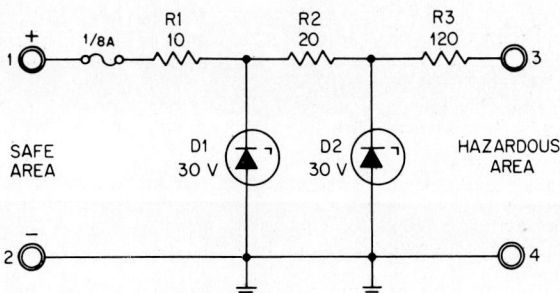

Fig. 6   Schematic of zener diode barrier, positive type.

will continue to protect the inductor. Both must be connected close to the inductor being protected. Connection to the inductor must be especially reliable so that a fault between the inductor and the protective shunt diodes can be assumed not to occur. If such a fault occurs, ignition is possible because of the release of energy from the inductor. The purpose of the shunt diodes is to absorb energy stored in the inductor if the circuit external to the protected inductor opens.

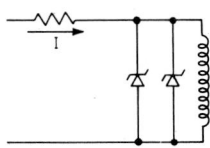

Fig. 7 Shunt diodes reduce incendivity of inductor.

The Analytical Method

The outline presented here can be considered both as a method of assessing a circuit and a mechanical design which already exists, or as a means for both analyzing the circuit and designing the layout. Only a slight difference in point of view is required. The steps in the analysis are essentially the same in both cases.

The *first step* is to identify the portion of the circuit which is to be intrinsically safe. Only the circuit in a Division 1 location need be intrinsically safe. A fault occurring in a nonhazardous or Division 2 location is of no consequence from the standpoint of the energy that may be released at that location. The fault is of concern if it affects the amount of energy which can be released at a Division 1 location.

Second, review the circuit or the hardware for isolating constructions which will allow one to assume that certain interconnections will not occur. If the hardware exists, review the mechanical layout to determine whether the use of adequate spacings or partitions allows one to avoid considering an interconnection between some nonintrinsically safe circuits and the intrinsically safe circuit. Usually the transformer must be one of the special protective constructions. Otherwise it is unlikely that the circuits can be both functional and safe if one assumes a line voltage connection to the transformer secondary. If the transformer is of special construction, then the wiring or the printed wiring board layout must ensure that the primary leads of the transformer are separated from the secondary leads by sufficient space or by a partition, so that connection between the two can be ignored. If the hardware has not been designed, one must determine where intrinsically safe portions of the circuit must be separated by appropriate spacing or partition from the higher energy portion of the circuits. The specific nature of the spacings must be determined from further detailed analysis.

Having reviewed for isolating construction, one should assume normal operation of the circuit. Compute the current and voltage in the circuit and compare it to the reference curves to determine whether the appropriate test factor has been observed. If not, adjust the circuit constants until the requirements are met. In this step, and in subsequent steps, it is essential that an orderly approach to record keeping be adopted. Even a relatively simple circuit requires the consideration of many steps. It is essential to record each combination of open, short, or ground of field wiring (these are not counted as faults) and component failures, so that when the analysis is complete one can verify that the *worst case situation* has been analyzed. If one is submitting the circuit for later review by certifying agencies, the availability of a well-organized fault table will ease and expedite the assessment process.

Third, if the hardware already exists, one considers the layout and spacing to determine what circuits must be assumed to be shorted together, what connections can be considered to be a fault, and what connections can be considered never to occur. After identifying these, recompute the current and voltage under fault conditions. This must be done for a single fault in combination with open, grounding, and shorting of the external wires—and also for two faults. One cannot assume that two-fault situations will be the most hazardous, because of the difference in test factor required when only a single fault is considered. Adjust the circuit constants until the voltages and currents are suitable.

After the analysis for arc ignition, consider whether current flowing in the circuit under fault conditions may produce a high surface temperature on resistors, transistors, etc. The temperature rise of small components is typically 50 to $100°C/W$. If the hardware exists, of course, one can measure the temperature of components which appear to be suspect under fault conditions. In some approval laboratories, because a barrier or other intrinsically safe device powering the device in the Division 1 location is marked with its maximum voltage and maximum current output, it is assumed that the power supply has a rectangular VI characteristic, so that maximum voltage and maximum current can occur together. The temperature rise must be assessed, therefore, at the wattage of the maximum voltage and current occurring simultaneously. At other laboratories, if the characteristics of the voltage and current limiting device are known, then the appropriate power to be used is $V_{max}I_{max}/4$ if the voltage limiting circuit has a resistive source impedance. The limiting temperature to be considered must, of course, be determined from the standard being used as a criterion for design.

Simplifying Assumptions

If one is analyzing a circuit, the validity of the analysis is only as good as the data to which the analysis can be compared. In general, the available data are limited to simple *RL, RC,* or resistive circuits. It may not be possible to analyze the effect of shunt elements included for some functional purpose. For example, an inductive force coil might be shunted with a variable resistor for a span adjustment. There is no reliable way to assess the additional safety provided by the shunt resistor. In general, one ignores such a parallel component in the analysis. Similarly, a capacitor might be wired in shunt with the force coil to provide damping of the electromechanical system. This, too, is ignored in the analysis.

It is assumed by almost all experts that an iron core inductor is less efficient in releasing stored energy than an air core inductor, because some of the energy, rather than being released to the arc, is dissipated in eddy-current and hysteresis losses. Therefore, if one is analyzing a circuit in which the inductor has a ferromagnetic core and the circuit will be safe with an air core inductor of the same value, then it will certainly be safe. The converse is not true. Inductance is a measure of the slope of the B-H curve of the core material. Many small inductors, especially those with ferrite cores, have high inductance because the core material has a very high initial permeability. However, if the volume of the material and the level at which it saturates are low, the amount of stored energy in the inductor may be considerably less than that calculated from a measured inductance value.

Another simplifying assumption, which in many circuits reduces the amount of analysis considerably, is to determine the highest possible power supply voltage which may ever exist under fault conditions and with a high line voltage. This value is then used to determine the stored energy in all the capacitors or to determine the current and the resulting stored energy in all the inductors. If all the calculations are safe, one need not calculate the actual circuit currents and voltages.

Another simplification, illustrated in the example, results from the need to prevent the discharge of a large capacitor in a hazardous location. Although curves giving the value of voltage on a capacitor discharge through a resistor (which causes ignition) are available, it is still common to assume that the capacitor is a battery charged to the fault voltage and to select a series resistor based on the resistive circuit ignition characteristics. The resistor selected will be higher than that based on ignition tests of capacitors because the voltage on the capacitor decays when current flows through the resistor. The connection between the resistor and the capacitor must be prevented from contacting any surrounding circuit. The capacitor must not discharge except through the resistor. Therefore spacing, potting, or some other technique must be used to ensure that no fault to that point in the circuit can exist.

Testing of Special Cases

It is not possible to determine the safety of all circuits by analysis alone. For example, some experts feel that one should verify the safety of diode-shunted inductors by conducting ignition tests. Inductors, especially small ones with ferromagnetic cores, may require testing to verify that they are safe despite high measured inductance.

Another common piece of apparatus which may have to be tested to determine safety is a regulated power supply. The reference curves of open-circuit voltage and short-circuit current for ignition in resistive circuits assume that the source impedance of the circuit, i.e., the Thevenin equivalent impedance, is resistive. If the power supply is regulated, the voltage will remain essentially constant until a critical level of current is reached, beyond which the voltage will drop off with a further increase in current. The safety of maximum voltage and maximum current from such a supply cannot be determined from the reference curves for resistive circuits. In general, safety must be established by test, although some special methods of analysis may be used.

Transmission lines are another special case. The common method of assessing safety, i.e., multiplying capacitance per foot (meter) or inductance per foot (meter) by the number of feet (meters) of cable and comparing these values with limit values for the voltage and current in the cable, yields conservative results. It is well known that because the resistance, capacitance, and inductance are distributed, the actual cable will be safer than this lumped constant analysis suggests. However, there are not sufficiently good reference data available to allow one to analyze on a more scientific basis. If answers from lumped constant approaches are not satisfactory, then the cable must be tested.

However, it is recognized that the L/R ratio of the cable, if it is sufficiently low, may be such that no hazard exists, even though the total inductance exceeds that which would be safe if the inductance were lumped. If one assumes a cable of resistance R_x and inductance L_x per foot (meter) operating from a circuit of maximum voltage V_{max} and source resistance R, then maximum energy will be stored in the cable when the cable resistance is equal to the source impedance. The current will be $V_{max}/2R$. Therefore, the maximum inductance permitted will be 4 times that permitted at the short-circuit current of the source. If the ratio L_x/R_x does not exceed $4L_{max}/R$, where L_{max} is the maximum permitted connected inductance, the cable will be safe regardless of its length.

Summary

The design techniques used in all commercially available systems are similar. In this section we noted that the fundamental techniques are very few in number. Manufacturers may introduce variants of the basic techniques, some quite imaginative, but the fundamental design techniques are similar in all commercially available systems. Although some systems have used current limiting resistors and put voltage limiting in the power supply, some have used active barrier isolators, and some have used Redding-type barriers, there is no difference in safety among the various approaches. Any of the techniques properly applied will yield a safe system.

System Design Using Commercially Available Hardware

The objective of the system design produced by an engineering consultant or a user is to ensure that the voltage and currents in a hazardous location are limited to recognized safe values. Depending on the jurisdiction under which the plant is being built, the objective may be to make the system safe or to have the system safe and certified. This discussion assumes that the system designer has the option of at least a partially uncertified system. Even in the United States where government regulations apply, for special combinations of apparatus there is allowance for the use of uncertified apparatus. See 29CFR 1910.307. The most attractive option for most plants and most jurisdictions is to select a completely certified system. It is often possible to specify that all the hardware in a measuring and control system be procured from one vendor or to select hardware which, although manufactured by different vendors, has been listed as an intrinsically safe combination. The latter situation exists for many varieties of transmitters, valve positioners and current-to-pressure transducers. The user often has a relatively wide choice of vendors because the vendors have already secured certification of various combinations of intrinsic safety interface elements, such as barriers, and their field-mounted products. For less common field-mounted devices, however, the hardware selection may be very restricted if certification is an essential criterion.

Even if the products of two manufacturers are not already certified together, it may be possible to specify these products if the procurement has a lead time of 6 to 12 months so that certification can be obtained by one of the manufacturers. It is also necessary that both products be certified in some other system so that certification of the combination is a paper exercise rather than a complete assessment of either piece of apparatus.

Specifying a completely certified system is often impractical if the lead times are short and if a system requires products from many vendors. Certification as a primary criterion may also fail if highly specialized devices such as analyzers are needed.

The user or contractor may occasionally feel that the responsibility for obtaining listings can be assigned to the major supplier of hardware for the system. If only cross-listings, i.e., certification of combinations of devices which have already been otherwise certified, are involved, this may be practical. If any of the products have not been certified as intrinsically safe, however, only the vendor of that product can obtain the basic intrinsic safety listing. If such products are involved, then the time required to obtain certifications may be 18 to 24 months.

Self-Certification by the Manufacturer

Under some circumstances, the supplier of the major portion of hardware for a system may be willing to assume total responsibility for safety of the system. A user contemplating such an arrangement should insist on written justification of safety from the supplier. The user also must allow the manufacturer veto power on equipment selection in order to ensure safety.

If all the pieces have been certified in some way, then the system may perhaps be procured from many sources. If uncertified apparatus is intended for a Division 1 location, it may be impossible for anyone but the vendor of the apparatus to self-certify unless it is very simple in construction. Certification of Division 1 apparatus is usually time-consuming and requires complete documentation, and it always involves an assumption that the design evaluated is indeed the design which will be subsequently produced. Third-party certification always imposes a responsibility on the manufacturer to ensure that this is indeed the case, and in some countries there is also a periodic audit to check that the products shipped are the same as those whose design was certified.

Self-Certification by the User or Contractor

This option is not often a viable one. It is necessary to ensure that the required expertise is available. However, it is advantageous because it retains responsibility for the safety and selection of apparatus in the same organization and yields maximum flexibility. Techniques for design by the user or contractor are discussed for three situations.

Field-Mounted Equipment Certified, but Control House Apparatus Not Certified

This situation commonly exists if the user wishes to use panel board equipment which has not been certified for intrinsic safety. The common solution is to purchase a barrier which as been certified with the field-mounted device and install it between the field-mounted device and the panel-mounted device. If the field-mounted device is a low-energy device like a thermocouple, most barrier vendors can supply a suitable barrier because specific thermocouple designs need not be certified with a barrier.

In the case of a field-mounted transmitter one should purchase barriers whose ratings meet the certification data of that field-mounted transmitter. In this case the user must ensure that a barrier which provides safety also provides the desired function. The user should make certain that the sum of the voltage drops through the load and barrier and the voltage drop across the transmitter do not exceed the available minimum voltage from the power supply. An obvious step, often overlooked, is to make certain that the voltage ratings of the zener diodes in the protective barrier are higher than the nominal voltage of the power supply. If not, the diodes may fire, blowing out the protective fuse the first time the system is energized. In some cases it may be necessary also to verify that current leakage through the zeners does not impair the desired accuracy of the system.

After ensuring that the system will function, it is also necessary to make certain that no voltage in any of the apparatus connected behind the barrier exceeds the rating of the barrier, usually 250 V root mean square (rms) but occasionally higher. The voltage of the resonant winding in ferroresonant power supplies, voltages in switching power supplies, and voltages in cathode ray tubes (CRTs) commonly exceed the rating of barriers. Most vendors of such apparatus who also supply barriers have had the pieces of apparatus containing high voltages especially certified by their certifying agency to be appropriate for use behind barriers. If such certification exists, it implies safety as well for any other barrier of the same input rating. However, this is the author's opinion and not necessarily the position of the approval bodies.

After determining that the barrier selected and the field-mounted apparatus both function and provide a safe system within the limits of the certification of the barrier, it is necessary to design the wiring system to ensure that the parameters of the cabling and any storage elements in the field-mounted devices do not exceed the levels permitted by the barrier certification. It is essential also to design the wiring system to maintain isolation of intrinsically safe wiring from all nonintrinsically safe wiring. ISA RP12.6 should be used as a guide.

Field-Mounted and Control House-Mounted Equipment Both Certified, but Not as a Connected System

Apparatus which has been marked in accordance with the marking system used for many years by PTB, and adopted in CENELEC standards, is in theory easy to mix and match simply by reading the labels of the devices to be interconnected. This is so because the control house equipment is marked with the maximum current and voltage it can deliver and with the permissible values of connected inductance and capacitance. Similarly, the field equipment is marked with the values of inductance and capacitance which can be seen at its terminals and the maximum values of current and voltage which may be safely applied. In theory, one simply compares the marked voltage and current values on the field and associated apparatus. If the field apparatus values are equal to or greater than those of the control house apparatus, and if the inductance and capacitance of the field apparatus are less than those marked on the control house apparatus, the combination is safe.

This system worked very well in Germany because of a constraint on the marking which for many years was not recognized as being one of the factors which made the system usable. Most German apparatus has been marked so that the maximum power which can be delivered by or to the apparatus is 3 W. Therefore many field-mounted pieces of equipment are marked for a maximum of 20 V and 150 mA, or a maximum of 30 V and 100 mA.

Although in North America some transmitters and barriers have been certified and are being marked in a similar manner, the great variation among the maximum current and voltage levels for which barriers have been designed in the United States makes it difficult to simply read the data and arrive at an immediate conclusion. Although a few field-mounted devices are being certified with barriers of various electrical characteristics, in general some use of the basic ignition curve is necessary to determine whether a particular combination is safe.

An additional complication arises because of the question whether a transmitter certified with a barrier whose maximum voltage is 24 V is also safe relative to hot wire ignition or hot surface ignition if it is used with a barrier certified for the same group of gases but with a maximum voltage of 30 V and a lower current. If both barriers are passive so that their output V-I characteristic is resistive, then one can be certain that safety against ignition by hot surfaces will be achieved if the maximum power output of the second barrier is less than that of the barrier with which the device was originally certified.

If the apparatus is loop-certified with a barrier, as has often been the case in the United States, Canada, and the United Kingdom, one must first secure from the manufacturer of the control house apparatus the maximum current and voltage values and the permissible connected inductance and capacitance values.

The second step is to determine from the manufacturer of the field-mounted apparatus the values of current, voltage, inductance, and capacitance of the control house apparatus with which certifications are held.

One then compares the second set of data with the first. The analysis is similar to that described above but may be more complicated because there has been relatively little commonality of barrier design or field-mounted apparatus design in North America.

Field-Mounted Equipment Not Certified

Unless the field-mounted equipment is very simple, it is usually impractical for anyone to certify it except the manufacturer or a certifying agency working with the manufacturer. It is necessary to perform a detailed analysis of the faults, construction, and stored energies, jsut as the manufacturer must do. This requires a complete set of manufacturing documentation, as well as a commitment on the part of the manufacturer not to deviate from the documentation supplied. In addition, the investment in time required to perform such an assessment is large, and a user or contractor is not likely to attempt this task.

Installation of Intrinsically Safe Systems

Because an intrinsically safe system is incapable of igniting a flammable atmosphere even under fault conditions, cables of special construction or conduit are unnecessary. It is, however, necessary to install intrinsically safe systems so that ignition-capable energies will not intrude from another circuit.

ISA RP12.6 assumes that an intrinsically safe system will be safe *after installation* if the installation

1. Conforms to the limiting parameters and installation requirements on which approval was based

2. Prevents intrusion of nonintrinsically safe energy on intrinsically safe circuits

3. Prevents power system faults or differences in ground potential from making the circuit ignition-capable

To minimize the probability of interconnections of circuits in a way not envisioned when the system was approved, RP12.6 mandates separate cables for systems supplied from power supplies of a different voltage or with different ground reference points. Circuits supplied from the same supply may be run in cable if all the conductors have a minimum insulation thickness of 0.25 mm.

To prevent the intrusion of other circuits, intrinsically safe circuits must be run in separate cables, wireways, or conduits from other circuits. Terminals of intrinsically safe circuits must be separated from other circuits by spacing (50 mm) or partitions.

The grounding of intrinsically safe circuits must be separate from the grounding of power systems except at one point.

Nonincendive Equipment and Wiring

It is not necessary to provide intrinsically safe equipment in Division 2 locations. The equipment need only be nonincendive, i.e., incapable in its normal operating condition of releasing sufficient energy to ignite a specific hazardous atmospheric mixture. Such equipment has been recognized without specific definition in the NEC. In Division 2 locations, equipment without make-or-break or sliding contacts and without hot surfaces may be housed in general purpose enclosures.

Requirements for apparatus suitable for use in Division 2 and Zone 2 locations are nearing completion by both ISA and IEC (as of early 1983). Both documents provide more detail than is found in the NEC, partially because of a trend toward certification especially of nonincendive circuits which may be normally sparking but release insufficient energy in normal operation to cause ignition. Both documents also define tests for sealed devices which are needed by industry. The ISA and IEC documents are likely to be very similar, except that the ISA standard will not include restricted breathing. CSA 22.2 No. 157 includes requirements for nonincendive circuits and sealed devices, as well as requirements for intrinsically safe systems.

REFERENCES

Editorial Note. The cooperation of the Instrument Society of America in permitting extractions and adaptations of text and illustrations from *Electrical Instruments in Hazardous Locations,* 3d ed., 1979, is greatly appreciated.

ANSI/NFPA; *ANSI/NFPA 70,* National Fire Protection Association, Boston, Massachusetts, 1981.

API: "Classification of Areas for Electrical Installations in Petroleum Refineries," *API RP 500A,* American Petroleum Institute, New York.

BSI: "Electrical Apparatus for Potentially Explosive Atmospheres," *BS5501,* British Standards Institution, London, 1978. Part 1: "General Requirements" *(EN 50014);* Part 2: "Oil Immersion 'o' " *(EN50015);* Part 3: "Pressurized Apparatus 'p' " *(EN50016);* Part 4: "Powder Filling 'q' " *(EN50011);* Part 5: "Flameproof Enclosures 'd' " *(EN 50018);* Part 6: "Increased Safety 'e' " *(EN 50019);* Part 7: "Intrinsic Safety 'I' " *(EN 50020).*

ISA: "Electrical Instruments in Hazardous Atmospheres," *ISA RP12.1,* Instrument Society of America, Research Triangle Park, North Carolina, 1965.

ISA: "Instrument Purging for Reduction of Hazardous Area Classifications," *ISA S12.4,* Instrument Society of America, Research Triangle Park, N.C., 1970.

ISA: "Electrical Safety Practices," *ISA Monogr.* 110–113, Instrument Society of America, Research Triangle Park, N.C., 1972.

ISA: "Installation of Intrinsically Safe Instrument Systems in Class I Hazardous Locations," *ISA RP12.6,* Instrument Society of America, Research Triangle Park, N.C., 1976.

Magison, E. C.: *Electrical Instruments in Hazardous Locations,* 3d ed., Instrument Society of America, Research Triangle Park, N.C., 1978.

NEC: *National Electrical Code,* National Fire Protection Association, Quincy, Massachusetts, 1984, Art. 500–503.

NFPA: "Classification of Gases, Vapors and Dusts for Electrical Equipment," NFPA 497M, National Fire Protection Association, Quincy, Massachusetts, 1983.

NFPA: "Intrinsically Safe Process Control Equipment for Use in Hazardous Locations," *NFPA 493,* National Fire Protection Association, Quincy, Mass., 1978.

NFPA: "Purged Enclosures for Electrical Equipment," *NFPA 496,* National Fire Protection Association, Quincy, Mass., 1982.

UL: "Industrial Control Equipment for Use in Hazardous Locations," *UL 698,* Underwriters Laboratories, Chicago, 1966.

Information and Display Systems*

Phillip J. Agard. *Bowmar/ALI, Inc., Division of Bowmar Instrument Corporation, Acton, Massachusetts. (Information Display and Storage Systems)*

Richard W. Borut. *Engineering Department, M. W. Kellogg Company, Houston, Texas. (Control Centers)*

Richard A. Brown. *Honeywell Inc., Test Instruments Division, Denver, Colorado. (Information Display and Storage Systems)*

Susan Bruijnes. *National Semiconductor Corporation, Santa Clara, California. (Voice Recognition and Speech Synthesis)*

Earle D. Duncan. *Chessell Corporation, Newtown, Pennsylvania. (Information Display and Storage Systems)*

Engineering Staff. *Bell Laboratories, Murray Hill, New Jersey. (Voice Recognition and Speech Synthesis)*

Alexandra B. Guldberg. *Kaye Instruments Inc., Bedford, Massachusetts. (Information Display and Storage Systems)*

R. L. Hanes. *Rochester Instrument Systems, Rochester, New York. (Alarms and Annunciators)*

Nels S. Hanssen. *Aquatrol Corporation, St. Paul, Minnesota. (Information Display and Storage Systems)*

Richard E. Kehrli. *Tektronix, Inc., Beaverton, Oregon. (Information Display and Storage Systems)*

---

*Persons who authored complete articles or subsections of articles, or who otherwise cooperated in an outstanding manner in furnishing information and helpful counsel to the editorial staff.

Dave Postetter. *John Fluke Manufacturing Company, Inc., Everett, Washington. (Information Display and Storage Systems)*

Robert B. Ross. *Gould Inc., Industrial Controls Division, Westminster, Maryland. (Relays and Contactors)*

Walter A. Skowron. *Hewlett-Packard, Loveland Instrument Division, Loveland, Colorado. (Information Display and Storage Systems)*

Richard F. Speicher. *Micro Switch Division, Honeywell Inc., Freeport, Illinois. (Panel, Keyboard, and Control Switches)*

Information Display and Storage Systems

This article is divided into two major sections—Indicating Means and Recording Means. Descriptions under indicating means include moving-pointer indicators; moving-scale indicators; multipoint and multirange indicators; digital indicators; illuminative display elements, including light emitting diodes (LEDs), liquid-crystal displays (LCDs), and cathode ray tubes (CRTs); alarms and annunciators; and graphic displays, including computer graphics. Descriptions under recording means include graphic recorders; strip chart recorders; circular chart recorders; xy recorders; alarm information recorders; oscillographic recorders; and loggers, plotters, and printers. Electronic storage means are described in the article Digital Technology in Sec. 14.

INDICATING MEANS

Indicating instruments are used for displaying the following:

1. **Essentially Instantaneous, Real-Time Changes in a Process Variable.**
   a. *Analog indicators* require a calibrated scale and a "pointer"; the pointer may move with reference to a stationary scale, or the scale may move with reference to a stationary pointer. The pointer need not be a pointer in the literal sense, but may take another form, such as a moving ribbon which provides indication in the form of a bar graph. The analog approach requires skill in reading, particularly in determining values which lie between two scale graduations. Although various means have been used to eliminate reading errors and to assist in interpolation (e.g., the vernier scale), these factors remain inherent disadvantages of analog indicators.
   b. *Digital indicators* display discrete values of the measured variable at any given instant, but this display may not show how the instantaneous value relates to the full-scale or desired operating range of the instrument. In most cases, this desirable reference is confined to an adjacent digital display of the set point. No interpolative skills are required, but failure to relate the instantaneous value to the general backdrop of the measurement is an inherent disadvantage of digital indicators.
   c. *Hybrid indicators* employ the advantages of both analog and digital indicators and appear in various formats and combinations.

2. **Alarm Conditions.** Alarm conditions are displayed visually, audibly, and frequently in combination—depending on the degree of importance of the alarm. By nature, an alarm condition is a discrete value, but it is important for the operator to know from which direction the *alarm value* has been approached, i.e., whether it is a *high-limit alarm* in which the process variable is moving *up-scale* or a *low-limit alarm* in which the process variable is moving *down-scale*. This knowledge can be imparted through the use of only two discrete indicators. Actuation of alarms may be effected by instruments used for routine indication, recording, or control, or for reasons of safety and security, entirely separate alarming circuitry may be used.

3. **Essentially Static Conditions of a System.** Illuminated pushbuttons and keys indicate that the system may be under automatic or manual control, or a certain algorithm may be in effect, or lights may be used to indicate the key points on a graphic diagram, among many other indications of conditions and situations. By nature, the indication of these conditions involves the display of discrete information—essentially steady state unless altered by the operator. Such indicators may be likened to illuminated "signs."

4. Instructions, Troubleshooting, and Operating Procedures. At one time this material was confined to notebooks or files, and now can be called up from electronic storage for display as "pages" on a CRT.

Other Characteristics

Indicating instruments may be classified in terms of speed which, unless associated with a rapid recording means, must fall within the capabilities of human identification and resolution. Although these characteristics vary considerably from one person to another, the standard of 24 frames per second as projected by motion picture equipment is indicative of the human limitation in time-sensing individual events. In selecting the most appropriate (including economic justification) indicating mechanism for an instrument, the designer must consider the overall response of the measuring equipment as well as the time-related importance of the measured data and the operator-instrument interface—and, in so doing, not overengineer or underdesign the mode of indication.

There are several forms of indication, reasonably uncommon except for specific purposes, which illustrate the ingenuity that instrument engineers over the years have applied to data display problems. Pyrometric cones and colored crayons and paints have indicated (historically) whether a certain temperature range has been exceeded (as in the firing of ceramics in a kiln). Manually operated, color matching optical pyrometers also present an unusual and interesting indicating format. Several of these highly specialized indicating methodologies are described where appropriate in various articles throughout this Handbook.

Moving-Pointer Indicators

Fixed-scale movable-pointer indicators vary in appearance, form of scale, and the plane in which the indicating pointer moves. In all cases, it is essential that the scale be graduated and mounted so

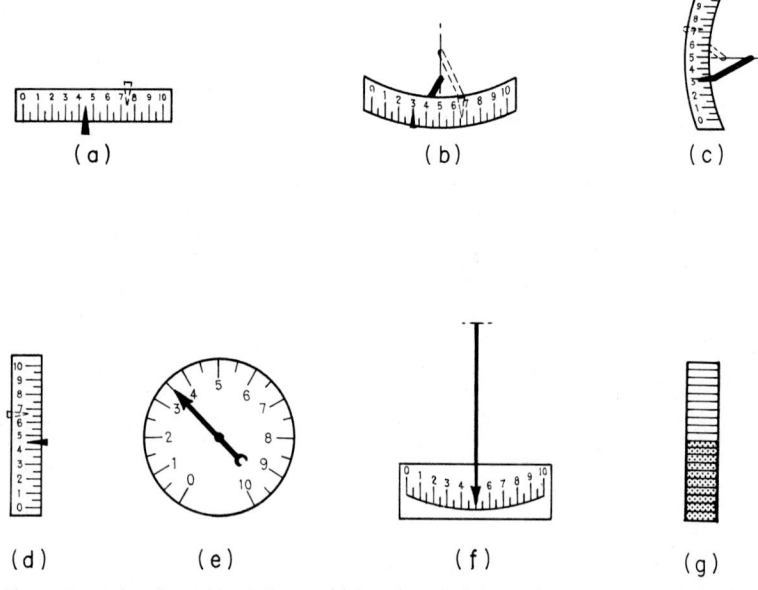

Fig.1 Types of moving-pointer indicators. (*a*) Straight scale (horizontal), (*b*) arc scale (horizontal), (*c*) arc scale (vertical), (*d*) straight scale (vertical), (*e*) circular scale, (*f*) segmental scale, (*g*) bar graph where, instead of a moving pointer, there is a moving, semitransparent ribbon (usually colored), the top of which indicates the present value, or several individually illuminated, vertically arranged solid-state display elements, such as liquid crystals.

as to minimize reading errors. To avoid parallax errors, a precision indicating instrument, which usually is read at very close range, often uses a mirror mounted behind the pointer and adjacent to the scale graduations. In another arrangement, the graduated portion of the scale is raised nearly to the level of the pointer tip, thus allowing accurate readings from greater angles of observation. The index may be in the form of a thin edgewise strip, a hairline scratched on each side of a transparent member, or a sharp, arrow-shaped tip.

Illumination, external or internal, is extremely important. Indicators intended for distant reading have entirely different design criteria, but usually have fewer graduations and the pointers have proportionally larger tip areas.

Several types of movable-pointer indicators are shown in Fig. 1a through g. Frequently, the scales shown are used with two indicating pointers. The second pointer (shown as a broken line) may indicate the value of a second variable, or it may function as a manually set index or reference spotter. These forms are essentially purely mechanical in nature and are among the means most widely used. The incorporation of these principles in several contemporary instruments is shown in Figs. 2 to 4. The application of a

Fig. 2 Circular scale used on a pressure gage. *(Robertshaw Controls.)*

light beam in an indicating instrument is shown by the projected moving-pointer instrument of Fig. 5. The scale is fixed, and a pointer of light moves along the scale. The scale markings are printed on the front surface of a translucent scale plate which is also the light diffusing surface.

Bar Graph Indicators

This format is considerably more recent in concept, but nevertheless has been available for a number of years. A vertical bar (as on a graph) may rise or fall with changes in the process variable. The bar may consist of a moving ribbon or a series of solid-state LCD or LED elements. See Figs. 6 to 8.

Fig. 3 Vertical edgewise arc scale used in a contemporary indicating controller. *(Robertshaw Controls.)*

Moving-Scale Indicators

Several forms of movable-scale fixed-pointer configurations are shown in Fig. 9. A flat disk scale is shown in Fig. 9a. In practice, the angle of rotation seldom exceeds 340° with a maximum calibrated scale length of ~30 in (~76 cm). Shown in Fig. 9c is an arrangement sometimes referred to as a *precision indicator,* which is a variation of the configuration in Fig. 9a, the chief difference being that the scale is finely divided (from 300 to 600 divisions) in a ~24-in (~61-cm) length. The stationary index generally is in the form of a sharp tip or hairline. Shown in Fig. 9d is, in essence, a moving-scale indicator of early (mechanical) format. Such indicators are still widely used on mechanical counters. The projected moving-scale instrument shown in Fig. 10 illustrates one way of increasing the scale length.

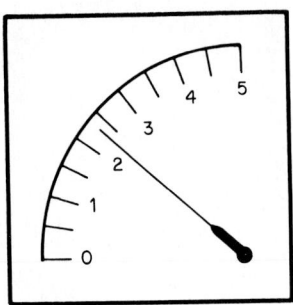

Fig. 4   Moving-pointer and stationary-arc scale with the center of the meter movement located in a corner of the instrument, as contrasted with the center of the scale being located at the symmetrical center of the instrument. It has been claimed that such designs are easier to read than conventional styles. An innovation was introduced in the early 1980s in which this principle is applied in connection with a series of electric meters made by Goerz Metrawatt.

Multipoint and Multirange Indicators

A multipoint indicator consists of a single measuring and indicating instrument used in conjunction with a selector switch, permitting a number of different input signals to be read, e.g., from a group of thermocouples. The switch may be either manually or automatically operated (as in a scanner). A variety of buttons and switches, including dial systems, are used for point selection. A multiple-point indicator should not be confused with a multiple-pointer instrument.

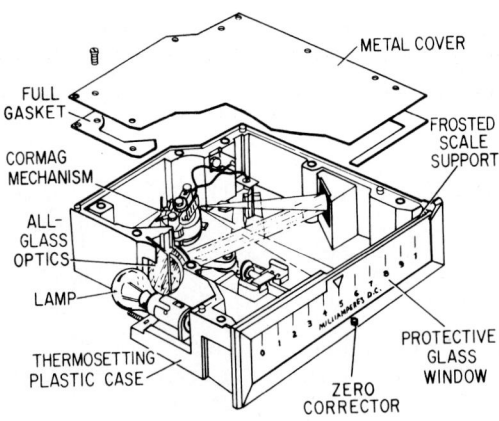

Fig. 5   Panel meter in which the moving pointer is optically projected. *(Weston.)*

The utility of an instrument can be increased by incorporating calibration for several ranges in one instrument, each range having a separate scale. Correctly identifying the range selected with its associated scale is of obvious importance and can be accomplished by using a scale range pointer automatically tied to the range selector switch; or a mask may be used to cover all the scales except the one selected.

Digital Indicators

The popularity of digital indicators expanded immensely during the 1970s, setting a trend for continuance in the 1980s and beyond. Digital indication, which is highly compatible with modern control systems, has been made a practical reality largely through the development of relatively new illuminative display elements, such as LEDs and LCDs. These devices are described in more detail later in this article.

Where earlier digital displays depended mainly on glowing filaments shaped as numerals and letters, a majority of digital indicators today utilize individual display elements, arranged in an orderly matrix, combinations of which form numerals, letters, and various symbols. For common numerical displays, the popular matrix comprises seven elements, all elements being required at one time only in the case of the figure "eight" (8). See Fig. 11. The individual display elements usually are not narrow rectangles but have chamfered corners to provide an optically satisfactory fit for all the numbers. The actuation sequence of seven separate elements to achieve nine numerals plus zero is shown in Fig. 12. A few of the many contemporary instruments utilizing digital displays of this nature are shown in Figs. 13 to 15. In another approach, a matrix may be made up of circularly configured display elements, comparable on a small scale to a large sign made up of incandescent lamps connected to a selector circuit. See Fig. 16. Of course, a CRT display unit can serve as a digital indicator as well as a recorder, logger, and plotter when provided with the necessary electronic circuitry.

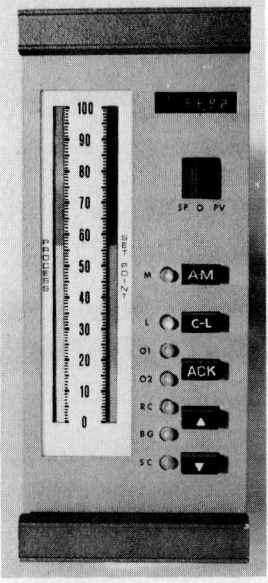

Fig. 6 Indicating controller in which moving-ribbon bar graph displays are used for the process variable and the set point. *(Robertshaw Controls.)*

Illuminative Display Elements

Modern illuminative display elements fall into the following general classifications: (1) elements based on the phenomenon of *luminescence,* of which *fluorescence* and *phosphorescence* are particular aspects; (2) LEDs; and (3) LCDs. Some of these elements are quite small, requiring a number of them to be activated to form alpha and numeric characters. In contrast, the CRT provides a means for displaying information, often simultaneously, of various shapes, sizes, and colors.

Luminescence

Instrument displays have made much progress during the past decade or two, advancing from essentially exclusive dependence on incandescent lamps. However, through ultraminiaturization and special bulb design, these lamps still play an important role in certain types of displays. Recent advances and changes in instrument displays largely have resulted from exploitation of the phenomenon of luminescence. In 1888, E. Wiedemann defined *luminescence* as "all those phenomena of light not solely conditioned by the rise in temperature." Luminescence may be better defined today as a characteristic nonthermal emission of electromagnetic radiation by a material upon some form of excitation.

Whereas the output from blackbody radiation consists of broadband emissions which follow Stefan-Boltzmann temperature relationships, luminescence emission from phosphors consists of relatively narrow bands which do not follow the blackbody laws. Thus, light emission due solely to the temperature of a source is referred to as *incandescence,* while luminescence, unlike incandescence, is

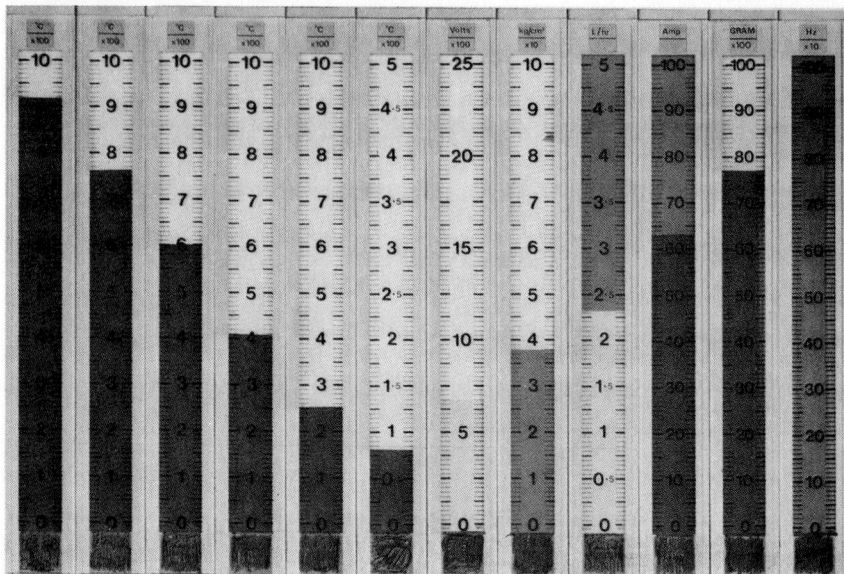

Fig. 7 Series of moving-ribbon bar graph indicators mounted side by side to provide a profile of the process being monitored. The effect is similar to that of mounting liquid manometers side by side as traditionally used in engine testing setups. Individual scales may be mounted horizontally to achieve a similar effect. The scales are 230 mm long. Customer-adjustable set points cover the length of the scale. High and/or low alarms with relays are also available to provide an on-off output. The ribbon drive is an aluminum-bronze lead screw. In the standard case version, a fully enclosed panel mounting case has a 288 × 36 mm bezel measuring 282 mm high, 33 mm wide, and 25 mm deep. Red, blue, green, and white ribbon colors are available for additional highlighting of variables. Shading indicates how colors can be combined. *(Chessell, Eurotherm International.)*

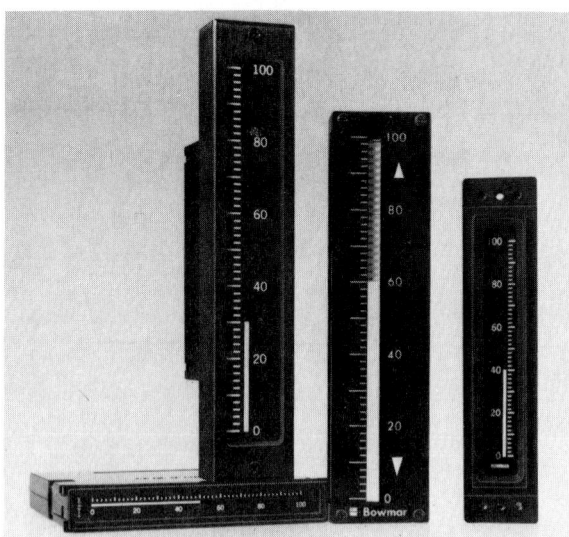

Fig. 8 Group of solid-state analog panel meters which utilize the bar graph display principle. The bar graph may run vertically or horizontally. Each bar graph display incorporates 51 segments (LEDs). They are available in red, green, yellow, and orange. Meters in the series include both voltage and current types and are optionally available in center-zero, voltmeter, and ammeter models with up to two set points and/or control (or alarm) outputs. In another series of analog bar graph panel meters, 50 segments, or up to 100 segments (LCD elements), are used. It is claimed that meters of this type are easy to read in direct sunlight or in high-intensity ambient light. *(Bowmar/ALI.)*

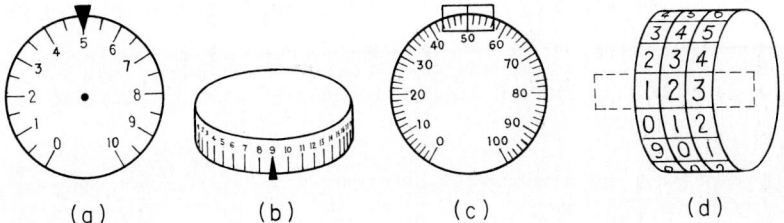

Fig. 9 Types of moving-scale indicators. (*a*) Moving dial, (*b*) moving drum, (*c*) precision indicator (moving dial), and (*d*) drum counter.

Fig. 10 Method of optically projecting the scale to achieve greater effective scale length and hence readability. Although the indicating space measures only about 2½ × 3 in (6.4 × 7.6 cm), the equivalent scale length of a 7 × 9 in (17.8 × 23 cm) pointer-type panel meter can be achieved. This is accomplished by mounting a scale of film on the moving coil of a core magnet mechanism in place of the usual pointer. A lamp, lens system, and mirror project this moving scale onto a coated window. The optical system expands the scale to 10 times its original value, and the portion to be read appears on both sides of a hairline on the coated window. This design avoids the usual problems of parallax in reading. (*Weston.*)

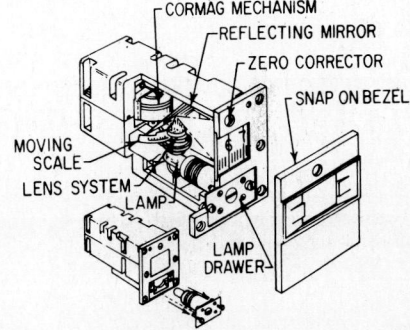

a function of the specific material involved. Although "fluorescence" and "phosphorescence" are sometimes used synonymously with "luminescence," *fluorescence* can be more rigidly defined as luminescence having a persistence (afterglow) shorter than about 10^{-8} s, and *phosphorescence* as being longer than 10^{-8} s.

The luminescence process itself involves (1) absorption of energy, (2) excitation, and (3) emission of energy, usually in the form of radiation in the visible portion of the spectrum. The type of luminescence is usually identified by the means of excitation, e.g., *cathodoluminescence* refers to excitation by cathode rays. The most commonly encountered types of luminescence are listed in Table 1. The luminescence material may be considered a transformer of energy, e.g., from ultraviolet photons to photons of lower energy, from cathode rays to photons, and from electric fields to photons. An inorganic luminescent material or phosphor usually consists of a crystalline host material to which has been added a trace of impurity (an activator or coactivator).

Phosphors

Initially, natural substances were used in CRTs for converting the energy of the scanning electron beam into light. It was during this early period that the word *phosphor* was coined. Synthetic phosphors have been used for many years. They are usually zinc, cadmium, calcium, and magnesium compounds (as sulfides, selenides, silicates, and tungstates). The materials must withstand "bakeout" temperatures of 400°C (752°F) or higher. They must have a low vapor pressure and an ability to hold up over long periods of time against the bombardment of electrons. Variation in the quality of specific phosphors is obtained through the use of accelerators, notably copper, silver, magnesium, chromium, and bismuth, among others. The activators permit the selection of efficiency, color of luminescence, and decay time.

1 2 3 4 5 6 7 8 9 0

Fig. 11 Numbers displayed by individual elements, such as LEDs, are frequently constructed from a possible total of seven elements, all elements being required at one time only in the case of the figure 8.

There are several scores of commercial phosphors from which to select. A significant percentage of these compounds is made up of group II and group VI elements of the periodic table. Zinc sulfide activated by magnesium produces a blue emission, whereas zinc and/or cadmium sulfide activated by copper or aluminum produces a green emission. Zinc sulfides activated by silver or copper can convert up to 20% of electron beam energy to light. These compounds are important in commercial CRTs. Where particularly long periods of electron bombardment are involved, $ZnSiO_4$ activated by manganese is well suited and thus finds application in oscilloscopes and aircraft instruments which require bright displays. A green luminescence is produced. See Table 2.

In recent years, a number of rare-earth elements, such as terbium and europium, have found use in color tubes. Compounds containing these elements emit a red color that is comparable in efficiency to the color emitted by well-established green and blue emitting compounds. Other rare-earth element compounds include La_2O_2S activated by terbium (green emission) and Y_2O_2S, also activated by terbium (white emission). An outstanding advantage of rare-earth phosphors is that they do not become saturated at high power levels. Confinement of their emission to rather narrow bands is also advantageous in providing image with high contrast even in the presence of high ambient light levels.

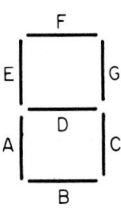

Digit	Elements activated		Elements not activated	
1	C, G	(2)	A, B, D, E, F	(5)
2	A, B, D, F, G	(5)	C, E	(2)
3	B, C, D, F, G	(5)	A, E	(2)
4	C, D, E, G	(4)	A, B, F	(3)
5	B, C, D, E, F	(5)	A, G	(2)
6	A, B, C, D, E, F	(6)	G	(1)
7	C, F, G	(3)	A, B, D, E	(4)
8	A, B, C, D, E, F, G	(7)		(4)
9	B, C, D, E, F, G	(6)	A	(1)
0	A, B, C, E, F, G	(6)	D	(1)

Fig. 12 Sequencing of seven display elements to achieve nine digits and zero.

The desired persistence time of phosphors varies with application. Whereas a time of 30 to 40 ms is satisfactory for many commercial needs, including television, a longer time (up to a second or even longer) is desirable for radar displays. Zinc-cadmium sulfide activated by copper persists for a number of seconds with a yellowish-orange color. For extremely short persistence, as required in flying-spot scanners, a material such as calcium-magnesium silicate, with persistence in terms of a fraction of a microsecond, is desirable. This compound emits in the violet-ultraviolet range.

Phosphor particles range in diameter from 1 to 10 μm. Image resolution varies inversely with the diameter of the particle, but efficiency decreases when particles are too small. CRTs are discussed later in this article.

LED Display Elements

Recombination or injection electroluminescence was first observed in 1923 by Lossew, who found that when point electrodes were placed on certain silicon carbide crystals and current passed through them, light was often emitted. An explanation of this emission had to await the development of semiconductor theory. If minority charge carriers are injected into a semiconductor, i.e., if electrons are injected into *p*-type material or "positive holes" into *n*-type materials, they recombine spontaneously with the majority carriers existing in the material. If some of these recombinations result in the emission of radiation, *electroluminescence* results. Minority carrier injection may occur not only at point contacts but also at broad area rectifying junctions; in this case, the junction must be biased in the forward or "easy flow" direction, and the electric field in the junction is lower when the voltage

Fig. 13 A wide variety of digital indicating instruments, ranging from very small hand-held indicators to other portable and bench-top indicators, are available for accepting inputs from thermocouples, resistance temperature detectors (RTDs), strain gages, and other transducers. Commonly used display elements are LCDs and LEDs. (Doric Trendicator electronic thermometers, *Doric Scientific, Division of Emerson Electric.*).

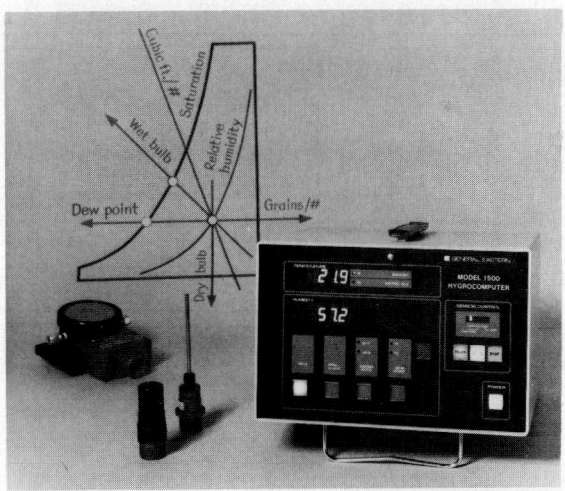

Fig. 14 Digital indication is frequently associated with microprocessor-based computing. Shown here is a fully automatic general purpose optical condensation hygrometer with remote dew-point and temperature sensors and a built-in computer to provide readout and output signals for any of the generally used humidity parameters. The instrument provides two digital displays, one for temperature and the other for one of the following selectable units: dew point, percent relative humidity, H_2O concentration in parts per million (volume), parts per million (weight), and grains per pound. The instrument displays metric or English units, which are switch-selectable. The display uses LED elements. *(General Eastern Instruments.)*

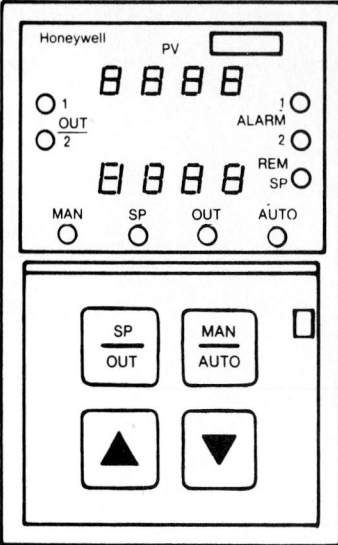

Fig. 15 Digital indication in connection with a controller—in this case a digital controller. The upper digital display shows the value of a process variable in engineering units, and the lower display normally indicates either the set-point value in engineering units or the output in percent, as selected. At any time, the operator can preempt these normal displays to show an operating parameter, tuning constant, or configuration word. The unit holds and updates the preempted value in memory and redisplays it 1 min after the first key stroke or when another key is pressed. When the manual mode is selected, the lower display automatically shows the output value. Eleven LEDs provide pertinent operating information: two indicate alarms; two indicate control relay outputs; two indicate minus signs for process variables and set-point displays; one indicates the set point is being displayed; one indicates the output is being displayed; one indicates remote set-point input; one indicates the manual mode is active; and one indicates the automatic mode is active. The microprocessor-based unit is provided with five control equations (algorithms). The indicating controller measures $4 \times 6 \times \sim 12$ in (depth) ($100 \times 150 \times \sim 325$ mm). (UDC 500, *Honeywell.*)

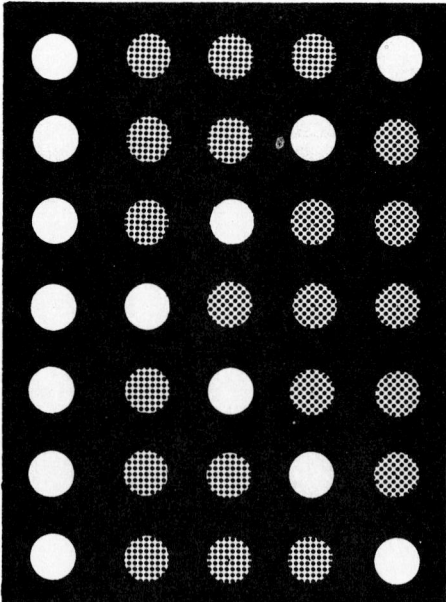

Fig. 16 Alphanumerics can be displayed by matrices made up of display elements, such as LEDs. In this case, the matrix is 5×7. In a system developed by the General Instrument Corporation, the dots are ~ 0.415 in (~ 10.5 mm) in diameter and thus useful for distant viewing.

15.14

Table 1 Principal Types of Luminescence

Type of luminescence	Excitation source	Example
Photoluminescence	Photons	$ZnS \cdot Ag$
Cathodoluminescence	Cathode rays	$Zn_2SiO_4 \cdot Mn$
Electroluminescence	Electric fields	$Zn \cdot (S \cdot Se) \cdot Cu$
Chemiluminescence	Chemical reactions	Oxidation of luminol (lightning bugs)
Bioluminescence	Biochemical reactions	Luciferin
Triboluminescence	Mechanical disruption	$ZnS \cdot Mn$

Table 2 Typical Phosphor Data Characteristics

Type of phosphor	Fluorescence	Relative luminance, %*	Relative photographic writing speed, %†	Comments
P-2	Bluish green	55	40	Good compromise for high- and low-speed applications
P-4	White	50	40	Television displays
P-7	Blue	35	75	Long decay, double-layer screen
P-11	Purplish blue	15	100	For photographic applications
P-15	Bluish green	15	15	Very short decay for flying-spot scanner use
P-31	Yellowish green	100	50	General purpose, brightest available phosphor

*Taken with a Spectra brightness spot meter which incorporates a CIE standard eye filter. Representative of 10-kV aluminized screens. P-31 as reference.

†P-11 as reference with Polaroid 410 film. Representative of 10-kV aluminized screens.

is applied than in its absence. This type of emission has been observed in several materials, including diamond, silicon, germanium, SiC, CdS, ZnS, ZnSe, ZnO, and some group III to group V compounds, such as GaSb, GaAs, GaP, InP, and InSb, among others. The emission of many of these materials lies in the infrared region of the spectrum. For radiation in the visible region, the energy difference between the holes and the electrons (the band gap of the semiconductor) must be more than 1.8 eV. Numerous materials satisfy this requirement, notably those used for CRT phosphors, but the materials present difficulties in fabricating p-n junctions.

The comparatively limited but growing list of materials used in LEDs includes GaP, GaAsP, GaAlAs, GaN, and SiC. Early commercial LEDs were made from $GaAs_{0.6}P_{0.4}$, deposited epitaxially as a thin layer on a GaAs crystal substrate. With these, p-n junctions were made, using diffusion techniques similar to those used in making silicon diodes. The band gap is 1.92 eV. There is an emission of red light with a peak at ~650 nm, resulting from direct recombination of electrons and holes.

GaAsP has a high index of refraction, and consequently only light emitted toward the surface (4%) is usable—the remainder is reflected back. A diode can be encapsulated in epoxy material to take on the shape and form of a lens. These diodes are particularly effective when a number are fabricated in close proximity on a single-crystal chip.

Diodes that emit light in shorter wavelengths (green, yellow, etc.) can be made by increasing the phosphorus content, but only up to ~40% because of a rapid decrease in efficiency. Efficiency can be increased by incorporating nitrogen atoms into crystals. Nitrogen atoms act as isoelectronic centers, trapping electron-hole pairs in an excited state. Three types of nitrogen-doped diodes have gained some importance: $GaAs_{0.65}P_{0.35}$ (orange light), $GaAs_{0.85}P_{0.15}$ (yellow light), and GaP (green light). Zinc and oxygen doping are also used. Diodes operate more efficiently if driven with periodic pulses of high current rather than with constant current. The short response time of junction diodes to current pulses (a fraction of a microsecond) and their rectifying property (they block current flow in the reverse, nonemitting direction) combine to make diodes a good choice for xy addressing arrangements.

As a result of intense proprietary research and development, the utility of LEDs has improved much during the past few years. In a patented process involving the overlaying of GaP with GaAsP, LED arrays with a luminous intensity of 2000 μcd at 20 mA forward current have been developed. One firm has introduced bicolor light bars for light bar and bar graph arrays. In this approach, four LEDs of each of two colors are used to uniformly illuminate the light emitting surface of a device with an emitting surface of 0.35 × 0.35 in (~9 × 9 mm). The units provide tristate indication, with a choice of off, red, and green, or off, red, and yellow. Another manufacturer, using the dual-color capability, packages a green and an amber LED in a common case with a diffuse top viewing plane. By mixing the two colors, intermediate shades of green-yellow, yellow, and light orange, in addition to the basic amber, can be achieved.

In the comparatively young LED technology, it is evident that many more innovative steps will be taken during the next few years.

LCD Elements

The technology of liquid crystals, particularly that of their practical application, is relatively recent. Even though LCDs have been in very limited use since the mid-1970s, as of the mid-1980s the field continues to be research and development intensive. Although much progress toward the improvement of LCDs has occurred during the last few years, overcoming several of the original complications and shortcomings, it is reasonable to assume that further improvements will proceed apace. Prior to describing practical applications—because the technology is recent—it is in order to review the fundamentals.

Basic Nature and Classes of Liquid Crystals

Liquid crystals are *liquids* that have the structural character of cybotactic liquids,* but that are considerably more viscous, with viscosities extending from that of a light glue to that of a glassy solid. They also exhibit much more definite evidence of structure than cybotactic liquids.

Liquid crystals must be geometrically highly anisotropic†—usually long and narrow—and revert to an isotropic liquid through thermal action *(thermotropic mesomorphism)* or through the influence of a solvent *(lyotropic mesomorphism)*. Several thousand organic compounds are known which meet these criteria, but significant molecular features found in thermotropic liquid crystals include the following. (1) The molecule is elongated and rectilinear. (2) When "flat segments" (e.g., benzene rings) are present, liquid crystallinity is enhanced. The molecule is rigid along its long axis, and double bonds are common in this direction. The simultaneous existence is seen in the molecule of strong dipoles and easily polarizable groups. Of lesser importance are weak dipolar groups at the extremities of the molecule.

Some authorities classify thermotropic liquid crystals into three groups:

1. *Smectic liquid crystals,* such as *p*-ethyl azoxybenzoate, have their molecules arranged in definite strata, a variety of molecular arrangements being possible within each stratification. In *smectic type A* crystals, the molecules may be considered to "stand on end" with their long axes perpen-

---

Cybotaxis is a condition in which certain liquids, under x-ray examination, give evidence of structure resembling that of crystals. Diffraction maxima or halos, somewhat like the diffraction rings produced by powdered crystals, are observed. These suggest that molecules are temporarily arranged in rows, layers, or stacks like bricks in a pile and that they have one, two, or even three different dimensions or spacings, corresponding, in accordance with Bragg's law, to the different angles of diffraction observed. A closely related property is exhibited by certain substances known as liquid crystals, which appear to be intermediate between merely cybotactic liquids and true cyrstals. In these materials there appear to be large groups of molecules which, though able to move and turn about, retain their structural arrangement. Such mesomorphic substances manifest some of the properties of optical crystals, which the former type do not.

†An isotropic medium has different optical or other physical properties in different directions. Calcite crystals, for example, are anisotropic, while fully annealed glass and, in general, fluids at rest are isotropic.

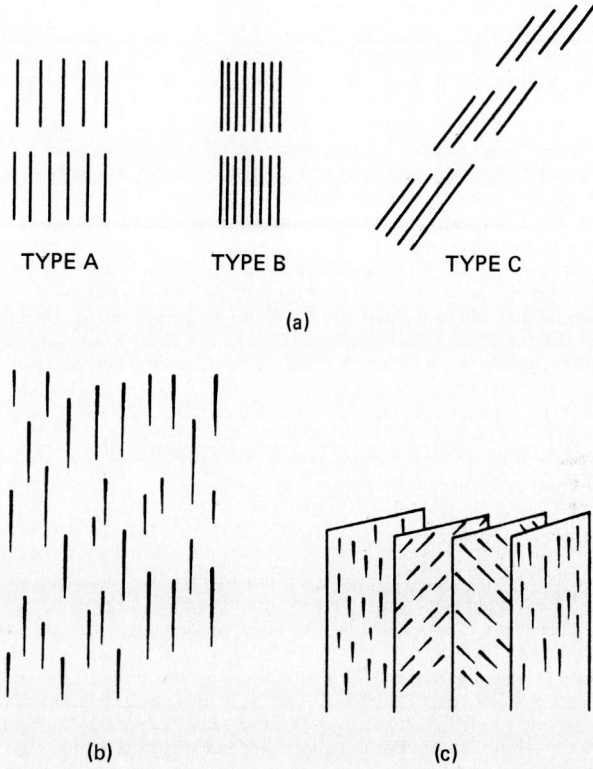

Fig. 17 Classes of liquid crystals. (*a*) Smectic liquid crystals, types A, B, and C; (*b*) nematic crystals; (*c*) cholesteric crystals.

dicular to the plane of the layer, but with their centers irregularly spaced. When the molecular centers adopt hexagonal close packing, the crystals are considered *smectic type B*, and when they adopt a tilted form of type A, they are classified as *smectic type C* crystals. See Fig. 17*a*.

2. In *nematic liquid crystals,* the molecular structures possess a high degree of long-range orientation order, but no long-range translational order. The molecules are spontaneously oriented with their long axes approximately parallel, but without the stratification seen in smectic crystals. Nematic liquid crystals like *p*-azoxyanisole are generally optically uniaxial, positive, and strongly bire-fringent, and some are composed of hundreds of molecules (cytotactic groups), with the molecular centers in each group arranged in layers. See Fig. 17*b*.

3. *Lyotropic* or *cholesteric liquid crystals* possess at least two components. One of these is water and the other is amphible (a polar head group attached to one or more long hydrocarbon chains). In the lamellar form, water molecules are sandwiched between the polar heads of adjacent layers, while the hydrocarbon tails lie in a nonpolar environment. These crystals have very complex structures, but occur abundantly in nature, particularly in living systems. See Fig. 17*c*. The name "cholesteric" stems from the fact that these substances sometimes contain cholesterol, although cholesterol itself is not known to exist as a liquid crystal.

Polarized light is the most powerful tool for investigating liquid crystals, all of which exhibit characteristic optical properties. A smectic liquid crystal transmits light more slowly when perpen-dicular to the layers than when parallel to them. Such substances are said to be optically positive. Nematic liquid crystals are also optically positive, but their action is less definite than that of smectic liquid crystals. However, the application of a magnetic field to nematic liquid crystals lines up their molecules, changing their optical properties and even their viscosity.

Both smectic and nematic crystals split a beam of ordinary light into two polarized components whose transverse vibrations are at right angles to each other. This is the well-known phenomenon of double refraction. Cholesteric liquid crystals exhibit the phenomenon of *circular dichroism*. That is, they break a beam of ordinary light into two components, one with the electric vector rotating clockwise and the other with it rotating counterclockwise. The first is usually transmitted, and the second is the one reflected. It is this property which gives cholesteric crystals their characteristic iridescent colors when illuminated by white light.

The ability to exhibit colors is one of the most useful attributes of liquid crystals. Many cholesteric substances behave as liquid crystals only within a certain temperature range. Above that, they are colorless, but as they are cooled they assume a succession of colors, running down the spectrum from red to violet and ultimately to the point where they become colorless. In this final stage, they still retain their molecular orientation, but it is that of smectic liquid crystals rather than cholesteric crystals. Some cholesteric liquid crystals do not exhibit all the colors mentioned, and others which are naturally colored simply change to another color on heating or cooling. Since the exact temperatures at which these color changes occur are known, the substances can be used for measuring temperature—with combinations of them covering a range from −20 to 250°C (−4 to 482°F). However, it is not this characteristic, but rather the electrooptical qualities of liquid crystals that make them attractive for use as display elements. It is interesting to note that, when used as display elements, liquid crystals consume far less energy than LEDs.

Advances in LCD Technology

Initially, to obtain color in an LCD, one had to use either polarized filters or provide colored background lighting. The viewing angle was restricted with these approaches and, unless the observer was close to the center line of the element, the display degraded to black. The addition of polarizing filters resulted in the elements becoming sensitive to humidity. The turn-off response time of LCDs generally increases in a nonlinear way with a decrease in temperature. This meant that, without special means taken in LCD design, the elements would be very sluggish in instrument displays at lower temperatures, such as 0°C (32°F). Some LCD suppliers have tackled this problem through better control of the fabrication process—e.g., by using supports between the glass to reduce the thickness of the LCD film and thus improving the response time. Currently, there is much better uniformity in this regard among LCDs from various suppliers than was originally the case.

Early LCDs had relatively short lives. Initially, the operating life was quoted by most suppliers to be in the 10,000-h range. However, the experience of users tended to confirm that the useful life was actually considerably shorter. Through the use of more reliable sealing methods and, in particular, as a result of investigating numerous LCD materials, the lifetime of LCDs has been increased manyfold.

The humidity sensitivity problem also has been partially solved—through better sealing and through the use of dip-coating of the element. With little reduction in polarizing efficiency, sensitivity to relative humidity has been significantly lessened. Further, in the early 1980s, dichroic dyes,* which function on the basis of light absorption rather than light polarization and reflection, were introduced. These dyes are opaque in the "off" state, but on application of an electric field become light-transmissive and thus permit the color of the reflector to be observed. Thus, there is a phase change—from the light absorbing (cholesteric) state to the light absorbing (nematic) state. The nature of the dye is such that there is a high absorbance limit for light polarized parallel to the major axis of the dye molecules and, in contrast, a low absorbance of light polarized perpendicularly to the major axis. This comparatively recent technique (sometimes called "host-guest") is claimed to provide a 180° viewing angle of essentially constant contrast, as well as the elimination of polarizing filters and their associated problems.

By the mid-1980s, LCDs have generated significant market acceptance and are found in a wide variety of instrument displays. The principal advantages of LCDs have been found to be a low power requirement, visibility over a wide range of ambient light, and flexibility in formatting.

Dichroic is a term used in crystallography to denote crystals which refract incident light in two directions, thus displaying two colors when observed from different angles.

CRTs—Displays

In the instrumentation field, CRTs were initially and for a long time mainly associated with oscilloscopes and industrial closed-circuit television. The role of the CRT in what might be called mainline instrumentation and control systems changed markedly, particularly during the 1970s, and is continuing at a rapid pace into the mid-1980s. Although, as previously described, numerous other display techniques are widely used (these are sometimes grouped together and called *nonvideo*), it has been evident for a number of years that the CRT has markedly altered the field. The CRT, along with some disadvantages, has many advantages—possibly its foremost advantage being its compatibility with computers and digital electronics. The CRT has a universal quality. Basically, it is an indicator, but it can serve other information display functions—recording, logging, monitoring, alarming, instructing, storing, among others—when it is regarded as a system appropriately supported by other electronics rather than simply as a display component. CRT displays are used singly, as they are in a variety of data terminals (Figs. 18 and 19), or in groups (two, three, four or more), as they are applied in the central control rooms of large, complex processes. Most contemporary distributed control systems make use of one or several CRTs. See Fig. 8 in Control Centers—From Spec to Job Site later in this Handbook section.

The variety of information that can be displayed on the face of a CRT seems almost unlimited. Some examples include: (1) a diagrammatic overview of an entire plant, showing all vital status and alarm information; (2) high-resolution and coordinated bar graphs for displaying analog information; (3) an overview of the alarm status of many different devices shown in a single display; (4) the status of analog points sampled at specified intervals and the results plotted on demand; (5) for utilities, power distribution systems displayed complete with all measured parameters and the status of all switchgear; (6) management information reports of many kinds presented in tabular form—for a single shift, day, or week; (7) diagnostic displays generated which utilize data in a programmable controller, simplifying troubleshooting by plant personnel; and (8) secondary, detailed displays

Fig. 18 CRT terminal used in connection with an electronic weighing system (designed for use by scrap dealers and processors) provides information on the amount of material moving through a facility and an accounting by customer for each type of material—by weight and value—for each transaction, day, week, and/or month. Associated units shown are a page printer, a disk drive and storage unit, a digital weight indicator, and a tape printer. *(StreeterAmet Division, Mangood Corporation.)*

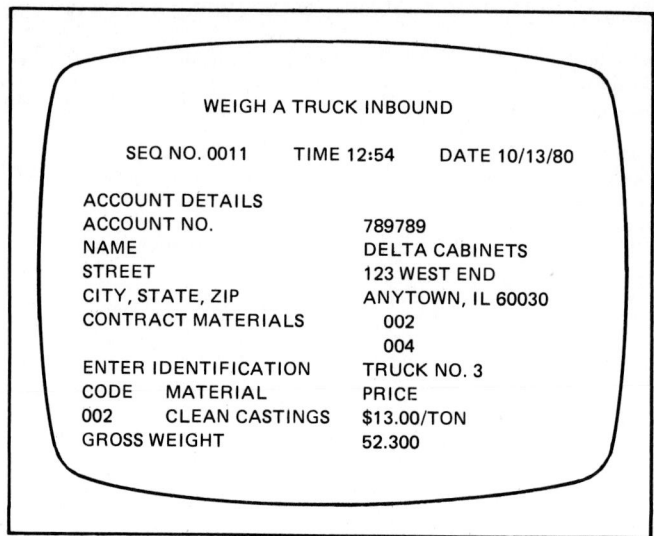

WEIGH A TRUCK INBOUND

SEQ NO. 0011   TIME 12:54   DATE 10/13/80

ACCOUNT DETAILS
ACCOUNT NO.   789789
NAME   DELTA CABINETS
STREET   123 WEST END
CITY, STATE, ZIP   ANYTOWN, IL 60030
CONTRACT MATERIALS   002
                           004

ENTER IDENTIFICATION   TRUCK NO. 3
CODE   MATERIAL   PRICE
002   CLEAN CASTINGS $13.00/TON
GROSS WEIGHT   52.300

Fig. 19 Typical CRT screen readout produced by the CRT terminal illustrated in Fig. 18.

shown in a "zoom-in" fashion. Views of such displays are not shown here because much of their meaningful effect cannot be captured in black and white.

The complete role, the advantages, and the limitations of CRT displays remain to be worked out, a process that probably will go on for another decade or two. For example, what is the role of the CRT in an alarm situation? Should it predominate or replace conventional annunciators and alarm systems? Or should the CRT display be secondary to other alarm systems but perform the important support function of furnishing the operator with emergency instructions drawn from storage rather than from a notebook? There is insufficient space here to ponder the probable enhanced role of the CRT in the future, but several references are cited at the end of this article.

CRT Fundamentals*

CRTs convert electron beam energy into light energy. Figure 20 shows the basic CRT divided into five sections. In operation, electrons are emitted from a thermionic cathode and controlled by the

*Prepared by Richard E. Kehrli, Tektronix, Inc., Beaverton, Oregon.

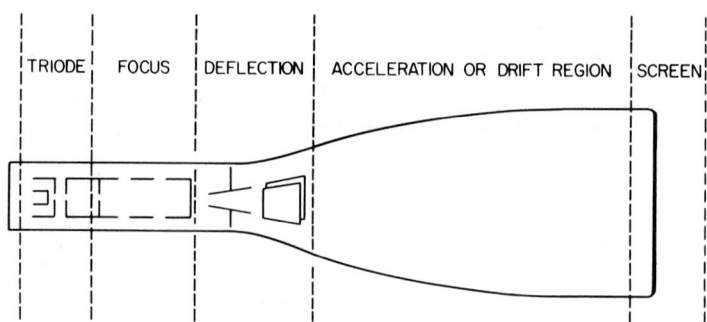

TRIODE | FOCUS | DEFLECTION | ACCELERATION OR DRIFT REGION | SCREEN

Fig. 20 Major sections of a CRT.

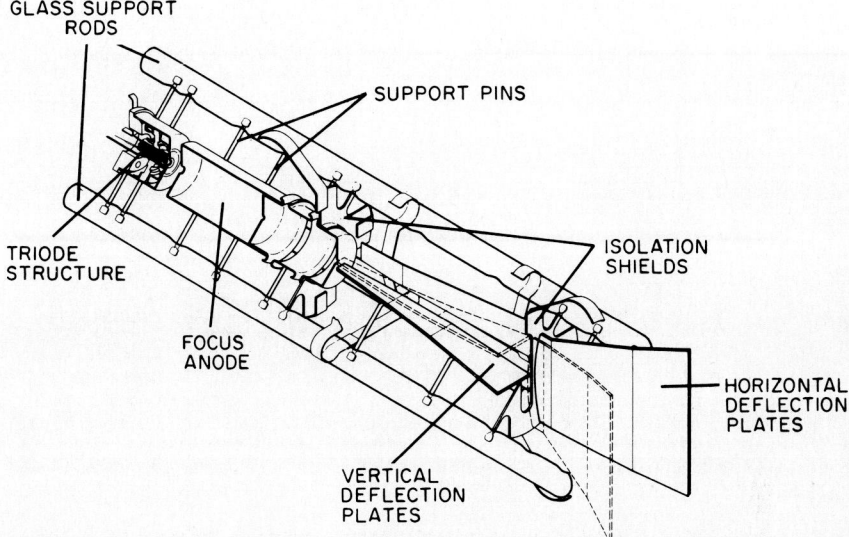

GLASS SUPPORT
RODS

SUPPORT PINS

ISOLATION
SHIELDS

TRIODE
STRUCTURE

FOCUS
ANODE

HORIZONTAL
DEFLECTION
PLATES

VERTICAL
DEFLECTION
PLATES

Fig. 21   Typical CRT electron gun.

triode section. They are then formed into a beam and accelerated in the focus section. The deflection section deflects the beam (typically on vertical and horizontal axes) by internal electrostatic deflection plates or by external electromagnetic deflection coils. The acceleration or drift area controls (and often accelerates further) the electron beam until the energy reaches the CRT screen. The electrons strike a light emitting phosphor on the inside face of the CRT screen, which fluoresces and emits visible light. CRT phosphors possess the characteristic of phosphorescence, emitting light energy after the electron beam has been removed. This image persistence allows a repetitive pattern to appear as a stationary display. See Table 2 for some of the phosphors used in CRTs. In addition to presenting x and y information on the deflection plates, a CRT uses the cathode or grid of the CRT gun to present z-axis intensity information. A typical gun structure is shown in Fig. 21.

CRTs are classified into two major subgroups, monoaccelerators and postaccelerators. *Monoaccelerators* generally apply a high voltage of 3 to 4 kV to the second (focus) anode, whereas *postdeflection acceleration* (PDA) tubes typically apply 10 to 14 kV to a high-voltage electrode near the CRT screen. PDA tubes typically have higher light output, since the light output from a phosphor increases with the voltage through which the beam electrons have been accelerated. This also allows the deflection region to be kept at a relatively low voltage and thus aids deflection sensitivity.

PDA tubes are usually aluminized with a thin coating of aluminum, which acts as a mirror and reflects to the screen light energy that would otherwise be lost. This also acts as a heat sink and helps prevent burning of the phosphor.

Storage CRTs

These tubes have two electron sources. A *writing gun* provides the electrons for writing, and a *flood gun* provides broad coverage of low-velocity electrons that bombard the storage screen uniformly. This flood of electrons holds the writing gun information in the "written" mode by means of secondary emission electrons and maintains the stored image after the writing beam has been cut off.

Storage CRTs are useful in displaying signals that occur only once (transients) or have low repetition rates. They eliminate much of the necessity of photographing transients, since this information may be obtained from the CRT screen. High-resolution storage is also very useful in presenting graphic and alphanumeric displays in computer readout applications. This eliminates much of the local storage normally required for continually refreshing displays and provides a flicker-free display. Storage displays may typically be retained from 15 min to an hour or longer. See Fig. 22.

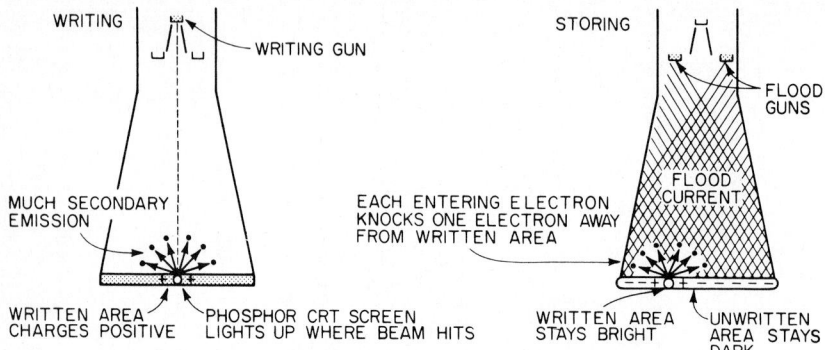

WRITING

WRITING GUN

MUCH SECONDARY
EMISSION

WRITTEN AREA
CHARGES POSITIVE

PHOSPHOR CRT SCREEN
LIGHTS UP WHERE BEAM HITS

STORING

FLOOD
GUNS

FLOOD
CURRENT

EACH ENTERING ELECTRON
KNOCKS ONE ELECTRON AWAY
FROM WRITTEN AREA

WRITTEN AREA
STAYS BRIGHT

UNWRITTEN
AREA STAYS
DARK

Fig. 22 Direct view bistable storage CRT. The action of the writing gun is shown at the left, and that of the flood gun at the right. The writing gun bombards the screen. High-energy electrons light the phosphor and also knock loose many secondaries. The written area, losing electrons, develops a positive charge. Flood gun electrons hit unwritten areas too slowly to light the phosphor. They merely accumulate, making the area negative. But the written area (positively charged) attracts electrons at high speed, keeping the phosphor lit, as well as knocking enough secondaries away to keep the area positive.

Storage CRTs are classified as bistable or halftone tubes. The stored display on a *bistable tube* has one level of intensity, whereas a *halftone tube* may display a stored signal at different levels of intensity. The intensity of a halftone tube is dependent on beam current and the time the beam remains on a particular phosphor particle. A bistable tube either stores or does not store, with all stored events having the same intensity.

Figure 23 shows a typical display of a high-resolution direct view bistable storage CRT.

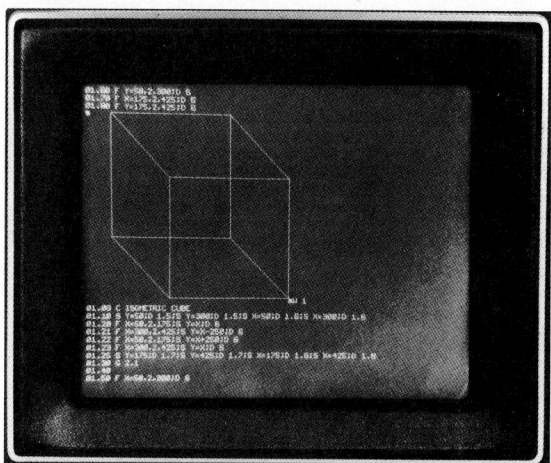

Fig. 23 Direct view storage CRT display. *(Tektronix.)*

Character Generation Tubes

In these tubes alphabetic, numeric, alphanumeric letters, and various forms of symbols are generated by passing the electron beam through an aperture of appropriate shape. Three basic methods are used to generate character information:

1. The *raster scan* technique involves controlling the intensity of the electron beam during the sweep. The process is similar to facsimile recording where the characters are generated in segments.

2. In the *Lissajous* technique the electron beam serves as a pencil.

3. In the *shaped-beam* teachnique the tube incorporates a number of stencil-type openings that are used to shape the electron beam.

Scan Conversion Tubes

Even though these tubes are not display tubes per se, they provide an important link in a display system. A scan conversion tube converts radar blips to television signals for viewing on a television screen. The scan conversion tube allows information to be put in at one rate and taken out at another rate, thus providing some storage. The technique is particularly useful for retaining aircraft locations in an air traffic control instrument so that the path of an aircraft can appear as a dotted line.

Other Electronic Displays

In classifying electronic display elements, it is interesting to note that they may be (1) emissive or (2) passive. The majority are emissive. In the emissive class, there are (*a*) electroluminescent devices, typified by LEDs, (*b*) gas discharge lamps, (*c*) cathodoluminescent devices, typified by CRTs as well as vacuum fluorescent devices, and (*d*) incandescent lamps, the latter depending on blackbody radiation (thermal). In the passive class, there are (*a*) liquid crystal devices and (*b*) possible future devices which may be electrochromic, electrophoretic, or electromagnetic. Most of the schemes used on a broad scale in process instrumentation already have been described in some detail. The other schemes, while widely accepted for certain applications, are frequently associated with customized systems.

Impressive research-and-development progress is underway in connection with planar gas discharge displays. Available are multiregister gas plasma displays that have a microprocessor interface design. These displays can provide a data terminal with an editing capability and a keyboard option. One design (Burroughs Self-Scan) has an array of 480 alphanumeric capacity in 12 registers of 40 characters in 5 x 7 matrices. Each register has 7 rows of display anodes and 283 cathode columns, the latter consisting of 2 reset cathodes, 1 scan monitor cathode, and 280 display cathodes.

The use of incandescent displays has not been overlooked. A display system has been developed to alter colors from green through orange, amber, and yellow to red in a continuous, smooth warning transition. Rather than warning an operator of a possible forthcoming emergency or alarm condition by way of sudden discrete steps, the psychological effect is akin to noting a small flame growing into a fire.

Alarms and Annunciators

An *alarm* is a device that signals the existence of an abnormal condition by means of an audible or visible discrete change, or both, intended to attract attention (Instrument Society of America). Commonly, alarm displays (audible or visible) are placed where the process or machine operator(s) is located, as in a central control room for a large, complex process. There has been considerable study in recent years pertaining to the human engineering factors involved in the design and location of alarm displays, particularly as these factors relate to the regular control loop instrumentation. In some cases, where a complex process is spread out over a large area, local audible alarms adjacent to processing units or machines may be used to alert personnel in specific areas.

The choice of an alarm system is affected by the degree to which a process or machine may be potentially dangerous—to personnel, equipment, and materials in process. There are wide differences in process complexity, including the manner in which several variables may interact, in the innate stability of the process or machine, and, most importantly, in the consequences of various faults, including not only the effects on personnel, equipment, and materials, but of downtime as well. Thus it is obvious that classifying and evaluating risks is difficult, but one fact is certain—the control engineer responsible should analyze each process separately from the perspective of safe operation and then base the choice of an alarm system on this analysis. As pointed out later, alarm systems are, in essence, *accident prevention systems*.

In the Hanes (1978) reference listed, five types of alarms are recommended for inclusion in an optimized plant monitoring system:

1. **Most Critical Alarms.** Alarms that require prompt operator action in order to maintain the unit on-line (conditions resulting in a complete loss of load), protection of major equipment, or safety of personnel

2. **Less Critical Alarms.** Alarms that require prompt operator action in order to maintain the unit load (conditions resulting in a partial loss of load) or to protect equipment

3. **Noncritical Alarms.** Alarms that require corrective action, but not directly by the operator

4. **Status Information.** Displays that indicate the status of events and which do not require corrective action

5. **Trip Analysis Information.** Conditions directly related to, or that can lead up to, tripping of the unit, this category also including information from special pretrip or posttrip logs in order to more readily determine the actual unit condition before a restart

It is further suggested that the foregoing different categories be displayed on different devices. Most critical alarms (red) and less critical alarms (white) would be displayed on visual annunciators, and noncritical information possibly displayed on a CRT. Status information would be available as hard copy or possibly on a separate CRT. Trip analysis would be hard copy.

Further observations (Hanes, 1978) on an optimized alarm system include: (1) Visual flashing lights accompanied by an audible warming may be the best way to convey the urgency of critical alarms to operating personnel; (2) plant operators require many more data on the events that indicate the status of the plant that can be, or should be, displayed with critical alarms; and (3) alarm and event data should be permanently logged, along with the exact time and sequence for an operations event analysis and safety review. These matters are further pursued under Analysis of Alarm Condition Information later in this article.

Alarm Actuation

Each alarm unit is actuated by either making or breaking a contact in its alarm circuit. The latter method has the advantage of alarming if for any reason a break occurs in the alarm circuit. Traditionally, three approaches have been taken, depending on the alarm system designer's evaluation of potentially hazardous conditions as previously mentioned.

1. **Actuation from a Contact within an Instrument.** A contact within an instrument that indicates, records, transmits, or controls the variable to be alarmed. This approach, of course, does not provide the advantages of a separate, *dedicated* alarm system. Specific methods (Fig. 24) that have been used include:

   a. Low-voltage contacts may be mounted on both the set pointer and the indicating pointer and so arranged to break contact when deviation occurs.

   b. A vane may be mounted on the indicating pointer so as to pass between two sensing coils mounted on the set pointer, thereby detuning an oscillating circuit which in turn operates a relay.

   c. High voltage with a limiting current may be used to jump the gap between the set pointer and the indicating pointer, thereby operating a relay.

   d. Commutator-type rotating contacts can be positioned by the indicating pen, and the segments positioned by the set pointer.

   e. A cam can be positioned by the indicating pointer through a servo motor, and the contacts positioned by the set pointer. The contact device may be a miniature switch or a tilting-type mercury switch.

   f. A tilting-type mercury switch may be positioned by the indicating pointer through a servo motor, and the contacts positioned by the set pointer. The contact device may be a miniature switch or a tilting-type mercury switch.

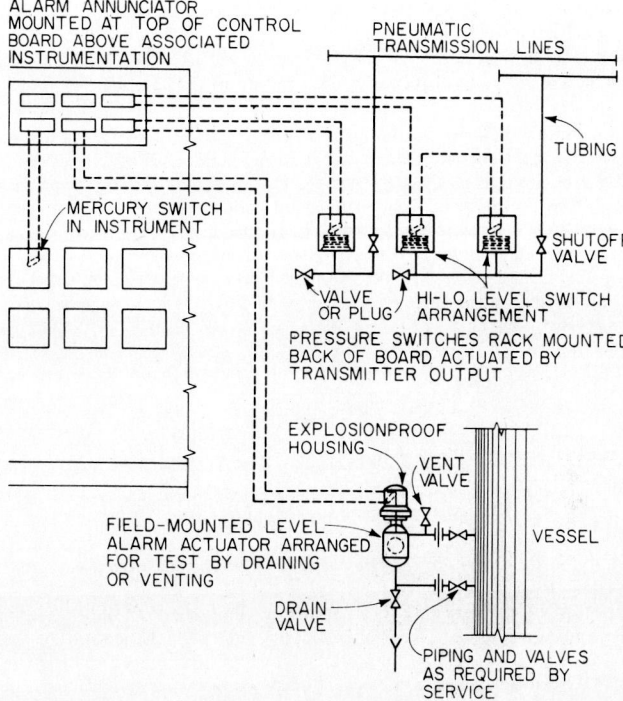

Fig. 24 Representative alarm actuator installation. *(American Petroleum Institute.)*

  g. A magnetic mercury switch may be used with the set pointer positioning the mercury switch and the indicating pointer positioning the magnet.

  h. A flapper and a nozzle sometimes are used in a pneumatic circuit, one set by the set pointer and the other by the indicating pointer. By using a bellows in this circuit, a set of contacts in a miniature or a mercury-type switch can be actuated.

2. *Actuation from Transmitter Output.* In a fully dedicated alarm system, the system would have independent signal transmitters used only for the alarm system.

3. *Operation by Direct Acting Actuators.* These devices might include a thermostat or temperature switch; a pressure switch; a level switch; a flow switch; a rotation or speed switch that detects the rotational speed of a shaft; a vibration switch whose contacts are actuated by an increase in the amplitude of vibrations in rotating equipment, such as motors and compressors; an auxiliary contact, typically on a motor starting device, which indicates electrical failure of a motor circuit; and, in chemical-composition determinations, a device, usually an analyzer, with contacts actuated at a predetermined percentage of the variable monitored, such as carbon monoxide, hydrogen sulfide, or hydrocarbon vapors.

Alarm Inputs

Input signals to alarm units usually consist of one of the following.

1. Relay contacts, normally closed or normally open. Isolated 24 or 125 V dc is normally furnished for use on field contacts; 12, 48, or 250 V dc also may be used.

2. Computer or other instrument-type logic input voltages.

3. Separate mounted electronic field transmitters or other input signal conversion devices, the usual ranges of which are 4 to 20 mA, 1 to 5 V, and 0 to 15 V. On such units, the alarm point is adjustable over a specified range, and high or low alarms are available. These transmitters, separate from regular process instrumentation, provide a fully dedicated alarm system.

The trend is toward two-wire transmitters. In selecting resistance temperature detector (RTD) transmitters, the distance between the measuring element and the transmitter determines whether a two-wire or a three-wire configuration is necessary. For applications where the distance between the sensor and the transmitter is short, a two-wire configuration is usually acceptable because distance introduces only a small error in measurement. For greater distances, a three-wire configuration is required. In this case, errors introduced on the leads can be electronically compensated for by the transmitter. Slidewire transmitters are also available. Alarm system equipment, of course, is available in explosion-proof housings where specified. Many alarm system components used today are designed for *intrinsic safety*.

Alarm Information Displays

Traditionally, alarm displays have been called *annunciators,* the majority of which over many years have been of the window format, illuminated from behind the window. Various colors have been used to indicate the "degree" of the alarm. A legend on each window identifies the particular situation being alarmed. The unit shown in Fig. 25 monitors 24 alarm points. The unit shown in Fig. 26 monitors 48 alarm points and is of a different configuration. LEDs are used as the display elements.

In addition to the traditional configurations, the CRT has gained an important role in the alarming function, as discussed later.

Some alarm display suppliers have incorporated the advantages of solid-state and other technological developments over the past decade or so, but have generally maintained the original physical format of the annunciator. In addition to using the more recent illuminative display elements, such as LEDs and LCDs, advantage has been taken of microprocessors for annunciator logic. The annun-

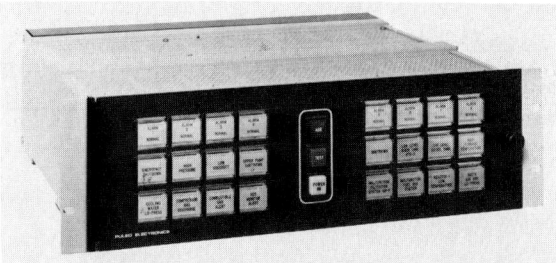

Fig. 25   Alarm annunciator for monitoring 24 alarm points features legend window displays. *(Puleo Electronics.)*

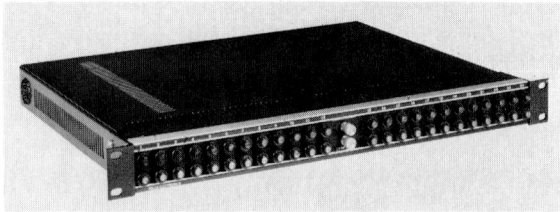

Fig. 26   Alarm annunciator for monitoring 48 alarm points features LED displays. *(Puleo Electronics.)*

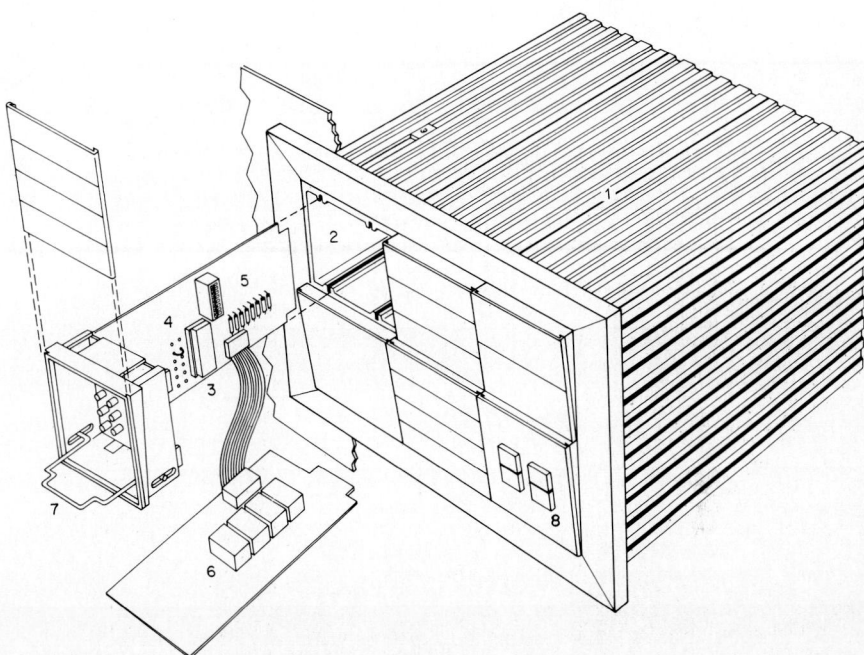

Fig. 27 Annunciator featuring a modular design. Individual cells (six are shown) are mechanically interlocked on a building-block basis, permitting various formats (in width, height, and number of displays). (1) Extruded aluminum housing. (2) Individual cell space within housing. (3) Microprocessor-based logic (one for each cell) made up of a single chip capable of handling up to four inputs—thus there is a capacity for four windows per cell even though only one, two, or three may be used in any given format. (4) Jumper matrix which permits in-the-field selection of the sequence on a per-cell basis. Several standard sequences, including three first-out modes, are available. (5) Field-selectable input operation—a dip switch bank permits selection of operation from normally closed (NC) or normally open (NO) inputs on a per-point basis and designation of critical or noncritical or audible group 1 or audible group 2 alarms; this is accomplished without wiring changes. (6) Repeat relays (optional)—addition of this board permits repeating the relay function per point, with eight different relay operating modes. As explained in Fig. 28, a single modular cell can accommodate one-, two-, three-, or four-point displays. This is accomplished by divided lamps and windows. Sizes may be mixed or matched in any combination. No additional logic or internal rewiring, or increase in the physical size of the unit, is required to make these changes. Redundant lamping is provided for all points. (7) A retractable bail permits front-of-panel removal of the lamp assembly and logic board. (8) Operator pushbuttons. (MicroLarm, *Rochester Instrument Systems*.)

ciator shown in Figs. 27 and 28 is illustrative of the incorporation of technological change, while still retaining much of the appearance of the earlier window-type instruments.

CRT Alarm Displays

Not all authorities agree on the most effective role of the CRT in displaying alarm information. One school considers that the CRT can play a very effective backup role in furnishing details and instructions pertaining to an alarm condition; others would give the CRT the major role, but only a minority would eliminate the traditional annunciators. A video display brings several new dimensions to the monitoring function that are not duplicated in conventional devices, including the graphics panel. A CRT display can provide a greater depth of status information and, for many operators, in a more comprehensible format. It can provide the exact sequence of all alarms rather than a few selected first-out alarms, and it can centralize the display so that alarms are always shown in the same place. It can show the real time of occurrence for each event, with a detailed language description of the associated field device or function. The CRT offers valuable sort and grouping functions and has a

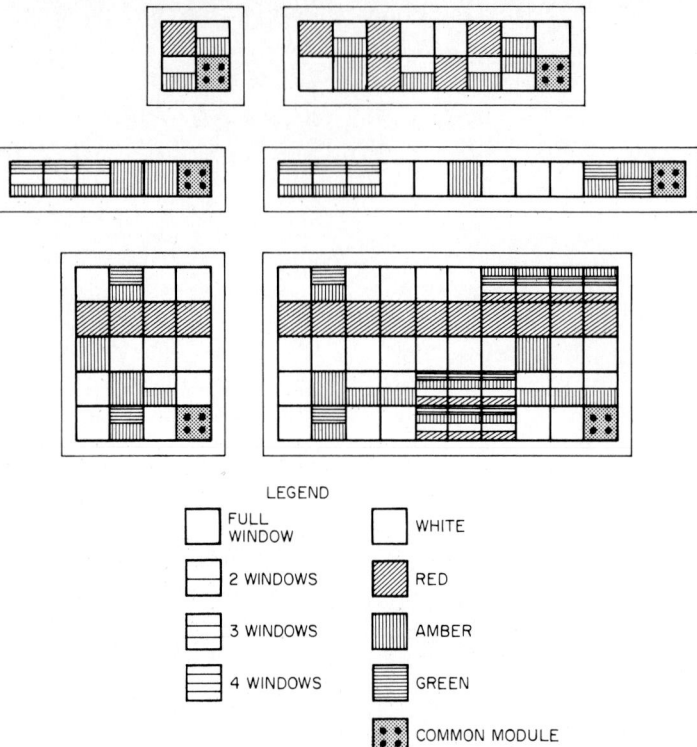

LEGEND

	FULL WINDOW		WHITE
	2 WINDOWS		RED
	3 WINDOWS		AMBER
	4 WINDOWS		GREEN
			COMMON MODULE

Fig. 28 Examples of the flexibility in annunciator formatting obtainable through modular design. Command module for all annunciator assemblies is located in bottom right-hand corner of device. This module has four pushbuttons: "ack," "test," "reset," or "first-out reset," and "silence." The nominal size of a full window is 3 × 3 in (76 × 76 mm). Where two windows are used in one modular space, the height of each window is one-half that of full modular size; and one-third and one-fourth for three and four windows, respectively. As indicated in the legend, white plus three colors are obtainable for all window sizes. (MicroLarm, *Rochester Instrument Systems.*)

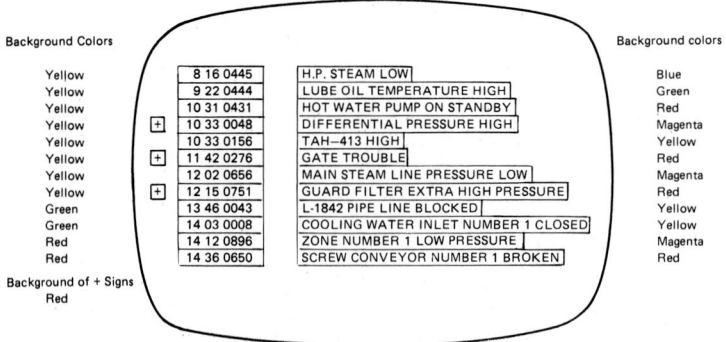

Background Colors   Background colors

Yellow		8 16 0445	H.P. STEAM LOW	Blue
Yellow		9 22 0444	LUBE OIL TEMPERATURE HIGH	Green
Yellow		10 31 0431	HOT WATER PUMP ON STANDBY	Red
Yellow	+	10 33 0048	DIFFERENTIAL PRESSURE HIGH	Magenta
Yellow		10 33 0156	TAH—413 HIGH	Yellow
Yellow	+	11 42 0276	GATE TROUBLE	Red
Yellow		12 02 0656	MAIN STEAM LINE PRESSURE LOW	Magenta
Yellow	+	12 15 0751	GUARD FILTER EXTRA HIGH PRESSURE	Red
Green		13 46 0043	L-1842 PIPE LINE BLOCKED	Yellow
Green		14 03 0008	COOLING WATER INLET NUMBER 1 CLOSED	Yellow
Red		14 12 0896	ZONE NUMBER 1 LOW PRESSURE	Magenta
Red		14 36 0650	SCREW CONVEYOR NUMBER 1 BROKEN	Red

Background of + Signs
Red

Fig. 29 Example of a color-coded CRT and alarm grouping display. In addition to the yellow status colors shown with the time and point number, each 64 character-capacity legend may have a background color to show the function of the point. The colors red, green, yellow, blue, magenta, cyan, white, and black can be used. A plus sign may be inserted to the left of any event to alert the operator to a critical event; the background color follows the status. (Videolarm, *Rochester Instrument Systems.*)

Fig. 30   Control room in which both window-type and CRT annunciators are used. *(Rochester Instrument Systems.)*

large capacity for color coding. A prime advantage of the CRT is its compatibility with computer-based process control systems.

The CRT annunciator shown in Fig. 29 can provide a 64-character legend and is available with ISA Sequence A or R-12. A combination of window-type annunciators and a CRT annunciator is shown in Fig. 30.

Alarm System and Annunciator Logic

Over the years, various schemes have been developed for the sequence used in actuating displays and in deactuating them. See Table 3 for a brief explanation.

Analysis of Alarm Condition Information

Traditionally, alarm systems have been dedicated to a single purpose—alerting human operators that one or more conditions in a process or machine may lead to personnel injury and damage to equipment and material in process. Alarm condition information, augmented by operator training and judgment, leads to immediate or very short-term corrective action. In recent years, considerable attention has been given to an important additional use of alarm condition information for post-emergency analysis of faults and fault patterns—in a manner that parallels the use of a flight data recorder after an aircraft accident. To be fully meaningful, alarm condition information should document not only the process or machine conditions precisely related to time, but also the effects of operator interaction with the process during the period of emergency or near emergency.

Table 3 Annunciator Operational Sequences*

ISA Designation A

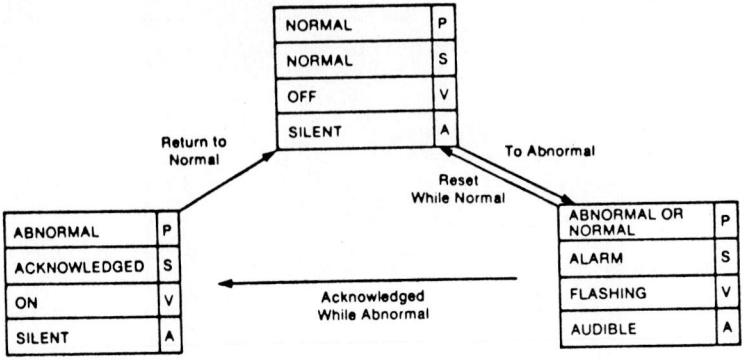

Sequence highlights: (1) "Acknowledge" and "Test" pushbuttons, (2) alarm audible device, (3) lock-in of momentary alarms until acknowledged, (3) audible device silenced and flashing stopped when acknowledged, (4) automatic reset of acknowledged alarm indications when process conditions return to normal, (5) operational test

ISA Designation M

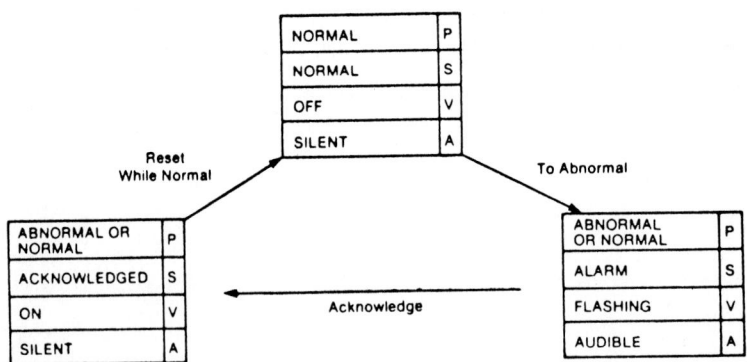

Sequence highlights: (1) "Acknowledge," "Reset," and "Test" pushbuttons, (2) alarm audible device, (3) lock-in of momentary alarms until acknowledged, (4) audible device silenced and flashing stopped when acknowledged, (5) manual reset of acknowledged alarm indications after process conditions return to normal, (5) operational test

Table 3 Annunciator Operational Sequences* (*continued*)

ISA Designation R-12

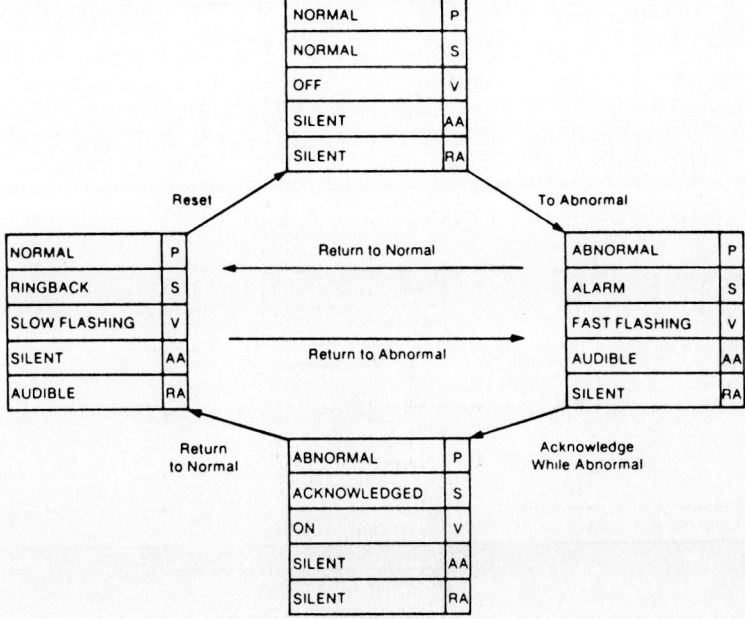

Sequence highlights: (1) "Acknowledge," "Reset," and "Test" pushbuttons, (2) alarm and ring-back audible devices, (3) audible device silenced and fast flashing stopped when acknowledged, (4) ring-back visual and audible indications when process conditions return to normal, (5) manual reset of ring-back indications, (6) operational test

ISA Designation F3A

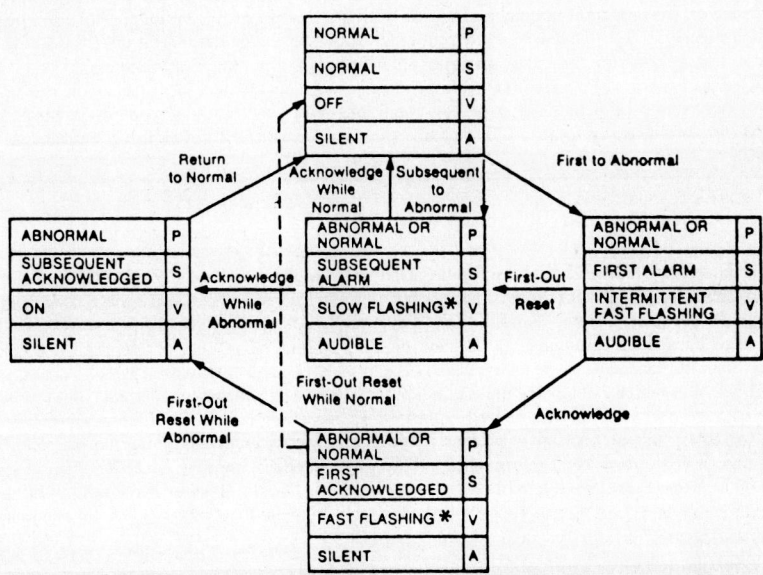

Table 3 Annunciator Operational Sequences* (*continued*)

Sequence highlights: (1) "Acknowledge," "First-Out Reset," and "Test" pushbuttons, (2) alarm audible device, (3) lock-in of momentary alarms until acknowledged, (4) "First-Out" flashing differing from subsequent flashing, (5) "First-Out" pushbutton for changing the "First-Out" visual indication to be the same as subsequent visual indications, (6) automatic reset of acknowledged alarm indications when process conditions return to normal, (7) operational test

RIS† Designation F2M-1 or FFAM2

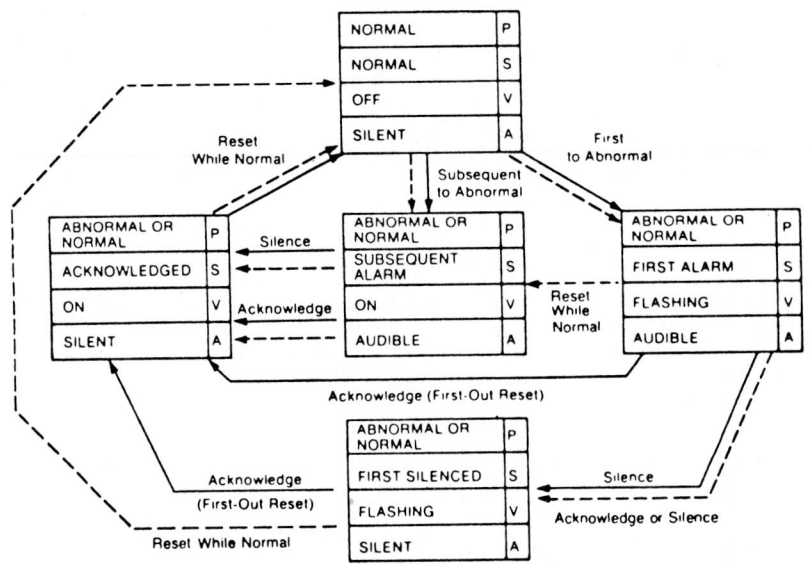

Sequence highlights: (1) "Silence," "Acknowledge," "Reset," and "Test" pushbuttons, (2) alarm audible device, (3) lock-in of momentary alarms until acknowledged, (4) option 1: "Silence" pushbutton to silence the audible device while retaining "First-Out" flashing indication, (5) flashing indication for first alarm only, new subsequent alarms having the same visual indication as acknowledged alarms, (6) "First-Out" indication reset when acknowledged, (7) manual reset of acknowledged alarm indications after process conditions return to normal, (8) operational test. (FFAM2 is similar, except that the "First-Out" point can be reset only after it has returned to normal.)

*P, Process; S, sequence; V, visual; A, audible; AA, alarm audible; RA, ring-back audible.
†Rochester Instrument Systems. Dotted lines indicate FFAM2.

It is obvious, of course, that alarm conditions arise from one or more causes (faults). Faults may develop slowly (e.g., heat exchanger fouling) but ultimately lead to a critical unsafe condition; or faults may occur in an instant (pump failure). There are hundreds of potential equipment faults that can lead to hazardous situations. Faults may develop within the process per se (e.g., distillation column flooding, inadequate reaction rates), the causes of which may be several—changes in throughput rates which fall outside the original process design parameters, inadequate initial process design, changes in the properties of feed materials, among many others. Routine measurements of single variables, unless potential hazardous conditions are fully understood, often do not reveal an impending crisis (a pressure or temperature a bit high), but two or more variables (a high pressure and a high temperature) may reliably forecast an unsafe condition. Two or more faults, acting in concert, may severely complicate not only the operator's immediate actions in a time of emergency, but postemergency fault analysis as well.

Thus modern thinking is leading to the development of alarm systems that not only fulfill their traditional objectives but also serve as *fault diagnostic tools*. Through careful and accurate documentation of preemergency, emergency, and postemergency conditions, including the effects of operator-process interactions, symptoms of faults (and impending faults) can be delineated on a scientifically sound basis. Special sequential event recorders (see section on recorders later in this article) have been developed as one important tool for obtaining the required documentation. The immediate value of determining patterns of symptoms and fault diagnosis is obvious—important corrections in process equipment and management can be made so that the probability of an emergency of like nature is significantly reduced. The long-term value is somewhat less obvious. By dedicating instrumentation to symptom and fault analysis (in essence, developing an in-depth understanding of process performance including the idiosyncrasies and eccentricities of the total system), the likelihood of future "surprises" in process behavior that could lead to an emergency situation is decreased. Modern computer technology can be very helpful in developing this understanding. Then, based on this knowledge, analytical and diagnostic procedures can be automated. An appropriate software program can be developed. With this enhanced role, the term "alarm system" no longer seems adequate—perhaps "accident prevention system" would more clearly define the system objective.

Related articles on this topic include those by Bristol and Wade (1976), Hanes (1980), and Hollands (1980), listed in the references at the end of this article.

Graphic Displays

Preamble

Control engineers two or three generations ago recognized the need to cut back on the ever-increasing size of control panels. The advent of vacuum tube electronics made it possible to increase display

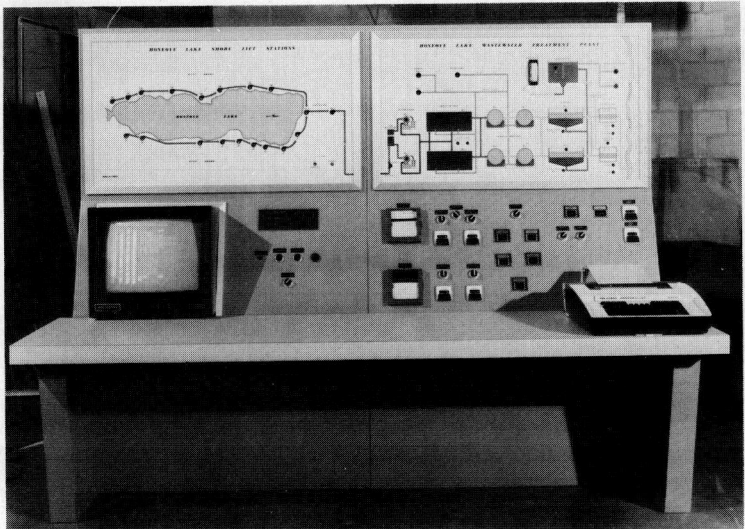

Fig. 31 Semitraditional color graphics panel installed at a wastewater treatment plant (Honeoye Lake, N.Y.). Two graphics panels are provided. One displays the remote lift station and pump station status (locations along the lake). The other graphics panel displays the local treatment plant status. A color CRT is provided to display complete remote station data. A solid-state teletype is provided to log any alarm condition as it occurs and to print out remote station data on demand. A common alarm panel is provided for treatment plant alarm monitoring from the fire hall when desired. See Fig. 32. *(Aquatrol.)*

density, but not to a significant degree. It remained for instrument manufacturers to become serious and competitive in an effort to reduce component and subsystem size (of mechanical, pneumatic, and electron tube instruments) through intensive research and development. Out of these efforts emerged the so-called miniature indicators, recorders, and controllers for panel mounting. Information density increased markedly, making it possible to use shorter panels for complex processes. Concurrently, new appraisals concerning the value of information developed, which led to a further increase in information density through the much greater use of multipoint recorders, loggers, and printers in connection with noncritical variables. Then, during a period of about a decade, thinking was directed toward the use of color and of graphics—techniques that went far beyond simply grouping instruments in an orderly fashion that roughly matched the process architecture per se. Bold steps were taken to "insert" the smaller format instruments into reasonable facsimiles of process flowcharts—to provide the operator with information traditionally available to installers and maintenance personnel, namely, the process instrument flow diagram.

Solid-state electronics permitted a further reduction in the size of display elements (a process that continues into the mid-1980s). This development has proved helpful because, as more knowledge was gained concerning human factors engineering, the concern switched from data too widely spread over a large area to too much information in a small space that could overwhelm an operator. When colors and graphics were first introduced in instrument panels, the panels were highly customized and graphics patterns were firmly a part of the panel (difficult to change). This concept did not allow the relocation of parts of a diagram or of instruments to accommodate the changes which are almost inevitable after a process has been on-stream for about a year. The result was many panels with a poor appearance that did not impress visiting VIPs very much. During the period prior to the advent of the CRT display, there were many innovations, but there remained a general feeling that the ultimate in graphic panels had not been achieved. It is true today, however, that well-designed graphics panels of this type persist, particularly in connection with processes that do not have a large number of control loops, processes that are reasonably stable with time so that panels need not be changed, and where there are no complex computer interfaces. See Figs. 31 through 35.

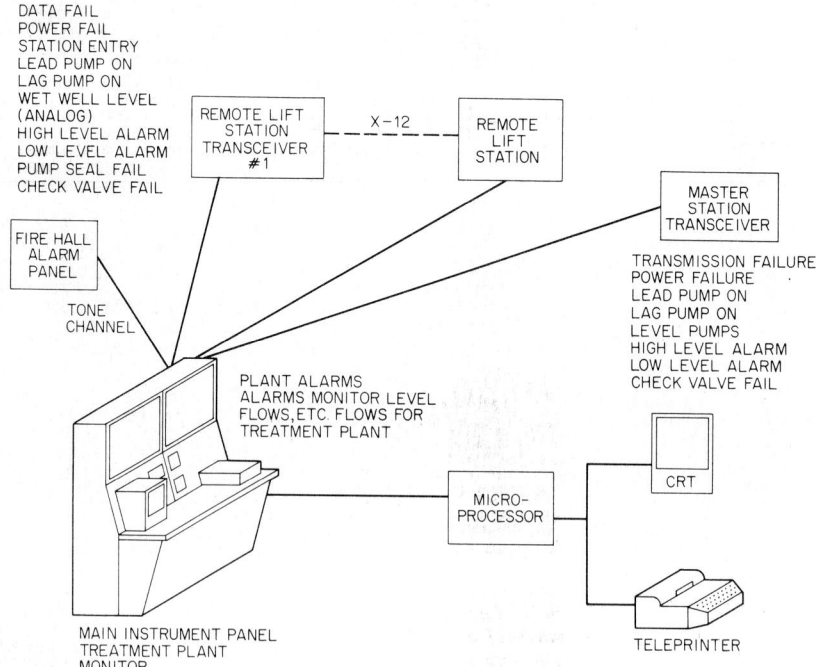

Fig. 32   Ties to the color-graphics panel installation illustrated in Fig. 31. *(Aquatrol.)*

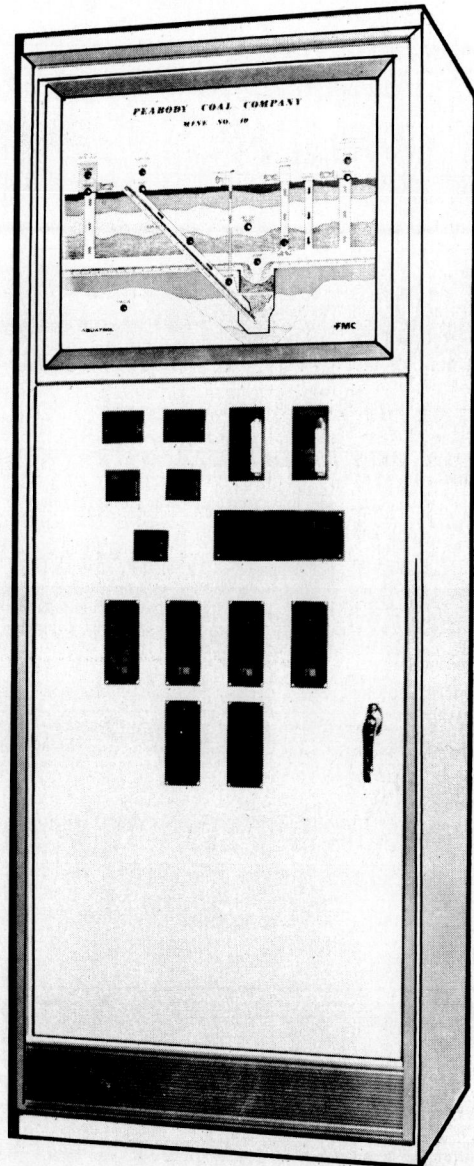

Fig. 33 Semitraditional graphics panel and control cabinet for monitoring a coal mining operation. *(Aquatrol.)*

It then followed, commencing in earnest in the early 1970s, that CRT displays were no longer considered limited to oscilloscopes and industrial television—particularly when highly reliable, large color tubes became a reality. As will be brought out later, the CRT has advantages not currently available from other types of displays. However, from the numerous discussions—pro and con—concerning the probable complete or ultimate role of the CRT in process instrumentation graphics,

Fig. 34 For some types of applications, users may prefer a neatly arranged grouping of indicating and alarm instruments with an associated separate CRT display. The system shown here continuously monitors 81 remote stations at a trona mine (FMC). The remote stations supervise 68 conveyor belt lines, 5 main power tie breakers, 4 hoists, and 4 shafts. Signals are digitally telemetered by a frequency-shift keying (FSK) tone to a master control panel over a single pair of telephone lines. The microprocessor drives a digital data logger which prints changes in status and alarms with the address and time. The master station displays and logs a total of 1100 status and alarm signals from a total of 81 remotes. The operator can transmit 239 command signals to the remote locations for start, stop, or reset of equipment. See also Fig. 35. *(Aquatrol.)*

it can only be concluded that the majority of users and operators, as of the mid-1980s, still do not feel that a utopian solution has been found. The present is a time of much experimentation and consequent growth in practical knowledge—hence, unlike the situation in an old technology, firm rules cannot be laid down. Obviously, a balance between CRTs and other display elements remains to be established. As in the case of annunciators, previously discussed, it remains to be determined during the years ahead whether the CRT should play the dominant role in an alarm situation with the traditional annunciators being subordinated, or vice versa; or whether there is an optimum use of both displays yet to be developed. As some authorities have observed, as of the present, we are just beginning to learn how to use the CRT display.

Computer Graphics

In some instrumentation systems and notably in connection with computer-aided design or simulation, a graphic input to a computer is the best or possibly the only way to communicate. In chemical research, for example, new compounds may be studied visually; molecular structures can be computer-generated with certain parameters controlled by the operator. Studies can be conducted in three dimensions; a new automobile design can be drawn on a CRT and then rotated so that it can be looked at from all sides. Generating visual data requires considerable software as well as hardware. One way to ease the computer load and also to provide a more realistic picture is to project slide or film images onto the face of a CRT and superimpose computer-generated information. This technique was pioneered for military command and control systems for displaying terrain maps and superimposing troop concentrations and equipment locations. The technique has scores of uses, including, for example, aircraft flight plan analysis. Much simplification has been accomplished in recent years through the incorporation of microprocessors in such systems.

A generalized block diagram for a CRT display system is shown in Fig. 36. Entry into the system is at the operator's console, which contains an alphanumeric keyboard for writing instructions of

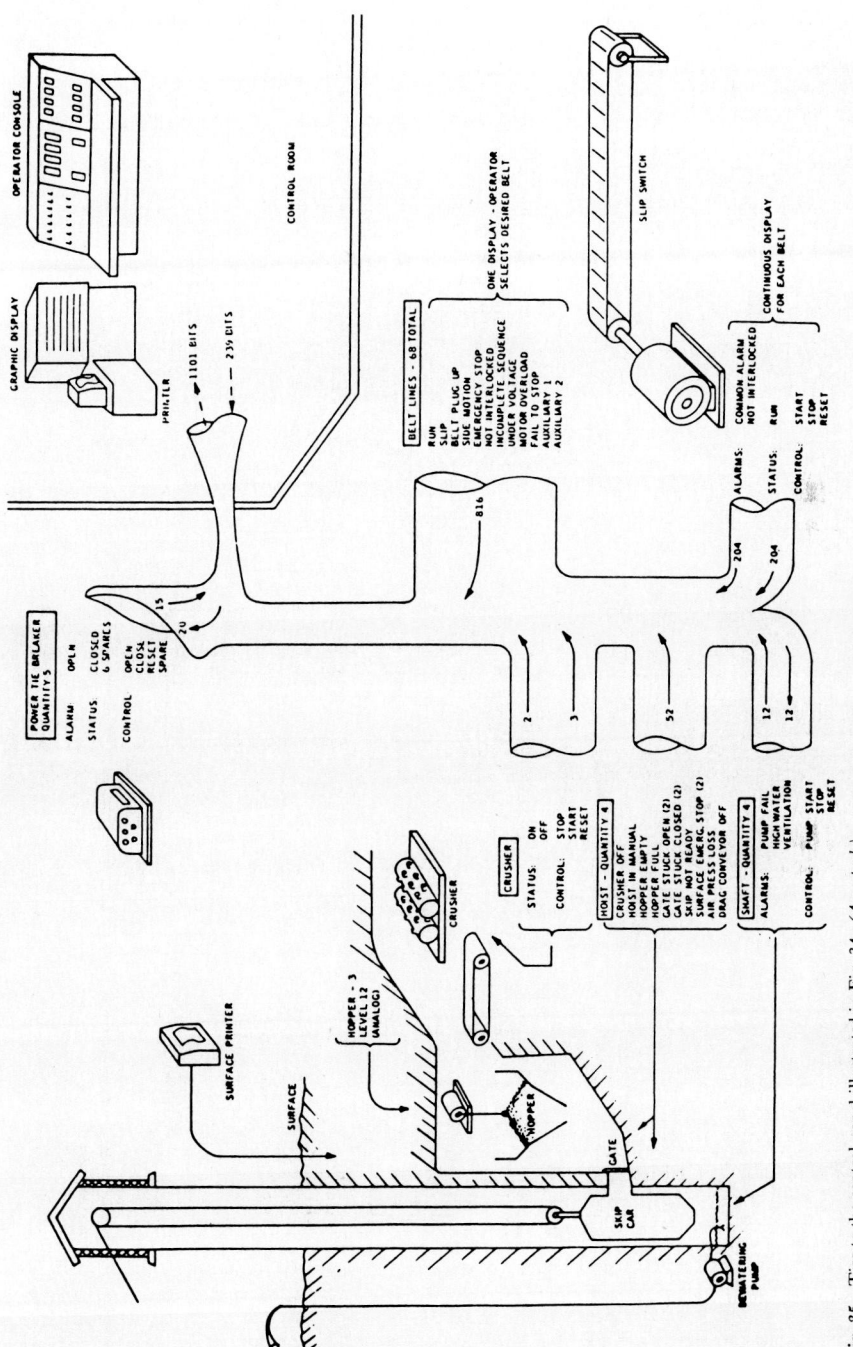

Fig. 35 Ties to the control panel illustrated in Fig. 34. (*Aquatrol.*)

15.37

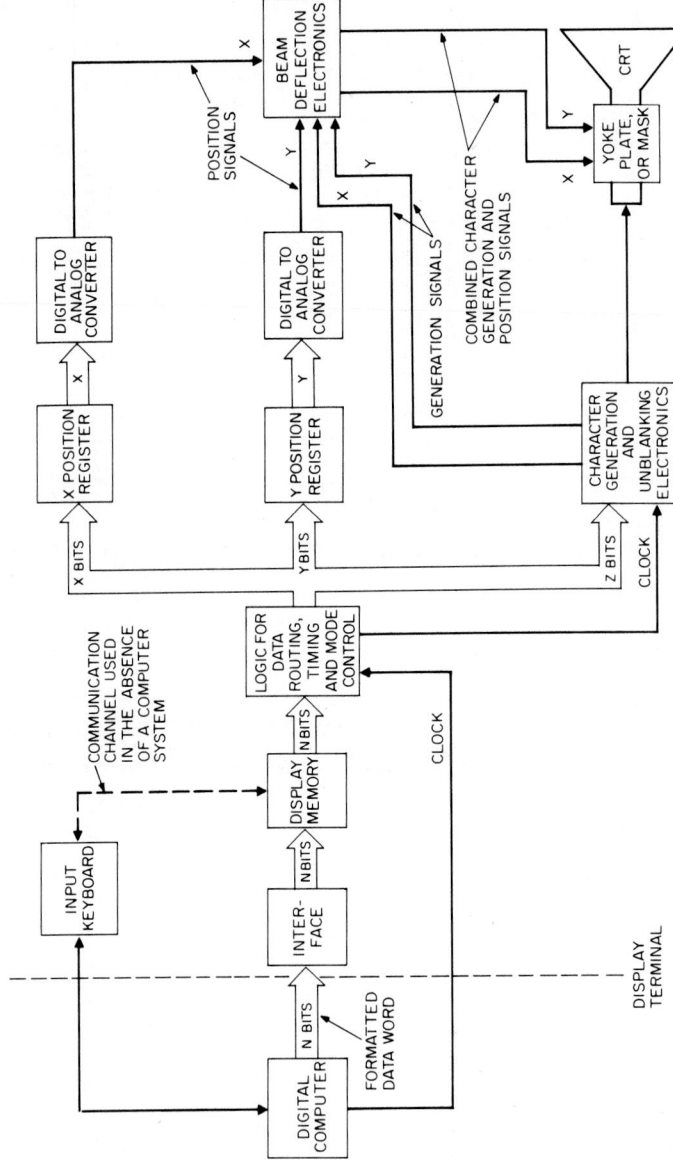

Fig. 36 Block diagram of a CRT display system. The system can be simplified to provide local computing with the use of a microprocessor-based data system.

entering data, a function selector to make inquiries or select operating modes, and a communications device. Light pens are sometimes used. With the proper device, the operator can point to information on the screen, designate a location at which information should appear, enter information, or possibly draw a line.

The operator can communicate with a digital computer that is part of the console or one that is located remotely. Simple data retrieval systems require only a memory bank from which data are extracted, edited, and reentered. Computer information enters the display console through an interface unit which conditions the input to ensure compatible logic levels and data word lengths. Additional signals are provided by the computer for sequencing computer signals with display generation circuits. Timing and mode control ensures the proper positioning of characters and symbols on the screen.

Among the types of displays that can be generated are point plotting, alphanumeric characters, and graphic or pictorial views. The simplest function is point plotting. This requires binary x and y addressing to position the CRT beam. Two output functions are needed to represent the plot adequately: position generation and intensity generation (blanking).

GRAPHIC RECORDERS

A *graphic recorder* can be traditionally defined as an instrument which, in some automatic fashion and with little if any manual intervention, draws a *graph*—a graph that may relate two or more variables to time, or to each other (typified by an *xy* recorder). A graph essentially is an analog representation of discrete information. Although, in the majority of contemporary recorders, the graph is an analog representation of data received in the form of analog signals (which vary with the measured variable), the chart, when read, is interpreted in terms of discrete values. The resolution of these values is determined by the fineness of the record produced and the coordinate background, coupled with the visual acuity and skills of the reader. Thus, the chart reader in essence performs the functions of an analog-to-digital (A/D) converter. In the case of a trend recorder, this manual A/D conversion usually is not important because the chart reader is essentially interested in the steadiness, upward trend, or downward trend or a variable and not in actual discrete values.

Large numbers of graphic recorders have been used for so long because practical experience indicates that the majority of persons find them compatible with their own nature as humans. A majority of individuals are inclined, probably more by intuition than training, to reach for a pad of paper and a pencil to sketch a crude curve when first confronted with a variable relationship that is somewhat difficult to comprehend immediately. These situations arise in business and in technology. When one thinks of a rising temperature or an increasing deficit, the mind's eye tends to see a curve rather than a tabular summary of data, whether hard copy or on the face of a CRT.

Because graphic recorders have been used for decades, the technology is highly advanced. Fortunately, too, most manufacturers of graphic recorders have kept pace and have adapted the latest developments of solid-state electronics, materials research, and other technologies to improve the performance of their instruments which have been refined over the years. Even more significant, manufacturers have added new features which greatly expand the fundamental curve drawing role of these instruments.

For many years, the selection of the best graphic recorder for a given application has required considerable analysis. But, with modern recorders, the numerous optional features have increased this task. A partial checklist of the factors involved in classifying and selecting recorders is given in Table 4.

From a commercial cataloging standpoint (as in a buyer's guide), graphic recorders are generally classified as (1) strip chart records, (2) circular chart recorders, and (3) special purpose recorders, such as *xy* recorders, event recorders, and oscillographs, among others.

Balancing means and circuits for recorders, notably strip chart recorders, are described in the article Self-Balancing Electrical Instruments in Sec. 14.

Graphic Coordinates

The majority of hand-drawn charts that relate variables one to another or to time are *rectilinear,* consisting of curves plotted on a square or rectangular coordinate matrix. So it is with instrumental

Table 4 Factors in the Classification and Selection of Information Storage Devices

Electronic information storage
  See the article Digital Technology in Sec. 14
Hard copy record making instruments
  Type of record
    Format
      Graphic coordinates
        Rectilinear
        Curvilinear
        Circular
      Tabular
      Log
    Marking
      Line
        Continuous
        Broken
          Dot
          Symbol
          Segmental record
  Number of records
    One (single-point)
    Multiple
      Simultaneous recording
      Time-shared recording
      Distinction between records
        Different colors
        Different symbols
  Relationship recorded
    Variable(s) versus time
    Variable versus one or more variables:
      xy, xx_1y, xyy_1
  Marker-medium system
    Ink on paper
      Pen
        Bucket
        V pen
        Ballpoint
        Cartridge
      Dot or symbol printer
        Fixed cycle
        Synchronized balance
      Pressure bar on inked ribbon
      Ink supply
        Manual
        Capillary feed
    Stylus on treated surface
    Moving beam and radiation-sensitive
      paper
    Mechanical printer
    Electronic printer
    Laser printer
  Calibrated scale
    Precision
    Readability
      Zoom device
    Trend recording
  Signal input, measuring and balancing
    system, range
    Analog
    Digital
    Electric
      Common- and normal-mode rejection
    Pneumatic
    Mechanical

Normal recording period
  Short (seconds or minutes)
  Intermediate (hours)
  Long (days and weeks)
  Varying requirements
    Flexibility of chart drive timing
Performance requirements
  Accuracy
  Dead band
  Response time
  Recording speed
    Flexibility of chart drive timing
Auxiliary functions
  Indication
  Alarms
  Event marking
  Control
    Electric
    Pneumatic
  Integration
  Totalizing
Case size
  Small (miniature)
  Intermediate
  Large
Mounting requirements
  Desk or bench top
  Panel
    Front of board
    Flush
Power requirements
  Servo system, chart drive, and illumination
    Electric
      Line
      Battery
    Pneumatic
    Spring
Application
  Centralized process instrumentation
    Large, complex process (many loops)
      Size constraint
        Graphic panel requirements
    Small, simple process (few loops)
  Uncentralized process instrumentation
    Local setting
  Testing (machines and equipment)
  Component of integrated measuring system
    (chromatograph, spectrometer, etc.)
  Uses requiring portability
    Field installations (pipelines, etc.)
    Aircraft and vehicles
  Accounting, inventory control, and other
    management information
  Sequential event recording
Environmental constraints
  Corrosive, dirty atmosphere
  Explosive condition
  Vibration and shock
  Position sensitivity
  Excessive temperature
  Excessive humidity
  Radiation exposure

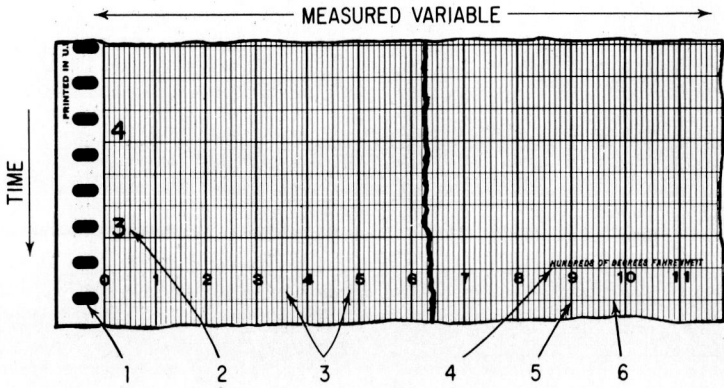

Fig. 37 Rectilinear coordinates. Shown here (reduced) is a section of an 11-in (279-mm) strip chart. (1) Humidity compensating holes, (2) major time markings, (3) minor graduations, (4) identifying calibration, (5) major graduations, (6) minor time markings.

recorders, the strip chart recorder using rectilinear coordinates found in numerous applications ranging from laboratory research to industrial process control. Because of geometric constraints imposed by certain meter movements, e.g., in a galvanometer, the *curvilinear* format was developed and is still used in some electric meters and laboratory instruments. This coordinate system matches the circular arc movement of the meter pointer, along curvilinear lines, with straight lines in the other

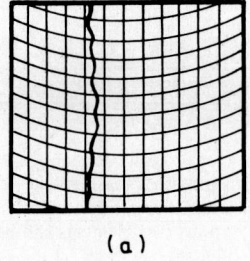

(a)

Fig. 38 Curvilinear coordinates. (*a*) Close-up of a curvilinear record, (*b*) barograph in which a drum- or cylinder-type recorder is used. The record is curvilinear. *(Bendix.)*

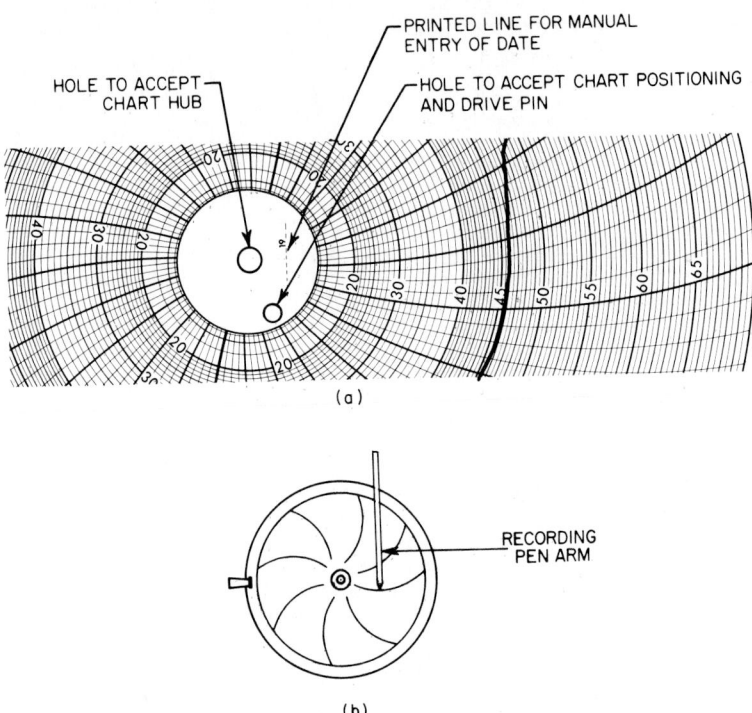

Fig. 39 Circular chart coordinates. (*a*) Section of a 12-in (305-mm) circular chart (reduced); (*b*) highly generalized representation of a round or circular chart recorder.

direction. *Circular chart* coordinates were probably developed early in recorder history to take advantage of the availability and convenience of spring-wound clock and synchronous motor movements to drive the chart in a circular direction. Although they continue to be popular for certain applications, particularly where mechanical measuring elements can be directly connected to the recorder as signal inputs, the popularity of circular chart or round chart recorders has diminished considerably during the past several decades. Example of the three coordinate systems are given in Figs. 37 to 39.

Strip Chart Recorders

From a commercial cataloging standpoint, strip chart recorders are frequently grouped by (1) the number of points recorded (single or multiple); (2) the type of marking-medium combination used (pen-ink-paper, ink-printwheel-symbols, stylus–specially treated surfaces, among others) or the type of marking made (continuous or broken lines, dots, or symbols); (3) the length of the calibrated scale (short, long), which is closely tied to (4) the case size (large, intermediate, small, or miniature); (5) the performance in terms of accuracy, speed, and other parameters (highly precise, fast, or standard and satisfactory for most uses of lower quality, and low cost for routine, noncritical applications); (6) the availability of auxiliary functions (separate indications, alarm switches, event markers, ability to integrate or totalize, and a full control ability—pneumatic or electric). Usually in the latter case, the instrument is identified as a controller rather than a recorder, with the term "recorder-controller" commonly used. (7) Further, recorders may be classified in terms of their final end use—for local mounting, for process control panels, for incorporation with other instrument systems (a chromatograph or other analytical instrument), or for portable use. Categories of applications and environmental constraints, among other recorder parameters, are listed in Table 4.

Marking-Medium Combinations

At one time, strip chart recorders were almost exclusively of the inked-pen-and-paper type. Over the years, many advancements have been made in the design of and the materials used for pens, and inks, and chart papers have also been improved. Concurrent with these developments, there have been innovations in the manufacture of specially treated papers that are sensitive to an electric current, to thermal stimuli, or to a radiation beam—so that today a significant number, but not the majority, of recorders are inkless.

Pen and Ink

Early pen-and-ink recorders replaced the even earlier use of instruments which scratched records on a smoked surface, particularly in seismographic recorders. There are several types of pens, including the bucket pen, the V pen, the fiber-tipped pen, and the ballpoint pen. Long popular were various types of capillary feeding systems, both pressurized and gravimetric. Refilling has been made more convenient through the use of ink cartridges and replacement pen assemblies. Disposable ink cartridges and pens are widely used today.

Impact Printing

This is a special variation of the ink-paper system. In very early versions, a moving pointer was clamped in position for a slightly prolonged instant, and a presser bar above the pointer mechanism pressed down on a carbon ribbon located between the paper chart and the pointer. Carbon markings served as the record. The system was once very popular for use in millivolt pyrometers. Impact printing was refined into considerably sophisticated systems several decades ago when the print wheel was first adapted to this use. In this system, an engraved wheel prints a number or symbol that identifies each variable being measured. The print wheel, like a pointer, travels from one end of the scale to the other. Accompanying the print wheel is an ink pad wheel which keeps the wheel adequately inked. The system* is available in large-format recorders with an 11-in (279-mm)-wide strip chart. In any process, up to 20 variables, such as temperature, that can be converted to dc millivolt signals, can be monitored and recorded essentially simultaneously. The record point selection and printing function are mechanically driven. A commutator switch connects the circuits in sequence. Records can be of a single color (purple) or multicolored (purple, red, black, blue, green, and brown).

In *fixed-cycle* printing, the printing mechanism produces a sequential recording of process variables in a fixed time cycle. In *synchro-balance* printing, a special clutch governs the printing cycle of each point in relation to the time required for the servo system to reach a balance at that point. This reduces the time required to record rapidly changing variables because there is no appreciable delay in printing action after the balance is reached at each point.

Thermal Writing

One of the earliest inkless techniques to be introduced, this method involves thermal writing elements which are actuated by electric currents and produce output traces on heat-sensitive paper. Although this technique was initially slow to be accepted, improvements have reached a state where this method of writing is offered by a number of manufacturers. Like other inkless methods, thermal writing advancements have tended to follow progress made in the calculator and business machine field.

In one contemporary system,† the printing element is a fixed, solid-state thermal metal and ceramic bar, 4.2 in (107 mm) wide. The bar contains 420 printing elements through which heat is applied to sensitized paper. Each element, separately controlled, generates a 100-dots per inch (~4-dots per millimeter) trace on a 4.5-in (114-mm)-wide paper strip. It is claimed that the special heat-sensitive paper is not prone to gumminess and does not smear when normally handled. In the recorder adapted for this system, multipoint traces up to six channels are identified by letter characters. There are also periodic printings of the date and time.

---

*Exemplified by the Honeywell Electronik 15 multipoint strip chart recorder.
†Tigraph, Texas Instruments Geophysical and Scientific Division.

Electric Writing

In this system, the electric writing paper is in the form of a dense, black substrate coated with aluminum. A tungsten wire stylus is in very light contact with the aluminum surface. As energizing voltage is applied to the stylus, the aluminum is etched, thus exposing the black substrate. A voltage of 35 V dc minimum is required. The energy is derived from a free-running oscillator driving a voltage tripler. To avoid possible grounding through the aluminized chart, the supply current is continuously interrupted at approximately 8 kHz, a frequency high enough to avoid observable effects on the chart even at the fastest combined pen and chart speeds. The pattern is initiated by movement of either the stylus or the chart; when the chart is in motion or when the slidewire voltage changes by more than 1% full-scale, the pattern is activated. See Fig. 40. In multiple-point recorders, the line patterns are varied as shown in Fig. 41.

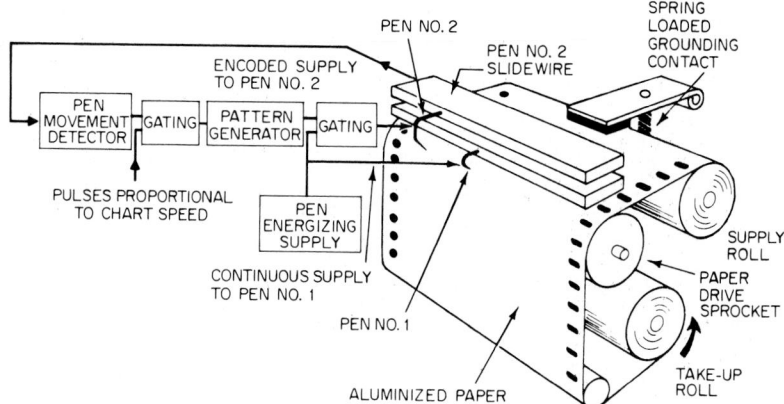

Fig. 40 Trace encoding in a two-pen electric writing strip chart recorder and message printer. *(Chessell, Eurotherm International.)*

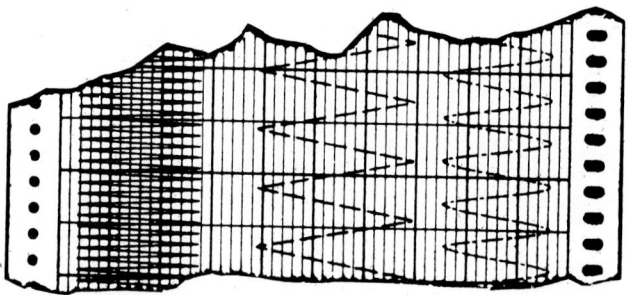

Fig. 41 Sample of a record produced by three-channel electric writing. Note how the channels are distinguished from each other by the use of solid lines, dashed lines, and dashed-dotted lines. Additionally, instruments are available which print alphanumeric messages, the date, and the time. The messages, written by the user and stored in programmed read-only memory (PROM), summarize key events in the monitored process, such as "hi pres alarm," "fuel low," "wash cycle ends." Up to 16 event messages, each of up to 15 characters, can be recorded in the nonvolatile storage. When initiated by an external command, the selected message is printed, together with the date and time, if required, by a scanning process which develops alphanumeric characters in dot matrix form. *(Chessell, Eurotherm International.)*

Light Beam Methods

The use of a mirror mounted on a galvanometer to amplify the scale by a large factor (depending on the distance between the mirror and the calibrated scale), particularly useful for the type of precision required in a laboratory, has been known for a number of decades and, in fact, similar principles are used in contemporary oscillographic recorders, as will be described later. Just one step is required in going from optically projected indication to recording. Some recorders are offered today that depend on ultraviolet rather than visible radiation, essentially eliminating the effects of ambient light. In one system, a programmable light gate array is used to record directly on ultraviolet-sensitive paper. The system* has been claimed to free the recorder from the problems of linearity, beam deflection, tangential error, overshoot, and inertia, among other factors which have been recurring problems in the case of galvanometric and CRT oscilloscope devices.

Electrostatic Writing

In this system,† there are three elements in the writing system—an imaging head, a toning head, and a vacuum knife. As reported, the imaging head is a linear array of 1000 wire elements, spaced 4 wires per millimeter for a total length of 250 mm. On each side of the array, there are 32 copper bars, referred to as "shoes." As the paper moves over the image head, a negative voltage is applied to selected wire elements and a positive voltage is applied to the closest shoes. This results in a positive point charge on the paper at the "written" points. The paper then passes the toner head, and negatively charged ink particles adhere wherever the paper had a positive charge. A vacuum knife then removes all excess toner and particles, leaving the charged-particle image. On exposure to air, the adhesive-coated particles permanently bond to the paper and the record emerges dry.

Recorder Servo Systems

The nature and quality of the pen, print wheel, or stylus driving system largely determine the performance parameters of a recorder. Most recorders incorporate some form of a self-balancing bridge or potentiometer circuit with pen-position feedback control. Systems of this type are described in some detail in Self-Balancing Electrical Instruments in Sec. 14. Useful information, particularly concerning unusual drive schemes, can be found in the Hulshizer and Kurimski (1980) and Morris (1980) references listed at the end of this article.

Standard Strip Chart Recorders

Recorders for usual applications are described in the next few paragraphs. Recorders for special applications are described later.

Shown in Fig. 42 is a grouping of contemporary single- and dual-pen strip chart recorders,‡ some with 6-in (152-mm) calibrated charts, particularly suited for mounting on panels with a high density of information display, and some with 11-in (279-mm) charts, to be used where readability is important, and space is not a constraint. Design highlights are as follows. (1) Servo electronics, input cables, and measuring circuits are enclosed in a full floating Faraday shield to protect against the effects of common-mode stray on the input circuit. (2) The servo amplifier is solid state and includes a transistor-type chopper. Components are mounted on a single printed wiring board. Gain adjustment is provided on all amplifiers. (3) The servo potentiometer has a conductive plastic slide-wire and a collector element consisting of four-fingered gold-alloy contacts. (4) A cascade zener diode network supplies a constant voltage for the precision thick-film resistor network on the measuring circuit. Adjustable zero and span adjustments for calibration are accessible from the front of the instrument. (5) The chart transport unlatches and rotates forward for loading while the instrument is operating. (6) The cartridge pen used is designed to write a minimum of 3000 ft (914 m) of line. (7) The control error signal generator senses any difference between pen (process variable) and index (set-point) position and generates an error signal proportional to the differences for the control mode used.

*Bell & Howell, CEC Division.
†Gould Inc., Instrument System Division.
‡Honeywell 111 and 112 strip chart recorders.

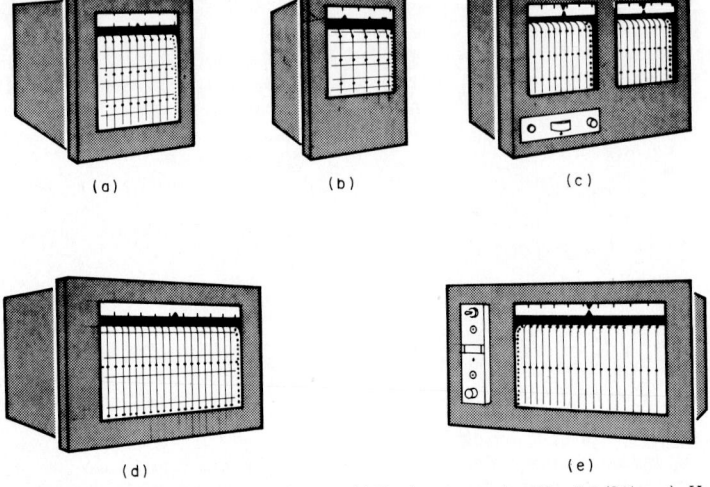

Fig. 42 Various strip chart recorder formats. (*a*) Single-pen recorder: W = 9.5 (241 mm); H = 10.5 in (267 mm); D = 13.6 in (346 mm). (*b*) Two-pen or single-pen dual-range recorder: W = 9.5 in (241 mm); H = 14 in (346 mm); D = 13.6 in (346 mm). (*c*) Dual-case (combined strip and circular chart configuration also available): W = 17.5 in (444 mm); H = 13.4 in (341 mm); D = 13.6 in (346 mm). (*d*) Single-pen single- or dual-range recorder: W = 15.5 in (394 mm); H = 10.5 in (267 mm); D = 13.6 in (346 mm). (*e*) Single-pen recorder with a three-mode electric control unit: W = 19 in (482 mm); H = 10.5 in (267 mm); D = 13.6 in (346 mm). W = width; H = height; D = depth. (The depth dimension given is that of the back of the panel. Disposable fiber-tiped pens and ink cartridge (blue or red) pens are used. Chart widths: (*a* to *c*) 6 in (152 mm) calibrated; (*d* and *e*) 11 in (279 mm). All formats are available with electric contact, electric proportional, or pneumatic proportional control. (Patterned after Honeywell Models 111 and 112.)

Specifications include:

Accuracy:

Spans 5 mV or greater: $\pm(0.3 + 0.1 \times$ suppression ratio)% of span, or 0.20 mV, whichever is greater

Spans less than 5 mV: $\pm(0.3 + 0.1 \times$ suppression ratio)% of span, plus 0.005 mV

Resistance thermometer: $\pm(0.3 + 0.01 \times$ suppression ratio)% of span (ohms)

Thermocouple input: $\pm(0.35 \pm 0.05 \times$ suppression ratio)% of span or $\pm[4.5/\text{span (mV)} + 0.05 \times$ suppression ratio]% of span, whichever is greater (for types J, K, T, E, and W5W26 thermocouples)

Thermocouple input: $\pm[3.0/\text{span (mV)} \pm 0.05 \times$ suppression ratio]% of span or $\pm[3.0/\text{span (mV)} + 0.05 \times$ suppression ratio]% of span, whichever is greater.

Reproducibility: $\pm0.15\%$ of span (2r hour)

Dead band: 0.1% of span

Actuation and input range:

Thermocouple, 1 to 100 mV (single range); 5 to 100 mV (dual range), suppression, 2 times span. Universal reference junction compensated within ±0.014 mV for types E, J, K, T, and W5W26; ±0.0025 mV for types R and S Radiation pyrometer, 1 to 100 mV, suppression, 2 times span

DC current: 5 to 100 mA (single range); 2 to 100 mA (dual range), suppression, 2 times spand

DC volts: 0.1 to 200 V (single range); 0.1 to 10 V (dual range), suppression, 2 times span

RTD: 3 to 1000 Ω, suppression, 20:1 ratio

Span and zero adjustment: $\pm10\%$ of span

Set-point travel: 4 to 96% of scale

Chart speeds (½ and 2 times change gears furnished):

Single speeds (in/h)	(in/min)
1, 2, 6, 10, 12, 20, 30, 60	2, 4, 6, 10, 12, 20, 30

Two speeds (in/h)
1 and 10, 2 and 20, 6 and 60, 12 and 120, 0.5 and 30, 1 and 60, 2 and 20

Note: 1 in/min = 25.4 mm/min; 1 in/h = 25.4 mm/h

Shown in Fig. 43 is a miniature strip chart recorder, available in two widths—wide format, 6 × 6 in (150 × 150 mm), and narrow format, 3 × 6 in (78 × 150 mm). Each format is furnished with

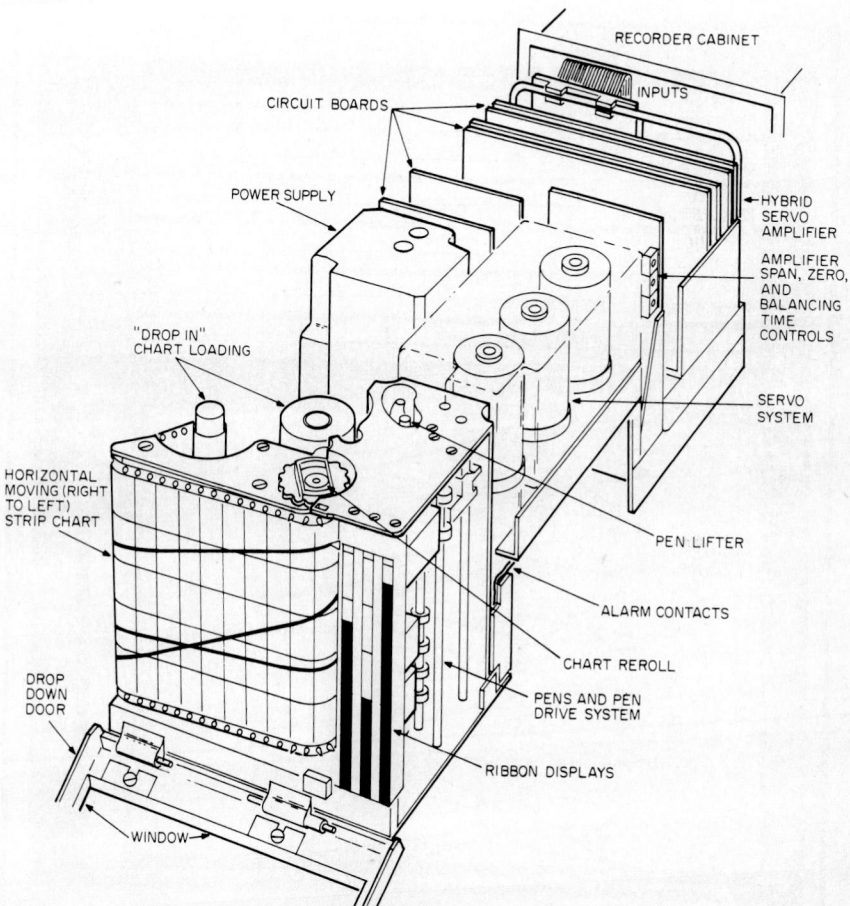

Fig. 43 Simplified diagram of the front and chassis of a miniature strip chart recorder available in a wide or a narrow format for panel mounting. The wide-case format is shown here: W = 6 in (150 mm); H = 6 in (150 mm); D = 14 in (347 mm). Narrow case: W = 3 in (78 mm); H = 6 in (150 mm); D = 24 in (610 mm). W = width; H = height; D = depth. Calibrated scale is 4 in (100 mm) (vertical). Up to three variables may be recorded, and indicated by vertical ribbon displays located to the right of the recording chart. Completely isolated channels accept inputs from thermocouples, resistance temperature detectors (RTDs) (thermohms), thermal radiation detectors, conductivity cells, and a variety of other millivolt sources and standard high-level current and voltage signals. (Diagram based on Leeds & Northrup Speedomax 100 Series.)

three pens and three ribbon-type vertical indicators located just to right of the recorder portion of the instrument. Design highlights are as follows. (1) The servo assembly and amplifier couple a precision dc motor to permanently lubricated drive gears. There is a permanently lubricated conductive-plastic slidewire in an enclosed module. Mechanical end stops are built into the servo module, and the amplifier includes electronic off-scale protection. Up to two alarm contacts may be mounted on the servo module's output shaft. (2) The key to the servo amplifier is a hybrid package which is subjected to a 168-h 140°C "burn-in" period in the interest of long-term performance reliability. (3) Charts are approximately 20 ft (exactly 16 m) long to permit continuous recording for 33 days at a speed of ¾ in (2 cm)/h. Markers are disposable and are furnished in three colors (red, blue, and green).

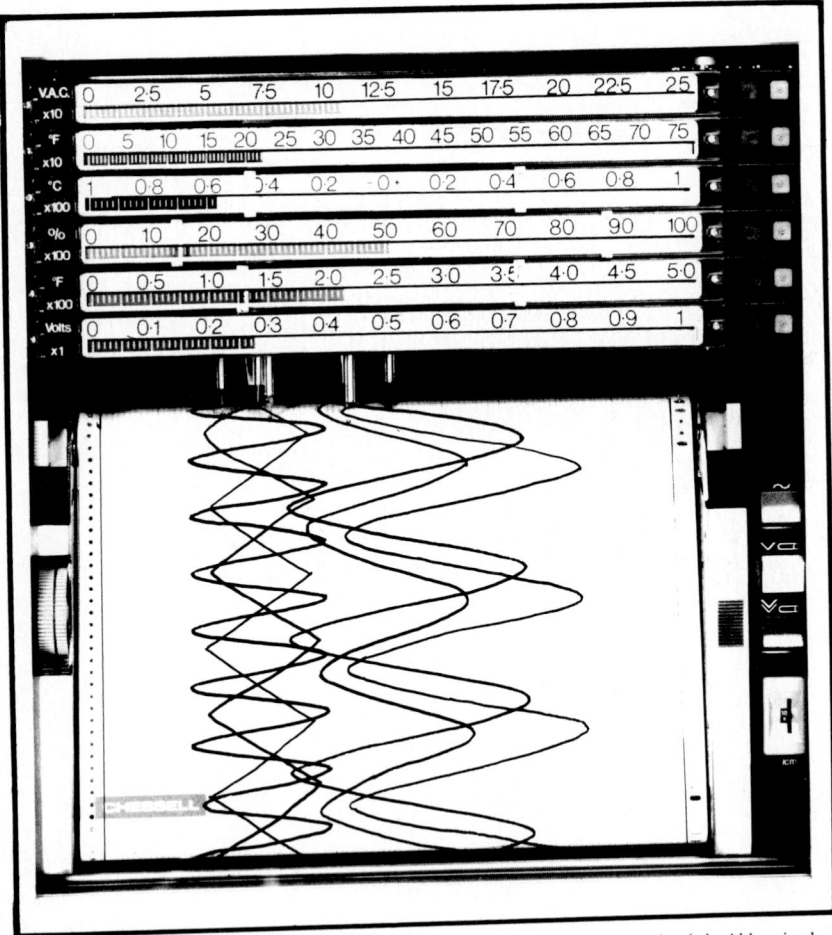

Fig. 44 Multipen recorder with up to six fully overlapping pens on a 10-in (250-mm) ruled-width strip chart. Indicators are colored moving ribbons. Matching ribbon and ink colors are green, red, blue, black, violet, and brown. The writing system is capillary ink or a disposable cartridge with individual pen lifters. The chart drive is line-synchronized electronic. Low speed (LS) and high speed (HS) versions each provide up to 10 switch-selectable speeds. The chart drive is roll or Z-fold. Electronic alarm relays can be fitted to each recording channel. Event markers are optional. (Diagram based on Chessell Model 320 recorder.)

Specifications include:

Accuracy: ±0.5%. High-level circuits without a preamplifier and with spans below 500 mV have ±1% accuracy

Dead band: 0.2% maximum

Response time: 1 to 10 s, adjustable

Chart speeds: 1 or 2 synchronous motor speeds, or a 14-speed stepping motor drive

A multipen recorder with up to six fully overlapping pens on a 10-inch (250-mm) ruled-width strip chart is shown in Fig. 44.

Hybrid Recorders

A microprocessor-based high speed recorder which accepts a variety of analog signals up to 30 channels and converts them to digital signals for printing as dot traces, plus incorporating a digital display, is shown in Figs. 45 through 48.

Circular Chart Recorders

A representative single-pen, direct acting circular chart recorder is shown in Fig. 49. Because they have been used for several decades, circular or round chart recorders have reached an advanced state of engineering. The format is generally restricted to 12-in (305-mm)-diameter charts, although instruments that use somewhat smaller charts are available. Circular chart recorders were much

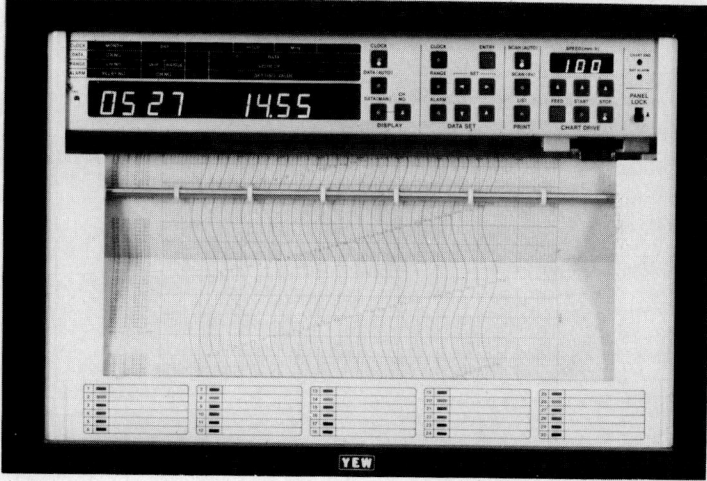

Fig. 45 Identified as a hybrid recorder by the manufacturer, this instrument is a microprocessor-based high speed recorder which accepts a variety of analog signals up to 30 channels. Inputs include dc voltage, thermocouples, and resistance temperature detectors (RTDs). Analog inputs are converted to digital signals and printed as dot traces and digital data in six colors by way of a stepping motor and a high speed wire dot printer. The full-scale range, chart speed, alarm set points, skip, and clock are programmed by means of the keyboard located on the front panel. The effective recording span is ~9.8 in (250 mm). The basic accuracy claimed in ±0.25% of span. The print cycle time (scan) is 8 maximum, and the printing rate is 30 channels (8 s). The chart is Z-fold, marked in 100 uniform divisions. The instrument is 17½ in (444 mm) wide and 11⅜ in (288 mm) high and extends 13¾ in (350 mm) in back of the panel. The left margin of the chart provides space for digital printout of channel numbers and data; similarly the right margin provides space for digital printout of alarms. (ER 250, *Yokogawa Corporation of America.*)

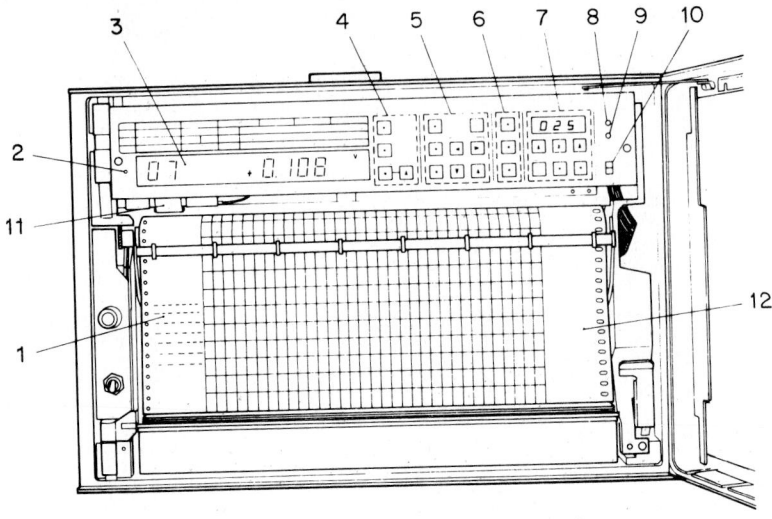

Fig. 46 Operator-instrument interface of a hybrid recorder. The principal functions of the displays and the keyboard used in observing and programming the instrument include the following. (1) Digital measurement data are printed in the left margin of the chart. Printing is executed automatically at a fixed time interval (coupled with selection of the chart speed). (2) Fail lamp indicates a fail condition of the recorder. A fail condition also produces a contact output (open signal) on the rear of the panel. (3) Digital display. (4) Display selector buttons for each of three displays: (a) clock; (b) channel number, alarm sign H (high) or L (low), polarity sign (+ or −); (c) manual change of display sequentially from the first to the last channel. (5) Data set buttons: (a) clock, (b) range, (c) alarm, (d) set. (6) Scan selector and list buttons: (a) scan (auto)—the scan interval is automatically changed in response to the programmed chart speed; the chart advances 0.25 mm in a single scan; (b) scan (8-s)—the scan interval is fixed at 8 s; (c) list—all program memory is listed on the chart. (7) Chart speed selector buttons and chart-drive buttons. (8) Chart end lamp—indicating an end-of-chart condition by actuating an LED indicator. (9) Battery alarm lamp—indicating low battery power. (10) Key lock selector—when selector is in down position, program entries are not accepted from the keyboard. (11) Printer. (12) Right-hand margin of chart is used to print alarm conditions, including relay number, H or L sign, and time of out-of-limit occurrence. Operator may call up a program list printout as shown in Fig. 48. (ER 250, *Yokogawa Corporation of America.*)

more popular during the period prior to the consideration of graphics panels and the desire for greater information density on panel boards. Standard circular chart recorders are rather large, on the average measuring about 15.5 in (394 mm) in width, 18.5 in (470 mm) in height, and 5.5 in (140 mm) in depth.

Circular chart recorders are particularly suitable for direct actuation by a number of mechanical-type sensors, such as bellows, bourdon tubes, filled-system thermometer bulbs, and differential-pressure elements as used in flow and liquid-level measurements—all accomplished *without* the need for transducers which generate electric signals. Because of the lower cost and relative ease of installation in certain applications, circular chart recorders continue to find acceptance for incorporation into small process control panels where information density is not a problem, for local mounting on machines and in processes in situations where there is no centralized control, for pipelines, and in other similar field situations. Circular chart recorders are available with auxiliary electric switches and as pneumatic controllers for all control modes. For isolated situations, circular chart recorders can operate on spring power.

The recorder shown in Fig. 49 has a helical pressure tube as the measuring element. The operating mechanism in this particular design comprises levers *a* and *b* and link *c*. These parts convey motion from the measuring element to the recording device (pen arm and pen). Numerous other designs have been developed. The recording of two or more separate variables on one chart by a single instrument poses particularly difficult design problems—because of the need to maintain a correct relationship between each pen and the chart movement (which is not alterable). Space con-

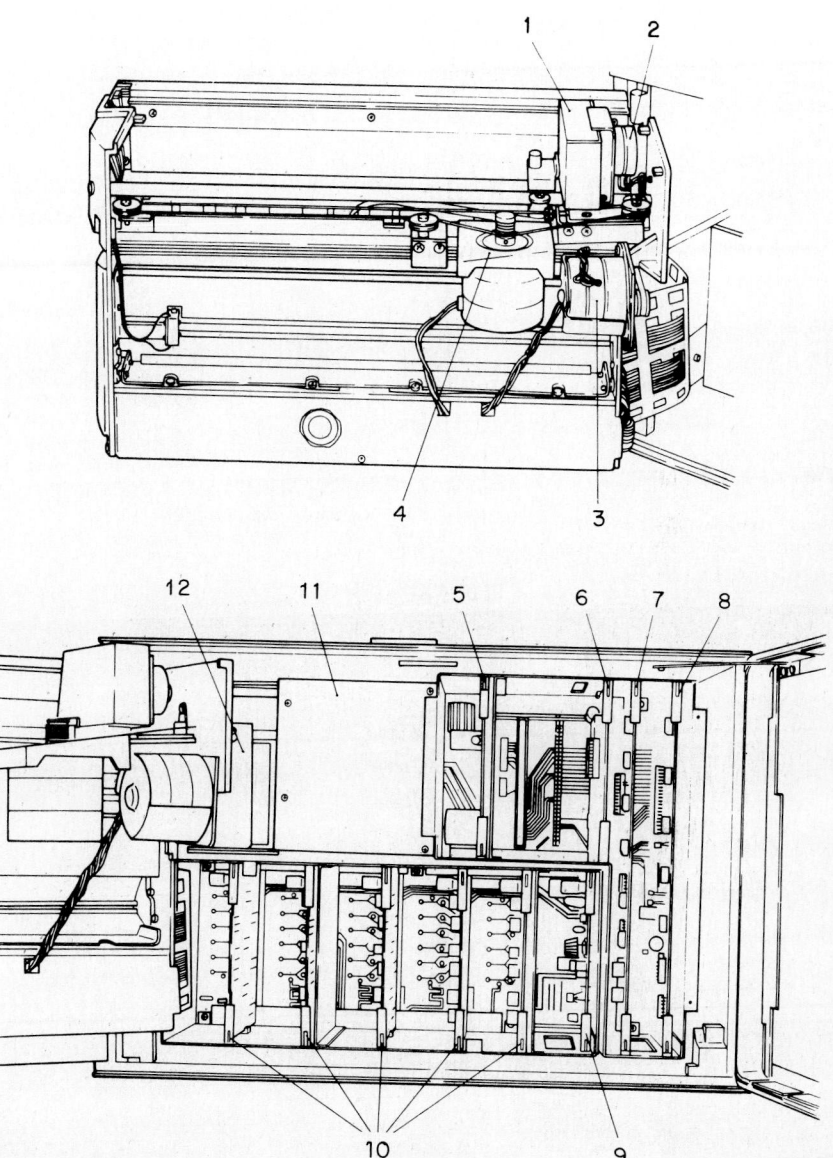

Fig. 47 Principal components of a hybrid recorder: (1) ribbon cassette containing a six-color ribbon which is serviceable for ~6 months when the recorder is continuously operated at 25 mm/h; (2) ribbon-drive motor; (3) chart-drive motor; (4) carriage drive motor; (5) power card; (6) memory card; (7) interface card; (8) central processing unit (CPU) card; (9) analog-to-digital (A/D) card; (10) scanner card; (11) power transformer; (12) battery. (ER 250, *Yokogawa Corporation of America*.)

RELAY NO.	CH	L/H	SET VALUE		RELAY NO.	CH	L/H	SET VALUE
01	01	H	65.0°C		02	02	H	-499.0
03	03	H	-1.00MV		31	01	L	60.0°C
32	02	L	+500.0		33	03	L	-2.00MV

DATE 06/03 18:26   CHART SPEED 23 MM/HOUR

CH RANGE	LEFT END	RIGHT END		CH RANGE	LEFT END	RIGHT END
01 K	-100.0°C	100.0°C		02 0-10MV	-500.0	500.0
03 20MV	+1.50MV	14.50MV		04 20MV	-2.00MV	14.00MV
05 20MV	+2.50MV	13.50MV		06 20MV	-3.00MV	13.00MV
07 20MV	+3.50MV	12.50MV		08 20MV	-4.00MV	12.00MV
09 20MV	+4.50MV	11.50MV		10 20MV	-5.00MV	11.00MV
11 20MV	+5.50MV	10.50MV		12 20MV	-6.00MV	10.00MV
13 200MV	+65.0MV	95.0MV		14 200MV	-70.0MV	90.0MV
15 200MV	+75.0MV	85.0MV		16 200MV	-80.0MV	80.0MV
17 200MV	+85.0MV	75.0MV		18 200MV	-90.0MV	70.0MV
19 20V	-9.50V	6.50V		20 20V	-10.00V	6.00V
21 20V	-10.50V	5.50V		22 20V	-11.00V	5.00V
23 20V	-11.50V	4.50V		24 20V	-12.00V	4.00V
25 20V	-12.50V	3.50V		26 20V	-13.00V	3.00V
27 20V	-13.50V	2.50V		28 20V	-14.00V	2.00V
29 20V	-14.50V	1.50V		30 20V	-15.00V	1.00V

Fig. 48 Recorder program can be printed out on a chart when called for by the operator. (ER 250, *Yokogawa Corporation of America.*)

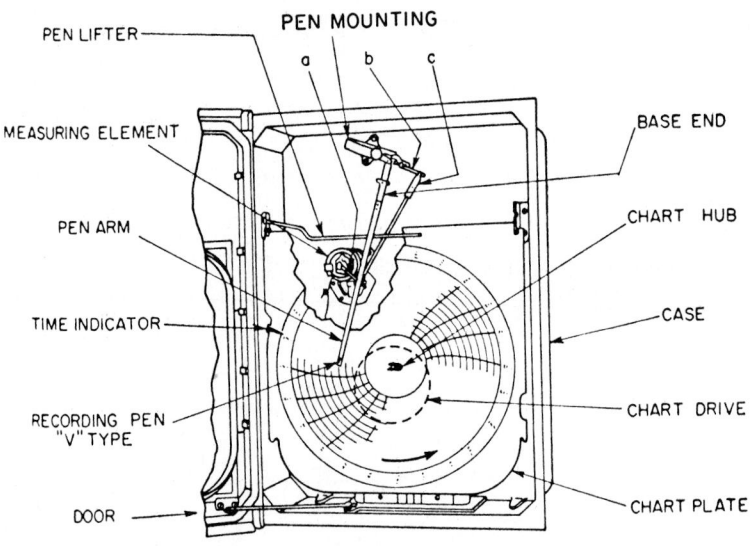

Fig. 49 Single-pen direct acting circular chart recorder.

straints in the usually shallow case also are severe. There are two basic approaches: (1) concentric shafts, where all pen arms pivot about a common center, as shown in Fig. 50, and (2) nonconcentric designs, where the pen arms are pivoted from different locations but still maintain their proper spatial relation with the chart, as shown in Fig. 51. In either design, obviously there must be sufficient clearance between the individual pen arms and pens. Because of these problems, over the years, the circular chart recorder has generally been limited to recording a maximum of four variables, although there have been exceptions. Sometimes segmental charts, as shown in Fig. 52, have been used to handle several variables with one instrument.

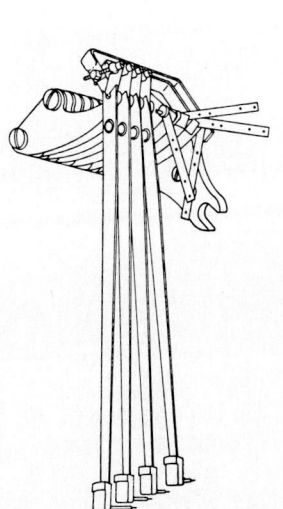

Fig. 50 Concentric pen-arm assembly sometimes used in multiple-pen circular chart recorders.

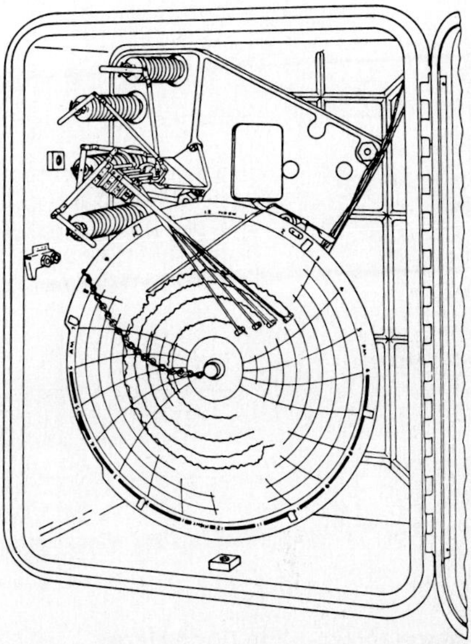

Fig. 51 Nonconcentric pen-arm assembly sometimes used in multiple-pen circular chart recorders.

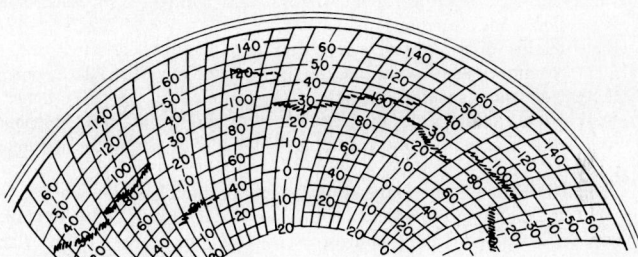

Fig. 52 Circular segmental chart.

xy Recorders

These instruments are used where it is desired to plot the relationship between two variables, $y = f(x)$, instead of plotting each variable separately as a function of time. An xy recorder closely resembles and functions like a single-pen recorder except that the chart (y axis) is moved in response to changes in a variable instead of at a uniform time rate. The chart may be driven by a separate servoactuated measuring element similar to that used in positioning the recorder pen, or by a self-synchronous motor in a remote control system. In most applications for the xy recorder a square-shaped graph sheet is employed. Thus, the full-scale chart travel (y axis) is made equal to the full-scale pen travel (x axis). The addition of a second servoactuated measuring element and recording pen to an xy recorder yields an xx_1y or an xyy_1 recorder in which the relationship among three associated variables can be plotted.

For laboratory use of an xy recorder, it may be desirable to have the horizontal or x axis driven as a function of time. Operated in this way, an xy recorder may be thought of as a low-frequency, direct writing oscillograph. A circuit for time base operation is shown in Fig. 53.

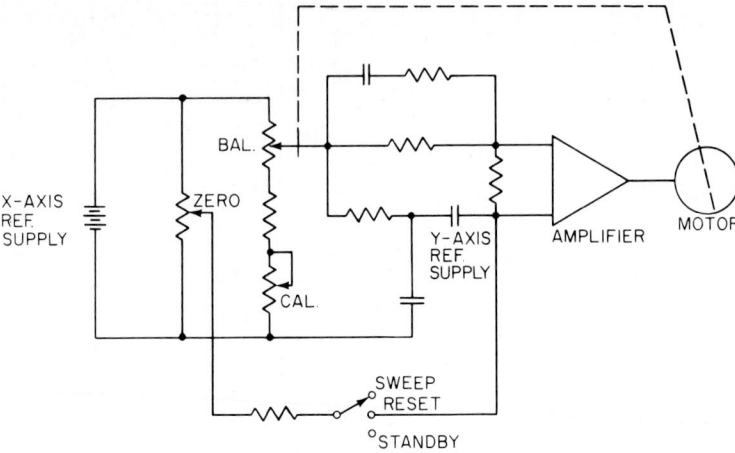

Fig. 53 Time base circuit for a *XY* recorder.

Alarm Information Recorders

As pointed out earlier in this article under Alarms and Annunciators, there is growing interest in the analysis of alarm condition information. Recognizing this need, some instrument manufacturers have approached the problem either by adaptating standard multipoint recorders or by designing specifically dedicated instrumentation for this purpose.

As an optional feature of a multipoint recorder,* a microprocessor-based unit replaces the mechanical switch-type alarm, and with the resulting expanded capability, development of a more flexible and extensive alarming system is possible. The instrument can be programmed to alarm or signal at up to four separate levels per point. All alarm programming, including the point and alarm assignments, the types of alarms, and the output relay designations, is field-programmable by the user.

The instrument consists of a data entry keyboard with a digital display, an interface with the measuring circuit of the multipoint recorder, a central processing unit (CPU) with a microprocessor for alarm processing and programming, and a relay rack for the alarm outputs. There is also a battery backup for memory storage. The instrument is furnished for use with all listed recorder ranges and all standard thermocouple types. Necessary linearization data are stored in the CPU so that alarm levels are programmed directly in engineering units rather than percent-of-scale figures.

The system provides for the programming of four independent levels of alarm, each with its set point and alarm action (high or low). Thus, up to 120 separate alarm points can be selected on a 30-point recorder. An additional capability involves an alarm which is activated when the *rate of change* of a programmed point exceeds a preset limit. For instance, if an alarm has been programmed for a maximum temperature increase of $10°$/min and the temperature rises at a rate of $12°$/min, alarming will occur. This limit can be assigned to operate in either a positive or negative direction.

Alarm logic can be configured to group points on an AND or an OR basis to provide more comprehensive information regarding process status.

Up to 60 lock-in alarm relays are available for alarm signaling. These devices are usually mounted in separate relay racks. The assignment of relays and alarm levels is programmed through the instrument's data entry keyboard, which is located inside the recorder on a slide-out chassis. The data entry keyboard is shown in Fig. 54. A four-digit security code is required to enter data or to alter information.

---

*Alarm Processor, an option available with the Leeds & Northrup Speedomax 165 or 250 multipoint recorder.

There are five program modes: (1) *security mode;* (2) *range mode,* which allows entry into the processor of existing ranges of the recorder and the points associated with each range; also used to enter new range data if a range (or range type) is changed after the initial programming; (3) *alarm mode,* for choosing the level, set point, type of alarm, and relay number for each recorder point to be alarmed; (4) *link mode,* which enables the operator to link alarms in any combination of AND-OR modes; and (5) *time mode,* for entering the correct time and date into the clock function.

Each of the relays may be programmed to alarm (1) at any point and on any of the four levels; (2) at any one of a number of point-level combinations in an OR linkup, where one point causes the relay to be activated; (3) at all point-level combinations in an AND linkup, where all selected points within a group must be in alarm status before the relay closes; and (4) on any combination of AND groups linked to other AND groups on an OR basis. Only one of the groups must be in alarm status for the relay to operate.

Sequential Events Recorder

This is an instrument designed for and dedicated to recording the sequential events which occur during an alarm condition. The instrument has the following features: (1) up to 128 inputs in a compact, self-contained system that is expandable with the simple addition of a modular external input chassis; (2) a matching 128 levels of memory with each 128 inputs, which ensures that every input change is time-tagged to within 1 ms of its occurrence regardless of how many changes take place at the same time; (3) an integral printer that provides both black (normal events) and red (abnormal events) printing for alarm highlights; (4) availability of up to 64 characters per legend in the language printout version for full event description; (5) an optional portable programming keyboard for convenience and security in legend entry and modification; (6) automatic and manual on-line diagnostics to simplify troubleshooting and maintenance; (7) a wide range of operator-direct input, modification, verification, and data request capability through front panel access and control facilities.

The basic architecture of the instrument (Fig. 55) is that of a distributed processing system. It consists of a master microprocessor module (MMM) controlling a number of slave microprocessor modules in combination with a "motherboard" data bus in which a variety of special function modules may be installed. The simplest configuration is made up of the MMM, a time keeper module (TKM), a printer control module (PCM), and one or more contact processing modules (CPM). Each CPM is dedicated to processing 128 inputs. The field contact is interfaced with the instrument via input cards containing optical isolation. Each input card has 16 inputs with deliberately simplified logic to ensure maximum immunity to external process noise. A status change detected by the optical coupler is transferred to the CPM where it is filtered via digital filtering techniques. If the input change is stable for five samples, the CPM stores it in sequence memory with the precise time the change was detected.

The MMM continuously polls the CPMs, searching for data in the sequence memory. When data are detected, the information is transferred to the printer, the oldest data being transferred first. As each event is removed from memory, space is available for any new event data.

If the CPM memory is completely filled because of a major process or system upset, there are provisions for one additional event per input to be stored by each CPM. However, these events are printed without time information—a positive indication to operating personnel that events beyond the 128-level capacity of each CPM have occurred.

The front panel of a sequential events recorder is shown in Fig. 56.

Oscillographic Recorders

Oscillographs are instruments for recording analog or digital information on paper or film. The information may be recorded with ink, powder, electrostatic charge, liquid, electric arc, or light. Technically, the following recorders may be considered oscillographs: (1) galvanometer recorders, (2) CRT recorders, (3) pen recorders, (4) *xy* graphic recorders, and (5) strip chart recorders. But, modern usage of the word "oscillograph" implies either a galvanometer or a CRT recorder—instruments that have bandwidths in excess of 20 kHz.

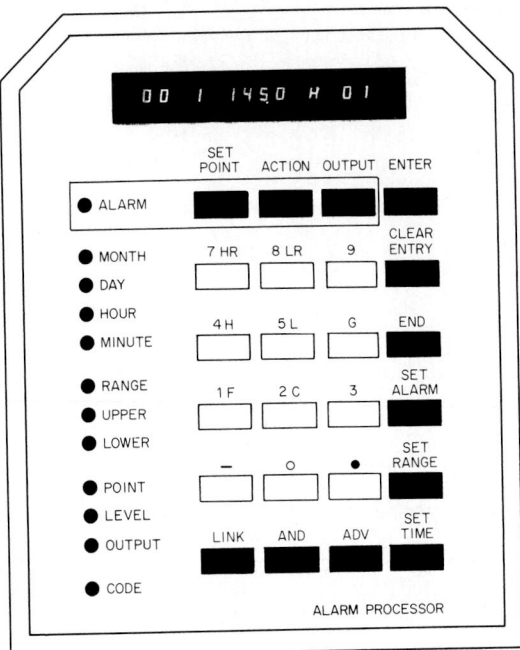

Fig. 54 Arrangement of a data entry keyboard for an alarm processor. Grouped in the center of the keyboard are buttons 0 to 9, plus a decimal button and one used to enter a negative quantity. Several of these buttons also have a secondary function. Buttons 7 and 8 are used to select either a high rate (HR) or a low rate (LR) type of alarm action. Buttons 4 and 5 permit selection of high alarm (H) or low alarm (L) action. Button 1 is used to enter degrees Fahrenheit when setting temperature ranges and button 2 to enter degrees Celsius. The decimal-point button is used to start the security programming.

The "Enter" button is used to enter previously keyed data into memory but is also used to indicate an OR condition when linking alarms. The "clear entry" button clears data incorrectly keyed before the data are entered in memory (it does not alter a function or calculation previously entered). The "end" button is used to signify the end of a programming effort. Pressed twice in succession, it takes the processor out of a programming mode. The "set alarm" button starts the alarm programming function, and "set range" button starts the clock programming function. The "adv" button advances the program to the next step in the programming sequence. The "link" button initiates the link programming function, and the "and" button enters an AND link in the alarm link program.

The "set-point," "action," and "output" pushbuttons are activated only when the "alarm" light immediately to their left is on. When lit, it signifies that the next entry is to be either the set-point value, the action high, low, high rate, low rate or the relay number associated with the set point and action previously entered.

The LED indicator lights arranged vertically to the left of the pushbuttons function as follows:

"ALARM"............................... indicates that set-point, action, and output buttons are to be activated

"month," "day," "hour," and "minute" lights call for their corresponding entries to be made in the "set time" program mode

"RANGE"............................. calls for the number of the range (1, 2, or 3) to be entered, and the "upper" and "lower" lights go on when the range limits of the recorder calibration are to be entered

"POINT"................................ is used in the range setting, alarm setting, and linking programs; when lit, it calls for the next entry to be a point number

15.56

Galvanometer Recorders

The principles of the galvanometer recorder date back many decades, and the instrument was available many years before introduction of the CRT. Galvanometer recorders are two-dimensional display-recording devices which contain a light source, mirror galvanometers and an optical system, light-sensitive recording media, and a system for transporting the record. The basic operation of a galvanometer recorder is shown in Fig. 57. Part of the light source is reflected from the galvanometer mirror and then off a recording mirror; it is then passed through a recording lens and onto the recording medium. When the galvanometer is connected to an external source, the electric signal conducts through the coil and causes the galvanometer mirror to rotate. Thus, the light beam moves in a straight line across the record, in proportion to the amplitude and to the rate of change in the input signal. The movement of the record provides a time history. Thus, the record provides an amplitude (*y* axis) and time (*x* axis) recording of the input signal. This configuration is known as *longitudinal recording*.

In the past, some oscillographic recorders used a high-pressure mercury-vapor lamp as an ultraviolet light source. When ultraviolet-sensitive paper is used as the recording medium, an instantly readable, permanent record can be obtained without the need for chemical processing.

More than one galvanometer may be installed and used in a recorder, thus providing many time-coincident channels of recording data. The galvanometers share the same light source and may be electrically or mechanically positioned.

With several galvanometers making simultaneous and overlapping recordings, specific identification of what trace goes with what galvanometer can be difficult. The traces can be identified by a momentary, sequential interruption. The trace interruption is caused by pins located in a moving endless belt positioned so that they pass laterally in front of the galvanometers. The pins interrupt the light beam from the recording lamp as it enters the galvanometer lens, and as a pin casts a shadow on a particular galvanometer mirror, the recording trace for that galvanometer is momentarily interrupted. As these breaks occur in the traces on the record, they correspond to the relative positions of the galvanometers in their magnet assemblies, regardless of whether the traces do or do not overlap.

Some recorders identify the trace interruptions by printing a series of numbers corresponding to the galvanometer positions on the edge of the record. This system consists of a numbered film-wheel assembly and the optics necessary to transmit a portion of the light from the recording lamp through the wheel to the paper.

Reference grid lines may be recorded by passing a portion of light via mirrors from the recording lamp through a grid line aperture. This aperture blocks out all but a series of fine bars of light running in the same direction as the record travel, which are exposed on the record simultaneously with the galvanometer traces.

Reference timing lines may be recorded across the record, perpendicular to the grid lines. The timing lines may be controlled externally or internally. A timing circuit activates a flash tube which directs lights to the recording mirror, though the recording lens, and onto the record.

A typical optical system for the galvanometer trace, grid line trace, and time lines is shown in Fig. 58.

"LEVEL"	is associated with the alarm setting and linking programs and indicates that the next entry identifies which of the previously selected point's alarm levels is being assigned an alarm level
"OUTPUT"	is used in linking and calls for designation of the relay on which points and levels are being linked
"CODE"	signifies that the security code number must be entered before programming may proceed; should the operator wish to review the information in memory, depressing the "enter" button will provide it; the security code entry is not required for data review

(Alarm processor, *Leeds & Northrup, Unit of General Signal.*)

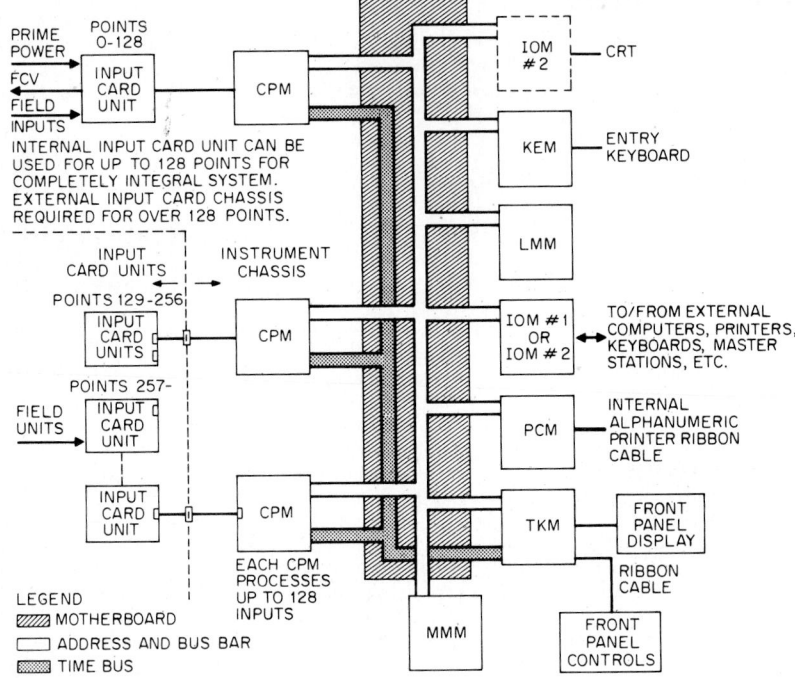

Fig. 55 Basic architecture of a sequential events recorder.

MMM = Master Microprocessor Module—controls flow of data through system. Contains solid-state memory that stores user-assigned data and parameter. Other master program data are also stored in this module.

TKM = Time Keeper Module—provides the basic time and date data for use by other modules and also interfaces the front panel touch pad keys into the system.

CPM = Contact Processing Modules—each module is dedicated to processing 128 inputs, allowing for one millisecond resolution regardless of number of alarms that may occur at any given instant. The modules detect input changes and store related data until it is transferred to the MMM.

PCM = Printer Control Module—provides an interface to the internal printer.

LMM = Legend Memory Module—solid-state memory for storing descriptive language legends to be printed with each event. EAROM (electrically alterable programmable ROM) and magnetic bubble versions are available.

IOM #1— Input/Output Module, Type 1—provides a serial data port to communicate with external printer or other devices, such as computers and data modems.

IOM #2— Input/Output Module, Type 2—microprocessor-based interface with both serial and parallel capabilities. This interface is used with an optional CRT or to accomplish other special interfacing requirements.

KEM = Keyboard Entry Module—provides interface capabilities for the external keyboard used for user inputs and alphanumeric legend information.

(RA-2800, *Rochester Instrument Systems.*)

15.58

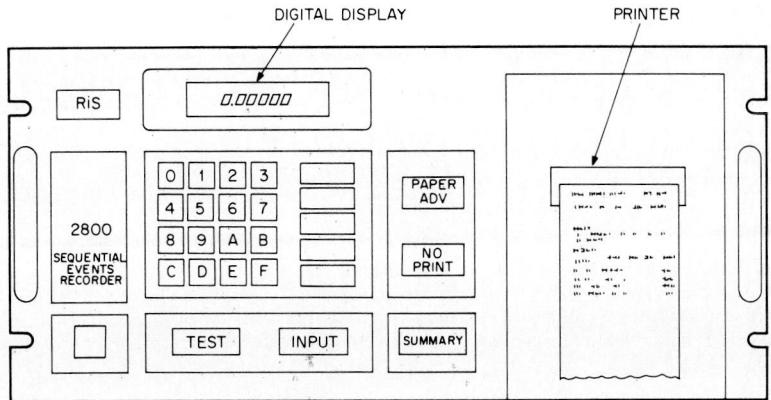

Fig. 56 Front panel of a sequential events recorder made up of three basic sections: (1) a LED display to provide status and system operating information, (2) an integral numeric or full language printer with appropriate controls and alarms, and (3) touch pad membrane switches incorporating both operational and functional controls. In the operation of the instrument, the output of the alphanumeric printer may be multiline-per-event printout (64 characters in length), with normal events printed in black and alarms printed in red and containing the following: (1) A, N, **, or S status characters corresponding to alarm, normal, test, and alarm summary, respectively; (2) the point number on which the event occurred; (3) the time of event occurrence in hours, minutes, seconds, and milliseconds. The digital display at the top of the instrument panel has eight characters. When not displaying alarm data, the display indicates real time in hours, minutes, and seconds. The "test" button triggers special diagnostics that exercise both hardware and software while searching for a problem. Test results are printed. "Input" control initiates a check of all input modules, including an optical coupler used for isolation. Testing includes comprehensive cycling of each input through both the open and the closed state. The input check is completed with a printout, indicating the correct operation of all inputs or listing any inputs that are faulty or inactive. The Summary function initiates a printed list of all points currently in the alarm or abnormal condition. The printout also includes a copy of all input points which have been deliberately removed from the active scanning mode. A second grouping on the front panel consists of number switches 0 through 9 and letter switches A through F, plus "time," "date," "function," "thru," and "enter" control switches. The numbered and lettered pushbuttons involve the functional aspects of the instrument. When used in varying combinations, these controls provide the operator with an array of system parameter changes, verifications, and data call-ups. A security code prevents unauthorized changes.

The "function" key is the initial control that gains access to the system and prepares it for a program change. This key is always followed by one of several two-digit funciton codes: the "thru" control key permits changes in groups of consecutively numbered points without a need to program in the same change for each point; the "enter" key is depressed only when new or modified data are to be finally entered into the system. It is the last command given to the system. If it is not activated within a predetermined time after the function codes have been initiated, the system will return to the previous state. In any program or special function procedure, it is necessary only to select the proper digit code for the function desired, put in the appropriate point number(s), and depress the "enter" key to finally execute the change.

The parameters which can be defined or changed through combination of function controls are (1) time, (2) date, (3) normally open or normally closed input select, (4) delete point from scan, (5) delay return to normal, (6) inhibit return to normal, (7) identify critical inputs, (8) early critical printout, (9) call-up of point parameters, and (10) assorted housekeeping functions. (Sequential events recorder Model RA-2800, *Rochester Instrument Systems*.)

CRT Recorders

These may be three- or four-dimensional display-recording devices. A four-dimensional device is simply a continuously recording oscilloscope, while a three-dimensional device is a line-scan recorder.

The four axes of a recording oscilloscope are the x axis (time base or horizontal), y axis (amplitude or vertical), z axis (spot intensity), and y' axis (movement of recording medium). See Fig. 59.

The phosphor on the faceplate of the CRT is selected to have a spectral output which matches that of the recording medium. This provides an efficient transfer of energy to the light-sensitive record. The light output from the CRT is focused on the record either by an optical lens system or by providing a fiber-optic faceplate on the CRT and placing the record in intimate contact with the fiber optics.

The four axes provide the capability of recording continuous or transient signals that previously required magnetic tape or oscilloscope-camera techniques. In the continuous mode, signals may be recorded longitudinally, xy axes (such as the galvanometer recorder), or transversely, xyy' axes. In the transverse mode, high-frequency data may be recorded without utilizing high record speeds. See Fig. 60.

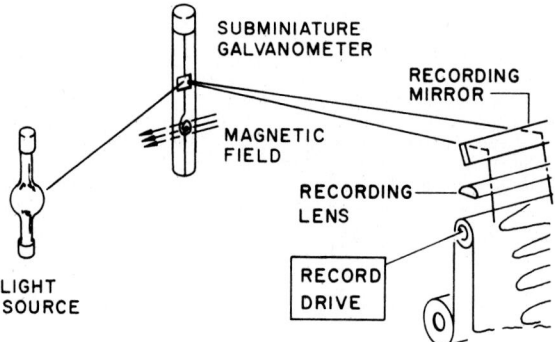

Fig. 57   Essential elements of a galvanometer recorder.

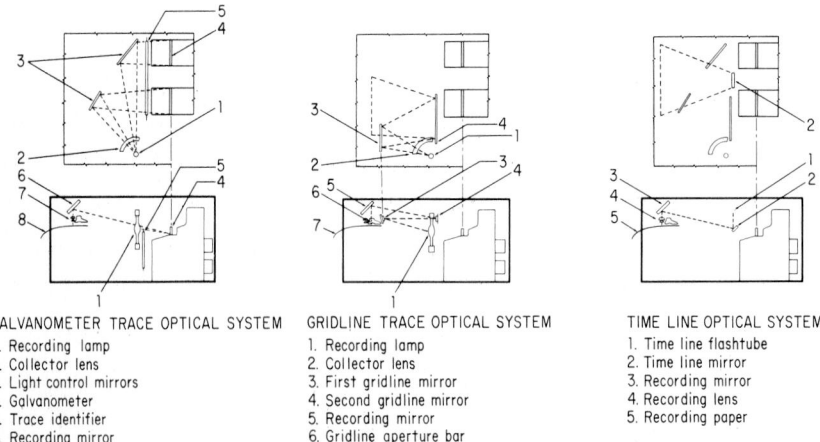

GALVANOMETER TRACE OPTICAL SYSTEM
1. Recording lamp
2. Collector lens
3. Light control mirrors
4. Galvanometer
5. Trace identifier
6. Recording mirror
7. Recording lens
8. Recording paper

GRIDLINE TRACE OPTICAL SYSTEM
1. Recording lamp
2. Collector lens
3. First gridline mirror
4. Second gridline mirror
5. Recording mirror
6. Gridline aperture bar
7. Recording paper

TIME LINE OPTICAL SYSTEM
1. Time line flashtube
2. Time line mirror
3. Recording mirror
4. Recording lens
5. Recording paper

Fig. 58   Recorder optical system for galvanometer, grid line, and timing traces.

Because of the large combination of horizontal sweep times (typically 2.5 s/cm to 1 μs/cm) and record speeds (typically 0.1 to 250 cm/s) the transverse signal may be skewed, i.e., not perpendicular to the edge of the record. Electrical or mechanical compensation must be provided in order to minimize this skew error.

Reference grid lines in the horizontal and vertical direction are recorded on the record by means of light from the flash tube passing through an aperture. The timing of the flash tube can be controlled externally or internally. Internal control may be from a self-contained timer or synchronized off the record drive system.

The four axes provide a recorder which is very versatile. Besides transverse and longitudinal recordings, video pictures from television or facsimile may be recorded by use of the zx axes and either the y or the y' axis. The quality of the pictures depends on the sensitivity of the recording medium, the bandwidth of the z-axis amplifier, and the spot size that can be recorded.

The recording medium sensitivity and the spot size also determine the writing speed with which data may be recorded. Characteristically, recording oscilloscopes have been capable of recording data 100 to 1000 times faster than galvanometer recorders.

A *line-scan recorder* is a three-axis recording oscilloscope. It does not have a vertical amplifier channel and therefore loses the capability of y-axis recording.

Alphanumeric characters can be printed out, thus making the recorder a high speed printer. The characters can be recorded by the use of xyz axes.

Fiber-Optic CRT Recorders

The instrument* shown in Fig. 61 provides a selection of up to 18 recording channels, but when combined with an accessory housing, the channel capacity can be increased to 32. Records that are relatively insensitive to normal room light are produced by ultraviolet light generated by a CRT and directed through a special fiber-optic faceplate onto photosensitive direct print paper. Recordings are produced immediately without the delays associated with wet processing or toner application.

A fiber-optic CRT faceplate consists of about 10 million glass fibers, each only a few micrometers

*Honeywell 1858 Visicorder.

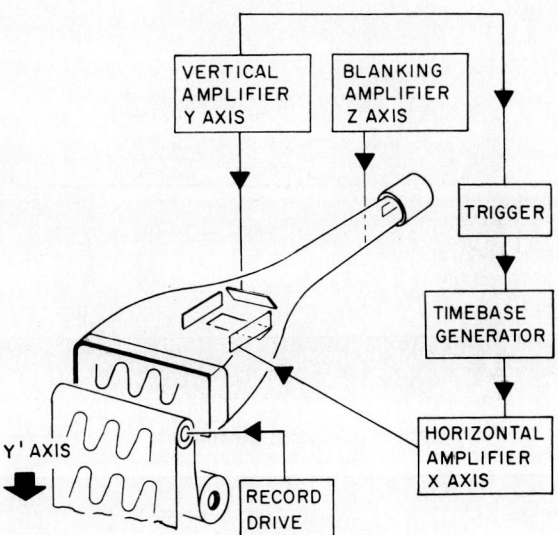

Fig. 59   Recording oscilloscope block diagram.

in diameter, fused together within a 0.2 × 8 in (5 × 200 mm) area. These fibers provide efficient transfer of ultraviolet light from the phosphor inside the faceplate to the photosensitive paper, enhancing the writing speed and trace resolution. Since the recording paper is in intimate contact with the faceplate, additional optical elements are not required. The intensity and focus are factory-preset. To prolong phosphor life, the recording area is gradually moved in both axes. The movement is so slow that it does not detract from the accuracy or readability of the recording.

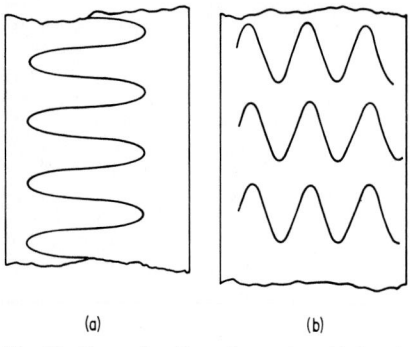

(a)   (b)

Fig. 60 Types of oscillograph recorders. (a) Longitudinal, (b) transverse.

Grid Lines

To assist in measuring signal amplitudes in calibrated increments, the instrument shown in Fig. 61 can place longitudinal lines each 0.2 in (5 mm) across the paper. The lines are accentuated at 1-in (25-mm) intervals and appear across 7.2 in (183 mm) of the record.

The grid lines may be continuous or interrupted, as selected via a front panel switch. The switch also allows the grid lines to be turned off for applications where signal levels are not important.

Interrupted lines appear for approximately 1/16 in (1.6 mm) following each time line, retaining amplitude references without cluttering the record. The interrupted grid line feature is operable only when the timing line option is included and is in use. See Fig. 62.

Recorder Drive

Recording paper is driven over a 1200:1 speed range by a servo system consisting of a dc drive motor, tachometer generator, and associated electronics. There are no gears and clutches. Speed selection is

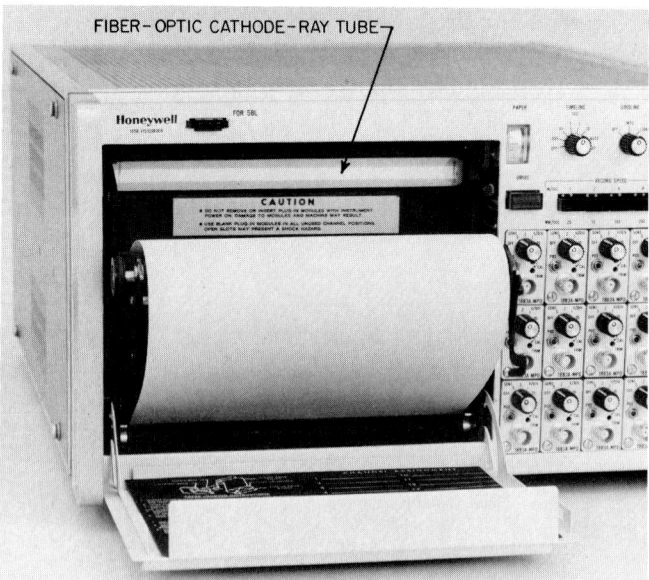

Fig. 61 Fiber-optic CRT recorder. (Model 1858 Visicorder, *Honeywell.*)

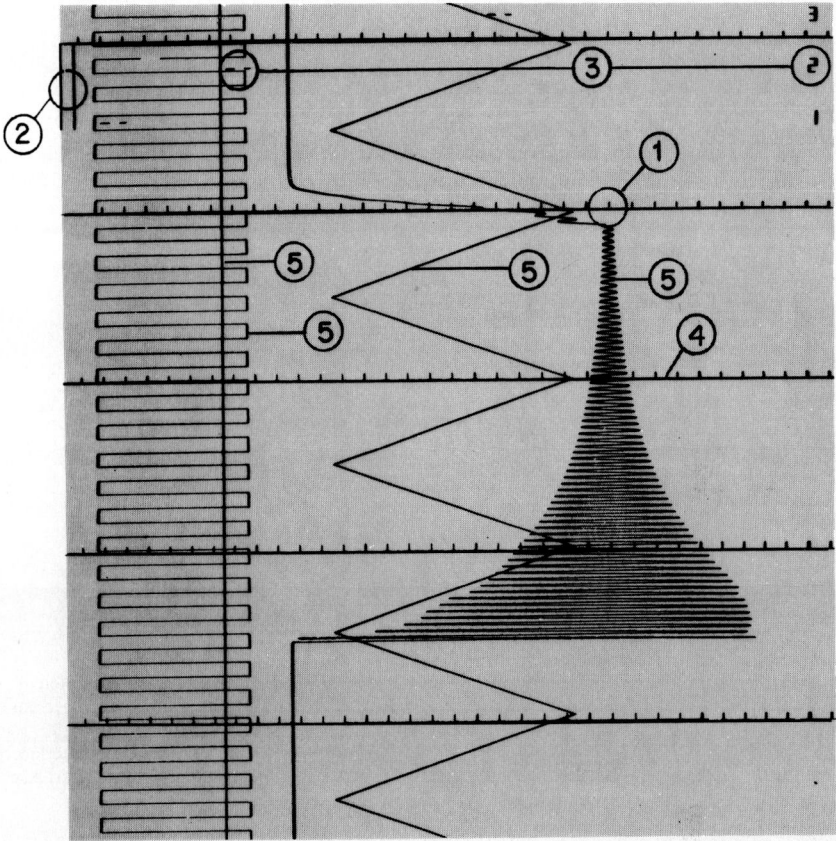

Fig. 62   Grid lines: (1) interrupted grid line, (2) time interval marker, (3) numbered trace identification, (4) time lines, (5) data traces. The graph as shown is approximately 55% of the original size. (Model 1858 Visicorder, *Honeywell.*)

controlled from the front panel by five pushbuttons of 1, 2, 4, 8, and 16 in/s (25, 50, 100, 200, and 400 mm/s). By depressing the switches in binary addition, 42 discrete speeds can be selected.

Remote control of drive "on-off" and CRT unblanking is provided by a rear panel connector. The local or remote selector switch on the rear panel locks out actuation by other than the selected mode. Remote actuation is transistor-transistor-logic (TTL)-compatible or through a single-pole single-throw switch closure. By applying 0.008 to 10 V to a rear panel connector, a speed range of 0.1 to 120 in/s (2.5 to 3000 mm/s) can be remotely programmed in continuously variable or discrete steps.

Signal Conditioning Modules

A variety of plug-in modules are available to accept input signals from 100 μV to 200 V per major division from voltage and transducer sources with common-mode voltages to 300 V. Operating power is drawn from supplies within the electronic housing. Modules are compact—up to 18 single-width, 9 dual-width, or a combination of both can be accommodated, and up to 32 single-width or 16 dual-width with additional housing. See Fig. 63.

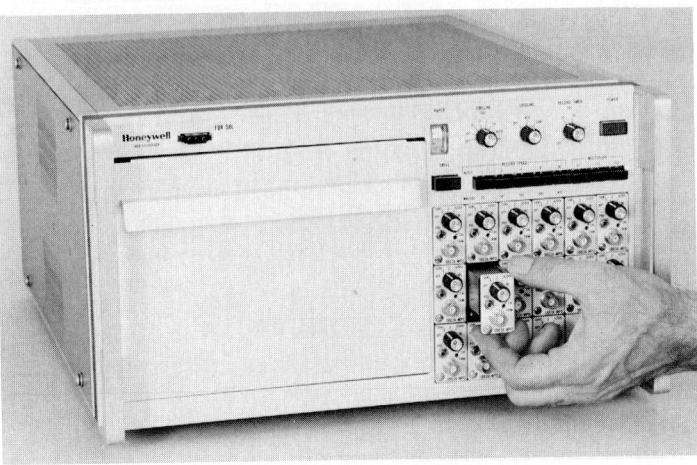

Fig. 63 Oscillographic recorder with signal conditioning modules shown at the right. (Model 1858 Visicorder, *Honeywell*.)

Sensitivity calibrations are in volts per division. Each module has a switch to turn its display on or off without disturbing the established baseline position. A position control allows the baseline of the trace to be placed anywhere across the entire width of the paper. Each module is capable of producing a recording at any amplitude up to 7.2 in. (18.3 cm) and at frequencies from direct current to 5 kHz.

Signal conditioner modules include high-gain, low-gain, and medium-gain amplifiers, interface modules, strain gage amplifiers, thermocouple and microvolt amplifiers, tape recorder amplifiers, and frequency-to-voltage converters.

Loggers

For many years, the logger was essentially a passive instrument, competing with graphic recorders for the presentation and storage of information. The early loggers with their digital tabulation of information, when viewed now in perspective, were ahead of the times. On the parts of some users in earlier years, there was considerable resistance to tabular information formats—a resistance that has diminished because of the digitalization of so many aspects of instrumentation, including control. But, as in the past, the manner in which information can be presented and stored to the best advantage of the user remains a matter of preference in specific applications.

Hybridization, when used in a general way, has typified instrumentation developments since the early 1970s and certainly has affected the kinds of logging systems available today. Particularly during the early 1980s, there were strong trends (which continue) toward markedly enhancing the capabilities of loggers—in fact up to the point of making a central logger not only a complete data acquisition system but one with some controlling capabilities as well. Thus, it is indeed very difficult as of the mid-1980s to classify loggers, except to identify them as quite simple and straightforward and ranging through a series of sophisticated features and enhanced capabilities to systems that, to the average instrument engineer, are no longer simply printers and loggers, but rather complete data acquisition systems, some with control abilities; hence the logging capability of some systems appears as a secondary role.

The mid-1980s are characterized by several diverging opinions as to what loggers and logging systems should comprise. There are not only differing opinions among users, but among suppliers

Fig. 64 A printer, microprocessor-based and utilizing needle printing, for use with an electronic scale. Up to 72 characters can be printed on a single line, or information can be configured on multiple lines. The printing speed is approximately 64 characters per second at 10 characters per inch (4 characters per centimeter). Characters are 0.13 in (3.3 mm) high. Examples of five print styles are shown in Fig. 65.

The input is 20 mA ASCII at 300 bauds, even parity; the output is 20 mA ASCII, selectable at 300 or 1200 bauds. The unit can print right-side-up or inverted, providing a capability for handling a wide variety of weigh tickets and forms. Data can be viewed as printed. A dual-register accumulator and keyboard entry capability provides totalization of weights for a shift, a day, or longer, as well as automatic recording of the time and date and consecutive numbering. The unit incorporates self-diagnostics.

Four LEDs in the upper right corner of the panel indicate the presence of paper, if the consecutive numbering is set at 1, if any data are stored in the operating register, and if any data are stored in the accumulating register. When an error occurs, a specific LED flashes to indicate the particular type of error.

The printer can be provided with either a numeric or an alphanumeric keyboard as shown under digital display. Up to 30 characters (3 fields of 10) can be used to identify weight information.

There are 20 color-coded pushbuttons in the lower right corner of the panel. Their function is explained below. (Printweigh, *Toledo Scale Division, Reliance Electric*.)

Key	Functions of control keys
+	Used to make a plus entry into the operating register of the printer
−	Used to make a minus entry into the operating register of the printer
T	Totalizes and clears the contents of the operating register and transfers the total to the accumulating register
Δ	Prints information from the weight indicator and/or from the printer keyboard but does not affect the contents of either register
ST	Initiates printing of a subtotal from the accumulating register but does not clear that register
GT	Totalizes and clears the accumulating register
∧, ∨	Initiates a single-line or multiline paper feed
Repeat print	Initiates a second printing of information just printed, except totals
Test verify	Artificially induces an error during a test print, which is indicated by flashing LEDs for circuit testing purposes. The error must be cleared using the "clear error" key.
CN reset	The consecutive number is reset to 1 when this key is continuously held for 3 s and the (CN = 1) LED is illuminated
Clamp release	Releases paper clamp to allow positioning of form
Gross weight entry	Used in conjunction with the +, −, or Δ button. Computes a net weight based on a printer-keyboard-entered gross weight and a weight transmitted from the digital indicator; on pressing the +, −, or Δ button, the gross, tare, and net weights are printed and the appropriate entry is made in the operating register
Tare weight entry	Procedure is similar to that just described, but the information from the weight indicator is used as the gross weight and the keyboard-entered value is the tare weight
Time/date set	Used to display and set the time and date
Space	Generates a space within a selective numbering field
Clear error	Used to reset printer electronics after an error has been corrected
Keyboard entry	Used with the + or − button to make an entry into the operating register; also used with the Δ button
Clear	Dual-function pushbutton—used to clear an entry from the display and memory; also used to recall and set a consecutive number
Enter	Dual-function pushbutton—used to recall and set a selective number and to return to the time display

```
318.8LB  SELECTIVE NUMBERING ON 1 LINE    JUN.08.81 A 10:21
                        (1)
```

```
   000001      JUN.25.81 A 08:34           119.6LB
   119.6LB     SELECTIVE                     19.6LB T
    19.6LB T   NUMBERING                   100.0LB N
   100.0LB N   ON 3 LINES                  SELECTIVE
                                           NUMBERING
                                           ON 3 LINES

         (2)                                    (3)
```

```
JUN.25.81                              000002
SELECTIVE                              199.4LB
NUMBERING                                9.4LB  T
ON 3 LINES                             190.0LB  N
                        (4)
```

```
000003  JUN.25.81   PART NO.  159.4LB    9.4LB T    150.0LB N+

000004  JUN.25.81   PART NO.  199.2LB    9.2LB T    190.0LB N+

                                                    340.0LBM*

                                                   1090.0LB  ◇

                                                   1090.0LB  *
                        (5)
```

Fig. 65 Examples of print formats which can be provided by the unit shown in Fig. 64. (1) Double-width gross weight, single-width ID, time, and date. (2) Single-width consecutive number, gross-tare-net, time, and date; ID on three lines. (3) Single-width gross-tare-net; ID on separate lines. (4) Same as (2) except that date only, double width, and fields are formatted in the extreme left and right margins. (5) Totalization providing single-width consecutive number, date, ID, gross-tare-net with total, subtotal, and grand total of net weights. (Printweigh, *Toledo Scale Division, Reliance Electric.*)

as well. Because attempts to classify loggers are rather impractical at this time, the following paragraphs present what might be called a potpourri of current systems. A few of the less complex systems are shown in Figs. 64 through 67.

Recording, Logging, and Alarming

According to the manufacturer, the instrument* shown in Fig. 68 has the trending and large-page readout of a multipoint recorder, the accuracy and speed of a premium data logger, and the alarm, math, and format flexibility of a minicomputer system. Additional models with greater capacities are available. See also Fig. 69. Features and characteristics include:

Printer

(1) Full-width printer using 8½ × 11 in fanfold impact paper; (2) programmable headers that automatically print at the top of each new page; (3) 32 programmable alarm message outputs; (4) a

---

*Kaye Instruments Digistrip.

Fig. 66 Data logger that collects up to 100 analog inputs including milli-volts from low-level sensors, plus voltage, current, and RTDs and other resistance signals—at speeds up to 14 points per second. The unit also accommodates up to 96 digital inputs and 250 alarm limits. Each logging output can be directed to from one to five recording and display devices, including a front-panel LED readout, built-in thermal printer, magnetic tape recorder, paper tape punch, modem, terminal, or computer. To the left of the unit in view is shown a front-end module which permits measurements without exposing the mainframe of the unit to an unfavorable environment. These modules can be located up to 1 mi (1.6 km) from the mainframe. Instrument shown is furnished with five such modules. (Digitrend 235, *Doric Scientific Division, Emerson Electric.*)

Fig. 67 Printed circuit assemblies used in data logger shown in Fig. 66. The data logger can be tested from the front. Each printed circuit assembly is a complete subsystem. (Digitrend, *Doric Scientific Division, Emerson Electric.*)

15.67

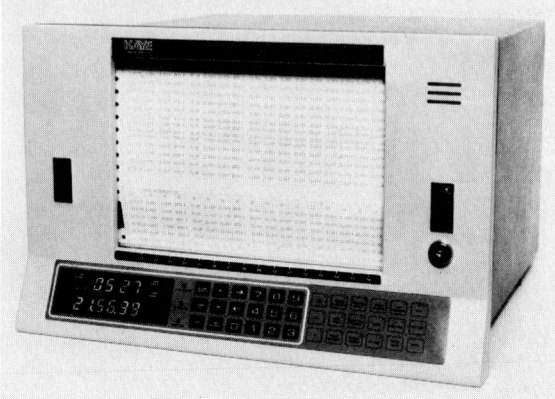

Fig. 68 Instrument that records, logs, and alarms. (Digistrip, *Kaye Instruments.*)

clock-calendar, battery-protected; (5) day and time printout; (6) an integral paper supply and take-up tray; (7) asterisk indication of a channel in alarm; and (8) an "A" indication of alarm log printout.

The scanning capacity is up to 128 channels, and restart occurs automatically after a power failure. RS232C and passive current loop ports are provided for computer connection.

Operational Keys

(1) "Alarm Search" automatically displays all channels in an alarm condition, sequencing at the rate of one channel per second; (2) "Alarm Log" causes a single log of data to be printed on every new alarm event; (3) log automatically prints a single log of data at the programmed log interval; (4) "Limit," when pressed sequentially, causes a programmed value of limit 1 and then limit 2 to be displayed alternatively; (5) "Time" displays days, hours, minutes, and seconds; and (6) "Time interval" displays the log interval in minutes.

Programming

(1) No programming language is required; (2) engineering conversion for E, J K, R, S, and T thermocouples in degress Celsius or Fahrenheit and 100-Ω platinum RTDs; (3) type T, 0.01°C resolution; (4) 65,000-mV, 650.00-mV, 6.5000-V programmable ranges; (4) 32 programmable linear scaling functions; (5) autocalibration; (6) 16 programmable accumulation functions; (7) high, low, and rate-of-change alarms independently programmable on every channel; (8) alarm closure contacts; (9) channel skip capability; and (10) cassette program loading capability.

Accessories

These include a frequency scanner (a remote satellite scanner), a digital signal scanner, a bridge conditioner, an analog output module for control and/or trending, a relay drive module, and a relay closure module. A grouping of optional accessories is shown in Fig. 70.

Data Logger and Monitor

The front panel of a data logger* is shown in Fig. 71. More detail of the operator controls and display section is shown in Fig. 72. A block diagram of the system is shown in Fig. 73.

*Fluke 2240C data logger.

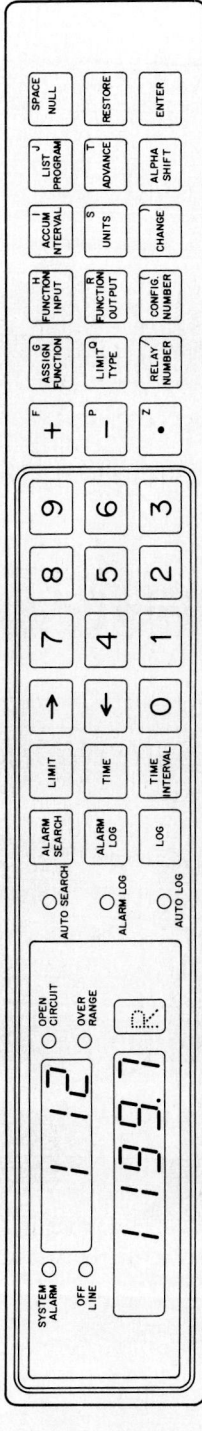

Fig. 69 Details of operational keys and displays. (Digistrip, *Kaye Instruments.*)

Fig. 70 Grouping of optional accessories. (Digistrip, *Kaye Instruments.*)

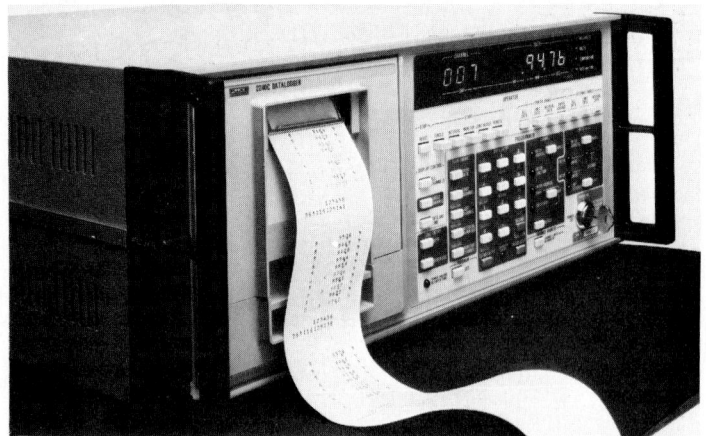

Fig. 71 Front panel of a data logger and monitor. (Model 2240C data logger, *Fluke.*)

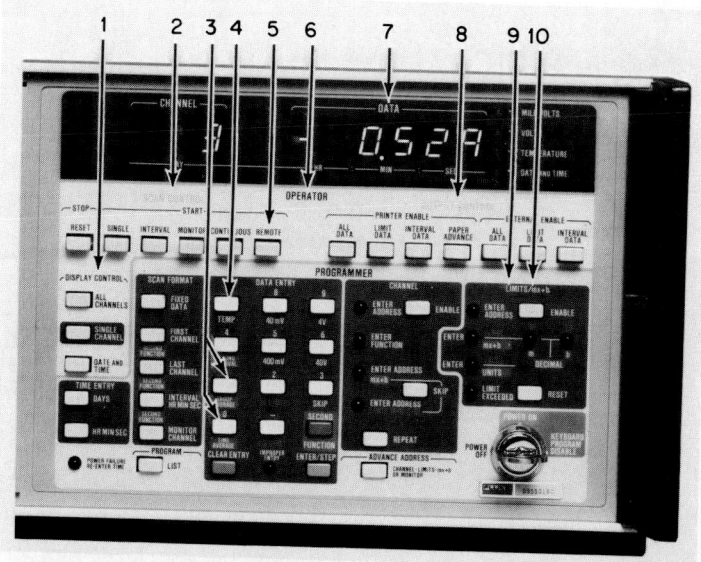

Fig. 72 Detail of the operator controls and display section of the instrument shown in Fig. 71. Note that the left-hand portion of the instrument containing the printer is not shown in this partial view. (1) Display controls, (2) scan controls, (3) data averaging—time average from 1 to 30 individual channels, (4) temperature linearization, (5) remote programming, (6) operator control, (7) high-performance A/D standard, (8) output selection, (9) alarm limits, and (10) $mx + b$ scaling. (Model 2240C data logger, *Fluke.*)

Some of the features of this instrument include:

Temperature Linearization

Eleven front-panel programmable thermocouple and/or RTD linearizations are provided from a wide selection of linearizations based on NBS, DIN, and JIS standards.

$mx + b$ Scaling

Thirty linear scaling functions can be programmed by the user. These functions are utilized in scaling analog inputs for a wide variety of transducers to their appropriate engineering units. This option includes a table of engineering units for printer and external output notation.

Data Averaging

This function operates in two modes—time averaging and group averaging. The *time averaging mode* allows the averaging of up to 30 individual channels over a selectable interval of 2 to 99 scans. In the *group averaging mode,* up to 30 groups of channels may be averaged during each scan. The group size is user-selectable and may include up to 99 channels.

Alarms

Up to four limit set points per channel may be programmed and coupled with alarm outputs for versatile alarming capability and control.

Dual-Scan Interval

Two completely independent scan intervals may be programmed. The "second" interval may contain, in relation to the primary interval, an identical, overlapping, or entirely different set of channels.

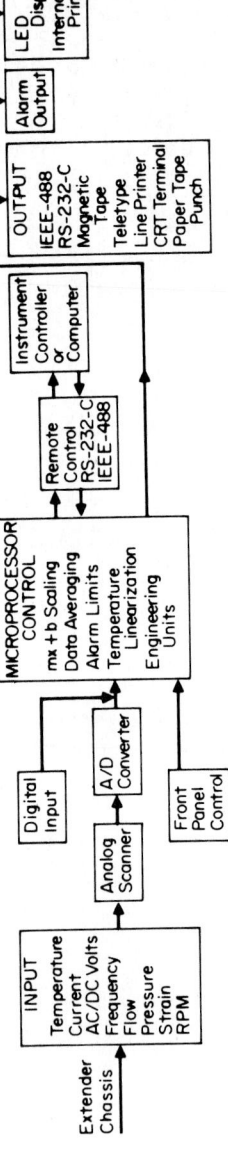

Fig. 73 Block diagram of a data logging system. (Model 2240C data logger, *Fluke*.)

Fig. 74   Data logging system designed for permanent mounting or portable use off a 12 V dc supply. (Model 2280A data logging system, *Fluke.*)

Independent control of the scan mode and output mode permits scanning continuously for alarms, yet outputting data on only a specified interval to an external recorder.

Inputs

These include a wide range of analog inputs, including ac voltage, dc voltage, direct current, and outputs from thermistors, RTDs, and thermocouples.

Remote Programming

Front panel programming and operational commands may be implemented over RS232C or IEEE 488 from a remote terminal, computer, or instrument controller. Remote programming is switch-selectable.

Program Memory

A 5-year nonvolatile memory prevents program loss if the power fails or if the unit is switched off.

Printer

A two-color printer prints data in black with alarm data in red. The printer, on command, provides a program list—a concise summary of programmed data logger parameters stored in memory.

Scanning

The unit can sample inputs at either 15 readings per second or 3 readings per second. Scan modes include single scan, monitor, interval scan, and continuous scan.

Portable Data Logger

The instrument shown in Fig. 74 is designed for permanent mounting or for portability. It can operate from a 12-V dc supply in a car, truck, ship, etc., where ac power is not available. The logging system measures a variety of inputs, including dc volts, megavolts, and current; thermocouples, binary-coded decimal, binary and contact closure, RTDs with linearization, ac volts [true root mean square (rms)], resistance, ¼, ½, and full-bridge strain gage, and frequency and totalizing (pulse). The instrument features a cartridge tape drive providing 500K bytes of storage for programs and data.

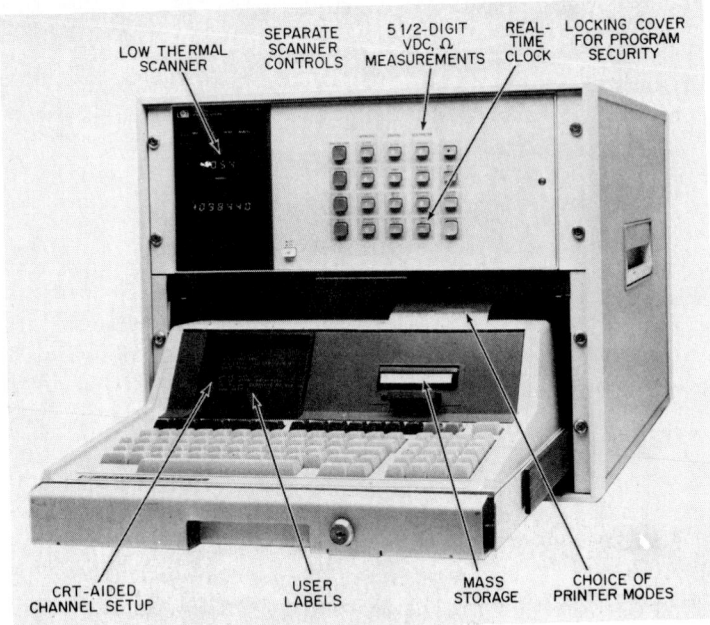

LOW THERMAL SCANNER · SEPARATE SCANNER CONTROLS · 5 1/2-DIGIT VDC, Ω MEASUREMENTS · REAL-TIME CLOCK · LOCKING COVER FOR PROGRAM SECURITY

CRT-AIDED CHANNEL SETUP · USER LABELS · MASS STORAGE · CHOICE OF PRINTER MODES

Fig. 75 Data logger with graduated software and a computer. (3054DL data logger, *Hewlett-Packard.*)

Data may be recorded at high speed for transfer later to the internal printer or to slower peripheral devices.

Data Logger with Computer and Graduated Software

The instrument* shown in Fig. 75 is a data logger with full computational and analysis capabilities. In addition to traditional transducer measurements, the system is designed for applications in research and development and in production testing. Three-level graduated data logging software enables the operator to start logging data immediately, without programming or computer experience. The device consists of a 5½-digit scanning unit, low thermal scanner (multiplexer), ohmmeter, current source, real-time clock, and interface (IEEE 488). The unit is contained in a 16-in (406-mm)-high cabinet. The computer is accommodated in a lower sliding drawer.

Instead of utilizing an internal microprocessor, the device utilizes a computer (HP 85F) to control the data logger. The HP 85F can be used as a separate computer when not logging data. Applications software for the HP 85F is available for mathematics, general statistics, electrical engineering, linear programming, and regression analysis. When logging data, the printer can show trend information or the results of a test. Graphic plots and histograms can be made to gain insight into the process that is being logged. Knowledge of a computer language is not required.

Measurement Performance

The autoranging voltmeter is designed for precision low-level measurements in the presence of noise. The resolution is 10 ppm, accuracy is $\pm 0.007\% + 1$ count, and sensitivity is 1 μV. The resistance resolution is 1 mΩ, and the temperature resolution is $0.1\,^\circ$C.

*Hewlett-Packard 3054DL data logger.

Low Thermal Scanner (Multiplexer)

Up to 100 analog channels (20 channels per card) or up to 80 digital input-output slots (16 channels per slot) are available with plug-in assemblies. An autoranging frequency counter is also available for logging frequency and/or totalizing data.

Graduated Data Logging Software

Three graduated levels of data logging software are supplied with each instrument. Level-1 software is a "menu-data" entry method which allows the user to begin data logging immediately. Level-2 software offers line entry instructions from the operator to test for limits, print the logged data, and/or close a relay if a limit does not meet specifications. Level-3 software differs from levels 1 and 2 in that it requires the operator to write a computer program. Most of the level-3 software is prewritten in the form of subroutines. Each subroutine performs a specific task and converts the data logger to a data acquisition system. Transducers can be linearized, and logged data can be stored on a tape cartridge. Two-variable graphs of the data can then be plotted using statistical histogram routines to determine the distribution of the logged data. Level-3 software comes with a comprehensive set of documentation capable of handling complex measuring situations without requiring the operator to become a computer expert.

Data Logger and Personal Computer

The instrument* shown in Fig. 76 combines a personal computer with a logger to provide low-cost solutions in connection with data collection problems.

*Hewlett-Packard Model 3421A with a HP-85 personal computer.

Fig. 76 Data acquisition system combined with personal computer. (Model 3421A with HP-85 personal computer, *Hewlett-Packard.*)

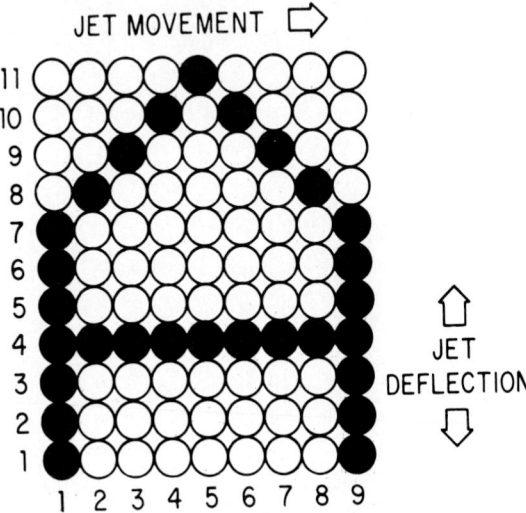

Fig. 77 Dot matrix (9 × 11) used in one type of ink jet printer. Ink jet printers utilize a stream of controlled ink droplets to form characters on a printing medium. In the printer, ink under pressure is forced through a nozzle energized with an ultrasonic signal (66 kHz) to cause velocity modulation of the ink stream. This modulation causes the ink stream to break up into over 65,000 discrete droplets per second. The droplets are of uniform size and spacing, both being a function of pressure at the nozzle, viscosity of the ink, nozzle diameter, and vibration frequency at the nozzle. Each droplet may be given a precise electrostatic charge as it forms. The voltage present at the time the droplet breaks away from the ink stream induces the charge. Once it breaks away, the charge is trapped, staying with the droplet until it reaches its destination. Because the droplets pass between a pair of high-voltage deflection plates, the charge of each droplet determines the precise path of travel and thus the final position of the droplet on the medium being printed. The nozzle is moved in a horizontal plane, driven by a servomechanism-controlled high speed low-inertia motor. The movement of the nozzle provides horizontal deflection of droplets; the high-voltage plates provide vertical deflection.

Other Printing Techniques

In addition to the printing methods previously described, ink-jet printing has been available for a number of years. See Fig. 77. A more recent development is laser printing. See Fig. 78. Both of these methods are widely used in various business and information processing machines but also find application in process instrumentation.

REFERENCES

Andreiev, N.: "Data Loggers Edge into Controller Market," *Control Eng.*, vol. 28, no. 2, 1981, p. 84.

Andreiev, N.: "Data Loggers Show Two Faces—Data Acquisition and PID Control," *Control Eng.*, vol. 28, no. 3, 1981, p. 84.

Bader, F. P.: "Custom CRT Graphics Enhance Operator-Process Interaction," *Control Eng.*, vol. 29, no. 2, 1982, p. 75.

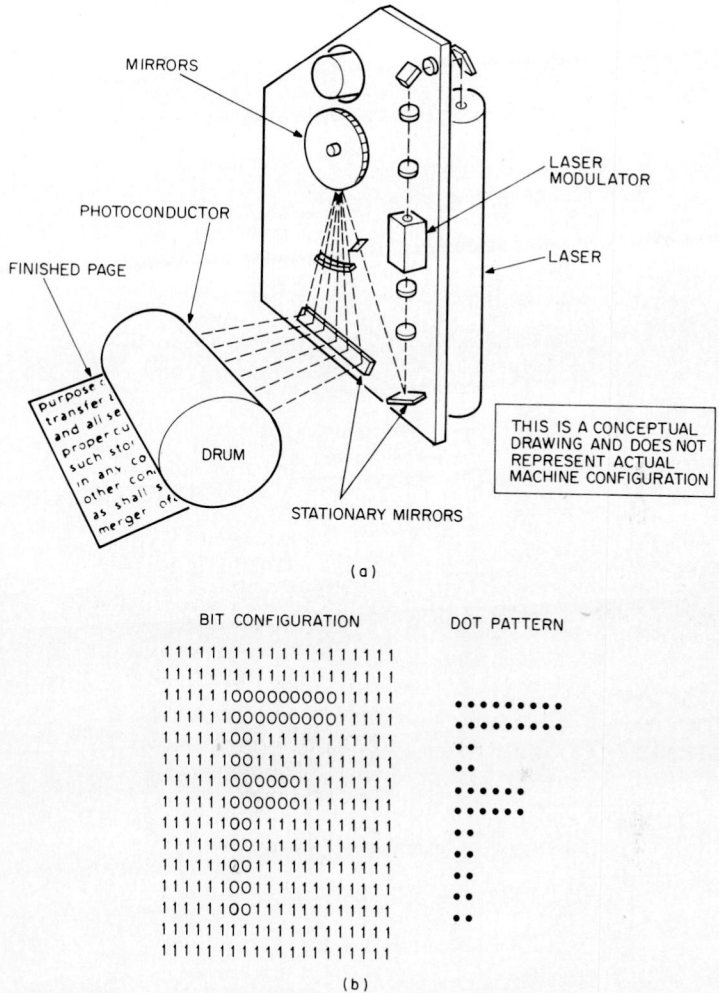

MIRRORS

LASER
MODULATOR

PHOTOCONDUCTOR

FINISHED PAGE

LASER

DRUM

THIS IS A CONCEPTUAL
DRAWING AND DOES NOT
REPRESENT ACTUAL
MACHINE CONFIGURATION

STATIONARY MIRRORS

(a)

BIT CONFIGURATION　　　　DOT PATTERN

```
111111111111111111111
111111111111111111111
11111100000000011111
11111100000000011111
11111100111111111111
11111100111111111111
11111100000011111111
11111100000011111111
11111100111111111111
11111100111111111111
11111100111111111111
11111100111111111111
11111100111111111111
111111111111111111111
111111111111111111111
```

(b)

Fig. 78 Creating an image using a laser. (a) A rotating surface having several mirrors reflects a low-powered laser beam and scans it across the length of a photoconductor. One scan represents one column of the dot pattern shown in (b). As the beam scans, signals from the processor to the modulator in effect turn the beam on and off. Each additional scan creates an additional column of the dot pattern until the entire matrix of information (one page) becomes a latent image on the photoconductor. The latent image is then developed, transferred to paper, and fused as in a copying process. *(IBM.)*

Bailey, S. J.: "Status Displays Still Essential in Man-Computer Interface," *Control Eng.*, vol. 24, no. 4, 1977, pp. 38–41.

Bailey, S. J.: "Non-Video Display Rounds Out Vital Man-Machine Interfaces," *Control Eng.*, vol. 27, no. 5, 1980, p. 75.

Bailey, S. J.: "Intelligent Terminals Guide the Control System Operator," *Control Eng.*, vol. 27, no. 11, 1980, p. 87.

Bailey, S. J.: "Data Acquisition Systems Mature into Control Loop Decision Makers," *Control Eng.*, vol. 29, no. 2, 1982, p. 66.

Beaverstock, M. C., Stassen, H. G., and H. W. Schneider: "Modeling the Operator: A Needed Approach to Interface Design," *Instrum. Technol.*, vol. 26, no. 4, 1979, pp. 49–51.

Bylander, E. G.: *Electronic Displays*, McGraw-Hill, New York, 1979.

Crook, K.: "CRT Touch Panels Provide Maximum Flexibility in Computer Interaction," *Control Eng.*, vol. 23, no. 7, 1976, p. 33.

Dallimonti, R.: "Human Factors in Control Center Design," *Instrum. Technol.*, vol. 23, no. 5, 1976, p. 39.

Dobrowolski, M.: "Guide to Selecting Strip and Circular Chart Recorders," *Instrum. Technol.*, vol. 28, no. 10, 1981, p. 35.

Dobrowolski, M.: "Alarms and Annunciators: Technology Alert," *Instrum. Technol.*, vol. 28, no. 9, 1981, p. 16.

Gregory, S. K.: "Interactive Television for Control Systems," in *Advances in Instrumentation*, vol. 34, part 2, Instrument Society of America, Research Triangle Park, N.C., 1979, pp. 195–201.

Hanes, R. L.: "Design Alarm Systems to Favor Critical Information," *Control Eng.*, vol. 25, no. 2, 1978, pp. 122–128.

Hollands, D. H.: "Why Use Microprocessors in Something as Simple as an Annunciator?" *Control Eng.*, vol. 27, no. 2, 1980, pp. 49–51.

Hulshizer, M. L., and M. J. Kurimski: "Getting Quick Pen Response in an Accurate Recorder," *Control Eng.*, vol. 27, no. 3, 1980, pp. 136–140.

Jutila, J. M.: "New Directions for Instrumentation," *Instrum. Technol.*, vol. 27, no. 12, 1980, p. 9.

Jutila, J. M.: "Guide to Selecting Alarms and Annunciators, *Instrum. Technol.*, vol. 28, no. 3, 1981, p. 35.

Kreiss, D.: "SERs Move into Industry," *Control Eng.*, vol. 28, no. 11, 1981, p. 122.

Lundstrom, J. E.: "CRT Terminal Uses Mosaics to Build Process Graphics," *Control Eng.*, vol. 28, no. 2, 1981, p. 63.

Miller, T. J.: "Displays and Annunciators Acquire Color and Intelligence to Keep Pace with Electronic Control," *Control Eng.*, vol. 28, no. 10, 1981, pp. 111–114.

Miller, T. J.: "Digital Data Logging or Analog Recording? Microprocessors Blur the Distinction," *Control Eng.*, vol. 29, no. 4, 1982, p. 61.

Morris, H. M.: "Breakthroughs Bring More Colors, Wider Applications to Electronic Displays," *Control Eng.*, vol. 28, no. 6, 1981, p. 6.

Pluhar, K.: "Alarms and Annunciators Choices Range from Simple to Complex," *Control Eng.*, vol. 27, no. 10, 1980, p. 87.

Pluhar, K.: "Tracing Drawings Produces Custom Process Graphics on CRT," *Control Eng.*, vol. 28, no. 2, 1981, p. 64.

Pluhar, K.: "Recorder or Logger? Microcomputing Makes It Hard to Tell," *Control Eng.*, vol. 28, no. 7, 1981, p. 70.

Popovic, J. R., Ashewell, R. E., and J. E. Smith: "CRT Man-Machine Communication in Nuclear Power Stations," *IEEE Trans. Nucl. Sci.*, vol. NS-26, no. 1, February 1979.

Shirley, R. S., Campbell, B. D., and W. B. Robinson: "What's Needed at the Man/Machine Interface?" *Instrum. Technol.*, vol. 28, no. 3, 1981, p. 59–61.

Staff: "Selectable Sound Patterns Aid Alarm Identification," *Control Eng.*, vol. 24, no. 5, 1977, p. 28.

Staff: "Selecting CRT-Based Process Interfaces," *Instrum. Technol.*, vol. 26, no. 2, 1979, p. 28.

Staff: "Alarms and Annunciators," *Instrum. Technol.*, vol. 28, no. 9, 1981, p. 16–30.

Tucker, T. W.: "How to Organize Process Information for CRT Display," *Instrum. Technol.*, vol. 28, no. 11, 1981, p. 59.

Yamashita, T., and F. Hara: "The Use of Face Graphs in Man-Machine Communication," Proceedings of an International Conference on Cybernetics and Society, IEEE, New York, November 1979.

Zey, R. B.: "Using Interactive Color CRTs as Operator Interfaces," *Instrum. Technol.*, vol. 25, no. 12, 1978, p. 49.

Panel, Keyboard, and Control Switches

During the past decade, in addition to the continuing refinement in the design of traditional switching formats, developments and trends include (1) greater emphasis on human engineering factors (e.g., the tactile response of the operator in manipulating a switch), (2) the application of solid-state switching circuitry and the incorporation of such principles as the Hall effect, (3) expanding the role of switches in the total system (e.g., encoding switches), and (4) the integration of switches in large groupings in complex data input and data output systems (e.g., microcomputer-based touch panel keyboards), among others. During the recent past, *membrane switches* have undergone several generations of development.

Instrumentation and control panels and consoles require numerous manually operated controls for start-up, shutdown, troubleshooting, and emergency procedures, as well as to provide for semi-automatic operation of various equipment and processes. In addition to the use of switches on complex, centralized control panels, there are many hundreds of different applications for manually operated switches on the production floor, particularly in the discrete-piece manufacturing and assembly industries. The joystick switch, for example, is representative of a special switch designed to initiate movement in various directions under manual control. In addition to manually operated switches, light-capacity mechanically actuated switches are described briefly in this article. Proximity, limit, and other heavy-duty switches which are instrumentally controlled are described in the article Position, Displacement, Motion, and Object Detection Systems in Sec. 9. Mercury switches are described briefly in this immediate article, and a brief summary on relay trends is presented.

Guidelines

Well-established guidelines pertaining to switch design and application have been developed over the years. These include:

1. Legends should be easily read and controls located within reach of the operator's normal position.

2. The amount of operator effort required to actuate the control should be consistent with the speed and duration of actuation. For example, rapid-repeat actuations for long periods dictate use of a light-force pushbutton to lessen operator fatigue.

3. There should be a visual or touch feedback (or both) to indicate that the contacts have changed position. Examples of visual feedback are change in display state on a lighted pushbutton and change in actuator position on a toggle or rotary switch.

4. To ensure prompt and correct identification of a lighted display, the choice of colors should be limited to six—red, green, blue, white, yellow, and amber. Amber should be used in place of yellow when it is near white, or if lamps are used at derated voltage.

5. Adequate safeguards should be provided in cases where accidental actuation endangers the operator or equipment. Toggle switches can be furnished with pull-to-unlock levers. Toggle switches with low-profile rocker-type buttons provide snagproof actuation. Pushbutton switches can have hinged guards over the buttons or raised barriers between units that prevent two buttons from being pushed at once.

6. Selection of manual switches should complement normal habit-pattern reflexes. For example, where necessary, provision should be made for jogging a machine or machine component into position. Pushbutton switches commonly are used here because the operator can retain complete control while intermittently operating the button. Shifting a quantity from one point to another

is a common application for a two-position toggle switch or an alternate-action pushbutton with a lighted two-section screen.

7. Matte or nonreflecting panel finishes aid contrast and visibility. Matte dark gray or black yields good contrast to a display of any color.

8. The contrast of a lighted display increases as the ambient light level is diminished. This is especially appreciated with low-output lamps and projected color effects. The angle of a panel to the dominant source of ambient light is a major factor in contrast and visibility. Glare becomes a factor when ambient light is derived from undiffused or concentrated sources.

Traditional Switch Configurations

Because switches have been an intimate part of instrumentation and control for many decades, it is not surprising that there is a standard switch available to suit each practical, everyday need of the control field. Because of this vast variety, it is quite impractical to cover detailed specifications in this Handbook. Figures 1 to 13, some of which have extensive captions, are presented as a brief summation of available contemporary switch configurations. The more recent membrane-type switches are described in more detail later.

The Hall effect, which is being used in increasing numbers of manually operated switches, is described in some detail in the article Position, Displacement, Motion, and Object Detection Systems in Sec. 9, as previously mentioned. Light emitting diodes (LEDs) are also being used with increasing frequency in manual switches. In LED-illuminated switches, internal resistors control the LED current to a nominal 20 mA on 5- and 15-V dc devices. For devices without internal current limiting resistors, suitable external control of the LED current must be provided. It is suggested that a minimum of 5 V dc open-circuit voltage with an appropriate series resistance be used to drive LED devices. This minimizes the effect of temperature (current variation) on the forward voltage of the LED. The reverse breakdown voltage of the LED is 5 V minimum. In such instances, the following example illustrates a simple dc drive circuit and the equation used to determine the value of the series resistance.

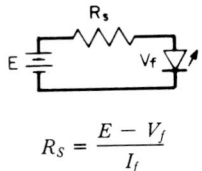

$$R_S = \frac{E - V_f}{I_f}$$

where R_S = series resistance
   E = supply voltage
   V_f = forward voltage of LED
   I_f = circuit current

If a diode is added in series for reverse polarity protection, then

$$R_S = \frac{E - V_f - V_{PD}}{I_f}$$

where V_{PD} = forward voltage of protection diode.

Checklist of Specifications

Factors which make up the specifications for a control switch, considering the large variety of styles and configurations, include:

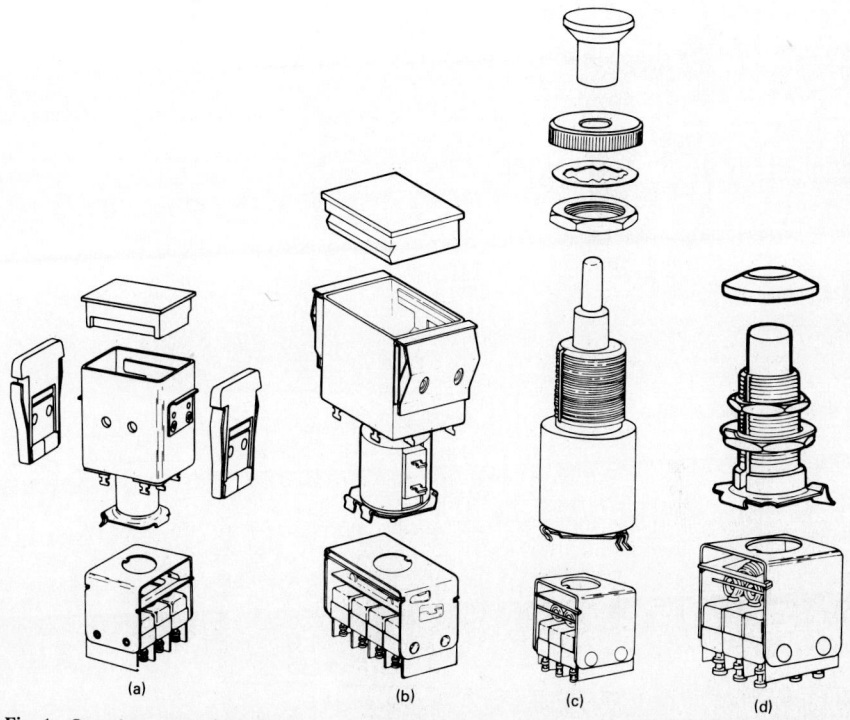

Fig. 1 One of a variety of pushbutton switches and indicators. Modules are designed for ease of assembly. Switches use up to four incandescent lamps and feature a one-, two-, three-, or four-section display and transmitted or projected color. An optional integral hold-in coil provides remote-released contacts. On the flange mount only, pull-in coils permit remote actuation. (*a*) Lighted square buttons with barriers that serve as mounting devices and separate the display screens to prevent accidental operation. (*b*) Same type of device, flange-mounted; flange-mounted units have mounting clips attached to the housing, and groupings can be separated by optional spacing barriers. (*c*) Heavy-duty sealed pushbutton switch. (*d*) Pushbutton switch with face nut fastening. *(Micro Switch.)*

1. **Panel Area and Switch Area Requirements.** Total square inches (millimeters) the switch, including all mounting brackets and other hardware, will occupy.

2. **Effect Display Area.** The actual size of the switch and bezel that will be viewed by the operator, considering placement of the switch (distance from the operator).

3. **Switching Action and Circuitry.** Number of poles and positions; momentary or maintained action. The circuit is usually defined as the contact arrangement with the switch actuator and contacts in their normal positions.

   a. **Single-pole single-throw (SPST).** A switch with only one moving and one stationary contact. Available either normally open (NO) or normally closed (NC).

   b. **Single-pole double-throw (SPDT).** A switch which may either make or break a circuit, depending on how it is wired.

   c. **Double-pole double-throw (DPDT).** A switch which makes and breaks two separate circuits; this circuit provides a normally open and a normally closed contact for each pole.

   d. **Bifurcated contact.** A movable contact which is forked to provide two contact mating surfaces in parallel, for more reliable contact.

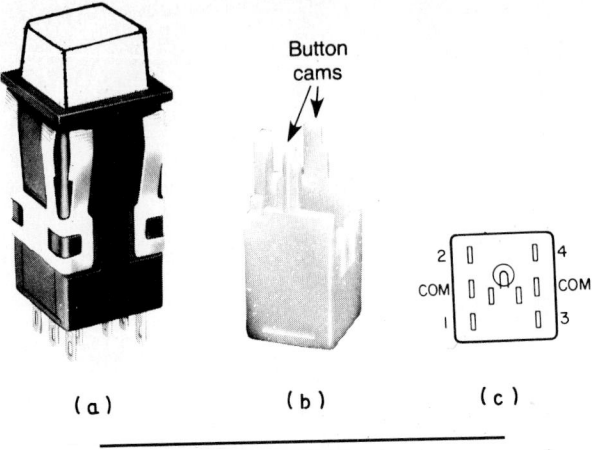

(a)   (b)   (c)

	Circuit			
Cam code	1	2	3	4
0	X	O	O	O
1	X	X	O	O
2	X	O	X	O
3	X	X	X	O
4	X	O	O	X
5	X	X	O	X
6	X	O	X	X
7	X	X	X	X

X, Circuit closure; O, circuit open.

Fig. 2 Electronic control pushbuttons have silver or gold contacts and handle up to 3 A 125 V dc, one, two, or four poles. (*a*) The particular switch shown here is an encoded switch. (*b*) Buttons on these switches each have a different plunger to perform the encoding function. The cams accomplish various contact closure arrangements to generate a binary-coded decimal output in the switch. (*c*) Circuit connection arrangement. Representative cam codes (others are available) are shown in the table. (*Micro Switch.*)

  e. *Maintained contact switch.* A switch designed for applications that require sustained contact after the plunger has been released, but with provision for resetting.

  f. *Momentary switch.* A switch with contacts that return from operated condition to normal condition when the actuating force is removed.

  g. *Pulse switch.* A switch that provides a single pulse of current for each cycle of operation.

  h. *Two-circuit switch.* A switch which, when in one position, moving contacts complete one circuit, and when in the other position, contacts complete another separate circuit.

  i. *Break distance.* The minimum open gap distance between stationary and movable objects.

 4. **Contact Rating.** Because of so many switch configurations and applications, standard contact ratings vary widely. Some of these are:

  10 μA to 0.15 A, 125 V ac, 28 V dc
  1 mA to 0.1 A, 5 to 28 V dc
  1 mA to 0.1 A, 50 V ac, 28 V dc
  0.1 to 5 A, 125 V ac
  0.1 to 15 A, 125 V ac
  3 A, 250 V ac
  10 A, 125 V dc
  10 A, 6 to 600 V ac
  15 A, 250 V ac

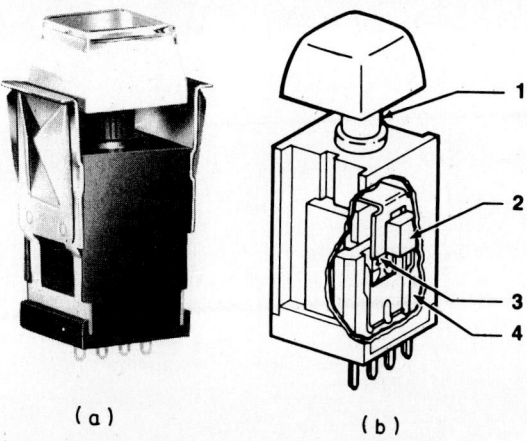

(a)   (b)

Fig. 3 Solid-state pushbutton switches are available in many types. They incorporate Hall-effect solid-state switching with bounce-free output. They are available for mounting on programmable controller boards or for snap-in panel mounting. (a) Snap-in panel mount, (b) sectional view of switch with a truncated button. (1) Button plunger assembly (the only moving part) which is made of durable plastic, (2) stable barium-ferrite-filled PVC magnet, (3) integrated circuit switching element, (4) switch outputs extending through frame. Depending on the style, the switches may be nonlighted or lighted with incandescent lamps or LEDs. Logic-scan modules with one isolated input and one isolated output are available. These modules are designed to readily interface with logic devices for microcomputer control. The output signal is valid when the interrogation signal (clock pulse) is low and the key is depressed. The input signal (at ground) will enable a logic gate; the output will reflect the normal performance of a level sinking output. When the input is at a high level, the output is inhibited by forcing the output transistor into the "off" state. *(Micro Switch.)*

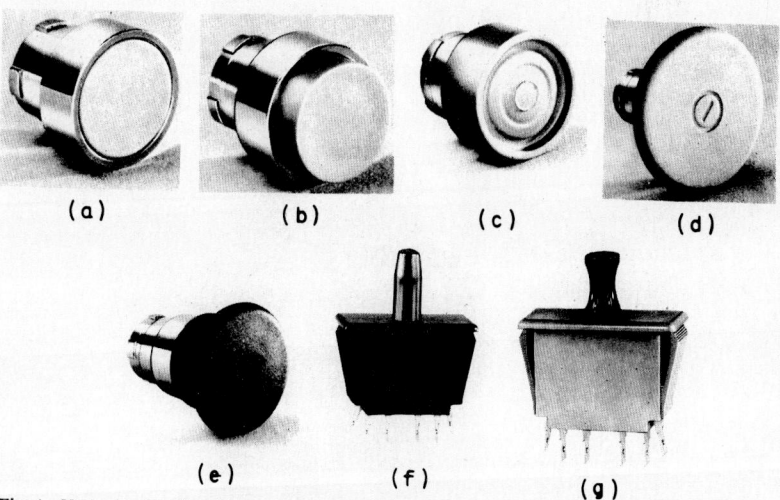

Fig. 4 Various styles of pushbutton heads. (a) Flush, (b) extended, (c) booted, (d) mushroom, and (e) push-and-pull-to-release. *(Telemecanique.)* (f) Bullet nose plunger, and (g) finger grip plunger. *(Micro Switch.)*

Fig. 5 High-voltage-rated control buttons. From left to right, pushbutton with pilot light, illuminated pushbutton, and wing lever. One to six NO or NC contact blocks can be fitted on pushbuttons. Operating range is from 6 to 600 V. Contact blocks are identical for the whole range. They are double-break, self-wiping, and cleaning. Contacts are nickel-silver with copper contact carriers. The nominal current is 10 A, and the terminals are captive screws and saddle clamps. *(Telemecanique.)*

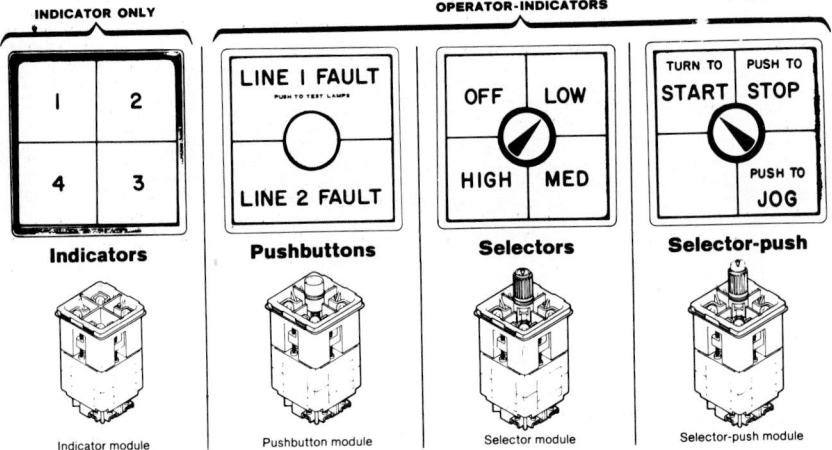

Fig. 6 Modular indicators and operator-indicators. The switches at the right illustrate the combination of push-buttons and selector buttons. *(Micro Switch.)*

Fig. 7 Three-position rotary selector switch with a wing lever. *(Micro Switch.)*

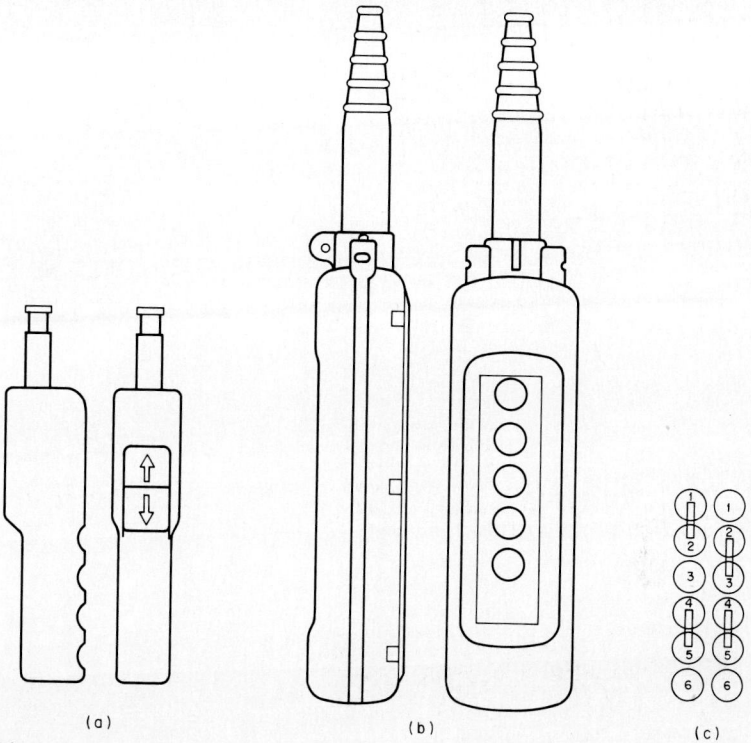

Fig. 8 Pendant pushbutton stations for the control of overhead and tower cranes, fixed and beam hoists, and other moving equipment in a production plant or warehouse. (*a*) Pistol-grip type with sliding switch; (*b*) multielement type (2-, 4-, 6-, 8-, and 12-element versions); (*c*) interlocking of switches is possible with the arrangement indicated. (*Telemecanique.*)

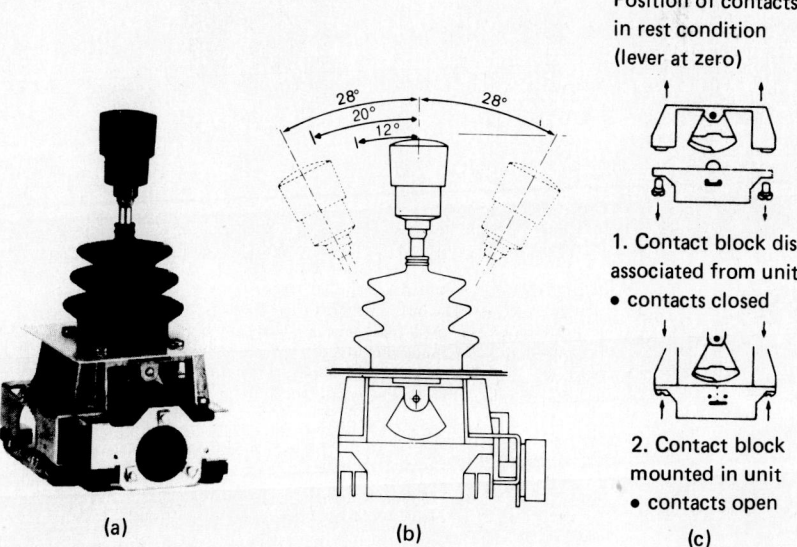

Position of contacts in rest condition (lever at zero)

1. Contact block disassociated from unit
   • contacts closed

2. Contact block mounted in unit
   • contacts open

Fig. 9 Joystick controller widely used in connection with light lifting equipment, cranes, and various mechanical handling machines. (*a*) External view of control switch, (*b*) sectional view, (*c*) contact action. (*Telemecanique.*)

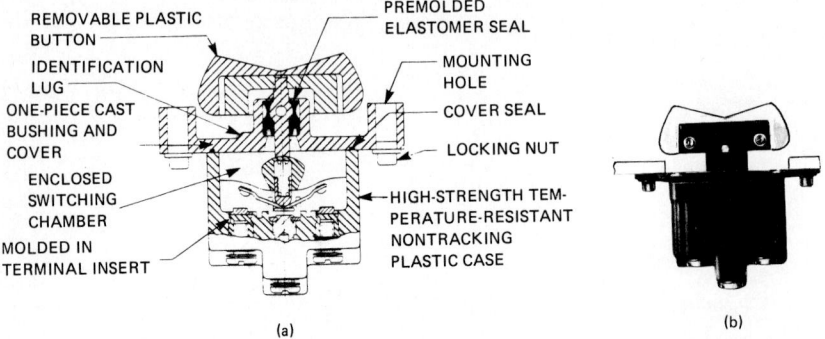

REMOVABLE PLASTIC BUTTON

IDENTIFICATION LUG

ONE-PIECE CAST BUSHING AND COVER

ENCLOSED SWITCHING CHAMBER

MOLDED IN TERMINAL INSERT

PREMOLDED ELASTOMER SEAL

MOUNTING HOLE

COVER SEAL

LOCKING NUT

HIGH-STRENGTH TEMPERATURE-RESISTANT NONTRACKING PLASTIC CASE

(a)

(b)

Fig. 10 Rocker button switch. Switches of this type are available with two- and three-position pushbutton action. Buttons are removable and interchangeable and are of two styles: *transparent* (colored plastic), which accept under-the-surface legend inserts for station and function identification; and *translucent* (white plastic), which have a clear appearance when unlighted and are effective for edge lighting. Colored buttons are effective for color coding switch functions. (*a*) Sectional view of switch, showing principal elements. (*b*) External view of swtich. Buttons may be mounted above or flush with the panel. The switch illustrated is for above-panel mounting. *(Micro Switch.)*

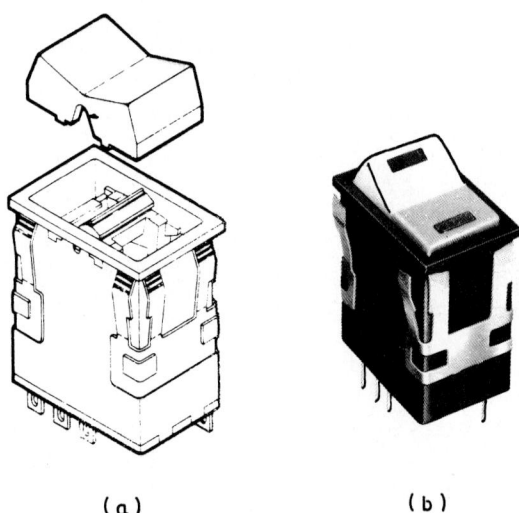

(a)

(b)

Fig. 11 Solid-state rocker button switch. (*a*) Detail of module and rocker, (*b*) external view of switch. The switch incorporates the Hall effect (described in the text) and LEDs which are flush with the rocker surface, providing wide-angle indication. Optional diode protection for the LEDs is available. The 5- and 15-V dc devices have an internal resistor to maintain the LED current at a nominal 20 mA. *(Micro Switch.)*

15.86

5. Service Life. From 1×10^6 to 3×10^7 operations

6. Relamping. Designs vary. This is an important consideration in the instrument maintenance program.

7. Termination. Quick-connect, solder terminals, screw terminals, programmable controller (PC) board, push-on.

8. Housing and Sealing. Panel seal, hermetically sealed, bushing and cover-to-cover seals:

   a. **Enclosed switch.** A switch housed in a durable metal case with the enclosure designed to protect the switching unit and to provide mounting means and fittings for a connection.

   b. **Environmental-proof switch.** A completely sealed switch for ensuring constant operating characteristics. Sealing may include an O ring on the actuator shaft and fused glass-to-metal terminal seals, or complete potting and an elastomer plunger-case seal.

   c. **Explosion-proof switch.** An Underwriters' Laboratories or otherwise listed switch capable of withstanding an internal explosion of a specified gas without igniting surrounding gases.

   d. **Hermetically sealed switch.** A switch completely sealed to provide constant operating characteristics. All junctures are made with metal-to-metal or glass-to-metal fusion.

   e. **Magnetic blowout switch.** A switch that contains a small permanent magnet which provides a means of switching high dc loads. The magnet deflects an arc to quench it.

   f. **Terminal enclosure.** A housing that fits over switch terminals to protect against electric shock and accidental shorting and to facilitate wiring.

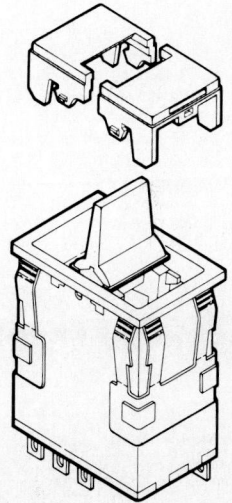

Fig. 12 The paddle configuration (so named because of the shape of the actuator) is a popular switch format. A wide variety of specifications are obtainable with this design, just as in the case of rocker button and toggle switches. *(Micro Switch.)*

9. Buttons and Actuators. Pushbutton, toggle, paddle, rocker, joystick, push-and-pull-to-release. Type of heads: flush, extended, boot, mushroom, bullet nose, finger grip, among others. Color and material used: plastic and/or metal. Color display options include:

   a. **Transmitted color.** Color can be distinguished whether the lamp is on or off.

   b. **Dead front.** Display appears black until illumination causes the legend and color to appear.

   c. **Projected color.** White display is diffused with color when illuminated.

10. Behind-Panel Depth Required. For example, standard versus low-profile designs.

11. Accessory Hardware. These items include mounting brackets, barriers, and switch guards.

   a. **Strip and matrix mount assemblies.** Where several switches are required in the same panel area, special mounting hardware can simplify installation. This hardware allows sub-panel mounting of devices in factory-assembled metal "cans" which are welded together in strips or matrices. Hardware also ensures accurate alignment and allows prewiring. Hardware also simplifies panel fabrication, since only one large cutout is required. Such hardware also facilitates printed circuit board mounting. The operating force is transmitted to the mounting hardware rather than to the PC board. See Fig. 14.

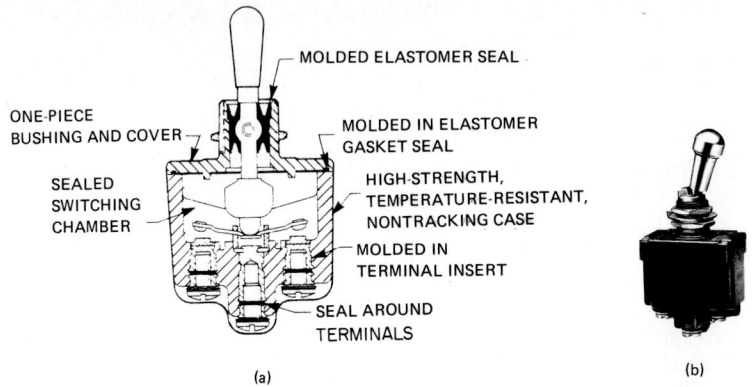

ONE-PIECE
BUSHING AND COVER

MOLDED ELASTOMER SEAL

MOLDED IN ELASTOMER
GASKET SEAL

SEALED
SWITCHING
CHAMBER

HIGH-STRENGTH,
TEMPERATURE-RESISTANT,
NONTRACKING CASE

MOLDED IN
TERMINAL INSERT

SEAL AROUND
TERMINALS

(a)

(b)

Fig. 13 Toggle switch with environment-proof sealing. Switches of this type are available with one-, two- and four-pole circuitry and two- and three-position, maintained, and momentary toggle action. The switches have an operating temperature range from −85 to 160°F (−65 to 71°C). Contacts are silver-cadmium oxide. (*a*) Sectional view of switch, showing principal elements. The standard toggle lever operates on a direct action spring-loaded toggle mechanism to provide tactile feedback in both the momentary and maintained toggle positions. The toggle lever is approximately 0.68 in (16 mm) long and has a nonglare matte nickel-plated finish. (*b*) External view of switch. Pull-to-unlock toggle levers prevent accidental toggle movement. The knobbed toggle lever must be pulled out approximately 0.09 in (2.3 mm) to change positions. (*Micro Switch.*)

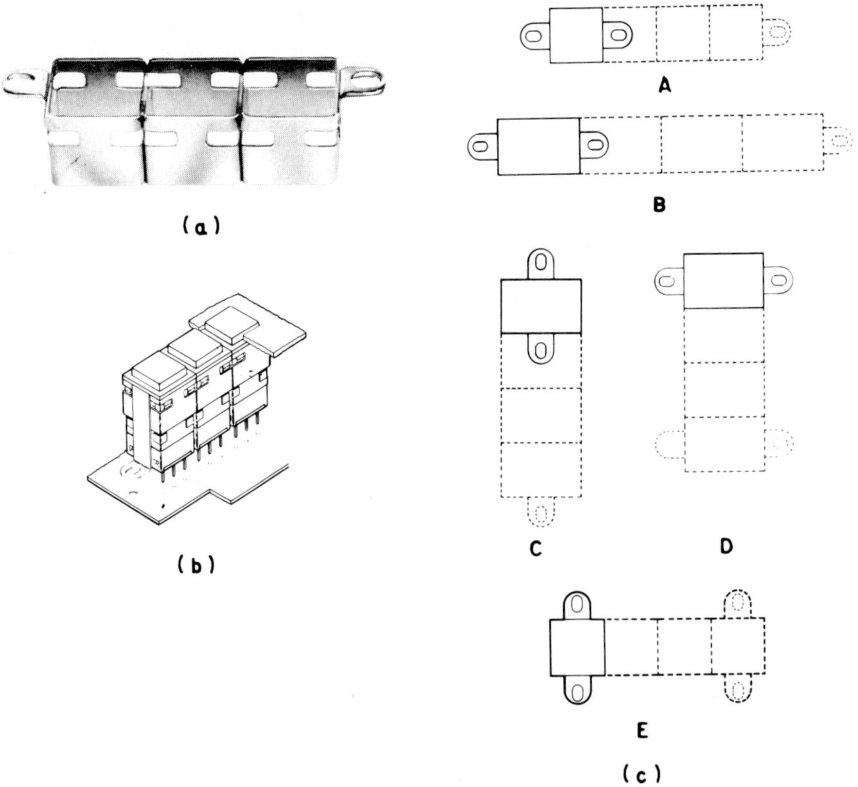

(a)

(b)

A

B

C

D

E

(c)

Fig. 14 Strip and matrix mounting hardware. (*a*) The strip shown can accommodate three rectangular switches. (*b*) The various strip configurations available are indicated. Note that both rectangular and square cans (receptacles) can be furnished. Configurations *A, C, D,* and *E* can accommodate up to 12 cans. The limit for configuration *B* is 8 cans. (*c*) Strip designed for PC board mounting. (*Micro Switch.*)

b. **Barriers.** These are separate, individually mounted switches and indicators; this helps to prevent inadvertent actuation of two pushbutton switches with a single push. Front-of-panel mounting simplifies installation. See Fig. 15.

c. **Panel Seals.** These provide protection from contamination from accidental beverage spills, dust, and dirt. They are easy to install (without tools), and have no effect on display color, light intensity, or legend quality. Seals can be conveniently replaced or removed for relamping without removing the switch from the panel.

d. **Switch Guards.** With a guard in place, a button cannot be operated, thus preventing accidental operation. In critical situations, a wire lock-down feature further prevents unintentional actuation of the switch. Lamps can be replaced with the switch guard attached. Switch guards can be installed initially or retrofitted onto existing switches. See Fig. 16.

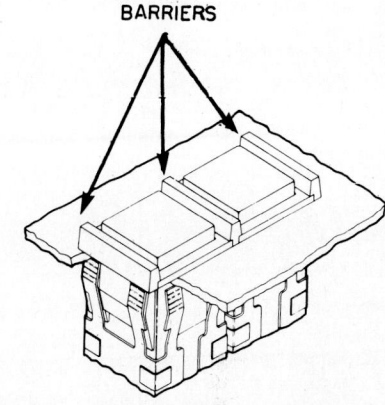

Fig. 15 Two pushbutton switches equipped with barriers which assist in preventing inadvertent actuation of adjacent switches. *(Micro Switch.)*

Membrane Switches

In recent years, membrane switches (sometimes called *touch panels*) have gained rather wide acceptance for certain applications in the fields of electronics and instrumentation and, notably, have been used frequently in peripherals, keyboards, data loggers, and, in fact, in a large number of consumer electronic appliances and products. Their acceptance is attributed to an inherent ruggedness and an immunity to dirty environments, as well as the manner in which graphics can be enhanced through their use. Costs and prices have ranged widely. Particularly in connection with more simplistic, lower quality membrane switches, the relatively low cost has been attractive to both switch manufacturer and user. For membrane switches of high quality designed to meet the needs of demanding equipment specifications, the cost advantage is less marked.

The cross section of a high-performance membrane switch is shown in Fig. 17a. The figure

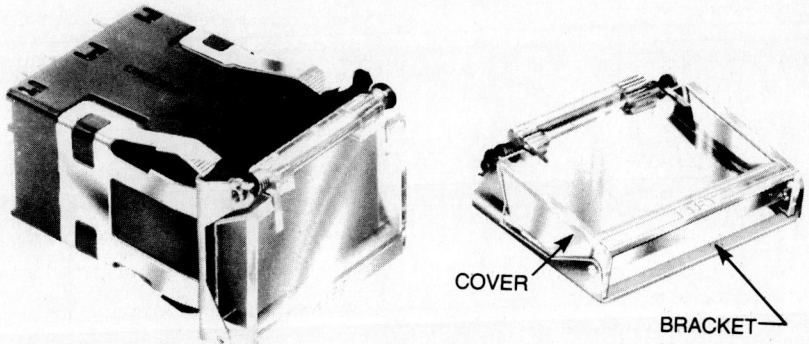

Fig. 16 Switch guard. (*a*) Top and side view of a pushbutton switch with a guard shown in the closed position; (*b*) view of cover and mounting bracket. *(Micro Switch.)*

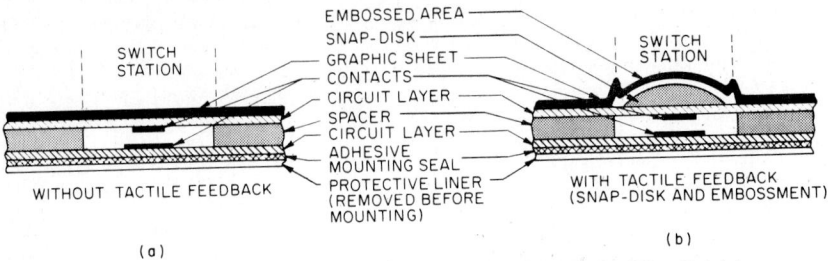

Fig. 17 Membrane switch. (*a*) Without tactile feedback, (*b*) with tactile feedback. *(Micro Switch.)*

Operating Voltage 0.5 to 30 V ac/dc

Operating Current 10 μA to 100 mA (resistive)

Operating Power 1 W (maximum)

Operating Force 12 oz (3.34 N), with tactile feedback
6 oz (1.67 N), without tactile feedback

Operating Temperature −20 to +65°C (−4 to +149°F)

Operating Life 2 million operations, with tactile feedback
5 million operations, without tactile feedback

Circuitry *xy* or common bus

Closed-circuit resistance 100 MΩ (typical), depending on size and layout

Open-circuit resistance 0 MΩ

Capacitance < 20 pF per station

Dimensions 0.5 × 0.5 in (12.7 × 12.7 mm), minimum station size

Travel 0.02 in (0.51 mm), with tactile feedback
0.01 in (0.25 mm), without tactile feedback

Storage Temperature −40 to +85°C (−40 to +185°F)

Altitude −100 to 10,000 ft (−30 to 3050 m)

illustrates the sandwichlike construction of a membrane switch. Contacts and circuitry paths are screened with conductive ink on two polyester films (circuit layers). They are separated by an insulating sheet (spacer) with holes that space the contact points. When a force is applied to a switch station, contact points "make" to close an SPST normally open circuit (SPST-NO). As shown in Fig. 17*b*, to impart tactile feedback from the switching action, a snap disk is located between the graphic sheet and the first circuit layer. Where tactile feedback is not desired, this feature can be eliminated.

The graphic sheet (touch surface) is shaped by embossing to provide tactile feedback from the switch station locations. There is a built-in provision for membrane venting, without disturbing the adhesive mounting seal. Prior to mounting the touch panel, the installer makes a small hole in the mounting surface which links with a matrix of channels in the membrane and a louvered opening in the lower circuit layer, allowing the internal switching chambers to "breathe," i.e., to be vented to the atmosphere. Thus, air pressure variations which can result from altitude and temperature changes are equalized to ensure continuous uniformity of operating characteristics. This is an important consideration with portable equipment that may be subjected to altitude changes, and it also has been found advantageous in lengthening the shelf life of membrane switches.

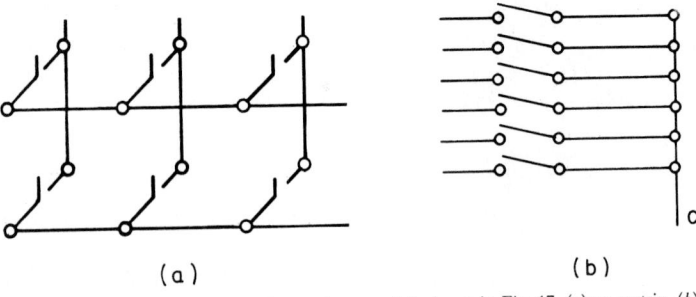

Fig. 18 Circuitry available with the membrane switch shown in Fig. 17. (*a*) *xy* matrix, (*b*) common bus.

Membrane switches essentially offer the same flexibility available with the more traditional switches. For example, multiple-contact configurations and interconnections are possible and frequently used in panels that incorporate large numbers of membrane switches. Such interconnections may be in the form of an xy matrix (Fig. 18a) or a common bus (Fig. 18b). The common bus actuates the function directly with no additional decoding required, but becomes increasingly difficult to use as the number of switch positions increases. In the xy matrix configuration, all the switches in each row are connected to an x output, and all the switches in each column are connected to a y output. Thus, each contact closure on the panel causes a distinct combination of xy outputs. These require a decoding device to interpret the xy outputs.

Keyboard Applications

Both traditional Hall-effect plunger snap-acting switches and contact or capacitance membrane switches are used in modern keyboard assemblies. Other switching principles used include inductive resistance. Commencing in the mid-1970s and continuing into the mid-1980s, the applications for keyboards in essentially all forms of electronic information handling systems—minicomputers, microprocessor-based devices, and other computer interfaces—have expanded severalfold.

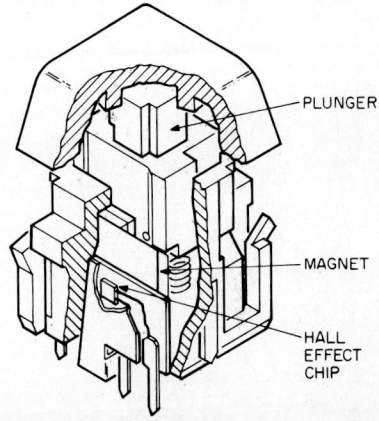

Fig. 19 Hall-effect snap-acting device. *(Micro Swtich.)*

Snap-Acting Devices

Introduced in the late-1960s for use in keyboards, the Hall-effect snap-acting switch was responsible for development of the first entirely solid-state keyboard. See Fig. 19. A Hall generator and a trigger circuit are integrated on a single silicon chip. The Hall element is sensitive to magnetic field strength

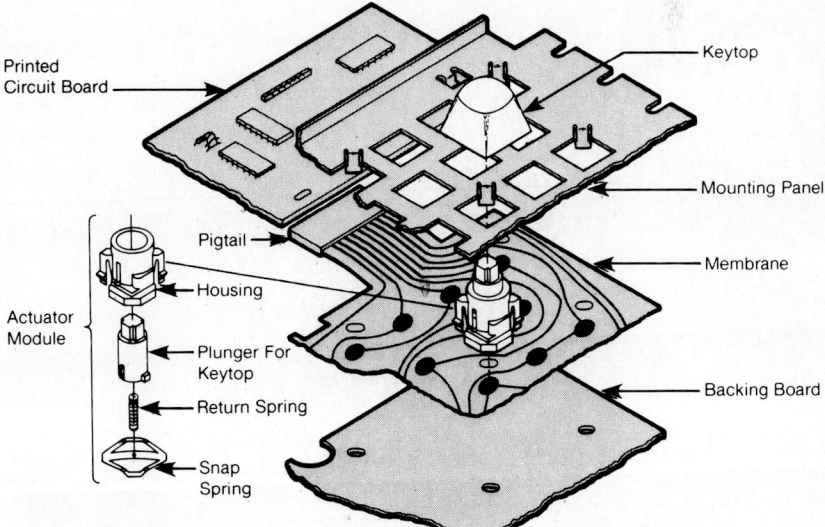

Fig. 20 Mounting configuration used for switch modules on a communications-oriented contact membrane keyboard. *(Micro Swtich.)*

and direction. A signal is generated as a magnet mounted on a plunger passes by the chip during actuation. In construction, the switching elements are sealed within the actuator module to protect them from the environment. Hall-effect keyboards are capable of handling very high throughput applications and continue to enjoy an impressive competitive position in the keyboard switching field.

Contact-Type Membrane Switches

Application of a contact-type membrane switch of the configuration previously described in a keyboard is shown in Fig. 20. Individual actuator module housings have a plunger for the key top, a compression return spring, and a snap spring actuator. When the key top is depressed, force is transmitted from the plunger through the compression spring to flex the hysteresis snap spring to close an SPST-NO circuit in the switching membrane, imparting tactile feedback from the switching action. When finger pressure is released, the snap spring relaxes, opening the circuit, and the key top self-returns.

Mechanical hysteresis is 0.010 in (0.25 mm) minimum (the difference between the operate and

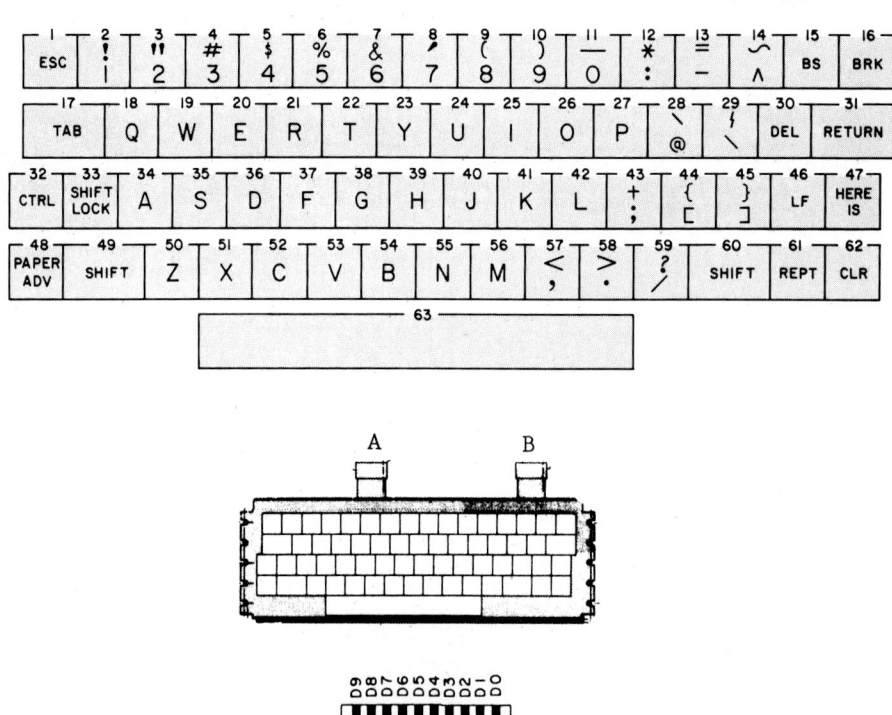

Termination Detail A

Termination Detail B

Fig. 21 Full-travel wired-only communications-oriented contact membrane keyboard. *(Micro Switch.)*

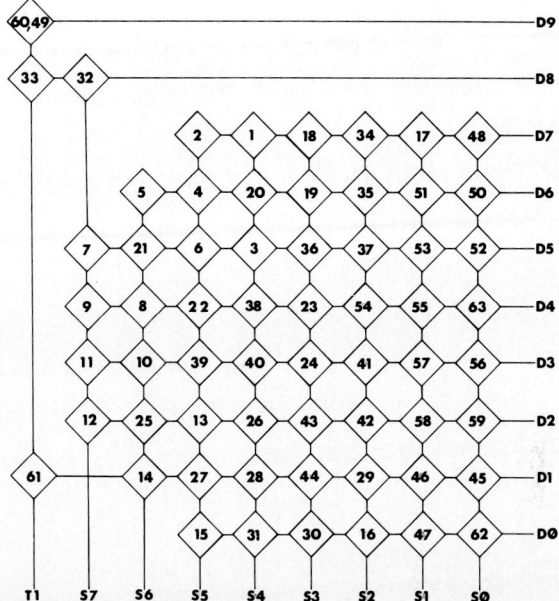

Fig. 22   Wiring termination matrix for the communications-oriented contact membrane keyboard shown in Fig. 21. *(Micro Switch.)*

release points). Combined with the tactile feedback feature, hysteresis helps to prevent slight hesitation or extra finger movement from causing erroneous signals to be generated. This combination of tactile feedback, good feel, and hysteresis promotes operator acceptance and reduces fatigue.

A representative keyboard employing a contact-type membrane switch is shown in Fig. 21. Actuator module plungers are stepped, with a ⅜-, ³⁄₁₆-, and ⅜-in (9.5-, 4.75-, and 9.5-mm) offset between rows within the staggered array. Key tops are spaced on ¾-in (19-mm) centers. The wiring termination matrix is shown in Fig. 22. This schematic provides a matrix of the circuitry terminations for each of the 63 key stations. Designations D0 through D9 terminate moving contacts, while T1 and S0 through S7 terminate stationary contacts. These designations are shown in termination details A and B given in Fig. 21.

Capacitance-Type Membrane Switches

Although similar in appearance to the membrane sandwich constructions previously described, the capacitance membrane switch utilizes capacitance rather than direct contact action. As shown in Fig.

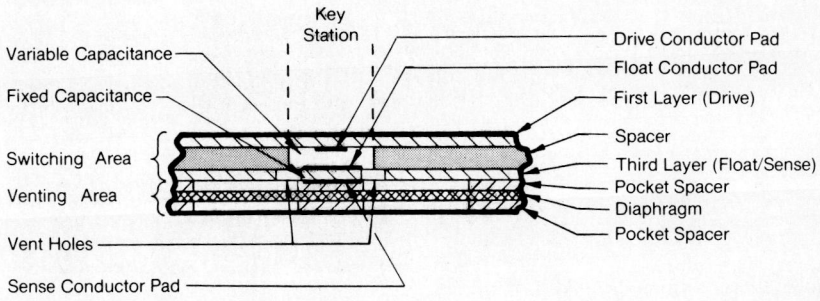

Fig. 23   Capacitance-type membrane switch. *(Micro Switch.)*

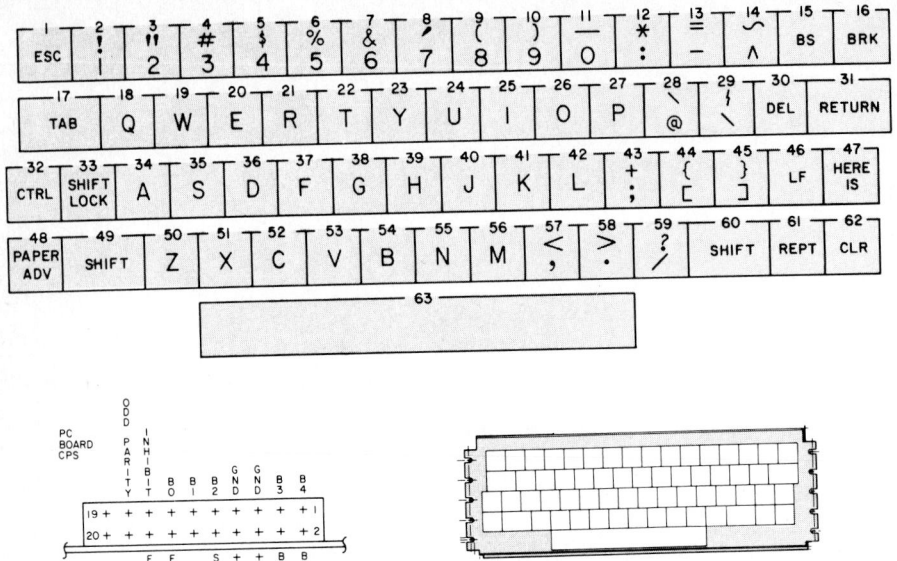

PC BOARD CPS

Termination detail

USASCII CODE ASSIGNMENT

Key #	Mode 1 Unshifted 87654 3210	2 Shifted 87654 3210	3 Control 87654 3210	4 Control Shifted 87654 3210
1	1001 1011	1001 1011	1001 1011	1001 1011
2	0011 0001	1010 0001	0011 0001	0011 0001
3	0011 0010	1010 0010	0011 0010	0011 0010
4	1011 0011	0010 0011	1011 0011	1011 0011
5	0011 0100	1010 0100	0011 0100	0011 0100
6	1011 0101	0010 0101	1011 0101	1011 0101
7	1011 0110	0010 0110	1011 0110	1011 0110
8	0011 0111	1010 0111	0011 0111	0011 0111
9	0011 1000	1010 1000	0011 1000	0011 1000
10	1011 1001	0010 1001	1011 1001	1011 1001
11	1011 0000	1101 1111	1011 0000	1011 0000
12	1011 1010	0010 1010	1011 1010	1011 1010
13	1010 1101	0011 1101	1010 1101	1010 1101
14	0101 1110	1111 1110	1001 1110	0101 1110
15	0000 1000	0000 1000	0000 1000	0000 1000
16	FUNC. 1	FUNC. 1	FUNC. 1	FUNC. 1
17	1000 1001	1000 1001	1000 1001	1000 1001
18	1111 0001	0101 0001	1001 0001	0101 0001
19	1111 0111	0101 0111	1001 0111	0101 0111
20	1110 0101	0100 0101	1000 0101	0100 0101
21	1111 0010	0101 0010	1001 0010	0101 0010
22	1111 0100	0101 0100	1001 0100	0101 0100
23	0111 1001	1101 1001	0001 1001	1101 1001
24	0111 0101	1101 0101	0001 0101	1101 0101
25	1110 1001	0100 1001	1000 1001	0100 1001
26	1110 1111	0100 1111	1000 1111	0100 1111
27	0111 0000	1101 0000	0001 0000	1101 0000
28	0100 0000	1110 0000	1000 0000	0100 0000
29	1101 1100	0111 1100	0001 1100	1101 1100
30	0111 1111	0111 1111	0111 1111	0111 1111
31	0000 1101	0000 1101	0000 1101	0000 1101
32	CTRL	CTRL	CTRL	CTRL

Key #	Mode 1 Unshifted 87654 3210	2 Shifted 87654 3210	3 Control 87654 3210	4 Control Shifted 87654 3210
33	SHIFT LOCK	SHIFT LOCK	SHIFT LOCK	SHIFT LOCK
34	0110 0001	1100 0001	0000 0001	1100 0001
35	0111 0011	1101 0011	0001 0011	1101 0011
36	0110 0100	1100 0100	0000 0100	1100 0100
37	1110 0110	0100 0110	1000 0110	0100 0110
38	0110 0111	1100 0111	0000 0111	1100 0111
39	0110 1000	1100 1000	0000 1000	1100 1000
40	1110 1010	0100 1010	1000 1010	0100 1010
41	0110 1011	1100 1011	0000 1011	1100 1011
42	1110 1100	0100 1100	1000 1100	0100 1100
43	0011 1011	1010 1011	0011 1011	0011 1011
44	0101 1011	1111 1011	1001 1011	1111 1011
45	0101 1101	1111 1101	1001 1101	1111 1101
46	1000 1010	1000 1010	1000 1010	1000 1010
47	FUNC. 2	FUNC. 2	FUNC. 2	FUNC. 2
48	0000 1011	0000 1011	0000 1011	0000 1011
49	SHIFT	SHIFT	SHIFT	SHIFT
50	0111 1010	1101 1010	0001 1010	1101 1010
51	1111 1000	0101 1000	1001 1000	0101 1000
52	1110 0011	0100 0011	1000 0011	0100 0011
53	0111 0110	1101 0110	0001 0110	1101 0110
54	0110 0010	1100 0010	0000 0010	1100 0010
55	0110 1110	1100 1110	0000 1110	1100 1110
56	0110 1101	1100 1101	0000 1101	1100 1101
57	0010 1100	1011 1100	0010 1100	0010 1100
58	1010 1110	0011 1110	1010 1110	1010 1110
59	0010 1111	1011 1111	0010 1111	0010 1111
60	SHIFT	SHIFT	SHIFT	SHIFT
61	REPT	REPT	REPT	REPT
62	0001 1111	0001 1111	0001 1111	0001 1111
63	0010 0000	0010 0000	0010 0000	0010 0000

Fig. 24 Keyboard arrangement for a microcomputer-based capacitance membrane communications keyboard, showing termination detail and USACII code assignments. *(Micro Switch.)*

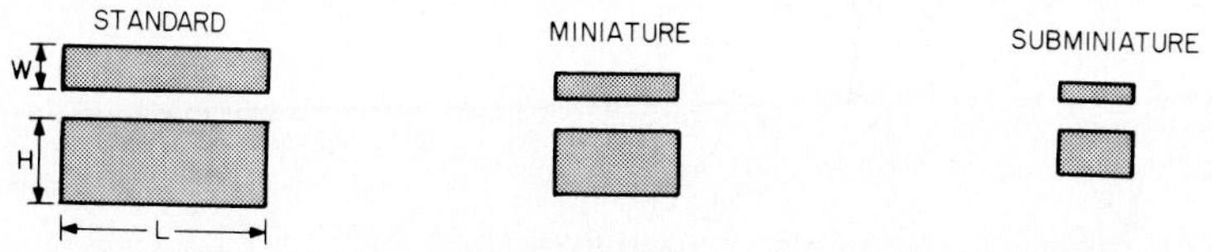

STANDARD	MINIATURE	SUBMINIATURE

Approximate Package Size

H 0.95–1.3 in (24–33 mm)	H 0.35–0.63 in (9–16 mm)	H 0.35 in (9 mm)
W 0.68–1.38 in (17–35 mm)	W 0.25–0.40 in (6–10 mm)	W 0.20 in (5 mm)
L 1.94–2.50 in (49–64 mm)	L 0.50–1.09 in (13–28 mm)	L 0.50 in (13 mm)

Representative Electrical Data

15–22 A; 20–25 A 125, 250, 480 V ac 10 A 125 V ac or V dc 10 A 125, 250 V ac 15 A 125, 240, 250, 480, 600 V ac	11A–15 A 125, 250, 480 V ac	28 V dc, 250 V ac

Minimum Operating Force

1–14 oz (28–396 g)	15–20 g (maximum)	6 g (maximum)

Minimum Differential Travel

0.004–0.109 in (0.10–2.8 mm)	0.001–0.002 in (maximum) (0.25–0.05 mm)	0.001 in (maximum) (0.25 mm)

Circuitry

SPST, SPDT, DPDT	SPDT	SPDT

Temperature Range

Upper: 250–400°F (121–204°C) Nominal: 180°F (82°C) Lower: −65°F (−53°C)	400–600°F (204–316°C) 180–250°F (82–121°C) −65°F (−53°C)	400°F (204°C) 180°F (82°C) −65°F (−53°C)

Fig. 25 Comparison of the sizes of standard, miniature, and subminiature basic switches, along with key specifications. The latter have been averaged to cover numerous designs and configurations.

23, sensing pads and conductive paths are screened on polyester films which form the first and third layers of the switch. Note that the third layer has pads on both sides. They are separated by an insulating layer with cutouts for each key location. The thickness of the third layer acts as a dielectric and very closely defines the closed-key capacitance for tighter control. There are two distinct capacitor circuits at every key location, namely, a fixed capacitance between the float and sense pads and a variable capacitance between the drive and float pads. When the drive pad approaches the sense pad (key plunger depressed), the variable capacitance increases to a level which triggers the detector circuit. Follow-through and touching of the pads are inconsequential. No second output occurs if the pads touch. This permits electrical and physical overtravel with no bounce or teasing, thus providing a clean, clear output.

The drive and sense lines terminate in pigtails which connect to the printed circuit board electronics. When a key is depressed, one sense circuit and one drive circuit are energized. The unused circuits are automatically grounded, shielding the keyboard from stray capacitance, cross-coupling of electrical noise that could generate erroneous codes.

Internal Venting

The lower three polyester film layers permit internal venting. There are two pocket spacer sheets with a thin diaphragm between them. Together they provide buffer chambers to stabilize air pressure in the switching area. When a key is depressed or changes in ambient temperature or pressure occur, the air passes through vent holes in the switching area and pushes against the diaphragm. When the key is released or the air cools and contracts, the diaphragm resumes its normal position. This occurs while the entire membrane remains completely sealed.

Mounting Arrangement

The manner in which capacitance membrane switches are mounted in a keyboard assembly is similar to that shown in Fig. 20. The individual actuator module consists of the housing, plunger, snap spring, and compression (return) spring. When the key top is depressed, force is transmitted from the plunger through the compression spring to the snap spring. The snap of the spring causes the capacitive pads to move toward each other and provides the operator positive feedback from the switching operation. The change in capacitance is detected, and an output results. When finger pressure is released, the snap spring returns, opening the circuit. The compression spring self-returns the key top. This combination of instantaneous operation, tactile feedback, and mechanical hysteresis reduces the possibility of teasing on the downstroke and multiple outputs.

The USACII code assignment and keyboard for a microcomputer-based capacitance membrane communications keyboard is shown in Fig. 24.

BASIC SWITCHES

Between panel and keyboard switches just described and the heavy-duty proximity, limit, and other position sensing switches described in Sec. 9 of this Handbook, there is a broad category of switches, sometimes aptly termed *basic switches*. These switches are often incorporated within instruments and controllers. Although applicable for manual actuation, they are usually actuated mechanically. Consequently there is a variety of designs available. These switches are characterized by their relatively small size. They are sometimes cataloged by manufacturers as standard, miniature, or subminiature. A size comparison, along with key specifications, of these three classifications is given in Fig. 25. In addition to the switches listed, there are other designs for handling high temperatures and severe industrial environments. Types of actuators commonly used with basic switches are illustrated in Fig. 26.

Standard basic switches are used for power load switching in numerous applications. Where there is a requirement for switching high-capacity systems, magnetic blowout designs are available. One design has double-break switching that is used in limit or control mechanisms in machine tools, presses, and other industrial equipment. Also, assemblies are available that put the control of two tandem-mounted basic switches under a common actuator.

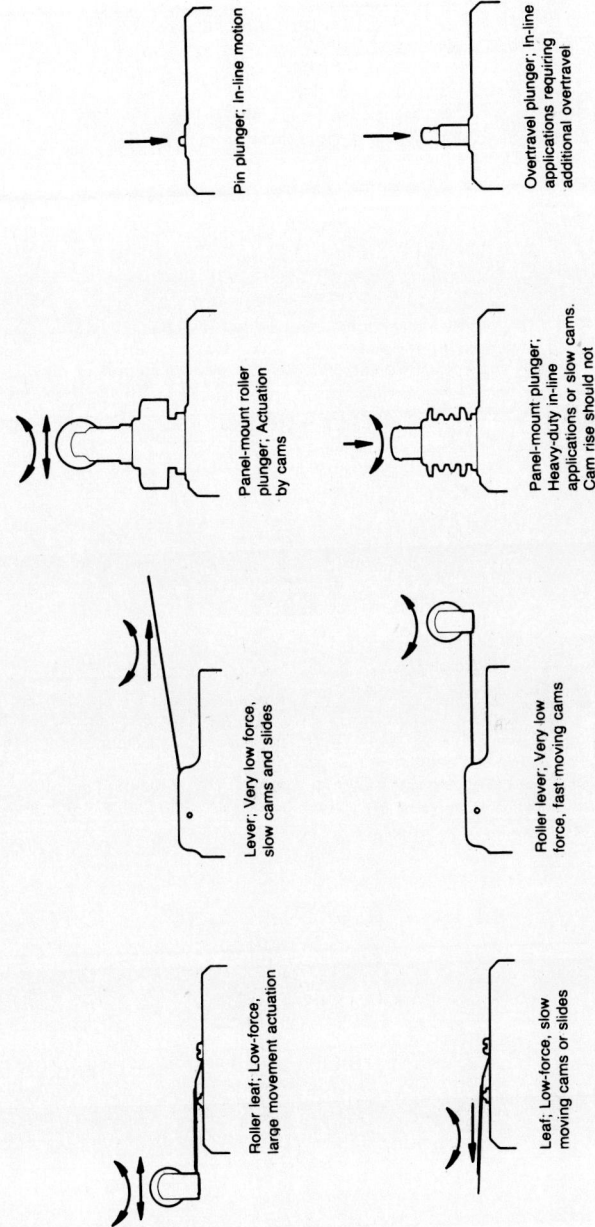

Fig. 26 Types of actuators used with basic switches. (*Micro Switch.*)

Pin plunger; In-line motion

Overtravel plunger; In-line applications requiring additional overtravel

Panel-mount roller plunger; Actuation by cams

Panel-mount plunger; Heavy-duty in-line applications or slow cams. Cam rise should not exceed 30°

Lever; Very low force, slow cams and slides

Roller lever; Very low force, fast moving cams

Roller leaf; Low-force, large movement actuation

Leaf; Low-force, slow moving cams or slides

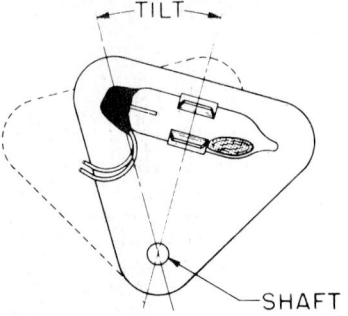

Over Center

"Over center" mounting. If the support mechanism has a small free motion, an extra "over center" snap-action can occur as the mercury will over-balance the device, picking up this free motion.

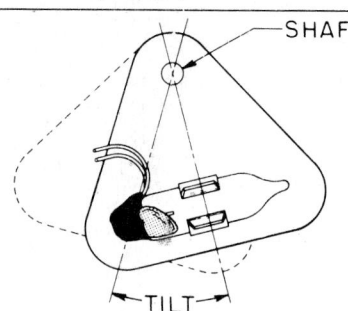

Below Center

If only minimum operating force is available, "below center" mounting could be used. Lowest operating force can be secured by counter-balancing the switch and its support. The actuating mechanism and switch tube move while the mercury almost remains stationary.

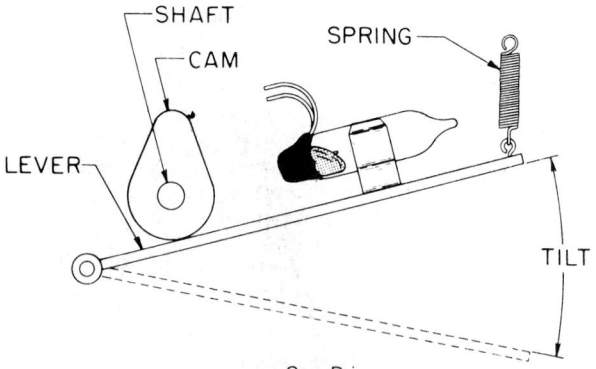

Cam Driven

A mercury switch can be actuated by a cam-driven lever.

Fig. 27 Mounting and actuation methods for mercury switches. *(Micro Switch.)*

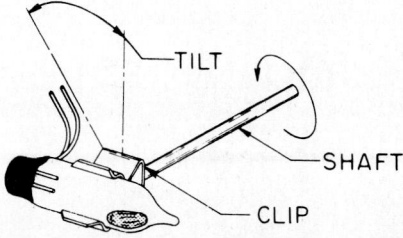

On Center

Where minimum operating space is a requirement, "on center"
mounting results in a minimum displacement of each end of the tube.

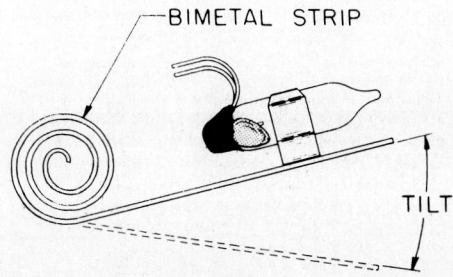

Temperature Control

For temperature control, the switch may be mounted on bimetal strip,
the angle varying with the temperature.

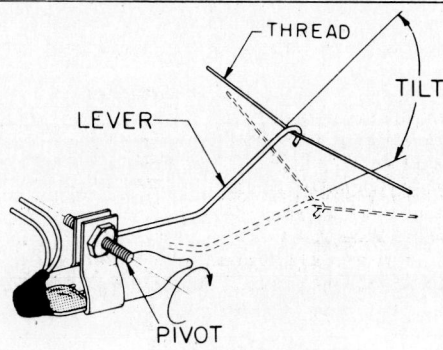

Pivoted Actuator

A pivoted actuator can be used with a mercury switch to detect a break
in thread or wire tension. When the strand parts or slacks, the actuator
arm drops and tilts the switch interrupting current to the reel motor.

Fig. 27 (*Continued*).

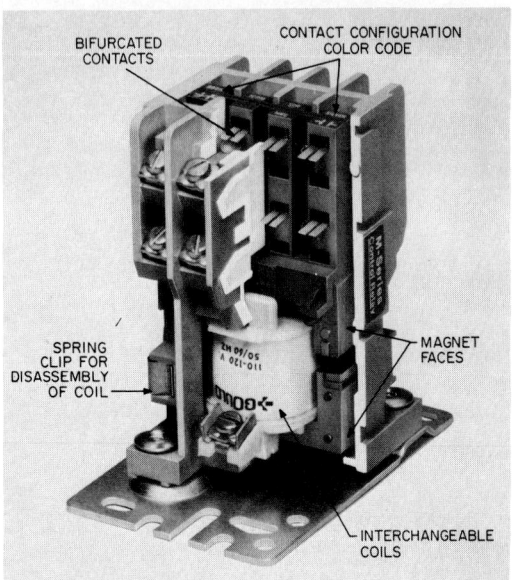

BIFURCATED CONTACTS

CONTACT CONFIGURATION COLOR CODE

SPRING CLIP FOR DISASSEMBLY OF COIL

MAGNET FACES

INTERCHANGEABLE COILS

Fig. 28 Modern electromagnetic control relay featuring bifurcated self-wiping contacts. Contact bounce has been reduced to less than 1 ms, which is particularly important where solid-state device interfaces or signal generating equipment may be included in a system. Part of the design simplicity claimed for the device shown is based on a study indicating that convertible, replaceable relay contacts are very infrequently changed or replaced once a relay has been installed. Relays are available in either four- or eight-pole versions, in any combination of contact configurations, but once they are set at the factory, the contact cannot be converted from normally open to normally closed, or vice versa. This simplication contributes to lowered costs. The relays are available with eight different ac coil voltages and six different dc coil voltages. Coils can be interchanged by disassembling the relay with a screwdriver. *(Gould.)*

Subminiature switches are small in size but have ample electrical capacities, including 11-A power load handling and a ¼ hp motor load. Most of these switches have gold and bifurcated gold contacts.

Mercury Switches

A conventional mercury switch uses stationary electrodes and a pool of mercury within a hermetically sealed glass tube and is particularly suited for very severe atmospheres and where the available operating force is extremely low. Actuation and release are accomplished by tilting the switch. Mounting and actuation methods for mercury switches are illustrated in Fig. 27. Mercury switches have a high electrical capacity—up to 20 A, with low contact resistance. They can operate in a temperature range from −65 to 500°F (−54 to 260°C). Installation is easy, operation is silent, and life is extremely long.

ELECTROMECHANICAL RELAYS

As of the mid-1980s, electromechanical relays (EMRs) are continuing to enjoy wide usage in industry even though developments during the past decade, including programmable controllers (PCs),

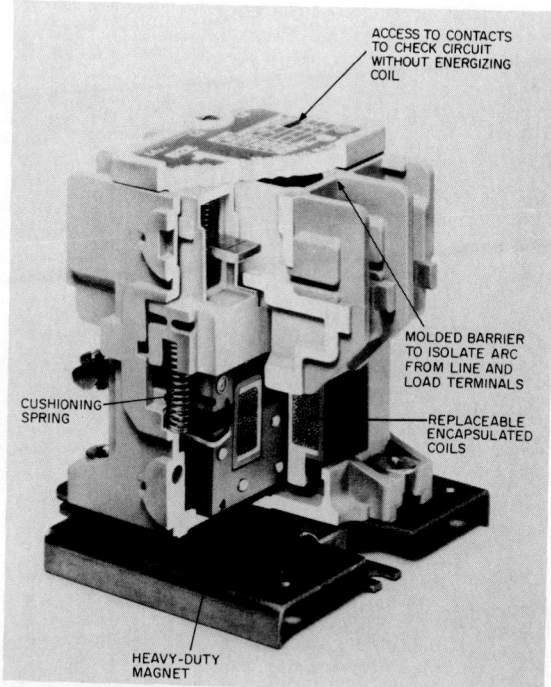

ACCESS TO CONTACTS
TO CHECK CIRCUIT
WITHOUT ENERGIZING
COIL

MOLDED BARRIER
TO ISOLATE ARC
FROM LINE AND
LOAD TERMINALS

CUSHIONING
SPRING

REPLACEABLE
ENCAPSULATED
COILS

HEAVY-DUTY
MAGNET

Fig. 29 Magnetic contactor with a 600-V rating for motor loads and resistive loads. The mounting area has been reduced to 11.5 in² (74 cm²) for small units and 14.6 in² (94 cm²) for larger units, allowing a significant reduction in the panel space required. The coil burden has been reduced to 3 W for small sizes and 4 W for larger sizes, permitting more contactors in a given panel because of the lower amount of heat generated. This also contributes to an improved energy efficiency ratio (EER). Ten different ac coil voltages and four different dc coil voltages are available. Coils can be replaced by disassembling the contactor. *(Gould.)*

were targeted to a large degree to displace EMRs. As pointed out in the article Programmable Controllers in Sec. 18, the first PC was designed in the late 1960s to replace relays and relay logic (but not the ladder diagram). Of course, as that article points out, the PC has grown in function and flexibility far beyond the original expectations. Relay manufacturers have taken full advantage of advancements in electronic technology and have streamlined their designs in the interest of lowering costs and increasing their competitive position. The relative economics of PCs and relays has tended to become more complex rather than simpler as EMRs improve and PCs continue along their evolutionary design path. Thus, it appears that the EMR will continue to be an important control component for many years.

Possibly the greatest impact of PCs on EMRs has occurred in connection with machine tool relays, although relays continue to be used in large numbers by machine tool builders. It has been reported that it is not the *same number* of relays per se required by a control and monitoring system that determines the economic breakpoint for relays and PCs. Rather, it is the amount of logic performed by the relays. For example, an economic study of one installation involving about 500 relays costed out at about one-half the price of a PC. The basic reason was that the system was used mostly for input-output of indication and alarm signals and for discrete signals to and from a computer, and there was very little logic. It must be recognized that PCs must have the same number of input-

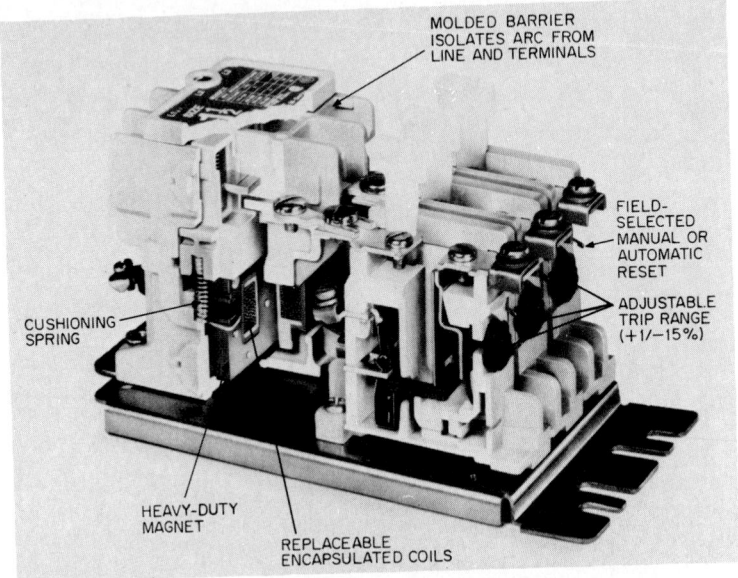

Fig. 30   Modern motor starter rated at 600 V with a choice of overload relays. Starters are available in six different sizes. Motor starters can be specified in a full-voltage nonreversible (FNVR) or in a full-voltage reversing (FVR) configuration. There are two basic types of overload relays—one that trips within 20 s at 600% of motor full load current, and one that trips within 10 s under the same conditions. The starter coil burden has been reduced to 3 W for smaller sizes and 4 W for larger sizes. Coils are available with 10 different ac voltages and 4 different dc voltage ratings. *(Gould.)*

outputs and, if they are solid state, costs run higher. Further, increasing relay logic is now possible with smaller, less expensive relays. It is also important to point out that the output of the solid-state relays (SSRs) of PCs generally require interposing EMRs in order to operate motor contactors.

A few examples of modern relay technology are illustrated and briefly described in Figs. 28 through 30. Engineering details of control relays are well covered in *Engineers' Relay Handbook,* 3d ed., National Association of Relay Manufacturers, Elkhart, Ind., 1981.

Voice Recognition and Speech Synthesis

The possible use of machines to recognize human voice commands has been under investigation for several decades, and considerable progress has been made since the mid-1970s. Although serious consideration has been given and some applications have gone somewhat beyond the experimental stage, voice recognition techniques have not yet made serious inroads in the process control field. The reverse process, speech synthesis, where machines "talk" to people, is a more recent technology dating back only a few decades. The primary applications of speech synthesis to date are found largely outside the process control field, in elevators, trains, subway systems, consumer appliances, toys, and games, and in automotive, nautical, and aeronautical instrumentation annunciators. Voice-interactive computer terminals, of course, are used for a large variety of applications.

Because voice recognition and speech synthesis may make important inroads in the process field during the next several years, these topics are described briefly here.

Voice Recognition

Speech sounds fall into two classifications: (1) *voiced* sounds which make use of the regular "buzz" of the vocal cords as their source, and (2) *unvoiced* sounds which are derived from the hiss of the breath passing through various narrow constrictions formed by the lips, teeth, and palate. Both these sounds are modulated by the adjustment of resonance chambers in the throat and mouth cavities. These sounds produce complex sound waves for carrying information which is encoded in frequency, intensity, and time. The human ear abstracts these messages generally in terms of spoken words.

Speech information has two parts: (1) a message content (what is said), which is, in terms of the receiver, usually referred to as *word recognition,* and (2) a source content (who said it), usually referred to as *speaker identification.*

A representative word recognition system, shown in Fig. 1, consists of a frequency analyzer, a feature extractor, sequential feature storage, and a word categorizer. The output is a word code. The code may locate the correctly spelled word in an auxiliary storage and cause it to be typed out. The purpose of the frequency analyzer is to subdivide complex voice waves into frequency components. In one type, there is a bank of frequency selectors or filters, usually arranged so that the filter characteristics overlap somewhat in order to cover a band of frequencies in the desired portion of the

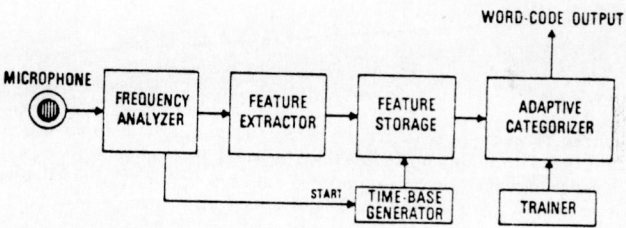

Fig. 1   Block diagram of a word recognition system.

frequency range from 100 to 10,000 Hz. For word recognition, a relatively small number of filters having desirable broad-bandpass characteristics can be used. A minimum practical filter bank is shown in Fig. 2. The feature extractor separates out identifiable elements of the frequency components. Normally, it includes some type of threshold, such as the floating-differential threshold system shown in Fig. 3. Detected features appear on output lines $H1$ to $H6$. The analog differentiators are integrating pulse shapers that operate when the voltage at the A input is greater than the threshold point E. The number of units that can be operated at any given time is set by the constant current source.

Serial sound elements for simultaneous observation are temporarily held in feature storage. This is usually a matrix of storage units as shown in Fig. 4. Each storage unit shown in the 6×6 matrix represents one of six features in a given time period $T1$ to $T6$. A bilateral flip-flop may be used for the storage unit. The time base generator determines the order of features being stored. Often it is provided by a gated multivibrator driving an open-ended ring. The ring is reset to the "talk" position and moves through stages $T1$ to $T6$, after a rise in the automatic-gain-controlled voltage indicates the start of a word.

Representative patterns are shown in Fig. 5 for short words. With wideband filters, speaker differences are minimized. However, note that speaker *A* has a midwestern background and speaker *B*, a southern accent. The word categorizer identifies spoken short words or word segments by pattern matching.

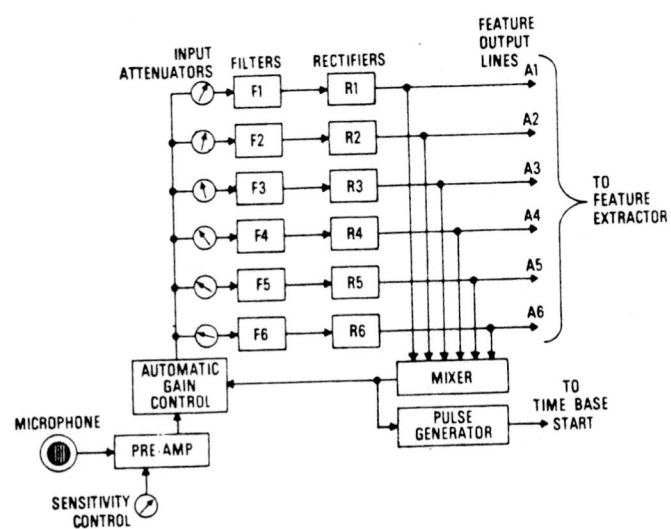

Fig. 2   Word recognition system frequency analyzer.

Speaker Identification

Differences that exist between speakers may be used to determine who is speaking. A frequency analysis is made in greater detail than that described for word recognition. A finer screen is needed to separate the voices of different speakers enunciating the same phrase. This is followed by feature summation, feature ranking and encoding, and adaptive categorization of the resultant speaker pattern. When a system has been set for a given speaker, that speaker alone will be able to match the pattern to the necessary degree and thus gain admittance, for example, to a bank vault. The range of frequencies analyzed for speaker identification should include the lower frequencies of the male fundamental pitch—70 to 80 Hz and upward. Also, an absolute rather than a floating threshold is used to identify the features and control summation. Another difference is that the time sequence is retained and used in the pattern for word recognition, whereas integration is used to remove sequential information for speaker identification, inasmuch as the identification of speech characteristics is more important than when they occurred.

Speech Synthesis

As pointed out by Smith and Weinrich (1980), the basic phonological element of speech is the phoneme. The *phoneme* represents a simple sound that by itself cannot distinguish different words. Phonemes, together with speaker inflection, volume, and emphasis, among other factors, are the fundamental building blocks of speech. The overall quality of any speech synthesizer, therefore, is directly controlled by its ability to faithfully reproduce all the necessary speech attributes and not just phoneme reproduction. The common American English language consists of approximately 38

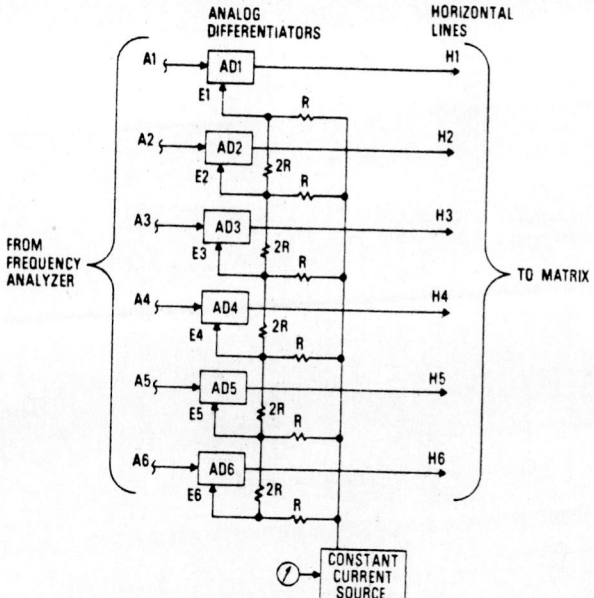

Fig. 3 Word recognition system feature extractor.

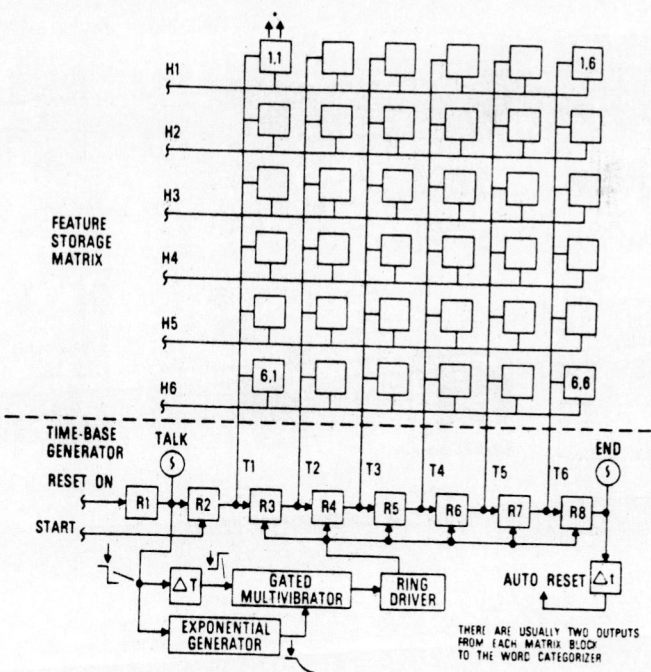

Fig. 4 Word recognition system feature storage.

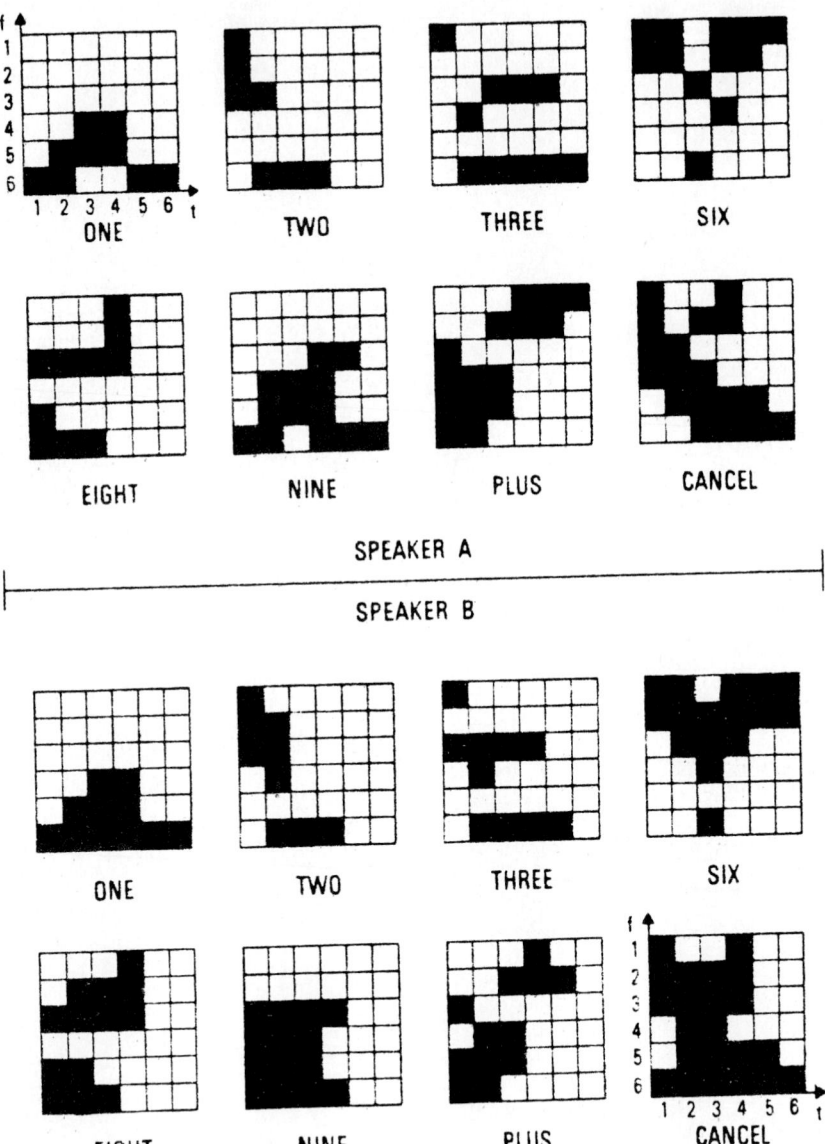

Fig. 5 Digitized word patterns stored in a word recognition system matrix.

to 40 phonemes—14 to 16 vowel sounds and 24 consonant sounds. Each phoneme is generated with either a voiced sound, as in "eye," or an unvoiced sound like the "sh" in "shy." This difference between a voiced and an unvoiced sound is important because unvoiced sounds are generally fewer in number and less dependent on the physiological characteristics of the speaker. A speech synthesizer can exploit this important difference. Normal speech rates are approximately 10 to 15 phonemes per second (including silence intervals). Since 38 to 40 phonemes can be coded using 6 bits, the normal bit rate for phoneme reproduction is approximately 60 to 90 bits/s. This bit rate, of course, refers only to phoneme information, which is only one of the many important speech attributes.

Some speech synthesis models use two driving functions—an impulse source for voiced sounds and a noise source (hiss noise) for unvoiced sounds. Each of these driving signals is filtered into specific frequency bands or formants by time varying filters.

Interest in talking machines stems from as far back as the ancient Greek and Roman civilizations. Only with the advent of electronics did speech synthesis pass from a curiosity to a potentially useful communication technique. Computers in partnership with the theory of sampled-data systems expanded the sophistication of speech synthesis and established a new application, namely, voice readout of computer-stored data. The ingredients of a computer voice response system are shown in Fig. 6. The machine is required to speak a message, typically specified in ordinary English, French, German, Spanish, etc., text. A synthesis program must access from a stored vocabulary a description of the required sequence of words. It must put them together head to tail, with the proper duration,

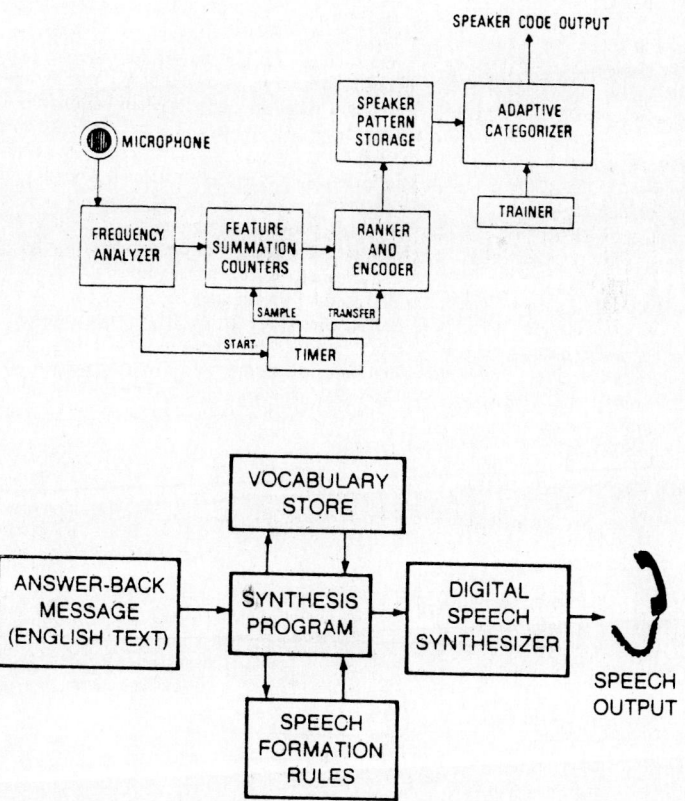

Fig. 6 Block diagram of a computer voice-response system. *(Bell Laboratories.)*

intensity, and inflection for the prescribed context. This description of the connected utterance is then given to a synthesizer device which generates a signal for transmission over a voice circuit.

Although many different approaches to speech synthesis exist, the techniques* are distinguished largely by the storage capacity needed for the vocabulary and by the complexity of the control rules for generating the speech. Three different techniques serve to illustrate the range of complexity: (1) *adaptive differential pulse code modulation* (ADPCM), (2) *formant synthesis,* and (3) *text synthesis.* Their typical bit rates for vocabulary storage are compared in Table 1.

The first and simplest technique utilizes a vocabulary composed of human-spoken words whose waveforms are digitally coded. The coding method is chosen to effect an advantageous compromise among signal quality, the storage capacity required for the word vocabulary, and the simplicity of the program for assembling the output messages. This coding is exemplified by ADPCM, typically at 24K bits/s.

Adaptive quantization applied in differential coding provides dynamic adjustment of quantizer step size according to signal characteristics, and it achieves a quality improvement to ~ 2 bits of quantization. As shown in Fig. 7a, the coder and decoder have the conventional DPCM arrangement, with the exception of a logic L. This logic observes the bit stream produced by the coder and adjusts the quantizer step size to minimize slope overload and granular distortion. Algorithms for adaption can involve memory for one or more past code words, and the rate of adjustment can vary between instantaneous and syllabic. Signal coding at 3 bits per sample, by an exponential logic that has one code word of memory, is shown in Fig. 7b. ADPCM is seen to provide a better estimate of the signal than conventional DPCM.

For message assembly, the synthesis program merely pulls the digitally coded words from disk storage in the required sequence and supplies them to an ADPCM decoder which produces an analog output signal. In this case, then, the digital synthesis device of Fig. 6 is a simple ADPCM decoder. The only control of *prosody* (accent, voice modulation) that can be effected in assembling the message is the adjustment of time intervals between successive words and possibly their relative intensities. No control of voice pitch or merging of vocal resonance at word boundaries is possible. This technique is more appropriate, therefore, for messages requiring a modest vocabulary size and having very lax semantic constraints. Typical messages are voice readout of numerals (such as telephone numbers), instructions for sequential operations (such as equipment assembly or wiring), stock price quotations, and inventory reporting.

This answer-back technique is currently being used experimentally for automated directory assistance, for voice-directed wiring, and for travel scheduling information. Because the instructions for tasks, such as point-to-point wiring in the assembly of electronic equipment, are not highly contextual and since the wiring list is often designed by a computer and supplied as a card deck or magnetic tape, a convenient means for converting the wiring list directly to spoken form is shown in Fig. 8. In this case, the computer voice output is recorded automatically on a cassette tape and taken to the work location. For example, in custom-wiring the main distributing frame of an electronic switching system, $\sim 13,000$ point-to-point connections (read automatically from as many punched cards representing the computer-designed wire list) may be used—representing a total of ~ 30 h of conversation (one-way).

*Source of the remainder of this article is Bell Laboratories.

Table 1 Comparison of Data Rates for Storage of Speech Vocabularies

Coding	Data rate, bits/s	Duration of speech in 10^6 bits of storage, min
ADPCM	20K	1
Formant synthesis	500	30
Text synthesis	70	240

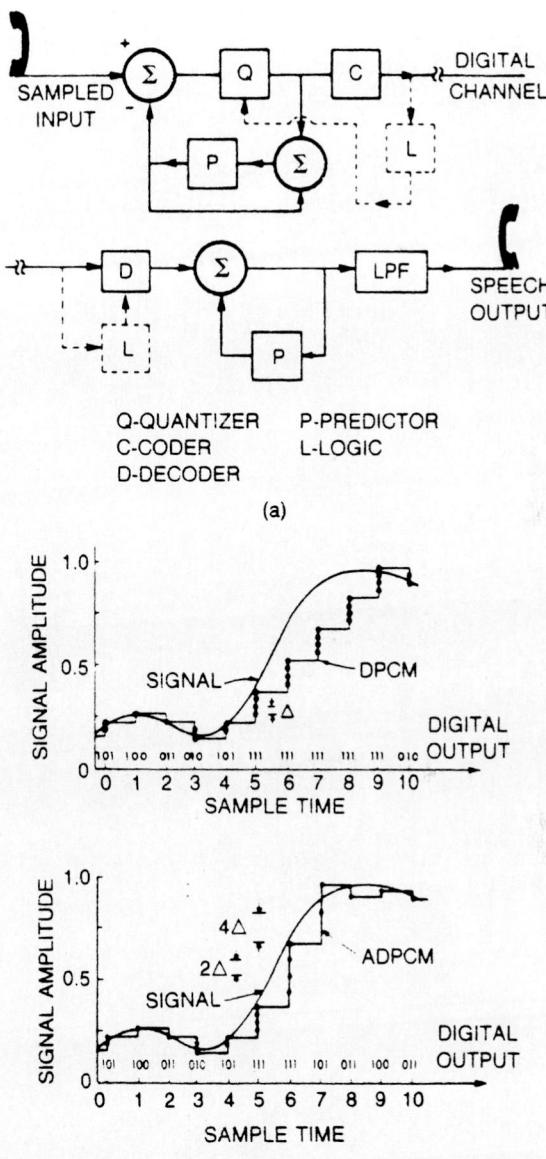

Fig. 7 Adaptive differential coding of speech. (*a*) Block diagram of a coder (adapted from U.S. Pat. 3 772 682). (*b*) Comparison of waveforms coded by 3-bit DPCM and ADPCM. *(Bell Laboratories.)*

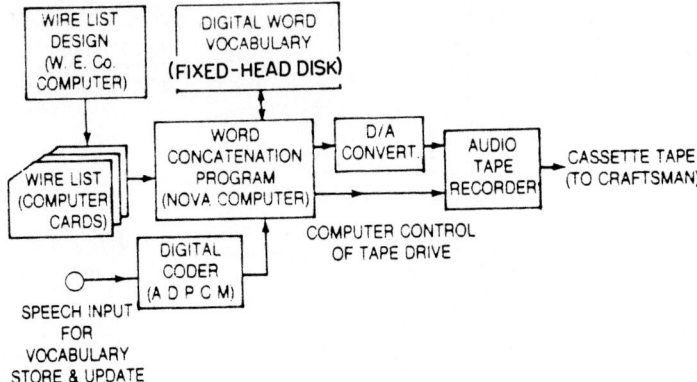

Fig. 8 Computer system for automatically coverting a wire list for telephone equipment into spoken instructions. *(Bell Laboratories.)*

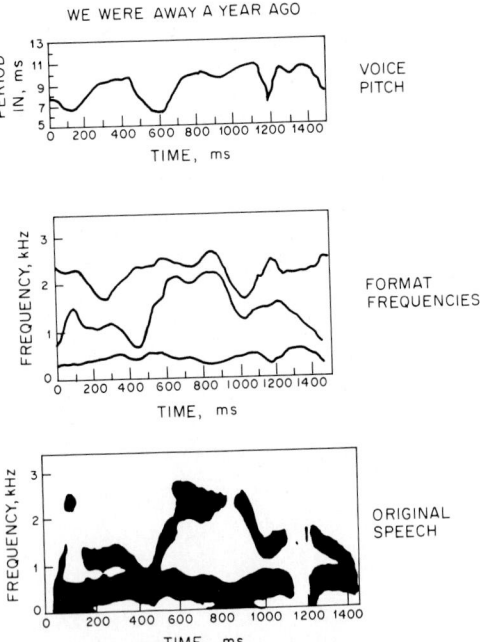

Fig. 9 Analysis of the sentence, "We were away a year ago." The sound spectrogram in the bottom panel shows the time variation of vocal tract resonances, or formants (dark areas). Computer-derived estimates of the formant frequencies and the voice pitch period are shown in the top two panels. *(Bell Laboratories.)*

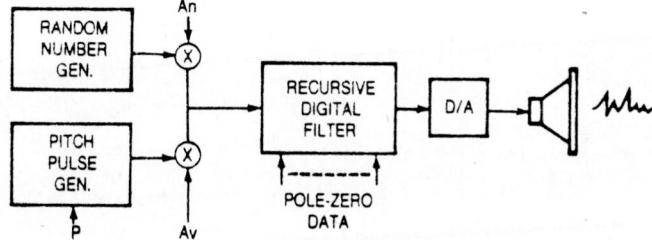

Fig. 10   Block diagram of a digital formant synthesizer. *(Bell Laboratories.)*

The second voice response technique, *formant synthesis,* utilizes a synthesis model in which a word library (again spoken initially by a human) is analyzed and parametrically stored as the time variations of vocal tract resonances, or formants. Speech formants are represented by the dark bars on the sound spectrogram in the bottom panel of Fig. 9. These frequency variations as well as the fundamental frequency, or voice pitch, can be computer-analyzed, as shown in the upper two panels of Fig. 9. These data can be represented by bit rates as low as 500 bits/s.

In formant synthesis, the synthesizer device is a digital filter, as shown in Fig. 10. The filter's transmission resonances and antiresonances (poles and zeroes) are controlled by the computer, and its input excitation is derived based on programmed rules for voice pitch for the amplitudes of voiced sounds and of noiselike unvoiced sounds. Because of the parametric description of the speech information, this synthesis method can control and modify the prosodic characteristics (pitch, duration, intensity) of synthetic speech, and it can produce smooth formant transitions at successive word boundaries. Like the ADPCM method, this technique has also been used experimentally for generating automatic-intercept messages (telephone numbers) and for producing wiring instructions by voice.

The third and most forward-looking voice response technique is called *text synthesis.* It strikes at the problem of storing and voice-accessing voluminous amounts of information or of converting any printed message to spoken form. It generates speech from an information input at a rate corresponding to that of a typewriter, ∼75 bits/s. In one design, the vocabulary is literally a stored pronoucing dictionary, each entry of which (just as in an ordinary dictionary) includes the word, a phonetic transcription of the word, word stress marks, and rudimentary syntax information. The synthesis program includes a syntax analyzer which examines the message to be generated and determines the role of each word in the sentence. Using the results of parsing by the syntax analyzer, the stored rules for prosody then calculate the sound intensity, sound duration, and voice pitch approx-

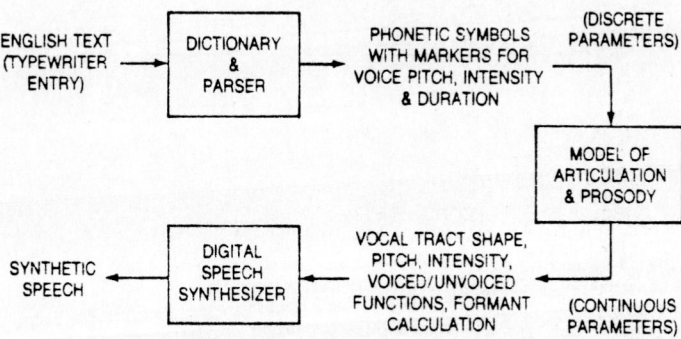

Fig. 11   System for synthesizing speech directly from printed text—functional diagram of system. *(Bell Laboratories.)*

imate for each phoneme in the particular context. See Fig. 11. At the heart of the synthesis program is a dynamic articulatory model of the human vocal tract. The calculated controls cause the vocal tract model to execute a sequence of "motions." These motions are described in terms of changes in the coefficients of a wave equation for sound propagation in a nonuniform tube. From the time varying equation, the eigenfrequencies (or formants) of the deforming tract are computed iteratively, 100 times per second. These resonant frequencies, along with the calculated pitch and intensity information, are given to the digital formant synthesizer shown in Fig. 10. This system, therefore, generates its synthetic speech completely from stored rules and without resorting to any vestige of human-produced speech. Language text can be typed directly into the system and the message synthesized on-line.

Since the system, through its model of articulation, already calculates physiologically based parameters in deriving controls for the formant synthesizer, the possibility exists of controlling a synthesis device which is more directly based on the human speech mechanism. A system of this type is shown in Fig. 12. This synthesizer approximates the human vocal system (schematicized at the top of Fig. 12) by the bilateral transmission line (shown at bottom of Fig. 12). Included in the transmission line formulation is an aerodynamic oscillator which models the behavior of human vocal cords. As input controls, the synthesizer requires physiologically based parameters corresponding to subglottal lung pressure, vocal cord tension and the rest area of opening, the cross-sectional shape of the vocal tract, and the area of coupling to the nasal tract.

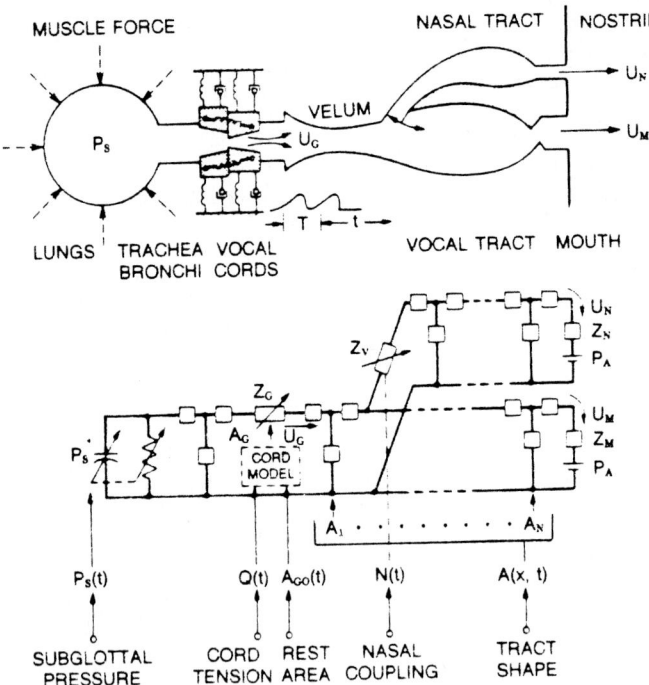

Fig. 12 Speech synthesizer based on computer models for vocal cord vibration, sound propagation in a yielding-wall tube, and turbulent flow generation at places of constriction. The control inputs are analogous to human physiology and represent subglottal air pressure in the lungs (P_s), vocal cord tension (Q) and the area of opening at rest (A_{GO}), the cross-sectional shape of the vocal tract $[A(x)]$, and the area of coupling to the nasal tract (N). *(Bell Laboratories.)*.

REFERENCES

Andreiev, N.: "Speech Synthesis," *Control Eng.,* vol. 28, no. 9, 1981, p. 95.

Atal, B. S.: "Effectiveness of Linear Prediction Characteristics of the Speech Wave for Automatic Speaker Identification and Verification," *J. Acoust. Soc. Am.,* vol. 55, 1974, pp. 1304–1312.

Flanagan, J. L.: "Computers That Talk and Listen," *Proc. IEEE,* vol. 64, no. 4, 1976, pp. 405–415.

Grunza, E. F.: "Voice Data Entry Can Improve Operator Performance," *Instrum. Technol.,* vol. 28, no. 11, 1981, p. 63.

Itakura, F.: "Minimum Prediction Residual Principle Applied to Speech Recognition," *IEEE Trans. Acoust. Speech Signal Process.* vol. ASSP-23, 1975, pp. 67–72.

National Semiconductor: Series of *Application Notes:* "MM54104 Digitalker Speech Synthesis System," December 1980; "DT100 Digitalker Speech Synthesis Evaluation Board," December 1980; "DT1050 Digitalker Standard Vocabulary Kit," December 1980; "DT10151 Digitalker Speech Evaulation Kit," June 1981; "DT1052 Digitalker Basic Numbers Kit," June 1981; "DT1056/DT 1057 Digitalker Standard Vocabulary Kit," August 1981; National Semiconductor, Santa Clara, Calif.

Olive, J. P.: "Fundamental Frequency Rules for the Synthesis of Simple Declarative English Sentences," *J. Acoust. Soc. Am.,* vol. 57, 1975, pp. 476–482.

Pfauth, M., and W. M. Fisher: "Voice Recognition Enters the Control Room," *Control Eng.,* vol. 30, no. 9, 1983, pp. 147–150.

Robinson, A. L.: "Communicating with Computers by Voice," *Science,* vol. 203, 1979, pp. 634–638, 734–736.

Smith, J., and D. Weinrich: "Speech Synthesis," *Bull. AN-252,* National Semiconductor, Santa Clara, Calif., 1980.

Control Centers—From Spec to Job Site

by
Richard W. Borut*

The effectiveness of an otherwise carefully engineered process control system can be thwarted if inadequate attention is given to the design, specification, and construction of the control center. The control center, whether a single panel of simplified design for local mounting or part of an extensive control system console, literally is the interface between the operator and the process.

Control Center Design Objectives

Among the more important of control center design objectives are:

1. To simplify the operator-process interface so that:

    a. The critical, most important process conditions are immediately available to the operator without a conscious effort.

*Section Manager, Instrument Engineering Division, M. W. Kellogg Company, Houston, Tex.

b. Less important information can be called on by the operator in accordance with operating schedules in a routine fashion, but not in any way interfere with the supervision of key variables. The center should be designed to effectively sort out and separate information elements so that the operator's attention during any time of emergency or special situations will not be diverted to information of lesser importance, or worse, to data of complete irrelevance to the situation.

2. To accommodate the special control situations which arise periodically, such as process start-up, shutdown, and major maintenance of either process or control equipment and instrumentation.

3. To provide information for cost accounting and other factors of managerial importance that may or may not be reflected immediately in the day-to-day adjustments of operating conditions.

4. To allow for expansion and modernization of the process. Spare capacity must be available in the existing instrument system to permit added instrumentation.

5. To allow for eventual upgrading of the instrument system. For example, the conversion from basic, simplified pneumatic controls to electronic direct digital controls. When present plants were originally built, in most cases no significant design features were included for later upgrading of the instrumentation. In a plant today, if a process computer is not tied in at the time of design, there is usually a plan for its later tie-in. Consideration must be given to the cost of providing for future improvements and the probability of later enhancement.

6. Provide a continuous record of data on unit operations for optimization analyses.

7. To permit convenience of installation and ease of maintenance of all panel-associated equipment.

8. To provide an aesthetically pleasing appearance (particularly in the case of control panels), because the control is essentially the "window to the process" and usually is the center of visits by VIPs. Further, an attractive control center, well lighted and air-conditioned, contributes to the efficiency of the operators and protection of the equipment.

All the foregoing objectives and more are subject, of course, to tradeoffs wherein cost must be weighed against projected factors, such as:

1. Selection of instruments in terms of their display and controllability features as well as size and maintainability.

2. Functional arrangements of the instruments, including annunciators and all other accessories. For example, pilot lights and selector switches must be located in accordance with the anthropometric principles developed in the study of human engineering to minimize mental and physical fatigue.

3. Lighting of the panel and control room and other environmental factors, including air-conditioning.

4. The most effective means of routing the field cables and leads to the control center must be chosen—either overhead in a hung ceiling or under a computer-type raised floor.

5. The point of transition between field wiring and control center wiring must also be considered, i.e., whether to wire into an interposing junction box or wire directly into the input-output cabinets furnished by the instrument vendor. This decision will be heavily influenced by the unit construction schedules. Experience has shown that the use of prefabricated plug-in cables between and interconnecting the vendor-furnished cabinets and the input-output cabinet is the most viable approach. The choice of who identifies the input-output points must also be made—the designer or the manufacturer. If the schedule permits, an orderly payout of instrument leads from the field cables can be achieved by the designer by specifying the location of the tie-ins in the I/O cabinets.

Control center design, with so many objectives to be achieved, including economic objectives, involves scores of tradeoff decisions. Only a few decades ago, panel design was largely determined by what was available in standard, unsophisticated instrumentation. Usually, the instruments were contained in large cases not particularly designed for remotely controlling a complex process from a central control panel. Earlier instruments were not designed with their mounting on a panel as a factor of ultimate importance. The situation has radically changed during the past 20 years, but not without going through several steps in panel design evolution, including full-graphic and semigraphic panel designs. Numerous panel design experiments have succeeded and are now commonplace; others have failed and have been dropped.

An early major step in the improved design of panels was the direct result of miniaturized electronic, electrical, and mechanical instrument components which, in turn, allowed the design of practical miniature instruments. Now, of course, there are analog-digital-analog control and direct digital control systems where the operator interface is through cathode ray tubes (CRTs). While the physical concept is somewhat different, the basic concepts of design are similar and may be generalized in a great many ways.

A thorough exploration of the many details that go into control center planning and engineering is not attempted in this article, but a representative example of modern control center design specification is described. In this instance, the major procedures and paperwork involved in the specification, construction, and installation of a control center are viewed from the standpoint of the engineering firm. Except for differences in some of the procedures, however, the fundamental factors remain the same—whether the user handles the control center engineering and installation fully as an in-house project, or whether the responsibility is turned over to a firm that specializes in control center engineering or to an instrument builder who has a special control center engineering department.

Basic Specifications

Preferably a company standard, this primary specification sums up all the company's control center requirements. In addition to the design criteria, this primary document is also a vehicle for informing the manufacturer of commercial procedures and other activities that must be undertaken to complete the contract. The standard specification should be general enough to cover all types of panels and control centers. It must also be sufficiently detailed to indicate requirements for each type of center normally used. The document is one of the most valuable assists in control center design and, ideally, should answer all the manufacturer's questions regarding design and material, from nameplates to wire and tubing material to switch application.

While allowing flexibility in points of little concern, the specification must be explicit in all points that are important. In one document of this type,* it is noted that panel stiffeners must keep the panel straight and true, but that the location of stiffeners is left to the manufacturer. It is noted that flanged spade lugs must be on the ends of all wires, but that the brand of lug to be used is selected by the panel fabricator.

The standard control center specification referred to has about 16 pages of fine print. See Table 1 for an outline of the table of contents. Although the initial drafting of a standard specification is a major effort, the document need be written only once, aside from periodic updating. This adaptive specification should be updated at least once a year to keep up with new ideas and the state of the art—and to tighten up areas of design that have resulted in past problems.

In addition to containing instructions to the control center manufacturer, the specification tells the engineering firm's clients exactly what will be furnished for their plant.

The Addendum

As the standard specification applies to the overall design and materials picture, it cannot cover all the specific requirements of each control center for each job. Thus, an overriding specification addendum is normally attached to each bid package.

The specification addendum takes into account any special requirements for a particular job. For example, test tees at all instruments on a pneumatic control center, or a change in the number of recorders per electric circuit, or a particular type of wire or wire identification method. These are typical deviations from the standard specification which may occur on a job. In this way, the rigid standard specification is made flexible by the addendum.

*M. W. Kellogg Company, Houston, Tex.

Table 1 Typical Table of Contents of a Standard Control Center Specification*

Section title	Comment
1.1 Scope	Defines drawings, codes, standards, and specifications to which the control center is to be manufactured
1.2 Bids	Instructs bidders on information to be submitted in their proposals
1.3 Drawings	Describes drawings the manufacturer is to develop and furnish; also states requirements for "as-built" documentation
2.1 General	Describes types of control centers and extent of fabrication
2.2 Mechanical	Specifies mechanical details of construction, namely, finish, joint stiffeners, mounting details, and shipping bolts
2.3 Graphic presentation	Specifies graphics colors, lines, and symbols
2.4 Nameplates	Specifies letter sizes, colors, engraving, and label attachment
2.5 Piping systems	Details pipe and tubing, supports, air header size and takeoff valves, air sets, tubing installation, bulkheads, and materials
2.6 Electrical systems	Details wiring specifications and methods, primary power distribution, and materials
2.7 Electrical instruments	Describes requirements of secondary power supplies, distribution method, wiring of chart drives, disconnects, and fusing
2.8 Electronic instruments	Covers voltage stabilization, wire layout, low voltage, dc power and signal distribution, auxiliary device installation, and fusing
2.9 Alarm and binary control systems	Details wiring and mounting of alarm switches, solenoid valves, relays, and annunciators
3.1 Inspection	Details tests prior to shipping, depth of functional tests, and test equipment to be furnished by control center manufacturer
3.2 Marking and shipping	Specifies crating, cargo type, bracing, weight limitations, and skids
3.3 Transportation by ship	Describes utilization of sealed containers and specifies above- or below-decks cargo
3.4 Rejection	Sets conditions of partial and complete rejection and extent of responsibility

*M. W. Kellogg Company, Houston, Tex.

Control centers are custom-engineered for every job. Experience has shown that allowances for client preferences, comments, changes, additions, and deletions must be included in the specification. The addendum also accomplishes these factors in a simple fashion.

Bid Request

The bid request and purchase order packages include a number of other documents, as indicated in Table 1, Item. 1.1. In addition to the standard control center specification and addendum, other forms and drawings must be sent to the panel manufacturer. When design changes occur, these drawings must be rigorously updated and promptly distributed as part of the general liaison effort.

Table of Instruments

All instruments and instrument components (being free-issued to the manufacturer) are listed by the instrument item number or tag number. The form also lists the manufacturer, component identification (recorder, controller, thermocouple-millivolt signal converter, model numbers, and the like), requisition or purchase order number, and check columns for case and drawing requirements. With this list, the panel manufacturer is able to report on deliveries of equipment, thereby lightening the purchaser's expediting efforts. Instruments reported as delivered to the panel manufacturer can be removed from the expediter's work list, but are again accounted for when the panel is inspected in the manufacturer's plant and when they are received in the field.

General Arrangement Drawing

The drawing in Fig. 1 is the key drawing for panel design. It locates the instruments, sets the panel size and shape, and goes into other details. Nameplate legends, for example, are listed on the arrangement drawing if space permits.

A comprehensive nameplate description of the specific instrument services (on the front as well as on all back-of-panel components) is important for smooth unit operation and convenient panel maintenance. All flush-mounted instruments are identified as to the instrument item number, service description, and multiplying factor in engineering units. Each back-of-panel component is identified with its own instrument item number or the item number of the instrument loop it serves. Back-of-panel nameplates should be durable, either engraved or of the raised-print type; they should not be fastened to removable covers that may be replaced on the wrong item. For distributed control systems (Fig. 2), the general layout drawing of the control center identifies the type and location of the various instruments and cabinets in the control center. Normally this drawing is fully dimensional and drawn to scale. It may also be used to indicate the limits of various vendor responsibilities. For instance, the alarm and binary control cabinets and the annunciator semigraphics may be furnished by a vendor other than the instrument control system manufacturer.

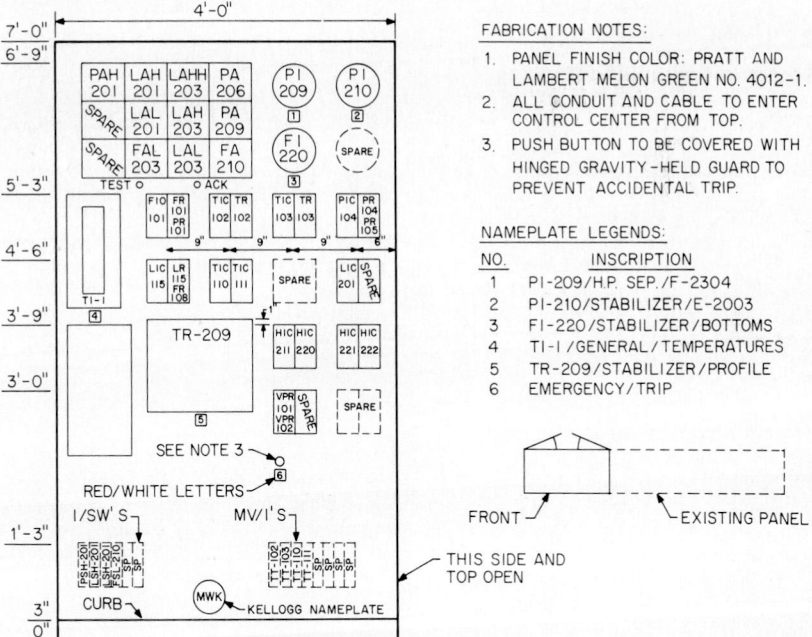

Fig. 1 General arrangement drawing locates the instruments, sets the panel size and shape, and presents other important details. Nameplate legends, for example, are listed on the arrangement drawing if space permits. Depending on preference and custom, English and/or metric dimensions may be used. *(M. W. Kellogg.)*

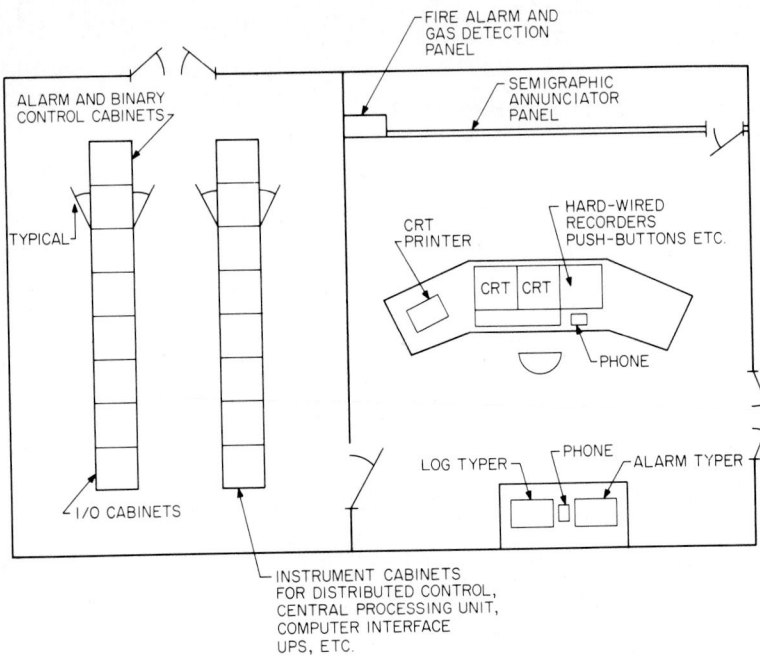

Fig. 2   A possible layout of control and peripheral equipment rooms. *(M. W. Kellogg.)*

Standard Fabrication Drawing

The field installation, the mounting, and the general fabrication of the panel should be shown. A standard detail drawing that specifies the panel height, method of fabrication, and typical assembly details is convenient to use. There is no need to standardize on panel widths. The fewer the number of separate panel sections, the better. This reduces the number of interconnecting joints in the panel steel, piping, and wiring.

This approach also reduces panel cost, as well as overall panel length. Construction crews can normally handle and install panels up to ~20 ft (~6 m) long without difficulty, and within their allocated installation work-hour budgets. Standards are not set up for local control centers, as local panels are furnished only when unavoidable. If a local panel is an absolute requirement, it is designed for the job and is seldom, if ever, duplicated for another one.

Schematics of the Piping

The standard specification describes the air supply distribution system—so a diagram is not needed. For example, it may state: "Air supply header shall be provided with a separate ¼-in NPS connection with an isolating valve for each air-consuming instrument, with an additional number of spare connections and valves totaling 10% of the air consuming instruments." The location of the connection for the field tie-in of the instrument air header is shown on the arrangement drawing.

Loop Diagrams

Single-line loop diagrams are drawn for all electronic instrumentation, but not for all the pneumatic loops. However, a loop diagram for a complicated circuit will be furnished. Pneumatic instrumentation has some standardization, such as connection sizes, air supplies, and signal pressures, whereas electronic instrumentation has relatively little standardization. Most clients accept loop descriptions for pneumatic instruments but insist on loop diagrams for electronic instruments.

The control center manufacturer requires only a simplified single-line loop diagram and a complete electronic component tabulation to design the electronic instrument installation for conventional electronic instruments.

For distributive- or computer-type instrumentation, a different single-line diagram is required. This diagram must indicate not only the hardware components of the loop but the software as well. This is particularly important in a system that has some common arithmetic functions generated externally.

The Instrument Society of America (ISA) is now working on ISA SP5.2, "Flow Diagram Symbols for Distributed Control/Shared Display Instrumentation Logic and Computer Systems." For two-wire loop diagrams, an arrangement can be made with the instrument manufacturer to furnish the loop interconnection diagrams with a blank area at the input-output interface. This allows the instrument installation designer to complete the loop by showing the field cabling and junction box tie-ins in the blank area. There is no need for both parties to draw physical details identifying instrument terminals and bulkheads at the control center interconnections. The control center manufacturer has designed the routing of the tubing and cabling on the back of the panel and, therefore, can readily identify and locate the field tie-ins.

Control systems should be designed so that most of the instrument loop components and interconnections are at the panel. When this is done, the field crew need only consider the transmitted signals in and the controlled signals out.

Alarms and trip circuits must be correct. To reduce the possibility of misinterpretation, ladder-type functional diagrams are developed which delineate the emergency trip devices and their place in the operational scheme. To assist the panel manufacturer and any future troubleshooter, the trip point (signal level) and the location of each of the various components are given by means of a small symbol; i.e., a small triangle adjacent to an alarm switch indicates that the switch is mounted behind the main board. If the triangle is filled in, it indicates that the device is located behind a local control center.

A recommended philosophy is that annunciator wiring for emergency trips or operational overrides be hard-wired, i.e., wired separately from the distributive control or computer system software and hardware. Additionally, they should operate on an electrical and logic system completely isolated from the computer.

Physical wiring diagrams of the annunciator and trip circuits are not furnished to the panel fabricator. The actual routing of the wires is left to the manufacturer's discretion. Similar to control loops, systems are designed so that most of the interconnecting logic circuits are within the control center unit. The relatively few field-mounted components minimize the number of field tie-in points to solenoid valves and actuating switches. The control center manufacturer has already been directed by the standard specification and ladder diagrams to identify these tie-in points by instrument item number or wire numbers on panel terminal blocks and physical wiring diagrams.

Electric power distribution, circuit breakers, and fuses are described in the standard specification. A tabulation identifies the items controlled in the primary and secondary power supply circuit. A one-line diagram is used if the distribution circuit is particularly complicated or unusual.

Inquiries and Changes

Thus far in this article it has been assumed that all the information given to the panel manufacturer and panel designer is complete at the time of the inquiry and that changes in the design will be minimal. However, schedules often demand that bids be issued on the basis of preliminary or incomplete data. Cost control of a panel purchased under these conditions is extremely difficult. To minimize this problem, a Class O Board Quotation (OBQ) form which fixes the value of any design change after the final quote may be used. See Fig. 3. This prevents unlimited future cost additions and allows the panel manufacturer to justify any price changes in the base contract.

The form for unit price additions and deductions is attached to each inquiry—to be completed and returned by the control center manufacturer, who must comply to be eligible for consideration. The bidder may add items to or amend this form with permission. One of the few exceptions is the escalation clause. For example, the bidder may state that the prices shown are to be discounted 10%

KELLOGG	JOB _____
CLASS "O" **BOARD QUOTATIONS** *BIDDER SHALL COMPLETE THIS FORM AND RETURN WITH QUOTE*	CLASS _____
	DATE: _____
	PAGE _____ OF _____
	□ INDOOR □ OUTDOOR
	N.E.C. □ GENERAL PURPOSE
	CLASS 1 GROUP "C" OR "D"
	□ 1 □ 2

	UNIT PRICES	
DESCRIPTION OF ITEMS ADDED OR DELECTD	ADD	DEDUCT
MINIATUARE INSTRUMENTS □ ELECTRONIC □ PNEUMATIC		
RECORDER		
RECORDER CONTROLLER		
SECOND PEN		
RATIO OR INDEX SET		
INDICATOR		
INDICATING CONTROLLER		
MANUAL CONTROLLER		
COVERPLATES		
ALARM AND 120 VOLT COMPONENTS		
ANNUNICIATOR ALARM (ASSUMING NO CABINET CHANGE)		
PILOT LIGHT		
PUSH BUTTON SWITCH (DPDT)		
SELECTOR SWITCH (SPOT)		
DISCONNECT DEVICE (FUSED)*		
DISCONNECT DEVICE (UNFUSED)*		
120 VOLT RELAY COIL WITH DPDT CONTACTS		
ADDITIONAL POLES FOR ALL OF ABOVE DEVICES		
ADDITIONAL WIRE TERMINATIONS (PER WIRE)		
SOLENOID VALVE		
AUXILIARY BACK OF PANEL MOUNTED INSTRUMENTS		
ALARM SWITCH (PNEUMATIC)		
ALARM SWITCH (ELECTRONIC)		
EMF CONVERTER		
INTEGRATOR		
MINIATURE INSTRUMENT CUT-OUT (ON PANEL ASS'Y FLOOR)		
NAMEPLATES - FRONT OF PANEL (WITH INSCRIPTION)		
NAMEPLATES - BACK OF PANEL (ITEM NO. ONLY)		
AIR SWITCHES 3-WAY (INCLUDING NAMEPLATE)		
AIR SWITCHES 4-WAY (INCLUDING NAMEPLATE)		
ADDITIONAL BLANK PANEL PER LIN. FT. (PAINTED NO CUTOUTS)		

NOTES:
1. UNIT INSTALLATION PRICE FOR ALL ITEMS INCLUDES THE COMPLETE INSTALLATION. I.E.: PANEL CUT-OUT, INSTRUMENT MOUNTING, PIPING, WIRING AND/OR POWER SUPPLY, ENGINEERING', ETC. AND WHERE NOTED THUS ( ) IS TO INCLUDE THE COST OF THE INDICATED EQUIPMENT.

2. PRICES SET FORTH HEREIN, WILL REMAIN FIRM FOR THE DURATION OF THIS PROJECT ONLY.

3. BIDDER SHALL COMPLETE ONE SHEET FOR EACH TYPE OF CONTROL PANEL SPECIFIED, I.E. OUTDOOR-INDOOR, GENERAL PURPOSE – DIVISION 2

4. EFFECT ON DELIVERY PROMISE TO BE ESTABLISHED AT THE TIME THE CHANGE IS MADE AND PURCHASER IS NOTIFIED.

5. NO DEVIATIONS ARE PERMITTED FROM THIS FORM WITHOUT THE PERMISSION OF PULLMAN KELLOGG

Fig. 3 OBQ form is a vehicle for establishing unit pricing at a particular control center. *(M. W. Kellogg.)*

before steel fabrication is started. Thereafter, the listed price is valid. In addition, the bidder may state that the prices shown will increase by a specified percentage after the panel has been painted or assembly has begun. This document enables the vendor to justify cost increases.

Every type of change need not be listed on the form. The key changes listed can be adapted or generalized to cover a multitude of other modifications.

Evaluation of the Quotation

Inquiries are sent to several panel manufacturers who have previously been accepted as eligible bidders. These firms have either successfully completed panel jobs for the buyer or have been investigated as to their capability, experience, and solvency.

The engineering firm's instrument design group determines not only the best price but also the best value. If the quotation states, "One (1) panel in accordance with your drawings and specifications," then there is nothing to evaluate. Vendors must explain furnished items in detail. This approach provides a basis for judgment and proves that the panel vendor understands the requirements and intends to honor a quotation in detail as well as in spirit. Factors, in decreasing order of importance, judged in a panel quote include:

1. *Total Price.* The total adds up to one control center.

2. *Bill of Material.* The bidder must produce a complete description of furnished equipment, including amount, type, manufacturer, and model numbers.

3. *Consistent Amounts.* For example, a bidder may quote 18 circuit breakers instead of the 24 units that are required and quoted by other bidders.

4. *Acceptable Hardware.* Hardware must be of industrial quality.

5. *Delivery Date.* The date must be realistic and fit into the job schedule.

6. *OBQ Form.* See Fig. 3. The bidder must complete the form; the price should be reasonable in comparison to those of other bidders. This is particularly important when significant numbers of changes are expected because of early ordering in the system.

7. *Estimated (or Firm) Freight Charges and Crating Costs.* The latter applies to overseas shipment, or if shipment is not made by a special "air ride" van.

8. *Past Performance.* This involves a comparison of relative merits. Although slowness on requotes is a nuisance, this might not be considered a valid reason for dropping a vendor.

Inspection and the Functional Test

Over the past several years, the scope of control center inspection has burgeoned to a complete functional testing of every loop in the panel manufacturer's shop. If a functional test is not possible, each tube or wire is checked or "rung out." Every item is checked out from instrument model numbers and scale ranges to jewel colors on pilot lights. The checkout follows a preconceived system.

During the functional test, the buyer's inspector lists all fabrication mistakes, additional requirements, missing components, incorrect nameplates, and other errors. At the conclusion of the tests, the inspector uses the list as a *back-check*. Before the inspector leaves the shop, a "punch list" is developed—including items to be completed before the equipment is shipped. One copy of the punch list is left with the manufacturer; another copy goes to the field instrument supervisor. The field supervisor then checks the control center to make sure all the items on the list have been taken care of. If they have not, the field supervisor then initiates a "back-charge" file.

The vendor's drawings are marked up during the inspection so that they reflect the "as-built" status. Two additional sets of the manufacturer's drawings are marked up with all the as-built notations: One set is included with the panel shipment (if the updated drawings are not ready when the center is shipped), and the other is kept by the inspector for reference. The first, or original working set, is for the use of the manufacturer in updating the originals prior to making the official drawing issue. Requirements vary with each job as to the number of sets of as-built prints and the type of reproducible drawings.

The extensive panel checkout program has proved its worth in many ways:

1. *Increases Efficiency.* The functional test is usually performed by the process engineering firm's designer who developed the control center design. The designer works with technicians who were the original installers of the wiring and piping in the panel. This combination efficiently troubleshoots errors likely to be found within the control center.

2. *Improves Workmanship.* Any wiring or piping rework found necessary as a result of the test is done by the vendor's technicians who are usually trained for this specialized type of installation. Experience has shown that the workmanship and appearance of rework done in the shop is far

superior to that done in the field. Field mechanics are capable, but they work in all phases of their craft and cannot obtain specialized experience.

3. *Saves Time.* A complete check of the instruments at the shop gives a few weeks' margin for changes involving wrong instrument models, scales, and charts. With present construction schedules, control center installation occurs during the period of the job near the start-up date. Any opportunity to save time during this critical stage should not be overlooked.

4. *Minimizes Back-Charges.* Although the manufacturer is legally required to meet specifications, back-charges based on field trouble reports are always troublesome and time-consuming and usually never come close to the actual cost of materials or lost time. For example, a field electrical crew, consisting of perhaps two electricians, a foreman, and a supervisor from the engineering firm, may spend several exhaustive hours tracking down a wiring error in the panel only to find a *single* mistermination. It is virtually impossible to justify a 24-work-hour back-charge for a one-wire error.

5. *Simplifies Field Troubleshooting.* Once the panel has been installed and all the field connections made, malfunctions in the system can be considered external to the panel.

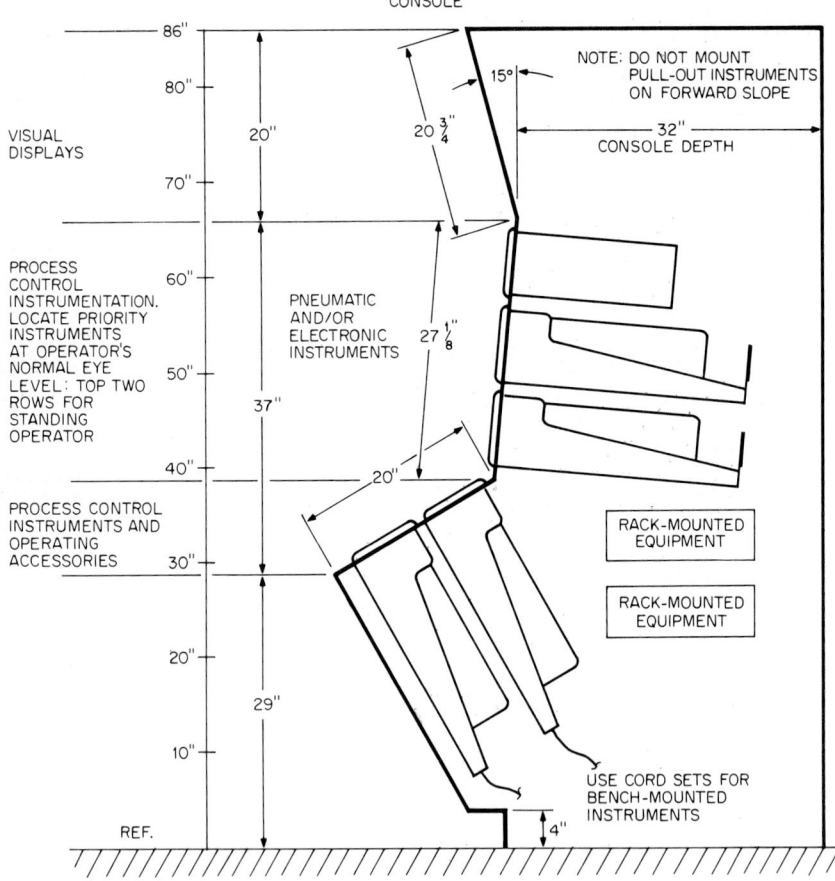

Fig. 4   Console designed to meet anthropometric requirements.

At the Job Site

According to the engineering firm's critical path scheduling, the control center may be scheduled to arrive at the job site before completion of the control room roof. The equipment can then be lifted from the van and put directly in place. Field handling and unnecessary jostling thus can be minimized. Conferences during the early stages of design ensure that doorways accessing the control room are sufficiently high and wide to admit the control center in case of late delivery. Special protection should be considered to guard the control center from weather and the construction environment. The construction instrument supervisor visually inspects the control panel immediately on its arrival for damage incurred in transit. The supervisor also utilizes the inspector's previously mentioned punch list to be sure that all the listed corrections and changes have been completed.

As the job progresses and connections are made, the construction supervisor keeps a strict accounting of all back-chargeable material and work-hour expenditures related to fabricator errors. It is expected that these errors will have been minimized by the shop inspection. Serious problems are immediately referred back to the vendor through the engineering firm's purchasing department.

Proper field handling is important in completing the installation. Abuse of the equipment and careless wiring and tubing hook-up can delay the start-up and nullify the time, experience, and interest that produce a good job.

Nameplate Data

From an instrument designer's viewpoint, nameplate design can consume a great many work-hours. It is a subject that engenders a multiplicity of opinions in every associated parameter. It is best that approvals be obtained from the client management, process, operating, and every other interested party before issuance to the control center manufacturer. The inscriptions, when issued, must not be changed by the manufacturer without explicit approval of the purchaser. Although nameplates cannot replace sound operating judgment and a thorough knowledge of the process, it has been found that a comprehensive description of the specific instrument service on the front, as well as instrument number identification on all back-of-panel components, is an important adjunct to smooth unit operation and convenient panel maintenance. See Table 1.

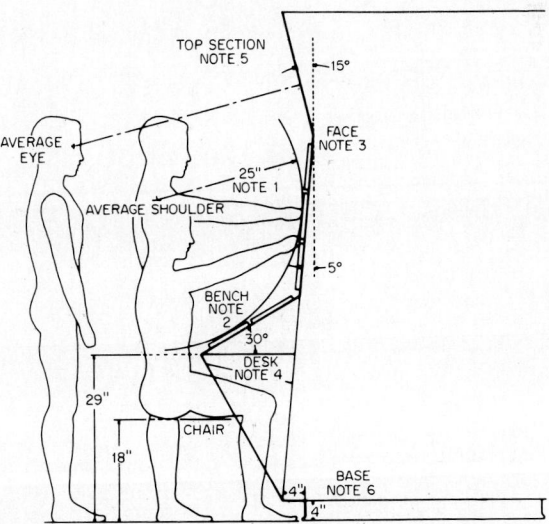

Fig. 5 Combination console and panel showing recommended relationships at the human and control center interface.

Console-Type Control Center

A typical arrangement of a console-type center is shown in Figs. 4 and 5. The dimensions shown are those believed to be optimum for this type of center. However, the angles and lengths may be adjusted to suit the particular instruments and equipment used.

Fig. 6   Modern control center following many human engineering dictates. *(Shell Chemical.)*

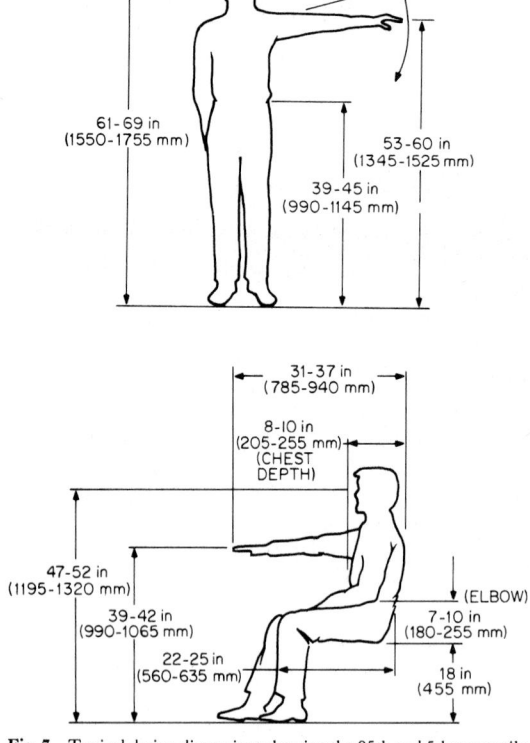

Fig. 7   Typical design dimensions showing the 95th and 5th percentile ranges.

Panel Design Philosophy

The panel illustrated in Fig. 6 contains several examples of modern control center design philosophy: (1) Controllers are mounted at good angles for ease of reading; (2) manual adjustments are mounted within the shoulder arm arc for ease of manipulations and to minimize back strain; (3) selector switches, pushbuttons, and pilot lights are easily accessible; (4) instruments are grouped into logical operational sections with spaces between entities; and (5) controllers are located where process trends may readily be noticed.

The installation shown utilizes the anthropometric principles of human engineering. The consoles are designed based on the 95th to 5th percentile of the American male; i.e., 5% of the male population may be too short and 5% too tall for the design provisions. Serious consideration must be given to including women in the percentages, although this tends to reduce to some degree the convenient working area. If the control center is to be used outside the United States, the working 95th to 5th percentile may vary significantly—the variations in dimensions encountered in Japan, Sweden, and Malaysia, are examples. See Fig. 7 for acceptable anthropometric size ranges for use in the United States.

Distributive or Computer Control

A typical distributive control system is shown in Fig. 8. A CRT with multicolor displays and its associated keyboard combines the main human engineering features offered in conventional control centers, i.e., the overviews, the semigraphic with a simplified diagram of the process, and the advantage of a full graphics panel with the process values included in the graphics picture next to its point of measurement. In most cases, the distributive control systems presently available have been designed in accordance with the latest human engineering principles. It is important for all engineers concerned to design control centers so that these principles are upheld.

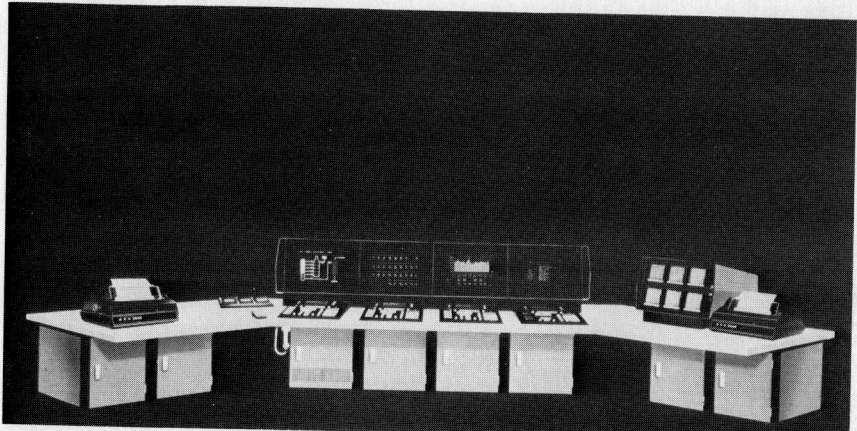

Fig. 8 A modern distributive control system layout.

REFERENCES

EPRI: "Human Factors Review of Nuclear Power Plant Control Room Design," *EPRI NP-309*, Electric Power Research Institute, Palo Alto, Calif., March 1977.

ISA: "Control Center Design," *ISA RP60.1* through *60.11* (various publication dates); "Symbol Standards," *S5.1,* 1982; "Flow Diagram Graphic Symbols for Distributed Control/Shared Display Instrumentation Logic and Computer System," Draft Standard *dS5.3*, 1981, Instrument Society of America, Research Triangle Park, N.C.

MIL-STD: "Human Engineering Design Criteria for Military Systems, Equipment and Facilities," *MIL-STD 14728-1974* (and later addenda).

NUREG: "Human Factors Evaluation of Control Room Design and Operator Performance at Three Mile Island-2," *NUREG/CR-2170,* vols. 1–3, U.S. Dept. of Commerce, Washington, D.C., January 1980.

SECTION 16

Data Communications Systems*

Staff. *Bell Laboratories, Murray Hill, New Jersey. (Optical Transmission Systems)*

J. D. Warnock. *Manager, Systems Engineering, Moore Products Company, Spring House, Pennsylvania. (Pneumatic Transmission and Control)*

*Persons who authored complete articles or subsections of articles, or who otherwise cooperated in an outstanding manner in furnishing information and helpful counsel to the editorial staff.

Telemetry

Telemetry may be defined as the science of measuring at a distance and is employed in industrial, commercial, military, and space operations. Telemetry systems may be classified on the basis of the characteristics of the electric signal, namely, voltage, current, position, frequency, and pulse; or as analog or digital, where the signal, respectively, has been transmitted as an analog measurement or has been converted to a code representing that measurement. Voltage, current, and position telemetering operations require a physical connection between the transmitter and the receiver. This physical connection, usually referred to as a *channel,* consists of one, two, or more wires, depending on the system design. Frequency and pulse systems not only can operate over a physical wire channel, but also can utilize other channels, such as telegraph, teletype, telephone, radio, or microwave. It is this ability and flexibility that allows these methods to be used for long-distance telemetering.

Voltage Telemetering

A voltage telemetering system transmits the measurement as a function of an ac or dc voltage. Most systems use primary elements which produce a voltage as an intrinsic part of their measuring function, such as thermocouples, tachometers, and differential transformers. Applications are mainly over industrial plant distances of about 1000 ft (300 m) or less. Self-balancing potentiometers are sometimes used as receivers for such systems. Deflectional indicators can be used if they are calibrated for the line resistance involved. Thermal converters are used in the electric power industry to transmit power measurements to a central control office. They are used over lines as long as 50 mi (80 km) but require good quality cable circuits for such distances.

Voltage telemetering systems require higher quality circuits than current systems, especially if they employ low voltages. The signal-to-noise ratio must be relatively high, i.e., on the order of 2 or greater. *Signal* refers to the power or energy level of the transmitted intelligence; *noise* refers to the energy level of the spurious interference which accompanies the signal. In voltage telemetering the current is quite low, and therefore the signal power, which is the product of the voltage and the current, is extremely small. Consequently, the transmission circuits must be protected from sources of interference which are on the same order of magnitude as the signal power.

A voltage telemetering system is suitable for adding several output voltages in series, provided the measurement is linear; but it usually requires a more costly receiving instrument and generally is not adaptable to the use of several receivers at the same time. See Fig. 1*a*.

Current Telemetering

Early current telemetering systems were similar to that shown in Fig. 1*b*. A slidewire potentiometer is connected in series to a battery with a sliding contact positioned by a pressure-sensitive bourdon tube. The telemetering channel consists of a pair of wires connected to a current measuring device. As the measured pressure changes, the bourdon tube moves and positions the sliding contact, thereby changing the current in the telemetering channel. The milliammeter measures the current and positions a pointer on a scale graduated in terms of the pressure being measured. In a force-balance current telemetering system, a position transducer, such as a differential transformer or a variable capacitor, replaces the slidewire (Fig. 1*c*). The motion of the bourdon tube varies the position of the core of the differential transformer on the movable portion of the variable capacitor. In the differential transformer, the mechanical motion of the core varies the amplitude of the voltage on the second winding, which is amplified and rectified to provide a dc milliampere signal (typically 4 to 20 mA dc) to the transmission channel. It will be noted from Fig. 1*c* that part of the current output is fed back to oppose the motion of the input variable.

Current telemetering systems can develop higher voltages than most voltage systems and, consequently, can be made more immune to the effects of thermal and inductive voltages in the interconnecting leads—a; well as line resistance—but are susceptible to errors due to leakage currents to ground. Current tra; smitters usually cost more than voltage transmitters because of the amplifier required, but inexpensive milliammeters can be used as receivers without calibration for line resistance. Current telemetering systems can be adapted to adding the output of several transmitters, operating several receivers at once, and operating multipoint receivers. The speed of response is almost instantaneous with many types. Most transmitters deliver some definite minimum current when the measured variable is at its lower limit and thus provide a *live zero*, a means of detecting an open circuit in the channel at the receiver.

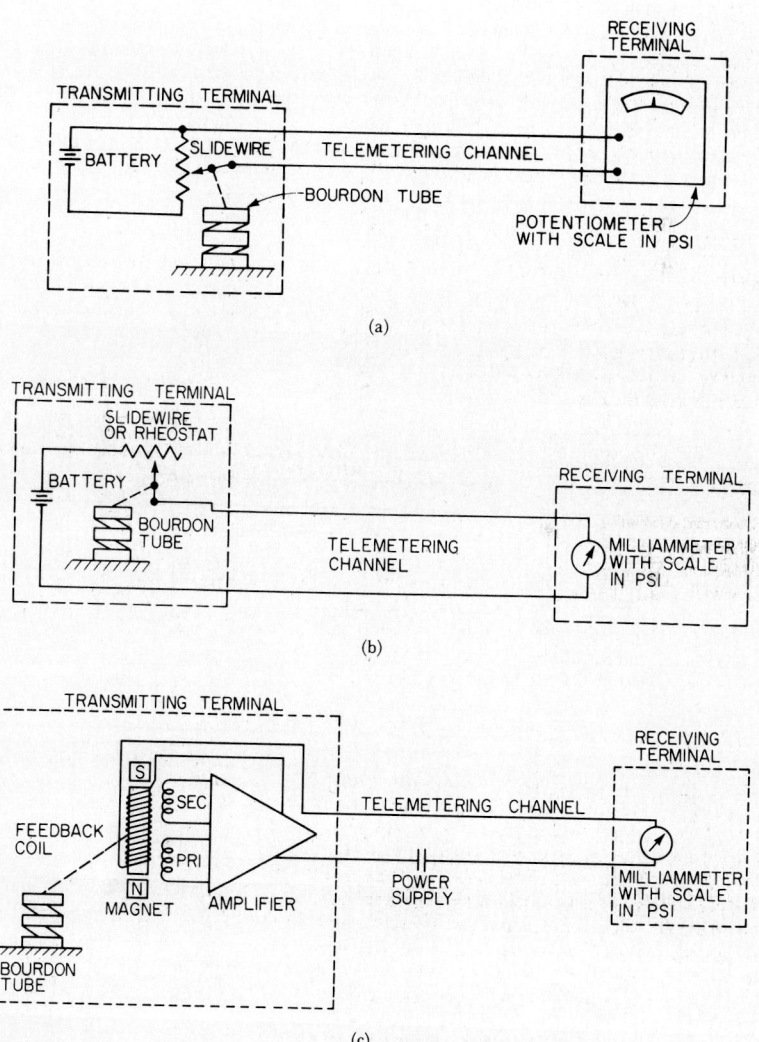

(a)

(b)

(c)

Fig. 1 Traditional voltage and current telemetry systems. (*a*) Voltage, (*b*) current, and (*c*) force-balance current.

Position Telemetering

A position telemetering system transmits and reproduces the measured variable by positioning variable resistors or other electrical components in a bridge circuit configuration so as to produce proportional changes at both the transmitter and the receiver. With reference to the traditional system shown in Fig. 2, the potentiometers at both the transmitting and receiving locations are energized by a common power supply. The sliding contact at the transmitting end is positioned by a bourdon tube as the pressure varies. If the sliding contact at the receiving terminal is positioned until the meter needle indicates zero current, the position of the contact will assume the same position as the contact at the transmitter. The receiving contact moves the pointer, which indicates on a scale graduated in units of pressure. This is the same principle as that of a Wheatstone bridge.

The *synchromotor,* used for many years, is similar to a three-phase induction motor but has two-pole fields carrying a single-phase excitation winding (Fig. 3). As the torque transmitter field winding rotates the relative magnitude of the voltages of the three stator windings is varied. If the rotor in the receiver is in the same position as that of the torque transmitter, the receiver stator windings will generate voltages equal to those of the corresponding torque transmitter's windings. However, if the receiver rotor is not in the same position, the receiver stator voltages will be different and current will flow in the three interconnecting leads. This current induces a torque on both the transmitter and receiver rotors, tending to make their positions coincide.

Although various types of position telemetering devices may appear to be quite different, they usually have much in common. The quantities whose relative magnitudes are changed in accordance with the measured variable can be resistances, inductances, ac voltages, and alternate half-cycles of a single ac voltage, among others. Resistance slidewire bridges, inductance bridges, and ac self-synchronous motors are typical examples. It is difficult to generalize about this broad class, but most

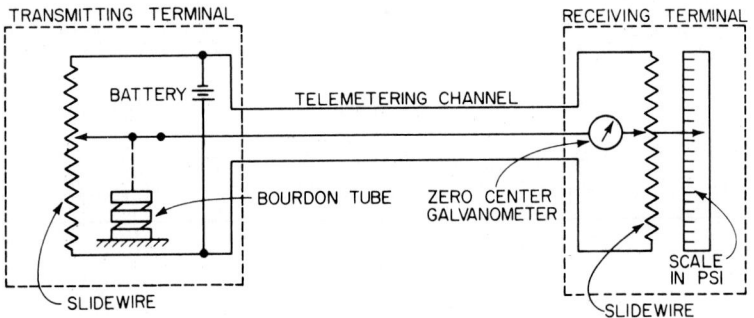

Fig. 2 Traditional position telemetering system.

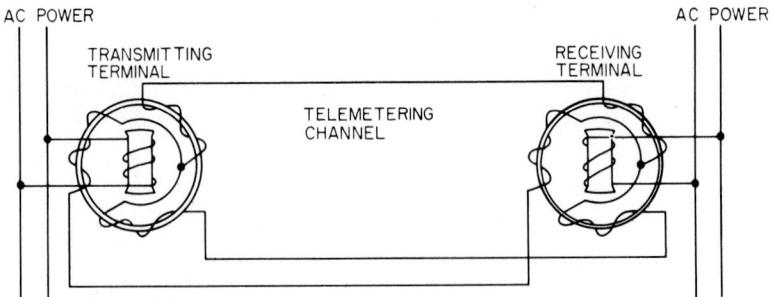

Fig. 3 Synchromotor telemetering system in which the angular input displacement is converted into the relative magnitude of three ac voltages.

ratio systems are intended for distances less than a few miles or kilometers, and many for distances less than a few hundred feet (meters). Most systems, but not all, require more than two interconnecting wires. The transmitters usually are inexpensive and require only low forces from the primary measuring elements. Many systems can be used with either low-cost deflectional receivers or servoactuated receivers if higher torque is required.

Frequency Telemetering

A system of this type varies the frequency of an electric signal in accordance with the information to be transmitted: 0% measurement is represented by x hertz, and 100% by y hertz. The frequencies used in commercially available systems for x and y are 7.5 to 15, 5 to 15, 9 to 15, 5 to 25, 6 to 27, 10 to 30, and 18 to 30 Hz.

Frequency telemetering was developed to provide a system that was independent of the electrical characteristics and variations of the telemetering channel. The frequency ranges selected take advantage of the low-cost telegraph and teletype communication channels commercially available.

As shown in Fig. 4, a frequency telemetering system consists of a transducer and transmitter at the transmitting location and a receiver and readout device at the receiving location. The pressure-to-current transducer converts the bourdon tube motion to a 4- to 20-mA dc proportional signal as previously described. This 4- to 20-mA dc signal then is converted to a 9- to 15-Hz signal by the current-to-frequency transmitter—utilizing a relay output (as shown in Fig. 4), or solid-state electric keyed output. At the receiving location, the frequency is converted to a current or voltage and displayed on an ammeter, a voltmeter, or a similar instrument with a scale calibrated in terms of pressure. A conventional frequency telemetering transmitter and receiver are shown in Figs. 5 and 6, respectively.

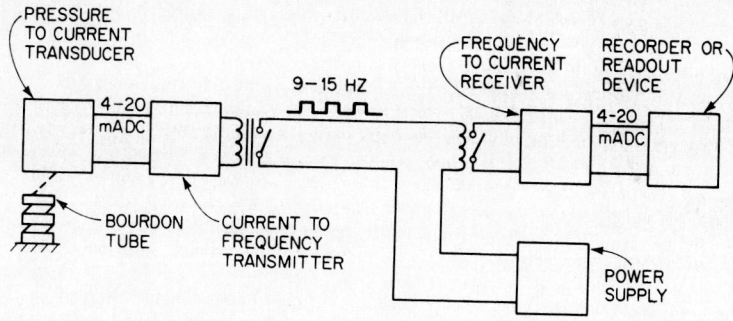

Fig. 4   Conventional frequency telemetering system.

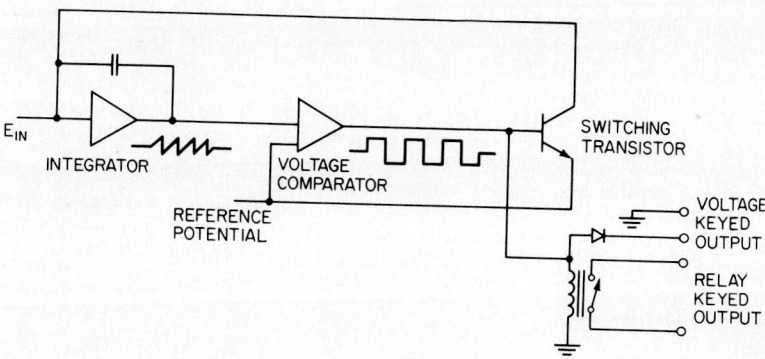

Fig. 5   Frequency transmitter.

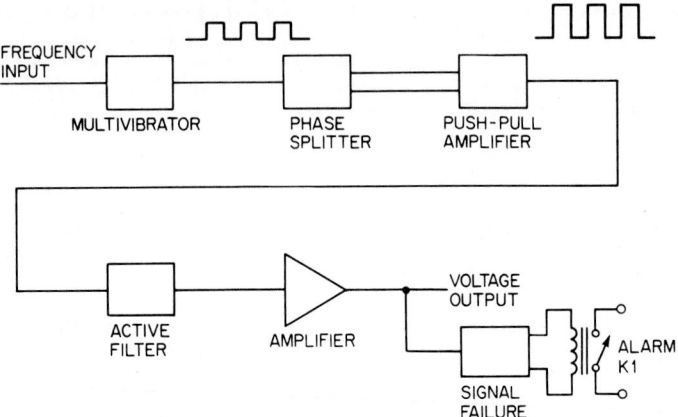

Fig. 6 Frequency receiver.

With reference to Fig. 5, a frequency transmitter is designed to convert dc input signals to proportional pulse frequency information suitable for transmission to a remote receiver. Whenever a dc signal is applied to the input E_{in}, the output of the integrator circuit ramps from zero to a predetermined value. The ramp voltage developed by the integrator circuit is coupled to a voltage comparator and compared, in level, to a dc reference potential. The comparator is a device whose output switches rapidly from one state to another whenever its two input voltages are equal. When the input ramp voltage reaches a level that is equal to the reference voltage, the comparator output changes from one state to another, driving the switching transistor into current saturation.

Since a saturated transistor approximates a short circuit, the reference voltage becomes zero. Therefore, the capacitor, which has already charged to the reference voltage, now can begin discharging through the switching transistor. As it discharges, current is drawn very rapidly from the integrator output. As a result of this action, the integrator output is driven back to zero. Once the condition is reached where the two input voltages of the comparator are both at zero level, that circuit switches its output back to the original state. As such, the switching transistor cuts off, thus resetting the integrator circuit for the next cycle. The total time required for the integrator output to change from zero to the reference voltage is inversely proportional to the instantaneous level of the dc input signal.

The output of the voltage comparator consists of a series of pulses. These pulses, which are used to trigger the switching transistor, provide a voltage-keyed output or relay-keyed output.

The basic operation of a frequency receiver is illustrated in Fig. 6. The frequency input signal triggers a multivibrator. Normally, this device is resting in a stable state when the input signal is at zero level and no output is provided. However, when an incoming pulse appears, the multivibrator switches to an unstable state and generates a square wave pulse output. Such pulses, which are of a constant duration, are connected to a phase splitter.

The phase splitter stage provides the necessary positive and negative signals to drive the push-pull amplifier. The amplified pulses at the output of the push-pull amplifier are connected to an active filter circuit. The filter subjects the incoming pulses to a high degree of ac attenuation—so as to provide a dc voltage equal to the average value of the signal pulses. This signal, after amplification, provides the dc voltage output proportional to the frequency input.

A signal failure circuit, also shown in Fig. 6, is used to alert the operator in the event that the input signal is cut off. This circuit activates the relay whenever the output level drops below 0%. To achieve the desired warning, the contacts of $K1$ are connected to an external alarm circuit.

Pulse Telemetering

These systems make use of various types of electric pulses and transmit them as a function of time, independently of electrical variations in the transmission channel. Telegraph and teletype systems employ this principle. Pulse telemetering systems can be classified as pulse duration, pulse count, and pulse code or digital telemetering systems.

Pulse Duration Systems

These systems may comprise electromechanical or electronic devices. The input quantity can be a mechanical position of an instrument pen or pointer, or a current or voltage at the transmitting end. The primary measuring element can be a low-torque device, such as a flowmeter or pressure gage.

Total cycle times of 15 and 5 s are common with mechanical and electronic systems. This means that a system is visibly intermittent in action and consequently does not give the instantaneous response required by many applications. For this reason, it is sometimes undesirable to include a system of this type within the control loop of an automatic control system. Equipment is made available for local control that is capable of telemetering the reading and receiving a telemetered set point.

Pulse duration systems are readily adaptable to the use of several receivers in parallel or in series and may include a suppressed zero signal to indicate channel failure. Pulse duration signals are not basically suitable for adding, but auxiliary equipment is available for accomplishing this by mechanical or electrical means. Pulse duration is adaptable to direct integration by applying the signal directly to a synchronous motor driving a counter.

Pulse duration systems have been applied extensively to wire circuits. They can use either ac or dc impulses—so it is possible, with the aid of filter circuits, to operate two systems over the same pair of wires, or to send ac power and a dc signal over the same pair of wires, thus allowing telemetering where power is not readily available at the transmitting end. There is no basic limitation in distance, since telegraph repeaters may be used where circuit resistances are higher than a few thousand ohms. Ground return is permissible in many applications.

An electronic pulse duration system is shown in Fig. 7. The dc voltage input signal is connected to a switching comparator and compared with the output from a ramp generator. The ramp generator output ramps from a minimum to a maximum voltage during one cycle, i.e., 15 s; it then reverts back to the original minimum output by a step change and repeats the cycle. Each time the ramp output reaches the same value as the dc voltage input signal, the comparator output deenergizes the output relay coil. The coil is energized each time the ramp generator output reaches the minimum voltage corresponding to the 3-s time. The 3-s pulse corresponds to a 0% input, and a 12-s pulse corresponds to a 100% input.

An electronic pulse duration receiver circuit is shown in Fig. 8. Basically, an "on" pulse from a pulse duration transmitter energizes relay $K1$. Contact $B1$ of the relay $K1$ makes contact to connect capacitor $C1$ to the constant current source; and capacitor $C1$ charges to a voltage level determined by the duration of the "on" signal and the amount of current delivered by the constant current storage circuit.

Once the "on" signal is completed, the current through the coil of $K1$ drops to zero and the relay deenergizes. When this occurs, contact $B1$ switches over to connect capacitor $C1$ to the operational

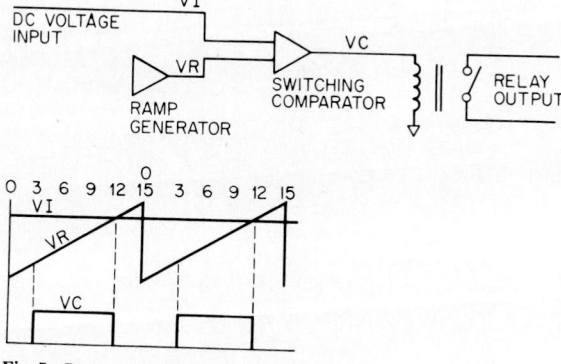

Fig. 7 Basic electronic pulse duration transmitter.

amplifier. The voltage stored in $C1$ (which is proportional to the duration of the "on" signal) is compared with a dc reference voltage applied to the input of the operational amplifier.

During the time period when relay $K1$ is deenergized ("off" segment of the input signal), contact $B2$ connects capacitor $C2$ to the reference input of the operational amplifier. Since the voltage stored in $C2$ is the same as the voltage present at the reference input, the amplifier output remains unchanged for the remainder of the "on" pulse cycle.

Because the operational amplifier is supplied with a balanced input during both the "on" and "off" portions of the input cycle, steady dc signals appear at the output E_o.

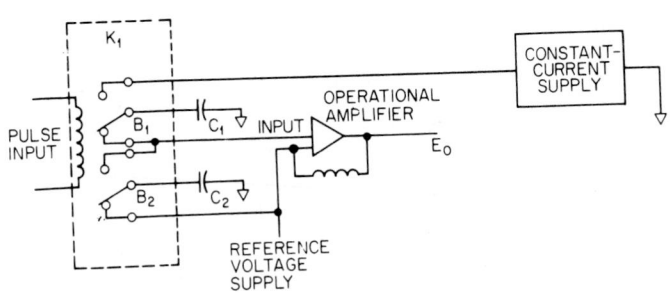

Fig. 8 Basic electronic pulse duration receiver.

Pulse Count Telemetering Systems

Pulse count or pulse rate telemetering systems are similar to frequency systems. The difference is that a frequency system provides a sine wave or square wave signal, whereas pulse count systems provide an "on-off" switching signal. In either case, the number of signals per second represents the measurement.

Pulse Code (Digital) Telemetering Systems

These systems use the binary numbering system to transmit information. The binary system is uniquely suited to telemetering because it is necessary to recognize only two conditions, represented by the symbols 0 and 1. These symbols are binary digits, or bits, and can be arranged in various configurations to represent any decimal number. Examination of the binary system (Table 1) shows how the quantity and location of the two available binary digits represent the decimal numbers 0 through 36. For example, the decimal number 9 is represented by 1001, or $1 \times 2^3 + 0 \times 2^2 + 0 \times 2^1 + 1 \times 2^0$.

Binary-Coded Decimal (BCD) Codes

To simplify the hardware requirements of a digital telemetering system, a BCD code is commonly used. This code represents each digit in a decimal number by 4 bits (Table 1). For example, decimal number 22 is represented by BCD 0010-0010 instead of 10110 in a straight binary code. This code simplifies the receiving equipment design, since only a total of 4 bits at a time must be decoded to find the decimal number.

8421 Coding

It will be noted that the positional weights or place values of the binary digit 1 change from 2^3 to 2^2 to 2^1 to 2^0, or 8 to 4 to 2 to 1, dependent on the column location. This is known as the 8421 code.

Baudot Code

This code consists of three parts. See Fig. 9. A *start pulse* that is always a space, for synchronization of the transmitter and receiver; five *data* or *information pulses,* whose combination designates the

Table 1 Comparison of Decimal, Binary, and Binary-Coded Decimal Number Counting

Decimal		Binary						Binary-coded decimal
10's	1's	32's	16's	8's	4's	2's	1's	
	0						0	0000
	1						1	0001
	2					1	0	0010
	3					1	1	0011
	4				1	0	0	0100
	5				1	0	1	0101
	6				1	1	0	0110
	7				1	1	1	0111
	8			1	0	0	0	1000
	9			1	0	0	1	1001
1	0			1	0	1	0	0001 0000
1	1			1	0	1	1	0001 0001
1	2			1	1	0	0	0001 0010
1	3			1	1	0	1	0001 0011
1	4			1	1	1	0	0001 0100
1	5			1	1	1	1	0001 0101
1	6		1	0	0	0	0	0001 0110
1	7		1	0	0	0	1	0001 0111
1	8		1	0	0	1	0	0001 1000
1	9		1	0	0	1	1	0001 1001
2	0		1	0	1	0	0	0010 0000
2	1		1	0	1	0	1	0010 0001
2	2		1	0	1	1	0	0010 0010
2	3		1	0	1	1	1	0010 0011
2	4		1	1	0	0	0	0010 0100
2	5		1	1	0	0	1	0010 0101
2	6		1	1	0	1	0	0010 0110
2	7		1	1	0	1	1	0010 0111
2	8		1	1	1	0	0	0010 1000
2	9		1	1	1	0	1	0010 1001
3	0		1	1	1	1	0	0011 0000
3	1		1	1	1	1	1	0011 0001
3	2	1	0	0	0	0	0	0011 0010
3	3	1	0	0	0	0	1	0011 0011
3	4	1	0	0	0	1	0	0011 0100
3	5	1	0	0	0	1	1	0011 0101
3	6	1	0	0	1	0	0	0011 0110

letters of the alphabet, the numbers 0 through 9, and various symbols; and a *stop pulse,* in the mark position to signal the end of the character. The combination of information pulses shown in Fig. 9 represents the letter "D." All pulses are of equal length except the stop pulse, which is 1.42 times the length of the others. A *word* is defined as being equivalent to six characters of the Baudot code and is considered to be a group of characters treated as a unit.

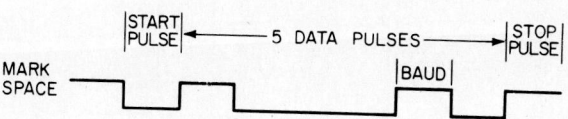

Fig. 9 Baudot teletype code.

Bauds versus Bits

All pulses are commonly called *bits,* but only the five that represent the character carry information. The start and stop pulses are *noninformation* bits, as are pulses that provide a means to detect and correct errors and bits are used to achieve and maintain synchronization.

The *bits-per-second rate* is the time required to transmit the five data pulses. In a 60-word-per-minute teletypewriter transmission, each bit is 21.95 ms in length except the stop bit which is 2.95 × 1.42 = 31.17 ms. The total character time is 6 × 21.95 + 31.17, or 162.9 ms.

The *baud* is a measure of signaling speed and is the reciprocal of the length (in seconds) of the shortest pulse transmitted. Thus,

$$\text{Bauds} = \frac{1}{0.02195} = 45$$

The *bits-per-second speed* is defined as the number of *information pulses* transmitted per second. In this case, a total of 5 information bits is transmitted in 162.9 ms, giving a bit per second rate of 30.

As will be noted, delays between characters and additional bits added for security do not change the baud rate, but they do alter the bit rate. The bit rate is primarily of interest to the computer, where it determines processing speed, whereas the baud rate is of interest in determining the communication channel requirements.

American Standard Code for Information Interchange (ASCII)

The ASCII code uses a 7-bit-plus-parity code. In a binary code, the base 2 raised to the number of information bits used gives the number of characters that can be formed. In this case, a 7-bit code allows transmission of 2^7, or 128, characters.

Parity

Parity is a method used to check for errors by adding an extra bit to make either an odd number of 1's (odd parity), or an even number of 1's (even parity). Table 2 illustrates the 8421 code with an odd-parity bit. If an error due to electrical noise occurs in the coded message and inverts a bit, the receiver will detect that the message does not contain an odd number of 1's and the message will be rejected. Of course, two exactly compensating errors cannot be detected in this manner if they occur within a very short interval of each other.

Two-out-of-Five Code

This code incorporates parity so each character includes two 1's and three 0's. See Table 3.

Table 2 The 8421 Code with Even Parity

Decimal	Binary				Parity
	8	4	2	1	
0	0	0	0	0	1
1	0	0	0	1	0
2	0	0	1	0	0
3	0	0	1	1	1
4	0	1	0	0	0
5	0	1	0	1	1
6	0	1	1	0	1
7	0	1	1	1	0
8	1	0	0	0	0
9	1	0	0	1	1

Table 3 Two-out-of-Five Code

Decimal	Two-out-of-five code				
	7	4	2	1	0
0	1	1	0	0	0
1	0	0	0	1	1
2	0	0	1	0	1
3	0	0	1	1	0
4	0	1	0	0	1
5	0	1	0	1	0
6	0	1	1	0	0
7	1	0	0	0	1
8	1	0	0	1	0
9	1	0	1	0	0

Extended Binary-Coded Decimal Interchange Code (EBCDIC)

This is a packed decimal code which uses two sets of BCD codes, giving a total of 8 bits. No parity is provided.

Transmission Modes

Parallel and Serial Transmission

In *parallel transmission*, a separate channel is used to transmit each bit making up the coded word. This can be done over short distances by using a separate wire for each bit, or for long distances by frequency division multiplexing. In *serial transmission*, a method of identifying the bits is required and is accomplished by some form of synchronization between the transmitter and the receiver. Although parallel transmission is faster, it is also more expensive.

Serial and parallel transmission are often implemented synchronously, with data transmitted at fixed intervals determined by a timing signal. Asynchronous transmission is also used, with control signals preceding predetermined blocks of data. Asynchronous transfers are slower than those for synchronous systems because of the necessary receiver-transmitter coordination, but hardware costs are lower. Digital communication normally requires a pair of wires to send and receive each bit, plus a line to serve as a common or ground. Duplex transmission can be implemented to lower wiring costs. In a full-duplex system, sending and receiving are performed on the same wires simultaneously. Half-duplex transmissions allow sending and receiving on the same wires, one at a time. Half-duplex synchronous serial transmission is often used in distributed systems.

Synchronous transmission transmits bits at an exact time interval and requires synchronization of the transmitter with the receiver. At the beginning of a pulse train, a synchronizing pulse (usually longer than a 1 or 0 pulse) is transmitted to start an electronic timing circuit in the receiver. The receiver then samples the transmission channel at fixed intervals to determine whether a mark or space is present.

Asynchronous transmission sends bits by adding time periods between each 2 bits, so that each bit carries its own synchronizing information. Because synchronous transmission does not have to return to zero between each 2 bits, theoretically twice as many bits can be transmitted synchronously as contrasted with the asynchronous method.

Interface Standards

Early distributed control systems used proprietary hardware interfaces, protocols, modems, and transmission parameters. The interconnection of elements from different manufacturers usually was impractical. Much effort has been made in recent years to achieve some standardization. Standards define factors, such as circuit board designs, cable connector pin assignments, and signal levels. For example, IEEE Standards 488 defines electromechanical aspects of parallel transmission and is used in a number of systems. The computer-automated measurement and control (CAMAC) standard, published as IEEE 583, 595, and 596, was developed to provide interfaces for parallel, bit serial, and byte serial communication and found major use in nuclear, test, and some process control applications. Electronic Industries Association (EIA) Standards RS232C, RS422, and RS449 define bit serial synchronous and asynchronous environments and are used in data processing as well as in a number of process control applications.

Although rigidly adhered to standards are needed, many communication protocols are proprietary. Specific protocols are associated with IEEE 448 and CAMAC. Serial systems use protocols derived from synchronous data link control (SDLC) and high-level data link control (HDLC) for communications within the network. ASCII is often used for external serial communication. See Hertanu (1980) reference at the end of this article.

Code Security and Error Detection

Characteristics of the transmission channel, plus external influences, can cause errors in the transmitted code. Various methods are used to detect such errors.

Bit Counting

This is commonly employed when the receiver counts the number of bits in a message to determine whether they total the required number.

Redundant Transmission

Each message is transmitted twice and compared at the receiver.

Check Digits

This is a method of adding digits that bear some relationship to the message being transmitted. A message may be the number 4786 to which a fifth digit can be added for checking purposes. This fifth digit must bear a particular relationship to the original number. One method is to multiply each digit of the original number by a particular number and then add the products and subtract this sum from another particular number. For example:

$$\frac{\begin{array}{cccc} 4 & 7 & 8 & 6 \\ \times 1 \times & 2 \times & 3 \times & 4 \end{array}}{4 + 14 + 24 + 24} = 66$$

$$70 - 66 = 4 \quad \text{use for check number}$$

The number actually transmitted would be 47864.

This procedure is known as modulus-n checking. In this case, since the sum of the products (66) is subtracted from the next higher product of 10 (70), it is called a modulus-10 system.

Once an error in a message has been detected, it is a simple matter for the receiver to reject that input and wait for the next transmission. Therefore, error correction by retransmission is the simplest method of error correction. Codes also can be designed so that errors can be detected and corrected at the receiver without retransmission.

Hamming Code

If a code consists of characters that differ from one another by a change in 1 bit, then any error produces a legitimate character. By adding a parity bit, a single-bit error can be detected (as previously described for the two-out-of-five code). With parity, every character differs from all other characters by 2 bits. The number of bits that must be changed to produce a legitimate character is called the *Hamming distance*, after R. W. Hamming, who developed the first error correcting codes.

A Hamming distance of 3 allows a single-bit error to be detected and corrected at the receiving terminal. For example, if the letter "A" is represented by the code 000, and "B" by 111, then a single-bit error in either character can be easily detected, since the erroneous character is still closer to the original than to any other. Suitable programming at the receiver allows the original character to be reproduced.

> A—000—Transmitted Character A
> 001—with single-bit error
> 010—with single-bit error
> 100—with single-bit error

> B—111—Transmitted Character B
> 110—with single-bit error
> 101—with single-bit error
> 011—with single-bit error

Telemetering Communication Channels

Commonly used channels for telemetering are (1) two-wire privately owned channels, (2) leased commercial telegraph channels, (3) leased commercial teletype channels, (4) leased commercial telephone channels, (5) leased wideband channels, (6) microwave channels, (7) radio channels, and (8) power line carrier channels. The major difference between the channels is the range of frequencies (bandwidth) that each channel can transmit.

Two-Wire Privately Owned Channels*

In-plant telemetering is usually accomplished over hard wire channels installed by plant personnel or electrical contractors. Table 4 gives the resistances of the most used conductors according to wire gage number. For a loop resistance value, multiply by 2 for every 1000 ft (300 m) or 1 mi (1.6 km) of conductor. Where multiconductor cable is used (25 or 50 pairs), no. 22 or 26 gage wires are usually furnished.

Such hard wire circuits are suitable only for dc or low-frequency signals, and the problems involved in the transmission of such signals over these circuits are similar to those encountered in dc telegraph transmission. The same engineering considerations frequently are applied.

Telemetering systems ordinarily transmit in only one direction and require two to five conductors—with two-conductor systems predominating. Line characteristics that may affect the operating range of low-frequency systems are (1) linear resistance, (2) to a lesser extent, leakage conductance (or insulation resistance), and (3) occasionally, linear capacitance and inductance. The insulation resistance of cable circuits ordinarily is so high that it is not an important factor in the operation of telemetering systems. Insulation resistance between the wires of a cable pair seldom drops below ~1000 mΩ-mi (~0.001 μmho/mi) and ranges from this value upward to 10,000 mΩ-mi or higher. The insulation resistance from a wire to ground is usually about two-thirds of the value between wires. The capacitance between two conductors or a pair ranges from ~0.0060 to 0.085 μF/mi; and the inductance of a nonloaded pair is ~1 mH/mi for the types of cable commonly used.

In cases where the circuits are routed through distribution boxes, substantially lower values of insulation may be encountered. In some instances, therefore, it may be desirable to avoid routing conductors through such boxes when circuits are used for telemetering.

The capacitance between two wires (12-in spacing) is ~0.009 μF/mi, and the inductance of such open wire pairs is approximately 3.4 mH/mi.

The insulation values obtained on open-wire circuits are, in general, very much lower than those for cable circuits. Consequently, the leakage is frequently the limiting factor in determining the range over which telemetering systems may be operated on open-wire facilities. The leakage conductance varies over wide limits, dependent chiefly on weather conditions and the type and maintenance of the line—and, to a smaller extent, on the insulators used.

---

*Information in this section has been adapted with permission from the IEEE publication, "Telemetering, Supervisory Control, and Associated Circuits," pp. 6–10, sec. 4.

Table 4 Wire Resistances

Gage Conductivity Resistance at Ohms per conductor	Solid copper AWG or B&S 100% 20°C or 68°F		Solid aluminum AWG or B&S 61% 20°C or 68°F		Solid steel Am. steel wire 12.5% 20°C or 68°F		Copper-clad steel AWG or B&S 40% 23.9°C or 75°F	
Gage no.	Per 1000 ft	Per mile	Per 1000 ft	Per mile	Per 1000 ft	Per mile	Per 1000 ft	Per mile
0	0.098	0.519	0.161	0.851	0.883	4.66	0.246	1.30
1	0.1239	0.654	0.203	1.073	1.036	5.47	0.310	1.64
2	0.1563	0.825	0.256	1.353	1.204	6.36	0.390	2.06
3	0.1970	1.04	0.323	1.706	1.397	7.38	0.492	2.60
4	0.2485	1.31	0.408	2.15	1.635	8.63	0.622	3.28
5	0.3133	1.65	0.514	2.71	1.936	10.22	0.782	4.13
6	0.3951	2.09	0.648	3.42	2.25	11.88	0.987	5.21
7	0.4982	2.63	0.817	4.32	2.65	13.98	1.25	6.60
8	0.6282	3.32	1.03	5.44	3.16	16.69	1.57	8.29
9	—	—	—	—	3.77	19.92	1.98	10.5
10	0.9989	5.28	1.64	8.65	4.55	24.0	2.50	13.2
11	—	—	—	—	5.71	30.2	3.15	16.2
12	1.588	8.38	2.61	13.76	7.45	39.4	3.97	21.0
13	—	—	—	—	9.91	52.3	5.00	26.4
14	2.525	13.3	4.14	21.9	12.96	68.5	6.31	33.3
15	3.184	16.8	5.22	27.6	16.01	84.5		
16	4.015	21.2	6.59	34.8	21.2	112.1		
17	5.064	26.7	8.31	43.8	28.5	150.2		
18	6.385	33.7	10.5	55.3	36.8	194.2		
19	8.051	42.5	13.2	69.7	49.4	261		
20	10.15	53.6	16.7	87.9	68.5	362		
21	12.80	67.6	21.0	110.9	82.3	435		
22	16.14	85.2	26.5	139.8	101.4	536		
23	20.36	108	33.4	176.3	124.6	658		
24	25.67	135	42.1	222	156.8	828		
25	32.37	171	53.1	280	199.4	1053		
26	40.82	216	67.0	353	253	1337		
27	51.46	272	84.4	446	277	1464		
28	64.90	343	106	562	316	1669		

It is common practice to engineer systems on the basis of wet-weather leakage values. It is convenient to employ a single leakage value which may be considered representative for any open-wire circuit. On this basis, the value of wet-weather leakage conductance between one wire and ground may be taken as 5 μmho/mi (0.20 MΩ-mi). On the average, this value probably will not be exceeded on more than ~1 day in 100, and then for only part of a day. Individual sections of the United States may vary materially in this respect. The wet-weather leakage conductance between wires probably will not be more than 2.5 μmho/mi.

For determining the effect of uniformly distributed leakage conductance on the accuracy of a dc telemetering system of the *current type,* the percentage error usually may be taken as equal to 50 times the ratio of loop resistance (in ohms) to total insulation resistance (in ohms). For example, assume a 20-mi open-wire grounded circuit having a 200-Ω resistance, including terminals. With the leakage figures previously given, the resistance between the wire and ground under wet-weather conditions would be 0.01 MΩ (0.20 MΩ divided by 20 mi). The ratio of the circuit resistance to the shunt resistance, therefore, would be 0.02. This value, multiplied by 50, would indicate that the error would be about 1% under wet-weather conditions.

For a dc *voltage-type* telemetering system an estimate of the error caused by leakage should take into account the dc resistance of the telemetering equipment itself. An estimate of the error might be made in each case by simple computations of the change in voltage across the receiver caused by a shunt resistance equivalent to the leakage.

Circuits may be divided into two classes: (1) metallic circuits in which a metallic conducting path is used throughout, and (2) grounded circuits in which the ground is used as one side. For most applications metallic circuits are preferred. Grounded circuits, however, are satisfactory in many instances and are more economical.

Leased Channels

These channels are available to industry from communications common carriers (telegraph, teletype, telephone, etc.). The principal channels are:

30-baud private lines

45- to 55-baud and 75-baud private lines

150-baud private lines

Voice-grade private line data

Data communications, using a switched telecommunications network

The 30-baud channel is commonly called a dc telegraph (or measuring) circuit because of its original use in telegraph communications. In its simplest form it consists of a single conductor with a return path through earth. Because of the difficulties that arise from using earth ground, a second conductor generally is employed to eliminate electrical interference caused by differences in ground potential between the transmitting and receiving locations.

The terms *mark* and *space* originated with the telegraph system. When the transmitter key was closed, a current flowed in the loop, causing a mark to be made by a pen on a recorder. When the telegraph key was open, no current flowed and thus a space appeared on the recorder.

As shown in Table 5, the 30-, 45-, 55-, 75-, and 150-baud channels are limited to telemetering speeds of 15, 22.8, 28.4, 37.5, and 75 Hz, respectively. An increase in transmission rate requires additional line conditioning, and increased cost results. Therefore, the lowest speed telemetering system should be selected consistent with expected measured variable excursions and criticality. For example, with reference to Fig. 3, an 18- to 30-Hz frequency telemetering system will require the use of a 75-baud channel and could not use the lower cost 30-, 45-, or 55-baud channels. However, a 9- to 15-Hz system could make use of the lowest cost 30-baud channel.

For short distances, within the boundaries of a central telephone office, such as 10 mi, a telegraph channel could consist of direct connection of transmitter and receiver by a physical pair of wires. However, the common arrangement is for the channel not to have direct continuity between the two locations, but instead to be connected through repeaters.

Although telegraph- and teletype-grade circuits are lowest in cost, they are limited in speed of transmission and do not allow efficient time division multiplex operation to be used. Also, frequency multiplexing is not possible.

Telephone circuits available for data transmission transmit frequencies in the audible voice range of 300 to 3200 Hz. Mark-space data must be in this frequency band to be transmitted by a telephone-grade circuit. All telephone-grade circuits are not identical, but have different transmission characteristics dependent on the line conditioning used.

Principal channels of U.S. communication networks are shown in Table 6.

Table 5 Telegraph and Teletype Transmission Channels

Classification, bauds	30	45	55	75	150
Minimum pulse length, ms	33.3	21.95	17.59	13.33	6.67
Equivalent teletypewriter speed, words/min		60	75	100	
Maximum frequency, Hz	15	22.8	28.4	37.5	75

Table 6 Principal Channels of National Communication Networks (United States)

Series 1000 (sub-voice-grade channels)

These are unconditioned channels capable of transmitting dc, mark-space, or binary signals at rates up to 150 bauds. These channels are not suitable for the transmission of ac tones.

Type 1001. Up to 30 bauds for remote metering, supervisory control, and miscellaneous signaling.

Type 1002. Up to 55 bauds for Morse, teletypewriter, teletypesetter, data, or remote metering, supervisory control, and miscellaneous signaling.

Type 1003. Up to 55 bauds for remote operation of radiotelegraph.

Type 1005. Up to 75 bauds for teletypewriter, data, or remote metering, supervisory control, and miscellaneous signaling.

Type 1006. Up to 150 bauds for teletypewriter, data, or remote metering, supervisory control, and miscellaneous signaling.

Series 2000 (voice-grade channels)

These are channels with a bandwidth not exceeding 4000 Hz. They are similar to series 3000 channels, but separated for engineering considerations to designate primary use for voice rather than for data transmission.

Type 2001. Approximate bandwidth of 300 to 3000 Hz. Furnished for voice transmission, for half-duplex (or for duplex on a two-point basis). Channels may be used to connect user-provided terminal equipment or communication systems to other user-provided terminal equipment or communication systems, or to connect such equipment or systems to a private branch exchange (PBX) system, Centrex (a type of PBX in which incoming calls can be dialed direct to any extension without an operator's assistance; outgoing and intercom calls are dialed direct by the extension users). System or other switching equipment or station equipment furnished to the user by the telephone company.

Type 2002. Approximate bandwidth of 300 to 3000 Hz. Furnished for voice transmission or combined voice transmission and control purposes for half-duplex (or for duplex on a two-point basis) in connection with remote operation of mobile radiotelephone systems.

Type 2003. Approximate bandwidth of 300 to 3000 Hz. Furnished for transmission of single-frequency tones in connection with remote operation of mobile radiotelegraph systems.

Type 2004. Specially adapted for operation and tone signal control between the frequencies of ~300 and ~3000 Hz. Furnished in connection with remote operation of a high-frequency point-to-point radiotelephone system for civil defense purposes.

Type 2006. Approximate bandwidth of 300 to 3000 Hz. Furnished in connection with foreign service exchange.

Special United States government configurations are not included in the foregoing.

Series 3000 (voice-grade channels)

These are channels with a bandwidth not exceeding 4000 Hz and are designed for data transmission, remote metering, supervisory control, and miscellaneous signaling. These channels are not suitable for the transmission of dc pulses. The number of stations that may be connected and the distance over which satisfactory transmission is possible may be limited by operating and transmission factors on channels furnished for data transmission. Channel conditioning at extra cost is available.

Type 3001. Approximate bandwidth of 300 to 3000 Hz. Furnished for remote metering, supervisory control, and miscellaneous signaling.

Type 3002. Approximate bandwidth of 300 to 3000 Hz. Furnished for data transmission.

Series 4000 (voice-grade channels)

These are conditioned channels with a bandwidth not exceeding 4000 Hz and are furnished for three-level data transmission or transmission of telephotograph (facsimile) material. These channels are not suitable for transmission of dc pulses.

Table 6 Principal Channels of National Communication Networks (United States) (*Continued*)

Type 4001. Approximate bandwidth of 300 to 3000 Hz. Furnished for three-level data transmission at a rate of 1300 and 1600 bits/s and at a normal error rate of not more than 1 in 100,000 bits. The channels are specially conditioned and maintained with regard to steady and impulse noise and to envelope delay characteristics and net loss as follows:

The envelope delay distortion shall not exceed between 1000 and 2500 Hz, a maximum difference of 500 μs.

The loss deviation with frequency (from 1000-Hz reference) shall not exceed between 500 and 1000 Hz, -3 to $+3$ dB ($+$ means more loss).

Two-point services are furnished for half-duplex operation in one direction only or for duplex operation. Multipoint services are furnished for half-duplex operation in one direction only. A data set is required to condition signals generated by user's equipment to signals suitable for transmission on the channel, and to condition signals received from such a channel to signals for delivery to the apparatus furnished by the user.

Type 4002. For telephotograph transmission. Specially adapted for the transmission of picture material between the frequencies of approximately 1200 Hz and approximately 2600 Hz.

Two-point services are furnished for half-duplex or duplex operation. Multipoint services are furnished for half-duplex or duplex operation. Duplex operation is available at the terminal stations of multipoint services, with intermediate service points limited to receiving-only operation, and will be furnished where (*a*) there are not more than four service points on the channel and conditioning arrangements are provided for, and (*b*) there are not more than seven service points on the channel and conditioning arrangements are *not* provided for.

Series 5000 (Telpak channels)

Many users have a requirement for bulk communications between specified locations. Telpak is the designation for a service offered for the leasing of wideband channels between two or more points. Channels under this tariff are provided through the use of Telpak service terminals, or connecting arrangements provided by the telephone company.

Telpak service terminals are furnished under the following classifications: (1) teletypewriter grade; (2) telephone grade, including telephotograph and data channels; and (3) wideband services.

Series 6000 (program transmission)

These channels are available for (1) program transmission services and (2) the transmission of educational television or radio programs—as further defined by regulations.

Series 7000 (television)

These channels are for television transmission services. They carry video and the associated audio for television.

Series 8000 (wideband)

This series of channels provides services suitable for high speed data and facsimile transmission or for use as individual voice-grade channels up to a maximum of 12 or for use as channels in connection with a 50-kilobit foreign exchange switched service. Additionally, when provided for high-speed data or facsimile transmission, the channels can be arranged for alternate use as individual channels of voice grade up to a maximum of 12.

Series 1000 (voice)

These channels are furnished for the purpose of extending user-provided communication systems to a premise of the user. Channels are furnished for half-duplex or duplex operation on a two-point basis. Approximate bandwidth is 300 to 3000 Hz. Furnished to the extent permitted by the normal transmission characteristics of the grade of channel, for types of transmission similar to those set forth for series 1000, 2000, and 3000 channels. The user's premises must be located 25 air miles or less from the point at which the user-provided communication channel is connected to the telephone company extrance facility.

Electrical Protection

Whenever telemetry equipment is operated over leased or privately owned channels, protection against the following hazards is required: (1) direct or induced voltages due to lightning, (2) voltages due to accidental contact of a transmission circuit with adjacent high-voltage power circuits, and (3) voltages induced electromagnetically or electrostatically by high-voltage power circuits that parallel the transmission circuits.

Leased telephone and telegraph circuits are usually provided with adequate protection against lightning and high voltage. However, this protection may not limit voltages at user-owned terminals to values that are safe for the connected equipment. Such equipment should be protected separately.

Privately owned transmission circuits fall into three categories, dependent on the location of transmission wires with respect to high-voltage wires: (1) Transmission circuits do not cross and are not adjacent to high-voltage power circuits. Protection against lightning only is required. (2) Transmission circuits are adjacent to high-voltage power lines, thus requiring both lightning and accidental contact protection. (3) Transmission circuits are parallel to high-voltage power lines. Protection is necessary against lightning and against electromagnetic and electrostatically induced voltages. Various devices for opening circuits to the transmitters and receivers and grounding the transmission lines are available for the protection of both mechanical and electronic equipment.

Wideband Channels

In addition to the basic voice circuit which transmits frequencies between 300 and 3000 Hz, circuits are available from licensed carrier companies which permit the transmission of much higher frequencies. For example, a single coaxial cable can transmit a number of 3000-Hz frequency bands, one being the basic 200- to 3200-Hz band. The second 4200 to 7200 Hz band is obtained by modulation of a basic 200- to 3200-Hz band by a 4000-Hz carrier. This process can be duplicated until 1800 channels are being transmitted simultaneously. A pair of open wires can transmit up to 150,000 Hz and as many as sixteen 3000-Hz bands.

Microwave Channels

This is a special case of radio transmission. The need for a large number of communication channels in such industries as power and gas transmission and the increased crowding of the spectrum in the radio range led to rapid growth in the use of microwave radio transmission. Several bands have been allocated in the range 890 MHz to 30 GHz. These frequencies are similar to television and radar in that they are line-of-sight and are beamed directly from one antenna to another so that terrain considerations become important. The equipment is rugged and reliable and can provide a large number of channels. In general, the basic cost of a transmitter and receiver is high, but the cost of adding extra channels is relatively small per channel. Consequently, microwave equipment becomes more economically competitive for multichannel requirements.

Power Line Carrier Channels

Power line carriers have been used for many years to communicate between the central generating station and local distribution stations by utilities supplying electric power. An ac carrier is coupled to a high-voltage transmission line, and the telemetered information is superimposed on the transmitted power.

Optical transmission systems are described in a separate article later in this section.

Multiplexing

Multiplexing is a method of transmitting more than one measurement over a single transmission channel. As the distance increases between transmitting and receiving points, the cost of multiconducting cable with separate wires for each measurement (installation, maintenance, and periodic replacement) becomes economically prohibitive. In addition, the cost of future expansion cannot be justified on the basis of separate transmission channels for each measurement.

Multiplexing commonly is accomplished by either time or frequency sharing of the transmission channel between the individual measurements.

Time Division Multiplexing

In this system, measurements are transmitted in sequence until all have been sampled—at which time the cycle is repeated. The time required to complete the transmission of all the measurements may be sufficiently short so that any one measurement cannot change excessively between samplings.

A pulse duration telemetering system using time division multiplexing is shown in Fig. 10. Each transmitter is sequentially connected to its corresponding receiver. The system connects transmitter 1 to receiver 1 for a specified length of time—then transmitter 2 to receiver 2—on so on.

To ensure that the rotary scanning switches do not get out of step because of power failure or other failure in the transmission channel, the two rotary switches must be synchronized. This is

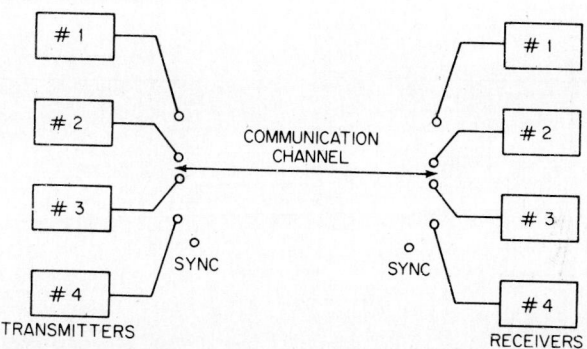

Fig. 10 Time division multiplexing.

accomplished by additional circuitry that stops the transmitter switch at the synchronization point until the receiver switch also is at its synchronization point. As previously described (under Pulse Code), time division multiplexing synchronization also can be accomplished by coding the message being transmitted when digital telemetering is utilized.

Frequency Division Multiplexing

This system uses electrical filters to separate the transmission channels into many frequency bands. These frequencies can be used to transmit data by amplitude modulation (AM), frequency modulation (FM), or phase modulation (PM).

As previously described (telemetering communications channels), a leased telephone channel transmits frequencies in the range 300 to 3000 Hz. Therefore, it is possible to transmit several distinct frequencies or tones over this channel at the same time. See Fig. 11. Approximately 24 separate frequencies can be transmitted over a single telephone-grade circuit.

Each transmitter consists of a tuned oscillator, a line driver, and an output network. The oscillator is on continuously and is connected to the transmission line by the line driver each time the carrier key closes. Therefore, as each carrier key operates in accordance with the requirements of a pulse duration, pulse position, pulse code, or frequency transmitter, the particular frequency assigned to that measurement is applied to the transmission channel.

At the receiver, the filter accepts only its particular frequency, and the output contacts duplicate the carrier key contacts at the transmitter. For example, a frequency telemetering transmitter with a range of 9 to 15 Hz would turn the carrier frequency 365 Hz on channel 1 on and off in the range 9 to 15 Hz proportional to the pressure measurements.

Another frequency telemetering transmitter with a range of 9 to 15 Hz would turn the carrier frequency 3500 Hz on channel 24 on and off in the range 9 to 15 Hz proportional to a different measurement. Similar operation of the intermediate 24 channels also would occur. Table 7 lists specifications of frequency-shift keying (FSK) frequencies.

The frequencies that would be telemetered when transmitting nine measurements using 120-Hz channel spacing and ±30-Hz frequency shifts are shown in Fig. 12. Each mark frequency is +30 Hz below the carrier frequency. The baud rate indicates the maximum possible transmission speed.

As shown in Table 7, each channel spacing has a baud figure which is one-half the channel

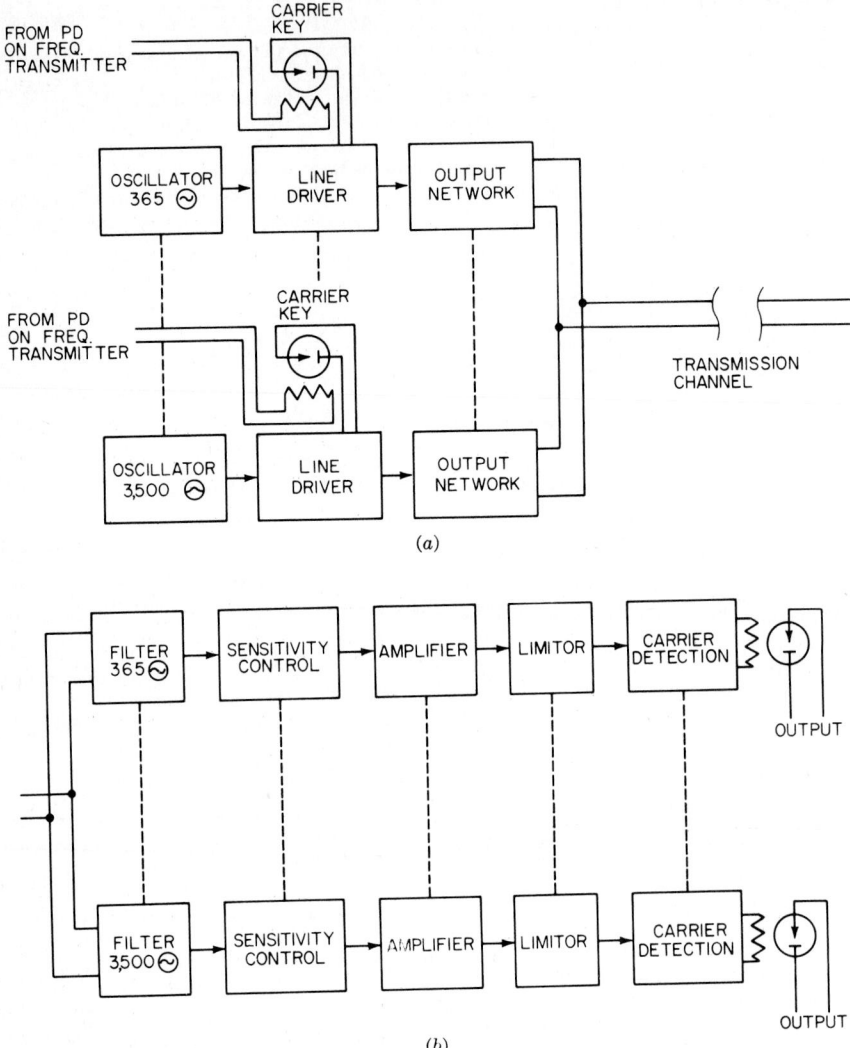

Fig. 11 Frequency division multiplexing. (*a*) Transmission; (*b*) receiving.

spacing and equal to the bandwidth. The keying speed is being modulated between mark and space at a second frequency less than or equal to half the bandwidth. The result is a sum of complex frequencies. The usable total energy is contained within the bandpass. When the keying frequency exceeds half the bandwidth, or the baud rate exceeds the bandwidth, sideband frequencies with significant energy levels outside the bandpass are generated. These are rejected by the receiver bandpass filter, information is lost, and errors in transmission occur.

The complex transmitted frequencies are passed through the tuned filter in a receiver designed to accept a particular frequency band only. In order to reject signals from adjacent channels, a guard band is provided between adjacent mark and space frequencies which limits the channel spacing. The guard band is generally equal to the bandpass.

The adjacent channel frequency rejection is typically 35 dB. See Fig. 13. To avoid cross talk or

Table 7 Typical FSK Audio Channel Frequencies

Channel spacing, Hz	100		120		170	300	600	1200
Bauds	50		60		85	160	300	600
Maximum frequency shift, Hz	±25		±30		±42.5	±75	±150	±300
Frequency channels								
	365	2000	426	1980	495	465	915	1815
	465	2100	540	2100	595	765	1515	
	565	2200	660	2220	765	1065	2115	
	665	2300	780	2340	935	1365	2715	
	765	2400	900	2460	1105	1665	3315	
	865	2500	1020	2580	1275	1965		
	965	2600	1140	2700	1445	2265		
	1075	2700	1260	2820	1615	2565		
	1175	2800	1380	2940	1785	2865		
	1275	2900	1500	3060	1955	3165		
	1375	3000	1620	3180	2125	3465		
	1475	3100	1740	3300	2295			
	1575	3200	1860		2465			
	1675	3300			2635			
	1775	3400			2805			
	1875	3500			2975			
					3145			
					3315			
					3485			
					3655			

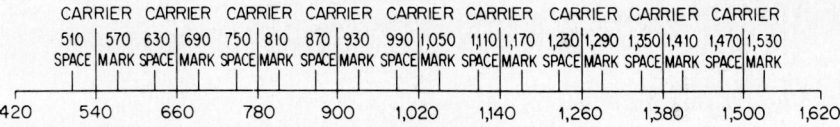

Fig. 12 Frequency-shift keying (FSK) channel frequency spacing.

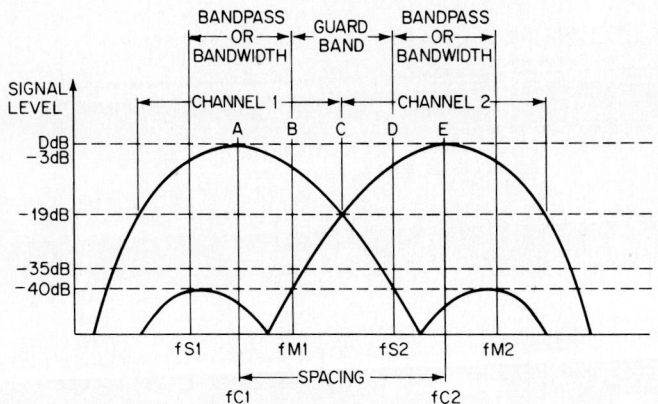

Fig. 13 Frequency-shift keying (FSK) filter characteristics. fC, Adjacent channel carrier frequencies; fS, space frequency; fM, mark frequency.

Table 8 Decibel Values versus Power and Current or Voltage Gains, and Decibel-Milliwatts versus Voltage Relationships

dB	Power ratio	Voltage or current ratio	dB·mW	dB·mW versus voltage*
0	1.00	1.00		
0.5	1.12	1.06		
1.0	1.26	1.12	−10	0.24
1.5	1.41	1.19	−9	0.27
2.0	1.58	1.26	−8	0.31
3.0	2.00	1.41	−7	0.35
4.0	2.51	1.58	−6	0.39
5.0	3.16	1.78	−5	0.44
6.0	3.98	2.00	−4	0.49
7.0	5.01	2.24	−3	0.55
8.0	6.31	2.51	−2	0.62
9.0	7.94	2.82	−1	0.69
10	10.0	3.16	0	0.77
15	31.6	5.62	+1	0.87
20	100	10	+2	0.98
25	316	17.8	+3	1.09
30	1,000	31.6	+4	1.23
40	10,000	100	+5	1.38
50	10^5	316	+6	1.55
60	10^6	1,000	+7	1.73
70	10^7	3,162	+8	1.95
80	10^8	10,000	+9	2.18
90	10^9	31,620	+10	2.45
100	10^{10}	10^5		

*Ac volts (based on 1 mA in 600 Ω = 0 level.)

interference from signals on adjacent channels, the sum of both filter attenuations is usually required to be equal to or less than −35 dB, as shown for points a, b, c, and d in Fig. 13. For a comparison of decibel values to power or voltage and current gains, refer to Table 8. Table 9 shows recommended voltage settings for increasing numbers of tone transmitters to avoid cross talk.

Table 9 Voltage Output Level Settings for Multiple-Tone Transmitters

Number of tone transmitters	Suggested output level setting*	
	AC volts	dB·mW
1	0.77	0
2	0.54	−3
3	0.44	−5
4	0.39	−6
5	0.34	7.5
6 or 7	0.31	−8
8 to 10	0.26	9.5
12 to 16	0.21	−13
17 to 25	0.18	−25

*These levels are for reference only. The local telephone company communications manager should be contacted for local requirements.

Modulation

Modulation is the process of varying the characteristics of an electric signal called a *carrier*. An electric signal can be turned on and off as in the mark-space system of a telegraphy system, in which case the signal is amplitude-modulated (AM). An electric signal also can be varied in frequency (FM) or shifted in pulse. See Fig. 14.

In a frequency telemetering system where the modulating signal is a frequency in the range of

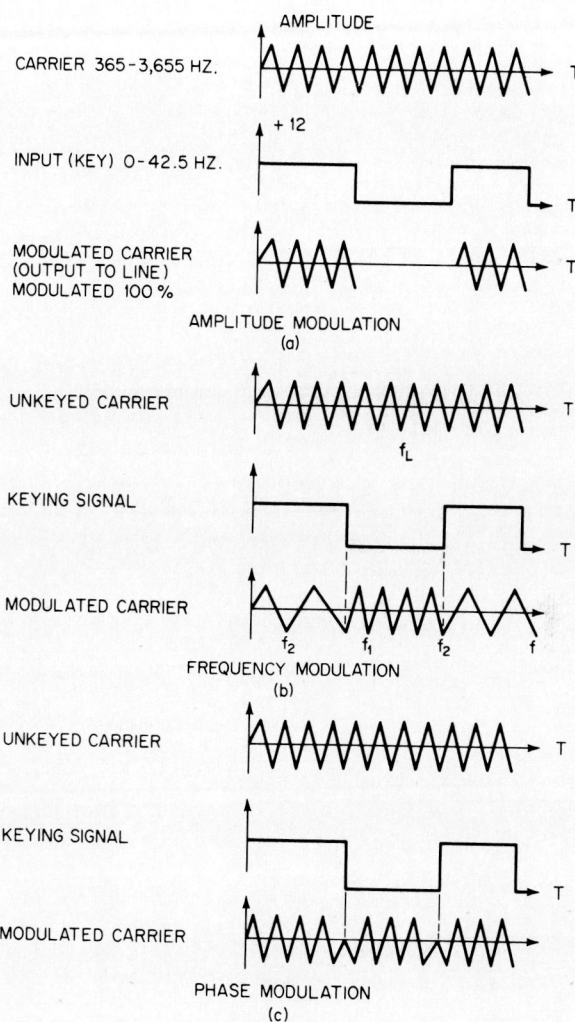

Fig. 14 Comparison of amplitude, frequency, and phase modulation. (*a*) Amplitude modulation—varying the amplitude of a sinusoidal carrier in step with the input (keying) signal. (*b*) Frequency modulation—varying (modulating) the frequency of a carrier in step with the input signal; the amplitude stays constant. (*c*) Phase modulation—varying the phase of a carrier in step with the input signal.

18 to 30 Hz, which is amplitude-modulating a carrier, the carrier is turned on and off at a rate equal to the 18- to 30-Hz modulating signal.

In a pulse code telemetering system, the carrier is on when a mark is being transmitted, and off when a space is being transmitted.

In FM or FSK, the carrier is shifted between two frequencies at a rate proportional to the 18- to 30-Hz modulating signal. In a pulse code telemetering system, one frequency is transmitted for the mark condition, and another for the space condition.

Phase modulation (PM) is similar to FSK. The phase of the carrier is shifted 180° at a rate proportional to the 18- to 30-Hz modulating signal, or shifted 180° each time the code changes from a mark to a space.

Of the three methods of modulation (AM, FM, and PM), many industrial telemetering applications use FM. AM is the most susceptible to noise which may change the amplitude of the signal, but the cost of AM is the lowest. FM is not susceptible to amplitude variations and, therefore, is less susceptible to noise. Since PM is independent of both amplitude and frequency, it is the most secure. In addition, PM has the advantage of permitting higher speeds of transmission, because each cycle can convey data. However, PM has the disadvantage of being the most expensive in initial costs, requiring a better grade and a more expensive carrier channel (than is normally available in the telephone grade) and being more difficult to maintain.

REFERENCES

Editor's Note: Closely related topics are described in this Handbook in the article Digital Technology, in Sec. 14, the article Information Theory, in Sec. 16, and the article Distributed Control in Sec. 18.

Andreiev, N.: "In Quest of a Common Data Bus," *Control Eng.,* vol. 24, no. 4, 1977, p. 50.

Andreiev, N.: "A Closer Look at Data Bus Systems," *Control Eng.,* vol. 24, no. 7, 1977, p. 33.

Bailey, S. J.: "Signal Transmission: Electronic and Optical Options," *Control Eng.,* vol. 23, no. 3, 1976, p. 32.

Bennett, W. R., and J. R. Davey: *Data Transmission,* McGraw-Hill, New York, 1965, pp. 295–300.

Chase, D.: "Consider Every Error Source in Data Acquisition Design," *Control Eng.,* vol. 27, no. 6, 1980, p. 65.

Deshon, W. E.: "Microprocessors Improve Serial Communications," *Control Eng.,* vol. 25, no. 3, 1978, p. 127.

Fling, J.: "Multiplexing," *Control Eng.,* vol. 23, no. 4, 1976, p. 87.

Gawlowicz, D. J.: "Use Existing In-Plant Telephone Lines for Data Transmission," *Control Eng.,* vol. 25, no. 5, 1978, p. 54.

Gooze, M., and G. Nelson: "What Do You Have to Know about Serial Data Communication?" *Instrum. Technol.,* vol. 28, no. 6, 1981, p. 65.

Hagelbarger, D. W.: "Recurrent Codes: Easily Mechanized, Burst-Correcting Binary Codes," *Bell Syst. Tech. J.,* vol. 38, 1959, pp. 969–984.

Hamming, R. W.: "Error Detecting and Error Correcting Codes," *Bell Syst. Tech. J.,* vol. 26, no. 2, April 1950, pp. 147–160.

Hertanu, H.: "Using Communication Techniques in Distributed Control," *Instrum. Technol.,* vol. 27, no. 12, 1980, p. 45.

Kemp, R. E.: "Close-Coupled Telemetry for Measurements on Gas Turbines," *Instrum. Technol.,* vol. 25, no. 9, 1978, p. 105.

Muller, D. E.: "Application of Boolean Algebra to Switching Circuit Design and to Error Detection," *IRE Trans. Electron. Comput.,* vol. EC-3, no. 3, 1954, pp. 6–12.

Peterson, W. W.: "Encoding and Error-Correction Procedure for the Bose-Chaudhuri Codes," *IRE Trans. Inform. Theory,* vol. IT-6, no. 4, September 1960, pp. 459–470.

Reed, I. S.: "A Class of Multiple-Error-Corrective Codes and the Decoding Scheme," *IRE Trans. Inform. Theory,* vol. PGIT-4, September 1954, pp. 38–49.

Reed, I.S., and G. Solomon: "Polynomial Codes over Certain Finite Fields," *J. Soc. Ind. Appl. Math.,* vol. 8, no. 2, June 1960, pp. 300–304.

Sliger, E. J., and G. S. Osgood, Jr.: "Communications Protocol Is Aimed at Low-Cost I/O," *Control Eng.,* vol. 27, no. 6, 1980, p. 83.

Staff: "Control Needs a Standard Data Highway," *Control Eng.*, vol. 28, no. 3, 1981, p. 144.

Taylor, J. L.: "Improving Data Acquisition in Dynamic Measurements," *Instrum. Technol.*, vol. 26, no. 8, 1979, p. 51.

Wozencraft, J. M., and B. Reiffen: *Sequential Coding,* John Wiley, New York, 1961.

Pneumatic Transmission and Control

by
J. D. Warnock*

Pneumatic transmission is a telemetering method in which analog information is transmitted as air pressure. In a typical pneumatic transmission system, a penumatic transmitter located in the field converts a process measurement, such as a level or temperature, to a proportional pneumatic pressure. The pressure signal is transmitted to a pneumatic indicator, recorder, or controller, usually located in the control room. In the case of a pneumatic control loop, the controller, responding to the transmitted measurement signal, sends a return signal to the field to position a final control element, such as a pneumatically operated control valve. See Fig. 1.

PNEUMATIC TRANSMISSION SYSTEMS

Pneumatic transmission systems, using ¼- or ⅜-in (~6.4- or 9.5-mm) plastic or metal tubing, with transmission distances up to 1000 ft (~305 m), are widely employed in the process industries. Transmission distances over 1000 ft are not common because of the effect of transmission lags on closed-loop automatic control.

Most of the measurement and control systems used in the process industries are based on either pneumatic or electric signal transmission. The basic concept of separate measuring and receiving instruments communicating with a standard transmission signal offers a number of advantages that are particularly significant in the process industries.

Advantages of Transmission Systems

Whether pneumatic or electrical, the advantages of transmission systems include:

1. *Control Operations Can Be Consolidated.* Measurements from a large plant can be transmitted to a central control room for more efficient operator supervision.

---

*Manager, Systems Engineering, Moore Products Company, Spring House, Pa.

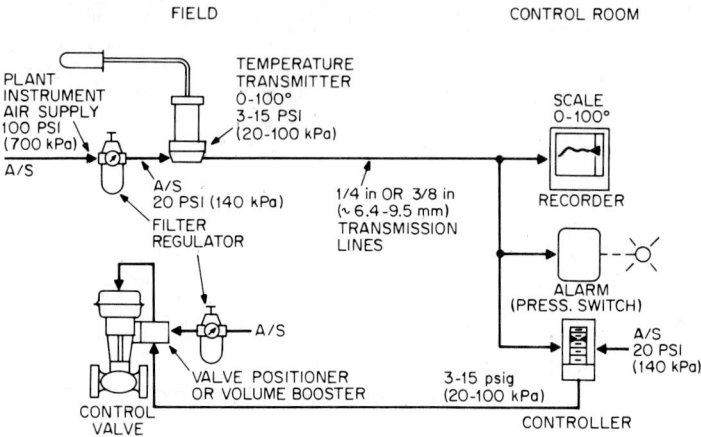

Fig. 1 Typical pneumatic transmission system. A/S, Air supply.

2. *Receiving Instruments Are Not Exposed to Field Conditions.* Only the transmitter is exposed to process fluids and field atmospheres. Controllers, indicators, and recorders operate more reliably in a clean control room environment.

3. *Receiving Instruments Can Be Standardized.* The same kind of controller, indicator, or recorder can be used for all process variables (flow, temperature, level, etc.).

Generally, pneumatic or electrical instruments can provide the same functions with equal accuracy. Because of the delay in pneumatic transmission, electrical systems may be preferred for installations with long transmission distances. However, penumatic instruments offer certain other advantages, again, of particular interest in the process industries.

Advantages of Pneumatic Transmission

Among these advantages are:

1. *Pneumatic Instruments Are Inherently Explosion-Proof.* Electrical instruments must meet stringent safety specifications for use in hazardous areas. See the article Safety in Instrumentation and Control Systems in Sec. 14.

2. *Pneumatic Systems Provide Protection against Power Failure.* With compressed air storage or turbine-driven compressors, a pneumatic system can maintain process control even during an electric power failure.

3. *Pneumatic Systems Are Directly Operated with Air-Operated Valves.* Air-operated valves are by far the most common final control elements in the process industries. Electrical control systems require an electric-to-pneumatic transducer for pneumatic valve operation.

In the early development of transmission-based process control systems, pneumatic systems predominated, simply because of the greater availability of suitable pneumatic instruments. At that time, additional advantages cited for pneumatic instruments were better reliability and maintainability. With the advance of solid-state electronics and, particularly, integrated circuits, equivalent electrical instruments were developed. As of 1982, a new installation of an extensive analog instrumentation system is more likely to be electronic than pneumatic. A large, centralized control system can be implemented with either pneumatic or electrical transmission. The importance of the differences between the two methods is determined by the specifics of each installation; an intelligent choice requires a knowledge of the essential differences and an evaluation of how these differences apply to particular installations.

Transmission Signal Range

Standard Signal Range

In any transmission system, a standard signal range, familiar to all users, is desirable. One important reason for this is that instruments from different manufacturers can be used in the same system. The Instrument Society of America (ISA) specifies two standard pneumatic transmission ranges: 3 to 15 psig (20 to 100 kPa gage)* and 3 to 27 psig [20 to 180 kPa].* The 3-psig [20-kPa] pressure represents 0% of the measured variable range or valve stroke, and the 15- or 27-psig [100- or 180-kPa] pressure represents 100% of the measured variable. These signal ranges are used by manufacturers for calibrating transmitters, receivers, and final control elements. The 3 to 15 psig range is by far the most common range used in the United States. Signal ranges used in other countries are similar and are shown below with SI equivalents for comparison. Since there is little difference between these ranges, a transmitter or receiver designed for the 3 to 15 psig range can usually be calibrated for any of the other ranges shown:

Pneumatic Signal Ranges

3 to	15 psig	20.68 to 103.42 N/m^2	(U.S. preference)
0.2 to	1.0 kg/cm^2	19.61 to 98.04 N/m^2	
0.2 to	1.0 bar	19.99 to 99.97 N/m^2	
0.2 to	100 kPa	19.99 to 99.97 N/m^2	

Live Zero

The 3-psig [20-kPa] pressure in the 3 to 15 psig [20 to 100 kPa] range and the equivalent pressure in the other ranges is called the *live zero*. Although 0 psig or atmospheric pressure may seem an obvious starting point for a pneumatic signal range, a live zero is commonly used for this reference for a number of reasons:

1. Instruments can be checked and calibrated more accurately when a signal below the zero reference can be read on test equipment.
2. The dynamic response of a decreasing signal is improved with a live zero. Pneumatic transmitters decrease system pressure by exhausting air to the atmosphere. With a live zero, substantial differential pressure is available to produce exhaust flow, even at minimum output.

Other Ranges

In addition to 3 to 15 psig [20 to 100 kPa] and 3 to 27 psig [20 to 180 kPa], a number of other ranges are also in use. These nonstandard ranges are encountered for the most part in older equipment and need not be considered except when an older system is updated to include new equipment. Fortunately, inexpensive pneumatic relays are available for interfacing between different signal ranges, and this situation presents no serious problems. In some applications, higher pressure ranges, such as 6 to 30 psig [40 to 200 kPa], are preferred because the higher pressure gives more force to operate a control valve. In most applications, however, the controller output signal is applied to a valve positioner or booster relay, rather than being sent directly to the valve actuator, and this consideration is not pertinent.

*Throughout this article, SI equivalents of English units are shown in parentheses. A comparison of similar pneumatic signals is shown in brackets, with the numbers in brackets representing the same percentage of full scale.

Air Supply Requirements

Air Quality

To ensure accurate and reliable operation of pneumatic instruments, a clean, dry, regulated air supply must be provided. Solid or liquid contaminants in compressed air can clog the small passages that are a part of all pneumatic instruments, causing loss of accuracy or even complete failure. Moisture in the air, condensing and freezing, can cause the same problems. Corrosive contaminants and oil in compressed air can damage metal and plastic instrument components, resulting in unreliable operation and excessive maintenance costs. For these reasons, considerable care is given to the design and maintenance of instrument air supply systems. In most plants, separate air supply systems, including compressors, filters, and moisture removal equipment, are used for instrument air.

ISA Standard S7.3 recommends the following minimum standards for instrument air:

Moisture. The dew point at the line pressure should be 10°C (18°F) below the minimum ambient temperature, but in no case should it be above 2°C (35°F).

Particle Size. The maximum particle size in the air stream at the instrument should be 3 μm (0.0001 in).

Oil Content. The maximum oil content should be as close to zero as possible, but under no circumstances should it exceed 1 ppm.

Air Supply Pressure

The instrument air supply pressure must be regulated to maintain accurate operation. With a typical pneumatic transmitter, a change of 5 psi (33 kPa) in supply pressure will change the transmitter output by about 1%, causing a measurement uncertainty. As shown in Fig. 1, a separate filter-regulator is used for each field instrument. This provides one final stage of filtering and moisture removal, as well as a stable instrument supply pressure. In a control panel, a single high-capacity regulator and filter set is used to supply all instruments through an air supply header. A second regulator and filter set is usually provided for standby service.

The instrument air supply pressure must be high enough to provide an adequate flow capacity at maximum output, but not so high that receiving instruments can be damaged by overranging. Most manufacturers in the United States have standardized on a 20-psig [140-kPa] supply pressure for instruments with a 3 to 15 psig [20 to 100 kPa] output (ISA Standard S7.4 allows supply pressures from 18 to 20 psig for 3 to 15 psig instruments). This means that the manufacturer calibrates instruments with a 20-psig [140-kPa] supply pressure and that instruments from different manufacturers can be used on a common 20-psig [140-kPa] panel supply without recalibration.

Transmission Dynamics

System Response

The principal difference in performance between pneumatic and electrical analog transmission is the slower response of the pneumatic system caused by lag in the transmission tubing. Steady-state system accuracy is determined only by the accuracy of the transmitting and receiving instruments and is about equal with electronic and penumatic process control instruments. The dynamic system response, i.e., how closely the system output follows a changing system input, is of most concern in closed-loop automatic control but also affects other system functions, such as indication, recording, and alarms.

The dynamic performance of a pneumatic transmission system is affected by the characteristics of the transmitting instrument, the receiving device, and the transmission tubing. The overall system response includes the responses of the individual elements. The frequency response of the system, for example, can be determined by combining the frequency response characteristics of the transmitter, receiver, and transmission tubing. However, the response of the tubing is affected by the flow capacity of the transmitter and the inlet volume of the receiver, and these factors must be considered in determining the response of the system. See Fig. 2.

Step Response of Pneumatic Tubing

Step response tests for 500 ft (152.4 m) of ¼-in (6.35-mm) and ⅜-in (9.5 mm) tubing are shown in Fig. 3. These output responses to a step input show a dead time, related to the sonic velocity, and a high-order completion curve characteristic of a distributed parameter system. The pneumatic resistance to flow and the pneumatic capacitance (volume) are distributed along the length of the transmission line. The 63.2% completion times shown are of no analytic significance for a complex response of this kind, but this figure is often used to compare step responses in all kinds of systems. The 1 and 95% times are similarly arbitrary and are used simply to show approximately when the output response starts to change and when it is nearly complete.

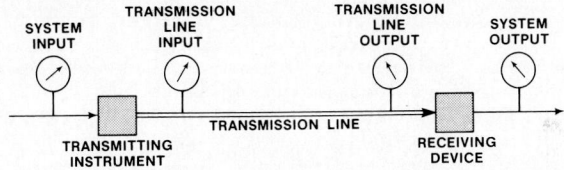

Fig. 2 Dynamic response of a pneumatic transmission system includes responses of the transmitter, receiver, and transmission line.

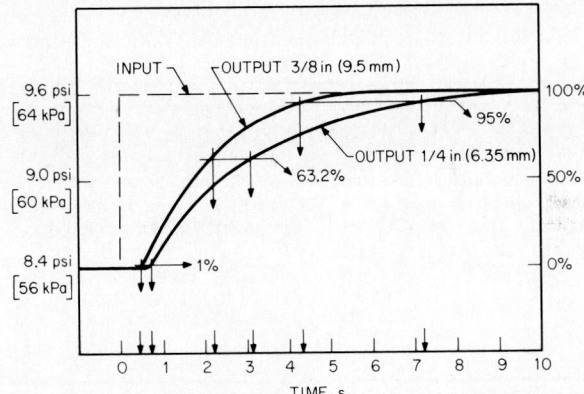

Fig. 3 Step-response tests for 500 ft (152.4 m) of ¼-in (6.35-mm)- and ⅜-in (9.5-mm)-OD plastic tubing. Tests were conducted with a high-capacity source and a 1.2-in³ (0.02-L) receiver volume.

Factors Affecting Tubing Response

A number of factors, most of them indicated in Fig. 3, affect the dynamic response of a pneumatic transmission line:

Pressure Level

Transmission line response is faster at higher pressures. It is possible to obtain faster response by using higher pressure, but most systems are operated over the standard 3 to 15 psig [20 to 100 kPa] range. Response is somewhat faster at 15 psig [100 kPa] than at 3 psig [20 kPa], so tests are normally run near the 9-psig [60-kPa] midscale value.

Signal Amplitude

Transmission line response is somewhat slower for large signal changes. Tests are usually run at ±5% amplitude near the midscale value (ISA Standard S26), because this kind of change is considered more typical than a full-scale change in process control systems.

Air Temperature

Transmission line response is slightly faster at lower temperatures. The effect of temperature over a range of 20 to 120°F (−6.7 to 49°C) is small and can be considered negligible, particularly when compared to the pressure effect from 3 to 15 psig [20 to 100 kPa].

Tubing Diameter

The response time of ¼-in (6.35-mm) tubing is nearly double the response time of the same length of ⅜-in (9.5-mm) tubing. Tubing smaller than ¼ in (6.35 mm) or larger than ⅜ in (9.5 mm) is seldom used for pneumatic transmission. To minimize transmission lag, ⅜-in (9.5-mm) tubing is often used between the control room and the field instruments. However, shorter lengths inside the control room and control panel have little effect on the total system response, and ¼-in (6.35-mm) tubing is more commonly used. Figure 3 shows responses for plastic tubing. Metal tubing is slightly larger in inside diameter and slightly faster in response. Inside diameters of plastic and metal tubing are:

	Inside diameter	
Outside diameter	Plastic	Metal
¼ in (6.35 mm)	0.170 in (4.32 mm)	0.190 in (4.83 mm)
⅜ in (9.5 mm)	0.250 in (6.35 mm)	0.305 in (7.75 mm)

Tubing Length

Doubling the transmission line length approximately quadruples the response time. The reason for this is that increasing the length increases both the line resistance and the line volume. Figure 4 shows the response time for step changes in lines up to 2000 ft (608 m) in length.

Transmitter Flow Capacity

Transmission line response is slower if the transmitter or controller does not have an adequate flow capacity. Figure 3 indicates a high-capacity source, meaning that any increase in the source capacity would not improve the response. An inadequate capacity can retard system response more or less seriously, depending on the line length and diameter. A greater flow capacity is necessary to prevent degradation of system response with a larger diameter or shorter length tubing. The flow capacity requirement is determined by the maximum instantaneous flow into the tubing. This flow is higher with large diameters because of the larger volume, and higher with short lengths because the disproportionate increase in speed more than offsets the reduction in volume.

Transmitter capacity should be considered when selecting the tubing diameter. There is no point in using larger tubing for faster response if the transmitter does not have a flow capacity sufficient for the larger tubing. The flow capacity requirement for short lengths, however, is not a practical concern because the effect of short tubing on fast response can usually be considered negligible. If the transmitter flow capacity is high enough for a 200-ft (61-m) length of tubing of a particular diameter, it can be considered sufficient for any length.

Most transmitters and controllers have a flow capacity of at least 1 sd ft^3/min (28.2 L/min), which is considered adequate for ¼-in (6.35-mm) and ⅜-in (9.5-mm) tubing. Manufacturers usually specify the flow capacity, but the method of testing is not standardized. One method is to specify the flow that will cause an output change of 1 psi (6.9 kPa), with a constant input. Since the transmitter must both fill and exhaust the transmission line, the supply and exhaust flow should be measured. Figure 5 shows the flow characteristic of a typical pneumatic transmitter.

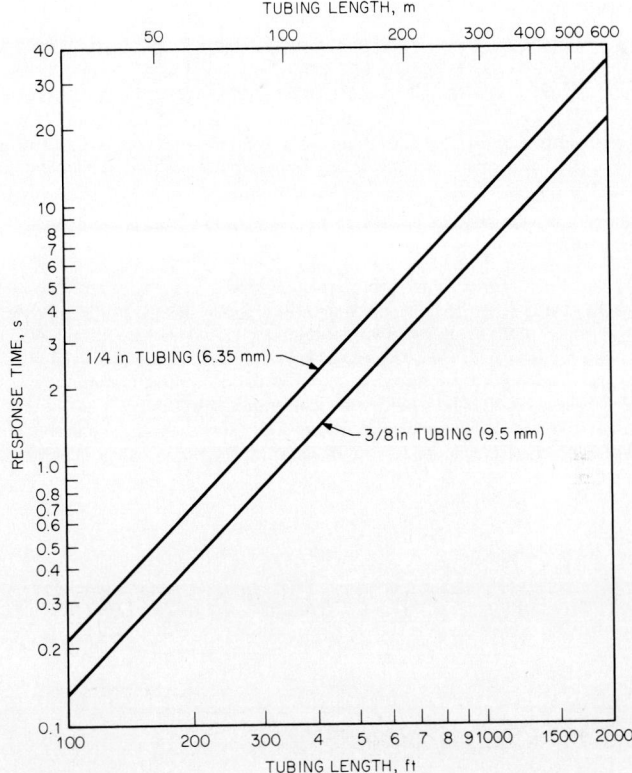

Fig. 4 Response times for a 63.2% response for a 8.4 to 9.6 psi [56 to 64 kPa] step input—for ¼-in (6.35-mm)- and ⅜-in (9.5-mm)-OD plastic tubing.

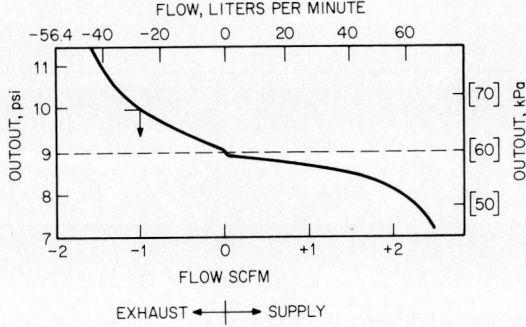

Fig. 5 Typical airflow characteristics of a pneumatic transmitter. The flow that will cause an output change of 1 psi (6.9 kPa) is often stated as the flow capacity. Note that the exhaust flow capacity is less than the supply flow capacity.

Receiver Volume

A large receiver volume seriously increases the response time in pneumatic transmission. The problem with a large receiver volume is that of filling a large volume through a length of restrictive tubing, rather than the transmission requirement of simply filling the tubing. Typical indicators, recorders, controllers, and valve positioners have inlet volumes on the order of the 1.2-in³ (0.02-L) standard test volume indicated in Fig. 3. The inlet volume of a control valve (without a valve positioner) is much greater than this. Where fast response is important, a valve positioner or a booster relay should be used to terminate the transmission line in a small volume.

Use of Volume Boosters

Most process control instruments have a sufficient flow capacity and a suitably small inlet volume so that no special precautions need be taken to maintain the best possible transmission system response. Furthermore, in many installations optimum response speed is not necessary to meet process objectives, and the effects of a low flow capacity or a large termination volume are not objectionable. In applications where fast response is required and equipment limitations cause slow transmission, volume booster relays can be used to improve response.

Figure 6 shows the use of volume booster relays (see also Fig. 11) to correct for equipment limitations. The booster at the output of the process transmitter is used to compenste for a low

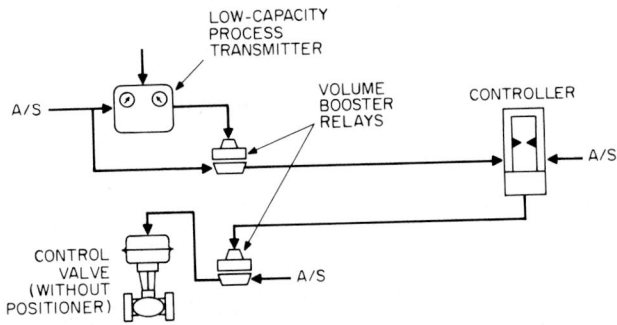

Fig. 6   Use of volume booster relays to improve system response. A/S, Air supply.

transmitter flow capacity. This booster is not necessary and will, in fact, introduce a small additional lag if the transmitter flow capacity is 1 std ft³/min (28.2 L/min) or more. The booster is located as close to the transmitter as practical and must be accurate enough that transmission accuracy is not seriously affected. The primary accuracy requirment in the booster is linearity, since the transmitter and booster can be calibrated as a system that compensates for minor gain and zero errors. The booster should, of course, have a flow capacity of at least 1 std ft³/min (28.2 L/min).

The booster relay at the input to the control valve is used to isolate the transmission line from the large volume of the air motor. This booster should be located as close to the valve as practical, so that piping losses do not restrict the flow to the valve actuator. A booster relay used for valve service need not be as accurate as a booster used for the measurement signal, but should have a high flow capacity for fast response. A volume booster with a pressure gain of 2 is sometimes used to interface a 3 to 15 psig (20 to 100 kPa gage) transmission system with a 6 to 30 psig (40 to 200 kPa gage) valve actuator.

Transmission Lags and Automatic Control

Figure 3 shows that, with a 500-ft (152.4-m) ¼-in (6.35-mm) transmission line, the full effect of a step change in a measured variable is not seen in the control room until 10 s later. This delay is not serious if the measurement is only indicated or recorded, although it should be noted that the full magnitude of a transient disturbance with a duration of much less than 10 s is not seen in the control room. The delay can be more serious if the transmission signal is used for alarm or interlock functions. However, the main concern in process control systems is with the effect of transmission lag on the response of an automatic-control loop, specifically with the delay in return to the set point after a load disturbance.

Figure 1 illustrates that a change in measurement caused by a load change is delayed in transmission to the controller, and the controller response, in turn, is delayed in transmission to the valve. The effect on the response of the control loop is then determined by the characteristics of two separate lines, each the length of the transmission distance from the field to the control room. The lags in longer transmission lines and the resultant effect on controller tuning can be expected to slow the response of a control loop to set-point changes and load disturbances. It is not possible to determine the effect of two separate transmission lines directly from step response data, nor would this be of any particular use in predicting the dynamic performance of a control loop. Step response data are useful mainly to give an intuitive picture of the magnitude of transmission lags and to show the effect of changing length and diameter.

Frequency response data on transmission lines, showing output phase shift and amplitude changes for sinusoidal input pressure signals at different frequencies, can be used to show the resultant response of two separate transmission lines. This information can be combined with frequency response data on other loop components to show the effect of transmission lines on a control loop. A limited amount of frequency response data on pneumatic transmission lines is available in the literature (Refs. 2, 4, and 10).

Figure 7 shows the results of tests on a liquid-flow control loop with different tubing lengths and

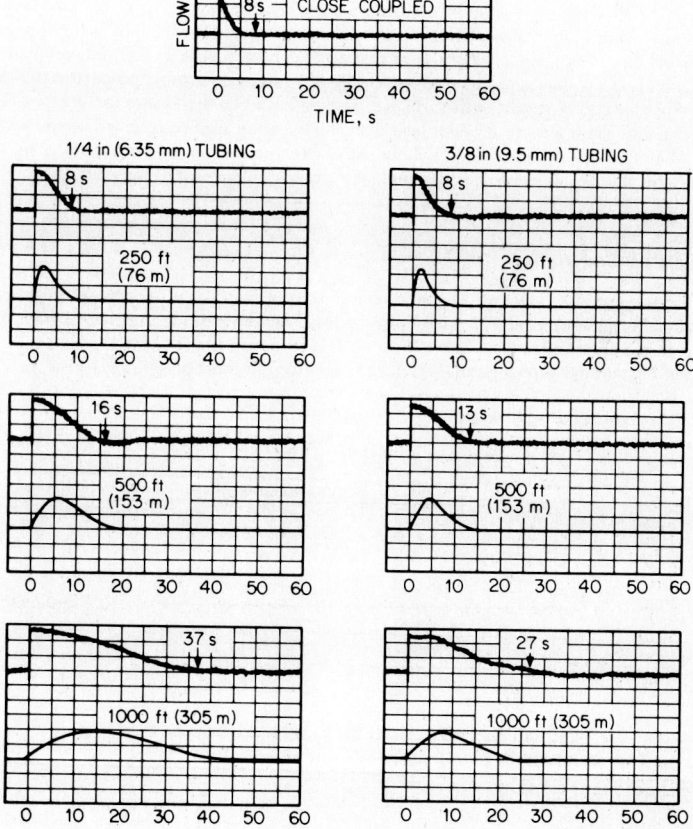

Fig. 7 Effect of transmission distance on flow control after a step load change. The upper curve in each graph shows the flow recorded at the transmitter. The lower curve is displaced to show the flow recorded in the control room.

diameters. Settling times, the times required for the flow to return almost completely to the set point after a load disturbance, are shown for comparison. The tests were conducted by applying a step change in liquid pressure at the control valve. The change in pressure required a 10% change in valve position to return the flow to the set point. The flow controller was retuned for each test to give the fastest return to the set point with the same degree of stability. The proportional gain and reset time increased as the line length increased. These tests were conducted on what may be considered a typical flow loop with an orifice plate, differential-pressure transmitter, proportional-plus-integral (PI) controller, and air-operated control valve. The transmitter and controller airflow capacities were high enough and the terminating volumes were small enough that equipment limitations did not affect the response of the transmission lines. The results apply specifically only to the loop tested, but certain general conclusions can be drawn:

1. Control response is not much affected at 250 ft (76 m). This is because the relatively small transmission lags are similar to the lags of the other loop elements—the valve, transmitter, and controller.

2. There is little advantage in using ⅜-in (9.5-mm) tubing at a 250-ft (76-m) transmission distance. Again, this is because transmission lags at this distance are relatively small.

3. At longer transmission distances, the full magnitude of a brief disturbance is not recorded in the control room.

Tests were conducted on a liquid-flow control loop because flow loop response is among the fastest in process control and shows the effect of transmission lags clearly. This is not to imply that all flow control loops require fast response or that other kinds of loops do not. However, the effect of transmission lags on slower loops, such as temperature control loops, is proportionately smaller and causes less concern. A step load disturbance, as used in these tests, causes the greatest deviation from the set point. With a more gradual load disturbance, controller reset action would reduce the deviation. Whether automatic control will be satisfactory at a particular transmission distance depends on the nature of the loop, the kind of load disturbance and, most important, process requirements.

Field-Mounted Controllers

Since the increase in response time of an automatic control loop is caused by lags in transmitting signals to and from the controller, response can be improved by locating the controller in the field, close to the transmitter and valve. Figure 8 shows this arrangement. The system is complicated by the provision for manual valve operation, which is included in most process control loops. Manual

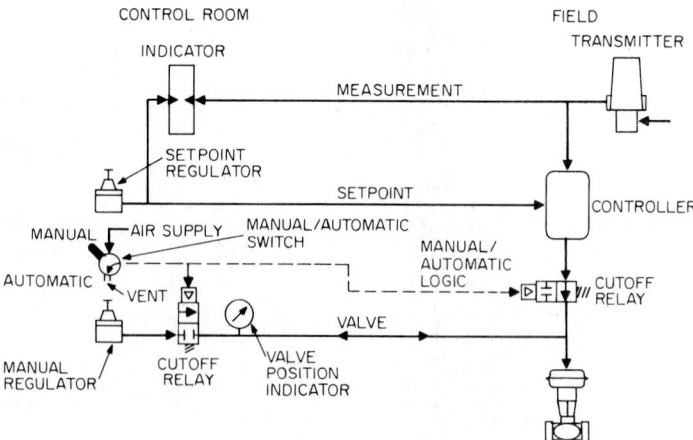

Fig. 8   Field-mounted controller used for fast control response with long transmission lines. Manual or automatic switching is accomplished with air-operated relays.

operation is accomplished by using a logic signal to shut off the controller and then operating the valve from a manual loading device in the control room. The closed-loop response to a load disturbance with this arrangement is about the same as that of a close-coupled controller without transmission lags. The branch lines connecting the transmitter and controller to the control room present an additional load on these outputs, but this is not a serious effect if the instruments have an adequate flow capacity.

The field-mounted controller arrangement in Fig. 8 is called a *four-pipe system*. The obvious disadvantage is the cost of the two additional transmission lines required for sending the set-point and logic signals to the field. Another disadvantage is that the controller is not accessible for tuning adjustments. Although the closed-loop response to load disturbances is improved, transmission lags still affect indication of the measurement and manual operation of the valve. In the early application of pneumatic transmission systems, four-pipe systems were commonly specified for all flow control loops, regardless of the transmission distance. As a result of wider understanding of transmission dynamics and more realistic evaluation of process requirements, the use of a four-pipe system is now likely to be restricted to loops where fast response is critical.

PNEUMATIC DEVICES

The design of pneumatic transmitters, controllers, and all other pneumatic instruments is based on a number of relatively simple pneumatic devices which are combined in different mechanical arrangements to give a specific output-input relationship. These pneumatic devices can be considered the equivalent of certain passive or active electrical elements, and a useful analogy between a pneumatic system and electric circuits is often apparent.

Baffle-Nozzle Amplifier

A baffle-nozzle amplifier is the primary detector in almost all pneumatic transmitters and controllers. Its function is to convert mechanical motion to a pneumatic signal. Figure 9 shows a baffle-nozzle

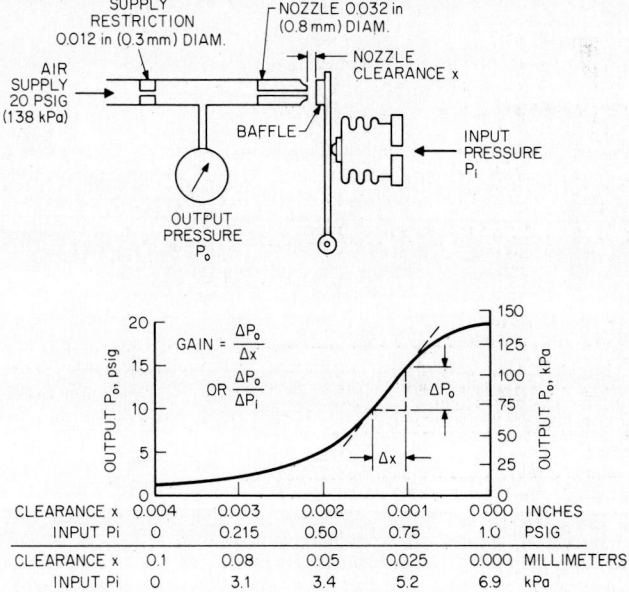

Fig. 9 Baffle-nozzle amplifier, the primary detector in many pneumatic instruments.

actuated by pressure in an input bellows. Because the baffle is often mounted on a pivoting element, this instrument is also called a *flapper-nozzle amplifier.*

In principle, the operation of a baffle-nozzle is quite simple. The output increases from a minimum value to supply pressure as decreasing nozzle clearance blocks the flow of air.

The minimum output pressure is a function of the supply pressure and the relative sizes of the restriction and nozzle, and is usually less than 1 psig (6.9 kPa gage). Certain operating characteristics and limitations of the baffle-nozzle are of particular interest in the design of pneumatic transmission instruments.

1. The gain of a baffle-nozzle, stated as the output pressure change per unit change in nozzle clearance, is high. In Fig. 9, a 3 to 15 psig (20 to 100 kPa gage) output requires only a 0.002-in (0.05-mm) change in nozzle clearance. This gain is determined by the nozzle size and the ratio of the nozzle size to the restriction size. The gain can also be stated as the ratio of the output pressure change to the input pressure change. The pressure gain depends on the characteristics of the input element and can be made as high as necessary for a particular application. Increasing the baffle-nozzle gain improves the accuracy of the instrument in which it is used. Pressure gains as high as 1000 are used in pneumatic instruments.

2. The restriction and nozzle diameters shown in Fig. 9 are typical. These diameters are a compromise to meet different operating requirements. Small diameters increase the gain but also increase the danger of clogging and the difficulty of aligning the baffle and nozzle. Large diameters increase the continuous air consumption, which is equivalent to power consumption in an electrical device.

3. The output of a baffle-nozzle amplifier is nonlinear and subject to drift. Gradual accumulation of foreign matter in the supply restriction reduces the output for a given input and changes the gain. The output is also sensitive to changes in air supply pressure. For these reasons, a simple baffle-nozzle amplifier cannot be used as an accurate transmission instrument.

4. The output flow capacity of a baffle-nozzle is low and is, therefore, not suitable for use as a transmission signal. All output flow is through the small supply restriction.

A baffle-nozzle amplifier, then, is a high-gain, nonlinear, drift-prone, low-capacity element which is not by itself useful as an accurate transmission device. It is equivalent to the active elements of electronic circuits—transistors and, in some respects, operational amplifiers. As with the electronic equivalents, negative feedback techniques are used with a baffle-nozzle amplifier to design accurate linear instruments.

Baffle-Nozzle Controller

Despite its limitations, a baffle-nozzle amplifier is quite suitable for use as a simple proportional-only controller. Figure 10 shows a temperature controller with an adjustable gain. In pneumatic controllers, the gain is usually described in terms of proportional band percentage and in this kind of controller may be adjustable to give a full-scale output change for 2 to 10% of the set-point adjustment span. Typical temperature set-point spans are from 100 to 1000°F (38 to 538°C). The action of this controller is reversible to suit the valve action (air-to-open or air-to-close).

Normal baffle-nozzle drift errors are not serious in this kind of control because the controller acts over such a narrow temperature range; small output pressure changes represent very small temperature errors. Since this device is not intended for use as an accurate transmitter, linearity is not a primary concern. Controller output linearity, in any case, is not critical because of the larger nonlinearities in most control valve characteristics. Even the limited flow capacity of a baffle-nozzle may be acceptable for short transmission lines and slow control loops. Where flow capacity is a problem, an inexpensive volume booster can be used.

Pilot Relays

These relays, also called *pilot valves* and *booster relays,* are used with baffle-nozzle amplifiers to provide a good flow capacity for transmission service. In addition to increased flow capacity, pilot relays usually provide some pressure gain. The pressure gain serves to increase the gain of the baffle-nozzle system for improvement in instrument accuracy. Additional gain permits the baffle-nozzle to

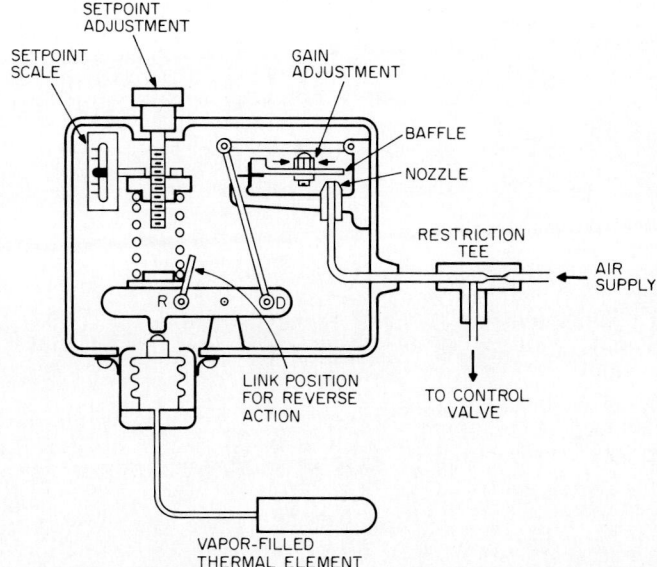

Fig. 10 Baffle-nozzle temperature controller.

operate over a narrow, more linear portion of the pressure characteristic and, by decreasing baffle-nozzle pressure, reduces air consumption.

In all the pilot relays shown in Fig. 11, an increase in nozzle pressure acts on the input diaphragm to push the valve plunger down, opening the supply port to increase the output pressure. A decrease in nozzle pressure causes the diaphragm to rise, opening the exhaust port to decrease the output pressure.

In the high-gain relay shown in Fig. 11*a*, for any output less than the supply pressure, both the supply and exhaust ports must be open, and there is a continuous flow of air to the atmosphere. This kind of pilot is also called a *bleed-type relay*. Maximum steady-state air consumption occurs when the output is about half the supply pressure. It can be seen that increasing the size of the supply and exhaust ports to increase the flow capacity also increases air consumption. Thus a high-gain relay with a flow capacity of 2 std ft^3/min (56.4 L/min) may have a steady-state air consumption of 1 std ft^3/min (28.2 L/min) at midscale output. A typical pressure gain for this kind of relay is 10, and the output-input linearity is moderate. The gain is determined primarily by the stiffness of the input diaphragm and spring and by the travel of the plunger between the supply and exhaust seats. Stiffer elements and longer travel decrease the gain. Since a pilot relay is usually enclosed in a feedback circuit with a baffle-nozzle, the exact gain and linearity are not critical. The spring below the input diaphragm determines the nozzle pressure required for a particular output and is used to fix the operating range of the baffle-nozzle on a more linear part of its characteristic curve. With a high-gain relay, very little change in nozzle pressure is required to provide a 3 to 15 psi (20 to 100 kPa) output.

The feedback pilot relay shown in Fig. 11*b* differs from a high-gain relay because an output feedback diaphragm is used and the supply and exhaust ports can be closed at the same time. The gain of the relay is determined by the ratio of the input to the output diaphragm area. Gains of from 2 to 6 are commonly used. Higher gains can be obtained by reducing the area of the output diaphragm but are not often used because of problems in maintaining sensitivity and adequate valve plunger travel with smaller diaphragms. With a diaphragm ratio of 2, a 1-unit increase in nozzle pressure opens the supply port until the output increases to 2 units to close it. Similarly, a 1-unit decrease in nozzle pressure opens the exhaust port until the output decreases 2 units. In effect, the feedback diaphragm is used to close the supply and exhaust ports when the output reaches the correct value. Theoretically, there is no flow in the steady state, and this is called a *nonbleed* relay. Most

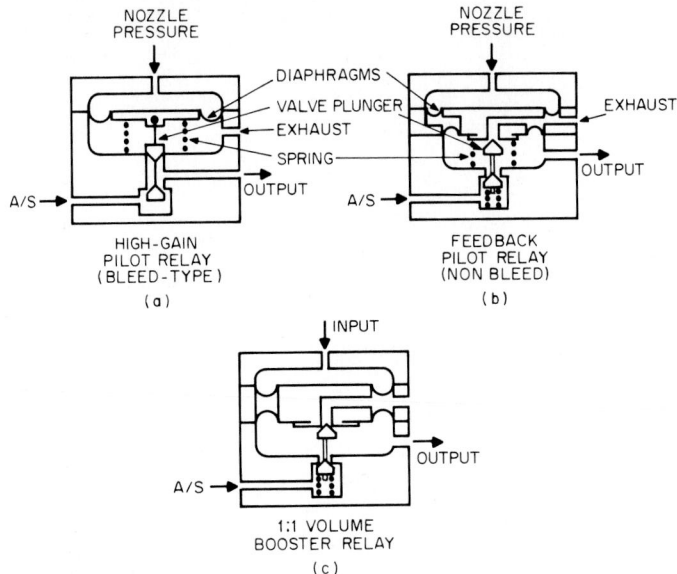

Fig. 11 Pilot relays. (*a*) and (*b*) are used with baffle-nozzle amplifiers for increased flow capacity and pressure gain, and (*c*) is used as a volume booster on transmission lines. A/S, Air supply.

manufacturers design a small bleed into the supply or exhaust port so that the other port is always slightly open. This prevents a dead spot or hysteresis in responding to small input changes. Actual air consumption is on the order of 0.1 std ft^3/min (2.8 L/min) for a typical feedback relay.

Another result of using output feedback is that the output-input pressure relationship is linear. This is not important when used in a baffle-nozzle circuit but means that this kind of relay can be used in other applications as a linear amplifier. As with a high-gain relay, a bias spring is used to determine the optimum nozzle pressure range for the 3 to 15 psig [20 to 100 kPa] output. This spring can be made adjustable to give specific output-input relationships when the relay is used as a linear amplifier.

The 1:1 *booster relay* shown in Fig. 11*c* is not often used in baffle-nozzle circuits but is shown here as another example of a feedback relay. A booster relay provides increased flow capacity, but no pressure gain. The relay is used with low-capacity transmitters and for small-volume termination on transmission lines, as previously discussed under Transmission Dynamics.

Pressure Divider Circuit

In pneumatic instruments, a pressure divider circuit is used for simple gain adjustments. A graduated needle valve is used to adjust the output as a fraction of the input. In Fig. 12*a*, closing the needle valve gives a 0 gain, and opening the needle valve fully gives a gain of nearly 1. The electrical equivalent of a pressure divider is sometimes used for analysis, but the analogy is not exact. Because airflow is not a linear function of differential pressure, the output of a pressure divider is nonlinear. The nonlinearity is not severe [with a nominal gain of 0.5, the nonlinearity for a 3 to 15 psig (20 to 100 kPa gage) input range is ~4%], but a pressure divider is normally not used in transmitter circuits where better linearity and drift stability are required. A pressure divider is employed primarily in the gain adjustment circuit of a controller. Note that a pressure divider is a low-capacity device; any flow from the output causes a drop in output pressure. Another kind of pressure divider in which both the input and vent restrictions are adjusted is shown in Fig. 12*b*.

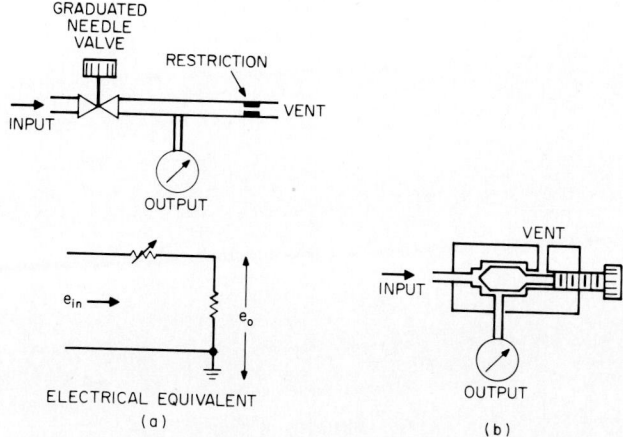

Fig. 12 Pressure divider circuits used for gain adjustments in pneumatic instruments. (*a*) A graduated needle valve is used to adjust the output as a fraction of the input. The electrical equivalent is shown in the lower diagram. (*b*) A pressure divider in which both the input and vent restrictions are adjusted.

First-Order Lag

Pneumatic resistance and capacitance respond as a first-order lag in a manner closely analogous to the behavior of an electrical resistance-capacitance circuit. The first-order lag is used to generate reset and derivative responses and as a stabilizing element in feedback devices. The response of a first-order lag is completely characterized by a time constant which can be measured as the 63.2% completion time for response to a step input. In the electric circuit, the time constant is equal to the RC product (ohms × farads = seconds). In pneumatics, there are no universal units of resistance and capacitance, but the time constant is proportional to the product of resistance to flow and volume of the chamber. The pneumatic time constant can be calculated if consistent units are used. For example, resistance expressed as a differential in bars per standard cubic feet per minute of flow, multiplied by capacitance in cubic feet, gives the time constant in minutes. Appropriate metric substitutions will yield the same result. Pneumatic time constants, however, are usually determined experimentally.

With a constant volume, an adjustable resistance can be used to change the time constant. The needle valve can be graduated in seconds or minutes to indicate the time constant. In a pneumatic controller, reset action is generated with a first-order lag, and the time constant is the reset time. See Fig. 13.

Because airflow, at a fixed pressure differential, increases with pressure level, pneumatic resistance decreases with operating pressure. Pneumatic capacitance is proportional to the chamber volume and is not affected by pressure. As a result of the resistance change, the reset time in a pneumatic controller is slightly longer at a 3-psig [20-kPa] than at a 15-psig [100-kPa] output. Resistance is considered to be reasonably constant over the 3 to 15 psig [20 to 100 kPa] range but should be determined at 9 psig [60 kPa] midscale to minimize errors at higher and lower pressures. For small signal changes near a specific operating pressure, resistance is almost constant, and the electrical analogy can be used to calculate pneumatic step response and frequency response with good accuracy.

The response graphs of Fig. 13 show the basic property of the first-order lag. The rate of change of the output is proportional to the differential across the resistance. This is intuitively apparent in the step response of the pneumatic system, where the pressure in the chamber changes more and more slowly as the differential and the airflow decrease. This also suggests a method of measuring the rate of change which will be useful for a derivative response in a pneumatic controller. The ramp response shows clearly that a differential-pressure measurement across the resistance can be used as a meaurement of the rate of change.

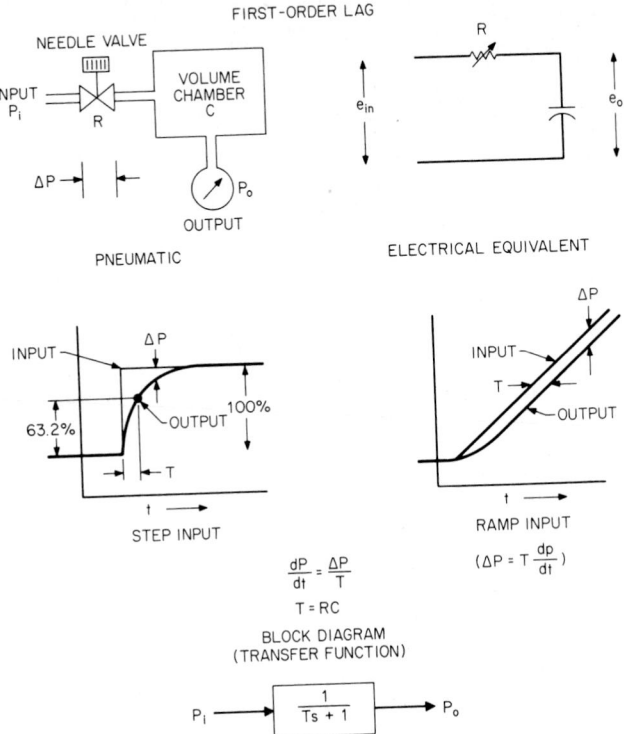

Fig. 13 Pneumatic first-order lag used for reset and derivative modes in pneumatic controllers.

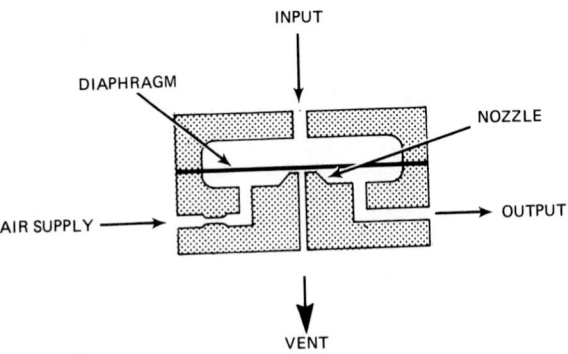

Fig. 14 A 1:1 repeater used as a buffer on the output of a first-order lag or pressure divider.

The transfer function shown represents a first-order lag in a pneumatic, electric, or any other kind of system. The transfer function can be used for system analysis, and it is commonly employed simply to designate the first-order lag in a block diagram.

1:1 Repeater

No airflow can be taken from the output of a pressure divider or a first-order lag without changing the output pressure. Where some flow capacity is required on these outputs, a 1:1 repeater is used. In electrical terms, a 1:1 repeater is equivalent to a buffer amplifier and is said to have a high input impedance.

Airflow to the repeater is through a supply restriction. The diaphragm, acting as a baffle, controls airflow through the nozzle so that the output pressure matches the input pressure. See Fig. 14. On an increase in input, the diaphragm reduces the nozzle clearance to increase the output. With a thin, flexible diaphragm, only a small difference between the input and output pressure is required to change the nozzle clearance over the operating range, and the accuracy of the 1:1 ratio is excellent. Since all output flow is through the small supply restriction, a repeater has a limited flow capacity, and its primary use is as an internal component of pneumatic controllers. It is also used in this simple form as a level transmitter, with the process liquid acting directly on the diaphragm.

Feedback Amplifier—Force Balance

If the baffle-nozzle is enclosed in a feedback circuit, the problems associated with nonlinearity and drift are eliminated. This is the principle on which most pneumatic transmitters and controllers operate. See Fig. 15.

This device is called a *force-balance amplifier* because the baffle-nozzle changes the pressure in the feedback bellows to balance the force generated by the input, whether the input force is generated by another bellows or by any other kind of input element. An increase in input force reduces the nozzle clearance, and increases the nozzle pressure and the output of the pilot relay until the feedback

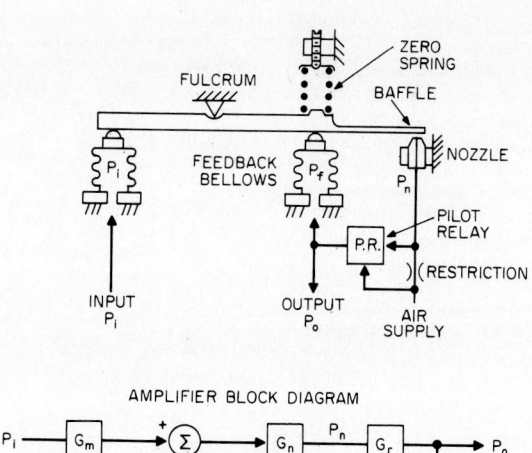

AMPLIFIER BLOCK DIAGRAM

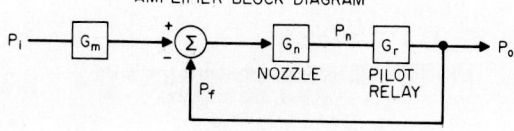

SIMPLIFIED BLOCK DIAGRAM

Fig. 15 Force-balance feedback amplifier.

pressure, which is also the output, balances the input force. The output pressure, being then proportional to the input, can be used as an accurate transmission signal for measurement and control.

The block diagram of Fig. 15 shows the operation of a feedback amplifier schematically. The negative sign in the summing junction shows that an increase in feedback bellows pressure causes a decrease in nozzle pressure and that this is a negative feedback device. From the diagram, the output-input relationship can be determined:

$$Ga = \frac{\Delta Po}{\Delta Pi} = Gm \frac{GnGr}{1 + GnGr}$$

where Ga = amplifier gain (ratio of output pressure change to input pressure change of amplifier)

Gm = mechanical gain of amplifier determined by bellows areas and location of fulcrum (representing the feedback pressure change necessary to exactly balance the force of an input pressure change)

Gn = nozzle gain (ratio of nozzle pressure change to feedback pressure change)

Gp = pilot relay gain (ratio of relay output pressure change to nozzle pressure change)

In a linear amplifier, the gain Ga should be constant for all values of output and input. The mechanical gain Gm is determined by the geometry of the amplifier and is not affected by pressure. However, the nozzle and pilot responses are nonlinear, and gains Gn and Gp do change with pressure. With a combined nozzle and pilot gain $GpGm$ which changes from 50 to 100 for a 3- to 15-psig (20- to 100-kPa gage) output, the effect on the amplifier gain is

$$GpGn = 50 \qquad Ga = Gm(50/51) = 0.98 \ Gm$$

$$GpGn = 100 \qquad Ga = Gm(100/101) = 0.99 \ Gm$$

This shows that a markedly nonlinear element with a gain varying by a factor of 2 can, when enclosed in a feedback circuit, produce a linear response. The amplifier gain changes only from 0.98 to 0.99, indicating about a 1% nonlinearity. Increasing the pilot or nozzle gain, even with the same degree of nonlinearity, further improves amplifier linearity—with $GpGn$ changing from 500 to 1000, Ga changes only from 0.998 to 0.999, indicating $\sim$0.1% nonlinearity. Specifications for pneumatic transmitters and controllers show nonlinearity in the range of 0.1 to 0.5%. Good linearity is important in transmitters, more so than in controllers, because transmitted measurements are indicated on linear scales and are used in analog computations.

The feedback circuit also compensates for output drift and supply pressure variations in the baffle-nozzle and pilot relay circuits. When the supply restriction in the baffle-nozzle circuit is partially obstructed or the air supply pressure changes, the gain of this part of the circuit is changed. As previously mentioned, these changes have little effect on the output span of the amplifier.

Note that the equation for amplifier gain Ga applies only to changes in input and output and does not define the actual output level. As shown in Fig. 15, a zero-adjusting spring can be used to adjust the output level, e.g., to give a 3-psi (20-kPa) output at zero input. The mechanical gain Gm in the amplifier can easily be designed for a wide range of inputs by proper selection of input and feedback elements. Force-balance amplifiers are used in transmitters for measuring pressure ranges as low as 0.1 psig (0.7 kPa gage) and as high as 10,000 psig (70 MPa gage). They are also used with other input elements for the measurement of temperature, level, and other process variables.

Feedback Amplifier—Motion Balance

Some input elements produce a displacement, rather than a change in force, and a *motion-* or *displacement-balance amplifier* is used to convert the displacement to a proportional pneumatic pressure. An advantage of this arrangement is that the input displacement also can be used for direct, mechanical indication of the measurement.

Increasing pressure in the bourdon tube (Fig. 16) raises the left end of the baffle lever and reduces the nozzle clearance. Increasing the nozzle pressure through the pilot valve increases the feedback pressure to reposition the nozzle lever. Since the nozzle clearance changes only a few thousandths of an inch (fractions of a millimeter) over the full output range, the baffle lever and nozzle lever remain parallel. The feedback bellows displacement and output pressure are, therefore, proportional to the

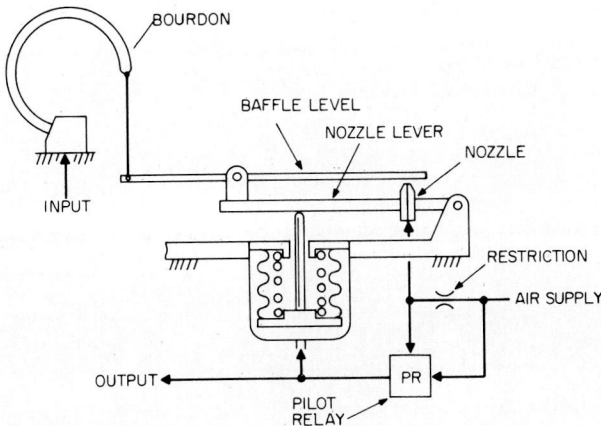

Fig. 16 Motion-balance feedback amplifier.

input displacement. In effect, the nozzle is moved to follow the input motion, and the pressure change required to reposition the nozzle is used to measure the input motion. Use of the feedback circuit results in a linear, stable output signal—as in a force-balance amplifier.

The force required to reposition the motion-balance mechanism is derived primarily from the output pressure. Little force is required from the input element, so very sensitive input elements can be used without loss of accuracy. The input displacement for a motion-balance amplifier ranges from 0.02 to 3.00 in (0.5 mm to 7.6 cm). With different input elements, motion-balance amplifiers are used in pressure, temperature, and level transmitters, for direct dimensional measurement, and in some kinds of pneumatic controllers.

PNEUMATIC CONTROLLERS

Pneumatic controllers can be conveniently divided into two classifications: (1) receiver controllers and (2) direct-connected controllers. *Receiver controllers* act on pneumatic input signals from transmitters and can be located in a control room some distance from the point of measurement. *Direct-connected controllers* include measuring elements directly connected to the process; they are usually located in the field. In the following portions of this article, force-balance receiver controllers are used to describe operating principles of pneumatic controllers—these principles are also used in direct-connected controllers.

Proportional Action

The proportional controller shown in Fig. 17 is a force-balance feedback amplifier with a differential-input bellows for the measurement and the set point. The resultant input to the controller is the difference between the measurement and the set point, called the *error* or *deviation*. The gain of the controller is changed by adjusting the pivot position. For a high gain, the pivot is at the right-hand end of the lever, so that a large change in output is required to balance a small change in input.

In most pneumatic controllers, the gain adjustment is graduated in proportional band units. The *proportional band* is the percent input change that will produce a 100% (usually 3 to 15 psi, 20 to 100 kPa) output change. The proportional band can be viewed as a band, on the indicator or recorder, through which a measurement must change to produce a full-scale valve position change. It should be noted that a receiver-controller has no inherent pressure range. The pressure range of the system is determined by the calibration of the valve and transmitter. A receiver-controller can be

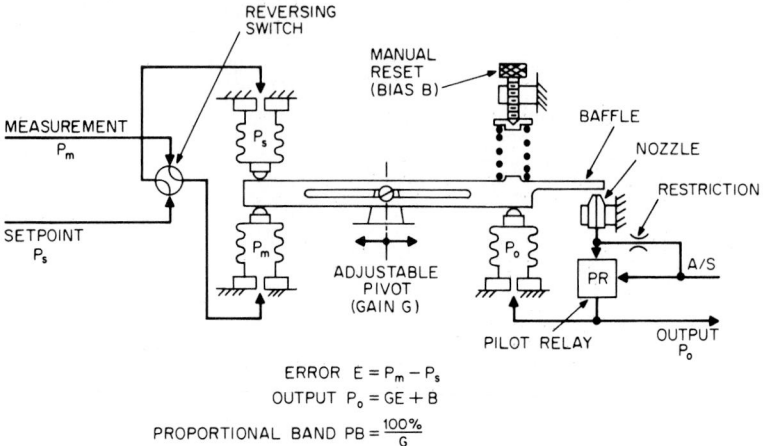

$$\text{ERROR} \quad E = P_m - P_s$$
$$\text{OUTPUT} \quad P_o = GE + B$$
$$\text{PROPORTIONAL BAND} \quad PB = \frac{100\%}{G}$$

Fig. 17 Force-balance proportional controller. A/S, Air supply.

used for any pressure range up to the pressure limits of the input and output elements and can, in fact, be used for different input and output ranges.

As shown in Fig. 17, the controller is *direct acting;* i.e., the output increases as the measurement increases. The reversing switch permits interchanging of the position of the measurement and set-point bellows to achieve *reverse action.* The controller action must be set for each control loop so that an increase in the measurement changes the controller output in the direction required to decrease the measurement—as is required in any negative feedback loop. Because of operating considerations in selecting the valve action (most control valves are air-to-open), most process control loops are set for reverse action.

The manual reset or bias adjustment determines the controller output when the measurement is at the set point (error = 0). This adjustment is used to set the valve opening required to bring the measurement to the set point. With proportional control, a load disturbance (after the manual reset has been set) will cause an offset between the set point; i.e., a different valve opening is then required to bring the measurement to the set point. If the offset is large enough to be objectionable, the manual reset must be readjusted for the new load condition. The offset can be reduced by reducing the proportional band (increasing the controller gain), but the stability of the control loop determines the minimum proportional band that can be used. A proportional band too narrow for the loop will cause control oscillation.

Proportional-plus-Reset Action

Automatic reset action causes the controller output to change as long as there is an error between the measurement and the set point. This eliminates the offset, which is characteristic of simple proportional control. Proportional-plus-reset action is also called *proportional-plus-integral* (PI) action.

In a proportional-plus-reset controller, as shown in Fig. 18, the output is connected to an opposing reset bellows through a needle valve and volume. From the output pressure equation for a proportional controller, it can be seen that the output is equal to the proportional output GE plus the pressure Pr in the reset volume. How this arrangement causes a continuous output change can be seen by considering the differential pressure across the reset needle valve. From the output pressure equation, the needle valve differential is equal to the product of the gain and the error, GE. Therefore, as long as there is an error, there is airflow into or out of the reset volume, which changes the reset pressure Pr and the controller output. The controller output continues to change until the error returns to zero. At zero error, the controller output is equal to the reset pressure Pr.

The pressure in the reset volume is proportional to the integral of the airflow into the volume. If the airflow is, in turn, proportional to the differential pressure (which is approximately so for

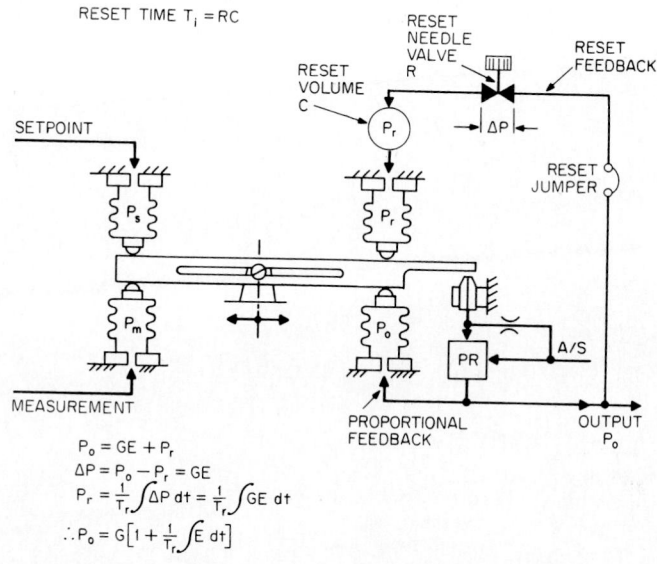

Fig. 18   Force-balance proportional-plus-reset controller. A/S, Air supply.

small differentials and for the small flows in the controller), then the reset pressure is proportional to the integral of GE. The response of the proportional-plus-reset controller can then be expressed in the equation showing integration of the error.

Reset action is adjusted by changing the reset time Ti with a graduated needle valve. Closing the needle valve increases Ti and slows the reset response. The normal range of the reset adjustment in pneumatic controllers is from 0.01 to 60 min. The reset time is commonly designated as *minutes per repeat,* meaning the time required for the controller output to change by an amount equal to the proportional response GE. The reset adjustment on some controllers is graduated to show the reset rate rather than the reset time, in which case the units are *repeats per minute.* The reset rate is simply the reciprocal of the reset time.

Reset action, then, is generated by connecting the controller output back into the controller through a first-order lag. This is a positive feedback because the reset pressure is added to the controller output. This is shown in the block diagram (Fig. 19) where the first-order lag is represented by its transfer function. Simplifying the block diagram results in the transfer function for a proportional-plus-reset controller, which can be recognized as equivalent to the output pressure equation in Fig. 18.

Reset Windup

In a controller with reset action, the output continues to change as long as there is an error. If the error persists, the output and the reset pressure Pr will increase to full supply pressure or decrease to zero—a condition known as *reset windup* or *reset saturation*. When this happens, the controller output is beyond the 3 to 15 psig [20 to 100 kPa] valve operating range.

Reset windup occurs when, for an extended period, the controller is not able to bring the measurement to the set point. Reset windup can be a problem because the measurement must cross the set point (the sign of the error must change) before the reset pressure and controller output can change. It takes some time to bring the reset pressure back to the valve operating range after crossing the set point, so the measurement moves well past the set point before settling back. The measurement overshoots the set point when control is resumed after reset windup.

In many situations, it is desirable to interrupt reset action to avoid the problems of reset windup. Since reset action is generated by positive feedback of the controller output, it is possible to interrupt

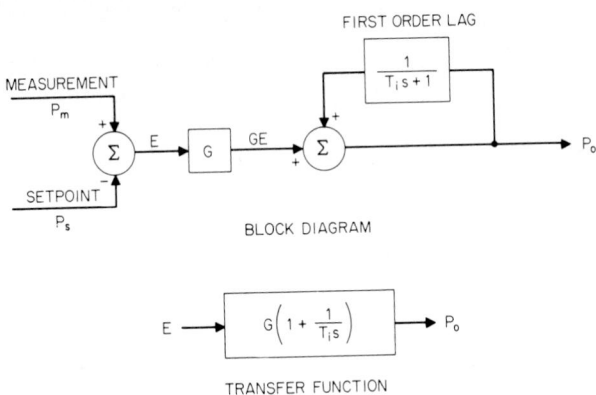

FIRST ORDER LAG

BLOCK DIAGRAM

TRANSFER FUNCTION

Fig. 19 Block diagram of Fig. 18 reduces to the transfer function for a proportional-plus-reset controller.

it by disconnecting the reset feedback from the output. The feedback jumper shown schematically in Fig. 18 is employed for this purpose in systems where alternate feedback signals are used. For example, when there is a switch between the controller and the valve, the jumper is removed and the reset feedback signal is taken from the output of the switch (the valve loading). Reset action is then interrupted when the controller output is not connected to the valve. A separate feedback connection is used very frequently in loops with manual/automatic switching. It is also used in other kinds of loops where the controller output is interrupted, such as limiting and override circuits.

Mechanical Gain Adjustment

In Fig. 18, the controller gain or proportional band is changed by repositioning the pivot. Another design for changing the gain mechanically is shown in Fig. 20. In this design, a floating disk acts as the balance beam and as a nozzle-baffle. Rotating the adjusting lever changes the position of the fulcrum rollers for a 5 to 500% proportional band adjustment. This arrangement permits compact design of the controller and gives smooth adjustment of the proportional band while the controller is in operation. Although mechanical details differ from those shown in Fig. 18, the principles of operation are the same.

Pneumatic Gain Adjustment—Stack Controller

A proportional-plus-reset controller in which the proportional band adjustment is made with a needle valve is shown in Fig. 21. Input elements are elastomer diaphragms separated by metal rings. This arrangement is referred to as a *diaphragm stack controller*.

In the error detector section, the area of the large diaphragms and the area of the smaller center diaphragm are in a ratio of 5:4. The measurement pressure, acting downward on a large diaphragm and upward on the small diaphragm, produces a resultant force downward, which is balanced by the output. Because the output acts on the full diaphragm area and the measurement acts only on the difference in areas, a change of 5 psi (34.5 kPa) in the measurement can be balanced by a change of 1 psi (6.9 kPa) in the output. Equation 1 in Fig. 21 shows the output required to balance the two forces produced by the other three pressures acting on the different diaphragm areas.

Gain adjustment is achieved by feedback of the output through a pressure divider. The gain g of the pressure divider is adjustable from 0 to 0.99 by means of the needle valve. This is a positive feedback and acts to increase the overall gain G of the controller. Substitution of the expression for P_c—the output of the pressure divider—into Eq. 1 and rearrangment gives Eq. 3. This shows the controller output as a function of the pressure divider gain g. As g is adjusted from 0 to 0.99, the controller gain changes from 0.2 to 20 and the proportional band from 500 to 5%. With the proportional needle valve closed, there is no positive feedback. The gain is 0.2, as determined by the ratio of the diaphragm areas.

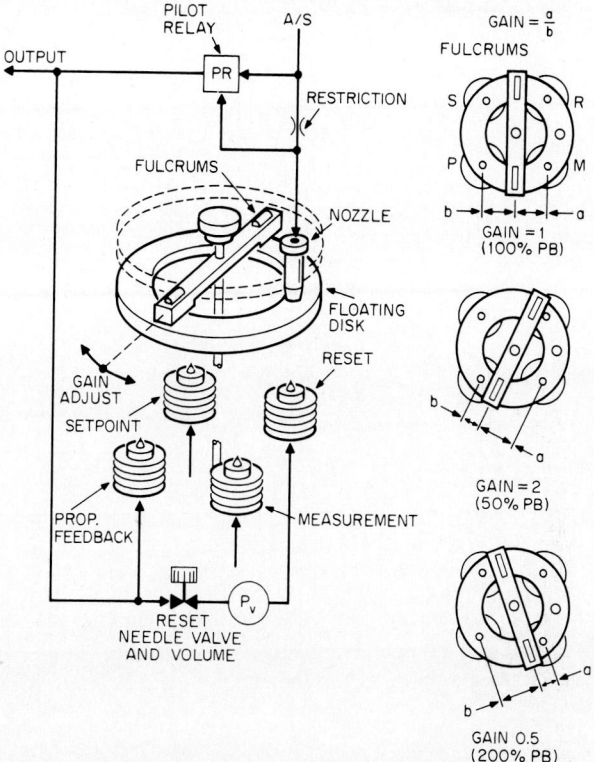

Fig. 20 Floating disk force-balance controller (proportional-plus-reset).

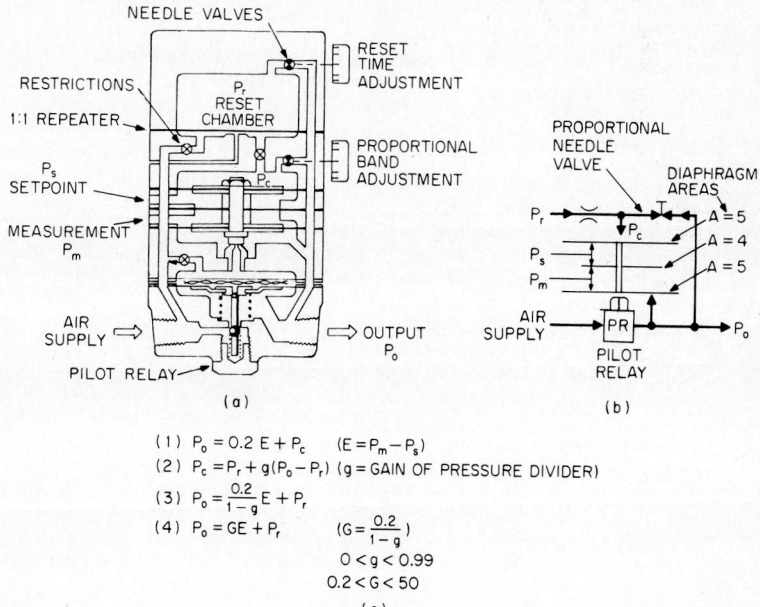

$$(1)\quad P_o = 0.2\,E + P_c \quad (E = P_m - P_s)$$

$$(2)\quad P_c = P_r + g(P_o - P_r) \quad (g = \text{GAIN OF PRESSURE DIVIDER})$$

$$(3)\quad P_o = \frac{0.2}{1-g}E + P_r$$

$$(4)\quad P_o = GE + P_r \qquad \left(G = \frac{0.2}{1-g}\right)$$
$$0 < g < 0.99$$
$$0.2 < G < 50$$

(c)

Fig. 21 Diaphragm stack controller with pneumatic gain adjustment. (a) Sectional view, (b) schematic of a proportional section, (c) relationship.

Since Eq. 4 is the same as the output equation for the PI controller in Fig. 18, reset can be obtained in the same way. The reset pressure Pr is generated by the output through the first-order lag of the reset needle valve and volume. Note that a 1:1 repeater is used on the reset chamber pressure to provide flow capacity for the pressure divider. A block diagram of a stack diaphragm controller is shown in Fig. 22.

The advantage of a stack controller is elimination of the pivot or fulcrums as potential points of wear and friction. The linearity of the proportional response is not as good as in controllers with a mechanical gain adjustment, but the linearity of the controller output is not critical. The accuracy of

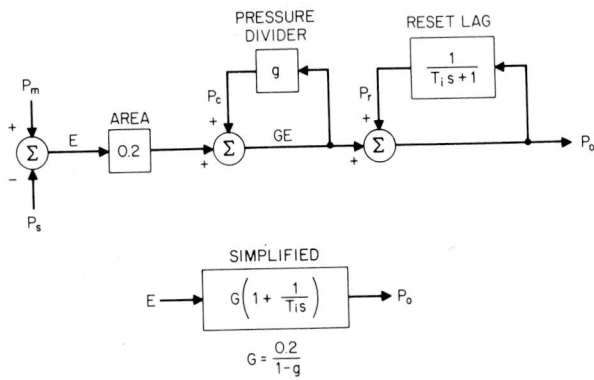

Fig. 22 Block diagram of the diaphragm stack proportional-plus-reset controller shown in Fig. 21.

the reset action is not affected by the linearity of the proportional response. Reset action continues until the error is reduced to zero, at which point the controller output is the same as the reset pressure and there is no flow in the pressure divider.

Proportional-plus-Derivative Action

Derivative action adds to the proportional response a component that is proportional to the rate of change in input. The derivative time Td is usually defined as the time separating the proportional and proportional-plus-derivative responses—with a straight-line ramp input, after transients have subsided. This is the time with which the derivative adjustment is graduated. A proportional-plus-derivative response is often described simply as a derivative action, although, in fact, a derivative response without the proportional component is seldom used in process control. Derivative action is also called *rate action* or *preact.*

One method of generating derivative action is by restricting the feedback pressure Pf in a force-balance amplifier as shown in Fig. 23. The amplifier is designed so that $Pf = Pi$ when the amplifier is in balance. With a ramp input Pi, the output must change to maintain the same pressure ramp in Pf. Since Pf is the output of a first-order lag, output Po must lead the input by the lag time constant, as indicated in Fig. 23. The derivative component $(Po - Pi)$ is then equal to the slope or derivative of Pi multiplied by the derivative time.

The foregoing explanation is simplified by the assumption that output Po can instantaneously drive the feedback Pf to match the input. This would require an infinite nozzle gain and would produce the theoretical derivative response shown. The nozzle gain, although high, is limited, and the actual derivative response shows a transient before the steady state is reached.

The block diagram (Fig. 24) represents the feedback amplifier of Fig. 23 with a finite gain a and a first-order time constant D. The gain a represents the product of the nozzle and pilot valve gains and is called the *derivative gain.* Simplifying the block diagram and considering that gain a is much greater than 1 yields the transfer function shown. This can be recognized as the transfer function for an actual proportional-plus-derivative response. The expression $Ds + 1$ represents theoretical pro-

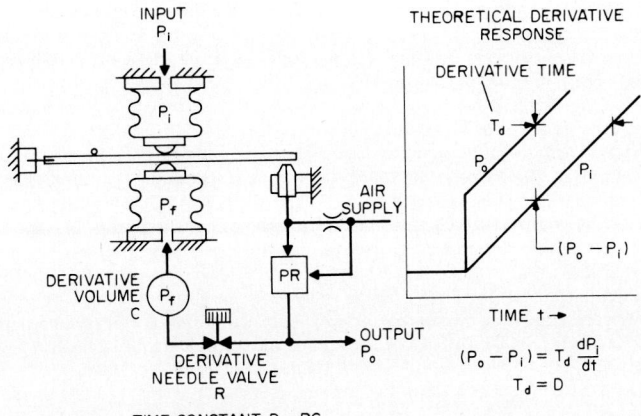

Fig. 23 Proportional-plus-derivative amplifier showing a theoretical derivative response with infinite nozzle gain.

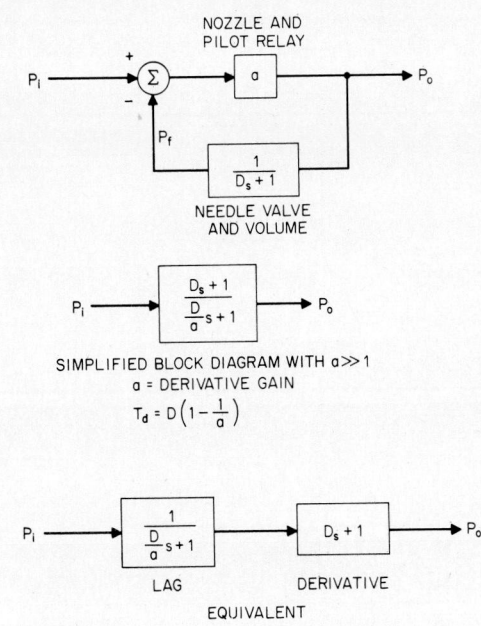

Fig. 24 Block diagram of a proportional-plus-derivative amplifier (Fig. 23) with finite nozzle gain reduces to the transfer function for the actual proportional-plus-derivative response. A simplified block diagram and equivalent are also shown.

16.51

portional-plus-derivative action. The time constant D/a is the time constant of the transient in the actual response. It can be seen that the actual derivative response is the resultant of the ideal response and a first-order lag. The derivative time Td is a function of time constant D and gain a. In the theoretical response, $Td = D$.

The actual derivative response can be made as close to the theoretical response as is necessary by increasing the derivative gain a. In practice, this is not done. The derivative gain is purposely limited to prevent the amplification of noise in the input signal. Actual derivative gains in pneumatic controllers range from 6 to 30. The specific value of a derivative gain is not critical. It should not be so low that the transient response is too slow, nor so high that noise is a problem. It is useful to think of the first-order lag, caused by the limited derivative gain, as a noise filter.

The derivative gain is limited by providing a direct proportional feedback from the output, as shown in Fig. 25. A change in input is balanced immediately by a change in the small proportional feedback bellows. Since Pc is the first-order lag response of the output, the pressure equation can be represented by the feedback block diagram, which can be simplified to show the proportional-plus-derivative transfer function. The derivative gain is determined by the ratio of the bellows areas. The step response graph shows clearly the effect of a limited gain.

A diaphragm stack proportional-plus-derivative unit is shown in Fig. 26. The pressure equation is the same as that shown in Fig. 25, and the output response is the same. This device is available as a separate unit for adding the derivative response to an existing controller or for dynamic com-

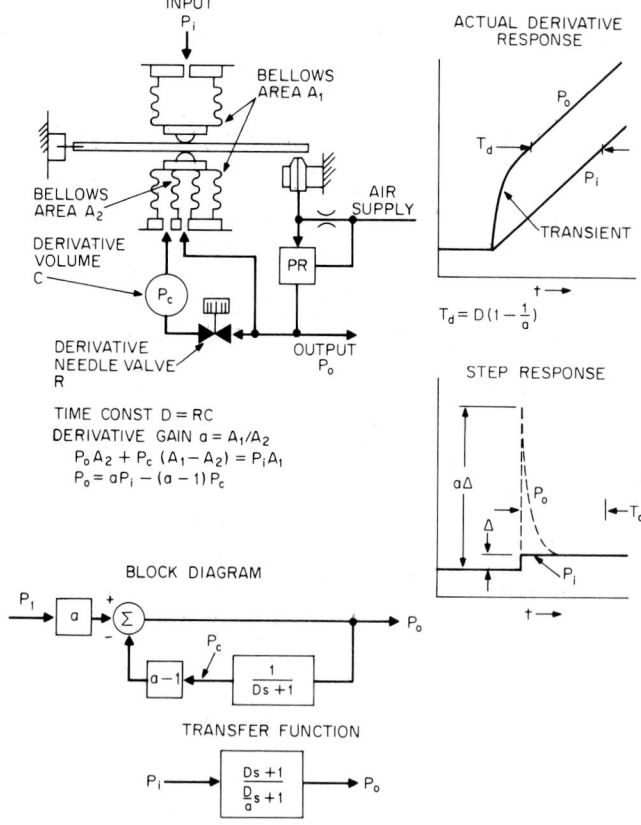

Fig. 25 Proportional-plus-derivative amplifier with a limited derivative sign.

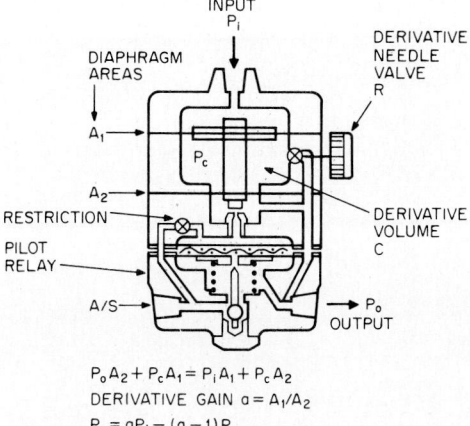

$$P_o A_2 + P_c A_1 = P_i A_1 + P_c A_2$$

DERIVATIVE GAIN $a = A_1/A_2$

$$P_o = aP_i - (a-1)P_c$$

Fig. 26 Diaphragm stack proportional-plus-derivative unit. Note that this device has the same pressure equation as that shown in Fig. 25.

pensation in feedforward systems. It is more commonly used as an integral component in a proportional-plus-reset-plus-derivative controller.

Combination of Control Modes—Derivative Position

The derivative technique shown in Fig. 25 is easily applied to the output section of the controllers of Figs. 17 or 18 to make a proportional-plus-derivative (PD) or a proportional-plus-reset-plus-derivative (PID) controller. When this is done, the derivative action is said to be on the controller *output*. Alternatively, the derivative unit is used on the measurement signal as the *input* to the proportional or PI controller. One disadvantage of a derivative on the output is that derivative action applies on set-point as well as measurement changes, which makes it difficult to make gradual set-point changes. In addition, an output derivative is not as effective under reset windup conditions because it cannot respond until the measurement crosses the set point. Although a controller with an input derivative section is somewhat more complex, this construction is common in current designs. The block diagram (Fig. 27) shows the different derivative positions. The product of the two responses shows the transfer function for a three-mode controller. Note that the response to a measurement change is the same for both kinds of controllers. Note also that a simplified transfer function $(TdS + 1)$ is usually used to denote derivative action. The actual transfer function of Fig. 25 is necessary only for a detailed loop analysis.

Control Stations—Manual or Automatic Switching

Pneumatic receiver controllers are used with panel mounting control stations, which include provisions for indicating the measurement and adjusting the set point and for manual operation of the control valve. Manual operation by the operator is useful for process start-up and for emergency conditions, such as removing the controller for maintenance. Manual operation requires a switch to bypass the controller and a manual loading device. Manual loading and set-point adjusting devices can be simple pressure regulators. In some control stations, motion-balance amplifiers, operating on the principle shown in Fig. 16, are used to convert the set-point indicator position to a pneumatic set-point signal.

In switching between manual and automatic control, it is desirable that the transition be made without disturbing the valve position. This is called *bumpless switching*. Before switching to manual control, the operator must adjust the manual output to match the controller output. Switching back

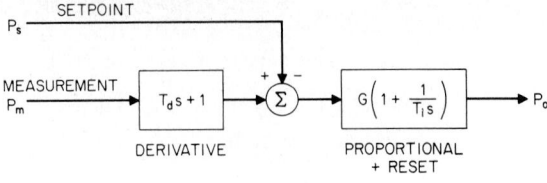

DERIVATIVE ON INPUT

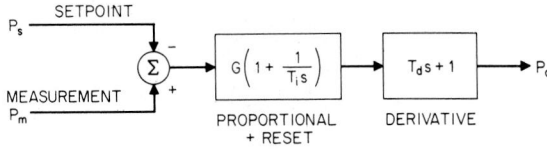

DERIVATIVE ON OUTPUT

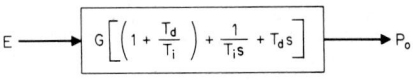

TRANSFER FUNCTION — PID CONTROLLER

Fig. 27 The derivative can be on the input or the output of a controller. A derivative on the output responds to set-point changes.

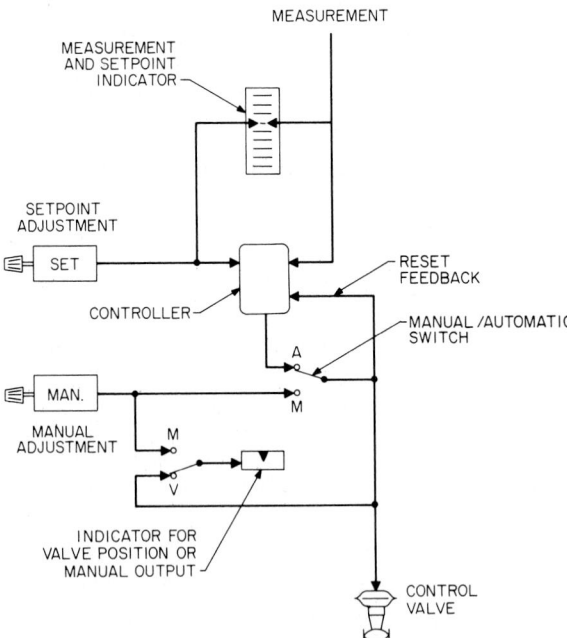

Fig. 28 Control station with a proportional-plus-reset or a proportional-plus-reset-plus-derivative controller and manual or automatic switching. The set point is adjusted to match the measurement before switching to automatic control.

to automatic control requires that the controller output be made to match the valve loading. This is accomplished by adjusting the set point to match the measurement. See Fig. 28. When this is done, the controller output is equal to the reset feedback ($Po = GE + Pr$, as indicated in Fig. 18), which is the valve loading pressure. In some controllers, the reset feedback signal is connected directly to the reset volume during manual operation so that the reset pressure Pr can follow the valve loading without the delay imposed by the reset needle valve. Since this method of switching from manual to automatic control depends on the reset pressure, it does not apply to a proportional controller without automatic reset. The manual reset of a proportional controller must be adjusted for bumpless switching to automatic control.

Control stations of recent design include automatic balancing circuits which eliminate the need for operator adjustments before switching. This is called *procedureless switching*. The manual loading device in a procedureless control station tracks the controller output in an automatic operation so that the transfer to manual control can be made at any time. There are two methods for procedureless switching from automatic to manual control. In one method of procedureless swtiching from manual to automatic control (see Fig. 29), the set-point device tracks the measurement in manual operation, and the principle of matching controller output to the manual loading is the same as just described. Both set-point and manual devices can track an input. In automatic operation, the manual device tracks the controller output. Switching is accomplished with pneumatic logic signals and air-operated cutoff relays. In automatic control, the A logic signal blocks the output of the manual loader and causes this output to track the valve position. In manual control the M logic signal blocks the controller output and causes the set point to track the measurement. In manual operation (no A signal), the valve loading is connected directly to the reset volume, bypassing the reset needle valve so that there is no delay in the controller output following the valve pressure.

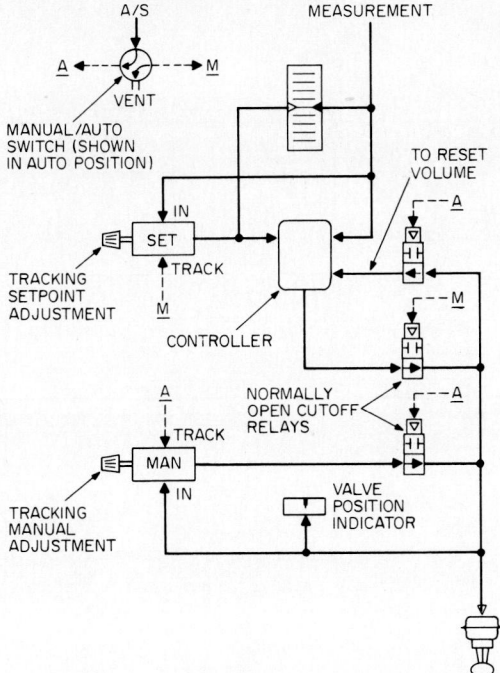

Fig. 29 Procedureless switching control station with set-point tracking. Switching is accomplished with air-operated cutoff relays. In manual operation, the set-point adjusting device tracks the measurement.

In the second method of procedureless switching from manual to automatic (Fig. 30), the set-point device does not track the measurement. In manual operation, a small balancing controller compares the controller output to the valve loading and adjusts the reset pressure to cause the controller output to match. The balancing controller may have high-gain proportional action or integral action. Switching is accomplished with pneumatic logic signals, as shown in Fig. 29. In manual operation (no A logic signal), the balancing controller connects directly to the reset volume. On switching to automatic control, the controller output is equal to the valve loading, even though the measurement is not at the set point, and normal proportional-plus-reset action begins.

In addition to the single-loop control station with manual or automatic switching, control stations with additional switching circuits are available for multivariable control. These include:

1. *Remote-set stations,* with external set-point and remote or local set-point switching. These are used in cascade systems and systems with computed set points.

2. *Cascade stations,* with primary and secondary controllers and switching for cascade or single-loop control.

3. *Ratio stations,* with gain adjustment on external set-point and remote or local set-point switching. These stations are used for flow ratio control.

4. *Computer-set stations,* with electrically driven set-point or valve loading and computer or local switching. These are used for supervisory and direct digital computer control (DDC).

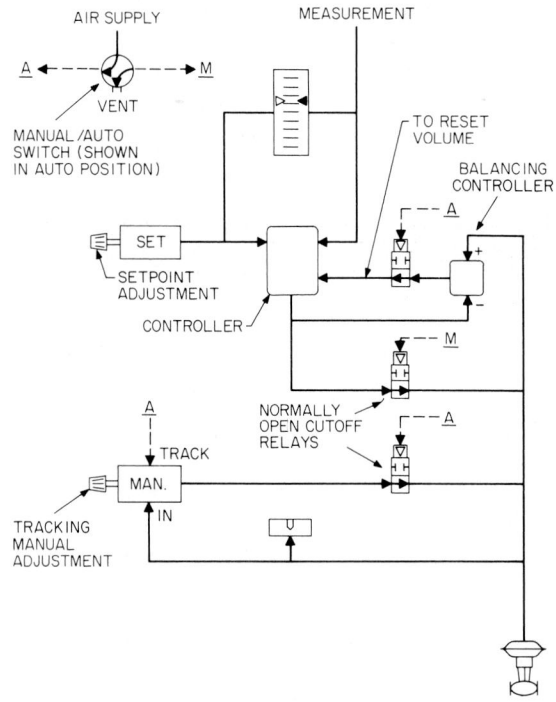

Fig. 30 Procedureless switching control station with a nontracking set point. In manual operation, the automatic balancing controller adjusts the reset pressure to make the controller output equal to the valve loading.

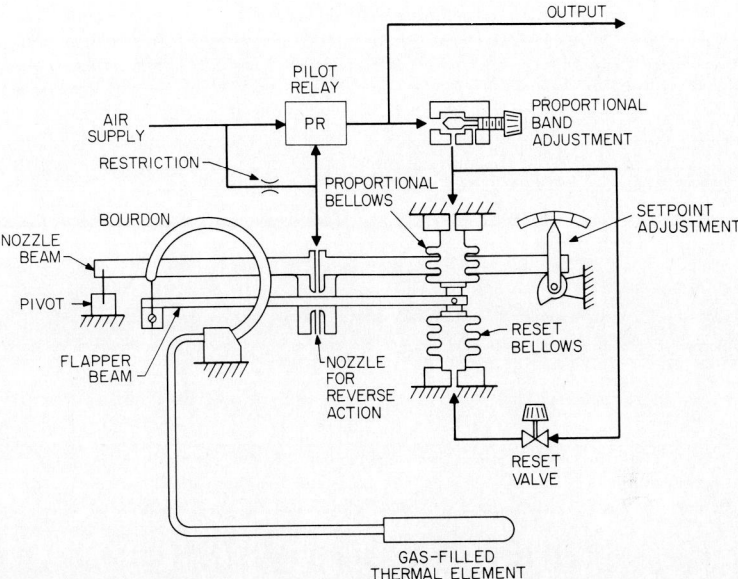

Fig. 31   Direct-connected temperature controller.

Direct-Connected Controllers

The direct-connected temperature controller shown as an example in Fig. 31 is a motion-balance instrument with proportional-plus-reset action. An increase in temperature in the gas-filled thermal element increases pressure in the bourdon tube, raising the left end of the flapper beam and increasing the nozzle pressure. Increasing the nozzle pressure, through the pilot valve, increases the output. The output is fed back to the proportional bellows, which lowers the left end of the flapper beam to restore nozzle clearance. Pneumatic gain adjustment is by a pressure divider, which feeds back some fraction of the output change to the proportional bellows. The proportional gain is increased by reducing the feedback to the proportional bellows so that a larger change in output is required to reposition the flapper beam. Reset action is generated, as shown in Fig. 18, by connecting the proportional feedback to an opposing bellows through a first-order lag.

The controller shown is direct acting. Action can be reversed by relocating the nozzle as shown and by interchanging the positions of the proportional and reset bellows. The set point is adjusted by changing the position of the nozzle with the cam supporting the nozzle beam.

REFERENCES

1. Buckley, P. S.: "Dynamic Design of Pneumatic Control Loops," *Instrum. Technol.,* vol. 22, nos. 4 and 5, 1975.

2. Caldwell, W. I., Coon, G. A., and L. M. Zoss: *Frequency Response for Process Control,* McGraw-Hill, New York, 1959.

3. Considine D. M., and S. D. Ross: *Handbook of Applied Instrumentation,* Robert E. Krieger, Melbourne, Fl., 1964, pp. 17-82 to 17-91.

4. Hougen, J. O., and O. R. Martin: "Dynamics of Pneumatic Transmission Lines," *Control Eng.,* vol. 10, no. 9, 1963.

5. ISA: "Air Pressure for Pneumatic Controllers for Transmission Systems," *Standard S7.4,* Instrument Society of America, Research Triangle Park, N.C., 1981.

6. ISA: "Dynamic Testing of Process Control Instrumentation," *Standard S26,* Instrument Society of America, Research Triangle Park, N.C., 1975.

7. ISA: "Quality Standard for Instrument Air," *Standard S7.3,* Instrument Society of America, Research Triangle Park, N.C., 1976.

8. Liptak, B. G. (Ed.): *Instrument Engineers Handbook,* vol. II, sec. 4.1: "Pneumatic vs. Electronic"; sec. 4.3; "Pneumatic Controllers," Chilton, Philadelphia, 1970.

9. Moore: "Reset Feedback in Pneumatic Controllers," *Tech. Bull. AD50-7,* Moore Products Company, Spring House, Pa., 1980.

10. Sandell, R. P., and N. H. Ceaglske: "Frequency Response of Pneumatic Transmission Lines," *ISA J.,* December 1956.

11. Tomkins, J.: "Pneumatic or Electronic Instrumentation," *Instrum. Control Syst.,* January 1979.

12. Warnock, J. D.: "How Pneumatic Tubing Influences Controllability," *Instrum. Technol.,* vol. 14, no. 2, 1967.

Optical Transmission Systems*

Transmission systems and circuits based on fiber-optics technology are attractive for instrumentation and process control applications for a number of reasons, including the following. (1) They are immune to electromagnetic interference (EMI), inclusive of radio frequency interference (RFI) and noise arising from electromagnetic pulses (EMPs). (2) They have a high information carrying capacity. Optical fibers developed as of the mid-1980s have 10,000 times the capacity of copper circuits of comparable size, with the attendant savings in weight and space—factors which contribute to lower installed costs. (3) Fiber-optics systems are effective in alleviating ground loop problems. (4) Optical fibers enhance the safety of systems that must operate in hazardous environments. (5) Optical systems are particularly suited for handling digital data. Additional benefits of this technology will accrue as more experience is gained from their use in the instrument field. Early incentives for the development of fiber-optics technology stemmed from telecommunication (telephony) and military applications. One of the early major examples of a process control application is shown in Fig. 1.

Light Wave Transmission

As in electrical transmission systems, the transmission sequence of a light wave system begins with an electric signal. This signal is converted to a light signal by a light source, such as a light emitting diode (LED) or a laser. The source couples the light into a glass fiber for transmission. Periodically, along the fiber, the light signal may be renewed or regenerated by a *light wave repeater unit.* At its destination, the light is sensed by a special receiver and converted back to an electric signal. Then it is processed like any signal that has been transmitted in electrical form.

A simplified schematic illustration of a light wave telecommunication system is shown in Fig. 2. The *transmitter circuitry* (1) modulates or pulses in code the light from a *light source* (2). An *optical fiber waveguide* (3) conducts the light signal over the prescribed distance, selected because of its

*Information for this article was largely supplied by L. V. Pfaender, Owens, Illinois; and staff members of Bell Laboratories.

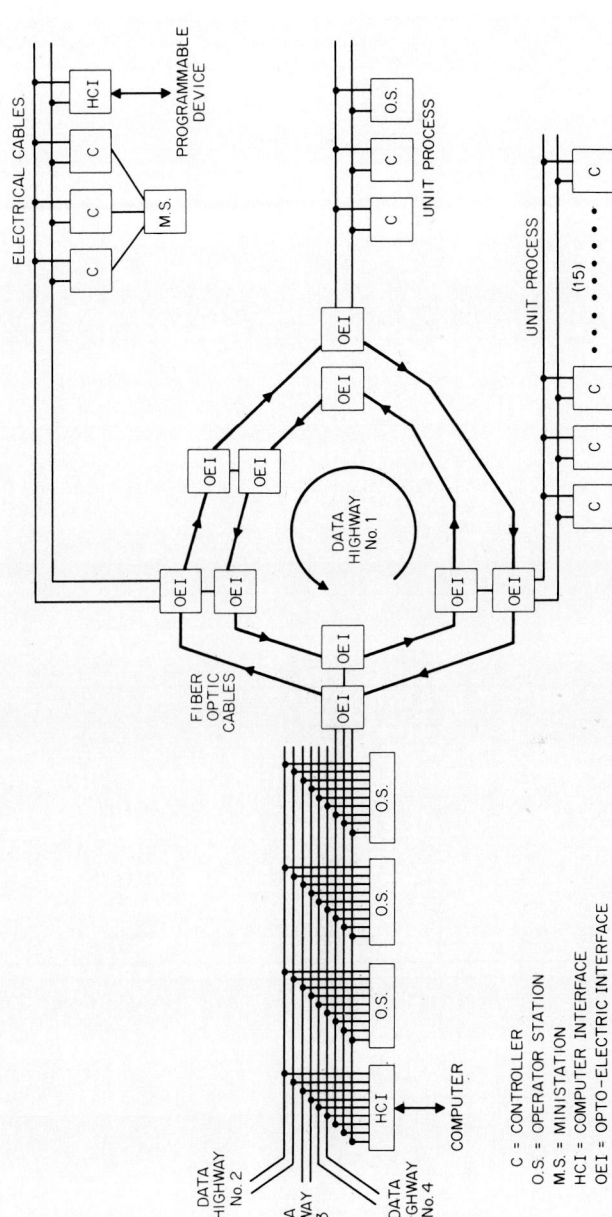

Fig. 1 Fiber-optic data highway for plant-wide distributed control introduced by Leeds & Northrup in mid-1982. Four data highways communicate with a central control room with a data rate of 500 kilobaud. Each data highway supports 30 stations, at least one of which must be an operator station or computer interface. Access to the optic loop is by way of dual optical-electrical interfaces. Provided that the total circumference of the loops does not exceed 20,000 ft (6096 m), the distance between optical-electrical couplers may be as long as 7000 ft (2134 m). The system has an eight-loop capacity in each control and over 900 loops, along with over 14,000 digital inputs and outputs. Individual stations are tied into the network by traditional electrical highways may stand alone if the large distances provided by the optical loop are not required.

16.59

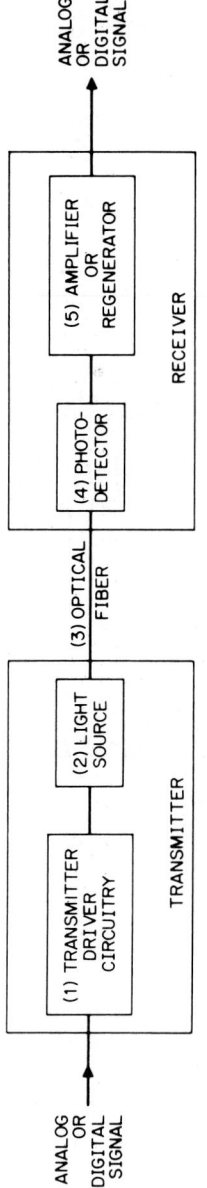

Fig. 2 Schematic of a light wave telecommunications system.

particularly good transmission capability at the wavelength of the light source. The terminal end of the waveguide is attached to a *detector* (4), which may be a *p-n* junction semiconductor diode or an avalanche photodiode, to accept the light and change the signal into an electromagnetic form for the *receiver circuitry* (5) which decodes the signal, making it available as useful electronic analog or digital output. When two-way communication is needed, the system is fully duplexed and two circuit links of the type shown are required.

Optical Fibers

Glasses of many compositions can be used for optical fibers, but for intermediate- and low-loss applications the options become increasingly limited. Multicomponent glasses containing a number of oxides are adequately suited for all but very low-loss fibers, which are usually made from pure fused silica doped with other minor constituents. Multicomponent glasses are prepared by fairly standard optical melting procedures, with special attention given to details for increasing transmission and controlling defects resulting from later fiber drawing steps. In contrast, doped fused silica glasses are produced by very special techniques that place them almost directly in a form from which fibers may be drawn. A detailed description of the principal processes, such as the *rod-in-tube method,* the *double-crucible method,* and the *chemical vapor deposition method,* can be found in *Van Nostrand's Scientific Encyclopedia* (D. M. Considine, ed.), 6th ed., Van Nostrand Reinhold, New York, 1982.

Digital Light Wave Systems

Much research has been directed toward light wave systems that are digital. In a digital system, the light source emits pulses of light of equal intensity, rather than a continuous beam of varying intensity (analog approach). Each second is divided into millions of slices of time. In the Bell System's FT3 system used in telephony, for example, there are 44.7 million such slices. The light source inserts 1 bit of information into each time slot, which flashes on briefly or remains off.

The telecommunications receiver looks for 1 bit in each slot. If the receiver senses a pulse, it registers a 1; if the absence of a pulse, a 0. Eight such bits of information make up a digital word. From a series of such words, other elements of the transmission system can reconstruct the original signal. For more fundamental digital background information, see the article Digital Technology in Sec. 14.

The capacity of a digital light wave system is the maximum rate at which pulses can be sent and received. The maximum pulse rate is limited by how much the signal is distorted by dispersion as it travels along the fiber. *Dispersion* means that a pulse is spread out in time, so that some of the pulses arrive in the wrong time slot. If enough is lost from the proper slot, the receiver may not sense a pulse that was sent. If enough is received in an adjoining slot, the receiver may sense a pulse when none was sent. The greater the dispersion, the longer the time slots must be for the receiver to sense accurately.

Basic Types of Fibers

Dispersion is of two kinds: (1) modal and (2) chromatic. *Modal dispersion* is the spreading of light as it traverses a length of fiber along different paths or modes. See Fig. 3. Each path is a different length, and thus light takes a different time to travel through each. The highest capacity fiber has only a *single mode,* so it has no modal dispersion. However, such fibers are much smaller, more difficult to couple light into, and harder to splice and connect with other types of fibers.

The more common type of fiber is *multimode,* either step index or graded index. These fibers have wider diameter cores than single-mode fibers and accept light at a variety of angles. As light enters at these different angles, it travels through the fiber along different paths. A *step index optical fiber* design is shown in Fig. 4. A light beam passing through a step index fiber travels through its central glass core and in the process ricochets off the interface of the cladding adhering to and surrounding the core. The core-cladding interface acts as a cylindrical mirror that turns light back into the core by a process known as *total internal reflection.* To ensure that total internal reflection occurs, fibers are usually made from two glasses: core glass, which has a relatively higher refractive index,

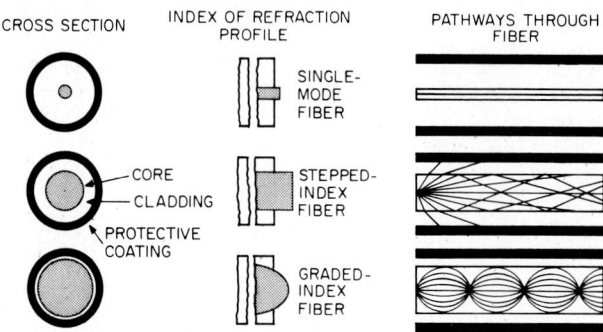

Fig. 3 Dispersion. The structure of the fiber determines whether and how the light signal is affected by *modal dispersion*. A single-mode fiber permits light to travel along only one path—therefore, there is no model dispersion. In contrast, a step index fiber provides a number of pathways of different lengths—and only one index-of-refraction boundary between layers, which bends the light back toward the center. Here, modal dispersion is high. A graded index fiber has many layers. The resulting series of graded boundaries bends the various possible light rays along paths of nominally equal delays, thus reducing modal dispersion. *(Bell Laboratories.)*

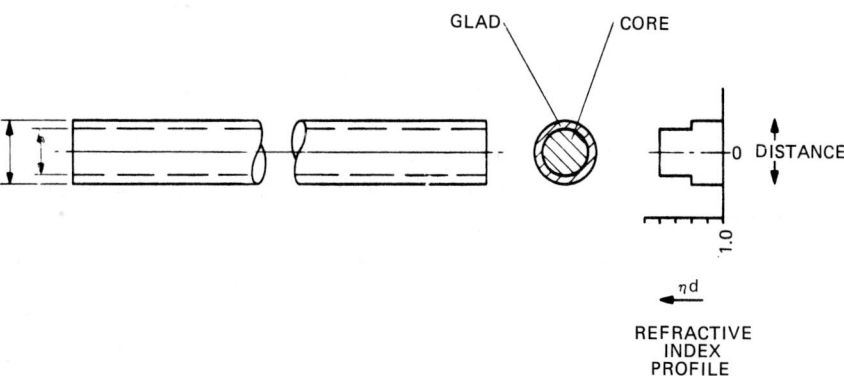

Fig. 4 Step index optical fiber design.

and clad glass, or possibly a plastic layer surrounding the core, which has a somewhat lower refractive index. When the seal interface between the core and clad is essentially free of imperfections and the relative refractive indices of the glasses used are correct, many millions of internal reflections are possible and light can travel through many kilometers of fiber and emerge from the far end with only a modest loss in brightness or intensity. A step index fiber has just a single composition inside the cladding. Light must travel to this boundary before it is bent toward the center. The paths in this type of fiber disperse the pulse more than in a graded index fiber.

A fiber with a *graded index profile* is shown in Fig. 5. Light is guided through it by means of refraction or bending which refocuses it about the center axis of the fiber core. Here each layer of glass from the center of the fiber to the outside has a *slightly* decreased refractive index compared to that of the layer preceding it. This type of fiber construction causes the light ray to move through it in the form of a sinusoidal curve rather than in the zigzag fashion of the step index variety. With

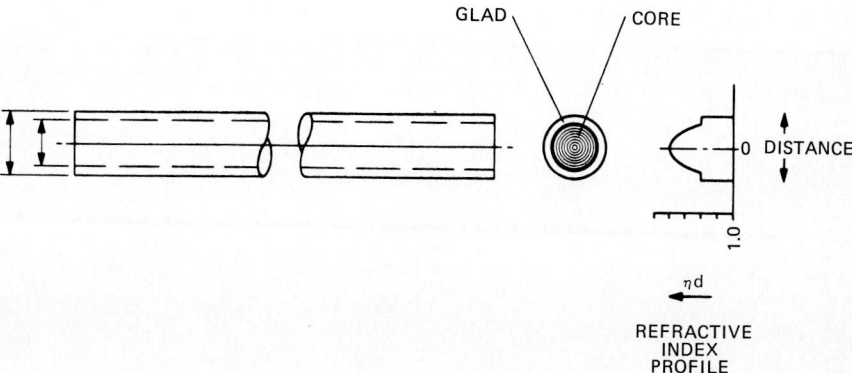

Fig. 5 Graded index (GRIN) optical fiber design.

this type of fiber, when the physical design is correct and the glass flaws are limited, light can also be conducted over very long distances without severe loss because it is trapped inside and guided in an efficient manner.

Optical Fiber Performance

The fiber core (Figs. 4 and 5) is the portion of an optical fiber that conducts the light from one end of the fiber to the other. Fiber core diameters range from 6 to ~250 μm.

Fiber Cladding

To help retain the light being conducted within the core, a layer surrounding the core of an optical fiber is required. Glass is the preferred material for the cladding, although pastic-clad silica fibers are common in less demanding applications. The cladding thickness may vary from 10 to ~150 μm, depending on the particular design.

Index of Refraction

This is the ratio of the velocity of light passing through a transparent material to the velocity of light passing through a vacuum using light at the sodium D line as a reference. The higher the refractive index of a material, the lower the velocity of light passing through the material and the more the ray of light is bent on entering it from an air medium.

Numerical Aperture (NA)

For an optical fiber, this is a measure of the light capture angle and describes the maximum core angle of light rays reflected down the fiber by total reflection. The formula from Snell's law governing the NA number for a fiber is

$$NA = \sin \theta = \sqrt{n_1^2 - n_2^2}$$

where n_1 is refractive index of the core and n_2 is refractive index of the clad glass.

Most optical fibers have NAs between 0.15 and 0.4, and these correspond to light acceptance half-angles of about 8 and 23°. Typically, fibers having high NAs exhibit greater light losses and lower bandwidth capabilities.

Light Loss or Attenuation through a Fiber

This is expressed in decibels per kilometer (dB/km). It is a relative power unit according to the formula

$$\text{Decibels} = 10 \log \frac{I}{I_0}$$

where I/I_0 is the ratio of light intensity at the source to that at the extremity of the fiber. A comparison of light transmission with light loss in decibels through 1 km of fiber is as follows:

80% transmission per kilometer $\simeq$ a loss of $\sim$1 dB/km

10% transmission per kilometer $\simeq$ a loss of $\sim$10 dB/km

1% transmission per kilometer $\simeq$ a loss of $\sim$20 dB/km

Bandwidth

This is a rating of the information carrying capacity of an optical fiber and is given either as pulse dispersion in nanoseconds per kilometer (ns/km) or bandwidth length in megahertz-kilometers (Mhz·km). Light pulses spread or broaden as they pass through a fiber, depending on the material used and its design. These factors limit the rate at which light carrier pulses can be transmitted and decoded without error at the terminal end of the optical fiber. In general, a large bandwidth and low losses favor optical fibers with a small core diameter and a low NA.

The longer the fiber the more the dispersion. Thus, modal dispersion limits the product of the pulse rate and distance. A step index fiber can transmit a maximum of 20 megabits of information per second for 1 km, and a graded index fiber, more than 1000 megabits. The process for making very low-loss fibers is essentially the same whether the fiber is step or graded index. Consequently, nearly all multimode fibers presenty used or contemplated for high-quality systems are of the higher capacity graded index type. Possibly for very high-capacity installations of the future, single-mode fibers may be attractive.

A typical optical fiber loss spectrum versus wavelength is shown in Fig. 6 for a graded index fiber made by the chemical vapor deposition method. While it is normal for losses in all optical fibers to vary with the wavelength of the transmitted light used, the curve clearly illustrates that a typical fiber meets an objective of achieving a relatively low loss at about 0.82 μm, where available laser diode emitters function. Even lower losses are possible at longer wavelengths if the hydroxyl ion absorption peak is controlled. However, adequately reliable emitters have not yet been developed for longer wavelength applications.

An idea of the relative performance of optical fibers made by various processes is given in Table 1. Since improvements are continually being made in fiber making processes, a range of values is shown, the best values being for fibers made in controlled laboratory endeavors rather than for fibers available from commercial operations.

Signal distortion is also a function of the wavelength range of a light pulse. See Fig. 7. These curves are based on experience with an installed telecommunications system. Each transmitter emits light primarily at a certain wavelength, such as 0.82 or 0.88 μm, but the light is not pure—it contains other wavelengths (or colors). Each travels at a different speed, so the broader the range emitted, the more the pulse disperses as it traverses a length of fiber. This is called *chromatic dispersion* because it depends on the colors of the light. This dispersion also depends on certain characteristics of the glass from which the fibers were made.

Light Sources and Detectors

LEDs produce a relatively broad range of wavelengths and, in the 0.8-μm-wavelength range, this limits present systems to $\sim$140 megabits/s for a 1-km path. Semiconductor lasers emit light with a much narrower range of wavelengths. Chromatic dispersion is comparatively low, so $\sim$2500 megabits/s may be transmitted for a 1-km path.

Another factor affecting system capacity is the response time of the sources and detectors. In

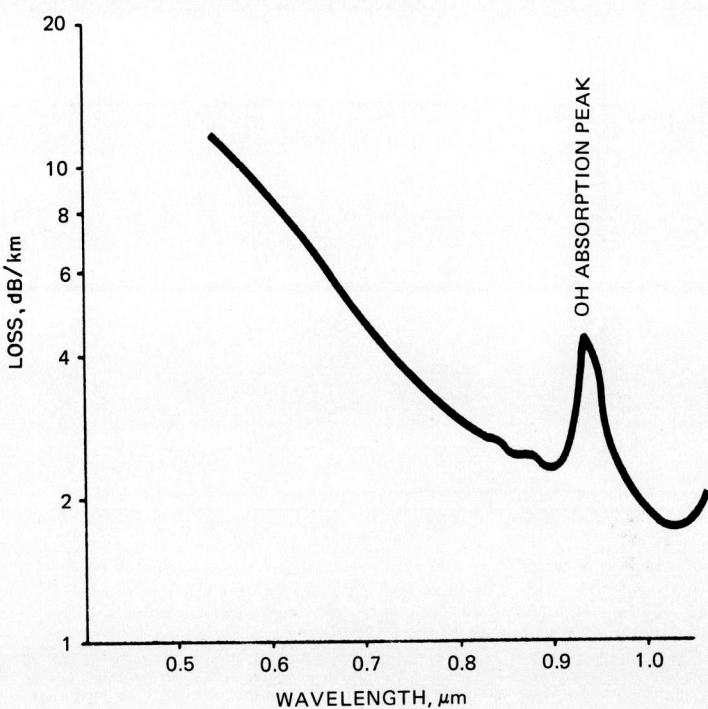

Fig. 6 Typical optical-fiber loss spectrum (graded index fibers made by a chemical vapor deposition process.)

Table 1 Relative Performance of Optical Waveguide Glasses and Fibers*

	Multicomponent glass core or tube redraw process		Multicomponent glass double-crucible process		Doped fused silica CVD process	
	Typical core, wt%	Typical clad, wt%	Typical core, wt%	Typical clad, wt%	Typical core, wt%	Typical clad, wt%
SiO_2	40	51	$\simeq 25$	$\simeq 60$	$\simeq 96$	$\simeq 92$
GeO_2					$\simeq 3$	
B_2O_3	12	14	$\simeq 50$	$\simeq 13$	$\simeq 1$	$\simeq 8$
Al_2O_3	9	19				
CaO	17	5				
MgO		8				
SrO	10					
La_2O_3	9					
Na_2O	3	3	$\simeq 25$	$\simeq 27$		
	100%	100%	100%	100%	100%	100%

	Step index fiber	Graded index fiber	Step index fiber
Loss at 820 μm, dB/km	50–1000	5–50	2–5
Bandwidth, MHz·km	70–200	20–200	500–1500
NA	0.3–0.66	0.1–0.3	0.1–0.25
Distance limitations	Short (0.01–0.1 km)	Medium (0.1–1 km)	Long (>1 km)

*Owens-Illinois.

16.65

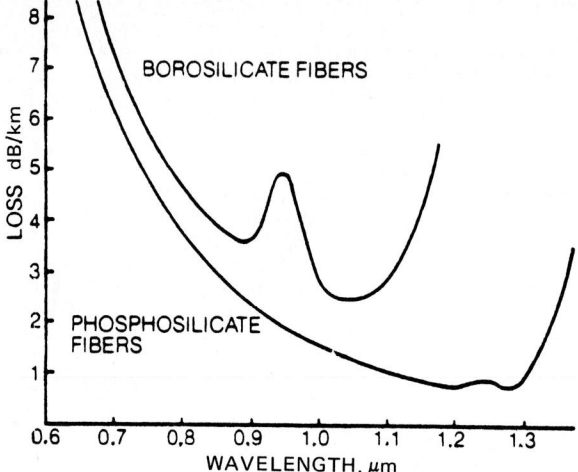

Fig. 7 Fiber loss. Signal loss in fibers decreases as the wavelength increases in the range shown. The upper curve represents the loss in the types of fibers used in earlier telecommunications systems. The lower curve indicates the lower loss achieved in newer fibers. *(Bell Laboratories.)*

general, it is possible to build sources and detectors with sufficiently short response times that the fiber, rather than the devices, becomes the capacity limiting factor. With single-mode fibers, lasers, and high speed detectors, transmission rates of more than 10^9 bits/s have been achieved experimentally. This corresponds to more than 15,000 digital voice channels. Although it is interesting to learn how fast a rate can be achieved, in practice the system designer must balance other technical, operational, and economic constraints in deciding how much capacity to require of an individual fiber.

The capacity of a transmission system helps determine whether it is practical. Another factor is repeater spacing. How long can the segments of cable be before the signal must be regenerated? This interval depends on the power, or strength, of the signal as it travels through the system, as well as

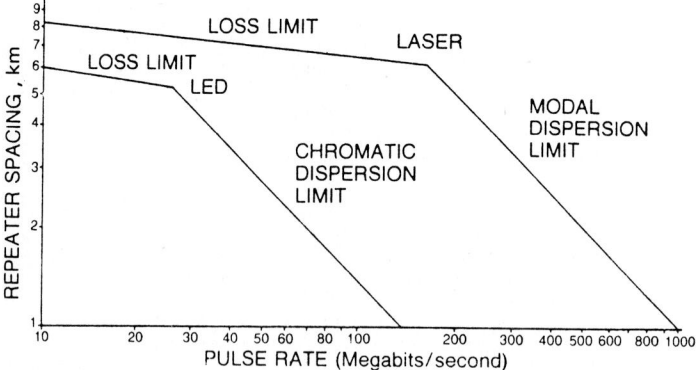

Fig. 8 Repeater spacing. The spacing of repeaters in light wave systems depends on the extent of signal dispersion and on the strength of the signal. At lower bit rates, the signal loss is limiting; at higher bit rates, the dispersion is limiting. A LED couples a weaker signal into the fiber as contrasted with a laser, and chromatic dispersion then limits the pulse rate. Laser transmission is so little affected by chromatic dispersion that modal dispersion is the limiting factor. This diagram shows typical spacing for 0.8-μm graded index fiber systems. *(Bell Laboratories.)*

on the maximum power produced by the transmitter, the minimum to which the receiver is sensitive, and the loss per unit length of transmission medium. See Fig. 8. The signal that arrives at a repeater or receiver must be sufficiently strong to trigger the acknowledgment of pulses in the correct time slots. If so much signal strength is lost that the signal has to be regenerated every few hundred meters, the system will not be practical for many uses. The components of present systems, however, maintain a signal that must be renewed only every few miles (several kilometers). In telephony, this is important because many urban central offices are in closer proximity than this and thus light wave cables can connect them without the use of repeaters. See Fig. 9.

Present semiconductor lasers made of aluminum-gallium-arsenide can couple ∼1 mW of optical power into a fiber. On the decibel scale, this is expressed as 0 dBm, meaning 0 dB above a reference power of 1 mW. Although some increase in power is possible, the small size and temperature sensitivity of these lasers make them inherently low-power devices. LEDs can be made that emit as much power as lasers, but since they project light over a wide angle, much of it is lost just coupling it into the fiber. This loss is typically ∼10 to 20 dB. Lasers are more complex and require more control circuitry than LEDs, but they are the light source of choice when repeaters must be far apart and the desired capacity is high.

Light wave receivers contain photodiodes which convert incoming light to an electric current. The receivers used in the FT3 Bell telecommunications system are *avalanche photodiodes* (APDs) made of silicon. They are called avalanche devices because the electric current is amplified inside the

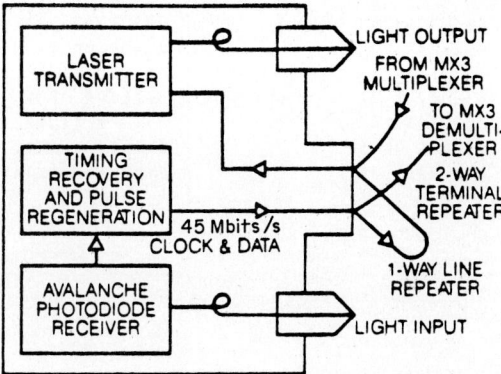

Fig. 9 Plug-in circuit used as a two-way terminal repeated within the light wave terminal. The incoming light wave signal is detected by the avalanche photodiode (APD), which converts it into an electric signal. The electric signal is further amplified, a timing signal is recovered, and in each time slot the presence or absence of a pulse is determined. The regenerated pulses and timing signals are then routed to demultiplexing circuits contained in other plug-ins. For the opposite direction of transmission, the electrical output from the multiplexer plug-ins is used as the input to the laser transmitter. The same light wave regenerator plug-in serves as a one-way repeater at intermediate repeater locations. Here, the regenerated electric pulses serve as the input to the laser transmitter.

Thus, the two functions (two-way terminal repeater and one-way line repeater) are achieved with the same plug-in, the different functions being obtained by differences in the wiring of the equipment into which the plug-in is inserted. Thus, the number of plug-in types to be manufactured is reduced.

The optical signal enters the regenerator through the connector plug at the top right of the unit shown in the diagram. It is carried by a fiber to the avalanche photodiode receiver. The signal, now electric, leaves the receiver, passes through other circuitry, and then is sent to the laser transmitter. The renewed optical signal is carried by another fiber to the plug connector. *(Bell Laboratories.)*

diode. This results in a more sensitive receiver than photodiodes without internal amplification. Again, this improved performance is achieved at the expense of added complexity. APDs require high-voltage power supplies, but they are the detectors of choice when high performance is desired.

Even with APDs, light wave receivers are less sensitive than the best electrical ones; they require a larger minimum received power. This is a consequence of the random fluctuations in optical signal intensity known as *shot noise*.

Light wave systems can compensate for this. They can carry a much wider bandwidth than electrical systems, and bandwidth can be used to offset noise.

Advantages of Digital Transmission

A particularly efficient way of trading bandwidth for immunity to noise is to transmit a signal in digital format. In the digital transmission of voice, for example, the analog signal is sampled 8000 times per second, and each sample is encoded as an 8-bit digital word. These 8000 analog samples per second become 64,000 bits/s of digital data. A digital system must transmit 8 times as many bits as there are analog samples. Because their bandwidth is wide, light wave systems readily provide the capability for high pulse rates. See Fig. 10.

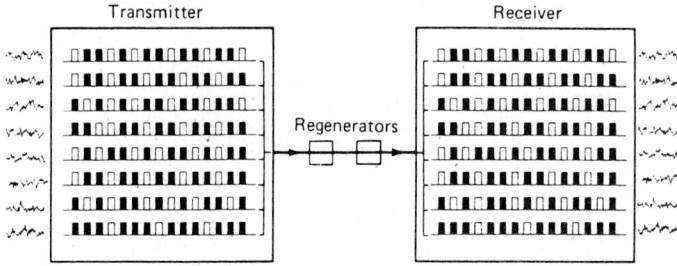

Fig. 10   Principle of digital transmission. Acoustic waves enter the transmitter at the left. The transmitter converts sound to digitized electric signals which are transmitted, sometimes through many regenerators to maintain signal strength, to a receiver where they are reconverted to acoustic waves. In a light wave transmission system, the digitized signals transmitted are in the form of light pulses.

Since noise alters the amplitude of an analog signal, analog transmission systems are susceptible to noise. The digital bits are 1's and 0's—the receiver must sense only the presence or absence of a light pulse. Noise matters less. With digital transmission, maximum repeater spacing can be achieved, links may be added in tandem without degradations accumulating, and linearity requirements for the transmitter and receiver are less severe than in analog systems.

With a light source such as a semiconductor laser and a receiver such as an avalanche photodiode, a digital light wave transmission system typically can accommodate ~40 dB of signal loss in the fiber between repeaters. Although some of the signal is lost in splices and connectors, most is lost to the fiber itself.

Cabling and Connections

Although optical fibers are very strong (having a tensile strength in excess of 500,000 psi, 3450 MPa), a fiber with a diameter of 0.005 in (0.1 mm), including the light guide core and cladding, has a maximum tensile strength of only ~10 psi (0.07 MPa). Unlike metallic conductors, which serve as their own strength members, fiber cables must contain added strength members to withstand the required forces. Also, pulling forces on unprotected fibers may increase their losses, as the result of bending or being under tension. Sometimes this is called *microbending loss*.

In one system (Bell Laboratories), ribbon cable to protect fibers from pulling tension has been developed. This configuration also permits precise alignment for connecting and splicing. Twelve fibers are embedded between two strips of plastic in a flat ribbon, and as many as 12 ribbons can be stacked in a cable. A 12-ribbon 144-fiber cable enclosed in a polyethylene-and-wire sheath is ~0.5 in (12.5 mm) in diameter and can carry more than 40,000 voice channels in the Bell FT3 system.

Imaging Requirements

In imaging, both ends of the group of fibers must maintain the exact same orientation one to another so that a coherent image is transmitted from the source to the receiver. Flexible coherent bundles of optical fibers having only their terminal ends secured in coherent arrays are used primarily in endoscopes to examine the inside of cavities with limited access, e.g., body cavities such as the stomach, bowel, and urinary tract.

Rigid fiber bundles fused tightly together along their entire length can be made to form a solid glass block of parallel fibers. Slices from the block with polished surfaces are sometimes used as fiber-optic faceplates to transmit an image from inside a vacuum to the atmosphere. A typical application is the cathode ray tube (CRT) used for photorecording. The requirement for this type of application is for both image coherence and vacuum integrity, so that when the fiber-optic array is sealed to the tube, the vacuum required for the tube's operation is maintained. However, any image formed electronically by phosphor films on the inside surface of the fiber-optic face is clearly transmitted to the outside surface of the tube's face. High-resolution CRT images can easily be captured on photographic film through fiber-optic faceplates.

REFERENCES

Allard, F. C.: "Wide Dynamic Range, Audio Frequency, Fiber Optic Transmission Line," Electro-Optical Systems Design Conference and International Laser Exposition, November 11–13, 1975.

Allard, F. C.: "Fiber Optics Sonar Link," *Chem. Eng.* vol. 63, May 10, 1976.

Andreiev, N.: "Industrial Fiber Optics," *Control Eng.,* vol. 24, no. 3, 1977, p. 36.

Andreiev, N.: "New Fiber Optic Data Highway Is a Masterless Token-Passer," *Control Eng.,* vol. 29, no. 6, 1982, p. 98.

Bailey, S. J.: "Signal Transmission: Electronic and Optical Options," *Control Eng.,* vol. 23, no. 3, 1976, p. 32.

Bailey, S. J.: "Fiber Optics Firms Tailor to Suit Plant Floor Needs," *Control Eng.,* vol. 27, no. 8, 1980, p. 53.

Baues, P.: "The Anatomy of a Fiber Optic Link," *Control Eng.,* vol. 26, no. 8, 1979, p. 46.

Baumbick, R. J., and J. Alexander: "Fiber Optics Sense Process Variables," *Control Eng.,* vol. 27, no. 3, 1980, p. 75.

Brambley, J. H.: "Fiber Optics: A Bright Future," *Instrum. Technol.,* vol. 29, no. 5, 1982, p. 9.

Brown, W. W., et al.: "System and Circuit Considerations for Integrated Industrial Fiber Optic Data Links," *IEEE Trans. Commun.,* July 1978.

Cowen, S. J.: "A High Performance Fiber Optic Pressure Penetrator for Use in the Deep Ocean," *NTIS Rep. AD-A100 732/7,* 1981.

Elion, G. R., and H. A. Elion: *Fiber Optics in Communications Systems,* Dekker, New York, 1978.

Faust, G.: "Programmable Controller Offers Fiber Optic Data Link for Remote I/O," *Control Eng.,* vol. 26, no. 10, 1979, p. 53.

Fayfield, R. W.: "Role of Fiber Optics in Photoelectric Sensing Applications," 33rd IEEE Conference, Record of Annual Conference on Electrical Engineering Problems in the Rubber and Plastics Industries, April 6–7, 1981.

Gawlowicz, D. J.: "Asynchronous Transmission in Electronic, Fiber Optic Systems," *Control Eng.,* vol. 25, no. 2, 1978, p. 67.

Greenwell, R. A.: "Fiber Optics Cost Models for the A-7 Aircraft," *Fiber Integrated Opt.,* vol. 1, 1977.

Hansberry, J., and R. Vanzetti: "Fiber Optics—A New Approach to Monitor and Control Process Temperature," *Ind. Heating* May 1981.

Hanson, D. C.: "Fiber Optic Data Links for Distributed Computer Communication," 28th IEEE Computer Society International Conference, March 1979.

Husbands, C. R.: "Airborne Integrated Communication Network Utilizing Fiber Optics," NTC Conference Board, National Telecommunications Conference, November 27–29, 1979.

Intrieri, A. J.: "Optical Fibers Look around Corners to Measure Temperature," *Control Eng.,* vol. 24, no. 12, 1977, p. 42.

ITC: *International Telemetering Conference Proceedings,* vol. XVI (1980), vol. XVII (1981), vol. XVIII (1982), vol. XIX (1983), available from Instrument Society of America, Research Triangle Park, N.C.

James, K. A., Quick, W. H., and V. H. Strahan: "Fiber Optics: The Way to True Digital Sensors?" *Control Eng.,* vol. 26, no. 2, 1979, p. 30.

Klein, R., and P. Onorato: "Glass Fibers for Optical Communication," *GTE Profile,* vol. 4, 1979, p. 8.

Krueger, A. H.: "Applying Fiber Optics to Photoelectric Switches," *Control Eng.,* vol. 27, no. 8, 1980, p. 161.

Levy, M. S.: "Optically-Coupled Triac Drivers Reduce Parts Count," *Control Eng.,* vol. 27, no. 9, 1980, p. 134.

McGowan, M. J.: "Fiber-Optic Link Runs 5 km at 150 Megabaud," *Control Eng.,* vol. 27, no. 8, 1980, p. 65.

Makuch, J. A.: "Connector Standards for Fiber Optics Application," *Insulated Circuits,* June 1981.

Matsui, T., et al.: "Optical Components for Fiber Optic Communications," *Mitsubishi Denki Ghio (Japan),* May 1980.

Miller, S. E.: "Photons in Fibers for Telecommunication," *Science,* vol. 195, 1977, p. 1211.

Morris, H. M.: "The Status of Fiber Optics for Industrial Control," *Control Eng.,* vol. 26, no. 10, 1979, p. 49.

O'Neil, V. P.: "Monolithic Integrated Circuit Optical Detector/Pre-Amplifier for Fiber Optics Systems," 2nd International Fiber Optics and Communications Exposition, 1979.

Ong, R. H.: "Application of Fiber Optics for a Peripheral Controller to a High Speed Printer," 3rd International Fiber Optics and Communications Exposition, 1980.

Pitt, D. A., et al.: "Optobundle—A Unique Fiber Optic Multiplier," *NTIS Rep. AD-A044 599/9ST,* June 1977.

Reese, I.: "Fiber Optic Data Bus for Control," *Control Eng.,* vol. 24, no. 7, 1977, p. 43.

Staff: "Voices in the Light," *Sci. Am.,* vol. 242, no. 3, 1980, p. 96.

Taylor, H. F.: "Navy Applications of Fiber Optics Communications," *Soc. Photo-Opt. Instrum. Eng., Semin. Proc.,* vol. 63, 1975.

Weik, M. H.: *Fiber Optics and Lightwave Communications Standard Dictionary,* Van Nostrand Reinhold, New York, 1980.

Principles of Automatic Control

E. H. Bristol *Research Department, The Foxboro Company, Foxboro, Massachusetts. (Control Algorithms)*

G. A. Hall, Jr. *Westinghouse Electric Corporation, Pittsburgh, Pennsylvania. (Fundamentals of Automatic Process Control)*

Stephen P. Higgins, Jr. *Retired. Senior Principal Software Engineer, Advanced Systems, Honeywell Inc., Process Management Systems Division, Phoenix, Arizona. (Process Control Techniques)*

Richard H. Kennedy *The Foxboro Company, Foxboro, Massachusetts. (Fundamentals of Automatic Process Control)*

Joe M. Nelson *Section Head, Applications Systems Division, Honeywell Information Systems, Honeywell Inc., Billerica, Massachusetts. (Process Control Techniques)*

Robert L. Osborne *Manager, Diagnostics and Controls, Westinghouse Electric Corporation, Orlando, Florida. (Fundamentals of Automatic Process Control)*

Fundamentals of Automatic Process Control

by
Robert L. Osborne*

Basically, automatic control is relatively easy to comprehend. In many ways it is much like manual control. However, the automatic controller does not necessarily duplicate what the human operator does by hand. Automatic equipment gives continuous, minute attention to the control application. Automatic controllers can compute and remember, but they cannot reason from new conditions, nor can they forecast beyond the data which are built into them. Thus, these are the basic differences between human and automatic-control means—factors that govern automatic controller design and use.

The terminology used here conforms largely with the Scientific Apparatus Makers Association, Process Measurement and Control Section SAMA Standard PMC20-2-1970, "Process Measurement and Control Terminology," which was adopted as an Instrument Society of America standard in 1976 (ISA-S51.1) and is available from the Instrument Society of America, Research Triangle Park, N.C.

Typical Process

To illustrate this presentation clearly, the simple heat-exchange process of Fig. 1 is used.

The term *process,* as employed here, signifies the function and operations utilized in treatment of material. In Fig. 1, for example, the operation of adding heat energy to water is a process. The steam coils, tank, pipes, and valves comprise the plant within which the heating process is accomplished. The parts and characteristic behavior of this typical process are analyzed in the following paragraphs to portray what effect these factors have on the *controllability* of the process. Although a very simple process, this example serves as a good fundamental starting point in the development of automatic-control principles.

The Energy Exchanger

The water heater of Fig. 1, like many processes, can be considered an energy exchanger. In many other processes, the exchange of materials alone, or of materials and energy, may be involved. In reference to Fig. 1, energy is put into the process, passes through a series of exchanges, and emerges as energy output. The rate of energy output equals the rate of energy input minus (1) the rate of energy lost and (2) the rate of energy storage in the process.

Editor's Note. This section comprises a nonmathematical introduction to automatic-control principles and is written especially for technicians, engineers, and scientists who are *not* expert in the field. Even though one may not be concerned daily with the solution to automatic-control problems, the fundamentals outlined here will prove helpful in the selection, application, and adjustment of automatic-control equipment. A more advanced discussion of the subject, especially as related to the mathematical techniques used in solving automatic control problems, will be found in the article Process Control Techniques later in this Handbook section.

*Manager, Diagnostics and Control, Westinghouse Electric Corporation, Orlando, Fla. Excerpts from article in the prior edition of the Handbook have been used throughout this revision. Prior authors were George A. Hall, Jr., Westinghouse Electric Corporation, and Richard H. Kennedy, The Foxboro Company, Foxboro, Mass.

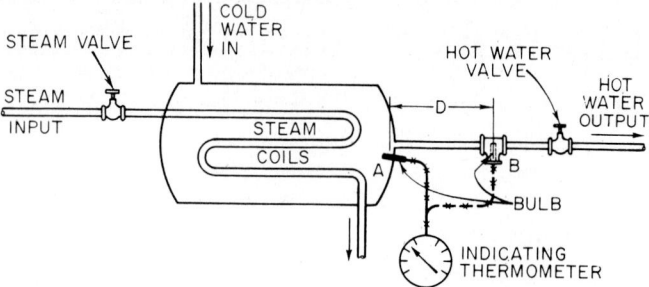

Fig. 1 Typical heat exchange process (used as an example throughout this section on automatic process control fundamentals).

In the heat exchanger, the rate of energy *output* depends on (1) the rate of water flow as regulated by the hot water valve, (2) the temperature of the incoming cold water, and (3) the heat energy losses, such as through the tank walls. The rate of energy *input* depends on (1) the rate of steam flow and (2) the quality and pressure of the steam supply. Therefore, whether the variables of the process are steady or are changing depends on whether the rate of energy input equals the rate of energy output.

Balanced Condition

If left to itself, the temperature of the output water would ultimately come to a steady value such that the output energy would equal the input energy. When input energy equals output energy, the process is in a *steady-state* condition; i.e., it is in balance. Any disturbance to either the input or the output of energy will upset this balance and consequently cause a change in the values of the process variables. In this example, one of the process variables is output-water temperature. When the heat energy output balances the heat energy input, the output-water temperature remains at a constant value until this heat energy relationship is upset.

Self-Regulation

Some processes possess an inherent characteristic which aids in limiting deviation of the controlled variable. In Fig. 1, when the steam input increases, the water temperature reaches a point of balance at some new, higher value; i.e., the water temperature does not increase indefinitely. This inherent ability of a process to balance its energy (or material) output and input is termed *self-regulation*.

In the self-regulating process of Fig. 2a, the flow out through resistance *R* tends to equal the flow in through valve *A*. If valve *A* is opened wider, the level in the tank will increase until the

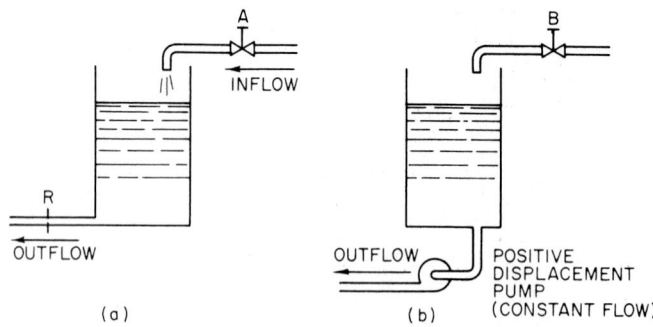

Fig. 2 Examples of self-regulating processes. (*a*) Self-regulation, (*b*) non-self-regulation.

outflow through R equals the new inflow. Thus, through wide limits, the process will self-regulate its outflow to equal its inflow. The limits in this instance depend on the depth of the tank.

Contrast the self-regulating process of Fig. 2a with the non-self-regulating process of Fig. 2b. In the latter the outflow is held constant by a fixed-speed positive displacement pump. Unless the inflow exactly equals this fixed outflow, the tank will either empty or overflow. There is no tendency for this process to balance input and output—hence the term "non-self-regulation."

Although self-regulation is very helpful in applying automatic control to a process, a negative self-regulating characteristic may make automatic control difficult or even impossible. Negative self-regulation may be defined as the tendency of a process (of itself) always to go in the direction of imbalance.

Function of Automatic Control

Most processes, whether controlled manually or automatically, perform well and efficiently (if at all) only when the values of certain process variables are held within given limits. These process variables will change unless energy (or material) input equals output. Thus, the fundamental function of process control is to manipulate the energy (or material) input-output relationship so as to keep the process variables within desired limits.

An automatic controller can be defined as a system that measures the value of a process variable and operates to limit the deviation of that variable from the set point. A process variable held within limits is termed a *controlled variable*. The automatic controller regulates the controlled variable by making corrections in another variable of the process, which is termed the *manipulated variable*. In Figure 1, the controlled variable is the temperature of the hot water. The set point is the desired hot water temperature. The manipulated variable is the rate of steam flow. Any change in the steam valve opening as dictated by the automatic controller comprises a correction in the manipulated variable. Thus, in Fig. 1, it is possible to hold or change the output hot water temperature by manipulating the balance of energy input and energy output.

Process Time Lags

At first glance, control of the water temperature in Fig. 1 may seem quite easy. All that appears necessary is to observe the hot water thermometer and to correct the steam valve opening accordingly so as to hold or change the water temperature to the set point. However, processes have the characteristic of delaying and retarding changes in the values of the process variables. This process characteristic greatly increases the difficulty of control. These delays and retardations generally are termed *process time lags*.

In addition to inertia, time lags are caused by three properties of the process namely, capacitance, resistance, and dead time (transportation lag).

Capacity and Resistance

The parts of a process that have the ability to store energy (or a quantity of material) are termed *capacities*. In Fig. 1, for example, the walls of the steam coils and the water in the tank can store heat energy. This energy storing property gives these capacities the ability to retard change. For instance, if the incoming steam temperature increases, it requires some time for more energy to be added to the water in the tank so as to bring it up to a new, higher temperature.

The parts of a process that resist transfer of energy (or material) between capacities are termed *resistance*. In Fig. 1 the walls of the steam coils and the insulating effect of the layers of steam and water on either side of them resist the transfer of heat energy between the steam in the coils and the water on the outside of the coils. The combined effect of supplying a capacity through a resistance produces time retardations in the transfer of energy between capacities. Such resistance-capacity (RC) time retardations are often called *capacity lags* or *transfer lags*.

It is desirable to classify processes by the number and arrangement of their resistance-capacity pairs, since processes containing a similar number and arrangement of RC pairs tend to exhibit similar behavior and controllability.

In order to classify processes in this manner, it is necessary to assume that their resistance-capacities are separated into discrete units. This way of analyzing processes is called the *lumped-capacity* (lumped-parameter) method. In some processes resistance-capacity is distributed. These are termed *distributed-capacity* (distributed-parameter) processes. A natural gas pipeline is an example where resistance and capacity are both distributed over long lengths of pipe.

Dead Time

A third property which contributes to time lag is the time required to carry a change from one point to another in the process. If the temperature (see Fig. 1) of the incoming cold water drops, some time will elapse before this colder water can travel through the tank and reach the thermometer bulb at point *A*. This time lag is not just a slowing down or retardation of a change, but a discrete time *delay* (or dead time) during which no change whatever occurs at the output. The length of this dead time depends on both the velocity with which the change is transported and the distance over which it is carried. Thus, dead time is also termed *distance-velocity lag* or *transportation lag*.

Figure 3 is a block diagram of the water heater shown in Fig. 1, with the capacities for heat storage, resistances to heat flow, and dead time identified. This shows how the amount of dead time depends on the distance and the velocity with which temperature changes are transmitted. For example, when the hot water flows with a velocity of 1 ft (0.3 m)/s with the bulb at point *A* (2 ft or 0.6 m from the tank), the dead time is 2 s. With the bulb at point *B* (10 ft or 3 m from the tank), the dead time is 10 s. But, if the water velocity is dropped to 0.5 ft (0.15 m)/s in the aforementioned examples, the dead times become 4 and 20 s, respectively.

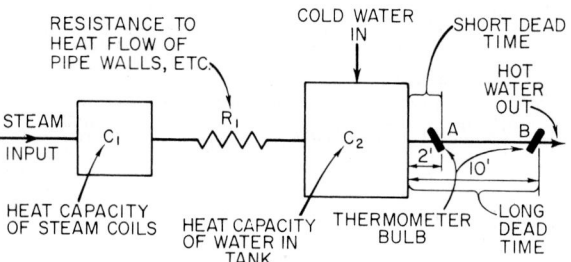

Fig. 3 Block diagram of the water heater shown in Fig. 1, illustrating the source of time lags.

Types of Process Disturbances

In analyzing a process from the standpoint of automatic control, it is well to give particular consideration to two of the several types of process disturbances that can occur:

1. **Supply Disturbances.** This is a change in the energy (or materials) *input* to a process. In the heat exchanger of Fig. 1, changes in steam quality, steam pressure, or steam valve opening are supply disturbances.
2. **Demand Disturbances.** This is a change in the energy (or material) *output* from a process. In Fig. 1, changes in cold water temperature and rate of water flow are demand disturbances.

These disturbances are usually called *supply-load* and *demand-load changes,* respectively. There are important differences in the reaction of a process to these two types of load changes.

Process Reaction Curves

By studying the reaction of process variables to load changes under uncontrolled conditions, much can be learned about the characteristics of a process that determine its controllability. In the discussion that follows, the process represented in Fig. 1 can be assumed to be in a steady state. The effects of sudden *step* load changes of both the supply and demand types are shown. Reaction curves are given for several combinations of resistance-capacity and dead time.

Single-Capacity Processes

The heat exchanger of Figs. 1 and 3 can be considered a single-capacity process (approximately), provided the heat capacity C_1 of the steam coils, tank walls, and thermometer bulb is assumed to be insignificant when compared with the heat capacity C_2 of the water in the tank. Thus, in this assumption, one of the resistance-capacity pairs of the process is so large as to dominate all others. Under these conditions, how does the output water temperature behave following sudden demand- and supply-load changes?

Effect of a Sudden Supply-Load Change

Figure 4 gives the uncontrolled reaction curves following the sudden supply-load change caused by increasing the steam valve opening at time zero. Good mixing of the water is assumed in these illustrations. Each curve indicates how the temperature starts to increase *immediately* after the load is changed—and then how it increases at a slower and slower rate until it finally rises to a new steady-state value. Worthy of note is how the full response of the temperature is retarded in time; i.e., as the heat storing capacity of the process is increased, more time is required for the temperature to reach its final value. This is an excellent example of how the heat capacity of the water and the resistance to heat flow retard the rise in temperature. This retardation is the capacity (transfer) lag.

Effect of Both Supply- and Demand-Load Changes

Figure 5 gives the uncontrolled reaction curves for the heat exchanger of Fig. 1 in response to both supply- and demand-load changes. Curve *a* shows the effect of a sudden demand-load change made at time zero by increasing the opening of the hot water valve. The significant point to note in curve *a* is that the temperature *immediately* begins to change as soon as the demand disturbance occurs. Curve *b* shows the effect of a sudden supply-load change made at time zero and represents the increased steam supply just sufficient to *exactly* correct for the demand disturbance represented by curve *a*. Curve *c* shows the effect of *simultaneously* applying the same demand-load change (curve *a*) and the exact supply correction for it (curve *b*). This would be theoretically possible if the hot water and steam valve openings were simultaneously increased by the same amounts used in curves

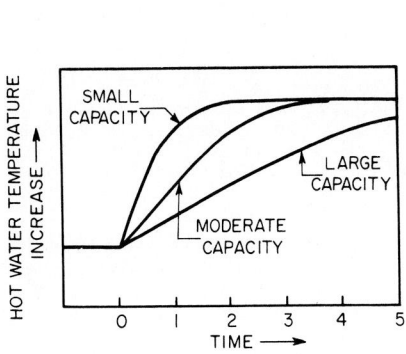

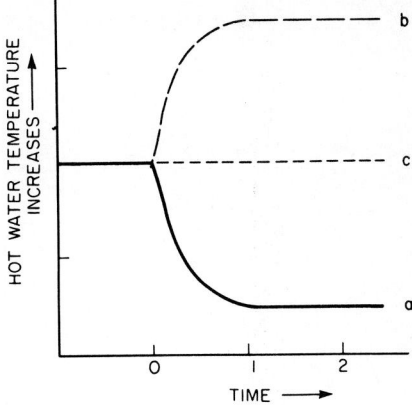

Fig. 4 Reaction curves for a single-capacity process, showing the effect of three different-sized capacities on the time lag. The supply load increased at time zero.

Fig. 5 Reaction curves for a single-capacity process (curves are superimposed). (*a*) Demand disturbance, (*b*) exact supply correction, (*c*) simultaneous demand disturbance and exact correction.

a and *b*. Note from curve *c* that, in this single-capacity process, the exact supply correction, when applied simultaneously with a demand disturbance, prevents the temperature from changing at all. This can be true only where the time constants are equal.

The reaction curves of Figs. 4 and 5 are typical for all processes that are approximately single-capacity in nature and have no dead time. However, *true* single-capacity processes are practically impossible to produce.

Two-Capacity Processes

If it is assumed that the steam heating coils of Fig. 3 are sufficiently large to have a heat capacity C_1 which is quite significant when compared with the capacity of the water in the tank C_2, the process can be considered a two-capacity process. Then, the resistance R_1 between capacities C_1 and C_2 is the resistance to heat transfer offered by the walls of the steam coils and the insulating film of water at their inner and outer surfaces.

Figure 6 gives the uncontrolled reaction curves for this two-capacity process after a sudden *supply-load* change is caused by increasing the steam valve opening at time zero. Comparison of Figs. 4 and 6 illustrates a very significant difference between single- and two-capacity processes. In the two-capacity process the temperature, instead of changing *immediately* (as in the single-capacity process of Fig. 4), begins to rise slowly, then faster, then slowly again, gradually rebalancing at a new steady-state value. This S-shaped reaction curve is characteristic of the effects of supply-load changes on a process with two or more relatively equal capacity-resistance pairs, that is, a *multicapacity* process. Resistance R_1 (Fig. 3) to the transfer of energy between the steam coil heat capacity C_1 and the water heat capacity C_2 causes this retardation *capacity lag* in temperature. Figure 6 shows that, as the capacity lag of the process is increased, more time is required for the temperature to reach its final value.

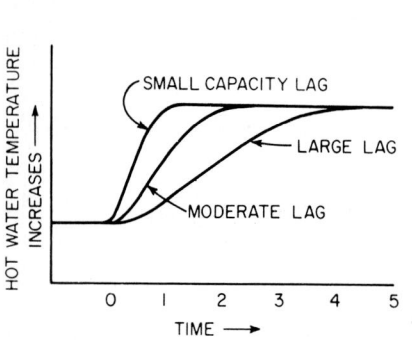

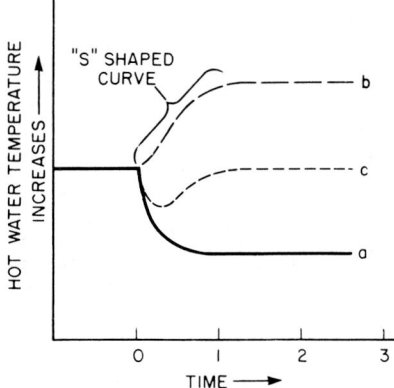

Fig. 6 Reaction curves for a two-capacity process, showing the effect of three different-sized capacity lags. The supply load increased at time zero.

Fig. 7 Reaction curves for a two-capacity process (curves are superimposed). (*a*) Demand disturbance, (*b*) exact supply correction, (*c*) simultaneous demand disturbance and exact correction.

Figure 7 shows, for a two-capacity process, the same demand and exact supply correction reaction curves shown in Fig. 5 for a single-capacity process. Curve *b* shows the effect of an *exact* supply correction by the steam valve at time zero. Worthy of note is the S-shaped response characteristic of a supply-load change in a multicapacity process. Curve *c* shows the result of simultaneously applying the demand-load change of curve *a* and the *exact* supply correction of curve *b*. The difference in behavior between the single-capacity process of Fig. 5 and the two-capacity process curve of Fig. 7 should be noted. Even though an *exact* correction is *immediately* applied, the temperature in the two-capacity process changes widely and does not return to its initial value for some time.

Multicapacity Processes

Although many processes have more than two resistance-capacity pairs of relatively equal size, the behavior of these multicapacity processes is similar to that for the two-capacity process as shown by the reaction curves in Figs. 6 and 7. Thus these reaction curves can be considered typical for all processes (that have no dead time) with two or more RC pairs of relatively equal size. In general, inclusion in a process of two or more capacities of about the same size greatly increases the difficulty of control.

Effect of Dead Time in Multicapacity Processes

As pointed out previously, the equipment shown in Fig. 1 will have considerable dead time if the thermometer bulb is installed at point B instead of at point A; i.e., more time will be required to carry a temperature change over the greater distance to point B. Figure 8 shows the effect of dead time on a two-capacity process. The reaction curves are simply shifted along the time scale by the amount of dead time. The addition of dead time does not change the shape or size of the reaction curves. The transportation of process energy or material through a distance can cause dead time wherever it occurs, and, therefore, dead time can exist in any part of a single-capacity or multicapacity process.

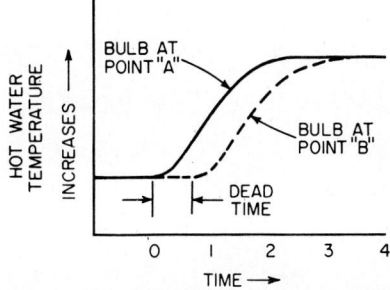

Fig. 8   Effect of dead time on the reaction curves for a two-capacity process. The supply load increased at time zero. Dead time due to moving the thermometer bulb from A to B (see Fig. 1) is shown. The curve is shifted along the time axis by the amount of dead time.

Manual Control

Figure 9 shows the heat exchanger process of Fig. 1 when under control by a human operator. The operator's job is to feel the output-water temperature and to turn the steam valve so as to keep the water at the desired value. If it is assumed that the process is balanced so that the output-water temperature is at the desired value, what will happen in this manual-control system if there is an increase in the rate of water flow?

Because of time lags of the process, some time will elapse before cooler water reaches the operator's right hand. When the operator feels this temperature drop, the operator must compare it with the temperature desired—then mentally compute how much and in what direction the steam valve must be repositioned—and then manually make this correction in the valve opening. Some time is required, of course, to make these decisions and to correct the valve position.

It is also true that some time will elapse before the effect of the valve correction on the output-water temperature can reach the output and thus be sensed by the operator. It is only then that the operator will be able to know whether the first correction was too little or too much. At that time, a second correction is made which will, after a time, cause another output-temperature change. The results of the second correction will be observed, and a third correction will be made, and so on.

This series of measuring, comparing, computing, and correcting actions will go round and round through the operator and through the process in a closed chain of actions until such time that the water temperature is finally balanced again at the value desired by the operator. This type of control is aptly termed *feedback control*. The heavy dashed loop of Fig. 9 shows the direction and path of this closed series of control actions. *This closed-loop concept is fundamental to a full understanding of automatic control.*

Problem of Control

Study of the manually controlled process of Fig. 9 shows the basic problem of control.

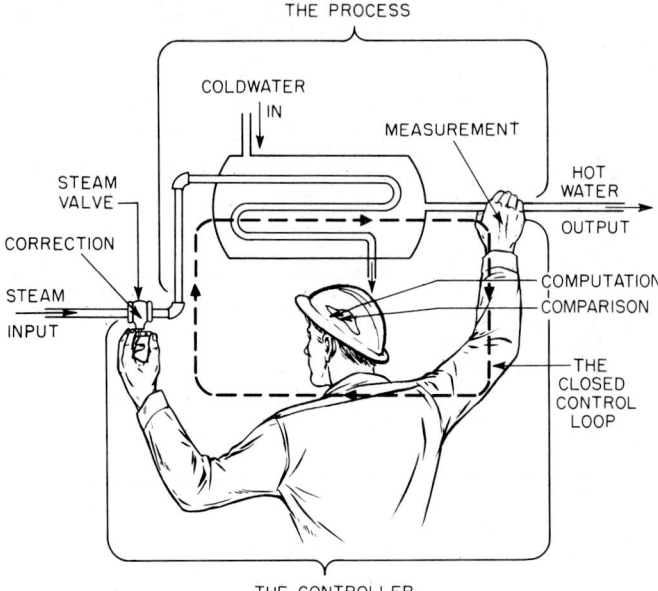

THE PROCESS

COLDWATER
IN

MEASUREMENT

STEAM
VALVE

HOT
WATER

CORRECTION

OUTPUT

STEAM
INPUT

COMPUTATION
COMPARISON

THE
CLOSED
CONTROL
LOOP

THE CONTROLLER

Fig. 9   Heat exchange process of Fig. 1 when under manual control. The controller and process together form a closed loop of control (represented by the heavy dashed line).

Feedback Control

A correction for a disturbance cannot be made until the effect of the disturbance is known. But process time lags require that some time must elapse before this can be ascertained. Also, some time is needed to measure the deviation and make the correction. Then, in turn, process time lags require again that time pass before the effect of the correction can be known. Meanwhile, the controlled variable continues to deviate from the desired value for some time. In short, *the problem of control is to overcome the effect of time lags that occur around the closed loop of control.*

In Fig. 5 it was shown that, in a single-capacity process, an exact correction applied simultaneously with a load change prevents any deviation of the controlled variable whatsoever. How does this process react when an exact correction is applied some time *after* a disturbance occurs? Figure 10 shows reaction curves for the single-capacity process of Fig. 9. Curve *a* gives the hot water temperature, and curve *b* the steam valve openings. It must be stressed that this process is *not* under automatic control. At time zero, a demand-load change occurs, caused by an increased flow of hot water. Curve *a* indicates how the temperature reacts. At time unit 2, curve *b* (solid line) shows an exact supply correction made to the steam valve. Curve *a* shows how the temperature rises again to its initial value after some time. But because this exact correction is *not* applied at the instant when the load change is applied, the temperature deviates widely (and for some time) from the desired value. In any process having time lags, exact corrections cannot be applied simultaneously with load changes because the time lags of the process prevent the effect of the disturbances from being known for some time. Since all processes have time lags of one magnitude or another, this situation is typical of the general automatic-control problem.

Excess Correction

In the above example (Fig. 10), if the steam valve had been opened *wide* at time 2, steam would have been supplied much in excess of this exact correction and the temperature would have been returned to its initial value much more rapidly. The dashed curve *Y* (Fig. 10) shows such an excess correction to the steam valve applied at time 2 and then reduced to the exact correction at the moment when the temperature was returned to its initial value. Dashed curve X shows that this *excess* correction returned the temperature to its initial value by time T before the exact correction did. Thus, an excess correction properly applied and withdrawn can return a variable to a desired value faster than an exact correction alone.

The energy supplied in excess of the exact correction is represented by the cross-hatched area under curve Y. It can be concluded that a controller which can produce a reaction like curve X is better than one which can produce curve a. Thus, a desirable function of a controller is to apply excess corrections as large as the process will permit and to reduce these to the exact corrections at the proper time. Excess corrections enable the controller to partially "catch up" for the time lost due to time lags around the control loop. In other words, they provide a partial solution to the basic problem of control, but they cannot be applied to processes of very small capacity, such as most rate-of-flow control problems.

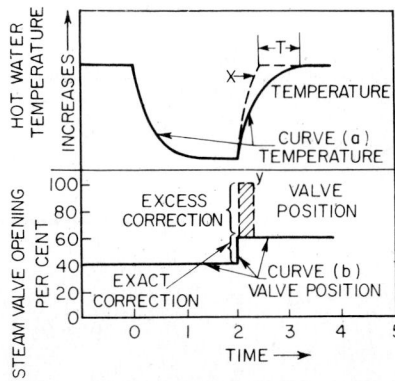

Fig. 10 Reaction curves for the single-capacity process of Fig. 9, showing advantages of excess correction over exact correction.

Practical Meaning of Process Reaction Curves

1. **Single-Capacity Process.** Processes dominated by one resistance-capacity pair are easier to control because:

   a. They begin to react immediately to load changes—so deviations can be known and corrections applied without delay (Fig. 4).

   b. Corrections are immediately effective in reducing deviations (Fig. 5).

2. **Multicapacity Processes.** Processes having two or more resistance-capacity pairs of *comparable* size are more difficult to control because:

   a. They do not begin to react to load changes immediately (Fig. 6), so deviations do occur and corrections can be applied only after some time has passed.

   b. Corrections are *not* immediately effective in reducing deviations (Fig. 7).

3. **Dead Time.** Dead time (transportation lag) increases control difficulty wherever it is found.

Basic Functions of Control

In the manually controlled process of Fig. 9, the operator *measures* the output temperature, *compares* it to the desired value, computes *how much* to open the steam valve, and makes that *correction* to the steam supply. Thus, the basic functions performed by the operator in manual control are (1) measurement, (2) comparison, (3) computation, and (4) correction. These, then, are the basic control functions that must be provided in any automatic-control system if it is to perform the function of a human operator.

Elements of Automatic Control

The functional elements of an automatic-control system in relation to the clsoed loop of control are shown in Fig. 11. A comparison of Fig. 11 with Fig. 9 shows that the automatic controller performs the same basic functions (and in the same order) as the human operator of a process.

The measuring elements (Fig. 11) perform the measurement function; they sense and evaluate an output variable of the process.

The function of comparing the measured value of the output variable with its desired value is accomplished by the *error detector,* which produces a signal when there is a deviation between the

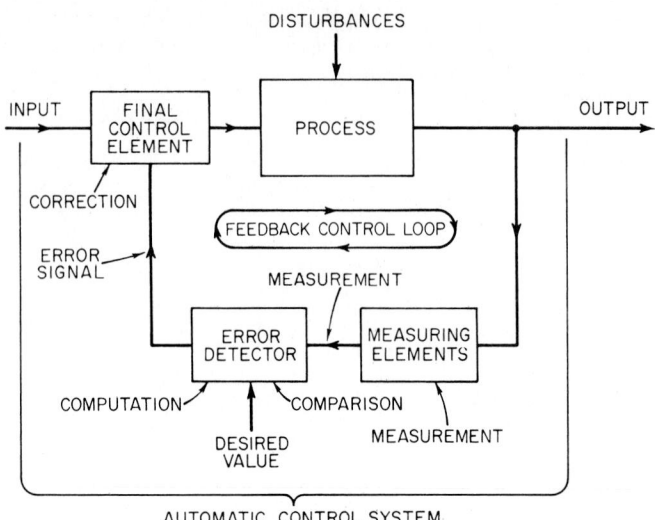

Fig. 11   Relation of four basic control functions and basic automatic controller elements in a feedback control system.

measured and desired values. This *actuating signal* is often called an *error signal*. This signal bears some size relation to the deviation.

The correction of an input to the process is made by the *final control element* which is actuated by the error signal.

Thus, an automatic controller is an error-sensitive self-correcting device. It takes a signal (energy) from an output of a process and *feeds it back* into a process input. Therefore, closed-loop control is also commonly referred to as *feedback control*.

A Typical Control System

The foregoing theoretical discussion can be clarified by considering how the basic control functions are performed by an *actual* automatic-control system. As a typical example, Fig. 12 shows schematically an air-operated control system applied to control of the water heater process of Fig. 1. The parts of the control system, arranged in block diagram form to illustrate the basic control function that each part performs, are shown in Fig. 13.

Measuring Means

The measuring system of this typical controller consists of three parts:

1. ***Thermometer Bulb (Primary or Sensing Element).*** This is the part of the measuring system that is directly sensitive to the controlled variable (temperature). The primary element converts energy from the controlled medium (hot water) to a measurable signal (fluid pressure).
2. ***The Capillary Tubing (Transmitting Means).*** This part of the system carries the signal from the primary element to the receiving element.
3. ***Bourdon Pressure Element (Receiving Element).*** This part of the system evaluates the signal from the primary element and converts it to scale readings, chart records, and actuation for the error detector.

In some controllers, the aforementioned three parts of the measuring system are combined into one or two devices.

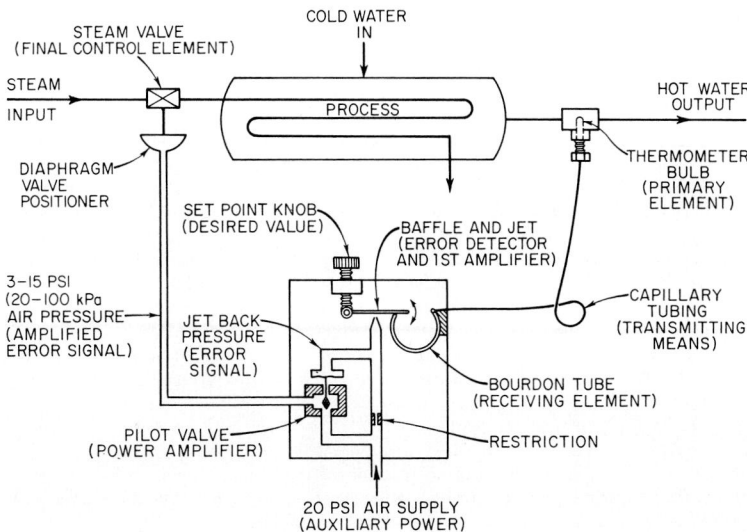

Fig. 12   Typical air-operated automatic control system applied to the heat exchange process of Fig.1.

Error Detector Elements (Unbalanced Detector, Summing Point, or Primary Relay)

In Fig. 12, the error detector is the baffle and jet. The error detector compares the measured value of the controlled variable with its desired value and produces an error signal when a deviation exists. A brief description of how the error detector of Fig. 12 operates is given in the following paragraphs.

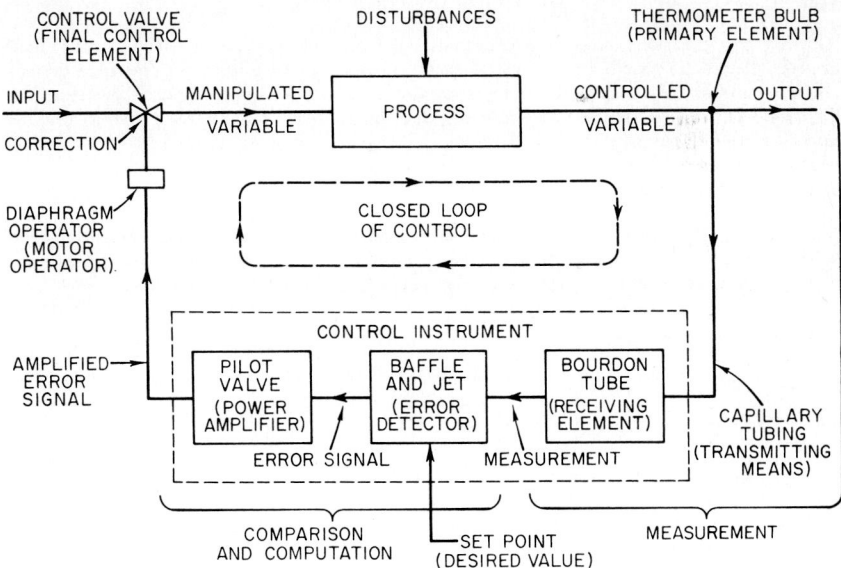

Fig. 13   Block diagram resolving parts of the control system of Fig. 12 into basic control functions.

The set point is represented by the position of the left end of the baffle, as determined by manual adjustments to the set-point knob. The *measured* value of the controlled variable is represented by the position of the right end of the baffle as determined by the deflection of the bourdon tube. Thus, the baffle is a differential lever; the position of its center (at the jet) represents the deviation (difference) or error between the desired and measured values of the controlled variable. In order to be usable, this very weak, small deviation or error signal, represented by the position of the baffle midpoint, must be both measured and amplified. This is done (in this particular controller) by the baffle and jet error detector.

The jet back-pressure air system is continuously furnished with air which bleeds in from the air supply through the restriction. This air bleeds out from the jet back-pressure system through the jet to the atmosphere. As the center of the baffle (representing the error) is moved closer to or farther from the air jet, the resistance to the flow of air through the jet changes. This changes the back-pressure in the jet air system. This back-pressure is proportional to the baffled position through a small range of baffle motion. Thus, the amount of deviation or error is measured and converted to an air jet back-pressure, which is an amplified error signal. The error detector is the heart of any automatic controller, for it is the part that senses deviation and first instigates corrective action.

Amplifier

In order not to restrict the sensitivity or accuracy of the measuring system, an error detector must take very little power from the measuring system. Thus, error signals usually are very weak. In order to operate most final control elements, the error signal must be greatly amplified in power. Therefore, an automatic controller usually contains a power amplifier which employs auxiliary power to increase the strength of the error signal. In the typical air-operated controller of Fig. 12, some amplification is obtained from the jet-and-baffle error detector, which might be called a *first-stage amplifier*. Additional power amplification is produced by the pilot valve, which might be called a *second-stage* or *power amplifier*. Both stages use auxiliary power furnished by the 20-psi (133-kPa) air supply. The jet back-pressure (error signal) from the error detector is applied to the diaphragm of the pilot valve. This causes the pilot-valve plug to move up and down, which delivers more or less 20-psi (133-kPa) supply air to form an amplified error signal or *output* from the air controller that is still proportional to the error or deviation represented by the baffle midpoint.

Motor Operator

The error signal must be converted to corrections in the manipulated variable of the process. In most control systems, this requires some form of motor to operate the final control element. In the air-operated control system of Fig. 12, the motor operator that positions the steam valve is the diaphragm air motor. The output air pressure (amplified error signal) from the pilot valve is applied to the motor-operator diaphragm.

Final Control Element

The final control element corrects the value of the manipulated variable. In the control systems of Figs. 12 and 13, the final control element is the steam valve which is in direct contact with the control agent (steam) and makes corrections in the manipulated variable (rate of steam flow).

Self-Operated Controllers

Some control systems obtain all power for operating the error detector and final control element from the controlled medium of the process via the primary element and measuring means. Such control systems are termed self-operated controllers. See Fig. 14.

Control-System Time Lags

Automatic-control systems have time lags that can have a serious effect on the performance of fast-acting control loops. The same kinds of lags—resistance-capacity lags and dead time—which are found in processes are also encountered in control systems. Further, the lags found in controllers are caused by the same properties—inertia, capacity, resistance, and transportation.

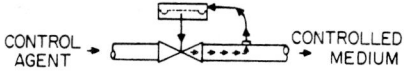

CONTROL AGENT →   → CONTROLLED MEDIUM

Fig. 14 Self-operated controller using only energy from a controlled medium through a primary element.

Measuring-Means Lags

Much of the time lag in automatic controllers occurs in the measuring system parts. For example (Fig. 12), the fluid in the thermometer bulb has a heat capacity and the shell of the bulb has a resistance to heat transfer. Together, they form a resistance-capacity with the same kind of lag found in processes. Inclusion of a thermometer well would greatly increase the time lag of this primary element. Thus, thermometer wells, thermocouple sockets, or anything that slows the speed of response of the primary element should be avoided wherever practical in control systems.

The resistance to fluid flow in the capillary tubing and the volumetric capacity of the tubing and bourdon spring also form a resistance-capacity pair which has a time lag. Therefore, long lengths of thermometer capillary tubing should be avoided in control systems. For the same reason, any pneumatic or hydraulic transmitting means used between the primary element and the receiving element in any control system should be as short as possible. Other examples include piping between pressure gages and the process, between flow manometers and orifices, and between pneumatic transmitters and receivers.

Error Detector and Amplifier Lags

The time lags in pneumatic error detectors and amplifiers, such as in Fig. 12, are usually insignificant. However, lost motion and friction in bearings and linkages constitute dead time. These factors can be considerable in poorly made or badly worn controllers. In controllers of the electric contact type, dead time is commonly found. For example, the differential gap between the high and low contacts of the controller in Fig. 17a creates dead-time lag.

Transmission Lags

In the output air system of Fig. 12, the resistance to airflow and the volumetric capacity of the pipe and diaphragm motor operator cause a time lag. Therefore, the length of output air pipe and the volume of the motor operator should be as small as possible.

Final-Control-Element Lags

The physical inertia and frictional resistance in the stem and seat of the control valve create time lags. In addition, if valve stem friction is large, no valve motion whatever may occur during a considerable change in output air pressure. Then suddenly, the standing friction overcome, the valve jumps to a new position. This constitutes dead time in the valve which has a very bad effect on control-loop stability. Therefore, where valve stem packing must be tight, valve positioners, steam lubricators, and careful valve maintenance are all required.

In general, *every part of an automatic-control system must be designed, selected, installed, and maintained to avoid time lags as much as possible.* It must be stressed that a given time lag has just as bad an effect on control-loop stability when it occurs in parts of the automatic-control system as when it occurs in the process itself. This can be illustrated by examining how the time lags of the control system shown in Fig. 12 can be reduced by use of solid-state electronics. This, in turn, will allow the reader to obtain some insight into the tradeoffs a designer must make.

Microprocessor-Based System and Time Lags

A microprocessor-based system, which could be used to control the heat exchanger of Fig. 1, is shown in Fig. 15. The effects of such a system on time lags are described as follows.

Measuring Means

The thermometer bulb (primary or sensing element) and the capillary tubing (transmitting means) have been replaced by a pair of thermocouple wires. This has eliminated the RC lag of the ther-

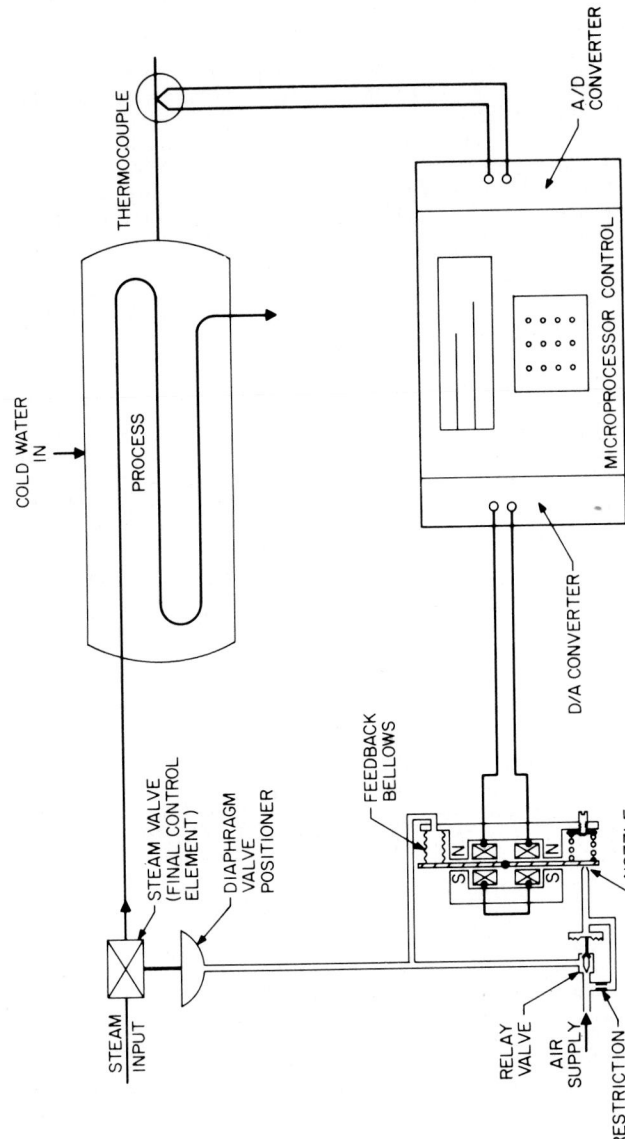

Fig. 15 Microprocessor-based control system.

mometer bulb and capillary tubing. The bourdon pressure element (receiving element) has been replaced by a cold junction thermocouple and an analog-to-digital (A/D) converter in order to transform the signal into a digital form for temperature display, recording, and calculation of the error signal. Thus, the capacitance and inertia delay of the bourdon tube have been eliminated, since the A/D conversion can be made essentially instantaneously relative to all other system delays.

Error Detector Elements

The unbalanced detector, summing point, or primary relay of Fig. 12 has been replaced by the microprocessor shown in Fig. 15. The set point is entered by the operator through the keyboard. The microprocessor subtracts the measured temperature from the set-point temperature to obtain the temperature error signal. This digital error signal is converted to an analog signal in the A/D converter. This process can be carried out as many times per second as is required to make this time delay small with respect to all other system time delays. In addition, the microprocessor is capable, by software program changes, of implementing all the control actions described in the next portion of this article.

Amplifier

The pilot valve (power amplifier) in Fig. 12 is replaced by a somewhat similar device located next to the diaphragm air motor (motor operator). This eliminates the time delay caused by the 3 to 15 psi (20 to 100 kPa) air pressure line, since the signal is now transmitted electrically. The electric signal is converted to a mechanical movement of the baffle and jet by a torque motor. The response of the torque motor can be made very high because of the high power level of the electric signal.

Final Control Element

This is the same as in Fig. 12. Thus the number and size of the time lags have been reduced, and the number of control functions which can be used has been increased.

Control Actions

In manual control (Fig. 9), when the operator senses that the temperature is deviating below the desired value, he can make steam valve corrections in several ways:

1. Instantly open the valve wide.
2. Open the valve slowly at constant speed as long as deviation continues.
3. Open the valve more when the deviation occurs rapidly.
4. Open the valve a constant amount for each unit of deviation.

The operator also can use other methods or combinations of methods of valve manipulation.

The *control action* is the manner in which a control system makes corrections relative to deviations. When applied to only the control instrument itself, the term *control action* describes the manner in which that particular controller changes its output relative to an error signal input (deviation). But in practice, it is often assumed, even by instrument engineers, that the nominal controller action is the control action actually being applied to regulation of the process. It must be stressed that the control action *actually applied* to correction of the process input results from the *combined operating characteristics of all functional elements* comprising the control system.

Modern industrial control instruments are usually made to produce one (or a combination) of the following control actions (modes of control):

1. Two-position (on-off)
2. Time proportioning (average position, proportional input)
3. Floating

4. Integral (reset, proportional speed floating)
5. Proportional (throttling)
6. Proportional plus integral
7. Proportional plus derivative
8. Proportional plus integral plus derivative

Each control action has characteristic advantages and limitations. Therefore, it is important that the characteristics of each be fully understood. In general, the more difficult a process is to control, the farther down in the foregoing list one must go to find the best-suited control action.

Two-Position Action

In this mode of control, the final control element is moved relatively quickly from one of two fixed positions to the other at a single value of the controlled variable. Inasmuch as these two positions of the final control element are usually open and closed, this mode is often termed *on-off* control. An example of a typical two-position electric control system is shown in Fig. 16a. When the temperature is at or above the set-point value, the contact is closed and the valve closes; when the temperature is below the set point, the contact is opened and the valve opens. Figure 16b shows the characteristic valve position corrections as the temperature deviates above and below this set point. It is seen that two-position control cannot make an exact correction; its correction must be either greater or less than exact. Therefore, no stable balanced condition of input and output energy is ever achieved, and the controlled variable must continue to cycle up and down, as shown in Fig. 16b.

Two-Position Differential-Gap Control Action

This is a common variation of the two-position action. Here the final control element is moved relatively quickly from its first position to its second when the controlled variable reaches a set value from one direction, and can return to its first position only after the variable has passed in the opposite direction through a range of values (called the *differential gap*) to a second value. Figure 17a shows a typical two-position differential-gap electrically operated control system, and Fig. 17b shows its characteristic valve corrections as the temperature passes through the differential gap. Note that no valve action occurs while the variable is within this differential gap.

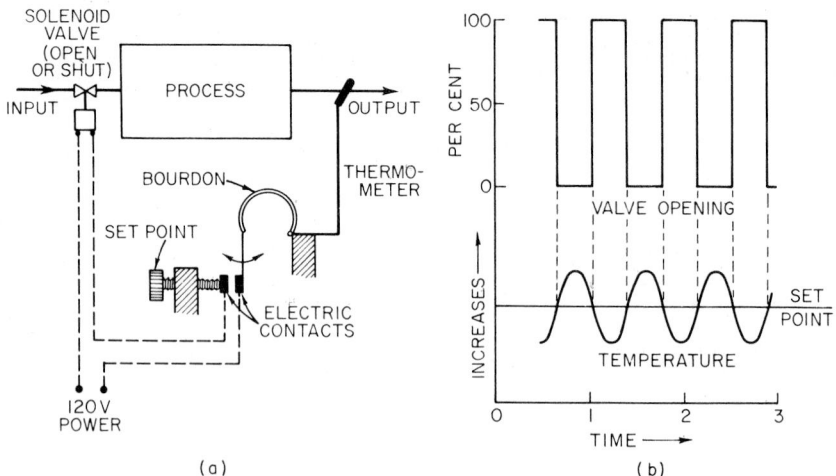

Fig. 16   (a) Typical two-position electrically operated control system. (b) Relation between valve position and controlled variable in two-position action.

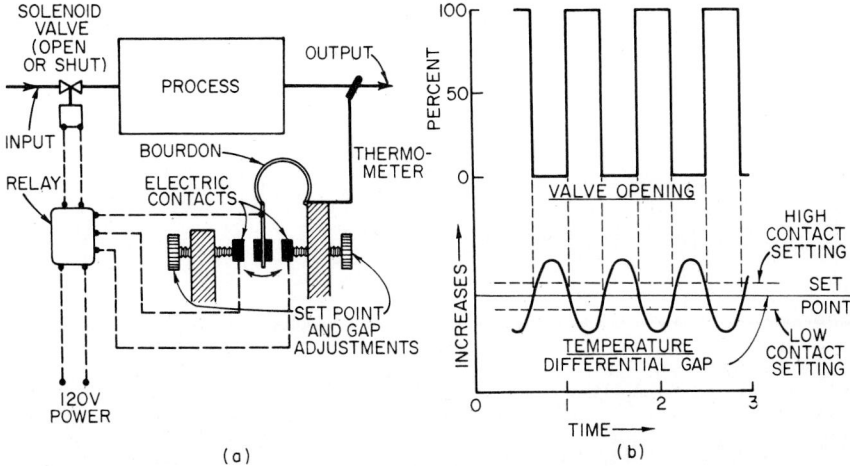

Fig. 17 (a) Typical two-position differential-gap electrically operated control system. (b) Relation between valve position and controlled variable in two-position differential-gap control action.

Multiposition Control Action

In this action the final control element is moved to one of *three* or more fixed positions, each corresponding to a definite range of values of the controlled variable. The control instrument of Fig. 17b can produce a three-position action if it is used with a final control element that takes a third position when the variable value is within the differential gap. The characteristic valve position corrections of three-position control are shown in Fig. 18. Since multiposition control is capable of a limited number of corrections, this control action rarely produces an exact correction for any load condition and, therefore, produces continuously cyclic control.

Time Proportioning Control Action

In this action there is a predetermined relation between the value of the controlled variable and the time-average position of a two-position final control element. Figure 19a shows schematically an average-position action temperature control-

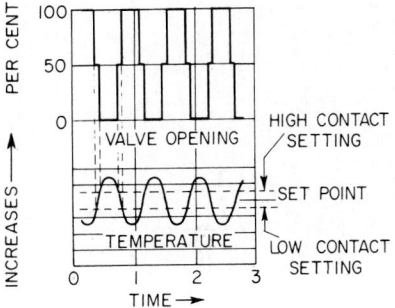

Fig. 18 Relation between valve position and controlled variable in three-position (multiposition) action.

ler. A constant speed motor rotates a cam that causes the right-hand electric contact to swing back and forth continuously. The bourdon tube positions the left-hand contact in accordance with the value of the temperature. Suppose that contact settings are such that the set point is 500° and the proportional band is 20° wide, extending from 490 to 510°. When the temperature is below the proportional band (below 490°), the contacts are open *all the time* and the valve is *open* all the time. When the temperature is within the proportional band (between 490 and 510°), the contacts are closed and opened each time the cam rotates. The percentage of valve-open time is proportional to the position of the temperature within the proportional band. For example (Fig. 19b), when the temperature is at 505°, the valve is open 25% of the time and closed 75% of the time. If the cam rotation is one per minute, at 505° the valve will be open 15 s (25%) and closed 45 s (75%) out of each minute. In other words, the input of energy to the process is proportional to the position of the controlled variable within the proportional band. When the temperature is above the proportional band (above 510°F), the contacts are closed all the time and the valve is closed all the time.

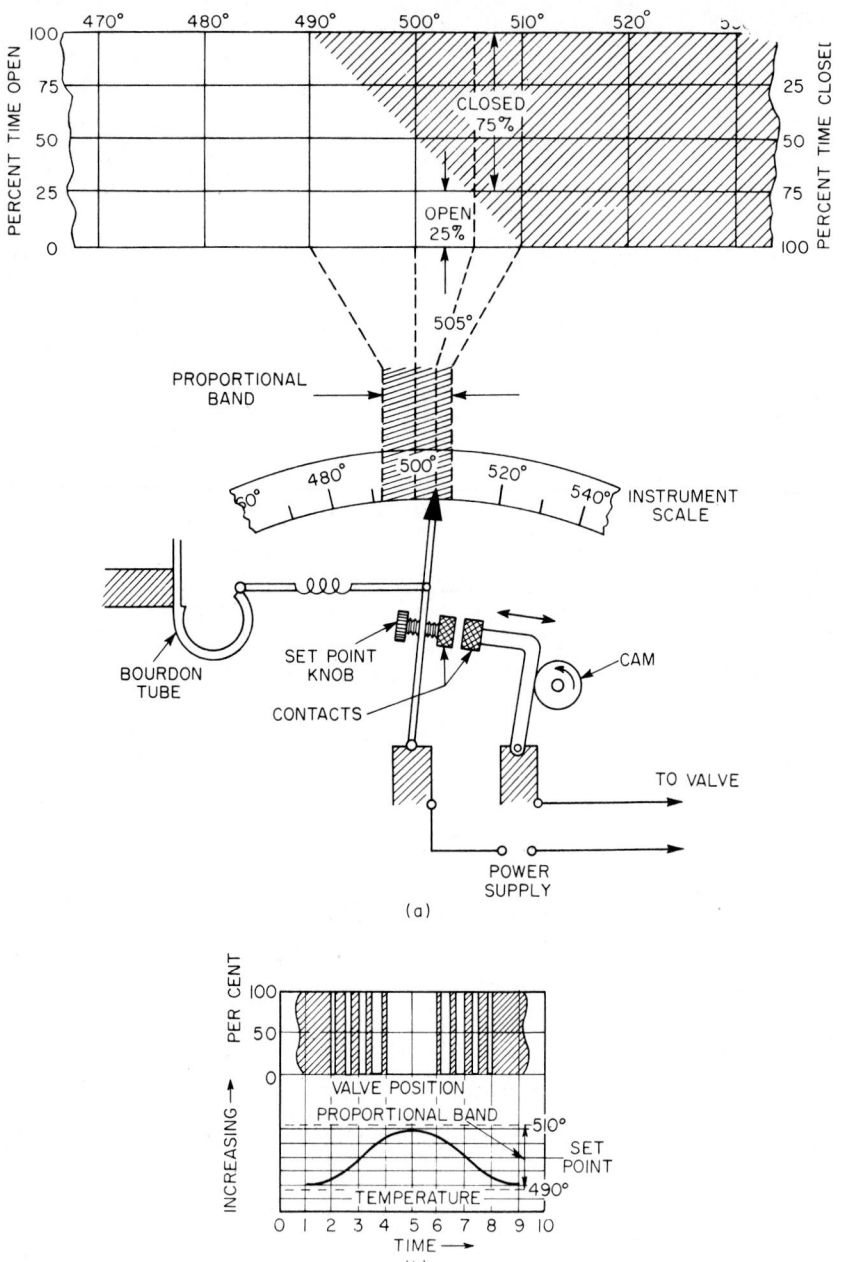

Fig. 19 (*a*) Typical time proportioning electrically operated controller. (*b*) Relation between valve position and controlled variable in time proportioning control action. Hatched area represents valve-open time, i.e., energy input to process.

Floating Control Action

In this control action there is a predetermined relation between the deviation and the rate of travel of the final control element. The final control element moves relatively slowly toward either one or the other of its two extreme positions, depending on whether the controlled variable is above or below the set point. The three common types of floating actions are described in the following paragraphs.

Single-Speed Floating Control Action

In this control action the final control element moves at a single rate regardless of the amount of deviation. The single-contact on-off controller of Fig. 16b can produce single-speed floating action if used with a slow-running reversible electric-motor-driven valve. When the temperature is at or above the set point, the valve runs toward its closed position at a single speed (see Fig. 20a). When the temperature is below the contact setting, the valve runs toward its open position at the same single speed.

   In actual practice, true single-speed floating control is little used because, like two-position control, it produces continuous cycling.

Single-Speed Floating Control Action with a Neutral Zone

In actual practice, single-speed floating control is usually employed with a *neutral zone*. The two-contact controller (Fig. 17a) when used with a relatively slow reversible-motor valve produces such neutral-zone action (see Fig. 20b). When the value of the controlled variable is in the neutral zone between the contact setting, no contact is made and the valve remains motionless—its rate of travel is zero. It is possible for this control system to produce almost an exact correction for any load condition. Thus, this scheme can reduce deviation almost to zero and can, if correctly adjusted, eliminate valve motion, except following load or set-point changes.

Multispeed Floating Control Action

In this control action the final control element is moved at two or more rates, each rate corresponding to a definite range of values of deviation. See Fig. 20c. In other words, at which of several speeds the final control element moves is dependent on the amount of deviation.

Proportional-Speed Floating Control Action

In this control action the position of the final control element is changed at a rate that is proportional to deviation. In other words, the greater the deviation becomes, the faster the valve moves, as shown in Fig. 20d. This is also called *integral control action*.

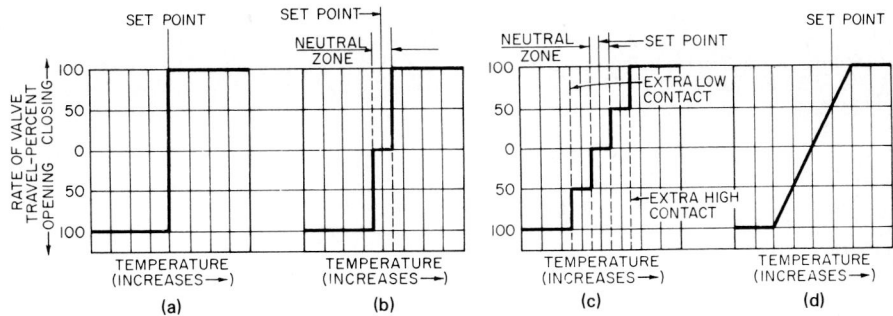

Fig. 20   Relation between rate of valve travel and controlled variable in floating control actions. (*a*) Single speed, (*b*) single speed with neutral zone, (*c*) multispeed, (*d*) integral (proportional speed).

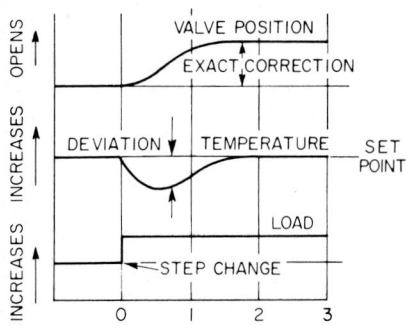

Fig. 21 Response of a process under proportional-speed floating control to a step change in load. Note that proportional-speed floating control action produces an *exact* correction.

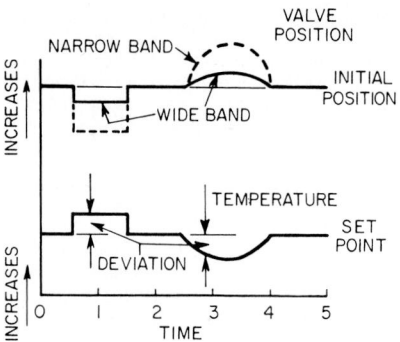

Fig. 22 Relation between valve position and controlled variable under proportional-position control action.

Integral Control Action

Figure 21 shows the temperature-to-valve-position relation characteristic of integral (proportional-speed) floating control action following a step change in load at time zero. Note that the rate of valve travel is proportional to the deviation. For example, when the deviation is maximum (time ½ min), the speed of valve travel is also maximum. The factor of primary importance in connection with integral control action is that *as long as deviation continues, the correction to the valve position continues.* Integral control responds to both the amount and time duration of the deviation. Thus, this mode of control continues to operate until it produces an *exact correction* for any load change. This is a unique advantage of integral control action over any other action.

An understanding of the advantages of integral control action is important because, when combined with proportional control action, it is so often used to produce the popular proportional-plus-integral control action.

Proportional Control Action

In proportional control action (throttling or modulating control action), there is a continuous linear relation between the value of the controlled variable and the position of the final control element (within the proportional band). In other words, the valve moves the same amount for each unit of deviation of the controlled variable from the set point.

Figure 22 shows the relation between valve position and controlled variable characteristic of proportional-control action. Note that the valve position changes in exact proportion to deviation. Figure 22 also shows that proportional-control action responds only to the *amount* of deviation and is insensitive to the rate or duration of deviation. Note that at 2 min and 4 min, when the temperature was returned to its initial value, the valve also returned to its initial value; in other words, there was *no valve correction without deviation.*

Proportional Band (Throttling Range)

This is the change in value of the controlled variable that is necessary to cause full travel of the final control element. Figure 22 shows the relation between valve position and temperature for two different proportional bands.

The proportional band of a particular instrument is usually expressed as a percent of its full range. For example, in Fig. 12, if the full range of the instrument is 200° and it takes a 50° change in temperature to cause full valve travel, the percent proportional band is 50° in 200°, or 25%. In practical controllers the proportional band may cover less than 1% to much over 200%. Figure 23 helps to explain the concept of percent proportional band. Note that proportional bands over 100% cannot cause full valve travel even for a full-range change in a controlled variable.

Gain (Sensitivity)

Another concept for expressing proportionality is *gain* or *sensitivity*. These terms describe the ratio between the input and output to and from a control device. Mathematically, gain and sensitivity are reciprocal to proportional band. The proportionality adjustment of some control instruments is calibrated in gain or sensitivity instead of in proportional band.

Offset (Droop)

It is fundamental that any change in process load requires a new valve position to correct for it. But, as shown in Fig 22, proportional control action requires a *change in deviation* in order to produce a new valve position. Figure 24 shows the load, temperature, and valve position for the heat exchanger process of Fig. 1 under proportional control action. Initially, the set point is at the desired value of 100°. At a time of 1 min, a step load change occurs. Note that the temperature is not returned to the desired value of 100° by proportional control, but balances out at a new value of 90°.

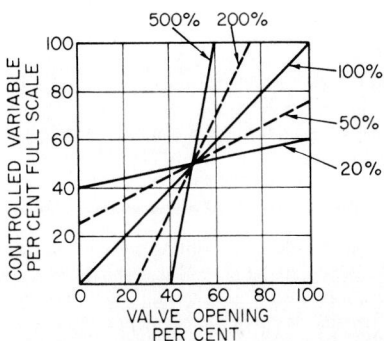

Fig. 23   Relation between valve travel and controlled variable under several percentages of proportional band in proportional control action.

Proportional control action can produce an *exact* correction for only one load condition; at all other loads, there must always be some deviation (error) left. This error is called *offset* (droop) and is an inescapable characteristic of proportional control action. Note in Fig. 24 that the offset from time units of 3 to 8 min was 10°F.

Manual Reset

This offset deviation can be eliminated by manually resetting the set point. In Fig. 24, at 8 min the set point was raised manually to bring the temperature to the desired value of 100°.

It must be emphasized, however, that such manual reset eliminates offset only for a single value of load. The factor of primary importance in connection with *proportional control action is that it is a powerful stabilizing control action capable of wide adjustment and application, but it has the undesirable characteristic of offset error.*

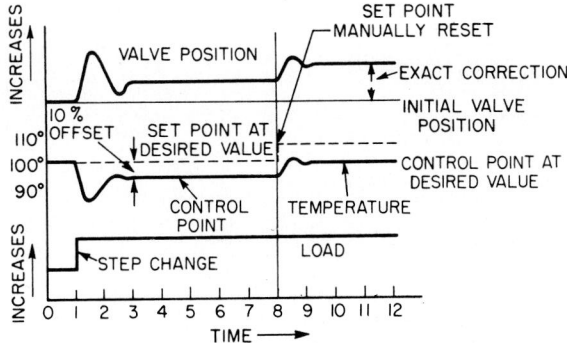

Fig. 24   Illustration of offset characteristic of proportional control action following a load change and how offset can be eliminated by manually resetting the set point.

Proportional-plus-Integral Control Action

It has been noted that floating control actions have the unique advantage of continuing to correct valve position until no deviation remains. Thus, by adding integral control action to proportional control action, the undesirable offset characteristic or proportional control action can be overcome. The addition of integral control action does automatically what manual reset does by hand. Thus, integral control is sometimes called *reset control*. The best way to explain the operation of these combined control actions is to picture separately the components of valve motion due to each control action and observe what each contributes to the resultant valve position.

Figure 25 shows the component of valve movement due to proportional control action combined with that due to integral control action (open-loop system). At time zero, a step change deviation occurs. The valve motion component due to proportional action occurs instantly, just as in Fig. 22. But, because of the integral component, the valve continues to move at a constant rate.

Figure 26 shows a similar analysis of valve position components under *actual process control*. At time zero, a step increase in load occurs. Because of proportional action, a large valve correction is imposed as the temperature deviates, and later is withdrawn completely as the temperature is returned to the set point. But it should be noted that the final *exact correction* is applied wholly by the integral action component. The cross-hatched area *A*, under the proportional component curve, represents the energy added by proportional action. Area *B* represents the energy added by integral action. Area *C*, under the resultant valve position, represents the excess correction, i.e., the correction in excess of the exact correction) that was applied and withdrawn by proportional action.

The facts of primary importance in connection with proportional-plus-integral control action are that control without offset error is possible under all load conditions, but integral action itself does not contribute to the stability of the control loop.

Integral Rate

The integral rate adjustment changes the rate of change of the integral component. Integral rate usually is expressed in repeats per minute.* This term denotes the number of times per minute that

---

*Applies only to controllers where integral speed is affected by proportional-band width.

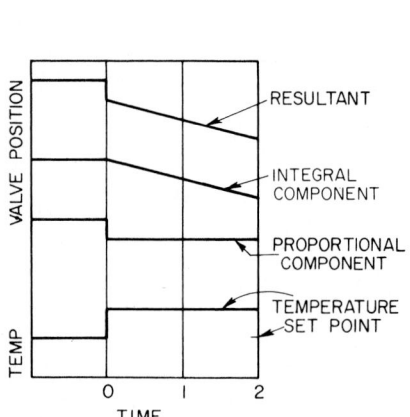

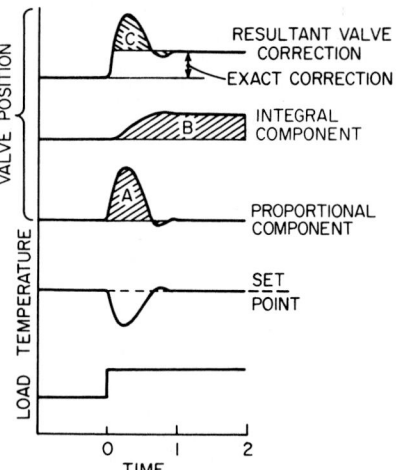

Fig. 25 Relation between valve position and deviation under proportional control action (process not on automatic control). Components of the valve position due to proportional action and integral action are shown separately.

Fig. 26 Relation of valve position to temperature under proportional-plus-integral action control of a process following a load change at time zero. Valve motion components due to proportional action and integral action are shown separately.

valve motion due to proportional action is repeated by integral action. Figure 27 explains this graphically (open-loop system). At time zero, a step change in temperature occurs. The valve immediately changes position by amount A because of the proportional component. The valve also begins to move at a constant rate because of the integral action. Figure 27 gives curves for the integral component at three different integral rates. The integral rate can be easily calculated. For example, in curve a, the proportional action valve motion A was repeated by integral action in ½ min. Thus, this integral rate is 2.0 repeats per minute. In curve b, valve motion A was repeated by integral action in 1 min. Thus, this integral rate is 1.0 repeat per minute. Some industrial control instruments are calibrated in *minutes per repeat*, which is the mathematical reciprocal of repeats per minute.

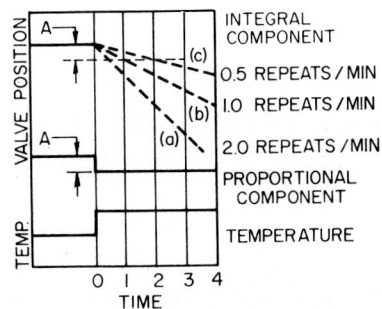

Fig. 27 Graphical representation of the integral rate in repeats per minute (process not on automatic control). The valve motion A due to a proportional component is repeated by the integral component.

Start-Up of Processes under Integral Action Control

The start-up of processes under control incorporating any form of automatic reset action usually results in a large overshoot above the set point by the controlled variable. This behavior is explained by Fig. 28a and b. Figure 28a shows start-up under proportional action only. At a time unit of 1 min, the controller and control agent are turned on. The control valve remains wide open until the controlled variables rises into the proportional band at a time unit of 2 min. Then the valve begins to close, and, after a small overshoot and cycling, the variable balances somewhat below the set point (offset error).

Figure 28b shows start-up under proportional-plus-integral action. At a time unit of 1 min, when the controller and control agent are turned on, the control valve is opened wide and the upper edge of the proportional band is at the set point. Now, at start-up, the deviation is great. Therefore, as soon as the controller is turned on, integral action begins to shift the proportional band upward at a rapid rate. Thus, the band is soon shifted completely until its lower edge is at the set point. But this band shifting by integral action accomplishes nothing, since the control valve is already wide open.

But then, because of this band shifting, the variable *rises all the way to the set point* before proportional action can even *begin* to close the control valve. This start-up should be compared with that under proportional action (Fig. 28a). Note that this band shifting by integral action cannot even begin to be withdrawn until the variable rises above the set point (i.e., polarity of deviation is

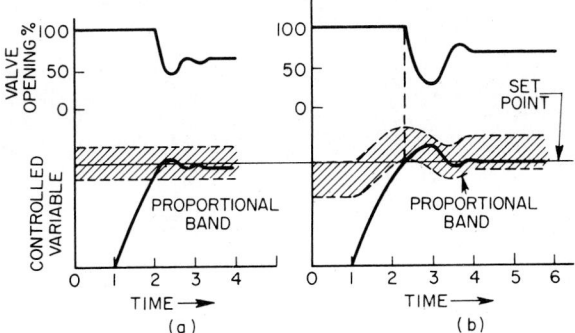

Fig. 28 (a) Start-up of a process under automatic control by proportional action. (b) Start-up of a process under automatic control by proportional-plus-integral action. Note the large initial overshoot of the set point due to proportional-band shifting by integral action.

reversed). Thus, the result is overshoot and cycling much greater than under proportional action alone. This same excessive proportional-band shifting by integral action can occur under any large, long-sustained load change where the controlled variable passes out of the proportional band.*

Several manufacturers have developed controllers that partly overcome this difficulty automatically. In these controllers, the output from a proportional-plus-derivative action control element is connected as the input to a proportional-plus-integral action control element. It will be shown later (Fig. 30) that in proportional-plus-derivative action the output leads the input. Thus, the integral control element is made to begin withdrawal of its extreme proportional-band shifting *before* the controlled variable actually reaches the set point (i.e., before the deviation changes polarity). Such integral controllers can be adjusted to produce satisfactory stability during both start-up and normal operation.

Proportional-plus-Derivative Control Action

In this action (sometimes referred to as *rate action*), there is a continuous linear relation between the rate of change (i.e., first derivative) of the controlled variable and the position of the final control element. In other words, the valve motion is proportional to the speed at which the temperature is changing; the faster the temperature changes, the more the valve is moved. Again, the easiest way to explain proportional-plus-derivative action is to picture separately the components of valve motion due to each, as in Fig. 29.

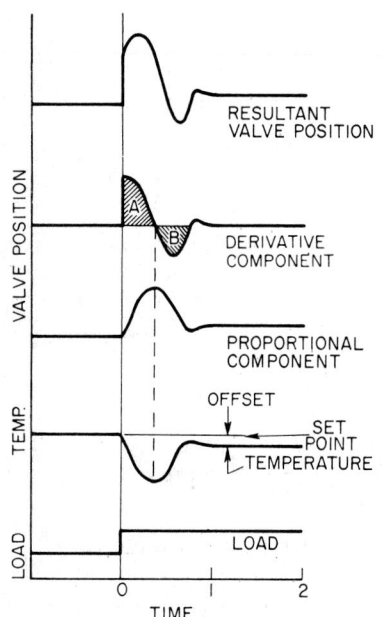

Note that the size of the derivative action correction is proportional to the rate of change (slope) of the controlled-variable curve. When the variable is changing fastest (time zero), the derivative action correction is largest. When the variable is reversed (time 0.4), its rate of change is zero; so the derivative action component is also zero. When the variable is changing *away* from the set point, derivative action *adds* the energy represented by area *A* to oppose this change. When the variable is changing *toward* the set point, derivative action subtracts the energy represented by area *B* to oppose this change. Thus, derivative action has a great stabilizing effect on control. Note, however, that after the variable stops changing (time 1.0), only the valve correction due to proportional action is left. Thus, derivative action has no direct effect on offset error.

The factors of primary importance in connection with derivative action are that, by opposing all change, derivative action has a great stabilizing effect on control, but it does not eliminate the undesirable offset characteristic of proportional action.

Fig. 29 Relation of valve position to temperature under proportional-plus-derivative action control of a process following a load change at time zero. Valve motion components due to proportional action and derivative action are shown separately.

Derivative Time

The derivative adjustment is expressed in a derivation time, which is the time in minutes by which derivative action advances the effect of proportional action on the final control element. Figure 30 shows this graphically. Proportional action alone would produce the valve position shown by the solid curve. But proportional-plus-derivative action

---

*Several manufacturers offer devices that minimize or eliminate this proportional-band shifting on start-up of integral action controllers.

would produce this same valve position by time T *earlier* than proportional action alone, as shown by the dotted curve. This time T, in minutes, is the derivative time. In Fig. 30, the derivative time is 0.6 min.

Proportional-plus-Integral-plus-Derivative Control Action

All three previously described control actions can be combined in one control instrument to obtain all their advantages. Figure 31 shows separately the components of valve motion produced by each action after a step change in load at time zero. Figure 31 also sums up the previous discussion.

1. The proportional action component corrects the valve position by an amount proportional to the deviation and thus produces a temporary increased energy input represented by area A.

2. The integral action (reset) component corrects the valve position at a rate proportional to deviation and thus produces a *permanent* increased energy input represented by area D.

3. The derivative action rate component corrects the valve position by an amount proportional to the rate of change of the controlled variable. Derivative correction first added the energy represented by area B and then subtracted the energy represented by area C.

The resultant valve position curve shows that, first, an excess correction is applied to oppose the change in the variable away from the desired value. This excess energy input is represented by area E. Then less than an exact correction is applied as the variable is returned to the desired value. This less-than-exact energy input is represented by area F. Note that ultimately neither proportional action nor derivative action produced any correction; it was the integral action only which produced the new valve correction that exactly satisfied the new load condition.

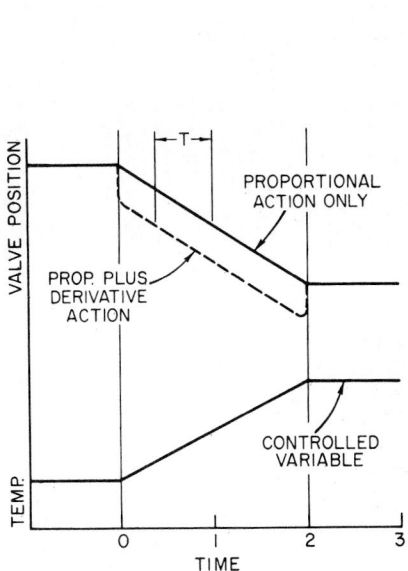

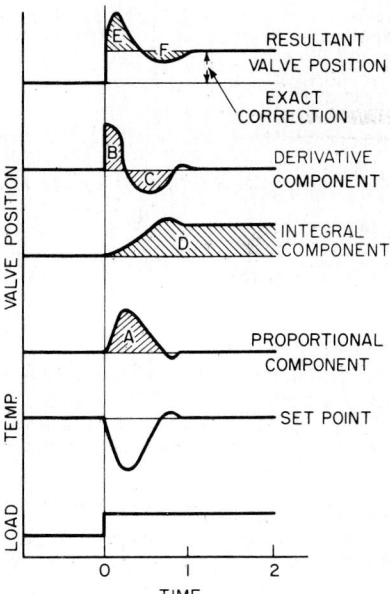

Fig. 30 Graphical representation of derivative time (process not on automatic control). Proportional-plus-derivative action produces a given valve position by time T before proportional action alone can do so. A derivative time of 0.6 min is shown.

Fig. 31 Relation of valve position to temperature under proportional-plus-integral-plus-derivative action control of a process following a load change at time zero. Valve motion components due to each mode are shown separately.

Selection of Control Action

Probably the most troublesome decision in designing a control system is selection of an adequate yet economical control action. The solution is usually a compromise between the quality of control obtained and the cost of the control system. The control system must be sufficient to meet the tolerance of the process, but it should not include refinements beyond those required, or its cost will be excessive. But, if there is any doubt, completely adequate instrumentation should be chosen, because the economic loss due to "overinstrumentation" is slight compared to product quality and quantity loss resulting from penny pinching on the control equipment.

Tables 1 and 2 give the forms of control suited to processes of various characteristics. These tables should be considered general guides only and must not be construed as more than approximating aids in selecting control actions. In such selection, there is no substitute for actual experience with identical or related processes. It should be recalled that *single-capacity* denotes a process dominated by one of its resistance-capacity pairs, while *multicapacity* denotes a process with two or more RC pairs of significant size.

Table 1 Process Characteristics versus Control Action

Number of process capacities	Process reaction rate	Resistance-capacity (RC)	Dead time (transportation)	Load changes Size	Load changes Speed	Suitable control action
		Process time lags		Load changes		
Single	Slow	Moderate to large	Small	Any	Any	Two-position, two-position with differential gap
				Moderate	Slow	Multiposition, time proportioning
Single (self-regulating)	Fast	Small	Small	Any	Slow	Floating actions: single-speed, multispeed
					Moderate	Integral
Multiple	Slow to moderate	Moderate	Small	Small	Moderate	Proportional
Multiple	Moderate	Any	Small	Small	Any	Proportional plus derivative
Multiple	Any	Any	Small to moderate	Large	Slow to moderate	Proportional plus integral
Multiple	Any	Any	Small	Large	Fast	Proportional plus integral plus derivative
Any	Faster than that of the control system	Small or nearly zero	Small to moderate	Any	Any	Wideband proportional plus fast integral

Table 2 Types of Processes versus Control Action

Process	Gain (proportional)	Integral	Derivative
Flow and liquid pressure	0.2–50 (500–2% PB)	Required	Not required
Gas pressure	20–∞ (5–0 % PB)	Not required	Not required
Liquid level	2–20 (50–5 % PB)	Occasionally required	Not required
Temperature	1–10 (100–10 % PB)	Required	Required

PB, Proportional band.

Processes with a very small capacity, notably flow processes, are successfully controlled by proportional-plus-integral action alone, or by wideband proportional control combined with very fast integral action.

Stability of Control Loops

Feedback control loops can be either stable or unstable. A feedback control loop is said to be *stable* when it is in such a condition that after a disturbance its controlled variable eventually reaches a steady, noncyclic value. It is the purpose of the following discussion to explain in a simple, practical way why and when control loops will be stable or unstable and to give some practical criteria for expressing their stability. (For a mathematical discussion of stability see the article Process Control Techniques later in this Handbook section.

Automatic control consists of a chain of energy exchanges—measurement, comparison, correction, reaction—going around and around in a closed loop through the control system and the process. Both the process and the control system can retard and delay these energy exchanges. The simplest explanation of control-loop stability is in terms of the time retardations and energy loss or gain around the loop. This closed loop can be simplified into two elements—the process and the controller—as shown in Fig. 32. A signal (energy) from a process output is fed back by the control system into an input to the process.

Corrections to process input must be *opposed* to changes in process output if these corrections are to restore the desired value of output. For example, if the process output value is increasing (Fig. 32), the correction to process input must be in the direction which will oppose this increase. This means that the correction to process input must be negative compared to the change in process output. In other words, the signal (energy) fed back must be in such a direction as to change the process input *negatively* with respect to the change in process output. Thus, control systems are made with a built-in 180-angular-degree phase lag or shift ($-180°$), that is, *negative feedback*.

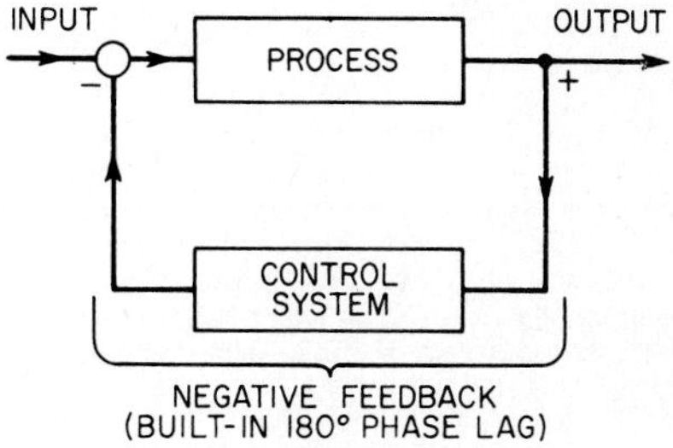

From a study of Figs. 4 to 8, it is known that the process capacity, resistance, and transportation cause lags or negative phase shifts between changes in process input and resultant changes in output. It has been pointed out, too, that these same properties cause negative phase shifts in the control system itself. These process and control-system phase shifts *add* to the negative 180° phase shift built into the controller. And, if these several phase shifts are large enough, the result can be a *total* phase shift around the control loop that approaches 360 angular degrees, that is, *positive* feedback. In other words, if the total of phase shifts (time lags) in the process and the control system is large enough, the controller corrections applied to process input will be *in phase* with process output. Under this condition, the controller corrections will actually increase output changes instead of opposing them.

Fig. 32 Closed loop of control. The control system must incorporate a 180° phase lag in order to oppose process output changes.

In the series of energy exchanges around the control loop, energy is both added and consumed, that is, gained and lost. If the energy added is just equal to that used, it can be stated that the "gain is one." If more energy is added than used, the "gain is more than one," and so on. These two properties—phase shift and gain—provide a simple yardstick for measuring and defining control-loop stability. *A control loop will be stable if, at the frequency of oscillation that gives a total phase shift of 360° around the loop, the gain around the loop is less than one.* However, if at the frequency of oscillation that gives 360° total phase shift, the gain around the loop is one or more, the loop is unstable. Any disturbance causes the loop to "break" into endless oscillations. These oscillations continue, because at gains over one, more energy is being added than is being used, and it is being fed back in phase with itself; so oscillations are self-sustaining or self-excited. This tendency toward self-sustaining oscillation is the greatest limitation to feedback control.

Frequency-Response Analysis

A technique for analyzing the stability of control loops and of their components has been developed by servomechanisms engineers. See Claridge and Berk (1952) and Higgins and McKnight (1952) references listed at the end of this article. The method consists of feeding sinusoidal inputs into the part of the loop concerned and measuring the phase shift and gain between its input and output, through a suitable range of frequencies. For example, in Fig. 33a the control loop of Fig. 32 has been "opened" between controller output O_c and process input I_p. A sinusoidal input I_p, curve b in Fig. 33, is fed into the process input. Curve c in Fig. 33 represents the resulting process output O_p. Note that it lags the input I_p by phase angle θ_p. Also note that the amplitude ratio or gain between I_p and O_p is less than one.

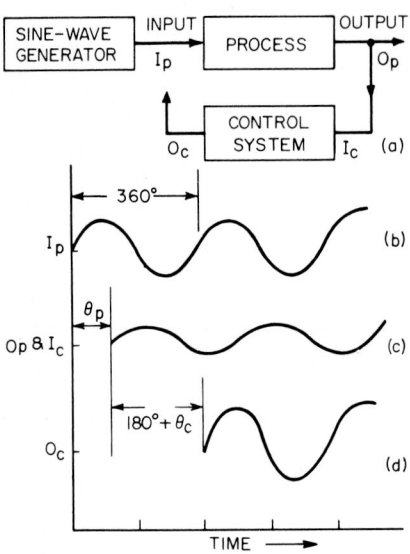

The output from the process O_p is also the input to the control system I_c. Curve d in Fig. 33 represents the resulting controller output O_c. Note that the controller output O_c lags behind its input I_c by $-180°$ *plus* phase angle θ_c, that is, by the 180° phase lag built into the controller plus the total θ_c of other phase lags occurring through the control system. Note also that the overall amplitude ratio or gain through the control loop is over one, because of the large gain of the control system. These are typical frequency-response curves for an industrial control loop.

If this loop is closed again, controller output O_c will become process input I_p. Such a closed system will produce self-sustained oscillations, because the amplitude of oscillation increases around the loop (gain over one) and the total phase shift is 360°. (Note that for all loops there is a frequency that produces a 360° total phase shift. This frequency is called the *crossover frequency*.)

Fig. 33   Curves showing the response of a typical open loop to sinusoidal inputs. Phase shifts around the loop total 180°, which, when added to the 180° phase shift built into the controller, totals 360°. θ_p = phase shift through process; θ_c = phase shift through control system.

In frequency analysis, the gain and phase shift are measured and plotted over a suitable range of frequencies. Frequency analysis can be applied to a complete control loop or to a component or group of components from any part of the loop, as, for example, to measuring systems, power units, final control elements, and parts of a process.

The phase shift and gain of each component can be combined mathematically to give the stability characteristic of the whole loop. Frequency-response curves, together with reaction curves to step changes (Figs. 4 and 6), provide valuable data for plant and control-system design and adjustment, as well as for expected performance. To make the control more stable, more feedback (higher gains) must be used. On the other hand, more feedback increases the tendency to oscillate. The only remedy is to design process and control-system parts with the smallest possible phase shifts relative to amplitude ratio and to each other or to use means that correct for phase lag (negative phase shift) by introducing phase lead (positive phase shift).

Types of Oscillation

Feedback control loops can oscillate or "cycle" in three ways.

1. Increasing Amplitude. If at the frequency that gives a 360° phase shift the gain around the loop is *over one,* the amplitude of oscillation will *increase.* This condition is shown in Fig. 34a.

2. *Constant Amplitude.* If at the frequency that gives a 360° phase shift the gain around the loop is *just one*, the oscillation will continuously sustain itself at *constant* amplitude (see Fig. 34*b*.).

3. *Decreasing Amplitude.* If at the frequency that gives a 360° phase shift the gain around the loop is *less than one*, the oscillations will *decrease* in amplitude and eventually stop (Fig. 34*c*). This is the stable condition sought in automatic control.

Following a disturbance, there are in turn three responses possible under stable conditions.

1. Figure 35*a* shows *cyclic* recovery to eventual stability.
2. Figure 35*b* shows *aperiodic* recovery. Here, the variable returns to a constant value in the shortest time possible without overshooting.
3. Figure 35*c* shows *overdamped* recovery. Here, the variable returns to a constant value in a time longer than the minimum without overshooting.

Effect of Control Actions on Stability

Proportional Action

The problem in feedback control is to improve control performance through the use of more feedback, but this tends to produce instability and self-excited oscillation. Therefore, proportionality (or gain)

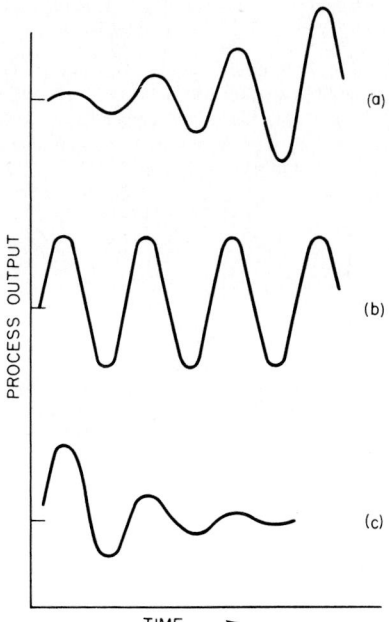

Fig. 34 Three types of oscillation or cycling. At the frequency which produces a total loop phase shift of 360°, the gain can be curve *a* (greater than one), curve *b* (equal to one), or curve *c* (less than one).

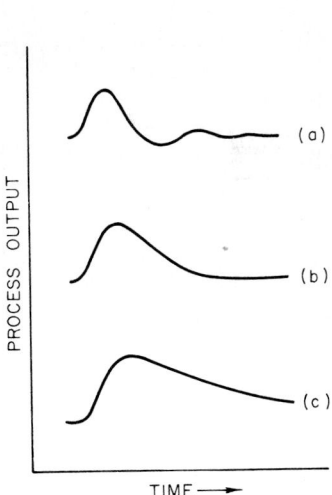

Fig. 35 Three possible types of recovery of a stable control loop following a disturbance. (*a*) Cyclic recovery to eventual balance, (*b*) aperiodic recovery, (*c*) overdamped recovery.

in most automatic controllers is made adjustable. This provides means for increasing the gain of energy around the loop (by narrowing the proportional band) until the control loop approaches instability, i.e., until the loop gain approaches unity at the frequency that yields a 360° total phase shift (crossover frequency). This explains, in terms of frequency response, the well-known fact that there is a minimum stable proportional band (maximum gain) for each control loop. Narrower bands cannot be used without producing constant cycling. It also explains why processes with dead time and multicapacity lags (large relative phase shifts) require wide proportional bands (low gain) for stability.

Integral Action

The addition of integral action (reset) to proportional action causes a negative phase shift in the response of the controller, as shown in Fig. 36. Note that the proportional component of output lags just 180° behind the input variable because of the built-in negative feedback (i.e., point B lags point A by 180°). Also, the integral action component lags behind the proportional component by 90° (point C lags point B by 90°). Thus, because of integral action, the resultant controller output lags behind its input by θ degrees *more* than the built-in negative 180° (point D lags point A by 180° + θ).

As the integral rate adjustment is increased, the amplitude of this integral component is increased relative to the amplitude of the proportional component. Thus, as the integral rate is increased, the phase lag θ between controller output and input increases. The amplitude of the integral component is large at low frequencies of cycling, resulting in high gain and negligible offset error. This explains, in terms of phase shift and gain, the commonly observed fact that loop instability increases as the integral rate is increased, until, finally, continuous oscillation results; i.e., the total loop phase shift approaches 360°. In short, *increasing the integral rate increases instability becuase it increases negative phase shift.*

Derivative Action

It has been stated that one remedy for oscillatory tendencies in feedback loops is to use means that correct for phase lag by creating phase lead. Such a means is drivative or rate action. Note that, while the integral component *lagged* the proportional by 90° (Fig. 36), the derivative component *leads* the

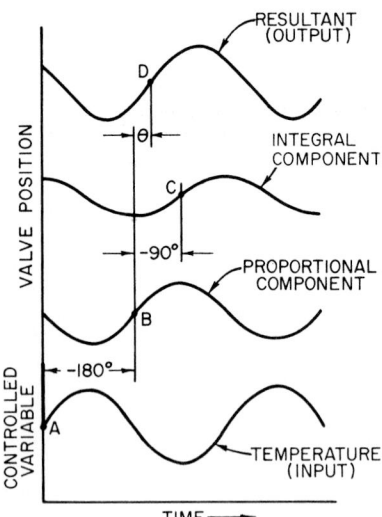

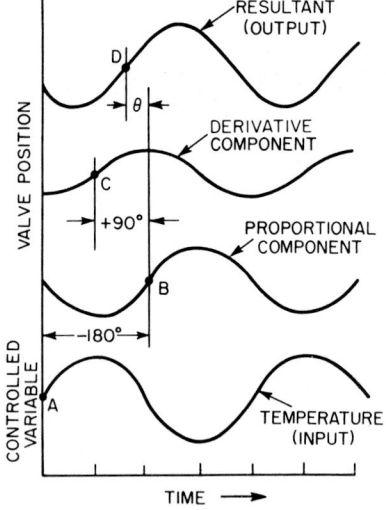

Fig. 36 Phase lag between sinusoidal input to and output from a proportional-plus-integral action controller caused by its integral action.

Fig. 37 Phase lead between sinusoidal input to and output from a proportional-plus-derivative action controller caused by its derivative action.

proportional by 90° in Fig. 37 (point C leads point B by 90°). Thus, because of derivative action, the resultant controller output lags behind its input by θ degrees *less* than the built-in negative 180° (point D lags point A by 180° less θ).

As the derivative time adjustment is increased, the amplitude of the derivative component is increased relative to the amplitude of the proportional component.

Thus, as the derivative time is increased, the phase *lead θ* between controller output and input increases. This explains the commonly observed fact that the use of derivative action has a great stabilizing effect on control loops, sufficiently so that, after inclusion of derivative action, the proportional band usually can be narrowed somewhat. Or, in other words, derivative action partly corrects for phase lags around the loop, thereby permitting the use of greater gain without increasing instability. Note that the amplitude of the derivative component is large at high frequencies of cycling. Thus, derivative time adjustments which are too large will cause instability due to excessive gain. In short, *derivative action increases stability by producing a positive phase shift, but derivative time which is too large produces instability due to excessive gain.*

Criteria of Control Quality

What is good control? There are three criteria for rating the quality of controller performance. The choice of criterion depends on the process concerned. *The best performance for one process may not be the best for another.* Which criterion to use in a given case is discussed in the following paragraphs. All these criteria refer to the shape and duration of the reaction curve following a disturbance.

Minimum-Area or Damping-Ratio Criterion

According to this criterion, the area under the recovery curve should be minimum (see Fig. 38a). When this area is minimum, the deviation averages the smallest amplitude for the shortest time. It has been found that this area is minimum when the amplitude ratio between peaks of succeeding cycles is 0.25. That is, each wave is only one-fourth as big as the wave before it. In terms of phase shift, this minimum area will be obtained with a phase margin of about 30°, when the loop gain is one, i.e., a total loop phase shift of 360° $-$ 30° $=$ 330°. This is the most generally used control quality or stability criterion. It applies especially to processes where the *duration* of the deviation is as important as the *size* of the deviation. For example, in a given process, any deviation above or below a narrow range may result in an off-specification product. Here, the best control is that which permits departures from this range for the minimum time.

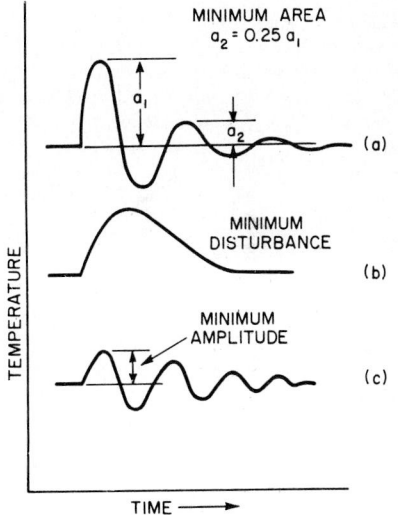

Minimum-Disturbance Criterion

According to this criterion, control actions should create the minimum disturbance to both the control-agent supply to the process and the output from the process. This usually requires noncyclic recovery curves similar to curve b of Fig. 38. This criterion applies to control loops where corrective actions (especially cyclic ones) constitute disturbances to associated controlled processes. For example, sudden or cyclic corrections to a steam control valve may upset steam-supply pressures and seriously disturb other controlled processes served by the same steam supply. Or, in the common condition where the output of one controlled process is the input to a second, sudden or cyclic changes in the first output may create intolerable load changes for the second.

Fig. 38 Relation curves following a disturbance, showing the three criteria for control quality or stability.

Harmonic Cycling

Inasmuch as each control loop has a definite natural frequency of cycling, these periodic supply or output changes from one loop may happen to resonate with the natural frequency of associated control loops. This condition, where the self-excited oscillation frequencies of two interacting control loops are nearly alike, can result in wide sympathetic oscillation, generally called *harmonic cycling*. A remedy for harmonic cycling consists of changing the frequency of either loop so that the two loops can no longer resonate. This can be done by increasing or decreasing the capacity lag in one process—the best remedy, but also the most expensive. The easiest solution is to change the controller adjustment in one loop to obtain noncyclic recovery, as in Fig. 38*b*. However, such adjustments result in lowered control accuracy for that loop. The minimum disturbance criterion is the justification for averaging liquid-level controls and other systems designed to produce minimum disturbance to associated controlled processes.

Minimum-Amplitude Criterion

According to this criterion, the amplitude of deviation should be minimum (see Fig. 38*c*). This criterion applies especially to processes where plant equipment or product may be damaged by even momentary excessive deviations. Here, the magnitude of deviation is more important than its duration. For example, in the melting of some metallic alloys, notably aluminum, even a temporary overshoot of the temperature above the set point can "burn" the metal, greatly reducing its quality. Another process of this kind is the nitration of toluene in manufacturing TNT explosives. Here, if temperatures are allowed to exceed the set point by as little as $5°F$ ($2.8°C$), great acceleration of the exothermic reaction occurs, which can result in total destruction of the plant. For such processes, control actions must be selected and adjusted to produce minimum amplitude deviations.

Effect of Load Changes on Stability

Anything that changes the balance between energy or (material) flow of the input and output of a control loop can be termed a *load change*. The *size, rate, frequency,* and *duration* of load changes greatly affect the quality of control obtained. For example, a control loop with a controller adjustment that gives stable recovery from a *slowly* applied load change can cycle badly if the same load change is *rapidly* applied. Or, if the frequency of load change happens to be in phase with the loop natural frequency, serious cycling may result (see Harmonic Cycling, above). In addition, load changes of large *size* can cause changes in phase shift or gain in nonlinear components of the control loop that can greatly affect its stability. Also, *large* load changes of long *duration* can cause severe cycling when integral action is used. See Start-Up under Integral Action earlier in this article.

Kinds of Load Changes

Load changes are of four general kinds:

1. Supply-load changes
2. Demand-load changes
3. Set-point changes
4. Ambient variable changes

The effects of supply- and demand-load changes are discussed under Process Reaction Curves earlier in this article. See also Fig. 7.

Set-point changes are usually difficult for several reasons.

1. They are generally applied very suddenly.
2. They ordinarily are disturbances to the supply and so must pass through the whole process before they can be measured and corrected.

Ambient variable load changes are those occurring under variable conditions related to the process, such as a change in air temperature surrounding the hot water tank (Fig. 1).

Nonlinearity of Control-Loop Components

Thus far, control loops have been considered as though they were linear, i.e., as though the loop phase shift and gain were constant for all conditions of load and set point. This is an oversimplified concept, because many control-loop components do change their phase shift or gain under different conditions. For example, the current flow through an electrical resistor changes proportionally with the voltage drop across it. Thus, it is a linear device. However, the fluid flow through the resistance of a pipe or an orifice changes exponentially with the pressure drop across it. Thus, it is a nonlinear device. Some sources of nonlinearity are described in the following paragraphs.

Nonlinear Phase Shift

Many control-loop components produce different amounts of phase shift (lag) under different conditions. The following are some practical examples: (1) In liquid-level control, the capacity lag of a tank with nonparallel sides changes as the set point is moved up and down; (2) in flow-rate control processes, the transfer lag changes with the set point, viscosity, and density; (3) in many processes, the dead time (phase lag) changes greatly with load, as in Fig. 1, where the transportation time from tank to thermometer bulb changes directly with the demand for hot water.

Nonlinear Gain

Control-loop components can produce different gain under different conditions. As an example, many controllers have nonlinear (such as square root) scales, so that their proportional band (or gain) is different in every part of their range.

One component of control loops that commonly has a nonlinear gain is the final control element. For example, if a constant pressure drop across the steam valve (Fig. 12) is assumed, the control system can produce proportional changes in flow for each unit change in temperature—strictly linear, proportional, control action. But the pressure differential across control valves is seldom constant; the upstream pressure usually drops and the downstream pressure usually rises with increased valve opening. Thus, the amount of flow change per unit change in the controlled variable usually varies with the valve opening—the proportionality (or gain) of the control system changes with load.

Note that the effect of these nonlinearities in phase and gain accumulates around the loop. Therefore, nonlinearities in one part of the loop can either add to or compensate for nonlinearities in another part. For example, the nonlinearity of the lift-versus-flow characteristics of a valve as shown in curve a of Fig. 39 can be compensated for by a second element in the control loop with the characteristics shown in curve b. This will result in the linear relationship in curve c. The reader is referred to Sec. 19 of this Handbook for a more detailed discussion of valve characteristics. Implementation of curve b is facilitated in the microprocessor control of Fig. 15 by reprogramming.

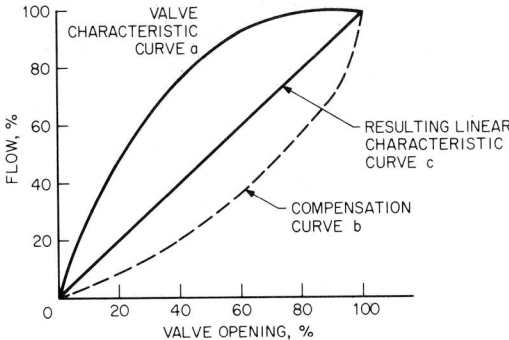

Fig. 39 Valve nonlinearity compensation.

Practical Results of Nonlinearity

Load Changes

As previously described, load changes can affect the dead time (phase shift) and can change the loop gain. Also, note that the phase shift and gain under increasing load conditions may be different than under decreasing load conditions. These facts explain the common observation that controller adjustments which may produce good stability at one load may produce severe cycling at an increased or decreased load.

Set-Point Changes

As explained previously, set-point changes can vary the resistance-capacity lags (phase shift) in processes and also vary the gain. This explains the common observation that controller adjustments which produce good stability at one set point may produce severe cycling at higher or lower set points. The only solution is to compromise, i.e., to select control adjustments which give slightly cyclic recovery under the most unstable load and set-point conditions. This procedure will provide somewhat less than the fastest recovery under the most stable load and set-point conditions.

Stability Changes with Time

With the passage of time, the phase shift and gain of a control loop can gradually change. For example, in Fig. 12 the resistance-capacity lag (phase shift) can slowly increase because of rust and scale accumulations inside and outside the steam coils or on the thermometer bulb. Or friction and lost motion can develop in parts of the controller, diaphragm operator, and valve because of wear, corrosion, and dirt. Particularly common is the development of static friction in control-valve stems, bearings, and packings. Excessive wear of the control-vlave ports or leaks in the output-air system (of a pneumatic controller) can change the proportionality (gain). Thus, it may be found that controller adjustments which produced optimum stability at one time may need to be readjusted at a later time because of slow changes in the process and controller.

Problem of Noise

Random, meaningless signals can occur in many parts of control loops. These signals, often referred to as *noise,* can interfere with the intelligence of the signal. For example, in Fig. 12 the cold water and heated water may not be completely intermixed by the time they reach the thermometer bulb. Slugs of cold water may alternate with hot water to give a rapidly fluctuating, wholly meaningless temperature signal at the bulb. If such a noise bearing signal is allowed to reach the controller, it may result in wild and meaningless corrections to the process, which may cause fluctuating or completely unstable automatic control. Similar noise problems can occur in connection with most signals, e.g., (1) random pulsations in pressure signals, (2) waves in liquid-level signals, (3) turbulence in differential-measured flow signals, and (4) induced alternating or direct currents in thermocouple circuits.

Solutions to Noise Problem

Derivative action produces difficulties where noise exists and, therefore, it should generally not be used in such instances. Noise in the signal also creates difficulties in narrowband proportional control. Therefore, the effect of noise can be partially overcome by the use of wide proportional bands plus integral action.

Two other practical solutions are available:

1. Reduction or elimination of the noise at its source.
2. Filtering or averaging the noise out of the signal. For example, in Fig. 12 the source of the thermal noise can be eliminated by better mixing of the hot and cold water in the tank or by using an

averaging-type thermometer bulb that measures temperature over a considerable length instead of at one point. Other common examples of noise filters include (a) pulsation dampers for pressure signals, (b) stilling wells for liquid-level signals, (c) damping restrictions on mercury manometers for flow signals, and (d) input filters for thermocouples. Integral action "averages out" the noise by its time integrating action and so is useful on "zero-capacity" processes where noise is a problem, such as liquid flow-rate control.

As the signal-to-noise ratio decreases, however, the problem of filtering or averaging out the noise becomes more difficult, if not impossible. Then, the only solution is to reduce or eliminate the noise at its source, e.g., by the use of (1) rotary instead of reciprocating pumps to avoid pulsating pressures, (2) larger mixing tanks or surge tanks, (3) stirrers to obtain a uniform signal, (4) longer pipe runs and straightening vanes in flow measurement, (5) shielding of thermocouple wires against stray voltages, and (6) feedforward, described in more detail later in this article.

It has developed here that a feedback control system must have a built-in negative 180° phase shift (½ cycle) so that its corrections to the input of the process will oppose changes in the output from the process, and also that resistance-capacity lags and dead times around the loop can add phase shifts. Thus, any system with negative feedback will continuously oscillate when disturbed if:

1. Its time lags add up to 180° (½ cycle, giving a total of 360° —one full cycle).
2. The energy gain around the loop is one or more.

It should now be clear:

1. Why any component anywhere in the loop that adds to these time lags lowers the quality of control and, therefore, should be avoided
2. That control components in general should be designed and selected to produce a linear gain through the desired operating range of load and set point.

Worthy of emphasis is that time lags can occur anywhere around a control loop: A given time lag can have just as bad an effect on control quality when it occurs in the control system as when it occurs in the process. Ignorance of this fact has caused many control failures, not because the process was difficult to control but because the control system was.

Adjustment of Automatic Controllers

The quality of regulation obtained from automatic-control systems depends greatly on the adjustments made in their various control actions. To obtain the best control, a *systematic* adjustment method must be used; haphazard adjustments rarely find the best settings. In general, the best settings are those that produce the desired stability. What stability is desirable depends, in turn, on the process involved. See Criteria of Control Quality, described earlier in this article.

Any method of adjustment cannot be expected to produce satisfactory control unless the action employed was correctly selected for the process concerned (see Table 1). The effects of nonlinearities on stability also cannot be overlooked. See Nonlinearity of Control-Loop Components, described earlier in this article. Adjustments should be made under the most unstable conditions of load and set point that normally can occur. The three most widely used methods of adjusting controllers are:

1. Systematic trial method
2. Ultimate-sensitivity calculated method*
3. Reaction-curve calculated method†

---

*Applicable only for proportional, integral, or derivative action controllers.
†See the Ray (1981), Shinskey (1977), and Bristol (1966) references listed at the end of this article.

Systematic Trial Method

This method of controller adjustment requires that the controller and the process be completely installed and operating in their normal manner. The general procedure is to start with broad adjustments of all actions and to narrow them a little at a time, in turn, until the desired stability is obtained. This method uses the closed-loop reactions of the entire controlled system.

Creating Load Changes

In order that the effects of trial adjustments can be known and compared, uniform, repeatable disturbances to the process must be made. This is best done by making small changes in the set point. During this time, all other load changes to the process should be stopped, if possible, to prevent them from producing false results. Set-point load changes should be made as follows:

1. In both the up- and down-scale directions *from the normal set point*
2. To a degree great enough to provide a considerable disturbance, but not so great as to spoil the product, damage the plant, or create intolerable disturbances to associated processes
3. In exactly the same amount each time

Sufficient time must be allowed after each set-point change to observe the full effect of the last adjustment. On very slow-reacting processes, this may require up to 2 or 3 h. A systematic trial adjustment procedure for each action is described as follows.

Adjustment of Two-Position Differential-Gap Control

Start with a wide gap and gradually narrow it a step at a time. Observe the amplitude and frequency of cycling. As the gap is narrowed, the amplitude of cycling decreases and the frequency increases. It must be stressed that constant cycling is characteristic of two-position action controllers. No matter how they are adjusted, they will always cycle.

Adjustment of Time Proportioning Control

Most time proportional controllers have two adjustments: (1) a length of cycle or *pulse-time* adjustment and (2) a width of proportional-band adjustment. The correct pulse time depends on the reaction rate of the process. The pulse time must be short enough so that the controlled variable does not rise or fall too much while the input to the process is on or off. Thus, shorter pulse times must be used on fast-reacting processes. Longer pulse times can be used on slow-reacting processes. The width of the proportional band is adjusted to give the desired stability of control and recovery from load changes. Start with the widest available band adjustment and narrow it in small steps, while creating set-point load changes, until cyclic recovery is approached.

Adjustment of Floating Controllers

Many floating controllers have two mode adjustments: (1) speed of floating and (2) width of neutral zone. Starting with a slow floating speed, increase the speed in small steps, each time creating set-point load changes, until the desired stability is obtained. In general, use slow floating speeds with slow-changing processes, and fast floating speeds with fast-changing processes. Increasing the floating speed increases instability (tendency to cycle). However, decreasing the floating speed permits long-duration deviations following rapid load changes.

Adjustment of Proportional Controllers

Most proportional-action controllers have adjustable proportionality (gain or sensitivity). As the proportional band is narrowed (gain increased), instability increases. However, as the band is widened (gain decreased), the amount of offset error is increased. Start with a wideband (low gain) and narrow it gradually step by step, while observing the behavior following set-point load changes, until

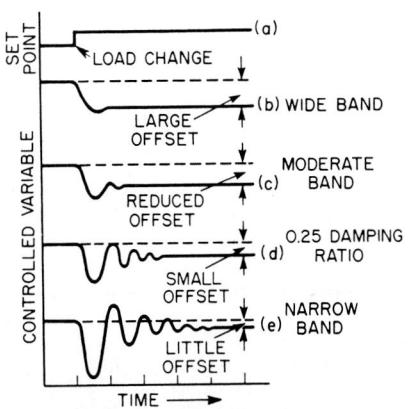

Fig. 40 Effect of narrowing the proportional band on recovery curves following a set-point load change, showing how proportional-band adjustment is a compromise between instability and offset error.

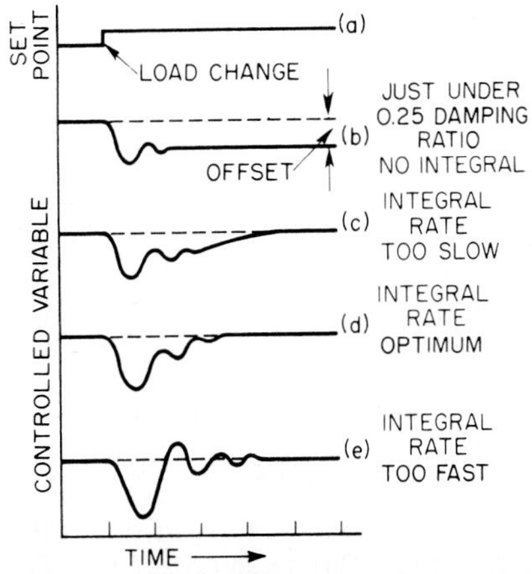

Fig. 41 Effect of increasing the integral rate on recovery curves following a set-point load change. Note the elimination of offset error, and that the period of integral cycling (curve *e*) is shorter than that of proportional cycling (curve *e* of Fig. 40).

the desired stability is obtained, as shown in Fig. 38. (Changes of less than 50% in control adjustments are seldom worth trying.) Figure 40 shows characteristic recovery curves as the proportional band is narrowed.

Adjustment of Proportional-plus-Integral Controllers

Most integral controllers have two adjustments: (1) proportional-band width and (2) integral rate. The trial adjustment involves two steps:

Step 1. With the integral rate at zero (or its lowest value), follow the procedure given above for proportional controllers, narrowing the proportional band until a damping ratio of 0.25 is obtained (Fig. 40*d*). Then increase the bandwidth slightly to obtain a curve like Fig. 40*c*.

Step 2. Allowing the proportional band to remain at this setting, begin with the slowest integral rate and increase the integral rate in small steps, while creating set-point load changes, until cyclic behavior begins to increase. Then reduce the integral rate slightly. (Wide proportional bands and fast integral rates sometimes are used to average out noise.) Figure 41 shows a characteristic recovery curve as the integral rate is increased.

Adjustment of Proportional-plus-Derivative Controllers

The trial adjustment of derivative action controllers involves two steps:

Step 1. With the derivative time at zero, follow the foregoing procedure given for proportional controllers, narrowing the band until a 0.25 damping ratio is obtained (see Fig. 40*d*).

Step 2. Allowing the proportional band to remain at this setting, increase the derivative time in small steps, while creating set-point load changes, until cyclic behavior begins to increase. Then reduce the derivative time slightly. Often the proportional band can then be narrowed a little without decreasing stability. Figure 42 shows characteristic recovery curves as the derivative time is increased.

Adjustment of Proportional-plus-Integral-plus-Derivative Controllers

The trial adjustment of "three-term controllers involves three steps.

Step 1. With the integral rate and derivative time at zero (or minimum settings), follow the foregoing procedure given for proportional controllers, narrowing the band until just less than a 0.25 damping ratio is obtained (see Fig. 40c).

Step 2. Allowing the proportional band to remain at this setting, slowly increase the integral rate as previously described until the point of instability is approached.

Step 3. Allowing the proportional band and integral rate to remain at these settings, slowly increase the derivative time as previously described. Again the proportional band often can be narrowed with improved results after the addition of derivative action.

Excessive Adjustments

An excessive adjustment to any control action—proportional, integral, or derivative—will cause cyclic recovery. The practical difficulty is to know which control action is in excess. Such cycling may be termed *proportional, integral,* or *derivative cycling.* The following general rules help to distinguish them:

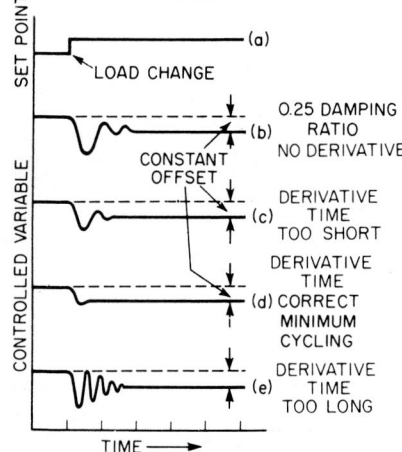

Fig. 42 Effect of increasing the derivative time on recovery curves following a set-point load change. Note that offset error is not affected, and that the period of "derivative cycling" (curve e) is shorter than that of "proportional cycling" (curve e of Fig. 40).

1. Integral cycling has a relatively long period.
2. Proportional cycling has a relatively moderate period.
3. Derivative cycling has a relatively short period.

Compare curves *e* in Figs. 40 to 42.

Ultimate-Sensitivity Method

A more exact way of using this self-excited closed-loop analysis is the ultimate-senstivity method developed by Ziegler and Nichols (1942). This method permits calculating all three controller adjustments from data obtained in a simple quick test of control-loop characteristics. With integral and derivative adjustments at their lowest value, the proportional band is narrowed (gain or sensitivity increased), while creating small set-point changes, until the process *just begins* to cycle continuously. The period of cycling P_u (in minutes) of the oscillations at this ultimate proportional band PB_u is noted. The *controller adjustments that will produce approximately* a 0.25 amplitude ratio (Fig. 38a) are calculated as follows:

1. Proportional-action controllers:

$$\text{Proportional band (\%)} = 2PB_u \tag{1}$$

2. Proportional-plus-integral-action controllers:

$$\text{Proportional band (\%)} = 2.2PB_u \tag{2}$$

$$\text{Integral rate (R/min)} = \frac{1.2}{P_u} \tag{3}$$

3. Proportional-plus-integral-plus-derivative-action controllers:

$$\text{Proportional band } (\%) = 1.6PB_u \qquad (4)$$

$$\text{Integral rate } (R/\text{min}) = \frac{2}{P_u} \qquad (5)$$

$$\text{Derivative time (min)} = \frac{P_u}{8} \qquad (6)$$

Reaction-Curve Method

The general procedure in the reaction-curve method of controller adjustment is to open the control loop just before the final control element and to create a small, sudden step change in process input. From the shape of the resulting reaction curve are obtained two factors characteristic of that process, from which controller adjustments are calculated by simple equations.

Creating Step-Input Changes

The method of making a reaction-curve adjustment can best be explained by reference to the typical temperature control process of Fig. 43. The control loop is opened by disconnecting the controller output air pipe from the diaphragm-valve operator. In its place are connected an air pressure regulator and a pressure gage.

While the process is being operated in balance at its normal set-point value under manual control, a sudden, small, sustained increase or decrease in air pressure to the diaphragm operator is made. The resulting reaction curve of the process is taken, preferably by a recording-type measuring instrument.

Analyzing the Reaction Curve

Figure 44 shows a typical reaction curve resulting from such a step input change. The controllability of the loop can be assessed with reasonable accuracy from two characteristics of its reaction curve: (1) the reaction rate R and (2) the time lag L. Figure 44 shows how these two factors are obtained

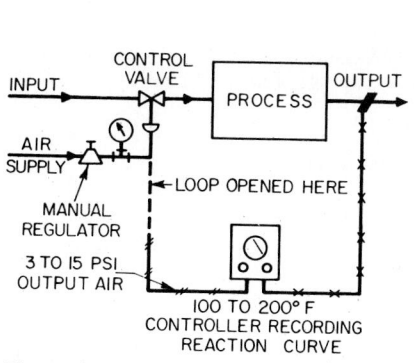

Fig. 43 Equipment connections for producing open-loop reaction curves for a typical air-controlled process. A step change in the process input is created by changing the air pressure supplied to the valve motor.

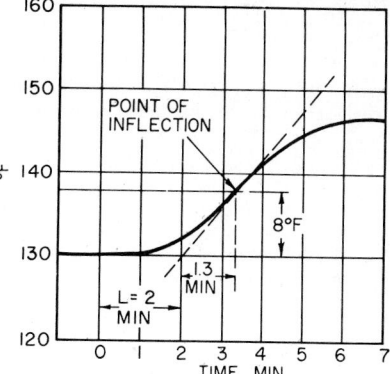

Fig. 44 Typical reaction curve produced by a step change in input, showing how time lag L and reaction rate R are derived from the curve.

from the reaction curve. A straight line is drawn tangent to the point of inflection, as shown. R is then the slope of this tangent straight line.

$$R = \frac{\% \text{ change in variable}}{\text{time (min)}}$$

The second factor, the time lag L, is the time in minutes between the step change and the point where this tangent straight line crosses the initial value of the controlled variable. Calculations are as follows:

1. Proportional-action controllers:

$$\text{Proportional band (\%)} = \frac{100RL}{\Delta P} \tag{7}$$

where ΔP is the percent change in the final-control-element (control-valve) position used to produce the raction curve.

2. Proportional-plus-integral-action controllers:

$$\text{Proportional band (\%)} = \frac{110RL}{\Delta P} \tag{8}$$

$$\text{Integral rate (repeats per min)} = \frac{0.3}{L} \tag{9}$$

3. Proportional-plus-integral-plus-derivative-action controllers:

$$\text{Proportional band (\%)} = \frac{83RL}{\Delta P} \tag{10}$$

$$\text{Derivative time (min)} = 0.5L \tag{11}$$

$$\text{Integral rate (repeats per min} = \frac{0.5}{L} \tag{12}$$

Sample Calculations

Suppose that in the air-operated controlled process of Fig. 43, after the temperature has been balanced at 130° under manual control, the manual pressure regulator is changed suddenly at time zero to increase the pressure on the valve motor by 2 psi. Then ΔP, the percent change in valve position, will be 2 psi in 12 psi (3 to 15 psi) [14 kPa in 84 kPa (20 to 100 kPa)] or 16.6% of full valve travel. The straight line drawn tangent to the point of inflection (Fig. 44) (point of steepest slope) intersects the 130° initial temperature at a time of 2 min. Thus, the time lag L is 2 min. The reaction rate R is the slope of this straight line. The temperature rose 8° or 8% of the 100° full-scale in 1.3 min. Thus,

$$R = \frac{8\%}{1.3 \text{ min}} = 6.15\% \text{ per min}$$

Then, from Eqs. 8 and 9 the correct settings for a proportional-plus-integral-action controller on this process are

$$\text{Proportional band} = 110 \times 6.15 \times 2/16.6 = 82$$

$$\text{Integral rate} = 0.3/2 = 0.15 \text{ repeats/min}$$

Limitations of the Reaction-Curve Method

Ultimate sensitivity and the foregoing equations were empirically derived from the characteristics of many industrial control loops. Some control systems have characteristics making these equations inapplicable. Some processes cannot tolerate cycle conditions even during tests. The adjustment figures obtained are valid only for load and set-point conditions existing when the cyclic or reaction curves were made. Thus, the curves should be taken at the most unstable load and set-point conditions of the process. On many processes, it is difficult to avoid disturbances to the process long enough to obtain respresentative curves resulting from the step change only.

Multielement Control (Multivariable, Multiple-Loop)

The previous discussion covered feedback control loops where one manipulated variable is controlled from one measured variable. Multielement control occurs when two or more input signals jointly affect the action of the control system. Examples of multielement control are:

1. Cascade control
2. Ratio control
3. Auctioneering control (override control, limiting control)

Cascade Control

In cascade control, the output of one controller is the set point for another controller. Each controller has its own measured variable with only the primary controller having an independent set point and only the secondary controller providing an output to the process. Both the primary and secondary controllers can be packaged together in one case. Figure 45 illustrates cascade control of a heat exchanger.

When the output of the primary controller is connected directly to the control valve, the control system consists of single-loop control. When a long time lag exists between a change in the manipulated variable and its effect on the measured variable, control is difficult, if not impossible, if significant supply-side load changes occur. By the time the effects of a steam flow change influence the hot water temperature, a considerable energy change has occurred within the heat exchanger, which will continue to drive the hot water temperature away from the set point long after a correction is made by the steam flow valve. This results in cycling of the measured variable around the set point and/or unacceptable delays in the return of the measured variable to the set point.

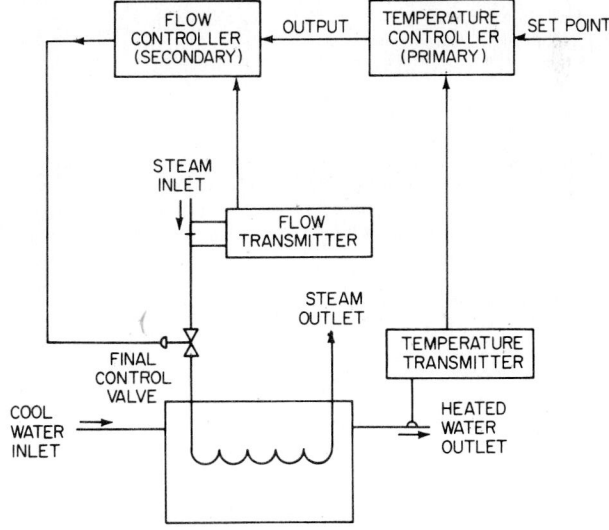

Fig. 45 Cascade control of a heat exchange process.

The addition of a second (secondary) controller whose measured variable is steam flow allows steam-flow variations to be corrected for immediately before they can affect the hot water temperature. By connecting the output of the first (primary controller) to provide the set point to the secondary controller, gradual excursions of the hot water temperature from the set point are also corrected.

Therefore, in cascade control, at least two control loops exist: the primary (slow or outer loop), consisting of the primary (master) temperature controller, the process temperature, the control valve, and the process; and a secondary (fast or inner loop) consisting of the secondary (slave) steam-flow controller, its process variable (steam flow), the control valve, and the process.

As shown in Fig. 46, the process is separated into two parts: A and B. The A part of the process contains the external disturbances, while B contains the long time constants. The slave or secondary

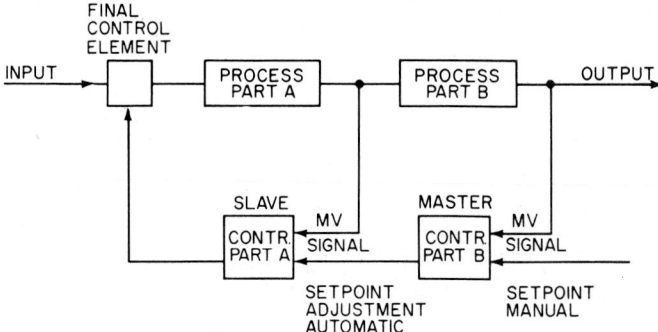

Fig. 46   Cascade control system showing the two loops involved.

loop is used to reduce the effects of supply-side disturbances in part A. Usually only proportional control action is required in the secondary controller. The master or primary controller senses the desired controlled variable and usually includes proportional-plus-integral control action, plus derivative where required.

When a cascade system is placed in operation, both controllers are intially set on manual during the start-up. After the process stabilizes, the secondary controller is placed on automatic and the correct control action settings are determined as previously described. When the secondary controller operation on automatic is satisfactory, the primary controller is then placed on automatic and its optimum control action settings are determined.

Ratio Control

In ratio control, a predetermined ratio is maintained between two or more variables. Each controller has its own measured variable and output to a separate final control element. However, all set points are from a master primary signal that is modified by individual ratio settings.

A typical application of ratio control is the control of the fuel flow/airflow ratio in a combustion control system such as that shown in Fig. 47. Other applications are cement kiln speed versus slurry flow control, propane gas-versus-airflow mixing controls, natural gas flow versus bottled gas flow mixing controls, steam flow versus airflow boiler control, and miscellaneous liquid blending process control.

With reference to Fig. 47, the gas-flow signal, in British thermal units per hour, is multiplied by the desired ratio and compared to the airflow signal. Any difference is indicated on the airflow controller. This difference is characterized in accordance with the proportional, integral, and derivative adjustments and produces an output that controls the airflow to eliminate any deviation from the desired ratio.

The ratio multiplying circuitry may be separate from the airflow controller as shown by the dashed outline in Fig. 47. This allows both fuel and airflow to be placed on manual or automatic control independent of each other, as well as in ratio control.

A second application of ratio control is shown in Fig. 48. In the manufacture of lead a sinter

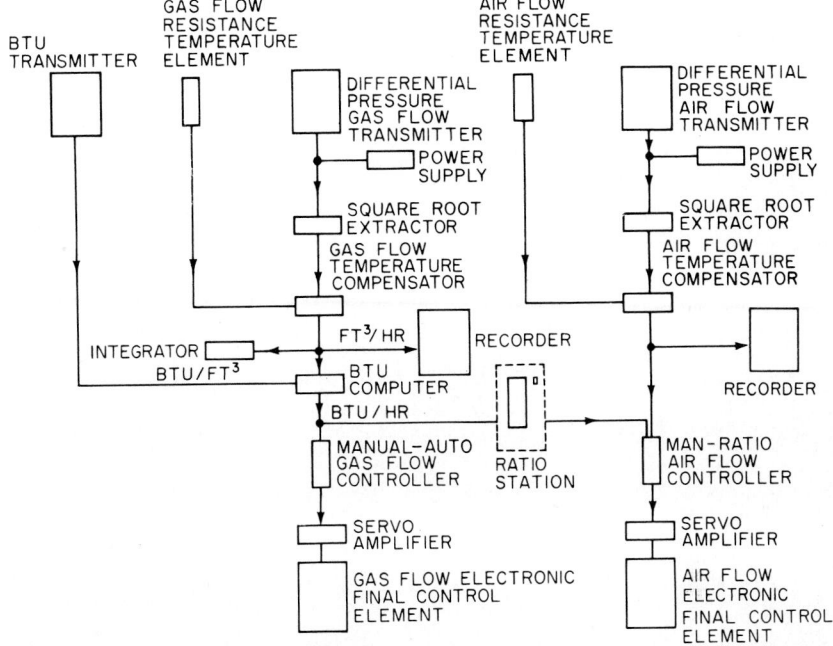

Fig. 47   Gas flow/air flow ratio combustion control system.

must be prepared by mixing lead concentrate, produced by crushing, grinding, and flotation of the raw ore, with slag forming fluxes such as iron ore, limestone, and silica. This charge is fed to a sintering furnace where sulfur is burned off and the charge fuses into a hard homogeneous sinter. This sinter is crushed into lumps and fed into the blast furnace along with coke and slag. The molten lead from the blast furnace goes to the refining plant where copper, silver, and slag are removed, leaving 99.9% pure lead. After refining, the lead is formed into 100-lb pigs or 1-ton blocks by continuous casting methods.

In order to properly proportion the six ingredients making up the charge to the sintering furnace and the three ingredients making up the charge to the blast furnace, ratio control is used.

The sinter charge demand is set from a manual loading station located on the sinter furnace panel and calibrated in tons per hour. As the sinter furnace demand changes, the operator alters the setting of the sinter charge demand loading station. This signal is the set point for the transfer belt loading controller where it is compared to the tons-per-hour measured variable signal received from the no. 1 transfer belt weight rate transmitter. The weight rate signal is obtained from a tons-per-foot weight measurement and a feet-per-hour rate measurement, producing a weight rate signal proportional to tons per hour which is the product of these two signals.

The output of the transfer belt loading controller is a function of the desired rate of charge delivered to the sinter furnace and the actual rate. As the sinter furnace demand increases, the output of the transfer belt loading controller also increases, causing an increase in the speed of all the ingredient feed belts. Increasing the speed of the ingredient feed belts loads more charge on the transfer belts and therefore delivers more charge to the sinter furnace.

The output of the transfer belt loading controller is the set point for the recycle feed controller. The recycle sinter is the primary feed, and all the other ingredients in the total charge are added as percentages of the recycle sinter feed rate at any given instant. Each of the feeders, except the recycle sinter, has a ratio controller whose scale is in terms of percentage of the recycle sinter instantaneous feed rate.

Assuming an increase in the sinter charge demand, the set point for the transfer belt loading controller is changed by the sinter furnace operator. The output of the transfer belt loading controller

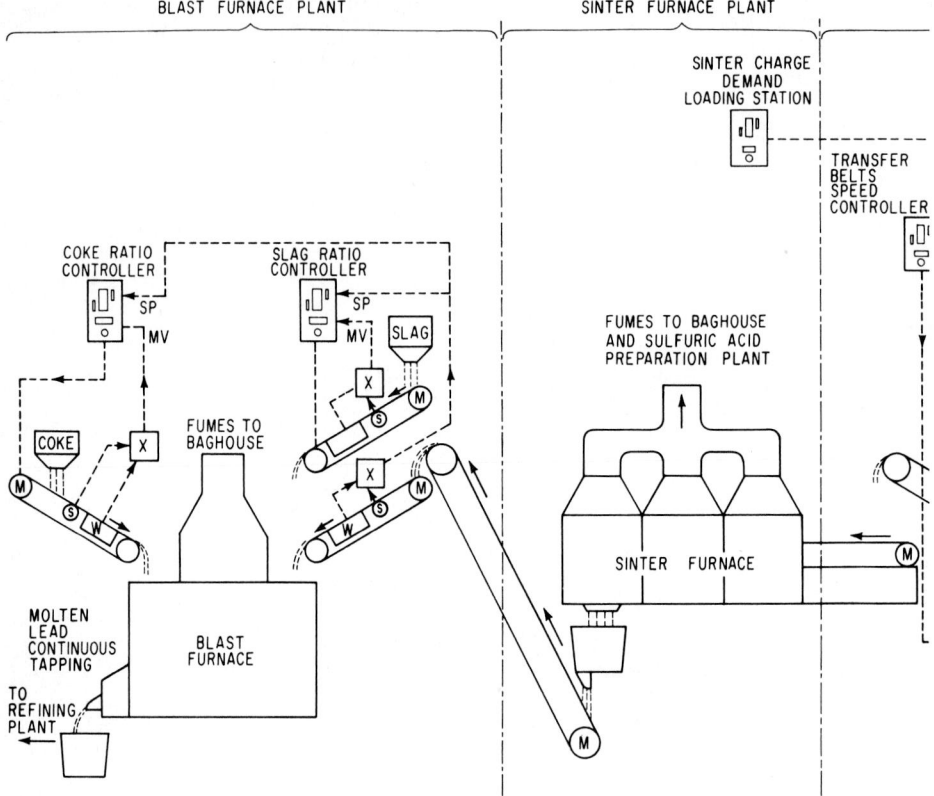

Fig. 48 Use of ratio control in a lead refining process.

increases, thereby increasing the set point of the recycle sinter feed controller. This causes the recycle sinter feed to increase so as to increase the charge being delivered to the no. 1 transfer belt weighing system. As the recycle sinter feed rate is the set point for the ratio controllers, the other ingredient feeds increase at the same rate, keeping the proportion of each at the desired percentage. The transfer belt speed controller holds the transfer belt speed constant, regardless of the belt loading.

Auctioneering Control (Override Control, Limiting Control)

In auctioneering control, either the highest or the lowest signal from two or more input signals is automatically selected. Common examples include the following:

1. *Suction and Discharge Pressure Compressor Control.* The discharge control valve is normally regulated from the discharge pressure (Fig. 49). However, if the suction pressure drops below its set point, control is transferred to the suction pressure controller. This prevents excessive suction on the supply side, from demand exceeding supply, with resultant compressor damage.

2. *Flow and Pressure Control of Gas Distribution Systems.* The distribution control valve is normally regulated from the discharge pressure (Fig. 50). However, under high-demand conditions, control is transferred to the flow controller. This arrangement limits the maximum flow rate and the maximum pressure in a distribution system.

3. *Temperature or Vacuum and Differential Pressure Control for a Packed Distillation Column* (Fig. 51). The column is controlled either from a vacuum set point, with the steam flow

SINTER CHARGE PREPARATION PLANT

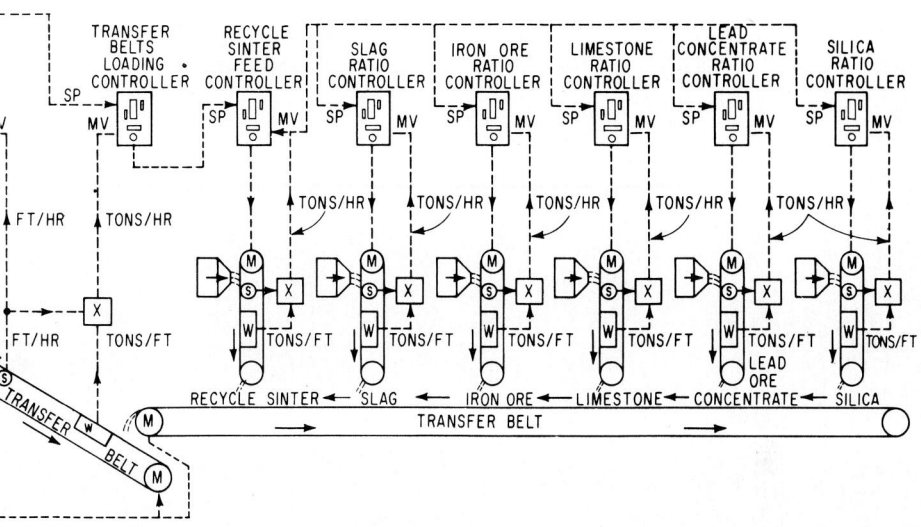

on manual for constant heat input, and a differential-pressure override to prevent flooding; or regulation is from a temperature set point with the vacuum on manual for constant vacuum with a differential-pressure override to prevent flooding.

Feedforward Control

This is control action whereby information concerning one or more conditions that can disturb the controlled variable are converted into corrective action to minimize deviations of the controlled variable.

Feedback control, by definition, requires that a deviation exist between the measured variable and the set point before control action can take place. The feedback controller changes its output until the deviation (error signal) between the set point and the measured variable is negligible, which is essentially a trial-and-error process characterized by the oscillatory nature that feedback control often exhibits.

The cascade-control system shown by Fig. 45 would be adequate until such time as the inlet water pressure and temperature fluctuate. This condition requires the measurement of both the inlet water temperature and flow with both variables combined to control the steam flow rate—Fig. 52.

The computer is designed to determine the heat energy required to be added to the cool water to bring it to the desired temperature at the outlet and to allow the correct amount of steam to enter the heating coils. The two supply-side load changes, flow and temperature, are measured and provide

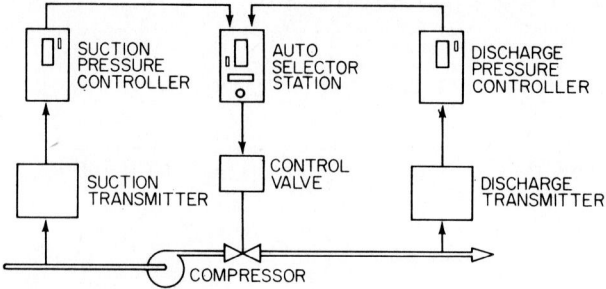

Fig. 49 Suction and discharge pressure compressor control system.

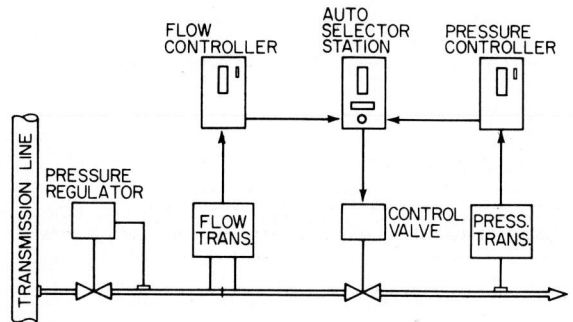

Fig. 50 Flow and pressure control of a gas distribution system.

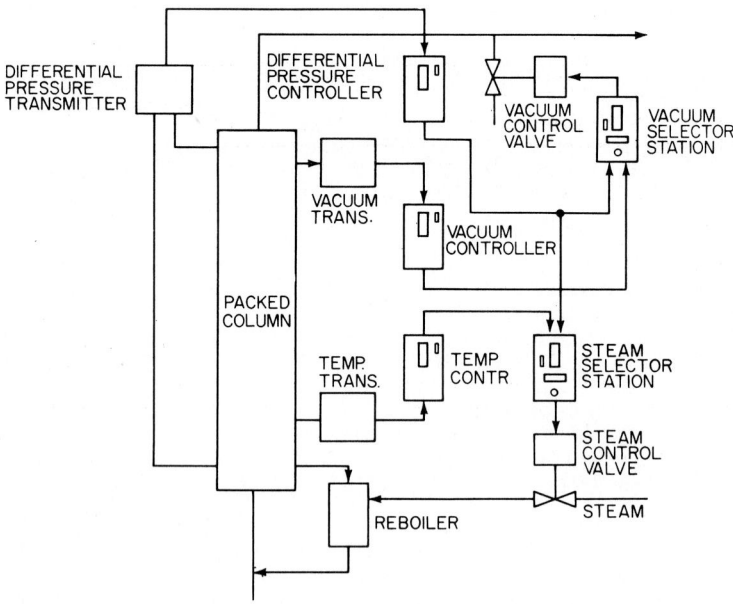

Fig. 51 Temperature or vacuum and differential-pressure control system for a packed distillation column.

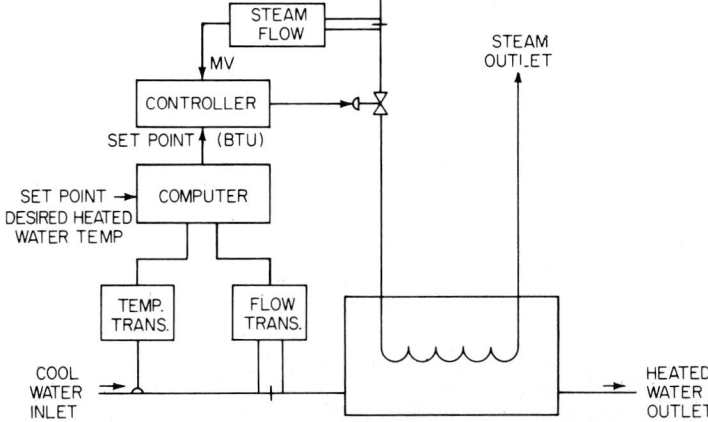

Fig. 52   Feedforward control of a heat exchange process.

inputs to the computer controller. The computer controller calculates the difference between the desired manually set heated water temperature and the measured cool inlet water temperature. This temperature difference is multiplied by the inlet water flow, thereby providing the measure of British thermal units required. The output control signal to the valve is computed to allow the correct amount of steam (British-thermal-unit input) to enter the heating coils.

   In a true feedforward control system the controller variable is not used as one of the inputs. However, it is usually difficult, as well as not practical, for the computer-controller to include corrections for all variables such as radiation losses, changes in heat-transfer characteristics caused by scaling, and inaccuracies in temperature and flow measurement. Therefore, it is customary to include a feedback loop to prevent inaccuracies in the feedforward loop from adversely affecting the process. Refer to Fig. 53.

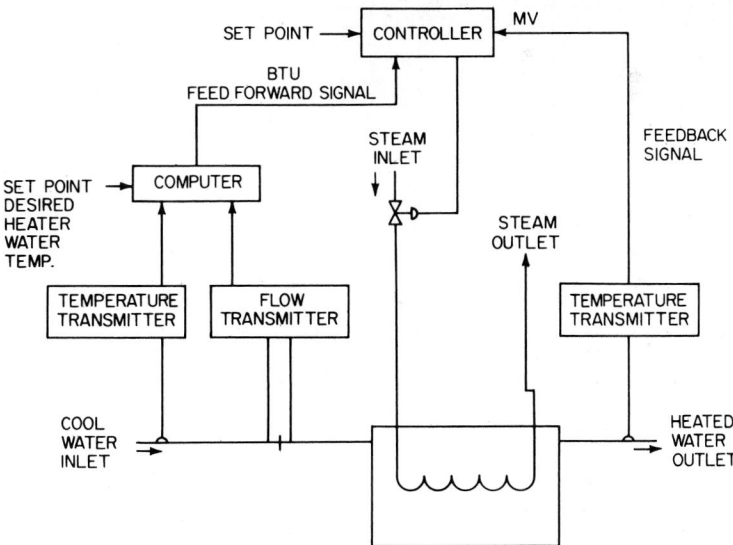

Fig. 53   Feedforward control with feedback of a heat exchange process.

The feedforward signal is scaled by gain and bias adjustment in the controller and combined with the controller output signal which has previously been characterized by the control action—proportional, integral, derivative—on the error, or deviation between the set point and the measured variable.

Multiple-Variable Interactive Control

Many processes have two or more output variables to be controlled. Normally the control system designer can choose one input variable which can be manipulated to control one of the output variables. By repeating this for each output variable a control loop is created for each output variable. If these loops do not have significant interactions, each loop can be designed and tuned by the methods described earlier in this article.

A method for evaluating the degree of interaction between control loops involves use of the relative gain concept or the Bristol array.* A two-output-variable interactive system is shown in Fig. 54. The Bristol array for this process is given in Fig. 55. The concept is similar for a system of any size. Figure 55 is filled out from either a system test or an analytical analysis. In each square of the table, the numerator is the ratio of the change in an output variable to the change in an input variable with all other *input* variables constant. This ratio is divided by the ratio of the same pair of output and input variables with all other *output* variables held constant (i.e., all other loops under perfect control). Thus, when there is zero interaction between the inputs of one loop and the outputs of another loop, the ratios in all loops on the diagonal, box (1, 1) and box (2, 2), will be one. The off-diagonal terms, boxes (1, 2) and (2, 1), will be zero. No matter how many boxes there are in Fig. 55, each row and each column will add to one. Thus, in a system with any number of loops, if the diagonal terms are one, the off-diagonal terms must all be zero.

The relative gain approach is based on steady-state gains. Although there is no generally accepted measure of dynamic interactions, the Ray (1981) reference contains a good discussion of noninteracting control.

A practical problem with the implementation of an interacting control is that it is difficult to operate the system on manual control when one of the loops fails. It is generally preferable, if possible, to use noninteracting control loops for this reason.

*See the Ray (1981), Shinskey (1977), and Bristol (1966) references listed at the end of this article.

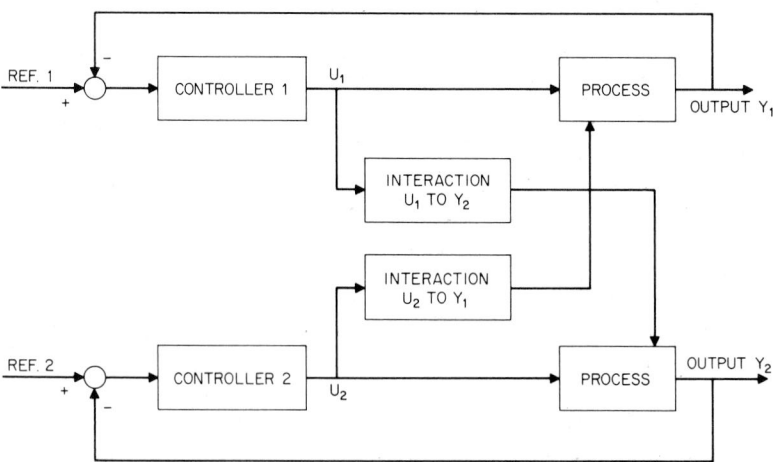

Fig. 54 A two-output variable interactive system.

Box (1,1)	Box (1,2)
$$\frac{\text{change in } Y_1}{\text{change in } U_1}\ \left(U_2 = \text{constant}\right)$$ $$\frac{\text{change in } Y_1}{\text{change in } U_1}\ \left(Y_2 = \text{constant}\right)$$	$$\frac{\text{change in } Y_1}{\text{change in } U_2}\ \left(U_1 = \text{constant}\right)$$ $$\frac{\text{change in } Y_1}{\text{change in } U_2}\ \left(Y_2 = \text{constant}\right)$$
Box (2,1)	Box (2,2)
$$\frac{\text{change in } Y_2}{\text{change in } U_1}\ \left(U_2 = \text{constant}\right)$$ $$\frac{\text{change in } Y_2}{\text{change in } U_1}\ \left(Y_1 = \text{constant}\right)$$	$$\frac{\text{change in } Y_2}{\text{change in } U_2}\ \left(U_1 = \text{constant}\right)$$ $$\frac{\text{change in } Y_2}{\text{change in } U_2}\ \left(Y_1 = \text{constant}\right)$$

Fig. 55   Bristol array for a two-loop system.

Self-Tuning Controllers

In many applications it is desirable to have the control system automatically adjust its parameters, such as gain, reset time, and rate, in order to give an optimal response. Such a control system is called *self-tuning*. Applications where it is particularly desirable are processes where the parameters change significantly with different operating conditions or where plant personnel are inexperienced at tuning control systems.

Self-tuning controllers adjust their parameters like a trained person would in order to obtain an exponential-type response to a plant disturbance. Figure 56 demonstrates the operation of a self-tuning controller. Between points t_1 and t_4, the system responded to two small, deliberately introduced square wave inputs in order to identify the control-loop dynamics. At t_4 the process model was completed; the operator transferred the parameters to the control loop and initiated the self-adaptive mode which automatically enters the updated parameters as they are calculated.

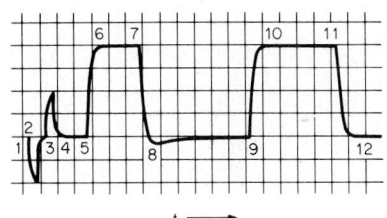

Fig. 56   Facsimile of an actual chart record demonstrating the operation of a self-tuning controller, with intentional upsets at t_2 and t_3 used to identify process characteristics.

At t_5 the set point was manually lowered, and the process variable responded smoothly. At t_6 the process response time increased, causing an overshoot at t_8. Before t_9, the control parameters were again updated, resulting in the smooth process response at t_{10} and t_{12}. For a more detailed description of self-tuning controllers, reference is made to the Andreiev (1981) articles listed at end of this article.

GENERAL CONTROL-SYSTEM DESIGN PROCEDURE

Thus far in this article, specific design procedures have been covered for most types of equipment being controlled. In actual practice, many other considerations must be covered to develop a well-designed control system. These considerations include the effects of the environment in which the control system must operate, the basic reliability and availability of both the process being controlled and the control system itself, the adequacy of manual operation when the control system fails, protective functions to ensure that the process is shut down before specific process variables exceed critical values, and sufficient recording equipment to determine which variables deviated from normal and in what order.

To help ensure that all necessary considerations are included in control-system design, the following procedure is suggested. This procedure is built around the generation of three key documents:

1. Description of equipment operation
2. Control system specification
3. Control system test specification

For the successful design of a control system, a complete understanding of both normal and abnormal equipment operation is required. Modern equipment and plants of any significant size are designed by a team of engineers, each specializing in a different area. A practical method for the control engineer to have both a qualitative and quantitative measure of the operation and limitations of the equipment is to generate a *description of equipment operation*. This document should be written and approved by the plant designers or jointly written by the control designer and the plant designers. It describes the start-up from a cold plant condition through normal operation to plant shutdown. *All* abnormal operating conditions and the limits on *all* critical variables must be described so that the control designer is aware of the acceptable operating boundaries of the plant.

The *control system specification* spells out and specifies the functional characteristics of the system and to some extent the control hardware. Utilizing the *description of equipment operation,* the control system designer determines what control functions are needed to implement safe, effective operation of the plant. The designer may employ simulation if the plant is complex, or simple analytical analysis if the plant is simple. This specification normally includes not only the control functions but all necessary protection and monitoring systems. Specifically the following would be included:

Controlled variables, measurement accuracies required, and the required scanning rate if the system is digital

Manipulated variables

Monitored variables

Functional description of system

Operator interfaces (panels, cathode ray tube, etc.)

Interfaces with other systems

Start-up and shutdown control

Protection systems

Code requirements for the system

Environmental requirements

Reliability (MTDF, MTTR, etc.) targets

The foregoing items should be approved by the equipment user. This helps to ensure acceptance of the system by the ultimate user.

In order to verify that a control system meets the *control system specification,* it is advisable to test the system before it leaves the vendor's factory, as well as in the field. This may involve utilizing a model of the equipment for pre-start-up tests, since there usually is only a limited number of tests which can be made once the plant starts up—because of physical and business constraints. A good control system test specification should ensure that requirements for all the functions are met by the actual control system.

Utilizing the foregoing documents, the control system designer will have a road map which, when used together with the techniques described in this and the next article, will result in a system that meets the expectations of the user.

REFERENCES

Andreiev, N.: "Autotuning Temperature Controller Optimally Tunes Up to Four Loops," *Control Eng.,* vol. 28, no. 10, 1981, p. 107.

Andreiev, N.: "A New Dimension: A Self-Tuning Controller That Continually Optimizes PID Constants," *Control Eng.,* vol. 28, no. 8, 1981, p. 84.

Bibbero, R. J.: *Microprocessors in Industrial Control,* Instrument Society of America, Research Triangle Park, N.C., 1982.

Borut, R. W. (Ed.): "Centralized Control," 1st ISA Control Centers Symposium, New Brunswick, N.J., 1970, Instrument Society of America, Research Triangle Park, N.C.

Bristol, E. H.: "On a New Measure for Multivariable Process Control," *IEEE Trans. Autom. Control,* vol. AC-11, p. 133, 1966.

Bristol, E. H.; "Recent Results on Interaction in Multivariable Process Control," Seminar on Recent Developments in Control and Estimation, American Institute of Chemical Engineers, Newark, N.J., April 1980.

Bristol, E. H.: "Pinned Zeros: An Introduction to Multivariable Zeros and Dymanic Interaction Analysis," ISA Chemical Symposium, Newark, N.J. April 1980.

Bristol, E. H.: *Strategic Design: A Practical Chapter in a Textbook on Control,* Joint Automatic Control Conference, San Francisco, 1980.

Bublitz, A. T., Mouly, R. J., and R. L. Thomas: "Statistical Feedback Squeezes Product Variations," ISA J., vol. 13, no. 11, 1966, p. 55.

Claridge, R. E., and W. J. Berk: "New Tools for the Instrument Engineer," *Chem. Eng.,* May 1952, p. 173.

Corripio, A. B.: *Tuning Techniques for Industrial Systems,* Instrument Society of America, Research Triangle Park, N.C., 1982.

Eckman, D. P.: *Principles of Industrial Process Control,* Wiley, New York, 1945.

Gould, J. K.: *Controllers and Control Elements,* Instrument Society of America, Research Triangle Park, N.C., 1983.

Hammond, P. H. (ed).: "Theory of Self-Adaptive Control Systems," Proceedings of the 2d IFAC Symposium on Theory of Self-Adapting Control Systems, Teddington, England, 1965.

Higgins, S. P., Jr., and G. W. McKnight: "Practical Application of Servomechanism Techniques to a Process Control Problem," *Instruments,* vol. 25, no. 6, June 1952.

ISA: "Process Instrumentation Terminology," *Standard ISA-S51.1,* 1976; "Theory of Automatic Control in Simple Language," *Tape No. U70AC,* 1983; "Feedforward—New Technologies," *Tape No. U70FNT2,* 1983; "Automatic Process Control," *Film No. U101OR,* 1983; "Principles of Automatic Control," *Publ. 1108-7,* 1968; "Understanding Process Control: An Overview," *IRP110,* 1983; "Introduction to Process Control," *ITTP Training Program Tape,* 1983. Foregoing available from Instrument Society of America, Research Triangle Park, N.C.

Llewellyn, J. A., and R. Gilbert: *Principles and Applications of Digital Devices in Measurement Control Systems,* Instrument Society of America, Research Triangle Park, N.C., 1983.

MacMillen, R. H.: *An Introduction to the Theory of Control in Mechanical Engineering,* Cambridge University Press, New York, 1951.

Minar, E. J., Bollinger, L. E., and J. G. Truxal (eds.): "Optimizing and Adaptive Control," Proceedings of the 1st International Symposium on Optimizing and Adpative Control, Rome, 1962.

Mollenkamp, R. A.: "Introduction to Automatic Control," *ISA IRP400,* Instrument Society of America, Research Triangle Park, N.C., 1982.

Mollenkamp, R. A.: "Practical Control Applications," *ISA IRP420,* Instrument Society of America, Research Triangle Park, N.C., 1983.

Murrill, P. W.: *Fundamentals of Process Control Theory (1981),* Instrument Society of America, Research Triangle Park, N.C., 1982.

Peters, J. C., and T. R. Olive: "Fundamental Principles of Automatic Control," *Chem. Met. Eng.,* May 1943, p. 97.

Raven, F. H.: *Automatic Control Engineering,* McGraw-Hill, New York, 1961.

Ray, W. H.: *Advanced Process Control,* McGraw-Hill, New York, 1981.

Rosenbrock, H. H.: "The Future of Control," *Automatica,* vol. 13, no. 4, 1977, p. 389.

Shinskey, F. G.: *Distillation Control,* McGraw-Hill, New York, 1977.

Shinskey, F. G.: "Energy Management Systems for Cogeneration," *Control Instrum.,* July–August, 1978.

Shinskey, F. G.: *Process Control Systems,* 2d ed., McGraw-Hill, New York, 1979.

Shinskey, F. G.: "Controlling Multivariable Processes (1981)," *ISA Publ. 1529-5,* Instrument Society of America, Research Triangle Park, N.C., 1982.

Staff: *Instruments and Process Control,* New York State Vocational Arts, Delmar Publishing, Albany, N.Y., 1945.

Thomas, R. H., and P. P. Tong: "Discrete-State Feedback: A Route to Minimum Control," *Instrum. Technol.,* vol. 14, no. 4, 1967, p. 59

Truxal, J. G., (ed.): *Control Engineer's Handbook,* McGraw-Hill, New York, 1958.

Tucker, G. K., and D. M. Wills: *A Simplified Technique of Control Systems Engineering,* Honeywell Inc., Fort Wasington, Pa., 1958.

Williams, T. J.: *Use of Digital Computers in Process Control,* Instrument Society of America, Research Triangle Park, N.C., 1983.

Process Control Techniques

by
Stephen P. Higgins, Jr.,* and Joe M. Nelson†

This article summarizes the basic control concepts and techniques which have been successfully applied to the design and analysis of analog and digital process control systems.

The presentation is in five parts covering the following topics:

1. System characteristics and representation
2. Basic control concepts
3. Continuous and sampled-data control systems
4. Stability
5. System design and control strategies

Generally useful information is presented, and sources of more detailed treatments and application examples are given.

Excellent texts are available covering practical applications of process control techniques at var-

*Retired. Previously Senior Principal Software Engineer, Advanced Systems, Honeywell Inc., Process Management Systems Division, Phoenix, Ariz.

†Section Head, Applications Systems Division, Honeywell Information Systems, Honeywell Inc., Billerica, Mass.

ious levels of sophistication. (See, for example, Refs. 1 to 4.) Several important topics are best covered by specific examples and are not discussed here. They are:

1. Batch process control (see Ref. 5)
2. Reset windup (see Refs. 6 and 7)
3. Protective controls—(override and cutback) (see Refs. 8 and 9)
4. Distillation column control (see Refs. 10 and 11)
5. pH control (see Ref. 12)

Reference 5 also describes specific instrumentation for a wide range of process applications.

SYSTEM DEFINITION

The first step in the design of a control system is to develop a thorough understanding of process operation and control objectives. The next step is the development of a model for the system. System boundaries are defined and inputs and outputs are specified. An example is shown in Fig. 1; the formal representation is shown in Fig. 2. The relationships are often expressible in terms of algebraic equations and Laplace transforms, or z transforms in the case of sampled-data systems.

When changes occur in a manipulated variable u or in a disturbance v, one or more of the measured variables x will be affected. The effect may be only a temporary transient, or it may also result in a sustained (steady-state) change in x. The *state* of the process for control purposes is defined by the set of measured variables x_i ($i = 1, 2, \ldots, n$). The process model within the system boundaries specifies the dynamic and steady-state effect of manipulated variables u and disturbances v on the measured variables x_i. A generalized diagram of the overall system with the control and the measurement subsystems added is shown in Fig. 2.

The reference inputs r are also disturbances in the sense that they command a new state for the process. In the simplest case, the r's are *set points* for specific controlled variables. The k's represent parameter adjustments to the controller. In the simplest case, they are manually set and represent

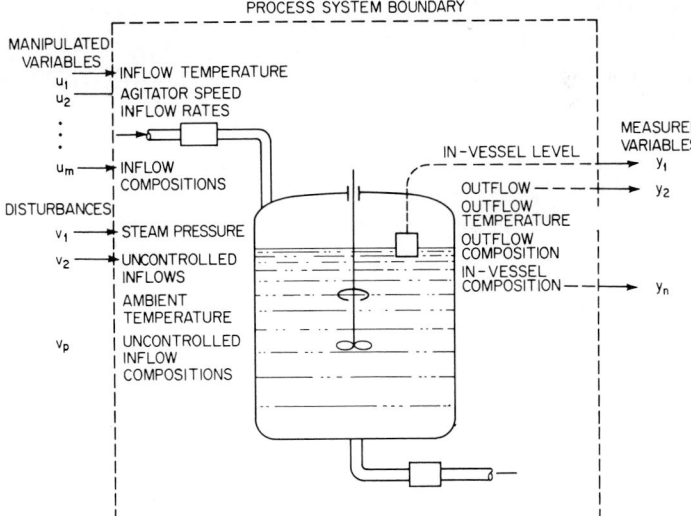

Fig. 1 System definition for a blending system.

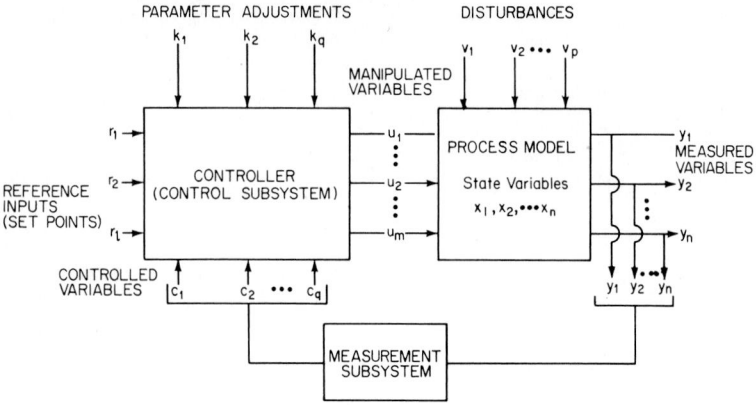

Fig. 2   Control system diagram for the system of Fig. 1. (Symbols follow ANSI Standard MC85.1M-1981, Ref. 13.)

proportional gain and time-parameter adjustments. In more sophisticated systems, it may be necessary to adjust reference inputs and/or the parameters k by means of optimizing or adaptive control functions.

Measurements and Manipulated Variables

The nonideal aspects of measured variables and final control elements are often important. Sensor input-output nonlinearity causes undesired changes in feedback loop gain and complicates feedforward design. Noise on sensor inputs represents an unusable component and, if in the frequency range of control signals, often causes excessive valve activity and creates maintenance problems. Correction often requires filtering or special nonlinear controller gain functions. (See Ref. 14 for an example.) Noise is usually from one of three sources: (1) uncontrollable process disturbances too rapid to be reduced by control action; (2) measurement noise due to flow turbulence around flow sensors, sensor vibration, and sensor electrical noise; or (3) stray electrical pickup from power lines, power switching, etc. For sampled-data control systems, noise frequency components at or near the sampling frequency translate into low-frequency components.

Since the manipulated variable usually involves a control valve, the following nonideal characteristics are possible:

1. Minimum, as well as maximum, flow limits may exist.

2. Flow-versus-signal characteristics, as well as the limits, are dependent on changes in upstream and downstream pressures.

3. Significant rate-of-change limits may exist for flow changes with large valves.

4. Accuracy and resolution of valve positioning may affect control, and hysteresis in positioning may be noticeable.

Once the problems of measurement and manipulated variables are understood, control techniques permit the engineer to make the best use of existing capabilities.

SYSTEM CHARACTERISTICS

System characteristics may have a significant effect on the techniques used for the control system design. Figure 3 shows some of the important aspects of the characterization and their effect on the difficulty of analysis and design.

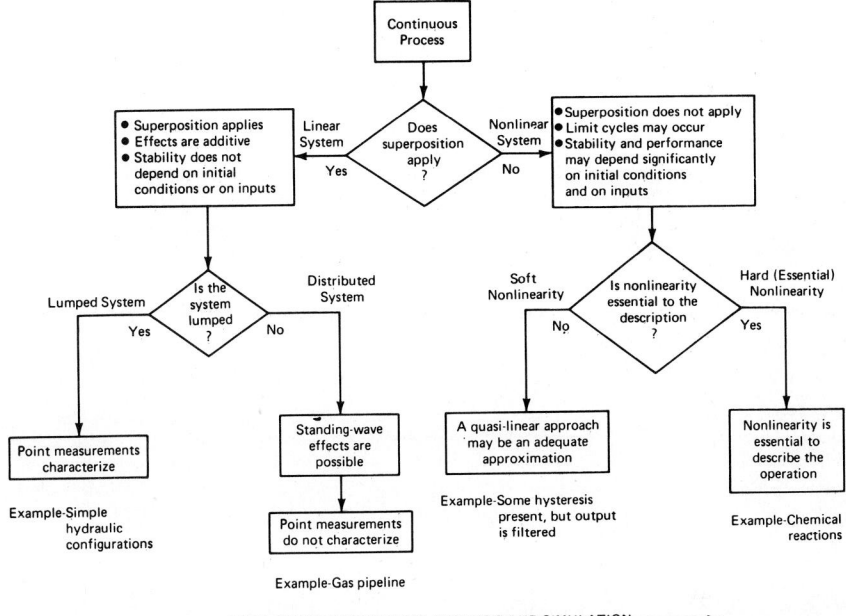

Fig. 3 Process system characteristics.

Linearity

The defining characteristics of a system in an engineering sense are dependent on the system inputs and the intended use. In many cases, if the system is to be subjected only to small disturbances, the assumption of *linearity* may be made. In a *linear system*, the output signal does not vary in form or in relative magnitude with varying size, level, or direction of the input. For instance, the slope of the flow-versus-lift curve of a control valve varies with the *size* and *level* of change. However, it is often permissible to consider the characteristic about a mean position and to assume a constant slope flow-versus-lift curve having an average slope within a certain region. If a system can be assumed to be linear, the *principle of superposition* can then be applied. The principle of superposition permits the total response of a system to be determined directly by summing the responses due to each input applied separately.

A linear system may have distance as a parameter, e.g., the resistance and capacitance per unit length of a pneumatic transmission line. Such systems are termed *distributed systems*. In practice it is usually convenient and permissible to approximate a distributed-parameter system by an equivalent *lumped system*, e.g., representation of a pneumatic line by two resistances and capacitances. This may be done except where standing waves, such as may occur in short gas-filled lines, affect the system dynamics. In such cases, a more detailed definition is necessary.

Two types of processes can be treated by linear methods, although they may not appear to be linear: (1) linear systems containing dead time and (2) otherwise linear systems which have parameters that vary slowly with time. The second type of system is linear provided that the variation in parameters is slow compared to the dynamics of interest.

Linear systems that contain sampled-data components, such as samplers, may often be considered *piecewise linear*. This considerably simplifies the analysis and design.

Nonlinearities

When significant nonlinearities are present, the analysis of stability and performance becomes more complicated. There are two notable cases where nonlinearities cannot be ignored. First, in chemical reactions the reaction rate is a strong function of temperature as indicated by the Arrhenius law

(Refs. 15 to 17). A second case is found in control of pH and is clearly described in Ref. 18. The system is characterized by nonlinear measurement and by wide-range operation. In some practical cases of nonlinear operation, useful information can be gained by system analysis and test, although the approach requires some ingenuity. Even where simulation must ultimately be employed to obtain a solution, some preliminary analysis, preferably quantitative, is useful to ensure that the problem is properly formulated for the simulation. For this work, extensions of the linear theory are particularly advantageous.

In many practical cases, the effect of the nonlinearities is small and can be treated as though they were superimposed on dominant linear effects. As an example, small-amplitude limit cycles may occur because of control valve hysteresis in an otherwise linear system. The amplitude and frequency of the limit cycles can often be estimated using *describing-function methods*.

SYSTEM REPRESENTATION

Two types of diagrammatic representations are common in the literature on control systems. See Table 1. Figure 4 shows a simple feedback control system represented in two forms. Both types permit complex systems to be represented in compact, meaningful form. By following set rules, the diagrams permit system characteristics to be studied and investigated. Table 2 shows some typical configurations. The block diagram offers the most physical representation. The signal-flow diagram is easily related to system equations, and manipulations of the diagram are closely related to manipulations of the equations.

Continuous-Data System Diagrams

Transfer functions for complex systems can be obtained using the transformation rules of Table 3 for block diagrams. The basic method for reducing signal-flow graphs to obtain transfer functions is suggested by equivalences shown in Table 1. For linear, continuous-data systems, transfer functions can be obtained directly from the signal-flow graph using Mason's gain formula (Ref. 19).

Table 1 Conventions for Control System Diagrams

Algebraic Expression	Block Diagram	Signal Flow Graph
Variable x		
Transfer Function $x_2 = gx_1$		
Summing Point $x_3 = x_1 - x_2$		
Multiply Function $x_3 = x_1 \cdot x_2$		No Equivalent
Divide Function $x_3 = x_1 / x_2$		No Equivalent

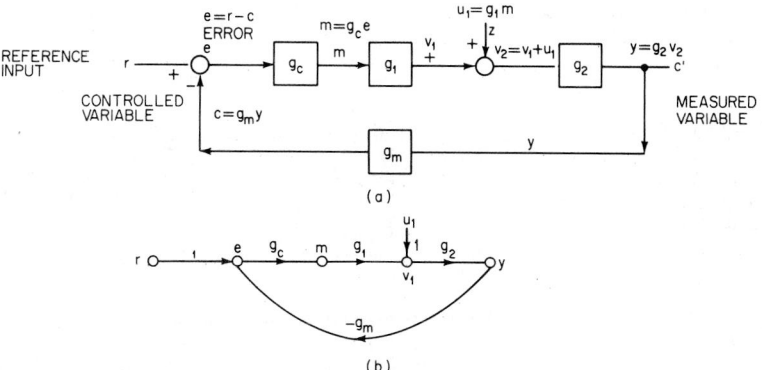

Fig. 4 Representations of a simple feedback control system. (*a*) Block diagram, (*b*) signal-flow graph.

Table 2 Comparative Representations

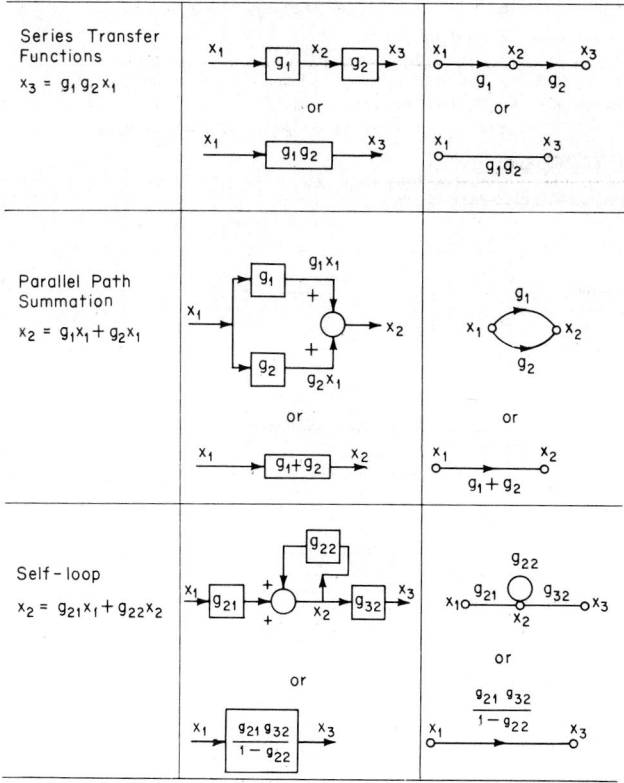

Table 3 Manipulation of Block Diagrams

Equivalent diagrams		Rule
1		Blocks in cascade are multiplied: $z = ABx$
2		Parallel operators add: $z = (A + B)x$
3		Blocks may be moved past branch and summing points if diagram algebra is preserved: $z = A(x + y)$
4		$z = Ax + y$
5		$z = Ax$
6		$z = Ax$
7		A simple single-loop feedback control system is equivalent to a single block and vice versa, $\dfrac{C}{R} = \dfrac{G}{1 + GH}$

provides the equivalent of a very high (infinite) gain, although it may require some time to do so. We also noted that derivative functions provide a prompt large gain in response to a rate of change on the input.

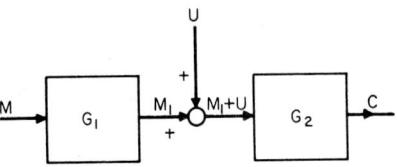

Fig. 9 Block diagram of a simple system subjected to a single disturbance.

By use of the block diagram representation, some essential characteristics of systems with and without feedback can now be examined. Figure 9 shows the block diagram of a simple system without feedback, subjected to a single disturbance U. This might represent, for instance, a process on manual control where M is the manipulated variable, C is the controlled variable, and U is a disturbance. From Eq. 2, it can be seen that the effect of the disturbance is only modified by G_2. Furthermore, even if there is no disturbance ($U = 0$), a change in G_1 or G_2 will cause a corresponding change in the controlled variable C. Since $C = G_1G_2M$ when $U = 0$, nonlinearities in the gains G_1 and G_2 as they affect C versus M_1 can be made self-canceling to some extent.

$$C = G_1G_2M + G_2U \qquad (2)$$

Gain Calculations

Consider the case where G_1 represents a control valve and G_2 represents a flow transmitter. If G_1 is a 1-in ($\sim$2.5-cm) valve with a C_v of 10 and a pressure drop of 9 psi ($\sim$60 kPa), the steady-state gain is

$$G_1 = \frac{C_v \sqrt{\Delta p}}{100\%} = \frac{10 \sqrt{9}}{100} = 0.30 \text{ gal/min/per \% change in valve opening}$$

If the flow transmitter is an orifice-type flowmeter, the output will be proportional to the square of the flow. Differentiating,

$$G_2 = 2 \frac{\text{flow}}{\text{max. flow}} = 2 \times \frac{9}{10} = 1.80\% \text{ per gal/min}$$

Note that the steady-state gains have been calculated under the operating conditions and will change as the operating conditions change.

If the disturbance U is due to a 0.5-psi decrease in the pressure drop across the valve, the flow disturbance U can be calculated as

$$Q = C_V \sqrt{\Delta p} \text{ gal/min}$$

$$Q_1 = 10 \sqrt{9} = 30 \text{ gal/min}$$

$$Q_2 = 10 \sqrt{8.5} = 29.2 \text{ gal/min}$$

$$U = -0.8 \text{ gal/min}$$

Feedback Control Equations

Since the block diagram conventions are based on linear operations, superposition applies, and Eqs. 3 to 7 can be easily developed algebraically by considering each input separately with the other inputs equal to zero. An example of the algebra for the system of Fig. 5 is

$$C = \frac{G}{1 + GH} R + \frac{G_2}{1 + GH} U_1 - \frac{GH_4}{1 + GH} U_2 \qquad (3)$$

3. The control-ratio response C/R depends primarily on the feedback elements H. In terms of control and measurement devices, this means that the linearity and drift of a feedback device may be made satisfactory by using a high value of loop ratio GH and quality feedback elements H. This is useful, since G_1 may be a pneumatic pilot relay or electronic amplifier stage which must supply power and may not have the required linearity and low-drift characteristics.

4. For unit feedback (H equal to unity) and high forward gain G the output G will reproduce the input R. If H is then increased above unity, C will be less than R. This points out, in a general way, the effect of imperfections in feedback elements, whether they involve dynamic or merely static deficiencies.

5. If one of the forward elements G_2 contains nonlinearities, the effect of feedback may be to shift the resulting distortion from C, the output of G_2, back toward $M + U_1$, the input of G_2.

Control and Regulation

The algebraic analysis above shows several important characteristics of feedback control systems. A high *loop gain* (large value of loop ratio GH), e.g., due to a narrow proportional band controller provides:

1. Accurate set-point tracking (C/R approaches 1 when GH is large and $H = 1$).

2. Effect of disturbances on the controlled variable C is small, as desired in a regulator.

Integrators have a high (ideally infinite) gain and therefore produce the same two effects. However, they act less promptly than the proportional action. Derivative response in the controller produces a prompt, but not sustained, high-gain response to rapidly changing disturbances. If it were not for stability considerations, many control requirements could be met by a proportional controller with a very narrow proportional band, as indicated below.

$$\frac{C}{R} = \frac{G}{1 + GH} \qquad \frac{C}{U_1} = \frac{G_2}{1 + GH}$$

For $G = 1000$, $H = 1$, and $G_2 = 1$,

$$\frac{C}{R} = \frac{1000}{1001} \approx 1 \qquad \frac{C}{U_1} = \frac{1}{1001}$$

Stability

Note that, if $GH = -1$, the denominator goes to zero and the expression blows up. This occurs when the signal goes out of phase as it traverses the loop and is a result of dynamic elements in the control loop, as discussed later. Unfortunately, this limits the maximum value of loop gain which can be used.

Feedforward Control

Closed-loop feedback control tends to effectively reduce disturbances and to force the variable to the new command level. This is true, however, only when a high "gain" can be attained without instability. Processes that contain significant dead time require both low proportional gain and slow integral gain for stability. The result is a slow response to set-point changes and low attenuation of disturbances with significant high-frequency components. Provided that the disturbance can be measured or calculated, feedforward control can often be applied.

Figure 10 shows three basic configurations for feedforward control. Figure 10a shows the simplest (nondynamic) compensation. In Fig. 10b, simple dynamic compensation requires a lead (derivative term) to compensate for the lag g_1. In Fig. 10c a lead or lag is required since the disturbance cannot be measured directly at the summing point.

Figure 11a indicates how the output of the feedforward controller can be applied to improve response to set-point changes. Figure 11b illustrates the way in which feedforward may be used to

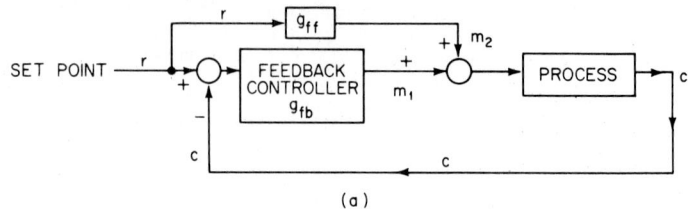

(a)

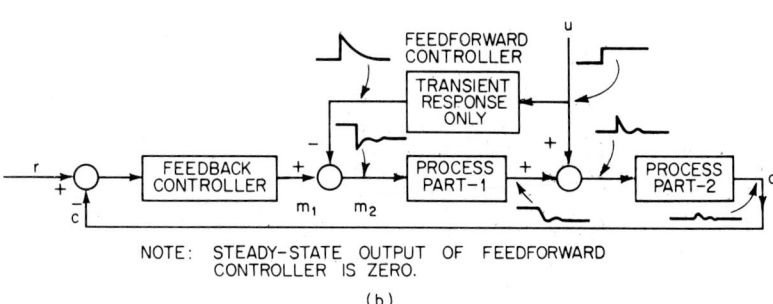

NOTE: STEADY-STATE OUTPUT OF FEEDFORWARD
CONTROLLER IS ZERO.

(b)

Fig. 11 Additional feedforward techniques. (a) Improving set-point response using feedforward control, (b) feedforward for transient correction only.

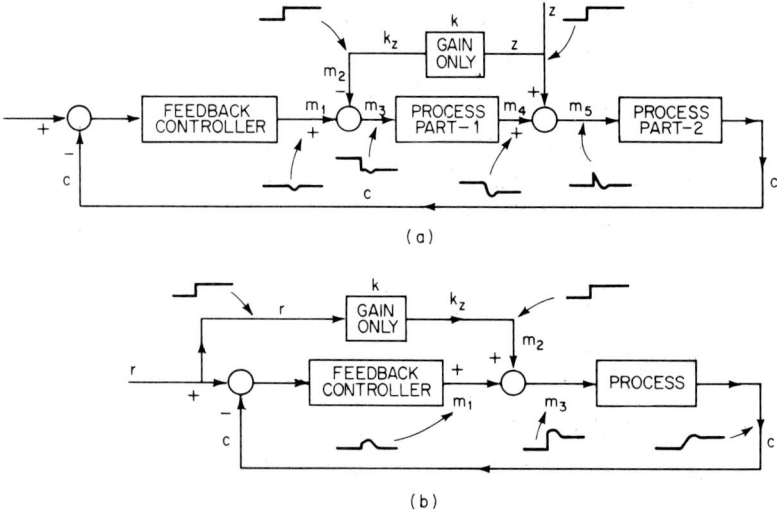

(a)

(b)

Fig. 12 Feedforward control with steady-state-only response. (a) Steady-state-only feedforward for disturbances, (b) steady-state-only feedforward for set-point response.

in the model, with attendant problems due to noise. Simple models are often used. The feedforward concept is obviously extendable to multiple variables and to cases where the variables are calculated. An important fact to note is that a properly designed feedforward system *will not affect the stability of feedback controllers involved.* A detailed and practical treatment of feedforward control is given by Shinskey (Ref. 1). Fast models can also be used for feedforward control. This is discussed under Modeling and Simulation.

CONTINUOUS SYSTEM ANALYSIS

A thorough understanding of the dynamic response of linear systems is essential if one is to design control systems. The topics below cover linear system dynamics and summarize basic techniques for the analysis of continuous linear systems. Linear discrete-data systems are piecewise linear systems and are discussed in the next subsection.

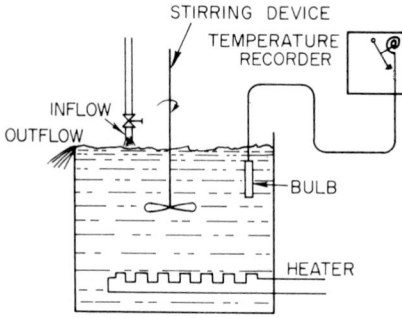

Fig. 14 Temperature measuring system.

Differential Equations

System differential equations are fundamental in describing the system dynamics and may be arrived at by an analysis based on experience and physical laws. For instance, the temperature measuring system shown in Fig. 14 may be considered for many purposes to be a lumped-parameter linear system. Assuming that the rate of heat flow to the bulb is entirely controlled by the surrounding fluid film and the bulb thickness is much less than the bulb diameter, Fourier's equation may be written as in Eq. 10.

$$\frac{dq}{dt} = -kA \frac{d\theta}{dx} \tag{10}$$

where θ = temperature
t = time
x = distance from inside of film
A = area normal to heat transfer
k = thermal conductivity of film
$d\theta/dx$ = temperature gradient in film (assumed constant)
θ_0 = bulb temperature
θ_f = fluid temperature
L = thickness of film
C = thermal "capacity" of bulb
M = mass of bulb
h = k/L, film heat-transfer coefficient
T = CM/hA, time constant
dq/dt = rate of heat flow to bulb

Assuming that $d\theta/dx$ is constant,

$$\frac{d\theta}{dx} = \frac{\theta_0 - \theta_f}{L} \tag{11a}$$

and

$$\frac{dq}{dt} = -\frac{kA}{L}(\theta_0 - \theta_f) \tag{11b}$$

Assuming a uniform bulb temperature,

$$\frac{dq}{dt} = CM \frac{d\theta_0}{dt} \tag{12}$$

Substituting Eq. 11a in Eq. 12 and simplifying yields

$$\frac{CM}{hA} \frac{d\theta_0}{dt} + \theta_0 = \theta_f \tag{13a}$$

Although mathematical restrictions exist on the use of Eq. 14, it may be applied without qualification to problems involving only linear lumped-parameter systems.

In addition to Eq. 14, a number of useful theorems exist which facilitate the use of Laplace transform methods; the following are some of the most generally useful theorems:

1. Linearity Theorem

$$\mathcal{L}[Af(t)] = A\mathcal{L}[f(t)] = AF(s) \tag{15}$$

where A = a constant.

$$\mathcal{L}^{-1}[AF(s)] = Af(t) \quad \text{for } t \geq 0 \tag{16}*$$

$$\mathcal{L}[f_1(t) \pm f_2(t)] = F_1(s) \pm F_2(s) \tag{17a}$$

$$\mathcal{L}^{-1}[F_1(s) \pm F_2(s)] = f_1(t) \pm f_2(t) \quad \text{for } t \geq 0 \tag{17b}$$

2. Real Differentiation Theorem

$$\mathcal{L}\left[\frac{d}{dt}f(t)\right] = sF(s) - f(0+) \tag{18a}$$

and

$$\mathcal{L}\left[\frac{d^n}{dt^n}f(t)\right] = \mathcal{L}[f^n(t)]$$
$$= s^n F(s) - s^{n-1}[f(t)]_{t=0+} - \cdots - s[f^{n-2}(t)]_{t=0+} - [f^{n-1}(t)]_{t=0+} \tag{18b}$$

where $f(0+)$ = value of $f(t)$ immediately following the disturbance.

3. Real Integration Theorem

$$\mathcal{L}\left[\int f(t)\, dt\right] = \frac{F(s)}{s} + \frac{1}{s}[\int f(t)\, dt]_{t=0+} \tag{19}$$

where the last term designates the value of the integral at $t = 0+$ (immediately after the disturbance is applied).

4. Initial-Value Theorem

$$\lim_{s\to\infty} sF(s) = \lim_{t\to 0} f(t) \tag{20}$$

This theorem is useful in establishing the initial value $f(0+)$ following an input to a system or component. The disturbance is included in $F(s)$.

5. Final-Value Theorem

$$\lim_{s\to 0} sF(s) = \lim_{t\to\infty} f(t) \tag{21}$$

This theorem is useful in establishing the final value of the system response $f(t)$ following an input disturbance. Use of the theorem makes it unnecessary to solve for the complete time response where only the final value is required. The disturbance is included in $F(s)$.

Although the direct and inverse Laplace transformations can be obtained by integration, fairly complete tables of transforms are available in most modern control texts. A limited transform table is given in Table 6.

*$\mathcal{L}^{-1}$ indicates the inverse Laplace transform.

where $Q(s) = K_0 s^{m-n} + K_1 s^{m-n-1} + \cdots$
$A(s) = s^r + A_1 s^{r-1} + \cdots + A_{r-1} s + A_r$
$B(s) = s^n + B_1 s^{n-1} + \cdots + B_{n-1} s + B_n$
K, A, B = constants
m, n, r = positive integers

$Q(s)$ is easily transformed, since it involves only terms in positive powers of s. Considering the fraction $A(s)/B(s)$ when $B(s)$ is factored,

$$F(s) = \frac{A(s)}{B(s)} = \frac{A(s)}{(s - s_1)(s - s_2) \cdots (s - s_k) \cdots (s - s_n)} \tag{22}$$

where $s_1, s_2, \ldots, s_k \ldots, s_n$ = roots of $B(s)$.

$F(s)$ is the quotient of two polynomials in s, where there are no common factors between the numerator and denominator and the order of the numerator is less than that of the denominator. The *partial fraction expansion* can therefore be used to obtain the inverse Laplace transform as follows:

If $B(s)$ in Eq. 22 contains no repeated roots,

$$F(s) = \frac{C_1}{s - s_1} + \frac{C_2}{s - s_2} + \cdots + \frac{C_k}{s - s_k} + \cdots + \frac{C_n}{s - s_n} \tag{23}$$

$$f(t) = C_1 e^{s_1 t} + C_2 e^{s_2 t} + \cdots + C_k e^{s_k t} + \cdots + C_n e^{s_n t} \tag{24}$$

where

$$C_k = \lim_{s \to s_k} \left[\frac{(s - s_k) A(s)}{B(s)} \right]$$

$$C_1 = \frac{(s - s_1) A(s_k)}{B(s)} \bigg|_{s = s_1} \tag{25}$$

$$C_2 = \frac{(s - s_2) A(s)}{B(s)} \bigg|_{s = s_2}$$

If $B(s)$ in Eq. 22 contains repeated roots,

$$F(s) = \frac{A(s)}{B(s)} = \frac{A(s)}{(s - s_1)^q (s - s_2) \cdots (s - s_k) \cdots (s - s_n)} \tag{26}$$

where q = a positive integer and s_1 = a typical repeated root.

$$[f(t) = e^{s_1 t}(C_{11} + C_{12} t + \cdots + C_{1j} t^{(j-1)} + \cdots + C_{1q} t^{(q-1)}$$

$$+ C_2 e^{s_2 t} + \cdots + C_k e^{s_k t} + \cdots + C_n e^{s_n t} \tag{27}$$

where

$$C_{1j} = \left\{ \frac{1}{(j-1)!} \frac{d^{(j-1)}}{ds^{(j-1)}} \left[\frac{(s - s_1)^q A(s)}{B(s)} \right] \right\} \bigg|_{s = s_1}$$

$$C_{11} = \frac{(s - s_1)^q A(s)}{B(s)} \bigg|_{s = s_1} \tag{28}$$

$$C_{12} = \left\{ \frac{d}{ds} \left[\frac{(s - s_1)^q A(s)}{B(s)} \right] \right\} \bigg|_{s = s_1}$$

The complete time response of linear lumped-parameter systems due to most disturbances is relatively simple to derive using the theorems, formulas, and tables just discussed. It is, however, necessary that the denominator of the response equation be factored.

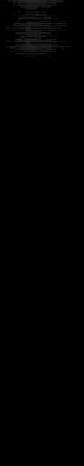

Table 7 Step-Response Equations for Some Simple Systems

Description of system	Transfer function	Step response
Integrator	$\dfrac{1}{T_1 s}$	$\theta_0(t) = \dfrac{t}{T_i}\,\Delta\theta_i$
First-order lag	$\dfrac{1}{Ts + 1}$	$\theta_0(t) = \Delta\theta_i(1 - \epsilon^{-1/T})$
Second-order system (overdamped)	$\dfrac{1}{(T_1 s + 1)(T_2 s + 1)}$	$\theta_0(t) = \Delta\theta_i\left(1 - \dfrac{T_1}{T_1 - T_2}\epsilon^{-1/T_1} + \dfrac{T_2}{T_1 - T_2}\epsilon^{-1/T_2}\right)$
Second-order system (underdamped)	$\dfrac{1}{\left(\dfrac{s}{\omega_n}\right)^2 + \dfrac{2\zeta}{\omega_n}s + 1}$ where $\zeta < 1$	$\theta_0(t) = \Delta\theta_i\left[1 + \dfrac{\epsilon^{-\zeta\omega_n t}}{\sqrt{1 - \zeta^2}}\sin(\omega_0 t + \phi)\right]$ where $\omega_0 = \sqrt{1 - \zeta^2}\,\omega_n$ $\phi = \tan^{-1}\left(\dfrac{\sqrt{1 - \zeta^2}}{-\zeta}\right)$

NOTE: $_i(t) = 0$ when $t < 0$; $\theta_i(t) = \Delta\theta_i$ when $t > 0$. ζ is actual damping/critical damping; $\omega_n = \zeta\pi \times$ (undamped natural frequency).

sion that two factors influence the effect of any root s_k on the transient response: the real part of the root and the magnitude of the corresponding coefficient C_k multiplying the exponential.* Since the exponential factor $e^s k^t$ decreases rapidly, the real part of s_k is usually the most important factor. Thus, if $s_k = -1/T$, the effect of this term on the transient response is usually small by the time $t = 4T$ or $5T(\epsilon^{-4} \approx 0.018,\ \epsilon^{-5} \approx 0.007)$. This means that, if one wishes to measure the step response of a system which has two time constants of $0.1T$ and T seconds, the transient effect of the first time constant $0.1T$ will be negligible after $0.4T$, whereas the effect of the time constant T will not be negligible until about $4T$ (see Fig. 16, curve b). Curve a shows the step response of a first-order lag with time constant T. Note that after 0.2 to $0.3T$ the second-order curve lags the first-order curve by an amount $0.1T$, the magnitude of the smaller time constant. In determining the value of the smaller time constant from test results, the first part of the curve must be examined closely. Since the total response in the region is small, the measurement of time zero (and output zero) must be good. Any noise or insensitivity of the measuring instrument will reduce or destroy the possibility of determining the magnitude of the smaller lag accurately. From Fig. 16 it can be seen that the step response of this two-time-constant system can be approximated by an initial delay, or dead time, plus a first-order lag response. In this case, the "initial delay" is equal to the smaller time constant $0.1T$, and the time constant of the slower

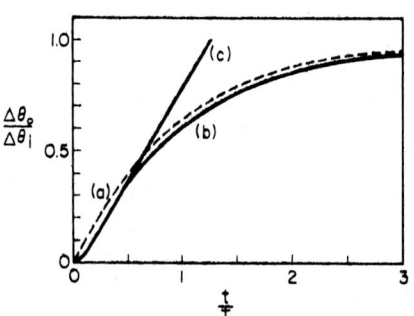

Fig. 16 Step response of a second-order lag with time constants T and $\frac{1}{10}T$.

response is equal to the time constant T. If the response b represents the step response of a process, an even rougher approximation may be permissible. Since the initial portion of the curve is generally

---

*In stable systems the roots s_k always have negative real parts where $s_k = -s_k + j\omega k$; $-sk$ is the real part.

Closed-Loop Frequency Response

If the loop is closed around components that have a frequency-response transfer function $G(j\omega)$, then

$$\frac{C}{R}(j\omega) = \frac{G(j\omega)}{1 + G(j\omega)} = \frac{M(j\omega)/\theta(j\omega)}{1 + M(j\omega)/\theta(j\omega)} \tag{33}$$

where $(j\omega)$ = a function of frequency
$M(j\omega)$ = magnitude ratio
$\theta(j\omega)$ = phase

The graphical significance of Eq. 33 is indicated in Fig. 19.

Suppose that $G(j\omega)$ represents the composite transfer function of a controller and a process. The *polar plot* of $G(j\omega)$ is the locus of the tips of vectors, each of which represents the magnitude at a given frequency. Such a plot is shown in Fig. 20 for a proportional controller and a multilag process. Using the graphical relations of Fig. 19, we note the following relationships between $G(j\omega)$ and the closed-loop response for the solid-line plot ($K = 1$).

1. At low frequencies below ω_1, the magnitude of the control ratio is less than unity, but the control-ratio phase is less negative than the phase of $G(j\omega)$.

2. Over the intermediate-frequency range between ω_1 and ω_2 the magnitude of the control ratio increases from unity to about 1.35 and then decreases to unity, while the phase of the control ratio decreases rapidly to about $-163°$. There is little improvement over the phase of $G(j\omega)$.

3. At high frequencies above ω_2, the magnitude of the control ratio decreases rapidly and the control-ratio phase is more negative than the phase of $G(j\omega)$.

The effect of a change in gain [multiplication of $G(j\omega)$ by a constant] is to change the scale of the plot. The plot of $G(j\omega)$ with $K = 1.25$ is shown by the dotted plot. Note that the peak value of the magnitude of the control ratio has been increased to ~ 2, and the zero-frequency magnitude increased from 0.5 to 0.556 (less offset). To improve the system of Fig. 20 a further increase in gain K results in a large control ratio at intermediate frequencies unless the $G(j\omega)$ plot is distorted by the addition of other dynamic functions. If, however, the low-frequency portion of $G(j\omega)$ is caused to fall along the negative $\mathcal{J}$ axis as shown, the low-frequency C/R response is improved.

This can be accomplished by using a proportional-plus-integral (PI) controller instead of a proportional controller. See Table 8, item 6. The PI controller moves the zero frequency point of the locus $G(j\omega)$ to $0 - j\infty$, along the negative $\mathcal{J}$ axis as shown by the dash-dot line in Fig. 20. The

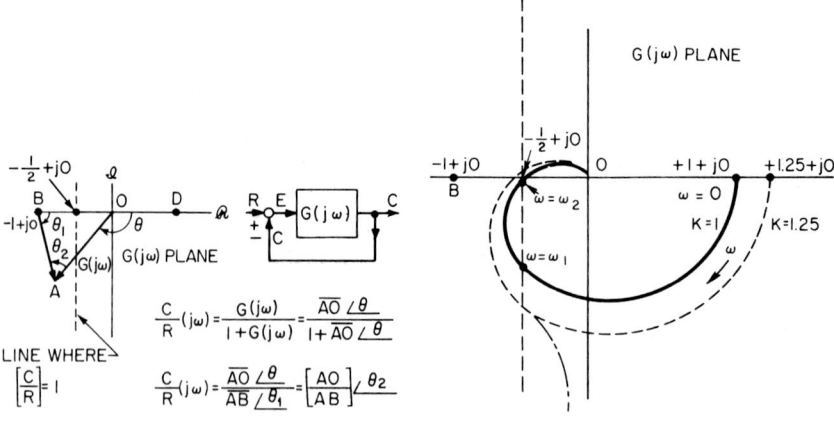

Fig. 19   Vector relations for a single-loop feedback control system.

Fig. 20   Polar plot of $G(jw)$.

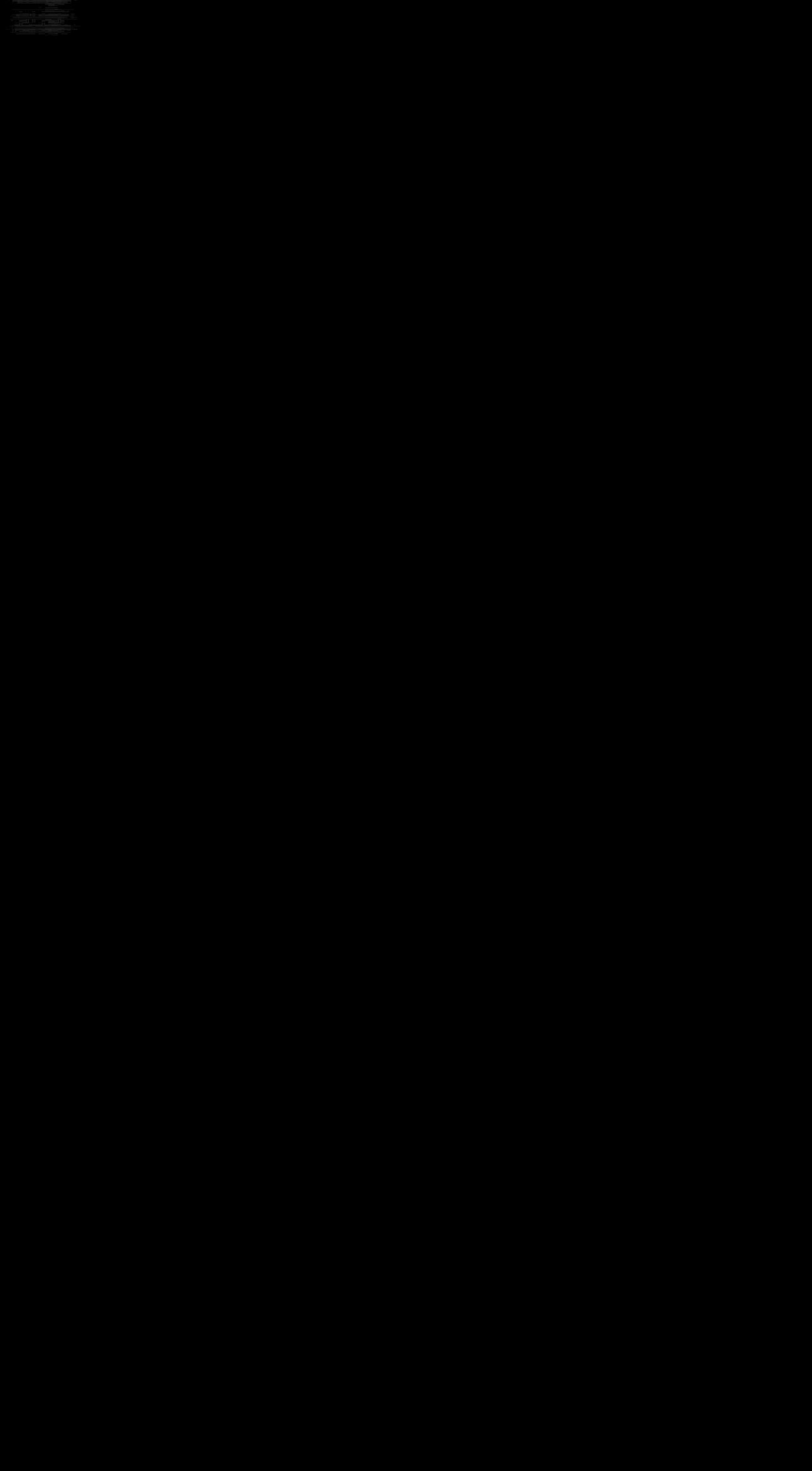

Table 8 Frequency-Response Equations for Some Common Control-System Elements

Description	Transfer function $G(s)$	Frequency response $G(j\omega)$	Magnitude ratio	Phase angle
1. Dead time $T = $ dead time	ϵ^{-Ts}	$\epsilon^{-j\omega T}L$	1	$-\omega T$ radians
2. First-order lag	$\dfrac{1}{Ts+1}$	$\dfrac{1}{j\omega T + 1}$	$\dfrac{1}{\sqrt{\omega^2 T^2 + !}}$	$-\tan^{-1}(\omega T)$
3. Second-order lag	$\dfrac{1}{(Ts+1)(aTs+1)}$	$\dfrac{1}{-a\omega^2 T^2 + j(1+a)\omega T + 1}$	$\dfrac{1}{\sqrt{(1-a\omega^2 T^2)^2 + (1+a)^2\omega^2 T^2}}$	$-\tan^{-1}\left[\dfrac{(1+a)\omega T}{1 - aT^2\omega^2}\right]$
4. Quadratic (underdamped)	$\dfrac{1}{\left(\dfrac{s}{\omega_n}\right)^2 + \dfrac{2\zeta}{\omega_n}s + 1}$	$\dfrac{1}{-\left(\dfrac{\omega}{\omega_n}\right)^2 + j2\zeta\dfrac{\omega}{\omega_n} + 1}$	$\dfrac{1}{\sqrt{\left(1-\dfrac{\omega^2}{\omega_n^2}\right)^2 + 4\zeta^2\left(\dfrac{\omega}{\omega_n}\right)^2}}$	$-\tan^{-1}\dfrac{2\zeta\dfrac{\omega}{\omega_n}}{1-\left(\dfrac{\omega}{\omega_n}\right)^2}$
5. Ideal proportional controller	K	K	K	0

Table 9 Vector Notation and Vector Operations

Vector operation	Complex number form	Polar form
Vector addition: $\mathbf{z} = \mathbf{x} + \mathbf{y}$	$z = x + jy$	$s = re^{j\phi}$
x and **y** are perpendicular vectors	where $j = \sqrt{-1}$ $-j$ denotes clockwise rotation of a vector; $+j$ denotes counterclockwise rotation	where $r = \sqrt{x^2 + y^2}$ $\phi = \tan^{-1}\left(\dfrac{y}{x}\right)$ or $\quad z = r\underline{/+\phi}$
Multiplication of vectors: $\qquad \mathbf{z_1} \cdot \mathbf{z_2}$ where $z_1 = x_1 + jy_1$ $\qquad z_2 = x_2 + jy_2$ $\qquad z_1 = r_1 e^{j\phi_1}$ $\qquad z^2 = r^2 e^{j\phi_2}$	$z_1 \cdot z_2 = (x_1 + jy_1)(x_2 + jy_2)$ $\qquad = x_1x_2 - y_1y_2 + j(x_1y_2 + x_2y_1)$	$z_1 \cdot z_2 = r_1 e^{j\phi_1} \cdot r_2 e^{j\phi_2}$ $\qquad = r_1 r_2 e^{(j\phi_1 + \phi_2)}$ or $z_1 \cdot z_2 = r_1 r_2 \underline{/\phi_1 + \phi_2}$
Vector division: $\qquad \dfrac{\mathbf{z_1}}{\mathbf{z_2}}$	$\dfrac{z_1}{z_2} = \dfrac{x_1 + jy_1}{x_3 + jy_2}$ $\qquad = \dfrac{(x_1 + jy_1)(x_2 - jy_2)}{x_1^2 + y_2^2}$ $\qquad = \dfrac{x_1x_2 + y_1y_2 + j(y_1^\ast x_2 - y_2x_1)}{x_1^2 + y_2^2}$	$\dfrac{z_1}{z_2} = \dfrac{r_1 e^{j\phi_1}}{r_2 e^{j\phi_2}}$ $\qquad = \dfrac{r_1}{r_2} e^{j(\phi_1 - \phi_2)}$ or $\dfrac{z_1}{z_2} = \dfrac{r_1}{r_2} \underline{/\phi_1 - \phi_2}$

gins, information in this form is very useful. In treating systems involving three or more components, the tolerance on component data is particularly important. Standards have been developed by the Instrument Society of America (ISA) for testing system components (see Ref. 29).

Polar Plots of Frequency Response

Graphical techniques exist for the design of feedback control systems using polar frequency-response plots. However, polar plots are chiefly useful in stability analysis (see Nyquist Criterion). The log magnitude and phase plots described below are generally used for system design.

Log Magnitude and Phase Plots

If separate plots are made of the logarithm of the magnitude ratio versus the logarithm of frequency and of the (linear) phase versus the logarithm of frequency, several advantages are gained. Since the magnitude ratios are in logarithmic form, the overall log magnitude ratio of a series of noninteracting components at each frequency is obtained by addition of the values. Graphically, this means merely adding distance on a plot. Another advantage is that the frequency response of the more common component forms can be easily sketched with sufficient accuracy for simple calculations.

Some definitions involved in the use of this type of plot are as follows:

One octave = frequency changes by a factor of 2

One decade = frequency changes of a factor of 10

Corner (or breakpoint) frequency = frequency at intersection of two asymptotes

Magnitude ratio in decibels (dB) = $\log_{10}$ (magnitude ratio)* (Fig. 21)

---

*Since most of the literature on feedback control uses this form of log magnitude and many standard plots are in this form, the decibel notation is used here. A conversion plot of decibels versus magnitude ratio is given in Fig. 21.

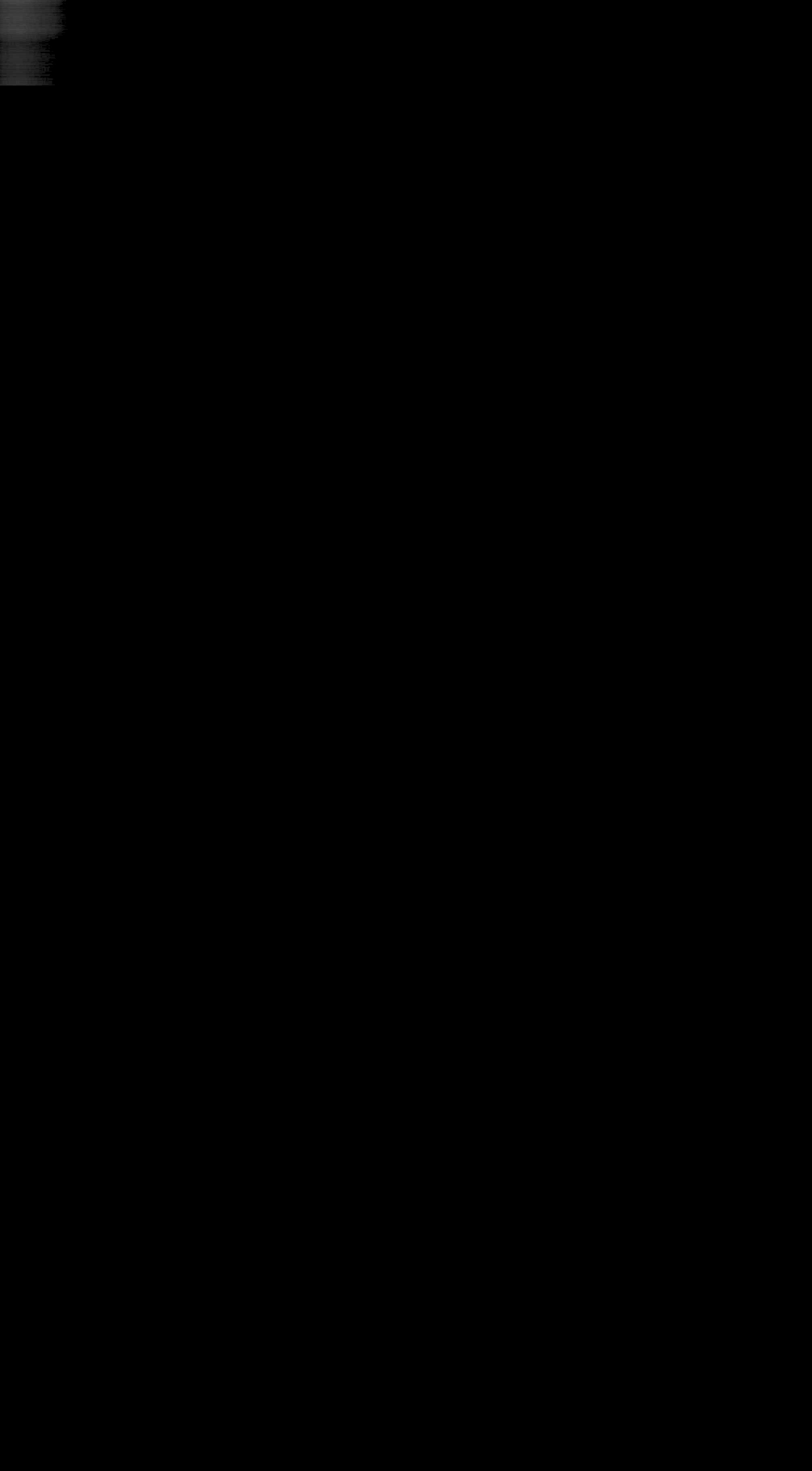

Table 10 Log Magnitude and Phase Plots for Some Common Control-System Components

Component	Transfer function	Log magnitude and phase plots
Proportional-plus-integral (PI) controller	$G(j\omega) = K\left(1 + \dfrac{1}{j\omega T_i}\right)$	
Second-oder lag	$G(j\omega) = \dfrac{1}{(j\omega T_1 + 1)(j\omega T_2 + 1)}$	

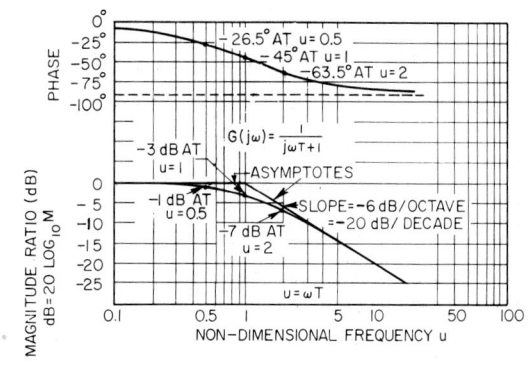

Fig. 22 Log magnitude and phase plot for a first-order lag.

K. This is generally done by using a transparency. For the plot of Fig. 24, $K = 9$ dB* or 2.8, and the control ratio has a maximum peak value M_P of ~ 2 dB or 1.25, occurring at $\underline{/G(j\omega)} = -140°$. With reference to Fig. 23, this corresponds to $\omega T = 1.5$.

The required reshaping of the $G(j\omega)$ plot to improve the frequency response is evident from the plot of Fig. 24. Improved low-frequency response will be obtained, if at low frequencies $G(j\omega)$ approaches the $M = 0$ dB ($M = 1$) contour. This can be accomplished by using a PI controller (see Table 10). With some practice, the required reshaping can be obtained directly from the log magnitude and phase plot, but a Nichols chart replot is often desirable as a check.

Thus, the log magnitude and phase plots permit simple, rapid plotting and useful asymptotic approximations in the construction of plots, but require a replot for conversion to the closed-loop response. The Nichols plot, however, shows immediately and numerically the effect that changes in the plotted open-loop response have on the closed-loop response. This is of practical importance,

---

*Thus, the zero-frequency phase, $\underline{/G(j0)} = 0°$, is at $G(j0) = 9$ dB.

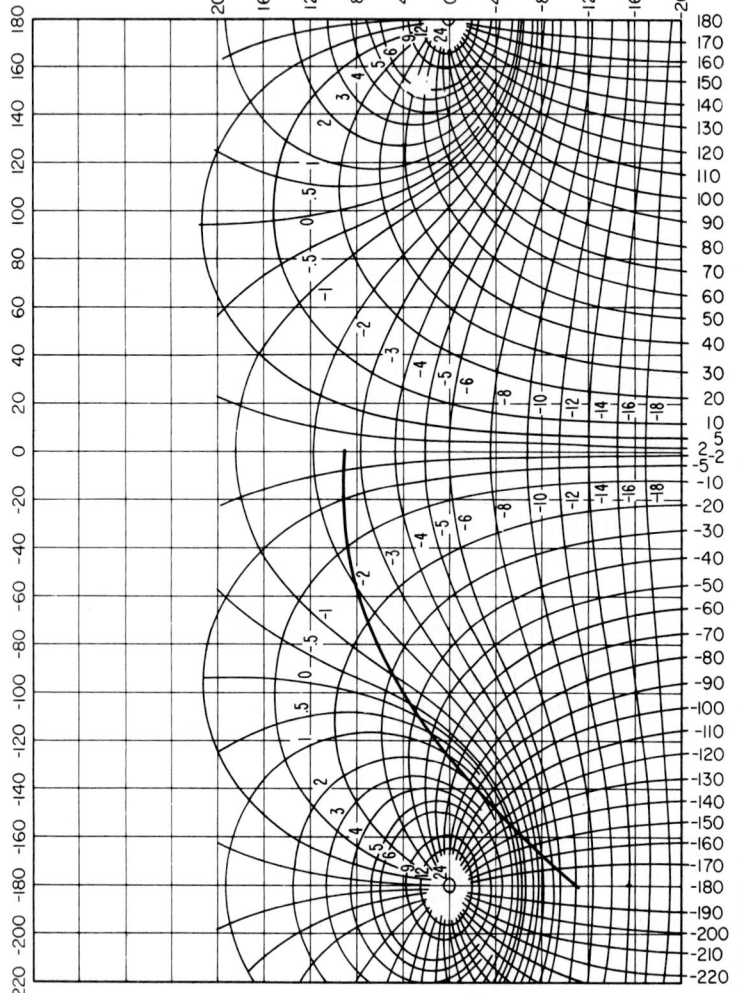

Fig. 24 Nichols chart.

When this is done, the corresponding **A** matrix becomes the *Jacobian matrix:*

$$
\mathbf{A} \equiv
\begin{bmatrix}
\dfrac{\partial f_1}{\partial x_1} & \cdots & \dfrac{\partial f_1}{\partial x_n} \\[2ex]
\dfrac{\partial f_2}{\partial x_1} & \cdots & \dfrac{\partial f_2}{\partial x_n} \\[2ex]
\cdots & \cdots & \cdots \\[1ex]
\dfrac{\partial f_n}{\partial x_1} & \cdots & \dfrac{\partial f_n}{\partial x_n}
\end{bmatrix}
\tag{42}
$$

Similar definitions apply for **B, C, D,** and **F,** and Eq. 39 becomes

$$
\delta \dot{x} = A\,\delta x + B\,\delta u + F\,\delta z
\tag{39b}
$$

although the δ's are often dropped.

Time-Invariant Linear Systems

The state equations solutions can be developed by analogy to the solution of the component *scalar equations* as indicated below:

$$
\dot{x} = ax + bu \qquad \text{and} \qquad x(0) = x_0
\tag{43}
$$

$$
x = x_h + x_p
\tag{44a}
$$

where x_h = homogeneous (free, undriven) solution
$\quad x_p$ = particular (driven) solution
a and b = constants

Assume

$$
x = a_0 + a_1 t + a_2 t^2 + \cdots
\tag{45}
$$

Substituting in Eq. 43 with $u = 0$ and solving for coefficients leads to

$$
x_h = (1 + at + \tfrac{1}{2}a^2 t^2 + \cdots)x_0
\tag{46a}
$$

$$
x_h = e^{at}x_0
\tag{46b}
$$

$$
x_p = e^{at}\int_0^t e^{-a\tau}bu(\tau)d\tau
\tag{47}
$$

$$
x = e^{at}x_0 + e^{at}\int_0^t e^{-a\tau}bu(\tau)\,d\tau
\tag{44b}
$$

Similarly, for the vector case,

$$
\dot{\mathbf{x}} = \mathbf{A}\mathbf{x}
\tag{48}
$$

$$
\mathbf{x}(t) = e^{\mathbf{A}t}\mathbf{x}_0 + e^{\mathbf{A}t}\int_0^t e^{-\mathbf{A}t}\mathbf{B}u(\tau)\,d\tau
$$

where

$$
e^{\mathbf{A}t} = \mathbf{I} + \mathbf{A}t + \tfrac{1}{2}\mathbf{A}^2 t^2 + \cdots
\tag{49}
$$

and $\quad \mathbf{I}$ = identity matrix with diagonal element unity and other elements zero

$$
\mathbf{A}^2 \triangleq \mathbf{A}\mathbf{A}
$$

and α_{ij} has a partial fraction expansion of the form

$$\alpha_{ij} = \frac{a_1}{s - \lambda_1} + \frac{a_2}{s - \lambda_2} + \cdots + \frac{a_n}{s - \lambda_n} \tag{55}$$

where the a's are constants and the λ's are the roots of the *characteristic equation*

$$|s\mathbf{I} - \mathbf{A}| = 0$$

The λ's are the *eigenvalues* of the system and correspond to the exponential terms

$$e^{\lambda_1 t}, e^{\lambda_2 t}, \ldots, e^{\lambda_n t}$$

of the homogeneous solution. These are the familiar transient terms covered previously. Eigenvalues may be real, imaginary, or complex. If complex, they occur as complex conjugate pairs.

If Eq. 55 contains terms of the form $a_k/(s - \lambda_k)^l$, where k and l are integers and $l > 1$, the system has *multiple eigenvalues,* and terms of the form $e^{\lambda_k t}, te^{\lambda_k t}, \ldots,$ occur.

If no multiple eigenvalues occur, the system is said to have *distinct eigenvalues.* The existence of distinct eigenvalues greatly increases the ease of analysis.

Inverse Laplace Transform

Applying the inverse Laplace transform to the first term of Eq. 53, with $\mathbf{U}(s) = 0$,

$$\mathbf{x}(t) = \mathcal{L}^{-1}[(s\mathbf{I} - \mathbf{A})^{-1}\mathbf{x}_0] = \mathcal{L}^{-1}[s\mathbf{I} - \mathbf{A})^{-1}]\mathbf{x}_0$$

Comparing with Eq. 48 for $\mathbf{u}(\tau) = 0$ gives a new and explicit expression for the *matrix exponential:*

$$e^{\mathbf{A}t} = \mathcal{L}^{-1}[(s\mathbf{I} - \mathbf{A})^{-1}]$$

State Variables from Transfer Functions

State equations can be obtained from transfer functions by various methods. Of particular interest is that based on expanding the transfer function into partial fractions as shown below.

Given the transfer function

$$\frac{C(s)}{U(s)} = \frac{k}{s^3 + a_1 s^2 + a_2 s + a_3} = \frac{k_1}{s + s_1} + \frac{k_2}{s + s_2} + \frac{k_3}{s + s_3} \tag{56}$$

where the k's and a's are constant,

$$C(s) = \frac{k_1}{s + s_1} U(s) + \frac{k_2}{s + s_2} U(s) + \frac{k_3}{s + s_3} U(s)$$

Define the state variables:

$$X_1(s) \triangleq K_1(s + s_1)^{-1}U(s)$$

$$X_2(s) \triangleq K_2(s + s_2)^{-1}U(s)$$

$$X_3(s) \triangleq K_3(s + s_3)^{-1}U(s)$$

$$c(t) = x_1(t) + x_2(t) + x_3(t)$$

leading to

$$\mathbf{A} = \begin{bmatrix} -s_1 & 0 & 0 \\ 0 & -s_2 & 0 \\ 0 & 0 & -s_3 \end{bmatrix} \quad \mathbf{B} = \begin{bmatrix} k_1 \\ k_2 \\ k_3 \end{bmatrix} \quad \mathbf{C} = \begin{bmatrix} 1 & 1 & 1 \end{bmatrix}$$

where $\mathbf{C} = $ a *row vector.*

Eigenvectors

The free response of the system defined by

$$\dot{x} = Ax \quad \text{and} \quad x(0) = x_0$$

can be written

$$x(t) = Ee^{\lambda t}$$

where

$$E = \begin{bmatrix} \epsilon_{11} & \cdots & \epsilon_{1n} \\ & \cdot & \\ & \cdot & \\ & \cdot & \\ \epsilon_{n1} & \cdots & \epsilon_{nn} \end{bmatrix} \quad \text{(the eigenvector matrix)}$$

$$e^{\lambda t} = \begin{bmatrix} e^{\lambda_1 t} \\ \cdot \\ \cdot \\ \cdot \\ e^{\lambda_n t} \end{bmatrix}$$

Since there are n degrees of freedom and $n \times n = n^2$ coefficients ϵ_{ij}, a variety of linearly independent groups of the coefficients could be used to express the response $x(t)$. Taking the eigenvector matrix by rows,

$$x(t) = \epsilon_1 e^{\lambda_1 t} + \cdots + \epsilon_n e^{\lambda_n t}$$

The vector term $\epsilon_j e^\lambda_{jt}$ is referred to as the jth *natural mode*.

where

$$\epsilon_i = \begin{bmatrix} \epsilon_{1i} \\ \cdot \\ \cdot \\ \cdot \\ \epsilon_{ni} \end{bmatrix}$$

Substituting in the original equation,

$$\lambda_i \epsilon_i = A\epsilon_i \quad i = 1, 2, \ldots, n$$

This gives the eigenvalues within a proportionality factor. The actual values are calculated based on the initial conditions.

$$x(t) = \alpha_1 \hat{\epsilon}_1 e^{\lambda_1 t} + \alpha_2 \hat{\epsilon}_2 e^{\lambda_2 t} + \cdots \tag{57}$$

where carets indicate unit vectors and α's are scalar magnitudes. By properly choosing the initial values $x(0)$, selected modes can be suppressed. The contributions of the modes to the total solution are functions of the initial state only.

Controllability

If the state equations can be put in *canonical form*, no coupling exists between the variables, and a necessary condition for controllability is that the **B** matrix for the canonical form have no zero rows. Variables having zero rows are referred to as *noncontrollable*. Note that it is *possible to control any finite number of state variables using only one control force* if the eigenvalues are distinct. Consider the scalar response equations,

$$x_i(t_f) = e^{\lambda_i t_f} x_i(0) + e^{\lambda_i t_f} \int_0^{t_f} e^{-\lambda_i \tau} b_i u(\tau) \, d\tau \quad i = 1, 2, \ldots, n$$

where t_f is the final time, and the initial conditions $x_i(0)$ and the final time t_f are fixed.

SAMPLED-DATA SYSTEMS

Sampled-data techniques apply when a control system involves a digital computer, a microprocessor, or an analyzer with a sampled output. A brief summary of techniques is given below. More detailed discussions and derivations are given in Ragazzini and Franklin (Ref. 36), in Kuo (Ref. 37), and Tou (Ref. 38). For a treatment oriented to digital process control systems, see Ash (Ref. 4).

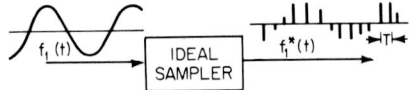

Fig. 26   Response of an ideal sampler to sine wave input.

The Sampling Process

Figure 26 shows an ideal sampler with a sinusoidal input. The frequency response of the sampler is given by Eq. 58.

$$F^*(j\omega) = \frac{1}{T} \sum_{k=-\infty}^{\infty} F\left[j\left(\omega + \frac{2\pi k}{T} \right) \right] \tag{58}$$

where $F^*(j\omega)$ = Fourier transforms of sample output
F = Fourier transform of the continuous function (input)
k = an integer ($\ldots, -2, -1, 0, 1, 2, \ldots$)
T = sampling interval

Figure 27 shows an input cosine wave and the corresponding Fourier transform. The output Fourier spectrum is shown in Fig. 28 (solid lines). The output magnitudes, indicated by the arrows, are $A/2T$.

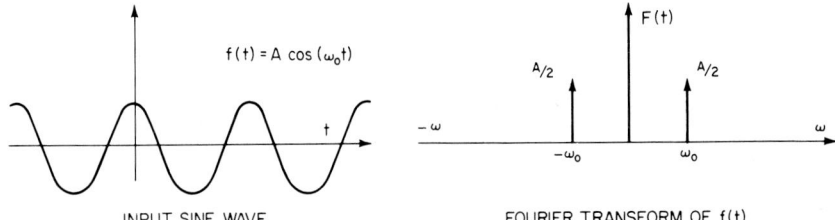

Fig. 27   Input to an ideal sampler.

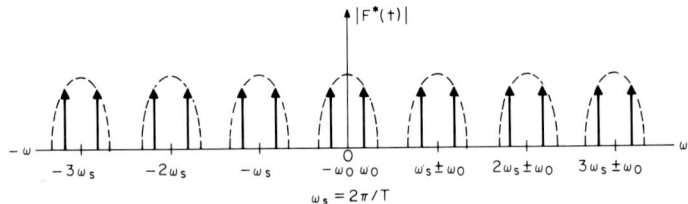

Fig. 28   Spectrum of ideal sampler output when driven by a cosine (solid lines) and by a spectrum (dashed lines).

In order to reconstruct the input from the sampled output, the sampling frequency must provide at least two samples per cycle for the highest frequency present in the input *(Shannon's sampling*

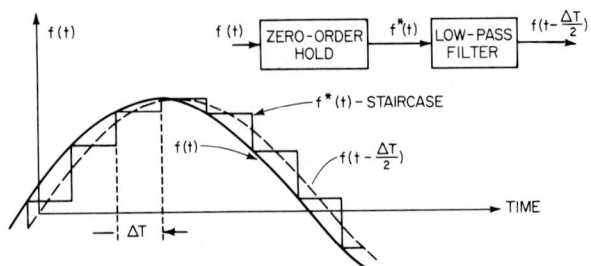

Fig. 31   Time response of a zero-order hold and filter.

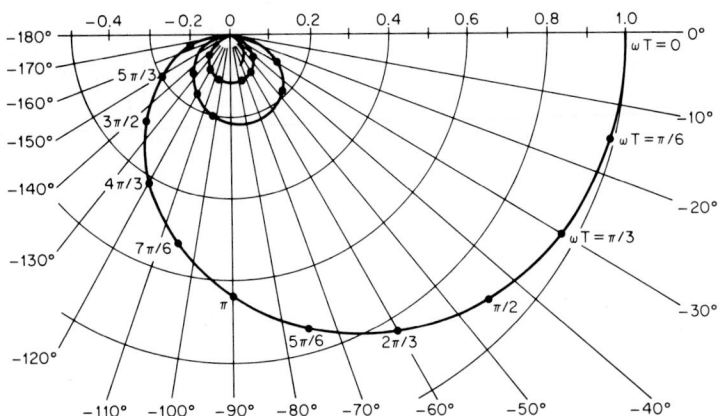

Fig. 32   Polar plot of frequency-response transfer function of the zero-order hold. *(After J. T. Tou, "Digital and Sampled-Data Control Systems," McGraw-Hill, 1959.)*

z Transforms

The sampler can be defined in terms of the Laplace transform

$$X^*(s) = \mathcal{L}[x^*(t)] = \sum_{n=0}^{\infty} x(nT)e^{-nTs} \tag{60}$$

where T = sample interval, n = 0, 1, 2, . . . , an integer, and * indicates the sampled function. A more convenient notation is to use

$$z = e^{Ts} \tag{61}$$

$$X(s) = X^*\left(\frac{1}{T}\ln z\right) = \sum_{n=0}^{\infty} x(nT)z^{-n} \tag{62}$$

and

$$X(z) = z\{x(t)\} \tag{63}$$

where the z transform of $x(t)$ is indicated.*

---

* The z transform is used for discrete-data systems in the same way that the Laplace transform is used for continuous-data systems.

Table 11 z Transforms

No.	$G(s)$	$g(t)$	$G(z)$	$g(nT)$
1	1	$\delta(t)$	1 or z^0	$\delta(nT)$
2	e^{-k_s}	$\delta(t-kt)$	z^{-k}	$\delta(nT-kT)$
3	$\dfrac{1}{s}$	$u(t)$	$\dfrac{1}{1-z^{-1}}$	$u(nT)$
4	$\dfrac{1}{s^2}$	t	$\dfrac{Tz^{-1}}{(1-z^{-1})^2}$	nT
5	$\dfrac{1}{s+a}$	e^{-at}	$\dfrac{1}{1-z^{-1}e^{-aT}}$	e^{-anT}
6	$\dfrac{1}{(s-a)^2}$	te^{-at}	$\dfrac{Te^{-aT}}{(1-z^{-1}e^{-sT})^2}$	nTe^{-anT}
7	$\dfrac{a}{s(s+a)}$	$1-e^{-at}$	$\dfrac{(1-e^{-aT})z^{-1}}{(1-z^{-1})(1-z^{-1}e^{-aT})}$	$1-e^{-anT}$
8	$\dfrac{a}{s^2(s+a)}$	$t-\dfrac{1-e^{-at}}{a}$	$\dfrac{Tz^{-1}}{(1-z^{-1})^2}$ $-\dfrac{(1-z^{-1})}{a(1-z^{-1})(1-z^{-1}e^{-aT})}$	$\dfrac{1}{a}\left(anT-1+e^{-anT}\right)$
9	$\dfrac{ab}{s(s+a)(s+b)}$	$1-\dfrac{be^{-at}-ae^{-bt}}{b-a}$	$\dfrac{1}{1-z^{-1}}$ $+\dfrac{z^{-1}(be^{-bT}-ae^{-aT})-(b-a)}{(b-a)(1-e^{-aT}z^{-1})(1-e^{-bTz^{-1}})}$	$1-\dfrac{be^{-anT}-ae^{-bnT}}{b-a}$
10	$\dfrac{a}{s^2+a^2}$	$\sin at$	$\dfrac{z^{-1}\sin aT}{1-2z^{-1}\cos aT+z^{-2}}$	$\sin anT$
11	$\dfrac{s}{s^2+a^2}$	$\cos at$	$\dfrac{1-z^{-1}\cos aT}{1-2z^{-1}\cos aT+z^{-2}}$	$\cos anT$

may depend on the region or set of states in which the system is operating. Thus, stability may be a local phenomenon, and the system may be stable for one set of initial conditions and disturbances and yet quite unstable for another closely related set.

Linear System Stability

A *stable linear system or element* is one in which the system response is always bounded (does not increase without limit) for any bounded system input. An example of a stable system is $G(s) = (1 + Ts)^{-1}$; the system $G(s) = (1 - Ts)^{-1}$ is an example of an unstable system. A step input is a

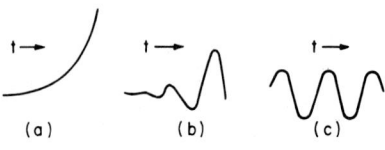

(a)   (b)   (c)

bounded input. A ramp input is not a bounded input, since it increases without limit with increasing time. Limits exist in all physical systems, and inputs or outputs beyond certain ranges cause the system to become nonlinear. In applying the linear system theory, care must be taken that the system remains substantially linear. If this is not so, extensions of linear theory may be useful. See Describing Functions.

For a stable linear system the response to any bounded input will not be of the form shown in Fig. 33a or b. Although responses of the form shown in Fig. 32c and d are mathematically stable responses, such oscillatory behavior is often not permissible. Figure 32e shows a stable response.

(d)   (e)

Fig. 33   Stable and unstable responses.

The stability of a discrete-data system is ensured if the poles (eigenvalues) of the system lie inside the unit circle about the origin in the z plane. Stability may be tested by means of the *Schur-Cohn criterion*. The methods for determining stability of a continuous linear system depend on the type of system. For lumped-parameter systems, it is ensured if the poles (eigenvalues) of the characteristic equation lie in the left-half plane. Stability may be tested with the *Routh-Hurwitz criterion* or the *Nyquist criterion*. For distributed-parameter systems, or systems with dead time, a lumped approximation can be employed; otherwise the Nyquist criterion can be applied.

Systems Equations and Eigenvalues

Linear system stability depends on the eigenvalues or poles of the closed-loop system equations. When the state variable equations are used, computer solutions may provide the eigenvalues. For simpler configurations where the transfer function approach is employed, it is necessary to evaluate stability based on the characteristic equation. Digital computer programs are available to solve for the roots of higher order algebraic equations. These are generally used to factor characteristic equations. An easy-to-remember method for hand calculation or desk calculator use is the *synthetic division method*, as follows (Ref. 44):

Given the characteristic equation

$$A_0 s^n + A_1 s^{n-1} + \cdots + A_{n-2} s^2 + A_{n-1} s + A_n = 0$$

divide through by A_0, obtaining

$$s^n + b s^{n-1} + \cdots + g s^2 + h s + k = 0$$

where

$$b = \frac{A_1}{A_0}, \ \ldots, h = \frac{A_{n-1}}{A_0}, k = \frac{A_n}{A_0}$$

If n is odd, the first approximation is $s + (k/h)$. If n is even, then the first approximation is $s^2 + (h/g)s + k/g$. Use synthetic division:

$$s^2 + \frac{h}{g}s + \frac{k}{g} \ \overline{\left) \begin{array}{l} s^{n-2} + \cdots \\ s^n + b s^{n-1} + \cdots + g s^2 + h s + k \end{array} \right.}$$

P = number of poles of $1 + G(s)H(s)$ in right-half s plane
Z = number of zeros of $1 + G(s)H(s)$ in right-half s plane

The application of the criterion to a case where $G(j0)$ equals a constant is shown in Fig. 34a. If the gain of $G(j\omega)$ were increased sufficiently, the plot would encircle the $-1 + j0$ point and the system would be unstable.

Figure 34b shows a plot for a case where $G(j0)$ approaches $1/j\omega$, and Fig. 34c shows a plot where $G(j0)$ approaches $(1/j\omega)^2$. In these cases a closure must be made with (infinite) circular contours as shown. The closure is made clockwise as ω goes from -0 to $+0$. Both the systems of Fig. 34b and c are conditionally stable, since either an increase or a decrease in gain will cause instability. In case instability occurs, the change in shape required to obtain stability is evident from the $G(j\omega)H(j\omega)$ plot.

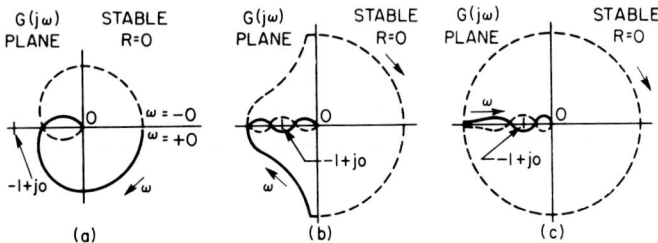

Fig. 34   Examples of the use of the Nyquist criterion.

It now remains to determine the zeros and poles when they exist in the right-half s plane. The method is as follows:

1. For open-loop stable single-loop systems with unit feedback, $R = 0$.

2. For a system with multiple loops, the loops are opened, starting with the innermost loop first, and the poles and zeros of the final main feedback loop are known. This may be accomplished either mathematically or by successive application of the criterion.

A more complete discussion of the Nyquist criterion can be found in Refs. 44 to 46.

Sampled-Data Systems

Stability considerations for linear systems containing sampling or other discrete operations are somewhat different from stability considerations for similar systems which omit these operations. In the linear continuous system, poles of the characteristic equation in the right-half s plane indicated instability. Since $z = e^{Ts}$, the left-half s plane (location of poles for stable systems) maps into a unit circle about the origin in the z plane. By substituting $s = \sigma + j\omega$, the magnitude of z becomes e^{Ts}. As $\sigma \to -\infty$, the magnitude approaches zero. For $\sigma = 0$, magnitude equals unity. $e^{j\omega T}$ is the equation for a circle of unit radius.

A direct method available for determining absolute stability of discrete-data systems is the *Schur-Cohn stability criterion* (Table 13). A second method which is sometimes useful is to map the interior of the unit circle into the left-half plane by means of a bilinear transformation:

$$z = \frac{1 + w}{1 - w} \tag{70}$$

where w = a complex variable.

The Routh-Hurwitz criterion can then be applied in the w plane. Also, the general stability methods (Stability of Nonlinear Systems) may be applied.

Stability of Nonlinear Systems

Since there are no general methods for analysis and design of nonlinear control systems, the designer usually avoids nonlinear stability problems where possible. However, nonlinear effects must some-

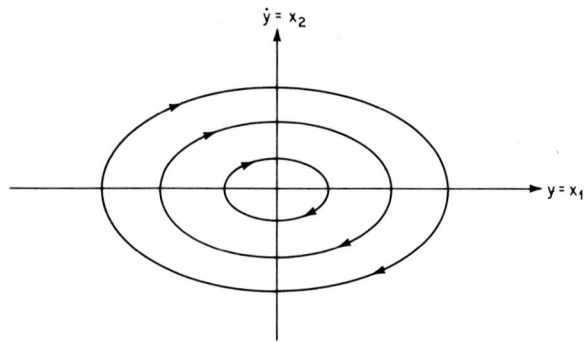

Fig. 35 Phase-plane trajectories for a center.

Table 14 Characteristics of Selected Types of Singular Points

$$\ddot{y} + 2\zeta\omega\dot{y} + \omega^2 y = 0$$

Singular Point	Eigenvalues	Phase Plane Plot
Center $\zeta = 0$ $\omega^2 > 0$		
Stable Focus $0 < \zeta < 1$ $\omega^2 > 0$		
Unstable Focus $-1 < \zeta < 0$ $\omega^2 > 0$		
Saddle Point $\omega^2 < 0$	S_1 S_2	

17.108

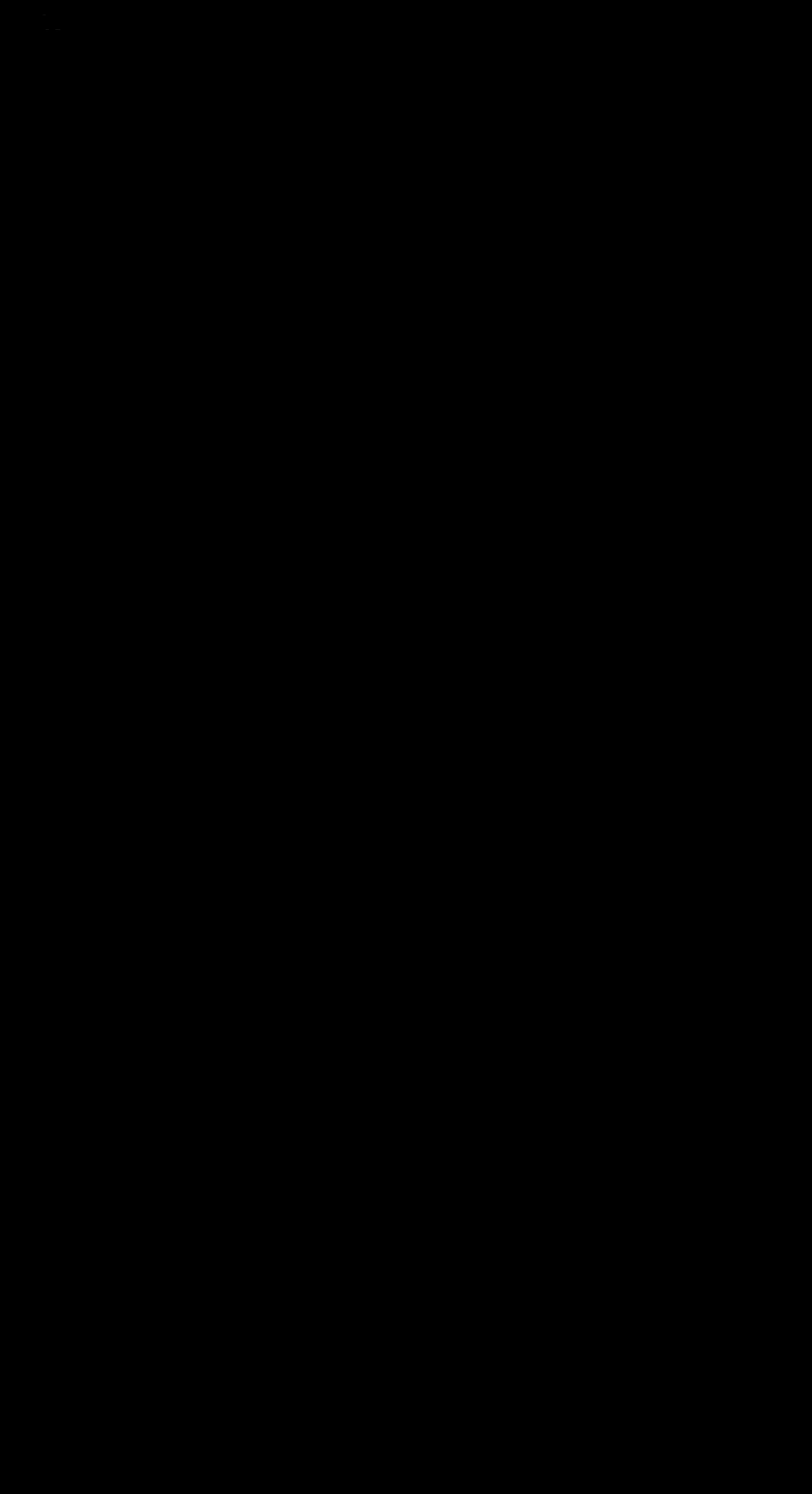

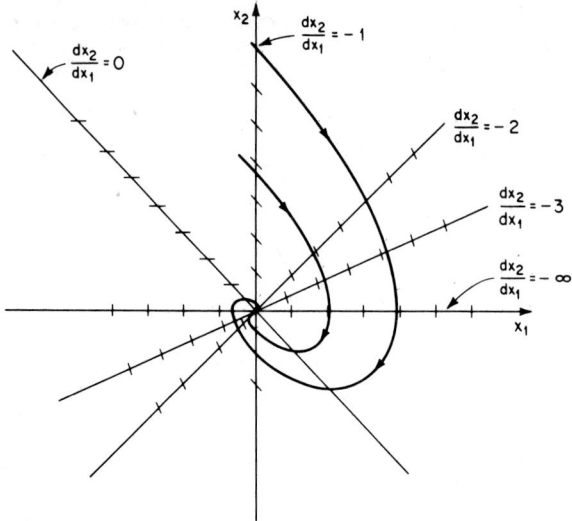

Fig. 36   Phase-plane portrait of the isoclines for $\zeta = 0.5$.

where $\mathbf{J}(\mathbf{x}_{is})$ is the *Jacobian matrix:*

$$\mathbf{J}(\mathbf{x}_{is}) = \begin{bmatrix} \dfrac{\partial f_1}{\partial x_1} & \dfrac{\partial f_1}{\partial x_2} & \cdots & \dfrac{\partial f_1}{\partial x_n} \\[2ex] \dfrac{\partial f_2}{\partial x_1} & \dfrac{\partial f_2}{\partial x_2} & \cdots & \dfrac{\partial f_2}{\partial x_n} \\[2ex] \cdots\cdots\cdots\cdots\cdots\cdots \\[1ex] \dfrac{\partial f_n}{\partial x_1} & \dfrac{\partial f_n}{\partial x_2} & \cdots & \dfrac{\partial f_n}{\partial x_n} \end{bmatrix} \tag{80}$$

Substituting $\mathbf{u} = \mathbf{x} - \mathbf{x}_{is}$, Eq. 81 is the *variational equation:*

$$\dot{\mathbf{u}} = \mathbf{J}(\mathbf{x})\mathbf{u} \tag{81}$$

Variational equations are used to investigate the stability of system behavior in the region of a singularity. The variational equation often reduces the original nonlinear equation to a familiar form with known solutions, thus greatly simplifying the analysis.

Describing-Function Techniques*

If a nonlinear system or component is subjected to a sinusoidal input, the output will not necessarily be a sine wave. However, in many practical cases, the output response will contain the input frequency as the fundamental Fourier-series components. In this case, the magnitude ratio and phase angle of the Fourier fundamental versus input amplitude, the frequency, and other factors form the *describing function*. The describing function may be determined either analytically or experimentally.

To use the describing-function techniques, the nonlinear components are separated from the "linear" components of the system. The transfer function of the linear component is its frequency

---

*Sometimes called *harmonic linearization*. For a more detailed discussion, see Refs. 30 and 47.

Table 15 Describing Functions

Type	Characteristics	Describing function*
Dead space		$a_1 = 0$ $\dfrac{b_1}{M} = 1 - \dfrac{2}{\pi}\theta_1 - \dfrac{D}{\pi M}\cos\theta_1$ where $\theta_1 = \sin^{-1}\dfrac{2M}{D}$
Hysteresis		$\dfrac{a_1}{M} = \dfrac{H}{\pi M}\left(\dfrac{H}{M} - 2\right)$ $\dfrac{b_1}{M} = \dfrac{1}{\pi}\left(\dfrac{\pi}{2} + \theta_1 + \dfrac{M-H}{M}\cos\theta_1\right)$ where $\theta_1 = \sin^{-1}\dfrac{M-H}{M}$
Saturation		$a_1 = 0$ $\dfrac{b_1}{M} = \dfrac{2}{\pi}\left(\theta_1 + \dfrac{B}{M}\cos\theta_1\right)$ where $\theta_1 = \sin^{-1}\dfrac{B}{M}$

$*\theta = M\sin\omega t$

$\theta_0 = \dfrac{a_0}{2} + a_1\cos\omega t + b_1\sin\omega t + a_2\cos 2\omega t + b_2\sin 2\omega t + \cdots$

$\theta = \dfrac{c_0}{2} + c_1\sin(\omega t + \phi_1) + c_2\sin(2\omega t + \phi_2) + \cdots$

where $c_1 = \sqrt{a_1^2 + b_1^2}$

$\phi_1 = \tan^{-1}\dfrac{a_1}{b_1}$

2. **Global Stability or Stability in the Large.** Applies to the entire state space of the system.

3. **Finite Stability.** Applies to a region larger than that of definition 1 but smaller than that of definition 2.

   Ordinarily definitions 1 to 3 do not exclude the possibility of limit cycles.

4. **Asymptotic Stability.** For any initial conditions within the region, the state point approaches arbitrarily close to the singularity as time approaches infinity. Dynamic equilibrium such as a limit cycle is excluded.

5. **Monotonic Stability.** Similar to asymptotic stability for *fixed-parameter systems*. The above generally apply only to autonomous (undriven) systems. *Total stability* for a nonautonomous (driven) system indicates that for every input within a reasonably small bound, the output is bounded.

Liapunov's Second Method

See De Russo et al., Ref. 31 and Gibson, Ref. 47. This provides several theorems which when applied to the so-called Liapunov V functions will indicate stability or instability. The result is obtained without solving the system equations, and the method is sometimes called the *indirect method*. It applies to systems that may be autonomous or forced, linear or nonlinear, stationary or time-varying, or deterministic or stochastic (Ref. 31, p. 498). The method is analogous to the rate-

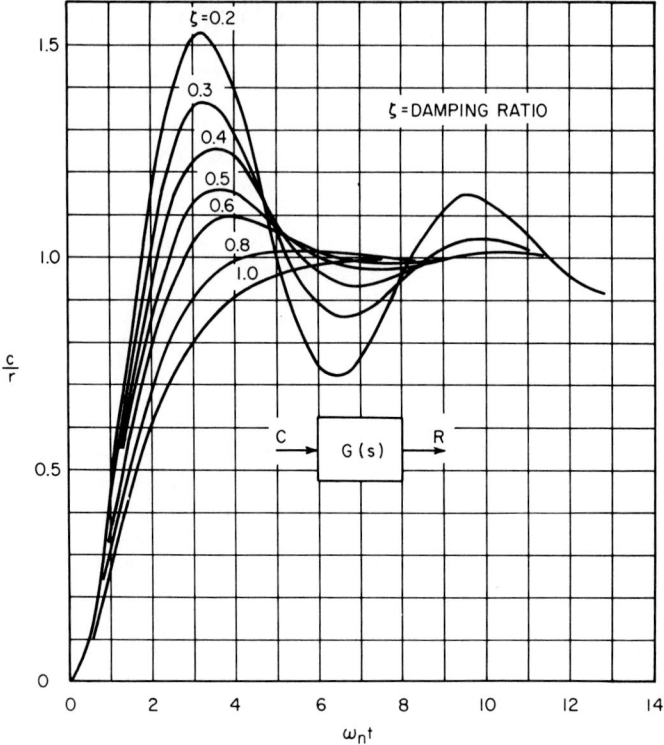

Fig. 37 Relative step response of element $G(s) = \omega_n^2(s^2 + 2\zeta\omega_n s + \omega_n^2)^{-1}$ versus ζ.

The step response for this system is shown in Fig. 37. The time response for Fig. 37 is given by Eqs. 90:

$$y(t) = 1 - \frac{e^{-\zeta\omega_n t} \sin (\omega_n \sqrt{1 - \zeta^2}\, t - \psi)}{\sqrt{1 - \zeta^2}} \tag{90a}$$

$$\psi = \tan^{-1} \frac{\sqrt{1 - \zeta^2}}{-\zeta} \tag{90b}$$

The parameters ζ and ω_n of the response are determined by the s-plane locations of the conjugate complex poles of $G(s)$, as shown in Fig. 38. The transient response to other types of input still contains terms involving the exponential envelope $e^{-\zeta\omega_n t}$ and the damped sinusoidal response with an oscillating frequency $\omega_n \sqrt{1 - \zeta^2}$ for $\zeta < 1$. For other types of inputs, the transient parameters persist as shown in Fig. 39.

Actually, the presence of additional poles or additional zeros distorts the response somewhat. The effect of a zero in $G(s)$ is indicated by comparing numbers 10 and 11 in Table 6. If the denominator of $G(s)$ contains an additional term $s + \lambda$, where λ is real and positive, an additional term $e^{-\lambda t}$ appears in the transient response and the coefficient of the second term of Eq. 90a will contain a term in λ. For a dominant second-order system of the type shown, the specification of ζ and ω_n is often a useful performance criterion.

Related performance criteria are sometimes specified. Some typical definitions of system performance following a set-point change are shown in Fig. 40.

Newton et al. (Ref. 58) use Parseval's theorem,

$$\int_{-\infty}^{\infty} x_1(t)x_2(t)\ dt = \frac{1}{2\pi} j \int_{-j\infty}^{j\infty} x_1(s)x_2(-s)\ ds$$

to provide tables for evaluating the integrals where $x(s)$ is a rational function of s. A limited table is given in Gould's book (Ref. 15, appendix B).

An easily programmed method has also been proposed by Pemberton (Ref. 59) for evaluating ISE and integral time squared error (ITSE) for systems with distinct poles (eigenvalues).

Continuous-Control System Design

Complete specification of system performance generally includes a definition of *steady-state errors* as well as a description of dynamic performance. Systems with nonzero steady-state errors include certain systems with proportional control and feedforward control systems without feedback. Given the performance requirements, the basic design problem is to meet these requirements by applying (1) compensating control functions (lead and lag functions), and (2) feedback paths. The process generally consists of altering the open-loop poles of the system to achieve an acceptable pole-zero configuration for the (closed-loop) system. The design of feedforward systems is usually a matter of modeling and simulation and is often based on transient response methods (Refs. 60 and 61). See also Control Concepts. If both feedforward and feedback are used, the feedforward design is often performed first. The discussions below are concerned primarily with feedback control.

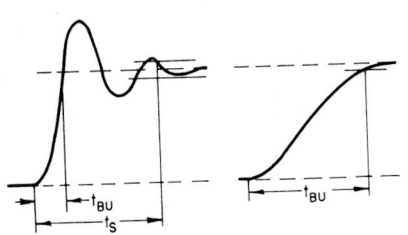

Fig. 40   Performance definitions.

Root-Locus Techniques

The performance of a lumped linear system is determined by the poles and zeros of the system and the zeros of the transfer function. The root-locus method of Evans (Ref. 48) permits the rapid estimation of closed-loop poles and zeros of the system. Consider the transfer function of the single-loop control system:

$$\frac{C}{R}(s) = \frac{G(s)}{1 + G(s)} = \frac{KA(s)}{KA(s) + B(s)} \tag{91}$$

where $G(s) = KA(s)/B(s)$. $A(s)$ and $B(s)$ are polynominals in s. K is a constant steady-state gain of $G(s)$. The zeros of C/R are the zeros of $G(s)$; the poles of C/R are the poles of $1 + G(s)$. Therefore, the poles are obtained by solving

$$1 + G(s) = 1 + K\frac{A(s)}{B(s)} = 0 \tag{92}$$

or

$$\frac{A(s)}{B(s)} = -\frac{1}{K} \tag{93}$$

Vectorially this can be written as

magnitude

$$\frac{A(s)}{B(s)} = \frac{1}{K} \tag{94}$$

and angle

$$\frac{A(s)}{B(s)} = \pm n180° \tag{95}$$

Frequency-Response Techniques

For a system in which the set-point response is of concern, a dominant second-order system is often assumed for the closed-loop response. The performance specification can then be made in terms of ζ and ω_n, as discussed previously. Frequency-response plots for the second-order system are shown in Fig. 43.

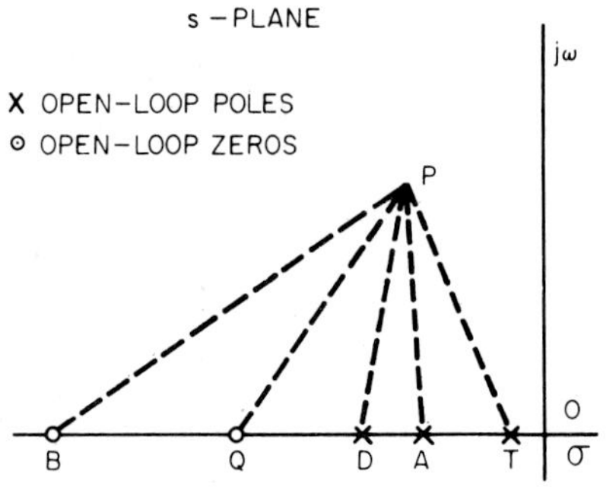

s − PLANE

X OPEN−LOOP POLES

o OPEN−LOOP ZEROS

Fig. 42   Root location for a system with three poles and two zeros.

Another method frequently employed is the M_p *criterion*, which uses the peak value of magnitude ratio M_p as a measure of the relative stability of the system. Since underdamped second-order systems with ζ less than 0.7 are commonly used to obtain fast response, a peak value exists, and this method may be applied. The speed of response is indicated by the resonant frequency at which the peak value occurs. Generally, values of M_p between 1.3 and 1.5 (2.28 to 3.52 dB) give satisfactory stability. Figure 44 shows M_p versus ζ for a second-order system.

Frequency-response techniques use "lead" or "lead/lag" functions to distort the open-loop transfer function $G(jw)$ in the critical frequency range near $-180°$ phase shift. The log magnitude and phase functions of a lead or lag function are shown in Fig. 45. Depending on the value of α, useful phase leads of about $+35$ to $+60°$ can be obtained. Although a pure "lead" might be preferred, noise sensitivity and excessive gain at higher frequencies dictate values of α between 5 and 15. The effect of lead or lag compensation is shown in Fig. 46.

Regulation and Disturbances

The effect of closed-loop control on disturbances is important, since many control loops function as regulators. Equation 97 expresses the ratio of the error amplitude to the amplitude of a sinusoidal

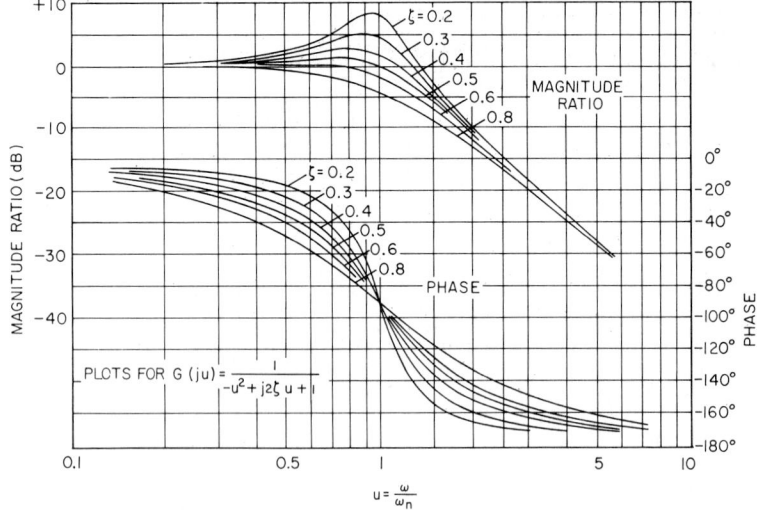

Fig. 43   Log magnitude and phase plots for $G(ju) = 1/(-u^2 + j2\zeta u + 1)$.

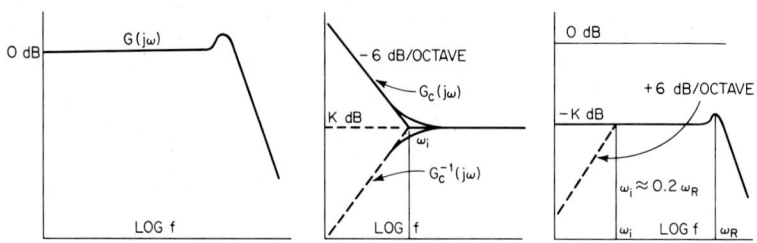

Fig. 47   Log magnitude plots for error versus disturbance.

different from unity at phase crossover. *Phase crossover* is a point on the plot of loop ratio $G(j\omega)H(j\omega)$ at which its phase angle is $-180°$. *Phase margin* is the angle by which the phase of the loop ratio $G(j\omega)$ of a stable system differs from $-180°$ at gain crossover. *Gain crossover* is a point in the plot of loop ratio $G(j\omega)$ at which the magnitude of the loop ratio is unity. See Fig. 48. Some judgment is necessary in applying either of the two methods, since the rules tend to break down for systems that exhibit certain types of closed-loop frequency responses.

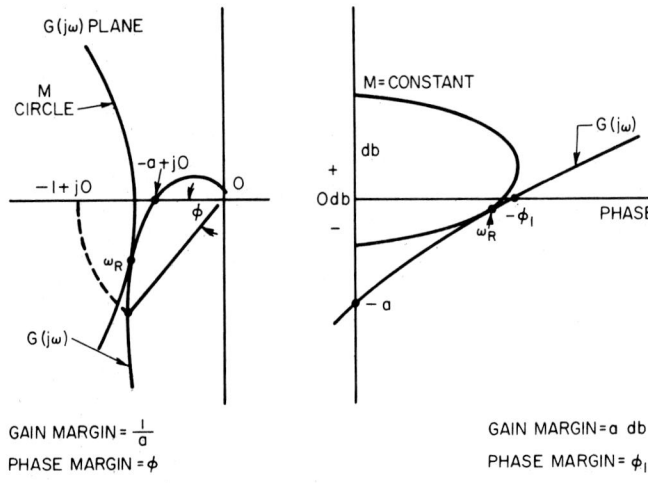

GAIN MARGIN = $\dfrac{1}{a}$
PHASE MARGIN = ϕ

GAIN MARGIN = a db
PHASE MARGIN = ϕ_1

Fig. 48   Gain margin and phase margin.

Sampled-Data System Design

Sampled-data control systems can be designed using approximations based on continuous control techniques—however, there are significant advantages in designs based on purely sampled-data techniques.

1. A wider range of control functions can be realized. (In particular, realistic simulation of dead time.)
2. Input and disturbance time functions can be specified.
3. Computations and logic are more easily accommodated, and special control algorithms are not difficult to develop, when necessary.
4. Digital computer design aids and simulation programs make it feasible to design in the time domain.

Setting

$$D(z) = \frac{M(z)}{E(z)}$$

where $M(z)$ = transform of the manipulated variable and $E(z) = R(z) - C(z)$.
  Solving,

$$M(z) = \frac{1-q}{K(1-a)}(e_n - ae_{n-1}) + qM_{n-1} + (1-q)M_{n-N-1} \tag{101}$$

where $q = e^{-T/\lambda}$.
  Krivoshein and Corripio (Ref. 64) present a detailed analysis of the Dahlin direct synthesis controller. They also discuss the causes and avoidance of "ringing" which can occur when the process must be approximated by two time constants plus dead time, rather than one time constant plus dead time. Ringing causes excessive activity of the final control element. Simulation results are presented.

Control with Dead Time

The Smith linear predictor has been discussed earlier. Another approach using prediction based on a model has been proposed by Moore. Applications have been summarized by Meyer et al. in Ref. 22. A shortcoming of systems for control with significant dead time has been the relatively poor response to unmeasured disturbances. A controller proposed by Gautaman and Mutharassan (Ref. 65) is designed to overcome this deficiency. The above systems are summarized by Ash and Deshpande in Ref. 4, including summaries of test results. Vogel and Edgar (Ref. 66) describe an adaptive control for dealing with variable dead time.

Design Aids

Design aids are available to assist in the design and simulation of sampled-data control systems. For example, Ash and Deshpande (Ref. 4) present examples of FORTRAN programs for (1) converting $C(z)$ as a ratio of polynomials in z to a time series in z (z-transform inversion), and (2) a closed-loop simulation of a sampled-data PID control system with a user-specific set point or disturbance inputs and a process simulation of dead time plus one or two first-order lags.

CONTROL STRATEGIES

The selection of a control strategy requires that two types of information be available: (1) the characterization of the process, including a dynamic model with a definition of disturbances; and (2) a clear understanding of control objectives and criteria for evaluating performance. The strategies for various types of process model characteristics are discussed below. However, in applying some of the more sophisticated techniques it is important to realize that the control parameters to be adjusted must usually have some rational physical interpretation. There are two alternatives: (1) The control parameters can be calculated and presented to an operator based on some criteria. This method was used by Adkison and Kohler (Ref. 67). (2) Adaptive techniques can be used.

Process Characteristics

The selection of a control strategy depends on the type of process model, as shown in Fig. 49. The assumption has been made here that normal operation involves significant changes in process parameters. For this reason, a preference for feedback control is indicated, even where feedforward control is used. With this approach, the feedback control will usually compensate for the effect of process parameter variations.

Control of Distributed Systems

The treatment of distributed systems in Fig. 49 requires some explanation. In a *shunt configuration,* the measurement and the final control element are in close proximity at "one end" of the system, e.g., a local pipeline pressure control system. In a *series configuration* the measurement and the final control element are at "opposite ends" of the distributed system, e.g., pipeline pressure control from a remote pressure measurement. See Buckley (Ref. 2) for details and examples.

Single-Variable Control with Process Dead Time

If the process model indicates a significant amount of dead time, performance of feedback control may be inadequate. One of three approaches can be followed: First and simplest, feedforward control can be used. Second, a special dead time controller can be used, such as the one described previously. Figure 49 covers only single-loop and simple strategies. If interaction between control loops is expected, this can be checked and remedied as discussed under Multivariable Control. Noise is also a special problem and is discussed below.

Cascade Control

It can be seen from Fig. 50 that the secondary loop appears to the primary as its closed-loop characteristic. The method of analysis is to design the secondary loop first. Franks and Worley (Ref. 68) present the results of an analog computer study indicating improvements which are possible using cascade control versus single-loop control. The results compare the performance of the two configurations using the ratio of ITAE for single-loop to ITAE for cascade versus the ratio of dominant lags in the primary and in the secondary. Step disturbances were used. For primary set-point disturbances, relative performance improvements ranged from about 3:1 to 5:1 as the ratio of primary-to-secondary time constants ranged from 10:1 down to 1:1. The corresponding improvement in response to primary process disturbances ranged between 6:1 and 8:1. The response improvement for secondary disturbances ranged from very large values down to about 80:1 when the lags were equal. Schork and Despande (Ref. 69) discuss the analysis and tuning of a double cascade.

Noise

Noise is a factor in the design and the tuning of some control systems. Most random noise found in industrial control applications can be characterized as exponentially correlated noise. Fertig (Ref. 14) describes a method for characterizing noise amplitude and noise bandwidth in terms of easily measured parameters. Avoidance of excessive final-control-element motion using filtering, reduced gain, and nonlinear gain are analyzed. Filter parameters, scan frequencies, and PID control parameters are recommended for sampled-data control. Tuning charts are included, and a furnace pressure control application is described.

Multivariable Control

In the design of multivariable control systems, an adequate process model is crucial. The most useful applications of the theory have been in the selection of variables in applications involving interaction (Refs. 1, 70, and 71).

The relative gain array (RGA) method of Bristol (Ref. 72) is very useful for this purpose. The method is described below with a simple example. The general method for larger arrays is summarized in Table 16. McAvoy (Ref. 142) explores the technique in detail with a variety of application examples and provides extensive references.

Blending System

Figure 51 shows a simple blending system. Composition X and total flow C are to be controlled using control valves on streams A and B. Interaction is to be minimized. System equations are:

$$\text{Total flow} \quad C = M_1 + M_2$$

$$\text{Composition} \quad X = \frac{M_i}{M_1 + M_2}$$

Table 16 Interaction Analysis

1. Form array $\mathbb{A}$ of measured variables and manipulated variables and calculate the *uncontrolled response* for each pair.

Manipulated variables

		m_1	m_2	$\cdots$	m_n
Controlled variables	C_1	a_{11}	a_{12}	$\cdots$	a_{1n}
	C_2	a_{21}	a_{22}	$\cdots$	a_{2n}
	.	.	.		.
	.	.	.		.
	.	.	.		.
	C_n	a_{n1}	a_{n2}	$\cdots$	a_{nn}

2. Calculate the inverse $\mathbb{A}$ matrix $\mathbb{A}^{-1}$ and transpose $\mathbb{A}^{-1}$. This is the $\mathbb{B}$ matrix.

$$\mathbb{B} = (\ \mathbb{A}^{-1})^T$$

3. Form the relative gain array $\mathbb{\Lambda}$ by multiplying corresponding terms of $\mathbb{A}$ and $\mathbb{B}$. (Note: This is not conventional matrix multiplication.)
4. Select manipulated-variable–controlled-value pairs by selecting those with positive relative gain closest to 1.0.
5. Properties of the relative gain array (RGA):
   a. Rows and columns of $\mathbb{\Lambda}$ sum to 1.0.
   b. If $a_{ij} = 0$, then $\lambda_{ij} = 0$.
   c. Pairing on a negative RGA element results in either an unstable system or an inverse responding system.

The RGA is

$$\mathbb{\Lambda} = \begin{vmatrix} \lambda_{11} & \lambda_{12} \\ \lambda_{21} & \lambda_{22} \end{vmatrix}$$

where $\lambda_{ij} = \left. \dfrac{\partial x_i}{\partial M_j} \right|_{\text{all loops manual}} \Big/ \left. \dfrac{\partial X_i}{\partial M_j} \right|_{\text{all loops but } M_j - X_i \text{ pair perfect}}$

$X_1 = C$
$X_2 = X$

"Perfect" means that the controlled variable is maintained at the design value.

The arrangement of controlled variables and manipulated variables in the array is shown below.

	M_1	M_2
C	λ_{11}	λ_{12}
X	λ_{21}	λ_{22}

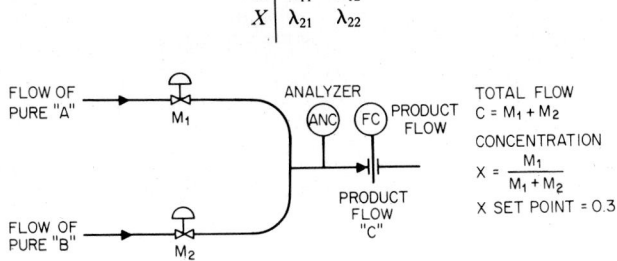

Fig. 51 Simple blending system.

RGA elements are:

$$\lambda_{11} = \cfrac{\left.\cfrac{\partial C}{\partial M_1}\right|_{M_2 = \text{const}}}{\left.\cfrac{\partial C}{\partial M_1}\right|_{X = X_0}} \qquad \lambda_{12} = \cfrac{\left.\cfrac{\partial C}{\partial M_2}\right|_{M_1 = \text{const}}}{\left.\cfrac{\partial C}{\partial M_2}\right|_{X = X_0}}$$

$$\lambda_{21} = \cfrac{\left.\cfrac{\partial X}{\partial M_1}\right|_{M_2 = \text{const}}}{\left.\cfrac{\partial X}{\partial M_1}\right|_{C = C_0}} \qquad \lambda_{22} = \cfrac{\left.\cfrac{\partial X}{\partial M_2}\right|_{M_1 = \text{const}}}{\left.\cfrac{\partial X}{\partial M_2}\right|_{C = C_0}}$$

Solving for λ_{11}

Differentiate the C equation with constant M_2 (Fig. 51)

$$\left.\frac{\partial C}{\partial M_1}\right|_{M_2} = 1$$

From the X equation (Fig. 51)

$$X(M_1 + M_2) = M_1$$

Differentiating with $X = X_0$

$$X_0\left(1 + \left.\frac{\partial M_2}{\partial M_1}\right|_{X = X_0}\right) = 1$$

$$1 + \left.\frac{\partial M_2}{\partial M_1}\right|_{X = X_0} = \frac{1}{X_0}$$

Differentiating the C equation with $X = X_0$

$$\left.\frac{\partial C}{\partial M_1}\right|_{X = X_0} = 1 + \left.\frac{\partial M_2}{\partial X_1}\right|_{X = X_0}$$

Substituting for $1 + \left.\dfrac{\partial M_2}{\partial X_1}\right|_{X = X_0}$

$$\left.\frac{\partial C}{\partial M_1}\right|_{X = X_0} = \frac{1}{X_0}$$

and

$$\lambda_{11} = \left.\frac{\partial C}{\partial M_1}\right|_{M_2 = \text{const}} \Bigg/ \left.\frac{\partial C}{\partial M_1}\right|_{X = X_0} = X_0$$

Other elements can be similarly derived. However, rows and columns of the RGA Λ sum to 1.0, and the RGA can be written down immediately for the 2 × 2 array.

$$\Lambda = \begin{bmatrix} X_0 & 1 - X_0 \\ 1 - X_0 & X_0 \end{bmatrix} = \begin{bmatrix} \lambda_{11} & \lambda_{12} \\ \lambda_{21} & \lambda_{22} \end{bmatrix}$$

Substituting for X_0 (Fig. 51)

$$\Lambda = \begin{bmatrix} 0.3 & 0.7 \\ 0.7 & 0.3 \end{bmatrix}$$

The least interactive pairs are selected as those with the λ's closest to 1.0, as shown by the circled elements below.

$$
\begin{array}{c c}
 & M_1 \quad M_2 \\
C & \boxed{\begin{array}{cc} 0.3 & \textcircled{0.7} \end{array}} \\
X & \boxed{\begin{array}{cc} \textcircled{0.7} & 0.3 \end{array}}
\end{array}
$$

Stream B flow is selected to control total flow C; stream A flow is selected to control concentration X.

Based on this analysis, if interactions appear to be a problem, control loops may be decoupled (Refs. 15, 30, and 73).

Matrix methods are sometimes used for the design of multivariable control systems, as discussed under State Variable Theory and under System Design. However, where noninteraction is sought, the number of controller functions can become large. In a two-input–two-output system, a total of four controller functions is required. Two of these functions are needed to cancel interactions.

Practical Considerations

Good system design involves many practical considerations. Three examples are cited. Shunta (Ref. 74) describes an application requiring sampled-data control of level with a dead time. The storage tank receives product from a distillation train and maintains the level within tank limits by controlling feed to the train. Luyben sampled-data design charts (Ref. 75) are used to design a proportional-level control system to a specified relative damping factor. The analysis includes estimating the vessel size based on expected conditions. Two papers give a good idea of the practical factors involved in the design of pH control systems. A paper by Hoyle (Ref. 76) describes the influence of tank volume, mixing, and inlet-outlet locations on process dead time. An article by Shinskey (Ref. 77) describes experience with an adaptive control system and a difficult pH application.

MODELING AND SIMULATION

Process models are necessary for control-system design and are fundamental to all practical types of optimal or adaptive control. The appropriate type of model may be quantitative or qualitative. For most applications a quantitative model is necessary. (For an example of a qualitative model see Hill Climbing.) It is convenient to classify the method of derivation of quantitative models as *fundamental* or *empirical models*. The resulting model may be composed of algebraic or differential equations and may be either linear or nonlinear.

To achieve a particular degree of control the model should represent the state of the process over the expected range of operating conditions to the accuracy with which the objective function is expected to approach the true maximum. It therefore follows that a good model is necessary for a good optimal or adaptive control system.

Fundamental Model

This can be derived from basic principles and knowledge of the process, is the most accurate, and gives the best control system. It can, however, lead to computational problems if it is too detailed. By means of a fundamental model, parameters can be calculated from basic physical data or from process design information. The fundamental derivation of the model should be used whenever possible.

Empirical Model

This is generally based on a gross assumption about the nature of the process so as to give a set of equations that is easy to solve. Techniques such as multiple regression have a model, often a linear one, built in. With an empirical model it is not so easy to include theoretical information. An example of an empirical model is the assumption that the process can be represented by a second-order lag with a constant gain.

Model Formulation

The first step in the formulation of any model is to establish the boundaries of the portion of the process to be modeled. The choice of the boundaries should be made so as to:

1. Exclude as many variables and equations as possible
2. Eliminate as many unmeasured disturbances as possible
3. Hold measured (uncontrolled) disturbances to a minimum
4. Allow explicit relationships between measured variables and factors that disturb the process
5. Include all significant variables

These requirements are often conflicting, and compromises must be made. A good starting point is often a reasonably simple, hence manageable and understandable, model. The process characteristics described by Gould (Ref. 15), Lee et al. (Ref. 78), Franks (Ref. 79), Smith et al. (Ref. 80), Luyben (Ref. 81), and Kim and Friedly (Ref. 82), and in the references given at the end of this article often provide basic equations which can be used. Design equations can be applied in many cases if the physical dimensions are treated as constants and the flows, temperatures, etc., treated as variables.

For example, in control investigations, process dynamics are often simulated by a dead time and two first-order lags. Reference 82 shows that this approximation can be justified for large staged systems. Three different methods of analysis are shown to lead to the same form and to approximately the same parameter values.

Economic models are a second level of fundamental models. They are derived by applying cost-and-profit relationships to the various physical operations and then employing a specific accounting method (return on investment, discounted cash flow, etc.) to establish a *profit objective function.* These models are often useful over and above the process model. Any model should meet the following criteria: (1) It should be capable of representing the process to the accuracy desired over the range of operating conditions required. (2) The equations should be solvable for the desired control function. (3) Data must be available to evaluate the constants in the equations.

Some methods do not require the explicit statement of a model; however, inevitably there is a model which is implicit in the method. For example, in the experimental gradient hill climbing method, the general assumptions are usually of the following form: (1) The process variables are continuous and have continuous first and second derivatives in the vicinity of the current operating point. (2) The dynamics of the process are fast compared to the interval for each experimental point. (3) The process is stable and does not change significantly during the running of a large number of experiments. (4) The noise level (unmeasured disturbances) is low enough, the measurements accurate enough, and the allowable perturbations large enough to establish gradients.

Steps in Building a Model

1. The operating procedures and practices for start-up, shutdown, and steady-state operation, changes in operation, raw material changes, grade changes, etc., must be written down and described in a quantitative manner. For existing plants the actual operating practice should be closely observed for subtleties in operation that may be critical to the real operation of the process (Weiss, Ref. 83).

2. After the boundaries of the process have been determined, all the inputs and outputs that cross these boundaries must be examined and reasonably defined. These include measurements, disturbances, and manipulated variables. All three of these should be examined for range, noise, dynamic response, nonlinearities, resolution, hysteresis, maximum slew rate, sensitivity, and bias. The nature, form, and location of the various disturbances should be established.

3. Internal to the process, the range of parameters which describe the system should be established as a function of the operating point over the range through which the operating point is moved. This includes steady-state and dynamic characteristics (dead time, lags, and gains). Given the model of a mixing operation in an open tank, the lags are a function of the liquid level of the

tank, and hence the range of the liquid levels of the tank establishes a range for the lags in the process.

4. Use the available data and run the experiments necessary to evaluate the parameters in the model; then verify the model by making sure that it matches the process. Data on various parameters such as absolute reaction rates or thermal conductivity may well be obtained from batch, laboratory, pilot plant semiworks, or from the literature or from tests on existing plants or equipment. For example, the heat-transfer characteristics of a similar heat exchanger in some other type of operation may be directly applicable. Information for a particular parameter may be available from more than one source. The value chosen may be based on the accuracy of the data available. Several methods of data anlysis such as multiple regression analysis and multiple regression after filtering can then be applied to the experimental data and the values for individual parameters chosen on the basis of the least probable error (see Dahlin and Brewster, Ref. 84).

5. When the parameters in the process model are being estimated, the error in the estimation should also be evaluated. There are several important reasons for this. First, errors can propagate through the process and the propagation of these errors can become significant. Mickley et al. (Ref. 16) give the mathematics of the propagation of the errors. Second, it indicates the reliance which can be placed on the results and hence how close to the limits it is reasonable to operate. Third, it provides criteria for dropping terms. Eliminate the parameters for which the values cannot be established with sufficient accuracy (see Ref. 85).

Model Strategies

In many types of systems the difficulty of dealing with a large model having many nonlinearities can be circumvented by splitting the model into two or more parts. One approach is to build a simple linearized model including the dynamics that represent a portion of the system at a particular operating point. The other portion of the model would include little or no dynamics and could be very complex and detailed. A hierarchical system may make this type of structure feasible. The models may in fact be in two different computers, at two different levels in the hierarchy. This follows through to the optimization where a simple model, hence a simple optimization, could be used to obtain tight first-level control. A more complicated, but nondynamic, non-differential-equation model could be used for the optimization operation which is less dynamic in nature and may be performed infrequently.

Contingency

For any process with a reasonable amount of equipment involved, there always exists the possibility of equipment, sensor, and actuator failures, open wires, short circuits, or out-of-normal-range signals. A capability for coping with these conditions may be the reason for selecting one model over another, and the important characteristics to consider are:

1. Detection of the problem situation and signaling the occurrence

2. A strategy for correcting the situation, that is to say, keeping the operation going if possible in spite of the failure. This should cause a minimum degradation of the system performance, and the system should "fail soft."

3. There should be a recovery procedure for getting back on line after the difficulty has been corrected. This includes proper initialization, etc.

For any or all three of these steps operator intervention may be required; hence a good operator interface may be needed. For example, suppose that a model is good only over a certain range and a signal is received that is outside the range; whether it is a true signal or an error condition, this must be detected and appropriate action must be taken. Under certain circumstances such as a missing signal, a control system should not be started up or placed in operation. Interlocks may be desired to ensure that all the required conditions are met before initiation. Testing may be either by automatic checking through the computer or by detailed operator responses to queries, indicating that all conditions have been met. Usual engineering practice is to provide for first-level failures. Design for outage beyond first-level and backup equipment is seldom profitable. It may be necessary to provide for manually overriding a single interlock, in case a contact fails, or something of this nature. These characteristics should be examined carefully in terms of reliability, safety, and cost. Any adaptive or optimizing scheme should be checked by simulation for these characteristics before it is placed in operation.

Linear and Nonlinear Algebraic Equations

Multiple regression techniques are often applicable in evaluating constants in equations of the form

$$x_i = \sum_{j=1}^{m} a_{ij}u_j \quad \text{or} \quad \mathbf{x} = \mathbf{Au} \tag{102}$$

where $\mathbf{x}$ = state vector and $\mathbf{u}$ = control vector.

A number of techniques exist for control based on this type of model, particularly with quadratic objective functions. Because the coefficients generally do not bear a simple relationship to identifiable physical or other known characteristics of the process, great care must be taken to make sure that the model is a true representation of the process. Some types of nonlinearities can be introduced by treating the nonlinear terms as another independent variable.

Although this overcomes some of the problems of the linear model (small range of applicablility, no indication of a curvature, etc.), the difficulty in verifying the model is increased along with other computational problems.

Statistical Models

A statistical model deals chiefly with future probabilistic behavior, irrespective of the past history of the process. Statistical models are in general most useful in describing the environment around the process being controlled. When they are applicable in a process model it is frequently for some type of discrete operation or filtering as in a Kalman filter.

As an example, if the noise in the measurement of the state of a process is equivalent to the output of a linear dynamic filter with gaussian noise as input (Gauss-Markov process), then a Kalman filter will give an optimal estimate of the state. On the other hand, statistical methods such as multiple regression, autocorrelation, and cross-correlation are usful in determining the parameters in a deterministic model. Bryson and Ho (Ref. 86) describe the mathematics involved in differential games, probability, random processes, and Gauss-Markov sequences and processes.

Differential Equations or Transfer Functions

The general state vector approach

$$\dot{\mathbf{x}} = \mathbf{f}(\mathbf{x}, \mathbf{u}, \mathbf{z}, t) \tag{103}$$

is suitable if the functions can be defined and the resulting control problems solved.

This standard form of state equations is directly translatable into an analog or digital computer simulation. The analog simulation is developed as follows:

1. Set up an integrator for each state variable. The output of the ith integrator is x_i.
2. Set up a simulation of the various other terms in the equations. Set up noise generators for noise inputs, etc.
3. Set up the functions of x_i, u_i, t for $\dot{x}_i$. Connect each of these functions to the input of the corresponding integrator.
4. Set the initial conditions on the integrators to the desired state of the system at time zero.

For example, if the equations are of the form

$$\dot{x}_1 = -(ax_1 + bx_2)cx_3 + dx_n \tag{104}$$

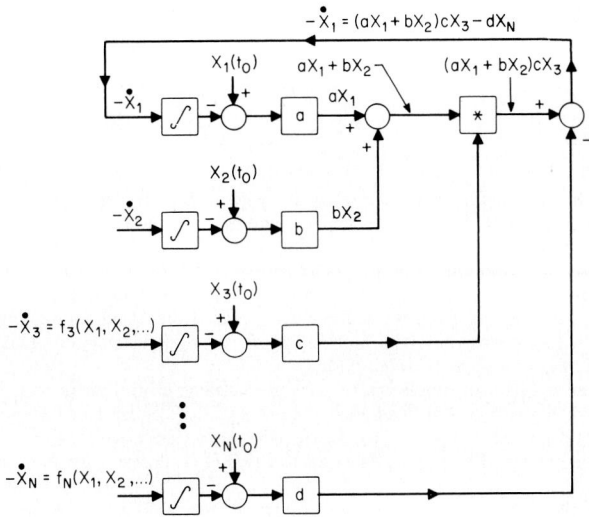

Fig. 52 Simulation of state equations.

$$\dot{x}_2 = -f_2(x_1, x_2, \ldots) \tag{105}$$

$$\dot{x}_3 = -f_3(x_1, x_2, \ldots) \tag{106}$$

. . . .

$$\dot{x}_n = -f_n(x_1, x_2, \ldots) \tag{107}$$

then the form of the simulation diagram is as shown in Fig. 52. Only the first equation is shown completely; the other equations would be treated in the same manner.

Typically the simpler form

$$\dot{\mathbf{x}} = \mathbf{A}\mathbf{x} + \mathbf{B}\mathbf{u} \tag{108}$$

is used for many analytical techniques. Some nonlinearities can be accomodated by transformation, as in the linear and nonlinear algebraic models above.

Finite Difference Equations

Finite difference equations are convenient for digital computer operations. Differential equations, or transfer functions, can be transformed into finite difference equations (Ref. 87).

A general form of the finite difference equations employing a fixed sample interval is

$$x_i = \sum_{j=1}^{m} \sum_{k=1}^{n} a_{jk} x_j Z^{-k} + \sum_{j=1}^{m} \sum_{k=0}^{n} b_{jk} u_j Z^{-k} \tag{109}$$

This form is applicable both as a model and as a form for a sampled-data controller.

The transfer function $x/u = 1/(\tau s + 1)$ or the equivalent differential equation $\dot{x} = -(1/\tau)(u + x)$ can be approximated using forward differences by

$$x(t) = \left(1 - \frac{\Delta t}{\tau}\right) x(t - 1) - \left(\frac{\Delta t}{\tau}\right) u(t) \tag{110}$$

or in the z-transform notation,

$$x = \left(1 - \frac{\Delta t}{\tau}\right) xz^{-1} - \left(\frac{\Delta t}{\tau}\right) u \tag{111}$$

Simulation

Simulation permits testing and validation of control schemes and devices under a variety of conditions, some of which would be difficult, costly, or dangerous to duplicate on the actual process. Simulation also permits a comparison of alternative strategies under identical conditions. Time can often be compressed in the simulation, thus decreasing test time. Simulation is often used for (1) testing control strategies prior to installation, with or without the control hardware; (2) validating a model for optimization or for adaptive control under a wide range of conditions and parameter mismatch; and (3) operator training, especially in the handling of contingency situations. This is useful both for preinstallation and for ongoing training.

For a routine checkout of simple control strategies, analog control modules or microprocessor algorithms can often be connected to simulate the process. Lead or lag modules or algorithms are often sufficient to represent the process dynamics. Microprocessors usually contain an algorithm to simulate dead time using a delay table. Analog simulation of dead time is usually not attempted for simple cases.

Three major types of electronic equipment have been successfully used for general large-scale process control system simulations: (1) analog, (2) digital, and (3) hybrid. Analog simulation and hybrid simulation were used extensively from 1950 to 1960 but have been largely replaced by digital simulation.

Analog Simulation

This is convenient and effective for systems of ordinary differential equations. Most kinds of nonlinearities are easily accommodated. The exceptions are large delays and large numbers of discrete-data operations. If the system of equations is entirely discrete data, then digital simulation is indicated. Modest numbers of logic and/or sampled data operations can be handled on an analog computer.

Analog computers have a fixed limit on the accuracy of simulation of $\sim$0.1 to 0.01%, which is adequate for most process control simulations. Detailed information on analog simulation can be found in Refs. 88 to 90.

The principal restriction in analog computation is the size of the problem. It is limited by the number of amplifier and integrators, and in general by the number of computational elements available in the hardware because of the parallel operation. Dead time is not directly simulated with normal analog elements. It must be approximated or simulated with special dead time simulation equipment or hybrid simulation.

Dead Time Approximations

A commonly used approximation for small amounts of dead time (0 to 0.4 radian) is the Padé approximation. The most typically used formulas are:

$$e^{-st} \approx \frac{s^2t^2 - 6st + 12}{s^2t^2 + 6st + 12} \tag{112}$$

$$e^{-st} \approx \frac{(st)^4 - 20(st)^3 + 180(st)^2 - 840st + 1680}{(st)^4 + 20(st)^3 + 180(st)^2 + 840st + 1680} \tag{113}$$

The derivation and other details are available (Refs. 2, 91, and 92). Because the highest exponent in the numerator does not exceed the highest exponent in the denominator, differentiation is not required. The step response of the Padé approximation is very poor (the full-amplitude step comes straight through immediately as a spike); hence the sampled-data solution should be used for this type of operation.

Digital Simulation

The two types of digital simulation, continuous and discrete, are both done by discrete methods. There are a number of simulation programs of the continuous variety which hopefully simplify this operation: Madas, Pactolus, Mimic, DSL/90, and CSSI, to mention a few. There are also a number of languages that facilitate discrete simulations: SIMULA, GASP, ASTRA, SIMSCRIP, and GERT.

Digital Simulation of Continuous Systems

This implies the solution of differential equations by various methods such as the trapezoidal rule or the Runge-Kutta-Gill techniques. This method of simulation attacks the same general class of equations that are handled by analog simulation (Ref. 93 to 96).

In some of the continuous-simulation methods, programming is from simulation diagrams and is very analogous to analog patching. Digital simulation can simulate dead time effectively by storing a large queue of sampled points.

Digital simulation can provide extremely high accuracy when this is necessary; an example is the integration of rocket trajectories to the moon. It also permits the methods to be varied to suit the problem: e.g., changing the steps used for dead time simulation. Very complex digital evaluations and logic can be handled. Situations that cannot be managed by the simulation language can be treated with another programming language such as FORTRAN. Once a program has been completed and checked out, it can be set up to run again very quickly.

The principal limitation of digital simulation is the speed restriction for equivalent accuracy. This is described by Giloi (Ref. 96) in the *ACTES Proceedings*. References 97 and 98 describe CSMP, a widely used simulation language which permits formulation in terms of Laplace transforms and from analog computer diagrams. It offers a number of alternative integration options and can present results in terms of transient and frequency response. Reference 98 includes program listings and presents simulation examples for a polymer reactor, an ammonia process, and an ethylene furnace.

Discrete Simulation

Discrete-time simulation involves the breaking up of time into discrete intervals which represent the time between interactions among different elements of a system. The use of logical instead of mathematical relationships moves the system from one interaction point to another. This type of simulation enables the user to gain some insight into the performance, critical parameters, and potential trouble spots of a discrete system (Refs. 99 and 100).

Hybrid Simulation

This type of simulation employs a combination of analog and digital techniques. It may consist of specially designed equipment or simply interconnected analog and digital computers. In the case of the specialized hybrid simulation equipment, the analog computer can incorporate a sample-and-hold function and adjustable initial condition settings under control of the digital computer. The user is free to split the simulation between the analog and the digital portions. For instance, dead time can be simulated in the digital computer and the analog system can supply very high speed simulation of other process dynamic functions, thus taking advantage of the best capabilities of each technique. Descriptions of hybrid simulation equipment and hybrid techniques are presented in Refs. 92, 96, 101, and 102.

Partial differential equations, which are generally a problem with any technique, can often be simulated by employing the analog computer to integrate with respect to one independent variable and utilizing the digital computer to store the data and step in discrete increments along the other independent variables. This technique uses a lumped-parameter representation for all but one of the independent variables.

Programming for the digital computer may be done in FORTRAN or in a special simulation language. Programming for the analog computer is the same as always.

The principal limitation of hybrid simulation is the cost and availability of the hybrid computer system.

OPTIMIZATION

Optimal control is a control system that maximizes (or minimizes) some performance index. As a practical engineering consideration, a less than theoretically perfect example (suboptimal control) is often used. In some instances, especially those involving large processes, simple controllers, and simple interconnections of control devices (when considered as a single loop) do not provide the most economic operation. In these cases, advanced control strategies should be investigated. These specialized techniques are classified as (1) *optimization,* trying to obtain the best results from a known process; and (2) *adaptive control,* in which part of the process is not known or is highly variable and the identification problem is combined with the optimization problem. See Ref. 103 for a summary description of an industrial plant utilities optimization system.

Steady-state optimization considers the process stationary and hence is concerned only with the operating point. When the process is stationary, the resulting optimum operating point can often be maintained with conventional control; e.g., setting precalculated control set points and precalculated control parameters depending on process feed, level of operations, etc. If the process changes from time to time, new optimum set points must be computed for each change.

If the process is changing dynamically all the time or if the dynamics of changing from one steady state to another are important, then dynamic optimization is in order.

Incentives and Constraints

There are numerous incentives for optimizing the performance of a system: improved quality, increased production, decreased waste, greater efficiency, and safety. Most, if not all, of the incentives can be expressed on the common basis of economics. If nothing else, the examination of optimum performance shows the difference between current or proposed performance and the ultimate performance possible.

Constraints are the limitations of the system. These include such items as safe operating conditions, miminum and maximum power available, valve stroke limits, storage capacity, maximum (or minimum) allowable pressures, temperatures, and speed, composition, and in general any hard limits. Imposed constraints include the practicality of measuring certain state variables and certain control actions.

Steady-State versus Dynamic Optimization

If the process is steady state, a simple control strategy may be able to maintain close to optimum performance once the optimum is established. If slow search techniques are used, the steady-state system can provide a stationary system long enough to find the optimum. If the method of finding the optimum is long and difficult, the steady-state system will provide good results with only very infrequent updating. Adaptive or optimizing control may be able to track slowly varying conditions. Dynamic systems, on the other hand, provide for the measurement of variables over a much wider range and hence produce of better signal-to-noise ratio and range of applicability. To find out how a process alters with modifications in conditions it is necessary to change the conditions. In steady-state systems optimization usually implies controlling state variables to fixed conditions; dynamic systems usually imply the path of state variables as a function of time. Most methods for optimizing dynamic systems use a finite time interval, e.g., one cycle of a batch reactor. If the system is continuous but not steady state, e.g., a hydroelectric power system, the problems are more difficult to solve.

Performance Index

The *performance index* or the *objective function* is the measure of the system performance that is to be optimized. In most of the discussions in this section a profit type of performance index is assumed; hence a maximum will be sought. If a cost type of performance index is wanted, a simple reversal of signs will make any of the maximum methods produce a minimum. Hence the two problems are basically the same.

The ultimate goal in most industrial environments is economic; therefore the optimizing system should strive for an economic goal. Return on investment and discounted cash flow are perfectly good economic goals depending on the individual company's accounting system. Because long-range marketing plans, process modifications, etc., change very slowly compared with the dynamics of the process, many of these items can be treated as constants in the optimization operations. If a plant is throughput-limited, maximum throughput might be called for. This assumes that there is, and will continue to be, a market for everything produced, and that the incremental cost of increased production is less than the incremental value of increased production.

On the other hand, if maximum production is not required, then the per-unit or total cost of production should be minimized. The cost or value of all the various factors such as quality, rework, raw materials, power consumption, waste disposal, and inventory should be included. Many simplifying assumptions can be made, but care must be taken to be sure that goal being strived for is at least close to the true goal. The error-squared type of performance index is often chosen, not because it is the optimum form but rather because it is easier to use and it exhibits a monotonic increase for deviations in either direction.

For state variables the quadratic performance index

$$J = \mathbf{E}^T\mathbf{QE} + \mathbf{u}^T\mathbf{Wu} \tag{114}$$

where $\mathbf{E}$ = error vector
$\mathbf{u}$ = control vector
$\mathbf{J}$ = the objective function
$\mathbf{Q}$ = weighting factor matrix
$\mathbf{W}$ = weighting factor matrix

is often chosen for the same reason.

Linear Programming

If the performance index is a linear algebraic function of the controlled variables and the constraints are also linear functions of these variables, then linear programming is a practical method for finding the optimum (Ref. 104). The *simplex method* can be used with digital computers or analog computers to solve the linear programming problem. The solution is always in a "corner," that is to say, up against some of the constraints.

Linear programming is widely used for blending and other systems which meet the linearity requirements and the necessity for steady-state operation (no dynamics involved).

Nonlinear Programming

When the system of algebraic equations is not too nonlinear, the solution can sometimes be obtained by nonlinear programming, particularly if the solution is in a corner. An analog computer solution is often applicable. The problems involved in the analog solution are in the implementation and in the occurrence of an occasionally local maximum. Another technique is to approximate the nonlinear equations with a number of linear subregions. Another method is to linearize the equations about a point. Search for the maximum, then linearize about the resulting point, and repeat until the iterations do not yield further improvement. Other variations are available (Abadie, Ref. 105; Dantsig, Ref. 104).

Mixed-Integer Optimization

Mixed-integer optimization permits recommendations for equipment start-up and shutdown to be included in the optimization. Application to an industrial plant utilities system is described in Ref. 106.

Hill Climbing

If the problem of finding the largest value of a function of two variables is given, the geometric interpretation of finding the highest point on a surface, the name *hill climbing* is apropos. Other names are *gradient method* and *method of steepest ascent (descent)*. The following examples and discussion illustrate some of the techniques. For additional information refer to the work of Box and Wilson (Ref. 107) and others (Refs. 108 to 110). Although the geometric significance is lost in a

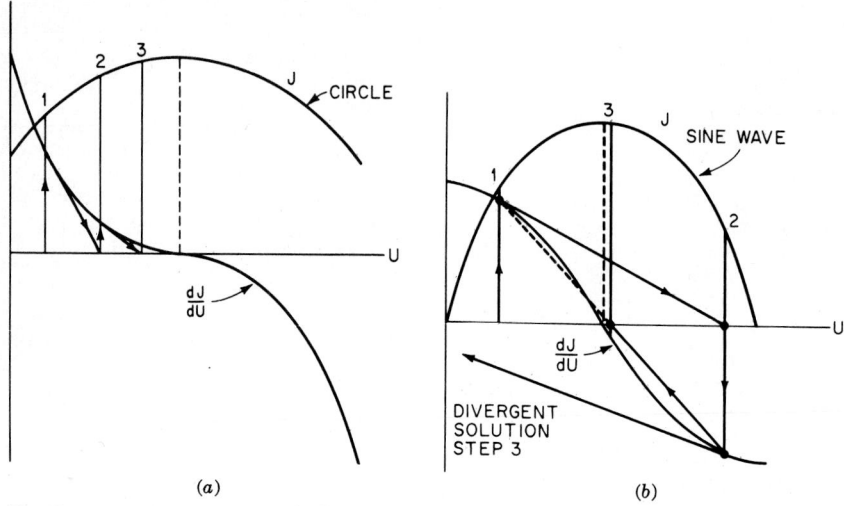

Fig. 53   Hill climbing—a geometric interpretation.

multidimensional hyperspace, hill climbing is a very useful tool for finding a maximum of a function of several variables.

Hill climbing is essentially a trial-and-error method. The real art is to make a good guess. Consider a two-dimensional case; and also assume for a moment that the slope, curvature, and height can be measured at each point. To find the maximum, assume a quadratic form and use the second derivative to estimate the point at which the first derivative would be zero. Use this point for the next try. Repeat this procedure until the first derivative changes sign, then interpolate between the points on each side of the maximum. The reason for this change in procedure is shown in Fig. 53. If the third step in Fig. 53b is taken, using $d^2 J/du^2$, the solution will be divergent. Higher-order curves will lead to faster convergence in some instances, but they should be employed with extreme care since they can project multiple zero crossings or no zero crossing at all. They are also much more sensitive to noise.

Step Size

From the geometric interpretation it can be seen that the ideal step size for a quadratic curve is:

$$\Delta u \approx \frac{dJ/du}{-d^2 J/du^2} = \frac{\dot{J}}{-\ddot{J}} \tag{115}$$

This has two properties which are frequently used: (1) The step size decreases as the optimum is approached (2) for a quadratic curve, e.g., $d^2 J/du^2 =$ a constant, and the optimum is reached in one step. The step also is proportional to the first derivative. It should also be remembered that a small step size leads to a slow stable solution which is equivalent to an overdamped solution. Obviously, if $\ddot{J} \geq 0$, a negative value must be substituted for $\ddot{J}$.

Now consider the case where only the first derivative is available. After the first step the second derivative at the midpoint between the steps is about $(\dot{J}_2 - \dot{J}_1)/(u_2 - u_1)$. Hence the next guess is

$$u_2 = -K \frac{\dot{J}_1 + \dot{J}_2}{2} \frac{u_2 - u_1}{\dot{J}_2 - \dot{J}_1} + \frac{u_1 + u_2}{2} \tag{116}$$

where the gain K is the reciprocal of the damping factor. Again, for the one-dimensional case, when $\dot{J}$ changes sign, interpolation instead of extrapolation should be employed. For stability in a multidimensional hyperspace, where the overshoot is not obvious, the gain K should be less than one. When the term $(u_2 - u_1)/(\dot{J}_2 - \dot{J}_1)$ is not negative, it cannot be used and a negative value must be substituted: $\Delta u = +K\dot{J}$. As the optimum is approached, both Δu and $\Delta \dot{J}$ approach zero, which causes computational or measurement problems, and the use of a constant is indicated. Consider the more realistic and difficult case where the derivative is not directly computable. The two steps will give three points which can be fitted with a quadratic curve to give the derivatives. After the first overshoot, the highest point and the two adjacent points should be used.

For simplicity of presentation, hill climbing will be described for one or two independent variables, but the methods are directly applicable to multiple variables. A random walk algorithm is as follows:

1. Evaluate the objective function at the base point.
2. Move in a random direction to a new point near the base point.
3. Evaluate the objective function at the new point.
4. Select the point with the highest value as the new base.
5. Go to step 2.

This method is simple, but it converges very slowly because many of the steps are in the wrong direction. One improvement is to modify the probabilities of the random step directions based on the uphill direction found in the previous steps.

If the gradient (steepest slope) can be computed, a simple algorithm is as follows:

1. Compute the gradient.
2. Move to a new point in the direction of steepest ascent.
3. Go to step 1.

If the gradient must be evaluated experimentally, or if it is difficult to compute, it may be expedient to make several steps along the gradient before reevaluating it.

Moving along the gradient can be treated as a one-dimensional optimization problem. There is normally little value in searching for a close approach to this intermediate point before establishing a new gradient. On the other hand, the search must be continued until an increase in the objective function is achieved. For convergence and stability, each base point should have a higher objective function than the preceding point. This applies to all hill climbing methods.

If the value of the gradient is not easily obtained from an experimental evaluation, it may be expedient to increase the step size by, say, 10% for consecutive steps in the same direction and then decrease the step size by, say, 25% for each reversal of direction on a given component, as shown below.

Sequential Stepping

A variation of random walk which does not involve explicit gradient determinations is shown in Fig. 54.

Here, Gain-1 is a number slightly greater than one, Gain-2 is a number less than one, and (Gain-1)(Gain-2) is less than one. The flowchart does not show the assumption of the original values of the control vector **u** and Old-u, the direction vector DIR, or the step size vector "step." **J** is the value of the objective function at the last step; Old-u is the previous value of the control vector **u**. The subscript i implies the ith component of a vector.

If a boundary is encountered during a search, move along the boundary until either the maximum is reached or the gradient points away from the boundary.

Figure 55 illustrates the results of eight steps with each of these methods.

If more than one peak or maximum exists, hill climbing and many other methods may yield a local optimum instead of the global optimum. With hill climbing, in order to climb the highest mountain one must start in the vicinity of the highest mountain.

Evolutionary Optimization (EVOP)

EVOP is a technique for systematically and repeatedly perturbing system parameters in order to find and maintain an optimum operating state. A parameter is perturbed according to a preset strat-

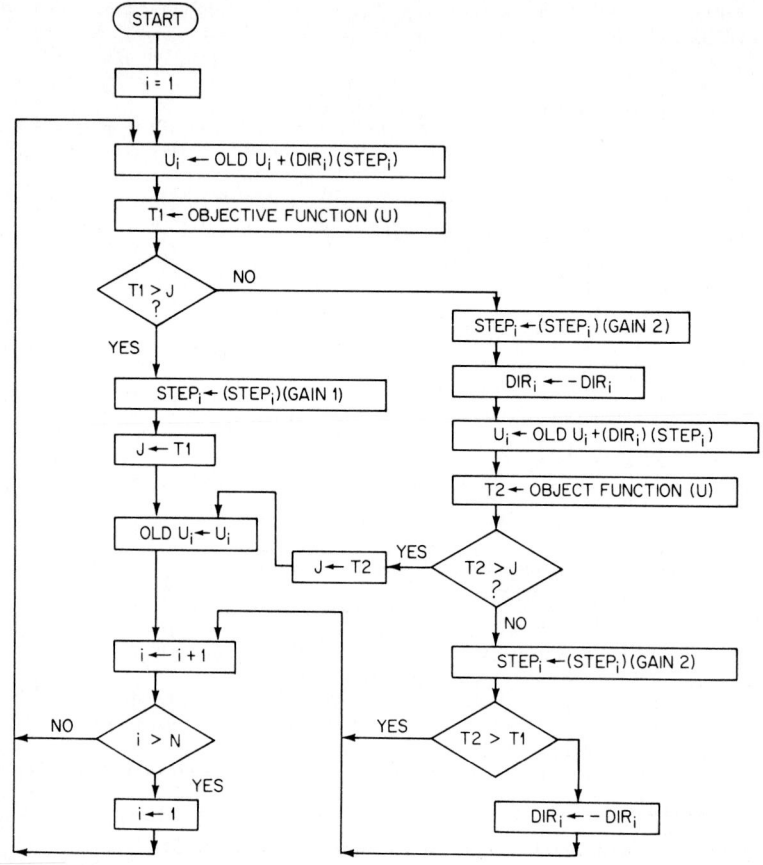

Fig. 54 Hill climbing flowchart.

egy, and after a suitable delay measurements are made and the objective function is calculated. This procedure is repeated until no further improvement is found. No model is required. (See Ref. 111.)

Mathematical Model Methods

If a good model of the system is available, the optimization methods may be used on the model to find the optimum, instead of on the actual process. In fact, most methods require models of some type anyway. The limitations on the cost of experimentation on the actual process indicate that most of the work should be done with the model.

Dynamic Programming

This technique, developed by R. E. Bellman et al. (Refs. 112 to 114), is particularly well adapted to digital computers. Continuous systems are sampled and treated as data sequences; hence the optimum control is treated as a multistage decision process—the *principle of optimality*. A policy that is optimal over the interval $0 \rightarrow N - 1$ is necessarily optimal over any subinterval $n \rightarrow N - 1$, where $0 \leq n < N - 1$. This means that for each step there is only one point that must be retained from the preceding step, and hence the objective function must be evaluated only for the increments from the points on the preceding step. This technique can be illustrated with the following example from Elgerd (Ref. 30).

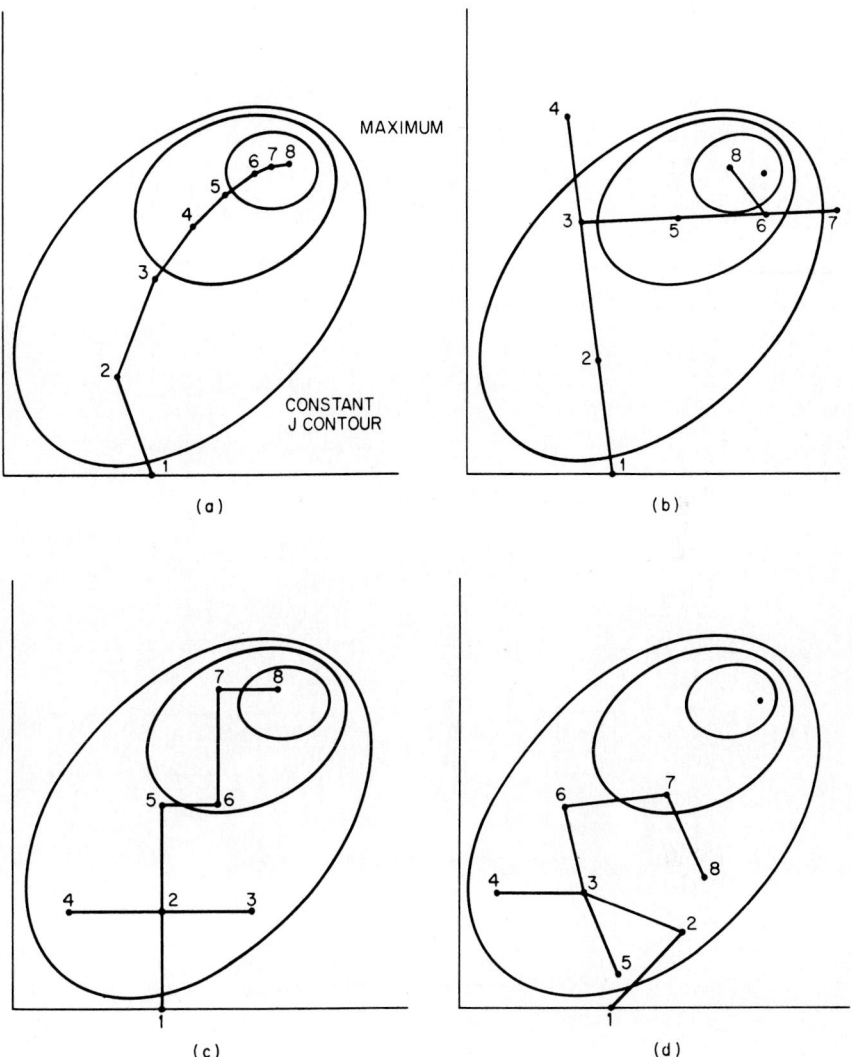

Fig. 55   Hill climbing examples. (a) Gradient, (b) modified gradient, (c) sequential steps, (d) random walk.

Consider the one-dimensional decision process shown in Fig. 56. The increments of the objective function between different stages are shown in the graph. We desire to find the path that minimizes the total objective function between I and II. Where

$$N = 6 \qquad r = 3 \qquad n = 1$$

There are $3^5 = 243$ different combinations to compare. Instead of doing this, we proceed as follows:

> **Step 1.**   Go from stage 0 to stage 1 and memorize all the $r = 3$ increments (framed numbers in Fig. 57). Note that we do not know at this time whether the optimum policy will take us through state $1a$, $1b$, or $1c$, and so we *memorize all possible cost functions (invariant embedding)*.

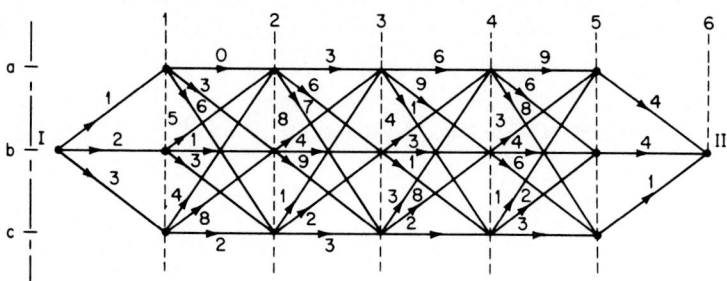

Fig. 56 Dynamic programming example problem.

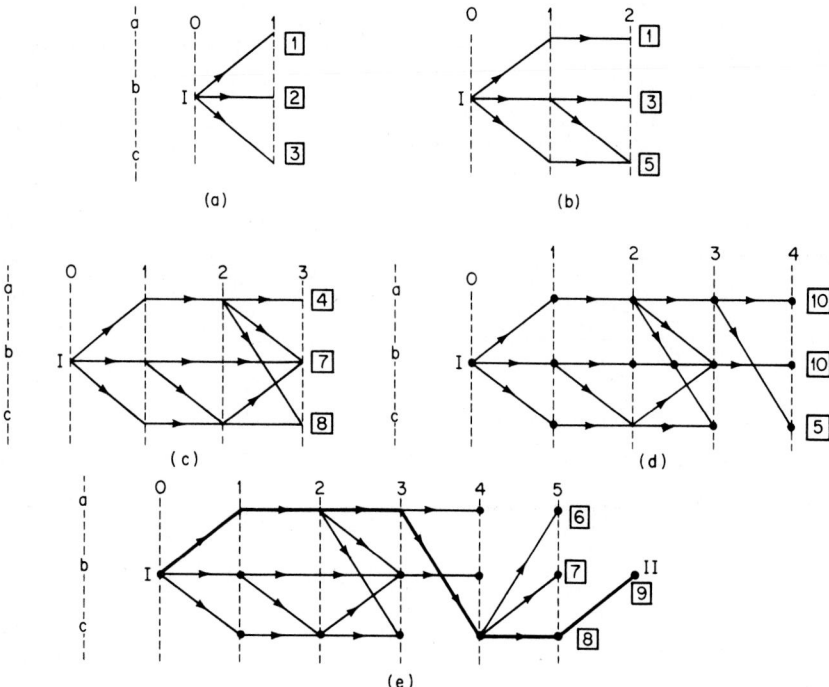

Fig. 57 Dynamic programming example solution.

Step 2. Proceed from stage 1 to stage 2. Memorize again the total cost function. Again we check all possible states 2a, 2b, and 2c. But in addition we invoke the principle of optimality: If it later turns out that the optimum policy will take us through state 2a, then we *certainly* must get there from state 1a. (This conclusion was reached after a comparison of r numbers.) The principle is then applied to *each* of the states in stage 2. A total of nine ($= r^2$) comparisons must be performed. The result is shown in Fig. 57b. Note that the principle of optimality in effect has permitted us at this point to *discard* two-thirds of all possible path combinations leading from stage 1 onward, as they cannot possibly be future candidates for the optimum path.

Steps 3 to 6. By successive comparisons and the application of the principle of optimality we obtain the results shown in Fig. 57c to e. At each step we discard two-thirds of all remaining possibilities.

The optimum policy from I to II can be determined only in the last step. The result is indicated boldface in Fig. 57*e*.

There are two basic problems with dynamic programming. They are "the curse of dimensionality" and "the menace of the expanding grid." That is to say, if there is a large number of independent variables, or a large number of intervals for each independent variable, the number of points and hence both the memory requirements and computation time get out of hand. For the expanding grid one method of attack is to use a coarse grid, and then a smaller grid only in the vicinity of the optimum indicated by the first try.

Dynamic programming has the distinct advantage of having very few limitations on the form of the system equations or on the objective function. Discontinuous nonlinear systems with all kinds of constraints can be solved with this technique. The principal advantage of dynamic programming over a full search is the massive reduction in the amount of memory and computation required.

Calculus of Variations

Whereas linear programming looked for optimal solutions which were in the corner of a linear system, calculus of variations attacks nonlinear systems with nonlinear constraints (boundary conditions). This may also be thought of as an extension of optimizing systems without constraints to systems with constraints.

Calculus of variations is a collection of analytical techniques which are described in many textbooks such as those by Pars (Ref. 115) and Elsgolc (Ref. 116). As an illustration, consider a dynamic system operating over some time interval $t_0 \leq t \leq t_f$.

$$\dot{\mathbf{x}} = f[\mathbf{x},\mathbf{u},i] \tag{117}$$

where the initial state $\mathbf{x}_0$ is given.

The system has n state and m control variables. A scalar objective function to be optimized is

$$J = k[\mathbf{x}_f,t_f] + \int_{t_0}^{t_f} \mathbf{L}[\mathbf{x},\mathbf{u},t] \; dt \tag{118}$$

Define a *Hamiltonian* scaler:

$$H = H[\mathbf{x},\mathbf{u},\lambda t] = L[\mathbf{x},\mathbf{u},t] + \sum_{i=1}^{n} \lambda_i f_i \tag{119}$$

λ is known as the Lagrange multiplier. Define a modified objective function:

$$J = \mathbf{k}[\mathbf{x}_1,t_1] + \int_{t_0}^{t_f} H \; dt \tag{120}$$

The resulting solution is optimal when

$$\dot{\lambda} = -\frac{\partial H}{\partial \mathbf{x}} = -\frac{\partial L}{\partial \mathbf{x}} - \sum_{i=1}^{n} \lambda_i \frac{\partial f_i}{\partial \mathbf{x}} \tag{121}$$

$$\lambda_f = \frac{\partial \mathbf{k}}{\partial \mathbf{x}} \tag{122}$$

$$\frac{\partial H}{\partial \mathbf{u}} = \frac{\partial L}{\partial \mathbf{u}} + \sum_{i=1}^{n} \lambda_i \frac{\partial f_i}{\partial \mathbf{u}} = 0 \qquad t_0 \leq t \leq t_1 \tag{123}$$

The applicability of these techniques is limited because of the difficulty often encountered in solving equations such as Eqs. 121 to 123 for complex systems.

Pontryagin's Maximum Principle

Pontryagin's maximum principle is an advancement over the calculus of variations in that it expands the range of applicability to bounded control signals (Refs. 110 and 117). The most useful feature of these techniques is that they determine the general form of the optimal control. The maximum principle can be illustrated by the following example.

Given the system

$$\dot{x}_i = f_i(\mathbf{x},\mathbf{u},t) \tag{124}$$

and an objective function

$$J = \sum_{i=1}^{n} k_i x_i(t_f) \tag{125}$$

The maximum principle states that, if the control vector $\mathbf{u}$ is optimum, then the Hamiltonian

$$H = \sum_{i=1}^{n} p_i f_i \tag{126}$$

is maximized with respect to the control vector over the interval.

$$\dot{p}_i = -\sum_{j=1}^{n} p_j \frac{\partial f_j}{\partial x_j} \tag{127}$$

$$p_i(t_f) = -k_i \tag{128}$$

p is the *adjoint* or *costate vector*.

$$\frac{\partial H}{\partial p_i} = f_i(\mathbf{x},\mathbf{u},t) = \dot{x}_i \tag{129}$$

$$\frac{\partial H}{\partial p_i} = \sum_{j=1}^{n} p_i \frac{\partial f_i}{\partial \mathbf{x}_i} = -\dot{p}_i \tag{130}$$

The initial values of $\mathbf{x}$ provide the remaining constants necessary to solve these equations for the optimum control vector. Methods are described in the literature for the solution of systems involving:

1. Objective functions of the form

$$J = \mathbf{k}[\mathbf{x}(t_f), t_f] + \int_{t_0}^{t_f} \mathbf{L}[\mathbf{x}(t), \mathbf{u}(t), t] \, dt \tag{131}$$

2. Various constraints on $\mathbf{x}$ and $\mathbf{u}$
3. Various initial and final conditions
4. Transformation between state variables and observed variables

$$\mathbf{y} = \mathbf{Cx} \tag{132}$$

5. Applications of the method to sampled-data systems

The equations that result from the calculus of variations or the maximum principle generally include differential equations with split boundary values. Split boundary value problems are usually difficult. In many cases the optimum solution can be found by taking the form of the solution indicated and solving for the switching points by trial-and-error (hill climbing) methods.

For linear systems with bounds on the control vector the optimal solution is "bang-bang" control, for example:

For the system,

$$\dot{\mathbf{x}}(t) = \mathbf{ax}(t) + \mathbf{bu}(t) \tag{133}$$

with control signals constrained to

$$|u_i| \leq u_i \text{ maximum} \tag{134}$$

the optimum control which minimizes the performance index

$$J = \mathbf{k}[\mathbf{x}(t_f)] \tag{135}$$

will switch between $+u_i$ maximum and $-u_i$ maximum.

The maximum principle is discussed in many books on control (Refs. 21, 30, 38, 109, and 118).

The "minimum time problem" of moving from one state (position and velocity) to another state with limited acceleration in either direction has a bang-bang solution. This is to apply maximum acceleration up to some point in time and then switch to maximum acceleration in the opposite direction for the remainder of the time. A velocity-versus-position plot for the solutions to this problem is shown in Fig. 58. The optimal (minimum time) path from point 1 to the origin is a maximum acceleration along path A to point 2 and then a maximum deceleration along path B to the origin.

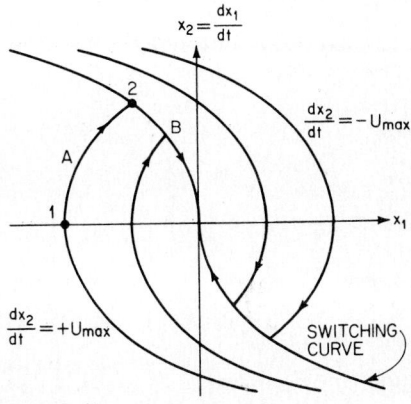

Fig. 58 Optimal phase-plane plot.

Kalman Filter

The Kalman filter is a method for providing an optimal estimate of variables in the presence of noise. Because of the similarity of a filter design to a controller design, Kalman also provides a design method for an *optimal controller* for some types of systems. One example from Kalman's paper gives the solution to the Wiener problem (Ref. 119).

For the discrete-data system,

$$\mathbf{x}(t + 1) = \boldsymbol{\phi}(t)\mathbf{x}(t) + \mathbf{u}(t) \tag{136}$$

$$\mathbf{y}(t) = \mathbf{M}(t)\mathbf{x}(t) \tag{137}$$

where terms are defined in the table below.

Variable	Definition	Matrix size
$\mathbf{x}(t)$	State vector at time t	$n \times 1$
$\phi(t + 1; t)$	Transition matrix from time t to time $t + 1$	$n \times n$
$\mathbf{u}(t)$	Independent gaussian random input vector with *zero mean* at time t	$n \times 1$
$\mathbf{M}(t)$	Matrix effect of state on observation at time t. $p \le n$	$p \times n$
$\mathbf{y}(t)$	Observed output vector at time t	$n \times 1$
t	Time	
$\mathbf{x}^*(t_1 \mid t)$	Optimal estimate of $\mathbf{x}(t_1)$, given $\mathbf{y}(0), \dots, \mathbf{y}(t)$	$n \times 1$
$\mathbf{x}(t_1 \mid t)$	Error in optimal estimate of $\mathbf{x}(t)$, given $\mathbf{y}(0), \dots, \mathbf{y}(t)$	$n \times 1$
$\mathbf{Q}(t)$	Covariance of random excitation $\mathbf{u}(t)$ at time t	$n \times n$
$P(t)$	Optimized estimation of error at time t	$n \times n$
$\mathbf{P}^*(t)$	Covariance of optimized estimation error at time t	$n \times n$
$\phi^*(t + 1; t)$	Transition matrix, for optimal estimation of error, from time t to time $t + 1$	$n \times n$
$\Delta^*(t)$	Weighting matrix of observation for optimal estimation	$n \times n$
E	Conditional expectation	
L	Loss function	
e	Estimation error	

Given the observed values of $y(t_0), \ldots, y(t)$, the estimate $\mathbf{x}^*(t_1 | t)$ of $\mathbf{x}(t_1)$ which minimizes the expected loss $L(e)$ is

$$\mathbf{x}^*(t + 1 | t) = \boldsymbol{\phi}^*(t)\mathbf{x}^*(t | t - 1) + \boldsymbol{\Delta}^*(t)\mathbf{y}(t) \tag{138}$$

The estimation error is given by

$$\tilde{\mathbf{x}}(t + 1 | t) = \boldsymbol{\phi}^*(t)\tilde{\mathbf{x}}(t | t - 1) + \mathbf{u}(t) \tag{139}$$

The covariance matrix of the estimation error is

$$E\tilde{\mathbf{x}}(t | t - 1)\mathbf{x}^T(t | t - 1) = \mathbf{P}^T(t) \tag{140}$$

The expected quadratic loss is

$$\sum_{i=1}^{n} E\tilde{x}_i^2(t | t - 1) = \text{trace } \mathbf{P}^*(t) \tag{141}$$

The matrices $\boldsymbol{\Delta}^*(t)$, $\boldsymbol{\phi}^*(t)$, $\mathbf{P}^*(t)$ are generated by the recursion relations

$$\boldsymbol{\Delta}^*(t) = \boldsymbol{\phi}(t + 1; t)\mathbf{P}^*(t)\mathbf{M}^T(t)[\mathbf{M}(t)\mathbf{P}^*(t)\mathbf{M}^T(t)]^{-1} \tag{142}$$

$$\boldsymbol{\phi}^*(t) = \boldsymbol{\phi}(t) - \boldsymbol{\Delta}(t)\mathbf{P}(t) \tag{143}$$

$$\mathbf{P}(t + 1) = \boldsymbol{\phi}^*(t)\mathbf{M}(t)\boldsymbol{\phi}^T(t) + \mathbf{Q}(t) \qquad t \geqq t_0 \tag{144}$$

In order to carry out the iterations one must specify the covariance $\mathbf{P}^*(0)$ of $\mathbf{x}(0)$ and the covariance $\mathbf{Q}(t)$ of $\mathbf{u}(t)$. Finally, for any $s \geqq 0$, if $\boldsymbol{\phi}$ is invertible:

$$\mathbf{X}^*(t + s | t) = \boldsymbol{\phi}(t + s; t + 1)\mathbf{x}^*(t + 1 | t)$$

$$= \boldsymbol{\phi}(t + s; t + 1)\boldsymbol{\phi}^*(t + 1; t)\boldsymbol{\phi}(t; t + s - 1)\mathbf{x}^*(t + s - 1 | t - 1)$$

$$+ \boldsymbol{\phi}(t + s; t + 1)\boldsymbol{\Delta}^*(t)y(t) \tag{145}$$

Combining the recursion relations gives the Riccati equation,

$$\mathbf{P}^*(t + 1) = \boldsymbol{\phi}(t + 1, t)\{\mathbf{P}^*(t)$$
$$- \mathbf{P}^*(t)\mathbf{M}^T(t)[\mathbf{M}(t)\mathbf{P}^*(t)\mathbf{M}^T(t)]^{-1}\mathbf{P}^*(t)\mathbf{M}(t)\}\boldsymbol{\phi}^T(t + 1; t) + \mathbf{Q}(t) \qquad t \geqq 0 \tag{146}$$

Figure 59 shows the sampled-data computation for the Kalman filter. As can be seen from the diagram, the matrices needed for the Kalman filter are $\boldsymbol{\Delta}^*(t)$, $\boldsymbol{\phi}(t + 1; t)$, $\mathbf{M}(t)$, $\mathbf{x}^*(0)$, $\mathbf{P}^*(0)$, and $\mathbf{Q}(t)$.

By combining Eqs. (138) and (142) the optimal estimate can be expressed as

$$\mathbf{x}^*(t + 1 | t) = \boldsymbol{\phi}(t + 1; t)\mathbf{x}^*(t | t - 1) + \boldsymbol{\Delta}^*(t)[\mathbf{y}(t) - \mathbf{M}(t)\mathbf{x}^*(t | t - 1)] \tag{147}$$

The estimate of the observed variable s is

$$\overline{\mathbf{y}}(t | t - 1) = \mathbf{M}(t)\mathbf{x}^*(t | t - 1) \tag{148}$$

The error in the estimate of $\mathbf{y}(t)$ is

$$\tilde{\mathbf{y}}(t | t - 1) = \mathbf{y}(t) - \overline{\mathbf{y}}(t | t - 1) = \mathbf{y}(t) - \mathbf{M}(t)\mathbf{x}^*(t | t - 1) \tag{149}$$

For extrapolation (prediction) to time $t + s$,

$$\mathbf{x}^*(t + s | t) = \boldsymbol{\phi}(t + s; t + 1)\mathbf{x}^*(t + 1 | t) \qquad s \geqq 1 \tag{150}$$

For additional information see Refs. 21, 38, 109, 118, and 120. Boxenhorn (Ref. 121) gives a simple example.

Kalman's technique generates recursion formulas which are convenient to use in computer solutions. His method in general attacks the systems that have either a quadratic objective function (loss) or uncorrelated gaussian noise as an input. Most other types of noise can be treated if a filter can be

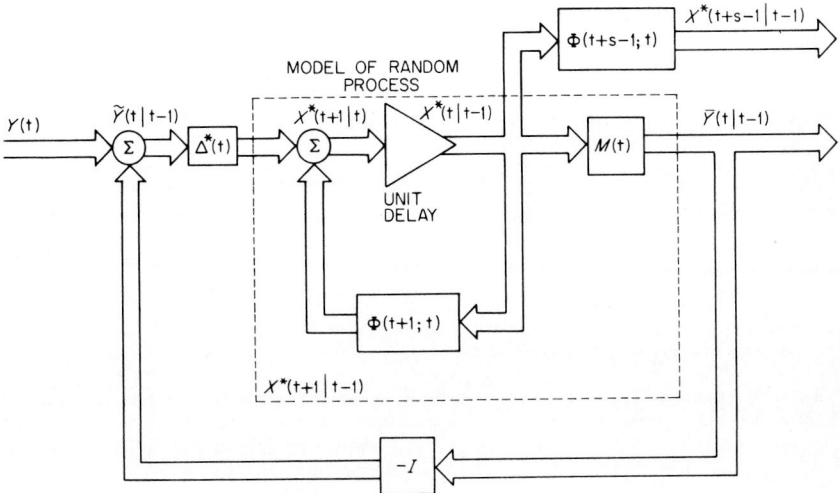

Fig. 59 Optimization matrix block diagram of a Kalman filter.

designed that will translate the gaussian noise into the observed noise. A means of applying the Kalman filter to a nonlinear system is to find a good estimate of the system and then use it to define a new set of linear state equations which approximates the nonlinear system at the normal operating point. The filter can then be applied to this set of linearized equations. It can be utilized for systems with control signals, forcing disturbances, and more than one noise source. Performance criteria other than mean square error can be employed for the Kalman filter. It also provides an estimate of the error. One point to be carefully noted is the assumption of zero mean for the noise. A bias in the measured signal will not be filtered out and will not show up in the computed error. For a comparison of the Kalman filter with the classical digital filter, see Kneile and Luecki (Ref. 122).

Applications of Optimization Techniques

In addition to the references given earlier, the following references present data on experimental results with various optimization techniques. Lynch and Ramirez (Ref. 123) present results for a stirred-tank reactor. Seborg and Fisher (Ref. 124) summarize the results of a 10-year program for the experimental evaluation of multivariable control techniques. The techniques evaluated include optimal feedback control, time optimal control, modal control, eigenvalue assignment, Kalman filters and Luenberger observers, and identification of state space models.

ADAPTIVE CONTROL

Adaptive control is on-line identification of process gains and time parameters and the subsequent use of this information to improve performance of a control loop or strategy. In the simplest form, this is analogous to control loop tuning; in more complex cases, it is analogous to on-line design of a strategy.

Adaptive control is similar to optimal control in that various parameters are changed as a function of time. It differs from optimal control in that parameters in the model of the process are evaluated on-line. In other words, adaptive control is the combination of control with the solution of the identification problem. The control itself may or may not be optimal control. Adaptive control is particularly applicable in cases where the process dynamics are not well defined or change with time.

In the simplest form, adaptive control involves the scheduling of control parameters based on the measurement of throughput, load, product, etc., using fixed logic. Another simple example is the use of multiplicative feedforward with feedback trim. In this case, the feedback controller sets the gain of the feedforward path. Bristol (Ref. 125) provides a good, easy-to-read overview of the subject.

Aerospace applications such as rockets and aircraft where control response changes with fuel consumption, load, altitude, etc., have provided considerable literature on adaptive control.

The basic criteria for applying adaptive control are:

1. Enough improvement in performance must be possible to justify the extra computation.

2. The state variables or the model parameters to be evaluated must be observable either directly or indirectly.

3. If the parameters are not directly computable from the measured signals, then there must be enough variation on inputs to allow the estimation of these parameters by their influence.

4. The dynamics of the system must be such that the parameters can be evaluated in a time that is reasonable based on their rate of change.

The use of semiautomatic adaptive control in the form of tuning aids can be helpful (Adkison and Kohler, Ref. 126), especially for slow control loops, which are time-consuming to tune. Batch chemical reactors, where the characteristics change with temperature, catalyst activity, etc., are also good candidates for adaptive control. (See Adams and Schooley, Ref. 127.) Shinskey (Ref. 77) provides an excellent description of the practicalities of applying a special purpose adaptive control system to a difficult pH control application. Kalman (Ref. 128) provides an in-depth analysis of the general adaptive control problem. A commercially available adaptive self-adjusting controller is described in Ref. 129. System identification, especially the selection of test signals and the treatment of noisy signals, is an important aspect of adaptive control and is discussed below.

The basic operation of adaptive control is shown in Fig. 60.

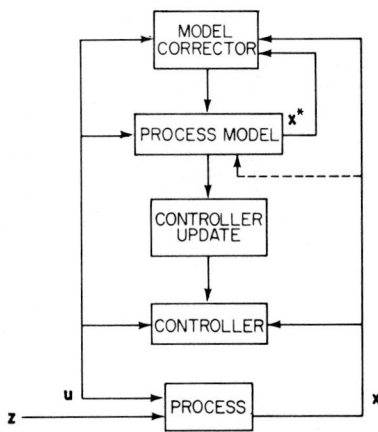

Fig. 60 Hierarchical control structure.

System Identification

Adaptive control is only one application of identification theory. Others are controller tuning (Ref. 27), developing off-line models (Ref. 2), and troubleshooting (Ref. 130). The basic operations involved are shown in Fig. 61 for the simple case of two inputs and two outputs.

Process inputs and outputs are measured and analyzed to produce outputs which characterize the process as to steady-state gains K and dynamic response G. If the form of the model is assumed, this greatly simplifies the problem, and this is the approach most generally employed.

If small disturbances can be introduced into the system, then the step-response or frequency-response techniques for estimating the system parameters can be utilized. An important characteristic is the signal-to-noise ratio. If the ratio is sufficiently high, a relatively small perturbation may quickly identify the dominant term in the process model. Higher signals and averaging the results of repeated experiments will be necessary to identify other terms in the equations.

Mathematical model techniques and dynamic analysis techniques can be applied to the steady-state system if there is a sufficiently high signal-to-noise ratio. Particular care must be taken to avoid estimation of nonlinear characteristics outside the range of actual operation. Multiple regression analysis must be carefully checked for the effect of noise.

Perturbation Methods

Both the theory and the practical aspects of identification are simplified if test signals can be applied to the system. An engineering evaluation of various test signals and techniques is given by Hougen

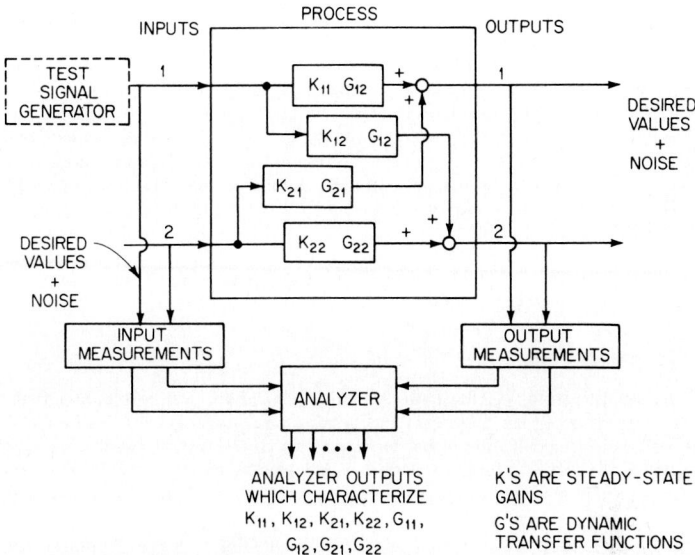

Fig. 61 Basic operations of identification.

and Walsh (Ref. 131). The basic problem is to provide adequate signal amplitude at the output without using large input signal that is so large that it disturbs process operation. Table 17 compares various test signals used in process identification.

As indicated in Table 17, sine wave (frequency-response) testing is the least convenient. Testing must cover a wide range of frequencies and consumes considerable time. However, results are definitive in that a frequency-response plot (or table) can be used to evaluate "best fit" model parameters and to indicate the tolerance of the model approximation over the dynamic range (Ref. 131). Even when a model is derived by other methods, conversion to the frequency-response form may be useful during the analysis. Computer programs are available to provide frequency response from pulse input data and corresponding output response data. (See Ash and Deshpande, Ref. 4.) Most processes are at least somewhat nonlinear, and the sine wave input generally provides a more accurate dynamic estimate of steady-state operating gain. The output sine wave is easier to separate from

Table 17 Comparison of Various Test Signals for Identification

Test signal	Advantages	Disadvantages
Sine wave	Definitive results, measures dynamic gain Simple data reduction, reasonably immune to noise	Long test time, special equipment usually required
Pulse	Simple to generate Returns to initial value, short test time	Selection of pulse width and amplitude not simple, subject to drift, analysis of results not simple
Step	Simplest to generate Relatively simple analysis of results	Long test time compared to pulse, leaves variable offset from initial value, subject to drift.
Normal operating signals	Does not disturb normal process operation	Analysis of results usually complex, noise a problem
Random signals	Minimizes process disturbances	Analysis of results complex

noise and drift than the step or pulse. Where a low signal-to-noise ratio is unavoidable, special techniques have been developed. (See Cowley, Ref. 132.)

In most instances, in order to obtain adequate test data, it is necessary to disable automatic controls which may distort the response. In the simplest case this involves placing a control loop on manual control for the duration of each response test. For some processes, such as those which include an integration, the use of an open-loop step test signal may not be permissible. Sometimes the use of a pulse signal may be preferable because of the shorter duration of the test and because the input is restored to the starting value at the end of the test. A modified step test has been suggested by Bakke (Ref. 57). The step is terminated and the input restored to the initial value once a maximum rate of change has been achieved. Identification using normal operating records is based on a discrete-data formulation and is discussed below. The use of random signals depends on the noise transmission properties of linear systems and utilizes autocorrelation and cross-correlation functions. This is also discussed below.

Analysis for Identification

Table 18 summarizes various identification techniques which have been employed. Often, a simple model such as a first-order lag plus dead time is adequate. Otherwise, a two-lag-plus-dead-time model is usually satisfactory. If a low signal-to-noise ratio is present, special methods are useful. The techniques utilized by Dahlin (Table 18, item 8) ensure high immunity to noise and are suitable for a wide range of models and input signals. The process is a paper machine which poses a particularly difficult measurement problem because of noise. The method proposed by Pemberton (Table 18, item 3) minimizes the data reduction requirements by limiting the number of measurements to three definite points on the pulse-response curve. The method proposed by Bristol et al. does not use a specifically stated model or test input and operates based on the shape of the response curve as defined by the measuremnt of specific preset points on the transient-response curve. Inputs are normal set-point changes or disturbances.

A number of other methods are found in the literature for the determination of model parameters based on step response. However, there is a basic problem with these methods in accurately locating the point of inflection. This makes accurate determination of the smaller time constant difficult. Ash and Deshpande (Ref. 4) list a representative set of references. They also give a detailed description of a method by Sundaresan based on the method of moments, which appears to overcome the problem of earlier methods. The method is applicable to overdamped and underdamped responses. Ash and Deshpande also discuss step and pulse testing in detail, including the selection of pulse test signals and data reduction. Programs for data reduction are included, together with the results.

The *random signals method* referred to in Table 17 is a basic technique using white noise as a test signal. A simple example is shown in Fig. 62. (See Gibson, Ref. 47, p. 502.)

Since $n(t)$ and $x(t)$ are uncorrelated, the average output of the multiplier or low-pass filter is

$$\overline{z(t)} = \phi_{xy}(\tau_i) = \int_0^\infty \phi_x(\tau_i - \lambda_1)g(\lambda_1)\,d\lambda_1$$

Since the bandwidth of $x(t)$ is much greater than the passband of $g(\lambda)$ (white noise),

$$\phi_{xy} \approx \Phi_x g(\tau_i)$$

where Φ_x is the area under $\phi_x(\tau_i)$, or

$$\phi_x(\tau) = \Phi_x\,\delta(\tau)$$

Complete identification requires either a number of correlation channels in parallel or sequential processing, and summation. For a more complete description of correlation functions, see Truxal (Ref. 137).

Adaptive Tuning of Controllers

A basic and thorough treatment of the subject is given by Kalman (Ref. 128). Most results are for PI or PID controllers, either DDC or continuous. Surveys and evaluations of criteria and methods are available in references by Scheib (Ref. 138) and Miller et al. (Ref. 27). Tuning maps developed

Table 18 Summary of System Identification Techniques

Authors	References	Application	Test signal	Type of test	Form of model	Analysis method	Remarks
1. Ziegler and Nichols	24 133	P, PI, PID Controller tuning	Step Controller-induced sine wave	Open loop Closed loop	Dead time and first-order lag	Empirical based on theory	Simple and graphical
2. Gallier and Otto	56	Operator-initiated tuning for DDC	Pulse	Open loop	Dead time and second-order lag	Least-squares criteria. Nonlinear regression to fit coefficients	Similar to Kalman method
3. Pemberton	134	General method	Pulse (single-sided or two-sided)	Open loop	Dead time plus second-order lag	Three-point measurement Formulas based on theory	Reduced data storage and processing
4. Ash and Deshpande	4	General method	Step and pulse	Open loop	Dead time plus second-order lag	Computer program	See reference for test data
5. Kalman	128	General method	Pulse	Open loop	Generalized discrete-data (z transform)	Weighted mean-square fit of coefficients	Extensive calculation and memory required
6. Bakke	57	On-line tuning on DDC at operator request	Controller-induced cycling Modified step	Part closed loop Part open loop	Dead time and first-order lag	Formulas based on theory	Validity test included
7. Stanton	130	Process trouble-shooting	Sine wave	Open loop	Dead time and second-order lag	Cowley (Ref. 132) and graphical	Off-line analysis
8. Dahlin	135 136	On-line process identification for control model	Normal operating signals	Closed loop	Dead time and first-order lag	Model reference Least-squares method with data filtering Two methods Best results selected	Two-input/two-output system with strong interaction and low signal-to-noise ratio
9. Bristol, Inaloglu, and Steadman	53	General controller tuning method	Normal operating signals	Closed loop	No explicit model (pattern recognition based on transient)	Empirical tuning equations	May be triggered automatically on set point or disturbance magnitude
10. Schork and Deshpande	69	General method	Pulse	Closed loop	None	Frequency response	Sampled-data control on outer loop (double cascade)

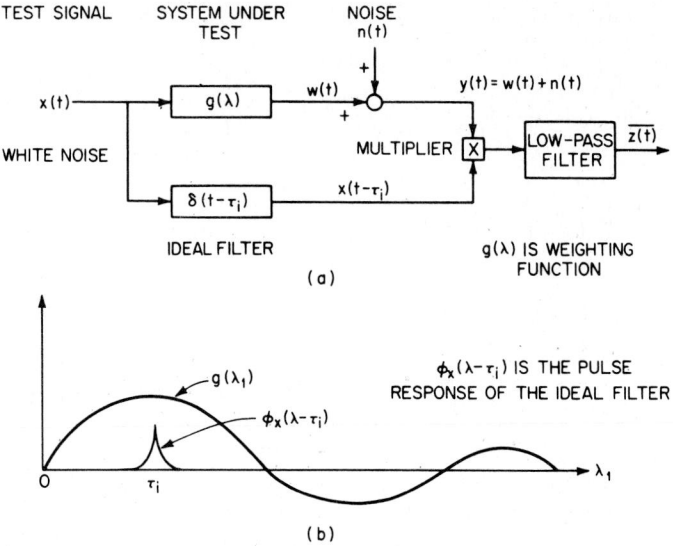

Fig. 62   Cross-correlation identification method.

by Wills (Ref. 26) indicate the sensitivity of tuning, etc. The system used by Adkison and Kohler (Ref. 126) is an off-line method using ITAE criteria and a modified Hook-Jeeves search to optimize closed-loop response. Gallier and Otto (Ref. 56) consider DDC full-value and incremental PI or PID algorithms as well as a special minimum-settling-time algorithm. The criterion is IAE, and the search method of Fletcher and Powell (Ref. 239) is employed. Nondimensional parameters are used, and the effect of noise on tuning is evaluated. Bakke (Ref. 57) uses the ITAE criteria to tune a PI controller, and a protective reasonability check is provided.

Hill Climbing

The principal addition to the hill climbing methods described previously is the control of step size. If perturbations or variations in the system are used to obtain information for adaptation (see Lee et al., Ref. 78), the rate of adaptation must be controlled by the amount, or quality, of information available. If, for example, the signal-to-noise ratio on the gradient measurement is too low, one should cut the step size down to zero, e.g., stop. On the other hand, the step size cannot be set at zero when the maximum is reached, because the maximum may move.

Model Reference Adaptive Control

The model reference adaptive control compares the process output **x** with the model output **x*** to estimate the error in the model parameters. In Fig. 60 the model corrector updates the model based on the inputs and outputs of both the real and the simulated processes.

The controller updater utilizes the model parameters to update the controller. The disturbances **x** are assumed to be unmeasured. If they are measured, they should also be fed into the model and the model corrector.

The model corrector can employ any of the control or identification techniques to update the model parameters. Care must be taken to avoid making corrections when the signal-to-noise ratio is too low. The parameters in the model should be constrained to a reasonable range.

Model Reference Adaptive Example

If the errors in the model parameters **p** can be clearly identified as a function of the errors in the estimation of the process variables **x**, a simple integral of this function multiplied by the signal level may be a suitable generator for **p**. It may be desirable to apply the current values of **x** to update the state of the model occasionally.

The key advantage of this technique over using the model equations and the plant input-output to compute the parameters directly is the effect of noise. In the model reference case lags and integration tend to suppress noise, whereas in the other case differentiation or lead functions tend to magnify noise.

One reason why model reference adaptive control is a desirable method is that the model and many of the other computed quantities are easily related to the physical process.

As an illustration of this technique (from a presentation by J. M. Nelson at the First Winter Institute on Optimal and Adaptive Control, University of Florida, February 1962), consider a process:

$$\dot{x} = - \left[\left(\frac{1}{\tau} \right) x + \left(\frac{G}{\tau} \right) u + z \right]$$

where both $1/\tau$ and G/τ are functions of time. Assume that there is a controller acting on x to generate u and that there is something to update the controller based on the estimate A of $1/\tau$ and the estimate B of G/τ. One possible solution to this problem is shown as a simulation diagram in Fig. 63.

Integrator 1, which is the model of the system, generates an estimate x^* of the state variable x. This value is then subtracted from x to give the error. This error has components due to three sources; therefore it is necessary to separate it at least approximately into these components.

$E1$. The component due to the error in estimation of A

$E2$. The component due to the error in estimation of B

$E3$. The component due to noise

Signals u which generate a rate of change in x are the ones that provide information about A. A signal level $S1$ and A is defined as $S1 = u - u/(\tau s + 1)$. In order to avoid adjusting A when the signal-to-noise ratio is too low, only the excess signal above some base level $N1$ is used as a multiplier for the error to generate $E1$.

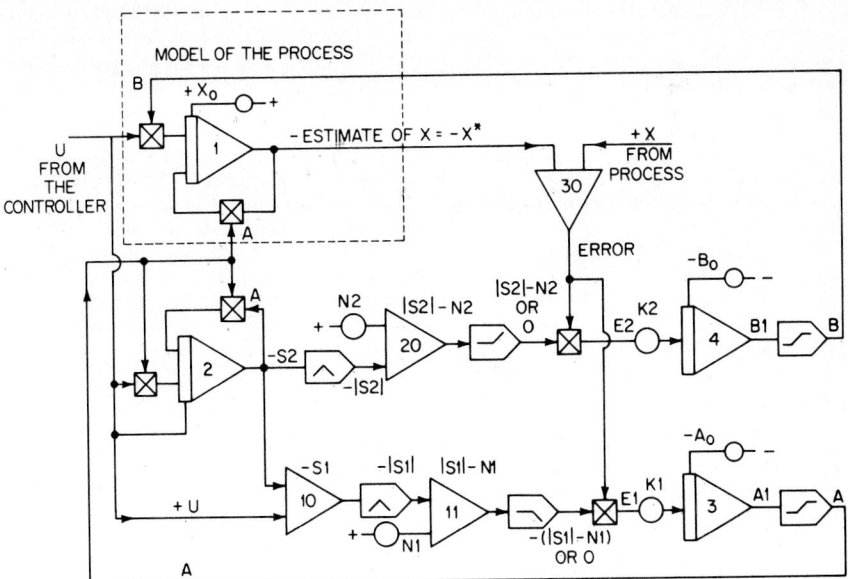

Fig. 63 Model reference adaptive example.

$$E1 = (|S1| - N1)E \qquad \text{for } |S1| > N1$$

$$E1 = 0 \qquad\qquad \text{for } |S1| \leq N1$$

The signal $E1$ is multiplied by a gain $K1$ and integrated (integrator 3) to produce $A1$ and the estimate of A. $A1$ is then limited to the maximum or minimum values that are reasonable for A.

$$A = A \text{ maximum} \qquad \text{for } A1 > A \text{ maximum}$$

$$A = A1 \qquad\qquad \text{for } A \text{ minimun} \leq A1 \leq A \text{ maximum}$$

$$A = A \text{ minimum} \qquad \text{for } A1 < A \text{ minimum}$$

The value of A thus generated is the one employed in the model and also in defining the controller characteristics. The gain $K1$ is in effect the reset rate of a control loop which adjusts A to minimize the error. Because this is a closed loop, the sloppy split of the error between $E1$ and $E2$ results in only minor fluctuations in the estimate of A. Increasing the gain $K1$ causes the system to adapt more rapidly at the expense of noise in A.

A similar argument can be utilized to derive the control loop shown for B.

This example problem did not illustrate the use of optimal filtering, signal separation, identification, etc. The following features were illustrated.

1. The differentiation of a noisy signal was not required.
2. Corrections were made only when a good signal was available.
3. If the process is at steady state, no adjustments will be made on A.
4. The parameters in the model were held to reasonable values.
5. Additional control loops with their own response and stability characteristics were added to the system.
6. Optimal separation of the signals, etc., was not required for good operation.

The preceding example was presented as an analog solution. The corresponding digital solution is similar and equally applicable.

The components of the error(s) due to the various parameters in the model are equivalent to the gradients required for the gradient hill climbing method. The gradients can be derived by taking the partial derivative of the summation of errors squared with respect to each of the model parameters to be identified.

Predictive Control

Given a model of the process and some measure of the state of the process, it is possible to predict what will happen to the process for a given set of control actions. Because of noise and uncertainties, this prediction is normally limited to a relatively short time. Predictive control can be described as utilizing this information to anticipate and hence avoid undesired deviations. The first portion of the "best control action" found in the fast simulation is then employed while the procedure is repeated. The combination of adaptive and predictive control can be advantageous in some situations, particularly for unstable processes or reactors (Ref. 78). See Adams and Schooley (Ref. 127) for an example showing a dramatic improvement in performance.

Restrictions

The adaptive and optimizing techniques may be thought of as on-line automatic redesign of the control system. It therefore follows that the limitations imposed on design must also be imposed on redesign, using these techniques (Refs. 83 and 141).

Stability

In addition to stability of the individual loops, it is necessary also to have stability of the overall system and of the computational procedures.

In multivariable systems the interaction among various loops may tend to increase or decrease stability of the individual loops. If the result is a decrease in stability of the loop, this requires additional design work to stabilize the system. Even if the interaction tends to stabilize the system, there may still be a problem when the gains in individual loops are increased as a result of the interconnection. Under these conditions special action may have to be taken to avoid instabilities when interconnections are broken. Such a problem may arise during start-up or during partial manual operation of the system.

The techniques described earlier for analyzing stability must be applied over the complete range of operating conditions. These must include start-up, shutdown, loops on manual control, switching conditions, and nonlinear regions of operation.

REFERENCES

1. Shinskey, F. G.: *Process Control Systems: Applications, Design, Adjustment,* McGraw-Hill, New York, 1979.

2. Buckley, P. S.: *Techniques of Process Control,* Wiley, New York, 1964.

3. Luyben, W. L.: *Process Modeling, Simulation and Control for Chemical Engineers,* McGraw-Hill, New York, 1973.

4. Ash, R. H., and P. B. Deshpande: *Elements of Computer Process Control,* Instrument Society of America, Research Triangle Park, N.C., 1981.

5. Liptak, B. G.: *Instrumentation in the Processing Industries,* Chilton, Philadelphia, 1973.

6. Khandheria, J., and W. L. Luyben: "Experimental Evaluation of Digital Algorithms for Antireset Windup," *Ind. Eng. Chem., Process Des. Dev.,* vol. 15, no. 2, 1976, pp. 278–285.

7. Cox, R. K., and J. P. Shunta: "Tracking Action Improves Continuous Control," *Chem. Eng. Prog.,* vol. 69, no. 9, 1973, pp. 56–61.

8. Buckley, P. S.: "Override Controls for Distillation Columns," *Instrum. Technol.,* vol. 25, no. 8, 1978, pp. 51–58.

9. Buckley, P. S.: "Protective Controls for a Chemical Reactor," *Chem. Eng.,* April 20, 1970, pp. 145–150.

10. Buckley, P. S.: "Distillation Column Design Using Multivariable Control, Part I: Process and Control Design," *Instrum. Technol.,* vol. 25, no. 9, 1978, pp. 115–122; "Part II: Economics, Energy, and Equipment," *Instrum. Technol.,* vol. 25, no. 10, 1978, pp. 49–53.

11. Shinskey, F. G.: *Distillation Control for Productivity and Energy Conservation,* McGraw-Hill, New York, 1977.

12. Shinskey, F. G.: *pH and pIon Control in Process and Waste Streams,* Wiley, New York, 1973.

13. ANSI: "Terminology for Automatic Control," *ANSI MC85.1M-1981,* ASME, United Engineering Center, New York, 1981.

14. Fertig, H. A.: "Tuning Controllers for Noisy Processes," *ISA Trans.,* vol. 14, no. 4, Instrument Society of America, Research Triangle Park, N.C. 1975, pp. 292–304.

15. Gould, L. A.: *Chemical Process Control, Theory, and Applications,* Addison-Wesley, Reading, Mass., 1969.

16. Mickley, H. S., Sherwood, T. K., and C. E. Reed: *Applied Mathematics in Chemical Engineering,* 2d ed., McGraw-Hill, New York, 1957.

17. Perry, R. H., Chilton, C. H., and S. D. Kirkpatrick (eds.): *Chemical Engineers' Handbook,* 5th ed., McGraw-Hill, New York, 1975.

18. Field, W. B.: "Design of a pH Control System by Analog Simulation," *ISA J.,* January 1959.

19. Mason, S. J.: "Feedback Theory: Further Properties of Signal Flow Graphs," *Proc. IRE,* vol. 44, July 1956, pp. 920–926.

20. SAMA: "Functional Diagramming of Instrument and Control Systems," *SAMA Stand RC2211-1966* (Reviewed and reconfirmed, April 1973), Recorder-Controller Section, Scientific Apparatus Makers Association, New York, 1973.

21. Kuo, B. B.: *Discrete-Data Control Systems,* Prentice-Hall, Englewood Cliffs, N.J., 1970.

22. Meyer, C., et al.: "An Experimental Application of Time Delay Compensation Techniques to Distillation Column Control," *Ind. Eng. Chem., Process Des. Dev.,* vol. 17, no. 1, 1978, pp. 62–67.

23. Gardner, M. F., and J. L. Barnes: *Transients in Linear Systems,* Wiley, New York, 1942.

24. Ziegler, J. G., and N. B. Nichols: "Optimum Settings for Automatic Controllers," *Trans. ASME,* vol. 64, December 1942, 759–765.

25. Oldenbourgh, R. C., and H. Sartorius: *The Dynamics of Automatic Controls,* H. L. Mason (trans.), American Society of Mechanical Engineers, New York, 1948.

26. Wills, D. M.: "Tuning Maps for Three-Mode Controllers," *Control Eng.,* April 1962, pp. 104–108.

27. Miller, J. A., et al.: "A Comparison of Controller Tuning Techniques, *Control Eng.,* December 1967, 72–75.

28. James, H. M., Nichols, N. B., and R. S. Phillips: *Theory of Servomechanisms,* McGraw-Hill, New York, 1947.

29. ISA: "Dynamic Response Testing of Process Control Instrumentation," *ISA-S26 Std.,* Instrument Society of America, Research Triangle Park, N.C., 1968.

30. Elgerd, O. I.: *Control Systems Engineering,* McGraw-Hill, New York, 1967.

31. DeRusso, P. M., Roy R. J., and C. M. Close: *State Variables for Engineers,* Wiley, New York, 1965.

32. Ellis, J. K., and O. W. T. White: "An Introduction to Modal Analysis and Control," *Control,* June 1965, pp. 193–197.

33. Kalman, R. E.: "On the General Theory of Control Systems," *Proc. Int. Congr. Autom. Control (London),* vol. 1, 1961, pp. 481–492.

34. Kalman, R. E.: "Mathematical Description of Linear Dynamical Systems," *J. Control (SIAM),* ser. A, vol. 1, no. 2, 1963, pp. 152–192.

35. Gilbert, E. G.: "Controllability and Observability in Multivariable Control Systems," *J. Control (SIAM),* ser. A, vol. 1, no. 2, 1963, pp. 128–151.

36. Ragazzini, J. R., and G. F. Franklin: *Sampled-Data Control Systems,* McGraw-Hill, New York, 1958.

37. Kuo, B. B.: *Discrete-Data Control Systems,* Prentice-Hall, Englewood Cliffs, N.J., 1970.

38. Tou, J. T.: *Digital and Sampled-Data Control Systems,* McGraw-Hill, New York, 1959.

39. DeBolt, R. R., and B. E. Powell: "A 'Natural' Three-Mode Algorithm for DDC," *ISA J.,* vol. 13, no. 9, 1966, pp. 43–47.

40. Cox, J. B., Hellums, L. F., and T. J. Williams: "A Practical Spectrum of DDC Chemical Process Control Algorithms," *ISA J.* vol. 13, no. 10, 1966, pp. 65–72.

41. Lane, D. W.: "Digital Control Algorithms for Slow Sampling Rates," *Instrum. Technol.,* July 1970, pp. 47–52.

42. Dahlin, E. B.: "Designing and Tuning Digital Controllers," in *DDC Tuning Reference Book,* Rimbach Publications, Philadephia, 1959.

43. Brockett, R. W.: "The Status of Stability Theory for Deterministic Systems," *IRE Trans. Autom. Control,* vol. AC-11, July 1966, pp. 596–606.

44. Brown, G. S., and D. P. Campbell: *Principles of Servomechanisms,* Wiley, New York, 1948.

45. Chestnut, H., and R. W. Mayer: *Servomechanisms and Regulating Systems Design,* vol. 1, Wiley, New York, 1951.

46. Ahrendt, W. R., and J. F. Taplin: *Automatic Feedback Control,* McGraw-Hill, New York, 1951.

47. Gibson, J. E.: *Nonlinear Automatic Control,* McGraw-Hill, New York, 1963.

48. Evans, W. R.: "Control Systems Synthesis by Root Locus Method," *Trans. AIEE* vol. 69, part 1, 1950, pp. 66–69.

49. Greif, H. D.: "Describing Function Method of Servomechanism Analysis Applied to Most Commonly Encountered Nonlinearities," *AIEE Tech. Pap. 53–254,* June 1953.

50. Kochenberger, R. J.: "A Frequency Response Method for Analyzing and Synthesizing Contractor Servomechanisms," *Trans. AIEE,* vol. 69, 1950, pp. 270–283.

51. Shinners, S. M.: "Dual Input Describing Functions," *Control Eng.,* February 1971, pp. 53–55.

52. Shapiro, E. Y.: "Rx: Circle Criterion for Nonlinear System Instability," *Control Eng.,* March 1970, pp. 81–83.

53. Bristol, E. H., Inaloglu, G. F., and J. F. Steadman: "Adaptive Process Control by Pattern Recognition," *Instrum. Control Syst.,* March 1970, pp. 101–105.

54. Schultz, W. C., and V. C. Rideout: "Control System Performance Measures: Past, Present, and Future," *IRE Trans.,* February 1961, pp. 22–35.

55. Wills, D. M.: "Tuning Maps for Three-Mode Controllers," *Control Eng.,* April 1962, pp. 104–108.

56. Gallier, P. W., and R. E. Otto: "Self-Tuning Computer Adapts DDC Algorithms," *Instrum. Technol.*, February 1968, pp. 65–70.

57. Bakke, R. M.: "A Straightforward Approach Allowing Industrial Operators to Adjust Continuous and Digital Controllers with the Aid of a Digital Computer," IFAC/IFIP Symposium on Digital Control of Large Industrial Systems, Toronto, June 17–19, 1968.

58. Newton, G. C., Gould, L. A., and J. Kaiser: *Analytical Design of Linear Feedback Controls*, Wiley, New York, 1957.

59. Pemberton, T. J.: "Simplified Method of Evaluating ISE and ITSE Performance Criteria," *Control Eng.*, April 1970, pp. 89–91.

60. Shinskey, F. G.: *Process Control Systems*, 2nd ed., McGraw-Hill, New York, 1979.

61. Woolverton, P. W., and P. W. Murril: "Evaluation of Four Ideas in Feedforward Control," *Instrum. Technol.*, January 1967, pp. 35–40.

62. Bristol, E. H.: "Designing and Programming Control Algorithms for DDC Systems," *Control Eng.*, January 1977, pp. 24–26.

63. Dahlin, E. B.: "Designing and Tuning Digital Controllers," *Instrum. Control Syst.*, June 1968, pp. 77–83.

64. Krivoshein, K. D., and A. B. Corripio: "A Study of Ringing in Dahlin's Digital Control Algorithms," *ISA Trans*, vol. 20, no. 4, 1982, pp. 1–11.

65. Gautaman, R., and R. Mutharassan: "A General Direct Digital Control Algorithm for a Class of Linear Systems," *AIChe J.*, vol. 24, no. 2, 1978, pp. 360–364.

66. Vogel, E. F., and T. F. Edgar: "An Adaptive Dead Time Compensator for Process Control," *ASME Pap. 80-WA/DSC-10*, 1980.

67. Adkison, B. M., and G. K. Kohler: "Optimum Tuning of a Generalized Three-Mode Digital Control Algorithm," *ISA Pap. 512-70*, Instrument Society of America, Research Triangle Park, N.C., 1970.

68. Franks, R. G., and C. Worley: "Quantitative Analysis of Cascade Control," *Ind. Eng. Chem.*, vol. 48, no. 6, 1956, pp. 1074–1079.

69. Schork, F. J., and P. B. Deshpande: "Double Cascade Controller Tested," *Hydrocarbon Process.*, June 1978, pp. 113–117.

70. Nisenfield, A. E., and H. M. Schultz: "Interaction Analysis Applied to Control System Design," *Instrum. Technol.*, April 1971, pp. 52–57.

71. Nisenfeld, A. E., and D. L. Hoyle: "Dynamic Feedforward Control of Multi-Effect Evaporators," *Instrum. Technol.*, February 1970, pp. 49–54.

72. Bristol, E. H.: "On a New Measure of Interaction for Multivariable Process Control," *IEEE Trans. Autom. Control*, vol. AC-11, January 1966, pp. 133–134.

73. Akatsuka, T.: "The Decoupling of Discrete Multivariable Systems by State Variable Feedback," *SRC 70-3*, Case-Western Reserve University, Cleveland, Ohio, June 1970.

74. Shunta, J. P.: "Sampling Control of Level with Deadtime," *Control Eng.*, June 1974, pp. 65–67.

75. Luyben, W. L.: "Damping Coefficient Design Charts for Sampled-Data Control of Processes with Deadtime," *AIChE J.*, vol. 18, no. 5, pp. 1048–1052.

76. Hoyle, D. L.: "Designing for pH Control," *Chem. Eng.*, November 8, 1976, 121–126.

77. Shinskey, F. G.: "Adaptive pH Controller Monitors Nonlinear Process," *Control Eng.*, February 1974, pp. 57–59.

78. Lee, T. H., Adams, G. E., and W. M. Gaines: "Computer Process Control," in *Modeling and Simulation*, Wiley, New York, 1968.

79. Franks, R. G. E.: *Modeling and Simulation for Chemical Engineers*, Wiley, New York, 1972.

80. Smith, C. L., Pike, R. W., and P. W. Murrill: *Formulation and Optimization of Mathematical Models*, International Textbook, Scranton, Pa., 1970.

81. Luyben, W. L.: *Modeling, Simulation, and Control for Chemical Engineers*, McGraw-Hill, New York, 1973.

82. Kim, C., and J. C. Friedly: "Approximate Dynamic Modeling of Large Staged Systems," *Ind. Eng. Chem. Process. Des. Dev.*, vol. 13, no. 2, 1974.

83. Weiss, M. D.: "Problems in the Application of Digital Computers to Batch Reactors," IFAC/IFIP Symposium on Digital Control of Large Industrial Systems, Toronto, June 17–19, 1968.

84. Dahlin, E. B., and D. B. Brewster: "Process Identification for Control System Design and Tuning," *Control Eng.*, April 1969, pp. 81–84.

85. Fisher, R. A.: *Statistical Methods for Research Workers*, Oliver and Boyd London, 1956.

86. Bryson, A. E., and Y.-C. Ho: *Applied Optimal Control,* Ginn, Waltham, Mass., 1969.

87. Rosenbruck, H. H., and C. S. Torey: *Computational Techniques for Chemical Engineers,* Pergamon, New York, 1966.

88. Rogers, A. E., and T. W. Connolly: *Analog Computation in Engineering Design,* McGraw-Hill, New York, 1960.

89. Jackson, A. S.: *Analog Computation,* McGraw-Hill, New York, 1960.

90. James, M. L., Smith G. M., and J. C. Wolford: *Simulation of Engineering Systems,* International Textbook, Scranton, Pa., 1966.

91. Smith, O. J. M.: *Feedback Control Systems,* McGraw-Hill, New York, 1958.

92. Korn, G. A., and T. M. Korn: *Electronic Analog and Hybrid Computers,* McGraw-Hill, New York, 1964.

93. Chu, Y. F., Sansom, J., and H. E. Peterson: *Digital Simulation of Continuous Systems,* McGraw-Hill, New York, 1969.

94. Clancy, J. J., and M. S. Fineberg: "Digital Simulation Languages: A Critique and a Guide," *Proceedings of Fall Joint Computer Conference,* Spartan, New York, 1965, pp. 23–36.

95. Rennan, R. D., and R. N. Linebarger: "A Survey of Digital Analog Simulation Programs," *Simulation,* vol. 3, no. 6, 1964, pp. 22–35.

96. Giloi, W.: "Future Developments in Hybrid Computation," International Analogue Computation Meeting, Lausanne, Switzerland, August 28–September 2, 1967, in *Proc. ACTES,* Presses Academiques Européenes, Brussels, 1967.

97. Speckhart, F. H., and W. L. Green: *A Guide to Using CSMP,* Prentice-Hall, Englewood Cliffs, N.J., 1976.

98. Shah, M. J.: *Engineering Simulation Using Small Scientific Computers,* Prentice-Hall, Englewood Cliffs, N.J., 1976.

99. Evans, G. W., II, Wallace, G. F., and G. L. Sutherland: *Simulation Using Digital Computers,* Prentice-Hall, Englewood Cliffs, N.J., 1967.

100. Tocher, K. D.: *The Art of Simulation,* Van Nostrand, Princeton, N.J., 1963.

101. Huskey, H. D., and G. A. Korn: *Computer Handbook,* McGraw-Hill, New York, 1961.

102. Nelson, J. M., and E. B. Dahlin: "Hybrid Computer Facility Provides Higher Speed and Accuracy," *Res. Dev.,* March 1963.

103. Horn, B. C.: "On-Line Optimization of Plant Utilities," *Chem. Eng. Prog.,* June 1978, pp. 76–79.

104. Dantzig, G. E.: *Linear Programming and Extensions,* Princeton University Press, Princeton, N.J., 1963.

105. Abadie, J.: *On Linear Programming,* Wiley, New York, 1967.

106. Krall, R. A., and B. C. Horn: "On-Line, Mixed Integer Optimization of an Industrial Plant's Utilities System," IFAC Meeting, Houston, Texas, October 1980.

107. Box, G. E. P., and K. B. Wilson: "On the Experimental Attainment of Optimum Conditions," *J. Roy. Statistical Soc.,* vol. 13(B), 1951, pp. 1–45.

108. Lee, T. H., Adams, G. E., and W. M. Gaines: *Computer Process Control in Modeling and Simulation,* Wiley, New York, 1968.

109. Bryson, A. E., and Y.-C. Ho: *Applied Optimal Control,* Ginn, Waltham, Mass., 1969.

110. Leitman, G., Ed.: *Optimization Techniques: With Applications to Aerospace Systems,* Academic Press, New York, 1962.

111. Laspe, C. G.: "Recent Experiences in On-Line Optimizing Control of Industrial Processes," 5th Annual Control Conference, Purdue University, West Lafayette, Ind., April 9–12, 1979.

112. Bellman, R.: *Dynamic Programming,* Princeton University Press, Princeton, N.J., 1957.

113. Bellman, R., and R. Kalaba: *Dynamic Programming and Modern Control Theory,* Academic Press, New York, 1965.

114. Bellman, R. E., and J. E. Dreyfus: *Applied Dynamic Programming,* Princeton University Press, Princeton, N.J., 1962.

115. Pars, L. A.: *An Introduction to the Calculus of Variations,* Wiley, New York, 1962.

116. Elsgolc, L. E.: *Calculus of Variations,* Addison-Wesley, Reading, Mass., 1962.

117. Pontryagin, L. S., et al: *The Mathematical Theory of Optimal Processes,* Wiley, New York, 1962.

118. Lee, R. C. K.: "Optimal Estimation, Identification, and Control," *Research Monogr. 28,* MIT Press, Cambridge, Mass., 1964.

119. Kalman, R. E.: "A New Approach to Linear Filtering and Prediction Problems," *ASME J. Basic Eng.*, March 1960, pp. 35–44.

120. Leondes, C. T. (ed.): *Advances in Control System Theory and Applications*, vol. 3, Academic Press, New York, 1966.

121. Boxenhorn, B.: "Using Kalman Filtering to Estimate Control Parameters," *Control Eng.*, July 1969, pp. 69–72.

122. Kneile, R. G., and R. H. Luecki: "Comparison of the Kalman Filter with a Classical Digital Filter," *AIChE J.*, vol. 20, no. 3, 1974, pp. 598–600.

123. Lynch, E. B., and W. F. Ramirez: "Real-Time Optimal Control of a Stirred Tank Reactor Using Kalman Filtering for State Estimation," *AIChE J.*, vol. 21, no. 4, 1975, 799–804.

124. Seborg, D. E., and D. G. Fisher: "Experience with Experimental Applications of Multivariable Computer Control," *Trans. ASME*, vol. 101, June 1979, pp. 108–116.

125. Bristol, E. H.: "Adaptive Process Control: Versatile On-Line Tool," part I, *Control Eng.*, April 1973, pp. 41–44; part II, *Control Eng.*, June 1973, pp. 41–43.

126. Adkison, B. M., and G. K. Kohler: "Optimum Tuning of a Generalized Three-Mode Digital Control Algorithm," *ISA Pap. 512-70*, Instrument Society of America, Research Triangle Park, N.C., 1970.

127. Adams, P. G., and A. T. Schooley: "Adaptive-Predictive Control of a Batch Reaction," *ISA Pap. 68-818*, Instrument Society of America, Research Triangle Park, N.C., 1968.

128. Kalman, R. E.: "Design of a Self-Optimizing Control System," *Trans. ASME*, vol. 80, February 1968, pp. 468–478.

129. Andriev, N.: "A New Dimension: A Self-Tuning Controller That Continually Optimizes PID Constants," *Control Eng.*, August 1981, pp. 84–85.

130. Stanton, B. D.: "Application of Systems Engineering in Trouble Shooting an Existing Process," *ISA Conf. Pap. 5.3-3-63*, New York, October, 12–15, 1964, Instrument Society of America, Research Triangle Park, N.C., 1964.

131. Hougen, J. O., and R. A. Walsh: "Pulse Testing Method," *Chem. Eng. Prog.*, March 1961, pp. 69–79.

132. Cowley, P. E. A.: "An Active Filter for the Measurement of Process Dynamics," Proceedings of ISA Engineering Conference, 1964, Instrument Society of America, Research Triangle Park, N.C., 1964.

133. Ziegler, J. G., and N. B. Nichols: "Process Lags in Automatic Control Circuits," *Trans. ASME*, vol. 65, July 1943, 433–444.

134. Pemberton, T. J.: "An Improved Pulse Testing Method," *Instrum. Technol.*, December 1970, pp. 61–67.

135. Dahlin, E. B.: "On-Line Identification of Process Dynamics," *IBM J. Res.*, July 1967, pp. 406–426.

136. Dahlin, E. B.: "Interactive Control of Paper Machines," *Control Eng.*, January 1971, pp. 76–81.

137. Truxal, J. G.: *Control System Synthesis*, McGraw-Hill, New York, 1955.

138. Scheib, T. J.: "Automatic Tuning of DDC Loops," *SRC-70-4*, Case-Western Reserve University, Cleveland, Ohio, September 1970.

139. Fletcher, R., and M. J. D. Powell: *Comput. J.*, vol. 6, 1963, p. 163.

140. Ramaker, B. L., Smith, C. L., and P. W. Murrill: "Digital Predictor Controls Noisy Continuous Reactor," *Instrum. Technol.*, June 1970, pp. 61–66.

141. Rosenbrock, H. H.: "Distinctive Problems of Process Control," *Chem. Eng. Prog.*, 1962, pp. 43–55.

142. McAvoy, T. J.: *Interaction Analysis: Principles and Applications*, Instrument Society of America, Research Triangle Park, North Carolina, 1983.

Control Algorithms

by
E. H. Bristol*

The implementation of a control function on a process requires either an analog control system, natural to the process but with its own problems, or a digital computation. The process of sampling an analog value, converting the resulting value to a (quantized) number, computing a control action, and outputting that value involves issues remote from the abstractions of design control theory. This section develops one view of these issues (Ref. 1). Other attitudes toward practical algorithm design may be of interest (Refs. 2 to 4).

The effect of sampling is covered in standard texts (Ref. 5) as a natural branch of control theory and so will not be addressed here. Nevertheless, it should be pointed out that the normal theoretical development of sampled-data control deals with neutral parameters not clearly related to time constants, loop gains, and other such similarly intuitive forms. The position taken here is that the operation of the control system will be more effective if algorithms can be set or tuned in terms of natural process-related values and control behavior, which are at least qualitatively related to analog practice. This means, of course, that control functions not normally available in analog systems, such as Smith predictor controllers, will also be expressed similarly. This in no way restricts the use of formal design methods, because for every computed set of neutral control coefficients in reasonable designs, the corresponding gains and time constants can be directly computed. The algorithms given will include the process time and sampling time as explicit parameters. As a result, they are, as much as possible, compensated for sample time or even for irregular sample times.

The one aspect of sampling which should be mentioned is *aliasing*. This is a bizarre effect where high-frequency noise looks like an offset (Ref. 5). It is not normally a problem because most controllers sample frequently compared to process time constants, but it should be understood. Because of sampling, the minimum width in time of a disturbance (pulse) as seen by the system must be one sample period; pulses are either ignored or treated as being at least that long. But, less obvious, when the noise is rapidly varying compared to the sampling frequency, the noise may synchronize one-sidedly with the sampling so that the noise looks to the digital system like a constant offset, or an aliased low-frequency variation. The control system then takes inappropriate corrective action. For this reason, filtered (before sampling) measurements or fast sampling is preferred. On the other hand, fast sampling requires greater data precision and computer time to accommodate the wider dynamic range of the resulting computations. These costs of unnecessarily fast sample time are real and significant to control.

Except for aliasing, the choice of a sampling time is today a nonissue. If the process is sampled faster than the *dominant closed-loop time constants* of the *economically important process variables,* one will rapidly approach a point of diminishing returns. At the sample times universally available commercially, the sampling effect with *well-designed* algorithms is normally, at worst, barely noticeable. If the sole economical effect of a fast flow loop were on an end process quantity which took an hour to respond, one would in principle see *no difference* if the *flow loop were sampled once every 5 min*. Of course, many "housekeeping functions," like flow loops, may have constraining side effects whose violation would involve real costs if this logic were actually implemented. Sampling times faster than 1 s, well filtered, *rarely matter* in practice even when the local process dynamics are faster. As with other hardware considerations, the inexpensive microprocessor has caused a conservative design tradeoff to favor faster sample times because the cost is slight. Nevertheless a 10 to 1 reduction in sample time corresponds to a 10 to 1 reduction in potential capability if one has the tools and imagination to use the capability.

---

*Research Department, The Foxboro Company, Foxboro, Mass.

The main emphasis of this section is a refined view of the properties of algorithms. A casual experimenter or designer of a one-use algorithm is justified in implementing the algorithm as a basic controller in FORTRAN, using floating-point computation. In this case the normal effort and debugging which take place are usually adequate to discover any serious deficiencies.

But the designer, connoisseur, or critic-evaluator of serious algorithm packages should understand the deficiencies of present-day floating-point higher level language implementation and be prepared to consider the advantages of fixed-point implementation, not only in terms of speed but, most important, in terms of precision and control of the algorithm. The difficulties of working in fixed-point arithmetic or even machine language are overstated; the real problem is the lack of understanding of the issues, which suggests clear ways of specifying the details of an algorithm and testing it. A later section will address this problem. A grasp of good methods here is the difference between good workmanship and mediocrity.

In discussing error and precision in control algorithms, one must recognize that the purpose of a control system is control, not simulation or computation. Rarely is the precise duplication of control parameters and tuned performance (say, to better than 10%) important *even* when the control parameters are computed or adapted. There are many control situations where a 2:1 or greater error in parameters is unimportant. In contrast, there are situations *within* a calculation where a quantization error of 1 part in 100,000 or more will affect the qualitative input-output character of the algorithm significantly or cause the control to completely change character or fail. These internal computational issues are the most important ones. Thus perfection of the end computed value is unimportant; only control effects matter. (But where the effect of an error is uncertain it is better to be perfect.)

The rest of the section details the different kinds of numerical representation and arithmetic and relates these effects to proper algorithm design. The performance of algorithms can be more easily guaranteed if the proposed method of specification is used. Several sections address the proper use of floating-point calculations where their limitations are unavoidable, as in matrix-oriented control packages. A common thread throughout this article is the effect of differences between large values in exaggerating error and of multiplication in creating these large values. The emphasis is on the design of high-quality dynamic algorithms for basic operation and control. The user interested in the computation of nonlinear compensating functions should consult the references (Refs. 6 and 7). Large computations peripheral to on-line dynamic control such as optimization are also ignored.

NUMBER SYSTEMS AND THE EFFECTS OF THE BASIC ARITHMETIC OPERATIONS

There are two basic formats for representing control data as numbers, fixed point and floating point. The fixed-point format represents the data as an integer, with a fixed range of possible values which must be scaled (painfully) in much the same way analog control systems and simulations were scaled. The scale factors, as well as multiple-precision computations and conversions, are the explicit responsibility of the programmer. Floating-point formats are automatically scaled in that they have a "fraction" or "mantissa" part (f), corresponding to the integer value of the fixed point, and an exponent (e), which defines as a scaling factor a base value (b) raised to the power (e) (see Ref. 6).

Until recently floating-point formats were not standardized (Ref. 8). Moreover, the floating-point format inherently involves arbitrary truncations and uncertain relationships between single- and double-precision computation. For this reason high-quality control algorithms are best written and analyzed in a fixed-point format.

Fixed-Point Format

A fixed-point format represents an integer as a binary number stored in a *word* containing a fixed number of bits, usually in what is called two's complement format. This binary format causes the range of positive numbers to be 0 to $2^n - 1$ and signed numbers to be -2^{n-1} to $2^{n-1} - 1$, n being the number of bits in the word. In two's complement arithmetic, negative numbers are represented

SIGN AND VALUE		TWO'S COMPLEMENT
		DECIMAL
DECIMAL	BINARY	EQUIVALENT
0	000	0
1	001	1
2	010	2
3	011	3
-4	100	4
-3	101	5
-2	110	6
-1	111	7
(3)+(-3)	1000	(3)+(5)=8

⤒ CARRY BIT

Fig. 1   Mapping two's complement numbers onto decimal integers ($N = 3$).

V.S	V.2	V.1	V.0
123456	12	34	56

Fig. 2   Decimal multiple-precision (two digits per word).

in binary notation as very large numbers, and addition is "around the clock." Thus, in Fig. 1 (with $n = 3$), negative 3 (-3) is represented by a binary number whose decimal equivalent is 5. And 5 added to 3 becomes 8 which is also represented as 0 in binary notation with an ignored carry. Thus 5 is a perfectly good negative 3.

Because of the around-the-clock nature of the arithmetic operations, the effect of combining numbers into a result too large to fit into a word must be taken care of in one of three ways: (1) One may produce a carry from the result used in a multiple-precision computation (using two or more words to represent the data). (2) One may saturate the result. Saturation means to take the value as computed and replace it by the largest possible value if the theoretical result is too large, or the smallest possible value if the theoretical result is too small. (3) One may ignore the problem if it is clear that the result will never overflow (i.e., become too large or too small).

This section will not examine the details of binary arithmetic further, decimal arithmetic being adequate to explain all effects. Suffice it to say that a good fixed-point hardware design includes the basic operations as well as carry bits, overflow bits, and other features which simplify the creation of multiple-precision operations when they are needed. Also, by the nature of things, multiplication and division or scaling by powers of 2 is particularly easy and fast. This explains the perfectly reasonable tendency of designers to scale by powers of 2 when they are adequate.

Throughout this section, whenever referring to multiple-precision scaled representation of a variable (say one named V) or a calculation, we will append special symbols to define different aspects of that scaled variable. Thus $V.S$ will become the scaled value taken as a whole. $V.0, V.1, V.2, \ldots$ will be different storage words which make up the representation of $V.S$. Thus, in Fig. 2, $V.0$ is the rightmost word and $V.1$ is the next rightmost word, etc.

Fixed-Point Scaling

In this notation $V.S$ will be the result of scaling V by a fixed-point fraction with numerator $V.N$ and denominator $V.D$ [either of which can in principle be multiple-precision integers as $V.N = (V.N.1, V.N.0), V.D = (V.D.1, V.D.0)$]. Thus

$$V.S. = \frac{V * V.N}{V.D}$$

Just as we will not be concerned by binary and decimal representation, since they come up only when conversion is made for data entry or display, so we will generally ignore V, $V.N$, and $V.D$, working entirely with $V.S$ (but referring to it as V when no confusion arises). Rarely, other scalings may be used. The conversion between binary and decimal numbers (radix conversion) for input and display is straightforward (Ref. 6) and need not concern us.

A multiple-precision scaled value $V.S$ can be considered the sum of the values $V.0$, $V.1$, $V.2$, etc., each with its own scaling factor:

$$V.S = V.0 + V.1 * B + V.2 * B^2$$

The value B is one greater than the largest value in the range of each storage word ($B = 2^n$ for positive data or $B = 2^{n-1}$ for signed data). The powers of B are not explicitly needed.

In scaling the parameters of a control algorithm one chooses scalings which are representative of the actual need, convenient to compute with, and otherwise make the best use of the available data range without forcing a naturally single-precision calculation to double precision. For a controller measurement or value on a 16-bit word machine, ½% precision would be minimally adequate. If both quantities are positive, the 1 part in 255 is better than ½%, and 256 is 2^8, corresponding to 1 byte (8 bits) of storage. Scaling by powers of 2 is particularly efficient. One might scale 0 to 100% into 0 to 256 with $V.N = 256$ and $V.D = 100$.

On the other hand, for a 16-bit microprocessor there are 8 more bits in the word which could be used to give a smoother valve action and simplify calculation. Working only with positive values, 16 bits corresponds to the unnecessary accuracy of 1 part in 65,535. But the controller calculation will need a signed value to represent error. This leads to a positive scaling of 1 part in 32,767 or, for safety, 1 part in 16,384. This scaling is well above minimum accuracy but does not force multiple precision. The required 14-bit analog-to-digital (A/D) and digital-to-analog (D/A) converters are reasonable.

Control parameters may have different natural data ranges or scalings. A controller gain might be scaled so that the minimum gain is ½₅₆ and the maximum gain is 256. In this way the range of values spans equally high-gain or low-gain processes about a nominal unity gain. For time constants the resolution may be important. When a minimal reset time of 1 s is adequate, then 1 h corresponds to 3600 s and 2^{16} corresponds to more than 18 h. On the other hand, using the full range of data storage for the control parameters may require creating arithmetic routines for which the arguments are mixed signed and unsigned. The additional convenience of avoiding this may be worth reducing the range, so that all data are signed, even when negative numbers are prevented from occurring.

Error in Fixed-Point Arithmetic

The objective of design decisions relating to quantization and multiple precision is to avoid poor control arising from inaccuracies not inherent in the real process. All the algorithms likely to be used are made up of combinations of additions, subtractions, multiplications, and divisions; they define rational functions of their data. It would be unimportant how calculations were ordered if all calculations were carried out with as much precision as needed. But, in practice this is not done; thus careful understanding of the effects of the basic arithmetic operations is needed. First, one should examine how each operation affects the numerical range of the result and how errors accumulate. As will be seen, range effects are typically more important.

Error can be considered in terms of absolute error, i.e., the worst case of actual error, or relative error, i.e., the error as a percentage of the value in error with respect to the worst case error. Usually the relative error is important in final results and products (or quotients), whereas the absolute error is important in determining how error accumulates in a sum.

When two numbers are added, their potential range doubles, either by requiring a carry when two large positive numbers are added or by requiring a sign when two large positive numbers are subtracted (see Fig. 3). Additions and subtractions cause the absolute error to accumulate (add up). Addition (of numbers of the same sign) can never increase the relative error. But subtraction increases the relative error. In Fig. 3, two 10% relative error numbers, when added, result in a 10% relative error. But when subtracted, a large relative error of 30% can arise. In Figs. 3 and 4, the bar over the number indicates an error term added to (or subtracted from) the "ideal" value of a calcu-

$$99+99=198 \qquad\qquad (20-\bar{2})+(10-\bar{1}) = 27 = (30-\bar{3})$$

$$99-(-99)=198 \qquad\qquad \frac{(20-\bar{2})+(10-\bar{1})}{20+10} = \left(\frac{30}{30} - \frac{\bar{3}}{30}\right)$$

$$99-0 \quad =99$$

$$0-99 \quad =-99 \qquad\qquad \frac{(20-\bar{2})-(10+\bar{1})}{20-10} = \left(\frac{10}{10} - \frac{\bar{3}}{10}\right)$$

Fig. 3   Addition and subtraction.

$$99*99 = 980 \qquad\qquad (10-\bar{1})(10-\bar{1})=81=(100-1\bar{9})$$

$$9999/99 = 101$$
$$9800/99 = 98 \text{ WITH REMAINDER } 98 \qquad \frac{(100+\bar{8})}{(10-\bar{1})} = 12=(10+\bar{2})$$
$$\text{BUT } 9800/1 = 9800$$

Fig. 4   Multiplication and division.

lation input or output. In this way the difference between large numbers explains most computing accuracy problems.

Fixed-point multiplication not only increases the range of the result but also doubles the required storage of the result, as shown in Fig. 4. In hardware implementations most single-precision multipliers return a double-register result. By analogy most hardware dividers consider division the reverse of multiplication, taking a double-precision quantity and dividing it by a single-precision value to obtain a single-precision quotient and a single-precision remainder. But, as Fig. 4 shows, the double-precision dividend can give rise to a double-precision quotient.

The dividend storage requirement (number of digits or bits) equals the sum of the quotient and divisor requirements. In the hardware implementation the double-precision quotient shows up as an overflow (division by zero should always be reflected as an overflow). Of course, one can always generalize the division in software, but one must allow for this character of hardware dividers if they are used.

A basic strength of fixed-point arithmetic, with its remainders, overflow and carry bits, and multiple-precision results, is that no information is lost to truncation except by choice. As a consequence, it is easy to build multiple-precision arithmetic out of single-precision primitives. Further, it is easier to alter precision in the course of a calculation.

Fixed-Point Multiplication and Division

The extensive expansion of range of a multiplication is the most problematic aspect of multiplication. Among other consequences it generates large numbers whose differences may cause large errors. It makes smaller numbers smaller still, compared to the errors caused by the large numbers. The important issue is the ratio of the large numbers, which make the errors, to the small numbers, which in the end constitute the result. It is usually desirable to avoid repeated multiplications. Often multiplications and divisions occur together in a manner in which they may be alternated, the multiplication generating a double-precision value and the division returning that value to a single-precision quotient. The basic proportional controller calculation is a good example:

$$\text{Output} = \frac{100 \times \text{error}}{\text{proportional band}} + \text{bias}$$

Figure 4 also shows that the multiplication and division *always* increase the potential relative error and, in the case of fixed-point multiplication, the absolute error.

The expansion of range caused by multiplication forces a choice: One can choose to preserve the precision throughout the calculation, or one can force intermediate values back into the earlier data range. Ultimately, it is necessary to do this, because D/A converters have a limited range. Range compression operations are natural at the machine language level, since they make use of overflow bits and different kinds of masking and bit shifting operations. Conceptually the operations are equivalent to appropriate combinations of division after scaling and saturation on overflow.

There are combinations of the basic operations that are of common occurrence and can be specially designed for. In general, one should try to maintain a constant data range throughout the calculation. Often a product is naturally intended to return a value in the same range as the principal output. There are two natural cases: either a value is multiplied by a gain or by a proper fraction. In the above controller calculation, the value 100/(proportional band) is likely to be scaled to fall between $\frac{1}{256}$ and 256. If the gain is expressed as a single scaled integer rather than a fraction, one can still perceive a single-precision result by taking the middle single-precision set of digits out of the double-precision scaled result as in Fig. 5.

One way of calculating the output value Y of a lag is as a weighted average of the past output and the current input X:

$$Y \leqq \frac{1}{T+1} X + \frac{T}{T+1} Y = Y + \frac{1}{T+1}(X - Y)$$

In this case, if the two proper fractions are scaled as integers, they may be multiplied as integers and the leftmost word of the double-precision result returned as a single-precision product. It is often appropriate to combine constants or parameters into a common effective parameter which is this kind of gain or fraction. The combined parameters can be made to act consistently on the data even if in error as calculated from the original parameters. Any errors can be entirely explained as errors in the effective values of the original parameters. Note that the final expression above involves a feedback between the Y and the difference between X and Y. Often such a feedback can be used to improve an otherwise error-prone algorithm, making it self-corrective like any other feedback system.

There is a parallel between the issues presented here and traditional dimensional analysis in that sums must be of values of identical units and result in the same units, whereas multiplications and divisions result in different units (Ref. 9). Quantities with similar units are likely to have similar ranges. Similarly, as indicated above, a multiplication by a dimensionless unit is unlikely to change the natural range of the data.

All these calculations generate rational functions of their inputs, i.e., ratios of sums of products. Figure 6 shows an explicit example of this arising out of solving equations where there is a real problem. The range of the quotient is likely to be limited by the physical nature of the process in ways not limiting the range of the numerator and denominator and not apparent in the data being

$$\frac{100}{\text{P.B. } (=50)} \times \text{ERROR}(40\%) = 80\%$$

SCALE FACTORS: 10 1 1

SCALED VALUE: 20 x 40 = 0800 = 80

(THE MIDDLE WORD)

(a)

0.7 x 50 = 35

SCALE FACTORS: 100 1 1

SCALED VALUE: 70 x 50 = 3500 = 35

(THE LEFT WORD)

(b)

Fig. 5 Multiplication by (*a*) a gain and (*b*) a fraction.

$$AX = B$$

$$A_{11} X_1 + A_{12} X_2 = B_1$$

$$A_{21} X_1 + A_{22} X_2 = B_2$$

$$X = \frac{B}{A}$$

$$X_1 = \frac{A_{22} B_1 - A_{12} B_2}{A_{11} A_{22} - A_{12} A_{21}}$$

Fig. 6   Solution of equations.

computed. Thus, a process gain may be equal to one, whether calculated as a ratio of an input change of 0.01 to an output change of 0.01, or as a ratio of 100 to 100.

Digital Integration for Control

Figure 7 shows a different kind of division situation common to process control integration. In this figure, the object is to integrate a process error as in a reset calculation. This can be done by computing a common fractional multiplier $\Delta t/T$ which has been scaled to give a good range and which might also include the proportional action P, as in $100 * \Delta t/(P*T)$. This result would be multiplied by the error and added in double precision to the previous value of the sum to obtain the current value.

Double precision is essential here, since a large value of T which corresponds to a small fraction $(\Delta t/T)$ can prevent the cumulative integration of significant errors. For example, suppose that the error and the sum are scaled as a signed integer $-10,000$ to $+10,000$. Suppose $\Delta t = 1$ and the minimum $T = 1$ also. The corresponding maximum $(\Delta t/T)$ which equals 1 must be scaled to a value of 10,000. Then the natural scaling of the controller output can be achieved by dividing by 10,000. (See the scaling discussions elsewhere.) That is, the scaled equations should be (remembering that "Sum.S" means the scaled value of "sum")

$$\text{Sum.}S = \text{old sum.}S + \left[\frac{(\Delta t/T).S * \text{error.}S]}{10,000} \right.$$

In the division only the integer quotient is considered. Thus, if the product $[(\Delta t/t).S * \text{error.}S]$ is less than 10,000, it is treated as zero and the error becomes a permanent offset. For example, if $(\Delta t/T).S = 1$, or rather $T = 10,000$ (≈ 3 h), any error less than 100% (scaled to 10,000) will be lost. Under control this could cause a 100% offset.

The better answer shown in Fig. 7 preserves the division and achieves an exact result with the same storage as double precision. In this case the product of the error and the sample time (the sample time may equal 1) is formed as a double-precision value added to a remainder of the previous

$$\sum_{i=0}^{n} \text{ERROR}_i \frac{\Delta t}{T} \qquad \frac{\sum_{i=0}^{n} \text{ERROR}_i \, \Delta t}{T} = \frac{\sum_{i=0}^{n-1} \text{ERROR}_i \, \Delta t}{T} + \frac{\text{ERROR}_n \, \Delta}{T}$$

$$= \frac{\sum_{i=0}^{n-1} \text{ERROR}_i \, \Delta t}{T} + \left(\text{QUOTIENT} \left(\frac{\text{ERROR}_n \, \Delta t + \text{REMAINDER}}{T} \right) \right)$$

$$\text{NOTE: ERROR}_n \, \Delta t = T \left(\text{QUOTIENT} \left(\frac{\text{ERROR}_n \, \Delta t + \text{REMAINDER}}{T} \right) \right)$$

Fig. 7   An integration trick.

sampled calculation. This net value is divided by T to get back a single-precision quotient to be added to the old sum, and a remainder to be added back in the next time. All truncated data are accumulated in the remainder.

Floating-Point Format

With the floating-point format, all remainders, carries, and multiple-precision results of single-precision inputs disappear. The new floating-point standard (Ref. 8) deals with this problem by using an awkward combination of different levels of precision and so-called *guard bits*. For floating-point data the errors introduced through differences of large numbers become more severe, and at the same time less obvious because their development is automated. For example, when a large number L is added to a small number S the small number may totally disappear. The sum $L - L + S$ should equal S. If the calculation is carried out in the natural ordering, $(L - L) + S$, then S will result. But $(L + S) - L$ will generate zero.

There are a number of intricate ways of avoiding this problem:

1. Convert all floating-point numbers to ratios of integers and operate in the fixed-point format. This is a good way to study the properties of an algorithm.

2. Use adequate precision. Practically, this is unpredictable, and theoretically it is impossible because repeating decimals require infinite data. This suggests that one might sometimes prefer to avoid division until the last operation.

3. Reorder the calculation for the most favorable computation, either statically based on algorithm properties or by using an arithmetic package which continuously reorders the operands. It is useful in the following discussion to consider every list of terms to be added (subtraction being taken as addition of a negated number) to be ordered by magnitude. The individual operations are carried out according to these rules:

   a. Out of an ordered list of values or products to be added, take the smallest pair of adjacent values whose difference in magnitude is to be computed and replace them by their difference. Thus $10{,}023 + 33 - 26 + 1$ is replaced by $10{,}023 + 7 + 1$. (See rules c and d below.) Repeat until all entries are of the same sign.

   b. Out of an ordered list of values or products to be added, all having the same sign, replace the smallest two by their sum. Thus, $10{,}023 + 33 + 26 + 1$ is replaced by $10{,}023 + 33 + 27$. (See rules c and d below.) Repeat until only one entry remains.

   c. Out of a list of product terms (or quotients) to be added, any pair, with a common factor whose difference is to be computed by rule a above, is factored and the differentiating terms subtracted first to create a new product term. Thus $10{,}023 + 11 * 3 - 3 * 9 + 1$ becomes $10{,}023 + 3 * 2 + 1$.

   d. Out of a list of values to be multiplied or divided (forming a product or quotient of products to be added by rule a later), cancel or divide out, in order, the most nearly equal numerator and denominator terms. When all denominator terms are gone, then the remaining multiplications are carried out, the largest element with the smallest element. This minimizes the chance of floating-point overflow, ensures any cancellations, and ensures that (since a product of values is evaluated the same way, independently of original order) any pair of products, each made from "identical" sets of values, will give identical results ($2 \times 3 \times 4$ will produce the same result as $4 \times 3 \times 2$). (As a result, additive cancellations of terms will occur properly.) Thus $100 * {}^{23}\!/_{00} * 50 * 6$ becomes, successively, $100 * 23 * 1.2 * 6$, $120 * 23 * 6$, $720 * 23$, $16{,}560$. Repeat until there are no products to be evaluated.

Generalized Multiple-Precision Floating Point

Normally the multiple-precision floating-point format involves using a larger fraction and exponent. The author has experimented with a different kind of floating-point format, illustrated in Fig. 8.

$$12340.36789 \Rightarrow 1234 \times 10^1 + 3679 \times 10^{-4} - 1000 \times 10^{-8}$$

$$12349.995678 \Rightarrow 1235 \times 10^1 - 4322 \times 10^{-6}$$

ADDITION:

$$A = 3057 \times 10^6 \qquad\qquad B = 4263 \times 10^4$$
$$A + B = 309963 \times 10^4 = 3100 \times 10^6 - 3700 \times 10^2$$

Fig. 8   A generalized multiple-precision floating-point ($M = 4$).

Here the multiple-precision floating-point value is represented by a set or sum of floating-point numbers, each of whose fractional parts has up to m digits (just as a multiple-precision fixed-point number is represented by a set of fixed-point numbers). Note that, where appropriate, the members in the set or sum may be rounded up. The rules for converting a large number N to this format are as follows:

1. Start with an empty set and N.
2. If N has a leading decimal value >4 with decimal position n (9876.5 has such a position with $n = 4$) and the set is not empty and the smallest member in the set is less in magnitude than 10^{m+n}, add 10^n to the magnitude of that smallest member, replace N by a value of opposite sign and of magnitude $10^n - |N|$, and continue.
3. Find the largest integer l such that $N * 10^l$ has a signed integer part M less in magnitude than 10^m, place the floating-point number number $M * 10^{-l}$ in the set, and replace N by $N - M * 10^{-l}$.
4. Continue steps 2 to 4 until N is reduced to zero.

The resulting set is a best representation of the sum of the values in the set (the original N), in the sense that the largest number in the set (and in each subset) is rounded and contains the maximum precision possible. One can, in the same sense, formulate standard operations (add, subtract, multiply, and divide) for single-precision inputs, which return a "best" rounded single- or multiple-precision value as well as a remainder so that no data are lost. Figure 8 illustrates the addition. The four operations can be structured so that:

1. Addition and subtraction take two single-precision values and return double-precision results (which can be viewed as a single-precision result plus a remainder). The two values are analogous to the fixed-point single-precision result plus carry.

2. Multiplication takes two single-precision values and returns a double-precision result, as with fixed-point arithmetic.

3. Division takes two single-precision values and returns a single-precision quotient and a double-precision remainder. This allows the earlier integration procedure to be applied to floating-point operations.

The development of these routines is straightforward from the earlier definitions (and Ref. 6), and they are as easily used to make multiple-precision (infinite precision) routines as in fixed-point arithmetic. If one prefers the floating-point format for the freedom from scaling it offers and the extra cost in time is not unreasonable, then some form of infinite-precision floating point should be the answer. It can automatically provide the accuracy to make any mathematically sound algorithm work without special care.

SPECIFICATION OF FIXED-POINT ALGORITHMS

The key to providing a clear fixed-point specification is to superpose on a conventional algebraic notation the specification of the sequential order for the successive calculations, and the scaling of the variables and the intermediate results. For instance, specification of the calculation of the proportional controller calculation previously given might take the format

$$\text{Output}_1, \text{ error}_2, \text{ proportional band}_3, \text{ bias}_4$$

$$\text{Output} = \left(\frac{100 * \text{error}}{\text{proportional band}} \right)_5 + \text{bias}_6$$

The parentheses define the order of computation. Where a normal left-to-right order of processing of a single-precision addition is appropriate to the specification internal parentheses are left out. The subscripts define scaling, as indicated later.

Where several constants are to be added, the specification assumes that they will be combined into a single constant in the working program. Where direct alternation of multiplication and division from left to right in a term is appropriate the parentheses can be left out. In the latter case, all constants in the numerator and denominator are also premultiplied into a single numerator-denominator pair for the working program. When the equation is scaled, as below, the scaling constants should be combined with the existing ones (and multiplied and divided first) into the premultiplied pair of constants. Within the term in parentheses, multiplications and divisions will alternate left to right until only all multiplications and divisions occur. These are then carried out from left to right.

The subscripts above refer to scaling specifications for the variables and parenthesized expressions, entered in a separate table shown as Fig. 9. Such a table should include all scaling, saturations, and conversions appropriate to the designer's system. In Fig. 9, the entries indicate whether saturation should occur on data input or output, what the high and low limits should be ("yes" means a limit on the overflow of the data precision). The scaling conversions define the fraction used to convert to an internal format. The sign indicates whether data are signed or simply positive, and the precision indicates the number of words to be used to represent the result. The actual calculation is carried out on the internal format. One should remember that for each variable V the internal format is calculated as

$$V.S = \frac{V.N * V}{V.D}$$

The scaled calculation used for actual computation is developed by replacing V by its expression in terms of $V.S$.

$$V = \frac{V.D * V.S}{V.N}$$

SPECIFICATION NUMBER	IN/OUT	SATURATION HI	SATURATION LOW	SEALING NUMERATOR	DENOMINATOR	SIGN	PRECISION
1	IN	100	0	16384	100	+	1
2	IN	100	-100	16384	100	±	1
3	IN	16384	0	2	1	+	1
4	IN	100	-100	16384	100	±	1
5	OUT	YES		16384	50	±	1
6	OUT	100	0	16384	100	+	1

Fig. 9   Fixed-point specification.

The above proportional calculation then becomes

$$\text{Output.}S * \frac{100}{16{,}384} = \frac{100 * \dfrac{100}{16{,}384} \text{ error.}S}{\dfrac{2}{1} \times PB.S} + \frac{100}{16{,}384} * \text{bias.}S$$

or

$$\text{Output.}S = \frac{50 * \text{error.}S}{PB.S} + \text{bias.}S$$

where PB = proportional band.

The final equation above, with scaling factors combined, would be the programmed calculation, except that each operation would be saturated according to its specification. Usually, the terms in a sum are scaled the same as the sum and as the natural cumulative scaling of factors in any one term.

Thus, the net effect of all scaling in the term subscripted 5 is equal to the ratio of the two component scalings: $(16{,}384/100)/(1/2) = 16{,}384/50 = 8192/25$. If this scaling were not the same as the actual scaling specification numbered 5, then an error would be recognized. On the other hand, the saturation and precision specification may be different for different parts of a calculation. For that part of the specification, the local specification (5 in the example) would prevail. It may occasionally be useful to use double parentheses, e.g., when the calculation should be carried out in double precision and then limited to single precision.

The above form of specification completely defines the intended calculation with the underlying calculations clearly stated. Implementation of the algorithm would have three parts: programming of the necessary arithmetic and saturation routines, manual development of the scaled calculation, and implementation using the routines. Clearly, the fewer scaling specifications used, the fewer the corresponding arithmetic routines needed.

As a matter of practice an algorithm should be tried first in floating-point FORTRAN (language form or equivalent) and then scaled to fixed-point FORTRAN using the earlier techniques before any attempt at machine language is made. When all three forms can be run comparatively, then an extremely effective checkout-debugging tool has been created. The FORTRAN versions naturally lend themselves to a display of intermediate results and should differ only as a consequence of the fixed-point–floating-point difference. With such tools the final implementation of good algorithms should not take significantly more time, effort, and design argument whether written in the fixed-point or floating-point format.

BASIC CONTROL ALGORITHMS

This section considers three basic algorithms as examples: the lag calculation, the lead-lag calculation, and the proportional-plus-integral-plus-derivative (PID) controller.

The Lag Calculation

The lag calculation corresponds to the following continuous transfer function:

$$\frac{L(s)}{I(s)} = \frac{1}{\tau s + 1}$$

where L-output and I-input, and the differential equation

$$\tau \frac{dL}{dt} + L = I$$

The usual practice of going to z transforms of the corresponding sampled-data forms should not be overemphasized. No exact translation is possible. But one practical direct sampled-data approximation of the above differential equation is

$$I(t) = \tau \frac{\Delta L}{\Delta t} + L(t) = \frac{\tau}{\Delta t} [L(t) - L(t - \Delta t)] + L(t)$$

or

$$L(t) = \frac{I(t) + (\tau/\Delta t)L(t - \Delta t)}{1 + (\tau/\Delta t)} = \frac{\Delta t * I(t) + \tau * L(t - \Delta t)}{\tau + \Delta t}$$

This approximation amounts to a weighted average of new input and old output. Note that from a scaling point of view each of the products can be viewed as having the same range as the sum; scaling is simple. The calculation is stable for all positive τ, accurate for large τ, and qualitatively natural as τ approaches 0 with $L(t)$ equaling $I(t)$ if $\tau = 0$, as intended.

The calculation is usable even in single precision if the term $\Delta t * I$ is truncated up (and τ is not too large). In this case the product is never truncated to a magnitude less than one unless the product is truly zero. This guarantees that the output will always settle out above any steady-state input value. Normal truncation and the output always would leave both quotients truncated below their "theoretical" value, a result similar to the integral offset defined earlier.

But a trick, similar to the one used with integrators before, can be applied to calculating lags:

$$L(t) = \text{quotient} \left[\frac{\Delta t * (t) + \tau * L(t - \Delta t) + \text{remainder}}{\tau + \Delta t} \right]$$

with the remainder being saved for the next calculation.

Lead or Lag Calculation

The easiest way of calculating lead or lag or a filtered derivative is by analogy with the transfer function calculations:

$$\frac{O(s)}{I(s)} = \frac{\tau s}{k\tau s + 1} = \frac{1}{k} \left[1 - \frac{1}{(k\tau)s + 1} \right]$$

and then

$$\frac{O(s)}{I(s)} = \frac{\tau s + 1}{k\tau s + 1} = \frac{1}{k} \left[1 - \frac{1}{(k\tau)s + 1} \right] + \frac{1}{(k\tau)s + 1} = \frac{1}{k} \left[1 + \frac{k - 1}{(k\tau)s + 1} \right]$$

or

$$\frac{O(s)}{I(s)} = \frac{As + 1}{Bs + 1} = \frac{1}{B} \left(A + \frac{B - A}{Bs + 1} \right)$$

Proceeding in reverse order, the output of a lead or lag would be calculated from the output of the lag (with lag time constant B):

$$O(t) = [A * I(t) + (B - A)L(t)]/B$$

A basic filtered derivative can be calculated based on a lag value L_D, assuming $k = 0.1$, and the lag time constant $0.1\tau_D$, related to the derivative time constant τ_D:

$$D(t) = 10 * [I(t) - L_D(t)]$$

with L_D calculated according to the above lag calculation.

PID Controller Calculation

A standard PID transfer function takes the form

$$O(s) = \frac{100}{PB}\left(S - M - \frac{\tau_D s M}{0.1\tau_D s + 1}\right)\left(1 + \frac{1}{\tau_I s}\right) \qquad \text{version 1}'$$

where M = measurement and S = set point.

An equivalent form is

$$B = \frac{O(t)}{T_I S + 1}$$

$$O(t) = \frac{100}{PB}\left(S - M - \frac{\tau_D s M}{(0.1\tau_D s + 1)}\right) + B \qquad \text{version 2}'$$

It turns out that the sampled equivalents of these two forms are not equivalent to each other. The common terms are combined to simplify analysis:

$$X = \frac{100}{PB}\left(S - M - \frac{\tau_D s M}{0.1\tau_D s + 1}\right)$$

The scaling and saturation of this calculation is identical to that of the proportional controller discussed earlier. Then, the first form of controller becomes, in sampled form:

$$O(t) = X(n \cdot \Delta t) + \sum_{i=1}^{n} \frac{\Delta t \cdot X(i \cdot \Delta t)}{\tau_I} \qquad \text{version 1}''$$

This controller action has the property that the integrating gain becomes arbitrarily large if τ_I becomes small or Δt becomes large, whereas a more natural integrating gain would limit for large $\Delta t/\tau_I$. Referring to the form of the lag calculation, one might use an integrating gain equal to $\Delta t/(\tau_I + \Delta t)$, which has the desired limiting property, approaching 1 as $\Delta t/\tau_I \to \infty$.

The first algorithm has another practical problem. In this same limit the best control is a pure floating sampled integration. If the proportional term is modified by a factor $\tau_I/(\tau_I + \Delta t)$, then the proportional action decreases naturally as $\Delta t/\tau_I \to \infty$. The resulting controller tunes naturally over its entire operating range and does not prevent control theory calculated settings:

$$O(t) = \frac{\tau_I}{\tau_I + \Delta t} X(n \cdot \Delta t) + \sum_{i=1}^{n} \frac{\Delta t}{\tau_I + t} X(i \cdot \Delta t) \qquad \text{version 1}$$

$$= X(n \cdot \Delta t) + \frac{\Delta t}{\tau_I + \Delta t} \sum_{i=1}^{n-1} X(i \cdot \Delta t)$$

Version 2′ translates directly into an algorithm:

$$F(t) = O(t)$$

$$Q(t) = X(t) + B(t)$$

$$B(t) = \text{quotient}\left[\frac{\Delta t \cdot X(t) + \tau_I * F(t - \Delta t) + \text{remainder } (t - \Delta t)}{t_I + \Delta t}\right] \qquad \text{version 2}$$

Normally F is equivalent to the output. But, for reasons discussed later, F and O are distinguished. It can be shown that version 1 and version 2 compute equivalent actions.

Quantization and Saturation Effects

All the calculations can now be carried out. The discussion has not addressed the effects of truncation on derivative but, as developed here, they are not more serious than for the proportional control. In both cases, truncation will cause a very small limit cycle, on the order of the minimum quantization of the D/A converter, when the loop operates about the steady state. However (see Ref. 1), if the derivative is not carefully filtered after a derivative calculation, or filtering is not integrated into the calculation, as shown above, severe problems arise. An approximate unfiltered derivative calculation would have the form

$$D(t) = \tau_D \frac{\Delta M}{\Delta t} = \frac{\tau_D}{\Delta t} \frac{\Delta M}{\Delta t}$$

Quantization has the awkward effect of forcing a minimum nonzero value of change in M for any sampled calculation. This value is multiplied by $\tau_D/\Delta t$ which usually represents a large gain. The result is very large pulses on the output of the controller. For ΔM quantized to 1 part in 1000 (0.1%), $\Delta t = 1$ s, and $\tau_D = 10$ min $= 600$ s, the minimum nonzero derivative signal equals

$$D(t) = 600 \times 0.1 = 60\%$$

These derivative problems can be made far worse by internal saturations in the algorithm after integration, particularly in incremental algorithms (algorithms which output the desired change in valve position rather than the desired valve position and which therefore include an extra differentiation). The problem arises because the saturation is likely to be unsymmetrical. A differentiated pulse, unsymmetrically saturated, can give rise to an offset of the same order as the original pulse when reintegrated. For example, suppose that in the previous example M went up to 0.1% and down the next sample time. The result of differentiation would be a pair of 60% pulses, the first up and the second down. Suppose the output saturates on the negative pulse so that only the positive pulse passes to the integrator. This would be integrated (without the compensating second pulse) into a 60% offset step.

OPERATIONAL EFFECTS

One of the classic controller effects is integration wind-up. It is an indefinite continuation of internal integrating action after the output has reached a limiting value. Each of the above digital algorithms can be made to limit the integrating action if the value hits a bound. If the limit is known, then one strategy for limiting further integration is to saturate the output and the integration directly. But a more refined possibility exists. Suppose O_L is a known limit. Then, for the two versions and whenever $O(t) > O_L$ (and Σ, the integrating variable in version 1), the algorithm can be modified as follows.

For version 1, compute Σ as follows:

$$\Sigma = \Sigma + \text{quotient} \left[\frac{X(i \cdot \Delta t) * \Delta t + \text{remainder}}{\tau + \Delta t} - (\text{output} - O_L) \right]$$

For version 2, set F to O_L as long as the limit is exceeded. In each case the integration naturally self-limits.

When the constraint on the output is not fixed, it is often possible to feed back the effective current (limited) value of the output (as distinct from the output intended by the controller) and limit windup. One term for this technique is *external feedback* (F above). When this technique is used, it is important that all computational elements which might be used between a controller and the final actuator be *back-calculated*. This operation involves computing for every input of an element the value which corresponds to the actual effective output, taking into account limiting at the device output. In this way the effect of any limit is propagated back to any controller external feedback connection in the same terms (scaling, etc.) in which the output was originally generated.

There are many different approaches to arranging a PID controller (as in versions 1 and 2). One common view is the concept of *bumpless transfer*. If the above algorithms with external feedback were continuously calculated but otherwise disconnected from the valve, then their output would always tend to match the actual output if no measurement error existed. One can arrange these calculations so that, when the controller is reconnected to the valve, the value does not change rapidly (bump), even when a measurement error exists. In this case, one back-calculates the integrating variable whenever a manual-to-automatic transfer is made. For the two versions this one-time-only calculation is:

$$\Sigma = O(t) - X(n \cdot \Delta t) \qquad \text{version 1}$$

$$B = O(t) - X)n \cdot \Delta t) \qquad \text{version 2}$$

In writing other algorithms, one should devote as much time to these considerations as to the others discussed earlier.

IDENTIFICATION AND MATRIX-ORIENTED OPERATIONS

Current academic control theory emphasizes different matrix-oriented formulations. These are becoming more common as more engineers are trained in them. From the author's point of view the traditional control elements and attitudes (including the Smith predictor and decouplers) are more effective for normal control application efforts, but there are aspects of advanced control algorithm design where these newer methods are appropriate. The current interest in adaptive control is an example of such methods. Space permits only an introduction to the appropriate techniques and a reference to suitable literature (Refs. 10 to 14).

A typical equation from the class being considered is the normal equation for the least squares fit, used in filtering and adaptive control:

$$A^T A \mathbf{x} = A^T \mathbf{y}$$

In this equation $\mathbf{y}$ is a vector of data values, $\mathbf{x}$ a parameter vector, and A a matrix of data vectors which occur in a set of equations of the following form:

$$a_{i1}\mathbf{x}_1 + a_{i2}\mathbf{x}_2 + \cdots = \mathbf{y}_i$$

The problem is to find the best fit for the parameters $\mathbf{x}$, given the known A and $\mathbf{y}$. In principle, the first equation has the solution

$$\mathbf{x} = (A^T A)^{-1} A^T \mathbf{y}$$

This equation can be solved directly or recursively (Ricatti equation, Kalman filter); however, recent work has indicated that the direct solution and the usual recursive techniques are unnecessarily sensitive to numerical rounding and truncation errors. The related problems of solving for eigenvalues have been equally difficult, and only recently have good solutions been understood, particularly in the control community (Refs. 10 and 13).

In essence the problem is similar to the problems discussed earlier involving individual numbers: If one computes differences with large floating-point numbers, one creates large errors. And every time one multiplies, one doubles the number of digits of the data range, unless one multiplies by a number close to 1.

In matrix calculations the concept of a matrix being "close to 1" is examined several ways. The determinant is one-way. Is $|A|$ close to 1? But because matrices involve several "directions," the determinant is misleading. Another measure of a matrix is its norm:

$$\|A\| = \max_{x=0} \frac{\sqrt{\mathbf{x}^T A^T A \mathbf{x}}}{\sqrt{\mathbf{x}^T \mathbf{x}}}$$

Underlying the concept of the norm are orthogonal matrices and singular values (these two concepts will now be defined) (Ref. 10). An *orthogonal matrix* is one whose inverse equals its transpose: $Q^TQ = I$. (The letter Q will designate an orthogonal matrix.) In essence an orthogonal matrix is "equal to 1" in nearly every possible way except that it is *not* an identity (or permutation) matrix. Orthogonal matrices have several important properties which make then equal to 1:

1. Their determinant equals 1.
2. They do not affect the length of a vector when they transform it.
3. Therefore (from property 2), their norm equals 1.
4. The products of orthogonal matrices are orthogonal.
5. Every element of Q has magnitude ≤ 1.

The singular values of σ_i of A are the square roots of the eigenvalues of A^TA. More interestingly, every matrix (square or not) obeys the singular-value decomposition theorem (Ref. 9):

$$A \equiv Q_1 \Sigma Q_2$$

where the Q's are orthogonal and Σ is a diagonal matrix of the singular values σ_i. From this and property 4 and the definition of the orthogonal matrix above:

6. The singular values of Q are all 1.
7. For those concerned, from property 5 above, the elements of the relative gain array or interaction measure are all between 0 and 1 (Ref. 15). Since the case involving 0 to 1 represents easy control, the relation between computation and feedback should make this result natural.

Because of the form of the singular-value decomposition, one can think of the orthogonal matrices as twisting any vector operated on by A (without changing its length) into an orientation where each component is multiplied by only one of the singular values. Thus the vector most amplified in length by A is one oriented so that it is multiplied by the largest σ_i. For this reason the norm of A equals the largest σ_i. The vector most diminished in length by A is the one multiplied by the smallest σ_i. Generally, a matrix is hard to compute with (is "much larger than one") if there are significant off-diagonal terms and the ratio of the largest to the smallest σ_i is much larger than 1.

With reference to the normal equation above, multiplying A^TA squares this ratio, making computation more difficult. If the normal equations could be solved without squaring A, or if multiplications could be limited to multiplications of orthogonal matrices, better results would be achieved. And that is how effective algorithms are designed.

As a simple example of the use of orthogonal transformations, consider the solution of the equation $A\mathbf{x} = b$, where A is the matrix in Fig. 10. The conventional gaussian solving procedure (without pivoting, which is an orthogonal operation) amounts to factoring A into two triangular matrices L and U (as in Fig. 10), which are then directly invertible by "back-solving." Note that in this case the numbers are large (to be avoided). There is another decomposition based on orthogonal matrices which may also be used to solve $A\mathbf{x} = b$ by decomposing A into two matrices, $A = QR$. Here R is triangular and Q is orthogonal and, in this case, a symmetrical form called a *Householder transformation*. In each case, L, U, Q, R are inverted easily to perform the solution. Only the first case

GAUSSIAN: $A = \begin{bmatrix} 0.1 & 1 \\ 1 & 1 \end{bmatrix} = LU = \begin{bmatrix} 1 & 0 \\ 10 & 1 \end{bmatrix} \times \begin{bmatrix} 0.1 & 1 \\ 0 & -9 \end{bmatrix}$

Q.R.: $A = \begin{bmatrix} 0.1 & 1 \\ 1 & 1 \end{bmatrix} = QR = \begin{bmatrix} -0.0995 & -0.9950 \\ 0.9950 & 0.0995 \end{bmatrix} \times \begin{bmatrix} -1.0050 & -1.0945 \\ 0 & -0.8955 \end{bmatrix}$

Fig. 10 Equation solution: gaussian versus QR.

presents a problem. The whole thrust of the new set of algorithms can be summed up as follows: Avoid matrix multiplication if possible (it expands the range) and try to restrict necessary multiplications to those involving orthogonal matrices.

SUMMARY

Digital control algorithms can be designed for experimental or single applications using the easiest tools: FORTRAN and floating-point implementation. In this case, there is a reasonable hope that normal commissioning debugging will weed out all problems. But, if a sense of workmanship motivates one, or many people intend to use the algorithm, then attention to refinement and foolproofing are important.

Among the issues addressed in this article are:

1. The causes of numerical problems in fixed-point and floating-point algorithms
2. The documentation, control, and testing of detailed scaling, precision, and saturation within the algorithm
3. Design for natural tuning and qualitative behavior predictable from an analog perspective
4. Nasty surprises due to quantization
5. Windup and other accommodations of the limits to control
6. Bumpless transfer and operational considerations

With the current state of programming languages, the proper design of a control algorithm is no more covered by the usual FORTRAN textbook equation than is an automobile design covered by the styling artist's sales renditions.

REFERENCES

1. Bristol, E. H.: "On the Design and Programming of Control Algorithms for DDC Systems," *Control Eng.,* January 1977.
2. Claggett, E. H.: "Keep a Notebook of Digital Control Algorithms," *Control Eng.,* October 1980, pp. 81–84.
3. Fehevari, W.: "Asymmetric Algorithm Tightens Compressor Surge Control," *Control Eng.,* October 1977, pp. 63–66.
4. Tu, F. C. Y., and J. Y. H. Tsing: "Synthesizing a Digital Algorithm for Optimized Control," *Instrum. Technol.,* May 1979, pp. 52–56.
5. Franklin G. F., and J. D. Powell: *Digital Control of Dynamic Systems,* Addison-Wesley, Reading, Mass., 1980.
6. Knuth, D. E.: *The Art of Computer Programming,* vol. 2, Addison-Wesley, Reading, Mass., 1981.
7. Collected Algorithms from ACM, 1960–1976.
8. Stevenson, D.: "A Proposed Standard for Floating Point Arithmetic, Draft 8 of IEEE Task P.754," *IEEE Comput.,* vol. 14, no. 3, March 1981, pp. 51–62.
9. *Chemical Engineers' Handbook,* 5th ed., McGraw-Hill, New York, 1973, pp. 2–81 to 2–85.
10. Strang, G.: *Linear Algebra and Its Applications,* Academic Press, New York, 1976.
11. Bierman, G. J.: *Factorization Methods for Discrete Sequential Estimation,* Academic Press, New York, 1977.
12. Lawson, C. L., and R. L. Hansen: *Solving Least Squares Problems,* Prentice-Hall, Englewood Cliffs, N.J., 1974.
13. Laub, A. J., and V. C. Klema: "The Singular Value Decomposition: Its Computation and Some Applications," *IEEE Trans. Autom. Control,* vol. AC-25, no. 2, April 1980, pp. 164–176.

14. MacFarlane, A. G. J., and Y. S. Hung: "A Quasi-Classical Approach to Multivariable Feedback Systems Design," 2nd IFAC Symposium, Computer Aided Design of Multivariable Technological Systems, Purdue University, West Lafayette, Ind., September, 1982, pp. 39–48.

15. Bristol, E. H.: "On a New Measure of Interaction for Multivariable Control," *IEEE-PTGAC,* vol. AC-11, no. 1, January 1966, pp. 133–134.

Automatic Controllers *

C. E. Bernstein. *Process Control Division, Honeywell Inc., Fort Washington, Pennsylvania. (Digital Control Programmers)*

R. G. Crane. *Division Manager, Research, Inc., Minneapolis, Minnesota. (Programmers)*

James H. Koegel. *Robertshaw Controls Company, Industrial Instrumentation Division, Anaheim, California. (Electronic Controllers)*

Richard J. Kotalik. *Robertshaw Controls Company, Industrial Instrumentation Division, Anaheim, California. (Electronic Controllers)*

C. L. Mamzic. *Moore Products Company, Spring House, Pennsylvania. (Fluidics)*

Ardmore M. Miller. *Love Controls Corporation, Wheeling, Illinois. (Electric Controllers)*

John P. King. *The Foxboro Company, Springfield, New Jersey. (Distributed Instrument Systems)*

Chris Perkins. *Automatic Timing & Controls Company, King of Prussia, Pennsylvania. (Timers)*

R. G. Reip. *Senior Development Engineer, GPE Controls, Division of Brunswick Corporation, Morton Grove, Illinois. (Hydraulic Controllers)*

Bruce R. Rusch. *President and General Manager, Gould Inc., Programmable Control Division, Andover, Massachusetts. (Programmable Controllers)*

George Willis. *The Bristol Saybrook Company, Old Saybrook, Connecticut. (Timers)*

*Persons who authored complete articles or subsections of articles, or who otherwise cooperated in an outstanding manner in furnishing information and helpful counsel to the editorial staff.

Electronic Controllers

by
James H. Koegel* and Richard J. Kotalik*

The first process controller was a person, adjusting a valve to maintain a constant level in a tank or a constant flow out of a pipe. Somewhere in history, it was learned that this control could be accomplished more easily and accurately by using air pressure in conjunction with actuators, orifices, and bellows, among other devices. By having pneumatic time delays built in and controlling the mechanical advantage (gain) of the system, the pneumatic analog process controller evolved and was used exclusively to control critical industrial process situations for many years. Pneumatic controllers continue to be preferred for some applications. See the article Pneumatic Transmission and Control in Sec. 16.

In the late 1950s, the state of the art was such that time delays and gain could more easily and more accurately be configured using electronic technology. The first electronic controllers utilized vacuum tubes and/or magnetic amplifiers as forward amplifiers. These were three-mode proportional-integral-derivative (PID) controllers where voltage signals were used to transmit process variable information to the controller input, and the controller output was a voltage signal for actuating a final control element. Eventually, current signals evolved as the signal standard (*originally,* 1 to 5 mA; *now,* 4 to 20 mA) because of its inherent immunity to noise and ability to drive a multiplicity of series loads without affecting signal accuracy.

In effect, a process controller is a special purpose analog computer. When properly tuned, it continuously solves a differential equation required for stable control of a process.

Controller Requirements

Set Point

The controller set point is an independent variable of the controller and its reference. An internal set-point adjustment is always provided and must be derived from a stable voltage source. The adjustment is usually controlled from a potentiometer adjustment or a pushbutton ramp adjustment on the front of the instrument. A precision dial or other electrical indicator is employed to display the set point on the front of the instrument.

Provisions are included to switch the set-point input of a controller to an external connection for remote set-point applications or cascade set-point control.

Reverse-Direct Action

Controllers have means to switch the relationship between the process variable and output from direct action to reverse action. This is normally done by reversing the process variable input and the set-point input. When a controller is set in the direct action mode, an increase in the process variable results in an increase in the output signal. The reverse-direct switch is always set to the position which will provide a stable loop and is dependent on the action of the final control element.

*Robertshaw Controls Company, Industrial Instrumentation Division, Anaheim, Calif.

Manual-Automatic Operation

What has been described thus far relates to automatic operation of a controller. Under process start-up conditions, or any time the process loop is opened, it is necessary to operate under manual control. There is usually a switch on the front of the instrument which accomplishes this.

Under manual control, the output of a controller is directly coupled from an adjustable voltage source to the final control element. The manual adjustment is made from a potentiometer or from pushbuttons at the front of the instrument. This enables the operator to manually adjust the final control element to any desired value.

When switching from manual to automatic control, a means must be provided to do so without upsetting the process. It is, therefore, necessary to either balance the set-point value with the process variable value or balance the output amplifier prior to transferring from manual to automatic control. Most modern electronic controllers automatically balance the amplifier circuit. Unbalanced conditions at the time of transfer cause what is called *output bump* and can result in a serious process upset. It is, therefore, important to accomplish a "bumpless" transfer when switching from manual to automatic control, and vice versa.

Indicators

The variables of a process must be monitored by the control room operator to ensure continued satisfactory control of the process. There are normally three indicators: set point, process variable, and output; or set point, deviation, and output.

Set Point

The indicator displays either the internal or external set-point value. The scale is either graduated in 0 to 100% or in units of the process variable, such as pounds per square inch, degrees Celsius, degrees Fahrenheit, gallons per minute, and so on.

Process Variable

The indicator displays the process variable value. The scale is graduated either in 0 to 100% or in units of the process variable. The process variable indicator normally displays the same units as the set-point indicator. This indicator is usually positioned so that the process variable value can easily be compared with the set-point value.

Deviation

This type of indicator is often used in lieu of a process variable indicator. It receives its input from the controller error signal directly and displays the deviation between the process variable and set-point values. The scale is graduated in percent deviation, usually $0 \pm 10\%$.

Output

This indicator is used to monitor the position of the final control element and is graduated in 0 to 100%. Since control valves and other final control elements can be either direct or reverse acting, it is common practice to include an output reference indicator associated with the output indicator. This is usually some sort of movable marker which tells the control room operator if the final control element is open at 0 or 100%.

Proportional-Integral-Derivative (PID)

The primary purpose of a controller is stable and efficient control of a process loop. To accomplish this, controllers have gain and frequency-response characteristics which are adjustable to best match the dynamic characteristics of the process. The classical differential equation which a typical controller solves can be expressed as

$$E_0 = K_G e + K_D \frac{de}{dt} + K_I \int e \, dt$$

where K_G = gain constant
    K_D = derivative factor
    K_I = integration factor
    e = input error voltage

To implement this solution, controllers are designed with the transfer characteristics as shown in the following Laplace equation:

$$G_S = K \frac{S + \tau_I}{S} \frac{S + \tau_D}{S + 10\tau_D}$$

The gain and frequency characteristics of this equation can be plotted on a Bode chart as shown in Fig. 1.

The low-frequency characteristic of the controller is a negative sloping response of -20 dB per decade until the first break frequency (F_I) is reached. F_I corresponds to $1/2\pi\tau_I$, where τ_I is as shown in the equation for G_S. τ_I is a RC time constant commonly called *reset*. It defines the *lag time constant* of the controller and provides the integral (I) action. Reset can be expressed in minutes or its reciprocal, repeats per minute. The end result is the same, but the units used are according to the manufacturer's preference. A typical controller has a reset range of 0.05 to 50 min or 20 to 0.02 repeats per minute.

The bottom horizontal portion of the curve represents the *proportional gain*. When the *proportional band* is set to 100%, the gain is 1 or 0 dB. Proportional band can be expressed as a percentage (%) or a dimensionless number called *gain*. When expressed as a percentage, the number represents the percent input change required to provide a 100% output change. When expressed as gain, the number represents the ratio of the resulting output change to the corresponding input change. A typical controller has a proportional-band range of 2 to 3000%, or a gain of 50 to 0.033. In effect, gain is the reciprocal of the percentage proportional band multiplied by 100.

F_D represents the frequency at which the characteristic breaks, upward at 20 dB per decade. F_D corresponds to $1/2\pi\tau_D$, where τ_D is shown in the equation for G_S and represents an RC time constant commonly called the *rate*. τ_D defines the lead corner frequency and provides the *derivative (D) action* of the controller. The curve slopes upward at this rate for 20 dB and then flattens at the final breakpoint. The 20-dB relationship at the high-frequency end of the characteristics is commonly called the *rate gain* of the controller.

Rate action is expressed in minutes or seconds, and the range of adjustment of a typical controller is 0.01 to 12 min. Rate is normally used when the process loop is very fast and a quick reaction to the process is required.

The most common controllers used are two-mode (proportional-reset action). This means there is a lag in the frequency response (time delay), plus a gain factor, such that the output varies in

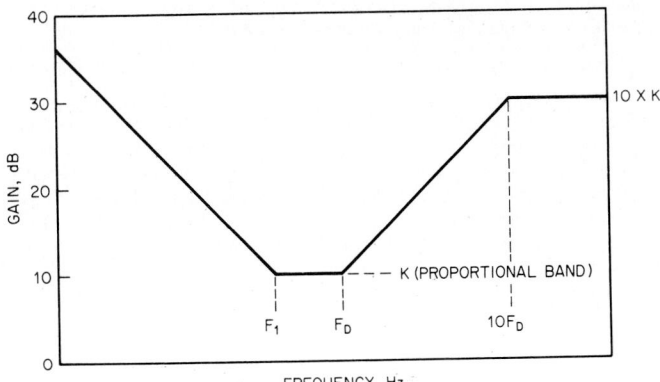

Fig. 1   Bode frequency-response curve for a three-mode controller.

proportion to the process variable signal. Rate action is seldom used without reset action and is a lead or lag in the frequency-response characteristics of the controller.

Proportional-Only

For some special applications, it is desired to operate a controller with a flat frequency response and a straight gain relationship between the output and the error signal at the input. Because of the direct-coupled relationship between input and output, an intolerable offset can result, especially at higher gain settings. Therefore, proportional-only controllers are designed with an output bias adjustment. This adjustment is called *manual reset*. Proportional-only controllers are sometimes used in place of ratio amplifiers in cascade loops where deviation control is required but an accumulation of reset cannot be tolerated.

Controller Applications

Basic Control Loop

The most common configuration of a process control loop is shown in Fig. 2. The sensor interfaces the controller with the process and may be sensing pressure, temperature, level, or any dependent variable.

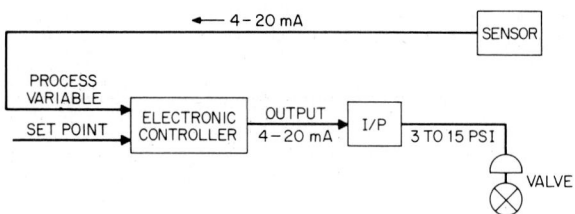

Fig. 2 Typical process control loop.

The set point is the independent variable and may be internal to the controller or derived from an external source. In any case, the value of the set point determines the value about which the process variable is to operate.

A current-to-pneumatic (I-to-P) converter is commonly used to interface the controller with a final control element. There are many applications, however, where other types of converters are used. The final control element is often an air pressure-actuated valve but can be any means used to regulate the process according to the controller output current.

Tuning

The controller's tuning parameters (proportional, integral, and derivative) must be adjusted to match and compensate the characteristics of the process. It is usually desirable to tune the controller somewhere between a slightly *underdamped* to a slightly *overdamped* process control loop.

The tuning parameters often can be approximated from the known characteristics of the system. This method is sometimes used prior to start-up. During start-up there are several methods of optimizing the tuning parameters other than trial and error. Two of these are the ultimate-sensitivity and reaction curve methods. These methods are described in considerable detail in the Ziegler-Nichols reference listed at the end of this article.

Ultimate-Sensitivity Method

This method requires that the controller gain be increased to the point of instability of the process loop. Both the derivative (rate) and integral (reset) must be set to "off." The controller is then set to minimum gain (widest percentage of proportional band) and is switched from 50% manual control to automatic control. While monitoring the process variable signal, the operator slowly decreases the controller percentage proportional band until the process oscillates at a constant amplitude. See Fig. 3c.

The conditions at which the amplitude remains constant are defined as the *ultimate sensitivity.* The proportional band required to obtain this condition is the *ultimate proportional band* (PB_U), and the resulting period of oscillation is the *ultimate period* (P_U). When these two values are known, the approximate tuning parameters can be determined from the following calculations:

Proportional-only control (single-mode controller):

$$\text{Proportional band} = 2PB_U \text{ percent}$$

Proportional-plus-reset control (two-mode controller):

$$\text{Proportional band} = 2.2PB_U \text{ percent}$$

$$\text{Reset} = 1.2/P_U \text{ repeats per minute}$$

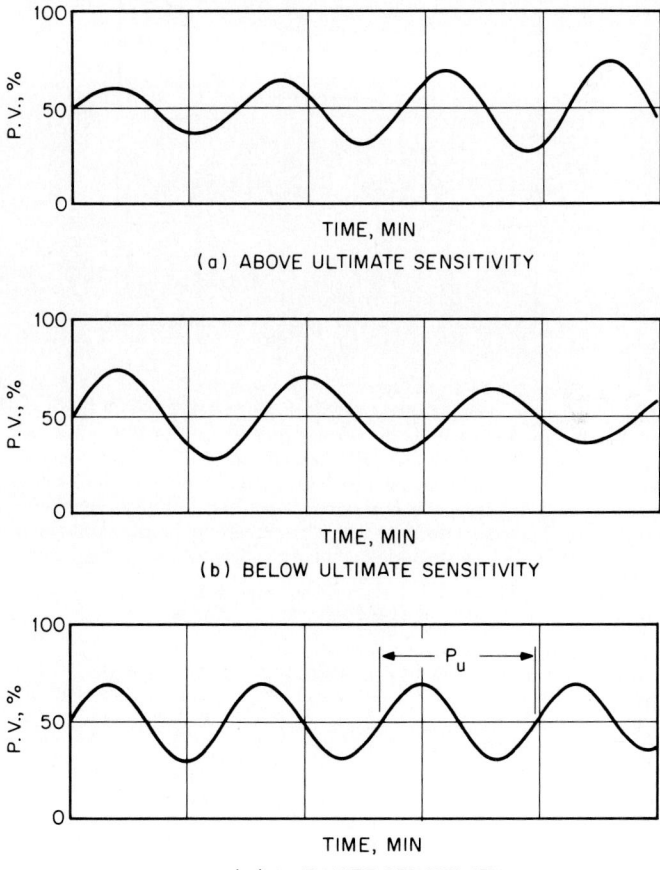

Fig. 3 Ultimate-sensitivity tuning curves. P.V., Process variable signal.

Proportional-plus-reset-plus-rate control (three-mode controller):

$$\text{Proportional band} = 1.6PB_U \text{ percent}$$

$$\text{Reset} = 2/PB_U \text{ repeats per minute}$$

$$\text{Rate} = P_U/8 \text{ minutes}$$

Reaction Curve Method

This method is more widely used and is the safest because it does not require large upsets in the process. It is a measurement of the process reaction and does not involve the controller parameters as such.

   With the controller in manual operation, the process variable input is monitored with an instrument recorder, as is done with the ultimate-sensitivity method. The process must be allowed to stabilize to a steady horizontal line. At time zero, a small step change is made in the controller output signal. This change must be carefully executed and of a known increment. The reaction of the process variable signal is then observed as shown in Fig. 4. After the process stabilizes, an asymptotic line

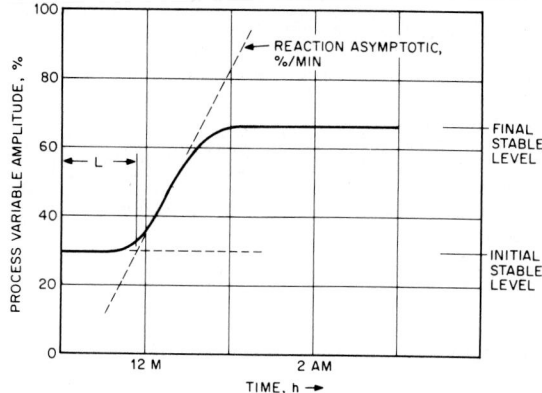

Fig. 4   Process reaction curve.

is drawn tangential to the steep slope of the reaction curve. The slope (R) of this asymptote is then read from the curve in percent per minute. The reaction lag (L) is also read from the curve as the elapsed time from time zero to the intersection of asymptote R and the initial stability level of the process variable value. With knowledge of these two values and the initial percentage step change in the controller output (V), the approximate controller settings can be calculated as follows:

   Proportional-only control (single-mode controller):

$$\text{Proportional band} = \frac{100 \times R \times L}{V} \text{ percent}$$

Proportional-plus-reset control (two-mode controller):

$$\text{Proportional band} = \frac{90 \times R \times L}{V} \text{ percent}$$

$$\text{Reset} = \frac{0.3}{L} \text{ repeats per minute}$$

Proportional-plus-reset-plus-rate (three-mode controller):

$$\text{Proportional band} = \frac{120 \times R \times L}{V} \text{ percent}$$

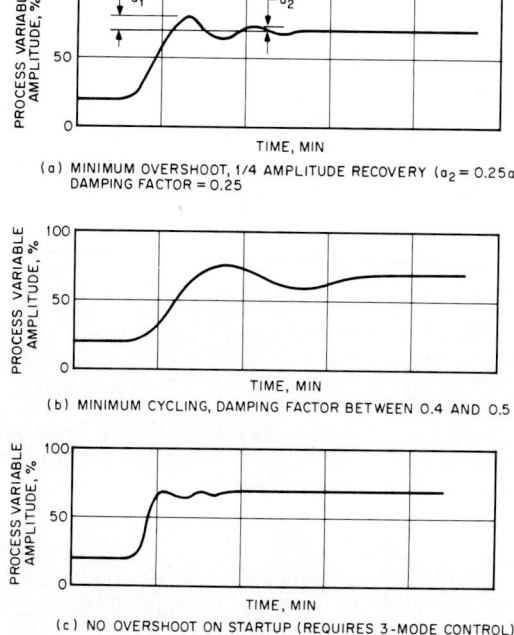

(a) MINIMUM OVERSHOOT, 1/4 AMPLITUDE RECOVERY ($a_2 = 0.25a_1$), DAMPING FACTOR = 0.25

(b) MINIMUM CYCLING, DAMPING FACTOR BETWEEN 0.4 AND 0.5

(c) NO OVERSHOOT ON STARTUP (REQUIRES 3-MODE CONTROL)

Fig. 5   Reaction recovery curves.

$$\text{Reset} = \frac{0.5}{L} \text{ repeats per minute}$$

$$\text{Rate} = 0.5L \text{ min}$$

After the process is started with the approximate setting for either of these methods, some fine-tuning is usually required to optimize the process. This must be done by trial and error. Figure 5 illustrates typical process reaction curves for a step set-point change with the controller under automatic control.

Deviation Controllers

In general, a *deviation controller* is one in which any deviation between set-point and process variable values is amplified, independent of the absolute values. A display of this difference is made on a deviation meter with a range of $0 \pm 10\%$.

Temperature Controllers

When the sensor shown in Fig. 2 provides a measurement of temperature, the process loop is called a *temperature control loop*. Often a controller is designed to accept a temperature signal as direct input, such as from a thermocouple or a resistance temperature detector (RTD)—in which case the unit is simply called a *temperature controller*.

Ratio Controller

There are applications where a common source is used for the external set point of several dependent control loops. In such cases, one or more of the set-point signals must be modified, and sometimes biased, to a different value in relation to individual loop parameters. A *ratio amplifier* can be used

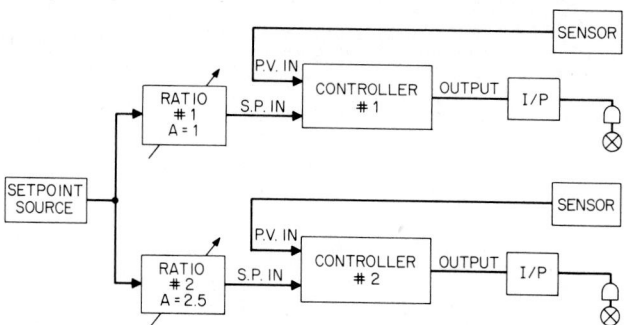

Fig. 6 Ratio control loop. P.V., Process variable signal; S.P., set point; I/P, integral-plus-proportional.

to interface the primary set-point signal with the controller's set point-signal. Figure 6 illustrates a ratio control loop with two controllers. If a controller is designed with the ability to ratio and bias the external set-point input, it is called a *ratio controller.*

Cascade Control

This is used in complex control applications where the controller on the primary variable is a master controller whose output provides the set-point input for one or more slave controllers. A typical example is a steam boiler control system where the process variable input to the master controller provides the set point to a flow controller providing air for the combustion system. Often, the fuel-

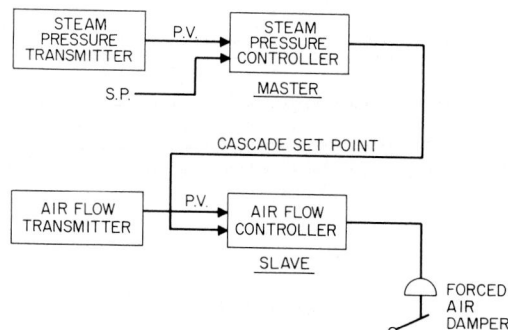

Fig. 7 Cascade control loop. P.V., Process variable; S.P., set point.

flow controller set point is a ratio of the airflow signal. An example of a steam or air cascade control loop is shown in Fig. 7.

Underride-Override Control

In some instances, the final control element may be actuated by one of two (or more) independent controllers, depending on the relation between the output of these controllers. An example is control of a pipeline pumping station. It is possible to control by suction pressure or discharge pressure of the pump. It is desired to maintain control on the lowest of the two pressures to prevent cavitation of the pump. In this case, two controllers are used as shown in Fig. 8.

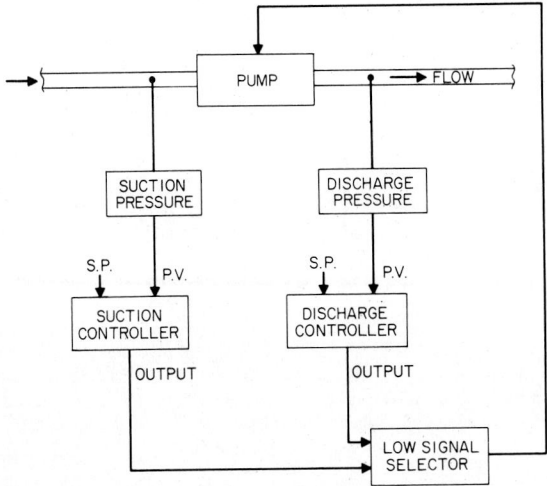

Fig. 8 Suction-discharge and override-underride control loops. P.V., Process variable; S.P., set point.

ANALOG CONTROLLER THEORY AND DESIGN

Amplifier Design

There are a number of ways of implementing controller functions into an electronic design. High-impedance circuits are required, with very low leakage characteristics, because of the long time constants involved. To achieve this in the past, several techniques have been employed, including vacuum tubes, electrometer tubes, self-oscillating varactor* bridges, and more recently, dc amplifiers employing field effect transistors.

The most common functional block diagram of a modern analog controller is shown in Fig. 9. Input can be either the process variable or the set point. If input 1 is the process variable, then the controller is direct acting. The "direct-reverse" switch merely interchanges these two inputs.

Amplifier A_1 is a deviation amplifier which amplifies the difference between the process variable and *set point*. This amplifier normally has a gain of 1 or 2, depending on the input voltage range. The gain is determined by the ratios R_1/R_2 and R_3/R_4. R_1 must equal R_3, and R_2 must equal R_4.

If the controller is lined out (the process variable equals the set point), the output of the deviation, point A, is equal to the reference voltage.

*Voltage variable capacitance derived from a back-biased semiconductor diode.

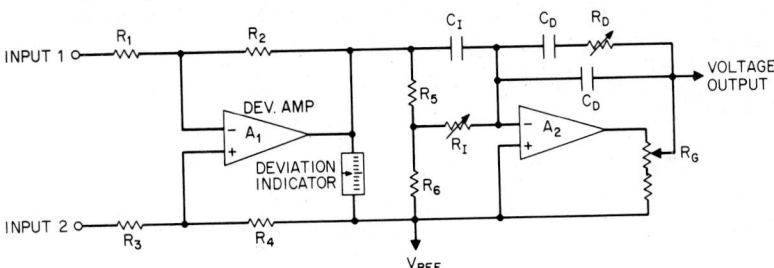

Fig. 9 Functional block diagram of a typical analog controller.

Amplifier A_2 is a high-impedance operational amplifier. The input impedance of this stage must be greater than 1×10^7 Ω. This amplifier configuration has a transfer characteristic as expressed in Eq. 2. R_I and C_I provide the reset time constant which determines F_I on the Bode plot of Fig. 1. R_5 and R_6 divide the deviation voltage by a factor of 10 ($R_5 = 9R_6$). This allows R_I to be $\frac{1}{10}$ of the value required to obtain the same time constant if this divide were not used. $\Upsilon_I = 10R_IC_I$.

The feedback of this amplifier is through the combination of R_D, C_D, and C_D' which provide the rate time constant for the controller. These components determine the lead or lag at F_D and $10F_D$ on the Bode plot of Fig. 1. The initial lead at F_I is determined by the time constant of R_D and C_D; $\Upsilon_D = R_DC_D$.

C_D is selected for a value of $\frac{1}{10}C_D'$, the resulting combination results in a lag at $10F_D$. $0.1\Upsilon_D = R_DC_D[C_D/(C_D + C_D')]$.

The feedback voltage is adjusted by the voltage divide R_G, which is the proportional-band adjustment. With the minimum feedback voltage, the amplifier gain is highest, usually a gain of 50 or a proportional band of 2%. Some manufacturers utilize a capacitance charge divider for the proportional-band feedback adjustment, rather than resistive voltage divides. The advantage of this is that there is no dc shift when an adjustment is made, eliminating any disturbance to the process. The disadvantage is increased output drift due to capacitive leakage.

The voltage output of this operational amplifier is normally fed to a current amplifier which provides the standard output process signals, usually 4 to 20 mA.

Common Problems

Stability

There are several areas where stability can be a problem in analog circuit design. The primary area of concern is drift due to component leakage. Since extremely large time constants are required, accompanied by amplifiers with very high input impedance, leakage currents in the nanoampere vicinity can cause significant output drift. For this reason, it is important to carefully select components associated with the operational amplifier.

Since this is a feedback amplifier, care must also be taken in designing the amplifiers and actual feedback network so that the circuit will remain stable under all operating conditions.

Controller-Caused Disturbances (Bump)

When a controller is switched from automatic to manual control, and vice versa, the output signal should remain constant. A sudden disturbance to the process resulting from improper switching from automatic to manual control is called a *bump*.

The manual signal can be derived in a number of ways, including a separate current amplifier to which the output can be switched or utilization of the operational amplifier itself. If the manual signal is derived from a separate current amplifier, special provisions for balancing the manual output with the automatic controller output must be provided. If the operational amplifier is used, it is possible to have the automatic feedback network (C_D and C_I) track the manual output so that bumpless transfer can be achieved. By using capacitive feedback for manual control, it is possible to achieve bumpless transfer in both directions, but then a practical manual adjustment scheme is difficult to achieve. This difficulty is commonly overcome by using slewing buttons rather than a potentiometer for the manual adjustment.

Reset Windup

Another common problem associated with analog controllers is *reset windup*. This occurs when the output of the controller is saturated or cut off for any extended period of time. It may be caused by part of the process being shut down or by excessive dead time in the process. While this occurs, the reset capacitor in the controller continues to change in an attempt to correct the process variable error. When the process variable does respond, the charge on the reset capacitor cannot be dissipated immediately, resulting in an extended out-of-control situation. The problem can be minimized by limiting the maximum deviation and the summation point of the operational amplifier. It is difficult, however, to eliminate the problem completely.

DIGITAL CONTROLLER DESIGN AND THEORY

The concept and attraction of a process controller which is digital in nature have existed since computers were considered for process control applications. Some of the earliest digital control systems began to appear in the mid-1950s. However, these systems were composed of large mainframe computers which performed a control system function rather than discrete process control. As semiconductor devices displaced vacuum tubes, digital controllers became increasingly realizable. It was not until the development of the microprocessor, however, that the digital controller was feasible.

Advantages and Disadvantages of a Digital Controller

Two major advantages have been the driving force in the development of digital controllers:

1. The ability to communicate with other instruments and/or a computer
2. The performance of a digital versus an analog controller

The development of a digital controller had been hindered by the lack of a technology which would permit the complex functions to be implemented at a reasonable cost.

A controller which can communicate with a computer is a prerequisite in implementing a distributed control system. Unless a controller can conveniently inform a supervisory computer system of its process parameters and receive operating commands from the supervisory computer system, a distributed system cannot be implemented. Communication via analog signals is not impossible, but more information may be communicated faster, with fewer errors, using digital techniques. A controller which is digital-based lends itself to communication with a digital computer.

The performance of a digital controller is superior to that of an analog controller in several ways. The accuracy and stability of digital hardware is fundamentally unlimited. Overall system accuracies of less than 0.1% can be easily achieved. Errors such as nonlinearity, hysteresis, and thermal or long-term drift are not present in a digital system. However, the overall system performance is usually limited by the performance of the analog-to-digital (A/D) and digital-to-analog (D/A) converters. These elements, because of their partial analog nature, can exhibit all the errors associated with analog instruments.

The performance of a digital controller is also superior to that of an analog controller because of the complexity of the control function it can provide. Digital controllers can assume, with a minimum of additional hardware, many of the functions associated with additional instruments in an analog controller. Such functions include alarms, square root extraction, totalizing, complex control algorithms, and special displays. The availability of a variety of control algorithms permits special processes which are nonlinear, for example, to be controlled. The result is that better control may be exercised in closer proximity to the process than previously possible.

Digital controllers are more complex, hence more expensive, than analog controllers. Much of the complexity results from the need to convert analog input signals to digital signals before they may be processed and then to return the output digital signals to analog signals. The task of performing these conversions is not especially difficult, but it requires hardware not needed in a completely analog system.

Another consideration influencing complexity is the need for digital controls and displays. Digital displays and controls are a mixed blessing. The displays or controls themselves offer increased accuracy and reliability because they are usually solid-state components or contain a minimum of moving parts. However, interfacing these controls with a digital controller often requires additional hardware, compared to interfacing with an analog control or display. The net result is that a digital controller usually is more costly to produce than an analog controller for a minimum control requirement. However, additional functions and features may be offered at very little additional cost. Controllers with options such as alarms, totalizers, square root extractors, etc., are usually priced lower than comparable analog controllers.

Brief Chronology of Digital Controller Development

The realization of a digitally implemented controller has been marginally feasible for many years. As practical solid-state components became readily available, some of the functions of analog controllers, such as the alarm logic circuitry, were provided by digital techniques. These "hybrid" controllers, composed of analog and digital portions, led the transition to a fully digital controller, but they did not become an established technology because improvements in digital components were occurring rapidly. At least one manufacturer attempted to design a digital controller implemented with custom digital integrated circuits (ICs). As this approach became viable, however, the microprocessor emerged as a component.

Microprocessors were expensive devices when they were first introduced. However, they offered the promise of implementing relatively complex requirements at a reasonable cost within a comparatively short time. The first microprocessor-based controllers were developed to control more than one loop (usually four loops). This allowed the cost of the hardware to be spread over a number of loops. When the cost of microprocessors decreased, single-loop digital controllers were produced. These controllers offered the long-sought advantages of digital control on a single-loop basis.

Multiloop controllers evolved into a split architecture configuration. In this scheme, the controller is provided as a back-of-panel instrument with a separate display module or cathode ray tube (CRT).

Microprocessor-Based Controller

Microprocessor-based controllers greatly increased the performance capability of controllers. The ability to process data and perform mathematical computations added a new dimension to controllers. Of course, this capability is not unlimited, but it continues to grow with new microprocessor developments. A typical digital controller is shown in Fig. 10.

The features currently provided in controllers are more a matter of what a broad base of users

Fig. 10 Digital controller. *(Robertshaw Controls, Industrial Instrument Division.)*

desire, rather than what the equipment can do. Many of the standard features in early controllers included alarms, for example. All conceivable types are provided: process variable, set point, deviation, output, and rate of change in the process variable. These may be configured as high, low, high-high, or low-low alarms, with or without a clamp at the alarm level. Some of the other functions which are options for analog controllers are square root extractors, totalizers, and filters on the process variable, and they are usually included in the standard digital controller.

The functional block diagram of a typical controller in Fig. 11 shows the relation of functions implemented by hardware and software. It provides an understanding of the instrument in more conventional instrument terms.

The real significance of digital controllers is demonstrated with features such as linearization, choice of algorithms, and batch control. Linearization of the process variable input for thermocouples, RTDs, irregularly shaped vessels, transmitter characteristics, etc., may be easily accomplished with digital controllers. The specific linearization function may be implemented in the field by the user or may require a factory modification, depending on the controller and the complexity of the function. Some controllers also offer the ability to compensate the output for the characteristics of a control actuator or valve.

Several controllers offer a choice of control algorithms for the user to select. The standard control algorithm is usually PID. However, this is often implemented with noninteracting gain, reset, and rate coefficients. This means that the reset coefficient indicates the ramp rate of the output with a given error independent of the gain coefficient. The ranges of adjustment of the coefficients have been increased to be suitable for a variety of processes. The hardware limitations of adjustment resolution are no longer a factor.

Nonlinear control algorithms are provided to optimize various control functions. For example, linear PID with a dead band may be used in a noisy process to control a valve without stroking it unnecessarily and without the time delay of a filter on the process variable. Nonlinear algorithms which control using the square or the square root of the error instead of the error may be used for nonlinear reactions such as the control of pH. The effective gain of the loop when the error is large is different than the effective gain when the error is small. This permits the loop to be stable with no error, yet correct large errors quickly. Typical gain functions are shown in Fig. 12. User-definable functions of the control parameters relative to error, process variable, and output are also available. Specialized controllers for specific applications offer control algorithms optimized for the control requirements.

Some manufacturers offer controllers which implement control of a batch process. Usually these are multiloop controllers with the capability to accept contact closure inputs. The algorithms of the controllers are determined by the relation of the contacts. For example, a controller might control a process which requires filling a container with reagents, mixing the reagents, cooking the mixture, and then emptying the container. The loop controllers within the multiloop unit may be used to ratio and control the flow of the reagents until the correct total accumulation is determined by the built-in totalizers. The mixer may then be run for a fixed time or until another condition is satisfied, and then the temperature of the container controlled until the desired characteristics are obtained. Finally, a valve and pumps are controlled to remove the processed material.

As can be seen in this example, the distinction between a multiloop controller and what was considered a multitask system is becoming very vague.

Operation of Digital Controllers

Digital controllers implement control equations and functions in a different manner than analog controllers. Digital controllers process data in a batch mode, rather than on a continuous basis as in an analog unit. This is of little significance to the user for routine operations, but a brief explanation of the fundamentals is in order here.

The microprocessor can perform only one task or one computation at a time. This means that first the process data must be input. Then the data must be checked for alarm conditions and linearized, if necessary, prior to calculation of the control algorithm.

The control algorithm is in the form of a difference equation:

$$\text{Output} = Q + K_G(E_2 - E_1) + K_I E_2 T + K_D(E_2 - 2E_1 + E_0)$$

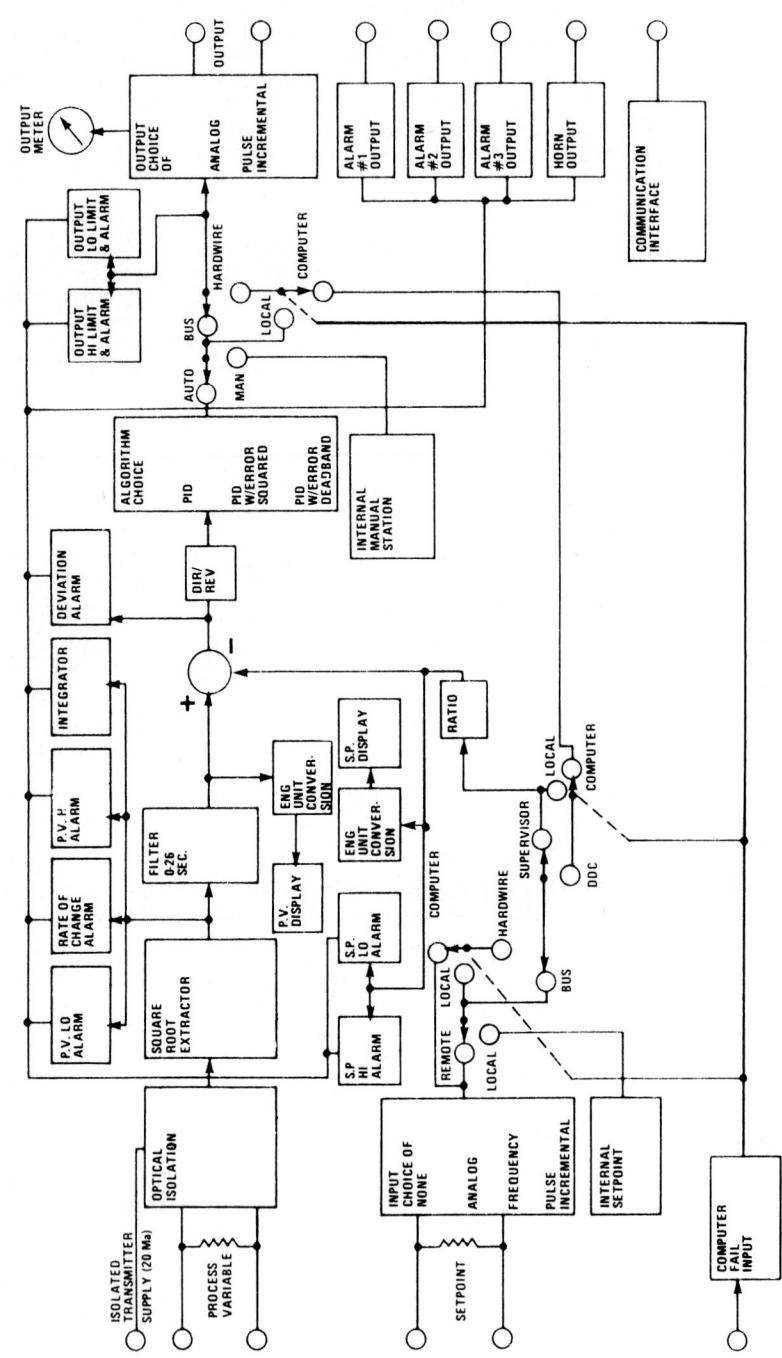

Fig. 11 Functional block diagram of a typical controller. P.V., Process variable; S.P., set point.

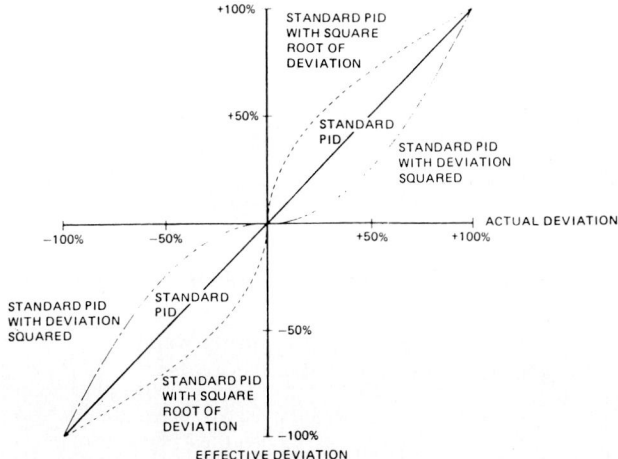

Fig. 12   Illustration of the effective deviation for three control algorithms. PID, Proportional-plus-integral-plus-derivative.

where output = revised value of output of controller
Q = previous value of output
K_G = gain constant
K_I = reset constant
K_D = rate constant
E_2 = current value of error
E_1 = last value of error
E_0 = second last value of error
T = time interval since last calculation or data input

The calculations required to determine the output are straightforward. They involve calculation of the changes in the error and the change in error weighted by the appropriate constants. The analog integral function (reset) is replaced by a "rectangular" approximation.

When the new value of the output has been calculated, the alarm functions and limits relative to the new value of the output are checked, and then the output value is converted to an analog value.

The mircoprocessor then may accomplish certain display functions, such as calculating the engineering value of the process variable and set point. After a prescribed interval of time, usually 0.05 to 0.5 s since the data were last inputted, the microprocessor begins the cycle anew. During any particular part of the cycle, the microprocessor (and the controller) are not cognizant of any abnormal or alarm conditions which may have developed. It is not until the processor performs a task related to an abnormal condition that it is noted. Therefore, conditions which must be monitored on a continuous basis are implemented with discrete hardware, so that a short-term abnormal or special condition can be noted and latched for review by the microprocessor at a later time.

Communications with other instruments are handled on an interrupt basis. A special purpose IC performs initial processing on the communicated message being received. When the data have been converted to a form suitable for the microprocessor, the IC signals the microprocessor to input the data.

Hardware

The block diagram for a typical controller is shown in Fig. 13. The controller consists of A/D and D/A converters, a microprocessor section, a controls and displays section, and a communications port. The microprocessor section, detailed in Fig. 14, includes not only the computational element, but also the program storage element or read-only-memory (ROM) and the operating storage "scratch pad" memory or random-access memory (RAM).

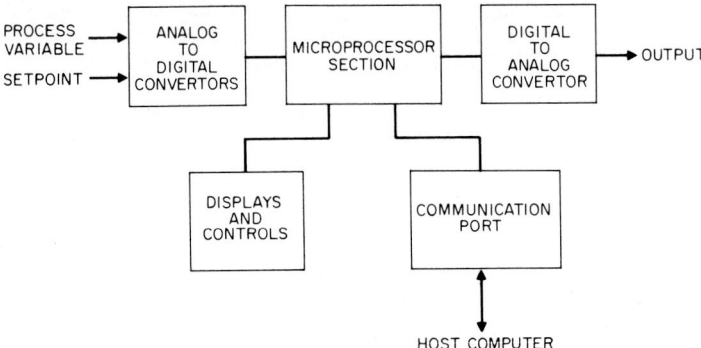

Fig. 13 Typical hardware block diagram of a digital controller.

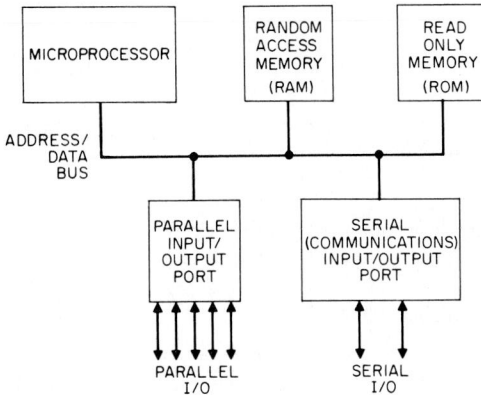

Fig. 14 Block diagram of a microprocessor section.

There are many ways to implement the A/D converters which convert the analog process variable signal to a digital format. The primary requirements for a process control application are accurate, relatively high-resolution conversions, with a large tolerance for noise and a medium to slow conversion rate. For such requirements averaging converters such as dual-slope integrators and voltage-to-frequency converters with counters are especially attractive. The latter offer the additional advantage of easily providing optical isolation of the digitized signal.

The D/A converter for the output signal has similar requirements. It is usually implemented with a straightforward resistor ladder network.

Some controllers offer other semidigitized inputs such as a pulse train. The frequency of the pulse train may indicate the input value. For example, a frequency of 10 to 50 kHz might be used instead of a 1 to 5 V dc signal. The input may also be an incremental pulse train. In this case, each pulse of a series increments or decrements the input register from the current value, depending on the status of an additional direct input. Several controllers with a pulse incremental input may be used with a controller with a pulse incremental output to implement a ratio control where the output of the master controller will not saturate. The input may also be time proportional pulse duration. In this case, a counter is gated for the duration of the input pulse.

Because of the digital format of the controller, the displays and controls are digitally oriented. Ideally, a D'Arsonval meter movement is replaced by a digital display. Unfortunately, digital displays do not convey the innate value of the variable that an analog display does. An operator can scan a panel of analog meters and notice an abnormal parameter by the position of the pointer faster than he or she can read a series of numeric displays and note the same problem.

There has been some development of digital displays which have an inherent analog readout. The most widespread are gas discharge, light emitting diode (LED), and liquid crystal bar graphs. These devices not only display a quantitized image but also have the added capability of flashing to draw the attention of the operator. A bar graph display augmented with a digital readout which provides the variable in engineering units provides the best of both analog and digital worlds.

The potentiometers found on analog controllers have been replaced by switches, pushbuttons, and keyboards. While it is possible to determine the position of a potentiometer wiper, it is easier and more accurate to read the position of a switch in a digital system. Thus, parameters are entered as numbers via keyboards or incremented or decremented by slewing pushbuttons. Both techniques allow a much wider range of parameter values to be entered with more resolution than a potentiometer could provide. A switch matrix and/or keyboard combination is often used to display and modify the controller parameters on a limited number of displays. A typical operator's panel is shown in Fig. 15.

There are many microprocessors and associated components. They operate at various speeds, and each has advantages and disadvantages relative to others. The use of a particular microprocessor in a controller is of minor significance to the user from a performance point of view. The only consideration that is significant to the user is the long-term availability of the components. A popular standard component should be available as a spare part for a longer time than a custom component.

A digital controller usually has many more operating parameters than an analog controller. Parameters such as alarm trips, timing constants, etc., are each set by a dedicated potentiometer on an analog controller. However, these parameters are entered via a switch matrix and retained in the microprocessor memory (RAM) in a digital system. Provision must be made to retain the memory in the absence of supply power so that a parameter need not be reentered after a power interruption. One technique for implementing a nonvolatile RAM is to incorporate a long-life or rechargeable

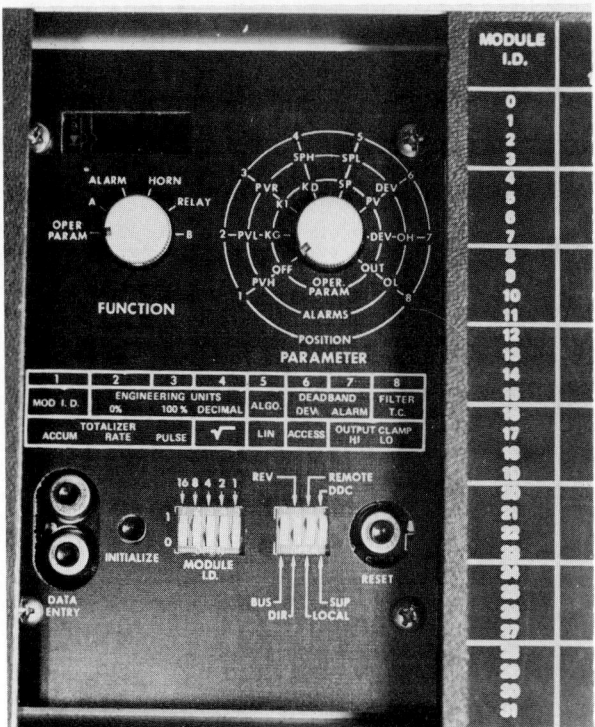

Fig. 15 Operator control panel of a digital controller.

battery in the controller to maintain power to the RAM during loss of primary power. Such a system can maintain parameters for periods of months to years without a primary power supply.

Another technique which is becoming more popular is the use of an electrically alterable or erasable programmable ROM (EAPROM or EEPROM). These components can maintain parameters without power for years. However, they are capable of accepting only a limited number of parameter charges. At the present time, this may limit their useful life to a year or two.

Communications

One of the most significant advantages of a digital controller is its ability to communicate with other equipment. The equipment can be other control equipment or a computer. Communications can consist of process data sent to the computer, control information sent to the controller, or both.

There are many communication formats which may be used, such as RS232, RS422, and IEEE 488, among others. And the data may be coded in many ways—binary, binary-coded decimal, ASCII, etc. With the combinations of standards and variations available, it is virtually assured that equipment from various manufacturers *cannot be interconnected without very special provisions.* Most of the communication formats, however, are serial rather than parallel to minimize wiring; but the data rates vary from a few kilobaud to several megabaud, depending on the manufacturer. Some of the communications are unidirectional on a single conductor; others are bidirectional on two conductors.

Several manufacturers offer controllers compatible with standard hardware interfaces and sufficient information to allow a user to establish communication with a computer. However, this requires the user to perform the programming required by the computer to code and decode the commands and data communicated by the controller.

Further Developments

The full potential of the digital controller has not been reached, and there are many opportunities for further development. Most of the capabilities of the digital controller are a direct result of microprocessors. The future of the controller is expected to parallel that of the microprocessor. As of the early 1980s, more powerful and lower cost microprocessors are being introduced at a rapid rate. Digital controllers will take advantage of these devices in two ways. First, more complex functions will be implemented. The user will have more flexibility in programming and in tailoring a standard controller for a given application. Specialized controllers for multivariable operations, such as boiler control, rather than single-variable flow control, will become commonplace. Some of these controllers will begin to utilize microprocessors.

Second, lower cost single-loop controllers will become available. These will eventually displace most analog controllers except for minimum-requirement situations.

Another feature expected to become more widespread is self-optimization. One form of self-optimization is self-tuning of the control parameters by the controller. All types of self-optimization require a definition of the performance criteria to be optimized and access to the performance data.

Interest in digital sensors and final control elements has existed for many years without significant results. It can only be hoped that the appearance of more digitally oriented controllers will encourage the development of related sensors and control elements.

Third, standards for communication interfaces and systems programming need to be encouraged. Progress in this area has been slow and painful in the past and probably will continue to be. However, more uniform standards will permit distributed systems to be implemented using components from a variety of manufacturers.

RELATED TOPICS

There are several articles related in some way to the general topic of electronic controllers. See Self-Balancing Electrical Instruments and Digital Technology in Sec. 14. See the articles in Sec. 15 on Information Display and Storage; in Sec. 17 on Fundamentals of Automatic Process Control; and in this section (18) on Programmable Controllers and Distributed Instrument Systems.

REFERENCES

Editor's Note: A very long list of journal articles could be given on this very active topic, but the majority of these—because of a rapidly developing technology—do not have long-term usefulness.

Bibbero, R. J.: *Microprocessors in Industrial Control,* Instrument Society of America, Research Triangle Park, N.C., 1982.

Bristol, E. H.: *Proceedings of Advanced Control Conference,* Purdue University, West Lafayette, Ind., April 1974.

Pryor, R.: "Microprocessor Fundamentals," *IRP630,* Instrument Society of America, Research Triangle Park, N.C., 1982.

Pryor, R.: "Microprocessors in Control Systems," *IRP640,* Instrument Society of America, Research Triangle Park, N.C., 1982.

SAMA: "Process Measurement and Control Terminology," *SAMA Stand. PMC20-Z-1970,* Scientific Apparatus Makers Association, New York, 1970.

Skrokov, R.: "The Benefits of Microprocessor Control," *Chem. Eng.,* October 11, 1976.

Tucker, G. K., and D. M. Wills: *A Simplified Technique of Control System Engineering,* Honeywell Inc., Fort Washington, Pa., 1958.

Ziegler, J. G., and N. R. Nichols: *Optimum Setting for Automatic Controllers,* American Society of Mechanical Engineers, New York, December 1941.

Electric Controllers (Traditional)

The distinction between electric and electronic controllers is not sharp. Their differences tend to be characterized as much by chronology as by technology. Modern electronic controllers are described in the prior article in this Handbook. The traditional electric controllers used in industry for several decades remain in place in many installations and, in fact, are currently supplied and are described in this article.

Electric controllers may be classified by control mode:

1. Two-position
2. Two-position time proportioning
3. Three-position
4. Proportional (stepless)
5. Proportional-reset
6. Proportional-reset-rate

Self-Operated Controllers*

In self-operated instruments, the energy output (usually mechanical or hydraulic) of the primary element directly actuates a measuring element. Operation of the final control element then is provided

*The remainder of this article furnished by Ardmore M. Miller, Love Controls Corporation, Wheeling, Ill.

by electric energy. Typically, such instruments are of the bimetallic or expansion-bulb type for temperature measurement, and of the bourdon tube, diaphragm, bellows, or float type for pressure, level, or flow measurements. In such instruments, the measured variable may or may not be indicated and/or recorded. The controlling means may be actuated directly from the primary element, as in a typical room thermostat, without indication of the measured variable. In an indicating or recording instrument, the controlling means (final element) is actuated on movement of the measuring system, which usually includes the indicating pointer or recording pen.

In the use of self-actuated instruments for electric control, two considerations are important: (1) sensitivity of the measuring element, i.e., the frictional drag and accompanying hysteresis introduced by the measuring and/or recording system, and (2) additional hysteresis introduced by the associated control elements.

Because of the small amount of power usually available in self-operated instruments, which may in some cases limit their accuracy and control sensitivity, instruments using electric or electronic output sensors often are required where better performance is required.

Electric Controllers

Although self-operated controllers usually are less expensive, electric controllers are used widely because of their versatility and sensitivity. Of primary concern in this description are electronic instruments which receive an electric signal from the primary element and, from that signal, record or indicate and control a final element. Selected nonindicating, nonrecording electronic instruments also are described briefly.

In considering controllers, one of the most important factors is control sensitivity, i.e., the ability to respond to small changes in the measured variable and to produce a corresponding change in the controller output. A second factor is response time, which may be of vital importance for certain applications involving rapidly changing process conditions.

A variety of control modes has evolved because of the manner in which control characteristics differ from one process to the next and because of variations in the requirements for closeness of regulation.

Modes of Control

Two-Position (On-Off) Control

This mode of electric control may be used satisfactorily if some cyclic fluctuation in the controlled variable is permissible. These fluctuations may be small or large, depending on process characteristics and controller sensitivity. Two-position control generally is available in both self-operated and electronic types of controllers. A highly simplified diagram of an electric thermostat, useful for room temperature control as well as in some industrial heating and refrigeration applications, is given in Fig. 1. Other typical forms of mechanical and hydraulic two-position devices are shown in Figs. 2 and 3. In Fig. 2, the contact points are made to close or open as the expanding shell increases or decreases in length in relation to the nonexpanding strut. A disadvantage of the type of control shown in Fig. 3 is the need for capillary tubing which is vulnerable to damage unless carefully handled and installed. Electronic two-position controllers, which incorporate amplifiers, provide higher accuracy than self-operated controllers, are more conveniently adjusted, and permit longer distance (remote) between the sensor and the controller.

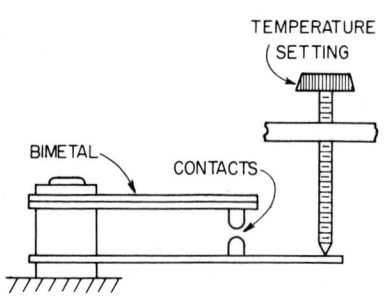

TEMPERATURE
(SETTING

BIMETAL
CONTACTS

Fig. 1 Highly simplified diagram of an electric thermostat.

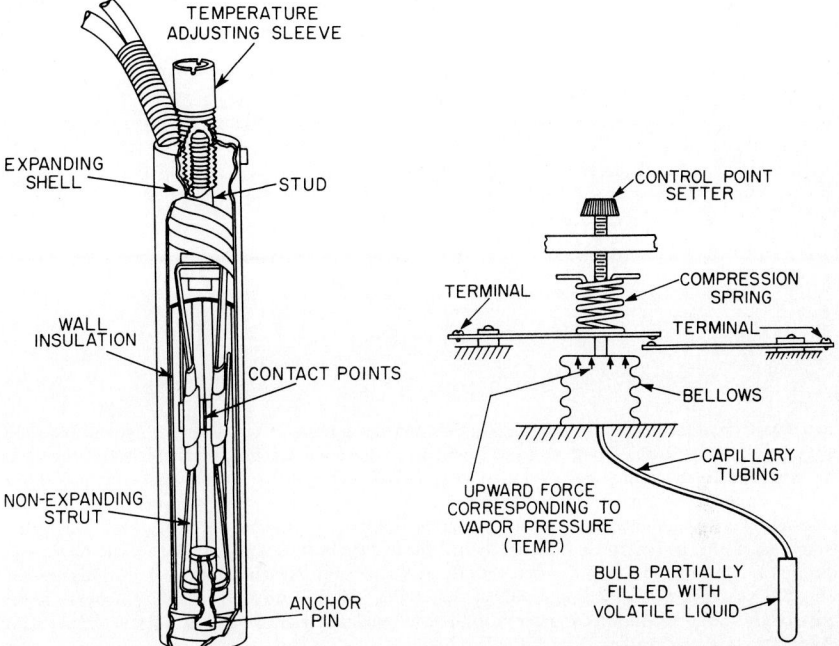

Fig. 2   Sectional view of a thermoswitch thermostat.

Fig. 3   Self-operated two-position electric controller, using an expansion-bulb primary element.

A bridge-type two-position controller is shown in Fig. 4. A bridge circuit containing, on one arm, a temperature-sensitive resistor is supplied with voltage across potential points $P1$ and $P2$. Bridge imbalance is detected at points $P3$ and $P4$. The set point is determined by adjusting the position of contacts on slidewires $S1$ and $S2$. Typically, a three-lead resistance bulb is used for compensation of the lead resistance, with the resistance of lead A being balanced by an equal resistance of lead B in the opposition arm of the bridge. Slidewire $S2$ serves to maintain a 1:1 ratio in the upper arms of the bridge for various set points to match the 1:1 ratio in the lower arms of the bridge and thus maintain exact resistance-bulb lead compensation throughout the range of set ponts.

A *millivoltmeter-type two-position controller* is shown in Fig. 5. This controller is designed for

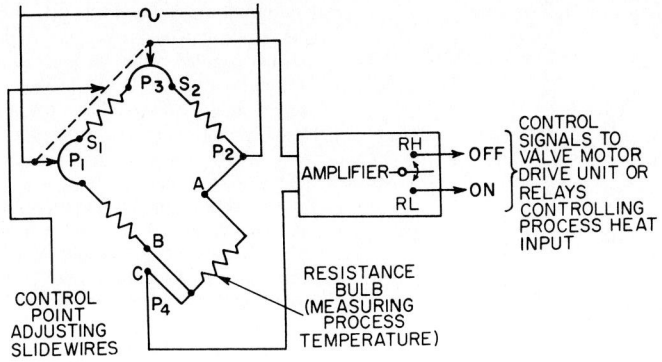

Fig. 4   Bridge-type two-position controller.

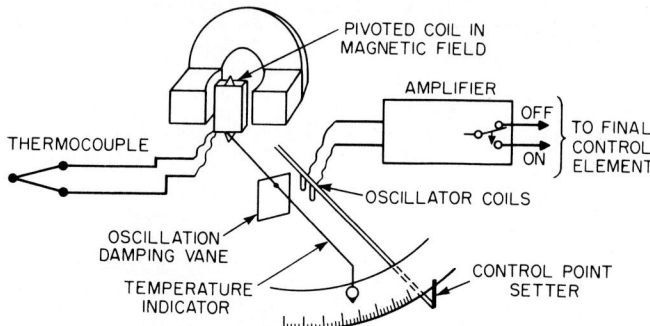

Fig. 5 Millivoltmeter-type two-position controller.

actuation from a thermocouple or radiation thermometer output. It is a deflection-type instrument, usually with an indicating pointer and scale and a suitable error detector, adjustable across the scale. By means of an electronic oscillator, amplifier, and associated relay, the instrument regulates the process input to maintain the temperature at the control point. Some designs of these controllers are plug-in, whereby the galvanometer and the error detector are built as a replaceable plug-in unit which plugs into the control amplifier section. This feature facilitates replacement of the rather delicate galvanometer. Some manufacturers of millivoltmeter controllers replace the oscillation damping vane and oscillator coils with a light source, photocell, and light obscuring vane. The signal from the photocell is amplified and used to operate an output relay. Resistance changes in the external sensor circuit can introduce reading errors in millivoltmeters.

In a *potentiometric-type two-position controller,* the voltage produced by the sensor is electronically compared to a known set-point voltage. The resulting differential voltage is amplified and used to operate a relay. This type of control is rugged and reliable because of the absence of moving parts. The resistance in the sensor circuit has no effect on accuracy (usually up to 100 Ω).

Two-Position Time Proportioning Control

This is a widely used mode of control and is a major improvement over on-off control. Time proportioning controls basically turn the controlled variable on and off, but the on-time and off-time are proportioned in accordance with the demand of the process. Essentially, all proportioning controllers anticipate the control action required—by changing the relay position from "on" to "off" before the controlled variable reaches the set point. As the controlled variable overshoots slightly and begins to drop below the set point, the anticipating action again is automatically introduced as the set point is approached.

The proportioning action occurs within a band around the set point. Within this band, various on-time to off-time cycles occur. At the very edge of the band below the set point, the controlled variable will be on 100% of the time. At the set point, on-time will be 50% and decrease steadily until the upper edge of the band is reached, where on-time will be zero. See Fig. 6.

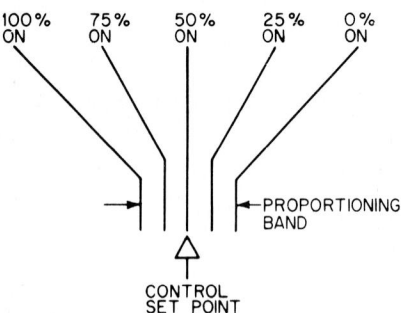

Fig. 6 Controlled variable output versus band position.

There are several means for introducing proportioning action in an on-off controller. In general, all designs require a false signal to be introduced in a positive direction below the set point and in a negative direction above the set point. This allows the controller to reach the set point from either direction before it would without the false signal, and thus provides anticipation of control action.

Table 1 Switching Possibilities of a Three-Position Controller

Application	Control function by relationships of controlled variable to set points		
	Below both set points	Between set points	Above both set points
Heating with automatic cooling	Heat on	Heat off	Cooling
High-low-off control	High input	Low input	Off
High-medium-low control	High input	Medium input	Low input
Process on-off control with high-limit alarm	Process on	Process off	High-limit alarm
Floating control	Process input increasing	Process input steady	Process input decreasing

Three-Position (Floating) Control

This control mode is quite versatile and, in a sense, is almost the equivalent of two on-off controllers. A three-position controller contains two relays that can be set to operate at two different set points within the range of the controller. The various switching possibilities of a three-position controller are listed in Table 1.

In addition, it is possible to incorporate proportioning action on either relay position, thus providing solutions to more complex control tasks. For example, the proportioning action configuration is widely used for temperature control in manufacturing thermoplastic extruders, where frictional heat is a problem in the process. The extruder barrel is divided into zones and is controlled with three-position controllers that incorporate proportioning action on the lower temperature control set point. In the absence of frictional heat, the temperature is controlled normally by the proportioning control set point. Should frictional heat cause an increase in temperature beyond the normal operating value, the instrument automatically becomes a cooling controller.

The instrument switches from no heat to cooling, either by operating air cooling blowers or by solenoid actuation of a water or vapor cooling system. Because of the unpredictability of frictional heat and because excessive temperatures can be damaging, the advantages of this type of control are evident.

A three-position controller is commonly used to control the position of gas or oil valves on fuel-fired furnaces. The controller operates an actuator which opens the valve if the controlled temperature is below the set-point value, stops the valve at the set point, and closes the valve above the set

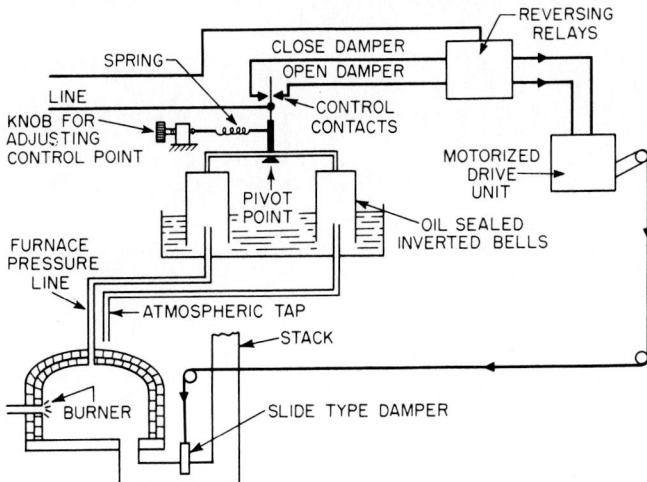

Fig. 7   Single-speed floating-action controller for controlling furnace pressure.

point. This is termed *floating control* because the valve has the ability to stop or move in either direction around the set point.

Floating-type controls often are used in the regulation of furnace pressure. A typical single-speed furnace pressure controller and its associated process are shown in Fig. 7.

An increase in furnace pressure above the control point, for example, deflects the pivoted pressure bell system and results in closure of the right-hand contact, actuating the motorized damper drive unit to open the stack damper slowly until the pressure drops to the control point, whereupon the controller contact opens and corrective action ceases. The motorized drive is geared down so that the rate of damper movement is relatively slow. Accordingly, pressure correction takes place after only partial movement of the damper has occurred. As a result of the prompt process response and the relatively slow damper movement, the damper position typically floats in small increments between its two extremes in response to small corrective impulses from the controller contacts.

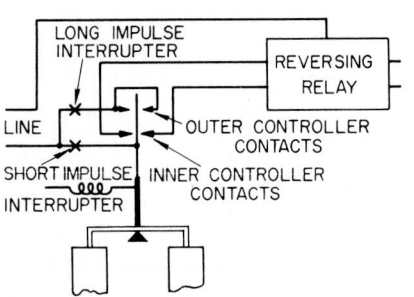

Fig. 8 Two-speed floating-action controller for controlling furnace pressure.

Two-speed floating control may be achieved by an arrangement of the type shown in Fig. 8. For small deviations in pressure, the inner controller contacts are closed which, through the short impulse interrupted, move the drive unit at a relatively slow average rate which does not overcorrect for small pressure deviations. For larger deviations in pressure, the outer controller contacts close, bringing into play the long impulse interrupter to move the drive unit at a faster average rate and thus correct more quickly for larger pressure changes. This faster action is discontinued when the pressure is brought back closer to the desired condition, and thus overcorrection is avoided.

Proportional (Stepless) Control

This is the simplest form of modulating control and is widely used and available in a variety of configurations. As the controlled variable changes, a proportional controller alters the process input by an amount proportional to the magnitude of the change and in a direction to restore process conditions to the desired value. As the controlled variable moves through a range, termed the *proportional band,* the final control element is proportionately adjusted from minimum to maximum input. This mode also is called *stepless control.*

Proportional controllers generally provide a linear current signal from their output stages instead of actuating a relay. This current signal can be fed to electropneumatic transducers, silicon controlled rectifier (SCR) circuits, saturable core reactor circuits, or a variety of transmitters and converters to perform a multitude of control tasks.

For temperature control, two popular final elements have emerged: the saturable core reactor and the SCR.

The *saturable core reactor* is the simplest and is a very reliable method for modulation of electric heating processes. Three disadvantages of saturable core reactors are:

1. Because the reactor is always in the live circuit, it relies on the impedance of its core to attain the shutoff of power, and the saturation of its core to give full power capability. In practical use, the shutoff condition still yields a minimum of 3% of power, and the saturated condition (turned on) usually yields no more than 90% of power.

2. The reactor must match the load wattage and voltage in order to stay within the 3 to 90% power control range. In applications requiring the controller to operate on several different loads, this may present a problem. Reactors, tapped for several ratings, are available to provide some versatility.

3. The weight and size of reactors can be disadvantageous. The appearance of a typical reactor is similar to that of a transformer. Depending on its rating, the reactor can become quite large.

A saturable reactor control system is shown in Fig. 9.

The reactor system offers a large safety advantage. It is impossible, even by direct short circuit of the electric heater, to burn out a reactor. Since it is a current limiting device, it will not permit enough current to flow to burn it out under any circumstances. A stepless controller utilizing an SCR has certain advantages over the reactor type and can control any size load which is at or below its rating without the necessity of load matching. There also is a substantial saving in space and weight. Complete shutoff to the electric heater is attained, and practically full power output is secured.

Manufacturers of stepless controllers must provide protection for the SCR, which is a semiconductor device. The controller must be protected from both fast and slow transient surges, high inrush currents, load short-circuiting, con-

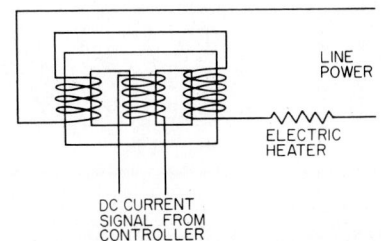

Fig. 9   Saturable reactor temperature control system.

trol circuit component failure, and controller component overtemperature. Fast-blow fuses and spike suppressing circuitry are employed.

SCR-type stepless controllers generally use two SCR semiconductors in order to control the full wave of ac power. A triac, which actually is two SCRs in one package, may also be used with equal results.

SCR stepless controllers turn the ac line on in very short or long bursts to apply the proper power to the electric heating load. Two types of circuits have been developed to fire the SCR at the proper level. One of these, called *phase angle firing,* turns each ac cycle on for the portion of its cycle which produces the proper amount of power to maintain the heating condition required. This type of firing generates radio frequency interference and also causes power line spikes. These effects can cause interaction between two or more controllers when they are used on the same line. Waveforms of the power line using phase angle firing are shown in Fig. 10.

A firing method has been developed in which the switching of the SCR occurs only as the ac voltage passes through zero. This is termed *zero-crossover firing* or *zero-voltage firing.* The term "proportioning" also may be used to describe this method. With zero-crossover firing, full cycles of power pass to the electric heater load. For example: one cycle on and four cycles off (20% power); one cycle on and one cycle off (50% power); or four cycles on and one cycle off (80% power). With this system, line spikes and radio frequency interference are reduced to an absolute minimum. The

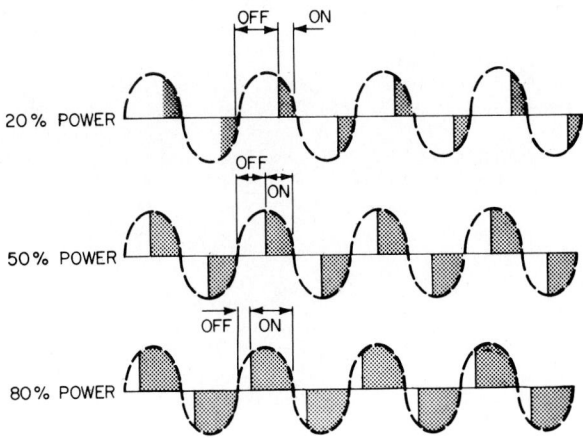

Fig. 10   Waveform of power being supplied to electric heaters using phase angle firing of silicon controlled rectifiers.

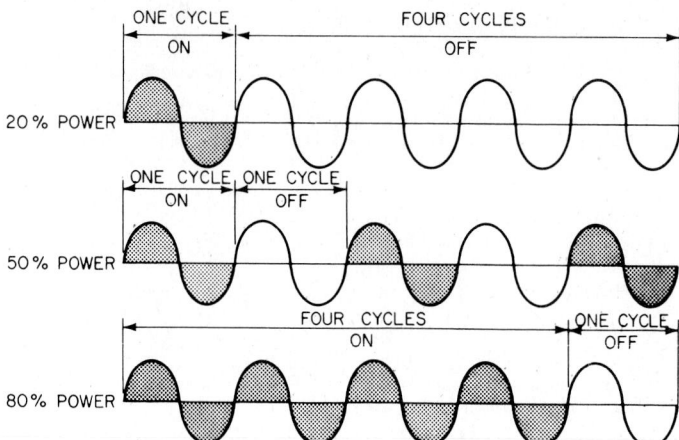

Fig. 11 Waveform of power being supplied to electric heaters using zero-crossover firing of silicon controlled rectifiers.

waveform of power being supplied to electric heaters using the zero-crossover firing method is illustrated in Fig. 11.

A comparison of phase angle firing with zero-crossover firing can be made by using a simple analogy. Phase angle firing can be compared to the opening or closing of a switch while the power is on. Zero-crossover firing, on the other hand, opens or closes a switch while the power is off.

Proportional control also is applied to regulate fuel-fired temperature-controlled processes by using the current signal to position valves. In one method, an electropneumatic transducer into which the current signal is fed is employed. The transducer converts this signal to an equivalent air signal to operate an air-operated valve. See Fig. 12.

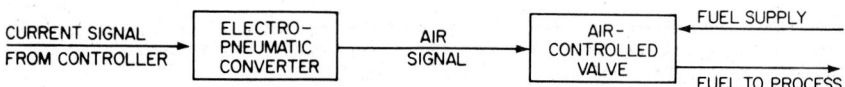

Fig. 12 Proportional control for an air-operated valve.

In another type of proportional control, a slidewire balancing system is used to position a motorized drive unit which positions a valve. See Fig. 13. A drive-unit slidewire is mounted in and actuated by the motorized drive unit. This unit, together with the control slidewire, reset slidewires (if used), and associated end resistances, constitutes the control bridge which is fed with a source of voltage. The relay detector is connected to the control and drive-unit slidewire contacts. On balance of the control-slidewire contact with respect to the drive-unit slidewire contact, an error signal is fed into the detector relay. This causes the output relay to close the appropriate contact to run the motorized unit in a direction to restore the bridge to balance, and also to operate the final control element in the proper direction to adjust the process input correctly.

For each change in the measured variable within the proportional band, a corresponding change is produced in the drive-unit position. The proportional-band adjuster is a shunt on the drive-unit slidewire, arranged so that an adjustable ratio is secured between control-slidewire movement and drive-unit travel. The two manual-reset rheostats, when turned in a given direction, change their resistances in an opposite sense to each other and accordingly shift the relationship between the control slidewire and drive-unit slidewire. This provides the proper nominal final control element position required to hold the controlled variable at the set point for the average load condition encountered.

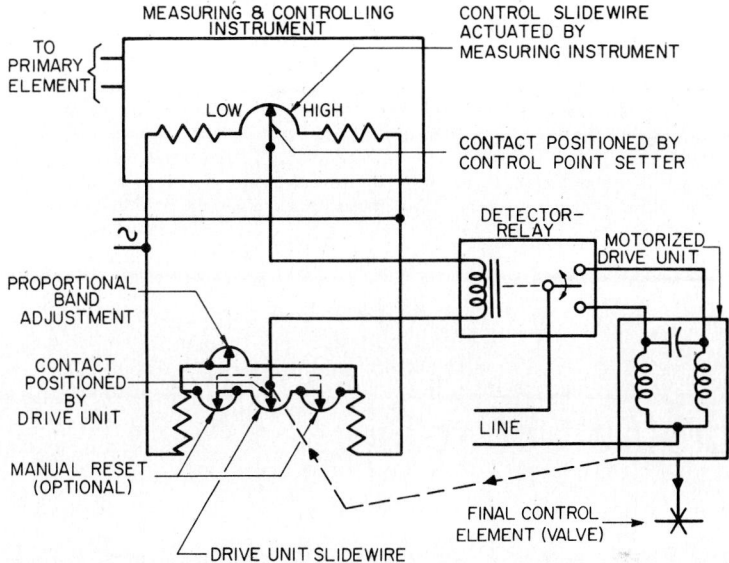

Fig. 13   Slidewire balancing controller.

Proportional-Reset Control

Two-position time proportioning and proportional controllers perform their control functions within a band, as previously shown in Fig. 6. Normally, when setting controllers having these modes, the set point represents a setting made in the center of the control band. For example, for a temperature control having a bandwidth of 2% of scale range and a range 0 to 800°, the bandwidth would be 16°. This means that, if the control point were set at 400°, the bandwidth would extend from 392 to 408°. In the course of control, it is possible for the temperature at the process to settle out anywhere within the control band. If the process temperature settled out at 404°, then the offset from the desired control point would be +4°. Obviously, this could be corrected by placing the set point at 396° so that proper control could be obtained at 400°. This would have to be done manually after determining the offset. See Fig. 14.

To avoid this situation, automatic reset was developed—to eliminate offset or droop. Special circuits in the control settle out at a point either above or below the desired set point and electronically shift the control band up or down to remove the error. In the foregoing example, the band would shift down by 4°. See Fig. 15.

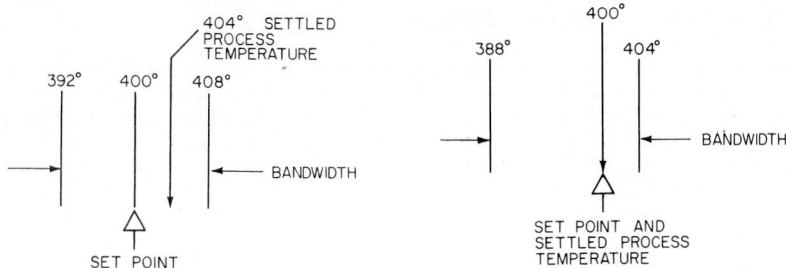

Fig. 14   Controller without reset, indicating a settled temperature that is 4° high.

Fig. 15   Controller with the reset band shifted down 4° so that the settled temperature is at the desired set point.

Proportional-Reset-Rate Control

Should a fast upset to the process occur, the input signal usually must drop to the lower edge of the proportional bandwidth before full power is applied. If full power could be applied instantaneously on sensing a fast upset, the process variable (ideally) would move only slightly before returning to the true control point. This is accomplished through the use of derivative circuits and is termed *rate action*. It is described in detail in other portions of this volume. Please consult the alphabetical index.

Basically, if the process upset is fast, the resulting correction will be fast (essentially narrowing the proportional band) and, conversely, if the process upset is slow, the resulting correction will be slow.

Hydraulic Controllers

by
Raymond G. Reip*

A hydraulic controller is a device that uses a liquid control medium to provide an output signal which is a function of an input error signal. The error signal is the difference between a measured variable signal and a reference or set-point signal. Self-contained closed-loop hydraulic controllers continue to be used for certain types of process control problems, but as the use of the computer, with its electrical output, expands in process control applications, the electrohydraulic servo valve gains in usage. The combination adds the advantages of hydraulic control to the versatility of the computer. Also contributing to the expanding use of hydraulics is the steady improvement in fire-resistant fluids.

Since the servo valve does not accept low-level digital input directly, a digital-to-analog (D/A) converter and servo amplifier are required. So, where it can be used, a hydraulic controller which senses a controlled variable directly is preferred in the industrial environment for its easy maintainability.

Elements of Hydraulic Controllers

The major elements include (1) a hydraulic relay or amplifier, (2) an error detector, and (3) a signal sensing section.

As in all control loop applications, a measured variable must be sensed and compared with a set point, and the difference or error signal manipulated by the controller to cause some final control element to correct the process. Depending on the particular construction of the hydraulic controller, these functions often can be incorporated physically into the hydraulic controller per se.

The hydraulic relay is common to all hydraulic controllers. The relay allows a small mechanical displacement to control the output from a hydraulic power supply in a fashion that will drive a work load. The jet pipe and spool valve are examples of these relays. See Fig. 1. If the process variable is being sensed directly, such as pressure by means of a bourdon tube or differential pressure with a

*Senior Development Engineer, GPE Controls, Division of Brunswick Corporation, Morton Grove, Ill.

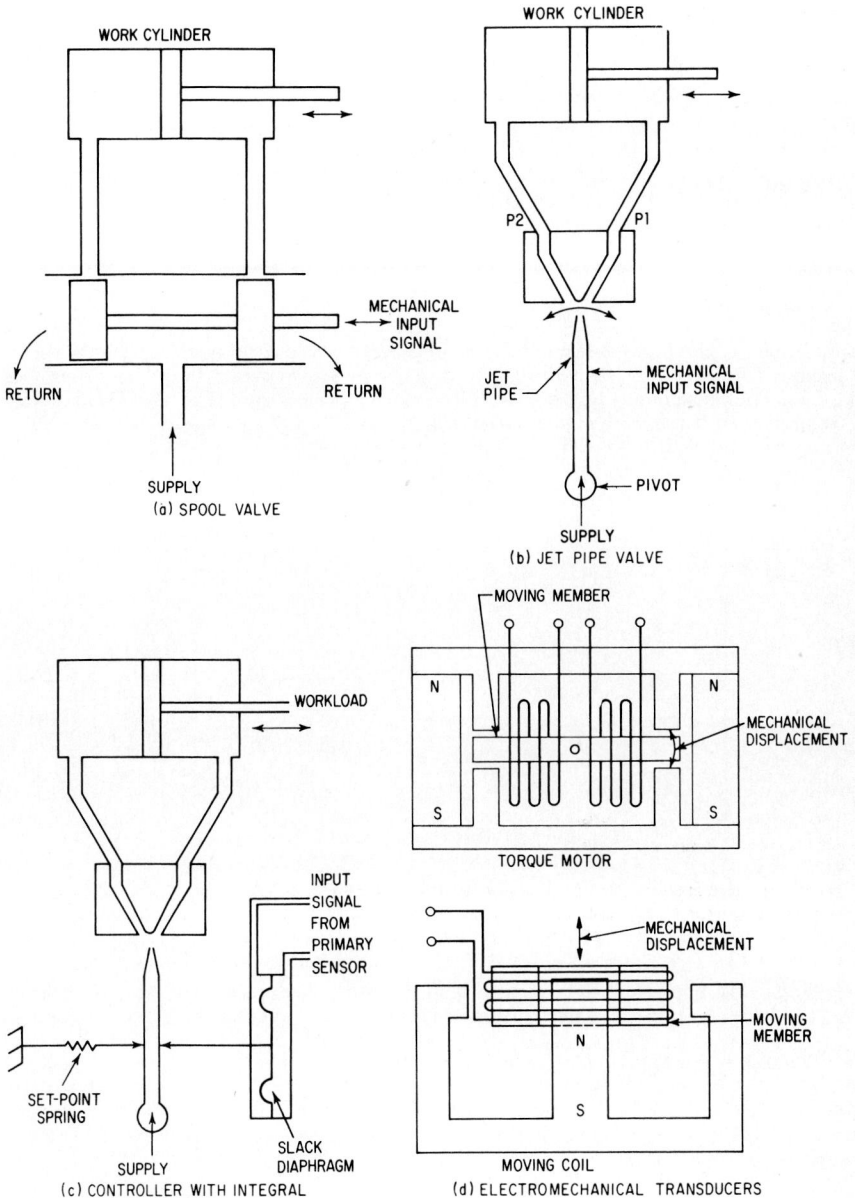

Fig. 1 Elements of hydraulic controllers.

WORK CYLINDER

MECHANICAL
INPUT
SIGNAL

RETURN

RETURN

SUPPLY
(a) SPOOL VALVE

WORK CYLINDER

P2   P1

JET
PIPE

MECHANICAL
INPUT SIGNAL

PIVOT

SUPPLY
(b) JET PIPE VALVE

WORKLOAD

INPUT
SIGNAL
FROM
PRIMARY
SENSOR

SET-POINT
SPRING

SUPPLY
(c) CONTROLLER WITH INTEGRAL
SIGNAL SENSING

SLACK
DIAPHRAGM

MOVING MEMBER

N   N

MECHANICAL
DISPLACEMENT

O

S   S

TORQUE MOTOR

MECHANICAL
DISPLACEMENT

MOVING
MEMBER

N

S

MOVING COIL
(d) ELECTROMECHANICAL TRANSDUCERS

diaphragm, the signal sensing and error detection can become an integral part of the hydraulic controller. As shown in Fig. 1c, a force is generated from the signal system and compared with a setpoint force operated by the spring, and the resulting mechanical displacement (error signal) moves the hydraulic relay (in this case, a jet pipe) to control the work cylinder.

Hydraulic Relays

The three major types of hydraulic relays are the (1) jet pipe valve, (2) flapper valve, and (3) spool valve.

Jet Pipe Valve

By pivoting a jet pipe, as shown in Fig. 1b, a fluid jet can be directed from one recovery port to another. The fluid energy is converted entirely to a velocity head as it leaves the jet pipe tip and then is reconverted to a pressure head as it is recovered by the recovery ports. The relationship between jet pipe motion and recovery pressure is shown in Fig. 2b. Although a jet pipe can be used at higher

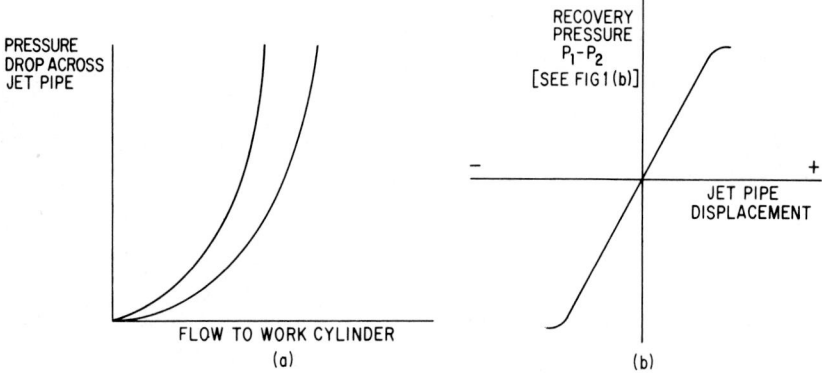

Fig. 2 Jet pipe characteristics. (a) Jet pipe flow versus differential pressure. (b) Jet pipe recovery pressure versus displacement.

pressures, most applications are less than 800 psi (5.5 MPa). The proportional operation of the jet pipe makes it very useful in proportional-speed floating systems (integral control), as shown in Fig. 3a.

Position feedback can be provided by rebalancing the jet pipe from the work cylinder, as shown in Fig. 3b. A proportional-plus-reset arrangement is shown in Fig. 3c. In the latter case, the proportional feedback is reduced to zero as the oil bleeds through the needle valve. The hydraulic flow obtainable from a jet pipe is a function of the pressure drop across the jet pipe. Typical flows for jet pipes are shown in Fig. 2a.

Flapper Valve

This device makes use of two orifices in series, one of which is fixed, and the other variable. The variable orifice consists of a flapper and a nozzle. The nozzle restriction is changed as the flapper is positioned closer to or farther from the nozzle. When the variable restriction is changed, the pressure drop across the fixed orifice changes, thus producing a variable output pressure with flapper position. Single- and double-flapper arrangements are available, as diagramed in Fig. 4a and b.

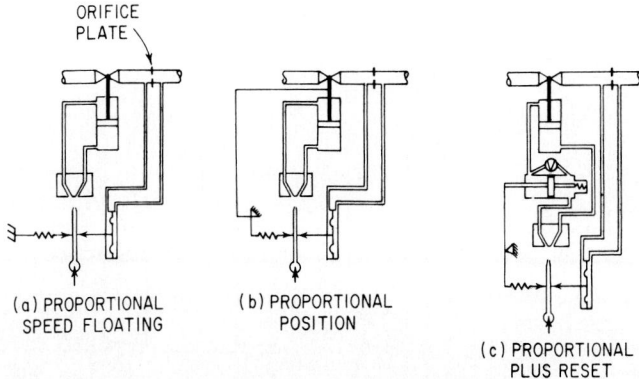

Fig. 3 Hydraulic controllers arranged in different control modes.

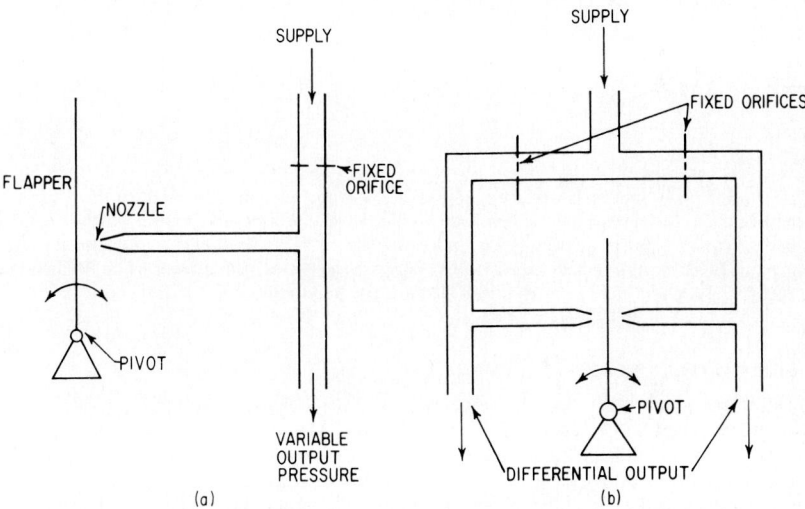

Fig. 4 Flapper valves. (*a*) Single; (*b*) double.

Spool Valve

This device (Fig. 5), when used as a hydraulic relay, generally is constructed in either a three- or a four-way valve porting arrangement. In both cases, the mechanical displacement of the spool allows the hydraulic pressure supply to be ported in a fashion that displaces the work cylinder in either direction, depending on the spool displacement. A four-way valve can put full supply pressure on one side of the work cylinder and drain the other side, or vice versa, while a three-way valve requires one side of the work cylinder to be pressurized from a constant pressure source. If the widths of the spool lands are greater than those of their respective ports, the valve is overlapped and is referred to as a *closed-center valve*. An underlapped valve is referred to as an *open-center valve*. A *line-to-line valve* is one in which the spool lands are the same width as the porting. Flow through the spool valve is proportional to the square root of the differential pressure across the valve, as is the case with a jet pipe.

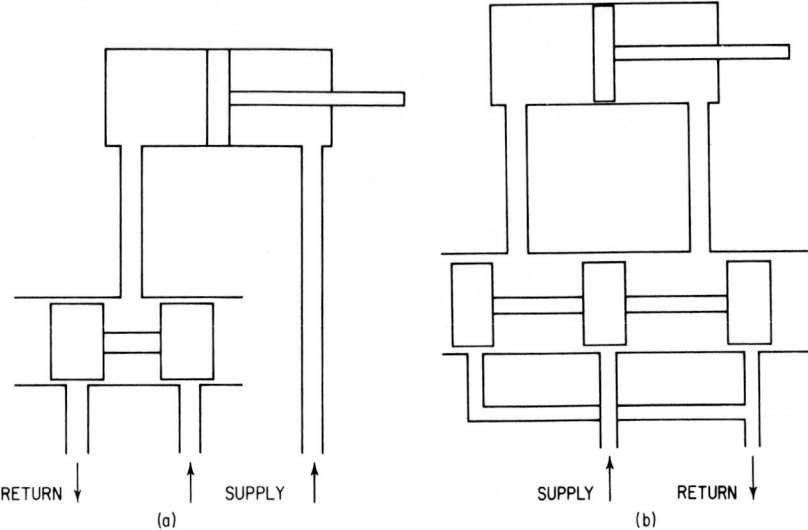

Fig. 5   Spool valves. (*a*) Three-way; (*b*) four-way.

Two-Stage Valves

When higher flows and pressures are required, a two-stage valve may be used. In these cases, the second stage usually is a spool valve, while the first stage may be a jet pipe, a spool valve, or a flapper valve. Different combinations of two-stage valves are shown in Fig. 6. In two-stage valves, the second stage must be either positioned, such as with springs, or provided with a feedback to the first stage as required. Feedback may be mechanical, electrical, or hydraulic.

Electrohydraulic Relay

See Fig. 6. Frequently, the process variable is converted to an electric signal by an appropriate transducer. In these cases generation of the set point and the error detection is performed electrically, and the error signal fed to the hydraulic relay is an electric signal. When this is done, an electromechanical transducer is required to generate the mechanical displacement. Two of the more popular schemes are shown in Fig. 1*d*: the moving coil and the torque motor. The principal difference between these two transducers is in the direction of the mechanical displacement. The moving coil produces a linear mechanical displacement, while the torque motor output is rotational. Electromechanical transducers mechanically packaged together with a hydraulic relay commonly are referred to as *electrohydraulic servo valves* or just *servo valves*.

Servo Valves

Figure 6*b* illustrates one of the most widely used servo valve designs—a two-stage valve. One and three stages are also available. The single-stage design is used for the highest response rate. Three-stage designs are used for the highest power output.

Typical dimensions and clearances for servo valves are given in Table 1. These figures apply to valves manufactured for aerospace or industrial use. The larger numbers shown apply to larger valves.

Most servo valve failures are due to sticking spools, usually directly traceable to contamination. Reliable servo valve systems depend on maintaining clean systems.

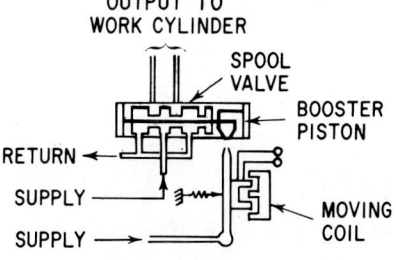

FIRST STAGE: JET PIPE
SECOND STAGE: SPOOL VALVE
FEEDBACK: HYDRAULIC

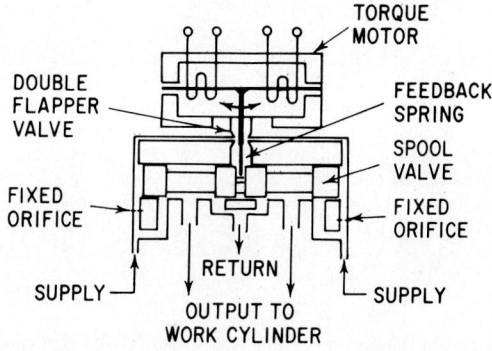

FIRST STAGE: DOUBLE FLAPPER
SECOND STAGE: SPOOL VALVE
FEEDBACK: MECHANICAL

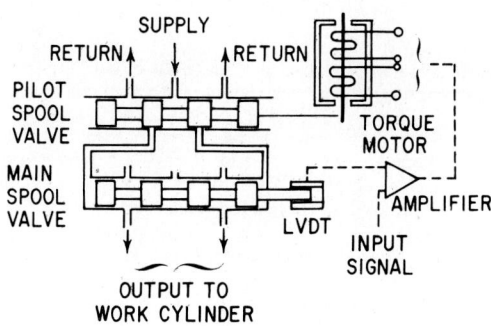

FIRST STAGE: SPOOL VALVE
SECOND STAGE: SPOOL VALVE
FEEDBACK: ELECTRIC

Fig. 6   Various combinations of two-stage valves.

18.37

Table 1 Typical Dimensions and Clearances in a Servo Valve

Parameter	Dimension	
	Inches	Millimeters
Air gap spacing (each gap)	0.010–0.020	0.254–0.508
Maximum armature motion in air gap	0.003–0.006	0.076–0.152
Inlet orifice diameter	0.005–0.015	0.127–0.381
Nozzle diameter	0.010–0.025	0.254–0.635
Nozzle-flapper maximum opening	0.002–0.006	0.051–0.152
Drain orifice diameter	0.010–0.022	0.254–0.559
Spool stroke	$\pm 0.010 - \pm 0.060$	$\pm 0.254 - \pm 1.524$
Spool or bushing radial clearance	0.0001–0.00015	0.0025–0.0038

Hydraulic Fluids

Although much increased in cost over the past several years, petroleum oil is still the optimum hydraulic fluid unless fire resistance is required. With additives, petroleum oil does not have major shortcomings of the magnitude of those found in all fire-resistant fluids. In designing a hydraulic control system, the choice of fluid must be made early because a number of features of the system must be tailored to the fluid used.

Although there are numerous petroleum-based hydraulic oils designed specifically for various types of systems, automatic transmission fluid has been found to be a good engineering compromise for many systems because (1) it is available worldwide to the same specifications; (2) it contains viscosity index improvers which reduce the effects of viscosity change with temperature (like 10W-30 motor oil); (3) like other quality hydraulic oils, it has rust, oxidation, and foam inhibitors; and (4) it contains an antiwear zinc compound additive, although not as much as is contained in extreme pressure (EP) hydraulic oils.

Synthetic hydrocarbon oils have all the advantages of natural oils, plus complete freedom from wax content. This permits synthetic oils to remain fluid and thus be pumpable down to $-65°F$ ($-54°C$). Although the cost of synthetic oils is relatively high, their use makes it possible to install hydraulic controls in adverse outside environments.

Fire-Resistant Fluids

The principal available fire-resistant fluids are described briefly in Table 2. Their characteristics are compared with those of petroleum oils in Table 3.

Table 2 Representative Types of Fire-Resistant Hydraulic Fluids*

Letter designation	Fluid description
HF-A	High-water-content emulsions and solutions are composed of formulations containing high percentages of water, typically greater than 80%. They include oil-in-water emulsions and solutions which are blends of selected additives in water.
HF-B	Water-in-oil emulsion fluids consist of petroleum oil, water emulsifiers, and selected additives.
HF-C	Water-glycol fluids are solutions of water, glycols, thickeners, and additives to provide viscosity and other properties desirable for satisfactory use in hydraulic systems.
HF-D	Synthetic fluids are nonwater fluids, such as phosphate esters or blends of phosphate esters with petroleum oils.

*Extracted from ANSI B93.5M-1979.

Table 3 Comparison of Hydraulic Fluids

| Characteristic | Petroleum oil | HF-A, high water content | | HF-B, water emulsion | HF-C, water glycol | HF-D, phosphate ester |
		Low viscosity	Thickened			
Fire resistance	Poor	Excellent	Excellent	Fair	Good	Very good
Cost	1	0.1	0.5	1	2	4
Lubricity	Excellent	Fair	Good	Good	Good	Excellent
High temperature	Excellent	Fair	Fair	Fair	Fair	Very good
Low temperature	Very good	Poor	Poor	Poor	Good	Fair
Corrosion protection	Excellent	Good	Good	Good	Good	Very good
Standard hardware compatibility	Excellent	Fair	Good	Fair	Fair	Very good
Toxicity	Good	Very good	Very good	Good	Excellent	Fair

Power Considerations

Hydraulic controllers are generally applied where either high-power or high-frequency response is required. The two requirements may or may not be needed for the same application. Frequency-response requirements are a function of the control loop dynamics and can be determined by appropriate automatic control theory techniques discussed elsewhere. See Process Control Techniques in Sec. 17 of this volume. Power considerations will be discussed here.

Because hydraulic fluids can generally be considered incompressible, they can be handled under higher pressure conditions than would be considered safe or practical with pneumatics. The amount of power delivered to the load is a function of the hydraulic power supply size, the pressure drop required by the hydraulic controller, and the pressure drop losses in the lines. With reference to Fig.

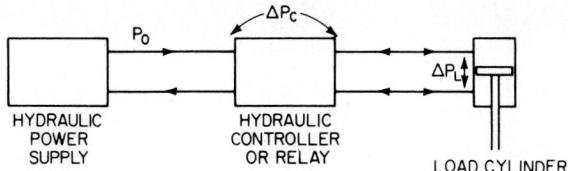

Fig. 7 Hydraulic flow loop.

7 and neglecting any line losses, it can be said that the pressure drop available for the load is equal to the supply pressure less the hydraulic controller pressure drop. In equation form,

$$\Delta P_L = P_o - \Delta P_c$$

If the oil flow under these conditions is Q, then

$$\text{Power available at the load} = \Delta P_L Q$$

Q_{max} is determined by the hydraulic power supply capacity. As Q increases, the term $\Delta P_L Q$ tends to increase. However, increasing Q increases ΔP_c, thus reducing ΔP_L. It can be shown that for most hydraulic controllers the maximum $\Delta P_L Q$ occurs when $\Delta P_L = 2\Delta P_c$.

Figure 8 shows the force and power ranges normally encountered in hydraulics contrasted with

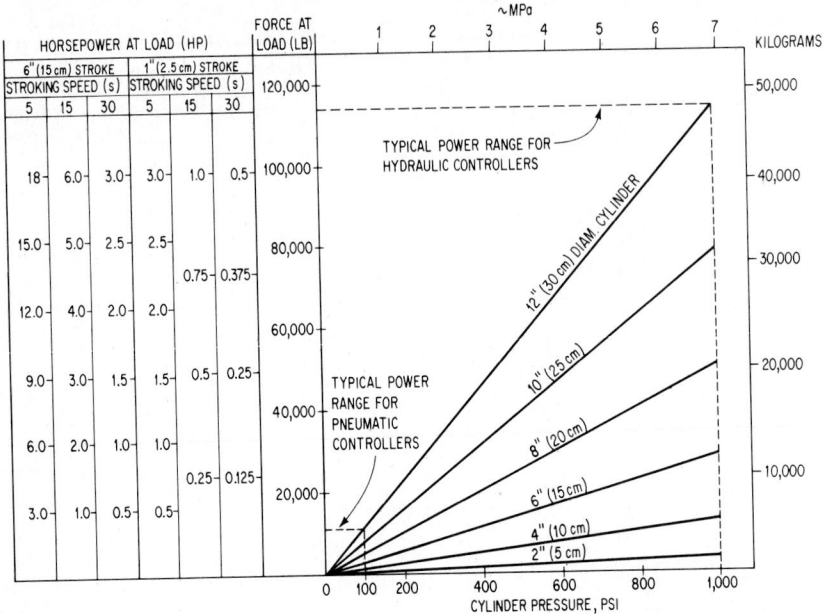

Fig. 8 Power relationships. Force and power ranges normally encountered in hydraulic systems contrasted with the ranges commonly used in pneumatic systems.

the ranges commonly used with pneumatics. The load horsepower scales were calculated by selecting two representative load stroke lengths and three different full-stroke speeds, all corresponding to the force scale. The hydraulic power supply horsepower requirement is greater, however, since it must in addition supply the power for the hydraulic controller, line losses, and pump inefficiency. This can be expressed as

$$ HP_{ps} = \frac{HP_L + HP_c + HP_{LL}}{\text{pump efficiency}} $$

The power lost across the hydraulic controller is a function of the controller capacity. This is generally expressed in terms of a flow–pressure drop curve. Figure 9 gives some typical examples.

Conservation

Increasing energy costs have resulted in greater use of systems that load the pump motor only when there is a demand for hydraulic control power. Methods include (1) open-center control valves that bypass flow at low pressure; and (2) pressure-compensated variable volume pumps that pump only a volume sufficient to maintain pressure in the system.

Applications

Generally hydraulic control systems are selected based on more demanding control requirements rather than on lower initial costs. However, as power requirements increase, even the initial cost may favor hydraulics. Three typical hydraulic controller applications are shown in Figs. 10 to 12.

The approximate power and performance requirements versus applications of servo valve control systems are illustrated in Fig. 13.

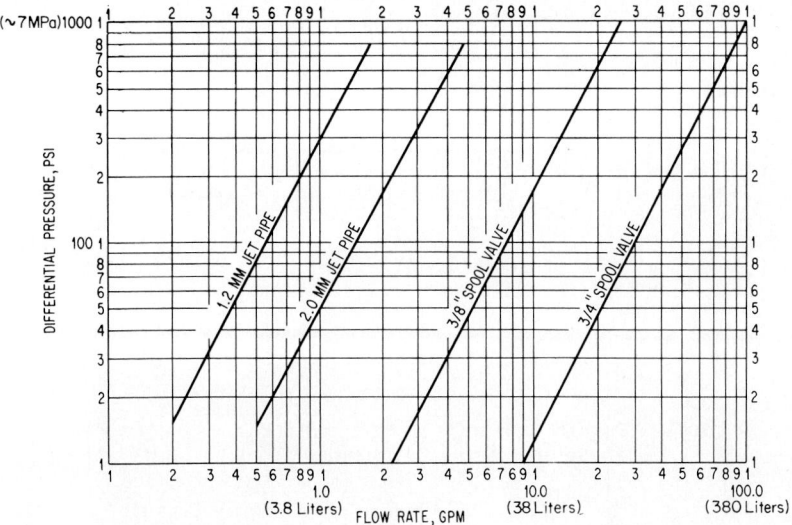

Fig. 9 Flow–pressure drop curves for hydraulic controllers.

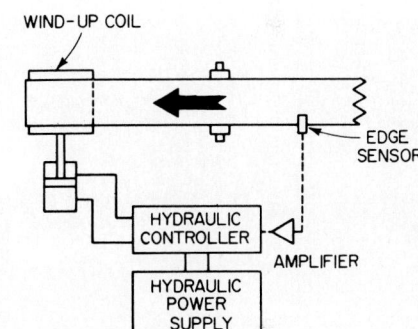

Fig. 10 Hydraulic edge-guide control system. The windup coil is shifted in accordance with an edge sensor signal to provide an even coil.

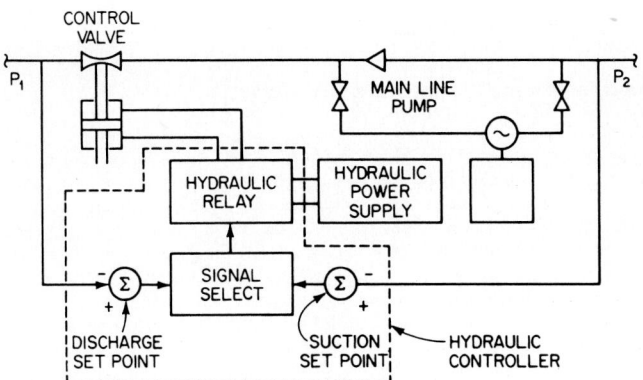

Fig. 11 Hydraulic pipeline control system. If discharge pressure (P_1) exceeds the set point, or if the suction pressure (P_2) goes below the set point, the control valve is throttled closed to correct the situation.

18.41

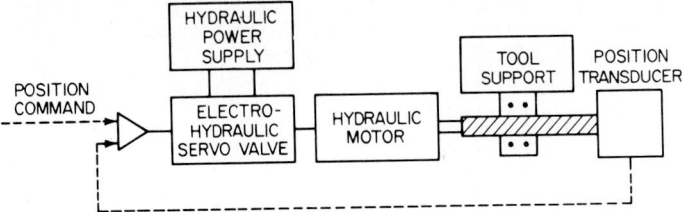

Fig. 12   Hydraulic machine tool control system. The tool support is positioned in accordance with a command.

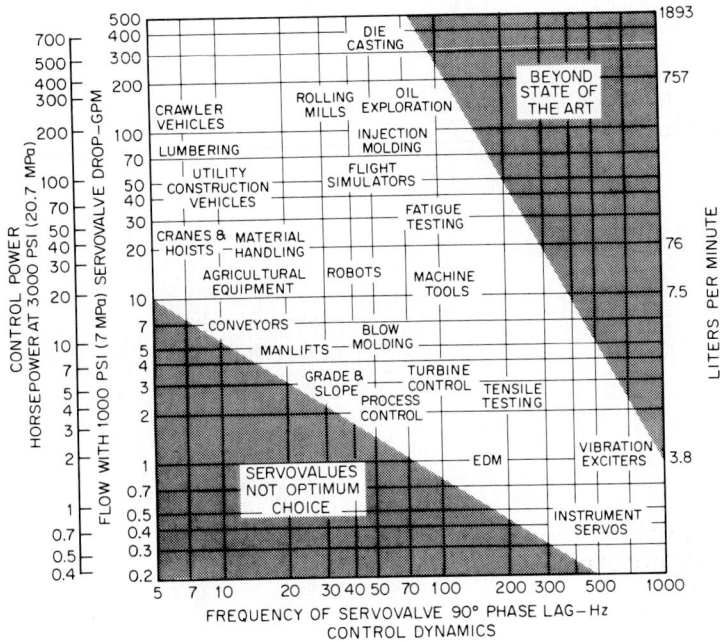

Fig. 13   Spectrum of industrial applications for hydraulic servo valves.

Relative Advantages and Limitations of Hydraulic Controllers

Advantages include the following. (1) High speed of response—the liquid control medium, being effectively incompressible, makes it possible for a load, such as a work cylinder, to respond very quickly to output changes from the hydraulic controller. (2) High power gain—since liquids can be readily converted to high pressures or flows through the use of various types of pumps, hydraulic controllers can be built to pilot this high-energy fluid. (3) Simplicity of the final actuator—most hydraulic controller outputs are two hydraulic lines that can be connected directly to a straight-type cylinder to provide a linear mechanical output. (4) Long life—the self-lubricating properties of most hydraulic controls are conducive to a long, useful life. (5) Relatively easy maintainability.

Some limitations include the following. (1) Maintenance of the hydraulic fluids—depending on the type of hydraulic hardware used, filtration is required to keep the fluid clean; the use of fire-resistant fluids, because of poorer lubrication and possible corrosive effects, requires more careful maintenance. (2) Leakage—care must be taken with seals and connections to prevent leakage of the hydraulic fluid. (3) Power supply—new hydraulic power supplies usually must accompany new hydraulic controller installations; this is contrasted with pneumatic and electrical or electronic systems where such power normally is available.

REFERENCES

Atland, G.: "Creating Motions Hydraulically," part I, *Automation,* vol. 14, no. 5, May 1967, p. 77; part II, vol. 14, no. 7, July 1967, p. 80.

Bailey, S. J.: "Fluid Power Control Provides A Wide Torque/Force Choice," *Control Eng.,* vol. 29, no. 4, 1982, p. 69.

Dushkes, S. Z., and S. L. Cahn: "Analysis of Some Hydraulic Components Used in Regulators and Servomechanisms," *Paper 41-A22,* Atlantic City ASME Meeting, 1951, American Society of Mechanical Engineers, New York.

Gebben, V. D.: "Unconventional New Four-Way Spool Valve Equation Holds Up in Dynamic Simulations," *Control Eng.,* vol. 24, no. 1, 1977, p. 36.

Henke, R.: "Speed and Displacement: Keys to Comparing LSHT (Low speed, high torque) Hydraulic Motors," *Control Eng.,* vol. 24, no. 2, 1977, p. 37.

Henke, R.: "Energy Costs Forcing Fluid Power Trends," *Control Eng.,* vol. 25, no. 3, August 1978, p. 37.

Henke, R.: "Proportional Hydraulic Valves Offer Power, Flexibility," *Control Eng.,* vol. 28, no. 4, 1981, p. 68.

Holzbock, W. G.: "Comparing Electromagnetic Servovalve Actuators for Hydraulic Power Control," *Control Eng.,* vol. 23, no. 5, 1976, p. 41.

Kandelman, A., and D. J. Nelson: "Simplified Model Eases Hydraulic System Simulation," *Control Eng.,* vol. 25, no. 4, 1978, p. 65.

Maskrey, R. H., and W. J. Thayer: "A Brief History of Electrohydraulic Servomechanisms," *ASME J. Dyn. Syst. Meas. Control,* June 1978.

Niemas, F. J., Jr.: "Understanding Servo Valves and Using Them Properly," *Hydraul. Pneum.,* vol. 31, no. 13, October 1977, p. 152.

Pluhar, K.: "Computer Control in Fluid Power—An Emerging Technology," *Control Eng.,* vol. 26, no. 10, 1979, p. 63.

Pluhar, K.: "High Performance Hydraulics Interface Electronics in Metal Forming and Simultaion," *Control Eng.,* vol. 27, no. 7, 1980, p. 77.

Smith, R. H.: "Direction-Control and Servovalves," *Mach. Des.,* vol. 42, no. 22, September 1970, p. 34.

Staff: "High Water Content Systems for Profit-Making Designs," *Hydraul. Pneum.,* vol. 35, no. 4, April 1982, pp. HP1–HP18.

Stegner, J. C.: "Servovalves for Mobile Hydraulics," Moog Inc., East Aurora, N.Y., 1980.

Walters, R.: "Hydraulic and Electro-Hydraulic Servo Systems," CRC Press, Boca Raton, Fla., 1967.

Ziebolz, H. W.: "Designing Hydraulic Servos," *Mach. Des.,* vol. 19, no. 7, July 1947, p. 123.

Fluidics

by
C. L. Mamzic*

Fluidics is the technology of sensing, controlling, and information processing with devices that use a fluid medium and whose operation is based solely on the interaction between fluid streams. The particular function of each device, none of which has moving parts, is dependent on the geometric shape of the device.

Fluidics is considered to have commenced in 1959 with the work of B. M. Horton, R. E. Bowles, and R. Warren, scientists at the Diamond Ordnance Fuze Laboratories (U.S. Army), Washington, D.C. Some of the principles employed in fluidics involve earlier discoveries by scientists such as Henri Coanda, a Rumanian engineer who in 1926 identified the wall attachment phenomenon for fluid jets—a principle now known as the Coanda effect.

Fluidic principles have been applied in the construction of amplifiers, oscillators, computing and logic elements, analog controllers, flowmeters, temperature proximity and dimensional gaging sensors, process control valves, and level controllers.

Characteristics

Some of the important characteristics of these devices include:

1. *Use of Any Fluid.* Almost any gas, liquid, or slurry may be utilized in fluidic devices. Any material that can be conducted through pipe is potentially amenable to control through these principles. Combinations of fluids may be employed in a single device, e.g., the use of a gas stream to control a liquid stream.

2. *High Amplification.* As an example of the high amplification obtainable with fluidic devices, the entire flow of liquid in a 4-in (~10-cm) diverting valve can be switched from one outlet to the other by opening (or closing) to the atmosphere a small control orifice slightly greater than ⅛ in (~3.2-mm) in diameter.

3. *No Practical Size Limitation.* Fluidic devices can be built in practically any size. There is no reasonable limit as to how large a process control valve can be. Conversely, miniaturized units, having fluid passageways on the order of 0.010 in (0.254 mm), are used in making high speed switches and logic elements.

4. *Wide Choice of Materials of Construction.* Almost any material that can be cast, molded, machined, formed, or etched can be used to make a fluidic device.

5. *High Response Speeds.* Oscillators have been made to operate at frequencies of 10,000 Hz. High speed dimensional gaging switches can detect and actuate a memory switch in 2 ms. The flow in a large 4-in diverting valve can be switched completely in 0.1 s.

6. *No Shock or Water Hammer.* A significant characteristic, particularly in a device used as a diverting valve, is the absence of shock or water hammer when the fluid stream is switched from one port to the other—even though this switching occurs at high speeds.

7. *Unaffected by Environment.* Since fluidic devices can be built of practically any material, the devices can operate under extremes of temperature, vibration, and radiation.

*Manager, Systems and Application Engineering, Moore Products Company, Spring House, Pa.

Principle of Operation

The simplest form of fluidic amplifier (Fig. 1) uses cross-directed fluid jets. The main jet can be deflected in proportion to the strength of a cross-directed control jet. If receiving apertures are placed downstream, the deflection manifests itself as a relative change in flow and pressure at the apertures. This type of amplifier works entirely on the momentum exchange principle and, therefore, is not very efficient.

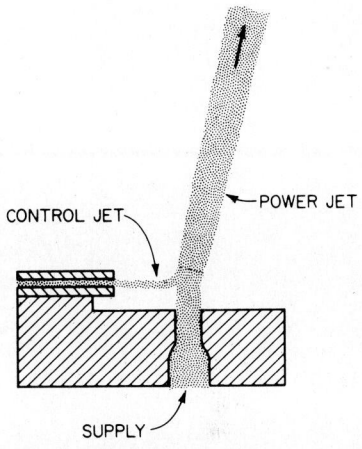

Fig. 1 Basic fluidic amplifier.

An additional principle employed in most fluidic devices is the *Coanda effect*, sometimes referred to as the *wall attachment principle* or *boundary layer effect*. To illustrate this principle, assume a device having a cross section as shown in Fig. 2 and top and bottom plates so that the downstream end and left side are open. An end view is shown in the figure. Assume this device is operating open to the atmosphere and that compressed air is supplied to the power nozzle. At the moment the supply air is turned on, the main jet tends to project in a straight line, as in Fig. 2a. Note that the jet, being rectangular in cross section, seals against the top and bottom plates and has the effect of a barrier or movable wall running parallel to the sidewall. Whenever a jet flows into a body of stagnant fluid, whether the jet flow is laminar or turbulent, it entrains some of the surrounding stagnant fluid and starts it into motion. In this case, as the ambient air is entrained and ejected along both sides of the main jet, replenishing air continuously moves into this region. Along the open wall, the replenishing air moves in, unimpeded, and the average pressure along this side is essentially atmospheric. However, along the right side of the jet, the replenishing air must flow down through the restricted opening between the wall and the jet boundary. The average pressure on the wall side, therefore, is somewhat below atmospheric.

The resultant differential in pressure across the two sides of the jet causes the jet to move closer to the wall (Fig. 2b). This, in turn, further restricts the passage down through which the replenishing air must move, making the pressure along the wall decrease further, while the differential across the jet correspondingly increases.

This action is regenerative and continues until it terminates with the jet attached to the wall (Fig. 2c). Under these conditions, a vortex forms in the region between the inside boundary of the jet and the wall. This region is also referred to as a *separation bubble*. The pressure in this region would be at a high vacuum in the example given, and the jet would remain attached to the wall because of the differential pressure impressed on it.

To detach the jet from the wall, it would be necessary to supply sufficient replenishing air to the separation bubble. This could be done by supplying the air through a control port, as will be shown in succeeding examples.

The main jet and the ambient fluid in this example need not be air. These principles apply even if the jet and ambient fluids are not air. Likewise, the jet and ambient fluids can be different from each other. In fluidic control valves, for example, it is common to control a liquid jet by means of ambient air. The Coanda effect, therefore, is the basis of the "memory" function obtainable in these devices, and the positive feedback it provides results in high amplification and much more efficient operation.

Bistable Amplifier

Figure 3a shows a cross section of a bistable amplifier, or flip-flop, with the main jet attached to the right outlet and both control ports closed to atmosphere. This unit is made in a sandwich form with top and bottom plates for sealing the jet. All openings are rectangular in cross section. In order to switch the jet, the right control port is opened, allowing replenishing air to enter. Immediately, the point of attachment of the jet begins to move outward (Fig. 3b), and the jet shifts to the left until it attaches to the left wall (Fig. 3c). Once it is attached to the left wall, the right control port can be

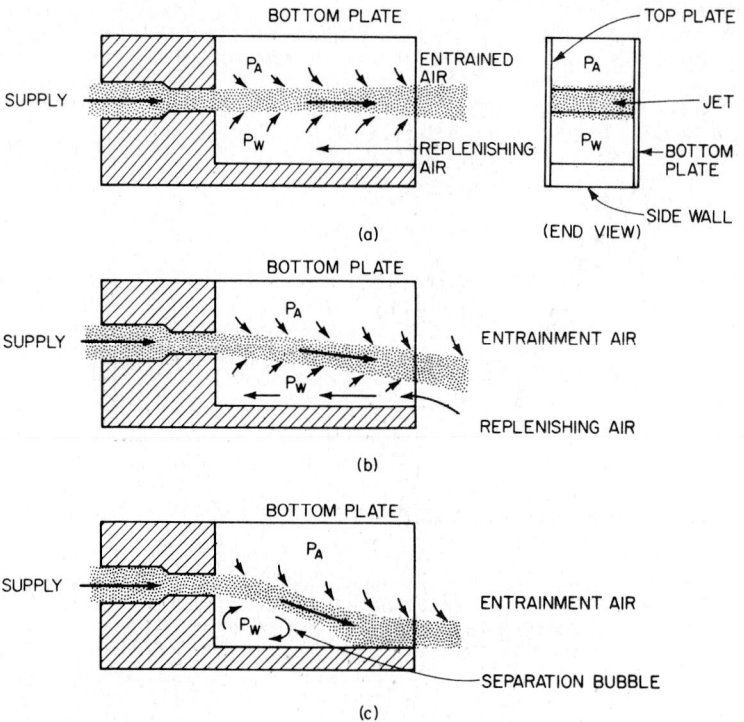

Fig. 2 Principle of the Coanda effect. (*a*) At the instant of start-up, the jet tends to move straight; (*b*) entrainment makes P_W drop, and the differential ($P_A - P_W$) begins to move the jet toward the wall; (*c*) the jet attaches to the wall and remains attached because of the difference between P_A and P_W.

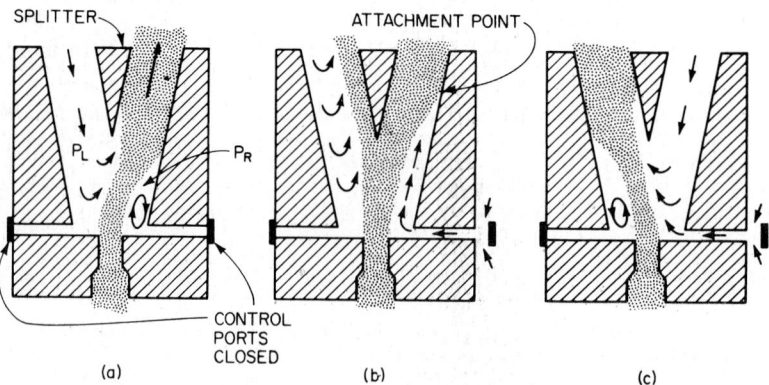

Fig. 3 Operation of a bistable amplifier. (*a*) A flip-flop is locked to right outlet; (*b*) the right control port is opened, and diversion of the power jet begins; (*c*) diversion is completed.

closed again and the main jet remains locked to the left wall because of the Coanda effect. This construction is the basis of a two-position diverting valve or a bistable logic element with memory.

Proportional Amplifier

To make a proportional amplifier, the divider of the unit shown in Fig. 3 is moved closer to the nozzle. This allows the operation to stabilize with the flows at the two outlets at any proportion to each other. Such a unit is controlled by varying the relative opening of the two control ports.

Oscillator

In the oscillator shown in Fig. 4, the instant the power jet switches to the right outlet, ambient air begins to flow counterclockwise through the circular feedback path. This airflow is preceded by propagation of a shock wave which moves at the speed of sound. When the shock wave front collides with the main jet, sufficient energy is transferred to cause the main jet to be diverted to the left outlet. The identical action then occurs when the succeeding shock wave travels down the right outlet and clockwise through the circular feedback path. The frequency of oscillation in this type of unit is a function of the length of the feedback path. As mentioned previously, oscillators of this type have been made to operate at frequencies above 10,000 Hz.

Low-frequency oscillators can be made by a different design in which part of the power jet flow through the outlets is alternately collected and fed back to a control port. The feedback flow causes the separation bubble to increase in size, shifting the point of attachment outward until the main jet switches.

Pulse Counter

This device, also known as a *flip-flop* or an *alternator,* is basic for counting and binary computation. In a pulse counter, each time an input pulse is received, the main jet flow switches to an alternate outlet. In Fig. 5, when the main jet is switched to the right outlet, a slight priming draft is set up, running counterclockwise through the heart-shaped feedback path. When a pulse signal appears, the pulse signal flow follows the path of least resistance and joins the priming draft flow. This, in turn, switches the main jet from the right to the left outlet. As long as the pulse signal is maintained, it remains locked to the right wall of the feedback path because of the Coanda effect. When the pulse signal is removed, the main jet remains locked to the left outlet, also by the Coanda effect. Once the main jet is switched to the left outlet, a priming draft develops in a clockwise direction, thus setting up the unit for the next pulse.

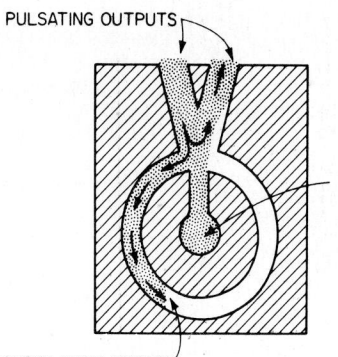

PULSATING OUTPUTS

SHOCK WAVE FRONT

Fig. 4 Fluidic oscillator.

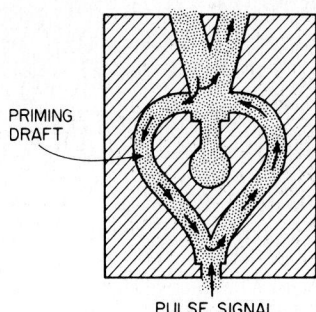

PRIMING DRAFT

PULSE SIGNAL

Fig. 5 Fluidic pulse counter.

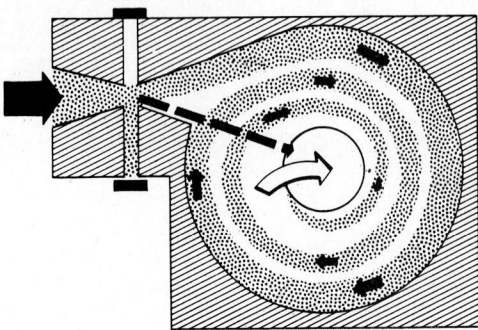

Fig. 6   Vortex amplifier.

Vortex Amplifier

In this device, shown in Fig. 6, if the upper control port is open, the power jet flow is diverted radially toward the outlet (dashed line) and the resistance to flow is minimal. If the lower control port is opened and the upper port closed, the power jet is aimed tangentially toward the cylinder wall and the fluid resists moving through the outlet because of the law of conservation of momentum. The flow, however, is forced to spiral inward toward the outlet and, as the radius decreases, the velocity increases, shear stresses in the fluid increase, and the pressure drop increases. Hence, the flow rate and pressure drop can be made to vary by changing the direction of the power jet.

Applications

In the early 1960s, much attention was directed toward fluidic logic systems, but interest waned in favor of electronic logic as extraordinary advancements were made through IC technology. Nevertheless, fluidic logic systems and proximity sensors have been applied in the control of machine tools and sequential manufacturing operations. Pulse counters are used in shift registers and counters. In areas other than logic, oscillators have been employed as temperature sensors and to provide timing functions. A fluidic oscillating spray nozzle is standard for the windshield washer system of several makes of automobiles. Among consumer products is a pulsating shower head. Major areas of development and reduction to practice have been in the field of process control valves and flow metering.

Fluidic Control Valves

These are available as two-position diverting and proportional diverting valves. They are similar in construction and operation to the bistable amplifier shown in Fig. 3. They can be applied to control gas, liquid, or slurry streams. The low-energy controlling fluid can be atmospheric air, gas, or liquid.

Figure 7 shows a 4-in (10.2-cm) two-position liquid diverting valve. It operates by opening control ports to the atmosphere. The control ports can be as small as $\%_4$ in (3.6 mm). All the flow can be made to switch from one outlet to the other in $\%_0$ s, and the switchover occurs with no shock or water hammer.

The most common application for fluidic valves has been in controlling the flow of cooling water to the radiators on diesel-electric locomotives.

Fluidic Flowmeter

A fluidic oscillator forms the basis of a unique in-line flowmeter which has the following advantages:

1. Linear output with the frequency directly proportional to the flow rate
2. Rangeability up to 30:1
3. Unaffected by shock, vibration, or field ambient temperature changes

Fig. 7   A 4-in two-position diverting valve. *(Moore Products.)*

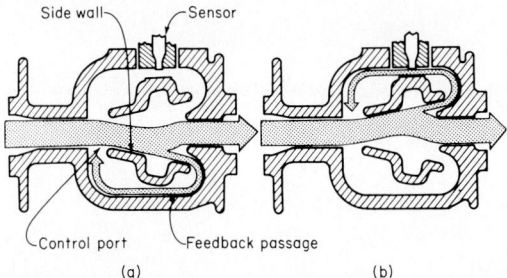

Fig. 8   Operating principle of a fluidic flowmeter. *(Moore Products.)*

4. Calibration in terms of volume flow unaffected by changes in fluid density

5. No moving parts, and no impulse lines

   A cross section of the meter is shown in Fig 8. The geometric shape of the meter body is such that when flow is initiated the flowing stream attaches to one of the side walls as a result of the Coanda effect. A portion of the main flow is diverted through a feedback passage to a control port. The feedback flow increases the size of the separation bubble. This peels the main flow stream away from the wall until it diverts and locks onto the opposite wall, where the feedback action is similar. The frequency of the self-induced oscillation is a function of the feedback flow rate, which in turn is directly proportional to the flow rate of the mainstream.

   A heated thermistor sensor is located in one of the feedback passages. When the mainstream is diverted toward the sensor, the feedback flow, and hence the removal of heat, is maximum. When the mainstream is diverted to the opposite wall, the fluid in the area around the sensor is stagnant. The fluctuation in voltage required across the sensor thus is synchronized with the oscillating frequency of the meter. Integrating these electric pulses gives a signal proportional to the instantaneous rate of flow. The pulses can also be totalized to give an accurate reading of the total flow.

   A similar flowmeter, but without diverging sidewalls, using momentum exchange instead of the Coanda effect, has proved effective in metering viscous fluids such as heavy fuel oils.

REFERENCES

ASME: *Fluid Jet Control Devices,* American Society of Mechanical Engineers Symposium, New York, 1962.

ASME: *Advances in Fluidics,* American Society of Mechanical Engineers Fluidics Symposium, Fluid Amplifier Associates, Ann Arbor, Mich., 1967.

ASME: *Proceedings of the 20th Anniversary of Fluidics,* American Society of Mechanical Engineers Symposium, Chicago, Ill., 1980.

BHRA: *Proceedings of the Cranfield Fluidics Conference,* British Hydromechanics Research Association, 2d Conference, Cambridge, England, January 3-5, 1967.

Humphreys, E. F., and D. H. Taramoto: *Fluidics,* Fluid Amplifier Associates, Ann Arbor, Michigan, 1967.

Kirshner, J. K.: *Fluid Amplifiers,* Fluid Amplifier Associates, Ann Arbor, Mich., 1967.

Kirshner, J. K.: *Design Theory of Fluidic Components,* Academic Press, New York, 1976.

Schaedel: "Fluidic Elements and Networks" (in German), Vieweg & Son, Braunschweig and Wiesbaden, 1979.

U.S. Army: *Proceedings of the Fluid Amplification Symposium,* Diamond Ordnance Fuze Laboratories, Washington, D.C., October 1967.

U.S. Army: *Fluidic State of the Art Symposium Proceedings,* Harry Diamond Laboratories, Washington, D.C., 1974.

Programmable Controllers

by
Bruce R. Rusch*

Programmable controllers (PCs) have become increasingly popular as control tools for industrial applications since their inception in 1969.† Because of their power, flexibility, and ease of use they have won acceptance by design engineers, operators, and maintenance personnel.

Functions of a PC

According to National Electrical Manufacturers' Association (NEMA) standards, a PC (Fig. 1) is a digital electronic device that uses a programmable memory to store instructions and to implement specific functions such as logic, sequence, timing, counting, and arithmetic to control machines and processes.

For example, unlike numerical control (NC) and computer numerical control (CNC) units, which are used to control position, the PC is used for sequence control. And, unlike the computer, the PC is designed specifically to interact directly with the industrial environment. In operations that require periodic tool changes, the PC eliminates the need for a separate control system. To change assembly line operations controlled by a PC, new instructions are written into the PC's internal memory by pushing a few buttons on a portable programming panel (Fig. 2). The task requires no special computer or computer programming skills.

PCs have other advantages: (1) They are rugged, solid-state equipment that can endure the industrial environment; (2) they have no moving parts, which eliminates many maintenance prob-

*President and General Manager, Gould Inc., Programmable Control Division, Andover, Mass.

†*Editor's Note:* Since the major thrust of programmable controllers occurred after publication of the last edition of this Handbook and because of the exceptional acceptance of PCs during the past decade, it is fitting to include a concise history of the origination and early years of the PC. This is given at the end of this article.

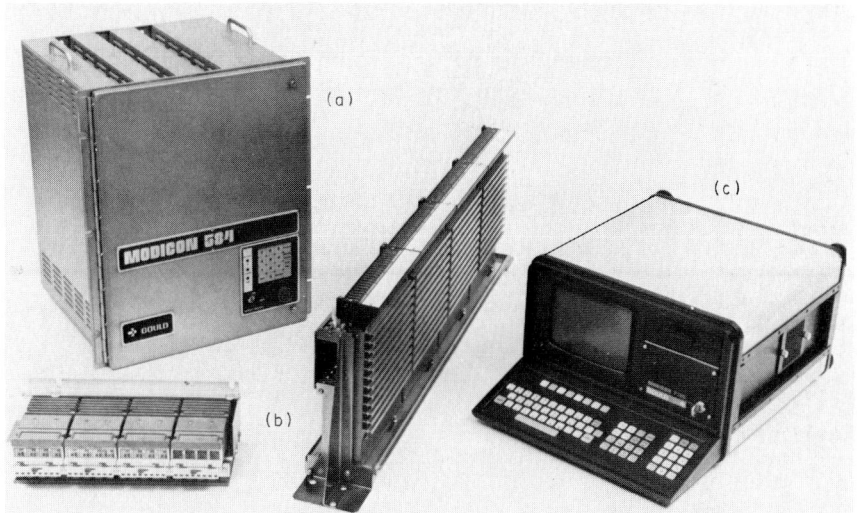

Fig. 1 Principal elements of a programmable controller. (*a*) Processor unit; (*b*) input-output modules; and (*c*) programming panel. *(Modicon 584.)*

Fig. 2 Portable programming panel. *(Modicon.)*

lems; and (3) in an era of intense pressure on profit margins, they are cost-effective. Moreover, PC technology does not require production management and maintenance personnel to learn a computer language. Ladder logic, used for years by electricians and electrical maintenance personnel, has been adapted for PC programming and troubleshooting.

While the PCs available in the early 1980s are designed for a wide variety of tasks, the most powerful models offer capabilities such as:

Relay logic	Arithmetics
Latches	Comparison
Timing	Computer interface
Counting	Matrix manipulation
ASCII interface	Binary-coded decimal (BCD)
Proportional-integral-derivative (PID) loops	Binary conversion
Shift register	Analog data manipulation
Data highway communications	Printer interface

Principle of Operation

All PCs contain a central processing unit (CPU), a memory or established program, a power supply, input-output modules (Fig. 3), and a programming device for programming, editing, and troubleshooting.

On receiving instructions from the memory, along with the feedback on the status of the input-output devices, the central processing unit (CPU) generates commands to the output module to control devices such as relay coils, solenoid valves, indicator lamps, and motor starters.

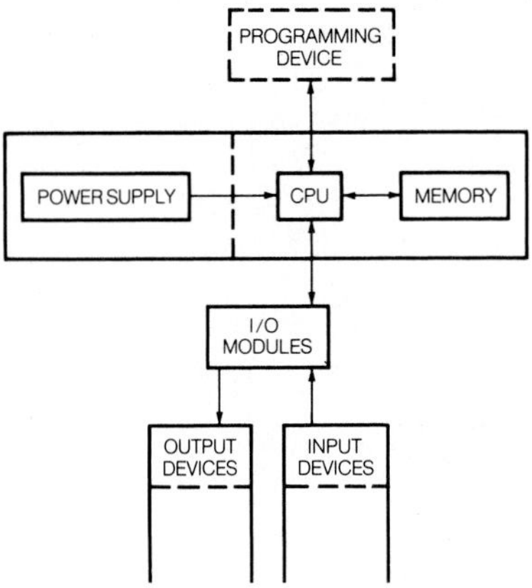

Fig. 3   Basic elements of a programmable controller.

PC versus Relays

In today's industrial environment, PCs compare very favorably with computers and other types of controllers, especially in their originally intended application of strictly replacing relays and relay logic. Before development of the PC, the user designed a relay panel or panels and had them built, wired, and checked out. Any modifications in the relay logic involved redrawing, rewiring of relays, and rechecking. In the event of failure, troubleshooting was difficult.

For example, in the automobile industry, at the beginning of each year's production, model changes required changes in the relay logic used with production machines. Auto manufacturers actually found it less costly to build new cabinets than to attempt to save the older ones. In contrast, at the beginning of a new model year—with a system using PCs—only the ladder diagram has to be reprogrammed in most cases. The PC can be used again and again with different logic.

PCs can be programmed with the user's ladder diagram logic after the wiring is completed because the program is not highly dependent on the wiring. The program also can be stored on magnetic or paper tape for direct insertion into several systems or into a spare system. In the case of a malfunction, status lights on the PC inform the operator which part of the system is malfunctioning.

Evolution of PCs

A second generation of PCs appeared in 1973, and since that time additional models have been developed. Each generation of PCs has incorporated new design features, including higher speeds and a greater logic capacity, among other features for making the PC increasingly competitive with modern, sophisticated control hardware.

The PC has encroached significantly on applications that once were considered exclusive to the minicomputer. Although the fast number crunching and powerful text handling capabilities of the minicomputer continue to make it attractive for some applications, the PC enjoys unique advantages even in large systems. The logic solving ability of the PC easily outstrips that of the minicomputer. Also, it is much easier to interface discrete external devices, such as limit switches, with a PC than with a minicomputer.

PC-Computer Comparison

A few key characteristics make the PC unlike a computer or other types of controllers. First, the PC is designed to communicate directly with the process to be controlled (i.e., inputs from the process and controlling outputs to the process are wired directly to the PC system), and the PC recognizes these inputs and outputs as part of its system at a fixed address. The inputs are made up of limit switches, pushbuttons, thumbwheels, pulses, analog signals, and ASCII serial data. The outputs are voltage or current levels which drive end devices such as solenoids, motor starters, relays, and lights, among others. Other outputs are to analog devices, digital BCD displays, ASCII-compatible devices, and computers.

The second basic difference between PCs and computers or other controllers is the ease of programming. The PC uses simple programming techniques that are easily understood by a plant technician or electrician. Simple relay ladder diagram programming does not require a knowledge of FORTRAN, BASIC, or any other computer language, nor does it require the use of Boolean or other logical expressions—although these can be easily converted into a relay ladder diagram format if desired. See Fig. 4.

The PC can be programmed and reprogrammed on-line while the process is running and without hardware modifications.

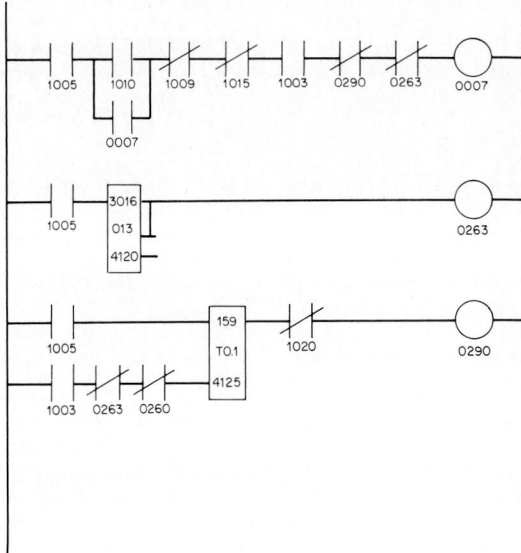

Fig. 4 Typical ladder logic programming language used in a programmable controller (PC) circuit design employs the same language used daily by control engineers and designers. Similarly, PC maintenance and troubleshooting involve logic diagrams already familiar to electricians and electrical maintenance personnel. Thus, there is no requirement for personnel to learn complex programming procedures or new languages.

The third, possibly most important, difference between PCs and computers is that PCs are designed for the demanding industrial environment. A well-designed unit allows location of the PC in a high-noise high-vibration high-temperature high-humidity (noncondensing) environment without affecting its operation.

It is suggested that the three foregoing factors are largely responsible for the wide acceptance of PCs—they require no special programmers, no air-conditioned rooms, and no special input-output systems to be designed.

Components and Configuration

The basic configuration of the PC has undergone relatively little change in the past decade. On smaller PCs, the processor unit and the modular input-output subsystem have been combined into a single, compact unit.

Processor Unit

In a PC, the processor unit contains the CPU, memory, peripheral interfaces, and often the power supply. To protect these components from electromagnetic interference (EMI), the processor is frequently housed in a metal enclosure.

The CPU is responsible for solving relay ladder logic as well as performing timing and counting, arithmetic, PID loop control, and other advanced functions. In accomplishing these functions, the PC uses a scanning technique, i.e., it sequentially executes the operations represented by the rungs of the relay ladder. After inputs are read, the information is used in the solution of the relay ladder logic, and outputs are set accordingly.

Modern PCs usually employ a microprocessor-based microcomputer as a CPU. To obtain increased performance, a multiprocessor design—one incorporating multiple microprocessors (often two)—is sometimes used. Here, the PC's tasks are divided between the two processors. For example, one may be used to solve relay ladder logic and special program functions, while the other is assigned the job of communicating with the input-output. A few PCs utilize CPUs implemented with bit-slice designs, which is another technique for obtaining higher performance than can usually be achieved with a CPU based on a discrete microprocessor.

PC Memory*

Memory is an important part of a PC. It stores the PC's executive program as well as the user's applications program. The latter instructs the PC what to do; the former how to do it.

In designing a PC memory system, the objective is to provide the user with a programmable nonvolatile memory. This means that the contents of memory (i.e., the applications program) can be repeatedly changed by the user, but memory contents are retained, even if electric power is lost, until they are deliberately changed.

There are basically three types of memory that can be used to provide these characteristics: (1) magnetic core, (2) battery backed-up random-access memory (RAM), and programmable read-only memory (PROM).

Magnetic Core Memory

In a magnetic core memory, information is stored in tiny ferrite cores that are magnetized with one of two polarities representing binary digits. Core memory is nonvolatile—memory contents remain unchanged when electric power is removed. However, memory contents can be changed easily by the user, since core is also a read-write memory. This means that memory can be repeatedly read or written to. Originally used more extensively than it is today, magnetic core memory is being displaced by a variety of semiconductor memory types.

---

*See also the article Digital Technology in Sec. 14 for descriptions of various types of memory.

RAM

This feature is a type of semiconductor memory and, like core, offers a read-write capability. Unlike core, however, RAM is usually volatile—if electric power is removed from the memory, its contents are lost. To prevent this, RAM is usually available with a battery back-up power system. If the main power is lost, the memory contents are supported by battery power, with switchover occurring automatically. Low cost and small size are the main advantages of RAM.

PROM

This is another type of semiconductor memory. It is nonvolatile, like core. However, as the name implies, it has a read-only capability. Generally, after they have been programmed, memory contents cannot be changed.

Some types of PROM can be erased and reprogrammed. Erasure of memory contents usually entails exposing the memory to ultraviolet light or the application of a special electric signal. After erasing, memory contents can be programmed anew. Like RAM, PROM is relatively inexpensive and compact, but reprogramming is not necessary, as it is part of the executive instructions.

Memory Capacity

The memory capacities of early PCs were small, while some large contemporary PCs can support up to 128K (1K = 1024 words) words of memory. Typically, memory size can be expanded by adding memory modules in increments of ¼K, ½K, 1K, 2K, 4K, or 8K. The amount of memory required for a particular application, of course, depends on the size of the applications program and the amount of input-output used.

Memory Protection

To protect memory contents from being changed by unauthorized personnel, PC processor modules are frequently equipped with a memory-protect key switch. When activated, the memory-protect feature prevents memory contents from being altered, except as a result of execution of the applications program.

Operator Interface

Users must have a means of interacting with the PC. Methods available include a (1) programming panel, (2) cassette tape loader, (3) hand-held monitor, and (4) register-access panel (RAP).

Programming Panel

As shown in Fig. 2, programming panels are relatively compact devices that allow the user to enter the applications program into the PC by using relay ladder logic language. Since it is a highly visual language, a programming panel provides a means to view the relay ladder logic program. Some designs incorporate a video display which allows approximately 70 contacts to be viewed at one time. Other programming panels are equipped with simple displays that allow a single rung of the relay ladder to be displayed at one time.

There is a keyboard on the programming panel for the operator to enter the elements of the relay ladder logic. The panel also usually supports additional functions. For example, it can display the amount of unused memory left in the PC and the number of output coils that have been programmed, as well as other information. Often, a programming panel can be used to monitor the operation of a PC. Some PCs allow the programming panel to change an applications program in the PC's memory while it is running.

Cassettes

To avoid the chore of manually entering a PC program repeatedly, most PC systems allow the applications program to be recorded on cassette tape. Special cassette tape player–recorder systems are available for this purpose; some programming panels have tape facilities built into them.

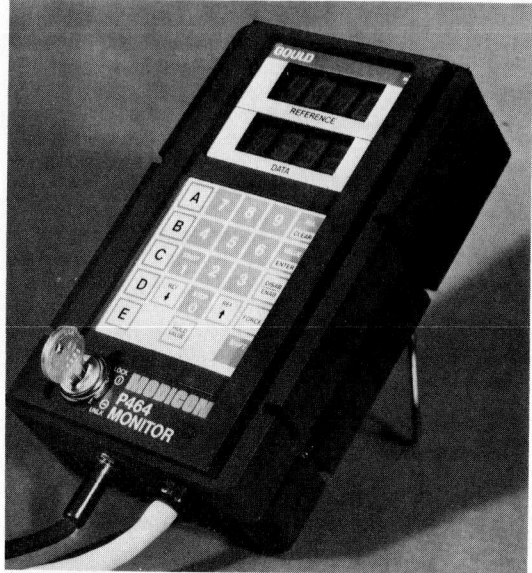

Fig. 5 Hand-held monitor for a programmable controller. *(Modicon.)*

Hand-Held Monitor and RAP

Two relatively recent developments in the area of user-PC interfacing are the hand-held monitor (Fig. 5) and the RAP. These devices perform basically the same functions. The RAP is built into the PC, while the hand-held monitor is a separate device that plugs into the PC.

These devices allow certain basic operations to be performed without the use of a programming panel. For example, they can be used to verify the operation of input-output modules and field devices. They also provide the operator with access to the dynamic status of inputs, outputs, and registers and allow discrete input-output to be forced into a desired state. And, if allowed by the memory-protect switch, they can be used to alter the contents of holding registers.

Function Block Programming

Electronic processing elements within the PC have the capability to support advanced functions. However, a simple technique is required to program these functions and to interface them with the relay ladder logic structure. Function block programming technique is a simple means of representing functions (often complex) that are compatible with the relay ladder logic format.

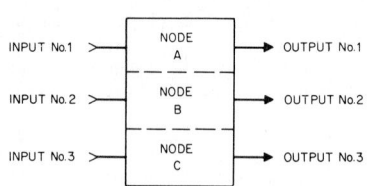

Fig. 6 General architecture of a function block. The numbers of nodes, inputs, and outputs vary with the type of function block.

The general architecture of a function block is shown in Fig. 6. The number of nodes, inputs, and outputs vary with the type of function block and are generally independent of one another. The type of function block is usually identified by a legend that appears within the body of the function block when it is shown on the video display of a PC programming panel.

Each node of a function block is used to specify a numerical value, although these values may have different representations, depending on the type of function block and on the particular node within the function block. Specifically, the value represented by a node can designate (1) a numerical constant to be used as an operand in the function, (2) the address of a register in which the result of the operation is to be stored or in which an operand for use in the function is stored, or (3) a numerical parameter that governs a certain characteristic of the function block's operation.

Relay ladder logic interfaces with the function block via the function block's inputs and outputs. The inputs of a function block are typically activated by relay contacts and are used to control various aspects of the function block's operation that have two states. For example, the function block's inputs can represent the following:

"On" state	"Off" state
Perform function	Do not perform function
Reinitialize function	Do not reinitialize function
Do not increment parameter	Increment parameter

As a result of the operation of a function block, its outputs are set to either the "on" or the "off" state. The function block's outputs can be interfaced with contacts, coils, or even the inputs of other function blocks. The latter technique can be used to implement quite complex functions. See Fig. 7.

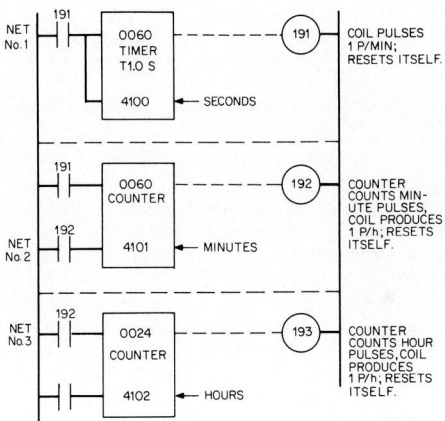

Fig. 7 Sophisticated functions can be implemented by cascading function blocks. Shown here is the relay ladder logic for a real-time clock with signals for minutes, hours, and days and register addresses where these values are stored.

The status of the outputs of a function block take on special meanings in association with the control application, although they have very definite meanings with regard to the function block itself. For example, the outputs of a function block can signify:

"On" state	"Off" state
Function complete	Function not complete
Zero result obtained	Nonzero result obtained
High alarm limit reached	High alarm limit not reached

Table 1 Function Block Types Supported by Modern PCs

Timers and counters
    Timers (1.0-, 0.1-, 0.01-s time base)
    Counters (count up, count down)
Arithmetic
    Addition
    Subtraction (numerical comparison)
    Multiplication
    Division
Data transfer
    Register to table
    Table to table
    Table to register
    Block move
    First-in–first-out (FIFO) queuing
    Table search
Matrix
    Logical AND, OR, XOR, NOT
    Comparison
    Bit sense, set, clear, rotate
Special functions
    Read or write ASCII messages
    Proportional-integral-derivative (PID) analog loop control

---

Over the years, a reasonably comprehensive library of function blocks has been assembled for the PC. A typical large, modern PC offers a variety of function blocks. See Table 1. The assortment of available function blocks will increase as the manufacturers of PCs enhance the capabilities of their products, particularly in connection with new kinds of applications.

Many of the functions listed in Table 1 represent relatively complex operations. When the PC performs these operations, it actually executes a series of simpler operations. The combination of these simpler operations is the equivalent of a function block. Thus, like relay ladder logic, function block programming is a true high-level language.

The nature of function block programming is also intuitively understandable; a function block can be regarded as a "black box" operation by the user. That is, the user specifies what function is to be performed—but not how—and provides the required operands and function-activation conditions. After the operation is performed by the PC, the result is made available for further processing.

On the whole, the function block programming technique has proven to be a powerful yet concise and simple means for extending the utility of relay ladder logic. Importantly, it does so without sacrificing the traditional virtues—most notably ease of use—that have made the PC so popular. As a result, the versatility of the PC has been enhanced, allowing it to be used in a wide variety of new applications. Of these, the process industries probably represent the largest potential area for new uses of PCs.

Process Control Applications

Prior to the entry of the PC into process control, two basic techniques of process control were available. These involved the use of (1) dedicated analog controllers and (2) computer-based controls. The former are relatively easy to use, but somewhat limited in capability. Computers offer more flexibility and computational ability, but they are more difficult to use. It is the PC's unique combination of bit- and number-oriented functions that makes it a prime candidate for process control applications. The PC's ability to handle operations that combine binary (open-close, on-off, in-out) and analog functions (time, set point, position, level) offers great flexibility in the implementation of process control systems.

PID functions supported by PCs can be used to regulate temperature, flow rate, volume, velocity, or position in process control applications. Notably, nearly all process parameters used by a PID control loop can be modified by the PC during the course of the process. Other features of a PC's PID capability include bumpless transfer, reverse and direct acting modes of control, antiwindup protection, high- and low-alarm output, and the use of engineering units with programmable scales. The processing time for a loop can be as little as 1.5 ms.

ASCII Message Handling

Some contemporary PCs have the ability to store and transmit ASCII data that represent alphanumeric characters. Thus, the PC user is provided with a means of configuring an applications program that can generate English language messages based on some predefined condition and shown on a video display or printout.

Suitable applications for this capability are displaying instructions or prompts for the operator of a PC-controlled machine, generating alarm messages, and printing out tags for defective parts that specify the nature of the fault, among others. Almost any application where information derived from the process being controlled must be communicated to people can benefit from the PC's ASCII message handling capability.

PCs make extensive use of numerical data bases for storing and calculating values for use with timers, counters, and analog functions, among others. Depending on the sophistication of the data functions available, the data base may also be used to construct production-oriented messages for management.

A large variety of data can be accumulated in the PC's memory and retrieved periodically for use in management and production reports, such as the number of batch cycles, quantities of products, profiles of temperature cycles, operating times for motor and valves, the number of on-line parameter changes, and so on. Using the PC's data transfer capabilities, this information can be routed to a printer to produce a hard copy report, transferred to a computer for further processing, or communicated to another PC for use in another part of the process being controlled.

Outlook

It is envisioned that PCs of the future will continue to incorporate retentive memory, the relay ladder diagram format, a separate input-output section, modular maintenance, a friendly interface with people, and logic solved in scans. It is expected that PCs will gain in sophisticated capabilities and compete effectively against minicomputers in the industrial environment. Stock inventories, maintenance diagnostics, report generating, and maintenance scheduling will most likely become more commonplace functions for PCs. Already available are PCs that can use ASCII devices, such as the teletype, line printer, cathode ray tube (CRT) terminal, and so on. Some PCs incorporate parallel processing and mass memory, as well as the ability to dynamically alter the logic. These features provide minicomputer capabilities in a relay ladder diagram PC. New devices, such as microprocessors, will continue to be incorporated into PCs, but it is unlikely that the basic structure will change. There will be increased use of complementary metal oxide semiconductor (CMOS) memory, and acceptance will be based on longer battery back-up (2 to 5 years) with lower costs and faster scan rates. New capabilities will include local communications networks to connect PCs in an industrial environment, including peer-to-peer arrangements and master-to-slave arrangements with a computer as the "master" polling the "slave" PCs.

Historical Footnote

The original PC was built by hand at Bedford Associates. It was a "hard hat" special configuration of a computer built for specific jobs. Its language was not developed until a decision was made to manufacture the unit commercially. In 1969, a large U.S. auto manufacturer (General Motors) purchased the first PC. It was designated the "084" because it was the eighty-fourth project undertaken at Bedford Associates. The original PC embodied many features still found today on most PC units.

At that time, a capacity of 256 words was more than adequate, but later the PC was expanded to a 1000-word machine. It had a noninterrupt structure that was quite novel at the time and a 16-bit word, and was a dedicated machine. Its input-output structure was directly accessible by user software. Seeking a language that would be compatible with industry, the developers decided to use ladder listing, a language started in Germany and known throughout the world by people of all languages. Another decision the original PC designers faced involved the type of memory. Vendors were attempting to sell high speed, highly susceptible memory planes, but the application called for slow memories with big cores, as they had to be immune to electrical interference. Core memory initially solved the problem.

The early machine with 1000 words had a relatively low speed, and it could not do arithmetic—only relay logic. A program panel was added, along with higher speed and improved input-output structure. The most difficult part of the design involved the logic solver. When the program first started, the logic solver and the PC consisted of a computer and memory—input-output. But, the computer had difficulty executing ladder logic quickly. In contrast, present PCs are essentially divided into three major parts—a general purpose computer, a logic solver, which is a special purpose piece of hardware used to solve and scan the ladder listing program, and a memory.

Another early challenge was the input-output modules, which had to be reliable and directly connected to the program. Much effort was expended in designing the input-output structure, including its input-output cards and input sensors and output triacs for driving 110 V. The objective was to have the user perceive the PC as a device just as reliable as relays but substantially easier to program and start up.

REFERENCES

Jannota, K. L.: "What is a PC?" *Instrum. Controls,* February 1980.

Loomis, R. M.: "Programmable Controllers in Automated Industrial Systems," *Ind. Eng.,* September 1982.

Staff: "Programmable Controllers from A to Z," *Assemb. Eng.,* May, June, July, and August 1982.

Time-Base Controllers

A very significant percentage of the instruments and control systems used for process and machine control, communications and information processing, appliance operation, navigation and traffic control systems, geophysical and other scientific measurements, and medical, biological, and other areas of scientific research require a time-base reference (1) certainly for all phenomena that are time-related, (2) to ensure proper time coordination of the information or control system per se, and (3) to operate equipment in a time-sequenced manner where each of a series of operations must be performed in its proper order and for a specified interval of time. In many instances the time base of the instrument is practically taken for granted, as in a recording instrument where one axis almost always is time. The *clock,* the device that generates periodic signals used for coordination of a synchronous computer, exemplifies the need for a time base by the instrumentation system proper.

Sequencing does not always connote the need for a time base, however, as in the use of a large transfer machine for machining automotive parts or in the operation of various conveyor systems. An *event-oriented* system can proceed without a time base through the use of transducers and limit switches which cause action on the completion of various tasks and movements. Systems also may be *count-oriented,* whereby a next sequence of events may occur on satisfying a counter without reference to a time base. Even in such systems, however, it is not unusual to introduce a time base somewhere along the line to ensure that not only *quantity* but also *rate* objectives are being satisfied and, in other instances, to protect equipment. See the article Programmable Controllers in this Handbook section.

Some processes may operate quite satisfactorily without a time base, as for instance a domestic hot water supply system which may be analogous to some industrial systems. But unlike the householder who simply may watch the monthly utility billings, industry desires to calculate costs on a much finer basis, requiring the generation of measured information on relatively short time-base intervals.

With the exception of the more obvious cases, the presence of time measurement and time actuation in instruments and control systems often is passed over lightly because the feature is built into the original equipment. This section will cover the more obvious situations in which timers and time-base controllers are considered *separate,* important elements of the total instrumentation systems.

Functional Classification of Timing Devices

From an overall functional viewpoint timing devices fall into two basic groups:

1. *Simple Measurement of Elapsed Time.* These devices measure and indicate and/or record the passage of time. Watches, clocks, and chronometers of all types fall into this classification. The configurations are numerous—as, for example, the stopwatch which can have an industrialized counterpart.

2. *Time-Base Controllers.* These *control* one or more functions on the basis of the passage of time. They require, of course, time measurement devices of the type described in category 1. As will be noted later, time measurement devices may be quite different from what one normally associates with conventional clocks and timepieces.

Basic Elements of Timing Devices

As shown in Fig. 1, there are four basic elements of a complete timing device for accomplishing the foregoing functions. A fifth element, display, usually is optional.

TIME MEASUREMENT SYSTEMS

Time reference systems for clocks include the *frequency rate* or *cycles* of (1) the earth, (2) a pendulum, (3) a balance wheel as in a watch, (4) controlled alternating electric current, (5) a tuning fork, (6) radioactive decay, and (7) the oscillations of an atom of cesium 133. *Time is what a clock reads.* There is no absolute time standard.

Atomic Clocks

Atomic time is based on the atomic second, officially defined in 1967 by the International Bureau of Weights and Measures as 9 192 631 770 oscillations of an atom of cesium 133. The value expresses the ephemeris time second as closely as possible in terms of an atomic standard and was derived by a joint experiment performed by the United States Naval Observatory and Great Britain's National Physical Laboratory, using a dual-rate moon position camera and a cesium beam clock. Corrections to the frequencies transmitted by the National Bureau of Standards (NBS) ratio stations are determined with regard to the NBS atomic standard and are published monthly *(Proc. IEEE).* The stations' carrier and modulation frequencies are offset intentionally from the atomic frequency stan-

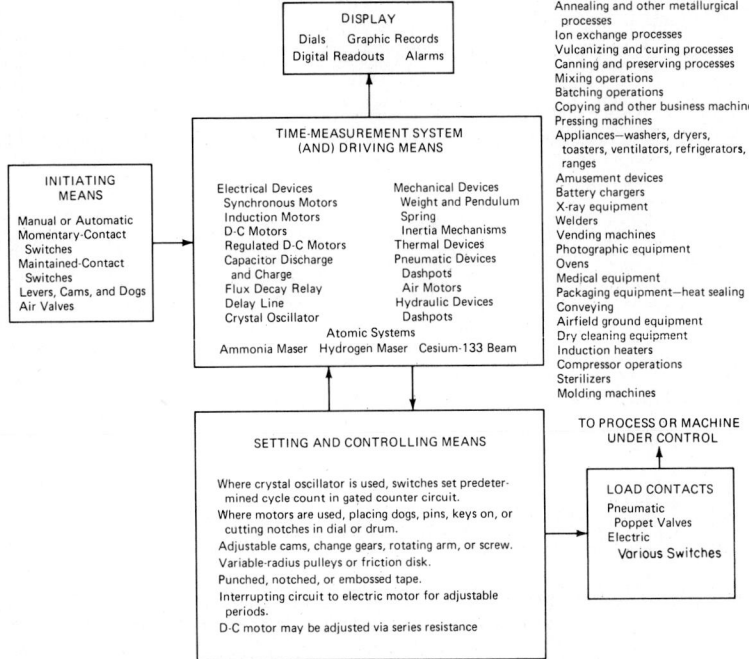

DISPLAY
Dials   Graphic Records
Digital Readouts   Alarms

**TIME-MEASUREMENT SYSTEM
(AND) DRIVING MEANS**

Electrical Devices	Mechanical Devices
Synchronous Motors	Weight and Pendulum
Induction Motors	Spring
D-C Motors	Inertia Mechanisms
Regulated D-C Motors	Thermal Devices
Capacitor Discharge	Pneumatic Devices
and Charge	Dashpots
Flux Decay Relay	Air Motors
Delay Line	Hydraulic Devices
Crystal Oscillator	Dashpots

Atomic Systems
Ammonia Maser   Hydrogen Maser   Cesium-133 Beam

**INITIATING
MEANS**

Manual or Automatic
Momentary-Contact
  Switches
Maintained-Contact
  Switches
Levers, Cams, and Dogs
Air Valves

Annealing and other metallurgical
  processes
Ion exchange processes
Vulcanizing and curing processes
Canning and preserving processes
Mixing operations
Batching operations
Copying and other business machines
Pressing machines
Appliances—washers, dryers,
  toasters, ventilators, refrigerators,
  ranges
Amusement devices
Battery chargers
X-ray equipment
Welders
Vending machines
Photographic equipment
Ovens
Medical equipment
Packaging equipment—heat sealing
Conveying
Airfield ground equipment
Dry cleaning equipment
Induction heaters
Compressor operations
Sterilizers
Molding machines

SETTING AND CONTROLLING MEANS

Where crystal oscillator is used, switches set predeter-
  mined cycle count in gated counter circuit.
Where motors are used, placing dogs, pins, keys on, or
  cutting notches in dial or drum.
Adjustable cams, change gears, rotating arm, or screw.
Variable-radius pulleys or friction disk.
Punched, notched, or embossed tape.
Interrupting circuit to electric motor for adjustable
  periods.
D-C motor may be adjusted via series resistance

TO PROCESS OR MACHINE
UNDER CONTROL

LOAD CONTACTS
Pneumatic
  Poppet Valves
Electric
  Various Switches

Fig. 1   Basic elements of timing devices. The initiating or starting means may be manual or automatic from an external source. The time measurement and driving means determine the limits of the range and the general accuracy of the timer. The setting and controlling means are the parts of the device which actuate the load contacts and provide the adjustments for obtaining the desired time intervals and complexity of cycling and sequencing operations. The load contacts provide the link between the timer and the process, machine, or other operation it may be controlling. Aside from a setting dial, many timers used industrially and for appliances do not incorporate a separate display.

dard by a precisely known amount so that, on average, the time signals broadcast remain as close as possible to universal time (UT). In 1966, 1967, and 1968, the offset was -300 parts in 10^{10}.

All atomic clock systems involve the use of atoms or molecules in certain energy states that are exposed to microwave energy to permit measurement of the quantum changes in the energy states of atoms. Even though highly accurate, such systems do not maintain strictly uniform time under all conditions because their *running rates* (as predicted by the theory of relativity) are affected by their position relative to the sun and moon, latitude, and height, and by the gravitational field in which they are placed or through which they move.

Rubidium gas cell systems now serve as secondary standards, like quartz crystal oscillators, because they must be referenced to more accurate systems. Rubidium systems have a characteristic resonance at 6835 MHz and are useful because, unlike other atomic frequency standards, they require little power and are relatively compact. Portable rubidium atomic clocks weighing as little as 44 lb (20 kg) and occupying about 1 ft^3 (0.028 m^3) were introduced by the U.S. Army Electronics Command in 1965. These clocks operate on standard 110-V current, the 24-V output of military vehicles, or both. They are used to set precise radio broadcasting frequencies, to synchronize radar nets, and to assist in the accurate tracking of missiles and satellites. Exceptional accuracy is required by tracking stations following a satellite moving at 25,000 ft(7620 m)/s in an earth orbit, because a time error of $\frac{1}{10}$ s is equivalent to a position error of about $\frac{1}{2}$ mi (0.8 km). Other atomic frequency standards include (1) gas cell frequency standards using alkali metals such as sodium, cesium, and rubidium, (2) thallium beam devices, and (3) the hydrogen maser, first developed at Harvard University in 1960.

Crystal Oscillator

The most precise *mechanical* resonator and frequency generator available is the crystal oscillator. The quest for a stable, accurate, low-cost frequency generator for the precise control of commercial radio and other higher communications frequencies led to the development of piezoelectric resonators—notably quartz crystals for a wide variety of applications. Also, the quartz crystal has become highly developed as a frequency standard for timekeeping and as a time signal generator. Piezoelectricity was discovered by the Curie brothers in 1880. The term "piezo" is derived from the Greek word meaning "to press." Piezoelectricity causes a crystal to exhibit electric polarity when it is subjected to mechanical pressure. Conversely, it is physically deformed when subjected to an electric potential. Specifically, piezoelectricity is a property of nonconducting solids that have a crystal lattice which does not have a center of symmetry.

Quartz, tourmaline, Rochelle salts, and such synthetic crystals as ethylenediamine tartrate (EDT), dipotassium tartrate (DKT), and ammonium dihydrogen phosphate (ADP) have varying suitability as piezoelectric elements. Tourmaline, an expensive material used mainly in hydrostatic pressure measuring devices, is more durable than quartz and, for a given frequency, normally is more rugged than quartz. Rochelle salt has a greater piezoelectric effect than any other crystal but has the disadvantage of a greater sensitivity to temperature changes than quartz. EDT has an advantage over quartz when used in frequency-modulated oscillators because of the wide gap between its resonant and antiresonant frequencies.

All piezoelectric crystals should have a good temperature coefficient, that is, show as little change in resonant frequency as possible under large variations in temperature. Ideally, the piezoelectric constant of proportionality between the mechanical and electrical variables must be the same for both direct (pressure-to-electricity) and converse effects.

Natural quartz crystal and new synthetic quartz crystal have been found to best meet the properties required of a piezoelectric element. Quartz is hard (7 on the Moh scale as compared with a diamond which is 10) and is relatively abundant and stable. Because of the anisotropic structure of quartz, cutting in different orientations makes it possible to obtain crystals for the widest range of applications, including frequency control, filters, resonators, and electromechanical transducers. Each of these uses depends on the orientation of the crystal cut with respect to the crystallographic axes. As shown in Fig. 2, the principal axes in quartz are identified as the optic Z, the mechanical Y, and the electrical X axes. By use of polarized light and x-ray diffraction techniques, the various axes may be properly located, permitting a crystal plate to be cut from the quartz with the performance characteristics desired. After cutting and extensive processing, the crystal plate is mounted at the nodal points of its normal vibration. These points, which also serve as electric connecting points, allow the crystal to vibrate freely with a minimum of interference or damping. Finally, the mounted crystal is hermetically sealed in a dry, inert atmosphere within a crystal holder.

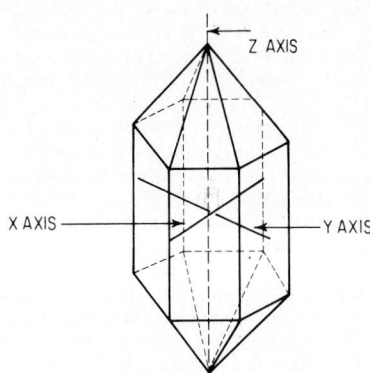

Fig. 2   Principal axes in a quartz crystal.

A stability of one part in a billion is routinely possible with the use of proportional ovens that can regulate the crystal temperature within $\pm 0.001°C$. By employing multiplying or binary techniques, frequency generating devices in the range from less than 1 Hz to more than 600 MHz are available. In a temperature-compensated crystal oscillator (TCXO), stability over a wide range of temperatures is obtained by a compensating network that shifts the frequency of the crystal an approximately equal and opposite amount to the shift of frequency caused by the temperature change. This network eliminates the need for an oven. TCXOs are available with temperature stabilities of ± 0.5 ppm over a temperature range of -40 to $70°C$. See also Fig. 3. Somewhat allied in principle is the tuning fork which is illustrated and described in Fig. 4.

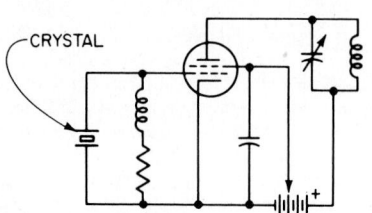

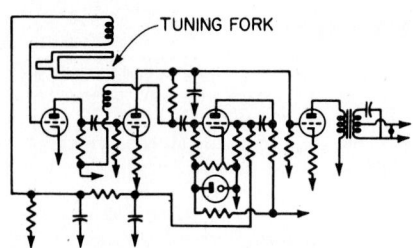

Fig. 3 Basic crystal oscillator circuit. Electronic multiplying and dividing circuits may be added to provide higher and lower frequencies, and output may be amplified to produce power for other functions. Accuracy may be as high as 1 part in 10^8 if stable electronic components are selected and close control of temperature, humidity, barometric pressure, voltage, and loading is maintained.

Fig. 4 Tuning-fork oscillator circuit. The fork is driven by one electromagnet, and its output is delivered to the other electromagnet, which is excited by the vibrating tines. The frequency may be changed, as in the crystal oscillator generator or by clamping the fork at different points. The output frequency may be constant within 1 part in 10^5 if the ambient conditions mentioned in the legend for Fig. 3 are kept constant.

Pendulum Clock

Some clocks use pendulums as their frequency standard. Galileo discovered the principle of *isochronism* of a pendulum in 1582, when he observed experimentally that a pendulum's frequency or period of swing seemed independent of its arc. Not until 1641 did he begin to adapt the pendulum to clocks. In 1656 Christian Huygens substituted a pendulum for the foliot bar in the then-existent weight-driven clocks. Theoretically, Huygens noted that the period of swing of the pendulum, unlike the oscillation rate of the foliot bar, was practically independent of the clock's drive system. In Huygens' clock, built in 1657 by Salomon Coster, a pendulum was weight-driven through a verge and foliot escapement. However, the pendulum was required to maintain an arc of ~40°.

Huygens recognized that a pendulum with such a large arc is not isochronous and determined that its bob should swing in a cycloidal (more U-shaped) curve. An English contemporary, William Clement, solved the problem by inventing an anchor escapement that permits the use of a small arc of 3 or 4°. Only within such a small arc is a pendulum practically isochronous and, therefore, more accurate. The small arc also minimizes power requirements and made practical narrower clock cases.

A later development, the *deadbeat* escapement, was invented by George Graham in England in 1715. This perfected the anchor escapement and was employed in many observatory pendulum clocks for more than 200 years.

The first electrically driven pendulum clock was built in 1843 by Alexander Bain in Scotland. The bob swung between two permanent magnets. In 1873 George Airy, British Astronomer Royal, designed the first electric pendulum clock to compensate for variations in barometric pressure—which otherwise affect the period of swing. Airy's clock became recognized as the standard time reference throughout the world until 1922, when the Shortt electric pendulum clock replaced it as the primary standard at the Royal Greenwich Observatory outside London.

In 1898 James Rudd of England built the first *free pendulum* clock that used two pendulums: a master or free pendulum that drove no mechanism, and a slave pendulum that drove an anchor escapement. The Shortt clock was an adaptation of this system. The slave pendulum always ran slow compared with the master, with a maximum difference of only ½₄₀ s, as reported at the Royal Greenwich Observatory. The NBS also used the Shortt pendulum clock as its primary standard. Quartz crystal clocks replaced the Shortt pendulum as the primary NBS standard during the period 1946 to 1950. The accuracy of the Shortt clock ranged to 1 part in 4 000 000, equivalent to a variation of 0.020 s/day.

After replacement of pendulum clocks as primary standards, a new Riefler electric pendulum appeared that eliminated the slave pendulum by substituting a diaphragm and pinhole on the master pendulum. A pencil beam fixed on the diaphragm activates a photocell that periodically penetrates the pinhole as the pendulum swings, creating a precise frequency standard without burdening the pendulum with a mechanical escapement. Electric *free pendulum* clocks must be set on ultrastable mountings and operate in a controlled atmosphere environment, hence are too delicate and large for common use.

Electric Clocks

An automobile clock generally uses current from a battery to automatically rewind a mainspring. In other electric clocks the mainspring usually is eliminated, and the battery current drives the balance wheel through electromechanical contacts. An early line current system employed an auxiliary mainspring that was kept fully wound by the current for use should the line power fail. Later a synchronous motor controlled by alternating current replaced the balance wheel system in electric plug-in clocks.

The electromechanically driven pendulum system (Bain) probably was the first electric clock. More recently, the so-called Western Union-type clock utilized two 1.5-V cells to wind a mainspring that maintained a pendulum while accepting an hourly time-pulse signal by telegraph to synchronize its time display on the hour with a remote master clock system. The time-pulse signal corrected for accumulated errors from all sources.

A consumer cordless electric clock may use a 1.5-V dry cell to drive the balance wheel electromagnetically by means of an electronic circuit. Because a transistorized circuit is employed as a switch to drive the balance wheel equipped with a tiny magnet, electromechanical contacts are eliminated. Another type of clock has as its frequency standard an electromagnetically driven 360-Hz tuning fork powered by a single aspirin-size 1.3-V mercury oxide cell. An electronic control circuit is required to drive the tuning fork. Accuracy averages ± 2 s/day. If three of the tuning fork systems are electrically interconnected, as in a marine navigation clock, the daily deviation in rate drops to less than 1 s.

In the United States, plug-in electric clocks generally use self-starting shaded-pole synchronous motors controlled by the 60-Hz alternating line current. Thus, their accuracy is dependent on the line frequency. Throughout the United States, line frequency normally is properly stabilized and corrected several times daily, thus eliminating accumulated error and permitting an accuracy of up to ± 4 s. The systems used by utilities to effect precise frequency control are not described here.

Synchronous Motors

Most timers are electrically driven, and those involving intervals greater than 1 min are usually driven by synchronous motors because of their inherent timing accuracy. A synchronous motor is essentially a cycle counting device and is as accurate as the frequency of its power source. The instantaneous frequency error of a typical 60-Hz utility generating plant in the United States is ± 0.1 cycle.

The torque required of timing motors places them in the "flea-power" range. The motors used are miniature single-phase self-starting types.

Hysteresis Type

A common synchronous timer motor is shown schematically in Fig. 5. It is referred to as a hysteresis type. It is nonreversible and may be stalled.

Permanent-Magnet Type

A more efficient higher torque type is a salient-pole motor with a permanent-magnet field. The permanent magnet also greatly reduces coasting, which will affect timing accuracy unless a quick-acting clutch is interposed between the motor and the driven mechanism. Coasting may be prevented in other motors by some form of magnetic or dynamic braking or by large gear reductions between the rotor and the output shaft of the motor. Another higher torque motor uses a capacitor to produce the phase lag necessary for starting.

The synchronous speed of these motors is not affected by ordinary voltage fluctuations. One inherent timing error is due to the delay in reaching synchronous speed from a stopped condition, which may amount to 2 cycles. For this reason it may be desirable to measure elapsed time with the

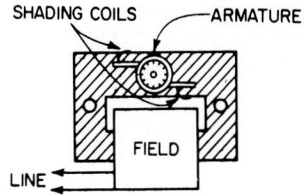

Fig. 5 Shaded-pole synchronous motor.

motor running at the start of the interval (using the aforementioned clutch), particularly if the interval is short.

Induction Motors

If controlled-frequency alternating current is not available, or if the load requires a high starting torque, an induction motor may be employed. Suitable induction motors, though having twice the torque of synchronous motors of equal size and number of poles, vary in speed from -3 to -10% of the synchronous speed, depending on the local situation. Capacitor-start, capacitor-run, and shaded-pole types are most often used.

DC Motors

If the power source is direct current or if it is desired to control timing remotely (accomplished with a rheostat), dc motors may be the driving means. Since constant speed is the first consideration, a shunt-wound motor is a logical choice. Actually, permanent-magnet field motors are more practical, since they are smaller and less costly. The speed varies with voltage, load, and temperature. Voltage ranges from 4.5 to 115 are customary. Series resistances, calibrated under load, may be employed to produce required speeds. It is difficult to hold speeds closer than $\pm 10\%$ without auxiliary governing devices, because of changing loads in many timers.

Dc motor life is shorter than that of an ac motor because of the inevitable wear on commutators and brushes.

Regulated DC Motor

Relatively constant speed may be produced by the use of a centrifugal governor on a dc motor. Line contacts are opened by the governor when the motor overspeeds and are set so that at normal full-load speed they are open part of the time, for better regulation. Contacts operate at a high rate to minimize pulsation. Accuracy may be held to within 1 and 5% but is affected by contact wear or fouling.

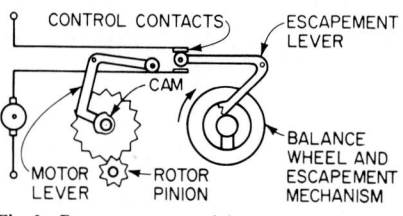

Fig. 6   Escapement-governed dc motor.

A governed motor which employs a balance wheel and escapement, but which can handle a greater load with less error than the ordinary escapement-regulated movement because it does not drive through the escapement, is shown in Fig. 6.

The motor is geared to a cam which oscillates a lever, causing control contacts to open. When the balance wheel returns from its free swing, it releases the escapement lever and closes the motor contacts. Thus pulses of line voltage are applied at regular intervals.

When jeweled bearings and a temperature-compensated hairspring are used, the speed of this motor at 1 rpm remains constant to $\pm 0.1\%$.

Spring-Actuated Devices

A spring clock uses a mainspring to power its balance wheel. Until the fifteenth century, all European clocks were weight-driven. The early spring clock, developed before the seventeenth century invention of the balance wheel and hairspring by Huygens, had no effective control over the mainspring's progressive loss of power as it unwound. Consequently, first the *fusee** and later the *stackfreed†* were developed to equalize the spring's declining force.

Though not accurate, the resulting clocks were useful because, unlike weight-driven clocks, spring clocks worked continuously while and after being moved; i.e., they were portable. Huygens used a hairspring to control the periodic oscillation of the balance wheel which, in turn, determined the periodic freeing of the pallet from the escape wheel. The latter rotated to work the clock's wheel train which operated the hands on the dial.

The eighteenth century English horologist John Harrison designed and built the first accurate spring clock for use as a ship's chronometer. Harrison used a mainspring-powered balance wheel movement (incorporating a fusee) set in a watchlike case 5 in (127 mm) in diameter. In 1770 Thomas Mudge of England invented a detached lever escapement that freed the balance wheel from direct contact with the escape wheel, thus completing the basic mechanical clock movement. This basic movement (without a fusee), versions of which still are employed in clocks today, was miniaturized in the twentieth century for use in small jeweled-lever travel clocks and wristwatches. A mainspring-powered balance wheel normally oscillates 2½ times a second, causing the jeweled lever escapement to *tick* 5 times a second.

Spring-operated timing devices cover a range of intervals which may vary from a fractional second to a week or a month or more. A typical movement uses a lever-type escapement (Fig. 7), and its accuracy depends on precision of manufacture, temperature compensation, and the effect of external loads.

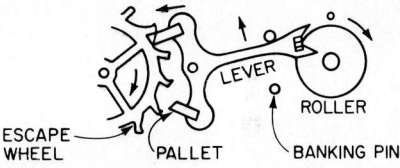

Fig. 7   Lever escapement.

Pendulum escapements are occasionally employed, but they are bulky and suitable only for stationary service.

Spring movements are limited to driving light loads, because of low torque and lack of constant torque. Accuracy may be within 0.2 s abs in a good stopwatch and 5 s/day in a good clock movement.

Air Motors

Flea-power air motors for timing are used principally where the introduction of electric power would be hazardous. Usual construction provides a jet of air impinging on a turbine wheel which revolves at high speed and drives a gear train regulated by an escapement.

Capacitor Discharge and Charge—*RC* Circuits

A simple method for obtaining short intervals is to utilize the time required for transient voltage and current changes in a resistance-capacitance *(RC)* circuit. If a capacitor in parallel with a resistor is charged to a definite voltage and the source of charge removed, the capacitor will then discharge through the resistor at a definite rate. This rate is determined by the time constant of the circuit, or *RC* product: after an elapsed time equal to *RC*, the charging voltage will have dropped to 37% of its initial value:

$$\text{Time (s)} = R \ (\text{M}\Omega) \times C \ (\mu\text{F}) \qquad \text{for decay to 37\% of charge}$$

---

*Derived from the Latin *fusata*—a cone-shaped device spirally wound with a thread, attached to the spring, designed to compensate through leverage for the declining power of the unwinding clock spring.

†Probably derived from German—a device for compensating through leverage for the declining force of an unwinding clock or watch spring, employing a pinion fixed to the arbor around which the spring is coiled and a second spring with an eccentric cam.

By the same token, charging a capacitor follows a similar law:

$$V_c = E(1 - e^{-t/RC})$$

where E is the impressed dc voltage and V_c the capacitor voltage.
Figure 8 shows the basic circuits and the exponential charging and relaxation curves.

$$V_R = \text{voltage across } R = E(e^{-t/RC})$$

The RC circuit is usually applied to the grid of an electron tube or solid-state component of a type which becomes conducting when V_c reaches a critical value.

Timing adjustment is obtained by varying the resistance or by switching to different values of R or C. Accuracy depends on the constancy of the charging voltage and the characteristics of the components, such as the temperature coefficient of the resistor, leakage, and the hysteresis effect of the capacitor. The physical size and reliability of large capacitors are limiting factors in the upper time range of timers using this measuring means. With a sensitive relay operated directly from a capacitor, it would take $\sim$200 μF to provide a 50-s delay. If, however, the capacitor voltage is used to "fire" a gas tube at its critical value, a 50-s delay becomes more practical, since the tube operates the relay and the capacitor need not be so large.

Most electronic timers are in the fractional-second to 1-min range, though longer intervals are available.

Flux-Decay Relay

A time delay may be produced in a dc electromagnet by placing a heavy copper ring or highly inductive short-circuited coil around the core, inside the main coil. When the circuit is opened, the dropout of the magnet is retarded by current induced in this copper "slug," which flows in a direction to oppose the decrease in flux in the core. The time constant L (inductance)/R (resistance) of the slug determines the delay. Adjustment may be made by using a movable copper sleeve between the core and coil or by a rheostat and neutralizing coil. The neutralizing coil operates the main flux and can cut the dropout time. This arrangement is shown schematically in Fig. 9.

Timing accuracy can be maintained within 10%; the delay produced ranges from 0.1 to 4 s.

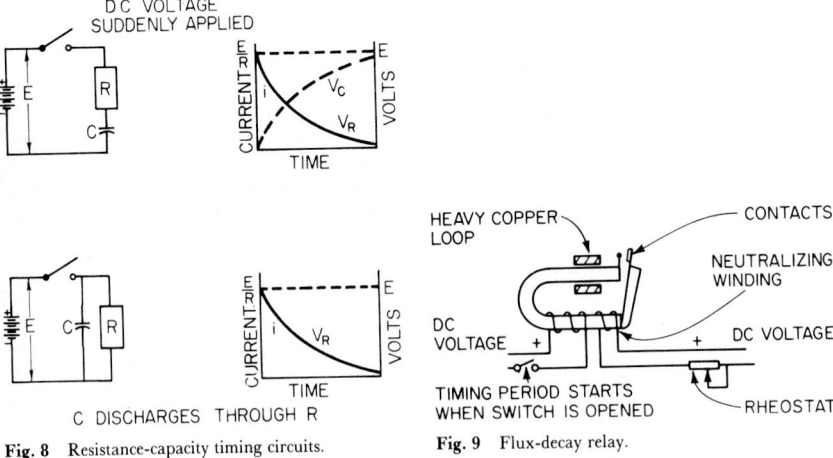

Fig. 8   Resistance-capacity timing circuits.

Fig. 9   Flux-decay relay.

Delay Lines

Although not associated with control timers, delay lines are important elements of certain instrumentation systems. A delay line is used to hold a signal for a discrete period and then retransmit it with a minimum of distortion.

The *magnetostrictive type* converts electric pulses into acoustical pulses for transmission along a wire and then reconverts to electric pulses after a delay depending on the length of the wire. In its simplest form, two coils, biased with small permanent magnets, are placed around a strip of magnetostrictive material. A current pulse in the input coil produces a longitudinal stress wave that travels along the wire at $\sim 2 \times 10^5$ in/s ($\sim 5.08 \times 10^5$ cm/s). When the wave reaches the output coil, it introduces a corresponding pulse. Absorption pads are clamped at the ends of the magnetostrictive strip to eliminate reflected pulses. This device is limited to ~ 50 μs, since this length of delay occupies a 15-in (~ 381-mm) package and longer delays become unwieldy. Maximum signal frequency is ~ 4 MHz, and the delay is ~ 5 μs/in (~ 127 mm) of transmission material.

The *torsional type* is another form of magnetostrictive delay line. Longitudinal stress waves generated in magnetostrictive material are converted to torsional waves in a transmission wire. The torsional pulse induced in the transmission wire travels along in helical fashion at $\sim 1.2 \times 10^5$ in/s (~ 3.05 cm/s). Delays may be as long as 20 ms. Bandpass characteristics are linear over a frequency band equal in width to the center frequency of operation. Thus the 3-dB bandpass for a line designed to operate at 1 MHz will extend from 0.5 to 1.5 MHz. Lines can be designed for frequencies ranging from 200 kHz to 4 MHz. Principal uses are in electronic memories and recirculating buffers in display systems.

Blocks of glass or quartz also can be used as a medium to transmit acoustical pulses. An electric pulse is converted to an acoustical signal by a transducer on one face of a multifaced block. The signal is totally internally reflected in the block many times before it reaches the receiving transducer on another face. Transducers may be piezoelectric-ceramic or crystalline quartz according to application. Ceramic transducers produce a low loss of ~ 40 dB at frequencies up to ~ 10 MHz. Crystalline quartz transducers produce a 60- to 80-dB loss but can be used at up to 60 MHz. Delays of up to 5 ms can be generated in fused quartz lines. If ferromagnetic metal is used for the transmission block, the frequency of signals is limited to ~ 5 or 10 MHz because of the absorption of higher frequencies in the metal domains. A glass delay line can be mounted on a circuit board together with integrated circuitry to utilize it as a digital memory. Applications include buffers for computer printers, readouts, and typewriters, as well as short-time memories for small computers, numerical controls, and electronic calculators.

Signals passing along a conductor are delayed according to the length of the conductor. For electric signals, this delay is about 1 ns/ft, which is of practically no use as a time delay, although it must be considered in the design of integrated control circuits. For pneumatic signals, this delay is closer to 1 ms/ft (depending on the propagation velocity of sound waves), and it is employed extensively in fluidic circuits to generate time-based functions. Usable delays can be produced in an electric lumped-constant transmission line. A practical limit to the length of the delay is ~ 1 μs, since size becomes cumbersome if the lines are built for longer delays. Lumped-constant delay lines are utilized extensively to adjust the output of glass delay lines.

If a signal is recorded on a moving tape or drum and then read again some distance away from the writing head, it will be delayed by the time it takes the tape or drum to move between the reading and writing heads. This method provides long, easily adjustable delays, but because of the cost and complexity of the driving equipment, it is seldom employed in commercially available equipment. Various types of long mechanical-type delays have been used in experimental dead-time process controllers.

Thermal Devices

Thermal timing devices, suitable for ac or dc operation, make use of the rate of expansion of metal when heated or of the differential expansion of a bimetallic element. There are several methods, which vary in accuracy, adjustability, repeatability, and time-delay range.

A quick-acting wire-wound heater placed near a bimetallic strip will cause the strip to bend and, after a time interval, operate a contact (Fig. 10). If two strips are used, temperature compensation is introduced. The two bimetals, each with a contact, move backward and forward with ambient temperature change but remain the same distance apart. When one of them is heated by an element wrapped around it, the contacts are made or broken.

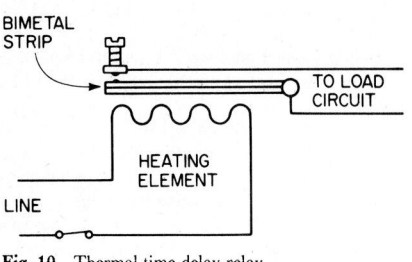

Fig. 10 Thermal time-delay relay.

Another type, suitable for the 6 to 14 V range, utilizes a hot wire. The wire is attached to a bent strip which carries a contact, and on being heated straightens the bow to close or open the circuit. The small thermal mass of the heater makes rapid operation possible.

Accuracy depends to a considerable degree on voltage fluctuations and ambient temperature changes and on whether sufficient time has elapsed between successive operations for cooling of the element. Arrangements employing both the heating and cooling times are more accurate for repeated cycles. Time intervals range from a few tenths of a second to 3 s for the hot wire type and from 5 s to 10 min for bimetal types. Error varies from 5 to 20%.

Thermal timers are simple in design, low in cost, silent, and unaffected by frequency or position change.

Dashpots

Fluid dashpots are useful in providing delays in the 1 s to 2 min range. Devices employing dashpots are generally ac or dc solenoids in which the armature is the plunger. Armature motion is retarded by the fluid having to escape past the plunger as it moves (see Fig. 11). Accurate timing requires constant viscosity in the fluid; therefore it is affected by ambient temperature changes. If silicone oils are utilized, this effect is greatly reduced at temperatures between -30 and $120°F$ (-34 and $49°C$).

When longer time intervals or rapid recycling are required, a quick-escape check valve is put in the plunger to permit rapid return to the bottom of the dashpot. Speed adjustment may be obtained by opening or closing bypass holes in the piston. Expected accuracy is 5 to 10%.

Pneumatic Devices

Use of the time required for a fixed volume of air to pass through a restriction is made in relay devices offering time delays of 0.1 s to 5 min. Customary construction provides an adjustable orifice which meters the volume of air passing into a diaphragm chamber. The diaphragm (Fig. 12) may be attached to a rod which is raised by the action of a solenoid, forcing air from the chamber through a check valve. With the solenoid deenergized, a spring forces the diaphragm downward, its movement being retarded by the rate at which the air can flow through the needle valve. Accuracy is within 5 and 10% and is little affected by ambient temperature. The air must be kept free of dirt, however.

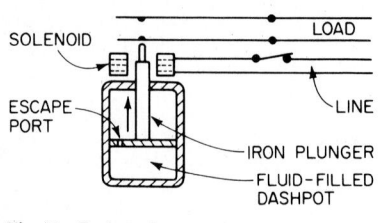

Fig. 11 Dashpot timer.

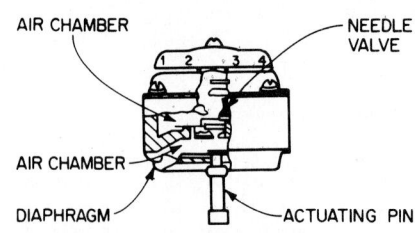

Fig. 12 Pneumatic time delay.

Inertia Devices

In an ac or a dc magnetic relay a weight may be added to achieve brief time delays (0.08 to 0.12 s). The weight is added to the armature and produces a delay by increasing the inertia it must overcome.

If shock, vibration, or change in position might affect the operation of a weighted device, the armature may be coupled to a flywheel. The flywheel is rotated as the armature moves, and the inertia effect may be varied by changing its travel with movable stop pins. An accuracy of 5% is possible, with a maximum delay of 0.2 s.

Electronic Timers

Because of the frequent mingling of electrical, electronic, and mechanical principles in one timing device, the term *electronic timer* sometimes is used rather loosely. The generally accepted classification considers crystal-controlled, tuning-fork-controlled, and *RC*-time-constant devices as qualifying as electronic timers. Actually, to fully qualify as an electronic timer, the last-mentioned device should incorporate a trigger circuit and a solid-state output. It is principally the advantages of this device (generally unaffected by shock, vibration, and contaminating atmospheres) that in an overall way are attributed to electronic timers.

As will be evident later in this article, solid-state timers have become widely available in recent years and are preferred for many applications. A relatively recent development is the inclusion of minicomputers and microprocessors in timing controllers to achieve what are sometimes called "smart" devices.

TIME-BASE CONTROLLER CONFIGURATIONS

Time-base controllers may (1) directly relate time to specific actions which must be started or stopped at definite times (where the length of time between these actions or time steps is critical to the objectives of the operation or process being controlled), or (2) relate a specific process variable to the passage of time (e.g., a time-temperature relationship). Time-base controllers, then, can be roughly classified on the basis of these two general functions. For category 1, the simple word "timer" (with appropriate modifiers) can be used. Lumped into this category are devices sometimes called *time-cycle controllers* or *step programmers*. For category 2, *time-schedule controller* is the currently acceptable term. The remainder of this article is based on these two general classifications.

TIMERS

This broad category of devices includes time switches, time-delay relays, interval timers, percentage timers, repeat-cycle timers, reset timers, and time-cycle controllers and step programmers. Two subcategories of these types of timers of notable importance are (1) electromechanical devices and (2) solid-state devices.

Time Switches

The time switch was one of the first time-base instruments to be adopted for industrial and commercial applications, dating back many decades. A time switch is a 24-h (usually) timing device used to open and close electric circuits in accordance with the time of day, on a continually repeating day-after-day schedule. Such timers are applied to daily automatic control of plant lighting, heating, and ventilating equipment; factory work signals, pumps, and compressors; and also for the commencement of heating processes prior to the start of the workday—as may apply to ovens and furnaces, lead and glue pots, dip tanks, soldering apparatus, molding press dies, and even testing equipment.

The greatest usefulness of time switches has been in the saving of electric power (lighting, for example) and labor costs (automatic versus manual connecting of numerous circuits). Time switches were adapted to electric time clocks for employee "punching in and out" before the turn of the century and, in fact, the manufacture of these time clocks spawned one of today's leading computer and office automation firms.

In a typical time switch (Fig. 13), a disk, graduated in time of day (and night), is rotated by a clock mechanism; attached to its periphery are adjustable dogs for tripping contacts open or closed at the desired time.

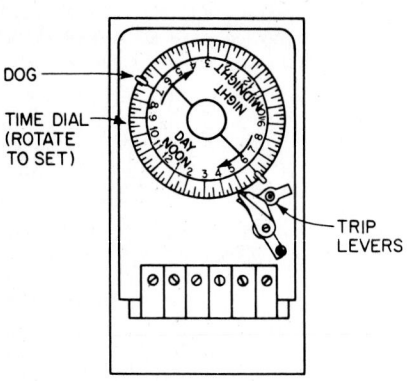

DOG

TIME DIAL
(ROTATE
TO SET)

TRIP
LEVERS

Fig. 13 Traditional time switch.

Time-Delay Relays

A time-delay relay is an electronic (solid-state), electromechanical, or mechanical contact making device in which there is a time lag between the energizing or deenergizing of the control circuit and the subsequent opening or closing of the load circuit or circuits. A major function of this type of controller is to delay the start-up, continuation, recycling, or shutdown of an operation until optimum conditions have been produced. Time-delay relays are commonly used for such applications as in electronic equipment warm-up and overload protection where the interval is relatively short, requiring no external adjustment once set, and repeatability within 5 to 10% is normally adequate. Hence, simple electromagnets with delayed armature movement created by copper slugs, capacitors, thermal devices, dashpots, and weights have been among the traditional delay methods used for many decades. In more modern configurations, a delay time may be obtained by using a complementary metal oxide semiconductor (CMOS) integrated circuit, an internal potentiometer-controlled oscillator, a programmable binary counter, and output-controlled logic. Although usually considered in terms of relatively short time ranges, i.e., in terms of seconds or minutes, delay action can be extended to 4 or 4.5 h, depending on the requirements and relay design. Where adjustment of the time delay is provided, the indication of time values to which the device can be set is usually quite coarse. Where greater accuracy and flexibility are required, the user generally turns to an interval timer rather than a time-delay relay.

The life expectancy of a time-delay relay varies with the type and design. For mechanical types, it is generally in excess of 20 million operations. For electrical types, a minimum of 100,000 operations at full rated load is usually claimed. The operating temperature range is frequently stated by suppliers as between 14 and 140°F (-10 and 60°C). Reset time is typically 100 ms.

A flasher may be classified as a form of time-delay relay. Such a device is used in electrical and electronic equipment as a flasher for lamp, relay, and solenoid circuits, and in electronic systems as a power pulser or system clock-driver. For dc voltage, units have a fixed flash rate ranging from 10,000 flashes per second to 10 flashes per minute; ac voltage units have a range from 6 to 300 flashes per minute.

Delay-on-Make Timers

Functionally a delay-on-make timer is essentially the same as a time-delay relay; i.e., after an initiating signal is received by the device, a timing action is commenced for a fixed (or adjustable) period during which no energy flows to the machine or process being controlled. On expiration of the timed period, current is permitted to flow through the timer contacts; it continues until such time that power to the timer is cut off and remains so for a short period prior to resetting for another cycle of operation. See Fig. 14.

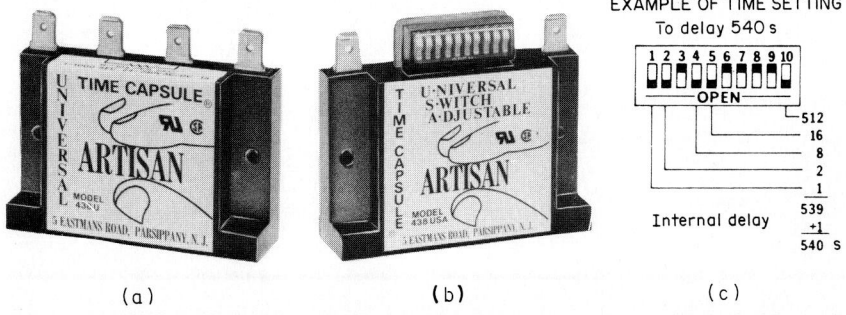

EXAMPLE OF TIME SETTING
To delay 540 s

	512
	16
	8
	2
	1
	539
Internal delay	+1
	540 S

(a)　　　　　(b)　　　　　(c)

Fig. 14 Solid-state delay-on-make timing module. (*a*) In this unit, the timing period ranges from 1 to 1000 s and is preset by connecting a resistor across the two center terminals. This timing period is essentially linear from 1 to 1000 and is achieved by connecting 10,000 Ω for each second of delay required. Should a delay period of 100 s be required, a 1-MΩ resistor connected across the two center terminals will provide approximately 100 s. Each time power is applied, the device remains in the "off" state, permitting only leakage current to flow through the load device for the present delay period. At the end of this preset delay, the unit switches on and permits full current to flow. The unit remains on for as long as power is applied. If power is removed for at least 50 ms, the unit will reset, turn off, and run through another timing period. (*b*) In this unit, 10 switches are incorporated to permit the setting of the timing period from 1 s, with all switches closed, to 1024 s, with all switches open. (*c*) Commencing with 1 s (minimum of range), opening the switches adds 1, 2, 4, 8, 16, 32, 64, 128, 256, and 512 s to the time delay in progressive order from switch 1 to switch 10 to give a total of 1024 s. Other units of this general design are available for fixed timing periods ranging from 100 m to 10,000 s. Potentiometer-adjusted units are also available. *(Artisan Electronics.)*

Interval Timers

Most timers are interval timers, though they may differ widely in such respects as means of initiating timing, range of intervals, contact operation, driving means, arrangement for resetting, and whether or not timing is continuously repeated. An interval timer holds circuits open or closed for a predetermined interval and then returns them to their original position. When the timer is energized, load contacts or air pilots (if pneumatic) close or open and remain so until it has timed out. Thus, it controls the duration of an operation or process.

A simple interval timer is shown in Fig. 15. Timing is started by turning the knob to the desired interval value, which also winds the spring motor and closes the mercury load switch. At the end of the timed period, the unwinding spring has returned the knob to its original position and the mercury tube tilts to open the load circuit. A device of this type (more often electrically driven and started by making a pushbutton contact—or completely solid state) may be used to control the operating time of a variety of processes, such as those involving plastic molding presses, die casting machines, machine tools, conveyors, centrifuges, and pumps, among many others. Interval timers are also frequently used with bottle filling machines, photoprinters, mixers, bake ovens,

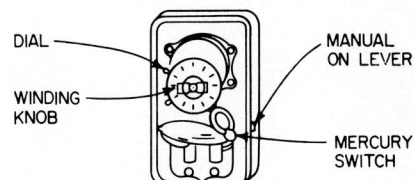

Fig. 15 Spring-driven interval timer.

resistance welding equipment, induction heaters, chemical feeders, and x-ray equipment; and to cooking, sterilizing, distillation, rubber curing, washing, electroplating, and heat-treating operations. In considering the many applications of control timers, one develops an enhanced appreciation of the great numbers of manufacturing operations that are time-dependent. The interdependence of time and temperature in a bake oven is one of the most obvious—time and temperature, taken together, actually define the variable of concern, namely, the quantity of heat available to be absorbed by the materials flowing through the oven. Within a fairly wide margin, crackers moving fast through a very hot oven can be satisfactorily browned or toasted; but the same result can be achieved by moving them more slowly through a less hot oven.

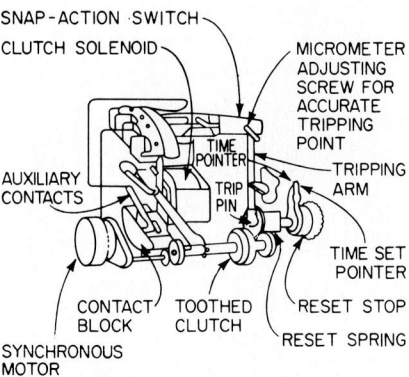

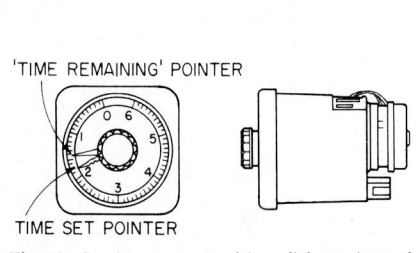

Fig. 16 Synchronous-motor-driven dial-type interval timer.

Fig. 17 Internal view of the synchronous-motor-driven dial-type interval timer of Fig. 16.

A synchronous motor-drive interval timer is shown in outline in Fig. 16, and schematically in Fig. 17. The control means shown in this example is a pin which operates a tripping arm. Its initial position, which determines the time interval, is set by the adjusting knob with a time-set pointer. A second pointer, which is set by the knob but not firmly attached to it, rotates during timing to indicate the time remaining. At time out, the pin hits the tripping arm, which is connected to the contact finger of a snap-acting single-pole double-throw (SPDT) load switch. A solenoid clutch starts the timing and permits the spring reset to operate when it is deenergized. An auxiliary set of load contacts is operated by the solenoid itself. This description applies to what may be termed traditional designs. There are other actuation configurations, including, of course, solid-state designs. Although the latter cover a wide timing range, they are particularly applicable for intervals of less than 1 min. For intervals in the microsecond range (10 μs to 1 s or more), cycle counters operating from crystal frequency generators can be adapted to control purposes. A series of switches, one for each decade stage, is set up to control the number of counts, which are indicated on a light emitting diode (LED), liquid crystal display (LCD), or a neon tube readout. Output pulses for the operation of load contacts are supplied at the start and finish of each interval.

An interval timer has a set of contacts, switches, or a relay that transfers at the start of the time cycle, holds through the cycle, and transfers back at the end of the cycle. Interval timers usually reset at the end of the timed interval. The operating cycle of a representative interval timer is diagramed in Fig. 18. Interval timers are typified by the devices shown in Figs. 19 and 20.

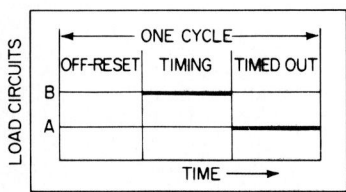

Fig. 18 Timing diagram for an interval timer.

The close relationship between interval timers and time-delay relays has been mentioned previously. Many interval timers may be used as time-delay relays by arranging the contacts so that they operate at the end of an interval only, instead of at both the beginning and end. These are sometimes referred to as *delay-interval timers*. Unlike most time-delay relays, interval timers are frequently used for close control (within 1%) and where they are likely to be adjusted to different time settings. As compared with the run-of-mill time-delay relays, these timers frequently include an indicating dial and knob, often with vernier scales, for ease in changing the timing adjustment.

Delay-Interval Timers

These are interval timers which are electrically reset to delay the energization or deenergization of a circuit. Electromechanical types are furnished with a direct or reverse clutch. A representative grouping of delay-interval timers is described and illustrated in Fig. 21.

Fig. 19 Pushbutton-type interval timer for panel mounting. The device features a manually controlled adjustable timer control and a separate progress indication pointer. A pilot light is available for visually monitoring the control circuit operation. The standard version is one single-pole double-throw (SPDT) switch, although a second SPDT can be provided where isolated load control is required. When two switches are provided, the load switch operates at the end of the preset time and the second switch turns off the timer motor ~1% of full time scale later. Single-switch timers are wired to deenergize at the end of the cycle. The timer remains in the timed-out position after each cycle but may be wired to turn a load on or off at the end of the timed period. Manually depressing the reset button after the unit has timed out resets the timer for another cycle. Depressing the reset button during the timing cycle resets the mechanism without transferring the switches. Releasing the pushbutton starts another full cycle. Timers like this are available in several standard time ranges: 6, 15, 30, 60, 120 s; 3, 6, 15, 30, 60, 120 min; 3, 6, 15, 30 h; 3 days. Contact ratings are 10 A at 120 and 240 V ac, ⅙ hp at 120 and 240 V ac, 0.5 A at 120 V dc, and 0.25 A at 240 V dc. Contacts are silver. Power consumption is approximately 3 W. The life expectancy of a timer is claimed to be in excess of 1 million operations, and that of switches, over 1 million cycles at 80% of rated load. The dimensions of the unit are approximately 2⅞ × 2⅞ × 4¼ in deep (72 × 72 × 108 mm). (Series N10, *Bristol Saybrook.*)

Fig. 20 Manually set motor-driven interval timer for rear panel mounting. This unit is designed for simplicity, small size, and low cost. The unit is 2 × 2¼ × 2 in deep (51 × 57 × 51 mm). The motor is wired in parallel with the controlled load, through the normally open pole of the control switch. The timer is started by turning the knob shaft from the zero time position to the desired setting. A hysteresis synchronous motor drives the cam back to the zero time position and turns itself off along with the controlled load. Time ranges from 15 s to 24 h are available, with a 300° scale. The switch rating is 10 A. The unit is mainly used by original equipment manufacturers for such items as medical and dental equipment, washing and drying machinery, and commercial cooking and food processing machinery. (Type 40162, *Bristol Saybrook.*)

A solid-state timer has been designed to protect compressor and motor circuits from recycling instantaneously. When the power is removed from a compressor or motor circuit, the device turns off. After power has been restored, it will not turn back on until a given delay period has passed. The delay is fixed and specified by the user, ranging from 1 s to 10 min. The circuit is shown in Fig. 22.

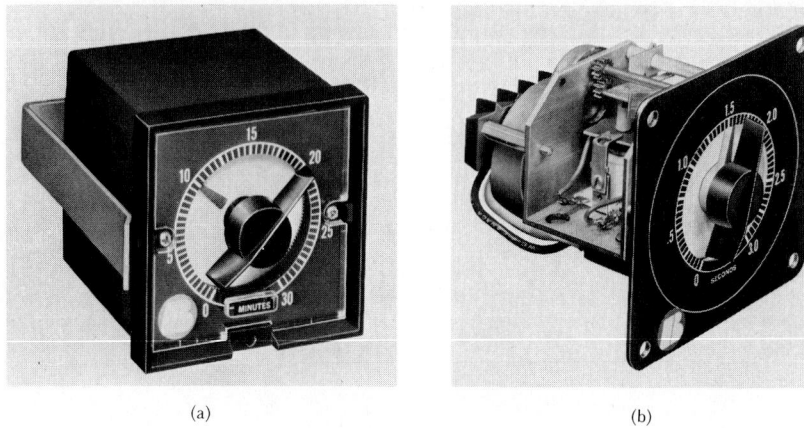

(a)

(b)

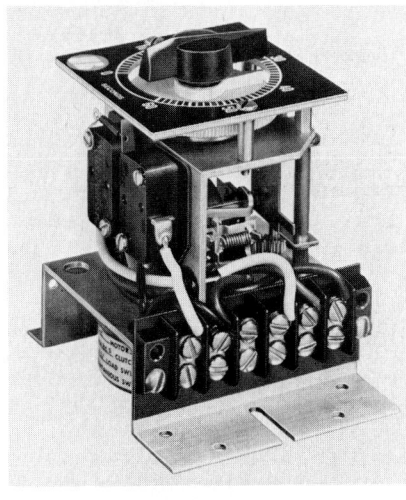

(c)

Fig. 21 Delay interval timers which are electrically reset to delay energization or deenergization of a circuit. (*a*) A panel-mounted timer with a scale, adjusting knob, and separate pointer for visual indication of timing progress. A pilot light is also available for visually monitoring the control circuit operation. The unit has a direct clutch which engages the timing mechanism when power is applied and disengages the mechanism, permitting the timer to reset when power is removed. Two single-pole double-throw (SPDT) switches are provided. The isolated load switch operates at the end of the preset time, whereas the motor control switch, which is used to control the time motor, operates ~1% of the time range later. (*b*) A design similar to that of (*a*) except that both the timer motor and load are controlled by a single switch, requiring that the load voltage be the same as the timer operating voltage. (*c*) A design generally similar to that of (*a*) and (*b*), showing more detail of the operating mechanism. Units are also available with a reverse clutch which engages the timing mechanism when the circuit is deenergized. The timer is reset by applying a short electric pulse to the clutch coil. These timers do not reset in the event of a power failure and resume timing when power is restored to the timer motor.

All units illustrated are available with time ranges of 6, 15, 30, 60, 120 s; 3, 6, 15, 30, 60, 120 min; 3, 6, 15, 30 h; and 3 days. The scale length is 5 in (127 mm), with 300° travel. The repeat accuracy claimed is ±½% of full-scale range. (*Bristol Saybrook.*)

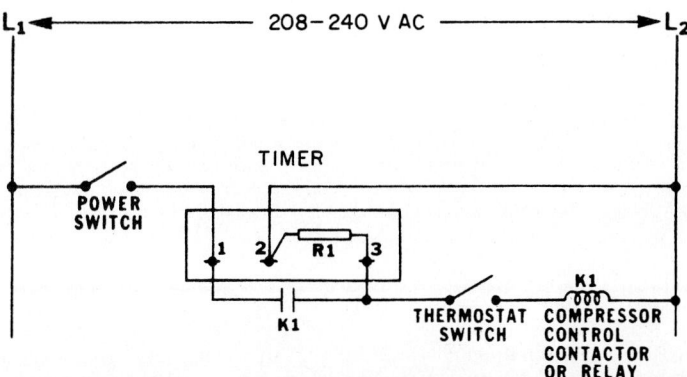

Fig. 22 Circuit of a timer designed to protect compressors and motors. As soon as the main power switch is closed and power is applied to the system, the timer acts like a large resistor. This causes the voltage at terminal 3 to remain below the level where closing the thermostat will not operate the compressor contactor. If the thermostat is in the open mode, the resistor $R1$, connected across the timer, will cause it to begin timing. At the end of the timing period, the timer turns on and applies $L1$ to terminal 3. If the thermostat is closed at this time, the compressor contactor $K1$ will pick up and hold itself through holding contacts $K1$. Other contacts of $K1$ will then be permitted to operate the compressor or motor circuits. The action of contacts $K1$ across the timer causes it to reset. If the thermostat is opened and then closed quickly, the contactor $K1$ will drop out and the timer will repeat another cycle. This wiring arrangement ensures that, if the compressor has been off for the time period of the timer delay, the closing of the thermostat will not require a waiting period and the compressor will operate. If, however, the compressor has not been off for the preset delay of the timer, the compressor control contactor $K1$ will not operate until the desired delay has expired. This ensures that, if the thermostat is operated from "off" to "on" and then this process is repeated, the compressor will not operate until the timer permits it to do so. (Cycletector, *Artisan Electronics*.)

Fig. 23 Percentage timer. The desired percentage of the total time range is set by manually turning the setting knob in either direction to the appropriate percent on the dial. The setting may be changed during operation. However, the new percentage setting may not occur until the next full cycle. By applying power to the motor, the timer starts and a load switch is actuated for the desired percentage of time. The selection of wiring connection—normally open or normally closed—on the load switch determines whether the dial percentage is related to the "on" or the "off" time. Operation of the timer is continuously repeated as long as power is applied to the motor. When power to the motor is interrupted, the timer stops and then continues from that point when power is reapplied. A visual indication of timing progress is shown by the rotation of the cycle progress indicator in the center of the dial. (Series 5750, *Bristol Saybrook*.)

18.77

Percentage Timers

Percentage timers are designed for applications that require a continuously cycling time that is remotely controlled and has an easily adjusted "on" or "off" setting as a percentage of full cycle time. These units cycle continuously whenever power is applied to the motor. When power to the motor is interrupted, the timer stops and will continue from that point when power is reestablished. Wiring through the common and normally open contacts of the switch gives the percentage of "on" with respect to the dial setting. Wiring through the common and the normally closed contacts of a switch gives the percentage "off" with respect to the dial setting. See Fig. 23.

Repeat-Cycle Timers

A repeat-cycle timer performs its function over and over until it is stopped. A timer of this type may be motor-driven or solid state. In the electromechanical type, the control means may be a disk with trippers or a shaft with cams. Generally, a repeat-cycle timer is not designed for frequent changes because switch tripping means usually are not easily adjusted in the field. For frequent changes, split cams and other means designed for flexibility of adjustment are preferred. However, some units marketed as repeat-cycle timers incorporate a considerable degree of adjustment. Electromechanical repeat-cycle timers have been used for decades and are technically mature. They continue to be widely used. Electronic solid-state timers are also available to provide the repeat-cycle function, and many timer manufacturers offer both electromechanical and solid-state designs. A repeat-cycle timer can be wired, of course, to perform one cycle of the program and stop, but in such a case it would no longer meet the definition of repeat-cycle timer.

A number of representative repeat-cycle timers are illustrated and described in some detail in

Fig. 24   Multirange repeat-cycle timer. A plug-in double-pole double-throw (DPDT) relay can be energized by the unit shown during the first timing cycle (T_1) or during the second cycle (T_2) by moving a push-on connector from one programming pin to another on the circuit board. This allows the load operation to be changed from one set of contacts to the other (normally open to normally closed) without changing the wiring. The timer operates continuously through its two timing ranges (T_1 and T_2), one after the other, transferring the relay contacts as it times out each range. There is no start circuit, and the timer resets on power interruption. There are six continuously adjustable switch-selected ranges for each timing circuit: 1 s, 10 s, 1 min, 10 min, 1 h, and 10 h. The minimum setting is 3% of range, except for 50 ms for the 1-s range. The unit incorporates a light emitting diode (LED) progress annunciator for each timing circuit. Except in the 1- and 10-s ranges, the blinking rate of the LED increases as the cycle progresses. In the 1- and 10-s ranges, the pilot light is off before the timing cycle starts and stays on during timing. Contact ratings are 5 A (resistive) at 120 or 240 V ac. (Series 342, *Automatic Timing & Controls*.)

Fig. 25 By selecting the appropriate time ranges for two timers, which are internally wired to recycle each other, it is possible to control an electrical load for a very short period of time during a long total time cycle. For example, it is possible to energize a load for 5 s every 24 h, this being an extreme case. As shown here, the timer with the shorter time range is in the left position and operates first. The minimum setting of the timer located at the right is 1% of the time range of the left-hand timer, plus ¼ s. Both timers may have the same time range if required, but the left-hand timer always operates first. The timers are operated by motors of ∼3 W each, and the clutches by motors of ∼2 W each. Both motors and clutches for both timers are on simultaneously for ∼1% of the fastest time range. The time ranges available are given in Table 1. Automatic recycling of the two timers continues as long as the unit is energized. Removal of power at any point in the cycle completely resets the unit, and the cycle again restarts with the left-hand timer when the unit is reenergized. Timing progress is indicated by a pointer on each timer. The life expectancy of each timer is claimed to be in excess of 1 million operations, and of switches, in excess of 1 million cycles at 80% of the rated load. The switch contact rating is 5 A at 120 or 240 V (resistive). Much more complex cascaded types of multiple control functions can be achieved through various circuitry configurations using dual-range units. (Series T-10, *Bristol Saybrook*.)

Figs. 24 through 28. Representative time ranges for the dual-range recycling timer shown in Fig. 25 are given in Table 1.

Reset Timers

A reset timer is designed to automatically reset itself to the preselected setting once it has timed out. The *reset time* is the interval required for a timer to return its pointer to the point where it started, or how long it takes to prepare a solid-state timing circuit for a new cycle. It determines the closeness of timing cycles.

An adjustable plug-in solid-state ac-dc timer is shown in Fig. 29. Timing begins when the "start" switch is closed. At the same instant, the timing LED goes on and a relaxation oscillator starts to run at a gate determined by the dial adjustment. The device times out and the timing LED turns off when the oscillator count is equal to the level set by the range switch. At time-out, the load relay is energized, transferring its contacts, and the timing circuit is automatically deenergized. Reset occurs when the start switch is opened or when power is interrupted. A second LED (labeled "timed out") turns on when the load relay is energized at time-out, and turns off when the start switch is opened or power is interrupted.

A multirange timer (also called a time-delay relay) incorporating MOS digital circuitry is shown in Fig. 30. This unit comprises a variable frequency oscillator (VFO) whose frequency depends on the settings of the range selector and dial adjustment and whose output feeds a digital counter. The frequency of the VFO determines how fast the counter reaches the limit of its fixed range and,

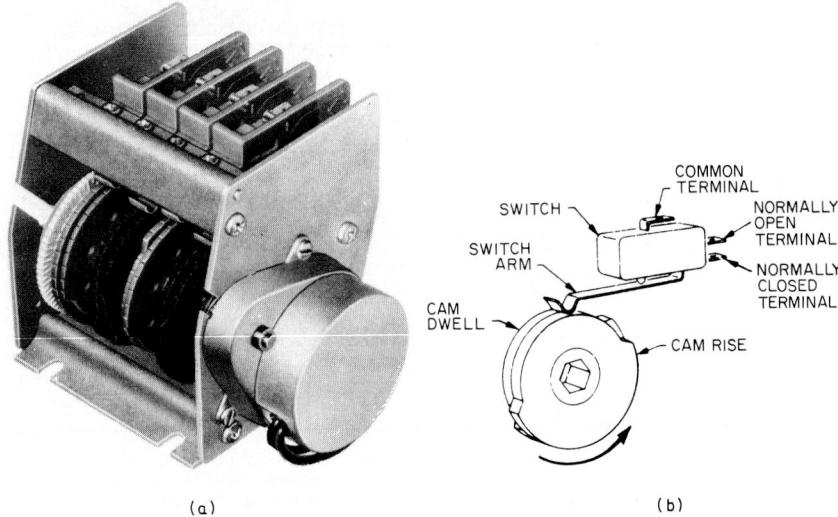

(a)   (b)

Fig. 26 Cam-type repeat-cycle timer. (*a*) Standard cams are of the random-set dual adjustable type. A degree wheel is provided for guidance in programming. Precision-cut cams for fixed or multiple-switch actuations are also available. The standard motor is a permanent-magnet type, but hysteresis motors are also available. (*b*) Arrangement of a cam and switch. When the switch arm is in the dwell of the cam, the circuit is from the "common" terminal to the "normally closed" terminal. When the switch arm is on the rise, the circuit is from the "common" to the "normally open." A number of time cycles can be accommodated for operation on different supply voltages and frequencies. Timers of this type are widely used in process controls, commercial dishwashing machines, air driers, and food vending machines, among other equipment. Standard cycle times are 1 revolution in 5, 6, 10, 12, 15, 20, 30, 40, 45 s; 1, 1.5, 1, 2.5, 3, 4, 5, 6, 7.5, 10, 15, 20, 30, 40, 45 min; and 1, 1.5, 2, 2.5, 3, 4, 5, 6, 7.5, 10, 12, 15, 24 h.

Switch ratings are 10 A at 120 and 240 V ac (resistive), ⅛ hp at 120 and 240 V, 0.5 A at 120 V dc, and 0.25 A at 240 V dc. Higher switch ratings are obtainable. Power consumption is ~3 W. The repeat accuracy is ±0.5% of cycle on continuous cycling; the setting accuracy is ±3° at any one operating point and ±6° between any two operating points. The life expectancy of mechanical timers is in excess of 5 million operations, and of electrical timers, in excess of 1 million cycles at 80% of the rated load. (C-10 and C-11, *Bristol Saybrook*.)

therefore, how fast the device times out. In *delay-on-make* operation, timing begins when the start switch is closed. The load relay contacts transfer at the end of the time period. Reset occurs when the start switch is opened or when there is a power interruption.

In *delay-on-break* operation, timing begins when the start switch is opened. The load relay contacts transfer at the end of the timed period and back again at reset. Reset occurs when the start switch is closed or when there is a power interruption. Control action of all loads is delayed, either closed-closed-open or open-open-closed.

In interval control, timing begins when the "start" switch is closed. The load relay contacts transfer at the beginning and at the end of the timed period, thus providing true interval control, either open-open-closed or closed-open-closed. The start signal may be either sustained or momentary. In the latter case, the signal is "latched in" by wiring it to one of the load relay's two sets of contacts. Power interruption resets the timer. Because of the interface circuitry, it is feasible to connect external loads in parallel with the "start" switch. Typical installation circuitry for this unit is given in Fig. 31. A motor-driven pushbutton digital reset timer is shown and described in Fig. 32, and a computing timer with a built-in microcomputer is shown in Fig. 33.

Flexible Functionality through Design

It is obvious from the previous descriptions that the variations in timing range, size, etc., that—for equipment manufacturers who make a number of products that require built-in timers—this can

Fig. 27 Dual-range repeat-cycle timer. This device has two separately adjustable timing ranges. The unit continuously switches back and forth between the two ranges and is typically used to interrupt a process, turning it on and off repeatedly for dial-set intervals, e.g., to cycle two variables or to reverse a motor, operating it in each direction for a dial-set interval. The dial is equipped with two set-point hands. The zero position between the two separate ranges is shown at the bottom center of the dial. As the pointer passes through the zero position, the two single-pole double-throw (SPDT) load switches individually transfer from one set of contacts to the other. Each switch is thus actuated during one timing range and not actuated during the other. The two load switches are adjusted to transfer one before the other, providing a dwell or dead zone when the cycle progress pointer passes through zero. The switching differential can be used to provide the dwell, which is especially useful for reversing a motor. The dwell gives the motor sufficient time to coast to a stop before it is reversed. The factory-set dwell is 1% of the timing range but may be made as much as 4% if specified. There is a choice of nine ranges from 30 s to 240 min. The minimum setting is $\frac{1}{60}$ of the range. A repeat accuracy of $\pm 0.33\%$ of range is claimed. For manual override, a friction clutch allows shortening of either time cycle. Contact ratings (noninductive) are 10 A at 120 V ac and 5 A at 240 V ac. (Series 305, *Automatic Timing & Controls.*)

pose a serious inventory problem. Exemplary of the flexibility that can be obtained from one basic design configuration and in size reduction for original equipment manufacturer (OEM) applications are the diagrams and explanations given in Fig. 34.

Time-Cycle Controllers and Step Programmers

A time-cycle controller opens and closes one or more circuits during a timed interval. Because it can control a predetermined series of related events in a preselected interval, it, more than any other class of timer, may be called a *program controller*. Since a time-cycle controller may have to perform many functions, it is often designed with a greater degree of flexibility than other timers. It may be of either the repeat-cycle or stop-cycle type and may have means for readily changing the sequence and duration of operations, adding or subtracting operators, changing the interval between operations, changing the overall time range, and changing the form (NO or NC) and type (air or electric) of the operator. See Fig. 35.

The time-cycle controller, frequently combined with auxiliary timers which control long intervals or process variable schedules, can be the control center of an automatic machine or process. In such complex applications, the device usually is referred to as a *program controller*. A typical time-cycle controller program is diagramed in Fig. 36. This is contrasted with the timing diagram for a simpler interval timer as previously shown in Fig. 18. For some industrial operations, the program shown in Fig. 36 may represent only a relatively small portion of the total complexity of the program.

(a)

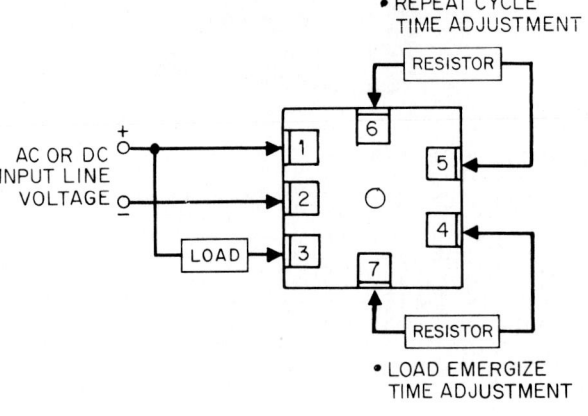

• REPEAT CYCLE
  TIME ADJUSTMENT

RESISTOR

AC OR DC
INPUT LINE
VOLTAGE

LOAD

RESISTOR

• LOAD EMERGIZE
  TIME ADJUSTMENT

(b)

Fig. 28 Solid-state repeat-cycle timers require no moving parts, such as a motor drive and clutch. The unit shown is fully encapsulated for protection against dirty environments. Fixed or resistor-adjustable full-cycle timing periods range from 200 ms to 24 h, with fixed or adjustable energized periods ranging from 0.1 s to 90% of the full-cycle time. Solid-state design permits minaturization. The unit illustrated measures 2 × 2 × 1.25 in (5 × 5 × 3.2 cm). Available operating voltages are 12, 14, and 48 V dc and 24, 48, 120, and 240 V ac. *(Artisan Electronics.)*

Table 1 Representative Time Ranges Available with a Dual-Range Recycling Timer*

Time range	Minimum setting	Time range	Minimum setting
6 s	0.2 s	30 min	30.0 s
15 s	0.25 s	60 min	60.0 s
30 s	0.5 s	120 min	120.0 s
60 s	1.0 s	3 h	3 min
120 s	2.0 s	6 h	6 min
3 min	3.0 s	15 h	15 min
6 min	6.0 s	30 h	30 min
15 min	15.0 s	3 days	3 h

*Bristol Saybrook Series T-10.

The manner of relating the basic timing driving means to events to be time-controlled is subject to many configurations. Only a representation of the setting and control means used in connection with motor-driven timers is given in Fig. 37. In addition to gear-and-clutch arrangements for speeding up or slowing down the basic drive, many arrangements, including cams, notched disks, and drums with pins and rollers, are used to operate in proper sequence either various forms of electric switches or pneumatic poppet valves. The configuration of Fig. 35 is only one of many possibilities. Punched cards also have been used as information input to time-cycle controllers.

Interrupters

A simple form of single-cam time-cycle controller, known variously as a percentage timer, intermitter, or interrupter, continually opens and closes a single contact. The portion of the period of rotation (usually 1 rpm) during which the contact is open or closed depends on the adjustable dwell time of the cam. Such a device is often employed to change the rate of another timer by interrupting its motor-drive circuit.

Fig. 29 Adjustable plug-in solid-state ac-dc timer. (Series 319, *Automatic Timing & Controls.*)

Variants of this type of timer have been used to control the elements in electric furnaces or other heating devices, where they are known as *input controllers.* The "on" and "off" time of the heating element may be set manually or controlled automatically from deviation contacts in a temperature controller. In each case the load contact is operated for timed periods on a repeat-cycle basis.

A Note Regarding PCs

During the past decade with the wide acceptance of the PC and its continuing adaptation to new uses, there has been an impact on time-base program controllers. See the article *Programmable Con-*

Fig. 30 Multirange timer (also called a time-delay relay) incorporating metal oxide semiconductor digital circuitry. The timing ranges available are 1 and 10 s, 1 and 10 min, and 1 and 10 h. Reset times are: on delay, 100 ms maximum; off delay, 50 ms maximum; interval, 100 ms maximum. The load relay is double-pole double-throw (DPDT) hard-wired. The repeat accuracy is claimed to be ±1% of the setting or 15 ms under constant voltage and temperature. The setting accuracy is ±10% of the range, and the operating temperature 0 to 150°F (−18 to 66°C). The contact ratings are 5 A (resistive) at 120 or 240 V ac; ⅒ hp at 120 V ac, and 5 A at 24 V dc. (Series 328, *Automatic Timing & Controls.*)

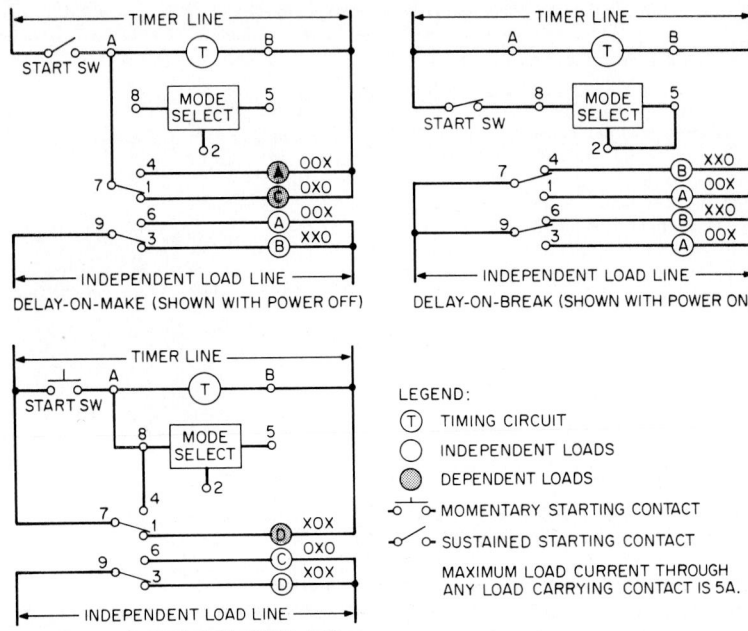

Fig. 31 Typical installation circuitry for various modes of the multirange timer shown in Fig. 30. *(Automatic Timing & Controls.)*

trollers in this Handbook section. The final selection of instrumentation for a given application still revolves around the time-honored consideration of such factors as simplicity versus complexity and performance versus cost.

Fig. 32 Motor-driven timer with a pushbutton digital setting with a choice of On Delay or Off Delay operation. A majority of industrial applications can be handled by one of the three standard ranges: 0 to 999.9 s, 0 to 999.9 min, and 0 to 999.9 h. The claimed repeat accuracy is 0.005%, the reset time is 200 ms minimum, and the minimum setting is 1 part in 10,000. By pushing the reset button at the left, the setting can be changed during a timing cycle. (Series 325, *Automatic Timing & Controls.*)

Step Programmers

Advancing a step at a time in response to external input pulses, the motor-driven step programmer shown in Fig. 38 closes and opens up to 20 load switches at each step in a predetermined but fully adjustable pattern of 2 to 20 interrelated steps. Available with two optional tap switches, the device can respond to a separate external impulse at each step of the program, thus providing interlocked sequence control of complex operations without the use of relays. The unit consists of 20 sliding segments. To actuate a load switch at any step in the program, the appropriate segment is moved to the right to actuate a load switch, or to the left if the switch is not to be actuated. The segments are held securely in position but can be readily changed to the opposite position at any time. The cam shaft assembly can be removed from the unit without tools and without disturbing the

Fig. 33 Computing timer which covers an overall span of 0.01 s to 999 h in nine switch-selected ranges of 0 to 9.99, 99.9, or 999 s, minutes, or hours. The three-digit cycle progress display (top) times up or down from the set point. After timing out, the display either stops or proceeds to display the time elapsed after timing out. To the right of the three-digit display, a timing bar blinks once per second during the timing cycle and more rapidly after timing out. At the left of the digital display, an inverted triangle marker turns on when the delayed relay is energized. The device is equipped with a self-diagnostics program. Through an internal microcomputer, the unit keeps track of the set point throughout the time cycle. Whenever there is a change in the set point, even during a cycle, the microcomputer instantly recomputes the time remaining and determines the time-out. This capability is used in time-down modes, allowing the cycle to be shortened without a loss of accuracy. Digitally clocked by the microcomputer, the timer's reset time is consistently of the same duration, unaffected by line voltage variations. The unit is not subject to false reset by momentary power interruptions (less than 30 ms).

When power is applied to the timer, the parameters of one group of input-output ports—for the range switches, display mode, and 50/60 Hz jumpers—are read into random-access memory (RAM) registers and stored for use during the timing cycle. The parameters at the other input-output ports—for the set-point switches and clock output—are continuously sampled so that any change is immediately reflected in the appropriate RAM registers.

When the microprocessor receives a clock pulse, it increments the time already stored in RAM. In the time-down mode, it subtracts the pulse from the stored time interval, displaying the amount of time remaining before time-out. In the time-up mode, it adds the pulse to the stored total, displaying the amount of time that has elapsed since the start.

At each computation, the microprocessor compares the new incremented total with the current set-point switch settings. As soon as the total is equal to or greater than the switch settings, the microprocessor times out, energizing the delayed relay and the inverted-triangle display marker.

In the up-and-go mode, the microprocessor continues to count clock pulses after the delayed relay is energized, displaying the time accumulated after time-out until the timer is reset. In the down-and-go mode, it reverses direction at time-out and counts up from zero, giving a direct overshoot reading. (Series 365, *Automatic Timing & Controls*.)

program—and replaced with a separate assembly for a different program. In this case, the cam shaft assembly provides a storable memory of a complete program. The circuit for the programmer is given in Fig. 39.

The unit is available with a range of steps, numbering from 2 to 20. Six, 12, or 20 cams can be specified. The cams are made of high-impact plastic and are 1.75 in (44.5 mm) in diameter. There is a load switch for each cam. The contact rating is 10 A at 115 V ac (noninductive), and the contact action is SPDT. Contact life expectancy is 200,000 operations at 10 A, and 1.5 million at 5 A (average). The operating temperature range is 0 to 125°F (−18 to 52°C), and the power requirement is 15 W.

TIME-SCHEDULE CONTROLLERS

A time-schedule controller links time to the control of a process variable. The value of the quantity or condition involved (such as temperature) is regulated in accordance with a predetermined time schedule. Devices for accomplishing this action have three basic functions: (1) automatic control of

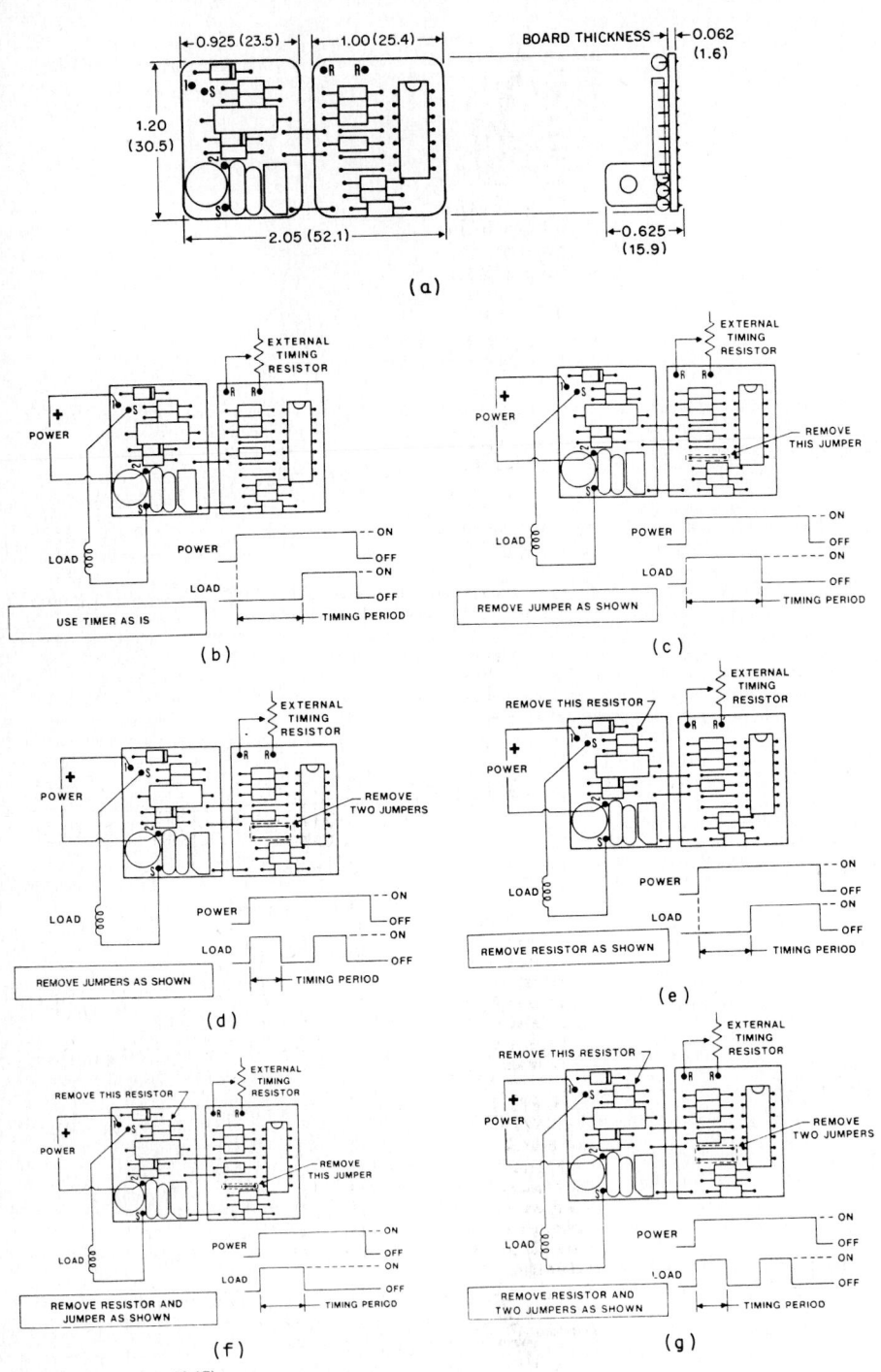

(a)

(b)
EXTERNAL TIMING RESISTOR
POWER
LOAD
POWER
LOAD
USE TIMER AS IS
ON
OFF
ON
OFF
TIMING PERIOD

(c)
EXTERNAL TIMING RESISTOR
POWER
LOAD
REMOVE THIS JUMPER
POWER
LOAD
REMOVE JUMPER AS SHOWN
ON
OFF
ON
OFF
TIMING PERIOD

(d)
EXTERNAL TIMING RESISTOR
POWER
LOAD
REMOVE TWO JUMPERS
POWER
LOAD
REMOVE JUMPERS AS SHOWN
ON
OFF
ON
OFF
TIMING PERIOD

(e)
REMOVE THIS RESISTOR
EXTERNAL TIMING RESISTOR
POWER
LOAD
POWER
LOAD
REMOVE RESISTOR AS SHOWN
ON
OFF
ON
OFF
TIMING PERIOD

(f)
REMOVE THIS RESISTOR
EXTERNAL TIMING RESISTOR
POWER
LOAD
REMOVE THIS JUMPER
POWER
LOAD
REMOVE RESISTOR AND JUMPER AS SHOWN
ON
OFF
ON
OFF
TIMING PERIOD

(g)
REMOVE THIS RESISTOR
EXTERNAL TIMING RESISTOR
POWER
LOAD
REMOVE TWO JUMPERS
POWER
LOAD
REMOVE RESISTOR AND TWO JUMPERS AS SHOWN
ON
OFF
ON
OFF
TIMING PERIOD

(See caption on page 18.87)

18.86

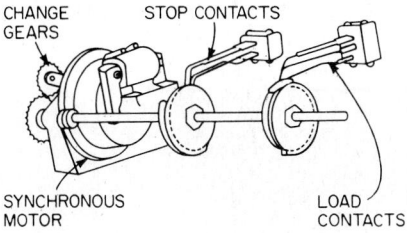

Fig. 35 One form of mechanism used in a time-cycle controller.

Fig. 36 Timing diagram for a time-cycle controller.

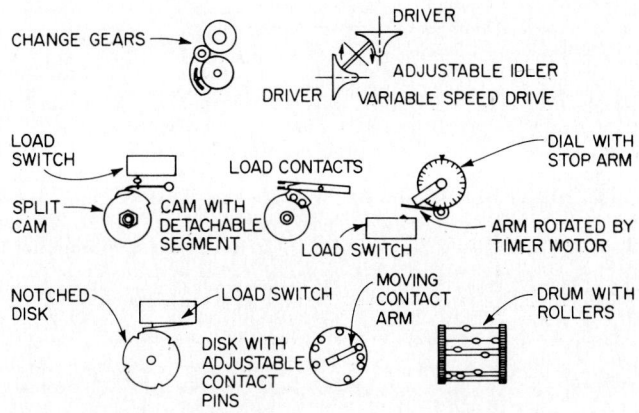

Fig. 37 Some setting and controlling means of motor-driven timers.

Fig. 34 Design of a solid-state multifunctional timer for maximum flexibility. (*a*) Reproduction is close to actual size. Dimensions are given in inches (millimeters in parentheses). Two printed circuit boards, connected by three wires, are used. This permits bending of the timer for operation in very cramped spaces. Five timing ranges are available on the two basic models (12 V ac-dc and 120 V ac): 0.1 to 10, 1 to 30, 2 to 100, 10 to 500, and 30 to 900 s. As shown by the diagram of fundamental models, the timers can be set for fixed timing periods or connected to variable resistors for variable timing period control. To wire either of the basic models [shown in (*b*)], if the timer is to operate as a fixed timing component, only four wires must be connected (two wires for the power and two wires for the load). If the timer is to be made adjustable by means of a potentiomer, the two *R* wires must be connected to the potentiometer as shown.

Delay-on-make: To create a delay-on-make operation, both jumpers must be left on the circuit board as in (*b*) and (*e*). When power is applied to the timer, the load remains off for the preset delay period. At the end of the delay period, the timer turns on the load. To recycle, the power must be interrupted for at least 50 ms.

Interval timer: To create interval operation, the jumper shown in (*c*) and (*f*) must be cut. When power is applied to the timer, the load energizes immediately and remains energized for the length of the preset delay period. At the end of the preset delay period, the load deenergizes. To recycle, the power must be removed and reapplied.

Repeat-cycle: To create repeat-cycle timing, both jumpers as shown in (*d*) and (*g*) must be removed. When power is applied to the timer, the load turns on and off in a repeat-cycle manner for as long as power is applied. The length of time the load is turned on always equals the length of time the load is turned off and is determined by a timing resistor, which may be fixed or variable.

To change the voltage setting: The basic model (12 V ac-dc) shown in (*b*) is wired to operate on a 12-V source, and no changes are required. To use the same model on a 24-V source, the resistor is cut off the circuit board as shown in (*e*). (Series 4784 and 4785, *Artisan Electronics.*)

18.87

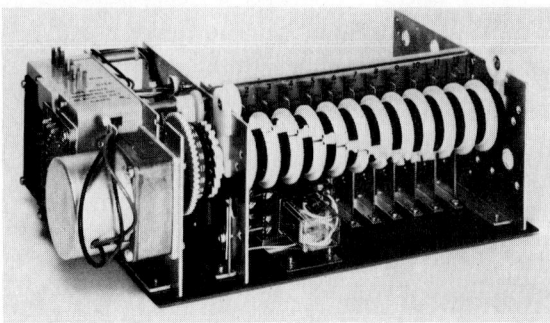

Fig. 38 Motor-driven step programmer. The unit shown has 12 cams. Up to 20 cams are available on larger units. The device provides interlocked sequence control of complex operations without the use of relays. (Series 1800, *Automatic Timing & Controls*.)

the variable, (2) timing, and (3) changing the set point of the controller. In essence, *a time-schedule controller* is an automatic controller whose set point is automatically positioned for a preselected time interval. It may control a variable at a number of rates of change, as well as at different fixed values, for a series of time intervals. In other words, a record of the variable may show a series of smooth rises and falls, or a relatively staccato, steplike variation. Other names for time-schedule controllers include *contour, time-variable, time-pattern,* and *program controllers.*

Time-schedule is the designation officially adopted by the Scientific Apparatus Makers Association (SAMA PMC20.1-1973), wherein the device is defined as *"a controller in which the set point or the reference-input signal automatically adheres to a predetermined time schedule."*

The function of a time-schedule controller is required by numerous processing and manufacturing applications, including annealing and other metallurgical and ceramics operations; food processing operations, including pasteurizing, cooking, and canning; oven and furnace operation; environmental test chambers; dyeing processes; air conditioning; polymerizing; vulcanizing; and curing, among many others.

The basic elements of a time-schedule controller are (1) a timing means, (2) a controller to maintain the process variable at the desired value, and (3) a link between the timing element and the controller, i.e., a means of changing the set point of the controller in accordance with a prescribed time plan. Like other instrumentation systems, time-schedule arrangements can be quite simple, as well as quite complex. For example, if a material needs to be maintained at a given temperature for a given length of time and where the temperature rate of rise is not critical (e.g., allowing a furnace to heat up at the maximum possible rate) and where the temperature rate of fall after the timed period is not critical (natural cooling), then all one needs is a very simple interval timer that cuts in the temperature controller at a certain time and later cuts the controller out. Although this is not what is usually meant by the term "time-schedule controller," the situation does qualify under the definition.

In earlier approaches to time-schedule control, which are, incidentally, methods still considered quite acceptable, the timing mechanism of a recording instrument served not only to drive the chart in the traditional manner but also as the timed driver of the link to the set point of a controller. This permitted the time-schedule controller to be housed in one case along with the recording, indicating, and process variable controlling means, or in adjacent housings.

Contoured Cams and Followers

This concept was developed for use in large-case circular chart mechanical-type instruments. A metal cam blank 12 in (305 mm) in diameter and printed in the fashion of a typical circular chart is cut to the program contour desired. See Fig. 40. A roller mounted on a lever rides along the edge of the cam and via linkages continuously adjusts the set point of the controller, which may be electric or pneumatic. Where a record is required, two instruments may be needed, one being the conventional recorder-controller, mounted either above or to the side of the instrument that houses the cam. Where only a single variable is required, all functions can be in one housing.

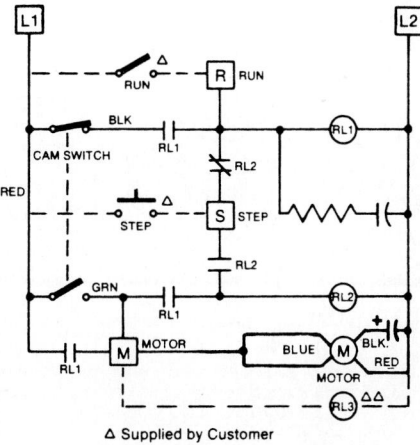

Fig. 39 Circuit of the motor-driven step programmer shown in Fig. 38. The unit advances a step every time it receives a signal from an external device, such as a limit switch, temperature switch, timer, or pushbutton. Without an optional tap switch, the device usually receives the step input signal from the same device; with a tap switch, as many as 20 such devices may be connected to different steps in a 20-step program so that the unit advances only when it receives a signal from the desired device.

Power is applied continuously to the $L1$ and $L2$ terminals, but the unit does not step until power is applied to the "step" terminal for at least 0.030 s. The pulse motor advances a single step on the make of the step signal, regardless of how long the signal remains on.

At each step, each single-pole double-throw (SPDT) load switch is actuated whenever the same segment is in the right-hand position, and not actuated when it is on the left. *Homing,* i.e., running without a pause through unused positions in the 20-step program, is accomplished by applying power to the "run" terminal. This can be done through a separate homing switch, through one of the load switches selected for this purpose, or through one of the tap switches.

When the programmer is specified with an optional tap switch, a chassis-mounted relay is provided. Its coil is factory-wired between the "motor" and $L2$ terminals, and its normally closed contact is wired in series with the wiper contact of the tap switch to protect it against arcing. All terminals are factory-wired to the unit's control module and motor. The unit measures approximately $9 \times 5 \times 19$ in deep ($229 \times 127 \times 483$ mm deep). *(Automatic Timing & Controls.)*

△ Supplied by Customer
△△ Supplied with Tap Switch Models

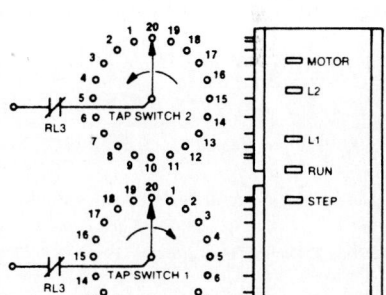

In addition to using a roller to track a cam, other means have been used, such as a photoelectric pickup that follows the markings of the desired program on a regular paper circular chart. With suitable circuitry, the output of the optical pickup can adjust the set point of pneumatic or electronic instruments.

Strip Chart Recorder-Controller with a Motor-Driven Set Point

This general type of system has been used for many years for temperature-time programming in the heat-treating and metalworking industry but has enjoyed acceptance in other fields, including environmental test chamber control. A block diagram of such a system is shown in Fig. 41. In this instrument, there are two motor-driven set-point variations:

1. *Synchronous Drive.* The set-point mechanism is positioned by a digitally pulsed motor which operates at a fixed (synchronous) speed with respect to power line frequency.

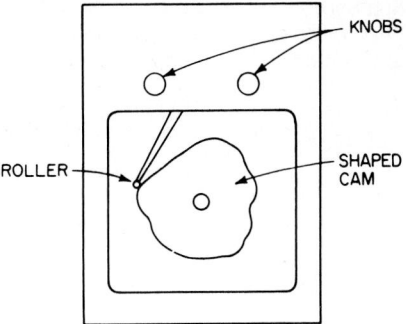

Fig. 40 Cam cut to the contour of a desired program for a single variable.

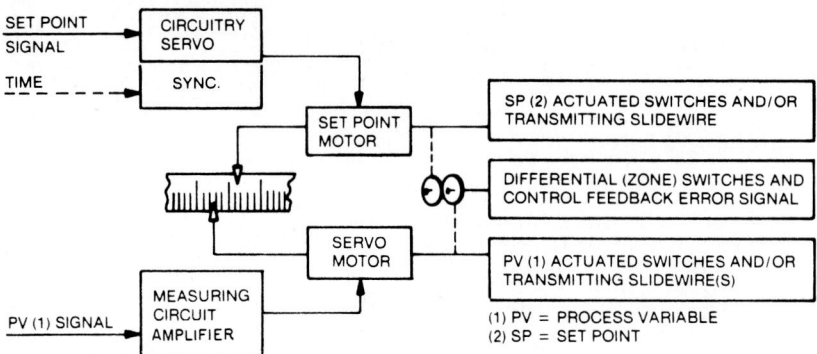

Fig. 41 Block diagram of a strip chart recorder-controller with a motor-driven set point for time-schedule control. *(Honeywell.)*

Drive speeds are programmed using plug-in printed circuit wiring boards. The basic up- and down-scale drive speeds may be different. Either or both basic rates may be slowed by percentage timers.

2. **Servo Drive.** This design is for master-slave control system requirements in applications such as cascade-ratio control systems and multiple-zone programming. The set-point mechanism is driven by a null-balance servoamplifier-motor system in response to a remotely generated signal. This signal may be a current (4 to 20 mA dc) or a voltage (1 to 5 V dc) from a remote slidewire (500, 1000, or 5000 Ω).

Both of the motor-driven set-point variations just described have three types of switching actions:

1. **Process Variable-Actuated (Backset).** Operates from the process variable servo shaft. Up to four SPDT contact switches, or two transmitting slidewires and one switch, or two switches with a transmitting slidewire can be set in relation to any point on the instrument scale. They are actuated by the relative pen or indicator position along the scale.

2. **Set-point Actuated (Backset).** Operates from the shaft of the set-point drive. The combination of switches and slidewires are the same as those of the process variable, except that they are actuated in relation to the set-point position along the scale.

3. **Differential Actuated (Zone).** Operates from the process variable set-point differential mechanism. Up to four SPDT switches are actuated by the position of the process variable relative to the position of the set point. Each switch can be independently adjusted to operate anywhere in a zone ±10% of full-scale span around the set point, and this zone will "track" the set-point position.

Digital Control Programmer

In an effort to reduce space requirements and to further utilize the advantages of solid-state technology, the microprocessor-based instrument shown in Fig. 42 was developed in 1982. The instrument provides a means for storing set-point programs. The programs are used with an integral electric proportional controller to control batch processes such as heat-treating and environmental testing. The set points are combined with the lengths of time they are to be held (soak segments), and the rates or times required to change from one set point to another (ramp segments), to make up a program. Programs can be as simple as one soak segment or as complex as 99 segments. A typical program with 3 soak segments and 2 ramp segments is shown in Fig. 43. The microprocessor performs several functions. In one application concerned with wet- and dry-bulb temperature measurements, the microprocessor subtracts the wet-bulb temperature from the dry-bulb temperature to obtain the wet-bulb depression, converts the wet-bulb depression to percent relative humidity

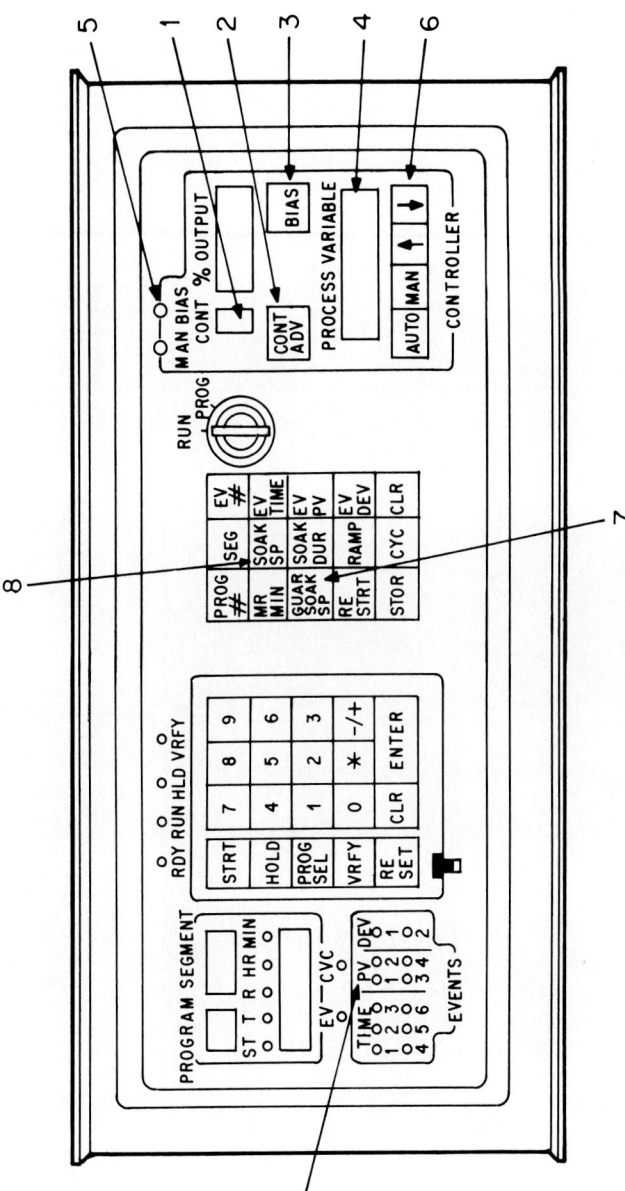

Fig. 42 Control panel of a digital control programmer. In this microprocessor-based unit, all the functions required for temperature-versus-set-point programming and three-mode control for up to nine separate programs involving up to three separate zones or process parameters can be handled. (1) The controller number display indicates the controller whose values are being displayed in the "1% output" and "process variable" windows and whose bias, tuning constants, and output limits may be changed. (2) The controller advance pushbutton advances numbers in the controller number display sequentially. (3) The bias pushbutton is pushed once to display (4) the amount of bias (if any) on the set point of the controller whose number is displayed. The bias LED (5) flashes, and the amount of bias may be changed by using the increment-decrement buttons (6). (7) The soak set-point pushbutton is used, when programming, if the soak is to be delayed until process variable (PV) no. 1 is within 5° (or 10° as desired) of controller no. 1 set point. (8) The soak set-point pushbutton is used, when programming, if the soak timing begins as soon as the set point reaches the soak temperature. (9) The PV and deviation events are common to all channels. For example, if PV event no. 4 is set at 500°, it will have an output if any PV exceeds 500°. (DCP 7700, *Honeywell*.)

18.91

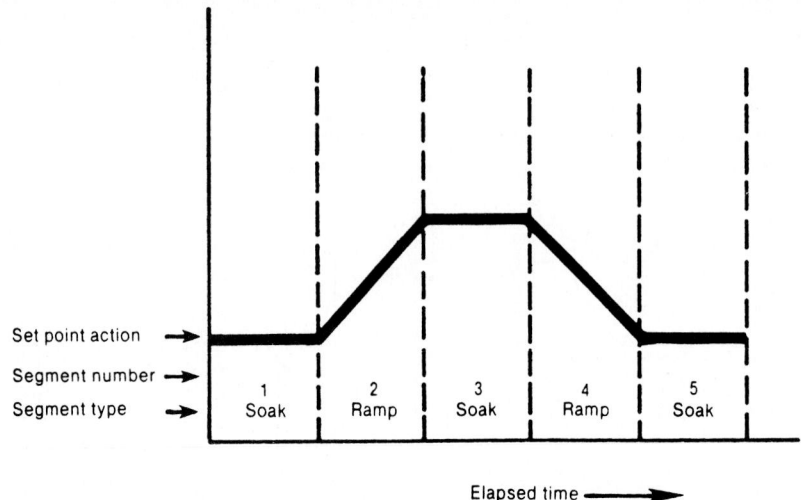

Fig. 43 Typical five-segment program.

Fig. 44 Front panel of a microprocessor-based device, one function of which can provide time-schedule control. (MicRicon, *Research*.)

(%RH), and then compensates the %RH value for atmospheric pressure. This value of %RH is then compared to the programmed set point, and the output of the controller is acted on accordingly.

The microprocessor-based device shown in Fig. 44 combines a dual-profile programmer with three-mode controllers. The basic system provides two three-mode controllers and two profiles (one time-base)* of variable set-point-versus-time programming. The unit can be expanded up to eight dual-profile programmers and/or up to 16 control loops. See also Fig. 45.

*To effect time-schedule control.

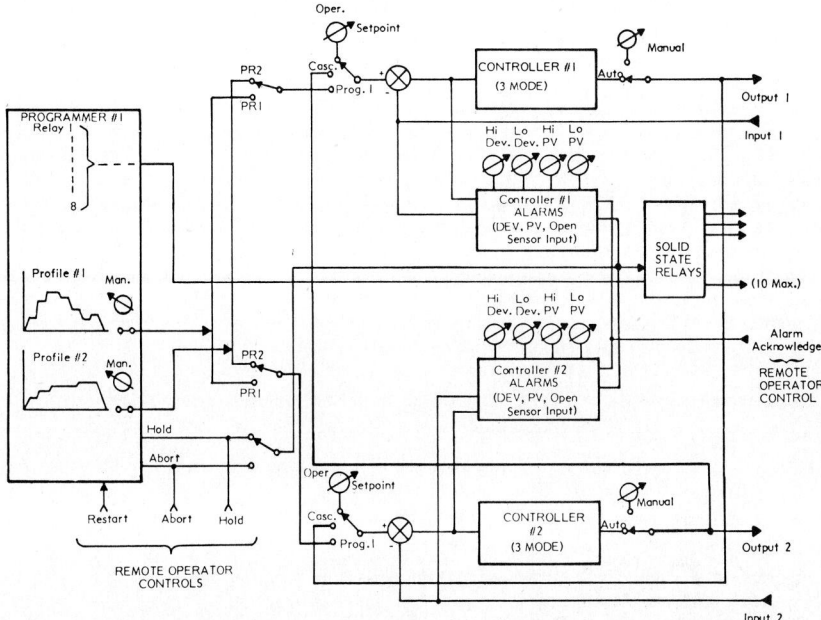

Fig. 45 Input-output controller representation of the unit shown in Fig. 44. (MicRicon, *Research.*)

A Practical Look at Distributed Instrument Systems

by
John P. King*

The microprocessor revolution has made its mark on the process control industry. Video-based operations work centers are displacing conventional panels in control rooms. Shared-function modules are displacing discrete components (recorders, indicators, totalizers, and controllers). High speed communications networks are linking once independent control rooms. Process computers are becoming commonplace.

The revolution has not occurred without generating certain transitional problems. These changes are taxing the abilities of almost everyone affected. Classic job skills within user organizations and

*The Foxboro Company, Springfield, New Jersey.

among contractors and vendors are being strained severely. Even the basic vocabulary required to communicate is causing confusion.

Why are these systems being accepted so enthusiastically? Given the complexities and uncertainties of the competitive market of the early- and mid-1980s (much of which, ironically, can be attributed to the use of microprocessors and computers in fields such as telecommunications and banking), the tools of the past are no longer adequate. Prior to the early 1970s, when energy was cheap, feedstocks plentiful, markets stable, and competition predictable, the role of instrumentation was to ensure stable operation of a process within the equipment design constraints. Feedforward and other optimizing techniques were being installed, but these were the innovators, the pioneers, the exceptions. The 1980s demand much greater productivity in the face of increasingly uncertain operating conditions and objectives. *Flexibility* has become a key word.

The demands for greater efficiency, more flexibility, and better product quality are almost insatiable. These needs require "working smarter" by taking advantage of the power and speed of the microprocessor, its calculation and logic capabilities, its information storage and retrieval, and its ability to follow complex instructions. Objectives of these systems include:

1. Reduced operating costs, particularly fuel savings, via optimizing control strategies

2. Reduced product giveaway through more uniform operation and tighter control enforcement

3. Reduced capital investment through improved scheduling of capital equipment (or increased production from existing equipment)

4. Improved responsiveness to operating philosophy changes, e.g., changes in product distribution or raw material specifications

5. Increased production uptime through the incorporation of contingency strategies which anticipate various plant failure conditions (e.g., a feed pump out of service)

6. Improved plant safety and plant equipment protection via similar contingency strategies

7. Availability of better and more timely information to plant operations and maintenance managers to enable them to keep a plant running longer and more efficiently, and to plant managers to enable them to run more profitable operations (inventory management, quality audits)

8. Better integration of plant operations through improved information and more flexible control practices

Architecture

A typical microprocessor-based distributed instrument system is composed of five major subsystems:

1. High speed, long-distance network communications

2. Operations work centers

3. Microprocessor-based controllers

4. Auxiliary data collection stations

5. Information management-supervisory stations

These subsystems are arranged in the modular architecture shown in Fig. 1.

The most obvious benefit of a systems architecture of this type is its contribution to the design of an integrated plantwide information and control system through the distribution of various functions throughout the plant. Thus, operators and unit supervisors can be made aware of conditions outside their domains (e.g., upstream or downstream bottlenecks or plant utilities constraints) and take corrective action *before* these factors adversely affect their operations.

Distributed architecture also can provide functionally dedicated stations to various disciplines, e.g., maintenance and engineering, so that system and process information can be dispersed throughout the plant without interfering with operations. The operator can check pressure drops across heat exchangers, for example, to determine the degree of fouling without leaving the maintenance shop.

Communications

Certain functional characteristics of the communications subsystem are required to accomplish these objectives. First, the communications subsystem must allow a large amount of information to be passed in a very brief time interval. It must also be extremely responsive to individual requests, returning information in some cases in fractions of a second.

The communications subsystem must be capable of supporting a large number of stations of various types, each having its own specialized needs for information transfer. A moderate number of these stations must be capable of initiating communication as well as responding to requests, e.g., operations or maintenance work stations.

The communications subsystem must be capable of transmitting over long distances to stations or clusters of stations located along the network. Because control rooms may be in environmentally harsh, dangerous, or electrically noisy surroundings, the communications line should be protected and the transmitted information rigorously checked. Finally, because of the dependence of everything else in the system on its proper operation, the communications subsystem must be extremely fault-tolerant; i.e., it must continue to function when damaged. Of course, a corollary to this requirement is the ability to be repaired rapidly while in operation.

Network communications of the type described are extremely complex—so much so that the International Standards Organization (ISO) has separated the system into seven specifications (called *layers*) and developed a mechanism to ensure compatibility. The ISO approach was to develop a model called the *open systems interconnection* (OSI) reference model. The separate layers are:

1. *Physical.* Electrical, mechanical, and packaging specifications of circuits

2. *Data Line.* Transmission of data in the network—message framing

3. *Network.* System addressing and routing

4. *Transport.* Security issues

5. *Session.* Communication and transaction management

6. *Presentation.* Transformation of various types of information, e.g., file transfers

7. *Application.* The actual information exchange process

While this work is in its infancy, a number of interesting approaches to message framing and obtaining control of the network are currently being used. Regarding message framing, two techniques are in use.

The first is a structured message format which consists of a fixed length frame or serial information subdivided into a number of functional elements, each of a fixed length. A typical fixed-frame construction is shown in Fig. 2.

The second approach, which has been adopted by most of the distributed instrument systems in the United States, is the *high-level data link communications* (HDLC) protocol. This method offers significantly more flexibility and more easily accommodates such functions as computer-to-computer communications. The disadvantage is that the length of the frame must be expanded to support the enhanced flexibility, which could result in lower data throughput for a given communications speed. To overcome this condition, particularly with multifunction controllers, a single frame is used to communicate many pieces of information, e.g., all measurements at given controller stations. A typical HDLC frame construction is shown in Fig. 3.

Control of the network as of the mid-1980s is being accomplished by three distinctly different arbitration mechanisms:

External Arbitrator

A mechanism wherein a single (or redundant) device grants permission to one of the various member stations to initiate communication with another station. The station is usually chosen on a first-come basis, with request messages queued. One novel approach for this request-grant communciation is to use subdata channel frequencies on the communications network. See Fig. 4a.

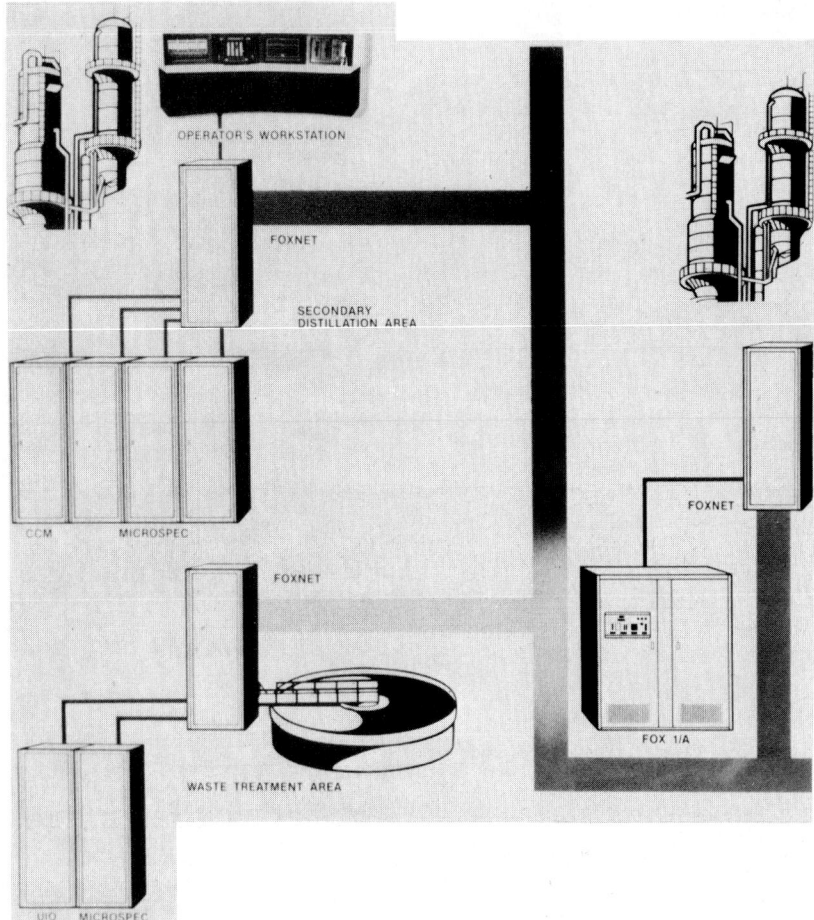

Fig. 1 Total process management and control. *(Foxboro.)*

Carrier Sense Multiple Access with Collision Detection (CSMA/CD)

All stations that need to initiate communication "listen for a lull" on the line and try to assume control once a prior communication transaction is completed. If two stations begin transmitting simultaneously, the resultant message is garbled, a condition sensed by the transmitters. They cease transmission for a random period, and the process for assuming control is repeated by all interested stations. See Fig. 4*b*.

Peer Group Token Passing

Each station in the network uses an embedded communications algorithm to determine if it is the next to assume control. This eliminates the need for an arbitrator, while providing a more orderly method for succession of control than is possible with CSMA/CD. Early designs employed a simple ring network where control was passed to adjacent stations. See Fig. 4*c*. More advanced algorithms are emerging which account for various failure modes, e.g., token passing through a failed station, high-priority urgent multiway communications needs; and recovery from a lost token.

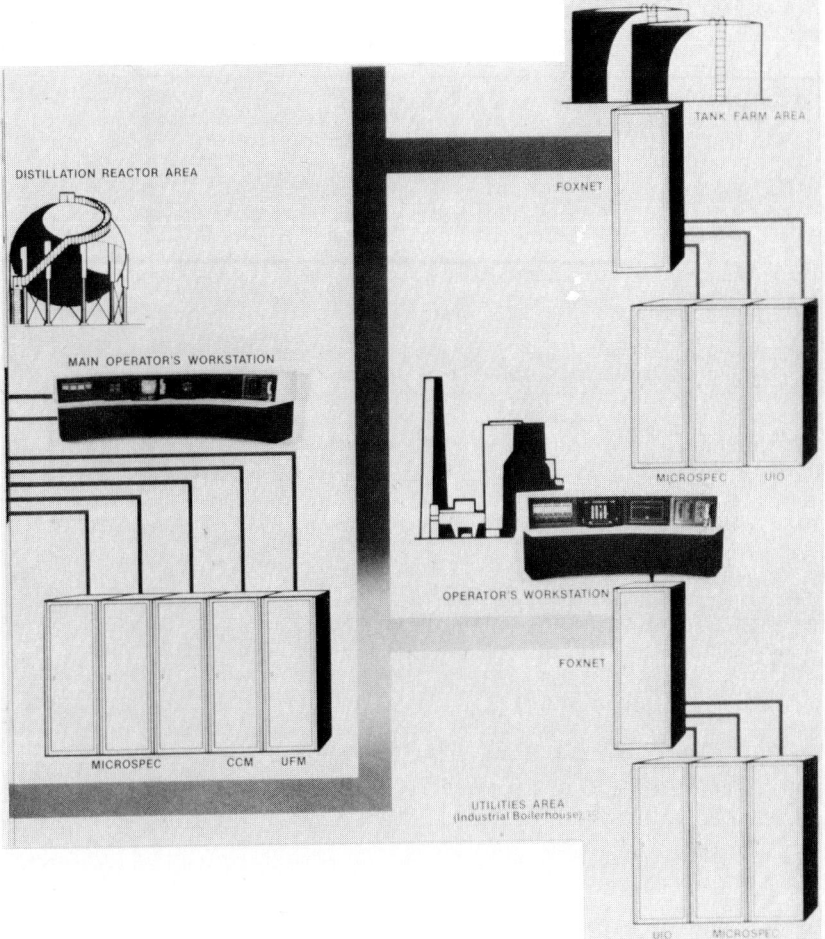

Fig. 1. (*Continued*)

A variety of technologies also exist which affect the media of communication:

1. *Twisted pair* configured in a daisy-chained manner with a speed of up to 250,000 bits/s and distances of about ∼1 m (1.6 km)

2. *Multiconductor cable* (e.g., IEE 488) for high speed (1 million bits/s), short-distance (250 ft, 75 m) communication

3. *Coaxial cable* communication for high speed (10 million bits/s), long-distance [several miles (kilometers)] multistation (as many as 1000) requirements

4. *Fiber optics,* where light pulses are transmitted in lieu of electromagnetic signals. Theoretically, this method offers the best potential. The line has no *RC* characteristics and therefore can transmit at much greater speeds. Fiber-optic cable is immune to electromagnetic interference. Since it is not a conductor of electricity, it can be run through hazardous areas. Further development work is required to make it commercially feasible in applications where multiple stations are required. One hurdle is the high degree of signal attenuation at the tee connection at each station. In the

START	TYPE CODE	SOURCE ADDRESS	DESTINATION ADDRESS	INFORMATION	FRAME CHECK

ELEMENT	FUNCTION
START	Identify beginning of frame.
TYPE CODE	Identify frame type (i.e., read, command, data reply...).
SOURCE ADDRESS	Station address of the requestor (return address).
DESTINATION ADDRESS	Address of station to which the message is directed.
INFORMATION	Functional or literal data which are meaningful only to the station.
FRAME CHECK	An integrity check on the contents of the frame.

Fig. 2 Structured message format which consists of a fixed-length frame or serial information subdivided into a number of functional elements, each of a fixed length.

FLAG	SOURCE ADDRESS	DESTINATION ADDRESS (OPTIONAL)	CONTROL CODE	INFORMATION FIELD LENGTH	INFORMATION FIELD	SECURITY FLAG CHECK

ELEMENT	FUNCTION
FLAG	Start/stop code.
SOURCE ADDRESS	Station address of the message frame sender.
DESTINATION ADDRESS	Address of station to which the message is directed.
CONTROL CODE	Identifies network level message type (i.e., broadcast to all stations, chained message, etc.).
INFORMATION FIELD	Station level protocol message (as opposed to communication network level) identifying length of field, data typing, and use.

Fig. 3 Message framing approach adopted by most of the distributed instrument systems used in the United States—the high level data link communications (HDLC) protocol.

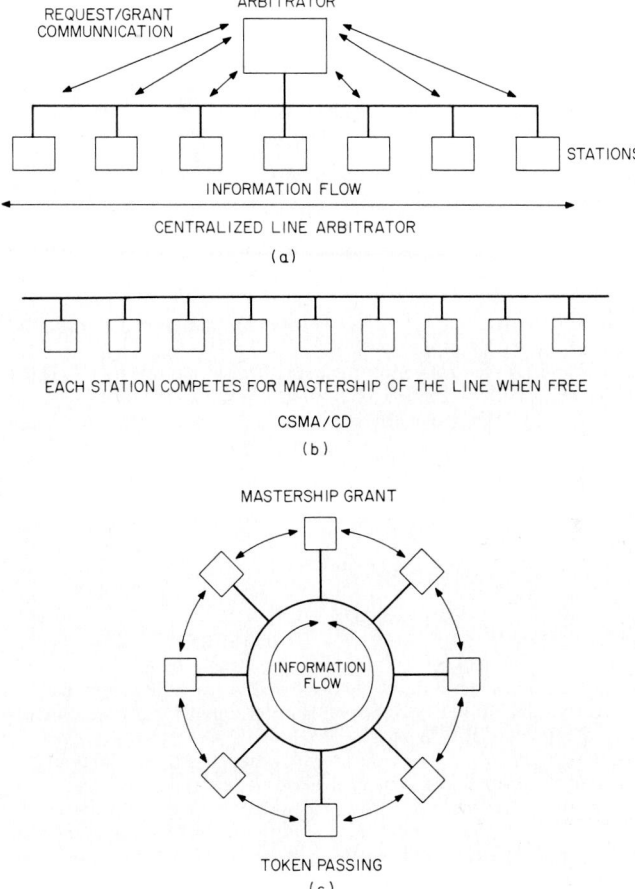

Fig. 4 Different arbitration methods.

early 1980s, each station must either boost or retransmit information, adding to the cost and reducing overall reliability.

Operations Work Centers

Another major benefit associated with distributed systems is the ability to keep various plant personnel better informed about plant operating conditions and therefore increasing efficiency. A major factor is the timely and accurate translation of raw data into meaningful patterns and trends in a manner that clearly portrays a plant situation requiring a decision or action. The intelligent work station (sometimes referred to as a man-machine interface) serves this purpose. Thus, it is a useful tool for everyone from a shift operator attempting to recover from a process upset to a plant manager attempting to balance production rates of various products against current inventories and capital utilizations.

There are several salient characteristics of these human-engineered intelligent work stations. The essential functions that must be performed by the critically important process operations work stations are:

1. Monitor plant operations
2. Call attention to process upsets and alarms
3. Provide a facility for taking corrective action when necessary
4. Present various trends occurring within the process
5. Interact with plant personnel performing functions outside the scope of the control system
6. Maintain good records on plant operating characteristics
7. Ensure that the control system is functioning properly

These functions are packaged into console-based operations work centers typified by the work center shown in Fig. 5.

Work stations typically contain between two and five video displays (although 1 is reasonable and as many as 20 have been packed into some panel-oriented work centers). The stations are supplemented by a variety of information devices, including:

1. Alarm printers
2. High speed report writers
3. Video copiers
4. User-assignable multipen trend recorders
5. Annunciator lamps
6. Audible signals (messages)

The primary information device is the videoscreen(s). Because they are intended to replace and, in fact, extend the capabilities of a conventional panel, a considerable amount of ergonomic design has been involved in producing these units.

The most difficult challenge, of course, is providing mechanisms for monitoring hundreds (sometimes thousands) of process parameters in an effective manner with a very limited display area (the combined surface area of the video screens at the work center). Fundamental to all designs is the concept that a human being can assimilate only a relatively small amount of information at a time. Also, in practical terms, with a large panel, the operator must stand so far back that detailed infor-

Fig. 5   Representative console-based operations work center. *(Foxboro.)*

mation becomes indistinguishable. For example, considering that, at a 20° peripheral view angle, a person must stand more than 68 ft (21 m) from the center of a 50-ft (15-m) panel to see it all. The work station approach, then, is to create a set of display hierarchies which will enable plant operators to view the process at the equivalent of four or more distinct distances from a conventional panel with better and more detailed information available at each level. Figure 6 depicts one such hierarchy.

To date, there has not been full agreement within the process control field as to the best format for representing these data. There are two competing approaches: (1) panelboard emulation and (2) process graphic representations. As with most well-conceived approaches, each type offers strong advantages. Most commercially available systems include elements of both approaches. There is far less agreement with regard to the method for moving among these displays. Among the techniques employed are:

Keyboard
1. Dedicated function keys
2. Group numbering addressing
3. Tag referenced addressing
4. Display-dependent function keys
5. Cursor positioning, using dedicated directional keys

Other Techniques
1. Light pen
2. Mouse (auxiliary cursor positioning technique)
3. Touch-sensitive screens
4. Joystick (pan and zoom)
5. Voice actuation (a possibility for the future)

Again, each of these techniques has advantages and disadvantages. Some represent first-generation attempts; others are still in the experimentation stage. The acceptable approaches include a number of common characteristics:

1. Simplicity of operation
2. Reinforcement of relatively few patterns of use
3. Fault tolerance (minimization of the occurrence of inadvertent actions; ease of recovery from improper actions)
4. Reliable design requiring relatively little maintenance

The most popular current approach employs a self-prompting or *various function keyboard* with keys which relate directly to the display on the video screen. This technique segments the screen into a number of partitions which are directly related to an equal number of function keys. Depressing a key selects the associated screen segment and expands its information content in the subsequent display, providing a sort of telescoping effect. See Fig. 7. Other variable-function keys provide the operator with the ability to move up or across the display hierarchy in a similar associative manner.

Another approach which currently appears to be gaining acceptance is the *joystick concept*. This is functionally similar to the variable function key approach, except that the entire variable-function keyboard is replaced by a multidirectional lever which enables the operator to move up, down, or across the display hierarchy. The operator can move closer to the center of the display, thus expanding the detail of the area immediately surrounding the center at the expense of losing the information at the edges altogether. This is called *zooming*. The operator can also translate the display in a number of directions to view adjacent information at the same level of detail. This is called *panning*.

Microprocessor-based systems really begin to excel when compared with conventional approaches in such areas as comparing, sorting, calculating, logical decision making, storing infor-

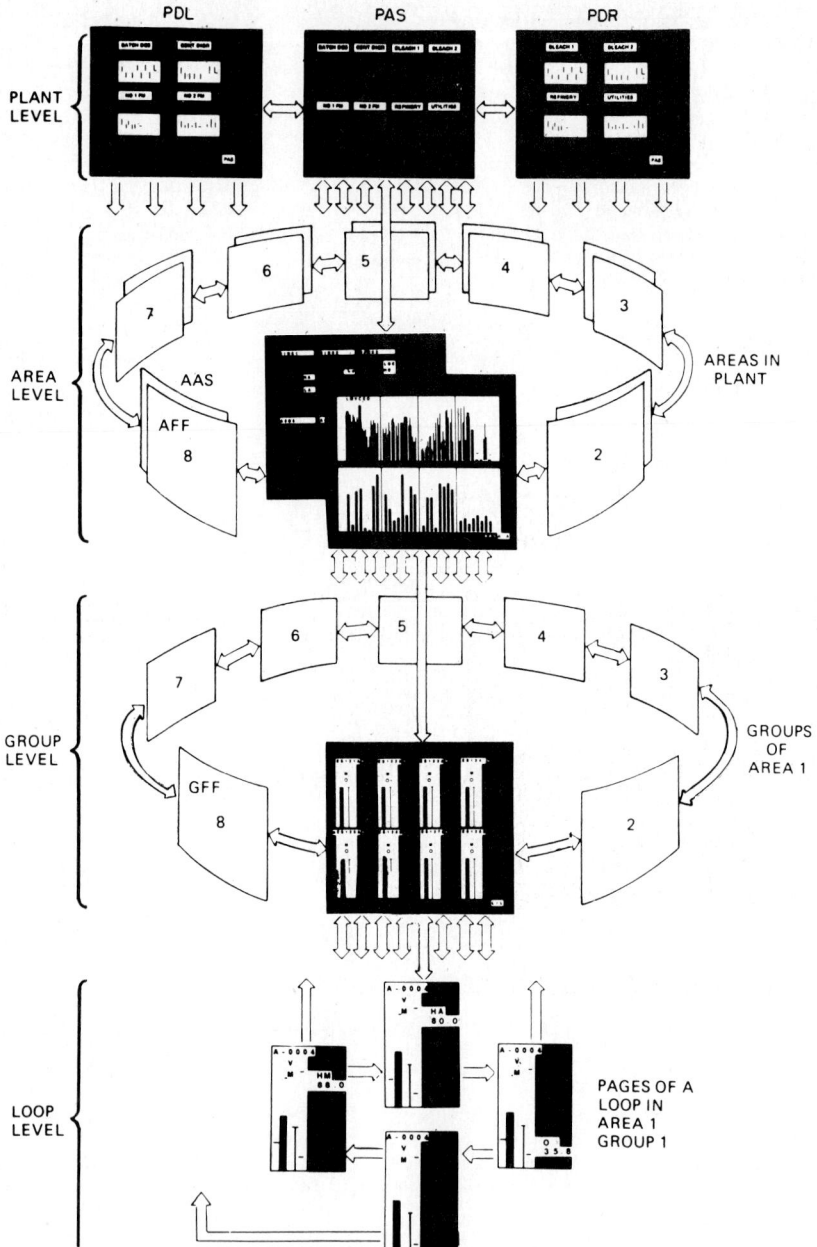

Fig. 6   Process display hierarchy.

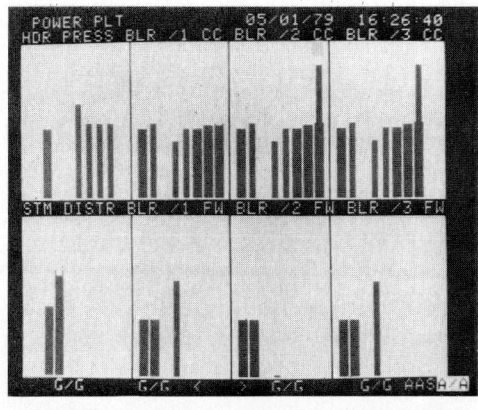

AREA LEVEL DISPLAY PORTRAYING
8 SCREEN SEGMENTS OF 8 LOOPS
EACH FOR A TOTAL OF 64 CONTROL LOOPS.

KEYBOARD WITH 8 FUNCTION BUTTONS—
EACH BUTTON IS ASSOCIATED WITH A
DISPLAY SEGMENT

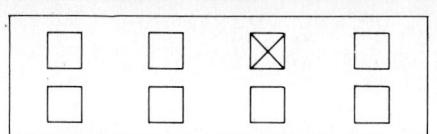

DEPRESSING THE THIRD KEY FROM THE LEFT,
TOP ROW EXPANDS THE VIEW OF THE EIGHT LOOPS
ASSOCIATED WITH THAT SEGMENT TO FILL THE
SCREEN

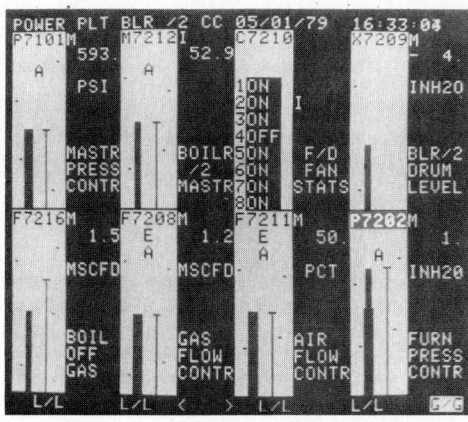

GROUP LEVEL DISPLAY PORTRAYING
8 SCREEN SEGMENTS CONTAINING
1 LOOP EACH

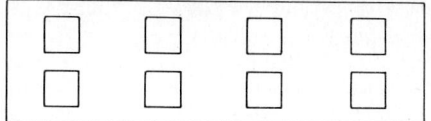

FURTHER DEPRESSION OF ONE OF THE EIGHT
FUNCTION KEYS ENABLES AN INDIVIDUAL
LOOP TO BE SELECTED

Fig. 7   Example of a variable function keyboard.

mation, retrieving information, reporting, and displaying. This enhances the utility of operations work stations for such functions as:

1. Alarm management (compare, sort)
2. Trending (store, retrieve)
3. Record keeping (store, sort, calculate)
4. Self-diagnostics (compare, calculate)

For these functions, the utility is substantially greater than any cost-effective alternative. The constraint on the design of these systems is performance. Tradition has established an economic value for control systems which is translated into cost targets for operations work centers. Because of this and because of the demanding performance requirements, most commercially available systems are forced to compromise between dynamic responsiveness (e.g., how often a process variable is checked for alarm conditions) and the scope of its functional capability (e.g., how many process variables there are per work station). An example of a performance-related systems design issue is explored later in this article.

Microprocessor-Based Controllers

Microprocessor-based controller subsystems provide a substantial tool for implementing advanced regulatory control techniques, such as cascade, ratio, and feedforward control, as well as dynamic modeling and optimizing control strategies. Also, many systems incorporate interlock, sequencing, timing, switching, and other logic functions which can be combined with process control functions to provide truly integrated control systems.

There are currently three major categories of controller subsystems:

1. *Dedicated Loop Controller.* This is quite similar in function to a conventional controller, with additional features such as the ability to configure the controller to suit the particular application and the ability to append a limited number of additional functions within the controller, such as:

   a. Process variable signal conditioning

   b. Specialized alarming and external contact actuation

   c. Reconfiguration of the control algorithm

   d. Auxiliary calculations, such as totalization

   Also, because it is an integral member of the distributed instrument system, its parameters, such as measurement, set point, output, automatic or manual transfer switch, alarm limits, and tuning parameters can be viewed and modified from workstations.

2. *Programmable Controller.* This is functionally similar to stand-alone PCs with added network communication for displaying and manipulating logic elements, such as those for starting and stopping motors or engaging permissives. The major advantage of the stand-alone variety is the ability to supervise multiple-PC devices in concert with dedicated loop controllers from a single work station.

3. *Multifunction Multiloop Controller.* This is functionally similar to a small dedicated purpose process computer. It combines the process control functions of a dedicated loop controller with many of the logic functions of a programmable logic controller and adds capability in areas such as calculation, decision making, dynamic and static characterization, and stored memory. This type of subsystem in a distributed network represents the most dramatic break with the past. It permits the existence of a control strategy for an entire unit operation—including the electrical requirements (e.g., motor control)—within a single controller and thereby encourages interactive control strategies.

However, there are potential drawbacks associated with this approach, the most significant being a higher exposure to station failure. To achieve true unit operations control, the station would have to contain a number of process loops and many motor control loops. If the station were to fail, the process upset could be substantial. To overcome this, many of the systems currently available offer redundant component entry, particularly where the component is critical to station operation. Figure 8 shows one such example.

Another potential drawback is complexity. The predecessor of the multifunction controller is the direct digital control (DDC) computer. Current systems have evolved, however, which include highly

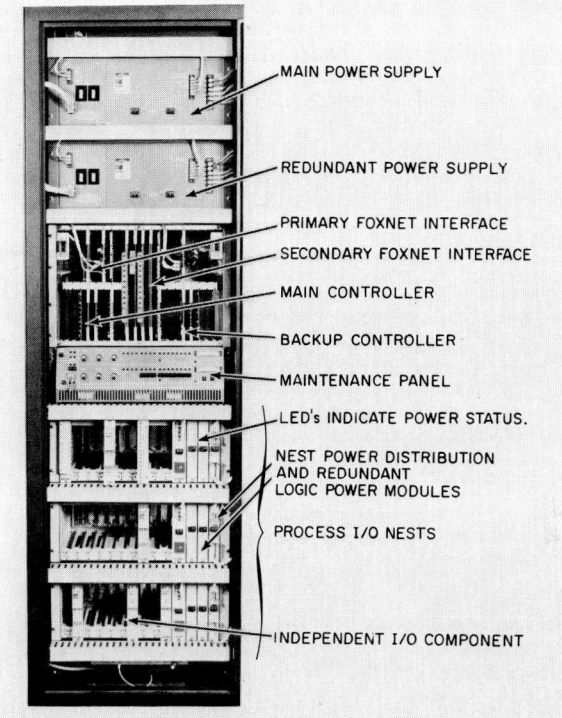

MAIN POWER SUPPLY

REDUNDANT POWER SUPPLY

PRIMARY FOXNET INTERFACE

SECONDARY FOXNET INTERFACE

MAIN CONTROLLER

BACKUP CONTROLLER

MAINTENANCE PANEL

LED's INDICATE POWER STATUS.

NEST POWER DISTRIBUTION
AND REDUNDANT
LOGIC POWER MODULES

PROCESS I/O NESTS

INDEPENDENT I/O COMPONENT

Fig. 8 Some systems offer redundant component entry, particularly where the component is critical to station operation. Shown here is one such example. *(Foxboro.)*

refined sets of application tools, such as a block-structured format with the blocks being functional analogs of conventional control tools such as:

1. Proportional, integral, and derivative controller (PID block)
2. Timer block
3. Autoselector block
4. Summer block
5. Logic block
6. Sequencer block
7. General purpose calculation block

The foregoing list is usually quite extensive, and the arrangement of the blocks, including parameter assignment and block interconnection, is usually handled via a menu-driven question-and-answer format. Also, an engineer's utility function is typically provided so that portions of the configuration (or all of it) can be modified as process needs change.

Another concern is modularity, i.e., the size of the station itself. This concern is manifested in two ways. The station may be either too small or too large for operation of the unit. At this writing, modularity appears to be at the 8- to 32-loop level.

Finally, the question of performance arises. Unlike a dedicated loop controller, which has a microprocessor performing a limited number of functions, or a PC, which executes relatively simple logic and compare instructions, a multifunction controller requires an enormous amount of work to complete each execution cycle. This device typically runs much more slowly than its counterparts, having a range of between 0.25 and 1 s/cycle, as compared to the other two, which have a typical sample interval range between 0.03 and 0.1 s.

Recognizing that all microprocessor-based controllers have sampled-data intervals, it becomes problematical (depending mostly on the application needs) as to what represents a reasonable sample interval.

Auxiliary Data Collection

Auxiliary data collection stations have existed in stand-alone form for a decade or more. The major advantage of combining them into a distributed system is the ability to consolidate their data with microprocessor controllers to create a better presentation of information at operators' work stations. Their function is to gather noncontrol data from various areas throughout the complex. One of the economic incentives for using such stations is their ability to support satellites which can be located in close proximity to a process unit, subsantially reducing cabling costs. See Fig. 9.

Microprocessor technology has allowed these devices to support quite a variety of functions and features, including:

1. Diversity of signal handling capabilities, including current inputs, thermocouples, resistance temperature detectors (RTDs), contacts, and sequence-of-events monitoring

2. Sophisticated alarming, including absolute, deviation from target values, rate of change, and contact change of state

3. Self-checking autodiagnostic features and redundancy of operation (particularly where a large concentration of process variables are resident at a single station)

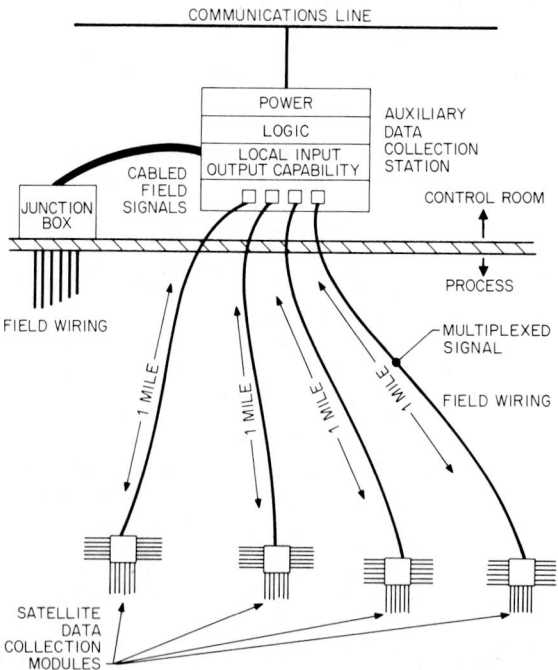

Fig. 9   Auxiliary data collection system.

Information Management

Information management requirements may benefit most from a distributed instrument system architecture. A station in a network has access to virtually all process information on a real-time basis. Although many front-office decisions are based on trends which develop over weeks or perhaps months, some have an activity cycle of hours. The shorter time cycle is typically related to operational planning or management. Other examples include opportunity decision making, such as spot-buying or spot-selling situations.

Operations management ordinarily looks for patterns which will improve the overall throughput or efficiency through immediate action. Typical examples include daily or shift reports containing information on

1. Material balances around process units
2. Energy balances around process units
3. Conversion of any yield calculations on chemical reactors
4. Storage tank utilization
5. Production rate by product
6. Process unit equipment allocation
7. Equipment out of service and the planned maintenance schedule

Another typical use involves operational data review and trend evaluation in an attempt to identify the causes of process upsets which have occurred in the last 72 h. By reconstructing the events leading to a major upset, technical services or supervisory personnel often can correct the situation or avoid a repetition of the upset.

The distributed system benefits management by delivering dispatch files of compressed data once a day. They are sent to a front-office plant accounting system or a corporate management information system.

Occasionally, plant accounting and operations management functions reside in the same system. If this approach is chosen, the planners should fully grasp the real-time nature of operations management and reconcile this with the need of many business machines to be taken off-line to add a new accounting program or data base management package.

Systems Considerations

When considering a microprocessor-based distributed instrument system, it is important to recognize that such a system is a highly sophisticated network of complex subsystems which must interact at extremely high speeds. Each depends on the proper operation of the whole. Also, the whole is definitely more complex than the sum of its parts.

Traditionally, an instrument and control engineer worked with process, operations, and project engineers to complete piping and instrument drawings. The control engineer then compiled a loop index and prepared loop sheets. Simultaneously, the engineer issued technical specifications for procurement. Typical categories included

1. Control and indicating instrumentation
2. Annunciator subsystems
3. Recorders
4. Pushbutton and selector switch assemblies
5. Auxiliaries
6. Panel construction and design specification

Traditionally, the control engineer, basically having designed the system, felt in total control.

Application of this same methodology to microprocessor-based system projects does not work successfully—for a number of reasons:

1. Build-it-yourself design strategies aimed at a single project are difficult to execute because microprocessor manufacturers require a very high volume to be profitable and direct their marketing efforts toward original equipment manufacturers (OEMs) with large technical staffs. Therefore, a significant investment is required to apply the equipment to a specific end use.

2. Vertical integration of a microprocessor-based network (distributed system) requires a thorough understanding of quite complex and diverse disciplines, such as:

    a. Microprocessor bus architecture

    b. Interface engineering

    c. Systems packaging for high availability

    d. Assembly language programming

    e. Real-time operating systems

    f. Design of support services and utilities

    g. Digital control theory

    h. Human factors engineering

    i. Communications theory

    j. Data structures and data base management

    k. Operational and performance testing

3. Systems furnished by integrated systems vendors, such as process control companies, are preengineered to perform in a particular manner. Although they can be configured to meet the specifics of a given project, the fundamental workings of the system (e.g., how a display is called up) cannot be modified without customization.

4. Each integrated systems supplier offers different and basically incompatible equipment, making it virtually impossible to combine them; e.g., a work station from vendor A would be incompatible with a multifunction controller from vendor B.

5. The rapidly changing technology precludes any comprehensive standardization for the foreseeable future. Completely new generations of equipment will emerge from work currently underway. Areas such as very large-scale integrated circuitry resulting in dramatic price-performance shifts in memories and microprocessor speeds, other fields such as artificial intelligence, and new substrate materials promise even more marked effects at some later date.

Faced with this situation, the prudent instrument and control engineer must change the approach from designer to selector. The control engineer now has entered the somewhat uncomfortable world of complex systems. In this period of transition, some control engineers may be torn between two emotional responses—first, a sense of euphoria, an impulse to relax and enjoy the learning experience, and, second, an unsettling anxiety.

It is in order here to introduce a useful parascientific rule regarding complex systems. This concept, called the *generalized uncertainty principle,* is eloquently explained by John Gall,* who examines such diverse systems as garbage collection in large cities and the space program. Succinctly put, the principle states: "Complex systems fail [to meet the expectations of their planners/users] in highly unpredictable ways."

In the case of the control engineer who is selecting "the best" microprocessor-based system, the rule can be interpreted as follows: "A failure to fully appreciate the subtleties of the selected system will inevitably result in surprises—some pleasant, some inconsequential, some correctable (at some cost) and some not."

To demonstrate this point and add credence to the subtlety of many of the issues involved, two examples will be investigated. The first is a relatively simple operator command (chainging a set point); the second, a control problem.

---

*John Gall, *Systematics: How Systems Work and Especially How They Fail,* Times Books, New York, 1977.

Performance Consideration Example

After careful study, it has been determined that the preferred way to adjust a process parameter is via a ramp key. This enables the operator to control the rate as well as the magnitude of change. Also, fewer errors occur than with a numeric entry technique—where the operator key strokes a value onto a numeric pad and then depresses an "enter" key. It is worth noting that this is an action which should be totally anticipated by vendor and user alike. Also, there are no variations where action may be taken in an unexpected manner.

This approach, although simplest for the operator, presents a relatively complex systems problem. In order to ensure accurate and safe operation, feedback mechanisms should be implemented to the greatest extent possible. In our example, this means that, before an incremental change is displayed to the process operator as a result of depressing the ramp key, the change action must be accepted by the subsystem using the set point for control purposes. See Fig. 10. In fact, the display of the set point on the video screen should always be the actual value in the controller. Also, for a ramp to perform properly, it must be responsive to operator action and appear to perform in a continuous manner. Studies have shown that small incremental changes occurring faster than 4 times per second appear to change continuously.

This means that the system must accept an interrupt from a keyboard and execute a program in the video processor to request a change in one of the process variables of a controller at one of the

PROCEDURE FOR AN OPERATOR TO RAMP A SET POINT UP OR DOWN – SYSTEM
ELEMENTS INCLUDE OPERATOR KEYBOARD, OPERATOR WORKSTATION,
MICROPROCESSOR, COMMUNICATIONS SUBSYSTEM, CONTROLLER SUBSYSTEM,
AND DISPLAY CRT.

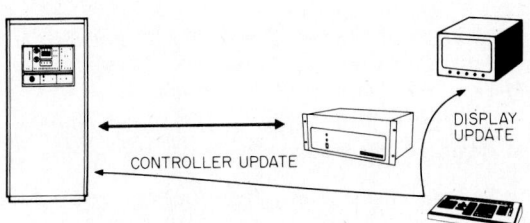

DISPLAY UPDATE

CONTROLLER UPDATE

INCORRECT METHOD – BECAUSE THE DISPLAY IS UPDATED
INDEPENDENT OF THE CONTROL SUBSYSTEM, THERE IS NO
ASSURANCE THE DISPLAYED VALUE AND THE CONTROLLER
SET POINT ARE THE SAME.

(a)

DISPLAY UPDATE

CONTROLLER UPDATE

CORRECT METHOD – INFORMATION IS FIRST SENT TO THE
CONTROLLER. THEN THE DISPLAY IS UPDATED BY THE
ACTUAL CONTROLLER SET POINT.

(b)

Fig. 10 Procedure by which an operator ramps a set point up or down. System elements include an operator keyboard, operator work station, processor, communications subsystem, controller subsystem, and display cathode ray tube.

stations attached to the communications network. The communications subsystem must then route the information to the proper place and ensure that it arrives correctly. The controller processor must update the process variable and feed back the new value. The communications subsystem must then route the response message back to the sender and ensure that it arrives correctly. The video processor then must take the information and update the information on the videoscreen.

All this must occur at least every quarter of a second for as long as the the operator keeps a finger on the ramp key. Also, all this must happen simultaneously with all the other tasks being performed by the process executing the set-point ramp function. The communications subsystem, for instance, might be to various subsystems in the network which require concurrent use of the communications subsystem, such as polling all control devices and auxiliaries for alarm information (both system and process) and communicating files of data to network members.

Microprocessors are very fast, and systems analyses of issues such as these are typically quite well considered. However, there is a point in all these systems where communications throughput limitations begin to affect response time adversely. In our example, this means that our set point will not ramp continuously. When the communications subsystem is seriously overloaded, the process variable appears to jump erratically. System constraints which bring on this phenomenon certainly are related to such factors as:

1. Transmission speed (bits per second)
2. Number of control loops
3. Number of stations supported and their function

Unfortunately, the problem is not so simple. Other factors involved include:

4. Application scope—computer-based direct digital control, for instance, which adds a heavy communications load
5. Protocol efficiency for various communications requirements
6. Queuing theory and message queuing policies
7. Communications loads during abnormal conditions, e.g., plant start-up and major process upsets

Many of the factors in the latter category are project-specific. Some, like upset conditions, are situational and cannot be predicted in advance. This may lead to the extremely unfortunate circumstance of having the system work very well except when it is really needed.

Advanced Control

The issues are not associated simply with taxing the system to its limits, nor are they restricted to communications matters. Consider the controller subsystem. When control is accomplished via the processor, all instructions are executed in serial fashion. This introduces a number of subtle control theory problems, including the following.

Sampled-Data Control

Microprocessor-based controllers periodically sample the process variable and execute a control function, then remain idle for the remainder of the sample time interval while holding the last output. This essentially adds dead time to the loop, but this is not noticeable as long as the interval is much shorter than the effective loop time constant.*

---

*For an excellent treatment of the effects of dead time, the reader is referred to F. Shinskey, ed., *Process Control Systems*, 2d ed, McGraw-Hill, New York, 1979.

Synchronization

Because multiloop controllers execute each function in serial fashion, care must be taken to ensure that the data-flow path coincides with the control function execution path. This is particularly important when two parallel paths join to form a common path or when downstream controllers feed information to upstream blocks [a specific example of this, called a multiple-outlet control system (MOCS), is analyzed later in this article].

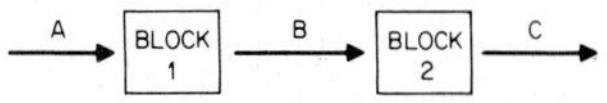

A	0	1	2	3	4	5	6	7	8	9	10	11	12
B	0	0	2	2	4	4	6	6	8	8	10	10	12
C	0	0	0	2	2	2	6	6	6	8	8	8	12

Fig. 11 Effect of a simple ramp (represented by consecutive ascending values in line A) on the output of block B (line C).

Phasing

In some systems, the sample intervals for various controller blocks can be individually adjusted, creating a variable pass-through rate. Consider two blocks connected in series, with the first block processed every *two* sample intervals and the second processed every *three* intervals. Figure 11 shows the effect of a simple ramp represented by consecutive ascending values in line A on the output of block B (line C). It will be noted that B characterized the ramp quite accurately, but that C is distorted. When parallel paths of different lengths are joined, phasing can result in serious control anomalies.

Initialization

Many control functions rely on the memory of past events for proper operation. A simple example is a first-order digital filter. The equation is

$$X_F = \mathcal{L}X_0 + (1 - \mathcal{L})X$$

where X_F = updated filtered input value
$\mathcal{L}$ = filter constant
X_0 = previous updated filtered value
X = updated unfiltered value

Note: When $\mathcal{L} = 0$, $X_F = X$.

For this equation to work properly, some value for X_0 must exist, which is not the case at the start. Therefore, some artificial initial condition must be inserted to provide a place to start (in the case of the digital filter, X_0 is set to X initially).

Initialization schemes which work for most control schemes create difficulties when the assumptions associated with the initialization philosophy are not valid. A classic example occurs in a proportional-plus-integral controller on transferring control from manual to automatic. A popular initialization technique is to force the set point equal to the measurement, *assuming* the controller output is in the correct position.

Again, most systems suppliers recognize the consequences of these issues on at least simple loops and preengineer their systems to address the potential problems. However, problems typically begin to arise when these products are used to implement complex control strategies. This is because most vendors cannot anticipate all the control applications for which the product can be used. Therefore, the user cannot be assured that the vendor has preengineered the product for the particular problem at hand. Once again, we may be confronted with a dilemma—the system works well except when we try to use it for the reasons we selected it (advanced control applications).

Advanced Control Example

To illustrate the problem, consider a parallel output problem where one process variable is manipulated by more than one controller output, as shown in Fig. 12a. This situation is typically encoun-

MULTIPLE OUTPUT CONTROL SYSTEM

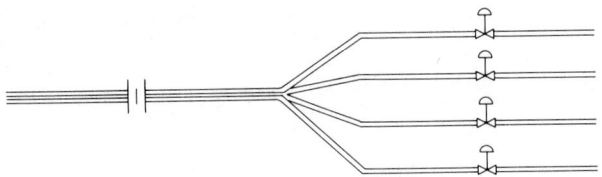

MOCS EXAMPLE – TOTAL FLOW IS CONTROLLED BY MANIPULATING
ALL FOUR CONTROL VALVES IN PARALLEL.

(a)

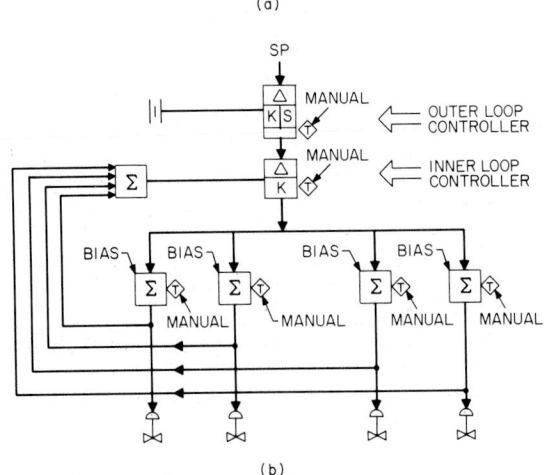

(b)

Fig. 12 Problem where one process variable is manipulated by more than one
controller.

tered in parallel metering systems. Forced-draft fans in combustion control applications and boiler
dispatching are other examples.

These applications share a common element which differentiates them from single-input–single-
output problems. Each manipulated variable can be individually biased, put on manual control, or
taken out of service. The classical approach to this problem is to employ a MOCS—see Fig. 12b—
which interposes a high-gain proportional controller between the process variable controller and the
manipulated variables and also furnishes automatic-manual control with a bias station on each of
the manipulated variables.

The high-gain proportional controller receives its set point from the process variable controller
and its measurement from a scaled summer, which in turn receives its inputs from the outputs of
each of the automatic-manual bias stations.

The purpose of this inner loop (high-gain controller, scaled summer, and automatic-manual bias
stations) is to ensure that the control action of the outer loop (process variable controller) functions
exactly like a single-input–single-output controller. Thus, if one of the outputs is biased, the inner
loop will automatically compensate by adjusting the other outputs. Also, if one of the outputs is put
on manual control or taken out of service, the inner loop will drive the other outputs "harder" so as
to provide the same dynamic response without having to retune the process variable controller. For
continuous control systems, this approach provides an elegant solution.

The MOCS strategy described, when implemented with microprocessor controllers (sampled-
data control), would result in nothing short of disaster. The slightest action taken by the process

variable controller (outer loop) would cause the inner loop to oscillate wildly out of control. The reason for this is simple. The stability of the inner loop is dependent on the fact that it has no dynamic elements. Thus, the output of the scaled summer provides instantaneous feedback to the high-gain controller to limit control action to that required to force the measurement to equal the set point. The microprocessor-based system introduces a dynamic element—the sampled-data interval. The net effect is that the measurement to the high-gain controller, because it is generated by the scaled summer which is executed *after* the high-gain controller, is representative of the controller output of the prior sample interval.

Introducing this dead time into the high-gain controller makes the output cycle wildly, as shown in Fig. 13a. Also, any settling of the proportional band less than 100% will result in limit cycling, while any settling greater than 100%, will result in undamped oscillation. Both will cycle at a period of twice the sampled data period, as shown in Fig. 13b.

The previous example illustrates the pitfalls of assuming that the area of concern is limited to specialized systems issues. Microprocessor-based systems, like all complex systems, change the nature of the problem. In the case of the MOCS example, the microprocessor approach would work well if an integral-only controller were substituted for the high-gain proportional controller, so long as the sample data interval is significantly shorter than the time constant of the outer loop.

The integral controller divides the sampled-data interval time by the integral time constant (scaled to sampled-data intervals per repeat), multiplies the result by the error, and adds the result to the output. Because its effect is additive (integrated), the oscillation encountered with the proportional-only controller is prevented. In fact, any setting resulting in a dimensionless value of τ_s/R between 0.0 and 2.0 would be stable, with a 1.0 setting being ideal, as shown in Fig. 13c.

A Systematic Approach

What is a methodology for avoiding the type of pitfalls just described? *First,* the engineer must realize that both the anticipated benefits and the nature of the tools are quite different from those of conventional instruments. Therefore, the selection criteria and methodology are fundamentally different—not just a different slant on the same approach.

Second, the engineer must determine which departments and disciplines within the plant may potentially benefit from the system, considering a long-term view of the installation. The engineering department may decide to institute a series of energy recovery schemes, while the maintenance department may want to use the system for plant fault locating. The possibilities are many.

Third, all parties with a stake in the system should meet to establish a list of objectives in as detailed a fashion as possible, quantifying the expected benefit for each. The list should include topics such as:

1. Piping and instrumentation drawing (P & ID) review
2. Control strategy objectives
3. Process alarming requirements
   a. Types (absolute, deviation, output, contact, logic, calculated, system, etc.)
   b. Operator alert, access, and acknowledge
   c. Alarm record keeping
   d. Priority structure
4. Process operations and supervision
   a. Shift reports
   b. Process trends
   c. Operator actions
5. Process analysis
   a. Material and energy balances
   b. Conversions and yields

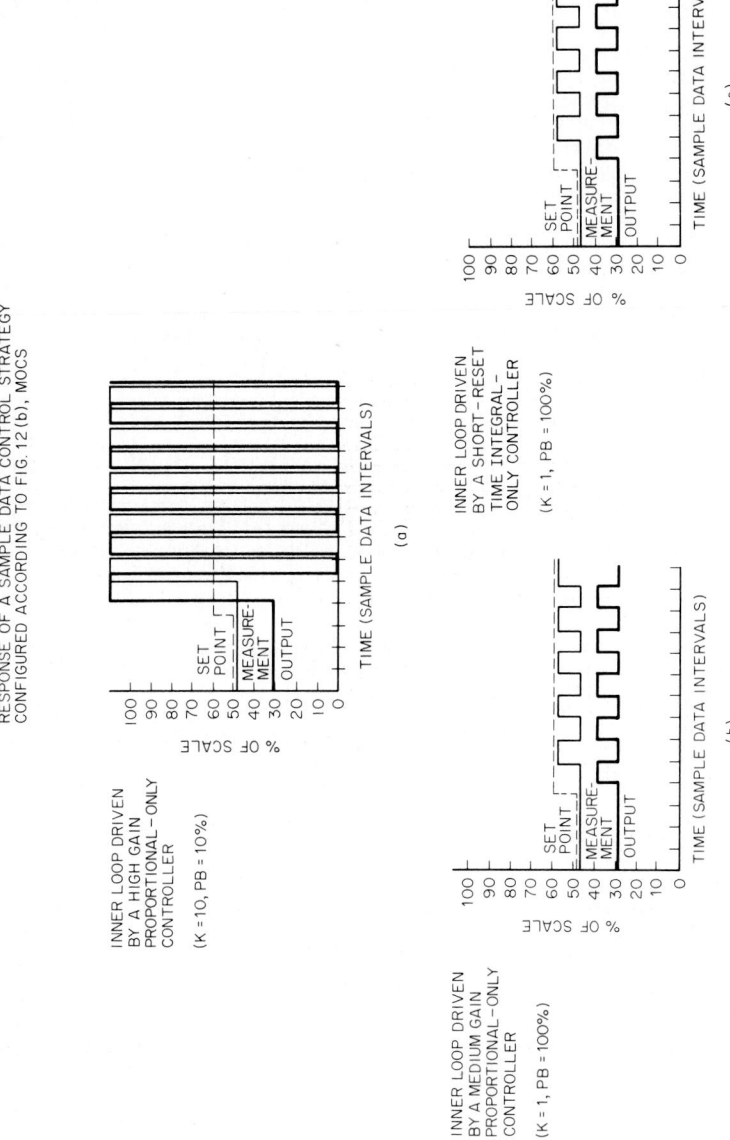

PROCESS VARIABLE AND MANIPULATED VARIABLE
RESPONSE OF A SAMPLE DATA CONTROL STRATEGY
CONFIGURED ACCORDING TO FIG. 12 (b), MOCS

INNER LOOP DRIVEN
BY A HIGH GAIN
PROPORTIONAL-ONLY
CONTROLLER

(K = 10, PB = 10%)

100
90
80
70
60 — SET
50 — POINT
40 — MEASURE-
30 — MENT
20 — OUTPUT
10
0

% OF SCALE

TIME (SAMPLE DATA INTERVALS)

(a)

INNER LOOP DRIVEN
BY A MEDIUM GAIN
PROPORTIONAL-ONLY
CONTROLLER

(K = 1, PB = 100%)

100
90
80
70
60 — SET
50 — POINT
40 — MEASURE-
30 — MENT
20 — OUTPUT
10
0

% OF SCALE

TIME (SAMPLE DATA INTERVALS)

(b)

INNER LOOP DRIVEN
BY A SHORT-RESET
TIME INTEGRAL-
ONLY CONTROLLER

(K = 1, PB = 100%)

100
90
80
70
60 — SET
50 — POINT
40 — MEASURE-
30 — MENT
20 — OUTPUT
10
0

% OF SCALE

TIME (SAMPLE DATA INTERVALS)

(c)

Fig. 13 Comparison of response using a high-grain proportional-only controller (a), a medium-gain proportional-only controller (b), and a short-reset-time integral-only controller (c).

    c.   Process study requirements

    d.   Fault locating needs

6. Interprocess communications

    a.   Display of process upset condition in upstream or downstream units

    b.   Utility and inventory information

7. Process management

    a.   Accounting needs

    b.   Forecasting needs

8. Maintenance needs

    a.   Plant equipment and systems diagnostics

9. Long-range plans

    a.   Expansions

    b.   Plant equipment modifications

10. Security issues

    a.   Operation during various failure modes
        i.   Power
        ii. System
        iii. Plant equipment, e.g., feed pump

11. Protection against unauthorized tampering

    a.   Invalid (incorrect) operator actions

    b.   Other

Fourth, a functional definition of the attributes of the system should be developed. This should focus on preferred methods for achieving the objectives previously established. A conscious effort should be made to avoid any system-specific or technology-associated requirements. This section should be fleshed out to incorporate as much detail as is practical. If reports are desired, the number, frequency, data, calculations, and format should be spelled out.

If complex control strategies are envisioned, a detailed description of the problem, if not the solution technique, should be included. If plant technical services personnel intend to use the system for process upset reviews, the data, analytical tools, and presentation format should be explained. If multiple variable-trend data are required, for instance, the format sample interval and time frame should be specified, but the device (e.g., a trend recorder) should not be called out.

Fifth, and a key step in the overall evaluation, is a series of detailed interviews with various vendors. The objective is to determine whether the vendor is qualified. Care must be taken here to avoid extending the requirements list to include an "interesting feature" of a particular vendor, unless the attendant benefit was obviously overlooked at an earlier stage and is considered important to the project. Otherwise, the requirements may be frivolously extended to exclude everyone, or may be unnecessarily biased against the better system. The primary objective should be to determine how well the vendors can satisfy the functional requirements, with particular emphasis on specialized needs such as a unique control problem. Each vendor should be asked to present a recommended approach to the functional requirements and to describe in detail how each of the objectives will be accomplished.

Vendors who do not seem to understand the requirement may not have anticipated the problem in their design. If these vendors are felt to be qualified because of strengths in other areas, their systems should be scrutinized in detail in those questionable areas to ensure minimum acceptable conformance.

Once a preliminary vendor qualification list has been established, plant visits should be scheduled to evaluate the unquantifiables, such as operator acceptance, impact of system load on overall performance, etc. Proper planning in this task is crucial. The plant selected for a visit should have similar application requirements, operations philosophy, and scope. A checklist of questions should be developed to ensure an accurate assessment of the visit. After all, plant personnel visited will probably respond spontaneously, and areas of importance may be overlooked.

Once the list of qualified vendors is finalized, a formal specification should be issued reiterating the functional requirements with expanded system selection criteria in areas such as

1. Expandability
2. Reliability and security
3. Project execution, including project documentation, e.g., block connection drawings, dimensions, weights, power, configuration layout, etc.
4. Standard manuals (operations, maintenance, initial setup)
5. System acceptance criteria
6. Contractual issues, warranties, etc.
7. Auxiliary equipment required for process interface, power, and earthing or grounding, intrinsic safety requirements, etc.
8. Environmental issues—temperature, humidity, etc.
9. Commercial terms—price, delivery, etc.

Note: Because the topic of this article represents such a fast-moving technology as of the mid-1980s, specific references are not cited. However, in terms of supporting fundamental details, as contrasted with specific system architectures, the lists of references appended to the articles in Sec. 17 of this Handbook may be found of value.

SECTION 19

Final Control Elements *

Allen C. Fagerlund. *Senior Research Specialist, Fisher Controls International, Inc., Marshalltown, Iowa. (Control Valve Noise)*

H. C. Gery. *Application Consultant, Leeds & Northrup (A Unit of General Signal), North Wales, Pennsylvania. (Electric Actuators)*

Alexander Kusko. *President, Alexander Kusko, Inc., Needham Hills, Massachusetts. (Electric Motor Drive Controls)*

John J. Murray. *The Superior Electric Company, Bristol, Connecticut. (Stepping Motors)*

Marc L. Riveland. *Senior Research Engineer, Fisher Controls International, Inc., Marshalltown, Iowa. (Control Valve Cavitation)*

Paul Wing. *Consultant, Hingham, Massachusetts. (Control Valves) (Control Valve Sizing) (Control Valve Characteristics)*

*Persons who authored complete articles or subsections of articles, or who otherwise cooperated in an outstanding manner in furnishing information and helpful counsel to the editorial staff.

19.1

Control Valves

by
Paul Wing*

Since the purpose of this section is to summarize the physical characteristics and function of control valves, it is important at the outset to define the subject. The following definition of a *control valve* is used by the Process Measurement and Control Section of the Scientific Apparatus Makers Association (SAMA): "A valve with a pneumatic, hydraulic, electric (excluding solenoids) or other externally powered actuator that automatically, fully or partially opens or closes the valve to a position dictated by signals transmitted from controlling instruments."

This definition is not incorrect, but there may be too much stress on the combination of a valve body assembly and an actuator. Many control valves may easily be recognized as such, even though the actuator is manual. Control valve body assemblies differ from those of shutoff (stop) valves in several basic ways, and many potential field problems can be avoided if the following criteria are kept in mind:

1. Control valves are used primarily to throttle energy in a fluid system and not for shutoff purposes.

2. Valve internals must withstand high fluid velocity and turbulence for long periods without maintenance.

3. Even when designed for shutoff as well as control, the degree of tightness can seldom be expected to be that obtainable in a property designed shutoff valve.

The actuator-to-plug connection in a control valve, whether manual or automatic, must have absolutely no lost motion. A stop valve generally has a loose connection such as a tee head to allow self-alignment for optimum shutoff. The internal moving parts of a control valve generally must have

*Consultant, Hingham, Mass.

heavy guiding and be more precisely aligned. The stem connection reaching the outside of the valve body, whether rotary or linear, must allow operation often at high pressures and temperatures with a minimum of friction. This means that a specially designed stem seal or packing box is required. Many control valves are designed to maintain one of a number of mathematical relations between the valve stroke and effective port area known as the *flow characteristic*. Others, such as the common butterfly valve and the full-port ball valve, have flow characteristics fixed by the basic valve construction.

The actuator used in throttling service moves the valve stem, either sliding or rotary, to establish the desired port area through the valve. The motive power may be pneumatic, hydraulic, electric, or some combination of these. On-off actuators such as simple solenoids and pneumatic or hydraulic cylinders are not treated here. It is assumed that the control valve actuator is capable of positioning and maintaining a valve position in response to the signals from a proportional controlling instrument. The most common design is the spring-opposed pneumatic diaphragm actuator. The range of application of actuators can be extended and operation often improved through the use of a servo device known as a *valve positioner,* which through appropriate relays and an independent operating supply matches the stem position to the controller signal. The valve positioner may also convert the controller output from electric to pneumatic or hydraulic, raise the actuator pressure level, and perform other useful secondary tasks. It is also widely employed to convert simple on-off actuators such as pneumatic or hydraulic cylinders for use in throttling control.

In summary, the control valve body assembly and the actuator can be considered quite separately. There are many valve body and actuator designs specifically for control service. These are complemented by modifications of standard shutoff valves and on-off actuators. The modifications, although important, are often easily made, and many useful control valves appear at first glance to be automated hand valves.

BODY STYLES

Because of the extremely wide range of flowing conditions, a large variety of valves have been especially developed or adapted from shutoff or block valve designs. The process continues, but there is an underlying trend toward standardization. The result has been a substantial realignment of the control valve selection process. The major change has been a shift toward sliding stem single-port balanced and unbalanced globe valves and greatly increased usage of improved rotary stem valves, such as eccentric disk globe, butterfly, and ball valves.

When the term "control valve" is mentioned, most engineers think of a globe valve. The term "globe" simply refers to the general shape of this style of body. The end connections are machined to suit the piping. They are generally straight through on the horizontal center line but may be offset or of angle configuration.

From the standpoint of the number used, the sliding stem globe valve is still the most common. A number of design modifications of single-port globe valves have been introduced which, in combination with more powerful actuators, have virtually displaced the double-port semibalanced version of the valve on which the industry was founded.

The second major control valve classification embraces rotary stem valves which now include general purpose versions as well as the familiar butterfly valve and the relatively recent ball valve. Properly designed, this class of valves can handle all but very small flows and very high pressure drops. Although an accurate figure is not possible, a reasonable estimate is that ~35% of all control valves sold are of this type, with some additional penetration to be expected in the next few years.

For service involving very corrosive applications, for abrasive slurry service, for very low flows or very high pressure drops, and for either high or low temperature extremes and other special applications, a variety of special valves are available. Typical examples will be described in the following discussion. For the more widely used valve body styles, a summary under each heading has been prepared as a general application guide. Valve capacity is shown as a multiple of the nominal diameter squared (d^2), which allows a comparison of the relative capacity of different valve styles.

Sliding Stem Globes

SINGLE-PORT TOP-ENTRY GLOBE (See Fig. 1)

Size: ¾ to 10 in (∼19 to 254 mm)
Pressure Rating: all ratings through ANSI Class 1500
Pressure Drop: >1000 psi (∼7 MPa) or as limited by actuator or body rating
Temperature: −320 to 1000°F (∼−196 to 538°C)
Body Materials: cast carbon and alloy steels plus corrosion-resistant alloys
Relative Capacity: $C_v = 14d^2$
C_f: 0.85

A general purpose design illustrated by a top-guided percentage-contoured lathe-turned plug and threaded seat ring, this valve style is widely used in ¾ and 1 in (∼19 and 25 mm) sizes with full or reduced diameter trim for relatively small flows and high pressures. Because of the simple construction and easy accessibility for servicing, the range of sizes extends to 10 in (∼254 mm), and with powerful pneumatic actuators this type of valve has made heavy inroads on the classic double-port design. In the larger sizes, it is being displaced by the two following modifications.

SINGLE-PORT GLOBE—TOP-ENTRY QUICK-CHANGE (See Fig. 2)

Size: 1 to 10 in (∼25 to 254 mm)
Pressure Rating: all ratings through ANSI Class 1500
Pressure Drop: <1000 psi (∼7 MPa) or as limited by actuator or body rating
Temperature: −320 to 1000°F (∼−196 to 538°C)
Body Materials: cast carbon and alloy steels plus corrosion-resistant alloys
Relative Capacity: $C_v = 12d^2$ to $14d^2$
C_f;   0.85

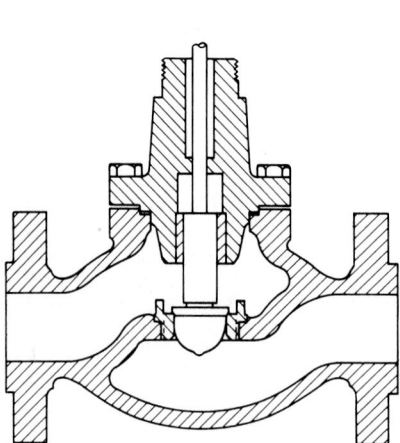

Fig. 1   Top-entry top-guided single-seat globe valve.

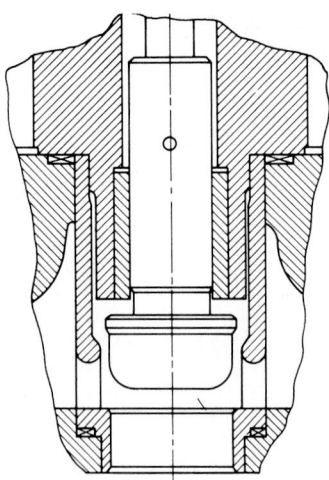

Fig. 2   Single-seat single-port top-entry valve with a threadless gasketed seat ring and hold-down spacer.

In the previously mentioned design, the replaceable seat ring is threaded into the body and is often difficult to remove; it also may develop leaks behind the ring, particularly in steam or flashing liquid service. This modification was developed to overcome these objections using a cylindrical hold-down spacer and a threadless controlled-compression gasketed seat ring. Machining tolerances are so adjusted that the bonnet closure joint and the gasketed seat ring are clamped simultaneously, generally using spiral-wound gaskets. This permits easy trim replacement and also the use of more exotic seat ring materials where required. The price premium is small, and this style is gaining in popularity.

SINGLE-PORT—TOP-ENTRY CAGE-GUIDED (See Figs. 3 and 4)

Size: 1 to 10 in ($\sim$25 to 254 mm)
Pressure Rating: all ratings through ANSI Class 2500
Pressure Drop: >1000 psi ($\sim$7 MPa) or as limited by actuator or body rating
Temperature: -320 to $750°F$ (~-196 to $400°C$)
Body Materials: cast carbon and alloy steels
Relative Capacity: $C_v = 12d^2$ to $15d^2$
C_f (Flow-to-Close) Balanced: 0.90

As a further extension of the quick-change idea illustrated in Fig. 3, the spacer or cage, made of an antigalling corrosion-resistant material, may serve as a heavy-duty plug guide in addition to retaining the seat ring. The tip of the plug may be contoured for flow characteristics or, as in the more popular version, the flow ports may be in the cage itself, the plug becoming essentially a flat-headed piston. This minimizes changes in dynamic force across the plug during the stroke, allowing higher operating pressure drops without plug instability. In the valve in Fig. 4, holes brought up through the plug result in a balanced plug basically similar in function to the once popular double-port valve. As shown, an auxiliary seal ring is provided in the guide portion. A relatively large port diameter results in a high flow capacity. Both static and dynamic forces are low, reducing actuator thrust requirements. This balanced version of the top-entry valve is superior to the old double-port valve and is a major factor in present control valve usage, particularly at moderate to high pressure drops and in sizes above 2 in ($\sim$5 cm).

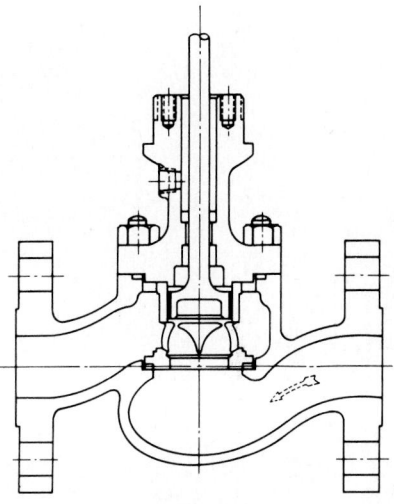

Fig. 3   Unbalanced single-port valve with a piston plug and ported cage guiding.

SINGLE-PORT HORIZONTALLY SPLIT GLOBE (See Fig. 5)

Size: ½ to 6 in ($\sim$13 to 152 mm)
Pressure Rating: all steel ratings through ANSI Class 600
Pressure Drop: as limited by actuator or body rating
Temperature: -450 to $1000°F$ (~-268 to $538°C$)
Body Materials: cast carbon and alloy steels plus corrosion-resistant alloys
Relative Capacity: $C_v = 12d^2$
C_f: 0.80

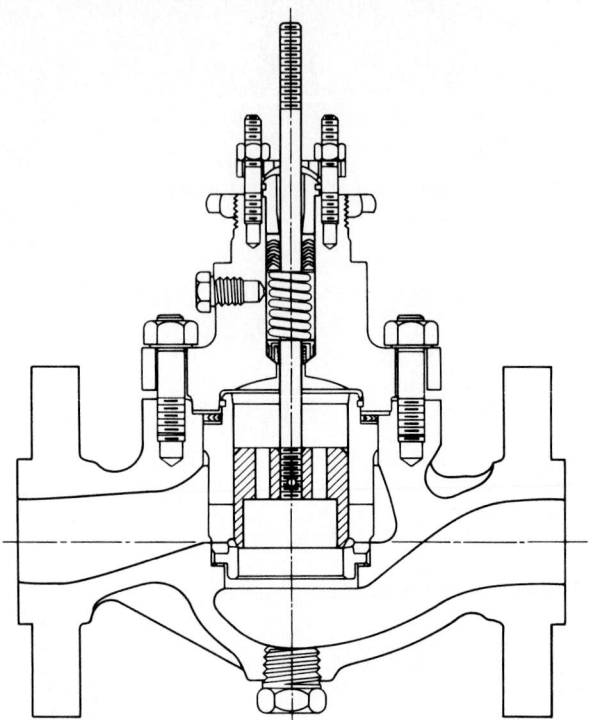

Fig. 4 Piston-balanced single-seat single-port globe valve with a ported quick-change seat ring retainer.

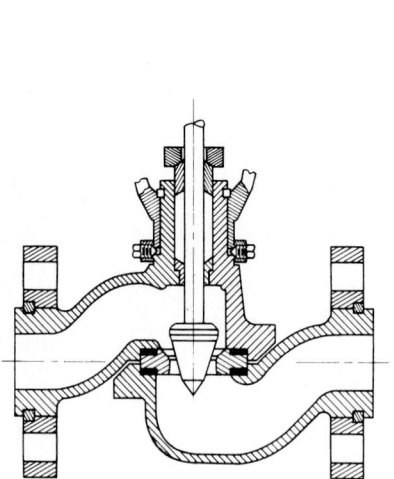

Fig. 5 Split-globe control valve body.

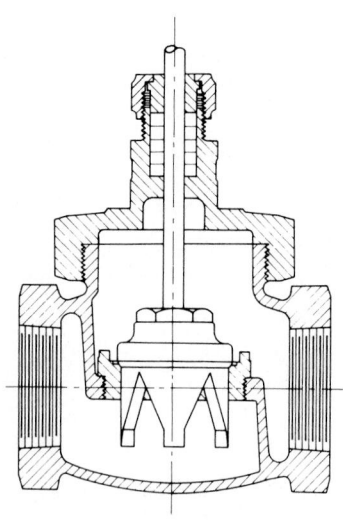

Fig. 6 Single-port nonreversible globe body with a threaded bonnet and threaded ends.

This general purpose body style features a threadless easily replaceable seat ring and a relatively large-diameter plug stem which doubles as a plug guide. Symmetrical bolting at the split body joint (not shown) permits assembly either as a straight-through or a right-angle valve or the substitution of an outlet piece pointing straight down for the conventional angle configuration. The version shown has removable flanges which allow a common body assembly to be used for ANSI Class 150, 300, and 600 ratings to reduce the cost of alloy bodies by using carbon steel flanges.

This valve is generally more popular in sizes through 4 in ($\sim$102 mm), ANSI Class 300, and in alloy construction, but is available in ratings through 2500 lb ($\sim$17.2 MPa). The unbalanced single-port design requires a relatively large actuator thrust. It is not suited for weld end construction. The use of face-to-face separable flanges does not match ANSI standards, but integrally cast flanged ANSI-dimensioned bodies are generally available through 4 in ($\sim$102 mm).

SINGLE-PORT GLOBE—THREADED BONNET (see Fig. 6)

Size: ½ to 2 in ($\sim$13 to 51 mm)
Pressure Rating: ANSI Class 300 maximum
Pressure Drop: generally not over 150 psi ($\sim$1 MPa)
Temperature: 0 to 450°F ($\sim$ −17.8 to 232°C)
Body Materials: bronze for general noncorrosive service, stainless steel for chemicals
Relative Capacity: $C_v = 8d^2$
C_f: 0.98

This is an inexpensive low-pressure valve widely used in building heating and for low-pressure process applications. It is generally furnished as shown, with a seat-ring-to-plug skirt guide and a simple V-port plug.

BOTTOM-ENTRY GLOBE (See Fig. 7)

Size: 1 to 4 in ($\sim$25 to 102 mm)
Pressure Rating: steel ratings through ANSI Class 300
Pressure Drop: up to 300 psi ($\sim$2.1 MPa)
Temperature: 0 to 500°F ($\sim$ −17.8 to 260°C)
Body Materials: steel and alloys
Relative Capacity: $C_v = 7d^2$
C_f: 0.85

This variation on the quick-change-type globe is supplied in a flangeless version, most often with a barstock body, and finds wide usage in corrosive service, hence is produced in a variety of alloys. The weight of the alloy material is kept to a minimum through flangeless construction and use of carbon steel separable flanges to retain the other body closures.

Similar designs are also made in chemically resistant plastic for highly corrosive low-temperature applications.

TOP-AND-BOTTOM-GUIDED INVERTIBLE GLOBE (See Figs. 8 and 9)

Size: 1 to 16 in ($\sim$25 to 406 mm), double port; 1 to 6 in ($\sim$25 to 152 mm), single port
Pressure Rating: all ratings through ANSI Class 2500
Pressure Drop: up to 1000 psi ($\sim$7 MPa)
Temperature: 320 to 1000°F ($\sim$160 to 538°C)
Body Materials: cast iron, cast steel, all castable alloys
Relative Capacity: $C_v = 12d^2$
C_f: 0.98, V port; 0.85, lathe-contoured

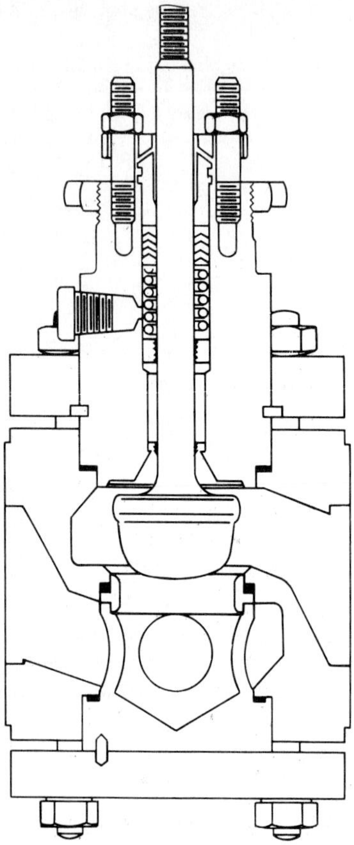

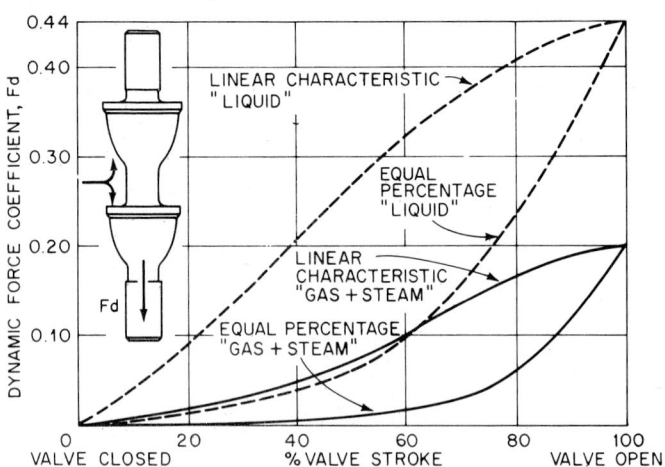

Fig. 7 Barstock flangeless quick-change design for trim removal through a bottom flange.

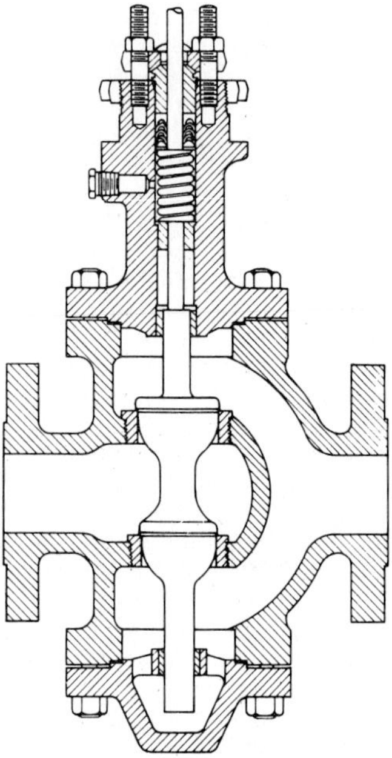

Fig. 8 Typical double-port reversible globe control valve body with a lathe-turned plug.

Fig. 9 Dynamic force coefficient for a double-port lathe-turned valve as a fraction of the total upper port area.

The double-ported invertible globe body with a lathe-contoured "parabolic" plug dates back over 50 years and is an outgrowth of "balanced" valves developed for direct acting pressure regulation. It filled the need for a valve that could be positioned with relatively little actuator power. Most designs have one port slightly larger than the other for ease of assembly. This produces a static imbalance in the closed position, which must be taken into account when determining the net valve stem thrust.

In service, dynamic force reactions from the flowing fluids create unbalanced forces that may be greater than those resulting from the difference in port area. This is particularly true for the lathe-contoured plug illustrated. With the valve installed as in Fig. 8, with the flow from left to right, dynamic force reactions tend to close the valve. Figure 9 shows the expected order of magnitude of these forces for a lathe-contoured plug in both liquid and compressible fluid service. F_d, the dynamic force coefficient, is the fraction of the total unbalanced force on the upper port area A_1 (the total dynamic force equals $F_d A_1 \, \Delta P$). Note that the maximum is almost half the total imbalance.

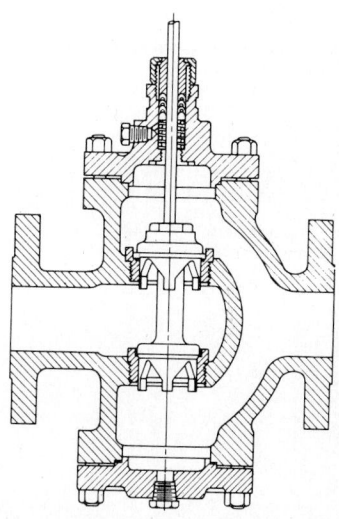

Figure 10 shows the same valve equipped with a skirt-guided V-port plug. These plugs are subject to torsional forces that may be of serious consequence and are generally limited to relatively low-pressure-drop service even though the vertical dynamic force F_d is less than one-half that of the lathe-contoured design. Both the torsional forces in the V-port design and the dynamic suction on the lathe-contoured plugs can be minimized, but not eliminated, by altering the plug configuration. Tight shutoff is not practical, and the accepted leakage figure is ½% of the rated maximum flow.

Similar top-and-bottom-guided designs were

Fig. 10 Double-port control valve with a skirt guide.

also produced in a single-port unbalanced configuration, generally in sizes up through 2 in (~51 mm) but extending through 6 in (~152 mm). Top and bottom guiding was well accepted, but the principal feature of this valve class was invertibility. The design was a consequence of a requirement for direct acting pressure regulators where back-pressure and reducing service required opposite action because an increase in pressure on the actuator always extended the stem which then had to close or open the valve port for reducing and back-pressure service, respectively. For many years reverse acting actuators (increasing air supply to retract the stem) have been readily available, and this valve class is no longer a significant factor.

Rotary Stem Valves

ECCENTRIC DISK GLOBE (See Fig. 11)

Size: 1 to 12 in (~25 to 305 mm)
Pressure Rating: ANSI Classes 150 to 600
Pressure Drop: up to 1000 psi (~7 MPa)
Temperature: −320 to 750°F (~−196 to 400°C)
Body Materials: cast steel, stainless steel
Relative Capacity: $C_v = 13d^2$
C_f: 0.85

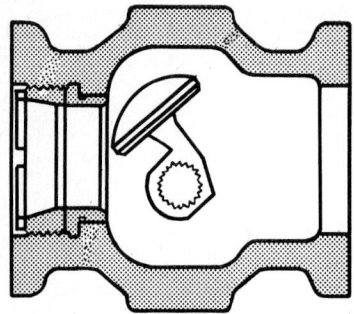

Fig. 11 Eccentric disk globe.

The rotary stem valve was designed for general control service. A spherically faced plug segment, eccentrically mounted, rotates to engage the in-line seat ring as shown. Shaft rotation is generally limited to 50°, resulting in a flow capacity similar to that for general purpose globe-style control valves and a linear inherent flow characteristic. Standardized materials of construction, a built-in rotary shaft extension for temperature dissipation, and a flangeless body permit a wide range of applications with a minimum of design modification. This valve is directly applicable as a substitute for valves of conventional globe design.

The operating torque is low, and the change in torque with valve opening at a constant pressure drop is small, permitting either flow-to-close or flow-to-open action to suit process requirements.

ROTARY STEM CONTROL BALL (See Fig. 12)

Size: 2 to 16 in (∼51 to 406 mm)*
Pressure Rating: ANSI Classes 150 to 300
Pressure Drop: up to 720 psi (∼5 MPa)
Temperature: −320 to 500°F (∼−196 to 260°C)
Body Materials: cast carbon and stainless steel
Relative Capacity: $C_v = 22d^2$
C_f: 0.60

---

*Full-bore designs to 30 in (∼72 mm) in diameter in ratings through ANSI Class 600.

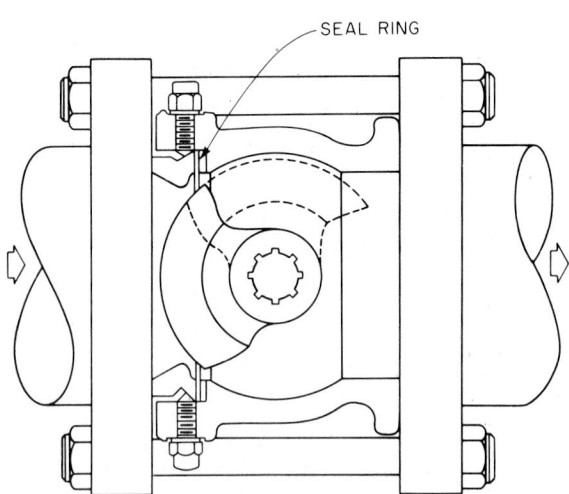

SEAL RING

Fig. 12 Flangeless control ball design using partial (half) ball design and a flexible metallic seal ring for a wide flow range and minimum leakage.

These valves are a special adaptation of the general purpose ball valve developed originally for shutoff service. In this case, a segmental or half-ball is trunnion-mounted and sweeps through 90° for full opening. The surface of the ball is hard-faced and is usually in sliding contact with a seal ring which may be a tempered grade of stainless steel or any one of a number of plastics. When tight shutoff is not required, the sliding seal may be replaced by a hard-faced ring for a higher pressure drop or for erosive service. The leading edge of the ball may be shaped to improve the flow characteristic.

Dynamic forces are relatively low, but friction forces are high and a relatively powerful actuator is required. The valves have a wide usable flow range and an extremely high capacity. They are widely used for the control of paper stock and in modified erosion-resistant versions for general slurry services. They are also useful in general applications at ratings up to ANSI Class 300. The operating temperature limit is generally imposed by the type of seal ring used. The valve has a high degree of pressure recovery as it nears the full open position, and this must be taken into account, particularly in liquid applications where severe cavitation may result.

In addition to the specialized design illustrated, full-ball valves, described previously for shutoff service, have been adapted for control by eliminating lost motion in the rotary-stem-to-ball connection. Operating characteristics are similar to those of the control ball valve. In heavy-duty designs, operating pressure drops to 1000 psi ($\sim$7 MPa) are possible, and sizes through 30 in ($\sim$762 mm) or more are available.

GENERAL PURPOSE BUTTERFLY (See Figs. 13 and 14)

Size: 2 to 48 in ($\sim$51 to 1220 mm)
Pressure Rating: generally through ANSI Class 300
Pressure Drop: shutoff to 150 to 200 psi ($\sim$1 to 1.4 MPa), 60° open as limited by actuator and strength of parts
Temperature: -450 to $1000°F$ (~ -268 to $538°C$)
Body Materials: iron, carbon and alloy steels plus cast and flame-cut alloys
Relative Capacity: $C_v = 18d^2$ to $20d^2$ for 60°, $25d^2$ to $30d^2$ for 75°, $30d^2$ to $35d^2$ for 90°; line size/valve size = 1.0
C_f: 0.68 for 60°, 0.65 for 75°, 0.60 for 90°

The modern butterfly valve used for control is a far cry from the sheet metal damper used for centuries to control draft. It combines compactness and economy with a very high relative capacity

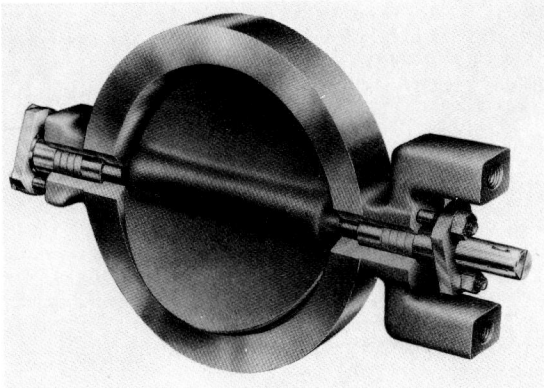

Fig. 13   Flangeless swing-through butterfly valve.

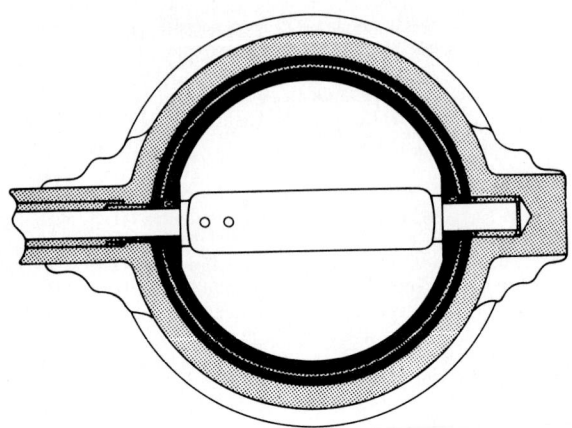

Fig. 14   Elastomer-lined butterfly valve for tight shutoff service.

and is widely used in the process control field, the most common size range being 4 to 12 in (~102 to 305 mm). Valves as small as 1 in (~25 mm) are available, and many up to 48 in (~1219 mm) in diameter are used, the upper limit being over 100 in (~2540 mm). Figure 13 shows a butterfly valve of the swing-through design in a flangeless body designed for bolting between mating line flanges. The metal disk is pinned to a rotary shaft extending out through a stem seal assembly, in this case a conventional packing box. Swing-through valves with all-metal internals are widely used for control where leakage flow between the disk and body can be tolerated. This leakage is generally less than 1% of the maximum rating.

Second in popularity is a modification of the swing-through type made by inserting an elastomer liner in the body as shown in Fig. 14. The liner may be a reinforced molded replaceable cylinder, as shown, or vulcanized directly to the body. Buna N, neoprene, Nordel, and Viton are among the available elastomers, and shutoff ratings up to 150 psi (~1 MPa) are obtainable at temperatures up to 350°F (~177°C).

HIGH-PERFORMANCE BUTTERFLY (See Fig. 15)

Size: 4 to 16 in (~102 to 406 mm)
Pressure Rating: ANSI Class 600 maximum (for installation purposes only)
Pressure Drop: shutoff to 720 psi (~5 MPa), throttling as dictated by actuator selection
Temperature: −320 to 450°F (~−196 to 232°C)
Body Materials: cast steel, stainless steel
Relative Capacity: $C_v = 20d^2$
C_f: 0.7 at rated opening

A butterfly valve with a major improvement in design, using a pressure-activated seal and an offset disk, is making large inroads in shutoff service, replacing gate, globe, and ball valves. The disk rotates in and out of the in-line seal, making contact only at the moment of closure. The breakaway torque on opening is low, and shutoff is reliable. The design is also readily made fire-safe. Standard designs have a relatively large rotary shaft for a high shutoff pressure capability.

With the rotary shaft downstream, static imbalance created by the offset construction tends to cancel the dynamic torque induced by the flowing fluid. Consequently, this valve is gaining favor for many control applications, particularly in the larger sizes.

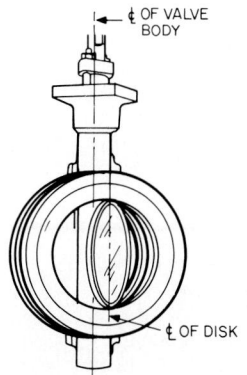

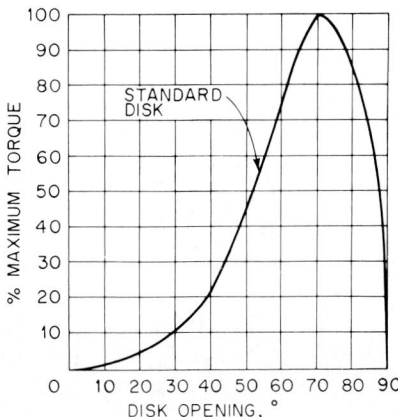

Fig. 15 Offset disk design and special seals provide good shutoff and low operating torque in a high-performance butterfly valve.

Fig. 16 Dynamic torque in closing direction on symmetrical disk butterfly peaks at a 70° opening.

General Butterfly Valve Design Considerations

Because of simplicity of construction, a large variety of subdesigns are available. End connections may be flanged, flangeless, or butt-welded. The valves may be refractory-lined for service up to 2000°F (1093°C) or equipped with shaft extensions for cryogenic service. Improved but not tight shutoff is possible through the use of a disk that contacts the body bore at an angle on closure. The basic behavior is similar for all classes.

When used for control, the disk is generally roated through 60° rather than 90°. The result is a modified equal percentage characteristic, which is quite suitable for general application.

Rotary Shaft Torque

The torque required to actuate a butterfly valve is based on three factors:

1. Side thrust acting on the shaft bearings
2. Dynamic torque generated by the flowing fluid
3. Torque required for seating in tight shutoff versions

The seat load allowance is based on laboratory tests. Bearing friction is readily computed and verified in the laboratory, but the dynamic torque requires more careful consideration. A butterfly disk, even though symmetrical, accelerates the fluid flow on one side more than on the other, and the result is a torque tending to close the valve, as shown in Fig. 16. Only at 0 and 90° is the torque zero for the symmetrical disk. It is proportional to the pressure drop times the cube of the diameter and at ∼70° open in a symmetrical design amounts to as much as 20% more than the equivalent torque on one-half of the disk in the closed position. Fortunately, in most applications, the pressure differential across the disk diminishes with increase in flow to the point where the net torque requirement for actuation is generally larger with the valve closed than at maximum flow.

If the pressure drop is maintained as the valve opens, the dynamic torque severely limits the application and, as a consequence, low-torque vane designs have been introduced, one of which is shown in Fig. 17 with the accompanying torque curve. This permits use at wider openings and consequently higher capacity ratings at a lower net torque with a given shaft diameter and actuator size. The high-performance butterfly design also has a favorably low net operating torque.

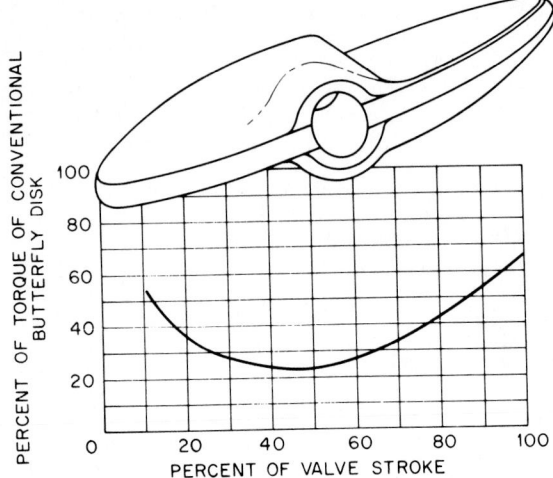

Fig. 17 Reduction in dynamic torque through use of a special vane design.

These valves are generally used at moderate pressure drops for throttling service. Heavy-duty valves with large-diameter rotary shafts are made for higher operating drops, and this may alter the flow pattern slightly, particularly for larger openings and smaller sizes. At very high static pressures, the maximum pressure drop in the closed position is generally limited to a small percentage of the static pressure. A typical high-pressure application might be a bypass around a high-pressure heat exchanger operating at 2000 psi ($\sim$13.8 MPa), where the maximum valve pressure drop does not exceed 50 psi ($\sim$345 kPa).

The pressure recovery factor C_f is low, and in liquid flow applications care must be taken to avoid cavitation. The high relative capacity may also result in a valve substantially less than line size, and the reduction in net capacity can be substantial, as shown in the article Valve Sizing in this Handbook section.

Special Valve Design and Application

The valve designs just described cover all but a small percentage of the requirements of process control applications. This by no means ends the possibilities of selection. Many variations on the basic designs and valves based on entirely different concepts are available. Rather than attempting to present a comprehensive list of mechanical designs, the following discussion concerns applications, using only representative illustrations.

Small Flow

A small percentage of control valve applications in the laboratory and in commercial plants involve extremely small flowing quantities. These needs are filled by a number of special designs. Figure 18 shows a scaled-down version of a single-ported valve where the port diameter is usually ¼ in ($\sim$64 mm) or less and a cylindrical fluted plug is used. The stroke is relatively long, and the plug and seat are machined to fit with extremely small clearances and are made of special hard antigalling materials.

For these extremely small flows, often at a high differential pressure, a number of proprietary designs are available. One approach is shown in Fig. 19. A flat-faced cylindrical plug, substantially larger in diameter than the valve orifice, operates with a very short stroke converted from a longer stroke actuator through a hydraulic reduction capsule. Fluid emerging from the valve orifice is forced

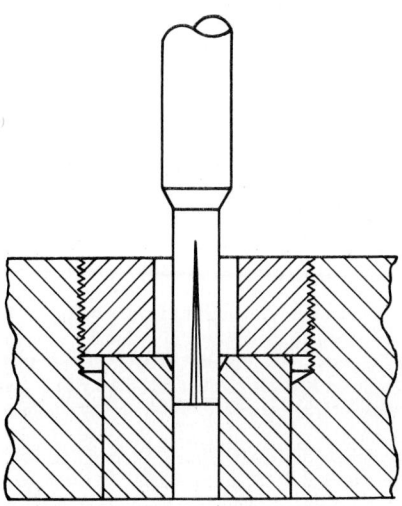

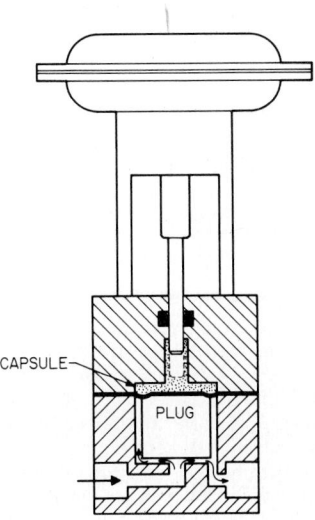

Fig. 18 Typical seat and valve plug construction of a control valve used for extremely small flows.

Fig. 19 Small-flow design utilizes laminar flow principle.

radially between the plug and seat with a clearance such that laminar flow is obtained. An extremely wide flow range is obtainable in this manner.

A high-pressure reciprocating stem design is shown in Fig. 20. Here the springless piston actuator travel of $\frac{3}{4}$ in ($\sim$19 mm) is reduced to an adjustable stroke of between 0.010 and 0.150 in ($\sim$0.25 and 3.8 mm). With suitable orifice diameters and tapered needle trim material, extremely minute flows may be controlled.

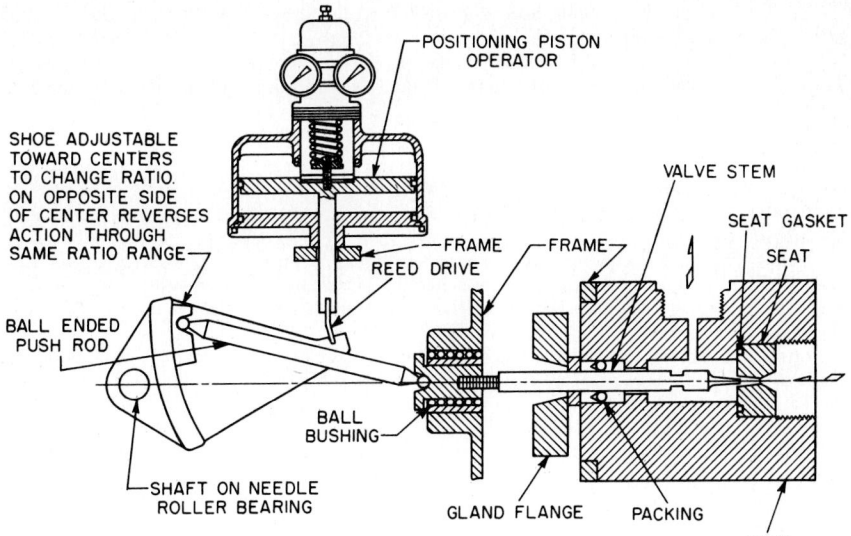

Fig. 20 Control valve design for small flows, where operator travel is reduced through a system of levers to a valve plug stroke that is adjustable between 0.010 and 0.150 in ($\sim$0.25 and 3.8 mm).

C_v ratings below approximately 0.01 must be used for reference only, and trim selection is of necessity partly trial and error. This is because the flow pattern in many cases may be laminar rather than turbulent, or some mixture of the two. Fortunately, this is not a serious problem because of the large flow range of most designs, particularly the variable short-stroke types.

Cryogenic Service

The term *cryogenic* applies to processes operating at temperatures below $-150°F$ ($-101°C$). Down to this temperature, the primary objective is to reduce heat inflow to the system, which can be done by a simple extension on the valve body where the actuator stem or rotary shaft is brought out.

Valves at cryogenic temperatures are generally required to handle extremely cold gases or liquefied gases. Heat in-leakage must be minimized. Thus, the valve is heavily insulated. All equipment in the cold section of the plant may be in a "cold box." The valve may be equipped with an exceptionally long plain extension bonnet as shown in Fig. 21, where a large-diameter thin-walled section, fabricated from stainless steel, is brought out through the wall of the cold box. The valve plug and seat ring may be removed without disturbing the body which is welded into the system to minimize the possibility of leakage.

In liquid hydrogen or helium service, even more elaborate precautions are required, including the use of vacuum jackets and fabricated designs to minimize the cool-down weight. The body configuration may be a Y pattern to reduce head loss to a minimum. A combination of tight shutoff and control is generally required to reduce the number of components in the piping system. Special butterfly valves with an extension bonnet with an offset vane and an elastomer lip seal are also used.

High Pressure

A special class of applications in the process industry are the high-pressure services, particularly involving the outlet of catalytic conversion units where the pressure drop may range from 1000 to 30,000 psi ($\sim$7 to 207 MPa) or more. The possibility of corrosion sometimes complicated by erosion due to entrainment of catalyst fines is always present. Design details are generally of angle configuration and produced from barstock or hammer forgings for body wall integrity. A typical design is shown in Fig. 22.

The valve actuator must handle high thrust and often must stroke the valve at high speed as part of the plant safety system. For these valves, a high-performance electrohydraulic actuator may be required. Valve plug and seat design are simple, permitting fabrication from the widest possible range of erosion- and corrosion-resistant alloys. Special high-pressure end connections are required.

Slurries

This is a difficult area of valve application where a number of special designs are used. Factors to be considered are particle size and shape, operating pressure drop, temperature, corrosiveness, and shutoff requirements.

Mildly abrasive slurries or mixtures, such as paper stock, often may be handled with conventional control valves, using abrasion-resistant trim materials. The control ball valve, previously shown in Fig. 12, was originally designed for paper stock, since the seal ring in tight contact with the ball shears the stock to give good shutoff to prevent dewatering.

Elastomers are very abrasion-resistant at moderate pressure drops, and a number of designs originally created for shutoff service are available. Reasonable to excellent service life is obtained with a low first cost within the pressure and temperature limits.

A Saunders patent valve is shown in Fig. 23. Here a heavy fabric-reinforced diaphragm serves both as a seal and a closure member when forced down against a weir in the body. The simple body configuration and absence of a stuffing box make it especially attractive for a variety of chemical and slurry services within the limitations imposed by the flexible diaphragm itself. The bodies come in a wide variety of alloys and may be lined with tantalum, glass, Teflon, etc. The diaphragm may be faced with Teflon or made of elastomers. The imbalance force in the opening direction and the load for seating are high. Consequently, usage is generally limited to 8 in ($\sim$203 mm) maximum on control, with occasional applications for larger sizes using powerful piston actuators. The normal

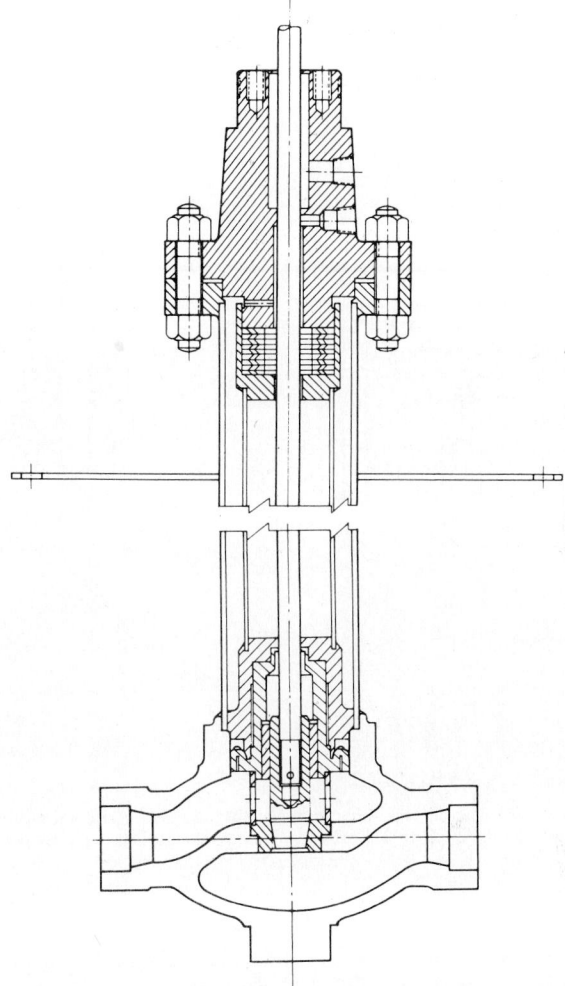

Fig. 21 Large-diameter thin-walled bonnet section reduces heat in-leak and permits trim replacement at the site.

flow characteristic is semi-quick-opening, and an improved version is available using a two-part thruster. In other modifications, a straight-through body and a deeply molded diaphragm are used for straight-through flow in the open position. Design limitations are otherwise similar.

Another useful design is the mechanical pinch valve, which is simply a flexible tube compressed mechanically or pneumatically like a rubber tube and a laboratory clamp. It can handle very coarse or extremely viscous materials. A typical enclosed mechanical design is shown in Fig. 24.

Mixing or Proportioning

For splitting of flow streams or combining two streams in some desired proportion, a single three-way valve may be practical. Two basic forms as adapted from either single- or double-port globe bodies are shown in Figs. 25 and 26. To minimize the effect of off-balance plug forces, it is generally

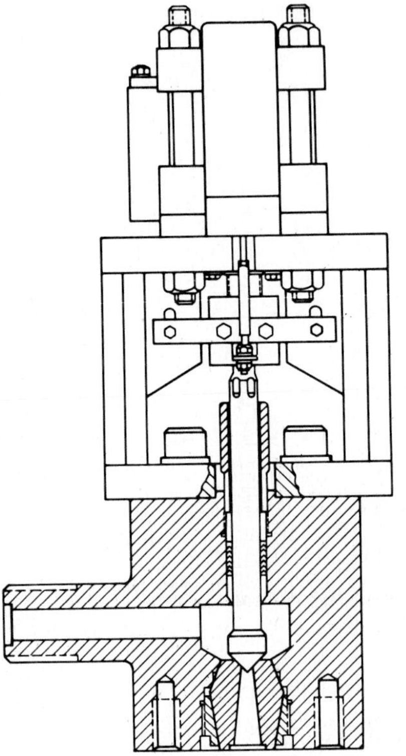

Fig. 22 High-pressure forged angle valve with an electrohydraulic actuator for 30,000 psi (∼207 MPa) service.

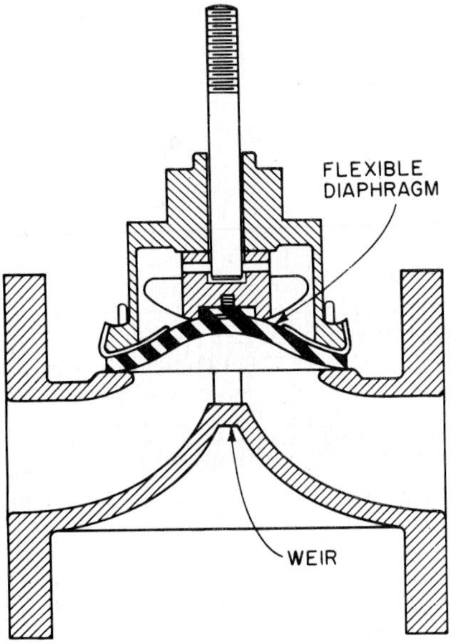

Fig. 23 Saunders patent valve using a fabric-reinforced diaphragm as both a seal and closure member.

FLEXIBLE
DIAPHRAGM

WEIR

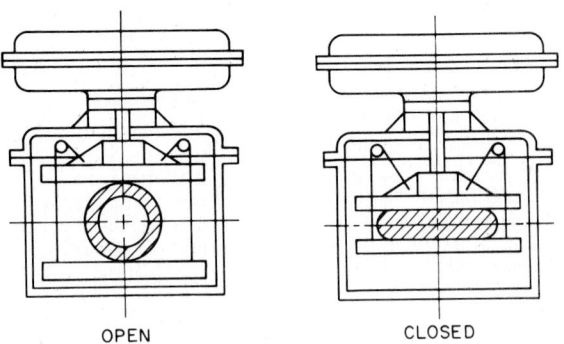

OPEN

CLOSED

Fig. 24 Flexible tube pinch valve for low-pressure low-temperature slurry service.

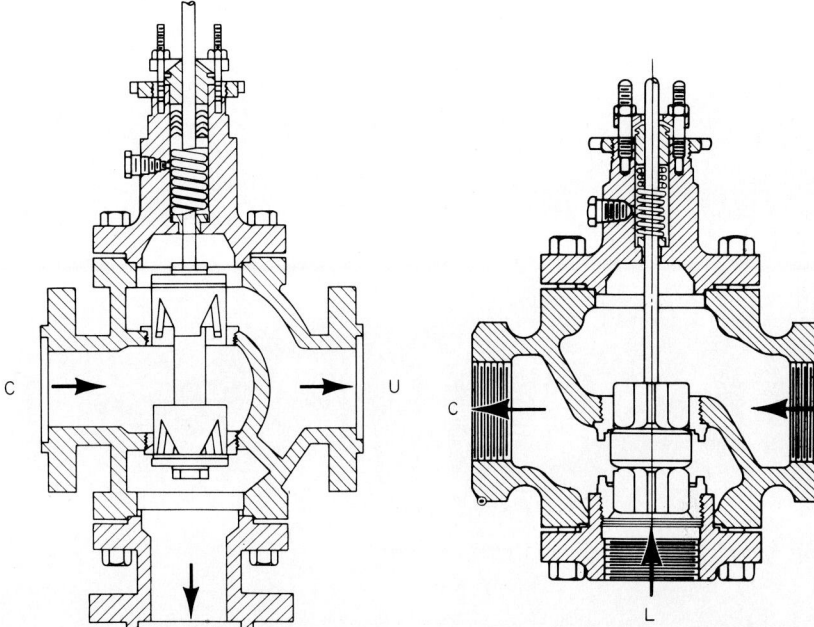

Fig. 25   Outside seating gives flow-to-open stability in flow diverting service.

Fig. 26   Three-way valve for combining two flows (inside seating for flow-to-open stability).

desirable to have the flow tend to open the valve ports. Hence, the design of Fig. 25 is generally used with a single flow brought into the common port and split into two streams, whereas the design shown in Fig. 26 is used to combine two separate streams into a common outlet.

Two common applications used for temperature control are heat exchanger bypass service and splitting a larger flow into two equal smaller ones. For large flows at lower pressures, a pair of butterfly valves mounted on a tee and mechanically linked to a common operator is sometimes used.

Angle and Y-Pattern Valves

One may be confused regarding the application of angle valves because the design is based on two unrelated factors. From the manufacturing standpoint, an angle-style body can easily be machined from a cylindrical or rectangular block with a full-sized interior passage. This feature is independent of any application requirements.

In cast form, angle-style valves were once used to eliminate a fitting and conserve space, but the primary field of application has been for special erosion-resistant designs or for those in which a greatly expanded outlet section is required for velocity and noise reduction. (See the article Control Valve Noise later in this Handbook section.)

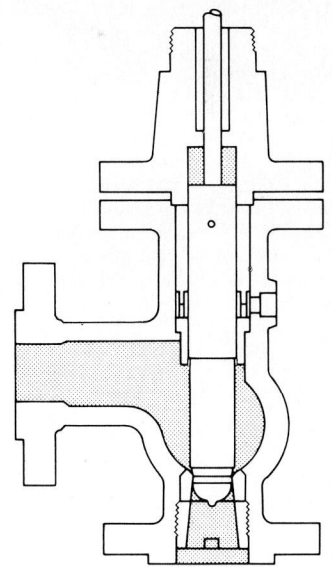

Fig. 27   High-pressure streamlined angle valve with a venturi outlet.

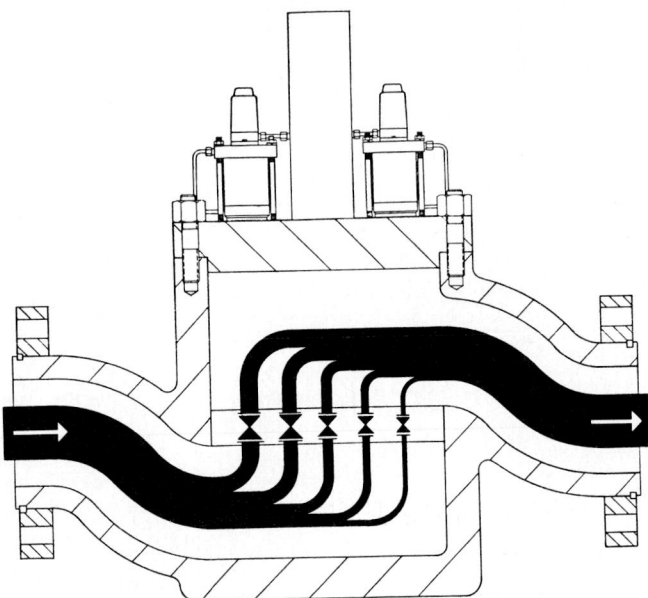

Fig. 28 Binary sequence of on-off solenoid-actuated valves provides wide variable flow on digital control.

The classic "streamlined" angle valve is shown in Fig. 27. This valve is normally installed with flow in over the plug and out through the venturi-shaped seat ring. This results in undesirable flow-to-close action, coupled with a very low C_f. (See the article Valve Sizing later in this Handbook section.) This makes it difficult to obtain stable operation and to avoid cavitation in liquid flow, severely restricting general application.

Digital Valves

With rapidly increasing acceptance of electronic process control systems, there is renewed interest in all electric valve actuation and particularly the possibility of direct digital drive. For the foreseeable future, however, there is strong reason to believe that conventional valves and pneumatic actuators will continue to dominate. Developments will come in conversion units which take computer output signals and converts them into the required pneumatic signals.

One digital valve design is shown in Fig. 28. The body casing contains a number of separate solenoid-operated on-off valves with a capacity varying in the binary sequence (1, 2, 4, 8, 16, . . .). An eight-step unit has a resolution of 1 part in 256. Applications are generally limited to very clean fluids at moderate temperature and pressure. For gases at a critical pressure drop, precise control of flow, coupled with very rapid response, is possible.

Valve Plug Design Details

In control valves of the sliding stem globe type, the flow-versus-lift relationship may be altered by changing the shape of the plug and/or seat ring. A large variety of physical combinations are possible, some of the most common being illustrated here. The actual shape is greatly influenced by the total port area exposed, the ratio of valve travel to port diameter, and other secondary factors such as flow efficiency.

Figure 29 shows two forms of equal percentage trim as used in top-guided single-port globes.

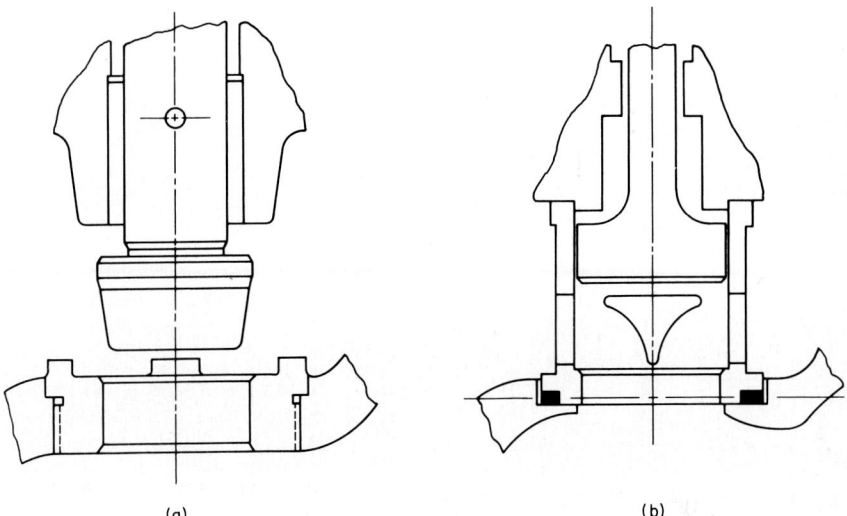

(a)   (b)

Fig. 29 Top-guided and cage-guided single-seat equal percentage valve trim.

The plug in Fig. 29*a* is a top-guided equal percentage lathe-turned plug, and the plug in Fig. 29*b* is essentially a flat-surfaced cylinder which cooperates with ports pierced through the supporting cage.

Figure 30 indicates four common styles of top-and-bottom-guided double-seat plugs.

For sizes 1 in (~25 mm) and under, smaller C_v ratings are obtained by reducing the seat ring bore and using a series of plugs with a common rated stroke, as shown in Fig. 31.

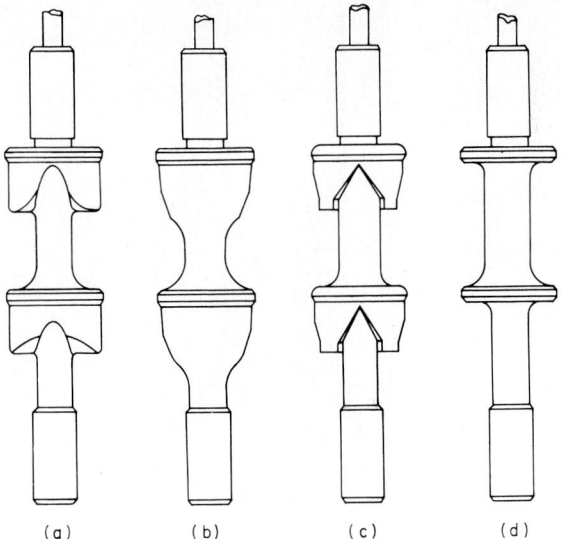

(a)   (b)   (c)   (d)

Fig. 30 Types of top-and-bottom-guided double-port plugs: (*a*) V-port plug with staggered ports, (*b*) equivalent lathe-turned unit, (*c*) straight-sided V-port design, (*d*) quick-opening disk.

Plug Guiding

With the control valve normally operating partially open under extremely turbulent flow conditions, a special support for the valve plug is required. For reciprocating stem valves, the most common form is shown in Figs. 1 and 2 in the earlier part of this article. A hardened stainless steel guide bushing is inserted in the valve bonnet. Design criteria include antigalling properties of the metal and proper fit to provide alignment yet prevent lateral vibration with an allowance for temperature extremes. The guide diameter and length must be chosen for good service life.

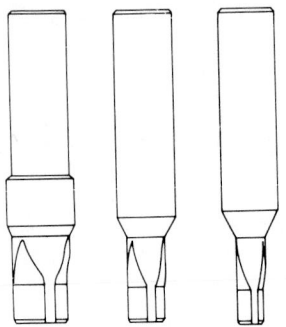

In Fig. 5, the guide bushing is shown as part of the packing box assembly, and the guide as the plug stem itself. A stem diameter large in proportion to the valve port size is used.

Seat ring guiding is commonly used for small plugs [1 in ($\sim$25 mm) and less]. One design using a plug in the form of a shaped piston is shown in Fig. 31. A skirted V port generally used in low-pressure steam, water, and air service is shown in Fig. 6. The various cage-guided designs, as shown in Fig. 4, provide exceptionally heavy-duty guiding.

Fig. 31 Part of a typical set of top-guided single-port trim. A range of orifice diameters between ¼ in and 1 in ($\sim$6.4 mm and 25.4 mm) is used in bodies as large as 2 in ($\sim$50.8 mm).

Other trim styles are included in the description of individual valve classes. In reciprocating stem designs, the determining criteria are adequate guiding and a satisfactory combination of flow characteristics and valve capacity. For rotary stem designs, the guides are the supporting trunnions, which must be made to withstand the side thrust without galling or showing excessive wear.

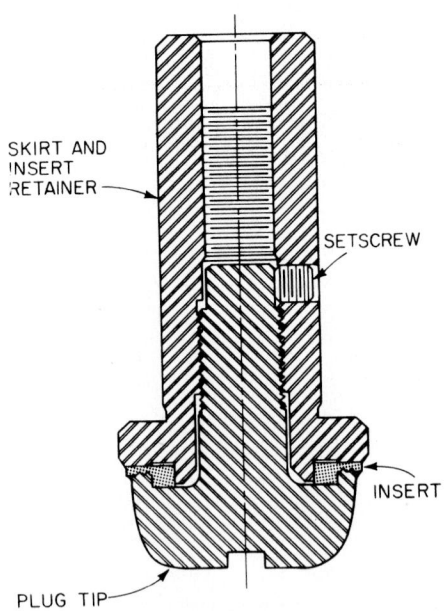

SKIRT AND INSERT RETAINER

SETSCREW

INSERT

PLUG TIP

Fig. 32 Top-guided single-port soft seat design.

Seat Leakage

All the plugs shown in the illustrations just mentioned are entirely of metal. For most control valves in continuous throttling services, tight shutoff is not a requirement and, in fact, can be obtained only through special designs. Both single- and double-port valves are available with elastomer inserts, such as Buna N and Teflon. In the double-port design, special provision must be made to allow adjustment of the seat-to-seat dimension, both in assembly and in the field.

A typical top-guided single-port soft seat design is shown in Fig. 32. Other designs may use O rings, and the soft insert may be in the seat ring rather than in the plug.

Control Valve Seat Leakage

Standard ANSI B16.104-1976 now sets forth five classes of leakage in order of increasing seat tightness:

Class II*:   0.5% of rated capacity
Class III*: 0.1% of rated capacity
Class IV*:   0.01% of rated capacity
Class V†: 5×10^{-4} ml/min of water per 1 in ($\sim$25 mm) of orifice diameter per 1 psi (7 kPa) differential
Class VI‡:   a special class for resilient seated control valves

*Tests conducted using air or water at 45 to 60 psig ($\sim$6.5 to 8.7 kPa).

†Used on metal-seated valves requiring exceptional tightness. The test is conducted using water at the maximum operating differential.

‡Test conducted at 50 psig ($\sim$7.3 kPa) or at maximum rated differential pressure, whichever is less, using air.

It is important to understand that absolute tightness is not practical for control valves because of the limited seating thrust attainable with conventional pneumatic actuators and because of the wear at the sealing surface which may occur during long periods of throttling service.

For soft seat designs, leakage generally approaches, but not quite reaches, zero. The seat load provided by the actuator must be sufficient. A figure of 25 lb per lineal inch ($\sim$4.5 kg per lineal centimeter) of seat circumference for Buna N seats and double this amount for Teflon are suggested in addition to the normal seat load allowance. Additional seat loads are also required for metal-to-metal seat valves requiring exceptional tightness.

Body Design Details

Globe body design details have become quite standardized through a long process of evolution. Valve body wall sections, end connections, and face-to-face dimensions are in accordance with recognized standards.

The bonnet and bottom flanges, if they contain plug guides or the packing assembly for the valve stem, must be accurately aligned. A double-offset joint which completely retains the gasket and also provides a suitable alignment guide is shown in Fig. 33. Stud bolting is generally used through ANSI Class 600 ratings, with through bolting or special mechanical designs more common for higher pressure ratings.

Threaded bonnet joints, such as shown in Fig. 34, are used for reasons of economy on low-pressure control valves through 2 in ($\sim$50 mm).

Packing Box

The function of the packing box is to provide a pressuretight seal for the valve stem, either sliding or rotary, which must transmit motion from the valve actuator to the plug.

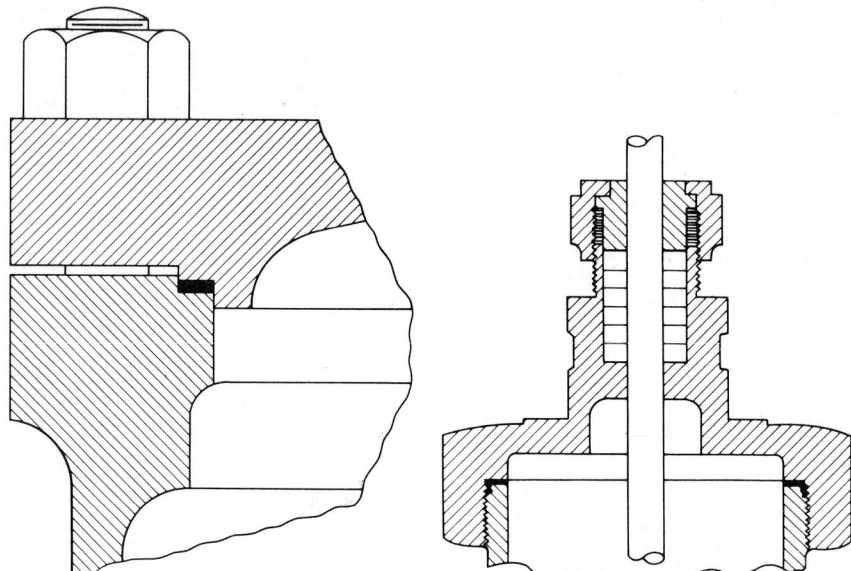

Fig. 33 Double-offset bonnet joint retains gasket and provides alignment.

Fig. 34 Threaded bonnet joint for economy in low-pressure service.

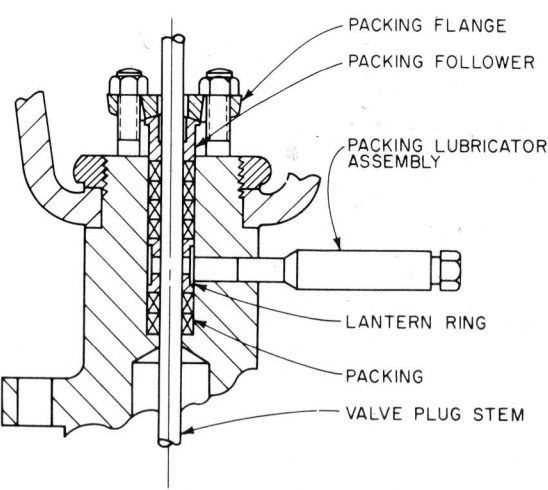

PACKING FLANGE

PACKING FOLLOWER

PACKING LUBRICATOR ASSEMBLY

LANTERN RING

PACKING

VALVE PLUG STEM

Fig. 35 Conventional sliding-stem control valve packing box.

The conventional control valve packing box shown in Fig. 35 is made up of packing to a depth of 1 to 1½ stem diameters, followed by a lantern ring or spacer and then additional packing to a total depth of 5 to 6 times the stem diameter. The packing is adjusted by squeezing down on the packing or follower through the bolted packing gland. For low-pressure services a threaded packing nut may be used and the packing depth may be reduced (see Fig. 6).

A number of packings are available which may be used with a wide variety of process fluids and maintain a tight seal with low friction. The most common are based on Teflon, since this material is inert to most chemicals, is usable at temperatures from -150 to $450°F$ (-101 to $232°C$), and has unique self-lubricating characteristics. It may be combined with shredded asbestos or asbestos yarn made up in the form of split or premolded solid rings or in shaped pure Teflon rings. In one widely used construction (Fig. 36) molded Teflon rings are preloaded by a corrosion-resistant spring and the packing follower is brought down against a shoulder so that no adjustment is needed.

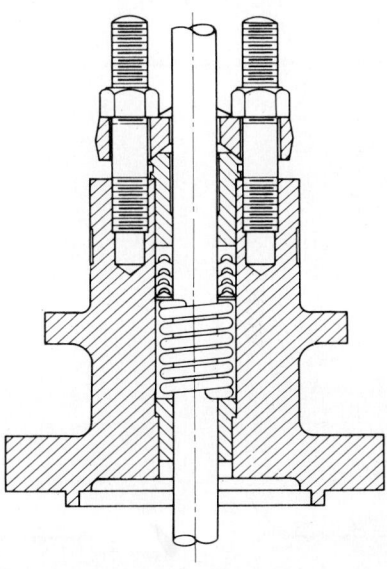

For temperatures over $450°F$ ($232°C$) or below $0°F$ ($-17.8°C$), an extension bonnet with or without radiating fins, as shown in Fig. 37, may be used. Where this construction is not practical, special high-temperature packings are available with the disadvantage of higher stem friction. For extremely high temperatures [$>850°F$ ($>454°C$)], various proprietary packing ring combinations based on a special form of pure carbon known as Graphoil are used.

When very toxic or valuable fluids are handled, a metallic bellows seal may be used as shown in Fig. 38. The most common bellows material is AISI Type 316 stainless steel, and pressure ratings are usually limited to 600 psi (~4.1 MPa) maximum, although bellows suitable for 2000 psi (~13.8 MPa) have been designed.

Fig. 36 Spring-loaded molded solid Teflon V rings permit nonadjustable packing box assembly.

Body Wall Thickness

Body wall sections are usually based on the values for flanged fittings in ANSI B16.5. An allowance of from ⅟₁₆ in (~1.6 mm) to ¼ in (~6.4 mm) is generally added to the tabular values to compensate for pattern core shift. Section VIII of the American Society of Mechanical Engineers' *Boiler and Pressure Vessel Code* is also used as a reference. Designs for large valves or complex shapes are carried out on an individual basis.

End Connections

The end connections for control valves are normally flanged. Threaded connections may be used through 1½ in (~38 mm) in size, particularly in air, water, and steam service. For general power plant application, particularly at higher pressures, weld end construction is used. Steel flanged end valves are generally furnished with dimensions to ANSI Class 600, and ring joint facings at higher pressures.

Cast iron flanges 1 in (~25 mm) and up are available in 125- and 250-psi (~0.9- and 1.7-MPa) ratings per ANSI B16.1. Bronze flanges are generally made based on ANSI B16.24. This standard includes Class 150 and 300 flanges.

Proprietary designs are available to meet special operating conditions. One of these, the Grayloc coupling, is shown in Fig. 39. It combines a compact size with ease of assembly and disassembly

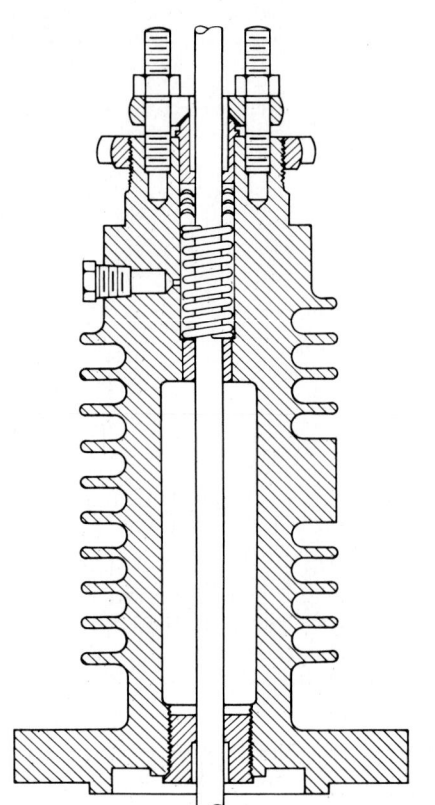

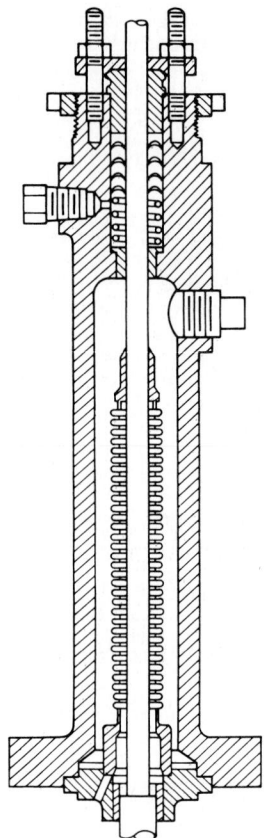

Fig. 37 Extension bonnet with radiating fins for high-temperature service.

Fig. 38 Bellows seal design for handling toxic or valuable fluids.

while minimizing reliance on good piping alignment at pressures to 10,000 psi (~69 MPa) and above. Another special design is the lens ring joint shown in cross section in Fig. 40. This applies to high-pressure valves above ANSI Class 2500, which are generally built to individual user specification.

For standard flanged globe bodies, 1 to 8 in (~25 to 203 mm) in ANSI Classes 125 and 250 for iron and Classes 150, 300, and 600 for steel are furnished with face-to-face dimensions per ANSI B16.10. For valves 10 to 16 in (~254 to 406 mm), the face-to-face dimensions for similar ratings are covered by FCI Standard 65.2. These dimensions are given in Table 1.

Face-to-face dimensions for flangeless control valves covered by ISA-S75.04 are given in Table 2.

Valve Body Materials

Since control valves are called on to handle all kinds of fluids from clean, dry air to corrosive chemicals at service temperatures from near absolute zero to well over 1000°F (538°C) and at pressures from almost a complete vacuum to 40,000 psi (~276 MPa) or higher, proper material specifications are of the utmost importance. For the pressure containing parts—the body, bonnet, and bottom

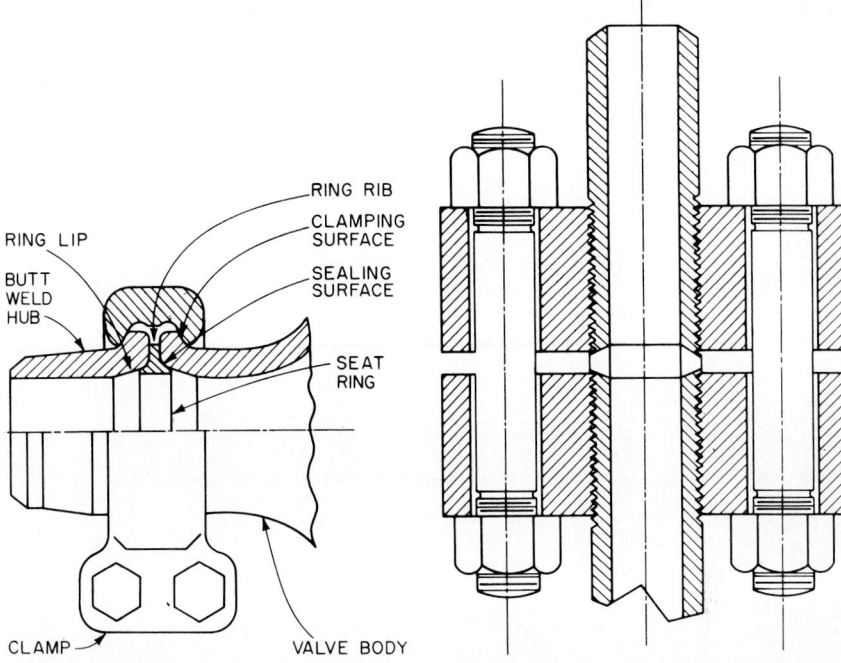

Fig. 39 Grayloc coupling used in high-pressure and high-temperature service.

Fig. 40 Lens ring joint. One of many similar forms used for high-pressure service.

Table 1 Face-to-Face Dimensions for Standard Flanged Globe Bodies

Nominal valve size		ANSI primary service pressure rating					
		Iron, 125 psi (862 kPa), steel, 150 psi (1034 kPa)		Iron, 250 psi (1724 kPa), steel, 800 psi (2068 kPa)		Steel, 300 psi (1068 kPa)	
Inches	Millimeters	Inches	Millimeters	Inches	Millimeters	Inches	Millimeters
½	13	—	—	7½	190	8	203
¾	20	—	—	7⅝	194	8⅛	206
1	25	7¼	185	7¾	197	8¼	210
1½	40	8¾	220	9¼	235	9⅞	251
2	50	10	255	10½	267	11¼	286
2½	65	10⅞	275	11½	292	12¼	311
3	80	11¾	300	12½	318	13¼	337
4	100	13⅞	350	14½	368	15½	394
6	150	17¾	450	18⅝	473	20	508
8	200	21⅜	545	22⅜	568	24	610
10	250	26½	675	27⅛	689	29⅝	753
12	300	29	735	30½	775	32¼	819
14	350	35	890	36½	927	38¼	972
16	400	40	1015	41⅝	1057	43⅝	1108

Table 2   Face-to-Face Dimensions of Flangless Control Valves

Nominal valve size		ANSI classes 150 through 600			
Millimeters	Inches	Millimeters (L)	Inches (L)	Tolerance, mm	Tolerance, in
20	0.750	76	3.00	±1.6	±0.0625
25	1.000	102	4.00	±1.6	±0.0625
40	1.500	114	4.50	±1.6	±0.0625
50	2.000	124	4.88	±1.6	±0.0625
80	3.000	165	6.50	±1.6	±0.0625
100	4.000	194	7.62	±1.6	±0.0625
150	6.000	229	9.00	±1.6	±0.0625
200	8.000	243	9.56	±1.6	±0.0625
250	10.000	297	11.69	±1.6	±0.0625
300	12.000	338	13.31	±3.2	±0.125
400	16.000	400	15.75	±3.2	±0.125

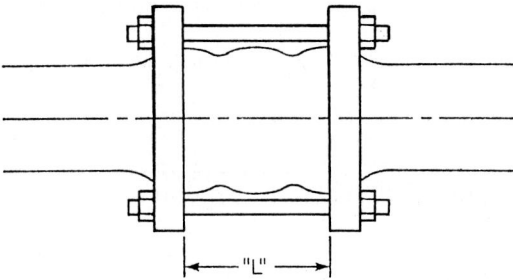

flange—cast iron or carbon steel is the most common material, since the majority of valve applications are relatively noncorrosive at reasonable temperatures and pressures.

Cast Iron

Cast iron is usually specified as ASTM A126, Class B or C. Class B is a 31,000-psi (~214-MPa) tensile-strength iron, and Class C has a 41,000-psi (~283-MPa) minimum tensile strength. In the major process industries, steel may be specified instead of iron for uniformity and for added protection against the hazard of fire.

Carbon Steel

Carbon steel is normally ASTM A216, Grade WCB. This steel is a welding grade and has a maximum carbon content of 0.35 and a minimum tensile strength of 70,000 psi (~483 MPa). WCB steel, as it is commonly called, meets the requirements for pressure and temperature ratings listed under carbon steel in ANSI B16.34 (Table 2).

Alloy Steels

For handling steam, petroleum products, and other relatively noncorrosive fluids at temperatures between 800 and 1050°F (427 and 566°C) alloy castings per ASTM A217 are usually specified for body components. This specification covers nine grades, the choice of which depends on the physical, temperature resistance, and corrosion resistance qualities required. These alloys meet the pressure and temperature ratings listed under "Carbon Moly and Chrome Moly Steels" in Table 2 of ANSI

B16.34 as explained in Table 1 of the same standard. Grade WC1, a carbon-molybdenum alloy, Grade C5, a 5-chrome-½-molybdenum alloy, and WC9, a 2¼-chrome-1-molybdenum alloy, are the most common. Grades C5 and WC9 are often used on valve bodies handling steam condensate and boiler feedwater in power plants, even though the pressures and temperatures are low. They resist the washing action of oxygen-free hot flashing water. Molybdenum provides creep strength at high temperatures, and chromium adds corrosion resistance and strength. These steels are all of welding quality but should be stress-relieved after welding. Grades C5 and WC9 require preheating.

For control valves handling fluids at temperatures between −50 and −150°F (−45.6 and − 101°C), a 3½% nickel alloy carbon steel is sometimes specified. A typical analysis is covered by ASTM A352, Grade LC3. This steel has a minimum Charpy impact value of 15 ft·lb (20.3 N·m) at −150°F (−101°C).

Stainless Steel

Austenitic stainless steel bodies per ASTM A351, Grades CF8M and CF8C, are used for many corrosive applications. Both grades also resist scaling at quite high temperatures. Grade CF8M (AISI Type 316), which contains molybdenum, retains considerable strength to temperatures as high as 1250°F (677°C) and is usable to 1500°F (816°C). Pressure and temperature ratings for stainless steels are given in Table 2 of ANSI B16.34. Grade CF8C (AISI Type 347) is recommended for corrosion resistance at high temperatures. Other stainless steel alloys that are castable and machinable may also be specified on a custom basis to meet particular corrosion requirements.

None of the austenitic stainless steels lose impact strength at low temperatures. Consequently, one of these steels is usually specified for temperatures below −150°F (−101°C).

Bronze

Case bronze bodies are used on direct-operated pressure and temperature regulators and in small sizes for low-pressure applications handling air, water, and steam. In the process industries, above the 2-in (50-mm) size, cast iron or steel is generally adequate and more economical unless corrosive conditions specifically warrant bronze. A bronze per ASTM B62 (85% copper, 5% tin, 5% lead, and 5% zinc) is most commonly used for body castings, although zinc-free bronze may be specified for its acid resisting properties. Bronze bodies may also be used for low-temperature applications because of their high impact strength. Bronze is also suitable for cryogenic temperatures and is specified for oxygen gas service at ambient temperatures as a safety measure, since it will not support combustion in the event of a fire.

Corrosion-Resistant Alloys

Many other alloys are available on special order. These include Monel, Hastelloy B or C, nickel, and special grades of austenitic stainless steels, such as AISI 329 and Alloy 20. They are used primarily for specific corrosive chemical applications. Such alloys are inherently expensive, but in many cases their use may be mandatory for satisfactory service life.

Bolting Materials

Carbon steel cap screws are generally considered adequate for body closure bolting on cast iron and low-pressure bronze bodies.

The bolting used on carbon steel and alloy steel bodies is generally per ASTM A193. The grade most commonly used is B7 up to 850°F (454°C). Above this temperature, Grade B16 is used and, as an option, strain-hardened austentic stainless Grade B8. This latter material is also recommended for temperatures below −20°F (−29°C).

Nuts used on studs for steel and alloy steel bodies are generally per ASTM A194, Grade 2H, for temperatures to 850°F (454°C), and Grade 4 or Grade 8 for higher temperatures. Grade 8 nuts (AISI Type 304 stainless) are also used with Grade B8 studs for low-temperature service.

Stainless steel and other corrosion-resistant bodies may be bolted with the steel studs and nuts just described, although if the bolting will be exposed to corrosive fluids or gases, it may be necessary to use materials similar to those used in the body.

Gasket Materials

The most common gasket material for flanged joints, either for body closures or line flanges, is asbestos fiber sheet in ⅟₃₂ or ⅟₁₆ (0.8 or 1.6 mm) thickness. This material is inexpensive, readily available, resists many forms of corrosion, and can withstand temperatures as high as 800°F (427°C). Metal-clad asbestos-filled gaskets of various designs are becoming increasingly popular.

On valves subject to thermal cycling, particularly in steam service, spiral-wound metal gaskets with asbestos as a filler material are recommended. These gaskets are compressed to a specified thickness as controlled by the body and bonnet machining or by a special spacer ring as an integral part of the gasket. Ring-type joints are not common for body closures but are sometimes used for line flange connections. They are covered by ANSI B16.20, and the rings are commonly made of soft iron or an alloy similar to the body material. Other solid metal gaskets including metal O rings are used for special designs, particularly for high-pressure service.

Research is under way to develop substitute materials for asbestos because of its hazardous nature.

Valve Trim Materials

While there is not complete agreement, the term *valve trim* for present purposes will include the valve plug, seat rings, guide bushings, and valve stem and internal packing box parts.

Standard Trim Materials

By far the most common trim material is austenitic 18-8 molybdenum stainless steel ASTM A351, Grade CF8M (AISI Type 316). This material is used for all trim parts except guide bushings. The latter should be of a material that does not gall or seize when in sliding contact with the plug guides. AISI Type 416, a 12 to 14% chromium stainless steel, hardened to 38 to 42 Rockwell C, is widely used, as is AISI Type 440-C, a higher chrome stainless steel with excellent wear resistance. 17-4 PH, a precipitation hardening stainless steel, is also used, since it combines satisfactory resistance to galling and wear with excellent corrosion resistance.

High-Pressure-Drop Service

Both plugs and seat rings may be made of the 400 series stainless steels when required for resistance to erosion. AISI Type 440-C stainless steel, heat-treated to 55 to 58 Rockwell C, may be used for plugs and seat rings as well as bushings. It gives excellent resistance to erosion at temperatures to 850°F (454°C) and has satisfactory corrosion resistance for many applications in power plants and refineries. However, since it is not readily castable, its use is limited to sizes and shapes that can be machined from barstock.

The most common high-pressure-drop trim is AISI Type 316 stainless steel hard-faced at the points of expected wear using hard-facing alloys of the cobalt base variety such as Stellite or nickel-chrome-boron types such as Colmonoy. Stellite 6 may be machined with only moderate difficulty, and this has led to an increase in the use of the solid alloy as a precision casting or barstock for valve plugs, seat rings, and guide bushings. The hard-facing alloy is occasionally laid directly in the body in lieu of seat rings for power plant or other high-temperature applications.

The valve stem may require special consideration if the applied force is high. AISI Type 17-4 PH (H-1075) is often used for increased strength at temperatures to 650°F (343°C).

High-Temperature Service

Special consideration must be given to materials for plugs, stems, plug guides, and guide bushings at elevated temperatures. Guide bushings for service above 950°F (510°C) may be Stellite or Colmonoy, faced or made of the solid alloy.

The plug guide should be hard-faced, using a different grade of alloy to minimize any tendency toward galling. A high-strength plug stem may also be required, and when the 650°F (343°C) limit of 17-4 PH stainless is exceeded, a more exotic alloy, such as Inconel X750 may be used, thus extending the temperature range to 1150°F (621°C).

Special Corrosion Resistance

A wide range of corrosion-resistant materials, such as Hastelloy, Monel, nickel, and Alloy 20, are used for valve trim in chemical service. Trim parts may also be coated with or made from solid or reinforced plastics, such as Teflon and Delrin, among others. Improvements in solid metallic coating techniques, such as vapor deposition of tantalum and other refractory or chemically resistant alloys, are sometimes used but have not gained wide acceptance.

Pressure and Temperature Ratings

Maximum pressure and temperature ratings for control valves are generally listed in accordance with recognized ANSI standards.

Cast Iron Control Valves

Pressure and temperature ratings for cast iron control valves are listed in ANSI B16.1. Most manufacturers and users follow the dictates of these standards.

Steel and Stainless Steel Control Valve Bodies

Most control valve manufacturers limit flanged steel valves to the pressure and temperature ratings prescribed in ANSI B16.5. This standard gives pressure and temperature ratings for steel, alloy steel, and stainless steel pipe flanges or flanged fittings for ANSI Classes 150, 300, 400, 600, 900, 1500, and 2500. It also prescribes hydrostatic test pressures for these primary ratings.

Bronze Control Valve Bodies

Cast bronze flanged control valves are normally rated according to the values listed in Table 1 of ANSI B16.24. This standard covers Class 150 and 300 bronze flanges and flanged fittings.

Corrosion-Resistant Alloy Bodies

For maximum pressure and temperature ratings of corrosion-resistant alloy bodies, such as Hastelloy and Monel, the purchaser should consult the manufacturer.

Low-Temperature Ratings

It is quite an accepted rule of thumb that control valve body materials should have a Charpy impact value of 15 ft·lb (20.3 N·m) at the minimum operating temperature.

Ordinary carbon steel bodies may be expected to exhibit this Charpy impact value, or more, at temperatures as low as $-50°F$ ($-45.6°C$). For the temperature range between $-50°F$ ($-45.6°C$) and $-150°F$ ($-101°C$), a 3½% nickel steel per ASTM A352, Grade LC3, is usually specified. For temperatures below $-150°F$ ($-101°C$), austenitic stainless steels or bronze materials should be used.

Actuator Design and Applications

For a control valve to perform its basic function, it must be provided with an actuator capable of positioning the valve plug in response to the demands of a controller. The design problem is primarily one of meeting requirements for positioning accuracy, speed of response, adequate force or thrust level, and stable operation. Secondary factors include behavior on loss of signal or auxiliary power supply, also known as *failure action*, long service life, and economical first cost and operating cost. An important distinction is that the actuator continuously positions the valve between open and closed. On-off actuators, or those operating in discrete steps, are not included in this article. See the article Stepping Motors in Sec. 19.

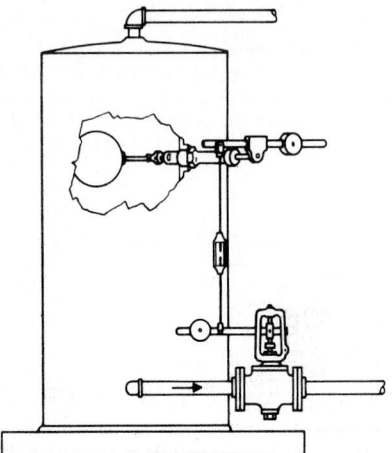

Fig. 41 Direct-connected level control with sliding-stem lever valve.

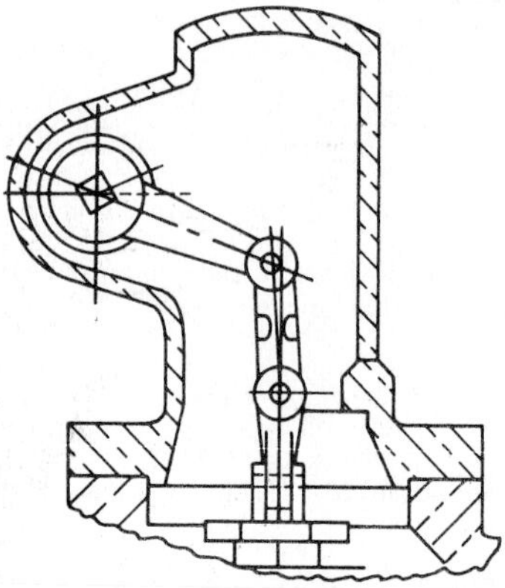

Fig. 42 Rotary shaft mechanical valve actuator.

Mechanical Actuation

A few control valves (primarily for level service) may be mechanically linked to the primary element. A level controller using a lever-operated valve with a sliding stem construction is shown in Fig. 41. In another version, the valve bonnet may contain a linkage for converting reciprocating motion to rotary motion (see Fig. 42), or a rotary valve such as a butterfly valve may be used. Despite the low cost, application is limited by the requirement for close location of the primary element and valve, limited power, and lack of flexibility in control adjustment.

Self-actuated pressure, flow, and temperature regulators are basically similar in function and are subject to the same limitations, although they are widely used, particularly in air, steam, and water service. A self-actuated pressure regulator generally utilizes as a primary element a spring-opposed fabric-reinforced flexible diaphragm assembly directly connected to the valve plug. By interposing a pneumatic controller and using a larger effective area for the sensing diaphragm assembly and a separate compressed air supply, the basic limitations of direct-connected regulator-type control were overcome. The result was the pneumatic spring diaphragm actuator which is still used on the majority of control valves with only minor modifications.

Spring-Opposed Diaphragm Actuator

The modern spring diaphragm actuator is an inexpensive rugged sensitive device. Basically, it consists of a pressure-tight housing sealed by a flexible fabric-reinforced elastomer diaphragm. A diaphragm plate is held against the diaphragm by a heavy compression spring. A typical construction is shown in Fig. 43. In this version, signal air pressure from a controlling instrument or a manual pressure loading device is applied to the upper diaphragm case. This, in turn, applies a load to the diaphragm plate and actuator stem assembly. By selecting the proper spring rate or stiffness, load carrying capacity, and initial compression, the desired no-load stroke can be obtained for any air-input signal (e.g., 3 to 15 psi; 20 to 100 kPa). In this actuator, an increase in air pressure extends the stem—referred to as *direct action*. In the design shown in Fig. 44, the actuator is reversed so that an increase in air pressure retracts the stem *(reverse action)*.

The diaphragm is generally molded to shape. Buna N and neoprene with a synthetic fabric reinforcement are common materials. Service life is exceptionally long at normal operating pressures [generally up to 50 psi ($\sim$345 kPa) maximum]. The diaphragm is clamped between the upper and lower cases of the actuator, forming its own gasket. The temperature limit for Buna N is -20 to $150°F$ (-28.9 to $66°C$) in this service. Special elastomers are available, as well as fabric reinforce-

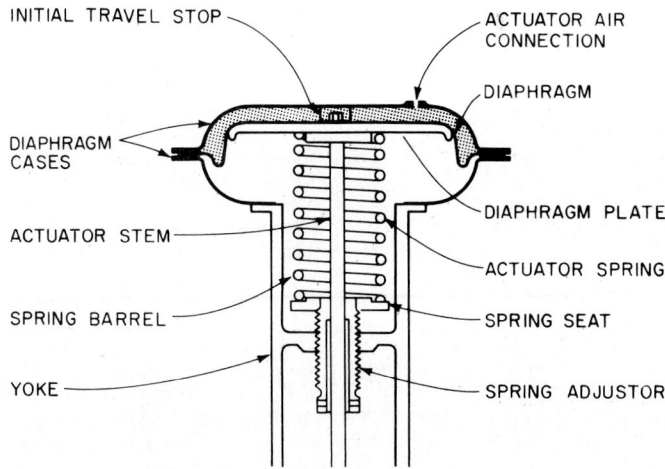

INITIAL TRAVEL STOP

ACTUATOR AIR CONNECTION

DIAPHRAGM

DIAPHRAGM CASES

DIAPHRAGM PLATE

ACTUATOR STEM

ACTUATOR SPRING

SPRING BARREL

SPRING SEAT

YOKE

SPRING ADJUSTOR

Fig. 43 Direct acting spring diaphragm actuator.

ments for exceptionally high ambient temperatures and also for low temperatures down to −60°F (−51.1°C).

The effective area of this actuator design is based on a diameter somewhere between the case ID and the diaphragm plate OD. In most modern designs, the area remains nearly constant with the stroke. In earlier designs made to accept flat sheet diaphragms or shallow molded types, the area varied with the stroke. This is unimportant from the practical standpoint, but the area-versus-stroke relationship must be known in order to predict net actuator thrust and to select proper springs to produce the desired pressure-stroke characteristic.

Bench Range

This is the no-load stroke-versus-air-pressure characteristic of the actuator. The most common ranges are 3 to 15 psi (20 to 100 kPa) and 9 to 15 psi (60 to 100 kPa) for split-range operation of two valves and ranges for higher operating pressures such as 6 to 30 psi (40 to 200 kPa). When used without a valve positioner, the bench range must not exceed the air-output range of the controller, which is zero to just under the supply pressure and not the commonly quoted 3 to 15 psi (20 to 100 kPa) which is primarily a standard for pneumatic transmission or used as the input signal to a valve positioner.

The actuator must oppose and cancel the valve stem forces, which may be large and which often vary substantially with the valve opening. The bench range, therefore, is really only a reference. As a simple example, for a 100-in² (∼645-cm²) actuator opposing a constant 100-lb (45.4-kg) thrust, Fig. 45 shows a bench range of 3 to 15 psi (20 to 100 kPa) and an installed range of 4 to 16 psi (∼27 to 107 kPa). The permissible offset between the bench and installed range is limited by the initial spring compression at the start of the stroke and the maximum available air supply pressure at the end of the stroke.

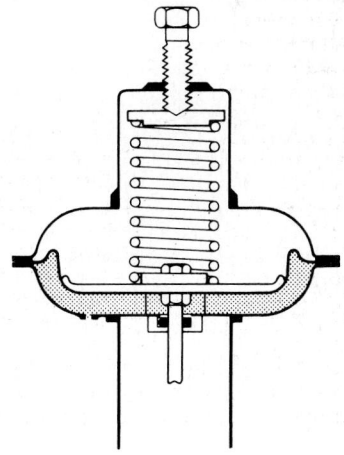

Fig. 44 Reverse acting spring diaphragm actuator.

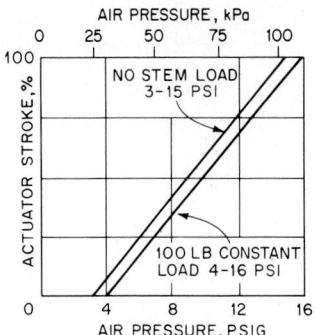

Fig. 45 Offset in actuator stroke versus air pressure caused by constant 100 psi (~690 kPa).

Actuator Thrust

The *maximum net actuator thrust* at the start of the stroke is the effective area times the actuator bench initial air pressure. The *final thrust* is the actuator effective area times the difference between the bench final air pressure and the maximum applied air pressure. With a bench final air pressure of 15 psi (100 kPa) and an applied air pressure of 20 psi (~138 kPa), a value of 5 psi (~35 kPa) times the effective area is available for doing work.

When an independent air supply is provided and a valve positioner (described later in this article) is used, the actuator bench range is limited only by the capacity of the compression spring and the maximum actuator pressure rating. For applications requiring high thrust or air failure, e.g., a normally closed flow-to-open single-port valve, a typical bench range might be 15 to 30 psi (100 to 200 kPa). If a high final thrust is required, an efficient bench range might be 3 to 8 psi (20 to ~53 kPa) with a 35-psi (~242-kPa) supply.

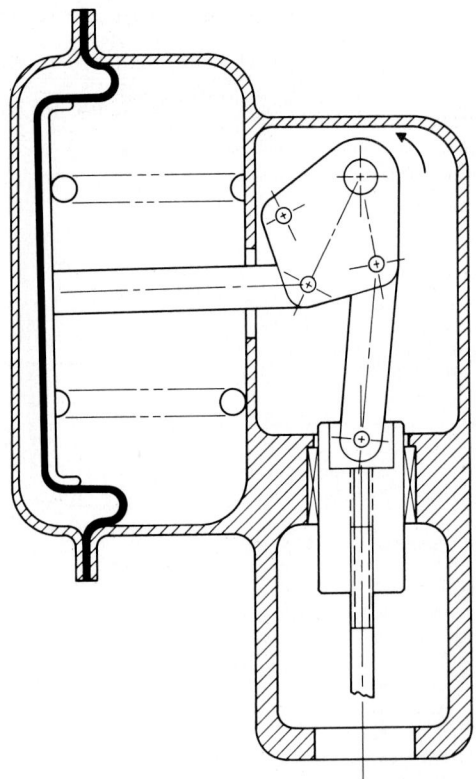

Fig. 46 Rotating bell crank linkage varies actuator output thrust to match valve requirements.

Actuator Stiffness

A factor in actuator design of fundamental importance is actuator stiffness. A spring-opposed dia-phragm actuator is a spring-mass system. The "spring" is the combined rate of the actual compres-sion spring and that imparted by the air trapped in the diaphragm case. The "mass" includes the weight of the moving actuator parts plus the plug-and-stem assembly. Dynamic force changes under service conditions can set up unwanted vertical oscillation. Also, the rate of change in force on the valve plug as it moves may exceed that of the actuator, causing the valve to fly open or closed.

When applying large valves at a high operating differential pressure, this factor must be given special attention. No simple rules have been formulated for all cases, and the valve manufacturer should be consulted. In many instances, the bench range has to be much higher than a simple static calculation would indicate in order to ensure operating stability.

Variable Thrust Actuator

By interposing a simple linkage system between the spring diaphragm power unit and the valve stem, the actuator output thrust may be altered to more closely match the valve requirements. In the design shown in Fig. 46, a spring diaphragm power unit of relatively long stroke rotates a plate-and-linkage assembly which drives the valve stem. As the connecting link approaches the vertical center line, the output thrust vector increases.

The unit, as shown, is in the reverse mode (an increase in air pressure retracts the stem). Direct action is obtained by simply moving the upper end of the connecting link to the left-hand position in the rotating plate assembly. This design is particularly well adapted for flow-to-open single-port control valves, since the imbalance force acting on the valve plug increases substantially as the valve closes. The inherent flow characteristic of the valve-actuator combination is also altered. For exam-ple, an inherently linear valve will produce a modified percentage characteristic.

Other designs are made in both hydraulic and pneumatic versions for special applications, such as high-performance butterfly valves and large ball valves in liquid pipeline service.

Piston Actuator

A large variety of piston actuators are used for throttling control. Figure 47 shows one such unit made for application in control valve service. Air pressure applied to the underside of the piston drives it upward, retracting the stem. In this ver-sion, a low-capacity spring installed above the piston provides a biasing force in the downward direction.

By itself, the piston actuator would be suit-able only for on-off service. It is converted into a positioning unit through the use of a valve positioner. One approach is to apply a constant air loading pressure to one side of the piston. The output of the positioner is then brought to the opposite side. By adjusting the loading pres-sure, the output thrust of the actuator may be biased in either direction. For example, if equal thrust is desired in either direction, the loading pressure must be set at one-half the supply pres-sure. This design has the advantage of simplic-ity, but is disadvantageous in certain applica-tions, because the net stem thrust is limited by the constant and opposing loading pressure.

A second approach (see Fig. 48) uses a posi-tioner with a four-way air relay so that the air pressure is increased on one side of the piston and simultaneously decreased on the other. The

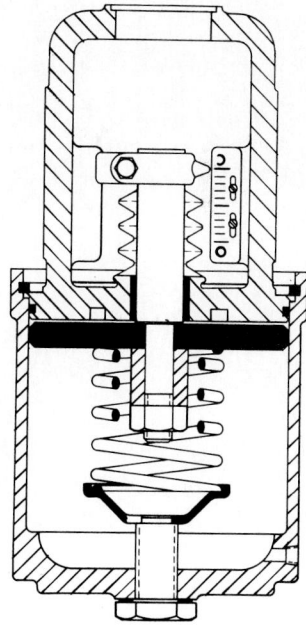

Fig. 47 Piston actuator equipped with low-capacity motion biasing spring.

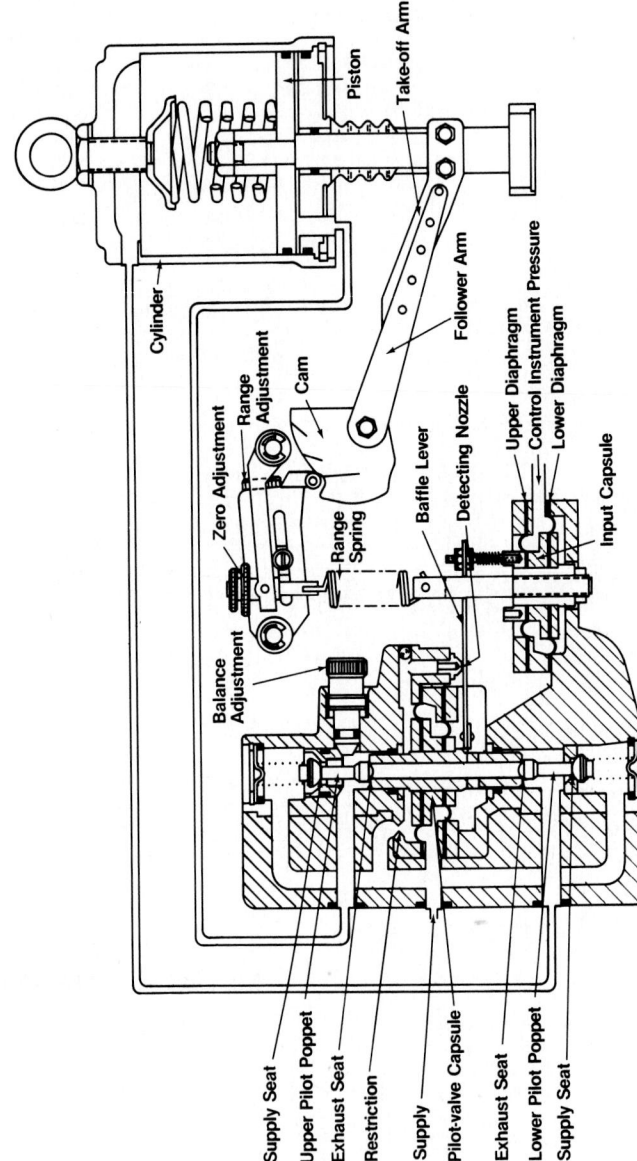

Fig. 48 Force-balance four-way valve positioner for piston actuation.

Take-off Arm

Piston

Cylinder

Zero Adjustment

Range Adjustment

Cam

Range Spring

Balance Adjustment

Follower Arm

Baffle Lever

Detecting Nozzle

Upper Diaphragm

Control Instrument Pressure

Lower Diaphragm

Input Capsule

Supply Seat

Upper Pilot Poppet

Exhaust Seat

Restriction

Supply

Pilot-valve Capsule

Exhaust Seat

Lower Pilot Poppet

Supply Seat

19.38

full potential thrust of the actuator is then obtained in either direction. This is particularly advantageous for valve designs, such as control ball valves where friction may represent a substantial portion of the load.

For a given size, a piston actuator is more expensive, but it offers the following advantages:

1. Long stroke versus piston area
2. An operating air pressure to 150 psi ($\sim$103 kPa) for high thrust
3. Compact size and weight

Fail-safe action is sometimes obtained by locking the actuating air pressure on both sides of the piston by means of auxiliary relays. The possibility of a leak past the piston makes this procedure less reliable than the use of an essentially hermetically sealed diaphragm in a spring diaphragm actuator.

The increasing popularity of rotary valves for shutoff service has resulted in a number of interesting piston actuators which are of compact design and are specially adapted for rotary action. Such actuators may be adapted for control valve use by the addition of a conventional valve positioner. In these adaptations, the design must eliminate any significant lost motion that may exist in the two-position (on-off) version.

Valve Positioners, Transducers, and Pneumatic Relays

A valve positioner is a device designed either integrally or as an attachment to a positioning-type valve actuator. The actuator stem motion is accurately compared with the signal from a controlling instrument or manual loader. Any deviation from the desired position results in an error signal which activates a pneumatic relay having an independent air supply. Air is then either admitted or exhausted from the actuator to drive the stem to the desired position. Positioners were first developed for use with spring diaphragm actuators over 40 years ago and are now produced in a variety of models, all based on the simple operating principles described here.

Pneumatic positioners can always be classified as force-balance or motion-balance. A *force-balance* positioner with a four-way pilot for piston operation was previously shown schematically in Fig. 48. The 3 to 15 psi (20 to 100 kPa) control instrument pressure creates a force through the input capsule that is balanced by the range spring. A change in control air pressure temporarily upsets the balance, covering or uncovering the detecting nozzle which, in turn, through the pilot valve capsule, simultaneously opens and closes alternate inlet and exhaust ports and drives the piston in the direction that restores the system equilibrium. The net motion of the detecting assembly is very small, hence the term "force-balance."

A cam interposed in the stem motion feedback system permits alteration of the inherent flow characteristic of a valve. It can also be used to alter the input air pressure span for split-range operation. Multiple cam surfaces are sometimes provided for the most popular options, such as 3 to 9 and 9 to 15 psi (20 to 60 and 60 to 100 kPa). One design at a premium price provides a larger cam which can be altered in the field to match a particular service.

In motion-balance designs, the 3 to 15 psi (20 to 100 kPa) input signal is brought into a spring-opposed capsule which moves one end of a lever an amount proportional to the applied air pressure. This movement covers or uncovers a flapper-nozzle assembly which through a pneumatic relay applies air to or exhausts air from the actuator. The motion of the actuator stem is fed back through a mechanical linkage to the other end of the lever position by the input capsule. The position of the lever is detected by a flapper-nozzle assembly which drives the pneumatic relay until it is balanced at the point of tangency. Here again, the motion feedback linkage may include a cam and means for varying the ratio between the stem motion and input signal.

Steady-state air consumption for a well-designed positioner should be low and is generally in the range of 10 to 30 std ft^3/h, approximately equivalent to 283 to 850 L/h. The maximum output of the relay, however, is generally in the range 150 to 250 std ft^3/h (4250 to 7080 L/h), since one purpose of the relay is to provide faster operation. Relay capacities and general performance characteristics of different designs vary substantially and should be carefully evaluated.

Typical Positioner Performance

Representative characteristics include:

> Pneumatic Signal Input Ranges: 3 to 9, 3 to 15, 9 to 15, 6 to 30 psi (20 to 60, 20 to 100, 60 to 100, 40 to 200 kPa)
> Air Supply Pressure: 20 to 100 psig (138 to 690 kPa)
> Supply Pressure Influence: less than 0.25% per 1 psi (7 kPa) supply change
> Conformity of Span: ±2% of span
> Dead Band: less than 0.25% of span (based on valve-positioner combination)
> Open-Loop Gain and Dynamic Response: (described in following paragraphs)

On the assumption that a valve positioner has only a single time constant, its response is given by

$$\xrightarrow{I} \boxed{\dfrac{KG}{I + KG}} \xrightarrow{O} \tag{1}$$

where G = positioner and actuator transfer function
I = controller input signal, %
K = static gain (open loop)
O = stem position, %

$$G = \frac{1}{1 + Ts} \tag{2}$$

where T = open-loop time constant (seconds) and s = Laplace operator. Therefore

$$\frac{O}{I}(s) = \frac{K}{K + 1 + Ts} \tag{3}$$

The break frequency ω (radians per second) is useful in evaluating the positioner response. It is defined as

$$\omega = \frac{K + 1}{T} \tag{4}$$

The open-loop gain K ranges from 40 to over 200 in various commercial positioners.

Positioner Applications

Valve positioners are now specified for a majority of spring diaphragm actuators, even though they may not be necessary and may even cause stability problems on fast loops, such as in flow control. Many plants are 100% positioner-equipped. From the standpoint of control dynamics, the positioner adds a time constant to the loop. The magnitude of the change can be relatively large for certain designs employing relays, which can be considered two-stage, i.e., a separate nozzle and pneumatic amplifying relays.

Advantages of a Positioner

The desirable features of a positioner include:

1. A positioner helps to overcome valve stem friction through its high open-loop gain.
2. Regardless of the bench range or imbalance forces on the valve stem, the positioner closely matches the input signal to the valve stroke.
3. Through the use of a cam, the *stroke-versus-input relationship* can be conveniently varied.
4. The controller signal is dead-ended in a small chamber, greatly increasing the speed of response when the distance between the controller and valve is large.
5. A faster stroking speed is generally attainable.
6. The signal input can be one-half or one-third of the full signal, e.g., for split-ranging without changing the bench range of the actuators.
7. Reverse action can provide an increase in output pressure for a decrease in signal pressure.
8. Subject to the pressure limitations of the actuator and available air supply, the output of a positioner is independent of the input signal, permitting much more efficient use of the actuator and greater net thrust.

Electropneumatic Positioner

This class of positioner is identical in function to the pneumatic force-balance type. A schematic diagram is shown in Fig. 49, where a low-level electric signal (commonly 4 to 20 or 10 to 50 mA dc) is fed through a coil, producing a variable force that is balanced by a feedback spring sensing the control valve stem position. Since the unit is valve-mounted, it must be resistant to vibration and shock.

Electropneumatic Transducer

These devices convert analog electronic controller signals into proportional air signals, and they are widely utilized. See Fig. 50. Since there is no feedback from the valve stem position, the electropneumatic transducer can be mounted remotely where it is less subject to vibration. All electronic

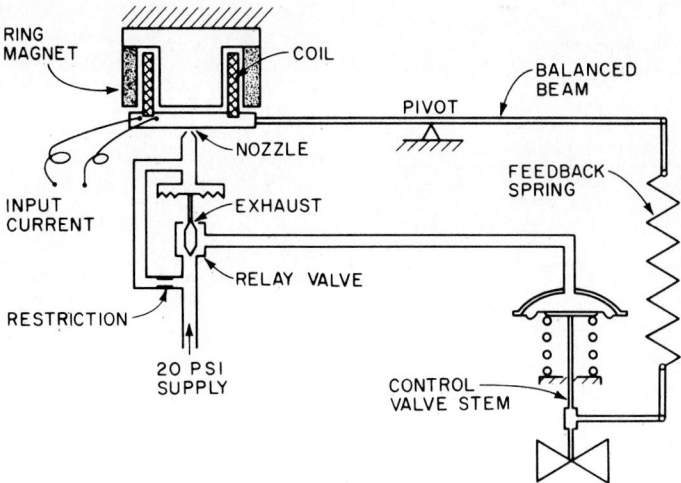

Fig. 49   Force-balance electropneumatic positioner.

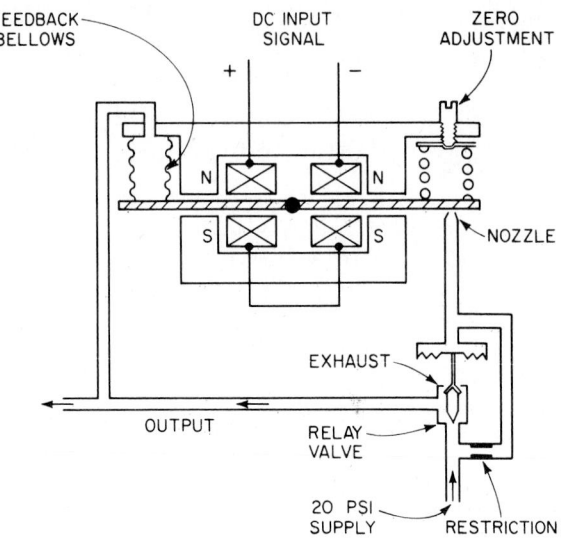

Fig. 50 Transducer which converts low-level electric signal to proportional air signal for valve actuation.

equipment located within the process area is generally built to meet explosion-proof requirements for Class 1, Group D, Division 1. Both of these low-level devices may also meet the requirements for intrinsic safety when employed with specific controlling instruments. The output from the transducer may be fed directly to the control valve or may serve as the input to a pneumatic positioner or an amplifying air relay. (See the article Safety in Instrumentation and Control systems in Sec. 14.)

Auxiliary Air Relays

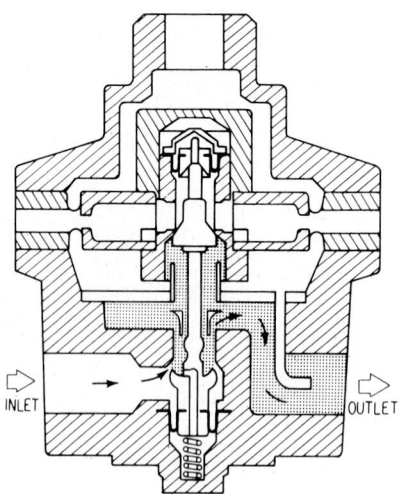

Fig. 51 One-to-one ratio balanced high-capacity air relay.

A number of air-loaded self-contained pressure regulators are available for use with control valve actuators. The most common requirement is for increased speed of operation. This is achieved through the use of a high-capacity 1:1 regulator known as a *booster*, as shown in Fig. 51. Both inlet and outlet seats have balancing rolling diaphragm assemblies—minimizing plug imbalance forces—so there is a very small pressure differential between the supply and exhaust at any pressure setting. These versatile relays come in a range of sizes capable of almost any reasonable speed and also in ratio diaphragm models for 1:2, 1:3, 2:1, and 3:1. The maximum flow capacity ranges from 10 to 150 std ft^3/min ($\sim$285 to 4250 L/min). This type of relay may be used with a spring diaphragm actuator without a positioner.

When the relay capacity of a positioner is not adequate, a modified version of the 1:1 volume booster relay, shown in Fig. 52, can be used. The positioner output is the signal to this relay. The output from the relay fed from an

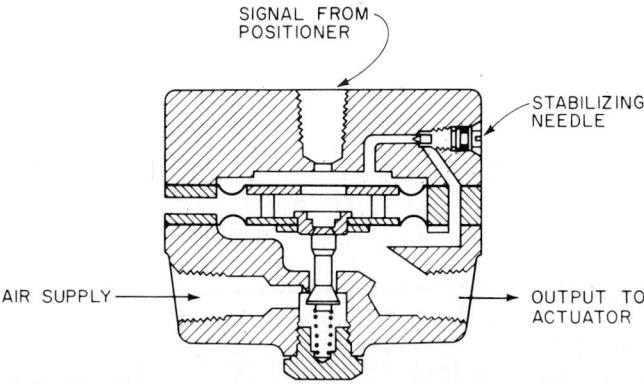

SIGNAL FROM
POSITIONER

STABILIZING
NEEDLE

AIR SUPPLY

OUTPUT TO
ACTUATOR

Fig. 52 One-to-one ratio amplifying relay with a stabilizing bypass for use with pneumatic valve positioners.

independent air supply is piped back to the signal connection through a stabilizing needle valve which is essential for operation.

RELATED REFERENCES

ANSI: "Cast Iron Pipe Flanges and Flanged Fittings," *ANSI B16.1-1975;* "Pipe Flanges and Flanged Fittings, Steel, Nickel Alloy, and Other Special Alloys," *B16.10-1973;* "Face-to-Face and End-to-End Dimensions of Ferrous Valves," *B16.10-1973;* "Bronze Pipe Flanges and Flanged Fittings, Class 150 and 300," *B16.24-1981;* "Valves, Flanges and Butt-Welding End—Steel, Nickel Alloy, and Other Special Alloys," *B16.34-1981;* "Control Valve Seat Leakage, Quality Control Standard for," *B16.104-1976;* American National Standards Institute, New York.

ASTM: *Annual Book of ASTM Standards,* issued annually: Part 1: Steel Piping, Tubing, and Fittings. Part 2: Ferrous Castings; Ferroalloys. Part 6: Copper and Copper Alloys. Part 8: Nonferrous Metals. ASTM, Philadelphia.

ISA: "Control Valve Sizing Equations," *ANSI/ISA S75.01,* 1981; "Uniform Face-to-Face Dimensions for Flanged Globe Style Control Valve Bodies," *ANSI/ISA S75.03, 1979;* "Face-to-Face Dimensions of Flangeless Control Valves," *ANSI/ISA S75.04, 1979;* "Control Valve Manifold Designs," *ISA RP75.06, 1981;* "Control Valve Standards" (containing all of the foregoing ANSI/ISA standards), 1982; "Control Valve Terminology," *ISA dS75.05,* 1982. ISA Audio Cassette Packages: "Control Valve Study Package," *UCVOO;* Herb Simon, "General Review of Factors Influencing Valve Selection," *UCVOI-I;* Mathew Freeman, "Valve Bodies and Valve Trim," *UCVO2-I;* L. R. Driskell, "Control Valve Sizing," *UCVO3-I;* R. L. Moore, "Flow Characteristics," *UCVO4-I;* Jerry L. Lyons, "Actuators and Accessories," *UCVO5-I;* "Control Valve Maintenance," *UCVO6-I;* Carl L. Askland, "Flow Control, Rangeability, Characteristics, and Leakage," *UCVO7-I;* "Pressure Drop, Cavitation, Noise and Two-Phase Flow," *UCVO8-I;* "Suspension and Slurry Service Conditions," *UCVO9-I;* Mathew Freeman, "High Temperature, Pressure and Mass Flow: Low Flow Control," *UCVIO-I;* Ross Forman, "Cost, Connections and Geometry Considerations," *UCVII-I;* Jerry L. Lyons, "Special Process Applications," *UCV12-I;* Instrument Society of America, Research Triangle Park, N. C.

Miller, T. J.: "Pneumatic Valve Actuators Continue to Dominate as Electronics Move into Process Control," *Control Eng.,* vol. 28, no. 10, 1981, p. 88.

Control Valve Sizing

by
Paul Wing*

The selection of proper port and body size for a control valve is based on calculation of the correct valve flow coefficient C_v and consideration of a number of secondary factors discussed in this article. An accurate prediction of actual valve capacity for most services is now possible as a result of the efforts of the Instrument Society of America (ISA), culminating in ANSI/ISA S75.01. This standard sets forth basic equations with which control valve flow determinations can be made for a wide range of operating conditions. Since the equations in the standard are not in a convenient form for everyday use, a proprietary set, easier to use, is outlined here. The results do not differ materially from those obtained when using the aforementioned standard.

The actual sizing process is still largely dependent on good analysis of the data available as a basis for computation. Usually one or more of the flowing conditions is an approximation. The engineering evaluation of these data is the most important single factor in determining the final valve size. For this reason, an explanation of the relative importance of the data used in sizing is presented before outlining a detailed procedure.

Pressure Drop across the Valve

Pressure drop is perhaps the most arbitrary factor in valve sizing. For a simple back-pressure or reducing application, the drop across the valve may be known quite accurately. This is also true for a liquid-level control installation where the liquid passes from one vessel to another at constant pressure. If the valve is installed in a long piping system or a system including heat exchange equipment, the pressure drop across the valve should be estimated at the maximum flow condition with reasonable allowance for the pressure losses in series with the valve.

Pressure drop across a control valve is often expressed as a percentage of the total friction drop in the system. A good working rule is that at least one-third of the total system drop should be absorbed by the control valve at maximum flow. This rule may, of necessity, be relaxed for extremely long or high-pressure-drop systems, particularly when the actual desired flow is known reasonably accurately and does not vary substantially. Reasonably good control can then be attained with as little as 15% of the total system drop across the valve. An interesting exception to this rule is encountered in long liquid pipeline transportation systems where the fully open control valve may absorb even less than 1% of the system pressure drop. The important point in this case is that the expected flow variation is very small and the valve functions primarily to balance the hydraulic gradient in the system. Even so, special attention must be paid to the valve characteristics and to the control system as a whole.

In many instances, it is necessary to make an arbitrary choice of pressure drop because of meager process data. For valves in a pump discharge, a drop of 10 psi (70 kPa) minimum or 10% of the discharge pressure may be assumed to be sufficient if the discharge system is not extremly long or complicated by large pressure drops through other equipment.

In systems involving centrifugal pumps, the change in pump head with change in flow may become a major factor and should be taken into account as discussed in the subsection Installed Flow Characteristic in the article Control Valve Characteristics in this Handbook section.

*Consultant, Hingham, Mass.

Flowing Quantity

The three figures to be considered are the normal, maximum, and minimum flow. The maximum figure for valve sizing should be chosen realistically and must be carefully related to the available pressure drop. It should be the required maximum flow, *not* the full capacity of the valve. The valve capacity is then generally set 25 to 60% above the required or normal maximum flow. An alternative approach may be to double the expected normal flow. Unrealistic combinations of flow and pressure drop should be avoided.

Specific Gravity

The actual specific gravity of the flowing medium at operating conditions is relatively unimportant. If the actual gravity is not known, a reasonable assumption can usually be made since the use of 0.9 sp gr, for example, instead of 0.8 would cause an error of less than 5% in the capacity determination.

Flow Coefficient C_v

After a proper evaluation of flowing conditions, the required valve flow coefficient C_v can be determined. By definition, this flow coefficient is "the number of gallons per minute of water which will pass through any flow restriction with a pressure drop of 1 psi." For example, a control valve with a maximum flow coefficient C_v of 12 has an effective port area in the full-open position such that it passes 12 gal/min of water with a 1-psi pressure drop. Basically, it is a universally accepted capacity index by which an engineer is able to rapidly and accurately estimate the required size of restriction in any fluid system. This definition, of course, can be converted to suitable metric units.

While the coefficient is defined in terms of water flow, it can be used conveniently and with sufficient accuracy for determination of the flow of a compressible fluid as well. In view of the enormous range of flowing conditions and the variety of valve styles, it is not surprising to find that there is still much to be done toward improving the accuracy of the sizing formulas. A more rigorous approach for compressible fluid service is set forth in ANSI/ISA S75.01, but the equations in this article give comparable results.

The first control valve sizing standards based on the flow coefficient C_v, such as the Fluid Controls Institute Voluntary Standards 58-1 (1958) and 61-1 (1962) did not take into account the variation in predicted capacity caused by pressure recovery effects. A substantial variation in flow prediction occurs because of differences in the contouring of the valve flow passages. The current ANSI/ISA standard and the equations discussed in this article take this important factor into account.

Critical Flow Factor C_f

Every valve or flow restriction exhibits some degree of pressure recovery downstream of the principal restriction. This means that, at some point within the valve body, the static pressure will be lower than the valve outlet pressure. The zone of minimum pressure is known as the *vena contracta,* and whenever the static pressure at this point is lowered to the vapor pressure of a liquid medium, or to the critical pressure of a gas causing sonic velocity, no further flow will be obtained as the valve outlet pressure decreases.

The effect of pressure recovery in subcritical compressible fluid flow is that critical flow (choked flow) is achieved at a differential between valve upstream and downstream pressure that is less than would be predicted by the earlier simplified formulas based on the actual pressure ratio from valve inlet to outlet. For subcooled liquid service, the vapor pressure of the liquid may be reached at the vena contracta, causing localized flashing. The bubbles formed in this region collapse just downstream as the pressure rapidly increases. With pure liquids, such as water, this collapse causes high localized stresses with attendant noise and vibration known as *cavitation.* This phenomenon is discussed later in this article and in considerable detail in a later article in this Handbook section.

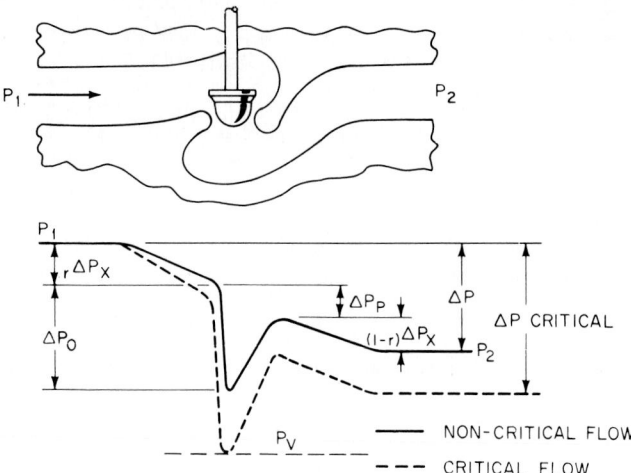

Fig. 1 Pressure gradient through a control valve.

The critical flow factor C_f is a dimensionless expression of the ratio between the flow coefficient C_v obtained under critical conditions, such as liquid vaporization or gas sonic velocity at the vena contracta, and the C_v measured under normal pressure recovery conditions. The pressure gradient in a control valve is shown in Fig. 1. The pressure loss due to the valve body restriction has been included as ΔPX, with the inlet loss shown as $R\,\Delta PX$ and the outlet loss as $(1 - R)\,\Delta PX$. These losses in general are small in comparison to the loss across the principal restriction. If the pressure at the vena contracta does not fall below the vapor pressure of a liquid or the critical pressure of a gas with respect to the inlet pressure, the ratio between the pressure drop across the valve and the pressure drop from the inlet to the vena contracta remains quite constant for each valve style and plug position. In another current method, the ratio between the two pressure drops is used rather than the corresponding flow coefficients, and the results are, of course, similar in practical application. A good correlation has been obtained in the laboratory for both flashing liquid and critical compressible flow.

For maximum accuracy, C_f factors should be obtained from the valve manufacturer. Good estimates for several basic valve styles can be obtained by using the values shown in Table 1. The critical flow factor may vary substantially with the valve stroke, as shown for a number of valve styles in Fig. 2, and this must be considered in predicting flashing or cavitation.

The installation of a valve with pipe reducers affects the pressure recovery, particularly in the case of high-capacity valves such as control ball and butterfly valves. Values for the combined coefficient C_{fr} are also shown in Table 1, and the method of determination is discussed later in this article under Effect of Pipe Reducers.

Control Valve Flow Formulas

The formulas listed in Table 2 (liquids) and Table 3 (gases and vapors) with the supplementary information that follows can be used in calculating the flow coefficient C_v or, conversely, the capacity of a control valve for the majority of applications. Control valve manufacturers also offer computer sizing programs of varying sophistication. Such programs automatically check for potential cavitation in liquid service or excessive velocity in gas service. Noise-level estimation is easily included even when small hand calculators are used. Longer programs may include selection by valve type and initial cost or other similar refinements. The particular manufacturer should be consulted if a large number of valves are involved.

Table 1 Flow Factors at Full Opening

Valve type	Trim size*	Flow to	C_f	K_c^*	C_{fr} $D/d = 1.5$ or greater	$D/d = 1.5$ R	C_{fr}/R	$D/d = 2.0$ R	C_{fr}/R
Single-port top-entry globe	A	Close	0.85	0.58	0.81	0.96	0.84	0.94	0.86
		Open	0.90	0.65	0.86	0.96	0.89	0.94	0.91
	B	Close	0.80	0.52	0.80	1.0	0.80	1.0	0.80
		Open	0.90	0.65	0.90	1.0	0.90	1.0	0.90
Single-port balanced quick-change	A	Close	0.90	0.65	0.86	0.96	0.89	0.94	0.91
	B	Close	0.90	0.65	0.90	1.0	0.90	1.0	0.90
Top-entry angle	A	Close	0.81	0.53	0.78	0.96	0.81	0.94	0.82
		Open	0.89	0.64	0.85	0.96	0.88	0.94	0.90
	B	Close	0.80	0.52	0.80	1.0	0.80	1.0	0.80
		Open	0.90	0.65	0.90	1.0	0.90	1.0	0.90
Streamlined angle	A	Close	0.48	0.17	0.45	0.85	0.53	0.77	0.57
		Open	0.90	0.65	0.84	0.95	0.89	0.91	0.91
	B	Close	0.55	0.23	0.54	1.0	0.54	1.0	0.54
		Open	0.95	0.72	0.93	1.0	0.93	1.0	0.93
Split body globe	A	Close	0.80	0.51	0.77	0.96	0.80	0.94	0.81
		Open	0.75	0.46	0.72	0.96	0.75	0.94	0.77
	B	Close	0.80	0.52	0.80	1.0	0.80	1.0	0.80
		Open	0.90	0.65	0.89	1.0	0.89	1.0	0.90
Top-and-bottom-guided double-port	A	Contoured	0.90	0.70	0.86	0.96	0.90	0.94	0.92
		V port	0.98	0.80	0.94	0.96	0.98	0.94	1.0
	B	Contoured	0.80	0.31	0.80	1.0	0.80	1.0	0.80
		V port	0.95	0.73	0.94	1.0	0.94	1.0	0.94
Single-port eccentric rotating	A	Close	0.68	0.35	0.65	0.95	0.68	0.92	0.71
		Open	0.85	0.60	0.80	0.95	0.84	0.92	0.86
	B	Close	0.70	0.39	0.70	1.0	0.70	1.0	0.70
		Open	0.88	0.62	0.87	1.0	.87	1.0	0.87
Butterfly	A	60°	0.68	0.35	0.63	0.91	0.69	0.85	0.73
		90°	0.58	0.25	0.51	0.77	0.66	0.67	0.74
Control ball	A	—	0.60	0.24	0.55	0.87	0.63	0.80	0.68

*A, full capacity trim, orifice diameter ≈ 0.8 valve size; B, reduced capacity trim 50% of A and below.

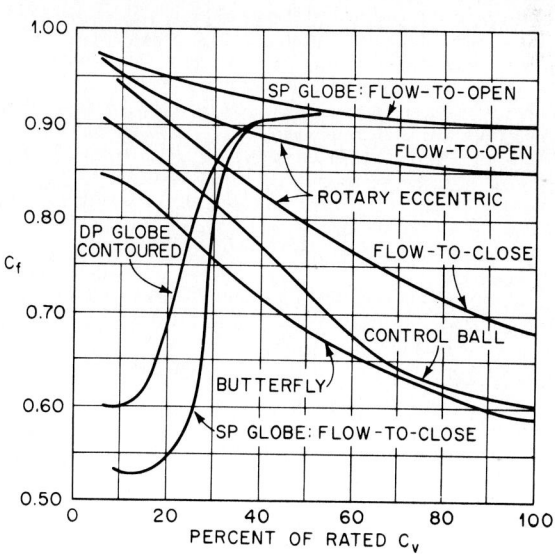

Fig. 2 Variation in the critical flow factor C_f with valve stroke.

Table 2 Valve Sizing Formulas—Liquid Service

Subcritical flow $\Delta P < C_f^2(\Delta P_S)*$	Critical flow, cavitation or flashing $\Delta P \geq C_f^2(\Delta P_s)*$
Volumetric flow:	
$C_v = q \sqrt{\dfrac{G_f}{\Delta P}}$	$C_v = \dfrac{q}{C_f} \sqrt{\dfrac{G_f}{\Delta P_s}}$
Flow by weight:	
$C_v = \dfrac{W}{500\sqrt{G_f\,\Delta P}}$	$C_v = \dfrac{W}{500 C_f \sqrt{G_f\,\Delta P_s}}$

where C_v = valve flow coefficient
   C_f = critical flow factor
   G_f = specific gravity at flowing temperature (water = 1 at 60° F)
   P_1 = upstream pressure, psia
   P_2 = downstream pressure, psia
   P_c = pressure at thermodynamic critical point, psia
   P_v = vapor pressure of liquid at flowing temperature, psia
   ΔP = actual pressure drop $P_1 - P_2$, psi
   q = liquid flow rate, U.S. gal/min
   W = liquid flow rate, lb/h

---

*$\Delta P_s = P_1 - [0.96 - 0.28 \sqrt{(P_v/P_c)}]P_v$ or for simplicity, if $P_v < 0.5P_1$, $\Delta P_s = P_1 - P_v$. $C_f^2 \Delta P_s$ is the maximum ΔP for sizing purposes. A valve is not limited in application to this pressure drop, but at a higher pressure differential, choked flow will occur without an increase in flow rate. This formula for ΔP_s is sufficiently accurate for general use on liquids.

Special Considerations

Cavitation

This is an undesirable phenomenon which can occur in control valves in liquid service. It is caused by transformation of a portion of liquid into the vapor phase during rapid acceleration of the fluid as it passes the valve orifice and subsequent collapse of these vapor bubbles downstream. The collapsing vapor bubbles can cause localized pressures of up to 100,000 psi (690 MPa) and can result in rapid wear of the valve trim, body, or outlet piping. While generally associated with high-pressure-drop conditions, many valve styles, particularly with a low C_f factor (high efficiency), may cavitate severely at relatively low pressure drops. Not only is there potential for serious damage to the valve trim or body, but there is often a corollary severe noise and vibration problem.

The potential for cavitation exists in a valve once the static pressure at the vena contracta reaches the vapor pressure of the fluid. At this point, vapor bubbles are formed which will recollapse just downstream, even before reaching the valve outlet, if the pressure at this point is above the vapor pressure of the liquid. Valves with a low critical flow factor C_f are most likely to cavitate and, in general, this includes the class of valves commonly referred to as *streamlined*.

The pressure drop condition (ΔP_{crit}) at which full cavitation obtains can be defined as follows:

$$\Delta P_{\text{crit}} = C_f^2\,\Delta P_s$$

$$\Delta P_{\text{crit}} = \left(\frac{C_{fr}}{R}\right)^2 \Delta P_s$$

where R = subcritical flow capacity correction factor. See Table 1.

Table 3 Valve Sizing Formulas—Gas and Vapor Service

Subcritical flow $\Delta P < 0.5C_f^2 P_1$	Critical flow $\Delta P \geq 0.5C_f^2 P_1$

Volumetric flow:

$$C_v = \frac{Q}{963}\sqrt{\frac{GT}{\Delta P(P_1 + P_2)}} \qquad\qquad C_v = \frac{Q\sqrt{GT}}{834 C_f P_1}$$

Flow by weight:

$$C_v = \frac{W}{3.22\sqrt{\Delta P(P_1 + P_2)G_f}} \qquad\qquad C_v = \frac{W}{2.8 C_f P_1 \sqrt{G_f}}$$

For saturated steam:

$$C_v = \frac{W}{2.1\sqrt{\Delta P(P_1 + P_2)}} \qquad\qquad C_v = \frac{W}{1.83 C_f P_1}$$

For superheated steam:

$$C_v = \frac{W(1 + 0.0007 T_{sh})}{2.1\sqrt{\Delta P(P_1 + P_2)}} \qquad\qquad C_v = \frac{W(1 + 0.0007 T_{sh})}{1.83 C_f P_1}$$

where C_v = valve flow coefficient
$\quad\quad\; C_f$ = critical flow factor
$\quad\quad\; G$ = gas specific gravity (air = 1.0)
$\quad\quad\; G_f$ = specific gravity at flowing temperature
$\quad\quad\; P_1$ = upstream pressure, psia
$\quad\quad\; P_2$ = downstream pressure, psia

ΔP = pressure drop $P_1 - P_2$, psi
Q = gas flow rate at 14.7 psia and 60°F, std ft^3/h
T = flowing temperature, °R, (460 + °F)
T_{sh} = steam superheat, °F
W = flow rate, lb/h

Special considerations:

a. Effect of pipe reducers   **b.** Limitation of outlet velocity

The valve sizing formulas for gas and vapor tabulated above introduce some error between flow at a low pressure drop ($\sim 0.2 P_1$) and critical flow. This error approaches 12% under the most unfavorable conditions. For a more precise determination of C_v, especially within the transition zone from subcritical to critical flow and for computerized valve sizing, the following unified gas sizing formula should be used.

For gas volumetric flow:

$$C_v = \frac{Q\sqrt{GT}}{834 C_f P_1 (Y - 0.148 Y^3)}$$

Flow by weight:

$$C_v = \frac{W}{2.8 C_f P_1 \sqrt{G_f}\,(Y - 0.148 Y^3)}$$

For saturated steam:

$$C_v = \frac{W}{1.83 C_f P_1 (Y - 0.148 Y^3)}$$

For superheated steam:

$$C_v = \frac{W(1 + 0.0007 T_{sh})}{1.83 C_f P_1 (Y - 0.148 Y^3)}$$

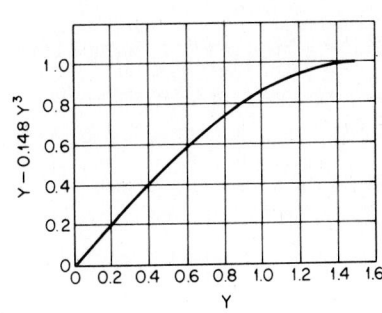

where

$$Y = \frac{1.63}{C_f}\sqrt{\frac{\Delta P}{P_1}}$$

with a maximum value of $Y = 1.50$. At this value, $Y - 0.148 Y^3 = 1.0$. The value $Y = 0.148 Y^3$ may be taken from the accompanying graph. All nomenclature same as previously defined.

In practical application, the onset of cavitation is somewhat gradual, and where no cavitation can be tolerated the coefficient of incipient cavitation K_c should be employed in place of C_f^2. Values of K_c are listed in Table 1 and may be used with or without reducers. For the differential pressure causing incipient cavitation,

$$\text{Incipient cavitation} = K_c \, \Delta P_s$$

To avoid cavitation, it is necessary to reduce the pressure drop across the valve to below ΔP_{crit}. This can be done, for example, by increasing the inlet pressure P_1 through the selection of a valve location at a lower elevation in the piping system. Another possibility is the selection of a valve type with a larger C_f or K_c factor. For example, one might choose a V port instead of a contoured plug. In extreme cases, two identical control valves may be installed in series. The combined C_f factor is then estimated as follows:

$$C_f \text{ total} = \sqrt{C_f} \text{ of the single valve}$$

For extremly high pressure drops, special valves with low-efficiency multiple-series restrictions are available. See the article Control Valves earlier in this Handbook section.

Effect of Pipe Reducers

When valves are mounted between pipe reducers, there is a decrease in actual valve capacity. The reducers create an additional pressure drop in the system by acting as contractions or enlargements in series with the valve. A flow capacity correction factor R is used to compensate for this reduction. The required C_v is found simply by dividing the calculated C_v by R, tabulated in Table 1 for line size/valve size ratios of 1.5:1 and 2:1.

The factor R is based on the velocity head loss of an abrupt contraction and enlargement. This represents the maximum loss and allows calculation of valve size on the safe side.

$R = $ *subcritical flow* capacity correction factor for inlet and outlet reducers

$$R = \sqrt{1 - 1.5\left(1 - \frac{d^2}{D^2}\right)^2 \left(\frac{C_v}{30d^2}\right)^2}$$

$R' = $ *subcritical flow* capacity correction factor for outlet reducer only or inlet reducer with entrance angle less than 40° and outlet reducer

$$R' = \sqrt{1 - \left(1 - \frac{d^2}{D^2}\right)^2 \left(\frac{C_v}{30d^2}\right)^2}$$

where d = valve size (estimated before C_v calculation), in
D = line size, in
C_v = required valve flow coefficient

[See Table 2 (liquids) or Table 3 (gases and vapors).]

The flow capacity of control valves under critical pressure drop conditions (choked flow) is also reduced when the valve is smaller than line size, but the reduction tends to be small because the outlet reducer does not influence the flow. The critical flow capacity may be calculated using C_{fr} from Table 1.

The effect on capacity is largest in high-recovery valves, such as control ball and butterfly valves. Here the high relative capacity often results in a valve size substantially less than the nominal line size. The critical flow factor for reducer-valve combinations C_{fr} for line size/valve size ratios of 1.5 or greater is also shown in Table 1.

To calculate the onset of cavitation, it is necessary to find the installed pressure recovery factor of the valve-reducer combination by dividing the installed critical coefficient $C_{fr}C_v$ by the installed subcritical coefficient RC_v. This factor is listed in Table 1 as C_{fr}/R.

C_{fr} for other ratios may be calculated from the formula

$$C_{fr} = \left[\frac{1}{C_f^2} + \left(\frac{C_v}{30d^2} \right)^2 \left(1 - \frac{d^4}{D^4} \right) \right]^{-1/2}$$

where C_f = valve critical flow factor
$\quad C_{fr}$ = valve-reducer critical flow factor
$\quad C_v$ = valve flow coefficient
$\quad d$ = valve size, in
$\quad D$ = line size, in

$\dfrac{C_{fr}}{R}$ = Installed recovery factor of valve-reducer section used to find ΔP for sizing.

$$\text{Maximum } \Delta P \text{ for sizing (valve-reducer section)} = \left(\frac{C_{fr}}{R} \right)^2 (\Delta P_s)$$

To compensate for reducer losses at *subcritical flow*, divide C_v calculated from the subcritical sizing formula by R:

$$\text{Valve } C_v \text{ required if subcritical} = \frac{\text{calculated } C_v}{R}$$

To compensate for reducer losses at *critical flow*, use C_{fr} in place of C_f in the appropriate critical flow formula without further adjustment of C_v.

High-Viscosity Laminar Flow of Liquids

Laminar flow may result when the liquid viscosity is high or when the value ΔP or C_v is small.
In calculating turbulent flow C_v and laminar flow C_v, the larger value should be used as the required valve C_v.

$$\text{Turbulent flow } C_v = q \sqrt{\frac{G_f}{\Delta P}}$$

$$\text{Laminar flow } C_v = 0.072 \sqrt[3]{\left(\frac{\mu q}{\Delta P} \right)^2}$$

where q = liquid flow rate, U.S. gal/min
$\quad G_f$ = specific gravity at flowing temperature
$\quad \Delta P$ = actual pressure drop, psi
$\quad \mu$ = viscosity, CP

Two-Phase Flow

For a liquid and noncondensable gas entering a valve, if there is no vaporization of the liquid and if the flow velocity ensures a turbulent well-mixed stream:

$$C_v = \frac{W}{44.8\sqrt{\Delta P(w_1 + w_2)}}$$

For a liquid and its vapor entering a valve where additional vaporization of liquid occurs; if the flow velocity ensures a turbulent, well-mixed stream:

$$C_v = \frac{W}{63.3\sqrt{\Delta P w_1}}$$

$$\text{Maximum } \Delta P \text{ for sizing} = C_f^2\left(\frac{P_1}{2}\right)$$

where W = flow rate, lb/h

ΔP = actual pressure drop, psi

w_1 = upstream specific weight, lb/ft^3 calculated from weight fraction of gas or vapor in the stream X_g and specific volumes of gas or vapor and liquid at upstream pressure V_{g1} and V_f

$$w_1 = \frac{1}{V_f} = \frac{1}{X_g(V_{g1} - V_f) + V_f}$$

REFERENCES

ANSI/ISA: "Control Valve Sizing Equations," *Stand. S75.01* (revised periodically). Available from Instrument Society of America, Research Triangle Park, N.C., or American National Standards Institute, New York.

ANSI/ISA: "Control Valve Seat Leakage," *Stand. B16,104* (revised periodically). Available from Instrument Society of America, Research Triangle Park, N.C., or American National Standards Institute, New York.

Hutchison, J. W.: *ISA Handbook of Control Valves,* 2d ed., Instrument Society of America, Research Triangle Park, N.C., 1976.

Keith, G. A.: "Linear Installed Characteristics Simplify Control Valve Selection," *Control Eng.,* vol. 24, no. 11, November 1977, p. 48.

Masoneilan: "Masoneilan Handbook for Control Valve Sizing," *Bull. OZ-100,* Masoneilan Division, McGraw-Edison Company, Norwood, Mass., 1981.

Miller, S. F.: "Computer Aided Design Helps Produce Custom Valves," *Control Eng.,* vol. 28, no. 4, April 1981, p. 50.

Stiles, G. F.: "Liquid Viscosity Effects on Control Valve Sizing," *Texas A&M Symposium on Instrumentation for the Process Industries,* 1964.

Usry, J. D.: "Stepping Motors for Valve Actuation," *Instrum. Technol.,* vol. 24, no. 3, March 1977, p. 58.

Wing, P.: "Plain Talk on Valve Rangeability," *Instrum. Technol.,* vol. 25, no. 4, April 1978, p. 53.

Wing, P.: "Determining and Using the Control Valve Pressure Recovery Factor," *Instrum. Technol.,* vol. 26, no. 8, August 1979, p. 55.

Wolter, D. G.: "Control Valve Selection," *Instrum. Technol.,* vol. 24, no. 10, October 1977, p. 55.

Control Valve Characteristics

by
Paul Wing[*]

The flow characteristic of a valve defines the flow behavior as the valve operates through the rated stroke. To take on full meaning, the definition must be considered from two viewpoints: (1) the inherent flow characteristic and (2) the installed flow characteristic.

The *inherent flow characteristic* is based on a constant pressure drop across the valve body throughout the stroke. It is generally idealized and applied to a whole family of valves. This must be taken into consideration in analyzing problems, since the difference between actual and published curves may be fairly large.

The *installed flow characteristic* is the actual relationship between valve stroke and flow with the valve applied to some particular flow system. Many factors in addition to the inherent characteristic influence the installed characteristic. The most common are a restriction such as line loss in series with a valve, a change in total pressure drop in the system as a function of flow as may be caused by a pump, and an open bypass around the valve. The effect is often substantial and must always be considered in a complete analysis of any control problem.

Inherent Flow Characteristics

The inherent flow characteristic is the only parameter convenient for presentation in catalog form. In describing the various combinations currently available, two general classifications can be made.

1. Flow characteristics derived mathematically. These include the linear and equal percentage characteristics.

2. Flow characteristics resulting from various basic constructions. In the conventional sliding stem valve, this category includes the quick-opening or poppet type and the straight-sided V port. Also included are special valve constructions, such as the butterfly, Saunders patent, and rubber pinch, and other proprietary designs.

Linear Characteristic

The linear characteristic is shown in Fig. 1. The relationship between valve opening and flow at a constant drop plots as a straight line. The mathematical expression is

$$Q = ky \tag{1}$$

where Q = flow at constant pressure drop
  y = valve opening
  k = constant

*Consultant, Hingham, Mass.

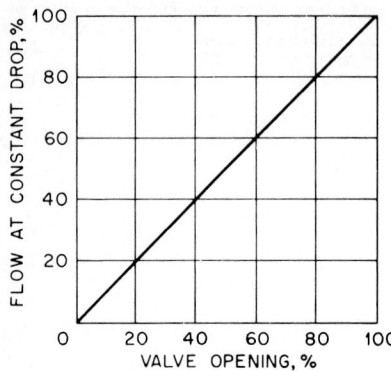

Fig. 1   Linear characteristics.

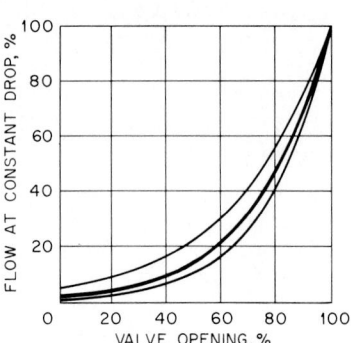

Fig. 2   Equal percentage characteristic.

Equal Percentage Characteristic

Figure 2 gives a series of idealized equal percentage curves differing in leakage flow values. The common property of these curves is that equal increments of stem motion produce equal percentage changes in flow at a constant pressure drop based on the flow just before the change is made. The most usual one is emphasized on the graph, and the rate of change in flow is ∼4% per percent change in valve stroke. This particular characteristic is known as 50-to-1 equal percentage. Unfortunately, this definition overemphasizes the possible flow range of the valve, which is an entirely independent characteristic. The important property of an equal percentage valve is a constant percentage rate of flow change at constant drop per unit change in valve opening through the major part of the stroke. The resultant flow range is an entirely separate consideration discussed under Rangeability.

The general equation for flow versus opening for any valve having equal percentage characteristics is

$$Q = be^{ay} \tag{2}$$

where Q = flow at constant pressure drop
$\quad y$ = valve opening
a and b = constants
$\quad e$ = base of natural logarithms

Constants a and b can be evaluated to give the more convenient form

$$Q = Q_0 e(\log R/y_{max})y \tag{3}$$

where Q_0 = flow at constant drop at zero stroke
$\quad R$ = flow range of valve, maximum to minimum at constant drop
y_{max} = maximum rated valve opening

For a flow range of 50 to 1, Eq. 3 becomes

$$Q = 2e^{(3.91/100)y} = 2e^{0.0391y} \tag{4}$$

This can also be written as

$$y = 25.6 \log \frac{Q}{2} \tag{5}$$

Figure 3 shows the same family of curves plotted as straight lines on semilogarithmic paper.

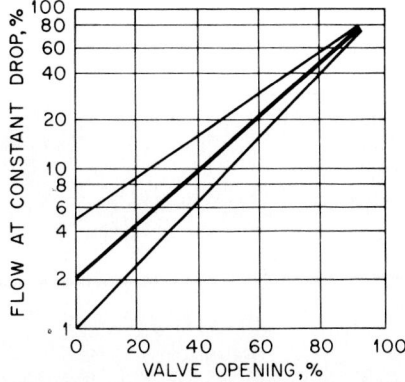

Fig. 3 Equal percentage characteristic plotted in semilogarithmic form.

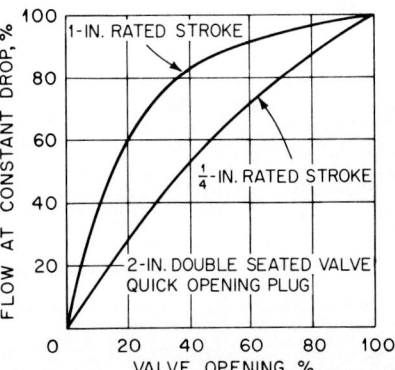

Fig. 4 Inherent flow characteristic of a typical 2-in (50-mm) double-seated valve at 1 in (25 mm) rated stroke and ¼ in (6.4 mm) rated stroke.

Quick-Opening Characteristic

This is the characteristic of the bevel-seated disk or plain flat disk type of plug. It is not generally defined mathematically. The flow-versus-stroke relationship is approximately linear up to a valve opening equal to one-quarter of the port diameter or to 60 to 70% of the body flow passage area. The inherent flow characteristic therefore may be considered linear so long as the rated stroke is chosen to limit the maximum valve port area and body friction losses do not seriously alter the relationship. Figure 4 shows the inherent flow characteristic of a typical 2-in (50-mm) valve with a quick-opening disk-type plug instead of a characterized plug. One curve is based on retaining a rated stroke of 1 in (25 mm), and the other on a stroke of 0.25 in (6.4 mm). This brings out the importance of considering the maximum flow coefficient and rated stroke in discussing quick-opening valves.

Straight-Sided V-Port (Parabolic) Characteristic

Many control valve plugs still in common use have port openings approximating a straight-sided "V." The port-area-versus-valve-opening relationship of such a valve can be expressed by

$$Q = ky^2 \tag{6}$$

where Q = flow at constant pressure drop
y = valve opening
k = constant

This is the equation for a parabola, and the characteristic is therefore sometimes designated parabolic. It should not be confused with the solid lathe-turned type of plug once called "parabolic" but now generally referred to as "lathe-contoured."

For practical purposes, the characteristic approaches the equal percentage. Figure 5 shows one idealized parabolic curve plotted on semilog paper along with an equivalent equal percentage curve.

Three-Way Valve Characteristic

The inherent characteristic of a three-way valve is difficult to define. The valves are commonly used for on-off diverting service and then are equipped with a quick-opening disk which merely shuts off

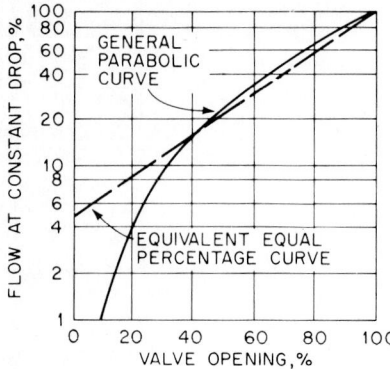
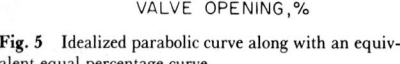
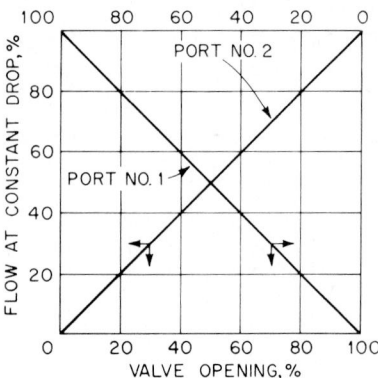

Fig. 5 Idealized parabolic curve along with an equivalent equal percentage curve.

Fig. 6 Theoretical inherent flow characteristic of a linear three-way valve with equal exchange of port areas.

the alternate port. For throttling service, the theoretical inherent characteristic would be linear, as shown in Fig. 6. This characteristic curve is produced by a plug designed so that the port areas are equally interchanged and is based on an equal pressure drop through both ports, a condition that very seldom exists in practice.

This is the only characteristic which could theoretically be obtained in a three-way valve without seriously violating the principle that equal areas be interchanged between the two ports. In the practical sense, a three-way valve operating, for example, as a bypass around a heat exchange system, usually behaves like a single control valve, throttling the flow in either the main or bypass line and with relatively little restriction in the alternate port. As a result, several combinations of porting are available which are satisfactory though difficult to analyze.

For more flexibility, a pair of control valves is often used; one valve is normally open, and the other normally closed. Such a system provides more flexibility in the field. For large flows, two butterfly valves linked to a common actuator may be used.

Rotary Valve Characteristics

The two most common rotary stem valves adapted from shutoff valves for control service are the butterfly and ball valves. Fortunately, both of these exhibit a characteristic which is approximately equal percentage, as shown in Fig. 7. An intermediate line also shows the inherent flow characteristic of a ball valve with a characterized port for general purpose control applications. In comparing this curve with that of a conventional full-bore ball, it must be remembered that the flow coefficient at full opening is substantially larger for the full-bore ball, and consequently the installed flow characteristic is generally much more severely altered.

The actual inherent characteristic of a particular butterfly valve is somewhat dependent on the thickness of the disk with respect to its diameter, particularly at large openings. The actual shape of the disk, which may be especially designed to minimize dynamic torque, also has a small effect. The characteristic also varies with the maximum angle of opening, which for conventional symmetrical disk valves has generally been 60° and for the new low-torque designs may range from 75° to the full 90° opening. These differences are not significant in a general discussion of characteristics. The important point in all cases with high capacity valves of this nature is to consider carefully the installed characteristic, including the effect of reducers and line loss in series with the valve.

Installed Flow Characteristics

The most common factor altering the inherent characteristic is variation in pressure drop across the valve with changes in flow. Figure 8 shows a control valve in series with a fixed resistance. The total

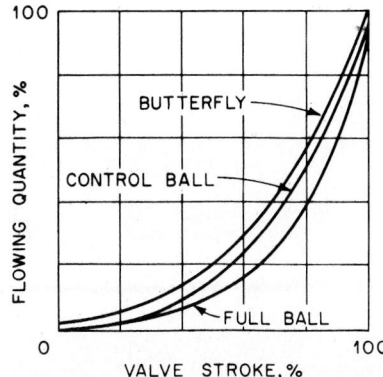

Fig. 7 Inherent flow characteristic of 60° rotation butterfly and 90° rotation full-bore and characterized ball valves.

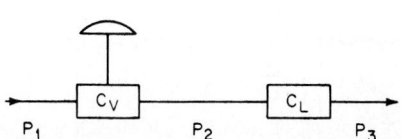

Fig. 8 Valve in series with a fixed resistance which alters the inherent flow characteristic.

pressure drop in the system, $P_1 - P_3$, is assumed a constant. The fixed resistance may be a length of pipe or be made up of pipe, heat exchangers, and mixing nozzles. For analysis, it can be lumped into a common resistance having a flow coefficient C_L. If the inherent flow coefficient of the valve at a given stroke is designated by C_v, the installed flow coefficient C_E is determined as follows:

$$\frac{1}{C_v^2} + \frac{1}{C_L^2} = \frac{1}{C_E^2} \tag{7}$$

or

$$C_E = C_v \frac{C_L^2}{C_L^2 + C_v^2} \tag{8}$$

where C_L = system flow coefficient exclusive of valve
C_v = valve flow coefficient
C_E = installed flow coefficient

By expressing the pressure drop across the valve at the rated capacity in terms of percentage of total system loss, one can draw a family of curves giving the installed characteristic for either the 50-to-1 equal percentage or the linear valve. These are shown in Figs. 9 and 10 for a 100% drop across

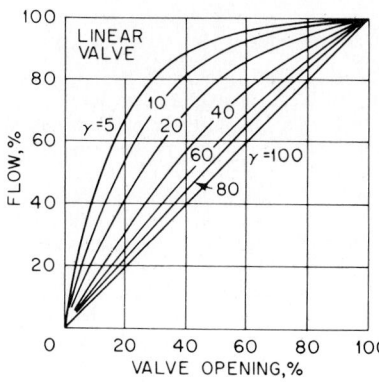

Fig. 9 Effective characteristic of a 50-to-1 equal percentage valve against the percent system drop across a fully open valve.

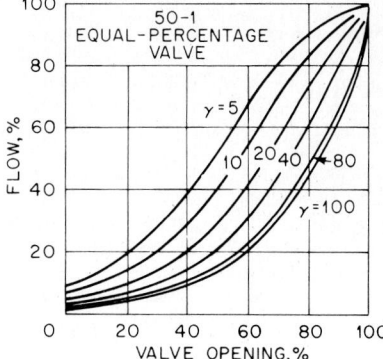

Fig. 10 Effective characteristic of linear valve against percent system drop across a fully open valve.

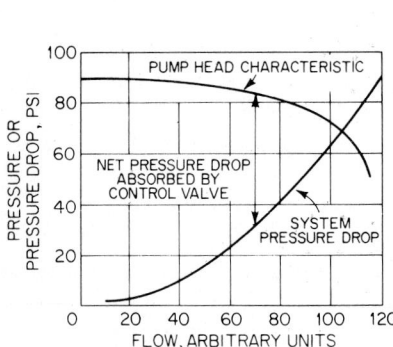

Fig. 11 Type of plot for estimating effective flow characteristic in a pump discharge system.

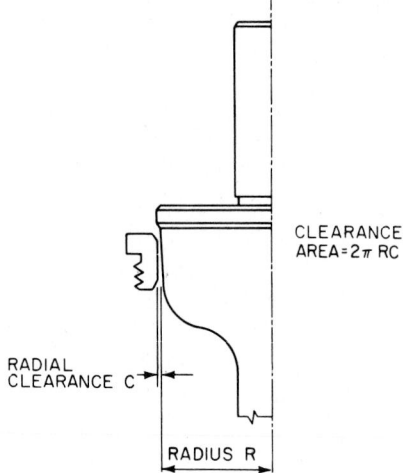

Fig. 12 Minimum controllable flow in some valve designs is directly proportional to the clearance between the plug and ring.

the valve (γ = 100 down to 5%). γ = 5, where γ is the percentage of the total system drop across the valve at the rated capacity. Note how the linear valve characteristic falls off in the direction of the typical quick-opening characteristic, as shown in Fig. 4, as the percentage line loss increases. The equal percentage valve tends to approach a linear installed characteristic as the percentage of the system loss across it is reduced. This is an important advantage of the equal percentage characteristic, since valves are often installed in flow systems with as little as 10 to 20% of the total drop available at the maximum rated flow.

For a more complicated system, the installed characteristic may be calculated from a plot similar to that in Fig. 11. Here a valve is installed in the discharge of a centrifugal pump with a pump head characteristic as shown. The system pressure drop, which is a square root function, is plotted as a function of flow. The drop to be absorbed by the control valve regardless of the inherent characteristic is the difference between the two curves as indicated. The pressure drop increases with a decrease in flow, and the inherent characteristic is altered in a manner similar to that shown for the constant head case (Figs. 9 and 10).

Rangeability

The term *rangeability* is still widely used to describe the flow range of a control valve. One definition of rangeability is the ratio of maximum to minimum controllable flow. The problem lies in the definition of the two limits. Another definition of the rangeability of a control valve is the flow range through which a particular inherent characteristic is maintained within prescribed limits.

Neither definition is satisfactory nor are they of any value in the valve selection process. As a result, the term is loosely used, primarily in advertising and sales promotion. The usable flow range of a control valve is naturally of interest, and a brief discussion of the factors involved follows.

For conventional globe-style control valve plugs, either solid lathe-turned or with ported skirts, the minimum flow area is directly proportional to the clearance between the plug and the seat ring (Fig. 12). This clearance is determined largely by the mechanical problem of alignment between the various body assembly components with an allowance to take care of operating temperature extremes. With the plug just off the seat, the flow passage is an annular ring which is very narrow with respect to the seat ring diameter. In commercial designs, a minimum leakage flow of 2% of maximum is within practical attainment. A practical range for standard single-seated valves might be from ~3% minimum for 1 in (25 mm) and under down to 1% for 10 in (250 mm) and over.

This leakage flow is generally accepted as the minimum controllable flow used for determining rangeability. In some valve designs, such as the control ball and the eccentric rotating plug valve, there is no point of abrupt discontinuity, and the theoretical rangeability can be much greater even though it seldom is possible to operate that closely to the closed position for other reasons.

At the other end of the scale the maximum controllable flow is equally difficult to determine. Use of the manufacturer's published C_v ratings is the only practical approach. The flow range is further affected by the pressure recovery factor C_f which varies with valve stroke.

Compounding these problems is the fact that no standard has been established for limiting deviation from the theoretical flow characteristic between these limits.

The candid fact is that the term "rangeability" is very much overworked and of negligible practical value. The "flow range" of all properly designed control valves is automatically large.

Selection of Characteristics to Suit the Process

Arguments on the selection of proper flow characteristics for a control valve are far from well defined. The ideal valve for most applications would have an installed characteristic such that the control loop would have equal stability through full load change. Careful interpretation of this statement is the key to the problem. The principal controversy lies in the selection of linear-versus-equal-percentage characteristics, since these are the two inherent characteristics generally available for continuous process control.

Most control systems give optimum performance with linear installed characteristics. For example, the stability of gas pressure control systems is largely dependent on the storage volume being controlled, not the flow rate. Liquid-level control stability is determined by the surface area and level range, not the flow rate. There are very few cases where the process itself will specifically indicate the equal percentage characteristic for optimum control. It is used primarily to help compensate for other system friction losses. Paradoxically, the use of equal percentage valves is greater than that of linear valves. For more critical applications, the practical problem of selecting the valve characteristic should be considered in light of the following factors:

1. *Load Variation (Constant Pressure Drop).* Over a load range of as much as 3 to 1, the effective characteristics of the 50-to-1 equal percentage or the linear valve do not differ materially. For this condition, which covers a majority of all applications, either valve can be used, although the sizing of the linear valve is more critical.

2. *Variable Pressure Drop.* Almost all valves installed in pump discharge lines, or in series with other process equipment, have an effective characteristic differing substantially from the inherent. In this case, the equal percentage valve tends to become linear, and the linear to become quick-opening. The equal percentage valve is definitely the better choice.

   If the load change is small, the linear valve can be employed, but it becomes increasingly difficult to choose the proper size. Too large a potential maximum flow and too small a pressure drop are often used for valve sizing. An oversized linear valve is definitely disadvantageous.

3. *System Stability.* If the control system is simple, allowing very high gain settings on the controller, valve characteristics become relatively unimportant. The principal question is whether to use on-off or proportional control. As the division point between proportional and on-off control is approached, quick-opening low-lift valves may be used with a substantial reduction in cost.

4. *System Nonlinearities.* In theory, the effective characteristic of a control valve should produce a constant gain in the control loop at any load. The loop includes the primary measuring element itself, which may be nonlinear. Two common examples are the differential-pressure cell utilized for measuring flow as a function of the square root of the pressure drop and vapor-pressure-type filled thermal systems. These can be shown to require special characteristics for the best performance over a wide setting range. In the practical sense, however, the process requirements seldom justify a custom design.

5. *Special Characteristics.* It is possible to produce valves with characteristics to fit special systems. The alteration may be in the basic valve design or may be accomplished through cam-equipped positioners. It is justifiable only for systems which are exceptionally complex or will be duplicated in large quantity.

SUMMARY

1. For most processes, particularly level control and pressure control applications, valve characteristics are relatively unimportant. Selection should be based on suitable construction and interchangeability for standardization.

2. Over a load change of up to 3 to 1, the performance of both the linear (properly sized) or the equal percentage valve is almost identical.

3. An oversized linear valve can be definitely inferior to an equally large equal percentage valve.

4. If the pressure drop across the valve under maximum flow conditions is less than 25% of the system drop and a linear installed characteristic is desired, an equal percentage inherent characteristic or an approximation thereof should be used.

5. The majority of all characterized valves sold are currently equal percentage. The choice is primarily based on practical grounds. The use of linear characteristics is increasing and will continue to do so as more accurate process data for sizing and system analysis become more readily available.

REFERENCES

Please consult preceding sections of this Handbook.

Control Valve Noise

by
Allen C. Fagerlund*

Fluid and transmission systems are a major source of industrial noise. Elements within the system that contribute to the noise are control valves, abrupt expansions of high-velocity flow streams, compressors, and pumps. This article deals with control valve noise, a principal source of noise in many fluid systems. Control valve noise is a result of the turbulence introduced into the flow stream in producing the permanent head loss required to fulfill the basic function of the valve. Specific topics covered are a brief review of noise terminology, identification of sources of valve noise, relation of gross flow parameters to noise potential of the valve, techniques for prediction of control valve noise, and methods of noise control.

*Senior Research Specialist, Fisher Controls International, Inc., Marshalltown, Iowa.

NOISE TERMINOLOGY

Noise is commonly defined as unwanted or annoying sound. Because of its vague definition, neither an absolute method of measurement nor specific dimensions of noise have been determined. For lack of a better method, noise is frequently described or specified by the physical characteristics of sound. The definitive properties of sound are the magnitude of sound pressure and the frequency of pressure fluctuation, as illustrated by Fig. 1.

Sound pressure (P_s) measurements are normally root-mean-square (rms) values of sound pressure expressed in microbars. Because the range of sound pressure of interest in noise measurements is $\sim 10^8$ to 1, it is customary to deal with sound pressure level (SPL) instead of sound pressure. SPL is a logarithmic function of the relative amplitude of sound pressure and is expressed mathematically as

$$SPL = 20 \log_{10} \frac{P_s}{0.0002 \; \mu bars} \; dB$$

The arbitrary selected reference sound pressure of 0.0002 μbars is approximately the sound pressure required at 1000 Hz to produce the faintest sound that the average young person with normal hearing can detect. The characteristic of the SPL scale is such that each change of 6 dB in level represents a change in the amplitude of sound pressure by a factor of 2.

The apparent loudness of a sound varies not only with the amplitude of sound pressure but also as a function of frequency. The human ear responds to sounds in the frequency range between 20,000 and 18,000 Hz. The normal ear is most sensitive to pressure fluctuations in the neighborhood of 3000 to 4000 Hz; therefore, the degree of annoyance created by a specific sound is a function of both sound pressure and frequency.

SPL measurements are often weighted to adjust the frequency response. Weighting that attenuates the lower frequencies to approximate the response of the human ear is called *A-weighting*. Figure 2 shows SPL correction as a function of frequency for A-weighted octave band analysis.

Sound intensity (I) is defined as the acoustic sound power transmitted per unit area, perpendicular to a specified direction. The common unit of measurement for sound intensity is watts per square centimeter. Sound intensity for a plane wave is given by the relation

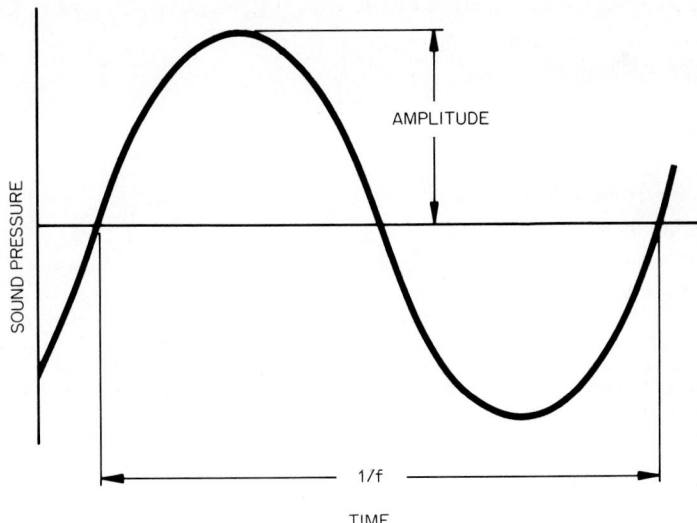

Fig. 1   Properties of sound.

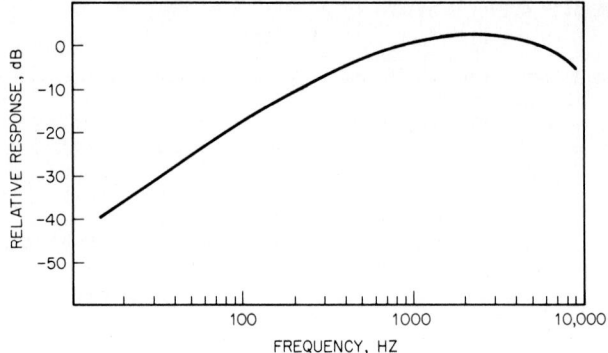

Fig. 2   A-weighting curve.

$$I = \frac{P_s^2}{\rho c} \tag{1}$$

where the product of the density (ρ) and sonic velocity (c) of the transmitting medium represents the characteristic impedance.

For measurement and comparison of sound intensities, it is more convenient to deal with *sound intensity levels* than with absolute values of sound intensity. Sound intensity level is commonly defined by the following relation:

$$\text{Sound intensity level} = 10 \log_{10} \frac{I}{10^{-16}} \text{ dB} \tag{2}$$

The reference sound intensity is selected as 10^{-16} W/cm^2. This is approximately the minimum intensity audible to the average human ear at 1000 Hz/s for standard air.

Table 1 presents the approximate overall sound levels of some familiar sound environments.

SOURCES OF VALVE NOISE

The major sources of control valve noise are:

1. Mechanical vibration of valve components
2. Fluid generated noise
   a.   Hydrodynamic noise
   b.   Aerodynamic noise

Mechanical Noise

The vibration of valve components is a result of random pressure fluctuations within the valve body or fluid impingement on movable or flexible parts. The most prevalent source of noise resulting from mechanical vibration is the lateral movement of the valve plug relative to the guide surfaces. Sound produced by this type of vibration normally has a frequency of less than 1500 Hz and is often described as a metallic rattling. The physical damage incurred by the valve plug and/or associated surfaces is generally of more concern than the noise emitted.

A second source of mechanical noise is a valve component resonating at its natural frequency.

Table 1 Approximate Sound Level of Familiar Sounds

Sound	Sound level, dB
Pneumatic rock drill	130
Jet takeoff (at 200 ft, 61 m)	120
Boiler factory	110
Electric furnace (area)	100
Heavy street traffic	90
Tabulation machine room	80
Vacuum cleaner (at 10 ft, 3 m)	70
Conversation	60
Quiet residence	50
Electric clock	20

Resonant vibration of valve components produces a single pitched tone, normally having a frequency between 3000 and 7000 Hz. This type of vibration produces high levels of stress that may ultimately produce fatigue failure of the vibrating part.

Noise resulting from mechanical vibration has, for the most part, been eliminated by improved valve design and is generally considered a structural problem rather than a noise problem.

Hydrodynamic Noise

Control valves handling liquid flow streams can be a substantial source of noise. The flow noise produced is referred to as *hydrodynamic noise* and may be categorized with respect to the specific flow classification or characteristic from which it is generated. Liquid flow can be divided into three general classifications: (1) noncavitating, (2) cavitating, and (3) flashing.

Noncavitating liquid flow generally results in very low ambient noise levels. It is generally accepted that the mechanism by which the noise is generated is a function of the turbulent velocity fluctuations of the fluid stream, which occur as a result of rapid deceleration of the fluid downstream of the vena contracta as the result of an abrupt area change.

The major source of hydrodynamic noise is cavitation. This noise is caused by the implosion of vapor bubbles formed in the cavitation process. Cavitation occurs in valves controlling liquids when the service conditions are such that the static pressure downstream of the valve is greater than the vapor pressure and at some point within the valve the local static pressure, either because of the high velocity and/or intense turbulence, is less than or equal to the liquid vapor pressure.

Figure 3 depicts the pressure profile of a cavitating flow stream as a function of distance along

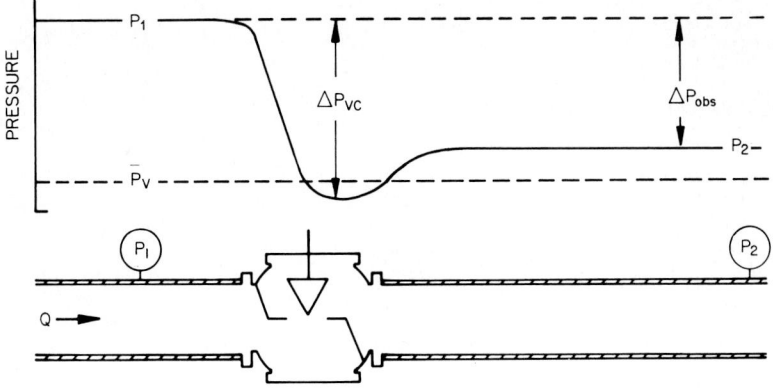

Fig. 3   Static pressure along a stream line of cavitating flow.

the stream. Vapor bubbles are found in the region of minimum static pressure and subsequently are collapsed or imploded as they pass downstream into an area of higher static pressure. Noise produced by cavitation has a broad frequency range and is frequently described as a rattling sound similar to that which would be anticipated if gravel were in the fluid stream.

Since cavitation may produce severe damage to the solid boundary surfaces that confine the cavitating fluid, noise produced by cavitation is of secondary concern.

Pertaining to the design of quiet valves for liquid application, the problem resolves itself into one of designing to reduce cavitation. Service conditions that will produce cavitation can readily be calculated. The use of staged or series reductions provides a viable solution to cavitation and hence hydrodynamic noise.

Flashing is a phenomenon that occurs in liquid flow when the differential pressure across a restriction is greater than the differential between the absolute static and vapor pressures at the inlet to the restriction:

$$\Delta P > P_1 - P_v$$

The resulting flow stream is a mixture of the liquid and gas phases of the fluid. Noise resulting from a valve handling a flashing fluid is a result of the deceleration and expansion of the two-phase flow stream.

Test results supported by field experience indicate that noise levels in noncavitating liquid applications are quite low and generally are not considered a noise problem.

Aerodynamic Noise

The major source of valve noise is aerodynamic noise. *Aerodynamic noise* is a result of the turbulence created in a flow stream as a result of deceleration, or impingement. The principal area of noise generation in a control valve is the recovery region immediately downstream of the vena contracta, where the flow field is characterized by intense turbulence and mixing.

NOISE CONTROL

Noise control employs either one or both of the following basic approaches:

1. *Source Treatment.* Prevention or attenuation of the acoustic power at the source (quiet valves)
2. *Path Treatment.* Reduction of noise transmitted from a source to a receiver

QUIET VALVES

Based on the preceding discussion, the parameters that determine the level of noise generated by compressible flow through a control valve for a given application are the geometry of the restrictions exposed to the flow stream, the total valve flow coefficient, the differential pressure across the valve, and the ratio of the differential pressure to the absolute inlet pressure.

It is conceivable that a valve could be designed that utilizes viscous losses to produce the permanent head loss required. Such an approach would require valve trim with a very high equivalent length, which becomes impractical from the standpoint of both economics and physical size.

The noise characteristic or noise potential of a regulator increases as a function of the differential pressure (ΔP) and the ratio of the differential pressure to the absolute static pressure at the inlet $(\Delta P/P_1)$. Thus for high pressure ratio applications $(\Delta P/P_1 > 0.7)$ an appreciable reduction in noise

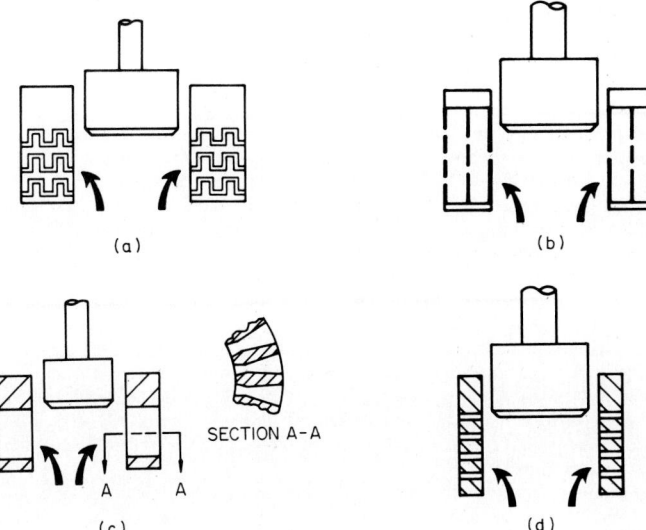

Fig. 4 Valve trim designed for noise attenuation.

can be effected by staging the pressure loss through a series of restrictions to produce the total pressure head loss required.

Generally, in control valves noise generation is reduced by dividing the flow area into a multiplicity of smaller restrictions. This is readily accomplished with a cage-style trim as shown in Fig. 4.

Critical to the total noise reduction that can be derived from the utilization of many small restrictions versus a single or a few large restrictions is the proper size and spacing of restrictions such that the noise generated by jet interaction is not greater than the summation of the noise generated by the jets individually. It has been found that the optimum size and spacing is very sensitive to the pressure ratio $\Delta P/P_1$.

It should be noted that, when multiple small-hole restrictions are used in series within the valve to distribute the pressure drop, they can act as strainers and are very susceptible to plugging as a result of either solid particles in the gas stream (dirty gas) or as a result of hydrate formation prior to the last stage.

However, for control valve applications operating at high pressure ratios ($\Delta P/P_1 \geqq 0.7$) the series restriction approach, splitting the total pressure drop between the control valve and a fixed restriction (diffuser) downstream of the valve can be very effective in minimizing the noise. In order to optimize the effectiveness of a diffuser, it must be designed (special shape and sizing) for each given installation so that the noise levels generated by the valve and diffuser are equal. Figure 5 depicts a typical valve-plus-diffuser installation.

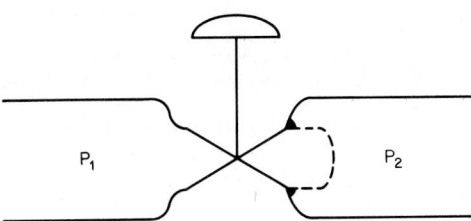

Fig. 5 Two-stage pressure reduction with a diffuser.

Path Treatment

A second approach to noise control is path treatment. Sound is transmitted through the medium that separates the source from the receiver. The speed and efficiency of sound transmission is dependent on the properties of the medium through which it is propagated. *Path treatment* consists of regulating the impedance of the transmission path to reduce the acoustic energy communicated to the receiver.

In any path treatment approach to control valve noise abatement, consideration must be given to the amplitude of noise radiated by both the upstream and downstream piping.

Since, when all else is equal, an increase in static pressure reduces the noise transmitted through a pipe, the upstream noise levels are always less than those downstream. Also, the fluid propagation path is less efficient moving back through the valve.

Dissipation of acoustic energy by the use of acoustical absorbent materials is one of the most effective methods of path treatment. Whenever possible the acoustical material should be located in the flow stream either at or immediately downstream of the noise source. This approach to abatement of aerodynamic noise is accommodated by in-line silencers. In-line silencers effectively dissipate the noise within the fluid stream and attenuate the noise level transmitted to the solid boundaries. Where high mass-flow rates and/or high pressure ratios across the valve exist, in-line silencers are often the most realistic and economical approach to noise control. The use of absorption-type in-line silencers can provide almost any degree of attenuation desired. However, economic considerations generally limit the insertion loss to ~25 dB.

Noise that cannot be eliminated within the boundaries of the flow stream must be eliminated by external treatment or isolation. This approach to the abatement of control valve noise includes the use of heavy-walled piping, acoustical insulation of the exposed solid boundaries of the fluid stream, and the use of insulated boxes, rooms, and buildings to isolate the noise source.

In closed systems (not vented to the atmosphere) any noise produced in the process becomes airborne only by transmission through the solid boundaries that contain the flow stream. The sound field in the contained flow stream forces the solid boundaries to vibrate, which in turn causes pressure disturbances in the ambient atmosphere that are propagated as sound to the receiver. Because of the relative mass of most valve bodies, the primary surface of noise radiation to the atmosphere is the piping adjacent to the valve. An understanding of the relative noise transmission loss as a function of the physical properties of the solid boundaries of the flow stream can be beneficial in noise control for fluid transmission systems.

A detailed analysis of noise transmission loss is beyond the scope of this chapter. However, it should be recognized that the spectrum of the noise radiated by the pipe has been shaped by the transmission loss characteristic of the pipe and is not that of the noise field within the confined flow stream. For a comprehensive analysis of pipe transmission loss, see Ref. 4.

Acoustic insulation of the exposed solid boundaries of a fluid stream is an effective means of noise abatement for localized areas. Test results indicate that ambient noise levels can be attenuated 10 dB/in of insulation thickness.

Path treatment such as the use of heavy-walled pipe or external acoustical insulation can be a very economical and effective technique for localized noise abatement. However, it should be pointed out that noise is propagated for long distances via a fluid streams and that the effectiveness of heavy-walled pipe or external insulation terminates where the treatment is terminated.

A simple sound survey of a given area will establish compliance or noncompliance with the governing noise criterion, but not necessarily either identify the primary source of noise or quantify the contribution of individual sources. Frequently piping systems are installed in environments where the background noise due to highly reflective surfaces and other sources of noise in the area make it impossible to use a sound survey to measure the contribution a single source makes to the overall ambient noise level.

A study of sound transmission loss through the walls of commercial piping indicated the feasibility of converting pipe wall vibrations to sound levels. Further study resulted in a valid conversion technique as developed in Ref. 5.

The vibration levels may be measured on the piping downstream of a control valve or other potential noise source. Sound pressure levels expected are then calculated based on the characteristics of the piping. Judgment can then be made as to the relative contribution of each source to the total sound field as measured with a microphone. The use of vibration measurements effectively isolates a source from its environment.

VALVE NOISE PREDICTION—AERODYNAMIC NOISE

To date, most theories on aerodynamic noise have been developed with special reference to the noise of free jets as opposed to a gas stream restrained by solid walls. A standard reference on this topic is Lighthill's paper (Ref. 1).

Development of a complete analytical model which accurately describes the precise mechanism of aerodynamic noise generation by control valves is very unlikely, in view of the complexity of the geometry and the lack of knowledge concerning the absolute values of the relevant parameters. It seems obvious, therefore, that an accurate technique for predicting valve noise must be based on empiricism. Such an approach requires the noise characteristic for each valve style and adjacent piping configuration to be established by actual test. Obviously, it is neither practical nor feasible to determine noise levels for all valve applications by laboratory tests. Dimensional analysis is a logical tool for correlation and extrapolation of the test data. This approach has led to the following expedient and accurate technique for predicting ambient noise levels resulting from the flow of compressible fluids through control valves:

$$SPL = SPL_{\Delta P} + \Delta SPL_{C_g} + \Delta SPL \frac{\Delta P}{P_1} + \Delta SPL_K + \Delta SPL_{P_2} \qquad (3)$$

where SPL = overall noise level at a predetermined point (1 m downstresam from valve outlet and 1 m from pipe surface), dB

$SPL_{\Delta P}$ = base SPL determined as a function of ΔP, dB
$\Delta SPLCg$ = correction for gas sizing coefficient as defined in Ref. 2, dB
$\Delta SPL_{\Delta P/P_1}$ = correction for valve style and pressure ratio, dB
ΔSPL_K = correction for acoustical treatment (heavy-walled pipe, insulation, in-line silencers, etc.), dB
ΔSPL_{P_2} = correction for effect of downstream pressure, dB

The terms ΔSPL_K and ΔSPL_{P_2} were developed from theoretical considerations (Ref. 4), the rest being empirical.

The following serves as a guide for using the noise prediction technique. Information presented in Figs. 6 to 9 facilitates the prediction technique.

Given a 300-lb cage-style globe body (ANSI Fisher Design ES) installed in an 8-in schedule 40 pipeline.

$$P_1 = 615 \text{ psia}$$

$$\Delta P = 348 \text{ psi}$$

$$C_g = 4000 \qquad \text{calculated as required}$$

$$SPL_{\Delta P} = 51 \text{ dB} \qquad \text{Fig. 6}$$

$$\Delta SPL_{C_g} = 72 \text{ dB} \qquad \text{Fig. } 7a$$

$$\Delta SPL_{\Delta P/P_1} = 15.5 \text{ dB} \qquad \text{Fig. } 8a$$

$$\Delta SPL_K = -27.1 \text{ dB} \qquad \text{Table 2}$$

$$\Delta SPL_{P_2} = -3.2 \text{ dB} \qquad \text{Fig. 9}$$

$$SPL = 51 + 72 + 15.5 + (-27.1) + (-3.2) \qquad \text{Eq. 3}$$

$$= 108.2 \text{ dBA}$$

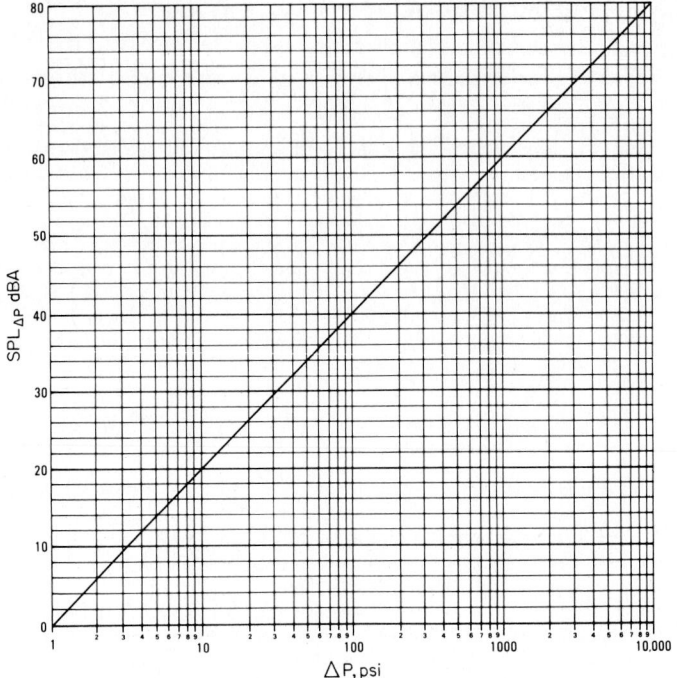

Fig. 6 $SPL_{\Delta P}$ versus ΔP for all valve styles.

If we were to use schedule 80 pipe from Table 2,

$$\Delta SPL_K = -30.9 \text{ dB}$$

$$SPL = 51 + 72 + (15.5) + (-30.9) + (-3.2)$$

$$= 104.4 \text{ dBA}$$

As a possible alternate valve selected to minimize noise, we will repeat the above problem using a cage-style globe body (Design EW) with Whisper Trim I. *Note:* the only change in our calculation will be in the $\Delta SPL_{\Delta P/P_1}$ value.

$$\Delta SPL_{\Delta P/P_1} = -2 \text{ dB} \hspace{3cm} \text{Fig. 8}b$$

$$SPL = 51 + 72 + (-2) + (-27.1) + (-3.2)$$

$$= 90.7 \text{ dBA}$$

Further reductions would be possible with a Whisper III type of cage.

Summary

From the foregoing discussion it should be obvious that a substantial amount of progress in the area of control valve noise has been achieved in the past few years. Prior to about 1968, the control valve industry as a whole had done little work on this extremely important subject. Perhaps the consensus

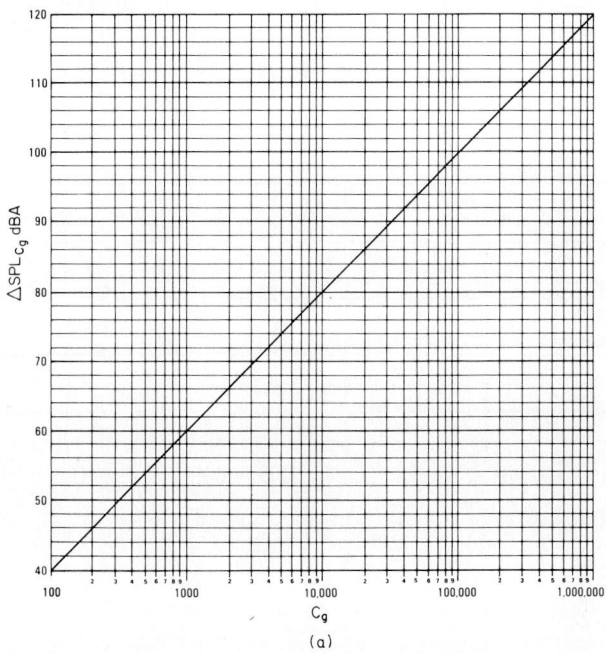

(a)

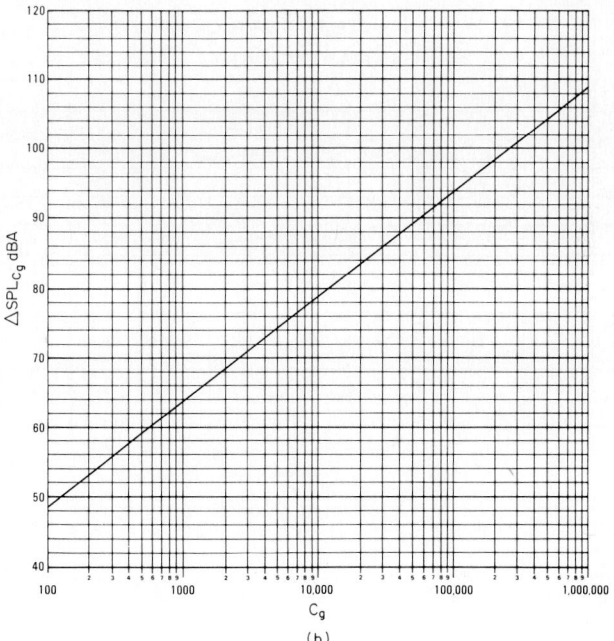

(b)

Fig. 7 (a) ΔSPL_{C_g} versus C_g for a line-of-sight valve or a globe valve with standard or a Whisper Trim I cage. (b) ΔSPL_{C_g} versus C_g for an in-line diffuser or a globe valve with a Whisper Trim III cage.

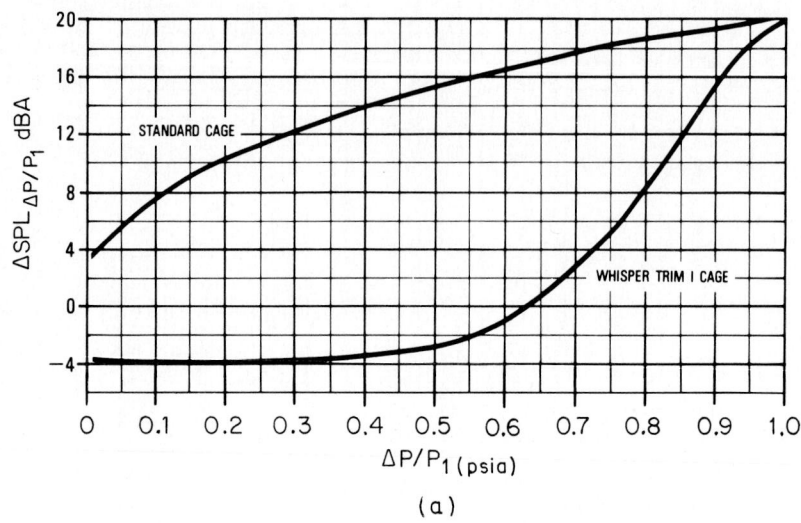

(a)

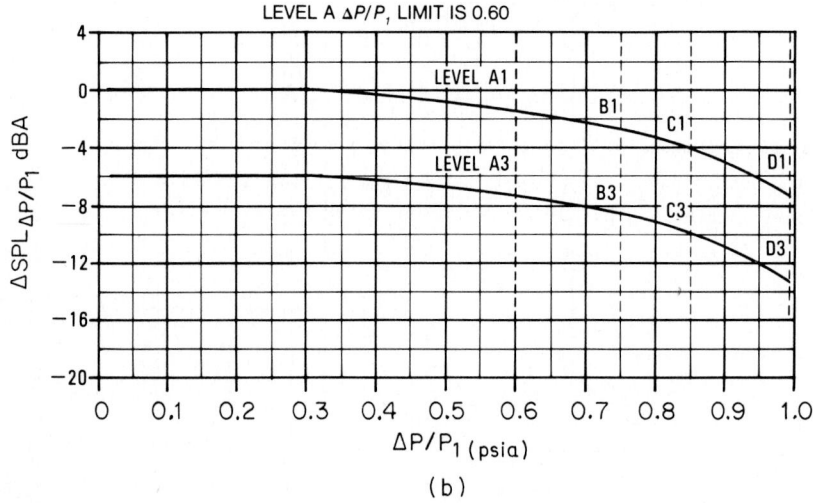

(b)

Fig. 8 (a) $\Delta SPL_{\Delta P/P1}$ versus $\Delta P/P_{1(\text{psia})}$ for a globe valve with a standard or Whisper Trim I cage. (b) $\Delta SPL_{\Delta P/P1}$ versus $\Delta P/P_{1(\text{psia})}$ for a globe valve with a Whisper Trim III cage.

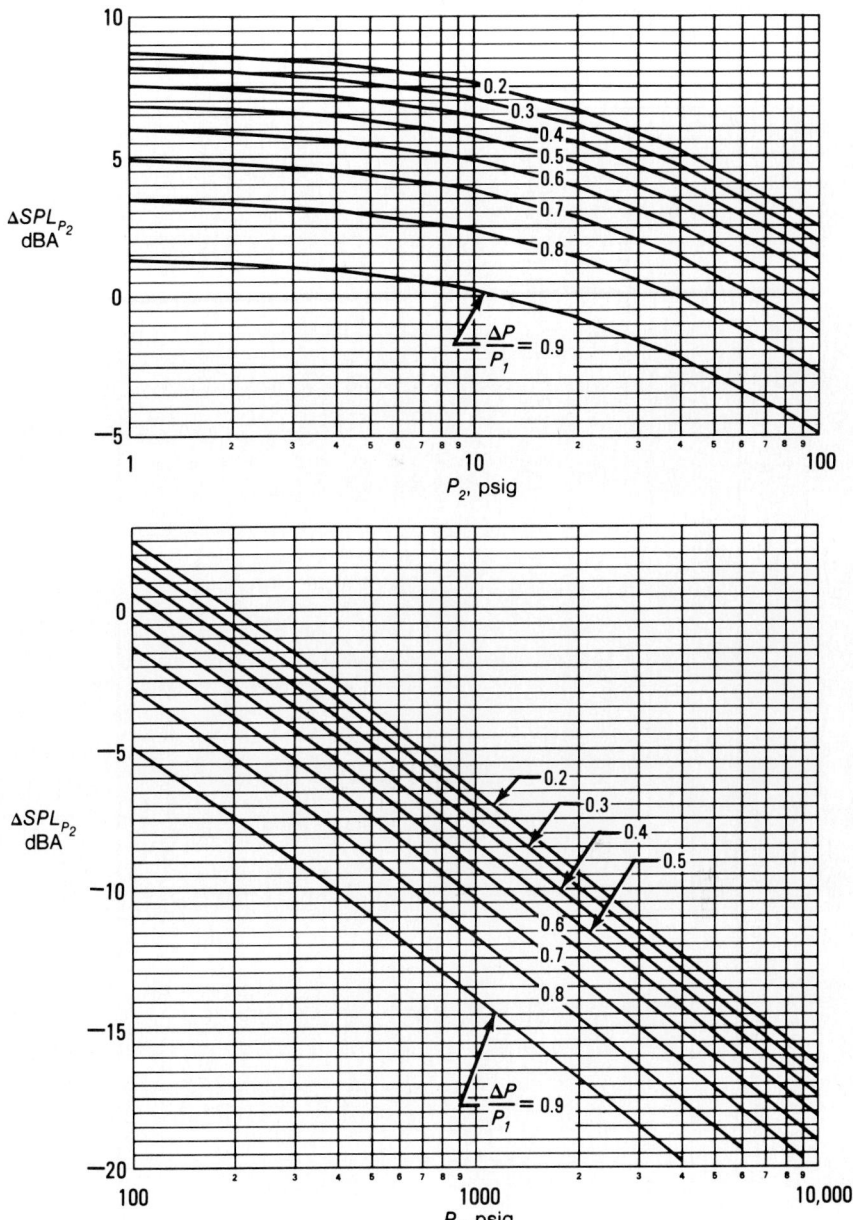

Fig. 9 ΔSPL_{P2} versus P_2 for all valve styles.

Table 2 ΔSPL_K Correction for Steel Pipe Wall Attenuation, dBA

Pipe size, in	Steel schedule												
	10	20	30	40	60	80	100	120	140	160	STD	XS	XXS
1	—	—	—	−19.0[a]	—	−21.6	—	—	—	−24.5	−19.0	−21.6	−27.6
1½	—	—	—	−19.8[a]	—	−22.6	—	—	—	−25.6	−19.8	−22.6	−28.6
2	—	—	—	−20.4[a]	—	−23.4	—	—	—	−27.3	−20.4	−23.4	−29.4
3	—	—	—	−23.4[a]	—	−26.2	—	−29.5	—	−29.5	−23.4	−26.2	−32.3
4	—	—	—	−24.2[a]	—	−27.2	—	−31.8	−35.1	−31.2	−24.2	−27.2	−33.2
6	—	—	—	−25.8[a]	−29.1	−29.5	−32.4	−34.1	−37.0	−33.9	−25.8	−29.5	−35.5
8	—	−24.9	−25.8	−27.1[a]	−31.0	−30.9	−34.1	−35.5	−38.1	−36.1	−27.1	−30.9	−35.8
10	—	−24.9	−26.7	−28.2[a]	−32.1	−32.4	−35.6	−37.1	−39.1	−38.0	−28.2	−31.0	—
12	—	−25.1	−27.5	−29.3	−32.6	−33.8	−36.6	−38.0	−40.4	−39.5	−28.6	−31.1	—
14	−25.4	−27.1	−28.7[a]	−30.0	−33.6	−34.7	−37.6	−39.0	−41.2	−40.1	−28.7	−31.2	—
16	−25.3	−27.2	−28.8[a]	−31.3	−34.8	−35.8	−37.6	−40.1	−42.3	−41.3	−28.8	−31.3	—
18	−25.3	−27.2	−30.1	−32.3	−35.6	−36.7	−38.6	−40.9	−43.9	−42.3	−28.8	−31.3	—
20	−25.4	−28.9[a]	−31.4	−32.9	−37.4	−37.7	−39.1	−42.8	—	−43.3	−28.9	−31.4	—
24	−25.6	−29.1[a]	−32.6	−34.4	—	−39.4	−41.3	—	—	−45.0	−29.1	−31.6	—
30	−27.7	−31.8	−33.7	—	—	—	—	—	—	—	−29.3	−31.8	—
36	−28.0	−32.1	−34.1	−35.6	—	—	—	—	—	—	−29.6	−32.1	—
42	—	—	—	—	—	—	—	—	—	—	−29.8	−32.3	—
44	—	—	—	—	—	—	—	—	—	—	−29.9	−32.4	—
48	—	—	—	—	—	—	—	—	—	—	−30.0	−32.5	—
52	—	—	—	—	—	—	—	—	—	—	−30.2	−32.7	—
56	—	—	—	—	—	—	—	—	—	—	−30.3	−32.8	—
60	—	—	—	—	—	—	—	—	—	—	−30.4	−32.9	—

[a] Standard schedule.

was that the noise these devices generated in performing their jobs was just a price which had to be paid. More probably, however, the real reason was simply the absence of a sufficient reason for the industry to spend the time and money required. Although the control valve buying public may have, by its occasional complaints regarding excessive noise, helped provide the incentive to attack the problem, the major impetus was furnished by the federal government in the form of legislation designed to protect people from noise pollution.

Based strictly on noise considerations, there are very few control valve installations which can be considered truly standard. They are unique from the standpoint of installation geometry, service conditions, noise attenuation requirements, or some combination of these. With so many possible installation variables and the numerous pieces of control valve noise abatement equipment, it becomes extremely important to consult persons knowledgeable in the application of this equipment. For example, several approaches may be taken to the same problem. One approach might produce the quietest installation but at a prohibitive cost, whereas another approach could meet the required noise specifications and at a substantial savings. Without the ability to predict noise levels and without the choices of equipment, optimizing a given installation would not be possible from the noise and cost standpoints.

Where do we go from here? Comparing current noise technology with other important control valve technologies, such as systems analysis and valve sizing, indicates that noise technology is in a growing state and progress can be expected in the future. Studies are in progress which are intended to increase the understanding of noise generation mechanisms and identify parameters not presently being considered. These studies and others should generate new and more efficient items of equipment and result in more precise techniques for the prediction of control valve noise.

REFERENCES

1. Lighthill, M. J.: "On Sound Generated Aerodynamically," *Proc. Roy. Soc.,* part 1, 1952; part 2, 1954.

2. Buresh, J. F., and C. B. Schuder: "Development of a Universal Gas Sizing Equation for Control Valves," *Publ. TM-15,* Fisher Controls Company, Marshalltown, Iowa.

3. Stiles, G. F.: "Development of a Valve Sizing Relationship for Flashing and Cavitating Flow," ISA Final Control Elements Symposium, Wilmington, Del., 1970.

4. Fagerlund, A. C.: "Sound Transmission through a Cylindrical Pipe Wall," *J. Eng. Ind., Trans. ASME,* vol. 103, 1981, pp. 355–360.

5. Fagerlund, A. C.: "Conversion of Vibration Measurements to Sound Pressure Levels," *Publ. TM-33,* Fisher Controls Company, Marshalltown, Iowa.

Cavitation in Control Valves

by
Marc L. Riveland*

Cavitation is a hydrodynamic phenomenon which began to receive recognition as a technical engineering problem in the early 1900s. At that time, observations led to the conclusion that vaporization of water in the vicinity of highspeed propellers was responsible for a decrease in their effectiveness. It is of current concern to the control valve industry not only because of a decrease in the effectiveness of control valves, but also because of the other attendant side effects—noise, vibration, and material damage.

Cavitation involves more complex phenomena than simple phase changes of a liquid. Innumerable factors can affect the hydrodynamic process as well as the response or reaction of the environment to the process. As a result, single indices of cavitation do not suffice as universal similarity parameters and have numerous scale effects associated with them. In fact, several authors state outright that the influences are so abundant that the study of cavitation defies classical modeling methods.

Knapp et al. (Ref. 1) feel that a definition of cavitation is too restrictive and resorts to describing the process as a liquid phenomenon which concerns growth and collapse of cavities resulting from pressure dynamics.

No restrictions are placed on the contents of these cavities; i.e., the contents are not necessarily exclusively liquid vapor. This leads to a consideration of the types of cavitation, of which three kinds are generally recognized (Holl, Ref. 2).

1. *Vaporous cavitation* is probably the most familiar form and is generally implied by the term "cavitation." It consists of explosive cavity growth resulting from local liquid vaporization, followed by rapid cavity collapse and vapor condensation. The phase change sequence is caused by local pressure fluctuations in the liquid. This form of cavitation is more detrimental than the other forms.

2. *Pseudocavitation* involves the growth and contraction of a fixed mass of noncondensible gas. This volume fluctuation is in direct response to local pressure changes. In other words, a small bubble expands as the pressure in the liquid surrounding it is decreased. Conversely, it decreases in size as pressure is increased.

3. *Gaseous cavitation* refers to the rectified diffusion of noncondensible gas entrained in the liquid into the cavity. It differs from pseudocavitation in that growth occurs because of an increase in mass in the cavity and is often called *outgassing*.

All three types of cavitation can and, in many instances, do occur simultaneously, making study of the individual types complicated.

Two phenomena which are related to cavitation, but which differ in mechanics and effects, need to be distinguished from cavitation.

The boiling and condensation process is similar to cavitation in that the liquid undergoes a liquid-vapor-liquid phase change sequence. It differs in that these changes result primarily from thermal dynamics as opposed to pressure dynamics. As a result, these transients generally are much slower than cavitation phase changes and thus less detrimental.

Flashing, like cavitation, consists of a phase change which results from pressure dynamics. Specifically, it is the change of a liquid to its vapor as a result of reduced pressure. In this circumstance,

*Senior Research Engineer, Fisher Controls International, Inc., Marshalltown, Iowa.

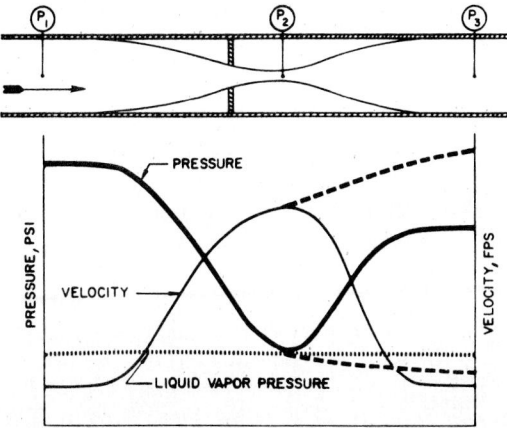

Fig. 1   Pressure profiles for flashing and cavitating flows through a simple restriction.

however, the fluid pressure remains below the pressure at which the liquid changed to vapor. The vapor thus never returns to the liquid state. Furthermore, it is the vapor-to-liquid phase change which has been shown to be associated with the production of noise, damage, and significant vibration. Hence, elimination of this phase change eliminates the damage potential, and flashing, like boiling, is considered less of a threat than cavitation, although choked flow can present sizing limitations.

Fluid Dynamics Which Cause Cavitation

To understand the conditions in a control valve which can cause cavitation, consider the simple restriction shown in Fig. 1. In this figure the local mean pressure of the liquid P is plotted as a function of the axial distance x through the simple restriction shown at the top of the figure. *Note:* For symbols used in this article, see Table 1.

If the flowing fluid is an incompressible liquid, it is realized that, in order to maintain a constant flow rate through the restriction, the mean flow velocity must increase to offset the effect of the decreased cross-sectional flow area. (This is basically a statement of the continuity equation.) Likewise, energy must be conserved. As the energy due to fluid velocity (kinetic energy) increases, other forms of fluid energy must be offset accordingly. The point of minimum cross-sectional area (and for this example the point of maximum velocity) is known as the vena contracta.

Experience has shown that this particular process of increasing the velocity through a geometrically simple restriction is nearly isentropic, that is, little or no energy is irreversibly lost (Streeter and Wylie, Ref. 4, pp. 243–244). Neglecting the work terms (no shaft work is done across a fixed restriction or valve) and changes in elevation, the energy equation between point 1 and the vena contracta may be written as

$$P_1 + \frac{\rho V_1^2}{2g_c} = \frac{\rho V_{vc}^2}{2g_c} + P_{vc} \tag{1}$$

Thus, the increase in kinetic energy must be offset by a decrease in static pressure.

As the cross-sectional flow area returns to its original value, the velocity decreases to its original value. However, for real fluids this recovery process is not reversible—fluid friction becomes significant and converts some of the mechanical energy into heat which may either be dissipated to the surroundings or stored in the form of internal energy. Some of the mechanical energy may also be transferred to the surroundings in the form of acoustic energy. Again, neglecting the work terms and

Table 1 Nomenclature

C_v	Flow coefficient for valves
g_c	Dimensional conversion factor
G	Specific gravity of fluid
H_l	Available energy loss—energy converted into heat
K_m	Valve pressure recovery coefficient
P	Pressure
q	Heat transferred into or out of the fluid
Q	Flow rate, gal/min
r_c	Critical pressure ratio
U	Internal energy of fluid
v	Velocity of fluid
Z	Elevation of center line of flow stream above a given datum

Greek letters
Δ	Change in
ρ	Density of fluid

Subscripts
1	Upstream conditions
2	Downstream conditions
c	Choked flow condition
m_v	Maximum velocity
v_c	Vena contracta

changes in elevation, this equation may be written as follows:

$$P_{vc} + \frac{\rho V_{vc}^2}{2g_c} + U_{vc} = P_2 + \frac{\rho V_2^2}{2g_c} + U_2 - q \tag{2}$$

Further combining the thermal terms into a single term H_l results in

$$P_{vc} + \frac{\rho V_{vc}^2}{2g_c} = P_2 + \frac{\rho V_2^2}{2g_c} + H_l \tag{3}$$

Combining Eqs. 1 and 3:

$$P_1 = P_2 + H_l \tag{4}$$

Thus, the original pressure is not restored; rather some lower value is attained.

If the restriction is insulated or the process is relatively fast, little or no heat is transferred to the surroundings. In such a case the internal energy (and temperature) of the liquid rises accordingly (although actual amounts are small). Thus, while little or no actual energy is lost, there can be an appreciable decrease in the available energy.

This entire process is shown by the curve in Fig. 1. As the fluid enters the area of decreased cross-sectional flow, the local mean pressure decreases from its original value P_1. After reaching a minimum value it increases to some final value P_2. This final value is less than the original value because of the conversion of available energy into heat.

It is these pressure dynamics in combination with the fluid characteristics which result in cavitation. As the fluid passes through this low-pressure region it may encounter a pressure level at which the liquid phase is not stable at the given inlet temperature (temperature changes very little from the original value). If these conditions prevail for a sufficient period of time, the liquid will change to a vapor very rapidly. Experience has shown the local mean pressure at which this occurs to be very near the vapor pressure of the fluid at this given temperature (the process of initiating bubble formation is known as *nucleation*).

Once a bubble has formed and proceeds to traverse the reduced pressure region, it grows in response to the continually decreasing pressure and liquid vaporization. Eventually, the pressure recovery halts the growth of the bubble and forces it to collapse. Under certain circumstances several regrowth-collapse cycles known as *rebounds* can occur.

As mentioned earlier, flashing occurs if the pressure of the fluid remains below the vapor pressure of the fluid. The pressure conditions for both cavitation and flashing in a simple restriction are shown in Fig. 1.

There are four primary negative side effects: excessive noise, excessive vibration, material damage, and deterioration of the flow relationship or *effectiveness*. The latter side effect concerns the relationship between the flow rate and the pressure drop. As implied by the energy equation, the flow through the restriction is normally proportional to the square root of the pressure drop (for a liquid flow). The *constant of proportionality* is the liquid-flow coefficient of the restriction C_v (Ref. 3) divided by the square root of the specific gravity G. Therefore,

$$Q = C_v \left(\frac{\Delta P}{G}\right)^{1/2}$$

(5)

This relation suggests that the flow rate can be continually increased by increasing the pressure differential across the restriction. However, in practice this relation begins to break down when a sufficient amount of the vapor phase (produced in the cavitation process) is formed. Less of a flow increase is realized for the same pressure differential increase (at a given P_1) until finally the flow remains constant despite an increase in the pressure drop. Figure 2 portrays this liquid "choking" phenomenon graphically.

The exact mechanisms of liquid choking are not fully confirmed, although there are parallels between it and critical flow in gas applications. In gas flows, the flow chokes at a point where the fluid velocity is equal to the acoustic wave speed. For pure liquids (incompressible fluids) the acoustic wave speed is so high that it is "unattainable" for all practical intents and purposes. In liquids which partially vaporize, however, the fluid is actually a two-phase mixture and typically has a very low acoustic wave speed (actually lower than that of a pure gas). Thus it is possible for the mixture velocity to become equal to the sonic velocity and choke the flow.

While all the aforementioned phenomena are side effects having the same origin (the formation and collapse of cavities), Robertson (Ref. 5) points out that they exhibit different trends with respect to cavitation intensity. Thus, it is apparent that inferring the degree of one effect by observation of another can have undesirable consequences. For example, if it were desired to evaluate the damage or damage potential of cavitating flow through a valve, it would be misleading to examine noise behavior. The noise level increases to a point and then decreases. This lower level of noise may create the false impression of safe operation, even though damage may already be occurring.

Noise, vibration, and loss of flow efficiency are certainly real concerns in sizing valves. However, physical damage to the valve is probably the most frequent concern because of its cost, inconvenience, and unpredictable nature. To aid in understanding the causes and conditions behind this phenomena, the next section discusses cavity dynamics and damage mechanisms in more detail.

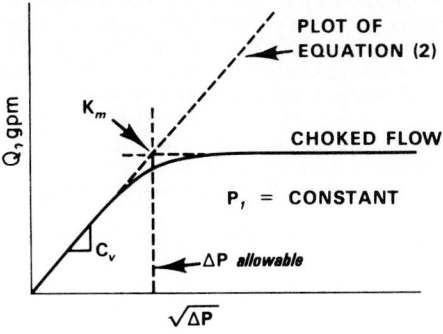

Fig. 2 Flow curve showing choked flow.

Cavity Mechanics

The behavior of cavities, both individually and collectively, has a direct bearing on the degree and extent of the negative side effects that will result, particularly damage.

As mentioned previously, there are four primary events in the bubble cycle: nucleation, growth, collapse, and rebound.

In order for a liquid to cavitate at or near the vapor pressure of the liquid it is necessary to have stabilized nuclei (such as an entrained noncondensible gas) available in the liquid. Furthermore, these nuclei must be of a certain minimum size in order to explosively grow or cavitate. This critical minimum size is determined from the prevailing conditions, and any nuclei below this size will not result in cavitation.

A nucleus of "subcritical" size is capable of growing to the critical threshold if ample time in a reduced pressure region is afforded. For example, a subcritical nucleus may enter a reduced pressure region, grow to the critical size, and then grow explosively by vaporization of the surrounding liquid. The degree and extent of the reduced pressure are established by the geometry of the flow and velocity of the fluid. The time a nucleus spends in the reduced pressure region is established by the nucleus velocity and the extent of the low-pressure region. If conditions are such that the reduced pressure region is relatively short and the nucleus is of subcritical size and is traveling at a relatively fast velocity, it is possible that the cavitation process may be suppressed entirely, even though the minimum pressure in the flow path is below the vapor pressure.

A good experimental study particularly illustrative of bubble mechanics was conducted by Knapp and Hollander (Ref. 6). Briefly, they employed very high speed photographic techniques to observe the behavior of a single bubble in external flow over a bluff body.

Their data showed the initial growth phase to be relatively slow and symmetrical as compared to the ensuing collapse.

The collapse stage, on the other hand, was fast and highly asymmetrical. Close examination of the cavity photographs indicated that the velocity of the leading edge of the bubble actually reversed direction. Other experimental studies (see Hammitt, Ref. 7, pp. 16, 17, 162–183) have revealed that a small "microjet" forms during this asymmetric collapse.

After the first growth-collapse cycle several additional cycles follow. This regrowth and collapse is known as rebound and may or may not occur. Rebound results from the compression of noncondensed vapors or gases, during which mechanical energy is stored and subsequently "re-released"; i.e., a "spring" effect occurs. The collapse of a rebound cavity is generally slower, more symmetrical, and marked by the absence of the microjet formed in the first collapse. In addition, the intensity of the pressure of the shock waves generated on collapse generally increases for the rebound.

The preceding discussion briefly outlines the behavior or life cycle of a single "model" bubble. It is important to realize that a number of "real" flow effects can alter this behavior.

One such effect has already been mentioned, namely, the absence of cavitation, even though the minimum mean pressure in the flow path may drop below the vapor pressure of the fluid. This condition results from subcritical-sized nuclei traversing the reduced pressure region in a time insufficient to grow to critical size.

Another effect may cause the opposite trend, namely, cavitation occurring when the minimum mean pressure is greater than the vapor pressure of the fluid. Thus far in this discussion, the fluid pressure at any given point in the flow path has been taken as the *mean* pressure of the cross section. Velocity gradients, highly turbulent free shear zones, separated flows, and irregularly shaped flow passages are all capable of producing localized regions where conditions differ substantially from the mean condition. Of concern are regions where the local instantaneous pressure may be considerably less than the mean pressure. In such instances the mean pressure may be greater than the vapor pressure of the fluid, suggesting that cavitation will not occur. However, the pressure at a given point in the liquid may momentarily drop below the vapor pressure, causing some localized cavitation. Thus, highly turbulent flow may be more prone to cavitate at pressures slightly above the vapor pressure.

Also associated with nucleation are effects associated with the entrained air content. Since an entrained air bubble serves as a good nucleus for cavitation, it logically follows that the degree of cavitation can be affected by the amount of air entrained. Over an indefinite range an increase in the amount of entrained air has the undesirable effect of creating an environment more conducive to cavitation. However, increasing the air content past a certain volume fraction impedes the growth-and-collapse process.

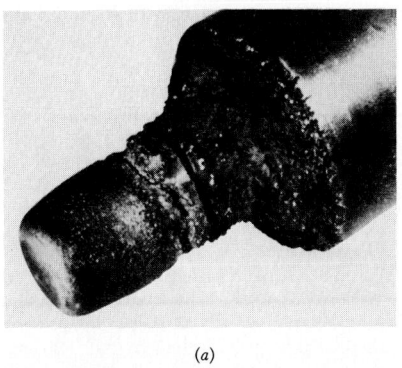

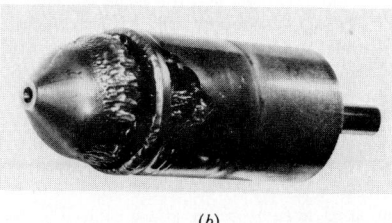

(a)   (b)

Fig. 3 Typical appearance of (a) cavitation damage, and (b) flashing damage.

Cavitation damage is usually the most troublesome negative side effect plaguing the valve industry. It does not take many examples of such damage to appreciate the destructive capabilities of cavitation.

Typically, cavitation damage is characterized by a very irregular, rough surface. The phrase "cinderlike appearance" has been used to describe it. It is discernible from other types of flow damage such as erosion and flashing damage, which are usually very smooth and shiny in appearance. Figure 3 illustrates these differences.

While the results of cavitation damage are all too familiar, the events and mechanisms of the cavitation damage process are not completely known or understood in spite of extensive study over the years. There is general agreement, however, on a number of aspects of the process and a consistency in certain observations. [*Note:* Much of the following discussion is based on the very thorough coverage of the subject given by Hammitt (Ref. 7, Secs. 5.2 and 11).]

Cavitation damage has been consistently observed to be associated with the collapse stage of the bubble dynamics. Furthermore, this damage consists of two primary events or phases: (1) an attack on a material surface as a result of cavitation in the liquid, and (2) the response or reaction of the material to the attack. Any factor which influences either of these events will have some sort of final effect on the overall damage characteristics. Note that "damage" in this context is defined as any permanent deformation or loss of material—the degree of this deformation or loss is not defined.

The attack stage of the damage process has been attributed to various single mechanisms in the past, but none of them have successfully accounted for all the observed results. It currently appears that this attack consists not of a single effect, but primarily involves two factors which interact in a positive reinforcing manner: (1) mechanical attack and (2) chemical attack.

There is evidence indicating the almost universal presence of a mechanical attack component which can occur in either of two forms:

1. Erosion resulting from a high-velocity microjet impinging on the material surface

2. Material deformation and failure resulting from shock waves impinging on the material surface

In the first type of mechanical attack a small, high-velocity liquid jet is formed during the asymmetrical collapse of a vapor bubble. Obviously, the orientation and proximity of this microjet with respect to the material surface are critical. If these two conditions are favorable, a damaging attack will occur on the surface; otherwise, no adverse material effect results. Consequently, only a relatively small percentage of the total number of existing vapor bubbles are damage producing. This is the most probable form of mechanical attack, and its presence is supported by high speed cinematography, liquid drop impingement comparisons, and various analytical studies.

The second type of mechanical attack—shock wave impingement—does not appear to be as dominant. Analytical estimations of vapor bubble collapse pressures do not suggest that the shock waves produced are on a damaging order of magnitude—at least during the initial collapse. Experimental studies bear this out but also reveal resulting collapse pressures to increase in magnitude with subsequent rebound collapses and hence become potentially damaging. The intensity of this shock wave dissipates quite rapidly with propagation distance, again necessitating close proximity of the bubble to a wall in order to impart significant damage.

The other primary component of attack—corrosion—is perhaps more significant, as it interacts with the mechanical component rather than acting by itself. After a period of mechanical attack many of the protective coatings (films, oxides, etc.) are physically removed, making the base material more vulnerable to chemical attack. Material then subjected to corrosive action is more easily removed by continued mechanical attack, and so on. Furthermore, corrosion can create pits which can serve as mechanical wave guides, thus concentrating the mechanical attack and resulting in an apparent preferential attack.

One can conclude therefore that cavitation attack is primarily mechanical, but often compounded by a chemical (corrosion) attack, and, furthermore, that this mechanical attack is produced by collapsing bubbles in *close proximity* to a wall.

The response or reaction of the wall to the cavitation attack is equally diverse and depends on the exact makeup of the attack as well as the characteristic properties of the material. No single material property has been identified which consistently correlates with a material's ability to resist cavitation atatck (although hardness, strain energy to failure, and ultimate resilience show some degree of association).

Material damage results from such phenomena as high-velocity erosion, overstressing, cold-working and fatigue and is further aggravated by factors such as chemical environment, temperature, and electrochemical effects. The multiplicity of types of attack and material responses and their various interactions results in a nonconstant, nonlinear damage rate.

Just as a number of influential variables have an effect on the behavior of individual cavities, so too there are influences affecting the degree and extent of material damage. Once again, many details are presently unknown, but primary influences include such variables as air content, pressure, velocity, and temperature.

Air content impacts cavitation damage primarily through its effect on cavity mechanics, as previously discussed. Again, two opposing trends are evident on increasing the amount of air. Additional air supplies more entrained air nuclei which produce more cavities and thus can increase the total damage. After a point, however, additional increases in air content disrupt the mechanical attack component and effectively reduce the total damage.

Pressure effects also exhibit two opposing trends. Given a fixed inlet pressure P_1, decreasing the back-pressure P_2 tends to increase the number of cavities formed, thus creating a worse situation. However, a lower back-pressure also creates a lower collapse pressure differential ($P_1 - P_{vc}$), resulting in a decrease in the intensity of the cavitation. Knapp et al. (Ref. 1, Sec. 9.3) cite experimental work confirming this dual nature of pressure effects.

An additional pressure effect unrelated to the above concerns the location of damage. As the back-pressure is changed, the pressure required to collapse the cavities moves up- or downstream, depending on whether the pressure is increased or decreased, respectively. Thus, in addition to a change in the severity of the total damage, there may be an accompanying change in the physical location of the damage when pressure conditions are altered.

In internal flow situations, such as those involving a control valve, the pressure and velocity of the fluid are usually not independent quantities. That is, a velocity change usually accompanies a pressure change. Thus, it is not easy to identify the individual effects of each in such situations. Controlled studies (see Hammitt, Ref. 7, Sec. 5:2:3) have indicated, however, that damage rates are proportional to a high power of the velocity (exponents in the range of 3 to 10 have been observed).

It should now be apparent that the cavitation damage process is a complex function of:

1. Intensity and degree of cavitation (cavitation attack)

2. Material of construction (material response)

3. Time of exposure

Thus, while the above-mentioned influences have been observed, they remain to be quantified. The consolidation of all these influences into a simple, methodical model or prediction algorithm (much less a single parameter) transcends current analytical and even experimental capabilities, resulting in a lack of ability to accurately predict cavitation damage in the field.

Cavitation Control

While the damage rate and the total damage are not highly predictable, a number of strategies may be employed to successfully reduce the risk of hardware damage. For convenience, they may be classified into three groups: (1) system design, (2) material selection, and (3) anticavitation products.

Consideration of potentially damaging cavitation conditions at the time a system is designed is a primary and preferred strategy. Awareness and avoidance of unfavorable conditions are highly effective means of reducing the risk of damage. The placement of control valves in high-back-pressure locations, when possible, reduces the tendency of a valve to cavitate. On occasion, when placement is not flexible, downstream "breakdown" orifices may sometimes be inserted to "artificially" increase the back-pressure. However, this is not a preferred method for two reasons. First, the effective flow range of such an installation is very limited. If the valve is throttling at a very low flow rate, fluid velocities through the breakdown orifice may be so low as to eliminate its effectiveness (i.e., the bulk of the pressure drop will again be "seen" by the valve). At higher flow rates the orifice plate may become the primary restriction and in turn limit or completely choke off the flow.

Second, if properly sized to a particular valve, the downstream orifice may prevent the valve from cavitating; however, it may cavitate itself. Thus, the cavitation is not eliminated—merely relocated. Critical components immediately downstream of the orifice may still sustain damage.

A third strategy may prove feasible within the context of system design considerations—the use of a sacrificial member. In some situations, economics may warrant the installation of a lower cost valve which is allowed to cavitate (e.g., a butterfly valve). Immediately downstream of such a valve, where cavitation damage is likely to occur, a comparatively inexpensive pipe or fitting is installed and periodically replaced. This method does not afford relief from the other side effects (noise, vibration, and choking), but in certain instances allows the most economical control of damage.

If the process under control can tolerate the injection of a noncondensible gas, cavitation damage can potentially be reduced. Although no general recommendation regarding how much gas is required has been made, noise and vibration can also be reduced with this technique.

The second category, material selection, also does not offer one "best" solution, since no material is completely immune to cavitation damage. Mechanical attack interacts synergistically with corrosion to create a different reaction to each set of application conditions. Currently, qualitative force ranking of material resistance to cavitation attack, in combination with experience based on rules of thumb, governs material selection. In general, metals with greater hardness, ultimate resilience, or strain energy to failure offer better resistance to cavitation (e.g., certain stainless steels, tool steels, and alloy steels).

Elastomers, and compliant surfaces in general, exhibit an ability to withstand levels of cavitation attack greater than standard structural indicators would suggest. This apparently results from an interaction between the surface and the bubble which orients the microjet away from the surface, thus eliminating the mechanical attack. While such behavior is appealing from a damage control standpoint, Sanderson (Ref. 8) points out that bonding difficulties, as well as potential pressure and temperature limitations, have curbed widespread use of such materials in industry.

Finally, perhaps the most effective means of reducing the risk of cavitation damage is the use of cavitation control equipment. Virtually all such valve products on the market employ two basic strategies: (1) control energy transformations, and (2) isolate cavities.

The former is usually attained by *staging*. This approach to damage control routes flow through several restrictions in series, as opposed to a single restriction. Each restriction dissipates a certain amount of available energy and "presents" a lower inlet pressure to the next stage. This effectively reduces the efficiency of the device and results in lower pressure recovery. Thus, in a well-designed device, the valve will be able to take a large pressure differential, yet maintain the minimum fluid pressure above the vapor pressure of the liquid; this prevents the liquid from cavitating. Conventional trim, on the other hand, because of its higher efficiency, will have a larger degree of pressure recovery. For the same pressure differential then, the minimum fluid pressure will be less than for the staged trim, and the liquid will be more prone to cavitate. See Fig. 4.

Trims which dissipate available energy have an additional advantage. If the design pressure differential is exceeded and cavitation does occur, the intensity will be less. This is because the pressure which causes the collapse of cavities (i.e., the recovered pressure) will be less.

The second design strategy listed does not key on preventing cavitation as much as it does controlling the cavitation that occurs. As pointed out earlier, a mechanical component of attack is always

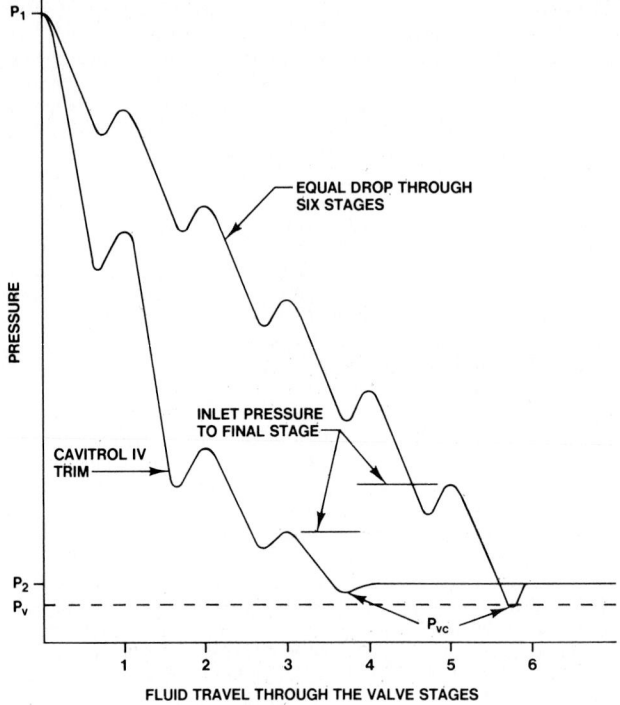

Fig. 4 Pressure profile for staged trim.

present in cavitation damage. Further, this mechanical attack occurs in two forms—microjet impact and shock wave impact—both of which need to occur in close proximity to the surface in order to be effective in producing damage. Thus, it follows that, if they occur far enough away from the surface, no mechanical attack will take place and no damage will occur. Some cavitation control trims, therefore, try to strategically place the minimum pressure regions (the most probable regions of cavitation) away from critical surfaces. While this approach does not necessarily eliminate the cavitation, it does minimize the possibility of cavitation damage to these surfaces.

REFERENCES

1. Knapp, R. T., J. W. Daily, and F. G. Hammitt: *Cavitation,* McGraw-Hill, New York, 1970.

2. Holl, J. W.: "Limited Cavitation," *Cavitation State of Knowledge,* Evanston, Ill., June 16–18, 1969, ASME, New York, 1969.

3. Hutchison, J. W.: *ISA Handbook of Control Valves,* 2d ed., Instrument Society of America, Pittsburgh, Pa., 1976, p. 9.

4. Streeter, V. L., and E. B. Wylie: *Fluid Mechanics,* 7th ed., McGraw-Hill, New York, 1979.

5. Robertson, J. M.: "Cavitation Today—An Introduction," *Cavitation State of Knowledge,* Evanston, Ill., June 16–18, 1969, ASME, New York, 1969.

6. Knapp, R. T., and A. Hollander: "Laboratory Investigations of the Mechanism of Cavitation," *Trans. ASME,* vol. 70, 1948, pp. 419–435.

7. Hammitt, F. G.: *Cavitation and Multiphase Flow Phenomena,* McGraw-Hill, New York, 1980.

8. Sanderson, R. L.: "Elastomers for Cavitation Damage Resistance," Instrument Society of America, International Conference and Exhibit, C.I.82-908, Philadelphia, October 1982.

Electric Actuators

by
H. C. Gery*

The selection of a control actuator depends on the characteristics of the process, the manner in which the process inputs can be regulated, and the required performance of the overall system. The selection of the appropriate control actuator determines the type of controller output. Other process-oriented decisions, such as failure mode analysis, may favor one type of actuator over another. For example, protection of the process may require that the actuator (1) remain in position, (2) move to a closed (or minimum) position, and (3) move to an open (or maximum) position when power is lost. Abbreviated diagrams of typical electric controller output connections are given in Fig. 1. There are many variations of these arrangements designed to meet specific process needs. See Table 1.

Electromechanical devices, such as relays and contactors, or solid-state devices, such as triacs and silicon controlled rectifiers, may be operated as two-position or on-off controllers with a dead band between the open and closed circuits. Batch processes with moderate to long process lags, such as the heating and cooling of habitable spaces, lend themselves to this type of control.

The switching may be operated by a proportional-plus-integral action controller which modulates the ratio of open to closed time of a relay or solenoid. If the process lag is long, the on-off pulses tend to be integrated and the effect is the same as if the energy were continuously modulated. Because loss of controller power opens the relay, an open relay must correspond to a safe process condition. If restoration to automatic control on return of power is undesirable, a lockout relay should be provided.

The current-output section provides a direct current proportional to the required output. This current is derived from a proportional-plus-integral action controller. The controller may also include derivative action. On loss of controller power, the valve moves to the position corresponding to zero current; so this must correspond to a safe process condition.

In the dual-relay output section, the controller switches are interlocked to prevent the simultaneous energizing of both motor fields. They may be mechanically or electrically operated in an on-off controller for two-position control. A time-pulse switch is in the common lead to the motor for floating control. The switches may be in a proportional action or a proportional-plus-integral-plus-derivative controller. On loss of controller power, the motor will remain in its last position.

*Application Consultant, Leeds & Northrup Company, A Unit of General Signal, North Wales, Pa.

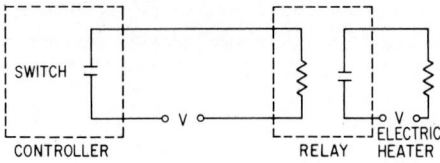

(a) SINGLE-RELAY OUTPUT SECTION

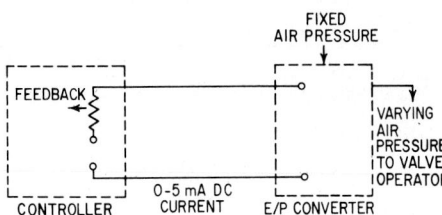

(b) CURRENT OUTPUT SECTION

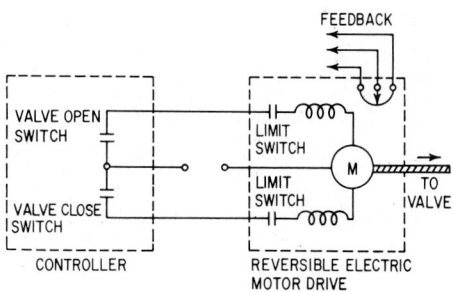

(c) DUAL RELAY OUTPUT SECTION

Fig. 1 Electric controller output sections: (*a*) Single relay, (*b*) current output, (*c*) dual-relay output.

Table 1 Electric Controller Ouptut Section versus Actuator Used

Output section	Actuator
Single-relay output section	Relay, solenoid, and other two-position actuators
Current output section and internally controlled feedback	Variable speed electric motor drive, electropneumatic converter, silicon controlled rectifier, saturable-core reactor
Dual-relay output section usually including provisions for a direct position feedback from the drive	Reversible electric motor drive

Classification

Electric actuators fall into two basic categories:

1. Inherently two-position actuators
   - a. Solenoid
   - b. Relay
2. Inherently infinite-position actuators
   - a. Reversible electric motor drive
   - b. Saturable-core reactor
   - c. Silicon controlled rectifier
   - d. Variable-speed electric motor drive
   - e. Electropneumatic converter

In some situations, an infinite-position actuator is capable of two-position actuation.

Reversible Electric Motor Drive

The actuator shown in Fig. 2 is suitable for operating valves, dampers, and similar lever-operated process regulators. Models are available with torques ranging from 5 to 500 lb·ft (6.8 to 680 N·m) and with full-stroke speeds ranging from 10 to 60 s. The operating voltage usually is 120 V, single-phase, 50 to 60 Hz—with a maximum running current of 1 A. A handwheel is provided for local manual operation. Internal worm and worm gear prevent back-drives by unbalanced loads. The linkage may be arranged for characterizing the process. Limit switches and position feedback transmitters can be provided for complex control.

Reversible Electric Motor Actuator with High Torque

The actuator shown in Fig. 3 is similar to that of Fig. 2 except that it is designed with torques ranging from 150 to 10,000 lb·ft (203 to 13,558 N·m)—for very large valves and dampers. Full-

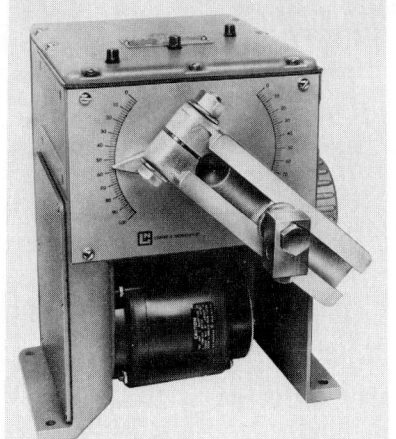 

Fig. 2 Reversible electric motor actuator with medium torque. *(Leeds & Northrup, Unit of General Signal.)*

Fig. 3 Reversible electric motor actuator with high torque. *(Leeds & Northrup, Unit of General Signal.)*

Fig. 4 Saturable-core reactor used with electric controller. *(Leeds & Northrup, Unit of General Signal.)*

stroke speeds range from 30 to 300 s. Operating voltages for larger sizes are 208 or 460 V, three-phase, 50 to 60 Hz. Power consumption at full running load ranges from 0.5 to 5.0 kV-A.

Saturable-Core Reactor

A saturable-core reactor can be used with electric controllers to modulate power input to an electrically heated process. See Fig. 4. Models are available for single- and three-phase power in sizes ranging from a few volt-amperes to several thousand kilovolt-amperes.

Fig. 5 Electropneumatic converter used with electric controller. *(Leeds & Northrup, Unit of General Signal.)*

Electropneumatic Converter

This type of unit changes electric output from the controller to a proportional pneumatic signal for operating pneumatic valve actuators. See Fig. 5. The most common range of transmission is 4 to 20 mA. Other current ranges and, occasionally, voltages ranges may be encountered.

Solid-State Power Switch

An electronic solid-state power switch of the type used with electric controllers is shown in Fig. 6. The device utilizes *silicon controlled rectifier* (SCR) *switches* for running, reversing, and braking high-torque reversible electric motor actuators. When used with high-inertia control devices, means are usually provided for dynamic braking. Tachometer feedback or plug stopping is frequently employed.

Electric motor drives are described in the

Fig. 6 Solid-state power switch used with electric controller. *(Leeds & Northrup, Unit of General Signal.)*

Table 2 Major Classes of Applications for Electric Actuators

Application	Actuator
Dampers, feeders, governors, pumps, valves, and other motion-actuated process regulators to modulate the flow of gases, liquids, and solids	Solenoid; reversible electric motor drive; variable-speed electric motor drive; electropneumatic converter
Modulation of flow of electric energy for heating or magnetic excitation	Relay; silicon-controlled rectifier; saturable-core reactor

next article of this Handbook. Saturable-core reactors and SCRs are described in further detail in the article Electric Controllers (Traditional) in Sec. 18.

Applications of the various kinds of electric actuators are delineated in Table 2.

REFERENCES

Driskell, L. R.: "Introduction to Control Valves and Other Final Control Devices," *ISA IRP300,* Instrument Society of America, Research Triangle Park, N.C., 1982.

Driskell, L. R.: "Selection of Control Valves and Other Final Control Devices," *ISA IRP310,* Instrument Society of America, Research Triangle Park, N.C., 1982.

Hordeski, M. F.: "Adapting Electric Actuators to Digital Control," *Instrum. Technol.,* vol. 24, no. 3, 1977, p. 53.

Electric Motor Drive Controls

by
Alexander Kusko*

A classification of electric motor drives is shown in Fig. 1. Constant-speed motor drives are supplied directly from the line. Adjustable-speed motor drives are supplied with processed power or use mechanical or eddy-current transmission to adjust the motor speed.

Constant-Speed Drives

Induction motors and synchronous motors are used for constant-speed applications, where speed is determined by their number of poles. These motors are started preferably at full line voltage but can be started at reduced voltage if the ac line voltage dips below 80 to 85%. A wound-rotor induction motor with secondary resistor control is suitable for starting high-inertia loads with a low starting line current. Solid-state adjustable-voltage or adjustable-frequency converters can be used to limit the starting line current. The block diagrams and ratings of constant-speed drives are given in Figs. 2 to 5.

Adjustable-Speed Drives

Direct current adjustable-speed drives employ phase-controlled thyristor rectifiers (silicon controlled rectifiers) to control the armature voltage and the speed of independent-field dc motors, as shown in Fig. 6. Field weakening control can extend the speed up to 3 times (depending on the dc motor) above base speed under constant horsepower loading constraints. Four-quadrant operation requires either an additional rectifier to reverse the armature voltage or a field reversing contactor. The armature voltage and speed of the dc motor can also be controlled with an ac-motor-driven dc generator.
    Alternating-current adjustable-speed drives employ solid-state inverters to control the frequency and the speed of induction motors or synchronous motors as shown in Fig. 7. The inverters use transistors up to ~40 hp and thyristors (SCRs) up to several thousand horsepower. The motor voltage and frequency are kept proportional up to the base speed of the motor. The motor voltage is kept constant up to 3 times the base speed under a constant-horsepower loading constraint. Large ac drives using synchronous motors employ one thyristor converter as a rectifier and a second as an inverter, as shown in Fig. 8. Other types include wound-rotor induction motors and fixed-speed ac motors with eddy-current couplings as shown in Fig. 9.

Drive Control

Speed regulators for dc drives employ analog or digital circuits. A typical analog regulator is shown in Fig. 10. The speed feedback signal is taken from the armature of the motor and corrected to the back electromotive force (emf) (proportional to speed), or from a tachometer. The speed reference

*President, Alexander Kusko, Inc., Consulting Engineers, Needham Heights, Mass.

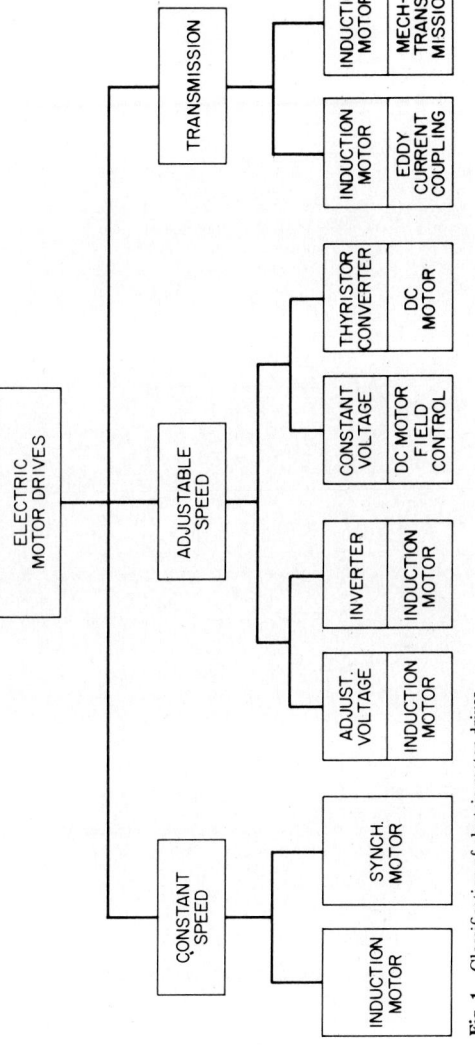

Fig. 1 Classification of electric motor drives.

FULL-VOLTAGE STARTER

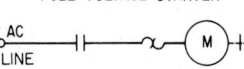

Fig. 2 Induction motor constant-speed drive block diagram. Standard drive systems are from ¼ to 5000 hp, with larger horsepower available as specials. Sixty-hertz synchronous speed of 3600, 1800, 1200, 900, 720, and 600 rpm. Single speed or multispeed.

FULL-VOLTAGE STARTER

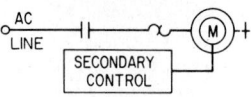

Fig. 3 Wound-rotor induction motor constant-speed drive block diagram. Standard drive systems are from 2 to 5000 hp, with larger horsepower available as specials. Sixty-hertz synchronous speed of 1800, 1200, 900, 720, and 600 rpm.

FULL-VOLTAGE STARTER

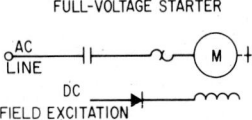

Fig. 4 Synchronous motor constant-speed drive block diagram. Standard drives are from 20 to 5000 hp, with larger horsepower available as specials. Sixty-hertz synchronous speed of 1800, 1200, 900, 720, 600, and 514 rpm.

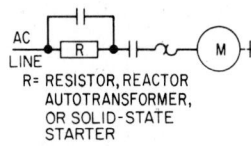

Fig. 5 Ac motor (with reduced in-rush starter) constant-speed drive block diagram. Standard drives are from 5 to 1500 hp, with larger horsepower available as specials. Sixty-hertz synchronous speed of 3600, 1800, 1200, 900, 720, and 600 rpm.

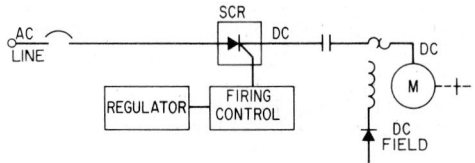

Fig. 6 Thyristor (SCR) rectifier adjustable-speed dc drive block diagram. Available standard drives from ¼ to 2000 hp with larger horsepower available as specials. The speed range is up to 20:1.

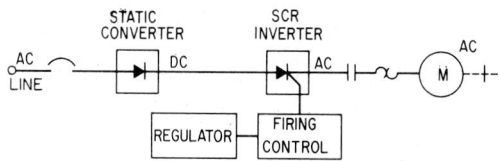

Fig. 7 Thyristor (SCR) inverter adjustable-frequency ac drive block diagram. Standard drives are from ¼ to 500 hp, with larger horsepower available as specials. The speed range is up to 20:1.

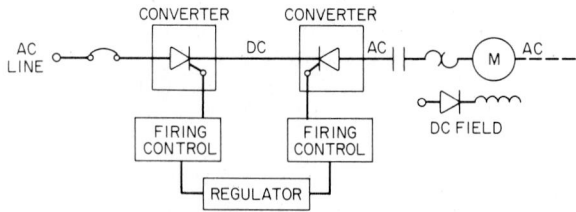

Fig. 8 Dual thyristor (SCR) converter adjustable-frequency ac drive with synchronous motor. Available drives are from 1000 to 10,000 hp. Speed range up to 20:1.

signal, typically 0 to 10 V, comes from a panel potentiometer or a remote control source. Such regulators typically employ an inner armature current loop to cancel the nonlinearity of the rectifier and increase the bandwidth of the overall drive.

Speed regulators for ac drives are analog or digital. Analog regulators respond to a reference voltage to control the inverter frequency and the speed of an ac motor open loop. The slip of the induction motor may be compensated. As shown in Fig. 11, digital regulators use microprocessors to compare the speed reference with feedback from a digital tachometer and to generate firing signals for the inverter. The speed reference signal can be an analog voltage or a serial or parallel digital signal delivered by a programmable controller or computer.

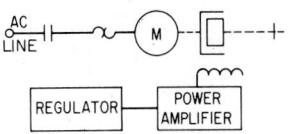

Fig. 9   Eddy-current coupling-type adjustable-speed ac drive block diagram. Standard drives are from ¼ to 750 hp, with larger horsepower available as specials. The speed range is up to 20:1.

Servo Drives

Electric motor servo drives are used to position mechanical loads in response to feedback signals from liner or rotary encoders. Two-phase ac servomotors operate at up to ~1 hp. Stepper motors, which typically move 1.8°, per step, operate open or closed loop. Direct current servomotors driven by transistor amplifiers deliver up to 10 hp; when driven by thyristor converters, they can deliver up to 500 hp. Brushless servo drives using permanent-magnet synchronous motors with position sensors operate similarly to dc servo drives as shown in Fig. 12, with velocity bandwidths up to 50 Hz. To maximize the torque/inertia ratio, dc servomotors use moving-coil shell or printed circuit armatures.

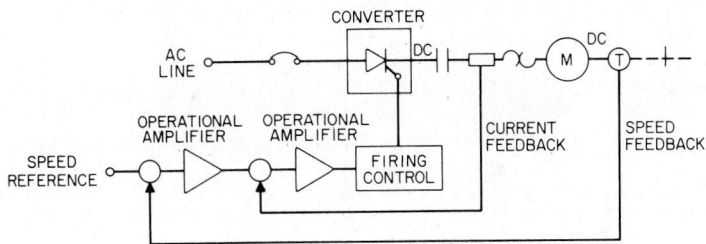

Fig. 10   Block diagram of analog speed regulator for dc drive.

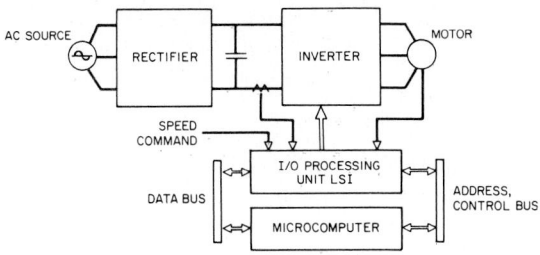

Fig. 11   Block diagram of digital microcomputer controllers for an adjustable-frequency pulse-width-modulated ac drive. *(From A. Kusko and D. Galler, "Survey of Microprocessors in Industrial Motor Drive Systems," paper presented at the IEEE-IAS 1982 Annual Meeting.)*

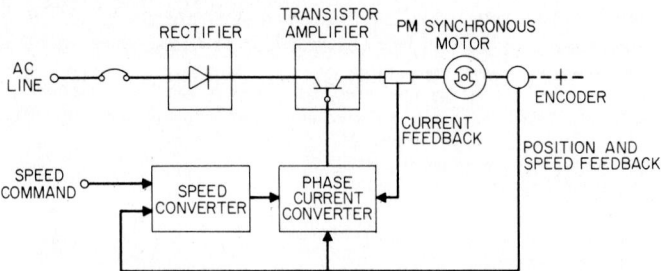

Fig. 12   Block diagram of brushless ac servo drive using a permanent-magnet synchronous motor.

FURTHER READING

Andrieiev, N.: "Solid State Devices Promise a Bright Future for Switching AC Motor Drives," *Control Eng.,* vol. 28, no. 8, 1981, p. 79.

Andrieiev, N.: "Software Driven LSI Chips Are Key to Future AC Drives," *Control Eng.,* vol. 28, no. 12, 1981, p. 62.

Bailey, S. J.: "Motor Drives Move toward Newer Technologies," *Control Eng.,* vol. 26, no. 11, 1979, p. 41.

Bailey, S. J.: "Variable Speed Motor Drives," *Control Eng.,* vol. 28, no. 1, 1981, p. 61.

Kervin, D.: "Zero-Crossing Triac Drivers Simplify Circuit Design," *Control Eng.,* vol. 29, no. 3, 1982, p. 76.

Kintigh, S.: "Reliable High-Power SCR Controller Design," *Control Eng.,* vol. 26, no. 5, 1979, p. 61.

Kompass, E. J.: "The Usage of Electrical Motors and Motor Controls in Industrial Control," *Control Eng.,* vol. 27, No. 6, 1980, p. 55.

Kusko, A., and T. Knutrud: "High Frequency SCR Chopper Circuit Improves DC Servo Motor Response," *Control Eng.,* vol. 24, no. 3, 1977, p. 44.

Lockert, C. A.: "Low Load Technique Controls Motor Energy Losses," *Control Eng.,* vol. 29, no. 2, 1982, p. 142.

Mawla, K. A.: "How Magnet Design Affects Linear Motor Performance," *Control Eng.,* vol. 29, no. 2, 1982, p. 78.

Miller, T. J.: "DC Drives Shrink in Size and Cost through Electronic Improvements," *Control Eng.,* vol. 28, no. 12, 1981, p. 68.

Morris, H. M.: "Unusual Motors for Special Design," *Control Eng.,* vol. 26, no. 4, 1979, p. 37.

Oachs, T. A.: "When to Use Variable Frequency AC Drives," *Control Eng.,* vol. 24 No. 9, 1977, p. 58.

Pluhar, K.: "Microelectronics Spur Changes in Motor Selection," *Control Eng.,* vol. 27, no. 6, 1980, p. 58.

Schroeder, E. G.: "Choose Variable Speed Drives for Pump and Fan Efficiency," *Instrum. Technol.,* vol. 27, no. 9, 1980, p. 83.

Staff: Linear-Induction Motor Positions Machine Tool Slides," *Control Eng.,* vol. 27, no. 5, 1980.

Staff: "AC and DC Motor Drives," *Control Eng.,* vol. 28, no. 1, 1981, p. 61.

Walker, P.: "Low-Inertia DC Servo Motor Has Wide Bandwidth Operating Range," *Control Eng.,* vol. 24, no. 10, 1977, p. 53.

Stepping Motors

by
John J. Murray*

A stepping motor is an electromechanical device which rotates a discrete step angle when energized electrically. The step angle usually is fixed for a particular motor and thus provides a means for accurately positioning in a repeatable, uniform manner. Typical step angles vary from as small as 0.72° to as large as 90°. Several means for electrically energizing stepping motors include dc pulses, square waves, and fixed logic sequence or multiple-phase square waves. Basic design types are (1) solenoid-operated ratchet, (2) permanent magnet, and (3) variable reluctance. Variations of these basic types may be combined with gears or hydraulic amplifiers to provide increased-output-torque stepping motors. The characteristics of a limited representation of hardware are given in Table 1.

Solenoid-Operated Ratchet

A solenoid is used to operate a ratchet device or star wheel. The main advantages of this configuration are low cost and high nonenergized holding torque. Limitations include short life and low stepping rates.

Permanent Magnet

Most stepping motors in use are of this type. Permanent-magnet motors offer high-torque capabilities—up to 2700 oz·in (19.1 N·m); for fast step response, 2 ms is possible; and for a high stepping rate, up to 20,000 steps per second. The accuracy of the step may be within 3% of one step, any error being nonaccumulative. Permanent-magnet types may be operated from pulses or square waves, dependent on motor design. The permanent magnet, which may be in the rotor or the stator—working with electrically produced flux—causes poles and teeth to align and effect rotation. Operation of two of the more common types is shown in Figs. 1 and 2.

Variable Reluctance

The variable reluctance type of stepping motor offers essentially the same advantages as permanent-magnet types. However, the torque capabilities have not been developed to the level of the permanent-magnet types. In the variable reluctance motor, the rotor-stator poles or teeth are aligned by the electric fields. Therefore, when the motor is deenergized, there is no residual holding torque. This is not a real disadvantage, however, in that most permanent-magnet types, as well as variable reluctance types, require that power remain on the motor between step movements. The same basic drive logic used for permanent-magnet steppers also applies to variable reluctance steppers. A multiphase variable reluctance stepping motor is shown in Fig. 3.

Drives

Although most stepping motors can be driven from switches or relays, most drive circuits incorporate solid-state devices which permit high-powered fast operation. Simple drives convert low-level pulses to power pulses or correctly phased power. More complex drives adjust power levels to allow running at very high rates.

*The Superior Electric Company, Bristol, Conn.

Table 1 Stepping Motor Characteristics*

Normal operation speed at rated load, pps	Maximum speed, no load, pps	Torque, oz·in		Rotor inertia, g·cm²	DC volts	Input current/power		Steps per revolution
		Operation speed	Detent, zero speed			Phase	Watts	
Solenoid-operated ratchet								
To 25	—	160	—	—	12, 28	1.0 A	84–210	18
600	1,000	1.4–6.1	—	—	6–220	—	6	10 or 12
400	750	3.0–15.7	—	—	6–220	—	9	10 or 12
300	600	6.0–24.2	—	—	6–220	—	11	10 or 12
Electrohydraulic								
5,000	5,000	10,000	10,000	—	2	5.4 A	10.8	200/500
4,000	8,000	1,440	1,440	—	28	3.0 A	10	200/250
4,000	8,000	2,880	2,900	—	28	3.0 A	10	200/250
4,000	4,000	8,000	8,200	—	28	3.0 A	10	200/250
Variable reluctance								
10	725	2.4	4.5	1.0	28	6.0 A	15	24
18	—	12	45	8.4	28	2.7 A	75	24
	900	—	8.0	0.74	28	1.93 A	100	24
400	5,000	25	40	180	5.4	1.5 A	—	200/72
250	3,000	1.0	2.7	0.004	28	355 mA	20	24
300	660	2.7	6.5	3.11	28	35 mA	20	24
450	6,000	0.5	0.95	0.004	28	330 mA	18	24
2,000	6,000	640	1,920	—	28	20.0 A	60	240
2,000	8,000	480	900	—	28	10.0 A	30	240
2,000	10,000	20	30	—	28	3.0 A	10	160
2,000	16,000	—	30	18.6	—	6.0 A	36	4
Permanent magnet								
10	130	0.1	0.22	0.15	28	90 mA	2.5	4
20	80	0.2	0.2	3.7	27	150 mA	4	24
40	320	0.47	0.97	2.6	24	150 mA	3	48
50	—	—	4.0	0.024	26	465 mA	12	8
75	190	1.5	5.5	4.02	28	43 mA	13	4
100	260	1.9	4.3	4.93	28	47 mA	13	8
120	145	0.6	0.3	0.77	28	280 mA	7.8	3
150	250	0.45	0.13	0.45	28	93 mA	5.2	4
175	200	5.0	9/7.4	7.05	28	1.25 A	35	4
200	320	1.0	0.3	0.77	28	93 mA	5.2	8
225	260	8.5	17/13.3	22.4	28	1.65 A	46	8
250	300	2.2	3.8/3.6	1.36	28	860 mA	24.1	4
300	500	0.3	0.7	0.003	26	104 mA	5.4	4
360/600	—	1.3/1.8	—	34.8	20–50	—	2.0/1.8	12/24
400	—	12	40	180	5.4	1.5 mA	—	72
500	—	200	600	7,310	—	—	—	200
600	—	100	260	1,520	4.5	3.5 mA	—	200
45	—	35	90	877	3.0	4.0 mA	—	500
45	—	1,200	2,500	38,010	12	3.8 mA	—	200

*A limited representation of commercially available hardware. Characteristics that fall between parameters shown are probably also available.

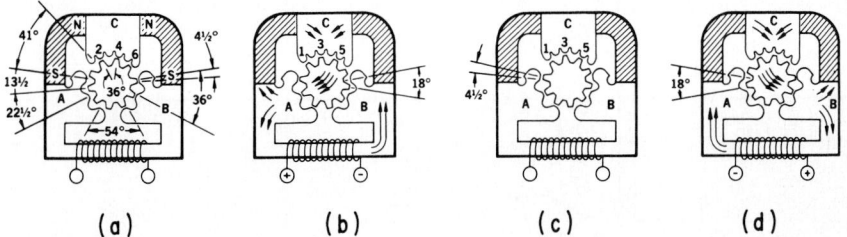

(a)   (b)   (c)   (d)

Fig. 1 Operation of a stepping motor (electromagnetic rotary incremental actuator) which converts digital pulse inputs to analog shaft output motion. The motor diagrammed starts and stops in discrete angular excursions of uniform magnitude (15°) to produce output steps always equal in number to the pulse inputs. Two permanent magnets, connected in parallel for uniform flux distribution, link two portent poles (*A* and *B*) and one detent pole (*C*). Each portent pole has three sets of teeth spaced 36° apart and is positioned so that, when one set of portent teeth aligns with three rotor teeth, the other set aligns with spaces between rotor teeth. Also aligned with three rotor teeth are three of six detent-pole teeth spaced 18° apart. Rotor position establishes alternate flux paths. Each pulse of alternating polarity applied to the coil advances the rotor one step by switching the magnetic flux from one path to the other. Dc pulses of plus or minus polarity overcome the permanent magnetic detent and produce torque to advance the rotor from one detent to the next. A similar 15° motor, designed to step through 24 fixed shaft positions in one revolution, operates in a similar manner. (Cyclonome, *Sigma Instruments.*)

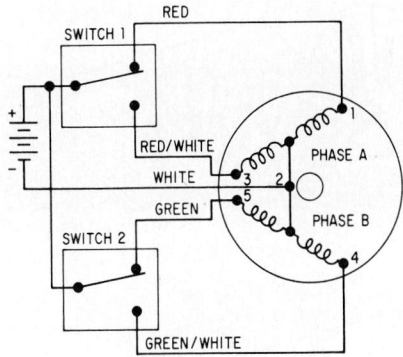

SWITCHING SEQUENCE *

STEP	SWITCH 1	SWITCH 2
1	1	5
2	1	4
3	3	4
4	3	5
1	1	5

* TO REVERSE DIRECTION,
READ CHART UP FROM BOTTOM.

Fig. 2 Connection diagram and switching sequence of a typical phase-switched stepping motor. Diagram is for a motor having a permanent magnet stator. *(Superior Electric.)*

19.95

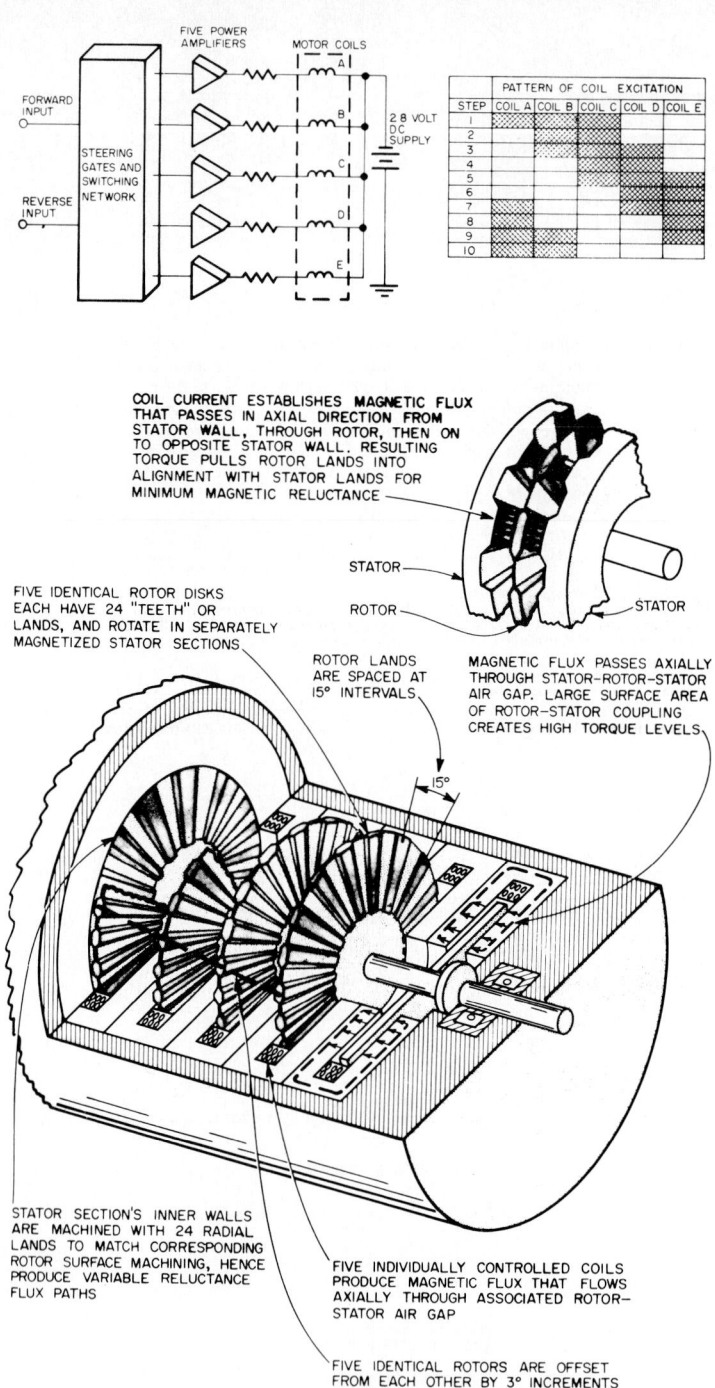

Fig. 3 Variable reluctance stepping motor. *(ICON.)*

Step Response

At very low rates, the step movement resembles the classic damped oscillation curve. Damping can be added to modify the curve to provide critical damping. This may be done mechanically, electrically, or with a viscous fluid. A typical step function for both underdamped and approaching critical damping conditions is shown in Fig. 4.

Applications

Stepping motors provide an economical means for accurate and reliable positioning. Since steppers move in discrete steps, they avoid the problems of stability and feedback inherent in most servo devices by allowing direct digital control in an open-loop mode of operation. Although not required, pulse feedback can be readily coupled to a stepping motor to provide closed-loop operation. Applications range from simple indexing positioners to complex numerical contouring controls. Large as well as small inertial loads can be handled by controlling the acceleration and deceleration rates of the input pulses. The most common applications for stepping motors include printers, tape readers, memory devices, valve actuators, positioning tables, counters, and indexers.

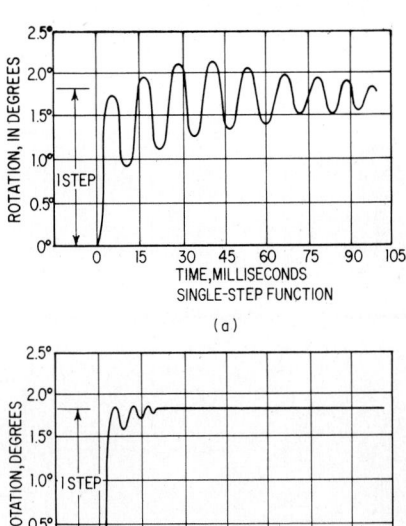

Fig. 4 Typical step curves for a stepping motor in (*a*) the undamped condition and (*b*) approaching critical damping.

REFERENCES

Bailey, S. J.: "Incremental Actuation: Stepper Torque and Force Moving Up," *Control Eng.*, vol. 25, no. 5, 1978, p. 47.

Bailey, S. J.: "Stepping Controls Mature as Digital Actuators," *Control Eng.*, vol. 26, no. 8, 1979, p. 36.

Berris, R., D. Hazony, and R. Resch: "Discrete Pulses Put Induction Motors into the Stepping Mode," *Control Eng.*, vol. 29, no. 1, 1982, p. 85.

Budzilovich, P. N.: "Electrohydraulic Stepping Motors," *Control Eng.*, vol. 17, no. 1, 1970, p. 82.

Kompass, E. J.: "Five-Phase Stepping Motor," *Control Eng.*, vol. 25, no. 3, 1978, p. 63.

Leenhouts, A. C.: "Techniques for Microstepping Control of Step Motors," *Control Eng.*, vol. 26, no. 3, 1979, p. 58.

Usry, J. D.: "Stepping Motors for Valve Actuation," *Instrum. Technol.*, vol. 24, no. 3, 1977, p. 58.

Index